The

ENCYCLOPEDIA

of

BIOCHEMISTRY

The

ENCYCLOPEDIA

of

BIOCHEMISTRY

Edited by

ROGER J. WILLIAMS

Professor, The University of Texas, Austin, and formerly Director of the Clayton Foundation Biochemical Institute

and

EDWIN M. LANSFORD, Jr.

Research Scientist Associate, Clayton Foundation Biochemical Institute, The University of Texas, Austin

ROBERT E. KRIEGER PUBLISHING COMPANY
HUNTINGTON, NEW YORK
1977

Original Edition 1967
Reprint 1977

Printed and Published by
ROBERT E. KRIEGER PUBLISHING CO., INC.
645 NEW YORK AVENUE
HUNTINGTON, NEW YORK 11743

Printed in the United States of America

Library of Congress Cataloging in Publication Data

Williams, Roger John, 1893-
The encyclopedia of biochemistry.

Reprint of the edition published by Reinhold, New York.
Includes bibliographies.
1. Biological chemistry—Dictionaries. I. Lansford, Edwin M., joint author. II. Title.
[QP512.W5 1977] 574.1'92'03 77-23257
ISBN 0-88275-534-X

CONTRIBUTORS

LEO G. ABOOD, Center for Brain Research and Department of Biochemistry, University of Rochester, Rochester, New York, 14627. MICROSOMES AND THE ENDOPLASMIC RETICULUM.

EUGENE ACKERMAN, Section of Biophysics, Mayo Clinic, Rochester, Minnesota. BIOCHEMICAL THERMODYNAMICS.

W. WILBUR ACKERMANN, Department of Epidemiology, University of Michigan, Ann Arbor, Michigan. VIRUSES (REPLICATION).

LUCILE F. ADAMSON, Department of Medicine, Harvard Medical School, Boston, Massachusetts. MALNUTRITION.

R. P. AHLQUIST, Medical College of Georgia, Augusta, Georgia. ADRENALINE AND NORADRENALINE.

ANTHONY A. ALBANESE, Nutrition and Metabolic Research Division, The Burke Rehabilitation Center, White Plains, New York, and Geriatic Nutrition Laboratory, Miriam Osborne Memorial Home, Rye, New York. AMINO ACIDS, NUTRITIONALLY ESSENTIAL.

NICHOLAS M. ALEXANDER, Department of Internal Medicine, Yale University School of Medicine, New Haven, Connecticut, 06520. GOITER.

CONSTANTINE J. ALEXOPOULOS, Department of Botany, University of Texas, Austin, Texas. SLIME MOLDS.

ROSLYN B. ALFIN-SLATER, School of Public Health, University of California, Los Angeles, California. FATTY ACIDS, ESSENTIAL.

CARL ALPER, Department of Biological Chemistry, Hahnemann Medical College and Hospital, Philadelphia, Pennsylvania. BLOOD (CHEMICAL COMPOSITION).

R. E. ALSTON, Department of Botany, University of Texas, Austin, Texas. CHEMOTAXONOMY.

CARL E. ANDERSON, Department of Biochemistry, School of Medicine, University of North Carolina, Chapel Hill, North Carolina. FATTY ACID METABOLISM.

H. VASKEN APOSHIAN, Department of Microbiology, Tufts University School of Medicine, Boston, Massachusetts, 02111. MICROBIOLOGY.

JOSEPH C. ARCOS, Department of Medicine, Tulane University School of Medicine, New Orleans, Louisiana. CARCINOGENIC SUBSTANCES.

FRANK B. ARMSTRONG, Department of Genetics, North Carolina State of the University of North Carolina at Raleigh, North Carolina. PROTEINS (BIOSYNTHESIS).

W. D. ARMSTRONG, Department of Biochemistry, College of Medical Sciences, University of Minnesota, Minneapolis, Minnesota. FLUORIDE IN METABOLISM.

DANIEL I. ARNON, Department of Cell Physiology, University of California, Berkeley, California. LIGHT PHASE OF PHOTOSYNTHESIS.

CAMILLO ARTOM, Bowman Gray School of Medicine, Wake Forest College, Winston-Salem, North Carolina. METHYLATIONS IN METABOLISM.

W. ROSS ASHBY, Departments of Electrical Engineering and Biophysics, University of Illinois, Urbana, Illinois. HOMEOSTASIS.

BERNARD D. ASTILL, Laboratory of Industrial Medicine, Eastman Kodak Company, Rochester, New York. ANTIOXIDANTS.

LAZARUS ASTRACHAN, School of Medicine, Western Reserve University, Cleveland, Ohio. MESSENGER RIBONUCLEIC ACIDS.

DANIEL E. ATKINSON, Department of Chemistry, University of California, Los Angeles, California. REGULATORY SITES OF ENZYMES.

VICTOR H. AUERBACH, Departments of Pediatrics and Biochemistry, Temple University School of Medicine, and St. Christopher's Hospital for Children, Philadelphia, Pennsylvania. METABOLIC CONTROLS (FEEDBACK MECHANISMS).

HAROLD BAER, Division of Biologics Standards, National Institutes of Health, Bethesda, Maryland, 20014. BLOOD GROUPS.

ROBERT C. BALDRIDGE, Department of Biochemistry, Temple University School of Medicine, Philadelphia, Pennsylvania, 19140. SULFUR METABOLISM.

ROBERT BARER, Department of Human Biology and Anatomy, University of Sheffield, England. PHASE MICROSCOPY.

S. B. BARKER, University of Alabama Medical Center, Birmingham, Alabama, 35233. IODINE.

ALLAN D. BASS, Department of Pharmacology, Vanderbilt University Medical School, Nashville, Tennessee. CYANIDE POISONING.

RICHARD S. BEAR, Graduate School, Boston University, Boston, Massachusetts, 02215. NUCLEOPROTEINS.

ERNEST BEERSTECHER, University of Texas Dental Branch, Houston, Texas, 77025. KWASHIORKOR.

HELMUT BEINERT, Institute for Enzyme Research, University of Wisconsin, Madison, Wisconsin. FLAVINS AND FLAVOPROTEINS.

L. L. BENNETT, JR., Kettering-Meyer Laboratory, Southern Research Institute, Birmingham, Alabama. CANCER CHEMOTHERAPY.

A. A. BENSON, Department of Marine Biology, Scripps Institution of Oceanography, University of California, La Jolla, California. CARBON REDUCTION CYCLE (IN PHOTOSYNTHESIS).

BRUCE BERMAN, Office of Research Information, National Institutes of Health, Bethesda, Maryland, 20014. NATIONAL INSTITUTES OF HEALTH.

I. A. BERNSTEIN, Department of Dermatology, University of Michigan, Ann Arbor, Michigan, 48104. CARBOHYDRATE METABOLISM.

ROBERT E. BEYER, Institute of Science and Technology and Department of Zoology, University of Michigan, Ann Arbor, Michigan. PHOSPHORYLATION, OXIDATIVE.

JOHN G. BIERI, Nutritional Biochemistry Section, Laboratory of Nutrition and Endocrinology, National Institute of Arthritis and Metabolic Disease, Bethesda, Maryland. VITAMIN A.

JOHN J. BIESELE, Department of Zoology, University of Texas, Austin, Texas. CYTOTOXIC CHEMICALS.

RICHARD J. BING, Department of Medicine, Wayne State University School of Medicine, Detroit, Michigan. HEART METABOLISM.

HENRY BIRNBAUM, Deputy Administrative Manager, National Science Foundation, Washington 25, D.C. NATIONAL SCIENCE FOUNDATION.

PAUL V. BLAIR, Department of Biochemistry, Indiana University School of Medicine, Indianapolis, Indiana, 46207. MITOCHONDRIA.

ERIC BLOCH, Department of Gynecology and Obstetrics, Albert Einstein College of Medicine, Yeshiva University, Bronx, New York, 10461. SEX HORMONES.

BEN BLOOM, 6202 Bannockburn Drive, Bethesda, Maryland, 20034. LIPIDS (CLASSIFICATION AND DISTRIBUTION).

STANLEY M. BLOOM, Polaroid Corporation, Cambridge, Massachusetts, 02139. POLYPEPTIDES, SYNTHETIC.

LARRY E. BOCKSTAHLER, Max-Planck-Institut für Molekulare Genetik, Tübingen, Germany. TOBACCO MOSAIC VIRUS.

F. J. BOLLUM, Department of Biochemistry, University of Kentucky Medical School, Lexington, Kentucky. RIBONUCLEIC ACIDS (STRUCTURE AND FUNCTIONS).

NESTOR R. BOTTINO, Department of Biochemistry and Biophysics, Texas A and M University, College Station, Texas, 77843. PHOSPHOLIPIDS.

M. JOHN BOYD, Department of Biological Chemistry, Hahnemann Medical College and Hospital, Philadelphia, Pennsylvania. BLOOD (CHEMICAL COMPOSITION).

EMIL BOZLER, Department of Physiology, Ohio State University, Columbus, Ohio. MUSCLE (STRUCTURE AND FUNCTION).

DALE C. BRAUNGART, Department of Biology, Catholic University of America, Washington, D.C. ULTRASONIC VIBRATIONS.

G. E. BRIGGS, Department of Botany, University of Cambridge, Cambridge, England. OSMOSIS AND OSMOTIC PRESSURE.

GEORGE M. BRIGGS, Agricultural Experiment Station, and Department of Nutritional Sciences, University of California, Berkeley, California. NUTRITIONAL REQUIREMENTS (MAMMALS).

ARNOLD F. BRODIE, Department of Microbiology, University of Southern California School of Medicine, Los Angeles, California. OXIDATIVE METABOLISM AND ENERGETICS OF BACTERIA.

FELIX BRONNER, Department of Physiology and Biophysics, University of Louisville School of Medicine, Louisville, Kentucky, 40202. CALCIUM.

MARION A. BROOKS, University of Minnesota, St. Paul, Minnesota. SYMBIOSIS.

NATHAN BROT, National Heart Institute, National Institutes of Health, Bethesda, Maryland, 20014. COBALAMINS AND COBAMIDE COENZYMES.

F. CHRISTINE BROWN, Brain Research Institute, University of Tennessee, and the Tennessee Department of Mental Health, Memphis, Tennessee. HALLUCINOGENIC DRUGS.

GENE M. BROWN, Division of Biochemistry,

Department of Biology, Massachusetts Institute of Technology, Cambridge, Massachusetts, 02139. FOLIC ACID.

H. P. BURCHFIELD, Pesticide Research Laboratory, U.S. Public Health Service, Perrine, Florida, 33157. CHROMATOGRAPHY, GAS.

DEAN BURK, Laboratory of Biochemistry, National Cancer Institute, Bethesda, Maryland, 20014. GLYCOLYSIS.

SIR MACFARLANE BURNET, School of Microbiology, University of Melbourne, Melbourne, Australia. ANTIBODIES.

DENNIS CALDWELL, Department of Chemistry, University of Utah, Salt Lake City, Utah. OPTICAL ACTIVITY AND ISOMERISM.

MELVIN CALVIN, Department of Chemistry, University of California, Berkeley, California, 94720. CHELATION AND METAL CHELATE COMPOUNDS.

JOHN R. CANN, Department of Biophysics, University of Colorado Medical Center, Denver, Colorado, 80220. CENTRIFUGATION, DIFFERENTIAL.

M. KATHLEEN CARTER, Department of Pharmacology, Tulane University School of Medicine, New Orleans, Louisiana, 70112. GLOMERULAR FILTRATION.

JOHN E. CASIDA, Department of Entomology and Parasitology, University of California, Berkeley, California, 94720. INSECTICIDES.

HAROLD G. CASSIDY, Department of Chemistry, Yale University, New Haven, Connecticut. CHROMATOGRAPHY (GENERAL); OXIDATION-REDUCTION POLYMERS.

WINSLOW S. CAUGHEY, Department of Physiological Chemistry, The Johns Hopkins University School of Medicine, Baltimore, Maryland, 21205. PORPHYRINS.

ALFRED CHANUTIN, Department of Biochemistry, University of Virginia School of Medicine, Charlottesville, Virginia. IRON.

VERNON H. CHELDELIN, deceased; formerly, Science Research Institute, Oregon State University, Corvallis, Oregon. MICROBIOLOGICAL ASSAY METHODS.

GREGORY R. CHOPPIN, Department of Chemistry, Florida State University, Tallahassee, Florida. ISOTOPIC TRACERS.

B. CINADER, Department of Medical Biophysics, University of Toronto, Toronto, Canada. IMMUNOCHEMISTRY.

WILLIAM C. CLARK, Department of Pharmacology, Indiana University Medical Center, Indianapolis, Indiana, 46207. GASTRIN.

SEYMOUR S. COHEN, Department of Therapeutic Research, University of Pennsylvania School of Medicine, Philadelphia, Pennsylvania, 19104. COMPARATIVE BIOCHEMISTRY.

STANLEY COHEN, Department of Biochemistry, Vanderbilt University School of Medicine, Nashville, Tennessee. MORPHOGENIC INDUCTION.

MARCELLUS T. COLTHARP, Department of Chemistry, University of Texas, Austin, Texas. ADSORPTION.

JOHN G. CONIGLIO, Department of Biochemistry, Vanderbilt University School of Medicine, Nashville, Tennessee. FAT MOBILIZATION AND TRANSPORT.

C. FRANK CONSOLAZIO, Bioenergetics Division, Fitzsimons General Hospital, Denver, Colorado, 80240. CALORIC INTAKE AND REQUIREMENTS.

N. E. COOK, Boston District, United States Food and Drug Administration, Boston, Massachusetts. U.S. FOOD AND DRUG ADMINISTRATION.

CECIL COOPER, Department of Biochemistry, Western Reserve University School of Medicine, Cleveland, Ohio. THYROID HORMONES.

M. S. COOPER, Lederle Laboratories, Pearl River, New York. ANTITOXINS.

JACK M. COOPERMAN, Department of Pediatrics, New York Medical College, New York, N.Y., 10029. ANEMIAS.

WILLIAM M. CORBETT, Purdue University, Lafayette, Indiana. STARCHES.

MILTON J. CORMIER, Department of Biochemistry, University of Georgia, Athens, Georgia. BIOLUMINESCENCE.

W. E. CORNATZER, Department of Biochemistry, Ireland Research Laboratory, University of North Dakota School of Medicine, Grand Forks, North Dakota. IONIZING RADIATION.

ALSOPH H. CORWIN, Department of Chemistry, The Johns Hopkins University, Baltimore, Maryland. CHLOROPHYLLS.

GEORGE C. COTZIAS, Medical Research Center, Brookhaven National Laboratory, Upton, New York, 11973. MANGANESE.

J. R. COUCH, Department of Poultry Science, Texas A&M University, College Station, Texas. NUTRITIONAL REQUIREMENTS (VERTEBRATES OTHER THAN MAMMALS).

DAVID J. COX, Clayton Foundation Biochemical Institute and Department of Chemistry, The University of Texas, Austin, Texas, 78712. MULTIENZYME COMPLEXES.

GERALD J. COX, 4731 Stanton Avenue, Pittsburgh, Pennsylvania, 15201. CARIES, DENTAL.

DANA I. CRANDALL, Department of Biological Chemistry, University of Cincinnati College of Medicine, Cincinnati, Ohio, 45219. ALKAPTONURIA.

FREDERICK L. CRANE, Department of Biological Sciences, Purdue University, Lafayette, Indiana, 47907. COENZYME Q.

ALFRED J. CROWLE, Division of Immunology,

Webb-Waring Institute for Medical Research, Denver, Colorado. IMMUNODIFFUSION.

VELIMIR CUBRILOVIC, Pharmacia Fine Chemicals, Inc., Monterey Park, California. GEL FILTRATION.

LEON CUNNINGHAM, Department of Biochemistry, Vanderbilt University School of Medicine, Nashville, Tennessee. PROTEOLYTIC ENZYMES.

GEORGE L. CURRAN, College of Physicians and Surgeons, Columbia University, Welfare Island, New York, N.Y., 10017. VANADIUM.

WINDSOR C. CUTTING, School of Medicine, University of Hawaii, Honolulu, Hawaii. ADRENOCORTICOTROPIC HORMONE.

R. DUNCAN DALLAM, Department of Biochemistry, University of Louisville School of Medicine, Louisville, Kentucky. VITAMIN K.

ISIDORE DANISHEFSKY, Department of Biochemistry, New York Medical College, New York, N.Y. 10029. CARBOHYDRATES (CLASSIFICATION AND STRUCTURE).

R. E. DAVIES, University of Pennsylvania School of Veterinary Medicine, Philadelphia, Pennsylvania, 19104. GASTRIC JUICE.

FRANK F. DAVIS, Department of Physiology and Biochemistry, Rutgers, The State University, New Brunswick, New Jersey. SOLUBLE RIBONUCLEIC ACIDS.

NEIL C. DAVIS, Department of Pediatrics, University of Cincinnati College of Medicine, Children's Hospital Research Foundation, Cincinnati, Ohio. PEPTIDASES.

ROBERT E. DELAVAULT, Department of Geology, University of British Columbia, Vancouver, B.C. GEOBIOCHEMISTRY.

J. F. DIEHL, Bundes-Forschungs-Anstalt, 75 Karlsruhe, Engesserstr. 20, Germany. KERATIN.

E. A. DOISY, JR., Saint Louis University School of Medicine, St. Louis, Missouri, 63104. INOSITOL.

RALPH I. DORFMAN, Institute of Hormone Biology, Syntex Research, Palo Alto, California. HORMONES, STEROID.

H. H. DRAPER, Department of Animal Science, University of Illinois, Urbana, Illinois. BONE FORMATION.

KENNETH P. DUBOIS, Department of Pharmacology, University of Chicago, Chicago, Illinois. ACETYLCHOLINE AND CHOLINESTERASE.

ROBERT E. EAKIN, Department of Chemistry, and Clayton Foundation Biochemical Institute, University of Texas, Austin, Texas. AVIDIN; ORIGIN OF LIFE.

BERNICE E. EDDY, Section on Experimental Virology, Division of Biologics Standards, National Institutes of Health, Bethesda, Maryland, 20014. VIRUS DISEASES.

JOHN T. EDSALL, Biological Laboratories, Harvard University, Cambridge, Massachusetts. PROTEINS.

ELIZABETH ROBOZ EINSTEIN, Departments of Neurology and Biochemistry, University of California School of Medicine, San Francisco, California. MULTIPLE SCLEROSIS.

LEWIS L. ENGEL, Harvard Medical School, Massachusetts General Hospital, Boston, Massachusetts. ANDROGENS.

HAROLD J. EVANS, Department of Botany and Plant Pathology, Oregon State University, Corvallis, Oregon, 97331. COBALT.

HENRY EYRING, Graduate School, University of Utah, Salt Lake City, Utah. OPTICAL ACTIVITY AND ISOMERISM.

GARY FELSENFELD, Laboratory of Molecular Biology, National Institute of Arthritis and Metabolic Diseases, National Institutes of Health, Bethesda, Maryland, 20014. HEMOCYANINS.

WALLACE O. FENN, School of Medicine and Dentistry, University of Rochester, Rochester, New York, 14620. GASEOUS EXCHANGES IN BREATHING.

EDMOND H. FISCHER, Department of Biochemistry, University of Washington, Seattle, Washington, 98105. GLYCOGEN AND ITS METABOLISM.

WILLIAM H. FISHMAN, Tufts University School of Medicine, New England Medical Center Hospitals, Boston, Massachusetts. GLUCURONIDASES; TESTOSTERONE.

JAMES W. FISHER, Department of Pharmacology, University of Tennessee, Memphis, Tennessee, 38103. KETOSTEROIDS.

JOHN FISSEKIS, Sloan-Kettering Institute for Cancer Research, Rye, New York. PURINES.

WARNER H. FLORSHEIM, Veterans Administration Hospital, Long Beach, California, 90804. THYROTROPIC HORMONE.

R. M. FORBES, Department of Animal Science, University of Illinois, Urbana, Illinois, 61801. MINERAL REQUIREMENTS.

DARHL FOREMAN, Department of Biology, Western Reserve University, Cleveland, Ohio, 44106. HORMONES, ANIMAL.

HUGH S. FORREST, Department of Zoology and Genetics Foundation, University of Texas, Austin, Texas. PTERIDINES.

DANIEL W. FOSTER, Department of Internal Medicine, University of Texas Southwestern Medical School, Dallas, Texas, 75235. LIPIDS (CLASSIFICATION AND DISTRIBUTION).

JOSEPH F. FOSTER, Department of Chemistry, Purdue University, Lafayette, Indiana, 47907. ALBUMINS.

ERNEST C. FOULKES, University of Cincinnati College of Medicine, Cincinnati, Ohio. METALLOPROTEINS.

L. FOWDEN, Department of Botany, University College London, London, England. AMINO ACIDS, NON-PROTEIN.

C. FRED FOX, Ben May Laboratory for Cancer Research, The University of Chicago, Chicago, Illinois. RIBONUCLEIC ACIDS (BIOSYNTHESIS).

JACK J. FOX, Sloan-Kettering Institute, Rye, New York. PYRIMIDINES.

IRWIN FRIDOVICH, Department of Biochemistry, Duke University Medical School, Durham, North Carolina, 27706. XANTHINE OXIDASE.

MELVIN FRIED, Department of Biochemistry, University of Florida College of Medicine, Gainesville, Florida, 32601. PLASMA AND PLASMA PROTEINS.

EARL FRIEDEN, Department of Chemistry, Florida State University, Tallahassee, Florida, 32306. METAMORPHOSIS, AMPHIBIAN.

EDWARD H. FRIEDEN, Department of Chemistry, Kent State University, Kent, Ohio, 44240. RELAXIN.

ARMAND J. FULCO, Department of Biophysics and Nuclear Medicine, and Department of Biological Chemistry, University of California School of Medicine, Los Angeles, California, 90024. UNSATURATED FATTY ACIDS.

SIDNEY FUTTERMAN, Department of Ophthalmology, University of Washington, Seattle, Washington. LIGHT ADAPTATION.

WILLIAM L. GABY, Department of Health Sciences, East Tennessee State University, Johnson City, Tennessee. CEPHALINS AND LECITHINS.

O. H. GAEBLER, Edsel B. Ford Institute for Medical Research, Detroit, Michigan, 48202. CREATINE AND CREATININE.

HANS GAFFRON, Institute of Molecular Biophysics, Florida State University, Tallahassee, Florida, 32306. PHOTOSYNTHESIS.

ARTHUR W. GALSTON, Department of Biology, Yale University, New Haven, Connecticut. PHOTOTROPISM.

R. L. GARNER, Department of Biochemistry, New Jersey College of Medicine and Dentistry, Jersey City, New Jersey, 07304. GLUCONEOGENESIS.

JAMES L. GAYLOR, Graduate School of Nutrition, Cornell University, Ithaca, New York, 14850. ISOPRENOID BIOSYNTHESIS.

T. A. GEISSMAN, Department of Chemistry, University of California, Los Angeles, California, 90024. PLANT PIGMENTS.

DAVID M. GIBSON, Department of Biochemistry, Indiana University School of Medicine, Indianapolis, Indiana, 46207. CARBOXYLATION ENZYMES.

FRANK GIBSON, School of Microbiology, University of Melbourne, Parkville N. 2. Victoria, Australia. AROMATIC AMINO ACIDS.

LEWIS I. GIDEZ, Departments of Biochemistry and Medicine, Albert Einstein College of Medicine, Yeshiva University, Bronx, New York, 10461. LIPOPROTEINS.

JAMES B. GILBERT, Department of Biochemistry, West Virginia School of Medicine, Morgantown, West Virginia, 26506. CARBONIC ANHYDRASE.

THOMAS J. GILL III, Laboratory of Chemical Pathology, Harvard Medical School, Boston, Massachusetts. ANTIGENS.

T. W. GOODWIN, Department of Biochemistry, The University, Liverpool 3, England. CAROTENOIDS.

MANUEL J. GORDON, Beckman Instruments, Inc., Spinco Division, Palo Alto, California. ULTRACENTRIFUGATION.

NICHOLAS M. GREENE, Division of Anesthesiology, Yale University School of Medicine, New Haven, Connecticut. ANESTHESIA.

ROBERT R. GRINSTEAD, Research Laboratories, Dow Chemical Company, Walnut Creek, California. COPPER.

GEROLD M. GRODSKY, Metabolic Research Unit, University of California Medical Center, San Francisco, California, 94122. BILE AND BILE PIGMENTS.

ARTHUR GROLLMAN, Department of Experimental Medicine, University of Texas Southwestern Medical School, Dallas, Texas, 75235. ADRENAL CORTICAL HORMONES.

SIGMUNDUR GUDBJARNASON, Wayne State University School of Medicine, Detroit, Michigan. HEART METABOLISM.

WALTER R. GUILD, Department of Biochemistry, Genetics Division, Duke University Medical Center, Durham, North Carolina, 27706. TRANSFORMING FACTORS.

FRANK R. N. GURD, Department of Chemistry, Indiana University, Bloomington, Indiana, 47401. MYOGLOBINS.

M. H. HACK, Department of Medicine, Tulane University School of Medicine, New Orleans, Louisiana, 70112. GANGLIOSIDES.

W. KNOWLTON HALL, Department of Biochemistry, Medical College of Georgia, Augusta, Georgia. POTASSIUM AND SODIUM.

R. S. HANNAN, T. Wall & Sons (Meat and Handy Foods) Ltd., London, N.W. 10., England. STERILIZATION.

BOYD A. HARDESTY, Department of Chemistry and Clayton Foundation Biochemical Institute, University of Texas, Austin, Texas, 78712. BIOCHEMICAL GENETICS.

ROBERT A. HARTE, Executive Officer, American Society of Biological Chemists, Bethesda, Maryland, 20014. AMERICAN SOCIETY OF BIOLOGICAL CHEMISTS.

PHILIP E. HARTMAN, Department of Biology, Johns Hopkins University, Baltimore, Maryland, 21218. BACTERIAL GENETICS.

STANDISH C. HARTMAN, Department of Biological Chemistry, Harvard University Medical School, Boston, Massachusetts, 02115. PURINE BIOSYNTHESIS.

W. STANLEY HARTROFT, The Research Institute of the Hospital for Sick Children, Toronto 2, Canada. ATHEROSCLEROSIS.

OSAMU HAYAISHI, Department of Medical Chemistry, Kyoto University Faculty of Medicine, Kyoto, Japan. OXYGENASES.

PETER HEMMERICH, Institut für Anorganische Chemie, Universität Basel, Basel, Switzerland. FLAVINS AND FLAVOPROTEINS.

L. M. HENDERSON, Department of Biochemistry, Institute of Agriculture, University of Minnesota, St. Paul, Minesota. NICOTINIC ACID.

RICHARD W. HENDLER, Laboratory of Biochemistry, Section on Cellular Physiology, National Heart Institute, Bethesda, Maryland, 20014. LIPOAMINO ACIDS AND LIPOPEPTIDES.

VICTOR HERBERT, Department of Medicine, Mount Sinai School of Medicine, New York, N.Y. VITAMIN B_{12}.

EDWARD J. HERBST, Department of Biochemistry, University of New Hampshire, Durham, New Hampshire, 03824. POLYAMINES.

ROGER M. HERRIOTT, Department of Biochemistry, Johns Hopkins University, School of Hygiene and Public Health, Baltimore, Maryland, 21205. PEPSINS AND PEPSINOGENS.

IRA D. HILL, Sun Oil Company, Research and Engineering, Marcus Hook, Pennsylvania, 19061. HYDROCARBON OXIDATION.

DAVID I. HITCHCOCK, Burpee Hill Road, New London, New Hampshire, 03257. BUFFERS; MEMBRANE EQUILIBRIUM; pH.

GEORGE H. HITCHINGS, Chemotherapy Division, Burroughs Wellcome and Co. (U.S.A.) Inc., Tuckahoe, New York, 10707. SULFONAMIDES.

D. S. HOARE, Department of Microbiology, University of Texas, Austin, Texas. PHOTOSYNTHETIC BACTERIA.

ROBERT E. HODGES, Department of Internal Medicine, University Hospitals, University of Iowa, Iowa City, Iowa, 52240. VITAMIN E.

EDWARD S. HODGSON, Department of Biological Sciences, Columbia University, New York, N.Y., 10027. TASTE RECEPTORS.

PHILIP HOFFMAN, Research Institute for Skeletomuscular Diseases, Hospital for Joint Diseases, New York, N.Y., 10035. MUCOPOLYSACCHARIDES.

ROBERT W. HORNE, Institute of Animal Physiology, Agricultural Research Council, Babraham, Cambridge, England. VIRUSES (COMPOSITION AND STRUCTURE).

H. L. HOUSE, Research Institute, Research Branch, Canada Department of Agriculture, Belleville, Ontario, Canada. NUTRITIONAL REQUIREMENTS (INVERTEBRATES).

T. C. HSU, Section of Cytology, The University of Texas M. D. Anderson Hospital and Tumor Institute, Houston, Texas. CHROMOSOMES.

CLYDE G. HUGGINS, Department of Biochemistry, Tulane University School of Medicine, New Orleans, Louisiana, 70112. VASOPRESSIN.

ROBERT B. HURLBERT, University of Texas, M. D. Anderson Hospital and Tumor Institute, Houston, Texas, 77025. PYRIMIDINE BIOSYNTHESIS.

ROSS B. INMAN, Departments of Biophysics and Biochemistry, University of Wisconsin, Madison, Wisconsin. POLYNUCLEOTIDES, SYNTHETIC.

A. F. ISBELL, Department of Chemistry, Texas A&M University, College Station, Texas. AMINO PHOSPHONIC ACIDS.

KURT J. ISSELBACHER, Harvard Medical School, Massachusetts General Hospital, Boston, Massachusetts. GALACTOSE AND GALACTOSEMIA.

L. JAENICKE, Institut für Biochemie der Universität Köln, Köln, Germany. FOLIC ACID COENZYMES.

ANDRE T. JAGENDORF, Division of Biological Sciences, Cornell University, Ithaca, New York, 14850. PHOSPHORYLATION, PHOTOSYNTHETIC.

W. T. JENKINS, Department of Chemistry, Indiana University, Bloomington, Indiana, 47405. TRANSAMINATION ENZYMES.

ELWOOD V. JENSEN, The Ben May Laboratory for Cancer Research, The University of Chicago, Chicago 37, Illinois. OVARIAN HORMONES.

BRUNO JIRGENSONS, The University of Texas M. D. Anderson Hospital and Tumor Institute, Houston, Texas. BENCE-JONES PROTEINS.

C. M. JOHNSON, Department of Soils and Plant Nutrition, University of California, Berkeley, California, 94720. NUTRITIONAL REQUIREMENTS (PLANTS).

MARY ELLEN JONES, Graduate Department of Biochemistry, Brandeis University, Waltham, Massachusetts. CARBAMYL PHOSPHATE.

JEREMIAS H. R. KÄGI, Biophysics Research Laboratory, Peter Bent Brigham Hospital,

Harvard Medical School, Boston, Massachusetts. ZINC.

HERMAN M. KALCKAR, Harvard Medical School, Massachusetts General Hospital, Boston, Massachusetts. GALACTOSE AND GALACTOSEMIA.

JULIAN N. KANFER, Section on Lipid Chemistry, National Institute of Neurological Diseases and Blindness, Bethesda, Maryland, 20014. BRAIN COMPOSITION.

FREDERICK H. KASTEN, Department of Ultrastructural Cytochemistry, Pasadena Foundation for Medical Research, Pasadena, California, 91101. HISTOCHEMICAL METHODS.

WILBUR KAYE, Scientific Instruments Operation, Beckman Instruments, Inc., Fullerton, California, 92634. INFRARED AND ULTRAVIOLET SPECTROSCOPY.

LOUIS A. KAZAL, Cardeza Foundation for Hematologic Research, Jefferson Medical College, Philadelphia, Pennsylvania, 19107. ION EXCHANGE POLYMERS.

JAMES E. KINDRED, Department of Anatomy, University of Virginia School of Medicine, Charlottesville, Virginia. BLOOD (CELLULAR COMPOSITION); HEMOPOIESIS.

C. G. KING, Institute of Nutrition Sciences, School of Public Health and Administrative Medicine, Columbia University, New York, N. Y., 10032. ASCORBIC ACID.

CHARLES R. KLEEMAN, Divisions of Medicine, Cedars-Sinai Medical Center, University of California Medical School, Los Angeles, California, 90048. ELECTROLYTE AND WATER REGULATION.

ISRAEL S. KLEINER, deceased. Formerly Department of Biochemistry, New York Medical College, Flower and Fifth Avenue Hospitals, New York, N.Y. METABOLIC DISEASES.

STEPHEN KROP, Division of Pharmacology, Bureau of Science, Food and Drug Administration, U.S. Department of Health, Education, and Welfare, Washington, D.C. 20204. TOXICOLOGY.

ABEL LAJTHA, New York State Research Institute for Neurochemistry, Ward's Island, New York, N.Y., 10035. BRAIN METABOLISM.

KOLOMAN LAKI, Laboratory of Biophysical Chemistry, National Institute of Arthritis and Metabolic Diseases, Bethesda, Maryland, 20014. BLOOD CLOTTING.

WILLIAM E. M. LANDS, Department of Biological Chemistry, The University of Michigan Medical School, Ann Arbor, Michigan. PHOSPHATIDES AND PHOSPHATIDIC ACIDS.

EDWIN M. LANSFORD, JR., Clayton Foundation Biochemical Institute, The University of Texas, Austin, Texas, 78712. AMINO ACIDS, PROTEIN-DERIVED; COENZYMES; ELECTRON TRANSPORT SYSTEMS; GENETIC CODE.

B. L. LARSON, Department of Dairy Science, University of Illinois, Urbana, Illinois. MILK (COMPOSITION AND BIOSYNTHESIS).

HANS LAUFER, Department of Zoology and Entomology, University of Connecticut, Storrs, Connecticut, 06268. BALBIANI RINGS.

A. J. LEHMAN, Division of Pharmacology, Bureau of Science, Food and Drug Administration, U.S. Department of Health, Education, and Welfare, Washington, D.C., 20204. TOXICOLOGY.

I. R. LEHMAN, Department of Biochemistry, Stanford University School of Medicine, Palo Alto, California, 94304. NUCLEASES.

AARON B. LERNER, Section of Dermatology, Yale University School of Medicine, New Haven, Connecticut. SKIN PIGMENTATION.

RACHMIEL LEVINE, Department of Medicine, New York Medical College, Flower and Fifth Avenue Hospitals, New York, N.Y. INSULIN (FUNCTION).

CHOH HAO LI, Hormone Research Laboratory, University of California, Berkeley, California, 94720. GROWTH HORMONE.

TING-KAI LI, Department of Medicine, Harvard Medical School, Peter Bent Brigham Hospital, Boston, Massachusetts, 02115. ALCOHOL DEHYDROGENASE.

CHIUN T. LING, Department of Physiological Chemistry, Johns Hopkins University School of Medicine, Baltimore, Maryland, 21205. CARCINOMA.

FRITZ LIPMANN, The Rockefeller University, New York, N.Y. 10021. COENZYME A.

ROYCE Z. LOCKART, JR., Central Research Department, E. I. Du Pont de Nemours and Co., Wilmington, Delaware. INTERFERONS.

JOHN M. LOWENSTEIN, Graduate Department of Biochemistry, Brandeis University, Waltham, Massachusetts, 02154. CITRIC ACID CYCLE.

TOM J. MABRY, Department of Botany, University of Texas, Austin, Texas. CHEMOTAXONOMY.

CATHERINE F. C. MACPHERSON, Allan Memorial Institute of Psychiatry, and Department of Biochemistry, McGill University, Montreal 2, Quebec, Canada. CEREBROSPINAL FLUID.

H. GEORGE MANDEL, Department of Pharmacology, George Washington University School of Medicine, Washington, D.C., 20005. ANTIBIOTICS.

E. MARGOLIASH, Abbott Laboratories, Scientific Divisions, North Chicago, Illinois, 60064. CYTOCHROMES.

AMEDEO S. MARRAZZI, Department of Pharmacology, University of Minnesota Medical School, Minneapolis, Minnesota, 55455. NERVE IMPULSE CONDUCTION.

A. E. MARTELL, Department of Chemistry, Illinois Institute of Technology, Chicago, Illinois, 60616. CHELATION AND METAL CHELATE COMPOUNDS.

WALTER MARX, Department of Biochemistry, University of Southern California School of Medicine, Los Angeles, California. HEPARIN.

HAROLD L. MASON, Lovelace Foundation, Albuquerque, New Mexico, 87108. GLUTATHIONE.

HOWARD S. MASON, Department of Biochemistry, University of Oregon Medical School, Portland, Oregon, 97201. BIOLOGICAL OXIDATION-REDUCTION.

H. S. MAYERSON, Department of Physiology, Tulane University School of Medicine, New Orleans, Louisiana. LYMPH.

HUGH J. MCDONALD, Department of Biochemistry and Biophysics, Loyola University, Stritch School of Medicine, Chicago, Illinois, 60612. ELECTROPHORESIS.

A. D. MCLAREN, College of Agriculture, University of California, Berkeley, California. ULTRAVIOLET RADIATION.

J. M. MCLAUGHLAN, Nutrition Division, Food and Drug Research Laboratories, Department of National Health and Welfare, Ottawa 3, Ontario, Canada. PROTEINS (NUTRITIONAL QUALITY).

WALTER MERTZ, Department of Biological Chemistry, Walter Reed Army Institute of Research, Washington, D.C., 20012. CHROMIUM.

DAVID E. METZLER, Department of Biochemistry and Biophysics, Iowa State University of Science and Technology, Ames, Iowa, 50010. PYRIDOXINE FAMILY OF VITAMINS.

ADE T. MILHORAT, Institute for Muscle Disease, Inc., New York, N.Y. 10021. MUSCULAR DYSTROPHY.

ERSTON V. MILLER, Department of Biology, University of Pittsburgh, Pittsburgh, Pennsylvania, 15213. CARBOHYDRATES (ACCUMULATION IN FRUITS AND SEEDS).

GORDON C. MILLS, Department of Biochemistry, The University of Texas Medical Branch, Galveston, Texas. PEROXIDASES.

URSULA MITTWOCH, The Galton Laboratory, University College London, London, England. CHROMATIN.

RICHARD D. MOCHRIE, School of Agriculture and Life Sciences, North Carolina State University, Raleigh, North Carolina, 27607. RUMINANT NUTRITION AND METABOLISM.

KIVIE MOLDAVE, Department of Biochemistry, University of Pittsburgh School of Medicine, Pittsburgh, Pennsylvania, 15213. TRANSFER FACTORS IN PROTEIN SYNTHESIS.

WILFRIED F. H. M. MOMMAERTS, Department of Physiology, and Cardiovascular Research Laboratory, University of California, Los Angeles, California, 90024. CONTRACTILE PROTEINS.

L. O. MORGAN, Department of Chemistry, University of Texas, Austin, Texas. CARBON (RADIOCARBON) DATING.

L. E. MORTENSON, Department of Biological Sciences, Purdue University, Lafayette, Indiana, 47907. NITROGEN FIXATION.

JACK MYERS, Laboratory of Algal Physiology, Department of Zoology, University of Texas, Austin, Texas. ALGAE.

AMOS NEIDLE, New York State Psychiatric Institute, Department of Pharmacology, New York, N.Y., 10032. HISTONES.

J. B. NEILANDS, Department of Biochemistry, University of California, Berkeley, California, 94720. ENZYMES (KINETICS).

GEORGE NÉMETHY, The Rockefeller University, New York, N.Y., 10021. PROTEINS (BINDING FORCES IN SECONDARY AND TERTIARY STRUCTURE).

MUTSUKO NISHIHARA, Department of Microbiology, Tufts University School of Medicine, Boston, Massachusetts, 02111. MICROBIOLOGY.

SUSAN M. OACE, Department of Nutritional Sciences, University of California, Berkeley, California, 94720. VITAMIN B GROUP.

JOHN S. O'BRIEN, Department of Pathology, Division of Chemical Pathology, University of Southern California School of Medicine, Los Angeles, California, 90033. MYELIN.

BOYD L. O'DELL, Department of Agricultural Chemistry, University of Missouri, Columbia, Missouri, 65202. MAGNESIUM.

JAMES OFENGAND, Department of Biochemistry, University of California School of Medicine, San Francisco, California, 94122. ACTIVATING ENZYMES.

JOHN R. OLIVE, American Institute of Biological Sciences, Washington, D.C., 20016. AMERICAN INSTITUTE OF BIOLOGICAL SCIENCES.

W. H. ORME-JOHNSON, Institute for Enzyme Research, University of Wisconsin, Madison, Wisconsin, 53706. OXIDATIVE DEMETHYLATION.

ALINE U. ORTEN, Wayne State University College of Medicine, Detroit, Michigan, 48207. ABSORPTION.

IRVINE H. PAGE, Research Division, The Cleveland Clinic Foundation, Cleveland, Ohio. HYPERTENSION.

T. S. PAINTER, JR., 2706 Rio Grande St., Austin, Texas. ALLERGIES.

GEORGE E. PALADE, The Rockefeller University, New York, N.Y., 10021. ELECTRON MICROSCOPY.

A. M. PAPPENHEIMER, JR., Biological Laboratories, Harvard University, Cambridge, Massachusetts, 02138. TOXINS, ANIMAL AND BACTERIAL.

S. W. PELLETIER, Deartment of Chemistry, University of Georgia, Athens, Georgia. ALKALOIDS.

JOHN G. PIERCE, Department of Biological Chemistry, University of California School of Medicine, Los Angeles, California, 90024. COUNTERCURRENT DISTRIBUTION.

G. W. E. PLAUT, Department of Biochemistry, Rutgers, The State University, Medical School, New Brunswick, New Jersey, 08903. RIBOFLAVIN.

HANS POPPER, The Mount Sinai Hospital, New York, N.Y., 10029. LIVER FUNCTIONS.

OSCAR W. PORTMAN, Oregon Regional Primate Research Center, Beaverton, Oregon, 97006. INTESTINAL FLORA.

VAN R. POTTER, McArdle Laboratory for Cancer Research, Medical Center, University of Wisconsin, Madison, Wisconsin, 53706. CANCER CELL METABOLISM.

MORTON D. PRAGER, Wadley Research Institute and Blood Bank, Dallas, Texas, 75246. LEUKOCYTES AND LEUKEMIA.

J. M. PRESCOTT, Department of Biochemistry and Biophysics, Texas A&M University, College Station, Texas, 77843. SNAKE VENOMS.

R. D. PRESTON, Astbury Department of Biophysics, University of Leeds, Leeds 2, England. CELLULOSE.

I. D. RAACKE, Department of Biology, Boston University, Boston, Massachusetts, 02215. NUCLEOPROTEINS.

JESSE C. RABINOWITZ, Department of Biochemistry, University of California, Berkeley, California, 94720. SINGLE CARBON UNIT METABOLISM.

JOSEPH L. RABINOWITZ, Radioioisotope Research, Veterans Administration Hospital, Philadelphia, Pennsylvania, 19104. ISOTOPE DILUTION METHODS.

NORMAN S. RADIN, Mental Health Research Institute, The University of Michigan, Ann Arbor, Michigan, 48104. CEREBROSIDES.

T. V. RAJKUMAR, Department of Biochemistry, University of Washington, Seattle, Washington, 98105. ISOZYMES.

MAURICE M. RAPPORT, Department of Biochemistry, Albert Einstein College of Medicine, Yeshiva University, Bronx, New York, 10461. PLASMALOGENS.

LESTER J. REED, Clayton Foundation Biochemical Institute and Department of Chemistry, University of Texas, Austin, Texas, 78712. LIPOIC ACID; MULTIENZYME COMPLEXES.

JAMES F. REGER, Department of Anatomy, University of Tennessee Medical Units, Memphis, Tennessee, 38103. NERVE CELL COMPOSITION; SYNAPSES AND NEUROMUSCULAR JUNCTIONS.

RAYMOND REISER, Department of Biochemistry and Biophysics, Texas A&M University, College Station, Texas, 77843. PHOSPHOLIPIDS.

LLOYD H. REYERSON, 5 Huckleberry Lane, Ridgefield, Connecticut. HYDRATION OF MACROMOLECULES.

A. GLENN RICHARDS, Department of Entomology, Fisheries, and Wildlife, University of Minnesota, St. Paul, Minnesota, 55101. CHITIN.

DAN A. RICHERT, Department of Biochemistry, State University of New York, Upstate Medical Center, Syracuse, New York, 13210. MOLYBDENUM.

G. W. RICHTER, Department of Pathology, Cornell University Medical College, New York, N.Y., 10021. FERRITIN.

WILLIAM RIEMAN III, School of Chemistry, Rutgers, State University, New Brunswick, New Jersey, 08903. CHROMATOGRAPHY, COLUMN.

AUSTIN F. RIGGS, II, Department of Zoology, University of Texas, Austin, Texas. HEMOGLOBINS.

PHILLIPS W. ROBBINS, Division of Biochemistry, Department of Biology, Massachusetts Institute of Technology, Cambridge, Massachusetts, 02139. SULFATE AND ORGANIC SULFATES.

J. W. ROBINSON, Department of Chemistry, Louisiana State University, Baton Rouge, Louisiana, 70803. FLAME PHOTOMETRY.

WILLIAM G. ROBINSON, Department of Biochemistry, New York University School of Medicine, New York, N.Y. AMINO ACID METABOLISM.

L. J. RODE, JR., Department of Microbiology, University of Texas, Austin, Texas, 78712. SPORES, BACTERIAL.

LORENE L. ROGERS, Graduate School, University of Texas, Austin, Texas, 78712. MENTAL RETARDATION.

ANTHONY H. ROSE, Department of Microbiology, The University of Newcastle Upon Tyne, Newcastle Upon Tyne 1, England. ALCOHOLIC FERMENTATION.

IRWIN A. ROSE, The Institute for Cancer Research, Philadelphia, Pennsylvania, 19111. ENZYMES (ACTIVATION AND INHIBITION).

SAUL ROSEMAN, The Rackham Arthritis Research Unit, The University of Michigan, University Hospital, Ann Arbor, Michigan. HEXOSAMINES AND HIGHER AMINO SUGARS.

DOROTHEA RUDNICK, Department of Biology, Albertus Magnus College, and Yale University, New Haven, Connecticut. ORGANIZER, IN CELLULAR DIFFERENTIATION.

WILLIAM J. RUTTER, Department of Biochemistry, University of Washington, Seattle, Washington, 98105. ISOZYMES.

B. V. RAMA SASTRY, Department of Pharmacology, Vanderbilt University Medical School, Nashville, Tennessee. CYANIDE POISONING.

GERHARD SCHMIDT, The Boston Dispensary, Boston, Massachusetts. PHOSPHATASES.

KNUT SCHMIDT-NIELSEN, Department of Zoology, Duke University, Durham, North Carolina, 27706. BASAL METABOLISM.

W. A. SCHROEDER, Division of Chemistry and Chemical Engineering, California Institute of Technology, Pasadena, California, 91109. PROTEINS (END GROUP AND SEQUENCE ANALYSIS).

JACK SCHUBERT, Graduate School of Public Health, University of Pittsburgh, Pittsburgh, Pennsylvania, 15213. BERYLLIUM AND BERYLLIOSIS.

V. T. SCHUHARDT, Department of Microbiology, University of Texas, Austin, Texas. LYSOSTAPHIN.

RICHARD EVANS SCHULTES, Botanical Museum of Harvard University, Cambridge, Massachusetts, 02138. TOXINS, PLANT.

IRWIN H. SEGEL, Department of Biochemistry and Biophysics, University of California, Davis, California, 95616. PHOSPHATE BOND ENERGIES.

EWALD E. SELKURT, Department of Physiology, Indiana University School of Medicine, Indianapolis, Indiana, 46207. RENAL TUBULAR FUNCTIONS AND URINE COMPOSITION; SWEAT.

ELLIOTT N. SHAW, Department of Biology, Brookhaven National Laboratory, Upton, New York, 11973. ENZYMES (ACTIVE SITES).

WILLIAM SHIVE, Clayton Foundation Biochemical Institute and Department of Chemistry, University of Texas, Austin, Texas. ANTIMETABOLITES.

N. W. SHOCK, Gerontology Branch, Baltimore City Hospital, Baltimore, Maryland, 21224. GERONTOLOGY.

BEN A. SHOULDERS, Department of Chemistry, University of Texas, Austin, Texas. NUCLEAR MAGNETIC RESONANCE.

PHILIP SIEKEVITZ, The Rockefeller University, New York, N.Y., 10021. RIBOSOMES.

DANIEL H. SIMMONS, Medical Research Institute, Cedars-Sinai Medical Center, Los Angeles, California, 90048. ACID-BASE REGULATION.

LEON SINGER, Department of Biochemistry, College of Medical Sciences, University of Minnesota, Minneapolis, Minnesota. FLUORIDE METABOLISM.

C. G. SKINNER, Department of Chemistry, North Texas State University, Denton, Texas. GERMINATION (PLANT SEED).

R. M. S. SMELLIE, Department of Biochemistry, University of Glasgow, Glasgow, W.2, Scotland. DEOXYRIBONUCLEIC ACIDS (REPLICATION).

HARRY SOBOTKA, deceased. Formerly The Mount Sinai Hospital, New York, N.Y. BILE ACIDS; GLYCOSIDES, STEROID; STEROIDS.

ERNEST SONDHEIMER, State University College of Forestry, Syracuse University, Syracuse, New York. FLAVONOIDS AND RELATED COMPOUNDS.

DARREL H. SPACKMAN, Department of Obstetrics and Gynecology, School of Medicine, University of Washington, Seattle, Washington, 98105. PROTEINS (AMINO ACID ANALYSIS AND CONTENT).

B. R. STANERSON, American Chemical Society, Washington, D.C., 20036. AMERICAN CHEMICAL SOCIETY.

E. L. R. STOKSTAD, Department of Nutritional Sciences, University of California, Berkeley, California, 94720. VITAMIN B GROUP.

ROGER L. STORCK, Department of Biology, Rice University, Houston, Texas, 77001. DEOXYRIBONUCLEIC ACIDS (STRUCTURE).

ELEANOR E. STORRS, Pesticide Research Laboratory, U.S. Public Health Service, Perrine, Florida, 33157. CHROMATOGRAPHY, GAS.

A. STRACHER, State University of New York College of Medicine, Downstate Medical Center, Brooklyn, New York. MYOSINS.

JACK L. STROMINGER, Department of Pharmacology and Toxicology, The University of Wisconsin, Madison, Wisconsin, 53706. BACTERIAL CELL WALLS.

H. ELDON SUTTON, Department of Zoology, University of Texas, Austin, Texas, 78712. HEMOGLOBINS IN HUMAN GENETICS.

PATRICIA SWAN, Department of Biochemistry, University of Minnesota, St. Paul, Minnesota. NICOTINIC ACID.

A. E. TAKEMORI, Department of Pharmacology, University of Minnesota College of Medical Sciences, Minneapolis, Minnesota, 55455. NARCOTIC DRUGS.

ROY V. TALMAGE, Department of Biology, Rice University, Houston, Texas, 77001. PARATHYROID HORMONES.

J. HERBERT TAYLOR, Institute of Molecular Biophysics, Florida State University, Tallahassee, Florida. NUCLEUS AND NUCLEOLUS.

RAYMOND L. TAYLOR, American Association for the Advancement of Science, Washington, D.C., 20005. AMERICAN ASSOCIATION FOR THE ADVANCEMENT OF SCIENCE.

T. T. TCHEN, Department of Chemistry, Wayne State University, Detroit, Michigan, 48202. STEROLS (BIOGENESIS AND METABOLISM).

KENNETH V. THIMANN, Department of Biology, University of California, Santa Cruz, California, 95060. HORMONES, PLANT.

OSCAR TOUSTER, Department of Molecular Biology, Vanderbilt University, Nashville, Tennessee, 37203. PENTOSES IN METABOLISM.

ROBERT D. TSCHIRGI, Office of the Chancellor, University of California, La Jolla, California, 92038. BLOOD-BRAIN BARRIER.

WALTER TULECKE, Boyce Thompson Institute for Plant Research, Inc., Yonkers, New York, 10701. TISSUE CULTURE, PLANT.

B. L. TURNER, Department of Botany, University of Texas, Austin, Texas. CHEMOTAXONOMY.

GEORGE K. TURNER, G. K. Turner Associates, Palo Alto, California, 94303. FLUORESCENCE AND FLUOROMETRY.

JOHN T. TYPPO, Food and Nutrition Section, School of Home Economics, University of Missouri, Columbia, Missouri. NUTRITIONAL REQUIREMENTS (MAMMALS).

W. W. UMBREIT, Department of Bacteriology, Rutgers, State University, New Brunswick, New Jersey, 08903. MANOMETRIC METHODS.

JOHN T. VAN BRUGGEN, Department of Biochemistry, University of Oregon Medical School, Portland, Oregon, 97201. ACTIVE TRANSPORT.

N. L. VANDEMARK, Department of Dairy Science, Ohio State University, Columbus, Ohio, 43210. REPRODUCTION, FERTILITY, AND STERILITY.

JOSEPH J. VITALE, Department of Nutrition, University of Wisconsin, Madison, Wisconsin, 53706. DIGESTION AND DIGESTIVE ENZYMES.

PETER H. VON HIPPEL, Department of Biochemistry, Dartmouth Medical School, Hanover, New Hampshire, 03775. COLLAGENS.

HEINRICH WAELSCH, deceased. Formerly New York State Psychiatric Institute, Department of Pharmacology, New York, N.Y. HISTONES.

ROBERT P. WAGNER, Department of Zoology, University of Texas, Austin, Texas, 78712. GENES.

GEORGE WALD, Department of Biology, Harvard University, Cambridge, Massachusetts, 02138. VISUAL PIGMENTS.

JAMES B. WALKER, Department of Biology, Rice University, Houston, Texas, 77001. ARGININE-UREA CYCLE.

DARRELL N. WARD, Department of Biochemistry, The University of Texas, M. D. Anderson Hospital and Tumor Institute, Houston, Texas. GONADOTROPIC HORMONES.

HARRY V. WARREN, Department of Geology, University of British Columbia, Vancouver, B.C. GEOBIOCHEMISTRY.

ALBIN H. WARTH, 29 York Court, Baltimore, Maryland, 21218. WAXES.

R. H. WASSERMAN, Department of Physical Biology, New York State Veterinary College, Cornell University, Ithaca, New York. PHOSPHATE (IN ANIMAL NUTRITION).

E. R. WAYGOOD, Department of Botany, The University of Manitoba, Winnipeg, Canada. RESPIRATION.

CHARITY WAYMOUTH, The Jackson Laboratory, Bar Harbor, Maine, 04609. TISSUE CULTURE, ANIMAL.

EDWIN C. WEBB, Department of Biochemistry, The University of Queensland, Brisbane, Australia. ENZYME CLASSIFICATION AND NOMENCLATURE.

LESLIE T. WEBSTER, JR., Department of Pharmacology, Western Reserve University School of Medicine, Cleveland, Ohio, 44106. AMMONIA METABOLISM.

T. ELLIOT WEIER, Department of Botany, University of California, Davis, California. CHLOROPLASTS.

HERBERT WEISSBACH, National Heart Institute, National Institutes of Health, Bethesda, Maryland, 20014. COBALAMINS AND COBAMIDE COENZYMES.

W. W. WESTERFELD, Department of Biochemistry, State University of New York, Upstate Medical Center, Syracuse, New York. ALCOHOL METABOLISM.

ROY L. WHISTLER, Department of Biochemistry, Purdue University, Lafayette, Indiana, 47907. STARCHES.

PHILIP R. WHITE, The Jackson Laboratory, Bar Harbor, Maine, 04609. NUTRIENT SOLUTIONS AND NUTRITION OF CELLS.

H. R. WHITELEY, Department of Microbiology, University of Washington, Seattle, Washington, 98105. FERMENTATIONS.

CHARLES G. WILBER, Marine Laboratories, University of Delaware, Newark, Delaware, 19711. CARBON DIOXIDE; CARBON MONOXIDE.

ROBERT R. WILLIAMS, deceased. Formerly with The Research Corporation, New York, N.Y. THIAMINE AND BERIBERI.

ROGER J. WILLIAMS, Clayton Foundation Biochemical Institute and Department of Chemistry, The University of Texas, Austin, Texas,

78712. ALCOHOLISM; APPETITES; BIOCHEMICAL INDIVIDUALITY; CHROMATOGRAPHY, PAPER; GENETOTROPHIC PRINCIPLE; HISTORY OF BIOCHEMISTRY; PANTHOTHENIC ACID; VITAMIN D GROUP.

WALTER L. WILSON, Department of Biology, Oakland University, Rochester, Michigan, 48063. CELL DIVISION.

MILTON WINITZ, Life Sciences Laboratory, United Technology Center, Sunnyvale, California, 94086. AMINO ACIDS.

H. G. WITTMANN, Max-Planck-Institut für Molekulare Genetik, Tübingen, Germany. TOBBACO MOSAIC VIRUS.

HARLAND G. WOOD, Department of Biochemistry, Western Reserve University School of Medicine, Cleveland, Ohio, 44106. BIOTIN.

MARK W. WOODS, Laboratory of Biochemistry, National Cancer Institute, Bethesda, Maryland, 20014. GLYCOLYSIS.

VAL W. WOODWARD, Department of Genetics, University of Minnesota, St. Paul, Minnesota. NEUROSPORA.

WALTER D. WOSILAIT, Department of Pharmacology, University of Missouri School of Medicine, Columbia, Missouri, 65201. NICOTINAMIDE ADENINE DINUCLEOTIDES.

BERNARD S. WOSTMANN, Lobund Laboratory, University of Notre Dame, Notre Dame, Indiana, 46556. GERM-FREE ANIMALS.

BARBARA WRIGHT, John Collins Warren Laboratories, Huntington Memorial Hospital of Harvard University, Boston, Massachusetts, 02114. DIFFERENTIATION AND MORPHOGENESIS.

LEMUEL D. WRIGHT, Graduate School of Nutrition, Cornell University, Ithaca, New York, 14850. ISOPRENOID BIOSYNTHESIS.

J. LYNDAL YORK, Department of Biochemistry, University of Tennessee Medical Units, Memphis, Tennessee. PORPHYRINS.

STEPHEN ZAMENHOF, Department of Medical Microbiology and Immunology, Department of Biological Chemistry and Molecular Biology Institute, University of California, Los Angeles, California, 90024. MUTAGENIC AGENTS.

E. A. ZELLER, Department of Biochemistry, Northwestern University Medical School, Chicago, Illinois, 60611. AMINE OXIDASES AND AMINO ACID OXIDASES.

PREFACE

This Encyclopedia of Biochemistry may provide the *first* word, but not necessarily the *last* word, about biochemical matters for readers who have some general scientific background but who may have had no previous information, or perhaps no previous interest, in some biochemical topic about which they find themselves in need of explanatory information. To non-biochemists, this book may usefully bring their first word about relatively broad topics in biochemistry; to chemists and biochemists, the Encyclopedia may be useful in certain more specialized areas outside their own customary fields of specialty. The relatively broadly defined articles of this book presuppose no more than a general education of their readers. The more narrowly defined topics, especially those articles written by currently active research specialists, tend to be more technical and detailed even though these also, by intention, and also because of the brevity of space available, serve partly as introductions to a much larger body of literature. Nearly all the articles are followed by a short list of references, including in most cases general references and in some cases references to selected current research reports.

This Encyclopedia is not a dictionary or a handbook of detailed data. Readers in need of detailed information about particular compounds and substances and their physical properties are referred to such excellent general sources as the *Handbook of Chemistry and Physics*, Chemical Rubber Publishing Company; *Lange's Handbook*, Handbook Publishers, Inc.; and, especially for biologically and pharmacologically significant materials, *The Merck Index* (Seventh Edition), Merck and Company, Inc.

Readers will notice that many terms in this Encyclopedia have been printed in SMALL CAPITALS. This is our device for indicating a cross-reference to another article or topic within this volume. Our policy has been to introduce a suggestion for cross-reference only when we thought such additional reading would add something to the reader's understanding of the topic at hand and the point being made about it—something that would usefully and helpfully extend the discussion from which the cross-reference originates. We have not been slavish to the precise form of speech of the cross-referenced title in every case; for example, a mention of "MITOCHONDRIAL structure and function" may be exploited to indicate cross-reference to the article whose precise title is MITOCHONDRIA.

The Editors suggest that this Encyclopedia can be useful as an aid to a planned study of biochemistry by readers whose lack of previous acquaintance with biochemistry might leave them feeling lost in a bewildering maze of alphabetical entries. For the possible help of those who might use this book for self-study, we present a short outline that we believe relates this book to certain fundamental questions to which the science of biochemistry has found answers (or is still attempting to find them). For many years, one of us has, in his lectures to students, expressed the broadest aims of biochemistry as attempts to answer these central questions: (1) What are living organisms made of, chemically? (2) What do living cells require from their environment? (3) How do organisms build and maintain themselves from the nutrients they take in? Included in this third question is a fourth broad one, applicable to higher, multicellular organisms; it has to do with the intercellular relationships and dependencies characteristic of such complex communities of cells, tissues and organs. Certain articles in this Encyclopedia are directly appropriate to these central questions, as indicated in the following outline:

I. **What are the chemical constituents of living organisms?** (A question of chemical *analysis.*) The principal organic and macromolecular constituents are discussed in the following:

A. CARBOHYDRATES (CLASSIFICATION AND STRUCTURAL INTERRELATIONS).

B. LIPIDS (CLASSIFICATION AND DISTRIBUTION).

C. PROTEINS, and their subunits, AMINO ACIDS (OCCURRENCE, STRUCTURE AND SYNTHESIS).

D. Nucleic acids, discussed in two subdivisions, DEOXYRIBONUCLEIC ACIDS (DISTRIBUTION AND STRUCTURE), and RIBONUCLEIC ACIDS (STRUCTURE AND FUNCTIONS).

E. Cellular inorganic constituents and their roles are treated in MINERAL REQUIREMENTS, as well as in separate articles for many individual elements, for example CALCIUM (IN BIOLOGICAL SYSTEMS); see also CHELATION.

II. **What substances do living cells need from their environment?** (The central question of the field of *nutrition.*)

A. At the level of intact organisms, nutritional needs are discussed in the articles AMINO ACIDS, NUTRITIONALLY ESSENTIAL; FATTY ACIDS, ESSENTIAL; MINERAL REQUIREMENTS; in the several articles on NUTRITIONAL REQUIREMENTS, subdivided into the areas of plants, invertebrate animals, vertebrates below mammals, and mammals; and in MALNUTRITION. Vitamins are discussed in separate articles for each vitamin and in VITAMIN B GROUP; for cross-references, see COENZYMES and VITAMINS (GENERAL).

B. At the cellular level, nutritional problems are introduced in the article NUTRIENT SOLUTIONS AND NUTRITION OF CELLS.

III. **How do cells build and maintain themselves from their nutrients?** (A question including the area traditionally termed *intermediary metabolism*).

A. The cellular catalysts, the *enzymes*, are centrally discussed in ENZYME CLASSIFICATION AND NOMENCLATURE.

B. Pathways for consumption of the principal energy-yielding cellular fuels are discussed in CARBOHYDRATE METABOLISM and in additional detail in terms of its subdivisions, GLYCOLYSIS and CITRIC ACID CYCLE. Energy production by cells is generally considered in RESPIRATIONS and in OXIDATIVE METABOLISM AND ENERGETICS OF BACTERIA. Energy storage in the form of "high-energy" compounds is discussed in PHOSPHATE BOND ENERGIES. More detailed aspects of the mechanism of cellular energy production are discussed in many related articles including BIOLOGICAL OXIDATION-REDUCTION; PHOSPHORYLATION, OXIDATIVE; MITOCHONDRIA.

C. Utilization and interconversion of protein constituents are treated in AMINO ACID METABOLISM and in AROMATIC AMINO ACIDS (BIOSYNTHESIS AND METABOLISM).

D. Other major areas of metabolic interconversion are treated in the articles FATTY ACID METABOLISM (OXIDATION AND BIOSYNTHESIS) and those concerning the nucleic acids and their constituents, including DEOXYRIBONUCLEIC ACIDS (REPLICATION); RIBONUCLEIC ACIDS (BIOSYNTHESIS); PURINE BIOSYNTHESIS; PYRIMIDINE BIOSYNTHESIS.

E. How are organisms faithfully replicated in subsequent generations? Several articles contribute to an account of the burgeoning area of "molecular biology" devoted to this question, including BIOCHEMICAL GENETICS; GENES; PROTEINS (BIOSYNTHESIS);

and BACTERIAL GENETICS; as well as the articles on nucleic acid biosynthesis mentioned in III.D.

IV. **What, in chemical terms, are the special problems and their natural solutions in multicellular, complex higher organisms?** Some major aspects include the following:

A. How do the intercellular distinctions arise in the development of the organisms? See DIFFERENTIATION AND MORPHOGENESIS, and METAMORPHOSIS.

B. How are these distinctive cellular forms and functions among the tissues and organs of higher animals and plants regulated and kept in states of harmonious adjustment? Biological controls and their mechanisms are active areas of current scientific study. A central discussion is given in METABOLIC CONTROLS (FEEDBACK MECHANISMS). Failure of normal biological controls may be implicated in CANCER METABOLISM. The special "chemical messengers" (humoral agents, hormones) that contribute to intercellular and inter-organ regulations are centrally considered in HORMONES, ANIMAL, and HORMONES, PLANT.

C. The broadest type of interdependence among living organisms is possibly that between plants and animals; the photosynthetic role of plants as captors of sunlight energy (to the subsequent benefit of animals) and as liberators of oxygen (to be consumed as oxidant by animals) is discussed in PHOTOSYNTHESIS; CARBON REDUCTION CYCLE (IN PHOTOSYNTHESIS); LIGHT PHASE IN PHOTOSYNTHESIS; and PHOSPHORYLATION, PHOTOSYNTHETIC.

This outline is by no means intended to depreciate the importance and interest of the many other topics ably contributed to this Encyclopedia, but rather to suggest that the articles named in this outline are some of those *most directly* related to the broad biochemical questions listed here and are articles from which the reader can conveniently radiate, with the aid of cross-references and his own intuition, into other more specialized areas. This outline does no justice, obviously, to the numerous sections treating various biochemical *methodologies*, for example, to be found in this Encyclopedia. Those readers interested primarily in relatively detailed and technically specialized articles not outlined here will in general be readers who need no outline or guide.

Readers versed in enzymology will undoubtedly realize that the articles on enzymological topics to be found in this volume constitute merely an arbitrary selection, for which the Editors must be responsible, from a vastly larger number of possibilities, among a catalog of known enzymes numbering close to a thousand—a selection that may acquaint the reader with some general aspects of enzymes, as well as with details of some few representative individual enzymes.

Our gratitude is expressed to the many contributors of signed articles in this volume, whose helpfulness was impressive, extensive and essential. We gratefully acknowledge also much capable stenographic aid, and assistance in handling of manuscripts and in preparation of drawings, by Mrs. Frances Sanders, Miss Sandra Panzarella, and Mr. James Konvicka, and the very capable work of the Reinhold production staff.

Austin, Texas
September, 1966

ROGER J. WILLIAMS
EDWIN M. LANSFORD, JR.

A

ABBREVIATIONS

Abbreviations have come to be used widely in biochemistry. Sometimes, used out of context, they may be puzzling. In the following list are some of the more common abbreviations and what they stand for. (Synonymous meanings, after a given abbreviation, are separated by a comma; independent, distinct meanings of the same abbreviation are separated by a semicolon.)

ACTH	ADRENOCORTICOTROPHIC HORMONE
AcCoA, AcSCoA	acetyl coenzyme A
$[\alpha]_D$	specific rotation (light source, sodium D line)
ADH	ALCOHOL DEHYDROGENASE
ADP	adenosine diphosphate
AICA, AICAR	aminoimidazolecarboxamide (ribotide)
Ala	ALANINE, alanyl
AMP	adenosine monophosphate
Arg	ARGININE, arginyl
Asp	ASPARTIC ACID, aspartyl
Asp-NH_2, Asg, Asn	ASPARAGINE, asparaginyl
ATP	adenosine triphosphate
ATPase	adenosine triphosphatase
CCD	COUNTERCURRENT DISTRIBUTION
CDP	cytidine diphosphate
CDPC	cytidine diphosphate choline
CF	citrovorum factor (see FOLIC ACID COENZYMES)
CMP	cytidine monophosphate
CoA	COENZYME A
CoI	coenzyme I (NICOTINAMIDE ADENINE DINUCLEOTIDE)
CoII	coenzyme II (nicotinamide adenine dinucleotide phosphate)
CTP	cytidine triphosphate
Cys, CySH	CYSTEINE, cysteinyl
$CySO_3H$	cysteic acid, cysteyl
d-	dextrorotatory
D-	related in absolute configuration to a standard reference compound, D-glyceraldehyde (see CARBOHYDRATES)
2,4-D	2,4-dichlorophenoxyacetic acid (see HORMONES, PLANT)
dADP, dAMP, dATP	deoxyadenosine di-, mono-, triphosphate (respectively)
dCDP, dCMP, dCTP	deoxycytidine di, mono-, triphosphate (respectively)
dGDP, dGMP, dGTP	deoxyguanosine di-, mono-, triphosphate (respectively)
dTDP, dTMP, dTTP	thymidine di-, mono-, triphosphate (respectively)
dUMP	deoxyuridine monophosphate
DDT	dichlorodiphenyltrichloroethane
DFP	diisopropylfluorophosphate
DNA	DEOXYRIBONUCLEIC ACIDS
DNase	deoxyribonuclease (see NUCLEASES)
DNP	dinitrophenol; dinitrophenyl (see PROTEINS, END GROUP AND SEQUENCE ANALYSIS)
DOPA	dioxy- or DIHYDROXYPHENYLALANINE
DPN, DPN^+	diphosphopyridine nucleotide (alternative name for CoI or NICOTINAMIDE ADENINE DINUCLEOTIDE)
DPNH	diphosphopyridine nucleotide (reduced form)
EDTA	ethylenediaminetetraacetic acid
EMF	electromotive force; erythrocyte maturation factor (VITAMIN B_{12})
ES	enzyme-substrate complex
ESR	electron spin resonance (see FREE RADICALS)
ETP	electron transport particles
FA	FOLIC ACID
FAD	flavin adenine dinucleotide, oxidized form (see FLAVINS)
$FADH_2$	flavin adenine dinucleotide, reduced form (see FLAVINS)
FFA	free fatty acids (also NEFA)
FGAR	formylglycinamide ribotide (see PURINE BIOSYNTHESIS)
FH_4	tetrahydrofolic acid (see FOLIC ACID)
f^5-FH_4	N^5-formyltetrahydrofolic acid, CF, folinic acid, leucovorin (see FOLIC ACID; FOLIC ACID COENZYMES)
$f^{5,10}$-FH_4	$N^{5,10}$-methenyltetrahydrofolic acid

f^{10}-FH_4	N^{10}-formyltetrahydrofolic acid
FIGLU	formiminoglutamic acid
FMN	flavin mononucleotide (see FLAVINS)
FSH	follicle-stimulating hormone (see GONADOTROPIC HORMONES)
γ	"gamma" = 1 μg (1 microgram)
GA	glyceric acid; GLUTAMIC ACID; etc.
GAR	glycinamide ribotide (see PURINE BIOSYNTHESIS)
GDP	guanosine diphosphate
GDPM	guanosine diphosphate mannose
GH	GROWTH HORMONE, somatotrophin, STH
Glu	GLUTAMIC ACID, glutamyl
Glu-NH_2, Glm, Gln	GLUTAMINE, glutaminyl
Gly	GLYCINE, glycyl
GMP	guanosine monophosphate
GSH	GLUTATHIONE, reduced form
GSSG	GLUTATHIONE, oxidized form
GTP	guanosine triphosphate
HCG	human chorionic GONADOTROPIN
His	HISTIDINE, histidinyl
HMG	hydroxymethylglutaric acid (see ISOPRENOID BIOSYNTHESIS)
HMP	hexose monophosphate (see CARBOHYDRATE METABOLISM)
Hylys	hydroxylysine
Hypro	hydroxyproline
IAA	indoleacetic acid (see HORMONES, PLANT)
ICSH	interstitial cell stimulating hormone, LH (see GONADOTROPIC HORMONES)
IDP	inosine diphosphate
Ileu	ISOLEUCINE, isoleucyl
IMP	inosine monophosphate (see PURINE BIOSYNTHESIS)
INH	isonicotinic acid hydrazide
IR	infrared
ITP	inosine triphosphate
l-	levorotatory
L-	related in absolute configuration to a standard reference compound, L-glyceraldehyde
LBF	*Lactobacillus bulgaricus* factor (PANTETHEINE)
LDH	lactate dehydrogenase
Leu	LEUCINE, leucyl
LH	luteinizing hormone, ICSH (see GONADOTROPIC HORMONES)
LTH	luteotrophic hormone (see GONADOTROPIC HORMONES)
Lys	LYSINE, lysyl
Met	METHIONINE, methionyl
mμ	millimicron
mRNA	MESSENGER RIBONUCLEIC ACID
MSH	MELANOCYTE STIMULATING HORMONE
MW	molecular weight
μg	microgram (10^{-6} gram)
NAD, NAD^+	NICOTINAMIDE ADENINE DINUCLEOTIDE
NADH	nicotinamide adenine dinucleotide, reduced form
NADP, $NADP^+$	nicotinamide adenine dinucleotide phosphate
NADPH	nicotinamide adenine dinucleotide phosphate, reduced form
NEFA	non-esterified fatty acids, FFA
NMN	nicotinamide mononucleotide
NMR	nuclear magnetic resonance
NPN	non-protein nitrogen
ODP, OMP, OTP	orotidine di-, mono-, triphosphate (see PYRIMIDINE BIOSYNTHESIS)
~P	energy-rich phosphate (see PHOSPHATE BOND ENERGIES)
PAB, PABA	*para*-aminobenzoic acid (see FOLIC ACID; SULFONAMIDES)
PBI	protein-bound iodine (see IODINE)
PGA	phosphoglyceric acid; pteroylglutamic acid (FOLIC ACID)
Phe	PHENYLALANINE, phenylalanyl
P_i	inorganic phosphate, orthophosphate
PNA	pentose nucleic acid (RIBONUCLEIC ACID)
P:O	moles ATP formed per atom oxygen consumed
PP, PP_i	inorganic pyrophosphate
PRA	phosphoribosylamine (see PURINE BIOSYNTHESIS)
Pro	PROLINE, prolyl
PRPP	5-phosphoribosyl-1-pyrophosphate
PTH	PARATHYROID HORMONE
Q_{10}	enzyme activity in relation to temperature
Q_7, Q_8, Q_9, etc.	forms of COENZYME Q
Q_{275}	form of COENZYME Q
R-1,5-DP	ribulose-1,5-diphosphate
R-5-P	ribose-5-phosphate
RNA	RIBONUCLEIC ACIDS
RNase	ribonuclease (see NUCLEASES)
RQ	respiratory quotient, ratio of CO_2 produced to O_2 utilized
rRNA	ribosomal ribonucleic acids
Ser	SERINE, seryl
SPCA	serum prothrombin conversion accelerator (see BLOOD CLOTTING)
STH	somatotrophic hormone, somatotrophin, GROWTH HORMONE
TCA	trichloracetic acid; tricarboxylic acid
TEPP	tetraethyl pyrophosphate
Thr	THREONINE, threonyl

TMV	TOBACCO MOSAIC VIRUS
TPN, TPN^+	triphosphopyridine nucleotide (alternative name for NADP, $NADP^+$)
TPNH	triphosphopyridine nucleotide (alternative name for NADPH)
TPP	thiamine pyrophosphate
Tris	tris(hydroxymethyl)aminomethane
Try, Trp	TRYPTOPHAN, tryptophanyl
TSH	thyroid stimulating hormone, THYROTROPHIC HORMONE
Tyr	TYROSINE, tyrosinyl
UDP	uridine diphosphate
UDPG	uridine diphosphate glucose
UMP	uridine monophosphate
USP	United States Pharmacopoeia
UTP	uridine triphosphate
UV	ultraviolet
Val	VALINE, valyl
VPC	vapor phase chromatography (see CHROMATOGRAPHY, GAS)
XO	XANTHINE OXIDASE

A number of these abbreviations are included in a list given at the beginning of each issue of the *Journal of Biological Chemistry* and are thereby recognized as currently acceptable for use without definition in articles submitted to the *Journal*. For discussion of the nomenclature of nicotinamide coenzymes (NAD, NADP, etc.), see DIXON, M., *Science*, **132**, 1548 (1960).

ABSORPTION (OF SUBSTANCES INTO CELLS)

Absorption may be regarded as the passage of molecules from an external environment, as from the lumen of the intestine or from interstitial fluid, through the cell membrane into the interior of cells. The process is vital to the transfer of the products of DIGESTION from the intestine into the blood plasma and to the uptake of nutrients and metabolites from body fluids by all cells. Intestinal absorption has been aptly termed the "first common pathway" (Hogben), since all food nutrients in the digestive tract are topologically still outside the body. They must traverse the barrier of the intestinal absorptive cell membrane before it can be said that they are actually "inside" the body and available for metabolic purposes.

The cell membrane plays a vital role in selectively screening, admitting or rejecting substances for absorption, as well as serving as a retaining wall for the essential cell constituents. It is a boundary which confines and maintains an internal aqueous medium differing markedly in composition from that of the surrounding external environment; it also presents a variety of barriers to absorption. To accomplish these functions, the membrane is adapted in its physicochemical structure and is provided with numerous biochemical absorptive mechanisms.

ELECTRON MICROSCOPY has revealed numerous structural cellular details formerly not visible. The membrane is now looked upon as a complex organized mosaic comprised of protein and lipid which form a sandwich-like structure about 80 Å thick. It appears to be a monolayer of protein molecules on both sides with an inner bimolecular layered lipid core. At intervals, the membrane becomes discontinuous, with *polar pores* extending through it. The pore consists of a succession of polar sites on polypeptide chains extending through the bimolecular layer of lipid. This structure permits relatively free passage of some solvents and of hydrated ions less than 8 Å in diameter; *e.g.*, K^+ and Cl^- are transferred freely but Na^+, $SO_4^=$, and Ca^{++} are not. Other areas of the membrane seem to be solvent-like surfaces which present a variety of barriers to the absorption of larger ions and molecules. Thus, in order to be absorbed, molecules must penetrate the membrane either by pinocytosis or by some type of diffusion.

Pinocytosis is a process in which regions of the cell membrane invaginate, surround droplets of, or particles from, the surrounding medium, and "eat or drink" them into the cell interior. In this way the cell membrane participates in the absorption of substances into cells with an active turnover of the membrane itself. It has been shown that the absorption of dietary fat from the intestine can take place in this way. Presently, pinocytosis is the only known mechanism capable of translocating large protein molecules across cell membranes, particularly those of the absorbing epithelial cells of the intestinal mucosa.

Simple diffusion ("passive") through a membrane is rather a slow procedure except for smaller ions and molecules. However, specific biological mechanisms are operative which not only enhance the rate of absorption of larger molecules such as various nutrients, but also concentrate them within the cell. Two interrelated accelerating mechanisms appear to be involved, "carriers" and a "pump."

According to the concepts of Danielli (1954), there are three sites of resistance to the free diffusion of solutes through cell membranes: (1) the water-membrane interface for diffusion into the membrane, (2) the lipid interior of the membrane, and (3) the membrane-water interface for diffusion from the membrane into the aqueous interior of the cell. A molecule that has polar groups, *e.g.*, hydroxyl, which are water soluble, forms at least one hydrogen bond with water for each polar group. Such hydrogen bonds must be fractured by an energy-requiring process if the molecule being absorbed is to penetrate the water-membrane barriers. Thus, molecules having a large number of polar groups, *e.g.*, glucose, require a relatively large amount of kinetic energy to effect penetration. Conversely, substances with few polar groups, *e.g.*, ethanol, penetrate easily. Adenosine triphosphate (ATP) serves as the source of energy. The more rapid absorption of amino acids with longer nonpolar groups, isoleucine, leucine, and methionine, from mixtures in the human intestine, may in part be explained on the basis of their greater solubility in the lipid core of the membrane. Amino acids of a more polar nature, threonine, serine, and glycine, are more slowly absorbed from such mixtures.

Penetration of the lipid layer of the cell membrane by water-soluble substances is accomplished through *carrier* mechanisms that render them temporarily lipid soluble. The precise chemical nature of carriers is uncertain. However, there are suggestions that weak association complexes between glucose and phospholipid, sodium and phosphatidic acid may be a part of their respective carriers.

Carriers in the cell membrane are highly specific, *e.g.*, there is a stereospecific carrier for the D-sugars which in turn have varying degrees of affinity for the carrier. Such a carrier has not been demonstrated for the L-sugars. Insulin, by attaching to pores in certain cell membranes, may serve as a stereospecific carrier of D-glucose, transporting it through the membrane into the cytoplasm of the cell for utilization. Insulin does not act in this manner in the intestinal mucosal cell. There are likewise specific amino acid carriers. However, it does not appear that individual carrier mechanisms are operative for the individual amino acids, but rather that there are several which serve groups of similar amino acids. Pyridoxal phosphate is involved in the carrier mechanism in the intestinal absorption of some amino acids. With solubility in the lipid layer as a factor, a type of diffusion referred to as *carrier-facilitated diffusion* can thus take place.

In order to account for the absorption of substances from an environment of lower concentration into one of a higher concentration along with the concentration of the substance in the cell, an additional mechanism within the membrane, a *biological pump*, is postulated. The expression "pump" denotes an ACTIVE TRANSPORT system of a biological membrane which is capable of transporting solutes by an energy-requiring process from a lower into a higher concentration. The pump derives its energy for operation from the breakdown of ATP which is derived from metabolic processes within the cell. The existence and operation of such a pump is indicated by a variety of lines of evidence, *e.g.*, the accumulation of glucose and amino acids in the cell against concentration gradients, and the reverse, the maintenance of low concentration of Na^+ in the cell as compared with the surrounding interstitial fluid. The analogy to a mechanical pump is appropriate since energy is utilized to move the substance in question against an opposing gradient. The fact that the required energy is supplied by the oxidative metabolism in the cell is supported by the fact that the process is inhibited by anoxia, cyanide and other respiratory inhibitors. Involved in the operation of the pump is an enzyme in the cell membrane which has the properties of adenosine triphosphatase (ATPase) and which splits the ATP high-energy bond, thus making energy available for the operation of the pump. Na^+ and K^+ ions are essential for the action of the enzyme. The relative rate of absorption will depend on (1) the degree of affinity the substance in question has for the carrier and (2) the affinity of the carrier complex for the pump.

No concentrating mechanisms have as yet been demonstrated for the water-soluble vitamins, thiamine, riboflavin, niacin, pyridoxine, pantothenic acid, biotin, and folic acid. These apparently pass across the intestinal mucosa by simple diffusion. The absorption of VITAMIN B_{12}, however, is enhanced in an unknown manner by a factor, the "intrinsic factor" secreted by the gastric mucosa. The fat-soluble vitamins, A, D, E, and K, appear to be absorbed from the intestine by mechanisms similar to those of other lipids, *i.e.*, diffusion by solution in the lipid core of the membrane. There is no evidence for a specialized transport system for these compounds. Bile salts (see BILE ACIDS) are essential for their adequate absorption, presumably by facilitating emulsification. Vitamin D, possibly by regulating the genetic formation of a "carrier transport protein" in the mucosal cell, increases the absorption of CALCIUM and PHOSPHATE from the intestine.

Thus, numerous biochemical mechanisms are implicated in the transfer of substances into cells. They are necessary for the absorption of organic nutrients, metabolites, and certain inorganic anions and cations. These mechanisms represent adaptations of basic cellular transport devices occurring at a molecular level.

References

1. GRAY, E. G., "Electron Microscopy of the Cell Surface," Endeavour, **23**, 61 (1964).
2. BOURNE J., "Division of Labour in the Cell," New York, Academic Press, 1962.
3. WILSON, T. H., "Intestinal Absorption," Philadelphia, W. B. Saunders Co., 1962.
4. PALAY, S. L., AND KARLIN, L. J., "An Electron Microscopic Study of the Intestinal Villus. II. The Pathway of Fat Absorption," *J. Biophys. Biochem. Cytol.* **5**, 373 (1959).
5. CRANE, R. K., "Intestinal Absorption of Sugars," *Physiol. Rev.*, **40**, 789 (1960).
6. CHRISTENSEN, H. N., "Biological Transport," New York, W. A. Benjamin Inc., 1962.
7. ORTEN, A. U., "Intestinal Phase of Amino Acid Nutrition," *Symposium, Federation Proc.*, **22**, 1103 (1963).

ALINE UNDERHILL ORTEN

ABSORPTION SPECTRA. See FLAME PHOTOMETRY AND ATOMIC ABSORPTION SPECTROSCOPY; INFRARED AND ULTRAVIOLET SPECTROSCOPY.

ACCEPTOR RIBONUCLEIC ACIDS (ACCEPTOR RNA)

Acceptor ribonucleic acids (synonymous with transfer RNA) are components of the group of SOLUBLE RIBONUCLEIC ACIDS, each of which functions by alternately accepting a specific amino acyl group and then transferring the amino acyl group into a growing polypeptide chain in the process of PROTEIN BIOSYNTHESIS.

ACETALDEHYDE

Acetaldehyde, $CH_3—\overset{O}{\overset{\|}{C}}H$, (ethanal; acetic aldehyde), b.p. 21°C, is produced by oxidation of ethyl alcohol (ethanol) chemically or enzymatically (see ALCOHOL METABOLISM).

ACETIC ACID AND ACETATE METABOLISM

Acetic acid, CH_3COOH (ethanoic acid), is obtained by the destructive distillation of wood or by oxidation of ethyl alcohol (ethanol) or of acetaldehyde (see ALCOHOL METABOLISM). Pure "glacial" acetic acid melts at 17°C, boils at 118°C. Acetic acid (or acetate ion) plays a central role in metabolism, being a product of oxidative CARBOHYDRATE METABOLISM; it is produced from pyruvate initially as acetyl coenzyme A by the action of the pyruvate oxidative decarboxylation complex (see MULTIENZYME COMPLEXES). Free acetate may be activated (see ACTIVATING ENZYMES) to form acetyl coenzyme A. In the latter bound form (see COENZYME A), acetate may then serve as the precursor for fatty acid chain lengthening (see FATTY ACID METABOLISM), for ISOPRENOID BIOSYNTHESIS (and subsequent STEROL BIOSYNTHESIS), for energy-yielding oxidative metabolism (see CITRIC ACID CYCLE), and for other biosyntheses. In a few microorganisms, acetate is a final excreted product of oxidative metabolism; some of these organisms are exploited industrially for such production.

ACETOACETIC ACID

Acetoacetic acid, $CH_3—\overset{O}{\overset{\|}{C}}—CH_2—COOH$, is an intermediate product in the metabolism of fatty acids and the ketogenic amino acids. It contributes to the metabolic acidosis which accompanies diabetes, starvation, and other conditions, and is one of the three "ketone bodies" (acetone bodies, acetone substances). It may be reversibly reduced enzymatically to β-hydroxybutyric acid (another so-called acetone substance) and readily undergoes spontaneous decarboxylation to form acetone and carbon dioxide [see FATTY ACID METABOLISM (OXIDATION AND BIOSYNTHESIS); METABOLIC DISEASES].

ACETONE

Acetone, $CH_3—\overset{O}{\overset{\|}{C}}—CH_3$, b.p. 56°C, was first recognized as a metabolic product by its presence in urine and in the breath, particularly of diabetic patients. ACETOACETIC ACID, β-hydroxybutyric acid, and acetone are the "acetone bodies" referred to in early biochemical literature. Acetone is readily produced from acetoacetic acid by spontaneous decarboxylation; acetoacetic acid is produced as an intermediate product in the biological combustion of fatty acids. Acetone is a product of a number of bacterial FERMENTATIONS.

ACETYLCHOLINE AND CHOLINESTERASE

Acetylcholine is a substance of great importance to the medical sciences. It has an essential role in transmitting the effects of stimulation of certain nerves into functional changes in organs supplied by the so-called cholinergic nerves. Thus when the cholinergic nerves to the muscles and glands are stimulated, they release acetylcholine which acts upon those structures to produce increased muscular or glandular activity (see SYNAPSES AND NEUROMUSCULAR JUNCTIONS). The release of acetylcholine from the vagus nerves causes slowing of the heart rate. Acetylcholine also serves as the chemical transmitter of the nerve impulse in ganglia which function as relay stations to transmit impulses from one nerve to a second nerve in an autonomic chain. It apparently plays a similar role in transmission of impulses from one nerve to another in the brain. Normal activity of various organs such as the intestine, the smooth and skeletal muscles, the heart and bladder, which is referred to as the normal tone of these structures, is maintained by constant release of acetylcholine from cholinergic nerves.[1]

Acetylcholine was first synthesized in 1867. It consists of a combination of choline and acetic acid in an ester linkage. The component parts of the acetylcholine molecule are both normal constituents of the body. Acetylcholine has the following chemical structure:

$$(CH_3)_3\overset{+}{N}—CH_2—CH_2—O—\overset{O}{\overset{\|}{C}}—CH_3 \quad OH^-$$

Acetylcholine assumed no importance to biologists until 1899 when Hunt identified the presence of choline in extracts of the adrenal glands and suggested that some derivative of choline was capable of causing a fall in blood pressure. This suggestion stimulated interest in studying the physiological effects of various choline derivatives, and in 1906 Hunt and Taveau found[2] that acetylcholine was 100,000 times more effective than choline in causing a fall in blood pressure. Soon afterward acetylcholine was identified in extracts of ergot, a fungus that grows on rye and other cereal grains. This constituted the first demonstration of the presence of acetylcholine in biological materials. However, the first real proof of the role of acetylcholine in transmitting the effects of nerve stimulation did not come until 1921 when Loewi[3] showed that when the vagus nerve to a frog heart was stimulated, a substance appeared in the fluid bathing the heart which was capable of slowing the second heart and thus mimicking the effects of nerve stimulation. His experiments were duplicated with warm-blooded animals and identification of the substance as acetylcholine was subsequently made. Following this discovery it was seen that nerves act by releasing chemical compounds which, in turn, act on the organs supplied by these nerves and that acetylcholine is one important mediator, while epinephrine (see ADRENALINE AND NORADRENALINE) and related compounds are other important mediators with an opposite action on various organs. Thus stimulation of activity of an organ is produced by one mediator and inhibition of activity can be produced by another mediator through stimulation of other nerves. Balance or tone of various structures is thus maintained by the two opposing systems. This has led to the concept of classifying nerves on the basis of the chemical substance released from them.

Acetylcholine is stored in the tissues in a physiologically inactive form. It is bound to some constituent of the nerve cell, probably to fat or protein. Stimulation of nerves containing acetylcholine causes a release of acetylcholine from its bound form. Liberation of acetylcholine from its bound form normally occurs only under the influence of nerve stimulation.

The supply of acetylcholine in the nervous system is maintained by synthesis from choline and acetate. This synthesis is controlled by a catalyst which is an enzyme called *choline acetylase*. This protein catalyst acts in the presence of an energy-yielding compound, adenosine triphosphate, and coenzyme A to form acetylcholine from its component parts. *Choline acetylase* is present in the brain and in all of the nerves that liberate acetylcholine upon stimulation.

After acetylcholine is liberated from the bound form and exerts its action on cells, it is then quickly rendered inactive. This inactivation consists of breakdown by hydrolysis to choline and acetic acid. The inactivation or detoxification is catalyzed by enzymes called *cholinesterases*.

The exact manner by which acetylcholine causes a response in various cells is not known. There have been a number of theories advanced which serve as a basis for further research on this subject. One possible mode of action involves reaction with a protein in the cell membrane to stimulate the activity of an enzyme or to facilitate the transport of ions through the cell wall.

As early as 1914, it was suggested that an enzyme is present in the blood which brings about destruction of acetylcholine. During the following years, a great deal of attention was devoted to a study of the enzymatic destruction of acetylcholine. It is now well established[4] that there is a specific enzyme located in the nervous tissue that catalyzes the rapid hydrolysis of acetylcholine. It is called *true* or *specific cholinesterase* and its concentration is highest at the nerve endings where acetylcholine is liberated. The rapid hydrolysis of acetylcholine catalyzed by cholinesterase is the reason that the acetylcholine liberated when cholinergic nerves are stimulated exerts only a very short-lasting effect upon the cells. Furthermore, its rapid destruction prevents acetylcholine liberated at one site from being transported by the blood to other organs. This mechanism thus restricts the action of acetylcholine to a localized area of the body at the immediate site of its liberation from the nerve endings. In addition to the specific cholinesterase present in the nervous system, a variety of tissues contain enzymes capable of hydrolyzing acetylcholine. These enzymes are referred to as *pseudocholinesterases*. They are present in highest concentrations in the serum, the intestinal mucosa and the liver. They hydrolyze acetylcholine and a variety of other esters. Normally they are not involved in the detoxification of acetylcholine. However, they do have an important role in protecting against acetylcholine poisoning under abnormal circumstances. For example, the pseudocholinesterase of the gastrointestinal tract rapidly hydrolyzes acetylcholine before it can be absorbed. This serves as a protective mechanism against poisoning from ingestion of foods containing acetylcholine. The pseudocholinesterase of the serum also has a protective role in those instances where the specific cholinesterase is inhibited. Under these circumstances, the nonspecific enzymes in the serum and liver hydrolyze acetylcholine and thus decrease its general effects.

The acetylcholine-cholinesterase system has served as the basis for the development of a number of drugs needed to alter the activity of the autonomic nervous system in certain disease states.[1] The action of acetylcholine is too transient to have important therapeutic applications due to its rapid breakdown. Several more stable derivatives of acetylcholine are used as substitutes. They are less rapidly hydrolyzed by cholinesterase. In cases where it is desirable to prolong the action of acetylcholine, another means of achieving this objective involves *inhibitors* of *cholinesterase*. The first effective cholinesterase inhibitor to be discovered was the alkaloid physostigmine which is commonly known as eserine. By temporarily preventing acetylcholine hydrolysis, physostigmine raises the acetylcholine level of the tissues. A number of synthetic compounds with a reversible inhibitory action on cholinesterase are available for therapeutic use. Inhibitors of cholinesterase are now being widely used as INSECTICIDES. The particular class of compounds that are generally employed for this purpose consists of organic esters of phosphoric acid. Parathion and malathion are examples of organic phosphate cholinesterase inhibitors that are widely used as agricultural insecticides. They produce a reversible inhibition of cholinesterase activity.[5] When the cholinesterase activity is sufficiently inhibited, the accumulated acetylcholine causes marked and excessive stimulation of a variety of structures in the body. This effect is responsible for the lethal action of these compounds. High doses of these and other cholinesterase inhibitors are also capable of producing poisoning and death in man and domestic animals by the same mechanism. The symptoms of poisoning consist of bronchoconstriction, sweating, salivation and other increased glandular secretions, anorexia, nausea, vomiting, diarrhea, muscular twitching, convulsions, and paralysis of respiration. These effects can be antidoted by administration of atropine, an alkaloid which acts by blocking many of the actions of acetylcholine.

References

1. GOODMAN, L., AND GILMAN, A., "The Pharmacological Basis of Therapeutics," Second edition, pp. 389–475, New York, The Macmillan Co., 1955.
2. HUNT, R., AND TAVEAU, R., "On the Physiological Action of Certain Choline Derivatives and New Methods for Detecting Choline," *Brit. Med. J.*, **2**, 1788–1791 (1906).
3. LOEWI, O., "Uber humorale Uberstragbarkeit der Herzenervenwirkung," *Arch. Ges. Physiol.*, **189**, 239–242 (1921).
4. AUGUSTINSSON, K., "Cholinesterases and Anticholinesterase Agents," in "Handbuch der Experimentellen Pharmakologie," pp. 89–128, Berlin, Springer-Verlag, 1963.
5. DUBOIS, K. P., in "Cholinesterases and Anticholinesterase Agents," in "Handbuch der Experimentellen Pharmakologie," pp. 833–859, Berlin, Springer-Verlag, 1963.

KENNETH P. DUBOIS

ACHROMOTRICHIA

Achromotrichia is the greying or loss of pigmentation of hair. This is due to metabolic failure which may be associated with deficiency of pantothenic acid, *p*-aminobenzoic acid or thiamine. In human beings, it is often due to unknown causes, probably not uniform for different individuals.

ACID-BASE REGULATION

Most data on acid-base regulation have been obtained from studies on mammals. Therefore, this discussion will basically be an outline of mammalian acid-base physiology. With relatively few exceptions, studies which have been done on birds, reptiles, and fish suggest that mechanisms for acid-base regulation in these groups are generally similar to those in mammals.

Problems of acid-base physiology largely revolve about (a) terminology and methodology of the description of the acid-base status of various body compartments, and (b) mechanism of production and disposition of acid or base loads, including internal storage and distribution, and involving buffering capacity and acid-base balance (*i.e.*, difference between intake or production of acid and its excretion).

The classical Brønstedt-Lowry method of classifying materials as acids (*i.e.*, hydrogen ion or proton donors) and bases (*i.e.*, hydrogen ion acceptors) is currently accepted. Disturbances of normal acid-base status are broadly classified as *respiratory* (Figs. 1 and 2) if changes in the concentration of the volatile acid CO_2 (or H_2CO_3) change body acidity, since the concentration of CO_2 is primarily determined by respiration (see GASEOUS EXCHANGE IN BREATHING). Changes in acidity due to accumulation or deficit of nonvolatile acids or bases are called nonrespiratory. Since the mechanisms involved can be broadly called "metabolic" processes, these acid-base disturbances are also often labeled *metabolic* (Fig. 2).

Arterial Blood pCO_2 (liquid phase) = Alveolar pCO_2 (gas phase) =

$$K \frac{CO_2 \text{ Production}}{\text{Alveolar Ventilation}}$$

FIG. 1.

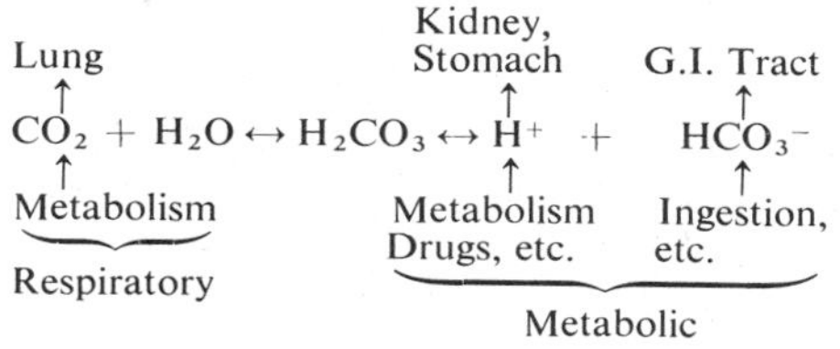

$$\underbrace{\overset{\text{Lung}}{\uparrow}\, CO_2 \underset{\text{Metabolism}}{\uparrow}}_{\text{Respiratory}} + H_2O \leftrightarrow H_2CO_3 \leftrightarrow \underbrace{\overset{\text{Kidney, Stomach}}{\uparrow}\, H^+ \underset{\text{Metabolism Drugs, etc.}}{\uparrow} + \overset{\text{G.I. Tract}}{\uparrow}\, HCO_3^- \underset{\text{Ingestion, etc.}}{\uparrow}}_{\text{Metabolic}}$$

FIG. 2.

Description of the acid-base status of the body is based on the status of the blood (usually arterial), which is the fluid compartment most easily sampled. Deviations of blood pH from normal are often due to many distinct contributing factors, both respiratory and metabolic. Each of these individual factors is usually called an "acidosis" or "alkalosis," depending on whether it tends by itself to lower or raise blood pH. These factors are also classified as (1) primary or (2) secondary, or compensatory, *i.e.*, physiologic mechanisms which tend to return blood pH to normal in response to primary acid-base disturbances. For example, primary respiratory acidosis, *i.e.*, CO_2 accumulation (Fig. 2) resulting from inadequate ventilation in lung disease (Fig. 1), causes the kidney to excrete an increased amount of acid (or retain increased amounts of bicarbonate), resulting in a secondary or compensatory metabolic alkalosis.

Blood pH (normal value, 7.38–7.42 in arterial blood) is a measure of over-all acid-base status. The contribution of respiratory abnormalities to the over-all status is quantitatively estimated by the pCO_2 of arterial blood (normal 38–42 mm Hg), an elevated value respresenting respiratory acidosis and *vice versa* (Fig. 2). Commonly used measurements for quantitative estimation of the non-respiratory components of an acid-base disturbance include: (1) whole blood buffer base (concentration of all buffer anions in blood in milliequivalents per liter; normal 45–55 meq/liter); and (2) base excess (milliequivalents of excess base per liter of blood; equals abnormality in whole blood buffer base; normal −2 to +3 meq/liter). These measurements are independent of the pCO_2 of the blood; therefore, changes from normal quantitate non-respiratory acid-base disturbances.

These values in themselves provide no indication of whether the factor is primary or secondary. This requires either a knowledge of the precipitating factor or a knowledge of the normal compensatory responses to a primary disturbance of given degree, or both.

Normally, approximately 2400 mmoles of the volatile acid CO_2 (or H_2CO_3) is produced daily by the adult human and excreted through the lungs. In contrast, only approximately 50–75 meq of nonvolatile acid is produced daily to be excreted by the kidneys. (The normal fecal contribution to acid excretion is not definitely established, but appears to be both small and at a fixed rate.) Nonvolatile acid is produced primarily by metabolism of ingested sulfur-containing compounds (principally the methionine of protein) to end products including H_2SO_4. Metabolism of phosphorus-containing compounds (to phosphoric acid) and formation of organic acids make smaller contributions to the normal acid load.

Compensatory Responses. When the normal acid-base status is disturbed, compensatory responses return blood pH toward normal in each major type of disturbance as follows:

(*I*) *Metabolic Acidosis.* Sudden addition of nonvolatile acid in excess of the normal rate of production results in:

(1) Increased acidity of the urine and an increased rate of excretion of buffered hydrogen ion. This response begins within minutes. Ultimately, the entire acid load must be eliminated by the kidney, but this takes at least 3 days. In the meantime, the excess acid is buffered within the body, as below.

(2) Buffering by the extracellular fluid (principally hemoglobin, bicarbonate, and plasma proteins). Initially, this must account for disposition of the entire acid load, but within 2–12 hours, approximately one-half the load is transferred into cells by diffusion.

(3) Buffering by intracellular buffers (including bone). Transfer of hydrogen ion into the cell results in transmembrane exchange, primarily with the cations Na^+ and K^+ approximately in the ratio 2:1. Since intracellular buffers combine with approximately one-half an acid load, and since the intracellular space is approximately 2.5 times the extracellular space, intracellular buffering capacity per unit volume is only roughly 40% of extracellular. This may account in part for the fact that intracellular pH is, on the average, 0.4–0.6 unit lower than that of blood.

(4) Increased ventilation due to stimulation of chemoreceptors in large arteries and in the central nervous system probably in fairly direct contact with the cerebro-spinal fluid (CSF), whose acidity is affected by the acid load. Ventilation increases approximately fourfold when CSF pH decreases 0.1 unit. This creates a respiratory alkalosis returning pH only in part toward normal. The response begins within minutes but is probably maximal only after hours or days. It accounts for approximately 40% of the "protection" against change in pH, while buffering, as described above, accounts for the other 60%.

(*II*) *Metabolic Alkalosis.* Sudden addition of base to the extracellular fluid results in the same types of responses as does acidosis (but of course in the opposite direction), with the following differences:

(1) Renal excretion of the base load is more rapid than the excretion of a comparable acid load.

(2) The respiratory response is less, since other overriding factors prevent the depression of respiration and respiratory acidosis which would normally compensate for metabolic alkalosis. Arterial pCO_2 does not rise above 50 mm Hg with even the most extreme alkalosis.

(3) Increased metabolic production of organic acids, primarily due to slowing of Krebs' cycle reactions. This is associated with increased urinary excretion of organic anions.

(*III*) *Respiratory Acidosis.* An acute increase in alveolar and blood pCO_2 results in the following compensatory mechanisms:

(1) Increased renal excretion of acid. This response is initially much less marked than to a comparable degree of metabolic acidosis (*i.e.*, per unit change in blood pH), but in spite of this, renal compensation may return the pH completely to normal over weeks if the arterial pCO_2 is less than 70.

(2) Extracellular and intracellular buffering of carbonic acid as for nonvolatile acids. Since CO_2 diffuses across cell membranes very readily, this results in greater intracellular acidosis than does extracellular metabolic acidosis.

(*IV*) *Respiratory Alkalosis.* An acute lowering of pCO_2 results in the following compensatory mechanisms:

(1) Decreased renal excretion of acid or increased excretion of base. This response not only is not marked initially, as in respiratory acidosis, but in addition is only transient, the urine returning to its normal acid state after approximately an hour. This appears teleologically appropriate, since the kidney is obligated to excrete nonvolatile acid end products of metabolism, which can be done only by acidifying the urine.

(2) Increased production of organic acids (particularly lactic, pyruvic, and citric) which may compensate for 30% of the potential change in blood pH. This results in increased renal excretion of citrate (metabolically equivalent to hydroxyl ion) which accelerates renal compensation.

(3) Extra- and intracellular buffering as in respiratory acidosis.

Acid-base regulation in normal metabolic states is presumed to be similar qualitatively to that under abnormal conditions of acid loading, etc., described above.

References

1. Pappenheimer, J. R., Fend, V., and Heisy, S. R., "Role of Cerebral Fluids in Control of Respiration as Studied in Unanesthetized Goats," *Am. J. Physiol.*, **208**, 436–450 (1965).
2. Simmons, D. H., "Regulation of pH of Body Fluids," in "Clinical Disorders of Fluid and Electrolyte Metabolism" (Maxwell, M. H., and Kleeman, C. R., editors), Ch. 3, pp. 71–114, New York, McGraw-Hill Book Co., 1961.
3. Astrup, P., Sigaard-Andersen, O., Jørgensen, K., and Engel, K., "The Acid-Base Metabolism:

A New Approach," *Lancet*, 1035–1039 (May 14, 1960).

4. SIGAARD-ANDERSEN, O., "Acute Experimental Acid-Base Disturbances in Dogs: An Investigation of the Acid-Base and Electrolyte Content of Blood and Urine," *Scand. J. Clin. Lab. Invest.*, **14**, Suppl. 66 (1961).
5. GIEBISCH, G., BERGER, C., AND PITTS, R. F., "The Extra-renal Response to Acute Acid-Base Disturbances of Respiratory Origin," *J. Clin. Invest.*, **34**, 231 (1955).
6. HUNT, J. N., "The Influence of Dietary Sulfur on The Urinary Output of Acid in Man," *Clin. Sci.*, **15**, 119 (1956).
7. SCHWARTZ, W. B., ORNING, K. J., AND PORTER, R., "The Internal Distribution of Hydrogen Ions with Varying Degrees of Metabolic Acidosis," *J. Clin. Invest.*, **36**, 373 (1957).

DANIEL H. SIMMONS

ACIDOSIS AND ALKALOSIS. See ACID-BASE REGULATION; METABOLIC DISEASES.

ACTH. See ADRENOCORTICOTROPIC HORMONE; HORMONES, ANIMAL.

ACTIVATING ENZYMES (AMINO ACID AND CARBOXYL ACTIVATION)

Acids (and amino acids) exist as the carboxylate anion at physiological pH in the cell and as such are unreactive both thermodynamically and kinetically to nucleophilic substitution reactions. The negative charge of the carboxylate anion repels nucleophiles, and the resonance stabilization of the carboxylate group does not favor synthetic reactions such as the condensation with alcohols or amines. For example, the synthesis of peptide bonds at physiological pH requires a standard free energy input of 0.4–4 kcal/mole. In order to circumvent this unfavorable situation, mechanisms have been developed in nature for the conversion of the carboxylate anion to a kinetically and thermodynamically more reactive derivative. The reactions involved in these mechanisms are termed "activation" reactions and the enzymes catalyzing them are called activating enzymes.

In this series of reactions, the splitting of a special "high-energy" class of compound (A ~ B) is coupled to the desired condensation reaction according to Eqs. (*1*) + (*2*) or (*1*) + (*3*) + (*4*).

$$RCOO^- + A \sim B \rightleftharpoons RCOOA + B^- \quad (1)$$
$$RCOOA + HY \rightleftharpoons RCOY + HOA \quad (2)$$
$$RCOOA + HX \rightleftharpoons RCOX + HOA \quad (3)$$
$$RCOX + HY \rightleftharpoons RCOY + HX \quad (4)$$

The active derivative, RCOOA, may react directly with the compound to be condensed (HY) as in Eqs. (*1*) + (*2*) (Mechanism I) or it may be converted to a second more stable derivative, RCOX, with a somewhat lower free energy of hydrolysis yet still sufficiently reactive kinetically for subsequent condensation [Eqs. (*1*) + (*3*) + (*4*); Mechanism II]. The equilibrium concentration of product, RCOY, will be the same by either mechanism and will be appreciable in spite of the unfavorable thermodynamics of condensation because of the high free energy of hydrolysis of A ~ B under physiological conditions.

Examples of Mechanism I in which RCOOA is a free intermediate [Eq. (*5*)] are the reactions catalyzed by (a) acetate and butyrate kinases resulting in the formation of acetyl and butyryl phosphate and (b) β-aspartyl kinase which promotes the synthesis of β-aspartyl phosphate.

$$RCOOH + ATP \rightleftharpoons R\overset{\overset{O}{\|}}{C}O{-}PO_3H_2 + ADP \quad (5)$$

More commonly, however, when Mechanism I is used in nature, the active derivative is not free but remains enzyme-bound and reacts on the enzyme surface. Such binding stabilizes the reactive intermediate, although the mechanism by which this occurs is unclear. Examples of this variant of Mechanism I are (a) the synthesis of carnosine and of pantothenic acid (Fig. 1A), and (b) all of the reactions involving cleavage of ATP to ADP and inorganic phosphate (see below), with the possible exception of the succinic thiokinase reaction.

Activation by Mechanism II, in which RCOOA is converted to a more stable RCOX which still retains a high group potential, is by far the most widespread mechanism for activation used in the cell (Fig. 1B and 1C). This mechanism possesses the advantages of (a) a more stable carboxyl derivative without the necessity of binding to a specific enzyme as in Fig. 1A, (b) a sufficiently high group potential (free energy of hydrolysis), and (c) mobility, the ability to be synthesized in one part of the cytoplasm and utilized at a different location.

The general features of this mechanism can be summarized as follows. The first reaction is an attack of the nucleophilic carboxyl group on the innermost phosphate of ATP, splitting out pyrophosphate, and producing a mixed anhydride between the carboxyl group and the 5′-phosphate group of AMP [Fig. 1(a)]. This intermediate, corresponding to RCOOA of Eq. (*1*), remains enzyme-bound and cannot be found free in solution. The enzyme-bound acyl adenylate is then attacked by a second nucleophilic species at the carboxyl carbon, to give the product plus AMP and free enzyme. The second nucleophilic species may be the final acceptor (Fig. 1A), coenzyme A (Fig. 1B), or transfer RNA (Fig. 1C) (see SOLUBLE RIBONUCLEIC ACIDS). In the latter two cases, another reaction [Fig. 1(e), 1(g)] must still take place with the final acceptor to give the desired end product.

Case A of Fig. 1 is really a variant of activation by Mechanism I where the reactive intermediate remains enzyme bound. Some examples of this are (a) the synthesis of carnosine in which β-alanyl-adenylate-enzyme reacts with histidine to form the dipeptide product, β-alanylhistidine; (b) pantothenic acid synthesis in which pantoyl-adenylate-enzyme reacts with β-alanine to yield pantothenic acid; and (c) a number of other less well-defined

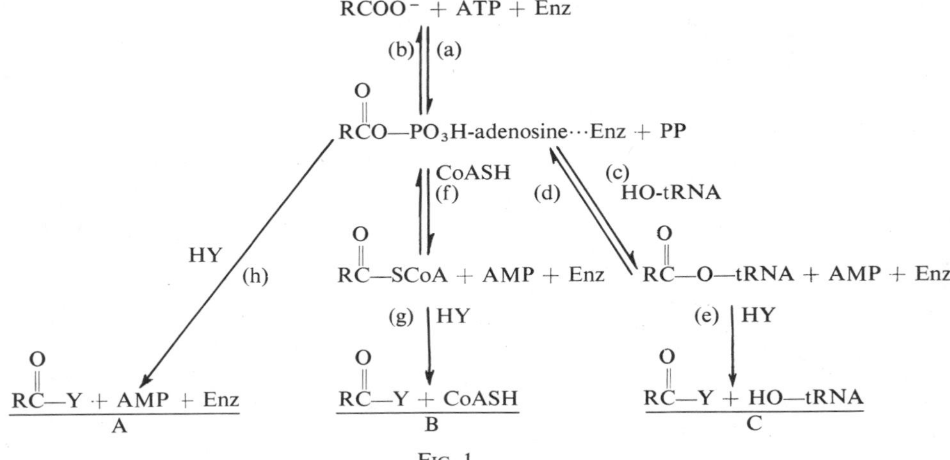

FIG. 1.

reactions in which both pyrophosphate and AMP are produced and where O^{18} is transferred from RCOOH to AMP. These latter reactions include the amination of the nucleotide ring and the synthesis of argininosuccinic acid.

Case B of Fig. 1 illustrates the usual means by which carboxyl acids are activated, and the only one known in which ATP is split to AMP and pyrophosphate. The intermediate CoA derivatives formed possess a group potential as high as ATP (as shown by the fact that the equilibrium constant for reaction (a) + (b) + (f) at pH 7 is near unity) and thus react readily with a variety of acceptors. Since acyl CoA compounds are not enzyme-bound they serve as a mobile, reactive form of carboxyl acid. Moreover, the common presence of pyrophosphatase in cells probably acts to pull this reaction toward acyl CoA synthesis even in otherwise unfavorable situations, and may account for the strong preference shown in nature for this mechanism of carboxyl activation over that depicted in Eq. (5).

There are three well-known enzymes catalyzing acyl CoA synthesis according to reactions (a) + (f). (1) Acetate thiokinase is specific for acetate, ATP and CoASH, and as mentioned above, the equilibrium constant is approximately unity. (2) Fatty acids, C_4–C_{12} units long, are activated by a second fatty acyl thiokinase which is also able to utilize benzoate, phenylacetate, *p*-aminobenzoate, and similar aromatic acids. (3) For long-chain fatty acids, C_{10}–C_{18}, still another enzyme is known. The final acceptors of these activated acyl compounds are varied. As an example, HY may be an amine to yield an acylated amine as product or HY may be an alcohol such as glycerol in which case the product would be a lipid.

As mentioned above, thermodynamic measurements have shown that the synthesis of the peptide bond is an endergonic process. It is now known that the mechanism of amino acid activation for the synthesis of peptide bonds in protein proceeds exclusively by reactions (a)–(e) of Fig. 1. Another mechanism involving the hydrolysis of ATP to ADP and inorganic phosphate exists for the activation of amino acids but appears to be used only for the synthesis of small peptides (see below).

All of the comments made above regarding the general features of the reaction in Fig. 1 apply with equal force when $RCOO^-$ is an amino acid. In addition, the following points can be made. There are separate amino acid activating enzymes, more properly termed amino acyl RNA synthetases, for each of the 20 naturally occurring AMINO ACIDS. Present evidence indicates that there is no more than one synthetase for each amino acid. Most, but not all, of the enzymes require reduced SH groups for the maintenance of their biological activity, and a number of them show a marked stimulation by K^+. While most of these enzymes are localized in the cytoplasm of the cell, they have also been found in the nucleus and in mitochondria. The K_m for amino acids varies widely in magnitude but is quite low, 10^{-4}–$10^{-6}M$. The rate of reaction (a) is in all cases greater than reaction (c), the ratio of rates ranging from 20–500. Reactions (a) and (b) always require Mg^{++} or another divalent cation for activity, while reaction (c) may or may not be dependent on such cations depending on the species. The requirement for reaction (d) has not yet been tested. As was the case for acyl CoA, the amino acyl RNA product is a compound with a group potential as high as ATP. This follows from the fact that the equilibrium constant for the reaction (a)–(d) measured at pH 7 ranges from 0.3–0.7 depending on the amino acid. In all cases, the amino acid is linked via its carboxyl group to the 3′(2′)-hydroxyl of the terminal adenosine end of the transfer RNA and thus there is only one amino acid bound per tRNA molecule (see SOLUBLE RIBONUCLEIC ACIDS).

The final acceptor of amino acyl RNA compounds is the α-amino group of another amino acyl RNA, resulting in peptide bond formation. Strictly speaking, the donor is usually a peptidyl tRNA rather than amino acyl RNA although the principle is the same (see PROTEIN BIOSYNTHESIS).

It is now clear that by virtue of linking amino acids to specific transfer RNA molecules, the specificity once contained in the amino acid side chain has been converted into a specificity of nucleotide sequences (anticodon) contained in the tRNA. It is this anticodon which is used to detect the codon sequence in the messenger RNA specifying that particular amino acid (see also GENES). Since the position that a given amino acid will occupy in a polypeptide depends exclusively on the particular tRNA molecule to which it is linked, it is clear that reactions (a)–(c) must be extremely precise if proteins of unique sequence are to be obtained. Experimental measurements of less than 0.03% error in the amino acid sequence of the finished protein illustrate the high specificity achieved by amino acyl RNA synthetases. The way such specificity is achieved can be best described by considering each reaction in turn.

Reaction (a) is highly specific for a single L-amino acid, and requires the presence of a free α-amino group (where tested). In a few cases, the synthetase is not completely amino acid specific. For example, a synthetase for isoleucine will also carry out reactions (a) and (b) with valine, and a valine-specific enzyme will function with threonine. However, although the maximum rate with the "wrong" amino acid is one-third that of the correct amino acid or greater, the K_m is 100 times greater so that at cellular amino acid concentrations the discrimination is much better (95–99% selective). Amino acid analogues can be used in this reaction in many instances, but there are no general rules as to which types of analogues will or will not work. Obviously those analogues known to be incorporated into proteins in place of the normal amino acid must react according to reactions (a)–(e). Still others may react only part way in the sequence.

Reaction (b), tested with synthetic amino acyl adenylates, is much less specific than reaction (a). A free α-amino group is required, but "wrong" amino acyl adenylates and D-isomers will readily yield ATP, and kinetic experiments have shown that this reaction occurs at the same enzyme site as reaction (a).

However, the "wrong" amino acids are selected against at the next step, reaction (c). Thus, synthetic "wrong" amino acyl adenylates are not transferred to tRNA, and even in the case described above, where valine reacts in (a) with an isoleucine-specific enzyme, no valine is transferred to RNA. This increased specificity in reaction (c), shown by the same enzyme, is a second line of defense, as it were, to keep "wrong" amino acids from the committed step of being linked to tRNA and hence into polypeptide.

Although amino acyl RNA synthetases are highly specific for forming only the correct amino acyl RNA when tRNA from the same species is used, a somewhat different result is found when enzyme of one species is tested with RNA of another species. The possible combinations and results of such experiments are summarized in Table 1 which includes the results of *intra* species reaction for comparison. Note the complication which arises because there can be more than one distinct tRNA for a single amino acid even in the same organism. Since many of the possible combinations of amino acid and species have not yet been tested, some of the (−) may become (+) in future. Despite this caution, it is remarkable that of the many examples tested so far, only one case of the enzymatic incorporation of a "wrong" amino acid onto RNA has been found.

TABLE 1. AMINO ACYL RNA FORMATION IN INTERSPECIES REACTION

RNA*†	aa_1–AMP–E^1_x *	aa_1–AMP–E^2_x *	
	aa-RNA formation	aa-RNA formation	Complex breakdown‡
RNA^{1a}_x	+	−	−
RNA^{1b}_x	+	−	−
RNA^{2a}_x	−**	−	+
RNA^{2b}_x	−**	−	?
	(a)***(b)***(c)***		
RNA^{1a}_y	− + +	−	−
RNA^{1b}_y	− − +	−	−
RNA^{2a}_y	+††	−	−
RNA^{2b}_y	?	−	−

* Subscript denotes species, superscript denotes amino acid.

† Only two RNA chains specific for a single amino acid are shown (i.e., 1a, 1b) although there may be more in some cases.

** Not rigorously rested for every amino acid but the specificity of protein synthesis predicts it.

‡ Catalysis of the reaction: $aa_1\text{-AMP-}E^2_x \rightarrow aa_1 + AMP + E^2_x$.

*** All three cases are known with appropriate combination of x, y, and amino acid. In some negative cases, interaction may occur although amino acyl RNA formation fails.

†† Only one example is known so far.

? Has not been tested.

Relatively little is known of the specificity of the reverse of this step, namely reaction (d), except that if a "wrong" amino acyl RNA is made (by a chemical method), then neither the enzyme specific for the amino acid nor the one specific for the RNA will catalyze the reaction.

There are a few examples of amino acid and carboxyl acid activation and condensation where a mechanism other than that of Fig. 1 is involved. These condensations are readily identifiable by the fact that they invariably result in the hydrolysis of ATP to ADP and inorganic phosphate. The reaction sequence (Fig. 2) is an example of Mechanism I activation where the active intermediate, RCOOA, remains enzyme bound.

Examples of this type include (a) the condensation of glutamic acid with ammonia to give GLUTAMINE catalyzed by glutamine synthetase, (b) the synthesis of GLUTATHIONE from glutamic acid,

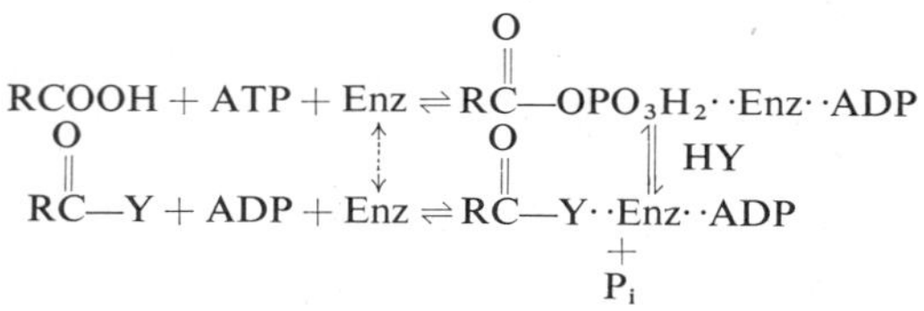

$$\mathrm{RCOOH} + \mathrm{ATP} + \mathrm{Enz} \rightleftharpoons \mathrm{R\overset{O}{\overset{\|}{C}}{-}OPO_3H_2\cdot\cdot Enz\cdot\cdot ADP}$$

$$\Big\updownarrow \qquad \Big\downarrow \mathrm{HY}$$

$$\mathrm{R\overset{O}{\overset{\|}{C}}{-}Y} + \mathrm{ADP} + \mathrm{Enz} \rightleftharpoons \mathrm{R\overset{O}{\overset{\|}{C}}{-}Y\cdot\cdot Enz\cdot\cdot ADP} + \mathrm{P_i}$$

Fig. 2.

cysteine, and glycine catalyzed by two enzymes, (c) a series of less well-studied enzyme reactions which activate both D- and L-amino acids for incorporation into cell-wall peptides, and (d) an enzyme reaction which activates glycine for PURINE BIOSYNTHESIS. Only the first two examples have been well studied. Essentially all that is known about the other reactions is that ADP and phosphate are produced in the process of condensation.

There is one other well-known carboxyl activating enzyme of this type, succinic thiokinase, which catalyzes the conversion of succinate, ATP and CoA to succinyl CoA, ADP and inorganic phosphate without producing any free intermediates. The mechanism of this reaction is not yet well understood. However, it does not appear to fit exactly into the format of Fig. 2.

References

1. Moldave, K., *Ann. Rev. Biochem.*, **34**, 419–449 (1965).
2. Berg P., *Ann. Rev. Biochem.*, **30**, 293–395 (1961).
3. Meister, A., "Biochemistry of the Amino Acids," Vol. I., pp. 439–518, New York, Academic Press, 1965.
4. Brown, G., "Progress in Nucleic Acid Research, Vol. 2, pp. 260–273, New York, Academic Press, 1963.
5. Jencks, W. P., "The Enzymes," Vol. 6, p. 373, New York, Academic Press, 1962.
6. Stulberg, M., and Novelli, G. D., "The Enzymes," Vol. 6, p. 401, New York, Academic Press, 1962.

James Ofengand

ACTIVATORS OF ENZYMES. See ENZYMES (ACTIVATION AND INHIBITION).

ACTIVE TRANSPORT

The field of active transport is one in which much research is taking place but also one in which, because of a lack of final answers to many problems, the literature is voluminous and often contradictory. This article will serve to introduce the subject but will not attempt to present a conclusive picture of any of the divisions of the field.

The animal organism is made up of many cell types and although these cells differ in shape, size and function they have certain characteristics in common. These generally include a cell membrane, a nucleus, and variety of cell organelles distributed throughout the cytoplasm. Common to certain of the organelles and to the cell itself are structures referred to as membranes (see CELL WALL; MICROSOMES; MYELIN). At this point, it will only be stated that cellular membranes are generally composed of lipid and protein molecules spatially oriented so that the inner part of the membrane is an area of interdigitated phospholipid and cholesterol ester molecules and the inside and outside coatings of the membrane are largely protein. The thickness of biomembranes varies, but many membranes appear microscopically as two dark lines of ca. 10 Å thickness separated by a lighter band of about 50 Å. (One Ångstrom is one hundred-millionth of a centimeter.) Myelin, the covering of nerve axons, is a repeating structure of lipid-protein layers of 140–170 Å. The molecules making up these thin structures serve to isolate the cell contents from the environment, however, they do not fit tightly together as a solid wall or surface but have areas in which there are "holes." These holes or spaces through bio-membranes are limited, however, in size and serve to effectively prevent the passage of a variety of large molecules, thus giving to most bio-membranes the property of semipermeability.

Most cells contain considerably more protein than is present in the fluids bathing the cells. The presence of a high concentration of cellular protein, together with the high concentration of salts associated with the charged protein structure, causes an osmotic gradient to exist and results in the flow of solvent into the cell in an attempt to compensate for the OSMOTIC PRESSURE difference. Plant cell walls have rigid structural features that prevent cell wall rupture and cell death when osmotic or hydrostatic pressures are imposed. Animal cells, however, lack supportive wall structures and must depend on other mechanisms to restore water and solute balance (and thus osmotic balance). The cell membrane then becomes a dynamic focal point of fluid and solute flow rather than just a static barrier unassociated with the life process.

Some mention has been made of membrane structure because the movement of materials through cell membranes can be accomplished by a variety of means, each of which may depend on structural characteristics of the membrane.

Diffusion is that motion which is imparted to solutes by the random molecular movements of materials in solution. The diffusion movement of solute is increased with increasing temperature and is directly dependent on its concentration. In dilute solutions, the diffusion of one species or particle is independent of the diffusion of another species provided there is no interaction between the species. For a small solute molecule (or the solvent itself) the movement through the small distance involved in the thickness of bio-membranes may be of a similar magnitude to the movement in free solution. Water moves rather quickly across many cell membranes. A solute may cross a cell membrane at a greater rate than it would by simple diffusion in water if, in the process of flowing rapidly through the membrane, water "drags" solute with it. This process is known as a "solvent drag."

The movement of certain molecules through cell membranes may be restricted or aided because of the lipid layers in the membrane. If a material is not small enough to diffuse through the solvent phase of the membrane, or is hydrophobic in nature, then it may still pass through the membrane by first dissolving in the membrane phase, then later leaving this phase and entering the cell. Since diffusion is related directly to concentration, high solubility of a solute in the membrane should lead to a high probability of its crossing the membrane.

Charged solutes (cations and anions) may be subjected to additional forces in their movements. If one side of the membrane, *i.e.*, the outside of the cell, has a positive charge on it and the inside of the cell has a negative charge, a potential difference exists. Now a charged solute in the vicinity of the inner or outer environment of the cell will move with greater or less speed depending on the nature of its own charge. An area of opposite charge then can be an attracting force and cause the solute to move at a speed greater than that expected by simple diffusion. Conversely, the electrical field may serve as a barrier to an ion of the same charge as the field.

The net movement of material across a biological membrane at a rate greater than that predicted by simple diffusion or by electrical gradients is considered to be *facilitated diffusion* or *active transport*. Facilitated diffusion does not directly involve cell metabolism and will be discussed below. The descriptive term "uphill transport" has been used synonymously with "active transport" to lend credence to the concept of transport being a work process. A work process requires energy, and the energy requirements of a biological system associated with active transport is, as expected, closely associated with high-energy phosphate compounds, substrate utilization and O_2 consumption. The term active or uphill transport then is used to describe a process by which cells may control the qualitative and quantitative aspects of their intracellular environment. There is no intent in this definition of active or uphill transport to imply a mechanism.

The field of active transport is growing and it is not known with certainty just how many substances are actively transported. The greatest amount of work in this field is concerned with carbohydrate, amino acid, water and ion transport. Ion and carbohydrate transport will be described briefly as illustrative systems.

Cation Transport. Epithelial "membranes" have in common the ability to move (transport) alkali metal ions in the face of osmotic, electrical and concentration gradients. By mechanisms not as yet understood, cells may maintain a high internal concentration of K^+ and a low concentration of Na^+ while existing in an environment that has a low K^+ and a high Na^+ concentration. The advent of radioisotopes of these ions revealed that this unlikely distribution is not due to a lack of movement or passage across the cell membranes but is due to selective processes that extruded ("pumped") ions from an area of low to an area of high concentration. In certain cell systems, the outward pumping mechanism for Na^+ is coupled to the reverse movement of K^+. The use of the descriptive term "pump" for the mechanism that causes ion movement is a useful analogy for pump implies a mechanism, a source of energy or fuel, and the fuel a source of oxygen for its combustion.

Considerable data supports the concept that the driving force of the pump is closely linked to reactions involving the use of ATP (adenosine triphosphate). This prime energy source is supplied by the catabolism of foodstuffs and the union of food hydrogen with environmental O_2 (see RESPIRATION). At various points in this interlinked, metabolic, active transport system, it is possible to interrupt the sequence of reactions. Such interruption results in a cessation of the active transport of Na^+.

The active transport of Na^+ can be demonstrated and measured in several ways. If a test membrane such as frog skin or toad bladder is used as a barrier to separate two identical salt solutions bathing the outside and inside surfaces of the tissue, then the addition of radioactive Na^+ to one of the solutions will allow the measurement of Na^+ movement through the tissue barrier into the opposite solution. If active transport is not present then it is to be expected that the movement of Na^+ will be equal in both directions across the membrane. If an active process is present, the movement can be expected to be greater in the direction of the active transport. It is not difficult to show that the ratio of movements (outside → inside)/(inside → outside) is greater than 1 and is often 30/1. Research in this field was greatly stimulated by the development of a simple but ingenious electrical device to measure the amount of active Na^+ transport. A diagram of this apparatus is shown in Fig. 1. With the use of this "short-circuit" technique, it is possible to show that the current flowing in the outer circuit is essentially identical to the net inward movement of Na^+ through frog skin.

It is not possible to describe with certainty the nature of the mechanism whereby Na^+ is pumped from the cell. Among the various theories now prevalent, the one receiving the greatest support is the ATPase theory. The enzyme, adenosine triphosphatase, along with other enzymes, is located in the cell membrane structure as a part of the protein component. The very special cation requirements and spatial relations of the enzyme and its associated ions lend considerable plausibility to the theory.

Sugar Transport. A variety of mechanisms operate for the movement of sugars into cells. Sugars may enter by simple diffusion, but this is a slow process without great structural specificity. This movement is always away from the region of highest concentration, however. Transport from a high to a low concentration is often called "downhill" transport. In the case of many cells, the movement of certain sugars appears to be much more rapid than expected from simple diffusion. In those instances where there is rapid movement, but the movement is still "downhill," the transport is

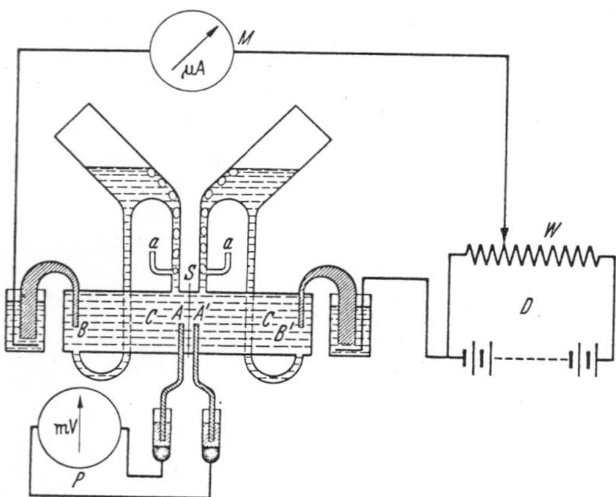

FIG. 1. Diagram used for the determination of short-circuit current and Na fluxes through an isolated membrane. The membrane, S, separates the two chambers, C. Two agar-Ringer bridges, A and A′, connect the bathing solutions with calomel electrodes. The potential difference across the skin is measured by the potentiometer, P. By means of the potential divider, W, a variable current can be passed through the skin via the agar-bridges B and B′. The current passing the circuit is read on the microammeter, M. The bathing solutions are aerated and mixed by air from the inlets, a. (Reprinted with permission from Ussing, H. H., "The Alkali Metal Ions in Isolated Systems and Tissues," in "Handbuch der experimentellen Parmakologie," Bd. 13, S. 1, Berlin-Gottingen-Heidelberg, Springer, 1960).

called facilitated transport. Facilitation of solute movement is presumed to result from the interaction of the sugar with a carrier substance in the membrane. The complex moves across the membrane, the sugar is discharged on the far side and the carrier returns for another cycle. In the case of glucose uptake by human red cells, the exchange is up to 100 times faster than that predicted by simple diffusion. Although this facilitated transport shows specificity for structures and demonstrates saturation phenomena, the process does not seem to directly require metabolic energy.

The absorption of sugars in the small intestine is an "uphill" process for sugar does accumulate in the cell, and metabolic energy is required for the transport, thus the intestinal absorption of glucose and certain other sugars is an active transport process. Many other sugars that are absorbed are not actively transported.

The active transport process is apparently localized near the luminal border of the epithelial cells. The sugars transported have certain common structural features which include a pyran ring, a methyl group or substituted methyl group at carbon-5 of the ring, and a hydroxyl group in the glucose configuration at carbon-2. This active transport requires energy and the process exhibits Michaelis-Menten kinetics. If the process involves a chemical reaction, it probably takes place with the hydroxyl at carbon-2 but the reaction does not lead to an exchange of O^{18}. One additional characteristic of this active transport should be mentioned. It does not take place in the absence of Na^+ in the bathing solutions. Active Na^+ transport is well known, and it is difficult to separate out the two active processes if they do coexist. It is interesting that even the non-transported sugars that are absorbed seem to require the simultaneous transport of Na^+.

A proposed mechanism for the active glucose transport involves the formation of a glucose-Na-carrier complex in the brush border. Movement of the complex and release of glucose into the cytoplasm of the cell also releases Na^+ which can be actively extruded again to the lumen.

Since all bio-membrane phenomena are associated with water solutions it would be expected that water transport itself should have received considerable study. It is beyond the scope of this article to consider this subject in any detail, but it should be recognized that water movement itself has been considered to be "active." Since it is known that water movement is usually associated with solute movement, it is difficult to dissociate these two so as to establish a claim for active water transport.

The final resolution of the phenomenon known as active transport promises to be an exciting endeavor.

References

1. USSING, H. H., in "The Alkali Metal Ions in Biology," Part I, in "Handbuch der Experimentellen Pharmakologie," Berlin, Springer-Verlag, 1960; KRUHØFFER, P., THAYSEN, J. H., AND THORN, N. A., *ibid.*, Part II.
2. KLEINZELLER, A., AND KOTYK, A. (EDITORS), "Membrane Transport and Metabolism," New York, Academic Press, 1961.
3. CHRISTENSEN, H. N., "Biological Transport," New York, W. A. Benjamin Inc., 1962.
4. DAVSON, H., "A Textbook of General Physiology," Boston, Little, Brown & Co., 1959.

J. T. VAN BRUGGEN

ACTOMYOSIN. See MYOSIN; MUSCLE (STRUCTURE AND FUNCTION); CONTRACTILE PROTEINS.

ADDISON'S DISEASE

This condition which is ascribed to insufficiency of the adrenal cortex may be mild or severe. It is characterized by anemia, weakness, low blood pressure, feeble heart action, bronze-like melanin pigmentation of the skin and mucous membranes. In severe cases, prostration, diarrhea and acute digestive disturbances occur. Unless appropriate therapy is instituted, there is serious loss of sodium and retention of potassium which in turn may precipitate many difficulties. ADRENAL CORTICAL steroids alleviate the condition which otherwise may cause death.

ADENINE

Adenine, 6-aminopurine,

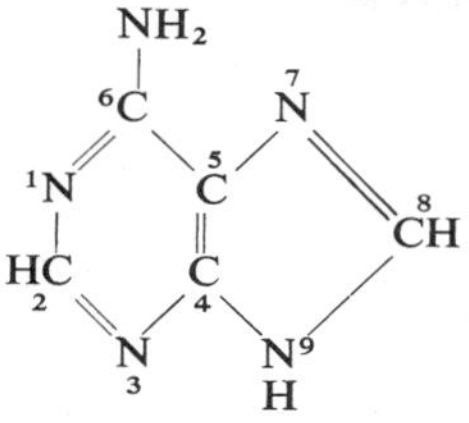

```
           NH2
            |
         6C      7N
       //   \   /   \\
    1N        5C       8CH
     |        ||       /
    HC        C       /
    2 \\    / 4 \    /
        N          N9
        3          H
```

is a prominent member of the family of naturally occurring PURINES. It occurs not only in RIBONUCLEIC ACIDS (RNA), and DEOXYRIBONUCLEIC ACIDS (DNA) but in nucleosides such as ADENOSINE, and nucleotides, such as adenylic acid, which may be linked with enzymatic functions quite apart from nucleic acids. Adenine, in the form of its ribonucleotide, is produced in mammals and fowls endogenously (see PURINE BIOSYNTHESIS) from smaller molecules and no nutritional essentiality is ascribed to it. In the nucleosides, nucleotides and nucleic acids, the attachment or the sugar moiety is at position 9.

The purines and pyrimidines absorb ultraviolet light readily, with absorption peaks at characteristic frequencies; this has aided in their identification and quantitative determination.

ADENOSINE

Adenosine, an important nucleoside, has the following formula:

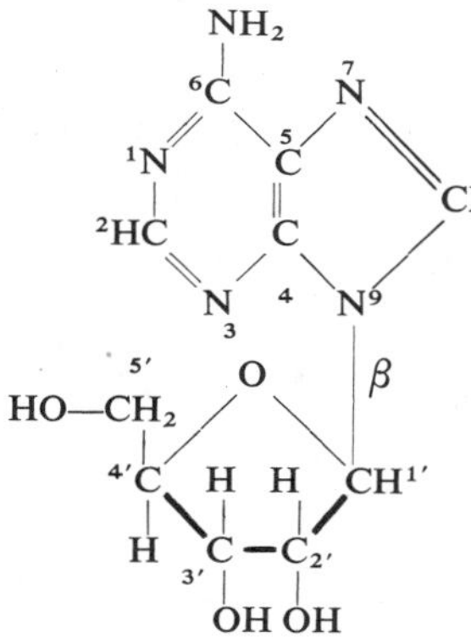

```
              NH2
               |
            6C      7N
          //   \   /   \\
       1N        5C       CH8
        |        ||       /
      2HC        C       /
         \\    /   \    /
           N   4     N9
           3          |
    5'        O       | β
 HO—CH2    /     \    |
      |   /        \  |
     4'C   H   H    CH1'
      | \  |   |   /
      H   C —— C2'
         3' |   |
            OH  OH
```

The upper portion represents the adenine moiety, and the lower portion the pentose, D-ribose (see PURINES).

ADENOSINE PHOSPHATES

Adenosine phosphates include *adenylic acid* (adenosine monophosphate, AMP) in which adenosine is esterified with phosphoric acid at the 5′-position; *adenosine diphosphate* (ADP) in which esterification at the same position is with pyrophosphoric acid,

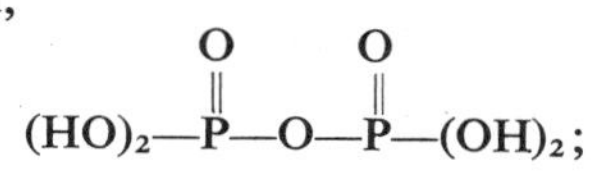

$$(HO)_2\text{—}\overset{\overset{\displaystyle O}{\|}}{P}\text{—O—}\overset{\overset{\displaystyle O}{\|}}{P}\text{—}(OH)_2;$$

and *adenosine triphosphate* (ATP) in which three phosphate residues

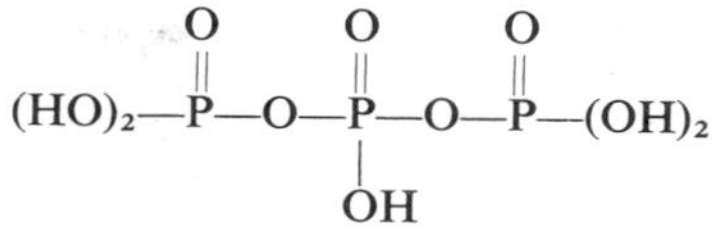

$$(HO)_2\text{—}\overset{\overset{\displaystyle O}{\|}}{P}\text{—O—}\underset{\underset{\displaystyle OH}{|}}{\overset{\overset{\displaystyle O}{\|}}{P}}\text{—O—}\overset{\overset{\displaystyle O}{\|}}{P}\text{—}(OH)_2$$

are attached at the 5′-position. Adenosine-3′-phosphate is an isomer of adenylic acid, and adenosine-2′,3′-phosphate is esterified in two positions with the same molecule of phosphoric acid and contains the radical

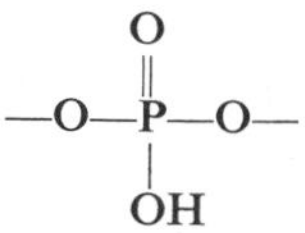

$$\text{—O—}\underset{\underset{\displaystyle OH}{|}}{\overset{\overset{\displaystyle O}{\|}}{P}}\text{—O—}$$

(see PHOSPHATE BOND ENERGIES; PURINES; RESPIRATION).

ADRENAL CORTICAL HORMONES

The hormones elaborated by the adrenal cortex are steroidal derivatives of cyclopentanoperhydrophenanthrene related to the SEX HORMONES. The structural formulas of the important members of the group are shown in Fig. 1. With the exception of *aldosterone*, the compounds may be considered as derivatives of *corticosterone*, the first of the series to be identified and named. The C_{21} steroids derived from the adrenal cortex and their metabolites are designated collectively as *corticosteroids*. They belong to two principal groups: (1) those possessing an O or OH substituent at C_{11} (corticosterone) and an OH group at C_{17} (cortisone and cortisol) exert their chief action on organic metabolism and are designated as *glucocorticoids*; (2) those lacking the oxygenated group at C_{17} (desoxycorticosterone and aldosterone) act primarily on electrolyte and water metabolism and are designated as *mineralocorticoids* (see STEROIDS for numbering system). In the human being, the chief glucocorticoid is cortisol; in the rodent, it is corticosterone. The chief mineralocorticoid in both species is aldosterone. Desoxycorticosterone (DOCA), the first of the corticosteroids to be synthesized (by Reichstein in 1938), is present in only small amounts in the adrenal and it is unlikely that it acts in the organism except as an intermediate in the biosynthesis of corticosterone and aldosterone.

Functions of the Corticosteroids. The glucocorticoids are concerned in organic metabolism and in the organism's response to stress. They accelerate the rate of catabolism and inhibit the rate of anabolism of protein; reduce the utilization of carbohydrate and increase the rate of gluconeogenesis from protein; and exert a lipogenic as well as a lipolytic action, potentiating the release of fatty acids from adipose tissue (see FAT MOBILIZATION). In addition to these effects on the organic metabolism of the basic foodstuffs, the

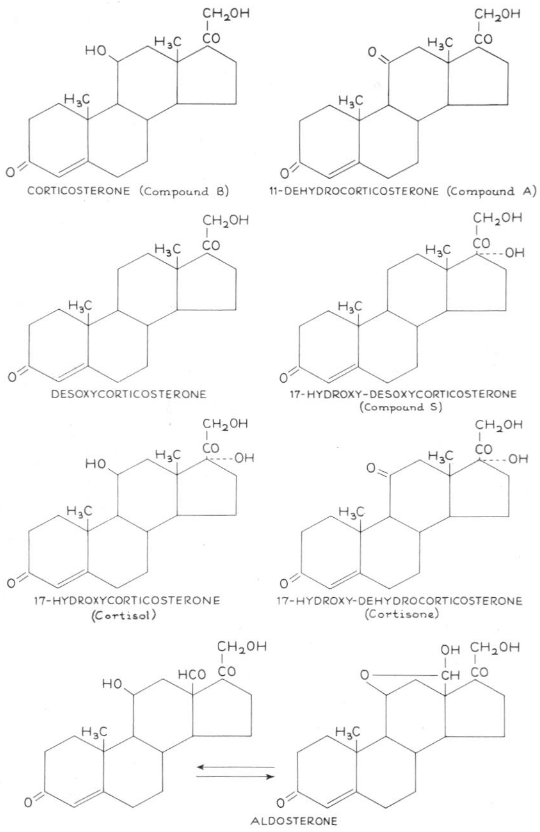

CORTICOSTERONE (Compound B)

11-DEHYDROCORTICOSTERONE (Compound A)

DESOXYCORTICOSTERONE

17-HYDROXY-DESOXYCORTICOSTERONE (Compound S)

17-HYDROXYCORTICOSTERONE (Cortisol)

17-HYDROXY-DEHYDROCORTICOSTERONE (Cortisone)

ALDOSTERONE

FIG. 1.

glucocorticoids in some unknown manner, affect such fundamental bodily functions as the allergic, immune, inflammatory, antibody, anamnestic and general responses of the organism to environmental disturbances (see IMMUNOCHEMISTRY). It is these reactions which are the basis for the wide use of the corticosteroids therapeutically.

Aldosterone exerts its main action in controlling the water and electrolyte metabolism. Its presence is essential for the reabsorption of sodium by the renal tubule, and it is the loss of salt and water which is responsible for the acute manifestations of adrenocortical insufficiency. The action of aldosterone is not limited to the kidney but is manifested on the cells generally, this hormone affecting the distribution of sodium, potassium, water and hydrogen ions between the cellular and extracellular fluids independently of its action on the kidney.

The differentiation in action of the glucocorticoids and the mineralocorticoids is not an absolute one. Aldosterone is about 500 times as effective as cortisol in its salt and water retaining activity but is one-third as active in its capacity to restore liver glycogen in the adrenalectomized animal. Cortisol in large doses, on the other hand, exerts a water and salt retaining action. Corticosterone is less active than cortisol as a glucocorticoid but exerts a more pronounced mineralocorticoid action than does the latter.

In addition to the above-described corticosteroidal hormones, the adrenal produces several 17-oxysteroids, small amounts of testosterone and other ANDROGENS, estrogens, progesterone and their metabolites. In this biosynthetic activity, the adrenal cortex reflects its common embryologic origin from the same anlage as the gonads. Specialization in hormone production is determined by the predominant enzyme systems present in the several tissues. In certain pathogenic states the adrenal cortex may produce sufficient androgen or estrogen to induce striking heterosexual effects which are apparent clinically. Normally, the secretion of small amounts of the sex hormones by the adrenal contributes to the appearance and the maintenance of the axillary and pubic hair.

Biosynthesis. The adrenal cortical hormones are synthesized in the gland from cholesterol by way of at least two metabolic pathways. Cholesterol may be converted through three intermediates to pregnenolone, a C_{21}-steroid, or to dehydroepiandrosterone which in turn is transformed to pregnenolone. This, in turn, is transformed by successive hydroxylation to the hormones. Aldosterone is formed from desoxycorticosterone by the introduction of an aldehyde group at C_{18} and 11-hydroxylation. The congenital absence of certain enzymes concerned with the biosynthesis of the adrenocortical hormones results in the hormonal derangements responsible for the manifestations of a group of disorders designated as the adrenogenital syndrome.

The normal human adrenal secretes about 20 mg of cortisol and about 0.14 mg of aldosterone daily. The production of cortisol is regulated by the ADRENOCORTICOTROPIC HORMONE (ACTH) of the pituitary; that of aldosterone appears to be regulated primarily by ANGIOTENSIN which is formed by renin produced in the juxtaglomerular apparatus of the kidney. Aldosterone has been synthesized but is not available for clinical use. Cortisol and its dehydro derivative, cortisone, are used as substitution therapy in adrenocortical insufficiency. A number of derivatives have been synthesized which are widely used in clinical practice in a variety of conditions to modify the response of the organism in certain diseases.

References

1. DEANE, H. W. (EDITOR), "The Adrenocortical Hormones: Their Origin, Chemistry, Physiology and Pharmacology," *Handb. der Exptl. Pharmakol.*, **14**, 1 (1962).
2. GROLLMAN, A., "The Adrenals," Baltimore, Williams and Wilkins, 1936.
3. GROLLMAN, A., "Clinical Endocrinology," Third edition, Philadelphia, J. B. Lippincott Co., 1964.
4. HECHTER, O., AND PINCUS, G., "Genesis of Adrenocortical Secretion," *Physiol. Rev.*, **34**, 459 (1954).

5. Muller, A. F., and O'Connor, C. M. (Editors), "Aldosterone: Ciba Symposium," Boston, Little, Brown & Co., 1958.
6. Neher, R., "Chemie der Corticosteroids," *Antibiot. Chemotherapy*, **7**, 1 (1960).

Arthur Grollman

ADRENALINE AND NORADRENALINE

Adrenaline (epinephrine) and its immediate biological precursor noradrenaline (norepinephrine, levarterenol) are the principal hormones of the adult, mammalian adrenal medulla. Noradrenaline is also considered to be the principal transmitter substance in sympathetic, adrenergic neuro-effector junctions. These, and related catecholamines, have been found also in toads, insects and plants.

$(HO)_2C_6H_3{-}CH(OH){-}CH_2{-}NH_2$

Noradrenaline

$(HO)_2C_6H_3{-}CH(OH){-}CH_2{-}NH{-}CH_3$

Adrenaline

A principal mammalian biosynthetic pathway is: phenylalanine → dihydroxyphenylalanine (DOPA) → dopamine → noradrenaline → adrenaline. The enzymes involved are stereospecific; and the end products are levorotatory. Synthetic dextrorotatory isomers are relatively inactive.

Several different pathways and nonspecific enzymes are involved in the degradation and inactivation of the catecholamines. Two of the more important enzymes are: *ortho*-methyltransferase (see METHYLATIONS IN METABOLISM) and monamine oxidase (see AMINE OXIDASES). Regardless of the sequence of action of these enzymes, the common metabolite of the catecholamines is 3-methoxy-4-hydroxymandelic acid. Other mechanisms of inactivation are tissue binding, sulfate conjugation and quinone formation.

The site of physiological action for the catecholamines, whether released from adrenergic nerves or the adrenal medulla or exogenously administered, is the adrenergic receptor. This receptor is defined for present purposes as the postulated specific molecular site or structure in (or on) the effector cell with which molecules of an agonist, in this case adrenaline or some of its analogues, must react to elicit the characteristic response of the cell (see SYNAPSES AND NEUROMUSCULAR JUNCTIONS). Some of the effector tissues that have adrenergic receptors are smooth muscle, cardiac muscle, some glands, and possibly an enzyme.

Some of the physiological effects produced by adrenaline are: contraction of the dilator muscle of the pupil of the eye (mydriasis); relaxation of the smooth muscle of the bronchi; constriction of most small blood vessels; dilation of some blood vessels, notably those in skeletal muscle; increase in heart rate and force of ventricular contraction; relaxation of the smooth muscle of the intestinal tract; and either contraction or relaxation, or both, of uterine smooth muscle depending on the species tested. Electrical stimulation of appropriate sympathetic (adrenergic) nerves can produce all of the above effects with the exception of the vasodilation in skeletal muscle.

Noradrenaline, when administered, produces the same general effects as adrenaline. However, noradrenaline is less potent than adrenaline. A synthetic analogue of noradrenaline, isoproterenol, is more potent than adrenaline in relaxing some smooth muscle, producing vasodilation and increasing the rate and force of cardiac contraction.

$(HO)_2C_6H_3{-}CH(OH){-}CH_2{-}NH{-}CH(CH_3)_2$

Isoproterenol

Adrenaline is the single most potent catecholamine; it seems probable, therefore, that the adrenergic receptor is designed specifically to make the best "fit" with this catecholamine. There appear to be two types of adrenergic receptor. These are differentiated by relative responsiveness to the adrenaline analogue and by being subject to competitive block by two different types of compounds.

The *alpha* receptor is most responsive to adrenaline, least responsive to isoproterenol, blocked by such substances as phentolamine ("Regitine") or phenoxybenzamine ("Dibenzyline"), and associated with vasoconstriction, mydriasis, myometrial contraction and intestinal relaxation.

The *beta* adrenergic receptor is most responsive to isoproterenol and adrenaline, less responsive to noradrenaline, blocked by certain isoproterenol analogues such as the dichloro or naphthyl derivatives, and associated with vasodilation, relaxation of smooth muscle, and myocardial stimulation.

Adrenaline and noradrenaline have two different normal functions in the living organism. The first, as described above, is to serve as the chemical transmission link between certain motor nerves and their effector cells. The second function is in the control of carbohydrate metabolism. Adrenaline produces hepatic glyogenolysis presumably by increasing accumulation of cyclic 3,5-AMP (see GLYCOGEN AND ITS METABOLISM).

At the present time, receptors are invoked to help explain the physiological responses produced by the catecholamines. This is true only because the real site of action is unknown.

References

1. "Adrenergic Mechanisms," Ciba Foundation Symposium, Boston, Little, Brown & Co., 1960.
2. von Euler, V. S., "Noradrenaline," Springfield, Ill., Charles C. Thomas, 1956.
3. Hamilton, W. F., and Dow, P. (Editors), "Handbook of Physiology," Part 3 of Section on Circulation, Baltimore, American Physiological Society, 1965.
4. Sutherland, E. W., and Rall, T. W., *Pharmacol. Rev.*, **12**, 265–299 (1960).
5. Ahlquist, R. P., *Am. J. Pharm. Education*, **28**, 708–710 (1964).

R. P. Ahlquist

ADRENOCORTICOTROPIC HORMONE (ACTH)

The history of the adrenocorticotropic hormone began when Evans prepared adrenocorticotropic extracts from the anterior pituitary in 1932. Through the years since then, work on the structure of the hormone has been continued in the same laboratory at the University of California by Li, who isolated pure active polypeptides. Shephard in 1956 determined the structure of ACTH, and Hofmann in 1960 and Schwyzer in 1963 synthesized active molecules.

Structure. ACTH is a single-chain polypeptide of 39 amino acids and a molecular weight of about 4500. The structure may be written as follows:

Ser-Tyr-Ser-Met-Glu-His-Phe-Arg-Try-Gly-Lys-Pro-Val-Gly-Lys-Lys-Arg-Arg-Pro-Val-Lys-Val-Tyr-Pro-
1 2 3 4 5 6 7 8 9 10 11 12 13 14 15 16 17 18 19 20 21 22 23 24
(*melanocyte activity* 1–13) (*ACTH activity* 1–24)

NH_2 (on Glu 33)

-Asp-Gly-Glu-Ala-Glu-Asp-Ser-Ala-Glu-Ala-Phe-Pro-Leu-Glu-Phe
25 26 27 28 29 30 31 32 33 34 35 36 37 38 39
(*species differences* 25–33) (*MSH suppression* 14–39)

Beef ACTH

The first 13 amino acids are identical with alpha melanocyte stimulating hormone (MSH); the first 23 or 24 appear to be essentially responsible for the ACTH action; species differences occur between 24 and 33; and it has been suggested that the last 26 amino acids largely suppress the melanocyte effect of the first 13 in complete ACTH.

Biological Activities. The primary action of ACTH is to stimulate the adrenal cortex to produce various ADRENAL CORTICAL HORMONES, especially the glycosteroids. The production of hydrocortisone and cortisone is specifically controlled, but there may be a sustaining element in the production of aldosterone and other steroids as well. The cortical steroids carry on peripheral actions in the tissues and also inhibit further liberation of ACTH.

ACTH is actually only one member in a chain of actions which results in the production of adrenocortical hormones. The sequence begins with the liberation of corticotropin releasing factor (CRF) from the hypothalamus as the result of stimulation from several sources. These include trauma (stress), which influences CRF via neural pathways; drugs, such as barbiturates and anesthetics, which act via the reticular activating center; and influences from other parts of the brain which may be transmitted through the hypothalamus. The releasing factor appears to be a polypeptide, the first 10 amino acids of which are similar to FSH (see GONADOTROPIC HORMONES). This substance is transported to the anterior pituitary with resultant release of ACTH into the blood stream.

In man and experimental animals, the most obvious manifestations from the administration of ACTH are similar to those from the administration of cortisone. Thus there are increased protein breakdown, increased gluconeogenesis, and a number of secondary effects. In diseased states, inflammation may be inhibited and allergic phenomena suppressed.

The fundamental mechanism by which ACTH acts has not been determined. It has been suggested that ACTH may promote the accumulation of adenosine-3′,5′-monophosphate which contributes to the energy requirements for the synthesis of corticosteroids from cholesterol.

ACTH appears to have some actions in addition to those on the adrenal cortex. A lipolytic action has been demonstrated in cultured cells of rat fat, possibly the result of activation of a lipolytic enzyme (see FAT MOBILIZATION). ACTH in the absence of the adrenals may cause darkening of the skin, presumably a manifestation of the melanocyte stimulating effect of the first portion of the molecule.

Little is known about the fate of ACTH in the body. It is destroyed in the alimentary tract and therefore must be administered parenterally. Its effects rarely last more than 6 hours.

Toxicity. Large doses may stimulate the formation of levels of cortical hormones adequate to produce the characteristic toxicity of these hormones, but such effects are less common than with the potent steroids. There may be moderate salt and water retention, hypertension, darkening of the skin, and slight androgenic effects. The asthenia upon withdrawal is usually shorter and less severe than that from cortisone.

Uses. ACTH has much the same indications as the many cortical steroids, natural and synthetic, which stem from cortisone. ACTH has the theoretical advantage of producing less atrophy of the adrenal cortex, but this superiority is largely overbalanced by the greater potency available from the natural and synthetic cortical steroids. Also, the latter may be administered by mouth in many cases, and allergic sensitivity is unlikely. Thus ACTH, though a useful agent, is eclipsed clinically in most instances by the more available cortical steroids.

The most frequent indications for use are inflammatory states, such as rheumatoid arthritis, ulcerative colitis, and exfoliative determatitis, and various hypersensitivity reactions. The following preparations of ACTH are available commercially:

(a) Corticotropin (ACTH; Acthar; Cortrophin). Dose: 10–20 units, 4 times a day, intramuscularly or intravenously.

(b) Corticotropin, Purified (ACTH, purified; Depo-ACTH; and others), gel suspensions which give prolonged effects. Dose: 40–80 units, once daily, subcutaneously or intramuscularly.

(c) Corticotropin-Zinc Hydroxide (Cortrophin-Zinc). Dose: 40–80 units, once daily, intramuscularly (long acting).

References

1. Smelik, P., and Sawyer, C., "Pharmacological Control of Adrenocorticoid and Gonadal Secretions," *Ann. Rev. Pharmacol.*, **2**, 313 (1962).
2. Li, C., "The ACTH Molecule," *Sci. Am.* **209**, 46 (July 1963).
3. Michael, M., "Uses and Misuses of Adrenal Corticosteroids," *J. Am. Med. Assoc.*, **185**, 280 (1963).

Windsor C. Cutting

ADSORPTION

Adsorption is the concentrating or holding of a chemical species at a surface. The concentration changes due to adsorption from a bulk phase (such as a solution or a gas) onto a solid surface are detectable only if the surface area is very large, as with activated charcoal, silica gel, ION EXCHANGE RESINS, or any material in a finely divided state. Several different types of adsorption isotherms (plots of amount adsorbed *vs* concentration in the bulk phase) have been predicted theoretically. Of these, the Langmuir isotherm describes many practical situations where the amount adsorbed increases linearly with concentration up to a plateau region, above which little further adsorption occurs. This behavior is explained by the holding of the adsorbed species at special locations or sites on the surface. A molecule so held is frequently favorably oriented for reaction with other molecules at nearby sites or in the bulk phase adjacent to the bound molecule. This catalytic effect of finely divided or colloidal materials is of importance in many processes, ranging from the industrial cracking of petroleum to enzyme-substrate interactions [see ENZYMES (KINETICS)].

The variety of the forces involved in adsorption make the separations obtained in gas, column, and other types of CHROMATOGRAPHY possible, and are important in the intermolecular associations of cytoplasmic components, particularly on the surfaces of membranes. These forces range from covalent through ionic to van der Waals, resulting in very specific interactions between a surface and the molecules of the bulk phase. This selectivity is evident in the dye-membrane reactions of histological staining (see HISTOCHEMICAL METHODS) and in differential drug activity. The interaction between the surface and *all* components of a solution is illustrated particularly well in the phenomenon of negative adsorption, in which the concentration of a solute species increases as adsorption takes place; this behavior is understood when it is realized that the solvent has been more strongly adsorbed than the solute.

Detailed information on adsorption can be found in references 1–4. Useful information on biochemical aspects of the subject may also be found in references 5 and 6.

References

1. "Advances in Catalysis and Related Subjects," New York, Academic Press (published annually).
2. Adamson, A. W., "Physical Chemistry of Surfaces," New York, Interscience Publishers, 1960.
3. Young, D. M., and Crowell, A. D., "Physical Adsorption of Gases," London, Butterworths, 1962.
4. Hayward, D. O., and Trapnell, B. M. W., "Chemisorption," Second Edition, Washington, Butterworths, 1964.
5. Boyer, P. D., Lardy, H., and Myrbäck, K. (Editors), "The Enzymes," Vol. 1, p. 193, New York, Academic Press, 1959.
6. West, E. S., and Todd, W. R., "Textbook of Biochemistry," Ch. 5, New York, The Macmillan Co., 1961.

Marcellus T. Coltharp

AGING (BIOCHEMICAL ASPECTS). See GERONTOLOGY.

α-ALANINE

α-Alanine or simply alanine, $CH_3CH(NH_2)$—COOH, is the simplest optically active amino acid (see ASYMMETRY) obtained by hydrolysis of PROTEINS. The most convenient protein source is silk fibroin which yields about 30% of its weight in this one amino acid. Like other protein-derived AMINO ACIDS, alanine from protein sources belongs to the L-family. Its rotatory power in 6*N* HCl solution is $[\alpha]_D = +14.5°$. Alanine is nonessential in the nutrition of mammals and birds.

```
    COOH              COOH
     |                 |
H2N—C—H            H—C—NH2
     |                 |
    CH3               CH3
 L-Alanine         D-Alanine
```

β-ALANINE

β-Alanine, $CH_2(NH_2)$—CH_2—COOH, is not derived from proteins. It was known as a constituent of CARNOSINE and ANSERINE before its importance as a constituent of PANTOTHENIC ACID and COENZYME A was discovered. β-Alanine possesses no asymmetric carbon atom.

ALBINISM

Albinism is a condition of absence of SKIN PIGMENTATION in animals, resulting from the genetic absence of the enzyme tyrosinase which normally acts in the conversion of TYROSINE through the intermediate dihydroxyphenylalanine (DOPA) to the pigment melanin (see AROMATIC AMINO ACIDS). Albinism is one of a group of metabolic disorders termed "inborn errors of metabolism" by Garrod in 1908. The study of such diseases, which included ALKAPTONURIA, phenylketonuria, and the porphyrias (see METABOLIC DISEASES) gave early support to the concept that one GENE, or unit of inheritance, exerts close and direct control over one enzyme, catalyzing a single step of metabolic conversion (see also BIOCHEMICAL GENETICS).

ALBUMINS

In its classical connotation, the term albumin referred to any unconjugated heat-coagulable PROTEIN which is soluble in pure water, as distinct from such water-insoluble proteins as globulins (salt soluble), glutelins (alkali soluble) and prolamins (soluble in ethanol-water mixtures). A further point of distinction between albumins and globulins was based on the failure of albumins to precipitate on half-saturation of their aqueous solutions with ammonium sulfate. This classification scheme has lost favor because many proteins do not fit unambiguously into any one of these classes, but several important proteins retain names based on their solubility behavior. The term albumin must be distinguished from the older form "albumen" which was applied in an even broader connotation to water-soluble protein systems such as egg white.

The egg white contains a number of proteins of which *ovalbumin*, the major component, is water soluble, readily crystallizable and rather well characterized. Ovalbumin contains a well-balanced complement of the dietarily essential amino acids, in accord with its presumed physiological role as a source of nutrient amino acids for the developing embryo. It also contains a small amount of covalently bonded carbohydrate as well as esterified phosphate groups so that it is a conjugated rather than a simple protein. Another water-soluble protein of egg white, *conalbumin*, is present in relatively small amounts, crystallizes with more difficulty, and has a powerful affinity for ferric iron. Limited digestion of ovalbumin with a bacterial proteinase, subtilisin, gives rise to a new crystallizable protein named plakalbumin (in recognition of the fact that it crystallizes as platelets rather than the typical needle-like crystals formed by ovalbumin).

Lactalbumin, a name applied earlier to a crystalline protein fraction isolated from milk whey, is now known to consist of a number of protein fractions of which the two major ones are *α-lactalbumin* and *β-lactoglobulin.* In addition, it contains some serum albumin and a number of enzymes.

By far the best known and most studied albumins are the *serum* (or *plasma*) *albumins* which constitute the major protein component of the blood serum of vertebrates (see PLASMA PROTEINS). Electrophoretic analysis of the serums of ten mammals show the albumin component to comprise approximately 30–60% of the total protein. The albumin component of equine serum was first crystallized by ammonium sulfate precipitation. A better method, applicable to many or most mammalian serum albumins, is crystallization from aqueous ethanol. Crystallization is markedly improved by addition of trace amounts of decanol or other higher alcohols or of fatty acids. The most often studied serum albumins are obtained from bovine and human blood. These albumins can be prepared virtually free of carbohydrate and are essentially unconjugated proteins. Generally they do contain small amounts of various lipids, particularly fatty acids, which are considered to be contaminants rather than essential structural components. The serum albumins are extremely soluble in water, are not precipitated by half-saturated ammonium sulfate and are heat coagulable, so they fit the classical definition in all respects.

There is a close similarity between the serum albumins of various mammals. All appear to contain a single peptide chain of approximately 550–600 amino acid residues (molecular weight 65,000–70,000), which chain, in the cases studied, terminates in an aspartic acid residue at the amino end and in either leucine, alanine or valine at the carboxyl terminus. The resemblance between human and bovine albumins is apparent in the amino acid compositions reported in Table 1. It will be seen that each of these proteins contains a very large number of titratable groups, namely approximately 100 carboxyl groups (aspartic plus glutamic acid residues less amide ammonia) and 100 basic groups (arginine, lysine and histidine) plus the phenolic groups of tyrosine. All of the half-cystine residues except one exist in the oxidized state as cystine, thus introducing numerous (17–18) disulfide cross-linkages into the protein molecule. The odd half-cystine residue is reactive in only a part of the protein molecules; thus even crystallized samples exhibit less than one reactive sulfhydryl group per molecule, typically 0.5–0.6. This is probably due in part to a masking of some sulfhydryl groups by the three-dimensional folding of the protein and in part to mixed disulfide formation with cysteine or glutathione. That fraction of the total albumin possessing a reactive sulfhydryl group can be separated from whole albumin by crystallization as a mercury dimer and is named mercaptalbumin. In fact, the analyses reported in Table 1 were performed on mercaptalbumins rather than whole albumins.

TABLE 1. AMINO ACID COMPOSITION OF BOVINE AND HUMAN SERUM ALBUMINS[a]

Amino Acid	Number of Residues per molecule[b] Human	Bovine
Glycine	12	16
Alanine	62	46
Valine	41–42	36–37
Leucine	60	62
Isoleucine	8	14
Proline	26	30
Phenylalanine	30	27
Tyrosine	17–18	20
Tryptophan	1	2
Serine	19	26
Threonine	24–25	34
Half-cystine	37	36
Methionine	6	4
Arginine	24	22–23
Histidine	16	17
Lysine	62	62
Aspartic acid	51	54
Glutamic acid	78–79	77–78
Amide NH_3	34–35	32–33

[a] Reprinted with permission from Spahr, P. F., and Edsall, J. T., *J. Biol. Chem.*, **239**, 850 (1964).
[b] To nearest integer, based on an assumed molecular weight of 66,000.

Other than the difference in reactivity of sulfhydryl groups, no differences between the two components of plasma albumins are known.

The physical chemical behavior of the plasma albumins in solution has been studied in great detail and emphasizes two dramatic properties of the protein, namely the ability to bind a large variety of substances, both ionic and non-ionic, and the ability of the protein to undergo extensive reversible alterations in three-dimensional folded structure. Probably the two types of phenomena are interrelated. There is evidence that the single polypeptide chain may be folded into two similar globular units which, in the native protein in neutral solution, are closely associated with one another. Probably the association is due mainly to the existence of largely hydrophobic surfaces which tend to associate as a means of escaping the aqueous environment (hydrophobic interaction). In acid solution, these units separate to some extent although they are still bonded through the common peptide chain (N—F transformation). The resultant form has strikingly different properties from the native protein in that it is less soluble in water and has a markedly reduced affinity for detergent ions, lipids, and even hydrocarbons. It seems probable that the binding of some or all of these substances may take place in the hydrophobic interface in the native form of the protein. The acid structural change is also accompanied by a normalization of the titration behavior of a large number of carboxyl groups which are masked or stabilized in the anionic form in the native protein. There is also evidence for the exposure of several tyrosyl residues and possibly even peptide amide groups in the N—F transformation. At lower pH, an even more drastic uncoiling of the protein molecule takes place (see PROTEINS, BINDING FORCES IN SECONDARY AND TERTIARY STRUCTURE).

Very recent evidence suggests that even crystallized mercaptalbumin is far from homogeneous. The protein probably consists of a continuum of molecular forms which differ in the ease with which they undergo the reversible acid transformations as well in their lability toward irreversible thermal denaturation. The structural basis for this "microheterogeneity" is not yet clear, but it may reflect some randomization of the pairing of the large number of half-cystine residues in the disulfide cross-linkages.

As the major protein of plasma, and in view of its relatively low molecular weight and high net negative electrical charge (as compared with other plasma proteins as they exist at normal physiological pH), the plasma albumins account for a large fraction of the colloid osmotic pressure of blood. They are thought, therefore, to play an important role in the maintenance of plasma volume. During World War II, intensive efforts (at Harvard Medical School under the late Professor Edwin J. Cohn) were directed toward the production of crystallized human plasma albumin for clinical use in the treatment of shock. Such preparations also proved effective in the control of edema resulting from certain kidney disorders. The unusual binding ability of the plasma albumins suggests that these proteins may also serve a vital role as transport agents for a variety of low molecular weight compounds in the blood. Specific roles which have been proposed are the mobilization of lipids (see FAT MOBILIZATION), the transport of hormones between one tissue and another, and the transport of waste materials to the kidney. Many drugs, *e.g.*, sulfonamide derivatives, are preferentially bound by the albumin component.

For further information on the preparation and properties of albumins and especially plasma albumins, the articles listed in the References are recommended.

References

1. FEVOLD, H. L., "Egg Proteins," *Advan. Protein Chem.*, **6**, 188 (1951).
2. MCMEEKIN, T. L., AND POLIS, D. B., "Milk Proteins," *Advan. Protein Chem.*, **5**, 201 (1949).
3. EDSALL, J. T., "The Plasma Proteins," *Advan. Protein Chem.*, **3**, 384 (1947).
4. PHELPS, R. A., AND PUTNAM, F. W., "Chemical Composition and Molecular Parameters of Purified Plasma Proteins" in "The Plasma Proteins" (PUTNAM, F. W., ED.), Ch. 5, New York, Academic Press, 1960.
5. FOSTER, J. F., "Plasma Albumin," in "The Plasma Proteins" (PUTNAM, F. W. ED.), Ch. 6, New York, Academic Press, 1960.

JOSEPH F. FOSTER

ALCOHOL

This term, unless otherwise qualified, refers to *ethyl alcohol*, CH_3CH_2OH (ethanol). As a product of ALCOHOLIC FERMENTATION it has claimed the attention of biochemists for many decades. Its production led to the discovery of the *enzymes* (meaning "in yeast") which, as biological catalysts, play central roles in modern biochemistry. Many alcohols other than ethyl alcohol are also biochemical products. Ethyl alcohol can be produced synthetically from acetylene or ethylene. An azeotropic ethyl alcohol–water mixture, containing 94.9% alcohol by volume (commonly called "95% alcohol") is the product obtained by distillation from alcohol–water mixtures of other compositions. Pure anhydrous ethyl alcohol must be produced by special procedures, such as by distillation of "95% alcohol" with appropriate amounts of benzene, so as to remove initially a ternary (alcohol–benzene–water) azeotropic mixture.

ALCOHOL DEHYDROGENASE

Alcohol dehydrogenase (ADH) catalyzes the reversible oxidation of alcohols to aldehydes, employing the coenzyme diphosphopyridine nucleotide (DPN; alternatively termed NICOTINAMIDE ADENINE DINUCLEOTIDE, NAD) as the principal hydrogen acceptor. The general reaction is expressed by the equation:

$$RCH_2OH + DPN^+ \rightleftharpoons RCHO + DPNH + H^+$$

The enzyme has been isolated in crystalline form from yeast and equine liver. It has been purified partially from the livers of humans, fish and rats, and from wheat germ and *Leuconostoc mesenteroides*. ADH activity has also been found in a large number of animal and plant tissues as well as in many microorganisms, demonstrating the general importance of the enzyme in metabolic processes.

In plants and microorganisms, ADH participates in the *anaerobic* breakdown or FERMENTATION of sugars. Glucose is metabolized to acetaldehyde which is then converted to ethanol by the action of this enzyme. Simultaneously, DPN is regenerated from DPNH and reutilized in the glycolytic cycle. Organisms which depend upon *aerobic* pathways of glucose metabolism may utilize ethanol as a source of carbon and energy. Ethanol is oxidized to acetaldehyde which is then converted to carbon dioxide and water by the enzymes of the Krebs tricarboxylic acid cycle (see CITRIC ACID CYCLE). The enzyme also plays an important role in the visual process. It catalyzes the reversible oxidation of vitamin A_1, which is an alcohol, to retinene or vitamin A_1 aldehyde which is a component of the visual pigment, rhodopsin (see VISUAL PIGMENTS). Moreover, ADH is involved in the degradation of fructose: it reduces glyceraldehyde, formed through the cleavage of fructose, to glycerol.

The enzymatic, physical and chemical properties of the two crystalline enzymes from *yeast* and *horse liver* have been studied and show many similarities. However, the enzymes also differ significantly in many respects, reflecting genetic variation as well as perhaps a difference in the function they serve in metabolism, since yeast depends in large measure upon anaerobic processes while mammals are aerobic. Both enzymes contain stoichiometric amounts of the element ZINC as an integral part of the molecule and, hence, are zinc METALLOENZYMES. Both catalyze the reversible transfer of hydrogen, as the hydride ion, from substrate to coenzyme directly, without participation of the solvent. In the process, a ternary complex consisting of enzyme-coenzyme-substrate is presumably formed. Moreover, the transfer of hydrogen is stereospecific, indicating that the substrate and coenzyme molecules bind to the enzyme surface in such a manner that they are spatially oriented in a fixed position with respect to each other.

Yeast ADH is a globular protein containing 4 gram atoms of zinc per 150,000 molecular weight of protein. The zinc atoms are firmly incorporated into the protein matrix during the biosynthesis of the enzyme and are essential for its catalytic function. The enzyme readily oxidizes a number of primary straight-chain alcohols, but attacks branched-chain alcohols, secondary alcohols and polyalcohols very slowly. It is inhibited by metal complexing agents, a number of metals from the first transition and group IIB of the periodic table, reagents which react with sulfhydryl groups, pyridine and purine bases, and by molecules which are structurally similar to the coenzyme, DPN(H). The enzyme has four independently functioning active catalytic centers since it binds four moles of DPN(H) per mole of protein. Zinc is a component of the active centers, and the coenzymes bind at or near the metal site. Removal of the zinc atoms by chelating agents leads to dissociation of the protein molecule into four inactive subunits, indicating that the enzyme consists of four polypeptide chains held together in part by the interaction of zinc with these chains. In addition to the four zinc atoms, four sulfhydryl groups per mole of protein have been found to be essential for the catalytic function of this enzyme.

In contrast, *horse liver ADH* has a molecular weight of only 84,000 and contains two functionally essential zinc atoms per molecule of enzyme. The zinc atoms are also firmly bound to the protein but their removal brings about concomitant denaturation and aggregation of the protein. Its catalytic activity toward ethanol is approximately one-tenth that of the yeast enzyme. However, it has broader substrate specificity. Liver ADH readily oxidizes a variety of primary, secondary and aromatic alcohols and polyalcohols, including vitamin A and ethylene glycol. It also oxidizes formaldehyde (to formic acid) and other aldehydes, and can employ triphosphopyridine nucleotide (TPN, or NADP) as coenzyme. The enzyme is inhibited by metal chelating agents, metals from groups IB and IIB of the periodic table, sulfhydryl-specific reagents, purine and pyrimidine nucleotides, and substrate homologues such as fatty acids and fatty acid amides. The enzyme binds two moles of DPN(H) per mole of protein and has two active catalytic centers as is also apparent from tryptic hydrolysis and fingerprinting of the protein. The coenzyme binds through the adenine ribonucleotide portion of the molecule at or near the catalytically essential zinc atoms of the enzyme. As with yeast ADH, sulfhydryl groups are essential for the catalytic function of liver ADH, one for each of the two active catalytic centers. The amino acid sequence surrounding these essential —SH groups of liver ADH has been identified. It is similar but not identical to that of yeast ADH.

Recent studies of the partially purified enzyme from *human livers* show that it is more similar structurally and enzymatically to horse liver ADH than to the yeast enzyme. It has all the characteristics of a zinc metalloenzyme and has broad substrate specificity, including methanol and ethylene glycol. Such studies are significant not only in relation to biochemical evolution but also medically. They provide a means for studying the molecular basis of the toxicity of alcohols, such as methanol and ethylene glycol, as well as normal and abnormal ethanol metabolism in man.

References

1. VALLEE, B. L., in *Proc. Intern. Congr. Biochem., 4th* (NEURATH, H., AND TUPPY, H., EDITORS), **8**, 138 (1960).
2. SUND, H., AND THEORELL, H., in "THE Enzymes" (BOYER, P. D., LARDY, H., AND MYRBÄCK, K., EDITORS), Second edition, Vol. 7, p. 25, New York, Academic Press, 1963.
3. HARRIS, I., *Nature*, **203**, 30 (1964).
4. LI, T. K., AND VALLEE, B. L., *Biochemistry*, **3**, 869 (1964).
5. VON WARTBURG, J.-P., BETHUNE, J. L., AND VALLEE, B. L., *Biochemistry*, **3**, 1775 (1964).

TING-KAI LI

ALCOHOL METABOLISM

The generally accepted pathway for alcohol metabolism involves an initial oxidation to acetaldehyde and a subsequent conversion to acetic acid and acetyl CoA. Once the 2-carbon unit of alcohol has been converted to acetyl CoA, it is in the "mainstream" of metabolic reactions, and at that point, it is chemically no different from the acetyl CoA which is formed from carbohydrate (see CARBOHYDRATE METABOLISM), fat (see FATTY ACID METABOLISM) or protein. Since the major oxidative pathway for acetyl CoA is via the CITRIC ACID CYCLE, the major oxidative pathway for alcohol beyond the acetyl CoA stage is also via the citric acid cycle. Since acetyl CoA can be converted to other substances (fatty acids, cholesterol, ketone bodies, acetylated amines, etc.), the carbons from alcohol can also be found in these other substances. Alcohol is, therefore, a foodstuff from which the animal can derive energy by means of these metabolic oxidations; it differs from the usual foodstuffs inasmuch as it also produces pharmacological effects.

Sites and Pathway of Alcohol Metabolism. Only liver and kidney are capable of oxidizing alcohol to acetaldehyde at a significant rate. Liver is quantitatively the most important organ in the body for both the initial oxidation of alcohol and the subsequent oxidation of the acetaldehyde to acetic acid. The enzyme ALCOHOL DEHYDROGENASE, which is responsible for the oxidation of alcohol to acetaldehyde, is present in liver and kidney but absent from muscle and brain. Blood normally contains little or no acetaldehyde, but this intermediate appears in the blood of most species at somewhat less than 1% of the alcohol concentration during alcohol metabolism. Blood acetaldehyde increases to much higher levels during alcohol metabolism when the animal has been pretreated with a drug such as Antabuse or cyanamide, and the "Antabuse reaction" is caused by, or at least is related to, such high levels of blood acetaldehyde. There are a number of enzymes in mammalian tissues which are capable of utilizing acetaldehyde as a substrate. The one which appears to be responsible for much of the acetaldehyde metabolism *in vivo* is a DPN-requiring dehydrogenase which converts it to acetic acid. Blood normally contains little or no free acetic acid, but detectable amounts are found in the blood leaving the liver during alcohol metabolism. Appreciable quantities of the acetic acid which is formed from alcohol in the liver pass via the blood stream to other organs for ultimate oxidation. Muscle, for example, cannot utilize alcohol *per se* at a significant rate, but can utilize the acetic acid and the ketone bodies which are formed from the alcohol by the liver. Brain tissue is also unable to oxidize alcohol directly, but does receive some acetaldehyde and acetic acid via the blood-stream during alcohol metabolism. *In vitro* experiments have demonstrated the incorporation of ethanol carbons into the components of the citric acid cycle in the same way that acetate is metabolized, and the citric acid cycle pathway is presumably utilized *in vivo* for the final oxidation of the acetate derived from the alcohol.

While there is no evidence which directly contradicts this scheme as the major pathway of alcohol metabolism, there are two observations which require further explanation: (1) alcohol is a better precursor than acetate for the synthesis of fatty acids, cholesterol, and the acetyl group of an acetylated amine; (2) the rate at which each of the two carbons of ethanol is converted to CO_2 *in vivo* is somewhat different from the corresponding rates obtained with acetate. Such differences suggest a possible divergence in the pathways of alcohol and acetate metabolism, and such a divergence could theoretically take place at the acetaldehyde step, *i.e.*, acetaldehyde could theoretically give rise to an intermediate other than acetate or acetyl CoA. While such a possibility is not precluded, all intermediates currently identified as arising from alcohol can also be obtained from acetate by known metabolic pathways. Another possible explanation for the observed metabolic differences between alcohol and acetate is that the conversion of alcohol to acetyl CoA takes place in the soluble cytoplasm of the cell, while a larger percentage of the acetate is metabolized in the mitochondria. Hence, a larger percentage of the labeled acetyl CoA derived from alcohol would be available for synthetic reactions in the soluble or microsomal fractions, while a larger percentage of the labeled acetyl CoA derived from acetate would be oxidized through the citric acid cycle in the mitochondria.

If there are no unique intermediates in the metabolism of alcohol yet to be discovered, then alcohol and acetaldehyde are the only substances present in the body during alcohol metabolism which could not also be derived from the usual foodstuffs. Traces of alcohol and acetaldehyde are found in the body in the absence of any ingested alcohol, but they apparently have no physiological significance. The phenomenon of intoxication appears to be due to the alcohol *per se*, rather than to the acetaldehyde, since the level of acetaldehyde which is found in the blood during alcohol metabolism is too low to produce the symptoms of intoxication.

Rate of Alcohol Metabolism. Alcohol is absorbed passively by diffusion from the gastrointestinal tract without any upper limit to its rate of absorption. Food delays the emptying time of the stomach and thereby slows the absorption of any alcohol taken simultaneously, but there is no specific mechanism available in the intestine to control the rate of absorption of alcohol. Once absorbed, it is distributed throughout the body (including the brain and cerebrospinal fluid) in proportion to the water content of the individual tissues. Following its distribution throughout the body, the rate of disappearance of alcohol from the blood can be used to measure the rate of alcohol metabolism. There are no selective excretory mechanisms available to remove alcohol from the body fluids. It cannot be stored as such, and only a relatively small percentage is converted to fat. Once ingested, it can be removed only by metabolic reactions, and the bulk of it is simply oxidized to CO_2 while other substrates are spared. Alcohol is present in urine and respiratory air in proportion to the blood alcohol level, but the total amount excreted by these routes is only a few per cent of the amount ingested.

The blood alcohol disappearance curve is unusual among metabolites because it is linear with time. This results from a saturation of the alcohol dehydrogenase with substrate at the very low level of 0.01%, and blood alcohol concentrations higher than this do not increase the reaction rate. The slowest step in the sequence of reactions by which alcohol is metabolized is the initial oxidation to acetaldehyde. The oxidation-reduction potentials for the two systems involved are very unfavorable for the oxidation of alcohol to acetaldehyde by the DPN-enzyme complex, and the reaction stops *in vitro* at less than 1% completion. This reaction effectively oxidizes alcohol *in vivo* only because the end products (acetaldehyde and DPNH) are removed from the sphere of action by other metabolic reactions. Under these conditions, the rate at which the DPNH is reoxidized is probably a major factor in determining the rate of alcohol metabolism. Since the oxidation of alcohol to acetaldehyde is carried out in the soluble cytoplasm of the cell and is not influenced by the metabolic needs of the animal, the rate of alcohol metabolism is not influenced by hyperthyroidism, exposure to cold, or muscular exercise.

DPNH can be reoxidized by the mitochondrial electron transport chain, or by a coupled oxidation-reduction reaction with another substrate, such as pyruvate. A number of substances have been reported to increase the rate of alcohol metabolism when administered to the intact animal. These include insulin plus glucose, fructose, pyruvate and alanine. This has been a controversial area because such substances give an effect only when the initial rate of alcohol metabolism is sluggish. Presumably these substances increase the rate of alcohol metabolism by increasing the rate of reoxidation of DPNH.

During alcohol metabolism, there is more DPNH and a more reducing atmosphere in the liver cell, and this affects other metabolism reactions taking place simultaneously. Norepinephrine is deaminated in the usual way during alcohol metabolism, but is then reduced to the corresponding glycol instead of being oxidized to the usual acid metabolite, 3-methoxy-4-hydroxymandelic acid. The increased DPNH concentration is also responsible for the lowered galactose tolerance observed during alcohol metabolism.

The human metabolizes from 100–250 mg of alcohol per kg body weight per hour; this is equivalent to a decrease in blood alcohol concentration of 10–25 mg/100 ml/hr. The larger value is equivalent to a total metabolism of from one-fifth of a gallon to one quart of 100-proof whiskey per 24 hours for a 70-kg man, and this is usually considered to be the maximum amount of alcohol which can be metabolized per day. However, it is known that a high blood alcohol level triggers off a high rate of metabolism. The above rates were determined at moderate blood levels, so it is possible that higher rates of metabolism and a larger maximum might be observed at high blood levels. The rate is not altered by the habitual use of alcohol by the chronic alcoholic.

A maximum blood level of about 0.1% is reached in 2 hours after the consumption of 1 ml/kg (4–5 ounces of whiskey), and 1 oz/hr will maintain it at that level thereafter. The combustion of alcohol yields 7 cal/g. One ounce of 100-proof whiskey or 12 ounces of 4% beer or 4 ounces of 12% wine all supply approximately the same amount of alcohol and the same 81–84 calories. The consumption of one-fifth of whiskey per day supplies 2240 calories, but such a distilled beverage supplies nothing but calories. When taken in small amounts, the calories from alcohol appear to be utilized by the animal in a normal way, but when alcohol supplies a major portion of the daily calories, it is not as good for the economy of the animal and the calories are not utilized as efficiently as an equal number of calories derived from carbohydrate or fat.

References

1. Westerfeld, W. W., *Am. J. Clin. Nutr.*, **9**, 426 (1961); *Texas Rep. Biol. Med.*, **13**, 559 (1955).
2. Westerfeld, W. W., and Schulman, M. P., *J. Am. Med. Assoc.*, **170**, 197 (1959).
3. Jacobsen, E., *Pharmacol. Rev.*, **4**, 107 (1952).

W. W. Westerfeld

ALCOHOLIC FERMENTATION

Alcoholic fermentation is the name given to the process by which certain microorganisms, particularly strains of yeasts, convert sugars into a mixture of ethanol (ethyl alcohol) and carbon dioxide (see FERMENTATIONS; GLYCOLYSIS). Brewers, wine-makers, distillers and bakers have for centuries been exploiting this ability of yeasts to ferment sugars to alcohol and carbon dioxide, and the term "fermentation" (Latin, *fermentare*, to boil) was first used to describe the frothing appearance of worts and musts during the making of beers and wines. The alcoholic fermentation of sugars was almost certainly discovered by accident, probably as a result of some cereal infusion becoming infected with yeasts and other microorganisms. This discovery is thought to have taken place well over 10,000 years ago among the ancient civilizations inhabiting the Nile Valley. Although the alcoholic fermentation industries have progressed considerably over the years, they are still nevertheless very empirical in their approach and only during the past few decades has there been any appreciable incursion of the sciences of biochemistry and microbiology. This is mainly because so little is known about the biochemical basis of the quality of an alcoholic beverage so that manufacturers are loth to introduce variations into processes that have for centuries provided organoleptically acceptable products. The ethanol produced by alcoholic fermentation is responsible for much of the desired potency in an alcoholic beverage, but many of the more subtle qualities of taste and aroma are caused by qualitatively minor products of the yeast fermentation, particularly higher alcohols and esters of these alcohols.[5] The recent introduction of the sensitive gas chromatographic technique has made possible an assessment of these organoleptically important volatile products in alcoholic beverages.

Beer. The essential step in the brewing of beer involves adding yeast to a water extract of germinated barley (called malt wort) that has been flavored with hops.[3] Brewing yeasts, which are specially selected strains of *Saccharomyces cerevisiae* and *S. carlsbergensis*, ferment the principal sugars in the wort (maltose and glucose) to give beer. The alcohol content of beers ranges from 3–5%. But the fermentation of malt wort comes only as a climax to a whole series of operations, and the taste, aroma and appearance of the beer depend to a large extent on the stages leading to the fermentation, namely, the preparation of the malt wort.[1,4] Two main types of brewing yeasts are recognized. Some strains tend to rise to the top of the wort during fermentation and so are termed top-fermentation yeasts. These are widely used in Britain and in some parts of the United States for brewing ales. But the bulk of the world's beer is produced by bottom-fermentation yeasts.

The brewer looks for two main qualities in selecting a yeast. The first of these is the ability to bring about a vigorous fermentation in malt wort. Presumably these yeasts are particularly well endowed with the enzymes of the Embden-Meyerhof-Parnas scheme (see CARBOHYDRATE METABOLISM; GLYCOLYSIS) as well as with efficient mechanisms for transporting sugars into the yeast. A brewer also requires that the yeast be capable of flocculating or depositing itself from the beer at the end of the fermentation. The flocculating ability of a yeast depends on the chemical and physical properties of the outer layers of the cell wall, and there is some evidence that flocculating yeasts differ from non-flocculating strains in having a greater proportion of the polysaccharide mannan in their cell walls.[2]

Wine. Wine-makers in the United States resemble brewers in that they use pure cultures of selected strains of yeast (*S. cerevisiae* var. *ellipsoideus*) to ferment the grape juice sugars. European vintners, however, rely on a spontaneous fermentation of the juice brought about by some of the yeasts that occur naturally as a "bloom" on the surface of the grapes. The juice expressed from the grapes contains 15–20% sugar, chief among which are fructose and glucose and, although this is ample for the yeast, extra sucrose is occasionally added to the juice. The alcohol content of the fermented juice or wine ranges from 6–10%. Those wineries which rely on a spontaneous fermentation of the juice usually treat it with sulfur dioxide in amounts ranging from 100–450 ppm. This treatment kills off most of the microorganisms in the bloom with the exception of certain strains of *S. cerevisiae* var. *ellipsoideus* that are resistant to sulfur dioxide.

Spirits. Manufacturers of potable spirits such as whiskey, brandy, gin and rum also use yeasts to carry out an alcoholic fermentation of sugars. The fermented liquid is distilled at least once to give the potable spirit. Whiskey is obtained by the distillation of a liquid produced by the alcoholic fermentation of an extract of malted barley, corn or rye. Brandy results from the distillation of fermented grape juice, while gin is made by distilling fermented maize or rye worts and flavoring the distillate with herbs. Rum is the alcoholic distillate from the fermented juices of the sugar cane or sugar molasses. Distillery fermentations are brought about by selected strains of *Saccharomyces cerevisiae* although rum is occasionally fermented by strains of the related yeast, *Schizosaccharomyces*. The distiller aims to convert as much sugar as possible into alcohol during the fermentation and this is reflected in the choice of yeast strains. In fermenting grain worts, for example, strains are chosen that are capable of hydrolyzing the oligosaccharides that remain after α- and β-amylases have attacked the grain STARCH during germination of the barley. Compared with brewer's and wine yeasts, distiller's strains of *S. cerevisiae* are also more alcohol-tolerant.

Ethanol ranks second only to water as a solvent in the chemical industry. Some industrial ethanol is made by an alcoholic fermentation of inexpensive sugar-containing raw materials such as molasses and sulfite waste liquid. The fermentation process is similar in many ways to that used to produce potable spirits, and the strains of *S. cerevisiae* used have the same general characteristics as those used by the distiller. In the United States and much of Western Europe, however, the manufacture of fermentation alcohol has been almost completely replaced by a synthetic process using raw materials from the petrochemical industry. But in other areas of the world where petrochemicals are scarce and saccharine raw materials are readily available, alcohol is still produced by fermentation.

The baker also exploits the ability of yeasts to carry out an alcoholic fermentation of sugars, but with the aim of using the carbon dioxide rather than the ethanol. When baker's yeasts (selected strains of *S. cerevisiae*) are mixed with the dough, there is an alcoholic fermentation of the sugars present in the dough and subsequently of those sugars liberated by the action of the diastatic enzymes in the flour. The carbon dioxide evolved is retained in the dough and leavens or aerates it such that, on baking, the loaf has the honeycomb structure characteristic of well-baked bread. The alcohol is driven off from the dough during baking and is lost in the steam.

References

1. DE CLERCK, J., "A Textbook of Brewing," London, Chapman & Hall, 1958, 2 vols.
2. MASSCHELEIN, C. A., JEUNEHOMME-RAMOS, C., CASTIAU, C., AND DEVREUX, A., *J. Inst. Brewing*, **69**, 332 (1963).
3. ROSE, A. H., *Sci. Am.* **200**, 90 (June, 1959).
4. ROSE, A. H., "Industrial Microbiology," London, Butterworths, 1961.
5. Webb, A. D., AND INGRAHAM, J. L., *Advan. appl. Microbiol.*, **5**, 317 (1963).

A. H. ROSE

ALCOHOLISM

Alcoholism, the compulsive uncontrolled drinking of alcoholic liquor, is now recognized as a progressive and often fatal disease which, however, has roots that are uncertain.

Sociologists are concerned with social factors, affluence, family life, and cultural customs which undoubtedly influence the incidence of the disease. Those who hold to the general tenets of Freudian psychology emphasize such causative factors as "insecurity" which may have started in early infancy. Biochemists on the other hand are likely to ask questions such as the following: Exactly how and in what organs and tissues is alcohol metabolized in the body? What enzymes are involved? Are there alternative pathways? What role does the liver play in this metabolism, and is there extra-hepatic metabolism? What relation exists between alcoholic cirrhosis and alcoholism? Is there a relationship between thyroid activity and alcoholism? How does alcohol affect brain metabolism and functioning, and what kinds of toxicities does it exhibit? What is the role of glutamine metabolism in relation to alcohol metabolism? How does alcohol affect the various endocrine glands? Is every person susceptible to alcoholism or are some innately vulnerable? If there are vulnerable individuals, what constitutes their vulnerability?

The answers to many of the above questions would not directly answer the query, "Biochemically, what is alcoholism?", but they would furnish basic information. None of these questions presently can be answered satisfactorily in a definitive way; hence, there is vast room for further investigation.

These biochemical investigations need not be completely shut off from all other considerations. There is a growing appreciation of the fact that one's family life certainly may affect one's body chemistry and that insecurity may influence the functioning of one's endocrine system. Accordingly the investigation of the biochemical aspects of alcoholism need not involve a denial of the pertinence of psychological stresses. The question whether alcoholism should be regarded as a "mental" disease tends to be eclipsed by a growing realization that probably every mental disease has its physiological and biochemical concomitants.

Two ideas have been emphasized by the writer and his associates in connection with their investigations of alcoholism. One is that the disease is related to BIOCHEMICAL INDIVIDUALITY; the other is that it is related to nutrition and the GENETOTROPHIC PRINCIPLE. These ideas do not constitute a definitive solution to the alcoholism problem as much as they point to areas of interest for further exploration.

The intricacies of biochemical individuality are such that if one accepts this concept, one rejects the possibility of finding one universal specific cause or one specific agent which will cure or ameliorate the condition in all individuals. The application of the genetotrophic principle is in one sense unsatisfying inasmuch as it postulates uncertainty; it allows for the possibility that any nutritional deficiency (involving 35 or more nutrients) might be involved in any specific case. From the practical standpoint, however, it is by no means impossible at present to move in the direction of complete nutritional adequacy for alcoholics and those threatened with alcoholic difficulties.

Certain biochemical aspects of alcohol and its metabolism are discussed in the separate articles on ALCOHOL METABOLISM, ALCOHOL DEHYDROGENASE, and ALCOHOLIC FERMENTATION.

There is a voluminous literature on alcoholism. The *Quarterly Journal of Studies on Alcohol*,

Rutgers University, New Brunswick, New Jersey, deals solely with material in this field.

References

1. KALANT, H., "Some Recent Physiological and Biochemical Investigations on Alcohol and Alcoholism—A Review," *Quart. J. Studies Alc.*, **23** 52–93 (1962).
2. "Symposium on Biochemical and Nutritional Aspects of Alcoholism," sponsored by the Christopher Smithers Foundation and the Clayton Foundation Biochemical Institute, October 2, 1964, Austin, University of Texas.
3. MANN, M., "Primer on Alcoholism," New York, Holt, Rinehart and Winston, 1953.
4. WILLIAMS, R. J., "Alcoholism: The Nutritional Approach," Austin, University of Texas Press, 1959.

ROGER J. WILLIAMS

ALDOLASES

In one of the steps of the sequence of enzyme-catalyzed reactions termed GLYCOLYSIS, an enzyme known as aldolase catalyzes the cleavage of fructose-1,6-diphosphate into two triose phosphate molecules (see also CARBOHYDRATE METABOLISM, Fig. 1). As in the cases of many other enzymes, aldolases occur in variant forms; see ISOZYMES.

ALDOSES

The aldoses are sugars which contain the aldehyde group or a potential aldehyde group that can be formed upon opening of an internal hemiacetal ring structure. The most important are the aldohexoses (six carbon atoms) and aldopentoses (five carbons). Usually the term is reserved for the simple, non-hydrolyzable, sugars (monosaccharides), but such hydrolyzable disaccharides as maltose and lactose also have aldose properties [see CARBOHYDRATES (CLASSIFICATION AND STRUCTURAL INTERRELATIONS)].

ALGAE

The algae comprise a heterogeneous assemblage of simple plants. The only common feature is a general lack of cell specialization. Great diversity in structure, reproduction, metabolism, and size is reflected in a taxonomy which recognizes as many as 11 separate phyla or divisions. Considered in this discussion are the salient features only of those algae treatable as microorganisms, the microalgae.

Within the algae there is a complete nutritional spectrum from obligate heterotrophy (requirement for organic compounds) in colorless forms, to obligate autotrophy (utilization of CO_2). Close to the middle of the spectrum is the most commonly studied genus, *Chlorella*. Some species of *Chlorella* can make the transition between dark assimilation of glucose and photosynthetic assimilation of carbon dioxide (and *vice versa*) without any of the lag expected of adaptive enzyme formation (see CARBON REDUCTION CYCLE IN PHOTOSYNTHESIS). A further versatile aspect of metabolism is found in some related genera (*e.g.*, *Scenedesmus*) in which anaerobic activation of hydrogenase permits photoreduction of carbon dioxide with hydrogen rather than water as the hydrogen donor.

Growth in most algae is a consequence of a photosynthetic metabolism in which bulk cellular constituents are synthesized from carbon dioxide and inorganic nitrogen. Growth rates are not as high as those of many bacteria but are the highest (up to 0.4 $hour^{-1}$) observed in autotrophic organisms. Although excretory products of algae are held to be of ecological significance, excretion of organic materials is not a notable characteristic. In balance studies most ($>90\%$) of the carbon and nitrogen assimilated have been recovered as cell product. As in other microorganisms, the algae typically have a high (*ca.* 50%) protein content. Hence their over-all metabolism is largely describable in terms of *protein* synthesis. In this respect, they are quite different from higher plants in which, partly because of large amounts of cellulose, metabolism is more closely based upon a carbohydrate economy.

Considerable variability in metabolism of any one alga can be induced by special conditions. In *Chlorella*, photosynthetic metabolism can be switched to almost exclusive carbohydrate synthesis by a preceding period of dark starvation or light-limited growth. Nitrogen deficiency leads to shunt or overflow metabolism and accumulation of reserve materials such as starch, various other exotic polysaccharides, and fats.

The high nutritional requirement for nitrogen is provided most commonly by ammonia, nitrate, or urea. Ability to use amino acids is highly variable between different species. Nitrogen fixation occurs in certain species of blue-green algae. It has been suggested that the blue-green species include a higher proportion of nitrogen-fixing species than any other group of microorganisms.

Most of the algal species in culture do not have growth factor requirements. However, it has been suggested that cultures have been selected arbitrarily by isolation procedures, that in nature growth factor requirements are more common, and that distribution of marine algae is essentially a biochemical geography because it is partially determined by locally available growth factors. The most common growth factors are thiamine, cobalamin, and biotin.

Mineral requirements provided by culture media usually include millimolar concentrations of potassium, magnesium, phosphate, and sulfate. At micromolar or lower concentrations, iron, manganese, zinc, molybdenum, copper, cobalt, and vanadium have been demonstrated as essential in one or more species.

In photosynthetic forms, chlorophyll a and carotenoids are found universally. Other pigments, active in photosynthesis, are characteristic of certain phyla: other chlorophylls, b, c, and d; a special carotenoid, fucoxanthin; biloproteins, the phycoerythrins and phycocyanins.

The algae have provided experimental organisms for the study of many cellular processes, notably PHOTOSYNTHESIS. They have been used as assay organisms for growth factors (*e.g.*, COBALAMIN). They provide means of synthesis of a number of biochemicals, either in bulk or isotopically labeled: porphyrins, carotenoids, quinones, pteridines, sterols, glycerides, amino acids.

Culture collections of algae are maintained in the Botany Departments of the University of Indiana, Bloomington, Indiana, and Cambridge University, Cambridge, England.

Reference

LEWIN, R. A. (EDITOR), "Physiology and Biochemistry of the Algae," New York, Academic Press, 1962.

JACK MYERS

ALKALINE PHOSPHATASES

Alkaline phosphatases are those of the group of PHOSPHATASE enzymes that show optimal catalytic activity at relatively high pH values (in the pH range 9–10). Alkaline phosphatase is present in blood and in osteoblasts and may play a role in BONE FORMATION.

ALKALOIDS

Introduction and Definition. From ancient times man has utilized alkaloids as medicines, poisons and magical potions. Only relatively recently has he gained precise knowledge about the chemical structures of many of these interesting compounds. The term, *alkaloid*, which was first proposed by the pharmacist, W. Meissner, in 1819 and means "alkali-like," is applied to basic, nitrogen-containing compounds of plant origin. Two further qualifications are usually added to this definition—the compounds have complex molecular structures and manifest significant pharmacological activity. Such compounds occur only in certain genera and families, rarely being universally distributed in larger groups of plants. Chemical, pharmacological and botanical properties must all be considered when classifying a compound as an alkaloid. Examples of well-known alkaloids are *morphine* (opium poppy), *nicotine* (tobacco), *quinine* (cinchona bark), *reserpine* (rauwolfia) and *strychnine* (strychnos nux-vomica).

It should be emphasized that many widely distributed bases of plant origin, such as methyl-, trimethyl- and other open-chain simple alkylamines, the cholines and the phenylalkylamines, are not classed as alkaloids. Alkaloids usually have a rather complex structure with the nitrogen atom involved in a heterocyclic ring. Yet, THIAMINE, a heterocyclic nitrogenous base, is not regarded as an alkaloid mainly because of its almost universal distribution in living matter. On the other hand, *colchicine* is classed as an alkaloid even though it is not basic and its nitrogen atom is not incorporated into a heterocyclic ring, because of its particular pharmacological activity and limited distribution in the plant world.

Occurrence and Distribution. Today a little over 2000 alkaloids are known, and it is estimated that they are present in only 10–15% of all vascular plants. They are rarely found in cryptogamia (exception—ergot alkaloids), gymnosperms or monocotyledons. They occur abundantly in certain dicotyledons and particularly in the families: **Apocynaceae** (dogbane, quebracho, pereiro bark), **Papaveraceae** (poppies, chelidonium), **Papilionaceae** (lupins, butterfly-shaped flowers), **Ranunculaceae** (aconitum, delphinium), **Rubiaceae** (cinchona bark, ipecacuanha), **Rutaceae** (citrus, fagara) and **Solanaceae** (tobacco, deadly nightshade, tomato, potato, thorn apple). Well-characterized alkaloids have been isolated from the roots, seeds, leaves or bark of some 40 plant families, and it is probable that the remaining families will provide only an occasional alkaloid-bearing plant. **Papaveraceae** is an unusual family in that all of its species contain alkaloids. The majority of plant families occupy an intermediate position in which most species within a genus or closely related genera either do or do not contain alkaloids. Thus all *Aconitum* and *Delphinium* species elaborate alkaloids whereas most of the other genera (*Anemone*, *Ranunculus*, *Trocleus*) in the family **Ranunculaceae** do not. It is generally true that a given genus or related genera yield the same or structurally related alkaloids; *e.g.*, seven different genera of **Solanaceae** contain hyoscyamine. It is also true that simple alkaloids often occur in numerous, botanically unrelated plants while the more complicated ones such as quinine, nicotine and colchicine are usually limited to one species or genus of plant and form a distinguishing characteristic of it.

Nomenclature and Classification. The nomenclature of alkaloids has not been systematized—both because of the complexity of the compounds involved and for historical reasons. The two commonly used systems classify alkaloids either according to the plant genera in which they occur or on the basis of similarity of molecular structure. Important classes of alkaloids containing generically related members are the aconitum, cinchona, ephedra, lupin, opium, rauwolfia, senecio, solanum and strychnos alkaloids. Chemically derived alkaloid names are based on the skeletal feature which members of a group possess in common. Thus indole alkaloids (*e.g.*, psilocybin, the active principle of Mexican hallucinogenic mushrooms) contain an indole or modified indole nucleus, and pyrrolidine alkaloids (*e.g.*, hygrine) contain the pyrrolidine ring system. Other examples of this classification include the pyridine, quinoline (see quinine), isoquinoline, imidazole, pyridine-pyrrolidine (see nicotine), and piperidine-pyrrolidine type alkaloids. A large number of important alkaloids have received names derived directly from those of plants—papaverine, hydrastine, berberine. A few have been named for their physiological action, as morphine (German,

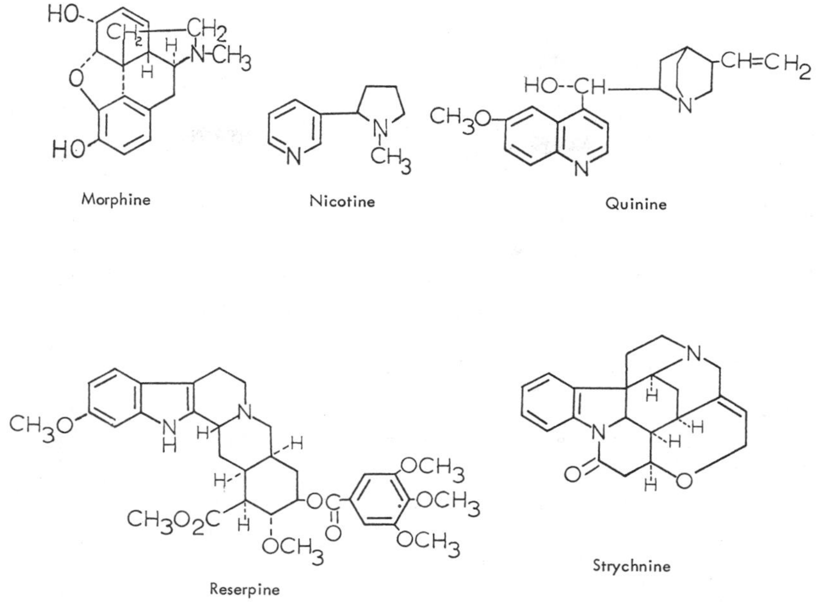

Morphine

Nicotine

Quinine

Reserpine

Strychnine

FIG. 1.

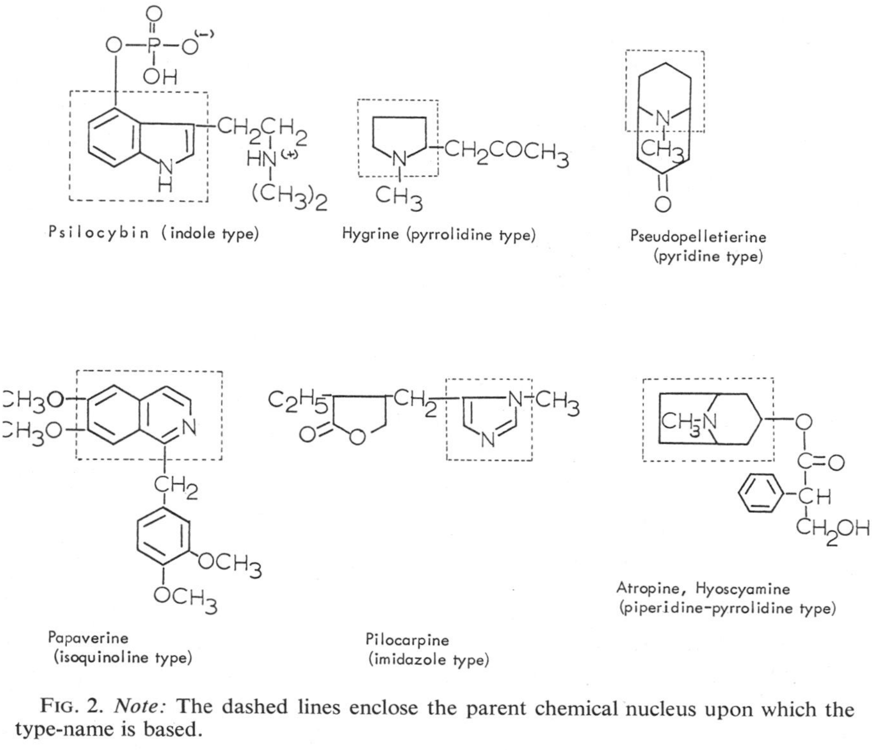

Psilocybin (indole type)

Hygrine (pyrrolidine type)

Pseudopelletierine
(pyridine type)

Papaverine
(isoquinoline type)

Pilocarpine
(imidazole type)

Atropine, Hyoscyamine
(piperidine-pyrrolidine type)

FIG. 2. *Note:* The dashed lines enclose the parent chemical nucleus upon which the type-name is based.

morphin, God of dreams), narcotine (Greek, *narkoō*, to benumb) and emetine (Greek, *emetikos*, to vomit). Only one, the pelletierine group, has been named after a chemist—Pierre Joseph Pelletier, the discoverer of emetine (1817), colchicine (1819), quinine (1820), cinchonine (1820), strychnine (1820), brucine (1820), caffeine (1820), piperine (1821), thebaine (1835) and also chlorophyll.

History. The beginning of alkaloid chemistry is usually dated back about 160 years when F. W. Sertürner announced the isolation of morphine in 1805. He prepared several salts of morphine and demonstrated that it was the principle responsible for the physiological effect of opium. Later (1810) Gomes treated an alcoholic extract of cinchona bark with alkali and obtained a crystalline precipitate when he named "cinchonino." Subsequent studies (1820) by P. J. Pelletier and J. B. Caventou of the Faculty of Pharmacy, Paris, showed that "cinchonino" was a mixture which they separated into two new alkaloids named quinine and cinchonine. Subsequently various investigators isolated more than two dozen additional bases from species of *Cinchona* and *Remijia*. Between 1820 and 1850, investigations were intensified and a large number of alkaloids of new and varied types were isolated and characterized. Among the important representatives discovered during this period were aconitine, one of the most toxic materials of plant origin known to man; atropine, a powerful mydriatic agent (4.3×10^{-6} gram will cause dilation of the pupil of the eye); colchicine, the alkaloid of the meadow saffron, which is used extensively in the treatment of gout; coniine, the principle responsible for the death of Socrates when he drank the cup of poison hemlock; codeine, a close relative of morphine, and a valuable pain killer and cough repressant; hyoscyamine, the optically active form of atropine; piperine, an alkaloid of pepper plants; berberine of barberry root; strychnine, a highly poisonous alkaloid used in certain cardiac disorders and for exterminating rodents; emetine, a powerful emetic sometimes used for treating amoebic dysentery.

Determination of Molecular Structure. Because of the complex molecular architecture of these compounds, with few exceptions, little progress was made in the elucidation of their structures during the nineteenth century. Only within the past few years have certain of the more complex alkaloids yielded the secrets of their molecular structure to the chemist.

The first step in the determination of the structure of a pure alkaloid consists in ascertaining the molecular formula. This information can be obtained by a combustion analysis for the elements present (always carbon, hydrogen and nitrogen and often oxygen) together with a molecular weight determination or, where high-resolution mass spectrometry is available, by this modern technique. The chemist next proceeds to ascertain the function of oxygen and nitrogen atoms in the molecule. Oxygen is most frequently present in the form of hydroxyl or phenol (—OH), methoxyl ($—OCH_3$), acetoxyl ($—OCOCH_3$), benzoxyl ($—OCOC_6H_5$), carboxyl ($—CO_2H$) or ketone ($>C{=}O$) groups. Oxygen in the form of a phenolic group can be recognized by alkali solubility, a color reaction with ferric chloride, acylation to an ester or alkylation to an ether; in the form of an alcohol by acetylation, dehydration or oxidation and by characteristic absorption in the infrared in the 3.0-μ region. Carboxyl groups are suggested by solubility in weak base, by esterification and by specific absorption in the infrared. Methoxyl groups can be determined quantitatively by the method of Zeisel which involves boiling the alkaloid with hydriodic acid and determining the quantity of methyl iodide formed. The estimation of the methylenedioxyl group is accomplished by reactions in which formaldehyde is split out by means of sulfuric acid. Carbonyl groups such as ketones or aldehydes may be detected by spectroscopic means (infrared, ultraviolet, nuclear magnetic resonance) as well as by standard chemical tests.

The determination of alkyl groups on nitrogen is carried out by heating the alkaloid hydriodide salt at 200–300° and measuring the amount of alkyl iodide produced. Thus an alkaloid bearing an $N—CH_3$ group (*e.g.*, nicotine) will produce methyl iodide and one containing an $N—CH_2CH_3$ group (*e.g.*, aconitine) will liberate ethyl iodide. In most cases, the nitrogen in an alkaloid is involved in a ring structure and will usually be secondary or tertiary. It is sometimes difficult to distinguish between these forms though several tests are available. The most widely used method for ascertaining the environment about the nitrogen is exhaustive methylation, also known as the Hofmann degradation. This method is based on the property of quaternary ammonium hydroxides, when heated, of decomposing with loss of water and cleavage of a carbon-nitrogen linkage to give an olefin. Thus with the acyclic tertiary amine shown in Fig. 3, a single methylation and decomposition eliminates the nitrogen as trimethylamine.

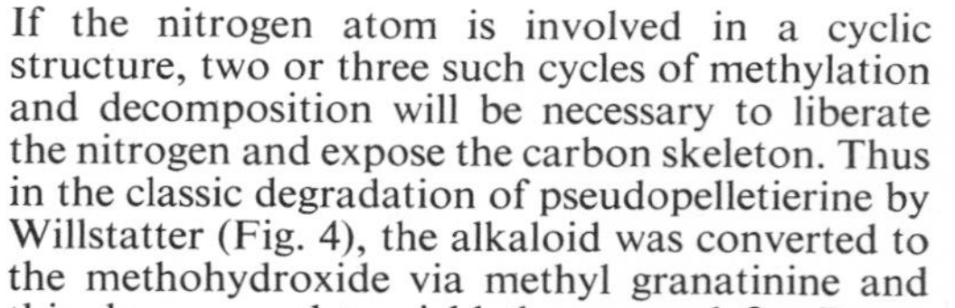

$$CH_3CH_2CH_2N(CH_3)_2 \xrightarrow{CH_3I} CH_3CH_2CH_2\overset{(+)}{N}(CH_3)_3I^{(-)}$$

$$\downarrow Ag_2O$$

$$H_2O + CH_3CH{=}CH_2 + (CH_3)_3N \xleftarrow{heat} CH_3CH_2CH_2\overset{(+)}{N}(CH_3)_3\overset{(-)}{O}H$$

FIG. 3.

If the nitrogen atom is involved in a cyclic structure, two or three such cycles of methylation and decomposition will be necessary to liberate the nitrogen and expose the carbon skeleton. Thus in the classic degradation of pseudopelletierine by Willstatter (Fig. 4), the alkaloid was converted to the methohydroxide via methyl granatinine and this decomposed to yield the monoolefin. Repetion of the methylation and decomposition yielded 1,5-cyclooctadiene. Where nitrogen is linked in ring structures through three valencies, three methylations and decompositions are necessary to eliminate the nitrogen.

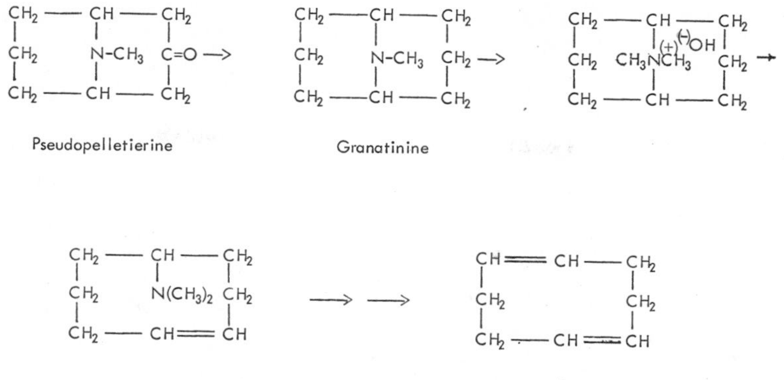

Pseudopelletierine Granatinine

1,5-Cyclooctadiene

FIG. 4.

Numerous other degradative methods are available which transform the alkaloid into stable, easily recognized compounds. Thus distillation over hot zinc dust will sometimes degrade an alkaloid to a stable aromatic derivative, *e.g.*, morphine gives phenanthrene, and cinchonine yields quinoline and picoline. Another valuable method involves dehydrogenating the alkaloid with a catalyst such as sulfur, selenium or palladium. During dehydrogenation, peripheral groups such as hydroxyls and C-methyls are eliminated. From the dehydrogenation reaction, relatively simple, easily recognized products can sometimes be isolated. These may provide a ready clue to the gross skeleton of the alkaloid. A good example of the value of this method is seen in the dehydrogenation of atisine, an alkaloid of *Aconitum heterophyllum*. The products, 1-methyl-6-ethylphenanthrene and 1-methyl-6-ethyl-3-azaphenanthrene (Fig. 5), provided early information as to the gross skeleton of atisine and thereby saved much valuable time in the elucidation of the structure of this alkaloid. Another interesting example of the use of this technique is the dehydrogenation of isorubijervine, an alkaloid found in a number of *Veratrum* species. A crystalline hydrocarbon, cyclopentanophenanthrene, and a liquid aromatic amine, 2-ethyl-5-methylpyridine are the main products of the reaction and provide immediate insight into the gross structure of isorubijervine (Fig. 6).

Other methods of degradation are available which allow the removal of particular groups or the cleavage of the molecule at specific points. The

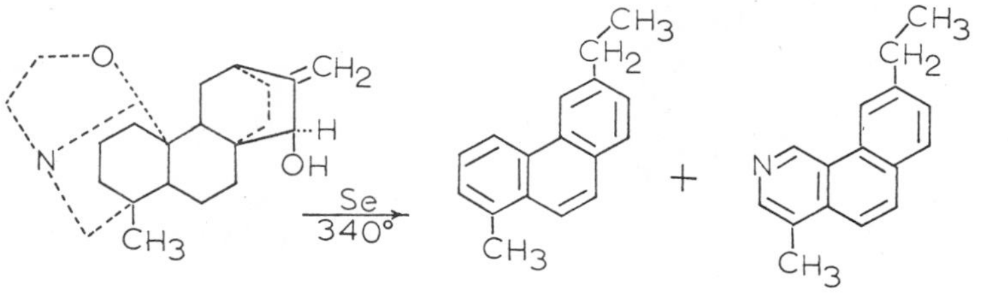

Atisine 1-Methyl-6-ethylphenanthrene 1-Methyl-6-ethyl-3-azaphenanthrene

FIG. 5.

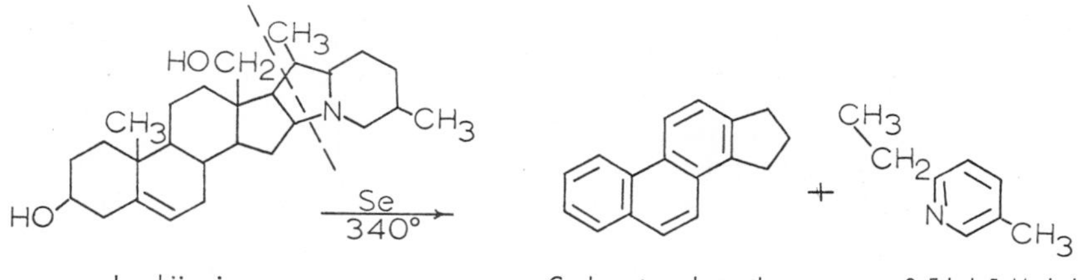

Isorubijervine Cyclopentanophenanthrene 2-Ethyl-5-Methyl pyridine

FIG. 6.

aim is to break the molecule into smaller fragments which can be identified and furnish clues which will allow the chemist to visualize the complete structure. The final part of the elucidation of an unknown alkaloid is usually concerned with the stereochemistry of the molecule, *viz.*, the determination of the complete three-dimensional representation of all the atoms in the molecule. Obviously, in the case of complex alkaloids this degradation and detective work can often take years to complete.

Biogenesis of Alkaloids. One of the most exciting and fascinating subjects pertaining to alkaloids is the mode of their synthesis in plants. Over the past few decades, chemists have proposed many biogenetic schemes for the synthesis of individual alkaloids. Most of these schemes have been based upon the idea that alkaloids are derived from relatively simple precursors such as phenylalanine, tryptophane, "acetate" units, terpene units, methionine and a few other amino acids such as ornithine. On paper at least, the structures of most alkaloids can be derived from such simple precursors using a few well-known chemical reactions. As a matter of fact, several simple alkaloids have been synthesized in the laboratory from amino acid derivatives under physiological conditions using just such biogenetic concepts. With the advent of isotopically labeled compounds these theories have been subjected to experimental test.

The modern approach to biosynthetic studies of alkaloids involves the administration of labeled precursors to selected plants and, after a suitable period of growth, the isolation of the alkaloids. These are then degraded in a systematic fashion to determine the position of the labeled atoms. Using this technique, it has been demonstrated that morphine and its companions are produced in the plant from TYROSINE. Thus when tyrosine-$2C^{14}$ was fed to *Papaver somniferum* plants, it was incorporated into morphine, codeine and thebaine. Degradation of the morphine revealed a distribution of radioactivity in the molecule which can be accounted for by the construction of morphine in the plant from two molecules of tyrosine (or a close biological equivalent) by way of norlaudanosoline. A study of the comparative rates of $C^{14}O_2$ and tyrosine-$2C^{14}$ incorporation indicates that thebaine is the first alkaloid synthesized and that it is converted by demethylation successively into codeine and morphine. By similar experiments, many other alkaloids (*e.g.*, nicotine, hyoscyamine, pellotine, papaverine, colchicine, gramine) have been shown to be synthesized from amino acids. The present state of research on alkaloid biogenesis represents a dramatic breakthrough. Not only are plants known to incorporate amino acids, acetate and mevalonolactone, but in specific cases large intermediates have been successfully introduced into the plant's biosynthetic system.

Function of Alkaloids. Unlike biogenesis, the function of alkaloids in the plant remains a subject for speculation. Many authorities regard them as by-products of plant metabolism. Still others conceive of alkaloids as reservoirs for protein synthesis; as protective materials discouraging animal or insect attacks; as plant stimulants or regulators in such activities as growth, metabolism and reproduction; as detoxicating agents, which render harmless, by processes such as methylation, condensation, and ring closure, substances whose accumulation might otherwise cause damage to the plant. While a particular explanation may have application to a given plant, it should be remembered that 85–90% of all plants manage very well without elaborating any alkaloids. In conclusion one might say that while much has been learned about the biogenesis and metabolism of alkaloids, their functions in the plant, if any, are still largely unknown.

References

1. SMALL, L., "Alkaloids," in GILMAN'S "Organic Chemistry—An Advanced Treatise," second edition, Vol. 2, pp.1166–1258 New York, John Wiley, 1947.
2. HENRY, T. A., "The Plant Alkaloids," Fourth edition, Philadelphia, Blakiston, 1949.
3. BENTLEY, K.W., "The Chemistry of the Morphine Alkaloids," London, Oxford University Press, 1954.
4. BENTLEY, K. W., "The Alkaloids," New York, Interscience Publishers, 1957.
5. Manske, R. H. F., AND HOLMES, H. L., "The Alkaloids: Chemistry and Physiology," Vols. 1–8, New York, Academic Press, 1950–1960.
6. BOIT, H. G., "Ergebnisse der Alkaloid-Chemie bis 1960," Berlin, Akademie-Verlag, 1961.
7. MORGAN, K. J., AND BARLTROP, J. A., "Veratrum Alkaloids, "*Quart. Rev.* **12**, 34–60 (1958).
8. BATTERSBY, A. R., AND HODSON, H. F., "Alkaloids of Calabash—Curare and Strychnos Species," *Quart. Rev.*, **14**, 77–103, (1960).
9. BATTERSBY, A. R., "Alkaloid Biogenesis," *Quart. Rev.*, **15**, 259–286 (1961).
10. PELLETIER, S. W., "Recent Developments in the Chemistry of the Atisine-Type Diterpene Alkaloids," *Tetrahedron*, **14**, 76–112 (1961).
11. PELLETIER, S. W., "The Chemistry of Certain Imines Related to Diterpene Alkaloids,"*Experientia*, **20**, 1–10 (1964).

S. W. PELLETIER

ALKALOSIS. See ACID-BASE REGULATION; METABOLIC DISEASES.

ALKAPTONURIA

Alkaptonuria is a rare hereditary disorder of metabolism in which the stepwise oxidation of the amino acid TYROSINE to carbon dioxide and water is interrupted at the level of homogentisic acid oxidation [see AROMATIC AMINO ACIDS (BIOSYNTHESIS AND METABOLISM)]. This acid is a metabolic derivative of tyrosine, arising in liver and to a lesser extent in the kidney, in which the aromatic ring is still intact. Normally it undergoes immediate oxygenative ring fission to form maleylacetoacetate which is readily transformed into

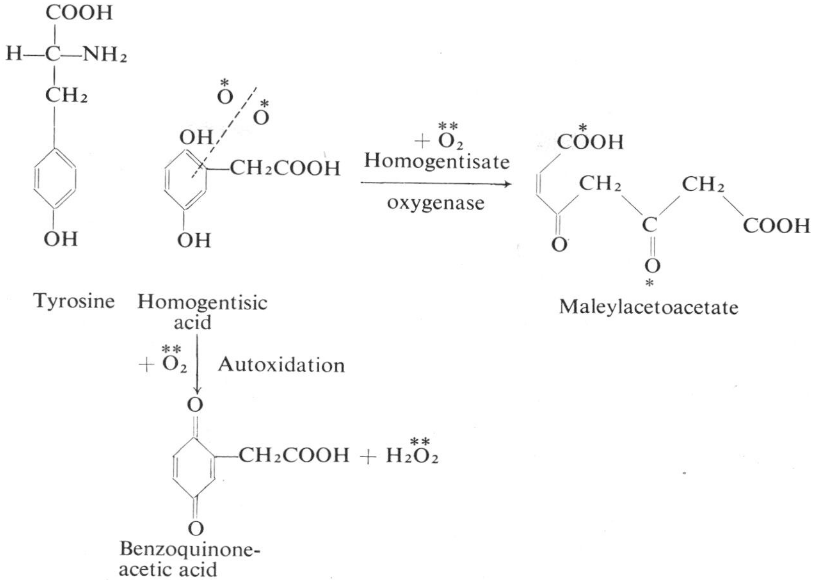

FIG. 1.

metabolites of the CITRIC ACID CYCLE. In the alkaptonuric, however, it accumulates due to the complete absence of the enzyme, homogentisate oxygenase.

This abnormal accumulation of homogentisate in liver and kidney causes it to enter the circulation and to be excreted in the urine. The urine then turns black on standing due to autoxidation of homogentisate to the corresponding quinone (benzoquinoneacetic acid) and polymerization of the latter into a melanin-like pigment. The autoxidative blackening of the urine is hastened by making it alkaline, and the term *alkapton* was coined from the Arabic word *alkali* and the Greek *kaptein*. The latter means to swallow eagerly and refers to the increased rate of absorption of oxygen caused by the addition of alkali to alkapton urine. The structure of tyrosine and the reactions undergone by homogentisic acid are indicated below (Fig. 1).

From its familial distribution, the disease appears to be inherited as a single Mendelian recessive trait which, in accord with modern BIOCHEMICAL GENETIC theory, would lead to the prediction that the alkaptonuric is deficient in a single enzyme, namely homogentisate oxygenase. The absence of this enzyme has been demonstrated in surgical specimens obtained from both the liver and the kidney of alkaptonurics.

The blackening of the urine is observed from birth and persists throughout life. With the exception of this and of a pigmentation of cartilages, tendons, and ligaments, termed *ochronosis*, which occurs in middle life, the alkaptonuric appears to be essentially normal. Ochronosis results from the gradual selective accumulation of homogentisate in these tissues followed by autoxidation to the quinone and polymerization of the latter to form the pigments. Although very striking on postmortem examination, the pigmentation may be manifested in life by a bluish discoloration of the knuckles, ears, and nose.

References

1. HARRIS, H., AND MILNE, M. D., in "Biochemical Disorders in Human Disease" (THOMPSON, R. H. S., AND KING, E. A., EDITORS), pp. 792–797, New York-London, Academic Press, 1964.
2. GARROD, A. E., "Inborn Errors of Metabolism," second edition, London, University Press, 1923.
3. BOEDEKER, *Henle's Zeitschrift*, **7**, 130 (1859).

DANA I. CRANDALL

ALLERGIES

The term allergy often is defined as an acquired state of specifically altered reactivity to some chemical or physical agent. It implies an allergic *reaction*, an active union of an ANTIGEN and an ANTIBODY resulting in harm to the host. Development of allergy depends on previous *exposure* to a substance, a period of *sensitization* during which there is formation of antibodies; *re-exposure* to the sensitizing agent (the antigen) results in an antigen-antibody reaction which produces adverse effects.

The evidence of antigen-antibody reaction in diseases of allergy is not always easily found, but new antigen-antibody detection techniques are giving more proof of this concept. It is evident that allergy is closely related to immunity, many aspects of which also depend on antigen-antibody reactions. Indeed, the same fundamental processes are believed to be involved in certain types of allergic and immune reactions, but in the former the host is likely to be injured, while in the latter the host is protected (see also IMMUNOCHEMISTRY).

Types of Allergy. A useful classification is to divide allergy according to the location and type of antibody found after the sensitization period: (1) Circulating antibody ("immediate," "anaphylactic," and "wheal-reacting allergy" or "histamine release allergy"). (2) Cellular antibody ("delayed allergy," "bacterial allergy," "tuberculin-type allergy").

The allergic antibody and "allergy" can be passively transferred temporarily to another individual by transfer of serum in the first type and by transfer of lymphocytic white blood cells in the second. The antibodies are *specific*; *i.e.*, allergic antibody to egg, for example, reacts selectively with egg antigen.

Circulating Antibody Type. Anaphylactic shock in animals and the "atopic" diseases of man exemplify the circulating antibody type of allergy. The atopic diseases in man occur on natural exposure to various environmental substances; they seem to have a hereditary basis and are the type of reaction which is commonly thought of as "allergy." In addition to the hereditary factor, to what substances allergies develop depends on exposure. Common agents (usually PROTEIN in nature but polysaccharides, polypeptides, fats, and other chemicals can sensitize) include air-borne substances (pollens from wind-pollinated plants, mold spores, house dust, animal danders, insect particles), ingested food or drug, things injected (drugs, serums, insect stings), and physical agents (light, temperature, pressure). The allergic reaction or union of antigen and antibody results in the release locally and systematically of "mediators" (histamine, bradykinin, "slow reacting substance," acetylcholine, and possibly others). These chemicals cause changes according to their properties and the location where they are released and/or act. The reaction is acute or *immediate* and in general produces swelling or whealing (from capillary dilation and increased vascular permeability), smooth muscle contraction, and increased mucus formation. When this occurs in the mucous membrane of the nose, *allergic rhinitis*, or hay fever, is produced; when in the bronchi, *asthma*; and if in the skin or other mucous membrane, *urticaria* (hives) or *angioedema*. The gastrointestinal tract, genitourinary tract, and central nervous system also may be affected, though this is less common. The reaction may be generalized and produce profound vasodilation, shock, and death. In general, however, the effects are temporary and usually reversible.

Much has been made of some secondary factors such as emotional influences, weather, physical exertion, endocrine state, etc. These things do influence allergic reactions, and in some individuals they seem conspicuous. However, they may be acting simply as precipitating or secondary factors modifying an individual's response to an underlying antigen-antibody reaction.

These serum-transferable or circulating allergic antibodies are still under investigation in regard to their physical and chemical characteristics, but they seem different from the normal immune antibodies and seem to be formed only in individuals with the inherited tendency.

Cellular Antibody Type. The second type of allergy is less well understood. Hereditary influences are not so important, and apparently anyone may develop this type. The reaction is delayed 24–48 hours or more after exposure and results from a local inflammatory process rather than from release of chemicals into the bloodstream. This is not to infer that at some time during the sensitization process dispersion of the antibody does not take place systematically but only that at the time of the allergic reaction, the antibody seems to be in or on the cells rather than in the serum.

Clinically, delayed allergy may be separated into: (1) Contact allergy; (2) delayed allergy associated with reaction to bacteria, bacterial products, and body tissues.

An example of contact allergy is the reaction to poison ivy plant oil. The oil contacts the skin of a sensitive individual and reacts with the antibody to it, previous sensitization having occurred. As far as is known, no chemicals are released into the bloodstream. Other substances can cause contact allergy, such as metals, dyes, many chemicals, and other plant oils.

The subject of delayed allergic reactions to bacteria, bacterial products, and body tissues is a field about which little is known. It is called "tuberculin type" because following an infection with tubercle bacilli, a test with tuberculin protein results in a delayed inflammatory response at the site of the test. Some of the clinical manifestations of many infectious diseases no doubt are due to this type of hypersensitivity. This mechanism is believed to be important also in the "autoimmune" diseases, a large group of conditions in which it is speculated that one may become sensitized to certain of his own tissues. Delayed allergy also is thought to be important in the graft rejection which occurs when tissue from a nonidentical source is transplanted into the body. This is a developing field in which much new information is expected which is certain to change our concepts and treatment of these conditions.

The classification and discussion of allergy here presented is convenient and accurate as far as is now known. However, some investigators suggest that there is not as sharp a dividing line between circulating-type and cellular-type allergic antibodies as presented and that these types represent stages in an individual's response to antigenic substances, the type depending possibly on amount of exposure, inherited factors, or other unknowns.

References

1. *The American Journal of Medicine*, Seminar on Allergy, reprinted from Vol. XX (January to June, 1956).
2. Ciba Foundation Symposium, "Cellular Aspects of Immunity," Boston, Little, Brown & Co., 1959.

3. Feinberg, S. M., "Allergy in Practice," Chicago, The Year Book Publishers, 1949.
4. Henry Ford International Symposium, "Mechanisms of Hypersensitivity," Boston & Toronto, Little, Brown & Co., 1959.
5. Sheldon, J. M., Lovell, R. G., and Mathews, K. P., "A Manual of Clinical Allergy," Philadelphia, W. B. Saunders Co., 1953.
6. Sherman, W. B., and Kessler, W. R., "Allergy in Pediatric Practice," St. Louis, C. V. Mosby Co., 1957.
7. Talmage, D. W., and Cann, J. R., "The Chemistry of Immunity," Springfield, Ill., Charles C. Thomas, 1961.

T. S. Painter, Jr.

ALLOSTERIC SITES (OF ENZYMES). See REGULATORY SITES (OF ENZYMES).

ALLOXAN DIABETES

Alloxan is a PYRIMIDINE derivative of the following structure:

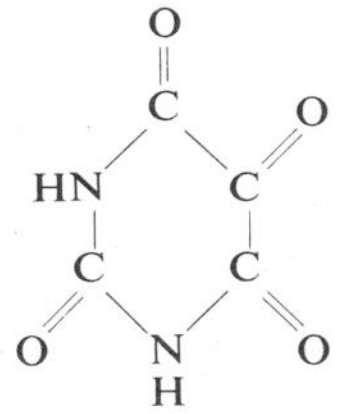

```
          O
          ‖
          C     O
        /   \  //
     HN       C
      |       |
      C       C
    //  \   /  \\
   O      N      O
          H
```

It has been used experimentally in animals to produce damage to the pancreatic β cells, which normally secrete the hormone, INSULIN. Alloxan thus induces an artificial diabetic state, or state of hypoinsulinism [see GLUCONEOGENESIS; GLYCOGEN AND ITS METABOLISM; INSULIN (FUNCTION IN METABOLISM)].

ALVEOLAR AIR. See GASEOUS EXCHANGE IN BREATHING.

AMERICAN ASSOCIATION FOR THE ADVANCEMENT OF SCIENCE

Formed in 1848, by a reorganization of the Association of American Geologists and Naturalists, founded in 1840, the Association has as its objects "to further the work of scientists, to facilitate cooperation among them, to improve the effectiveness of science in the promotion of human welfare, and to increase public understanding and appreciation of the importance and promise of the methods of science in human progress." It was the first, and still is the only, American scientific society national in scope, with interests extending into all the fields of the natural and social sciences, and open to all scientists.

For the first third of its history, the Association served as the focal point for the special fields of science. As science grew, the need for more specialized societies and meetings became imperative, and many of the large thriving scientific organizations of today grew out of the sections of the Association and then became affiliates of the parent group. There are now 297 societies and academies affiliated or associated with the AAAS, and individual membership has grown to over 98,000.

The first meeting in September, 1848, was organized around two sections: "Natural History, Geology, etc." and "General Physics, etc." These two sections have evolved by a process of subdivision into 20 sections, each with an appointed secretary who serves for four years and an annually elected chairman who also holds the title of vice-president of the Association. The program of the section is developed by a section committee composed of the Council representatives of the societies affiliated with that section, the secretary and vice-president of the section, plus four members-at-large. The Council as a whole elects the Board of Directors and officers and determines general policy.

Chemistry first appeared as a specific part of the program at the 1850 meeting in New Haven. The Section of Chemistry and Mineralogy scheduled five sessions with 29 chemical papers. At the Hartford, Connecticut, meeting in 1874, a request from the previous year calling for a subsection on chemistry under Section A was approved, and S. W. Johnson served as the first chairman. The Cincinnati meeting in 1881 marked the establishment of the full-fledged section on chemistry, Section C.

Although the American Chemical Society (founded in 1876) did not actually spring from Section C, the Section Committee spearheaded by F. W. Clarke and H. W. Wiley played a key role in putting the floundering young society on a firm national foundation. Although the ACS has long since outgrown the possibility of holding joint meetings with the AAAS, the Section C Committee and the Association itself have enjoyed continuous close cooperation with both the local sections and the national office of the American Chemical Society, not only in conducting meetings but in many other efforts toward the advancement of chemistry and science as a whole. Two examples are the Cooperative Committee on the Teaching of Science and Mathematics with members from both the ACS and AAAS and the separate affiliation of the Division of Chemical Literature in the Association's Section on Information and Communication (T).

Publications. The publications of the Association have always played an important role in the dissemination of chemical knowledge. In the early years, before the meetings became so large as to make it impossible, the publication of the "Proceedings" supplied an annual review of significant advances in chemistry as well as other fields.

From the time of its establishment by Thomas A. Edison in 1880, *Science* has published a continuous account of important technical developments in all major fields of science. It became an official journal of the AAAS in 1900, although ownership and responsibility for its publication did not come until 1938. *Science* is distributed weekly on a world-wide basis and contains news of scientists, scientific developments and technical papers reporting the results of current research. In recent years, important review articles, coverage of national news of import to all scientists and other features have resulted in 4900 pages of text per year. Present circulation is 115,000.

The Association has published some 81 symposium volumes, most of them based on symposia presented at its meetings. "Recent Advances in Surface Chemistry and Chemical Physics" (1939) and "Monomolecular Layers" (1954) are typical of those in chemistry.

Gordon Research Conferences. The Gordon Research Conferences now represent the Association's most extensive activity in chemistry. Dr. Neil E. Gordon organized the first summer conference of the series at Johns Hopkins University in 1931. From this small meeting of 25 faculty members and students of the University, the movement has grown to a total of 60 conferences with some 6000 participants from around the world. Far more important than numbers has been the increasingly significant stimulus to chemical research afforded by these informal week-long meetings.

After four years on the Hopkins campus, the conferences were moved to Gibson Island, Maryland, on Chesapeake Bay, a spot more conducive to leisurely informality and recreation than a city university could provide. By 1947 the facilities at Gibson Island could no longer accommodate the conferences, and they were moved to Colby Junior College, New London, New Hampshire. Subsequent expansion has required the establishment of four additional centers at New Hampton School, New Hampton, N.H.; Kimbal Union Academy, Meriden, N.H.; Tilton School, Tilton, N.H.; and Proctor Academy, Andover, N.H. The program for 1965 included 54 Conferences. There is also a West Coast series in Santa Barbara, California.

Each conference is a continuing autonomous entity on a specialized but fairly general subject, *e.g.*, polymers, catalysis. The group attending one year selects the special phase of the subject to be covered the next year and elects a chairman to arrange the program (unless a previously elected vice-chairman becomes chairman). The chairman, with the assistance of the Director of the GRC, selects those privileged to attend, a maximum of about 100, from the applicants. Those selected must be active mature researchers who can contribute both information and ideas. Every effort is made to distribute attendance from among academic, governmental, industrial, and foreign laboratories. Funds are available to assist foreigners and academic scientists.

The Annual Meeting. Each annual meeting of the Association brings together scientists from all fields of science—including chemistry. The meeting includes sessions sponsored by the Association as a whole, by AAAS committees, by all 20 sections, and by some 40 or 50 participating societies. The emphasis is on symposia, especially those interdisciplinary in scope, and on lectures and programs of interest to all scientists. There are also, however, specialized symposia and some opportunities for contributed papers. Programs of particular interest to chemists—and typical of any AAAS meeting—included "Moving Frontiers of Science" lectures by Norman F. Ness on "A New Look at the Earth's Magnetic Field," "The Physiological Basis of Mental Activity" by Jerome Y. Lettvin, and "Some Aspects of Low Temperature Physics" by William M. Fairbank. Symposia included: "Proteins and Nucleic Acids" (arranged by Wendell M. Stanley); "Behavior, Brain, and Biochemistry" (David Krech); "Nonprotein Neurotoxins" (Harry S. Mosher); and "Recent Developments in the Study of Energy Transfer" (George C. Pimentel and Harmon W. Brown). Several sessions were for short papers in analytical, physical, and inorganic chemistry. The entire program in chemistry was co-sponsored by the California Section of the ACS.

RAYMOND L. TAYLOR

AMERICAN CHEMICAL SOCIETY

In April 1876, 35 chemists met in New York City to form an American Chemical Society. The first published list of members included 133 names. Originally the Society was incorporated in New York State. By action of the seventy-fifth Congress of the United States, a national charter was granted effective January 1, 1938.

The objects of the Society always have been "to encourage in the broadest and most liberal manner the advancement of chemistry in all its branches; the promotion of research in chemical science and industry; the improvement of the qualifications and usefulness of chemists . . .; the increase and diffusion of chemical knowledge; and . . . to promote scientific interests and inquiry; thereby fostering public welfare and education, aiding the development of our country's industries, and adding to the material prosperity and happiness of our people."

To serve specialized fields of chemistry, subject matter divisions are authorized. The first five were created in 1908; today there are 25. Geographic organization was conceived as a means of providing to the members in a relatively small area a program of activities beneficial to the science and the profession and readily available to participants. The first local section was formed in 1890; at the end of 1965 there were 166.

The Society as a whole has grown tremendously. At the end of 1965, there were about 104,000 members, all meeting minimum standards of professional education and training. In addition, there were nearly 9000 student affiliates. At the first national meeting in 1890, 43 chemists were in attendance. Registration at the National Meetings in 1965 was 17,499.

The American Chemical Society owns and publishes the following: *Analytical Chemistry*; *Biochemistry*; *Chemical and Engineering News*; *Chemical Reviews*; *Chemistry*; *Industrial and Engineering Chemistry*, in combination with quarterly supplements: *Process Design and*

Development, *Product Research and Development*, *Fundamentals*; *Inorganic Chemistry*; *Journal of Agricultural and Food Chemistry*; *Journal of the American Chemical Society*; *Journal of Chemical Documentation*; *Journal of Chemical and Engineering Data*; *Journal of Medicinal Chemistry*; *The Journal of Organic Chemistry*; *The Journal of Physical Chemistry*; Chemical Abstracts Service, including *Chemical Abstracts*, *CA-Applied Chemistry Sections*, *CA-Biochemical Sections*, *CA-Macromolecular Sections*, *CA-Organic Chemistry Sections*, *CA-Physical Chemistry Sections*, *Chemical-Biological Activities*, and *Chemical Titles*. The Division of Chemical Education owns the *Journal of Chemical Education*, and the Division of Rubber Chemistry owns and publishes *Rubber Chemistry and Technology*.

The usefulness of *Chemical Abstracts* has been expanded through periodic publication of collective indexes of various kinds: Author, Subject, Formula, Numerical Patent Index, and numerous special services. In the *Advances in Chemistry Series*, the Society publishes specialized symposia and compilations of useful data. In 1943 Atherton Seidell assigned to the ACS the copyright of his "Solubilities of Inorganic and Organic Compounds." Subsequently the ACS publishes supplements, and a complete revision was finished in 1965. The work of the Committee on Analytical Reagents is reflected from time to time in a publication presenting specifications for reagent chemicals.

The first serious attempt to build up a chemical literature in English without primary regard to commercial considerations was begun in 1920 when the series of American Chemical Society Monographs was inaugurated. These books are published by the Reinhold Publishing Corporation, 430 Park Ave., New York 10022.

The American Chemical Society is active in the educational field. Its first contact with students is with vocational counseling work at the high-school level. For the employed individual, continuation courses and special lecture series on specialized topics are sponsored by local sections. Between these two extremes are many other educational programs including the approval of departments in colleges and universities as qualified to give professional training.

Literally thousands of ACS meetings are held annually, including national, regional, local, divisional, and cooperative with other professional societies.

Professional matters are of great importance in the Society's program. Beginning in 1917, the work of the News Service has expanded gradually until today it is considered as an authoritative source of information for all those who disseminate information to the public—press, radio, television, magazines, books, etc.

The Society has increased the professional and economic status of chemists and chemical engineers through various procedures. It has extended opportunities for the profession; new jobs have resulted from its efforts. It maintains various kinds of effective employment aids. It has protected the chemist and chemical engineer against legal encroachments on their rights and has acted positively to gain recognition for their proper status.

The Society maintains effective liaison with all branches of the federal government. In certain cases, advisory bodies to government agencies are created. As a federally chartered body, it has executed many assignments for the government within its proper field.

The Society administers 32 awards, and its local sections and divisions sponsor others.

The popular deliberative assembly of the Society is its Council. Each local section and each division is represented therein. Ex officio members consist of the President, the President-Elect, the Past Presidents, the Executive Secretary, the Treasurer, and the Directors. Total membership approximates 450.

The Board of Directors is composed of the President, the President-Elect, the most recent Past President, six Regional Directors and four Directors-at-Large, and is the legal representative of the Society. The Constitution provides that it "shall have, hold, and administer all the property, funds, and affairs of the Society."

The Society is administered from headquarters at 1155 Sixteenth St. N.W., Washington, D.C. Some of its journals also are edited there. The staff of *Chemical Abstracts* occupies a building designed specifically for its use in Columbus, Ohio.

At the end of 1965, the full-time staff of the ACS, including its publications, numbered about 1000. Other persons are employed during peak periods or in special assignments. These figures do not include abstractors for CA, contributing editors, advisory boards, personnel in the advertising department, or those engaged in printing; the last two are handled under contract. Local section, division, and committee personnel contribute hours of service equal in the aggregate to substantial staff increases.

B. R. STANERSON

AMERICAN INSTITUTE OF BIOLOGICAL SCIENCES

The American Institute of Biological Sciences was organized February 20, 1948, within the National Academy of Sciences–National Research Council.

In September 1954, the Member Societies of AIBS voted to have the Institute become independent of the National Research Council, and in January 1955, the Institute became incorporated in the District of Columbia as a nonprofit, national organization of professional biological societies and biologists.

The purposes of the Institute have always been "the advancement of the biological, medical, and agricultural sciences and their applications to human welfare and to foster, encourage and conduct research in the biological sciences."

To serve these purposes, the AIBS assists societies, other organizations, and biologists in such matters of common concern as can be dealt with effectively by united action; cooperates with local, national, and international organizations concerned with the biological sciences; promotes unity and effectiveness of effort among all those who are devoting themselves to the biological sciences and their applications by research, by teaching, and by study; holds and sponsors scientific meetings and fosters the relations of the biological sciences to other sciences, to the arts and industries, and to the public good.

Prior to January 1, 1964, members of Member Societies of the Institute automatically became members of the AIBS. By Governing Board action in August 1963, the constitution and bylaws of the Institute were amended and, effective January 1, 1964, membership in the AIBS is on a direct member basis. There are five classes of direct membership: Life, Honorary, Sustaining, Individual and Student. Effective liaison is maintained with professional biological societies and industry. In 1965 there were 42 Adherent Societies and 22 Industrial Members affiliated with the AIBS.

The official journal of the Institute, *The AIBS Bulletin*, initiated in January 1951, and published five times a year, was mailed to 14,000 members of the 16 Member Societies of the AIBS. With the growth of the Institute, over 40,000 issues to members of 27 Member Societies were mailed in 1963. Effective January 1, 1964, the bimonthly *AIBS Bulletin* was published as a monthly journal and the name was changed to *BioScience*. The constitution and bylaws of the Institute were amended, and effective with the January 1964 issue of *BioScience* only direct members of the Institute received the official journal.

In addition to *BioScience* and *The Quarterly Review of Biology*, the AIBS publishes, in cooperation with professional societies, government agencies, and individuals, books covering all disciplines of the biological sciences. Of importance among these in recent years are: "Style Manual for Biological Journals," "Sharks and Survival," "World Directory of Hydrobiological and Fisheries Institutions," and "Reprography and Copyright Law." The AIBS has also published many monographs, proceedings of symposia, and conference programs.

The first AIBS Meeting of Biological Societies was held on the campus of Ohio State University in Columbus, in September 1950. At this first meeting, 14 societies participated with 1556 individuals registered. Many of the attending societies requested a similar 1951 meeting and as a result the Annual AIBS Meeting of Biological Societies was initiated. The 16th Annual Meeting was held August 15–20, 1965 at the University of Illinois at Urbana, with 21 participating societies and an attendance of over 4000. These joint meetings, at which around 2000 scientific papers are presented, offer biologists an opportunity to expand their knowledge and learn of advances in diversified fields.

The AIBS has always maintained a Placement Service. Originally this service was available to members of Member Societies of the Institute at no cost but with a small fee to Affiliates. Effective January 1, 1964, Placement Service is available only to direct members of the AIBS.

Through its Committee on Education, AIBS has led in establishing programs to provide better curricula for students in biology. AIBS initiative launched both the Biological Sciences Curriculum Study (BSCS), which is now administered at the University of Colorado at Boulder, and the Commission on Undergraduate Education in the Biological Sciences (CUEBS), which is administered by George Washington University, Washington, D.C. AIBS has also established a Sub-Committee on Facilities and Standards to compile recommendations as guidelines for biology departments in self-evaluation and as standards for college administrations establishing new departments.

To help solve the national shortage of scientists by making use of the talents and knowledge of senior biologists retired from their lifetime work, the AIBS, in 1964, initiated the Emeritus Biologist Program. The AIBS maintains a register of qualified and willing emeritus biologists and matches them against needs for various services in universities, government and industry.

The Bioinstrumentation Advisory Council, initiated in 1965, was set up to meet a recognized need for interdisciplinary communications to smooth the flow of knowledge of instrumentation technique and theory from the engineering and physical sciences to the life sciences and also to enrich engineering and physics with concepts from biology.

The AIBS provides panels and committees to advise various government agencies. One such group, the Shark Research Panel, has worked with the Office of Naval Research since 1958. With NASA, AIBS has set up six Study Councils to recommend appropriately valuable biological experiments to be performed in space. For the AEC, AIBS publishes a *Newsletter* to inform biology teachers and researchers of advances in the use of radiation as a tool in the study of biology. For the NSF and the AEC, the Visiting Biologists Program offers high schools, colleges and universities the opportunity to have as guest lecturers nationally known biologists. For the National Science Foundation, the AIBS maintains the National Register of Scientific and Technical Personnel in the field of biology.

The business and activities of the AIBS are managed by the Governing Board composed of board members and officers. The Governing Board is comprised of one board member selected by each Adherent Society, one board member-at-large per each 1000 direct members elected by member ballot, the President, the retiring President, Vice President, Secretary-Treasurer, and the Executive Director. The Executive Committee consists of the President as Chairman, the Vice President, the Secretary-Treasurer, the Executive Director as Secretary, not more than eight or less than three members of the Governing Board appointed by the Governing Board, and the immediate past President as ex officio.

The headquarters of the Institute is maintained at 3900 Wisconsin Ave., NW, Washington, D.C.

JOHN R. OLIVE

AMERICAN SOCIETY OF BIOLOGICAL CHEMISTS

The American Society of Biological Chemists was formed by the action of 81 charter members in 1906. According to its Articles of Incorporation and its Constitution, "the purposes for which this corporation was formed are to further the extension of biochemical knowledge and to facilitate personal intercourse between American investigators in biological chemistry." The Society was incorporated under the laws of the State of New York in 1919.

The Society is the owner and publisher of the *Journal of Biological Chemistry*. It has appeared continuously since 1906 and continues as one of the major media for the communication of biochemical research results through the world. Currently membership in the Society number 2276, including 44 honorary members. Honorary members are, by definition, eminent biochemists living outside of America who would not otherwise be eligible for membership.

Representatives of the American Society of Biological Chemistry are among the members of the U.S. National Committee for the International Union of Biochemistry, and are also active in national and international affairs through the aegis of the National Academy of Sciences–National Research Council, and of the International Union of Biochemistry. In addition, the Society maintains an active Educational Affairs Committee, publishes a brochure on the subject of careers in biochemistry, and participates generally in the affairs of biochemistry.

The Society holds a scientific meeting each year as a part of the annual meeting of the Federation of American Societies for Experimental Biology of which it, together with the American Physiological Society, were original founders. This annual meeting constitutes a forum for the communication of results from the research institutes, universities, and industrial laboratories engaged in biochemical investigation. In addition, an important feature of this annual meeting is a series of symposia by outstanding scientists on subjects of current importance in biochemistry.

Membership in the American Society of Biological Chemists is by election following nomination of candidates by existing members. The standards for membership require that sponsors of candidates produce evidence that their candidates have in fact published meritorious original investigations.

The Society maintains executive offices in Bethesda, Maryland.

ROBERT A. HARTE

AMIDATION AND AMINATION REACTIONS.

See AMINO ACID METABOLISM; AMMONIA METABOLISM; TRANSAMINATION ENZYMES.

AMINE OXIDASES AND AMINO ACID OXIDASES

The process of converting the amino group into the corresponding carbonyl residue [Eq. (*1*)]

$$R\text{—}\underset{R'}{\overset{H}{C}}\text{—}NH_2 \rightarrow R\text{—}\underset{R'}{C}\text{=}O \qquad (1)$$

is quite a common one throughout the living kingdom and many different mechanisms are involved in it, *e.g.*, (a) dehydrogenation of L-glutamic acid, L-alanine, and L-proline with the aid of NAD and NADP enzymes; (b) transaminations carried out in the presence of pyridoxal phosphate enzymes; (c) deamination of hydroxy- and mercaptoamino acids by pyridoxal enzymes with concomitant replacement of the hydroxyl and sulfhydryl groups by hydrogen; and (d) oxidative deamination in a narrower sense, the only one considered in this article and summarized in Eq. (*2*).

$$R\text{—}\underset{R_1}{\overset{H}{C}}\text{—}NH_2 + O_2 + H_2O \rightarrow R\text{—}\underset{R_1}{C}\text{=}O + NH_3 + H_2O_2 \qquad (2)$$

Either pyridoxal phosphate with copper, or riboflavin is employed in the electron transfer to molecular oxygen. Amines (R_1 = H) and amino acids (R_1 = COOH) are the principal substrates of this type of oxidative deamination.

Amine Oxidases. Amine oxidases have been found in many living organisms, from bacteria to mammals. Their action can be described by Eq. (*2*) (R_1 = H). They can be divided into two distinct classes according to their sensitivity toward unsubstituted acylhydrazides (*e.g.*, semicarbazide) and by other characteristics listed in Table 1. The acylhydrazide-insensitive oxidases ("monoamine oxidases") form a group of closely related enzymes, while the acylhydrazide-sensitive amine oxidases comprise a set of fairly different enzymic entities. Some of the sufficiently characterized members of the latter class are the classical

TABLE 1. CHARACTERIZATION OF AMINE OXIDASES

	Acyl-hydrazide Resistant	Acyl-hydrazide Sensitive
Degradation of N-methylated amines	+	−
Preferred ring-substitution in benzylamines	*meta*	*para*
Sensitivity toward phenyl-cyclo-propylamine and N-benzyl-N-methylpropynyl-amine	+	−
Intracellular distribution	Particulates	Cytoplasm[a]

[a] Exception: rabbit liver DAO.

hog kidney diamine oxidase, the diamine oxidases of mycobacteria, germinating plant seeds, human placenta, and rabbit liver mitochondria, and the blood plasma amine oxidases. Transaminases, in contrast to their importance in AMINO ACID METABOLISM, play a minor role in amine synthesis and degradation.

Acylhydrazide-insensitive Amine Oxidases: Monoamine Oxidases.[1–4] [Monoamine:O_2 oxidoreductases (deaminating), 1.4.3.4]. This group of enzymes—discovered in 1928 by M. L. C. Hare-Bernheim—catalyzes the oxidative deamination of many primary and a few secondary and tertiary amines (Table 1). Since partial purification has been accomplished only recently,[5,6] the composition has not been determined as yet, and little is known about the mechanism of action. It seems likely, however, that only one of the two α-hydrogens, selected by its stereospecific position, is removed and transferred to oxygen.[7,8]

The range of amines which are easily attacked by MAO is considerable. They comprise (a) aliphatic amines from C_1 to C_8 and beyond, the optimal chain length varying from species to species, (b) arylalkylamines, including phenylalkylamines from C_1 to C_5, the optimal chain length again depending on the species from which the enzyme was prepared,[9] (c) hydroxylated arylakylamines, *e.g.*, tyramine, octopamine, and catecholamines, and (d) "heterocyclic" amines such as tryptamine and 5-hydroxytryptamine. In benzylamine, MAO prefers *meta*-substitution over other ring positions. Among the monohalogenated benzylamines, *m*-iodobenzylamine was found to be the most interesting substrate.[10] Of considerable theoretical and practical interest is a series of powerful inhibitors which mimic the structure of typical substrates, *e.g.*, iproniazid, phenylcyclopropylamines (tranylcypromine), N-benzyl-N-methylpropynylamine (pargyline), phenylalkylhydrazines, phenylalkylguanidines, harmaline, and α-ethyltryptamine.[11–16]

According to Eq. (*2*), the rate of MAO reaction can be measured (a) by following the consumption of oxygen or the liberation of ammonia, (b) by measuring the disappearance of the substrate by biological or sensitive physicochemical methods, (c) by following spectrophotometrically the transition from substrate to product with kynuramine[17] or *m*-iodobenzylamine,[18] and (d) by taking advantage of the difference in solubility between substrate and product, extracting the isotopically labeled product into an appropriate solvent.[19,20]

MAO is found in cells derived from all three germ layers. It is quite unevenly distributed in the various parts of the brain and kidney. In the course of ontogenesis marked changes occur, and extremely low values are often found at birth. The MAO activity in mouse liver is depressed by THYROXINE and SEX HORMONES, particularly testosterone.[21] Within the cell, the enzyme is located almost exclusively in the MITOCHONDRIA and to a small degree in microsomes.

Some facets of the biological function of MAO are beginning to be understood due to the availability of the powerful *in vivo* inhibitors mentioned above. The enzyme, present in the intestinal mucosa and in the liver and placenta, protects the organism against the invasion of powerfully active biogenic amines produced by intestinal bacteria or occurring in some foods, *e.g.*, cheese. MAO also participates in the regulation of the intracellular concentration and concentration gradients of biogenic amines and is thus involved in the process of neurotransmission.[22] In previous investigations, a number of biogenic amines could not be found in the mammalian cell because they were obviously destroyed by MAO before they could be isolated. However, after treating animals *in vivo* with powerful MAO inhibitors, compounds such as octopamine, considered to be a "false neurotransmitter," became detectable. Similarly, the role of MAO in the development of certain forms of edema became apparent when it could be shown that MAO inhibitors prevented the accumulation of fluid. Finally, these inhibitors are in use as therapeutic agents in the treatment of mental depressions, hypertension, angina pectoris and Parkinson's disease. The development of this new class of pharmaceuticals is a direct consequence of certain enzymological studies carried out in 1952 in this laboratory. Although several thousand effective inhibitors have been synthesized and millions of patients have been treated with them, many questions about the future role of these drugs remain unanswered.[11–16,23]

Acylhydrazide-sensitive Amine Oxidases: Diamine Oxidases.[1,24–27] [Diamine:O_2 oxidoreductase (deaminating), 1.4.3.6]. In 1929, C. H. Best discovered a histamine-destroying enzyme which he and E. W. McHenry called histaminase and which later was shown by E. A. Zeller to attack aliphatic diamines as well and, therefore, was renamed diamine oxidase.

This group of enzymes catalyzes the oxidation of primary amines only (Table 1). Since the hog kidney DAO is by far the most extensively studied member of this group, the subsequent discussion is devoted mostly to this entity. DAO has been obtained in fairly pure state[26,27]; the presence of copper, pyridoxal phosphate (see PYRIDOXINE), and possibly of FAD (see FLAVINS) has been ascertained. The occurrence of pyridoxal has been suspected ever since it was found that all carbonyl reagents, *e.g.*, acylhydrazides, are powerful inhibitors of these enzymes. Little precise information regarding the mechanism of the reaction is known other than the suggestion that the primary amino group may form a Schiff base with the aldehyde residue of pyridoxal.

A large number of aliphatic diamines have been observed to be attacked by DAO, cadaverine (C_5) being the optimal substrate. Only one of the amino residues is removed from the diamine. The aldehyde forms a Schiff base with the second basic group [Δ^1-pyrroline in the case of putrescine (C_4)], as indicated by the production of the yellow or orange 1,2-dihydroquinazolinium derivatives with *o*-aminobenzaldehyde. The second basic group is replaceable by a secondary or a tertiary, but not by a quaternary, amino residue, by the guanidino group as in agmatine, or by the imidazole ring as in histamine, or it may be left out altogether.[28] The identity of "histaminase" and DAO, although established in 1938, was not accepted by some authors. The difference in opinion was reconciled when it was found recently that the indigo method used to measure histaminase activity was more complex than anticipated.[29] The inhibitors of DAO are comprised of chelating agents and acylhydrazides, *e.g.*, semicarbazide and aminoguanidine, molecules which mimic the substrate (various hydrazines and amines).

In principle, the same array of methods as described for MAO (a, b, d) can be used for the determination of DAO activity. In addition, the disappearance of the blue color of indigodisulfonate, due to the hydrogen peroxide formation by the DAO reaction,[29] or the appearance of a yellow or orange color in the presence of *o*-aminobenzaldehyde (see above), can be photometrically followed.

While diamine oxidases seem to occur more widely throughout the living kingdom than do monoamine oxidases, the latter are found in a greater variety of tissues than DAO. The highest DAO activity generally is found in the kidney, intestinal mucosa, lung, and placenta, the last of which appears to be the source of the high DAO level found in the blood of pregnant women. It is now well established that DAO, while absent from the mammalian brain, is present in the fish brain.[30]

DAO in the intestinal mucosa and in the placenta probably exerts a protective function against biologically active amines similar to that of MAO in the same tissues. Since histamine is supposedly involved in allergic reactions, DAO is thought to play a role in these reactions by destroying toxic amounts of this amine. Many questions raised in this field of research, however, remain unanswered—mainly because other enzymic systems, including MAO, are involved in the termination of histamine action.[31] In plants, DAO appears to be a part of the chemical sequence which leads to the synthesis of ALKALOIDS. Certain observations suggest that amine oxidases are engaged in the production of analogous basic systems whithin the mammalian organism.[32]

Plasma Amine Oxidases.[25,33] [Spermine:O_2 oxidoreductase (donor cleaving), 1.5.3.3]. To this set of enzymes belongs the first amine oxidase ever obtained in the crystalline state. It was isolated from beef plasma, has a molecular weight of 255,000 and appears to contain pyridoxal and 3.7 gram atoms of copper.[34] A similar enzyme was crystallized from hog plasma.[35] Its pyridoxal phosphate content was demonstrated by biological tests and by the reactivation of the apoenzyme upon addition of the coenzyme. The mechanism of the reaction of plasma amine oxidases is not yet clear. With benzylamine as a substrate, benzaldehyde and ammonia are formed, while the products of spermidine [$H_2N(CH_2)_4NH(CH_2)_3NH_2$] degradation are putrescine, ammonia and acrolein.[36] The substrate pattern of this amine oxidase ranges from monoamines to long-chain diamines to POLYAMINES. Unfortunately, these enzymes have been called plasma *mono* amine oxidases, a trend which has created considerable confusion among non-enzymologists. The inhibitors are the same as for other members of the group of acylhydrazide-sensitive amine oxidases.

Amine oxidases of this type are found in the blood of ruminants. Closely related enzymes with a slightly different substrate pattern occur in other species such as hog, horse, and man. In contrast to the ordinary DAO, no increase of activity during human pregnancy is observed.[37] The origin of this enzyme is unknown. However, an amine oxidase of similar properties has been partially purified from the microsomes of beef liver and kidney.[38]

Substrates of plasma amine oxidases, particularly spermidine and spermine, occur widely in nature and may be involved in many biological processes. Notably, interaction with membranes and nucleic acids have been reported. Spermine oxidase, by affecting the POLYAMINE levels, may thus be of substantial importance for the regulation of the action of nucleic acids and transport. More specifically, plasma amine oxidase converts spermine into products toxic for the kidney, tubercle bacilli,[16,33,39] and virus particles.[40]

Amino Acid Oxidases.[41–43] Although amino acid oxidases are found in many animal cells, their importance for the degradation of AMINO ACIDS is overshadowed by the coupled action of TRANSAMINASES and glutamic dehydrogenase (see also AMINO ACID METABOLISM). Sarcosine oxidase and other demethylases, although representing true oxidative deamination with regard to the methylamino residue, leave the amino acid intact. Apparently, glycine oxidase is identical with D-amino acid oxidase (D-AAO). Particular attention will be given to the ophidian L-AAO and to the D-AAO of mammalian kidney, since these enzymes have been brought to a relatively high state of purity and have, therefore, been available for studies pertaining to the mechanisms of these reactions.

L-*Amino Acid Oxidases.* [L-Amino acid:O_2 oxidoreductase (deaminating) 1.4.3.2]. L-Amino acid oxidases have been found in microorganisms, invertebrates, and vertebrates. In general, they react best with neutral aliphatic and aromatic amino acids as substrates. The L-AAO occurring in certain mollusks, however, exhibits a preference for basic amino acids.

An L-AAO from rat kidney has been isolated which displays a wide substrate spectrum, including N-substituted L-amino acids, *e.g.*, proline. No similar oxidase could be located in other species. This distribution is not as peculiar as it may appear. The rat kidney enzyme, an FMN protein, was observed to catalyze the oxidation of α-hydroxy acids better than that of the corresponding amino acids. A similar enzymic action toward α-hydroxy acids is found in the kidney of other species, *e.g.*, hog kidney, which is not able to attack amino acids. The rat kidney oxidase, therefore, differs from related α-hydroxy acid oxidases by extending its substrate range into the L-amino acids.[44] Other L-amino acid oxidases, *e.g.*, a microsomal enzyme in chicken liver[45] and the ophidian L-AAO, are without effect on the α-hydroxy acids.

Ophidian L-*Amino Acid Oxidase.*[42,43,46,47] This oxidase was discovered in many snake venoms in 1944 by E. A. Zeller and A. Maritz and obtained in the crystalline state by A. Meister *et al.* from the venom of *Crotalus adamanteus*. It has a molecular weight ranging from 120,000–150,000 and possesses two moles of FAD. It has been suggested that this oxidase has an active site containing two closely associated flavins, each of which accepts one hydrogen atom from the substrate to yield the half-reduced enzyme.

The substrate pattern is markedly different from that of other L-amino acid oxidases. It prefers aliphatic amino acids of medium chain length, such as methionine and leucine, and aromatic amino acids, *e.g.*, phenylalanine and tryptophan. N-Methylated amino acids, proline, and D-amino acids are excluded. Ring substitution in phenylalanine and tryptophan markedly affects the constants of the Michaelis-Menten relationship, whereby localization and size of the residue rather than its chemical nature, as expressed by the Hammett equation, are of importance. These data suggested a new form of the relationship between substrate concentration and reaction rate.[48]

Most of the assays described are based upon measurement of oxygen uptake, ammonia production, keto acid production in the presence of added catalase, or reduction of a dye or other electron acceptor.

The enzyme occurs not only in the venoms of most poisonous snakes, but also in their livers. Its biological function is far from being understood. Unlike other proteinaceous components of venoms, the oxidase is not inhibited by antisera (see also SNAKE VENOMS).

Since the preference of the ophidian enzyme for L-amino acids seems to be absolute, this oxidase has been extensively used to (a) obtain D-amino acids free from L-amino acids, (b) destroy completely the L-component in racemates, (c) prepare keto acids, which are otherwise difficult to obtain, from available amino acids, and (d) assay L-phenylalanine in the blood of children suffering from oligophrenia phenylpyruvica[50] (see MENTAL RETARDATION).

D-*Amino Acid Oxidases.* [D-Amino acid:O_2 oxidoreductase (deaminating), 1.4.3.3]. Mammalian D-AAO was discovered by H. A. Krebs and subsequently purified in O. Warburg's laboratory. Kubo obtained the first crystalline material. This enzyme exists in a series of polymeric forms of unit molecular weight of 54,700, each unit of which can combine with one FLAVIN group; it dissociates into the monomer upon dilution. Recently, a crystalline preparation of molecular weight 108,000 was isolated containing 1.5 molecules of FAD. Upon addition of FAD, the apoprotein was observed to accommodate two molecules.[51] Crystals of the apoenzyme, of the flavoprotein, and of the enzyme-substrate complex have been prepared. Although the mechanism of action is still under investigation, the following scheme may be proposed: (a) binding of the amino acid to the enzyme, (b) electron transfer to FAD, converting the former to a semiquinone and the latter to a free radical, (c) binding of oxygen to the enzyme-substrate complex, (d) formation of hydrogen peroxide, and (e) dissociation of the imino acid to regenerate the free oxidized enzyme.

The enzymic rate is measured in the same manner as for the ophidian oxidase. In addition, some especially sensitive methods have been developed for the D-AAO. As an example, a procedure based in principle upon the catalytic amplification method is given: 3-Amino-1,2,4-triazole destroys catalase in the presence of hydrogen peroxide, which is produced by the catalytic action of the enzyme [Eq. (*2*)]; this method, therefore, consists in measuring the catalase activity of the system after incubation.[52] A considerable number of D-amino acids, including N-substituted compounds, are attacked by the oxidase. Until recently, D-AAO was considered to be absolutely specific for D-amino acids. The increase in measuring sensitivity, however, permitted the detection of a small, but definite, degradation of the L-enantiomer.[52]

Inhibitors of D-AAO (a) act upon the proteinaceous part of the enzyme, *e.g.*, silver ions and other reagents blocking sulfhydryl residues, (b) compete with the substrate, *e.g.*, benzoic acid, or (c) compete with FAD for the apoenzyme, *e.g.*, the drugs chlorpromazine and chloramphenicol.

D-AAO activity has been found in the kidney and liver of nearly all mammals studied. A small amount of activity is also consistently observed in the central nervous system. The role of D-AAO in metabolism has yet to be elucidated.

Some of the data presented in this review resulted from work supported by a Public Health Research Career Award K6–GM–14,000 and by Public Health Research Grants GM 05927–05, from the division of General Medical Sciences, and HD 00979–01, from the Division of Child Health and Human Development. In this article, the following abbreviations have been used: MAO (monoamine oxidase); DAO (diamine oxidase); L–AAO and D–AAO (L– and D–amino acid oxidase, respectively); NAD and NADP (diphosphopyridine nucleotide and triphosphopyridine nucleotide); FMN and FAD (flavin mononucleotide and flavin-adenine dinucleotide). Individual papers are not cited unless they are not mentioned in the review given in the list of references.

References

1. ZELLER, E. A., "Oxidation of Amines," in "The Enzymes" (SUMNER, J. B., AND MYRBÄCK, K., EDITORS), Vol. II, Part 1, pp. 536–558, New York, Academic Press, 1951.

2. BLASCHKO, H., *Pharmacol. Rev.*, **4**, 415–458 (1952).
3. ZELLER, E. A., "Monoamine Oxidase," in "Monoamines et système nerveux central" (DE AJURIAGUERRA, J., EDITOR), pp. 51–57, Geneva, Georg, and Paris, Masson, 1962.
4. BLASCHKO, H., "Amine Oxidase," in "The Enzymes" (BOYER, P. D., LARDY, H., AND MYRBÄCK, K., EDITORS), second edition, Vol. VIII, pp. 337–351, New York, Academic Press, 1963.
5. GANROT, P. O., AND ROSENGREN, E., *Med. Exptl.*, **6**, 315–619 (1962).
6. GUHA, S. R., AND KRISHNA MURTI, C. R., *Biochim. Biophys. Res. Commun.*, **18**, 396–401 (1965).
7. BELLEAU, B., FANG, M., BURBA, J., AND MORAN, J., *J. Am. Chem. Soc.*, **82**, 5752–5754 (1960).
8. KAUFMAN, K., MAYERS, D., RAMACHANDER, G., AND ZELLER, E. A., Kinetics of Degradation of Stereospecifically Deuterized *m*-Methoxybenzylamine," unpublished data.
9. SARKAR, S., AND ZELLER, E. A., *Federation Proc.*, **20**, 238 (1961).
10. ZELLER, E. A., *Biochem. Z.*, **339**, 13–22 (1963).
11. "Amine Oxidase Inhibitors" ZELLER, E. A., EDITOR), *Ann. N. Y. Acad. Sci.*, **80**, 551–1045 (1959).
12. PLETSCHER, A., GEY, K. F., AND ZELLER, P., "Monoaminoxydase-Hemmer," *Progr. Drug Res.* **2**, 417–590 (1960).
13. "New Reflections on Monoamine Oxidase Inhibition" (ZELLER, E. A., EDITOR), *Ann. N. Y. Acad. Sci.*, **107**, 809–1158 (1963).
14. ZELLER, E. A., AND FOUTS, J. R., *Ann. Rev. Pharmacol.*, **3**, 9–32 (1963).
15. ZELLER, E. A., "Monoamine and Polyamine Analogs," in "Metabolic Inhibitors" HOCHSTER. R. M. AND QUASTEL, J. H., EDITORS), Vol. II, pp. 53–78, New York, Academic Press, 1963.
16. SOURKES, T. L., AND D'IORIO, A., "Inhibitors of Catechol Amine Metabolism," *Ibid.*, pp. 79–98.
17. WEISSBACH, H., SMITH, T. E., DALY, J. W., WITKOP, B., AND UDENFRIEND, S., *J. Biol. Chem.*, **235**, 1160–1163 (1960).
18. ZELLER, V., RAMACHANDER, G., AND ZELLER, E. A. *J. Med. Chem.*, **8**, 440–443 (1965).
19. WURTMAN, R. J., AND AXELROD, J., *Biochem. Pharmacol.*, **12**, 1439–1441 (1963).
20. OTSUKA, S., AND KOBAYASHI, Y., *Biochem. Pharmacol.*, **13**, 995–1006 (1964).
21. CHORDIKIAN, F., LANE, R. E., SCHWEPPE, J. S., STANICH, G., AND ZELLER, E. A., (in press).
22. KOPIN, I. J., *Pharmacol. Rev.*, **16**, 179–191 (1964).
23. SPECTOR, W. G., AND WILLOUGHBY, D. A., *J. Pathol. Bacter.*, **80**, 271–280 (1960).
24. ZELLER, E. A., *Advan. Enzymol.*, **2**, 93–112 (1942).
25. ZELLER, E. A., "DIAMINE Oxidases," in "The Enzymes" (BOYER, P. D., LARDY, H. AND MYRBÄCK, K., EDITORS), second edition, Vol. VIII, pp. 313–335, New York, Academic Press, 1963.
26. KAPELLER-ADLER, R., AND MCFARLANE, H., *Biochem. Biophys. Acta*, **67**, 542–565 (1963).
27. FOUTS, J. R., BLANKSMA, L. A., CARBON, J. A., AND ZELLER, E. A., *J. Biol. Chem.*, **225**, 1025–1031 (1957).
28. MONDOVI, B., ROTILIO, G., FINAZZI, A., AND SCIOSCIA-SANTORO, A., *Biochem. J.*, **91**, 408–416 (1964).
29. ZELLER, E. A., *Federation Proc.*, **24**, Part 1, 766–768 (1965).
30. BURKARD, W. P., GEY, K. F., AND PLETSCHER, A., *J. Neurochem.*, **10**, 183–186 (1963).
31. "Symposium on Histamine Metabolism," (UNGAR G., MODERATOR), *Federation Proc.*, **24**, Part 1, 757–780 (1965).
32. HOLTZ, P., in Papers of the Second International Catechol Amines Symposium, Milan, 1965, *Pharmacol. Rev.* (in press).
33. TABOR, H., TABOR, C.W., AND ROSENTHAL, S.M., *Ann. Rec. Biochem.*, **30**, 579–604 (1961).
34. BUFFONI, F., AND BLASCHKO, H., *Proc. Roy. Soc. London Ser. B*, **161**, 153–167 (1964).
35. YAMADA, H., GEE, P., EBATA, M., AND YASUNOBU, K. T., *Biochim. Biophys. Acta*, **81**, 165–171 (1964).
36. ALARCON, R. A., *Arch. Biochem. Biophys.*, **106**, 240–242 (1964).
37. MCEWEN, C. M., JR., *J. Lab. Clin. Med.*, **61**, 540–547 (1964).
38. GORKIN, V. Z., reviewed in *Chem. Abstr.*, **57**, 17050 (1962).
39. TABOR, H., AND TABOR, C. W., *Pharmacol. Rev.*, **16**, 245–300 (1964).
40. BACHRACH, U., *Biochem. Biophys. Res, Commun.*, **19**, 357–360 (1965).
41. FRANKE, W., "Die Dehydrasen der Aminosäuren und verwandte Enzyme," in "Die Chemie der Enzyme" (VON EULER, H., EDITOR), zweiter Teil, dritter Abschnitt, pp. 592–611, Munich, J. F. Bergmann, publishers, 1934.
42. KREBS, H. A., "Oxidation of Amino Acids," in "The Enzymes" (SUMNER, J. B., AND MYRBÄCK, K., EDITORS), Vol. II, Part 1, pp. 499–535, New York, Academic Press, 1951.
43. MEISTER, A., AND WELLNER, D., "Flavoprotein Amino Acid Oxidases," in "The Enzymes" (BOYER, P. D., LARDY, H., AND MYRBÄCK, K., EDITORS), Second edition, Vol. VII, pp. 609–648, New York, Academic Press, 1963.
44. ISELIN, B., AND ZELLER, E. A., *Helv. Chim. Acta*, **29**, 1508–1520 (1946).
45. STRUCK, J., JR., AND SIZER, J. W., *Arch. Biochem. Biophys.*, **90**, 22–30 (1960).
46. ZELLER, E. A., *Advan. Enzymol.*, **8**, 459–495 (1948).
47. ZELLER, E. A., "Enzymes as Essential Components of Bacterial and Animal Toxins," in "The Enzymes" SUMNER, J. B., AND MYRBÄCK, K., EDITORS), Vol. I, Part 2, pp. 986–1013, New York, Academic Press, 1951.
48. ZELLER, E. A., RAMACHANDER, G., FLEISHER, G. A., ISHIMARU, T., AND ZELLER, V., *Biochem. J.*, **95**, 262–269 (1965).
49. MEISTER, A., "Preparation of α-Keto Acids," in "Methods in Enzymology" (COLOWICK, S. P., AND KAPLAN, N. O., EDITORS), Vol. III, pp. 404–418, New York, Academic Press, 1957,
50. LADUE, B. N., HOWELL, R. R., MICHAEL, P. J., AND SOBER, E. K., *Pediatrics*, **31**, 39–46 (1963).
51. HELLERMAN, L., COFFEY, D. S., AND NEIMS, A. H., *J. Biol. Chem.*, **240**, 290–298 (1965).
52. SCANNONE, H., WELLNER, D., NOVOGRODSKY, A., Biochemistry, **3** 1742-1745 (1964).

E. ALBERT ZELLER

AMINO ACID CONTENT OF PROTEINS. See PROTEINS, AMINO ACID ANALYSIS AND CONTENT.

AMINO ACID METABOLISM

The major source of amino acids for the mammal is dietary protein which is hydrolyzed to amino acids and small peptides by PROTEOLYTIC ENZYMES in the stomach and intestine. The mixture of amino acids and peptides is then absorbed from the intestine and eventually distributed to all tissues. Since the metabolic fates of the 20 amino acids found in PROTEIN are so diverse, in this brief review only a few of the biochemical reactions involved in amino acid metabolism will be presented.

Throughout the first half of the twentieth century, intensive nutritional and biochemical research was devoted to determining whether all of the 20 amino acids are required in the diet. The development of synthetic diets, in which protein was replaced by a mixture of the amino acids, made an answer to this question possible. When young rats were fed the complete diet, they grew at a rapid rate. Omission of alanine from the diet produced no change in the growth rate. Therefore alanine is not essential. When threonine was omitted, however, the animals failed to grow, showing that threonine is essential. In this manner, ten of the amino acids were found to be required for the optimal growth of the rat. They are arginine, histidine, isoleucine, leucine, lysine, methionine, tryptophan, phenylalanine, valine, and threonine. The other ten amino acids are non-essential. Application of a modified experimental approach to humans showed, surprisingly, that arginine and histidine are not required by man, but the remaining eight amino acids in the list above are essential (see also the article on AMINO ACIDS, NUTRITIONALLY ESSENTIAL).

The most obvious conclusion to be drawn from these experiments is that higher animals possess biochemical mechanisms for synthesizing about half of the amino acids, but they are deficient in enzymes necessary for the synthesis of the *essential amino acids*. Plants and microorganisms are the ultimate source of the essential amino acids. After the amino acids have reached the tissues, they may be used for three major functions: (1) oxidation of the carbon skeleton to provide energy, (2) synthesis of special compounds such as glutathione, and (3) synthesis of protein. The remainder of this article will be devoted to functions (1) and (2). [Function (3) is discussed in the article PROTEINS (BIOSYNTHESIS).] A different metabolic pathway exists for the utilization of the carbon skeletons of each amino acid, but prior to the removal of the amino group, a few reactions common to several of the amino acids can occur.

Decarboxylation. Quantitatively, decarboxylation is not a major route of amino acid metabolism in animals. Nevertheless, it is important for the synthesis of certain primary amines such as histamine, a powerful vasodilator.

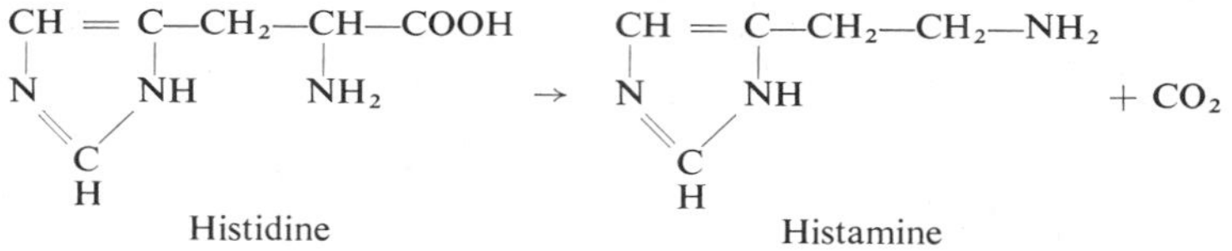

```
CH = C—CH2—CH—COOH          CH = C—CH2—CH2—NH2
|    |      |                |    |
N    NH     NH2       →      N    NH                 + CO2
 \\ /                         \\ /
  C                            C
  H                            H
      Histidine                    Histamine
```

Histidine decarboxylase, like other amino acid decarboxylases, requires pyridoxal-5-phosphate as coenzyme. Serotonin, ethanolamine, and α-aminobutyric acid are also produced by decarboxylation reactions.

Transamination. All of the amino acids except possibly glycine, threonine, and lysine can participate in transamination reactions of the following type:

$$\underset{\text{Glutamic acid}}{HOOC—CH_2—CH_2—CHNH_2—COOH} + R—CO—COOH$$
$$\rightleftharpoons \underset{\alpha\text{-Ketoglutaric acid}}{HOOC—CH_2—CH_2—CO—COOH} + R—CHNH_2—COOH$$

Pyridoxal-5-phosphate, the coenzyme for all transaminases, is an amino group carrier during this reaction. The amino group is transferred from the amino acid to the coenzyme to form pyridoxamine-5-phosphate which, in turn, transfers its amino group to a keto acid for the regeneration of pyridoxal-5-phosphate and the formation of a new amino acid. Glutamic acid (or α-ketoglutaric acid) can be replaced by other amino (or keto) acids, but the reaction with glutamic/α-ketoglutaric acid is particularly important in the eventual detachment of the amino group from the carbon chain (see PYRIDOXINE FAMILY OF VITAMINS; TRANSAMINATION ENZYMES).

Deamination. Four distinct types of reaction are known for the removal of the amino group: (1) direct elimination of ammonia, (2) dehydration followed by hydrolysis, (3) oxidative deamination, and (4) dehydrogenation. Superficially, the first of the reactions would appear to be the simplest, but it is also the rarest form of amino group removal in animal tissues, since it occurs only with histidine to yield ammonia and the corresponding α, β-unsaturated acid. A similar reaction, in which aspartic acid is converted to ammonia and fumaric acid, occurs only in bacteria.

Serine, threonine, and cysteine are deaminated by the second type of reaction as follows:

$$\underset{\text{Serine}}{\underset{\text{OH}\quad \text{NH}_2}{CH_2\text{—}CH\text{—}COOH}} \xrightarrow{-H_2O} \left[\underset{NH_2}{CH_2\text{=}C\text{—}COOH} \leftrightarrow \underset{NH}{CH_3\text{—}\overset{\|}{C}\text{—}COOH} \right] \xrightarrow{+H_2O} \underset{\text{Pyruvic acid}}{\underset{O}{CH_3\text{—}\overset{\|}{C}\text{—}COOH}} + NH_3$$

The dehydrases which catalyze this type of reaction require pyridoxal-5-phosphate. Threonine, as would be expected, yields α-ketobutyric acid rather than pyruvic acid, and cysteine, by the action of a desulfhydrase, gives hydrogen sulfide, ammonia, and pyruvic acid.

For many years, the oxidative deamination of amino acids has been something of an anomaly. Liver and kidney have a very active enzyme for the oxidation of the unnatural D-isomer of most of the amino acids, even though these compounds do not occur in animal tissues in significant amounts. A possible explanation for this curious observation may be that the deamination of the D-amino acids is catalyzed by the same enzyme which deaminates glycine, the only non-optically active amino acid. Glycine oxidase is known to be physiologically important for the conversion of glycine to ammonia and glyoxylic acid. The L-amino acid oxidase activity of kidney and liver is weak and plays a minor role in deamination. Both L-amino acid oxidase and glycine oxidase contain flavin adenine dinucleotide (FAD) as their prosthetic groups (see FLAVINS AND FLAVOPROTEINS). The general reaction for oxidative deamination is given by the equations:

$$R\text{—}CHNH_2\text{—}COOH \rightleftharpoons$$
$$R\text{—}CO\text{—}COOH + NH_3 + FADH_2$$
$$FADH_2 + O_2 \rightarrow FAD + H_2O_2$$

(see also AMINE OXIDASES AND AMINO ACID OXIDASES).

Although glutamic acid is one of the few amino acids not attacked by L-amino acid oxidase, liver contains large amounts of a very active nucleotide-dependent, glutamic acid dehydrogenase. As indicated by the equation below, the coupling of transamination and dehydrogenation by glutamic acid dehydrogenase accounts for most of the ammonia production from amino acids. (DPN is alternatively termed NAD, NICTOTINAMIDE ADENINE DINUCLEOTIDE.)

$$\text{α-Ketoglutaric Acid} + \text{Amino Acid} \rightleftharpoons \text{Glutamic Acid} + \text{α-Ketoacid}$$
$$\text{Glutamic Acid} + DPN^+ + H_2O \rightleftharpoons \text{α-Ketoglutaric Acid} + DPNH + H^+ + NH_3$$
$$\textit{Sum}\text{: Amino Acid} + DPN^+ + H_2O \rightleftharpoons \text{α-Ketoacid} + DPNH + H^+ + NH_3$$

Since ammonia is toxic, an animal would be in serious difficulty if there were no mechanism for its removal. In certain fish, the problem is solved very simply by excreting ammonia directly into the aquatic environment. Birds and reptiles remove ammonia by converting it to uric acid which they excrete. In mammals, however, urea is the chief end product of ammonia metabolism. Advantages of urea as an excretory product are: (1) it is neither acidic nor basic, (2) it is very soluble, and (3) it contains a high percentage of nitrogen.

An elaborate pathway, in which a number of amino acids participate, has been evolved for urea synthesis. Ornithine, an amino acid not found in protein hydrolysates, has a catalytic function as can be seen from the reactions below. N-Acetylglutamate is required for the synthesis of CARBAMYL PHOSPHATE, but its role in the reaction is not known (see equations at bottom of page).

Liver is the primary site of urea synthesis and contains all of the enzymes necessary for the operation of the urea cycle as depicted (see also ARGININE-UREA CYCLE).

Carbon Chain Conversions. Although a number of gaps still remain, the main routes for metabolism of the carbon chain of the individual amino acids are known. On the basis of their metabolic pathways, the amino acids can be divided into three groups: (1) glycogenic amino acids yield products which can be converted to glucose or glycogen, (2) ketogenic amino acids yield acetyl CoA or acetoacetic acid, and (3) glycogenic and ketogenic amino acids give products which could yield glucose and acetyl CoA or acetoacetic acid (see GLUCONEOGENESIS).

The glycogenic amino acids are alanine, cysteine, serine, threonine, glycine, aspartic acid, glutamic acid, proline, hydroxyproline, methionine, valine, histidine, arginine, and tryptophan. As indicated previously, dehydration of serine and desulfurization of cysteine yield pyruvic acid directly. By a reversal of glycolysis, pyruvic acid would yield glucose, but, normally, the bulk of it is oxidized to carbon dioxide and water by way of the tricarboxylic acid cycle (see CITRIC ACID CYCLE). Removal of the amino group from alanine also results in the formation of pyruvic

$$NH_3 + CO_2 + H_2O + 2ATP \rightarrow \underset{\text{Carbamyl phosphate}}{H_2N\text{—}CO\text{—}OPO_3H_2} + 2ADP + P_i + H^+$$

$$\underset{\text{Carbamyl phosphate}}{H_2N\text{—}CO\text{—}OPO_3H_2} + \underset{\text{Ornithine}}{H_2N\text{—}CH_2\text{—}CH_2\text{—}CH_2\text{—}\underset{NH_2}{CH}\text{—}COOH} \rightarrow$$

$$P_i + \underset{\text{Citrulline}}{H_2N\text{—}CO\text{—}NH\text{—}CH_2\text{—}CH_2\text{—}CH_2\text{—}\underset{NH_2}{CH}\text{—}COOH}$$

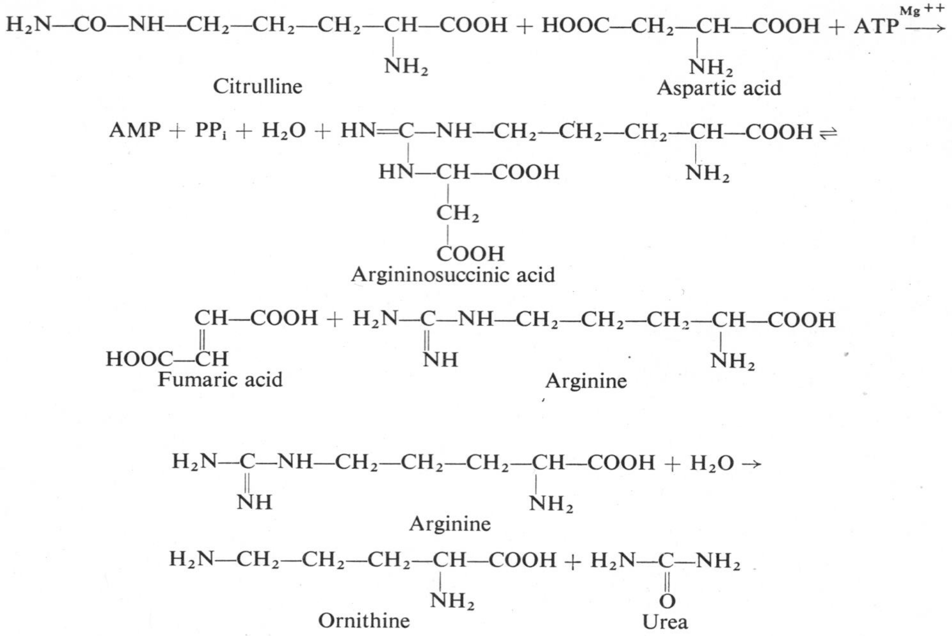

$$H_2N\text{—}CO\text{—}NH\text{—}CH_2\text{—}CH_2\text{—}CH_2\text{—}\underset{NH_2}{\underset{|}{CH}}\text{—}COOH + HOOC\text{—}CH_2\text{—}\underset{NH_2}{\underset{|}{CH}}\text{—}COOH + ATP \xrightarrow{Mg^{++}}$$

Citrulline — Aspartic acid

$$AMP + PP_i + H_2O + HN{=}\underset{HN\text{—}CH(CH_2COOH)\text{—}COOH}{\underset{|}{C}}\text{—}NH\text{—}CH_2\text{—}CH_2\text{—}CH_2\text{—}\underset{NH_2}{\underset{|}{CH}}\text{—}COOH \rightleftharpoons$$

Argininosuccinic acid

$$HOOC\text{—}CH{=}CH\text{—}COOH + H_2N\text{—}\underset{NH}{\underset{\|}{C}}\text{—}NH\text{—}CH_2\text{—}CH_2\text{—}CH_2\text{—}\underset{NH_2}{\underset{|}{CH}}\text{—}COOH$$

Fumaric acid — Arginine

$$H_2N\text{—}\underset{NH}{\underset{\|}{C}}\text{—}NH\text{—}CH_2\text{—}CH_2\text{—}CH_2\text{—}\underset{NH_2}{\underset{|}{CH}}\text{—}COOH + H_2O \rightarrow$$

Arginine

$$H_2N\text{—}CH_2\text{—}CH_2\text{—}CH_2\text{—}\underset{NH_2}{\underset{|}{CH}}\text{—}COOH + H_2N\text{—}\underset{O}{\underset{\|}{C}}\text{—}NH_2$$

Ornithine — Urea

acid. Oxaloacetic and α-ketoglutaric acids, intermediates in the tricarboxylic acid cycle, are formed from aspartic and glutamic acids. With glutamic acid as an intermediate, the ornithine derived from arginine is converted to α-ketoglutaric acid. Proline and ornithine are interconvertible through oxidation and reduction at the δ-carbon atom. Oxidation and opening of the heterocyclic ring of hydroxyproline forms γ-hydroxyglutamic acid which, after transamination followed by an aldol cleavage, yields glyoxylic and pyruvic acids. The α-ketobutyric acid derived from methionine and threonine is oxidatively decarboxylated to propionyl CoA which, by means of a series of reactions involving carbon dioxide fixation, yields succinyl CoA, another intermediate in the tricarboxylic acid cycle. Threonine aldolase also catalyzes the cleavage of threonine to acetaldehyde and glycine.

The pathway for the breakdown of valine is given below. The last two steps in the sequence are based on analogous reactions, but there is no direct experimental evidence that they occur as indicated:

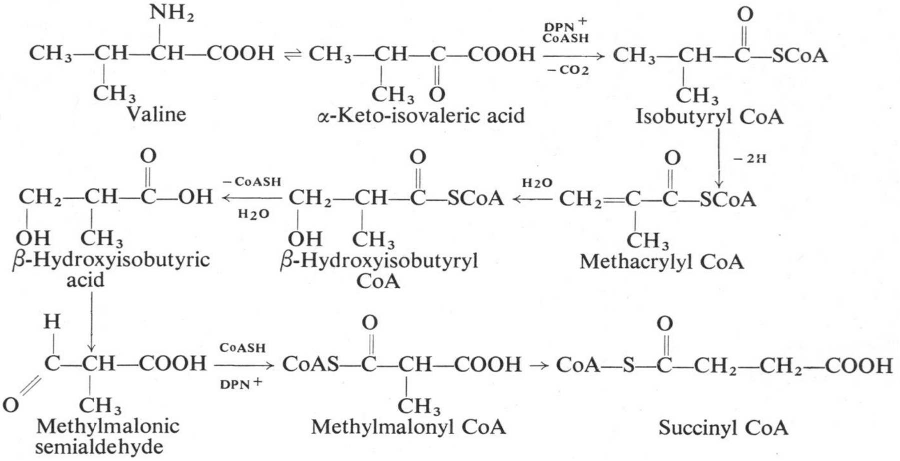

$$CH_3\text{—}\underset{CH_3}{\underset{|}{CH}}\text{—}\overset{NH_2}{\overset{|}{CH}}\text{—}COOH \rightleftharpoons CH_3\text{—}\underset{CH_3}{\underset{|}{CH}}\text{—}\underset{O}{\underset{\|}{C}}\text{—}COOH \xrightarrow[-CO_2]{DPN^+,\ CoASH} CH_3\text{—}\underset{CH_3}{\underset{|}{CH}}\text{—}\overset{O}{\overset{\|}{C}}\text{—}SCoA$$

Valine — α-Keto-isovaleric acid — Isobutyryl CoA

$$\xrightarrow{-2H} CH_2{=}\underset{CH_3}{\underset{|}{C}}\text{—}\overset{O}{\overset{\|}{C}}\text{—}SCoA \xrightarrow{H_2O} \underset{OH}{\underset{|}{CH_2}}\text{—}\underset{CH_3}{\underset{|}{CH}}\text{—}\overset{O}{\overset{\|}{C}}\text{—}SCoA \xrightarrow[H_2O]{-CoASH} \underset{OH}{\underset{|}{CH_2}}\text{—}\underset{CH_3}{\underset{|}{CH}}\text{—}\overset{O}{\overset{\|}{C}}\text{—}OH$$

Methacrylyl CoA — β-Hydroxyisobutyryl CoA — β-Hydroxyisobutyric acid

$$\rightarrow \overset{H}{\overset{|}{C}}({=}O)\text{—}\underset{CH_3}{\underset{|}{CH}}\text{—}COOH \xrightarrow[DPN^+]{CoASH} CoAS\text{—}\overset{O}{\overset{\|}{C}}\text{—}\underset{CH_3}{\underset{|}{CH}}\text{—}COOH \rightarrow CoA\text{—}S\text{—}\overset{O}{\overset{\|}{C}}\text{—}CH_2\text{—}CH_2\text{—}COOH$$

Methylmalonic semialdehyde — Methylmalonyl CoA — Succinyl CoA

The complex metabolic pathways for histidine and tryptophan metabolism will not be discussed except to mention that glutamic acid is the chief product of histidine degradation and that tryptophan eventually yields pyruvate from its side chain and the vitamin, NICOTINIC ACID, from the indole ring. In addition to its important role as a precursor of the PORPHYRINS, glycine is interconvertible with serine as follows:

$$\underset{\text{Glycine}}{CH_2(NH_2)\text{—}COOH} + [O] \rightarrow NH_3 + \underset{\text{Glyoxylic acid}}{OHC\text{—}COOH} \rightarrow \underset{\text{Oxalic acid}}{HOOC\text{—}COOH} \rightarrow CO_2 + \underset{\text{Formic acid}}{HCOOH}$$

$$\text{Formic acid} \xrightarrow{+\text{ATP},\ +\text{Tetrahydrofolic acid}} N^{10}\text{-Formyltetrahydrofolic acid}$$

$$\underset{\text{Serine}}{CH_2(OH)\text{—}CH(NH_2)\text{—}COOH} \underset{}{\overset{+\text{Glycine}}{\rightleftharpoons}} N^{10}\text{-Hydroxymethyl tetrahydrofolic acid} \xleftarrow{\text{DPNH}} N^{10}\text{-Formyltetrahydrofolic acid}$$

Leucine is the only amino acid which is entirely ketogenic. Its metabolic pathway is:

$$\underset{\text{Leucine}}{CH_3\text{—}CH(CH_3)\text{—}CH_2\text{—}CH(NH_2)\text{—}COOH} \rightleftharpoons \underset{\alpha\text{-Ketoisocaproic acid}}{CH_3\text{—}CH(CH_3)\text{—}CH_2\text{—}C(=O)\text{—}COOH} \xrightarrow{\text{DPN}^+,\ \text{CoASH}} \underset{\text{Isovaleryl CoA}}{CH_3\text{—}CH(CH_3)\text{—}CH_2\text{—}C(=O)\text{—}SCoA}$$

$$\text{Isovaleryl CoA} \xrightarrow{-2H} \underset{\beta\text{-Methylcrotonyl CoA}}{CH_3\text{—}C(CH_3){=}CH\text{—}C(=O)\text{—}SCoA} \xrightarrow[+CO_2]{\text{ATP}} \underset{\beta\text{-Methylglutaconyl CoA}}{HOOC\text{—}CH_2\text{—}C(CH_3){=}CH\text{—}C(=O)\text{—}SCoA}$$

$$\beta\text{-Methylglutaconyl CoA} \xrightarrow{H_2O} \underset{\beta\text{-Hydroxy-}\beta\text{-methylglutaryl CoA}}{HOOC\text{—}CH_2\text{—}C(OH)(CH_3)\text{—}CH_2\text{—}C(=O)\text{—}SCoA} \rightarrow \underset{\text{Acetoacetic acid}}{CH_3\text{—}C(=O)\text{—}CH_2\text{—}COOH} + \underset{\text{Acetyl CoA}}{CH_3\text{—}C(=O)\text{—}SCoA}$$

Isoleucine is both glycogenic and ketogenic since it yields equal amounts of acetyl CoA and propionyl CoA:

$$\underset{\text{Isoleucine}}{CH_3\text{—}CH_2\text{—}CH(CH_3)\text{—}CH(NH_2)\text{—}COOH} \rightleftharpoons \underset{\alpha\text{-Keto-}\beta\text{-methylvaleric acid}}{CH_3\text{—}CH_2\text{—}CH(CH_3)\text{—}C(=O)\text{—}COOH}$$

$$\alpha\text{-Keto-}\beta\text{-methylvaleric acid} \xrightarrow{-CO_2,\ \text{DPN}^+,\ \text{CoASH}} \underset{\alpha\text{-methylbutyryl CoA}}{CH_3\text{—}CH_2\text{—}CH(CH_3)\text{—}C(=O)\text{—}SCoA} \xrightarrow{-2H} \underset{\text{Tiglyl CoA}}{CH_3\text{—}CH{=}C(CH_3)\text{—}C(=O)\text{—}SCoA}$$

$$\text{Tiglyl CoA} \xrightarrow{+H_2O} \underset{\alpha\text{-Methyl-}\beta\text{-hydroxy butyryl CoA}}{CH_3\text{—}CH(OH)\text{—}CH(CH_3)\text{—}C(=O)\text{—}SCoA} \xrightarrow{\text{DPN}^+} \underset{\alpha\text{-Methylacetoacetyl CoA}}{CH_3\text{—}C(=O)\text{—}CH(CH_3)\text{—}C(=O)\text{—}SCoA} \rightarrow \underset{\text{Acetyl CoA}}{CH_3\text{—}C(=O)\text{—}SCoA} + \underset{\text{Propionyl CoA}}{CH_3\text{—}CH_2\text{—}C(=O)\text{—}SCoA}$$

Phenylalanine and tyrosine, which are also both glycogenic and ketogenic, have a common degradative pathway in that the first step in phenylalanine metabolism is oxidation to tyrosine. The final reaction products are fumaric and acetoacetic acids (see AROMATIC AMINO ACIDS). At present, lysine is difficult to classify, but judging from what is known of its metabolism, it should be ketogenic.

The biochemical reactions of amino acids discussed so far are largely degradative, but tissues also contain a large variety of small, nitrogen-containing compounds which are synthesized from amino acids. Only a few examples of this important class of compounds will be given. Histamine formation from histidine has already been mentioned.

CREATINE, which, as its phosphate ester, phosphocreatine, is a storage form of high-energy bonds, is synthesized from three different amino acids. In kidney, an enzyme catalyzes the transfer of the guanido group from arginine to glycine to yield ornithine and guanidoacetic acid:

$$\underset{\text{Arginine}}{H_2N{-}\underset{\|}{\overset{}{C}}(=NH){-}NH{-}CH_2{-}CH_2{-}CH_2{-}CH(NH_2){-}COOH} + \underset{\text{Glycine}}{H_2N{-}CH_2{-}COOH}$$

$$\rightleftharpoons \underset{\text{Ornithine}}{H_2N{-}CH_2{-}CH_2{-}CH_2{-}CH(NH_2){-}COOH} + \underset{\text{Guanidoacetic acid}}{H_2N{-}C(=NH){-}NH{-}CH_2{-}COOH}$$

Guanidoacetic acid is transported to the liver where it accepts a methyl group from S-adenosylmethionine:

$$\text{ATP} + \underset{\text{Methionine}}{CH_3{-}S{-}CH_2{-}CH_2{-}CH(NH_2){-}COOH} \rightarrow \underset{\text{S-Adenosylmethionine}}{\text{Adenine-ribose}{-}CH_2{-}\overset{+}{S}(CH_3){-}CH_2{-}CH_2{-}CH(NH_2){-}COOH} + PP_i + P_i$$

$$\underset{\text{Guanidoacetic acid}}{H_2N{-}C(=NH){-}NHCH_2COOH} + \text{S-Adenosylmethionine}$$

$$\rightarrow \underset{\text{Creatine}}{H_2N{-}C(=NH){-}N(CH_3){-}CH_2COOH} + \underset{\text{S-Adenosylhomocysteine}}{\text{Adenine-ribose}{-}CH_2{-}S{-}CH_2{-}CH_2{-}CH(NH_2){-}COOH}$$

Transmethylation reactions of this type are also involved in the synthesis of choline, epinephrine, and a number of other metabolites (see METHYLATIONS IN METABOLISM). The formation of S-adenosylmethionine is unusual in that it is the only known reaction of ATP in which all three phosphate groups are removed.

The tripeptide, GLUTATHIONE, is a coenzyme for some reactions and may even function as a hormone in some organisms. It is synthesized as follows:

$$\underset{\text{Glutamic acid}}{\mathrm{HOOC{-}\underset{NH_2}{\underset{|}{C}H}{-}CH_2{-}CH_2{-}COOH}} + \underset{\text{Cysteine}}{\mathrm{HS{-}CH_2{-}\underset{NH_2}{\underset{|}{C}H}{-}COOH}}$$

$$+\ \mathrm{ATP} \xrightarrow[\mathrm{K^+}]{\mathrm{Mg^{++}}} \underset{\gamma\text{-Glutamylcysteine}}{\mathrm{HOOC{-}\underset{NH_2}{\underset{|}{C}H}{-}CH_2{-}CH_2{-}\overset{O}{\overset{\|}{C}}{-}NH{-}\underset{SH}{\underset{|}{\underset{CH_2}{\underset{|}{C}H}}}{-}COOH}} + \mathrm{P_i}$$

$$+\ \mathrm{ADP} \xrightarrow[+\mathrm{ATP}]{+\mathrm{Glycine}} \underset{\gamma\text{-Glutamylcysteinylglycine (glutathione)}}{\mathrm{HOOC{-}\underset{NH_2}{\underset{|}{C}H}{-}CH_2{-}CH_2{-}\overset{O}{\overset{\|}{C}}{-}NH{-}\underset{SH}{\underset{|}{\underset{CH_2}{\underset{|}{C}H}}}{-}\overset{O}{\overset{\|}{C}}{-}NHCH_2{-}COOH}}$$

Routes for the biosynthesis of some of the nonessential amino acids in the mammal are fairly obvious. For example, alanine, glutamic acid, and aspartic acid are synthesized by amination of the corresponding keto acids. Serine is derived from phosphoglyceric acid. Tyrosine is not essential because it is the product of the enzymatic hydroxylation of phenylalanine, one of the essential amino acids. The biosynthesis of the essential amino acids by microoorganisms is a fascinating field involving a number of reactions with interesting control mechanisms, but that is another story.

References

1. Rose, W. C., *Physiol. Rev.*, **18**, 109 (1938).
2. Coon, M. L., and Robinson, W. G., *Ann. Rev. Biochem.*, **27**, 561 (1958).
3. Greenberg, D. M., (Editor), "Metabolic Pathways," Vol. 2, pp. 1–261, New York, Academic Press, 1961.
4. Sakami, W., and Harrington, H., *Ann. Rev. Biochem.*, **32**, 355 (1963).

William G. Robinson

AMINO ACID OXIDASES. See AMINE OXIDASES AND AMINO ACID OXIDASES.

AMINO ACID POOL

Amino acids are constantly added to body fluids by absorption from the intestine [see ABSORPTION (OF SUBSTANCES INTO CELLS)], by synthesis and by protein breakdown, and at the same time they are continually being removed for synthesis of protein (see AMINO ACID METABOLISM; PROTEINS, BIOSYNTHESIS) and other uses. A balance between these two processes results in a level of free amino acids in the body fluids and tissues which may be termed the amino acid pool. The study of this amino acid pool, and of the dynamic aspects of free amino acid exchange (and nitrogen exchange) with proteins, was one of the earliest applications of isotopic tracer methods in biochemistry, particularly as carried out by Schoenheimer and co-workers.[1-3] Such studies showed that in animals the free amino acid pool is small but turns over very rapidly.

Bacterial cells have been found capable of establishing intracellular concentrations of amino acids much higher than the external concentrations of the corresponding amino acids in the medium, and the term "pool" may be applied to the intracellular "accumulated" amino acids that are acid-extractable (cold 5% trichloracetic acid), in contrast to those intracellular amino acids that are covalently bound ("incorporated" or non-acid extractable).[4]

References

1. West, E. S., and Todd, W. R., "Protein Metabolism," in "Textbook of Biochemistry," Third edition, Ch. 25, New York, The Macmillan Co., 1961.
2. Fruton, J. S., and Simmonds, S., "Metabolic Breakdown and Synthesis of Proteins," in "General Biochemistry," Second edition, Ch. 30, p. 723, New York, John Wiley & Sons, 1958.
3. Schoenheimer, R., "The Dynamic State of Body Constituents," Cambridge, Mass., Harvard University Press, 1942.
4. Britten, R. J., and McClure, F. T., "The Amino Acid Pool in *Escherichia coli*," *Bacteriol. Rev.*, **26**, 292–335 (1962).

AMINO ACID SEQUENCE ANALYSIS AND END GROUP ANALYSIS. See PROTEINS, END GROUP AND SEQUENCE ANALYSIS.

AMINO ACIDS, ANALYTICAL METHODS. See PROTEINS, AMINO ACID ANALYSIS AND CONTENT.

AMINO ACIDS, AROMATIC (BIOSYNTHESIS AND METABOLISM). See AROMATIC AMINO ACIDS (BIOSYNTHESIS AND METABOLISM).

AMINO ACIDS, NON-PROTEIN

Other sections of this volume have discussed amino acids, numbering about 20, which are derived from PROTEIN and so are universal constituents of all living organism [see AMINO ACIDS (OCCURRENCE, STRUCTURE AND SYNTHESIS); AMINO ACIDS, PROTEIN-DERIVED]. A far larger number of naturally occurring amino acids show a more restricted distribution pattern and are rarely, if ever, introduced into protein molecules under normal circumstances so that a tendency to call them *non-protein amino acids* has developed. Normally, paper chromatography has provided the first indication of the presence of a "new" substance, and now about 150 such compounds have been chemically characterized. The majority of the compounds originate from higher plants, and seeds, especially, frequently accumulate large amounts of particular substances. Many others have a microbial origin, often as components of extracellular antibiotic peptides. Relatively few of the acids are characteristic of the animal kingdom.[2,3]

The present group of compounds exhibit many of the structural features encountered in the protein amino acids. For instance, in the formula $R\cdot CH(NH_2)\cdot CO_2H$, the radical R may include carboxyl, amino, amido and hydroxyl groups or it may be an aromatic or heterocyclic ring. Some examples of structural similarities are illustrated in Table 1; clearly, many of the non-protein amino acids are homologues, close isomers or simple substituted derivatives of the protein constituents. In addition, certain unique or rare chemical features may be distinguished, *e.g.*, the presence in a natural product of the unusual cyclopropyl ring system and the entirely novel occurrence of a pyrazole derivative.

Table 1 also includes the primary sources of the

TABLE 1. STRUCTURAL RELATIONSHIPS BETWEEN SOME PROTEIN AND NON-PROTEIN AMINO ACIDS

Protein Amino Acid	Related Non-Protein Compound	Occurrence
	Acidic	
$HO_2C\cdot CH_2\cdot CH_2\cdot CH(NH_2)\cdot CO_2H$ Glutamic Acid	$HO_2C\cdot C(=CH_2)\cdot CH_2\cdot CH(NH_2)\cdot CO_2H$ γ-Methyleneglutamic Acid	Peanuts, tulips and other random occurences.
	$HO_2C\cdot CH_2\cdot CH_2\cdot CH_2\cdot CH(NH_2)\cdot CO_2H$ α-Aminoadipic Acid	Widely distributed in animals, plants and microorganisms
	Amides	
$H_2N\cdot OC\cdot CH_2\cdot CH_2\cdot CH(NH_2)\cdot CO_2H$ Glutamine	$H_2N\cdot OC\cdot C(=CH_2)\cdot CH_2\cdot CH(NH_2)\cdot CO_2H$ γ-Methyleneglutamine	Peanuts, tulips and other random occurrences
	$(C_2H_5)HN\cdot OC\cdot CH_2\cdot CH(NH_2)\cdot CO_2H$ N-Ethylasparagine	Some cucurbits
	Basic	
$CH_2(NH_2)\cdot (CH_2)_3\cdot CH(NH_2)\cdot CO_2H$ Lysine	$CH_2(NH_2)\cdot CH_2\cdot CH(NH_2)\cdot CO_2H$ α,γ-Diaminobutyric Acid	Antibiotic peptides, some legumes, etc.
$(H_2N)(HN{=})C\cdot NH(CH_2)_3\cdot CH(NH_2)\cdot CO_2H$ Arginine	$(H_2N)(HN{=})C\cdot NH\cdot O\cdot CH_2\cdot CH_2\cdot CH(NH_2)\cdot CO_2H$ Canavanine	Jack bean and seed of some other legumes
	Hydroxy	
$CH_3\cdot CH(OH)\cdot CH(NH_2)\cdot CO_2H$ Threonine	$CH_2(OH)\cdot CH_2\cdot CH(NH_2)\cdot CO_2H$ Homoserine	Widely distributed among living organisms
	Aromatic and Heterocyclic	
HO–(benzene ring)–$CH_2\cdot CH(NH_2)\cdot CO_2H$ Tyrosine	O=(ring, HO)–$N\cdot CH_2\cdot CH(NH_2)\cdot CO_2H$ Mimosine	*Leucaena glauca* and *Mimosa* seed
HC=N, HN, C·$CH_2\cdot CH(NH_2)CO_2H$, CH (imidazole ring) Histidine	CH=N, CH, $N\cdot CH_2\cdot CH(NH_2)\cdot CO_2H$, CH (pyrazole ring) β-Pyrazol-1-ylalanine	Many members of Cucurbitaceae

	Sulfur-containing	
$HS{\cdot}CH_2{\cdot}CH(NH_2){\cdot}CO_2H$ Cysteine	$CH_2{=}CH{\cdot}CH_2{\cdot}\overset{O}{\overset{\uparrow}{S}}{\cdot}CH_2{\cdot}CH(NH_2){\cdot}CO_2H$ *S*-Allylcysteine Sulfoxide	Garlic
	$CH_3{\cdot}CH{=}CH{\cdot}\overset{O}{\overset{\uparrow}{S}}{\cdot}CH_2{\cdot}CH(NH_2){\cdot}CO_2H$ *S*-Propenylcysteine sulfoxide	Onion
	Imino Acid	
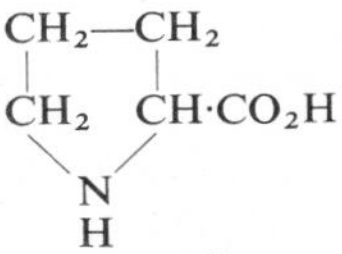 $CH_2{-}CH_2$, CH_2, $CH{\cdot}CO_2H$, NH (ring) Proline	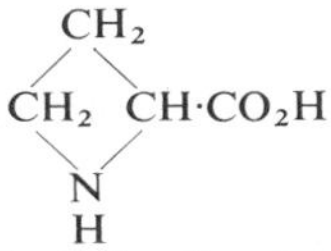 CH_2, CH_2, $CH{\cdot}CO_2H$, NH (ring) Azetidine-2-carboxylic Acid	Many Liliaceae
	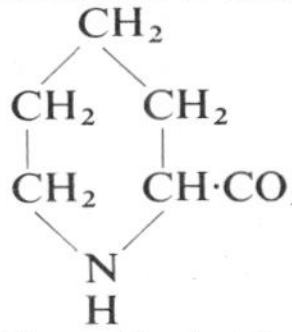 CH_2, CH_2, CH_2, CH_2, $CH{\cdot}CO_2H$, NH (ring) Pipecolic Acid	Widely distributed in animals, plants and microorganisms
	Miscellaneous	
No equivalent	$NC{\cdot}CH_2{\cdot}CH(NH_2){\cdot}CO_2H$ β-Cyanoalanine	Several species of *Vicia*
	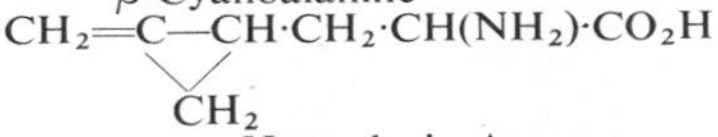 $CH_2{=}C{-}CH{\cdot}CH_2{\cdot}CH(NH_2){\cdot}CO_2H$ (C and CH bridged by CH_2) Hypoglycin A	Unripe fruit of *Blighia sapida*
	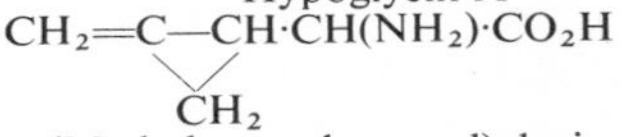 $CH_2{=}C{-}CH{\cdot}CH(NH_2){\cdot}CO_2H$ (C and CH bridged by CH_2) α-(Methylenecyclopropyl)glycine	Seed of *Litchi chinensis*

compounds. A few are distributed widely; indeed, compounds like homoserine, that are implicated as intermediates in biosynthetic pathways leading to protein amino acids, must be of universal occurrence. Most have a very limited distribution. Sometimes they form characteristic components of a few closely allied species, *e.g.*, azetidine-2-carboxylic acid in members of the plant family Liliaceae and β-pyrazol-1-ylalanine in many cucurbits, but other compounds, *e.g.*, γ-methyleneglutamic acid, show infrequent and haphazard distribution across members of the plant kingdom. It is noteworthy also that the two homologous cyclopropyl amino acids in Table 1 were isolated from members of the same, small plant family, the Sapindaceae.

In seeking a *raison d'etre* for this group of substances, no general convincing explanation for their biosynthesis has been advanced. Varied functions concerned with nitrogen assimilation, storage and transport in plants seem possible, and it is conceivable that certain compounds are endowed with growth-regulatory or metabolic-controlling action. Certainly, the accumulation in seeds, frequently of acids with high nitrogen content, endorses a storage role. Another view would envisage many of the compounds as by-products of an aberrant metabolism, perhaps even as evolutionary misfits not yet discarded.

The biogenesis of most compounds in this group remains to be studied. Preliminary work suggests that certain compounds homologous with protein constituents may be synthesized by the nonspecific action of universally distributed enzymes, *e.g.*, the dehydrogenase responsible for catalyzing the last step in proline biosynthesis also effects pipecolic acid formation, while glutamic-aspartic transaminase utilizes certain γ-substituted glutamic acids at a slow rate.

Another example of interaction at the level of protein amino acid metabolism is seen in PROTEIN BIOSYNTHESIS itself, where several of the newly recognized amino acids act as analogues of a normal constituent and become incorporated into protein in place of the usual residues after activation by the appropriate activating enzyme. Analogues must show a close stereochemical relationship to the normal component if they are to be activated. This type of analogue behaviour is illustrated by azetidine-2-carboxylic acid which can effectively replace proline residues during biosynthesis of a variety of proteins. Canavanine, a constituent of a group of legumes, similarly acts as an arginine analogue. The report that α-aminoadipic acid occurs in trace amount in maize seed protein perhaps may be ascribed to the nonspecific activation of the amino acid by the plant's glutamate activating enzyme.

This type of analogue behaviour may result in toxicity. For example, azetidine-2-carboxylic acid kills many seedlings and chick embryos presumably because the altered protein molecules then synthesized either lack, or possess decreased, enzymic activity. Plants normally producing the imino acid possess a proline activating enzyme that is unable to use the analogue as a substrate and by this means they have gained protection against their own toxic product. Other instances of toxicity caused by amino acids of this group include lathyrism and hypoglycemia. Lathyrism is caused by the ingestion of several species of legume seed (Bell, 1964) and the toxins have been shown to include α,γ-diaminobutyric acid, β-oxalylamino-α-aminopropionic acid, β-cyanoalanine and its γ-glutamyl peptide, all of which produce a neurolathyrism, while γ-glutamyl-β-aminopropionitrile (Lathyrus factor) is responsible for osteolathyrism. The toxicity of β-cyanoalanine may be due to interference with pyridoxal phosphate requiring enzymes, while γ-glutamyl-β-aminopropionitrile undoubtedly prevents normal COLLAGEN formation. Hair or wool growth is stopped when mimosine is fed to animals, and sheep have been observed to shed their fleece after consuming large quantities of the tropical legume, *Leucaena glauca*. Finally, hypoglycin A and α-(methylenecyclopropyl)glycine show definite hypoglycemic action, but in these cases it appears probable that the toxic agents are unsaturated fatty acids formed by catabolic oxidative deamination and decarboxylation of the original amino acids, which then inhibit the cycle of β-oxidation of fatty acids at the stage of crotonase action.

The lachrymatory properties of some plants may be attributed to their content of S-substituted cysteines. When garlic bulbs are crushed, the S-allylcysteine sulfoxide (alliin) present is rapidly split by a lyase enzyme to yield S-allyl sulfenic acid, pyruvate and ammonia: two molecules of the sulfenic acid then undergo condensation to yield allicin, the major lachrymatory principle. The isomeric S-propenylcysteine sulfoxide occurs in onion, and a similar lyase action yields S-propenyl sulfenic acid. This sulfenic acid does not undergo self-condensation and is considered to be the real lachrymator of onion (Virtanen, 1964).

Garlic

$$\underset{\text{Alliin}}{CH_2{=}CH{-}CH_2{-}\overset{O\leftarrow}{S}{\cdot}CH_2{\cdot}CH(NH_2){\cdot}CO_2H} \xrightarrow[\text{splitting}]{\text{Lyase}} \underset{\text{Allyl sulfenic acid}}{CH_2{=}CH{-}CH_2{-}\overset{O\leftarrow}{S}H} + \text{Pyruvate} + NH_3 \xrightarrow{\text{Self-condensation}} \underset{\text{Allicin}}{CH_2{=}CH{-}CH_2{-}\overset{O\leftarrow}{S}{-}S{-}CH_2{-}CH{=}CH_2}$$

References

1. Bell, E. A., *Nature*, **203**, 378 (1964).
2. Fowden, L., *Endeavour*, **21**, 35 (1962).
3. Fowden, L., *Ann. Rev. Biochem.*, **33**, 173 (1964).
4. Virtanen, A. I., "Studies on Organic Sulphur Compounds in Vegetables and Fodder Plants," Helsinki, Biochemical Institute, 1964.

L. Fowden

AMINO ACIDS, NUTRITIONALLY ESSENTIAL (FOR MAN AND OTHER VERTEBRATES)

Amino acids constitute the alphabet of the PROTEINS. At least 22 different letters make up the amino acid alphabet, and in various combinations these comprise the great multitude of proteins found in nature [see AMINO ACIDS (OCCURRENCE, STRUCTURE, AND SYNTHESIS)]. The qualitative and quantitative distribution of amino acids in a protein determines its chemical characteristics, its nutritive value, and its metabolic functions in the body. Eighteen different amino acids commonly occur in our food supply. The body can manufacture ten of them from materials derived from proteins, carbohydrates and fats in the food. The remaining eight cannot be made by the body from any of these materials, and they must be supplied preformed by the food. These are: valine, lysine, threonine, leucine, isoleucine, tryptophan, phenylalanine, and methionine. They are called the *indispensable amino acids* because it is essential to have them ready-made in the diet.

The presence in a protein of the nutritionally essential amino acids in significant amounts and in proportions fairly similar to those found in body proteins classifies it as a complete protein (see also PROTEINS, NUTRITIONAL QUALITY; NUTRITIONAL REQUIREMENTS). Apparently the body needs these amino acids to be available from foods in about the same ratios at each meal for use in body maintenance, repair, and growth. Most animal foods—meat, fish, poultry, eggs, milk, and cheese—contain complete proteins. However, very few vegetable foods contain complete protein; deficiencies of lysine, threonine, and tryptophan occur frequently.

The nutritional requirement for an amino acid is determined by changes in *body weight* or *nitrogen balance* in the test organism. The absence from the diet of an unessential amino acid will not hinder growth of young animals or reduce nitrogen retention of mature mammals. On the other hand, the lack of an adequate amount of an essential amino acid from the diet will cause a marked reduction in growth rate and loss of body nitrogen.[1] In weanling rats deficiencies of tryptophan and lysine have been shown to induce cataracts and corneal vascularization. In mature females a dietary deficiency of lysine causes an interruption of the estrous cycle; and a lack of tryptophan, resorption of the fetuses.

If the intake of any one of the essential amino acids is too small to meet the body's need, none of the other essential acids being fed can be used either for growth or maintenance of tissue. They will be deaminized, and the nitrogen will be excreted. Also, the body will be in negative nitrogen balance because it has to use some of its own tissue protein as a source of the needed amino acids. The goal in the studies with adults, then, is to find the amount of each essential amino acid that will maintain the person in nitrogen equilibrium when adequate amounts of the other essential ones are fed also.

The range of daily requirements of amino acids for nitrogen equilibrium is shown in Table 1. It will be noted that these amounts actually are very small compared with the amounts we eat in our protein foods every day.[2] This discrepancy remains to be explained. It may arise from the limitations of the nitrogen balance method or experimental designs.

The amino acids phenylalanine and methionine each have a special helper in a related but nonessential amino acid. Tyrosine can help phenylalanine so well that about three-fourths of the phenylalanine requirement can be met by tyrosine. There are still some functions that only phenylalanine can perform—but if there is only a limited supply of phenylalanine, it can be saved for these special functions by supplying plenty of tyrosine.

Cystine is the helper for methionine and is the only other common and plentiful amino acid that contains sulfur. At least three-fourths of the methionine requirement can be met by cystine. Because of this high degree of interchangeability, we are likely to refer to the requirement for total sulfur-containing amino acids, realizing, of course, that at least a small amount of methionine must be present.

The information on requirements and the amino acids in foods gives us a chance, albeit approximate, to check the food we eat to see how well it meets our needs as we know them now. We need not be concerned about the proportions of different amino acids when the supply of protein is generous and comes from a mixture of ordinary foods. In countries where the protein supply is small and perhaps inadequate, diets are low in calorie value, and the protein chiefly comes from a single food, usually cereal, in which the proportion or ratio of the amino acids may not be optimal.

It is possible that an oversupply of one amino acid may reduce the utilization of another amino acid so that a deficiency will occur. Also, an excess of one amino acid may increase the requirement of another acid. The high leucine content of corn protein, for example, may increase the requirement for isoleucine.

One application of these findings has been to translate the figures for requirements of amino acids into a pattern of desirable ratios for the eight essential ones. Such a pattern can be used as both a goal and a measuring stick in developing and supplying food supplements.

Examination of the amino acid content of proteins reveals that those occurring in animal foods are characterized by a lysine/tryptophan ratio of 5, or more; whereas in vegetable proteins the lysine/tryptophan ratios of 3, or less, predominate.[3] On the basis of this evidence and the results of rat experiments, Beach and Mitchell proposed that amino acid analyses of mammalian carcasses might prove to be an effective method for estimating the requirements of some, if not all, of the amino acids essential for growth.[4]

Available evidence gathered from feeding studies with animal proteins and various vegetable proteins supplemented with lysine indicates that the utilization of dietary proteins increases as their lysine and tryptophan content approaches that of muscle tissues. The importance of these relationships in attempts to improve the nutrition of populations of underdeveloped world areas, whose dietaries are comprised largely of wheat and other cereals, is obvious.

On the basis of limited data (see Table 1) on the minimum amino acid requirements determined experimentally with young adults fed purified amino acids diets, the FAO proposed the use of a pattern—1:3:3—for tryptophan, lysine, and methionine plus cystine, as ideal. Evaluation of this pattern by a number of investigators has shown it to be nutritionally inferior in young and older adults to diets which provide isonitrogenous

TABLE 1. COMPARISON OF TENTATIVE MINIMUM REQUIREMENTS OF AMINO ACIDS WITH AMINO ACID CONTENT OF AMERICAN DIETS

	Minimum Requirements		Amino Acid Content of Food Intakes			
	Women	Men	Reynolds	Futrell	Mertz	Wharton
Amino Acid	(g/day)		(g/day)			
Isoleucine	0.45	0.70	4.2	2.49–5.73	0.7–4.5	2.8–3.1
Leucine	0.62	1.10	6.5	3.28–7.35	1.3–7.8	4.4–4.9
Lysine	0.50	0.80	4.0	1.7 –8.6	1.3–5.6	3.5–4.0
Methionine } Cystine }	0.55	1.10	3.0	0.90–2.54	0.7–2.7	0.9–1.0
Phenylalanine	0.22	1.10	4.1	1.98–4.88	0.9–3.77	2.5–2.7
Tyrosine	0.90					
Threonine	0.31	0.50	2.8	1.68–3.44	0.9–3.8	2.6–2.9
Tryptophan	0.16	0.25	0.9	0.5 –1.28	—	0.4–0.5
Valine	0.65	0.80	4.2	2.85–5.39	0.8–4.9	3.1–3.4

quantities of animal protein.[5] Indeed, the conditions of the FAO pattern are met by the proteins of several cereals, notably rice and wheat, the staple of many population groups, which have long been known to be nutritonally inadequate.

Determination of protein requirements must also take into consideration the fact that essential amino acids constitute source substances for the formation of the biocatalysts, namely hormones and enzymes. Adrenalin, a powerful heart stimulant, and thyroxine, a vigorous metabolic activator, derive from tyrosine or phenylalanine. Insulin, which controls the utilization of sugars, contains 13% cystine. The adrenal and pituitary hormones have recently been shown to be composed of polypeptides having unique amino acid patterns. (See HORMONES, ANIMAL.)

Analyses of some enzymes essential to body function have also been shown to possess unusual amino acid composition. Cytochrome, for example, contains 22% lysine, and ribonuclease contains 10.4% lysine—respectively four and two times the lysine content of muscle tissues. It is clear from the foregoing that a diet high in animal proteins not only is necessary for normal growth and maintenance of the structural tissues of mammalian organisms, but also most fills the needs for biosynthesis of hormones and enzymes.

Obviously, the translation of information of minimal needs into practical dietary recommendations poses a number of problems arising from (a) sensitivity of the method, (b) artifacts of the test dietary, and (c) nutritional individuality of the test subject. The assumption that the human population is composed largely of individuals who have about average nutritional requirements has been questioned by some, notably Roger Williams,[6] and shown to be untenable in the light of the often observed large interindividual differences in requirements and metabolism of the essential amino acids.

References

1. ALBANESE, A. A., in "Protein and Amino Acid Nutrition" (ALBANESE, A. A., EDITOR), p. 297, New York, Academic Press, 1959.
2. LEVERTON, R. M., in "Protein and Amino Acid Nutrition" (ALBANESE, A. A., EDITOR), p. 477, New York, Academic Press, 1959.
3. ORR, J. B., AND LEITCH, I., *Nutr. Abstr. & Rev.*, **7**, 509 (1937–1939).
4. MITCHELL, H. H., in "Protein and Amino Acid Nutrition" (ALBANESE, A. A., EDITOR), p. 11, New York, Academic Press, 1959.
5. ALBANESE, A. A., AND ORTO, L. A., in "Modern Nutrition in Health and Disease" (WOHL AND GOODHART, EDITORS), Third edition, p. 125, Philadelphia, Lea & Febiger, 1964.
6. WILLIAMS, R. J., in "Protein and Amino Acid Nutrition" (ALBANESE, A. A., EDITOR), p. 45, New York, Academic Press, 1959.

ANTHONY A. ALBANESE

AMINO ACIDS (OCCURRENCE, STRUCTURE, SYNTHESIS)

Definition. An amino acid may be defined as any organic acid which incorporates one or more amino groups. This definition includes a multitude of substances of most diverse structure, such as β-alanine ($NH_2CH_2CH_2CO_2H$), taurine ($NH_2CH_2CH_2SO_3H$), carbamic acid (NH_2CO_2H), glycine ($NH_2CH_2CO_2H$), and aminomalonic [$NH_2CH(CO_2H)_2$], as well as a seemingly limitless variety of related compounds of differing molecular size and constitution which incorporate varying kinds and numbers of functional groups. Most extensive study has centered around the relatively small group of α-amino acids which are combined in amide linkage to form PROTEINS, in addition to a slightly larger number of related synthetic and naturally occurring compounds. With few exceptions, all of these possess the general structure NH_2CHRCO_2H, where the amino group occupies a position on the carbon atom *alpha* to that of the carboxyl group, and where the side chain R may be of diverse composition and structure.

General Characteristics. Few products of natural origin are as versatile in their behavior and properties as are the amino acids, and few have such a variety of biological duties to perform. They are at once:

(a) Water-soluble and amphoteric electrolytes, with the ability to form acid salts and basic salts and thus act as buffers over at least two ranges of pH;

(b) Dipolar ions of high electric moment with a considerable capacity to increase the dielectric constant of the medium in which they are dissolved;

(c) Compounds with reactive groups capable of a wide range of chemical alterations leading readily to a great variety of degradation, synthetic, and transformation products, such as esters, amides, amines, anhydrides, polymers, polypeptides, diketopiperazines, hydroxy acids, halogenated acids, keto acids, acylated acids, mercaptans, shorter- or longer-chained acids, and pyrrolidine and piperidine ring forms;

(d) Indispensable components of the diet of all animals including man (see AMINO ACIDS, NUTRITIONALLY ESSENTIAL);

(e) Participants in crucial metabolic reactions on which life depends, and substrates for a variety of specific enzymes *in vitro*;

(f) Binders of metals of many kinds;

(g) Absorbers of ultraviolet and infrared radiation within specific ranges of wavelength;

(h) Possessors with one exception of optical rotatory power related to the configuration of asymmetric centers; and

(i) Essential constituents of protein molecules whose biological and chemical specificities are determined in part by the number, distribution, and spatial interrelations of the amino acids of which they are composed.

They reveal at once uniformity and diversity; *uniformity*, because with rare exceptions they are α-amino acids with all the physical consequences which flow from this fact and because, for those that are constituents of proteins and hence of living tissues, the same optical configuration at the α-carbon atom is common to all; *diversity*, because each possesses a different side chain which confers upon it unique properties distinguishing it physically, chemically, and biologically from the others. In this duality, the array of the amino acids is a partial reflection of the larger biological world which is always the same and always different.

Optical Considerations. With the exceptions of glycine ($NH_2CH_2CO_2H$) and aminomalonic acid [$NH_2CH(CO_2H)_2$], all α-amino acids which are classifiable according to the general formula NH_2CHRCO_2H exist in at least two different optically isomeric forms. The optical isomers of a given amino acid possess identical empirical and structural formulas, and are indistinguishable from each other on the basis of their chemical and physical properties, with the singular exception of their effect on plane polarized light. This may be illustrated with the two optically active forms of alanine [$NH_2CH(CH_3)CO_2H$], one form of which exhibits the ability to rotate the plane of polarization of plane polarized light to the left (*levorotatory*) whereas the other rotates it to the right (*dextrorotatory*). Although the direction of optical rotation exhibited by these optically active forms is different, the magnitude of their respective rotations is the same. If equal amounts of the *dextro* and *levo* forms are admixed, the optical effect of each isomer is neutralized by the other, and an optically inactive product known as a *racemic modification* or *racemate* is secured.

The ability of the alanine molecule to exist in two stereoisomeric forms can be attributed to the fact that the α-carbon atom of this compound is attached to four different groups which may vary in their three-dimensional spatial arrangement. Compounds of this type do not possess complete symmetry when viewed from a purely geometrical standpoint and hence are generally referred to as *asymmetric*. As a consequence of this molecular asymmetry, the four covalent bonds of an asymmetric carbon atom can be aligned in a manner such that a regular tetrahedron is formed by the straight lines connecting their ends. Hence, two different tetrahedral arrangements of the groups about the asymmetric α-carbon atom of the alanine molecule may be devised so that these structures relate to one another as an object relates to its mirror image or as the right hand relates to the left. This relationship is depicted as follows;

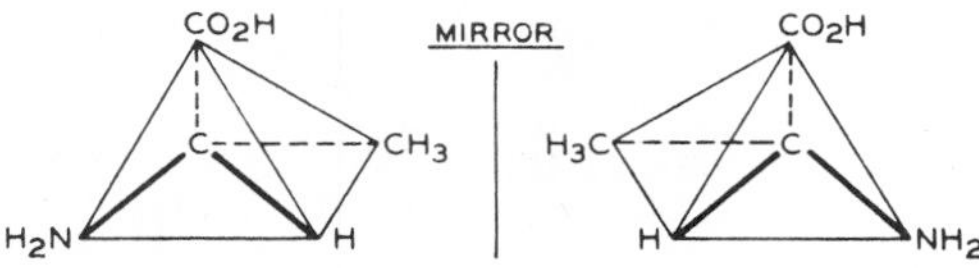

Molecules of this type are endowed with the property of optical activity and, together with their non-superimposable mirror images, are generally referred to as *enantiomorphs*, *enantiomers*, *antimers*, or *optical antipodes*.

In addition to an α-asymmetric center, amino acids may possess a second center of asymmetry and are therefore capable of existing in more than two stereoisomeric forms. Such a situation is illustrated by the threonine molecule, where I and II depict the optical antipodes of threonine, whereas III and IV depict the optical antipodes

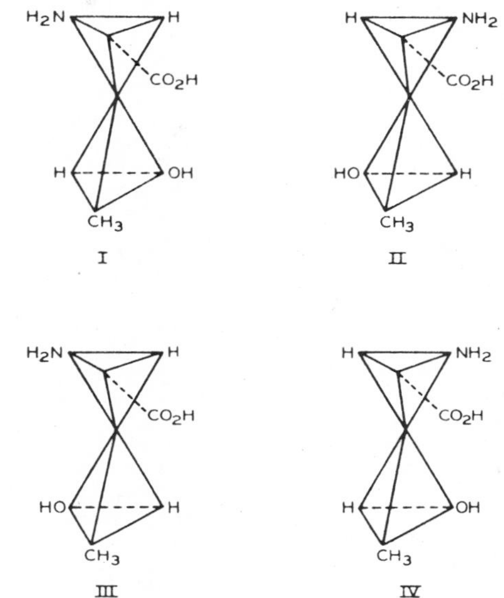

of allothreonine, its stereoisomeric forms. An equimolar mixture of form I and form II or of form III and form IV constitutes a racemic modification. However, the relationship between one or the other of the threonine antipodes (I and II) and one or the other of the allothreonine antipodes (III and IV), is quite another situation, since neither of the former two molecules is stereochemically identical with either of the latter molecules, nor are they mirror images of them. Stereoisomeric molecules which bear such a relation to one another are referred to as *diastereomers* or *epimers*. Unlike optical antipodes, diastereomers may exhibit differences in such physical properties as magnitude of optical rotation, solubility, infrared spectra, and melting points, while their chemical reactions, although frequently identical, may often proceed at markedly different rates.

Although certain diasymmetric amino acids, such as threonine, may exhibit four stereoisomeric forms, other diasymmetric amino acids, such as cystine [$HO_2C—CH(NH_2)—CH_2—S—S—CH_2—CH(NH_2)—CO_2H$], exist in only three stereoisomeric forms. Such situation is shown where V and VI, respectively, depict the levorotatory and dextrorotatory antipodes of cystine, whereas VII represents *meso*-cystine, an optically inactive diastereomer. The latter compound reveals no optical activity by virtue of the fact that a plane of

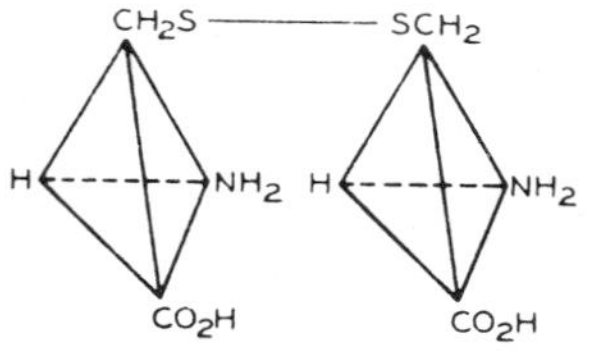

V 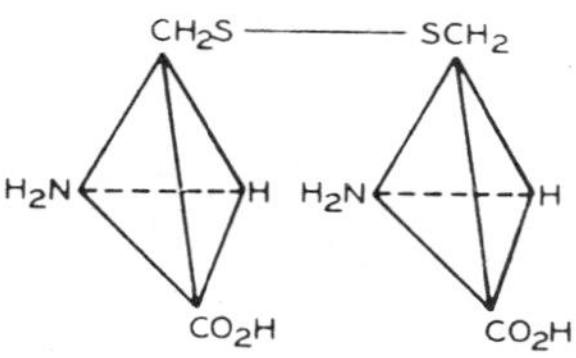 VI 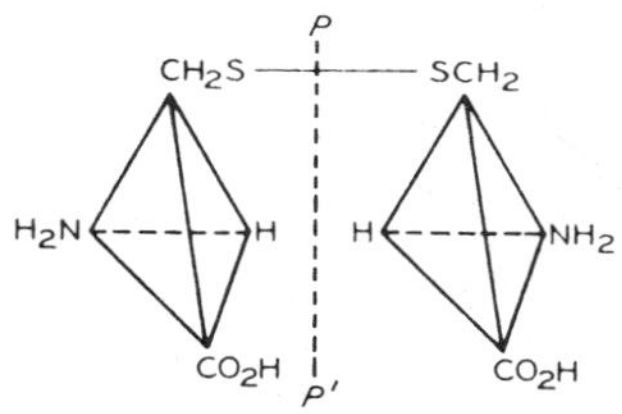 VII

symmetry, PP′, passes through the center of the molecule and theoretically divides the molecule into halves which are non-superimposable mirror images of each other. Although each half of the *meso*-cystine molecule rotates the plane of polarized light a like number of degrees, the direction of rotation of each half is the opposite of the other, thereby leading to a net rotation of zero for the molecule as a whole. Such compounds, which are devoid of optical activity as a consequence of internal compensation, are referred to as *meso* forms. Only a single form of *meso*-cystine exists since interchange of the hydrogen and amino groups of VII leads to an identical stereoisomeric form. As was the case with the other diastereomers discussed above, the physical and chemical properties of *meso*-cystine may differ from those of the optical antipodes of cystine or of the racemate secured by an equimolar mixture of these antipodes.

As three-dimensional representations of molecules, such as those just described, are too unwieldy for routine illustration on a plane surface, a more convenient scheme of representation of tetrahedral space relationships as plane projections was instituted in 1890 by Emil Fischer. Such a scheme is illustrated for dextrorotatory alanine as follows:

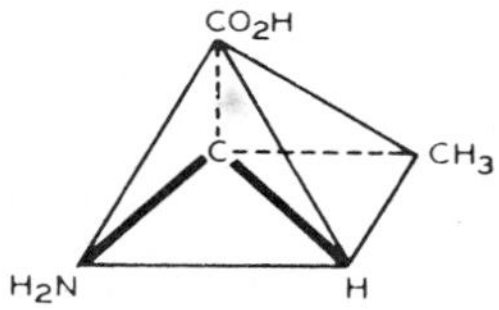

VIII 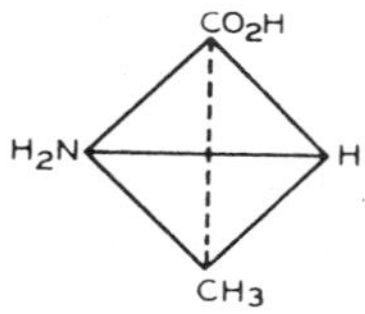 IX

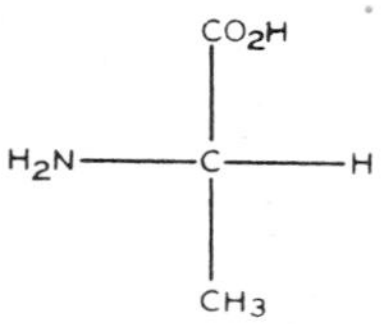

$$\begin{array}{c} CO_2H \\ | \\ H_2N—C—H \\ | \\ CH_3 \end{array}$$

X

Thus, if the C—H linkage is chosen as the apex of an opaque tetrahedron VIII, and if we look down from this apex on the other three groups, they will be found to occur in the clockwise order of methyl, amino, and carboxyl, as designated by formulation IX; the dotted line of IX denotes the hidden back edge of the tetrahedron VIII. According to the Fischer convention, formula X for dextrorotatory alanine implies derivation from the model specifically depicted by IX. The conventional formula for the enantiomorphic levorotatory form of alanine may be obtained in a like manner if the same line of reasoning is employed with the mirror image of VIII. Comparable simplified representations may be secured from three-dimensional models of amino acids which contain more than one center of asymmetry, for example,

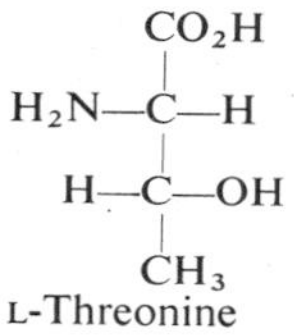

$$\begin{array}{c} CO_2H \\ | \\ H_2N—C—H \\ | \\ H—C—OH \\ | \\ CH_3 \end{array}$$

L-Threonine

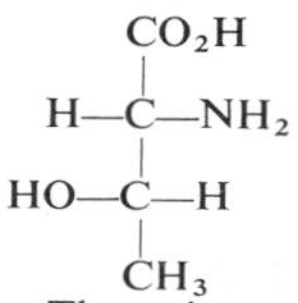

$$\begin{array}{c} CO_2H \\ | \\ H—C—NH_2 \\ | \\ HO—C—H \\ | \\ CH_3 \end{array}$$

D-Threonine

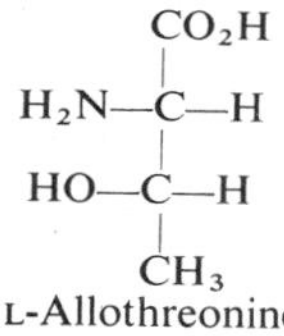

$$\begin{array}{c} CO_2H \\ | \\ H_2N—C—H \\ | \\ HO—C—H \\ | \\ CH_3 \end{array}$$

L-Allothreonine

$$\begin{array}{c} CO_2H \\ | \\ H—C—NH_2 \\ | \\ H—C—OH \\ | \\ CH_3 \end{array}$$

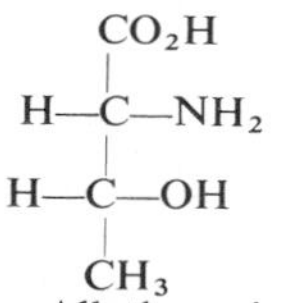

D-Allothreonine

Relative Configurations and Nomenclature. Plane projections of the type depicted by X provide a facile means for representing *relative* configurations. The establishment of a system of relative configurations among a family of optically active compounds, on the other hand, requires the initial arbitrary selection of some substance which, by definition, belongs to either the D- or L-category. Other optically active substances are then related to such a *standard of reference* through the use of appropriate physical, chemical or biological procedures. As the early standard of reference for the carbohydrate series, Emil Fischer selected natural *dextrorotatory glucose*. This compound, as the prototype of the D-configurational series, was thereupon designated as D-glucose, with the highest-numbered center of asymmetry chosen as the point of reference. A later suggestion, by Rosanoff, that *dextrorotatory glyceraldehyde* be employed in its stead resulted in

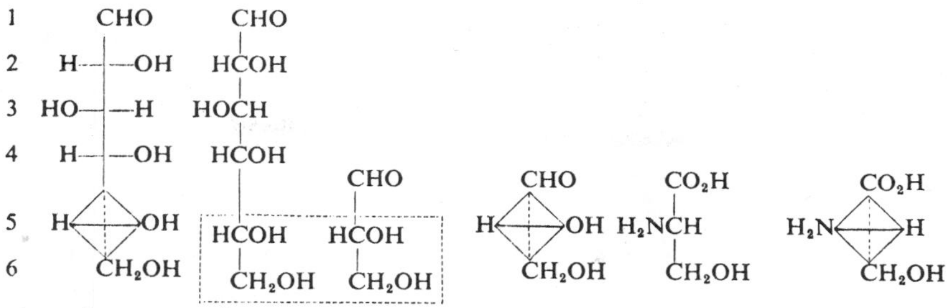

	XI	XII	XIII	XIV	XV	XVI
1	CHO	CHO				
2	H—|—OH	HCOH				
3	HO—|—H	HOCH				
4	H—|—OH	HCOH	CHO	CHO	CO_2H	CO_2H
5	H◇OH	HCOH	HCOH	H◇OH	H_2NCH	H_2N◇H
6	CH_2OH	CH_2OH	CH_2OH	CH_2OH	CH_2OH	CH_2OH

XI, XII: D-(dextro)-Glucose; XIII, XIV: D-(dextro)-Glyceraldehyde; XV, XVI: L-(levo)-Serine

the subsequent universal acceptance of this less complex compound as the standard of reference for the carbohydrate series.

The aforementioned standards of reference may be conveniently depicted either by simplified tetrahedral representations or by formulas XII and XIII. A similar representation obtains for L-serine (XV and XVI) which, by virtue of its structural resemblance to glyceraldehyde, was subsequently adopted as the standard of reference for the amino acid series. With but a single exception, the conventions employed for the assignment of configurations in both the amino acid and the carbohydrate series are identical in principle. For compounds of the former series which embody two or more centers of asymmetry, *e.g.*, isoleucine, threonine, and hydroxyproline, the configurational designation is based on the asymmetric carbon atom which is the *lowest numbered* according to conventional nomenclature practice, *i.e.*, the α-asymmetric carbon atom. For the carbohydrate series, on the other hand, this designation is governed by the configuration possessed by the asymmetric carbon atom *farthest* in the chain from the functional group of reference.

According to accepted rules of amino acid nomenclature, the configuration of the asymmetric α-carbon atom of an amino acid (or derivative thereof) should be indicated by a *small capital letter* D or L prefixed to the name, *e.g.*, L-alanine, D-alanine. It is important to note that the direction of optical rotation (*levorotation* or *dextrorotation*) of the amino acid, in a given solvent system, bears no relation to its optical configuration, *i.e.*, D or L designation. The prefix *meso* (or its abbreviation *ms*) in *lower case italic letters* is used to denote the isomers of amino acids and derivatives of these that are optically inactive because of internal compensation, *e.g.*, *meso*-cystine or *ms*-cystine.

Amino Acids of Protein Origin. PROTEINS may be viewed as high molecular weight polymers of amino acids which are bound, one to the other, in amide linkage through their respective α-amino and α-carboxyl groups. Cleavage of these amide linkages via the hydrolytic action of a strong mineral acid, alkali or a suitable PROTEOLYTIC ENZYME proceeds with the liberation of the constituent amino acids. This is exemplified as follows:

$$\text{—NHCHRCO—NHCHR}^1\text{CO—NHCHR}^2\text{CO—NHCHR}^3\text{CO—NHCHR}^4\text{CO—} \xrightarrow{+4H_2O}$$

$$\text{—NHCHRCO}_2\text{H} + \text{NH}_2\text{CHR}^1\text{CO}_2\text{H} + \text{NH}_2\text{CHR}^2\text{CO}_2\text{H} + \text{NH}_2\text{CHR}^3\text{CO}_2\text{H} + \text{NH}_2\text{CHR}^4\text{CO—}$$

At the present writing, some 25 different amino acids have been demonstrated to occur in the proteins of plant and animal tissues. All possess in common a primary amino substituent, a carboxyl moiety, and a hydrogen atom attached to the α-carbon atom. The imino acids, proline and hydroxyproline, may be considered as α-amino acids whose respective amino groups have become involved in the closure of a pyrrolidine ring. Table 1 lists the amino acids which have been isolated from hydrolysates of proteins and which have for the most part been so intensively investigated as to establish their existence in the parent protein beyond all reasonable doubt. Most of these are of widespread occurrence in proteins of various kinds. Some, however, like thyroxine, triiodothyronine, diiodotyrosine, dibromotyrosine, hydroxylysine, and hydroxyproline, reveal a more restricted distribution.

Special reference should here be made to the amino acid cysteine [$HSCH_2CH(NH_2)CO_2H$] which undergoes facile transformation into cystine upon mild oxidation. By virtue of this marked propensity toward oxidation, the hydrolysis of a protein wherein cysteine is normally present leads, under ordinary circumstances, to the ultimate isolation of cystine in lieu of the reduced form of the amino acid. There is nevertheless virtually no question of the actual presence of cysteine in many intact proteins. This can be attested to, for example, by the positive red color exhibited by proteins in native and in denatured forms, as well as by their hydrolysis products, when subjected to the action of sodium nitroprusside, a reagent for the detection of sulfhydryl groups.

The amino acids listed in Table 1 have been

TABLE 1. NATURALLY OCCURRING α-AMINO ACIDS FOUND IN PROTEINS

Class	Amino Acid		First Isolation as Product of Protein Hydrolysis		Protein Source of First Isolation
(I) Aliphatic (A) Monoamino-monocarboxylic acids	Glycine	$CH_2(NH_2)CO_2H$	Braconnot	1820	Gelatin
	L-Alanine	$CH_3CH(NH_2)CO_2H$	Weyl	1888	Silk fibroin
	L-Valine	$(CH_3)_2CHCH(NH_2)CO_2H$	Schützenberger	1879	
			Fischer	1901	Casein
	L-Leucine	$(CH_3)_2CHCH_2CH(NH_2)CO_2H$	Braconnot	1820	Muscle fiber, wool
	L-Isoleucine	$CH_3CH_2CH(CH_3)CH(NH_2)CO_2H$	Ehrlich	1904	Fibrin
	L-Serine	$CH_2(OH)CH(NH_2)CO_2H$	Cramer	1865	Sericine
	L-Threonine	$CH_3CH(OH)CH(NH_2)CO_2H$	Schryver and Buston	1925	Oat protein
			Gortner and Hoffmann		Teozein
(B) Diaminomono-carboxylic acids	L-Lysine	$CH_2(NH_2)CH_2CH_2CH_2CH(NH_2)CO_2H$	Dreschel	1889	Casein
	L-δ-Hydroxy-lysine	$CH_2(NH_2)CH(OH)CH_2CH_2CH(NH_2)CO_2H$	Schryver, Buston, and Mukherjee	1925	Fish gelatin
	L-Arginine	$H_2NC(=NH)NHCH_2CH_2CH_2CH(NH_2)CO_2H$	Hedin	1895	Horn
(C) Monoamino-dicarboxylic acids and amides	L-Aspartic acid	$CH(NH_2)CO_2H$ | CH_2CO_2H	Ritthausen	1868	Conglutin, legumin
	L-Asparagine	$CH(NH_2)CO_2H$ | CH_2CONH_2	Damodaran	1932	Edestin
	L-Glutamic acid	$CH(NH_2)CO_2H$ | $CH_2CH_2CO_2H$	Ritthausen	1866	Gluten-fibrin
	L-Glutamine	$CH(NH_2)CO_2H$ | $CH_2CH_2CONH_2$	Damodaran, Jaaback, and Chibnall	1932	Gliadin
(D) Sulfur-containing amino acids	L-Cystine	$SCH_2CH(NH_2)CO_2H$ | $SCH_2CH(NH_2)CO_2H$	Mörner	1899	Horn
	L-Methionine	$CH_3SCH_2CH_2CH(NH_2)CO_2H$	Mueller	1922	Casein
(II) Aromatic amino acids	L-Phenylalanine	[benzene ring]—$CH_2CH(NH_2)CO_2H$	Schulze and Barbieri	1881	Lupine seedlings
	L-Tyrosine	HO—[benzene ring]—$CH_2CH(NH_2)CO_2H$	Bopp	1849	Casein

	L-3 : 5-Dibromo-tyrosine	Br HO—[ring]—$CH_2CH(NH_2)CO_2H$ Br	Mörner	1913	Skeleton of *Primnoa lepadifera* (coral)
	L-3 : 5-Diiodo-tyrosine	I HO—[ring]—$CH_2CH(NH_2)CO_2H$ I	Drechsel	1896	Skeleton of *Gorgonia cavolinii* (coral)
	L-Triiodo-thyronine (3 : 5 : 3)	I I HO—[ring]—O—[ring]—$CH_2CH(NH_2)CO_2H$ I	Gross and Pitt-Rivers	1953	Thyroid tissue
	L-Thyroxine	I I HO—[ring]—O—[ring]—$CH_2CH(NH_2)CO_2H$ I I	Kendall	1915	Thyroid tissue
(III) Heterocyclic amino acids	L-Proline	CH_2—CH_2 CH_2 $CHCO_2H$ NH	Fischer	1901	Casein
	L-Hydroxyproline	HOCH—CH_2 CH_2 $CHCO_2H$ NH	Fischer	1902	Gelatin
	L-Tryptophan	[indole ring]—$CH_2CH(NH_2)CO_2H$ NH	Hopkins and Cole	1902	Casein
	L-Histidine	HC═$CCH_2CH(NH_2)CO_2H$ N NH CH	Kossel Hedin	1896	Sturine Various protein hydrolysates

categorized according to certain fundamental differences in structure, such as aliphatic, aromatic, and heterocyclic side chains. The aliphatic amino acids are further subdivided on the basis of the number of amino and carboxyl substituents that each incorporates. With the exception of glycine, all amino acids which have thus far been derived from proteins via the hydrolytic action of acids or proteolytic enzymes have been shown to be optically active. This is known because each of them has been isolated from various protein hydrolysates and, after extensive purification, each has been subjected to polarimetric analysis.

In the older literature the optical antipodes of amino acids were designated with the prefix *d* or *l* in order to show whether the compounds exhibited dextrorotation or levorotation in neutral aqueous solution. Hence, prefix designations, as exemplified by *d*-arginine, *l*-tryptophan, *l*-histidine and *d*-valine, were employed for the optical forms that were isolated from protein hydrolysates. These designations were abandoned when a considerable body of evidence accumulated which unequivocally demonstrated that all the purified, protein-bound amino acids possessed the same configurations about the asymmetric α-carbon atom. This evidence has been based largely on (a) the demonstration that the major portion of these amino acids could be converted either into each other or into structurally and configurationally identical compounds by reactions that did not implicate the asymmetric α-carbon atom, (b) like changes in the direction of optical rotation exhibited by the amino acids with changes in ionization or in wavelength (rotatory) dispersion, and (c) the susceptibility of these amino acids or of their derivatives to the action of the same optically specific enzymes and, conversely, their mutual resistance to other optically specific enzymes. All these experimental observations are consistent in themselves and compatible with the concept that the purified amino acids liberated by the hydrolysis of proteins possess an L-configuration, relative to L-serine as the standard of reference, about the asymmetric α-carbon atom.

To date, no unequivocal evidence for the presence of D-amino acids as constituents of proteins has been reported. However, the possibility that proteins of both animal and plant origin might contain D-amino acid constituents cannot be excluded. It should be emphasized that the amino acids which have hitherto been isolated from protein hydrolysates, and which appeared to be L in configuration, were essentially purified preparations. The purification procedures employed could in many instances be expected to free them not only from other amino acids but, conceivably, from appreciable amounts of their own racemic modifications as well. The presence of varying amounts of the racemic forms of several amino acids (glutamic acid, proline, histidine, hydroxylysine, etc.) in protein hydrolysates has long been known and has most frequently been explained on the basis of racemization of the L-forms during the period of hydrolysis. Although such racemization undoubtedly occurs to some extent under the frequently drastic hydrolytic conditions employed, the presence of preformed D-amino acid residues in proteins cannot be arbitrarily denied, and only a few such residues need be present to confer unique properties of spatial configuration and enzymic susceptibility upon the protein molecule of which they may be a part.

Experience has shown that certain amino acids may arise as a consequence of protein degradation, although they themselves are not constituents of the protein *per se*. Thus, although arginine can be isolated from acid hydrolysates of proteins, alkaline hydrolysis of these same proteins is accompanied by degradation of the amino acid to ornithine [$NH_2CH_2CH_2CH_2CH(NH_2)CO_2H$]. Another illustration of the formation of such artifacts is provided by the isolation of inactive lanthionine [$HO_2CCH(NH_2)CH_2$—S—$CH_2CH(NH_2)CO_2H$] from acid hydrolysates of wool, hair, lactalbumin, and several other proteins which had previously been treated with alkali. Presumably, treatment of the protein with dilute alkali causes partial desulfuration of the cystine residues with concomitant formation of lanthionine. The same proteins, not previously conditioned with alkali but subjected instead to direct acid hydrolysis, yield cystine but no lanthionine. Still another illustration of the phenomenon is the isolation of ornithine from hydrolysates of collagenous proteins previously treated with lime; the ornithine obviously is obtained at the expense of the arginine degraded by the alkaline treatment of the protein. Although many comparable examples are known, it should suffice to merely reemphasize that certain amino acids which are isolable from protein hydrolysates may arise as a consequence either of the hydrolytic conditions employed or of the previous history of the protein, and are not necessarily present as such in the intact, native protein.

In addition to questions relating to the validity of natural occurrence of certain amino acids introduced by artifacts, the intensive search for new amino acids in proteins has been, to a considerable degree, clouded by premature and unsubstantiated claims. For these reasons, in a brilliant review entitled "The History of the Discovery of the Amino Acids," Vickery and Schmidt proposed certain criteria that should be met before any report of the existence of a new amino acid as a constituent of proteins be generally accepted as valid. These criteria are given in what follows:

(1) In order that an amino acid shall be accepted as a definite product of the hydrolysis of proteins, it must also have been isolated by some worker other than its discoverer.

(2) Its constitution must have been established by synthesis and by demonstration of identity between the synthetic product and the natural racemized product, or by actual resolution of the synthetic product and preparation of the optically active natural isomer.

(3) The substance must be liberated by hydrolysis from a preparation of a protein of demon-

strated purity and must be adequately characterized by analysis of salts and typical derivatives.

The protein-derived amino acids selected in 1931 which were acceptable according to these criteria included all of those presently believed to occur in proteins with the exceptions of threonine, hydroxylysine, asparagine, glutamine and triiodothyronine, which had not yet been discovered, and also included β-hydroxyglutamic acid. Some two years later, norleucine was added to the list of accepted amino acids because of its apparent conformity with all of the required criteria. However, the seemingly irrefutable evidence upon which the acceptance of both β-hydroxyglutamic acid and norleucine rested was subsequently shown to be so tenuous that neither, at the present time, receives serious consideration as a constituent of proteins. Such exceptions notwithstanding, the Vickery-Schmidt criteria have served an invaluable role in that they prompted a generally more critical, as well as a more conservative, attitude toward the acceptance of a new amino acid. Thus, amino acids such as diaminoglutaric acid, hydroxyaspartic acid, sarcosine, α-aminoadipic acid, norvaline, and a host of others have, at one time or another, been reported to occur in proteins, but such claims either have subsequently been proved questionable or have not yet received the experimental support necessary for their general acceptance.

Amino Acids Not of Protein Origin. In addition to the amino acids which are known constituents of proteins, a number of amino acids have been found to occur in nature, either in the free state or in chemical combination in non-protein compounds of varying molecular size. Such amino acids have been detected in the biological fluids and tissues of various plants and animals, in the circulatory system, among the products of excretion, as intermediates in the metabolic processes, as components of antibiotics and as bacterial decomposition products. Reports in the chemical literature allude to well over a hundred different amino acids of non-protein derivation that have been isolated from such sources (see the separate article on AMINO ACIDS, NON-PROTEIN). By far the greater proportion of the α-amino acids known to occur in nature, whether as constituents of proteins, as constituents of smaller peptides, or in the free state, belong to the L-configurational family. In recent years, however, a constantly increasing number of observations has indicated the widespread natural occurrence of amino acids of the D-configuration, chiefly among the metabolic products of lower organisms, *e.g.*, D-serine, D-valine, D-leucine and D-phenylalanine in various of the polypeptide antibiotics.

Although some of the amino acids of non-protein origin have been unequivocally established as occurring in nature by a strong body of experimental data, somewhat more have been supported by data that would, at best, provide only strong presumptive evidence for their natural occurrence. With most of the amino acids, however, sufficient evidence has not yet been amassed to demonstrate their existence in nature with even a modicum of assurance. These considerations, in addition to the appearance of a continually increasing number of reports of new amino acids of both the D- and L-varieties in nature, point out the urgent need for criteria which, if met, would reliably establish the natural occurrence of these amino acids. Proposals for such criteria have been offered by Greenstein and Winitz as follows:

(1) The new amino acid should be physically isolated in sufficient quantity, and purified by several crystallizations from water, water-alcohol, or some other suitable solvent system, in order to obtain accurate and meaningful carbon, hydrogen, nitrogen and other possible elemental analyses.

(2) Chromatographic procedures, whether two-dimensional techniques on paper employing different solvent systems, or elution techniques, with columns employing various solvent mixtures should show it to possess a single spot or peak, respectively, in order to assure the possibility of molecular homogeneity.

(3) The molecule should be degraded by chemical organic procedures or unequivocal enzymatic procedures to identifiable products or fragments in order to obtain a knowledge of its structure.

(4) Verification of the structure ultimately assigned to the molecule should be achieved through synthesis of the material employing unequivocal chemical procedures.

(5) The synthetic material should be resolved into its optical antipodes, and the physical, chemical and biological properties of the antipodes so secured should be compared with those of the natural material; in the case of amino acids with more than one asymmetric center, separation of the diastereomers should precede the resolution of each racemate into its optical antipodes.

(6) Determination of the optical configuration of the molecule should be accomplished by any of a number of chemical, physical, optical or biological techniques.

(7) The natural occurrence of the new amino acid should be independently confirmed by some investigator other than the discoverer.

Methods of Synthesis. In recent years, a steadily increasing demand for sizable quantities of α-amino acids has developed because of their extensive employment in biochemical, nutritional, microbiological, and tissue culture research, in addition to their widespread use in pharmaceutical preparations and as supplements for natural and processed foods. Isolations from acidic, alkaline, or enzymic digests of proteins, or from other natural sources, have sufficed to supply a goodly portion of most of the protein-derived amino acids. In addition, a number of practicable chemical syntheses are presently known which permit ready access to the more common α-amino acids, on both an industrial and a laboratory scale.

Utilization of a particular method of preparation is often contingent on (a) whether the amino acid is to be produced industrially for commercial purposes, (b) whether the synthesis is directed toward a new or a commercially unavailable amino acid on a laboratory scale, or (c) whether the desired amino acid is to be isotopically labeled for tracer studies. In industry, where a premium is placed on low production costs, the employment, for example, of high-temperature, high-pressure reaction systems frequently permits the use of relatively inexpensive starting materials and markedly diminishes the time required to complete a reaction. While the routine use of such procedures on a laboratory scale would generally prove infeasible, the cost of reactants presents a far less critical problem as contrasted to the worth of the investigators' time, and the route of synthesis selected for the preparation of a new or a rare amino acid will generally weigh more in favor of convenience and economy of time than economy of materials. On the other hand, if the synthesis involves the incorporation of highly expensive isotopic material into the amino acid molecule, the reaction route of choice is generally that which does not necessarily lead to the best over-all yield for the entire reaction sequence, but rather to the best yield based on the point of introduction of the isotope; schemes are thereby sought wherein the isotope may be introduced as late as possible.

The synthetic procedures for α-amino acids may be considered as general insofar as the assumption is valid that the requisite starting materials are available and that the behavior revealed and techniques employed during each stage of the reaction sequence are comparable for all amino acids. Such is not invariably the case, however, for the side chains of amino acids may be possessed of a most diverse structure and may require the expenditure of considerable effort for their synthesis. The side chains of lysine, histidine, and arginine may be cited as illustrative of cases in point. Additional differences may arise from the necessity of masking the reactivity of certain functional groups (*e.g.*. hydroxyl), the prolonged reaction times which result from the steric hindrance imposed by highly branched side chains, the extreme lability of certain amino acids or their intermediates to acid or alkali, and the different solubilities which the various amino acids possess. It thus becomes apparent that the synthesis of a given amino acid by a general procedure may afford many problems unique unto itself. The following are methods upon which some of the more general of these syntheses are based:

(a) *Strecker synthesis*: This method involves the interaction of an aldehyde (XVII) with cyanide ion and ammonia or ammonium carbonate to give either an aminonitrile (XVIII) or an appropriately substituted hydantoin (XX) which, upon hydrolysis with acid or alkali, is converted to the desired amino acid (XXI).

(b) *Amination of α-halogen acids*: Treatment of a suitable α-chloro or α-bromo acid (XXV; X = Cl or Br) with ammonia or hexamethylenetetramine leads to the desired amino acid (XXVI) directly. The same amino acid may be obtained by condensation of the corresponding ester (XXIII) of the α-halogen acid with potassium phthalimide (XXII), followed by hydrazinolysis and subsequent acid hydrolysis of the intermediate phthalylamino acid ester (XXIV).

(c) *Reductive amination*: Here, the pertinent amino acid (XXX) is obtained via catalytic reduction of an α-keto acid (XXIX) in the presence of ammonia, or conversion of an α-keto acid (XXIX) or an appropriately substituted acetoacetic ester (XXXII) to a suitably constituted oximino (XXVII) or phenylhydrazone (XXXIII) derivative which, in turn, is subjected to chemical or catalytic reduction.

(d) *Amination via molecular rearrangement*. This method involves rearrangements of a carboxylic acid (XLIII), hydrazide (XXXVII), or amide (XLI) to the desired amino acid using the Schmidt, Curtius or Hofmann rearrangement. The general reaction scheme is shown in the sequence XXXV → XLIII.

(e) *Condensation of an aldehyde with an active methylene group*: This method which is used for the preparation of aromatic α-amino acids such as phenylalanine, tyrosine and tryptophan, or ring-substituted analogues thereof, is generally effected by the condensation of the pertinent aromatic aldehyde with an acylglycine, a hydantoin, a 2 : 5-diketopiperazine, a thiohydantoin, or a 2-mercaptothiazol-5-one, followed by reduction and subsequent hydrolysis of the intermediate so obtained. A typical reaction sequence is illustrated by XLIV → XLVIII.

(*f*) *Oxidation of aminoalcohols*: Oxidation of 2-aminoalcohols (NH_2CHRCH_2OH) with a suitable oxidizing agent leads to the desired amino acid. The method is limited by the variety of aminoalcohols which are commercially available.

(*g*) *Condensations with N-substituted aminomalonic esters*: This procedure involves the sodium ethoxide-mediated condensation either of acetylaminomalonic ester (XLIX; R = CH_3) or ethyl acetamidocyanoacetate (LIII; R = CH_3) both of which are commercially available, with a suitable alkyl or acyl halide (R′Br); acid or alkaline hydrolysis of the pertinent condensation product (L or LII) then yields the desired amino acid (LI).

If, in syntheses of the above type, an amino acid is prepared which possesses more than one asymmetric center, then a diastereoisomeric mixture of the racemates generally results. Separation of the racemates must precede their resolution into the optical antipodes. For this purpose, two general approaches for the separations of diastereomeric racemates have been generally employed, namely, differential solubility in various solvent systems and partition or ion-exchange chromatography.

Resolution Procedures. The term "resolution" may be defined as a procedure whereby the two optical isomerides which compose a racemic mixture or compound are separated and prepared in pure form. In one form or the other, all resolution procedures stem from the classical studies of Louis Pasteur, whose principles for the separation of optical antipodes were based essentially on the following: (a) the mechanical

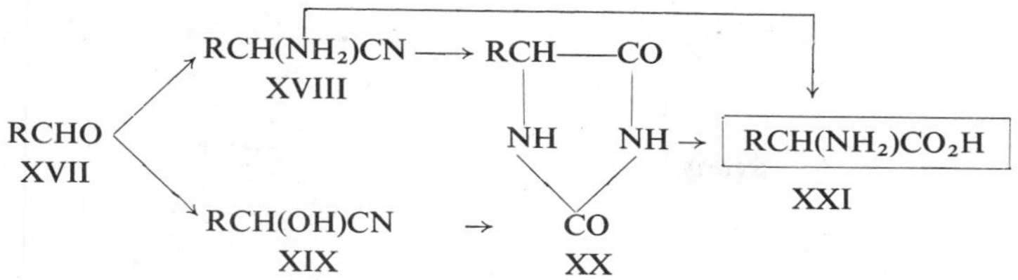

RCHO (XVII) → RCH(NH$_2$)CN (XVIII) → RCH—CO / NH NH / CO (XX) → RCH(NH$_2$)CO$_2$H (XXI)

RCHO (XVII) → RCH(OH)CN (XIX) → XX

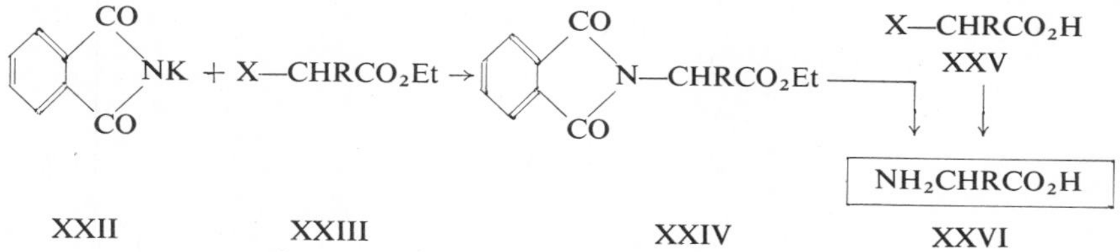

XVIII → XXI

Phthalimide potassium (CO)$_2$NK (XXII) + X—CHRCO$_2$Et (XXIII) → phthalimido (CO)$_2$N—CHRCO$_2$Et (XXIV) → NH$_2$CHRCO$_2$H (XXVI)

X—CHRCO$_2$H (XXV) → NH$_2$CHRCO$_2$H (XXVI)

RC(=NOH)CO$_2$H (XXVII) ← RC(=NOH)CO$_2$Et (XVIII)

RCOCO$_2$H (XXIX) → RCH(NH$_2$)CO$_2$H (XXX) ← RCH(NH$_2$)CO$_2$Et (XXXI)

XXVII → XXX; XVIII → XXXI

CH$_3$COCHRCO$_2$Et (XXXII) → XVIII

CH$_3$COCHRCO$_2$Et (XXXII) → RC(=NNHC$_6$H$_5$)CO$_2$Et (XXXIV)

RC(=NNHC$_6$H$_5$)CO$_2$H (XXXIII) ← RC(=NNHC$_6$H$_5$)CO$_2$Et (XXXIV)

XXXIII → XXX

CN—CH$_2$—CO$_2$Et (XXXV) → CN—CHR—CO$_2$Et (XXXVI) → CN—CHR—CONHNH$_2$ (XXXVII) → CN—CHR—CON$_3$ (XXXVIII) → CN—CHR—NHCO$_2$Et (XXXIX) → NH$_2$—CHR—CO$_2$H (XLII)

XXXVI → CN—CHR—CO$_2$H (XL) → CONH$_2$—CHR—CO$_2$H (XLI) → NH$_2$—CHR—CO$_2$H (XLII) ← CO$_2$H—CHR—CO$_2$H (XLIII)

RCHO (XLIV) + CH$_2$(CO—NH)(NH—CO) (XLV) → RCH=C(CO—NH)(NH—CO) (XLVI) → RCH$_2$CH(CO—NH)(NH—CO) (XLVII) → RCH$_2$CH(CO$_2$H)(NH$_2$) (XLVIII)

CO$_2$Et—CHNH—COR—CO$_2$Et (XLIX) —R′Br→ CO$_2$Et—C(R′)NH—COR—CO$_2$Et (L) → R′—CHNH$_2$—CO$_2$H (LI) ← CN—C(R′)NH—COR—CO$_2$Et (LII) ←R′Br— CN—CHNH—COR—CO$_2$Et (LIII)

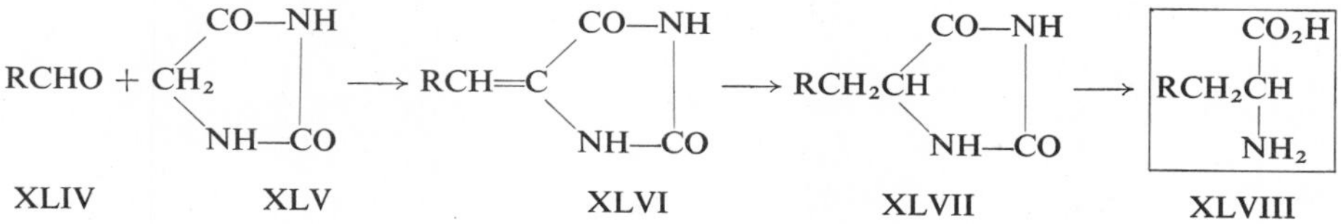

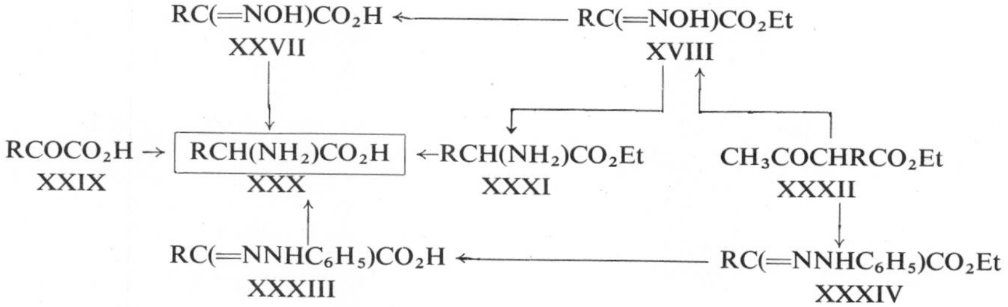

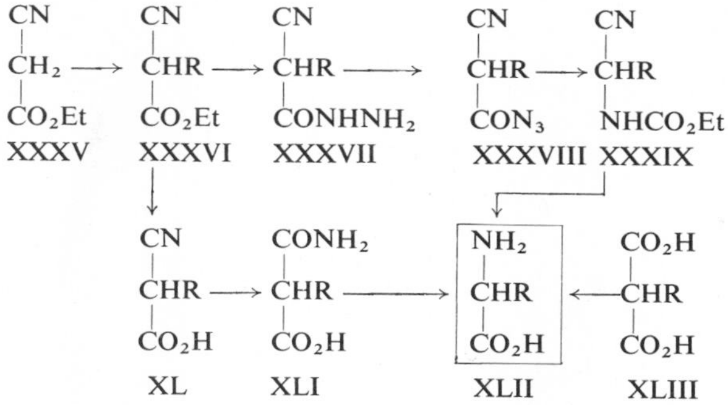

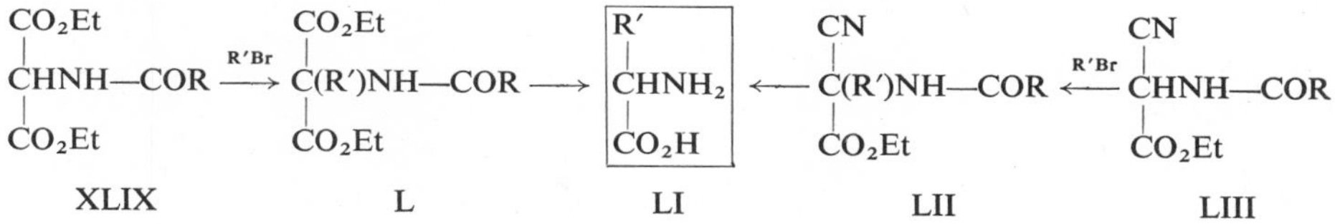

separation of crystals possessing requisite hemihedrism, (b) the differential solubility of diastereoisomeric salts of the racemate with optically active compounds, and (c) the action of living organisms or of enzymes derived from living organisms which utilize or attack preferentially one of the two optical antipodes or its derivatives.

(a) *Selective crystallization*: An occasionally useful, though often unsatisfactory means of effecting a direct separation of the enantiomers which compose a racemate is to induce one of the optical forms to preferentially crystallize from a saturated solution of the racemate which has been supersaturated with respect to this particular form. A typical example of such a procedure is the separation by successive crystallizations of a mixture of the L- and DL-forms of histidine hydrochloride into individual crops of the L-amino acid, the DL-amino acid and the D-amino acid.

(*b*) *Diastereoisomeric salts*: The formation of salts of optically active bases with racemic acids or of optically active acids with racemic bases leads to diastereomeric mixtures which may be resolved by the different solubility of the two components of such mixtures, *i.e.*,

$$(+)\text{B} + (dl)\text{A} \rightarrow (+)\text{B}\cdot(l)\text{A} + (+)\text{B}\cdot(d)\text{A}$$

or

$$(+)\text{A} + (dl)\text{B} \rightarrow (+)\text{A}\cdot(l)\text{B} + (+)\text{A}\cdot(d)\text{B}$$

The salts in turn may be decomposed by a metathetical reaction involving a stronger base than (+)B or a stronger acid than (+)A, to yield thereby the free, optically active isomerides of (*dl*)A or (*dl*)B. For the immediately successful resolution of a racemate to occur, the resolving agent must form a relatively tightly bound salt with at least one of the two isomers of the racemate, and this salt must be cleanly and quantitatively precipitable from that of the isomer under the conditions imposed. The procedure gives no clue to the optical configuration of the respective isomerides, and unless an isomeride of known configuration is available for comparison, the enantiomorphs so resolved must be subjected to further procedures designed to elicit this information.

Although most of these resolution procedures have employed α-amino acids acylated on the amino groups so as to cause them to be essentially acids and hence able to form salts with optically active bases, some have employed basic racemic amino acid esters or amides able to form salts with optically active acids. Such amino acids as lysine or histidine, which are already bases in the free state, have been resolved without conversion into any derivatives by simply mixing them with an optically active acid. Indeed, the resolution of DL-lysine with *d*-camphoric acid forms the basis for the resolution of this amino acid on the commercial scale.

(c) *Biological procedures*: The great virtue of resolution procedures involving biological materials is that of optical specificity. Once the specificity of a biological agent toward amino acids of known optical configuration has been established, its action toward amino acids of unknown optical structure can, with some confidence, be predicted. These biological procedures of resolution possess a marked advantage over purely chemical procedures since isomers are produced whose optical configurations are known. A further advantage of the use of biological agents is that they permit a more general approach and a more uniform resolution procedure.

The various biological procedures are as follows: (a) the use of the whole animal, *i.e.*, the feeding or injection of a racemic amino acid, followed by the isolation from the urine of one or the other antipode; (b) asymmetric oxidation or decarboxylation through the action of microorganisms or tissue fractions, whereby one of the enantiomorphs is unaffected and the other is metabolized to either the corresponding α-keto acid or amine; (c) asymmetric synthesis through the action of a protease on N-acylated racemic amino acids, whereby only the L-antipode in the most favorable cases partakes in a synthetic reaction with a base to form an insoluble and hence separable derivative, leaving most of if not all the D-antipode in solution; and (d) asymmetric hydrolysis through the action of amidases, esterases, and acylases on the appropriately substituted racemic amino acids, whereby only the L-antipode is hydrolyzed, leaving, with few exceptions, all products in solution to be separated by various devices.

Probably the most satisfactory of all presently available biological resolution procedures is that developed by Greenstein, who employed as the biological agents two separate enzyme systems isolated from hog kidney. One enzyme system, known as acylase I, is capable of acting asymmetrically only on the L-isomers of *N*-acylated DL-amino acids, while the other enzyme system, known as amidase, is capable of acting asymmetrically only on the L-isomers of DL-amino acid amides as follows:

$$\text{DL-RCO-NHCHR}'\text{CO}_2\text{H} \xrightarrow{\text{Acylase}} \text{L-NH}_2\text{CHR}'\text{CO}_2\text{H (organic solvent soluble)} + \text{RCO}_2\text{H} + \text{D-RCO-NHCHR}'\text{CO}_2\text{H (organic solvent insoluble)}$$

$$\text{DL-NH}_2\text{CHR}'\text{CONH}_2 \xrightarrow{\text{Amidase}} \text{L-NH}_2\text{CHR}'\text{CO}_2\text{H (organic solvent insoluble)} + \text{NH}_3 + \text{D-NH}_2\text{CHR}'\text{CONH}_2 \text{ (organic solvent soluble)}$$

References

1. Greenstein, J. P., and Winitz, M., "Chemistry of the Amino Acids," Vols. I, II, and III, New York, John Wiley & Sons, 1961.
2. Vickery, H. B., and Schmidt, C. L. A., "The History of the Discovery of the Amino Acids," *Chem. Reviews*, **9**, 169–318 (1931).

Milton Winitz

AMINO ACIDS, PROTEIN-DERIVED

Upon hydrolysis, PROTEINS yield mixtures (in proportions unique to a specific protein, but varying from one specific protein to another) of about 20 common "protein-derived" amino acids. These include α-ALANINE, ARGININE, ASPARAGINE, ASPARTIC ACID, CYSTEINE (or cystine), GLUTAMINE, GLUTAMIC ACID, GLYCINE, HISTIDINE, ISOLEUCINE, LEUCINE, LYSINE, METHIONINE, PHENYLALANINE, PROLINE, SERINE, THREONINE, TRYPTOPHAN, TYROSINE, and VALINE. [Enzymatic hydrolysis, as by PROTEOLYTIC ENZYMES, is usually employed to permit recovery of asparagine and glutamine since acid hydrolysis degrades (deamidates) these to yield ammonia plus aspartic acid and glutamic acid, respectively.] The sequence of these 20 amino acids within the protein macromolecule (linear polypeptide chain) is genetically controlled (see BIOCHEMICAL GENETICS; GENES; GENETIC CODE) during the process of PROTEIN BIOSYNTHESIS. Certain proteins, *e.g.*, COLLAGEN, contain in addition to members of the above group of amino acids, hydroxylysine and HYDROXYPROLINE, which may be formed by metabolic conversions subsequent to the genetically controlled incorporation of proline or lysine into the polypeptide chain. For certain specific proteins, the exact linear sequence of constituent amino acids is now known; these cases include INSULIN (the first to be determined), and also ADRENOCORTICOTROPIC HORMONE, some CYTOCHROMES, HEMOGLOBINS, MYOGLOBIN, RIBONUCLEASE, and TOBACCO MOSAIC VIRUS coat protein.

For man and certain other species, a definite subgroup (the members of which vary somewhat with the species) of the above protein-derived amino acids have been shown to be required in the diet for complete nutrition (see AMINO ACIDS, NUTRITIONALLY ESSENTIAL; NUTRITIONAL REQUIREMENTS).

E. M. LANSFORD, JR.

AMINO PHOSPHONIC ACIDS

For many years esters of phosphoric acid, such as adenosine triphosphate, triphosphopyridine nucleotide, cocarboxylase, nucleic acids, coenzyme A, lecithin and sphingomyelin, have been recognized as essential coenzymes and intermediates in the metabolism of biological substances. All such compounds are distinguished by the fact that they contain C—O—P bonds. Although many compounds of this type have been shown to be important biological substances, prior to 1959 no compound containing a direct C—P bond had ever been found in biological materials and it was generally believed that living organisms were incapable of synthesizing such compounds.

Naturally Occurring Amino Phosphonic Acids. In 1959, Horiguchi and Kandatsu[1] isolated from protozoa the first compound containing the C—P bond, 2-aminoethylphosphonic acid (2-AEP), the analogue of β-alanine. Additional papers by these authors reported further information, including the fact that in the organism *Tetrahymena*, 13–15% of the total phosphorus present was in the form of 2-AEP. They reported that this phosphonic acid was isolated after the acid hydrolysis of a lipid fraction. Kittredge, Roberts and Simonsen[2] reported the isolation of 2-AEP from a phospholipid of the sea anemone *Anthopleura elegantissima*. This unusual sphingolipid was isolated by Rouser and co-workers,[3] and a similar phospholipid from shell fish has been isolated by Hori and co-workers in Japan. Rosenberg[4] reported in 1964 that *Tetrahymena pyriformis* incorporated P^{32} into 2-AEP during the log phase growth of the cells and the lipid and non-lipid portions of the cells did not liberate free 2-AEP when treated with proteolytic enzymes. Quin[5] found 2-AEP in both the lipid and the protein fractions of a variety of marine organisms. The sea anemone *Tealia felina* contained 1.8% 2-AEP (dry basis), and the 2-AEP accounted for 50% of the total P present. The edible blue mussel, *Mytilus edulis*, and the littleneck clam, *Venus mercenaria*, also contained significant amounts. The manner in which this compound is incorporated into these proteins is not known at present, although Quin reported evidence that 2-AEP is attached through the amino group and the evidence did not exclude binding through the phosphonic acid group.

The second amino phosphonic acid to be found in living organisms was reported by Kittredge and Hughes.[6] They isolated 2-amino-3-phosphonopropionic acid, an aspartic acid analogue, from *Tetrahymena* and a zoanthid. The following formulas indicate the relationship between these two naturally occurring amino phosphonic acids and their relations to β-alanine and aspartic acid:

$$\begin{array}{ccc} PO(OH)_2 & & PO(OH)_2 \\ | & & | \\ CH_2 & \underset{-CO_2}{\overset{+CO_2}{\rightleftarrows}} & CH_2 \\ | & & | \\ CH_2 & & CHCO_2H \\ | & & | \\ NH_2 & & NH_2 \end{array}$$

2-Aminoethylphosphonic acid — 2-Amino-3-phosphonopropionic acid

$$\begin{array}{ccc} CO_2H & & CO_2H \\ | & & | \\ CH_2 & \underset{+CO_2}{\overset{+CO_2}{\rightleftarrows}} & CH_2 \\ | & & | \\ CH_2 & & CHCO_2H \\ | & & | \\ NH_2 & & NH_2 \end{array}$$

β-Alanine — Aspartic acid

Synthetic Amino Phosphonic Acids. A variety of amino phosphonic acids with the general formula

$$\begin{array}{l} RCH(CH_2)_xPO(OH)_2 \\ \;\;| \\ \;NH_2 \end{array} \qquad \text{I}$$

where R = H, alkyl or aryl groups, and $x = 0$, 1, 2, etc., have been synthesized in recent years. The first compound of this type, aminomethylphosphonic acid (I, R = H, $x = 0$), was described in 1942 in a patent issued to Engelmann and Pikl. The second and most important to date, the β-alanine analogue, 2-AEP (I, R = H, $x = 1$)

was first synthesized by Finkelstein[7] by subjecting diethyl 3-phosphonopropionamide to the Hofmann reaction.

It appears that around 1947 three groups of investigators began attempting to devise syntheses for compounds having the general formula (I). The Frenchman Chavane[8] and the American Kosolapoff[9] employed the Gabriel synthesis for producing aminomethylphosphonic acid (I, R = H, $x = 0$). Kosolapoff and the Russians Medved and Kabachnik[10] employed the aminolysis of halogen-substituted compounds to synthesize aminomethylphosphonic acid and 3-aminopropylphosphonic acid, (I, R = H, $x = 2$). In a series of papers starting in 1952, Kabachnik and Medved[10] described a Mannach-like reaction by which they synthesized a variety of 1-aminoalkylphosphonic acids, most of which were not closely related to the naturally occurring amino carboxylic acids. Chalmers and Kosolapoff[12] used this same condensation to prepare 1-aminoethylphosphonic acid (α-alanine analogue) (I, R = methyl, $x = 0$) and 1-amino-2-phenylethylphosphonic acid (phenylalanine analogue) (I, R = benzyl, $x = 0$).

What appears to be a more general synthesis for these compounds was described[13] by Chambers and Isbell. This synthesis involved the Curtius degradation of the carboxylic ester group in substituted phosphonoacetic esters [$RCH(CO_2Et)(CH_2)_xPO(OEt)_2$, R = H, alkyl or aryl groups, $x = 0, 1, 2$, etc.]. These esters are readily synthesized by three general routes, which make it possible to produce such esters possessing almost any configuration and arrangement of groups. By utilizing these syntheses, the phosphonic acid analogues of glycine, α-alanine, β-alanine, phenylalanine, aspartic acid, glutamic acid, γ-aminobutyric acid, valine, tyrosine, ornithine and approximately 20 additional related amino phosphonic acids have been prepared.

Future developments involving amino phosphonic acids will depend upon the determination of their biological roles. It seems likely that additional amino phosphonic acids and the related alkylphosphonic acids may be found in living organisms. It also appears that some of the amino phosphonic acids may be widely distributed in both the plant and animal kingdoms, including man. Whether they are incorporated in proteins through the phosphonic acid group, as well as the amino group, could have significant implications. How they are synthesized and metabolized by living organisms and the nature of the enzymes involved are fascinating questions still to be answered.

References

1. Horiguchi, M., and Kandatsu, M., *Nature*, **184**, 901 (1959); *Bull. Agr. Chem. Soc. Japan*, **24**, 565 (1960); *Agr. Biol. Chem. Tokyo*, **26**, 721 (1962).
2. Kittredge, J. S., Roberts. E., and Simonsen, D. G., *Biochemistry*, **1**, 624 (1962).
3. Rouser, G., Kritchevsky, G., Heller, D., and Lieber, E., *J. Am. Oil Chemists Soc.*, **40**, 425 (1963)
4. Rosenberg, H., *Nature*, **203**, 299 (1964).
5. Quin, L. D., *Science*, **144**, 1133 (1964); *Biochemistry*, **4**, 324 (1965).
6. Kittredge, J. S., and Hughes, R. R., *Biochemistry*, **3**, 991 (1964).
7. Finkelstein, J., *J. Am. Chem. Soc*,, **68**, 2397 (1946).
8. Chavane, V., *Bull. Soc. Chim.*, 774 (1948).
9. Kosolapoff, G. M., *J. Am. Chem. Soc.*, **69**, 2112 (1947).
10. Medved, T. Ya., and Kabachnik, M. I., *Akad. Nauk. SSSR, Inst. Organ. Khim. Sintezy Organ. Soedin. Sb.* **2**, 12 (1952).
11. Kabachnik, M. I., and Medved, T. Ya, *Dokl. Akad. SSSR*, **84**, 717 (1952); *Izv. Akad. Nauk SSSR, Otd. Khim. Nauk*, **1954**, 314; *Ibid.*, **1953**, 868.
12. Chalmers, M. E., and Kosolapoff, G. M., *J. Am. Chem. Soc.*, **75**, 5278 (1953).
13. Chambers, J. R., and Isbell, A. F., *J. Org. Chem.*, **29**, 832 (1964).

A. F. Isbell

AMMONIA METABOLISM

Ammonia (NH_3) is a gas which exists in tissue fluids near neutral pH predominantly as the monovalent ammonium ion (NH_4^+). This compound is widely distributed throughout biological systems where its metabolism is primarily associated with that of protein, amino acids, urea and uric acid. If ammonium salts made with the heavy isotope of nitrogen, N^{15}, are administered to experimental animals, most of the isotope appears not in amino acids or protein but in the major nitrogenous excretory product of the species tested, *e.g.*, *urea* in mammals, *uric acid* in birds, or *ammonium* in bony fishes. Urea is the chief nitrogenous excretory product of ammonium, amino acid and protein metabolism in most terrestrial vertebrates and it is for this reason that these animals are termed "ureotelic" organisms.

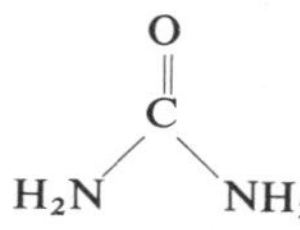

$H_2N{-}C({=}O){-}NH_2$

Urea

Urea is manufactured almost exclusively by the liver and, after entering the blood, most of it is rapidly excreted into the urine. However, a significant fraction of urea circulating in the blood enters the gastrointestinal tract where it has an important role in the metabolism of ammonium (*vide infra*). Birds and saurian reptiles are "uricotelic" vertebrates whereas mammals synthesize and excrete uric acid solely as an end product of

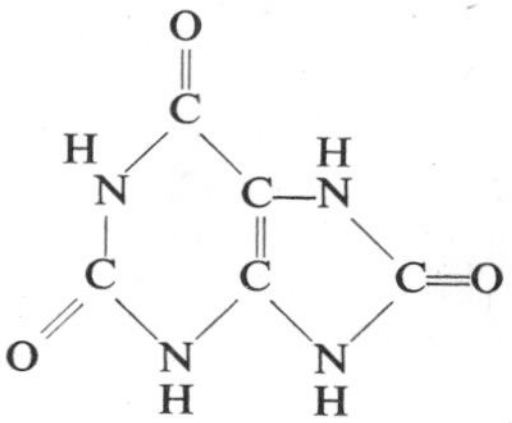

Uric acid

purine metabolism. A brief description of some of the major aspects of ammonium metabolism is given below (see also AMINO ACID METABOLISM; NITROGEN FIXATION).

Animals have essentially no capacity to convert atmospheric nitrogen into simple inorganic nitrogenous compounds such as nitrates or ammonia. This function must be carried out by plants and soil bacteria, with the result that animals then derive nitrogen from ingestion of plants or other animals. The nitrogen thus obtained is predominantly in the form of PROTEINS. These proteins are hydrolyzed to their constituent amino acids largely by the action of several PROTEOLYTIC ENZYMES and PEPTIDASES originating in the upper gastrointestinal tract or pancreas. During the process some ammonium is formed by deamination of amino acids or deamidation of glutamine (an amidated amino acid).

Deamination:

$$\underset{\text{Amino acid}}{R{-}\underset{\substack{|\\NH_2}}{CH}{-}COOH} \rightarrow \underset{\text{Keto acid}}{R{-}\underset{\substack{\|\\O}}{C}{-}COOH} + NH_3$$

Deamidation:

$$\underset{\text{Glutamine}}{HN{-}\underset{\substack{\|\\O}}{C}{-}CH_2{-}CH_2{-}\underset{\substack{|\\NH_2}}{CH}{-}COOH}$$

$$\rightarrow \underset{\text{Glutamic acid}}{HOOC{-}CH_2{-}CH_2{-}\underset{\substack{|\\NH_2}}{CH}{-}COOH} + NH_3$$

Transamination:

$$\underset{\text{Glutamic acid}}{HOOC{-}CH_2{-}CH_2{-}\underset{\substack{|\\NH_2}}{CH}{-}COOH} + \underset{\text{Oxaloacetic acid}}{HOOC{-}CH_2{-}\underset{\substack{\|\\O}}{C}{-}XOOH}$$

$$\rightleftharpoons \underset{\alpha\text{-Ketoglutaric acid}}{HOOC{-}CH_2{-}CH_2{-}\underset{\substack{\|\\O}}{C}{-}COOH} + \underset{\text{Aspartic acid}}{HOOC{-}CH_2{-}\underset{\substack{|\\NH_2}}{CH}{-}COOH}$$

However, most of the ammonium in the oral cavity, stomach, upper small intestine or large intestine of ureotelic non-ruminants is derived from endogenous urea circulating in the blood; urea enters the gastrointestinal tract where it is hydrolyzed to carbon dioxide and ammonium by the action of bacterial urease. The large intestine provides the major fraction of ammonium to the portal blood because it maintains a high content of urea-splitting bacteria. In ruminants, considerable ammonium is found in the upper gastrointestinal tract where it originates from the action of ruminal bacteria on dietary nitrogenous substrates (see RUMINANT NUTRITION AND METABOLISM).

Ammonium is absorbed from the upper and lower gastrointestinal tracts and appears in the portal blood in concentrations varying from 120 μg/100 ml to over 1000 μg/100 ml depending on the level of protein intake. The normal liver removes most of the portal blood ammonium so that the concentration of ammonium in hepatic venous blood drops to approximately 30 μg/100 ml.

The ammonium extracted from the portal blood can be considered with the larger quantities of ammonium produced in the hepatic cell. The latter are derived primarily by oxidative or nonoxidative deamination of amino acids which cannot be stored to any appreciable extent. The other product of the oxidative deamination reaction is an α-keto acid. The keto acid, among other functions, can accept the α-amino nitrogen of a number of amino acids by TRANSAMINATION. In this way, α-amino nitrogen is rapidly transferred among a variety of amino acids.

In tissues such as liver, smaller quantities of ammonium are generated by deaminating reactions where the original substrate is a protein or polypeptide, an amide (glutamine), a nucleotide (adenosine monophosphate), or an amine (5-hydroxytryptamine).

Ammonium is quite toxic and must be rapidly converted to nontoxic forms for storage and transport in order to keep the intracellular concentration low. Ammonium from dietary or endogenous sources is incorporated into organic compounds mainly by three reactions. In the first, which is present in both ureotelic and uricotelic vertebrates, ammonium combines with α-ketoglutarate and protons to form glutamate (see above for formulas). Glutamate is a dicarboxylic amino acid salt which plays a central role in ammonium metabolism in that its amino group readily interchanges with those of many other amino acids by transamination. In the second fixation reaction, glutamate is amidated by ammonium to form glutamine, which, along with glutamic acid, comprises a considerable portion of the non-protein nitrogen in vertebrate tissues. Glutamine is the major storage and transport form of ammonium in extrahepatic tissues. This amide also provides two of the four amino groups for the biosynthesis of uric acid (see PURINE BIOSYNTHESIS). The remaining two amino groups of uric acid are donated by aspartate, an amino acid formed from glutamate by transamination with oxaloacetate, and glycine. In ureotelic vertebrates, a third major ammonium fixation pathway is the reaction of ammonium, carbon dioxide and adenosine triphosphate to form CARBAMYL PHOSPHATE, a precursor of PYRIMIDINES and urea. The two nitrogens of urea are derived from carbamyl phosphate and aspartate (see ARGININE-UREA CYCLE).

$$H_2N{-}\overset{\overset{\large O}{\|}}{C}{-}O{-}PO_3H_2$$

Carbamyl phosphate

Thus, for practical purposes, the fate of ammonium in tissues is indistinguishable from that of nitrogen in glutamate, glutamine and carbamyl phosphate.

As mentioned above, most of the ammonium in portal blood is cleared by the normal liver. The concentration of ammonium in mixed venous or arterial blood is about 20–70 $\mu g/100$ ml which is slightly higher than that present in hepatic venous blood. The rise is due to the kidney which normally contributes ammonium to the circulation. However, under basal conditions, there is little net contribution or removal of blood ammonium by other tissues.

Ammonium is an important component of the urine of vertebrates, but only in the Teleostei (bony fishes) does it comprise the chief nitrogenous fraction. The cation is produced by the mammalian kidney from glutamine and other amino acids circulating in the blood. The excretion of ammonium is proportional primarily to the concentration of acids in the blood and is largely independent of the protein intake. Thus, when the organism receives or produces excessive acids, *e.g.*, in exercise, starvation or diabetes, more ammonium is manufactured through the action of renal glutaminase and amino acid oxidases. Excretion of acids with ammonium ions spares the elimination of metallic cations, sometimes called fixed bases, such as Na^+, K^+, Mg^{++}, Ca^{++}, which are essential for the economy of the organism. In severe acidosis in man, ammonium excretion may reach the high rate of 8.5 g/day. On the other hand, failure of ammonium production by the diseased kidney is the basis for the acidosis found in severe renal disease.

Ammonium metabolism may be altered in certain physiological or pathological states. For example, muscle contributes net amounts of ammonium to the circulation during exercise. Convulsions, coma and death may be produced in experimental animals and man by administration of excessive amounts of ammonium salts. Congenital defects of urea biosynthesis in man also can result in ammonium intoxication. In hepatic disease, systemic blood ammonium concentrations often become elevated because the liver fails to convert ammonium into less toxic compounds for storage or excretion. A serious and often fatal complication known as hepatic coma can be induced in patients with severe hepatic disease by the administration of compounds which either contain or liberate ammonium. A similar state may be observed in ruminants maintained on feeds rich in or supplemented with non-protein nitrogenous compounds; this results from the large quantities of ammonium produced by the ruminal bacteria. Raising the pH of the blood increases the toxicity of ammonium salts in animals, presumably because it favors formation of ammonia which more readily diffuses from the blood into the brain than the ammonium ion. However, the precise way in which excess ammonium produces central nervous system symptomatology is not understood.

References

1. FRUTON, JOSEPH S. AND SIMMONDS, SOFIA, "General Biochemistry," Second edition, New York, John Wiley & Sons, Inc., 1958.
2. COHEN, PHILIP P., AND BROWN, GEORGE W., JR., "Ammonia Metabolism and Urea Biosynthesis," in "Comparative Biochemistry" (FLORKIN, MARCEL AND MASON, HOWARD S., EDITORS), Vol. II, pp. 161–244, New York, Academic Press, 1960.
3. PITTS, ROBERT F., "Physiology of the Kidney and Body Fluids," Ch. 11, pp. 163–188, Chicago, Ill., Year Book Medical Publishers, Inc., 1963.
4. GABUZDA, GEORGE J., "Hepatic Coma: Clinical Considerations, Pathogenesis, and Management," *Advan. Internal Med.*, **11**, (DOCK, WILLIAM, AND SNAPPER, I., EDITORS),(1962).

LESLIE T. WEBSTER, JR.

AMPHOLYTES

An ampholyte is a substance each molecule of which contains simultaneously at least one portion that behaves as an acid and at least one portion that behaves as a base; *i.e.*, it is an amphoteric substance. AMINO ACIDS are ampholytes, each amino group being capable of bearing one positive charge, and each carboxyl group one negative charge. The fraction of such groups actually existing in the charged (ionized) state depends upon the acidity or alkalinity (hydrogen ion concentration; see pH) of their environment (see PROTEINS, section on Proteins as Acids and Bases).

AMYGDALIN

Amygdalin is a plant glycoside, first isolated from oil of bitter almonds in 1830, which on hydrolysis yields two moles of glucose, one mole of hydrogen cyanide, and one mole of benzaldehyde. The two glucose units are believed to be connected by a β-1,6-glucosidic linkage, and the cyanohydrin of benzaldehyde is linked to carbon-1 of one of the glucose units through a β-glucosidic linkage [see CARBOHYDRATES (CLASSIFICATION AND STRUCTURAL INTERRELATIONS)].

AMYLASES

Amylases are enzymes that catalyze the hydrolysis of STARCHES. Two classes of amylases have been recognized. The first type, formerly known as α-amylase, and currently also termed α-1,4-glucan 4-glucanohydrolase, is present in saliva and pancreatic juice, acts upon (a) amylose (a straight chain unbranched polysaccharide) at random locations to yield finally a mixture of glucose and maltose, and acts upon (b) amylopectin (a branched polysaccharide) to produce a mixture of branched and unbranched oligosaccharides containing many unattacked α-1,6-glycosidic bonds. The second type, formerly known as β-amylase, which occurs in barley malt among other sources, is currently termed α-1,4-glucan maltohydrolase, acts upon (a) amylose to yield almost exclusively maltose, and acts upon (b) amylopectin to attack only nonreducing chain ends, leaving unattacked a highly branched core polysaccharide or *dextrin* containing all the original α-1,6-bonds [see also CARBOHYDRATES (CLASSIFICATION AND STRUCTURAL INTERRELATIONS); DIGESTION AND DIGESTIVE ENZYMES].

References

1. WHITE, A., HANDLER, P., AND SMITH, E. L., "Chemistry of Carbohydrates," in "Principles of Biochemistry," Third edition, Ch. 4, pp. 49–53, New York, McGraw-Hill Book Co., 1964.
2. BERNFELD, P., "Amylases, α and β," in "Methods in Enzymology" (COLOWICK, S. P., AND KAPLAN, N. O., EDITORS), Vol. I, pp. 149–158, New York, Academic Press, 1955.

AMYLOPECTINS AND AMYLOSES. See STARCHES.

ANABOLISM

Anabolism is one of two broad types of *metabolism* (a term which encompasses all the numerous chemical and energy transformations that occur in living cells). *Anabolism* broadly refers to processes in which components of living cells (relatively complex molecules and structures) are biosynthesized from nutrient materials (relatively simple in structure). Anabolism thus contrasts with the other broad class of metabolic processes, *catabolism*, which implies degradation of relatively large, complex molecules to smaller ones. Anabolic processes, such, for example, as PROTEIN BIOSYNTHESIS, generally require energy, which is available in cells in the form of stored chemical energy in "high-energy phosphate" compounds (see PHOSPHATE BOND ENERGIES); these energy-rich compounds in turn are produced at the expense of energy released during degradative, catabolic processes (see GLYCOLYSIS; CITRIC ACID CYCLE; RESPIRATION). In non-growing living cells, anabolism and catabolism are balanced in a steady state far from equilibrium, and in growing cells, the anabolic processes predominate, although both are always active. In death of a cell or organism, anabolism ceases, but many catabolic processes may continue (see AUTOLYSIS).

ANAPHYLAXIS. See IMMUNOCHEMISTRY.

ANDROGENS

(From Greek ἀνήρ, a man, a male, and γείομαι, that which produces.) The relation between the testis and the male secondary sex characteristics has been recognized since antiquity. However, the first objective evidence that a chemical substance present in the testis could elicit androgenic effects was achieved in 1908 by Walker who prepared an aqueous glycerol extract of bull testis tissue that caused growth of the capon's comb. In 1927, McGee, in association with F. C. Koch, prepared much more active extracts from bull testes by using organic solvents. He assayed the extracts quantitatively by measuring the increase in area of the capon's comb. The discovery of androgenic activity in urine made possible the isolation of the first biologically active crystalline androgens by Butenandt and his co-workers in 1931–1934. These workers isolated androsterone (Fig. 1, structure 1) and dehydroisoandrosterone (3β-hydroxyandrost-5-en-17-one) (Fig. 1, structure 2) from human

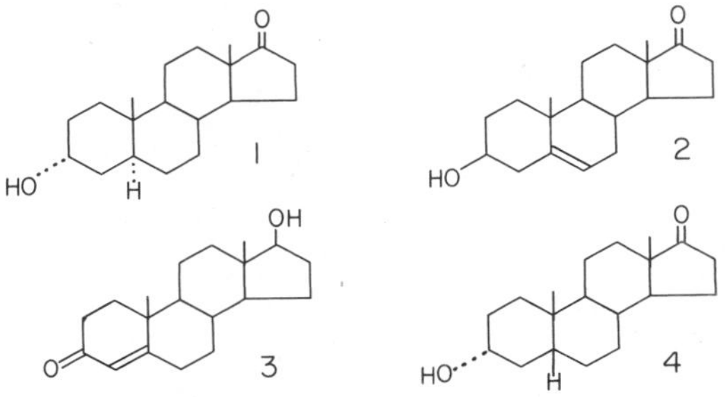

FIG. 1.

male urine (see STEROIDS for numbering system). In 1935, David and his collaborators isolated from bull testis extract a crystalline hormone having higher biological activity than either androsterone or dehydroisoandrosterone; they named it testosterone (Fig. 1, structure 3). Testosterone is the principal secretory product of the testis in most species. Androsterone and dehydroisoandrosterone were quickly shown to be steroids by their chemical synthesis from 5α-cholestan-3α-ol and cholesterol, respectively. Testosterone was also prepared from cholesterol within months of its isolation from testicular extract.

Androgens are detected and measured biologically by their profound effects upon the secondary sex tissues of male vertebrates. Quantitative biological assays for androgens are performed in mammals by measuring the increase in weight of the seminal vesicles and prostate glands of the immature or castrated male rat. Another very sensitive assay is based upon the increase in weight of the comb of the day-old chick under androgen stimulation. This test is derived from the classic capon comb test used by Koch.

The androgens stimulate the development of the male secondary sex structures, such as the penis, scrotum, seminal vesicles, prostate gland, vas deferens and epididymis. The deepening of the voice, the growth of pubic, axillary, body, and facial hair, as well as the development of the characteristic musculature of the human male, are also under the influence of testosterone. If the testes fail to develop or are removed prior to puberty, these changes do not occur. Thus, testosterone is essential for reproductive function in the male.

The testes are not the only site of androgen production; the adrenal cortex produces dehydroisoandrosterone (Fig. 1, structure 2) which is found in blood and urine largely conjugated as the sulfate ester. The amounts of androgen secreted by the normal adrenal cortex are, however, insufficient to maintain reproductive function in the male. The normal human ovary and placenta also produce small amounts of androgenic steroids that serve as precursors for the estrogens in these tissues. In the human, little testosterone is excreted into the urine and virtually none into the feces. The principal metabolic transformation products are androsterone (Fig. 1, structure 1) and 5β-androsterone (Fig. 1, structure 4) with small amounts of other reduced compounds. These substances are excreted in the urine in the form of esters with sulfuric acid or glycosides with glucuronic acid.

Androsterone, 5β-androsterone and dehydroisoandrosterone share the structural feature of a carbonyl group at C-17. They are therefore known as 17-ketosteroids. The measurement of urinary 17-ketosteroids by the Zimmermann reaction (a purple color appears upon treatment with *m*-dinitrobenzene and alkali) affords a method of assessing the production of androgens, although, of course, it does not distinguish between androgens produced by the gonads and those produced elsewhere (see also the separate article on KETOSTEROIDS).

Like all other classes of steroid hormones, the androgens are synthesized from acetyl coenzyme A *via* mevalonic acid, isopentenyl pyrophosphate, farnesyl pyrophosphate, squalene, lanosterol and cholesterol (see ISOPRENOID BIOSYNTHESIS; STEROL BIOSYNTHESIS). Enzyme systems in the testis then catalyze the cleavage of the sidechain of cholesterol to pregnenolone (3β-hydroxypregn-5-en-20-one) (Fig. 2, structure 5) which can give rise to testosterone by the two pathways shown in Fig. 2. The stages from cholesterol through dehydroisoandrosterone can also occur with the hydroxyl group at C-3 esterified as the sulfate.

Testosterone is formed by the interstitial or Leydig cells of the testes which develop under the influence of GONADOTROPHIC HORMONES discharged into the blood stream by the anterior pituitary gland. In pituitary insufficiency, this hormonal stimulus is lacking and, as a consequence, the Leydig cells do not secrete testosterone. In such cases, the male secondary sex characteristics fail to develop. On the other hand, interstitial cell tumors may occur, leading to excessive androgen production and precocious puberty. In women, tumors or excessive function of the adrenal cortex and, very rarely, of the ovary, result in the production of large amounts of androgens with associated virilization.

While androgens produce many effects in the whole organism that are reflected in biochemical changes, relatively little is known of the exact biochemical events that are set in motion by these substances. Since stimulation of growth of specific tissues is a characteristic of androgen action, it is

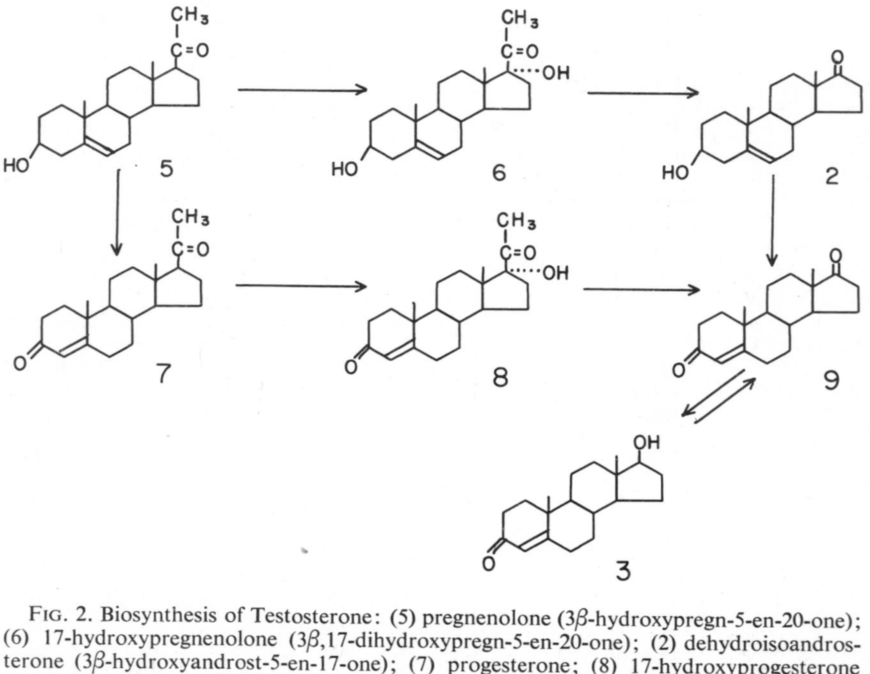

FIG. 2. Biosynthesis of Testosterone: (5) pregnenolone (3β-hydroxypregn-5-en-20-one); (6) 17-hydroxypregnenolone (3β,17-dihydroxypregn-5-en-20-one); (2) dehydroisoandrosterone (3β-hydroxyandrost-5-en-17-one); (7) progesterone; (8) 17-hydroxyprogesterone (17-hydroxypregn-4-ene-3,20-dione); (9) androstenedione (androst-4-ene-3,17-dione); (3) testosterone.

tempting to postulate an effect somewhere in the early stages of protein synthesis. There is experimental evidence suggesting that androgens affect the synthesis of messenger ribonucleic acid in the rat ventral prostate gland. However, the biochemical mechanism of androgen action is still an unsolved problem.

References

1. DORFMAN, R. I., AND SHIPLEY, R. A., "Androgens," New York, John Wiley & Sons, 1956.
2. RICHARDS, J. H., AND HENDRICKSON, J. B., "The Biosynthesis of Steroids, Terpenes and Acetogenins," New York, W. A. Benjamin, Inc., 1964.
3. FIESER, L. F., AND FIESER, M., "Steroids," New York, Reinhold Publishing Corp., 1959.

LEWIS L. ENGEL

ANEMIAS

Anemia occurs when there is a deficiency of red blood cells or a reduction in the hemoglobin content of the red blood cells. Since blood may easily be sampled and its constituents accurately measured, much of the information about anemias has been obtained from studies of human subjects. Some anemias occurring in man have not been duplicated in animals. Aside from blood loss due to hemorrhage, anemia occurs when (1) hemoglobin synthesis is blocked, (2) there is a block in the production of mature red cells, or (3) the red blood cells which are elaborated are defective and thus have a shortened survival time.

(1) Defects in Hemoglobin Synthesis. Hemoglobin is a red protein with a molecular weight of about 65,000. The main function of hemoglobin is to transport oxygen. This complex molecule does this efficiently because it can combine reversibly with oxygen. Normal globin contains two alpha and two beta polypeptide chains. A heme moiety is attached to each polypeptide. Heme is a protoporphyrin ring with an iron atom in the ferrous state at the center.

(a) *Heme Synthesis.* Heme is synthesized by the organism from glycine; the main steps in this scheme are:

$$\text{Glycine} + \text{Succinyl Coenzyme A} \xrightarrow{\text{Pyridoxal}\cdot\text{PO}_4}$$

$$\alpha\text{-Amino-}\beta\text{-ketoadipic Acid} \rightarrow \delta\text{-Aminolevulinic Acid}$$

$$\rightarrow \text{Porphobilinogen} \rightarrow \text{Protoporphyrin} \xrightarrow{\text{Fe}} \text{Heme}$$

The most common cause of anemia is IRON deficiency. In the absence of this mineral, hemoglobin production is markedly reduced, resulting in anemia. This anemia is characterized by small red blood cells, or microcytes, circulating in the blood.

COPPER deficiency, although rare, also results in anemia. Copper is essential for the utilization of iron and its incorporation into heme.

A deficiency of vitamin B_6 also results in a microcytic anemia. This vitamin (see PYRIDOXINE FAMILY OF VITAMINS), in its coenzyme form pyridoxal phosphate, is an essential cofactor in the biosynthesis of heme from glycine (see scheme above). In its absence, heme synthesis is effectively blocked.

Pantothenic acid, another B vitamin, is a constituent of COENZYME A, and it would be expected that a deficiency of this vitamin would also block heme synthesis. Because of its widespread distribution in foods, pantothenic acid deficiency in man is very rare. However, in experimental animals this deficiency results in such a block.

There is a series of inherited diseases called porphyrias which are characterized by excessive and abnormal synthesis of intermediates in heme formation, resulting in abnormal excretion of various heme precursors. These are inborn errors of metabolism, which are probably the result of excess formation of one or more enzymes necessary in the scheme outlined above.

Lead poisoning will result in mild to severe anemia. This is probably due to the fact that lead will block at least three steps in heme synthesis: (1) the incorporation of iron into protoporphyrin, (2) the formation of δ-aminolevulinic acid, and (3) the formation of porphobilinogen.

(*b*) *Globin Synthesis.* The protein globin normally consists of four polypeptide chains containing 19 amino acids in a specific sequence and, in combination with heme, is called hemoglobin A (see HEMOGLOBINS IN HUMAN GENETICS). Slight aberrations in this structure give rise to diseases called hemoglobinopathies, which may be fatal. For example, an anemia found mainly in negroes from West Africa is characterized by a sickle-shaped red blood cell.

The hemoglobin in these cells has different solubility properties and different electrophoretic mobility than normal hemoglobin. Extensive studies have shown that *sickle cell hemoglobin* differs from normal hemoglobin only by the substitution of a valine, a monoamino monocarboxylic acid, for a glutamic acid, a monoamino dicarboxylic acid, in the beta chain of the globin. The replacement of a negatively charged amino acid by a neutral amino acid accounts for the changes in the mobility of the protein in an electric field, which offers a convenient method for identifying it. When the red blood cells of patients with this disease are deoxygenated, they assume the sickle shape, perhaps due to the decreased solubility of the hemoglobin, which is called *hemoglobin S*. This anemia is inherited and is severe when the patients are homozygous for this abnormality. The heterozygous sickle cell trait results in mild anemia, and it probably confers protection against malaria in patients with this trait. This may account for the persistence of this disease in some areas of Africa.

Another hemoglobinopathy resulting in anemia is hemoglobin C disease. Hemoglobin C is a variant in which the diamino monocarboxylic amino acid lysine occurs in its beta chain in place of a glutamic acid in hemoglobin A. This changes the electrophoretic mobility of the hemoglobin, a characteristic used to diagnose the anemia (see

ELECTROPHORESIS). In recent times, a variety of hemoglobinopathies have been discovered from a comparison of the electrophoretic mobility of the abnormal hemoglobin with that of hemoglobin A. These abnormal hemoglobins are usually characterized by a single amino acid substitution in one of the polypeptide chains. When the amino acid substitution does not result in a change in the charge of the molecule, the mobility of the hemoglobin is unaffected and other means are necessary to identify it.

Thalassemia major is an anemia in which there is a diminished ability to make the beta chain of globin. No difference in the pattern of the amino acids in the polypeptide chains has been found. It is a genetic error of metabolism and results in severe anemia in patients with the homozygous disease. These patients usually require frequent transfusions of whole blood or red blood cells to survive. Since the mechanism for excretion of iron is limited, excessive iron resulting from these transfusions is deposited in the vital organs of these patients, resulting in hemochromatosis which may prove fatal. A thalassemia in which production of the alpha chain is reduced has also been detected.

(2) Defect in Maturation of Red Blood Cells. Megaloblastic anemia is characterized by the appearance of large nucleated immature red blood cells, or megaloblasts, in the bone marrow and peripheral circulation. These immature cells fail to mature properly. This is probably due to a deranged nucleoprotein metabolism and occurs as a result of a FOLIC ACID or VITAMIN B_{12} deficiency. PURINES and PYRIMIDINES, the building blocks of nucleoproteins, are synthesized *de novo* in the tissues; FOLIC ACID COENZYMES acting as one-carbon unit donors are involved in three reactions, the first two necessary for purine biosynthesis and the third for the biosynthesis of thymidylic acid, a pyrimidine:

(1) Glycinamide Ribotide + N^5, N^{10}-Methenyltetrahydrofolic Acid → Formylglycinamide Ribotide
(2) Aminoimidazolecarboxamide Ribotide + N^{10}-Formyltetrahydrofolic Acid → Formaminoimidazolecarboxamide Ribotide
(3) Deoxyuridylic Acid + N^5, N^{10}-Methylenetetrahydrofolic Acid → Thymidylic Acid

In folic acid deficiency, these reactions are presumably blocked, leading to deranged nucleoprotein synthesis.

The one-carbon units which the folate coenzymes donate to the above substrates are derived to a large extent from the degradation of histidine to glutamic acid by the following sequence:

$$\text{Histidine} \rightarrow \text{Urocanic Acid} \rightarrow \text{Formiminoglutamic Acid (FIGLU)} \xrightarrow{\text{Tetrahydrofolic acid}} \text{Glutamic Acid}$$

The step from FIGLU to glutamic acid requires tetrahydrofolic acid, which accepts the formimino group. In the absence of folic acid, FIGLU accumulates and is excreted in the urine. The measurement of urinary FIGLU after a histidine metabolic load is a specific test for folic acid deficiency.

Although vitamin B_{12} is undoubtedly involved in nucleoprotein synthesis, the specific site of action has not yet been determined (see also COBALAMINS AND COBAMIDE COENZYMES).

(3) Red Blood Cell Defects. The mature red blood cell is a delicately balanced living container for hemoglobin. The latter accounts for approximately 95% of the protein content of the red blood cell. The normal survival time of this cell is about 120 days. However, a metabolic defect may cause a reduction in the survival time, resulting in what is termed a hemolytic anemia.

The mature red blood cell obtains its energy from glucose by two pathways: (1) the Embden-Meyerhof anaerobic glycolysis, and (2) the pentose phosphate pathway. The former leads to the production of ATP and DPNH (NADH). The latter leads to the production of TPNH (NADPH) (see CARBOHYDRATE METABOLISM; RESPIRATION).

(*a*) *Embden-Meyerhof Pathway*. An anemia has been described in which there is reduction in the pyruvate kinase content of the red cell. This enzyme catalyzes the conversion of phosphoenol pyruvate to pyruvate with the production of ATP, the source of energy for the cell. In patients with a deficiency of this enzyme there is a reduction of ATP content and an accumulation in the red blood cell of intermediates of the conversion of glucose to pyruvate. The survival time of this defective erythrocyte is shortened, resulting in anemia.

(*b*) *Pentose Phosphate Pathway*. A hemolytic anemia has been observed to occur in patients receiving drugs such as primaquine, sulfanilamide, naphthalene, or synthetic vitamin K. The red blood cells of such subjects contain a decreased content of reduced GLUTATHIONE. Under normal conditions reduced glutathione, which is necessary for the integrity of the red blood cell, is regenerated by the following scheme:

$$\text{Glutathione} \xrightarrow[\text{Glutathione reductase}]{\text{TPNH}} \text{Glutathione·SH}$$

The erythrocyte produces TPNH from the oxidation of glucose-6-PO_4 to 6-phosphogluconolactone. The enzyme glucose-6-phosphate dehydrogenase which catalyzes this reaction was found to be sharply reduced in these patients, resulting in a decreased ability to make TPNH. The red blood cells of these subjects function normally until subjected to stress by the administration of the drugs above, at which time a hemolytic anemia occurs.

The biochemical studies of the metabolic reactions involved in a variety of anemias have provided a better rationale for a therapeutic approach to many of these conditions.

References

1. WINTROBE, M. M., "Clinical Hematology," Fifth edition, Philadelphia, Lea and Febiger, 1961.
2. INGRAM, V. M., "Hemoglobin and its Abnormalities," Springfield, Ill., Charles C. Thomas, 1961.
3. LEVINE, R., AND LUFT, R. (EDITORS), "Advances in Metabolic Disorders," Vol. 1, New York, Academic Press, 1964.

JACK M. COOPERMAN

ANESTHESIA

The nature of the biochemical alterations produced by anesthesia and the magnitude of these alterations depend upon the anesthetic agent employed, the circumstances under which it is employed and, because of major species differences in metabolic response, the type and age of the organism anesthetized. Because anesthesia has until recently been approached primarily from a pragmatic and technical point of view, surprisingly little basic information is presently available on its biochemical aspects. These aspects are, however, of major importance. From a clinical point of view, the nature and extent of the biochemical response to anesthesia in the 11,000,000 persons annually anesthetized in this country constitutes a major determinant of surgical morbidity and mortality. And, from a research point of view, the anesthesia so frequently administered to permit biologic experimentation to be carried out is in itself often a significant factor in the results of the experiments. Failure to recognize the influence of anesthesia in research studies has often and unnecessarily invalidated the results of research.

Despite a relative paucity of information on the biochemical effects of anesthesia, enough data are available to support four general statements. First, although the state of anesthesia may be associated with biochemical changes, it is not caused by these biochemical changes. The consensus is that the primary anesthetic site of action is the cell membrane. Here anesthetics stabilize the membrane (possibly by clathrate microcrystal formation), alter membrane permeability, or prevent membrane depolarization. But such membrane changes only secondarily influence cell metabolism. Changes in membrane function may cause metabolic alterations, but they are not the result of anesthetically induced metabolic alterations.

The second generalization on the biochemical aspects of anesthetics is that although their most obvious effect is a behavioral one mediated through their effects on the central nervous system, the metabolic and biochemical responses to anesthetics occur throughout the organism as a whole and are not limited solely to the brain. Indeed, the clinical efficacy and safety of anesthetics is more often determined by their effects on, for example, myocardial or hepatic metabolism than by their effects on nerve tissue metabolism.

Thirdly, it can be said that the frequently expressed and essentially unitarian approach to the effect that what one anesthetic does is characteristic of all anesthetics represents dangerous oversimplification and is not in accord with known facts. All drugs which are capable of producing central nervous system depression (*i.e.*, capable of producing "anesthesia") do not necessarily have the same physiologic or biochemical effects. True anesthetics agents (cyclopropane, halothane, diethyl ether, etc.) are nonmetabolized, inert gases or vapors of liquids which upon inhalation produce a readily reversible, nontoxic depression of tissue excitability. On the one hand, they must be differentiated on pharmacologic, physiologic, and biochemical grounds from local anesthetics, and on the other hand, from hypnotics such as barbiturates, narcotics such as morphine, and tranquilizers such as chlorpromazine. But even within such a classification, generalizations are often hazardous. The biochemical response to all inhalation anesthetics is not the same. The response to cyclopropane anesthesia is different from that to ether anesthesia, which in turn is different from that to halothane, methoxyflurane, nitrous oxide, or ethylene anesthesia.

And, finally, as the fourth generalization, it may be said that the biochemical effects of anesthetics may be either directly or indirectly mediated. The former represent those effects of the anesthetics which are due to their direct action on cell membranes, membrane carriers, or cell enzyme systems, etc. The latter are those effects of the anesthetics which may be ascribed to the fact that they concurrently alter pulmonary ventilation, organ blood flow, tissue gas tensions, circulating levels of hormones, etc., each one of which in turn has metabolic consequences. Direct biochemical effects of anesthetics are observable both *in vitro* and *in vivo*. Indirect effects are observable primarily *in vivo*.

Direct biochemical effects of anesthetics include the uncoupling effects of certain barbiturates whereby the normal relationship between inorganic phosphate uptake and oxygen consumption is disrupted. The uncoupling effect of barbiturates on oxidative PHOSPHORYLATION is such that inorganic phosphate uptake is depressed, with consequent decreased production of high-energy phosphate bonds in substances such as adenosine triphosphate, at a time when oxygen uptake is unaltered. This effect of barbiturates is not, however, related to their mode of action. As another direct biochemical effect of anesthetics, the inhibition by local anesthetics of membrane sodium pump mechanisms may be mentioned, although this effect of local anesthetics on sodium pump systems also is probably not related to their mode of action. As an example of the direct effects of inhalation anesthetics on metabolism, evidence may be mentioned which indicates that they act as competitive inhibitors of the carrier mechanisms whereby monosaccharides are transported across human erythrocyte membranes. Interestingly, the carrier for glucose is not influenced by anesthetics, although the carriers for other sugars are. Inhalation anesthetics have also been shown to have an observable but kinetically insignificant direct inhibitory effect (10–15% inhibition) on many enzyme systems *in vitro*.

The majority of the biochemical responses to anesthetics, especially to inhalation anesthetics, are indirectly mediated. The means by which anesthetics indirectly affect metabolism are illustrated by six events which may occur during anesthesia, events which are themselves associated with metabolic changes: (1) Many inhalation anesthetics, notably cyclopropane and, to a lesser extent, diethyl ether, are associated with a hypothalamically mediated increased activity of the sympathetic autonomic nervous system. One result of this is increased blood levels of metabolically active substances such as epinephrine (see ADRENALINE AND NORADRENALIN), but another result is hepatic glycogenolysis comparable to that observed after electrical stimulation of hepatic splanchnic nerves. This neurogenically induced glycogenolysis contributes to the hyperglycemia seen with certain anesthetic agents, although it is by no means the sole source of increased blood glucose levels. (2) Inhalation anesthetics may also be associated with reflex release of metabolically active hormones other than the neurogenically mediated increase in epinephrine (and norepinephrine). Diethyl ether anesthesia is accompanied, for example, by a release of pituitary ADRENOCORTICOTROPIC HORMONE with consequent increased blood levels of 17-hydroxycorticosteroids which in turn contribute to the abnormal blood levels of carbohydrate metabolites observed during ether anesthesia. The effect of anesthetics on the sympathetic nervous system and on hormonal activity with a resultant increase in metabolism is one of several reasons why general anesthesia is usually not associated with a decrease in oxygen consumption despite a decreased somatic activity (*i.e.*, truly basal conditions) (3) Changes in tissue gas tensions occur during anesthesia which in themselves result in changes in metabolism. The most important of these is the decrease in tissue oxygen tension which may occur during certain types of anesthesia. The decrease in tissue oxygen tension is only rarely due to decreases in alveolar or arterial oxygenation. It is instead usually due primarily to cardiovascularly mediated decreases in tissue blood flow (decreased cardiac output, arterial hypotension, peripheral vasoconstriction, etc., alone or in combination). A result of the decrease in tissue oxygen tension is an increase in blood lactic acid levels in excess of that which can be ascribed to any concurrent increase in blood pyruvic acid levels, this excess lactate being evidence of the fact that molecular oxygen is not available at the tissue level in amounts adequate to maintain the normal relationship between oxidized and reduced diphosphopyridine nucleotide. Such changes in lactate and pyruvate are especially prominent during ether or cyclopropane anesthesia. Increases in tissue carbon dioxide tension may also occur during anesthesia, secondary to inadequate alveolar ventilation. In fact, in modern anesthesia, respiratory acidosis (carbon dioxide retention) occurs more frequently than does hypoxia, for oxygenation can generally be more readily assured than can carbon dioxide removal. The metabolic consequences of increased carbon dioxide tension remain incompletely defined, however (see also GASEOUS EXCHANGE IN BREATHING). (4) The metabolic consequences of surgery and its attendant trauma during anesthesia are additive to the biochemical effects of anesthesia alone. It is frequently difficult to dissociate the metabolic responses to surgery from the effects of anesthesia if metabolic studies are made during surgery and anesthesia instead of during anesthesia alone. (5) If shock, especially hemorrhagic shock, occurs during anesthesia, the metabolic effects of shock are added to, and further accentuate, the metabolic response to anesthesia. (6) Renal function, especially renal blood flow and GLOMERULAR FILTRATION, are frequently depressed during anesthesia, the degree of depression depending on the type and depth of anesthesia. These changes in kidney function add to the metabolic changes observable during anesthesia by impairing renal excretory mechanisms. Renal vascular effects are further complicated by increased output of aldosterone and pituitary antidiuretic hormone during certain types of anesthesia. (7) The body's temperature may also change (in either direction) during anesthesia, with resultant biochemical consequences which become especially prominent if hypothermia is allowed to develop.

References

1. ANDERSEN, N. B., AND GRAVENSTEIN, J. S., "Effects of Local Anesthetics on Sodium and Potassium in Human Red Cells," *J. Pharmacol. Exp. Therap.* **147**, 40 (1965).
2. BREWSTER, W. R., JR., BUNKER, J. P., AND BEECHER, H. K., "Metabolic Effects of Anesthesia: Mechanism of Metabolic Acidosis and Hyperglycemia during Ether Anesthesia in the Dog," *Am. J. Physiol.*, **171**, 37 (1952).
3. GREENE, N. M., "Inhalation Anesthetics and Carbohydrate Metabolism," Baltimore, Williams and Wilkins Co., 1963.
4. GREENE, N. M., "Inhalation Anesthetics and Permeability of Human Erythrocytes to Monosaccharides," *Anesthesiology*, **26**, 731 (1965).
5. HENNEMAN, D. H., AND VANDAM, L. D., "Effect of Epinephrine, Insulin, and Tolbutamide on Carbohydrate Metabolism during Ether Anesthesia," *Clin. Pharmacol. Therap.*, **1**, 694 (1960).
6. PAULING, L., "A Molecular Theory of General Anesthesia," *Science*, **134**, 15 (1961).

NICHOLAS M. GREENE

ANGIOTENSIN

This term applies to two polypeptides, *angiotensin I* (also called hypertensin I or angiotonin) produced when a kidney enzyme (renin) cleaves the substrate protein angiotensinogen (or hypertensinogen), and *angiotensin II* (or hypertensin II), produced when a "converting" enzyme (an angiotensinase or hypertensinase) cleaves angiotensin I (see HYPERTENSION).

ANIMAL PROTEIN FACTOR

This is an older term for VITAMIN B_{12}, attributable to the fact that plants are relatively poor sources of Vitamin B_{12}, while many animal products were found to be good sources of this nutritional "factor" or vitamin (see also COBALAMINS AND COBAMIDE COENZYMES).

ANSERINE

Anserine (N-β-alanyl-1-methyl-L-histidine) is a dipeptide, occurring in mammalian muscle, which yields on hydrolysis two amino acids, neither of which is one of the common amino acids recognized as protein derived. The methyl group is located on the imidazole ring of the histidine.

ANTHOCYANINS

The anthocyanins, which occur naturally as plant pigments, are glycosides of anthocyanidins, a subdivision of the group of compounds broadly termed FLAVONOIDS (see also CHEMOTAXONOMY; PIGMENTS, PLANT).

ANTIBIOTICS

Antibiotic is a term first applied by Waksman in 1945 to a chemical compound of microbial origin which even in dilute solution possesses the ability to inhibit reproduction of, or cause the destruction of, microorganisms. In present usage the definition also encompasses close chemical derivatives of the natural product, even though they themselves may not be naturally occurring. The word is derived from *antibiosis*, coined in the latter half of the nineteenth century by DuBarry and Vuillemin to describe the competition in nature between species, such as one creature destroying the life of another in order to preserve its own. Most antibiotics are produced by fungi, actinomycetes and bacteria, many of which normally abound in the soil but are also present elsewhere. The compounds are usually prepared commercially from growing microorganisms, although several are synthesized. Historically, the first antibiotic cannot be established with certainty since many herbals have been used as antiseptics for centuries. The era of antibiotics was introduced by the now classical observation of Fleming in 1929 of the antagonistic action of *Penicillium notatum* upon growth of cultures of staphylococcus. Subsequently, Dubos isolated tyrothricin from *Bacillus brevis* (1939); Florey, Chain and colleagues, penicillin (*ca.* 1940); and Schatz, Bugie and Waksman, streptomycin, in 1944.

In the meantime, many new antibiotic agents have become available. The U.S. Tariff Commission estimates that 6.7 million pounds of antibiotics were produced in the U.S. in 1963, of which 4.2 million pounds were for human or veterinary use and 2.5 million pounds for animal feed supplements, food preservation and crop spraying. Almost 700,000 kg of penicillin, 500,000 kg of streptomycin and 200,000 kg of tetracycline were produced that year. Total annual sales were approximately $400 million. Table 1 lists some of the major antibiotics of clinical importance and their properties.

Action and Selectivity. Ideally, an antibiotic is a drug that is selectively toxic to a parasitic microorganism without producing any deleterious effect upon the infected host, such as man. Preferably, it also does not interfere with those microorganisms not responsible for the infection of the host.

Unlike most antiseptic compounds, an antibiotic inhibits growth of a characteristic spectrum of microorganisms. Generally the range of antibacterial action is predictable on the basis of microbial classification. For example, penicillin is more effective against Gram-positive bacteria, whereas streptomycin is more inhibitory to Gram-negative ones. The so-called broad-spectrum antibiotics, the tetracyclines and chloramphenicol, have a wider range of antibacterial activity. In spite of these generalizations, frequently the antibacterial action against a given strain of bacteria is unpredictable and requires empirical evaluation. This uncertainty is due to the presence of microbial strains which are normally resistant to a drug and the gradual development of resistance in other cell strains. Undoubtedly, the sensitivity of a bacterial cell is based on chemical or physical properties, but there is insufficient information available about differences in bacterial biochemistry at the moment to predict it in a rational manner.

Mechanism of Action. Although antibiotics have been receiving extensive clinical usage for several decades, an understanding regarding the more detailed mechanism by which they interfere with bacterial growth has become available only in the last few years. It is now believed that penicillin's major action is inhibition of the formation of CELL WALL of bacteria. This wall is very different from the membrane surrounding mammalian cells, and the probable reason for the selectively toxic effect on bacteria rather than mammalian cells rests on this difference. The bacterial cell wall protects the microbial cell from the osmotic difference that exists between the inside and outside of the bacterial cell. Thus, when the cell wall is sufficiently damaged, the cell will disrupt and die unless the osmotic strength of the medium is greatly increased to minimize the osmotic difference. Under these special conditions, cells free of cell wall (protoplasts) are formed which are no longer sensitive to penicillin. Further evidence for the mechanism of action of penicillin was obtained by Park and Strominger, who observed the accumulation of a conjugate of uridine diphosphate and N-acetylmuramyl peptide in the medium of penicillin-inhibited cells. The condensation into the cell walls of this latter compound, which is a constituent of bacterial cell walls but not of mammalian cell membranes, was prevented by the drug. Other antibiotics, such as bacitracin, cycloserine, novobiocin, vancomycin and the ristocetins, are believed to block specific steps in cell wall biosynthesis.

Several other antibiotics interfere with specific steps in PROTEIN BIOSYNTHESIS or nucleic acid biosynthesis (see RIBONUCLEIC ACIDS, BIOSYNTHESIS) of the microbial cell. Although much information is becoming available in this biochemical area, a more detailed pinpointing of the locus of action of many of these drugs must await further delineation of the steps of these reactions and an

Table 1. Properties of Some Clinically Used Antibiotics

Name	Structure	Source	Introduced	Useful Spectrum of Inhibition	Remarks
Penicillins Benzylpenicillin or penicillin G	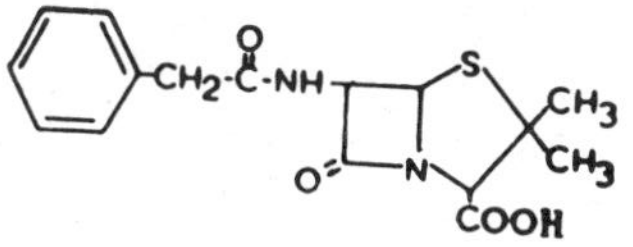	*Penicillium notatum; Penicillium chrysogenum*	1940, Fleming, Chain, Florey	Streptococci, pneumococci, some staphylococci, meningococci, gonococci, clostridia, spirochetes	Lactam ring split by penicillinase
Methicillin ("Staphcillin")	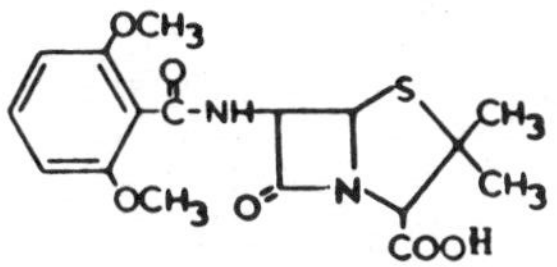	Synthesized from aminopenicillanic acid	1960, Rolinson *et al.*	Staphylococci resistant to penicillin G; otherwise spectrum like penicillin G	Stable to penicillinase; reserve for Staphylococcus aureus
Streptomycin	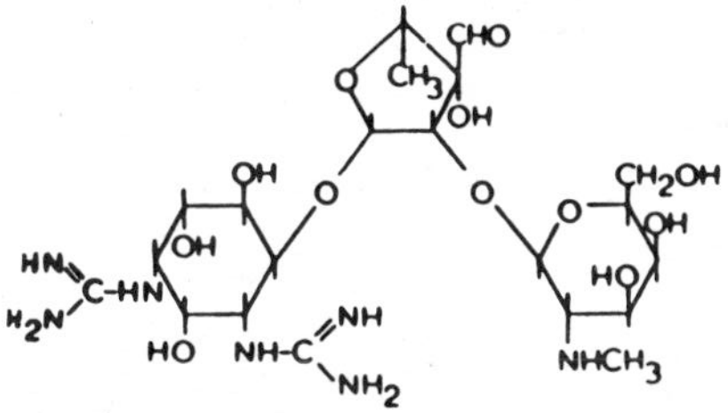	*Streptomyces griseus*	1944, Schatz, Waksman, *et al.*	Mycobacterium tuberculosis; streptococci; many Gram-negative bacteria	Closely related: neomycin, kanamycin
Chloramphenicol ("Chloromycetin")	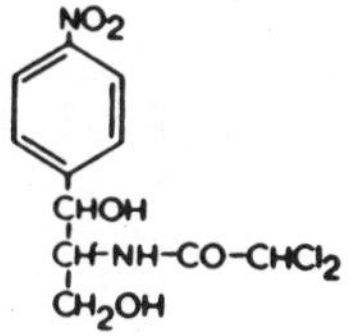	*Streptomyces venezuelae*	1947, Ehrlich *et al.*	Salmonellae, many others under special circumstances	"Broad-spectrum" drug

Tetracyclines Chlortetracycline ("Aureomycin")	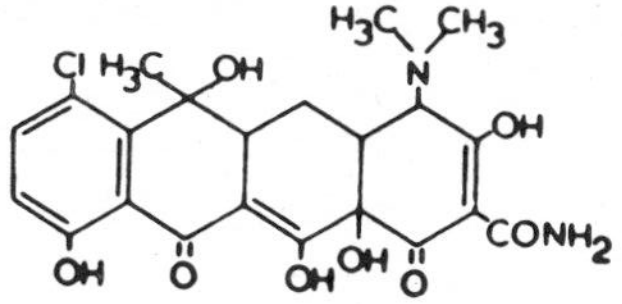 *Streptomyces aureofaciens*	1948, Duggar	Rickettsia, psittacosis virus, shigella, *E. coli* brucella, hemophilus, aerobacter, Klebsiella, pneumococci, etc.	"Broad-spectrum" drug
Oxytetracycline ("Terramycin")	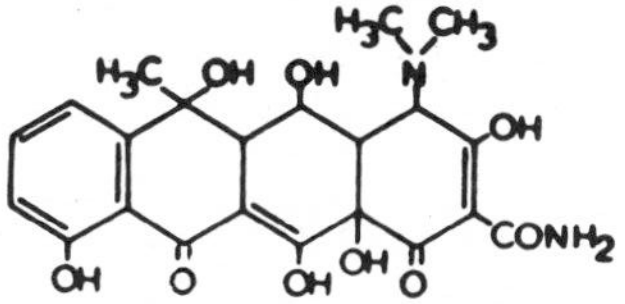 *Streptomyces rimosus*	1950, Finlay *et al.*		
Tetracycline ("Achromycin", "Tetracyn")	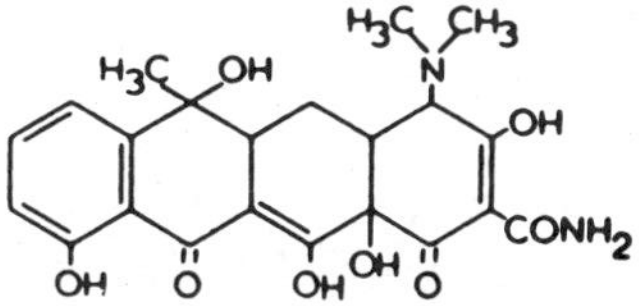 Dehalogenation of chlortetracycline	1953, Boothe, Conover *et al.*		
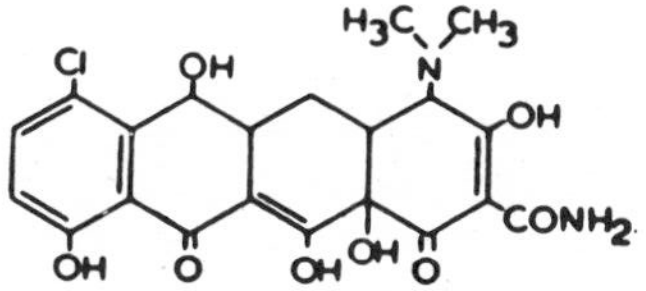 Demethylchlortetracycline	*Streptomyces aureofaciens*	1957, McCormick *et al.*		
Erythromycin ("Ilotycin," "Erythrocin")	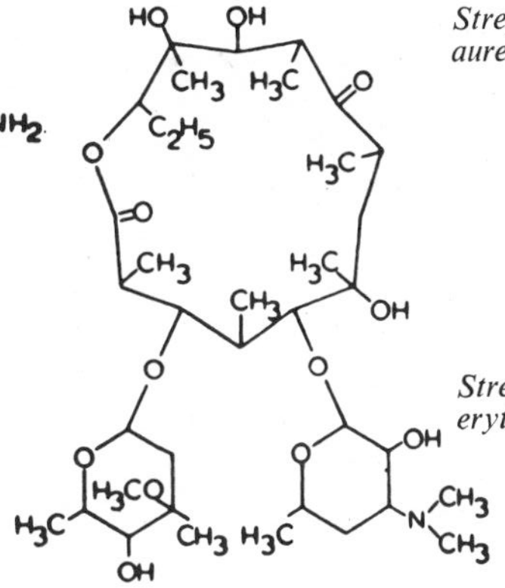 *Streptomyces erythreus*	1952, McGuire *et al.*	Streptococci, pneumococci, meningococci	Similar to penicillin in antibacterial spectrum; but penicillin preferred if possible; macrolide antibiotic; also as ester ("Ilosone")

TABLE 1—*continued*

Name	Structure	Source	Introduced	Useful Spectrum of Inhibition	Remarks
Bacitracin ("Baciguent")		*Bacillus subtilis*	1945, Johnson *et al.*	Wide spectrum, for use on skin; mainly against Gram-positive microorganisms	Serious systemic toxicity; usually used topically; closely related: tyrocidine, polymixin, colistin
Griseofulvin ("Grifulvin")		*Penicillium griseofulvum*	1939, Oxford *et al.*	Ringworm and related fungi	Given orally
Actinomycin D ("Dactinomycin")		*Streptomyces chrysomallus*	1952, Brockmann; Waksman and Pfennig	Wilms's tumor and other malignancies	Forms complex with DNA; also called actinomycin C_1

understanding of biochemical differences between these processes in bacterial and mammalian cells. Antibiotics believed to inhibit growth by interfering with bacterial protein synthesis include chloramphenicol, the tetracyclines, streptomycin, erythromycin and puromycin. Actinomycin D inhibits growth by preventing the normal functions of DNA in the synthesis of RNA. Nystatin, polymixin B and amphotericin B may inhibit susceptible microorganisms by altering permeability of the cell membrane.

Toxicity. Undesirable effects of antibiotics are well recognized. These responses may be due to interference with normal symbiotic relationships of microorganisms in the human intestine and may lead to overgrowth by certain microbial species (see INTESTINAL FLORA). Since normal gastrointestinal flora is essential for normal gastrointestinal function and for the synthesis of certain vitamins for the host, antibiotic-induced alterations of flora may lead to diarrhea, deranged absorption of foodstuffs and vitamin deficiencies. Because of their broad antibacterial spectrum, *tetracyclines* are particularly prone to cause this toxicity. As with all drugs, there is a chance of development of allergic reactions. This danger is particularly great with *penicillin* and may lead to death, in rare cases. Other toxic effects occasionally associated with specific antibiotics include aplastic anemia, after *chloramphenicol*; vertigo and deafness, after *streptomycin*; electrolyte changes and altered function of liver and kidney, following *tetracyclines*; enhanced tendency towards sunburn after *certain tetracyclines*. It is, therefore, extremely important that the agents be employed only when necessary and that the patient be watched for toxic drug effects. Local contact has a greater ability to sensitize the patient to the drugs than does their systemic use.

Drug Disposition. The action of any drug will depend on its delivery to the actual site in the body where it is needed, in a sufficient concentration and for at least a minimum period of time. Many factors affect this process, such as rate of absorption and excretion of the drug, its relative distribution in the tissues, and its inactivation by the microbial or host cells. Frequently, mechanical barriers may exist which prevent the drug from reaching its target tissue. These factors must be considered in formulating the desired dose required by the average individual for maximum therapeutic benefit with least toxicity. They also may explain differences frequently observed between *in vitro* microbial sensitivity tests to drugs and the effectiveness of the agents *in vivo*.

Resistance. Continuous exposure to an antibiotic may allow for the emergence of a cell population relatively insensitive to the drug. This selective survival leads to the development of resistance and may be a significant clinical problem. Resistance develops more readily to some antibiotics, such as streptomycin, than others; similarly, microorganisms such as *Staphylococcus aureus* are more likely to become insensitive to drugs than other strains. For example, in the early 1940's most such strains were sensitive to penicillin G, whereas now most strains are resistant. In general, shorter treatment with larger doses of an antibiotic will have less tendency to elicit resistance in a bacterial population than the prolonged use of smaller doses.

The most common mechanisms for the development of resistance to antibiotics include the enhanced ability by cells to produce an enzyme which catalyzes the drug's destruction, or an apparent decrease in permeability to prevent the drug's penetration into the cell and thereby aborting its inhibitory actions. For example, the enzyme penicillinase usually is elaborated by cells of *Staph. aureus* that have become resistant to penicillin. Some newer penicillin derivatives, synthesized from aminopenicillanic acid (*i.e.*, methicillin, oxacillin) are stable to penicillinase and thus are useful in the eradication of cells which have become resistant to penicillin G. Although cells resistant to other antibiotics frequently will accumulate less of the drug, compared to sensitive cells, it is not clear whether this differential effect determines sensitivity to the drug.

In a few instances it is possible to cultivate cells that have even become dependent on the presence of an antibiotic for optimum growth.

Uses and Scientific Contribution. The availability of antibiotic drugs has provided new and relatively specific tools for the control of large numbers of infectious diseases caused by microorganisms. Antibiotics have been most useful in the treatment of respiratory, gastrointestinal, genitourinary infections, as well as many others which previously defied medical treatment. Some antibiotics are inhibitory to certain types of malignant tumors, the psittacosis-lymphogranuloma group of viruses, and rickettsia. At the present time, none have been discovered which are effective against smaller viruses.

Certain antibiotics are used as supplements for stock feeds for poultry, swine and calves, where they increase the rate of growth and the efficiency of food utilization.

The agents have also served as tools for the delineation of biochemical pathways. By their selective inhibitory actions on cells they have revealed many of the intricate steps of biosynthesis and have provided information on the interrelationship of the synthesis and functions of DNA, RNA, proteins and cell wall that make up the process of growth.

References

1. *Historical*: a. WAKSMAN, S. A., "Microbial Antagonism and Antibiotic Substances," Commonwealth Fund, Cambridge, Mass., Harvard University Press, 1945. b. FLOREY H., "Antibiotics," Springfield, Ill., Charles C. Thomas, 1951.
2. *Medical*: MODELL, W., "General Considerations of Antibiotics and Antibiotic Therapy," Chapter 82, in "Pharmacology in Medicine" (DRILL, V. A., EDITOR), Second edition, New York, McGraw-Hill Book Co., 1958.
3. *Biochemical*: GALE, E. F., "Mechanisms of Antibiotic Action," *Pharmacol. Rev.* **15,** 481–530 (1963).

4. *General*: SCHNITZER, R. J., AND HAWKING, F., "Experimental Chemotherapy," Vols. II and III, New York, Academic Press, 1964.

H. GEORGE MANDEL

ANTIBODIES

For many years, an antibody could be defined as a serum protein produced in response to the injection of a foreign protein or other macromolecular substance and recognized by its capacity for specific union with the ANTIGEN that provoked its production. This is still a useful working approach. Virtually all experimental work on antibodies is initiated by immunizing an animal with an antigen which is often a highly purified protein such as bovine serum albumin (BSA). However, there is also much interest in "normal" antibodies which can be found in the serum of certain animals which, as far as can be judged, have never come into contact with the antigen the antibodies react with. For a number of experimental and theoretical reasons, any modern definition is concerned only with the nature of antibody as observed and without including, as the earlier one did, an implicit assumption about how the antibody was produced. An antibody is now considered to be a serum protein of one of a small number of physical types, the immunoglobulins, which reacts by selective union with a particular type of chemical structure, usually part of a macromolecule, which we call an antigenic determinant.

The Immunoglobulins. In mammalian serum, there are three types of protein which can function as antibody. In 1964, there was an international agreement to simplify older terms and call these immunoglobulins I.g.G., I.g.M and I.g.A. I.g.G. is the form taken by classical antibodies such as those produced when we immunize a rabbit against a simple protein like BSA or a virus like influenza. It has a molecular weight of about 160,000 and, in recent years, a great deal has been learned about its structure, much more than about the M and A types of immunoglobulin which can be passed over rather rapidly. The M type has a much larger molecular size, molecular weight about 1,000,000, and is characteristic of antibody against some polysaccharide antigens. The most interesting thing about I.g.M, however, is that when any *new* type of antibody is being produced, that found over the first few days is of the M type. Later on, the antibody, in most cases, is found almost wholly in the G form. The A type is more difficult to study because of its tendency to become attached to cells rather than to remain in the serum. It seems to be the basis of the antibodies that are responsible for allergies such as hay fever, and like G it has a molecular weight of about 160,000.

Modern work on the chemistry and genetic control of antibody has been almost confined to the G type, also called 7Sγ and γ2, and this "classical antibody" is the only form which will be discussed in the rest of this section.

For a large protein molecule of 160,000 molecular weight which is composed of about 1485 amino acid units, plus a small complex polysaccharide of 13 sugar equivalents, it is impossible to provide a detailed formula. The number and types of amino acids are known, so are the sugars in the polysaccharide but, in neither case, has the sequence been established. Study of the structure of the molecule has been largely confined to determining the number of polypeptide chains which can be clearly distinguished and showing how they are attached to one another and what part they play in giving "antibody" and "antigen' qualities to the molecule.

In general, the technical approach is to treat the purified antibody protein with either proteolytic enzymes, reagents (such as 2-mercaptoethanol) capable of breaking disulfide bonds, or detergents. After such treatment, the various fragments resulting are separated and characterized chemically and biologically. Such work is still actively going on, but in early 1964, there was an almost general acceptance of a model structure first clearly stated by R. R. Porter. According to this, γG antibody is composed of four separable polypeptide chains, two light (L) chains of about 25,000 molecular weight and two heavy (H) chains of 55,000 molecular weight. Each pair is made up of two identical chains, and each half of the symmetrical molecule is an L-H compound with the two elements held together by a single —S—S— bond. The two half molecules are likewise held together by a disulfide bond. In addition, there are weaker linkages between different parts of the various chains which hold the extremely complex intertwined mass of polypeptide chains into the form of a flattened cigar-shaped object approximately 250 × 60 × 20 Å. It is flexible and, at each end, has an "active patch" or "combining site" by which the antibody molecule unites with the antigenic determinant to which it corresponds (Fig. 1).

The active patch is quite small and seems to represent a configuration of 10–20 amino acid residues produced by close association of two

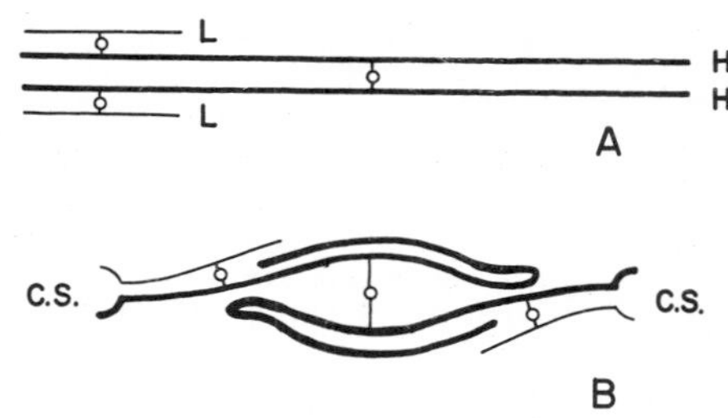

FIG. 1. A diagram to show the relationship of the 4 chains in an antibody molecule. (L and H are the light and heavy chains joined as shown by disulfide bonds. C. S. indicates combining site. A is the Porter diagram. B suggests in simplified form how a symmetrical molecule with two similar combining sites can be made on the same pattern.)

short lengths of L and H chains. There is no unanimity as to which chain is the more important, and there is even a minority opinion that the active patch may be essentially a third type of short chain attached specifically to appropriate points of the two L-H half molecules. Despite differences of opinion or detail, there is now virtual unanimity that the structure of antibody like that of every other protein is genetically determined and that the problem now is to find how the genetic information, probably from two or more genetic sources, flows together to give the highly specific antibody.

The Production of Antibody. If, again, we concentrate on the active production of 7S G type antibody, we can provide a fairly certain answer to the question, "What happens when an antigen like BSA is injected into a rabbit which, some months previously, has been injected with the same antigen and is therefore ready to produce rapidly a large amount of antibody?" The antigen BSA moves to the lymph nodes where it is in part taken up by phagocytic cells. After modification by the phagocytic cells or directly, the antigen meets lymphocytic (or reticulum) cells which, because they carry antibody-like receptor, can "recognize" and respond to the antigen. These cells respond by initially becoming more primitive in character and starting to divide very rapidly for 8–10 cell generations. As they divide, they gradually change character by forming, in their cytoplasm, the dense endoplasmic reticulum which will synthesize antibody. When this is fully formed, we have a plasma cell capable of synthesizing and liberating about 200 antibody molecules per second. The hundreds of cells descended from a single stimulated cell form a clone and, under most circumstances, all cells of the clone make the same antibody.

This is what we can call the standard situation, but most of the current interest of immunological research is in the vast number of questions which lie outside the standard situation. One of the most important, which cannot be discussed here, is simply "Why does an animal not produce antibody against its own constituents?"

There is, however, one difficult question which cannot be avoided, even in the briefest account of antibodies. If, as we have stated, antibody is determined by genetic information in the cells, how can the body produce the right sort of antibody appropriate for each of many thousand foreign substances that could be injected? The old answer was that genetic information produced standard globulin and that this globulin was molded into antibody against the actual antigen molecules. This is now excluded by, for instance, the evidence that there are amino acid differences in the constitution of physically similar antibodies reacting with different antigens.

The alternative is that the population of lymphocytic cells in the body must include an immense variety of patterns arising by processes still unknown, although somatic mutation is highly probable. In the human body, there are more than 10^{12} lymphocytes and, somewhere, appropriate cells for each type will be found. The function of the antigen is to *select* those of the right type to proliferate. This must hold for the first experience of a foreign antigen as well as for subsequent ones and, basically, the process is the same as we described for the standard production of antibody in an experienced rabbit. There are many different ways of elaborating such selection theories of antibody production, and we can probably define the most important current job of immunology as finding a way to describe an appropriate genetic mechanism which will give the flexibility and effectiveness of the immune reactions.

References

1. Burnet, F. M., "The Clonal Selection Theory of Acquired Immunity," Vanderbilt (and Cambridge) University Press, 1959.
2. Burnet, F. M., *Nature*, **203**, 451 (1964).
3. Edelman, G. M., and Gally, J. A., *Proc. Natl. Acad. Sci. U.S.* **51**, 846 (1964).
4. Fleischman, J. B., Porter, R. R., and Press, E. M. *Biochem. J.*, **88**, 220 (1963).
5. Nossal, G. J. V., *Sci. Am.*, **211**, No. 6, 106 (December 1964).

F. M. Burnet

ANTIGENS

Antigens are natural or synthetic macromolecules composed of materials found in living organisms. Non-biological macromolecules of similar size, such as vinyl polymers, are not antigenic. *Haptens* are small organic molecules capable of directing antibody specificity when coupled to a larger molecule (carrier), but not capable of eliciting antibody formation themselves. The term *antigenicity* encompasses the ability to elicit ANTIBODY formation and to react with antibody. *Immunogenicity* is a more restricted term that denotes just the ability to elicit antibody formation. In general, all molecules capable of eliciting an immune response can react with antibody.

An *adjuvant* is a substance that enhances the ability of an antigen to elicit antibody formation. It is usually either an emulsion of oil and mycobacterial waxes or material extracted from bacterial cell walls. Since the potency of a given antigen varies greatly among different species of animals and, to a lesser extent, among individuals of the same species, adjuvants are often necessary to elicit a good antibody response. In fact, a substance may be non-antigenic in one species and elicit a large antibody response in another species.

Haptens. The use of chemically well-defined, small, organic molecules as antigenic determinants (haptens) is well suited to the study of the intricate relationship between the stereochemistry of the antigen and the specificity of the antibody. The most intensively studied haptens are: *p*-azobenzenearsonate (Rp) and *p*-azobenzoate (Xp), which are negatively charged, *p*-azophenyltrimethylammonium (Ap), which is positively charged, and dinitrophenyl (DNP) and *p*-azophenyl-β-lactoside (Lac), which are uncharged.

The factors involved in the strength and specificity of the hapten-antibody interaction are electrostatic, hydrophobic, and dispersive (van der Waals) forces, hydrogen bonding, and conformation.

Studies with the complex hapten 5-azoisophthalyl-glycyl-leucine illustrate the heterogeneous nature of the antibody response: several populations of antibodies are formed, each complementary to a different portion of the large hapten. The specificity of the antibody for the hapten is not absolute, even for small haptens, and the closeness-of-fit of the antibody varies greatly with the hapten used. Cross-reactions among the *ortho*-, *meta*-, and *para*-derivatives of azobenzene-arsonate (R) show the latitude of antibody specificity. The amount of antibody precipitated with Ro-, Rm-, and Rp-ovalbumin is in the ratio of 1:1:0.5 with anti-Ro serum, 0.2:1:0.3 with anti-Rm serum, and 0:1:1 with anti-Rp serum; stronger anti-Rp sera show slight cross-reactivity with Ro-ovalbumin. The strength of the antibody-antigen interaction varies with the antigen used: the free energy of the hapten-antibody interaction ranges from −8 to −15 kcal/mole and that of the protein antigen-antibody interaction, from −5 to −8 kcal/mole.

Negatively charged haptens elicit antibodies with positively charged groups in the antibody combining sites, and positively charged haptens elicit antibodies with negatively charged groups in the antibody combining sites. The centers of charge on the antibody and the hapten are separated by approximately 8 Å, and the distance between the van der Waals contour of the antigen and antibody in the combining site is about 3 Å.

The optical configuration of the antigen is clearly reflected in the specificity of the antibody response. Antibody against the L-tartaric acid hapten will not cross-react with D-tartaric acid and will partially cross-react with *meso*-tartaric acid. This same pattern of reactivity is seen with D- and L-peptide haptens, the maleate and fumarate haptens, and the reaction between anthrax antiserum, which is directed against the poly-γ-D-glutamic acid of the anthrax cell wall, and synthetic γ-D-, γ-L-, and γ-*meso*-glutamic acid polymers.

Proteins. The polypeptide chains of the PROTEIN can contribute to the structure of its antigenic sites through four different structural modalities: the amino acid sequence (primary structure), the conformation of the polypeptide chain, *e.g.*, α-helix or β-conformation (secondary structure), the folding in space of the polypeptide chain (tertiary structure), and the association of subunits (quaternary structure).

The two fundamental approaches to the study of the antigenic structure of proteins are chemical modification of various functional groups and degradation of the native structure to varying extents. The object of chemical alteration or partial degradation is to assess the effect of modifying various groups on the ability of the modified protein to react with antibody to the native protein and to elicit antibody with a different specificity. The exact effects of the modification are difficult to assess in their totality, because they are not generally restricted to one group, but have profound repercussions throughout the entire molecular domain. Complete degradation of a protein antigen provides small peptides which can be used as inhibitors of the antibody-antigen reaction. The peptides that cause the greatest degree of inhibition probably contain part or all of an antigenic site, and hence by analyzing a series of inhibitors, one can deduce the structure of the antigenic sites on the protein. This approach is particularly useful for fibrous proteins, but with globular proteins it presents the problem of differences in spatial structure of the amino acid residues in the small peptides and in the native globular protein. The influence of the various structural features of proteins on their ability to elicit antibody formation is not well understood. In general, modification of the protein structure, including denaturation and degradation, qualitatively and quantitatively alters the ability of the protein to elicit antibody formation.

The role of primary structure in antigenicity is well illustrated by the studies on oxidized ribonuclease, which is essentially a polypeptide of known sequence. The peptide containing amino acids 39–61 is the most potent inhibitor and hence probably contains the major antigenic site; this peptide contains many nonpolar amino acids. Peptide 105–124, which also has many nonpolar amino acid residues, has a less potent antigenic site. The importance of primary structure can also be demonstrated by the alteration or loss of antigenic sites or of immunogencity by changes in amino acid sequence caused by genetic mutation, *e.g.*, in the enzymes trytophan synthetase from *E. coli* or lysozyme from bacteriophage.

The role of secondary structure in antigenicity is shown by the investigations on the highly helical molecules clam muscle paramyosin and bacteriophage lysozyme. Both are excellent antigens and have their antigenic sites on helical portions of the molecule; destruction of the helix destroys the antigenic sites. Therefore this modality of protein structure can be involved in eliciting antibody formation and in forming antigenic sites.

Tertiary structure is intimately involved in antigenicity as shown by studies on the limited enzymatic degradation of ribonuclease. The enzyme subtilisin splits ribonuclease into two fragments which subsequently can be separated by fractionation with trichloroacetic acid into the S-protein, which contains amino acids 21–124, and the S-peptide, which contains amino acids 1–20: these fragments can recombine to form enzymatically active ribonuclease. Antibody to native ribonuclease reacts completely with subtilisin-treated ribonuclease before separation into the two fragments and with ribonuclease reconstituted from the S-protein and the S-peptide, but only partially with the S-protein, and not at all with the S-peptide. Hence by altering the tertiary structure of the ribonuclease molecule, its antigenic sites were, to a large extent, destroyed.

Quaternary structure is also important in the formation of antigenic sites. Cross-reactions of the polymeric forms of ribonuclease with antibody to native ribonuclease indicate that twice as much of the dimer compared to the monomer is needed to precipitate a given amount of antibody and four times as much of the tetramer is needed to precipitate the same amount of antibody. Thus, in the formation of polymers some of the antigenic sites on the ribonuclease are covered and more ribonuclease is needed to react with a given amount of antibody. The enzyme tryptophan synthetase consists of two proteins subunits, which are easily dissociable, and the antigenic structure of the intact molecule depends upon the presence of both subunits. Separation of these subunits or genetic variations in their structure decreases the antigenic reactivity of the molecule with antibody to whole, wild-type tryptophan synthetase.

Two other studies which illustrate the importance and the interplay of various structural features are those with human serum ALBUMIN and sperm whale MYOGLOBIN. One of the antigenic sites on human serum albumin can be isolated in a peptide fragment of molecular weight of 7100 (60 amino acid residues) which contains 17% lysine and 15% glutamic acid. This finding indicates that an antigenic site from a globular protein can be isolated intact in a small peptide fragment. Considering the studies on ribonuclease and on human serum albumin, it is clear that both polar and nonpolar amino acid residues can participate in the formation of potent antigenic sites. The studies on myoglobin, which contains approximately 80% of its residues in the helical conformation, indicate that both linear and helical portions of the molecule are capable of forming antigenic sites. Some of the antigenic sites contain residues in both the linear and helical conformations while other sites have their residues mostly in the helical conformation. In addition, it appears that some of the sites lie within the molecule rather than on its surface. The structure and mode of action of these "internal determinants" are not as yet known.

Thus all the modalities of protein structure can enter into the composition of antigenic sites, and their interrelationships must be preserved in order to preserve the integrity of the antigenic sites.

Synthetic Polypeptides. The composition, the size, and, in some cases, the sequence of synthetic polypeptide antigens can be chemically controlled during their synthesis. By varying the composition and size of the molecule, its conformation (secondary structure) can be varied. In addition intramolecular cross-linking allows the construction of extended spatial (tertiary) structure (see POLYPEPTIDES, SYNTHETIC).

The fact that a completely synthetic macromolecule can be antigenic disposes of the notion that an antigen must be a "complex," biologically produced macromolecule. A polypeptide need not be rigid or contain any aromatic amino acids to be antigenic, although the latter considerably enhance antigenicity. Charge does not play a prominent role in antigenicity, with the exception that polypeptides containing greater than 80% net positive or negative charge are poor antigens and completely charged homopolymers are not antigenic. The uncharged homopolymer polyproline is very weakly antigenic and antibodies to forms I and II do not cross-react. An uncharged, desaminated multichain polypeptide is also antigenic. Linear-chain, branched-chain, and intramolecularly cross-linked synthetic polypeptides can be antigenic; hence, the shape does not have a crucial role in eliciting antibody formation. Polymers of molecular weight 4000 elicit antibody, and there is some evidence that even smaller peptides can be antigenic.

The antigenic site of a polypeptide must be on an accessible part of the molecule or else the polymer will not elicit an antibody response. Covering antigenic sites composed of tyrosine and glutamic acid in a branched-chain polypeptide with alanine residues destroys the ability of the polymer to elicit antibody formation. Removing the alanine residues restores the ability of the polymer to elicit antibody directed at the antigenic sites containing tyrosine and glutamic acid. There appears to be a hierarchy of antigenic potency among the amino acids. The antigenic sites of linear synthetic polypeptides reflect this fact and are composed of mosaics of such antigenically potent amino acids which do not necessarily reflect the same proportions of amino acids in the over-all composition of the polypeptide. Studies with intramolecularly cross-linked synthetic polypeptides show that antigenic sites involving spatially extended structures are more potent than those located on the linear portions of the polypeptide chain. The introduction of these spatially extended sites does not destroy the antigenic sites on the linear portion of the polypeptide chain—they merely supersede them.

Some, but not all, D-polypeptides can be antigenic, but they are very weak antigens. Antibodies to D- or L-polypeptides do not cross-react with the corresponding polymer of the opposite optical sense.

Carbohydrates. The most intensively studied CARBOHYDRATE antigens are the large, branched-chain pneumococcal polysaccharides, dextrans and BLOOD GROUP substances. Antibodies to the polysaccharide antigens are specific to the sugar residues involved and the type of linkage between them. Once the antibody against a carbohydrate has been characterized, it can be used to study the structure of unknown carbohydrates. One of the most extensively studied carbohydrates is dextran which is composed mainly of $\alpha(1 \rightarrow 6)$-linked glucose units. The dextran–anti-dextran reaction can be inhibited by oligosaccharides containing one to seven $\alpha(1 \rightarrow 6)$-linked glucose residues. Isomaltohexaose is generally the best inhibitor, and from this fact the maximal size of the antigenic site on dextran was inferred to be six glucose units.

Nucleic Acids. Purified nucleic acids and synthetic POLYNUCLEOTIDES are not antigenic. Antibody

TABLE 1. APPROXIMATE SIZE OF ANTIGENIC SITES AND IMMUNOLOGICAL CAPABILITIES OF VARIOUS ANTIGENS

Antigen	*Composition of the Antigenic Site*	*Approximate Size* (Å)	*Immunological Capability Complete Antigen*	*Hapten*
Hapten	1 small organic molecule	35–70	0	+
Protein	12–60 amino acid residues	40–100	+	+
Synthetic Polypeptide	6–10 amino acid residues	20–35	+	+
Carbohydrate	6–10 sugar residues	30–50	+	+
Nucleic acid	4–6 purines or pyrimidines	30–45	?[a]	+
Lipid	Not known		?	+
Cell	Largely carbohydrate or lipid + some protein	10^6 Å^2	+	

[a] Can elicit antibody when sonicated bacteriophage, DNA-methylated BSA., or ribosomes are used as the antigens

to DNA can be elicited by immunization with DNA from sonicated bacteriophage or by complexing DNA or small oligonucleotides with methylated bovine serum albumin. In addition, antibodies to a variety of thermally denatured DNA's are found in the sera of patients with systemic lupus erythematosus. Antibodies capable of reacting with DNA have been elicited by immunization with a purinoyl- or 5-acetyluracyl-protein conjugate, thus establishing that purines and pyrimidines can act as determinants when linked to protein carriers. Antisera to guanine, cytosine, uridine, thymidine, and adenine can be prepared by immunizing rabbits with the appropriate base coupled to bovine serum albumin or multichain poly-DL-alanine. The antibodies will react with thermally denatured DNA but not, generally, with RNA. The one exception is the small degree of cross-reaction between antisera to uridine-5′-carboxylic acid conjugated to multichain poly-DL-alanine and thermally denatured *E. coli* RNA and polyuridylic acid. Antibody to RNA can be produced by immunizing rabbits with RIBOSOMES.

Lipids and Steroids. It has not yet been clearly shown that lipids by themselves are antigenic. There is some evidence, however, that they can act as haptens. Steroids can act as haptens when coupled to bovine serum albumin.

Cellular Antigens. A wide variety of cells including bacterial cells, tumor cells, normal tissue cells (including histocompatibility antigens), and various blood cells are antigenic. The antigenic materials are located on the cell surface. They are, at present, not chemically well defined, but generally seem to be composed of a relatively large amount of polysaccharide or lipid and some protein. The structure of the antigens in the cells is under genetic control and consequently can be used to develop a system of taxonomy for microorganisms and a system of genetics for tumors (see CHEMOTAXONOMY).

Size of Antigenic Site. The size and composition of the antigenic sites of various types of macromolecules and cells can be estimated from data in the literature. These estimates are given in Table 1 along with a summary of the immunological capabilities of the various antigens.

References

1. LANDSTEINER, K., "Specificity of Serological Reactions," Cambridge, Harvard University Press, 1945.
2. KARUSH, E., "Immunologic Specificity and Molecular Structure," *Advan. Immunol.* **2,** 1 (1962).
3. PORTER, R. R., AND PRESS, E. M., "Immunochemistry," *Ann. Rev. Biochem.*, **31,** 625 (1962).
4. EISEN, H. N., AND PEARCE, J. H., "The Nature of Antibodies and Antigens," *Ann. Rev. Microbiol.*, **16,** 101 (1962).
5. "Antibodies to Enzymes," *Ann. N. Y. Acad. Sci.*, **103,** Art. 2 (1963).
6. SELA, M., "Immunological Studies with Synthetic Polypeptides," *Advan. Immunol.*, **5,** (1965).
7. "Symposium on Immunochemistry," *Federation Proc.*, **21,** 692 (1962).
8. HELLSTRÖM, K. E., AND MÖLLER, G., "Immunological and Immunogenetic Aspects of Tumor Transplantation," *Progr. Allergy*, **9,** 158 (1965).

THOMAS J. GILL III

ANTIMETABOLITES

Compounds structurally related to a particular biologically active compound have been observed to antagonize the biological action of the latter. Such antagonists have also been termed antimetabolites, or metabolic antagonists. Enzymology probably gave the first example of such action, in that the products of certain enzyme systems were reported to inhibit their formation to an extent greater than that attributable to mass action effects. In these enzyme reactions, the products were structurally similar to the starting material (substrate). Many such antagonistic effects were demonstrated in *pharmacology*, such as the effects of numerous structural analogues

OH

—OH

CHOH—CH$_2$—NH—CH$_3$

Epinephrine

$$\text{C}_6\text{H}_5\text{—CHOH—CH(NH}_2\text{)—CH}_3$$

Propadrine

of epinephrine (*e.g.*, propadrine) in antagonizing the action of epinephrine on smooth muscle. In *nutrition*, certain amino acids were found to inhibit growth, and a structurally related amino acid would prevent (reverse) this growth inhibition. Following the earlier postulation of many similar concepts in *chemotherapy*, the discovery of the biological antagonism between the SULFONAMIDE drugs (*e.g.*, sulfanilamide) and *p*-aminobenzoic acid was a major advance in the establishment of fundamental concepts of structural antagonisms and a rational approach to research in chemotherapy. The basic concepts of structural antagonism that have evolved from several areas of scientific research are the following.

(1) Each of the enzymes (biological catalysts for conversion of nutrients to other, often essential, products of cell metabolism) specifically catalyzes one type of reaction, with a high degree of specificity for the conversion of one particular substrate to a product. The enzyme, a large protein molecule, forms a complex with the substrate which is attached at a specific site on the protein molecule, and the complex then reacts further under normal conditions to form the product. The specificity of the enzyme for a particular substrate is believed to result from the existence on the enzyme of a structural pattern of intramolecular groups, into which the substrate molecule fits and is linked by a specific pattern of electrostatic and other interatomic forces.

(2) Compounds structurally related to the substrate can similarly interact with the enzyme, but the resulting complex either cannot undergo the normal reaction or can yield only a modified product, different in structure from the normal product. If such compounds structurally related to the substrate cannot perform the function of the normal substrate in a biological system, such structural analogues are called *inhibitors* (or inhibitory analogues, or antagonists of the substrate).

(3) The inhibitor may combine with the enzyme by reacting at the same site on the molecule as the normal substrate, competing with the latter and preventing its attachment at this site (competitive inhibition), or the mechanism of inhibition may involve attachment of the inhibitor to the enzyme at some other site on the molecule independent of the normal substrate attachment in such a manner that the inhibitor is still capable of forming a complex with the enzyme-substrate complex as well as the free enzyme (*non-competitive* inhibition).

(4) Coenzymes [cofactors of enzymes, which must combine with the enzyme protein molecule (apoenzyme) to form the active complete catalyst (or holoenzyme)] can be antagonized in a similar manner by structural analogues of the coenzymes, but in many such cases the equilibria of analogue-enzyme combination are not rapidly attained.

The fundamental concepts of structural antagonisms have been applied to a wide variety of biologically active compounds, such as vitamins, amino acids (the components of proteins), purines and pyrimidines (components of nucleic acids), hormones, etc (see CYTOTOXIC CHEMICALS).

The liberation of histamine is thought to be an important factor in anaphylactic shock and in many allergic conditions. While some substances counteract the effects of histamine by other mechanisms, certain structural analogues appear to be direct and specific histamine antagonists. Many of the antihistamines structurally resemble histamine.

Vitamin K or its probable precursor, 2-methyl-1,4-naphthoquinone, appears to function in the formation of prothrombin which affects the rate of blood coagulation, and appears to reverse the action of "Dicumarol," the hemorrhagic substance in spoiled sweet clover hay.

Most structural analogues of a biologically active compound are inert, many are antagonists, but some actually replace the natural active compound in performing the same biological function. For example, oxybiotin (*o*-heterobiotin) replaces the vitamin biotin for many organisms and is utilized without conversion to biotin. The analogue apparently is converted to an analogue of the coenzyme form of biotin, and this coenzyme analogue actually performs the normal function of biotin.

In other cases, an analogue replaces its corresponding substrate or growth factor for some organisms but inhibits growth of others. *p*-Aminosalicylic acid inhibits growth of *Mycobacterium tuberculosis* H37Rv, and *p*-aminobenzoic acid reverses the toxicity; however, for a certain mutant strain of *Escherichia coli*, *p*-aminosalicylic acid promotes growth in lieu of *p*-aminobenzoic acid. Cases are known in which an analogue replaces some of the biological functions of a vitamin but inhibits other functions.

Structural antagonisms, particularly in cases of competitive inhibition, offer a basis for the study of biochemistry in living organisms. Effects exerted upon systems which are inhibited by antagonists of a particular cell metabolite or nutrient include not only those exerted by the metabolite or nutrient itself, but also those exerted by the limiting precursors of the metabolite, by the products of the inhibited enzyme reaction, by substances which increase the effective enzyme concentration, and by compounds which influence the rates of destruction of either the metabolite or the inhibitory analogue. Testing techniques which elucidate the mechanisms of effects exerted by such secondary reversing agents of the inhibited biological system have been termed *inhibition analysis*, and offer a new method for the study of biochemistry.

References

1. "Metabolic Inhibitors," (HOCHSTER, R. M., AND QUASTEL, J. H., EDITORS), Vols. I and II, New York, Academic Press, 1963.

2. SHIVE, W., AND SKINNER, C. G., "Metabolic Antagonists," *Ann. Rev. Biochem.*, **27**, 643 (1958).
3. SHIVE, W., "Metabolic Antagonists," in "Proteins and Their Reactions" (SCHULTZ, H. W., AND ANGLEMIER, A. F., EDITORS), Westport, Conn., Avi Publishing Company, 1964.

WILLIAM SHIVE

ANTIOXIDANTS

Antioxidants are usually organic compounds of a wide variety of structures, which in catalytic quantities retard or inhibit the autoxidation of edible oils and fats, vitamins, soaps, gasolines, polymers, resins, and other such compounds. They act chiefly by breaking free radical chains, quenching electron mobility, sequestering trace metals and scavenging for free radicals. They may be classified as those which possess a direct antioxidant action, such as certain phenols, aromatic amines and the tocopherols, and those which by themselves have little effect on stability to autoxidation but act synergistically with direct antioxidants. Much interest has centered on the autoxidation *in vitro* of edible fats and oils, which leads to rancidity and deterioration and which is controlled by the extensive use of synthetic antioxidants.

The stability *in vivo* of many fats and oils of plant origin which are readily oxidized *in vitro* has long been known to indicate the existence of naturally occurring antioxidants in biological systems. Such antioxidants may also be direct or synergistic in their effect. Synthetic antioxidants which may be potent *in vitro* stabilizers frequently show little or no *in vivo* activity, or show activity which may be ascribed to an *in vivo* sparing effect on natural antioxidants.

Antioxidant effects *in vivo* are expressed mainly at the biological level, such as by the inhibition of peroxidation of highly unsaturated body fats, the prevention of a yellow-brown pigmentation in body fat and in the protection of VITAMIN A and carotene stored in the liver. Probably the most significant test of an *in vivo* antioxidant effect is in the alleviation or inhibition of VITAMIN E deficiency symptoms in animals. The principal natural antioxidants appear to be the *tocopherols*, which are of widespread natural occurrence in plants. High concentrations of tocopherols are found in the so-called inhibitol concentrates from wheat germ and cottonseed oils. Their role in stabilizing unsaturated lipids is shown by an enhanced Vitamin E requirement by both rats and humans when increased amounts of unsaturated fats are included in the diet (see FATTY ACIDS, ESSENTIAL; UNSATURATED FATTY ACIDS). Exudative diathesis and encephalomalacia, the well-known Vitamin E deficiency diseases in chicks, are believed to result from an abnormal oxidation of unsaturated fatty acids in the tissues and may be prevented by inclusion of tocopherols in the diet. Tocopherols inhibit lipid peroxidation in the depot fat of Vitamin E deficient chicks and in that of rats receiving diets high in cod liver oil, and also show an *in vivo* sparing effect on Vitamin A, carotene, and essential polyunsaturated acids.

Various tocopherols, designated as α, β, γ, δ, etc., from their chemical structure, have been isolated from plant sources, and their antioxygenic potency is assayed by various procedures. In general α-tocopherol has the greatest *in vitro* and *in vivo* potency. Other naturally occurring substances which possess antioxidant activity are lecithin, CEPHALIN, sesamol, gossypol, gallic acid, nordehydroguaretic acid, ASCORBIC, citric and oxalic acids and several FLAVONOIDS, although the *in vivo* effect may frequently be expressed as synergism with tocopherols.

The tocopherols are the principal *in vivo* antioxidants present in animal body fat. They must be entirely derived from the diet, although they are capable of storage for long periods with infrequent replenishment. Human daily dietary requirements have been placed at 7.6 mg (5.74 mg of the α-form). Although no selective distribution in body tissues is found, storage capacity is greatest in body fat, followed by the liver, with widespread distribution in other organs and tissues. Within the cell, the highest concentration is found in the MITOCHONDRIA, with a lesser quantity in the microsomal fraction and much lesser amounts in the nucleus.

In addition to naturally occurring antioxidants, some synthetic compounds exhibit antioxidant effects *in vivo*. Methylene blue prevents depot fat discoloration in rats, protects Vitamin A in the liver of chicks and rats, and inhibits depot fat peroxidation in chicks, while diphenyl-*p*-phenylenediamine prevents exudative diathesis and encephalomalacia in chicks. The effects of both these substances have been ascribed to the sparing of traces of tocopherols. Ethoxyquin may exhibit a direct *in vivo* antioxidant effect in the prevention of peroxide formation in the liver of Vitamin E deficient chicks. Experimental necrotic liver degeneration, a Vitamin E deficiency disease in rats, may be inhibited by ethoxyquin, diphenyl-*p*-phenylenediamine and di-*tert*-amylhydroquinone, and the associated respiratory decline may also be reversed. These are also considered to be *in vivo* antioxidant effects. Both experimental necrotic liver degeneration in rats and exudative diathesis in chicks can be reversed by the so-called Factor 3 selenium, which may originate as dietary selenium. Selenomethionine decomposes lipid peroxides and inhibits *in vivo* lipid peroxidation in tissues of Vitamin E deficient chicks; selenocystine catalyzes the decomposition of organic hydroperoxides; and selenoproteins show a high degree of inhibition of lipid peroxidation in livers of sheep, chickens and rats. Some forms of selenium thus exhibit *in vivo* antioxidant behavior.

An understanding of the mechanism of *in vivo* antioxidant action depends on a knowledge of the mechanism of *in vivo* autoxidation. This is at present believed to proceed in an analogous fashion to *in vitro* autoxidations or to be catalyzed by the widespread naturally occurring lipoxidase system. Conjugated peroxides and hydroperoxides are presumably formed which are converted to degradation products and polymers. In systems analogous to *in vitro* processes, antioxidants could act as inhibitors by the mechanisms outlined at the beginning of this article. The lipoxidase system may produce *in vivo* autoxidation by initiating free radical chain reactions, the enzyme catalyzing the initial abstraction of an electron. Alternatively an enzyme-unsaturated lipid-oxygen complex may be formed in which the enzyme accepts an electron, hydroperoxide, peroxide or polymer formation occurring on its surface by subsequent reaction of the lipid free radical. Antioxidants would then act to decrease lipoxidase capacity by donating electrons or hydrogen radicals to the enzyme, by protonating peroxide free lipid free radicals on the enzyme surface or by inhibiting free radical formation. The reversible formation of semiquinones by tocopherols and phenolic and aromatic amine antioxidants by facile electron donation is of considerable significance in their role as antioxidants.

The precise physiological function of the tocopherols and Factor 3 selenium is still a matter for speculation. They have been regarded as cofactors at specific sites of intermediary metabolism, and other workers have ascribed an essentially antioxidant function to them. Little is known of the action of biological antioxidants at the cellular level, their effects so far being expressed mostly in biological terms. They may be involved in maintaining the stability of biological membranes which contain unsaturated lipids, such as those of mitochondria, MICROSOMES and lysosomes. The protection of other nutritional factors, of proteins against damage by lipid peroxides, and of any enzyme system involving autoxidizable LIPOPROTEINS and PHOSPHOLIPIDS may also be important functions for *in vivo* antioxidants.

References

1. GREENBANK, G. R., in "Encyclopedia of Chemistry" (CLARK, G. L., AND HAWLEY, G. G. EDITORS), p. 96, New York, Reinhold Publishing Corp., 1957.
2. AAES-JØRGENSEN, E., "Auto-oxidation and Antioxidants *in Vivo*," in "Autoxidation and Antioxidants" (LUNDBERG, W. O., EDITOR), Vol. II, p. 1045, 1962.
3. "Biological Antioxidants," Transactions of Conferences of Josiah Macy Foundation, New York, 1946–1950.
4. "Interrelationships among Vitamin E, Coenzyme Q, and Selenium" (OLSON, R. E., Symposium Chairman), *Federation Proc.*, **24**, 55–92 (1965).

BERNARD D. ASTILL

ANTISEPTICS

Antiseptics are substances which, when applied to living tissues, kill or prevent the growth of microorganisms [see STERILIZATION in this volume; see also the article "Antiseptics" by G. H. Reddish, in "The Encyclopedia of Chemistry" (Clark, G. L., and Hawley, G. G., Editors), New York, Reinhold Publishing Corp., 1957].

ANTITOXINS

The history of the development of antitoxins in combating bacterial infection dates back to the early beginning of organized bacteriology. Behring was the first to show that animals that were immune to diphtheria contained, in their serum, factors which were capable of neutralizing the poisonous effect of the toxins derived from the diphtheria bacillus. While this work was carried out in 1890, prior to many of the great discoveries of mass immunization, and more recently the antibiotics, it is interesting to note that there is still an important place in medical treatment or prophylaxis for antitoxins for diseases such as tetanus and botulism.

The more important approach to immunity to infectious disease today is the development of *active immunity* by the injection of a vaccine. The vaccine may be either an attenuated live infectious agent or an inactivated or killed product. In either case, protective substances are generated in the bloodstream called ANTIBODIES which help to neutralize the infectious agent when it is introduced. The principle of *passive immunization*, on the other hand, involves the development of the antibodies in another host and most frequently a different species as well. The antiserum or antitoxin (from the other host) is employed in preventing the onset of the disease or in actual treatment of the active infection in subjects who have not had the advantage of becoming actively immunized due either to neglect or to the fact that an effective vaccine was not yet available. The use of antitoxins prepared in another species is not without some element of risk as will be shown later in this review.

Preparation of Antitoxins. The basic principle employed in preparing potent antitoxins is to inject the donor animals with frequent and increasing doses of toxin while maintaining a level at each injection that the animal can tolerate. The initial doses are critical since these toxins may be among the most poisonous agents known to man. In one of the current techniques in use, the toxin is diluted so that the first injection contains less than the minimum lethal dose. Other programs employ toxins that are inactivated with formaldehyde so that they are no longer poisonous but may still elicit an immune response and result in antitoxin that will neutralize the unaltered toxin. This method of inactivation was also developed prior to the twentieth century but is still the basic procedure in use today for preparing many of our important vaccines against diseases such as influenza, tetanus, and diphtheria.

The production of antitoxin in small laboratory animals is, of course, of great utility in carrying out research in this field, but in commercial production, larger animals are required. Perhaps the first species to be used for this purpose was the sheep, but it soon became evident that the horse was a very satisfactory animal for large-scale production of antitoxins. Horses are still used today for the major portion of antitoxin production in the United States, but a small amount of material is also derived from cattle. While the technique of producing antitoxins of high potency in horses has been actively studied and improved since 1895, the basic factors involved in obtaining the highest potencies and amounts of antitoxin from these animals have not been entirely elucidated even today. The management of the animals and the schedules and dosages are perhaps as much an art as an exact science. The selcetion of the horses that have the best potential as antitoxin producers is also a critical factor in producing these products. When a horse is receiving the maximum level of toxin in the hyperimmunization program he may be injected in a single dose with enough toxin to be fatal for 100 million mice if the material was suitably distributed.

One means of enhancing the potency of the antitoxin in horses is to employ an agent, usually mixed with the toxin, which is called an *adjuvant*. Many adjuvants have been used such as tapioca, mineral oil, and aluminum hydroxide. The mechanism by which these agents increase the intensity of the immune response is not entirely understood, but local inflammatory reaction and the resulting slower release of the injected material from the original site play a part in this phenomenon. Adjuvants are also frequently employed with vaccines for human use. A well-managed antitoxin production program allows horses to remain as active antitoxin producers for a considerable period. In one company, a horse remained an active producer for 11 years prior to his retirement. During his productive period, this animal gave 657 gallons of blood from which tetanus antitoxin and, at a different time, pneumococcus antiserum, was prepared.

One way in which the volume of serum removed from the horse can be increased, is by the return of the red cells after the removal of the plasma or liquid component. It has been shown that if the red cell level can be maintained in human donors of special serums, they can safely give as much as 1 liter of serum per week. If this can be applied to horses in the antitoxin program, it is possible that the antitoxin yield could be increased several fold over the usual amount collected by present methods.

The antitoxin component in horse serum is only a small fraction of the serum solids. The activity resides in a class of serum proteins called gamma globulins which make up about 3% of the total solids in serum. It is known that the concentration of this globulin component is considerably higher in an antitoxin producer than in a normal horse. Another area that has been investigated is the development of a new serum protein component in antitoxin horses. This so-called T component or gamma-1-globulin is detected by an instrumental technique called ELECTROPHORESIS. This technique measures the migration of the serum proteins through a liquid in which electrical current is passed. The purification of antitoxin by the removal of the non-gamma globulin fractions had been practiced even prior to 1900 when the precipitation of the antitoxic fraction was accomplished by the addition of a high concentration of one type of salt. Other procedures for separating the gamma globulin fraction have been developed such as treatment with alcohol in the cold, but the most practical methods used on a large scale today for horse origin antitoxin still employ the precipitation of the desirable fraction by the addition of a high concentration of salt.

Sensitivity to Foreign Serum Protein. The injection of serum components of another species into human patients has not been without problems. A condition known as serum sickness develops in an alarmingly high proportion of those treated in this way. Serum sickness is apparently due to a generalized sensitivity which develops to the foreign serum protein. The onset of illness is usually delayed for several days after the injection of antitoxin and the symptoms may be quite severe. The death rate due to this condition has been estimated as high as 1 in 50,000. In 1936, Parfentjev and somewhat later, Pope, developed methods for treating the antitoxin with the protein enzyme called pepsin. The objective here was to break down the protein molecule to the extent that serum sickness reactions would be reduced but to stop the digestion at a point where the modified antitoxin would still be effective. This procedure is in use today in many parts of the world and has unquestionably resulted in a significantly lower frequency of the serum sickness syndrome in treated patients. On the other hand, serum sickness may still occur and may be just as severe in the subjects where it develops.

A relatively recent development on a reasonably large scale has been the availability of antitoxins of human origin. Human donors or volunteers are injected with a course of vaccine to the extent that suitable levels of antitoxins are developed. This type of product can be used in human prophylaxis or therapy without the hazard of serum sickness. It has also been demonstrated in laboratory animals that antitoxin produced in the same species is effective in lower dosage and for a longer period of time because the serum protein from a different species is rejected within a short time since the recipient recognizes it as a foreign substance. The greater availability of antitoxin of human origin is due to the fact, as mentioned earlier, that techniques have been perfected for returning the blood cells to the donors and in this way it is not unusual to obtain one liter (about 1 quart) of serum per volunteer per week.

In summation, it is pointed out that in spite of the high effectiveness of the new chemotherapeutic agents and the vaccines which produce active immunity, there is still an important place in medicine for antitoxins against tetanus and botulism, and antiserums against other bacterial and virus diseases of man. It is not likely that the need for these antitoxins and antiserums will disappear in the near future.

References

1. WEINBERG, M., AND KREGUER, A., "Preparation of a Bivalent Anti-botulinum Serum by the Injection

in the Horse of Toxins in Lanolin," *Compt. Rend, Soc. Biol.*, **128,** 949 (1938).
2. Van Der Sheer, J., *et al.*, "The Electrophoretic Analysis of Several Hyperimmune Horse Sera," *J. Immunol.* **39,** 65 (1940).
3. Parfentjev, I. A., "Despeciated Protein Viewed as a Potential Plasma Substitute," *Ann. N. Y. Acad. Sci.*, **55,** 513 (1952).
4. Skudder, P. A., *et al.*, "The Incidence of Reactions Following the Administration of Tetanus Antitoxins," *The Journal of Trauma*, **1,** 41 (1961).
5. Reisman, R., *et al.*, "Serum Sickness," *J. Allergy*, **32,** 531 (1961).
6. Smolens, J., *et al.*, "The Persistence in the Human Circulation of Horse and Human Tetanus Antitoxins," *J. Pediat.* **59,** 899 (1961).
7. Deutsch, H. F., and Nichol, J. C., "Biophysical Studies of Blood Plasma Proteins," *J. Biol. Chem.*, **176,** 797 (1948).

M. S. Cooper

APOENZYMES

Many enzymes may be dissociated into a protein (polypeptide) component, called an *apoenzyme*, plus a nonprotein organic portion often called a *prosthetic* group, which may, for example, be a metal-porphyrin complex, or a coenzyme (containing a vitamin as part of its structure). The combination of apoenzyme with prosthetic portion forms the complete enzyme or *holoenzyme*.

See also COENZYMES.

APPETITES

This subject is not often dealt with in books on biochemistry. It is cited here because it requires investigation from the biochemical standpoint. It has been long known that THIAMIN deficiency, for example, causes loss of appetite in animals and man, but little is known of the mechanism and the extent to which thiamin is unique in this regard. It is also known that hypothalamic lesions may cause excessive eating. If human appetite for food was well understood and could be held in proper control, weight problems and many associated health problems could be banished.

Special appetites for salt, for sugar ("sweet tooth"), for fat, (Jack Spratt's wife), for calcium, for phosphate, and for specific vitamins, probably exist, and the appetite for alcohol which alcoholics (see ALCOHOLISM) possess may be thought of in the same connection.

The appetite for salt is reasonably well defined and its biochemical basis established. Damage to the adrenal cortex (which produce ADRENAL CORTICAL HORMONES known to be involved in salt retention and balance), whether surgically or by pantothenic acid deficiency, causes in rats large increases in their appetite for salt, which disappear if the impairment is restored, *e.g.*, by furnishing the deficient animal with the missing vitamin.

That appetite for sugar has a physiological basis (and may be pathological) is indicated by the fact that diabetics frequently have a strong craving for it. That appetite for alcohol likewise has a biochemical basis (in experimental animals at least) is shown by the fact that it can often be abolished by making the diet of the animals fully adequate (see GENETOTROPHIC PRINCIPLE).

Appetites for food are often strongly influenced by psychological suggestion. Many individuals might, for example, enjoy a stew of kitten or puppy meat if they were not aware of the source of the delicacy or had not been culturally conditioned against it. The importance of psychological factors, does not diminish the need for biochemical investigation. In human beings, biochemistry and psychology are closely intertwined.

Reference

1. Mayer, J., "Appetite and Obesity," *Sci. Am.*, **195,** 108–116 (Nov. 1956).

Roger J. Williams

ARABINOSE

Arabinose is one of the aldopentoses:

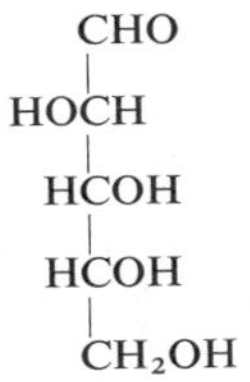

CHO
|
HOCH
|
HCOH
|
HCOH
|
CH_2OH

[see CARBOHYDRATES (CLASSIFICATION AND STRUCTURAL INTERRELATIONS)].

Arabinose got its name from its occurrence as a constituent of a complex plant carbohydrate, gum arabic (gum acacia). Like other aldoses it occurs principally in hemiacetal forms through interaction of the aldehyde group with a γ- or δ-OH group.

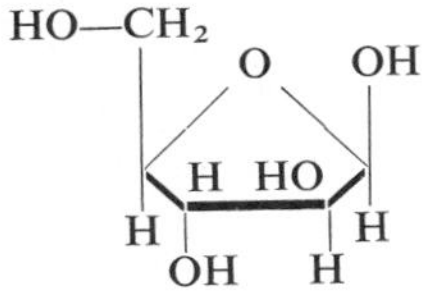

$HO-CH_2$ O   OH
H   HO
H   H
OH   H

β-D-Arabinofuranose (an internal hemiacetal form of arabinose)

ARGININE

L-Arginine, $H_2N-\overset{\overset{\displaystyle NH}{\|}}{C}-NHCH_2CH_2CH_2CH(NH_2)COOH$, is one of the three basic AMINO ACIDS derived from common PROTEINS. Some protamines on hydrolysis yield up to about 90% of their weight of this one amino acid. HISTONES also yield large amounts of arginine, and, like protamines, owe their basic properties largely to this content.

Arginine is regarded as nonessential nutritionally for adult mammals, but it is not produced rapidly enough by growing rats to permit maximum growth rate. Because of this, rat growth is favored if arginine is included in the diet. Its roles in metabolism are relatively well established (see also AMINO ACID METABOLISM; ARGININE-UREA CYCLE).

ARGININE-UREA CYCLE (ORNITHINE CYCLE)

In adult animals including man the characteristic tissue-specific levels of different enzymes are maintained by a dynamic balance between the independently controlled rates of biosynthesis and degradation of each enzyme. A dynamic rather than a static system was selected during evolution because it enables organisms to adapt to widely different nutritional conditions and other environmental changes. Depending upon the physiological state of the animal at a given moment, amino acids derived from hydrolysis of exogenous or endogenous protein may be predominantly *utilized for synthesis* of tissue-specific proteins, or their *carbon chains may be metabolized* further to provide energy (ATP) or intermediates for synthesis of other cellular constituents. When the carbon chains of amino acids are utilized to provide energy, some provision must be made for disposal of the reduced nitrogen components. Animal tissues in general cannot tolerate accumulation of ammonia. Aquatic animals, which are surrounded by a convenient diluent, can simply excrete ammonia as rapidly as it is formed. Land animals have had to devise other solutions to this problem; they convert amino acid nitrogen and ammonia into nitrogen-rich, nontoxic compounds such as *urea* and *uric acid*, which are then excreted at intervals. Synthesis of urea, the primary nitrogenous excretory product of mammals, is efficiently accomplished in the liver by combining a portion of the already established pathway of arginine biosynthesis with the hydrolytic degradative enzyme, arginase [Fig. 1, enzyme (4)]. This simple strategem transforms two linear enzymic sequences, a multi-step, reversible biosynthetic pathway and a single-step, irreversible degradative reaction, into a single, unidirectional cyclic pathway which provides the necessary coupling mechanism for converting large quantities of potentially harmful reduced nitrogen into innocuous urea.

In mammals ornithine carbamoyltransferase [Fig. 1, enzyme (1)] occurs only in the liver; hence the complete urea cycle occurs only in this organ (see CARBAMYL PHOSPHATE). The other enzymes are rather widely distributed in the various tissues. It is of interest that synthesis of citrulline from ornithine is also the step missing in birds and a number of lower vertebrates, with the consequence that these organisms cannot synthesize arginine and thus require it in their diet. In man, a hereditary deficiency of argininosuccinase [Fig. 1, enzyme (3)] leads to increased levels of blood argininosuccinate and urinary excretion of several grams of this compound each day. As is the case in a number of other inborn errors of amino acid metabolism (see AROMATIC AMINO ACIDS, section on phenylalanine), argininosuccinic aciduria is accompanied by mental deficiencies in individuals homozygous for this trait. Arginase [Fig. 1, enzyme (4)] catalyzes the primary pathway of degradation of excess arginine, forming urea and ornithine. The latter compound is a precursor of the biologically important amines, putrescine and spermidine (see POLYAMINES). During catabolism of arginine, the carbon chain of ornithine is converted to α-ketoglutarate, where it enters the CITRIC ACID CYCLE for further metabolism. In organisms such as higher plants which can synthesize arginine, α-ketoglutarate is a precursor of the ornithine moiety of arginine.

Experimentally it has been observed that above a certain basal level, the quantity of urea excreted is proportional to the amount of ingested protein. Over sixty years ago Folin proposed that excreted urea is an index of the body's "exogenous"

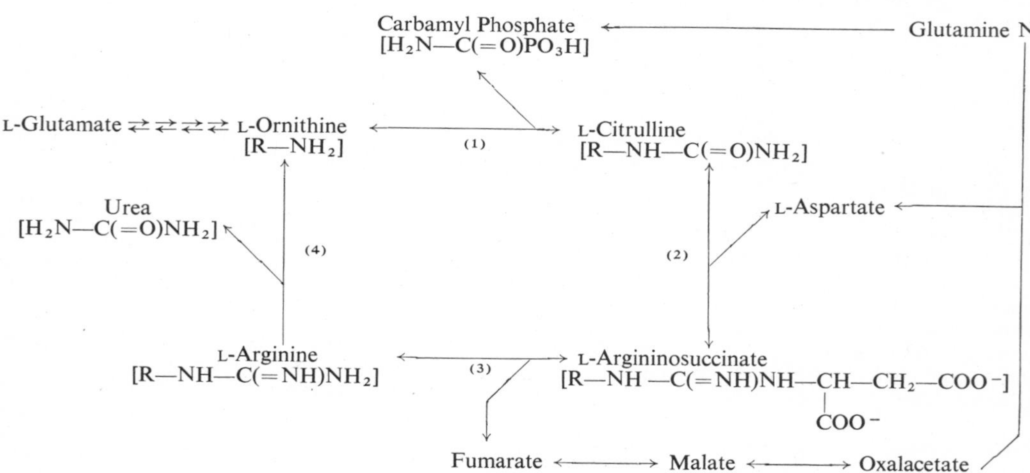

FIG. 1. The arginine-urea cycle.

metabolism of proteins, whereas excreted creatinine is an index of its "endogenous" metabolism. In contrast to urea, creatinine excretion is relatively constant for an individual, representing a first-order, irreversible, non-enzymic breakdown of muscle and brain phosphorylcreatine to creatinine. Of particular interest is the fact that arginine is a specific precursor of both of these medically important indices of metabolism (see CREATINE AND CREATININE).

In human liver a given molecule of arginine has four possible metabolic fates. Arginine can be converted to (a) argininosuccinic acid, (b) argininyl-sRNA, (c) ornithine plus urea, or (d) ornithine plus glycocyamine, the precursor of creatine. The flow along pathway (d) is regulated by feedback repression (see METABOLIC CONTROLS) in which the steady-state level of the enzyme involved, arginine: glycine amidinotransferase, is regulated by the concentration of liver creatine. In this fashion the organism is protected against runaway synthesis of creatine. If flow along this pathway were not regulated by supply and demand for its end product, creatine, excessive amounts of the three amino acid precursors of creatine (arginine, glycine, and methionine) would be diverted from biosynthesis of essential enzymes and special metabolites. The deleterious consequences of excessive creatine biosynthesis can readily be shown by feeding glycocyamine to experimental animals. Provision of this compound bypasses the physiologically controlled enzymic reaction, and large amounts of creatine are synthesized. The resulting fatty liver and other toxic symptoms attest to the importance of the bypassed control.

References

1. MEISTER, A., "Biochemistry of the Amino Acids," Second edition, New York, Academic Press, 1965.
2. WOLSTENHOLME, G. E. W., AND CAMERON, M. P., EDITORS, "Comparative Biochemistry of Arginine and Derivatives," Ciba Foundation Study Group No. 19, Boston, Little, Brown and Company, 1965.

JAMES B. WALKER

AROMATIC AMINO ACIDS (BIOSYNTHESIS AND METABOLISM)

The aromatic amino acids, phenylalanine, tyrosine and tryptophan, are essential components of most PROTEINS. Mammalian cells cannot synthesize aromatic rings and, therefore, must be supplied with aromatic compounds formed by plants or microorganisms. The pathways of biosynthesis of the aromatic amino acids have been worked out by experiments with microorganisms such as the bacterium *Escherichia coli* or the mold *Neurospora crassa* which grow on simple glucose-ammonium salts media.

That part of the pathway which is common to all the aromatic amino acids will be described first, and then the pathways to the specific amino acids.

The Common Pathway. This is set out in Fig. 1. The first specific reaction leading to aromatic biosynthesis is the condensation of two products of glucose metabolism, namely erythrose-4-phosphate and phosphoenolpyruvic acid. This condensation gives a seven-carbon compound which is then enzymatically cyclized to give 5-dehydroquinic acid. In this way, the six carbons destined to become the aromatic ring are joined. Removal of water gives 5-dehydroshikimic acid and subsequent reduction leads to shikimic acid. Shikimic acid is then phosphorylated to give shikimic acid-5-phosphate, and an enolpyruvic acid side chain is added. The removal of phosphoric acid leads to the production of the compound chorismic acid, from which the individual pathways of biosynthesis diverge. Chorismic acid is also a precursor of folic acid, ubiquinone and probably vitamin K.

Biosynthesis of Phenylalanine and Tyrosine. Figure 2 depicts the pathways of biosynthesis of these two amino acids. There are two enzymes capable of metabolizing chorismic acid through prephenic acid. One of these enzymes forms phenylpyruvic acid and the other 4-hydroxyphenylpyruvic acid. These keto acids are then converted by transamination to phenylalanine and tyrosine, respectively.

Formation of Tryptophan. Chorismic acid is converted into tryptophan by the series of reactions set out in Fig. 3. Chorismic acid is first converted to anthranilic acid in a rather complex reaction which nevertheless seems to be carried out by a single protein; glutamine appears to be the most efficient nitrogen source of this reaction. Phosphorylated ribose is added to anthranilic acid, then rearrangement and elimination of water and CO_2 follow, leading to indole-3-glycerolphosphate, the immediate precursor of tryptophan. Glyceraldehyde-3-phosphate is then eliminated and serine substituted to give tryptophan. Indole does not appear as a free intermediate.

The pathways set out in Figs. 1, 2 and 3 represent the results of experiments with microorganisms. The isolation of mutants which accumulate various intermediate compounds has allowed these pathways to be investigated much more readily in microorganisms than in higher organisms.

The main dietary source of aromatic amino acids for animals is plant protein. There is no reason to believe that the pathway of biosynthesis in plants differs very much from the pathway described above, although many aromatic compounds in plants, other than amino acids, are formed by the condensation of acetic acid units. In those sections of the biosynthetic pathway which have been investigated in plants, the reactions seem similar to those in microorganisms. It was once thought that yeast formed an intermediate compound involving fructose instead of ribose in the biosynthesis of tryptophan. While this is probably not so, some variations may exist on the pathway given, since it has already been found that the details of the pathway differ

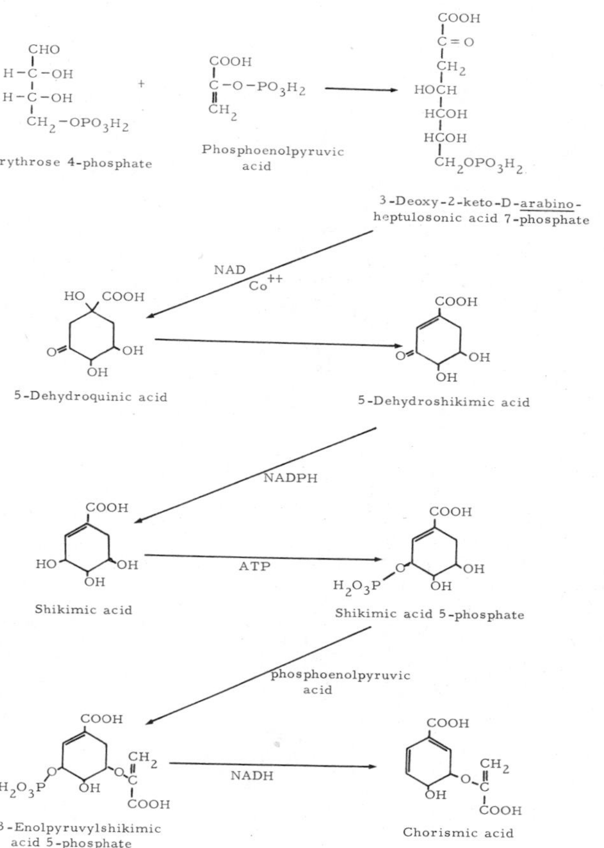

FIG. 1. The pathway of biosynthesis common to all the amino acids.

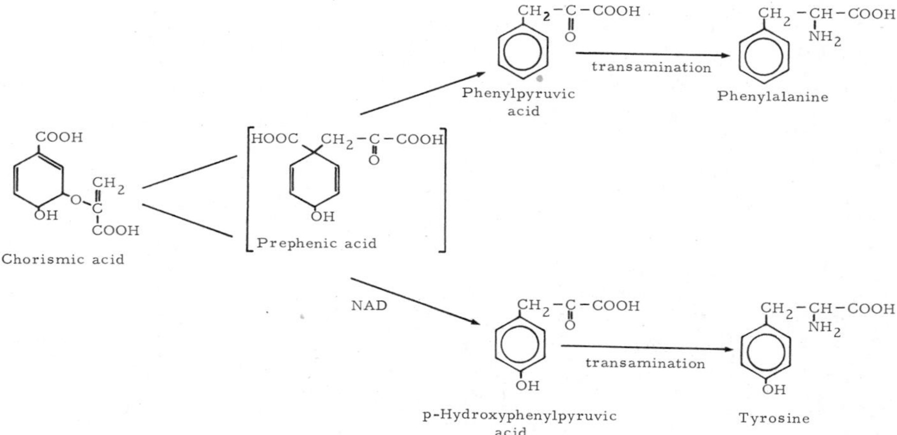

FIG. 2. Terminal stages in the biosynthesis of phenylalanine and tyrosine.

slightly in *Escherichia coli* and *Neurospora crassa.* Thus in the mold, two of the reactions shown in Fig. 2 (which outlines the pathway in the bacterium) appear to be carried out by a single enzyme. A number of the reactions shown have not yet been investigated in detail so a considerable amount of research remains to be done. With the more recent trends in biochemistry and genetics, some of the reactions in this pathway are receiving a great deal of attention. This branched biosynthetic pathway is suitable for studying how cells control the flow of metabolic intermediates by end-product inhibition and enzyme repression (see METABOLIC CONTROLS). The two likely points of metabolic control are the first reaction peculiar to aromatic biosynthesis (the condensation of erythrose-4-phosphate and phosphoenolpyruvic acid) and the conversion of

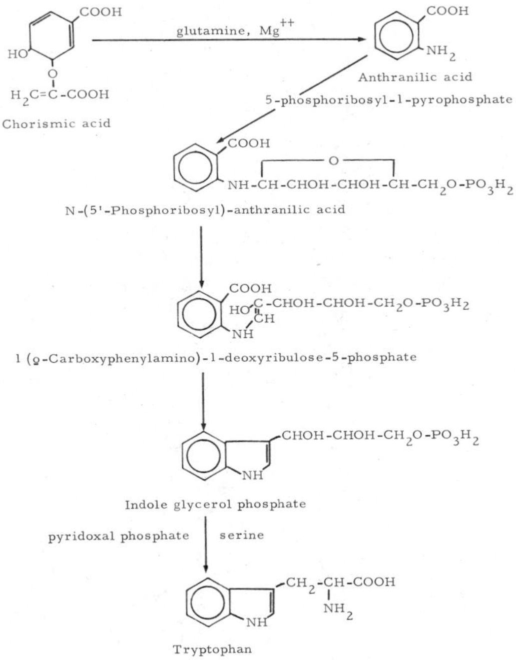

FIG. 3. The biosynthesis of tryptophan from chorismic acid.

chorismic acid into the first compound on each of the individual pathways. It appears likely that there are three enzymic activities in the cell capable of carrying out the formation of the seven-carbon compound and each of these is subject to some sort of control by a different aromatic amino acid. The enzymes leading from chorismic acid are found to be subject to end-product inhibition. The terminal enzyme of tryptophan biosynthesis (tryptophan synthetase) is being studied in great detail and is providing valuable information on the relationships between GENES and proteins.

Metabolism. Apart from PROTEINS, which are considered elsewhere, a wide range of naturally occurring aromatic compounds is derived from the amino acids. These range from complicated pigments (melanin; see SKIN PIGMENTATION) and plant structural materials (lignins) to relatively simple compounds such as ALKALOIDS, ANTIBIOTICS, HORMONES, vitamins and many other compounds formed by microorganisms and higher cells. The aromatic rings may also be broken and the resulting compounds used as sources of energy. The metabolism of each of the amino acids is outlined briefly below and varies with the organism and the environmental conditions.

Phenylalanine. Phenylalanine is normally converted irreversibly by animals to tyrosine, hence the latter amino acid is not essential in the diet of man. Most bacterial and plant cells cannot carry out this conversion, but phenylalanine may serve as an amino donor for transaminase reactions, leading to phenylpyruvic acid. Certain humans have inherited a genetic defect (phenylketonuria) resulting in an inability to convert phenylalanine to tyrosine. Such people form phenylpyruvic acid which is excreted in the urine as such, as phenyllactic acid, or other metabolic products. The presence of large quantities of phenylpyruvic acid (1–2 g/day may be excreted) leads to MENTAL RETARDATION.

Some of the products of phenylalanine metabolism in man are shown in Fig. 4. The intermediate compounds on these pathways are known for the most part and may be found in the references given below.

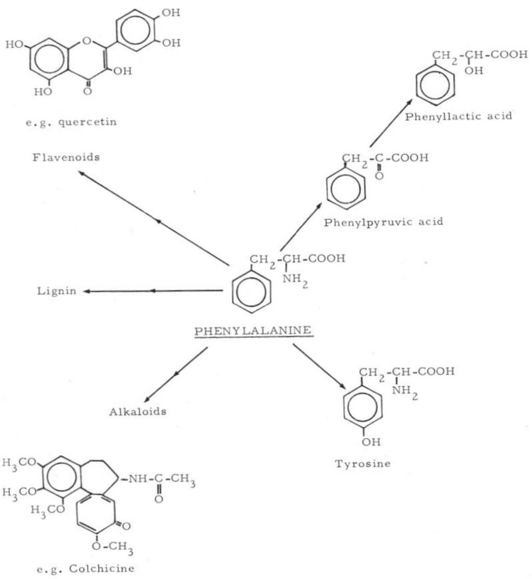

FIG. 4. Outline of metabolic routes from phenylalanine.

A wide variety of compounds found in plants appear to arise from phenylalanine: these include FLAVONOIDS, ALKALOIDS and lignins. The metabolic pathways for the formation of these compounds are not well established and are based on the observations that isotopically labeled phenylalanine is incorporated into their structures. These compounds are not formed by unicellular organisms and are not essential metabolites, even for the cells which do form them. Therefore, it has not yet been possible to apply many of the techniques which have been used so successfully in the studies of biosynthesis of the amino acids themselves. In the case of pigments, it is possible to search for non-pigmented mutants and then isolate accumulated intermediates. However, the isolation of such strains in diploid multicellular organisms is difficult.

Tyrosine. In man, tyrosine serves both as a source of energy and as a precursor of other physiologically important compounds, notably the hormones of the adrenal medulla (see ADRENALINE AND NORADRENALINE) and the pigment melanin. In the formation of the latter compounds an important intermediate is dihydroxyphenylalanine (DOPA) and this compound is formed in liver, in cells forming melanin, and in the adrenal medulla. An outline of the pathways concerned is shown in Fig. 5. The hormones epinephrine and norepinephrine are themselves rapidly metabolised to inactive compounds by methylation, oxidative deamination or conjugation, and excreted.

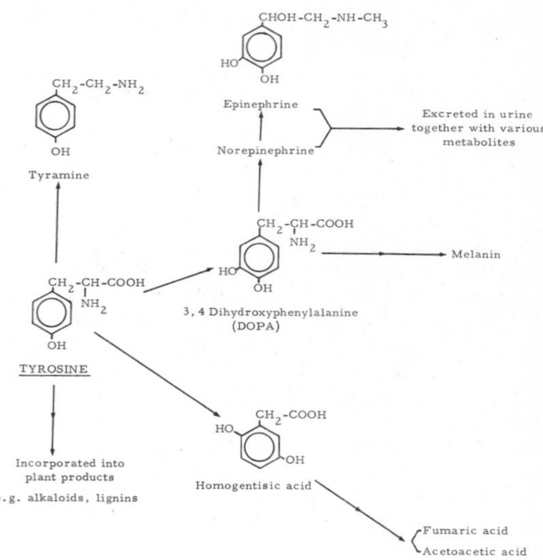

FIG. 5. Outline of main pathways of tyrosine metabolism.

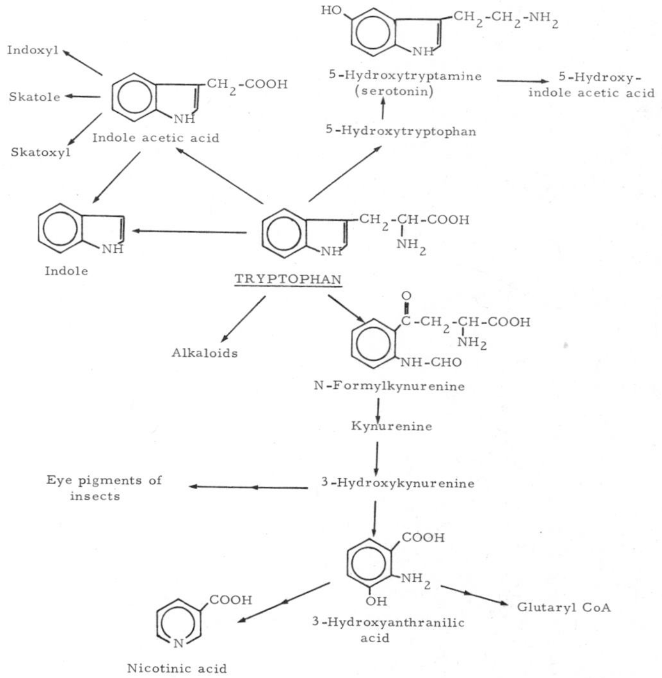

Melanin is formed in pigmented cells by a copper-containing enzyme which converts tyrosine through DOPA to a quinone which then undergoes a series of non-enzymic chemical transformations to give melanin. Tyrosine when bound in thyroglobulin is iodinated in the thyroid gland to give the hormone thyroxine [see IODINE (IN BIOLOGICAL SYSTEMS); THYROID HORMONES].

HO CH2-CH2-NH2 NH
5-Hydroxytryptamine (serotonin)
5-Hydroxy-indole acetic acid
5-Hydroxytryptophan
Indoxyl
Skatole
Skatoxyl
CH2-COOH NH
Indole acetic acid
NH
Indole
CH2-CH-COOH NH2 NH
TRYPTOPHAN
Alkaloids
O C-CH2-CH-COOH NH2 NH-CHO
N-Formylkynurenine
Kynurenine
Eye pigments of insects
3-Hydroxykynurenine
COOH NH2 OH
3-Hydroxyanthranilic acid
Glutaryl CoA
COOH N
Nicotinic acid

FIG. 6. Outline of some of the pathways of tryptophan metabolism.

The bulk of the dietary tyrosine (and phenylalanine) is metabolized by conversion through 4-hydroxyphenlpyruvic acid to homogentisic acid after which the benzene ring is opened. Fumaric acid and acetoacetic acid are eventually formed and metabolized by the usual pathways.

Tyrosine is deaminated by inducible enzymes in some bacterial cells (and to some extent in liver) to give tyramine. Various natural products including lignin, various alkaloids and the antibiotic novobiocin appear to be derived from tyrosine, but details of the pathways are not clear.

Tryptophan. Tryptophan can be metabolized in a number of ways leading to complete oxidation, to physiologically important compounds, or to complex natural products as outlined in Fig. 6. In man, for example, the heterocyclic ring may be broken by tryptophan pyrrolase, and a series of reactions follows which leads to the formation of glutaryl coenzyme A which is further metabolized. Two physiologically useful by-products of this series of reactions are eye pigments in insects and NICOTINIC ACID in man. The amount of nicotinic acid normally formed is not sufficient to replace the dietary requirement for this vitamin.

Tryptophan may also be converted to indoleacetic acid by bacteria, plants and animals. Indoleacetic acid is an important growth hormone (auxin) for plants (see HORMONES, PLANT), and bacteria can metabolize it further to give a number of products including indole and various indole derivatives (skatole, skatoxyl, indoxyl) found in feces. Many bacteria possess an enzyme (tryptophanase) which converts tryptophan directly to indole.

Another important product of tryptophan metabolism is 5-hydroxytryptamine (serotonin) a powerful vasoconstrictor widely distributed in the body. Some amphibia form acetylated or methylated derivatves of serotonin which are very toxic for man.

As with the other aromatic amino acids, tryptophan is incorporated by plants into complex compounds such as alkaloids (*e.g.*, lysergic acid).

Further details of the biosynthesis and metabolism of the aromatic amino acids may be found in the references cited below.

References

1. WHITE, A., HANDLER, P. AND SMITH, E. L., "Principles of Biochemistry," Third edition, New York, McGraw-Hill Book Co., 1964.
2. BERNFELD, P. (EDITOR), "Biogenesis of Natural Compounds," Oxford, Pergamon Press, 1963.
3. UMBARGER, E., AND DAVIS, B. D., "Pathways of Amino Acid Biosynthesis," in "The Bacteria" (GUNSALUS, I. C., AND STANIER, R. Y., editors), Vol. 3, New York, Academic Press, 1962.

FRANK GIBSON

ARSENIC

This is one of the elements commonly found in small amounts in the tissues of plants and animals (see MINERAL REQUIREMENTS). A human body may contain as much as 20 mg (As_2O_3). So far as is known, however, arsenic plays no role in natural biological phenomena. It is present in seawater to the extent of 0.006–0.03 ppm. In estuaries it may be ten times as high. Shellfish tend to accumulate the arsenic from the large amount of seawater with which they come in contact. Oysters may contain 3–10 ppm. However, shellfish of the same species grown in different localities show wide variations in arsenic content, suggesting that it is an accidental constituent which the organisms learn to tolerate. Its lack of function in the human body is suggested by the fact that it tends to accumulate in the hair and nails which are essentially non-living.

Arsenic is of course a poison (see TOXICOLOGY). The minimum fatal dose (As_2O_3) for humans is said to be from 60–180 mg, although individual cases appear to vary widely. Lower forms are more tolerant. Lead arsenate is extensively used in apple orchards, for example, to control the codlin moth. The spray residue is effectively removed from the fruit with dilute hydrochloric acid before commercial sale so that eating the fruit contributes very little arsenic to that which is inevitably present in our bodies. This small amount of arsenic, derived to a considerable extent from consuming fish, appears to do no harm. It conceivably may even play some biological role, although the evidence is against this supposition.

Certain organoarsenic compounds have been employed as haptens in the study of the stereochemistry of the antigen and the specificity of the antibody (see ANTIGENS).

Reference

Monier-Williams, G. W., "Trace Elements in Foods," New York, John Wiley and Sons, 1949.

ASCORBIC ACID (VITAMIN C) AND SCURVY

Vitamin C appears to have been the first nutrient to be associated with major outbreaks of disease. Even primitive tribes often knew that persons left without consumption of fresh food, particularly fresh fruit or vegetables for several months, would become sick and ultimately die, with the symptoms characteristic of *scurvy*. Although extremely unstable under most circumstances, this vitamin was also one of the first to be identified in crystalline form as a pure substance. It was then named *ascorbic acid* because of its function as a complete protective agent against scurvy in experimental animals and man. Commercial production with high yields from glucose followed rapidly.

Apparently all forms of life, both plant and animal, with the possible exception of simple forms such as bacteria that have not been studied thoroughly, either synthesize the vitamin from other nutrients or require it as a nutrient. Dormant seeds contain no measurable quantity of the vitamin, but after a few hours of soaking in water, one can easily demonstrate the formation of the vitamin, either by titration or by silver staining as the germ portion of the seed begins to sprout. Most fresh animal tissues also contain an easily determined amount of the vitamin, particularly in the glandular organs such as the pituitary, adrenals, corpus luteum, pancreas, kidneys, liver, and intestinal wall.

Scurvy. The only animals thus far shown not to have a capacity to synthesize the vitamin are man and other primates, guinea pigs, and two species recently studied in India by Guha *et al.*, a fruit-eating bat and the bulbul bird.[1] These are the only animal forms thus far identified as having lost genetically the presence of a specific enzyme (L-gulonolactone oxidase—see formulas) that normally permits other animals to form the vitamin readily from such sugars as D-glucose, and D-galactose. There is no evidence to indicate how many forms of animal life may have become extinct during past ages because of their having lost the capacity to form this nutrient from simple sugars or other nutrients.

In human history, however, it is clear that most primitive cultures had learned repeatedly and sometimes forgotten the requirement for fresh foods to furnish the quantities of vitamin C necessary to sustain life and health. Outbreaks of scurvy and frequent high death rates were observed during long ocean voyages, military sieges, long travels with grain as a chief foodstuff, long winter seasons, long periods of drought, or other circumstances that did not permit access to fresh foods. These early discoveries did not contribute as rapidly to advances in nutrition as one might expect, because there was no understanding of the nature of disease. In peoples' minds, there was no way to differentiate between diseases caused by witchcraft, infections, food deficiencies, or slowly acting poisons. However, they could recognize and pass on by word of mouth, through successive generations, that certain foods such as extracts of evergreen leaves, cabbage, sprouting grain, onion-type bulbs, watercress, oranges, or certain types of fresh meat, might restore to health those afflicted with the disease later to be called scurvy. The North American Indians recognized that fresh bulbs or root tips, or a hot-water extract of evergreen leaves would protect against the disease. The Canadian explorer, Cartier, reported such findings in 1535 after losing a large fraction of his encampment by scurvy.

The disease is characterized by swollen and bleeding gums, loosened teeth, tenderness to touch, weakness, swollen joints, beaded ribs, swelling of the feet and legs, and gradual development of small hemorrhages around the hair follicles or massive hemorrhages that appear much like bruises. Under moderate stress, even the ends of bone shafts may fracture.

The most famous of early experiments in studying the cause of scurvy was the controlled experimentation on shipboard by the British surgeon, James Lind, published in 1756. His studies of men in the same environment on shipboard, given specific medicine or food for the prevention and cure of scurvy, is often referred to as the first "controlled" experimentation in the history of nutrition or medicine. Dr. Lind was also a highly competent critic of the medical literature of his time. He pointed out the great number of misleading reports that had confused scurvy with other diseases, had falsely claimed to cure scurvy with such "medicine" as dilute sulfuric acid or cream of tartar, and had failed to observe the true relationship to nutrition. It was a dramatic observation at the time of the Lind Bicentennial celebration in 1956 to note that during Lind's service in the navy, scurvy was the most prominent cause of sickness and death among ocean voyagers, including the British Navy, but the retired Chief Surgeon who had served in the same naval hospital, and gave one of the bicentennial lectures, had never seen a single case of scurvy during his professional lifetime in the British Navy.[2]

A second major step that introduced experimental studies of scurvy on a much more active research basis was the discovery by Holst and Frohlich, in Norway, that guinea pigs not receiving fresh food in their diets also developed scurvy (1907). This discovery permitted laboratory studies of the vitamin, leading to its isolation and identification in 1932 by King and Waugh in the United States, and Svirbely and Szent-Györgyi in Hungary. Rapid progress followed in which Swiss, British, and German chemists established its molecular structure and developed low-cost methods of synthesis from D-glucose.

Stability of the Vitamin. The almost universal requirement for fresh foods as a source of vitamin C is readily explained by the extreme sensitivity of the vitamin to destruction by reaction with oxygen in the air. This tendency is greatly accelerated by the presence of minute quantities of enzymes that occur in most living tissues, in which copper or iron is combined with a protein to form a catalyst for the reaction of the vitamin with oxygen from the air. Traces of copper and iron compounds, other than enzymes, also act as catalysts in the reaction with oxygen. Many other chemicals such as quinones or high-valence salts of manganese, chromium, and iodine can also oxidize the vitamin rapidly in aqueous solutions. Most of these reactions increase rapidly in proportion to exposure to the air and rising temperature. In the dry crystalline state, however, and in many dried plant tissues, particularly if acidic in reaction, the vitamin is relatively stable at room temperature through periods of many months or even years. For example, a fresh green leaf nearly always contains a liberal quantity of the vitamin, but if permitted to wilt at room temperature it may lose most of its vitamin content within a few days. If the leaf is crushed, however, exposing the vitamin more vigorously to the air and to the catalysts contained in the leaf, the vitamin may nearly all disappear within a matter of minutes. On the other hand, if green leaves are dipped in boiling water or steamed to inactivate the enzymes, and then placed in a can or jar from which the air is removed, the content of the container may then be sterilized by heating under pressure and stored for many months with very little loss of the vitamin. Fresh-cut oranges or tomatoes or their juices may be exposed in an open glass for several hours without appreciable loss of the vitamin because of the protective effect of the acids present and the practical absence of enzymes that catalyze its destruction. In potatoes, when baked or boiled, there is only slight loss of the vitamin, but if they are whipped up with air while hot, as in the production of mashed potatoes, a very large fraction of the initial vitamin content may be lost during the normal preparation and consumption of a given meal.

In the freezing of foods for market, it is common practice to dip them in boiling water or to treat them briefly with steam to inactivate the enzymes, after which they are frozen and stored at very low temperatures. In this state the vitamin is reasonably stable. In a practical sense, foods marketed as fresh, frozen, or canned foods usually contain protective quantities of the vitamin in the quantities of food eaten in a mixed diet.

Biological Synthesis. By the use of glucose labeled with radiocarbon in known positions, it has been possible to trace the sugar through intermediate steps to the formation of ascorbic acid in plant and animal tissues, and then by using ascorbic acid with radiocarbon labeled in known positions it has been possible to determine with considerable accuracy the metabolic distribution, storage, and chemical changes characteristic of the vitamin molecule. Evidence of this kind was obtained by King *et al.* in 1950 and has been greatly extended by others using glucose labeled in position one or six, for example, followed by isolation of the vitamin formed in rats and determination of the position of radiocarbon in the product. These tests made it clear that the carbon atoms in glucose or galactose all retain their original positions along the carbon chain in the vitamin formed biologically. There was no rupture or replacement in the carbon chain during the conversion. The synthesis can be greatly increased by feeding the animals small amounts of chloretone or any one of dozens of organic compounds. The reactions are indicated by the following formulas:

Ascorbic acid is easily oxidized to dehydroascorbic acid. The latter is much less stable than ascorbic acid and tends to yield products such as oxalate, threonic acid, and carbon dioxide. When administered to animals or consumed in foods, dehydroascorbic acid has almost the same antiscorbutic activity as ascorbic acid, and it can be quantitatively reduced to ascorbic acid by hydrogen sulfide or many RSH compounds such as cysteine. It is not extremely unstable in weak acid solutions or at low temperatures.

Biochemical Function. Ascorbic acid clearly acts as a regulator in tissue respiration and may serve as an antioxidant *in vitro* by reducing oxidizing chemicals. Although ascorbic acid participates in a number of reaction systems, it has not been possible to establish its quantitative role compared with other respiratory sequences. In plant tissues, the related GLUTATHIONE system of oxidation and reduction is fairly widely distributed, and there is strong evidence that electron transfer reactions involving ascorbic acid are characteristic of animal systems.[1] PEROXIDASE systems may involve reactions with ascorbic acid also. In plants, either of two copper-protein enzymes are commonly involved in the oxidation of ascorbic acid. In one system, the vitamin is oxidized directly by molecular oxygen to dehydroascorbic acid without the formation of hydrogen peroxide. This enzyme is relatively specific for the exact molecular structure of ascorbic acid, although other closely related dienols of similar structure may be oxidized by the ascorbic acid oxidase. A second copper-protein enzyme, phenol oxidase, functions through the intermediate formation of quinones from native phenols such as catechol; the quinone in turn is reduced by ascorbic acid with formation of dehydroascorbic acid and the original phenol.

In animal tissues, it is easily demonstrated that as the vitamin content of tissues is depleted, many enzyme systems in the body are decreased in activity, but the specific nature of the decreased activity has not been clarified. In the total animal and in isolated tissues from animals with scurvy, there is an accelerated rate of oxygen consumption even though the animal becomes very weak in mechanical strength and many physiologic functions are disorganized.

```
                                               OH              OH              OH
                                               |               |               |
1   H-C=O           H-C=O                    H-C-H           H-C-H           H-C-H
      |               |                        |               |               |
2   H-C-OH          H-C-OH                   H-C-OH          H-C-OH          H-C-OH
      |               |                        |               |               |
3  HO-C-H    (O)   ┌─ C-H       (2H)      ┌── C-H    (O)   ┌─ C      (O)    ┌─ C-H
      |      -->   |  |         -->       |   |      -->   |  |     <==>    |  |
4   H-C-OH         O H-C-OH               O H-C-OH         O  C-OH   (2H)   O  C=O
      |            |  |                   |   |            |  ||            |  |
5   H-C-OH         | H-C-OH               | H-C-OH         |  C-OH          |  C=O
      |            |  |                   |   |            |  |             |  |
6   H-C-H          └─ C=O                 └── C=O          └─ C=O           └─ C=O
      |
      OH

 D-glucose      D-glucuronic          L-gulonic        L-ascorbic      L-dehydro-
                acid lactone          acid lactone       acid          ascorbic acid
```

With the onset of scurvy, the most conspicuous tissue change is the failure to maintain normal COLLAGEN. Sugar tolerance is decreased and lipid metabolism is altered. There is also marked structural disorganization in odonto-blast cells in the teeth and in bone-forming cells in skeletal structures. In parallel with the above changes, there is a decrease in many hydroxylation reactions (see OXYGENASES). The hydroxylation of organic compounds is one of the most characteristic features disturbed by the vitamin deficiency. These reactions are thought to constitute an important part of the vitamin's regulation of respiration, hormone formations, and control of collagen structure. Recent investigations also make it clear that ascorbic acid functions in the subcellular structures such as the mitochondria.[1]

In summary, the ascorbic acid $\rightleftharpoons$ dehydroascorbic acid system functions in an extensive number of enzyme reactions, including the regulation of tissue respiration, amino acid metabolism as in the case of tyrosine, in RSH systems, in cupric and ferric reactions, and in the hydroxylation of aromatic and aliphatic compounds.

Vitamin C Requirements. Estimation of the human vitamin C requirement can be approached in many ways: (a) by direct observation in human studies, (b) by analogy to experimentation with guinea pigs, (c) by analogy to experimental studies in monkeys and other primates, and (d) by analogy to animals such as the albino rat, that normally synthesize the vitamin in accordance with physiological need. In direct human observations, milk secreted by mothers when their diet furnishes the amount of vitamin C commonly found in mixed diets, supplies to the nursing infant about 20–25 mg of the vitamin per day. When thus nourished, the tissues of the infant maintain a vitamin C content in the blood, glandular organs, and muscle, that is about the same in concentration and distribution as observed in animals that maintain the vitamin under physiologic control. When the mother's diet is extremely low in vitamin C, there is a gradual reduction in the ascorbic acid content of the milk secreted. And when the mother's intake of vitamin C is unusually high, the concentration in breast milk remains about normal while the excess is excreted in the urine. Hence the over-all physiologic adjustments in human tissue appear to be comparable to those found in nearly all animals that have the vitamin under physiologic control, and the depletion sequences are comparable to those observed in the guinea pig and primates other than man. To maintain a "normal" concentration of ascorbic acid in the blood and tissues of growing children and adults, there is a gradual increment in the requirement parallel with growth, gestation, and lactation, as illustrated by the figures below, from the table of Recommended Dietary Allowances prepared by the Food and Nutrition Board of the National Academy of Sciences—National Research Council.[3]

Intakes appreciably higher than those indicated in the table result in increased urinary excretion and a moderate rise in blood and tissue content. Many forms of illness, trauma, infections, poisoning, or other tissue injuries, may accelerate depletion of the vitamin so that subsequent to many forms of stress, there is need for more than a normal intake until normal concentrations are restored. However, the above quantities are usually sufficient to serve in this respect when the imposed stresses are not extremely severe. Intakes of ascorbic acid in adults below about 10–25 mg/day result in severe depletion of the serum and tissue ascorbic acid content and may involve tissue changes coincident with the development of scurvy. Although quantities as low as 10 mg/day have maintained adults free from scurvy for fairly long periods under controlled conditions of living,[4] such low intakes are not regarded as either safe or satisfactory for the protection of health. In carefully controlled studies in experimental animals, it is evident that young, growing guinea pigs are protected from scurvy by intakes in the range of 0.5 mg of ascorbic acid per day, and 1 mg/day will permit a normal growth rate. However, gestation and lactation are distinctly more favorable when intakes are in the range of 5–10 mg/day. Guinea pigs with their normal diet of fresh forage receive intakes in this range or higher. Intakes in the range of 1 or 2 mg/day result in the animals being distinctly more sensitive to injury by diptheria toxin, for example, as shown both in the soft tissue structures and in injury to the odonto-blast structures in the growing incisors. Hence, the Recommended Dietary Allowances of the Food and Nutrition Board are believed to

	Age (years) from to	Ascorbic Acid (mg/day)		Age (years) from to	Ascorbic Acid (mg/day)
Men	18–35	70	Infants	0–1	30
	35–55	70	Children	1–3	40
	55–75	70		3–6	50
Women	18–35	70		6–9	60
	35–55	70	Boys	9–12	70
	55–75	70		12–15	80
Pregnant (2nd and				15–18	80
3rd trimester)		+30	Girls	9–12	80
				12–15	80
				15–18	70

be desirable and adequate for the maintenance of normal growth and other physiologic functions, and in addition, to meet the types of stress considered normal to daily living.

It is fairly easy to maintain intakes at the recommended levels by the use of mixed practical dietaries that include nominal quantities of fresh, canned, or frozen vegetables or fruits. Oranges, lemons, grapefruit, tomatoes, cantalopes, broccoli, currents, strawberries, and a great many common foods furnish quantities reliably in the above range or higher. In fact, baked or boiled potatoes, cabbage, parsnips, onions, and many root crops, supply protective quantities of the vitamin and make a substantial contribution toward the quantities indicated above. In the preparation of infant formulas, it is now common practice to add known quantities of the vitamin and to standardize the products to resemble the concentration in mother's milk.

References

1. BURNS, J. J., KING, C. G., *et al.*, "Vitamin C," *Ann. N.Y. Acad. Sci.*, **92**, 332pp. (1961).
2. STEWART, C. P., GUTHRIE, D., *et al.*, "Lind's Treatise on Scurvy," A bicentenary volume, Edinburgh University Press, 1953, 440pp.
3. "Recommended Dietary Allowances," Sixth Revised Edition, Food and Nutrition Board of the National Academy of Sciences—National Research Council, *Natl. Acad. Sci.—Natl. Res. Council Publ.*, **1146**, 59pp. (1964).
4. BARNES, A. E., *et al.*, "Vitamin C Requirements of Human Adults," *Med. Res. Council Spec. Rept. Ser.*, **280**, 179pp (1953).

C. G. KING

ASPARAGINE

L-Asparagine, $H_2N-\overset{\overset{O}{\|}}{C}-CH_2-CH(NH_2)-COOH$, the β-amide of aspartic acid, was named from asparagus from which it was first isolated. Asparagine is not commonly produced by the hydrolysis of proteins because it is cleaved by acid or alkaline hydrolysis into ASPARTIC ACID and ammonia. As a building unit of PROTEINS, asparagine is, however, important. It has been obtained from proteins by enzymatic hydrolysis and should be regarded as one of the constituent AMINO ACIDS derived from proteins.

Asparagine (like aspartic acid) is one of the nutritionally nonessential amino acids. Its specific rotatory power in $1N$ HCl is $[\alpha]_D = +20°$ $(c = 1M)$.

ASPARTIC ACID

L-Aspartic acid, $HOOC-CH_2-CH(NH_2)-COOH$, is one of the AMINO ACIDS derived from common PROTEINS by hydrolysis. It is interconvertible by TRANSAMINATION reactions with oxalacetic acid, an intermediate in the CITRIC ACID CYCLE; and it is nonessential nutritionally for mammals (see AMINO ACID METABOLISM; NUTRITIONAL REQUIREMENTS). For L-aspartic acid, $[\alpha]_D = +19.5°$ ($c = 1M$ in $1M$ HCl); $[\alpha]_D = -18.8°$ ($c = 1M$ in $1M$ NaOH).

ASYMMETRY

The recognition of the existence of asymmetric carbon atoms, *i.e.*, attached to four *different* groups, W, X, Y, and Z,

$$\begin{array}{c} W \\ | \\ Z-C-X \\ | \\ Y \end{array}$$

in organic compounds is closely related historically to biochemistry. In such a compound, it may be noted that the four different substituent groups W, X, Y, and Z, bonded to the carbon along directions corresponding to the apexes of a regular tetrahedron having the carbon at its center, may exist in two different spatial configurations. One of these configurations is the mirror image of the other, and they are not interconvertible unless bonds are broken. The two configurations represent a pair of antipodes (or enantiomorphs, or enantiomers.) A mixture of equal numbers of the two enantiomorphic molecular forms is called a racemic mixture. The "resolution" of a racemic mixture, *i.e.*, the separation of the mixture into its two enantiomorphic components, involves the use of biochemical products such as alkaloids, tartaric acid, etc. (see DIASTEREOISOMERS). The fundamental fact that enzymes are asymmetric in nature and selective in their action has been operative since life began on earth.

A discussion of considerations of optical asymmetry as they apply to amino acids is given in the article AMINO ACIDS (OCCURRENCE, STRUCTURE, SYNTHESIS); see also OPTICAL ISOMERISM.

Other atoms besides carbon may (rarely in biochemistry) when bonded to four different groups also act as centers of asymmetry.

By arbitrary convention, in the two-dimensional depiction of a tetrahedral atom bonded to four other atoms (such as an asymmetric carbon atom), the *horizontal lines* (*e.g.*, the bonds from C to Z and C to X in the diagram above) represent *bonds coming toward the reader* out of the plane of the paper, whereas the *vertical lines* (*e.g.*, the bonds from C to W and C to Y in the diagram) represent *bonds going away from the reader* behind the plane of the paper.

Reference

MORRISON, R. T., AND BOYD, R. N., "Organic Chemistry," pp. 318–331, Boston, Allyn and Bacon, 1959.

ATHEROSCLEROSIS

For the past several decades, the vital statistics of most countries of the western world have clearly reconfirmed the fact each year that the most common cause of death is a group of diseases of the heart and blood vessels. In the United

States, deaths from cardiovascular-renal disease not only outnumber deaths caused by all other killers, including cancer and accidents (the two other most common ones) but also are greater in number than deaths from *all* other causes lumped together. In 1962, 55% of total deaths in the United States resulted from arteriosclerosis (813,790 deaths) and other cardiovascular-renal diseases including hypertension and cerebrovascular disorders (197,590 "strokes"). Heart disease, which often strikes without warning in the form of the coronary attack, is the most dramatic of this group, and perhaps the most cruel, because it often cuts down relatively young men in the prime of their life. Atherosclerosis is the most important lesion of arteries which precedes the heart attack and sets the stage for the final *coup de grace*. But even in later decades, including the sixth and the seventh, deaths from attacks of myocardial infarction continue to occur with great frequency, outclassing all other causes. In these decades, unlike the earlier ones (third and fourth), the elderly woman is more frequently a victim than the male. If the frequency of deaths from heart disease is averaged over the entire life span, it is seen that women are affected just as often as men, a fact not realized until recently.

At the other extreme of life, deaths from heart attacks may occur in youngsters just out of their teens although not nearly so frequently at this stage of life as later. In fact, coronary attack as a cause of death in men in their twenties was not even recognized by the medical profession until autopsies on young American soldiers early in the 1940's demonstrated its presence. Perhaps even more important was the recognition of scars in the hearts of some of these boys caused by previous but non-fatal heart attacks which had not been recognized as such by either the patient or his physician. In a recent study (1963) of a group of young male Britishers (mean age of 27.3 years), significant functional narrowing of one or more coronary arteries was demonstrated in over 20% of them and visible gross narrowing (macroscopic) in over one-third. It is, therefore, probably no exaggeration to state that from the age of puberty onwards, ravages of diseases of the heart and blood vessels are already operative. In fact, evidence is now accumulating which leaves little doubt that the initial stages of damage to the coronary arteries and other vital blood vessels, which culminate eventually in these catastrophic events of later life, have their origin and beginnings in childhood. This concept is important because of its implication that prevention of heart disease may lie primarily within the pediatricians' sphere. Any nationwide attack on the problem must therefore not neglect possible prophylactic measures to be applied to our children which would be aimed at the objective of sparing them in their future decades from the same vascular disease which is now killing their fathers and mothers.

Etiology of Arteriosclerosis. Arteriosclerosis is the change in arteries which produces almost all the above serious clinical manifestations of sickness and death from vascular disease. This term means simply a thickening and hardening of the walls of major arteries of the body. There are only a few pathologic processes which may so alter these vessels, but the most common by far and the only one that need concern us here is that of atheroma. The name of this arterial disease is taken from the Greek term *athere*. Its literal translation is "porridge" or "gruel." It is a descriptive term referring to a late stage of the lesion in which the center of a thick plaque in the wall of the vessel has undergone a form of softening into a greasy porridge-like material. Like so many of the names of classical origin currently used in medicine, this particular feature of the lesion by which it happens to be known is a late one and, although probably important, is neither a cardinal nor an essential characteristic. Although the condition has been recognized in one or another of its forms since antiquity, its cause is still unknown. But several theories of its etiology have been developed and some have recently gained considerable strength through newly revealed evidence.

Aging is undoubtedly an important factor in the progress of the disease, because the lesions of atheroma are clearly more numerous, more advanced and more damaging in the old than in the young. But only a few decades ago many authorities shared the pessimistic view that atheroma was an inevitable consequence of the wear and tear that the years impose on vital arteries. In fact, this concept was propounded with great authority to the writer by his teachers when he was a medical student. Probably no other concept could have inhibited more effectively the imaginations of potential medical researchers in this field. It is probably safe to say that the once widespread acceptance of this view of atheroma is the single most important reason for the relative paucity of research in the field of cardiovascular disease up until World War II. Happily, we can say that this notion has now been completely discredited. Although age is still recognized as a factor which is important in the progression of the lesions, atheroma is no longer regarded by any responsible authority in the field as an inevitable consequence of growing old.

Heredity is currently recognized as another important factor controlling to some degree the rate of progression and distribution of lesions in the vascular system. But this view does not imply that some individuals are going to die of heart disease no matter what, because they come from a "coronary-family." The importance of heredity, including racial origins, in the pathogenesis of atheroma will be discussed more fully below. However, source of an individual's germ plasm is no longer regarded as the single, uncontrollable cause of heart disease. This defeatist view, once held by not a few eminent authorities is also now happily outmoded and no longer inhibits those who devote their lives to the search for ways of preventing or ameliorating the ravages of arterial damage.

Toxins, *poisons* and *infections* of various kinds have all been postulated as causes of atheroma from time to time. In the generally accepted sense of any of these terms, evidence to incriminate any specific environmental toxic agent or particular microorganism, either viral or bacterial, is now quite unconvincing. Although in both man and experimental animals arterial damage can be produced by these types of agents, the changes which result bear little specific resemblance to the lesion of atheroma in man.

Metabolic and *mechanical* concepts of the etiology of atheroma are most strongly supported by currently available evidence. Having dismissed the idea that atheroma is an inevitable result of aging, or is inherited in the form of some sort of hidden congenital anomaly, or is caused by some unidentified poison or organism, the remainder of this article will deal in somewhat more detail with notions which are favored by most authorities in the field nowadays. Probably most contemporary views, of which there are many, each differing from the other in various details, could be placed under one or the other of two chief headings or a combination of them. Although a great deal of new information relating to these two views has accumulated in only the past decade or two, they were each originally propounded many years ago in the nineteenth century.

The Lipid Imbibition Theory of Virchow. It was originally elaborated by this great German pathologist about a hundred years ago. Through the microscope he saw abnormal amounts of fat in the lesions of atheroma. This fat, unlike the normal lipid of the arterial wall, could be readily stained and visualized by suitable techniques. Virchow felt that the pathologically visible fat which he observed, entered the arterial wall from the bloodstream by a process of imbibition. This concept served to focus the minds of investigators then and now on factors that would elevate the levels of fat in the bloodstream—lipids of either a normal or an abnormal character. Virchow's hypotheses received an enormous impetus just after the turn of the century when another investigator, Anitschkow, discovered that he could significantly elevate the blood lipids of rabbits by feeding them large amounts of cholesterol either in the form of egg yolk or in a more purified state. In the arteries of these rabbits, Anitschkow found lesions not unlike those of atheroma in man. The reason he fed diets high in cholesterol (egg yolk) to his rabbits was because Virchow had observed that this particular form of lipid (cholesterol) was present in human atheroma. The resulting view that cholesterol in the diet or at least in the serum is the culprit of atheroma still dominates many researchers and clinicians in the field today. So strongly did these experiments of Anitschkow focus attention on the serum lipids that from that date until World War II, the limited research carried on in this field was restricted in a completely unbalanced way to the contents of the arterial lumen. The wall itself—the site of the lesion—was almost forgotten. This state of affairs persisted until the 1940's when an English pathologist of Scottish origin, Duguid, finally revived the opposite view of thrombogenesis to be described below, and thereby restored attention of researchers to the lesion itself instead of only to the serum bathing it.

The notion that atheroma is caused by abnormally elevated serum lipids, particularly cholesterol, which infiltrate the inner layers of the arteries is an attractive and relatively simple one. In an effort to explore this possibility further, much attention has been devoted in current approaches to the problem to find out what factors—dietary, hormonal, racial, those associated with stress, those under neural influence, etc.—may influence levels of circulating lipids. In this connotation, toxins and drugs have also been studied because of their potential effects on the level of fat in the bloodstream. Yet there is a respectable body of opinion today which feels that this entire concept involving blood lipids may be of only secondary importance in the etiology of arterial disease. There is no question that fat is increased in amount in atheromatous lesions affecting walls of arteries, and there is equally no question that cholesterol is one of the important fats concerned in this abnormal accumulation. Because this hypothesis is having such an important effect in directing the research efforts of so many, it is relevant here to present some of the data which support it. Some findings that do not fit the lipid imbibition theory will also be examined.

The very fact that abnormal amounts of fat, particularly cholesterol, are a constant and early component of the atheromatous lesion is responsible for the extensive studies that have been made on cholesterol in the diet and in the blood of both experimental animals and man. Once atheromatous-like lesions were produced in rabbits by feeding large amounts of dietary cholesterol, the results were speedily confirmed in innumerable laboratories the world over. Even today—more than a half-century later—there are few methods available to the experimentalist to reproduce atheroma in animals which do not involve cholesterol feeding or at least an abnormal elevation of cholesterol in the serum (produced by some means other than dietary, such as hormonal or pharmacologic). Until World War II, these approaches suffered from the criticism that atheroma had been produced by this method in only the rabbit and chick. It is now known that the rabbit's susceptibility is explained by the fact that it does not excrete cholesterol *via* its bile into the intestine as readily as do other species such as the rat, the dog or the monkey. As a result of this discovery, cholesterol feeding of other species was tried in combination with other procedures to depress similarly in them the excretion of cholesterol in the bile. These attempts (all in the past 10–15 years) were quite successful. Consequently, today the experimental atherologist is able to produce lesions for study not only in rabbits and chickens, but also in the rat, dog, pigeon, monkey, domestic pig, guinea pig, hamster, duck and others. An anatomist working in the prairies of Western Canada even found it was possible by this method to produce atherosclerosis in prairie gophers. But with all these species it has almost always been necessary to feed cholesterol in amounts greatly exceeding that ever consumed by man, accompanied often with equally abnormally large quantities of dietary fat or other agents which elevate the serum cholesterol and inhibit its excretion in the bile. These agents run from poisonous-like substances such as thiouracil, which adversely affects the thyroid gland, to adding bile itself to the diet. Probably the least extreme method so far devised has been that of feeding diets high in a saturated fat such as butter to domestic pigs. This procedure, which by itself does not produce lesions in most other species, unless cholesterol and other agents are also incorporated in the diet, after several months will produce atheromatous-like changes in the arteries of the pig. It should be noted that a number of biologists consider the domestic pig more closely related metabolically to man than many of the other experimental animals used in laboratories.

In man there are some diseases in which levels of serum cholesterol are abnormally elevated. The congenital disorder of familial hyperlipemia is an example. Serum cholesterols are very high in these individuals and they suffer more severe degrees of atheroma than the general population, thereby supporting the hypercholesteremic hypothesis here being discussed. Furthermore, many studies have now been completed in which the levels of animal fat and cholesterol in the diet as well as the levels of serum cholesterol have been compared with the amount of vascular disease as it varies in different parts of the world. Populations in China, South America, or Africa, as well as various ethnic groups throughout Europe and Asia, have been compared with population groups of England, the United States and Canada. In South America for example, the Guatemalans consume little animal fat or cholesterol, the amount of cholesterol circulating in their blood is low compared to North Americans and they suffer from much less disease of the heart and arteries than executives in New York. These geopathologic comparisons provide data that strongly support the notion that diets high in animal fat and cholesterol are atherogenic and may actually be the cause of heart disease in those socioeconomic groups which consume generous amounts of them in their diets as we in the United States do. But this evidence, which may at first seem almost overwhelming, is vulnerable to two rather serious criticisms. Philosophically, such studies yield results which, at best, can show only an *association* between the disease and these dietary habits; logically, they can never demonstrate a cause-and-effect relationship. An even more grave objection is the fact that in individual cases, a correlation between dietary habits, serum lipids and vascular disease frequently cannot be demonstrated. There have been a number of people studied in the United States who have died of heart attacks but who demonstrably never had exhibited abnormally high levels of cholesterol in their blood for as long as a decade prior to the time they died. In Canada, for instance, two carefully planned studies have been made with the object of looking for individual correlations between a high serum cholesterol and susceptibility to heart attack. One of these studies, now in its eleventh year, was carried out in a Veterans' hospital. The level of serum cholesterol in these men was measured four times a year. In a significant number of them who died of heart disease, there was no evidence that those with higher levels of serum cholesterol were any more susceptible to heart disease than those with levels falling within the normal range. In another study now in its fifteenth year, inmates of both sexes in a mental institution have been similarly studied. The numbers in this investigation are larger and, despite this fact, individual correlation of high levels of serum lipids before death and susceptibility of the patient to heart disease could not be established. But in New York City, a group of volunteers participating in what its originator, the late Professor Norman Joliffe, called "the coronary club" has been spared the expected rate of cardiovascular disease presumably because the dietary restrictions they follow (reduction of the amount of fat in the diet, correction of obesity, reduction of cholesterol and saturated fat in the diet, etc.) reduced their serum cholesterols and with it their susceptibility to heart attacks. However, this work, like another to be described, suffers from the serious criticism that the experimental designs did not provide adequate controls. In another "forward" project commonly referred to as the Framingham Study, which is now entering its tenth year, it has been demonstrated beyond statistical doubt that the expected incidence of heart disease in various members of this group is reduced if their levels of serum cholesterol are within the normal range rather than above. In this study, it has also been shown, however, that there are other factors of equal importance in predicting the rate of heart disease. These factors include obesity, moderate to heavy amounts of cigarette smoking, the existence of high levels of blood pressure, or a history in the individual's family of heart disease. The presence of diabetes enhances the effect of all these factors. These results emphasize the importance of a number of different influences rather than serum cholesterol *per se*.

Diet is probably the means most readily available to members of the general population who wish to reduce their levels of serum cholesterol. It has now been shown conclusively that it is perfectly feasible to lower the serum cholesterol in significant numbers of the normal adult population by manipulating the diet in two ways simultaneously. Firstly, foods that are rich in cholesterol itself, such as egg yolk and organ meats (e.g., sweetbreads, liver, kidney, etc.), are largely eliminated from the diet in order to reduce the total amount of exogenous cholesterol taken in by the individual. This method is an obvious outgrowth of the old experiments with rabbits initially reported more than half a century ago. In addition, a more recently devised and perhaps even more effective method of reducing the serum cholesterol by dietary means has been well established in man under a wide variety of conditions both within and outside of the laboratory. This method consists in reducing the total amount of fat in the diet from that commonly encountered in the average diet (where it supplies 40% of the calories) to a level that supplies only about 25% of the total caloric intake each day. In addition, the *kind* of dietary fat has to be altered. In the usual American diet, twice as much of the fat is of animal origin (saturated) as of vegetable origin (usually polyunsaturated; see FATTY ACIDS, ESSENTIAL). The ratio of different kinds of fat is usually expressed as the P:S ratio (polyunsaturated fat over saturated fat in the diet) and accordingly runs about 0.4 or 0.5% in the average American's food intake. By increasing the amount of vegetable oil consumed (*e.g.*, corn oil) and by reducing the amount of animal fat, it is possible to elevate this P:S ratio to 1.1 or even higher. When done even under quite free-living conditions, it has been found that the level of serum cholesterol often will be reduced by as much as 20% or even more in individual instances. Nevertheless, it must be emphasized that it still has not been shown that lowering the cholesterol in the blood by this amount will have any protective effect for the heart and vessels against the development of atheroma and the onset of its serious complications. A definitive (and very expensive) experiment which should provide such an answer has been planned and will be under consideration for support in the near future. Until data of this kind are available, it is a matter of considerable speculation to predict whether lowering the serum cholesterol by dietary means to this extent will protect the individual to any important degree from the risk of major heart disease. We have, however, set forth here the problem because it has been illustrated in no small measure how theoretical speculation concerning the cause of atheroma may influence practical decisions which seriously affect individual Americans and large Industries.

Mechanistic Theories. Mechanistic theories concerning the production of atheroma are concerned largely with the notion that the thickening of the wall and the formation of plaques projecting inwardly into the lumen are the result of precipitation and/or coagulation of one or more parts of the blood. It is generally conceded that the cause of death from heart disease results from one of the complications of atheroma, rather than atheroma itself, namely, occlusion of the lumen with obstruction of blood flow by a thrombus or clot forming within the vessel. The mechanistic theory of the etiology and pathogenesis of atheroma postulates not only that this complication of thrombotic occlusion of blood flow in affected arteries is the result of precipitation of blood by coagulation, but also that the atheromatous plaques *per se* represent smaller, non-occlusive thrombi flatted onto the inner side of the vessel. These mural thrombi, as they are usually called, according to some authorities, may have the same composition as ordinary blood clots containing precipitated red blood cells, fibrin, and the small, organized elements of the blood called platelets. Others feel that incorporation of red blood cells is minimal or absent and that the important constituents of such mural thrombi are fibrin and platelets. At first, one would think it was a matter of simple inspection under the miscroscope to determine whether the plaques of atheroma really resemble blood clots or not. But the matter is not so simple because it is generally recognized that the relatively high pressure within the arterial lumen may well alter the usual appearance of clots as seen in veins or other situations where pressure is not a major factor. Even the electron miscroscope has not settled this question conclusively to date, although it may be significant that it has so far provided little to corroborate the clot-theory. Even when it does seem quite clear in certain lesions that thrombi of one sort or another are present in the area of an atheromatous plaque, proponents of the lipid imbibition theory point out that such thrombi may be secondary in nature and have been superimposed on previously existent atheroma. The thrombogenic view was first proposed many years ago by a European pathologist, Rokitansky, but was almost forgotten when the consensus among students of the disease swung so largely in support of the lipid imbibition theory due to the influence of Virchow and particularly Anitschkow. In the early 1940's however, Duguid (already referred to) swung the pendulum back toward the Rokitansky theory by an ingenious experiment. He withdrew blood from rabbits and allowed it to clot in test tubes. He then broke the clot up into sufficiently small pieces that he could inject it back into the ear vein of the same animal. By using the same animal for the reinjection he thereby avoided sensitivity reactions. The injected clots were filtered out by arteries of the lung through which the clots were too big to pass. These entrapped thrombi underwent a series of changes, which Professor Duguid was able to follow in various animals that culminated eventually in lesions indistinguishable in most regards from those of recognized atheromatous plaques in man. Subsequent workers in several laboratories not only confirmed these experiments but also showed that if the animals were fed diets high in saturated fat (up to 40%) at the same time, there were more of the resulting plaques into which the injected and entrapped thrombi had been converted and they were bigger than if the diet contained small amounts of fat (2–10%). It has subsequently been shown that the mechanism whereby intravascular clots are normally dissolved is interfered with to some extent if the animals are fed abnormally high amounts of saturated fat. Polyunsaturated vegetable fat (40%) under similar conditions did not produce this enhancing effect on the lesions of rabbits injected with clots. It is apparent here that a dietary effect is also involved, but one operating in a different manner than that envisioned by the advocates of the lipid imbibition hypothesis of the etiology and pathogenesis of atheroma.

From a consideration of the above-mentioned findings it is apparent that a combination of the theories of lipid imbibition and the mechanistic theory of thrombus formation may actually be the hypothesis which will eventually best explain most of the available data.

The *distribution* of atheromatous plaques is by no means uniform throughout the arterial tree. The favourite sites for formation of plaques in most vessels are at their bifurcations or openings of tributaries of small diameter. At these places there is a certain amount of turbulence in the otherwise smooth flow of blood, a fact which has been demonstrated very beautifully not only in models of the cardiovascular system but also by direct observation in the living animal. Such sites of flow turbulence might well be expected to favor the precipitation of formed elements of the bloodstream in the form of mural thrombi. New evidence along these lines has recently been reported and is providing increasing support for the mechanistic thrombogenic theory. The fact that among Americans and Canadians, heart attacks are more frequent in those who suffer from abnormally high blood pressure (hypertension) also favors the thrombogenic theory. Against this point, however, it should be pointed out that in Africa, south of the Sahara, heart attacks among the native population are almost unknown, but hypertension is by no means rare. Again, even in the United States, the American Negro has a lower frequency of heart disease than the American White, but the comparative incidences of hypertension in the two races is the converse particularly in the third and fourth decades of life.

In summary, certain facts have been firmly established by the studies of atheroma in man and experimental animals so far referred to, but their interpretation remains difficult. Lipid is an important component of the lesion and, at least under some conditions, is associated with an elevation of similar lipids in the serum. Diet can influence the amount of fat in the bloodstream. But it is not clear in man that the amount of abnormal fat in the arterial wall reflects these variations in blood fat although animal experiments largely appear to support this view. Many factors have been found to influence the progression of the disease, but none has been established as an initiating event. Such factors include hormonal imbalances, stress, cigarette smoking, race and heredity. Other diseases such as hypertension and diabetes may aggravate the condition. There may be a single, important cause of atheroma, but it is clear that the lesions are also subject to the influence of many factors, and to this extent, atheroma can be regarded as a multifactorial condition. However, the very low incidence of heart disease in certain populations such as native Africans—even elderly ones—indicates clearly that if we knew enough we should surely be able to give Americans the same degree of protection as the Negro and some other races now enjoy for reasons not at all clear to us. It is obviously of paramount importance to us to find out exactly what these reasons are.

Reference

Reports of the Committee on Dietary Fat and Atheroma, The Food and Nutrition Board; Washington, D.C., The National Academy of Sciences—National Research Council, 1961 and later (in press).

W. Stanley Hartroft

ATOMIC ABSORPTION SPECTROSCOPY. See FLAME PHOTOMETRY AND ATOMIC ABSORPTION SPECTROSCOPY.

AUTOLYSIS

The energy derived from biological oxidations in living cells serves to promote ANABOLIC processes, *i.e.*, to produce relatively complex, highly ordered molecules and structures, and thus normally keeps living cells in a steady state remote from equilibrium. In organisms that lack cellular nutrients or oxygen (or in dead organisms or cells that have been disrupted so as to destroy much subcellular organization), the opposing catabolic tendency toward equilibrium, including the tendency toward degradation of macromolecules to simpler monomeric subunits, is not counterbalanced. These degradative processes, many of them enzymically catalyzed, are collectively termed autolysis. Autolytic processes may include, for example, hydrolysis of PROTEINS catalyzed by PROTEOLYTIC ENZYMES or hydrolysis of nucleic acids catalyzed by NUCLEASES. Autolysis of tissues (*e.g.*, liver homogenate) has sometimes been employed as a method for releasing bound molecules (*e.g.*, vitamins or coenzymes) into free soluble form.

AVIDIN

Avidin is a minor protein constituent of egg white (approximately 0.2% of the total protein) which has the distinctive property of combining firmly with BIOTIN (dissociation constant: $10^{-15}M$) and rendering it unavailable to organisms since PROTEOLYTIC ENZYMES do not destroy the avidin-biotin complex. It was discovered because of the observation that animals fed a diet containing large amounts of raw egg white (otherwise a nutritionally superior protein) suffered from "egg white injury," a condition which was proved eventually to be a biotin deficiency. Avidin looses its ability to combine with biotin when it is subjected to steam heat; hence, cooked egg white does not induce biotin deficiency. The isolation of avidin was facilitated because, in its presence, biotin is unavailable to yeast cells that normally require biotin for growth; such cells, unable to respond to biotin in the medium in the presence of avidin, were therefore used as convenient and rapidly responding test organisms for the bioassay of avidin.

The molecular weight range for avidin is 58,000–70,000, and the protein is composed of three or possibly four subunits, each subunit binding one molecule of biotin. Until recently, avidin was thought to occur only in the egg white of birds; however, a protein elaborated by *Streptomycetes* has biotin-binding properties similar to those of avidin, and the crystalline product (named streptavidin) was found to have many physical characteristics in common with avidin.

Although its biochemical function has not yet been established, avidin has proved to be an extremely useful tool (in addition to producing nutritional deficiencies) in establishing the role of biotin in enzymatic reactions. Unlike most of the other coenzymes, biotin is covalently bonded (amide linkage) to its apoenzyme and the biotin cannot be removed without complete destruction of the protein. However, avidin combines just as readily with biotin-containing enzymes as with free biotin and completely inhibits their enzymatic activities. Hence the susceptibility to avidin inhibition has been used as the best criterion for establishing whether or not an enzymatic process is biotin-dependent.

ROBERT E. EAKIN

AVOGADRO'S NUMBER

Avogadro's number is 6.02×10^{23}. It is the absolute number of molecules in a gram mole, whether the gram mole is of hydrogen (2.016 grams), oxygen (32.000 grams), sucrose (342.3 grams), etc., or a gram mole of an exceedingly complex virus weighing 10 metric tons.

B

BACTERIAL CELL WALLS (STRUCTURE AND BIOSYNTHESIS)

Bacterial cell walls are complex heteropolymeric molecules which envelop virtually all bacterial cells. They are well-defined morphological layers and have been visualized by staining methods as well as by electron microscopy. They are external to the cytoplasmic membrane (which is the permeability barrier of the cell) and, in encapsulated bacteria, they lie between the capsule and the membrane. Animal cells do not possess structures which are morphologically equivalent to the bacterial cell wall. Plant cells have a similar structure, which is chemically different from that of bacteria. L-forms or spheroplasts (protoplasts) of bacteria are organisms which have lost their cell walls or in which the mechanical rigidity of the cell wall has been lost. Such forms are produced by mutations in the biosynthesis of the cell wall or by treatment of bacterial cells with enzymes which hydrolyze bacterial cell walls, such as egg white lysozyme, or with agents which inhibit bacterial cell wall synthesis, such as penicillin or D-cycloserine. Most frequently, these spheroplasts are unstable and explode, since the cell membrane, in the absence of the cell wall, is unable to withstand the high internal OSMOTIC PRESSURE of most bacterial cells. These forms can be stabilized in media containing high concentrations of sucrose or salts. However, many bacterial species can form spheroplasts which are stable in ordinary media.

Cell walls can be prepared from bacteria after mechanical fragmentation of the cells with glass beads. The wall fraction, which is quite heavy compared to the other particulate components of the cell, can then be separated by differential centrifugation and subsequently purified with the aid of nucleolytic or proteolytic enzymes. Proteins which may be part of the cell wall are, however, removed by this purification procedure. When examined by electron microscopy, cell wall preparations have the shape and rigidity of the bacterial cell from which they were derived, *i.e.*, bacilli have rod-shaped cell walls and cocci have globoid cell walls. Physiological experiments with enzymes which hydrolyze cell walls or with agents which inhibit cell wall synthesis also led to the conclusion that it is the wall which has the form and shape of the microorganism from which it is derived.

Examination of the components of acid hydrolysates of cell walls has revealed that all bacterial cell walls contain a structure, termed a glycopeptide, which is very similar in different bacterial species. Its invariable constituents are two amino sugars (see also HEXOSAMINES) and four AMINO ACIDS. The amino sugars are N-acetylglucosamine and N-acetylmuramic acid (a 3-O-D-lactic acid ether of N-acetylglucosamine which was first isolated from a hydrolysate of spore peptide). The amino acids are L-alanine, D-alanine D-glutamic acid and either L-lysine or α, ε-diaminopimelic acid. Several of the three isomers of diaminopimelic acid (DD-, LL- or *meso-*) have been found in different bacterial species. The presence of acetylmuramic acid provides a chemical definition of a bacterial cell because no other type of cell is known to contain this sugar. D-Amino acids are found in very few other substances of natural origin and diaminopimelic acid has been found elsewhere only in a few blue-green algae. In addition to the components indicated above the cell wall glycopeptide in some bacteria may contain additional amino acids such as glycine or aspartic acid.

These components are put together in a three-dimensional network of exceedingly high molecular weight. Fundamentally, this network is composed of polysaccharide chains in which the two sugars alternate. A tetrapeptide, the sequence of which is L-ala-D-glu-L-lys-D-ala or L-ala-D-glu-DAP-D-ala, is linked to the carboxyl group of the muramic acid residues. Chains of these polysaccharide-tetrapeptide units are cross-linked to each other through peptide bridges to form the three-dimensional network. The nature of these peptide cross-links varies greatly among different bacteria; *e.g.*, in *Staphylococcus aureus* it consists of pentaglycine units which link the ε-amino group of lysine to the carboxyl group of the terminal D-alanine residue. In *Micrococcus lysodeikticus* the link appears to be a direct one between lysine and D-alanine with no intervening amino acid. The structure proposed for the glycopeptide component of one bacterial cell wall, that of *Staphylococcus aureus*, is shown in Fig. 1. At the present time, except for scanty information available on the nature of the cross-bridges, insufficient information is available on the detailed structure of a sufficient number of bacterial cell wall glycopeptides to state how much variation in structure may occur.

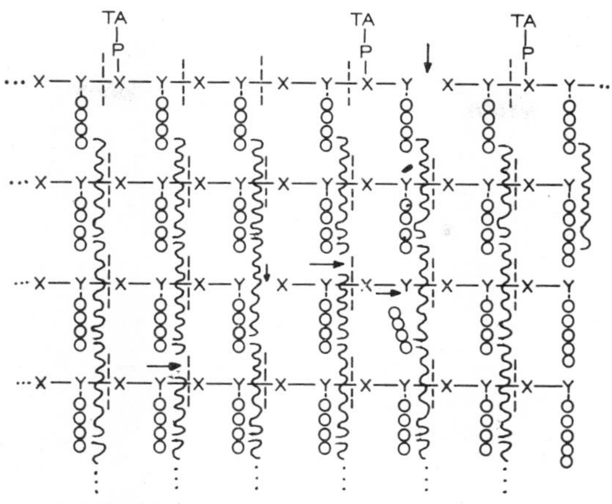

X = N-ACETYLGLUCOSAMINE
Y = N-ACETYLMURAMIC ACID OR N,O-DIACETYLMURAMIC ACID
O = AN AMINO ACID OF THE TETRAPEPTIDE OR PENTAPEPTIDE
∿∿∿ = A PENTAGLYCINE CROSS-BRIDGE
---- = BONDS CLEAVED BY ACETYLMURAMIDASE
→ = RANDOM BREAKS IN THE STRUCTURE

FIG. 1. A representation of the structure of the glycopeptide component of *Staphylococcus aureus* cell wall.

The glycopeptide is the rigid component of the bacterial cell wall and its integrity is essential for the maintenance of the life of the bacterial cell. Cell walls are solubilized with consequent lysis of the cell by bacteriolytic enzymes. The first example of this class of enzymes was egg white lysozyme discovered forty years ago by Sir Alexander Fleming. This enzyme is a glycosidase and it hydrolyzes the linkage between acetylmuramic acid and acetylglucosamine. Other glycosidases which hydrolyze bacterial cell walls have also been recognized, and in addition, an increasing number of peptidases have recently been discovered which selectively hydrolyze one or another linkage in the peptide network. Through the use of these enzymes, it has become clear that the insolubility of the bacterial cell wall is not due to the insolubility of any of the polymers which comprise it but is rather the consequence of a tight cross-linked network. Cleavage of any linkage within this network, whether a peptide link or a glycosidic link, results in solubilization of the wall. In addition to the glycopeptide, as far as is presently known all bacterial cell walls contain an additional structure, referred to as *special structure*. The nature of these materials varies greatly among different organisms. These substances are of great importance because they are often the mediators of responses between bacteria and infected animals. The specific immunological responses of animals to infection are directed against components of the special structure of bacterial cell walls. Similarly, toxic responses to infection (such as fever) are frequently also responses to chemical constituents within the special structure. The wide variety of substances found is evident from a listing of a few of the materials whose chemical structures have been investigated: the group-specific carbohydrates of streptococci and the M protein of streptococci, the teichoic acids of staphylococci and other Gram-positive bacteria, the lipopolysaccharide and protein of enterobacteria, the polysaccharide of corynebacteria, and the exceedingly complex lipid ("waxes") found in mycobacteria. The presence of lipids in the cell walls of Gram-negative bacteria is a striking feature which distinguishes them from the cell walls of Gram-positive bacteria. However, the precise basis for the Gram stain, probably related to cell wall structure, is presently unknown.

Studies of the biosynthesis of bacterial cell walls are inextricably woven with studies of mechanism of action of antibacterial agents (see ANTIBIOTICS). It was originally found that penicillin induced formation of spheroplasts in *Escherichia coli* and induced accumulation of uridine nucleotide in *Staphylococcus aureus*. Both of these phenomena are related to inhibition of bacterial cell wall synthesis by penicillin. The uridine nucleotide which accumulates in *S. aureus* is UDP-acetylmuramyl-pentapeptide. It is a precursor of the bacterial cell wall glycopeptide and its accumulation is a consequence of inhibition of wall synthesis by penicillin. A number of other antibacterial agents (vancomycin, ristocetin, bacitracin, cycloserine) are selective inhibitors of bacterial cell wall synthesis. *i.e.*, under conditions in which these substances inhibit cell wall synthesis no other metabolic process is known to be inhibited. By contrast, novobiocin and gentian violet are nonspecific inhibitors of cell wall synthesis.

In addition to inhibition by antibiotics, a number of other means are known by which nucleotide accumulation can be induced.

Different antibiotics and different treatments may lead to accumulation of different nucleotides. The nucleotides which accumulate represent a biosynthetic sequence (Fig. 2). One of the antibiotics mentioned above is a specific inhibitor of two of the enzymatic reactions required for this reaction cycle, namely, alanine racemase and D-alanyl-D-alanine synthetase. This is one of the few known examples in which the effect of an antibiotic on an enzymatic reaction has been demonstrated.

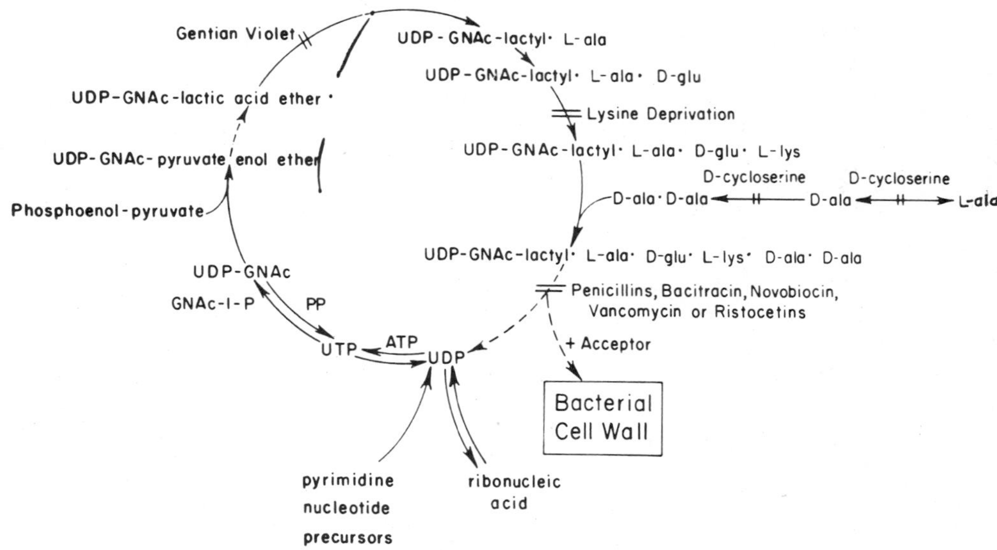

FIG. 2. Cell wall biosyntheis. The reaction cycle which leads to the synthesis of UDP-acetylmuramyl-pentapeptide in *Staphylococcus aureus*. Points of inhibition by antibiotics and other agents are indicated. GNAc represents N-acetylglucosamine; UDP, uridine diphosphate; UTP, uridine triphosphate.

The other antibiotics mentioned above interfere in the reaction sequence in which UDP-acetylglucosamine and UDP-acetylmuramyl-pentapeptide are utilized for the synthesis of the glycopeptide. In this sequence (Fig. 3), the two sugars which comprise the glycopeptide are transferred alternately from the nucleotide to a lipid carrier situated in the cell membrane. Repeated cycles of the type illustrated lead to the formation of a linear glycopeptide. In the case of *Staphylococcus aureus* and *Micrococcus lysodeikticus* and presumably in other bacteria, additional amino acids found in the glycopeptide are also transferred to the lipid intermediate. In *S. aureus*, for example, a pentaglycine chain is transferred by sequential transfer of glycine residues from glycyl-sRNA to lipid intermediates. The antibiotics vancomycin and ristocetin interfere specifically with the utilization of the lipid intermediates for glycopeptide synthesis, but at physiological concentrations of antibiotic they do not hinder the formation of these intermediates. Penicillin interferes with the terminal reaction, a transpeptidation in which linear glycopeptide strands are cross-linked to form the 3-dimensional network.

Considerable effort has also been devoted in recent years to studying the biosynthesis of several types of special structure. The teichoic acids have been extensively studied, as have the *lipopolysaccharides*, especially of *Salmonella*. In this genus, the polysaccharide is exceedingly complex and may contain eight or more sugars. Some of the terminal sugars are novel substances, 3,6-dideoxyhexoses, and these sugars have been

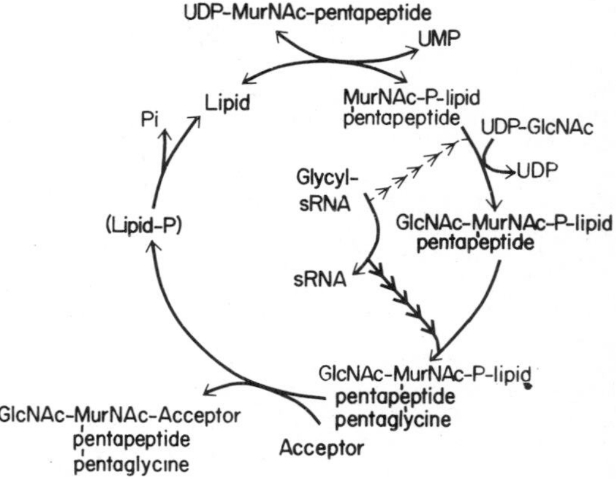

FIG. 3. The lipid cycle in bacterial cell wall glycopeptide synthesis. MurNAc represents an acetylmuramyl group.

shown to be important immunological determinants. The sequential transfer of sugar residues, activated as nucleotides, to form these complex polysaccharides is at present an area of extremely active investigation.

A great deal more work on the comparative biochemistry and biosynthesis of bacterial cell walls remains to be done.

References

1. SALTON, M. R. J., "The Bacterial Cell Wall," Amsterdam, Elsevier, 1964.
2. MCCARTY, M., AND MORSE, S. I., "Cell Wall Antigens of Gram-Positive Bacteria," *Advan. Immunol.*, **4** (DIXON, F. J., AND HUMPHREY, J. H., EDITORS), 249 (1964).
3. LÜDERITZ, O., STAUB, A. M., AND WESTPHAL, O., "Immunochemistry of O and R. antigens of *Salmonella* and Related Enterobacteriaceae," *Bacteriol. Rev.* (in press).
4. STROMINGER, J. L., "Biosynthesis of Bacterial Cell Walls," in "The Bacteria," Vol. III (GUNSALUS, I. C., AND STANIER, R. Y., Editors), p. 413, New York, Academic Press, 1962.
5. OSBORN, M. J., ROSEN, S. M., ROTHFIELD, L., ZELEZNICK, L. D., AND HORECKER, B. L., "Lipopolysaccharide of the Gram-negative Cell Wall: Biosynthesis of a Complex Polysaccharide Occurs by Successive Addition of Specific Sugar Residues," *Science*, **145**, 783 (1964).
6. IZAKI, K., MATSUHASHI, M., AND STROMINGER, J. L., "Glycopeptide Transpeptidase and D-Alanine Carboxypeptidase: Penicillin-Sensitive Enzymatic Reactions," *Proc. Natl. Acad. Sci. U.S.*, **55**, 656 (1966).

JACK L. STROMINGER

BACTERIAL GENETICS

Bacteria are microscopic, unicellular plant cells bounded by a membrane-wall complex and containing a variety of inclusions. Depending upon the species and cultural conditions, bacteria occur as individual cells or in clumps or chains of sister cells. Titration (assay) of the number of bacteria in a suspension is influenced by the pattern of growth as well as by random sampling errors. Quantitative enumeration of the number of bacteria in a suspension can be achieved by one of the following methods: (1) *Serial dilution and plating* of a known volume of the final dilution onto solid growth medium. Upon incubation of the plate, each *living* cell forms a visible group of descendants (called a clone if the organisms are descended from a single cell). (2) *Measurement of the optical density of the suspension* and calibration by comparison with some other method. (3) *Direct microscopic counts* in a known, small volume of suspension (Petroff-Hausser counting chamber). (4) *Direct count of the number of particles* in a small aliquot of the suspension with a Coulter counter. As each particle passes through a capillary tube, the resistance across the capillary is altered and the count is automatically recorded.

Nuclear Bodies. The DEOXYRIBONUCLEIC ACID (DNA) of bacteria is predominantly located in masses of variable shape, nuclear bodies, unbounded by a nuclear membrane. Bacteria thus are classified as procaryoids, in contrast to higher organisms containing nuclear membranes, the eucaryoids. In general, non-growing, stationary-phase bacteria contain one nuclear body per cell, whereas exponentially growing, log-phase bacteria contain two or more nuclear bodies per cell. These nuclei are the sister products of a preceding nuclear division.

When bacteria are inoculated into growth medium, there is a delay (lag phase) before division and exponential growth ensue. The rate of exponential growth is a characteristic of the bacterial strain, the temperature, and the nutritional environment. The amount of DNA per nuclear body remains constant at various growth rates, although cell mass and average number of nuclei per cell (and, consequently, the content of DNA per cell) are functions of the growth rate. Studies on induced mutations and results reported below show that the nuclear bodies bear most of the "genetic information" of the cell. Evidence cited below conclusively demonstrates that it is the DNA of the nucleus that contains this information. Cytoplasmic constituents, however, influence the expression of the genetic information contained in the DNA; they play a role in the hereditary capacities of the cell. For example, ability to manufacture typical cell wall material (see BACTERIAL CELL WALLS) relies on cell surface components (*e.g.*, possibly polysaccharide primers upon which additional constituents are laid) as well as on nuclear genetic factors. In another case, flagella, once synthesized, are passed from the parent bacteria to the daughter cells during vegetative multiplication.

Bacterial DNA. Chemical analyses and electron microscope observations indicate that bacterial DNA is *not* associated with *typical* histones; the secondary phosphoric acid groups of the DNA presumably are neutralized by POLYAMINES and inorganic cations. The DNA is composed of two, long, complementary deoxy-polynucleotide chains wound about each other in a Watson-Crick double helix. The genetic information contained in the DNA lies in the specific sequence in which the PURINE and PYRIMIDINE bases are ordered along the polynucleotide chain. The purine adenine (A) content equals the complementary pyrimidine thymine (T) content; the purine guanine (G) content equals the complementary pyrimidine cytosine (C) content. These two sets of bases are paired, by hydrogen bonding, in the double-stranded DNA (AT and GC base pairs). Minor quantities (less than 0.1%) of 5-methylcytosine and 6-methylaminopurine (methylated A), of unknown function, also are present. Certain analogues, *e.g.*, 5-bromodeoxyuridine, are readily taken up from the growth medium and incorporated into the DNA's of some bacteria.

The relative proportions of AT/GC vary from about 0.25–0.75 for different bacterial species, but DNA's of species classified together on morphological, biochemical, and immunological grounds resemble one another in composition. These latter DNA's also have similar nearest-neighbor frequencies (*e.g.*, the percentage content of dApdT) in their base sequences, whereas two DNA's from unrelated species may have DNA's similar in over-all base content but very dissimilar in nearest-neighbor frequencies. The non-covalent bonds holding together the two strands of the DNA can be melted out by heating *in vitro*, resulting in separation of the strands. The melting temperature in heat denaturation and the mean buoyant density of the DNA, as determined by centrifugation experiments, reflect its relative AT/GC content. Upon slow cooling, the two strands of the DNA, if present in adequate concentration, reanneal with each other. Separated strands of two different, but closely related, bacterial species form mixed, "hybrid" molecules upon reannealing. The proportion of hybrid molecules parallels the genetic relatedness of the two strains. Sensitive *in vitro* tests of genetic relatedness rely upon the base composition and sequence of the bases in the deoxypolynucleotide chains (see also CHEMOTAXONOMY).

Each nuclear body of most bacteria contains but a single complement of genetic information; *i.e.*, most bacteria are haploid organisms. The relative DNA contents per nucleus vary markedly among different bacteria. The haploid DNA complement ranges in molecular weight from approximately 5×10^7, as reported for some free-living, bacteria-like organisms called pleuropneumonia-like organisms (PPLO), through 2.8 $\times 10^9$ for the intestinal bacterium, *Escherichia coli*. The entire haploid content of DNA has been extracted intact from isotopically labeled *E. coli* cells. Autoradiographs show that the DNA complement exists in the form of a closed circle about 1100 μ in length. DNA of about 4×10^8 molecular weight, representing about one-half of the nuclear DNA complement, has been extracted from *Hemophilus influenzae* and subjected to physicochemical analysis. Such elongate molecules, being highly sensitive to shearing forces, are readily broken down to smaller molecules of 10^6–10^7 molecular weight, a size relatively resistant to the shearing forces encountered in standard laboratory procedures such as pipetting. There are few critical data to evaluate in the current debate as to whether the giant "DNA molecule" comprising the bacterial nucleus is composed of two continuous, uninterrupted deoxypolynucleotide chains or is composed of a series of shorter molecules of about 5×10^5 molecular weight joined by phosphoserine-containing peptides.

DNA Synthesis. DNA synthesis is initiated at a particular site on the giant bacterial DNA molecule and proceeds unidirectionally to completion. A single "growing point" proceeds sequentially around the circular molecule. Some process involving biosynthesis of a special protein is necessary for initiation of a new round of synthesis. A single round of DNA synthesis requires about the same length of time under different growth conditions; therefore, DNA replication occupies nearly the entire division cycle in bacteria dividing rapidly on rich mediums whereas smaller fractions of the generation time are consumed by DNA replication on poorer mediums where bacterial growth and division proceed more slowly. As the DNA replicates, longitudinally hybrid molecules are formed; each of these is composed of one conserved "old" unit and one newly synthesized unit. The integrity of the "old" unit, one of the two polydeoxynucleotide chains originally present, is maintained through the next replications ("semi-conservative replication").

Polynucleotide chains resembling bacterial DNA's have been synthesized *in vitro* by enzyme systems extracted from various bacteria. These syntheses, by DNA polymerases, require the presence of all four deoxynucleoside triphosphates (dATP, dGTP, dTTP, dCTP) and a DNA primer. The synthesized DNA reflects closely the composition, and at least many of the base sequences, of the primer DNA, indicating that the system contains the high degree of specificity required for the synthesis of genetic material. No conclusive evidence, however, has been presented for the test tube synthesis of biologically active DNA, although extensive attempts at this synthesis have been made. *In vitro*, DNA polymerase in the absence of primer can synthesize DNA-like polymers (*e.g.*, a deoxypolynucleotide chain of alternating A and T); such polymers have not been detected in bacteria but are present in certain tissues of some higher organisms [see also DEOXYRIBONUCLEIC ACIDS (REPLICATION)].

Exposure of DNA to ULTRAVIOLET RADIATION causes chemical changes in the DNA. Among the changes are a water-addition product of C, creation of T dimers, deamination of C to uracil (U), and possible formation of U dimers and mixed T-U dimers. One or more of these, or other unknown, changes block DNA synthesis, cause mutations, and lead to death of irradiated bacteria. Several enzymes in bacteria are able to repair some of the ultraviolet-induced lesions. One such enzyme acts by excising T dimers (along with one or several adjacent nucleotides); the "knick" in the affected DNA strand is then filled by Watson-Crick base pairing, and the continuity of the chain is restored by a second enzyme, probably DNA polymerase. A different repair enzyme, for which ultraviolet-irradiated DNA serves as substrate, will adsorb to the DNA in the dark, but consummation of the repair reaction requires irradiation with light in the visible region of the spectrum (photoreactivation). Recent data indicate that some enzymes involved in ultraviolet repair mechanism also participate in mediating recombination of genetic material carried on different DNA molecules ("crossing over," see below).

In summary, most of the genetic information of bacteria is contained in a single structure of fixed DNA content, a giant circular DNA molecule that replicates semi-conservatively. The enzymatic reactions involved in the biologically fundamental processes of DNA biosynthesis and genetic recombination are being elucidated in studies with bacterial systems.

Mutation. Mutations are genetic changes which occur suddenly and are thereafter heritable. Mutations arise through three general mechanisms: (1) chemical modification of preformed DNA, such as breakage and aberrant reunion of molecules or the changes elicited by ultraviolet light mentioned above, (2) errors in incorporation of the purine and pyrimidine bases, or additions and subtractions of bases, during DNA replication, and (3) unequal exchange between two identical or similar DNA molecules ("unequal crossing over") during recombination. These chemical changes normally occur with low frequency (spontaneous mutations), but the frequency can be greatly increased by means of various chemical and physical treatments (induced mutations). Even when so induced, the frequency of bacterial mutants for a particular trait usually is low—*e.g.*, one mutant in 10^4–10^{10} bacteria (see MUTAGENIC AGENTS).

The mode of action of a number of chemical mutagens is under investigation. Two examples of proposed mechanisms illustrate the approaches that have been used. Treatment of non-dividing bacteria with diethylsulfate results in appearance of mutations during subsequent growth. Chemical tests with cell-free DNA show that diethylsulfate ethylates A to 3-ethyladenine and G to 7-ethylguanine (EG). The presence of the ethyl group on EG weakens the glycoside linkage binding the base to the DNA backbone. This occasionally results in the release of EG from the DNA. The predominant (but not the only) cause of diethylsulfate-induced mutations appears to be the EG *still present* in DNA at the time of DNA replication; EG ionizes more readily than does G to a form that can base-pair with the forbidden partner T rather than the customary partner C. The mutation can therefore change the genetic code in the DNA through the following steps: (1) G/C is chemically altered to EG/C; (2) EG pairs with T during replication, giving rise to a molecule containing the EG/T base pair; (3) during the next round of replication of the molecule containing EG/T, T pairs with its normal partner, A. The over-all change is from G/C to A/T.

An example of the process of induction of mutations during DNA replication is the incorporation of 5-bromodeoxyuridine (BD) in place of the similar base T. BD contains a bromine atom in place of the methyl group found in T. Because of the difference in charge distribution on the base, BD pairs with the abnormal base C much more frequently than does T (pairing of BD with C is still a rare event). If the *initial* entry of BD takes place opposite a C in the DNA during replication, the "mistake in incorporation" leads to a BD/A base pair at the next and further rounds of replication. The over-all change, from a G/C base pair to an A/T base pair, is completed in two rounds of replication. The mutations occur and are expressed within several divisions of the bacteria following administration of BD. This early expression contrasts with the events when BD normally is taken up in place of T, creating BD/A base pairs. Most of the time these replicate as BD/A base pairs. Occasionally, however, BD makes a "mistake in replication," pairing with a G. Thus, the change in the genetic code is from an A/T base pair to a G/C base pair, and the mutations appear over a large number of divisions in the absence of exogenous BD.

The above mutational processes involve the replacement on the DNA of one nucleotide pair by another nucleotide pair. Diethylsulfate and BD largely cause point mutations, involving single sites or positions in the DNA molecule. Multisite mutations, involving several or many adjacent sites, are characterized through failure to recombine in genetic tests with a number of single-site mutations that freely recombine with one another. Multisite mutations are presumed to be deletions of portions of the DNA molecule since they never back-mutate to the wild-type form and since only the entire mutation may be replaced during recombination.

Several chromosomally localized mutator genes have been described in bacteria. Strains carrying a mutator gene show a high incidence of "spontaneous" mutation; the mutants are themselves unstable when in the presence of the mutator gene. One such mutator gene appears to lead to the accumulation in the cell of a base not normally present in amounts large enough to engender significant incorporation into DNA; the accumulated base may cause mutation by mechanisms analogous to that mentioned above for BD. The mode of action of other mutator genes awaits investigation. It is possible that some mutations are occasioned by the interaction of chromosomal DNA with DNA normally foreign to, or somehow rearranged in, the chromosome: *e.g.*, a bacterial virus (bacteriophage) has been described that penetrates bacteria, interacts in some fashion with various regions of the chromosome, and causes an alteration in the functioning of genetic material in its vicinity. This mechanism of "mutation" may have some specificity since it involves interactions between polynucleotide chains. Otherwise, mutation is a random process; no noncontroversial demonstration of locus-specific induced mutation has been made. That is, no chemical agents have been found that will induce mutations specifically in one small portion of the genetic material without influencing other regions almost equally.

Genetic Markers. Since 10^9 bacteria may be rapidly grown vegetatively from a single cell and spread onto a single plate of chemically defined medium, rare new phenotypes are readily detected. In many cases the precise phenotype presents clues important to the specification of the biochemical reaction(s) most directly influenced by the mutation.

Some mutants can be directly selected by plating large numbers of bacteria on a toxic medium which prevents growth of all but the rare resistant cells in the population:

(1) *Relative resistance to* ANTIBIOTICS such as streptomycin. Streptomycin resistance can be polygenically controlled (low-level, multi-step resistance), can be due to a single genetic change leading to high-level resistance, or can be due to dependence. While the precise mode of action of streptomycin, as of most antibiotics, is unknown, it is thought that streptomycin binds to RIBOSOMES and interferes with biosynthesis of legitimate proteins. The mutations to high-level resistance and to dependence would lead to production of resistant ribosomes through alterations in the structure of a ribosomal component.

(2) *Relative resistance to physical agents.* Resistance to irradiation with ultraviolet light is discussed above. Underlying biochemical mechanisms remain to be resolved for most physical agents.

(3) *Relative resistance to antimetabolites*, such as amino acid and purine analogues, nitrofurans, azide, and sulfonamides. Antimetabolite resistance has been attributed to different mechanisms in different situations: (a) Failure to take up the agent (*e.g.*, in rhamnose, azaserine, and 5-fluorophenylalanine resistance); (b) loss of an enzyme necessary for conversion of the inhibitor to an active substance (*e.g.*, nitrofuran reductase; fluorouracil riboside kinase); (c) alteration in an enzyme to a form with increased specificity for the natural substrate and decreased affinity for the analogue (*e.g.*, phenylalanyl-sRNA synthetase in 5-fluorophenylalanine-resistant bacteria); (d) alteration in an enzyme normally sensitive to inhibition of activity by an end product of the reaction sequence so that it is inhibited by neither the normal metabolite nor its analogue (*e.g.*, thiazolealanine-resistant mutants that excrete L-histidine); (e) loss of sensitivity to an analogue that serves to keep particular enzyme levels low through repression (*e.g.*, many triazolealanine-resistant mutants contain high levels of the enzymes involved in histidine biosynthesis); (f) gain in ability to detoxify the inhibitor.

(4) *Resistance or partial resistance to biological agents* such as bacteriophages or bacteriocins. Such mutants generally lack the specific receptors in their cell walls necessary for adsorption of the otherwise lethal agents.

Mutations affecting pigment production or colonial morphology and "dark" mutants of luminous bacteria also may be directly selected. Other mutants require enrichment techniques prior to their isolation. In the most widely utilized technique, the bacteria are placed in medium containing penicillin after a culture has been treated with a mutagen and allowed to grow for expression of mutations. A main site of penicillin action is prevention of new cell wall synthesis without much influence on protoplasmic growth. When the treated bacteria "outgrow" their cell walls and lyse, survivors are the viable cells that are unable to grow on the particular medium. In this general manner, one may select for several types of mutants: (1) Inability to utilize a compound as a carbon or nitrogen source. Such mutants are relatively or completely deficient in active uptake of a compound ("permease" mutants) or in one or several enzyme activities active in metabolizing the compound. (2) Inability to synthesize a metabolite required for growth (*e.g.*, amino acids, purines, pyrimidines, vitamins, hemin, chlorophyll). Many such mutants lack adequate amounts of a specific biosynthetic enzyme or of several enzymes or contain catalytically defective enzyme proteins.

Figure 1 depicts a genetic map of the chromosome of *Salmonella typhimurium*. The various genes, noted on the map and described in the legend indicate the wealth of traits available for genetic analysis in one bacterial species. Methods utilized in obtaining the genetic map are described in the sections below.

Recombination. Recombination in bacterial cultures results from a unidirectional transfer of genetic material from the donor parent to the recipient. The process involves three main steps: (1) mobilization of a portion of the genetic complement of the donor parent, (2) transfer from the donor bacterium to the recipient, and (3) integration of the transferred genetic material into the chromosome of the recipient bacterium. There are several different modes of transfer, each requiring its own characteristic process for the mobilization of donor material. These processes are roughly classified below under the categories of *transformation*, *infectious heredity*, and *conjugation*. Once transferred into the recipient bacterium, the genetic material from the donor may undergo one of the following, alternate fates: (a) the newly introduced DNA may persist and function without undergoing replication or integration; (b) it may be integrated in an unstable fashion and replicate more or less in synchrony with the chromosomal material of the recipient bacterium; (c) it may be stably integrated, replacing homologous material from the chromosome of the recipient.

The process of integration, of recombination at the chromosomal level, is just beginning to be understood. The process ("crossing over") appears to involve the breakage and reunion of DNA molecules, mediated by a series of bacterial enzymes.

Careful analysis of the mechanisms of genetic exchange in bacteria has been secured from test tube experiments. There is, however, ample evidence that the processes occur widely in nature, are important sources of bacterial variability, and lead to the origin of novel types of bacteria. Indeed, in environments containing complex swarms of antagonistic antibiotics and antibodies, the versatility of bacterial recombination processes often raises severe problems for physicians.

Transformation. Genetically active DNA may be extracted from donor bacteria and highly purified *in vitro* to contain less than 0.02% protein and no detectable uracil or serologically reactive substances such as capsular polysaccharides. During its extracellular existence, the DNA is sensitive to degradation by the enzyme deoxyribonuclease (see NUCLEASES). Genetic activities may be differentially inactivated by chemical and physical treatments of the cell-free DNA, and the DNA can be subjected effectively *in vitro* to chemical mutagens.

DNA is taken up by only those cells in a population which are "competent." The development of competence requires protein synthesis. In some cases, cell division and accessory factors, such as calcium and serum albumin, are required. Competent bacteria contain about 50 sites to which DNA reversibly adsorbs, but possibly only two or several sites are responsible for irreversible uptake. At high concentrations, DNA molecules compete for the DNA-adsorbing sites. A minimum molecular weight of about 10^6 is necessary for adsorption and uptake. Double-stranded DNA is much more readily taken up by most bacteria than is single-stranded DNA. The actual uptake of DNA from the adsorption sites is irreversible. Adsorption and uptake require no DNA or protein synthesis on the part of competent bacteria. (See also TRANSFORMING FACTORS).

The events following uptake seem to differ in different bacteria. In experiments with one bacterial species, biologically active DNA can be re-extracted quantitatively after uptake. The integration of the input DNA into the chromosomal material appears to involve some sort of breakage and reunion of double-stranded DNA molecules in the absence of extensive DNA synthesis. In other experiments with other bacteria, the biological activity of the DNA disappears upon uptake; one strand of the DNA is degraded and the second strand shortly thereafter finds its way into the continuity of the host chromosomal DNA. Again, this process does not appear to require extensive DNA synthesis. Work in progress on the enzymatic mechanisms involved should yield results of importance to interpretation of recombination processes at the molecular level.

Kinetic experiments show that irreversible uptake of one DNA molecule is sufficient for transformation of one bacterium. In suitable conditions, the number of transformed clones produced by a DNA preparation approximates the number of nuclei present in the bacteria from which the DNA was extracted. The expression of transformed characteristics requires RNA and protein synthesis and takes place in some cases before the transformed bacteria begin dividing. When DNA is denatured and the two complementary strands are partially separated by centrifugation techniques, the function of one of the strands is expressed very soon after uptake but expression of the second strand requires intervening DNA synthesis. An interpretation of this result is that only one of the two DNA strands functions in transcription (*i.e.*, as template for synthesis of the messenger RNA which serves as an intermediate during PROTEIN BIOSYNTHESIS —see Fig. 3).

The incorporation (integration) of genetic factors into the genome of the recipient bacteria involves replacement, in some of the daughter cells, of homologous material originally present in the recipient bacteria. Once the input DNA has been integrated into the replicating genome, a stably altered clone of descendents is built up by vegetative multiplication. DNA extracted from this clone is again able to transform the same trait in still further bacterial recipients.

Linked loci are sometimes transformed together by a single DNA molecule ("joint or linked transformations"). The extent of linkage depends upon the proximity of the two markers on the bacterial chromosome and the size of the DNA molecules used in the transformation experiment.

Analogous to transformations are infections of bacterial cells with DNA and RNA extracted from bacterial viruses, tissue culture infections with animal virus nucleic acids, and infection of plant cells with plant virus RNA. In these cases release of complete, infective virus particles ensues. These systems clearly demonstrate that genetic material is not solely DNA; some RNA's also possess genetic properties (*i.e.*, the RNA's are accurately replicated and transmitted to progeny whose properties they dictate). One biologically active, non-viral, messenger RNA has been extracted from sulfonamide-resistant bacteria or synthesized by RNA polymerase along DNA extracted from such cells. Such RNA phenotypically "transforms" genetically sensitive bacteria to partial sulfonamide resistance. This non-inherited, transient, resistant state presumably results from the direction, by the input RNA, of the synthesis of sulfonamide-resistant enzyme protein. This case is perhaps the first direct demonstration of the directive role in protein synthesis of a specific, non-viral RNA.

Infectious Heredity; Transduction. Bacteriophages inject their DNA (or their RNA, as the case may be) into bacteria. The nucleic acid is replicated, elicits the synthesis of phage-specific proteins, and eventually matures into infectious phage particles (the vegetative or lytic cycle). Alternatively, with "temperate" viruses, the DNA may remain in a noninfectious state (prophage state) and be replicated more or less in synchrony with the chromosome of the bacterial host (lysogenic response). Some prophages are associated with specific locations on the bacterial chromosome; their location may be mapped as for any other bacterial genes. The site of this integration is determined by just a portion of the total phage genome in conjunction with the specific gene locus (or loci) of the bacterial host. It now appears that the genetic material of the virus is inserted in the continuity of the chromosome. Other viruses may be integrated at one of many chromosomal locations. Still other prophages appear unassociated with the chromosome.

In the prophage state, the genetic material of the virus is presumably "naked"; it is noninfectious. Each cell containing the prophage, however, carries the hereditary potential to produce bacteriophage under suitable conditions. The prophage may mutate in any of a number of genes essential for its vegetative multiplication and maturation. Defective prophage may be genetically blocked in an early reaction, eliminating

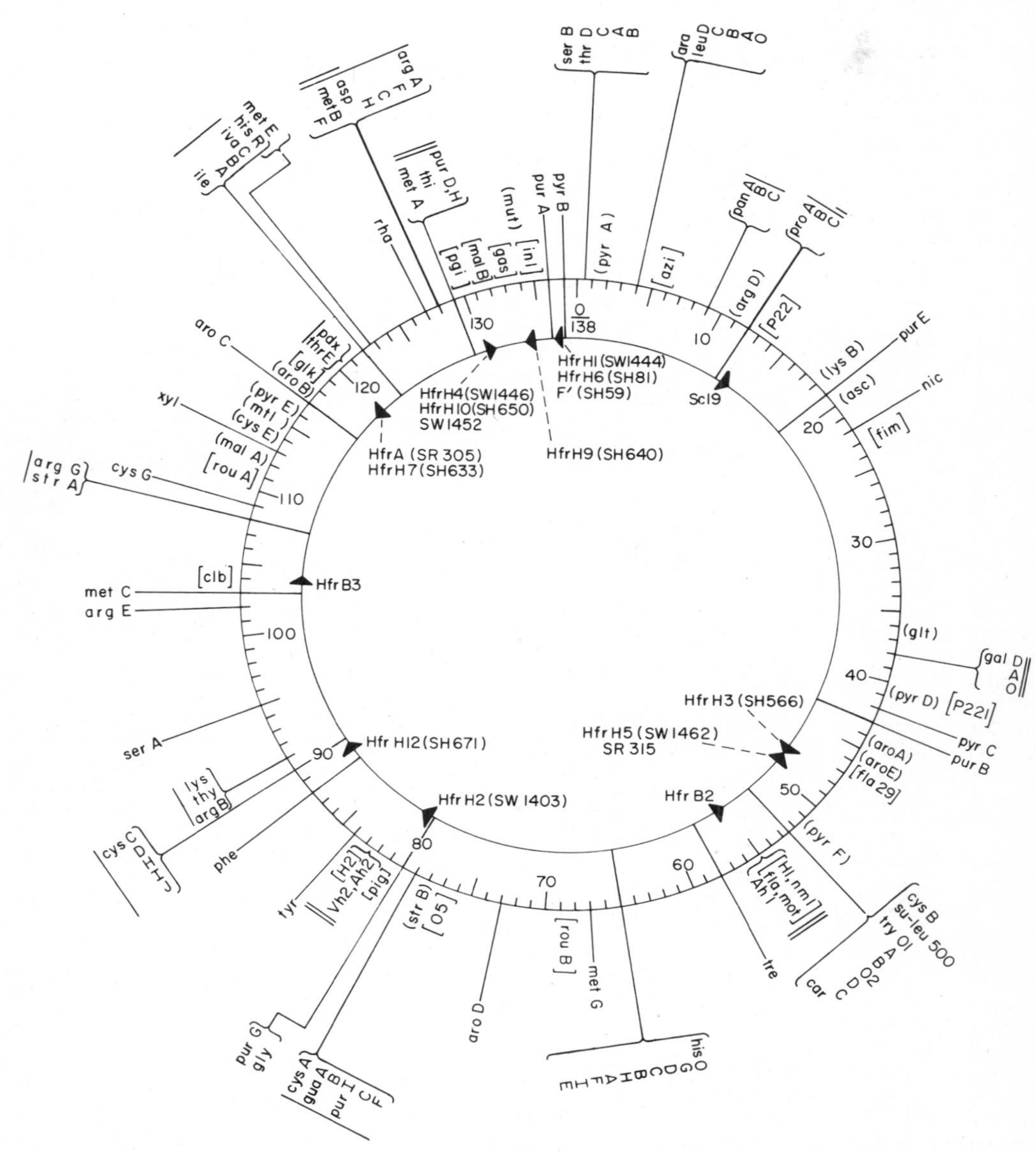

LINKAGE MAP OF SALMONELLA TYPHIMURIUM

} TRANSDUCING FRAGMENT; GENE ORDER & ORIENTATION KNOWN

| } TRANSDUCING FRAGMENT; GENE ORDER KNOWN, ORIENTATION NOT KNOWN

|| } TRANSDUCING FRAGMENT; NEITHER GENE ORDER NOR ORIENTATION KNOWN

( ) POSITION KNOWN APPROXIMATELY

[ ] MAPPED WITH col-FACTOR MEDIATED CONJUGATION

FIG. 1. Linkage map of *Salmonella typhimurium*, according to Sanderson and Demerec (published in *Genetics* 51: 897–913, 1965). Consult text for details on methodology.

The entire chromosome is about 5×10^6 nucleotide pairs in length. The numbers 0–138 are the relative times genes are first transferred during *Hfr*-mediated conjugation. The origin and direction of gene transfer for each *Hfr* strain is indicated on the inner circle. A key for the relation of various genes to chromosomal segments transduced by P22 bacteriophage (which incorporates 6.5×10^4 nucleotide pairs or less) is given under the figure.

H1 and *H2* denote loci dictating the structure of flagella of phase 1 and phase 2, respectively. At any given time, either *H1* or *H2* is expressed; the locus *Vh2* controls the rate of variation between the two phases, and the genes *Ah1* and *Ah2* determine which of the two genes, *H1* or *H2*, will function. A number of genes (*fla*) determine the presence or absence of flagella, and *mot* genes determine if bacteria possessing flagella will be actively motile or non-motile. Gene *nml* functions in the methylation of lysine residues specifically of flagellar protein to N-methyllysine. *05* in concerned in the production of a polysaccharide component of the cell surface, somatic antigen 5. Full expression of *05* and of other somatic antigen genes is influenced by *rou* genes ("rough)" which are involved in synthesis of cell wall substructure to which the polysaccharides comprising the somatic antigens are attached. Production of fine, fibrillar projections from the cell surface (fimbrae) is controlled by gene *fim. P22* and *P221* designate the sites of integration of two prophages. Genes concerned in the utilization of carbohydrates are: *car* (several carbohydrates), *clb* (cellobiase), *gal* (galactose), *gas* (gas production during glucose fermentation), *glk* (glycerol), *inl* (inositol), *mal* (maltose), *mtl* (mannitol), *pgi* (glucose), *rha* (rhamnose), *tre* (trehalose), and *xyl* (xylose). Genes involved in the synthesis of metabolites required for growth are: *arg* (arginine), *aro* (aromatic amino acids: phenylalanine, tyrosine and tryptophan), *asc* (ascorbate), *asp* (aspartate), *cys* (cysteine), *glt* (glutamate), *gly* (glycine), *gua* (guanine), *his* (histidine; *hisR* affects levels of enzymes elicited by the *his* genes); *ile* (isoleucine), *ilv* (isoleucine and valine), *lys* (lysine), *met* (methionine), *nic* (nicotinic acid), *pan* (pantothenate), *pdx* (pyridoxine), *phe* (phenylalanine), *pro* (proline), *pur* (purines), *pyr* (pyrimidines), *ser* (serine), *thi* (thiamine), *thr* (threonine), *thy* (thymine), *try* (tryptophan), and *tyr* (tyrosine). Gene *pig* affects pigmentation, and gene *mut* influences mutation rate throughout the genome. Resistance markers are genes *azi* (azide) and *str* (streptomycin). Gene *su-leu* is a suppressor gene that alleviates the effects of a mutation in one of the *leu* genes.

vegetative reproduction entirely, or in a late reaction, in which case replication occurs vegetatively, but the phage genomes fail to mature into infective particles or, once mature, fail to obtain release from the cell.

Similar to defective prophages in their behaviors are elements (F elements, below) allowing bacteria to undergo conjugation; elements engendering the hereditary potential for the lethal synthesis of proteins, termed bacteriocins, and other elements, termed resistance transfer factors. The above elements are collectively termed "episomes." *Episomes* are capable of replication either as integrated elements at chromosomal sites in synchrony with the bacterial host or in a vegetative state out of synchrony with chromosomal replication. The presence of an episome in a bacterium often results in changes in the immunological constitutions of its specific somatic (O) antigens (lipopolysaccharide-protein complexes), which are mainly present on the cell surface. Genes on the episomes influence the enzymatic constitution of the bacteria, eliminating biosynthesis of certain polysaccharide components of the cell surface and instigating the synthesis of others (antigen conversion or episomal conversion). In one case, the presence of a new protein, diphtheria toxin, is dependent upon the presence of prophage or vegetative phage in the bacteria.

Episomal elements occasionally (10^{-6} per chromosomal replication) incorporate genetic markers of the bacterium into their genetic continuity. This genetic material may be added to the preexistent episomal genome. In many cases, however, the bacterial genetic material replaces other genetic material originally present in the accessory (episomal) genetic fragment. The classic examples are the transducing derivatives of *lambda* prophage. Such phages exhibit "special transduction," *i.e.*, they contain only genes known by other criteria to be normally closely linked with the chromosal site at which *lambda* prophage is located. Lambda integrates in the chromosome of *E. coli* between the genes involved in galactose metabolism (gal, in Fig. 1) and a nearby gene concerned with biotin synthesis (not shown in Fig. 1). The chromosomes of *E. coli* and *Salmonella*, closely related species, are similar in gross genetic structure. Some lambda particles "pick up" the biotin gene while others, which are all defective in certain lambda functions, pick up the *gal* genes. A step necessary for the incorporation of the *gal* genes takes place only during replication in the prophage state, not during vegetative multiplication.

In contrast, other phages are capable of "general transduction." They incorporate various regions of the chromosome which collectively encompass the entire bacterial genome. In addition, they carry out this incorporation during vegetative multiplication. The genetic constitution of the bacterial fragment each transducing particle carries is characteristic of the last host upon which the phage was prepared. Particles competent in general transduction have lost at least one phage function; it is not yet clear if any phage genes are present along with the content of bacterial genetic material.

Transducing phages serve to transport ("transduce") relatively short pieces of genetic material from donor bacteria to recipient bacteria. When the genetic fragment is injected by the phage particle into a recipient bacterium, it may persist and function without replicating (abortive transduction), allowing the construction of partially heterozygous bacteria. In other cases, the input fragment undergoes rapid integration, replacing homologous genetic material in the recipient host chromosome (complete transduction). In still other cases, the transduced element persists and multiplies more or less in synchrony with the host chromosome (heterogenote). By growing up a culture of such heterogenotic bacteria, a relatively pure clone of transducing particles may be obtained (HFT—high-frequency transducing phage).

Genetic Fine Structure. Although the frequency of particles carrying a particular gene region may be rare (*e.g.*, 10^{-6} per infectious phage particle), the content of bacterial genes in a phage suspension still can be analyzed with sensitivity since one can easily plate on a single Petri dish 10^{10} particles. Transductions are useful in analysis of genetic fine structure, as is illustrated in the discussion of Fig. 2 in the legend. Figure 2 shows an enlargement of a portion of the *Salmonella* chromosome (located at 65 in the chromosomal map of Fig. 1).

Genes with related functions are clustered on the *Salmonella* chromosome (see section on Control of Gene Function, below). Each gene is a discrete element, a section of the DNA molecule, and is involved with dictating the structure of a single polypeptide chain through determination of its amino acid sequence. Since each gene is composed of many base pairs, it contains many sites which can mutate independently, and mutations at independent sites can recombine with with one another. Some enzymes appear to be composed of a single kind of polypeptide chain, dictated by a single gene, while other enzymes are composed of two (or more) polypeptide chains, the structure of each kind being dictated by a separate gene. Since the amino acid sequence is dictated by the gene through the intermediary of an RNA molecule (messenger RNA), all three of these structures are co-linear (Fig. 3). Elucidation of the mechanics of gene action now allows us to describe the genetic material through investigations of these molecules, in addition to the purely genetic tests utilized to date [see GENES; PROTEINS (BIOSYNTHESIS)].

The specificity of the protein appears to reside in its amino acid sequence. The composition and sequence of amino acids in the polypeptide chain, along with metabolites in the cell, determine the pattern of folding of the protein and the activity of the final form.

Conjugation. Recombination mediated by conjugation involves the following progressive sequence of events: (1) Random collision of donor bacteria with recipient bacteria. Donors

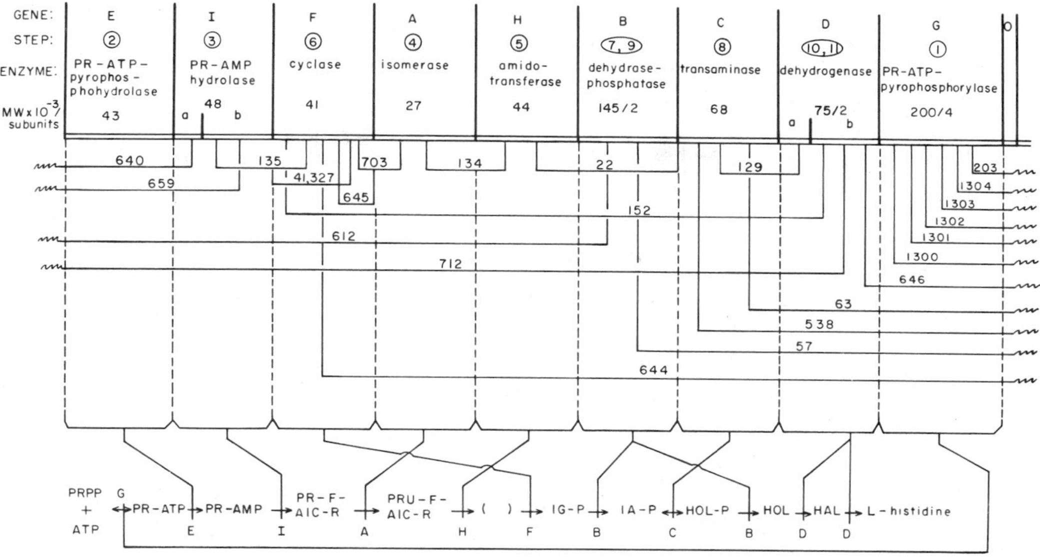

FIG. 2. Genes and proteins involved in histidine biosynthesis in *Salmonella*, after Loper *et al.* (*Brookhaven Symp. Biol.* 17,15, 1964).

The capital letters at the top of the diagram refer to the gene loci. In addition, genes *I* and *D* each may be subdivided into two genes, *a* and *b*; each subunit may dictate the structure of a separate polypeptide chain contained in the final enzyme. The steps in the reaction sequence controlled by each gene are circled, and the general name of the enzyme is listed. The approximate maximum molecular weight of each enzyme is listed in daltons $\times$ 10^{-3} directly beneath the enzyme name. The probable number of *identical* subunits comprising four of the enzymes is indicated along with the molecular weight. (In the case of gene *D*, the dehydrogenase with a total molecular weight of 75,000 may be composed of four polypeptide chains, two elicited by gene *Da* and two elicited by *Db*.) Thus, a rough idea of the molecular weights of the presumed monomers can be obtained from the numbers given. The extents of some multisite mutations, along with their isolation-stock numbers, are shown in the center of the diagram. Multisite mutations are useful since in genetic mapping, the presence or the absence of recombination, a qualitative test, (rather than the frequency of recombinants, a quantitive test) serves to localize sites of mutation. The data on multisite mapping also justify the construction of a one-dimensional scheme for the entire histidine region ("operon," see section on Control of Gene Function), since the chance of a deviation from linear topology among the overlapping multisite mutations is negligible. The histidine region appears to be a linear segment (about 11,000 nucleotide pairs) of a much larger linear structure, the bacterial chromosome (cf. Fig. 1).

At the bottom of the Figure are abbreviations for the intermediates in histidine biosynthesis, as elucidated in the studies of Ames and coworkers: ATP (adenosine triphosphate), PRPP (phosphoribosylpyrophosphate), PR-ATP (phosphoribosyl adenosine triphosphate), PR-AMP (phosphoribosyl adenosine monophosphate), PR-F-AIC-R (phosphoribosyl formamidino aminoimidazole carboxamide ribotide), PRU-F-AIC-R (phosphoribulosyl formamidino aminoimidazole carboxamide ribotide), ? (an unidentified compound), IG-P (imidazoleglycerol phosphate ester), IA-P (imidazoleacetol phosphate ester), HOL-P (L-histidinol phosphate ester), HOL (L-histidinol), and HAL (L-histidinal).

contain an episome which alters their surface properties through antigen conversion and allows them to stick to competent recipients. (2) Formation of an intimate union and a fusion bridge between bacteria. (3) Unidirectional genetic transfer from donor to recipient. (4) Gene function. (5) Integration of the transferred genetic material.

Different types of donors transfer different segments of the genetic complement. Donors, when F^+, contain a purportedly extra-chromosomal element; during conjugation only the F (fertility) element is transferred. In some cases, the F episome undergoes recombination with the host chromosome and becomes physically linked with bacterial genes (F′ element); F′ bacteria transfer both the F element and the bacterial genes associated with it but only very rarely transfer genes still in the bacterial chromosome. When such rare transfer of chromosomally located genes does occur, the transfer either may involve short, randomly distributed regions of the chromosome or may be highly oriented. In the latter instance, it is found that the orientation and chromosome transfer follow crossing over between the bacterial genes on the F′ element and homologous chromosomal genes. Similar to F′ episomes are resistance transfer factors (RTF) which transfer one to six different genes engendering resistance to various antibiotics. Some colicinogenic factors also elicit conjugation and their own transfer, and some even mobilize transfer of segments of the chromosome. Transfer of the above elements occurs with high frequency. Since the elements often replicate freely for a time in recipient bacteria and turn recipients into donors, the element can rapidly spread through an entire bacterial population.

A large number of positions on the chromosome are capable of mutation to an *Hfr* (high-frequency recombination) condition in the presence of the F episome. (Some sites of F integration are shown in Fig. 1.) In *Hfr* strains, the F element inserted into the continuity of the chromosome guides the transfer during conjugation of bacterial genes located to one side of it. The transfer of bacterial chromosome is oriented, starting with genes ajdacent to one side of the F element (origin) and proceeding away from the element which enters the recombinant progeny last. The transfer is slow and progressive, involving per minute only about 10^5 nucleotide pairs of the total 5×10^6 nucleotide pairs contained in the haploid *E. coli* or *Salmonella* chromosome. Some bacteria spontaneously break apart during the transfer process. Chromosome transfer may be experimentally interrupted by agitation of mating pairs in a Waring blender, or the mating donor bacteria may be killed with a drug or by infection with a virulent phage. Transfer also may be interrupted by allowing P^{32} decay to take place in labeled donor DNA before mating. Because mating pairs spontaneously break apart before transfer is completed, gene transfer most often involves only a portion of a haploid complement; in each case transfer starts at the crigin but ends at a position dependent upon the length of time transfer has taken place before the cells are severed. This unequal transfer is used as the basis for genetic mapping. (1) The earlier-transferred markers are present in highest frequency among the recombinants. (2) Usually few recombinants containing a proximally located marker also contain a distally located marker; however, those recombinant bacteria containing a distally located marker all have received the proximal markers and these are contained in a high (usually 50% or more) proportion of the recombinants. (3) The kinetics of appearance of markers in the recipient cell (time of first entry) allow measurement of the orders of markers that are not extremely closely linked. In experiments with P^{32}-labeled DNA, the time of entry is correlated with physical distance along the giant DNA

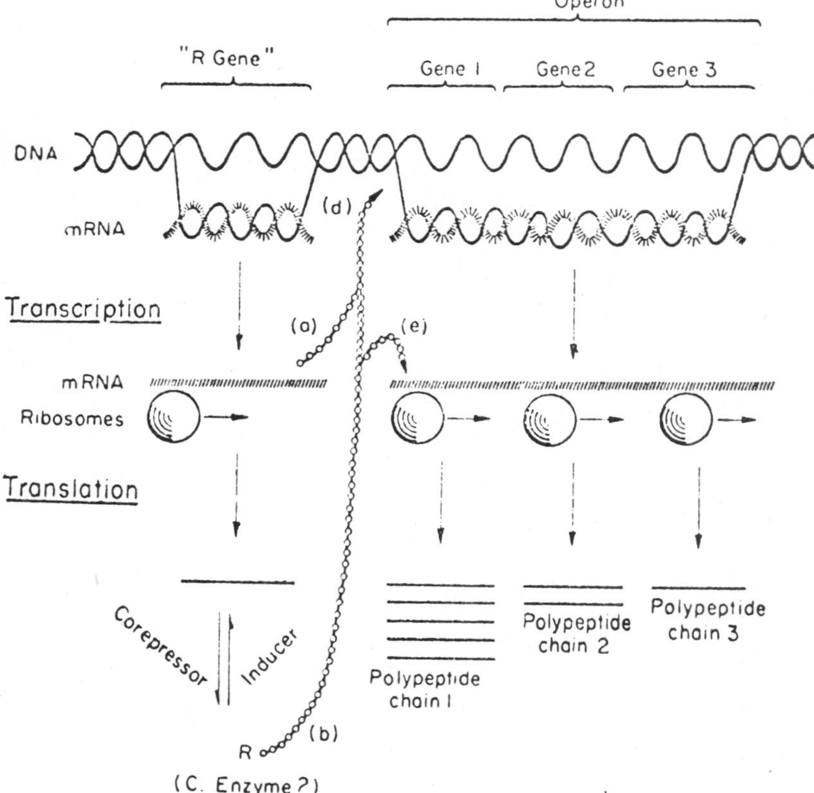

FIG. 3. Schematic representation of factors involved in genetic control of enzyme formation in bacteria, after Hartman [in "Genetics Today," Symposium No. 4: Gene Action: Control. *Proc. Intern. Congr. Genet., 11th The Hague* (September 1963)].

molecule of the donor bacteria. Through use of a number of *Hfr* strains, each transferring the chromosome in a unique orientation, a circular chromosome map containing all genetic markers for which transfer has been tested (Fig. 1) can be constructed. This map provides striking agreement with the cytological observation of a single, circular, giant DNA molecule, mentioned above in the section on Bacterial DNA. It shows that the DNA is the main, if not the exclusive, carrier of genetic information in bacteria.

Control of Gene Function. The genotype of a cell specifies the *limits* of the cell's potentialities; the environment of a cell plays an important role in the *expression* of the genetic information. The ultimate traits a cell possesses—its phenotype—are thus under dual control. Bacteria are being intensively studied by those interested in the regulation of gene action.

The genes of bacteria sometimes are clustered into groups with related functions (Fig. 1). In some of these groups (*e.g.*, the histidine cluster shown in Fig. 2), enzyme production by each of the member genes is coordinate. That is, when the bacteria are grown in one environment—*e.g.*, in the presence of L-histidine—the level of each enzyme activity is a certain value arbitrarily set at 1; when histidine is limiting, all the enzyme levels rise the same amount; if one enzyme increases tenfold, all of the others also rise tenfold. The levels of enzymes controlled by other genes in the cell are not influenced by the presence or absence of histidine. Thus, we say that L-histidine plays a specific role in repression of production of the enzymes involved in its biosynthesis. Examination of the RNA in bacteria manufacturing high levels of histidine enzymes (in comparison with bacteria not making these enzymes) reveals the presence of a giant messenger RNA (mRNA) molecule about 11,000 nucleotides long; since this is the same length estimated for the entire cluster of nine genes, it appears that one messenger is made for the entire cluster. Such functionally coordinated clusters of genes have been termed "operons" by Jacob and Monod.

Current evidence indicates that L-histidine serves as a "corepressor" (Fig. 3) changing the conformation of a protein (which may be an enzyme, C in Fig. 3) made by a repressor gene ("R gene"). It is thought that the active repressor protein either interacts at the gene level to shut down transcription of additional messenger RNA synthesis at the operon (d, in Fig. 3) or interacts with the messenger during translation of the genetic code at the time of protein synthesis (e, in Fig. 3). It is less likely, but possible, that messenger RNA made by the R gene is the active repressor (a, in Fig. 3).

References

1. Hayes, W., "The Genetics of Bacteria and Their Viruses," New York, John Wiley & Sons, 1964.
2. Gunsalus, I. C., and Stanier, R. Y. (Editors), "The Bacteria," Vol. V: "Heredity," New York, Academic Press, 1964.
3. Adelberg, E. A., "Papers on Bacterial Genetics," Boston, Little, Brown & Co., 1960.
4. Jacob, F., and Wollman, E. L., "Sexuality and the Genetics of Bacteria," New York, Academic Press, 1961.
5. Hartman, P. E., and Suskind, S., "Gene Action," Englewood Cliffs, N.J., Prentice-Hall, 1965.
6. Ames, B. N., and Martin, R. G., "Biochemical Aspects of Genetics: The Operon," *Ann. Rev. Biochem.*, **33**, 235 (1964).
7. Taylor, J. H. (Editor), "Molecular Genetics," Part I, New York, Academic Press, 1963.
8. Campbell, A. M., "Episomes," *Advan. Genet.*, **11,** 101 (1962).
9. Burdette, W. J., (Editor), "Methodology in Basic Genetics," San Francisco, Holden-Day, 1963.

Philip E. Hartman

BACTERIAL METABOLISM. See oxidative metabolism and energetics of bacteria.

BACTERIOPHAGES

Bacteriophages are viruses that invade bacterial cells and are replicated within the bacterial cells [see viruses (replication)]. The study of certain of these bacterial viruses has greatly advanced the knowledge of the nature and structure of genes (see bacterial genetics; also references 1 and 2.) Experimental studies using bacteriophages contributed to the evidence that a nucleotide *triplet* of deoxyribonucleic acid (DNA) is the coding unit (of the "genetic code") that specifies a particular amino acid to be built into the polypeptide chain in protein biosynthesis (see reference 3). See also the electron micrographs in the article viruses (composition and structure).

References

1. Jacob, F., and Wollman, E. L., "Viruses and Genes," *Sci. Am.*, **204,** No. 6 (June 1961).
2. Benzer, S., "The Fine Structure of the Gene," *Sci. Am.*, **206,** No. 1 (January 1962).
3. Crick, F. H. C., "The Genetic Code," *Sci. Am.*, **207,** No. 4 (October 1962).
4. Edgar, R. S., and Epstein, R. H., "The Genetics of a Bacterial Virus", *Sci. Am.*, **212,** No. 2, 70 (February 1965).

BALBIANI RINGS

One of the most fundamental problems of current biology is that of cellular differentiation. That is, how do the genes of the one-celled zygote act to give rise to a complete multicellular organism made up of specialized cells whose functions are integrated to constitute an individual? The primary evidence shedding light on this problem is derived from an analysis of Balbiani rings.

Balbiani rings were described in morphological detail by Balbiani in 1881 as curious bodies associated with nuclear threads or filaments within salivary gland cells of certain insects.

Such filaments were later shown to be polytenic CHROMOSOMES (Heitz and Bauer; Painter; King and Beams; see, for example, reference 2 for a complete review of this entire subject), which are actually multi-stranded complexes containing as many as several hundred or even thousands of times the chromosomal complement of the egg cell. Remaining in close association with their homologs, these chromosomes attain a highly organized superstructure giving the chromosome a banded appearance. Bands or discs seem to occur at particular sites along the length of giant chromosomes; these have been meticulously mapped.

The Balbiani rings are specialized regions or loci of polytenic chromosomes where the chromosomal superstructure, the highly condensed and coiled DNA, DEOXYRIBONUCLEIC ACID, has loosened during cell function. The Balbiani rings, as well as the rest of the chromosomes, take up both acidic and basic dyes, the basic stains binding nucleic acids, the acid stains binding basic proteins present. The rings are visible by phase contrast microscopy of even unfixed cells (see PHASE MICROSCOPY).

The synthetic activity and cytological dimensions of these rings change during development at precise stages and may be altered experimentally by changes in the environment. The dimensions are commonly two or, at times, three times the width of the unexpanded portion of the chromosome. The region has a particularly high concentration of RNA, RIBONUCLEIC ACID, as revealed by differential staining. It has also been shown to be synthesized there preferentially by autoradiographic visualization of isotope incorporation, when labeled RNA precursors are administered to the tissue, revealing that only the nucleoli are as active as the Balbiani rings. These loci function in RNA synthesis and they respond to changes in the environment by producing more or less of this nucleic acid. The particular type of RNA synthesized in the rings is thought to be specific messengers, mRNAs, which function in the synthesis of one or more proteins [see PROTEINS (BIOSYNTHESIS)]. The nature of these RNAs has been demonstrated by means of isotope tracer techniques and by base ratio analyses of RNA from localized chromosomal regions.

There are usually two or three active Balbiani rings in the genome of each *Chironomus thummi* or *Chironomus tentans* salivary gland cell, respectively. This number is usually constant, but the size and the number of rings may be decreased, depending upon both which lobe of the gland and which developmental stage are considered. The Balbiani rings are at their most expanded size when active during most of larval life; however, with the onset of metamorphosis and pupation, these loci regress. After pupation, the entire gland degenerates. Balbiani ring activity is therefore stage dependent. These particular Balbiani ring loci, which are active in the salivary glands, are not known to be active in other tissues, but other loci may be active. The activity of Balbiani loci is therefore considered also to be tissue-specific. The tissue specificity and stage specificity of Balbiani rings appear to reflect the specialized cellular function, characteristic of the tissue.

Electron micrographs of granules associated with the Balbiani rings, which are presumably synthesized there, indicate that the products of these large chromosomal structures differ from one another.

The Balbiani rings are active in RNA synthesis and seem to be a specialized case of the more general phenomenon known as chromosomal puffing. The loosening of the polytene chromosome at specific sites for increased RNA synthesis is called puffing. Other loci remain unpuffed and are relatively inactive in RNA synthesis. In its extreme form, puffing gives the appearance of a large chromosomal bulge or Balbiani ring.

Temperature changes, wounding, and ligature of the organism, which affect, among other things, the endocrine environment in the animal, alter the puffing pattern of the salivary gland cell. These changes reflect altered functional activity of the gland. Transplantation of the salivary gland to another animal with a different hormonal environment also alters puffing activity. It has been shown that puffing, particularly of the Balbiani rings, is controlled by hormones. Thus, impending metamorphosis seems to terminate synthetic activity at certain puffs. According to Clever, ecdysone, the molting hormone of insects (crystallized and characterized by Karlson), activates puffs. A particular locus, I-18-C seems to be activated as one of the first events observed in response to this hormone. The response can be seen 15–30 minutes after hormone injection. Other loci, such as one very near to I-18-C, regress after ecdysone treatment. Among those which regressed during subsequent development, following ecdysone stimulation, are the Balbiani rings. The hormone has recently been shown[3] to be a steroid with empirical formula $C_{27}H_{44}O_6$. Balbiani ring activity may be reduced or even eliminated by other treatments such as with the antibiotic actinomycin D, which presumably inhibits DNA-dependent RNA synthesis.[4]

Genetic evidence links the presence of one particular Balbiani ring with the appearance of a specific type of secretory granule.[1] This led Laufer and co-workers to examine the synthetic activity of salivary glands as well as the source of the salivary gland secretion. The evidence suggested that the proteins of the secretion of chironomids are of general distribution throughout the body and few if any appear to be synthesized by the gland. The gland was shown to transport proteins. It has been proposed[5] that the functional relationship between the Balbiani rings and the secretion lies in the synthesis of messenger RNA by the chromosomal loci. The RNA mediates the synthesis of proteins important in a transport system for the transfer of certain blood proteins as salivary secretion.

The general significance of chromosomal puffing, as exemplified by Balbiani ring activity, relates to its importance to the developmental concept of differential gene activity. This concept supposes that development of an organism from a single fertilized egg cell can be accomplished by an orderly sequential turning on and off of genetic loci. Thus the differentiated tissues of the various parts of the body all have the same genetic endowment, but different genes are active in different cells, and the complement of genes functional in any one cell may be altered during the course of the cell's existence. Thus the biochemical capacities of each cell are potentially like those of every other cell, but the expression of each cell type can be quite distinct, and it often is. The best evidence for this concept of cellular differentiation at the molecular level is derived from studies of chromosomal puffs, particularly of Balbiani rings.

References

1. BEERMANN, W., "Ein Balbiani-ring als Locus einer Speicheldrüsenmutation," *Chromosoma*, **12**, 1–25 (1961).
2. BEERMANN, W., "Riesenchromosomen," *Protoplasmatologia*, **6**, 1 (1962).
3. KARLSON, P., HOFFMEISTER, H., HOPPER, W., AND HUBER, R., "Zur Chemie des Ecdysons," *Liebigs Ann. Chem.*, **662**, 1–20 (1963).
4. LAUFER, H., NAKASE, Y., AND VANDERBERG, J., "Developmental Studies of the Dipteran Salivary Gland, I. The Effects of Actinomycin D on Larval Development, Enzyme Activity, and Chromosomal Differentiation in *Chironomus thummi*," *Develop. Biol.*, **9**, 367–384 (1964).
5. LAUFER, H., AND NAKASE, Y., "Salivary Gland Secretion and its Relation to the Chromosomal Puffing in the Dipteran, *Chironomus thummi*," *Proc. Natl. Acad. Sci. U.S.*, **53**, 511–516 (1965).

HANS LAUFER

BASAL METABOLISM

Metabolism of a living cell or organism refers to the total turnover of chemical material and energy. It consists of *anabolism*, or assimilation, mostly of substances of high potential energy (primarily protein, fat, and carbohydrate), and *katabolism* or dissimilation. In common speech, metabolism refers to the oxidation of major foodstuffs and the concomitant release of energy. Metabolic rate refers to the metabolism in a given period of time. The "basal" metabolic rate refers to the fundamental energy requirement for maintenance and continued functioning of the organism (aside from external muscular work and work of digestion), such as respiration, contraction of the heart, function of the kidney, the liver, and of all cells in general. Basal metabolic rate (BMR) in man refers to the determination of metabolic rate under certain standardized conditions, including complete physical rest (but not sleep), a fasting state, and an ambient temperature that does not require energy expenditure for physiological temperature regulation. Actually the BMR refers not to a "basal" rate but to a determination under these standard conditions. The BMR is below normal in sleep, starvation, anesthesia, and certain endocrine disturbances (hypothyroidism), and is elevated in fever, athletic training, under the influence of drugs (*e.g.*, caffeine) and endocrines (adrenalin, thyroid hormones).

In studies of animals it becomes technically difficult to make observations under standard conditions which include rest. Restraining an animal increases the metabolic rate and inactivation through anesthesia lowers it; ruminants and other plant eaters cannot be brought into a fasting state unless they are deprived of food for prolonged periods of time; small animals (*e.g.*, shrews) have such high metabolic rates that they must eat almost continuously to sustain a normal metabolic rate, etc.

Methods of Determination. In principle, the metabolic rate is determined in three different ways. (a) Determination of the energy value of all food less the energy value of excreta (mainly feces and urine) should give the energy turnover of the organism. However, the result must be corrected for any change in the composition of the body, mainly deposition or utilization of body fats. The method is cumbersome and is accurate only if the period of observation is sufficiently long. (b) Measurement of total heat production of the organism. This is fundamentally the most accurate method. The value obtained must be corrected for any external work performed, including such items as heating of the foodstuffs taken in, vaporization of water, etc. The determinations are made with the organism in a calorimeter, technically a rather difficult procedure, but it yields very accurate results. (c) The amount of oxygen used in oxidation processes can be used to determine the metabolic rate. (In theory, the carbon dioxide production could also be used, but it is less accurate, mainly because there is a large pool of carbon dioxide in the organism that undergoes changes relatively easily.) The reason that oxygen can be used is that similar amounts of heat are produced for each liter of oxygen, irrespective of whether fat, carbohydrate or protein is oxidized. The figures are: fat, 4.7 kcal; carbohydrate, 5.0 kcal; and protein, 4.5 kcal, per liter oxygen. It is customary to use an average value, 4.8 kcal/liter oxygen consumed. The use of oxygen consumption for the determination of metabolic rate is so common that the two concepts have become practically synonymous. Obviously, the oxygen consumption cannot be used for determinations of metabolic rate in, for example, anaerobic organisms.

Temperature Effects. Animals whose body temperature changes with that of the environment (poikilothermic or cold-blooded animals) have a metabolic rate which depends on their temperature. In general, the metabolic rate increases, within the range tolerated by the organism, some two- or three-fold for a temperature increase of 10°C. This change, designated as Q_{10}, is a term preferred by most physiologists and biologists over the use of the Arrhenius constant, which is a thermodynamically more correct way of expressing temperature dependence. Because of the temperature effect, information about metabolic rate in cold-blooded animals is meaningful only if the temperature is known.

Mammals and birds maintain a relatively constant body temperature within a wide range of ambient temperatures, and are called warm-blooded or homothermic animals. When the ambient temperature falls below a certain critical level, their metabolic rate increases so that the increased heat loss is balanced by increased heat production. Most of the increased heat production is due to involuntary muscle contractions (shivering).

Metabolic Rate in Relation to Body Size. If a uniform group of animals, such as mammals, is used for a comparison of metabolic rates, an interesting relationship is revealed. The smaller the animal, the higher is the metabolic rate per gram of body weight. If, on logarithmic coordinates, the metabolic rate is plotted against body size, we obtain a straight line (see Fig. 1) which corresponds to the equation: log metabolic rate $= k + 0.74$ log body weight (k being a constant whose numerical value depends on the units used). It has been suggested that this relationship expresses the need for a higher heat production in the smaller animal, which, because of its larger relative surface, must produce heat at a higher rate than a large animal in order to maintain its body temperature. However, similar relationships between metabolic rate and body size have been found in numerous groups of cold-blooded animals as well as plants, where the need for heat regulation cannot be the fundamental explanation of this interesting relationship.

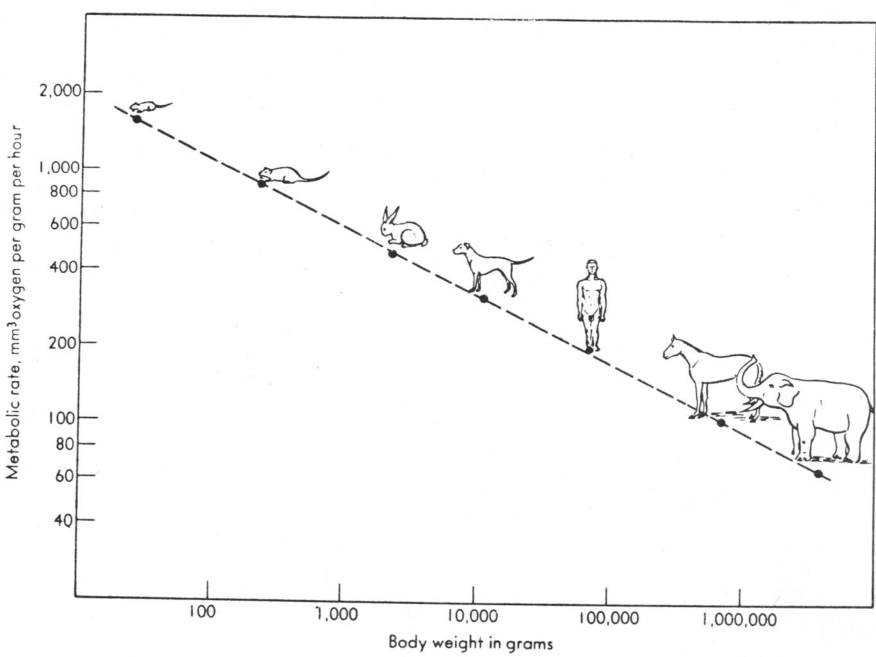

Fig. 1. The metabolic rate of animals, calculated per gram of body weight, increases with diminishing size in such a fashion that we obtain a straight line if the observations are plotted on logarithmic coordinates. (Fom Schmidt-Nielsen, K., "Animal Physiology," Second Edition, 1964. Reprinted by permission of Prentice-Hall, Inc., Englewood Cliffs, N.J.).

References

1. Brody, S., "Bioenergetics and Growth," New York, Reinhold Publishing Corp., 1945, 1023pp.
2. Hemmingsen, A. M., "The Relation of Standard (Basal) Energy Metabolism to Total Fresh Weight of Living Organisms," *Rept. Steno. Mem. Hosp. Nord. Insulin Lab.* (1950) 58pp.
3. Kleiber, M., "The Fire of Life, An Introduction to Animal Energetics," New York, John Wiley & Sons, 1961, 454pp.
4. Krebs, H. A., "Body Size and Tissue Respiration," *Biochim. Biophys. Acta*, **4**, 249–269 (1950).

Knut Schmidt-Nielsen

BEADLE, G. W.

George Wells Beadle (1903–) is an American geneticist who, using *Neurospora*, was able to demonstrate a direct relationship between genes and enzymes. With Edward Lawrie

TATUM he was the founder of BIOCHEMICAL GENETICS which is of growing interest today. For this work, Beadle and Tatum received shares of the Nobel Prize for physiology and medicine in 1958, 17 years after their classic work was done. JOSHUA LEDERBERG was the other recipient in 1958. George W. Beadle is now Chancellor of the University of Chicago.

BENCE-JONES PROTEINS

In about half of all cases of a malignant bone marrow disease known as multiple myeloma, the patients produce a strange protein which appears in urine. This protein has the peculiar property that, after being precipitated by heating to about 45–55°C, it redissolves on boiling. Upon cooling it reprecipitates again. This interesting protein was described more than a hundred years ago by H. Bence Jones[1] and since then has been known in the literature under the name of Bence-Jones protein. Considerable attention has been paid to this substance in the last decades for the following reasons: (1) its importance in the diagnosis of multiple myeloma, (2) its interesting physical and chemical properties, and (3) concern as to its origin. These studies have facilitated diagnosis and have yielded important information of the physicochemical properties, chemistry, and metabolism of this pathological protein.[2-4]

One of the most characteristic features of the Bence-Jones protein is individuality.[5] Even the redissolving on boiling is not shared by all samples, obtained from various individuals, to the same degree. In each individual case of multiple myeloma, the urinary Bence-Jones protein will have its own molecular weight, electrophoretic mobility, degree of homogeneity, amino acid composition, N-terminal amino acid, etc. For this reason of individual diversity, it is proper to speak about Bence-Jones proteins instead of a definite Bence-Jones protein.

The Bence-Jones proteins can be isolated from urine very easily. The urine is acidified with acetic acid to pH approximately 5 and, if necessary, filtered. The protein then is precipitated with saturated ammonium sulfate solution. The precipitant is added gradually during stirring. The precipitated protein is removed by filtration or centrifugation, dissolved in water, and dialyzed against water. The protein is obtained in dry state by freeze drying (lyophilization). The colorless protein material is readily soluble in water, dilute salt solutions, acids, and alkalies. It can be further purified by redissolving and chromatographic purification, such as GEL FILTRATION on Sephadex or ION-EXCHANGE chromatography on DEAE-cellulose. Several individual specimens of Bence-Jones proteins have been crystallized.

The homogeneity of the various individual specimens of Bence-Jones proteins varies considerably. Most of them, even if crystallized, appear to be inhomogeneous, *i.e.*, made up of several components. Although some of these components may be impurities, *e.g.*, small amounts of serum albumin may appear as contaminants, even highly purified samples usually contain several molecular species. For example, in one indvidual case reported, three components could be isolated from a Bence-Jones protein by ion-exchange chromatography.[6]

The sedimentation constants of Bence-Jones proteins are in the range of 2–4 Svedberg units. The molecular weights, calculated from the sedimentation, diffusion, and viscosity data, or from osmotic pressure, are between 20,000 and 48,000. In most cases, the molecular weight is close either to 20,000–22,000 or 40,000–44,000. The isoelectric point of the various individual specimens varies between pH 4.5 and 6.7, indicating that they are of a slightly acid nature. The electrophoretic mobility, by using $0.1M$ sodium barbiturate buffer of pH 8.6, has been found between -1.4 to -4.7 mobility units, *i.e.*, the mobility is in the range of the β- and γ-globulins. The viscosity of the Bence-Jones protein solutions is low, *i.e.*, the intrinsic viscosity of 0.03–0.04 dl/g has been found. This indicates that these proteins are globular and of a compact structure.

The amino acid composition of Bence-Jones proteins has been studied in detail,[4] and one of the remarkable features of this aspect is that chemically they all are similar to the serum γ-globulins and myeloma proteins. The serine and threonine content of these proteins is high and the methionine content very low. Examples of amino acid composition are shown in Table 1. The specimen Pu was homogeneous and could not be separated into subfractions; the other individual specimen

TABLE 1. AMINO ACID COMPOSITION OF INDIVIDUAL BENCE-JONES SPECIMEN PU AND OF THREE SUBFRACTIONS OF ANOTHER INDIVIDUAL SAMPLE T. THE NUMBERS INDICATE PER CENT NITROGEN OF TOTAL PROTEIN NITROGEN

	Pu	T1	T2	T3
Aspartic acid	7.64	6.86	7.34	7.15
Threonine	6.36	4.90	5.37	4.83
Serine	9.80	11.33	9.61	9.79
Glutamic acid	8.12	7.87	8.38	7.48
Proline	4.99	5.03	6.29	6.74
Glycine	4.72	4.88	5.44	4.78
Alanine	2.74	4.41	4.64	5.07
Cystine + cysteine	1.40	1.33	1.23	1.04
Valine	4.31	6.04	6.77	4.55
Methionine	0.21	0.83	0.61	0.56
Isoleucine	2.69	2.22	2.01	2.19
Leucine	6.27	7.66	7.09	7.12
Tyrosine	3.72	3.24	2.97	3.08
Phenylalanine	2.77	2.93	3.00	2.93
Lysine	8.90	6.90	8.76	8.26
Histidine	2.93	2.82	3.56	3.24
Arginine	10.41	10.78	9.90	10.75
Tryptophan		1.48	1.32	1.32
Amide N	8.46	6.56	8.46	8.25
Total	96.44	98.07	100.75	99.13

T could be separated into three components which appeared to be of somewhat different composition.[6]

Most of the well-purified Bence-Jones proteins do not contain any carbohydrate, lipid, or other material of non-protein nature, although a few cases of carbohydrate-containing samples have been reported.

With respect to immunological characteristics, all Bence-Jones proteins can be classified into two groups A and B. The B group has aspartic acid as the N-terminal, whereas the other group has some other amino acid at the end of the polypeptide chain. In several cases, no N-terminal group could be detected; this indicates that the terminus is somehow blocked, *e.g.*, the amino group may be acetylated. Knowledge of the molecular weight and amount of terminal amino acids makes it possible for one to conclude how many polypeptide chains the macromolecule contains. Recent studies have shown that either one or two polypeptide chains may be present, and that these are very similar or identical with the L (or B) chains of serum γ-globulins.[7]

The amino acid sequences in the various Bence-Jones protein specimens have been studied, but the complete sequence of any of the samples has not been unraveled as yet. The secondary and tertiary structure, *i.e.*, the conformation of the polypeptide chains in the macromolecules, has been studied by the optical rotatory dispersion method; and it has been found that the Bence-Jones proteins are non-helical.[8] In conformation, the Bence-Jones proteins are similar to serum γ-globulins and myeloma proteins.

Although metabolic studies with radioactive tracers have shown that Bence-Jones proteins are not breakdown products of plasma proteins but are synthesized from the amino acid pool,[2,3] the similarity of these proteins with the L chains of serum γ-globulins (ANTIBODIES) is striking. With respect to molecular size, electrophoretic mobility, amino acid composition, conformation, and even redissolving on boiling and immunochemical character, the Bence-Jones proteins can not be differentiated from the L chains which, in the monomeric form, have a molecular weight of 20,000–22,000. This indicates that the derailment in the biosynthetic mechanism is not in that a different polypeptide chain is produced but that the L chains are produced in excessive quantities and, because of their small size, are readily filtered out in the urine.

References

1. JONES, H. BENCE, *Lancet*, **ii**, 88 (1847).
2. PUTNAM, F. W., *Physiol. Rev.*, **37**, 512 (1957).
3. PUTNAM, F. W., in "The Plasma Proteins," Vol. II, pp. 377–386, New York, Academic Press, 1960.
4. VIS, H., AND CROKAERT, R., *Acta Clin. Belg.*, **11**, 455 (1956).
5. WILLIAMS, R. J., "Biochemical Individuality," New York, John Wiley & Sons, 1956.
6. JIRGENSONS, B., IKENAKA, T., AND GORGURAKI, V., *Clin. Chim. Acta*, **4**, 876 (1959).
7. GALLY, J. A., AND EDELMAN, G. M., *J. Exptl. Med.*, **119**, 817 (1964).
8. JIRGENSONS, B., *Arch. Biochem. Biophys.*, **85**, 89 (1959).

B. JIRGENSONS

BERIBERI. See THIAMINE AND BERIBERI.

BERNARD, C.

Claude Bernard (1813–1878) was a famous French physiologist who discovered GLYCOGEN, and also the digestive function of pancreatic juice. He made many other fundamental contributions relating to the maintenance within a body of a constant internal environment (see HOMEOSTASIS). His initial ambition was to become a playwright but through the good offices of a friend he was induced to study medicine instead. Physiology the world over advanced greatly through the impetus he gave it.

BERYLLIUM AND BERYLLIOSIS

Beryllium is a relatively rare and toxic metal with unique applications in nuclear energy, electronics, aircraft and missiles, and rocket fuel. It is most widely used in the form of a non-magnetic copper-beryllium alloy. It was formerly used in the preparation of fluorescent powders.

A single stable beryllium isotope of atomic weight 9 and several radioactive isotopes are known. The most useful radioisotopes for biological applications are: Be^7, half-life $= 53$ days, and Be^{10}, half-life $= 2.7 \times 10^6$ years. Beryllium is found in nature in the following average concentrations: air, 0.0002 $\mu g/m^3$; soil, 0.37 ppm; coal ash, 4 ppm; lung tissue, 0.33 $\mu g/100g$; urine and blood, 0.00 μg; and seawater, 0.0006 μg/liter. In persons exposed to beryllium, no significant correlations appear to exist between the excretion of beryllium, tissue levels, and the severity of the disease.

Beryllium disease includes acute and chronic manifestations which may follow inhalation of fumes or dusts of beryllium compounds. Symptoms of acute beryllium poisoning which include inflammatory lesions, chills, and fever, usually disappear within a year. Berylliosis refers to the chronic disease which may follow years after exposure, usually to insoluble beryllium compounds, and characterized by extensive lung damage, shortness of breath, and weight loss. Cortisone and related drugs are quite successful in treatment. Skin lesions also develop at the sites of beryllium-contaminated wounds. Berylliosis has been found in persons living near beryllium-processing plants, the so-called neighborhood cases. Modern standards of contamination control have drastically reduced the dangers from beryllium disease.

A variety of biological and biochemical effects

are produced by beryllium compounds. A single intravenous injection of beryllium sulfate at a dose level above 0.6 mg/kg as beryllium is lethal to many animal species. Beryllium salts stimulate the growth of algae and higher plants. Bone cancer eventually occurs in rabbits (but not in pigs or monkeys) following injection of insoluble salts, while inhalation of beryllium sulfate by rats has resulted in lung cancer. No cases of human cancer can be definitely attributed to beryllium. Addition of beryllium carbonate to the diet of young rats produces "beryllium rickets" since the beryllium ion interferes with phosphorus utilization by forming insoluble phosphates.

Beryllium combines with many proteins. Binding with bovine serum albumin is weak and appears to take place through carboxyl groups. Significant changes in the distribution of ribonucleic acid but not of deoxyribonucleic acid within the cytoplasmic components of lung tissue cells has been observed in rats following exposure to suspensions of beryllium oxide.

Two enzymes strongly inhibited by beryllium salts are alkaline PHOSPHATASE and MYOSIN. Less than 10^{-6} mole of beryllium inhibits alkaline phosphatase both *in vitro* and *in vivo*. The beryllium combines very strongly with the enzyme. The inhibition of alkaline phosphatase can be reversed either by metal ions which compete with beryllium ions for active sites or by chelating or complexing agents.

Binding by chelating agents in which sulfur or nitrogen atoms act as donors to the beryllium is relatively weak, while those containing phenolic oxygen as a donor are far stronger binding agents. Such compounds, including salicylic acid and the dye aurin tricarboxylic acid (ATA), are effective antidotes for animals acutely poisoned with soluble beryllium salts. Preliminary investigations indicate that ATA may not be effective against chronic beryllium disease.

References

1. WHITE, D. W., JR., AND BURKE, J. E. (EDITORS), "The Metal Beryllium," Cleveland, Ohio, The American Society for Metals, 1955.
2. SCHUBERT, J., "Beryllium and Berylliosis," *Sci. Am.*, **199**, 27–33 (August 1958).
3. "Conference on Beryllium Disease and Its Control," A.M.A. Arch. Ind. Health, **19** (February, 1959).
4. "Workshop on Beryllium," The Kettering Laboratory, University of Cincinnati, Cincinnati, Ohio, January, 1961.
5. TEPPER, L. B., HARDY, H. L., AND CHAMBERLIN, R. I., "Toxicity of Beryllium Compounds," New York, Elsevier Publishing Company, 1961.
6. KING, M. E., SHEFNER, A. M., AND EHRLICH, R., "Effectiveness of Aurin Tricarboxylic Acid (ATA) as an Antidote for Beryllium Poisoning," Ind. Med. Surg., **33**, 566–569 (1964).
7. EVEREST, D. A., "The Chemistry of Beryllium," New York, Elsevier Publishing Co., 1964.

JACK SCHUBERT

BERZELIUS, J. J.

Jons Jakob Berzelius (1779–1848), a Swedish scientist, was the most famous chemist of his time. Though chemists think of him because of his contributions to fundamental chemistry, including atomic weights, he had a medical degree and gave considerable attention to matters which are now regarded as biochemistry. He was a thoroughgoing mechanist. He grew to be an arbiter in his later years and not infrequently it turned out subsequently that he was wrong. He coined the words catalyst, isomer and protein.

BETAINE

Betaine, $(CH_3)_3N^+—CH_2—COO^-$, is trimethyl glycine which can exist only in the zwitterion form. It got its name from beets, the leaves of which may contain 3% betaine. It is present in many other plants but is not conspicuous in animal tissues. Betaine is related to choline (see METHYLATIONS IN METABOLISM) from which it may be produced by oxidation of the alcohol group to a carboxyl group.

BILE AND BILE PIGMENTS

Bile is produced in the polygonal cells of the LIVER and is secreted via the bile caniculi into the hepatic and cystic ducts. Its various components are listed in Table 1. Bile has a number of

TABLE 1. COMPONENTS OF BILE

	% of Total Bile		% of Total Bile
Water	97	Fatty acids	0.15
Bile salts	0.7	Lecithin	0.1
Bile pigments	0.2	Fat	0.1
Cholesterol	.06	Hormone conjugates	—
Inorganic salts	0.7	Drug conjugates	—

functions: (a) it facilitates the absorption of fat, fat-soluble vitamins and cholesterol by means of bile salts; (b) it provides an alkali reservoir for the neutralization of stomach acids; (c) it serves as the major route of excretion of drugs, hormones, toxins and bile pigments after their conversion to metabolic end products by the liver; (d) it provides the primary pathway of excretion of heavy metals, *e.g.*, copper, zinc, and mercury, concentrated by the liver.

The components of bile fall into two general categories: (a) substances such as sodium, potassium, chloride, creatinine, glucose and cholesterol, which are present in similar concentrations in the bile and plasma, and (b) substances such as bile salts (see BILE ACIDS), bile pigments and conjugates or metabolites of administered drugs and dyes which are present in greater concentration in the bile than in plasma.

Bile salts and some inorganic electrolytes enter

the bile by ACTIVE TRANSPORT resulting from stimulation by secretin, whereas water and most inorganic ions enter by passive diffusion, a result of the active secretion of the bile salts.

Compounds, once excreted into the bile, are are not necessarily eliminated in the feces, but to a varying degree may be reabsorbed from the gut into the portal circulation. They can reenter the liver, where they may be further metabolized to nonbiliary components or reexcreted into the bile. This process, known as the *enterohepatic circulation*, provides a conservation mechanism whereby important biliary components such as bile salts are continuously reutilized to facilitate absorption. Many substances ultimately excreted in the urine originally appear in the bile as metabolic intermediates which may or may not be further metabolized after reabsorption. ION-EXCHANGE RESINS such as cholestyramine are employed orally with variable success to reduce the concentration of cholesterol or bilirubin in the plasma by binding these substances in the gut and preventing their enterohepatic circulation.

Cholesterol. Part of the cholesterol newly synthesized in the liver is excreted into bile in a free non-esterified state (in constant amount). Cholesterol in bile is normally complexed with bile salts to form soluble choleic acids. Free cholesterol is not readily soluble and with bile stasis or decreased bile salt concentration may precipitate as gallstones. Most common gallstones are built of alternating layers of cholesterol and calcium bilirubin and consist mainly (80–90%) of cholesterol. Normally, 80% of hepatic cholesterol arising from blood or lymph is metabolized to cholic acids and is eventually excreted into the bile in the form of bile salts.

Bile Salts. The C_{24} bile acids arise from cholesterol in the liver after saturation of the steroid nucleus and reduction in length of the side chain to a five-carbon acid; they may differ in the number of hydroxyl groups on the sterol nucleus. The four acids isolated from human bile include *cholic acid* (3,7,12-trihydroxy) (Fig. 1),

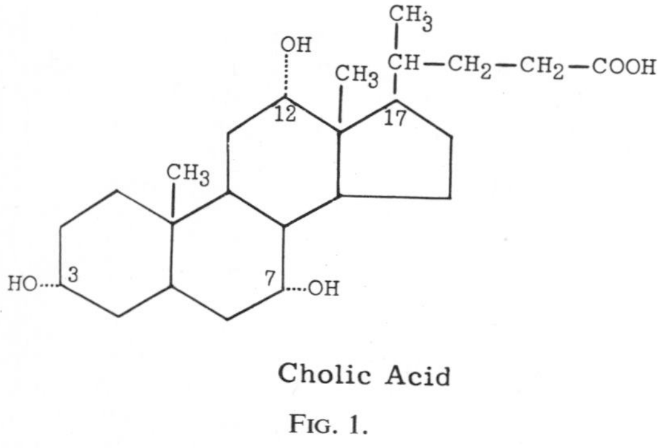

Cholic Acid

FIG. 1.

deoxycholic acid (3,12-dihydroxy), *chenodeoxycholic acid* (3,7-dihydroxy) and *lithocholic acid* (3-hydroxy). The bile acids are not excreted into the bile as such but are conjugated through the C_{24} carboxylic acid with glycine or taurine (NH_2—CH_2—CH_2—SO_3H). This esterification of the bile acids to soluble conjugates occurs in the MICROSOMES and requires COENZYME A, magnesium ion and ATP. Although taurocholic acid predominates at birth, the most abundant of the bile acids in adult man is glycocholic acid. In alkaline bile the conjugated bile acids exist in their ionized form as the bile salts, glycocholate or taurocholate. Bile salts can function as effective product feedback inhibitors of hepatic cholesterol synthesis [see METABOLIC CONTROLS (FEEDBACK MECHANISMS)]. Because of their detergent action, bile salts play an important role in the absorption of cholesterol, fats and fat-soluble vitamins. The bile salts are believed to facilitate absorption of these compounds by the formation of micelles or aggregates of low osmotic pressure. The bile salts themselves are not absorbed during this process; their absorption from the intestine occurs at a different site and at an entirely different rate than that of the lipids. Approximately 95% of the bile salts are reabsorbed, enter the enterohepatic circulation, and are ultimately reexcreted into the bile for further utilization in lipid absorption.

Hormones in Bile. Most of the hormones that are normally conjugated in the liver to form glucuronides or sulfates, such as the steroids, thyroxine, epinephrine and norepinephrine, are secreted into the bile but to a varying degree may be reabsorbed in the intestine and eventually excreted in the urine. The 17-hydroxysteroids, including cortisol (see ADRENAL CORTICAL HORMONES), are secreted into the bile primarily as reduced glycuronide conjugates. More than 70% of these conjugates enter the enterohepatic circulation and are eventually excreted into the urine; less than 30% are found in the feces. Progesterone, after its conversion to pregnanediol, is also excreted into the bile primarily as the glucuronide, some 75% of which is eventually excreted in the feces. Most ANDROGENS are excreted as sulfates in the urine, part of which may be of non-hepatic or non-biliary origin. Significant amounts of estrogens are excreted into bile as estriol glucuronide or estrone sulfates. Many derivatives of epinephrine and norepinephrine (see ADRENALINE) are eventually conjugated with either glucuronide or sulfate at the 4-hydroxy position and excreted into the bile. THYROXINE is predominantly conjugated with glucuronic acid and is excreted as such into the bile. The bile, however, is not a significant route for the net disposal of thyroxine, since this hormone is rapidly reabsorbed, enters the enterohepatic circulation and is eventually excreted as urinary metabolites.

Both steroids and thyroxine appear on the serosal side of the intestine as conjugates and so, in contrast to bilirubin, are not reabsorbed in the unconjugated form (see also GLYCOSIDES, STEROID).

Bile Pigments. The major components of bile, the bile pigments, can account for 15–20% of the total solids. *Bilirubin* comes primarily from the degradation of heme in the reticuloendothelial system in the spleen, bone marrow and, to a lesser extent, the liver. The initial step in the metabolism of heme is the cleavage of the PORPHYRIN ring and the elimination of the alpha methylene carbon to produce an open tetrapyrrole. This may exist as a complex with iron and globin called choleglobin. After removal of the iron and globin, the resulting tetrapyrrole, biliverdin, is rapidly reduced to bilirubin, the major pigment in human bile. Not all bilirubin results from the breakdown of hemoglobin from mature red cells. The early appearance of labeled bilirubin after injection of precursor glycine-C^{14} indicates that some bilirubin (approximately 10%) may arise from: (a) the rapid breakdown of immature red cells in the bone marrow, (b) from heme that had not entered hemoglobin, or (c) from destruction of newly formed red cells in the peripheral circulation. This "shunt" pathway for bilirubin formation may predominate in pernicious ANEMIA and some porphyrias. A small amount of bilirubin may also arise from other heme pigments such as myoglobin or the cytochromes. The bilirubin that enters the blood is rapidly and solely bound to ALBUMIN; the maximum binding is an estimated 2 moles of bilirubin/mole of albumin or approximately 65 mg of bilirubin/100 ml of serum. Normal circulating levels of bilirubin are less than 1 mg/100 ml. Bilirubin not bound to albumin (a result of high bilirubin concentration, low albumin concentration or the competition of organic acid drugs with bilirubin for the binding sites on albumin) appears more readily diffusible into the tissues. Free bilirubin, which readily crosses the BLOOD-BRAIN BARRIER in the newborn and to a lesser extent in the adult is an effective uncoupler of oxidative PHOSPHORYLATION in the brain and is highly toxic.

The hepatic transport of bilirubin from plasma to bile involves three independent but related mechanisms: uptake, conjugation and secretion. Plasma bilirubin is dissociated from plasma albumin in the liver and is rapidly concentrated in the cytoplasm of the hepatic cells by an unknown mechanism which precedes and is relatively independent of any subsequent hepatic conjugation. After concentration in the liver, bilirubin ("indirect" acting) is conjugated with 2 moles of glucuronic acid to form bilirubin diglucuronide ("direct" acting), the glucuronic acid moieties being attached in ester linkage to the carboxyl groups on the propionic acid side chains (Fig. 2). The reaction is shown in Fig. 3.

Glucuronyl transferase, the enzyme catalyzing the final step, is located in the smooth endoplasmic reticulum of liver and to a lesser extent in kidney and gastric mucosa, where a small amount of extrahepatic conjugation may occur. This enzyme has not been purified and it is unclear whether it nonspecifically catalyzes glucuronide conjugation of many non-bilirubinoid substrates or is bilirubin specific and a member of a large group of closely related glucuronyl transferases. Its activity can be induced by a variety of drugs and can be inhibited with steroids or steroid glucuronides found in plasma of pregnant women. "Bilirubin monoglucuronide," which is found in increased

STRUCTURE OF "DIRECT" BILIRUBIN

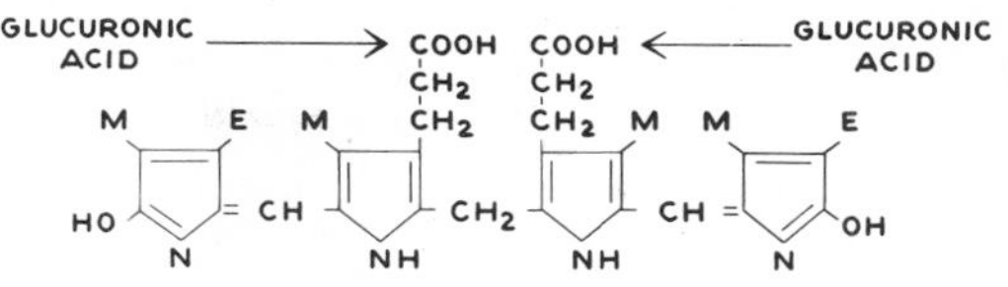

FIG. 2.

CONJUGATION OF BILIRUBIN

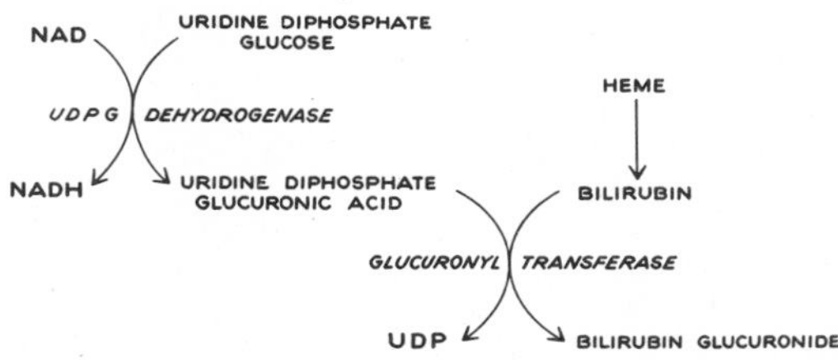

FIG. 3.

concentrations in the plasma in certain hepatic disease states, may represent incomplete conjugation or may be a complex of bilirubin with bilirubin diglucuronide. Small amounts of bilirubin may be excreted as the sulfate in the rat (not in man), but this pathway is inadequate for the elimination of bilirubin when glucuronide conjugation is impaired. Conjugated bilirubin is secreted into the bile (where it may be bound to lecithin or bile proteins) by a specific but unknown mechanism which does not favor secretion of the free pigment. Although some bilirubin glucuronide may be hydrolyzed and reabsorbed as the unconjugated pigment by the enterohepatic circulation, most of it is reduced and secreted via the feces.

Bilirubin entering the gut is reduced by bacterial action through a series of compounds to stercobilinogen (or L-urobilinogen). Since the methene and vinyl groups are reduced in urobilinogen, resonance is no longer possible; urobilinogen is therefore colorless and must be measured with Ehrlich's aldehyde reagent. When exposed to air, urobilinogen is partially oxidized to stercobilin (L-urobilin), the brown pigment that gives the color to the feces. In the newborn, or after oral administration of antibiotics in adults, intestinal flora are reduced and the feces are yellow because of their content of intact bilirubin. A small amount of urobilinogen (0.5–1%) may enter the enterohepatic circulation and eventually be excreted in urine. Measurement of urinary levels of urobilinogen is useful as an index of over-production of bilirubin and biliary bilirubin conjugates.

Clinical Significance and Biochemical Disorders. Classically, unconjugated and conjugated bilirubin in the serum of jaundiced patients are measured by the van den Bergh reaction, a reaction of diazosulfanilic acid with bilirubin to form azobilirubin. Water-soluble conjugated ("direct") bilirubin reacts immediately with the soluble diazo reagent, whereas nonpolar unconjugated ("indirect") bilirubin reacts only after addition of a solubilizing agent such as alcohol. In subjects with hepatic or extrahepatic biliary obstruction, the diglucuronide ("direct") bilirubin is regurgitated into the circulation and is the major bile pigment in the plasma. From plasma it is eventually excreted in the urine. In hepatitis and cirrhosis, hepatic uptake and conjugation are impaired and plasma "bilirubin monoglucuronide" as well as unconjugated ("indirect") bilirubin is high. Serum unconjugated bilirubin predominates in conditions characterized by bilirubin overproduction, such as hemolytic jaundice. Possibly because of its insolubility or strong binding to serum albumin, unconjugated bilirubin is not excreted in the urine regardless of its serum concentration.

Crigler-Najjar's disease in man and constitutional nonhemolytic hyperbilirubinemia in the Gunn rat are characterized by increased levels of unconjugated bilirubin in the serum. A genetic impairment of glucuronyl transferase, the enzyme responsible for the transfer of glucuronic acid from UDPGA (see Fig. 3), exists not only in the liver but in the kidney as well. Conjugative activity for *o*-aminophenol, steroids and phenolphthalein is also impaired, indicating a general defect of glucuronide conjugation rather than a specific defect of bilirubin metabolism (see GLUCURONIDASES). In these conditions, bilirubin is metabolized to unknown, polar, nonpigmented products which are excreted into the bile and eventually into the gut.

Gilbert's disease is characterized by a mild increase of unconjugated bilirubin in the plasma, which may result from a partial impairment of glucuronyl transferase, from a defect in bilirubin transport in the blood, or a defect in hepatic uptake. In subjects with Dubin-Sprinz (or Dubin-Johnson) disease, the serum contains high levels of both unconjugated and conjugated bilirubin, and an unindentified brown pigment is present in the liver. A defect in the secretion of the bilirubin conjugates from the hepatic cell is a probable causative factor. Rotor's disease is also characterized by increased serum levels of both unconjugated and conjugated bilirubin, but it differs from Dubin-Sprinz disease in that the hepatic brown pigment is not found.

The mild nonhemolytic jaundice often present in the newborn (physiological jaundice) or the more severe jaundice and kernicterus in premature infants may result in part from an inability of the immature liver to conjugate bilirubin; low hepatic levels of both glucuronyl transferase and UDPG dehydrogenase (the enzyme that catalyzes the synthesis of UDP glucuronic acid from UDP glucose) are found in fetus and newborn. Hepatic secretion of conjugated bilirubin may also be impaired. Bilirubin transport across the placenta, believed to be the major route of fetal bilirubin clearance, has been demonstrated in the guinea pig and the monkey but not in the rat.

References

1. WHITE, A., HANDLER, P., AND SMITH, E. L., "Principles of Biochemistry," New York, McGraw-Hill Book Co., 1964.
2. HARPER, H. A., "Review of Physiological Chemistry," Ninth edition, Los Altos, Calif., Lange Medical Publications, 1963.
3. BILLING, B., "Bilirubin Metabolism," *Postgrad. Med.*, **39**, 176 (1963).
4. CARBONE, J. V., GRODSKY, G. M., AND FANSKA, R., "Recent Advances in Bilirubin Metabolism," in "Current Concepts of Clinical Gastroenterology," (GAMBLE, J. R., EDITOR), p. 203, Boston, Little, Brown & Co., 1965.

GEROLD M. GRODSKY

BILE ACIDS

The bile acids are monocarboxylic acids of the steroid group with 24 carbon atoms and one to three secondary hydroxyl groups. They occur in the bile of all vertebrates from the teleosts upward, mostly in peptidic conjugation with glycine or taurine. The hydroxy groups are situated on C_3, C_0, C_7, and C_{12}, all in α-position (see Table 1; see STEROIDS for numbering system).

TABLE 1

Name	α-Hydroxy Groups in Position		
Lithocholic acid	3		
Hyodeoxycholic acid	3	6	
Chenodeoxycholic acid	3	7	
Deoxycholic acid	3	12	
Isodeoxycholic acid (synth.)	7	12	
Cholic acid	3	7	12
β-Phocacholic acid	3	7	23

The bile acids are the most important ingredient of the bile (see BILE AND BILE PIGMENTS), the secretion of the LIVER. The bile is stored in most vertebrates in a gall bladder; there, the bile is concentrated and its reaction changes from slightly alkaline to weakly acidic. Hence it is intermittently released into the duodenum where it emulsifies the alimentary lipides prior to their hydrolysis by pancreatic lipase. About 9/10 of the total amount of bile acids excreted is reabsorbed further down in the intestine. Thus, the liver has to replace only the portion lost with the feces and through catabolism. Oxidation of the bile acids leads as a first step to the conversion of the secondary hydroxyl groups to keto groups. Some such keto acids have been isolated from animal bile. The *in vitro* oxidation product of cholic acid, dehydrocholic acid (3,7,12-triketocholanic acid) is used as a choleretic, *i.e.*, as a stimulant of the bile flow.

Some of the bile acids, present in mammalian bile, are not synthesized by the animal itself, but are derived by reduction from originally present substances by the intestinal flora. The important deoxycholic acid is formed by elimination of the hydroxyl group on C_7 from cholic acid; lithocholic acid, by the analogous reaction from chenodeoxycholic acid.

Bile acids share with fatty acids the power to hemolyze red blood corpuscles. More specific is their ability to induce the lysis of live pneumococci, the so-called Neufeld phenomenon, which is widely utilized in bacteriological technique. Dehydrogenation of a bile acid derivative leads the the pentacyclic aromatic methylcholanthrene, one of the most potent chemical CARCINOGENS. A number of naturally occurring steroids also show minor carcinogenic activity. However, the hypothesis that aromatization of steroids takes place in the animal body and leads to actual carcinogenesis has been discarded.

Various color reactions are specific for bile acids; they go back to Pettenkofer's reaction with sucrose and concentrated sulfuric acid; in numerous modifications the sugar has been replaced by furfural, vanillin and other compounds, and concentrated sulfuric acid by more dilute sulfuric acid, phosphoric acid and hydrochloric acid. The ultraviolet absorption before and after treatment with sulfuric acid (65%) seems to give the best differentiation of the two most important bile acids, cholic acid and deoxycholic acid.

The chemical evolution of the bile acids with their structural and steric variations has been one of the most advanced chapters of comparative biochemistry. Snakes and lizards, being the more recently developed orders among the reptiles, form the same type of C_{24} bile acids as the higher classes of the vertebrates. The paleontologically older orders among the reptiles, the crocodiles and the turtles, have bile acids with 27 carbon atoms, derived from cholesterol by oxidation without shortening of the side chain. Among the amphibia, the toad, which shows peculiarities in steroid metabolism in connection with the elaboration of its poisonous secretion, also contains unusual bile acids with 27 carbon atoms.

The frogs among the amphibia, the cyprinidae (carp) among the teleostian fish, and all elasmobranchs (*e.g.*, the shark) use in lieu of bile acids the sulfate esters of neutral steroid alcohols with 27 carbon atoms, *e.g.*, ranol, cyprinol, scymnol, and, in the recently "resuscitated fossil fish" *Latimeria*, latimerol. Related neutral polyhydroxysteroids, presumably intermediate in the formation of bile acids, have been found in toad's bile, when synthetic processes are slowed down during "hibernation."

The steroids in general and the bile acids in particular exhibit a great tendency to form complexes with fatty acids, hydrocarbons, and other organic compounds. This property is most pronounced in deoxycholic acid, which forms well-defined coordination complexes, the "choleic acids," built on an axially symmetrical plan. Because of the optical activity of deoxycholic acid, choleic acids may serve to resolve racemates.

HARRY SOBOTKA

BILIRUBIN

Bilirubin, as the name suggests, is a red pigment found in bile. Chemically it is made up of four pyrrole rings with side chains, linked together, and is derived from heme released in the degradation of hemoglobin (see BILE AND BILE PIGMENTS). It is similar in structure to a green pigment biliverdin. Degradation of hemoglobin takes place principally in the reticulo-endothelial cells in the spleen, bone marrow, and liver. The bile pigments may accumulate in the blood in jaundice but are ordinarily excreted in the feces and urine.

BIOCHEMICAL GENETICS

Biochemical genetics has been broadly defined by Beadle[1] as "the branch of biology which seeks to define hereditary units in terms of the chemistry of their structures and of their functions." It has emerged into a rapidly expanding field through the fusion of many of the facts, concepts, ideas, and techniques of two classically independent branches of science until at the time of this writing, it encompasses many of the central problems of living systems. The fusion of genetics and biochemistry came about because of the scientific pressure to explain biological phenomena on a molecular level and typifies the trend in modern biology.

To fully understand the operational meaning of the term biochemical genetics, it is necessary to have some appreciation of the historical development of the field. Beadle's broad definition might well be assumed to include all work on the mechanisms of GENE action in which an underlying biochemical mechanism is assumed. It seems that Goldschmidt clearly envisioned a molecular basis for genetic phenomena in his book "Physiological Genetics"[2] in that he included a section on "the gene as an active molecule or group of molecules." Many other workers of the period sought to explain the phenomena of heredity on a biochemical level.[3,4] Yet these studies are generally considered to be during the era of "physiological genetics" or "developmental genetics." Similarly, in more recent times, the term molecular biology has come to be used to an increasing extent to indicate studies that most certainly should be included under Beadle's broad definition. Thus, to give a functional, limited definition of biochemical genetics that would distinguish it from other terms with similar meanings is undesirable if not impossible. It is a term that became very fashionable during a certain scientific era in the development of the basic understanding of heredity in terms of biochemistry. It has no exact limits or boundaries in time or in the types of experimental approaches or techniques used to study the underlying problem. It is a term currently used by some workers to indicate an area of investigation as broad as Beadle's definition. Others tend to limit its use to the area of study involving metabolic mutants similar to those employed by Beadle and Tatum[5] at the time the term biochemical genetics originally became popular. A brief outline of the development of the molecular basis of heredity will be provided in the hope that the reader may thus gain additional insight into the scope and meaning of this frequently used but nebulous term.

Perhaps the first clearly outlined work that would now be considered to be in the era of biochemical genetics was undertaken by Sir Archibald E. Garrod at about the turn of the century. By 1908 he and others had accumulated a considerable body of knowledge about alkaptonuria which he interpreted in two editions of his book "Inborn Errors of Metabolism."[6] Garrod clearly outlined his belief that alkaptonuria was the result of an abnormal inability of affected individuals to cleave the ring of homogentisic acid and that this was due to the absence or inactivation of the enzyme that normally catalyzes this reaction (see ALKAPTONURIA). This in turn was thought to be due to the absence of the normal form of a specific gene.

Garrod's interpretation of alkaptonuria and other inborn errors in metabolism as gene defects which resulted in inactivation of specific enzymes required for the specific biochemical reactions involved had little influence on the thinking of the biologists of his time. It remained for G. W. Beadle and those working with him in Paris, at the California Institute of Technology, and at Stanford University to outline clearly the relation between genes and enzymes. This work ultimately led Beadle to outline the "one-gene-one-enzyme" hypothesis[1] and introduced the era of biochemical genetics.

In the spring of 1935, Beadle joined forces with Borris Euphrussi at l'Institute de Biologie physio-chimique in Paris. After a period of classically discouraging and frustrating work,[7] they were able to transplant the imaginal disc of *Drosophila* larva of one genotype into the body of a larva of another genotype.[8] When imaginal discs from larva of mutant stocks with either cinnabar or vermilion eyes were transplanted into wild-type larva, wild-type eye color resulted in the transplanted tissue. Transplants of vermilion discs into cinnabar larva gave wild-type pigment; however, cinnabar discs transplanted into vermilion host produced cinnabar eye color. These results were explained by assuming that two discrete, diffusible substances were involved, one formed from the other. Beadle and Euphrussi gave a schematic representation of these results:[8]

$$\rightarrow \text{Precursor} \rightarrow \text{V}^+ \text{ Substance} \rightarrow \text{CN}^+ \text{ Substance} \rightarrow \text{Wild-type Pigment}$$

Later it was established that this sequence could be written in terms of specific chemical compounds involved:

→ Tryptophan —(V+)→ N-Formylkynurenine ⟶ Kynurenine —(CN+)→ Hydroxykynurenine → → Wild-type Pigment

Tryptophan: indole ring with $CH_2—CH(NH_2)—COOH$

N-Formylkynurenine: benzene ring with $CH_2—CH(NH_2)—COOH$ via C=O, and NH—CHO

Kynurenine: benzene ring with $CH_2—CH(NH_2)—COOH$ via C=O, and NH_2

Hydroxykynurenine: benzene ring with $CH_2—CH(NH_2)—COOH$ via C=O, NH_2, and OH

Vermilion mutants lack the enzyme to convert tryptophan to N-formylkynurenine. Cinnabar mutants lack the enzyme to convert kynurenine into hydroxykynurenine and accumulate kynurenine. When an imaginal disc from a vermilion mutant was transplanted into a cinnabar host, kynurenine would diffuse into the cinnabar tissue which contained normal enzyme for that portion of the synthetic pathway. The result was normal, wild-type eye pigment. *Drosophila* with vermilion- or cinnabar-colored eye pigment carried a mutation that led to the loss or inactivation of a single enzyme present in wild-type flies and necessary in the biosynthetic pathway leading to the normal, brown eye color of wild-type flies. With this relationship between mutations and enzymatic activity clearly in mind, the next step was to isolate more metabolic mutants. T. H. Morgan's choice of *Drosophilia* as an experimental organism had been very fortunate in the early days of genetics; however, it soon became obvious that a simple organism with simple nutritional requirements was needed if a large number of nutritionally deficient mutants were to be obtained.

Edward L. Tatum joined Beadle at Stanford University and soon *Neurospora crassa* (common bread mold) was chosen as an experimental organism. It had simple nutritional requirements (biotin, inorganic salts, and sugar), a haploid vegetative stage, and the feature of keeping all the spores resulting from a single meiotic cycle in order in an ascus, or spore pod, in a way that they could be dissected to yield all the progeny from a single meiosis. These features made *Neurospora* an almost ideal organism for the new approach to be undertaken (see also NEUROSPORA).

The technique was to irradiate asexual spores of *Neurospora*, cross them with a strain of the opposite mating type, allow sexual spores to be produced, then isolate them, grow the spores on a suitably supplemented medium and test them on an unsupplemented medium. Beadle has said that their concern at that time was not that mutations might not result in the loss of enzymatic activity necessary for the organism to convert sugar to vitamins, amino acids, and other simple compounds needed for growth, but rather that the mutations might occur so infrequently that they might become discouraged and give up before finding one. However, the 299th spore isolated gave a mutant strain requiring vitamin B_6 (pyridoxine) and the 1085th spore required vitamin B_1 (thiamine). Soon dozens of mutant strains with various nutritional requirements were available. They almost always involved single mutations that could be related to one enzyme responsible for a single biochemical step in the biosynthetic pathway leading to the required compound.[9] It was this relation that Beadle later put forth in the form of the "one-gene-one-enzyme" hypothesis.

The success with *Neurospora* lead Tatum to try essentially the same approach with bacteria to see if nutritional deficiencies followed their exposure to radiation. The first bacterial mutants of this type were successfully produced in *Acetobacter* and in *E. coli*.[10] These results in turn provided the basis for Joshua Lederberg's classical work on genetic recombination in bacteria.[11] Within a few years after the early work on *Neurospora*, *Acetobacter*, and *E. coli*, biochemical mutant strains of microorganisms had been isolated from almost every species tried, including other bacteria, yeasts, algae, and fungi. Ultimately the underlying relationship between genes and proteins was demonstrated for living organisms ranging from virus to man.

The recognition and isolation of nutritionally deficient strains of microorganisms had consequences for biochemistry beyond the "one-gene-one-enzyme" hypothesis. Nutritional mutants became a frequent tool in unraveling the detailed steps involved in the biosynthesis of vital cellular constituents. In a biosynthetic pathway, if a compound B is formed from another compound A through an intermediate X, according to the scheme:

$$A \xrightarrow{\text{Enzyme 1}} X \xrightarrow{\text{Enzyme 2}} B$$

usually metabolic mutants could be found that lacked the ability to carry out the over-all reaction from A to B. If a mutant lacked enzyme 2, the unknown intermediate X often accumulated to levels that greatly facilitated its isolation and characterization. If enzyme 1 were inactive in the mutant, compound A might be accumulated but the organism could still convert X to B so that frequently either X or B could be used to support growth.

Nutritional mutants used in this way were important in the elucidation of many biosynthetic pathways such as the sequence of reactions leading to the aromatic amino acids via dehydroshikimic and shikimic acids,[12] to phenylalanine by way of prephenic acid,[13] and to tryptophan via anthranilic acid, indole glycerol phosphate[14] and condensation of serine[15] and, of course, the conversion of tryptophan via kynurenine and 3-hydroxykynurenine to the eye pigments of insects and to niacin.[16,17] Mutant microorganisms played similar roles in elucidating the pathways leading to isoleucine, valine and leucine, arginine and proline, methionine and threonine, lysine, histidine, purines and pyrimidines, vitamins and many other compounds of biochemical importance. Wagner and Mitchell[18] present many such pathways and show some of the sites at which mutant blocks occur.

Molecular Basis of Genetic Information Transfer. The recognition of the relation between genes and enzymes that grew out of the work leading to the "one-gene-one-enzyme" hypothesis greatly stimulated the quest for the recognition and elucidation of the biochemical nature of the genetic material. By 1940 genetic and cytogenetic studies clearly pointed to the CHROMOSOMES of higher organisms as the structures within which genes were to be found. In 1936 the microspectrophotometric observations of Casperson[19] had made it clear that chromosomes contained large amounts of nucleic acid, and refinements of the Feulgen staining technique[20] (see HISTOCHEMICAL METHODS) made it clear that this nucleic acid was primarily DEOXYRIBONUCLEIC ACID (DNA). However, extraction from isolated nuclei and extension of Casperson's microspectrophotometric techniques indicated that the DNA probably existed primarily as a nucleoprotein complex in the living cells. The question, then, was what was the real genetic material? Genes might be made of DNA, protein, or a complex of both as nucleoproteins. The situation was complicated by the lack of knowledge of the chemical structure of DNA. It had been widely assumed during the 1930's that DNA might consist of a repetitive series of tetranucleotides, each of which contained adenylic, thymidylic, guanylic, and cytidylic nucleotides. If this were the case, all DNA would be identical or at least so similar that it was difficult to imagine how it could contain the information required for it to be the genetic material. It thus seemed likely that DNA might function merely as a structural element of chromosomes.

In the early 1940's evidence was presented from several sources on the chemical structure of DNA that made the repetitive tetranucleotide model untenable[21] and made it seem likely that there might be a very large number of different kinds of DNA. The idea that DNA itself might be the genetic material received its greatest support from the work of Avery, MacLeod, and McCarty[22] who succeeded in showing that the

TRANSFORMING FACTOR in pneumococcus was almost certainly DNA. Evidence accumulated from many sources,[23,24] until, by 1952, it seemed likely that DNA was the primary genetic material for living systems ranging from virus to mammals. In addition, during the same period, evidence was accumulating on the chemical structure of DNA that made it seem increasingly probable that it had the required chemical properties. Chargaff[25] pointed out the basic relation between the purine and pyrimidine bases (adenine equals thymine, and guanine equals cytosine) in DNAs from a wide variety of sources and with very divergent over-all base composition. Watson and Crick used this relation with X-ray diffraction data of the type provided by Wilkins[26] to work out the helical double-stranded structure of DNA[27] in which a highly specific hydrogen bonding occurs between the two chains of the polymer. This specific hydrogen bonding between adenine and thymine or guanine and cytosine accounted for the Chargaff rules of DNA base composition, and provided a structure that seemed to provide the required genetic stability.

As DNA began to be recognized as the primary genetic material, the central problem changed from what to how: how is the information contained in DNA and how is it transmitted into PROTEINS. In 1956, Ingram[28] provided some insight into this problem by demonstrating that sickle-cell hemoglobin isolated from humans carrying the mutant gene for sickle-cell anemia differed from normal hemoglobin by only one amino acid [see HEMOGLOBINS (IN HUMAN GENETICS)]. Work from many sources[29-31] led to the recognition that genes control the sequence of amino acids in polypeptides and that most mutations cause specific changes in this amino acid sequence. These changes may affect the properties of the protein formed from the mutant polypeptide, such as enzymatic activity, solubility or heat stability. This led Benzer[32] to refine and restate the "one-gene-one-enzyme" hypothesis as "one-gene-one-polypeptide chain."

The emerging concept held that DNA, the genetic material, contained the information necessary to position specific amino acids into sequences in a related polypeptide and that this information was somehow encoded into the sequence of bases of the DNA. It was soon discovered that a RIBONUCLEIC ACID (RNA) intermediate, messenger RNA, was involved and that the code consisted of a sequence of three nucleotides for each amino acid (see GENETIC CODE).

Further discussion of the molecular mechanisms whereby each nucleotide triplet of DNA is "transcribed" into a nucleotide triplet of messenger RNA, and in turn "translated" into a specific amino acid in the synthesized linear polypeptide chain, is found in the articles GENES; RIBONUCLEIC ACIDS (BIOSYNTHESIS); and PROTEINS (BIOSYNTHESIS). The mechanism of transfer of genetic information from the nucleotide triplet "language" of one generation to its daughter generation is dicussed in DEOXYRIBONUCLEIC ACIDS (REPLICATION).

References

1. BEADLE, G. W., *Chem. Rev.*, **37**, 15 (1945).
2. GOLDSCHMIDT, R., "Physiological Genetics," pp. i–ix, 1–375, New York, McGraw-Hill Book Co., 1938.
3. WADDINGTON, C. H., "Organizers and Genes," London, Cambridge University Press, 1940.
4. WRIGHT, S., *Physiol. Rev.*, **21**, 487 (1941).
5. BEADLE, G. W., AND TATUM, E. L., *Proc. Natl. Acad. Sci. U.S.*, **27**, 499 (1941).
6. GARROD, A. E., "Inborn Errors in Metabolism," Second edition, Oxford, Oxford University Press, 1923.
7. BEADLE, G. W., "Genes and Chemical Reactions in Neurospora," Nobel Lecture given December 11, 1958, *Science*, **129**, 1715 (1959).
8. BEADLE, G. W., AND EUPHRUSSI, B., *Genetics*, **29**, 291 (1937).
9. BEADLE, G. W., AND TATUM, E. L., *Proc. Natl. Acad. Sci. U.S.*, **27**, 499 (1941).
10. TATUM, E. L., *Cold Spring Harbor Symp. Quant. Biol.*, **11**, 278 (1946).
11. LEDERBERG, J., AND TATUM, E. L., *Nature*, **158**, 558 (1946).
12. TATUM, E. L., GROSS, S. R., EHRENSVÄRD, G., AND GARNJOBST, L., *Proc. Natl. Acad. Sci. U.S.*, **40**, 271 (1954).
13. METZENBERG, R. L., AND MICHELL, H. K., *Biochem. J.*, **68**, 168 (1958).
14. YANOFSKY, C., *J. Biol. Chem.*, **224**, 783 (1957).
15. TATUM, E. L. AND BONNER, D. M., *Proc. Natl. Acad. Sci. U.S.*, **30**, 30 (1944).
16. MITCHELL, H. K., AND NYE, J. F., *Proc. Natl. Acad. Sci. U.S.*, **34**, 1 (1948).
17. BONNER, D., *Proc. Natl. Acad. Sci. U.S.*, **34**, 5 (1948).
18. WAGNER, R. P., AND MITCHELL, H. K., "Genetics and Metabolism," New York, John Wiley & Sons, 1965.
19. CASPERSON, T., *Skand. Arch. Physiol.*, **73**, Suppl. No. 8, (1936).
20. FEULGEN, R., AND ROSSENBECK, H., *Z. Physiol. Chem.*, **135**, 203 (1924).
21. GULLAND, J. M., *Cold Spring Harbor Symp. Quant. Biol.*, **12**, 95 (1947).
22. AVERY, O. T., MACLEOD, C. M., AND MCCARTY, M., *J. Exptl. Biol. Med.*, **79**, 137 (1944).
23. HERSHEY, A. D., AND CHASE, M., *J. Gen. Physiol.*, **36**, 39 (1952).
24. ALLFREY, V. G., MIRSKY, A. E., AND STERN, H., *Advan. Enzymol.*, **16**, 411 (1955).
25. CHARGAFF, E., "The Nucleic Acids" (CHARGAFF, E., AND DAVIDSON, J. N., EDITORS), Vol. 1, New York, Academic Press, 1955.
26. WILKINS, M. H. F., STOKES, A. R., AND WILSON, H. R., *Nature*, **171**, 738 (1953).
27. WATSON, J. D., AND CRICK, F. H. C., *Cold Spring Harbor Symp. Quant. Biol.*, **18**, 123 (1953).
28. INGRAM, V. M., *Nature*, **178**, 792 (1956).
29. HOROWITZ, N., *Federation Proc.*, **15**, 818 (1956).
30. HELENSKI, D. R., AND YANOFSKY, C., *Proc. Natl. Acad. Sci. U.S.*, **48**, 173–183 (1962).
31. INGRAM, V. M., "The Hemoglobins in Genetics and Evolution," New York, Columbia University Press, 1963.

32. BENZER, S., "Elementary Units of Heredity" in "The Chemical Basis of Heredity" (MCELROY, W. D., AND GLASS, B., EDITORS), p.70, Baltimore, Md., Johns Hopkins Press, 1957.

BOYD HARDESTY

BIOCHEMICAL INDIVIDUALITY

Biochemical individuality is the possession of biochemical distinctiveness by individual members of a species, whether plant, animal or human. The primary interest in such distinctiveness has centered in the human family, and in the distinctiveness within animal species as it might throw light on human biochemistry.

While it has been known for centuries that bloodhounds, for example, can tell individuals apart even by the attenuated odors from their bodies left on a trail, the first scientific work which hinted at the existence of substantial biochemical distinctiveness in human specimens was the discovery of BLOOD GROUPS by Landsteiner about 1900.

A few years later Garrod noted what he called "inborn errors of metabolism"—rare instances where individuals gave evidence of being abnormal biochemically in that they were albinos (lack of ability to produce pigment in skin, hair and eyes), or excreted some unusual substance in the urine or feces. To Garrod these observations suggested the possibility that the biochemistry of all individuals might be distinctive (see BIOCHEMICAL GENETICS).

About fifty years later serious attention to the phenomenon of biochemical individuality resulted in the publication of several articles and a book on this subject. These reported evidence indicating that every human being, including all those designated as "normal," possesses a distinctive metabolic pattern which encompasses everything chemical that takes place in his or her body. That these patterns, like the abnormalities discussed by Garrod, have genetic roots is indicated by the pioneer explorations of Beadle and Tatum in the field of biochemical genetics in which they established the fact that the potentiality for producing enzymes resides in the genes.

Biochemical individuality, which is genetically determined, is accompanied by, and in a sense based upon, anatomical individuality, which must also have a genetic origin. Substantial differences, often of large magnitude, exist between the digestive tracts, the muscular systems, the circulatory systems, the skeletal sytems, the nervous systems, and the endocrine systems of so-called normal people. Similar distinctiveness is observed at the microscopic level, for example in the size, shape and distribution of neurons in the brain and in the morphological "blood pictures," *i.e.*, the numbers of the different types of cells in the blood.

Individuality in the biochemical realm is exhibited with respect to (1) the composition of blood, tissues, urine, digestive juices, cerebrospinal fluid, etc.; (2) the enzyme levels in tissues and in body fluids, particularly the blood; (3) the pharmacological responses to numerous specific drugs; (4) the quantitative needs for specific nutrients—minerals, amino acids, vitamins—and in miscellaneous other ways including reactions of taste and smell and the effects of heat, cold, electricity, etc. Each individual must possess a highly distinctive pattern, since the differences between individuals with respect to the measurable items in a potentially long list are by no means trifling. Often a specific value derived from one "normal" individual of a group will be several times as large as that derived from another.

The implications of this individuality are extremely broad. For medicine, they suggest that susceptibility to all disease—infective, metabolic, degenerative, mental or unclassifiable (including cancer)—probably has its roots in biochemical individuality and that the differences in responses to drugs including alcohol, caffeine, nicotine, carcinogens and morphine derivatives, which are well authenticated, have a sound and discoverable basis. It is a well-known fact that conditions which will produce disease in certain individuals will not do so in others. The basis for this observation has hitherto not been recognized; it doubtless has its roots in biochemical individuality—a development which has received little attention. People definitely possess what may be called "biochemical personalities," and it seems extremely likely that these are meaningful in connection with the numerous and increasing personality disorders and difficulties which afflict men and women in modern life.

Biochemical individuality offers a sound scientific basis for recognizing the existence in every individual of a unique makeup in the broadest sense. For centuries, "individuality" has been written about and its place in the scheme of things has been discussed, but the knowledge of what individuality consists of and how it manifests itself has indeed been scanty. Only in recent years have we had a basis for understanding it in a definitive way.

The understanding of individuality is basic to the understanding of human behavior. It is not enough to know the ways in which all human beings respond alike to certain stimuli; it is fully as important to know also why different people confronted with about the same stimulus react very differently. Since biochemistry underlies many of our moods and our reactions to different types of stimulus, and since it is in the area of biochemistry that individuality is most definitively recognized, biochemistry merits inclusion as one of the most important of the so-called behavioral sciences.

Because biochemical individuality points the way toward individuality in the broadest sense of the word, it has profound implications not only in medicine, psychiatry, and psychology but also in human relations, education, politics and even philosophy.

Reference

WILLIAMS, R. J., "Biochemical Individuality," New York, John Wiley & Sons, 1956.

ROGER J. WILLIAMS

BIOCHEMICAL THERMODYNAMICS

The concepts of energy and its conservation have played a central role in the development of physics. Energy is a mathematical, theoretical construct existing in the mind of man and in his writings. This construct has proved useful in man's physical descriptions of the universe. Physicists and chemists have come to believe that they do not fully understand a phenomenon unless they can describe it in terms of energy and energy changes. Studies dealing with changes of chemical energy and heat are often labeled chemical thermodynamics. A subset of this larger group are those which are concerned with biochemistry and which involve chemical reactions associated with living systems.

Thermodynamics. Thermodynamics is a macroview of the universe in which phenomenologic descriptions are developed for systems that are large compared to molecular sizes. By contrast, statistical mechanics interprets microscopic changes of energy in terms of the statistical behavior of molecules. (Both the macroscopic and the microscopic views will be used in this paper.)

In many textbooks, thermodynamics is based upon three laws.[1] Only a brief summary is included here. The *first law* of thermodynamics is a restatement of the law of conservation of energy; it emphasizes that if energy seems to disappear, this energy is stored as internal energy of the system. The internal energy is reflected in the temperature of the system.

The *second law* of thermodynamics states that unless external energy is supplied, heat always flows from a hotter body to a colder body. The second law of thermodynamics may be restated in terms of the change of *entropy*, which is defined as the heat energy put into a body divided by its absolute temperature. In molecular terms, entropy may be regarded as a measure of the randomness of the system. In these terms, the second law of thermodynamics states that the entropy of a completely isolated system always tends to increase to a maximum, *i.e.*, things become more and more disordered.

The *third law* of thermodynamics, in its strongest form, states that the entropies of all simple crystalline substances at a temperature of absolute zero are equal and may be chosen as zero. (If there is a random mixture in the crystal, such as some protons with their spin "up" and others with their spin "down," then the entropy at absolute zero may be greater than zero.)

Classical thermodynamics deals with reversible processes and can be applied to irreversible ones only by approximation. Most biochemical processes are irreversible. Accordingly, the body of knowledge concerned with the mathematical relationships between fluxes in irreversible processes[2] would appear to be the important one for biochemistry. However, to date, its applications have been extremely limited.

Chemical Thermodynamics. In chemical thermodynamics it is customary to define a number of quantities in addition to entropy. For example, the *enthalpy* (H) is a function defined as the sum of the internal energy plus the product of the pressure times the volume. In a system maintained at constant pressure, the change in enthalpy is equal to the heat put into the system. Enthalpies are useful in computing heats of reaction.

The thermodynamic quantity which has the most widespread application in biochemistry is called the *free energy*. Actually, there are two different quantities referred to by that name; the one called the Gibbs free energy (G) is the one which is convenient to use in biochemical thermodynamics. G is defined as the difference between the enthalpy and the product of the entropy times the absolute temperature. At equilibrium, G is a minimum for a system maintained at constant temperature and pressure. If the system is not at equilibrium, the difference between G and its minimum is equal to the maximum useful energy which can be obtained by letting the system approach equilibrium.

It is often convenient in chemical thermodynamics to describe how much of the free energy (or other quantities such as the entropy) can be associated with each of several molecular species in a mixture. If this is then divided by the number of moles of the substance in question, one obtains a partial molal quantity. Such quantities are used frequently in biochemical thermodynamics.

Equilibrium. There are many different types of applications of thermodynamics to biochemistry. The first of the restricted group considered here deals with equilibrium. Consider a bimolecular chemical reaction:

$$\mathrm{A + B \rightleftharpoons C + D} \qquad (1)$$

At equilibrium the concentrations of the reactants will be related in such a fashion that one may define an equilibrium constant by:

$$K = \frac{[\mathrm{C}]\,[\mathrm{D}]}{[\mathrm{A}]\,[\mathrm{B}]} \qquad (2)$$

The equilibrium constant is very closely related to the Gibbs free energy. For this relationship, one needs to know the partial molal free energy for each reactant in its standard state—*i.e.*, unit concentration and at the same temperature and pressure as the reaction mixture. These are found in standard tables.[3,4] These may be designated as $\tilde{G}_A{}^\circ$, $\tilde{G}_B{}^\circ$, $\tilde{G}_C{}^\circ$, $\tilde{G}_D{}^\circ$, respectively. The difference between these partial molal Gibbs free energies in the standard state for the right and left sides of the Eq. (*1*) is defined as:

$$\Delta G^\circ = \tilde{G}_C{}^\circ + \tilde{G}_D{}^\circ - \tilde{G}_A{}^\circ - \tilde{G}_B{}^\circ \qquad (3)$$

Then, by using only the relationship that G is a minimum at equilibrium, it can be shown that:

$$K = e^{-\Delta G^\circ/RT} \qquad (4)$$

where R is the universal gas constant (about 2 kcal/mole) and T is the absolute temperature. Thus, a knowledge of the Gibbs free energies allows the computation of equilibrium constants. Certain computer programs are reported to be easier to write if the data are supplied in terms of the partial molal free energies rather than in terms of equilibrium constants.[6]

Thermodynamic analysis relating equilibrium constants to the Gibbs free energy is valid for reversible reactions. However, in some examples it is not possible to observe the reactions in both directions. Consider, for example, the catalytic decomposition of H_2O_2 by the enzyme, catalase. According to Chance[7] the reaction follows the scheme:

$$H_2O_2 + \text{catalase} \underset{k_2}{\overset{k_1}{\rightleftharpoons}} H_2O_2 \cdot \text{catalase} \qquad (5)$$

$$H_2O_2 + H_2O_2 \cdot \text{catalase} \underset{k_4}{\overset{k_3}{\rightleftharpoons}} O_2 + 2H_2O + \text{catalase} \qquad (6)$$

It is possible to measure the reaction rates k_1, k_2, and k_3.[7] These allow a calculation of the equilibrium constant for Eq. (5). The equilibrium constant for Eq. (6) and the rate constant k_4, however, can only be found by calculations based on the partial molal Gibbs free energy.[5]

It would appear that, in examples such as the above, the use of irreversible thermodynamics would be fruitful. Irreversible thermodynamics also would seem to be preferable for describing steady-state concentrations of enzymes under the nonequilibrium conditions which characterize most living cells. However, these applications of irreversible thermodynamics have not been possible to date.

Heat of Reaction. Another aspect of biochemical thermodynamics is concerned with the energy changes which accompany biochemical reactions. As noted earlier, under the conditions of most biologic systems, the change in enthalpy is equal to the heat energy required for the reaction to proceed. The heat required (or liberated) per mole of product is called the heat of the reaction. By using the first and second laws of thermodynamics, one can readily show[1] that this heat of reaction is independent of the reaction pathway or intermediates. It is the same for spontaneous reactions and for enzyme-catalyzed reactions, depending only on the initial reactants (including their concentrations and temperature) and the final products (including their concentrations and temperature).

From this it follows that the energy changes are additive for a series of reaction steps. Accordingly, a definite energy or heat of dissociation may be assigned to any chemical bond. These heats of dissociation will vary somewhat for the bonds between a specific group and a variety of other groups, depending on the nature of the other group, but this variation tends to be small compared to the range of heats of dissociation for different types of bonds. An average value of the heat of dissociation for a given type of bond is called the bond energy.

The energy of bonds which can be oxidized or reduced is often expressed in terms of the equivalent electrical potential.[8] This is referred to as a redox potential. In electron transport, in which a series of oxidation-reduction reactions occurs, these reactions will be arranged in steps of increasing redox potential, from the reduced end of the change to the oxidized, if all of the substances are oxidized to about the same degree in the steady state. However, if the reactants in the chain are oxidized to quite varying degrees, electron transport may occur in the opposite direction from that which would be predicted from a knowledge of the redox potentials alone. (Chance and Williams[9] have shown that in the cytochrome system all of the molecules occur in comparable concentrations and, hence, the redox potentials are significant in indicating the direction of electron transport.)

Bond energy calculations are also used in other studies of metabolism; in general, the formulation is in terms of energy *per se* rather than redox potential. Such calculations allow the determination of the energy available from foodstuffs and from metabolic products. They also allow the determination of the energy required in the synthesis of new molecules. It is interesting to note that the efficiency of transferring energy in oxidative PHOSPHORYLATION apparently exceeds the efficiency of common heat engines such as the steam engine and the gasoline engine.

The energy released from the oxidation of foodstuffs is divided into small packets and stored as labile chemical bonds which can be readily used by biochemical and biophysical processes requiring energy. These bonds, such as the pyrophosphate bonds in adenosine triphosphate (ATP), are called high-energy bonds (see PHOSPHATE BOND ENERGIES). From the point of view of thermodynamics, there is nothing "high-energy" about these bonds. Rather, their significant property is the ease with which they may give up their energy to a variety of other reactions.

Membrane Transport. Thermodynamic analyses are used in studies of potentials at biologic membranes and of the transport of ions across these membranes. The potentials which exist at these membranes arise as a result of the unequal concentrations of ions on the two sides of the membrane. These concentrations in turn may be maintained by the transport of ionic species at the expense of metabolic energy. This latter process is called ionic pumping or ACTIVE TRANSPORT. In order to distinguish a steady-state maintained at the expense of metabolic energy from a true thermodynamic equilibrium, it is necessary to have a knowledge of the partial molal Gibbs free energies and of the concentrations of all the molecular and ionic species on each side of the membrane. For equilibrium to occur, the free energy must be a minimum; *i.e.*, transporting a small part of any molecular species present from one side of the membrane to the other side must introduce a negligible (but positive) change in the free energy of the system. (This is subject to various constraints such as keeping the net charge equal to zero.)

By requiring only that the Gibbs free energy be a minimum, it is possible to show that a potential (called a Donnan membrane potential) will develop across a dialysis membrane separating a mixture containing a salt and an ionized protein from distilled water and that the concentrations of the cation and of the anion in the salt will be unequal on the two sides of the membrane. (This is called Donnan equilibrium; see MEMBRANE EQUILIBRIUM.)

A quantity called the electrochemical potential (ϕ) is the same on both sides of the membrane for each species at equilibrium. ϕ is defined by

$$\phi = qV + RT \ln (c/c^\circ) \qquad (7)$$

where

q = charge on the ion expressed as coulombs per mole,
V = electrical potential expressed in volts relative to an arbitrary ground,
c = concentration,
c° = standard state concentration.

For equilibrium, ϕ on the inside (i) and outside (o) of the membrane must be equal, leading to the expression

$$\Delta V = V_i - V_o = \frac{RT}{q} \ln (c_o/c_i) \qquad (8)$$

The transmembrane potential, ΔV, must be the same for all molecular species at equilibrium.

For systems not at equilibrium, one may still assign to each ionic species a value of ΔV which would be the transmembrane potential if that ionic species were at equilibrium. The difference between this and the actual ΔV may be considered as an electrochemical potential difference driving that ionic species.

A test[10] for active transport is to compare the ratios of the unidirectional fluxes through the membrane for each ionic species with those computed from the electrochemical potential differences. An inequality establishes that active transport is occurring. Additional experiments are needed to determine whether the ion is being pumped, carried along as some other ion is pumped, or dragged by a solvent flux. In any of these cases, metabolic energy must be expended to maintain disequilibrium.

Membrane transport is one of the few subjects which have been studied by irreversible thermodynamics.[11]

Reaction Rate Theories. Another major aspect of biochemical thermodynamics deals with biochemical reaction rates. The rate at which the reaction proceeds is limited by the frequency with which the reactants collide with sufficient energy to react. Thus, in the catalase–hydrogen peroxide reaction discussed earlier, the rate constant k_1 will depend on the frequency of collision of a hydrogen peroxide molecule and an active site on the catalase molecule. In addition, there may be a probability factor, P, such that even collisions between molecules with sufficient energy to react will not always lead to a reaction.

Based on the statistical mechanics of gases, it is possible to write an equation for a rate constant k, namely:

$$k = PZe^{-\mu/RT} \qquad (9)$$

where

μ = the so-called Arrhenius constant or energy of activation,
Z = frequency of collision,
P = probability of a reaction as above.

In the case of gaseous reactions, μ represents an internal energy necessary for interaction; it is called the energy of activation. For reactions in liquids the intepretation of μ is not so clear-cut. However, if P, Z, and μ are all assumed to be independent of T, it is possible to approximate the temperature dependence of many reactions by Eq. (9). In this case, μ is called the Arrhenius constant.

In terms of Eq. (9), one may describe the role of enzymes as decreasing both μ and P. These values are decreased appropriately for both the forward and the backward reactions, so that the equilibrium concentrations will remain unaltered but the rate of reaching equilibrium will be greatly increased. It is also possible to show from Eq. (3) that in most biologic reactions the rate will be changed by a constant factor, Q_{10}, for a change in temperature of 10°C. For those reactions having a Q_{10} of about 2, one may readily show that μ is about 12 kcal/mole.

Since the interpretation of μ in biologic reactions is so unclear, an alternate interpretation known as absolute rate theory has been used.[12] This theory states that in a reaction such as Eq. (5), a small but not negligible percentage of the reactants (in this example, free catalase and hydrogen peroxide molecules) encounter each other and remain near each other sufficiently long to be regarded as a new, unstable molecular species called an activated complex. As in any other equilibrium, the concentration of this species will be determined by the difference of the Gibbs free energies in the standard state. The rate at which the activated complex is transformed to the product of the reaction can be approximated by an absolute quantity. In this fashion it can be shown that any rate constant k can be expressed as:

$$k = \frac{RT}{Nh} e^{-\Delta G^*/RT} \qquad (10)$$

where

N = Avogadro's number,
h = Planck's constant,
ΔG^* = the Gibbs free energy of activation.

This theory has been successfully applied to a wide variety of examples.[5,12] It can be used not only in discussions of reaction rates but also in the transport of molecules through a membrane.[12] It is interesting to note that values cannot be assigned to ΔG^* in most cases except by the use of the third law of thermodynamics.

Thermodynamics and Life. A more general biochemical application of thermodynamics describes the gross changes in energy and in entropy which occur in the life processes. Life implies a molecular ordering and a nonrandomness of biochemical molecules. Thus, the entropy of a living system must be lower than that of an equilibrium system. Since all systems on earth always tend toward equilibrium, there must be a continual supply of energy to earth in such a form as to reduce the entropy below the equilibrium value. This energy is supplied by the sun in the form of light. Thus, one may speak of the sun as supplying energy and "negative entropy" to the earth.

The "negative entropy" is lost if the light is converted into heat. On the other hand, green plants capture this "negative entropy" in the process of PHOTOSYNTHESIS by which free molecular oxygen and sugars are produced.

Some plants are eaten in turn by animals who also use the molecular oxygen in the atmosphere. The animals bring about a further degradation of the "negative entropy," but still preserve enough to make their life possible. These animals may be eaten by other animals or their bodies may be destroyed by bacterial decay. Eventually, after passing through many living systems, the energy received by the sun becomes evenly distributed as heat. The continuing life cycle demands continuing supplies of "negative entropy."

For a still more general point of view, thermodynamics implies that the entropy of the universe, and probably even of the solar system, is continually increasing toward equilibrium. Nonetheless, in a part of the universe (such as the earth) which is not thermodynamically isolated, the total entropy may decrease rather than increase. Similarly, within a living organism, particularly a growing one, the entropy may decrease continually when compared to the equilibrium value.

Information Theory. The last biochemical application of thermodynamics to be considered here deals with a branch of science called information theory.[13] Information theory attempts to resolve all information to a series of yes-no (or binary) decisions. It then measures information in bits as the base-2 logarithm of the probability of a yes or no before the "message" was received. The word "message" is interpreted in information theory to include messages conveyed by the structure and composition of biochemical molecules such as DEOXYRIBONUCLEIC ACID (DNA) and PROTEIN.

In information theory it is possible to compute a function which describes the average information received per message. This expression is formally identical to that for the negative of the entropy in statistical mechanics. Thus, it is tempting to associate information with negative entropy.

Whether the information in information theory is identical to the negative of entropy in thermodynamics has been the subject of many pages of journal articles and books. Suffice it to point out here that if the messages are biochemical molecules then the information content corresponds exactly to the negative of the entropy.[14] In fact, in these cases the justification for discussing the information coded in DNA rather than the negative of the entropy of the DNA is a matter of human convenience.

Summary. There are numerous applications of thermodynamics to biochemical systems. Those reviewed here include interpretations of equilibrium constants, heats of reaction, membrane transport, reaction rate theory, negative entropy and life, and information theory. All these and other applications of biochemical thermodynamics form a special subset of chemical thermodynamics which in turn is part of thermodynamics *per se*. Biochemical thermodynamics makes use of the three laws of thermodynamics as well as statistical mechanics to present a basic view of life in terms of energy changes.

References

1. KLOTZ, I. M., "Introduction to Thermal Thermodynamics," New York, W. A. Benjamin, 1964, 244pp.
2. PRIGOGINE, ILYA, "Introduction to Thermodynamics of Irreversible Processes," Second ed. New York, John Wiley & Sons, 1962, 119pp.
3. National Research Council, "International Critical Tables of Numerical Data, Physics, Chemistry and Technology," New York, McGraw-Hill Book Co.,
4. "Handbook of Chemistry and Physics," Ed. 45, Cleveland, Ohio, Chemical Rubber Publishing Company, 1964.
5. ACKERMAN, EUGENE, "Biophysical Science," Englewood Cliffs, N.J., Prentice-Hall, 1962, 626pp.
6. DAYHOFF, M. O., "Thermodynamic Equilibria in Prebiological Atmospheres," *Proc. 17th Ann. Conf. Eng. in Med. and Biol.*, **6**, 58 (1964).
7. CHANCE, BRITTON, "Reaction Kinetics of Enzyme-Substrate Compounds," in "Technique of Organic Chemistry: Investigation of Rates and Mechanisms of Reactions," (FRIESS, S. L., AND WEISSBERGER, A., EDITORS), Vol. 8, pp. 627–667, New York, Interscience Publishers, 1953.
8. WHITE, ABRAHAM, HANDLER, PHILIP, AND SMITH, E. L., "Principles of Biochemistry," Third edition, New York, McGraw-Hill Book Co., 1964, 1106pp.
9. CHANCE, BRITTON, AND WILLIAMS, G. R., "The Respiratory Chain and Oxidative Phosphorylation," in "Advances in Enzymology: And Related Subjects of Biochemistry" (NORD, F. F., EDITOR), Vol. 17, pp. 65–134, New York, Interscience Publishers, 1956.
10. USSING, HANS, "Active Transport of Inorganic Ions," in "Active Transport and Secretion" (BROWN, R., AND DANIELLI, J. F., EDITORS), pp. 407–422, New York, Academic Press, 1954.
11. KATCHALSKY, A., AND KEDEM, O., "Thermodynamics of Flow Processes in Biological Systems," *Biophys. J.*, **2**, 53–78 (March 1962).
12. JOHNSON, F. H., EYRING, HENRY, AND POLISSAR, M. J., "The Kinetic Basis of Molecular Biology," New York, John Wiley & Sons, 1954, 874pp.

13. GOLDMAN, STANFORD, "Information Theory," Englewood Cliffs, N.J., Prentice-Hall, 1953, 385pp.
14. QUASTLER, HENRY, "Essays on the Use of Information Theory in Biology," Urbana, Illinois, University of Illinois Press, 1953, 273pp.

EUGENE ACKERMAN

BIOCYTIN

Biocytin is a bound form of BIOTIN isolated from yeast. It is ε-N-biotinyl-L-lysine, and can serve as growth promoting substance for *Lactobacillus casei*.

Reference

WRIGHT, L. D., *et al.*, *Science*, **114**, 635 (1951).

BIOLOGICAL OXIDATION-REDUCTION

The expression "biological oxidation-reduction" refers to the movement of electrons in living organisms. This movement may take place with accompanying atoms or groups of atoms. The donor of reducing equivalents is called the reductant, and the acceptor of reducing equivalents is called the oxidant. Reduction equivalents pass from one molecule to another at biologically useful rates, providing the mechanism of the transfer does not involve improbable states. This condition is met in living organisms by means of enzymic catalysis.

Biological oxidation-reduction commonly takes the stoichiometric form of electron transfer, hydrogen atom transfer, or oxygen atom transfer between oxidant and reductant, catalyzed by "oxido-reductases." The details of the underlying processes are not known, and the elucidation of these mechanisms is a major objective of contemporary biochemistry and biophysics. The complexity of this problem is illustrated by RESPIRATION, during which molecular oxygen is reduced to water. The enzyme for this reaction, cytochrome *c* oxidase, catalyzes the transfer of four reducing equivalents to O_2 by a process which could be, among other possibilities, hydrogen atom transfer:

$$O_2 + 4H\cdot = 2H_2O$$

or electron and proton transfer:

$$O_2 + 4H^+ + 4e = 2H_2O$$

but no reaction intermediates have been detected, and the mechanism of this reaction, essential to aerobic life, is an enigma (in 1966).

Biological oxidation-reduction reactions support two major functions: transduction of energy and transformation of metabolites. Energy is required to perform muscular (mechanical), biosynthetic (chemical), transport (osmotic) and electrical work which life requires. The oxido-reductive transformation of metabolites produces many components of the metabolic network which is the chemical framework of life.

Oxidation-reduction Potentials. The equilibrium position of a redox reaction is a measure of the "pressure" of reducing equivalents from reductant toward oxidant. This pressure can also be measured as an electrical potential relative to a standard if the oxidant and reductant ("half-cells") are separated and connected through inert electrodes, a potentiometer, and a salt bridge. The standard is the normal hydrogen electrode, consisting of a platinum electrode under 1 atm H_2 pressure and immersed in a solution containing $1MH^+$. It is defined as having a potential of 0. The potential E of any half-cell measured against the standard hydrogen electrode is given by the Nernst-Peters equation,

$$E = E_0 + RT/nF \ln [ox]/[red]$$

where E is a measured potential in volts, R is 8.315 joules/degree (gas constant), T is absolute temperature, n is the valance change in the measured half-cell, F is 96,487 coulombs (the coulombic equivalent of the Faraday), and [ox] and [red] represent the concentrations of the oxidized and reduced forms of the substance comprising the half-cell. When these are equal to one another, the measured potential $E =$ the constant E_0, or standard redox potential. When the constant is represented as E_0', pH is specified. This is particularly useful for the redox potentials of biologically important half-cells which normally function near pH7 and are affected by pH. A selection of E_0' values of biochemical interest is given in Table 1. It is important to note that half-cells having electron pressures greater than that of the standard electrode have negative E' values, and those having electron pressures smaller than that of the standard electrode have positive E_0' values.

TABLE 1. E_0' VALUES (pH 7) FOR REDOX SYSTEMS OF BIOLOGICAL IMPORTANCE

System	E_0' pH 7. *Volts*
1/2 O_2/H_2O	+0.82
Cytochrome a_3^{2+}/cytochrome a_3^{3+}	+0.53
Cytochrome a^{2+}/cytochrome a^{3+}	+0.29
Cytochrome c^{2+}/cytochrome c^{3+}	+0.26
Ascorbate/dehydroascorbate	+0.08
Cytochrome b^{2+}/cytochrome b^{3+}	0.00
Flavoprotein$_{reduced}$/flavoprotein$_{oxidized}$	−0.06
DPNH/DPN	−0.32
Glutathione, G—S—S—G/ G—SH	−0.34
1/2 H_2/H^+	−0.42

During the oxidation-reduction of some classes of substances of biological importance, such as FLAVINS, diphenols (COENZYME Q) and ene-diols (ascorbate), an over-all two-equivalent transfer may consist of two overlapping one-equivalent steps. According to the Michaelis principle of compulsory univalent oxidation, *all* two-equivalent oxidation-reductions *must* proceed through single equivalent steps, but practically, the second step may take place so rapidly and completely after the first step that the univalent oxidation-reduction stage is not detectable. However, in some cases, it is observable by optical or electron spin resonance spectroscopy. In this case, three oxidation levels of a substance may be present at the same time: the completely reduced form R, the semioxidized form S, and the completely oxidized form T. These will be in equilibrium according to the following relationship,

$$R + T \rightleftarrows 2S$$

Thus, in any biological oxidation-reduction, free radicals may be detectable because (1) they are in equilibrium with R and T, but not formed by an enzyme-catalyzed one-equivalent step, or (2) the oxidation-reduction involves a one-equivalent step. Thus, the mere detection of free radicals does not establish mechanism.

Fundamental Principles of Biological Oxidation-reduction. In general, the enzymes which catalyze biological oxidation-reduction belong to one of the following classes: *dehydrogenases*, *electron carriers*, and *oxidases*. Many of these enzymes are organized in systems of enzymes which function as a terminal respiratory chain, *i.e.*, a group of enzymes which transfers reducing equivalents to O_2. Since there are many donors of reducing equivalents to the terminal respiratory chain, the first process consists of the activation of hydrogen atoms on certain metabolites by specific *dehydrogenases*. These activated hydrogen atoms are transferred as reducing equivalents through *electron carriers* and are then combined with O_2 to form respiratory water by a *terminal oxidase*.

Dehydrogenases. A very large number of dehydrogenases initiate the transfer of reducing equivalents to acceptors in the respiratory chain or other acceptors. Some dehydrogenases act with pyridine nucleotides (diphosphopyridine nucleotide, DPN or NAD, and triphosphopyridine nucleotide, TPN or NADP) (see NICOTINAMIDE ADENINE DINUCLEOTIDES). Examples are: alcohol dehydrogenase, lactate dehydrogenase, malate dehydrogenase, isocitrate dehydrogenase, glucose-6-phosphate dehydrogenase, and many others, which are listed comprehensively in "Enzymes" by Dixon and Webb. The binding of hydrogen atoms by the pyridine nucleotide in these reactions is alpha stereospecific with respect to the 4-position of the pyridine ring.

Other dehydrogenases act with FLAVIN or flavin nucleotides as acceptors. In these enzymes, the reaction of flavin can be readily observed to take place in one-equivalent steps with formation of flavin semiquinone, and some flavoprotein dehydrogenases appear to act by one-equivalent transfer mechanisms. Some flavoproteins contain metals, such as iron or molybdenum, which may play accessory roles such as stabilization of semiquinonoid flavins, or linkage in energy conservation. These enzymes can be classified in several ways, *i.e.*, (1) according to their donor and acceptor molecules, (2) according to the structure of the active site, and (3) according to the accessory components. Some examples of these types are the flavoprotein oxidases, glucose oxidase, D-amino acid oxidase, and lactic oxidative decarboxylase which utilize O_2 as an acceptor, milk xanthine oxidase which contains eight atoms of ion and one to two atoms of molybdenum as accessory factors along with two molecules of flavin (FAD, flavin adenine dinucleotide), and the respiratory chain dehydrogenases, succinic dehydrogenase and DPNH dehydrogenase, which are iron-flavoproteins.

Electron Carriers. The respiratory chain, the photosynthetic system, and other systems which require carriers of electrons contain CYTOCHROMES, or cellular pigments in which iron porphyrin compounds act as electron carriers. In the respiratory chain, these substances are reduced by flavoprotein dehydrogenases and oxidised ultimately by O_2.

Oxidases. The enzymes which catalyze the reactions of molecular oxygen are called "oxidases." When the reactions catalyzed by oxidases are studied, using $O_2{}^{18}$ as a tracer, it is found that they fall into three general categories: (1) oxygen transfer (oxygenation), $A + O_2{}^{18} \rightarrow AO_2{}^{18}$; the oxygen is incorporated into substrate, (2) mixed function oxidation, $AH + O_2{}^{18} + 2e \rightarrow AO^{18}H + O^{18--}$, one oxygen atom being incorporated into substrate, the other reduced by a two-equivalent donor to water; and (3) transfer of reducing equivalents only, such as two-equivalent transfer, $O_2 + 2e + 2H^+ \rightarrow 2H_2O_2$, or four-equivalent transfer, $O_2 + 4e + 4H^+ \rightarrow 2H_2O$. Several reviews of these enzymes may be consulted for additional details.[4,5]

Energy-linked Oxidation-reduction. The major part of respiratory oxidation-reduction is organized in MITOCHONDRIA in the form of a succession of oxidation-reduction reactions, at the beginning of which are the metabolites activated by specific dehydrogenases, and at the end of which is molecular oxygen. This organization serves to break the over-all oxidation of metabolites into small steps from which free energy may be efficiently recovered in the form of the terminal "energy-rich" phosphate linkage of adenosine triphosphate (ATP). The reaction by which this energy conservation is accomplished is called "oxidative phosphorylation" and consists in the reversal of the hydrolysis of ATP to ADP (adenosine diphosphate):

$$\text{ADP} + \text{Inorganic Phosphate} \rightleftarrows \text{ATP} + H_2O$$

The free energy of hydrolysis of ATP to ADP is approximately 7 kcal. There are three sites in the respiratory chain where oxidative PHOSPHORYLATION occurs. According to present observations, each pair of electrons which travels through the respiratory chain from metabolite to O_2 generates 21 kcal of biochemically useful free energy in the form of ATP. The respiratory chain now appears to consist of the following sequence of components: Metabolites — DPN — non-heme iron — flavoprotein — Coenzyme Q — cytochrome b — cytochrome c_1 — cytochrome c — copper — cytochrome a — cytochrome a_3 — oxygen. The sites of oxidative phosphorylation appear to lie between DPN and flavoprotein, cytochrome b and cytochrome c, and between cytochrome a and O_2. Both *high-energy intermediates* and *coupling factors* are involved, but because of the lability of these substances and the great dependence of oxidative phosphorylation upon the integrity of mitochondrial structure, they remain to be identified. The mechanism of oxidative phosphorylation remains very enigmatic.

References

1. West, E. S., Todd, W. R., Van Bruggen, J. T., and Mason, H. S., "A Textbook of Biochemistry," Fourth edition, N.Y., The Macmillan Co., 1966.
2. Clark, W. M., "Oxidation-Reduction Potentials of Organic Systems," Baltimore, Williams & Wilkins, 1960.
3. Dixon, M., and Webb, E. C., "Enzymes," Second edition, New York, Academic Press, 1964.
4. Mason, H. S., *Advan. Enzymol.*, **19**, 79 (1957).
5. Mason, H. S., *Ann. Rev. Biochem.*, **34**, 595 (1965).

Howard S. Mason

BIOLUMINESCENCE

Many living organisms exhibit the unique property of producing visible light, a phenomenon referred to as bioluminescence. Known light-emitting organisms have either oxidative or peroxidative enzymes that couple the chemical energy released from the enzyme reaction to give electronic excitation of a luminescent compound. The compound that is oxidized with subsequent light emission is usually referred to as *luciferin* and the enzyme which catalyzes the reaction as *luciferase*. Most luciferins and luciferases that have been isolated from unrelated species are different in molecular structure. With one known exception, combinations of luciferin and luciferase from different species do not exhibit bioluminescence.

The light-producing reaction in a number of organisms can be represented simply:

$$\text{Luciferin} + O_2 \xrightarrow{\text{Luciferase}} \text{Light}$$

Some luminous organisms catalyzing this reaction are: (1) *Cypridina* (a crustacean), (2) *Apogon* (a fish), and (3) *Gonyaulax* (a protozoan).[1] The latter organism is primarily responsible for the so-called phosphorescence of the sea.

On the other hand, some luciferins must first undergo a luciferase-catalyzed activation reaction prior to their being catalytically oxidized by the enzyme to produce light. There are two well-known cases:[1,2]

(1) The firefly:

$$\text{Luciferin} + \text{Adenosine Triphosphate (ATP)} \underset{}{\overset{\text{Luciferase; Mg}^{++}}{\rightleftarrows}} \text{Activated Luciferin}$$

$$\text{Activated Luciferin} + O_2 \xrightarrow{\text{Luciferase}} \text{Light}$$

(2) The sea pansy (*Renilla*):

$$\text{Luciferin} + 3',5'\text{-Diphosphoadenosine (DPA)} \overset{\text{Luciferase; Ca}^{++}}{\rightleftarrows} \text{Activated Luciferin}$$

$$\text{Activated Luciferin} + O_2 \xrightarrow{\text{Luciferase}} \text{Light}$$

Both of these activation reactions are linked to adenine-containing nucleotides of great biological importance. Since the measurement of light can be made an extremely sensitive and rapid technique, the most sensitive and rapid assays known have been developed for ATP and DPA using the above luminescent systems.[3,4] Nucleotide concentrations of less than $1 \times 10^{-9}M$ are easily detectable using commercially available electronic equipment. Firefly luciferase-luciferin preparations for ATP assays are now commercially available from a number of biochemical supply companies.

The structure of firefly luciferin has been confirmed by total synthesis.[2,5] The firefly emits a yellow-green luminescence, and luciferin in this case is a benzthiazole derivative. Activation of firefly luciferin involves the elimination of pyrophosphate from ATP with the formation of an acid anhydride linkage between the carboxyl group of luciferin and the phosphate group of adenylic acid forming luciferyl-adenylate.[2]

All other systems that have been extensively studied emit light in the blue-green region of the spectrum. In these cases the luciferins appear to be indole derivatives.

Some animals, such as the marine acorn worms (*Balanoglossus*), produce light via a peroxidation reaction and appear not to require molecular O_2 for luminescence.[6] The luciferase in this case is a PEROXIDASE of the classical type and catalyzes the following reaction:

$$2\ \text{Luciferin} + H_2O_2 \xrightarrow{\text{Luciferase}} \text{Light}$$

Commercially available horseradish peroxidase (crystalline) will substitute for luciferase in the above reaction. In addition, a compound of known structure, 5-amino-2,3-dihydro-1,4-phthalazinedione (also known as luminol), will substitute for luciferin. The mechanisms appear to be the same regardless of the way in which the crosses are made. Thus a model bioluminescent system is available and can be used as a sensitive assay for H_2O_2 at neutral pH.[6] The identification of luciferase as a peroxidase is of interest since this represents the only demonstration of a bioluminescent system in which the catalytic nature of a luciferase molecule has been defined.

Most of the luminescent systems mentioned above appear to be under some nerve control. Normally a luminous flash is observed after mechanical or electrical stimulation of most of the above-mentioned species. A number of these also exhibit a diurnal rhythm of luminescence.[1,2]

Among the lower forms of life there are two well-known examples of luminescence which are not under nerve control giving a continuous glow of visible light. These are the luminous bacteria, frequently found growing on dead fish, and luminous fungi which grow abundantly on rotting wood. These cells apparently depend on the oxidation of an organic molecule and hydrogen which is transferred through diphosphopyridine nucleotide (DPN; also termed NAD, NICOTINAMIDE ADENINE DINUCLEOTIDE) and the enzyme system to drive the luminescent reaction. Known details of these luminescent reactions are represented below. For bacteria:

$$\text{DPNH} + \text{H}^+ + \text{Flavin Mononucleotide (FMN)} \underset{}{\overset{\text{Oxidase}}{\rightleftharpoons}} \text{FMNH}_2 + \text{DPN}$$

$$\text{FMNH}_2 + \text{Long-chain Aliphatic Aldehyde} + \text{O}_2 \xrightarrow{\text{Luciferase}} \text{Light}$$

and for fungi:

$$\text{DPNH} + \text{H}^+ + \text{Unknown Compound (X)} \overset{\text{Oxidase}}{\rightleftharpoons} \text{XH}_2 + \text{DPN}$$

$$\text{XH}_2 + \text{O}_2 \xrightarrow{\text{Luciferase}} \text{Light}$$

Both of these systems are apparently closely linked to respiratory processes and in this sense are analogous to one another.[1,2] Luciferase from a luminous bacterium, *Photobacterium fischeri*, has recently been crystallized in high yield.[7]

References

1. CORMIER, M. J., AND TOTTER, J. R., *Ann. Rev. Biochem.*, **33**, 431–458 (1964).
2. MCELROY, W. D., AND SELIGER, H. H., *Advan. Enzymol.*, **25**, 119–162 (1963).
3. STREHLER, B. L., AND TOTTER, J. R., in *Methods Biochem. Anal.* (GLICK, D., EDITOR), **1**, 341–356 (1954).
4. CORMIER, M. J., *J. Biol. Chem.*, **237**, 2032–2037 (1962).
5. WHITE, E. H., MCCAPRA, F., AND FIELD, G. F., *J. Am. Chem. Soc.*, **85**, 337–343 (1963).
6. DURE, L. S., AND CORMIER, M. J., *J. Biol. Chem.*, **239**, 2351–2359 (1964).
7. KUWABARA, S., KREIS, P., AND CORMIER, M. J., *Federation Proc.*, **23**, No. 2, 163 (1964).

MILTON J. CORMIER

BIOS

Bios, which in Greek means life, was a word coined to describe a growth-promoting substance for yeast discovered by Wildiers in 1901. When added in small amounts to a sugar and salts medium, it permitted rapid growth of yeast even from a small seeding.

Subsequent investigations proved that there was not merely a single substance involved but that depending on the strain of yeast and the circumstances of testing, a number of different vitamins, etc., could act, often synergistically, to promote the rapid growth of yeast. PANTOTHENIC ACID, BIOTIN, INOSITOL, THIAMINE and PYRIDOXINE all have "bios" properties when tested appropriately. Even an amino acid may be a limiting factor for yeast growth when other needs are supplied. The term "bios" is not used in the current literature.

BIOTIN

Biotin is a vitamin which is required for growth and normal function by animals, yeast and many bacteria. The ill-effects caused by its deficiency in the diet are seldom seen in man or even in most laboratory animals, because the intestinal bacteria synthesize it in sufficient quantity to meet the needs of the animal (see INTESTINAL FLORA). Biotin deficiency does occur, however, in animals fed raw whites of eggs. The egg white contains a protein named AVIDIN which combines with biotin and this complex is not broken down by enzymes of the gastrointestinal tract, hence the deficiency develops. The symptoms in animals are lesions of the skin, retarded growth and nervous disorders which are cured if biotin is given in excess of the avidin.

Biotin is found in minute amounts in animal and plant tissue, usually in a combined form linked to the amino acid LYSINE. The most abundant sources are liver (less than 0.0001%), kidney, egg yolk, yeast and milk. Biotin was first isolated in pure form in 1936 by two Dutch chemists, Koegel and Tonnis,[4] who obtained 1.1 mg from 250 kg of dried egg yolk. They showed that the compound was necessary for the growth of yeast and gave it the name, biotin. Five years later in America, P. Gyorgy and co-workers[1] found that the same compound prevented the toxicity of raw egg white to animals, and in 1942 du Vigneaud and collaborators[2,7] determined the structure of the compound which is as follows:

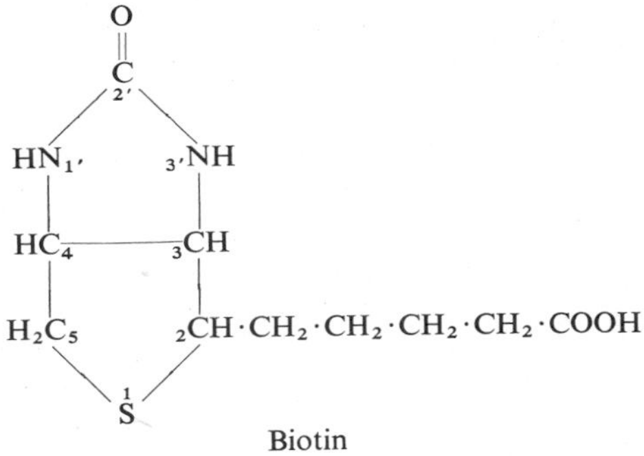

```
              O
              ‖
              C
             2'
           /    \
       HN1'      3'NH
        |          |
       HC4 ————— 3CH
        |          |
       H2C5       2CH·CH2·CH2·CH2·CH2·COOH
           \      /
              1
              S
```

Biotin

Biotin is essential because of its role in fixation of carbon dioxide which is a process by which carbon dioxide is combined with other compounds to form carboxylic acids[3,6,8] (see CABOXYLATION ENZYMES). Fixation of carbon dioxide is a required part of the metabolism of all forms of life and one type is involved in the formation of fatty acids (see FATTY ACID METABOLISM). The role of the biotin is as a coenzyme. The German biochemist, Lynen, winner of the Nobel Prize in 1964, as well as others, have shown that biotin occurs in certain of the enzymes which catalyze fixation of carbon dioxide and that the biotin is linked to lysine in these enzymes (see BIOCYTIN). The catalysis occurs by combination of carbon dioxide with the 1′-N of biotin which then transfers the carbon dioxide to the acceptor compound, thus forming the carboxylic acid. Biotin, therefore, serves as an essential agent for transfer of carbon dioxide during carbon dioxide fixation.

Biotin can be synthesized by some microorganisms and by plants. Apparently it is formed from a seven-carbon compound (pimelic acid), the amino acid cysteine, CO_2, and ammonia.[5]

References

1. GYORGY, P., ROSE, C. S., EAKIN, R. E., SNELL, E. E., AND WILLIAMS, R. J., *Science*, **93**, 477 (1941).
2. HOFMANN, K., "The Chemistry and Biochemistry of Biotin," *Advan. Enzymol.*, **3**, 289 (1943).
3. KAZIRO, K., AND OCHOA, S., "The Metabolism of Propionic Acid," *Advan. Enzymol.*, **26**, 378 (1964).
4. KOEGEL, F., AND TONNIS, B., *Z. Physiol. Chem.*, **242**, 43 (1936).
5. LEZIUS, A., RINGELMANN, E., AND LYNEN, F., *Biochem. Z.*, **336**, 510 (1963).
6. LYNEN, F., KNAPPE, J., AND LORCH, E., "Biotin, das Coenzym der Transcarboxylierung," *Proc. Intern. Congr. Biochem. 5th Moscow* (DESNUELLE, P. A. E., EDITOR), **4**, 225 (1961).
7. DU VIGNEAUD, V., MELVILLE, D. B., FOLKERS, K., WOLF, D. E., MOZINGO, R., KERESZTESY, J. C., AND HARRIS, S. A., *J. Biol. Chem.*, **146**, 475 (1942).
8. WOOD, H. G., AND UTTER, M. F., "The Role of CO_2 Fixation in Metabolism," in "Essays in Biochemistry," London, Academic Press, in press.

HARLAND G. WOOD

BLOCH, K. E.

Konrad Emil Bloch (1912–) is a German-born American biochemist whose research has had to do with the intermediary metabolism of amino acids and lipids and the biosynthesis of steroids. In 1964 he shared the Nobel Prize in physiology and medicine with FEODOR LYNEN.

BLOOD-BRAIN BARRIER

The term "blood-brain barrier" is the name given to the observation that many blood-borne solutes do not penetrate into central nervous tissue as rapidly as they penetrate into most other tissues. This phenomenon was first publicized by P. Ehrlich in 1885 who observed that certain aniline dyes, when injected into the bloodstream of mice, stained most tissues of the body rapidly, but left the nervous system largely uncoloured. During the ensuing half century, the slow permeation of the brain by dyes and other histologically identifiable subtsances (*e.g.*, ferricyanide and silver) was studied intensively.

When these materials were placed in the cerebrospinal fluid, they entered the brain without restriction by passive diffusion through the pial surface. These observations gave rise to the erroneous concept that all metabolic exchange between blood and brain occured via the cerebrospinal fluid. It is now recognized that metabolite transfer occurs throughout the central nervous system vasculature, but is subject to local controlling mechanisms not found in other tissues.

The availability of radioisotopes has enabled the study of rates of exchange for many physiologically significant substances and, with few exceptions, the exchange of blood-borne solutes with the central nervous system has been found to be significantly, often orders of magnitude, slower than with other tissues. Certain metabolites and metabolic products, such as glucose, oxygen, and carbon dioxide, as well as lipoid soluble substances and water itself, move rapidly between the blood and extravascular fluids of the central nervous system, but inorganic ions and most other highly dissociated compounds are very slow to equilibrate.

Many efforts have been made to evolve a general theory to explain the blood-brain barrier phenomenon. The most persistent approach has been the attempt to discover physicochemical properties of molecules which determine these rates of migration. This has led variously to explanations based on electric charge, molecular size, dissociation constant, protein binding, lipoid solubility, and combinations of these. None has been entirely satisfactory, although selected series of compounds can be found which behave quite predictably according to one or more of these criteria. There is considerable similarity between blood-brain barrier permeability and cell membrane permeability, and it appears that solutes, to pass from the plasma to the extravascular fluids of the central nervous system, must for the most part pass through and not between cells.

No agreement has yet been reached concerning the anatomical or physiological configuration of cells and fluid compartments within the central nervous system which determines this solute exchange. Assuming that a conventional interstitial space exists which is readily available to solutes from the cerebrospinal fluid, then it is necessary to postulate that capillary endothelium or its investing glial sheath represents a barrier to free access to this space from the blood.

Recently, electron micrographs have revealed the possibility that central nervous tissue may contain essentially no extracellular space. The blood-brain barrier would then reflect the permeability characteristics of glial and neuronal cell membranes since all extravascular substances would be within cytoplasmic compartments. Under these circumstances, the blood-brain barrier becomes a special case of inter-intracellular solute transfer and is not analogous to plasma-interstitial fluid transfer in other tissues.

Still another explanation for the blood-brain barrier phenomenon proposes that the metabolic activities of the central nervous tissue determine the rate of accumulation of blood-borne solutes. The central nervous system is assumed to contain compartments of low metabolic turnover, as in bone, which exchange slowly with the plasma.

Whatever may be the anatomical structures involved or the organization of the extravascular fluid compartments, evidence is accumulating that an active "secretory" process is involved in the transfer of some solutes across the blood-brain barrier (see ACTIVE TRANSPORT). These mechanisms appear to have similarities with analogous systems in the kidney tubules. Organic acids normally excreted by the kidney tubules may be actively transported into the plasma from central nervous tissue. Glucose may be actively transported from the plasma to metabolizing cells of the central nervous system by a process involving glucuronic acid and synthesis of glucosamine. Inorganic ions, particularly sodium and chloride accompanied by osmotic water, may be transported from plasma to extravascular fluids of the central nervous system in exchange for hydrogen and bicarbonate ions. These latter are presumed to be formed by hydrating metabolic carbon dioxide in the presence of the enzyme carbonic anhydrase. The mechanism for this ionic exchange is postulated to reside within the perivascular neuroglial membrane which contains a high concentration of carbonic anhydrase.

The functional significance of the blood-brain barrier mechanism is to buffer the neuronal microenvironment against changes in plasma concentrations of various important solutes and to regulate the composition of the neuronal "atmosphere" for optimum performance (see also CEREBROSPINAL FLUID; NERVE CELL COMPOSITION).

References

1. "Handbook of Physiology," (FIELD, J., EDITOR), Section 1, Vol. III, pp. 1761 and 1865, Washington, D.C., American Physiological Society, 1960.
2. TSCHIRGI, R. D., *Federation Proc.*, **21**, 665 (1962).
3. DOBBING, J., *Physiol. Rev.*, **41**, 130 (1961).
4. BAKAY, L., "The Blood-Brain Barrier," Springfield, Ill., Thomas, 1956.

ROBERT D. TSCHIRGI

BLOOD (CELLULAR COMPONENTS—IN HIGHER ANIMALS)

A brief description of the cellular components of the blood of the vertebrates, accompanied by two figures showing the types of cells and their relative sizes, may be found in reference 5. The blood cells or corpuscles are classified according to their morphology and tinctorial reactions into: (1) *erythrocytes* or *red blood corpuscles* (RBCs), which contain the respiratory pigment hemoglobin; (2) the *leukocytes* or *white blood corpuscles* (WBCs), which are motile and colorless; and (3) the *thrombocytes* or *blood platelets* which are specialized functionally to participate in the clotting of the blood.

In all of the vertebrates except the mammals, practically all of the erythrocytes of the circulating blood are nucleated. In mammals, the non-nucleated erythroplastid is derived from a nucleated precursor (see HEMOPOIESIS). Historically, as the erythrocytes or erythroplastids were studied, methods of measurement were invented by which the number of cells per cubic millimeter of blood could be counted; the amount of hemoglobin per 100 cc of blood could be measured; and the volume of the erythroplastids per 100 cc of whole blood (Hematocrit) could be estimated. By combining these data, certain information, known as the mean corpuscular volume and the mean corpuscular hemoglobin concentration, can be calculated and quantitative comparisons between species as regards hemoglobin function may be made [see also HEMOGLOBINS (COMPARATIVE BIOCHEMISTRY); HEMOGLOBINS (IN HUMAN GENETICS)].

The leukocytes have been classified by their granular content and by the tinctorial reactions of these granules to special blood stains containing a combination of acid and basic dyes, such as Wright's and Giemsa's stains (Romanofsky mixtures), which are widely used by hematologists. Supravital staining and other methods of a more detailed nature have been developed in the course of modern investigations. Because of their diverse contents, the leukocytes do not lend themselves to mass study as do the RBCs, although the count per cubic millimeter is often a source of pertinent information regarding physiologic or pathologic changes in the body.

In this essay, the cellular components of the blood of mammals are to be briefly considered with sampling information concerning the quantitative and qualitative characteristics of similar cells from members of the different orders, as revealed by the older methods noted above, and by newer methods including particularly HISTOCHEMISTRY, ELECTRON MICROSCOPY and autoradiography.

Some Quantitative Methods. Counts of the corpuscles: In mammals with large ears, blood may be drawn from ear vein; in the dog, from the external saphenous or jugular vein; in guinea pig, rat, mouse and other small animals, directly from the heart or from the cavernous sinus via the orbit. Granular *leukocytes* may be drawn from the peritoneal cavity after injection of irritating substances; *lymphocytes* may be collected by catheter from the thoracic duct, or from the buffy coat of centrifuged blood. In addition to the traditional method of counting blood corpuscles in a hemocytometer, photoelectric methods have been devised for counting cells in diluting fluids by opacity. Projection of cells by special television methods has also been used in counting as well as for studying changes in morphology. *Reticulocytes* are estimated by counting RBCs with standard methods in solutions of .05% Brilliant Cresyl Blue; for accuracy of count, many cells have to be counted and examination of suspensions in a dark field microscope has been recommended. The technique for counting blood *platelets* depends upon the staining of the platelets with Brilliant Cresyl Blue and counting them against the numbers of RBCs counted. A method for counting platelets in the phase microscope in a solution in which the RBCs are suspended has been used. Platelets have also been estimated by fluorescent methods after staining with auramine or acridine orange. RBCs may be collected by simple centrifuging. Platelets have been collected by withdrawing the thin plasma layer after rapid centrifuging of whole blood in dextran; the tubes must be clean and non-wettable. WBCs and platelets may be obtained by dipping glass slides, either plastic coated or not, directly into freshly collected blood. Blood may be examined without treatment by putting a drop of blood on a cover slip, inverting and sealing the edges with vaseline. This method is used with dark-field and phase microscopy for demonstration of small particles which do not react to the usual staining methods.

In Table 1 are shown some of the results of quantitative studies for representatives of the Class Mammalia, from the lowest to the highest orders. In some of the orders such as the Insectivora, Cheiroptera, Perissodactyla and Artiodactyla, there is a trend toward more RBCs per unit volume. The diameters of the cells as measured in the dried state show larger cells in the more primitive orders where locomotion is slow, as is the metabolic rate. The smallest RBCs are present in animals specialized for speed, such as horses; goats and deer (Artiodactyla) have even smaller RBCs, which may be related to their descent from animals which lived in higher altitudes where the RBCs are smaller and the counts higher as in the llama. The RBCs in all of the mammals, except the Camelidae, are circular biconcave disks; in the latter they are oval biconcave disks. These RBCs do not form rouleaux (piled up like coins) as do the RBCs of other mammals when they are examined in fresh plasma, but they become arranged in chains or strings. When the RBCs of camels are fixed in methyl alcohol, the hemoglobin is deposited in the center of the corpuscle which stains deeply in Romanofsky stains; this reaction may account for the erroneous idea that the RBCs of camels are nucleated. Counts of the RBCs of camel's blood ranges from 10–19.4 million/cu mm; the count in the llama is 11.3.

Making of blood smears or films seems easy to the uninitiated, but care must be taken to ensure good results. It is suggested that small drops of very *freshly* drawn blood about 2 mm in diameter be placed near the end of 1 × 3 inch glass slides and the edge of another narrower slide or cover slip be placed against this drop; after it has spread along the edge, it is drawn toward the other end of the slide. If drawn rapidly, a thick smear results; if slowly, a thin one. Rapid drying is necessary; moderate warmth is also necessary. A damp atmosphere is hemolytic. If the smears are collected in the field, as soon as they are dry they should be dipped in absolute methyl alcohol Pertinent information can be written in pencil on the dried blood at one end of the slide. Such slides can be wrapped in cellophane and stored indefinitely until they are stained by Wright's stain.

In most of the mammals, the shape is constant in the fresh drops, whether the cells are oxygenated or not; but in the deer (Cervidae of Artiodactyla; RBCs per cubic millimeter, 17–24 million; RBC volume, 50–58%; hemoglobin, 17–20 grams; diameter of RBCs, 3.5–4.5 μ) spontaneous sickling of the cells has been observed, and from electrophoretic and other studies the sickling has been thought to be caused by the biochemical nature of the hemoglobin.

Supravital (Intravital) Staining. This method has been used to stain intracellular structures while the cell is still alive. The drops of blood are mounted on an inverted coverglass which is placed upon a slide which has previously been covered with a thin layer of such dyes as Neutral Red, Janus Green B, Methyl Blue, Cresyl Blue,

Table 1. Ranges of Quantitative Values of Cellular Components of the Blood from Typical Representatives of Some Orders of the Mammalia

RBC, red blood corpuscles per cu mm in millions; D, diameter of RBC, in μ; Hb, hemoglobin, g per 100 g whole blood; Ht, hematocrit, or volume RBC per 100 cc whole blood; WBC, number of leukocytes per cu mm in thousands. Differential counts or percentage distribution as follows: L, lymphocytes; M, monocytes; N, neutrophils or heterophils; E, eosinophils; B, basophils; P, blood platelets, in 100,000's per cu mm (counts of cells from Jordan and Albritton, and recent literature; differential counts, data obtained by the writer).

Order	Animal	RBC	D	Hb	Ht	WBC	L	M	N	E	B	P
Marsupalia	Opossum	5–6	8–9	8–9	14–36	7–8	49–73	0–1	22–34	0–8	0–8	—
Insectivora	Hedgehog	9–11	7–9	—	—	5–6	14–30	5–6	56–77	7–13	2–5	—
Cheiroptera	Brown bat	7–14	5–7	—	—	1–7	30–44	3–5	8–20	—	0–5	—
Primates	Marmoset	5–7	5–8	12–13	42	5–13	21–47	5–8	31–55	0–2	0–1	1.7
Edentata	Anteater	—	8–13	—	—	—	63–95	0–2	12–26	Trace	Trace	—
Lagomorpha	Rabbit	5–7	6–8	12–13	37	6–14	51–91	3–6	11–27	0–4	Trace	1.7–11.1
Rodentia	Rat	6–7	5–7	13–14	46–49	7–13	69–82	Trace	18–22	0–2	0–5	2.0–10.0
Carnivora	Cat	6–8	5–7	8–12	28–41	15–21	13–26	2–8	69–86	3–11	0.2	1.6–7.6
Pinnepedia	Seal	6	7–8	—	48	—	9–22	3–5	47–53	17–22	3–4	—
Perissodactyla	Horse	7–11	4–7	11–15	27–47	6–14	7–44	0–7	44–83	2–8	0.5	2.4–5.6
Artiodactyla	Cow	5–10	4–7	11–14	33–48	5–12	54–72	4–8	14–50	4–11	0.4	5.4–9.7

Nile Blue, Dahlia Violet and Hemocyanin. The study must be carried on in a warm chamber, but the slides may be stored in a refrigerator for 12–24 hours. Very clean slides are needed for this method. The structures are difficult to see, but the chief object is to visualize the MITOCHONDRIA and Neutral Red vacuoles, which were thought to have specific arrangements in monocytes as compared with lymphocytes. The granules of the neutrophils are yellow; of the eosinophils, orange; of the basophils, deep red in the Neutral Red slides. Mitochondria are present in myeloblasts, myelocytes, lymphoblasts, lymphocytes and monocytes; they are rare in the granulocytes. RBCs stain pale yellow; 1% of the cells have a small Neutral Red vacuole. All precursors of the RBCs show mitochondria and Neutral Red vacuoles. Studies with the phase microscope and electron microscope have more or less replaced this method which is now rarely used, but study by phase microscopy can be used effectively in conjunction with supravital staining.

In order to test the storage capacity of the blood cells and their fate, examination of the organs of animals injected with Trypan Blue has been found to be very successful, since the cells obtained from the hemopoietic organs which are Trypan Blue positive can be studied in fresh preparations using Neutral Red and Janus Green B.

Phase Contrast Microscopy. This method is widely used, and the theory and methods of use are given in handbooks on optics (see PHASE MICROSCOPY). The chief advantage of this method is that the living cells can be studied for long periods of time; chromatin, mitochondria, centrosomes and specific granules can be seen and photographed at magnifications of 2500X. The method is excellent for the study of granules of the matrix of cells which is unseen in traditionally fixed and stained cells. It is an excellent aid for those who wish to use the electron microscope, because areas demonstrated by light can be compared with those visualized by the electron beam. Special objectives, condensers and filters are available for photography at high magnifications. Special methods for compression of the cells must be used to get good results. Spontaneous spreading of the cells is helpful. Dried and rehydrated materials may also be studied to advantage. Glycerin is the best medium for mounting rehydrated films.

Microcinematography. The study of blood by motion pictures has been used for many years. With the invention of the phase microscope, this approach to the study of the blood cells has been a very important tool. Studies of the movements of the lymphocytes in rats showed a softening of the membrane at the forward moving end, and pseudopod formation; contractions of the cell force the inner plasma forward, while the external plasmagel remains fixed except at the posterior end, then it becomes softer and passes through the stiffer ring of plasmagel to become more gelated at the anterior end. Changes in the shapes of the other WBCs are continual; they move by changes in a non-rigid plasma membrane, by amoeboid motion. The movements of the cells may be slowed down or accelerated by the action of irritating substances from the blood and tissues.

Using speed photography at 3200 frames/sec, the RBC has been observed to have an interior viscosity of 30 times that of water at 38°C. In the dog's mesentery, RBCs passing into capillaries from larger arterioles take the form of an inverted cap or parachute; when blood flow is stopped they become biconcave disks. The cup shape is suggested as bringing more surface close to the capillary endothelium.

Tissue Culture. This is a difficult field, because so many variables are involved. Young blood cells have been seen to develop further *in vitro* and to complete their maturation, but most of them undergo regression into fibroblasts and macrophages. Bone marrow can be cultivated in foreign plasma for transplantation in the production of chimeras. Of the cells studied *in vitro*, it has been shown that orthochromatic erythroblasts may mature into RBCs; granulocytes do not survive, but for the period of survival, eosinophils are most resistant; lymphocytes survive for five days; monocytes appear to change into macrophages; mesenchyme cells survive indefinitely as undifferentiated fibroblasts. In the rat, cells from the buffy coat of centrifuged blood, when placed in a culture of plasma, show the following activities: there is active migration of the neutrophils with a broad anterior end and thin, waving film-like pseudopodia, a blunt or sharp posterior end, which may be knoblike with long slender processes; contraction waves pass from the anterior end and may end in a partially detached knob; Brownian motion occurs within the cytoplasm. (See TISSUE CULTURE, ANIMAL.)

Chemical Agents. Many different inorganic and organic agents have been introduced into the bloodstream of animals, and the results have been assayed by the study of the volumetric and morphologic relations of the various types of cells. A few examples are as follows: injection of germanium has an erythropoietic effect; anemia has been produced in rabbits by injections of water; increase in the number of RBCs follows injections of extracts of beef kidney or pancreas; injection of pyridine, nitrobenzol, or pyrogallic acid causes curious electron-dense methyl-violet reacting granules (Heinz-Ehrlich) to appear in the RBCs; hemolytic agents produce changes in the colloids of the RBCs causing the formation of Cabot rings, which were once thought to be nuclear remnants such as occur in reticulocytes; repeated injections of RBCs, hemoglobin or egg white resulted in an eosinophilia of the peritoneal exudate; solutions of adrenal hormones injected also produced this effect; leukopenia is produced in dogs after injections of benzene or aminopyrine; injections of adrenocortical hormone caused lymphopenia by destruction of lymphoid tissue, inhibition of production of lymphocytes, redistribution of the lymphocytes, and reduced production of eosinophils in the bone marrow;

monocytes react to various agents, but lymphocytes do not; sodium citrate injected into the rabbit blood-stream caused rhythmic showers of degenerated leukocytes; carmine particles injected into the bloodstream of rabbits were found to be stored in the cells of the reticulo-endothelial system.

Physical Agents. The use of physical agents in study of blood is very old, but mechanical advances have supplanted more traditional primitive methods. Extreme heat of the external environment and absence of water in the diet produce sludging of the blood and finally stasis and death, which does not occur in camels because of the labile system of heat regulation. Frozen sections have always been used for rapid assay of tissues. Pinocytosis is the term applied to the activity of cells engaged in taking in fluid through their plasma membranes. It is thought to be effected by changes in surface tension and electrostatic charges on the membrane. Cyclic changes in eosinophilic counts have been related to varied effects of light via the optic pathways of the nervous system.

Ultracentrifugation. This method has been used to separate out parts of cells with regard to weight. In granulocytes, the nucleus is usually in the middle of the centrifuged cell, there is a vacuole at proximal end and granules distally; the nucleus becomes pearshaped; eosinophilic granules seem to be the heaviest of the granules; mitochondria lie just above the nucleus, and there is a layer of cytoplasm just above the nucleus which reacts to tinctorial tests for RNA. This layer appears in myeloid cells where none is apparent in smears. Thus traces of substances not seen by conventional methods can be visualized by this method. In plasma cells from the bone marrow, ribonuclear proteins can be separated from glycoproteins; differences are checked by digestive enzymes and tinctorial reactions to specific stains under controlled pH.

Cytochemistry. This method attempts to show the location of chemical substances which a cell contains. Accessory methods used in association with this technique are microincineration and digestion by enzymes. Books on biological techniques give details of these methods. A few examples will be cited to show the trends (see also HISTOCHEMICAL METHODS): DNA is demonstrated by the use of such specific stains as Toluidine Blue, and the reaction of the stained substances to deoxyribonuclease, applied before and after staining; the location of ribonucleic acid may be demonstrated in a similar way. Lipids are demonstrated by staining fresh tissues with Sudan Black; glycogen by iodine vapor; or by a special periodic acid method combined with controls run in saliva. Mucoproteins have certain reactions with Toluidine Blue or thionine, as in mast cell granules.

Enzymes. The enzymes first studied were the PEROXIDASES, which showed as blue-green areas in cells treated with special techniques utilizing benzidine and hydrogen peroxide. The granules of all of the myeloid cells reacted positively, but those of the lymphoid series were negative. The sites of alkaline and acid phosphatases are demonstrated by treating smears or sections by a series of substrates and fluids in which the final steps locate the enzyme sites with black or brown lead salts: *e.g.*, specific granules of the granulocytes are negative for alkaline phosphatase, but eosinophilic granules are always positive for acid phosphatase, and those of neutrophils only slightly so. Alkaline phosphatases have been demonstrated in the nucleoli. Hyaluronidase is said to dissolve selectively the granules of the neutrophils.

Microincineration. Such minerals as sodium, potassium and iron have been localized by this method in the RBCs, and their positional relationships in the cytoplasm and nucleus of other cells have been examined by electron microscopy. Leukocytes and platelets contain calcium. There seems to be no correspondence between the site of the inorganic ash and electron-dense materials in cell organelles. Special stains are used for the demonstration of materials in the RBCs such as ferric salts in the granules of siderocytes and methyl dyes for Heinz bodies. Shadowing of membranes or granules may be carried on by blowing metal particles over fixed films of blood, which have or have not been stained with Wright's stain.

Electron Microscopy. The following characteristics have been identified in thin sections of several types of blood cells from studies with the electron microscope. The erythroblast cytoplasm contains numerous MITOCHONDRIA and diffuse RIBOSOMES, and little or no endoplasmic reticulum; the nucleus is opaque, but the arrangement of slightly contrasting opacities suggests the "radkern" nucleus. In the lymphocyte, there are few mitochondria, few ribosomes, sparse endoplasmic reticulum, many RNA granules, and dense CHROMATIN in the nucleus. In the monocyte, there are a well-developed endoplasmic reticulum, many small mitochondria, and a few ribosomes. In the neutrophils, the endoplasmic reticulum is sparse; few granules are present; there are many coarse and fine opaque round or oval granules; the plasma membrane is very delicate and the nucleus is dense. The characteristic picture of the eosinophils is the presence of coarse, medium-opaque, banded, oval to rectangular shaped special granules (the acidophilic granules of the light microscope); there are a few small mitochondria (seen also in light microscope); the endoplasmic reticulum is not well marked; and there are few ribose granules. All indications of an inactive cell as regards proliferation are present.

The RBCs have been studied for a longer time, but it was realized quite early that for successful study of the contents, the RBCs had to be hemolyzed. Since complete hemolysis is very difficult, the opacity of remnants of hemoglobin caused some confusion. However, there are now methods for excellent hemolysis, and in preparations of RBCs made by these methods, it has been found that the surface of RBC is covered with electron-dense "plaques" or short cylinders, 100 × 500 Å wide and 50 Å thick. Since they react to pyrogallic acid which has been found to be a stain for proteins, these plaques are thought to be the protein components of the membrane, together with certain small fibrils which run internal to the plaques and at right angles to them; selected solvents have shown that these are fastened together by a lipid. The plaques have been called "elium" and the fibers "stromin." Pores have been suggested to lie between the fibers and plaques.

Blood platelets collected by centrifugation on "Formvar" films have been shown by electron microscopy to consist of a central group of tiny, electron-opaque granules, 50–100 per chromomere, surrounded by a less opaque matrix having a fibrillar and vacuolar nature, the hyalomere.

Autoradiography. This method promises to be one of the most fruitful tools for tracing the development and fate of the blood cells. Tritium (H^3) incorporated into selected amino acids is introduced into the animal, and its fate is traced by smears or sections or even by cell suspensions, when the materials are placed against photographic film in the dark. The site of the radioactive molecules imparts a black area to the film. This method supplants that in which P^{32} or C^{14} radioactive molecules were introduced into the bloodstream or tissue of animals and the sites of deposit located by radioactivity counters. In the original studies with P^{32}, it was found that this was incorporated into the sites of deoxyribonucleic acid formation, particularly in the thymus, spleen, lymph nodes, and bone marrow (of rats). Such a demonstration of DNA incorporation agreed with the more laborious methods of counting the populations and cells in mitosis in traditionally fixed and stained materials. With the use of radioactively labeled substances necessary for the formation of the contents of the blood cells, autoradiographic techniques can be applied with hope of an adequate answer to the problem of blood development. (See ISOTOPIC TRACERS.)

References

1. ALBRITTON, E. C., "Standard Values in Blood," Philadelphia, Saunders, 1952.
2. BESSIS, M., "Cytology of the Blood and Blood-forming Organ," New York, Grune and Stratton, 1956 (translated by E. PONDER).
3. HALL, B. E., "Evaluation of the Supravital Staining Method," in "Handbook of Hematology," Vol. I, p. 641, New York, Hoeber, 1938.
4. JORDAN, H. E., "Comparative Hematology," in "Handbook of Hematology," Vol. II, p. 700, New York, Hoeber, 1938.
5. KINDRED, J. E., "Blood, Vertebrates," in "Encyclopedia of Biological Sciences," New York, Reinhold Publishing Corp., 1961.
6. LOW, F. N., AND FREEMAN, J. A., "Electron Microscopic Atlas of Normal and Leukemic Human Blood," New York, McGraw-Hill Book Co., 1958.
7. MACFARLANE, R. T., AND ROBB-SMITH, A. H. T., "Functions of the Blood," New York, Academic Press, 1961.

JAMES E. KINDRED

BLOOD (CHEMICAL COMPOSITION—VERTEBRATES)

One may consider blood to be a liquid tissue lying within a system of tubes, the blood vessels (arteries and veins). The liquid part of the blood (the plasma) is regarded as the lifeless part, while the living part consists of white cells (leucocytes) and red cells (erythrocytes). The plasma delivers food and nutrient materials to the various tissues of the body and removes waste products from them. Metabolite levels in the plasma or serum are indications of dynamic equilibrium between formation and removal which go on at all times in the body.

The *white cells* have nuclei and are capable of spontaneous movement and reproduction. They are colorless, amoeboid cells which have the power of phagocytizing, or engulfing, solid particles that

TABLE 1. INORGANIC CONSTITUENTS OF HUMAN PLASMA OR SERUM

Constituent	Value/100 ml
Aluminum	45 μg
Bicarbonate	24–31 meq/liter
Bromine	0.7–1.0 μg
Calcium	9.8 (8.4–11.2) mg
Chloride	369 (337–400) mg
Cobalt	10 (3.7–16.6) μg
Copper	8–16 μg
Fluorine	109 (75–145) μg
Iodine, total	7.1 (4.8–8.6) μg
Protein bound I	6.0 (3.5–8.4) μg
Thyroxine I	4–8 μg
Iron	105 (39–170) μg
Lead	2.9 μg
Magnesium	2.1 (1.6–2.6) mg
Manganese	8.0 μg
Phosphorus, total	11.4 (10.7–12.1) mg
Inorganic P	3.5 (2.7–4.3) mg
Organic P	8.2 (7–9) mg
ATP P	0.16 (0–6.4) mg
Lipid P	9.2 (6–12) mg
Nucleic acid P	0.54 (0.44–0.65) mg
Potassium	16.0 (13–19) mg
Rubidium	0.11 mg
Silicon	0.79 mg
Sodium	325 (312–338) mg
Sulfur	
Ethereal S	0.1 (0–0.19) mg
Inorganic S	0.9 (0.8–1.1) mg
Non-protein S	2.8 (2.4–3.6) mg
Organic S	1.7 (1.4–2.6) mg
Sulfate S	1.1 (0.9–1.3) mg
Tin	4 μg
Zinc	300 (0–613) μg

TABLE 2. METABOLITES IN HUMAN BLOOD (VALUES ARE PER 100 ml)

Constituent	Plasma or Serum	Whole Blood
Acetoacetic acid	1.8 mg	0.08–0.40 mg
Adenosine	1.09 mg	
Adenosine triphosphate (total)		31–57 mg
,, ,, (P)		5–10 mg
Amino acids (total)		38–53 mg
Amino acid N	3.4–5.5 mg	4.6–6.8 mg
Ammonia N	0.1–1.1 mg	0.1–0.2 mg
Ascorbic acid	0.7–1.5 mg	0.1–1.3 mg
Base (total)	145–160 meq/liter	
Bicarbonate	25–30 meq/liter	19–23 meq/liter
Bile acids		0.2–3.0 mg
Bilirubin, 1 min. direct	0.1–0.5 mg	0.1–0.2 mg
,, , 30 min. indirect	0.1–1.0 mg	0.2–1.4 mg
Biotin	0.9–1.6 μg	0.7–1.7 μg
Blood volume		2990–6980 ml
Adult men	33.7–43.7 ml/kg	66.2–97.7 ml/kg
Adult women	32.0–42.0 ml/kg	46.3–85.5 ml/kg
Infants	36.3–46.3 ml/kg	79.7–89.7 ml/kg
Buffer base		45–52 meq/liter
Calcium	4.0–5.5 meq/liter	
Carotenoids	10–300 mg	24–231 mg
Carotenes	50–420 mg	
Carbon dioxide		
Arterial blood, total		45–55 vol % 20.3–24.7 meq/liter
Arterial blood, in solution		1.5–4 vol % 0.067–0.18 meq/liter
Venous blood, total	60–70 vol % 27–31.5 meq/liter	50–60 vol % 22.5–27 meq/liter
Venous blood, in solution	2.9–4.3 vol % 0.13–0.193 meq/liter	1.7–4.2 vol % 0.077–0.189 meq/liter
Cephalin	5–10 mg	30–115 mg
Chloride	96–108 meq/liter	77–86 meq/liter
Cholesterol, total	120–250 mg	115–225 mg
Cholesterol esters	75–150 mg	45–115 mg
Cholesterol, free	30–60 mg	82–113 mg
Choline, total	26–35 mg	11–31 mg
Choline, free	0.05–2.5 mg	1–4 mg
Citric acid (adult)	1.6–3.2 mg	1.3–2.5 mg
Cobalamine	0.01–0.07 μg	0.06–0.14 μg
Creatine	2.5–3.0 mg	2.9–4.9 mg
Creatinine	1–2 mg	0.6–1.2 mg
Deoxyribonucleic acid	0–1.6 mg	
Diphosphoglycerate P		8–16 mg
Ergothionine	0	1–20 mg
Fat, neutral	25–260 mg	85–235 mg
Fatty acids	190–450 mg	250–390 mg
Fatty acids, total esterified	7–20 mg	
Fatty acids, volatile		1.8 mg
Fibrinogen	200–400 mg	120–160 mg
Flavin adenine dinucleotide	8–12 μg	
Folic acid, total	1.6–20 μg	2.3–5.2 μg
Fructose (fetus)		0–1 mg
,, (adult)	7–8 mg	0–5 mg
Fumaric acid		<0.3 mg
Glucosamine (fetus)	42–55 mg	40–60 mg
,, (child)	52–69 mg	50–70 mg
,, (adult)	61–86 mg	60–80 mg
,, (aged)	70–89 mg	70–96 mg
Glucose (fetus)		35–115 mg
,, (newborn)		20–30 mg
,, (child)	65–105 mg	80–120 mg
,, (adult)	65–105 mg	80–120 mg
Glucuronic acid	0.8–1.1 mg	4.1–9.3 mg
Gluthathione, reduced		25–41 mg
Glycogen		1.2–16.2 mg
Glycoprotein acid	40–60 mg	
Guanidine		0.18–0.23 mg
Guanidoacetic acid	0.26 mg	

TABLE 2—*continued*

Constituent	Plasma or Serum	Whole Blood
Hemoglobin	Trace	14.8–15.8 g
Hexosephosphate P	0.0–0.2 mg	1.4–5.0 mg
Histamine		6.7–8.6 μg
β-Hydroxybutyric acid	0.1–0.9 mg	0.1–0.6 mg
Indican	0.08–0.50 mg	
Inositol	0.3–0.7 mg	
Iodine, protein bound	3.5–8.0 μg	4.0–8.5 μg
Ketone bodies, total	0.15–1.36 mg	0.23–1.00 mg
α-Ketonic acids (infant)		0.6–1.00 mg
α-Ketonic acids (adults)		0.1–3.00 mg
α-Ketoglutaric acid		0.8 mg
Lactic acid	30–40 mg	5–40 mg
Lactose	Trace	Trace
Lecithin	100–225 mg	110–120 mg
Lipids, total	400–700 mg	445–610 mg
Lipoprotein, S_f 12–20	10–100 mg	
Malic acid	0.1–0.9 mg	
Methylguanidine		0.2–0.3 mg
Mucopolysaccharides	175–225 mg	
Mucoproteins	86.5–96 mg	
Nicotinic acid	0.02–0.15 mg	0.5–0.8 mg
Nitrogen, total		3.0–3.7 g
Non-protein nitrogen	18–30 mg	25–50 mg
Nucleotide, total		31–52 mg
Nucleotide phosphorus		2–3 mg
Oxygen, arterial		17–22 vol %
Oxygen, venous		11–16 vol %
Pentose phosphate	2.0–2.3 mg	
pH	7.38–7.42	7.36–7.40
Pantothenic acid	6–35 μg	15–45 μg
p-Aminobenzoic acid		3–4 μg
Phenol, free		0.07–0.1 mg
Phenol, total	1.0–2.0	
Phospholipid	150–250 mg	225–285 mg
Phospholipid P	6–14 mg	11–14 mg
Polysaccharides, total	73–131 mg	
Protein, total	6.0–8.0 g	19–21 g
Protein, albumin	4.0–4.8 g	
Protein, globulin	1.5–3.0 g	
Pteroylglutamic acid, total	1.5–5.0 μg	2.3–5.3 μg
Pteroylglutamic acid, free	0.01–0.05 μg	0.5–0.13 μg
Purines, total		9.5–11.5 mg
Pyridine nucleotides	20–120 μg	2600–4600 μg
Pyridoxine	1–18 μg	5–20 μg
Pyruvic acid	0.7–1.2 mg	0.5–1.0 mg
Riboflavin	2.6–3.7 μg	15–60 μg
Ribonucleic acid	4–6 mg	50–80 mg
Sphingomyelin	10–47 mg	150–185 mg
Sulfate		0.9
Thiamine	1–9 μg	3–10.7 μg
Urea	28–40 mg	20–40 mg
Urea N	8–28 mg	5–28 mg
Uric acid, male	2.5–7.2 mg	0.6–4.9 mg
,, ,, , female	2.0–.60 mg	
Vitamin A, carotenol	15–60 μg	9–17 μg
Vitamin A, carotene	40–540 μg	20–300 μg
Vitamin B_{12} (cyanocobalamin)	0.01–0.07 μg	0.06–0.14 μg
Vitamin D_2	1.7–4.1 μg as calciferol	
Vitamin E	0.9–1.9 mg	
Water	93–95 g	81–86 g

enter the bloodstream. Their specific gravity is less than that of red cells, hence they collect as a thin layer (buffy coat) above the red cells when whole blood is centrifuged for a short time. They are variable in size and structure, their diameters varying from 4–13 μ. The total number per cubic millimeter appears to vary from 7000–10,000. Their composition is the same as that of typical cells consisting of albumins, globulins, phospholipids, sterols and nucleins, peptides, carbohydrates, etc., there being 3000–6000 different species of molecules present. (See LEUKOCYTES AND LEUKEMIA.)

The *red cells* are non-nucleated in mammals and appear in human blood in the form of biconcave disks. They are smaller than white cells, averaging 7–8 μ in diameter. In human blood the average number per cubic millimeter is between 5,000,000 and 6,000,000. They consist chiefly of a red-colored protein (HEMOGLOBIN) which makes up three-fourths to four-fifths of the total solids of the cells. The major function of hemoglobin is to carry oxygen from the lungs to the tissues and carbon dioxide from the tissues to the lungs. Lysis of the red cells, resulting in the removal of hemoglobin from the interior of the red cells, leaves corpuscular membranes called "ghosts," or stroma, which consist chiefly of LIPOPROTEIN.

The blood *platelets* (thrombocytes) are thin colorless disks found in the blood of mammals, but not of birds or other vertebrates lower than mammals. They vary in number from 300,000–1,000,000 per cubic millimeter of blood. In their chemical constitution, they appear to be similar to the stroma of red blood cells, phospholipid-protein constituents. They play a role in BLOOD CLOTTING.

Blood *plasma* is the fluid portion of the whole blood. It may be separated from whole blood by centrifugation. It is a straw-colored liquid representing 55–70% by volume of whole blood. In general, it contains water, electrolytes (ions), small organic molecules, and macromolecules (see PLASMA AND PLASMA PROTEINS).

Inorganic Constituents. A large number of inorganic analyses of vertebrate plasma and serum have been published during the last twenty years. Table 1 indicates some of the inorganic constituents of human plasma or serum.

Organic Constituents. The non-protein nitrogenous constituents of plasma and serum comprise such chemical compounds as urea, uric acid, creatine, creatinine, AMINO ACIDS, peptides, ammonia, etc. The chief analytical procedures for their determination have been colorimetric, chromatographic, and microbiological. Table 2 includes the analytical values for many of these constituents in human blood, as well as values for other organic constituents.

The protein constituents comprise a large number of low and high molecular weight compounds; their molecular sizes vary from a few thousand to several million (see PLASMA AND PLASMA PROTEINS; also ANTIBODIES; IMMUNOCHEMISTRY). The analytical procedures for the determination of these constituents are manifold. The chief chemical methods used in the past have involved fractionation by salting out with ammonium sulfate and precipitation with organic solvents, notably cold ethanol. Modern methods of separating the various components of plasma or serum are the use of ION-EXCHANGE RESINS, CHROMATOGRAPHY, ELECTROPHORESIS, molecular sieves (GEL FILTRATION), and ULTRACENTRIFUGATION.

In the present space, complete presentation cannot be made of all data available concerning other constituents of human blood, or of species differences among vertebrates. Much detailed information may be found in the references.

References

1. SPECTOR, W. S. (EDITOR), "Handbook of Biological Data," Philadelphia, Pa., W. B. Saunders and Co., 1958.
2. CORNELIUS, C. E., AND KANEKO, J. J. (EDITORS), "Clinical Biochemistry of Domestic Animals," New York, Academic Press, 1963.
3. DITTMER, D. S. (EDITOR), "Blood and Other Body Fluids," Washington, D.C., Federation of American Societies for Experimental Biology, 1965.
4. LONG, C. (EDITOR), "Biochemists Handbook," New York, D. Van Nostrand Co., 1961.
5. KUGELMASS, I., "Biochemistry of Blood in Health and Disease," Springfield, Ill., C. C. Thomas and Company, 1959.

M. JOHN BOYD AND CARL ALPER

BLOOD CLOTTING

Clotting proper takes place when the *fibrin* molecules join up with each other to form a network structure. This network structure, the clot, plugs the wound and thus prevents the loss of blood. Fibrin molecules are sparingly soluble and start forming a visible network structure at very low concentrations. In order to prevent their aggregation, nature furnished them with negatively charged peptides which, when placed on proper sites, prevent, with their repelling charges, the molecules from approaching each other so as to form the network. In this form, these "repellent" fibrin molecules are called fibrinogen molecules. When fibrin molecules are needed in abundance, the "repellent" peptides are removed by an enzyme, called *thrombin*. Obviously, thrombin cannot exist in quantities in the circulating blood. It is generated from its precursor, *prothrombin*, when needed.

In order to put the clotting in the proper perspective, let us go back in time hundreds of millions of years ago when life was developing in the oceans. Before "life" could venture out to dry land, it had to invent a scheme whereby it could carry the seawater with it so that its cells could still be bathed in this fluid. Apparently, blood plasma circulating in vessels was the solution. To make the system safe, a mechanism also had to be invented by which a wound could be sealed so that the precious fluid would not be lost. The clotting mechanism seals the wounded vessels.

It appears that in its simplest form, the sealing material (fibrin molecules) was locked up in circulating cells. In the case of emergency, the cells broke up and released the sealing material into the circulation. The remnants of this mechanism apparently still exist even in higher animals. Platelets still carry a fibrin-like material. Platelets also carry a contractile protein which is released to perform the last act in the clotting process, *i.e.*, the shrinkage of the clot.

In higher animals, however, a more sophisticated mechanism supplemented the original one. In these animals, the sealing material is distributed in the blood plasma in the form of *fibrinogen*. When loss of blood must be prevented, prothrombin is converted to thrombin at the site of the wound. Thrombin then splits off the peptides attached to the fibrin molecules which, devoid of the repelling peptides, associate together and form a gel-like network structure.

Otherwise, it is fibrinogen and prothrombin that circulate in the blood plasma. This is however, a leaky system and small amounts of thrombin and fibrin are present in the circulation. This leakiness is probably a deliberate act on the part of nature. The concentration of the fibrin molecule is kept on the edge of the gel point, ready to go into action when clotting is needed. This carefully balanced steady state is suddenly shifted when, at the surface of a wound, prothrombin is converted to thrombin.

As the demand during evolution became greater for more efficient hemostasis, this basic scheme became supplemented with a system that provided an amplification for the speedy conversion of prothrombin to thrombin. Figure 1 shows the events leading to a speedy conversion of prothrombin to thrombin.

First, a plasma component called the Hageman factor becomes activated. A glass surface can act as the activator. In the vessels, the uncoated surface of the capillaries probably activates this factor. Once this factor is activated, it acts on another plasma component called PTA (plasma thromboplastin antecedent). The activated PTA acts as an enzyme and converts its substrate, the Christmas factor into another enzyme. Active PTA needs Ca ions for its action; thus, complexing agents like citrate, oxalate, and the like can arrest the process at this stage. HEPARIN also acts apparently at this stage by forming an inactive complex with PTA. The enzymatic nature of PTA is revealed by the fact that DFP (diisopropylfluorophosphate), the well-known inhibitor of PROTEOLYTIC ENZYMES, also blocks the process at this stage.

Active Christmas factor is again an enzyme and converts its substrate the antihemophilic factor (AHF) into an active component. This reaction also requires Ca ions and is apparently blocked by heparin. The active AHF converts the so called Stuart factor into an active component which in turn activates proaccelerin into the principle that converts prothrombin to thrombin.

Most of these factors were discovered in patients in whose blood plasma one or another of these factors was missing, leading to serious bleeding conditions. The names like Christmas factor preserve names of the patients in whom its was first recognised. (These factors are also designated by roman numerals).

The process depicted above is called the intrinsic system. In case of injury, the broken tissue acts as the initiator by releasing thromboplastin, a lipoprotein which is not well characterized. This thromboplastin, with the intervention of another plasma component, called factor seven, activates the Stuart factor. The extrinsic system thus bypasses Christmas factor, PTA, and the Hageman factor.

The mechanism of amplification is clearly seen in the fact that as we move closer to fibrinogen on Fig. 1, the quantity of the various factors in plasma increases. While the Hageman factor represents only trace elements, prothrombin represents a few tenths of a milligram per milliliter. Fibrinogen occurs in milligram quantities. The scheme represented in Fig. 1 is probably not a final one. New factors may turn up in future investigations.

The details of the thrombin-fibrinogen interaction are much better known on the molecular level. Bovine fibrinogen contains three different peptide chains in duplicates. The average molecular weight of these chains is about 50,000. One of the three chains on the amino end terminates in glutamic acid, and one in tyrosine; in the third

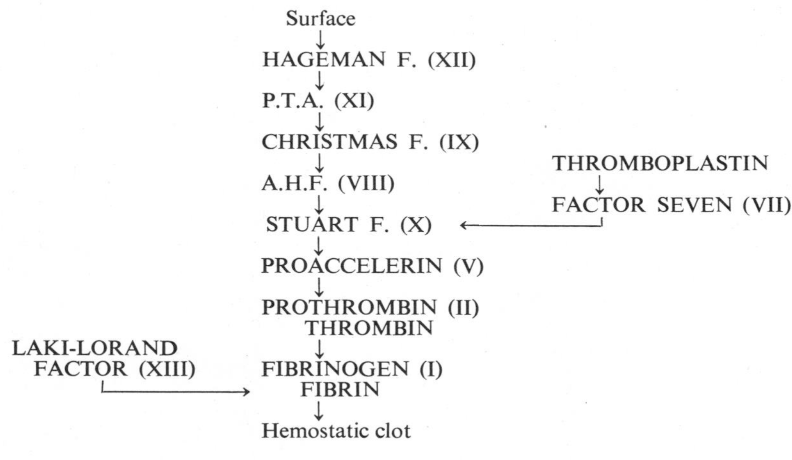

FIG. 1

chain, the amino group is masked. These peptide chains are folded in such a way that they form three globules that are connected like beads on a string.

Thrombin is much smaller than fibrinogen. The minimal molecular weight of bovine thrombin as judged from chemical and physical analysis is about 8000–9000. Thrombin turned out to be a highly specialized proteolytic enzyme. On synthetic substrates, like tosylarginine methyl ester, it acts like trypsin. On protein substrates however, thrombin proved to be a very poor enzyme except in the case of fibrinogen where it splits four bonds with great rapidity. These four bonds are the first arginine-glycyl bonds along two of the chains of the bovine fibrinogen.

The aggregation of the fibrin molecules is quite orderly, such that tyrosine residues on one molecule find the histidine residues on the other to form numerous hydrogen bonds. At this stage, the clot is not very firm, and a second enzyme is needed to strengthen the clot for its role as a hemostatic agent. If this strengthening fails due to lack of this enzyme, a serious bleeding condition arises.

The pattern discovered in the clotting of bovine fibrinogen repeats itself with other fibrinogens. In each case, the bond split is between arginine and glycine residues. Figs. 2 and 3 show the amino acid sequence of the peptides released from various fibrinogens.

The basic similarity of these peptides is striking. Although there are positions along the chain that are quite different, at certain places we always find the same amino acid. This indicates that certain of the amino acids are essential at a given position and cannot change during evolution.

It is understandable that the peptides terminate in arginine at one end because this is where

19 18 17 16 15 14 13 12 11 10 9 8 7 6 5 4 3 2 1

MAN H-Ala-Asp-Ser- Gly-Glu-Gly-Asp-Phe-Leu- Ala- Glu-Gly- Gly-Gly- Val- Arg-OH (Ser: OPO_3H_2)

OX H-Glu-Asp-Gly-Ser-Asp-Pro- Pro-Ser- Gly-Asp-Phe-Leu- Thr- Glu-Gly-Gly- Gly-Val- Arg-OH

SHEEP H-Ala-Asp-Asp-Ser- Asp-Pro- Val- Gly- Gly-Glu-Phe-Leu- Ala- Glu-Gly- Gly-(Gly,Val)-Arg-OH

GOAT H-Ala-Asp-Asp-Ser- Asp-Pro- Val- Gly-Gly-Glu- Phe-Leu- Ala- Glu-Gly- Gly-Gly- Val- Arg-OH

REINDEER H-Ala-Asp-Gly- Ser- Asp-Pro- Ala- Gly-Gly-Glu- Phe-(Leu,Ala, Glu, Gly, Gly, Gly,Val)-Arg-OH

PIG H-Ala- Glu-Val- Glu-Asp-Lys- Gly-Glu- Phe-Leu- Ala- Glu-(Gly,Gly,Gly, Val)-Arg-OH (Glu: NH_2)

RABBIT H-Val- Asp-Pro- Gly- Glu-Thr-Ser- Phe- Leu-(Thr, Glu,Gly,Gly)-Asp-Ala- Arg-OH

DOG H-Thr-Asp-Ser- Glu-Gly-Lys-Glu-Phe-Ileu- Ala- Glu-Gly- Gly- Gly-Val- Arg-OH (Glu: NH_2)

FIG. 2. Amino acid sequence of peptide A released from various fibrinogens.

21 20 19 18 17 16 15 14 13 12 11 10 9 8 7 6 5 4 3 2 1

MAN Pyr- Gly- Val- Asp-Asp- Asp-Glu-Glu-Gly- Phe- Phe- Ser- Ala-Arg-OH (Asp: NH_2; Asp: NH_2)

OX Pyr-Phe-Pro-Thr- Asp-Tyr-Asp-Glu- Gly- Glu- Asp-Asp- Arg-Pro- Lys- Val- Gly- Leu- Gly- Ala-Arg-OH (Tyr: OSO_3H; Glu: NH_2)

SHEEP H-Gly-Tyr- Leu-Asp-Tyr-Asp-Glu- Val- Asp- Asp-Asp- Arg-Ala- Lys- Leu- Pro- Leu- Asp-Ala-Arg-OH (Tyr: OSO_3H; Asp: NH_2)

GOAT H-Gly-Tyr- Leu-Asp-Tyr-Asp-(Glu,Val, Asp, Asp,Asp)-Arg-Ala- Lys- (Leu,Pro, Leu)-Asp-Ala-Arg-OH (Tyr: OSO_3H)

REINDEER Pyr- Leu-Ala- Asp-Tyr-Asp-Glu- Val- (Glu,His, Asp)-Arg-Ala- Lys-(Leu,His)-Leu- Asp-Ala-Arg-OH (Tyr: OSO_3H)

PIG H-Ala-Ileu-Asp- Tyr-Asp-Glu- Asp-Glu- Asp-Gly- Arg-Pro- Lys- Val- His- Val- Asp-Ala-Arg-OH (Tyr: OSO_3H)

FIG. 3. Amino acid sequence of peptide B released from various fibrinogens. (Pyr indicates pyrrolidone carboxyl, a cyclized form of glutamyl.)

thrombin carries out the splitting. Aspartic and glutamic acids with their negative charges are also essential to keep the fibrin molecules apart from each other. The role of other invariants like phenylalanine in position 9 is not clear.

From Figs. 2 and 3, it is quite obvious that the closer the animals are related, the more similar are the amino acid compositions of the peptides. The sequences in the goat and sheep peptides are identical. In the goat and reindeer A peptides, 90% of the amino acids are the same but only about 70% are common in the goat and ox peptides. Only about 45% of the amino acids are similar in the sheep and human peptides. Roughly, the correspondence between two peptides is greater the shorter the time they have separated away from a common ancestor.

References

1. MacFairlane, R. G., "An Enzyme Cascade in Blood Clotting Mechanism and Its Function as a Biochemical Amplifier," *Nature*, **202**, 499 (1964).
2. Laki, K., and Gladner, J. A., "Chemistry and Physiology of the Fibrinogen-fibrin Transition," *Physiol. Rev.*, **44**, 127 (1964).
3. Doolittle, R. F., and Blombäch, B., "Amino-acid Sequence Investigations of Fibrinopeptides from Various Mammals: Evolutionary Implications," *Nature*, **202**, 147 (1964).
4. Laki, K., "The Clotting of Fibrinogen," *Sci. Am.*, **206**, 60 (March 1962).
5. Laki, K., "Enzymatic Effects of Thrombin," *Federation Proc.*, **24**, 794 (1965).
6. Davie, E. W., and Ratnoff, O. D., "Waterfall Sequence for Intrinsic Blood Clotting," *Science*, **145**, 1310 (1964).

K. Laki

BLOOD GROUPS

The outer portion of the red blood cell is a very complex material composed of proteins, polysaccharides, and lipids, many of which are ANTIGENS referred to as blood group substances. The presence of most, if not all, of these antigens is genetically determined, and their number is such that there may be few, if any, individuals in the world with an identical set of antigens on his red cells—monozygotic twins excluded.

The nomenclature of the human blood groups is complex, each group is usually designated either by a letter or by the name of the individual or family in which the blood group was first observed. Some of the more important and best studied blood group systems and their symbols are: ABO, MNS, P, Rh. Lutheran (Lu), Lewis (Le), Kell (K), Duffy (Fy), Kidd (Jk), Diego (Di), Sutter (Js), Vel, I, and Xg. Each of these systems is composed of two or more allelic genes. Three general categories of blood groups have been defined: high-frequency groups such as Vel and I are present in more than 95% of humans thus far tested; low-frequency groups, such as Levay, are extremely rare and are found in one or few families; groups that occur with an intermediate frequency such as the ABO and MNS groups.

Blood groups are not limited to humans; they occur in every animal species thus far studied.

Methods. The presence of a blood group antigen may be determined by reacting erythrocytes with a specific antibody. The reaction is usually observed as a clumping or agglutination of the red cells. Sometimes special means may be required to bring about clumping, such as the use of a special suspending medium (*i.e.*, bovine albumin), pretreating the red cells with enzymes such as papain or ficin, or the addition of a second antibody, antiglobulin, to react with the specific antibody already on the cells; the latter is called the antiglobulin or Coombs test.

Specific antibodies for these tests may occur naturally, as in the case of antibodies to the A and B blood group antigens, or can be obtained from the sera of individuals who have received antigenic stimulus through blood transfusions or pregnancy. Antibodies may also be prepared by injecting animals with human red blood cells; antibodies to the M and N antigens are usually prepared in this manner in rabbits. It should be noted that the injection of human red cells into animals does not usually result in antibodies specific for a particular red cell antigen, but rather results in antibodies that agglutinate all human red cells; this procedure, therefore, is not regularly used.

Extracts of various plants, lectins, have been found to agglutinate human red cells possessing specific antigens. For example, an extract of the lima bean, *Phaseolus limnensis*, agglutinates cells containing the A antigen, while an extract of the seeds of the South American plant *Vicia graminea* most strongly agglutinates cells with the N antigen.

When an antigen can be obtained in soluble form, an inhibition test may be employed. The soluble antigen, *e.g.*, antigen A from saliva, is mixed with the anti-A antibody. The mixture is incapable of agglutinating red cells possessing the A antigen; the presence of the A antigen in saliva in soluble form is therefore inferred.

In this article, only a few of the many blood group systems can be discussed.

ABO System. This was the first blood group system to be defined and is still the most important one to be considered in selecting blood for transfusions. The system, consisting of three allelic genes, divides all humans into four groups, A, AB, B, and O. In a few rare individuals, the presence of a suppressor gene may prevent the expression of the A, B, O group character. The products of these genes are the A, B, and O antigens or substances; the O substance is still not clearly understood and is sometimes referred to as O (H) substance. These antigens not only are located on red cells, but are widely distributed in the body, occurring in the endothelium of capillaries, veins, and arteries; in cells of the Malpighian layers; in the mucus-secreting apparatus such as stomach lining and salivary glands. In addition to a cell-associated form, these antigens occur in soluble form in many body fluids, *i.e.*, saliva, gastric juice, urine, amniotic fluid, and in very high concentrations in pseudomucinous ovarian cyst fluid.

All individuals possess cell-associated A, B, O (H), antigens. The presence of the soluble form, however, is governed by a recessive gene called the secretor gene which exists as two alleles, Se and se. Individuals who possess at least one Se gene secrete the antigens, while those with two se genes do not.

The A, B, O (H) antigens are not uniquely human; in fact, they are widely distributed in nature. For example, they are found on primate erythrocytes, in the stomach lining of pigs and horses, and even in some strains of *Escherichia coli*, a bacterium that normally inhabits the intestinal tract.

Intensive investigation has produced a great deal of information concerning the chemical composition of these antigens. They are extremely stable substances, which is attested to by the fact that they can be extracted from Egyptian mummies, thus making it possible to obtain the blood groups of this ancient people.

Analysis of purified A, B, or O (H) substances reveals that about 75% of the weight is accounted for by four sugars, L-fucose, D-galactose, N-acetyl-D-glucosamine, and N-acetyl-D-galactosamine [see CARBOHYDRATES (CLASSIFICATION); HEXOSAMINES]. The remainder, consisting of AMINO ACIDS, has been given relatively little study, but all preparations are either free of, or have very low concentrations of, aromatic amino acids; most preparations have a high content of proline and the hydroxyamino acids threonine and serine.

Specific antigenic activity is associated with the carbohydrate moiety, and since the A, B and O (H) substances possess the same four sugars, the difference between them lies in their arrangement.

Mild acid hydrolysis at pH 1.5 for 2 hours results in a large non-dialyzable core, and dialyzable monosaccharides and oligosaccharides. From the blood group A substance, for example, was isolated galactose, fucose, N-acetylhexosamines and several di- and trisaccharides. N-acetylgalactosamine was the only simple sugar that displayed some A activity since, in high concentration, it could inhibit the agglutination of A cells by anti-A, and the precipitation of complete A substance by anti-A; this is undoubtedly the end sugar of the A substance carbohydrate. A trisaccharide with high A activity was also isolated from the dialysate and represents the last three sugars in the chain. It has the structure α-N-acetylgalactosaminoyl-(1 → 3)-β-galactosyl-(1 → 3)-N-acetylglucosamine.

The structure of blood group B substance is not as well elucidated, but it too contains a core and a side chain of sugars that carries the B specificity. The last two sugars on the side chain appear to be O-α-D-galactosyl-(1 → 3)-D-galactose.

Knowledge of the chemistry of blood group O (H) substance is very fragmentary. Fucose is unquestionably the end sugar important in determining specificity. Enzymes with fucosidase activity split fucose from blood group O (H) substance and simultaneously eliminate antigenic specificity; fucose also inhibits the agglutination of group O erythrocytes by anti-O antibody.

Lewis Groups. This blood group system consists of at least two alleles Le^a and Le^b and possibly a third, Le^c. The Lewis group is unusual in that the antigens are not really red cell antigens but are, rather, soluble antigens that occur in many body fluids including serum. Erythrocytes adsorb the antigen from serum and can then be agglutinated by the appropriate antiserum. In this respect they resemble the J antigen of cattle and R antigen of sheep.

Le^a antigen, the best studied of this group, occurs in the secretions of more than 90% of all humans, but the concentration is very variable. Those humans who are non-secretors of A, B, O (H) substances possess high concentrations of Le^a substance in their saliva and other fluids; their red cells agglutinate well with anti-Le^a serum. Secretors of A, B, O (H) substances have very low concentrations of Le^a substance in their fluids; their erythrocytes fail to agglutinate with anti-Le^a serum.

The Le^a substance contains the same four sugars as the A, B, O (H) substances. The end group that establishes specificity is branched and the final three sugars are:

β-D-galactosyl-(1 → 3)
　　　　　　　　　　　　>N-acetyl-D-glucosaminoyl-
α-L-fucosyl-(1 → 4)

Rh Group. The Rhesus, or Rh, group was so named because the antigen was first found in the red cells of Rhesus monkeys. It is now known, however, that the Rhesus antigen as it occurs in the monkey has certain similarities to the human Rh antigen but it is quite different.

This system is exceedingly complex consisting of perhaps 20 antigens. The $Rh_o(D)$ antigen is the most important since antibody to this antigen is the one most frequently involved in inducing hemolytic disease of the newborn.

Almost nothing is known of the chemistry of these antigens. They occur only on red cells and are presumed to be protein or to contain protein, but the evidence is scanty.

Distribution. While the erythrocytes of all people possess antigens of the various blood group systems, the individual antigens are not uniformly distributed over the world. Each of the aboriginal populations of the world tends to have a characteristic combination of blood group antigens. Thus, the B gene has its highest frequency in China, its lowest in Western Europe, and is completely lacking in the Indian population of North and South America. The A antigen, on the other hand, is absent in the South American Indians but present in Indians of northwest North America.

The Diego antigen is almost exclusively Mongolian, occurring in Chinese, Japanese, and American Indian populations but is curiously absent in Eskimos. Antigens V and Js^a are found almost exclusively in Negroes. These examples are merely illustrative of a very vast literature on blood group antigen distributions.

References

1. RACE, R. R., AND SANGER, RUTH, "Blood Groups in Man," Fourth edition, Blackwell Scientific Publications, 1962.
2. MOURANT, A. E., "The Distribution of the Human Blood Groups," Springfield, Ill., Charles C. Thomas, 1954.
3. KABAT, E. A., "Blood Group Substances," New York, Academic Press, 1956.
4. STONE, W. H., AND IRWIN, M. R., "Blood Groups in Animals Other than Man," *Advan. Immunol.*, **3**, (1963).
5. WATKINS, W. M., AND MORGAN, W. T. J., "Further Observations on the Inhibition of Blood Group Specific Serological Reactions by Simple Sugars of Known Structure," *Vox Sanguinis*, **7**, 129–150 (1962).
6. KABAT, E. A., "Antigenic Determinants of Dextrans and Blood Group Substances," *Federation Proc.*, **21**, 694–701 (1962).
7. WIENER, A. S., "Advances in Blood Grouping," Grune & Stratton, 1961.

HAROLD BAER

BLOOD ENZYMES

Blood plasma contains detectable amounts of many different enzymes, but the normal levels of activity are comparatively low, except for certain enzymes involved in BLOOD CLOTTING. The activities of certain enzymes in plasma may rise in particular pathological conditions, and these abnormal levels of activity may have diagnostic value. Examples include acid phosphatase and alkaline phosphatase (see PHOSPHATASES, section on Phosphomonoesterases), and glutamic-aspartic transaminase (see TRANSAMINATION ENZYMES) of blood plasma.

Reference

1. WROBLEWSKI, F., "Enzymes in Medical Diagnosis," *Sci. Am.*, **205**, 99–107 (August 1961).

BLOOD OSMOTIC PRESSURE

The presence of solute molecules and ions in relatively high concentrations in blood establishes an osmotic pressure which tends to transport water from the exterior, through the semi-permeable membranes of the blood vessel walls, into the bloodstream (see OSMOSIS AND OSMOTIC PRESSURE). This osmotic transport of water inward is opposed by the effect of hydrostatic pressure within the blood vessels, tending to force water (and solute substances) out through the capillary walls; the loss through "leakage" of some of these solutes is indirectly restored through the action of the LYMPHATIC SYSTEM. Among the blood constituents important in maintaining blood osmotic pressure (and thus in helping to regulate the volume of fluid in the bloodstream) are the blood proteins (see PLASMA AND PLASMA PROTEINS); of these, the protein fraction termed ALBUMINS, being relatively low in molecular weight, makes the greatest contribution to the total osmotic effect (see also ELECTROLYTE AND WATER REGULATION).

Reference

1. WHITE, A., HANDLER, P., AND SMITH, E. L., "Blood Plasma," in "Principles of Biochemistry," Third edition, Ch. 32, pp. 627–646, New York, McGraw-Hill Book Co., 1964.

BONE FORMATION

Bone formation is one of the earliest biological phenomena to be studied from a biochemical standpoint, yet its essential nature remains a subject of active research and speculation. Although the chief characteristic of bony tissue is its high content of inorganic salts, before considering the events associated with the formation of bone crystals some reference is necessary to the biogenesis of the organic portion within which the salts are deposited.

The bones originate from embryonic mesenchymal connective tissue cells which differentiate into bone-forming cells or osteoblasts. The formation of most embryonic bones occurs by calcification of a previously generated cartilaginous model, the remainder (intramembranous ossification) being the result of direct mineralization of connective tissue. In the former, the cartilage cells become hypertrophic and form centers of ossification from which cartilage is replaced centrifugally by bone cells. In the latter, calcification of the intercellular matrix occurs under the influence of osteoblasts which arise by transformation of connective tissue cells. The diaphysis, or shaft, elongates by calcification of the epiphyseal cartilage plate which is continuously regenerated by osteogenic mesenchymal cells. It increases in diameter by accretion beneath the layer of connective tissue covering the bone (the periosteum) and by concomitant removal from the endosteal surface of the marrow cavity.

Calcification results in the formation of the trabeculae of spongy bone, a form characterized by a high proportion of marrow and a profuse blood supply. Progressive deposition of new layers of bone, covered with osteoblasts, results in the generation of compact bone which is made up of units called haversion systems or osteones, each consisting of interwoven layers of bone oriented around a central vascular canal. The intercellular material is permeated by small spaces (lacunae) containing branched cells termed osteocytes. These cells are similar to osteoblasts, are rich in glycogen, and are necessary for the maintenance of bone cells. The osteones are subject to a continuing remodeling, apparently under the action of large, multi-nucleated osteoclasts located in tunnels which infiltrate the tissue prior to resorption.

The organic matrix of bone consists essentially of bundles of collagenous fibers imbedded in a ground substance. Although the general properties of bone collagen, which makes up over 90% of the dry fat-free organic matter, are similar to those of COLLAGEN derived from other forms of connective tissue throughout the body, the material present in osseous tissue apparently possesses some unique characteristic necessary for the nucleation of salt crystals. The ground substance is characterized chemically by the presence of MUCOPOLYSACCHARIDES, including chondroitin sulfate, hyaluronic acid and keratosulfate, the physiological significance of which is still obscure.

The process by which bone crystals are deposited in the organic matrix, their internal structure and their chemical constitution have been under investigation for many years. The mineral consists mainly of Ca^{++} and $PO_4^{\equiv}$ ions, with smaller amounts of $CO_3^{=}$, OH^-, Mg^{++}, Na^+, F^- and citrate$^{\equiv}$. However, the concentrations of the minor ions is uncertain owing to the occurrence of surface absorption phenomena, exchange with components of the fluid medium, and the possibility that some constituents (*e.g.*, citrate) are in a separate phase. Electron microscopy suggests that the crystals are rod-like with a diameter of about 50 Å and that they may be oriented in chains along the collagen fibrils. X-ray diffraction and chemical analysis indicate that bone mineral has the crystal lattice structure and composition of a substituted hydroxyapatite $[Ca_{10}(PO_4)_6(OH)_2]$. The architecture of the crystals provides for an enormous surface area in proportion to mass, thereby exposing the salts to intimate contact with constituents of the surrounding fluid. Exchange occurs actively, particularly in trabecular bone, not only between ions of the same species but also between dissimilar species: $CO_3^{=}$ for $PO_4^{\equiv}$, Sr^{++} for Ca^{++} and F^- for OH^-. The Ca:P molar ratio for bone is very close to the theoretical value for hydroxyapatite (1.67). (See also CALCIUM; FLUORIDE METABOLISM; PHOSPHATE.)

Studies on the formation of the bone salts have centered around the physicochemical concept that crystallization occurs when the concentration of Ca and P ions in the blood and circulating fluids exceeds the solubility product constant. Plasma P is present mainly as $HPO_4^{=}$, in a smaller amount as $H_2PO_4^-$ and in minute concentrations as $PO_4^{\equiv}$. Observations on the calcification of cartilage *in vitro* indicate that the product $[Ca^{++}] \times [HPO_4^{=}]$ is the critical ion relationship in crystal formation, and it has been proposed that whereas the serum and extracellular fluid are normally undersaturated with respect to $CaHPO_4$, they are supersaturated with respect to bone salts. This proposal suggests that an ion gradient exists between the interstitial bone fluid and the extracellular fluid which is maintained by cellular activity, and it stresses the importance of the minor ions, particularly citrate$^{\equiv}$, in determining the degree of saturation. The production of citrate$^{\equiv}$ by bone cells may determine whether the medium is undersaturated or supersaturated with respect to Ca^{++} and $HPO_4^{=}$, *i.e.*, whether dissolution or deposition of bone salts occurs. It is further suggested that VITAMIN D and parathormone may exert their influence on bone metabolism by regulating the metabolic activity of the cells and hence the production of citrate (see PARATHYROID HORMONES).

The mechanism by which crystal formation is initiated is still obscure. Following the discovery of phosphatase in calcifying cartilage, the view became prevalent that the local action of this enzyme on some organic phosphate ester produced a high concentration of phosphate ion which exceeded the solubility product constant of bone mineral. This theory has been largely discarded in favor of a "seeding" mechanism which assumes that some component of the organic material (presumably the collagen or the mucopolysaccharide) furnishes the seeding sites. Reconstituted collagen fibers have the ability to induce crystal formation *in vitro* from stable solutions of Ca^{++} and $HPO_4^{=}$ ions; however, it is difficult to prepare collagen that is completely free of mucopolysaccharides. Once the nuclei have been formed, crystallization proceeds spontaneously. Apart from the unexplained role of phosphatase, GLYCOLYSIS appears to be a necessary accompaniment of bone salt formation; inhibitors of glycolysis interrupt crystallization in a reversible manner. No integrated concept has been put forward to account for these various observations, and consequently no biochemical description of the bone mineralization process in terms of functional groups is as yet feasible.

References

1. McLean, F. C., and Urist, M. R., "Bone," Chicago, Ill., The University of Chicago Press, 1961.
2. Neuman, W. F., and Neuman, M. W., "The Chemical Dynamics of Bone Mineral," Chicago, Ill., The University of Chicago Press, 1958.

H. H. Draper

BORON

Boron is of biochemical interest because incontrovertible evidence has been obtained that it is essential for higher plants. That the need is low is indicated by the fact that it cannot be demonstrated in water culture if the experiment is carried out in Pyrex glass vessels; in this case enough boron dissolves from the glass to meet the demand.

Boron is inevitably a constituent of the tissues of animals that consume plants. Most animal tissues are said to contain less than 1 ppm. Several experiments have been performed to produce boron deficiency in rats without success. Based upon our knowledge of the unity of nature, one would suspect that eventually boron may be found to be essential for animal life.

References

1. Anderson, A. J., and Underwood, E. J., "Trace Element Deserts," *Sci. Am.*, **200,** 97 (January 1959).
2. Massey, A. G., "Boron," *Sci. Am.*, **210,** 88 (January 1964).

BRAIN COMPOSITION

The gross chemical composition of the brain represents the sum of the compositions of the various cell types. Because of the heterogeneity of the cell population, the tabulated values are undoubtedly not representative of the composition of individual brain cells. Anatomically, the brain can be divided into approximately equal gray and white matter portions. Although estimates are available for the composition of these two areas, the following values are given in terms of the fresh (wet) weight of the whole brain.

General Composition. The brain in common with all tissues is composed of organic and inorganic materials. The gross composition is the following:

	Approximate %
Water	77
Solids	23
Soluble organic substances	2
Inorganic salts	1
Carbohydrates	1
Protein	8
Lipids	13–14
Total nitrogen	1.6
Total phosphorous	0.3

Inorganic Ions. Although the distribution of inorganic ions is clearly different in intracellular and extracellular fluid spaces, the delineation of these compartments in the brain is still uncertain, so that no attempt at such a distinction has been made here.

	Approximate %
Calcium	0.01
Sodium	0.14
Potassium	0.32
Magnesium	0.02
Chloride	0.12
Bicarbonate	0.06
Phosphate	0.04

Lipids. Cerebral lipids qualitatively are similar to the types of lipids distributed throughout the body. Although ganglioside-like material has been reported to occur in other tissues, the structures of GANGLIOSIDES found in brain is unique to this organ. The lipid concentration of the white matter of brain is higher than that of any other organ.

	Approximate %
Total lipid	14
Cholesterol	3.3
Triglyceride	0.1
Cerebroside	4
Sulfatides	0.8
Gangliosides	0.1
Total phospholipids	6.5
Phosphatidyl choline	1.5
Phosphatidyl ethanolamine	2.0
Phosphatidyl serine	1.5
Phosphatidyl inositols	0.2
Sphingomyelin	1.2
Total plasmalogen	0.3

Proteins. The state of knowledge concerning the cerebral proteins leaves much to be desired. Data on proteins isolated from the brain provide only rough approximations of the relative amounts.

	Approximate %
Proteolipids	3.0
Globulins	4.0
Albumins	0.5
Collagen	0.2

Other Macromolecules. The brain contains approximately 0.1% glycogen, 0.12% RNA, and 0.09% DNA.

Vitamins and Cofactors. The following estimates are representative of certain biochemically important compounds.

	Approximate %
ATP + ADP	0.25
Phosphocreatine	0.06
DPN (NAD)	0.015
Coenzyme A	0.007
Pyridoxine	0.001
Riboflavin	0.0004
Ascorbic acid	0.04
Thiamine	0.0001

Amines and Related Substances. The brain contains several biologically active amines known or suspected to act as transmitters of impulses between nerve cells.

	Approximate %
Total acetylcholine	0.0003
Noradrenaline	0.00001
Adrenaline	0.000002
Histamine	0.00001
5-Hydroxytryptamine	0.00001
γ-Aminobutyric acid	0.03
Free amino acids	0.55

Tricarboxylic Acid and Glycolytic Intermediates. A continuous supply of nutrients is required by the brain in order to maintain its cellular functions (see GLYCOLYSIS; CITRIC ACID CYCLE). The levels of certain of these compounds are presented in the following table.

	Approximate %
Glucose	0.07
Lactic acid	0.02
Pyruvate	0.002
Citrate	0.006
α-Ketoglutarate	0.02
Fumarate	0.014
Malate	0.003
Oxaloacetate	0.009
Succinate	0.004

Many of the brain constituents mentioned in this summary are the subject of separate articles in this volume, including CEPHALINS AND LECITHINS; CEREBROSIDES; LIPIDS; MYELIN; PHOSPHOLIPIDS; PLASMALOGENS (see also CEREBROSPINAL

FLUID; BLOOD-BRAIN BARRIER; NERVE CELL COMPOSITION; NERVE IMPULSE CONDUCTION; SYNAPSES AND NEUROMUSCULAR JUNCTIONS).

References

1. McILWAIN, H., "Biochemistry and the Central Nervous System," Second edition, Boston, Little, Brown and Co., 1959.
2. LONG, C. (EDITOR), "Biochemists Handbook," Princeton, N.J., D. Van Nostrand Co., 1961.
3. ELLIOTT, K. A. C., PAGE, I. H., AND QUASTEL, J. H., "Neurochemistry," Second edition, Springfield, Ill., Charles C. Thomas, 1962.
4. TOWER, D. B., in "Handbook of Physiology: Neurophysiology," (FIELD, J., MAGOUN, H. W., AND HALL, V. E., EDITORS), Vol. 3, p. 1793, Washington, American Physiological Society, 1960.

JULIAN N. KANFER

BRAIN METABOLISM

The metabolism of the brain is usually measured either (1) in living organisms or (2) in isolated systems such as brain slices, brain homogenates, or extracts. In isolated systems, the different metabolic pathways can be closely observed, while measurements under living conditions show which of the possible capacities of the organ observed in isolated systems are operative, and at what rate, in the living brain. Such investigations have established that although, in general, the metabolic pathways that have been observed in the brain are similar to those observed for the rest of the organism, metabolic rates and the alterations of these show a characteristic behavior for brain.

The metabolism of living brain is usually measured by analyzing the arterial blood coming to, and venous blood coming from, this organ. Since cerebral circulation is extremely rapid (under normal conditions about 1/6 of the total circulation goes through the brain), this method can only measure those substances that are used or produced rapidly and in significant quantities by the brain. Such measurements of arterio-venous difference have established that GLUCOSE is the main metabolite of brain and that the oxidation of this compound can also account for most of the oxygen utilized by the brain. The utilization of oxygen in the brain is surprisingly high, higher than in most other organs, with about 1/5-1/4 of the oxygen taken up through the lung being utilized by the brain, which comprises only 2% of body weight. About 90% of the glucose is oxidized completely to carbon dioxide, the rest mostly to lactic acid, and a small portion to pyruvic acid. The whole brain of a normal young man consumes 46 ml oxygen and 76 mg of glucose per minute with a blood flow of 750 ml. These figures show that the metabolic rate of the brain is very considerable in the total body economics as far as energy and oxygen utilized, heat produced, etc., are concerned. With physical activity, the oxygen consumption of the whole body changes significantly, but brain metabolism does not. Oxygen uptake and glucose utilization do not change significantly during mental exercise or during sleep, but are altered only in extreme circumstances such as deep ANESTHESIA, which reduces metabolism, or high CO_2 content of the air, which increases cerebral blood flow.

The large amount of energy utilized by the brain is somewhat surprising since this organ does not perform work like that done by muscle and kidney for example. The most plausible explanation for the utilization of so much energy is that there is continuous very high activity in the brain. Such activity can be detected among other methods by the electroencephalogram. Continuous impulse conduction involves the movement of ions, since conduction by nerve is achieved by temporary change in ion permeability. Sodium ions penetrate the nerve as impulse travels, and according to the most plausible theory, energy by the brain is mostly utilized for pumping back these ions to the outside of the membrane (see NERVE IMPULSE CONDUCTION; also ACTIVE TRANSPORT).

Thus quantitatively the most important compound utilized is glucose, and it is mostly oxidized to CO_2. This metabolism accounts for more than 90% of the arterio-venous difference of the brain. Other reactions, however, also occur in the brain, and these may have great significance. Indeed, most present theories explaining brain function in biochemical terms attribute functions such as learning or the storage of memory not to glucose metabolism but to other reactions. Glucose is regarded as the basal metabolite of the brain, and quantitatively less significant metabolic reactions account for the functional metabolism of this organ.

It has been shown that a number of other components in addition to glucose are metabolized; these include LIPIDS, NUCLEIC ACIDS, and PROTEINS. Indeed, it seems that except for only a few constituents, such as cholesterol and deoxyribonucleic acids (DNA), most cerebral components are unstable. It has recently been found with the aid of radioactive amino acids, for example, that at least 98% of the cerebral proteins are in a dynamic equilibrium. Although in the adult brain the absolute quantity of the organ protein is not changed, it is constantly broken down and resynthesized. The average half-life of brain proteins (the time it takes for 50% to be broken down and resynthesized) is about 14 days in mice and rats. Almost no cell division in the adult brian can be observed; there is no significant cell regeneration either, and DNA is stable in the brain. Although the cells thus seem to be stable, the cell constituents themselves apparently are not, with most of them being in a dynamic equilibrium and being replaced at a rather high rate, since 14 days half-life means replacement of brain proteins several times per year. This is rather surprising if one visualizes the brain as depository of permanent information because memory storage then occurs in a structure that is constantly broken down and rebuilt.

There are two questions usually asked by investigators of cerebral metabolism, (1) whether there exist metabolic pathways that are specific for brain and not operative in other tissues and

(2) whether there are differences in metabolism due to activity or pathological alteration of brain. As far as the first question is concerned, no mechanisms have yet been found in the brain that are specific for this organ, although a number of compounds exist, such as γ-aminobutyric acid and acetyl aspartic acid, that do not seem to occur in organs other than brain. It seems likely that metabolic differences between brain and the rest of the organism are more quantitative than qualitative.

A property of brain that at least quantitatively distinguishes it from other organs is its permeability. Substances that are administered penetrate the brain to a lesser extent than they do other organs (see BLOOD-BRAIN BARRIER). Since penetration is restricted by membranes and active transport processes, the brain barrier system (as this phenomenon is called) in a way restricts cerebral metabolism, since even those substances for which metabolic machinery is present in the brain cannot be metabolized if they cannot get into the brain at significant rates. The greater restrictive power of cerebral membranes therefore restricts cerebral metabolism. This also explains the fact that cerebral constituents undergo less fluctuation than components of other organs, *e.g.*, in starvation. The high rate of oxygen and glucose consumption of the brain combined with the low rate of penetration of most other substances makes the brain very largely dependent on glucose for its basal metabolism; the oxygen reserves of the brain are enough for only about 6–10 seconds, and prolonged anoxia results in irreversible damage of the brain.

Cerebral metabolic rates change under a number of circumstances. There is a general decrease of oxygen utilization during development. It has been estimated that in newborn, cerebral respiration can amount to half of that of the whole organism. Metabolism reaches a plateau in adulthood, and further decrease after 50 years of age in humans has been suggested but not well confirmed. The metabolism of a number of compounds undergoes significant fluctuation under conditions such as convulsions, hibernation, demyelinating diseases, and other pathological states (see MULTIPLE SCLEROSIS; MYELIN). A lot of interest is presently centered on subtle changes in metabolism which are due to changes in mental activity, such as those shown by comparison of trained *vs* untrained animals, or animals reared in isolation *vs* others reared in a complex environment. It is of interest that metabolic changes that were found and which were interpreted as changes in functional metabolism are different in the two types of cells of the brain. It seems that in functional metabolism, neurons have preference and glial cells play a supporting role. Because of the high rate of supporting basal metabolism, extremely sensitive methods will have to be developed to measure functional alterations of metabolism in a meaningful manner. However, already much experimental evidence, although indirect, points to the participation of biochemical reactions in brain function. The effect of chemical compounds on brain function is well known. Alcohol could be mentioned as one of the oldest of such compounds, but a great many others are known, such as anti-depressants, drugs causing hallucinations, etc. Other evidence supporting the functional role of metabolism in the brain is that compounds that inhibit cerebral protein metabolism significantly also inhibit the deposition of short-term memory while they do not affect already deposited long-term memory.

Not only is metabolism different in various developmental stages and under conditions of increased or decreased activity, but there are differences between the various brain areas. The distribution of metabolites is heterogeneous, and the metabolic rate also changes from one brain area to another, showing that controls are highly specific for the various compounds as well as for the various areas involved.

References

1. QUASTEL, J. H., AND QUASTEL, D. M., "The Chemistry of Brain Metabolism in Health and Disease," Springfield, Ill., Charles C. Thomas, pp. 1–170, 1961.
2. MCILWAIN, H., "Chemical Exploration of the Brain," pp. 1–207, Amsterdam, Elsevier, 1963.
3. LAJTHA, A., "Protein Metabolism of the Nervous System," *Intern. Rev. Neurobiol.*, **6**, 1–98; **7**, 1–40 (1964).
4. "Neurochemistry," ELLIOTT, K. A. C., PAGE, I. H., AND QUASTEL, J. H. (EDITORS), pp. 1–1035, Springfield, Ill., Charles C. Thomas, 1962.

ABEL LAJTHA

BREATHING (BIOCHEMICAL ASPECTS).

See GASEOUS EXCHANGE IN BREATHING.

BROMINE

Bromine occurs in plant tissues and is thus indirectly introduced into plant-eating animals, but it has no known essential biochemical role, unlike other halogens (see MINERAL REQUIREMENTS). Certain synthetic bromine-containing compounds structurally similar to natural metabolites (see ANTIMETABOLITES) have been useful in biochemical studies; 5-bromouracil derivatives for example, can become incorporated by the action of DNA polymerase into polydeoxyribonucleotides (see POLYNUCLEOTIDES, SYNTHETIC). At least one bromine-containing amino acid occurs naturally; see Table 1, in AMINO ACIDS (OCCURRENCE, STRUCTURE, SYNTHESIS).

BUCHNER, E.

Eduard Buchner (1860–1917) was a German chemist noted because of his discovery that alcoholic fermentation could be brought about by cell-free yeast juice, which he carefully prepared by grinding the cells and expressing the juice. The existence and functioning of nonliving enzymes from within cells was thus demonstrated. For this work, Buchner was awarded the Nobel Prize in chemistry in 1907.

BUFFERS

When acid is added to an aqueous solution, the pH falls; when alkali is added, it rises. If the original solution contains only typical salts without acidic or basic properties, this rise or fall may be very great. There are, however, many other solutions which can receive such additions with only a slight change in pH. The solutes responsible for this resistance to change in pH, or the solutions themselves, are known as *buffers*. A weak acid becomes a buffer when alkali is added, and a weak base becomes a buffer on the addition of acid. A simple buffer may be defined, in Brönsted's terminology, as a solution containing both a weak acid and its conjugate weak base. Buffer action is explained by the mobile equilibrium of a reversible reaction:

$$A + H_2O \rightleftharpoons B + H_3O^+$$

in which the base B is formed by the loss of a proton from the corresponding acid A. The acid may be a cation such as NH_4^+, a neutral molecule such as CH_3COOH, or an anion such as $H_2PO_4^-$. When alkali is added, hydrogen ions are removed to form water, but, as long as the added alkali is not in excess of the buffer acid, many of the hydrogen ions are replaced by further ionization of A to maintain the equilibrium. When acid is added, this reaction is reversed as hydrogen ions combine with B to form A.

The pH of a buffer solution may be calculated by the mass law equation

$$\text{pH} = \text{p}K' + \log \frac{C_B}{C_A}$$

in which pK' is the negative logarithm of the apparent ionization constant of the buffer acid and the concentrations are those of the buffer base and its conjugate acid.

A striking illustration of effective buffer action may be found in a comparison of an unbuffered solution such as 0.1M NaCl with a neutral phosphate buffer. In the former case, 0.01 mole of HCl will change the pH of 1 liter from 7.0 to 2.0, while 0.01 mole of NaOH will change it from 7.0 to 12.0. In the latter case, if 1 liter contains 0.06 mole of Na_2HPO_4 and 0.04 mole of NaH_2PO_4, the initial pH is given by the equation:

$$\text{pH} = 6.80 + \log \frac{0.06}{0.04} = 6.80 + 0.18 = 6.98$$

After the addition of 0.01 mole of HCl, the equation becomes:

$$\text{pH} = 6.80 + \log \frac{0.05}{0.05} = 6.80$$

while after the addition of 0.01 mole of NaOH it is

$$\text{pH} = 6.80 + \log \frac{0.07}{0.03} = 6.80 + 0.37 = 7.17.$$

The buffer has reduced the change in pH from ± 5.0 to less than ± 0.2.

Figure 1 shows how the pH of a buffer varies with the fraction of the buffer in its more basic form. The buffer value is greatest where the slope of the curve is least. This is true at the midpoint, where $C_A = C_B$ and pH = pK'. The slope is practically the same within a range of 0.5 pH unit above and below this point, but the buffer value is slight at pH values more than 1 unit greater or less than pK'. The curve of Fig. 1 has nearly the same shape as the titration curve of a buffer acid with NaOH or the titration curve of a buffer base with HCl. Sometimes buffers are prepared by such partial titrations, instead of by mixing a weak acid or base with one of its salts. Certain "universal" buffers, consisting of mixed acids partly neutralized by NaOH, have titration curves which are straight over a much wider pH interval. This is also true of the titration curves of some polybasic acids, such as citric acid, with several pK' values not more than 1 or 2 units apart. Other polybasic acids, such as phosphoric acid, with pK' values farther apart, yield curves having several sections, each somewhat similar to the graph in Fig. 1. At any pH, the buffer value is proportional to the concentration of the effective buffer substances or groups.

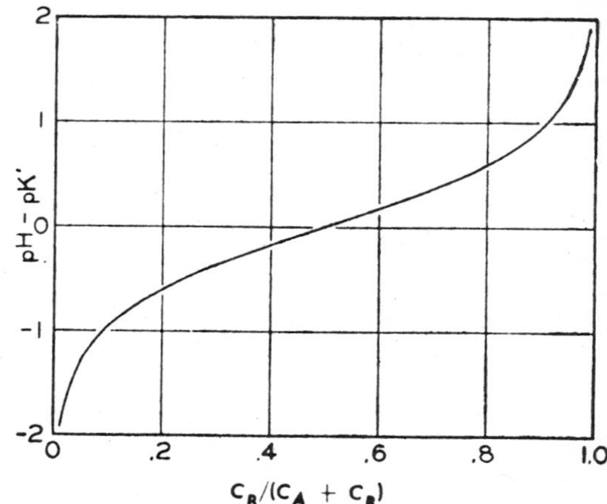

FIG. 1. pH of a simple buffer solution. Abscissas represent the fraction of the buffer in its more basic form. Ordinates are the difference between pH and pK'.

The following table gives approximate pK' values, obtained from data in the literature, for several buffer systems:

Constituents	pK'
H_3PO_4, KH_2PO_4	2.1
HCOOH, HCOONa	3.6
CH_3COOH, CH_3COONa	4.6
KH_2PO_4, Na_2HPO_4	6.8
HCl, $(CH_2OH)_3CNH_2$	8.1
$Na_2B_4O_7$, HCl or NaOH	9.2
NH_4Cl, NH_3	9.2
$NaHCO_3$, Na_2CO_3	10.0
Na_2HPO_4, NaOH	11.6

Buffer substances which occur in nature include phosphates, carbonates and ammonium salts in the earth, proteins of plant and animal tissues, and the carbonic acid–bicarbonate system in blood.

Buffer action is especially important in biochemistry and analytical chemistry, as well as in many large-scale processes of applied chemistry. Examples of the latter include the manufacture of leather and of photographic materials, electroplating, sewage disposal and scientific agriculture. See also ACID-BASE REGULATION; ELECTROLYTE AND WATER REGULATION; pH.

References

BATES, R. G., "Determination of pH," Ch. 5, New York, John Wiley & Sons, 1964.

CLARK, W. M., "The Determination of Hydrogen Ions," Chapters I, II, IX, Baltimore, Williams and Wilkins Co., 1928.

KOLTHOFF, I. M., AND LAITINEN, H. A., "pH and Electro Titrations," Chapters I, III, New York, John Wiley & Sons, 1941.

MACINNES, D. A., "Principles of Electrochemistry," pp. 275–278, New York, Reinhold Publishing Corp., 1939.

DAVID I. HITCHCOCK

BURNET, F. M.

Sir Frank Macfarlane Burnet (1899–) is an Australian physician whose extensive investigations of ANTIBODIES earned for him a share of the Nobel Award in Physiology and Medicine in 1960. The other recipient was PETER BRIAN MEDAWAR.

BUTENANDT, A. F. J.

Adolf Friedrich Johann Butenandt (1903–) is a German chemist who is distinguished for his isolation and synthesis of SEX HORMONES. In 1939 he shared with Leopold Ruzicka the Nobel Prize in chemistry. He was forced by Hitler to decline the award, but it was given him after World War II.

C

CALCIUM (IN BIOLOGICAL SYSTEMS)

Calcium, the fifth most abundant and the most important structural element, occurs in the human body largely in inorganic form. It is an essential nutrient, and is required for maintenance and reproduction, as well as for growth. In the mammalian body it is needed to insure the integrity and permeability of cell membranes, to regulate nerve and muscle excitability, to help maintain normal muscular contraction and to assure cardiac rhythmicity. It plays an essential role in several of the enzymatic steps involved in blood coagulation and also activates certain other enzyme-catalyzed reactions not involved in any of the preceding processes. Finally, calcium is the most important element of bone salt; together with phosphate and carbonate, it confers on bone most of its mechanical and structural properties.

The term *calcium metabolism* will be used to designate the aggregate of the various processes by which calcium enters and leaves the body and its various subsystems.

The *major pathways* of calcium metabolism are intake, digestion and absorption, transport within the body to various sites, deposition in and removal from bone, teeth and other calcified structures, and excretion in urine and stool. These pathways may be thought of as involving three subsystems of the body:

(a) The oral cavity (where ingestion occurs) and the gastro-intestinal tract where digestion and absorption take place and from which the feces is excreted.

(b) The body fluids, including blood, which transport calcium; the soft tissues and body organs to which calcium is transported and where many of its physiological functions are carried out. Some of the organs, like kidney, liver and sweat glands, are also responsible for calcium excretion.

(c) The skeleton, including the teeth, where calcium is deposited in the form of bone salt and from where it is removed (resorbed) after destruction of the bone salt.

Calcium intake varies in different populations and is related to the food supply and to the cultural and dietary patterns of a given population. The intake of a substantial fraction of the world population falls between 400 and 1100 mg/day, but a range encompassing 95% of all people would undoubtedly be even wider. Most populations derive half or more of their calcium intake from milk and dairy products. Calcium intakes of domestic and laboratory animals are higher than those of man; for example, rats typically ingest 250 mg Ca/kg body weight, and cattle 100 mg/kg, whereas man ingests only 10 mg/kg. Ingestion falls with age in all species (see also MINERAL REQUIREMENTS).

Absorption of calcium occurs mainly in the upper portion of the small intestine. The amount and therefore the fraction of calcium absorbed from the gut are a function of intake, age, nutritional status, and health. In general, the fraction absorbed decreases with age and intake and as the nutritional status improves; the absolute amount absorbed increases with intake and may or may not decrease with age. The mechanisms by which calcium is absorbed are not understood; ACTIVE TRANSPORT of the ion against an electrochemical gradient seems to be involved, but not all of the calcium appears to be absorbed by way of this process, as calcium absorption still occurs under conditions when active transport is severely depressed, as in vitamin D deficiency. Calcium absorption can be enhanced by the administration of large doses of VITAMIN D and is depressed in vitamin D deficiency. There is still dispute regarding the effect on calcium absorption of the PARATHYROID HORMONE, the major endocrine control of the blood calcium level (see below). Patients with hyperparathyroidism have been shown to have higher than normal absorption and patients with hypoparathyroidism have lower than normal absorption. Similar effects have also been observed in acute animal experiments, but in most of these a possible indirect effect has not been excluded.

The concentration of *calcium in the blood plasma* of most mammals and many vertebrates is remarkably constant at about 2.5 m*M* (10 mg/100 ml plasma). In the plasma, calcium exists in three forms: as the free ion, bound to proteins, and complexed with organic (*e.g.*, citrate) or inorganic (*e.g.*, phosphate) acids. The free ion accounts for about 47.5% of the plasma calcium; 46% is bound to proteins and 6.5% is in complexed form. Of the latter, phosphate and citrate account for half.

The mechanisms involved in the regulation of the plasma calcium level are not fully understood. The parathyroid glands regulate both level and constancy; when these glands are removed the plasma level drops and tends to stabilize at about 1.5 m*M*, but variations in calcium intake may induce fairly wide fluctuations in the plasma level. In the intact organism, on the other hand, wide variations in intake produce essentially no variations in the plasma calcium value stabilized at around 2.5 m*M*. The equilibrium between bone and plasma is thought to determine the level of the plasma calcium in parathyroidectomized animals, but this reasonable hypothesis still requires experimental support.

The problem of whether parathyroid regulation is due to a single hormone, with hypercalcemic properties, or to two hormones, one hypocalcemic, termed calcitonin, the other hypercalcemic, termed parathyroid hormone, is currently under active investigation. It has also been proposed that a hypocalcemic principle from the thyroid gland, termed thyrocalcitonin (a peptide), either assists or is identical with calcitonin.

When the *calcium* ion concentration is lowered *in* the fluids bathing *nerve* axons—fluids which are in very rapid equilibrium with the blood plasma—the electrical resistance of the axon membrane is lowered, there is increased movement of sodium ions to the inside, and the ability of the nerve to return to its normal state following a discharge is slowed. On the one hand there is thus hyperexcitability, but the ability for synaptic transmission is inhibited because the rate of acetylcholine liberation is a function of the calcium ion concentration. The neuromuscular junction is affected in a similar fashion; hence the end plate potential is lowered below the muscle membrane potential and the muscle membrane is in a hyperexcitable state. These events are reversed when the calcium ion concentration is raised above normal in the blood plasma and in the fluids bathing muscle and nerve. It is for these reasons that hypocalcemia is associated with hyperexcitability and ultimately tetany, and hypercalcemia with sluggishness and bradycardia. (See ACETYLCHOLINE; NERVE IMPULSE CONDUCTION; SYNAPSES AND NEUROMUSCULAR JUNCTIONS).

The role of calcium in *muscular contraction and relaxation* is currently under active investigation. It has been proposed that calcium is the link between the electrical and mechanical events in contraction. It has been shown *in vitro* that when calcium ions are applied locally, muscle fibers can be triggered to contract. It has further been postulated that relaxation of muscle fibers is brought about by an intracellular mechanism for reducing the concentration of calcium ions available to the muscle filaments. Others postulate that contraction occurs because calcium inactivates a relaxing substance which is released from the sarcoplasmic reticulum in the presence of ATP. At present, it is not possible to choose between these two hypotheses, but the central role of calcium in the contraction and relaxation of muscle is becoming increasingly clear [see MUSCLE (STRUCTURE AND FUNCTION); CONTRACTILE PROTEINS].

Bone is the most important reservoir of calcium in the body, accounting for well over 90% of the body's calcium. Moreover, calcium constitutes about one-fourth of the weight of fat-free, dried bones. Calcium occurs in bone mostly in the form of a complex, apatitic salt, so named for its structural resemblance to a family of calcium phosphates of which hydroxyapatite [$Ca_{10}(PO_4)_6(OH)_2$] is the best-known mineralogical example. Calcium occurs also as the carbonate and there has been considerable discussion as to whether bone salt contains the carbonate as a separate phase, whether some of the surface phosphate in apatite has been substituted for by carbonate, or whether bone mineral is in truth a carbonato-apatite, such as dahllite. Whatever the ultimate answer with regard to carbonate, it is important to recognize that the crystal lattice of the bone mineral, when first laid down, does not and probably cannot have all possible calcium positions occupied; whether stability is derived from hydrogen and/or organic bonds to which the mineral may be attached is still not known. Recently it has been proposed that bone salt is a lamellar mixture of octocalcium phosphate and hydroxyapatite; this hypothesis has to account for the amount of pyrophosphate formed when bone salt is heated and also for its evolution with age, *i.e.*, the increase with age in the calcification of bone and the corresponding drop in its induced pyrophosphate content, observations for which the apatitic structure can account. The proponents of the octocalcium phosphate hypothesis explain this by showing that octocalcium phosphate breaks down to apatite and anhydrous dicalcium phosphate which upon further heating give rise to pyrophosphate. Finally they adduce some evidence and calculations that strengthen the possibility that octocalcium phosphate may be present in young and presumably newly formed bone, whereas in older bone an apatitic phosphate admittedly dominates equilibrium. Clearly the final chapter in this story is still far from being written.

Calcium enters and remains in bone as a result of calcification processes (see BONE FORMATION). These involve two steps:

(a) Deposition of bone salt of a minimum calcium content and specific gravity. Deposition occurs by way of nucleation, probably an epitactic process on the collagen fibers, with the ground substance (mostly mucopolysaccharides) between the fibers exerting either a positive or an inhibitory effect on the nucleation process;

(b) Subsequent further mineralization of the bone mineral, leading to an increase in its calcium content and its specific gravity.

Calcium removal, on the other hand, involves destruction of the calcified structure *in toto*. There is no evidence that only particular structures are resorbed, *e.g.*, those with a given degree of mineralization.

The amount of calcium deposited in bone at any moment may be determined from experiments with radioactive calcium. In growing individuals, it exceeds the amount removed by bone destruction; in adults, it is about the same as the amount removed—hence such individuals are considered

TABLE 1. VARIOUS PARAMETERS OF CALCIUM METABOLISM: HUMAN SUBJECT (70 kg MALE)

Parameter	Magnitude	Units
Plasma calcium concentration	0.1	mg/ml
Total plasma calcium	350	mg
Soft tissue calcium concentration	0.05	mg/g
Total calcium pool[a]	5000	mg
Rate constant of pool	0.23	per day
Total calcium in bone	10^6	mg
Calcium intake	1000	mg/day
Food calcium absorbed	350	mg/day
Fecal endogenous calcium	150	mg/day
Total fecal output	800	mg/day
Urinary output	175	mg/day
Balance	+25	mg/day
Calcium deposition in bone	800	mg/day
Calcium removal from bone	775	mg/day

[a] The pool is defined as the totality of calcium that will undergo (rapid) isoionic exchange with radioactive calcium. Physiologically speaking, it comprises all of the calcium in the blood and all other body fluids, plus all calcium in all organs and other soft tissues, plus the rapidly exchangeable calcium of bone. The latter accounts for about half the calcium pool.

to be in "zero" calcium balance; in older persons, the amount deposited is less than the amount removed.

The principal routes of *excretion* are stool and urine. Calcium in the stool may be thought of as made up of unabsorbed food calcium and non-reabsorbed digestive juice calcium. The latter is termed the fecal endogenous calcium. The proportion of fecal endogenous to urinary calcium varies in different species; it is approximately 1:1 in man and 10:1 in the rat and in cattle. The calcium in the urine may have a dual origin—calcium that was filtered at the glomerulus and failed to get reabsorbed along the length of the nephron and calcium that may have originated from transtubular movement in certain regions of the nephron (see RENAL TUBULAR FUNCTIONS). The amount of calcium that may be lost in SWEAT can be great, but there is no convincing evidence that sweat is a habitual route of significant loss.

Table 1 presents a summary of the magnitude of various parameters of calcium metabolism in a typical man.

References

1. BRONNER, F., "Dynamics and Function of Calcium," in "Mineral Metabolism—An Advanced Treatise," (COMAR, C. L., AND BRONNER, F. EDITORS), Vol. IIA, Ch. 20, pp. 344–444, New York and London, Academic Press, 1964.
2. EICHLER, O., AND FARAH, A. (EDITORS), "Handbuch der Experimentellen Pharmakologie," Vol. 17 "Ions Alcalino-Terreux, Part 2, Organismes Entiers," (BACQ, Z. N., SUB-EDITOR), Berlin-New York, Springer-Verlag, 1964.

FELIX BRONNER

CALORIC INTAKE AND CALORIC REQUIREMENTS

The daily caloric requirements are affected by many variables including physical activity, environmental temperature, age, sex, radiation and body size and composition. The National Research Council's Food and Nutrition Board have utilized the "reference man or woman" in planning the daily requirements or minimal allowances for individuals.[1] The reference man is 25 years old, weighs 70 kg (154 pounds), lives in a temperate environment of 20°C and performs moderate physical activities. The NRC's daily allowance for the reference man has now been reduced from 3200 kcal/day[1] to 2900 kcal/day.[2] The reference woman is also 25 years of age, weighs 58 kg (128 pounds), is moderately active and lives in a temperate environment of 20°C. Her allowances are now reduced from 2300 to 2100 kcal/day. These allowances have been reduced due to the decreased physical activity of Americans and to the large segment of the population being overweight (Table 1).

Caloric allowances are increased in relation to increased body weight and decreased with an increase in age because of the decrease in metabolic rate and lessened physical activity as one gets older (Table 2). The new NRC allowance has reduced the energy requirement by 5% for each decade between the ages of 35 and 55 years, by 8% for each decade from 55–75 years, and by 10% in the decade after 75 years. As an example, at 55 years of age the allowances are decreased by 10%, and at 65 years they are 18% less than at age 25 years (Table 2).

The caloric requirements are increased with an increase in physical activity. The energy require-

TABLE 1. ARBITRARY EXAMPLES OF ENERGY EXPENDITURE BY REFERENCE MAN AND WOMAN[a]

Activity	Time (hr)	Man		Woman	
		Rate (kcals/min)	Total	Rate (kcals/min)	Total
Sleeping and Lying[b]	8	1.1	540	1.0	480
Sitting[c]	6	1.5	540	1.1	420
Standing[d]	6	2.5	900	1.5	540
Walking[e]	2	3.0	360	2.5	300
Other[f]	2	4.5	540	3.0	360
			2880		2100

[a] This is a reproduction of Table III from the NRC publication No. 1146, Recommended Dietary Allowances, 6th Revised Edition, 1964.
[b] Essentially basal metabolic rate plus some allowance for turning over or getting up or down.
[c] Includes normal activity carried on while sitting, *e.g.*, reading, driving automobile, eating, playing cards and desk or bench work.
[d] Includes normal indoor activities while standing and walking spasmodically in limited area, *e.g.*, personal toilet, moving from one room to another, etc.
[e] Includes purposeful walking, largely outdoors, *e.g.*, home to commuting station to work site, and other comparable activities.
[f] Includes spasmodic activities in occasional sports, exercises, limited stair climbing or occupational activities involving light physical work. This category may include weekend swimming, golf, tennis, or picnics using 5–20 cal/min for limited time.

TABLE 2. ADJUSTMENT OF CALORIE ALLOWANCES FOR ADULT INDIVIDUALS OF VARIOUS BODY WEIGHTS AND AGES[a]

[At a mean environmental temperature of 20°C (68°F) assuming average physical activity]

Desirable Weight		Calorie Allowance[b]		
kg	pounds	25 years	45 years	65 years
	Men	(1)	(2)	(3)
50	110	2300	2050	1750
55	121	2450	2200	1850
60	132	2600	2350	1950
65	143	2750	2500	2100
70	154	2900	2600	2200
75	165	3050	2750	2300
80	176	3200	2900	2450
85	187	3350	3050	2550
	Women	(4)	(5)	(6)
40	88	1600	1450	1200
45	99	1750	1600	1300
50	110	1900	1700	1450
55	121	2000	1800	1550
58	128	2100	1900	1600
60	132	2150	1950	1650
65	143	2300	2050	1750
70	154	2400	2200	1850

Formulas[c]

(1) $725 + 31W$ (2) $650 + 28W$ (3) $550 + 23.5W$
(4) $525 + 27W$ (5) $475 + 24.5W$ (6) $400 + 20.5W$

[a] This is a reproduction of Table II from the NRC publication No. 1146, Recommended Dietary Allowances, 6th Revised Edition, 1964.
[b] Values have been rounded to nearest 50 cal. To convert formulas for weight in pounds, divide factor by 2.2.
[c] W = weight in kg.

ments in a temperate environment for men performing various grades of physical activity can be classified in terms of kilocalories per kilogram of body weight. These values for sedentary to light physical activities range from 35–40 kcal; for light to moderate activities, 43–46 kcal; for moderate activities, 47–50 kcal; for moderate to heavy physical activities, 51–55 kcal; and for very heavy physical activities, 57–68 kcal/kg of body weight.

Effects of Environmental Temperature. Prior to the publication of the new National Research Council's Recommended Dietary Allowances,[2] the general concept had been that the caloric requirements were increased in a cold environment and decreased in a hot environment. This was based on the work of Johnson and Kark[3] who presented information showing that the energy requirements of men were inversely proportioned to the environmental temperature. They concluded that the calorie requirements of man performing moderate physical activities were 3000 kcal/day at 40°C and approximately 5000 kcal/day at −20°C. These authors[3] observed that the increase in requirements in the cold could not be explained in terms of size or different physical activities, but believed that the increased energy requirements in the arctic were due, in part, to the increased needs for maintaining thermal equilibrium and to the binding or "hobbling" effect of the clothing worn. This was later confirmed by Gray *et al.*[4] who measured the energy expenditure of men engaged in hard physical activity in three different environments while wearing standard arctic, temperate, and desert clothing. The results showed that the energy expenditure for a given amount of external work performed at a constant temperature increased approximately 5% when the clothing was changed from desert to temperate attire and approximately 5% more when the clothing was again changed from temperate to arctic wear. These authors regarded the heavy clothing as playing a major role in the increased requirements at the lower environmental temperatures but felt that the requirements under these conditions could not be increased by more than 100 kcal/day. Belding *et al.*[5] also observed that as the bulk of the clothing was increased, there was a greater increase in energy expenditure than could be accounted for by the increased weight of the uniform. As a result they also attributed the extra energy expenditure to the "hobbling" effect of the clothing worn.

On the basis of the study of Johnson and Kark,[3] the recommended dietary allowances of the 1958 NRC's Food and Nutrition Board[1] were adjusted for climatic differences. The old NRC formula[1] employed a base temperature of 20°C and increased the allowances at lower environmental temperatures by 5% for the first 10°C decrease in environmental temperature and an increase of 3% in allowances for each additional 10°C decrease in temperature. The caloric allowances were reduced for higher environmental temperatures by 5% for each 10°C increase in temperature above the base temperature of 20°C. These allowances were actually increased by 14% in men living at environmental temperatures of −20°C and decreased by 10% for men living at 40°C.

These allowances for climatic differences have been the subject of controversy for quite some time. As a result, Welch *et al.*[6] designed a study to obtain information concerning the maximal caloric requirements of 26 men existing on a self-sufficient basis in the arctic and performing hard physical work for extended periods. The men spent 21 days in the field where the mean outdoor temperature was −21.8°C.

During the prebivouac period, the daily caloric intake averaged 3355 kcal/man, or 48 kcal/kg of body weight, for moderate activity. During the 3-week bivouac period the intake was increased to an average of 4196 kcal/man/day (61.2 kcal/kg) (Table 3). This value may be regarded as a maximal value for sustained (more than 5 days) hard work in a cold environment. These men exercised vigorously, marching at least 8 miles/day, in addition to pulling sleds with their field equipment. Each morning the men broke camp and in the evening they made camp at a new site. It seems clear, however, that the food intakes recorded during the bivouac periods in this study were very similar to the actual energy requirement. Although the temperature increased 11.1°C between the prebivouac and bivouac periods, the food intake increased as the men worked harder. In other words, the food intake was not associated with the mean outdoor temperature, but with the activity level. The time per day spent outdoors was somewhat less in the prebivouac than in the bivouac situations since the men slept indoors during this period.

In view of these studies, it would seem that in an extremely cold environment, a caloric requirement of 4200–4500 kcal/man/day should be adequate for men who are working hard and living on a self-sufficient bivouac basis.

It is important to assess the physical activity level for an operation in a given environment prior to establishing the caloric (food) logistics. Although a cold environment has a well-known stimulating effect on the metabolic rate when the body is cooled, body cooling may occur during only a small fraction of the day in the well-clothed man, working and residing in the cold. His "macro" climate (outdoor ambient conditions) is far different from "micro" climate (inside his clothing). In the unusual situation where extensive body cooling results in shivering and involuntary muscular activity, an elevation in metabolic rate occurs, thereby increasing the daily energy requirements.

Welch *et al.*[6] concluded that there was no basis for the idea that the energy requirements should be increased in the cold unless (a) enough time is spent in the cold to make a man shiver, (b) a man is wearing extra-heavy clothing which would impose a resistance to body movement, or (c) an individual wears heavy footgear which would result in an increase in energy expenditure.

The decreased dietary allowances recommended by the NRC[1] made it increasingly apparent that a reevaluation of the energy requirements in

TABLE 3. FOOD CONSUMPTION IN AN EXTREMELY COLD ENVIRONMENT HEAVY PHYSICAL ACTIVITY[a]

Period	Food Consumption		Body Weight Changes
	kcal/day	kcal/kg body wt	
Prebivouac	3355	48.0	70.3
I	3902	56.6	70.2
II	4199	61.6	69.9
III	4488	65.5	69.1
Mean Bivouac Periods	4196	61.2	—

[a] Body weight loss of 1.2 kg for 21 days, which was apparently water, based on water balance and deuterium space calculations.[6]

extremely hot and cold environments was necessary. This was especially true since, due to lack of strict control of many factors, very few of the studies in the literature are comparable.

In 1959, Consolazio *et al.*[7] conducted a strictly controlled study in the extremely hot desert at Yuma, Arizona, in an attempt to answer some of the questions pertinent to the energy requirements of men living and working in the heat. The effects of solar radiation and extreme heat on the energy requirements of men performing a constant daily activity and on an *ad libitum* food and water intake were evaluated from (a) the daily energy expenditure and energy balance, (b) the fluid and nitrogen balances, (c) the sweat rates, (d) the body temperature changes, (e) the daily body weight changes, and (f) changes in body composition.

The study was divided into three 10-day experimental periods, using eight normal, healthy, young adults as test subjects. During Period I the men were kept outdoors in the direct sunlight. In Period II the men were also outdoors (40°C) but were kept in the shade under a large tarpaulin, and during Period III the men were moved indoors into an air-conditioned room (26°C). Food was *ad libitum* but measured, and the daily physical activity was constant. In this study food intake was increased significantly during the hot sun and hot shade periods, when compared to the cool shade period. These daily food intakes in a hot environment were considerably higher than the NRC[1] minimal allowances for men living and working in the heat (Table 4), being equivalent to 55.4, 56.4 and 36.6 kcal/kg body weight for the hot sun, hot shade and air-conditioned environment, respectively. In addition, significant increases were also observed in metabolic rate and body temperature.

Some of the main criticisms of this study[7] were whether the men in this experiment were fully acclimatized to the heat and whether the increase in energy requirements was due to insufficient training prior to the beginning of the experiment. A study was designed to rule out these two factors by measuring the metabolic rate of eight men under three strictly controlled levels of temperature and humidity.[8] The environmental temperatures for the three phases were 21.2, 29.4 and 37.8°C, each of 4 days' duration being repeated four times for a total of 48 days. The relative humidity averaged 30% during the entire study. The daily activity levels were controlled at a constant rate and comparisons of metabolic rates were performed at three activity levels: riding an ergostat at a fairly heavy level for 50 min/day, riding an exercycle at a moderate activity level for 50 min/day, and various resting activity levels.

This study again showed that there was a significant increase in the metabolic rate of men living in the heat. It is of interest that for each man the average at 37.8°C was higher than his average at 29.4°C (Table 5). The significant increases in metabolic rate averaged 7.8% for the light, 12.4% for the moderate, and 11.8% for the heavier activity. It is felt that these results were very important since they showed that neither acclimatization nor training were factors in the increased metabolism and increased energy requirements in the heat.

Other Environmental Effects. During the past year, this laboratory[9] completed an *altitude* study at 3475 meters (11,400 feet) in which food intake patterns and physical performance of young men acclimated to sea level and 1610 meters (5200 feet) were evaluated during a 21-day exposure to altitude. The food intake, corrected for body weight changes did not change appreciably at high altitude. These data suggest that the energy requirements of men living at 11,400-foot altitude are unchanged, in comparison to sea level requirements.

There seems to be no question as to a direct relationship between metabolic rate and *gravity*. As far back at 1935, Crowden[10] reported a marked decrease in oxygen consumption (almost 50%) for a designated exercise in water when compared with the same exercise in air. This decrease was later confirmed by Hill.[11] In another study, Goff *et al.*[12] measured oxygen consumption and observed that the isometric workloads for

TABLE 4. FOOD CONSUMPTION IN AN EXTREMELY HOT ENVIRONMENT, MODERATE PHYSICAL ACTIVITY

	Phase[a]		
	I	II	III
	Hot Sun	Hot Shade	Cool Shade
Food consumption	3560	3516	3156
Body weight change, g/day	+62	+36	+17
Water balance, g	+120	+121	−55
Weight change not due to water retention or loss, g	−58	−85	+38
Protein balance (N × 6.25)	−2.7	−4.2	+8.6
Weight change not due to water or protein (due to fat), g	−55	−81	+47
Caloric equivalent of body weight change, 9 kcal/g	−498	−729	+423
Energy requirement (food intake)	4058	4243	2733

[a] Three consecutive 10-day periods.[7]

TABLE 5. ENERGY EXPENDITURE IN RELATION TO ENVIRONMENTAL TEMPERATURE—OXYGEN USED, LITERS/MINUTE (STPD), MEAN OF 7 MEN

Activity	Temperature Phase		
	21.2°C	29.4°C	37.8°C[a]
Light	0.273	0.282	0.304
Moderate	0.521	0.525	0.590
Moderately heavy	1.422	1.404	1.570

[a] The oxygen consumption was highly significant when comparing the 37.8 to the 29.4°C phase. The light activity was significant at $P < 0.001$, moderate at $P < 0.005$, and moderately heavy at the $P < 0.025$ levels.

exercise and for rest were not significantly different in air and submerged. Data presented by Donald and Davidson[13] suggested that reduced postural effort may account for the near-basal oxygen uptake they observed in sitting submerged subjects in water at 70°F. McCally and Lawton,[14] in their summary, felt that "the relationship between metabolic rates and postural muscular activity suggests that the basal metabolic rate of weightlessness may be closely related to that of recumbency, inactivity or immersion."

Distribution of Calories. With the exception of the work of Sargent *et al.*,[15] very little factual information is available as to the best combination of protein, fat and carbohydrate consumption for maintaining maximum efficiency and well-being in extreme environments. The NRC's Committee on Dietary Allowances[2] has recommended daily allowances of 1 gram of protein per kilogram of body weight. No recommended allowances for fat and carbohydrate intakes are available, since only limited data are available on a reasonable fat allowance and the characteristics of a mixture of fatty acids that would be most favorable to promote good health.

In our studies,[16] we have found that the same proportions of protein, fat, and carbohydrate are consumed *ad libitum* in extreme heat, extreme cold, a temperate environment, and at high altitude (Table 6). It is the general feeling that this distribution of food calories is a matter of food habit and preparation. Americans are accustomed to a high fat intake because of the high economy level of individuals. In countries of lower economic level, the trend is toward a diet high in carbohydrate and low in fat. Rice and wheat products are relatively inexpensive and are the main food items in these countries [see also NUTRITIONAL REQUIREMENTS (MAMMALS)]. Regardless of the environmental temperature, humidity, and altitudes up to 11,400 feet, the distributions of protein, fat and carbohydrate calories in the diet appear to be relatively constant.

TABLE 6. DISTRIBUTION OF CALORIES CONSUMED IN VARIOUS ENVIRONMENTS

Environment	Total Calories Consumed (%)		
	Protein	Fat	Carbohydrate
Temperate	12.2	42.4	45.4
Extremely hot	13.0	38.6	48.4
Extremely cold	14.6	36.6	48.8
Altitude (11,400 ft)	13.1	36.5	50.4

Summary. The new NRC allowances for hot environments suggest that the energy requirements are increased in men performing prescribed work and living at high temperatures. They recommend little or no adjustment of requirements when living at environmental temperatures between 20 and 30°C, but with increased activity the requirements should be increased by at least 0.5% for each degree rise in environmental temperature between 30 and 40°C (5% increase for each 10°C). This means that the caloric requirements for a hot environment have changed from a decrease of 10% at 40°C to an increase of 5%, or an over-all increase of 15%.

The new NRC caloric allowances[2] for extremely cold environments, assuming an adequately clothed individual, are not increased except to compensate for the 2–5% increase in energy expenditure due to carrying or wearing of the cold weather clothing (hobbling effect). Under conditions where the individuals are not adequately clothed, the requirements are increased due to the increased metabolic rate of shivering.

References

1. National Academy of Sciences—National Research Council, Revised Edition, Publication No. 589, Washington, D.C., 1958.
2. National Research Council, "Recommended Dietary Allowances," 6th Revised Edition, Publication No. 1146, National Academy of Sciences, Washington, D.C., 1964.
3. JOHNSON, R. E., AND KARK, R. M., *Science*, **105,** 378–379 (1947).
4. GRAY, E. LEB., CONSOLAZIO, C. F., AND KARK, R. M., *J. Appl. Physiol.*, **4,** 270–275 (1951).
5. BELDING, H. S., RUSSELL, H. D., DARLING, R. C., AND FOLK, G. E., Report No. 37, Report from the Harvard Fatigue Laboratory to the Quartermaster General's Office, Washington, D.C., 1945.
6. WELCH, B. E., LEVY, L. M., CONSOLAZIO, C. F., BUSKIRK, E. R., AND DEE, T., USAMRNL Report No. 202, Denver, Colorado, 1957.
7. CONSOLAZIO, C. F., SHAPIRO, R., MASTERSON, J. E., AND MCKINZIE, P. S. L., *J. Nutr.*, **73,** 126 (1961).
8. CONSOLAZIO, C. F., MATOUSH, L. O., NELSON, R. A., TORRES, J. B., AND ISAAC, G. J., *J. Appl. Physiol.*, **18,** 65–68 (1963).
9. CONSOLAZIO, C. F., MATOUSH, L. O., AND NELSON, R. A., "The Energy Requirements at 3475 Meter Altitude," in press.
10. CROWDEN, G. P., "The Relative Energy Expenditure in Muscular Exercise Performed in Air and Water," (Abstract) *J. Physiol.*, **84,** 31 (1935).
11. HILL, L., "Exercises in Bath," *Brit. Med. J.*, **11,** 1153–1154 (1950).

12. Goff, L. G., *et al.*, "Effect of Total Immersion at Various Temperatures on Oxygen Uptake at Rest and During Exercise," *J. Appl. Physiol.*, **9**, 59–61 (July 1956).
13. Donald, K. W., and Davidson, W. N., "Oxygen Uptake of 'Booted' and 'Fin Swimming' Divers," *J. Appl. Physiol.*, **7**, 31–37 (July 1954).
14. McCally, M., and Lawton, R. W., "Pathophysiology of Disuse and the Problem of Prolonged Weightlessness, A Review," Tech. Doc. Rep. AMRL-TDR-63-3, US Air Force, June 1963.
15. Sargent. F., II, *et al.*, "The Physiological Basis for Various Constituents in Survival Rations," Part I, "The Efficiency of Young Men Under Temperate Conditions," WADC Tech. Rept. 53–484, Part I, US Air Force, 1954.
16. Consolazio, C. F., "Caloric Requirements of Long Flights, Conference on Nutrition in Space and Related Waste Problems," NASA SP70, Apr. 27–30, 1964.

C. Frank Consolazio

CALORIMETRY

Calorimetry, in biochemistry, usually refers to the measurement of rate of heat output from an animal, resulting from the metabolic combustion of food materials within the animal. In *direct calorimetry*, the entire animal is enclosed in the insulated chamber of a calorimeter. In the more convenient *indirect calorimetry*, the rate of O_2 consumption by the animal, or (less commonly) the rate of CO_2 output from the animal, may be measured, as an index of the rate of food oxidation; the exact relation between the measured value and the calories released in food oxidation then depends upon the type of food being utilized, *i.e.*, carbohydrate, lipid, or protein (see BASAL METABOLISM).

Reference

1. White, A., Handler, P., and Smith, E. L., "General Metabolism," in "Principles of Biochemistry," Third edition, Ch. 16, pp. 280–292, New York, McGraw-Hill Book Co., 1964.

CALVIN, M.

Melvin Calvin (1911–   ), University of California chemist, is noted primarily for his investigations of PHOTOSYNTHESIS for which he received the Nobel Award in Chemistry in 1961. He combines competence in organic chemistry, physical chemistry, and biochemistry.

CANCER CELL METABOLISM

The biochemistry of cancer has challenged the minds of scientists for over six decades, and although there is agreement as to the general nature of the chemical changes involved, there is no agreement as to the precise details. This is probably because the same can be said about the processes of CELL DIVISION and DIFFERENTIATION.

Cancer biochemistry has always been dependent upon two main streams of scientific development. One has been the experimental biology of cancer, in which the ability to produce experimental neoplasms by various chemicals, radiations, and viruses, and to transmit the resulting tumors from one animal to another has constantly enlarged the possible scope of chemical studies. The other necessary development has been the flow of new knowledge about the fundamental biochemical processes of metabolism, the chemistry and function of the nucleic acids, and the mechanisms of regulation of enzyme synthesis and enzyme activity.

The best-known biochemical theory of cancer was developed by Otto Warburg in Berlin in the period between 1924 and 1930. He stated categorically at that time and reiterated in 1956[1] his conclusion that cancer cells originate from normal cells as a result of an initial irreversible injury to respiration, which is followed by a changeover to an anaerobic or fermentative type of energy production. This conclusion was not really a theory of cancer since it never attempted to explain the connection between the proposed description of cancer cells and the malignant properties. It was based on what was believed to be a universal correlation between high rates of GLYCOLYSIS and the malignant property.

Opposition to the Warburg view was renewed when C^{14}-labeled substrates became available, and Weinhouse[2] in particular emphasized the lack of evidence to support the conclusion that any respiratory enzyme was damaged, altered, or causally related to increased glycolysis or to malignancy. It was generally agreed that malignant tumors had a high aerobic and anaerobic glycolysis but the mechanism of how this was related to preceding or following events was not explained, and indeed the obligation to explain the mechanism was denied.[3] It may be mentioned that the Warburg position was based on studies with tissue slices and more recently on suspensions of intact ascites or tissue culture cells and not on attempts to examine individual enzyme activities in cell extracts.

The latter approach was emphasized by Greenstein,[4] who noted a tendency for tumors to resemble each other more than a variety of normal cells resembled each other, and in fact to tend toward lower values for a variety of enzymes including the enzymes of aerobic processes. This was the so-called Convergence Theory and again it was not a theory of mechanism but an attempted description of cancer tissue in terms of individual enzymes. Insofar as it seemed to indicate a lowered amount of respiratory enzymes, it provided some support for the Warburg theory. Some clarification was provided in 1961 when Aisenberg[5] showed that the oxidative PHOSPHORYLATION in mitochondria from glycolyzing tumors was normal but the quantity of mitochondria per cell was lower than in some normal tissues.

By 1950 the expansion of knowledge of metabolic pathways led to the concept of alternative metabolic pathways and to the realization that no single network of converging and diverging metabolic pathways could be regarded as essential to life. Certain metabolites were essential but the manner in which they were obtained could vary from one cell type to another. Any given metabolite could be a substrate for more than one enzyme and the fate of the metabolite could be controlled by regulating the balance between competing enzymes.[6] This over-all view greatly enlarged the range of biochemical studies on cancer and permitted the elaboration of the Catabolic Enzyme Deletion Hypothesis of cancer formation, according to which the loss of strategic catabolic enzymes decreased the competition for certain building blocks for nucleic acids and thereby permitted growth to proceed.[7] Although numerous studies support the idea of the deletion of a variety of catabolic enzymes, it was realized that the problem of suitable control tissue for comparison with cancer cells had never been satisfactorily solved.[7] Accordingly, a search was made for hepatoma cells that might resemble normal parenchymal liver cells so closely that they could be compared with them.[7] This search was successful and led to the discovery of the Morris Hepatoma 5123, and subsequently to the Reuber Hepatoma H35, and a series of Morris Hepatomas, 7316, 7800, 7793, 7794, and 7787, among others.[8] These tumors have been referred to as "minimal deviation hepatomas" because they were found to contain all of the catabolic enzymes that had earlier been lacking in the fast-growing hepatomas such as the Novikoff hepatoma, even though this tumor was believed to be derived from parenchymal liver cells.

It now became possible to apply to minimal deviation hepatomas the earlier suggestion by Weber and Cantero[9] that many of the biochemical changes described in cancer literature might be concomitant and nonessential to the malignant property. It was found that many enzymes that were missing in some tumors were present in one or another minimal deviation tumor and could therefore not be regarded as a deletion that was essential to the cancer process.[8] By the same test, Aisenberg and Morris[10] concluded that neither aerobic nor anaerobic glycolysis nor the ability to incorporate amino acids into protein under anaerobic conditions could be considered as an essential feature of malignancy. These findings were of course relevant to the Warburg Theory and it would appear that the development of a respiratory defect can no longer be looked upon as the initial step in carcinogenesis, whether or not malignant cells develop an increasing rate of glycolysis as they progress from slower to faster growth rates.

Meanwhile it was recognized that the biochemistry of cancer involves not only the cancer itself but the host organism in which the cancer occurs. Jesse Greenstein[4] said, "The host-tumor relationship is the key to the cancer problem," and this may be generalized to the normal animal to suggest that the body: cell relationship is the key to the cancer problem. In a normal multicellular organism, every cell in the body is buttressed against its external environment by the presence of other cells and by a fluid medium that communicates with the blood, which in turn brings in nutrients and carries away waste materials. This is true even in the cells in the outermost layers of the skin or the lining of the intestine, since even these cells are backed up by other cells. In addition to carrying nutrients and waste materials the fluid medium is a communications channel, in which specific signals that are sent or received are in chemical terms, *i.e.*, in the form of precisely structured molecules that can increase or decrease specific chemical activities in the cells that they reach. Many of these information-carrying molecules are recognized HORMONES, but others are undoubtedly still uncharacterized. These considerations are important in the case of cancer cells because in contrast to the normal cells they do not cooperate in the normal organismic feedback processes that make the integrated activity of the organism possible. It seems clear that the outstanding feature of cancer cells is their failure to obey faithfully the signals coming in from the normal cells in the body. Furth[11] in particular has emphasized abnormal feedback relationships between tumors and host tissues. Feedback relationships can be studied in populations of cells in terms of changes in enzyme amount or enzyme activity and in individual cells by suitable autoradiographic techniques. Efforts are now being made to study the minimal deviation tumors in terms of a rather widely accepted hypothesis that there are defects in systems responsible for negative feedback on specific enzymes or enzyme-forming systems required for cell division and systems responsible for structural relations between contiguous cells.[12] It is still unknown whether the defects are at the genetic or the epigenetic level but the present rate and direction of research should provide an answer to this question. It appears that the feedback hypothesis can provide a focus for the effect of cancer-producing viruses as well as for the effect of chemical and physical carcinogens [see also METABOLIC CONTROLS (FEEDBACK MECHANISMS)].

Numerous studies show that there are many cell genomes that are malignant, and hopes for a single effective cancer cure have long since been modified because of the certainty that the evolution of drug resistance is possible in a population of cancer cells. Future research will involve both cancer prevention and cancer cure, and both chemical and immunologic studies will continue to be needed in order to build up as wide a variety of rational countermeasures as possible.

References

1. WARBURG, O., *Science*, **123**, 309 (1956); **124**, 269 (1956).
2. WEINHOUSE, S., *Science*, **124**, 267 (1956).
3. BURKE, D., AND SCHADE, A. L., *Science*, **124**, 270 (1956).
4. GREENSTEIN, J. P., *Cancer Res.*, **16**, 641 (1956).
5. AISENBERG, A. C., *Cancer Res.*, **21**, 295, 304 (1961).

6. POTTER, V. R., AND HEIDELBERGER, C., *Physiol. Rev.*, **30,** 487 (1950).
7. POTTER, VAN R., in "The Molecular Basis of Neoplasia," p. 367, The University of Texas M. D. Anderson Hospital and Tumor Institute, Austin, University of Texas Press, 1962.
8. POTTER, VAN R., in "Cellular Control Mechanisms and Cancer," (EMMELOT, P., AND MÜHLBOCK, O. EDITORS), p. 190, Amsterdam, Elsevier Publishing Co., 1964.
9. WEBER, G., AND CANTERO, A., *Cancer Res.*, **15,** 679 (1955).
10. AISENBERG, A. C., AND MORRIS, H. P., *Cancer Res.*, **23,** 566 (1963).
11. FURTH, J., *Cancer Res.*, **23,** 21 (1963).
12. POTTER, V. R., *Cancer Res.*, **24,** 1085 (1964).

VAN R. POTTER

CANCER CHEMOTHERAPY

Cancer chemotherapy has as its objective the control of cancer by the administration of specific chemical compounds to the tumor-bearing host. An effective cancer chemotherapeutic agent must then, when administered to the host, reach and penetrate the cancer cells in whatever anatomical site they may exist and destroy such cells or inhibit their growth without undue toxicity to the many types of normal cells of the host. The difficulties and problems peculiar to cancer chemotherapy are implicit in the following general statements: (a) Cancer is not a single disease but a number of closely related diseases; hence various types of neoplasms may differ markedly in their response to drugs. (b) Cure of the disease demands the destruction of all cancer cells [as evidenced by the demonstration in a model system (mouse leukemia) that one leukemic cell is ultimately lethal]. (c) The cancer cell is not an invading organism but is derived from a host cell and hence is generally similar biochemically to the cell type of origin.

Concerted effort in cancer chemotherapy has been largely a development of the last twenty years. Most research carried out in this area can be divided broadly into four parts: (a) biochemical comparisons of cancer cells and normal cells in a search for metabolic differences that might be exploited by the design of specific inhibitors; (b) the systematic synthesis and assay of potential antimetabolites based on present knowledge of biochemical pathways and of structure-activity relationships among inhibitors; (c) the empirical search for new classes of inhibitors among synthetic compounds and material of plant, animal, and microbial origin; (d) study of the mechanisms of action of known anticancer agents. A number of experimental systems are available for the assay of candidate anticancer agents; these include transplantable rodent tumors and leukemias, tumors that can be induced in animals by the administration of certain CARCINOGENS, strains of animals subject to high incidence of certain spontaneous tumors, and animal and human tumor cells in culture.

Classes of Anticancer Agents. A large number of compounds are known that have marked effects on one or more experimental tumors, and a considerable number of these have found some clinical use. The major classes of clinically useful agents, together with respresentative examples and some indication of the human neoplasms responding, are shown in Table 1. The greatest chemotherapeutic success has been with a relatively rare tumor, choriocarcinoma: about 40% of choriocarcinoma patients treated with methotrexate showed seemingly "permanent" regression. This tumor is the only example of a "cure" provided by chemotherapy, but impressive increases in the life span of leukemia patients have been obtained with cortisone, 6-mercaptopurine, and methotrexate, usually administered in sequence (see LEUKOCYTES AND LEUKEMIA). Note should be made of possible variations and special techniques in the use of chemotherapeutic agents such as (a) the use of combinations of agents in attempts to produce synergism or to prevent development of resistant cells; (b) the use of chemotherapeutic agents as an adjuvant to surgery or in combination with radiation; (c) regional chemotherapy by various techniques of infusion and perfusion.

Limitations of Cancer Chemotherapy. The following several factors limit or are suspected to limit the usefulness of presently known anticancer agents: (a) insufficient selective toxicity of the agent for cancer cells, (b) development of drug-resistant cell populations, (c) sequestration of cancer cells at anatomical sites at which they are not readily reached by the circulating agent. The first of these factors is probably the most important. The most effective known agents have at best only a slight differential toxicity to cancer cells and, at doses the same as or only slightly higher than the minimal effective tumor-inhibitory doses, produce marked effects on rapidly proliferating host tissues such as intestinal mucosa and bone marrow. The clinical significance of the possible development of resistant cell populations during chemotherapy cannot be estimated at present; in experimental animal systems, resistance to certain antitumor agents develops rapidly, and the usual clinical treatment, namely chronic therapy with low doses, is precisely that expected to lead to the development of resistant cell populations. An example of the operation of the third factor is the infiltration of leukemic cells into the central nervous system where they are not reached by most agents in clinical use because these agents do not pass the "blood-brain barrier."

Biochemical Aspects. Research in cancer chemotherapy has followed closely progress and trends in general biological sciences and particularly in biochemistry. Effective cancer chemotherapy obviously depends upon the existence of qualitative or significant quantitative metabolic differences between cancer cells and host normal cells. The search for metabolic profiles unique to, or characteristic of, cancer cells is complicated by incomplete knowledge of normal growth and its

TABLE 1. MAJOR CLASSES OF CANCER CHEMOTHERAPEUTIC AGENTS, WITH SELECTED EXAMPLES AND INDICATIONS OF CLINICAL USE.

I. Alkylating Agents (Hodgkins' Disease, leukemia, multiple myeloma, lymphosarcoma, carcinoma of lung, ovary, breast, and other tumors)

A. N-Bis(2-chloroethyl)compounds

$CH_3N(CH_2CH_2Cl)_2$ Nitrogen mustard

$HOOC(CH_2)_3$-pC_6H_4-$N(CH_2CH_2Cl)_2$ Chlorambucil

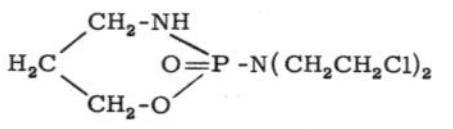

$H_2C(CH_2\text{-}NH)(CH_2\text{-}O)$ O=P-$N(CH_2CH_2Cl)_2$ Cyclophosphamide

B. Ethylenimines

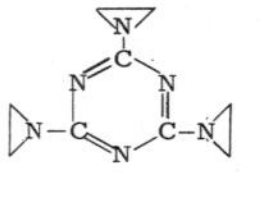

Triethylenemelamine (TEM)

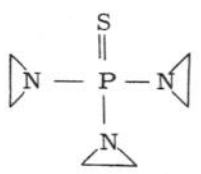

Triethylenethiophosphoramide (Thio-TEPA)

C. Bis Sulfonic Esters

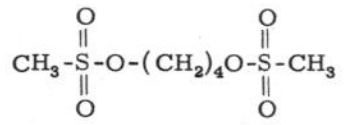

CH_3-S(=O)(=O)-O-$(CH_2)_4$O-S(=O)(=O)-CH_3

Myleran

D. N-nitrosoureas

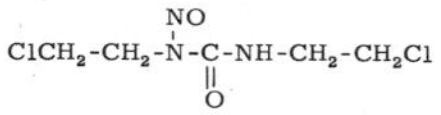

$ClCH_2$-CH_2-N(NO)-C(=O)-NH-CH_2-CH_2Cl

1,3-Bis(2-chloroethyl)-1-nitrosourea

II. Antimetabolites

A. Antifolics (Leukemia, choriocarcinoma)

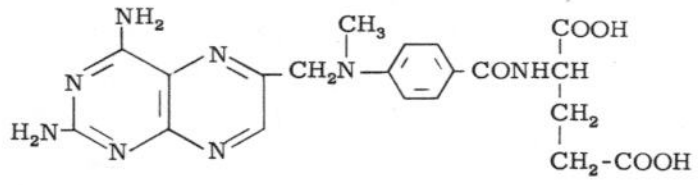

Methotrexate (A-methopterin)

B. Antipurines (Leukemia)

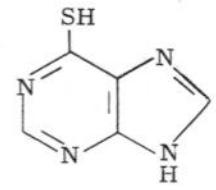

6-Mercaptopurine

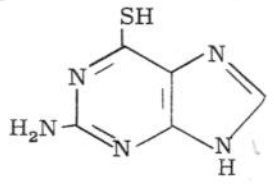

6-Thioguanine

C. Antipyrimidines (Various carcinomas and sarcomas)

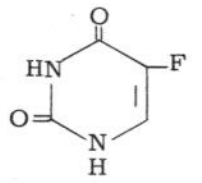

5-Fluorouracil

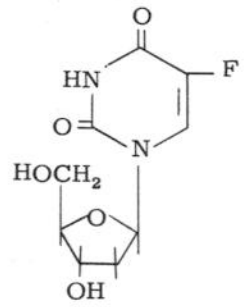

5-Fluorodeoxyuridine

III. Hormones

Prednisone, Cortisone (Leukemia)

Androgens, Estrogens (Various tumors of sex organs)

IV. Miscellaneous

A. Actinomycin (Wilm's Tumor)

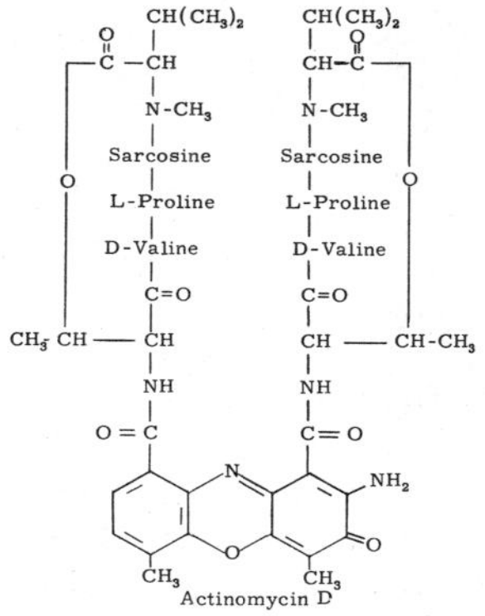

Actinomycin D

B. Vinca Alkaloids (Leukemia, Hodgkin's disease)

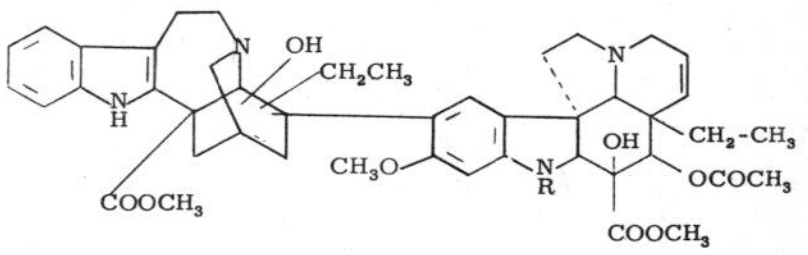

Vinblastine, R = CH_3; Vincristine, R = CHO

C. o,p'-DDP (Carcinoma of adrenal)

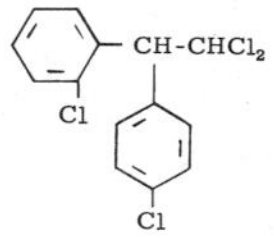

1,1-Dichloro-2-(o-chlorophenyl)-2-(p-chlorophenyl)ethane

D. Methylglyoxal Bis(guanylhydrazone) (Leukemia)

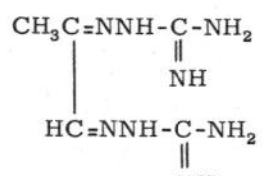

CH_3C=NNH-C(=NH)-NH_2 | HC=NNH-C(=NH)-NH_2

E. Urethan (Multiple myeloma, leukemia)

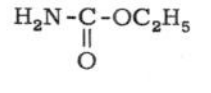

H_2N-C(=O)-OC_2H_5

regulation, the complexity of metabolism in all cells, the difficulty of determining which biochemical events represent the primary biochemical lesion and which secondary changes and, experimentally, the paucity of systems providing a biochemical comparison of the metabolism of a tumor cell and its normal cell of origin. Although the systematic biochemical comparison of cancer cells and normal cells appears to offer the only hope for establishment of a rational basis for chemotherapy, there is at present no knowledge of the primary biochemical lesion responsible for the initiation of any neoplasm (see also CANCER CELL METABOLISM).

Other biochemical problems germane to cancer chemotherapy are an understanding, in molecular terms, of the sites of action of known agents, the reasons for their selectivity, however slight, against neoplastic cells, and the biochemical basis for development of resistance. Compounds with activity against tumors embrace a variety of structures (Table 1); specific sites of action of many of these have been determined, while those of others are yet obscure. Alkylating agents are highly reactive compounds capable of alkylating a number of biologically important molecules; the alkylation of the 7-position of the guanine moiety of DNA has been stressed as a critical site of action. Methotrexate, by competition with folic acid for folic acid reductase, lowers the production of tetrahydrofolic acid, a requisite cofactor for many reactions including the synthesis of thymidylate and of purine nucleotides (see FOLIC ACID COENZYMES). 5-Fluorouracil and 5-fluorodeoxyuridine are converted to 5-fluorodeoxyuridylic acid, an inhibitor of thymidylate synthetase (see PYRIMIDINE BIOSYNTHESIS). 6-Mercaptopurine is converted intracellularly to the nucleotide; among the effects of the latter is the inhibition of synthesis *de novo* of purine nucleotides (see PURINE BIOSYNTHESIS). Actinomycin, by complex formation with the deoxyguanylic acid moiety of DNA, inhibits the replication of RNA on the DNA template; complex formation with macromolecules is also a suspected action of methylglyoxal bis(guanylhydrazone). All of these agents for which sites of action are known appear to affect the synthesis or function of nucleic acids; such action is consistent with the mutagenic activity of cancer chemotherapeutic agents in general. There is little or no information on the biochemical basis for the selective action of any of these agents against cancer cells.

Resistance mechanisms to several classes of agents have been elucidated in experimental systems. These include (a) failure to activate the inhibitor (purine antagonists); (b) degradation of inhibitor (purine antagonists); (c) increased production of an enzyme binding the inhibitor (methotrexate-folic acid reductase); (d) altered enzyme affinity for the inhibitor (methotrexate-folic acid reductase); and (e) decreased rate of entry into cell (methotrexate). The clinical significance of any of these mechanisms remains to be established.

Current Trends and Prospects. Because of the multiplicity of types of neoplasms and their differences in response to drugs, it is improbable that any single agent effective against all cancers will be discovered; it is rather to be anticipated that cancer chemotherapy will progress stepwise, with the control of one group of neoplasms at a time. In addition to the areas of research already mentioned, current concepts on the viral etiology of cancer, on feedback control and repression, and other developments of biochemistry and related areas, offer possibilities of new rationales for chemotherapy. It is also not unlikely that further study of the kinetics of the growth and death of neoplastic cells in the presence of drugs may result in new knowledge that will suggest more effective uses of presently known drugs in the control of cancer.

The following general references pertain in whole or (in the case of books) in selected chapters to cancer chemotherapy.

References

1. BUSCH, H., "An Introduction to the Biochemistry of the Cancer Cell," New York, Academic Press, 1962.
2. "Symposium on Problems Basic to Cancer Chemotherapy," *Cancer Res.*, **23,** 1181–1497 (1963).
3. "Advances in Cancer Research," Vols. 1–8, New York, Academic Press, 1953–1964.
4. KARNOFSKY, D., AND CLARKSON, B. D., *Ann. Rev. Pharmacol.*, **3,** 357–428 (1963).
5. GOLDIN, A., AND HAWKING, F., (EDITORS), "Advances in Chemotherapy," Vol. I, New York, Academic Press, 1964.
6. SKIPPER, H. E., "Perspectives in Cancer Chemotherapy: Therapeutic Design," *Cancer Res.*, **24,** 1295–1302 (1964).

L. L. BENNETT, JR.

CARBAMYL PHOSPHATE

Carbamyl phosphate, $NH_2\overset{\overset{\displaystyle O}{\|}}{C}OPO_3^{2-}$,[1] appears to be universally required for the biosynthesis of citrulline from ornithine and for the biosynthesis of carbamyl aspartate from aspartate.[1–3] Citrulline is a precursor of arginine which is utilized for protein synthesis as well as for urea production in ureotelic animals.[2] Carbamyl aspartate is required for the biosynthesis of the pyrimidine nucleotides via the orotic acid pathway[3] (see ARGININE-UREA CYCLE; PYRIMIDINE BIOSYNTHESIS).

Carbamyl phosphate is a labile compound whose half-life at 37°C is 40 minutes of physiological pH. This short half-life precludes using carbamyl phosphate in bacterial or tissue culture media unless results can be obtained after a short incubation (*i.e.*, 1 hour or less). The products of its decomposition at physiological pH are phosphate and cyanate.[4] Cyanate can react chemically with amines and alcohols so that it may be essential to have appropriate controls for nonenzymatic reactions.

Biosynthesis. The biosynthesis of carbamyl phosphate can be catalyzed by one of three distinct enzymes which utilize ATP as the source of the energy-rich phosphoryl group of carbamyl phosphate. These enzyme reactions are:

(1) Carbamyl phosphokinase (found in *Streptococcus faecalis*[4] and *Neurospora*[5]:

$$NH_2CO_2^- + ATP^{4-} \underset{}{\overset{Mg^{2+}}{\longleftarrow\!\!\longrightarrow}} NH_2CO_2PO_3^{2-} + ADP^{3-}$$

(2) Carbamyl phosphate synthetase I (ammonia-specific found in ureotelic vertebrates[2] and the earthworm):

$$NH_4^+ + HOCO_2^- + 2ATP^{4-} \xrightarrow{Mg^{2+},\ N\text{-acetyl-L-glutamate}} NH_2CO_2PO_3^{2-} + $$
$$+ 2ADP^{3-} + HOPO_3^{2-} + 2H^+$$

(3) Carbamyl phosphate synthetase II (glutamine enzyme found in basidiomycetes,[6] *Escherichia coli*, and yeast):

$$R\overset{O}{\overset{\|}{C}}NH_2 + HOCO_2^- + 2ATP^{4-} + H_2O \xrightarrow{Mg^{2+}} NH_2CO_2PO_3^{2-} + 2ADP^{3-} + HOPO_3^{2-} + RCOO^- + 2H^+$$
$$R = COOH\text{—}CHNH_2\text{—}(CH_2)_2\text{—}$$

The stoichiometry noted above has been demonstrated for reactions (1) and (2). For reaction (3) it has been found[6a] that two moles of ATP are required as in reaction (2). Ammonia can replace glutamine as N-donor for reaction (3), but it is a less efficient substrate.[6a,6b,10]

The carbamyl phosphokinase reaction is the simplest reaction of the three and is readily reversible. The substrate, carbamate, is formed from the gases NH_3 and CO_2, which are in equilibrium with ammonium bicarbonate in solution. The synthesis of carbamate from ammonium bicarbonate can be catalyzed by carbonic anhydrase. The long arrow, pointing to the left, indicates that the equilibrium position of the reaction favors ATP production. The equilibrium position is toward ATP and carbamate because the phosphoryl bond of carbamyl phosphate has a free energy of hydrolysis 1.8 kcal, greater than that for ATP. In *Streptococcus faecalis*, this enzyme is one of three induced enzymes formed when the organism is grown in the presence of arginine, which allow this microorganism to utilize arginine as an energy source (*i.e.*, a source of ATP).[1] In *Neurospora*,[5] this enzyme is utilized for the biosynthesis of arginine.

The carbamyl phosphate synthetase I utilizes bicarbonate as substrate[1] and efficiently fixes bicarbonate and, more particularly, ammonia at low concentrations. This efficient fixation of ammonia is important to vertebrates for whom ammonia is a toxic agent (see also AMMONIA METABOLISM). The over-all reaction is irreversible. The first mole of ATP is probably required for the dehydration of bicarbonate,[1,2,7] and the oxygen so removed enters the orthophosphate produced from this ATP. The second mole of ATP provides the phosphate of carbamyl phosphate. An unusual feature of the reaction is its nearly absolute requirement[8] for N-acetyl-L-glutamate which appears to be an allosteric cofactor (see discussion in reference 7). This enzyme, primarily present in the mitochondria of ureotelic vertebrate liver, increases in response to nitrogen intake of rats.[9]

The reaction catalyzed by glutamine-dependent carbamyl phosphate synthetase II is probably irreversible since the glutamine-amide bond is hydrolyzed. Preliminary evidence obtained from studies with intact mouse Ehrlich ascites cells[10] suggests that this enzyme (and not carbamyl phosphate synthetase I) is utilized to provide carbamyl phosphate for PYRIMIDINE BIOSYNTHESIS in this mammalian cell.

It would seem, therefore, that carbamyl phosphokinase would be found in organisms which degrade carbamyl or guanido compounds, such as arginine or citrulline, creatinine and allantoin,[1] to yield energy. Its importance for the biosynthesis of arginine and pyrimidine may be more limited, and it would not be expected to be so utilized for organisms where ammonia is a toxic substance. For biosynthesis, it is less obvious whether carbamyl phosphate synthetase I or II would be preferred, and information now available as to the distribution of these two enzymes is insufficient for generalization.

References

1. JONES, M. E., *Science*, **140**, 1373 (1963).
2. COHEN, P. P., in "The Enzymes" (BOYER, P. D., *et al.*, EDITORS), Vol. 6, p. 477, New York, Academic Press, 1962.
3. REICHARD, P., *Advan. Enzymol.*, **21**, 263 (1959).
4. JONES, M. E., AND LIPMANN, F., *Proc. Natl. Acad. Sci. U.S.*, **46**, 1194 (1960).
5. DAVIS, R. H., *Biochim. Biophys. Acta*, **107**, 44 (1965).
6. LEVENBERG, B., *J. Biol. Chem.*, **237**, 2590 (1962).
6a. ANDERSON, P. M., AND MEISTER, A., *Biochemistry*, **4**, 2803 (1965).
6b. KALMAN, S. M., DUFFIELD, P. H., AND BRZOZOWSKI, T., *Biochem. Biophys. Res. Comm.*, **18**, 530 (1965).
7. JONES, M. E., "Amino Acid Metabolism," *Ann. Rev. Biochem.*, **34** (1965).

8. GRISOLIA, S., AND COHEN, P. P., *J. Biol. Chem.*, **204**, 753 (1953).
9. SCHIMKE, R. T., *J. Biol. Chem.*, **237**, 459, 1921 (1962).
10. HAGER, S. E., AND JONES, M. E., *J. Biol. Chem.*, **240**, 4556 (1965).

MARY ELLEN JONES

CARBOHYDRATE METABOLISM

Cells utilize carbohydrates as a source of energy and as precursors for the manufacture of many of their structural and metabolic components. In the mammal, for example, D-glucose is the carbohydrate primarily used for this purpose. Certain microorganisms, on the other hand, can grow on a medium containing some other hexose or a pentose as the principal source of carbon. Green plants obtain their carbohydrates by photosynthesis while animals receive most of their carbohydrates by ingestion and digestion.

The complete oxidation of glucose to CO_2 and H_2O yields 689 kcal of heat per mole of glucose. When this oxidation occurs in a cell, the energy is not all dissipated as heat. Some of the evolved energy is conserved in biochemically utilizable form of "high-energy" phosphates such as adenosine triphosphate (ATP) and guanosine triphosphate (GTP) (see PHOSPHATE BOND ENERGIES; RESPIRATION). In addition to enzymes concerned with energy metabolism, there are enzymes in biological systems which catalyze the transformation of glucose into various carbohydrates, fatty acids, steroids, amino acids, nucleic acid components and other necessary biochemical substances. The entire network of reactions involving compounds which interconvert carbohydrates constitues *carbohydrate metabolism*. By convention, some reactions involving compounds which are not carbohydrates but which are derived from carbohydrates, are also included in this area of metabolism.

Anaerobic Oxidation of Glucose. Historically, the first system of carbohydrate metabolism to be studied was the conversion by yeast of glucose to alcohol (*fermentation*) according to the following equation: $C_6H_{12}O_6 \rightarrow 2CH_3CH_2OH + 2CO_2$. The biochemical process is complex involving the successive catalytic action of 12 enzymes (the *Embden-Meyerhof pathway*) as illustrated in Fig. 1.

In order for the cell to carry out a "controlled" oxidation of D-glucose and conserve some of the energy derived from the process, it appears necessary to first add phosphate to the hexose with the expenditure of energy. As shown, the necessary energy and, indeed, the phosphate itself is supplied by ATP in two separate reactions of the system. Since each molecule of glucose can yield two molecules of triose phosphate for oxidation, the conversion of glucose to pyruvic acid nets two molecules of ATP per molecule of hexose utilized. Approximately 30% of the evolved energy is conserved as ATP, but only about 8% of the total energy in glucose is made available in this anaerobic oxidation of glucose to pyruvic acid. Since NICOTINAMIDE ADENINE DINUCLEOTIDE (NAD^+),* which is involved in the oxidation of glyceraldehyde-3-phosphate, is present in the cell in small quantities only, this coenzyme must constantly be regenerated for the oxidative process to continue. This regeneration is accomplished by the reduction of *acetaldehyde* to *ethanol*. Since oxygen plays no role in this process, the system can obviously proceed anaerobically. In fact, the presence of oxygen decreases the net disappearance of glucose (the *Pasteur effect*). Fermentation occurs in many microorganisms but not all organisms reoxidize the reduced nicotinamide adenine dinucleotide (NADH) through the formation of ethanol. In certain organisms, for example, *pyruvic acid* is converted to *acetoin* which is then reduced with NADH to 2,3-butylene glycol. In other organisms and in animal tissues, NADH is oxidized in the reduction of *pyruvic acid* to *lactic acid*. In insects, and possibly in some animal tissues as well, the reduction of *dihydroxyacetone phosphate* to *α-glycerol phosphate* may serve to regenerate NAD^+. The conversion of glucose to lactic acid in animal tissues was named *glycolysis* in the mistaken impression that this process was markedly different from the microbial fermentative process. Fermentation and glycolysis are now known to differ primarily in the further anaerobic utilization of pyruvic acid (see ALCOHOLIC FERMENTATION; also the separate articles on FERMENTATIONS and GLYCOLYSIS).

Aerobic Oxidation of Pyruvic Acid. Pyruvic acid can be oxidized completely to CO_2 and H_2O in a cyclic enzymatic system called the *Krebs* CITRIC ACID CYCLE or *tricarboxylic acid cycle*. In this system (Fig. 2), a two-carbon unit in the form of acetyl coenzyme A, derived from the NAD^+-mediated oxidative decarboxylation of pyruvic acid (see MULTIENZYME COMPLEXES) in the presence of COENZYME A (CoA), is condensed with oxalacetic acid to form citric acid. This tricarboxylic acid is then converted back to oxalacetic acid in a stepwise manner with the formation of $2CO_2$ and $2H_2O$. In addition to this formation of CO_2, 1 reduced nicotinamide adenine dinucleotide phosphate (NADPH),† 2 NADH, 1 reduced flavin, and 1 GTP arise per two-carbon unit oxidized in the cycle. Since in the aerobic oxidation of the reduced flavin and the reduced nicotinamide adenine nucleotides, ATP is formed, the oxidation of a molecule of "acetate" results in the conservation of energy in the form of 12 molecules of triphosphate. In the complete oxidation of glucose through glycolysis and the citric acid cycle, about 40% of the energy originally present in the glucose can be retained as triphosphate. The ubiquitous distribution of this cycle in nature, suggests that the citric acid cycle is the major energy-yielding pathway in biological systems. Certain microorganisms have a modification of this cycle in which isocitric acid is cleaved to

* Another designation for this substance is diphosphopyridine nucleotide (DPN^+).

† Also called reduced triphosphopyridine nucleotide (TPNH).

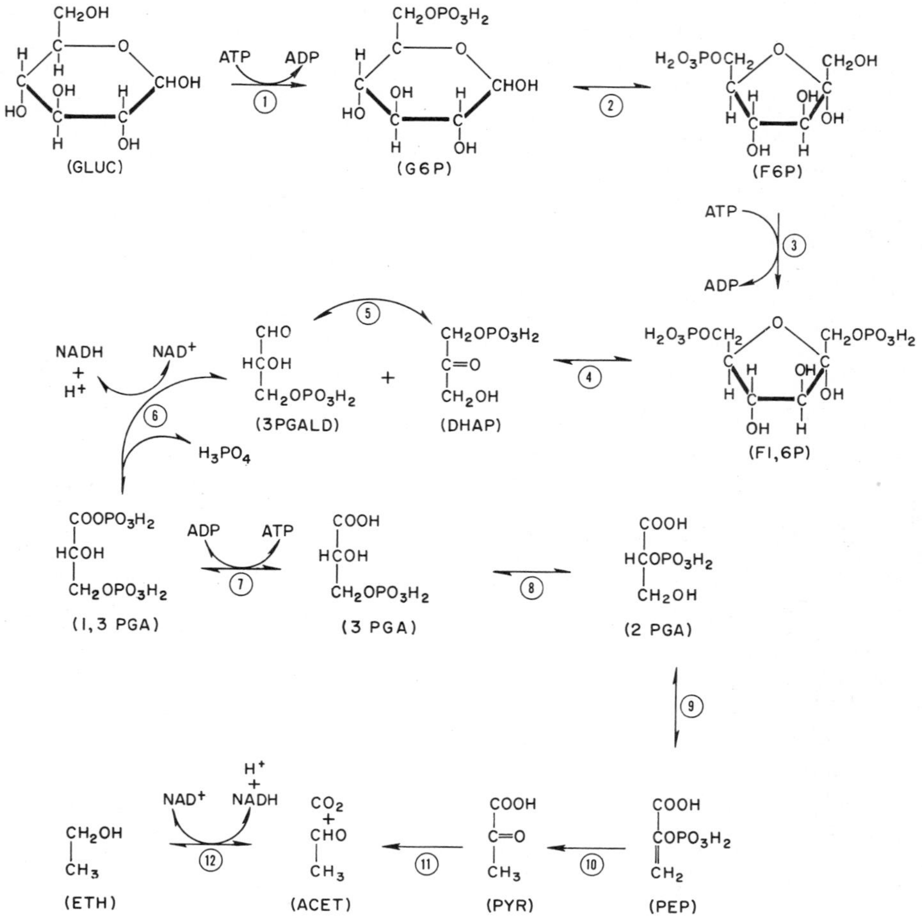

FIG. 1. Fermentation of glucose to ethanol. *Enzymes involved*: (1) hexokinase; (2) phosphohexoisomerase; (3) phosphofructokinase; (4) aldolase; (5) triosephosphate isomerase; (6) triose phosphate dehydrogenase; (7) 3-phosphoglyceric acid kinase; (8) phosphoglyceric acid mutase; (9) enolase; (10) phosphoenolpyruvic acid kinase; (11) carboxylase, and (12) alcohol dehydrogenase. *Abbreviations include*: gluc = glucose; G6P = glucose-6-phosphate; F6P = fructose-6-phosphate; F1,6P = fructose-1,6-diphosphate; DHAP = dihydroxy-acetone phosphate; 3 PGALD = glyceraldehyde-3-phosphate; 1,3PGA = 1,3-diphosphoglyceric acid; 3PGA = 3-phosphoglyceric acid; 2PGA = 2-phosphoglyceric acid; PEP = phosphoenolpyruvic acid; PYR = pyruvic acid; ACET = acetaldehyde; and ETH = ethanol.

succinic acid and glyoxylic acid. The latter acid is condensed with acetyl-CoA to form malic acid. In this modification (the *glyoxylic acid cycle*), oxalsuccinic acid and α-ketoglutaric acid are not involved. (This "glyoxylate shunt" pathway is indicated in more detail in the article CITRIC ACID CYCLE.)

Since in the citric acid cycle there is no net production of its intermediates, mechanisms must be available for their continual production. In the absence of a supply of oxalacetic acid, "acetate" cannot enter the cycle. Intermediates for the cycle can arise from the carboxylation of pyruvic acid with CO_2 (*e.g.*, to form malic acid), the addition of CO_2 to phosphoenolpyruvic acid to yield oxalacetic acid, the formation of succinic acid from propionic acid plus CO_2, and the conversion of glutamic acid and aspartic acid to α-keto-glutaric acid and oxalacetic acid, respectively (see CARBOXYLATION ENZYMES; TRANSAMINATION ENZYMES).

The utilization of carbohydrate intermediates for the biosynthesis of amino acids, fatty acids, steroids, etc., takes place at various stages of the cycle and its related reactions. Some of these transformations are indicated in Fig. 3.

Other Carbohydrate Interconversions. Two systems (Fig. 4) are available for the synthesis of ribose-5-phosphate, a precursor of the pentose moiety of RIBONUCLEIC ACID, ATP and other substances. The formation of ribose-5-phosphate from glucose-6-phosphate by formation and decarboxylation of 6-phosphogluconic acid and isomerization of the resulting ribulose-5-phosphate is called the *hexose monophosphate oxidative*

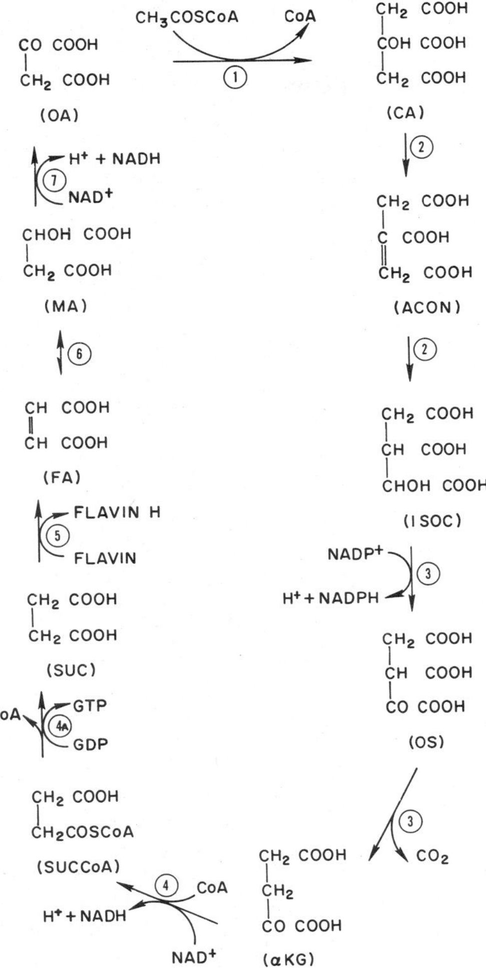

FIG. 2. Krebs citric acid cycle. *Enzymes involved*: (1) condensing enzyme; (2) aconitase; (3) isocitric acid dehydrogenase; (4) α-ketoglutaric acid dehydrogenase; (4A) succinic acid thiokinase; (5) succinic acid dehydrogenase; (6) fumarase, and (7) malic acid dehydrogenase. *Abbreviations include*: CA = citric acid; ACON = *cis*-aconitic acid; ISOC = isocitric acid; OS = oxalsuccinic acid; α KG = α-ketoglutaric acid; SUC = succinic acid; FA = fumaric acid, MA = malic acid, and OA = oxalacetic acid.

pathway. This scheme, together with the system involving the enzymes transketolase and transaldolase (which can also synthesize pentose as shown) which act to form hexose phosphate from pentose phosphate, is called the *pentose phosphate cycle*. This cycle represents an alternative pathway to glycolysis for the formation of triose phosphate from glucose-6-phosphate. The relative importance of the two pathways seems to be different among the various organisms and tissues (see also PENTOSES IN METABOLISM).

In a certain group of bacteria, still another pathway (the *Entner-Doudoroff pathway*) for the utilization of glucose has been found. Here glucose-6-phosphate is oxidized to 6-phosphogluconic acid which is dehydrated to 2-keto-3-deoxy-6-phosphogluconic acid. This substance is then split to pyruvic acid and glyceraldehyde-3-phosphate (which can also be converted to pyruvic acid).

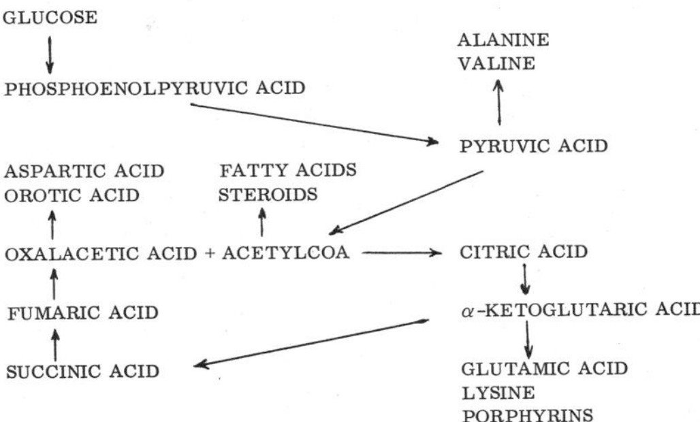

FIG. 3. Representative conversions of carbohydrates to other substances.

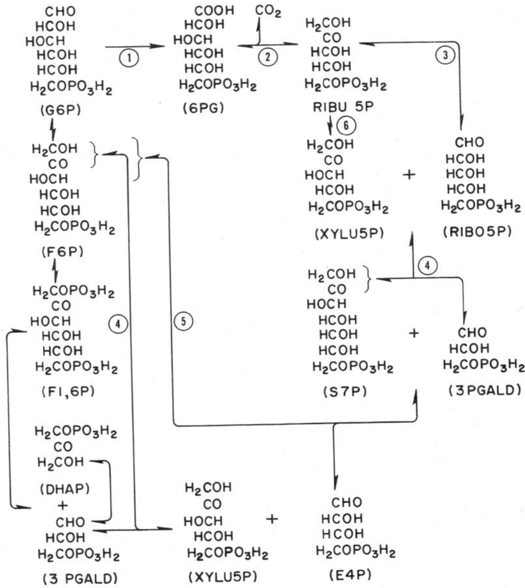

FIG. 4. Pentose phosphate cycle. *Enzymes involved include*: (1) glucose-6-phosphate dehydrogenase; (2) 6-phosphogluconic acid dehydrogenase; (3) pentose phosphate isomerase; (4) transketolase; (5) transaldolase, and (6) pentose phosphate epimerase. *Abbreviations include*: G6P = glucose-6-phosphate; 6PG = 6-phosphogluconic acid; RIBU5P = ribulose-5-phosphate; RIBO5P = ribose-5-phosphate; XYLU5P = xylulose-5-phosphate; S7P = sedoheptulose-7-phosphate; 3PGALD = glyceraldehyde-3-phosphate; E4P = erythrose-4-phosphate; F1,6P = fructose-1,6-diphosphate; DHAP = dihydroxyacetone phosphate, and F6P = fructose-6-phosphate. *Enzymes not named*, are those of glycolysis. $NADP^+$ is reduced in reactions (1) and (2).

The formation of deoxyribose, the pentose moiety of DEOXYRIBONUCLEIC ACID, can occur directly from ribose while the latter is in the form of a nucleotide diphosphate. Deoxyribose-5-phosphate can also be formed by condensation of acetaldehyde and glyceraldehyde-3-phosphate.

An enzymatic process called *transglycosylation* plays an important role in carbohydrate metabolism. Figure 5 represents the formation of the disaccharide, sucrose, as an example of this mechanism. In the upper reaction of the Figure, glucose-1-phosphate is the glycosyl donor and fructose is the acceptor. In the lower reaction, the sugar nucleotide, uridine diphosphoglucose (UDP-glucose), is the glycosyl donor. With UDP-glucose as donor and glucose-6-phosphate as acceptor, trehalose-6-phosphate may be formed. Polysaccharides may also be formed by this process. The donor residues provided by sugar nucleotides are added to preexisting polysaccharide chains (known as "primers") acting as glycosyl acceptors. In the formation of glycogen, for example, UDP-glucose donates the glucose moiety which is added to the end of a previously synthesized chain by a 1,4-linkage, thereby lengthening the chain by one glucose unit. [The branching which is found in glycogen is introduced by a different enzyme, amylo(1,4 → 1,6)transglucosylase, which forms 1,6-linkages between glucose residues on separate chains of 1,4-linked residues.] Transglycosylation seems to be a general mechanism for synthesis of polysaccharides. Different sugar nucleotides are used depending on the composition of the polysaccharide (*e.g.*, UDP-acetylglucosamine is used for CHITIN; UDP-glucuronic acid and UDP-acetylglucosamine, for hyaluronic acid; UDP-glucose or ADP-glucose, for starch, etc.; see also CELLULOSE; GLYCOGEN; MUCOPOLYSACCHARIDES; STARCHES).

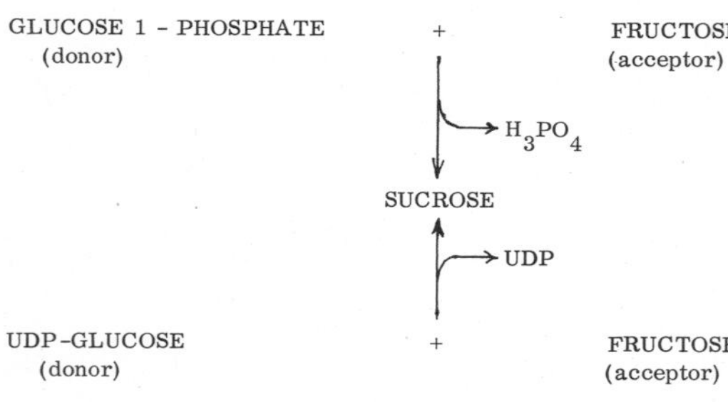

FIG. 5. Examples of transglycosylation.

The breakdown of polysaccharides appears to occur by mechanisms other than those used for synthesis. Starch and glycogen can be enzymatically hydrolyzed to maltose by amylase. Another degradative mechanism (phosphorolysis) leads to the formation of glucose-1-phosphate by addition of H_3PO_4 across the 1,4-glycosidic linkage in the polysaccharide. Glucose-1-phosphate can be converted to glucose-6-phosphate to enter glycolysis.

The sugar nucleotides are involved in a variety of carbohydrate interconversions as well as transglycosylation. For example, UDP-glucose can be *oxidized* to UDP-glucuronic acid with the concomitant reduction of NAD^+. By *decarboxylation*, this uronic acid derivative is converted to UDP-D-xylose. UDP-glucose can be *epimerized* to UDP-galactose; UDP-D-xylose, to UDP-L-arabinose; and UDP-N-acetylglucosamine, to the N-acetylgalactosamine derivative. The sugar moieties of sugar nucleotides may also undergo *exchange* (*e.g.*, UDP-galactose + glucose-1-phosphate → UDP-glucose + galactose-1-phosphate). By *reduction*, deoxysugar nucleotides can be formed from sugar nucleotides (*e.g.*, GDP-L-fucose or GDP-colitose from GDP-D-mannose; and thymidine diphospho-L-rhamnose from thymidine diphosphoglucose). The sugar nucleotides are themselves generally formed by reaction between a nucleotide triphosphate and a sugar-1-phosphate (*e.g.*, UTP + glucose-1-phosphate→ UDP-glucose + pyrophosphate). Alcohol nucleotides are also known. Cytidine diphosphoglycerol and cytidine diphosphoribitol are involved in the synthesis of teichoic acids, polymers of these polyols and phosphate, found in many microorganisms (see BACTERIAL CELL WALLS).

Glucosamine-6-phosphate can be formed by amination of fructose-6-phosphate. N-Acetylation of the hexosamine phosphate is accomplished with acetyl-CoA. N-Acetylmannosamine is a precursor of N-acetylneuraminic acid which is a constituent of the widely distributed sialic acids.

In the rat, glucose can serve as the precursor for the formation of ASCORBIC ACID through glucuronic acid, L-gulonic acid and 2-keto-L-gulonolactone. L-Gulonic acid can also be converted to fructose-6-phosphate through 3-keto-L-gulonic acid, L-xylulose, xylitol, D-xylulose and xylulose-5-phosphate. These reactions represent another pathway for the formation of pentose phosphate from glucose. Glucuronic acid can also arise from the oxidative cleavage of the cyclic polyol, *myo*-INOSITOL. The mechanism of synthesis of the inositols is uncertain as yet.

Glucose can be metabolized in some cases without the prior addition of phosphate. Enzymes which oxidize free glucose to gluconolactone exist in many organisms and tissues. After opening of the lactone ring, the gluconic acid may be phosphorylated on carbon 6 or may be further oxidized to various keto acids without phosphorylation. Galactose can apparently be oxidized similarly in a certain bacterium to form 2-keto-3-deoxygalactonic acid. Various non-phosphorylated aldohexoses and aldopentoses can be isomerized enzymatically. Enzymes for reducing certain hexoses to their corresponding polyols have been demonstrated.

Digestion, Absorption and Storage of Carbohydrate in the Mammal. In the mammal, complex polysaccharides which are susceptible to such treatment, are hydrolyzed by successive exposure to the amylase of the saliva, the acid of the stomach and the disaccharases (*e.g.*, maltase invertase, amylase, etc.) of the small intestine. The last mechanism seems to be the most important.

Absorption of the resulting monosaccharides takes place primarily in the upper part of the small intestine from which the sugars are carried to the liver by the portal system. The absorption across the intestinal mucosa occurs by a combination of ACTIVE TRANSPORT and diffusion. For glucose, the active transport mechanism seems to be the most important and appears to involve phosphorylation in an as yet undetermined manner. Agents which inhibit respiration (*e.g.*, azide, fluoracetic acid, etc.) and phosphorylation (*e.g.*, phlorizin) and those which uncouple oxidation from phosphorylation (*e.g.*, dinitrophenol) interfere with the absorption of glucose. Once the various monosaccharides pass through the mucosa, interconversion of the other sugars to glucose can begin, although the liver is probably the chief site for such conversions. Even though many organs and tissues store carbohydrate as GLYCOGEN for their own use, the liver provides the main source of glucose for all tissues through conversion of its glycogen (and other substances) to glucose-6-phosphate, hydrolysis of this ester by the specific liver glucose-6-phosphatase and transport of the free glucose in the bloodstream throughout the body (see also DIGESTION AND DIGESTIVE ENZYMES; STARCHES).

Endocrine Influences on Carbohydrate Metabolism. A number of HORMONES are known to influence carbohydrate metabolism in the mammal. INSULIN seems to increase oxidation of glucose, lipogenesis and glycogenesis. Its primary mode of action may be to facilitate the entry of glucose into the cell. Epinephrine and glucagon appear to accelerate the hydrolysis of liver glycogen by activating the phosphorylase which forms glucose-1-phosphate from the polysaccharide (glycogen) and inorganic phosphate. A number of other hormones from the adrenal cortex and thyroid also exert an influence indirectly on carbohydrate metabolism.

Vitamin Influences on Carbohydrate Metabolism. The involvement of NAD^+ and $NADP^+$ in many carbohydrate reactions explains the importance of nicotinamide in carbohydrate metabolism. Thiamine, in the form of thiamine pyrophosphate (cocarboxylase), is the cofactor necessary in the decarboxylation of pyruvic acid, in the transketolase-catalyzed reactions of the pentose phosphate cycle and in the decarboxylation of α-ketoglutaric acid in the citric acid cycle among other reactions. Biotin is a bound cofactor in the fixation of CO_2 to form oxalacetic acid from pyruvic acid. Pantothenic acid is a part of the CoA molecule. (See VITAMIN B GROUP.)

Photosynthesis. The formation of carbohydrates in green plants by the process of photosynthesis is discussed in the articles CARBON REDUCTION CYCLE (IN PHOTOSYNTHESIS) and PHOTOSYNTHESIS. The synthetic mechanism involves the addition of CO_2 to ribulose-1,5-diphosphate and the subsequent formation of two molecules of 3-phosphoglyceric acid which are reduced to glyceraldehyde-3-phosphate. The triose phosphates are utilized to again form ribulose-5-phosphate by enzymes of the pentose phosphate cycle. Phosphorylation of ribulose-5-phosphate with ATP regenerates ribulose-1,5-diphosphate to accept another molecule of CO_2.

References

1. AXELROD, B., "Glycolysis," in "Metabolic Pathways" (GREENBERG, D. M., EDITOR), Vol. I, p. 97, New York, Academic Press, 1960.
2. KREBS, H. A., AND LOWENSTEIN, J. M., "The Tricarboxylic Acid Cycle," in "Metabolic Pathways" (GREENBERG, D. M., EDITOR), Vol. 1, p. 129, New York, Academic Press, 1960.
3. AXELROD, B., "Other Pathways of Carbohydrate Metabolism," in "Metabolic Pathways" (GREENBERG, D. M., EDITOR), Vol. I, p. 205, New York, Academic Press, 1960.
4. HASSID, W. Z., "Biosynthesis of Complex Saccharides," in "Metabolic Pathways," (GREENBERG, D. M., EDITOR), Vol. I, p. 251, New York, Academic Press, 1960.
5. GLASER, L., "Biosynthesis of Deoxysugars," *Physiol. Rev.*, **43**, 215 (1963).
6. HORECKER, B. L., "Interdependent Pathways of Carbohydrate Metabolism," *Harvey Lectures*, **57**, 35 (1961–1962).
7. NEUFELD, E. F., AND HASSID, W. Z., "Biosynthesis of Saccharides from Sugar Nucleotides," *Advan. Carbohydrate Chem.*, **18**, 309 (1963).
8. MANNERS, D. J., "Enzymic Synthesis and Degradation of Starch and Glycogen," *Advan. Carbohydrate Chem.*, **17**, 371 (1962).
9. WOOD, H. G., "Significance of Alternate Pathways in the Metabolism of Glucose," *Physiol. Rev.*, **35**, 841 (1955).
10. STETTEN, D., AND STETTEN, M. R., "Glycogen Metabolism," *Physiol. Rev.*, **40**, 505 (1960).
11. HORECKER, B. L., "Pentose Metabolism in Bacteria," New York, John Wiley & Sons, 1962.
12. GUNSALUS, I. C., HORECKER, B. L., AND WOOD, W. A., "Pathways of Carbohydrate Metabolism in Microorganisms," *Bacteriol. Rev.*, **19**, 79 (1955).
13. CABIB, E., "Carbohydrate Metabolism," *Ann. Rev. Biochem.*, **32**, 321 (1963).
14. ASHWELL, G., "Carbohydrate Metabolism," *Ann. Rev. Biochem.*, **33**, 101 (1964).

I. A. BERNSTEIN

CARBOHYDRATES (ACCUMULATION IN FRUITS AND SEEDS)

Soluble carbohydrates are synthesized in the CHLOROPLASTS of green plants, and those not utilized immediately in respiration are translocated to other parts of the plant. These translocated carbohydrates may be utilized in RESPIRATION and growth, or may be stored as reserve foods. In many plants the most conspicuous site of stored foods is to be found in the fruit.

All fruits undergo four stages of development. (a) Following fertilization a fruit grows by cell division. (b) There follows a period of cell enlargement during which time sap-filled vacuoles are formed. Sugars accumulate in the vacuoles; the cytoplasm which, up to this stage, consisted chiefly of proteins, now contains starch. When a fruit has attained full growth, it may be considered *mature* though not necessarily *ripe*. (c) Ripening ensues during the third stage of development, during which period substances responsible for flavor and aroma are formed, acidity is reduced, sugars increase, and a certain amount of softening occurs. (d) The fourth stage of development, called senescence, begins where ripening leaves off and is not pertinent to this discussion.

It should be pointed out at this time that fruits may be loosely divided into two groups: those with a starch reserve and the ones without a starch reserve.

Fruits with a Starch Reserve. Typical of fruits with a STARCH reserve are the apple, banana, and pear. It has been shown, for example, that invert sugar and sucrose increase throughout the growing period of the apple fruit, but starch reaches its maximum when ripening processes begin. During the course of ripening, therefore, the starch is hydrolyzed to sugar. During the early stages of ripening, the only form of pectic substances present consist of protopectin (insoluble pectin). Soluble pectin develops during ripening, reaches its maximum when fully ripe and decreases thereafter. The sugars in a ripe apple consist mainly of glucose, fructose and sucrose.

Rather pronounced carbohydrate transformations take place in the banana during ripening. It has been shown, for example, that when the fruit changes from the green to the ripe stage, total carbohydrates drop from 26.56 to 19.00%, soluble carbohydrates increase from 1.30 to 17.02%, and insoluble carbohydrates decrease from 25.26 to 1.98%. In general, reducing sugars show a gradual increase during post-harvest ripening. The behavior of nonreducing sugars varies with the variety.

Another group of substances, known as tannins, are often classified as compound carbohydrates. These substances accumulate during the growth of certain fruits, accounting for astringency in the unripe stage. Unripe persimmons, olives, bananas, and dates are characterized by high tannin content. This is true to a lesser extent of certain varieties of pears. During the ripening of these fruits, astringency is reduced as tannins are converted to an insoluble form.

Fruits without a Starch Reserve. Fruits which do not accumulate a large carbohydrate reserve are typified by citrus fruits, blackberries, and raspberries, cherries, peaches, plums, strawberries and others. During ripening on the tree or bush, these fruits show an increase in sugars and a decrease in acids. Following harvest, fruits without a starch reserve may develop a characteristic color, soften (in some types), and lose a slight amount of acid through respiration, but they will not show any increase in sugar. A good variety of orange has been shown to contain 10.6% soluble solids (mainly sugars) and 0.85% acids when acceptable to consumers.

Several exceptions or variations from the general rule are to be found in this second group. Lemons, for example, do not undergo the same changes during ripening as those in oranges and grapefruit. The lemon fruit, during growth and maturation, does not increase in sugar. Free acids in the juice increase during ripening and predominate over sugars in the ripe fruit.

The avocado should be mentioned separately because in this instance, too, total sugar content decreases during maturation. With the loss of sugar there is a concomitant increase in oil.

Dates are unique not only because of the high sugar content in ripe fruits, but also because different varieties accumulate different kinds of sugar. Barhee, for example, accumulates mostly glucose and fructose and is therefore classed as an invert-sugar variety. Deglet Noor, on the other hand, contains mainly sucrose when ripe.

Other Fruits. Fruits borne on succulent vines, if of economic importance, are popularly known as vegetables, although, botanically speaking, they may be true fruits. In this category are included tomatoes, cantaloupes, watermelons, and pumpkins.

Mature green tomatoes contain a very slight amount of starch which disappears upon ripening. Reducing sugars increase with ripening, but only traces of sucrose have been found in these fruits in various stages of ripening.

Cantaloupes, honeydew and casaba melons undergo an increase in total solids, total sugar and sucrose during ripening and a decrease in invert sugar. The same general changes take place in the watermelon.

Both pumpkin and squash differ from related species in that immature fruits contain as much sugar as ripe ones.

Carbohydrate Accumulation in Seeds. Seeds may be grouped into three categories regarding accumulation of foods:

(1) Those in which carbohydrates represent the main food reserve. The cereal grains are typical of this group.

(2) Seeds that accumulate large quantitites of proteins. Many of the legumes, like peas and beans, fall into this group.

(3) Seeds in which large quantities of oil are stored. Sunflower, almond, tung, macadamia and castor bean seeds belong to this group.

According to Crocker and Barton, starch and hemicellulose predominate in seeds in which large quantities of carbohydrates are stored. Sucrose is generally present in smaller amounts. During the early stages of development of this type of seed, there occurs a gradual increase in sugar up to a maximum. Subsequently sugars decrease, and starch and other polysaccharides increase. It will be recalled that a number of seeds are utilized for food by man when the seeds are still rather succulent. Sweet corn is an example of this type. Appleman described four stages of maturity in sweet corn as indicated by the "nail test." The stage is determined by the nature of the exudate when a kernel is broken by pressure of the thumb nail. (1) The "pre-milk" stage, in which the exudate is opalescent. The ratio of sugar to starch at this stage was found to be 1.903. (2) Next is the "milk" stage, the sugar/starch ratio being 0.750.

(3) In the "early dough" stage, the sugar/starch ratio was equal to 0.239. (4) The "dough" stage had a sugar/starch ratio of 0.146.

It should be borne in mind that although some seeds, like peas, are rich in protein, they also accumulate carbohydrates, though to a lesser extent than do the carbohydrate-rich seeds, and the changes from sugar to polysaccharides proceed in the same order during ripening.

Oily seeds are like the avocado fruit, in that during growth, carbohydrates decrease and oil accumulates.

Seeds of the coconut will be considered separately because of two unique features—their large size and the presence of liquid endosperm (milk) during the maturing stages. Ripening stages of the coconut have been divided into three groups: (1) Before the formation of the endosperm, when invert sugar and amino acids accumulate in the milk; (2) when the loss of water from the nut takes place and sucrose appears in the milk; (3) when a sudden rise in the oil content of the endosperm and a loss of nutrients in the milk occur.

Some seeds have the carbohydrate reserves stored as hemicellulose in the tertiary, much thickened cell walls of the endosperm of the cotyledons, instead of in the interior of the cells as is the case with stored starch. The most striking seed of this kind is the seed of the ivory nut palm (*Phytelephas macrocarpa*) from South America. Seeds of the date palm also store carbohydrate in the form of hemicelluloses. Hydrolysis of this seed yields glucose, fructose, mannose, galactose, arabinose and xylose.

References

1. ALLEN, F. W., "Physical and Chemical Changes in the Ripening of Deciduous Fruits," *Hilgardia*, **6**, 381–441 (1932).
2. APPLEMAN, C. O., "Forecasting the Date and Duration of the Best Canning Stage for Sweet Corn," Univ. Maryland Expt. Sta. Bull. No. 254, 1923.
3. CROCKER, W., AND BARTON, L. V., "Physiology of Seeds. An Introduction to the Experimental Study of Seed and Germination Problems," Waltham, Mass., Chronica Botanica Co., 1953.
4. HARDING, P. L., WINSTON, J. R., AND FISHER, D. F., "Seasonal Changes in Florida Oranges," *U.S. Dept. Agr. Tech. Bull.*, **753** (1940).
5. JONES, W. W., AND SHAW, L., "The Process of Oil Formation and Accumulation in the *Macadamia*," *Plant Physiol.*, **18**, 1–7 (1943).
6. LOESECKE, H. W. VON, "Bananas. Chemistry, Physiology, Technology," New York, Interscience Publishers, 1949.
7. ROSA, J. T., "Changes in Composition during Ripening and Storage of Melons," *Hilgardia*, **3**, 421–442 (1928).
8. SANDO, C. E., "The Process of Ripening in the Tomato, Considered Especially from the Commercial Standpoint," *U.S. Dept. Agr. Bull.*, **859**, 1920.
9. WINTON, A. L., AND WINTON, K. B., "The Structure and Composition of Foods," Vols. I and II, New York, John Wiley & Sons, 1935.

ERSTON V. MILLER

CARBOHYDRATES (CLASSIFICATION AND STRUCTURAL INTERRELATIONS)

The class of compounds designated as carbohydrates includes polyhydroxyaldehydes, polyhydroxyketones and substances of higher molecular weight which yield these compounds upon hydrolysis. The monomeric polyhydroxyaldehydes and ketones are called *monosaccharides*; the dimers, *disaccharides*; the short chain polymers, *oligosaccharides*; and the high molecular weight polymers, *polysaccharides*. The name "carbohydrate" is based on the generalization that these compounds have the formula $C_m(H_2O)_n$. However, there are a number of substances in this class which do not have this formula and are essentially *derived* carbohydrates. These include uronic acids, deoxysugars and HEXOSAMINES. In addition to the above classification, carbohydrates are sometimes divided into *sugars* and *polysaccharides*. The sugars are sweet, water-soluble compounds, usually crystalline and having a well-defined molecular weight (*i.e.*, mono-, di-, trisaccharide). In contradistinction, the polysaccharides, like most polymers, are heterogeneous with regard to molecular weight, considerably less soluble or insoluble, and do not have a sweet taste.

Monosaccharides. The monosaccharides can be classified according to the number of carbons in the molecule (*e.g.*, *pentose* contains five carbons; *hexose*, six; *heptose*, seven) or according to the nature of the carbonyl group, whether aldehyde or ketone. In the former case, the compound is termed an *aldose* and in the latter a *ketose*. Examples of these and the method of numbering the carbons is shown below:

	An aldopentose		A ketohexose
1	CHO	1	CH_2OH
2	CHOH	2	CO
3	CHOH	3	CHOH
4	CHOH	4	CHOH
5	CH_2OH	5	CHOH
		6	CH_2OH

The generic name for the ketoses has the suffix "ulose." For example, a ketopentose is a pentulose and a ketohexose is a hexulose.

As a result of the presence of ASYMMETRIC carbon atoms in these molecules, a number of different structures are possible. Specifically, on the basis of the straight-chain structures for the aldoses shown above, there are four erythroses (four-carbon aldoses), eight pentoses, and sixteen hexoses. The designation of the configuration of these sugars is based on that of D- and L-glyceraldehyde. The aldoses in which the substituents on the *carbon adjacent to the primary alcoholic carbon* (—CH_2OH) are the same as in D-glyceraldehyde are classified as the D-series as shown below:

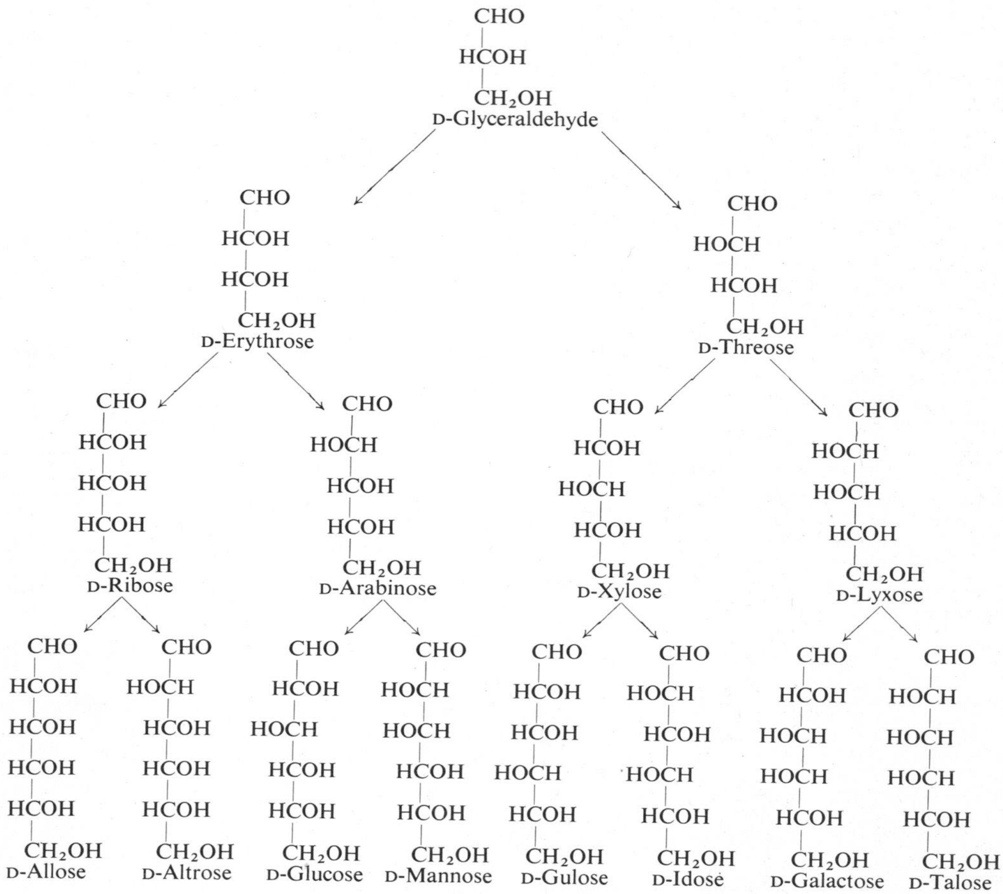

CHO
HCOH
CH_2OH
D-Glyceraldehyde

D-Erythrose	D-Threose
CHO	CHO
HCOH	HOCH
HCOH	HCOH
CH_2OH	CH_2OH

D-Ribose	D-Arabinose	D-Xylose	D-Lyxose
CHO	CHO	CHO	CHO
HCOH	HOCH	HCOH	HOCH
HCOH	HCOH	HOCH	HOCH
HCOH	HCOH	HCOH	HCOH
CH_2OH	CH_2OH	CH_2OH	CH_2OH

D-Allose	D-Altrose	D-Glucose	D-Mannose	D-Gulose	D-Idose	D-Galactose	D-Talose
CHO	CHO	CHO	CHO	CHO	CHO	CHO	CHO
HCOH	HOCH	HCOH	HOCH	HCOH	HOCH	HCOH	HOCH
HCOH	HCOH	HOCH	HOCH	HCOH	HCOH	HOCH	HOCH
HCOH	HCOH	HCOH	HCOH	HOCH	HOCH	HOCH	HOCH
HCOH	HCOH	HCOH	HCOH	HCOH	HCOH	HCOH	HCOH
CH_2OH	CH_2OH	CH_2OH	CH_2OH	CH_2OH	CH_2OH	CH_2OH	CH_2OH

Relationships of the D-aldoses.

The L-aldoses are the mirror images of the above structures, *e.g.*, L-glucose, L-mannose, etc.

The aldehyde group in these compounds is not free but interacts with a hydroxyl group on carbon 4 or carbon 5 to form a hemiacetal. In the former case, the resulting cyclic monosaccharide is called a *furanose* compound, and in the latter case, a *pyranose*. As a result of the formation of the ring, a new asymmetric center is created at carbon 1, thus giving rise to twice the original number of isomers. Carbon 1 in these instances is called the *anomeric* carbon and the monosaccharide is designated as *alpha* or *beta* according to the configuration of the anomeric carbon. Examples with complete names are shown below:

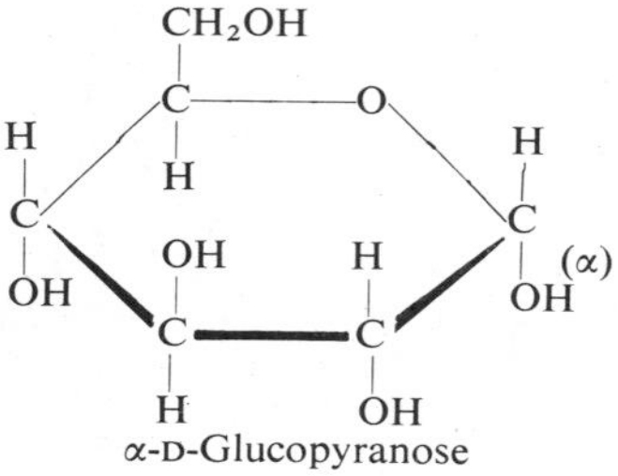

CH_2OH — C — O; H, C, OH; OH, H; C — C; H, OH; H, C, OH (α)

α-D-Glucopyranose

CH_2OH — C — O; H, C, OH; OH, H; C — C; H, OH; OH(β), C, H

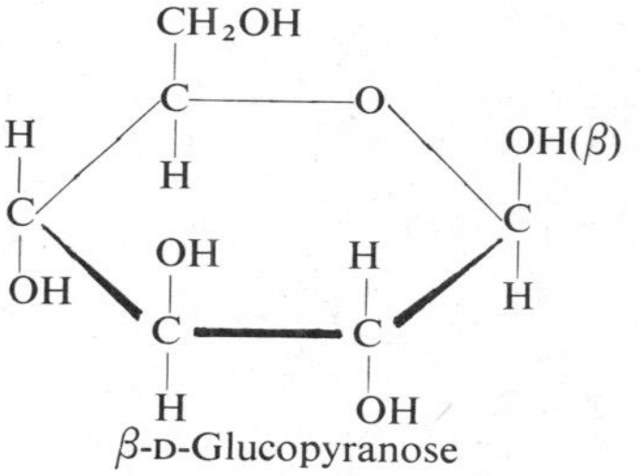

β-D-Glucopyranose

Ketohexose may also have a number of isomers, in addition to different ring structures. The best known of these is D-fructose.

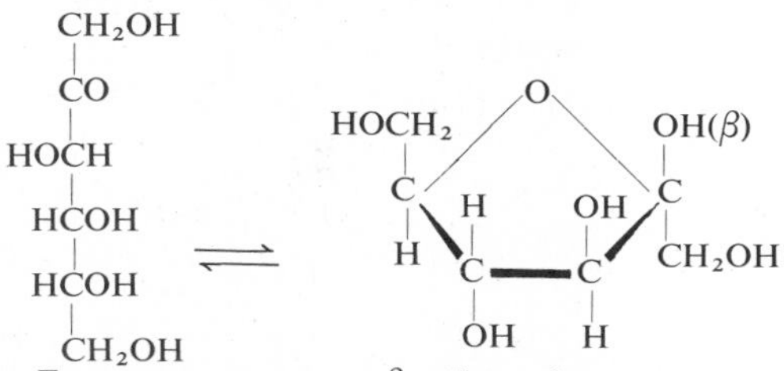

CH_2OH–CO–HOCH–HCOH–HCOH–CH_2OH ⇌ β-D-Fructofuranose

D-Fructose

β-D-Fructofuranose

In aqueous solution, the monosaccharides exist as equilibrium mixtures of the open chain (carbonyl) and different cyclic hemiacetal forms. The sugars will therefore undergo reactions characteristic for the carbonyl group. Examples of such reactions are: reduction of ammoniacal silver (Fehling's solution, Tollens' reagent); conversion of the monosaccharide by various reducing agents to the polyhydroxy alcohol (*e.g.*, glucose to sorbitol); oxidation to the *aldonic* and *aldaric* acid; reaction with phenylhydrazine to form phenylhydrazones and osazones; addition of hydrogen cyanide across the carbonyl group to form the cyanohydrin derivative.

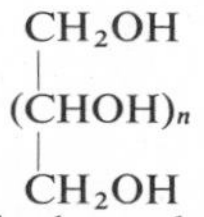

CH_2OH–$(CHOH)_n$–CH_2OH
Polyhydroxyalcohol

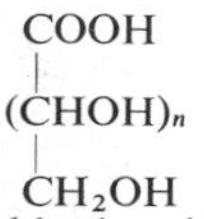

COOH–$(CHOH)_n$–CH_2OH
Aldonic acid

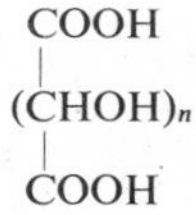

COOH–$(CHOH)_n$–COOH
Aldaric acid

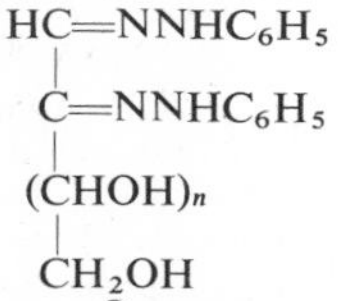

$HC{=}NNHC_6H_5$–$C{=}NNHC_6H_5$–$(CHOH)_n$–CH_2OH
Osazone

The derivatives obtained from glucose corresponding to the illustrated structures are, respectively, sorbitol, gluconic acid, saccharic acid and glucosazone.

As a result of the presence of the asymmetric carbons in the sugars, they are optically active and have characteristic specific rotations. These are specified as *plus* (+) or *minus* (−) depending on the direction of rotation. The measurement of specific rotation or polarimetry is thus a highly important procedure in carbohydrate chemistry (see also OPTICAL ACTIVITY AND ISOMERISM).

The carbonyl group in monosaccharides also reacts with alcohols or phenols to yield hemiacetals called glycosides. When this takes place, the carbonyl group is not oxidizable, and the cyclic form of the sugar is fixed. Thus the methyl glycoside of mannose has the following structure:

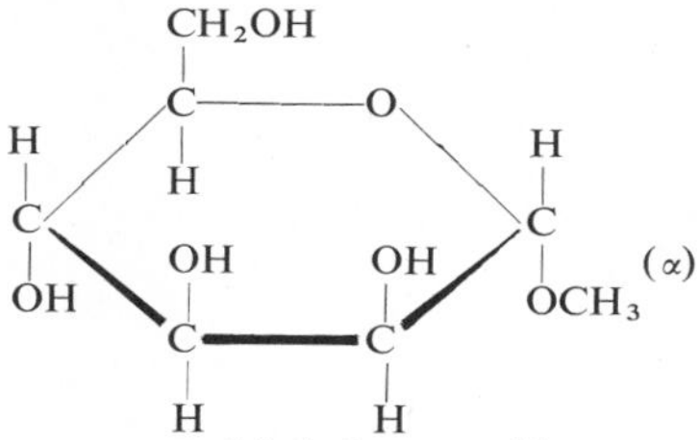

α-D-Methyl mannoside

The glycosidic linkage (to the methyl group) may also occur in the β-configuration, as in the case of α- and β-configurations of glucopyranose, illustrated above. The hydroxyl groups of sugars can be esterified to form acetates and benzoates. They can also be converted to ethers. Of the latter, the methyl ethers are highly important in the determination of the structures of carbohydrates.

In considering the cyclic structure of sugars, it was realized that the ring is not planar, but that it would exist in either *chair* or *boat* conformations. Modern studies on carbohydrates thus also include consideration of the possible conformations. The preferred form in most cases is the chair form and the situation where most of the bulkier hydroxyl groups are equatorial.

Disaccharides. Disaccharides are composed of two monosaccharide units linked through a glycosidic linkage. For example *maltose* is composed of two glucose units, *i.e.*, 4-O-(α-D-glucopyranosyl)-D-glucopyranose.

If the glycosidic linkage between the two units is *beta*, then the compound is *cellobiose*.

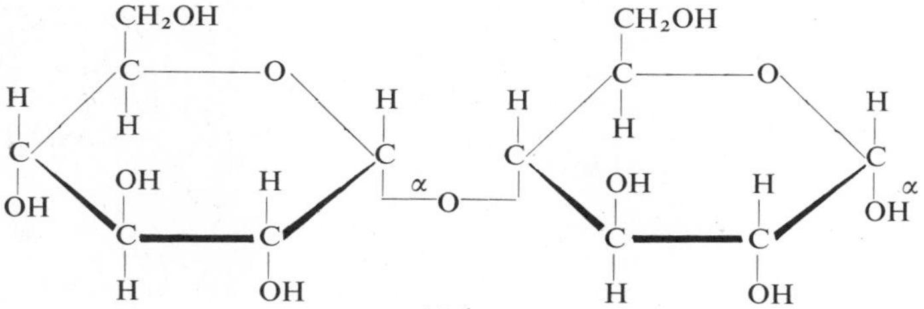

α-Maltose

Lactose, or milk sugar, is composed of glucose and galactose, 4-O-(β-D-galactopyranosyl)-D-glucopyranose (note that the right hand or glucopyranose ring is drawn inverted, relative to the previous structure):

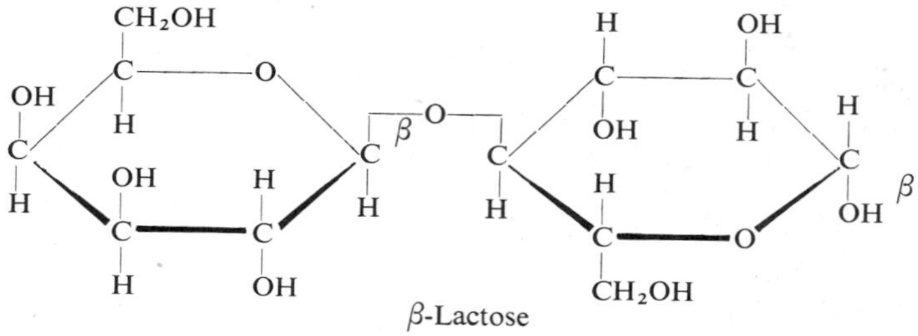

β-Lactose

Sucrose is a disaccharide composed of glucose and fructose as shown below:

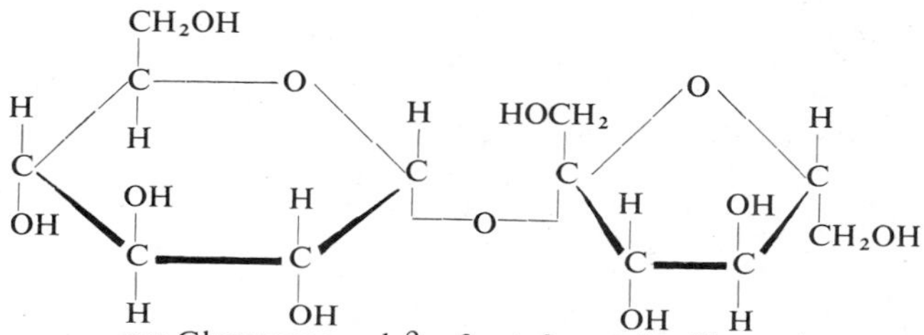

α-D-Glucopyranosyl-β-D-fructofuranoside (Sucrose)

In a manner analogous to disaccharides, three or more monosaccharides may be attached to each other through glucosidic linkages to yield oligosaccharides and polysaccharides. Among the oligosaccharides are *gentianose*, *raffinose* and *turanose*.

Polysaccharides. The polysaccharides include CELLULOSE, GLYCOGEN, STARCH, pectins and various plant gums. In addition to these monosaccharides and polysaccharides a number of derivatives of these compounds are classified as carbohydrates. These include monomers and polymers of glucosamine and uronic acids, *i.e.*, alginic acid, CHITIN and MUCOPOLYSACCHARIDES (see also HEXOSAMINES AND HIGHER AMINO SUGARS; BACTERIAL CELL WALLS).

Carbohydrates are major components of plant and animal tissues. The structural materials of higher plants are mainly cellulose and hemicellulose. In addition to this, plants contain considerable amounts of starch, pectins and lower molecular weight carbohydrates. The latter are found to a large extent in plant saps and include sucrose, glucose and fructose. Animal carbohydrates are primarily glucose and glycogen. The nucleic acids (see DEOXYRIBONUCLEIC ACIDS; RIBONUCLEIC ACIDS) of all cellular material contain the pentoses ribose and deoxyribose. Polysaccharides are also the major constituents of cell walls and capsules of various microorganisms.

Maltose is produced from starch during the brewing process and cellobiose is formed from cellulose. Sucrose, the sugar generally used in the diet, is found in all photosynthetic plants. However, its main sources are sugar cane and sugar beet.

References

1. PIGMAN, W. W. (EDITOR), "The Carbohydrates: Chemistry, Biochemistry and Physiology," New York, Academic Press, 1957.
2. STANEK, J., CERNY, M., KOCOUREK, J., AND PACAK, J., "The Monosaccharides," New York, Academic Press, 1963.
3. "Advances in Carbohydrate Chemistry," New York, Academic Press, published annually. Contains review articles by different authors.

I. DANISHEFSKY

CARBON (RADIOCARBON) DATING

The chemist Willard F. Libby was awarded the Nobel prize (1960) for discovering the occurrence of carbon 14 in nature and for developing its use in archaeological and geological chronology. Carbon 14 is an isotope of carbon which was first observed by Ruben and Kamen in 1939 as an artificially produced radionuclide. It is a beta-particle emitter with a half-life of 5570 years. That which occurs in nature is presumed to be continuously produced in the upper atmosphere by the interaction of cosmic ray neutrons with nitrogen 14 to yield a proton and carbon 14. The steady-state concentration of the isotope in atmospheric carbon (in the form of carbon dioxide) exhibits a radioactivity of 15.3 disintegrations per minute per gram of carbon. Plants which utilize the atmospheric carbon dioxide, and animals which eat the plants, are built up of carbon compounds having very nearly the same isotopic composition. In the normal course of events the carbon is returned to the atmospheric reservoir. However, if the plant or animal is buried, or otherwise prevented from decomposing, its carbon does not reenter the carbon cycle and does not maintain a steady-state concentration of carbon 14. Instead, radioactive decay leads to a steady decrease in the relative amount of carbon 14. Thus a determination of the isotopic composition of, say, the wood used in construction of the tomb of an Egyptian pharaoh leads to an estimation of its age.

The rate of radioactive decay of an unstable nuclide may be specified by its *half-life*, *i.e.*, the time required for one-half of the unstable nuclei to distintegrate, as for any first-order rate process. For carbon 14, with a half-life of 5570 years, a detectable amount remains in carbonaceous material for as long as 50,000 years, even though it is not replenished from the atmosphere. Assuming that no intrusion of other carbon has occurred, and that the modern equilibrium distintegration rate is the same as that prevailing at the time the carbonaceous material was alive, the specific activity of the carbon is expected to be 7.65 disintegrations per minute per gram if its age is 5570 years, 3.83 distintegrations per minute per gram for an age of 11,140 years, etc. Letting f be the fraction of the original specific activity remaining and t be the age (in years) of the carbon sample, one obtains

$$f = (\tfrac{1}{2})^{t/5570}$$

Apart from the possibility of contamination of the sample by carbon-containing materials of greater or lesser age, or by other radioactive substances in the environment, the principal problems associated with the use of the method result from the very low specific activity of the samples. Extreme precautions must be taken to screen out background activity from cosmic radiation and from natural radioactivity of the atmosphere, or the materials from which the detector is constructed. The low energy of the beta particles from carbon 14 makes them more difficult to detect and complicates the determination. The original detection method used by Libby was based on the use of a Geiger counter filled with gas containing carbon from the sample (carbon dioxide, methane, acetylene, etc.). More recently, the method of choice has utilized the proportional counter, also gas filled. Alternatively, liquid scintillation detectors have been used where ample amounts of carbon are available and the sample age is very great. In the last method, the sample carbon is converted to benzene (see also ISOTOPIC TRACERS).

Radiocarbon dating has been of major importance to archaeology and paleontology in establishing absolute chronology.

References

1. LIBBY, W. F., "Radiocarbon Dating," Second edition, Chicago, University of Chicago Press, 1955.
2. TAMERS, M. A., STIPP, J. J., AND COLLIER, J., *Geochim. Cosmochim. Acta*, **24**, 266 (1961).
3. The journal *Radiocarbon* (originally a supplement to *American Journal of Science*), and various journals of archaeology.

L. O. MORGAN

CARBON DIOXIDE

Carbon dioxide, which is a by-product of the metabolic activity of all cells, is one of the most important chemical regulators in the human body. It can be truly said that human life without carbon dioxide would be impossible. In less specialized forms of life, carbon dioxide is something to be gotten rid of; it is a waste product. In the more highly evolved animals such as man, nature has used the gas to regulate the activity of the heart, the blood vessels, and the respiratory system.

Normal air contains about 0.03% by volume of CO_2. A poorly ventilated room may contain as much as 1%. Concentrations of the gas from about 0.1–1% by volume induce lanquor and headaches; 8–10% concentrations bring about death by asphyxiation. High concentrations of the gas are toxic.

Resting men who are transferred from room air to a mixture of 4.1% by volume of CO_2 in room air show an increase in expiratory minute volume from an average of 8 to about 15 liters; the respiratory rate increases from 14 to about 18 per minute; and the tidal volume from 0.500 to almost 0.900 liters. If the amount of carbon dioxide is increased, much greater increases (percentagewise) in the above physiological measurements follow: expiratory minute volume, 420; tidal volume, 260; respiratory rate, 160.

As a general rule, the respiration of individual cells decreases as the concentration of carbon dioxide in the medium increases. Fish show a lessened capacity to extract oxygen from their environment with increasing amounts of carbon dioxide present. On the other hand, many invertebrates show marked increases in respiratory rate (or ventilation) with increased amounts of the gas in their surroundings.

Photosynthetic and autotrophic bacteria reduce carbon dioxide which is assimilated into complex molecules for use in synthesizing various cellular constituents. The gas is apparently assimilated, at least to a small extent, by the heterotrophic bacteria. Certainly it is required for any growth in these forms. Many pathogenic bacteria require increased carbon dioxide tension for growth immediately after they are isolated from the body. The production of hemolysins and like substances is greatly enhanced by adding 10–20% of CO_2 in the air which comes in contact with the cultures.

In men, the average amount of CO_2 in the alveolar air is about 5.5% by volume; during the breathing cycle this concentration varies only slightly. In women and children, somewhat lower mean values obtain.

The oxygen dissociation curve for blood is shifted to the right when the partial pressure of carbon dioxide in air is increased. This is referred to as the "Bohr Effect." It means that for a given partial pressure of oxygen, hemoglobin holds less oxygen at high concentration of carbon dioxide than at a lower. It is evident, then, that the production of carbon dioxide by actively metabolizing tissues favors the release of oxygen from the blood to the cells where it is urgently needed. Moreover, at the alveolar surfaces in the lungs, the blood is losing carbon dioxide rapidly, which loss favors the combination of oxygen with hemoglobin.

In every 100 ml. of arterial blood there is a total of 48 ml of free and combined CO_2. In venous blood of resting man there is about 5 ml more than this. Only about 1/20 of the carbon dioxide is uncombined, a fact which indicates that there is a specialized mechanism, aside from simple solution, for the transport of CO_2 in the blood.

About 20% of the CO_2 in the blood is carried in combination with hemoglobin as *carbaminohemoglobin*. The balance of the combined carbon dioxide is carried as bicarbonate. A CO_2 dissociation curve for blood can be prepared just as for oxygen, but the shape is not the same as for the latter. As the partial pressure of CO_2 in the air increases, the amount in the blood increases; the increase is practically linear in the higher ranges. Oxygen exerts a negative effect on the amount of CO_2 which can be taken up by the blood.

In working muscles large amounts of CO_2 are produced. This causes local vasodilation. The diffusion of some of the CO_2 into the bloodstream slightly raises the concentration there. It circulates through the body and the capillaries of the vasoconstrictor center, where it excites the cells of the center, resulting in an increase of constrictor discharges. If one recalls the stimulating effect of CO_2 on cardiac output, it is evident that a most effective mechanism exists for increasing circulation through active muscles. More blood is pumped by the heart per minute and the arterial pressure is increased by the general vasoconstriction; blood is forced from the inactive regions, under increased pressure, through the widely dilated vessels of the active muscles.

Narcosis due to CO_2 is characterized as follows: mental disturbances which may range from confusion, mania, or drowsiness to deep coma; headache; sweating; muscle twitching; increased intracranial pressure; bounding pulse; low blood pressure; hypothermia; and sometimes papilloedema. The basic mechanism by which carbon dioxide induces narcosis is probably through interference with the intracellular enzyme systems, which are all extremely sensitive to pH changes.

CHARLES G. WILBER

CARBON DIOXIDE FIXATION

Whenever carbon dioxide is used to build up more complex organic molecules, the carbon dioxide is said to be fixed. The outstanding example in nature is that which takes place in photosynthesis whereby enormous quantities of carbohydrates and other organic compounds are produced by green plants using carbon dioxide from the atmosphere as the source of carbon [see CARBON REDUCTION CYCLE (IN PHOTOSYNTHESIS)].

The ability to fix carbon dioxide is now known to be possessed by many tissues including those from animals. Isotopic studies have shown, for example, that in the liver of rats the carbon from carbon dioxide may be incorporated ultimately into glycogen. The first step involves its combination with pyruvic acid to produce oxaloacetic acid. Metabolic routes for carbon dioxide fixation are discussed in the article CARBOXYLATION ENZYMES.

CARBON MONOXIDE

Carbon monoxide is toxic to warm-blooded animals. Because of its extremely faint odor and taste, its lethal capacity can be insidious. The ordinary charcoal-filled gas mask is useless for filtering out carbon monoxide from contaminated air. Persons who are required to enter areas contaminated with carbon monoxide (firemen, rescue workers, maintenance men) must be provided with closed-circuit breathing apparatus which delivers oxygen through a mask to the wearer. This is essential in atmospheres which contain more than 2% by volume of carbon monoxide. In atmospheres which contain less, an ordinary gas mask can be used for short periods, if it is fitted with a special canister filled with hopcalite, a mixture of metallic oxides which serve to catalyze the oxidation of carbon monoxide to carbon dioxide. The reaction is exothermic and such canisters become very hot in use.

Carbon monoxide is physiologically quite inert, except for its strong combination with hemoglobin in the blood. It has no unique toxic action on any of the bodily tissues. As Henderson and Haggard point out: "Were it not for this one reaction carbon monoxide would be classified with nitrogen and hydrogen as a simple asphyxiant." The affinity of carbon monoxide for hemoglobin is about 300 times that of oxygen.

The reaction between carbon monoxide and hemoglobin is reversible:

$$HbO_2 + CO \rightleftharpoons HbCO + O_2$$

"Carbon monoxide displaces oxygen from hemoglobin, and in turn oxygen may displace carbon monoxide from its combination. Red corpuscles, in which the hemoglobin has been joined to carbon monoxide and then freed from the combination by means of oxygen, are not injured; they are capable of transporting oxygen as if they never had been exposed to the other gas. But so long as the combination with carbon monoxide continues they are incapable of fulfilling their respiratory function." Consequently, they cannot transport adequate oxygen to the various bodily tissues. Progressively severe anoxia results. Unfortunately, the victim is all too often unaware of his danger. Mechanical efficiency, *e.g.*, ability to drive a car, may persist until poisoning has advanced almost to the possibility of unconsciousness.

Death from inhalation of carbon monoxide can be summarized as follows: (1) reduction of the oxygen-carrying capacity of the blood due to the formation of HbCO; (2) tissue anoxia, especially in the brain, which is very sensitive to lack of oxygen; (3) consequent depression of respiratory center in the brain and decrease in respiration; (4) failure of the heart due to inadequate oxygen supply.

Carbon monoxide is absorbed into the body only through the alveoli in the lungs. It does not enter through the eyes, mucous membranes, cuts, or upper respiratory tract.

Tolerance Limits. On the basis of numerous experimental studies, the tolerance limits for the average man have been established. The following series of equations gives a ready method for estimating the safety of any carbon monoxide-air mixture under conditions of rest. Time is given in hours and concentration of carbon monoxide in parts per 10,000 of air:

(a) Time × Concentration = 3 (no perceptible effect).

(b) Time × Concentration = 6 (a just perceptible effect).

(c) Time × Concentration = 9 (headache and nausea).

(d) Time × Concentration = 15 (dangerous). Muscular activity or increased respiratory minute volume reduces the value in (a) to 1, 2, or less; it influences the other equations in like manner.

Drinker gives data on which the following table of allowable concentrations for carbon monoxide in air is based:

Concentration of Carbon Monoxide in		
Per Cent	Parts per 10,000	Effect
0.01	1	No symptoms for 2 hr
0.04	4	No symptoms for 1 hr
0.06–0.07	6–7	Headache and unpleasant symptoms in 1 hr
0.1 –0.12	10–12	Dangerous for 1 hr
0.35	35	Fatal in less than 1 hr

As a safe rule, based on sound experiments and experience, concentrations of carbon monoxide above 0.01%, or 1 part per 10,000, should not be permitted in houses, garages, laboratories, or industrial plants where prolonged exposure to the gas may be experienced.

Chronic Effects. There is no such physiological entity as "chronic carbon monoxide poisoning." The gas is not a cumulative poison; it is readily removed from the blood when the victim is exposed to pure air or oxygen.

After a severely acute exposure, the victim usually dies in about 36 hours or he recovers completely after a few days. The alleged chronic damage to man from carbon monoxide poisoning stems from prolonged cerebral anoxia which was severe enough to cause permanent brain damage but not severe enough to kill; it is not caused by retention of carbon monoxide in the body.

Treatment. The treatment of carbon monoxide poisoning depends on removal of the victim from the contaminated atmosphere, administration of artificial respiration, and inhalation of pure oxygen by the patient. If a good mechanical respirator is available, it can be used to advantage. In the absence of such device, air can be pumped into the victim's lungs by the Nielsen method of resuscitation, which is an arm lift-back pressure procedure. Rescue breathing or mouth-to-mouth resuscitation may also be used. If appropriate equipment is available, oxygen may be delivered to the victim under greater than atmospheric pressure.

The use of oxygen is essential for effective treatment. From the purely academic point of view, it might be argued that the addition of 5–7% carbon dioxide to the oxygen will result in more efficient resuscitation. On the practical level, however, there is little to justify the use of carbon dioxide during resuscitation.

Drugs are of little use and may even be dangerous. Under no circumstances should a patient who is recovering consciousness after carbon monoxide poisoning be permitted to arise and walk about. He must be kept in a prone or supine position; every effort must be made to keep his oxygen requirements at a minimum.

CHARLES G. WILBER

CARBON REDUCTION CYCLE (IN PHOTOSYNTHESIS)

Plants and certain bacteria convert ten billion tons of carbon dioxide each year to carbohydrates and more reduced compounds. The fixation of atmospheric CO_2 at a rate approximating that of a sodium hydroxide solution surface is catalyzed by a unique carboxylation enzyme system present in large quantities in the CHLOROPLASTS of green leaves and ALGAE and in the autotrophic bacteria. Rates as high as 0.2–0.4 cc of CO_2 per minute per gram of algae or leaf are observed.

The sequence of events in the reduction of carbon dioxide was established using $^{14}CO_2$ in the laboratories of S. Ruben and M. D. Kamen (1939–1943) and of Melvin Calvin and co-workers (1946–1954). The sugars produced during 30 seconds photosynthesis from radioactive $^{14}CO_2$ contained, at first ^{14}C in only carbon atoms C-3 and C-4 of the six-carbon glucose or fructose molecules (Fig. 1). When the time of photosynthesis in $^{14}CO_2$ was reduced still further, to 5 seconds, no sugar was formed and the major radioactive product was the three-carbon compound, phosphoglyceric acid (Fig. 2). It is the first stable product of the carboxylation reaction of photosynthesis. Degradation of its hydrolysis product, glyceric acid, by lead tetraacetate oxidation revealed that 95% of the ^{14}C was in the carboxyl (—COOH) group. These facts suggested that carbon dioxide was fixed by some two-carbon acceptor which therefore required a *cyclic* series of reactions for its continuous production of an acceptor for the CO_2 fixation.

To find the CO_2 acceptor, ^{14}C-labeled algae were flushed with nitrogen to remove $^{14}CO_2$. The 3-phosphoglyceric acid product of CO_2 fixation diminished and a five-carbon sugar, ribulose-1,5-diphosphate (Fig. 3), increased in concentration several fold. That it is the CO_2 acceptor was finally proved when a mixture of carboxylating enzyme from spinach, ribulose diphosphate, and $^{14}CO_2$ produced radioactive phosphoglyceric acid.

$$\text{Enzyme} + CO_2 + \text{Ribulose Diphosphate} = \text{Enzyme} + 2 \text{ Phosphoglyceric Acid}$$

i.e.,

$$C_1 + C_5 = 2C_3$$

The reaction therefore involved dismutation of the sugar, part of it (C-3 of the ribulose diphosphate) becoming oxidized to provide the free energy necessary to make the carboxylation proceed. The enzyme, ribulose diphosphate carboxylase (carboxydismutase), is part of the most abundant soluble protein in green leaves.

CHO	8%
HCOH	8%
HOCH	33%
HCOH	45%
HCOH	3%
H_2COH	3%

FIG. 1. ^{14}C-labeling pattern of glucose after 30 seconds. Photosynthesis in $^{14}CO_2$

COOH
|
HCOH
|
$H_2CO—PO_3H_2$

FIG. 2. 3-Phosphoglyceric acid, the first product of carboxylation.

$H_2CO—PO_3H_2$
|
C=O
|
HCOH
|
HCOH
|
$H_2CO—PO_3H_2$

FIG. 3. Ribulose-1,5-diphosphate, the CO_2 acceptor.

The cycle utilized for regeneration of ribulose diphosphate, Fig. 4, converts triose phosphates (C_3) to fructose-1,6-diphosphate (C_6), utilizing enzymes of GLYCOLYSIS. Fructose-6-phosphate is

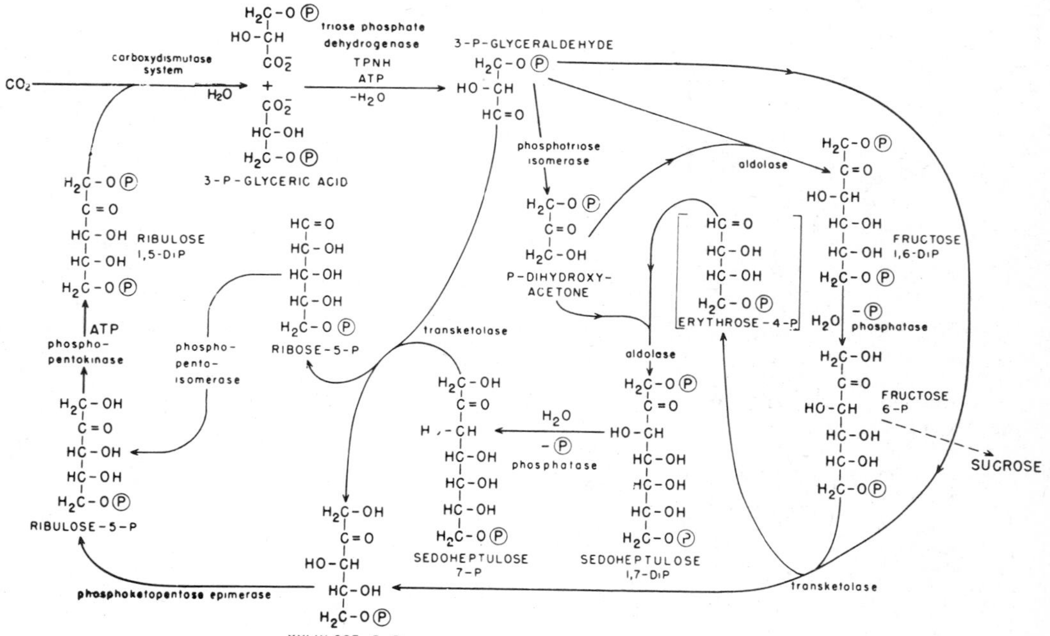

FIG. 4. The carbon reduction cycle of photosynthesis (after Calvin).

cleaved to give C_2 and C_4 fragments which undergo further combination with triose phosphate, C_3, to produce sedoheptulose mono- and diphosphates (C_7) and xylulose-5-phosphate (C_5) utilizing enzymes of glycolysis and transketolase. The latter transfers C-1,2 as a glycolyl thiamine pyrophosphate complex in a sequence leading to production of the C_5 sugar. Isomerization of the xylulose-5-phosphate yields ribulose-5-phosphate. These ketose phosphates differ from other sugar phosphates in that they are real carbonyl compounds and not the usual cyclic hemiketal or hemiacetal forms of sugars. Phosphorylation produces ribulose-1,5-diphosphate, the CO_2 acceptor.

The carbon reduction cycle requires three molecules of ATP and two of TPNH (NADPH) per molecule of CO_2 fixed. These are provided by the chloroplast and its electron transport (photophosphorylation) system (see PHOSPHORYLATION, PHOTOSYNTHETIC). The ATP contains a reactive phosphoric anhydride group with which ribulose-5-phosphate and 3-phosphoglyceric are phosphorylated, and subsequently the enzyme triose-phosphate dehydrogenase reduces 1,3-diphosphoglyceric acid to 3-phosphoglyceraldehyde, the first sugar derivative of the cycle.

Each of the enzymes catalyzing the steps of the carbon cycle has been isolated from plant tissue. Their coordinated function in the carbon reduction cycle produces sucrose and the many by-products. It has been possible to fix CO_2 and drive the cycle in cell-free homogenates of spinach leaf by adding ATP, TPNH, and extra ribulose diphosphate but neither the rates nor the distributions of products approximate those of the intact leaf. It appears that the enzymes are associated in multifunctional proteins which require the natural organization in the leaf (see also MULTI-ENZYME COMPLEXES).

The ultimate products of photosynthesis, new cells, derive their components from the intermediates of the cycle. Sucrose is formed from fructose-6-phosphate; many amino acids are formed from 3-phosphoglyceric acid; fats are produced from acetate derived from the same starting material; pentosans are formed from xylulose-5-phosphate, etc. The steady-state balance of concentrations of intermediates of the carbon reduction cycle depends on many factors such as the rates of production of ATP and TPNH and those of removal of the reduced intermediates of the cycle for use by the cell.

References

1. BASSHAM, J. A., BENSON, A. A., KAY, L. D., HARRIS, A. Z., WILSON, A. T., AND CALVIN, M., "The Path of Carbon in Photosynthesis, XXI. The Cyclic Regeneration of Carbon Dioxide Acceptor," *J. Am. Chem. Soc.*, **76**, 1760–1770 (1954).
2. CALVIN, M., AND BASSHAM, J. A., "The Photosynthesis of Carbon Compounds," New York, W. A. Benjamin, Inc., 1962.
3. BASSHAM, J. A., "The Path of Carbon in Photosynthesis," *Sci. Am.*, **206**, 89 (June 1962).
4. GEE, R., JOSHI, G., BILS, R. F., AND SALTMAN, P., "Light and Dark $C^{14}O_2$ Fixation by Spinach Leaf Systems," *Plant Physiol.*, **40**, 89 (1965).
5. RABIN, B. R., AND TROWN, P. W., "Mechanism of Action of Carboxydismutase," *Nature*, **202**, 1290–1293 (1964).

A. A. BENSON

CARBONIC ANHYDRASE

Carbonic anhydrase (CA), or carbonate hydro-lyase is an enzyme of somewhat broader substrate specificity than its name may imply. It catalyzes the reversible hydration of carbon dioxide.

$$CO_2 + H_2O \xrightleftharpoons{CA} HCO_3^- + H^+ \qquad (1)$$

This reaction also occurs spontaneously, *i.e.*, without enzyme. However, it occurs at a relatively slow rate, insufficient for many physiological requirements. Carbonic anhydrase also demonstrates esterase activity[1] (reaction 3 below) and aldehyde hydrase activity:[2]

$$CH_3CHO + H_2O \xrightleftharpoons{CA} CH_3CH(OH)_2 \qquad (2)$$

Since reactions (*2*) and (*3*) have only recently been described, the subsequent discussion will be oriented toward the role of carbonic anhydrase in the reversible hydration of carbon dioxide [reaction (*1*) above].

The enzyme has been found in animals, plants, algae, and bacteria. In general, it is associated with the soluble portion of various extracts and not with subcellular particulate fractions. In mammalian tissues, high concentrations are found in the erythrocytes, stomach, pancreas, kidney, brain, and eye. Carbonic anhydrase activity has been investigated in the reproductive tract. The level there varies with the amount of certain of the

$$\text{(naphthyl acetate, } CH_3C(=O){-}O{-}C_{10}H_7) + H_2O \xrightleftharpoons{CA} CH_3COOH + \text{(naphthol, } C_{10}H_7OH) \qquad (3)$$

steroid SEX HORMONES, *e.g.*, estrogen and progesterone.

The enzyme is necessary for the rapid removal of HCO_3^- as CO_2 from the arterial blood in the lung capillary to the alveolar air. This involves reversal of reaction (*1*) (see also GASEOUS EXCHANGES IN BREATHING). Carbonic anhydrase also possesses a fundamental role in the secretory process. Evidence for the latter function has been obtained indirectly by use of an inhibitor of carbonic anhydrase, namely, acetazoleamide. Administration of this compound has profound effects on the ionic composition and volume of urine. In the kidney, the inhibition of carbonic anhydrase impairs acid secretion in the proximal tubule and causes a more alkaline urine which contains more sodium, potassium and bicarbonate ions (see RENAL TUBULAR FUNCTIONS). Furthermore, the following effects on the secretory process are also observed after acetazoleamide administration: an impairment of gastric HCl secretion (see GASTRIC JUICE); a decrease in an elevated pressure in the anterior chamber of the eye; and a reduction in the bicarbonate concentration of the exocrine fluid of the pancreas. Carbonic anhydrase is, therefore, involved in the process of secretion in the above organs. In general, *in vivo* inhibition studies have revealed the presence of a considerable excess of enzyme. Consequently a large percentage inhibition of carbonic anhydrase must occur before function is impaired in a given organ or tissue. Other functions ascribed to carbonic anhydrase include eggshell formation, swim bladder gas concentration, and the possible adjustment of the pH of the endometrial site of implantation of a fertilized ovum.

The detailed physical, chemical, and compositional studies of carbonic anhydrase have dealt principally with the erythrocytic isoenzymes. These are two or three in number depending on species and in the human are known to show genetic variance. They are demonstrated best by gel ELECTROPHORESIS and column CHROMATOGRAPHY. The molecular weight reported for erythrocytic carbonic anhydrases approximates 30,000.[3] The isoelectric points of two human erythrocytic carbonic anhydrases are 5.7 and 7.3. One isoenzyme in the human red blood cell has 3–30 times the activity of another isoenzyme from the same source. The maximal rate of turnover in the hydration of carbon dioxide has been estimated for one isoenzyme as 620,000 molecules of carbon dioxide hydrated per molecule of enzyme per second. Erythrocytic carbonic anhydrases contain one atom of ZINC per molecule. The metal is essential for enzyme activity and can be reversibly removed. However, the plant enzyme does not contain zinc. Erythrocytic enzymes have been crystallized. The amino acid composition is known and reveals a low content of cysteine and methionine.

The study of the kinetics of carbonic anhydrase has been complicated by the use of impure preparations of enzymes (*e.g.*, presence of several isoenzymes), by effects dependent on the nature of the anions present, and by the rapidity of the reaction. Equations other than that of reaction (*1*) have been suggested in detailed reaction mechanisms.

References

1. TASHIAN, R. E., PLATO, C. C., AND SHOWS, T. B., *Science*, **140,** 53 (1963).
2. POCKER, Y., AND MEANY, J. E., *Abstr. Intern. Congr. Biochem. 6th, Washington, D.C.*, 327 (1964).
3. RICKLI, E. E., GHAZENFAR, S. A. S., GIBBONS, B. A., AND EDSALL, J. T., *J. Biol. Chem.*, **239,** 1065 (1964).
4. DAVENPORT, H. W., "Carbonic Anhydrase Inhibition and Physiological Function," *Ciba Found. Symp., Enzymes and Drug Action* (MONGAR, J. L., AND DE REUCK, A. V. S., EDITORS), 16 (1962).
5. DAVIS, R. P., "Carbonic Anhydrase," in "The Enzymes" (BOYER, P. D., LARDY, H., AND MYRBÄCK, K., EDITORS), Vol. 5, p. 545, New York, Academic Press, 1961.
6. DAVIS, R. P., "Measurement of Carbonic Anhydrase Activity," *Methods Biochem. Anal.* (GLICK, D., EDITOR), **11,** 307 (1963).
7. MALMSTROM, B. G., AND NEILANDS, J. B., "Metalloproteins," in *Ann. Rev. Biochem.*, **33,** 331 (1964).
8. PINCUS, G., AND BIALY, G., "Carbonic Anhydrase in Steroid-Responsive Tissues," *Recent Progr. Hormone Res.* (PINCUS, G., EDITOR), **19,** 201 (1963).

JAMES B. GILBERT

CARBOXYLATION ENZYMES

In living systems there are enzyme-catalyzed reactions by which carbon dioxide is added to another molecule. These may be defined as carboxylation enzyme processes. A new carbon-carbon (or a carbon-nitrogen) bond is formed, the CO_2 becoming a carboxyl group. Carboxylation reactions are ordinarily reversible, but in many instances the equilibrium is so far in the direction of free CO_2 that net CO_2 fixation is virtually impossible. Some of the latter instances are classified with decarboxylating enzymes (decarboxylases).

In most cells, the oxidation of carbon-containing metabolites to CO_2 and H_2O provides the energy for driving the mechanochemical manifestations of life (*e.g.*, muscular contraction). Many cells are consequently faced with the task of CO_2 elimination rather than CO_2 utilization. Nevertheless, CO_2 fixation is required for the maintenance of life since certain key metabolic transformations do involve a net addition of CO_2. The most important of these is clearly the process of PHOTOSYNTHESIS in which hexose is created from CO_2 and water [see also CARBON REDUCTION CYCLE (IN PHOTOSYNTHESIS)]. Indeed, this is the principal means by which carbon is brought into the biosphere.

The carboxylation enzymes may be placed into three major groups:[1] (1) reactions utilizing ATP, (2) systems coupled to the oxidation of reduced pyridine nucleotides, and (3) carboxylations having no additional "high-energy" cofactor requirement.

The Carboxylation Enzymes Requiring ATP. These enzymes contain the vitamin BIOTIN as a bound cofactor (prosthetic group) which participates in the carboxyl transfer. Although certain CO_2-fixing reactions were known to be impaired in biotin-deficient animals,[2] biotin has only recently been demonstrated to function as a true prosthetic group in an enzyme preparation.[3] AVIDIN, a protein from egg white that specifically binds biotin, blocks the enzyme system catalyzing the synthesis of fatty acids from acetyl coenzyme A (see FATTY ACID METABOLISM). The individual enzyme affected is acetyl-CoA carboxylase. Other enzymes carrying out similar reactions contain biotin[4,5] bound to the enzyme through the terminal (epsilon) amino group of the amino acid lysine. These enzymes are:

Pyruvate Carboxylase.

$$\text{ATP} + \overset{*}{\text{C}}\text{O}_2 + \underset{\text{Pyruvic acid}}{\text{CH}_3\text{COCOOH}} \rightleftharpoons \underset{\text{Oxaloacetic acid}}{\text{HOO}\overset{*}{\text{C}}\text{CH}_2\text{COCOOH}} + \text{ADP} + \text{P}_i$$

The net synthesis of oxaloacetic acid, a key intermediate in the CITRIC ACID (Krebs) CYCLE, is achieved by this carboxylase enzyme. A molecule of acetyl-CoA, bound to the enzyme, is necessary for catalytic activity, probably by influencing the three-dimensional structure (conformation) of the enzyme protein (see REGULATORY SITES OF ENZYMES). This is an example of a "positive feedback" control circuit, since oxaloacetate must be supplied to condense with (free) acetyl-CoA generated from other sources.

Acetyl-CoA Carboxylase.

$$\text{ATP} + \overset{*}{\text{C}}\text{O}_2 + \underset{\text{Acetyl-CoA}}{\text{CH}_3\text{COSCoA}} \rightleftharpoons \underset{\text{Malonyl-CoA}}{\text{HOO}\overset{*}{\text{C}}\text{CH}_2\text{COSCoA}} + \text{ADP} + \text{P}_i$$

This enzyme is essential for the synthesis of saturated fatty acids (*e.g.*, palmitic acid). The carboxylase product, malonyl-CoA, contributes two carbons (those which originated from acetyl-CoA) for progressive lengthening of the fatty acyl hydrocarbon chain. Long-chain fatty acids or fatty acyl-CoA inhibit the carboxylase. By this "negative feedback" circuit, fatty acids inhibit their own formation.

Propionyl-CoA Carboxylase.

$$\text{ATP} + \overset{*}{\text{C}}\text{O}_2 + \underset{\text{Propionyl-CoA}}{\text{CH}_3\text{CH}_2\text{COSCoA}} \rightleftharpoons \underset{\text{2-Methylmalonyl-CoA}}{\text{CH}_3\text{CH}\overset{*}{\text{C}}\text{OOHCOSCoA}} + \text{ADP} + \text{P}_i$$

Propionyl-CoA, a product of catabolism of certain amino acids and odd-numbered fatty acids, is oxidized to completion by way of methylmalonyl-CoA which is converted by other enzymes to succinyl-CoA, an intermediate in the citric acid cycle.

Methylcrotonyl-CoA Carboxylase.

$$\text{ATP} + \overset{*}{\text{C}}\text{O}_2 + \underset{\text{3-Methylcrotonyl-CoA}}{(\text{CH}_3)_2\text{C}{=}\text{CHCOSCoA}} \rightleftharpoons \underset{\text{3-Methylglutaconyl-CoA}}{\text{HOO}\overset{*}{\text{C}}\text{CH}_2(\text{CH}_3)\text{C}{=}\text{CHCOSCoA}} + \text{ADP} + \text{P}_i$$

The carboxylation product may be subsequently converted to 3-hydroxy-3-methylglutaryl-CoA (HMG CoA) which in turn leads to acetoacetate formation or cholesterol synthesis (see ISOPRENOID BIOSYNTHESIS).

The biotin enzymes just listed are especially important since they catalyze net CO_2 fixation. These systems share a common mechanism: The protein-bound biotin condenses with CO_2 in the presence of ATP. The carboxyl group attached to the ring nitrogen of biotin (*i.e.*, an "active CO_2") is then transferred to a specific acceptor[4] (see Fig. 1 and Equations 1 and 2).

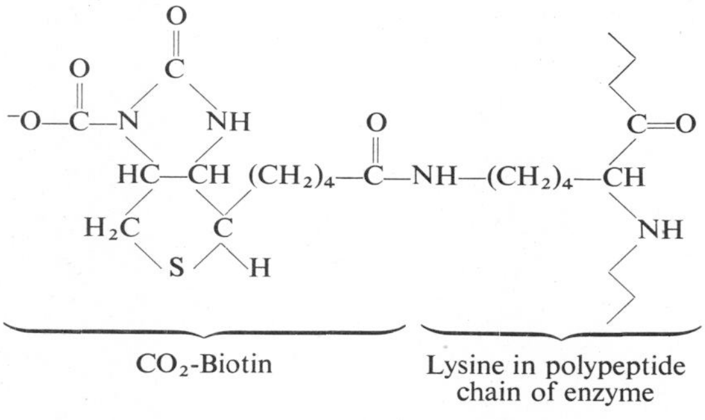

O
‖
O C
‖ / \
⁻O—C—N NH O C=O
\ / ‖ /
HC—CH (CH₂)₄—C—NH—(CH₂)₄—CH
/ \ / \
H₂C C NH
\ S / \ H

CO₂-Biotin | Lysine in polypeptide chain of enzyme

FIG. 1. The CO_2-biotin enzyme.

(1) $\text{ATP} + \text{CO}_2 + \text{Biotin-Enzyme} \rightleftharpoons {}^{-}\text{OOC-Biotin-Enzyme} + \text{ADP} + \text{P}_i$

(2) ${}^{-}\text{OOC-Biotin-Enzyme} + \text{Acceptor} \rightleftharpoons {}^{-}\text{OOC-Acceptor} + \text{Biotin-Enzyme}$

Except for a transcarboxylase enzyme, in which free CO_2 does not appear as an intermediate and ATP is not required, no other biotin-containing enzymes are known.

There is an enzyme system which depends on ATP to effect CO_2 incorporation, but which does not involve biotin. This is the carbamate kinase of the urea cycle. The product, CARBAMYL PHOSPHATE, is an active carbamyl donor in several reactions.

$$2ATP + \overset{*}{C}O_2 + NH_3 + H_2O \rightarrow H_2N\overset{*}{C}OOPO_3H_2 + 2ADP + P_i + H^+$$

The Carboxylation Enzymes Requiring Reduced Pyridine Nucleotides. These reactions must be coupled to the oxidation of reduced pyridine nucleotides (also termed NICOTINAMIDE ADENINE DINUCLEOTIDES) if any new carbon-carbon bonds are to be established. The group is technically classified with the oxido-reduction enzymes. In contrast to the biotin enzymes, these reactions ordinarily proceed in the direction of decarboxylation (and the formation of reduced pyridine nucleotides).

Malic Enzyme.

$$NADP + \underset{\text{Malic acid}}{HOO\overset{*}{C}CH_2COHCOOH} \rightleftharpoons NADPH + H^+ + \underset{\text{Pyruvic acid}}{CH_3COCOOH} + \overset{*}{C}O_2$$

Isocitric Dehydrogenase.

$$NAD + \underset{\text{Isocitric acid}}{HOOCCH_2OHCH(\overset{*}{C}OOH)CH_2COOH} \rightleftharpoons NADH + H^+ + \underset{\text{2-Ketoglutaric acid}}{HOOCCOCH_2CH_2COOH} + \overset{*}{C}O_2$$

Phosphogluconate Dehydrogenase.

$$NADP + \underset{\text{6-Phosphogluconic acid}}{HOO\overset{*}{C}—(CHOH)_4—OPO_3H_2} \rightleftharpoons NADPH + H^+ + \underset{\text{Ribulose-5-phosphate}}{CH_2OH—CO—(CHOH)_2—CH_2OPO_3H_2} + \overset{*}{C}O_2$$

The Carboxylation Enzymes Not Requiring Cofactors. In this group no high-energy cofactor is needed to initiate carboxylation since the carbon dioxide acceptor possesses sufficient potential to receive the CO_2 directly. There is no evidence to suggest that biotin is involved.

Diphosphoribulose Carboxylase.

$$\underset{\text{Ribulose-1,5-diphosphate}}{H_2PO_3OCH_2—CO—(CHOH)_2—CH_2OPO_3H_2} + \overset{*}{C}O_2 + H_2O \rightarrow \underset{\text{2 Molecules of 3-phosphoglyceric acid}}{H_2PO_3OCH_2—CHOH—\overset{*}{C}OOH + HOOC—CHOH—CH_2OPO_3H_2}$$

This is the initial step of the CARBON REDUCTION CYCLE (the "dark reaction") in photosynthesis. In this process, it is postulated that a six-carbon intermediate is formed and immediately splits into two molecules of 3-phosphoglyceric acid. The latter lead to the net production of one hexose for every six molecules of CO_2 fixed.

Phosphoribosyl-aminoimidazole Carboxylase.

$$\text{5'-Phosphoribosyl-5-aminoimidazole} + CO_2 \rightarrow \text{5'-Phosphoribosyl-5-amino-4-imidazole carboxylate}$$

One of the carbons of the purine ring enters by means of this reaction (see PURINE BIOSYNTHESIS).

Phosphopyruvate Carboxylase.

$$GTP + \underset{\text{Oxaloacetic acid}}{HOO\overset{*}{C}CH_2COCOOH} \rightleftharpoons \underset{\text{Phosphopyruvic acid}}{CH_2{=}CHOPO_3H_2COOH} + GDP + \overset{*}{C}O_2$$

This enzyme is important in gluconeogenesis since the carbons from the citric acid cycle intermediates may be converted to glucose by way of oxaloacetate and phosphopyruvate. The synthesis of oxaloacetate (the reverse carboxylation reaction) is not as favorable. Note that the guanosine nucleotides are involved in this reaction, but not as a source of energy in the direction of carboxylation.

References

1. Calvin, M., and Pon, N. G., *J. Cellular Comp. Physiol., Suppl. 1*, **54**, 51 (1959).
2. Lardy, H. A., and Peanasky, R., *Physiol. Rev.*, **33**, 560 (1953).
3. Wakil, S. J., and Gibson, D. M., *Biochim. Biophys. Acta*, **41**, 122 (1960).
4. Lynen, F., *Federation Proc.*, **20**, 941 (1961).
5. Vagelos, P. R., *Ann. Rev. Biochem.*, **33**, 139 (1964).

D. M. Gibson

CARBOXYPEPTIDASES. See PROTEOLYTIC ENZYMES.

CARCINOGENIC SUBSTANCES

Carcinogenic substances (or carcinogens) are chemical compounds of known or unknown structure which produce tumors in multicellular organisms.

Some Characteristics of Tumors. A tumor is an abnormal new tissue growth (neoplasm), which can originate in any tissue and which is generally not, or only partially, under the physiological control of the organism. A *benign* tumor has generally a very slow growth rate and sometimes regresses and disappears spontaneously; it does not invade the neighboring normal tissue from which it originates and in certain tissues it is sometimes encapsulated by connective tissue. A *malignant* tumor (commonly called a cancer) may have a slow or fast growth rate, but its tendency to grow is irreversible; it invades and destroys neighboring tissues. Malignant tumors often (but not always) have a tendency to *metastasize*, meaning that individual tumor cells or small clusters of cells break off from the tumor tissue (*primary tumor*), are carried away by the blood or lymph, and become lodged in distant regions of the animal body where they originate new tumors (*metastatic tumors*). Benign tumors sometimes turn malignant; in fact, some regard benign tumor cells as an intermediate stage in the malignant transformation of normal cells. The terminology of tumors is exemplified in the following. *Papilloma:* benign tumor of the skin or mucous surfaces. *Epithelioma:* malignant tumor originating from epithelial cells. *Fibroma:* (encapsulated) benign tumor originating from connective tissue. *Sarcoma:* malignant tumor originating from connective tissue. *Adenoma:* benign or malignant tumor of glandular origin. *Hepatoma:* malignant tumor originating from hepatic cells.

Testing Procedures. Because of their short life span (average 3 years), small rodents (mice, rats and hamsters) are most often used for the testing of chemicals for carcinogenic activity; occasionally testing is done in rabbits, dogs, fowls, monkeys, etc. While a great variety of ways of administration have been used, most commonly substances are tested in the following ways: (a) *skin "painting"*: small volumes of solution of the substance in an inactive solvent (*e.g.*, benzene) are applied to the shaved surface of the skin (generally of mice, in the interscapular region) daily or at longer intervals; (b) *subcutaneous injection* of the pure substance or its solution (once or at repeated intervals); (c) *feeding:* the substance is mixed in the diet at given levels, or dissolved in the drinking water. Testing of new substances for possible carcinogenic activity is conducted for a *minimum* of 1 year to be meaningful. At the end of the testing period all animals are autopsied, and all tumors and dubious tissues examined histopathologically.

A carcinogen which is highly active in one species may be totally inactive in another species, and *vice versa*. The *susceptibility* of a species to a given carcinogen also depends on the genetic strain, sex and dietary conditions. Moreover, carcinogenic substances generally show a rather selective *specificity toward certain target tissues; e.g.*, certain compounds produce exclusively hepatomas in the susceptible species. For these reasons, *no chemical compound may be stated safely to be devoid of carcinogenic activity toward man unless it has been found inactive when tested in a variety of mammalian species and by a variety of routes of administration for a length of time corresponding to half of the life span of each species*.

Some Types of Chemical Carcinogens. The high incidence of skin cancer in coal tar workers was recognized as early as 1880. The carcinogenic activity of coal tar was demonstrated in 1915, when Yamagiwa and Ichikawa obtained epitheliomas by its prolonged application to the ears of rabbits. Identification of the active material (in 1933) as the polycyclic aromatic hydrocarbon 3,4-benzopyrene (III, Fig. 1) is due to Cook, Kennaway, Hieger and their co-workers. This discovery was followed up by the synthesis and testing of a considerable variety of polycyclic aromatic hydrocarbons. All compounds of this class may be regarded as composed of condensed benzene rings. The arrangement of the hexagonal rings in various patterns results in a variety of compounds having different physical, chemical and biological properties. However, not all polycyclic aromatic hydrocarbons possess carcinogenic activity; certain requirements of molecular geometry must be met. For maximum activity the molecule must have (Fig. 1): (a) an optimum size; (b) a coplanar molecular conformation, meaning that all hexagonal rings must lie flatly in one plane; *in fact*, hydrogenation of many of the active hydrocarbons results in buckled molecular

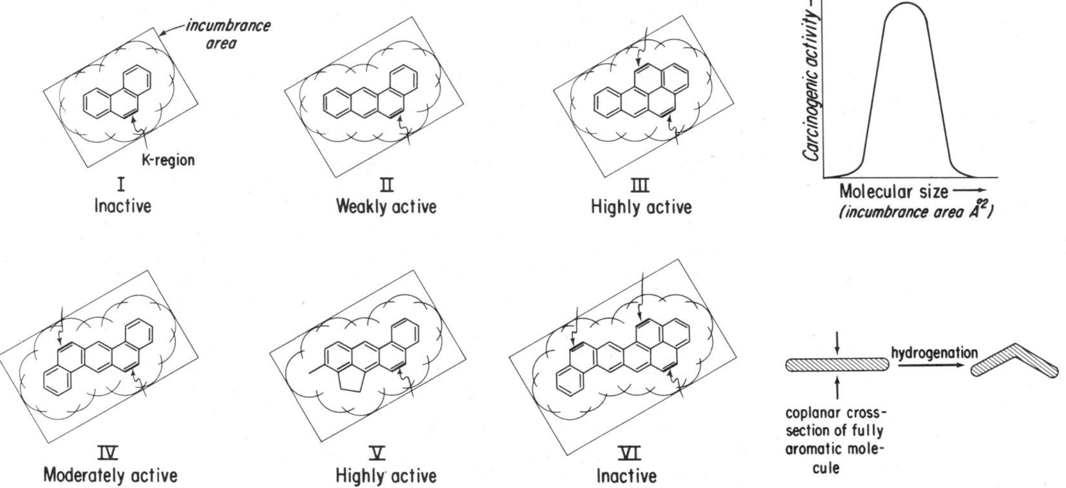

FIG. 1.

conformations and this is concomitant with partial or total loss of activity; (c) at least one meso-phenanthrenic double bond, also called K-region (see arrows, Fig. 1), of high π-electron density (*i.e.*, of high chemical reactivity). In addition to III, 1,2,5,6-dibenzanthracene (IV), and 20-methylcholanthrene (V) are commonly used to study the experimental induction of tumors. The activity of most hydrocarbon carcinogens was tested toward the skin of mice and the subcutaneous connective tissue of mice and rats. There is a vast body of evidence indicating that 3,4-benzopyrene and other carcinogenic hydrocarbons are formed during pyrogenation or incomplete burning of almost any kind of organic material. For example, carcinogenic hydrocarbons have been identified in overheated fats, broiled and smoked meats, coffee, burnt sugar, rubber, commercial paraffin oils and solids, soot, the tar contained in the exhaust fumes of internal combustion engines, cigarette smoke, etc.

Attention to the *carcinogenic aromatic amines* (Fig. 2) was drawn by the high incidence of urinary bladder tumors in dye factory workers exposed to 2-naphthylamine (VII) and benzidine (IX). The carcinogenic activity of VII, IX and 4-aminobiphenyl (X) toward the bladder of the dog and the mouse has been demonstrated. In the rat, however, there is a change in target specificity, and tumors are induced by IX and X in the liver, mammary gland, ear duct and small intestine. Carcinogenic activity is considerably heightened in 2-acetylaminofluorene (XI), without change of target specificity. Increased activity is due to the fact that XI is more coplanar than X, because of the internuclear methylene (—CH_2—) bridge in the former. 2-Acetylaminofluorene was proposed as an insecticide before its carcinogenic activity was accidentally discovered; it is a ubiquitous, potent carcinogen in a variety of species. Changing the internuclear bridge of XI to a —CH═CH—, as in 2-aminophenanthrene (XII), causes a shift in target specificity; thus, in the rat, XII is inactive toward the liver, but in addition to inducing tumors in the mammary gland, ear duct and small intestine, it produces leukemia (a malignant condition characterized by the production of abnormal and immature circulating LEUKOCYTES). Compound XII represents a structural link between the aromatic amine and polycyclic hydrocarbon carcinogens (compare XII to I); it is also interesting in this respect that 2-aminoanthracene (VIII), which is a higher homologue of VII, is inactive toward the bladder but potent to induce skin tumors in rats.

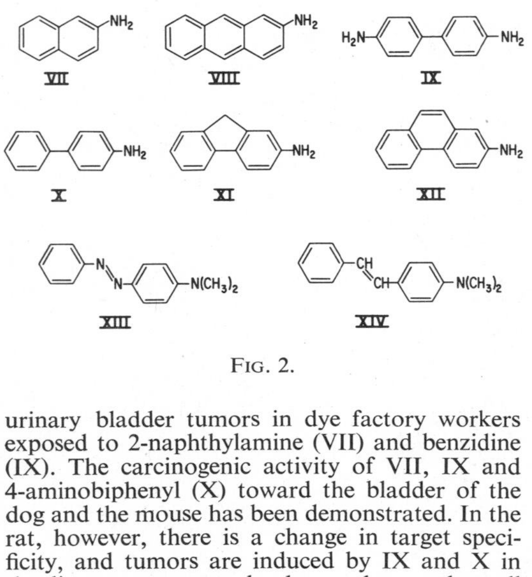

VII (NH2) VIII (NH2) IX (H2N, NH2)

X (NH2) XI (NH2) XII (NH2)

XIII (N, N, N(CH3)2) XIV (CH, CH, N(CH3)2)

FIG. 2.

4-Dimethylaminoazobenzene (XIII) is the parent compound of the *aminoazo dye carcinogens;* it is also known in the earlier literature as Butter Yellow, because it was used to color butter and vegetable oils before its carcinogenic activity was discovered. Many derivatives of XIII have been prepared and tested for carcinogenic activity. In the rat, the aminoazo dye carcinogens, administered in the diet, specifically induce hepatomas. Tumor induction by most of the aminoazo dyes is delayed or inhibited by high dietary levels of RIBOFLAVIN (vitamin B_2) or protein. Replacement of the —N═N—azo linkage by —CH═CH—, as in 4-dimethylaminostilbene (XIV), results in widening the target tissue spectrum; XIV induces tumors in the liver, mammary gland and ear duct. Mice are much more resistant than rats to the carcinogenic activity of both aminoazo dyes and aminostilbenes.

Figure 3 illustrates some *aliphatic carcinogens.* N-Methyl-bis-β-chloroethylamine (XV), a nitrogen mustard, produces local sarcomas, lung, mammary and hepatic tumors upon injection in mice; because of its tumor inhibitory properties, XV is also used in the therapeutic management of certain types of human cancers. Bisepoxybutane (XVI), β-propiolactone (XVII) and N-lauroylethyleneimine (XVIII) produce local sarcomas in rats upon injection. Ethylcarbamate (XIX), the parent compound of several hypnotic drugs used in humans, produces malignant lung adenomas in rats, mice and chickens. Dimethylnitrosamine (XX) is a potent carcinogen toward the liver, lung and kidney, and ethionine (XXI) toward the liver of the rat; the former is an intermediate in the manufacture of the rocket-fuel component, dimethylhydrazine $(CH_3)_2N—NH_2$, while the latter is the S-ethyl analogue of the natural amino acid, METHIONINE.

Classification of Carcinogenic Substances. Chemically identified carcinogens may be classified as inorganic ions and organic compounds. The *inorganic carcinogens* are the elements beryllium, cadmium, iron, cobalt, nickel, silver, lead, zinc and probably arsenic; these can form coordination compounds and/or react with sulfhydryl groups. Also asbestos powder is a powerful carcinogen toward the lung upon inhalation (asbestos cancer of miners). The organic carcinogens may be subdivided into synthetic aromatics and aliphatics, and naturally occurring substances. The main classes of *aromatic carcinogens* are: (a) Condensed polycyclic aromatic hydrocarbons and heteroaromatic polycyclic compounds; (b) aromatic amines and N-aryl hydroxylamines; (c) aminoazo dyes and diaryl-azo compounds; (d) aminostilbenes and stilbene analogues of sex hormones. The main classes of *aliphatic*

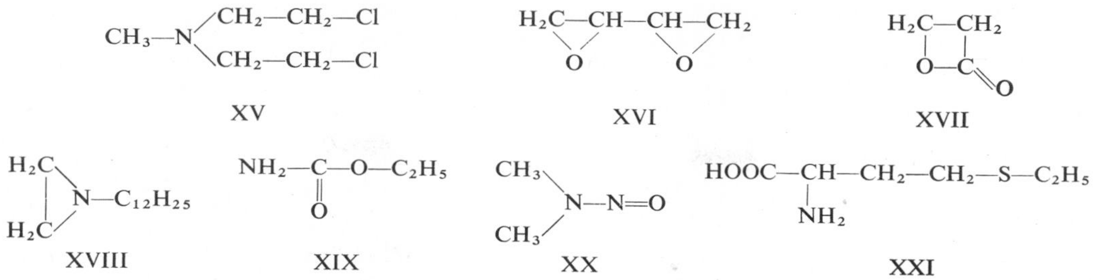

$CH_3—N(CH_2—CH_2—Cl)_2$

XV

$H_2C—CH—CH—CH_2$ (two epoxide rings, each bridged by O)

XVI

$H_2C—CH_2$, $O—C=O$ (four-membered ring)

XVII

H_2C, H_2C >N—$C_{12}H_{25}$

XVIII

$NH_2—C(=O)—O—C_2H_5$

XIX

$(CH_3)_2N—N=O$

XX

$HOOC—CH(NH_2)—CH_2—CH_2—S—C_2H_5$

XXI

FIG. 3

carcinogens are: (a) alkylating agents (such as sulfur and nitrogen mustards, derivatives of ethyleneimine, lactones, epoxides, alkane-α,ω-bis-methanesulfonates, certain dialkylnitrosamines, and ethionine); (b) lipophilic agents, detergent-like compounds and hydrogen-bond reactors; this class comprises a wide variety of agents, such as chlorinated hydrocarbons (chloroform, carbon tetrachloride, the compounds used as pesticides under the names of aldrin and dieldrin), bile acids, "Tween"-type detergents, certain water-soluble high polymers, certain phenols, urethane and some of its derivatives, thiocarbonyls, and cycloalkylnitrosamines; (c) naturally occurring carcinogens. Until about 1950, the idea prevailed that the activity of carcinogenic chemicals was somehow related to the fact that they were synthetic, "unnatural" substances which, since they were not present in a natural environment, were not factors of selection during the evolutionary process; hence, contemporary living organisms were not equipped for the effective metabolic "detoxication" of these compounds. *However, during the last 15 years a number of carcinogenic compounds of plant and fungal origin have been identified:* safrole in sassafras, capsaicine in chili peppers, various tannins, cycasin in the cycad groundnut, parasorbic acid in mountain ash berry, pyrrolizidine alkaloids in *Senecio* shrubbery, and patulin, griseofulvin, penicillin G, aflatoxin and actinomycin produced by various molds. The number and variety of identified naturally occurring carcinogens continue to increase rapidly.

Mechanisms of Action of Carcinogens. The biochemical pathways in the cell are closely interconnected and are in a state of dynamic equilibrium (HOMEOSTASIS). This equilibrium is maintained by feedback relationships existing between a great number of pathways [see METABOLIC CONTROLS (FEEDBACK MECHANISMS)]. Chemical "communication" between subcellular organelles, such as the *nucleus* (within which the chromosomes contain the genetic "blueprints" for cell reproduction and the synthetic processes of cell life), the *mitochondria* (the "powerhouse" of the cell, which assures the synthesis of the universal cellular fuel, ATP, through the metabolism of carbohydrates and fatty acids), and the *endoplasmic reticulum* (synthesizing the proteins of the cell and assuring the metabolic breakdown—"detoxication"—of a multitude of endogenous and foreign compounds), depends on the constant interchange of a large variety of metabolic products and inorganic ions between them. There are probably a very great number of loci (*receptor sites*) upon which these regulatory chemical "stimuli" act (see REGULATORY SITES OF ENZYMES). The receptor sites are of enzymic and nucleic acid nature. Other control points of protein character regulate the morphology of the intracellular lipoprotein membranes which serve as "floor space" to the organized arrangements of multienzyme systems (see MULTIENZYME COMPLEXES). The specificity of compounds of chemical control toward given receptor sites is due to three-dimensional geometric "fit" following the lock and key analogy. Such is the general scheme of functional interrelationships in monocellular organisms which, hence, in a favorable medium multiply unchecked to the limit of the availability of nutrients.

In multicellular organisms, the subordination of the individual cells to the whole is assured by the existence of additional receptor sites which enable the cells to be responsive to chemical "stimuli" emitted by neighboring cells in the tissue and to hormonal regulation by the endocrine system in higher organisms (see HORMONES). Hence, depending on the requirements of the moment, cells may remain stationary or may undergo cell division because of need for repair of tissue injury, may secrete different products, or may perform other specialized function depending on the nature of the particular tissue.

Carcinogenic substances are *nonspecific cell poisons* which cause the alterations and, hence, *functional deletion* of a large number *of metabolic control sites.* Present evidence suggests that these alterations are produced by the accumulation of the carcinogen in subcellular organelles, by covalent binding of the carcinogen to cellular macromolecules (proteins and nucleic acids) through metabolism, and by denaturation (*i.e.*, destruction of the three-dimensional geometry) of the control sites through secondary valence interactions (hydrogen bonds, hydrophobic bonding, etc.) with the carcinogen [see also CANCER CELL METABOLISM; CARCINOMA AND OTHER CANCERS (BIOCHEMICAL ASPECTS)]. Early stages of tumor induction generally coincide with extensive cell death (necrosis) in the target tissue because a number of the biochemical lesions cause the irreversible blocking of metabolic pathways essential for cell life. However, because of the random distribution of the biochemical lesions in the cell population, in a small number of cells vital pathways are only slightly damaged and the lesions involve those sites and pathways which are not essential for cell life proper but are necessary for organismic control. Thus, due to the action of the carcinogen, these cells escape physiological control and revert to a simpler, less specialized cell type (*i.e.*, dedifferentiate). Such cells respond to continuous nutrition with continuous growth, which is an essential characteristic of malignant tumor cells.

References

1. Wolf, G., "Chemical Induction of Cancer," Cambridge, Harvard University Press, 1952.
2. Greenstein, J. P., "Biochemistry of Cancer," New York, Academic Press, 1954.
3. Arcos, J. C., and Arcos, M., "Molecular Geometry and Mechanisms of Action of Chemical Carcinogens," *Prog. Drug. Res.* (Jucker, E., Editor), **4**, 407–581 (1962).
4. Clayson, D. B., "Chemical Carcinogenesis," Boston, Little, Brown and Co., 1962.
5. Boyland, E. (Editor), "Mechanisms of Carcinogenesis: Chemical, Physical and Viral," *Brit. Med. Bull.*, **20**, 87–170 (1964).

Joseph C. Arcos

CARCINOMA AND OTHER CANCER (BIOCHEMICAL ASPECTS)

The fundamental problem of cancer is to understand the biochemical mechanisms which regulate cell multiplication and growth. In the regenerating liver in a hepatectomized animal, cells may grow faster than those of hepatoma. When the regenerating liver attains approximately normal size, growth diminishes markedly, whereas in cancer tissues, cells continue to multiply without restraint. Is it due to lack of normal control mechanism in tumor tissues or is it that cancer cells become insensitive, nonresponsive to the control mechanism? Available experimental evidence points to both possibilities. Our knowledge on the control mechanism of cell multiplication and tissue growth is still meager: feedback regulation, hormonal control, tissue cell interaction; many possibilities exist but much remains to be learned. Feedback regulation is the inhibition of a biosynthetic reaction by its end product. One such example bearing on cell division and tissue growth is the feedback inhibition of thymidine kinase by thymidine triphosphate studied with materials from tumor cells by Smellie and by Van Potter.[1] This enzyme plays a key role in DEOXYRIBONUCLEIC ACID (DNA) biosynthesis. Certain defects in the control mechanism were found in tumor cells but not in normal liver, regenerating liver or the minimal deviation hepatomas (see also CANCER CELL METABOLISM; METABOLIC CONTROLS).

Hormonal Control. There is little doubt that HORMONES play an important role in the development and growth of neoplasms. Age and sex differences in the incidence of sarcoma and carcinoma point to the influence of endocrines on carcinogenesis. Animal experiments demonstrated that some hormones can cause cancer while others inhibit tumor growth or even induce profound regression. Steroid hormones have been used with some success in the management of prostatic and mammary cancers. Of special interest is the interaction between hormones. Huggins and Yang[2] observed that progesterone, an enhancer of breast cancer, becomes a suppressor when combined with minute quantity of estradiol-17β. Such demonstrated interaction suggests that changes in the balance between various hormones may either favor or prevent the growth of neoplasms.

Tissue cell interaction may be stimulatory or inhibitory on cell division and growth. The interdependence of epithelium and stroma for growth and function has been reviewed by Herrman, and specific interaction between thymus and spleen was reported by Auerbach. Dinning *et al.*[3] isolated a substance from rabbit liver which inhibited the incorporation of formate-C^{14} into DNA in bone marrow preparations. Contact inhibition[4] is another mechanism of cellular interaction which regulates cell division and movement in normal tissues. Studies revealed that many growing tumor cells are not subject to contact inhibition; changes in lipid composition of cell membrane with concomitant alteration in cell surface properties appear to be involved.

Studies on the deviation from the norm of cancer cells in chemical composition, metabolic pattern, enzyme activities, etc., have been carried out by many investigators.[1b] Some quantitative differences between certain types of cancer cells and their normal prototype have been found, but a common denominator characteristic of malignancy acceptable to everyone has yet to be discovered. Cancer cells resemble embryonic cells in carbohydrate metabolism, in cell structure characterizing immaturity,[1] and in their capacity to draw nutrients from the host regardless of existing deficiency. As a fetus can draw essential nutrients such as calcium from maternal blood to the extent that the mother's bones become softened and develop osteomalacia, cancer cells can absorb essential nutrients to the detriment of the host.

Abnormalities in lipid metabolism occur frequently in cancer tissues and the tumor-bearing host (hyperlipemia). Ambrose *et al.*[5] compared the effect of enzymes on growing cultures of normal kidney epithelia and of kidney tumors and found lipase showed the most strikingly selective inhibition on the tumor cultures. With the use of fluorescence labels, these authors further observed a characteristic difference between the reaction of the lipase with the normal and the tumor-cell membranes. The abnormal lipid composition of the cell membrane mentioned earlier may contribute to the invasiveness and freedom from contact inhibition of cancer cells. Rapport[6] isolated a lipid hapten, Cytolipin "H" from human tumor tissues, a cerebronyl-sphingosyl-glucosyl-galactoside, which they could not detect in normal tissues. Cytolipin "H" reacts with most rabbit sera immunized with human tumors (see CEREBROSIDES).

Carcinogenesis may be brought about by chemical or physical means or by virus infection; the cause may vary but the mechanism is probably the same—change in GENE structure. It has been known that a chemical as simple as nitrous acid can alter the structure of DNA; other carcinogens such as 3-methylcholanthrene, 2-acetylaminophenanthrene, benzo(α)pyrene, etc., probably act in the same manner (see CARCINOGENIC SUBSTANCES). The action of radiation or virus is known to change gene structure. However with virus infection, the mechanism is more complicated and variable than radiation [see VIRUSES (REPLICATION)]. Virus infection occurs only between certain viruses and specific types of cells (host specificity) and the resulting events vary according to the virulence of the virus and the response of the host cell. In some cases, the entire synthetic mechanism is converted to the formation of new viruses soon after infection, culminating in the destruction of the cell. In other instances, the viral DNA or RNA is believed to combine with cellular DNA forming hybrids or complexes altering the gene structure of the host cell. The virus disappear from the scene; the cell continues to multiply; a neoplasm may be formed. In other cases, like phage-infected lysogenic bacteria, the cell remains normal in appearance for a long time; it continues to replicate producing apparently healthy daughter cells but carries the virus inside. Later when environment changes, the cell may suddenly start to form viruses resulting in its own destruction (see BACTERIAL GENETICS). In this phenomenon known as "lysogeny," the viral DNA is unable to take command over the cell it has infected, but manages to live peaceably in delicate balance with the host and even to get itself replicated every time the cell DNA is replicated. During the latent period the virus seems not to do the host cell any harm and not a single complete virus particle can be found in the infected cell. But as soon as conditions become unfavorable to the cell, the precarious balance is upset, and the virus takes command and manufactures new viruses. Lysogeny has been the basis of argument for those advocating the theory that all malignant neoplasms have viral origin.

In the search for the cure of cancer, chemotherapy, antibiotics, plant alkaloids, hormonal and immunochemical control have all been pursued. Chemotherapy has not yet achieved its goal of selective action against cancer cells; since these cells arise from host cells, the biochemical difference between the tumor and the host tissues is not as much as that between the pathogenic bacteria or parasite and the host. Similar difficulties exist in the immunochemical approach for the cure of malignancy. Plant alkaloids from *Vinca rosea* have shown some promising effects on Hodgkins disease and certain types of leukemia.

An important factor in the use of antimetabolites is cell response; cancer cells often exhibit greater capacity to overcome an inhibitor than normal cells (a) by producing neutralizing substance or (b) by increasing enzyme activity. It has been shown that in normal and leukemic bone marrow cultures, cell division stopped soon after addition of a folinic acid antagonist; however, the leukemic cells began to divide actively after 24 hours, whereas normal bone marrow cells remained inhibited as long as they were exposed to the antagonist. Filtrates from leukemic cultures were no longer inhibitory to normal or leukemic cells, but filtrates from normal bone marrow cultures were still active. Huennekens *et al.*[7] observed that in leukemic patients, after treatment with the folic acid antagonist—amethopterin, dihydrofolic acid reductase activity in leukocytes increased 10–20 fold while the same drug had very little effect on the enzyme activity of leukocytes in normal individuals. (See also CANCER CHEMOTHERAPY; LEUKOCYTES AND LEUKEMIA).

Although many biochemists think that malignant transformation is irreversible, it is difficult to believe that changes in gene structure can only go in one direction. It is a well known fact that a certain number of patients bearing malignant tumors recover completely due to spontaneous regression. Similarly, in tissue culture, while cells of normal origin can undergo malignant transformation, those of tumorous origin often lose their capacity to produce tumor in susceptible animals. In the development of malignant neoplasms, the soil is as important as the seed, as exemplified by the negative result of a blood transfusion experiment from leukemic patients to volunteer prisoners. As Huggins and Yang have shown, the combination of two hormones can convert an enhancer of breast cancer to a suppressor, the balance between various hormones may play a major role on the "soil" condition—the *milieu interne*. Certain nutritional factors, especially some vitamins are known to affect the production and metabolism of hormones; hence their availability in the diet may alter the endocrine balance of the organism. With better understanding of the interaction among various hormones and between vitamins and hormones, it shall be possible to change the character of the *milieu interne* so that it becomes unsuitable for malignancy to develop and grow.

References

1. (a) Proceedings American Cancer Society Conference on Nucleus of the Cancer Cell, *Exptl. Cell Res. Suppl.* **9,** 245, 259 (1963).
(b) BRACHET, J., AND MIRSKY, A. E. (EDITORS), "The Cell," Vol. V, Chs. 7 and 8, pp. 405–496, 497–544, New York and London, Academic Press, 1961.
2. HUGGINS, C., AND YANG, N. C., "Induction and Extinction of Mammary Cancer," *Science*, **137,** 257 (1962).
3. DINNING, J., AND WILES, L., "A Substance in Liver which Inhibits Thymine Biosynthesis by Bone Marrow," *Science*, **129,** 336 (1959).
4. ABERCROMBIE, M., AND AMBROSE, F. J., "Surface Properties of Cancer Cells," *Cancer Res.*, **22,** 525 (1962).

5. Ambrose, E. J., *et al.*, "The Inhibition of Tumor Growth by Enzymes in Tissue Culture," *Exptl. Cell Res.*, **24**, 220 (1961); "Interaction of Enzyme with Tumor Cells," *Nature*, **190**, 1207 (1961).
6. Rapport, M. M., *et al.*, "Immunological Studies of Organ and Tumor Lipides. IV. Isolation and Properties of Cytolipin 'H'," *Cancer*, **12**, 438 (1959).
7. Bertino, J. R., *et al.*, and Huennekens, F. M., "Increased Levels of Dihydrofolic Reductase in Leucocytes of Patients Treated with Amethopterine," *Nature (London)*, **193**, 140 (1962).

Chiun T. Ling

CARIES, DENTAL (BIOCHEMICAL ASPECTS)

Dental caries, or tooth decay, is the pathological process of localized destruction of tooth tissues by microorganisms.

Heredity has been demonstrated to be a factor in susceptibility to dental caries. (a) Rats that showed early development of carious lesions and (b) those from the same strain with later caries development were bred. The same dietaries were used for all. The offspring in both series were used for succeeding generations. The susceptible strain would develop carious lesions within 40 days; the resistant strain generally went about 10 times as long with no detectable lesions. In communities with the same water supply and dietaries, Negro children generally have lower caries rates than white children.

Constituents of Teeth. *Enamel* is produced by ectodermal tissues. Human enamel contains approximately 0.4% protein. The protein of enamel is a eukeratin, based on the histidine: lysine:arginine ratio of 1:4:12 but possibly contains proline and hydroxyproline and a small amount of cystine, which is not typical of true keratins. *Dentin* and *cementum* are of endodermal origin. Human dentin contains approximately 20% protein. It is similar in composition to COLLAGEN.

The principal inorganic constituent of teeth is apatite, best shown as $3Ca_3(PO_4)_2{\cdot}Ca(OH)_2$. Magnesium replaces variable amounts of calcium, and carbonate replaces some of the hydroxyl. Minor replacements of hydroxyl by chlorine and fluorine occur, and trace amounts of lead, tin, manganese, iron, aluminum and strontium have been found.

The major elements of sound and carious enamel are as follows, expressed as per cent dry weight:

	Sound	Carious
Calcium	36.75 ± 0.17	35.95 ± 0.21
Magnesium	0.54 ± 0.01	0.40 ± 0.01
Phosphorus	17.41 ± 0.04	17.01 ± 0.06
CO_2	2.42 ± 0.02	1.56 ± 0.03
Ca/P	2.07 ± 0.02	2.08 ± 0.03
H_2O	2.02 ± 0.04	3.07 ± 0.05

Resistance to carious attack increases with age of the teeth. The density of enamel increases. There is an increase in phosphorus content of enamel and a marked increase in the fluorine content of the outer layer of enamel.

Analyses of sound and carious dentin show a tenfold increase in the fluoride content. Some other trace elements, such as zinc and tin also increase. This suggests remineralization of dentinal crystallites adjacent to those undergoing destruction. This hypothesis is supported by variation in morphology of crystallites in carious areas of dentin.

The organic matrix of carious dentin has a tendency to persist. This indicates that the initial destruction of dentin involves removal of the inorganic constituents.

Development of Caries. The process of the *initiation* of carious lesions probably differs from that of further development of cavities. For example, the process may be initiated in pits and fissures by acid attack on enamel. Smooth surface lesions may be initiated by microfracture and subsequent inoculation with microorganisms related to the development of the cavity. Thus formative fluoride is more effective in protection against smooth surface lesions than the pit and fissure type.

Carious lesions develop in both deciduous and permanent teeth and in specific locations, indicating possible difference in the sequence of factors that initiate the process. Pit and fissure lesions begin in areas of that description, principally in the occlusal surfaces of the molar and premolar teeth. Smooth surface lesions are generally in the interproximal areas of adjacent teeth. Supragingival cavities develop principally in buccal areas. Subgingival lesions, beginning below the enamel, namely, in the cementum, apparently originate because of gingival retraction.

After the carious process has been initiated, the development of cavities occurs because of the action on the tooth substances of products of metabolism of microorganisms that develop within the incipient lesion. The process may be arrested by change of conditions before any detectable lesion develops. The carious process may be reactivated, and this alternation of activity of specific lesions is probably common to all cavities, especially in the beginning stage.

Another type of carious lesion is recurrent caries, so-called because of its occurrence at the borders of fillings placed by a dentist.

Destruction of tooth substances in the carious process is by action of acids, chelating agents and PROTEOLYTIC ENZYMES. The predominant belief is that highly localized acid attack first removes the inorganic constituents and the organic substances are destroyed later by enzyme action. Chelation may occur in the process of destruction of both organic and inorganic material.

High saliva flow rates have been found associated with low rates of dental caries. Removal of the salivary glands of rats results in increased caries activity. Persons with conditions of injury to salivary glands, or absence of these glands, have high rates of dental caries. The effect of saliva may be because of more rapid removal of food substances from the mouth and also because of constituents of the saliva that inhibit the action of microorganisms.

The parts of the teeth that are attacked in dental caries are enamel, dentin, dentinoenamel junction and cementum. Accordingly theories of causation are based upon the inorganic and organic constituents, their locations and availability to the microorganisms of the oral cavity.

Enamel removed by the carious process is not replaced in any way by a vital process though there is evidence of some slight remineralization of enamel rods. Loss of dentin frequently results in the formation of secondary dentin on the pulp side of the dentin under carious attack.

Effects of Chemical and Microbial Agents. Hypotheses relative to the protection of teeth by fluoride against carious attack are as follows: (1) The solubility of enamel in acid is diminished by formation of fluorapatite. (2) The enamel is hardened by deposition of fluorapatite during formation. (3) The fluoride ion inhibits enzyme action of microorganisms. (This does not seem possible as it requires the fluoride ion in solution.) (4) The microstructure of enamel, as indicated by appearance, resists microfracture as a basis of initiation of the carious process. (5) Topical fluorides sterilize the incipient lesions with respect to decay producing microorganisms. (See also FLUORIDE METABOLISM.)

No evidence has been found of change in composition of the organic constituents of enamel by either formative or topically applied fluorides.

Acidogenic-aciduric microorganisms contribute principally to the caries process. Acids produced by microorganisms attack the inorganic constituents of the teeth as acids and as chelating salts.

Relative to the carious process, some microorganisms are protective since they metabolize substances which attack the teeth. There is antagonism between microorganisms as well as symbiosis. The aciduric aspect of caries-associated microorganisms is illustrated by *Lactobacillus acidophilus*. This microorganism contributes very little total acid, but because of survival in the acid conditions produced by the more numerous and more acidogenic cocci types of microorganisms, it is an index of caries activity in proportion to the number found in the saliva.

Dental caries is transmissive within the mouth as each lesion tends to maintain the microflora essential for the development of other lesions. Higher caries rates are found in rats caged together compared with those caged singly.

The presence of food, especially carbohydrates, is necessary for the carious process. Rats maintained by feeding by tube to the stomach develop no carious lesions as compared with litter mates on the same dietary by mouth, showing that there is no effective transfer of food to the mouth by saliva, soft tissues or teeth.

Carbohydrates contribute to the carious process in proportion to their availability to the microorganisms in contact with the teeth. Fermentable sugars, in solution, are retained on teeth to a lesser degree than sticky candies. Cooked STARCHES are more retained and, through action of salivary amylase, contribute to the carious process.

GERM-FREE ANIMALS, fed sterile foods, do not develop carious lesions, demonstrating that microorganisms play an essential part in dental caries. Such animals inoculated with specific microorganisms develop dental caries in varying degrees.

Mechanical removal of food debris from tooth surfaces, and especially from interproximal areas, tends to diminish the intensity of the carious process. Toothbrushing, especially immediately after consumption of foods, is the most effective of mechanical means. The use of dental floss and toothpicks is more effective in the removal of foods in interproximal areas. Fibrous foods may contribute to removal and less retention of food substances leading to the carious process.

The yellow-brown pigmentation of carious lesions is produced by a nonenzymatic browning reaction between exposed protein components of the teeth and carbonyl-containing fermentation products of carbohydrates, especially dihydroxyacetone, glyceraldehyde, furfural and 5-hydroxymethylfurfural.

Infection of the dental pulp can result from advanced active caries. Hyperemia of the pulp is produced with the occurrence of odontolagia, or toothache. Also there can be irritation of the pulp by products of metabolism of microorganisms associated with caries of the dentin. A response of the pulp to such irritants may be the formation of secondary dentin. If the carious lesion is not treated and a filling placed, the pulp will be destroyed by the carious process.

A preventive procedure used in dentistry is placing fillings in pits and fissures before decay of the same can be detected. Filling of cavities as early as they can be detected is the most effective procedure in the control of the carious process.

When decay has progressed to the stage beyond any possible repair, the affected tooth should be extracted as the resultant dental focal infection may cause a general systemic disease.

GERALD J. COX

CARNITINE

Carnitine, $(CH_3)_3{\equiv}N^+{-}CH_2CHOHCH_2COO^-$, is the betaine of β-hydroxy-γ-aminobutyric acid and has long been known as a constituent of muscle. The meal worm *Tenebrio molitor* (and several other insects) die before they undergo metamorphosis unless they are supplied carnitine as a constituent of their diet. Carnitine has been designated Vitamin B_t (Fraenkel), but the tendency now is to call all vitamins and other related compounds by noncommittal names that do not designate their functions. Carnitine is thought to be built up endogenously by mammals.

CARNOSINE

Carnosine (β-alanyl-L-histidine) occurs along with ANSERINE, its methylated derivative, in mammalian muscle. In snake muscle, another dipeptide, ophidine, is found. It has the same structure as carnosine except for the presence of a methyl group attached to the carbon atom between the two nitrogen atoms of the imidazole ring.

CAROTENOIDS

Distribution. Carotenoids are lipid-soluble, yellow to orange-red pigments universally present in the photosynthetic tissues of higher plants, ALGAE and the PHOTOSYNTHETIC BACTERIA. They are spasmodically distributed in flowers, fruit, and roots of higher plants, in fungi and in bacteria. They are synthesized *de novo* in plants and protists. Carotenoids are also widely but spasmodically distributed in animals, especially marine invertebrates, where they tend to accumulate in gonads, skin and feathers. All carotenoids found in animals are ultimately derived from plant or protistan carotenoids, although because of metabolic alteration of the ingested pigments some carotenoids found in animals are not found in plants or protista.

In photosynthetic tissues of higher plants and algae, carotenoids are located in the CHLOROPLASTS and localized in the grana, when these sub-organelles are present. In the photosynthetic bacteria, they are located in the chromatophores, which correspond to the grana of higher plants. In all cases they exist as LIPOPROTEINS in close association with the CHLOROPHYLLS. They can also exist as lipoproteins in other tissues of higher plants (*e.g.*, carrot root) and in fungi and bacteria (*e.g.*, *Corynebact.* spp.); but in others (*e.g.*, red palm, *Phycomyces* sp.) they are present as oil droplets. Caroteno-proteins are often differently colored from the parent pigments; the color of brown algae is largely due to a fucoxanthin-protein complex while the free pigment is yellow.

The carotenoids in the leaves of all higher plants are qualitatively the same, but variations occur in different classes of algae and in photosynthetic bacteria. Unique carotenoid distribution in higher plants occurs in flowers, fruit and roots, where numerous different and often highly specific

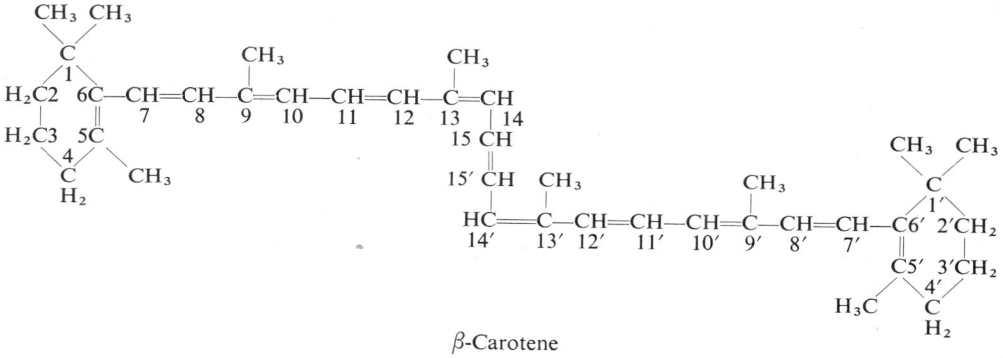

β-Carotene

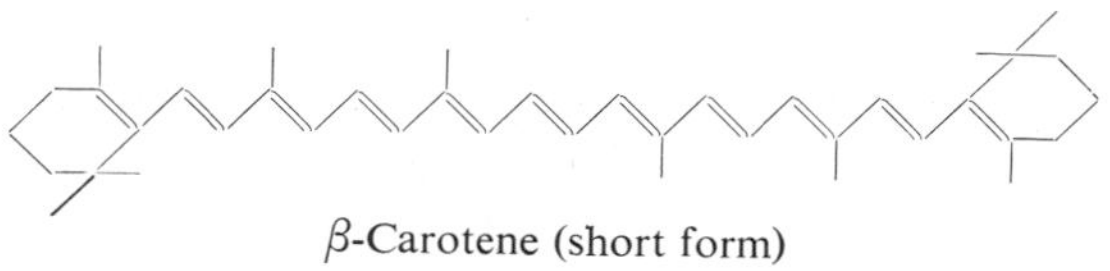

β-Carotene (short form)

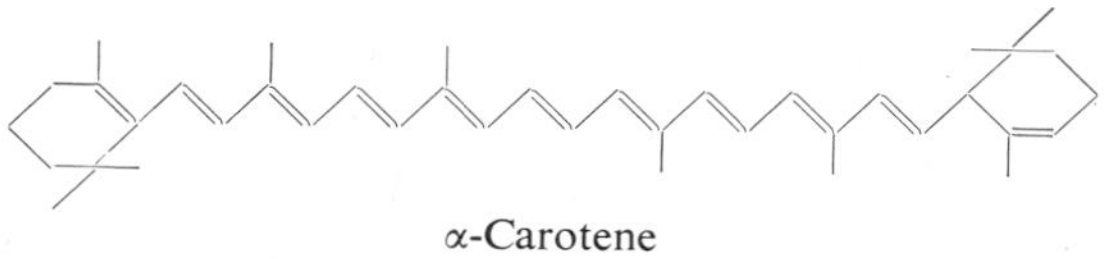

α-Carotene

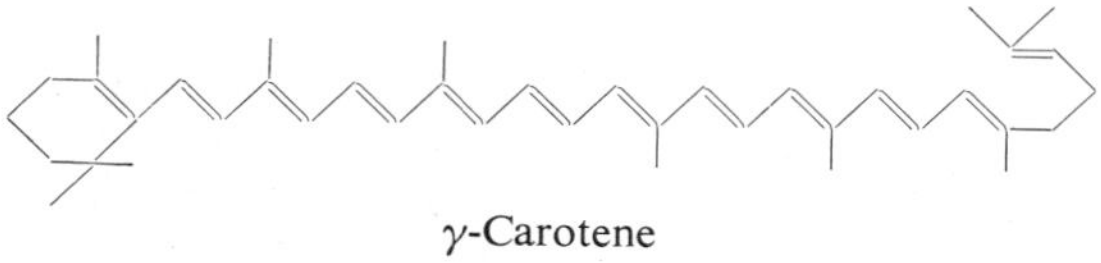

γ-Carotene

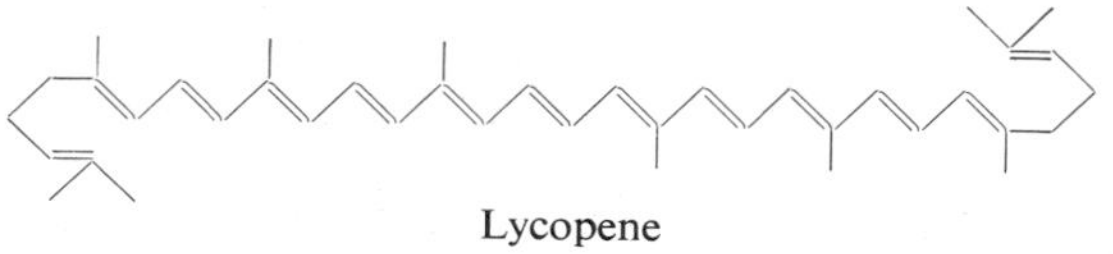

Lycopene

carotenoids are found. Carotenoids often specifically accumulate in reproductive regions of algae (*e.g.*, *Ulva*). About 0.2% of the dry matter of leaves is carotenoid; in non-photosynthetic tissues this value can frequently rise to 1%; in the brick-red coronas of the certain narcissi it is 16.5%.

In animals, carotenoids exist as lipid droplets in specialized cells (*e.g.*, xanthophores and erythrophores in trout skin), dissolved in body fat (locusts, cows) and as chromoproteins: green of lobster eggs and bluish-black of lobster carapace are both due to the same pigment, astaxanthin (which is red), attached to different proteins.

Structure. Carotenoids are tetraterpenoids, consisting of eight isoprenoid (C>C—C—C with C) residues, and can be regarded as being synthesized by the tail-to-tail dimerization of two 20-carbon units, themselves each produced by the head-to-tail condensation of four isoprenoid units. Hydrocarbon carotenoids are termed *carotenes*, and oxygenated carotenoids are known as *xanthophylls*. The structure of the best known carotene, β-carotene, is given in full and in the usual shorthand form. The other main carotenes are α-carotene, γ-carotene and lycopene.

α-Carotene is widely distributed in trace amounts together with β-carotene in leaves, γ-carotene is found in many fungi and lycopene is the main pigment of many fruits, *e.g.*, tomato.

The oxygen in xanthophylls can exist in various functions (hydroxy, keto, epoxy, methoxy, and carboxy) and typical examples are given in Table 1. Carotenoids with rather unexpected

TABLE 1. EXAMPLES OF DIFFERENT CLASSES OF XANTHOPHYLLS

Name	Structure	Major Distribution
Cryptoxanthin	3-Hydroxy-β-carotene	Fruit
Lutein	3,3′-Dihydroxy-α-carotene	Main xanthophyll in green leaves
Echinenone	4-Oxo-β-carotene	Blue-green algae
Violaxanthin	5,6,5′,6′-Diepoxy-3,3′-dihydroxy-β-carotene	Flower petals; Green leaves
Auroxanthin	5,8,5′,8′-Diepoxy-3,3′-dihydroxy-β-carotene	Flowers and fruit
Torularhodin	(See structure)	Red yeasts
Spirilloxanthin	(See structure)	Photosynthetic bacteria

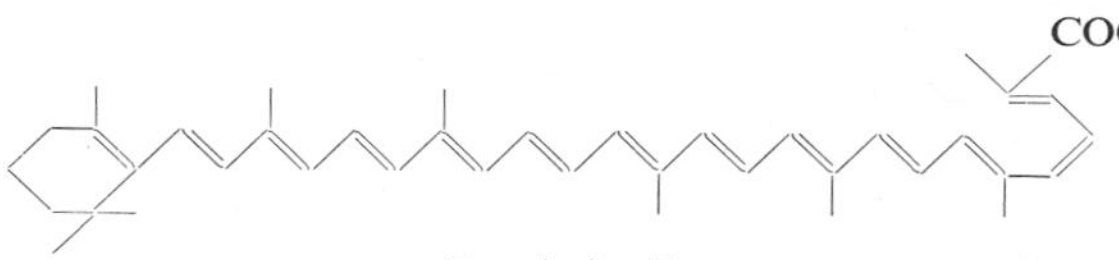

COOH

Torularhodin

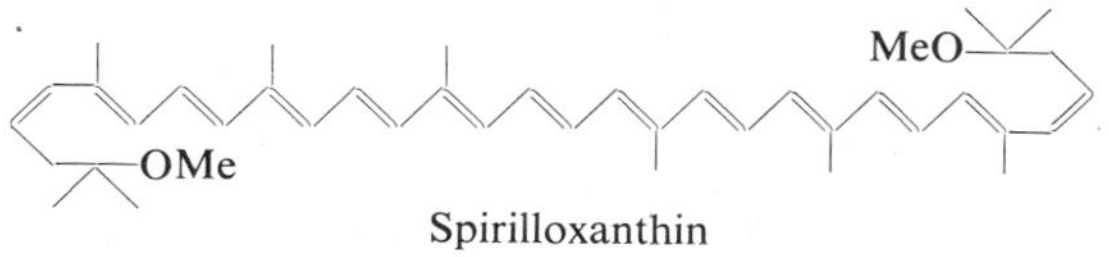

MeO— —OMe

Spirilloxanthin

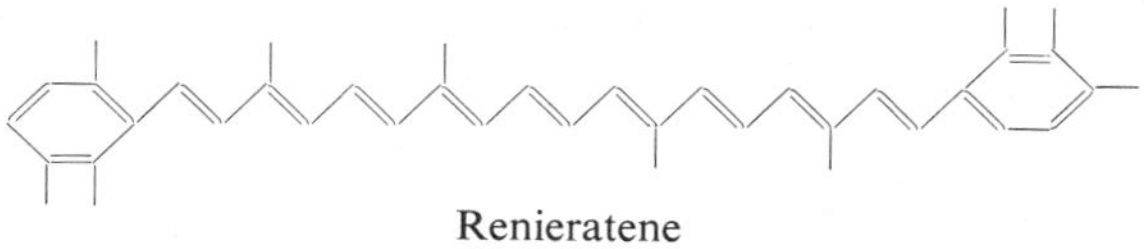

Renieratene

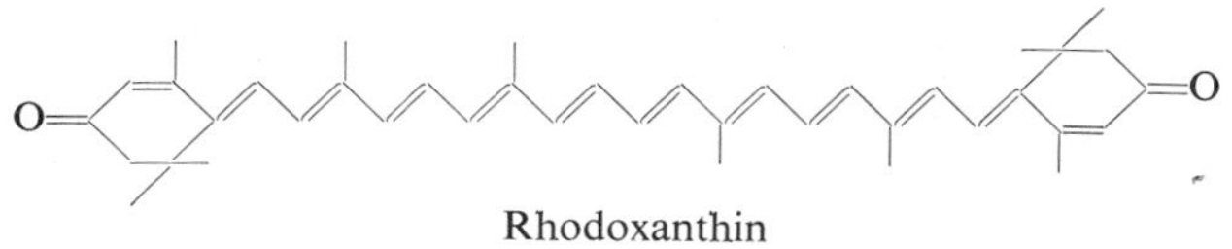

O= =O

Rhodoxanthin

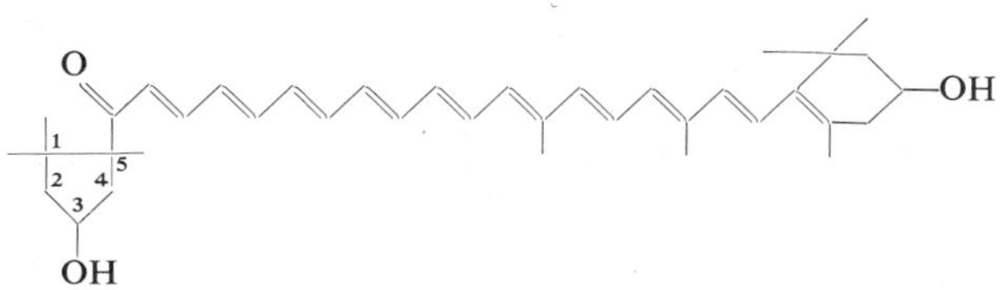

O —OH 1 2 3 4 5 OH

Capsanthin

structures are occasionally encountered, *e.g.*, aromatic carotenoids such as renieratene from sponges; retro carotenoids, such as rhodoxanthin from berries (retro carotenoids have the positions of the single and double bonds reversed compared with "normal" carotenoids), and carotenoids with five-membered rings, such as capsanthin from peppers.

Carotenes with some of the double bonds in the $C_{(7)}$-$C_{(7')}$ region saturated are also known and are biosynthetic precursors of the more fully unsaturated carotenoids (see next section). Carotenoids with fewer carbon atoms are known; *e.g.*, crocetin, C_{20}; these are possibly degradation products of C_{40} carotenoids.

HOOC–(polyene chain)–COOH

Crocetin

Biosynthesis. All terpenoids are derived from the basic isoprenoid precursor, isopentenyl pyrophosphate (IPP), which is synthesized from acetyl coenzyme A thus:

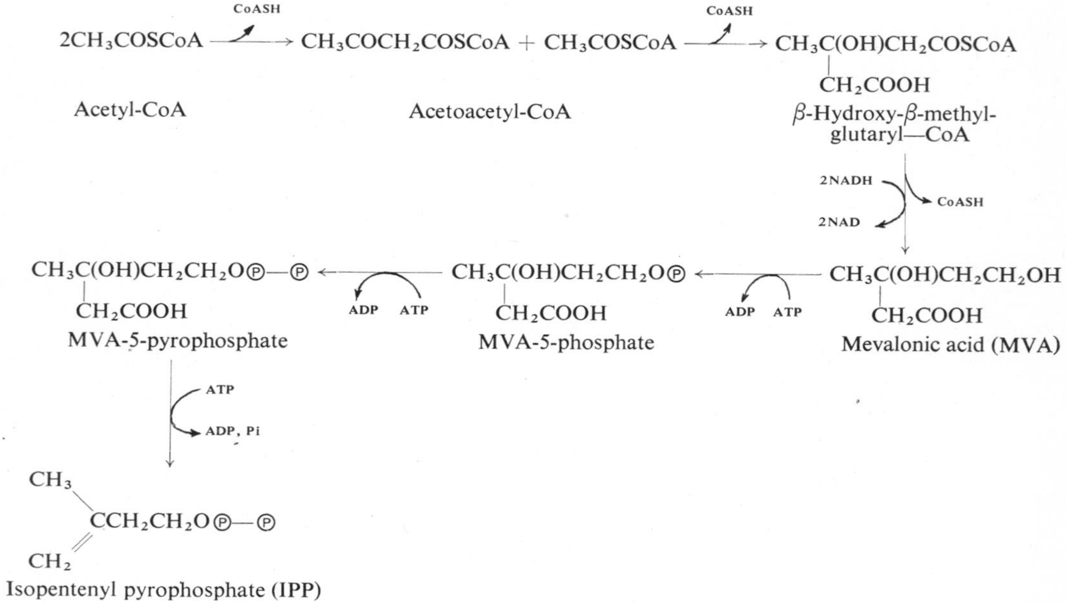

$$2CH_3COSCoA \xrightarrow{CoASH} CH_3COCH_2COSCoA + CH_3COSCoA \xrightarrow{CoASH} CH_3C(OH)(CH_2COOH)CH_2COSCoA$$

Acetyl-CoA; Acetoacetyl-CoA; β-Hydroxy-β-methyl-glutaryl—CoA

$$\xrightarrow{2NADH \rightarrow 2NAD;\ CoASH} CH_3C(OH)(CH_2COOH)CH_2CH_2OH$$

Mevalonic acid (MVA)

$$\xrightarrow{ATP \rightarrow ADP} CH_3C(OH)(CH_2COOH)CH_2CH_2O\text{Ⓟ}$$

MVA-5-phosphate

$$\xrightarrow{ATP \rightarrow ADP} CH_3C(OH)(CH_2COOH)CH_2CH_2O\text{Ⓟ—Ⓟ}$$

MVA-5-pyrophosphate

$$\xrightarrow{ATP \rightarrow ADP,\ Pi} CH_2{=}C(CH_3)CH_2CH_2O\text{Ⓟ—Ⓟ}$$

Isopentenyl pyrophosphate (IPP)

IPP is isomerized to dimethylallyl pyrophosphate, which acts as a starter for condensation with further molecules of IPP to form geranyl pyrophosphate (C_{10}), farnesyl pyrophosphate (C_{15}) and geranylgeranyl pyrophosphate (C_{20}). This mechanism, which is stereospecific for proton elimination, is indicated here as far as C_{10}:

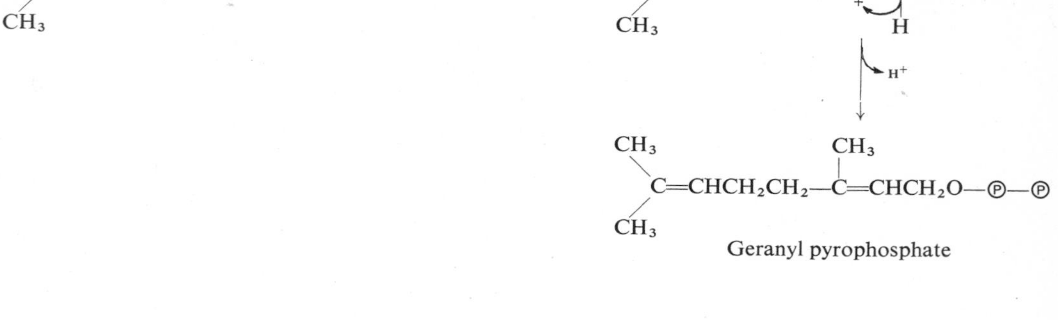

$$\underset{\text{IPP}}{CH_2{=}C(CH_3)CH_2CH_2O\text{—Ⓟ—Ⓟ}} \xrightarrow{H^+} (CH_3)_2\overset{+}{C}\text{—}CH(H)CH_2O\text{—Ⓟ—Ⓟ} \xrightarrow{-H^+} \underset{\text{Dimethylallyl pyrophosphate}}{(CH_3)_2C{=}CHCH_2O\text{—Ⓟ—Ⓟ}}$$

$$(CH_3)_2C{=}CHCH_2\text{—O—Ⓟ—P} + CH_2{=}C(CH_3)\text{—}CH_2CH_2O\text{—Ⓟ—Ⓟ} \xrightarrow{-P\text{—}P} (CH_3)_2C{=}CHCH_2\text{—}CH_2\text{—}\overset{+}{C}(CH_3)\text{—}CH(H)CH_2O\text{—Ⓟ—Ⓟ}$$

$$\xrightarrow{-H^+} (CH_3)_2C{=}CHCH_2CH_2\text{—}C(CH_3){=}CHCH_2O\text{—Ⓟ—Ⓟ}$$

Geranyl pyrophosphate

The addition of two further $C_{(5)}$ units to form geranylgeranyl pyrophosphate follows the same mechanism. Geranylgeranyl pyrophosphate then dimerizes tail to tail with the loss of two molecules of pyrophosphate to form the partly saturated polyene, phytoene. This is then stepwise dehydrogenated to neurosporene:

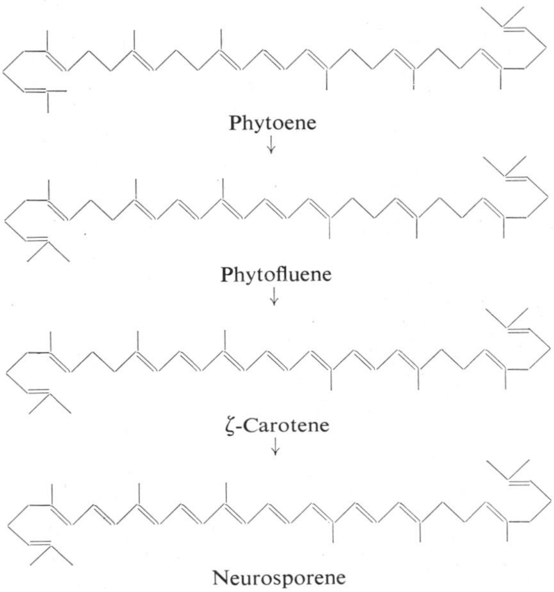

Insertion of oxygen occurs at the level of the fully unsaturated carotenoids, but the mechanism of the insertion of the various oxygen functions is unknown.

In purple photosynthetic bacteria, neurosporene is also the branch point; both pathways lead to the formation of methoxylated carotenoids, which

Phytoene

↓

Phytofluene

↓

ζ-Carotene

↓

Neurosporene

Neurosporene can then be channeled into two main pathways in higher plants, fungi and algae:

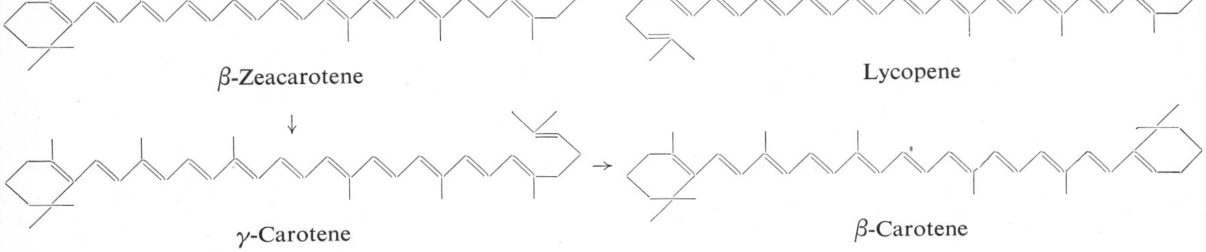

are not found outside this group of organisms. One pathway leads eventually to spirilloxanthin

Neurosporene

2H

β-Zeacarotene

Lycopene

↓

γ-Carotene

→

β-Carotene

(a) dehydrogenation to lycopene; (b) cyclization to β-carotene. The mechanisms for forming α-carotene, the aromatic carotenes and the retro carotenoids are now known.

(*Rhodospirillum rubrum*) and the other to spheroidene and spheroidenone (*Rhodopseudomonas spheroides*) and in some case also to spirilloxanthin:

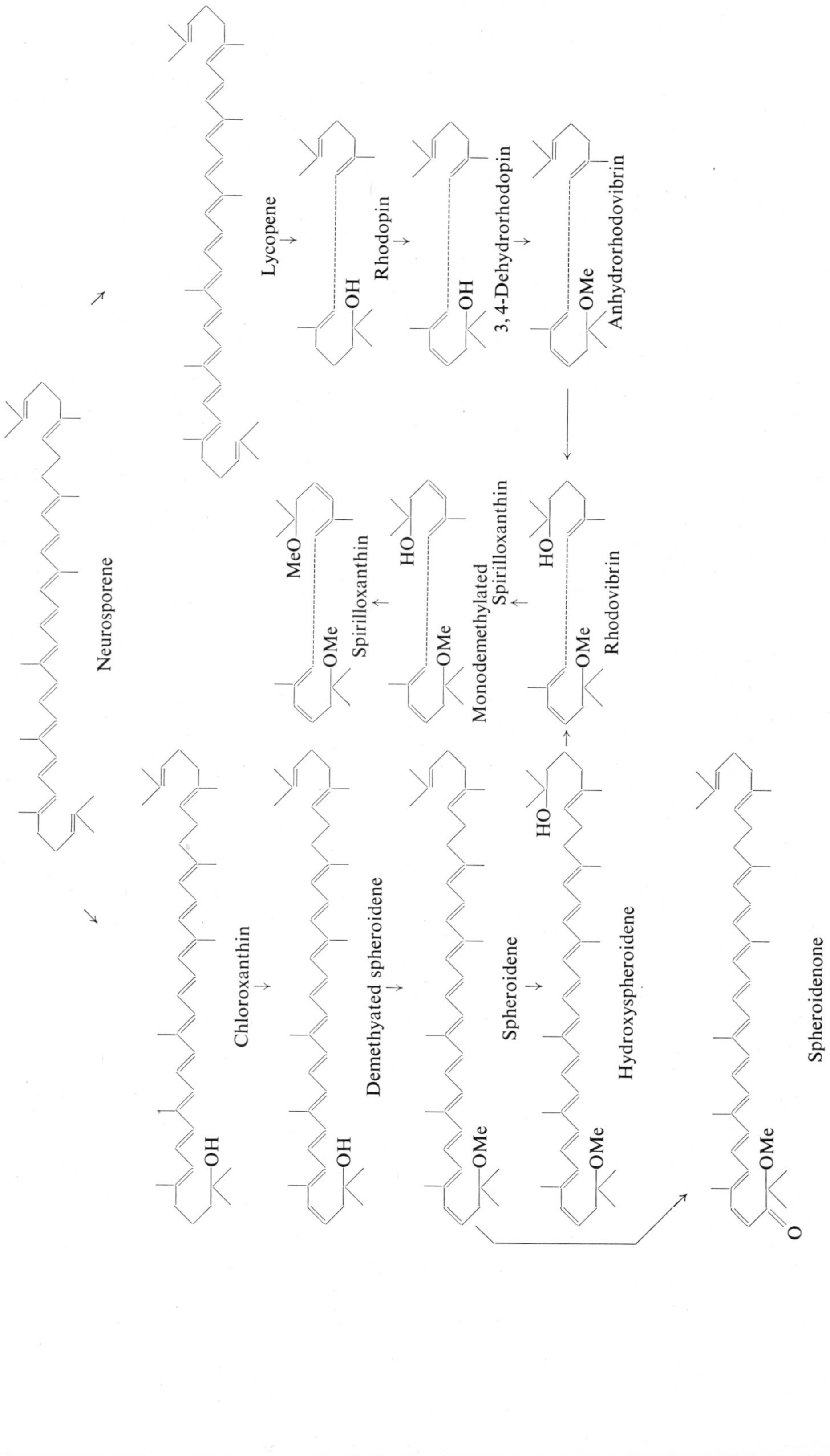
Neurosporene
Lycopene
OH
Rhodopin
OH
3, 4-Dehydrorhodopin
OMe
Anhydrorhodovibrin
MeO
OMe
Spirilloxanthin
HO
OMe
Monodemethylated
Spirilloxanthin
HO
OMe
Rhodovibrin
OH
Chloroxanthin
OH
Demethyated spheroidene
OMe
Spheroidene
HO
OMe
Hydroxyspheroidene
OMe
O
Spheroidenone

See also ISOPRENOID BIOSYNTHESIS.

References

1. *Chemistry*
   (a) KARRER, P., AND JUCKER, E., "Carotenoids," Amsterdam, Elsevier, 1950 (translated by E. A. BRAUDE).

   (b) WEEDON, B. C. L., in "Chemistry and Biochemistry of Plant Pigments" (GOODWIN, T. W., EDITOR), London, Academic Press, 1965.
2. *Distribution*
   (a) GOODWIN, T. W., "Comparative Biochemistry of Carotenoids," London, Chapman and Hall, 1952.
   (b) GOODWIN, T. W., in "Comparative Biochemistry," (FLORKIN, M., AND MASON, H. S., EDITORS), Vol. 4, New York, Academic Press, 1963.
   (c) GOODWIN, T. W., in "Chemistry and Biochemistry of Plant Pigments," (GOODWIN, T. W., EDITOR), London, Academic Press, 1965.
3. *Biosynthesis*
   GOODWIN, T. W., *Ibid.*, 1965.
4. *Identification*
   DAVIES, B. H., in "Chemistry and Biochemistry of Plant Pigments," (GOODWIN, T. W., EDITOR), London, Academic Press, 1965.

T. W. GOODWIN

CASEIN

Casein is the principal protein component, actually a mixture of a number of related proteins, of bovine milk [see MILK (COMPOSITION AND BIOSYNTHESIS); PROTEINS].

CATABOLISM. See ANABOLISM; BASAL METABOLISM.

CATALASE. See PEROXIDASES.

CATALYSIS

Catalysis is the process or phenomenon in which one substance, the catalyst, is able to alter (usually increase) the rate of chemical reaction among other substances. Important as catalysts in living organisms are the *enzymes*, a sub-class of the PROTEIN macromolecules [see ENZYME CLASSIFICATION AND NOMENCLATURE; ENZYMES (KINETICS); CHELATION AND METAL CHELATE COMPOUNDS]. Catalysis as a general chemical phenomenon is discussed in the reference below.

Reference

EMMETT, PAUL H., "Catalysis," in "The Encyclopedia of Chemistry" (CLARK, G. L. AND HAWLEY, G. G., EDITORS), New York, Reinhold Publishing Corp., 1957.

CELL DISRUPTION METHODS

Prior to the extraction and recovery of soluble components (*e.g.*, enzymes) or particulate components from tissues, cells, or bacteria, the cells may be disrupted by any of several methods. Commonly used methods include grinding of the tissue or cell suspension with abrasives, shaking or grinding with glass beads, subjection of the liquid or frozen suspension to shear stresses at high pressure, "homogenization" in a test tube containing the suspension and a close-fitting power-rotated pestle, and subjection of the suspension to supersonic vibrations generated by a magnetostriction oscillator (see ULTRASONIC VIBRATIONS).

Reference

POTTER, VAN R., "Tissue Homogenates," in "Methods in Enzymology" (COLOWICK, S. P., AND KAPLAN, N. O., EDITORS), Vol. I, pp. 10–15, New York, Academic Press, 1955.

CELL DIVISION

The division of one living cell into two is one of the most fascinating and most important of biological phenomena. By this process, continuity of a species is ensured and mutation of a species is made possible. Moreover, cell division plays an important role in the growth and DIFFERENTIATION of tissues in embryonic forms, in wound healing, in the formation of tumors, and in the normal replacement of old cells in certain tissues such as skin of humans. However, much about the chemistry of the division process remains to be elucidated.

A living cell consists of cytoplasm and a body within the cytoplasm, the nucleus. The division of a cell involves not only cleavage of cytoplasm but also a complex nuclear reorganization in which replicated genetic material, CHROMATIN, of the mother nucleus is distributed to each daughter nucleus.

Interphase. After cell cleavage and prior to visible nuclear changes of the next division, a cell is said to be in *interphase*. In this condition, the nucleus is bounded by a thin double-layered structure, the nuclear membrane. In the interphase nucleus, chromatin is present in the form of very fine, extended threads, and present also is at least one spherical body known as the nucleolus. Chromatin and nucleolus are immersed in a clear, homogeneous liquid, the nuclear sap, which has a viscosity only a few times greater than that of water. Within the cytoplasm of a cell in interphase are a number of components, such as mitochondria, lysosomes, plastids, vacuoles, endoplasmic reticulum with associated ribosomes and Golgi body, and centrioles (Fig. 1). Centrioles are always present in animal cells, but strangely they have not been detected in the cells of higher plants.

During interphase, a cell prepares for the ensuing division. During this period, replication of the chromatin threads occurs, forming sister threads. Chromatin consists essentially of two substances, a basic protein, HISTONE, and DEOXYRIBONUCLEIC ACID (DNA). (A nucleic acid is a substance of great molecular weight made up of

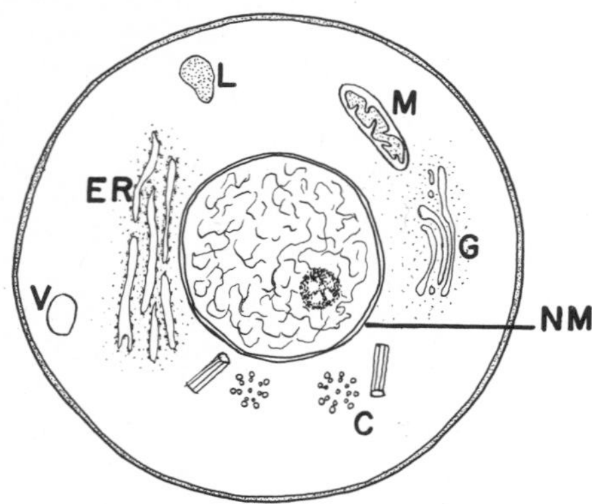

FIG. 1. Diagram of structures of interphase cell. C, centriole; ER, endoplasmic reticulum; G, Golgi body; L, lysosome; M, mitochondrion; NM, nuclear membrane surrounding chromatin threads and nucleolus; V, vacuole.

many units of nucleotides. A nucleotide consists of phosphoric acid, a five-carbon sugar, and an organic base, either a PURINE or a PYRIMIDINE. In the case of deoxyribonucleic acid, the five-carbon sugar is deoxyribose, the purine bases are adenine and guanine, and the pyrimidine bases are cytosine and thymine).

The synthesis of new DNA within an interphase nucleus can be marked by adding to the environment of a cell a precursor of DNA, thymidine, labeled with radioactive hydrogen (tritium). Radioactive thymidine passes into the cells and can be detected in the nucleus by means of autoradiography (a film is exposed to radiation emitted by the radioactive substance within the cell; Fig. 2). In a cell not destined to divide, DNA synthesis does not occur, and in a cell in which DNA synthesis does occur, division typically takes place. If DNA synthesis is blocked by treatment of cells with deuterium oxide, division is inhibited; when the block is removed, DNA synthesis proceeds and division occurs.[3]

Toward the end of interphase the adjacent two pairs of centrioles begin to move in opposite directions. When they come to rest, they will form the poles of a structure known as the spindle. The centrioles are by no means simple structures. From studies carried out with the aid of the electron microscope, it is known that a centriole is a cylindrical body made up of parallel, tubule-like structures. Often the centrioles of a pair lie at right angles to each other. Radiating from the region around a centriole pair is a system of fibers, the aster. Little is known about the chemistry of centrioles. There is some evidence that they contain RIBONUCLEIC ACID (RNA). (Ribonucleic acid differs from DNA in that the five-carbon sugar is ribose rather than deoxyribose and the pyrimidine base, thymine, is replaced by the base, uracil).

Prophase. About the time the centrioles begin to move, other events are initiated: dissolution of the nuclear membrane, disappearance of the nucleolus, condensation of the chromatin threads by coiling, and spindle formation. This stage of the division process is called *prophase* (Fig. 3).

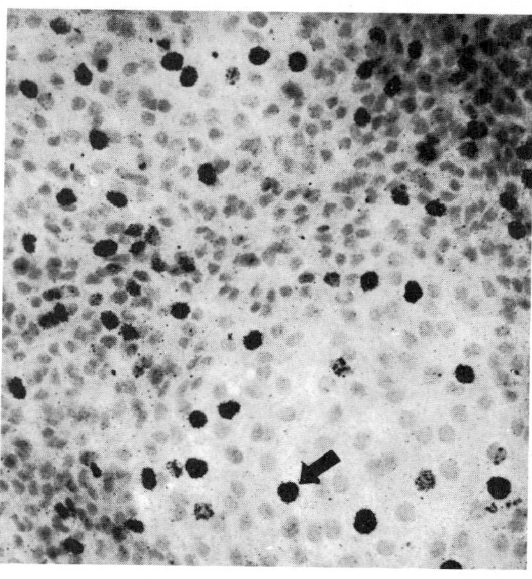

FIG. 2. Autoradiogram of epithelium of lens of rabbit eye. Radioactive nuclei and DNA synthesis are indicated by the black dots (arrow points to one radioactive nucleus). (Reprinted with permission, from reference 4.)

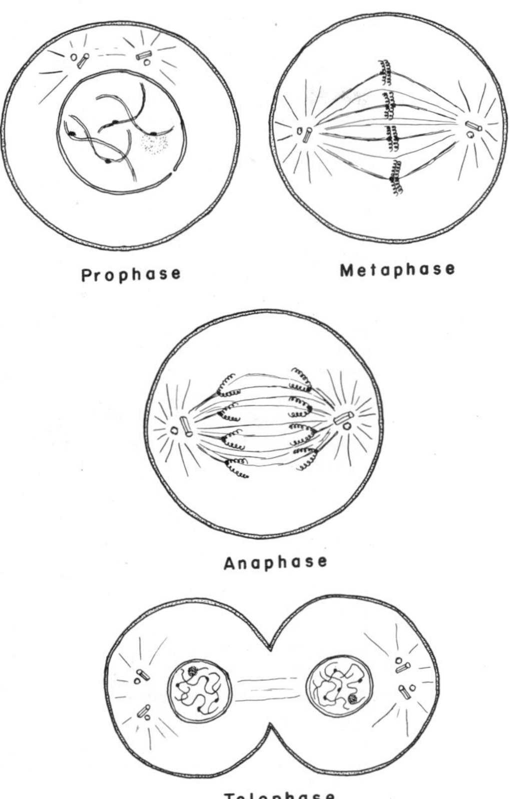

FIG. 3. Diagram representing the stages of cell division.

Presumably the nuclear membrane is a LIPOPROTEIN similar to that of the cell membrane. Dissolution of the nuclear membrane apparently is brought about by a calcium-ion activated proteolytic enzyme.[1]

The nucleolus contains protein and ribonucleic acid. RNA is not synthesized in the nucleolus but accumulates here after being synthesized by DNA.[2] During prophase, RNA accumulates on the coiling chromatin threads, the CHROMOSOMES, and it has been suggested that the accumulation of RNA on the chromosomes is the result of blockage of the transfer of RNA from the chromosomes to the nucleolus.[8] Late in prophase, the nucleolus disappears and its RNA passes into the cytoplasm.

During prophase, a dual system of fibers forms between the two poles from protein synthesized during interphase. One system, the primary spindle, consists of fibers extending from pole to pole; the other system, the chromosomal spindle, consists of fibers extending from the poles to the equatorial region between the poles.

Metaphase. Late in prophase the tightly coiled chromosomes begin a movement which will culminate in the alignment of the chromosomes in a narrow band in the equatorial plane of the spindle. When the chromosomes reach this position on the equatorial plate, the cell is said to be in *metaphase*. Each of the sister chromosomes of a pair is connected to a chromosomal spindle fiber of its pole by means of a body on the chromosome called the kinetochore. By this time, chromosomes, spindle fibers, centrioles, and asters have become surrounded by a gelatinous matrix. This whole structure, known as the mitotic apparatus, differs physically from the remainder of the cytoplasm. For example, the mitotic apparatus can be dislodged from its normal position and moved about the cell by means of ultrasonic vibration applied to the cell surface.[12]

In some respects, the chemical and physical properties of cytoplasm resemble those of blood. Like blood, cytoplasm can undergo changes in viscosity. In eggs of the marine invertebrates *Arbacia* and *Chaetopterus*, during prophase prior to the time spindle and mitotic apparatus are forming, cytoplasmic viscosity increases two- to threefold.[6] This viscosity increase is called the mitotic gelation. Then during the time that the spindle and mitotic apparatus are undergoing completion, the viscosity decreases, and by metaphase, the viscosity has decreased to its value prior to the gelation (Fig. 4). In these cells, cleavage is always preceded by a mitotic gelation, and this gelation appears to be related to the formation of the spindle. Just as HEPARIN prevents the gelation which occurs in the clotting of blood, heparin and heparin-like substances inhibit cell division by preventing the mitotic gelation.[5]

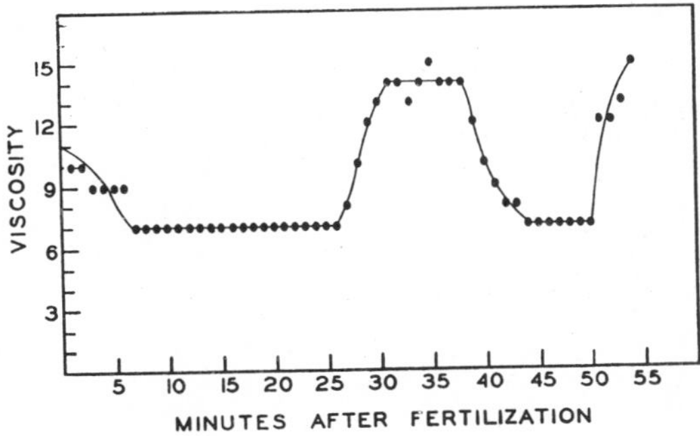

FIG. 4. Viscosity changes in the egg of *Chaetopterus* during the time between fertilization and first cleavage. The spindle forms in late prophase at about 40 minutes after fertilization during the time that the viscosity is decreasing. Metaphase occurs from 45–49 minutes after fertilization. (Reprinted with permission, from reference 6.)

The mitotic apparatus can be subjected to chemical analysis after being isolated from the cell.[8] The isolation can be accomplished by treating cells first with alcohol and then with digitonin or, preferably, by breaking cells in a medium containing sucrose or dextrose and dithiodiglycol ($OHCH_2CH_2SSCH_2CH_2OH$). Although a number of proteins are present, the mitotic apparatus is constructed largely of one protein. The proteins of the isolated mitotic apparatus (obtained by the alcohol-digitonin method) have an average molecular weight of 315,000 ± 20,000. RNA also is found in the mitotic apparatus and at least part of the RNA is present in the spindle. A considerable amount of the RNA is associated with the major protein component. Lipid is present and apparently is associated with ribonucleoprotein. With staining techniques, evidence has been obtained which indicates that the mitotic apparatus contains polysaccharide and zinc. Also, an enzyme has been isolated from the mitotic apparatus which can split adenosine triphosphate. The facts related above concerning the chemistry of the mitotic apparatus were compiled from Mazia.[8]

The importance of sulfur-containing groups in the division process has been emphasized by a number of investigators. However, the exact role these groups play is not fully understood. During its formation, the mitotic apparatus shows a high concentration of protein rich in sulfhydryl (SH) groups, whereas during later stages of the division process when the mitotic apparatus is undergoing disassembly, the protein SH groups are present in low concentration.[7] Presumably SH groups and disulfide groups (S-S) play a role in linking together the large protein molecules of the mitotic apparatus.[8]

Anaphase. The end of metaphase and the beginning of *anaphase* is marked by the separation of sister chromosomes and the beginning of movement of the sister chromosomes toward opposite poles. During anaphase, RNA is lost from the chromosomes, and toward the end of anaphase there is an accumulation of ribonucleoprotein in the equatorial region between the two chromosome groups.[9] The source of this equatorial RNA is not definitely known. Although the equatorial RNA might represent RNA lost from the chromosomes, it could originate from other sources.

Telophase. During the final stage of the division process, telophase, the cell body divides into two, a process called cytokinesis, and each daughter cell completes processes which restore it to an interphase cell. The mitotic apparatus disappears, the chromosomes become attenuated, the centrioles duplicate and split, the nuclear membrane becomes reconstituted, and the nucleolus reappears.

During anaphase, nucleolar material can be detected among the chromosomes, and by the end of telophase this material has been accumulated in the body known as the nucleolus. It is not known what fraction of RNA in the telophase nucleolus represents RNA lost to the cytoplasm from the prophase nucleolus. At least part of the RNA in the telophase nucleolus is newly synthesized.[13]

Although the centriole has the unusual capability of reproducing itself, little is known about its chemical makeup. There is some evidence which indicates that centrioles contain RNA.

Restitution of the nuclear membrane is brought about by interactions between chromosomes and elements of the endoplasmic reticulum.

Various schemes have been proposed to explain the furrowing or cleavage of the cell body. According to one hypothesis, cleavage is brought about by synthesis of new material in the region of the furrow.[10] Another hypothesis proposes that there is a contraction or constriction of cytoplasmic substance in the equatorial region of the cell. The "expanding membrane" hypothesis maintains that furrowing is caused by the expansion of folded protein molecules in the cortical region of the cell.[11] There is evidence in support of each of these ideas, and at present none can be rejected.

The events occurring at the furrowing cell surface and those preceding them in the interior of the cell are not unrelated. Furrowing always occurs in the equatorial plane between the poles of the spindle. Presumably the interior-to-surface messenger is a chemical substance (or substances), but its nature and source are not known.

Although the division process requires energy, the active phases of the mitotic cycle are not marked by great metabolic or respiratory activity. On the contrary, the period between active phases, interphase, is the time of high metabolic and respiratory activity. Thus many investigators believe that the energy required for division is obtained from a "reservoir" prepared during interphase.[8] The nature of the "reservoir" is not known. Conceivably a high-energy compound such as adenosine triphosphate is involved, but evidence for this is not strong.

The account presented above depicts the events in an ideal cell during the course of a somatic cell division. As a result of this division, each daughter cell has a full complement of chromosomes (diploid) just as the mother cell. In reproductive organs during the production of germ cells (sperm and eggs of animals and spores of plants), the cells undergo two divisions, called meiotic divisions, which result in gametes each having only half the chromosome complement (haploid) of a somatic cell.

References

1. Goldstein, L., *Biol. Bull.*, **105,** 87 (1953).
2. Goldstein, L., and Micou, J., *J. Biophys. Biochem. Cytol.*, **6,** 301 (1959).
3. Gross, P. R., and Harding, C. V., *Science*, **133,** 1131 (1961).
4. Harding, C. V., and Thayer, M. N., *Invest. Ophth.*, **3,** 302 (1964).
5. Heilbrunn, L. V., "The Dynamics of Living Protoplasm," New York, Academic Press, 1956.
6. Heilbrunn, L. V., and Wilson, W. L., *Biol. Bull.*, **95,** 57 (1948).
7. Kawamura, N., and Dan, K., *J. Biophys. Biochem. Cytol.*, **4,** 615 (1958).
8. Mazia, D., in "The Cell" (Brachet, J., and Mirsky, A. E., Editors), Vol. III, p. 77, New York, Academic Press, 1961.
9. Rustad, R., *Exptl. Cell. Res.*, **16,** 575 (1959).
10. Schectman, A. M., *Science*, **85,** 222 (1937).
11. Swann, M. M., and Mitchison, J. M., *Biol. Rev. Cambridge Phil. Soc.*, **33,** 103 (1958).

12. Wilson, W. L., Nyborg, W. L., Wiercinski, F. J., Schnitzler, R. M., and Sichel, F. J., Am. Inst. of Ultrasonics in Medicine, Proceedings of the Ninth Annual Conference, in press, 1965.
13. Woods, P. S., and Taylor, J. H., *Lab. Invest.*, **8**, 309 (1959).

Walter L. Wilson

CELL NUCLEUS COMPOSITION. See nucleus and nucleolus.

CELL WALL COMPOSITION. See bacterial cell walls.

CELLULOSE

Cellulose is the name given both to a specific polysaccharide—consisting of β-D-glucose residues joined end to end by linkage through —O— of C_1 of one residue to C_4 of the next (Fig. 1)—and to a resistant family of polysaccharides (containing cellulose in the strict sense) isolated from plants by specific chemical treatment. The

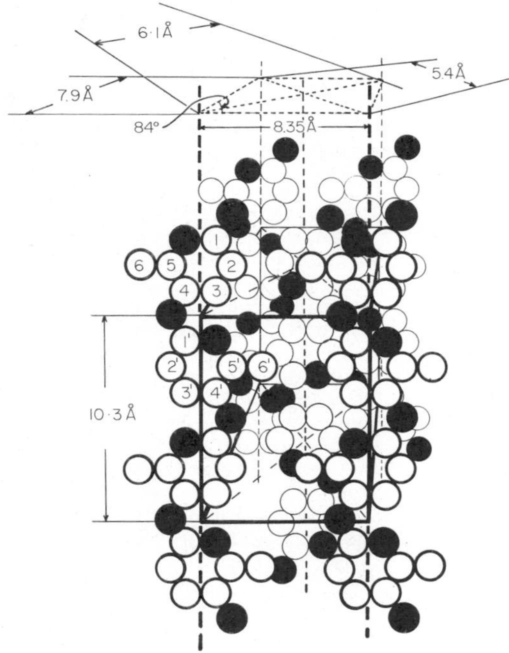

Fig. 1. Diagrammatic representation of the unit cell of cellulose. The monoclinic cell, dimensions 10.3 Å × 8.35 Å × 7.9 Å, β = 84°, is delineated by solid lines with one cellulose chain at each vertical edge and one (antiparallel) in the center. Open circles = carbon atoms; solid circles = oxygen atoms. For clearness of figure, the hydroxyl groups on carbons 2, 3 and 6 are omitted and hydrogen atoms are omitted. Two spacings at 6.1 Å and 5.4 Å are included since these are strongly represented in the X-ray diffraction diagram.

chemistry of this substance has been investigated in detail.[1] Cellulose is always associated in plants with other polysaccharides and polysaccharide derivatives such as mannan, xylan, araban, galactan, polygalacturonic acid, and in woody plants, with lignin. Except in many of the fungi, and in a few seaweeds,[2] it forms the skeletal polysaccharide of the walls of plant cells. Cellulose also occurs in some animals (*e.g.*, in the tunica of sea squirts) and is said to occur in mammalian, including human, skin. It is isolated from plant cell walls by removing other less resistant polysaccharides, *e.g.*, by boiling in water (to remove pectic compounds) followed by chlorination (thought to break association between polysaccharides), and treatment at room temperature with 4*N* KOH to remove alkali-soluble polysaccharides (hemicelluloses), *e.g.*, mannan, xylan, etc. The resultant cellulose contains non-glucose polysaccharides ranging up to 15% by weight in higher plants and 50% or more in some seaweeds[3] and sometimes pectic compounds and polyglucuronic acid, except in a few seaweeds (*Valonia*, *Cladophora*, *Chaetomorpha*) which appear to yield true cellulose in the crystallographic sense. The content of cellulose extracted in this way varies widely with cell type and plant species. It may be as low as 1.5% (some seaweeds, *e.g.*, *Pelvetia*) and as high as 70% or more (the seaweed *Valonia*) but is commonly around 40–50% (W/W) of the wall in mature plant cells.

The extracted cellulose takes the form of long, thin threads (the microfibrils) unless the alkali treatment has caused mercerization (see below). These vary in width from about 100 Å to about 300 Å and are about half as thick; these are said by some authorities to consist of elementary fibrils 35 Å in diameter but X-ray and electron diffraction analysis makes this seem unlikely. Cellulose is in part crystalline and it therefore yields X-ray diffraction diagrams; these correspond to a monoclinic unit cell $a = 8.35$ Å, $b = 10.3$ Å (fiber axis), $c = 7.9$ Å, $\beta = 84°$, in which the molecular chains lie parallel to each other (Fig. 1). Each microfibril consists of bundles of such chains in parallel crystalline array clothed in a cortex in which cellulose chains lie intermixed with chains of sugar residues other than glucose; the chains in this region are parallel but not regularly spaced and are said to be paracrystalline. Treatment with caustic soda solutions (of strength varying from about 10 to 20% according to plant species) causes swelling to an extent that when the alkali is washed out, the cellulose crystallizes into a new lattice ($a = 8.14$ Å; $b = 10.3$ Å; $c = 9.14$ Å; $\beta = 62°$). This is Cellulose II (as against the native Cellulose I) or mercerized cellulose. Only one plant is known (the seaweed *Halicystis*) in which it is suspected that this form of cellulose occurs in nature. Cellulose II can be converted to another crystalline form, Cellulose IV (closely similar to Cellulose I), by heating in water to 300°C. An older report that this form occurs in one plant (*Tussilago farfara*—coltsfoot) is probably in error since the X-ray diagram of glucuronic acid (which remains with the isolated cellulose in walls which contain this substance) is almost identical with that of cellulose IV.

Cellulose microfibrils are usually specifically oriented with respect to a cell axis, and the orientation can be determined by any of the standard methods for the determination of crystal axes such as X-ray or electron diffraction analysis, polarization microscopy or electron microscopy.[4-6] In elongating cells of higher plants, they tend to lie transversely, with considerable angular dispersion. In the thicker secondary walls, they often lie helically around the cell in lamellae in which the helical angle varies from one lamella to the next, except in isodiametric cells in which they lie at random. Cellulose is considered to be bonded in cell walls, by hydrogen bonds and through linkages with Ca^{++} and Mg^{++}, probably as phosphates, with the other polysaccharides in the wall. It has recently been shown that the cell walls of higher plants contain a specific protein which contains hydroxyproline residues and the possibility has been recognized that —S—S— bonds of this protein are significant for cell wall growth.[7] It is not known whether this protein is linked with cellulose, but the probability at the moment is that the linkage is through hemicelluloses and not to the cellulose itself.

The mechanism of cellulose synthesis and microfibril orientation is not known. Synthesis probably occurs from a glucose/phosphate precursor which may be guanosine diphosphate glucose.[8] The enzyme system involved can be extracted from the plant (*e.g.*, from *Acetobacter xylinum*[9]), and synthesis by cell-free extracts has been achieved. The synthetic mechanism in plants higher than bacteria is thought to be located on the cell surface, and both granular aggregates and microtubules seen in the electron microscope are considered as possible sites.

Cellulose is not digestible by mammals, utilization by ruminants resulting from the activity of the INTESTINAL FLORA (see also RUMINANT NUTRITION). Cellulose-digesting enzymes (cellulases) have been extracted from many fungi and other plants and from some animals, notably the snails. Isolated cellulase attacks cellulose rapidly only when the cellulose has been subjected to pretreatment such as swelling in alkali or mild acid hydrolysis. Cellulase-containing fungi, and perhaps bacteria, are responsible for the degradation of cellulose in the soil and therefore for the rotting of wood. In this sense, the rotting fungi are divided into three groups. The "brown rots" (*e.g.*, *Coniphora casebella*, *Poria monticola*) and the "white rots" (*e.g.*, *Polystictus versicola*) are both basidiomycetes. Both attack cellulose, but only the latter takes lignin to any large extent. "Soft rot fungi," recognized only comparatively recently as of importance in this regard, are members of the Ascomycetes and Fungi imperfecti (*e.g.*, *Chaetomium globosum*). They all attack the cellulose of wood, producing characteristic angular cavities. The evidence is that all these fungi attack the non- or paracrystalline component of cellulose more rapidly than the crystalline component.

References

1. OTT, E., AND SPURLIN, H. M., "Cellulose and Cellulose Derivatives," Vols. I–III, New York, Interscience Publishers, 1954–1955.
2. FREI, E., AND PRESTON, R. D., *Nature*, **192**, 939 (1961); *Proc. Roy. Soc. London Ser. B*, **160**, 293 (1964); *Proc. Roy. Soc. London Ser. B*, **160**, 314 (1964).
3. CRONSHAW, J., MYERS, A., AND PRESTON, R. D. *Biochim. Biophys. Acta*, **27**, 89 (1958).
4. PRESTON, R. D., "The Molecular Architecture of the Plant Cell Wall," London, Chapman and Hall, 1952.
5. FREY-WYSSLING, A., "Der Pflanzliche Zellward," Berlin, Spring-Verlag, 1959.
6. ROELOFSEN, P. A., "The Plant Cell Wall," Gebruder Borntraeger, 1959.
7. LAMPORTE, D. A., in "Advances in Botanical Research," Vol. II, New York, Academic Press, 1965.
8. BARBER, G. A., ELBEIN, A. D., AND HASSID, W. Z., *J. Biol. Chem.*, **239**, 4056 (1964).
9. COLVIN, J. ROSS, "The Biosynthesis of Cellulose," in "Formation of Wood in Forest Trees," (ZIMMERMANN, M. H., EDITOR), New York, Academic Press, 1964.

R. D. PRESTON

CENTRIFUGATION, DIFFERENTIAL*

Differential centrifugation refers to separation in the ultracentrifuge of macromolecules from each other and from small molecules due to differences in their rates of sedimentation or, as in the case of flotation analysis and sedimentation equilibrium in a density gradient, their densities. The rate of sedimentation of a solute molecule in a solvent of different density depends upon the magnitude of the centrifugal field, the molecular weight of the solute molecule and the frictional resistance it experiences in moving through the solvent. The rate of sedimentation is given by the expression, $dx/dt = \omega^2 xs$, where x is the distance (centimeters) from the center of rotation; t, the time (seconds); and ω, angular rotation (sec^{-1}) of the rotor. The sedimentation coefficient, s (seconds or Svedberg units, S, 10^{-13} second) is the rate per unit field and is related to molecular parameters by the Svedberg equation,

$$s = \frac{M(1 - \bar{V}\rho)}{Nf} = \frac{MD(1 - \bar{V}\rho)}{RT},$$

where M is the anhydrous molecular weight; f, the frictional coefficient which depends upon size, shape and state of hydration of the molecule; D, diffusion coefficient; $\bar{V}$, partial specific volume; ρ, density of the solvent; R, gas constant and T, absolute temperature. The Svedberg equation provides the theoretical basis for differential centrifugation of substances having different

* This work was supported by a grant from the National Institute of Arthritis and Metabolic Diseases, National Institutes of Health, AI 01482–14.

molecular weights and/or shape in a solvent less dense than the solute molecules (velocity sedimentation). It also shows that differences in solute density, $1/\bar{V}$, can be utilized to achieve separation; *e.g.*, consider a mixture of PROTEIN and LIPOPROTEIN in a salt solution whose density is less than that of the protein but greater than the lipoprotein. Under the action of a centrifugal field, the protein will sediment to the bottom of the centrifuge column while the lipoprotein floats to the top. If the density of the solvent is equal to that of the solute, the solute molecules neither sediment nor float. These principles are utilized in the separation of DEOXYRIBONUCLEIC ACIDS (DNAs) by the method of sedimentation equilibrium in a field-generated density gradient of cesium chloride.[1] Centrifugation of a concentrated cesium chloride solution results in an appreciable concentration gradient of cesium chloride and, thus, a density gradient. When DNA is present, the macromolecules redistribute themselves in the density gradient. The DNA molecules sediment, float upward, or remain stationary, depending upon the local density. At equilibrium, DNAs of different densities are narrowly banded at different positions in the ultracentrifuge column. Each DNA bands at a position where the density of the cesium chloride and the effective density of the macromolecule in that solution are equal. This method was originally devised to elucidate the mechanism of DNA replication and is finding important applications in nucleic acid chemistry. It has also been applied to the analysis of human plasma lipoproteins in sucrose gradients.

The abundant results obtained by differential centrifugation have contributed significantly to the rapid progress made in the past two decades in the various branches of biochemistry and medicine. Many important results[2] were obtained in the years following construction of the first oil-turbine ultracentrifugation by Svedberg and his co-workers in 1925–26, but progress[3,4] in both practice and theory has been dramatic since the advent about 15 years ago of commercially available electrically driven instruments. Like ELECTROPHORESIS, ultracentrifugation has become an indispensible research tool of the biochemist and molecular biologist. In clinical medicine, ultracentrifugal analysis of blood serum led to the recognition of macroglobulinemia and serves as an aid in the diagnosis of this disease. Ultracentrifugation is also playing an important role in the elucidation of the etiology of diseases such as rheumatoid arthritis and ATHEROSCLEROSIS.

The following discussion is concerned solely with velocity sedimentation. One distinguishes between analytical and preparative centrifugation depending upon whether analysis is carried out during sedimentation or on fractions removed from the centrifuge tube after cessation of centrifugation. Each may be practiced either as a moving-boundary or zone method.

Analytical Centrifugation. Analytical centrifugation is carried out in a sector-shaped cell using an instrument such as the Spinco Model E (see also ULTRACENTRIFUGATION). The principles of the moving boundary method are illustrated in Fig. 1. Initially the cell is filled with the solution of macromolecules to be analyzed, in this example a mixture of two kinds of molecules, A and B, having different sedimentation coefficients. The solution should contain a sufficient concentration of supporting electrolyte such as KCl to depress undesirable charge effects;[2,3] $0.1M$ suffices for most proteins at pH values near the isoelectric point. Differential centrifugation of the macromolecules at rotor speeds as high as 60,000 rpm (centrifugal force of the order of $200{,}000 \times g$) results in two moving boundaries, a slow one between pure solvent and solution of pure A and a faster one between the solution of pure A and one containing both A and B. The boundaries may be recorded photographically as a function of time using schlieren, interference or ultraviolet absorption optics. Thus, multiplication of the ordinate in Fig. 1(d) by an appropriate constant transforms the plot into a facsimile of a sedimentation pattern obtained with schlieren optics at a

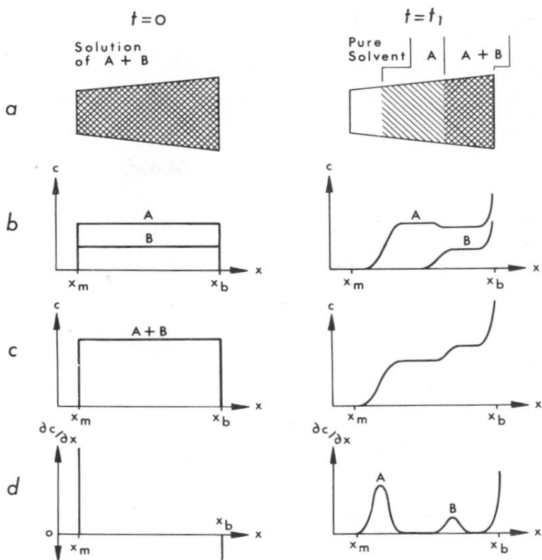

FIG. 1. Resolution of a mixture of two kinds of macromolecules, A, and B, with different sedimentation coefficients by the method of moving-boundary centrifugation: (a) diagrammatic representation of moving boundaries in a sector-shaped analytical cell; (b) and (c) plots of concentration, c, against x; and (d) plot of concentration gradient, dc/dx, against x. Original mixture contains two parts of A and one part of B. Sedimentation coefficient of B greater than A. Sedimentation proceeds to right. X_m and X_b are the distances from the center of rotation of the meniscus and the bottom of the cell, respectively. Broadening of the boundaries during sedimentation is due to diffusion. Only very high molecular weight substances like DNA and macroglobulins yield almost infinitely sharp boundaries. In general, a solution containing n macromolecules having sufficiently different sedimentation coefficients will give n moving boundaries.

particular instant of time, t_1, during sedimentation. The apparent composition is computed from the areas under the peaks in the pattern after correction for dilution in the sector-shaped cell by multiplying each area by $(\bar{X}/X_m)^2$. $\bar{X}$ is the distance of a particular peak from the center of rotation and is usually taken as either the position of the top of the peak or its first moment. The apparent relative proportion of each macromolecule is given by the relative contribution of the corresponding peak to the total corrected area. But because of the Johnson-Ogston effect,[3] these values are in error, the apparent relative proportion of the slower sedimenting macromolecule being too large and that of the faster one too small. The Johnson-Ogston effect arises from the dependence of the sedimentation coefficient of a given component upon the total macromolecular concentration of the solution through which it is sedimenting. Consequently, the concentration of the slower sedimenting molecule decreases across the faster moving boundary, as illustrated in Fig. 1(b). Several procedures are available for obtaining the correct composition, one of which makes use of the fact that the Johnson-Ogston effect is of negligible magnitude at very low concentrations. Analyses are made at progressively decreasing total macromolecular concentration, and the apparent compositions are extrapolated to infinite dilution. Finally, the sedimentation coefficient of each component is computed from the slope of a plot of log $\bar{x}$ *vs* t, corrected to water at 20°C as solvent[3] and then extrapolated to infinite dilution to obtain a coefficient which is a molecular parameter independent of solute-solute interactions. Estimates of molecular weight can be obtained from the extrapolated sedimentation coefficients by assuming the molecules to be spherical in shape. Thus, we see that moving-boundary ultracentrifugation provides information not only about composition but also about molecular size. Furthermore, fractionation can be achieved by the combined use of schlieren optics and a separation cell which permits isolation of pure slow-moving component. Also, with proper choice of solvent density, these techniques can be applied to the flotation analysis of lipoproteins.

The patterns of many highly purified proteins show a single boundary. The molecular weight is computed with the aid of the Svedberg equation using the extrapolated sedimentation coefficient (preferably computed from the second moment of the gradient curves and corrected for stretching of the rotor) and the extrapolated diffusion coefficient.*† The ratio of the frictional coefficient to that of a hypothetical sphere of the same mass can then be computed and interpreted in terms of the shape and state of hydration of equivalent hydrodynamic models. But, a single sedimenting boundary must not be taken as a sufficient criterion of ultracentrifugal homogeneity without first examining the shape of the boundary.[3] Excessive spreading during sedimentation may indicate heterogeneity. In that event, the distribution of sedimentation coefficients can be derived from a single ultracentrifuge experiment. Such analyses have shown that γ-pseudoglobulin is largely composed of molecules differing only slightly in size and shape and that its electrophoretic heterogeneity reflects a heterogeneity with respect to net charge, not size and shape.

An important application[3,6] of analytical centrifugation is the study of interactions of macromolecules with each other and with small molecules, *e.g.*, association-dissociation of proteins, antigen-antibody reactions, enzyme-coenzyme or enzyme-inhibitor interactions, and complexing of DNA with RIBOSOMES. Discussion of the significance of these investigations for our understanding of life processes is beyond the scope of this article. But, cognizance must be taken of the fact that reactions of the type $n\text{A} \rightleftharpoons \text{A}_n$ for $n > 2$, $\text{A} + \text{B} \rightleftharpoons \text{C}$, $\text{A} + n\text{X} \rightleftharpoons \text{B}$, and $2\text{A} + n\text{X} \rightleftharpoons \text{A}_2\text{X}_n$, where X is a small molecule, can give bimodal reaction boundaries despite instantaneous establishment of equilibrium. Also, isomerizing systems, $\text{A} \rightleftharpoons \text{B}$, can give trimodal boundaries for rates of interconversion of the same order as the rates of differential sedimentation of the isomers. Such reaction boundaries sometimes arise unexpectedly, and care must be exercised not to misinterpret them in terms of heterogeneity.

The schlieren method of recording moving boundaries is used with proteins, certain nucleic acid preparations, VIRUSES and other macromolecules at concentrations within the approximate range, 4–0.1 g/100 ml. Lower concentrations necessitate the use of the ultraviolet absorption method, which is particularly applicable to nucleic acids at 30–5 μg/ml. Unlike the schlieren method which records concentration gradients, the absorption method records concentration. Thus, multiplication of the ordinate in Fig. 1(c) by an appropriate constant transforms the plot into a facsimile of an inverted densitometer tracing of an absorption pattern. Interpretation is subject to all of the above considerations for schlieren patterns. In addition, the use of such low concentrations may require stabilization of the boundaries against convective disturbances. This can be achieved by sedimentation in a fairly small, preformed sucrose density gradient.

The principles of band centrifugation[7] (a zone analytical method so designated to distinguish it from zone sedimentation in a preformed density gradient) are illustrated in the rectangular approximation by Fig. 2. A thin lamella of a dilute solution of nucleic acid, protein or virus is layered onto a denser liquid (solvent containing more concentrated electrolyte or D_2O) under the

* Even if the extrapolated diffusion coefficient is not known, the molecular weight of a protein can still be computed from the Svedberg equation by using values of the sedimentation and diffusion coefficient measured at the same finite concentration and a relatively high ionic strength.

† The sedimentation-diffusion method for determining molecular weights is being rapidly supplanted by the Archibald approach-to-equilibrium method[3] and short-column equilibrium centrifugation.[5]

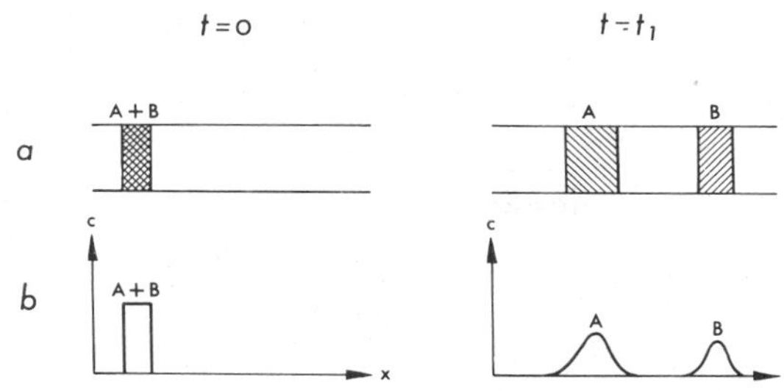

FIG. 2. Complete separation of two kinds of molecules, A and B, with different sedimentation coefficients by the methods of band and zone centrifugation in a hypothetical rectangular analytical cell or a cylindrical tube: (a) diagrammatic representation of zones in the sedimentation column and (b) plot of macromolecular concentration, c, against x. Original mixture contains two parts of A and one part of B. Sedimentation coefficient of B greater than A. Sedimentation proceeds to the right. In general, n components, having sufficiently different sedimentation coefficients, will give n zones.

influence of the centrifugal field in a special sector-shaped cell. Macromolecules of the same sedimentation coefficient then sediment through the liquid in a narrow band. Differential sedimentation of a mixture results in multiple bands. The density gradients necessary to stabilize the system against convection are generated during the experiment by diffusion of small molecules between the lamella and the bulk solution, and in some cases by sedimentation of the small molecules in the bulk solution. The bands are recorded using ultraviolet absorption optics. Multiplication of the ordinate in Fig. 2(b) by an appropriate constant transforms the plot into a facsimile of an inverted densitometer tracing of an absorption pattern obtained at time t_1. In contrast to the moving-boundary method, band sedimentation permits complete physical separation of different kinds of macromolecules, thereby permitting analysis without the complication of the Johnson-Ogston effect. Sedimentation coefficients may be evaluated from the motion of the bands. Band-centrifugation analyses are more appropriate than moving-boundary ones as pilot experiments for zone preparative centrifugation.

Preparative Centrifugation. The principles of preparative centrifugation[8,9] are essentially the same as those already described for analytical centrifugation. Experiments are performed in cylindrical Lusteroid tubes, using a swinging bucket rotor in an ultracentrifuge such as the Spinco Model L at rotor speeds and for lengths of time chosen to cause separation of the boundaries or zones before the most rapidly sedimenting one reaches the bottom of the tube. Upon cessation of centrifugation, the contents of the tube are fractioned by any one of several methods, *e.g.*, withdrawal of numerous successive layers of solution from the top of the fluid column with a Pasteur pipet and gentle suction, or piercing the bottom of the tube with a hypodermic needle and collecting drops. The fractions are then analyzed for radioactivity, biological activity, chemical properties, optical density, etc., and sedimentation patterns are constructed by plotting the results of the analyses against distance from the meniscus or center of rotation, volume, number of drops or fraction number. Thus, preparative sedimentation provides both analysis and fractionation of a mixture in a single operation. Furthermore, a particular biologically active substance in a crude extract may be localized by its activity and characterized at concentrations so low as to be undetectable by the most sensitive analytical centrifugation procedures.

Moving-boundary preparative centrifugation of proteins or DNA at a concentration of about 10 mg/ml requires no special precautions, but formation of sharp and stable boundaries at low concentrations, 1–0.1 mg/ml, necessitates sedimentation in a shallow, preformed sucrose density gradient. A continuous gradient can be produced by layering solutions of the same protein but varying sucrose concentration and twirling a sawtooth wire inserted into the tube. Sedimentation coefficients ranging from 4–2000S have been determined and found to be in excellent agreement with values obtained by analytical centrifugation.

In zone preparative centrifugation, a small volume of solution containing 5 mg to 5 μg macromolecule/ml is layered on top of a fairly small, continuous density gradient.* The usually linear gradient is preformed from light and heavy solutions with some mechanical device using gradient-forming materials such as sucrose, cesium chloride, glycerol or polyvinylpyrrolidone. Under the action of the centrifugal field, macromolecules with the same sedimentation coefficient sediment as a more-or-less narrow zone through the gradient, which stabilizes the system against convection. A mixture of several different kinds of molecules yields a corresponding number of discrete zones (Fig. 2). Sedimentation coefficients can be computed directly from the rates of migration of the zones and knowledge of the viscosity and density at each point in the gradient. Alternatively, a standard, well-characterized protein can be added to the original protein mixture. The sedimentation coefficient of the unknown is then given by the product of that of the standard and the ratio of the distances sedimented by the unknown and the standard. In the case of DNA, the molecular weight, M, of one species can be measured in terms of that of another from the ratio of distances, D, sedimented according to the relation, $D_2/D_1 = (M_2/M_1)^{0.35}$.

Zone centrifugation is used extensively to purify

* If the gradient is fairly steep, extending over the range of densities of the macromolecules or cellular particles to be separated, centrifugation proceeds until each macromolecule has reached a position corresponding to its own density. In this method, referred to as isopycnic gradient centrifugation, separation is due to differences in density rather than sedimentation coefficient.

and characterize viruses, cellular particles, enzymes, serum proteins such as the rheumatoid factor, and nucleic acids. An application of considerable importance is the characterization of DNA, RNA and ribosomes as to their size, biosynthesis and interactions. The significance of such experiments for elucidation of the molecular mechanism of information transfer from gene to protein is illustrated by the experimental discovery of messenger RNA. Finally, as in the case of analytical centrifugation, care must be exercised not to misinterpret possible reactions zones in terms of heterogeneity.

References

1. Meselson, M., Stahl, F. W., and Vinograd, J., *Proc. Natl. Acad. Sci. U.S.*, **43**, 581, (1957); **44**, 671 (1958).
2. Svedberg, T., and Pedersen, K. O., "The Ultracentrifuge," London and New York, Oxford University Press, 1940 (Johnson Reprint Corporation, New York).
3. Schachman, H. K., "Ultracentrifugation in Biochemistry," New York and London, Academic Press, 1959.
4. Fujita, H., "Mathematical Theory of Sedimentation Analysis," New York and London, Academic Press, 1962.
5. Yphantis, D. A., *Biochemistry*, **3**, 297 (1964).
6. Nichol, L. W., Bethune, J. L., Kegeles, G., and Hess, E. L., in "The Proteins: Composition, Structure and Function" (Neurath, H., Editor), Second edition, Vol. II, Ch. 9, New York and London, Academic Press, 1964.
7. Vinograd, J., Brunner, R., Kent, R., and Weigle, J., *Proc. Natl. Acad. Sci. U.S.*, **49**, 902 (1963).
8. de Duve, C., Berthet, J., and Beaufay, H., *Progr. Biophys. Biophys. Chem.*, **9**, 326 (1959).
9. Martin, R. G., and Ames, B. N., *J. Biol. Chem.*, **236**, 1372 (1961).

John R. Cann

CEPHALINS AND LECITHINS

Cephalin (*kephalē*, head), isolated from brain tissue by Thudichum in 1864, and lecithin (*lekithos*, egg yolk), isolated by Gabley in 1846 from egg yolk, are the best known examples of the large group of naturally occurring compound LIPIDS classified as phospholipids. They are found in all animal and plant cells and are lipid complexes consisting of glycerol to which are esterified ($R—C—O—\overset{O}{\overset{\|}{C}}—R'$) two fatty acids and a phosphoric acid which in turn is attached to the alcohol group of a nitrogenous base

$$(—O—\underset{O^-}{\underset{|}{\overset{O}{\overset{\|}{P}}}}—O—CH_2R'')$$

Like the fats they are present in nature as a mixture of several compounds containing different but related acyl groups, *i.e.*, palmityl, stearyl, oleyl, etc. Structurally these compounds are generally represented by the following formula where $—\overset{O}{\overset{\|}{C}}—R_1$ represents the fatty acid residues and $—CH_2—R_2$ the residual nitrogenous bases:

$$\begin{array}{l} \qquad\qquad\quad CH_2—O—\overset{O}{\overset{\|}{C}}—R_1 \\ \qquad\qquad\quad | \\ R_1—\overset{O}{\overset{\|}{C}}—OCH \\ \qquad\qquad\quad | \\ \qquad\qquad\quad CH_2—O—\underset{O^-}{\underset{|}{\overset{O}{\overset{\|}{P}}}}—O—CH_2—R_2 \end{array}$$

Hydrolytic Products. Acid hydrolysis of these complex lipids yields glycerol, fatty acids, phosphorous and nitrogenous bases. Lecithins contain choline ($—CH_2CH_2N(CH_3)_3$) as the nitrogenous base, whereas the cephalins may contain any one of several different nitrogenous bases, of which the best known are serine [$HO—CH_2—CH(NH_2)—COOH$] and ethanolamine ($HO—CH_2—CH_2—NH_2$). The carboxylesterases, lecithinase A and lecithinase B, hydrolyze the fatty acid residues (acyl esters) from lecithins and cephalins, producing the deacylated compounds, lysolecthins and lysocephalins. Of the phosphodiesterases, phospholipase (lecithinase) C is capable of splitting the nitrogenous bases from the phosphatides while phospholipase (lecithinase) D splits the nitrogenous bases at their ester linkage between glycerol and phosphoric acid.

Chemical Properties. Purified lecithin is an extremely hygroscopic, white, paraffin-like substance which on exposure to air turns dark brown, acquiring a disagreeable odor and taste. It is soluble in alcohol, benzene, ether, petroleum ether, chloroform, carbon tetrachloride, acetic acid, pyridine, glycerol, and many other organic solvents. One of the outstanding characteristics of both lecithins and cephalins is their insolubility in acetone. When mixed with water, lecithins and cephalins become opaque colloidal suspensions. The properties of cephalins are similar to those of the lecithins with the major exception that cephalins are less soluble in alcohol than are the lecithins. The reactions of cephalins differ from those of lecithins only when the nitrogenous bases are involved.

Since the terms lecithin and cephalin frequently have resulted in considerable confusion because they tend to connote single compounds, the use of a more consistent organic chemical nomenclature is recommended, *i.e.*, phosphatidylcholine (glycerylphosphorylcholine), phosphatidylserine (glycerylphosphorylserine), and phosphatidylethanolamine (glycerylphosphorylethanolamine).

Distribution. With few exceptions, these phospholipids comprise the bulk of the total lipids of animal and plant cells. The chemical composition of nerve tissue contains a high percentage of phospholipids (see NERVE CELL COMPOSITION). Approximately 40% of the dry weight of the white matter and 25% of the grey matter of beef brain are phospholipids (cephalins 55%, lecithins 25%, and sphingomyelins 20%). Normal plasma contains approximately 150 mg lecithins, 15 mg cephalins, and 30 mg sphingomyelins per 100 ml. The phospholipid content of erythrocytes is higher than that of plasma, with cephalins being present in greater proportions than lecithins. Skeletal muscle contains equal amounts of lecithins and cephalins. Phospholipids comprise approximately 15% of the dry weight of beef liver, with 6% cephalins, 8% lecithins, and 1% sphingomyelins. The intestine of beef contains approximately 7% phospholipids (dry weight) with 4% lecithins, 2% cephalins, and 1% sphingomyelins. The lipids of the Gram-negative bacillus *Pseudomonas aeruginosa* represent 6.8% of the dry weight of cells, with phosphatidylethanolamine comprising 90–95% and phosphatidylcholine 3–5% of the total bacterial lipids. Phosphatidylserine is present only in trace amounts. A complete chemical analysis of the lipids has been seriously hampered by the difficulty in separating and purifying these compounds, by their lability, and by the scarcity of pure compounds for study. Earlier investigators used the classical methods of separating cephalins on the basis of their insolubility in alcohol, lecithins by the formation of their metallic complexes with such salts as cadmium chloride, and sphingomyelins on the basis of their insolubility in ether. Only within recent years have newer techniques such as gas-liquid CHROMATOGRAPHY, solid-liquid chromatography and the concomitant use of isotopes become available for the isolation, purification and identification of the various lipids and their components. The more important chromatographic techniques which have made possible significant advances in the field of lipid chemistry are silicic acid impregnated filter paper, thin-layer, silicic acid column and gas-liquid chromatography. Fractionation of the polar lipids of *P. aeruginosa* on silicic acid columns with chloroform-methanol solvent mixtures and subsequent analysis of the column fractions give results of the following nature:

Chloroform in Methanol (%)	Major Nitrogenous Bases	Weight of Total lipids (%)	Total Phospholipids (%)
80	Ethanolamine, serine plus several ninhydrin positive spots	81	84
60	Ethanolamine, serine plus several ninhydrin positive spots	11.8	11.5
0	Choline, plus several ninhydrin positive spots	4	4.5

While such results demonstrate that complete separation of the cephalins and lecithins are not possible by silicic acid column chromatography, this technique is far superior to solvent fractionation. Hydrolysis of each of the above column fractions and analysis of the water-soluble hydrolysates by paper chromatography or by a Technicon Autoanalyser proved that these fractions contain several different amino acids (ninhydrin-positive compounds). Silicic acid impregnated paper and thin-layer chromatograms of phosphatidylethanolamine (PE), phosphatidylserine (PS), and phosphatidylcholine (PC) give the following R_f values when developed with appropriate solvents: paper = PE .63, PS .54, and PC .45; thin-layer = PE .50, PS .02, and PC .33.

The question of the possible biological function of cephalins and lecithins has resulted in much research and an even greater amount of speculation. The phospholipids have been implicated as a source of choline for nerve cells, as playing a role in the clotting of blood, as a transport for sodium and potassium ions (see ACTIVE TRANSPORT), as essential structural components of cells, as well as many other biological functions. It has been proved, however, that the cell membrane is rich in phospholipids, and recent experimental evidence has shown that the cephalin fraction of metabolically active animal and plant cells is capable of incorporating radioactive amino acids from the surrounding medium and donating the amino acids to the internal metabolic processes of the cell. These results implicate the phospholipids, and in particular the cephalins, as important intermediates in the metabolism of amino acids. Gaby *et al.*[1] have proposed that the phospholipid-amino acid complex may serve to transport amino acids across the cell membrane. Hendler,[2] and Hunter and Goodsall[3] have proposed models for protein synthesis in which the lipids play a major role (see LIPOAMINO ACIDS).

The chemical composition of such a phospholipid-amino acid complex was virtually unknown until Macfarlane[4] reported a glycerylphosphorylglycerol-alanine complex in the Gram-positive bacillus *Clostridium welchii*, and Houtsmuller and Van Deenen[5] described a glycerylphosphorylglycerol-ornithine complex in *Bacillus cereus*. On the basis of their chemical analysis,

Sinha and Gaby[6] have postulated the following structure of the labile phospholipid-amino acid complex consisting of two phospholipid molecules linked by glycerol to which an amino acid is bound by an O-ester linkage:

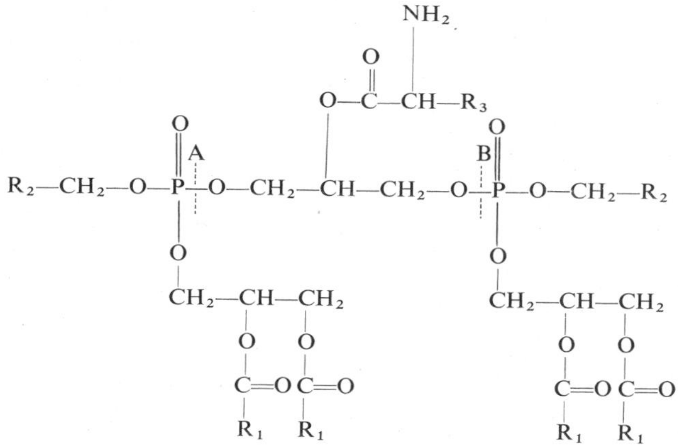

```
                                      NH2
                                      |
                              O       |
                              ‖       |
                            O—C———CH—R3
           O                |                 O
           ‖ A              |               B ‖
R2—CH2—O—P—O—CH2—CH—CH2—O—P—O—CH2—R2
           |                                  |
           O                                  O
           |                                  |
           CH2—CH—CH2                         CH2—CH—CH2
               |   |                              |   |
               O   O                              O   O
               |   |                              |   |
             C=O C=O                            C=O C=O
               |   |                              |   |
               R1  R1                             R1  R1
```

R_1 represents fatty acid residue; R_2 residual nitrogenous bases; R_3 amino acid.

If such a compound is cleaved at either position A or position B, 1 mole of the phospholipid amino acid complex and 1 mole of phospholipid results. A chromatographic fraction would either gain or lose 1 mole of aminoglycerol moiety attached to the phospholipid (cephalin) molecule. Chemical analysis as well as specific radioactivity of the phospholipid fractions supports such a postulation and explains the ability of phospholipids to transport amino acids.

References

1. Gaby, W. L., Naughten, R. N., and Logan, C., *Arch. Biochem. Biophys.*, **82**, 34 (1959).
2. Hendler, R. W., *Nature*, **193**, 821 (1962).
3. Hunter, G. D., and Goodsall, R. A., *Biochem. J.*, **74**, 34 (1960).
4. Macfarlane, M. G., *Nature*, **196**, 136 (1962).
5. Houtsmuller, U. M. T., and Van Deenen, L. L. M., *Biochim. Biophys. Acta*, **70**, 211 (1963).
6. Sinha, D. B., and Gaby, W. L., *J. Biol. Chem.*, **239**, 3668 (1964).

W. L. Gaby

CEREBROSIDES

Cerebrosides are substances which can be cleaved by heating with acid to yield three component parts: sphingosine, a fatty acid, and a sugar. The structure of a cerebroside is as shown:

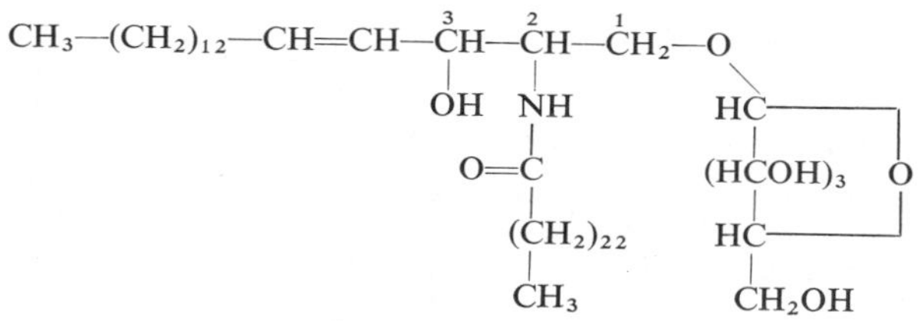

```
                           3    2    1
CH3—(CH2)12—CH=CH—CH—CH—CH2—O
                   |    |         \
                   OH   NH         HC———————
                        |          |        |
                      O=C          (HCOH)3  O
                        |          |        |
                      (CH2)22      HC———————
                        |          |
                        CH3        CH2OH
```

The sphingosine component is the horizontal part of the formula: its hydroxyl group on position 1 is combined with the sugar, its amine group on position 2 is combined in amide linkage with a fatty acid (lignoceric acid in this illustration), and its hydroxyl group on position 3 is free. Because of the many methylene groups (CH_2) in the sphingosine and fatty acid components, cerebroside is classed as a lipid. However, the presence of the many hydroxyl groups in the sugar component renders the compound rather insoluble in the usual lipid solvents (hexane, chloroform, alcohol). Solvents containing polar and nonpolar groupings are best: chloroform + alcohol; pyridine; tetrahydrofuran.

Distribution. Cerebrosides occur in nature, but with a limited distribution. The name is derived from "cerebral," pertaining to the brain, for this is their region of highest concentration and the source from which cerebrosides were isolated by Thudicum about 1870. In brain there are many cerebrosides, all containing galactose as the sugar (and thus termed *galactocerebrosides*) but differing in the nature of the fatty acid component. Two classes of fatty acids are present: ordinary fatty acids and fatty acids containing an hydroxyl group in the 2-position (next to the carboxyl group). Some of the fatty acids contain a double bond somewhere in the chain of methylene groups, in most cases nine carbons from the methyl end. The chain lengths of the acids vary, being mainly 24, 22, and 18 carbon atoms long. However appreciable amounts of 26-carbon acids occur, as well as 23- and 25-carbon acids. The relative amounts of the odd-numbered acids are unusually high (compared with other lipids), and they increase with the age of the animal. This increase with age is due in part to a one-carbon degradation process which attacks the fatty acids.

Some cerebrosides contain dihydrosphingosine as the amine component; in this case the double bond is absent from the 4,5-position. In animals, galactocerebrosides are known to occur (thus far) only in central nervous system and kidney; in spleen and possibly some other organs, the cerebrosides contain glucose as the sugar component. While galactocerebrosides are almost specific for brain, they probably do not occur in the brains of invertebrates. Cerebrosides probably do not occur in plants, although the earlier scientific literature indicates they do.

Biosynthesis and function. The biological function of cerebrosides is unknown. Their relatively high occurrence in the nervous system, particularly in MYELIN, suggests they have a structural role, perhaps helping to insulate the nerve axon from intraneural short circuits, or helping to increase the speed of NERVE IMPULSE CONDUCTION. However, cerebrosides occur in other regions of the brain cells and must have additional functions. Radioactivity studies in whole animals suggest the presence of relatively stable cerebroside molecules as well as molecules which undergo rapid degradation and rebuilding. Perhaps the stable molecules are in the myelin sheaths. In certain diseases involving progressive paralysis and loss of myelin (MULTIPLE SCLEROSIS), there is a concomitant loss of cerebroside.

Cerebroside appears to be formed enzymatically with the aid of 2 enzymes:

(1) Sphingosine + Uridine Diphosphogalactose → Uridine Diphosphate + Sphingosine Galactose (Psychosine)

(2) Psychosine + Fatty Acid-Coenzyme A → Cerebroside + Coenzyme A.

Galactocerebroside appears to undergo destruction in brain by two successive hydrolytic steps: removal of the galactose component (to form a ceramide) and removal of the fatty acid component (to yield a fatty acid + sphingosine).

A genetically controlled disease, Gaucher's disease, is characterized by a marked enlargement of the spleen and liver and particularly high accumulation of glucocerebroside in these organs and in bone marrow. The infantile form of the disease is fatal and is marked by severe maldevelopment of the brain.

A sulfuric acid ester of cerebroside ("sulfatide") exists in brain and kidney, and possibly elsewhere. The sulfate group is attached at the 3-position of the galactose. Once formed in brain, it undergoes very slow degradation; in kidney, however, the molecules are made and destroyed fairly rapidly. There is a fatal genetically determined disease, metachromatic leucodystrophy or sulfatidosis, in which considerable deposition of sulfatide occurs in brain and kidney and leakage of sulfatide into urine takes place. This lipidosis is accompanied by severe degeneration of the brain.

Cerebroside is the simplest member of a larger class of lipids, the *glycosphingolipids*, in which oligosaccharides (instead of glucose or galactose) occur bound to the 1-position of sphingosine. The trivial names assigned to some of these are cytolipin H, globoside, hematoside, and ganglioside.

References

1. DEUEL, H. J., JR., "The Lipids," pp. 473–506, New York, Interscience Publishers, 1951.
2. KISHIMOTO, Y., AND RADIN, N. S., "Composition Cerebroside Acids as a Function of Age," *J. Lipid Res.*, **1**, 79 (1959).
3. HAJRA, A. K., AND RADIN, N. S., "Isotopic Studies of the Cerebroside Fatty Acids in Rats," *J. Lipid Res.*, **4**, 270 (1963).
4. KOPACZYK, K. C., AND RADIN, N. S., "*In vivo* Conversions of Cerebroside and Ceramide in Rat Brain," *J. Lipid Res.*, **6**, 140 (1965).
5. HAGBERG, B., SOURANDER, P., AND SVENNERHOLM, L., "Sulfatide Lipidosis in Childhood," *Am. J. Diseases Children*, **104**, 644 (1962).
6. STANBURY, J. B., WYNGAARDEN, J. B., AND FREDRICKSON, D. S., (EDITORS), "The Metabolic Basis of Inherited Disease," pp. 603–633, New York, McGraw-Hill Book Co., 1960.

NORMAN S. RADIN

CEREBROSPINAL FLUID

The cerebrospinal fluid (CSF) is a clear watery liquid that surrounds the brain and spinal cord and fills the four cavities or ventricles of the brain. The bulk of the fluid is thought to originate in filamentous structures, the choroid plexuses, which are situated on the walls of the ventricles. The fluid flows out of the brain through openings in the roof of the fourth ventricle to circulate over the brain and around the spinal cord in the subarachnoid space. This area is simply the space between the membrane (pia mater) directly adjacent to the nervous tissue and the membrane (arachnoid) attached to the tough outer covering of the brain (dura mater). The liquid leaves the subarachnoid spaces to enter the blood by flowing through canals in the arachnoid villi which open onto large venous channels in the outermost membrane of the brain. There are similar structures related to veins in the arachnoid membrane covering the spinal cord. These spatial relationships are depicted schematically in Fig. 1.

In man, the entire volume of CSF is between 120 and 150 cc, about 30 cc of which is in the ventricles. The rate of formation is difficult to establish, as it appears to depend on the normal degree of absorption through the arachnoid villi. More than a liter may be formed in a day if drainage is brought about by artificial means or by spontaneous pathological processes. Ordinarily, daily production is estimated to be from one to three times the total volume.

One of the obvious functions of the CSF is to protect the brain and spinal cord against mechanical shock. The fluid also functions to maintain the intracranial pressure at a constant level. Thus, whenever the volume of brain tissue is altered because of differences in the volume of blood in

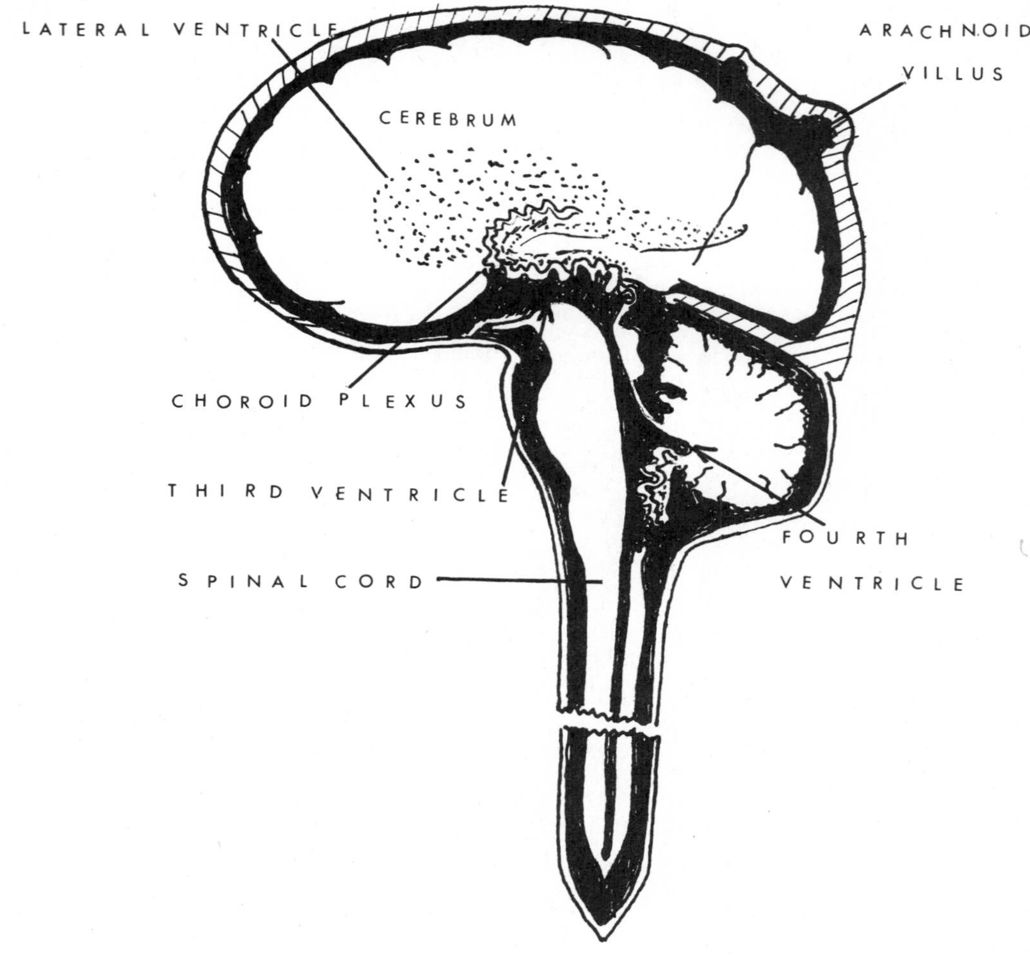

FIG. 1. The drawing represents a median section of the nervous system and shows the relation of the cerebrospinal fluid (in black) to the brain and spinal cord. The two lateral ventricles are not visible but their shape is denoted by the stippled area.

the brain or after surgical manipulation, the CSF volume is adjusted to maintain the normal intracranial pressure. At present, the fluid has no proven nutritive function and no unique biochemical action on the nervous tissue it bathes.

For many years there has been controversy as to whether the CSF was an ultrafiltrate of the blood plasma or whether the choroid plexuses formed the fluid by a process of active secretion. If the fluid were the result of ultrafiltration through the capillary walls, the concentrations of the various inorganic cations and anions should be the same as the concentrations of the same ions in a dialysate of plasma.

In Table 1 is shown a comparison of the concentrations of some ions and non-electrolytes in the CSF and in the plasma and plasma dialysate of a rabbit. The samples of CSF and blood were drawn from the same animal as nearly simultaneously as possible. The results in Table 1 indicate, for example, that the concentration of chloride ions is higher in CSF than in plasma or in a dialysate of plasma, while the concentration of glucose is markedly lower. While there may be numerical differences among various species, generally the experimental results lead to the conclusion that insofar as the inorganic ions and small uncharged molecules like glucose are concerned, the CSF is an active secretion of the choroid plexus and not a filtrate of the blood plasma.

The normal range of protein concentration is from 20–40 mg, with an average value of 25 mg, per 100 cc. This is roughly 1/250 of the protein concentration of blood plasma. The concentration of protein varies according to the level from which the fluid is withdrawn, being higher in the lumbar region than in the ventricles. The protein content also increases with age.

TABLE 1. SOME ELECTROLYTES AND GLUCOSE: COMPARISON OF THE CONCENTRATIONS IN THE CSF AND PLASMA DIALYSATE TO THOSE IN THE PLASMA, OF A RABBIT[a]

Substance	Concentration in CSF / Concentration in Plasma	Concentration in Plasma Dialysate / Concentration in Plasma
Cl	1.21	1.04
Na	1.03	0.95
Mg	0.80	0.80
K	0.52	0.96
HCO_3	0.94	1.04
Glucose	0.64	0.97
pH	7.27	7.46

[a] Values taken from Davson.[1]

In Table 2 are listed values for total protein at three different levels, as well as the percentage composition of normal lumbar fluid as determined by paper electrophoresis.

Although inorganic ions and some small molecules are believed to be actively secreted into the CSF, the plasma proteins enter the CSF partly by a process of filtration through the membranes lining the CSF space and partly by local synthesis in the choroid plexus and perhaps elsewhere in the brain. Pre-albumin is an example of a protein which comprises a negligible percentage of the total plasma proteins, but makes up 6% of the total protein of ventricular fluid. It may be concentrated from the plasma or synthesized locally by membranes in the ventricles, or both processes may occur.

It was not until 1961 that Clausen, and MacPherson and Cosgrove, showed that there were two globulins in human cerebrospinal fluid that were not demonstrable in normal human serum by ordinary techniques. One of these proteins was a gamma globulin, named the γ_c-globulin by MacPherson, and was found to compose about 5% of the proteins of normal CSF. It is present in human brain and is secreted continuously into the CSF. The nature and location of the neural cells which form it or in which it is stored are not known. Its function is not known at present. The concentration of protein remains at a relatively constant level, except in demyelinating diseases of the central nervous system, when the concentration falls, sometimes to very low levels. Recently, MacPherson and Saffran showed that bovine CSF also contains a γ_c-globulin and that this globulin is also present in bovine brain.

The other proteins which are components of the normal CSF are plasma proteins and these are found (in lumbar fluid) in concentrations which range from 1/250–1/300 of their concentrations in the plasma. The majority of the proteins in the CSF appear to be transposed unaltered from the serum. Exceptions are the iron-binding protein transferrin, which is converted into two active but electrophoretically different forms, and the copper containing oxidase, α_2-ceruloplasmin, which may lose its activity, although no other physicochemical changes are apparent. When one of the proteins of the plasma is found to be elevated, by disease or other means, a comparable increase in the concentration of the same protein will be reflected in the CSF. Only the plasma proteins with molecular weights lower than 200,000 usually pass the normal membranes.

TABLE 2. COMPARISON OF NORMAL CEREBROSPINAL FLUID AND SERUM (VALUES OF MAIN PROTEIN FRACTIONS DETERMINED BY PAPER ELECTROPHORESIS)

Fluid	Level	Total Protein (mg per 100 ml)	Percentage of Total Protein					
			Pre-albumin	Albumin	Globulins			
					α_1	α_2	β	γ
CSF	Ventricular	10						
	Cisternal	16						
	Lumbar	25						
			4.6	58	5.1	6.0	20	10
Serum		7250	—	56	7.0	9.0	13	15

When, however, the meninges or membranes covering the brain become infected, the protein concentration of the CSF may increase enormously and the largest protein molecules of the plasma (fibrinogen, γ_1-macroglobulin, α_2-lipoprotein) are then found in the CSF. These disappear and the protein concentration returns to normal when the patient recovers.

The maintenance of such a large differential in protein concentration between the plasma and CSF indicates that a highly selective membrane exists between the blood and the CSF. Indeed, the equilibration of substance between the blood and CSF does not occur with the same rapidity that it does between the blood and extracellular spaces of other body tissues. This impediment to rapid equilibration of substances between the blood and CSF has been called the blood-CSF barrier, although no anatomical structure can be discerned on which to explain its existence. It appears to reside in the nature of the cells of the choroid plexuses and the cells lining the surface of the brain. (See also BLOOD-BRAIN BARRIER.)

References

1. DAVSON, H., "Physiology of the Ocular and Cerebrospinal Fluids," London, J. & A. Churchill Ltd., 1956.
2. MILLEN, J. W., AND WOOLLAM, D. H. M., "The Anatomy of the Cerebrospinal Fluid," London, Oxford University Press, 1962.
3. "The Cerebrospinal Fluid," *Ciba Found. Symp.* (WOLSTENHOLME, G. E. W., AND O'CONNOR, C. M., EDITORS), (1958).
4. BOOIJ, J., "Pre-albumin in the Cerebrospinal Fluid," *Progr. Neurobiol*," ARIENS KAPPERS, J. EDITOR, 164 (1956).
5. HILL, N. C., GOLDSTEIN, N. P., MACKENZIE, B. F., MCGUCKIN, W. F., AND SVIEN, H. J., "Cerebrospinal Fluid Proteins, Glycoproteins and Lipoproteins in Obstructive Lesions of the Central Nervous System," *Brain*, **82,** 581 (1958).
6. BAUER, H., AND HEITMAN, R., "Chemische und Serologische Untersuchungen bei der multiplen Sklerose," *Deut. Z. Nervenheik.*, **178,** 47 (1958).
7. VAN SANDE, M., KARCHER, D., AND LOWENTHAL, A., "Acquisitions nouvelles par l'etude électrophorétique des proteins du liquid cephalorachidien dans la sclerose en plaques," *Acta Neurol. Belge.*, **59,** 572 (1959).
8. MACPHERSON, C. F. C., AND COSGROVE, J. B. R., "Immunochemical Evidence for a Protein Peculiar to Cerebrospinal Fluid," *Can. J. Biochem. Physiol.*, **39,** 1567 (1961).
9. CLAUSEN, J., "Proteins in Normal Cerebrospinal Fluid not Found in Serum," *Proc. Soc. Exptl. Biol. Med.*, **107,** 170 (1961).
10. MACPHERSON, C. F. C., "Quantitative Estimation of the γ_c Globulin in Normal and Pathological Cerebrospinal Fluids by an Immunochemical Method," *Clin. Chim. Acta*, **11,** 298 (1965).

CATHERINE F. C. MACPHERSON

CHELATION AND METAL CHELATE COMPOUNDS (IN BIOLOGICAL SYSTEMS)

Chemical reactions in biological systems are usually mediated by selective catalysts called *enzymes*. The high efficiencies and stereospecificities achieved require that enzymes have definite and characteristic geometries, whereby specific functional groups coordinated to the metal ion are held in definite spatial positions relative to each other and relative to the substances on which they exert their catalytic effects. The incorporation of metal ions into enzyme structures can assist in the maintaining of a definite geometrical relationship between ionic and polar groups, through the geometric requirements of the coordinate bonds of the metal ion. Certain metal ions may also participate in the catalytic properties of enzymes through ionic and coordinate bonding between the metal ion and electron donating groups of the enzyme and substrate, and through the ability of the metal ion to initiate oxidation-reduction reactions. Because of these chemical and steric effects, nature frequently employs coordinated metal ions in the complex compounds that catalyze biological reactions (see also METALLOPROTEINS).

Definition of Terms. Before proceeding to a discussion of biological catalysis by metals, a brief description of the types of compounds formed by metal ions is in order. The characteristic *coordination number* of a metal ion is the number of atoms that can combine with it by donation of an electron pair from a solvent molecule or by another molecule or anion having unshared pairs of electrons, as indicated by the following reactions for metal ions of coordination number four and six:

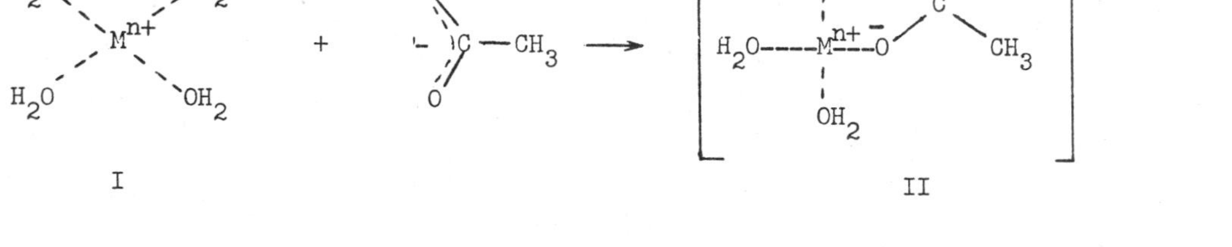

```
H2O     OH2                                      [      OH2        O       ] (n - 1)+
   \   /              O                          [       |         ||      ]
    Mn+        +    - C — CH3     ——>            [ H2O---Mn+---O — C       ]
   /   \              O                          [       |            \    ]
H2O     OH2                                      [      OH2           CH3  ]

     I                                                        II
```

Metal ion + ligand ⟶ metal complex
(in aqueous solution)

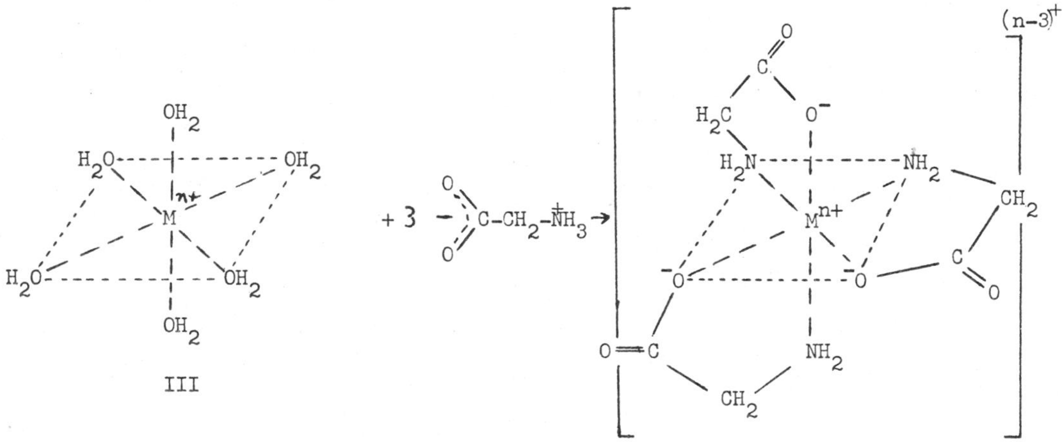

III

IV

When two or more donor groups are attached to the same molecule or anion, the complex compound formed is said to be a *chelate compound.* The chelate of glycine, formula IV, contains three metal chelate rings each coordinated through amino and carboxylate groups. Metal chelates have high solution stability, and the metal ions found in nature as part of a catalytic enzyme system are usually bound to donor groups of the enzyme through metal chelate rings.

Most of the metal ions that have biological functions have a coordination number of six, with the donor groups arranged in an octahedral fashion, as indicated by III and IV. There are a few metals, such as Mg^{2+} and Zn^{2+}, that frequently coordinate only four donor groups tetrahedrally, and Cu^{2+}, which has four coordinations directed to the corners of a square plane with the metal ion at the center of the plane.

Metal Ions and Complexes as Catalysts in Aqueous Systems. Many simple acid-base reactions are catalyzed by both metal ions and hydrogen ions. By virtue of its small size, the electronic interaction of the hydrogen ion with a substrate is much greater than that of a metal ion. The latter, on the other hand, has properties not possessed by hydrogen ions, which are useful in catalysis: the ability to coordinate a large number of electron donor groups simultaneously, the specific geometric orientation of the coordinate bonds of certain metal ions, and the ability of metal ions to undergo oxidation-reduction reactions. Many of these reactions are models of the more complex catalytic effects that occur in biological systems. Since these reactions of simple coordination compounds aid in the understanding of biological reactions, a few of the more common examples are listed in Table 1.

The function of the metal ions in the reactions summarized in Table 1 is to attract electrons from the substrate. When this effect takes the form of simple polarization of the functional groups of the substrate, charge variations and electron shifts in these groups facilitate the chemical reactions listed under solvolysis and acid catalysis. When the metal ion removes completely one or more electrons from the substrate, the first step in an oxidation reaction occurs. This type of catalysis can be accomplished only by metals capable of existing in more than one valence state.

Metal-Catalyzed Biological Reactions. The following is a description of various types of metal catalysis that have been observed in biological systems. In all cases, the metal is coordinated to, or "carried" by, the organic functional groups of the *enzyme.* In some cases, the metal ion is labile and dissociates readily from the enzyme. The binding energy of the metal ion in such systems may vary from weak to moderately strong, depending on the functional groups involved. In other cases, the metal ion is firmly bound in a specific molecular grouping, called a *prosthetic group.* Removal of the metal from enzymes of this type usually occurs in such a manner that it is still combined with the organic part of the prosthetic group.

A typical reaction scheme for a metal-catalyzed biological reaction is the following.

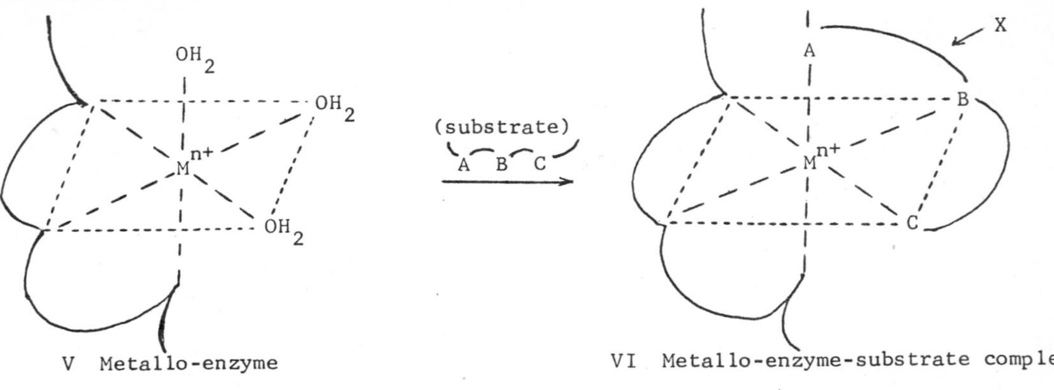

V Metallo-enzyme

VI Metallo-enzyme-substrate complex

↓

products ← { reactive intermediates + metallo-enzyme

As indicated above, the metal ion is "carried" by the donor groups of the enzyme in such a manner that one or more coordination positions of the metal are weakly coordinated to the solvent (or perhaps with auxiliary groups of the enzyme) and are available for interaction with the substrate. The electron attraction of the metal ion resulting from its positive charge and tendency to form electron pair bonds with the substrate results in the polarization of the donor groups (A, B and C) so as to make the substrate more susceptible to attack by a reactant X, or to the elimination of an atomic grouping already present in the substrate. As indicated by VI, the metal

TABLE 1. METAL ION AND METAL CHELATE CATALYSIS OF CHEMICAL REACTIONS

(A) Solvolysis and Other Reactions Involving Acid Catalysis by the Metal Ion

Reaction Type	Substrate	Catalyst
Solvolysis	Amino acid esters, peptides, and amides	Cu^{2+}, Co^{2+}, Mn^{2+}
	Phosphate esters	La^{3+}, Cu^{2+}, VO^{2+}
	Fluorophosphates	Cu^{2+}, $UO_2{}^{2+}$, diamine-Cu(II) complexes
	Polyphosphates	Ca^{2+}, Mg^{2+}
	Schiff bases	Cu^{2+}, Ni^{2+}
Transamination	Schiff bases of pyridoxal and α-amino acids	Fe^{3}, Cu^{2+}, Al^{3+}, Zn^{2+}, Ni^{2+}, Co^{2+}
Decarboxylation	α-Keto polycarboxylic acids (*e.g.*, oxalacetic and oxalsuccinic acids)	Cu^{2+}, Zn^{2+}, Ni^{2+}, Co^{2+}, Mn^{2+}, Fe^{2+}
Acylation	Acetylacetone	Co(III), Rh(III) or Cr(III) chelates of acetylacetone

(B) Catalysis of Oxidation Reactions by Electron Exchange with Metal Ions or Metal Complexes

Reaction	Substrate	Metal Ion or Complex
Oxidation by molecular oxygen	Ascorbic acid, catechols, quinoline, salicylic acid	Fe(III), Fe(III)-EDTA, Cu(II), Cu(II)-EDTA, V(IV)
Oxidation by hydrogen peroxide	Phenol, anisole	Fe(II) (Fenton's reagent), Fe(II)-hydroquinone, Fe(II)-EDTA-ascorbic acid
Formation of oxygen	Hydrogen peroxide	Fe^{3+}, Fe(III)-phthalocyanine chelate
Formation of disulfides from mercaptides	Thioglycolic acid	Fe^{3+}, Cu^{2+}

holds the substrate in a conformation which is favorable for the specific reaction catalyzed by the enzyme.

There is a saturation effect in the coordination of a metal ion by donor groups of both the enzyme and the substrate. Therefore one would expect that the interaction of a free metal ion with the substrate would be greater than that of the metalloenzyme (in which the metal is already partially coordinated). If this were true, the metal ion would have a greater catalytic effect than the metalloenzyme. The reverse is always the case; thus far no metal ions, or metal complex enzyme models, have ever been found to approach the catalytic activities of the corresponding enzyme. This high activity of the enzyme is ascribed to the special environment of the substrate around the active site of the enzyme, through which additional binding of the substrate by adjacent organic groups of the enzyme takes place [see also ENZYMES (ACTIVE CENTERS)].

Metalloenzymes. One of the best known proteinases is carboxypeptidase (see PROTEOLYTIC ENZYMES), which catalyzes the solvolysis of the peptide linkage at the carboxyl end of the protein or polypeptide, as indicated by VII:

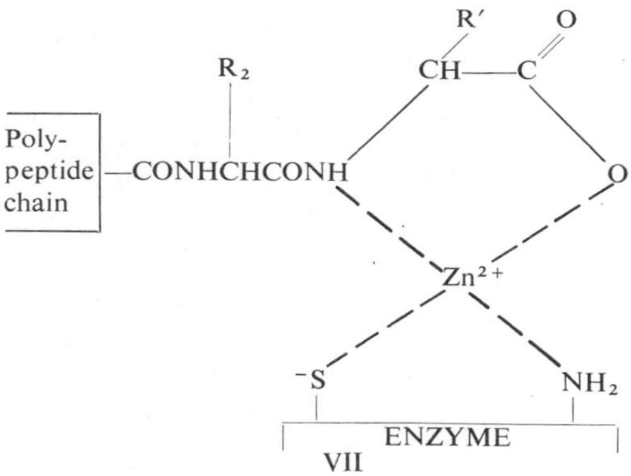

Poly-peptide chain —CONHCHCONH (R_2; R' CH—C(=O)O$^-$) ··· Zn^{2+} ··· $^-$S, NH_2 ENZYME

VII

The binding of the metal ion to this enzyme through amino and mercaptide groups was deduced by comparing the stability of the metal-enzyme complex with a number of simple complex models containing various combinations of possible donor groups.

It has been noted (Table 1) that metal ions catalyze the decarboxylation of oxalacetic acid and similar compounds, as indicated by the following reaction:

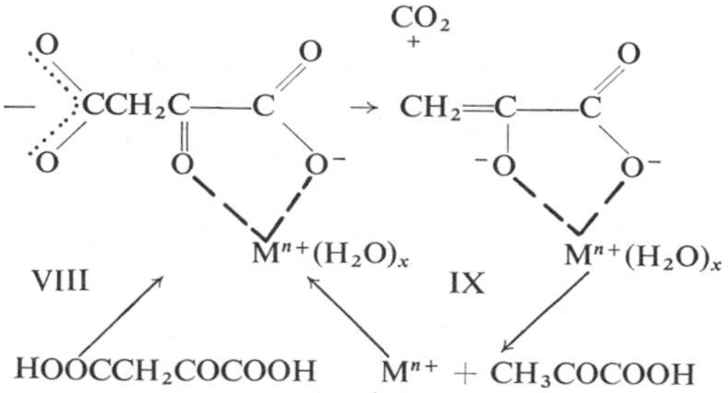

$$\text{—}\ddot{C}CH_2C(=O)\text{—}C(=O)O^- \cdots M^{n+}(H_2O)_x \;(\text{VIII}) \rightarrow CO_2 + CH_2{=}C(O^-)\text{—}C(=O)O^- \cdots M^{n+}(H_2O)_x \;(\text{IX})$$

HOOCCH$_2$COCOOH → VIII; IX → M^{n+} + $CH_3COCOOH$

This reaction is catalyzed by a wide variety of dipositive metal ions in the order of metal complex stability ($Cu^{2+} > Ni^{2+} > Zn^{2+} > Co^{2+} > Mn^{2+} > Mg^{2+} > Ca^{2+}$). The activation of the enzymatic decarboxylation reaction by metal ions, illustrated by Fig. 1,[6,pp.42–45] is much greater than the effect of metal ions alone and does not exhibit the regular behavior based on metal chelate stability shown in the non-enzymatic reactions. This difference is understandable on the basis of the fact that steric effects and special interactions not possible for simple metal ions probably occur at, and adjacent to, the reactive coordinating center of the enzyme.

The enzyme aconitase, which contains the Fe^{2+} ion at the reactive center, catalyzes the interconversion of citric, isocitric and aconitic acids (see CITRIC ACID CYCLE). The reaction has been shown to occur through the formation of a single intermediate carbonium ion structure, in which the Fe^{2+} ion is always bound to the same donor atoms, while the interconversion of the substrate occurs through the migration of only protons and electrons, as illustrated by the following scheme:

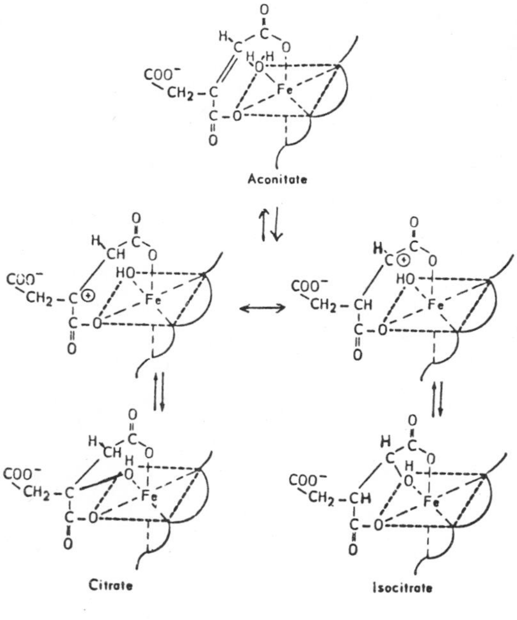

Aconitate

Citrate

Isocitrate

This is an example of the importance of stereochemistry and charge of the metal ion, as well as of chelate ring formation, in catalyzing chemical reactions by enzymes. The enzymatic cleavage of adenosine triphosphate (ATP) to adenosine diphosphate (ADP) is perhaps one of the most important reactions in biochemistry, since it provides the energy for a large number of processes, such as muscular contraction, and many chemical reactions. These reactions are always

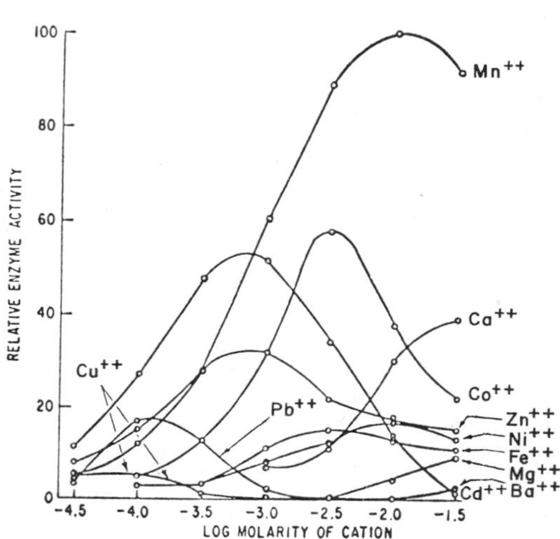

FIG. 1. Effect of metal ions on enzymatic decarboxylation of oxalacetic acid.

metal ion catalyzed, usually by Mg^{2+} or Mn^{2+}. The cleavage of ATP in biological systems can be understood on the basis of some "model" reactions carried out *in vitro*, as indicated by the following:

the coordinate bonds of the metal also attached to donor groups on the enzyme.

Metal Chelates Containing Prosthetic Groups. The structures of these stable metal chelate species, that catalyze essential biological reactions

XII Structure of coenzyme B_{12}

METAL CATALYZED REACTIONS OF ATP

X ATP

$\xrightarrow{Mg^{2+} \text{ or } Mn^{2+}}$

XI ADP

$$ATP + R{-}COOH \xrightarrow{M^{2+}} ADP + R{-}C(=O)OPO_3H$$

The cleavage is considered to take place after coordination with the metal ion, which brings another coordinated group (such as a water molecule, or a phosphate or carboxylate group shown above) in close proximity to the phosphate to facilitate the cleavage. The electrostatic attraction of the metal ion for the phosphate oxygens is also considered to favor the cleavage of the triphosphate to the diphosphate. The enzymic reactions are considered to involve similar metal ion-coordinated intermediates, with some of

as parts of larger (enzyme) systems, provide interesting examples of how the stereochemistry of the metal ion and of the coordinated donor groups of the ligand are well adapted to the specific catalytic function involved.

One such chelate compound for which the structure has been completely elucidated is the coenzyme vitamin B_{12} (see also COBALAMINS AND COBAMIDE COENZYMES), illustrated by XII. This chelate compound, with its metal-carbon bond and partially saturated pyrrole rings, is a unique structure without parallel in coordination catalysis. Its biological functions are believed to include the rearrangement of hydrogens and alkyl groups in various substrates. Known reactions are the isomerization of glutamic acid and β-methylaspartic acid, and the conversion of methylmalonic acid to succinic acid. The specific site of these catalytic effects is not known, but it may conceivably involve the unique metal-carbon bond, or positions 8 and 10 of the conjugated ring systems, which are known to undergo hydrogen atom or proton transfer.

The heme proteins provide an interesting example of variation of catalytic function with structure of the coordination sphere. All of these substances contain the Fe^{2+} ion strongly coordinated by the planar porphine ring, which is, in turn, bound to protein through substituents attached to the outer part of the ring. The most common of these structures is the Fe(II)-protoporphyrin complex, illustrated by XIII. This complex (protoheme) is the compound isolated by chemical degradation of hemoglobin, myoglobin, peroxidase, catalase and cytochrome c.

XIII Fe(II)-protoporphyrin complex.

The partial structure of cytochrome c, indicated by XIV, shows that the Fe(II) ion is completely and strongly coordinated by the PORPHYRIN ring, and by two imidazole groups derived from histidine units of the polypeptide chain of the protein. The porphyrin ring is firmly bound to the protein through sulfur bridges derived from

XIV Cytochrome c.

cysteine residues. The stereospecific structure of the protein holds the coordinating groups firmly in the correct position for strong coordination by all six donor groups. This stable arrangement is in accord with the fact that cytochrome c does not react with other molecules, but serves only as an electron transfer agent.

The donor groups in *catalase* and *peroxidase* are the same as for cytochrome c, but only one imidazole group is tightly coordinated to the metal, leaving the other position relatively labile and thus available for coordination of the peroxide substrate, as indicated by XV (see also CYTOCHROMES; PEROXIDASES).

XV Ligand attachment in peroxidase.

The biological functions of HEMOGLOBIN and MYOGLOBIN are to store and transport molecular oxygen by coordination of the oxygen molecule to the iron atom of the heme complex. Thus these compounds resemble catalase and peroxidase in that the iron atom is free to coordinate with an additional ligand, in this case molecular oxygen, as indicated by XVI.

Other Functions of Metal Chelates in Biological Systems. *DNA*. DNA is a good example of a biological catalyst that can be stabilized or

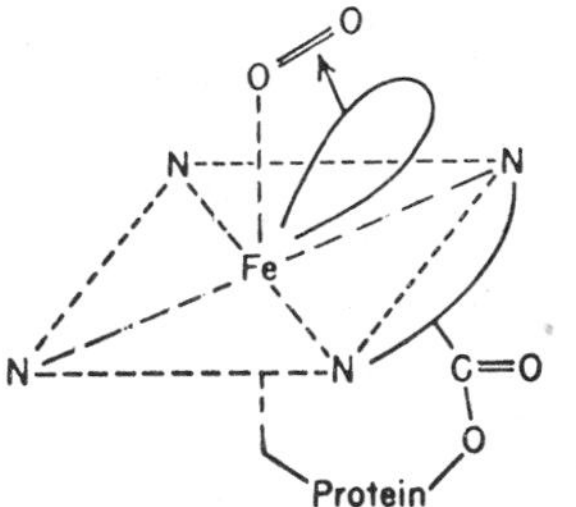

XVI Ligand attachment in hemoglobin.

destabilized by coordination with metal ions. The alkaline earth ions and most transition metal ions stabilize the DNA structure by coordinating the negative phosphate groups on the surface of DNA, since the mutual repulsions of these negative groups tend to disrupt the molecule. The stabilization effect is evidenced by an increase of the "melting" temperature of DNA in the presence of these metal ions. The Cu(II) ion, on the other hand, decreases the "melting" temperature and prevents reformation of the helical

TABLE 2. BIOLOGICALLY ACTIVE METAL CHELATES

Metal	Metalloenzyme	Other Biological Functions
Mg	Polynucleotide phosphorylase, ATPase, choline acylase, deoxyribonuclease, acetate kinase, adenosine phosphokinase, fructokinase, glyceric kinase, hexokinase	CHLOROPHYLL
Ca	α-Amylase, aldehyde dehydrogenase, lipase	
V		Green algae, blood of marine worm (ascidian)
Cr		Glucose tolerance factor
Mn	Arginase, carnosinase, prolinase, enolase, isocitricdehydrogenase, 3-phosphoglycerate kinase, glucose-1-P kinase	
Fe	Aconitase, formic hydrogenylase, phenylalanine hydroxylase, peroxidase, catalase, cytochromes	Hemoglobin, ferritin, hemosiderin, siderophilin
Co	Aspartase, acetylornithinase	Vitamin B_{12}
Cu	Laccase, phenolase, tyrosinase, uricase	Ceruloplasmin, cytochrome
Zn	Carbonic anhydrase, carboxypeptidase, alcohol dehydrogenase, glutamic dehydrogenase, acylase	
Mo	Nitrate reductase, xanthine oxidase	

structure on cooling. These effects are believed to be due to the coordination of Cu(II) to the nitrogen bases (adenine, guanine, cytosine and uracil), thus disrupting the hydrogen bonds that normally hold the double helix together [see also DEOXYRIBONUCLEIC ACIDS (STRUCTURE)].

Metal Ion Transport. The ferric ion is stored in liver, spleen and bone marrow as FERRITIN, containing approximately 20% iron, and as *hemosiderin*, which contains up to 55% iron by weight. In both substances, the iron is present as ferric hydroxide micelles, rather than as individually coordinated iron atoms. Transport of the iron in the plasma is accomplished by a truly specific protein chelate, called *siderophilin* or *transferrin*, a β-globulin that contains up to two iron atoms per molecule (see also IRON IN BIOLOGICAL SYSTEMS).

The nature of the transport of other metal ions is not well understood, but clearly takes place through formation of metal chelates or complexes.

Summary of Biological Reactions Requiring Metal Ions. Table 2 contains a list of the more important biological reactions that are catalyzed by metal ions.

References

1. "Biological Aspects of Metal Binding," *Federation Proc.*, **20,** No. 3, Part II (Supplement No. 10) (1961).
2. DWYER, F. P., AND MELLOR, D. P. (EDITORS), "Chelating Agents and Metal Chelates," New York, Academic Press, 1964.
3. BAILAR, J. C., JR., AND BUSCH, D. H. (EDITORS), "The Chemistry of the Coordination Compounds," Ch. 21, New York, Reinhold Publishing Corp., 1956.
4. MARTELL, A. E., AND CALVIN, M., "Chemistry of the Metal Chelate Compounds," Englewood Cliffs, N.J., Prentice-Hall Inc., 1952.
5. SEVEN, M. F. (EDITOR), "Metal Binding in Medicine," Philadelphia, Pa., Lippincott, 1960.
6. BUSCH, D. H. (EDITOR), "Reactions of Coordinated Ligands," *Advan. Chem. Ser.*, No. 37. (1963).
7. CHABERK, S., JR., AND MARTELL, A. E., "Organic Sequestering Agents," New York, John Wiley & Sons, 1959.

M. CALVIN AND A. E. MARTELL

CHEMOTAXONOMY

The application of chemical data to systematics represents a powerful new approach available to all biologists in their attempts to answer questions about relationships among organisms and in devising natural systems for the classification of organisms.[1] The chemical evidence may be equally well represented by macromolecular compounds such as DNA-RNA and proteins, or by smaller molecules such as the natural products of plants, *e.g.*, alkaloids, terpenoids and flavonoids. These latter classes of compounds have no clearly defined metabolic roles and are, therefore, often described as secondary compounds to distinguish them from the more vital substances as, for example, the protein amino acids.

The generally accepted theory for information transfer or "gene action" (see GENES) directly relates specific features of the chemical structures of secondary compounds, such as functional groups and number and size of rings, to segments of DNA (genes) *via* enzyme-controlled biosynthetic pathways. Since DNA is unique for each individual, we can infer that some enzymes and hence some secondary compounds may be unique for each organism. Unfortunately, for taxonomic purposes, the ability of two different organisms to synthesize the same secondary compounds can be interpreted in at least two different ways: it may reflect common ancestry or, alternatively, it may be a case of convergent evolution. Still, the careful application of chemical information to

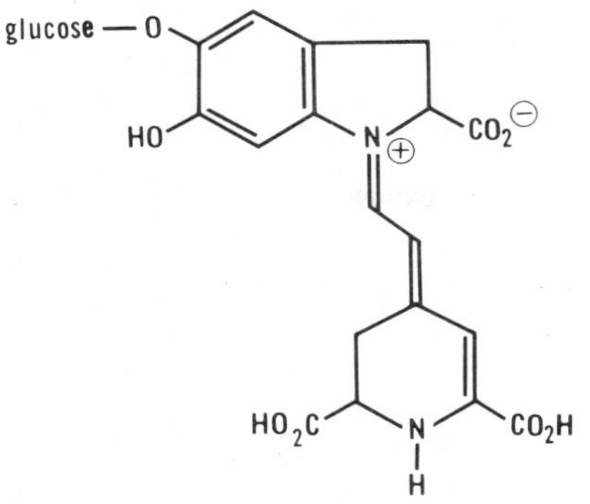

Betanin

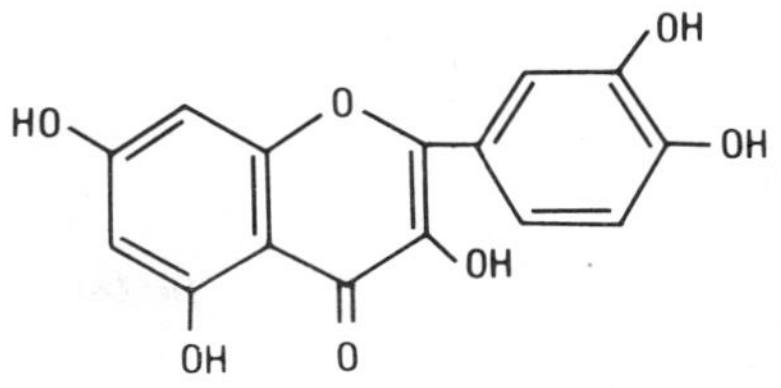

Quercetin

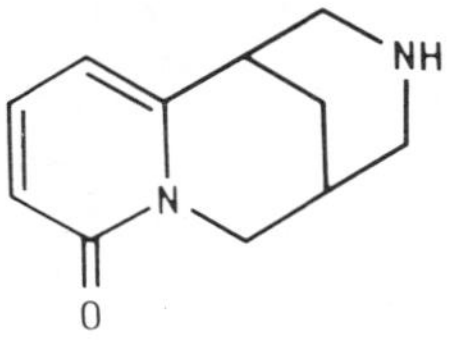

Cytisine

systematic problems can provide a level of insight into phylogeny not possible with morphological and cytological evidence alone. A few specific instances where such distributional chemical data have proved especially valuable will be mentioned.

Plant Pigments. *The Betacyanins-Betaxanthins and the Anthocyanins.* The betacyanins and betaxanthins are red-violet and yellow pigments, respectively, found, so far as is known, only in ten families of flowering plants belonging to the Order Centrospermae.[2] These families contain many well-known pigmented plants including the red beet, the *Bougainvillea*, the common four-o'clocks, *Portulaca* and the cacti. Typical members of the betacyanin and betaxanthin classes of pigments include, respectively, the red beet pigment, betanin, and the yellow cactus pigment, indicaxanthin (see structures). The more widely distributed anthocyanin pigments, which account for most of the red and blue pigments found in other families of flowering plants (*e.g.*, cyanin, the rose pigment) do not occur in any of the ten Centrospermae families. The positioning of the family Cactaceae in the Centrospermae by a few taxonomists purely on morphological grounds was strengthened by the presence of the betacyanin-betaxanthin pigments in cacti. Several workers have suggested, on the basis of the chemical as well as morphological evidence, that the Order Centrospermae might best be recognized as containing the ten betacyanin families. The totally different structures of the two classes of pigments, the betacyanins-betaxanthins and the anthocyanins, which indicate different biosynthetic pathways (hence different genetic material), their mutual exclusion, and the limited distribution of the betacyanins and betaxanthins suggest that the "betacyanin families" constitute an evolutionary line separate from those which gave rise to the other flowering plants (see also FLAVONOIDS).

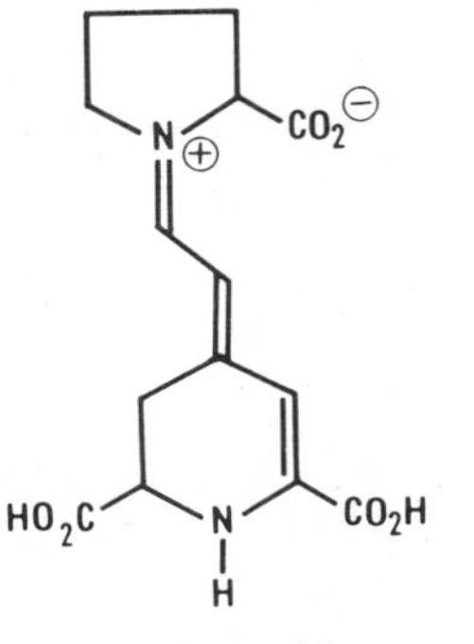

Indicaxanthin

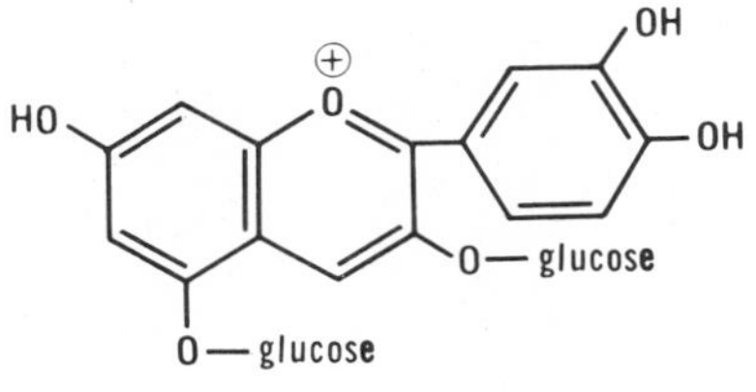

Cyanin

Secondary Compounds in Baptisia. The analysis by Alston, Turner, and Mabry[1a] of patterns of FLAVONOIDS (a typical one is quercetin), as disclosed by two-dimensional paper chromatography, has been successfully used for the recognition of the seventeen species of the genus *Baptisia* (family Leguminosae). Furthermore, the flavonoid patterns provided parental identification in the analysis and validation of hybrids among the species of *Baptisia* since the hybrid plants, in general, displayed the combined flavonoid chemistry of the parents. The availability of organic analytical and structural analysis tools, such as gas CHROMATOGRAPHY, NUCLEAR MAGNETIC RESONANCE (NMR) and mass spectrometry, has made possible the relatively rapid total structure analysis of hundreds of secondary compounds. Therefore, the early emphasis on paper chromatographic pattern analysis of plants has now been replaced, in many instances, by the total chemical identification of the substances detected by paper and gas chromatography.

For example, Cranmer and Mabry[3] recently identified and assessed the systematic value of most, if not all, of the lupine ALKALOIDS (*e.g.*, cytisine) in the genus *Baptisia*.

Macromolecular Chemistry and Systematics. The comparative chemistry of DNA-RNA (see DEOXYRIBONUCLEIC ACIDS and RIBONUCLEIC ACIDS) and PROTEINS is only now being applied to taxonomic problems, but they can be expected to increase in their application and importance as advances in macromolecular chemistry develop. Perhaps the most striking single case of the macromolecular approach to resolving phylogenetic problems for the higher categories, is in the results described by Hoyer, McCarthy and Bolton.[4] They demonstrated that the per cent of DNA similarities between organisms such as man, monkey, mouse and fish can be determined. In general, their procedures rests on the assumptions that double-stranded DNA can be separated into single strands and that these strands will form double strands again only with DNA of similar composition (*i.e.*, containing many similar genes at equivalent positions). Double-stranded, high molecular weight DNA, extracted from animal or plant tissue, is separated into single strands which are subsequently trapped in agar. The ability of sheared (or fragmented) DNA from the same organisms to form a duplex with the agar-embedded DNA is compared with the ability of another organism's sheared DNA to combine with the agar-embedded DNA. This gives directly a relative measure of the genetic similarities of the two organisms.

Much progress has been made in the analysis of primary structure of proteins, and a few proteins have been studied intensively on a comparative basis. Perhaps more comparative data exist for HEMOGLOBIN chains than for any other protein, and many abnormal human hemoglobins are now known to occur, each differing from normal hemoglobin in a single amino acid ascribable to a single nucleotide difference in the DNA governing the synthesis of the individual chain. Ingram[5] has developed an evolutionary scheme to account for the four types of chains typical of human hemoglobin. Other extremely promising comparative studies have been conducted on the primary structures of cytochrome c (see CYTOCHROMES). It is probable that the importance of protein structure in providing evolutionary insights is far greater than many biologists realize at this time.

In summary, the emerging discipline of chemotaxonomy reflects the efforts of chemists and biologists to bring all available evidence to bear on phylogenetic problems.

References

1. General References:
   (a) ALSTON, R. E., MABRY, T. J., AND TURNER, B. L., *Science*, **142**, 545 (1963).
   (b) ALSTON, R. E., AND TURNER, B. L., "Biochemical Systematics," Englewood Cliffs, N.J., Prentice-Hall, 1963.
   (c) SWAIN, T. (EDITOR), "Chemical Plant Taxonomy," New York, Academic Press, 1963.
   (d) LEONE, C. A. (EDITOR), "Taxonomic Biochemistry and Serology," New York, The Ronald Press Co., 1964.
2. MABRY, T. J., "The Betacyanins and Betaxanthins," in "Comparative Phytochemistry" (SWAIN, T., EDITOR), London, Academic Press, 1966.
3. CRANMER, M., "Biochemical and Biosystematic Investigations of the Lupine Alkaloids of the Legume Genus *Baptisia*," Doctoral Dissertation, The University of Texas, Austin, 1965.
4. HOYER, B. H., MCCARTHY, B. J., AND BOLTON, E. T., *Science*, **144**, 959 (1964).
5. INGRAM, V. M., "The Hemoglobins in Genetics and Evolution," New York, Columbia University Press, 1963.

TOM J. MABRY, R. E. ALSTON AND B. L. TURNER

CHEMOTHERAPY. See ANTIBIOTICS; CANCER CHEMOTHERAPY; SULFONAMIDES.

CHITIN

Chitin is a high molecular weight polymer composed of N-acetylglucosamine residues joined together by β-glycosidic linkages between carbon atoms 1 and 4 (Fig. 1). Modern chemical terminology would call it a polymer of 2-acetamido-2-deoxy-α-D-glucopyranose. The molecular chains are long and unbranched. Freed from the protein with which it is normally associated, the chitin chains are parallel to one another and adjacent chains run in opposite directions—an arrangement that maximizes the number of interchain hydrogen bonds. In many respects chitin is similar to CELLULOSE [see also CARBOHYDRATES (CLASSIFICATION)] from which it differs by the presence of an acetylamine group on the second carbon atom.

Chitin is perhaps best known as a characteristic chemical component in the skeletons of arthropods. It is also found in setae, jaws, and gut lining of annelids, in the radula and dorsal shield of mollusks, in the perisarc of medusoid coelenterates, in the stalk wall of bryozoans, in the egg shells of nematodes and acanthocephalans, and less certainly in a few other animal groups. It is also the common constituent of the walls of most but not all groups of fungi.

As usually prepared for chemical study, chitin is a colorless solid which is chemically stable and is insoluble and unaffected by most chemical reagents. Solutions in concentrated mineral acids quickly show degradation to short chain lengths, but some authors report that undegraded solutions can be made with lithium salts which are capable of strong hydration. Hot concentrated alkalies remove half or all of the acetyl side groups to give a product called chitosan which can be made to give a color reaction useful in identification of chitin in natural objects.

Birefringence and X-ray diffraction data show that chitin has crystalline areas, but there is not good agreement as to the details or dimensions of the unit cell of the crystal lattice. While chitin from diverse sources yields similar constituents on hydrolysis, X-ray diffraction studies reveal several crystalographic forms. According to Rudall, the commonest type is that found in arthropod skeletons, etc.; it has been termed α-chitin.

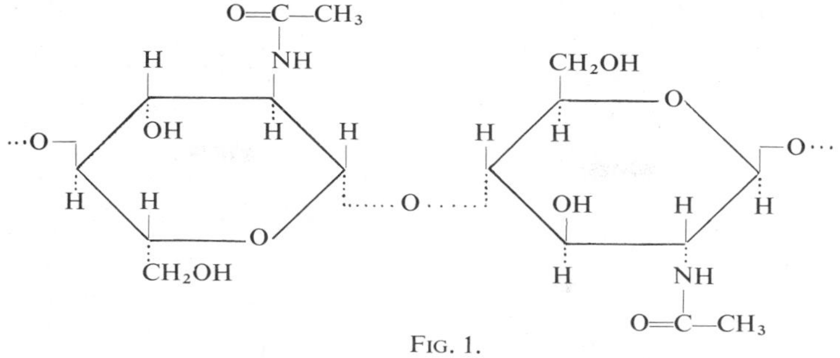

FIG. 1.

Another type with water incorporated in the lattice is β-chitin found in the "pen" of squid. Another type is found in the perisarc of coelenterates. What type occurs in fungi remains to be determined.

Chitin is not found in nature in a pure condition. Rudell reports that β-chitin is associated with COLLAGEN whereas α-chitin is not. At least in the arthropod skeleton, α-chitin is always associated with proteins called arthropodins. On the average, chitin accounts for only $\frac{1}{3}$–$\frac{1}{4}$ the dry weight of non-calcified arthropod cuticles.

It is now generally agreed that chitin is found in nature only linked to protein. The naturally occurring compound, then, is a glycoprotein from which the protein moiety is readily removed to yield what is called chitin. Purification is usually accomplished by prolonged heating in KOH or NaOH solutions followed by dilute acid and a strong oxidizing agent. Such treatment yields an ash-free product that was long considered to be pure chitin, but more recent analyses show all samples to contain some bound histidine and aspartic acid. Treatment of arthropod cuticles with solutions of "Versene" or lithium thiocyanate, however, results in the extraction of several chitin-containing glycoproteins. Analyses of these fractions indicate that chitin is bound to arthropodin by several kinds of bonds, including some covalent bonds; details of the bonding are not yet known but presumably will include some that are through histidyl and aspartyl residues. If, as some suspect, there is a small percentage of non-acetylated residues in the chains, these could also be involved in cross-bonding. Purified chitin can be made to react with pure arthropodins, peptides, and amino acids (especially tyrosine), but it is not known whether these *in vitro* reactions are the same as the bondings in nature.

Other recent evidence has been interpreted as indicating that in the cuticle of arthropods the chitin and protein chains form an interpenetrating lattice with the protein chains at a right angle to the chitin chains. Both of these sets of chains are parallel to the surface of the cuticle. It is well known that arthropod cuticles become sclerotized in certain areas, *i.e.*, the originally soft cuticle becomes first elastic, then hard, and usually more or less dark. Some chemical information on the process of sclerotization has been obtained but we do not yet know whether or not linkages to chitin chains are involved (see reference 1).

The metabolic source of chitin may be glycogen since there are numerous reports of glycogen decrease concurrent with chitin synthesis. However, no proof exists. Furthermore, no detailed knowledge is available on the metabolic steps involved in chitin synthesis. Since the naturally occurring substance is a glycoprotein, one would expect that monomers are added *in situ* to a mixed lattice.

The decomposition of chitin in nature is accomplished by chitinases. The main source of these enzymes is various soil bacteria, but chitinase activity has also been recorded for some fungi and even eelworms, earthworms and soil amoebas. The digestive juice of snails is a well-known source, but it is uncertain whether the enzyme is produced by the snail itself or by associated bacteria. Chitinase is also present in the molting fluid of insects, but whereas the entire cuticle disintegrates in nature, only the soft cuticle is dispersed during molting.

References

1. HACKMAN, R. H., "Chemistry of the Insect Cuticle," in "The Physiology of Insects," (ROCKSTEIN, EDITOR), Vol. 3, New York, Academic Press, 1965.
2. RICHARDS, A. G., "The Integument of Arthropods," Minneapolis, The University Press, 1951.
3. RUDALL, K. M., "The Distribution of Collagen and Chitin," *Symp. Soc. Exptl. Biol.*, **9**, 49–71 (1955).
4. RUDALL, K. M., "The Chitin/Protein Complexes of Insect Cuticles," in "Advances in Insect Physiology," (TREHERNE, BEAMENT, AND WIGGLESWORTH, EDITORS), Vol. 1, New York, Academic Press, 1963.

A. GLENN RICHARDS

CHLORIDE (IN METABOLISM). See ELECTROLYTE AND WATER REGULATION; MINERAL REQUIREMENTS; RENAL TUBULAR FUNCTIONS. The secretion of hydrochloric acid is discussed in the article GASTRIC JUICE.

CHLOROPHYLLS

Chlorophylls are a group of closely related green pigments occurring in leaves, bacteria and organisms capable of PHOTOSYNTHESIS. The major chlorophylls in land plants are designated *a* and *b*. Chlorophyll *c* occurs in certain marine organisms. Because of the overwhelming percentage of the total photosynthesis which is performed by marine organisms, it is possible that chlorophyll *c* is equivalent in importance to chlorophyll *b*. Chlorophyll *a* is several times as abundant as chlorophyll *b*. Bacteriochlorophyll contains two more hydrogens than the plant chlorophylls and has the vinyl group altered to an acetyl.

```
              CH3—C—C—CH=CH2
                  ‖  ‖
  CH3   CH———C  C———CH
   |    ‖     \ /    ‖      R
   |    ‖      N     ‖      |
  HC————C  +  -|-  + C——————CH
   |     \ N——Mg——N /       |
   |     //    |    \\      |
  HC————C      |     C——————CH
   |    |      N     |      |
  CH2   |     / \    |     C2H5
   |    C===C    C===CH
  CH2       |    |
   |        |    |
  C=O  HC   C====C—CH3
   |    | \  /
   O    |  C
   |    |  ‖
  C20H39|  O
        |
      COOCH3                   I. R=CH3
                              II. R=CHO
```

Formula I represents one of the canonical forms for chlorophyll *a*, II the similar form for chlorophyll *b*. These structures have been established by a long series of degradation studies mainly by R. Willstätter, Hans Fischer and their collaborators, and by synthetic studies in the laboratories of Fischer. The laboratories of Woodward and of Strell announced the complete synthesis of chlorophyll *a* nearly simultaneously.

The biological significance of the chlorophylls stems from their role in photosynthesis, the process by which plants fix the sun's energy in the form of organic matter. This process corresponds to the reversal of the combustion of hydrogen. The oxygen liberated is set free in the air. Under special conditions, some organisms are also capable of liberating the hydrogen, but usually this is used for chemical reductions in the plant. Atmospheric carbon dioxide is fixed enzymatically and is thus used as the source of the carbon in the synthetic process but is not reduced directly. The path of the carbon from carbon dioxide in photosynthesis has been elucidated largely by the studies of Calvin and his collaborators (see CARBON REDUCTION CYCLE). While it is known that most of the energy fixed in photosynthesis is absorbed originally by the chlorophylls, the exact reactions which they undergo to initiate the process of reduction are not yet understood. It is known, however, that the photosynthetic sequence requires a high degree of organization within the plant cells where it occurs and that destruction of the organization of the CHLOROPLASTS by processes like grinding are sufficient to bring photosynthesis to a stop, even when the chlorophyll and the soluble enzymes participating in the process are still presumably intact.

Weak acids remove the magnesium from the chlorophylls giving the pheophytins. Strong acids selectively hydrolyze the phytyl group, yielding the pheophorbides. Hydrolysis of the chlorophylls by the enzyme chlorophyllase in the absence of alcohols also removes the phytyl group, giving chlorophyllides. When the enzymatic process is conducted in the presence of alcohols, chlorophyllide esters, such as methyl or ethyl chlorophyllide, are produced by alcoholysis. The chlorophylls, because of the phytyl group, are microcrystalline waxes. The chlorophyllides, on the other hand, crystallize in visible crystals.

Hot, quick, alkaline saponification of the chlorophylls yields chlorophyllins, magnesium-containing pigments with three carboxylate ions. These result from the removal of the two alcohol groups and the cleavage of the five-membered isocyclic ring. The cleavage occurs readily because this ring contains a keto group β to a carboxylic ester group. Acidification of the product obtained in this manner from chlorophyll *a* removes the magnesium and gives chlorin e_6, one of the most readily obtainable and important degradation products of chlorophyll *a*. The corresponding degradation product of chlorophyll *b* is called rhodin g_7. The numerical subscripts in this and lower ranges refer to the number of oxygen atoms contained in the molecules. In the high ranges, for example pheopurpurin$_{18}$, the subscript refers to the so-called acid number, the percentage strength of hydrochloric acid which will remove two-thirds of the substance from an equal amount of its ethereal solution.

When an ethereal solution of a chlorphyll, or of a derivative containing an intact isocyclic ring, is treated with cold alcoholic KOH, the green color is momentarily discharged to a yellow or brown, depending upon the derivative, and the green color then reappears. This is known as the "phase test" and was discovered by Molisch. When chlorophyll derivatives are exposed to air in the presence of alkali, oxidation accompanies the hydrolysis and ring cleavage and the product is said to be "allomerized." Such a product will no longer give a positive phase test, nor will it crystallize readily. The prevention of allomerization is of great importance in securing high-quality chlorophyll derivatives. The complicated series of reactions taking place during allomerization was elucidated by the work of J. B. Conant and his collaborator. Equivalent oxidation can also be secured with quinone. Stronger oxidizing agents, such as ferricyanide or molybdicyanide, are capable of stripping off the extra hydrogens from the nucleus and converting members of the green chlorophyll series into the red porphyrins. Chromic acid degrades all these substances into maleic imide derivatives, together with other products.

Mild hydrogenation of chlorophyll derivatives saturates the vinyl group. Drastic hydrogenation is capable of discharging the color, with formation of leuco compounds, or even of cleaving the nucleus with the formation of pyrrole derivatives.

It is thus evident that the chlorophylls are sensitive to acid, alkali, reducing agents or oxidizing agents, whether weak or strong. This variety of sensitivities to chemical agents renders them hard to purify without change and makes chemical synthesis in the field difficult. In commercial practice, it complicates the problem of securing uniformity of product and gives rise to the possibility that each batch will contain different substances or different proportions of various products. The sensitivity of the *b* series to chemical reagents is generally greater than that of the *a* series.

Compared to the highly stable phthalocyanines, chlorophyll derivatives are pigments of relatively low stability to light, oxidizing agents and other common pigment destroyers. This property forms one of the serious limitations to their wider commercial exploitation. A moderate increase in stability is conferred by the substitution of copper for magnesium. This also produces a clearer shade that is frequently more desirable. Numerous metals which have square, planar bonds are quite tightly bound by chlorophyll derivatives. This is especially noticeable in the case of copper, which is so tightly held in this configuration that it is deprived of its usual catalytic action. Even these complexes possess relatively low stability to light and oxidizing agents, however. An additional difficulty is introduced by the fact that the destruction of the organic portion of the molecule renders the metal available for new combinations.

Commercially, chlorophyll derivatives are usually assayed spectrophotometrically or colorimetrically. These assays serve as an index of tinctorial power but do not accurately measure the chlorophyll derivatives because of the interferences of other plant pigments and decomposition products. For reliable assays, preliminary fractionations must be performed.

The special techniques most frequently used in the fractionation and purification of chlorophyll derivatives are acid fractionation, column chromatography and paper chromatography (see the separate articles on several types of CHROMATOGRAPHY). The acid fractionation procedure of Willstätter and Mieg relies upon the fact that porphyrins, chlorins and related substances are weak bases which can be extracted from ethereal solutions by varying concentrations of hydrochloric acid. The fractionations obtained are sharper than might be expected because small changes in acid concentration cause changes in the solubility of ether in acid and of aqueous acid in ether, thus producing greater effects than they would in a relatively invariant system. The method of chromatography was invented by Tswett expressly for the separation of chlorophylls and carotenoids but lay dormant in the literature for many years until rediscovered for the fractionation of carotenoids. Paper chromatography is a powerful tool for assay because of the small quantities of material needed for a satisfactory fractionation. Even these small quantities, however, are sufficient for quantitative purposes when eluted and assayed in a spectrophotometer.

Chlorophyll derivatives with the phytyl group intact are oil soluble and form a series of green dyes which have found wide commercial application in the coloring of oils and waxes. The chlorophyll soaps, resulting from combined saponification and cleavage of the isocyclic ring, form valuable "water-soluble" dyes, useful in the coloring of soaps and similar products.

Both the medical and the cosmetic literature are replete with claims of therapeutic or physiological activity of "chlorophyll." The substances utilized in this work range from partially purified chloroplasts to mixtures of materials which have undergone deep-seated chemical alteration. Some of the types of activity claimed can be shown to be due to incidental impurities. The field for investigation of the action of pure chemical individuals produced by the action of various reagents upon chlorophyll or its derivatives is unexplored. It is known, however, that neither chlorophyll nor hemoglobin in the diet is utilized by the body in the formation of the physiologically active pyrrole pigments. These are derived, instead, from such simple building blocks as glycine and acetate ion. Only the iron in dietary blood pigment can be utilized by the body.

The work of Granick has shown that, in the physiological processes of plants, chlorophyll is formed from protoporphyrin, which can be obtained in the laboratory by the removal of iron from hemin (see PORPHYRINS). The pathways to heme and to chlorophyll diverge at protoporphyrin. To form heme, an organism introduces iron into protoporphyrin. To form chlorophyll from protoporphyrin, an oxidation, a reduction, a ring closure and esterifications are performed and the magnesium is introduced. The end product of the enzymatic synthetic chain is presumably protochlorophyll, the magnesium derivative of the porphyrin corresponding in structure to chlorophyll. The addition of the two hydrogens necessary to convert protochlorophyll to chlorophyll is accomplished under the influence of light.

A. H. CORWIN

CHLOROPLASTS

The chloroplast is an organelle present in the cells of all green plants capable of PHOTOSYNTHESIS. It contains from one to several allied green pigments, the CHLOROPHYLLS, which are active in the absorption of radiant energy and its transfer to the chemical energy of the carbohydrates. The chloroplast then is the cellular structure in which occurs the synthesis of sugar, the basic compound from which the world's supply of food, fuel, clothing, and other organic materials is derived (see CARBON REDUCTION CYCLE IN PHOTOSYNTHESIS). Other pigments, the carotenes and xanthophylls, may also occur in chloroplasts. Chloroplasts are disk-shaped; they average about 5 μ in diameter and there are about 50 per cell. STARCH is the first visible product of photosynthesis; in most cases it is elaborated as small grains visible within the chloroplast. In the blue green algae and certain green algae, the pigment seems to be coextensive with the general cytoplasm of the cell. In some lower forms, the pigment is found in elaborate bands or nets usually termed *chromatophores*. In such chromatophores, the starch may accumulate about dense proteinaceous granules called *pyrenoids*, Fig. 1. This starch is known as pyrenoid starch, while the more usually occurring starch grains are called stroma starch. Both types may occur in the same chromatophore. In some red algae, starch grains regularly occur outside of the choroplasts in the general cytoplasm of the cell. In many plants, as the leaves or fruits mature, they change from green to red or brown. This is caused by the displacement of the chlorophyll within chloroplasts by carotenes or xanthophylls (see CAROTENOIDS), and the change is apparently not reversible. Plastids having a preponderance of these pigments are known as chromoplasts. In potatoes or iris roots, food may be stored in colorless plastids known as leucoplasts. In the great majority of plants, the chloroplast appears either as a homogeneous body or as a disk with a colorless background or stroma in which the grana are embedded (Fig. 2).

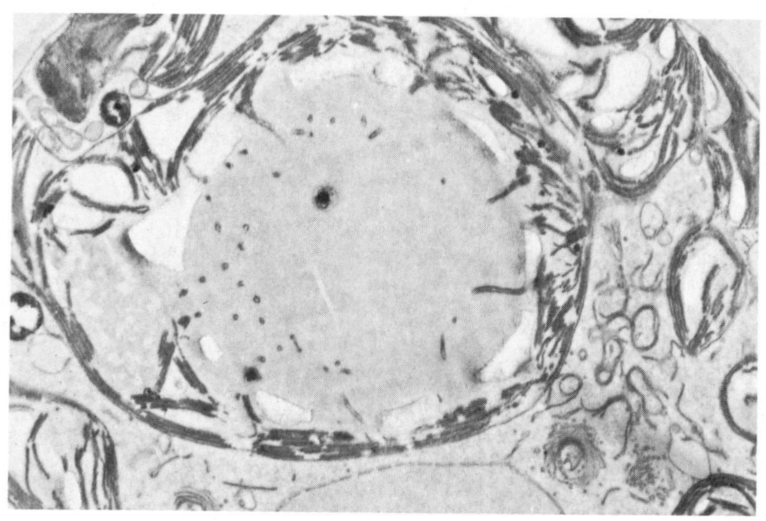

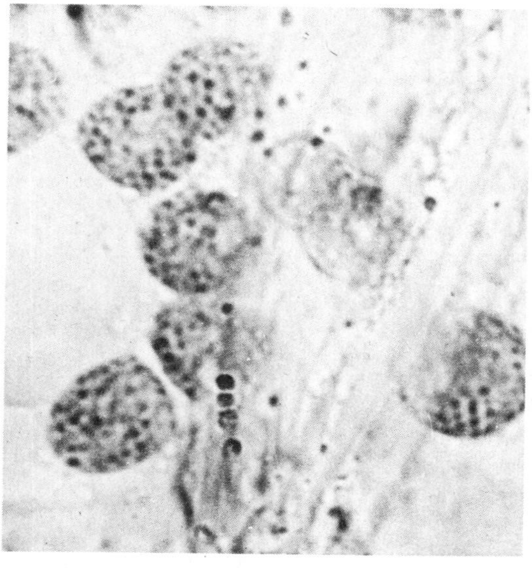

FIG. 1. Portion of a cell from the green alga *Scenedesmus* showing the spherical gray pyrenoid surrounded by the electron transparent starch grains. (× 8000)

FIG. 2. Living chloroplasts from a tobacco leaf cell showing the grana in the chloroplasts. (× 2500)

Under the electron microscope, the pigmented, photosynthetic area of the blue green algae may be seen to consist of an array of irregular membranes (Fig. 3). In higher plants, the chloroplasts appear as membrane-bounded organelles with an internal membrane system embedded in the stroma. These membranes are of two sorts. The grana appear as compartment cylinders formed by parallel electron-dense membranes, the partitions and margins, enclosing an electron-transparent space, the loculus (Fig. 4). The grana are interconnected by an irregular anastomosing system of channels, the frets (Figs. 4 and 5). Fret membranes and margins are formed by a single row of subunits having light cores and dark rims (Fig. 5). The average width of these membranes is 89 Å while the light cores average 37 Å in diameter. The partitions averaging 154 Å in width (Fig. 5) are formed of two rows of such subunits. Isolated chloroplasts remain photosynthetically active. They may be sonically ruptured into particles averaging 120 Å in diameter still capable of carrying on the Hill reaction. Particles of this size may be seen in isolated shadowed fragments of membranes. These units are called *quantasomes*, and measurements suggest that four of the subunits seen in sections of chloroplast

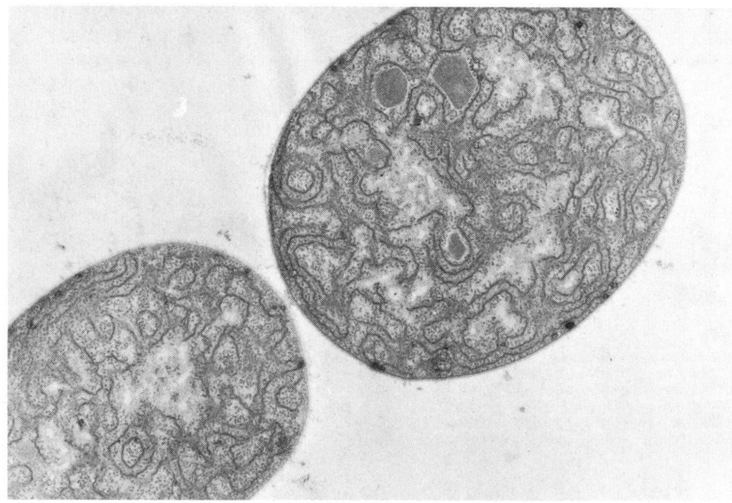

FIG. 3. Cells of the blue-green alga *Nostoc*. The peripheral arrays of irregular membranes are associated with the photosynthetic pigments. (× 6000)

FIG. 4. Chloroplast from a leaf cell of bean, *Phaseolus vulgaris*. (× 12,000)

membranes may be the equivalent of a quantasome. The bounding chloroplast envelope and the stroma may be washed away from the isolated chloroplasts leaving a system of naked membranes (Fig. 6). Grana may be separated from the frets

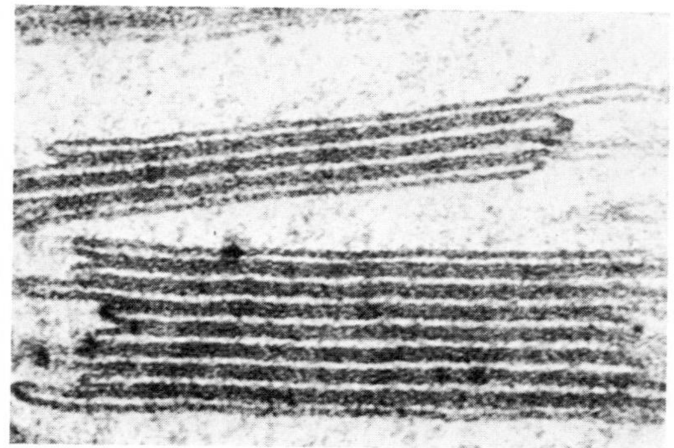

FIG. 5. Subunit structure of the membranes of chloroplast of bean, *Phaseolus vulgaris*. ($\times$ 100,000)

(Fig. 7). Sonic treatment ruptures margins, releasing individual partitions or groups of compartments into the homogenate (Fig. 8). When plastids are swollen in hypertonic solutions, some frets are ruptured and there also occurs a rearrangement of membranes and considerable swelling. The resulting swollen vesicles are composed of a complex union of the loculi and compartment of several grana with interconnecting frets (Fig. 9).

In etiolated seedlings there occurs a complex, crystalline-like body, the prolamellar body; a short treatment with light induces the rearrangement of the membranes of the prolamellar body into the typical grana-fret system accompanied by the transformation of protochlorophyll into chlorophyll (Fig. 10).

Studies on the inheritance of chloroplast characters indicate that GENES may bring about mutations in chloroplasts which may then be passed on to following plant generations independent of nuclear control. Some work with the evening primroses (*Oenothera*) furnishes evidence for complete independence of chloroplast inheritance from the nucleus. If this is generally so, it would suggest that chloroplasts in the higher plants, like those of the algae have a genetic continuity from one generation to the next. Ribonucleic acid (RNA) is definitely a plastid component and there is also strong evidence for the presence of DNA in many, if not all, chloroplasts.

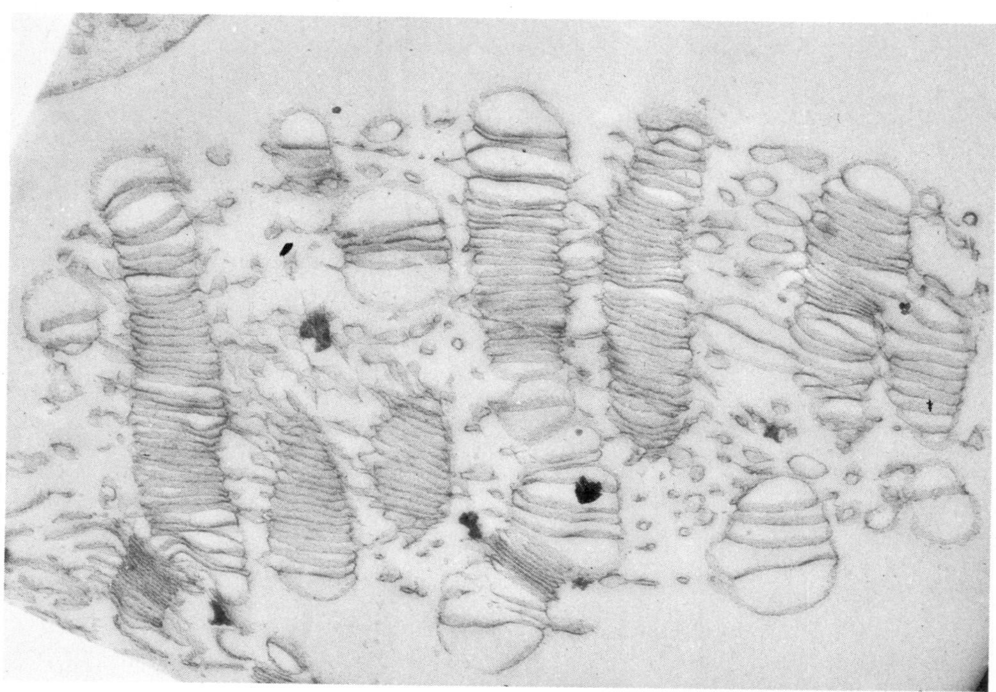

FIG. 6. Naked membrane system in a chloroplast isolated from a bean leaf. The outer envelope and stroma have been leached away. ($\times$ 12,000)

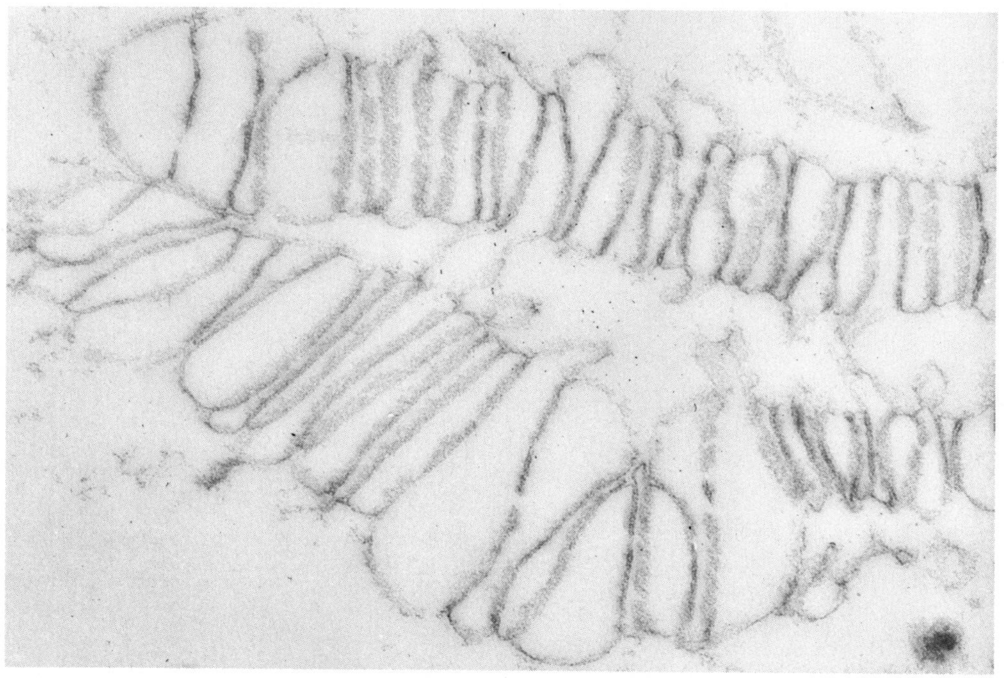

FIG. 7. Two grana isolated from a chloroplast of bean. ($\times$ 25,000)

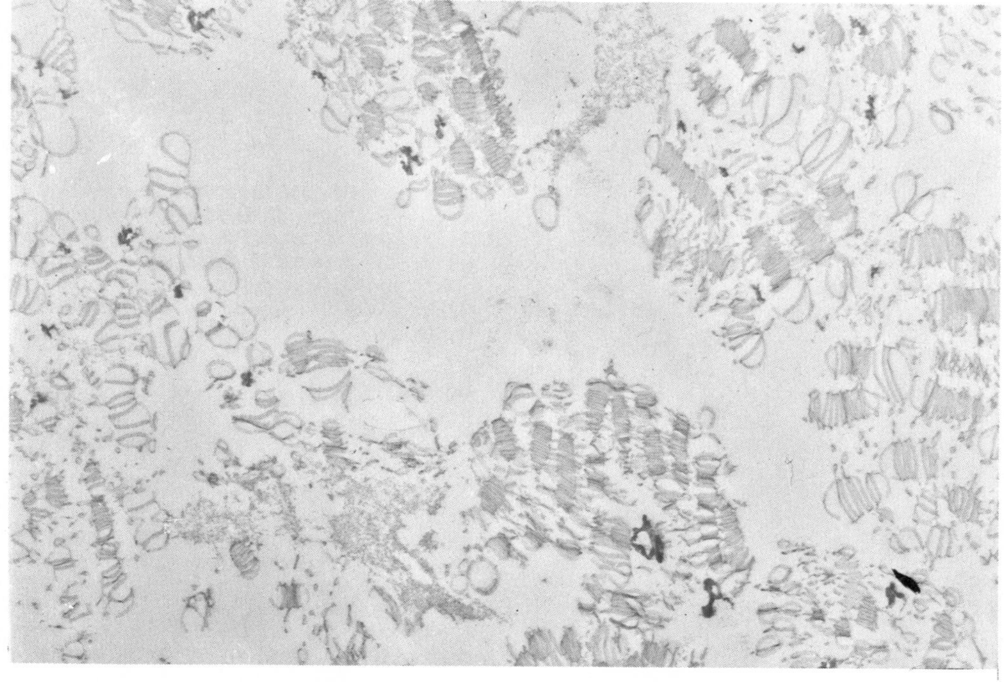

FIG. 8. Partial disruption of chloroplast membranes, by sonic rupture. ($\times$ 5000)

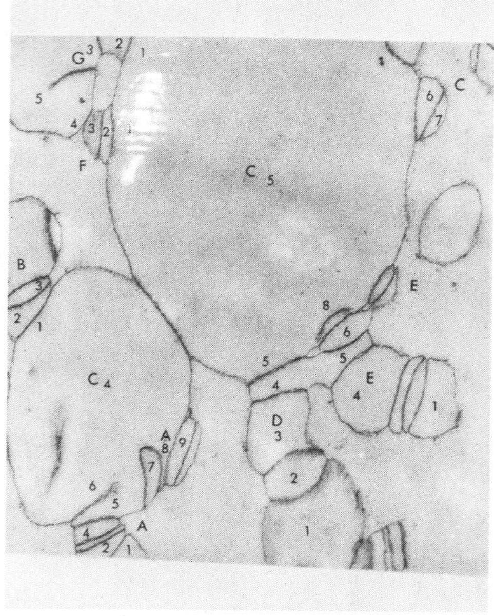

FIG. 9. Chloroplast, grana and interconnecting frets swollen in distilled water. (× 10,000)

References

1. GIBBS, S. P., "The Ultrastructure of the Chloroplasts of Algae," *J. Ultrastruct. Res.*, **7**, 418–435 (1962).
2. GIBOR, A., AND GRANICK, S., "Plastids and Mitochondria: Inheritable Systems," *Science*, **145**, 890–897 (1964).
3. JACOBSON, H. B., SWIFT, H. AND BOGORAD, L., "Cytochemical Studies Concerning the Occurrence and Distribution of RNA in Plastids of *Zea mays*," *J. Cell Biol.*, **17**, 557 (1963).
4. KLEIN, S., AND BOGORAD, L., "Early Stages in the Development of the Plastid Fine Structure in Red and Far-red Light," *J. Cell Biol.*, **22**, 433 (1964).
5. PANKRATZ, H. S., AND BOWER, C. C., "Cytology of Blue-green Algae. I. The Cells of *Symploca muscorum*," *Am. J. Botany*, **50**, 387–399 (1963).
6. PARK, R. B., AND BIGGINS, J., "Quantasome: Size and Composition," *Science*, **144**, 1009–1011 (1964).
7. WEIER, T. E., STOCKING, C. R., BRACHER, C. B., AND RISLEY, E. B., "The Structural Relationships of the Internal Membrane Systems of *in situ* and Isolated Chloroplast of *Hordeum vulgare*," *Am. J. Botany*, **52**, 339 (1965).
8. WEIER, T. E., ENGELBRECHT, A. H. P., HARRISON, A., AND RISLEY, E. B., "Subunits in Chloroplast Membranes," *J. Ultrastruct. Res.*, **13**, 92 (1965).

T. ELLIOT WEIER

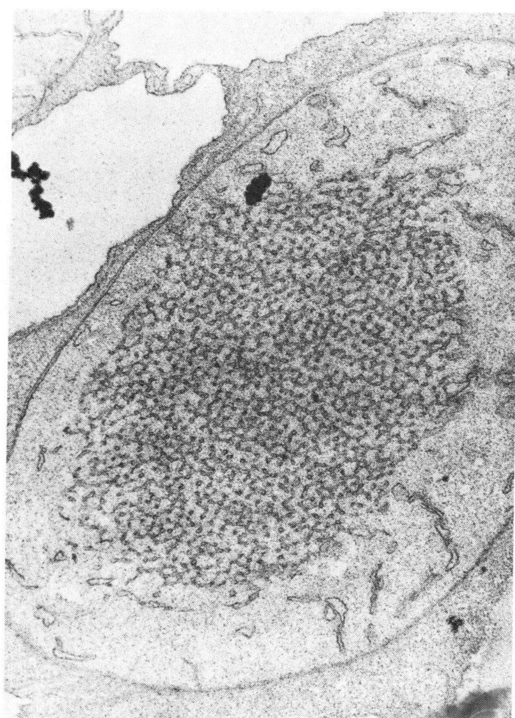

FIG. 10. The prolamellar body found in most chloroplasts of plants grown in darkness. (× 8000)

CHOLESTEROL

This important sterol (see STEROIDS for structure) occurs in blood to the extent of about 130–260 mg/100 ml. Its biosynthesis and metabolism are discussed in the articles HORMONES (STEROID); ISOPRENOID BIOSYNTHESIS; and STEROLS (BIOGENESIS AND METABOLISM). Its nutritional status and relation to cardiovascular disease are discussed in the article ATHEROSCLEROSIS. See also BILE; FATTY ACIDS, ESSENTIAL; LYMPH.

CHOLINE AND CHOLINESTERASE. See ACETYLCHOLINE AND CHOLINESTERASE. Choline also occurs as a constituent of certain PHOSPHOLIPIDS (see CEPHALINS AND LECITHINS; SPHINGOLIPIDS). The biosynthetic introduction of the methyl groups into choline is discussed in the article METHYLATIONS IN METABOLISM.

CHONDROITIN. See MUCOPOLYSACCHARIDES.

CHROMATIN

The name chromatin was given by Flemming[5] to denote that substance in cell nuclei which, in the usual treatment with nuclear dyes, takes up the color. In nondividing nuclei, chromatin is distributed throughout the entire nucleus, but in nuclei which are undergoing CELL DIVISION, chromatin is confined to the CHROMOSOMES. In Flemming's time, the chemistry of the nucleus was entirely unknown. In spite of the great advances which have since been made concerning the structure of DEOXYRIBONUCLEIC ACID and NUCLEOPROTEINS, it is still not known to what particular substance, or substances, the special affinity for dyes is due,[1] and the term "chromatin" is still being used in present-day scientific literature.

The constituents of chromatin, which was isolated from pea embryos by differential centrifugation and purified by sucrose gradient centrifugation,[7] were found to be as follows: deoxyribonucleic acid (DNA), 31%; RIBONUCLEIC ACID (RNA), 17.5%; HISTONE protein, 33%, nonhistone protein, 18%.

The majority of chromosomes are visible individually only in dividing nuclei, as they become greatly extended and dispersed in nondividing (interphase) nuclei. A minority of chromosomes, or parts of chromosomes, remain condensed in interphase nuclei. The state of condensation of some chromosome parts, while the rest of the chromosomes are extended, is called "heteropycnosis." On the basis of the morphological criterion of heteropycnosis, Heitz[6] has distinguished two classes of chromatin: (1) euchromatin, which becomes dispersed in interphase nuclei, and (2) heterochromatin, which remains condensed in interphase nuclei. The sex chromosomes of many animals are particularly rich in heterochromatin. In autosomes (*i.e.*, chromosomes other than sex chromosomes), smaller amounts of heterochromatin are usually present in the regions near the centromere, and sometimes in other regions.

It has recently been shown that heterochromatin differs from euchromatin in the time at which DEOXYRIBONUCLEIC ACID REPLICATION occurs. As a result of using radioactive markers, usually thymidine labeled with tritium, it has been found that chromosomes, and parts of chromosomes, which exhibit the morphological properties of heterochromatin, incorporate thymidine at a later stage in DNA synthesis than chromosomes consisting of euchromatin. This applies to the heterochromatin of sex chromosomes, as well as to heterochromatic regions of autosomes, and data are available from both animal and plant species.[4]

The sex chromatin body of mammals is an example of heterochromatin which has received much attention in recent years. Originally described in nerve cell nuclei of female cats,[2] sex chromatin was soon found to be present in the interphase nuclei of many cell types in other mammalian species, including man. In stained preparations, sex chromatin consists of an intensely stained body, of about 1 μ diameter (Fig. 1). Although sex chromatin is normally confined to the nuclei of females, it is not connected with anatomical or physiological sex, but merely reflects the number of X chromosomes present in the nucleus. If only a single X chromosome is present, as in males (whose sex chromosome constitution is XY), no sex chromatin is formed, but of the two X chromosomes of the female, one forms a sex chromatin body. A sex chromatin body is also present in male patients with XXY sex chromosomes (Klinefelter's syndrome), and it is absent from the nuclei of female patients with an XO sex chromosome constitution (Turner's syndrome). Wherever sex chromatin is present, one of the X chromosomes duplicates its DNA later than the other X chromosomes and the autosomes.[8] Clearly, a sex chromatin body represents a heterochromatic X chromosome.

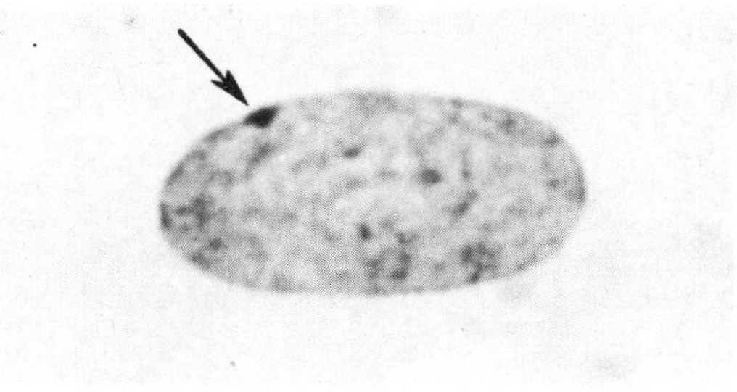

FIG. 1. Nucleus of fibroblast-like cell from human female, showing sex chromatin body; Feulgen stain, magnification ×2200.

Recent evidence[3] may confirm the long-held view that euchromatin and heterochromatin represent two different DNA control systems: euchromatin as the seat of Mendelian genes may be concerned with the production of biochemically specific products, while heterochromatin may control quantitative characteristics.

References

1. BAKER, J. R., "Principles of Biological Microtechnique," pp. 327–328, London, Methuen, 1958.
2. BARR, M. L., AND BERTRAM, E. G., "A Morphological Distinction between Neurones of the Male and Female, and the Behaviour of the Nucleolar Satellite during Accelerated Nucleoprotein Synthesis," *Nature*, **163**, 676–677 (1949).
3. COMMONER, B., "Roles of Deoxyribonucleic Acid in Inheritance," *Nature*, **202**, 960–968 (1964).
4. EVANS, H. J., "Uptake of ^{3}H-thymidine and Patterns of DNA Replication in Nuclei and Chromosomes of *Vicia faba*," *Exptl. Cell Res.*, **35**, 381–393 (1964).
5. FLEMMING, W., "Beiträge zur Kenntnis der Zelle und ihrer Lebenserscheinungen," *Arch. mikr. Anat.*, **18**, 151–259 (1880).
6. HEITZ, E., "Das Heterochromatin der Moose," *J. Jahrb. Wiss. Bot.*, **69**, 762–818 (1928).
7. HUANG, R. C., AND BONNER, J., "Histone, a Suppressor of Chromosomal RNA Synthesis," *Proc. Natl. Acad. Sci.*, **48**, 1216–1222 (1962).
8. MITTWOCH, U., "Sex Chromatin," *J. Med. Genet.*, **1**, 50–76 (1964).

URSULA MITTWOCH

CHROMATOGRAPHY, COLUMN

In 1906, Tswett, a Russian botanist, poured an extract of plant leaves in petroleum ether into a vertical tube, open at both ends and nearly filled with finely divided calcium carbonate supported by a plug of glass wool. After the addition of the extract, he passed more petroleum ether through the column. The various solutes were reversibly adsorbed by the calcium carbonate. Any individual solute molecule is adsorbed and redissolved many times in the process. The fraction of the time spent by such a molecule on the surface of the adsorbent depends on the affinity of solid surface for the solute. Since this affinity differs from one solute to another and since the solute molecules move downward through the column only when they are in solution, the various solute species move down the column at different rates. Thus they are separated into bands, the less adsorbed solutes being nearer the bottom of the column.

In Tswett's experiment, several distinct bands of yellow and green compounds were formed. Because of these colored bands, Tswett called this method of separation *chromatography*, a name that has endured in spite of the fact that chromatography is now applied to colorless compounds more frequently than to colored compounds.

The isolation of the several constituents of the sample may be accomplished by removing the adsorbent from the tube after the bands have been separated, cutting the cylindrical plug into several shorter cylinders so that each cylinder contains only one solute, and dissolving the solutes in suitable solvents. Another method, generally more advantageous, consists in continuing the passage of solvent through the column until all the solutes have been removed and collected in separate portions of the effluent or eluate.

Tswett's experiment is an example of *adsorption chromatography*. In *partition chromatography*, the stationary phase is water, held in the column by a porous solid such as silica gel or starch. The mobile phase is an immiscible liquid such as benzene or chloroform. In *reversed-phase partition chromatography*, the mobile phase is water, and the stationary phase is the immiscible organic liquid held in place by a porous hydrophobic solid such as beads of polystyrene resin, rubber, or silica gel coated with a silicone. Cations (or anions) can often be separated from each other by elution through a column of cation-exchange (or anion-exchange) resin with an aqueous salt solution as eluent. This is *ion-exchange chromatography*. ION-EXCHANGE RESINS can also be used for the separation of water-soluble organic compounds with water as the eluent. This is really a type of reversed-phase partition chromatography, the resin serving as the stationary organic phase. Such separations are greatly facilitated by using an aqueous salt solution as eluent. This method is called *salting-out chromatography*. Organic compounds of insufficient solubility in water can often be separated chromatographically with an ion-exchange resin as the stationary phase and an aqueous solution of an organic solvent such as alcohol, acetone, or acetic acid as eluent. This modification is called *solubilization chromatography*.

Each of the foregoing types of chromatography can be further subdivided as follows. In *elution chromatography*, a small amount of sample solution is added to the column, previously freed from tightly sorbed material. Then a suitable solvent is passed through the column, and the sample constituents are separated into bands as in Tswett's experiment. This type is most useful in analytical chemistry because the various sample constituents can be quantitatively separated from each other. In *frontal chromatography*, no eluent is used, but the solution of the sample is passed continuously into the column. The least sorbed of the sample constituents emerges first from the column, free from the other solutes of the sample. Then a mixture of this compound and the next least readily sorbed compound emerges, then a mixture of the three least readily sorbed, and so on until finally the effluent has the same composition as the sample. *Displacement chromatography* has some characteristics of both elution and frontal chromatography. A rather large sample is first added to the column, previously freed of tenaciously sorbed compounds. Then a solution of a vigorously sorbed compound is used as eluent or displacing agent. A band of the sample constituents moves down the column in front of the highly sorbed eluent. As it moves, it is gradually separated into several bands, each containing only one constituent of the sample. At the boundaries of these bands, there is always a

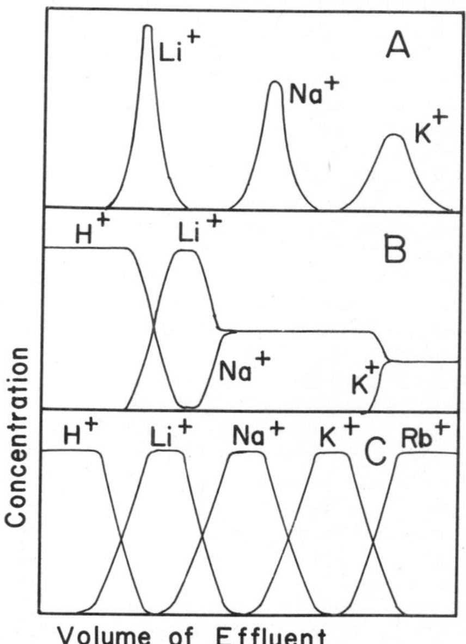

FIG. 1. Three types of ion-exchange chromatography. A hydrogen-form strong-acid cation exchanger is used in all cases. The sample is an equimolar mixture of salts of lithium, sodium, and potassium. (A) Elution chromatography with hydrochloric acid as eluent. (B) Frontal chromatography of acetates. (C) Displacement chromatography of acetates with rubidium acetate.

mixture of the solutes of the two bands. Since large quantities can be used and since a large fraction of each sample constituent can be recovered uncontaminated by the other solutes, this type of chromatography is used mostly in preparative work.

Figure 1 illustrates the behavior of a mixture of lithium, sodium, and potassium ions in the three types of ion-exchange chromatography.

Among the many books on chromatography, those by E. Heftmann, L. Zechmeister, H. H. Strain, and E. and M. Lederer may be mentioned. The topic is also discussed in more comprehensive works such as "Treatise on Analytical Chemistry," part 1, volume 3, edited by I. M. Kolthoff and P. J. Elving, and "Physical Methods in Chemical Analysis," edited by W. Berl.

References

1. Heftmann, E. (Editor), "Chromatography," New York, Reinhold Publishing Corp., 1961.
2. Lederer, E., and Lederer, M., "Chromatography, a Review of Principles and Applications," Second edition, Amsterdam and New York, Elsevier Publishing Co., 1957.
3. Strain, H. H., "Chromatographic Adsorption Analysis," New York, Interscience Publishers, 1942.
4. Zechmeister, L., and Cholnoky, L., "Principles and Practice of Chromatography," New York, John Wiley & Sons, 1951.
5. Kolthoff, I. M., and Elving, P. J., (Editors), "Treatise on Analytical Chemistry," Part 1, Vol. 3, New York, Interscience Publishers, 1961.
6. Berl, W. (Editor), "Physical Methods in Chemical Analysis," Second edition, New York, Academic Press, 1961.

William Rieman III

CHROMATOGRAPHY, GAS

Principles. The term gas chromatography is used to describe a process by which complex mixtures of chemical compounds are separated from one another by selective partition between a stationary liquid or solid phase and a mobile gas phase. The gas used as the moving phase is usually hydrogen, helium, nitrogen, or argon, depending upon the requirements of the detection system. Sometimes the compounds which are separated from one another are also gases, but more often they are vapors derived from solid or liquid samples. Thus this method can be used to resolve mixtures of permanent gases such as carbon dioxide, oxygen, and nitrogen at room temperature, or it can be employed to separate vapors of compounds such as steroids, fatty alcohols, and fatty acid esters at higher temperatures which are still far below their boiling points.

The stationary phase is usually contained within a glass or metal tube 0.2–1 cm in diameter and 0.5–10 meters long. When the stationary phase is an active solid such as molecular sieve or silica gel, the method is termed gas-solid chromatography. A very important advance was made in 1952 when James and Martin introduced gas-liquid partition chromatography.[3] In this technique, the stationary phase is a high-boiling liquid coated on an inert solid support such as a diatomaceous earth or small glass beads. Materials which are useful as stationary liquids include silicone polymers, polyesters, polyethylene glycols and high molecular weight esters of organic acids. The separations obtained depend upon the properties of the stationary liquid phase: hydrocarbons often separate sample components in order of boiling point, and oxygenated hydrocarbons and nitriles in order of polarity, other things being equal. The stationary phase must be essentially nonvolatile at the operating temperature of the column, which will generally be between −50° and 300°C. A well-prepared column, 5 mm in diameter and 1 meter in length, will often provide an efficiency of 1000–2000 theoretical plates.

Procedure. The equipment most generally used for gas chromatography consists of a heated injection block, a column (or several columns connected in series or parallel), and a device for measuring the concentrations of separated components in the effluent carrier gas. The temperature of the column must be thermostatically controlled, and may be held constant throughout the analysis, or programmed to permit the separation of mixtures having high boiling point ranges.

The sample (usually 0.001–100 μg dissolved in a few microliters of solvent) is injected with a syringe through a diaphragm into the heated block. (Many methods for sample introduction have been developed, but this is the one most generally used.) It is vaporized immediately and swept through the column by the carrier gas. The components are separated from one another in the column by selective partition or adsorption and, ideally, are detected separately in the effluent carrier gas as they emerge from the distal end. The output from the detector is picked up by a recorder which translates the impulses into a graph of response *vs* time. Thus a chromatogram is obtained having a number of peaks, each of which corresponds to a component (or unresolved components) of the mixture. The time that elapses between the injection of a sample and the appearance of a peak maximum on the chromatogram is called the retention time (uncorrected) of the component. Since this time will vary with gas flow rate, temperature, and other variables, it is often reported as relative retention, which is the ratio of the observed time to that obtained on a standard compound injected with the sample. The area under each peak is proportional to the concentration of the compound it represents. Thus the method can be used for quantitative analysis, as well as for separating the components of complex mixtures.

Detectors. The detection system which is employed will depend upon the nature of the problem. The detectors used most commonly for the analysis of inorganic gases or the isolation of relatively larger amounts of organic compounds from impurities are *thermal conductivity cells.*

These can be thermistors or catharometers. Thermal detectors can measure as little as 10^{-8} gram of each component of a complex mixture. For trace analysis and most biochemical work, the *flame ionization* and *argon detectors* are much more useful. The flame ionization detector responds to as little as 10^{-11} mole of each compound in the carrier gas, has a greater dynamic range than the argon detector, and is much more sensitive to organic compounds than the thermistor or catharometer. It is insensitive to water and most other inorganic compounds, and is highly sensitive to all organic compounds containing methylene groups. Therefore, it is the method of choice for the detection of compounds of biochemical interest when semi-universal response is desired. However, in the analysis of tissue extracts, it is sometimes desirable to employ more specific detection methods to eliminate interferences and reduce background noise. For example, the *electron capture detector* can be used to detect chlorinated hydrocarbons and many conjugated unsaturated compounds with a high degree of selectivity, since it is highly sensitive to these substances while being relatively insensitive to hydrocarbons. The *microcoulometer* provides even higher selectivity since it can be used in-stream for the quantitative analysis of chlorine, sulfur or phosphorus in organic compounds. In this method, organic compounds are oxidized or reduced directly in the gas stream with oxygen or hydrogen. The HCl, SO_2, PH_3, or H_2S formed during these processes is measured electrochemically.

Capillary Columns. Chromatography can also be carried out in long capillary columns, the stationary phase consisting of a thin film of organic material coated on the internal surface of the capillary tube. The tube thus serves as the solid support. A 50-meter column can be coiled so that it is very compact. It will often provide resolutions up to 150,000 theoretical plates. For some materials such as hydrocarbons, capillary columns far exceed packed columns in their capacities for separating complex mixtures of closely related compounds. However, they are not as generally applicable as packed columns to microanalytical work since retention values are sometimes difficult to reproduce, and the total amount of material which can be applied to the column is comparatively small.

Applications. The applications of gas chromatography to biochemical problems are many and varied.[1] Gas-solid chromatography can be used for measuring respiratory and photosynthetic gases, while gas-liquid chromatography can be used to measure the volatile compounds responsible for food aromas, terpenes and essential oils, lipids. Even nonvolatile compounds such as amino acids and oligosaccharides can be chromatographed after their conversion to volatile derivatives. Gas chromatography is sometimes used to analyze these materials directly. However, a more satisfactory approach is to first separate the sample into groups of related compounds by ancillary procedures such as distillation, liquid-liquid partition, liquid-liquid chromatography, ion-exchange chromatography, or thin-layer chromatography. Gas chromatography is then used for the final analysis. This provides for better resolution than when the method is used directly.

Permanent gases of interest to the biochemist which can be analyzed by this method include carbon dioxide, oxygen, and nitrogen. Oxygen and nitrogen are usually resolved from one another on molecular sieve. However, carbon dioxide is not eluted from this stationary phase except at high temperatures. Therefore, for a complete analysis of respiratory or photosynthetic gases, two columns are used—either in series or in parallel. Molecular sieve is used to separate oxygen from nitrogen, while silica gel is used to resolve carbon dioxide from a composite nitrogen-oxygen peak. The complete analysis can be carried out on molecular sieve alone if the temperature is programmed.

Organic compounds present in the atmosphere due to air pollution or exudation from vegetation may also be of interest to the biochemist. Air samples are usually not analyzed for trace components directly. Instead, the air is passed through a cold trap which contains an inert (or in some cases an active) solid. Trace organic components are condensed or adsorbed on the solid. The trap is attached to the chromatograph, and its contents are vaporized by heat and flushed into the column. Stationary phases which have been used for the analysis of air pollutants include di-*n*-butylphthalate, aluminum oxide modified with propylene carbonate, squalane, and solutions of silver nitrate in ethylene glycol. This latter liquid is particularly useful for separating alkenes from alkanes and for resolving isomeric olefins.

Volatile compounds present in foods and beverages are amenable to analysis by gas-liquid chromatography. These are defined arbitrarily as materials containing one, or rarely two, functional groups, and not more than eight to ten carbon atoms. They are found in the extracts or distillates of fruits, meats, and vegetables, or in the condensates obtained on freeze-drying these products. They are usually mixtures of lower fatty acids, amines, carbonyl compounds, thiols, sulfides, alcohols, and esters. Often they are chromatographed as complex mixtures, but more meaningful results can sometimes be obtained by prefractionating them. Thus, fatty acids can be isolated as a group by distilling off neutral and basic compounds from alkaline solutions, and recovering the free acids after acidification through distillation or extraction. Carbonyl compounds can be isolated as a group by precipitation with 2,4-dinitrophenylhydrazine, while thiols and sulfides can be separated by the formation of compounds with salts of heavy metals. After isolation, the various groups of related compounds are chromatographed separately. Complex mixtures of volatiles have been chromatographed on a wide variety of liquid substrates at temperatures ranging from $-80°$ to $200°C$. No single set of conditions is satisfactory for all samples. Usually it is best to chromatograph complex mixtures on both polar and nonpolar liquids to obtain the maximum numbers of peaks. Very often, additional information can be obtained by collecting unresolved fractions and rechromatographing them on a stationary liquid having characteristics different from the ones used for the primary separation.

Gas chromatography is also an excellent tool for the separation, characterization, and quantitative analysis of essential oils. Separations of components that formerly took days by tedious chemical and physical means, or were impossible by these older methods, can be accomplished in minutes. Terpene hydrocarbons and related oxygenated terpenoids are the main constituents of essential oils. The oils may be analyzed directly, or prefractionated into hydrocarbons and oxygenated hydrocarbons. Generally, the latter is accomplished by liquid-solid chromatography on silica gel, but preparative-scale gas chromagraphy can also be used since the terpenes as a group are eluted well ahead of the oxygenated compounds.

Gas chromatography has proved to be more successful for the analysis of lipids than for any other group of compounds. Here again, prefractionation is desirable before injection of a sample into the chromatograph. This is usually accomplished by liquid-solid chromatography on silica gel or thin-layer chromatography. These procedures separate the sample into fractions consisting of phospholipids, sterol esters, triglycerides, and higher fatty acids. Some lipids are chromatographed directly while others are converted to volatile derivatives. The following classes of compounds are usually modified chemically before injection: phospholipids, sterol esters, higher fatty acids, O-alkyl glycerols, and higher aldehydes. Sterols and higher fatty alcohols can be chromatographed either unchanged or as derivatives. Lipids are usually chromatographed on polar and nonpolar stationary liquids to obtain optimum resolution. Chromatography of the methyl esters of the higher fatty acids on nonpolar liquids such as "Apiezon" (a high-boiling hydrocarbon) results in separation according to boiling point. By contrast, chromatography on polar phases such as polyesters [*e.g.*, poly-(diethyleneglycol succinate)] leads to improved separation according to degree of unsaturation. Thus on polar liquids, saturated acids are eluted ahead of monoenes, monoenes ahead of the dienes, and diense ahead of trienes for compounds containing the same carbon skeleton.

Compounds which are ordinarily considered to be nonvolatile can be separated by gas chromatography by first converting them to volatile derivatives. Thus, AMINO ACIDS can be separated as N-acetyl alkyl esters, N-trifluoroacetyl alkyl esters, or trimethylsilyl derivatives. Di- and polybasic organic (Krebs cycle) acids can be chromatographed as their methyl esters, and oligosaccharides can be separated after converting their OH groups to OCH_3 groups. Excellent methods of separation have been developed for many of these compounds, but some problems still remain to be solved in quantitating results. It is certain that gas chromatography will continue to grow in usefulness in this field with continued advances in technology.

References

1. BURCHFIELD, H. P., AND STORRS, E. E., "Biochemical Applications of Gas Chromatography," New York and London, Academic Press, 1962.
2. GOLAY, M. J. E., *Anal. Chem.*, **29**, 928 (1957).
3. JAMES, A. T., AND MARTIN, A. J. T., *Biochem. J.*, **50**, 679 (1952).

H. P. BURCHFIELD AND ELEANOR E. STORRS

CHROMATOGRAPHY (GENERAL)

Chromatography is the generic name of a group of separation processes that have a certain characteristic in common. This common characteristic is that the separation depends upon the redistribution of the molecules of the mixture between *a thin phase in contact with one or more bulk phases*.[2] The chromatographies form a subclass of those separation processes which are applicable to molecular mixtures and which depend upon distribution between phases. This larger class includes distillation (vapor-liquid), solvent extraction, or absorption distributions (liquid-liquid), sublimation (vapor-solid), and crystallization (liquid-solid) distributions, in all of which the phases, indicated in parentheses are *bulk* phases. What makes the chromatographies different is that one of the phases is thin—often reaching molecular dimensions—and that for this reason molecular size and shape play an exaggerated role in the separation: in other words, extremely subtle separations become possible because the limited volume in the thin film emphasizes differences in molecular size, shape, packing and, hence, orientation.

A typical chromatographic separation (such as was described by Mikhael Tswett, the inventor of the method, 1906) is shown in Fig. 1, which illustrates the processes of adsorption chromatography.[2] A *column*, or *fixed bed*, of adsorbent, A, is formed in a tube by dry packing, or by pouring in a slurry from which the slurrying liquid is almost completely drained. This bulk phase will be the carrier of the thin phase, the adsorption layer. The mixture to be separated, B, dissolved in some poorly adsorbed solvent, is *applied* to the top of the column by letting the solution percolate into the bed. If the substances are adsorbed, they are removed from the solvent onto the surface of the adsorbent to form a *mixed zone*, MZ, while the solvent, emptied of solute, passes down the column. This behavior is the embodiment of the mass-action principle. The moving bulk liquid is referred to as the *mobile phase*, and the adsorbed layer as the thin *stationary phase*. The solute may move between the two in contact:

$$\text{Substances in Mobile Phase} \rightleftharpoons \text{Substances in Stationary Phase}$$

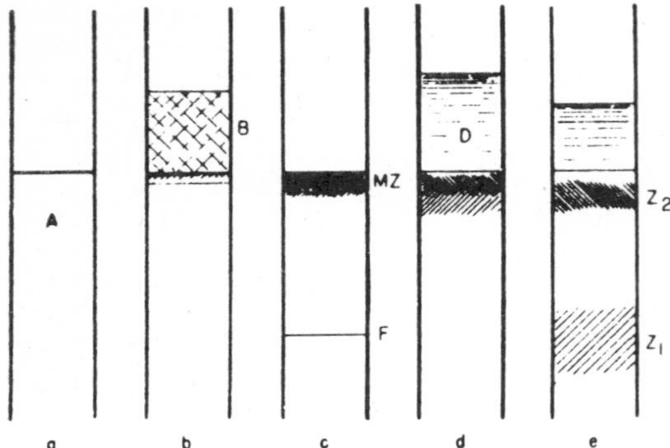

FIG. 1. Chromatography of a binary mixture. (A) Column of adsorbent in the chromatography tube. (B) Solution of mixture to be separated. MZ, mixed zone, applied to the column. F, front of empty solvent. D. developer, applied to the column. Z_1,Z_2, zones of separated components. To the packed column (a) the solution to be analyzed is applied (b), As it passes into the bed of adsorbent the components are retarded to form a mixed zone, while the solvent (also to some extent adsorbed) runs ahead (c). Developer is then applied (d), and as the development occurs the zones of separated material draw apart to produce the developed chromatogram (e)[1].

When the solution is first applied to the adsorbent, the latter is *empty* of solute; there are no "substances in the stationary phase"; thus, by the principle of Le Chatelier, solvent is adsorbed and substances move out of the mobile phase, accumulating and forming the stationary phase of adsorbed solutes and solvent. But the mobile phase flows on down the tube, being replaced by fresh mixture and continually coming into contact with empty adsorbent. Thus, because of the *differential countercurrent movement* of mobile and stationary phases relative to each other, the solute is *all* removed from the solution, and forms a *mixed zone*, or *band*, at or near the top of the column. The upper part of this zone may have reached equilibrium with the solution; the lower edge may be more dilute and may already contain more of the less strongly adsorbed solute, relative to the other solutes, than was present in the original mixture: in other words, some slight separation may already have taken place. It is evident that a poorly adsorbed solvent is desirable in this process, so that it will not compete with solute for the limited surface area of the adsorbent. The *front* of the advancing solvent is shown at F. If the adsorbent is uniformly packed, the front or face of the zone will be quite regular. *The top of the bed must never be allowed to become dry*, else it will crack.

The next step is to *develop* the chromatogram. A solvent, slightly more strongly adsorbed than before, is applied to the column. Once again the principle of mass action takes effect. Empty *developer*, D, takes up solute from the stationary phase as it flows down the bed, eventually, perhaps, reaching equilibrium with the adsorbed material, but then as it reaches the front of the zone, it meets empty adsorbent, and so deposition of solute occurs as before. The developer starts empty of solute and emerges from the bed empty of solute. But in the intervening period it will have picked up solute and transported it downward. Moreover, the *less strongly adsorbed* material will be carried preferentially, since the more strongly adsorbed will occupy preferentially the available sites, or space, in the thin film of the stationary phase. One says that the more strongly adsorbed substance, Z_2—the substance preferentially distributed into the stationary phase—is *retarded* more than the other, Z_1, which *runs on ahead*. As development progresses, the zones separate until, in favorable cases, they are well and completely separated.

If development is continued, the faster zones may be *washed* successively out of the bed. Alternatively, the development may be stopped, the developer sucked from the column, and the column of adsorbent *extruded* from the tube by means of a closely fitting plunger or dowel. The zones may then be dissected out with attendant adsorbent, and the pure substance *eluted*, or *desorbed*, from each by means of a good solvent for the substance which contains, if needed, a strongly adsorbed material which *displaces* the substance from the surface. The most efficient way to carry out this step is to pack each zone as a bed in a small chromatographic tube, and apply the *eluent*, or *displacer*, to it.

For control of the process it is convenient to characterize the zone by an R_f value: the velocity of movement of the front edge of the zone divided by the velocity of movement of developer in the tube above the column of adsorbent. (Other, analogous values, are used with other chromatographic techniques).

This process, described in detail for adsorption chromatography (but not with all the subtle ramifications of even that technique) describes also, in principle, all other chromatographic techniques. These differ in the nature of the bulk phases, whether liquid or gas; in the nature of the thin phase, whether it is immobile and fixed (as above) at the surface of a solid adsorbent, or is a "mobile" interfacial film between two liquid phases or between a liquid and a gel phase (often difficult to distinguish); and in the way the phases are brought into contact.

Thus in (a) *gas chromatography*, or (b) *vapor-liquid partition* chromatography (VPC), the mobile phase is a gas.[3,4] It flows through a tube packed with (a) adsorbent, or (b) an inert carrier coated with a film of an essentially nonvolatile liquid. In the latter case, the carrier may be dispensed with, and the film spread over the walls of a long capillary tube. The mixed zone is applied and development is carried out as described. The zones march successively out of the tube to be detected by some appropriate method. Because of the near-ideal behavior of most substances in VPC, excellent, highly quantitative results are possible with it. Minute amounts of mixtures may be handled. VPC was originated by A. J. P. Martin.

In (a) *column* or (b) *paper partition chromatography*,[1,2,5] the thin layer is a liquid film held by sorption in or on (a) silicic acid or other gel, or cellulose or related materials, or (b) paper strips or sheets; the bulk phase is a liquid that flows through the column, or down or up the paper by capillarity. This method, originated by A. J. P. Martin and R. L. M. Synge can handle minute amounts of mixtures and can be fast and quantitative in use (see also the separate articles on CHROMATOGRAPHY, COLUMN, GAS, and PAPER, respectively).

Adsorption chromatography was described in detail, above.[2,5] In *ion-exchange* chromatography, the "adsorbent" is an ion exchanger. This technique has been refined to a remarkable degree for the separation of rare earths. In principle, all the procedures are just as described for adsorption chromatography. *Thin-layer chromatography* utilizes a layer of solid adsorbent, or gel, on a glass strip or sheet, and the developer is moved by capillarity.[6] The mixed zone is applied as a spot or wide zone, and as development proceeds (the strips must be kept in a controlled atmosphere), it can readily be followed visually or by the use of reagents, since the zones are accessible, not being inside a tube, but arranged on the strip, or slab, of adsorbent.

Foam and emulsion fractionation are chromatographic methods in which the thin layer is at the surface of bubbles or droplets, and the mobile phase flows or drains between the bubbles or drops.[2] It is a difficult method to use because of problems of stabilizing the thin film, but it has several promising applications.

In each of these methods of chromatography (except the last), the literature is of vast proportions; the variations of technique are extraordinarily numerous; and the kinds of accessory apparatus, the detecting devices, the analytical machines, and so on, are legion. The principle, however, is that described. Chromatography may thus be epitomized as "*a separation process applicable to essentially molecular mixtures which relies on distribution of the mixture between an essentially two-dimensional, or thin, phase and one or more bulk phases which are brought into contact in a differential countercurrent manner.*"

References

1. BLOCK, R. J., DURRUM, E. L., AND ZWEIG, G., "A Manual of Paper Chromatography and Paper Electrophoresis," New York, Academic Press, 1955.
2. CASSIDY, HAROLD, G., "Fundamentals of Chromatography," Vol. 10 in "Techniques of Organic Chemistry" (WEISSBERGER, A., EDITOR), New York, Interscience Publishers, 1957.
3. DAL NOGARE, STEPHEN, AND JUVET, RICHARD S., JR., "Gas-Liquid Chromatography. Theory and Practice," New York, Interscience Publishers, 1962.
4. KAISER, RUDOLF, "Gas Phase Chromatography," Vols. I, II, III, Washington, Butterworths, 1963.
5. LEDERER, EDGAR, AND LEDERER, MICHAEL, "Chromatography, A Review of Principles and Applications," Second edition, Elsevier, 1957.
6. RANDERATH, KURT, "*Dunnschicht-Chromatographie*," Verlag Chemie, GMBH, Weinheim, 1962; also translation by Libman, D.D., "Thin-layer Chromatography," New York, Academic Press, 1964.

This is a very small selection from the many excellent monographs on chromatography.

HAROLD G. CASSIDY

CHROMATOGRAPHY, PAPER

Although paper chromatography as it is now conceived and practiced was antedated by Goppelsroeder's capillary analysis (1899) which is related to it, its modern beginnings date from a classical paper by Consden, Gordon and Martin in 1944, who successfully used paper chromatography to separate amino acids and to make possible analyses of amino acid mixtures.[1]

The first step in making a paper chromatogram consists in applying a small drop of the solution to be analyzed (in quantitative work 5 μl is often used) about 2–3 cm from the edge of a suitable piece of filter paper and allowing it to dry. Usually an organic solvent (or solvent mixture) containing water is then allowed to flow over and past the spot to a distance a few decimeters beyond. As this movement takes place, the various substances in the spot are picked up and transported by the moving solvent mixture. If the substances are extremely soluble in the solvent mixture used, they are carried along completely with the solvent front and no separation takes place. If they are extremely insoluble, the substances remain on the spot of origin; this, of course, results in no separation. If, however, the substances are soluble to an appropriate degree, each is transported to some intermediate position between the original spot and the edge of the solvent front. The choice of suitable solvent mixtures is thus crucial and depends on the nature of the substances to be separated.

Different substances have their distinctive affinities for water, for paper and for the constituents of the organic solvent mixture used. If in a given system a particular substance travels characteristically halfway to the solvent front, it is said to have a "R_F value" of 0.50. If it travels 1/4 or 3/4, respectively, of the distance traveled by the solvent, it is said to have an R_F value of 0.25 or 0.75.

In the original 1954 paper,[1] the authors used 18 different solvent systems employing phenol, collidine, *n*-butanol, *t*-amyl alcohol, benzyl alcohol, *o*-, *m*-, and *p*-cresols with various additions and modifications. In a phenol-NH_3 system, for example, the various amino acids had R_F values ranging from 0.12 (aspartic acid) to 0.90 (phenylalanine). Two or more amino acids may have R_F values about the same in a given solvent system, in which case they superimpose on one another. In this case a different solvent system must be used to separate them.

Paper chromatography is particularly well adapted to amino acid analysis because after the chromatograph is complete and the solvents have been removed by evaporation, the positions of the amino acids can be made visible by spraying with NINHYDRIN solution which produces a typical color (in most cases blue or purple) wherever the amino acids accumulate. Various means have been used, including visual comparison with standards, to obtain quantitative results. These have been rather surprisingly successful. For details, see references 1–3 below. Typical applications of paper chromatography to the separation and identification of amino acids and carbohydrates are illustrated in color in references 4 and 5.

In the pioneering work (1944), Consden and co-workers used an apparatus in which strips of paper were hung in a vertical position and the solvent mixture was allowed to flow downward over the spot from a trough. In 1948, the writer introduced a modification which involved using a cylinder of filter paper (made by clipping together the edges of a square or rectangular sheet) on which as many as 18 spots had been placed on a line about 3 cm from the bottom edge. The cylinder of filter paper (about 30 cm in height) was then placed upright in a shallow pool of the solvent mixture, and the mixture was allowed to ascend by capillary action over the various spots, carrying the substances with various R_F values upward in vertical lines above the corresponding spots.

An interesting variation used by Martin and his co-workers was that of two-dimensional paper chromatography. In this case, a single spot is placed about 3 cm from the corner of a square sheet and chromatographed in one direction along the edge of the sheet. After drying, the sheet is then placed so the solvent (a different mixture from that initially used) flows at right angles to the original flow. By this means, amino acids which were not separated the first time may be well separated the second time provided the two solvent mixtures are suitably chosen. This procedure can readily be used also in the ascending technique, in which case the first cylinder used is unfastened after chromatography and drying, and clipped together into a cylinder which has an axis at right angles to first cylinder. Regardless of the particular technique used, two-dimensional chromatographs require far more paper and time, and cut down the number of analyses that can be made in a restricted laboratory.

Small temperature changes sometimes have little effect on R_F values; in other cases they may affect these values markedly. Whatever apparatus or system is used, temperature control is essential for the best results. The carrying out of paper chromatographic investigations is subject to many ramifications and modifications. The particular technique which may prove most valuable will depend to a large degree upon the nature of the particular problem at hand.

Paper chromatography is applicable to analysis on a micro scale. A fraction of a microgram of an amino acid can be detected on a chromatographic sheet, and a solution to be analyzed need not contain more than 10 μg/ml of each of the amino acids. Since paper chromatography involves little apparatus, particularly when the ascending technique is used, it has been used with more or less success in a host of laboratories, not only for amino acid separations and analyses, but for many other analyses—amines, urea, proteins, sugars, polysaccharides, phosphate esters, aliphatic acids, lipids, steroids, purines, pyrimidines, phenols, aromatic acids, porphyrins, alkaloids, pigments, antibiotics, vitamins, and inorganic ions such as chloride, sulfate, phosphate and sodium [see also CHROMATOGRAPHY (GENERAL); PROTEINS (AMINO ACID ANALYSIS AND CONTENT)].

References

1. CONSDEN, R., GORDON, A. H., AND MARTIN, A. J. P., *Biochem. J.*, **38**, 224 (1944).
2. BLOCK, R. J., DURRUM, E. L., AND ZWEIG, G., "A Manual of Paper Chromatography and Paper Electrophoresis," New York, Academic Press, 1958.
3. "Biochemical Institute Studies IV. Individual Metabolic Patterns and Human Disease: An Exploratory Study Utilizing Predominantly Paper Chromatographic Methods," Publication No. 5109, Austin, University of Texas Press, 1951.
4. ABELSON, P. H., "Paleobiochemistry," *Sci. Am.*, **195**, No. 1, 83 (1956).
5. BASSHAM, J. A., "The Path of Carbon in Photosynthesis," *Sci. Am.*, **206**, No. 6, 88 (1962).

ROGER J. WILLIAMS

CHROMIUM (IN BIOLOGICAL SYSTEMS)

Chromium (atomic weight 52.01) represents 0.033% of the earth crust. It is present in trace amounts in seawater and in most organic matter. Higher concentrations of around 100 ppm have been observed in certain ascidians which concentrate the element from seawater (Levine, 1961) and in nucleic acid from beef liver. The highest known chromium content of biologic material was found in a beef liver fraction consisting of ribonucleic acid and protein in which chromium was present in a concentration of up to 830 ppm (Wacker and Vallee, 1959).

Of the three common valence states, the divalent is unstable because it is easily oxidized, and the hexavalent is known mainly for its toxicity. Trivalent chromium appears to be the physiologically active form, probably due to its strong tendency for coordination (coordination number 6), hydrolysis and olation (formation of polynuclear hydroxo complexes). Probably the oldest and best-understood application of chromium to biological materials is chromium tanning of leather. In this process, the trivalent form, in an olated state, coordinates with functional groups of the skin proteins, thus protecting these groups from enzymatic and other attacks which would cause hydrolysis. Here, as in other biological systems, the action is believed to depend on the state of olation as well as on the nature of the coordinated ligands (see CHELATION AND METAL CHELATE COMPOUNDS).

The first direct indication for a biological role of chromium in metabolism was derived from enzymatic studies (Horecker *et al.*, 1939). The succinate-cytochrome dehydrogenase system, an all-important enzyme system for the production of energy, requires certain inorganic cofactors. Of various elements tested, chromium produced the greatest increase of enzyme activity. A significant stimulation of the activity of phosphoglucomutase by chromium was described by Strickland in 1949. This system which has an important function in the early steps of carbohydrate metabolism requires magnesium and one other metal for optimal activity. Chromium was outstanding as "second metal" because it produced the highest enzyme activation and supported a measured amount of activity even when given alone.

Another effect of chromium, as described by Curran in 1954, is the stimulation of fatty acid and cholesterol synthesis from acetate in liver [see FATTY ACID METABOLISM; STEROLS (BIOGENESIS AND METABOLISM)]. The effect was also observed when the donor animals were injected with the element before the tissue was obtained. In this as in the other systems discussed, the action of chromium was not specific, and the amounts required for activity were above the levels normally found in nature.

The *in vivo* action of very small, physiological levels of chromium was established when trivalent chromium was identified as the active ingredient of "glucose tolerance factor," a dietary agent postulated as essential for maintenance of normal glucose tolerance in rats.[4] Whereas rats on chromium-sufficient diets are able to maintain an efficient rate of glucose removal from the blood throughout their life, animals on a low-chromium diet slowly develop an impairment of glucose utilization as they grow older. At the same time, isolated fat or muscle tissue of chromium-deficient animals responds to low doses of INSULIN significantly less than that of chromium-sufficient rats when glucose uptake from the medium, glucose utilization for lipogenesis or for oxidation to CO_2, is measured. Also, the effect of insulin on entry rates of a non-utilizable sugar, D-galactose, into cells is enhanced by chromium, thus defining the site of action at the cell membrane (Mertz and Roginski, 1963). Similar effects have been demonstrated on mitochondrial membranes (Campbell and Mertz, 1963), and on the basis of these and other findings (Christian *et al.*, 1963), the formation of a ternary complex between the sulfhydryl of membrane acceptor sites, chromium and the intrachain disulfide of insulin has been suggested as the mode of action.

That chromium is an essential cofactor for the action of insulin on the rat lens was shown by Farkas in 1964. In the absence of the element, no significant insulin effect on glucose utilization of lens can be demonstrated. Chromium supplementation to the donor animals results in a significant response of lens tissue to the hormone.

Further evidence for a biological role of chromium came independently from long-term experiments. When conducted in an environment which allows strict control of external metal contamination, supplementation of the drinking water with trace amounts of chromium results in a significantly lower mortality and in better growth rates in rats and mice, as compared to those of unsupplemented controls. While the mechanism responsible for the decrease of mortality remains to be established, one possible factor has been investigated: chromium supplementation protects against excess deposition of lipid in the aorta (Schroeder *et al.*, 1965; see also ATHEROSCLEROSIS). Despite the high amounts of chromium ingested by the animals during their lifetime, tissue levels are only moderately elevated at the termination of the experiments, indicating that the element is not accumulated to any significant degree. This may be explained on the basis of the poor absorption and the relatively efficient excretion mechanisms for *trivalent* chromium. Regardless of dietary history, only a small percentage, physiological or excessive, is absorbed into the organism. Of a physiological dose of chromium chloride injected intravenously into rats, approximately 90% is bound to transferrin, the iron-binding protein in blood. From this transport form, it is taken up by the tissues at different rates. Testes, kidney, spleen and bone accumulate chromium over the blood concentration, whereas heart and lung take it up at only a fraction of the blood concentration (Kraintz and Talmage, 1952; Hopkins, 1965). The excretion mechanism, as measured by determining total body activity upon intravenous injection of radioactive chromium, consists of at least 3 components with respective half-lives of 0.5, 5.7 and 78.5 days. Excretion rates are not influenced by dilution with additional chromium injections or by variations of dose given (Mertz *et al.*, 1965). Excretion is indirectly proportional to doses given within a wide range, and only a moderate increase in tissue levels results from increased intake. Perhaps because of these mechanisms of absorption, transport and excretion, *trivalent* chromium is considered to be relatively nontoxic. Long-term administration causes no adverse symptoms in a variety of species, and in rats a ratio of therapeutic to acute toxic doses of approximately 1:10,000 has been established. These observations sharply distinguish the trivalent from the *hexavalent* form. The latter, a strong oxidizing agent, is known for its toxicity. An increased incidence of lung cancer has been reported for people exposed to chromate dust, and local irritation reactions of skin and mucosae caused by chromate are well known.

Chromium levels in biological matter have been extensively studied.[2] In contrast to findings with other metals, chromium concentrations in the U.S. population are highest at the time of birth, with a pronounced decline during the lifetime, whereas they appear to remain high in some other countries (Thailand and the Phillipines). These findings suggest the possibility of a relative chromium deficiency in the United States. The relation of this to disturbances of carbohydrate metabolism in man, while under investigation, remains yet to be established.

References

1. UDY, M. J., "Chromium," ACS Monograph No. 132, Vol. 1, New York, Reinhold Publishing Co., 1956.
2. SCHROEDER, H. A., BALASSA, J. J., AND TIPTON, I. H., "Abnormal Trace Elements in Man: Chromium," *J. Chronic Disease*, **15**, 941 (1962).
3. SCHROEDER, H. A., VINTON, W. H., JR., AND BALASSA, J. J., "Effect of Chromium, Cadmium and Lead on the Growth and Survival of Rats," *J. Nutr.*, **80**, 48 (1963).
4. SCHWARZ, K. AND MERTZ, W., "Chromium (III) and the Glucose Tolerance Factor," *Arch. Biochem. Biophys.*, **85**, 292 (1959).
5. CHRISTIAN, G. D., KNOBLOCK, E. C., PURDY, W. C., AND MERTZ, W., "A Polarographic Study of Chromium-Insulin-Mitochondrial Interaction," *Biochim. Biophys. Acta*, **66**, 420 (1963).
6. FARKAS, T. G., AND ROBERSON, S. L., "The Effect of Chromium^{+++} on the Glucose Utilization of Isolated Lenses," *Exptl. Eye Research*, **4**, 124 (1965).

WALTER MERTZ

CHROMOSOME PUFFS

This term refers to ring-like enlargements of chromosomes (also called BALBIANI RINGS) visible, with the aid of suitable staining (see HISTOCHEMICAL METHODS) in the light microscope. Chromosome puffs are believed to be regions of active RNA synthesis [see RIBONUCLEIC ACIDS (BIOSYNTHESIS)], whereby portions of the genetic message stored as a linear sequence of the four bases (adenine, guanine, cytosine, and thymine) in DNA [see DEOXYRIBONUCLEIC ACIDS (DISTRIBUTION AND STRUCTURE)] are transcribed into a complementary linear sequence of the bases (uracil, cytosine, guanine, and adenine) in MESSENGER RNA molecules, which then migrate into the cytoplasm and control the amino acid sequence in PROTEIN BIOSYNTHESIS. The appearance of chromosome puffs at definite chromosomal locations has been correlated with certain events along the time course of DIFFERENTIATION AND MORPHOGENESIS, particularly in studies of insect metamorphosis.

Reference

1. BEERMAN, W., AND CLEVER, U., "Chromosome Puffs," *Sci. Am.*, **210**, 50 (April 1964).

CHROMOSOMES (BIOCHEMICAL ASPECTS)

Originally, "chromosome" was a term used by microscopists to describe the type of cellular organelles which could be observed microscopically during CELL DIVISON processes and which had strong affinity toward basic dyes. As genetic studies advanced, it was recognized that these cellular components represented the vehicles which carried the hereditary determinants (see GENES). During recent years, the term chromosomes has been used to describe the "gene carrier" of any cell, whether it is microscopically visible or not. Thus the term has lost its morphological connotation, because the chromosomes of most microorganisms, such as bacteria and viruses, are not always detectable by standard microscopic techniques. The chemical composition of these two types of chromosomes is also quite different. Unless specifically stated, the term chromosome henceforth refers to the classic definition, *viz.*, the chromosome of higher organisms.

Chemical Composition. Studies on the chemistry of chromosomes have not advanced very rapidly because of several factors. (1) A chromosome is not composed of a single type of compound, but is a composite of several species of complex macromolecules, whose interrelationships have not been well established. (2) No satisfactory method is available to isolate chromosomes in quantities to facilitate chemical analysis. (3) Chromosomes can be observed microscopically only when a cell enters division stages (mitosis or meiosis), which represent a small fraction of the life span of a cell. During the rest of time, the cell is in interphase when the chromosomes are invisible. The chromosomes are "decondensed" and are enclosed in a nuclear envelope. Isolation of chemical constituents can therefore be done only from isolated nuclei. It is not known that the deoxyribonucleoprotein so isolated equates to the chromosomes.

A certain amount of information on the chemical composition of chromosomes has been obtained by cytochemical analysis. The Feulgen reaction has sufficiently demonstrated that DEOXYRIBONUCLEIC ACID (DNA) is the major component of chromosomes. Alkaline fast green staining showed that there is also a basic PROTEIN component (histone) in the chromosomes. In some viruses, the genetic determinants are in the form of RIBONUCLEIC ACID (RNA) rather than DNA. However, DNA is the hereditary material of the overwhelming majority of life forms, including numerous viruses and bacteria, protozoa, and all higher plants and animals.

In bacteria and viruses, each "organism" possesses one chromosome, and each chromosome is a single, circular molecule. There is no evidence of proteins closely associated with DNA, as in the case of NUCLEOPROTEINS of higher life forms. A possibility exists, however, that amino acids or small peptides may be present to interrupt the continuity of the long DNA molecule. Bendich and collaborators postulate that amino acids may serve as punctuation points for the genetic messages.[1] In the chromosomes of higher forms, more than one DNA molecule per chromosome is probable.

Replication and RNA Synthesis. DNA molecules have two functions: replicating themselves and synthesizing RNA. Ample evidence has been accumulated to show that DNA replicates itself in a semiconservative manner. That is to say, each of the complementary strands of the double-helical structure synthesizes its complementary strand, resulting in two double helices each containing one old strand and one new strand. Apparently the chromosomes replicate the same way. Using tritiated thymidine as DNA precursor to study the mode of chromosome replication, Taylor and collaborators,[5] from the distribution of the radioactivity of the chromosomes in the ensuing mitoses, concluded that the replication of chromosomes follows the Watson-Crick scheme [see DEOXYRIBONUCLEIC ACIDS (REPLICATION); RIBONUCLEIC ACIDS (BIOSYNTHESIS)].

The bacterial chromosome[6] replicates itself from a predetermined point and proceeds around the circular chromosome with an apparent constant rate until the entire ring is duplicated. At that time, the two rings separate and each may begin another generation of replication (see BACTERIAL GENETICS). In higher plants and animals, the chromosomal replication may begin at multiple sites and is asynchronous among different chromosomes of the same cell. The asynchrony is not lost even after an artificial arrest of the DNA synthesis by analogues or inhibitors (*e.g.*, aminopterin, 2′-deoxy-5-fluorouridine). Under the influence of such agents, the DNA synthesis process ceases. When thymidine is later introduced to the arrested cells, DNA replication resumes at the point where it was stopped. The asynchronous pattern is not altered. This asynchrony of chromosome replication has much genetic significance.

As important as self-replication, the second function of DNA is to serve as template for RNA synthesis. The biosynthesized RNA molecule copies the base sequence of the DNA, so that it transcribes the information carried in the original genetic code. The function of RNA is to carry the genetic message from DNA to the cytoplasm to conduct PROTEIN BIOSYNTHESIS. Chromosomes synthesize RNA only when they are not condensed, *i.e.*, when they are in the interphase stage.

Control of Genetic Expression. All cells of an organism find their lineage from a single fertilized cell, the zygote. The original zygote thus contains all the genetic information required for the development and maintenance of this organism. A great deal of genetic information for a specific function necessary for one type of cell will be useless in other cells. In other words, not all the DNA molecules should actively synthesize RNA at one time. In fact, a specialized cell should have many inactivated DNAs whose information belongs to other cell types. Otherwise development cannot proceed in an orderly manner. Thus some mechanisms must be operative to control the RNA synthetic activity of the chromosomes.

In the chromosomes of higher forms, a considerable quantity of protein is present. One group of proteins is basic, known as the HISTONES. Histones are basic because of their high content of arginine and lysine. Stedman and Stedman[4] suggested that histones may be the agents which regulate the activity of genes. More recently, Huang and Bonner[3] found that histones inhibit the *in vitro* synthesis of DNA-dependent RNA. More recent data showed that various fractions of histones differ in their affinity to DNA.

Although the question of the role of histones is far from settled, it is attractive to hypothesize that histones are involved in the regulation of RNA synthesis. Presumably, histones bind with or detach from DNA molecules, thus competing for binding sites with RNA polymerase. Since histones can be extracted from the chromosomes by acid (HCl, for example) and the chromosomes do not lose their morphological characteristics, it is probably reasonable to infer that histones do not contribute to the "backbone" structure of the chromosomes. On the other hand, digestion with PROTEOLYTIC ENZYMES disintegrates the chromosomes. It is conceivable, therefore, that proteins other than histones contribute to the structural integrity of the chromosomes, while the histones associate with DNA as a secondary (disposable) structural component. Dounce and Hilgartner[2] suggested that the so-called residual protein molecules may situate perpendicularly to the axis of the DNA molecules, and there may be several such protein molecules linked by —S—S— bonds.

References

1. BENDICH, A., BORENFREUND, E., KORNGOLD, G. C., KRIM, M., AND BALIS, M. E., *in* "Acidi nucleici e loro funzione biologica," p. 214, Istituto Lombardo, Fondazione Baselli, 1964.
2. DOUNCE, A. L., AND HILGARTNER, C. A., *Exptl. Cell Res.*, **36**, 228 (1964).
3. HUANG, J. C., AND BONNER, J., *Proc. Natl. Acad. Sci. U.S.*, **48**, 1216 (1962).
4. STEDMAN, E., AND STEDMAN, E., *Nature*, **166**, 780 (1950).
5. TAYLOR, J. H., WOODS, P. S., AND HUGHES, W. L., *Proc. Natl. Acad. Sci. U.S.*, **43**, 122 (1957).
6. CAIRNS, J., "The Bacterial Chromosome," *Sci. Am.*, **214**, 36 (January 1966).

T. C. HSU

CHYMOTRYPSINS. See PROTEOLYTIC ENZYMES.

CITRIC ACID CYCLE

In most living organisms the citric acid cycle constitutes the final common pathway in the degradation of foodstuffs and cell constituents to carbon dioxide and water. An outline of the cycle is shown in Fig. 1. It can be imagined that the process starts when the two carbon atoms of acetic acid, in the form of acetyl coenzyme A, are condensed with oxaloacetate to form citrate. Each complete turn of the cycle results in the conversion of one molecule of acetic acid to carbon dioxide and water according to the balance shown in reactions (*1*)–(*3*).

$$CH_3\cdot COOH + 2H_2O \rightarrow 2CO_2 + 8[H] \quad (1)$$

$$8[H] + 2O_2 \rightarrow 4H_2O \quad (2)$$

$$\text{Sum:} \quad CH_3\cdot COOH + 2O_2 \rightarrow 2CO_2 + 2H_2O \quad (3)$$

Carbon dioxide arises directly as a result of the reactions of the cycle. The hydrogen atoms, which are shown in Fig. 1 as 2H, are transferred to coenzymes. In this and subsequent processes, the hydrogen atoms become dissociated into protons and electrons. The electrons pass through a complex chain of enzymes, called the electron transport system or respiratory chain and are ultimately used to reduce oxygen. This process occurs according to the balance shown in reactions (*4*)–(*7*).

$$4[H] \rightarrow 4H^+ + 4e \text{ (electrons)} \quad (4)$$

$$O_2 + 4e \rightarrow 2O^{2-} \quad (5)$$

$$4H^+ + 2O^{2-} \rightarrow 2H_2O \quad (6)$$

$$\text{Sum:} \quad 4[H] + O_2 \rightarrow 2H_2O \quad (7)$$

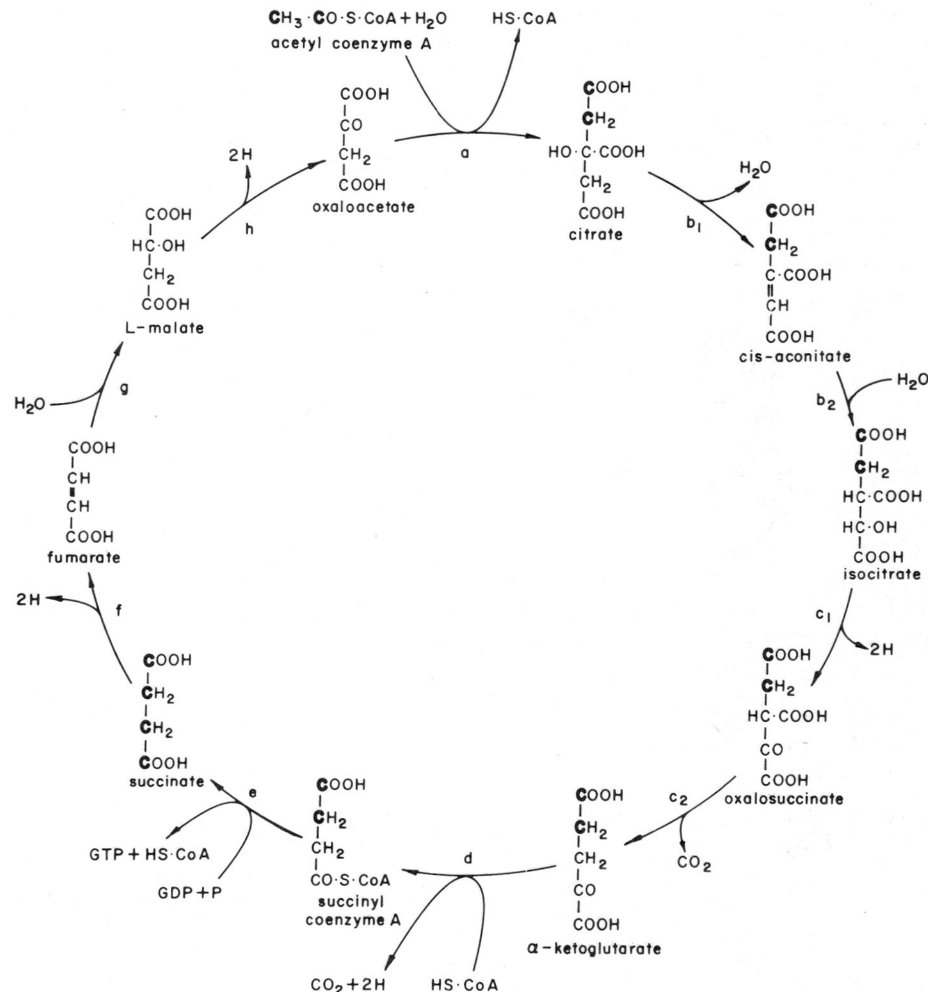

FIG. 1. Outline of the citric acid cycle. Operation of the cycle through one turn leads to the combustion of one acetic acid equivalent according to the balance shown in reaction (*3*). The two carbon atoms of acetate are fed into the cycle in the form of the acetyl group of acetyl coenzyme A. The fate of these carbon atoms is shown in heavy outline to the point where they reach succinate. The carbon atoms of an acetyl group fed into the cycle remain in the compounds of the cycle during its first turn. Under physiological conditions, the compounds that participate in the cycle exist as anions, and in the Figure they are named as such. However, for sake of clarity, the formulas shown are those of the free acids.

The enzymes catalyzing the reactions are: (a) condensing enzyme; (b_1) and (b_2) aconitase; (c_1) and (c_2) isocitrate dehydrogenase; (d) α-ketoglutarate dehydrogenase; (e) succinate thiokinase; (f) succinate dehydrogenase; (g) fumarase; (h) malate dehydrogenase.

Notes: (1) Citrate can also yield isocitrate without *cis*-aconitate occurring as free intermediate. (2) Oxalosuccinate may occur only as an enzyme-bound intermediate. (3) In brain α-ketoglutarate can also yield succinate via γ-aminobutyrate (see Fig. 2). (4) Succinyl coenzyme A can also yield succinate via reactions catalyzed by succinyl CoA deacylase [reaction (*14*)] or succinyl coenzyme A transferase [reaction (*15*)].

The function of the electron transport system is to convert the energy liberated in reaction (*7*) into chemical energy. This is achieved by condensation of orthophosphate and adenosine diphosphate (ADP) to adenosine triphosphate (ATP)

$$\text{Orthophosphate} + \text{ADP} \rightarrow \text{ATP} + H_2O \qquad (8)$$

The process is obligatorily linked to the oxidation of [H], and is referred to as oxidative PHOSPHORYLATION.

Degradation of Foodstuffs and Cell Constituents. Before a constituent of the cell or food enters the citric acid cycle, it is broken down to one of the compounds that lies on the cycle. Quantitatively the most important entry point is acetyl coenzyme A (see COENZYME A). Two-thirds of the carbon atoms of carbohydrate and all of the carbon atoms of fatty acids derived from fat are converted to acetyl coenzyme A before entering the cycle. About half of the carbon atoms of amino acids derived from proteins, as well as a number of other compounds, are also first converted to acetyl coenzyme A. However, some substances are degraded to one of the other members of the cycle. For example, aspartate gives rise to oxaloacetate, while glutamate, histidine, arginine, and proline give rise to α-ketoglutarate.

Synthesis of Cell Constituents via Reactions of the Cycle. Radioactive tracer experiments show that compounds of the citric acid cycle serve in a synthetic as well as a degradative capacity. Oxaloacetate serves as a precursor of hexoses such as glucose, of pentoses such as ribose, of pyrimidine bases of the nucleic acids, and of several amino acids. α-Ketoglutarate gives rise to several amino acids. Succinyl coenzyme A is a precursor of the porphyrins. Citrate is a source of carbon for the synthesis of fatty acids.

Enzymes of the Tricarboxylic Acid Cycle. In cells of higher animals, all the enzymes of the cycle occur in the MITOCHONDRIA. In addition some of these enzymes also occur outside of the mitochondria, in the non-particulate cytoplasm.

Step (a), the formation of citrate, is catalyzed by *condensing enzyme.*

$$\text{Acetyl Coenzyme A} + \text{Oxaloacetate} + H_2O \rightarrow \text{Citrate} + \text{Coenzyme A} \qquad (9)$$

The enzyme, which has been crystallized, is located chiefly in the mitochondria.

Steps (b_1) and (b_2) result in the conversion of citrate to isocitrate. They are catalyzed by the enzyme *aconitase.*

$$\text{Citrate} \underset{H_2O}{\longleftrightarrow} \textit{cis}\text{-Aconitate} \underset{H_2O}{\longleftrightarrow} \text{Isocitrate} \qquad (10)$$

The process is shown to proceed via *cis*-aconitate [reaction (*10*)]. Although the enzyme catalyzes the conversion of *cis*-aconitate to citrate and isocitrate, the reaction from citrate to isocitrate can occur without the formation of free *cis*-aconitate. The most remarkable feature of this enzyme reaction is that citrate does not react as a perfectly symmetrical compound. In the formulas below, the carbon atoms derived from the acetyl group of acetyl coenzyme A are shown in boldface. Experiments with isotopic tracers show that the enzyme catalyzes reaction x but not reaction y:

$$\begin{array}{ccccc}
\mathbf{C}OOH & & \mathbf{C}OOH & & \mathbf{C}OOH \\
HO\cdot\mathbf{C}\cdot H & & \mathbf{C}H_2 & & \mathbf{C}H_2 \\
H\cdot C\cdot COOH & \underset{y}{\longleftrightarrow} & HO\cdot C\cdot COOH & \underset{x}{\longleftrightarrow} & H\cdot C\cdot COOH \\
CH_2 & & CH_2 & & HO\cdot C\cdot H \\
COOH & & COOH & & COOH \\
\text{Isocitrate} & & \text{Citrate} & & \text{Isocitrate}
\end{array}$$

In other words, a symmetrical molecule, citrate, can behave asymmetrically when combined with an asymmetrical enzyme. This behavior is encountered only with compounds that do not contain elements of symmetry higher than a single plane of symmetry.

Step (c), the oxidation and decarboxylation of isocitrate to yield α-ketoglutarate is catalyzed by the *isocitrate dehydrogenases* [reaction (*11*)].

$$\underset{\substack{\text{(DPN)}\\ \text{Oxidized}\\ \text{coenzyme}}}{\text{Isocitrate} + \text{TPN}} \rightarrow \alpha\text{-Ketoglutarate} + CO_2 + \underset{\substack{(DPNH_2)\\ \text{Reduced}\\ \text{coenzyme}}}{TPNH_2} \qquad (11)$$

Cells of higher animals contain either two or three separate enzymes which carry out this reaction. The enzymes require magnesium or manganese ions for full activity. Mitochondria contain both a TPN- and a DPN-specific enzyme. Diphosphopyridine nucleotide (DPN) and triphosphopyridine nucleotide (TPN) are the principal hydrogen-carrying coenzymes of the cell. (DPN is also termed NAD, NICOTINAMIDE ADENINE DINUCLEOTIDE; and TPN is also termed NADP.) In addition, the nonparticulate cytoplasm of the cell contains a TPN-specific isocitrate dehydrogenase which is different from that found in the mitochondria. The DPN-specific enzyme from higher animals has the property of requiring catalytic quantities of ADP to activate the enzyme. In molds, the activator of the enzyme is AMP (adenosine monophosphate). It follows that in the living cell the levels of ADP or AMP determine to what extent the isocitrate dehydrogenase reaction results in the reduction of DPN or TPN.

Step (d), the oxidation and decarboxylation of α-ketoglutarate, is catalyzed by a complex set of reactions that involves four coenzymes, namely THIAMINE pyrophosphate, LIPOIC ACID, coenzyme A, and diphosphopyridine nucleotide. The complete reaction also requires at least three enzymes and magnesium ions:

$$\alpha\text{-Ketoglutarate} + \text{Coenzyme A} + \text{DPN} \rightarrow \text{Succinyl Coenzyme A} + DPNH_2 + CO_2 \qquad (12)$$

In cells of higher animals this enzyme system has been found only in the mitochondria (see MULTI-ENZYME COMPLEXES).

Step (e) results in the liberation of succinate from succinyl coenzyme A. Three alternative enzyme reactions exist for this purpose [reactions (*13*)–(*15*).]

Succinyl Coenzyme A + GDP + Orthophosphate
→ Succinate + Coenzyme A + GTP (*13*)

Succinyl Coenzyme A + H_2O
→ Succinate + Coenzyme A (*14*)

Succinyl Coenzyme A + Acetoacetate
→ Succinate + Acetoacetyl Coenzyme A (*15*)

The enzymes that catalyze reactions (*13*) and (*14*) occur in liver. Reaction (*15*) is catalyzed by a very active enzyme found in heart and skeletal muscle. It appears to be the only reaction whereby free acetoacetate formed in the liver can be activated for oxidation. This process is of importance in the regulation of the amounts of "ketone bodies" (acetoacetate, β-hydroxybutyrate and acetone) that circulate in the blood. Excessive amounts of these substances occur in starvation and diabetic ketosis; they are responsible in part for the toxic condition, ketoacidosis (see also FATTY ACID METABOLISM).

Step (f) is the dehydrogenation of succinate to fumarate. It occurs by the transfer of two electrons from succinate to ferricytochrome b, a member of the electron transport system. The enzyme that catalyzes the reaction, succinate dehydrogenase,

Succinate + 2 Cytochrome b (Fe^{3+})
→ Fumarate + 2 Cytochrome b (Fe^{2+}) + 2 H^+ (*16*)

contains a FLAVIN coenzyme as well as iron. In higher animals, the enzyme occurs in the mitochondria.

Step (g) is catalyzed by fumarase:

Fumarate + H_2O ↔ L-Malate (*17*)

The mammalian enzyme, which occurs both inside and outside of the mitochondria, has been crystallized. The enzyme is stereospecific for the addition of both OH and H of water to the double bond of fumarate.

Step (h), the formation of oxaloacetate, is catalyzed by malate dehydrogenases.

Malate + DPN ↔ Oxaloacetate + $DPNH_2$ (*18*)

Two separate enzymes occur in cells of higher animals, one being extramitochondrial, the other intramitochondrial.

The γ-Aminobutyrate Bypass. The reactions between α-ketoglutarate and succinate that are shown in Fig. 1 can be bypassed by the alternative sequence of reactions shown in Fig. 2. Tracer experiments show that this bypass is very active in brain, but its physiological significance is not clear.

The Glyoxylate Shunt. In some microorganisms as well as in certain plant seedlings, the reactions between isocitrate and malate that are shown in Fig. 1 are supplemented or replaced by the alternative sequence shown in Fig. 3. Acetyl coenzyme A now enters the cycle at the malate synthetase reaction as well as at the condensing enzyme reaction. One turn of the cycle operating via the glyoxylate shunt results in the formation of one molecule of succinate from two molecules of acetyl coenzyme A. Possession of this bypass enables certain microorganisms and plant seedlings to achieve a net conversion of acetyl groups to carbohydrate, a capacity lacked by higher animals.

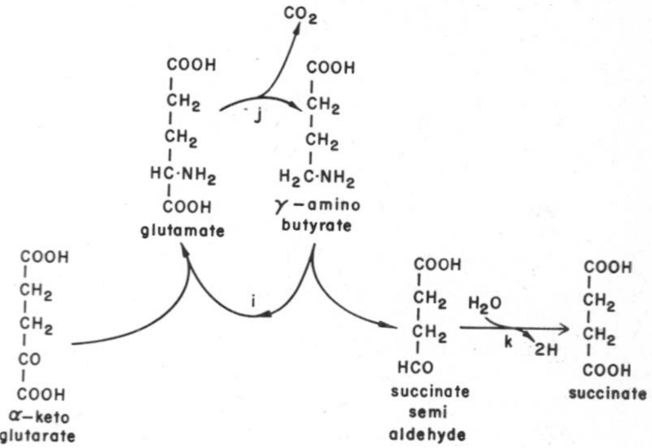

FIG. 2. The γ-aminobutyrate bypass. The reactions between α-ketoglutarate and succinate in Fig. 1 are bypassed by the alternative reactions shown. The enzymes catalyzing the reactions are: (i) α-ketoglutarate γ-aminobutyrate transaminase; (j) glutamate decarboxylase; (k) succinate semialdehyde dehydrogenase.

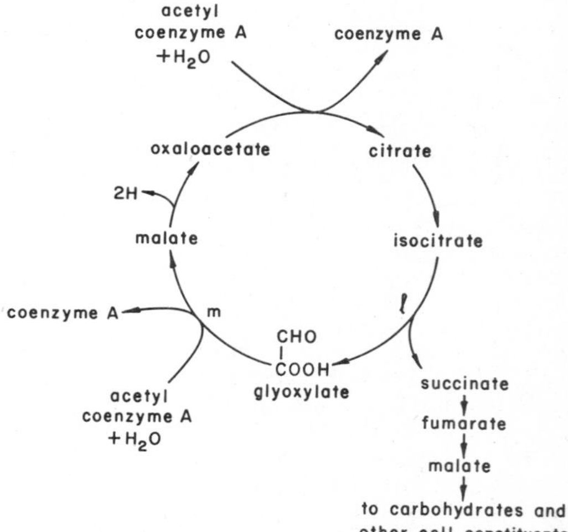

FIG. 3. The glyoxylate shunt. The reactions between isocitrate and malate in Fig. 1 are bypassed by the alternative reactions shown. Operation of this cycle through one turn leads to the condensation of two acetyl groups, supplied in the form of acetyl coenzyme A, to yield one molecule of succinate plus 2[H]. This cycle is found in certain microorganisms and seedlings. The enzymes catalyzing the reactions are: (*l*) isocitratase; (m) malate synthetase.

References

1. KREBS, H. A., AND JOHNSON, W. A., *Enzymologia*, **4**, 148 (1937).
2. KORNBERG, H. L., AND KREBS, H. A., *Nature*, **179**, 988 (1957).
3. KORNBERG, H. L., *Ann. Rev. Microbiol.* **13**, 49 (1959).
4. KREBS, H. A., AND LOWENSTEIN, J. M., "Metabolic Pathways" (GREENBERG, D. M., EDITOR), Vol. 1, p. 129, New York, Academic Press, 1960.
5. Recent references to specific topics dealt with in this article can be found by consulting the index of *Annual Reviews of Biochemistry*.

J. M. LOWENSTEIN

CITRULLINE

Citrulline, $H_2N—C(=O)—NH—(CH_2)_3CH(NH_2)COOH$, is a biologically important amino acid which, however, does not appear to be a constituent of common proteins. It was first isolated from watermelon juice and got its name from citron, a name sometimes applied to melons. It figures prominently in the urea cycle and is closely related structurally to ARGININE (see ARGININE-UREA CYCLE).

COBALAMINS AND COBAMIDE COENZYMES

The tedious search for anti-pernicious anemia factors led to the isolation of both FOLIC ACID and vitamin B_{12}. The latter vitamin was isolated independently in two laboratories in the late 1940's[2,3], although several additional years passed before its exact structure was elucidated (Fig. 1). Although the vitamin was indirectly implicated in a variety of biochemical reactions, it was not until 1958 that a clear-cut coenzymatic function for a vitamin B_{12} derivative was obtained. It was observed that cell-free extracts of *Clostridium tetanomorphum* decomposed glutamate to acetate and pyruvate. The first step in this reaction, the isomerization of glutamate to β-methylaspartic acid, was shown to require a charcoal adsorbable cofactor. This material was isolated and demonstrated to be a derivative of vitamin B_{12} as shown in Fig. 1. The coenzyme form of the vitamin contains a 5′-deoxyadenosyl moiety in place of cyanide which is found in the vitamin. Many types of vitamin derivatives are present in nature (varying in the nucleotide portion of the molecule), and similarly comparable forms of the B_{12} coenzymes have also been isolated. The unique feature of the coenzyme structure is the presence of the 5′-deoxyadenosyl group linked to the cobalt by a carbon-to-cobalt bond. It appears that the chemical and biochemical reactions of the cobamide coenzymes involve this ligand. The carbon-to-cobalt linkage is sensitive to visible light, cyanide and mild acid, and very likely is the active site in the enzymatic reactions requiring this coenzyme. [See also COBALT (IN BIOLOGICAL SYSTEMS)].

The biological synthesis of the carbon-to-cob bond has been studied in bacterial systems. The reaction as shown in extracts of *Cl. tetanomorphum* can be depicted as follows:

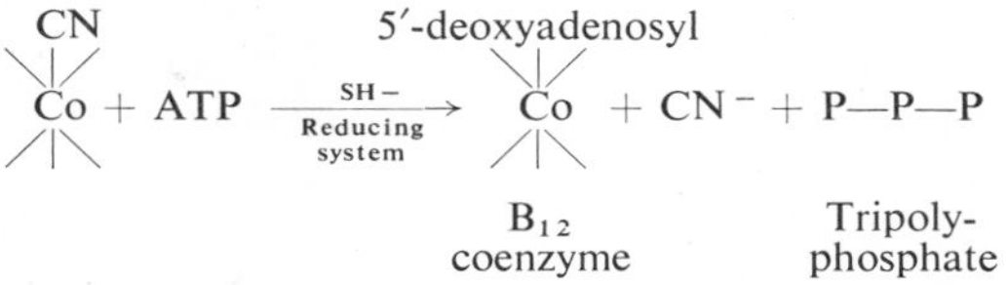

$$\text{CN—Co} + \text{ATP} \xrightarrow[\text{Reducing system}]{\text{SH}-} \underset{\text{B}_{12}\ \text{coenzyme}}{\text{5'-deoxyadenosyl—Co}} + \text{CN}^- + \underset{\text{Tripolyphosphate}}{\text{P—P—P}}$$

The enzyme catalyzes a transfer of a 5′-deoxyadenosyl group from ATP to the vitamin in the presence of a reducing system. Inorganic tripolyphosphate is liberated from the ATP in this system.

The chemical synthesis of the B_{12} coenzyme and related derivatives possessing a carbon-to-cobalt bond has been reported and involves the reduction of hydroxy-B_{12} (Co^{+3}) to hydrido-B_{12} (Co^{+1}), followed by a reaction with an appropriate alkyl halide. Thus the synthesis of methyl-B_{12} is shown below:

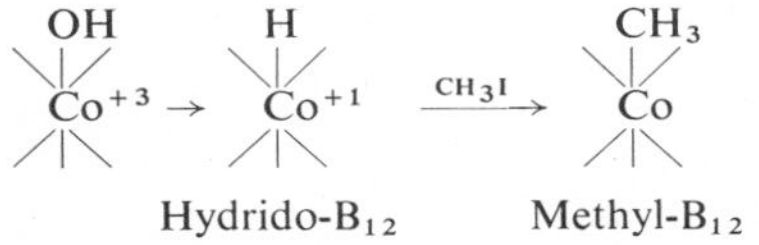

$$\text{OH—Co}^{+3} \rightarrow \underset{\text{Hydrido-B}_{12}}{\text{H—Co}^{+1}} \xrightarrow{\text{CH}_3\text{I}} \underset{\text{Methyl-B}_{12}}{\text{CH}_3\text{—Co}}$$

In an analogous manner, a whole series of alkyl cobamides has been chemically synthesized including deoxyadenosyl-B_{12} (B_{12} coenzyme). Thus both the chemical and the enzymatic syntheses of the carbon-to-cobalt bond require a reducing system.

Metabolic Functions of Cobamide Coenzyme. The cobamide coenzyme is required for four biochemical reactions listed below:

(*1*) *Glutamate isomerase*:

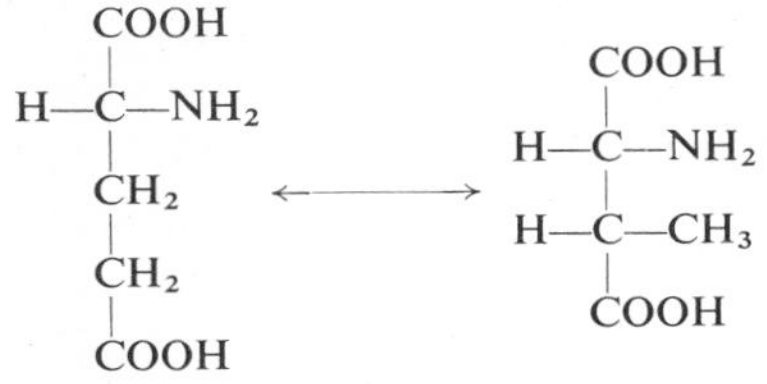

$$\text{HOOC—CH(NH}_2\text{)—CH}_2\text{—CH}_2\text{—COOH} \longleftrightarrow \text{HOOC—CH(NH}_2\text{)—CH(CH}_3\text{)—COOH}$$

Studies with purified enzyme preparations indicate that this carbon chain rearrangement does not involve such free intermediates as glycine, propionate or α-ketoglutarate. Carbon 3 of glutamate becomes the branched methyl of β-methylaspartate. The fermentation of β-methylaspartate in *Cl. tetanomorphum* eventually yields acetate and pyruvate.

(2) *Methylmalonyl CoA isomerase*:

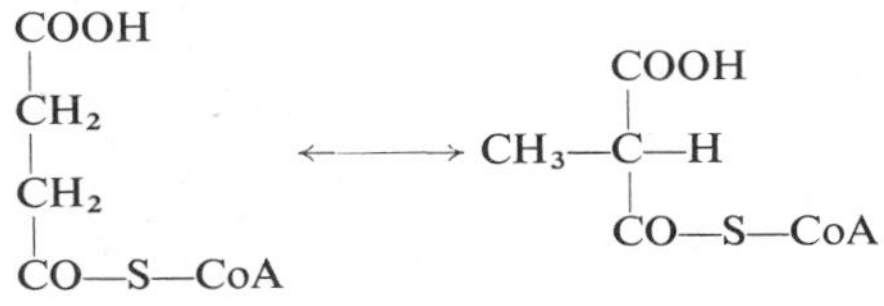

$$\text{HOOC—CH}_2\text{—CH}_2\text{—CO—S—CoA} \longleftrightarrow \text{CH}_3\text{—CH(COOH)—CO—S—CoA}$$

This isomerization, similar to the glutamate isomerase reaction, has been shown to involve a transfer of the thiol ester carboxyl group, not the free carboxyl group. The reaction is important in the metabolism of propionate in microorganisms and animals.

$$\text{Propionyl CoA} \xrightarrow{CO_2} \text{Methylmalonyl CoA} \xrightarrow{B_{12}} \text{Succinyl CoA}$$

(3) *Glycol dehydrase reaction*:

$$\begin{array}{c} H \\ | \\ H-C-OH \\ | \\ H-C-OH \\ | \\ CH_3 \end{array} \longrightarrow \begin{array}{c} HC{=}O \\ | \\ CH_2 \\ | \\ CH_3 \end{array}$$

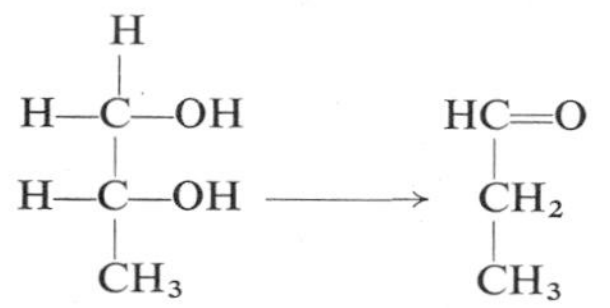

This reaction has been studied in extracts of *Aerobacter aerogenes* and a similar reaction, glycerol → β-hydroxypropionaldehyde, requiring a cobamide coenzyme has been reported in a species of a *Lactobacillus*.

(4) *Reduction of ribonucleotides*:

Cytidine Diphosphate → Deoxycytidine Diphosphate

This reaction, analogous to the glycol dehydrase reaction, has been shown to require a cobamide coenzyme in extracts of *Lactobacillus leichmannii*[4]. The cobamide requirement at the enzymatic level explains the known requirement that this organism has for either vitamin B_{12} or deoxynucleosides.

The above four reactions can be pictured as involving a hydrogen transfer as a common mechanism. Although the role of the cobamide coenzyme has not yet been elucidated, it could function as a hydrogen carrier in the above reactions.

Transfer of Methyl Groups. Besides the reactions which require the coenzyme form of the vitamin, a cobamide requirement (not the coenzyme) is also seen in the terminal reaction in methionine biosynthesis. This reaction involves a transfer of the methyl group from N^5-methyl-folate-H_4 to homocysteine. A cobamide-containing enzyme, S-adenosylmethionine (AMe) and a reducing system, are required.

$$N^5\text{-methyl-folate-}H_4 + \text{Homocysteine} \xrightarrow[B_{12}\text{ protein}]{\text{AMe, Reducing system}} \text{Methionine} + \text{Folate-}H_4$$

Although a satisfactory mechanism for the action of the coenzyme in the isomerase and glycol reactions is not yet known, there is more definitive data on the mode of action of the cobamide in methionine synthesis. It appears that a reduced derivative of the vitamin (enzyme-bound) accepts a methyl group from the folate derivative yielding a transient methyl-B_{12} enzyme. This reaction is similar to the chemical alkylation of a reduced cobamide which is the basis for the chemical synthesis of the carbon-to-cobalt bond. Subsequent methyl transfer to homocysteine to form methionine regenerates the reduced cobamide on the enzyme.

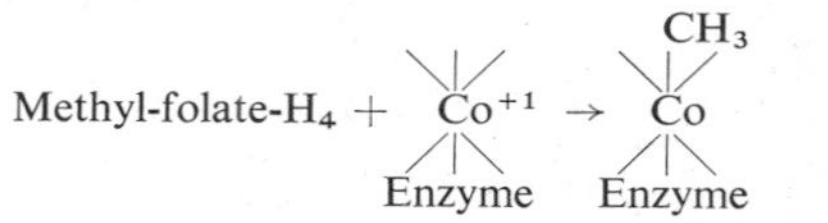

$$\text{Methyl-folate-}H_4 + \underset{\text{Enzyme}}{Co^{+1}} \rightarrow \underset{\text{Enzyme}}{\overset{CH_3}{Co}} \qquad (1)$$

$$\underset{\text{Enzyme}}{\overset{CH_3}{Co}} + \begin{array}{c} COOH \\ | \\ -C-NH_2 \\ | \\ -C- \\ | \\ -C-SH \\ | \end{array} \rightarrow \underset{\text{Enzyme}}{Co^{+1}} + \begin{array}{c} COOH \\ | \\ -C-NH_2 \\ | \\ -C- \\ | \\ -C-S-CH_3 \\ | \end{array} \qquad (2)$$

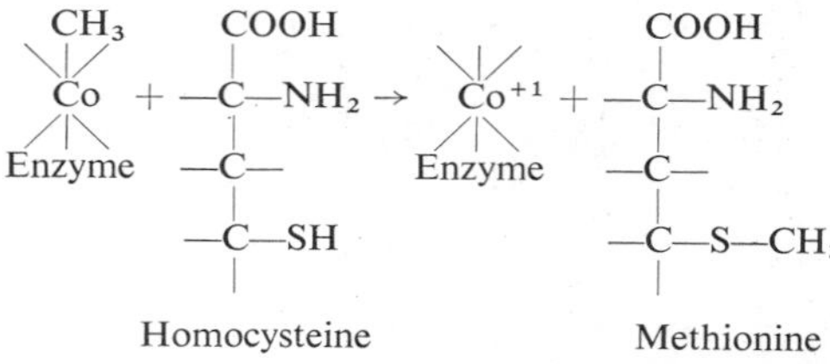

Homocysteine Methionine

The role of AMe is unknown. It functions catalytically in the reaction, possibly by methylating the enzyme or altering the protein structure. The reducing system is necessary to maintain the enzyme-bound cobamide in the reduced state. The ability of the enzyme to catalyze methyl transfer from methyl-B_{12} to homocysteine led to the initial postulate that enzyme-bound methyl-B_{12} is an intermediate in this reaction. (See also METHYLATIONS IN METABOLISM; SINGLE CARBON UNIT METABOLISM).

Of interest are the findings that methyl-B_{12} functions as a methyl donor in two other bacterial systems. The methyl group of methyl-B_{12} is enzymatically converted to methane by extracts of *Methanosarcina barkeri*, and to the methyl carbon of acetate in a system from *Clostridium thermoaceticum*[1b]. Whether a cobamide functions as a coenzyme in these reactions is not known although the ability to obtain methyl transfer from methyl-B_{12} is indicative of a cobamide involvement.

The requirement of a cobamide prosthetic group for methionine biosynthesis from N^5-methyl-folate-H_4 accounts, at the enzyme level, for the known nutritional interrelationship among vitamin B_{12}, folic acid and methionine. It should be noted that ample evidence is available, at the enzyme level, to show that the formation of methionine in animal tissues involves a cobamide enzyme. In most animals, a vitamin deficiency is obtained only when limiting amounts of methionine are present in the diet [see also VITAMIN B_{12} (ABSORPTION)]. The ability of folic acid to reverse some of the hematological aspects of pernicious anemia and the increased excretion of forminino-glutamic acid during vitamin-B_{12} deficiency indicate that an alteration in the folate partition occurs as a result of vitamin B_{12} deficiency. This could result from an inability to transfer the methyl group from N^5-methyl-folate-H_4 to homocysteine. Under such a situation, failure of the methyl transfer to homocysteine would result in an accumulation of N^5-methyl-folate-H_4 with a corresponding decrease in the rate of regeneration of folate-H_4. The need for folate-H_4 in purine and thymidylate synthesis and as an acceptor for the formyl group from formimino-glutamic acid could account for some of the

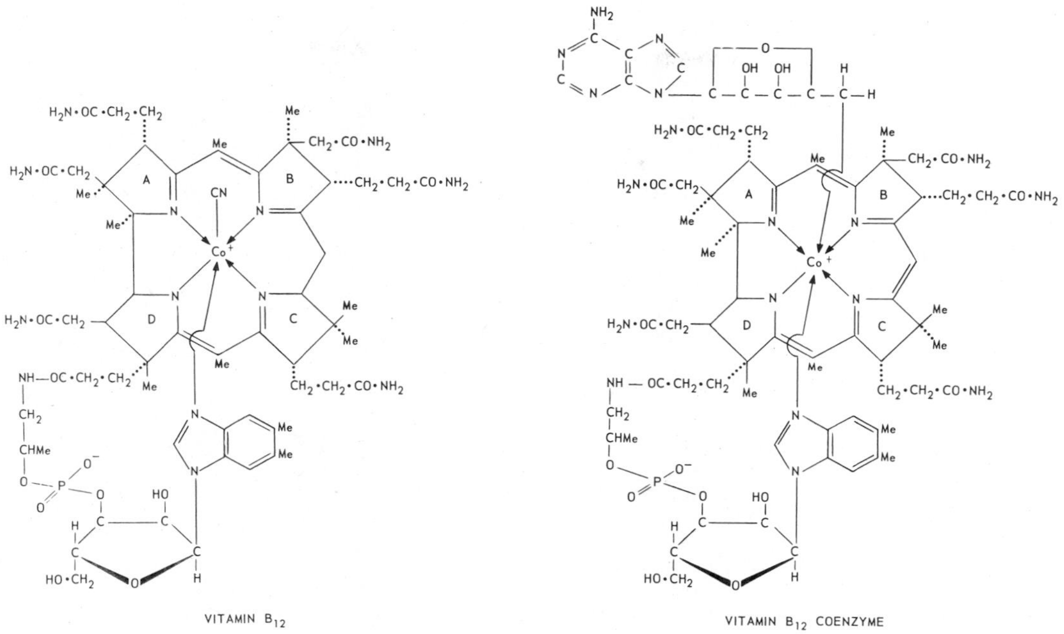

Fig. 1. Structure of vitamin B_{12} and vitamin B_{12} coenzyme.

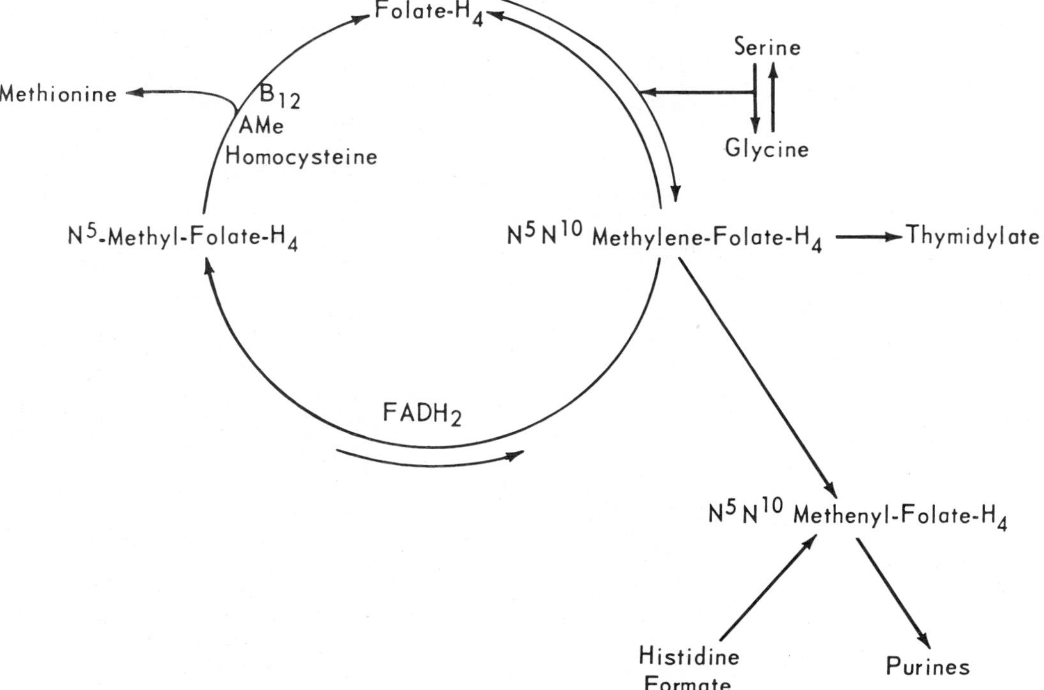

Fig. 2. Interrelationship among folic acid, vitamin B_{12} and purine and pyrimidine metabolism (see also PURINE BIOSYNTHESIS; PYRIMIDINE BIOSYNTHESIS).

observed effects of vitamin B_{12} deficiency (Fig. 2).

Abnormal nucleic acid metabolism in pernicious anemia may, therefore, result partly from altered folic acid metabolism, or as a direct consequence of diminished deoxyribose synthesis. Vitamin B_{12} has not yet been implicated in deoxyribose synthesis in animal tissue, although its involvement in this reaction in *L. leichmannii*[4] strongly suggests that a similar requirement will be seen elsewhere. At present, it appears that the participation of vitamin B_{12} in methionine synthesis and deoxyribose synthesis may account for the major physiological and biochemical alterations seen in vitamin B_{12} deficiency.

References

1. Several excellent reviews and symposia reports are available on this topic and cover most of the biochemical data presented. These include:
   (a) HEINRICH, H. C., (EDITOR), Second European Symposium on Vitamin B_{12} and Intrinsic Factor, Enke Stuttgart, Germany, 1962.
   (b) "Vitamin B_{12} Coenzymes," *Ann. N.Y. Acad. Sci.*, **112**, (1964).
2. RICKES, E. L., BRINK, N. G. KONIUSZY, F. R., WOOD, T. R., AND FOLKERS, K., *Science*, **107**, 396 (1948).
3. SMITH, E. L., AND PARKER, L. F. J., *Biochem. J.*, **43**, VIII (1948).
4. BLAKLEY, R. L., AND BARKER, H. A., *Biochem. Biophys. Res. Commun.*, **16**, 39 (1964).

HERBERT WEISSBACH AND NATHAN BROT

COBALT (IN BIOLOGICAL SYSTEMS)

Traces of cobalt are widely distributed in higher plants, animals and microorganisms. The concentration in normal plant materials ranges from less than 0.01 to 4.6 ppm of dry matter. When excesses of the element (0.02–29 ppm) are available in nutrient media, plants or microorganisms accumulate much higher concentrations, and symptoms of toxicity may by exhibited. Analyses of normal rat tissues indicate cobalt contents ranging from 0.001–0.006 ppm of dry matter. If excessive amounts of the element are supplied by injection into animals or by addition to their feed, toxicity symptoms referred to as polycythemia result.

Cobalt occurs in biological materials in a form loosely bound to protein and also as a component of VITAMIN B_{12}. In the ionic form, cobalt is known to function as a cofactor for several enzymes including certain pyrophosphatases and arginase, but it is not specifically needed for these enzymes and may be replaced by certain other divalent cations. Cobalt apparently plays its specific and indispensable role as a metal atom firmly attached by chelation to the corrin ring of vitamin B_{12} derivatives. All organisms that synthesize vitamin B_{12} and utilize a derivative of it as an indispensable component of metabolic processes require cobalt for life (see COBALAMINS AND COBAMIDE COENZYMES).

Animals and certain types of microorganisms lack the capacity to synthesize vitamin B_{12} and thus must depend upon an exogenous source of the vitamin. These organisms cannot utilize inorganic cobalt directly but must be supplied with the element as a constituent of vitamin B_{12}. Many microorganisms possess the capacity to synthesize vitamin B_{12} compounds and, therefore, require elemental cobalt in their nutrient media for this process. Microorganisms in the rumen of ruminant animals (see RUMINANT NUTRITION) and in the digestive tract of some non-ruminants (see INTESTINAL FLORA) require cobalt for the synthesis of vitamin B_{12}. When animals respond to inorganic cobalt, therefore, it is an indirect effect of the provision of the element to microorganisms for synthesis of the vitamin in the digestive system.

There is no convincing evidence that higher plants *per se* synthesize vitamin B_{12} and utilize derivatives of it in their metabolic processes. Certain plant tissues have been reported to contain traces of vitamin B_{12}, but it appears that the quantity detected was supplied by microbial contamination. Direct demonstrations of cobalt requirements of organisms, therefore, have been limited to microorganisms. All living organisms that require exogenous vitamin B_{12} are dependent upon microorganisms for the incorporation of elemental cobalt into the complex organic molecule of vitamin B_{12}. Some of the vitamin B_{12} needed by humans may be supplied by microorganisms in the digestive tract, but most of it is derived from meat or other animal products.

The importance of cobalt in biological processes was emphasized by the discovery in the early 1930's that cattle and sheep grazing on certain pastures in Australia and New Zealand suffered from severe deficiency symptoms that could be alleviated by the addition of traces of cobalt to feed. After vitamin B_{12} was isolated in 1948 and shown to contain cobalt, the symptoms of deficiency in ruminant animals were shown to be alleviated by injection of the vitamin. Inorganic cobalt salts supplied to deficient animals by injection were less effective. It seems clear now that cobalt deficiency in ruminants is essentially a deficiency of vitamin B_{12}. Cobalt-deficient soils in certain parts of the world produce pasture plants that contain insufficient amounts of the element. The microorganisms in the rumen or digestive tract of animals that consume cobalt-deficient feed fail to synthesize sufficient vitamin B_{12} because the cobalt supply is limiting. Under these circumstances vitamin B_{12} deficiency in the animals becomes apparent.

Only minute traces of cobalt are required for microorganisms; thus direct demonstrations of essentiality have been difficult. In 1954 Holm-Hansen and co-workers utilized a purified nutrient medium and precautions to avoid contamination and showed that four species of blue-green algae responded to the addition of cobalt. A concentration of 0.4 μg of the element per liter resulted in optimum growth and significant increases in the growth and nitrogen content of cells were observed upon the addition of 0.002 μg of the element per liter of medium. Those algal species that fixed atmospheric nitrogen appeared to exhibit a greater response to cobalt than those supplied with combined forms of nitrogen.

The next advance in an understanding of cobalt requirements developed from Burton's and Lochhead's observation that the symbiotic nodule bacteria from leguminous plants synthesized large amounts of vitamin B_{12}. Levin and Funk reported that nodules from legumes contained unusually high concentrations of the vitamin. These reports and the lack of positive evidence that non-leguminous plants requred cobalt induced several laboratories to investigate the possibility that cobalt may be required by leguminous species when cultured under conditions where symbiotic NITROGEN FIXATION was obligatory.

By the use of highly purified water and mineral salts and special techniques to prevent external cobalt contamination, it now has been proved that symbiotically cultured soybeans, alfalfa and subterranean clover require traces of cobalt for growth and nitrogen fixation. Cobalt is specific in its effects, and a concentration of 0.1 μg of the element per liter of medium is adequate. Cobalt-deficient soybean plants exhibit severe symptoms characteristic of nitrogen deficiency and show decreased contents of vitamin B_{12} and leghomoglobin in nodules. When leguminous plants were cultured in the absence of *Rhizobium* and were supplied with adequate nitrogen in the form of nitrate salts, no definite cobalt requirement was demonstrated. In contrast, Lowe and Evans have shown that various *Rhizobium* species cultured in media containing either ammonium or nitrate salts require cobalt for normal growth. A concentration of 0.01 μg per liter of medium is adequate. From this evidence, one may conclude that the cobalt needed for growth of inoculated legumes is associated with the *Rhizobium* cells in nodules rather than the leguminous plants *per se*. Since this research, it has been established that cobalt is required for *Azotobacter vinelandii*, *Clostridium pasteurianum* and symbiotically grown *Alnus glutinosa*, *Alnus rubra* and *Azolla filliculoides*. In those cases where sufficient analyses have been conducted, cobalt deficiency was shown to cause a decreased content of vitamin B_{12} in microorganisms. It seems apparent that the major function of the element is in the form of a component of vitamin B_{12} coenzymes. Now that methods have been developed for the elimination of cobalt contamination, undoubtedly many other microorganisms and symbionts will be shown to require the element.

References

1. AHMED, S., AND EVANS, H. J., "The Essentiality of Cobalt for Soybean Plants Grown under Symbiotic Conditions," *Proc. Natl. Acad. Sci.*, **47**, 24–36 (1961).
2. BURTON, M. O., AND LOCHHEAD, A. G., "Production of Vitamin B_{12} by *Rhizobium* Species," *Can. J. Botany*, **30**, 521 (1952).
3. EVANS, HAROLD J., AND MARK KLIEWER, "Vitamin B_{12} Compounds in Relation to the Requirements of Cobalt for Higher Plants and Nitrogen Fixing Organisms," *Ann. N.Y. Acad. Sci.*, **112**, Art. 2, 735–755 (1964).
4. HOLM-HANSEN, O., GERLOFF, G. C., AND FOLKE, S., "Cobalt as an Essential Element for Blue-green Algae," *Physiol. Plantarum*, **7**, 665–675 (1954).
5. SMITH, LESTER E., "Cobalt," in "Mineral Metabolism Volume II," CONNAR, C. L., AND BRONNER, FELIX, EDITORS), New York, Academic Press, 1962.
6. UNDERWOOD, E. J., "Mineral Metabolism," *Ann. Rev. Biochem.* (LUCK, J. MURRAY, EDITOR), 1959.
7. YOUNG, R. S., "Cobalt in Biology and Biochemistry," *Sci. Progress*, **44**, 16–37 (1956).

HAROLD J. EVANS

COCARBOXYLASE

Cocarboxylase is an older term for thiamine pyrophosphate (TPP), the coenzyme form of thiamine (see VITAMIN B GROUP, section on Thiamine Coenzyme). The term is not currently used, because of possible confusion with the coenzymatic roles, in carboxylation or decarboxylation reactions, of BIOTIN and of Vitamin B_6 (see PYRIDOXINE FAMILY OF VITAMINS).

COENZYME A

Discovery of Coenzyme A: The "Active" Acetate. *In vivo* experiments with isotopically marked acetate had shown that this compound is incorporated into a large number of metabolites including lipids and amino acids. This led to the assumption that an activated acetate exists which is a metabolic intermediary of prime importance. Therefore, when acetyl phosphate was isolated as the oxidation product of pyruvate using a soluble oxidase from *L. delbrueckii*,[1] it was hopefully considered to be this active acetate. However, it turned out to be a specific precursor in bacterial metabolism of a more general acetyl carrier, which was found to be a special coenzyme, coenzyme A (CoA). This was discovered by following the ATP-dependent acetylation of aromatic amines in pigeon liver extract.[2] In the course of these studies, a heat-stable cofactor was found to be involved. The enzyme system lost activity on autolysis or dialysis, but was reactivated by addition of boiled liver extract. This indicated the participation of a coenzyme for acetyl transfer. It proved to be identical to a heat-stable factor found by Berman and Nachmansohn to complement the ATP-dependent enzymatic acetylation of choline in brain extracts.

Structure. (See reference 3 for a general review of the chemistry of coenzyme A.) The coenzyme was first isolated from pigeon liver and later on from fermentation sources. One of the striking features of the new coenzyme was the presence of pantothenic acid,[4] the vitamin discovered many years earlier by R. J. Williams.[5] In addition, the coenzyme was found to contain adenylic acid, two moles of phosphate, and a —SH containing moiety which Snell's group identified as thioethanolamine.[6,7] They had found a growth factor, pantetheine, that was a derivative of pantothenic acid, in which the carboxyl group of pantothenic acid was linked peptidically to thioethanolamine. This turned out to be a fragment of CoA. The studies on the degradation and biosynthesis by Novelli[8] and its complete synthesis showed CoA to have the following structure:

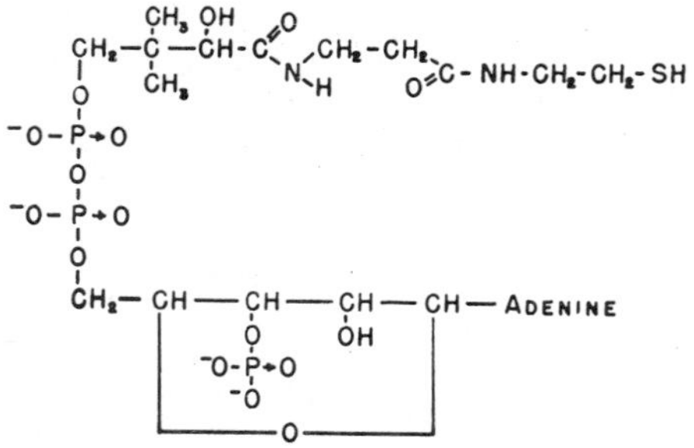

```
          CH3 OH    O
CH2—C —CH-C    CH2-CH2
    |         N         C - NH-CH2-CH2- SH
O   CH3       H      O
|
-O- P→O
|
O
|
-O- P→O
|
O
|
CH2— CH —— CH —— CH —— CH— ADENINE
            O      OH
        -O-P→O
           -O
     |________O________|
```

The pyrophosphate bridge between the terminal hydroxyl group of pantothenic acid and the 5′-hydroxyl on the ribose of adenosine, was verified largely through the work of Baddiley, who also successfully approached the partial synthesis of the coenzyme.[9] A most elegant complete synthesis of CoA has been worked out more recently by Khorana's group.[10] The manner in which the acetate was chemically linked to the coenzyme remained to be discovered by Lynen.[3,p.4] He isolated acetyl-CoA from yeast and identified it as the thioester of acetate linked to the —SH group of thioethanolamine in CoA.

The ATP-acetate-CoA Reaction.[11] Since the acetyl-carrying function of CoA was discovered in the ATP-dependent acetyl transfer reaction, it was obvious that the energy of the acetyl link to CoA could be derived from ATP. Berg[12] showed that the reaction occurs on one enzyme in two stages: (1) ATP + acetate to yield acetyl adenylate + pyrophosphate; (2) acetyl adenylate reacting with CoA to give the thioester of acetyl-thio-coenzyme A + adenylate. The equilibrium between ATP and acetyl CoA is near one, indicating that the group potential of the thioester link is nearly the same as that of the phosphoanhydride link in ATP. In *bacterial metabolism*, another mode of activation is prevalent. Here, acetate activation is transacted by two enzymes; the first, acetokinase, phosphorylates acetate to acetylphosphate: acetate + ATP → acetyl-P + ADP; the second enzyme, transacetylase, exchanges phosphate for CoA[13]: acetyl phosphate + CoASH ↔ acetyl-SCoA + P_i.

Metabolic Functions of CoA. The function of CoA as acetyl carrier was discovered through its participation in the acetylation of aromatic amines and of choline. Furthermore, synthesis of acetoacetate from 2 molecules of acetyl-CoA was demonstrated by an enzyme in pigeon liver extracts.[14]

The Role of Acetyl-CoA in Sugar Oxidation. The work of Krebs had shown a path of sugar oxidation by way of the CITRIC ACID CYCLE. The primary oxidation product, pyruvate, yields acetyl-CoA on further oxidation which, through the so-called condensing enzyme,[15] condenses with oxalacetate to citric acid:

$$\begin{matrix}\text{COOH} \\ | \\ \text{CO} \\ | \\ \text{CH}_2 \\ | \\ \text{COOH}\end{matrix} + \text{CH}_3\overset{\overset{\text{O}}{\|}}{\text{C}}\text{SCoA} \rightarrow \text{HO—}\begin{matrix}\text{COOH} \\ | \\ \text{C} \\ | \\ \text{CH}_2 \\ | \\ \text{COOH}\end{matrix}\text{—CH}_2\text{—COOH} + \text{CoASH} + \text{H}_2\text{O}$$

Oxalacetic acid   Acetyl-CoA   Citric acid

(See also MULTIENZYME COMPLEXES.)

Fatty Acid Synthesis. Early work by Knoop has shown that fatty acid oxidation occurs progressively by β-oxidation. On the other hand, already mentioned experiments had indicated that acetoacetate might be synthesized from 2 moles of acetyl-CoA. Lynen[16] showed that the aceto-acetyl-CoA formed in this reaction may be reversibly split into two acetyl-CoA's by an enzyme thiolase, the equilibrium, however, being toward lysis rather than synthesis. An important contribution now was the observation that the presence of carbon dioxide was essential for fatty acid synthesis. The role of carbon dioxide (see CARBOXYLATION ENZYMES) was explained by the discovery of malonyl-CoA, enzymatically derived from acetyl-CoA + CO_2, as the actual condensing reagent with acetyl-CoA to form β-ketoacyl-CoA + CO_2 in the condensing step:

$$CH_3COSCoA + \overset{\overset{COO^-}{|}}{C}H_2COSCoA \rightarrow CH_3COCH_2COSCoA + CoASH + CO_2$$

A path that appears to be an important metabolic source of acetyl CoA was found some time ago by Srere and Lipmann. In liver extracts they observed an enzymatic split of citrate, ATP, and CoA to acetyl CoA, oxalacetate, ADP, and phosphate. This derivation of acetyl CoA from citrate is now attracting much attention.

Most recent observations have shown that actually acyl-CoA in fatty acid synthesis transfers the acyl to —SH groups of proteins, in bacteria, to a polypeptidyl phosphopantetheine,[17,18] and in yeast to —SH groups on an enzyme which catalyzes the fatty acid synthesis without releasing an intermediary compound until the long chains of stearic or palmitic acid are formed. Finally, palmityl- and stearyl-CoA's are formed and react with glycerol derivatives to form PHOSPHOLIPID or neutral fat. (See also FATTY ACID METABOLISM.)

The Role of CoA in Steroid Synthesis. In a reaction sequence elaborated by Lynen,[19] two acetyl-CoA's condense to acetoacetyl-CoA, which then immediately reacts with another acetyl-CoA to form β-hydroxy-β-methylglutaryl-CoA:

$$\begin{array}{l} CH_3COCH_2COSCoA \rightarrow CH_3COHCH_2COSCoA \\ + \qquad\qquad\qquad\qquad\qquad | \\ CH_3COSCoA \qquad\qquad\quad CH_2COOH + SHCoA \end{array}$$

β-hydroxy-β-methylglutaryl-CoA then functions as a precursor of mevalonic acid, which, through a series of phosphorylations and condensation steps, yields isoprene structures (see ISOPRENOID BIOSYNTHESIS) such as squalene, the precursor of cholesterol and steroids (see STEROLS), or terpenes and rubber.

Succinyl-CoA Synthesis and Function. Succinyl-CoA may be derived by a series of reactions involving an ATP-linked carboxylation of propionate and CO_2 to methylmalonyl-CoA which rearranges to succinyl-CoA:

$$\begin{array}{l} CH_3 \qquad\qquad\qquad CH_3 \qquad\qquad CH_2COOH \\ | \qquad\qquad\qquad\qquad | \;\; COOH \leftrightarrow \;\; | \\ | \qquad + CO_2 \rightarrow \;\; | / \qquad\qquad\quad | \\ CH_2—COOH \qquad\quad CH \qquad\qquad CH_2COSCoA \\ \qquad\qquad\qquad\qquad\quad \backslash \\ \qquad\qquad\qquad\qquad\quad COSCoA \end{array}$$

Succinyl-CoA may also be directly derived by reaction between succinate, ATP or GTP, and CoA; this reaction is catalyzed by a complex enzyme in mitochondria. Succinyl-CoA is a building block of PORPHYRINS in HEME and CHLOROPHYLL. As Shemin showed, succinyl-CoA reacts with the methyl group of glycine to form α-amino-β-ketoadipic acid; this is decarboxylated to γ-aminolevulinic acid. Two of these condense to the monopyrrol precursor of porphyrin.

Summary. Coenzyme A participates as an acyl carrier in a great variety of biosynthetic reactions. The basic phenomenon in all these condensations is the versatility of an acyl thioester link to enter into condensations. This precursor function of the acyl thioester is, more generally, an example of the even more widespread functioning of carboxyl activation (see ACTIVATING ENYZMES) in biosynthetic mechanisms, such as found recently in polypeptide synthesis [see PROTEINS (BIOSYNTHESIS)] from the aminoacyloxy esters in aminoacyl sRNA's.

References

1. LIPMANN, F., *Cold Spring Harbor Symp. Quant. Biol.*, **7**, 248 (1939).
2. LIPMANN, F., KAPLAN, N. O., NOVELLI, G. D., TUTTLE, L. C., AND GUIRARD, B. M., *J. Biol. Chem.*, **167**, 869 (1947).
3. LIPMANN, F., *Bacteriol. Rev.*, **17**, 1 (1953).
4. DEVRIES, W. H., GOVIER, W. M., EVANS, J. S., GREGORY, J. D., NOVELLI, G. D., SOODAK, M., AND LIPMANN, F., *J. Am. Chem. Soc.*, **72**, 4838 (1950).
5. WILLIAMS, R. J., *Advan. Enzymol.*, **3**, 253 (1943).
6. BROWN, G. M., AND SNELL, E. E., *J. Am. Chem. Soc.*, **75**, 1691 (1953).
7. GREGORY, J. D., AND LIPMANN, F., *J. Am. Chem. Soc.*, **74**, 4017 (1952).
8. NOVELLI, G. D., *Physiol. Rev.*, **33**, 525 (1953).
9. BADDILEY, J., *J. Chem. Soc.*, **1951**, 3421.
10. MOFFATT, J. G., AND KHORANA, H. G., *J. Am. Chem. Soc.*, **83**, 663 (1961).
11. JONES, M. E., BLACK, S., FLYNN, R. M., AND LIPMANN, F., *Biochim. Biophys. Acta*, **12**, 141 (1953).
12. BERG, P., *J. Biol. Chem.*, **222**, 991 (1956).
13. STADTMAN, E. R., AND BARKER, H. A., *J. Biol. Chem.*, **180**, 1117 (1949).
14. STADTMAN, E. R., DOUDOROFF, M., LIPMANN, F., *J. Biol. Chem.*, **191**, 377 (1951).
15. STERN, J. R., AND OCHOA, S., *J. Biol. Chem.*, **191**, 161 (1951).
16. LYNEN, F., AND DECKER, K., *Ergeb. Physiol.*, **49**, 327 (1957).
17. MAGERUS, P. W., ALBERTS, A. W., AND VAGELOS, P. R., *Proc. Natl. Acad. Sci. U.S.*, **51**, 1231(1964).
18. MAGERUS, P. W., AND VAGELOS, P. R., *Federation Proc.*, **24**, 290 (1965).
19. LYNEN, F., *Proc. Intern. Congr. Biochem. 4th*, **13**, Colloquia, 267 (1961).

F. LIPMANN

COENZYME Q

Coenzyme Q designates a series of quinones which are widely distributed in animals, plants and microorganisms. These quinones have been shown to function in biological ELECTRON TRANSPORT SYSTEMS which are responsible for energy conversion within living cells. The nature and significance of coenzyme Q was first recognized in 1957. In structure, the coenzyme Q group closely resembles the members of the vitamin K group and the tocopherylquinones, which are derived from tocopherols (vitamin E), in that they all possess a quinone ring attached to a long hydrocarbon tail.

The quinones of the coenzyme Q series which are found in various biological species differ only slightly in chemical structure and form a group of related 2,3-dimethoxy-5-methyl-benzoquinones with a polyisoprenoid side chain in the 6-position which varies in length from 30–50 carbon atoms. Since each isoprenoid unit in the chain contains five carbon atoms, the number of isoprenoid units in the side chain varies from 6–10. The different members of the group have been designated by a subscript following the Q to denote the number of isoprenoid units in the side chain, as in coenzyme Q_{10}. The members of the group known to occur naturally are coenzyme Q_{10}, Q_9, Q_8, Q_7 and Q_6. These quinones have also been called *ubiquinones*, and the number of carbon atoms in the long side chain is indicated parenthetically. Thus ubiquinone (30) is the same as coenzyme Q_6. There are forms of coenzyme Q in which one of the isoprenoid units does not contain a double bond. For example, the designation coenzyme Q_{10} H-10 indicates that the last isoprenoid unit does not have a double bond. These forms have mostly been found in fungi.

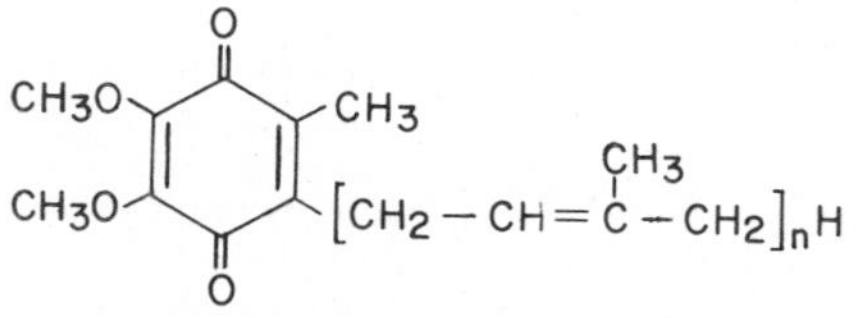

Coenzyme Q_n (quinone form), $n = 6$–10.

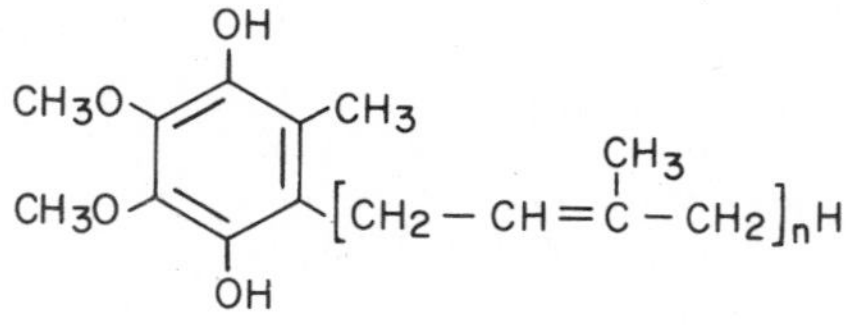

Reduced coenzyme Q_n (quinol form).

Coenzyme Q functions as an agent for carrying out oxidation and reduction within cells. Its primary site of function is in the terminal electron transport system where it acts as an electron or hydrogen carrier between the FLAVOPROTEINS (which catalyze the oxidation of succinate and reduced pyridine nucleotides) and the CYTOCHROMES. This process is carrried out in the MITOCHONDRIA of cells of higher organisms. It has been shown that during the oxidation of succinate, the quinone is reduced to the quinol, which is then reoxidized to the quinone by the cytochrome-containing part of the electron transport chain.

Since coenzyme Q is also found in other parts of cells in addition to mitochondria there may be other enzyme systems in which it functions. It has been found in the MICROSOMAL fraction from some cells especially in adrenal tissue, and it has been found in the soluble flavoprotein enzyme aldehyde oxidase.

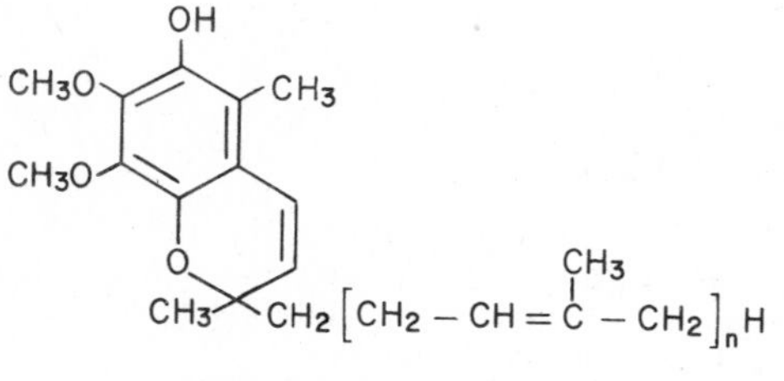

Ubichromenol, $n = 5$–9.

In some tissues a chromenol derivative of the quinone is found which has been called ubichromenol. It is possible that the conversion of coenzyme Q to ubichromenol in tissues may have some metabolic significance, but no evidence for a special function for this chromenol has been found. The structural relation between the ubichromenol and the tocopherols indicates that it may function as an ANTIOXIDANT (see also VITAMIN E.) Addition of ubichromenol to the suspending medium helps in the preservation of sperm cells during storage in a manner similar to the protective effect of tocopherols.

Certain bacteria and other lower organisms do not contain any coenzyme Q. It has been shown that many of these organisms contain vitamin K_2 instead (see VITAMIN K) and that this quinone functions in electron transport in much the same way as coenzyme Q. Similarly, plant CHLOROPLASTS do not contain coenzyme Q, but do contain plastoquinones which are structurally related to coenzyme Q. Plastoquinone functions in the electron transport processes involved in photosynthesis. In some organisms, coenzyme Q is present together with other quinones such as vitamin K, tocopherylquinones and plastoquinones, and each type of quinone can carry out different parts of the electron transport functions.

References

1. WOLSTENHOLME, G. E. W., AND O'CONNOR, C. M., (EDITORS), *Ciba Found. Symp. Quinones Electron Transport*, London, Churchill, 1961.
2. CRANE, F. L., "Quinones in Lipoprotein Electron Transport Systems," *Biochemistry*, **1**, 510–517 (1962).
3. HATEFI, Y., "Coenzyme Q (Ubiquinone)," *Advan. Enzymol.*, **25**, 275–328 (1963).
4. PAGE, A. C. JR., *et al.*, "Coenzyme Q XXVIII. Activity of the Coenzyme Q Group in Sperm Motility," *Biochem. Biophys. Res. Commun.*, **6**, 141–145 (1961).

FREDERICK L. CRANE

COENZYMES

Coenzymes are organic molecules, of a size intermediate between the small-molecule intermediary metabolites, which serve as the substrates of enzymatic reactions, and the macromolecular proteins. A coenzyme is an easily dissociable portion (sometimes called the prosthetic portion or group) attached to the protein component (apoenzyme) to form the complete, enzymatically active, conjugated protein (holoenzyme). Each coenzyme (or "cofactor") acts usually as acceptor or donor of some specific type of atom or group of atoms to be removed from or added to a small-molecule substrate in a reaction catalyzed by the holoenzyme. For example, the FOLIC ACID COENZYMES accept or donate single carbon units (at various states of oxidation) in a considerable number of enzymatic reactions grouped together as SINGLE CARBON UNIT METABOLISM.

Each of the B vitamins is a nutritionally assimilable form that is incorporated into, and exerts its essential biochemical function as part of, some coenzyme (excepting BIOTIN and LIPOIC ACID, which are covalently linked to enzyme protein). These respective coenzyme forms of the B vitamins are illustrated in the article VITAMIN B GROUP. In addition most of the B vitamins are discussed in this Encyclopedia in two articles for each vitamin: one emphasizing the "vitamin" or simplest nutritionally utilizable form and the other treating the enzymatic functions of the coenzyme form or forms. Thus the coenzyme forms of FOLIC ACID are discussed in FOLIC ACID COENZYMES; of NICOTINIC ACID, in NICOTINAMIDE ADENINE DINUCLEOTIDES; of PANTOTHENIC ACID, in COENZYME A; of RIBOFLAVIN, in FLAVINS AND FLAVOPROTEINS; of VITAMIN B_{12}, in COBALAMINS AND COBAMIDE COENZYMES. The coenzymatic roles of the PYRIDOXINE FAMILY OF VITAMINS are further discussed in the article TRANSAMINATION ENZYMES. Other types of molecules besides B-vitamin conjugates, *e.g.*, nucleotides, may also function as coenzymes in certain metabolic reactions.

E. M. LANSFORD, JR.

COLLAGENS

Collagen is the major protein component of connective tissue. In fact, in mammals as much as 60% of the total body protein is collagen. It comprises most of the organic matter of skin, tendons, bones and teeth, and occurs as fibrous inclusions in most other body structures.

Collagen fibers are easily identified on the basis of the following characteristic properties: they are quite inelastic; they swell markedly when immersed in acid, alkali or concentrated solutions of certain neutral salts and nonelectrolytes; they are quite resistant to most PROTEOLYTIC ENZYMES, but are specifically attacked by the collagenases; they undergo thermal shrinkage to a fraction of their original length at a temperature which is characteristic of the collagen from a given animal but varies from one species to another; and they are converted in large part to soluble *gelatin* by prolonged treatment at temperatures above the thermal shrinkage level. Collagen fibers are not unique to mammals: collagen has been identified in the tissues of almost all multicellular animals, ranging from the primitive porifera and coelenterates, through the annelids and echinoderms, and up to the vertebrates.

As a protein, collagen is unusual in both chemistry and structure. Close to 33% of its residues are GLYCINE, and an additional 20–25% are imino acids (PROLINE and HYDROXYPROLINE). It also contains hydroxylysyl residues. In terms of sequence, glycine occurs regularly in essentially every third position (indeed, this follows as a steric requirement of the secondary-tertiary structure—see below). Tripeptide sequences such as gly-pro-hypro and gly-pro-ala are quite common. Collagen also contains considerable quantities of polar residues: particularly ASPARTIC and GLUTAMIC ACIDS, LYSINE and ARGININE. These moieties seem to be concentrated in sequences about twenty residues in length, and alternate regularly with slightly longer, nonpolar regions containing most of the imino acids.

As a class, the collagens are most characteristically identified by a unique, wide-angle X-ray diffraction pattern, which is easily recognized by the strong 2.86 Å meridional spacing and by the 11–15 Å (hydration-sensitive) reflection on the equator. The unusual molecular geometry responsible for this pattern is now largely understood. In its essentials, the structure consists of three polypeptide chains, each wound around its own axis in a left-handed, three-residue per turn helix with a 9 Å pitch. Bundles of three such helices are then given a slight right-handed twist about a common axis to form a super-helix which repeats every 86 Å. The three chains of each bundle are held together by interchain hydrogen bonds (and an occasional covalent linkage—see below). Every third residue along each chain is required to be glycine, since no other residue will fit into the structure in these positions.

Collagen fibers will dissolve to a considerable extent in dilute acid or concentrated neutral salt solutions in the cold. The macromolecules obtained by such treatment are highly asymmetric rigid rods, about 2600 Å in length and about 15 Å in diameter. The molecular weight of these molecules is about 300,000 gms/mole. Each macromolecule contains three polypeptide chains, twisted about one another as indicated by the X-ray diffraction results. On gentle heating, these asymmetric, helical macromolecules suddenly collapse (denature) to a random-coil conformation. This helix-coil transition is the molecular analogue of the thermal shrinkage phenomenon observed with collagen fibers. Like the shrinkage temperature, the molecular transition temperature also varies from one collagen to another. Amino acid analyses show a direct correlation between this "melting" temperature and the total content of prolyl plus hydroxyprolyl residues, suggesting that the pyrrolidine ring is importantly involved in stabilizing the structure. On denaturation of most collagens, the random-coil gelatin chains which result do not all separate from one another. Rather they are partially covalently cross-linked by "ester-like" bonds which hold some of the chains together as two- and three-chain structures. The chemical nature of these bonds is not yet entirely clear. They may be of several types, perhaps involving the tightly bound carbohydrate or aldehyde components of collagen. Present evidence suggests that both intra- and interchain "ester-like" bonds occur, and that the peptide-linked subunits may have a weight-average molecular weight of only 20,000 gms/mole. Peptide bonds involving the ε-amino group of lysine and the γ-carboxyl group of glutamic acid have also been demonstrated in collagen.

The conversion of collagen to gelatin is at least partially reversed on cooling. Portions of the gelatin chains reform three-chain, helical units, separated by regions of the chain which remain random. Since individual chains may become involved in three-chain helices with several sets of partners along their length, the result in concentrated solution is to form a fibrous network interconnecting all the gelatin chains, which in turn results in the formation of characteristic low-temperature gelatin gels.

Collagen fibers have a very distinctive appearance in the electron microscope, showing distinct cross-striations which repeat regularly at about 660 Å intervals in the native material. Solutions of collagen macromolecules can be induced to precipitate by various means, forming fibrous aggregates which vary in appearance in the electron microscope depending on the precipitating conditions used. The macromolecules can be induced to line up side by side in end-to-end register, giving major striational repeats at 2600 Å intervals. They can also be precipitated under conditions which result in the reappearance of the native 660 Å repeat, which seems to be due to the collagen macromolecules in the fiber overlapping one another in an approximately quarter-staggered arrangement. The darkly staining striations observed in the electron microscope are due to the lateral alignment of the polar amino acid residue-containing sequences in the individual macromolecules. These sequences selectively adsorb the ionic electron-dense stains generally used and thus appear as dark "bands." The relatively unstained intervening regions are due to the non polar, imino acid residue-rich sequences and are called "interbands." It appears that specific side-chain interactions between polar residues on adjacent collagen macromolecules are largely responsible for ordering the macromolecules into fibers. Specific cooperative interactions between functional groups on appropriately oriented macromolecules seem to be involved in the heterogeneous nucleation of hydroxyapatite crystals, and thus in the initiation and control of mineralization in bones and teeth (see BONE FORMATION). As collagen fibers age *in vivo*, they seem to become progressively more intermolecularly cross-linked, perhaps by the "ester-like" bonds mentioned above. Little or no soluble collagen can be extracted from most mature connective tissue because of this extensive cross-linking, though this material can be converted into soluble gelatin by drastic thermal treatment.

Collagen and gelatin are of great commercial importance. As insoluble collagen, this material may be cross-linked further by tanning and thus converted to leather. The soluble gelatins are used in the manufacture of food, film emulsions, glue and so forth.

References

1. "Connective Tissue: Intercellular Macromolecules," Proceedings of a Symposium sponsored by the New York Heart Association, Boston, Little Brown and Company, 1964.
2. GUSTAVSON, K. H., "The Chemistry and Reactivity of Collagen," New York, Academic Press, 1956.
3. VEIS, A., "The Macromolecular Chemistry of Gelatin," New York, Academic Press, 1964.
4. HARRINGTON, W. F., AND VON HIPPEL, P. H., "The Structure of Collagen and Gelatin," *Advan. Protein Chem.*, **16**, 1 (1961).
5. HODGE, A. J., AND SCHMITT, F. O., "The Tropocollagen Macromolecule and its Properties of Ordered Interaction," in "Macromolecular Complexes" (EDDS, M. V., JR., EDITOR), New York, Ronald Press, 1961.

PETER H. VON HIPPEL

COLUMN CHROMATOGRAPHY. See CHROMATOGRAPHY, COLUMN; also CHROMATOGRAPHY (GENERAL).

COMPARATIVE BIOCHEMISTRY

Comparative biochemistry may be defined as the study of the nature, origin and control of biochemical diversity. Such a definition suggests that (1) biochemical differences are to be found among organisms, (2) such differences arise during the evolution of organisms, and (3) the biochemical properties of organisms are under a variety of controls in nature, and presumably may be modified when the biochemical properties and their natural controls are understood. These propositions are, in fact, consistent with the data of modern biochemistry; data pointing to the validity of the first two points are presented below.

In the approach to comparative biochemistry popular before the 1950's, exemplified by the books of Baldwin[1] and Florkin,[2] biochemists studied the similarities and differences among higher organisms, mainly among animals. The structures and distribution of certain significant compounds such as the various phosphagens, blood-transport pigments, CAROTENOIDS and the A vitamins, etc., were correlated with the postulated phylogenetic position of the animals possessing these substancse. The comparative aspects of nitrogenous excretion products and of salt and water balance have also been studied in great detail (see ELECTROLYTE AND WATER REGULATION). Such types of studies have indeed revealed metabolic differences among the Metazoa, differences related to their evolutionary and ecological niches. More recently the distributions of the ALKALOIDS and flavones of plants (see FLAVONOIDS), organic acids of lichens, and the pterins of *Drosophila* (see PTERIDINES), among other biochemical markers, have been analyzed in great detail to help in establishing genetic and evolutionary interrelations.

However, biochemistry since 1945 has concentrated very largely on cellular metabolism, and these studies, in their earliest phases at least, seemed to emphasize the uniformity of cellular biochemistry. Thus, almost all cells contained proteins built of the same 20 amino acids, RNA and DNA containing their characteristic constituents, common coenzymes ATP, DPN, coenzyme A, etc. Also, many similarities could be detected in the metabolic events relating to energy and biosynthesis in many cell types, *i.e.*, microorganisms, plants, and animals. This experience was soon interpreted (1) by the school of Kluyver and van Niel and their students to signify "the unity of biochemistry," a unity inferred to derive from a monophyletic evolution of organisms from a primitive cellular type. In this widely held view, biochemical differences among organisms are considered to reflect relatively late evolutionary divergences of which the metabolic differences among the Metazoa noted earlier, *e.g.*, patterns of nitrogen excretion such as ammonotelism, ureotelism, and uricotelism, are clear examples (see AMMONIA METABOLISM). It should be noted, however, that the biochemical differences recognized to exist between organisms are not only considered to be relatively late evolutionary events but are usually stated to represent minor alterations in the broad biochemical pattern common to many cells.

The evolution of the initial cells is widely thought, as formulated in the theory of Haldane, Oparin, and Urey, to have occurred in an anaerobic environment rich in organic compounds.[3] It has been suggested that as cellular growth and multiplication proceeded, various organic nutrients were depleted from the aqueous environment and only those cell types developed which were capable of synthesizing the essential missing organic compounds, *i.e.*, an initial heterotrophy preceded the development of autotrophy. According to an extension of this theory, modern cells have once again become heterotrophic by the loss of the ability to synthesize essential enzymes. Such a view has tended to be supported by the studies of bacterial nutrition by Knight and Lwoff indicating that highly parasitic organisms have lost more synthetic capabilities and have more extensive nutritional requirements than have free living forms. The experience of microbial genetics, in which the loss of function is far more common than is the acquisition of function, also tended to support this postulated pattern of biochemical evolution. Indeed, acquisitions of function at the cellular level have been detected to occur in the laboratory only as a result of a variety of sexual mechanisms.

Until recently, this hypothetical sequence of events was considered to be the only reasonable description of a monophyletic development of all the modern cellular types. However, it is now possible to ask if cells such as those of bacteria and blue-green algae (procaryotic cells) are really similar to higher (eucaryotic) cells, if, indeed, the range of structures in modern cells may not reflect a polyphyletic evolutionary development. Thus, the more recent detailed analyses of microbial and higher cells have revealed very significant differences in cell structure, *i.e.*, differences in nuclear, genetic, cytoplasmic, and wall structures as well as differences in metabolism between the two cellular types.[4] Such differences in cell structure reflect molecular differences; *e.g.*, substances such as muramic acid or diaminopimelic acid are found in procaryotic cell walls (see BACTERIAL CELL WALLS) but not in eucaryotic cells.

On the other hand, substances such as sialic acid are present in the mucoproteins of many higher cells but very rarely in microbes. Indeed, the biosynthetic path for synthesis of this component in higher cells is different from some other paths for biosynthesis of sialic acid found in bacteria. Thus, the sporadic distribution of this component in both procaryotic and eucaryotic cells appears to reflect convergent evolution to form a single substance via different paths. Similar conclusions may be drawn for the formation of other substances as diverse and important as bone salts, iodinated derivatives of tyrosine [see IODINE (IN BIOLOGICAL SYSTEMS)], CARBAMYL PHOSPHATE. Furthermore, the existence of these latter components appears to indicate the existence of relatively recent biochemical innovations during the evolution of multicellular forms.[5]

Recent studies of the distribution of lipids and of their biosynthetic paths have indicated additional distinctive differences between procaryotic and eucaryotic cells.[6] The former lack sterols and polyUNSATURATED FATTY ACIDS, compounds which require molecular oxygen for their biosynthesis. The development of atmospheric molecular oxygen itself is thought to be a consequence of the evolution of PHOTOSYNTHESIS from the anaerobic form of photosynthesis occurring in bacteria to the O_2-yielding type occurring in algae and higher plant cells. A rigorous correlation has been obtained between O_2 production in photosynthesis and the presence of α-linolenate in the chloroplasts of O_2-producing photosynthetic cells. Animal cells lack this polyunsaturated acid and possess γ-linolenate and arachidonate. Thus, these relatively late biochemical and still mysterious innovations of O_2 evolution and the formation of α-linolenate have permitted the development of new biochemical paths in higher cells for the formation of sterols, tyrosine, nicotinic acid, unsaturated fatty acids, etc., substances which have contributed in turn to the evolution of structure and function in these cells.

The existence of the sharp metabolic break in the biochemical record, represented by the presence of metabolically functional oxygen, has at least permitted the introduction of a hypothesis of polyphyletic origins of cell types. In the light of such a hypothesis, the common elements among cells may be explained as arising from the rigorous demands of convergent evolution selected from the enormous diversity provided by chemical evolution (see also ORIGIN OF LIFE). The dissection of PROTEIN BIOSYNTHESIS and the coding relations of trinucleotide sequences to amino acid insertion in a polypeptide chain have provided interesting data relevant to the problem of the unity of the organic world. Although this process is not yet entirely clear, present data on the mechanism of amino acid assembly suggest that convergent evolution would not be expected to develop the same coding relations in all cells. Nevertheless, despite many natural and experimental complications in the clarification of protein synthesis, present evidence tends to suggest a universality of the code, providing a potentially powerful argument for monophyletic evolution.

In summary, we may say that at the present time it is not quite clear whether biochemical evolution was, indeed, monophyletic, with structural and metabolic differences among organisms accounted for entirely by evolutionary divergence, or was polyphyletic with respect to several major streams of the biotic world, in which structural and metabolic similarities are accounted for by evolutionary convergence. This question may be formulated as the central problem of comparative biochemistry in its present phase of development. Compendia of relevant data are available[7] but have not yet been thoroughly analyzed in these terms.

Of course, many more significant data remain to be obtained. For example, it may be supposed that the exploration of Mars will shed light on this question since the discovery of organisms lacking DNA, RNA, proteins, or other substances thought to be critical in our cells will affirm at least the terrestrial origin of terrestrial cells. On the other hand, the presence on Mars of cells possessing all of the common components will suggest either that the survival of cells wherever they may arise generally requires these substances, tending to support the hypothesis of convergent evolution, or that cells, the Cosmozoa postulated by Arrhenius, have been flitting back and forth through outer space fecundating the Universe.

References

1. BALDWIN, E., "An Introduction to Comparative Biochemistry," Fourth edition, Cambridge, The University Press, 1964.
2. FLORKIN, M., "Biochemical Evolution," New York, Academic Press, 1949.
3. OPARIN, A. I., "The Origin of Life on the Earth," Third edition, New York, Academic Press, 1957.
4. STANIER, R. Y., AND VAN NIEL, C. B., *Archiv. Mikrodial.*, **42**, 17 (1962).
5. COHEN, S. S., *Science*, **139**, 1017 (1963).
6. BLOCH, K., "Symposium on Biochemical Evolution," Rutgers University, in press.
7. FLORKIN, M., AND MASON, H., "Comparative Biochemistry," Vols. 1–7, New York, Academic Press, 1960–1964.

SEYMOUR S. COHEN

COMPETITIVE INHIBITION. See ENZYMES (ACTIVATION AND INHIBITION).

CONTRACTILE PROTEINS

Proteins from the Contractile System in Muscle. The study of the fibrillar proteins of muscle invites interest for two reasons. On the one hand, they form the building stones of the machinery that performs a typical vital activity, movement, in its most specialized form; thus, the elucidation of their functions, and of the physical and chemical properties basic to it, represents one of the cardinal parts of molecular biology. On the other hand, the isolated proteins display such striking physical behavior that to the macromolecular physicist they are among the most fascinating materials. Unfortunately, they are difficult to obtain and exceedingly changeable and labile; thus, working with them is somewhat of an art, and requires much experience.

The muscle cell has a number of supportive functions besides the contractile mechanism itself. In addition to nuclei and other constituents common to all cells, mention must be made of the sytoplasm, here called sarcoplasm. This too consists mainly of proteins besides small-molecular solutes; these proteins are all enzymes subserving the anaerobic phases of metabolism (see GLYCOLYSIS), but they offer no special physicochemical interest other than that of enzymes in general. However, there may be more organization in this phase than is known at present. Then, there are mitochondria as in other cells, with an oxidative-PHOSPHORYLATIVE function. Finally, exceeding the usual endoplasmic reticulum in complexity and prominence, there is a sarcotubular reticulum, responsible for the conduction of the excitatory state from the surface membrane to the contractile fibril, and for the on-off control of contractile activity in the latter. Its modes of action are not understood, but its role in intracellular CALCIUM movements begins to be agreed upon, and isolated particulate components derived from it display a remarkable calcium uptake activity *in vitro*. All these systems control or support the workings of the contractile structures and supply its chemical energy in the form of ATP. It is the function of the contractile protein system to engage this ATP in chemomechanical transformations of such a nature that a linearly oriented motion results.

The major fibrous proteins are two: *myosin* and *actin*. The former has been known for some time, if not in a well-understood and reproducible form. Myosin was the subject of an outstanding classic of biochemistry in 1930, the work of Von Muralt and Edsall showing that it possessed the phenomenon of flow birefringence, indicative of the presence of elongated particles that can be oriented in the streaming field.[8] It had been suggested by studies on muscle fibers in the polarization microscope that parts of the fibrils must consist of long, perhaps rod-shaped, micelles in parallel arrangement, and for such a building plan myosin appeared to have suitable properties, as indicated by this study. Weber showed shortly afterward that myosin could be precipitated in the form of oriented fibers, the optical anisotropy of which was comparable to that of the birefringent part of the muscle fibrils. The second fundamental discovery was that by Engelhardt and Ljubimova,[2] who found myosin to be enzymatically active in splitting ATP. This observation bridged the gap between the structural and physical aspect of muscle function, and the biochemical aspect of the supply of chemical energy.

The myosin preparations used in earlier work were variable in their properties, but this was not realized until Szent-Györgyi, in 1941, could systematically control the preparation of either a moderately viscous or an extremely viscous and birefringent form of myosin. The latter showed striking physical changes in response to ATP, *e.g.*, a drastic drop in viscosity or turbidity. These properties are acquired by the combination of myosin proper (which predominates in the low-viscosity form) with actin, which was discovered in conjunction with these investigations by Straub (1942). Hence, the viscous responsive form is called *actomyosin*, and its physical changes in solution due to ATP are caused by the induced dissociation of the complex protein into its original constituents. Both myosin and actomyosin require a fairly high ionic strength to be soluble; neutral 0.5 *M* KCl has become the traditional solvent. At lower salt concentration they precipitate.

As discovered by Szent-Györgyi in 1942, such precipitated actomyosin forms contractile systems with ATP, undoubtedly one of the outstanding biological discoveries of the present epoch. A merely flocculent precipitate expells occluded water and settles to a smaller sediment volume; it is remarkable that, while wholly unorganized, such gel particles still display the essentials of contractility, proof that this depends on the molecular properties of these proteins. Actomyosin spun to a loose gelatinous thread shows contraction of the same undirected nature, becoming shorter and thinner, but threads internally oriented by various artifices become thicker when shortening. The best approach is to start not from separated proteins, but from muscle fibers from which all active constituents but actin and myosin are removed by prolonged extraction in aqueous glycerol at low temperature. These fiber preparations, likewise developed by Szent-Györgyi, preserve a great deal of structural integrity and, with ATP, show a degree of contraction and force development comparable to that of living muscle. All these contractile phenomena occur optimally in the presence of K and Mg ions in concentrations like those occurring intracellularly. They also appear to require trace amounts of calcium, and contraction can be prevented (or relaxation caused) by a relaxing factor from muscle, identical (at least in part) to the sarcotubular particles which, with their ability to concentrate calcium, can remove this essential ion from the fibrils or protein micelles.

Other proteins that may be involved in myofibrillar structure are *tropomyosin* and *paramyosin*. Their role is less obvious and there are no indications of a primary participation in the contraction process. They probably have modifier functions. This is clearly indicated for paramyosin, which occurs in invertebrate smooth muscles that can exert a holding function, *e.g.*, closing shells or attaching mussels to rocks. It is held that, once myosin and actin have effected a desired degree of contraction, another type of nerve stimulus now causes some structure formation on the part of paramyosin, so as to lock everything in place until released again.

Composition and Properties of the Contractile Proteins. All four proteins share certain chemical properties; among others they are all exceptionally rich in ionizing AMINO ACIDS, thus they are highly charged molecules. Tropomyosin is the extreme in this regard, with two-thirds of its residues classified as polar, and almost half anionic or cationic; in actin, these proportions would be somewhat over half polar, and over one-fourth ionic; myosin is in between. There is much similarity in composition between para- and tropomyosin in various species—a lower histidine content than in myosin and actin, and a near absence of proline and tryptophan being their distinctive features; the tropomyosins, furthermore, are especially rich in glutamic acid. Actin is especially rich in sulfur and aromatic amino acids. Generally speaking, the sulfur chemistry of these proteins is very interesting and not yet completely explored; a high SH content, and a variability in the reactivities of these groups invite further attention. There have generally been difficulties in identifying peptide chain end groups, and, apart from partial approaches, there has been no progress toward systematic structure and sequence studies; there are easier proteins for that.

While enzymatic digestion will of course degrade all these proteins, the digestion of myosin by trypsin goes through a relatively selective primary cleavage of the myosin molecule to form two subunits H and L (heavy and light) meromyosin. This entails a peculiar distribution of constituents: L may be somewhat more polar, H is much enriched in cysteine, proline and aromatic amino acids; L requires salt for its solubility, but H has become water soluble. The affinities toward ATP and actin are found in the H component.

Myosin is an adenosinetriphosphatase, and it retains this activity in various degrees of aggregation and in combination with actin, and also in the form of isolated myofibrils. The ATPase has a sharp pH optimum near pH6.4 and a strong increase of activity (until inactivated) in the alkaline range perhaps due to the use of OH^- instead of water in hydrolysis. Some electrolyte is needed for activity, but the specific effects of K^+, Mg^{++} and Ca^{++}, singly or in combination, are very complex and dependent on the presence or absence of actin; SH groups play a role in the ATPase activity; some of them can be substituted to cause a complete inactivation of myosin-ATPase in the absence of calcium, but a stimulation in its presence. There seems to be one enzymatically active site per molecule. However, the molecular weight is not accurately known; a number of various determinations cluster around 500,000, but others are about 650,000. Ultracentrifugation and light scattering have been the major methods used, the reasons for the variations are not clear. Light scattering dissymmetry suggests a mean molecular length of about 1600 Å, and electron microscopy suggests similar lengths, some less, some more, with the added information that the molecule has a thicker knob at one end, the whole being shaped like a lampbrush (see illustration in the separate article on MYOSINS). Of the meromyosins, L would be one end of the thinner rod, H the remainder of the rod with the knob. The latter part is supposed to make up the cross-bridges which are believed to link the thin rods described by Huxley and Hanson as extending from the I-bands of the myofibrils into the A-bands where they interdigitate with heavier rods restricted to the A-bands only. These A-band rods consist of myosin, the molecules of which are arranged into crystalloid regions which are about 1.5 μ long and 150 Å thick, from which the thick ends of the myosin molecules emerge according to a hexagonal screw pattern. The thinner I-band rods are actin [see MUSCLE (STRUCTURE AND FUNCTION)].

The main property of actin is that it can occur in a monomer or G- and a polymer or F-form. The former has a molecular weight in the 60,000 range, 57,000–62,000 being the values actually discussed. It has no enzymatic properties, but binds one mole of nucleotide, preferably ATP, without which it loses activity. Addition of salt initiates the transition to the fibrous polymer, the major reason probably being the elimination of negative charges which prevent molecular collisions. Originally it was thought that the G-F transition entails a stoichiometric breakdown of ATP to ADP with a strong binding of the latter to the polymer, but it has become obvious that while this may occur, it is not required; one can, for example, prepare G-actin in the exclusive presence of ADP and cause polymerization without change in the nucleotide. The best summing-up of the present knowledge would be as follows: when a monomer enters into the polymer, and in conjuction with the presence of salt (all ATPases seem to require salt activation), actin undergoes a change in conformation which gives it ATPase activity until it has become part of the polymer or has become inhibited by the ADP formed. Thus, the maximal nucleotide change is in a stoichiometric ratio, unless the transient form is maintained, which appears possible in an ultrasonic field. The aggregation is predominently end-to-end, but there is a dimerizing tendency as well because the F-form consists of a double strand of monomers. The arrangement is regular, F-actin shows ultraviolet dichroism due to transversal arrangement of the attached ADP. The resulting polymer forms a gel even at concentrations below 1%; this is thixotropic and highly birefringent, mere swirling in a beaker between crossed polaroids suffices to show interference colors. The molecular weight of the polymer is indeterminate, strands longer than 10 μ having been seen by electron microscopy. The I-band filaments from muscle can also be isolated, but have regular lengths of the order of 1 μ. In the polymerization process, 4–5 of the 6–7 SH groups of the monomer, which were slowly reacting with most reagents, become even less reactive. The other two react freely, in either G- or F-actin, with reagents such as mercurials or N-ethylmaleimide.

When myosin and F-actin are mixed they combine to form actomyosin. The optimal proportion corresponds to about 1 molecule of myosin per 2 molecules of actin monomer; thus, 1 myosin molecule is attached for every doublet along the actin filament. Actomyosin formation causes a great increase in turbidity because of the increased mass per unit length; it causes increased viscosity, partly because of the solvent occlusion by the voluminous structure. Dissociation by ATP reverses these changes, as long as the system is soluble. In actomyosin precipitates, syneresis and contraction phenomena occur, as described. These contraction phenomena can be prevented (and the actomyosin ATPase inhibited) by constituents from muscle known as the relaxation factor; this will also actually reverse the ATP-induced contraction of the glycerol-extracted muscle fiber. To a large extent, at least, this relaxing activity is due to the calcium concentrating activity of sarcotubular vesicles, inasmuch as trace amounts of calcium appear to be required for contraction under many experimental conditions. Research continues on the possibility of a relaxation substance besides this.

Tropomyosin has attracted attention because it is the easiest to crystallize. Its behavior, in a sense, is opposite to that of actin in that it polymerizes in the absence, and depolymerizes in the presence, of salts. Such a behavior would result if critical parts of the molecule held a pattern of charges responsible for aggregation. Its monomeric molecular weight has been measured as 53,000, and the figure of 62,000 has been suggested on the basis of its amino acid composition. Paramyosin, also called tropomyosin A, in distinction from the preceding which would be called tropomyosin B, occurs widely in invertebrate "catch" or tonus muscles, where it is responsible for a characteristic low-angle X-ray diffraction and electron optically demonstrable periodicity. It is soluble in strong salt, but unlike tropomyosin B, it forms needle-shaped crystals upon reduction of the ionic strength. Its molecular weight is reported around 135,000 with a length of 1400 Å, while some have found about double this weight value. The molecule is supposed to be completely helical. Neither tropomyosin A nor B has any demonstrated relation to ATP. But B, or proteins accompanying it, are reported to change some responses of actomyosin to ATP, and can accompany improperly prepared actin as tenacious impurity, while the special role of paramyosin in fixing muscle at a given state of contraction has already been mentioned. Some other proteins have been reported, one of which, delta protein, has an affinity to myosin.

The further study of the molecular and biochemical properties of these muscle proteins, as a part of the study of the mechanism of a vital process, will continue to deserve our attention among the fundamental analytical approaches to biology.

References

1. BANGA, I., ERDÖS, T., GERENDAS, M., MOMMAERTS, W. F. H. M., STRAUB, F. B. AND SZENT-GYÖRGYI, A., "Myosin and Muscular Contraction," Basel, New York, S. Karger, 1942.

2. Engelhardt, W. A., and Ljubimova, M. N., "Myosin and Adenosinetriphosphatase," *Nature*, **144**, 668–669 (1939).
3. Hanson, J., and Huxley, H. E., "The Structural Basis for Contraction in Striated Muscle," *Symp. Soc. Exptl. Biol.*, **9**, 228–264 (1955).
4. Kielley, W. W., "The Biochemistry of Muscle," *Ann. Rev. Biochem.*, **33**, 403 (1964).
5. Mommaerts, W. F. H. M., "Muscular Contraction, A Topic in Molecular Physiology," New York, Interscience Publishers, 1950.
6. Szent-Györgyi, A., "Chemistry of Muscular Contraction," New York, Academic Press, 1951.
7. Szent-Györgyi, A. G., "Proteins of the Myofibril," in "Structure and Function of Muscle" (Bourne, G. H., Editor), Vol. II, New York, Academic Press, 1960.
8. Von Muralt, A., and Edsall, J. T., "Studies on the Physical Chemistry of Muscle Globulin. IV. The Anisotrophy of Myosin and Double Refraction of Flow," *J. Biol. Chem.*, **89**, 351 (1930).
9. Weber, H. H., "Muskelphysiologie. Die Wirkung von Adenosintriphosphat auf die kontraktilen Proteine und die Kontraktion von Muskeln und Zellen," *Fortschr. Zoologie, N.F.*, **10**, 304 (1956).

Wilfried F. H. M. Mommaerts

COORDINATION COMPLEXES. See CHELATION AND METAL CHELATE COMPOUNDS.

COPPER (IN BIOLOGICAL SYSTEMS)

Plants. The activity of copper in plant metabolism manifests itself in two forms: synthesis of CHLOROPHYLL and activity of enzymes. In leaves, most of the copper occurs in close association with chlorophyll, but little is known of its role in chlorophyll synthesis, other than that the presence of copper is required.

Copper is a definite constituent of several enzymes catalyzing oxidation-reduction reactions (oxidases), in which the activity is believed to be due to the shuttling of copper between the +1 and +2 oxidation states. Ascorbic acid oxidase catalyzes the reaction between oxygen and ASCORBIC ACID to give dehydroascorbic acid. This oxidase occurs widely in plants, particularly in cucurbits and beans. Tyrosinase, also known as polyphenol oxidase or catechol oxidase occurs in potatoes, spinach, mushrooms, and other plants (see AROMATIC AMINO ACIDS; OXYGENASES.) It catalyzes the air oxidation of monophenols to *ortho* diphenols, and the oxidation of catechol to dark-colored compounds known as *melanins*. Laccase also catalyzes the oxidation of phenols and is fairly widely distributed. The cytochrome enzymes (see below) are also found in plants.

Traces of copper are required for the growth and reproduction of lower plant forms, such as algae and fungi, although larger amounts are toxic.

The effects of copper deficiency in plants are varied and include: die-back, inability to produce seed, chlorosis, and reduced photosynthetic activity. On the other hand, excesses of copper in the soil are toxic, as is the application of soluble copper salts to foliage. It is for this reason that copper fungicides are formulated with a relatively insoluble copper compound. Their toxicity to fungi arises from the fact that the latter produce compounds, primarily hydroxy and amino acids, which can dissolve the copper compounds from the fungicide.

Animals. Copper is also a necessary trace element in animal metabolism. The human adult requirement is 2 mg/day, and the adult human body contains 100–150 mg of copper, the greatest concentrations existing in the liver and bones. Blood contains a number of copper proteins, and copper is known to be necessary for the synthesis of HEMOGLOBIN, although there is no copper in the hemoglobin molecule.

Copper in PLASMA is mainly present in the blue protein ceruloplasmin, which is thought to be responsible for the transport of copper in the body. Copper has been shown to be a constituent of some of the CYTOCHROME enzymes. These proteins are the catalysts for the main respiratory reaction chains, involving transfer of electrons from various carbohydrates to oxygen. Copper is also required for the synthesis of a number of enzymes and is involved in the glycolysis or breakdown of sugars.

The blue copper protein HEMOCYANIN occurs in the blood of certain lower forms of animal life. This compound performs the oxygen-carrying function for these species. This protein is believed to be a polypeptide containing +1 copper. It is not, however, as efficient an oxygen carrier as hemoglobin. The enzyme tyrosinase is found in many animals, being mainly responsible for SKIN PIGMENTATION and for hardening of fresh tissue in molting species.

Copper is also found in bacteria; in the diphtheria bacillus, copper is necessary for the production of toxins.

ANEMIA can be induced in animals on a low copper diet, such as milk, and appears to be due to an impaired ability of the body to absorb iron. This anemia, however, is rare, because of the widespread occurrence of copper in foods. In some places, *e.g.*, Australia and Holland, diseases of cattle and sheep, involving diarrhea, anemia and nervous disorders, can be traced either to a lack of copper in the diet, or to excessive amounts of MOLYBDENUM, which inhibits the storage of copper in the liver.

Ingestion of copper sulfate by humans causes vomiting, cramps, convulsions, and as little as 27 grams of the compound may cause death. An important part of the toxicity of copper to both plants and animals is probably due to its combination with thiol groups of certain enzymes, thereby inactivating them. The effects of chronic exposure to copper in animals are cirrhosis of the liver, failure of growth, and jaundice.

References

1. Butts, A., "Copper," American Chemical Society Monograph No. 122, New York, Reinhold Publishing Corp., 1954.

2. Monier-Williams, G. W., "Trace Elements in Food," New York, John Wiley & Sons, 1950.
3. McElroy, W. D., and Glass, B., Editors, "Symposium on Copper Metabolism," Baltimore, Johns Hopkins Press, 1950.
4. Thompson, R. H. S., and Webster, G. R., section on "Copper Metabolism" in "Neurochemistry," *Ann. Rev. Biochem.*, **29**, 376 (1960).
5. Holmberg, C. G., and Blomstrand, R., section on "Copper Metabolism and Disease" in "Clinical Biochemistry," *Ann. Rev. Biochem.*, **28**, 336 (1959).
6. Florkin, M., and Stotz, E. H., Editors, "Comprehensive Biochemistry," Vol. 8, Pt. 2, p. 48, New York, Elsevier, 1963.

R. R. Grinstead

CORTISOL AND CORTISONE

Cortisol (hydrocortisone; 17-hydroxycorticosterone; 11β, 17α, 21-trihydroxy-4-pregnene-3, 20-dione) and cortisone (11-dehydro-17-hydroxycorticosterone; 17α, 21-dihydroxy-4-pregnene-3,11,20-trione) are STEROIDS, members of the group of ADRENAL CORTICAL HORMONES.

COUNTERCURRENT DISTRIBUTION

Countercurrent distribution is a multistep separation process in which differences in distribution (solubility) of components of a mixture are utilized for purification. The distributions are carried out between two immiscible liquid phases. Much of the development of both the technique and the apparatus has been done by L. C. Craig, O. Post and their colleagues at the Rockefeller Institute in New York City, and several complete reviews have been written.[1-3] The process is stepwise and discontinuous in operation. Fig. 1 illustrates the simplest method of operation; a single solute (with a partition coefficient, K, equal to one) is distributed through three transfers. The diagram shows that even after three transfers the larger proportion of the material is concentrated into the two middle tubes. As the process continues through an increasing number of transfers, the percentage of the *total number* of tubes containing *significant* amounts of the solute will decrease although the zone of tubes containing the material will spread. The distribution of the solute throughout a train of units (test tubes in Fig. 1) follows the binomial expansion $(x + y)^n = 1$, where x is the concentration in the lower phase and y is the concentration in the upper phase. If a second solute with a K (concentration in the upper phase/concentration in the lower phase) of, for example, 3 was present in the system shown in Fig. 1, the material would have begun to distribute in the units of the train ahead of the solute whose $K = 1$. Fig. 2 shows the extension of a distribution to 1000 transfers and the separation of a major component ($K = 0.31$) and a minor component ($K = 0.43$). K for each component can be measured experimentally by determining the concentrations in upper and lower phases of selected tubes or from the position of the tube containing the maximum amount of solute. The expression used is $M = nK/(K + 1)$, where M is the position of the maximum; n, the number of transfers; K, the partition coefficient. A correction for unequal volumes of upper and lower phases can be made, if necessary.

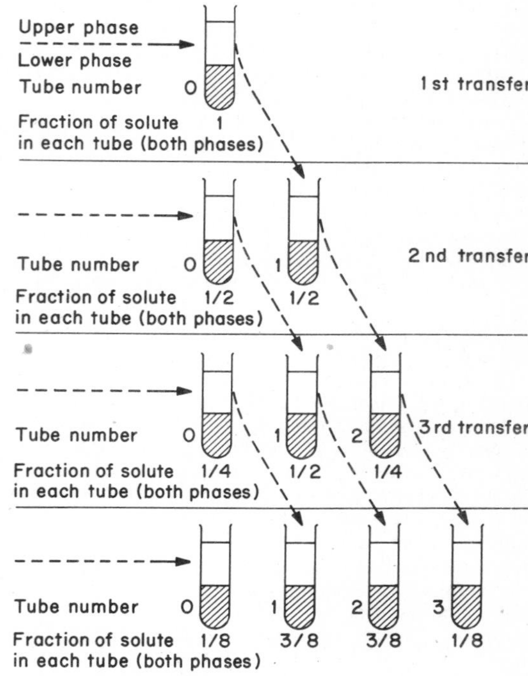

Fig. 1. Diagram illustrating the principle of a fundamental series in countercurrent distribution. Three transfers are illustrated and these proceed in sequence with a tube added to the series at each transfer. Distribution of the solute (whose partition coefficient K is 1) between the two phases in each tube is achieved by mixing the two phases and then allowing the phases to separate. The lower phase in each tube remains stationary during the transfer while each upper phase is moved to the adjoining tube in the series. Thus the upper phase moving into the new tube added to the series comes in contact with fresh lower phase while fresh upper phase is added to tube 0.

The distribution shown in Fig. 1 is the "fundamental procedure" of Craig and can easily be extended to 15 or 20 transfers in test tubes or in separatory funnels (in which the upper phase remains stationary and the lower phase is transferred to successive units). Convenient increase in the number of transfers beyond this point requires special apparatus and several designs are commercially available. Most common are apparatus, which are of glass construction, designed so that one operation permits simultaneous equilibration of the solute between the two phases in each of the tubes and followed by simultaneous transfer of the upper phase of each tube to the adjoining tube. These apparatus are often constructed with automatic devices for equilibration and transfer. Withdrawal or "recycling" procedures which are carried out following the "fundamental process" are useful in many instances and are described in the references. "Recycling" permitted the 1000-transfer distribution shown in Fig. 2 to be carried out in a train containing 200 tubes.

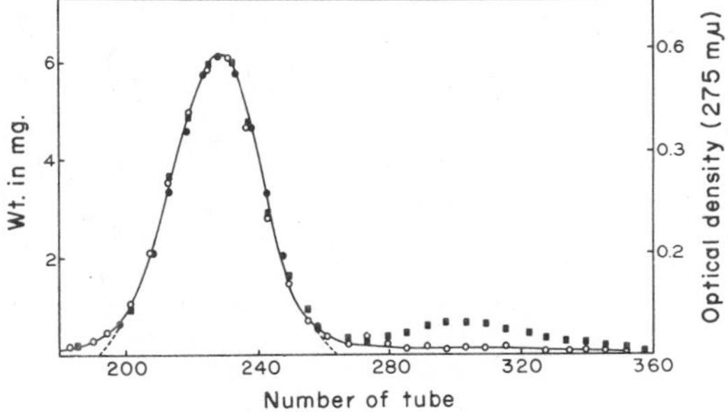

FIG. 2. The purification of the oxytocin hormone of the pituitary gland by countercurrent distribution through 1000 transfers. The hormone (an octapeptide) had previously been taken through several preliminary steps of purification including countercurrent distribution through fewer transfers. The figure shows the separation of a contaminant (the faster moving peak) with a relatively higher absorbancy in the ultraviolet. Absorbancy at 275 mμ is represented by ■; the weight of material in tubes by O. The theoretical distribution curve calculated for the major component (the hormone) is shown by ●. [Reprinted with permission from PIERCE, J. G., GORDON, S., AND DUVIGNEAUD, V., *J. Biol. Chem.*, **199**, 929 (1952).]

"Countercurrent distribution" has many characteristics in common with liquid-liquid partition chromatography in which solutes distribute themselves between an eluting liquid or mobile phase and a stationary liquid phase held on columns in which the packing material has, in itself, little attraction for the solute (see CHROMATOGRAPHY, COLUMN). Although in many cases such columns appear to be more convenient to operate, countercurrent distribution often has pronounced advantages for many problems in separation. The advantages include the fact that there is no question of adsorption of the solutes which thus can be recovered quantitatively following separation. The discontinuous nature of the process is of particular importance in that it allows complete equilibration of solute between the two phases before each transfer. A simple and rigorous mathematical analysis of the distribution curve can therefore be made when the concentrations of solutes are such that linear partition isotherms prevail. Theoretical curves for the distribution of a single substance can be calculated easily and fitted to the experimental curve. One is shown in Fig. 2 for the major component. The fit between experimental and theoretical distribution curves is a most rigorous criterion of purity for many substances. Theoretical curves can be calculated readily by several methods including direct calculation of each term of the binomial expansion.[4] For calculation of the curve after a distribution of many transfers, the mathematics of probability are easiest to apply because the distribution curve of a single solute is the plot of an ideal gaussian distribution. The equation

$$y = \frac{1}{\sqrt{2\pi nK/(K+1)^2}} e^{-x^2/[2nK/(K+1)^2]}$$

can be simplified[1] to allow calculation of y (values along the ordinate, Fig. 2) in terms of x (the number of tubes to the left or right of the maximum on the abscissa) and y_0 (the experimentally determined value of the maximum).

Applications. Separation and purification of many compounds within most of the major classes of biochemical substances have been accomplished by countercurrent distribution. Only limited success has been achieved with carbohydrates because of difficulty of finding suitable solvent systems. A description of many solvent systems and methods of operation is given by von Metzsch.[5] The process has been particularly successful in purification of peptides, including peptide ANTIBIOTICS, peptide HORMONES, the two types of polypeptide chains which constitute HEMOGLOBIN, and the peptides produced by enzymatic digestion of hemoglobin. The recent fractionation of low molecular weight RIBONUCLEIC ACID (transfer RNA) has led to the isolation of individual ribonucleic acids, each specific in accepting a single activated amino acid in *in vitro* systems which synthesize protein. R. W. Holley and associates have isolated transfer RNA specific for alanine and have determined the sequence of its component nucleotides.

Another significant development is the two-phase solvent system of P. A. Albertsson[6] which consists of aqueous solutions of certain polymers (*e.g.*, dextrans and polyglycols). These systems enable one to fractionate a number of biological macromolecules and particles including VIRUSES, nucleic acids, PROTEINS and whole cells. New modifications of the Craig apparatus have also appeared. One type is that of "counter double current distribution" which is effective for batchwise operation or for a "steady-state" operation where a continual feed of material occurs at the center of the train. Impurities are stripped from a major component at both ends. The term "countercurrent distribution" has been restricted by many authors to the discontinuous process. Numerous liquid-liquid distribution apparatus have been devised which permit continuous flow of mobile phase and which are valuable for many preparative purposes,[2] although equilibrium distribution of the solute between the two phases is usually not obtained.

References

1. Craig, L. C., and Craig, D., in "Technique of Organic Chemistry" (Weissberger, A., Editor), Second edition, Vol. 3, Part 1, p. 149, New York Interscience Publishers, 1956.
2. Morris, C. J. O. R., and Morris, P., in "Separation Methods in Biochemistry," p. 559, London, Pitman and Sons, 1963.
3. Craig, L. C., in "Comprehensive Biochemistry" (Florkin, M. and Stotz, E. H., Editors), Vol. 4, p. 1, Amsterdam, Elsevier Publishing Co., 1962.
4. Williamson, B., and Craig, L. C., *J. Biol. Chem.*, **168**, 687 (1947).
5. von Metzsch, F. A., in "Physical Methods in Chemical Analysis" (Berl W. G., Editor), Vol. IV, p. 317, New York, Academic Press, 1916.
6. Tiselius, A., Porath, J., and Albertsson, P. A., *Science*, **141**, 13 (1963).

J. G. Pierce

CREATINE AND CREATININE

Creatine was isolated from meat extract by Chevreul, who gave the compound its name, derived from the Greek word for flesh. He described its properties in 1835, and stated that it contained carbon, hydrogen, oxygen, and nitrogen. Liebig's classic researches on composition of meat extract, published in 1847, included quantitative analyses and formulas for creatine and compounds obtained from it. Creatine solutions, on evaporation with acid, yielded a base, creatinine, derived from creatine by loss of water. Another compound, sarcosine, obtained from creatine by the action of baryta, differed from creatine by one formula weight of urea. Volhard, in 1862, synthesized sarcosine from methylamine and monochloracetic acid, and in 1868 obtained a small yield of creatine from sarcosine and cyanamide, thus establishing the structure of creatine.

Structure and Biosynthesis. Structural formulas of compounds utilized for chemical synthesis of creatine and creatinine appear on the left in Fig. 1, and those of four compounds from which creatine is synthesized biologically appear on the right. Glycocyamine is formed in the kidney from glycine and arginine, by an enzymatic reaction known as transamidination. An amidine group from arginine is substituted for one of the amino hydrogens of glycine. By another enzymatic process known as transmethylation, occurring in the liver, glycocyamine is converted to creatine. The required methyl group is donated by a compound formed from methionine by an activating enzyme and ATP (see also methylations in metabolism). That the biological precursors of creatine are glycine, arginine, and glycocyamine derived from them, was established by Bloch and Schoenheimer with N^{15}-labeled substances. du Vigneaud and associates used methionine with deuterium in its methyl group to establish the source of this group in creatine (see also amino acid metabolism).

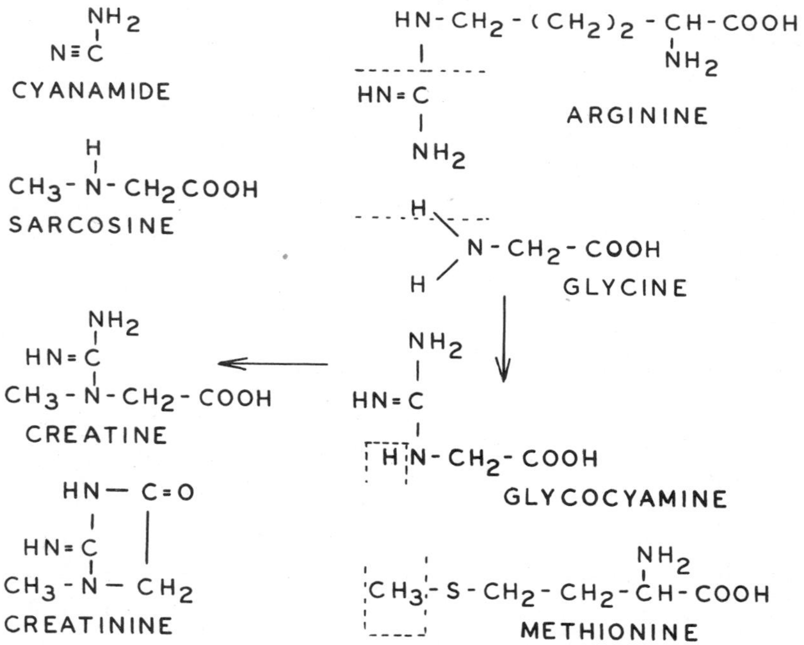

CYANAMIDE: $N{\equiv}C{-}NH_2$

SARCOSINE: $CH_3{-}NH{-}CH_2COOH$

CREATINE: $HN{=}C(NH_2){-}N(CH_3){-}CH_2{-}COOH$

CREATININE: $HN{=}C{-}N(CH_3){-}CH_2{-}C({=}O){-}NH$ (ring)

ARGININE: $HN{=}C(NH_2){-}HN{-}CH_2{-}(CH_2)_2{-}CH(NH_2){-}COOH$

GLYCINE: $H_2N{-}CH_2{-}COOH$

GLYCOCYAMINE: $HN{=}C(NH_2){-}HN{-}CH_2{-}COOH$

METHIONINE: $CH_3{-}S{-}CH_2{-}CH_2{-}CH(NH_2){-}COOH$

Fig. 1.

Determination Methods. In alkaline picrate solutions, creatinine develops a reddish-brown color like that of bichromate. This reaction, reported by Jaffé in 1886, was utilized by Folin in a quantitative method published in 1904. With alkaline 3,5-dinitrobenzoate solutions, creatinine yields a purplish rose color, utilized in quantitative methods of Benedict and Behre, Langley and Evans, and Bolliger. Creatine gives neither of these color reactions, but can be determined by either or them after conversion to creatinine. Several less widely used direct determinations of creatine have been described.

Non-specificity of the Jaffé reaction is a serious matter when concentrations of creatine or creatinine in a tissue or fluid are small and amounts of interfering substance are large. In such instances, isolation procedures and enzymatic methods are used to determine whether creatine and creatinine are present at all. After converting creatine to creatinine, Morris precipitated "total creatinine" as potassium creatinine picrate, from urines of children, fasting adults, diabetics, and women postpartum. Harding and Eagles isolated creatine, as creatinine potassium picrate, from brain. Gaebler and Abbott precipitated creatinine directly from serum ultrafiltrates as rubidium creatinine picrate, and converted this to creatinine zinc chloride for analysis. Linneweh confirmed earlier isolations of Gaebler and associates, and submitted the gold salt of isolated creatinine to elementary analysis. Enzymatic methods were introduced by Miller and Dubos, who used a creatinine-destroying enzyme, induced in soil bacteria by growing them on a medium in which nitrogen was supplied as creatinine.

Distribution. The body of an average adult contains about 100 grams of creatine. Over 90% of it is present in voluntary muscle, where its concentration, in different orders and species of vertebrates, varies from 350 to 500 mg/100 g. Cardiac muscle contains 250–300 mg/100 g. In smooth muscle and other tissues, the concentration is quite low. Thus the substance is abundant where extensive and rapid conversion of chemical energy to mechanical work occurs. The concentration of creatinine in blood and body fluids is normally very small—less than 1 mg/100 ml—but when renal failure occurs, it may rise above 20 mg/100 ml in blood.

Function. Eggleton and Eggleton observed that free inorganic phosphate appeared in increasing amounts in muscle during contraction under anaerobic conditions, and disappeared during recovery. Fiske and Subbarow observed that the inorganic phosphate content of acid filtrates of muscle increased steadily. The labile substance giving rise to phosphate was isolated and identified as phosphocreatine. This compound is an important reserve supply of PHOSPHATE BOND ENERGY, particularly under anaerobic conditions. An enzyme present in muscle, creatine kinase, catalyzes formation of ATP from ADP and phosphocreatine. The view has been widely held that ATP is the immediate source of energy for muscle contraction, and that it is resynthesized from ADP and phosphocreatine, which, in turn, is regenerated at the expense of energy derived from oxidation of lactic acid. Sachs has recently summarized evidence against the view that ATP is the initial and unique source of energy for contraction.

Excretion. Creatine normally appears in the urine of children and other young mammals. Adults on a creatine-free diet excrete only minute amounts, but creatinuria occurs in MUSCULAR DYSTROPHIES and many other diseases. Creatinine, on the other hand, is a major nitrogenous component of normal urine. The kidney concentrates it 100 times or more, so that urine may contain over 1 mg/ml.

That urinary creatinine originates from body creatine is evident from experiments of Bloch and Schoenheimer, in which body creatine and urinary creatinine were labeled to the same extent when N^{15}-labeled creatine was given. Administered labeled creatinine was simply excreted; it was not converted to body creatine. The amount of creatinine excreted per day is related to the amount of body creatine, and thus to the mass of muscle. Creatinine excretion per kilogram per day, known as the creatinine coefficient, is greater in muscular individuals than in obese ones.

Folin's discovery that the daily output of creatinine of a given individual is quite constant, and independent of protein intake, lent support to the concept that "exogenous" metabolism of food protein and "endogenous" metabolism of tissue protein could be sharply distinguished. Some 30 years later, Schoenheimer's studies with N^{15}-labeled amino acids demonstrated that dietary amino acids intermingle freely with those derived from tissue proteins, which are in a dynamic state of continuous breakdown and resynthesis.

References

1. FOLIN, O., "Laws Governing the Chemical Composition of Urine," *Am. J. Physiol.*, **13**, 66 (1905).
2. HUNTER, A., "Creatine and Creatinine" (monograph), London, Longmans Green and Co., 1928.
3. BLOCH, K., AND SCHOENHEIMER, R., "The Biological Precursors of Creatine," *J. Biol. Chem.*, **138**, 167 (1941).
4. GAEBLER, O. H., "Creatinine and Creatine" (review), in "Cyclopedia of Medicine, Surgery, and Specialties," Vol. 6, p. 466, Philadelphia, F. A. Davis and Co., 1950.
5. SACHS, J., "The Role of Adenosine Triphosphate in Muscular Contraction," *Perspectives Biol. Med.*, **7**, 285 (1964).

O. H. GAEBLER

CRETINISM

Cretinism or infantile myxedema is a disease state marked by dwarfism, an apathetic face, and delayed development, resulting from an inadequate production of THYROID HORMONES from the time of birth. It is associated also with a lowered body temperature and subnormal basal metabolic rate [see BASAL METABOLISM; see also GOITER; IODINE (IN BIOLOGICAL SYSTEMS)].

CRICK, F. H. C.

Francis Harry Compton Crick (1916–) is a British biochemist who, with JAMES DEWEY WATSON, published in 1953 a classic paper suggesting a double-helix structure for DNA molecules. For contributions in this field they jointly received a share in the 1962 Nobel Prize in physiology and medicine. The co-recipient was Maurice Wilkins, a British physicist.

CYANIDE POISONING

Occurrence. Hydrocyanic acid and alkali cyanides are found in fumigants, insectides, rodenticides, metal polishes, electroplating solutions and in various metallurgical and photographic processes. Nitriles occur in many plants. For example, amygdalin is a chemical compound of glucose, benzaldehyde, and hydrogen cyanide, and occurs in the bark, leaves, fruits and flowers of cherry laurel and in the seeds of cherry, plum, peach, apricot, apple and pear. Plants have been sources of cyanide poisoning in a small number of cases, but the majority of recognized poisonings in man have been traced to the accidental, suicidal or homicidal ingestion of cyanides.

Biological Action. Poisoning may arise from any substance which releases the cyanide ion. The lethality of most derivatives is proportional to the content of readily available cyanide. Cyanides reversibly inhibit oxidative enzymes, such as cytochrome oxidase (see CYTOCHROMES) and deprive tissues of necessary oxygen. Venous blood retains the bright red color of oxyhemoglobin because oxygen is not utilized by tissues. Cyanide itself does not react appreciably with hemoglobin. Brain is the organ most sensitive to lack of oxygen; therefore, during cyanide poisoning, a transient stage of central nervous system (CNS) stimulation is followed by CNS-depression, hypoxic convulsions, and death due to the respiratory arrest. Cardiac irregularites may be observed in some cases, but heartbeat invariably outlasts the breathing movements. Cyanides can produce toxic symptoms within a few minutes and death within one hour. The prognosis is fairly good if the victim survives for one hour after poisoning. The average oral lethal does is 60–90 mg of hydrogen cyanide or 200 mg of potassium cyanide. The mortality rate is high, but in non-fatal cases, the recovery is complete.

Antidotes and Treatment. Advantage has been taken of the metabolic pathways for the detoxification of cyanide *in vivo* in developing the antidotes or prophylactic measures agianst cyanide poisoning. A small amount of ingested cyanide is excreted by lungs, but most of it is converted to the comparatively nontoxic thiocyanate by the enzyme rhodanese (or transulfurase). The limiting factor for the above enzyme reaction is the availability of some source of sulfur. Administration of sulfur donors such as sodium thiosulfate and colloidal sulfur afford some protection against the toxic effects of cyanide. A quantitatively minor pathway of cyanide metabolism is the formation of cyanocobalamin (VITAMIN B_{12}). Administration of the precursors of Vitamin B_{12}, which lack the cyanide moiety, may accelerate this reaction. It has been found that both hydroxycobalamin and chlorocobalamin protect experimental animals against the toxic effects of cyanide. A third pathway for the detoxification of cyanide is its interaction with cystine with the formation of 2-imino-4-thiazolidine carboxylic acid, which is excreted in urine. This pathway has been demonstrated in the rat, but its clinical significance is not known.

Methemoglobin (MHgb) has greater affinity for cyanide than cytochrome oxidase, and therefore, it can compete effectively with cytochrome oxidase for cyanide ions. The cyanmethemoglobin (MHgb-CN) so formed is gradually changed to normal hemoglobin (Hgb) as the cyanide is converted to the relatively nontoxic thiocyanate [see also PORPHYRINS; IRON (IN BIOLOGICAL SYSTEMS)]. Therefore, when a part of the circulating Hgb is converted to MHgb *in vivo*, one can protect tissue cells until the cyanide ion is detoxified by rhodanese. There are many chemicals which produce MHgb *in vivo*. Inhalation of amylnitrite or intravenous administration of sodium nitrite produces appreciably high concentrations of circulating MHgb. Methylene blue is less satisfactory than nitrites for producing MHgb. *p*-Amino-propiophenone (PAPP) is more effective than sodium nitrite in promoting the formation of MHgb. However, PAPP is not a satisfactory antidote as MHgb formation is somewhat delayed.

Organic salts of cobalt (gluconate, glutamate) and the chelate with EDTA are effective antidotes against cyanide poisoning in animals. The exact mechanism of their action is not known, but it may possibly relate to the inhibition by cobalt of the enzymatic reduction of MHgb. Sodium cobaltinitrite is an effective antidote for cyanide poisoning in animals because it is both a former and sustainer of MHgb. As yet, the clinical usefulness of this compound has not been evaluated.

When combined with artificial respiration and general supportive therapy, the nitrite-thiosulfate regimen is the most effective treatment against cyanide toxicity. This regimen offers a combination of a sulfur donor with a compound which produces MHgb, a combination which is synergistic in its antidotal action. The procedure of treatment involves (1) inhalation of amylnitrite (one "Perle") by the patient for the rapid production of MHgb; (2) slow intravenous administation of sodium nitrite (3% solution, 2.5–5.0 ml/min, total 15 ml/adult) for sustained production of MHgb; (3) intravenous injection of sodium thiosulfate (12.5 grams in 50 ml) to enhance the conversion of cyanide to thiocyanate. The mechanisms of poisoning and detoxification can be depicted schematically as follows:

$$\text{Cyt (cytochrome oxidase)} \underset{}{\overset{\text{NaCN}}{\rightleftharpoons}} \text{Cyt—CN} \quad (1)$$

$$\text{Hgb (Fe}^{++}\text{) (hemoglobin)} \overset{\text{NaNO}_2}{\rightleftharpoons} \text{MHgb (Fe}^{+++}\text{) (methemoglobin)} \quad (2)$$

$$\text{MHgb} + \text{Cyt—CN} \longrightarrow \text{MHgb—CN (cyanmethemoglobin)} + \text{Cyt} \quad (3)$$

$$Na_2S_2O_3 + NaCN + [O] \underset{\text{SCN-oxidase}}{\overset{\text{Rhodanese}}{\rightleftharpoons}} NaSCN + Na_2SO_4 \quad (4)$$

Reaction (*4*) is slowly reversible through the action of thiocyanate oxidase, and this may account for the recurrence of symptoms of cyanide poisoning in some cases. If the symptoms were to reappear, the sodium nitrite and sodium thiosulfate should be repeated in one-half of the above does. Nitrites and thiocyanates are hypotensive agents, and consequently, the chief side effect is hypotension. Therefore, the blood pressure should be monitored carefully during the treatment.

References

1. CHEN, K., AND ROSE, C. L., "Treatment of Acute Cyanide Poisoning," *J. Am. Med. Assoc.*, **149**, 113 (1952).
2. DONE, A. K., "Clinical Pharmacology of Antidotes," *Clin. Pharmacol. Therap.*, **2**, 765 (1961).
3. GOSSELIN, R. E., AND SMITH, R. P., in "Clinical Toxicology of Commercial Products," (GLEASON, M. N., GOSSELIN, R. E., AND HODGE, H. C., EDITORS), Section III, pp. 54–56, Baltimore, The Williams & Wilkins Co., 1963.
4. HIRSCH, F. G., "Cyanide Poisoning," *Arch. Environ. Health*, **8**, 622 (1964).

ALLAN D. BASS AND B. V. RAMA SASTRY

CYSTEINE AND CYSTINE

These AMINO ACIDS are derived from many common PROTEINS by hydrolysis:

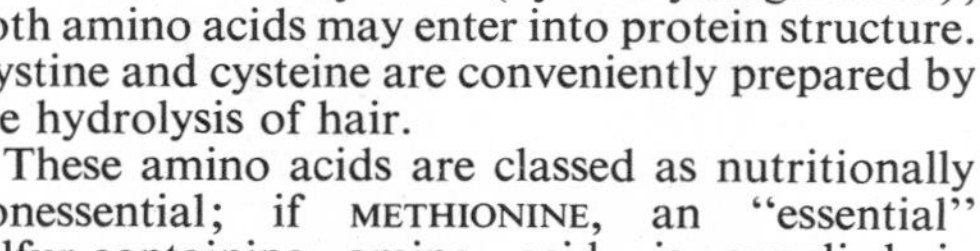

$$\begin{array}{c} COOH \\ | \\ H_2N—C—H \\ | \\ CH_2—SH \end{array} \qquad \begin{array}{ccc} COOH & & COOH \\ | & & | \\ H_2N—C—H & & H_2N—C—H \\ | & & | \\ CH_2—S & — & S—CH_2 \end{array}$$

L-Cysteine L-Cystine

Ordinarily unless precautions are taken to avoid oxidation, cysteine is not obtained as such but is converted into cystine (by dehydrogenation); both amino acids may enter into protein structure. Cystine and cysteine are conveniently prepared by the hydrolysis of hair.

These amino acids are classed as nutritionally nonessential; if METHIONINE, an "essential" sulfur-containing amino acid, is supplied in limited quantity, however, cystine can supplement it effectively in the diet. Part of the essential methionine is needed for the production of cysteine and cystine. Cysteine has a specific rotatory power $[\alpha]_D = +7.6°$ in *N* HCl solution. Cystine under similar acid conditions has a specific rotatory power of $-223°$. This contrast is unexpected.

CYTIDINE

Cytidine is a nucleoside consisting of cytosine attached through a β-glycosidic linkage to ribose, and is thus comparable to ADENOSINE. When esterified with phosphoric acid, it yields the corresponding nucleotides which may occur as monomeric building units of nucleic acids (see PYRIMIDINES).

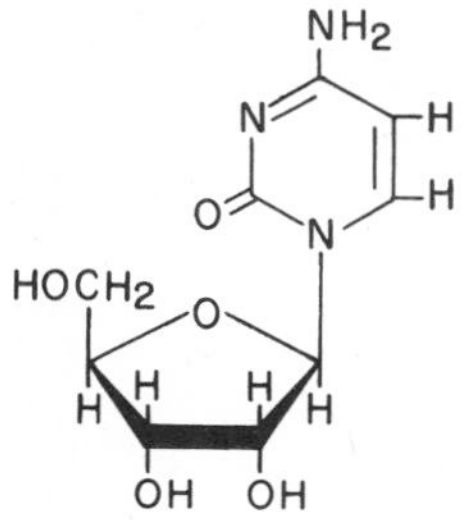

NH_2 N H H O N $HOCH_2$ O H H H H OH OH

CYTOCHROMES

Definition and History of "Mammalian-type" Cytochrome *c*–Cytochrome Oxidase System. The cytochrome *c*–cytochrome oxidase system represents the terminal segment of the respiratory chain common to the vast majority of organisms utilizing oxygen as the terminal oxidant in tissue RESPIRATION. The complete respiratory chain consists of a number of electron carriers, both protein and non-protein in nature, organized in a definite sequence within the walls and internal partitions of subcellular organelles known as MITOCHONDRIA. These structures carry the electrons which come from the substrates being oxidized and eventually react with oxygen. The energy released in several of the many steps of this series of reactions is utilized to make the high-energy compound adenosine triphosphate, a process known as oxidative PHOSPHORYLATION. The high-energy compound is, in turn, employed to drive the many reactions of metabolism which require chemical energy. Every component of the terminal respiratory chain is reduced by the component immediately preceeding it and then reduces the component immediately following it in the chain, itself becoming reoxidized. This article is concerned with the so-called mammalian-type of cytochrome *c*–cytochrome oxidase system which is common to all vertebrates and invertebrates, plants, as well as numerous microorganisms, and which must be distinquished from systems having similar functions but very different properties, which occur in numerous bacteria.

The cytochromes were first observed by MacMunn, as early as 1886, who described their spectral absorption bands in a large variety of organisms and tissues. His discovery was, however forgotten after a controversy with Hoppe-Seyler had raised doubts as to the validity of some of his conclusions, and it was not until 1925 that Keilin independently rediscovered the remarkable cytochrome spectrum in the flight muscles of a living insect.

Keilin's observations came at a time when the understanding of tissue respiration had advanced to the point of providing the foundations necessary for the unraveling of the physiological role and chemical nature of cytochromes. The first step had indeed been taken some forty years earlier by Ehrlich when he found that a variety of animal tissues could transform a mixture of α-naphthol and dimethyl-*p*-phenylenediamine to indophenol, in the presence of oxygen. A decade later, the enzyme responsible for this effect had been named indophenol oxidase, and it was shown that its activity was inhibited by cyanide. In the first decades of this century, Warburg, from studies of the catalysis of the oxidation of cysteine by iron-charcoal, considered as a "model" of cellular respiration, concluded that oxygen activation was the all-important process in cellular respiration and that an iron-containing enzyme, the "respiratory enzyme" or *atmungsferment*, is solely responsible for the transport of the oxidizing equivalents of oxygen to the substrates. An opposing view was taken by Thunberg, who had detected a large variety of dehydrogenases in tissues, and by Wieland who used palladium-hydrogen as a "model" of tissue respiration and believed that substrate-specific hydrogen activations were characteristic of all biological oxidation processes, the reaction with oxygen being nonspecific and relatively unimportant.

The controversy as to the respective roles and importance of hydrogen and oxygen activation faded into the background when, following his initial observations, Keilin demonstrated that the four-banded spectrum of cytochrome, observed in a large variety of tissues and organisms, was in fact the spectrum of the ferrous or reduced forms of three distinct cytochromes, cytochrome *a*, cytochrome *b* and cytochrome *c*. Keilin obtained a soluble preparation of cytochrome *c* from baker's yeast, and together with Hartree, in 1938–1939, showed that the indophenol oxidase activity of particulate tissue preparations was simply the result of a non-enzymic reduction of cytochrome *c* by dimethyl-*p*-phenylenediamine, the reduced heme protein being oxidized by indophenol oxidase in the presence of oxygen. Having established the nature of the final steps of tissue respiration, they renamed the enzyme "cytochrome oxidase," since its only function appeared to be the oxidation of cytochrome *c*. There had been no doubt of the overwhelming physiological importance of the system ever since 1934 when Haas found that in a number of tissues the rate of oxygen uptake was identical to that of cytochrome *c* reduction, demonstrating that nearly all of the oxidizing equivalents of oxygen were transmitted by the cytochrome *c*–cytochrome oxidase system.

That the material in tissues reacting directly with oxygen was in fact a HEME compound had been shown by the experiments of Warburg and collaborators on the effect of carbon monoxide on tissue respiration, carried out in the late 1920's. Warburg observed that carbon monoxide inhibits the uptake of oxygen by tissues and that this inhibition is reversed in bright light. Using this phenomenon, he succeeded in measuring the absorption spectrum of the carbon monoxide complex of the respiratory enzyme, a spectrum which turned out to be clearly that of a heme compound. Thus, when Keilin and Hartree in 1939 found that in the presence of carbon monoxide, cytochrome *a* showed up as two spectroscopic components, they were able to demonstrate that the new cytochrome, cytochrome a_3 was the substance responsible for the photochemical action spectrum of Warburg. Cytochrome a_3 was thus identified with the respiratory enzyme reacting directly with oxygen, and the system was considered to be composed of three entities, cytochromes *c*, *a*, and a_3, reacting consecutively, like all the other components of the respiratory chain.

Cytochrome *c*. *Primary Structure and Evolution.* Cytochrome *c* consists of a polypeptide chain, 104–108 amino acid residues in length. A single heme prosthetic group is attached by thioether bonds formed between the sulfhydryl side chains of two cysteine residues in the protein and the vinyl side chains of the PORPHYRIN ring (see Fig. 1). A microphotograph of horse heart cytochrome *c* crystals is shown in Fig. 2, the amino acid sequence of the protein is given in Fig. 3, and its spectrum in Fig. 4 (see ABBREVIATIONS, for amino acid residue abbreviations).

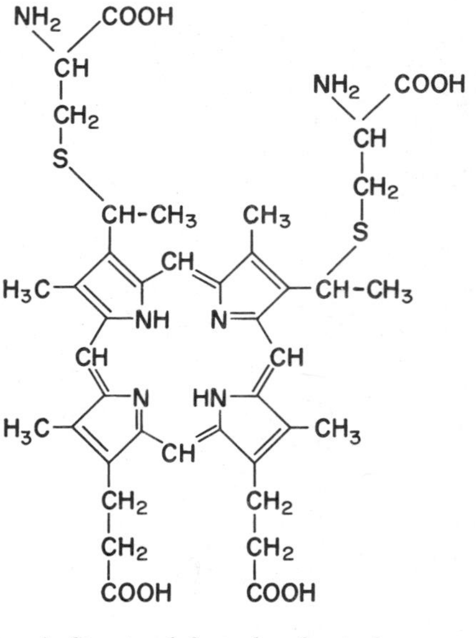

FIG. 1. Structural formula of cytochrome *c* porphyrin with its two attached cysteinyl residues (see top of Figure).

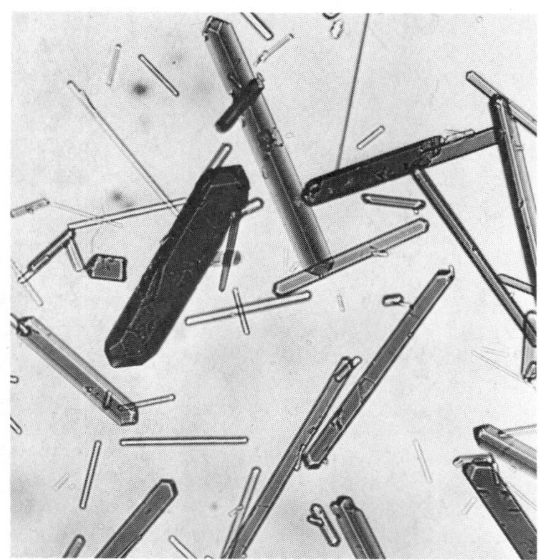

FIG. 2. Crystals of horse heart cytochrome *c*, magnified 75 times.

Acetyl-Gly-Asp-Val-Glu-Lys-Gly-Lys-Lys-Ile-Phe-Val-Gln-Lys-
10

CyS-Ala-Gln-CyS-His-Thr-Val-Glu-Lys-Gly-Gly-Lys-His-Lys-Thr-
20
HEME

Gly-Pro-Asn-Leu-His-Gly-Leu-Phe-Gly-Arg-Lys-Thr-Gly-Gln-
30 40

Ala-Pro-Gly-Phe-Thr-Tyr-Thr-Asp-Ala-Asn-Lys-Asn-Lys-Gly-
50

Ile-Thr-Trp-Lys-Glu-Glu-Thr-Leu-Met-Glu-Tyr-Leu-Glu-Asn-
60 70

Pro-Lys-Lys-Tyr-Ile-Pro-Gly-Thr-Lys-Met-Ile-Phe-Ala-Gly-Ile-
80

Lys-Lys-Lys-Thr-Glu-Arg-Asp-Leu-Ile-Ala-Tyr-Leu-Lys-Lys-
90 100

Ala-Thr-Asn-GluCOOH
104

FIG. 3. Amino acid sequence of horse heart cytochrome *c*.

Horse heart cytochrome *c* was the first for which a complete primary structure was determined in 1962, and it is quite typical of the large number of cytochromes *c* which have since been studied. These include the cytochromes *c* from man, pig, chicken, baker's yeast, cow, sheep, tuna, a rhesus monkey *Macacus mulatta*, the domestic rabbit, a saturnid moth, *Samia cynthia*, the dog, the great grey kangaroo, the rattlesnake, turkey, the domestic pigeon, the King penguin, Pekin duck, the screw worm fly, and the proteins from the kidney, liver, brain and skeletal muscles of the pig.

Like all cytochromes *c* from vertebrate species, the horse heart protein (see Fig. 3) lacks a free α-amino group at the amino-terminal residue, the chain starting with an N-acetylglycine. The heme is bound through the sulfhydryl groups of the two cysteinyl residues in positions 14 and 17, separated by two other amino acids. Studies with atomic models and more recent direct syntheses of this structure have shown that such a spacing is necessary for the establishment of the proper thioether bonds. Characteristically, a lysyl residue precedes the first cysteine, while a histidyl-threonine sequence follows the second cysteine.

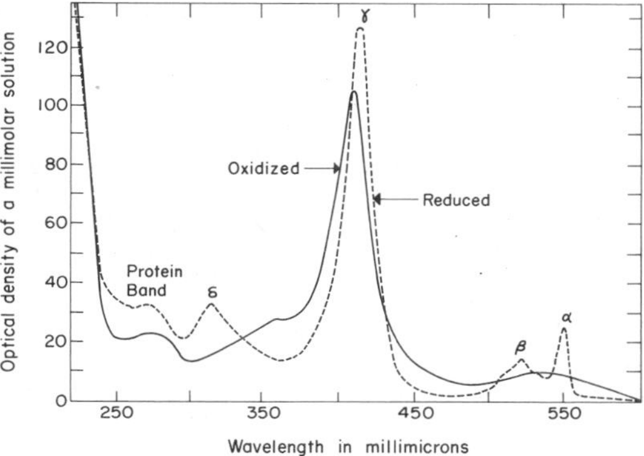

FIG. 4. Absorption spectrum of horse heart cytochrome *c* in the visible and ultraviolet regions.

This arrangement is universal among mammalian-type cytochromes *c* as well as in cytochromes with quite different properties which do not react with the same oxidase system, such as cytochrome c_2 of *Rhodospirillum rubrum* and the variant heme protein or cytochromoid of *Chromatium*. In some cytochromes *c*, such as the protein of baker's yeast, that of the silkworm and other insects, the lysine is replaced by an arginine. In all cases, however, there is a grouping of hydrophobic residues (see below) in the area immediately preceding this lysine or arginine, and there is a single hydrophobic residue immediately following the histidyl-threonine sequence.

In general, there is a marked tendency of hydrophobic and basic residues to occur in discrete groups along the polypeptide chain. Among the hydrophobic amino acids, one commonly includes valine, leucine, isoleucine, tyrosine, phenylalanine, tryptophan and methionine. The cytochromes *c* which have been examined to date show eight segments which contain from 23–27 of the 26–29 hydrophobic residues in the protein, and this hydrophobic territory covers a constant proportion of 32% of the entire polypeptide chain. Such a remarkable stability of distribution is probably related to the importance of these hydrophobic areas in the formation of intramolecular hydrophobic bonds, which are probably responsible for the formation and the maintenance of the particular spatial configuration of cytochrome *c*. A comparison of the structures of various cytochromes *c* reveals that even more strongly than the hydrophobic character of individual residues in certain positions, the entities which are conserved in evolution are the hydrophobic groupings themselves and their locations in particular areas on the polypeptide chain. The basic residues lysine, arginine, and histidine also tend to form clusters which are conserved, but in a manner which does not appear to be quite as rigid as for the hydrophobic segments. It is noteworthy that nearly every hydrophobic grouping or even each separate hydrophobic residue is contiguous to, or a single residue removed from, a single or a cluster of basic residues.

Another physicochemical character which appears to be rather strictly conserved during the evolution of the protein is its net charge and isoionic point. Notwithstanding a considerable degree of variability in the total number of basic and acidic residues, the net charge is quite constant, varying by no more than 1.5 cationic charges. Similarly, the isoionic point of all the cytochromes *c* which have been examined is, within experimental error, pH 10.04. Such a constancy of basic character may well be an expression of the fundamental significance of the cationic character of the protein in all of its physiological reactions, as will be discussed below. The intimate relationship between cationic and functional behavior could represent one of the cardinal selective forces guiding the evolutionary stability or flux of charged groups in the protein, possibly implying the occurrence of interrelated compensating evolutionary variations. It is indeed probable that the occurrence of a mutation leading to a variation in the charged groups of the molecule would tend to favor the retention of subsequent mutations that reestablish the most favorable balance of charges.

In the present set of mammalian type cytochromes *c*, variations in amino acid residues have been found to occur in no more than 55 positions along the polypeptide chain. The most numerous changes are observed at residues 89 and 92, in which 7 and 6 different amino acids have been detected, respectively, while there are 11 positions in which four different residues have been found, 14 in which 3 have been detected, and 28 in which 2 different amino acids have been observed. This 50% identity of amino acid sequence makes it probable that all of these proteins derive phylogenetically from a common ancestral form. This homology implies that the cytochrome *c* gene has survived as a distinct entity for some 1–2 billion years, essentially since organisms learned to use oxygen as the terminal oxidant of metabolism. Such a conclusion presupposes that the effective emergence of life on earth must have resulted from a unique occurrence. The only alternative hypothesis would be that these extensive similarities of primary structure arise as a result of convergent evolution, dictated by the functional requirements of the protein. If this were so, one would have to assume that on the average every second residue is an absolutely immutable structural requirement, a situation which is unlikely since it can be shown that protein functions and conformation are commonly compatible with a variety of amino acid sequences. Species known to be relatively closely related phylogenetically exhibit either few or no differences in their cytochromes *c*, while the proteins from more distantly related species show relatively larger degrees of variation. Thus, for example, the amino acid sequences of the cytochromes *c* from the pig and the cow are identical, and differ from that of the horse by only three residues. In general, cytochromes *c* from mammalian species differ from the fish proteins by approximately 20 amino acids, those from vertebrate species differ from the insect cytochromes *c* by approximately 30 amino acids, while at the other end of the scale, baker's yeast cytochrome *c* differs from all the other proteins examined by 43–49 residues. In addition to this qualitative connection between the taxonomic kinship of species and the number of variant residues, there appears to be a quantitative aspect to this relationship. The latter has been used to construct a curve relating the time elapsed since the divergence of species grouped in large categories of systematic classification and the number of variant residues in their cytochromes *c* (see Margoliash and Smith, 1965).

An area which has shown a particularly striking degree of variability is the sequence covered by the carboxyl-terminal four residues. Four or five different amino acids have been detected in each of these positions, and they range through acidic, basic and neutral residues, including cysteine and the amides of the dicarboxylic amino acids, but excluding all hydrophobic residues. The only structural requirement of this area appears to be absence of hydrophobic residues. Moreover, it has been shown that the removal of this terminal sequence by enzymic digestion results in no detectable change of the functional activity of the protein. Thus, this extensive variability can hardly be based on evolutionary selection for specific functional requirements, and it is likely that such variations have been fixed in the appropriate populations by mechanisms, such as genetic drift, that do not directly involve natural selection.

Strongly contrasting with such highly variable areas is the sequence which extends between residues 70 and 80, a segment which has remained strictly invariant in every one of the cytochromes *c* examined. This evolutionary constancy has recently been shown to be related to an unknown but nevertheless critical structural requirement. Indeed, alkylation of the methionyl residue in position 80 leads to a complete loss of functional activity, a similar but weaker effect being observed when one of the two lysyl residues in positions 72 or 73 is chemically changed by the addition of a trinitrophenylsulfonate group to the ε-amino function. The implications of such results with regard to the requirement for structural integrity of the invariant segment are clear when one considers, for example, that alkylation of the other methionine in position 65 has no effect whatsoever on cytochrome *c* function.

It is necessary to point out that notwithstanding the detailed information that has accumulated concerning the amino acid sequences of a large variety of cytochromes *c*, representing species at all levels of the phylogenetic scale, and the intriguing and possibly important relations that have been detected between the evolution of species and the structures of their cytochromes *c*, briefly discussed above, definitive interpretations in evolutionary terms of differences and similarities of primary structure still elude us. The difficulties stem mainly from the relative paucity of our understanding of the exact relationships of cytochrome *c* with its natural environment, the intra-mitochondrial respiratory chain. Whatever is known about this natural milieu, which in some evolutionary connotations may have to be extended to the entire organism, is discussed below.

Properties and Spatial Structure of the Protein. X-ray crystallographic work, which will eventually lead to a complete solution of the spatial structure of cytochrome *c*, is still at its very early stages. Some electron microscopic observations have indicated that the protein may have the general appearance of a letter "e" flattened in the vertical direction, thus consisting of three segments of about equal length running parallel to each other and connected by two turns. This model leads to an estimate of the dimensions of the molecule as approximately 40 × 28 Å. Because of the uncertainty of interpretations of structure based on negative contrast ELECTRON MICROSCOPY, it is difficult at this time to judge to what extent such a model does or does not approximate the actual spatial conformation of the protein. This model implies a very high content of α-helical structures, whereas direct estimations give cytochrome *c* a rather low helical content.

Whatever the three-dimensional structure of the protein will eventually turn out to be, there is little doubt that the prosthetic group is located in a so-called crevice which provides the specific environment required to impart to the cytochrome *c* heme its particular properties. Thus, the two amino acid side chains which must coordinate with the central heme iron atom to give the characteristic hemochrome spectrum form part of this environment. Similarly, the characteristic stability of the reduced or ferrous form of cytochrome *c* which does not react with molecular oxygen, or with ligands such as carbon monoxide, must also be a property of the environment of the prosthetic group. The identification of the amino acids which carry the hemochrome-forming side chains is still uncertain. There is general agreement that the imidazole group of the histidyl residue in position 18 is one of these groups, but what the other hemochrome-forming group is, has not been decided.

Cytochrome *c* readily forms polymers which have lost the typical enzymic activities, are autoxidizable, and react with carbon monoxide. At extremes of pH and in concentrated solutions of urea or guanidine hydrochloride, such polymers readily revert to the monomeric state. This reaction is accompanied by a complete recovery of the enzymic and hemochrome properties characteristic of the native molecule.

There is so far only one well-authenticated case of the occurrence of two molecular forms of the protein in a single organism. Baker's yeast contains two cytochromes *c*, isocytochrome-1 and isocytochrome-2, which are under independent genetic control. The amino acid sequence of isocytochrome-2 has not yet been completely determined. It differs from isocytochrome-1 by several residues, among which are a lysine replacing the carboxyl-terminal glutamic acid and glutaminyl and isoleucyl residues replacing the valine and the leucine in the area immediately adjacent to the attachments of the heme to the polypeptide chain. The two isocytochromes can be separated by cation-exchange chromatography, isocytochrome-2 being present in amounts varying from 1% to about 20% that of isocytochrome-1. An interesting hypothesis of the function of isocytochrome-2 states that the apoprotein acts as the genetic repressor of the synthesis of the main isocytochrome-1. Thus, when anaerobic yeast which has little if any cytochrome *c* is placed in an environment containing oxygen, heme is bound to the preformed apoprotein of isocytochrome-2, releasing the repression of synthesis of isocytochrome-1 and permitting the cell to develop its full complement of the protein.

The tissue concentrations of cytochrome *c* vary over a wide range. Muscles which act continuously over prolonged periods and require a relatively large expenditure of energy, such as the myocardium in most animals and the flight muscles of birds, contain a particularly large concentration. Conversely, various internal organs and embryonic tissues contain far less. The amount of cytochrome *c* in most tissues seems to be regulated, at least in part, by the THYROID HORMONE. Tissues of hyperthyroid animals contain increased concentrations of the protein while those of hypothyroid animals may have as little as half the normal amount. The antigenic activity of cytochrome *c* is low, and it is only quite recently that antibodies to the protein have been obtained.

Reaction of Cytochrome *c* with Cytochrome Oxidase. *Cationic Behavior of Cytochrome c in Relation to the Oxidase Reaction.* The kinetics of oxidation of soluble cytochrome *c* by cytochrome oxidase are different from those that are observed in most reactions between an enzyme and its substrate. A plot of the rate of oxidation of the reduced protein as a function of its concentration yields a hyperbolic curve, indicating that cytochrome *c* itself is an inhibitor of its oxidation by cytochrome oxidase. The most probable mechanism explaining this behavior involves the formation of a complex between cytochrome *c* and the oxidized or ferric form of cytochrome oxidase, then an internal oxido-reduction in which the oxidase heme is reduced and the cytochrome *c* oxidized, followed by the oxidation of the enzyme by molecular oxygen and the dissociation of the final complex between oxidized enzyme and ferricytochrome *c* into free enzyme and free cytochrome *c*. It is generally assumed that is is the complex between ferricytochrome *c* and the oxidase that inhibits the repetition of this cycle. Indeed, recent rapid spectrophotometric measurements of these kinetics have shown that after one mole of cytochrome *c* per mole of cytochrome oxidase heme is oxidized, a reaction which occurs particularly rapidly, the oxidation of the remainder of the ferrocytochrome is very much slower, taking place at the rates which are usually measured as the rate of oxidation of cytochrome *c* by cytochrome oxidase.

The forces which hold together the components of the inhibitory complex are both ionic and hydrophobic. The ionic forces, presumably due to electrostatic interaction between the basic groups on cytochrome *c* and oppositely charged groups on the oxidase, are affected by the degree of basicity of cytochrome *c*, the concentration and the valence of the inorganic cations in the solution, and the presence of polycationic substances. Thus, for example, if the lysyl residues in cytochrome *c* are guanidinated chemically, a reaction which leads to their replacement by homoarginyl residues, distinctly stronger bases than the original lysines, the maximal rates of oxidation occur at relatively low concentrations of guanidinated cytochrome *c* and are considerably lower than for the native protein. Similarly, the presence of other basic proteins such as ribonuclease, lysozyme and histone, or of synthetic polycationic substances such as polylysine or glucose polymers positively charged by the introduction of diethylaminohydroxylpropyl groups, strongly inhibits the cytochrome *c*–cytochrome oxidase reaction. Conversely, the addition of polyanionic substances to the reaction mixture inhibited by polycations reverses these inhibitions completely. Finally, inorganic cations, but not anions, compete with cytochrome *c* for attachment to preparations of the terminal oxidation chain deficient in cytochrome *c*. The higher the valency of the cation, the more effective is the competition. In general, the higher the cation concentration, the slower is the oxidase reaction.

Electrostatic forces are not the only ones responsible for the inhibition of cytochrome oxidase by cytochrome *c*. Indeed, partially acetylated cytochrome *c* which has lost its basic character can be shown to be quite effective in decreasing the rate constant for the oxidase reaction. Similarly, when synthetic polycations are used to inhibit the oxidase, the inhibition is not only a function of the charge density of the inhibitor but also of its hydrophobicity. This phenomenon is particularly obvious at relatively high ionic strengths. Under such conditions, copolymers of alanine and lysine have a considerably weakened inhibitory effect, while copolymers of phenylalanine and lysine exert their full effect, in direct relation to the proportion of phenylalanine in the compound.

Cytochrome c in Its Natural Milieu. In comparison to our detailed knowledge of the activities and physicochemical properties of cytochrome *c* in solution, our understanding of the reactions of cytochrome *c* in its natural environment, the mitochondrion, is very much less satisfactory. There is no question that the mitochondrion presents a highly organized structure of precise geometry having a high lipid content, and it is unlikely that in such an environment thermodynamic considerations derived from experiments in aqueous media are totally applicable. The picture obtained of the situation of cytochrome *c* in the mitochondrion is of a sort of cage in which the protein can shuttle between a reductase site providing the electrons which reduce it and an oxidase site at which these electrons are given up to the cytochrome oxidase complex. Cytochrome *c* is not extractable from intact mitochondria, and in order to obtain the protein in solution it is necessary to damage the mitochondrial structure. When this operation is carried out under well-controlled conditions, it is possible to remove all the cytochrome *c*, to separate the cytochrome *c*-deficient mitochondria and, at the appropriate ionic strength, to reincorporate cytochrome *c* in exactly the original concentration. Such reconstituted mitochondrial systems appear to carry out all the reactions in which cytochrome *c* is involved as well as, or nearly as well as, the original intact particles. There has nevertheless been some irreversible structural damage since a mere change of the cation concentration in the medium will release the protein, without going through the osmotic shock treatment required for native mitochondria. Whether or not cytochrome *c* is bound to lipid in its natural milieu has not yet been decided. The various lipid–cytochrome *c* complexes that have been obtained are artifacts of preparation and do not necessarily reflect the normal situation of the protein.

The major difficulty in attacking experimentally the relations of cytochrome *c* to its *in vivo* environment is that in strictly intact mitochondria the protein is out of contact with the medium and is thus not affected by the various conditions that have been shown to influence soluble cytochrome *c*, such as ionic strength or the presence of polycationic substances. All the observations which have been carried out so far do not appear to require mitochondrial binding sites, other than the ones on the enzymes with which the protein reacts. It is quite possible that the specific functions of mitochondrial structure in the region of cytochrome *c* are simply to keep the reductase and oxidase sites in the spatial relation that is most favorable for cytochrome *c* electron transfer activities, as well as to prevent, possibly mechanically, the diffusion of cytochrome *c* away from its locality of action.

Cytochrome *c* Oxidase and its Reaction with Oxygen. In contrast to the ease with which cytochrome *c* can be extracted from tissues, cytochrome oxidase appears to be firmly bound to the insoluble structural elements of mitochondria. It is only through the use of detergents, including bile salts, that so-called solublized preparations of the enzyme have been obtained.

Such preparations have been purified to the point that they do not contain any of the other members of the terminal respiratory chain, but it is as yet uncertain whether they are pure from a protein point of view. These preparations contain a considerable amount of lipid, approximately 20%, and the extensive removal of such lipid components causes the preparations to become insoluble and lose their enzymic activity. It is not possible to say whether these lipids are important, as such, in the functional activity of the enzyme, or whether they represent a means of maintaining the strongly hydrophobic protein in its native configuration.

Only a single heme, heme *a*, can be extracted from cytochrome oxidase preparations, though spectroscopically two varieties of heme protein are present, cytochrome *a* and cytochrome a_3. These are recognizable in that cytochrome a_3 reacts with ligands, such as carbon monoxide and cyanide, while cytochrome *a* exhibits no change of its spectrum in the presence of these reagents. Cytochrome a_3 is the component of the oxidase system which reacts directly with molecular oxygen, performing the final step of the long series of reactions carried out by the terminal oxidation chain, whereas cytochrome *a* appears to be a carrier, similar to other heme protein carriers of the terminal respiratory chain, transporting electrons between cytochrome *c* and cytochrome a_3. Since both the cytochrome components of the oxidase contain the same heme, it must be the specific environments of the prosthetic groups, provided by the protein, which induce the differences between cytochromes *a* and a_3. Whether a single protein molecule enfolds the two prosthetic groups, or whether there are two distinct protein molecules involved, is not known. The most recent estimates of the relative quantities of cytochromes *a* and a_3 indicate essentially equal amounts.

Although in their first papers on cytochrome oxidase, Keilin and Hartree, in 1939, found that their preparations contain a substantial quantity of non-dialyzable copper, the relation of this metal to cytochrome oxidase activity has become evident only during the last few years. There is today no doubt that copper is an essential portion of the enzyme, and undergoes reversible oxidation and reduction. It has been shown that one molecule of oxygen, containing four oxidizing equivalents, will oxidize only two molecules of heme *a*, the other two oxidizing equivalents presumably oxidizing the two copper atoms in the enzyme. Similarly, when cytochrome oxidase preparations are titrated with reduced coenzyme I, two electrons are accepted for each heme group, one presumably serving to reduce the heme and the other reducing the associated copper atom. The oxidation and reduction of the copper in cytochrome oxidase has been followed by electron paramagnetic resonance spectroscopy, as well as by the changes in a spectral adsorption band in the near-infrared region, at 820 mμ. This band appears to be a specific expression of the enzymically active copper related to cytochrome *a*, and its kinetic behavior in the reaction of reduced cytochrome oxidase with oxygen shows that the molecular species producing it is kinetically distinct from both cytochrome *a* and cytochrome a_3. On oxidation, the rate of the appearance of the 820 mμ band is intermediate between the rate of oxidation of cytochrome *a* and the very rapid maximal rate of oxidation of cytochrome a_3. These measurements are compatible with a model in which the copper producing the spectral band would be an obligatory intermediate between the *a* and a_3 components. It must be emphasized that the 820 mμ band appears to be associated with the copper related to cytochrome *a*, whereas the copper atom related to cytochrome a_3 is spectroscopically invisible. Whether the copper associated with cytochrome a_3, like the copper associated with cytochrome *a*, undergoes oxidation more rapidly than its accompanying heme is as yet unknown.

In summary, cytochrome oxidase consists of two heme components, contained within one or two proteins, each associated with a copper atom which is indispensible to the functional activity of the enzyme complex. Here again, as with cytochrome *c*, it is important to remember that our knowledge of cytochrome oxidase is largely based on preparations that have been forcibly removed from their natural environment; consequently, one cannot be certain that all *in vitro* observations apply completely to the *in vivo* situation. Major questions relating to the number of redox components involved in the activity of the oxidase, the pathways by which electrons are transferred within the oxidase complex, the actual mechanism of oxygen reduction, and the linkage of the oxidase to the respiratory chain as a whole, which permits the integration of its activities with those of the entire system, are yet to be completely answered.

References

1. KEILIN, D., "On Cytochrome, a Respiratory Pigment, Common to Animals, Yeast and Higher Plants," *Proc. Roy. Soc. London, Ser. B.*, **98**, 312 (1925).
2. FALK, J. E., LEMBERG, R., AND MORTON, R. K., (EDITORS), "Haematin Enzymes," London, Pergamon Press, 1961.
3. NICHOLLS, P. "Cytochromes, a Survey," in "The Enzymes" (BOYER, P. D., LARDY, H. AND MYRBÄCK K., EDITORS), Vol. 8, p. 3, New York, Academic Press, 1963.
4. YONETANI, T., "The a-Type Cytochromes," in "The Enzymes," (BOYER, P. D., LARDY, H. AND MYRBÄCK, K., EDITORS), Vol. 8, p. 41, New York, Academic Press, 1963.
5. PALEUS, S., AND PAUL, K. G., "Mammalian Cytochrome *c*," in "The Enzymes" (BOYER, P. D., LARDY, H. AND MYRBÄCK, K., EDITORS), Vol. 8, p. 97, New York, Academic Press, 1963.
6. MARGOLIASH, E., AND SCHEJTER, A., "Cytochrome *c*," *Advan. Protein Chem.*, **21**, 113 (1966).
7. MARGOLIASH, E., AND SMITH, E. L., "Structural and Functional Aspects of Cytochrome *c* in Relation to

Evolution," in "Evolving Genes and Proteins" (BRYSON, V., AND VOGEL, H. J., EDITORS), p. 221, New York, Academic Press, 1965.

8. GIBSON, Q. H., AND GREENWOOD, C., "Kinetic Observations on the Near Infrared Band of Cytochrome Oxidase," *J. Biol. Chem.*, **240**, 2694 (1965).

E. MARGOLIASH

CYTOSINE

Cytosine is a PYRIMIDINE base with the formula:

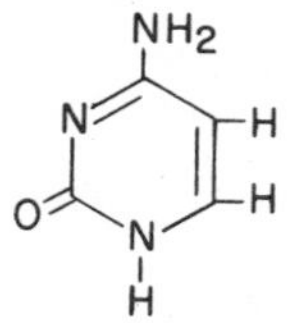

It enters into the composition of nucleic acids where it is bound to a pentose (ribose or deoxyribose) which in turn is esterified with phosphoric acid to form nucleotides such as cytidine-5′-phosphate (see DEOXYRIBONUCLEIC ACIDS; RIBONUCLEIC ACIDS).

CYTOTOXIC CHEMICALS

Cytotoxic chemicals may be defined as those chemical agents that damage cells to which they are applied. They are poisons, to which cells respond with injury, disease, or death. There are multitudes of cytotoxic chemicals; they act by a variety of mechanisms; and they have many different sorts of effects (see also TOXICOLOGY). An exposition of these matters calls for nothing less than a cellular pharmacology, for an elucidation of drug effects at the cellular level. Just as pharmacology based on multicellular organisms can be interpreted in terms of the responses of susceptible organ systems (*i.e.*, of certain sorts of cells), so is chemical cytotoxicity to be viewed as a polyphasic subject: not only is there a multiplicity of chemicals to consider, as well as a variety of mechanisms of action, but there are also many kinds of cells, differing significantly not only with respect to differentiated cell type within the organism but also with respect to the organism of which the cell is a part.

Specificity of Toxic Action. We must take the biological specificity of susceptible cells into account as well as chemical specificity of the cytotoxic agents. The biomedical research world has had this brought forcibly to its attention in the field of experimental cancer chemotherapy in the last two decades. The response of a particular sort of cancer cell in the human patient to a given chemical agent cannot be adequately predicted from the response of a mouse cancer cell, for example, whether of a similar or a different sort of neoplasm and whether grown in a mouse or in a test tube. It is notorious that similar tumors in different patients may respond differently to the same chemotherapeutic agent. Individual organisms vary in their responses to drugs, partly on a genetic basis, both within and between species.

Nevertheless, there are grounds for some generalizations. These we shall explore, bearing in mind this fundamental variability as well as the shifting sands of our concepts.

Types of Toxic Action. Let us begin with a simplified view of a generalized cell living in its native environment. Whether this be a free-living, single-celled protozoan or a particular sort of cell within a multicellular organism, it will have certain environmental requirements in terms of nutrition, inorganic ions, dissolved gases, perhaps radiant energy, and the possibilities of exchange of materials. Ordinarily, there are both upper and lower limits to the concentrations of these material components of the environment, between which our cell may live through its gene-programmed existence in a state of health.

Some of the chemicals of the native environment, when present in excessive amounts, either absolutely or relative to others, may cause what amounts to a nutritional imbalance. Others may damage solely on the basis of their OSMOTIC effects in solution: they cause an excessive withdrawal of water from the cell because the cell may not permit them free entry through its semipermeable plasma membrane. It is a question, of course, whether such components of the native environment can properly be called cytotoxic agents, but we will set no firm qualitative or quantitative boundaries in this discussion. Nor will we consider, on the other hand, VIRUSES and agents of immunologic mechanisms, such as ANTIBODIES, which are on the more complex side of our field of concern and verge on the nebulous domain of living matter.

Certain agents disturb the integrity of the cell's outer coat or its plasma membrane. This membrane is composed chiefly of proteins, lipids, and polysaccharides, and agents that attack these materials may be cytotoxic. Thus, various lipid solvents, such as benzene, chloroform, alcohols, and the like, may extract the lipids of the plasma membrane or of other membranes within the cell, possibly altering their permeabilities and special properties so drastically as to injure or kill the cell. Some agents of this sort, apparently acting through their physical presence rather than through any specific chemical reactivity, may narcotize the cell and depress cellular activity. Protein reagents, such as aldehydes or heavy metals, for instance, may combine with, denature, and even "fix", the proteins of biological membranes, possibly releasing lipids from combination with the proteins, with deleterious effects on biological activity. Agents that produce lipophanerosis act in this way through unmasking of bound lipid.

The special properties of biological membranes in life are maintained through the expenditure of energy by the cell. Energy is made available within cells chiefly through RESPIRATION and GLYCOLYSIS. Energy thus provided by the oxidation of carbohydrates or other substrates is captured in the form of adenosine triphosphate, which in turn is the common coin of energy supply used by the cell in its synthetic activities, movements, and exchanges with the environment. Obviously, interference in the energy supply may result in harm to the cell. Respiratory inhibitors such as CARBON MONOXIDE, sodium azide, and CYANIDE, which can react with metal-containing respiratory enzymes, thus form another class of cytotoxic chemicals. Related to these in effect is 2,4-dinitrophenol, which uncouples the phosphorylation of adenosine diphosphate from oxidation, thus preventing the formation of adenosine triphosphate (see PHOSPHORYLATION, OXIDATIVE).

Respiration is localized in MITOCHONDRIA, which are among the most sensitive and easily injured organelles of the cell. In health, they are long and filamentous, and a change to short, rounded forms, which may clump together or disintegrate, is one of the first morphological indications of cell injury.

Within the cell, and not only in mitochondria, are many vitally important enzymes whose functions can be inhibited by blockade of their thiol groups. There may be several score of these enzymes, and the blocking chemicals include the heavy-metal agents such as mercurials, arsenicals, and antimonials, oxidizing agents such as iodine, alkylating agents such as iodoacetamide and the mustards, arylating agents such as quinones, and miscellaneous other cytotoxic chemicals (see, for example, BERYLLIUM AND BERYLLIOSIS).

The alkylating and arylating agents deserve special mention, for they include some highly reactive compounds of great toxicity. In the field of experimental CANCER CHEMOTHERAPY, for example, numerous agents of this sort have been synthesized and tested for selective effect against tumor cells. Although these agents react with numerous constituent molecules within cells, including proteins and phosphate esters, the most biologically significant reaction of the mustards has been stated recently to be alkylation of the N_7-nitrogen of the guanine moiety of DEOXYRIBONUCLEIC ACIDS, DNA, of the nucleus (Lawley and Brookes). If this be so, it is an amazingly effective reaction, for cell death can follow on the alkylation of only one nucleotide for every twenty thousand or so. With higher concentrations of the mustards, many molecules of the cell must react, because a fixation of the cell can be brought about.

In common with a number of other sorts of agents, alkylating agents in sublethal doses are often mutagenic, giving rise to gene mutations and to visible chromosome breakage (see also MUTAGENIC AGENTS). They may also be carcinogenic. Mutagenicity and carcinogenicity of cytotoxic agents are properties usually displayed after a delay and do not appear to be consequences of acute toxicity. The highly planar nature of the carcinogic polycyclic hydrocarbons has led to suggestions of their intercalation between bases in DNA helices. Some planar molecules, such as acridine, also have a photodynamic effect in the presence of oxygen.

Antimetabolites. Another very interesting class of cytotoxic agents is made up of the antimetabolites, which are substances so similar to normal enzymatic substrate molecules or metabolites as to gain entry into the cellular machinery of intermediary metabolism, but once there they differ enough to cause enzymatic inhibition [see ENZYMES (ACTIVATION AND INHIBITION)] or, if incorporated into protein, nucleic acids, or coenzymes, for example, to diminish the biological worth of those substances. Spectacular agents of this sort include the antifolic acids (such as aminopterin and amethopterin), various other vitamin analogues, and analogues of the naturally occurring purines, pyrimidines, nucleosides, and amino acids. Effective action against the integrity of the cell appears to be exerted at a number of points of intermediary metabolism by these multifarious antimetabolites. Of particular interest with many of them is an interference in normal nucleic acid metabolism. 5-Fluoro-2′-deoxyuridine, for example, acts to inhibit the synthesis of thymidylate, a necessary precursor of DNA, and the related 5-bromo-2′-deoxyuridine is actually incorporated into new DNA in the place of thymidine (see PYRIMIDINE BIOSYNTHESIS). Both of these agents increase the frequency of chromosomal disturbances, in consequence, although by somewhat different means. Various other base analogues, if incorporated into DNA, can lead to GENE mutation by alteration of the normal sequence of nucleotides during replication through incorrect base pairing. 2-Aminopurine is an example of such a mutagen. 8-Azaguanine can be incorporated into RIBONUCLEIC ACIDS, which are thus rendered defective. Among the actions of 6-mercaptopurine is an interference in the biochemical activity of coenzyme A, with resultant mitochondrial damage. Such amino acid analogues as *p*-fluorophenylalanine can effectively halt cellular activities by being incorporated into new proteins, which thereupon fail to attain their proper enzymatic or other functions.

There are many ANTIBIOTIC substances of natural origin that can be classed as cytotoxic agents. Chemically, antibiotics may be of diverse sorts. The actions of several of these may be cited. Mitomycin C, for example, interferes in DNA synthesis, and azaserine inhibits PURINE BIOSYNTHESIS. Actinomycin D is particularly inhibitory to the transcription of the genetic code in chromosomal DNA into new messenger RNA being synthesized by RNA polymerase (see RIBONUCLEIC ACIDS, BIOSYNTHESIS). Hence the coded instruction for the synthesis of new enzyme or other protein molecules may be inhibited in its manufacture. Higher concentrations of actinomycin D are needed to inhibit DNA replication. Puromycin, on the other hand, also ultimately prevents the synthesis of new protein by the cell's ribosomes by competing with transfer RNA molecules bearing activated amino acids for the complex of ribosome and messenger RNA [see PROTEINS (BIOSYNTHESIS)]. Interference in possibly crucial protein syntheses at a particular point in the life of the cell may be damaging or lethal.

Many cytotoxic agents display antimitotic effects. Doses below the point of acute toxicity of many of these agents interfere in different syntheses (as of DNA and various proteins), intracellular movements, or energy supply needed for these processes, and in the over-all process of cellular replication and division (see CELL DIVISION). When the biological test object is a proliferating cell culture, growing root tip, or the like (see TISSUE CULTURE), interference in the normal pattern of mitosis is for many agents an obvious indication of their cytotoxic nature.

Reference

1. BIESELE, J. J., "Mitotic Poisons and the Cancer Problem," Amsterdam and New York, Elsevier Publishing Company, 1958.

JOHN J. BIESELE

D

DALE, H. H.

Henry Hallett Dale (1875–), an English scientist, first isolated ACETYLCHOLINE from ergot, and, based upon the work of OTTO LOEWI, demonstrated that it is a parasympathetic neurohormone. He and Loewi were jointly awarded the Nobel Prize in physiology and medicine in 1936.

DAM, H.

Henrik Dam (1895–) is a Danish biochemist who is noted for his discovery of VITAMIN K. He observed in chickens fed "synthetic" diets small hemorrhages which responded to no known nutrient. Eventually the chemical nature of the missing vitamin became known (see EDWARD ADELBERT DOISY), and in 1943 he and Doisy received the Nobel Prize in medicine and physiology.

DARK ADAPTATION (IN VISION). See LIGHT ADAPTATION.

DECARBOXYLASES AND DECARBOXYLATION REACTIONS. See AMINO ACID METABOLISM, section on Decarboxylation; CARBOXYLATION ENZYMES; PYRIDOXINE FAMILY OF VITAMINS, section on Metabolic Functions.

DEHYDROGENATION ENZYMES

This is a broad class of enzymes (approximately synonymous with the class of "oxidoreductases"; see ENZYME CLASSIFICATION) that catalyze the removal of hydrogen atoms from a substrate and the transfer of the hydrogen to an acceptor, most often one of the NICOTINAMIDE ADENINE DINUCLEOTIDE coenzymes or a flavin coenzyme (see FLAVINS AND FLAVOPROTEINS). See also BIOLOGICAL OXIDATION-REDUCTION, section on Dehydrogenases.

DENATURATION

Proteins in their natural or native state are long polymerized chains of amino acids wound up in a distinctive manner. In this form they can act as ANTIGENS, or if they are enzymes they show catalytic activity. By mild treatment with heat, acid, alkali, urea solutions or detergents, many native proteins are denatured so that they no longer act as antigens and lose whatever enzymatic properties they may have possessed [see PROTEINS, section on Denaturation; also PROTEINS (BINDING FORCES IN SECONDARY AND TERTIARY STRUCTURE)].

It is supposed that denaturation consists mainly of unwinding the polypeptide chains, not breaking the amino acid residues apart. If denaturation is carried out carefully, it is in many cases reversible. In these cases, the denatured protein if allowed to stand may wind itself up again spontaneously and recover a large part of its antigenic and/or enzymatic properties.

DENTAL CARIES. See CARIES, DENTAL (BIOCHEMICAL ASPECTS).

DEOXYRIBONUCLEIC ACIDS (DISTRIBUTION AND STRUCTURE)

Deoxyribonucleic acids (DNA) are high molecular weight linear polymers composed of the following nucleotides: adenylic, cytidylic, guanylic, and thymidylic acids, each of which contains deoxyribose. This sugar and thymine (5-methyluracil), which is the pyrimidine found only in thymidylic acid, serve to distinguish DNA from RIBONUCLEIC ACIDS (RNA) (see also PURINES; PYRIMIDINES). In the case of some bacterial viruses (bacteriophages), cytosine is replaced by 5-hydroxymethylcytosine, and in others, thymine is replaced by uracil or 5-hydroxymethyluracil. In addition to deoxyribose, the DNA from some bacteriophages also contains glucose and rhamnose. Polynucleotides composed of only two different types of deoxyribonucleotides, as for example, deoxyadenylate and deoxythymidylate have been synthesized *in vitro* with DNA polymerase (see POLYNUCLEOTIDES, SYNTHETIC).

In the polynucleotide chain, as shown in Fig. 1, the nucleotides are linked to each other by diester bonds formed by phosphoric acid between the 3′-hydroxyl group of one deoxyribose ring and the 5′-hydroxyl group of the neighbor deoxyribose ring. This covalent binding constitutes the backbone of the chain and accounts for the primary structure of DNA.

A convenient abbreviated way to represent polynucleotide chains which has been generally accepted reads as follows: pApCpTpG in the case of the example shown in Fig. 1. It is therefore easy to see that such chains have a polarity if one notices that the sequence from left to right (5′- to 3′-direction) differs from the sequence from right to left (3′- to 5′-direction). It is also worth

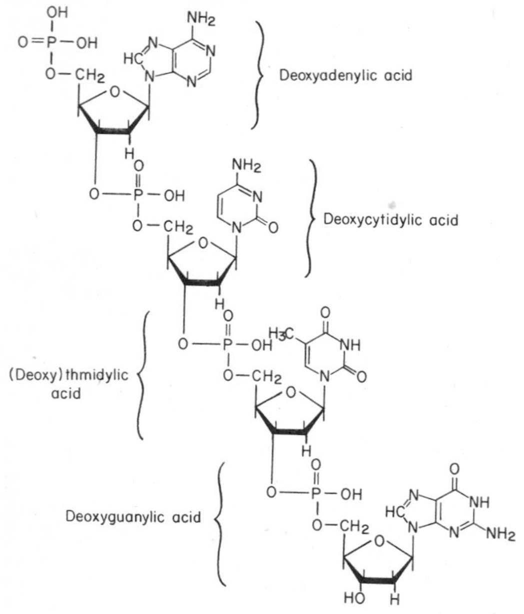

FIG. 1. A tetranucleotide, pApCpTpG.

while to note that without altering the sequence, one can distinguish, from the example shown in Fig. 1, molecules like ApCpTpG, where a terminal 5′-phosphate has been eliminated, or ApCpTpGp, where a terminal 3′-phosphate has been added.

Double Helical Structure. Study of X-ray diffraction patterns of DNA fibers led Watson and Crick (1953) to suggest that one molecule of DNA is in reality composed of two polynucleotide chains forming a double helix. In this model, the two chains are of opposite polarity and are held together by hydrogen bonds—two between adenine and thymine and three between guanine and cytosine (Fig. 2). This pairing of the bases is specific and constitutes a major feature of the secondary structure of a DNA molecule, and it is in agreement with E. Chargaff's earlier discovery that the molar ratios of adenine to thymine and guanine to cytosine were each equal to 1.0 in the hydrolysates of DNA extracted from animals, plants and microorganisms. This secondary structure probably persists almost over the entire length of the molecules. This fact would explain their rigidity and their rod shape when in solution.

A schematic representation of the double helix is shown in Fig. 3. The double helix is regular, has one axis and contains ten base pairs per each turn, which is equal to 34 Å. The bases are located in the core and are perpendicular to the axis, and the deoxyribosephosphate backbone is located at the periphery 10 Å from the axis. Each nucleotide pair is rotated relative to its neighbors by 36°.

The detailed atomic configuration is still under study and therefore it might be found that it departs somewhat from the model presented here. However, there is experimental support for the main features of the Watson and Crick model, namely, two complementary strands held together by specific hydrogen bonds. These bonds are weak and can be destroyed by heat. It has therefore been possible to separate the two

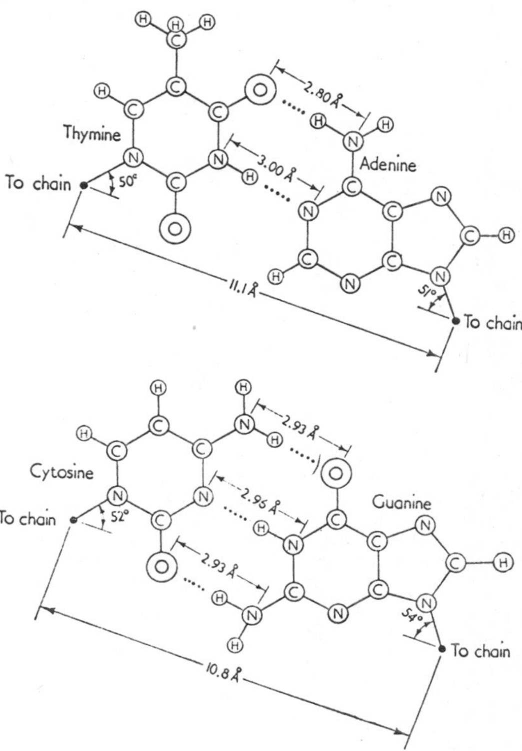

FIG. 2. Complementary base pairs of DNA double helix.

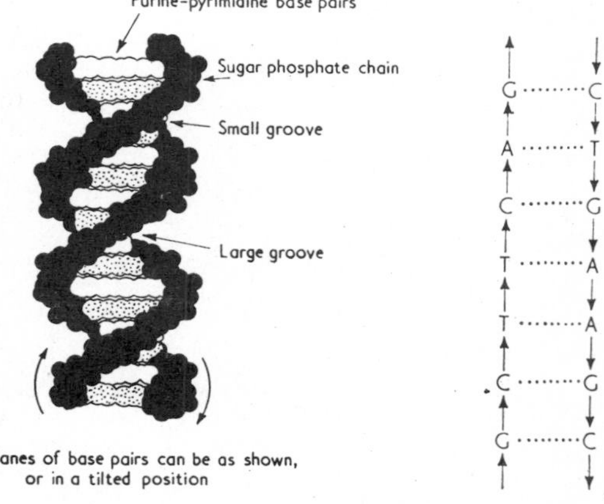

FIG. 3. Model of a portion of double-helical DNA molecule. (Reprinted with permission from A. M. MICHELSON, "The Chemistry of Nucleosides and Nucleotides," New York, Academic Press, 1963.)

strands by heating and to reanneal them by very slow cooling. Specific recombination of the two strands to reform native DNA can be proved by showing that the biological activity, such as the ability to "transform," is recovered. Further support in favor of the Watson-Crick model comes from the analysis of the polydeoxyribonucleotides synthesized *in vitro* with DNA polymerase by Kornberg and his associates [see DEOXYRIBONUCLEIC ACIDS (REPLICATION)], and also from the experiment of Meselson and Stahl. These two workers were able to show that during bacterial growth, DNA molecules are replicated by a semiconservative mechanism implying strand separation.

It must be indicated here that the DNA of some viruses exists in the form of a single strand. The best studied example is that of a bacteriophage discovered by Sinsheimer and called ϕX 174. In this case, molar ratios of adenine over thymine, and guanine over cytosine, are not equal to one.

Distribution. With the exception of some viruses which contain exclusively RNA, DNA is found in every form of life. It is present in a nuclear body almost centrally located in the cell. Bacteria and blue-green algae are called *procaryotic* organisms because, in contrast to the higher forms (*eucaryotic*), the nuclear body is not surrounded by a membrane. Furthermore, it appears that the structure and composition of the CHROMOSOMES in the two types of organisms are different. Whereas the chromosomes of the eucaryotics have a complex structure, as revealed by the electron microscope, and contain (in addition to DNA) RNA and protein, in procaryotics, there appears to be only one chromosome per nucleus which is composed exclusively of DNA. Furthermore, there is recent evidence, in the case of a few bacterial species, suggesting that this chromosome is in reality a unique, double-stranded DNA molecule. In one instance, *Escherichia coli*, the length of this molecule was estimated to be equal to about 1000 μ which would be equivalent to a molecular weight of approximately 3.0×10^9. If we assume that base pairs are separated by 3.4 Å (Fig. 3), such a molecule should contain about 3×10^6 such pairs. Similar estimates indicate that for viruses the molecular weight ranges from 10^7–10^8. Furthermore, in these instances, as well as in the case of *E. coli*, these macromolecules appear to be continuous and circular, but it cannot yet be concluded that this is true for all viral and bacterial chromosomes. It is necessary to emphasize that such intact molecules are extremely difficult to isolate in view of the fact that they are rigid, as indicated above, and therefore are easily broken by shearing stresses applied normally during extraction and purification procedures.

It has been recently demonstrated that mitochondria and chloroplasts which are found in the cells of eucaryotic organisms also contain DNA. Analyses indicate that this DNA has a chemical composition which differs from that of the nucleus.

Base Composition. Recent physicochemical methods have shown that any DNA sample is heterogeneous. From what was said above, we can expect to find molecules of different sizes, a fact which would account for physical heterogeneity; in addition, all these molecules do not have the same chemical compositon. The molar proportions of bases (base ratio) therefore represent a statistical average. This average is constant for a given species and independent of the tissue from which the DNA is extracted. It does not vary when the growth conditions are changed. The composition of DNA may thus be regarded as being very stable and therefore characteristic for the organism from which it is extracted.

Analyses of the base composition of DNA from microorganisms, plants and animals show that there is a compositional diversity within each group. Expressed in ranges of mole per cent G+C, the values 25–75, 37–64, 22–62, 35–48, 35–45 and 40–44 have been obtained, respectively, for bacteria, algae, protozoa, higher plants, invertebrates and vertebrates. Several hundreds of bacterial species have already been analyzed, and it was found that the compositional heterogeneity is small compared to plants and animals. Also, for all taxa analyzed, the species of the same genera have similar base composition. Similarity in base composition within a family is generally accompanied by genetic compatibility—transformation, conjugation, transduction (see BACTERIAL GENETICS). The determination of the base ratio of DNA was routinely obtained by chemical methods involving acid hydrolysis, chromatographic separation of the bases and spectrophotometric determination. More recently, two physicochemical methods have been developed which not only permit such determinations but, in addition, have clarified the concept of heterogeneity. The *first* method consists in the determination of the *buoyant density* of DNA in a linear gradient of CsCl generated in an ultracentrifugal field (see CENTRIFUGATION, DIFFERENTIAL). At equilibrium, the DNA molecules concentrate in a region of the gradient corresponding to their buoyant density. The distribution of the molecules is gaussian, and the mode, therefore, corresponds to an equilibrium position where centrifugal and thermal forces are equal. The variance of this distribution is thus a measure of the heterogeneity mentioned above. It has been shown that the buoyant density corresponding to the mode is directly proportional to the guanine plus cytosine molar content (G+C content). The *second* method is based directly on the Watson-Crick model. It consists in the determination of the relative increase in absorbance at 260 mμ as a result of increase in temperature. This increase in absorbance, or *hyperchromicity*, results from the separation of the two strands composing the DNA molecules. The temperature corresponding to the midpoint of this absorbance change or T_m is linearly related to GC content. Ample use is being made of these methods, not only for the determination of base ratios but also for the study of structural and biological properties of DNA solutions.

Genetic Information. It is now generally accepted that DNA molecules are the repository of the genetic information specifying the characteristics of all living forms (see GENES). It is believed that information is stored in the sequence of the bases. Although there are only four bases, it is easy to visualize that for long molecules a great number of permutations of the relative position of these bases is possible. This would thus constitute an alphabet, and the language would have a code composed, according to current views, of triplets of bases, each triplet required to specify one amino acid in the amino acid sequence specific for proteins. The major problem in molecular genetics is therefore to obtain knowledge of the sequence of the bases in a given DNA molecule, or part of it, which corresponds to the sequence of the amino acids in the polypeptide chain which it specifies. It is beyond the scope of this article to discuss the various approaches to the solution of this fascinating problem. However, it should be indicated that the sequences of bases in DNA molecules from various organisms show greater similarities between organisms which are genetically and systematically related than between organisms which are unrelated. Advantage of this is taken for systematic and phylogenetic studies (see CHEMOTAXONOMY). These studies on sequence homologies are made by separating the strands of DNA molecules and permitting annealing to take place between the strands of various origin.

References

1. DAVIDSON, J. N., AND COHN, W. E., "Progress in Nucleic Acid Research," Vol. 1 (1963), Vol. 2 (1963), Vol. 3 (1964), New York, Academic Press.
2. MICHELSON, A. M., "The Chemistry of Nucleosides and Nucleotides," New York, Academic Press, 1963.
3. STEINER, R. F., AND BEERS, R. F., JR., "Polynucleotides," New York, Elsevier Publishing Co., 1961.

R. STORCK

DEOXYRIBONUCLEIC ACIDS (REPLICATION)

DNA replication is the name given to the process by which two double-stranded molecules of deoxyribonucleic acid (DNA) are formed from one parent double-stranded molecule in such a way that each of the daughter molecules is physically, chemically and biologically identical to the parent. The available evidence suggests that in the process of replication *in vivo*, one strand of the parent DNA molecule is passed to each of the daughter molecules and that the second or complementary strand in each of these is synthesized from nucleotide units. Genetic and autoradiographic evidence indicates that synthesis of the two new strands of DNA *in vivo* proceeds simultaneously in one direction along the parent molecule from a single starting point (Fig. 1).

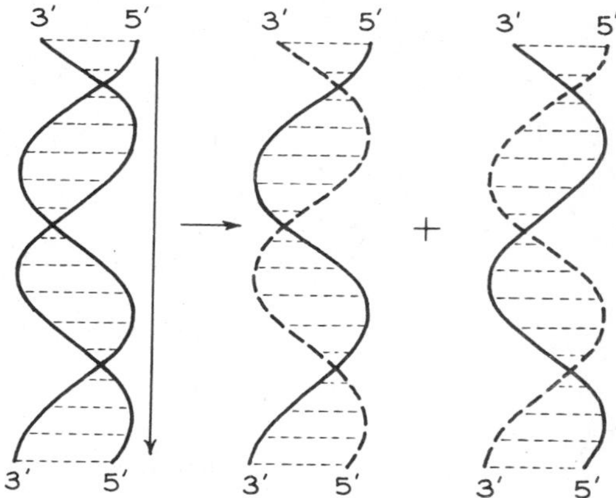

FIG. 1. Diagrammatic representation of the replication of DNA. The double helix on the left shows a parent DNA molecule with two strands of opposite polarity. Dotted lines represent hydrogen bonds between base pairs and the long vertical arrow indicates the direction in which replication may occur. As a result of this process, two new molecules of DNA are formed in each of which one strand (solid line) is derived from the parent DNA and one (interrupted line) is newly synthesized.

DNA Polymerases. Several enzymes have been identified in microbial and animal cells that catalyze the synthesis of DNA-like material. These fulfill many, although not necessarily all, of the requirements for an enzyme responsible for the replication of DNA. The enzymes are called DNA polymerases or DNA nucleotidyltransferases and they have been highly purified from extracts of *Escherichia coli* and *Bacillus subtilis* and less extensively purified from mouse ascites tumors and calf thymus.

DNA polymerases catalyze the formation of polydeoxyribonucleotides from the 5′-triphosphate of deoxyadenosine (dATP), deoxyguanosine (dGTP), deoxycytidine (dCTP) and deoxythymidine (dTTP) in the presence of Mg^{2+} ions and DNA. The reaction is completely dependent on the presence of DNA and Mg^{2+} ions, and the omission of one or more of the deoxyribonucleoside-5′-triphosphates greatly diminishes its rate and extent.

Highly purified preparations of DNA polymerase from *E. coli* bring about net synthesis of DNA that may exceed the amount of DNA added to the reaction mixture by a factor of about two, but less highly purified preparations that are contaminated with NUCLEASES which attack the added DNA may lead to net synthesis of twenty times the amount of added DNA. In contrast to this, DNA polymerases from mammalian sources do not normally give rise to net synthesis of DNA in excess of the amount of DNA added to the reaction mixture.

The nature of the DNA added to the reaction mixture, hereafter known as the *DNA primer*, is of some importance. DNAs from many different sources are equally effective primers of DNA polymerases. DNA polymerase from *E. coli* can utilize native, double-stranded DNA or the partially single-stranded material formed on denaturing DNA by heating, for example. It can also utilize the single-stranded DNA isolated from bacteriophage ϕX 174. Mammalian DNA polymerases, however, exhibit only low levels of activity when provided with native, double-stranded DNA and require either single-stranded DNA or denatured, partially single-stranded DNA.

When DNA is treated with nucleases to yield a partially single-stranded product of the type shown in Fig. 2, this material can be utilized by

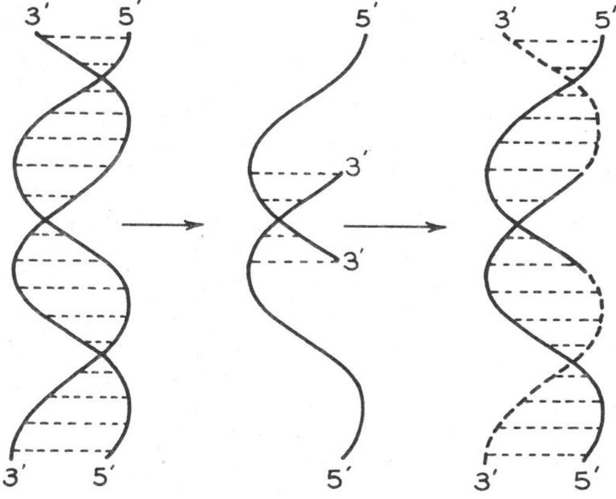

FIG. 2. Partial degradation of DNA by nuclease action followed by repair of the damaged molecule by DNA polymerase. The newly synthesized repaired section is indicated by the interrupted line.

bacterial DNA polymerases. The first product appears to be a repaired, double-stranded molecule with the newly synthesized strands covalently linked to the ends of the damaged DNA strands.

Products of Polymerase Action. The product of the action of DNA polymerase resembles native DNA in several respects. It has been shown to be a polymer of deoxyribonucleotide units with a molecular weight around 5 million that has many of the characteristics of a double-stranded helix. Chemical studies have shown that the deoxyribonucleoside units are linked together by phosphodiester bonds between the 3′-hydroxyl of one deoxyribonucleoside residue and the 5′-hydroxyl of the adjacent residue [see DEOXYRIBONUCLEIC ACIDS (DISTRIBUTION AND STRUCTURE)]. Comparison of the molar proportions of the four deoxyribonucleotides in the DNA primer and in the product of action of DNA polymerase has shown there to be close correspondence even when DNA primers of widely differing nucleotide compositions have been employed. The correspondence in composition between primer and products also holds when the proportions of the deoxyribonucleoside 5′-triphosphates supplied in the reaction mixture are varied extensively. It therefore appears likely that the nucleotide composition of the primer determines that of the product. This of course does not necessarily indicate identity between the DNA primer and product since this would require not merely that the molar proportions of the constituent deoxyribonucleotides were identical but that the sequences of these in both strands of the product were the same as those in the primer and that the physical structures of the primer and product were the same.

Nearest Neighbor Frequency Analysis. At the present time it is not possible to determine sequences of deoxyribonucleotides in a DNA molecule. This is an enormous task since each strand of a DNA molecule may contain many millions of deoxyribonucleotides. However, a procedure has been devised which enables measurements to be made of the frequency with which any given deoxyribonucleotide appears adjacent to any of the four possible residues in a DNA molecule. This procedure is known as nearest neighbor frequency analysis. Clearly these frequencies are determined by the sequence of deoxyribonucleotides in each strand of the DNA, and such measurements provide more information than do determinations of molar proportions of deoxyribonucleotides. It should be emphasized, however, that it would be quite possible for two DNA molecules to have identical base compositions and nearest neighbor frequencies but to possess very different nucleotide sequences.

The results of nearest neighbor frequency analyses on the products of DNA polymerase action have shown that in the reaction, both strands of the DNA primer have been copied and that the polarities of the two newly synthesized strands are opposite to one another. (Reference to Fig. 1 shows that one of the newly formed strands has a 3′-position at the top and a 5′-position at the bottom, whereas in the other strand this is reversed.) Furthermore, if different DNA primers are employed in several reactions, the nearest neighbor frequency distribution in the products are characteristic of the primers employed, and if the product of a DNA polymerase reaction of known nearest neighbor frequency is used as a primer in a reaction, the nearest neighbor frequency in the product of that reaction matches that in the primer.

Mechanism of Polymerase Action. All these features of DNA polymerase activity are consistent with the requirements of an enzyme for the replication of DNA, and it seems likely that the mechanism of DNA polymerase action can be described as shown in Fig. 3(a). Namely, deoxyribonucleoside-5′-triphosphates become aligned progressively to a single strand of the DNA primer by the specific hydrogen bonding that occurs between the base pairs adenine and thymine, on the one hand, and guanine and cytosine, on the other. The formation of internucleotide bonds then proceeds progressively along the strand by nucleophilic attack of the 3′-hydroxyl at the growing end of the polynucleotide chain on

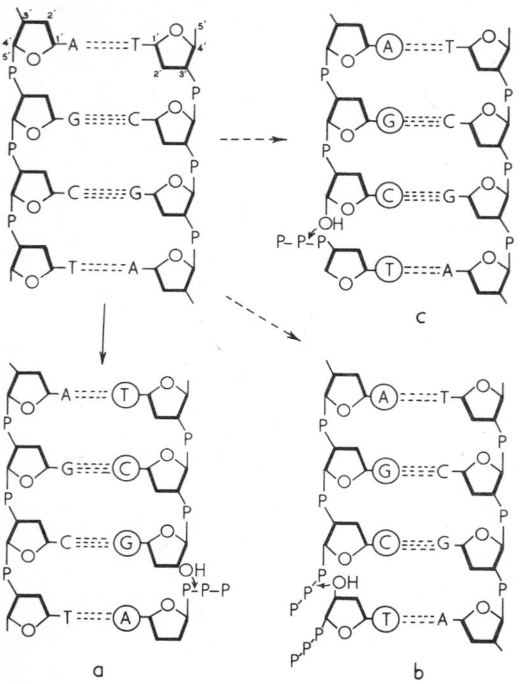

FIG. 3. Possible mechanisms for the replication of DNA. The figure at the top left represents the parent DNA molecule. Formation of a double-stranded molecule of DNA containing one parent strand and one newly formed strand is shown in (a), (b) and (c), and the new strand is shown with the bases in circles. The mechanism shown in (a) seems likely to be the reaction catalyzed by DNA polymerase; (b) and (c) are hypothetical mechanisms that might be required to account for the synthesis of the second daughter molecule of DNA if synthesis of both new strands proceeds simultaneously in one direction.

the pyrophosphate-activated deoxyribonculeoside-5′-triphosphate in the adjacent position. When one of the strands of the primer had given rise to a complete new chain in this way the resulting double-stranded molecule would be a replica of the parent molecule containing one parent strand and one newly synthesized strand.

Problems arise, however, when one considers how the second strand of the primer DNA might give rise to its complementary strand. If, as seems likely, synthesis of both new strands proceeds simultaneously in one direction along the parent molecule from a single starting point, the mechanism of formation of the second new strand cannot be the same as that for the first and one may have to postulate mechanisms of the type shown in Fig. 3(b) or 3(c) to account for this. As yet, there is no clear evidence for either of these proposed mechanisms.

A further problem that emerges from studies of DNA polymerase is that as yet there has been no unequivocal demonstration of enzymic synthesis of DNA with biological activity. In part, this is due to one of the anomalous physical properties of the product of DNA polymerase action. This is that it has not proved possible to separate physically the parent strand from the newly synthesized strand. Physical studies have shown that when denaturing procedures are applied to the product of action of DNA polymerase, separation of the strands occurs but the separated strands show a marked tendency to reassociate. This distinguishes this material from native DNA. Electron microscopy of the synthetic material indicates that branching occurs in the strands and this may account for its anomalous behavior. It remains to be seen whether this branching is a genuine function of DNA polymerase activity or whether it is an artifact of the *in vitro* assay conditions for the enzyme.

References

1. BESSMAN, M. J., in "Molecular Genetics" (TAYLOR, J. H., EDITOR), Part 1, p. 1, New York, Academic Press, 1963.
2. DAVIDSON, J. N., in "The Biochemistry of the Nucleic Acids," Fifth edition, London, Methuens, 1965.
3. KORNBERG, A., in "Enzymatic Synthesis of DNA," New York, John Wiley & Sons, 1961.
4. SMELLIE, R. M. S., *Brit. Med. Bull.*, **21**, (1965).
5. TAYLOR, J. H., in "Molecular Genetics" (TAYLOR, J. H., EDITOR), Part 1, p. 65, New York, Academic Press, 1963.

R. M. S. SMELLIE

DIABETES. See ALLOXAN DIABETES; INSULIN (FUNCTION); METABOLIC DISEASES, section on Melliturias.

DIASTEREOISOMERS (DIASTEREOMERS)

Diastereoisomers include stereoisomers (specifically, optical isomers) which are not antipodes.

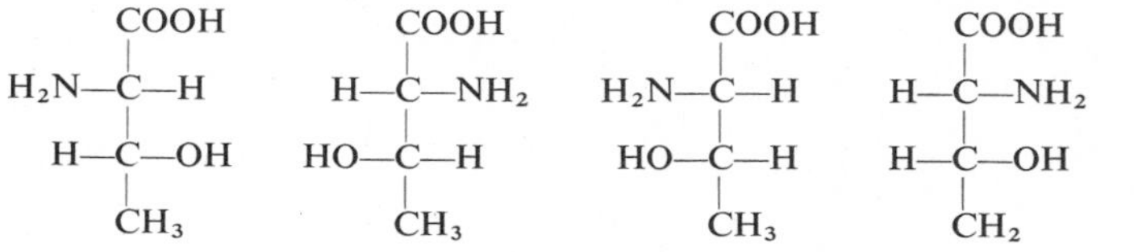

L-Threonine	D-Threonine	L-Allothreonine	D-Allothreonine
COOH	COOH	COOH	COOH
$H_2N—C—H$	$H—C—NH_2$	$H_2N—C—H$	$H—C—NH_2$
H—C—OH	HO—C—H	HO—C—H	H—C—OH
CH_3	CH_3	CH_3	CH_2

Thus, L-threonine and D-threonine are a pair of antipodes (related as mirror images to each other); also L-allothreonine and D-allothreonine are a pair of antipodes. L-Threonine and L-allothreonine are not antipodes but are a pair of diastereoisomers. As another example, D- and L-tartaric acid are antipodes, and have exactly the same solubilities and physical properties (except where asymmetric forces come into play). *Meso*tartaric acid, on the other hand, is a diastereoisomer of the other two and does not have the same solubility and physical properties. The resolution of optically active compounds depends upon the fact that diastereoisomers do not have the same solubilities, etc. If DL-tartaric acid is reacted with L-quinine, the two salts L-quinine-D-tartrate and L-quinine-L-tartrate are diastereoisomers (not antipodes) and hence can be separated by fractional crystallization. Biologically, all optical isomers are distinct molecular species, because enzymes are always asymmetric. In biological systems D-tartaric acid, L-tartaric acid and *meso*tartaric acid are all distinct in their behavior.

See also AMINO ACIDS (OCCURRENCE, STRUCTURE, AND SYNTHESIS), section on Optical Considerations; ASYMMETRY.

DIFFERENTIATION AND MORPHOGENESIS

Visible changes in the form and structure of an organism must be preceded and, in part, caused by invisible biochemical changes. Thus, morphogenesis usually implies the gradual accumulation or redistribution of structural materials, such as connective tissue, bone, or cell wall polysaccharides. This in turn implies alterations in the enzymatic activities responsible for the synthesis of these materials; the various mechanisms by which such changes might be initiated and controlled will be the subject of this article.

Figure 1 summarizes various ways in which the products characteristic of a differentiated cell might be made to accumulate during morphogenesis. The rate of product accumulation could be enhanced by an increase in the level of an enzyme, substrate, activator, or RNA template (used in the synthesis of an enzyme). The accumulation of any of these types of molecules implies nothing as to the mechanism; three possibilities are: (a) an increased rate of synthesis; (b) a decreased rate of destruction; and (c) an unmasking of a preformed, inactive form of the molecule. Thus, for each of the mechanisms listed in the Figure, the problem is simply pushed back to the level of an analysis of the underlying causes for changes in the level of the agent exerting the immediate effect on the rate of product accumulation. Although our present state of knowledge allows a discussion of these three possibilities only with respect to enzyme levels, as indicated in the Figure, levels of the other types of molecules could also be altered by similar mechanisms. Finally, it must be kept in mind that an observed increase in the level of any of these possible controlling factors will be critical to the formation of a product of differentiation only if it was already limiting this process in the cell. Such information is exceedingly difficult to obtain. Changing levels of an enzyme or substrate may only be correlated with, or an indirect result of, the morphogenic process observed, and can be due to causes quite unrelated to our naive and prejudiced interpretations!

Genetic Control. The details of the differentiation process in any given species are inherited. Studies with mutants have demonstrated that GENE-controlled events occur throughout the various stages of development. A mutation is lethal if the embryonic process concerned is essential to the survival of the organism (*e.g.*, failure to accomplish egg implantation, mesoderm formation or bone development in the mouse). Mutations affecting other characteristics such as

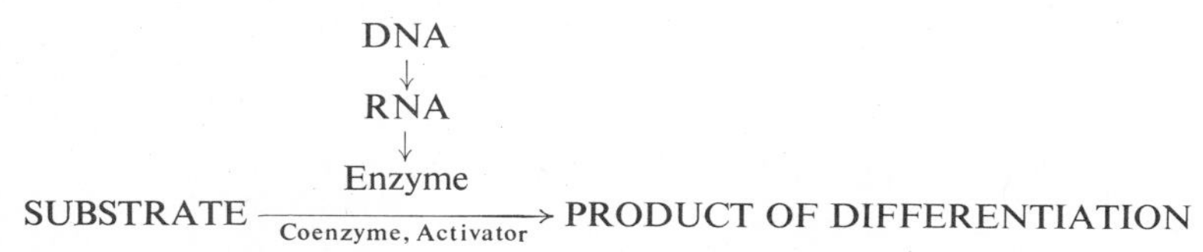

REACTION MAY BE STIMULATED BY:

(1) LEVEL OF ENZYME
  (a) INCREASED SYNTHESIS (RNA and/or DNA activity)
  (b) DECREASED DEGRADATION (stabilization)
  (c) UNMASKING OR ACTIVATION (of preformed protein)
(2) LEVEL OF SUBSTRATE
(3) LEVEL OF COENZYME, ACTIVATOR OR INHIBITOR

FIG. 1

eye color can occur without injury to the organism. The formation during differentiation of either vermillion or normal red eyes in *Drosophila* is a direct result of the (gene-dependent) presence or absence of a circulating substance, kynurenine. Abundant evidence exists that the *potential* for enzyme formation resides at the genetic level. However, although the presence of genes is essential for differentiation to occur, it does not follow that gene activity is primarily responsible for the initiation of morphogenesis or that it necessarily accompanies each step of morphogenesis. In the extreme case, genes may transmit their messages to RNA templates prior to the beginning of specific periods of differentiation, and the activation or selective use of these gene products then directs morphogenesis. The same genes are probably present in all presumptive cell types during morphogenesis and, for that matter, in all of the many cell types present in the fully differentiated organism. This can be strikingly illustrated by the fact that a single adult cell of a carrot plant can be stimulated to grow and differentiate completely into a new plant! We are, therefore, faced with the problem of explaining the pattern of unique genetic expression found in specific tissues. Depending upon cytoplasmic differences in various presumptive cell types, there may occur a differential activation of previously "masked" genetic activity. [As an alternative, all genes may be functioning at maximal rates in all cell types, but their products (RNA templates and eventually enzymes) are rapidly destroyed except when stabilized by specific substrates or particular environmental conditions.] Some of the most direct evidence for differential gene activation comes from studies of chromosome activity in the development of *Drosophila* larvae. Specific areas of the chromosome "puff" at various stages of development. RNA synthesis occurs actively in these areas of puffing, which may represent a pattern of gene activity. The mechanism of gene activation is unknown. With respect to the formation of a product of differentiation (see Fig. 1), the action of the gene is distant and indirect, being mediated through the action of RNA templates and enzymes. Since partial control of morphogenesis could occur independently at either the RNA or the enzyme level, the extent of their contribution must be better understood before we can clarify the role of selective gene activation.

RNA Control. Recent excitement over "MESSENGER RNA" (mRNA) has been largely responsible for stimulating research on the synthesis and behavior of various types of RNA during differentiation. PROTEIN BIOSYNTHESIS from activated amino acids presumably occurs on RIBOSOMES in the presence of mRNA. This template RNA was originally defined in microbial systems as a rapidly turning-over fraction with a base composition similar to that of DNA. It now appears that in some systems undergoing morphogenesis the mRNA fraction may be relatively stable. It is not yet possible to isolate a particular species of mRNA responsible for the synthesis of a specific enzyme necessary to differentiation. Thus the approach has been to correlate the presence and/or synthesis of various types of RNA with the process of differentiation.

One question currently being asked is whether the mRNA fraction is being actively made (from DNA) throughout differentiation or whether some "messages" may be already present in the unfertilized egg. Studies of amphibian morphogenesis have demonstrated that no significant increase in RNA levels occurs until the embryo has undergone extensive differentiation and reached the swimming stage (see METAMORPHOSIS). New ribosomal RNA is made only when growth begins; in fact, a mutant strain lacking the ability to synthesize ribosomes differentiates normally up to the swimming stage and then dies. There is no apparent increase in the amount of mRNA following fertilization or in the extent to which it becomes associated with the ribosomes. Thus some mRNA is apparently present but inactive in the unfertilized egg. A similar picture is emerging in studies of sea urchin egg development. Although fertilization triggers a burst of protein synthesis, this can also be induced by the artificial activation of anucleate fragments of sea urchin eggs. The stimulation of protein synthesis, and of differentiation as far as the blastula, does not appear to depend upon the concurrent formation of new mRNA. It may, rather, depend upon differential "unmasking" of mRNA activities by changes in the cellular environment. However, in the later stages of differentiation in both the amphibian and sea urchin, active synthesis of ribosomal and mRNA occurs, and is presumably necessary to direct the more advanced stages of morphogenesis.

In summary, current data suggest that although the presence of mRNA is of course necessary for differentiation, the synthesis of *new* mRNA may not be of primary importance to the initiation or control of certain early stages of development. In other words, the time of gene expression need not coincide with the time at which the gene product (mRNA) is used during development. However, in the later stages of morphogenesis, or in more complex organisms, such a correlation may well occur.

Enzyme Control. A study of enzyme levels brings us tantalizingly close to the actual product of differentiation. As yet, measurements of enzyme activity *in vivo* during development have rarely been attempted, and analyses *in vitro* are fraught with difficulties. One artifact which can occur is the destruction of an enzyme during its isolation at one stage of differentiation, yet not at another stage. Enzymes are usually measured under conditions of pH, ionic strength, substrate concentration, coenzyme, activator or inhibitor concentration which decidely do *not* reflect the conditions in the differentiating cell. Since these and other variables can (and do) change significantly *in vivo* during development, one is not justified in simply extrapolating from *in vitro* enzyme measurements made under optimal conditions to the activity of an enzyme in a differentiating cell. More data are needed in which enzyme activity is measured both *in vivo* and *in vitro*, and in which levels of the relevant substrates, coenzymes and activators of the reaction are determined *in vivo* at various stages of development. All these data, taken together, may then give a consistent picture of the activity of an enzyme in the differentiating cells.

There are many examples in which the *in vitro* level of an enzyme has been closely correlated with a differentiation process dependent upon the activity of the enzyme. Cholinesterase, an enzyme essential to the transmission of NERVE IMPULSES, accumulates specifically in the brain tissue of the honey bee. In the developing amphibian, cholinesterase is first detectable in the spinal cord, coinciding in time with the ability of the embryo to respond neurogenically to tactile stimuli. During subsequent development, enzyme activity appears sequentially from the posterior to anterior regions of the brain. This pattern coincides exactly in time with the functional maturation of the central nervous system, as indicated by the motor behavior of the developing animal. Another example in the amphibian concerns enzymes of the urea cycle. Metamorphosis represents, in part, a preparation for the transition from an aquatic to a terrestrial habitat. This conversion necessitates numerous anatomical and biochemical alterations, including a changeover from excreting ammonia to excreting urea (see ARGININE-UREA CYCLE). The enzymes involved in urea formation increase in activity during metamorphosis. It has recently been found that one of these enzymes, CARBAMYL PHOSPHATE synthetase, is formed from an enzymatically inactive macromolecular precursor. In another system, germinating castor beans, enhanced activity of enzymes of the "glyoxylate cycle" occurs during development. This series of reactions is essential to the transformation of seed fat to carbohydrates, a conversion which proceeds rapidly during germination. The SLIME MOLD offers an example of an enzyme which increases sevenfold in concentration during development, yet this change is apparently not reflected in its activity *in vivo*. The enzyme is an alkaline phosphatase, highly specific for AMP, which reaches a maximum *concentration* at the end of differentiation. However, inhibition of the enzyme by increasing levels of inorganic phosphate *in vivo* results in maximum *activity* in the intact cell, not at the end but in the middle of development. Thus, observed alterations in the concentration of an enzyme may not bear a direct relationship to its actual activity in the differentiating cell. This is probably the rule rather than the exception. Since the activity of an enzyme is entirely dependent upon levels of specific substrates, coenzymes, inhibitors, etc., knowledge of these variables in the intact cell is highly desirable in attempts to evaluate the significance of changing enzyme levels to a reaction important to differentiation.

Substrate Control. A number of instances of substrate control during differentiation has been found in the SLIME MOLD. The morphogenesis of this microorganism depends in part upon the breakdown of endogenous protein and its eventual conversion to carbohydrates. As protein degradation intensifies during differentiation, the intracellular concentration of glutamate increases an order of magnitude. Oxidation of this amino acid and its entry into the Krebs cycle are necessary to its eventual conversion to carbohydrate. The enzyme responsible for this oxidation, glutamic acid dehydrogenase, is very stable in extracts prepared throughout development. Although the *concentration* of the enzyme does not change, its *activity*, when measured *in vivo* (using radioactive glutamate), increases sevenfold during differentiation. The dehydrogenase was purified, and its affinity for glutamate (K_m) was determined. Knowing the effect of substrate concentration on the rate of this reaction, it was shown that the accumulation of glutamate *in vivo* could fully account for the enhanced rate of the reaction in differentiating cells. A comparable study has been carried out in the slime mold on cell wall synthesis. This highly resistant, insoluble material accumulates only at the end of the differentiation process and is composed of a mixture of CELLULOSE and GLYCOGEN polysaccharides in intimate association. An "activated" form of glucose, uridine diphosphoglucose (UDPG), has been shown to be the specific precursor donating glucose in the synthesis of this cell wall material. Studies *in vitro* demonstrated that glucose-6-phosphate (G6P), trehalose (a disaccharide) and magnesium also stimulate the synthesis of cell wall material from UDPG. Their effects are interdependent—*e.g.*, G6P stimulates strikingly at low but not at high UDPG levels; trehalose stimulates only in the presence of G6P. Both UDPG and G6P accumulate and reach a peak in concentration in the differentiating cells just prior to cell wall construction. Even at their peak concentrations in the cell, UDPG, G6P, trehalose and magnesium limit and interact in their effect on cell wall synthesis. Data on the accumulation of UDPG suggest similar complex circumstances.

It is apparent from these and other studies that the existence *in vivo* of multiple limiting factors in the synthesis of materials important to differentiation may be the rule rather than the exception. It is known that even in fully differentiated cells, enzymes are usually operating far below their potential activity due to substrate limitation. Since differentiating cells are frequently dependent entirely on endogenous metabolism, they have very limited resources from which to obtain the necessary energy and building blocks for the various synthetic reactions required for morphogenesis. Control of differentiation at the substrate (activator, inhibitor) level is relatively straightforward to analyze and interpret, since it can occur independently of changes at the genetic, RNA and enzyme level. Future studies may show that substrate control plays a more significant role in the morphogenesis of less complex organisms and plants, and that control at the hormonal and genetic level becomes very critical in the higher forms.

The relative contributions of substrate, enzyme, RNA and gene control, as well as the time at which each acts relative to the differentiation process, are problems of the future. The probable interactions and interdependence of all these mechanisms make this task challenging, to say the least!

References

1. WILLIER, B. H., WEISS, P. A., AND HAMBURGER, V., (EDITORS), "Analysis of Development," Philadelphia, W. B. Saunders Company, 1956.
2. "Developmental and Metabolic Control Mechanisms and Neoplasia," Nineteenth Annual Symposium on Fundamental Cancer Research, M. D. Anderson Hospital and Tumor Institute, Houston, Texas, March 1965; Austin, University of Texas Press (in press); see articles by D. D. BROWN, U. CLEVER, W. J. RUTTER, AND B. E. WRIGHT.
3. STEWARD, F. C., "The Control of Growth in Plant Cells," *Sci. Am.*, **209**, 104 (October 1963).
4. BEERMAN, W., AND CLEVER, U., "Chromosome Puffs," *Sci. Am.*, **210**, 50 (April 1964).
5. FLORKIN, M., AND MASON, H., (EDITORS), "Comparative Biochemistry," Vol. 6 (see articles by B. E. WRIGHT, P. KARLSON, and C. E. SEKERIS), New York, Academic Press, 1963.

BARBARA E. WRIGHT

DIGESTION AND DIGESTIVE ENZYMES

Digestion and digestive enzymes constitute the means by which food is hydrolyzed to such physical and chemical entities as are accessible to the mechanism governing absorption. Thus, CARBOHYDRATES (starch), PROTEINS and fats (see LIPIDS) are hydrolyzed to monosaccharides, amino acids and fatty acids, respectively, before they are absorbed. Water, many organic ions, and vitamins are liberated during the digestion of food and are also absorbed as such. The simplification of carbohydrates, proteins and fats is brought about by the action of several enzymes. Hydrolysis of foods is carried out in the digestive tract which includes the mouth, esophagus, stomach, small intestine and large intestine. There is little, if any, evidence that digestive processes of any sort are to be found in the esophagus or large bowel. In addition, secretions from the pancreas and LIVER (see also BILE) play an important role in the digestive processes. A discussion of the mechanisms involved in absorption of nutrients, however, is not within the scope of this brief review of digestion and digestive enzymes [see ABSORPTION (OF SUBSTANCES INTO CELLS)].

Carbohydrate Digestion. The first phase of normal digestion is not one involving enzymes but rather one involving a physical process. Mastication is the process by which food is chewed into size and consistency for swallowing, both of which are aided by the addition of saliva. Saliva is secreted by the salivary glands and contains a hydrolytic enzyme called α-amylase (ptyalin), the function of which is to split α-1,4-glucosidic linkages. (STARCH is probably the greatest source of carbohydrate for man. It consists of two types of molecules—a linear or non-branched polymer of glucose, called amylose, and a branched polymer, called amylopectin. Amylose is made up of linear chains of glucopyranose units joined by the first and fourth carbon atoms, *i.e.*, in a 1,4-linkage; amylopectin, the branched type, has branches joining the principal chain through 1,6-glycosidic linkages.) Stated more simply, ptyalin acts on the starch by making it more soluble and splitting part of the polysaccharide into mono- and disaccharides (glucose and maltose, respectively). About 50% of the starch may be broken down to glucose and maltose in the stomach, as a result of continued action of the salivary ptyalin, before reaching the duodenum (first portion of the small bowel). In the duodenum, the starch comes in contact with an enzyme liberated by the pancreas, pancreatic amylase, which completes the digestion of the starch to the end products, glucose (13%) and maltose (87%).

The ptyalin of saliva juice, like the amylase of the pancreas, belongs to the group of enzymes called α-amylases or diastases. Maltose is split into two molecules of glucose by an enzyme called maltase which is liberated within the small bowel. In the case of the other disaccharides which reach the small bowel, they are split into monosaccharides by intestinal enzymes liberated by the glands within the small bowel (Brunner's glands and the glands of Lieberkuhn). The disaccharide of milk, lactose, is split into glucose and galactose by the enzyme β-galactosidase or lactase; sucrose is hydrolyzed into glucose and fructose by the enzyme sucrase. There is good evidence that sucrose and lactose may also be first absorbed into the epithelial cells before being split into the monosacharides. Galactose and fructose are absorbed as such and then converted to glucose within the epithelial cells by various non-digestive enzyme systems. It is not entirely clear whether glucose or other monosaccharides have to be phosphorylated before absorption occurs. *In vitro* studies using isolated loops of rat's and hamster's small intestine suggest that phosphorylation is not a prerequisite for absorption of glucose and, further, that other monosaccharides may be absorbed to some extent without being converted to glucose [see CARBOHYDRATES (CLASSIFICATION) for structures].

Protein Digestion. The major portion of PROTEIN digestion occurs in the small intestine. Gastric digestion of protein is dispensable and individuals with complete gastrectomy can maintain nitrogen balance indefinitely while eating protein in its usual forms. At best, some 10–15% of ingested protein may be broken down to AMINO ACIDS within the stomach by the action of digestive enzymes liberated within the stomach.

Secretion of gastric juice probably begins before food actually reaches the stomach and is increased after ingestion of a meal. The smell, sight, or taste of food will activate the cephalic phase of gastric secretion, and the presence of food within the stomach, resulting in gastric distension, will cause mechanical stimulation of certain secretive factors.

There is apparently a chemical as well as a non-chemical influence on gastric secretion: an extract from the pylorus when injected into the blood causes a flow of GASTRIC JUICE. The term GASTRIN has been given to this substance which behaves like a hormone. The stomach secretes in 24 hours some 2–3 liters of gastric juice which is a conglomeration of many things such as ions, acid and enzymes. Pepsinogen, the inactive form of the PROTEOLYTIC ENZYME pepsin, is to be found in the gastric juice. Pepsinogen is converted into pepsin by HCl (hydrochloric acid of the gastric juice) and by PEPSIN itself. Gastric juice contains another proteolytic enzyme called *rennin* which acts on the casein of milk, changing it into a more soluble product to which the name paracasein has been given. Pepsin hydrolysis of protein produces rather ill-defined products called proteoses and peptones; usually no amino acids or dipeptides are formed during this phase of protein digestion. By peristaltic action, the food is moved into the small bowel in a more or less liquid form (chyme) where it is acted upon by intestinal juices, pancreatic juice, and BILE. Of the digestive juices, that of the pancreas is probably the most important. The presence of acid chyme in the small bowl stimulates the secretion of an intestinal hormone, secretin, which in turn stimulates the flow of pancreatic juice.

As in the case of pepsinogen, the PROTEOLYTIC ENZYME of the pancreatic juice is first secreted in the inactive (zymogen) form. Trypsinogen, a zymogen, is converted to the active form, trypsin, by enterokinase of the intestinal juice (succus entericus). Chymotrypsinogen, another proteolytic zymogen is converted to its active form, chymotrypsin, by trypsin. Both trypsin and chymotrypsin act on proteins, proteoses and peptones converting these into polypeptides and dipeptides (long- and short-chained linkages of amino acids, respectively). Other proteolytic enzymes are to be found in the intestinal juice.

The proteolytic enzymes have been classified as either *endopeptidases* or *exopeptidases*, depending on their specificity. Endopeptidases act on large protein molecules, whereas exopeptidases are enzymes which hydrolyze peptide linkages adjacent to a free polar group. Table 1 lists these enzymes, the type of substrate upon which they act and the types of amino acids involved. One aromatic ring in the side chain is particularly

TABLE 1. SPECIFICITY OF PROTEOLYTIC ENZYMES

Enzyme	Substrate	Preferred Side Chain (R)
Endopeptidases		
Pepsin	—CO—⋮—NHCH(R)CO—⋮—NH—	Tyrosine or phenylalanine
Trypsin	—CO—NHCH(R)CO—⋮—NH—	Arginine or lysine
Chymotrypsin	—CO—NHCH(R)CO—⋮—NH—	Tyrosine or phenylalanine
Exopeptidases		
Aminopeptidase	NH_2CH(R)CO—⋮—NH—	Leucine
Carboxypeptidase	—CO—⋮—NHCH(R)COOH	Variety of amino acids
Tripeptidase	NH_2CH(R)CO—⋮—NHCH(R′)CO—NHCH(R″)COOH	Variety of amino acids
Dipeptidases	NH_2CH(R)CO—⋮—NHCH(R′)COOH	Variety of amino acids

favorable for hydrolysis by pepsin; *e.g.*, pepsin splits bonds involving the amino group of tyrosine or phenylalanine or those involving the carboxyl group of an aromatic amino acid and those in which both residues or side chains involved are aromatic. Trypsin cleaves bonds involving the carboxyl group of the basic amino acids lysine and arginine. Esters and amides are split by trypsin also. Chymotrypsin behaves like pepsin in its specificity and attacks linkages in which the residues or side chains are aromatic. Like trypsin, chymotrypsin hydrolyzes esters and amides. Cathepsins are a small group of intracellular enzymes acting primarily on substrates with aromatic amino acids and in which there is a free α-amino group in the chain.

The exopeptidases include amino peptidases, carboxypeptidases, tripeptidases and dipeptidases. Amino peptidases split dipeptides, dipeptide amides or amino acid amides requiring a free α-amino group. Leucine amino-peptidase, found in the intestinal mucosal cells, splits for example, L-leucylglycine and leucylamide (Table 1). Carboxypeptidases split linkages of acylated dipeptides where the carboxyl group is free and the amino group has been blocked. Tripeptidases split tripeptides, and dipeptidases hydrolyze dipeptides only.

Once formed, the amino acids are absorbed as rapidly as they are split from polypeptide or smaller peptide linkages. Absorption of amino acids is an active process requiring energy derived from oxidative metabolism of the epithelial cells. Not all amino acids are absorbed against a concentration gradient. There is evidence also that some amino acids may compete for absorption sites and inhibit the absorption of others. Amino acid absorption has been studied *in vitro* and *in vivo*, and the results indicate a preferential order of absorption. As was stated earlier, small peptides may disappear from the lumen only to be hydrolyzed into amino acids within the mucosal cells; hydrolysis is essentially complete before the products of protein digestion, amino acids, reach the blood.

Lipid Digestion. There is little or no digestion of fat in the mouth, and none in the stomach. Most, if not all, of the digestion occurs in the small bowel. Almost all dietary fat, be it derived from animal or vegetable, consists of *triglycerides* (glycerol combined in ester linkage with three fatty acids all containing an even number of carbon atoms). PHOSPHOLIPIDS, in which one glycerol alcohol is linked to a phosphoric ester of an organic base, also occur in small quantities and have the same metabolic fate as triglycerides.

Lipids are first acted upon by pancreatic lipase and then need to be emulsified before absorption occurs. The major hydrolytic products which are produced by the action of pancreatic lipase are fatty acids and monoglycerides. Fatty acids are the predominant form in which fat is absorbed, and since 80–95% of the absorbed fat appears in the lymph as triglycerides, important changes in composition of the fat must occur as it passes through the mucosa. For the normal digestion and absorption of lipids, bile is essential. Bile formed in the liver is stored in the gall bladder and is released during fat ingestion. The hormone *cholecystokinin* secreted in the succus entericus brings about the contraction of the gall bladder and probably the relaxation of the common duct sphincter. BILE ACIDS, of which there may be several, are usually conjugates of cholic acid, an end product of cholesterol metabolism. In the conjugated form, the bile acids are water soluble. Taurine or glycine are the amino acids usually conjugated to the bile acid. Bile salts when mixed with fats make the fats more easily digested by pancreatic lipase and, when they combine with fatty acids, give rise to a complex usually referred to as bile salt micelles (fatty acids and glycerides) which are more soluble and more easily absorbed. Short- and medium-chain fatty acids (less than 14 carbon atoms) are absorbed into the circulation directly via the portal vein blood. There is some evidence obtained by electron microscopy that neutral lipids (triglycerides) and perhaps other nutrients may be absorbed directly from the gastrointestinal lumen by a process called pinocytosis. In this process, the fat droplet abuts the epithelial cell which invaginates and engulfs the droplet. The dietary lipids absorbed via the LYMPH are in the form of chylomicrons (low-density lipoproteins) and enter the circulation via the jugular vein. The chylomicrons are made up of the following approximate amounts (in per cent): neutral lipids, 86; cholesterol, 3; phospholipid, 8.5; protein, 2; and a small amount of carbohydrate. It is these chylomicrons which give the plasma its turbid appearance after the ingestion of a meal high in lipids. In defects where bile acids are lacking, there is a decrease in the absorption of fat and, as a result, a decrease in the absorption of other nutrients including the fat-soluble vitamins.

References

1. DAVENPORT, H. W., "Physiology of the Digestive Tract," Chicago, Year Book Medical Publishers Incorporated, 1961.
2. WILSON, T. H., "Intestinal Absorption," W. B. Saunders Company, Philadelphia, London, 1962.
3. ORTEN, A. U., "Intestinal Phase of Amino Acid Nutrition," *Federation Proc.*, **22,** 1103–1109 (1963).
4. HARROW, B., AND MAZUR, S., "Biochemistry," Philadelphia, London, W. B. Saunders Company, 1962.
5. ISSELBACHER, K. J., "Metabolism and Transport of Lipid by Intestinal Mucosa," *Federation Proc.*, **24,** 16–22 (1965).

J. J. VITALE

DIGITONIN AND DIGITOXIN

These constituents of the plant *Digitalis purpurea* are members of the group of steroid glycosides (see GLYCOSIDES, STEROID) in which a hydroxyl (alcoholic) group of the steroid moiety (which thus serves as the aglycone) is attached through a glycosidic linkage to one of an interlinked group of monosaccharide units. Digitonin is built from one digitogenin (the sterol *aglycone*) unit, two galactose units, two glucose units, and one xylose unit. Digitonin is useful as an agent that complexes with, and specifically precipitates, STEROIDS containing the 3β-hydroxy structure, such as cholesterol. Digitoxin consists of one digitoxigenin (the aglycone) unit and three digitoxose (monosaccharide) units.

DIHYDROXYACETONE

Dihydroxyacetone, $CH_2(OH)—C(=O)—CH_2(OH)$, is the simplest ketose; it is structurally an oxidation product of glycerol. It and glyceraldehyde, an aldose (another oxidation product of glycerol), are probably the simplest compounds which should be regarded as sugars. They do not have the acetal or hemiacetal structures of sugars, however, and in this respect are not typical. Dihydroxyacetone is optically inactive but like other α-hydroxy ketones it is a strong reducing agent (Fehling's solution, silver mirror test). Dihydroxyacetone in the form of its phosphate ester (dihydroxyacetone phosphate) is an intermediate in the "glycolytic sequence" of metabolic reactions that form a major pathway for glucose oxidation (see CARBOHYDRATE METABOLISM; GLYCOLYSIS).

DIHYDROXYPHENYLALANINE (DOPA)

3,4-Dihydroxyphenylalanine (DOPA) is closely related to TYROSINE in that it has two hydroxyl groups in the *ortho* position instead of the one present in tyrosine. Through the agency of an enzyme tyrosinase, tyrosine is oxidized to DOPA and then by a series of complex changes, melanin pigments of skin and hair are formed. Melanins can also be formed by similar processes starting with other phenolic substances such as adrenalin or homogentisic acid (see AROMATIC AMINO ACIDS; SKIN PIGMENTATION).

DIIODOTYROSINE

3,5-Diiodotyrosine is an amino acid with the following formula:

$$HO—C_6H_2I_2—CH_2—CH(NH_2)—COOH$$

It was first found by the alkaline hydrolysis of coral (horny skeleton of marine forms) but has since been found in many marine organisms. Sometimes it is accompanied by the corresponding dibromotyrosine. These facts along with the fact that tyrosine is easily iodinated or brominated in the 3,5-positions suggest that these amino acids may be formed in proteins by iodination or bromination.

Diiodotyrosine also occurs in the thyroid glands of animals and is related structurally to thyroxine and triiodothyronine both of which have THYROID HORMONE activity [see also IODINE (IN BIOLOGICAL SYSTEMS)].

DIKETOPIPERAZINES

Diketopiperazines are condensation products which may be formed even under such mild conditions as when amino acid solutions are evaporated to dryness. They are more easily formed from the esters of amino acids.

The diketopiperazine derived from glycine is formulated as follows:

$$\text{cyclo}(—NH—CH_2—C(=O)—NH—CH_2—C(=O)—)$$

Similar compounds may be produced by the condensation of other amino acids individually or by pairs. These compounds are not known to have biological significance.

DINITROFLUOROBENZENE

Dinitrofluorobenzene (1-fluoro-2,4-dinitrobenzene) is a reagent that reacts with free amino groups of amino acids, with free N-terminal amino groups of polypeptides, and with free ε-amino groups of polypeptide-bound lysine, to give yellow dinitrophenyl derivatives. These DNP derivatives are resistant to acid hydrolysis (which may be used to cleave the amino acid residues of a polypeptide); these yellow derivatives thus survive as identifiable "tagged" amino acids serving to indicate which amino acids bore free amino groups in the polypeptide state. See PROTEINS (END GROUP AND SEQUENCE ANALYSIS).

DIPEPTIDASES. See PEPTIDASES.

DIPHOSPHOPYRIDINE NUCLEOTIDE (DPN)

Diphosphopyridine nucleotide, also (and more recently) termed NICOTINAMIDE ADENINE DINUCLEOTIDE (NAD), is one of the coenzyme forms of the vitamin NICOTINIC ACID (niacin). This coenzyme acts as an oxidation-reduction catalyst (in association with any of a number of protein apoenzymes, each specific for certain substrates), and is the most commonly employed coenzymatic acceptor of electrons in biological oxidations. Dehydrogenase enzymes specifically requiring DPN as coenzyme are involved in many metabolic areas, *e.g.*, CARBOHYDRATE METABOLISM. Some dehydrogenases require the related coenzyme, triphosphopyridine nucleotide (TPN), also termed nicotinamide adenine dinucleotide phosphate (NADP).

Reference

1. White, A., Handler, P., and Smith, E. L., "Biological Oxidations," in "Principles of Biochemistry," Third edition, Ch. 19, New York, McGraw-Hill Book Co., 1964.

DISACCHARIDES. See CARBOHYDRATES (CLASSIFICATION AND STRUCTURAL INTERRELATIONS).

DISINFECTANTS

Disinfectants are agents used to kill or inactivate pathogenic or other microorganisms on inanimate objects; disinfectants are thus similar in action to ANTISEPTICS, a term used for agents applied to living organisms for the purpose of killing microorganisms. See also STERILIZATION.

Reference

1. Gadberry, H. M., "Disinfectants," in "The Encyclopedia of Chemistry" (Clark, G. L., and Hawley, G. G., Editors), New York, Reinhold Publishing Corp., 1957.

DOISY, E. A.

Edward Adelbert Doisy (1893–) is an American biochemist who did notable pioneer work on SEX HORMONES. His "discovery of the chemical nature of Vitamin K" was the basis for the Nobel Prize in physiology and medicine in 1943, shared with Henrik Dam.

DOMAGK, G.

Gerhard Domagk (1895–1964) was a physician-chemist in Germany who investigated the therapeutic effects of various synthetic dyes. His work finally led to the discovery that sulfanilamide, a well-known chemical, had an antibacterial effect (see SULFONAMIDES). Antibiotics which interfere with bacterial metabolism, including penicillin (see Alexander Fleming), came to the fore and ushered in a new era in the use of medicines. Domagk was awarded the Nobel Prize in physiology and medicine in 1939.

DONNAN EQUILIBRIUM. See ELECTROLYTE AND WATER REGULATION; MEMBRANE EQUILIBRIUM.

DU VIGNEAUD, V.

Vincent du Vigneaud (1901–), an American biochemist, has contributed widely to the chemistry of sulfur-containing compounds of biological importance: methionine, cystine, biotin, penicillin, oxytocin, vasopressin. He was awarded the Nobel Prize for chemistry in 1955 for the first synthesis of oxytocin and vasopressin, hormones from the posterior pituitary. These are octapeptides and their synthesis represents a milestone in protein chemistry.

DWARFISM

Dwarfism, a condition of subnormal or arrested skeletal development, may be associated with THYROID HORMONE deficiency (cretinism), or with a deficiency of the hormone, secreted by the adenohypophysis (anterior pituitary), termed GROWTH HORMONE or somatotrophin.

E

EGG WHITE INJURY. See AVIDIN; BIOTIN.

EIJKMAN, C.

Christiaan Eijkman (1858–1930) was a Dutch physician who first, in the 1890's, fed fowls polished rice and by this means produced beriberi (polyneuritis). At a time when physicians had finally fully accepted the germ theory of disease (and supposed that germs must be the cause of *all* disease) he did not know how to interpret his results. Eventually the correct interpretation emerged (see THIAMINE AND BERIBERI). He shared with FREDERICK GOWLAND HOPKINS the Nobel Prize for physiology and medicine in 1929.

ELECTROLYTE AND WATER REGULATION (VERTEBRATES)

Those vertebrates that now inhabit our land, seas, brackish waters and fresh waters have survived because they have developed homeostatic mechanisms that enable them to cope with considerable variation in the content and availability of *water*, *sodium*, *potassium* and *chloride* in their external environment. These mechanisms prevent life-threatening changes in their internal environment by (1) assuring that the cells are bathed by fluid with the same osmotic concentrations as themselves, and (2) by preventing major qualitative changes in the intra- and extracellular content of these ions or water. The lower the vertebrate on the phylogenetic scale, the better is it able to tolerate significant changes in the water and ionic content of its body fluids. As a corollary, as the vertebrate ascends the scale, the more complex and elaborate are its controlling systems. However, regardless of the species, or whether the animal lives primarily in the ocean, on land or in the air, one is impressed not by the differences but by the similarities in the ionic composition of their intra- and extracellular fluids.

The water content of the fat-free tissues of all vertebrates ranges between 70 and 80%. Sodium chloride is the major salt of the extracellular fluids (blood and interstitial fluid), while potassium salts of non-diffusible organic anions (primarily phosphates) comprise the major intracellular electrolytes. Despite the qualitative differences in composition of the intra- and extracellular fluids, their OSMOTIC PRESSURES are equal. Water diffuses freely along *its* concentration gradient (osmosis) throughout all body tissues. Therefore, any deviation of the osmotic pressure of intra- or extracellular fluids, by either withdrawal or addition of water, or osmotically active solute, causes an immediate movement of water from the more dilute to the more concentrated solution until osmotic equilibrium is reestablished.

The basic chemical mechanisms responsible for the movement of ions and water in all biologic and non-biologic systems are continuously acting to maintain the steady-state composition of the intra- and extracellular fluid. These are passive diffusion, osmosis, ultrafiltration, and active transport. The Donnan MEMBRANE EQUILIBRIUM, a special type of passive diffusion, should be singled out because the body fluid spaces (blood, interstitial and intracellular) contain widely different concentrations of non-diffusible anions (proteins, organic phosphates, etc.). In such systems, the diffusible ions distribute along electrochemical gradients governed by the concentration differentials of the non-diffusible ions. At steady-state equilibrium, the *products* of the concentrations of the *diffusible* cations and anions on both sides of a given membrane are equal, the diffusible cation concentration being higher and the diffusible anion concentration being lower on the side with the higher concentration of non-diffusible anions. Another important characteristic of this system is that at equilibrium the *sum* of the concentrations of *all* ions (diffusible and non-diffusible) is always greater on the side with the higher concentration of *non-diffusible* ions. Therefore, the *osmotic pressure* of the latter side will always be greater and water will tend to move to this side of the membrane to establish osmotic equilibrium. However, the movement of water changes the electrolyte concentrations and the electrolytes must now diffuse to establish a new Donnan equilibrium. It is immediately apparent from these comments that there must be some mechanism(s) acting across capillaries and across all cell membranes to oppose the osmotic effect of the Donnan equilibrium. In the capillaries, it is the hydrostatic pressure of the blood. Between the cells and extracellular fluid, it is the ACTIVE TRANSPORT of sodium against its electrochemical gradient.

When it first became known that the electrolyte composition of cells differed strikingly from their surrounding interstitial fluid, it was postulated that relative impermeability of the cell membranes accounted for these differences in concentration. However, radioisotope studies with labeled sodium, potassium, and chloride indicated that these ions move readily across all cell membranes and that the passive movements *in either direction* of a given ion are approximately equal. Therefore, although the Donnan membrane equilibria can contribute to the concentration differences observed across all body cells, some other mechanism must be present to prevent the cell from progressively increasing its water and NaCl content under the osmotic influence of the Donnan forces and to extrude sodium from the cell against its electrochemical gradient. Transmembrane potential measurements across vertebrate cells have almost invariably shown that the difference between intra- and extracellular electrical potential is approximately -70 to 90 mV, the inside of the cell being electrically negative to the "outside." This potential difference is the algebraic sum of the electrical forces caused by the differential rates of movements of the anions and cations across the cellular membrane; movement dependent on (1) the relative permeability of the membrane to the different ions; (2) the tendency of ions to diffuse along their concentration gradients while being opposed by different rates of movement of the ions of opposite charge (in some cases by the non-diffusibility of the latter); (3) the active transport of ions. Active electrolyte transport means that an ion can be transported "uphill" against the observed electrochemical gradient. This process requires energy derived from the aerobic and anaerobic metabolism of the cell, or more particularly its membranous components. There is increasing evidence that the energy derived from the conversion of adenosine triphosphate (ATP) to adenosine diphosphate (ADP) under the influence of a magnesium-dependent, sodium- and potassium-activated adenosine triphosphatase (ATPase) is specifically linked to the *active* transport of sodium and potassium between intra- and extracellular fluids and between the organism and its external environment. With few exceptions, chloride ion is not handled in an active manner but moves in and out of the body and across all membranes by diffusion along the electrochemical concentration gradient established by the transport of sodium and potassium and by the Donnan equilibrium [see also POTASSIUM AND SODIUM (IN BIOLOGICAL SYSTEMS)].

The nature and functions of the organs responsible for the maintenance of the sodium, potassium, chloride, and water content of the organism have certain basic similarities throughout the vertebrate kingdom. The differences derive from the adaptation of a given species to the habitat in which it lives and from the availability of water and electrolytes in this environment. The organs include the gills, kidneys, lungs, cloaca, skin, bladder, and rectal and nasal salt glands. They function to conserve, exclude, and eliminate ions and water so that the tonicity (osmolality), volume and specific electrolyte content of the body fluids can be maintained at optimum levels for the survival of the organism. Invariably, two or more of these structures exist in a given species and their respective functions are always integrated toward this end (see also RENAL TUBULAR FUNCTIONS).

A few examples of various types of homeostatic regulation in the vertebrates follow: All vertebrates, except the hagfish, bear the imprint of their earlier evolution in fresh water. Irrespective of the conditions in which they now live, they have NaCl concentrations in their blood (extracellular fluids) about 1/4–1/2 that of seawater. The hagfish has body fluids approximately isotonic with seawater, and the major part of its tonicity is derived from inorganic ions. Freshwater teleosts (body fish) must maintain the osmotic pressure and the sodium and chloride content of their extracellular fluids at a level distinctly hyperosmotic to the fresh water in which they live. Their kidneys and gills must conserve NaCl and eliminate the excess water that enters through their skin and gastrointestinal tract. Therefore, gill fluid and urine contain sodium and chloride at concentrations equal to or less than that in their freshwater habitat. The excess potassium, present in the water or organic nutrient they ingest, is eliminated through gills and kidneys. Marine teleosts, on the other hand, lose water through their skin to the hypertonic seawater, and they are forced to ingest the latter when obtaining food. Although the kidneys of the marine teleosts are unable to excrete a urine more concentrated in salt than their body fluids, their kidneys are essential for eliminating nitrogenous waste products and the large amounts of calcium and magnesium ions ingested. To their gills goes the main responsibility for maintaining the volume and salt content of their body fluids, and gill fluid may contain sodium and chloride at a concentration significantly higher than the concentration of these ions in seawater. For some time it has been known that the kidneys and gills of the elasmobranchs do not participate in the excretion of hypertonic NaCl, yet it is clear that they must ingest salt in hypertonic concentration when obtaining food. It was not until 1960 that it was discovered that the finger-like expansion of the gut, opening into the top of the rectum (the rectal gland) secreted a fluid that was almost pure NaCl, in concentrations exceeding that in seawater. This rectal gland secretion is probably adequate to account for the removal of all the excess salt in this animal's body. It is appropriate at this point to state that nasal glands with a comparable function exist in marine reptiles and birds. The kidneys of these latter two species are also unable to excrete a urine with a NaCl content above that of their body fluids. In these species, the elimination through the nasal gland of NaCl at concentrations >800 meq/liter allows for the conservation of water from the sea and the elimination of any excess salt ingested. Reptiles must adapt to seawater, fresh water, and arid lands. All reptiles have developed a thick nonporous coat or external covering, and very little fluid or electrolyte exchange occurs across this surface regardless of the environment.

For the most part, the amphibia (toads, salamanders, frogs) live in fresh water almost continuously or, like the toad, spend a good part of their existence in moist land areas very high in water content. The frogs and salamanders must be able to conserve NaCl and prevent overhydration while living in a markedly hypotonic environment (fresh water). They do this by active uptake of sodium chloride through their skin from concentrations in the fresh water as low as a fraction of a milliequivalent, by limiting the osmotic uptake of water through their skin, and by excreting a urine with an osmotic pressure less than 1/10 of that of their body fluids and almost free of sodium salts.

The terrestrial reptile has become more fully emancipated from water than the terrestrial amphibia, and this is due to its ability to restrict water loss from the body. The thick skins of reptiles decrease the rate of cutaneous evaporative water loss, and there is minimal, if any, loss of sodium or potassium salts through their thickened skin. Although the reptiles cannot produce a urine more concentrated in NaCl than their blood, the structure of their kidney is adapted to reduce over-all solute and water losses to very low levels. As mentioned, those marine reptiles (*i.e.*, turtles and lizards) that are forced to ingest NaCl in high concentration have nasal or lacrimal (tear) glands that can excrete NaCl at 800–900 meq/liter concentrations. These animals can therefore extract free water from seawater (47 meq/liter Na^+; 55 meq/liter Cl^-) by eliminating the excess salts.

Mammals, in general, have free access to sodium and potassium salts, and other minerals, in their diets, and they can ingest these at will (see also MINERAL REQUIREMENTS). Excesses of sodium, chloride, and potassium are readily excreted in the urine. Potassium, however, is much more ubiquitous in the food than NaCl. As the latter salt is the main electrolyte of the extracellular fluids of vertebrates, a deficiency of this salt will cause a contraction of both the blood and the interstitial volume and, if severe, will lead to failure of the circulation. Therefore all mammalian kidneys can maximally conserve NaCl and, if necessary, produce a NaCl free urine.

Water is lost from the body of mammals by the same routes as in other terrestrial vertebrates, namely, by evaporation across the skin and in the expired air, urine, and feces. The more arid the environment in which the mammal is forced to live, the more it is able to reduce water loss and tolerate longer periods of water dehydration and hypertonicity of its body fluids. It reduces water loss by (1) absence of sweating; (2) production of increasingly concentrated urine—in the desert rodent this may reach levels of 4000–5000 mOsm/liter (maximal urinary concentrations in man is 1400 mOsm/liter); (3) production of dry feces; (4) restriction of skin and respiratory losses.

It is apparent that all vertebrates have utilized similar basic transport mechanisms for regulating the movement of sodium, chloride, potassium, and water between the cells and extracellular fluids, and between the organism and the environment. These mechanisms are homeostatically regulated to maintain the volume, osmotic pressure, and electrolyte composition of the intra- and extracellular fluids. The functional and structural characteristics of the organ systems (gills, nasal glands, kidney, etc.) responsible for exchange with the environment are specifically adapted for optimum survival of the organism regardless of the habitat.

References

1. Maxwell, M. H., and Kleeman, C. R., "Clinical Disturbances of Fluid and Electrolyte Metabolism," New York, McGraw-Hill Book Company, 1962.
2. Christensen, H. N., "Body Fluids and their Neutrality," New York, Oxford University Press, 1963.
3. Schmidt-Nielsen, K., "Salt Glands," *Sci. Am.*, **200**, 109–116 (January 1959).

Charles R. Kleeman

ELECTRON MICROSCOPY

Electron microscopy is a body of techniques used to prepare specimens and to obtain and record their image in an electron microscope, *i.e.*, an instrument in which an electron beam and a set of 3–5 electrostatic or electromagnetic lenses are used to produce a highly magnified image of an object.

Instrumentation and Preparatory Techniques. The wavelength of the beam electrons depends on their acceleration which can be varied, according to instrumental design, from 30–100 kV (or more in instruments used primarily in metallurgical research). Although the corresponding wavelengths are short (~ 0.05 Å at 50 kV), the limit of resolution currently reached by the best microscopes is not better than ~ 4 Å. The limiting factor is the high spherical aberration and low numerical aperture of the electromagnetic lenses that can be produced at present. Most electron microscopes are provided with such lenses, since the spherical aberration of electrostatic lenses is even higher. Current work in electron microscope design and development aims at a limit of resolution of ~ 2 Å. Notwithstanding these limitations, the resolving power of modern electron microscopes reaches well within the range of molecular dimensions, but in many situations it cannot be used fully because of lack of adequate contrast in the object (see below).

Since high vacuum ($\gtrsim 1 \times 10^{-4}$ Torr) must be maintained on the path of the electron beam, biological specimens must be fixed (usually with OsO_4, $KMnO_4$ or various aldehyde solutions) and dried or embedded in a supporting matrix (methacrylic or epoxy resins) before being examined in the microscope.

In transmission electron microscopy, which is

the type of microscopy most commonly used at present, the specimen is mounted in the beam path supported by a thin (~200 Å) plastic membrane and/or a metal grid. The specimen itself must be extremely thin ($\gtrsim$ 1000 Å), because electrons are easily scattered or absorbed by matter. For this reason, specimens are comminuted into fine fragments, if of sufficient stability, or sectioned in 500–1000 Å thick slices after appropriate fixation and embedding. Special microtomes have been developed for obtaining such sections.

Image formation depends on electron scattering which is a function of the mass per unit volume of the specimen and of the atomic number of its components. Hence, contrast in the image is generally and inherently low in biological specimens. It can be increased in the instrument by introducing apertures in the objective lens, which prevent part of the scattered electrons from reaching the image plane, or it can be increased in the object by "staining" it with solutions of heavy metal salts. Most of the heavy ions now in use (phosphotungstate, plumbite, uranyl, lead, etc.) form salt linkages with the basic or acidic groups of the specimen. This is called "positive staining." In an alternative procedure, referred to as "negative staining," the specimen is embedded in an electron-opaque glass formed upon drying by a solution of a heavy metal salt (phosphotungstate or uranyl). Another procedure which obviates low contrast is "shadow casting," *i.e.*, deposition of a thin film on the surface of the specimen by metal evaporation in high vacuum. If the deposition is carried out at a low angle to the plane of the specimen, metal accumulates on one side of protruding structures (the side facing the metal source) and leaves a metal-free area (the "shadow") on the opposite side. If the specimen is continuously rotated during metal evaporation, the over-all dimensions of protruding structures are increased, but fine (diameter $\gtrsim$ 20 Å), long molecules (DNA) can be visualized in their entirety, irrespective of their convolutions.

Surface detail can also be studied on a replica of the object taken with a plastic film or a film of evaporated carbon. Relief in the replica can be made more visible by shadow casting.

Work at high magnifications ($>$ 20,000X) requires a beam of high intensity which can damage the specimen by irradiation and which leads to the rapid contamination of the surface by residual hydrocarbons in the microscope. At present, procedures are developed to reduce specimen damage and to eliminate contamination by introducing cold traps in the vicinity of the object.

The image of the specimen can be viewed directly on a fluorescent screen and can be recorded on photographic plates or films at initial magnifications up to 250,000X. These micrographs can be magnified further photographically.

Importance of Electron Microscopy in Biochemistry. During the last decade, electron microscopy has been used extensively in cytochemical and biochemical research, especially in correlated biochemical and morphological studies of subcellular components such as nuclei, mitochondria, lysosomes, microbodies (peroxysomes), microsomes (*i.e.*, fragments of the endoplasmic reticulum), ribosomes and others. Since most subcellular components do not lose their characteristic structural features upon tissue homogenization, electron microscopy can establish the cytological composition of various cell fractions and estimate their degree of heterogeneity or contamination. This is an important prerequisite before ascribing chemical identity and localizing enzymatic and biosynthetic activities in any subcellular component.

Electron microscopy also proved useful in monitoring the results of the chemical dissection of certain cell structures, as in the separation of RIBOSOMES and membranes from MICROSOMES by detergent treatment, or in the subfractionation of MITOCHONDRIA. It is also used in monitoring attempts to reconstitute cell structures (*e.g.*, ribosomes, mitochondria) from their isolated components.

Electron microscopy has been applied successfully to the study of MULTIENZYME COMPLEXES (the pyruvate dehydrogenase complex, for instance), natural and synthetic nucleic acids or POLYNUCLEOTIDES, and a wide variety of globular and fibrous PROTEINS, as well as filaments and fibrils, the latter form by lateral aggregation. Good examples are COLLAGEN fibrils, KERATIN filaments, and the actin, MYOSIN and paramyosin filaments of various muscles. The information obtained concerns primarily the size and general shape of such macromolecules and, in some cases, details of internal organization of macromolecular aggregates.

Electron microscopy has been extensively used in virology. In conjunction with X-ray diffraction studies, it proved extremely useful in the structural analysis of viral particles, while in conjunction with biochemical and biological assays, it helped elucidate virus-host cell interactions [see VIRUSES (COMPOSITION AND STRUCTURE)].

Recently, autoradiography and a number of HISTOCHEMICAL tests have been adapted for electron microscopy, thus further increasing the scope of its contributions to biochemistry.

References

1. HALL, C. E., "Introduction to Electron Microscopy," New York, McGraw-Hill Book Co., 1953.
2. HEIDENREICH, R. D., "Fundamentals of Transmission Electron Microscopy," New York, Interscience Publishers, 1964.
3. MAGNAN, C., "Traité de microscopie électronique," Paris, Herman, 1961.
4. PEASE, D. C., "Histological Techniques for Electron Microscopy," New York, Academic Press, 1964.
5. REIMER, L., "Electronenmikroskopische Untersuchungs- und Präparations-methoden," Berlin, Springer-Verlag, 1959.
6. Most of the common research in biological electron microscopy is published in the following journals:

Journal of Cell Biology, *Journal of Molecular-Biology*, *Journal of Ultrastructure Research*, *Virology*, *Journal de Microscopie* and *Zeitschrift für Zellforschung*.

GEORGE E. PALADE

ELECTRON TRANSPORT SYSTEMS

In biological oxidations, *i.e.*, in the removal of hydrogens or electrons from a substrate molecule, the intracellular carriers of the removed hydrogen atoms (or electrons) often act sequentially in a series or hierarchy of oxidizing agents of increasing oxidizing strength (*i.e.*, of increasing affinity for electrons), termed a "respiratory chain." Each oxidizing agent of the series is in turn reduced by the agent of next greater oxidizing potential in the sequence, the entire sequence forming a "bucket brigade" for electrons, *i.e.*, an electron transport system, in which the terminal and strongest oxidizing agent (in oxidative metabolism or "respiration") is usually O_2 (molecular oxygen). Certain of the electron transfer steps are coupled to reactions which produce adenosine triphosphate (ATP), the "energy-rich" compound which is the principal form in which some of the energy derived from the over-all oxidation reaction is stored for other metabolic use (see PHOSPHORYLATION, OXIDATIVE). The various catalytic oxido-reducing agents that act in such electron transport systems include the NICOTINAMIDE ADENINE DINUCLEOTIDES, the FLAVINS AND FLAVOPROTEINS, COENZYME Q, and the CYTOCHROMES. Their function in this electron transport role appears to depend upon complex and precise spatial organization of the component agents, together with various lipid and protein components, in such organized subcellular structures as the membranes of the MITOCHONDRIA, membranous structures of bacteria (see OXIDATIVE METABOLISM AND ENERGETICS OF BACTERIA), and possibly membranes of the endoplasmic reticulum (see MICROSOMES). The ultimate details of such spatial organization are currently under active study but are not yet completely understood. (See also BIOLOGICAL OXIDATION-REDUCTION).

References

1. LEHNINGER, A. L., "How Cells Transform Energy," *Sci. Am.*, **205**, 62–73 (September 1961).
2. GREEN, D. E., "The Mitochondrion," *Sci. Am.*, **210**, 63–74 (January 1964).
3. ERNSTER, L., AND LEE, C., "Biological Oxido-reductions," *Ann. Rev. Biochem.*, **33**, 729–788 (1964).

E. M. LANSFORD, JR.

ELECTROPHORESIS

Electrophoresis involves a study of the migration of charged particles, either colloidally dispersed substances or ions, through conducting solutions.

The decade beginning with the year 1950 marks the start of a golden era in the development and application of this technique. Prior to that time, two methods were generally employed for studying the electrophoretic behavior of charged particles in a liquid. In the microscopic method, the migration of particles is observed in a solution contained in a glass tube placed horizontally on the stage of a microscope. The method is suitable for the study of relatively large particles such as bacteria, blood cells or droplets of oil. Its usefulness was extended somewhat by finding that various finely divided inert materials such as tiny spheres of glass, quartz or plastic can, in some instances, be so completely covered with adsorbed protein that they act as if they were large protein particles and respond to an electrical field in terms of the charge on the protein. The method is, today, mainly of historical interest.

In the moving-boundary technique of electrophoresis, the movement of a *mass* of particles is measured, thus obviating the necessity of observing individual particles. The displacement of the particles in an electric field is recorded photographically as the movement of a boundary between a solution of a colloidal electrolyte, such as a protein, and the buffer against which it was dialyzed. The material to be studied is poured into the bottom of a U-tube, and on top of it, in each arm of the U-tube, a buffer solution is carefully layered so as to produce sharp boundaries between the two solutions. Electrodes, inserted in the top of each arm of the tube, are attached to a d-c electric source. If the material under study is a protein, bearing an excess of negative charges on its molecular surface, the boundary will move toward the positive electrode. Since the net electric charge on the protein molecule varies with the acidity of the buffer solution, the charge on the molecule, and hence its velocity, may be varied by varying the acidity of the buffer. As the acidity of the buffer is progressively increased, *i.e.*, as the hydrogen ion concentration is increased or the pH lowered, the velocity of the protein is reduced until a point on the pH scale is reached at which it fails to move (isoelectric point or pI). If the pH of the buffer is further reduced, the protein will acquire a net positive charge and will move toward the negative electrode. The migration velocity of any particular migrant is, of course, directly proportional to the applied voltage gradient. Other important factors which may affect the observed velocity include the molecular shape and structure of the specific substance under study, the concentration of the buffer solution, or more specifically its ionic strength, the temperature, and electroosmosis. Electroosmosis refers to the constant flow of liquid generally toward the negatively charged electrode.

Through the efforts of many investigators, but especially of Tiselius and co-workers in Sweden, the moving-boundary method was developed into a discriminating and rather accurate technique. Instead of round tubing, the U-tube has a narrow rectangular cross section which provides better cooling of the solutions and improved optical qualities. The U-tube is immersed in a water bath held at the temperature of the maximum density of the solution, approximately 2.8°C for a 0.1*N* sodium acetate solution. At this point, a change in temperature produces the least change in density and, therefore, minimum convection. The lowered temperature also increases the resolving power of the apparatus. The apparatus is equipped with a cylindrical lens system to render the boundaries visible as shadows, or "schlieren," from which this system is known as the schlieren method. The method is based on the fact that at a boundary between two transparent materials of different density, the light rays are refracted, thus casting shadows which mark the place of refraction. The instrument produces a photographic diagram in which the abscissa represents the refractive index gradients which can be related to concentration gradients. The areas under the curves are, therefore, proportional to the concentrations of the various components.

The third technique for carrying out electrophoretic separations is known variously as ionography, zone electrophoresis, electrochromatography, etc. Although the rootlets of the technique are discernible in the publications of Lodge dating back to 1886, they withered for all practical purposes until the 1920's when they were revived for a few years by Kendall and co-workers. In 1939, König and Klobusitzky, in Brazil, described a technique they used for the partial separation of a snake venom using paper-stabilized electrolytes. Shortly thereafter, Berraz, in Argentina, described quite a sophisticated apparatus and procedure for the separation of inorganic ions on a narrow strip of filter paper wetted with a conducting buffer solution. During 1950, several papers on electromigration in paper-stabilized electrolytes were published from a number of countries. From that time on, the number of reports on various applications, modifications and limitations of the technique has grown phenomenally, and this procedure has become one of the important tools in biochemical and clinical chemical research. It lends itself equally well to the study of the electromigration of ionic substances of low molecular weight such as AMINO ACIDS, peptides, nucleotides and inorganic ions, or of colloidal materials such as PROTEINS and LIPOPROTEINS. Under favorable conditions, substances not only are separated totally from mixtures but may be recovered almost completely. This aspect of the procedure is particularly valuable in work with radioactive materials. Rigid restrictions on the temperature, current and composition of the solutions are largely removed when electrophoresis is carried out in a solution stabilized with a material such as paper, cellulose acetate, starch, polyacrylamide gel or agar. In addition, only minute amounts of material are required, the equipment is relatively simple and inexpensive, and the method can be utilized over a wide range of temperature. Some of the many uses to which electrophoresis in stabilized media has been applied are identification of the individual components of a mixture, and establishing homogeneity, concentration and purification. In conjunction with other microanalytical techniques such as immunology and polarography, it may often be the means by which identification is ultimately established and quantitative assessment made. Adaptations of the technique, *e.g.*, curtain and planar electrophoresis, provide for

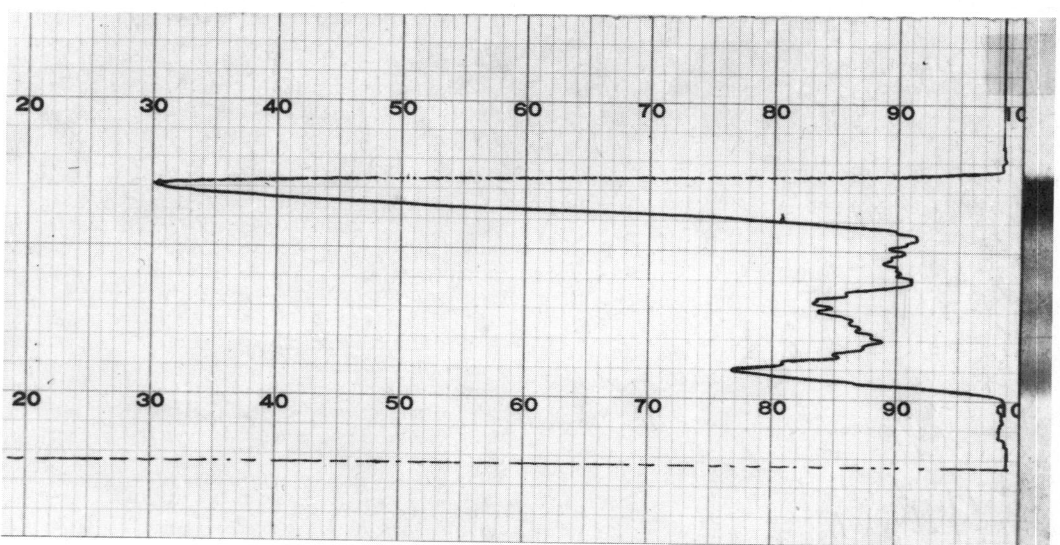

FIG. 1. Densitometer tracing of electrophoretically separated fractions of serum proteins from a child with recurrent acute rheumatic fever; run under following conditions: buffer, "Veronal"; ionic strength, 0.05; pH 8.6; potential gradient, 3 volts/cm; temperature, 25°C; time, 5 hours; monochromator set at 585 mμ. Thin line across ionogram from which tracings were made indicates initial point of application of plasma to paper strip. Proteins stained with bromophenol blue. Largest peak represents albumin fraction. Small peaks, in order of increasing size, represent α_1, α_2, β and γ globulin fractions.

continuous operation, thus permitting the separation of relatively large quantities of substances. In one form or another, electrophoresis in stabilized media is now the most widely used technique of electrophoresis.

Normally, a micro amount of the mixture to be separated is streaked across the midpoint of a horizontal column or strip of stabilizing agent, *e.g.*, a narrow strip of paper, which is saturated with a buffer solution, and a controlled d-c source of electric potential is applied to the ends of the column or strip. The substances under study, the migrants, begin to move and each rapidly reaches a constant velocity of electromigration through the stabilizing structure. The velocity depends, among other factors, upon the potential gradient, the charge on the substance, the ionic strength of the solution, the temperature, the barrier effect interposed by the stabilizing agent, the wetness, electroosmosis, and the hydrodynamic movement of the solvent through the stabilizing structure. In general, as the electromigration proceeds, the original zone separates into several discrete zones having different specific electromigration velocities. The distance of each separated spot, zone or band on the ionogram from the point of application provides a measure for determining the mobility of the particles making up the zone, and the density and the area of the spot provide an index to the quantity of material in the mixture. If the substances separated are colorless, the bands may be developed by the use of suitable dyeing reagents, *e.g.*, bromophenol blue for blood serum proteins.

Several methods have been utilized for the quantitative determination of the dyed zones (*e.g.*, protein fractions) on the ionogram. The strip may be cut into sections, the colored material eluted with suitable solvents and its concentration determined in a spectrophotometer fitted with small cuvettes. A key element in the various fractions, *e.g.*, nitrogen, may be determined by the micro Kjeldahl method, and the concentrations of the original components can be computed. The most common method in use today involves direct determination by a transmission densitometer. A motor-driven device moves the strip of paper, cellulose acetate, etc., past the exit slit of a monochromator. The light, after passing through the strip, impinges on a photoelectric cell. The impulse generated is amplified and, after passing through a log converter unit, is fed to a strip chart recorder. Movement of the chart paper on the recorder is synchronized with movement of the ionogram before the exit slit of the monochromator. The areas under the individual peaks of the graph on the chart are proportional to the amount of the component represented by the peak.

It was observed, many decades ago, that when electricity is passed through a solution containing colloidally dispersed particles, the negatively charged particles move toward the positive electrode, and positively charged particles move in the opposite direction. Moreover, the particles move at differing speeds, depending on such properties as their net electric charge, size and shape, thus making it possible to separate them from a mixture. The charge on a particle may arise from charged atoms or groups of atoms that are part of its structure, from ions which are adsorbed from the liquid medium, and from other causes. It soon became evident that the behavior of colloidal particles in an electric field, as compared to that of ions, differed in degree rather than in kind. Although a colloidal particle is much larger than an ion, it may also bear a much greater electrical charge with the result that the velocity in an electric field may be about the same, varying roughly from $0–20 \times 10^{-4}$ cm/sec in a potential gradient of 1 volt/cm.

To understand the phenomenon of electrophoresis, let us suppose, for simplicity, that a non-conducting particle, spherical in shape, of radius r, and bearing a net charge of Q coulombs, is immersed in a conducting fluid of dielectric constant D, having a viscosity of η poises. Suppose, further, that the particle moves with a velocity of v cm/sec under the influence of an electrical field having a potential gradient of x volts/cm. The force causing the particle to move, namely $Qx \times 10^7$ dynes, is opposed by the frictional resistance offered to its movement by the liquid medium. From Stokes' law, the latter is given by $6\pi\eta rv$. Under steady-state conditions, and by introducing the electrophoretic mobility $u = v/x$, rearrangement yields the expression $u = Q \times 10^7/6\pi\eta r$. It is evident that if the electrophoretic mobility of a particle can be computed, it should be possible to determine Q, the net charge on the particle. For the micro and moving-boundary techniques, a more rigorous treatment of the problem must take into account such complicating factors as electroosmosis, the actual size and shape of the moving particle, the electrolyte concentration in the solvent medium, and the conductivity of the particle itself. Although the ionographic technique is simple from an experimental standpoint, additional complex factors are introduced due to the presence of the stabilizing agent, namely paper, cellulose acetate, starch, etc. However, it is now possible to introduce suitable corrections for these factors and to arrive at mobility data of sufficient quality to be useful in physical chemical computations.

The most important applications of electrophoresis are to the analysis of naturally occurring mixtures of colloids such as various proteins, lipoproteins, polysaccharides, nucleic acid, carbohydrates, enzymes, hormones and vitamins. Electrophoresis often offers the only available method for the quantitative analysis and recovery of physiologically active substances in a relatively pure state. It provides the most convenient and dependable means of analyzing the protein content of body fluids and tissues, and it constitutes an important tool in most hospital laboratories. The marked differences between normal and pathological serum samples are useful in the diagnosis and understanding of disease. Such changes in the electrophoretic pattern of blood serum are evident in diseases characterized by marked protein abnormalities such as multiple myeloma, nephrosis, obstructive jaundice, liver cirrhosis and various parasitic disorders. Because of the small amount of fluid required, the method is applicable for the study of spinal fluids.

The electrophretic pattern is not to be interpreted as specific for a given disease, but rather as an index to the physiological condition of the patient, to be combined with other information for a more complete diagnosis. If the electrophoretic pattern of a patient's blood plasma shows an excess of γ-globulin, the inference is that the body may be suffering from an infection since most of the ANTIBODIES evoked by the presence of infectious microbes are γ-globulin-like proteins. An increase in the α-globulin, a result of the breakdown of tissue proteins, may herald a fever-producing disease, such as pneumonia or tuberculosis. When the blood shows a decrease in ALBUMIN, the clinician looks to the liver as a possible seat of the disease because it is the main site of albumin production. When the liver fails, other tissues will often react to the lower albumin level by producing an excess of globulins.

Because of the fact that the fractionation of mixtures by ionography and quantitative estimation of the components have been of such great practical value, the potentiality of the technique for theoretical electrochemistry and colloid chemistry is often overlooked. Although the presence of a stabilizer such as paper or cellulose acetate introduces many complications as far as mobility determinations are concerned, it is now possible to obtain reliable mobility data in most types of electrophoretic media. Since substances can be characterized by their electrophoretic mobilities, phenomena such as the binding of various ions to proteins, which produce changes in the net charge of a molecule, can, in principle, be evaluated from the changes in the mobility. It has been possible to arrive at some conclusions as to relative binding strength and the charge on the chemical site involved. The isolectric points of ampholytes such as proteins and lipoproteins can be determined, and initial steps have been made toward determining such factors as ionization constants and associated thermodynamic quantities. Since these determinations can be made with less than 1 mg of substance, the technique is often of considerable value for establishing the nature of functional groups.

References

1. Abramson, H. A., "Electrophoresis of Proteins," New York, Reinhold Publishing Corp., copyright 1942; reprinted by arrangement, New York, Hafner Publishing Co., 1964.
2. Bier, M., Editor, "Electrophoresis; Theory, Methods and Applications," New York, Academic Press, 1959.
3. McDonald, H. J., "Theoretical Basis of Electrophoresis and Electrochromatography," in "Chromatography" (Heftmann, E., Editor), Ch. 10, New York, Reinhold Publishing Corp., 1961.
4. McDonald, H. J., "Ionography; Electrophoresis in Stabilized Media," Chicago, Year Book Publishers, 1955.
5. Ribeiro, L. P., Mitidieri, E., and Affonso, O. R., "Paper Electrophoresis," Amsterdam, Elsevier Publishing Co., 1961.

Hugh J. McDonald

ELVEHJEM, C. A.

Conrad Arnold Elvehjem (1901–1962) was an American biochemist who, following a strong tradition at the University of Wisconsin, carried out a large number of studies dealing with nutrition. He was concerned broadly with all types of nutrients. The most significant single contribution made by him and his students was the discovery that NICOTINIC ACID would cure black tongue (pellagra) in dogs. He later became the President of the University of Wisconsin.

END GROUP ANALYSIS. See PROTEINS (END GROUP AND SEQUENCE ANALYSIS).

ENDOCRINE GLANDS. See HORMONES, ANIMAL; also, more detailed articles on ACETYLCHOLINE AND CHOLINESTERASE; ADRENOCORTICOTROPHIC HORMONE; GASTRIN; GONADOTROPIC HORMONES; GROWTH HORMONE; INSULIN; PARATHYROID HORMONES; SEX HORMONES; THYROID HORMONES; THYROTROPIC HORMONE; VASOPRESSIN.

ENDOPLASMIC RETICULUM

The endoplasmic reticulum is a term applied to a system of subcellular structures originally recovered from cells and studied in fragmented form as "microsomes" (see MICROSOMES). In many types of cells, ribonucleoprotein particles called RIBOSOMES are attached to the membranes of the endoplasmic reticulum. These ribosomes are the sites of PROTEIN BIOSYNTHESIS. The terms ergastoplasm and ergastoplasmic lamellae have sometimes been applied to the endoplasmic reticulum.

Reference

1. Brachet, J., "Biochemical Cytology," New York, Academic Press, 1957.

ENDOTOXINS

Endotoxins are lipopolysaccharide-protein complexes found in the cell walls of certain bacteria, and identical with their somatic or "O" antigens (see BACTERIAL CELL WALLS; TOXINS, ANIMAL AND BACTERIAL).

References

1. Stanier, R. Y., Doudoroff, M., and Adelberg, E. A., "The Microbial World," pp. 593–594, Englewood Cliffs, N.J., Prentice-Hall, 1963.
2. Braude, A. I., "Bacterial Endotoxins," *Sci. Am.*, **210,** 36 (March 1964).

ENTROPY. See BIOCHEMICAL THERMODYNAMICS.

ENZYME CLASSIFICATION AND NOMENCLATURE

Principles of Classification. The term *enzymes* refers to a group of PROTEINS which catalyze a variety of chemical reactions. Any systematic account of such a group (and any system of nomenclature) must be based on some property or quality common to them all, which varies sufficiently between individual members to be used as a distinguishing feature. There are three such properties which would appear to provide a basis for enzyme classification: the chemistry of the enzyme itself, the chemical nature of the substrate acted on, and the nature of the reaction catalyzed. These three systems will be briefly discussed.

In some fields of biochemistry (*e.g.*, in the study of coenzymes) interest has shifted from the functional side to the detailed *chemistry of the substance* involved, and a similar shift of emphasis will no doubt take place with enzymes. The functional differences between enzymes must result entirely from differences in the enzyme protein, and the most fundamental kind of classification would be based on these chemical differences. Unfortunately the information does not exist at this time to propose such a classification. Certain groups of enzymes can be distinguished on the basis of the chemistry of non-protein groups associated with the catalysis (*e.g.*, flavoproteins, hemoproteins, cuproproteins). In a very small number of cases it has been possible to show that certain amino acid residues or peptide groupings in the enzyme protein take part in the catalytic process, and it has been found that enzymes catalyzing somewhat diverse reactions may be similar in this respect. The term "serine proteinases" has been proposed for a group of enzymes catalyzing hydrolysis of peptides, amides or esters which are inhibited by organophosphorus compounds and depend upon formation of an acyl-serine derivative on the enzyme during catalysis. Information of this kind is available, however, for a very limited number of enzymes [see ENZYMES (ACTIVE CENTERS)], and a classification based on enzyme chemistry would be premature.

Many of the earlier attempts at enzyme classification were based on the *nature of the substrate*, and broad groups were described such as carbohydrases (enzymes acting on carbohydrates), proteinases (enzymes acting on proteins) and so on. Such groupings often obscure close functional similarities between enzymes in different groups; for example, enzymes hydrolyzing —CO·NH— bonds in proteins have close similarities to enzymes hydrolyzing —CO·NH_2 in simple amides, and flavoprotein from kidney bringing about the oxidation of an amino acid is similar in many respects to xanthine oxidase of liver or milk or to the flavoprotein in molds which brings about the oxidation of D-glucose. It is therefore generally agreed that the only practicable basis for a complete classification of enzymes is the nature of the reaction catalyzed.

In the system of classification described in this article, all enzymes are allocated to one of *six* main groups, each catalyzing one *reaction type*. Subdivision is based on more close definition of the reaction catalyzed and ultimately on the substrate(s) within which the reaction takes place.

Principles of Nomenclature. The first definite enzyme name was "diastase," put forward by Payen and Persoz in 1833. Duclaux proposed in 1898 that all enzymes should be named by adding "-ase" to a root indicating the nature of the substrate on which the enzyme acts. Most enzyme names which have been put forward use the "-ase" suffix, apart from a small group of PROTEOLYTIC ENZYMES: pepsin, trypsin, chymotrypsin, papain, etc.

A rational system of enzyme nomenclature should be clear and unambiguous; it should use the same kind of consideration for forming the names of all enzymes; it should be based on clearly understood principles so that it is readily extrapolated to provide names for newly discovered enzymes; it should, if possible, yield names which indicate a reasonable amount of information about the enzyme named; yet the names should be sufficiently short and euphonious for regular written and spoken use. The conflict between brevity and convenience on the one hand, and clarity and informativeness on the other, has been the greatest obstacle to universal agreement on enzyme nomenclature. In the Enzyme Commission system, discussed below, two names are given for each enzyme: a systematic name which is based on strict logical principles, but which is often long and cumbersome, and a "trivial" name which is short enough for general use, but not necessarily very exact or systematic.

A system of nomenclature, in contrast to an arbitrary list of individual names, must be based on a system of classification. For the reasons set out in the first section, therefore, enzyme names are based on the *reaction catalyzed*, a root indicating this being followed by -ase (*e.g.*, oxidase, hydrolase, transferase).

Enzyme Nomenclature: Recommendations (1964) of the International Union of Biochemistry. An International Commission, with Dr. M. Dixon as President, was set up by the International Union of Biochemistry in 1956 to consider the classification and nomenclature of enzymes and coenzymes; the Commission reported its results in 1962. Subsequently, a Standing Committee reviewed comments and criticisms received and submitted a revised Report, which was adopted by the Council of the International Union of Biochemistry in 1964 and published under the title "Enzyme Nomenclature: Recommendations (1964) of the International Union of Biochemistry." It contains the recommended classification scheme (see "Key" following this paragraph), a set of rules of nomenclature, a list of known enzymes (showing the systematic and trivial names for each), a code number, and the reaction, catalyzed by the enzyme, on which the systematic name is based.

Key to Numbering and Classification of Enzymes

1. Oxidoreductases

1.1 *Acting on the CH—OH group of donors*
- 1.1.1 With NAD or NADP as acceptor
- 1.1.2 With a cytochrome as an acceptor
- 1.1.3 With O_2 as acceptor
- 1.1.99 With other acceptors

1.2 *Acting on the aldehyde or keto group of donors*
- 1.2.1 With NAD or NADP as acceptor
- 1.2.2 With a cytochrome as an acceptor
- 1.2.3 With O_2 as acceptor
- 1.2.4 With lipoate as acceptor
- 1.2.99 With other acceptors

1.3 *Acting on the CH—CH group of donors*
- 1.3.1 With NAD or NADP as acceptors
- 1.3.2 With a cytochrome as an acceptor
- 1.3.3 With O_2 as acceptor
- 1.3.99 With other acceptors

1.4 *Acting on the CH—NH_2 group of donors*
- 1.4.1 With NAD or NADP as acceptor
- 1.4.3 With O_2 as acceptor

1.5 *Acting on the C—NH group of donors*
- 1.5.1 With NAD or NADP as acceptor
- 1.5.3 With O_2 as acceptor

1.6 *Acting on reduced NAD or NADP as donor*
- 1.1.6 With NAD or NADP as acceptor
- 1.6.2 With a cytochrome as an acceptor
- 1.6.4 With a disulfide compound as acceptor
- 1.6.5 With a quinone or related compound as acceptor
- 1.6.6 With a nitrogenous group as acceptor
- 1.6.99 With other acceptors

1.7 *Acting on other nitrogenous compounds as donors*
- 1.7.3 With O_2 as acceptor
- 1.7.99 With other acceptors

1.8 *Acting on sulfur groups of donors*
- 1.8.1 With NAD or NADP as acceptor
- 1.8.3 With O_2 as acceptor
- 1.8.4 With a disulfide compound as acceptor
- 1.8.5 With a quinone or related compound as acceptor
- 1.8.6 With a nitrogenous group as acceptor

1.9 *Acting on heme groups of donors*
- 1.9.3 With O_2 as acceptor
- 1.9.6 With a nitrogenous group as acceptor

1.10 *Acting on diphenols and related substances as donors*
- 1.10.3 With O_2 as acceptor

1.11 *Acting on H_2O_2 as acceptor*

1.12 *Acting on hydrogen as donor*

1.13 *Acting on single donors with incorporation of oxygen (oxygenases)*

1.14 *Acting on paired donors with incorporation of oxygen into one donor (hydroxylases)*
- 1.14.1 Using reduced NAD or NADP as one donor
- 1.14.2 Using ascorbate as one donor
- 1.14.3 Using reduced pteridine as one donor

2. Transferases

2.1 *Transferring one-carbon groups*
- 2.1.1 Methyltransferases
- 2.1.2 Hydroxymethyl-, formyl- and related transferases
- 2.1.3 Carboxyl- and carbamoyltransferases
- 2.1.4 Amidinotransferases

2.2 *Transferring aldehydic or ketonic residues*

2.3 *Acyltransferases*
- 2.3.1 Acyltransferases
- 2.3.2 Aminoacyltransferases

- 2.4 *Glycosyltransferases*
  - 2.4.1 Hexosyltransferases
  - 2.4.2 Pentosyltransferases
- 2.5 *Transferring alkyl or related groups*
- 2.6 *Transferring nitrogenous groups*
  - 2.6.1 Aminotransferases
  - 2.6.3 Oximinotransferases
- 2.7 *Transferring phosphorus-containing groups*
  - 2.7.1 Phosphotransferases with an alcohol group as acceptor
  - 2.7.2 Phosphotransferases with a carboxyl group as acceptor
  - 2.7.3 Phosphotransferases with a nitrogenous group as acceptor
  - 2.7.4 Phosphotransferases with a phospho-group as acceptor
  - 2.7.5 Phosphotransferases, apparently intramolecular
  - 2.7.6 Pyrophosphotransferases
  - 2.7.7 Nucleotidyltransferases
  - 2.7.8 Transferases for other substituted phospho-groups
- 2.8 *Transferring sulfur-containing groups*
  - 2.8.1 Sulfurtransferases
  - 2.8.2 Sulfotransferases
  - 2.8.3 CoA-transferases

3. Hydrolases

- 3.1 *Acting on ester bonds*
  - 3.1.1 Carboxylic ester hydrolases
  - 3.1.2 Thiolester hydrolases
  - 3.1.1 Phosphoric monoester hydrolases
  - 3.1.4 Phosphoric diester hydrolases
  - 3.1.5 Triphosphoric monoester hydrolases
  - 3.1.6 Sulfuric ester hydrolases
- 3.2 *Acting on glycosyl compounds*
  - 3.2.1 Glycoside hydrolases
  - 3.2.2 Hydrolyzing N-glycosyl compounds
  - 3.2.3 Hydrolyzing S-glycosyl compounds
- 3.3 *Acting on ether bonds*
  - 3.3.1 Thioether hydrolases
- 3.4 *Acting on peptide bonds (peptide hydrolases)*
  - 3.4.1 α-Aminoacyl-peptide hydrolases
  - 3.4.2 Peptidyl-aminoacid hydrolases
  - 3.4.3 Dipeptide hydrolases
  - 3.4.4 Peptidyl-peptide hydrolases
- 3.5 *Acting on C—N bonds other than peptide bonds*
  - 3.5.1 In linear amides
  - 3.5.2 In cyclic amides
  - 3.5.3 In linear amidines
  - 3.5.4 In cyclic amidines
  - 3.5.5 In cyanides
  - 3.5.99 In other compounds
- 3.6 *Acting on acid-anhydride bonds*
  - 3.6.1 In phosphoryl-containing anhydrides
- 3.7 *Acting on C—C bonds*
  - 3.7.1 In ketonic substances
- 3.8 *Acting on halide bonds*
  - 3.8.1 In C-halide compounds
  - 3.8.2 In P-halide compounds
- 3.9 *Acting on P—N bonds*

4. Lyases

- 4.1 *Carbon-carbon lyases*
  - 4.1.1 Carboxy-lyases
  - 4.1.2 Aldehyde-lyases
  - 4.1.3 Ketoacid-lyases
- 4.2 *Carbon-oxygen lyases*
  - 4.2.1 Hydro-lyases
  - 4.2.99 Other carbon-oxygen lyases

- 4.3 *Carbon-nitrogen lyases*
  - 4.3.1 Ammonia-lyases
  - 4.3.2 Amidine-lyases
- 4.4 *Carbon-sulfur lyases*
- 4.5 *Carbon-halide lyases*
- 4.99 *Other lyases*

5. Isomerases

- 5.1 *Racemases and epimerases*
  - 5.1.1 Acting on amino acids and derivatives
  - 5.1.2 Acting on hydroxy acids and derivatives
  - 5.1.3 Acting on carbohydrates and derivatives
  - 5.1.99 Acting on other compounds
- 5.2 *Cis-trans isomerases*
- 5.3 *Intramolecular oxidoreductases*
  - 5.3.1 Interconverting aldoses and ketoses
  - 5.3.2 Interconverting keto and enol groups
  - 5.3.3 Transposing C═C bonds
- 5.4 *Intramolecular transferases*
  - 5.4.1 Transferring acyl groups
  - 5.4.2 Transferring phosphoryl groups
  - 5.4.99 Transferring other groups
- 5.5 *Intramolecular lyases*
- 5.99 *Other isomerases*

6. Ligases or Synthetases

- 6.1 *Forming C—O bonds*
  - 6.1.1 Aminoacid-RNA ligases
- 6.2 *Forming C—S bonds*
  - 6.2.1 Acid-thiol ligases
- 6.3 *Forming C—N bonds*
  - 6.3.1 Acid-ammonia ligases (amide synthetases)
  - 6.3.2 Acid-aminoacid ligases (peptide synthetases)
  - 6.3.3 Cyclo-ligases
  - 6.3.4 Other C—N ligases
  - 6.3.5 C—N ligases with glutamine as N-donor
- 6.4 *Forming C—C bonds*

The main features of the Report are briefly set out in the following paragraphs.

Enzymes are divided into the following six groups:

1. Oxidoreductases
2. Transferases
3. Hydrolases
4. Lyases
5. Isomerases
6. Ligases or Synthetases.

The main group to which an enzyme belongs is indicated by the first figure of the code number. The second figure indicates the subclass; for the oxidoreductases it shows the type of group in the *donors* which undergoes oxidation; for the transferases, it indicates the nature of the group which is transferred; for the hydrolases it shows the type of bond hydrolyzed; for the lyases, the type of link which is broken between the group removed and the remainder; for the isomerases, the type of isomerization involved; and for the ligases, the type of bond formed. The third figure of the code number, indicating the sub-sub-class, shows for the oxidoreductases the type of acceptor involved; for the transferases and hydrolases, it shows more precisely the type of group transferred or bond hydrolyzed; for the lyases it shows the nature of the group removed; for the isomerases it shows, in more detail, the nature of the isomerization; and for the ligases it shows the nature of the substance formed. The enzyme nomenclature rules provide a detailed code for the sub-class and sub-sub-class numbers.

The complete Enzyme Commission number, which is commonly indicated by the prefix EC, gives fairly detailed information about the enzyme specified. For example, EC 1.2.3.1 indicates the first enzyme listed in sub-sub-class 1.2.3, which according to the rules set out includes all oxidoreductases acting on the aldehyde or keto groups of donors (sub-class 1.2), and using O_2 as acceptor (sub-sub-class 1.2.3). The enzyme is listed with a systematic name "aldehyde: oxygen oxidoreductase," and trivial name "aldehyde oxidase." Similarly, EC 4.1.3.6 refers to a lyase splitting a carbon-carbon bond (sub-class 4.1) by removing a keto acid (sub-sub-class 4.1.3); the enzyme is "citrate oxaloacetate-lyase," trivial name "citrate lyase," which cleaves citrate into acetate and oxaloacetate. It is recommended that at the beginning of each paper or abstract primarily dealing with an enzyme the code number and systematic name be given, although subsequently, for convenience, the trivial name may be used; many major biochemical journals have adopted this convention.

The systematic name of an enzyme is essentially a concise statement of the reaction catalyzed. In general the names have at least two parts: the second, ending in "-ase," indicates the type of reaction and often corresponds to the sub-class or sub-sub-class (phosphotransferase, carboxy-lyase, *cis-trans*-isomerase), while the first part gives some indication of the particular substrate(s) (as in ATP: D-glucose 6-phosphotransferase, acetoacetate carboxy-lyase, 4-maleylacetoacetate *cis-trans*-isomerase). While most enzymes present no difficulty, in certain cases it is necessary to make arbitrary decisions about the reaction on which the name is to be based. Some enzymes catalyze a number of different reactions, and a choice must be made if a single name is to be formed for each enzyme; with enzymes of wide specificity, the systematic name refers to one substrate only, usually the best natural substrate, unless there is a suitable word covering the whole group of substrates acted on, such as aldehyde or D-amino acid. The direction in which an enzyme-catalyzed reaction is written for the purposes of naming it is chosen so as to be the same for all the enzymes in a given class, and is not an indication of the equilibrium position.

Finally, it should be pointed out that a number of enzymes are known to exist which cannot be given a systematic name because the precise nature of the reaction catalyzed is not clear.

Enzyme Reactions. Nearly one thousand enzymes are known; 875 were listed and named in 1964 in the volume on "Enzyme Nomenclature." A casual survey of known enzymes suggests that they catalyze a staggering variety of chemical reactions which could be systematized with difficulty. A more careful analysis shows that this impression is due to the characteristically high specificity of many enzymes, so that they act only on one substrate, often of considerable complexity, although they bring about only one fairly simple transformation of one part of the substrate molecule. The comparative simplicity of the classification scheme described above bears testimony to the underlying unity of enzymatic catalysis; it is indeed surprising to find that all known enzymes can be allocated to one of the six main groups listed. In this section the main types of enzyme-catalyzed reactions will be rapidly surveyed.

The over-all reaction catalyzed by all the enzymes of Group 1 (*oxidoreductases*) can be written as hydrogen transfer, and these enzymes might be considered to be merely one section of the transferases (Group 2). A general reaction for Groups 1 and 2 can be written as:

$$AX + B \rightleftarrows A + BX$$

The oxidoreductases are classified separately because of their large number (221 are listed) and because of their great biological importance in bringing about the main energy-yielding reactions of living tissues. The main groups of *transferases* (Group 2) are concerned with transfer of one-carbon groups, acyl groups, glycosyl residues, amino- and other nitrogen-containing groups, phosphate, and sulfate. Oxidoreductases and transferases together represent over half of the enzymes at present recognized, with 454 entries in the Enzyme List in "Enzyme Nomenclature."

Hydrolases (Group 3, 211 entries) include esterases, glycosidases, peptidases, deaminases, and enzymes hydrolyzing acid anhydrides (such as the pyrophosphate group in adenosinetriphosphate). Many hydrolases have been shown to be able, under appropriate conditions, to catalyze transfer reactions; a high concentration of acceptor is usually necessary, since there is competition between the added acceptor and water for the group transferred. The detailed mechanism in these cases probably involves transfer of a part of the substrate on to a group on the enzyme, with subsequent transfer to an acceptor or hydrolysis, *e.g.*, for a hydrolase acting on a substrate AB to produce AOH and BH:

$$EH + AB \rightarrow E\text{—}A + BH$$

and

$$E\text{—}A + X \rightarrow E + AX$$

or

$$E\text{—}A + H_2O \rightarrow EH + AOH$$

These hydrolases (if not all) can therefore be regarded as transferases which include H_2O among their possible acceptors. Under normal conditions, in aqueous solution, hydrolysis will be the dominant reaction.

Lyases (Group 4) catalyze reactions of the type:

$$AX\text{—}BY \rightleftarrows A{=}B + X\text{—}Y$$

Molecules such as water, H_2S, NH_3 or aldehydes are added across the double bond of a second unsaturated molecule. Decarboxylases such as those acting on amino acids can be regarded as lyases (carboxy-lyases), assuming CO_2 and not H_2CO_3 to be the immediate product of decarboxylation. 116 lyases are known.

The *isomerases* (Group 5) include enzymes bringing about reactions similar to those in several other groups, but distinguished in that the reaction takes place entirely within one molecule, which is not cleaved, so that the overall reaction is:

$$A \rightleftarrows B$$

Thus there are intramolecular oxidoreductases (*e.g.*, ketol-isomerases), intramolecular transferases (*e.g.*, phosphomutases), and intramolecular lyases. Forty-seven isomerases are listed in "Enzyme Nomenclature."

The last group (*ligases*) catalyze reactions which are more complicated than those of the other groups and must involve at least two separate stages in the reaction. The over-all result is the synthesis of a molecule from two components with a coupled breakdown of adenosine triphosphate or some other nucleoside triphosphate; in general this may be written

$$X + Y + ATP \rightarrow XY + AMP + \text{Pyrophosphate}$$
$$(\text{or ADP} + \text{Phosphate})$$

These enzymes, of which about 50 are known, are of great importance in the conservation of chemical energy within the cell and in the coupling of synthetic processes with energy-yielding breakdown reactions.

References

1. "Enzyme Nomenclature: Recommendations (1964) of the International Union of Biochemistry," Amsterdam, Elsevier, 1965.
2. HOFFMANN-OSTENHOF, O., *Advan. Enzymol.*, **14**, 219 (1953).
3. DIXON, M., AND WEBB, E. C., "Enzymes," second edition, London, Longmans Green, and New York, Academic Press, 1964.

EDWIN C. WEBB

ENZYMES (ACTIVATION AND INHIBITION)

The response of an enzymatic rate to the addition of various and specific chemical entities (modifiers) represents an important approach to the study of enzyme reaction mechanisms and metabolic control mechanisms. The use of enzyme inhibitors has played an important part in the determination of metabolic pathways.[1] In general, the modifier interacts not with the reactants but with the enzmye or another catalytic component of the reaction (cofactor) so that there will be no effect on the equilibrium of the catalyzed reaction. Further, the interaction of the enzyme with modifier (M) is always reversible and can be represented by a simple equation: $E + nM \rightleftharpoons EM_n$ where n is usually a small whole number that can be easily determined from the effect of the concentration of M on the rate of reaction.[2] Generally, the rate of establishing the equilibrium between modifier and enzyme is very rapid as may be shown from studying the effect on the observed initial reaction rate of the order of addition of components to the reaction mixture. The rapidity and easy reversibility of interaction between enzyme and inhibitor distinguish inhibition from the effects of an inactivator which would be progressive and irreversible.

Many of the cases of metabolic modifiers that are known can be explained by interaction with specific enzymes or proteins. CARBON MONOXIDE, which causes asphyxiation of animals that rely on the oxygen carried by HEMOGLOBIN for respiration, is known to combine with the reduced iron of the hemoglobin, thus preventing combination with O_2. CO also interferes with the metabolism of O_2 by a similar reaction on the cytochrome oxidase molecule. As might be expected from the mode of inhibition, the effect of CO can be overcome by increasing the O_2 content of the solution. This example of competititive inhibition of cytochrome oxidase conforms to the general equation

$$v = \frac{V_{max}}{1 + \frac{K_s}{S}\left(1 + \frac{I}{K_i}\right)}$$

This equation describes the relation between the observed velocity, v, at a given concentration of substrate, S, and inhibitor, I, and the maximum velocity, V_{max}, which is approached as the substrate concentration becomes very large. The meaning of the term "very large" depends on the ease with which S displaces I from the enzyme (K_s lower than K_i) and the concentration of I. The K_i value is a dissociation constant,

$$K_i = \frac{(E)(I)}{(EI)}.$$

It can be determined from the kinetic measurements of v at different concentrations of I and S, and should agree with the direct measurement of binding of inhibitor to enzyme. The K_s value is not so simply defined in physical terms but, operationally, corresponds to that concentration of substrate which, in the absence of inhibitor, causes the reaction to proceed at one-half the maximum rate. The kinetics of inhibitor action that is not competitive with substrate will be described by an equation in which

$$V_{max}\Big/\left(1 + \frac{I}{K_i}\right)$$

appears in the numerator of the rate equation. If such a relation is observed, the modifier may or may not influence the apparent strength of interaction of the substrate with the enzyme, K_s, depending on the details of the process (for more information on the kinetics of enzyme reactions see references 1, 3 and 4). The equation for the hemoglobin case is much more complex because here 4 molecules of O_2 react successively with different affinities with the 4 hemes of the tetramer hemoglobin molecule. This effect is of great physiological importance because it allows the oxygenated hemoglobin to have a high affinity for O_2 when there is a high O_2 tension, at the lungs, and to give up the O_2 more readily at the tissues where the O_2 tension is lowered by metabolism. The explanation for this phenomenon probably resides not in the direct interaction of one O_2-heme group with a neighboring heme but rather in the effect of partial oxygenation on the conformation and ionic properties of the protein itself. A further complication arises from the observation that O_2 affinity may be dependent on hemoglobin concentration, and hence be a function of the dissociation of the tetramer protein into subunits that have a different O_2 affinity. Thus the ability of modifier to alter function through the physical and chemical state of the protein need not be restricted to the immediate site of modifier binding. This fact has been pointedly recognized by the use of the term "allosteric" to designate those agents that affect the function of the catalytic site of the enzyme although they are bound to some other site of the protein[5] (see REGULATORY SITES OF ENZYMES).

The product of a reaction will always be an inhibitor of the forward rate, which will be seen as competitive inhibition when in a single substrate, single product reaction the equilibrium is far toward product so that the effect of product is only to compete with substrate for the active site of the enzyme. In the case of two-substrate reactions $A + B \rightleftarrows C + D$, the effect of C on its own rate of production can give important information concerning the sequence in which reactants A and B interact with the enzyme. If C acts competitively with A, then it is presumed that A interacts with free enzyme and B reacts subsequently. This is confirmed if C is not competitive with B. It is well recognized that in catalyzing reactions, an enzyme forms many intermediary complexes with the substrates and their altered forms proceeding toward product formation. In general, the maximum velocity that can be obtained by increasing substrate concentration will not be affected by the presence of a given amount of modifier if the substrate and modifier interact with the same form of the enzyme, and will be altered if they interact with different forms.[6] This formalism says nothing about the sites on the enzyme with which the substrate and modifier interact, and indeed these need not interact at the same site in order to show competitive kinetics although this is more usually the situation.

Metabolic Regulations. Specific enzyme activation and inhibition at critical enzyme steps in the metabolizing and growing cell are of importance in the regulation and coordination of metabolic events. It is generally found, in bacteria, that the end product of a biosynthetic pathway is a strong and specific inhibitor of the earliest irreversible enzymatic reaction that leads to its formation[5] [see also METABOLIC CONTROLS (FEEDBACK MECHANISMS)]. Thus, L-isoleucine acts as such a feedback inhibitor to prevent the action of threonine deaminase in *E. coli.*[7] In this way, isoleucine synthesis for protein formation responds to the need for isoleucine by regulation of the mobilization of threonine. None of the four enzymatic steps that intervene between threonine deaminase and the formation of isoleucine are affected by isoleucine. Furthermore, another threonine deaminase in the same organism does not participate in the isoleucine biosynthetic path and is not affected by isoleucine. Other examples of enzymes that catalyze similar reactions (isozymes) in the same cell but are different in their response to modifiers are known, and these differences are surely good criteria of the different roles these enzymes have in the cell (see ISOZYMES).

Although the inhibitory effect of isoleucine for threonine deaminase can be overcome by high threonine concentration, the two compounds cannot be competing for the catalytically active or substrate site on the enzyme. Rather it seems that they may be competing for a distinct regulator site. This is indicated by several attributes that are often characteristic of such cases. Although it is quite unlikely that two molecules of threonine would be required in the deamination mechanism, the rate of reaction is proportional to S^2 rather than to S, as is generally the case, thus suggesting that a second threonine molecule must occupy a regulator site in order for the substrate reacting site to operate. Other compounds which cannot act as substrates for the enzyme are able to overcome the effect of the inhibitor and at the same time restore normal linear dependence on substrate concentration to the enzyme in the presence of isoleucine. That these modifier effects are probably acting through the conformational changes induced in the protein and not directly at the catalytic site is shown by the loss of these effects by treatments (urea, gentle heating, aging, mercuri-compounds and mutation) that may be expected to alter protein structure but are found not to prevent enzymatic reaction. Thus the modifier is no longer able to cause the transition in protein structure which is responsible for the influence on the reaction rate. This sensitivity of the effect to the state of the protein is especially well illustrated in a number of cases where a small change of pH is able to alter reversibly the susceptibility of the reaction to modifier effects.

The role of products and intermediates in controlling metabolic rate is certainly not restricted to biosynthetic systems in bacteria as can be illustrated by our current understanding of the control of glucose metabolism in the mammal. The product of the irreversible step catalyzed by hexokinase is glucose-6-P which is the immediate source compound of three alternative paths:

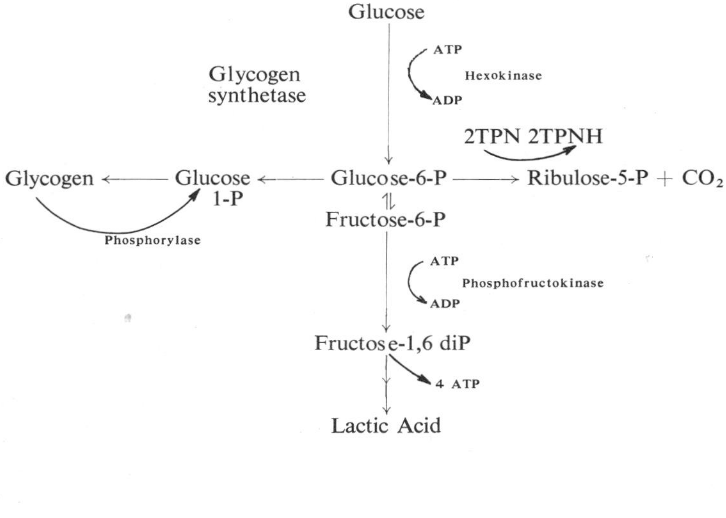

Glucose
ATP
Glycogen synthetase
Hexokinase
ADP
2TPN 2TPNH
Glycogen ← Glucose 1-P ← Glucose-6-P → Ribulose-5-P + CO_2
Phosphorylase
⇅
Fructose-6-P
ATP
Phosphofructokinase
ADP
Fructose-1,6 diP
4 ATP
Lactic Acid

one passing *via* the phosphofructokinase reaction to yield ATP and lactate. It is known that ATP is a strong inhibitor as well as being a substrate of phosphofructokinase. The inhibition is overcome by ADP which by its increase would activate the kinase, stimulating the production of ATP. If the kinase rate is depressed by a lessened demand for ATP, this would cause a rise in glucose-6-P level when the alternate paths are not using it rapidly. This increase in glucose-6-P has two effects: it is known that glucose-6-P activates glycogen synthetase so that storage of the energy source as GLYCOGEN would be expected. This occurs until a need for ATP is signaled through a decrease in the inhibition of phosphofructokinase leading to a fall in glucose-6-P and hence a decrease in glycogen synthesis. The second effect of an increase in glucose-6-P level is to inhibit its own synthesis, not by reversal of the hexokinase step, but by an action that seems to occur at an allosteric site. Thus glucose phosphorylation will be stopped under conditions that limit the need for glucose-6-P thus allowing it to rise to inhibiting concentrations. The glucose-6-P will have to be present since it is the controlling factor and hence it will remain available as an important source of reduced nicotinamide adenine dinuleotide-P (NADPH) through the third alternate path of glucose-6-P metabolism.

An interesting form of activation is involved in the regulation of glycogen breakdown in muscle. The process involves an enzyme called phosphorylase, which exists in an inactive unit called phosphorylase b. This moelcule may be activated in two ways that cause it to combine with itself to form a dimer. One way, which is very rapid and reversible, occurs in the presence of a common metabolite, adenylic acid-5′-P. The other way, which is relatively slow and irreversible, depends on another enzyme which transfers the terminal phosphate of ATP to phosphorylase b and causes the formation of a stable dimer made up of two molecules of the phosphorylase b-P. These two kinds of activation, one rapid and transient and hence responsive to the instantaneous changes in the metabolic network, and the other responding more slowly and dependent on the functioning of another enzyme which will itself be responsive to natural modifiers, exist side-by-side in skeletal muscle undergoing contraction. They are called forth under different physiological conditions depending on the demand for energy, supplied as ATP from glycogen breakdown. Clearly, however, the stimulation of glycogen conversion to glucose-P cannot, in itself, cause increased ATP synthesis unless subsequent steps are rapid enough. To provide for this, the same factors that stimulate phosphorylase b are known to overcome the inhibitory effects of ATP on phosphofructokinase. Hence a concerted activation of two rate-controlling enzymes causes the desired end. Although the details of these effects of inhibitors and activators have been established with the highly purified enzymes, the analysis of whole muscle caught in different states of physiological function tends to confirm this high degree of dependence of rate on the activators and inhibitors that are themselves under metabolic control.

References

1. WEBB, J. L., "Enzyme and Metabolic Inhibitors," Vol. 1, New York, Academic Press, 1963.
2. TAKETA, K., AND POGELL, B. M., *J. Biol. Chem.*, **240**, 651 (1965).
3. DIXON, M., AND WEBB, E. C., New York, Academic Press, 1964.
4. REINER, J. M., "Behavior of Enzyme Systems," Minneapolis, Burgess Publishing Co., 1959.
5. MONOD, J., CHANGEUX, J. P., AND JACOB, F., *J. Mol. Biol.*, **6**, 306 (1963).
6. CLELAND, W. W., *Biochim. Biophys. Acta*, **67**, 188 (1963).
7. UMBARGER, H. E., AND BROWN, B., *J. Biol. Chem.*, **233**, 415 (1958).

IRWIN A. ROSE

ENZYMES (ACTIVE CENTERS, OR ACTIVE SITES)

The active center of any enzyme is generally defined as that part of the protein structure coming into direct contact with the substrate acted upon. This functional region of an enzyme must be very small relative to its whole structure since the chemical change in the substrate is limited to a very small area undergoing alteration at a given time. A few amino acid side chains of the enzyme (and possibly the polypeptide backbone itself), groups not necessarily near one another in the primary sequence, are therefore visualized as coming into close spatial proximity and providing the essential features of the enzyme, *i.e.*, binding specificity and catalytic action. The binding specificity must be determined in part by the geometrical arrangement of the enzyme side chains in this region of the active center since the stereochemistry of the substrates plays such an important part in determining whether or not they are acted upon. The nature of these enzyme side chains is also of importance since in the absorptive phase of enzyme-substrate complex formation, they participate in non-covalent bond formation essential for binding. Such bonds may include ionic bonds, van der Waals forces, hydrogen bonds, hydrophobic bonds, dipole interactions, charge transfer complexes, and possibly others. Relatively little is known about the specificity features of enzyme active centers; more progress has been made in exposing their catalytically functioning groups.

Kinetic studies with substrates and inhibitors are of great value in studying enzyme active centers. Through the structural variation of such substances, it is possible to delineate the specificity of a given enzyme and to determine the probable forces involved in the formation of the enzyme-substrate complex, and their steric relationship. Through kinetic studies at various pH values, there have been deduced, in certain well-analyzed cases, the pH values of groups at the active center and whether they are important in substrate binding or at some later stage in the over-all process.[1,2] The nature of the amino acid residues cannot always be determined with certainty from such *p*K values because several groups may have a similar *p*K value, because the protein environment may have an effect on the ionizable group, shifting it from characteristic values, and, also, because the process studied may be a complex one. In studies of this kind, corroborative evidence from a variety of experimental sources must be brought together before a unified consistent picture is attained.

For some enzymes it is possible to demonstrate an intermediate enzyme derivative such as an acyl-enzyme in the case of PROTEOLYTIC ENZYMES[2] or a phospho-enzyme, in the case of PHOSPHATASES or transferases.[3] The functional serine residue generally found to be involved is clearly part of the active center and can be identified by a structural study of the intermediate. These are normal intermediates, but in most cases of enzymatic action it is not possible to characterize stable intermediates by protein degradative studies. However, a transitory state may be stabilized by chemical treatment. Thus several interesting examples of reducing the carbonyl group of a substrate (triosephosphate, in the case of aldolase) or a cofactor (pyridoxal phosphate in the case of phosphorylase) *in situ* have resulted in the stable reductive attachment of the small molecule to the enzyme and permitted the location of a particular, altered lysine residue.

Studies by Chemical Modification. The observation of the effect of chemical reagents on the activity of enzymes is a major way of studying active centers. For example, the loss of activity accompanying treatment with mercurials (—SH of cysteine), iodoacetate (—SH, imidazole ring of histidine, CH_3S— of methionine, or ε-amino group of lysine, according to pH), iodination (phenolic ring of tyrosine), photooxidation (H_3C—S— of methionine, rings of tryptophan and histidine), or dinitrofluorobenzene (—SH of cysteine, ε-amino group of lysine) may indicate the importance of the amino acid side chain indicated in parenthesis. Most of these reagents do not provide a specific group test. The interpretation of the results of such studies may be complicated since there is required a knowledge of the nature and number of altered sites within the enzyme and the effect on its properties. The possibility of altering the properties of an enzyme indirectly through a chemical change which is remote from the active center but alters it by distortion is an ever-present consideration. The scope of chemical modification is demonstrated in the particularly well-studied case of ribonuclease.[4]

A small number of reagents have an unusual specificity apparently because of pseudo-substrate character with resultant localization at enzyme active centers. Diisopropylfluorophosphate (DFP) is a classic example. It inactivates a number of hydrolytic enzymes by transferring a diisopropylphosphoryl group to a particular serine residue which is essential for catalytic activity and which can be identified in the protein structure by virtue of this label. DFP is not a group reagent for serine since it is inert to many other serines present in the susceptible enzymes. The advantage of designing specific reagents for discovering the essential amino acids in the active center of any enzyme by devising an inhibitor which combines substrate characteristics plus a chemically active grouping offers great promise for the study of active centers. One example of this approach is that of TPCK which was synthesized as an active center reagent for chymotrypsin by including phenylalanine since chymotrypsin acts hydrolytically at the carbonyl of phenylalanine in esters, amides, and proteins. In addition, TPCK includes a chloromethyl ketone function desired to achieve covalent binding with some functional group of the enzyme. TPCK inactivates chymotrypsin by alkylation of a histidine residue which thus becomes identifiable. In analogous fashion, TLCK (see structures) was expected to combine[1] at the active center of trypsin whose substrate preference has been incorporated in TLCK in the form of lysine; it was indeed found to inactivate trypsin by alkylation of an essential histidine in that enzyme. (However, TPCK has no effect on trypsin and TLCK, none on chymotrypsin, showing that the enzyme-specific structural feature determines whether or not action takes place.) These are not histidine reagents but active center probes for specific enzymes.

The elucidation of protein structure through X-ray crystallography has not yet been applied very much to the study of enzyme active centers. However, following the high-resolution study of hen egg-white lysozyme, the analysis of some lysozyme-inhibitor complexes became feasible and resulted in the localization of the binding site of the enzyme.[6] The identification of three tryptophan residues in this region agreed with prior chemical evidence of their functional significance. This important result indicates that major advances can be expected from the further application of this technique.

$(CH_3)_2CH{-}O{-}P(=O)(F){-}O{-}CH(CH_3)_2$

DFP

$C_6H_5{-}CH_2{-}CH(NH{-}SO_2{-}C_7H_7){-}C(=O){-}CH_2{-}Cl$

TPCK

$H_2N{-}CH_2{-}CH_2{-}CH_2{-}CH_2{-}CH(NH{-}SO_2{-}C_7H_7){-}C(=O){-}CH_2{-}Cl$

TLCK

The concept of the "active center" does not necessarily imply the nonessentiality of most of protein structure of enzymes. During the catalytic process, the realignment of large parts of the protein may provide essential changes at the active center and thus may be critical to the enzymic process.

References

1. DIXON, M., AND WEBB, E. C., "Enzymes," Second edition, New York, Academic Press, Ch. IV(c), 1964.
2. BENDER, M. L., *Ann. Rev. Biochem.*, **34**, 49 (1965).
3. HUMMEL, J. P., AND KALNITSKY, G., *Ann. Rev. Biochem.*, **33**, 15 (1964).
4. SCHERAGA, H. A., AND RUPLEY, J. A., *Advan. Enzymol.*, **24**, 161 (1962).
5. SCHOELLMANN, G., AND SHAW, E., *Biochemistry*, **2**, 252 (1963); SHAW, E., MARES-GUIA, M., AND COHEN, W., *ibid.*, **4**, 2219 (1965).
6. BLAKE, C. C. F., KOENIG, D. F., MAIR, G. A., NORTH, A. C. T., PHILLIPS, D. C., AND SARMA, U. R., *Nature*, **206**, 757 (1965); JOHNSON, L. N., AND PHILLIPS, D. C., *ibid.*, **206**, 761 (1965).

ELLIOTT N. SHAW

ENZYMES (INDUCTION AND REPRESSION).

See METABOLIC CONTROLS (FEEDBACK MECHANISMS).

ENZYMES (KINETICS)

Enzymes, the biological catalysts found in all living cells, generally occur in such trace levels that it is not possible to measure the amount present by a direct chemical or physical test for the enzyme molecule itself. Even in the most favorable circumstances imaginable—that where an enzyme of molecular weight 10,000 was present at a potency of 1%, the *molar concentration* of the catalyst would still be only 10^{-3}. Resort must therefore be made to measurement of the *catalytic activity* of the enzyme, and from these data estimates of the amount of enzyme present can be made. Once the enzyme has been obtained in pure form and the catalytic constants established, the actual grams of enzyme present in crude mixtures can be calculated. Such work requires an appreciation of reaction types and kinetics of enzymes.

The modern theory of enzyme action demands that an actual complex be formed between the biocatalyst and the compound acted upon (substrate). Numerous examples are now known in which the properties of the enzyme-substrate complex have been elucidated in rather minute detail, although it must be confessed that the precise mechanism of action of no enzyme is yet fully understood. Many years ago Michaelis and Menten,[1] through kinetic studies, predicted that intermediate compounds of enzyme and substrate were formed.

Reaction Types. *Zero Order*. When the enzyme molecule is saturated with substrate, the formation of product proceeds at a linear rate with time. Addition of more substrate does not augment the rate, and the reaction type is said to be *zero order* (Fig. 1). The rate is described by the equation

$$\frac{dp}{dt} = k^0$$

in which p is product and k^0 the first-order rate constant. The latter has the dimensions of amount per unit time, usually expressed as micromoles

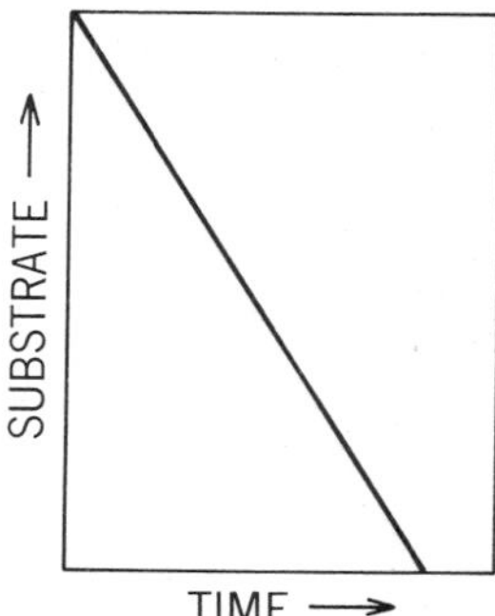

FIG. 1. Zero-order reaction.

per minute. Obviously, zero-order kinetics are desired when assays for enzyme concentration are being carried out in crude systems; one is then assured that a change in rate is actually caused by the enzyme and not by the adventitious addition of more substrate. A simple test for zero-order kinetics is to allow the reaction to run on for a double length of time; the amount of product formed should exactly double.

First Order. At less than saturating amounts of substrate the rate will be higher than zero order. A common type encountered is *first order*, in which the rate is proportional to the prevailing substrate concentration. This leads to a gradual loss in rate, with infinity being the time required for total completion of the reaction. The course of a first-order plot is given by the equation

$$\frac{dp}{dt} = k^1(a - p)$$

where a is the initial substrate concentration and p the amount of product formed in time t. This leads to the "die-away" curve shown in Fig. 2. Integration of the first-order equation affords

$$k^1t = 2.3 \log a/(a - p)$$

The first-order rate constant, k^1, has the dimensions of reciprocal time, *i.e.*, k^1/min. Although first-order kinetics are more cumbersome to work with than zero-order reactions, it is nonetheless sometimes necessary to use the former when dealing with insoluble, expensive or toxic substrates.

Higher Reaction Orders. As the substrate concentration is decreased, the rate will advance

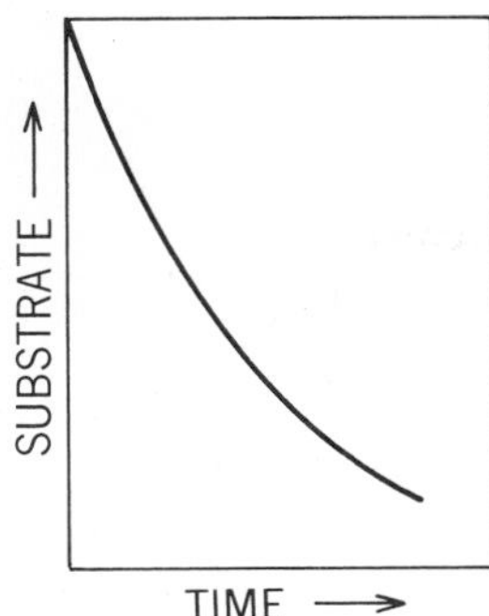

FIG. 2. First-order reaction.

through a complicated progression of reaction types to higher orders. The reader is referred elsewhere for further treatment of this subject.[2]

Application of Reaction Rates in Enzymology. The absolute term for the activity of an enzyme is the *turnover number*, usually defined as the moles of substrate converted per mole of enzyme per minute. The turnover number ranges from a few hundred to many thousands of moles. Where the molecular weight of the enzyme is not known, the activity is described in terms of the number of arbitrary units per milligram of protein. The *International Unit* is defined as one micromole per minute. For a zero-order reaction, the enzyme units present in a sample will be given by the ratio

$$\frac{k^0 \text{ (measured)}}{k^0 \text{ (1 unit)}}$$

A similar ratio applies for the first-order reaction. Irrespective of the reaction order, the rate will always be proportional to the enzyme concentration. From this it follows that

$$\text{Enzyme} \times \text{Time} \times \text{Product} = \text{Constant}$$

Effect of Substrate Concentration on Velocity. The quantity which describes the effect of substrate concentration on velocity is known as the Michaelis constant, K_m. This is a kinetic constant governing the rate of formation and decay of the intermediate enzyme-substrate complex, here denoted as A. The enzyme combines with the substrate to form A

$$E + S \underset{k_2}{\overset{k_1}{\rightleftharpoons}} A \overset{k_3}{\rightharpoonup} E + P$$

which itself breaks down into free enzyme and product (P). The rate of each part reaction is controlled by specific rate constants (k_1, k_2, k_3). At the start of the reaction, when P is very low, the amount of resynthesis of A will be negligible and may be disregarded. During the steady-state conditions, the rate of formation of A will equal its rate of decomposition and we may write

$$k_1[E][S] - k_2[A] = k_3[A]$$

By transposition, grouping and equating the constants $k_2 + k_3/k_1 = K_m$,

$$[E][S] = [A]\frac{(k_2 + k_3)}{k_1} = [A]K_m$$

$$[E]/[A] = K_m/[S]$$

Now the maximum velocity, V, will be proportional to the total enzyme concentration, $[E]_t$, and the actual velocity, v, will be proportional to the concentration of active enzyme, or [A]. Hence, since the free enzyme equals $[E]_t - [A]$

$$\frac{[E]_t - [A]}{[A]} = \frac{[E]_t}{[A]} - 1 = \frac{V}{v} - 1 = \frac{K_m}{[S]}$$

The usual form of the Michaelis equation is

$$v = \frac{V[S]}{K_m + [S]} = \frac{V}{1 + (K_m/[S])}$$

It should be noted that both K_m and V are constants for the enzyme although they may vary independently under various conditions.

The Michaelis equation shows that when $[S] = K_m$, the velocity is exactly half maximal. Similarly, when [S] is 0.1 and 10 times the K_m, the velocity will be 10 and 90% of maximum, respectively. K_m has the dimensions of concentration, *i.e.*, moles per liter. At high substrate levels, the Michaelis equation reduces to $v = V$; at low substrate concentration relative to K_m, the equation becomes

$$v = \frac{V[S]}{K_m}$$

and the velocity is dependent on [S].

The K_m is a kinetic constant comprised of the terms $(k_2 + k_3)/k_1$.

Individual Rate Constants. It has been shown that where K_m and V can be varied by altering values of k_3, the corresponding values for k_1 and k_2 can be determined.[3] Thus,

$$K_m = \frac{k_2[E]_t + k_3[E]_t}{k_1[E]_t} = \frac{k_2[E]_t + V}{k_1[E]_t}$$

From this equation it is seen that a straight line results from a plot of V against K_m, yielding k_2/k_1 as the intercept on the K_m axis. The slope of the line will be $1/k_1[E]_t$, and the point where it cuts the V axis will be $-k_2[E]_t$.

At the start of an enzyme reaction where all of the enzyme is present as A the velocity will be maximal and equal to $k_3[E]$. The turnover number is identical with k_3 if the concentrations of both enzyme and product are given in moles and the time unit is minutes.

The Haldane Relationship. The interdependence of the equilibrium constant and the rates of the forward and back reactions was first pointed out in 1930 by Haldane.[4] He showed that

$$K_{eq} = \frac{[P]}{[S]} = \frac{k_1k_3}{k_4k_2}$$

Multiplying the top and bottom of the latter equation by the term $[E]_t(k_2 + k_3)/k_1k_4$ affords

$$K_{eq} = \frac{k_1k_3[E_t]\left[\frac{k_2 + k_3}{k^1k_4}\right]}{k_4k_2[E]_t\left[\frac{k_2 + k_3}{k_1k_4}\right]} = \frac{V_fK_{m,r}}{V_rK_{m,f}}$$

Fumarase, an enzyme which has measurable equilibrium constants for both the forward and back reactions, was shown by Frieden and Alberty[5] to obey the Haldane relationship; several other examples could be cited.

Measurement of K_m and V. As increasing amounts of substrate are added to an enzyme reaction, the velocity will increase and gradually approach a maximum value as shown in Fig. 3. In

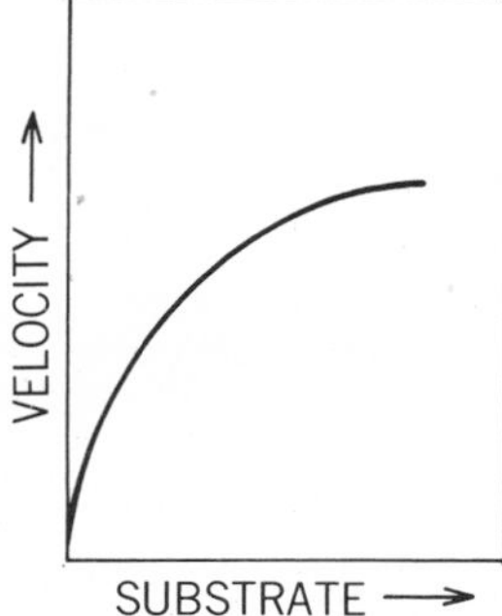

FIG. 3. Effect of substrate concentration on velocity.

practice it is not convenient to determine V and K_m in this type of experiment, however. Frequently very high levels of substrate are needed to saturate the enzyme, and before that point is reached substrate inhibition and solute effects may be encountered. Lineweaver and Burk[6] popularized the use of a straight-line form of the Michaelis equation for the laboratory estimation of K_m and V. Taking the reciprocal of each side of the Michaelis equation we have

$$\frac{1}{v} = \frac{K_m + [S]}{V[S]} = \frac{K_m}{V[S]} + \frac{[S]}{V[S]}$$

$$\frac{1}{v} = \frac{K_m}{V}\left(\frac{1}{[S]}\right) + \frac{1}{V}$$

This is the equation of a straight line of the form $y = ax + b$ in which a is the slope and b the intercept. The latter is $1/V$ while the slope of the line is K_m/V (Fig. 4). Extrapolation of the line across the $1/v$ axis afford K_m directly, as $-1/K_m$, at the point of intersection of the $1/S$ axis.

It should be noted in passing that the Michaelis equation is formally identical with two other expressions in common usage in biochemistry.[7] These are the Henderson-Hasselbalch equation for weak acids and bases and the fundamental electrode equation.

$$\text{pH} = pK + \log\frac{[\text{Salt}]}{[\text{Acid}]}$$

$$E_h = E_0 + \frac{0.06}{n}\log\frac{[\text{Oxidized Form}]}{[\text{Reduced Form}]}$$

$$\log [S] = \log K_m + \log\frac{v}{V - v}$$

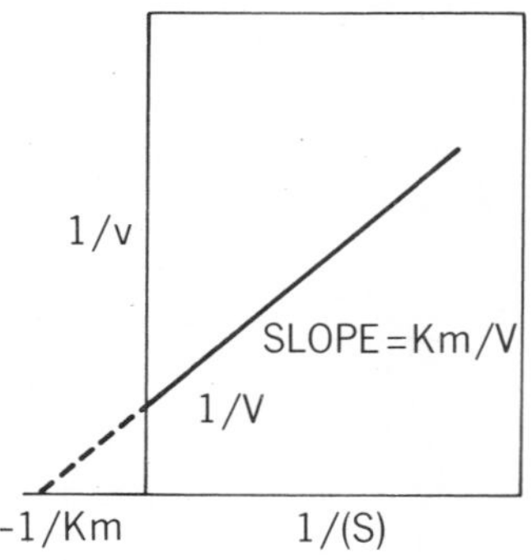

FIG. 4. Graphical determination of K_m and V.

The Value of K_m for Certain Enzymes. Actual numerical values of K_m range from over $0.1M$ to less than $10^{-6}M$.[3] If the dissociation constant of an enzyme-coenzyme complex is much less than about $10^{-8}M$, the compound is capable of isolation as a conjugated protein. The Michaelis constant for pyridine nucleotide coenzymes with a variety of dehydrogenases can be measured by the above methods and is of the order of 10^{-5}–$10^{-6}M$. In general, the K_m for hydrolytic enzymes is larger than that for oxidative enzymes. Thus the K_m for maltase acting on maltose is $0.21M$ while that for lactic dehydrogenase with pyruvate as substrate is $3.5 \times 10^{-5}M$.

Enzymes as Pure Catalysts. In general, enzymes are present in biological tissues at such low levels that they play no part in determining the thermodynamic equilibrium of a reaction. However, in cases where the enzyme is concentrated in cell particles, such as mitochondria, and there is an unusual affinity between either reactant or products for the enzyme, the latter can displace the equilibrium in the direction of the binding reaction. This is the case for the enzyme ALCOHOL DEHYDROGENASE from liver, a protein which displays a strong affinity for reduced diphosphopyridine nucleotide.[8] At high levels of the enzyme, the thermodynamic equilibrium constant becomes an *apparent* equilibrium constant and is proportional to the amount of enzyme added.

Effect of pH on Enzyme Activity. *Introduction*[9,10] Enzymes are so strongly affected by pH that it is this variable which is considered first in working with these substances in the laboratory. The most favored pH for activity is known as the "optimum" pH and may vary from values of less than 2 to very alkaline reactions. PEPSIN is an example of an enzyme which displays optimum activity in very acidic media while certain PHOSPHATASES and PEPTIDASES are most active in basic solution. In some cases, families of enzymes are identified according to the optimum pH, *i.e.*, "alkaline" and "acid" peptidases and phosphatases.

The pH vs Activity Curve. In determining the pH optimum for enzymes it is necessary to work over a wide range of acidic and basic reactions and this will involve the use of several buffer anions and cations. If possible, two buffers should be used at identical values of pH in order to be sure that the observed effect is caused by pH and not by the ions of the buffer solution. Usually acetate, phosphate, "tris" (trihydroxymethylaminomethane) and glycinate will suffice for the pH range 3.5–10.5.

Most enzymes have an optimum pH around neutrality and in most cases the pH *vs* activity curve is bell shaped (Fig. 5). Originally it was

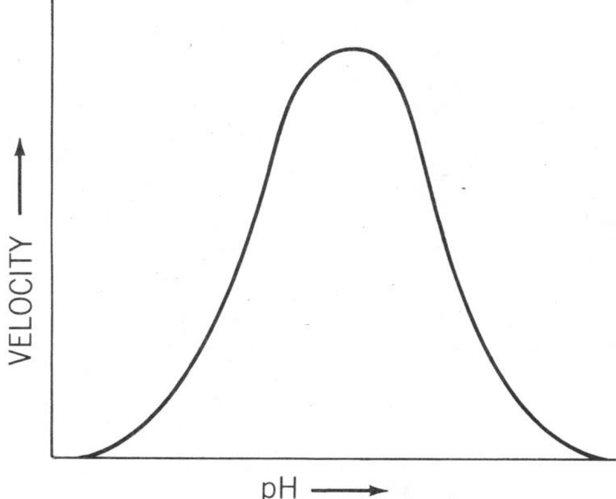

FIG. 5. A typical pH *vs* activity curve for an enzyme reaction.

believed that such a curve could be explained by assuming the isoelectric form of the enzyme to be the active species. Now, however, it is considered that the *active center* is capable of reversibly attaching a proton with only the intermediate form having activity.

$$EH_{2\,inactive} \rightleftharpoons HE_{active} \rightleftharpoons E_{inactive}$$

Obviously the effect of pH may be very complicated and may involve alterations in substrate and product as well as changes in the ionic state of buffer constituents and the enzyme protein itself. In addition, the effect of pH on the enzyme-substrate compounds needs to be considered.

Effect of pH on the Equilibrium of Enzyme-catalyzed Reactions. The substrates and products involved in enzyme reactions are very often capable of binding or discharging one or more protons. This means that the equilibrium point of the reaction will be dependent on pH. All of the very large number of reactions in which the pyridine nucleotides participate are affected in this way as a consequence of the conversion of a strongly basic quaternary nitrogen in the oxidized coenzyme into a weakly basic tertiary nitrogen in the reduced coenzyme.

Proton exchanges in the catalyzed reaction provide a useful means for following the activity of enzymes. Thus the removal of the terminal phosphate from adenosine triphosphate at pH 7.5 will be accompanied by the formation of one equivalent of hydrogen ion; at pH 6, however, no acid is produced in the hydrolysis. A similar situation exists for peptidases acting at pH values of 8.5 and 6.5, respectively.

Effect of Temperature on Enzymes. *Introduction.*[11] Temperature may affect enzyme activity in two ways. The enzyme protein may be modified in such a way that the catalytic activity is changed. The other means whereby enzyme activity is sensitive to heat changes is via the catalyzed reaction itself. These two effects will be considered separately.

Effect on the Enzyme Protein. All enzymes are PROTEINS and all proteins are to some extent altered by exposure to heat. In addition to the primary peptide bonds which hold together the amino acid building blocks, secondary and tertiary bonds exist within the giant protein molecules. When these are irreversibly broken, the enzyme protein is said to be denatured, a phenomenon which often involves precipitation of the protein from solution [see PROTEINS (BINDING FORCES IN SECONDARY AND TERTIARY STRUCTURE)].

Effect of Temperature on the Enzyme-catalyzed Reaction. Enzymes act as true catalysts in that they bypass at least a portion of the activation energy barrier (Fig. 6). This is achieved by formation of the intermediate enzyme-substrate compound, the conformation and configuration of which is such that the reaction can proceed at an unusually low temperature.

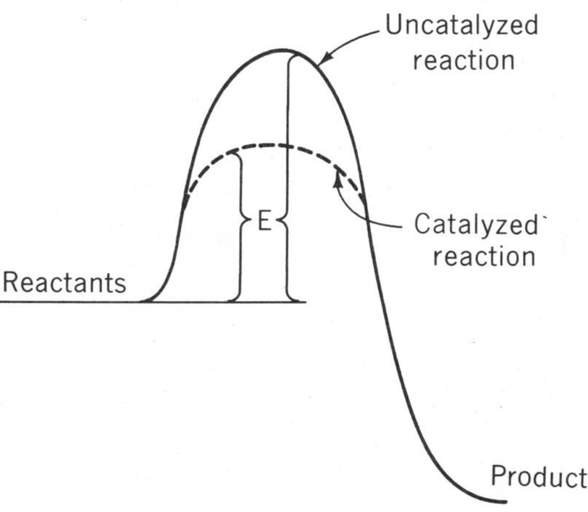

FIG. 6. The activation energy for catalyzed and uncatalyzed reactions.

As a general rule, the velocity of enzyme reactions is doubled for a 10° rise in temperature. Stated in another way, the activation energy (E) is of the order of 12,000 calories. Substitution of this value in the integrated form of the Arrhenius equation

$$\ln \frac{k_2}{k_1} = \frac{E(T_2 - T_1)}{RT_1T_2}$$

in which k_1 and k_2 are the rate constants at temperatures T_1 and T_2, respectively, and R is the gas constant, gives $\log(k_2/k_1)$ equal to 0.29 for a rise in temperature from 22–32°C, *i.e.*, k_2/k_1 is 2 and the reaction rate is doubled.[7]

References

1. Michaelis, L., and Menten, M. L., *Biochem. Z.*, **49**, 333 (1913).
2. Boyer, P. D., Lardy, H., and Myrbäck, K., (Editors), "The Enzymes," Second edition, Vol. 1, New York, Academic Press, 1959.
3. Dixon, M., and Webb, E. C., "Enzymes," second edition, New York, Academic Press, 1964.
4. Haldane, J. B. S., "Enzymes," London, Longmans, Green and Co., 1930.
5. Frieden, C., and Alberty, R. A., *J. Biol. Chem.*, **212**, 859 (1955).
6. Lineweaver, H., and Burk, D., *J. Am. Chem. Soc.*, **56**, 658 (1934).
7. Neilands, J. B., and Stumpf, P. K., "Outlines of Enzyme Chemistry," Second edition, New York, John Wiley & Sons, 1958.
8. Theorell, H., and Bonnichsen, R., *Acta Chem. Scand.*, **5**, 1105 (1951).
9. Alberty, R. A., *J. Cell. Comp. Physiol.*, **47**, Suppl. 1, 245 (1956).
10. Johnson, M. J., in "Respiratory Enzymes" (Lardy, H. A., Editor), Minneapolis, Minn., Burgess Publishing Co., 1949.
11. Moelwyn-Hughes, E. A., in "The Enzymes" (Sumner, J. B., and Myrbäck, K., Editors), New York, Academic Press, 1951.

J. B. Neilands

EPILEPSY

Epilepsy is a disease characterized by sudden convulsive seizures (grand mal) or by less severe attacks (petit mal) involving often a temporary loss of consciousness. It has been known since the time of Hippocrates and has afflicted many famous people including Julius Caesar and possibly Napoleon. Electroencephalography can be used to study the disease and anticonvulsants are widely used to treat it. Treatment is largely symptomatic, however, and the metabolic defect which gives rise to the trouble is unknown. Biochemical study will probably be able eventually to discover what goes wrong when a person has an attack and why only certain individuals are susceptible (see also MENTAL RETARDATION).

EPIMERIZATION

Epimerization is the conversion of configuration of one asymmetric center to its mirror image or enantiomeric configuration, while the remaining asymmetric center or centers within the molecule remain unchanged [see ASYMMETRY; AMINO ACIDS (OCCURRENCE STRUCTURE AND SYNTHESIS), section on Optical Considerations]. An example of epimerization is the conversion of D-glucose to D-galactose, through an enzyme-catalyzed inversion ("Walden inversion") of the optical configuration of carbon-4 of the sugar group [see GALACTOSE AND GALACTOSEMIA for this particular reaction; CARBOHYDRATES (CLASSIFICATION) for structures].

EPISOMES. See BACTERIAL GENETICS, section on Infectious Heredity.

ESSENTIAL AMINO ACIDS. See AMINO ACIDS, NUTRITIONALLY ESSENTIAL (FOR MAN AND OTHER VERTEBRATES).

ESSENTIAL FATTY ACIDS. See FATTY ACIDS, ESSENTIAL (IN ANIMAL NUTRITION).

ESTERASES

Esterases, a subclass of the hydrolases (see ENZYME CLASSIFICATION), are enzymes that catalyse the hydrolysis of ester linkages. Some PROTEOLYTIC ENZYMES can act also as esterases. Esterases include the lipases that hydrolyze dietary triglycerides into fatty acids and glycerol (see DIGESTION AND DIGESTIVE ENZYMES); the various NUCLEASES and PHOSPHATASES that hydrolyze phosphoric acid ester bonds; thioester and sulfate ester hydrolases; and a number of enzymes (such as occur in SNAKE VENOMS) that attack certain PHOSPHOLIPIDS. See also ACETYLCHOLINE AND CHOLINESTERASE.

ESTROGENS. See HORMONES, ANIMAL; OVARIAN HORMONES; SEX HORMONES.

ETHANOL. See ALCOHOL.

EULER-CHELPIN, H. V.

Hans von Euler-Chelpin (1873–1964) was a German-born Swedish biochemist whose pioneering work on vitamins and coenzymes was outstanding. For his investigations dealing with fermentation, he shared with Arthur Harden the Nobel Prize for chemistry in 1929.

EVOLUTION, BIOCHEMICAL. See ORIGIN OF LIFE. Certain aspects are discussed also in COMPARATIVE BIOCHEMISTRY.

EXTRINSIC FACTOR

Extrinsic factor is an alternative term for Vitamin B_{12}, originally used in relation to the "intrinsic factor" of Castle, which is a protein that combines with Vitamin B_{12} in the normal process of intestinal absorption of this vitamin [see VITAMIN B_{12} (ABSORPTION AND NUTRITIONAL ASPECTS); COBALAMINS AND COBAMIDE COENZYMES; ANEMIAS].

F

FAT MOBILIZATION AND TRANSPORT

Fats used by cells of the body for energy and other purposes originate either exogenously (dietary) or endogenously (synthesized by the body from other substances). In either case, it is necessary for these molecules, as well as those which have been stored in tissues, to be transported from one location in the body to others. The net movement of these materials, as well as their uptake or release from specific organs or tissues, is dependent on nervous, hormonal, and nutritional factors, some of which are not yet understood.

Transport of Fats in Blood. Dietary fat in the form of triglyceride [see LIPIDS (CLASSIFICATION)] may be directly incorporated into adipose tissue or it may first be routed to the liver where, after undergoing lipolysis and reesterification, it is released again into the circulation for further disposition.

Transport of fats from the intestine to various body sites takes place via the LYMPH and blood in the form of *chylomicrons*, colloidal particles of about 0.5–1.5 μ diameter composed largely of neutral fat (triglyceride) but containing also small amounts of lecithin and protein. Chylomicrons are removed rapidly from blood, with a half-life of about 10 minutes, and uptake has been shown to occur in several organs and tissues.

Chylomicron triglyceride may be acted upon by lipoprotein lipase ("clearing factor" lipase), converting the fat to lower glycerides or to glycerol and fatty acids, the latter being bound by ALBUMIN present in plasma. HEPARIN, injected intravenously, releases a lipolytic enzyme into blood plasma with subsequent increased hydrolysis of particulate triglyceride. If an inhibitor of the enzyme (*e.g.*, protamine) is used, removal of chylomicrons is delayed. Whether this lipolytic enzyme is important physiologically, however, is questionable since its appearance in the bloodstream is seen only after heparin injection. In the absence of heparin administration, only traces of enzyme activity are seen in plasma.

In addition to chylomicrons, which are present in plasma in large amounts during absorption of fat from the intestine, lipids in plasma exist as water-soluble complexes in the form of LIPOPROTEINS, which contain PHOSPHOLIPID, cholesterol, and glycerides associated with α- and β-globulins, or as unesterified fatty acids bound to albumin. Lipoproteins have been classified according to density: high-density lipoproteins (density >1.063 include α-lipoproteins and have a relatively large amount of protein), low-density lipoproteins (density <1.063 include β-lipoproteins and have relatively less protein associated with lipid), and chylomicrons. About three-fourths of the total blood lipids are present as β_1-lipoproteins, and about 8–12% of the total plasma proteins are involved in lipoprotein molecules. The low- and high-density lipoproteins are further subdivided into classes according to density. Lipoproteins of density 0.98 contain about 9% protein, 50% triglyceride, 18% phosphatide, and 22% cholesterol. Those of density 1.035 contain about 21% protein, 10% triglyceride, 22% phosphatide, and 46% cholesterol. The high-density lipoproteins include those of density 1.09 which contain about 33% protein, 8% triglyceride, 29% phosphatide and 30% cholesterol and those of density 1.14 which are composed of about 57% protein, 5% triglyceride, 21% phosphatide, and 17% cholesterol. The lipoproteins have also been characterized by flotation constants (S_f) which are analogous to sedimentation constants of ordinary proteins. The higher the S_f value, the lower is the density of the lipoprotein.

Mobilization and Transport of Adipose Tissue Lipids. Adipose tissue lipids, which amount to 10–20% of the body weight of normal adult animals and are largely triglycerides, are constantly being mobilized into the blood stream, and plasma fats are continuously being deposited in adipose tissue. Factors affecting the rates of these two directions are nutritional, hormonal, and nervous. The entry of triglycerides into adipose tissue is related to the activity of lipoprotein lipase, which is involved in the second of what may be a two-stage process: (1) non-enzymatic incorporation of triglyceride into adipose tissue; (2) rearrangement by lipolysis to free fatty acids followed by reesterification. The activity of the lipase, in turn, is related to diet and to insulin activity. Feeding (carbohydrate) and availability of INSULIN increase the activity of lipoprotein lipase and thus facilitate the entry of triglycerides into adipose tissue. Lack of insulin or decreased dietary carbohydrate decreases entry. Insulin is further involved since normal carbohydrate metabolism is needed to furnish α-glycerophosphate for the reesterification of the fatty acids in the adipose tissue.

Fat leaves adipose tissue as unesterified fatty acids bound to albumin, and this fatty acid-albumin complex represents the metabolic fuel of the body. Isotopically labeled unesterified fatty acids injected into the circulation have a half-life of about 2 minutes. Administration of glucose to a fasting animal decreases the concentration of plasma unesterified fatty acids. Similar results are produced by administration of tolbutamide, glucagon, and insulin, and by exercise. All of these increase the rate of utilization of glucose. Various factors are known to cause an increase in plasma concentration of unesterified fatty acids and among these are hormonal agents such as epinephrine and norepinephrine (see ADRENALINE), GROWTH HORMONE, ADRENOCORTICOTROPIC HORMONE, conditions of stress, fasting, and the diabetic state. The effect of most of these is due to action on adipose tissue.

Adipose tissue triglyceride is continuously hydrolyzed by an enzyme which appears to be distinct from the lipoprotein lipase involved in triglyceride entry into adipose tissue. Reesterification of the fatty acids resulting from the lipolysis requires the presence of α-glycerophosphate. Therefore, free fatty acids accumulate during deprivation of carbohydrate or decreased metabolism of carbohydrates. Decreased insulin or availability of carbohydrate thus results in increased release of unesterified fatty acid by adipose tissue into the bloodstream. The release of fatty acid from adipose tissue is stimulated by various hormones: epinephrine, norepinephrine, adrenocorticotropin, thyrotropin, growth hormone, THYROID HORMONE, and glucagon. Adipokinin, a factor isolated from the anterior pituitary, also causes mobilization of fat. The action of the hormones may be due to stimulation of the lipolytic process with a possible simultaneous inhibition of the reesterification mechanism. Excessive nerve stimulation causes loss of fat from adipose tissue (while paralysis or denervation leads to increased fat deposition). An intact sympathetic nerve supply is essential for the mobilization of lipid from adipose tissue in response to fasting. The effect of the sympathomimetic amines on release of fat from adipose tissue lends further support to the important role of the nervous system in the regulation of fat mobilization.

Mobilization and Transport of Liver Lipids. Part of the dietary fat absorbed from the intestine first enters the liver and then is released into the bloodstream for transport to peripheral tissues. Although dietary fat may be incorporated into adipose tissue without prior passage through the liver, it is believed that the principal route to adipose tissue is through liver. The dietary triglycerides are removed efficiently from blood by the liver, and the triglyceride molecule appears to enter the liver intact, hydrolysis and reesterification occurring after entry into the liver.

Uptake by liver of unesterified albumin-bound fatty acids from blood is very rapid and proportional to the concentration in the blood. Immediate esterification converts these to glycerides and phosphatides in the liver.

Fat leaves the liver in association with low-density lipoproteins. These lipoprotein complexes act as carriers of the triglycerides, and the surface-active components of the lipoprotein (protein, cholesterol, phospholipids) have a slower turnover rate than the triglyceride moiety. Some of the triglyceride in the low-density lipoproteins comes from sources other than the dietary fat, *e.g.*, from fatty acids produced in the liver by lipogenesis from dietary carbohydrate.

Blockage of triglyceride release by the liver may account for the cause of certain types of fatty livers induced by hepatotoxins such as carbon tetrachloride, white phosphorus, puromycin, and ethionine. Blockage of triglyceride release may be caused by interference with the synthesis by the liver of the proper protein moiety of the low-density lipoprotein essential as a carrier of the triglyceride.

It has also been found that there is in plasma a globulin protein ("lipid acceptor protein") which can combine with lipid in the liver with subsequent release into the plasma as lipoprotein. The relative contributions to lipoprotein formation of this circulating "lipid acceptor protein" and protein synthesized *de novo* in the liver are not yet known.

Uptake of Fat in Muscle and Other Tissue. Uptake of fat has been demonstrated also in other organs and tissues such as skeletal MUSCLE, heart (see HEART METABOLISM), diaphragm, and mammary gland. The activity of lipoprotein lipase is increased in muscle and in heart during fasting or in the diabetic state, both conditions being associated with a decreased entry of triglyceride into adipose tissue. Lipoprotein lipase activity of mammary gland increases at parturition to levels up to one hundred times greater than during pregnancy and remains at this level until suckling stops. These activities are apparently related to uptake of chylomicron triglyceride from the blood in the lactating animal. Increase in concentration of lipoprotein lipase during lactation supports the hypothesis that localized changes in the concentration of this enzyme play a role in the regulation of fat transport.

References

1. DOLE, V. P., AND HAMLIN, J. T., III, "Particulate Fat in Lymph and Blood," *Physiol. Rev.*, **42,** 674 (1962).
2. EVANS, J. R., "Cellular Transport of Long Chain Fatty Acids," *Can. J. Biochem.*, **42,** 955 (1964).
3. FREDRICKSON, D. S., AND GORDON, R. S., JR., "Transport of Fatty Acids," *Physiol. Rev.*, **38,** 585 (1958).
4. MENG, H. C., CONIGLIO, J. G., LEQUIRE, V. S., MANN, G. V., AND MERRILL, J. S., (EDITORS), "Lipid Transport," Springfield, Ill., Charles C. Thomas, 1964.
5. VAUGHAN, M., "The Metabolism of Adipose Tissue *in Vitro*," *J. Lipid. Res*, **2,** 293 (1961).

6. WINEGRAD, A. I., "Endocrine Effects on Adipose Tissue Metabolism," *Vitamins Hormones* (HARRIS, R. S., AND WOOL, I. G., EDITORS), **20**, 14 (1962).

JOHN G. CONIGLIO

FATTY ACID METABOLISM (OXIDATION AND BIOSYNTHESIS)

It has long been known through studies employing labeled fatty acids and fatty acid precursors that the oxidation of fatty acids leads to the formation of a "2-carbon fragment," also known as "active acetate," and now known to be acetyl coenzyme A (acetyl-CoA) (see COENZYME A). It also was believed that resynthesis to fatty acid could occur through reductive condensations of the "2-carbon fragment." Thus, the broad outlines of both fatty acid oxidation and biosynthesis were established. However, knowledge of the finer details of the individual enzymatic steps of the oxidative and synthetic pathways required waiting until the concept of activation of fatty acids through the formation of an energy-rich acyl thioester with coenzyme A could be established.[1]

A hypothesis regarding the manner in which fatty acids are oxidized in living tissues was first proposed by Knoop in 1904[2] and later extended by Dakin.[3] Earlier attempts had been made to trace the fate of fatty acids in tissues by feeding the lower members to animals and then analyzing the urine for end products. These attempts were unsuccessful because the fatty acids were either completely oxidized or excreted unchanged. Knoop found that ω-phenyl-substituted straight-chain fatty acids when fed to dogs were not oxidized completely to carbon dioxide. Phenyl derivatives of even-numbered fatty acids were excreted in the urine as phenylacetic acid whereas the phenyl derivatives of odd-numbered fatty acids were excreted as benzoic acid. These acids were present in the urine as the glycine conjugates phenylaceturic acid and hippuric acid, respectively, and as the unconjugated acid after hydrolysis. Knoop interpreted his results as arising from the successive removal of 2-carbon units following the oxidation of the β-methylene group to a β-keto group:

Even-numbered Fatty Acids

$$C_6H_5{-}CH_2{-}\overset{\beta}{C}H_2{-}\overset{\alpha}{C}H_2{-}COOH$$
$$\downarrow$$
$$C_6H_5{-}CH_2{-}\overset{O}{\overset{\|}{C}}{-}CH_2{-}COOH$$
$$\downarrow$$
$$C_6H_5{-}CH_2{-}COOH + CH_3COOH$$

Phenylacetic acid

Odd-numbered Fatty Acids

$$C_6H_5{-}CH_2{-}CH_2{-}COOH$$
$$\downarrow$$
$$C_6H_5{-}\overset{O}{\overset{\|}{C}}{-}CH_2COOH$$
$$\downarrow$$
$$C_6H_5{-}COOH + CH_3COOH$$

Benzoic acid

Catabolism. *β-OXIDATION. Even-numbered Saturated Fatty Acids.* Lipases and other lipolytic enzymes catalyze the hydrolysis of lipids to glycerol and fatty acids mainly. In animal tissues, the MITOCHONDRION is the major site of fatty acid oxidation.[4,5] Carnitine esters may serve as the transport form of fatty acids for entry into the mitochondrion.[6] In the mitochondrion, the activated fatty acids are oxidized successively (see Fig. 1) to acetyl-CoA and a fatty acyl-CoA shorter by two carbon atoms. Successive repetitions of this sequence of reactions results in the complete degradation of an even-numbered carbon fatty acid to acetyl-CoA.

Reaction 1—Activation (Fig. 1). The initial step in the oxidation of a fatty acid involves a reaction with adenosine triphosphate (ATP) and coenzyme A (CoA) in the presence of Mg^{++} to form adenylic acid (AMP), inorganic pyrophosphate (P—P) and a fatty acid thioester of CoA. (In the Figure, the symbol $\sim$ represents an "energy-rich" type of bond.) Fatty acids are relatively inert chemically. Their reactivity is raised when they become converted to thioesters. These esters are "energy rich." Three different ACTIVATING ENZYMES, or thiokinases, specific for short-, intermediate-, and long-chain fatty acids catalyze the formation of the fatty acyl-CoA ester. The thiokinases are named according to the length of the carbon chain of the acid with which they react most rapidly, *e.g.*, *acetic thiokinase* (for C_2 and C_3 fatty acids), *octanoic thiokinase* (C_4 to C_{12}), and *dodecanoic thiokinase* (C_{10} to C_{18}). The inorganic phosphate formed is known to undergo hydrolytic removal by an inorganic pyrophosphatase in mitochondria, thereby allowing the over-all reaction to proceed in the direction of the activated acyl-CoA.

Initiation of fatty acid activation in isolated mitochondria requires the presence of a CITRIC ACID CYCLE intermediate (oxaloacetate) as a priming agent. In liver mitochondria, where CoA is readily liberated from its ester by deacylases, the priming agent is replaceable by ATP alone. In animal tissues, activation of fatty acids occurs mainly by thiokinases, whereas in microorganisms reactions catalyzed by the thiophorases predominate. Acetic thiokinase has been demonstrated in a number of plants and has been partly purified from spinach leaves.[7]

Reaction 2—Dehydrogenation. The next step, a dehydrogenation, involves the removal of two hydrogens in the α,β-position of the fatty acyl-CoA. This step is catalyzed by FLAVOPROTEINS called *acyl-CoA dehydrogenases*. Three such enzymes containing flavin-adenine dinucleotide (FAD) as a prosthetic group have been isolated from pig liver[8] and named for the most sensitive substrate, *e.g.*, *butyryl-CoA dehydrogenase*, a green copper-containing protein, and *octanoyl-CoA dehydrogenase* and *hexadecanoyl-CoA dehydrogenase*, two yellow flavoproteins. The fatty acyl-CoA flavoproteins do not react directly with oxygen or with the cytochromes, but transfer hydrogen atoms to an "electron-transferring

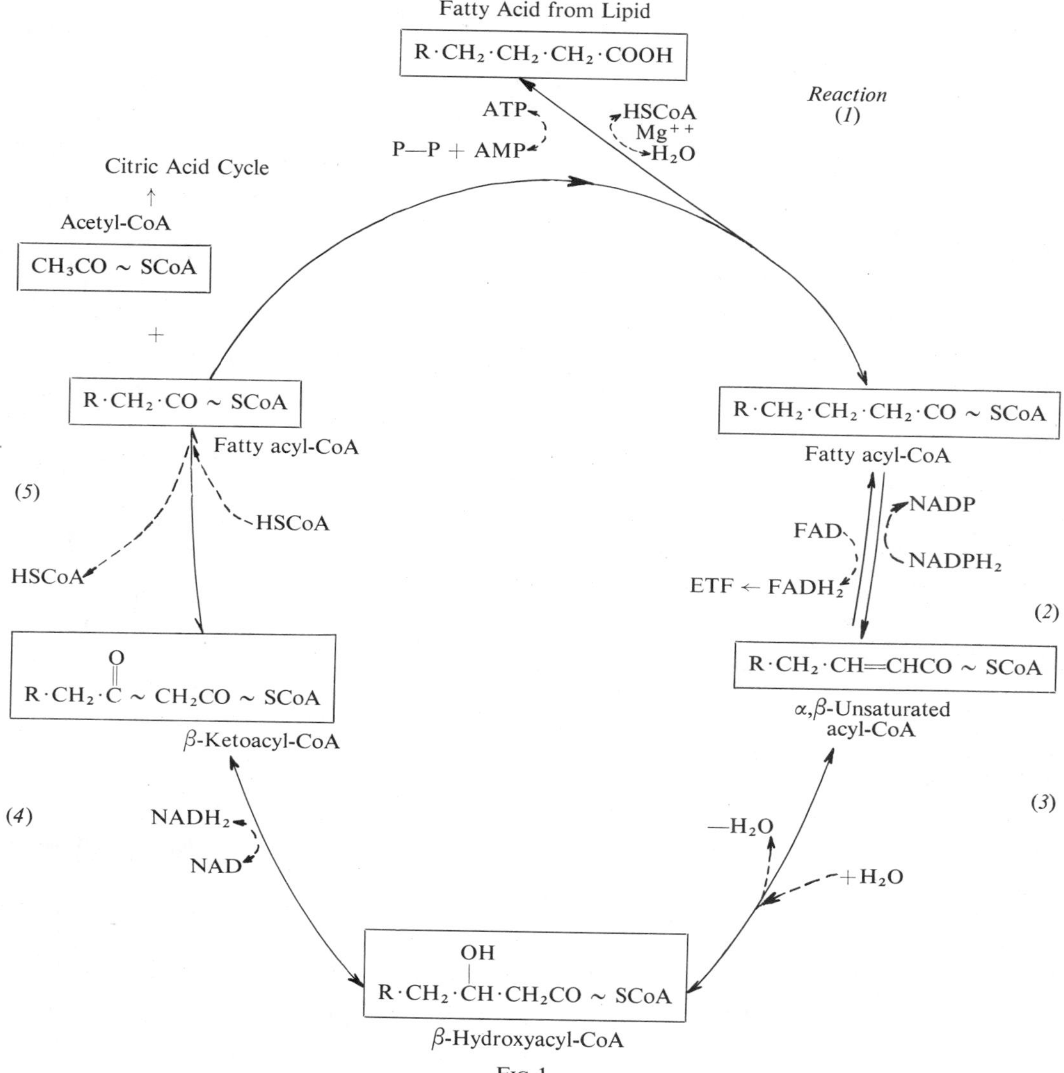

FIG 1.

flavoprotein" (ETF), which in turn appears to couple to the cytochrome system. Since the NICOTINAMIDE ADENINE DINUCLEOTIDE (NAD) coenzyme system is bypassed, two high-energy phosphate bonds are generated per hydrogen pair. This reaction is essentially irreversible and requires a separate enzyme, *crotonyl-CoA reductase*, utilizing reduced nicotinamide adenine dinulceotide phosphate ($NADPH_2$) as a coenzyme for synthesis.

Reaction 3—Hydration. In this reaction *enoyl-CoA hydrase* (*crotonase*), a sulfhydryl enzyme, catalyzes the addition of water across the double bond of the α,β-unsaturated acyl-CoA. The reaction is reversible and L-(+)β-hydroxyacyl-CoA is formed.

Reaction 4—Dehydrogenation. The second dehydrogenation is catalyzed by a *β-hydroxyacyl-CoA dehydrogenase*. This enzyme is specific for the L-stereoisomer of β-hydroxyacyl-CoA. The product is a β-ketoacyl-CoA. NAD^+ is the coenzyme and three high-energy phosphate bonds are produced as a result of oxidation through the respiratory chain.

Reaction 5—Thiolysis. In the final reaction of the β-oxidative cycle the β-ketoacyl-CoA undergoes thiolytic cleavage to form acetyl-CoA and an acyl-CoA shorter by two carbon atoms. The β-keto-linkage is energetically equivalent to the thioester or high-energy phosphate bond. Therefore, a reaction similar to hydrolysis but involving coenzyme A instead of water takes place. Although the reaction is reversible, the equilibrium position is greatly in the direction of cleavage.

The acetyl-CoA formed in the last reaction of each cycle sequence may enter the CITRIC ACID

CYCLE directly and become oxidized to CO_2 and water with the production of additional useful energy. Alternatively, in liver mitochondria and under conditions limiting entry into the citric acid cycle, acetyl-CoA may condense with itself, become deacylated to acetoacetate, reduced to β-hydroxybutyric acid and the acetoacetate can be spontaneously decarboxylated to acetone (the latter three substances are known as the "ketone bodies").

The other product of the reaction, the fatty acyl-CoA, now shorter by two carbon atoms, may now reenter the cycle and undergo β-oxidation with the removal of two additional carbon atoms. Successive cycling of the acyl-CoA gradually shortens the fatty acid to acetyl-CoA.

The complete oxidation of fat leads ultimately to CO_2 and water and the liberation of energy equivalent to 9 kcal/g. Because the fatty acid molecule is a highly reduced molecule, its oxidation yields more energy than can be obtained from an equivalent amount of carbohydrate or protein. The following illustrates the amount of energy for useful work that can be obtained by the oxidation of one mole of palmitic acid:

$$C_{16}H_{31}O_2 + 23O_2 \rightarrow 16CO_2 + 16H_2O + 2{,}339{,}700 \text{ calories}$$

The β-oxidation of palmitic acid requires 7 turns of the cycle and yields 8 molecules of acetyl coenzyme A.

Oxidation of 7 electron pairs in flavin system	$7 \times 2 =$	$14 \sim P$
Oxidation of 7 electron pairs in NAD system	$7 \times 3 =$	$21 \sim P$
	Total	$35 \sim P$
Since one high-energy phospate is required for activation		$-1 \sim P$
	Net gain $=$	$34 \sim P$
Oxidation in citric acid cycle	$8 \times 12 =$	$96 \sim P$
	Total gain $=$	$130 \sim P$

If each mole of ATP represents 7000 calories of free energy then:

$130 \times 7000 = 910{,}000$ calories of utilizable energy

at an efficiency of:

$$\frac{130 \times 7000}{2{,}339{,}700} \times 100 = \text{approximately } 39\%$$

Metabolism of Odd-carbon Fatty Acids.[9,10] Although the occurrence of odd-numbered fatty acids in both animal and plant tissues is rare, some are formed during the metabolism of such amino acids as ISOLEUCINE and VALINE and from bacterial degradation of carbohydrates in animals. These fatty acids undergo β-oxidation with the progressive splitting off of molecules of acetyl-CoA until the final residue of three carbons of the original chain forms propionyl-CoA. The metabolism of propionyl-CoA proceeds by various pathways through succinic acid, and involves the "fixation" of CO_2 by the biotin-containing enzyme *propionyl-CoA carboxylase* (see CARBOXYLATION ENZYMES):

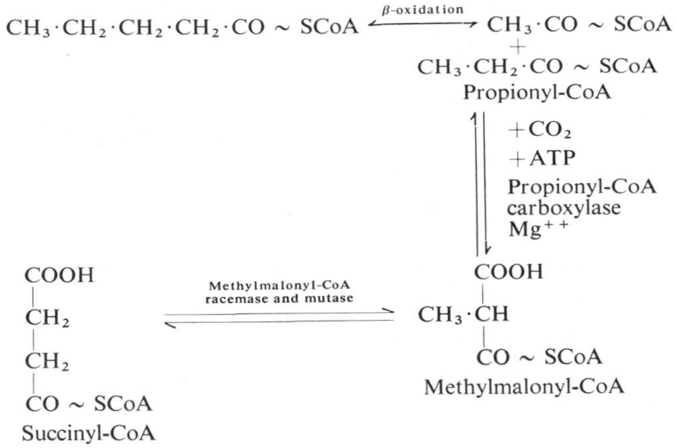

$$CH_3\cdot CH_2\cdot CH_2\cdot CH_2\cdot CO \sim SCoA \xrightarrow{\beta\text{-oxidation}} CH_3\cdot CO \sim SCoA + CH_3\cdot CH_2\cdot CO \sim SCoA$$

Propionyl-CoA

$\rightleftharpoons$ $+CO_2$, $+ATP$, Propionyl-CoA carboxylase, Mg^{++}

$$COOH\text{—}CH(CH_3)\text{—}CO \sim SCoA$$

Methylmalonyl-CoA

$\rightleftharpoons$ Methylmalonyl-CoA racemase and mutase

$$COOH\text{—}CH_2\text{—}CH_2\text{—}CO \sim SCoA$$

Succinyl-CoA

The conversion of methylmalonyl-CoA to succinyl-CoA appears to require two steps: first, the isomerization of the product of the carboxylase reaction by a specific racemase, and then the conversion of one of the optical isomers so formed to succinyl-CoA. Vitamin B_{12} (cyanocobalamine) is required as a coenzyme in the step catalyzed by *methylmalonyl-CoA mutase* (see also COBALAMINS).

The succinic acid formed can be oxidized in the citric acid cycle or, through pyruvate, converted to carbohydrate. It is utilized in the synthesis of PORPHYRINS and in introducing the succinyl group into such substances as sulfanilamide.

Metabolism of Branched-chain Fatty Acids.[10] Small amounts of acids with branched methyl groups are found in nature. For example, the catabolism of the amino acids, leucine, isoleucine and valine, after transamination and oxidative decarboxylation, yields isovaleryl-CoA, α-methylbutyryl-CoA, and isobutyryl-CoA, respectively (see AMINO ACID METABOLISM).

α-Methylbutyryl-CoA and isobutyryl-CoA are initially degraded by β-oxidation, or as in the case of isobutyryl-CoA by a slightly modified β-oxidation, to propionyl-CoA. This latter odd-carbon acid can be oxidized via propionate oxidation as described previously.

The degradation product of leucine, isovaleryl-CoA, appears to require a different pathway as follows:

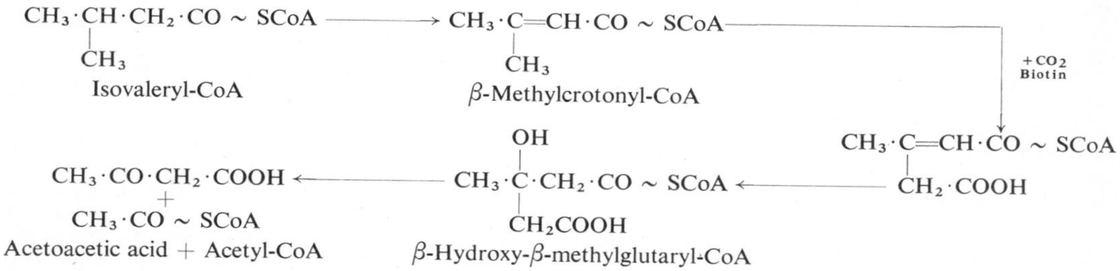

$$CH_3\cdot CH(CH_3)\cdot CH_2\cdot CO \sim SCoA \longrightarrow CH_3\cdot C(CH_3){=}CH\cdot CO \sim SCoA$$

Isovaleryl-CoA → β-Methylcrotonyl-CoA

$$\xrightarrow[\text{Biotin}]{+CO_2} CH_3\cdot C(CH_2\cdot COOH){=}CH\cdot CO \sim SCoA$$

$$\longrightarrow CH_3\cdot C(OH)(CH_2COOH)\cdot CH_2\cdot CO \sim SCoA \longrightarrow CH_3\cdot CO\cdot CH_2\cdot COOH + CH_3\cdot CO \sim SCoA$$

β-Hydroxy-β-methylglutaryl-CoA → Acetoacetic acid + Acetyl-CoA

chondrial extracts, $NADH_2$ as a coenzyme is required. This coenzyme is not necessary for fatty acid oxidation. Furthermore, fatty acid synthesis was found to proceed much more efficiently in non-mitochondrial fractions than in mitochondria. It is apparent, therefore, that two systems are involved: (1) a non-mitochondrial system

Metabolism of Unsaturated Fatty Acids.[10] Despite the widespread occurrence and abundance of unsaturated fatty acids such as oleic (9-octadenoic), linoleic (9, 12-octadecadienoic), linolenic (9,12,15-octadecatrenoic) and arachidonic (5,8,11,14-eicosatetroenoic) in animal and plant tissues, little is known of their metabolism. In animals, linoleic, linolenic, and arachidonic acid are essential for normal growth (see also FATTY ACIDS, ESSENTIAL). Studies of the synthesis of unsaturated fatty acids from sucrose-C^{14} in soybean pods suggest that sucrose is converted preferentially into oleic acid, and that linoleic and linolenic acid are formed by successive dehydrogenation of oleic acid. Interconversions between saturated and unsaturated fatty acids probably take place (see also UNSATURATED FATTY ACIDS).

Other Types of Oxidation. Although they may represent minor pathways of fatty acid oxidation both α- and ω-oxidation can take place in animal and plant tissues.

The methyl group most remote from the carbonyl end of the fatty acid molecule, the ω-carbon, undergoes oxidation to a carboxyl group. The dicarboxylic acid so formed can be further oxidized through the β-oxidative pathway.[10,11]

An α-oxidation system for long-chain fatty acids has been reported in both animals and plants.[10,12]

Anabolism. *Biosynthesis of Fatty Acids.*[9,10,13] Any substance capable of yielding acetyl-CoA can be considered a source of carbon atoms for fatty acid synthesis in living tissues. This process of synthesis is called *lipogenesis.*

Studies employing isotopes have shown that fatty acid synthesis and degradation occur at the highest rate in adipose tissue, liver, and intestinal mucosa and at a considerably slower rate in muscle, skin, and nervous tissue.

With the pathway of fatty acid oxidation in mitochondria established, it was assumed that a reversal of this pathway, employing the same enzyme system, would lead to the synthesis of fatty acids. However, it was found that such was not the case and that to accomplish fatty acid synthesis in isolated mitochondria or mito-

which, when supplemented with ATP, CO_2, Mn^{++}, and $NADH_2$, converts acetyl coenzyme A to long-chain fatty acids, and (2) a mitochondrial system which is virtually a reversal of the oxidative pathway, differing in that in the final dehydrogenation $NADPH_2$ is an obligatory coenzyme in order that this step may proceed.

The Mitochondrial System. This pathway involves successive additions of acetyl-CoA units by a reversal of the steps illustrated in the figure for the oxidative pathway. Evidence suggests that the steps involving the *thiolase*, *β-hydroxyacyl-CoA dehydrogenase*, and *enoyl-CoA hydrase* systems proceed but that the final dehydrogenation is accomplished by $NADPH_2$-dependent *enoyl-CoA reductase* rather than the fatty acyl dehydrogenase.

It is believed that the mitochondrial system is capable of only a limited synthesis of long-chain fatty acids. However, it must be noted that work on beef heart mitochondria[14] has indicated that mitochondria may be capable of rapid synthesis of long-chain fatty acids via the malonyl-CoA route rather than by the reverse action of the enzymes catalyzing fatty acid oxidation.

The Non-mitochondrial System. Although the sequence of reactions is not known it appears that a molecule of acetyl-CoA condenses with seven molecules of malonyl-CoA in a series of reactions comprising decarboxylation, dehydration, and the equivalent of a double reduction, forming a long-chain fatty acid (C_{16} in this example) in accordance with the following:

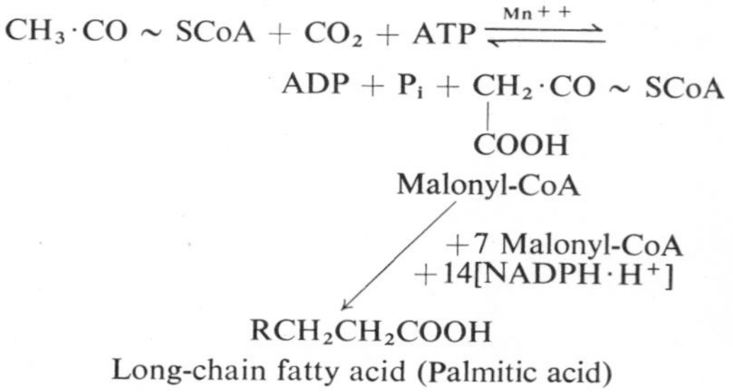

$$CH_3\cdot CO \sim SCoA + CO_2 + ATP \underset{}{\overset{Mn^{++}}{\rightleftharpoons}} ADP + P_i + CH_2(COOH)\cdot CO \sim SCoA$$

Malonyl-CoA

$$\xrightarrow{+7\ \text{Malonyl-CoA}\ +14[NADPH\cdot H^+]} RCH_2CH_2COOH$$

Long-chain fatty acid (Palmitic acid)

The enzyme catalyzing the formation of malonyl-CoA is the biotin-bound acetyl-CoA carboxylase. $NADH_2$ may replace $NADPH_2$ in the reductive step, but the synthesis then proceeds much more slowly. The pentose-shunt pathway of glucose metabolism provides the $NADPH_2$ and helps explain the absolute dependence of lipogenesis upon normal CARBOHYDRATE METABOLISM. It is believed that the steps of fatty acid biosynthesis involve an acyl carrier protein, and yield palmitate or stearate.[10] The factors which terminate synthesis at 16 or 18 carbons are unknown.

References

1. GREEN, D. E., AND GIBSON, D. M., "Fatty Acid Oxidation and Synthesis," in "Metabolic Pathways" (GREENBERG, D. M., EDITOR), Vol. I, p. 301, New York, Academic Press, 1960.
2. KNOOP, F., *Beitr. Chem. Physiol. Path.*, **6**, 150 (1904).
3. DAKIN, H. D., *J. Biol. Chem.*, **6**, 221 (1909); *Physiol. Rev.*, **1**, 394 (1921).
4. LEHNINGER, A. L., "Mitochondrion," New York, W. A. Benjamin, 1964.
5. BRADY, R. O., AND TRAMS, E. G., "The Chemistry of Lipids," *Ann. Rev. Biochem.*, **33**, 75 (1964).
6. BREMER, J., *J. Biol. Chem.*, **237**, 3628 (1963); **238**, 2774 (1963).
7. DAVIES, D. D., GIOVANELLI, J., AND REES, T. A., "Plant Biochemistry," p. 256, Philadelphia, F. A. Davis Co., 1964.
8. CRANE, F. L., HAUGE, J. G., AND BEINERT, H., *Biochim. Biophys. Acta*, **17**, 292 (1955).
9. WAKIL, S. J., "Lipid Metabolism," *Ann. Rev. Biochem.*, **31**, 369 (1962).
10. VAGELOS, P. R., "Lipid Metabolism," *Ann. Rev. Biochem.*, **33**, 139 (1964).
11. DEUEL, H. J., JR., "The Lipids," Vol. III, p. 87, New York, Interscience Publishers, 1957.
12. MEAD, J. F., AND LEVIS, G., *Biochem. Biophys. Res. Comm.*, **9**, 231 (1962).
13. MEAD, J. F., "Lipid Metabolism," *Ann. Rev. Biochem.*, **32**, 241 (1963).
14. HÜLSMANN, W. C., *Biochim. Biophys. Acta*, **58**, 147 (1962).

CARL E. ANDERSON

FATTY ACIDS. See LIPIDS (CLASSIFICATION AND DISTRIBUTION), section on Fatty Acids; also ACETIC ACID AND ACETATE METABOLISM; FATTY ACID METABOLISM; FATTY ACIDS, ESSENTIAL; PHOSPHATIDES AND PHOSPHATIDIC ACIDS; PHOSPHOLIPIDS; UNSATURATED FATTY ACIDS; and Table 1 following:

TABLE 1. COMMON SATURATED FATTY ACIDS

Common Name	Systematic Name	Number of Carbons	Formula	M.P.(°C)
Acetic	Ethanoic	2	CH_3COOH	17
Butyric	Butanoic	4	$CH_3(CH_2)_2COOH$	−6
Caproic	Hexanoic	6	$CH_3(CH_2)_4COOH$	−3
Caprylic	Octanoic	8	$CH_3(CH_2)_6COOH$	16
Capric	Decanoic	10	$CH_3(CH_2)_8COOH$	31
Lauric	Dodecanoic	12	$CH_3(CH_2)_{10}COOH$	44
Myristic	Tetradecanoic	14	$CH_3(CH_2)_{12}COOH$	54
Palmitic	Hexadecanoic	16	$CH_3(CH_2)_{14}COOH$	63
Stearic	Octadecanoic	18	$CH_3(CH_2)_{16}COOH$	70
Arachidic	Eicosanoic	20	$CH_3(CH_2)_{18}COOH$	76

FATTY ACIDS, ESSENTIAL (IN ANIMAL NUTRITION)

The fact that certain fatty acids are essential to the well-being of various species of animals was not recognized until 1929. Until this time, it was the feeling among nutritionists that as long as sufficient protein, vitamins, minerals and calories were supplied in the diet, any fat which the animal required could be provided through the enzymatic mechanisms available for fat synthesis in various tissues of the animal body; as precursor material, products resulting from the metabolism of proteins and carbohydrates could be used. The early work of Burr and Burr, and subsequently of other investigators showed that when rats and other species of animals were fed diets from which fat was excluded, growth and reproduction were impaired and survival was limited. A dermatitis, characterized by a scaliness of the tail and paws, soon developed. All these symptoms could be reversed by the addition of certain unsaturated fatty acids to the diet. These fatty acids are: linoleic (*cis, cis*-9,12-octadecadienoic), γ-linolenic (*cis, cis, cis*-9,12,15-octadecatrienoic), and arachidonic (*cis, cis, cis, cis*-5,8,11,14-eicosatetraenoic) acids. These fatty acids are called essential fatty acids (originally called vitamin F) since they are required by the animal and, obviously, the animal is unable to synthesize them from other materials (at least not in the quantities for which they are needed for the vital processes in which they are involved). For the health and well-being of the animal then, these essential fatty acids have to be supplied in the diet.

The essential fatty acids are not equally potent. Arachidonic acid is roughly about three times as effective as is linoleic in growth-promoting properties; linoleic acid is much more effective than is linolenic in relieving some of the essential fatty acid deficiency symptoms. Interrelationships between these fatty acids have been elucidated, and it is now known that linoleic acid is converted to arachidonic acid in the animal through enzymatic processes in which vitamin B_6, PYRIDOXINE, may be involved (see also UNSATURATED FATTY ACIDS).

It is now recognized that many animal species require essential fatty acids for proper nutrition: *e.g.*, rats, mice, guinea pigs, hamsters, dogs, calves, pigs, as well as the human infant. A need for essential fatty acids for the chick has also been reported. Apparently, however, either the young animal is able to store essential fatty acids (when these are supplied in the diet) for future use, or the requirement for these essential fatty acids is markedly decreased in older animals, since it is practically impossible to produce an essential fatty acid-deficiency syndrome in the adult of any species. The possibility does exist, however, that the utilization of essential fatty acids may be impaired in certain disease conditions which could increase the requirement for the adult animal. The amount of essential fatty acids required for optimum nutrition varies from species to species as well as from animal to animal within the same species. Other dietary constituents, *e.g.*, saturated fats, cholesterol, certain vitamins, may affect the utilization and, therefore, the requirement for essential fatty acids.

Deficiency Symptoms. Among the symptoms of essential fatty acid deficiency are depressed growth, interferences with reproduction and lactation, testicular and spermatogenic degeneration leading to sterility in the male, deposition of fat in the liver, capillary fragility, increased skin permeability resulting in a markedly elevated water consumption, a general elevation in BASAL METABOLIC rate, renal lesions and an accompanying hematuria, and the typical dermatitis involving, especially, the paws and tail. Although this eczematous-like dermatitis has been used as an index of the degree of essential fatty acid deficiency in animals (the "dermal score"), it is now known that the dermal symptoms are affected by humidity; a high humidity decreases the incidence and severity of the symptoms. Perhaps a more precise indication of the rate of development of essential fatty acid deficiency is the change in polyunsaturated fatty acid composition of animal tissues; a decrease in linoleic and arachidonic acids accompanied by a pronounced increase in the amount of a 20-carbon chain fatty acid with three unsaturated bonds (5,8,11-eicosatrienoic acid) in various tissues occurs long before the appearance of the external symptoms of the deficiency.

Distribution and Metabolic Functions. Essential fatty acids play a role in many metabolic processes. Essential fatty acids are widespread in their distribution in the animal body; they occur in various cell fractions, in mitochondria and microsomes of the cytoplasm of the cell, and in the cell wall. As early as 1942, Dr. Burr recognized the importance of essential fatty acids in cell structure and their involvement in lipid mobility. The importance of essential fatty acids in the maintenance of normal cell structure has been confirmed by various other investigators who have reported differences in morphology between normal and deficient MITOCHONDRIA; mitochondria of essential fatty acid-deficient animals are more fragile than those of normal animals and are more difficult to isolate in an intact condition. Since many enzyme systems are located in the mitochondria, it is possible that many aspects of intermediary metabolism not only of lipids but also of other nutrients may be altered in essential fatty acid-deficient animals. In the liver of the essential fatty acid-deficient animal, there develops an abnormal accumulation of cholesterol esters and neutral fat, containing, primarily, saturated fatty acids. Supplementation of the animal with linoleic acid decreases the amount of cholesterol and fat deposition, which confirms the fact that essential fatty acids are required for normal transport of lipid materials. Essential fatty acids, and fats containing these fatty acids, have been shown to reduce hypercholesterolemia in individuals fed these supplements in ample quantities—a fact which has strengthened the reported relationship between essential fatty acids and cholesterol metabolism (see ATHEROSCLEROSIS).

Sources. Essential fatty acids occur naturally and in considerable quantity, but not exclusively, in vegetable and seed oils. Cottonseed oil, corn oil and soybean oil contain approximately 50% linoleic acid as the triglyceride (a combination of three molecules of fatty acid with glycerol); walnut oil, safflower oil and wheat germ oil are considerably higher in linoleic acid content. The vegetable oil, olive oil, is rather low in linoleic acid (10%), whereas chicken fat contains approximately 20%. On the other hand, beef tallow and butter have, respectively, approximately 2 and 4% linoleic acid. Coconut oil, a vegetable product, contains only about 2% linoleic acid. Attempts to increase the linoleic acid content of the egg by manipulations in the diet of the chicken have resulted in an egg which is higher in total fat as well as in polyunsaturated fats. Although fish oils (*e.g.*, menhaden, tuna) are high in polyunsaturated fatty acids, these are not essential fatty acids.

Although it is now recognized that animals require essential fatty acids for good nutrition and protection against the development of many undesirable conditions, an understanding of the role of essential fatty acids in nutrition and in metabolic processes is far from complete.

References

Books

1. DEUEL, H. J., JR., "The Lipids," Vol. III, Ch. XIII, New York, Interscience Publishers, 1957.
2. SINCLAIR, H. M. (EDITOR), "Essential Fatty Acids," New York, Academic Press, 1958.

Review Articles

3. AAES-JØRGENSEN, E., "Essential Fatty Acids," *Physiol. Rev.*, **41,** 1 (1961).
4. HOLMAN, R. T., "Essential Fatty Acids," *Nutrition Reviews*, **16,** 33 (1958).

5. ALFIN-SLATER, R. B., "Newer Concepts of the Role of Essential Fatty Acids," *J. Am. Oil Chemists Soc.*, **34,** 574 (1957).

Research Articles

6. BURR, G. O., AND BURR, M. M., *J. Biol. Chem.*, **82,** 345 (1929).
7. BURR, G. O., AND BURR, M. M., *J. Biol. Chem.*, **86,** 587 (1930).

ROSLYN B. ALFIN-SLATER

FEEDBACK LOOPS IN METABOLISM. See METABOLIC CONTROLS (FEEDBACK MECHANISMS).

FERMENTATIONS

The process of fermentation has been used by man from prehistoric times in the preparation of foods and beverages, but the causative agents of fermentation were not recognized until the middle of the nineteenth century. The end products resulting from the natural fermentation of glucose, namely alcohol and carbon dioxide [Eq. (*1*)], were

$$C_6H_{12}O_6 \rightarrow 2CO_2 + 2C_2H_5OH \qquad (1)$$

identified by Gay-Lussac in 1810, but it was thought that this process resulted from contact catalysis and the decay of animal or vegetable materials. This explanation was refuted by the work of Pasteur (1857) on the lactic acid fermentation. In the course of this investigation, Pasteur determined that fermentation was caused by living cells, that different microbial species caused different fermentations, that the nitrogenous materials present served only to support the growth of the cells, that lactic acid was produced when cells (removed from the fermentation mixture) were added to a sugar solution, and that the natural fermentation yielded both alcohol and lactic acid but that the amount of each could be altered by changes in pH. In later studies, Pasteur showed that the conversion of glucose to alcohol [Eq. (*1*)] was caused by yeast cells growing under anaerobic conditions, thus leading to the definition that fermentation was "life without air."

For the purposes of this discussion, fermentations will be defined not as "life without air," but as those energy-yielding reactions in which *organic* compounds act as both oxidizable substrates and oxidizing agents. Anaerobic reactions in which *inorganic* compounds are utilized as electron acceptors may be termed "anaerobic respirations," whereas reactions in which oxygen serves as a terminal electron acceptor are RESPIRATIONS.

Almost any organic compound may be fermented provided it is neither too oxidized nor too reduced, since it must function as both electron donor and electron acceptor. In some fermentations, a compound is degraded via a series of reactions in which intermediates in the sequence act as electron donors and acceptors; in others, one molecule of the substrate may be oxidized while another molecule is reduced, or two different organic compounds may be degraded after a coupled oxidation-reduction reaction. These fermentations provide energy required for the growth of a variety of cells. In addition, many microorganisms can carry out, in appropriate conditions, a number of fermentative reactions (*e.g.*, oxidations, reductions, cleavages) which do not yield useful energy or do not yield sufficient energy for growth. The fermentations of carbon compounds and nitrogenous compounds discussed below will be concerned primarily with energy-yielding fermentations.

In view of the great variety of different compounds which may be fermented and the enzymatic capabilities of different microorganisms, it is not surprising that many compounds important in industry (*e.g.*, ethyl alcohol, butyl alcohol, acetone, 2,3-butylene glycol), in the production, preservation, and seasoning of food (*e.g.*, lactic, citric and glutamic acids), and in medicine (*e.g.*, VITAMINS are extracted from the yeast carrying out the alcoholic fermentation) may be produced most cheaply through microbial fermentations. In addition, fermentations continue to be important in the production of foods (*e.g.*, the lactic and propionic acid fermentations in the making of cheeses), beverages (*e.g.*, the ALCOHOLIC FERMENTATIONS in the making of wine and beer), and in the leavening of breads [by the CO_2 produced in Eq. (*1*)].

Fermentation of Carbon Compounds. *Fermentation of Carbohydrates to Pyruvate*. Carbohydrates are one of the chief sources of energy and cell components for the growth of plants, animals and microorganisms. Oligosaccharides or disaccharides are first split by enzymes to the component sugars (glucose, fructose, mannose, or galactose). If a phosphorolytic enzyme mediates this reaction, sugar phosphates are produced; if the enzyme is hydrolytic, the resulting free sugars are phosphorylated at the expense of the "energy-rich" compound, adenosine triphosphate (ATP). The subsequent metabolism of the phosphorylated sugars occurs by one of several pathways. In GLYCOLYSIS (the only fermentative pathway found in plant and animal cells and the most common pathway in microorganisms), fructose-1-6-diphosphate is produced from hexose monophosphates by a second phosphorylation with ATP and is degraded to pyruvate by a sequence of enzymatic reactions outlined in Fig. 1. In another pathway (discovered in the bacterium *Zymomonas lindneri*), glucose-6-phosphate is oxidized to phosphogluconic acid which is then split to phosphoglyceraldehyde and pyruvate. In a third pathway (found in heterofermentative lactic acid bacteria), phosphogluconic acid is oxidized and decarboxylated to yield ribulose phosphate which, in turn, undergoes cleavage to phosphoglyceraldehyde and ethanol. In all three pathways, phosphoglyceraldehyde is converted to pyruvate by the reactions shown in Fig. 1 and is metabolized further to end products characteristic of each type of organism.

In glycolysis, 2 "energy-rich" bonds (indicated in Fig. 1 as $\sim PO_3H_2$) are generated per molecule of triose oxidized, or 4 per molecule of fructose-1-6 diphosphate. These bonds are transferred to adenosinediphosphate (ADP), thereby yielding ATP which can be used as a source of energy for the

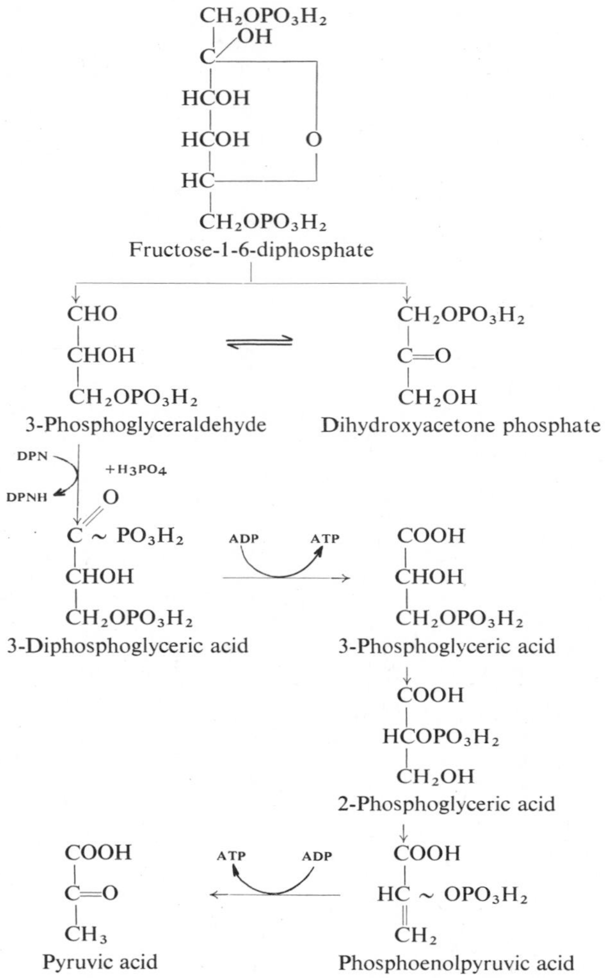

$CH_2OPO_3H_2$ — C(OH) — HCOH — HCOH — HC — $CH_2OPO_3H_2$ (ring through O)

Fructose-1-6-diphosphate

CHO — CHOH — $CH_2OPO_3H_2$ ⇌ $CH_2OPO_3H_2$ — C=O — CH_2OH

3-Phosphoglyceraldehyde ⇌ Dihydroxyacetone phosphate

DPN → DPNH; $+H_3PO_4$

$C(=O) \sim PO_3H_2$ — CHOH — $CH_2OPO_3H_2$ → (ADP → ATP) → COOH — CHOH — $CH_2OPO_3H_2$

3-Diphosphoglyceric acid → 3-Phosphoglyceric acid

COOH — $HCOPO_3H_2$ — CH_2OH

2-Phosphoglyceric acid

COOH — C=O — CH_3 ← (ADP → ATP) ← COOH — $HC \sim OPO_3H_2$ = CH_2

Pyruvic acid ← Phosphoenolpyruvic acid

FIG. 1. Formation of pyruvate from fructose-1-6-diphosphate.

synthesis of cell components (see PHOSPHATE BOND ENERGIES). Since the conversion of a hexose to its diphosphate derivative requires the expenditure of 2 molecules of ATP, glycolysis yields a net of 2 "energy-rich" bonds per molecule of hexose. The second and third pathways yield a net of 1 "energy-rich" bond, *i.e.*, 2 "energy-rich" bonds arise from the metabolism of phosphoglyceraldehyde but 1 bond is required to produce glucose-6-phosphate. In the respiration of glucose, 30 "energy-rich" bonds could be generated per molecule of glucose, if all of the carbon were oxidized to carbon dioxide. This accounts for the observation, first made by Pasteur, that a larger number of yeast cells are produced for a given amount of glucose when cultures are grown under conditions of respiration rather than fermentation.

Metabolism of Pyruvate. (1) Lactic and Alcoholic Fermentations. The reduction of pyruvate to lactate [Eq. (*2*)] may be carried out by a number of

$$CH_3COCOOH + 2[H] \rightleftharpoons CH_3CHOHCOOH \quad (2)$$

microorganisms and also by mammalian muscle. In certain bacteria (*e.g.*, homofermentative lactic acid bacteria), glucose is converted quantitatively to lactate; in others (heterofermentative lactic acid bacteria), ethanol and CO_2 are also produced [Eq. (*3*)]. In the latter fermentation, ethanol arises

$$C_6H_{12}O_6 \rightarrow CH_3CHOHCOOH + CH_3CH_2OH + CO_2 \quad (3)$$

by cleavage of ribulose phosphate (discussed above), but in the alcoholic fermentation of yeasts and certain bacteria (*Zymomonas lindneri*), ethanol arises from pyruvate. In this instance, pyruvate is decarboxylated to acetaldehyde [Eq. (*4a*)] and the latter is reduced to ethanol [Eq. (*4b*)]. The reduction of acetaldehyde to

$$CH_3COCOOH \rightarrow CH_3CHO + CO_2 \quad (4a)$$

$$CH_3CHO + 2[H] \rightarrow CH_3CH_2OH \quad (4b)$$

ethanol, or of pyruvate to lactate is linked to the

oxidation of 3-phosphoglyceraldehyde to 1,3-diphosphoglycerate (Fig. 1) and is mediated by the coenzyme diphosphopyridine nucleotide (DPN).

(2) Mixed Acid Fermentations. The fermentation of pyruvate may yield more than one organic acid. For example, lactate and acetate are produced under certain conditions by the bacterium *Escherichia coli* [Eq. (*5a*)]. This requires the oxidative decarboxylation of one molecule of pyruvate [Eq. (*5b*)] and reduction of another [Eq. (*5c*)]. Certain bacteria (*e.g.*, *Micrococcus lactilyticus*) do not form lactate from

$$2CH_3COCOOH + H_2O \rightarrow CH_3COOH + CO_2 + CH_3CHOHCOOH \quad (5a)$$

$$CH_3COCOOH + H_2O \rightarrow CH_3COOH + CO_2 + 2[H] \quad (5b)$$

$$CH_3COCOOH + 2[H] \rightarrow CH_3CHOHCOOH \quad (5c)$$

pyruvate, but carry out an oxidative decarboxylation in which the available hydrogen is released as hydrogen gas [Eq. (*6*)]. Others (*e.g.*,

$$CH_3COCOOH + H_2O \rightarrow CH_3COOH + CO_2 + H_2 \quad (6)$$

Clostridium butyricum) metabolize pyruvate to acetate and formate [Eq. (*7*)]; the latter compound

$$CH_3COCOOH + H_2O \rightarrow CH_3COOH + HCOOH \quad (7)$$

may be degraded to CO_2 and H_2 by other bacteria. Small amounts of succinate may also be formed in these fermentations (see section 5 below). The acetate produced in Eq. (*5b*) and Eq. (*6*) appears first not as free acetate but as the "energy-rich" COENZYME A derivative, acetyl coenzyme A (acetyl-CoA). Thus, the energy of oxidation is conserved and is made available [via Eqs. (*8a*) and (*8b*)] as

$$\text{Acetyl-CoA} \xrightleftharpoons{PO_4H} \text{Acetyl-phosphate} + \text{CoA} \xrightleftharpoons{ADP} \text{Acetate} + \text{ATP} \quad (8a)$$

$$\text{Acetyl-CoA} + \text{ADP} \rightleftharpoons \text{Acetate} + \text{ATP} + \text{CoA} \quad (8b)$$

ATP which may then be used for synthetic reactions. Equation (*8a*) occurs only in bacteria and Eq. (*8b*) has been found in plant and animal cells and in many microorganisms.

(3) Butylene Glycol Fermentation. In addition to variable amounts of formate, acetate, ethanol, CO_2 and H_2, some microorganisms (*e.g.*, *Aerobacter aerogenes*) produce 2,3-butylene glycol as the chief end product of glucose fermentation. This 4-carbon compound arises from the condensation of acetaldehyde [produced via Eq. (*4a*)] with pyruvate to form α-acetolactate:

$$\begin{array}{l} CH_3COHCOCH_3 \\ | \\ COOH \end{array}$$

The latter is decarboxylated to acetoin ($CH_3 \cdot CHOHCOCH_3$) which is subsequently reduced to 2,3-butylene glycol ($CH_3CHOHCHOHCH_3$). Acetaldehyde may also be condensed with longer compounds of similar structure thereby yielding more complex acyloins.

(4) Butyric Acid-Butanol-Acetone Fermentations. Certain anaerobic bacteria, notably species of clostridia, produce butyrate, butanol, acetone and isopropyl alcohol as the chief end products in the fermentation of glucose. The synthesis of butyrate requires the participation of the coenzyme A derivatives of several organic acids [Eqs. (*9a*)–(*9d*)]. In the synthesis of butyrate, 2 molecules of

$$2CH_3CO\text{—}CoA \rightleftharpoons CH_3COCH_2CO\text{—}CoA + CoA + H_2O \quad (9a)$$

$$CH_3COCH_2CO\text{—}CoA + 2[H] \rightleftharpoons CH_3CHOHCH_2CO\text{—}CoA \quad (9b)$$

$$CH_3CHOHCH_2CO\text{—}CoA \rightleftharpoons H_2O + CH_3CH{=}CHCO\text{—}CoA \text{ or } CH_2{=}CHCH_2CO\text{—}CoA \quad (9c)$$

$$CH_3CH{=}CHCO\text{—}CoA \text{ or } CH_2{=}CHCH_2CO\text{—}CoA + 2[H] \rightleftharpoons CH_3CH_2CH_2CO\text{—}CoA \quad (9d)$$

acetyl-CoA produced from pyruvate via Eq. (*5b*) are condensed to form acetoacetyl-CoA [Eq. (*9a*)] which is then reduced first to β-hydroxybutyryl-CoA [Eq. (*9b*)], then dehydrated to either crotonyl-CoA or vinylacetyl-CoA [Eq. (*9c*)]. The product of this reaction is reduced to butyryl-CoA [Eq. (*9d*)]. Deacylation and reduction of the latter yields butanol. Acetoacetyl-CoA formed in Eq. (*9a*) may be deacylated and decarboxylated to acetone [Eq. (*10a*) which may also be reduced to isopropyl alcohol [Eq. (*10b*)].

$$CH_3COCH_2COOH \rightarrow CH_3COCH_3 + CO_2 \quad (10a)$$

$$CH_3COCH_3 + 2[H] \rightarrow CH_3CHOHCH_3 \quad (10b)$$

(5) Propionic Acid Fermentation. Two different pathways have been described for the formation of propionate. The acetate and CO_2 produced by propionibacteria from glucose [Eq. (*11*)] are

$$3C_6H_{12}O_6 \rightarrow 4CH_3CH_2COOH + 2CH_3COOH + 2CO_2 + 2H_2O \quad (11)$$

formed via Eq. (*5b*) whereas propionate arises from a 4-carbon compound by the following reactions (the same reactions are responsible for the metabolism of propionate by a variety of animal cells). Pyruvate is condensed with CO_2 to yield oxaloacetate which is converted to malate → fumarate → succinate through the Krebs tricarboxylic acid cycle (see CITRIC ACID CYCLE). This pathway also accounts for the succinate found in the mixed acid fermentations (section 2 above). In propionate synthesis, succinate is isomerized to methylmalonate which is then decarboxylated to propionate and CO_2 [Eq. (*12*)]. The latter reactions

$$COOHCH_2CH_2COOH \rightleftharpoons \begin{array}{c} COOHCHCOOH \\ | \\ CH_3 \end{array} \rightleftharpoons CH_3CH_2COOH + CO_2 \quad (12)$$

require the participation of the coenzyme A derivatives of the acids rather than the free acids.

Another anaerobic bacterium (*Clostridium propionicum*) is unable to carry out the sequence shown in Eq. (*12*) although propionate is formed from a variety of substrates. It has been postulated that propionate is produced in this species by the dehydration of lactate [produced via Eq. (*2*)] to acrylate and reduction of the latter to propionate [Eq. (*13*)]

$$CH_3CHOHCOOH \xrightarrow{-H_2O} CH_2{=}CHCOOH \xrightarrow{+2[H]} CH_3CH_2COOH \quad (13)$$

Fermenation of Aldonic Acids, Polyhydric Alcohols and Pentoses. Bacteria adapted by growth on glucuronate, galacturonate, 2-ketogluconate and 5-ketogluconate are able to ferment these compounds to characteristic end products. Hexitols, glycerol and pentoses (ribose, xylose, arabinose and desoxyribose) may also be fermented by a variety of microorganisms. The pathways of fermentation of the latter group of compounds have not been completely established, but pentose fermentations are known to involve phosphorylation, before or after isomerization, followed by cleavage to 2-carbon and 3-carbon compounds.

Fermentation of Nitrogenous Compounds. *Fermentation of Amino Acids.* (1). Single Amino Acids. About 20 species of anaerobic bacteria (chiefly clostridia and anaerobic micrococci) are known to ferment single AMINO ACIDS. With the exception of proline, hydroxyproline and isoleucine, all of the common amino acids may be degraded by at least one species. As an example of this class of fermentation, the pathway for the fermentation of glutamate by the anaerobic bacterium, *Clostridium tetanomorphum*, will be outlined [Eqs. (*13a*)–(*13d*)]. Glutamate is isomerized

$$COOHCHNH_2CH_2CH_2COOH \rightleftharpoons COOHCHNH_2\underset{\displaystyle CH_3}{\underset{|}{C}}HCOOH \quad (13a)$$

$$COOHCHNH_2\underset{\displaystyle CH_3}{\underset{|}{C}}HCOOH \rightleftharpoons COOHCH{=}\underset{\displaystyle CH_3}{\underset{|}{C}}{-}COOH + NH_3 \quad (13b)$$

$$COOHCH{=}\underset{\displaystyle CH_3}{\underset{|}{C}}{-}COOH + H_2O \rightleftharpoons COOHCH_2\underset{\displaystyle CH_3}{\underset{|}{C}}OHCOOH \quad (13c)$$

$$COOHCH_2\underset{\displaystyle CH_3}{\underset{|}{C}}OHCOOH \rightleftharpoons CH_3COOH + CH_3COCOOH \quad (13d)$$

to β-methyl aspartate [Eq. (*13a*)] which is deaminated to mesaconate [Eq. (*13b*)]. Hydration of mesaconate yields citramalate [Eq. (*13c*)] which is cleaved to acetate and pyruvate, the latter giving rise to butyrate by the mechanism outlined earlier in Eqs. (*8a*)–(*8d*).

(2) Pairs of Amino Acids. Certain of the clostridia cannot ferment single amino acids, but are able to degrade appropriate pairs of amino acids (the "Stickland reaction"). One member of the pair is oxidized while another is reduced. For example, in the simultaneous fermentation of glycine and alanine [Eq. (*14a*)] alanine is oxidatively deaminated to pyruvate [Eq. (*14b*)] and pyruvate undergoes oxidative decarboxylation to acetate and CO_2 [Eq. (*14c*)]. These reactions are coupled to the reductive deamination of glycine [Eq. (*14d*)].

$$CH_3CHNH_2COOH + 2CH_2NH_2COOH + 2H_2O \rightarrow 3CH_3COOH + CO_2 + 3NH_3 \quad (14a)$$

$$CH_3CHNH_2COOH + H_2O \rightarrow CH_3COCOOH + NH_3 + 2[H] \quad (14b)$$

$$CH_3COCOOH + H_2O \rightarrow CH_3COOH + CO_2 + 2[H] \quad (14c)$$

$$2CH_2NH_2COOH + 4[H] \rightarrow 2CH_3COOH + 2NH_3 \quad (14d)$$

Fermentation of Heterocyclic and Other Compounds. Relatively few bacteria (for the most part clostridia and anaerobic micrococci) have been reported to ferment heterocyclic compounds. The variety of end products and differences in the intermediates formed in the fermentation of PURINES suggests that the degradation of these compounds by different bacteria does not proceed by the same pathway. Detailed studies on the degradation of xanthine by *Clostridium cylindrosporum* have shown that the following sequence is involved in this anaerobe: xanthine → 4-ureido-5-imidazole carboxylic acid → 4-amino-5-imidazole carboxylic acid (plus NH_3 and CO_2) → 4-amino imidazole (plus CO_2) → 4-imidazolone (plus NH_3) → formiminoglycine → glycine + NH_3 + HCOOH. PYRIMIDINES are fermented by three bacterial species, but only *Zymobacterium oroticum* can use these compounds as a source of energy for growth. In addition, the following compounds have been reported to be fermented by bacteria: allantoin, nicotinic acid, creatinine, and ergothionine.

References

1. BARKER, H. A., "Fermentations of Nitrogenous Organic Compounds," in *The Bacteria, II* (I. C. GUNSALUS AND R. Y. STANIER, EDITORS), pp. 151–208, New York, Academic Press, 1961.
2. ELSDEN, S. R., AND PEEL, J. L., "Metabolism of Carbohydrates and Related Compounds," *Ann. Rev. Microbiol.*, **12**, 145-202 (1958).
3. GUNSALUS, I. C., AND SHUSTER, C. W., "Energy-yielding Metabolism in Bacteria," in *The Bacteria, II* (I. C. GUNSALUS AND R. Y. STANIER, EDITORS), pp. 1–58, New York, Academic Press, 1961.
4. WOOD, W. A., "Fermentation of Carbohydrates and Related Compounds," in *The Bacteria, II* (I. C. GUNSALUS AND R. Y. STANIER, EDITORS), pp. 59–150, New York, Academic Press, 1961.
5. STANIER, R. Y., DOUDOROFF, M., AND ADELBERG, E. A., "The Microbial World," Second edition, Englewood Cliffs, N. J., Prentice-Hall, 1963.

H. R. WHITELEY

FERREDOXIN

Ferredoxin, an iron-containing metalloprotein, is one of the catalytic oxido-reducing agents or electron carriers in the ELECTRON TRANSPORT SYSTEM that accepts electrons from light-activated CHLOROPHYLL (see LIGHT PHASE OF PHOTOSYNTHESIS) and subsequently transfers the electrons, through other carriers, to carbon dioxide to form reduced organic compounds (see CARBON REDUCTION CYCLE IN PHOTOSYNTHESIS). Relative to other common electron carriers, ferredoxin has a rather negative redox potential, of about −0.5 volt; *i.e.*, in its reduced form, it is a relatively strong intracellular reducing agent (see also NITROGEN FIXATION; PHOSPHORYLATION, PHOTOSYNTHETIC).

FERRITIN

Ferritin, a protein rich in iron, was first isolated from horse spleen by Laufberger (1937), and has since been found in many animals and in some higher plants. It occurs in liver, spleen, kidneys, ovaries, lungs and other organs of vertebrates, and in certain tissues of some invertebrates as well (*e.g.*, in *Planorbis corneus*, *Arenicola marina* and in *Lumbricus* species). It has been isolated from pea and bean seedlings and cotyledons.

Chemical Composition and Properties. Molecules of ferritin consist of a protein moiety, *apoferritin*, with molecular weight of approximately 465,000 and of a micellar complex of iron hydroxide-phosphate with approximate empirical composition $[(FeOOH)_8 \cdot (FeOPO_3H_2)]$. The *iron content* of ferritin varies, depending on source and method of preparation, but averages about 20% of dry weight. The iron has an apparent paramagnetic susceptibility, corresponding to a magnetic moment of 3.8 Bohr magnetons, but has been shown to be antiferromagnetic. Crystals of ferritin are brown to yellow, and generally have cubic symmetry though non-cubic forms have been noted.

As demonstrated by electron microscopy, the iron hydroxide micelles form the core of the ferritin molecule. This core measures about 60 Å in diameter in dried preparations, or 75 Å when wet, and is surrounded by a shell of apoferritin.

Apoferritin has been well characterized. It is not denatured in aqueous solution at 80°C. For horse spleen apoferritin, the sedimentation coefficient, $S^0_{20,w}$, is 17.6. Isoelectric points vary. Values of pH 4.4 (horse spleen apoferritin) to pH 5.5 (human liver or spleen apoferritin) have been reported. Ultracentrifugation, X-ray data, and electron microscopy have indicated that the apoferritin shell of the ferritin molecule is close to spherical, with outer radius of 61 Å and inner radius of 37 Å when hydrated. This shell is probably composed of 20 equal subunits.

Electrophoretic mobilities of various ferritins are determined by the nature of the apoferritin moieties, not by the micelles that contain the iron. However, ferritins from various animal tissues and from cultured cells are inhomogeneous on ELECTROPHORESIS in gel systems.

The ANTIGENIC properties of each sort of ferritin are due to the corresponding apoferritin. Serological specificities vary with animal species. More than one antigenic determinant has been detected in highly purified ferritin (*e.g.*, from horse, rat, man).

Preparation. Tissue homogenates (*e.g.*, spleen, liver) in aqueous media are heated at 80°C; this precipitates out most other proteins. A crude product is then obtained from the aqueous supernatant by precipitation with ammonium sulfate. The precipitate is redissolved in water, and after further purification, ferritin is crystallized by addition of $CdSO_4$. Aqueous extracts of tissue homogenates also contain some "natural" apoferritin, which can be separated by density gradient ULTRACENTRIFUGATION. Apoferritin is prepared chemically from ferritin by removal of the iron hydroxide-phosphate complex through reduction, solubilization and chelation of the iron. The apoferritin so prepared can be crystallized with cadmium sulfate in the same habit as homologous ferritin.

Biological Properties. Ferritin is generally classified as an iron storage compound. Normally, the human body contains between 0.5 and 1.0 gram of ferritin, or 10–20% of its total iron as ferritin; most of this is located in liver and spleen, but there are functionally significant quantities in bone marrow and in the small intestine [see also IRON (IN BIOLOGICAL SYSTEMS)]. In higher animals, ferritin constitutes an important reserve of iron that is drawn on in emergencies, *e.g.*, after loss of blood and in various ANEMIAS. Iron from ferritin can be utilized for synthesis of hemoglobin by several metabolic pathways. The hypothesis, that absorption of iron from the gut is regulated via incorporation of Fe into, and release from, apoferritin in mucosal cells (duodenum, jejunum) has not been proved. Parenteral administration of iron stimulates synthesis of ferritin, into which much of the injected iron becomes incorporated. Diverse mammalian cells growing *in vitro* synthesize ferritin if excess iron is present in the culture medium. Similar observations have been made *in vitro* with surviving liver slices and with reticulocytes.

An "active" form of ferritin, in which the surface of the molecule has free SH groups, has been identified. These SH groups can bind and release iron in its ferrous state. Such labile Fe may be part of a chemical equilibrium between ionized (dissociated) iron in cells and micellar iron inside ferritin. Passage of Fe from SH-ferritin to intracellular PORPHYRIN (heme) groups or to extracellular transferrin has been postulated. Ferritin (=VDM of Shorr) exerts certain vasodepressor and antidiuretic activities that may depend on the presence of free SH groups on the molecule. In rats, small quantities of ferritin inhibit the response of smooth muscle cells of metarterioles and of precapillary sphincters in the meso-appendix to topical application of adrenalin. In dogs, antidiuretic hormone (see VASOPRESSIN) appears to be released from the neurohyphysis after intravenous injections of ferritin.

The presence of ferritin in diverse animal species as well as in some plants suggests that the capacity to synthesize this protein developed early in evolution, perhaps as an adaptive process connected with the regulation of cellular metabolism of iron.

References

1. Granick, S., *Chem. Rev.*, **38**, 379 (1946); *Physiol. Rev.*, **31**, 489 (1951).
2. Shorr, E., *Harvey Lectures Ser.*, **50**, 112 (1956).
3. Farrant, J., *Biochim. Biophys. Acta*, **13**, 569 (1954).
4. Bessis, M., in "The Kinetics of Cellular Proliferation" (Stohlman, F., Jr., Editor), p. 22, Grune & Stratton Inc., 1959.
5. Richter, G. W., *Lab. Invest.*, **12**, 1026 (1963).
6. Harrison, P. M., in "Iron Metabolism," an International Symposium (Gross, F., Editor), p. 40, Berlin-Göttingen-Heidelberg, Springer-Verlag, 1964.
7. Hyde, B. B., Hodge, A. J., Kahn, A., and Birnstiel, M. L., *J. Ultrastruct. Res.*, **9**, 248 (1963).
8. Mazur, A., and Carleton, A., *J. Biol. Chem.*, **238**, 1817 (1963).
9. Fischbach, F. A., and Anderegg, J. W., *J. Mol. Biol.*, **14**, 458 (1965).

G. W. Richter

FEULGEN REACTION. See histochemical methods.

FIBRIN AND FIBRINOGEN. See blood clotting

FISCHER, E. H.

Emil Herman Fischer (1852–1919) was an eminent German organic chemist whose studies on sugars and purines won him the Nobel Prize for chemistry in 1902. He is remembered by biochemists even more perhaps because of his fundamental work on amino acids, peptides and proteins. He was the first to isolate a number of amino acids and to determine that they are bound together in proteins by peptide bonds.

FISCHER, H.

Hans Fischer (1881–1945) was a German organic chemist who is noted for his elucidation of the structure of heme, for which contribution he received the Nobel Prize for chemistry in 1930.

FLAME PHOTOMETRY AND ATOMIC ABSORPTION SPECTROSCOPY

Flame photometry and atomic absorption spectroscopy both deal with the determination of trace metals in various types of samples. They are similar to each other in that both deal with the interaction of radiant energy and the atoms of the metal being examined. They differ from each other in that flame photometry deals with radiation *emitted* by the atom, atomic absorption deals with radiation *absorbed* by the atom. This is illustrated in Fig. 1.

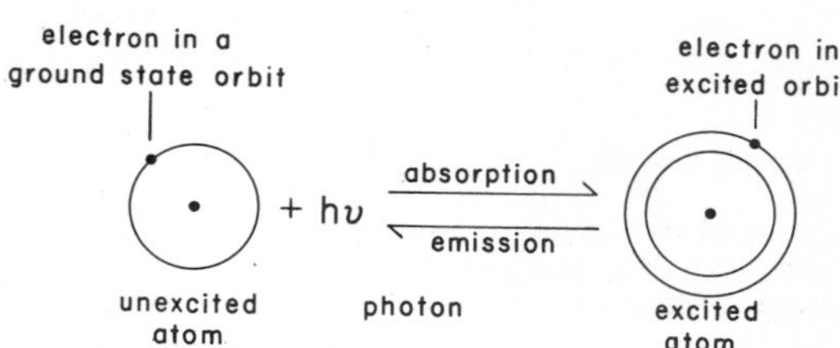

Fig. 1. Relationship between flame photometry and atomic absorption spectroscopy.

Any change which affects the total number of atoms in the system affects both phenomena. Any change which affects the distribution of excited and unexcited (ground state) atoms has opposite effects on the two phenomena.

Principles of Atomic Absorption Spectroscopy. A schematic diagram of the equipment is shown in Fig. 2. The sample is introduced into a flame atomizer. The metals present are reduced to *unexcited* atoms. Radiant energy from the source passes through the atomizer (and the metal atoms) to a monochromator, which selects light of the pertinent wavelength. The light then falls on a detector. The detector measures (a) the initial intensity of radiation leaving the source (by switching off the atomizer), and (b) the intensity of radiation after passing through the atomizer and the metal atoms. From these two readings the degree of absorption is measured and the analysis of the sample calculated. Radiation emitted by the atomizer at the same wavelength as the absorption wavelength is eliminated by modulating the equipment.

Flame Photometry. The schematic diagram for flame photometry is shown in Fig. 3. The sample is introduced into the hot flame. As before, reduction of the metals to the atomic state occurs,

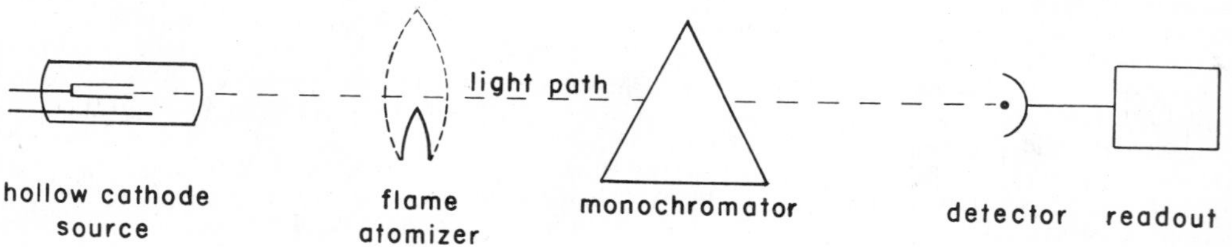

Fig. 2. Schematic diagram of atomic absorption equipment.

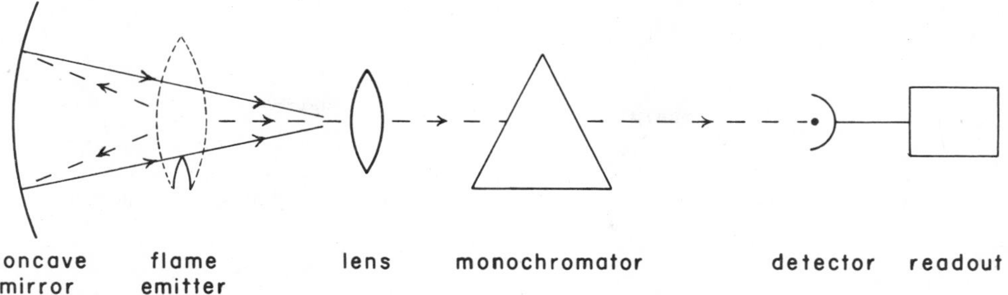

FIG. 3. Schematic diagram of a flame photometer.

with the flame operating to form *excited* atoms. These rapidly radiate light with a wavelength characteristic of the metal. This light is separated out by the monochromator, and its intensity is measured by the detector. Emission from excited metal oxide bands has also been studied extensively, but these bands are generally less sensitive and more troublesome than atomic lines.

In practice, atomic absorption is somewhat more sensitive than flame photometry and in general suffers from fewer interferences. In particular, high background radiation and excitation interferences can cause serious problems in flame photometry. These are only minor problems in atomic absorption. However, the equipment used for atomic absorption is more sophisticated and therefore costs more.

Applications. Both of these methods have found extensive applications in the biological field. Areas of particular interest have been body fluids in particular, *blood*, *urine* and *saliva*. Flame photometry has become a standard procedure for the direct determination of sodium, potassium, calcium and magnesium in blood and urine. Recently, however, atomic absorption has been used to an increasing degree for these analyses. *Body tissue* and *bone ash* have also been analyzed by both methods. Sample decomposition (combustion or chemical) and solution are necessary steps in these latter determinations.

A list of sample types and metals analyzed by these procedures is shown in Table 1.

References

1. WALSH, A., *Spectrochim, Acta*, **7**, 108 (1955).
2. WILLIS, J., *Clin. Chem.*, **11**, No. 2, 251 (1965).
3. ROBINSON, J. W., "Atomic Absorption Spectroscopy," New York, M. Dekker (in press).
4. DEAN, J. A., "Flame Photometry," New York, McGraw-Hill Book Co., 1960.
5 HERRMAN, R., AND ALKEMADE, C. T. J., "Flame Photometry," New York, Interscience Publishers, 1963 (translated by P. T. GILBERT).

J. W. ROBINSON

TABLE 1. SAMPLE TYPES AND ELEMENTS DETERMINED BY ATOMIC ABSORPTION SPECTROSCOPY

Sample Type	Element Determined	
	Atomic Absorption	*Flame Photometry*
Serum	Calcium, magnesium, sodium, potassium, cesium, chromium, manganese, molybdenum, selenium	Calcium, sodium, potassium, magnesium
Urine	Bismuth, cadmium, cesium, chromium, cobalt, copper, lead, magnesium, manganese	Calcium, magnesium, sodium, potassium
Bone Ash	Cesium, magnesium, potassium,, sodium	Copper, sodium, potassium
Tissue	Cesium, magnesium, Zinc	Copper, strontium, magnesium
Saliva	Cesium	
Fecal material	Magnesium	

FLAVIN ADENINE DINUCLEOTIDE (FAD) AND FLAVIN MONONUCLEOTIDE (FMN).

See FLAVINS AND FLAVOPROTEINS; RIBOFLAVIN.

FLAVINS AND FLAVOPROTEINS

The yellow prosthetic group of a class of oxidative enzymes, the flavoproteins, is called by the trivial name "*flavin*" (not to be confused with flavone).[1,2] This group is contained in a nonprotein molecular moiety, which in some cases may be dissociated from the protein. This moiety is called flavo- or flavin coenzyme or flavin nucleotide. There are two such coenzymes, the more frequently encountered *flavin adenine dinucleotide* (FAD) and *flavin mononucleotide* (FMN) (Fig. 1). Strictly speaking, these substances are *not* nucleotides, since they contain a ribityl, not a ribosyl, side chain. The correct name for FMN is therefore riboflavin-5′-monophosphate; and for FAD riboflavin adenosine diphosphate.

FLAVIN-MONONUCLEOTIDE (FMN)

ISOALLOXAZINE

RIBOFLAVIN (VIT. B_2) ADENOSINEDIPHOSPHATE

FLAVIN-ADENINE-DINUCLEOTIDE (FAD)

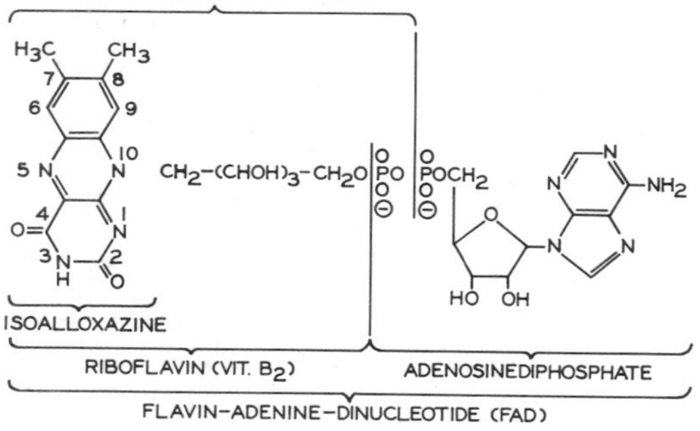

FIG. 1.

Acid hydrolysis of the flavocoenzymes produces RIBOFLAVIN. This was the first flavin to be isolated. Before its structure was established, it was known as vitamin B_2. Mainly in the form of its protein-bound derivatives, riboflavin is a ubiquitous constituent of living matter and biological fluids. The early names lactoflavin ("flavin from milk") ovoflavin ("flavin from egg white"), and hepatoflavin ("flavin from liver") for the same substance now called riboflavin attest to this.

In dilute solution, riboflavin as well as FMN are very *sensitive to light*,[2] whereas FAD is some twenty times more stable to light. Since flavin is an oxidant, it acquires the capability, when irradiated by light of the proper frequency (blue visible and ultraviolet light), of oxidizing its own side chain. This *photolysis* is irreversible. It can be prevented by addition of suitable electron-donating substances (tertiary amines, sulfides, etc.). Electrons so donated to photoactivated (triplet) flavin cause *photoreduction*, which may be reversed by oxygen.

In the pH range between 2 and 10, photolysis leads to complete oxidative splitting of the side chain. The product is a very pale yellow, extremely insoluble heterocyclic compound of strong bluish fluorescence: *lumichrome*. At pH 10, however, the ribityl side chain is not completely degraded. A methyl group remains at N(10). *Lumiflavin* is thus obtained, which, in contrast to lumichrome, retains all spectral, oxido-reductive and acid-base properties of riboflavin.

The first flavin synthesized was lumiflavin. This *synthesis* was analogous to the well-known condensation of alloxan and *o*-phenylenediamines leading to *alloxazines*. Hence, the flavin nucleus, obtained from alloxan and monoalkyl-*o*-phenylenediamines was named *isoalloxazine*. According to this, lumichrome is 7,8-dimethylalloxazine whereas lumiflavin is 7,8,10-trimethylisoalloxazine. The numbering system used here follows *Chemical Abstracts* (as shown in Fig. 1). An older system, analogous to that used for anthraquinone, is found in the earlier literature as well as in recent publications.[1] It appears more adequate to designate flavins as 10H-benzo-[g]-pteridine-2, 4-diones rather than isoalloxazines. This would also account for the *biosynthesis* of riboflavin which involves condensation of a benzene nucleus with a pteridine, rather than a quinoxaline with a pyrimidine, by the enzyme riboflavin synthetase.[4,4a] The biosynthesis of the coenzyme occurs in two steps: FMN is synthesized from riboflavin and ATP by flavokinase,[5] and FMN is converted to FAD by FAD pyrophosphorylase with ATP as donor of the AMP moiety.[6] The specificity of these enzymes for their substrates has been studied.[5,6]

The *technical synthesis* now in use is based on a rare type of splitting of an azogroup (Fig. 2).

FLAVIN SYNTHESIS

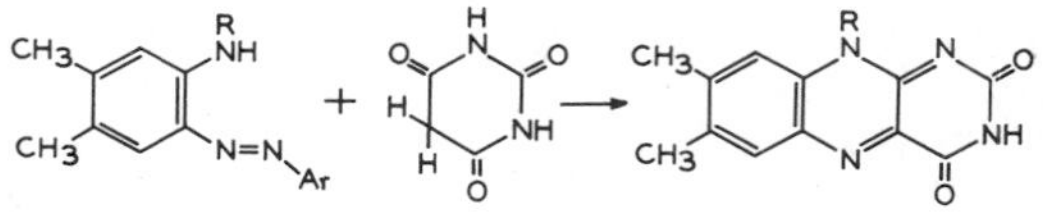

FIG. 2.

All free flavins exhibit a characteristic green *fluorescence*, which can be used for their detection at extremely low concentrations. There are many mechanisms, however, which interfere with flavin fluorescence: binding to proteins, uptake or loss of protons, reduction, charge transfer, and metal binding. The fluorescence of FAD is an order of magnitude lower than that of FMN because of intramolecular charge transfer.

Flavins can occur in three *states of oxidation*. The thermodynamic reversibility of this process allows the classification of flavin as a true quinoid system. Since the three states differ drastically in their physical and chemical properties, it should in every case be made clear which species is referred to. This can be accomplished, for instance, by a nomenclature system analogous to that of quinones: "flavoquinone" (=oxidized state), "flavosemiquinone" (=half-reduced or radical state), "flavohydroquinone" (=reduced or "leuco" state). Since the "leucoflavin" (=flavohydroquinone) is colorless only in very dilute solutions, the term "leuco" is misleading.

Flavoquinone is the only form stable under aerobic conditions. Despite this fact, however, the other more evasive redox forms of flavin should be given equal consideration from the standpoint of function. The neutral form of flavoquinone is stable over a wide range (pH 2–9) of pH. It shows the characteristic and well-known "flavin spectrum" with maxima at 375 and 446 mμ.[7,8] Below pH 2, a proton is added at N(1). The cation so formed shows only one maximum, at 239mμ. Dissociation of NH(3) at pH 10, however, affects the spectrum only slightly.

Flavoquinone is very stable to acid and moderately stable to alkali. Upon heating in 1*N* NaOH, the pyrimidine nucleus will open at CO(2). If dissociation of NH(3) is prevented (by alkylation or binding to protein), flavoquinone may become extremely alkali labile.

Flavoquinone is only slightly water soluble, if no ionic groups are present in the side chains (as in FMN, for instance). It dissolves in strong mineral acids and alkali, and can be purified most easily by reprecipitation from these solvents into hot aqueous buffer. When acetylated in the side chain, riboflavin becomes very soluble in chloroform.

Molecular complexes are readily formed between flavoquinone and any strong electron-donating agent, such as (hetero)aromatic electron-rich molecules, *e.g.*, tryptophan, phenothiazine, adenine, and "polarizable" inorganic ions (I^-, Hg^{2+}). In FAD a charge-transfer (sandwich) configuration between the two heteroaromatic moieties is adopted. Because of the electron deficiency in flavoquinone—obvious from the cited examples—no stable metal chelates exist for the flavoquinone state except with such cations as Ag^+, Cu^+ (donor-metal specificity).[9]

Chemical *reduction of flavins* can be carried out rapidly at neutral pH electrolytically, with H_2 and catalyst, or with $Na_2S_2O_4$, but only very slowly with NaSH, $NaBH_4$ or NADH.

Flavohydroquinone is pale yellow in the neutral form, in concentrated solutions orange-red, but nearly colorless in the protonated form ($pH < 1$) and as the anion ($pH > 6$). It has to be kept in mind that the acid dissociation of this form is within the physiological range of pH. Flavohydroquinone is not quite coplanar, in contrast to the other redox states. Flavohydroquinone reacts very rapidly with traces of oxygen, apparently in a two-electron step. H_2O_2 is liberated and undergoes further reduction only slowly.

In non-aqueous systems, the reduced flavin can be trapped by acetylation as colorless crystalline 5-acetylflavohydroquinone (Fig. 3). The 5-acetyl group can be hydrolyzed under very mild conditions.

Flavosemiquinone is the intermediate radical form in flavin oxido-reduction. It is present in equilibrium with flavoquinone and flavohydroquinone. The individual reactions in this disproportionation-comproportionation equilibrium are very rapid. This accounts for the fact that flavin is an efficient catalyst of one-electron and two-electron ("hydride") transfer. The equilibrium is displaced toward disproportionation for the free flavocoenzymes at physiological pH. It can be shifted toward the radical state by addition of mineral acid or complexing metal ions (Fig. 3) because flavosemiquinone has a higher proton as well as metal affinity than either flavoquinone or flavohydroquinone.

The neutral flavosemiquinone has blue color, but is not easily observed, since its absorption is hidden by the excess of flavoquinone and flavoquinhydrone (see below), which is always present. The cation and the metal chelates of flavosemiquinones are red. The only certain criterion for the presence of flavin radicals is therefore paramagnetism, as observed by electron paramagnetic resonance or susceptibility measurements.

At higher concentrations, the study of flavin oxido-reduction in aqueous medium is difficult because charge-transfer complexes form between closely related flavin species.[10] The most common complex between two flavin molecules is *flavoquinhydrone*. It has the structure of a "charge-transfer" complex of flavoquinone and flavohydroquinone, but can dissociate as well into two radicals. Flavoquinhydrone itself is diamagnetic. The charge-transfer absorption of flavoquinhydrone is very broad, extending from 600 mμ toward the infrared. The stability of flavin-flavin complexes is strongly diminished on addition of organic solvents and/or heating.

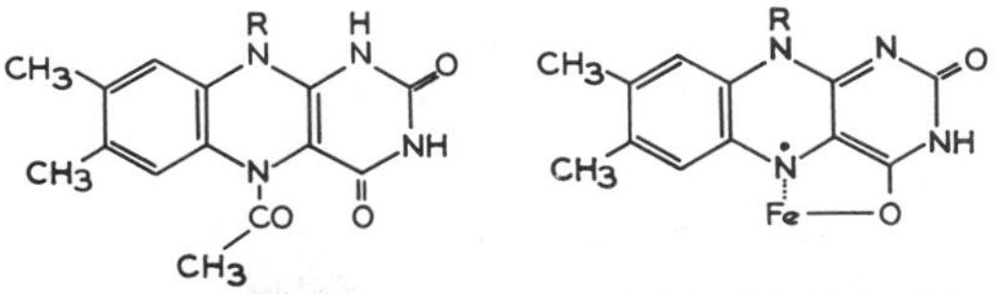

5-ACETYLFLAVO-HYDROQUINONE

FLAVOSEMIQUINONE METAL-CHELATE

FIG. 3.

Concerning *flavoproteins*, it shoud be emphasized at the outset that according to present knowledge, there are far-reaching differences in structure, properties,[11] function and reaction mechanism between individual members of this class. What has earlier been called "a typical flavoprotein" does not exist. There are, however, certain features which will be found to apply to a great number of cases. Most flavoproteins contain *two flavin groups* per molecule, both either FAD or FMN, with the exception of bacterial dihydroorotic dehydrogenase[12] which contains 1 FAD and 1 FMN. In most instances, a molecular weight equivalent of 60,000–80,000 is found *per flavin*. The *strength of binding* of the prosthetic flavin group to the protein varies widely. In extreme cases, it is dissociable at low pH and high ionic strength. Univalent cations expedite the dissociation.[13] A largely undenatured apoprotein may be obtained which will recombine with flavin. The specificity of these apoproteins for their prosthetic groups may thus be tested.[14] A quantitative assay for the combining flavin group is also available in this way.[15] With the majority of flavoproteins, the flavin is only released at a pH where the protein is denatured. In a few cases, proteolytic digestion is needed to liberate the flavin. The best-known example here is succinic dehydrogenase, where a FAD-hexapeptide can be obtained by proteolysis.[16] The place of attachment of the peptide chain to the flavin ring system is not known.

In addition to simple flavoproteins, where the only oxidation-reduction catalyst appears to be the flavin group, more *complex flavoproteins* exist. Strong evidence is available that sulfhydryl-disulfide systems may in some instances function in conjunction with flavin,[17] and there are possibly, in addition to this, *metal constituents* present in a small number of flavoproteins. Two *hemoflavoproteins* are known: yeast L(+)-lactic dehydrogenase[18] (cytochrome b_2) and formic dehydrogenase of *E. coli*;[19] 3 *molybdenum-iron-flavoproteins*: XANTHINE OXIDASE of milk and liver, aldehyde oxidase of liver, and the nitrate reductases from various sources; several *iron-flavoproteins*: the dehydrogenases of the respiratory chain for succinate and NADH and dihydro-orotic dehydrogenase from bacteria.[12] It should be emphasized that the iron component in the last two groups of flavoproteins is not in porphyrin linkage but in an as yet unknown complex probably involving sulfur.[20] Finally, a *zinc-flavoprotein*[21] has been found. In this last case, the metal functions in substrate binding, whereas in all the other cases, strong evidence is available from EPR spectroscopy that the metal components participate in oxidation-reduction. Some metalflavoproteins have a considerable excess of iron over flavin. It is not likely that all this iron participates directly in catalysis. It may be required for substrate binding or structural integrity.

Electron donors of flavoproteins can be grouped into the following classes: reduced pyridine nucleotides (→ oxidized pyridine nucleotides), α-amino and α-hydroxy acids (→ α-ketoacids), aldehydes or aldoses (→ acids or lactones), saturated carbon chains —CH_2—CH_2— (→ —$\overset{H}{C}$=$\overset{H}{C}$—) and azomethines, *e.g.*, xanthine, purine (→ ketimine).

More rarely encountered reactions catalyzed by flavoproteins are oxidative decarboxylation such as the oxidative decarboxylation of lactic acid (→ acetate and CO_2)[22] and oxalic acid (→ CO_2)[23] or oxidation of PYRIDOXINE and pyridoxamine phosphates to pyridoxal phosphate.[24]

A special case of a donor-acceptor system is found when *two flavoproteins* interact *in series* in oxidation-reduction, one flavoprotein being the donor and the other the acceptor. Several donor proteins in this type of system but only one acceptor-protein are known thus far. The latter has the provisional name "*electron-transferring flavoprotein*."[25] It is the only simple flavoprotein known that couples efficiently to the cytochrome system as it occurs in particles derived from mitochondria.

Only a few flavoproteins—those generally called *oxidases*—react efficiently with *oxygen as electron acceptor*. *Most flavoproteins* are, however, *autoxidizable* even if only at a slow rate. In some instances, the substrate or reaction product interferes with autoxidation which is readily observed after chemical reduction. Flavoproteins are, however, not oxidases (OXYGENASES) in the sense that oxygen is incorporated into the structure that undergoes oxidation. Oxygen is hydrogenated to hydrogen peroxide. In the iron-containing flavoproteins, the protein-bound iron complex may be the electron acceptor of the flavin.[26] The iron complex does not, however, necessarily react with oxygen at an appreciable rate.

For the flavoproteins that react poorly with oxygen, the natural electron acceptors are often not known. Artificial acceptors, such as dyes or ferricyanide, are widely used as acceptors in assays and in connection with isolation procedures. Such artificial systems, however, may be misleading in the sense that not the native enzyme but a modified form, more efficient in the assay system, is followed or produced in the isolation procedure. Examples for this are found in the history of the enzymes diaphorase-lipoic dehydrogenase[27] and succinic dehydrogenase.[28] Cytochrome *c*, although capable of reacting with some flavoproteins at a reasonable rate, cannot *a priori* be considered as a natural acceptor. Disulfides such as GLUTATHIONE[29] or LIPOIC ACID[27] are acceptors for specific flavoproteins. Hydrogen peroxide can serve as acceptor in rare instances, and some metalloflavoproteins are able to reduce nitrate and organic nitrogen compounds.[1]

No inhibitors specific for the flavin group are known, although this is sometimes assumed for atebrin (= atabrine = quinacrine = mepacrine). Conclusions as to the flavoprotein nature of an enzyme based on results with such inhibitors have to be considered with caution.[30]

The *optical absorption spectra* of the oxidized forms, even of those flavoproteins that are free of metals, vary widely and differ from those of the free prosthetic groups. Shifts of maxima to longer or shorter wavelengths may occur, and new peaks or shoulders appear.[31] A lipophilic environment of the prosthetic group is thought to produce the shoulder frequently seen with flavoproteins some 20 mμ toward longer wavelengths from the main absorption peak. Bound substrate or product may be responsible for a shoulder in this region as well as at shorter wavelengths (~ 420 mμ and ~ 315 mμ). Caution is necessary in the interpretation of absorptions at 410–430 mμ as these may also be due to small amounts of hemoprotein impurities. In rare instances, pure flavoproteins show absorption at 600-700 mμ and exhibit, therefore, a green color.

The spectra of metalloflavoproteins are even more complicated. It has been shown that the iron complex present in its oxidized state makes a definite and in some cases well-defined[32] contribution. It is not clear whether molybdenum contributes to the spectrum.

With so much variation encountered in the oxidized forms, it is not surprising that on interaction with different substrates and subsequent reduction a great number of divergent and often unexplained observations have been made.[11] Although easily measurable reduction of the prosthetic flavin, *i.e.*, decrease of absorption at ~ 450 mμ would be expected of an active substrate, even this criterion is not generally applicable. In many cases, particularly with metal-free flavoproteins, on addition of substrate or product new broad absorption bands appear at long wavelengths (>500 mμ), sometimes as transients but often also for periods of hours.[33] These bands can be due to charge-transfer interaction in the complexes formed between the flavoprotein and the added low molecular weight substance,[34] but they may also be due to true semiquinoid reaction intermediates. Great caution is advisable in the interpretation of these complex phenomena, and a definite assignment is only possible on the basis of extensive and *quantitative* studies by methods that are capable of measuring semiquinones directly (*e.g.*, EPR spectroscopy). The fully reduced state is not readily reached with many flavoproteins even when chemical reducing agents are used, and the proteins may be inactivated by complete reduction.

Most flavoproteins show no measurable *fluorescence* (see above). Contamination by traces of free flavins has to be ruled out in fluorescence studies on flavoproteins. Since the reduced forms of flavins are not fluorescent, fluorimetry can only be used with fluorescent flavoproteins in studies on early events involving the oxidized state and in studies on coenzyme-apoenzyme interactions.

Most flavoproteins are considerably less *sensitive to light* than the free coenzymes. With lactic oxidative decarboxylase and D-amino acid oxidase, however, semiquinone formation in ordinary room light has been reported. Reactions that are not normally occurring could thus be catalyzed by light. With flavoproteins of unknown properties caution is therefore advisable.

No unified picture of "the *reaction mechanism* of a flavoprotein" is available and such a mechanism is unlikely to be found. It appears rather that all chemical potentialities of the flavin molecule are utilized by nature but to a varied extent in different enzymes.

The flavin molecule is capable of one- or two-electron transfer (see above). With microsomal NADH cytochrome b_5 reductase, direct stereospecific transfer of hydrogen has been demonstrated.[35] This finding should, however, not be generalized before other flavoproteins have been similarly investigated. Evidence has been presented for catalytic mechanisms involving only the oxidized and semiquinone forms, where further reduction leads actually to inhibition[17,36] for a mechanism involving only the semiquinone and reduced forms,[37] and finally, in the reaction catalyzed by glucose oxidase,[36,38] no direct evidence for semiquinone participation has been found, so that a two-electron transfer from flavohydroquinone to flavoquinone is suggested. Unfortunately, except for microsomal NADPH cytochrome *c* reductase,[38] metal-free flavoproteins when reacting with substrate do not form semiquinone species detectable by EPR[11] so that only indirect spectrophotometric and kinetic evidence for *semiquinone formation* is available. Semiquinones are readily detected by EPR in metalloflavoproteins and appear to be kinetically significant intermediates in the reactions of these enzymes.[26] It is not certain whether in the iron flavoproteins electron transfer follows the path: substrate → flavin → iron complex → acceptor, or whether flavin and iron complex react as a unit in a two-electron (hydride ion) step; in the molybdenum-iron flavoproteins xanthine and aldehyde oxidase, molybdenum probably accepts electrons directly from substrate and donates to flavin. Direct *interaction between metal and flavin semiquinone* in metalloflavoproteins has been demonstrated by spin relaxation studies.[39] *Multiple forms* of molybdenum(V) have been observed in these two enzymes and multiple forms of radicals (probably flavin semiquinones) with dihydroorotic dehydrogenase.[26] These findings point to the likelihood that, at least with the more complex flavoproteins, mechanisms will be found to be more complicated than previously thought.

Lastly, attention should be drawn to the increasing number of purified and characterized flavoproteins[40-42] for which no obvious oxidoreductive function is apparent in the reaction that they catalyze. New facets of flavin function may become apparent from further work on such systems.

Many aspects of flavin and flavoprotein chemistry and function, which have been touched upon in this brief review, have been discussed from the (at the time of this writing) most advanced viewpoints at a symposium on "Flavins and Flavoproteins."[43] An advanced review of topics in this field will be published about the same time as this volume.[44]

References

1. Beinert, H., in Boyer, P. D., Lardy, H., and Myrbäck, K., (Editors), "The Enzymes," Vol. II, p. 339, New York, Academic Press, 1960.
2. Hemmerich, P., Veeger, C., and Wood, H. C. S., *Angew. Chem.* Internatl. Ed., **4**, 671 (1965).
3. Holmström, B., *Arkiv Kemi*, **22**, 329 (1964).
4. Plaut, G. W. E., in "Pteridine Chemistry" Pfleiderer, W., and Taylor, E. C., (Editors), New York, Macmillan, Pergamon Press, 1964.

4a. Winestock, C. H., and Plaut, G. W. E., in "Plant Biochemistry" Bonner, J., and Varner, J. E., (Editors), New York, Academic Press, 1965.

5. McCormick, D. B., and Butler, R. C., *Biochim. Biophys. Acta*, **65**, 326 (1962).
6. McCormick, D. B., *Biochem, Biophys. Res Commun.*, **14**, 493 (1964).
7. Beinert, H., *J. Am. Chem. Soc.*, **78**, 5323 (1956).
8. Dudley, K. H., Ehrenberg, A., Hemmerich, P., and Müller, F., *Helv. Chim. Acta*, **47**, 1354 (1964).
9. Hemmerich, P., Müller, F., and Ehrenberg, A., in "Oxidases and Related Redox Systems" (King, T. E., Mason, H. S., and Morrison, M., Editors), Vol. I, pp. 157–178, New York, John Wiley & Sons, 1965.
10. Gibson, Q. H., Massey, V., and Atherton, N. M., *Biochem. J.*, **85**, 369 (1962).
11. Beinert, H., *Proc. Intern. Congr. Biochem. 6th, Section IV, New York*, 285 (1964).
12. Friedmann, H. C., and Vennesland, B., *J. Biol. Chem.* **235**, 1526 (1960).
13. Strittmatter, P., *J. Biol, Chem.*, **236**, 2329 (1961).
14. Arsenis, C., and McCormick, D. B., *Biochim. Biophys. Acta*, **92**, 440 (1964).
15. Huennekens, F. M., and Felton, S. P., in "Methods of Enzymology" (Colowick, S. P., and Kaplan, N. O., Editors), Vol. III, pp. 950–959, New York, Academic Press.
16. Kearney, E. B., *J. Biol. Chem.*, **235**, 865 (1960).

17. Massey, V., and Veeger, C., *Biochim. Biophys. Acta*, **48**, 33 (1961).
18. Morton, R. K., and Shepley, K., *Biochem. J.*, **89**, 257 (1963).
19. Linnane, A. W., and Wrigley, C. W., *Biochim. Biophys. Acta*, **77**, 408 (1963).
20. Beinert, H., Der Vartanian, D., Hemmerich, P., Veeger, C., and Van Voorst, J. D. W., *Biochim. Biophys. Acta*, **93**, 530 (1965).
21. Cremona, T., and Singer, T. P., *J. Biol. Chem.*, **239**, 1466 (1964).
22. Sutton, W. B., *J. Biol. Chem.*, **226**, 395 (1957).
23. Datta, P. K., and Meeuse, B. J. D., *Biochim. Biophys. Acta*, **17**, 602 (1965).
24. Wada, H., and Snell, E. E., *J. Biol. Chem.*, **236**, 2089 (1961).
25. Beinert, H., in "The Enzymes" (Boyer, P. D., Lardy, H., and Myrbäck, K., Editors), Vol. VII, pp. 467–476, New York, Academic Press, 1963.
26. Beinert, H., and Palmer, G., in *Advan. Enzymol.* (Nord, F. F., Editor), **27**, 105 (1965).
27. Veeger, C., and Massey, V., *Biochim, Biophys. Acta*, **64**, 83 (1962).
28. King, T. E., *J. Biol. Chem.*, **238**, 4037 (1963).
29. Scott, E. M., Duncan, I. W., and Ekstrand, V., *J. Biol. Chem.*, **238**, 3928 (1963).
30. Hemker, H. C., and Hülsmann, W. C., *Biochim, Biophys. Acta*, **44**, 175 (1960).
31. Beinert, H., in "Light and Life" (McElroy, W. D., and Glass, B., Editors), pp. 163–169, Baltimore, Johns Hopkins Press, 1961.
32. Rajagopalan, K. V., and Handler, P., *J. Biol. Chem.*, **239**, 1509 (1964).
33. Beinert, H., and Sands, R. H., in "Free Radicals in Biological Systems" Blois, M. S., Jr., Brown, H. W., Lemmon, R. M., Lindblom, R. O., and Weissbluth, M., Editors), pp. 17–52, New York, Academic Press, 1961.
34. Massey, V., and Palmer, G., *J. Biol. Chem.*, **237**, 2347 (1962).
35. Strittmatter, P., *J. Biol. Chem.*, **239**, 3043 (1964).
36. Massey, V., and Gibson, Q. H., *Federation Proc.*, **23**, Part I, 18 (1964).
37. Masters, B. S. S., and Kamin, H., *J. Biol. Chem.*, **240**, 921 (1965).
38. Nakamura, T., and Ogura, Y., *J. Biochem.*, **52**, 214 (1962).
39. Beinert, H., and Hemmerich, P., *Biochem. Biophys. Res. Commun.*, **18**, 212 (1965).
40. Becker, W., Benthin, U., Eschenhof, E., and Pfeil, E., *Biochem. Z.*, **337**, 156 (1963).
41. Gupta, N. K., and Vennesland, B., *J. Biol. Chem.*, **239**, 3787 (1964).
42. Störmer, F. C., and Umbarger, H. E., *Biochem. Biophys. Res. Commun.*, **17**, 587 (1964).
43. Beinert, H., in "Flavins and Flavoproteins" (Slater, E. C., Editor), BBA Library Vol. 8, p. 49, Amsterdam, Elsevier, 1966.
44. Palmer, G., and Massey, V., in "Biological Oxidations" (T. P. Singer, Editor), New York, Wiley and Sons, in press.

Helmut Beinert and Peter Hemmerich

FLAVONOIDS AND RELATED COMPOUNDS

The flavonoids are a group of aromatic oxygen-containing, heterocyclic compounds, most of which have a 2-phenylbenzopyran skeleton as their basic ring system. The 2,3-dihydro derivative (I) is termed flavan. They are widely distributed

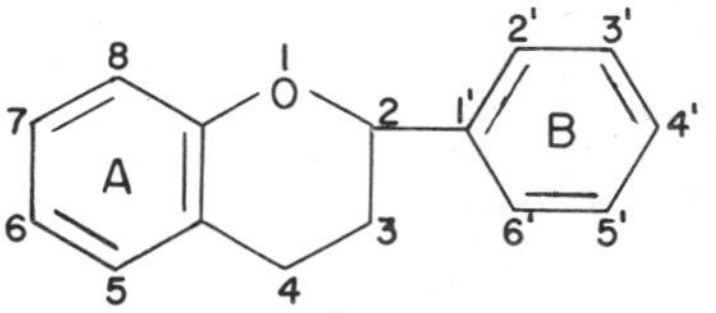

I

among higher plants including ferns and mosses but are not found in algae, fungi, bacteria or lichens. Many compounds belonging to this group of substances are highly colored. In fact, the majority of the yellow, red and blue pigments found in flowers and fruits, as well as other organs of higher plants, are flavonoids. Exceptions are the CAROTENOIDS which are water-insoluble pigments ranging from yellow to red and the "nitrogenous anthocyanins." The latter are water-soluble red pigments found in *Beta vulgaris* and *Bougainvillaea glabra*. These pigments are actually ALKALOIDS and their structure has only recently been elucidated[1] (see also CHEMOTAXONOMY).

Classification. The flavonoids are subdivided according to the oxidation state of the heterocyclic ring. The substances listed below are arranged in order of increasing oxidation state: (1) catechins; (2) leucoanthocyanidins and flavanones; (3) flavanols, flavones and anthocyanidins; (4) flavonols. Closely related substances which, however, do not have the 2-phenylbenzopyran structure are the isoflavones, 3-phenylbenzopyrone derivatives, the aurones (II) and chalcones, benzylideneacetophenone derivatives.

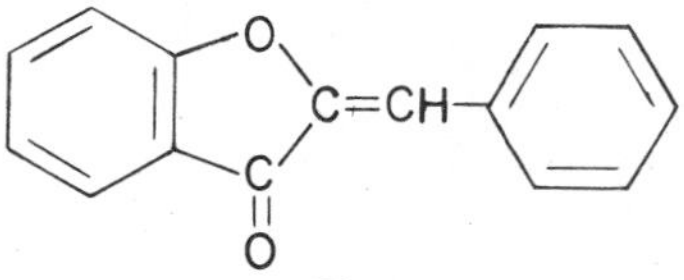

II

Catechin, a colorless, crystalline cold water-insoluble substance, is found primarily in woody tissue. A number of different catechins, all having the flavan-3-ol structure have been isolated. They differ in the number and positions of the additional hydroxyl or methoxyl groups as well as in regard to the stereochemistry around carbon-2 and carbon-3. Only rarely are glycosides of catechins found in nature. Catechins are frequently accompanied by large amounts of water-soluble, polymeric tannins. These have the ability to precipitate gelatin from solution. It seems quite probable that a large part of the natural tannins is derived from catechins by non-enzymatic reactions that occur in wood as postmortal changes. Enzymatic formation of tannins seems to occur only under non-physiological conditions such as are found as a result of injury, for example.

The *leucoanthocyanidins* are colorless substances which form intensely colored pigments, including anthocyanidins, when heated with acid. They are extremely widely distributed in the plant kingdom and although many have been shown to be flavan-3,4-diol derivatives, it is not at all certain that all have this structure. Since these substances are extremely susceptible to atmospheric oxidation and are very sensitive to acids and bases, their isolation is quite difficult. However, with CHROMATOGRAPHIC or COUNTERCURRENT DISTRIBUTION techniques, crystalline solids are frequently obtained.

The biogenetic relation between the leucoanthocyanidins and the naturally occurring anthocyanins is still highly uncertain. Although many of them have the same hydroxylation pattern as the naturally occurring anthocyanins, few occur as glycosides. The fact that they can readily be converted *in vitro* to anthocyanidins by oxidative dehydration should not be taken as proof that they are the biological precursors of the anthocyanins.

The *flavanones* are colorless substances that also appear to be widely distributed among higher plants either in the free form or as glycosides. About 25 different flavanones have been isolated. The carbonyl group is at the carbon-4 position, and these substances are therefore 2,3-dihydropyrone derivatives. The flavanones are readily converted in alkaline solution to chalcones. The absence of a highly conjugated system also accounts for the fact that they do not absorb light in the visible region. The *flavanonols*, 3-hydroxyflavanones, are chemically quite similar to the flavanones. Eleven differently substituted flavanonols are known. They are seldom found as glycosides and occur usually only in low concentration.

The *flavones* (2-phenyl-1,4-benzopyrones), *isoflavones* (3-phenyl-1,4-benzopyrones) and *flavonols* (flavon-3-ols) make up the largest group of ivory and yellow pigments found in higher plants. Of the approximately 100 different flavones and flavonols that have been isolated and characterized to date, many occur as β-glycosides of glucose, rhamnose, arabinose and galactose. There is usually a sugar in the 3-position. If a second position is glycosylated, it is commonly the 7-position. Biosides and triosides are also known with rutinose, α-L-rhamnosyl-(1 $\rightarrow$ 6)-D-glucose being the most widespread. The most common hydroxylation pattern is that of quercetin (5,7,3′,4′-tetrahydroxyflavon-3-ol), but flavones with fewer hydroxyl groups and with methoxyl substitutes are also found. In fact, flavone itself occurs as white flakes on the leaves and stems of a number of *Primula* species.

Biological Roles. An older use of flavones which is no longer of any economic significance is the application of these substances to cloth as vegetable dyes. Certain flavonoids, rutin, hesperidin, quercetin, for example, increase the tensile strength of capillary walls and decrease capillary permeability in experimental animals. This has led to claims that these substances act as vitamins. However, these suggestions do not seem to be well founded, and it may well be that the beneficial effects observed are due to a prolongation of the vasoconstrictor effects of adrenaline.[2] There do not appear to be any good clues concerning the possible functions of the flavones in the metabolism of higher plants.

The flavonoids with which the layman would be most familiar are the *anthocyanins*. These are the water-soluble pigments which account for most of the red, pink, purple and blue colors found in higher plants. The use of chromatographic procedures, which have greatly simplified the isolation and characterization of the anthocyanins, has shown that most plants contain more than one of these pigments and that they occur most prevalently as glycosides. The aglycons, *anthocyanidins*, have the structure of a flavylium cation (III) in acidic solution. The chlorides and

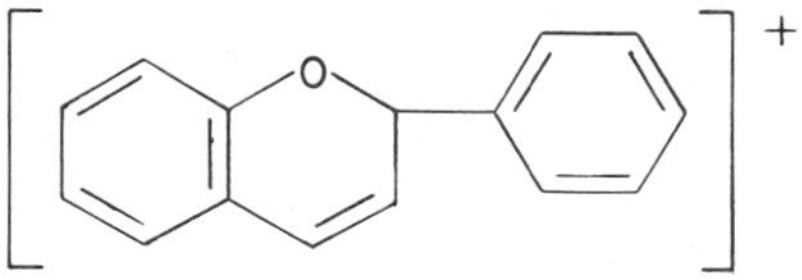

III

picrates frequently can be obtained as nicely crystalline compounds. The positive charge is distributed among various carbons and oxygens of this conjugated ring system, and it is this property that is responsible for the fact that these substances are colored. An increase in the number of hydroxyl or methoxyl groups found on the flavylium nucleus deepens the color tint. Other factors that can influence the observed color shades are the pH of the cell sap, co-pigments such as xanthones, metal ions (particularly aluminium), and the concentration of the anthocyanins. Although several hundred different anthocyanins are known, there are only 11 different anthocyanidins. The most widely distributed aglycon is cyanidin, 3,5,7,3′,4′-pentahydroxyflavylium salt. Apigenidin and luteolinidin are the only anthocyanidins without a hydroxyl at carbon-3. When present, the hydroxyl group at carbon-3 is always found to be glycosylated. This seems to increase the stability of the pigments. Also fairly common are 3,5- and 3,7-diglycosides. β-Glucosides are by far the most commonly found glycosides, but rhamnose, xylose and galactose derivatives are also known. Many of the latter occur together with glucose as biosides and triosides. In addition, anthocyanins have been isolated that are acylated with substituted cinnamic acids. The site of attachment of these acids to the anthocyanins is not known. It has not been possible to assign any specific metabolic role in plant metabolism to the anthocyanins. Possibly these pigments fulfill an ecological function in regard to pollination and seed dispersal through their ability to act as insect and bird attractants.

Biosynthesis. The information on the modes of biosynthesis of flavonoids is still rather meager. From tracer experiments it has been established that the A-ring is derived from acetate units by head-to-head addition. The B-ring and carbons-2, 3, and 4 are most likely synthesized via the shikimic acid pathway, probably through a phenylpropanoid unit [see AROMATIC AMINO ACIDS (BIOSYNTHESIS)]. Isoflavones seem to be formed from chalcones through a phenyl migration[3]. Very little is known concerning the enzymes involved in flavonoid synthesis. It has been shown that sugar nucleotides can participate in glycoside formation[4]. However, there is very little information concerning the interrelationships among the various anthocyanins. It is not known, for example, whether reductions of flavones to anthocyanins occur *in vivo*. A great deal of work has been done in the past on the genetics of color inheritance[6], and the availability of material of known genetic background should prove a great impetus to future studies on the biosynthesis and interrelations of the flavonoids. It might also be anticipated that such studies would yield valuable information on the mechanisms of genetic controls in higher plants. Harborne[5] has edited an excellent summary on the biochemistry of the flavonoids.

References

1. MABRY, T. J., WYLER, H., SASSU, G., MERCIER, M., PARIKH, I., AND DREIDING, A. S., *Helv. Chim. Acta*, **45**, 640 (1962).
2. DEEDS, F., in "Pharmacology of Plant Phenolics" (FAIRBAIRN, J. W., EDITOR), p. 91, New York, Academic Press, 1959.
3. GRISEBACH, H., AND BRANDNER, G., *Biochim. Biophys. Acta*, **60**, 51 (1962).
4. BARBER, G. A., *Biochemistry*, **1**, 463 (1962).
5. HARBORNE, J. B., "Biochemistry of Phenolic Compounds," New York, Academic Press, 1964.
6. ALSTON, R. E., in "Biochemistry of Phenolic Compounds" (HARBORNE, J. B., EDITOR), p. 171, New York, Academic Press, 1964.

E. SONDHEIMER

FLEMING, A.

Alexander Fleming (1881–1955), a British bacteriologist, discovered lysozyme, a protein enzyme (in tears, nasal secretions and egg white) which brings about the lysis of certain bacteria. Later, his outstanding discovery was penicillin, which led to his being awarded a share in the Nobel Prize in physiology and medicine in 1945, with Ernst Boris Chain and Howard Walter Florey.

FLOWER PIGMENTS.

See CAROTENOIDS; CHEMOTAXONOMY; FLAVONOIDS AND RELATED COMPOUNDS; PIGMENTS, PLANT.

Reference

1. CLEVENGER, S., "Flower Pigments," *Sci. Am.* **210**, 84–92 (June 1964).

FLUORESCENCE AND FLUOROMETRY

Fluorescence refers to luminescence of a substance following excitation by absorption of any form of electromagnetic radiation. The term is understood to embrace the range of wavelengths from X ray to near infrared. (X ray fluorescence, beyond the scope of this article, is useful primarily for elemental analysis.) This article is limited to the wavelength range from about 1900 Å in the ultraviolet to 1.2 μ in the infrared. Absorption of radiation of these wavelengths by a molecule involves elevation of an electron to an excited state, usually either the singlet (F) or indirectly to the triplet (P). Direct return from the singlet state (fluorescence) or from the triplet state (phosphorescence) to the ground state (N) involves emission of a photon, which because of loss of energy in radiationless transitions before return is generally of longer wavelength than the absorbed photon. Fluorescence takes place within about 10^{-8} second of absorption of the photon, whereas the related phenomenon, phosphorescence, may require several seconds, or even longer.

Fluorometry refers to the relative measurement of the efficiency of various wavelengths of absorbed light to cause fluorescence (activation spectrum), the relative measurement of the intensity of emitted light as a function of its wavelength (emission spectrum), and the measurement of the probability that an absorbed photon will generate an emitted photon (quantum efficiency). Such measurements have been utilized extensively in theoretical chemistry and physics and for the identification of compounds[1].

Fluorometry also refers to the quantitative determination of the concentration of a compound by the simple comparison of the intensity of its fluorescence with that of a known concentration of the same compound. The specificity and sensitivity of such measurements are such that analysis at concentrations of one part in 10^{11} is not uncommon. For this reason, fluorometry is widely used in trace metal analyses, biochemical analyses and medical laboratory analyses, and fluorescent dyes have achieved great popularity as tracers, particularly in oceanography[2]. A central library and clearing house for information on the analytical uses of fluorescence has now been established[3].

The fundamental relationship defining fluorescence in dilute solution may be expressed as:

$$F = 2.3 I_q E C D \phi$$

where F is the total fluorescence intensity, in quanta per second in all directions; I_q is the intensity of exciting light, in quanta per second; E is the molecular extinction coefficient as derived from spectrophotometric measurement; C is the concentration of fluorescent material; D is related to sample cell geometry and represents effective optical depth; ϕ is the quantum efficiency of the material under study.

There are two basic types of fluorometers, the filter fluorometer and the spectrofluorometer.

The *filter fluorometer* is widely used for quantitative analysis. Referring to the above equation, a filter fluorometer consists of: (1) a sensitive light detector which receives a fixed fraction of F, the light emitted by the sample; (2) a stable light source supplying exciting energy I_q; (3) a sample holder of fixed geometry; (4) filters to select the desired excitation and emission wavelengths. Calibration is established by use of a standard. Sensitivity may be increased by increasing light detector sensitivity or increasing the intensity of the light source. Specificity is high since the instrument will respond only to material with appreciable extinction coefficient at the chosen exciting wavelength and with appreciable quantum efficiency at the chosen band of light to which the light detector is sensitive.

The *spectrofluorometer* has similar components except the excitation and emission wavelengths are selected by monochromators, and generally it is used with an X-Y recorder to record wavelength *vs* fluorescence intensity. In spectrofluorometers, there are two subtypes, uncorrected and corrected. In the uncorrected spectrofluorometer, no provision is made for controlling the intensity of exciting light as the exciting wavelength is varied or for controlling the response of the photodetector as the emission wavelength is scanned. Such instruments, therefore, introduce artifacts into the spectra which are individual to the particular instrument used. They are in common use, and may be employed for limited qualitative (identification) data on compounds and for quantitative analysis with the convenience of dialing wavelengths. A number of commercially available filter fluorometers and uncorrected spectrofluorometers are discussed in reference 2.

The corrected (or absolute) spectrofluorometer may be used in the same way as the uncorrected instruments, but in addition, it eliminates instrumental artifacts, thereby permitting comparison of excitation spectra with absorption spectra (obtained with a spectrophotometer) and the determination of quantum efficiency. Such spectra are useful in gaining information on molecular structure.[1,2,4,5] Two absolute spectrofluorometers have recently become commercially available.[6,7]

A valuable use of corrected spectrofluorometers and, to a limited extent, of uncorrected spectrofluorometers lies in the determination of the purity of a compound without reference to a pure sample. In a molecule which has only one absorbing structure, or where all absorbing centers are electronically coupled, the excitation spectrum should coincide with the absorption spectrum, and the shape of the emission spectrum and the quantum efficiency should be independent of the exciting wavelength.

References

1. Pringsheim, P., "Fluorescence and Phosphorescence," New York, Interscience Publishers, 1949.
2. Udenfriend, S., "Fluorescence Assay in Biology and Medicine," New York, Academic Press, 1962.
3. G. K. Turner Associates; write to Dr. R. E. Phillips, 2524 Pulgas Avenue, Palo Alto, California 94303.
4. Weber, G., and Teale, F. W. J., *Trans. Faraday Soc.*, **53**, 646 (1957).
5. Parker, C. A. and Rees, W. T., *Analyst*, **85**, 587 (1960).
6. Available commercially from Perkin-Elmer Corporation, Norwalk, Connecticut; Slavin, W., Mooney, R. W., Palumbo, R. W., *J. Opt. Soc. Am.*, **51**, 93 (1961).
7. Available commercially from G. K. Turner Associates, 2524 Pulgas Avenue, Palo Alto, California 94303; Turner, G. K., *Science*, **146**, 183 (1964).

G. K. Turner

FLUORIDE METABOLISM

The element fluorine with atomic number 9 and atomic weight 19 is the most electronegative and reactive of all elements, with the result that elemental fluorine does not occur as such in nature, but only in chemical combinations. Fluoride is the seventeenth most abundant element in the earth's crust, and it has an ubiquitous distribution in rocks, soils, waters, and in plant and animal tissues. It is doubtful that any unrefined natural substance is completely devoid of fluoride. It occurs in varying quantities in all water supplies and foods. The average daily American diet, exclusive of water, probably contains 0.2 ± 0.3 mg of fluoride. Communal water supplies in the United States vary in fluoride content from a trace to over 5 ppm. About 292,000 people regularly use water with a fluoride content of 3 ppm or more, and water is the most variable natural source of fluoride intake.

Fluoride has not been shown to be essential for life since otherwise nutritionally adequate diets which are devoid of fluoride or which might be low enough to cause recognizable deficiency symptoms have not been prepared (see also MINERAL REQUIREMENTS). Fluoride is concentrated in the skeletal and dental hard tissues as fluorapatite, and it occurs only in small regulated concentrations in body fluids and soft tissues. Once the ion becomes part of a calcified tissue, it probably cannot be removed without resorption of the entire structure of the mineral phase. The previous fluoride exposure, skeletal fluoride concentration, and the age of the individual, influence the uptake of fluoride by the skeletal tissues and thus affect the further retention of additional amounts of fluoride[1]. There is evidence that an individual on a long-term relatively constant fluoride intake reaches an equilibrium between intake and retention at which time the fluoride uptake by the skeletal tissue is reduced and the concentration of fluoride in the urine, coincidentally, approximates that of the drinking water[2]. Thus an individual in a community with 1 ppm of fluoride in the drinking water can be expected eventually to excrete urine containing 1 ppm of fluoride.

Fluoride is readily absorbed from the gastrointestinal tract and there is only a slight transitory effect of ingested fluoride on the plasma fluoride concentration. Radioactive fluoride (F^{18}) tracer studies with human subjects have indicated that not more than 10% of an ingested fluoride dose is contained in the circulating plasma volume at one time. Once the fluoride is absorbed into the circulation there is a rapid adjustment of body fluid fluoride concentration by virtue of the uptake of fluoride by skeletal tissues and by urinary excretion.

The blood plasma fluoride contents of residents of five Midwest communities in which the communal waters were different in fluoride contents have indicated a constancy of plasma fluoride concentration (0.14 + 0.005 SE to 0.19 ± 0.0085 SE ppm) when the communal water varied between 0.15 and 2.5 ppm of fluoride[3]. When the drinking water contained 5.4 ppm of fluoride, the regulatory mechanisms did not operate perfectly and a slight but statistically significant rise in plasma fluoride content to 0.26 ± 0.0073 ppm was observed. The plasma fluoride contents of hospitalized individuals with various diseases were not found to be different from those of normal individuals of the same area.

Fluoride is being currently used in the treatment of metabolic bone diseases on the premise that fluoride promotes positive calcium balance and causes remineralization of the skeleton. The patients treated were not poisoned although their daily intake for 12–38 weeks was 50–75 mg of fluoride ion. In some cases, the plasma fluoride concentrations rose to 1.8 ppm but in no case was it increased in proportion to the fluoride intake[4].

Rats fed a diet containing 0.5 ppm fluoride had a mean plasma fluoride content of 0.17 ppm ± 0.053 (SD), but the plasma of animals receiving food containing 100 ppm fluoride was increased to 0.47 ppm ± 0.199 (SD). The muscle fluoride contents of the two groups of animals were, however, not different being 0.20 ± 0.082 and 0.21 ± 0.063 ppm, respectively. Nephrectomy resulted in an increase in the muscle fluoride content of the animals receiving the high fluoride diet, but it had no effect on the concentration of fluoride in the muscles of the animals of the control group, even though these animals received the high fluoride-containing food after they were nephrectomized.

Animals raised on widely varying fluoride intakes and then deprived of food for a week or given PARATHYROID extract sufficient to increase significantly the plasma CALCIUM content did not exhibit an alteration in the plasma fluoride content.

The kidney is an important organ in the regulation of the fluoride content of body fluids. The clearance of fluoride by the kidney exceeds chloride clearance by many fold and increases with urine volume. Fluoride clearance in normal human subjects is less than creatinine clearance and indicates that 51–63% of the fluoride in glomerular filtrate is resorbed[5]. Urinary fluoride excretion occurs by GLOMERULAR FILTRATION with a variable amount of tubular resorption (see RENAL TUBULAR FUNCTIONS). The dog kidney has been shown to concentrate the fluoride from the plasma by a factor of 10–20 times in producing urine, and this kidney effectively functions even when the plasma fluoride is elevated to more than 10 times the normal concentration.

Ingestion of a high amount of fluoride by rats was not found to interfere with the utilization of dietary calcium for BONE FORMATION and tooth formation[6] determined in experiments in which the dietary calcium was labeled with radioactive calcium. A high fluoride content of bone, enamel, and dentin was found to reduce the amount of calcium dissolved from the tissues by weakly acidic buffer solutions.

Fluoride, unlike chloride, occurs in intracellular fluid but at lower concentrations than in extracellular fluid. Approximately 75% of the blood fluoride is in the plasma. Less than 5% of the plasma fluoride is transported in protein bound form.

Solutions containing as much as 40 ppm of fluoride have been shown not to influence respiration of the rat gastrointestinal tissue, whereas 200 ppm of fluoride instantly decreased oxygen utilization. It thus appears that the oxidative processes would not be affected by far higher concentrations of fluoride than could conceivably result from the use of fluoridated water.

The growth of cultures of HeLa cells and human esophageal cells was not influenced by concentrations of fluoride of up to 4.5 ppm in the media. The generation time and protein synthesis by the HeLa cells was not significantly altered by up to 10 ppm of fluoride. DNA synthesis in rapidly growing rat metacarpal bone in tissue culture was not affected by up to 20 ppm of fluoride, whereas concentrations in excess of 10 ppm were required to reduce collagen synthesis. These results indicate that it is unlikely that fluoridation of water can lead to detrimental biological effects on cells since the concentrations of fluoride required to produce the effects mentioned above are far above the mean plasma fluoride of 0.26 ± 0.0073 ppm of individuals consuming 5.4 ppm of fluoride in their drinking water.

Dental decay (see CARIES, DENTAL) is less prevalent in regions where the drinking water contains a reasonable amount of fluoride as compared to those areas where only traces of fluoride are present. The full protection to the teeth is obtained when the water supply contains approximately 1.0 ppm of fluoride ion which results in a 50–60% reduction in the incidence of dental caries in children raised on this water. The benefits and safety of well-regulated fluoridation programs have been documented by the results of trials in this country, some of which have been in operation for 20 years, and in other countries.

Skeletal fluorosis which takes the form of hypercalcification of bones and tendons occurs from excessive fluoride intake in India. In the United States, roentgenographic evidence of increased skeletal calcification, without related symptoms, occurred in 10–15% of persons using water containing 8 ppm fluoride.

Epidemological, physical, and laboratory studies have been reported on persons using considerably more fluoride, over long periods of time, than the optimum amount in water. These reports have not revealed any increase in disease attack rates, or other abnormal findings, as a consequence of fluoride ingestion from water, with the exception of mottled enamel. The most critical index of excessive fluoride intake in a population is the appearance of mottled enamel of the permanent teeth of those persons who use the water during the period of tooth formation. Mottled teeth exhibit chalky white areas on the enamel surfaces and these may become secondarily stained to a brown color. Fortunately, the full dental benefits of water-borne fluoride can be obtained without producing unesthetic degrees of mottled enamel.

References

1. LARGENT, E. W., "Fluorosis," pp. 22–56, Columbus, Ohio, Ohio University Press, 1961.
2. ZIPKIN, I., LIKINS, R. C., MCCLURE, F. S., AND STEERE, A. C., "Urinary Fluoride Levels Associated with Use of Fluoridated Waters," *Public Health Rept.*, **71**, 767–772 (1956).
3. SINGER, L., AND ARMSTRONG, W. D., "Regulation of Human Plasma Fluoride Concentration," *J. Appl. Physiol.*, **15**, 508–510 (1960).
4. ARMSTRONG, W. D., SINGER, L., ENSINCK, J., AND RICH, C., "Plasma Fluoride Concentrations of Patients Treated with Sodium Fluoride," *J. Clin. Invest.*, **43**, 555 (1964).
5. CARLSON, C. H., ARMSTRONG, W. D., AND SINGER, L., "Distribution and Escretion of Radiofluoride in the Human," *Proc. Soc. Exptl. Biol. Med.*, **104**, 235–239 (1960).
6. SINGER, L., DALE, M. D., AND ARMSTRONG, W. D., "Effects of High Fluoride Intake on Utilization of Dietary Calcium and on Solubility of Calcified Tissues," *J. Dental Res.*, **44**, 582–586 (1965).

W. D. ARMSTRONG AND LEON SINGER

FOLIC ACID

Occurrence. The formula for folic acid, also known as pteroylglutamic acid, is given in Fig. 1. This vitamin contains three structural units: a pteridine compound, *p*-aminobenzoic acid, and glutamic acid. The formulas of two related compounds, pteroic acid and *p*-aminobenzoylglutamic acid, are also indicated in Fig. 1. Many derivatives of folic acid occur naturally. Pteroyltriglutamic acid is known to occur in materials of both animal and bacterial origin, and pteroylheptaglutamic acid also has been found in yeast. These compounds differ from folic acid in that they contain three and seven glutamic acid residues, respectively. The coenzyme form of the vitamin is 5,6,7,8-tetrahydrofolic acid or 5,6,7,8-tetrahydropteroyltriglutamic acid; however these compounds are so labile to oxidative degradation that they do not occur naturally in significant amounts. Various derivatives of these coenzyme

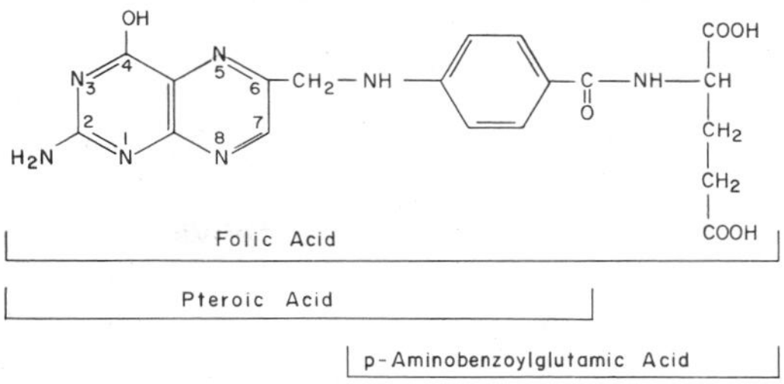

FIG. 1. Structure of folic acid.

forms are more stable and therefore are found widely distributed in nature. The best known of these derivatives is 5-formyltetrahydrofolic acid, also known as citrovorum factor (since it was discovered as a growth factor for *Leuconostoc citrovorum*), leucovorin, and folinic acid. Recently, 5-methyltetrahydrofolic acid has also been shown to occur in significant quantities in human blood serum. This compound provides the methyl groups for the biosynthesis of methionine. Rabinowitz and Himes[1] have claimed that most, if not all, of the folic acid compounds of bacteria can be accounted for as pteroyltriglutamic acid compounds. Although no quantitative assessment has been made of the amounts of 5-formyl and 5-methyl derivatives of tetrahydropteroyltriglutamic acid that might be present in various natural products, it seems possible, and even likely, that significant quantities of these compounds occur naturally. For a more detailed discussion of the occurrence of some of these compounds, the reader is directed to a review by Jukes and Stokstad[2] (see also PTERIDINES).

Biosynthesis of Folic Acid Compounds. 7,8-Dihydrofolic acid, and not folic acid, is the product formed enzymatically from precursors. The enzymatic reactions that account for the formation of dihydrofolic acid are shown in Fig. 2. The pteridine precursor has been established definitely as 2-amino-4-hydroxy-6-hydroxymethyl-7,8-dihydropteridine. This compound can be converted enzymatically in the presence of ATP to either dihydropteroic acid or dihydrofolic acid, depending on whether *p*-aminobenzoic acid or *p*-aminobenzoylglutamic acid is provided as substrate. These two compounds are not used equally well as substrate in all enzymatic systems tested. In the enzyme systems from *Lactobacillus plantarum*, *Neurospora crassa* and baker's yeast, the two are used approximately equally well; whereas systems from *Escherichia coli*, *Aerobacter aerogenes*, and *Micrococcus lysodeikticus* utilize *p*-aminobenzoic acid many times more effectively and the system from a species of *Corynebacterium* will not utilize *p*-aminobenzoylglutamic acid at all, but does use *p*-aminobenzoic acid very effectively. Thus, if *p*-aminobenzoylglutamic acid can be formed by the microorganisms, it can be used to make folic acid compounds (with the single possible exception of *Corynebacterium*). It appears more likely that *p*-aminobenzoic acid is the true substrate *in vivo*, since in no instance

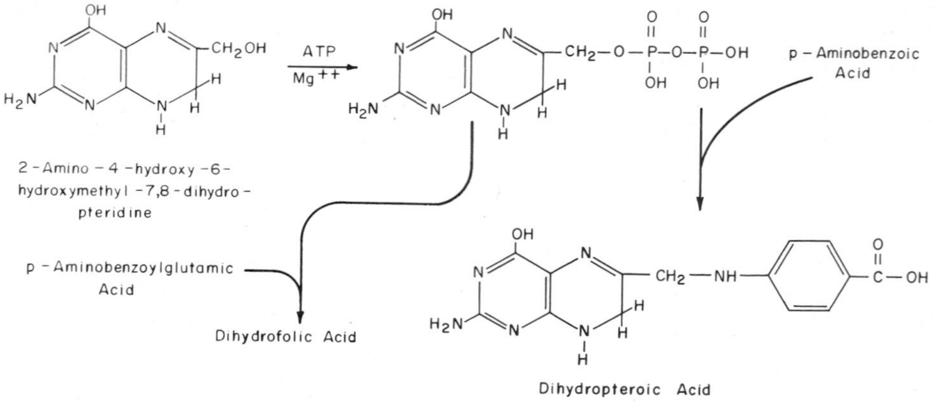

2-Amino-4-hydroxy-6-hydroxymethyl-7,8-dihydropteridine

ATP, Mg^{++}

p-Aminobenzoic Acid

p-Aminobenzoylglutamic Acid

Dihydrofolic Acid

Dihydropteroic Acid

FIG. 2. Enzymatic reactions that lead to the biosynthesis of dihydropteroic acid and dihydrofolic acid.

can an enzymatic system from any organism tested utilize *p*-aminobenzoylglutamic acid more effectively than *p*-aminobenzoic acid. Additional support for this view is the evidence to be discussed below which shows that the enzyme that catalyzes the formation of dihydrofolic acid from dihydropteroic acid and glutamic acid is widely distributed among a variety of microorganisms.

The conversion of 2-amino-4-hydroxy-6-hydroxymethyl-7,8-dihydropteridine to either dihydropteroic acid (in the presence of *p*-aminobenzoic acid) or dihydrofolic acid (in the presence of *p*-aminobenzoylglutamic acid) requires the presence of ATP. The enzyme system (from *E. coli*) that catalyzes this transformation has been separated into two protein fractions, both of which are necessary for synthesis of the product. One of the protein fractions catalyzes the formation of an ATP-dependent "activated" pteridine compound from 2-amino-4-hydroxy-6-hydroxymethyl-7,8-dihydropteridine. This "activated" compound is heat stable and is converted to dihydropteroic acid in the presence of the second protein fraction and *p*-aminobenzoic acid. The latter reaction is not dependent on the presence of ATP. The monophosphate and pyrophosphate esters of the pteridine compound were made synthetically and tested in the system. The resulting observations showed that the second protein fraction could use the pyrophosphate ester as substrate for the formation of dihydropteroic acid in place of the 6-hydroxymethyldihydropteridine compound and ATP, but that the monophosphate ester could not be converted to dihydropteroic acid even when both protein fractions and ATP were supplied. These observations indicate strongly that the pyrophosphate ester is an intermediate in the biosynthetic pathway, although it must be emphasized that this compound has not yet been synthesized enzymatically in amounts large enough to identify as a product. Details of the work described above can be found in research papers by Shiota *et al.*[3] and by Weisman and Brown.[4] This subject is reviewed in greater detail in reference 5.

The effect of SULFONAMIDES as inhibitors of folic acid production from *p*-aminobenzoic acid by bacterial cells has been known for several years. The discovery of the enzymes that are necessary for the incorporation of *p*-aminobenzoic acid into folic acid compounds provided the opportunity to study the effects of sulfonamides on the enzymatic synthesis of these compounds.[3,6,7] It was found that in *E. coli*, and probably also in *Viellonella*, sulfonamides compete with *p*-aminobenzoic acid as substrate for the enzyme that converts *p*-aminobenzoate to dihydropteroate; *i.e.*, in the presence of the sulfonamide, a product is formed that is apparently an analogue of dihydropteroic in that it contains a sulfonamide residue in place of the normal *p*-aminobenzoate residue.[6] Only indirect evidence exists for the formation of such an analogue, since not enough of it can be synthesized enzymatically to do proper identification studies.

The experimental observations described above strongly suggest that dihydropteroic acid is an intermediate in the biosynthesis of folic acid compounds. If this is true, the enzyme that catalyzes the formation of dihydrofolic acid from dihydropteroic acid and glutamic acid should be present in microorganisms. This enzyme has been found in extracts of a number of microorganisms, including *Escherichia coli*, *Corynebacterium* species, *Saccharomyces cerevisiae*, *Bacillus megaterium*, *Neurospora crassa*, *Streptococcus faecalis*, *Mycobacterium phlei*, and *Mycobacterium avium.*[8] These observations provide additional support for the view that dihydropteroic acid is an intermediate in the biosynthetic pathway.

The enzymatic reaction sequence for the addition of the three glutamic acid residues to form pteroyltriglutamic acid compounds has been shown in *E. coli* extracts to be as follows[8]:

(a) Dihydropteroic Acid + Glutamic Acid $\xrightarrow[Mg^{++}, K^{+}]{ATP}$ Dihydrofolic Acid

(b) Dihydrofolic Acid $\rightarrow$ Tetrahydrofolic Acid

(c) Tetrahydrofolic Acid + Glutamic Acid $\xrightarrow[Mg^{++}]{ATP}$ Tetrahydropteroyldiglutamic acid

(d) Tetrahydropteroyldiglutamic Acid + Glutamic Acid $\xrightarrow{ATP}$ Tetrahydropteroyltriglutamic Acid

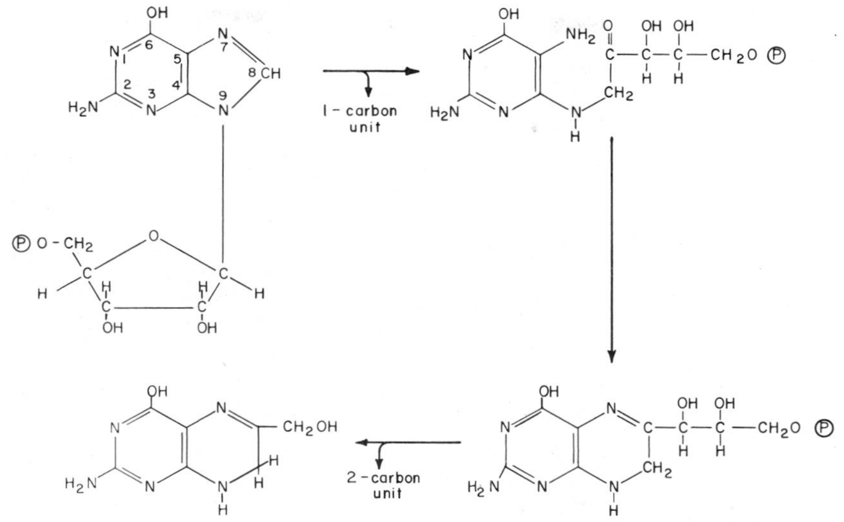

FIG. 3. Tentative biosynthetic pathway for the formation of 2-amino-4-hydroxy-6-hydroxymethyl-7,8-dihydropteridine from guanine nucleotide. Ⓟ represents a phosphate group.

Biosynthesis of the Pteridine Ring. A number of experimental observations have indicated that isotopically labeled purines can be converted to pteridine compounds by whole organisms. However it is only recently that a soluble enzyme system has been obtained that will carry out this transformation.[9] Extracts of *Escherichia coli* can convert guanine nucleotides (but no other purine compounds) to a pteridine compound that can then be utilized for the enzymatic synthesis of either dihydropteroic acid or dihydrofolic acid. Presumably this pteridine compound is 2-amino-4-hydroxy-6-hydroxymethyl-7,8-dihydropteridine, although the latter has not yet been identified directly as a product of the action of enzymes on guanine nucleotides. During this enzymatic transformation, carbon 8 (see Fig. 3 for numbering system) of guanine is lost and the ribose of the nucleotide provides the extra carbon atoms needed for completion of the pteridine ring. These facts were established experimentally with the use of substrates labeled appropriately with C^{14}. It seemed reasonable to think that during the transformation of the guanine nucleotide to a pteridine that could be used for the synthesis of folic acid compounds, an intermediate pteridine compound might be formed that would have a trihydroxypropyl side chain at the 6-position of the pteridine ring (this compound is shown in Fig. 3). The 3-carbon side chain would be derived from what originally were carbons-3, 4, and 5 of the ribose portion of the guanine nucleotide. This pteridine compound was synthesized and found to substitute very effectively for guanine nucleotide for the formation of dihydropteroic acid. A pathway for the enzymatic formation of 2-amino-4-hydroxy-6-hydroxymethyl-7,8-dihydropteridine that appears to be consistent with the experimental observations is shown in Fig. 3. It should be emphasized that this pathway is still tentative and is subject to possible revisions after further investigations.

References

1. RABINOWITZ, J. C., AND HIMES, R. H., *Federation Proc.*, **19**, 963 (1960).
2. JUKES, T. H., AND STOKSTAD, E. L. R., *Physiol. Rev.*, **28**, 51 (1948).
3. SHIOTA, T., DISRAELY, M. N., AND MCCANN, M. P., *J. Biol. Chem.*, **239**, 2259 (1964).
4. WEISMAN, R. A., AND BROWN, G. M., *J. Biol. Chem.*, **239**, 326 (1964).
5. BROWN, G. M., AND REYNOLDS, J. J., *Ann. Rev. Biochem.*, **32**, 419 (1963).
6. BROWN, G. M., *J. Biol. Chem.*, **237**, 536 (1962).
7. WOLF, B., AND HOTCHKISS, R. D., *Biochemistry*, **2**, 145 (1963).
8. GRIFFIN, M. J., AND BROWN, G. M., *J. Biol. Chem.*, **239**, 310 (1964).
9. REYNOLDS, J. J., AND BROWN, G. M., *J. Biol. Chem.*, **239**, 317 (1964).

GENE M. BROWN

FOLIC ACID COENZYMES (FOLATE COENZYMES)

Derivatives of pteroic acid (N-4[(2-amino-4-hydroxy-6- pteridyl)-methyl]-aminobenzoic acid), in which the carboxyl group of paraminobenzoic acid (PABA) is combined with one or several glutamic acid residues, are generically termed "folates" (see also FOLIC ACID). The latter compounds, carrying two to seven glutamyl residues as γ-peptides are the so-called conjugates.

The best known conjugates are the triglutamate (teropterin) and the heptaglutamate (vitamin B_c conjugate). Frequently it is assumed that these conjugates are the parent substances of the true cofactors, acting in metabolism as acceptors, donors or transformers of the biological one-carbon (C_1) unit.

In certain *Clostridia*, the triglutamate has been found to be the only folate; in others, only polyglutamyl derivatives are able to catalyze folate-specific enzymic reactions (serine/glycine transformation). There has even been evidence for natural conjugates, the side chain of which contains other amino acids besides glutamic acid. Also in higher organisms, polyglutamates may occur as native cofactors, since generally during their isolation or extraction no attention has been paid to the action of extra- or intra-cellular conjugases (γ-glutamyl peptidases). Such enzymes are found in several microorganisms, in serum, and in pancreas or kidney. Thus, in human serum, for example, folates seem to be present almost exclusively as monoglutamates. Chemically, the conjugates differ little from the parent compound, pteroylglutamic acid (PteGlu). They may be separated and analyzed by various chromatographic procedures, combined with microbiological assays. Generally, it may be said, that pteroyl monoglutamates and pteroic acid are growth substances for *Streptoccocus faecalis* R, whereas *Lactobacillus casei* uses monoglutamates as well as short-chain ($n \leq 2$) conjugates.

The problem of the conjugates being still unsolved, in the following only the non-conjugated compounds will be discussed, inasmuch as they show identical chemical behavior and were used successfully in all *in vitro* studies—at least as a model of the true coenzyme. Nevertheless, one may still keep in mind that the given effects actually may be due to more complex substances.

At the beginning, microbiological and pharmacological studies had demonstrated the role of PteGlu (and PABA) in the transfer of C_1 units. The first—if indirect—chemical evidence of this was the isolation of rhizopterin (10-formylpteroic acid) from culture media of the mold *Rhizopus*. However, enzymatic investigations soon showed that PteGlu itself has only indirect biological activity. As generally true in the B group, the vitamin isolated from natural sources, is only a precursor of the actual cofactor. In contrast to other vitamins of the B complex, however, the biologically active form of PteGlu is neither a phosphorylated nor a nucleotidic derivative, but 5, 6, 7, 8-tetrahydropteroylglutamic acid ($PteGluH_4$), shown in (I), and numbered accordingly (see also VITAMIN B GROUP).

So far, this very labile compound has been usable *in vitro* as the cofactor in all known enzymic folate-dependent reactions. In the molecule, the pyrazine ring is reduced. The hydrogen uptake occurs in two steps, passing through an intermediate dihydropteroylglutamic acid ($PteGluH_2$). Chemically, this reduction is accomplished by H_2/Pt, by dithionite, or by boranate. In alkaline milieu, the catalytic hydrogenation stops at the dihydro stage; in neutral solution or in glacial acetic acid, $PteGluH_4$ is obtained. On reduction, several additional features are introduced into the molecule, which are of prime relevance to its biochemical activity and chemical behavior. These are the weakly basic imino groups in nitrogens-5 and 8, and—in addition to the L-glutamyl residue—a second center of asymmetry at carbon-6 (asterisk). Thus, the synthetic tetrahydro compound (and also 5, 6-$PteGluH_2$!) occurs in two diastereomer antipodes, which can be resolved by the usual means. $PteGluH_4$ ($\lambda_{max} = 298$ mμ) is a quite unstable molecule. In alkaline solution, it is partly autoxidized to PteGlu, partly to unknown cleavage products. In acidic aqueous solution, it is decomposed by cleaving at the methylene bridge (carbon-9), yielding *p*-aminobenzoylglutamic acid, which may be used for quantitative assay. Reducing agents such as ascorbic acid or mercaptoethanol stabilize the biological activity. However, chemical changes are not excluded.

The biological reduction of PteGlu is catalyzed by the $NADP^+$-specific enzyme, folate reductase (folate and dihydrofolate : NAD(P)-oxidoreductase, 1.5.1.3/4). In this case, the (−) L-diastereomer ($\alpha_D = -16.9°$) is the product. The enzyme, which is found in avian and mammalian systems, and also in microorganisms, has the remarkable property of reducing PteGlu as well as $PteGluH_2$. It is therefore able to transform PteGlu to $PteGluH_2$ and the latter to $PteGluH_4$. Thus far,

Tetrahydropteroic acid

(I) Tetrahydropteroylglutamic acid (R = OH) $PteGluH_4$
Conjugates: R = (γ-L-glutamyl)$_n$, $n = 1 \ldots 6$

it has not been possible to separate the two activities. In several species of insects, however, specific dihydropterine reductases have been found; in *Cl. sticklandii*, a pteroylglutamate reductase was found which reduces PteGlu only to the 7,8-$PteGluH_2$ stage. Also in the systems from higher organisms, the first reduction product is the 7,8-isomer, symmetric in the pteridine moiety. Chemically prepared $PteGluH_2$ is identical with the natural 7,8-compound. Both give, upon stereospecific enzymic reduction, fully active (−)L-$PteGluH_4$ in 100% yield. This could not be the case, if the center at carbon-6 were reduced first. Folate reductase is blocked very

$$\text{PteGlu} \xrightarrow[(\text{NADPH}+\text{H}^+)]{2(\text{H})} 7,8\text{-PteGluH}_2 \xrightarrow[(\text{NADPH}+\text{H}^+)]{2(\text{H})} 5,6,7,8\text{-PteGluH}_4$$

specifically and pseudo-irreversibly by folate antagonists (Aminopterin = 4-desoxy-4-amino-PteGlu; Amethopterin = 4-desoxy-4-amino-10-methyl-PteGlu). This inhibition is the basis of the therapeutic use of antifolates in leukoses (see CANCER CHEMOTHERAPY and LEUKOCYTES).

According to our present knowledge, only nitrogens-5 and 10 and the ethylene group (6 and 9) flanked by them (bold-face type) are essential for the catalytic activity. On the other hand, the molecule cannot be varied much without being completely thrown out of the biological gear. Glycosidic compounds at nitrogen-8 have been synthesized; however, they seem to have no biological significance. Recently in insects, nucleotidic (?) pterine compounds have been isolated. There is nothing known yet of their metabolic role as cofactors or as intermediates in the biogenesis of pteroyl derivatives.

Coenzyme Forms Containing Single-Carbon Unit. At nitrogens-5 and 10, substitutions can occur. Interestingly, several of the reduced PteGlu derivatives thus substituted have later been rediscovered as cofactors of the C_1 cycles. The ethylenediamine grouping can be regarded as the active center of the cofactors, in a manner still somewhat obscure. So far, from living systems, there have been isolated six $PteGluH_4$-C_1 complexes. Three of them carry a C_1 unit at the oxidation level of formic acid, two of formaldehyde, and one of methanol. Table 1 demonstrates these relationships. Since in animal, plant, and bacterial metabolism, coenzymes containing PteGlu control the transfer and the transformation of C_1 units, the term "coenzyme F" (CoF) was coined early, and is still used.

As seen from Table 1, $PteGluH_4$ is acceptor coenzyme of the C_1 unit in the transformation of serine to glycine, in the splitting off of the formyl group of several formyl compounds, and in the biological activation of formic acid. It is the donor coenzyme of formylation, hydroxymethylation, and methylation reactions. Beyond that, the different oxidation levels of the folate-bound C_1

Table 1. Biochemical Activities of Folic Acid Coenzymes

One-carbon Unit	Coenzyme	Reaction[a]
Formyl residue (CHO—)	10-Formyl-$PteGluH_4$ (II)	Formate activation: Folate-H_4 + HCOOH + ATP ⇌ CHO-folate-H_4 + ADP + P_i Deacylation: CHO-folate-H_4 + H_2O → HCOOH + folate-H_4 Formylation of AICAR: CHO-folate-H_4 + AICAR ⇌ CHO-AICAR + folate-H_4
	5-Formyl-$PteGluH_4$ (III) (citrovorum factor)	Formylation of glutamate: CHO-folate-H_4 + Glu ⇌ CHO-Glu + Folate-H_4
	5, 10-Methenyl-$PteGluH_4$ (IV) (anhydrocitrovorum factor)	Formylation of GAR: CHO-folate-H_4 + GAR → CHO-GAR + Folate-H_4
Formimino residue	5-Formimino-$PteGluH_4$ (V)	Formimination of glycine and glutamate: CHNH-folate-H_4 + Gly (resp. Glu) ⇌ CHNH-Gly (resp.-Glu) + Folate-H_4
Hydroxymethyl residue ($HOCH_2$—)	5, 10-Methylene-$PteGluH_4$ (VI) (or 5(10)-CH_2OH-$PteGluH_4$)	Serine/glycine-conversion (transhydroximethylase, pyridoxal phosphate dependent): CH_2OH-Folate-H_4 + Gly ⇌ Ser + Folate-H_4 Thymidylate synthetase: CH_2OH-Folate-H_4 + dUMP → TMP + Folate-H_2
Methyl residue (CH_3—)	5-Methyl-$PteGluH_4$ (VII) (serum *L. casei* factor, prefolic A)	Methionine synthetase (vitamin B_{12} dependent): CH_3-Folate-H_4 + Homocysteine → Met + Folate-H_4 and other methylation reactions (?)

[a] AICAR = 4(5)-amino-5(4)-imidazole carboxamide ribotide; GAR = glycinamide ribotide.

units are interconnected by oxidoreductions. The folate-catalyzed conversions of C_1 fragments can be further grouped into:

(a) Reactions in which an unchanged C_1 unit is activated at the cofactor and transferred (transfer reactions);

(b) Reactions in which the oxidation level of the C_1 unit is first changed while bonded to the cofactor, and then transferred. In this case, the oxidation energy will be stored in an energy-rich compound (activated formaldehyde, activated formate). Here, two (or more) folate enzymes act together, a dehydrogenase and a transferase.

The various enzymic events are summarized in the reaction chains of Fig. 1, from which follow some long-known relationships.

Except for the functions of vitamin B_{12}—which cooperates in some folate reactions—the functions of folate cofactors, of "CoF", make up the most complicated chapter in the metabolism of vitamin-dependent processes. Quite reasonably, therefore, some details are still insufficiently known at the present time.

Not all of the $PteGluH_4$ coenzymes could be obtained from biological material, since only the compounds substituted at nitrogen-5 are stable enough to survive the steps of isolation without reoxidation or cleavage. Their solutions, however, can be kept for years at $-20°C$, even without addition of ascorbate. With 1% ascorbate (pH 6), they may be stored at 0°C for some length of time. Yet, large amounts of ascorbate or other antioxidants destroy the compounds altogether. Without additional reagents, the coenzyme forms can be tested microbiologically using the "aseptic technique," which avoids changes by sterilization. Quite universally, the folate coenzymes (reduced pteroylglutamates) are growth factors for *Pediococcus cerevisiae* 8081 (*Leuconostoc citrovorum*).

$PteGluH_4$ *per se* has not yet been demonstrated with certainty in biological material. In studies with isolated enzyme systems, however, it serves as C_1 acceptor in the folate formylase reaction (10 - formyl - $PteGluH_4$ - ligase (ATP), 6.3.4.3), the formation of 10-formyl-$PteGluH_4$ concomitant with the splitting of ATP; the formimino transferases (2.1.2.4. and 2.1.2.5); and a formaldehyde-activating enzyme. The product of the formate activation, 10-formyl-$PteGluH_4$ (II), is specific substrate of AICAR-transformylase (2.1.2.3) (carbon-2 of purines; see PURINE BIOSYNTHESIS). Like all 10-substituted $PteGluH_4$ derivatives, it is extremely susceptible to oxidation. In pure form it is colorless ($\lambda_{max} = 260$ mμ). From natural sources and enzyme runs, forming 10-formyl-$PteGluH_4$, commonly only 10-formyl-$PteGluH_2$ was isolated—or 5,10-anhydroformyl-$PteGluH_4$ (see below) from acidified solutions. Impure 10-formyl-$PteGluH_4$ is formed chemically by catalytic reduction of 10-formyl-PteGlu or by formylation of $PteGluH_4$. If such solutions are heated anaerobically, the formyl group migrates from nitrogen-10 to nitrogen-5, and 5-formyl-$PteGluH_4$ (III, citrovorum factor) is the product, which was found first as a vitamin for *L. citrovorum* (see above). For a time, 5-formyl-$PteGluH_4$ ($\lambda_{max} = 282$ mμ) was supposed to be the main form of biological folate compounds. It is stable enough to withstand autoclaving without destruction, even in alkaline solutions. By way of the calcium salts, the synthetic mixture of diastereomers was resolved. The biologically active form is the (−)L-isomer ($\alpha_D = -15.1°$). The formyl shift (from the less basic nitrogen-10 to the more basic nitrogen-5, where under physiological conditions the group remains firmly bound) probably goes *via* the cyclic 5, 10-anhydroformyl compound [5,10-methenyl-$PteGluH_4$ (IV)]. This substance is obtained both from 5-formyl- and 10-formyl-$PteGluH_4$ in an acidic milieu (pH < 1.3). First, a normal salt (isoleucovorinate) is formed, which in turn hydrolyzes, yielding the less soluble betain. This "anhydroleucovorin" is a yellow compound ($\lambda_{max} = 355$ mμ), fluorescing strongly and characteristically, suggesting the formation of a coplanar imidazolinium ring. In alkaline solution, this ring is opened, yielding 10 - formyl - $PteGluH_4$ — strikingly against the thermodynamic slope. This is the best way to obtain pure 10-formyl-$PteGluH_4$.

As a cofactor form of $PteGluH_4$, citrovorum factor is relatively unimportant, the only specific enzymic reaction being the formylation of glutamate. Much more essential is *anhydroleucovorin* as cofactor, as substrate, and as product of several enzymatic processes. It is the C_1 donor

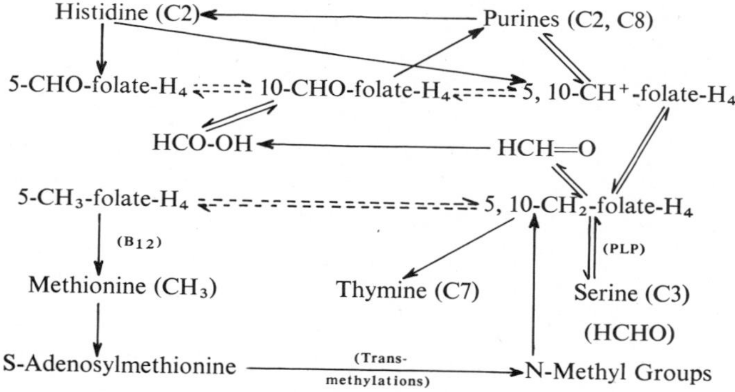

FIG. 1. Single-carbon unit transfer reactions.

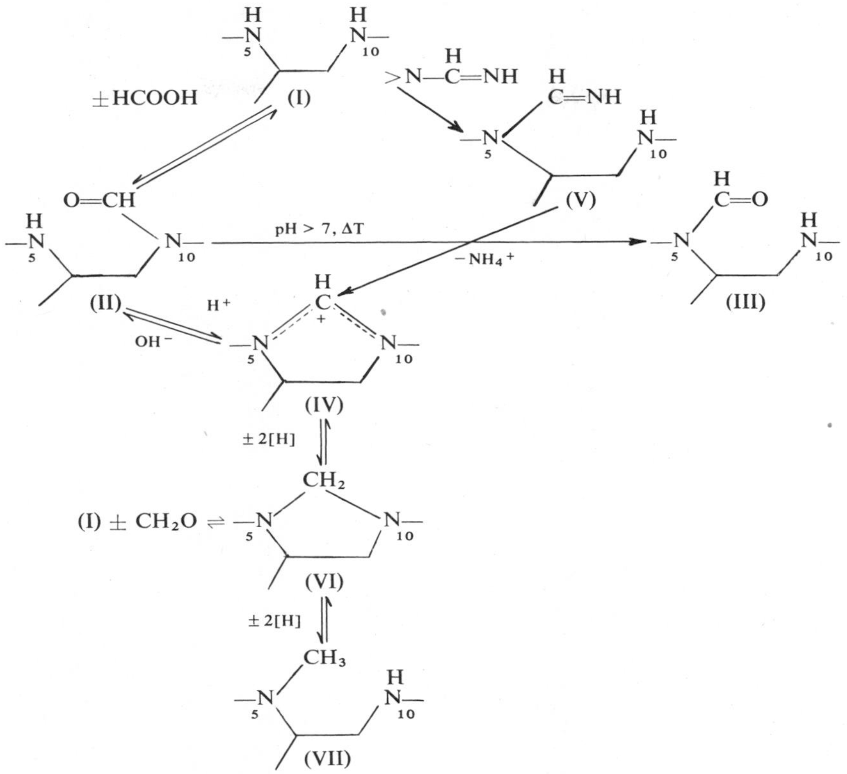

Fig. 2. Conversions of folate cofactors.

of GAR-transformylase (2.1.2.2) (carbon-8 of purines); the substrate of 5, 10-methylene-PteGluH$_4$ (NADP)-oxidoreductase (1.5.1.5) which reduces active formate to active formaldehyde [5, 10-methylene-PteGluH$_4$ (IV)]; and the product of the cyclodesaminase (4.3.1.4) and cyclohydrolase reactions. The latter is the enzymic analogue of the chemical ring opening reaction (IV $\rightarrow$ II). In the former, 5-formimino-PteGluH$_4$ (V)—as the triglutamate (?)—is cyclicized by splitting off ammonium ion. At pH 5–9 and 37°C this reaction proceeds nonenzymatically with a half-time of 1 hour. 5-Formimino-PteGluH$_4$ is produced by enzymatic transfer to PteGluH$_4$ of a formimino (—CH=NH) radical (see above), which is pre-formed in formiminoglycine and formiminoglutamate, catabolic products from xanthine and histidine, respectively.

In the transformation of serine to glycine, PteGluH$_4$ is acceptor coenzyme of the C_1 unit, yielding 5, 10-methylene-PteGluH$_4$ (VI). This compound also can arise by chemical reaction of PteGluH$_4$ with excess formaldehyde at pH 4.2. At this pH, nitrogen-5 is charged, nitrogen-10 uncharged, *i.e.*, 10-hydroximethyl-PteGluH$_4$ is formed, which, after dissociating OH^- ion, cyclizes to 5, 10-methylene-PteGluH$_4$ (λ_{max} = 294 mμ) in a concerted reaction. Evidence for the imidazolidine structure is manifold: *e.g.*, the unusual stability against hydrolysis and oxidation; the enzymic and chemical (with boranate) reduction of cyclic 5, 10-anhydroformyl-PteGluH$_4$ to cyclic 5, 10-methylene-PteGluH$_4$ (see above) but not of the non-cyclic compounds 5-formyl- or 10-formyl-PteGluH$_4$. Strong circumstantial evidence, furthermore, is the unusually high optical rotation of the natural compound (α_D = *ca.* + 165°), which also seems to indicate a bridge between the ring moieties of the molecule. In acidic but not in alkaline solutions, 5, 10-methylene-PteGluH$_4$ liberates formaldehyde. Among other things, 5, 10-methylene-PteGluH$_4$ is the donor of the hydroxymethyl group of 5-hydroxymethyl cytidylic acid and of the 5-methyl group of thymidylic acid (see PYRIMIDINE BIOSYNTHESIS). In the latter reaction, no primary reducing agent is necessary. The hydrogens stem from the tetrahydropyrazine part of the cofactor, which indeed, is oxidized to 7,8-PteGluH$_2$. Thus, the folate cofactors, besides being C_1 carriers, are even oxidation-reduction catalysts, resembling FLAVINS. The Mannich base, formed (formally) from 5-hydroximethyl-PteGluH$_4$ by dissociating OH^- (see above) seems to be the chemically reactive species of active formaldehyde.

By chemical (with boranate) or enzymic (with 5-methyl-PteGluH$_4$-reductase + FADH$_2$) reduction, this Mannich base gives 5-methyl-PteGluH$_4$ (VII, λ_{max} = 287 mμ). Uniquely among PteGluH$_4$ cofactors, the latter compound is only active for

L. casei. It seems to be the main biological storage form of folate cofactors in higher organisms and is identical with "serum folate" (serum *L. casei* factor) or "prefolic A". It stands out by its great stability. As mediate methyl donor in the synthesis of methionine from homocysteine, 5-methyl-$PteGluH_4$ is substrate in a vitamin B_{12}-dependent reaction. By this, the methyl group is transferred as $CH_3{}^+$ ion to a reduced B_{12} cofactor, and "methyl-B_{12}" is produced. This is apparently the immediate donor—now of a methyl radical (?)—in methionine formation in bacterial and mammalian systems. The detailed mechanism is not known yet (see also COBALAMINS AND COBAMIDE COENZYMES).

The different folate cofactors are not very firmly bound to their apoenzymes (pteroproteins). The Michaelis constants are in the range of $10^{-4}M$. It is not known which groups or forces connect the cofactor with the proteins. Inhibition experiments indicate that, at least in some instances, the substituents of the 2-amino-4-hydroxypyrimidine ring mainly participate in this interaction.

From Fig. 1, together with the reactions of Table 1, it becomes evident that the balance of several folate-dependent metabolic products will be an important clinical criterion of the folate status [see also ANEMIAS and VITAMIN B_{12} (ABSORPTION)]. Thus in the human, the amount of serum folate (5-methyl-$PteGluH_4$) is a sensitive indicator of the vitamin B_{12} level. In vitamin B_{12} deficiency, the methylation of homocysteine will be slowed down, and large amounts of the catalytically essential folate cofactors are not regenerated and are blocked for metabolic use. An accumulation of formyl folates will not be observed, since they can be cleaved by a deacylase, which not only recycles the limiting cofactors but also helps to transport formate through different cell compartments. In folate deficiency, formiminoglutamic acid excretion into the urine is increased (particularly after a histidine load), and so is that of aminoimidazole carboxamide. Both intermediates cannot be converted further to their respective end products.

Among the cofactors derived from "folates" may also be numbered the phenylalanine hydroxylation factor, mediating the electron transport in the reaction:

$$\text{Phenylalanine} + O_2 + \text{NADPH} + H^+ \xrightarrow[\text{Enzymes}]{\text{Factor}} \text{Tyrosine} + H_2O + \text{NADP}^+$$

This factor is found in liver extracts of various animals. While it may be replaced partially by $PteGluH_4$, it has been found that only the pteridine moiety of the latter is the essential portion. Among the various pteridines selected as models, 2-amino-4-hydroxy-6-methyl tetrahydropteridine was particularly active. Recently it was shown that the true cofactor is very closely related to unconjugated PTERIDINES of the type of biopterin (VIII), *e.g.*, sepiapterine, tetrahydrobiopterin. Mechanistically this pteridine factor strikingly resembles flavin. In the biological reaction—a mixed function hydroxylation—the tetrahydropteridine is transformed to an "oxidized pterine", which, in the absence of NADPH, isomerizes to 7,8-dihydropterine, inactive in supporting the hydroxylation reaction.

(a) $PteH_4$ + Phenylalanine + O_2 → "Oxidized Pterine" + Tyrosine + H_2O

(b) "Oxidized Pterine" + NADPH + H^+ ⇌ $PteH_4$ + $NADP^+$

(c) "Oxidized Pterine" → 7,8-$PteH_2$ (inactive)

At first, the "oxidized pterine," because of retaining its optical center at carbon-6, was believed to be the 7,8-dehydro compound. Very likely, however, it is the paraquinonoid 2′,5-dehydro compound (IX), tautomeric with the 7,8-dihydropteridine. The 2-amino group helps in shuttling the electrons. This makes the analogy to the flavin coenzymes still more striking. Possibly, the system is stabilized by chelation of the oxo group at carbon-4.

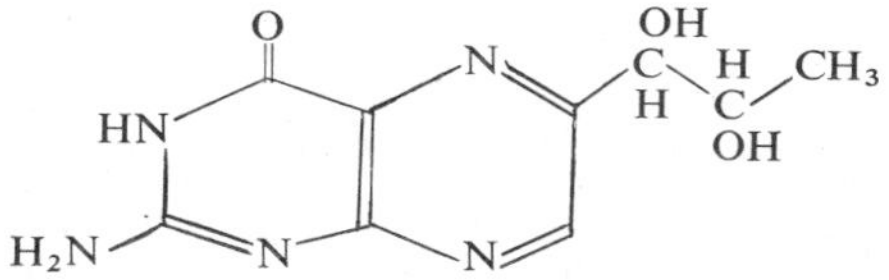

(VIII)

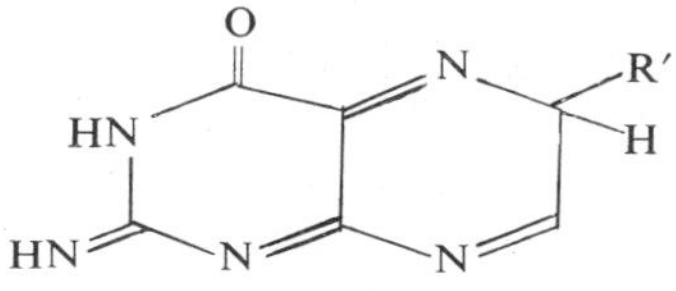

(IX)

References

1. Chapters on water-soluble vitamins in *Annual Review of Biochemistry* (Palo Alto).
2. JAENICKE, L., AND KUTZBACH, C., *Fortschr. Chem. Org. Naturstoffe*, **21**, 184 (1963).
3. STOKSTAD, E. L. R., in "The Vitamins" (SEBRELL AND HARRIS, EDITORS), Vol. III, p. 89, New York, Academic Press, 1954.

L. JAENICKE

FOLINIC ACID

Folinic acid, 5-formyl-5,6,7,8-tetrahydrofolic acid, also termed citrovorum factor and leucovorin, is a reduced form of FOLIC ACID in which the presence of the N^5-formyl group has the effect of stabilizing the derivative against air oxidation (see also FOLIC ACID COENZYMES; SINGLE CARBON UNIT METABOLISM).

FORMIC ACID

Formic acid (methanoic acid), H–COOH, is the simplest member of the monocarboxylic acid series, the even-carbon members of which (up to about C_{20}) are the common FATTY ACIDS. Formic acid got its name from the ant (Latin, *formica*, ant), since it was early observed in the products resulting from the distillation of ants. Formic acid may arise metabolically in a number of reactions including the oxidation of glycine and the oxidation of pyruvate to acetate and formate (see FERMENTATIONS, section on Mixed Acid Fermentations). In SINGLE CARBON UNIT METABOLISM, a single carbon unit at the oxidation level of formic acid, in the form of an N-formyl group carried on one of the FOLIC ACID COENZYMES, becomes incorporated during PURINE BIOSYNTHESIS into the carbon-8, and subsequently another is incorporated into the carbon-2, position of PURINES.

FORMOL TITRATION

When AMINO ACIDS, peptides or PROTEINS are treated with neutral concentrated formaldehyde solution, the free amino groups lose their basic properties with the result that carboxyl groups are freed and can be titrated as such. The chemistry of the formaldehyde reaction is complex and this method of study has lost much of its interest since it was discovered by Sorensen in 1912.

FRACTIONAL ELECTRICAL TRANSPORT

This is a technique whereby it is possible to determine whether an "active principle" (capable of being assayed by biological means) is a weak or strong acid (non-amphoteric), a weak or strong base (non-amphoteric), an amphoteric substance, a non-electrolyte or a combination of the above. If the agent is amphoteric, its isoelectric point can be determined at least approximately. High potentials up to 10,000 volts are used in the electrolysis of very dilute solutions in cells capable of being cut into a number of compartments; the use of membranes is avoided. Non-electrolytes such as sugar or nearly neutral ampholytes can, by the use of this technique, be exhaustively freed from anions and cations. In these cases, the technique can be used for preparative purposes. This technique was particularly useful in determining the nature of "BIOS" and of PANTOTHENIC ACID before its isolation was feasible.

Reference

1. WILLIAMS, R. J., *J. Biol. Chem.*, **110**, 589–597 (1935).

FREE ENERGY. See BIOCHEMICAL THERMODYNAMICS.

FREE RADICALS

The study of free radicals has not as yet had great impact on biochemistry, but this is a field that will doubtless develop since it has growing importance in organic chemistry. Electron spin resonance (ESR), also termed electron paramagnetic resonance (EPR), is a tool that is invaluable in the investigation of free radicals.

References

1. WALLING, C., "Free Radicals in Solution," New York, John Wiley & Sons, 1957.
2. PRYOR, W. A., "Introduction to Free Radical Chemistry," Englewood Cliffs, N. J., Prentice-Hall, 1965.

FRUCTOSE

D–Fructose (levulose) has the configuration represented by the formula:

$$\begin{array}{c} CH_2OH \\ | \\ C{=}O \\ | \\ HO{-}C{-}H \\ | \\ H{-}C{-}OH \\ | \\ H{-}C{-}OH \\ | \\ CH_2OH \end{array}$$

but exists, like other sugars, predominantly in inner acetal (ring) forms. The six-membered ring (pyranose) form predominates in free fructose which is strongly levorotatory (its specific rotatory power varies greatly with concentration and temperature). In glycosidic combination (*e.g.*, in sucrose), a furanose form of fructose is present which is dextrorotatory. *Inulin* is a polysaccharide, somewhat comparable to STARCH, which yields D-fructose on hydrolysis. Fructose is utilized in the body (see CARBOHYDRATE METABOLISM) and is conspicuously present in semen. [See also CARBOHYDRATES (CLASSIFICATION).]

FUMARIC ACID

Fumaric acid, represented by the formula

$$\begin{array}{c} H{-}C{-}C({=}O)OH \\ \| \\ HO(O{=})C{-}C{-}H \end{array}$$

is the *trans* isomer of maleic acid (the *cis* isomer). It is an important compound biochemically since it enters into the CITRIC ACID CYCLE. Many biochemically occurring derivatives are known. Fumarate is a by-product at certain stages in the ARGININE-UREA CYCLE and in PURINE BIOSYNTHESIS.

FUNK, C.

Casimir Funk (1884–), a Polish-born biochemist, gained distinction in 1911 by promulgating the "vitamine hypothesis," namely that beriberi, scurvy, rickets and pellagra were all deficiency diseases caused by the lack of "vitamine" (vital amine). While the "amine" ending was unfortunate (since we now know that the vitamins which prevent scurvy and rickets contain no nitrogen), the name *vitamin*, with altered spelling, has stuck, and has served to center attention on the problem. That Funk's idea was completely original cannot be accepted (see FREDERICK GOWLAND HOPKINS), but it was a bold guess which, except for the amine designation, was an excellent one. "Accessory substances" or "food accessories" or "water-soluble B" turned out to be poorer designations for these unknowns in nutrition than Funk's more imaginative one (see THIAMINE; VITAMIN B GROUP).

G

GALACTOSE AND GALACTOSEMIA

Congenital galactosemia is an hereditary disease due to a single recessive genetic defect.[1] The disease ensues if galactose [for structure, see CARBOHYDRATES (CLASSIFICATION)] or other galactose-containing ingredients are ingested. The pronounced clinical manifestation involves the eye, giving rise to the production of cataract as well as the development of necrotic and cirrhotic processes in the liver accompanied by fat infiltration. Moreover, in most cases, mental retardation develops. However, if a strict galactose-free diet is instituted early, it appears to prevent the beginning of development of tissue damage and mental retardation.[2,3] The exact relationship of galactose ingestion after birth and the development of mental retardation has yet to be defined.

It has been found that the galactosemic organism not only spills galactose in the urine and has elevated galactose in circulating blood (galactosemia), but that the red blood cell accumulates galactose-l-phosphate.[2-5]

Such an accumulation can also be observed in other tissues. The accumulation of galactose-l-phosphate indicates that a defect in the galactose metabolic pathway is blocked beyond the step catalyzed by galactokinase. The following steps catalyzed by specific enzyme constitute the major pathway of galactose metabolism in the human organism:[6,7]

(1) Galactose + ATP → Galactose-l-P + ADP

(2) Galactose-l-P + UDP-glucose ⇄ UDP-galactose + Glucose-l-P

(3) UDP-galactose ⇄ UDP-glucose

It has been generally believed that galactosemia is due to a block in "galactose waldenase." However, developments of specific enzymatic assays for steps (2) and (3), *i.e.*, transferase and epimerase (galactowaldenase) have made it possible to distinguish between galactose-l-phosphate accumulation due to defect in steps (2) or (3).[1,8,9] In the carefully studied case of hereditary galactosemia, it has always been found that the defect is due to a block in step (2), the transferase, and not in step (3). Quantitative enzymatic methods have been further developed so as to make it possible to study the level of these enzymes in parents and siblings of galactosemic children. It has been found that the heterozygotes have transferase levels which are not much more than half of that seen in normal individuals.[10,11] Various *in vitro* screen tests have been developed for detecting galactosemia or for following the status of galactosemia patients; most of them are based on the demonstration of the failure of respiration. The basis of other tests relates to the accumulation of galactose-l-phosphate.[12-15] The simple screening tests are of great value in that they may turn more attention to the detection of galactosemic families. However, it would seem most advisable once the first screening has been done to check these methods with the specific enzymatic tests. If this is not done, other defects such as epimerase defects or defects of UDP-glucose synthesis would not be recognized. There are ample examples of this from the study of mutants in bacteria.[6] If they were not recognized, infants maintained on a galactose-free diet, having defects in the two steps just mentioned, might develop gross abnormalities in the central nervous system; the galactolipids of the central nervous system are synthesized to a large extent after birth (see CEREBROSIDES; GANGLIOSIDES).

This biosynthetic pathway should be independent of supply of galactose or of transferase activity since the galactose is normally recruited from glucose metabolites. However, it does presumably depend on a normal operation of UDP-galactose synthetase and 4-epimerase since UDP-galactose has been shown to play a role as galactosyl donor in the biosynthesis of brain galactolipids.[6,16] It may, therefore, seem that galactose is not required in the development of the mammalian organisms. This may not mean, however, that it is not of selective value in the evolution and development of the normal mammalian organism. This is a problem of considerable interest. However, in regard to the simple single recessive transferase defect found in human hereditary galactosemia, galactose is not only a superfluous ingredient in the food but a harmful one. The disease is due mainly to the toxic effects of the accumulation of galactose-l-phosphate.[2,3] The cataract formation may, however, be related to the accumulation of dulcitol in the lens.[17,18] The transferase abnormality persists in any case. However, the lack of this enzyme is probably of very little consequence to normal development provided that the supply lines of exogenous galactose are blocked. This applies as well to leakage of galactose through the placenta from the maternal to the fetal organism.

References

1. Kalckar, H. M., Anderson, E. P., and Isselbacher, K. J., "Galactosemia, a Congenital Defect in a Nucleotide Transferase," *Biochim. Biophys. Acta*, **20**, 262 (1956).
2. Holzel, A., Komrower, G. M., and Schwarz, V., "Galactosemia," *Am. J. Med.*, **22**, 703 (1957).
3. Isselbacher, K. J., "Galactose Metabolism and Galactosemia," *Am. J. Med.*, **26**, 715 (1959).
4. Kirkman, H. N., "Symposium on Hereditary Metabolic Diseases, Galactosemia," *Metabolism*, **9**, 316 (1960).
5. Donnell, G. N., Bergren, W. R., and Cleland, R., "Galactosemia," *Ped. Clin. North Am.*, **7**, 315 (1960)
6. Kalckar, H. M., "Hereditary Defects in Galactose Metabolism in Man and Microorganisms," *Federation Proc.*, **19**, 984 (1960).
7. Kalckar, H. M., "Biochemical Genetics as Illustrated by Hereditary Galactosemia," *Am. J. Clin. Nutr.*, **9**, 676 (1961).
8. Maxwell, E. S., Kalckar, H. M., and Bynum, E., "A Specific Enzymatic Assay for the Diagnosis of Congenital Galactosemia. II. The Combined Test with 4-Epimerase," *J. Lab. Clin. Med.*, **50**, 478 (1957).
9. Kirkman, H. N., and Maxwell, E. S., "Enzymatic Estimation of Erythrocytic Galactose-1-phosphate," *J. Lab. Clin. Med.*, **56**, 161 (1960).
10. Kirkman, H. N., and Kalckar, H. M., "Enzymatic Deficiency in Congenital Galactosemia and Its Heterozygous Carriers," *Ann. N.Y. Acad. Sci.*, **75**, 274 (1958).
11. Donnell, G. N., Bergren, W. R., Bretthauer, R. K., and Hansen, R. G., "The Enzymatic Expression of Heterozygosity in Families of Children with Galactosemia," *Pediatrics*, **25**, 572 (1960).
12. Weinberg, A. N., "Detection of Congenital Galactosemia and the Carrier State using Galactose C^{14} and Blood Cells," *Metabolism*, **10**, 728 (1961).
13. Eggermont, E., and Hers, H. G., "Une nouvelle methode de detection de la galactosemie congenitale," *Clin. Chim. Acta*, **7**, 437 (1962).
14. Brewer, G. J., and Tarlov, A. R., "A Clinical Test for Galactosemia. The Galactose-methemoglobin Test," *Am. J. Clin. Pathol.*, **39**, 579 (1963).
15. Beutler, E., Baluda, M., and Donnell, G. N., "A New Method for the Detection of Galactosemia and Its Carrier State," *J. Lab. Clin. Med.*, **64**, 694 (1964).
16. Cleland, W. W., and Kennedy, E. P., "The Enzymatic Synthesis of Psychosine," *J. Biol. Chem.*, **235**, 45 (1960).
17. Van Heyningen, R., "Formation of Polyols by the Lens of the Rat with 'Sugar' Cataract," *Nature*, **184**, 194 (1959).
18. Kinoshita, J. H., Merola, L. O., and Dikmak, E., "The Accumulation of Dulcitol and Water in Rabbit Lens Incubated with Galactose," *Biochim. Biophys. Acta*, **62**, 176 (1962).

K. J. Isselbacher and H. M. Kalckar

GALL STONES. See bile and bile pigments, section on Cholesterol.

GAMMA GLOBULINS. See antibodies; immunochemistry; plasma and plasma proteins.

GANGLIOSIDES

Our knowledge of the gangliosides began with the recognition by E. Klenk[1] that an unknown sugar-containing lipid, which he called "substance x," was present in the protagon fraction obtained from the brain of a case of Niemann-Pick's disease. Two characteristic properties of "substance x" differentiated it from cerebrosides, *i.e.*, the production of humin by mineral acid and the formation of a red color with Bial's orcinol reagent. Continued work by Klenk with other lipidoses demonstrated that "substance x" was greatly increased in the brain of cases of Tay-Sach disease, and that the material responsible for the characteristic Bial reaction was a nine-carbon nitrogen-containing organic acid which he called *neuraminic acid.*[2] Development of a quantitative assay for neuraminic acid, based on the Bial reaction, permitted the demonstration that gangliosides were concentrated in the gray matter of brain. As a consequence of these facts, "substance x" was renamed *ganglioside*[3] and appears to have the general structure indicated by Fig. 1. Gangliosides are clearly the most complex of the known sphingolipids. Neuraminic acid has been shown to be present in gangliosides principally in the form of an N-acetyl derivative which is also known as *sialic acid.*

Gangliosides are white substances which crystallize from ethanol as spheroliths; they are soluble in mixtures of benzene : ethanol, chloroform : ethanol, pyridine, acetic acid, and water. They are insoluble in acetone, ethyl ether, and ethyl acetate. About 50% of the gangliosides from beef brain appear to contain a C_{20} homolog of sphingosine which has not yet been demonstrated in other sphingolipids. Stearic acid is present in uniquely high concentration. Galactose and glucose are in the only two hexoses which have been detected in gangliosides, and brain gangliosides contain galactosamine. Other hexosamines have been reported in gangliosides from other sources. There is now ample evidence that native gangliosides can be even more complicated than Fig. 1 indicates, *e.g.*, di- and tri-sialo gangliosides have been reported which contain 2-3 moles of a sialic acid per ganglioside molecule. Glycosidases (sialidase and neuraminidase) from *Vibrio cholerae* and influenza virus have been shown to hydrolyze gangliosides with the release of neuraminic acid. Ganglioside-like lipids and lipid complexes have been isolated from erythrocyte stroma and spleen, some of which are known as *globosides* and *hematosides*, and from brain which are known as *strandin* and *mucolipid.*

Application of modern chromatographic techniques[4] is rapidly expanding our understanding of the gangliosides; it was, however, the masterful application of classical organic chemistry which led to their dicsovery. The review by Svennerholm[4] provides considerable detail concerning the con-

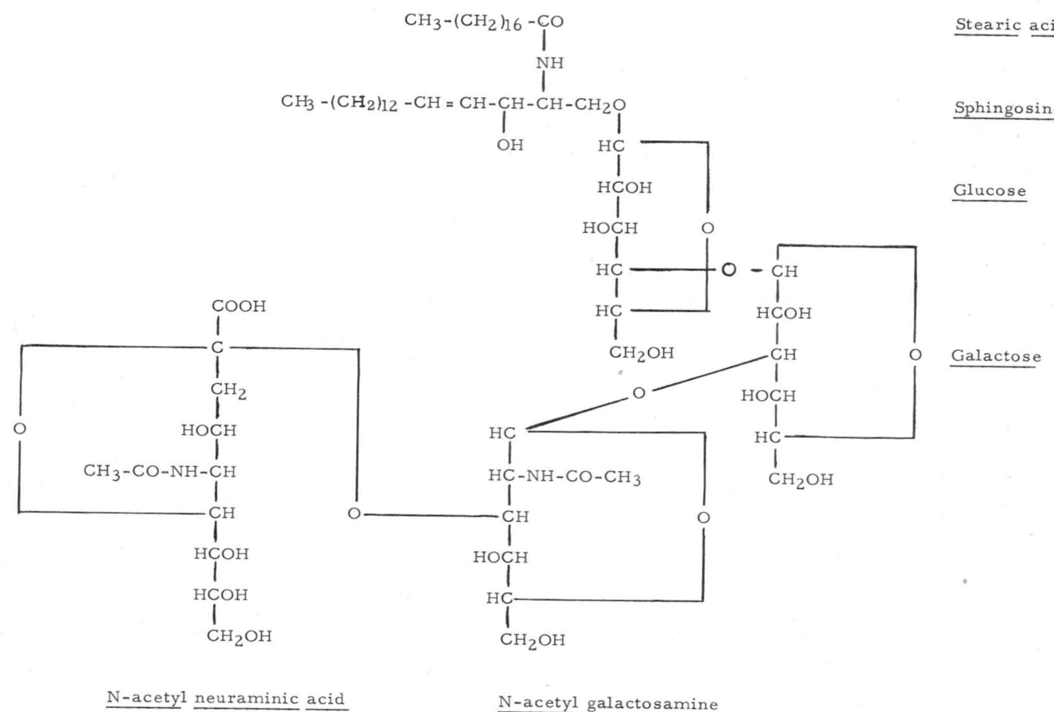

FIG. 1. Ganglioside structure.

tribution of a number of scientists currently active in this field.

References

1. KLENK, E., "Uber die Natur der Phosphatide und anderer Lipoide des Gehirns und der Leber bei der Niemann-Pickschen Krankheit," *Z. Physiol. Chem.*, **235**, 24–36 (1935).
2. KLENK, E., "Neuraminsaure, das Spaltprodukt eines neuen Gehirnlipoids," *Z. Physiol. Chem.*, **268**, 50–58 (1941).
3. KLENK, E., "Uber die Ganglioside, eine neue Gruppe von Zuckerhaltigen Gehirnlipoiden," *Z. Physiol. Chem.*, **273**, 76–86 (1942).
4. SVENNERHOLM, L., "The Gangliosides," *J. Lipid Res.* **5**, 145–155 (1964).

M. H. HACK

GAS CHROMATOGRAPHY. See CHROMATOGRAPHY, GAS.

GASEOUS EXCHANGE IN BREATHING (PULMONARY GAS EXCHANGE)

A 70-kg man consumes about 300 ml of O_2 per minute, and gives off about 250 ml of CO_2. These quantities of gas must diffuse each minute between the blood passing through the pulmonary capillaries and the gas contained in the lungs. This volume of gas called "alveolar air" has a volume of about 2500 ml at the end of expiration, and 3000 ml at the end of inspiration, if the "tidal air" at each breath equals 500 ml. In addition, there is about 150 ml of gas contained in the airway or dead space. Little or no gas exchange takes place in the dead space which includes the nasal passages, trachea, bronchi and bronchioles, down as far as the respiratory bronchioles, alveolar ducts and alveoli where the alveolar air is contained. Because of the dead space, the first 150 ml of each breath is nothing but alveolar air from the last expiration. The fresh room air which mixes with the alveolar air at each normal inspiration is only $500 - 150 = 350$ ml. This fresh air dilutes the partial pressure of the CO_2 of the alveolar air from 40 to 37 mm and increases the partial pressure of the O_2 from 100–103 mm Hg. The large volume of alveolar air, therefore, provides fairly constant partial pressures of O_2 and of CO_2 with which the blood gases can exchange during both inspiration and expiration.

The alveoli of the lung are tiny sacs about 280μ in diameter, which open into the alveolar ducts. There are about 3×10^8 alveoli in the lung, and their total surface area is about 95 square meters. The surface of each alveolus is covered by a hexagonal network of capillaries so that the blood flows over the alveolus almost in a thin sheet. The effective area of this blood-air barrier can be varied by changes in flow of both air and blood.[3]

The volume of inspired air reaching the alveoli with each breath ($V_A = 350$ ml) multiplied by the number of breaths per minute ($f = 12$) represents the "alveolar ventilation" ($\dot{V}_A = fV_A = 4.2$ liters/min), and equals the total ventilation minus the dead space ventilation. Since $\dot{V}_A \times F_{ACO_2}$ (fractional concentration of CO_2 in the alveolar air) equals the total CO_2 output per minute, $\dot{V}_{CO_2}$, the concentration of CO_2 in the alveolar air (F_{ACO_2}) remains constant so long as the $\dot{V}_A$ varies in proportion to the $\dot{V}_{CO_2}$. The rate of CO_2 output equals the metabolic rate of the body except in transition periods when $\dot{V}_A$ is increased or decreased. Thus, when $\dot{V}_A$ is increased, F_{ACO_2} falls, and CO_2 is given off from bicarbonate stores in the body, until a new equilibrium is attained (in about half an hour). Thus, the exchange ratios R, of CO_2 output to the O_2 intake may vary as the CO_2 stores of the body are increased or decreased. In the steady state, R equals the metabolic respiratory quotient, RQ, which varies from 0.7–1.0, depending on whether fat (0.7), protein (0.81) or carbohydrate (1.0) is being burned.

The volume of gas diffusing per minute, $\dot{V}$, is given by the equation

$$\dot{V} = K \times A/d \times S/\sqrt{M}\,\Delta P = D\,\Delta P.$$

In this equation, K is a constant, A is the area and d the thickness of the membrane between the alveolar air and blood, S is the solubility and M the molecular weight of the diffusing gas, and ΔP is the difference in partial pressure of the gas between the alveolar air and the mean partial pressure of the blood as it passes through the capillaries. D may be called the diffusing capacity of a given lung for a particular gas, and has the dimensions milliliters of gas per minute per mm Hg of pressure difference. It is difficult to measure D directly because the partial pressure changes so rapidly along the capillaries, and equilibrium is reached at different points. It is best estimated by adding small amounts of carbon monoxide to the inspired air, and measuring the amount of CO taken up by the blood per minute and the concentration of CO in the alveolar air. The latter may be taken as equal to ΔP (with some minor corrections), because the affinity for hemoglobin is so great that the partial pressure in the blood is nearly zero. This D_{CO} can then be used to calculate D_{O_2} or D_{CO_2} by correcting for the solubility and molecular weight of the gases concerned. In this way it is found that D_{CO}: D_{O_2}: $D_{CO_2} = 1 : 1.23 : 25.4$.

The high diffusing capacity for CO_2 as compared to O_2 is due to its much greater solubility (24.3 times). As a result of this, the ΔP as the blood enters the pulmonary capillary may be only 6 mm for CO_2 and 60 mm for O_2. The higher value of D_{CO_2} more than compensates for the much smaller initial ΔP and CO_2 reaches equilibrium much earlier than the O_2 as the blood traverses the capillary.

An average value for D_{O_2} in a normal man is 20 ml O_2/min per mm Hg. To transfer 300 ml/min, therefore, a mean pressure difference between blood and alveolar air must be maintained at an average value of 300/20 = 15 mm Hg. To increase the rate of O_2 intake from 250–3000 ml/min, as in muscular exercise, the value of $D_{O_2} \times \Delta P$ must be increased twelvefold. To do this, D_{O_2} may be increased threefold, by opening up more capillaries in the lung (increasing the area), and ΔP may be increased fourfold, by lowering the saturation of the venous blood and by increasing its velocity of flow through the capillaries of the lung.

If all the blood went to one lung, and all the air to the other lung, the capacity for gas exchange would be zero. Similarly, if some alveoli received relatively too little blood or too little air, the gaseous exchange would be impaired. The normal ratio of alveolar ventilation to blood flow (both in liters per minute) is about 0.8, and for maximum efficiency of gas exchange this same ratio should be maintained in every alveolus. Failure of this requirement is one of the chief causes of inadequate saturation of the arterial blood with oxygen in patients (uneven distribution of ventilation and blood flow). There are, apparently, some partially effective automatic mechanisms for maintaining a proper ratio of air flow to blood flow so that a decrease of air flow to a part of the lung brings about some decrease in blood flow, and vice versa.

Diffusion of gases across the alveolar membrane can also be impaired by an abnormal thickening of the membrane, as in pulmonary edema, or by filling of the alveoli with fluid. Escape of fluid from the blood is normally prevented by a colloid osmotic pressure of plasma proteins in excess of the hydrostatic pressure of the blood. Tendency of alveoli to draw fluid out of the blood by surface tension (as water is pulled into small capillary tubes) is minimized by a superficial layer of surfactant material, so that the air-water interface has a very low surface tension. Moreover, as the surface area diminishes in expiration, this surfactant material is concentrated, thus decreasing still further the surface tension, and thereby diminishing the tendency of the lung to collapse. The tendency to collapse is still further diminished by the presence of 80% nitrogen in the alveolar air. This gas does not exchange across the alveolar membrane under normal conditions, because the body is already in equilibrium with it. In an occluded lobule, however, the concentration of nitrogen becomes slightly increased by the absorption of oxygen, and all the gas is very slowly absorbed. Not until all the gas is absorbed can the lung be completely collapsed. Without any nitrogen, the lung will collapse in a few minutes, when all the oxygen is absorbed. When pure oxygen is breathed in an open normal lung, the nitrogen dissolved in the water and fat of the body will continue to escape exponentially, with a half-desaturation time of about 20 minutes or more, depending on the blood flow to different parts of the body.

The pulmonary gas exchange in the steady state is independent of the rate of blood flow in the lung, and independent of the rate of alveolar ventilation; it is dependent then only on the rate of metabolism in the tissues. If the alveolar ventilation is increased, there is a temporary increase in CO_2 output, because the concentration of CO_2 in the alveoli is decreased, and each unit of blood gives up more CO_2 as it passes through the lungs. Eventually, however, the partial pressure of CO_2 in the venous blood will diminish as the tissues give up their CO_2 stores, and the rate of CO_2 output will return to normal. An increase in alveolar ventilation will not affect the rate of oxygen uptake, even though the alveolar oxygen partial pressure is increased, because the arterial blood is already saturated with oxygen and cannot take up any more. If, in the steady state, the rate of blood flow is increased, there will be a temporary increase in the exchange of both oxygen and carbon dioxide in the lungs, but both will return to normal promptly, when the partial pressures of these gases in the venous blood readjust to the new rate. The pulmonary gas exchange cannot be permanently increased unless the metabolic rate in the tissues is also increased.

In general, the principles described here apply to all air-breathing vertebrates. The absolute amount of the pulmonary exchange varies, of course, with the size of the animal, and may be roughly estimated, for all mammals from mice to whales, from the equation,

$$B = 70W^{0.75}$$

where B is the basal metabolic rate in kilocalories per day and W is the body weight in kilograms. The sizes of the lung are roughly proportional to the body weight, but the diameters of the alveoli (d) are smaller in smaller animals where the oxygen consumption per kilogram per day, B/W, is high. Thus, according to Tenney and Remmers,

$$d = \text{Constant} \times (B/W)^{-.71}$$

Frogs having a moist skin absorb 37–70% of their oxygen through the skin, and some also through the buccal cavity. From the buccal cavity, air is further forced into the lungs by elevation of the floor of the mouth. Reptiles and amphibians (*i.e.*, frogs) breathe with the muscles of the throat and mouth and lack a diaphragm. Reptiles with a dry skin exchange all their gases through the lungs, as in the mammals, and they, likewise, have the pulmonary circulation quite separate from the systemic circulation as in man. Cold-blooded animals would be expected to have a lower metabolic rate for their size than mammals, but apparently, it is lower than can be accounted for by temperature alone.[7] The diffusing surface of the lungs of amphibians and reptiles is small compared to that of mammals of the same size. The lungs of amphibia vary from the single sac lung of *Proteus* to the lung of the frog with its complex septa.[8] The gas exchange of birds is apparently much like that of mammals, but their lungs are peculiar because of the presence of several large air sacs into which the inspired air may also enter.

Fish, of course, exchange gases through gills, except for the lungfishes, which breathe air. The metabolic rate is low in fish, compared to mammals, as it is in amphibians and reptiles. Fish are peculiar in having a very low partial pressure of carbon dioxide in the blood, of only 2–4 mm. Rahn has explained this as necessary for water-breathing.[5] The solubility of oxygen in water is so low compared to carbon dioxide, that an amount of water passing through the gills, which will provide sufficient oxygen necessarily carries away an excessive amount of CO_2. The lungs of rats and dogs have been successfully used for water breathing, but for that purpose, the water must be charged with oxygen at a pressure of several atmospheres, to provide enough oxygen and also to maintain a normal P_{CO_2}.

References

1. Comroe, J. H., Jr., Forster, R. E., II, DuBois, A. B., Briscoe, W. A., and Carlsen, E., "The Lung," second edition, Chicago, The Year Book publishers, 1962.
2. "Handbook of Physiology," Section III, "Respiration," Vol. I, (Fenn, W. O., and Rahn, H., Editors) Washington, D.C., *American Physiological Society*, 1964.
3. Weibel, E. R., "Morphometrics of the Lung," *ibid.*, 7.
4. Otis, A. B., "Quantitative Relationships in Steady State Gas Exchange," *ibid.*, Ch. 27.
5. Rahn, H., and Farhi, L. E., "Ventilation, Perfusion and Gas exchange," *ibid.*, Ch. 30.
6. Forster, R. E., "Diffusion of Gases," *ibid.*, Ch. 33.
7. Irving, L., "Comparative Physiology of Respiration," Ch. 5.
8. Krogh, A., "The Comparative Physiology of Respiratory Mechanisms," Philadelphia, University of Pennsylvania Press, 1941.
9. Kleiber, M., "The Fire of Life," New York, John Wiley & Sons, 1961.
10. Tenney, S. M., and Remmers, J. E., "Comparative Quantitative Morphology of Mammalian Lungs; Diffusing Areas," *Nature*, **197**, 54–56 (1963).
11. Fenn, W. O., "The Mechanism of Breathing," *Sci. Am.*, **202**, 138–148 (January 1960).

Wallace O. Fenn

GASTRIC JUICE

Gastric juice has a variable composition. It is a mixture of secretions evoked by many types of stimuli and contains organic and inorganic components. The most remarkable of these is free hydrochloric acid which can approach a concentration isotonic with blood, *i. e.*, 170 meq/liter. It is secreted by the parietal (oxyntic) cells of the gastric tubules. The surface epithelial cells of the stomach secrete a juice containing mainly sodium chloride together with potassium, calcium and bicarbonate ions. This juice contains much mucus. The chief cells of the gastric tubules make PEPSIN which is a proteolytic enzyme active in the presence of hydrochloric acid. It has optimum activity at about pH 2. The neck cells of the tubules make a mucoprotein and the "intrinsic factor" necessary for the absorption of VITAMIN B_{12} in the intestine.

Enzyme Content. Gastric juice contains a variety of enzymes which may be different in different species. Human gastric juice contains gelatinase and a proteolytic enzyme gastricsin with a pH optimum of 3.0. The pig has parapepsins while the calf has rennin. A gastric lipase is found in many species as are other less important enzymes.

In general, enzymes are made and stored in cells as inactive precursors such as pepsinogen. They are activated by the removal of a peptide. For example pepsinogen is a protein with a molecular weight of 42,500. It is converted into pepsin by acid or pepsin, and then has a molecular weight of 35,000. It hydrolyzes peptide bonds in which the aromatic amino acids L-tyrosine or L-phenylalanine provide the amino group for the peptide bond. Because of this specificity of attack, the products of peptic digestion are mixtures of peptides with a great range of size.

Secretion of Acid. A vast amount of research has been directed to the elucidation of the mechanism of formation of hydrochloric acid by the parietal cells. These cells contain intracellular canaliculi and experiments with acid-base indicators show that during acid secretion these canaliculi contain hydrochloric acid. Secretion can be initiated by stimulation of the vagus nerve, by gastrin and by histamine. It is interesting that the parietal cells secrete as much alkali as acid, because no net production of hydrochloric acid can be observed if a sheet of gastric mucosa is incubated in a physiological saline solution containing histamine. As much alkali is secreted from one side as acid from the other and they neutralize each other.

If the two sides face into two separate chambers, or if the whole isolated gastric mucosa is used, either normally or inside-out, and tied at both ends, as can be readily done with the lining of frog stomachs, then acid secretion can readily be observed.

The chloride of the hydrochloric acid comes mainly from the red cells of the blood by a reversed chloride shift. The hydrogen ions are made by oxidation of covalently bound hydrogen atoms from substrate and water molecules. For every hydrogen ion secreted there is an uptake into the parietal cells of one molecule of carbon dioxide which forms a bicarbonate ion which exchanges for chloride from the blood.

The actively acid-secreting stomach requires more carbon dioxide from the arterial blood than it requires oxygen. It has a negative respiratory quotient (carbon dioxide output/oxygen uptake). The requirements for carbon dioxide are so great that CARBONIC ANHYDRASE activity is required to catalyze the spontaneous reaction. The amount of the enzyme in the parietal cells is sufficient to cause the necessary rate of carbon dioxide uptake and when the enzyme is inhibited sufficiently, inhibition of acid secretion occurs.

Acid secretion requires calcium, potassium, magnesium, phosphate and bicarbonate ions together with an oxidizable substrate and oxygen as an energy supply. The transport of anions is not entirely specific since with isolated gastric mucosa it has been shown that bromide, iodide, sulfate and nitrate can be secreted along with the hydrogen ions.

The gastric mucosa is a source of electric power and maintains a potential difference of 30–80 mV across itself, depending on the animal and the conditions. The mucosa is negative to the serosa in an external circuit. This potential difference falls when acid secretion begins and rises when it stops, or when secretion is prevented by an inhibitor such as thiocyanate. If current from an external source of power is passed through the mucosa to enhance the natural potential difference, acid secretion is increased; if it is passed so as to oppose or reverse the natural potential difference, acid secretion is reduced or abolished.

If a current is passed so as to keep the potential difference just zero, the mucosa is effectively short circuited and the current it passes through itself is one of chloride ions alone. With chloride replaced by isethionate, the potential difference is reversed and a current of hydrogen ions can be observed. The secretion of hydrochloric acid is, of course, equivalent to the passage of two equal currents in opposite directions, so there is no net charge transfer.

2,4-Dinitrophenol (which uncouples oxidative PHOSPHORYLATION) inhibits acid secretion and abolishes the natural potential difference across the gastric mucosa. Measurements of changes in the rate of acid secretion and the content of adenosine triphosphate during the onset of anaerobiosis strongly suggest that adenosine triphosphate is used to drive hydrochloric acid production. One mole of pure parietal secretion requires about 10,000 calories of energy for its formation, and for the whole mucosa somewhat less than four hydrogen and chloride ions can be made and moved per oxygen molecule. The actual efficiency of the parietal cells is unknown, but these cells must be among the most energetically active in the body. It is probable that the hydrogen ions are created in an oxidation-reduction reaction associated with an electron transport system driven either directly by oxygen or indirectly from the energy of adenosine triphosphate.

References

1. JAMES, A. H., "The Physiology of Gastric Digestion," London, E. Arnold, Ltd., 1957.
2. MURPHY, Q. (EDITOR), "Metabolic Aspects of Transport Across Cell Membranes," Madison, The University of Wisconsin Press, 1957.
3. HIGHTOWER, N. C. (EDITOR), "Production of Hydrochloric Acid by the Gastric Mucosa," A Symposium, *Am. J. Digest. Disease, New Series*, **4** (March 1959).
4. SKORYNA, S. C. (EDITOR), "Pathophysiology of Peptic Ulcer," Montreal, McGill University Press, 1963.

R. E. DAVIES

GASTRIN

Discovery of secretin by Bayliss and Starling in 1902 led Edkins[1] to seek a similar hormone for stimulating the secretion of gastric acid. He prepared extracts from the gastric mucosa of the cat and pig and observed a stimulation of acid secretion following the intravenous injection of extracts made from cardiac and antral mucosa but not those from fundic area. He believed the active material was a hormone and named it gastrin. He reported that cold water extracts were activated by boiling or by preparing the extract with acidulated water. He also observed no decrease in response by the atropinized cat, and noted that both the active and inactive extracts caused a similar decrease in blood pressure. His extraction method did not eliminate the possible presence of histamine in the gastrin.

Keeton *et al.*[2] in 1920 demonstrated histamine to be a very powerful stimulus for gastric acid secretion, and a prolonged controversy arose over the possibility of gastrin and histamine being the same substance. Komarov[3] prepared an active, histamine-free gastrin from the protein fraction of the gastric mucosa. Similar histamine-free gastrin has been prepared by other workers. The gastrin literature is far too extensive to be reviewed in this short article. A comprehensive review by Babkin[4] and the review by Grossman[5] cover the literature up to 1950.

A study of the literature reflects the difficulties encountered by the investigators working in this area. These difficulties were associated with (1) the use of different extraction methods, (2) the use of impure extracts, (3) a possible interference in normal function as a result of surgery, (4) too rapid injection of gastrin, inhibiting gastric secretion obtained with the slow infusion, and (5) the augmentation of the response by parasympathetic nervous activity. Despite these handicaps, the biochemistry of gastrin is being elucidated.

Advances in the biochemistry of gastrin have been made possible by the development of (1) the guinea pig ilium histamine assay, (2) the removal of histamine from the extract by histaminase, (3) the use of portal vein injection route which inactivates histamine but not gastrin, (4) the development of a method for preparing a very highly purified gastrin, and (5) the recent development of an analytical method for demonstrating that gastrin depletes histamine in the gastric mucosa.

The PROTEIN-like nature of gastrin is apparent in the biochemical techniques employed in its purification. It is present in the protein fraction of the mucosal extracts, and it is precipitated by the protein precipitant, trichloracetic acid. It is subject to isoelectric point precipitation and is stable in acid or alkaline solutions. It is not dialyzable through cellophane.

The recent report by Haverback *et al.*[6] appears to confirm the hypothesis of Babkin[4] who suggested that gastrin probably brought about a release of histamine which then stimulated gastric secretion. Haverback and co-workers, using a highly purified gastrin which was free from histamine, demonstrated a mean reduction of histamine in the glandular stomach of the rat of approximately 50% two hours after the subcutaneous injection of the gastrin.

At the present time, it appears that either a mechanical stimulus, parasympathetic nervous activity, or the administration of gastrin can stimulate gastric acid secretion (see GASTRIC JUICE). It has been shown that parasympathetic stimulation after gastrin injection synergizes the secretory response. It is suggested that the normal physiologic secretion of gastric juice is largely, if not entirely, a response to mechanical or vagal stimulation which releases gastrin that in turn liberates histamine which then acts as the ultimate stimulus for the secretion of gastric acid. Since atropine blocks the vagal stimulated secretion but not that of gastrin, it seems unlikely that acetylcholine liberated by vagal stimulation may release histamine. Gastrin stimulates secretion in the cat, dog, guinea pig, rabbit, rat, pig, duck and frog.

References

1. EDKINS, J. S., "The Chemical Mechanism of Gastric Secretion," *Am. J. Physiol.*, **34**, 133 (1906).
2. KEETON, R. W., KOCH, F. C., AND LUCKHARDT, A. B., "Gastrin Studies III. The response of the Stomach Mucosa of Various Animals to Gastrin Bodies," *Am. J. Physiol.*, **51**, 454 (1920).
3. KOMAROV, S. A., "Gastrin," *Proc. Soc. Exptl. Biol. Med.*, **38**, 514 (1938).
4. BABKIN, B. P., "Secretory Mechanisms of Digestive Glands," Second edition, New York, P. B. Hoeber, Inc., 1950.
5. GROSSMAN, M. I., "Gastrointestinal Hormones," *Physiol. Rev.*, **30**, 33 (1950).
6. HAVERBACK, B. J., LECIMER, L. B., DYCE, B. J., COHEN, M., STURBIN, M. I., AND SANTA ANA, A. D., "The Effect of Gastrin on Stomach Histamine in the Rat," *Life Sciences*, **3**, 637 (1964).

WILLIAM C. CLARK

GEL FILTRATION

The Molecular Sieve Effect. Gel filtration is a chromatographic separation resulting from restricted molecular diffusion through a column of gel particles having suitable porosity and other properties. One type of gel used for this purpose (having the trade name "Sephadex") is a modified dextran obtained by fermentation of sugar; the linear macromolecules of dextran are cross-linked to produce a three-dimensional network of polysaccharide chains. When mixed with water or electrolyte solutions, this material swells considerably. Placed in a chromatographic column, it acts as a sieve for molecules of different sizes, since the porosity of the gel is determined by the amount of cross-linkage in the dextran network. A high degree of cross-linkage creates a compact structure with low porosity; a low cross-linkage produces a highly porous structure. Gels of a variety of degrees of cross-linkages are commercially available. The liquid imbibed by the gel particles is available as solvent to solute molecules of different sizes to a degree dependent on the porosity of the granules.

Volumes of a Gel Filtration Column. The total volume V_t of a gel filtration column is the sum of the volume outside the gel grains (V_o) plus the volume inside the gel grains (V_i) and the volume of the gel matrix or substances itself (V_g):

$$V_t = V_o + V_i + V_g \quad (1)$$

The outer volume V_o is the volume of liquid required to elute a substance through a column if the molecules are completely excluded from the gel particles. This volume is also known as the void volume. The inner volume V_i can be calculated from the known dry weight of the gel, a, and the water regain W_r:

$$V_i = a \cdot W_r \quad (2)$$

If the weight of the gel in the column is not known, the volume of the inner water V_i can be calculated from the wet density (d) of the swollen gel particles:

$$V_i = \frac{W_r \cdot d}{W_r + 1}(V_t - V_o) \quad (3)$$

The elution volume V_e of a substance depends on the volume external to the gel particles (V_o) and on the distribution coefficient K_d:

$$V_e = V_o + K_d \cdot V_i \quad (4)$$

Since the distribution coefficient, as defined in Eq. (*4*), can be determined from known volumes, K_d possesses a characteristic value for each substance and is independent of the geometry of the gel bed. Thus,

$$K_d = \frac{V_e - V_o}{V_i} = \frac{V_e - V_o}{a \cdot W_r} \quad (5)$$

For large molecules that cannot enter the stationary phase, $K_d = 0$, and for small molecules, where accessibility is complete, $K_d = 1$. For these extremes, complete exclusion and complete interchange, the K_d values are, in fact, identical with the distribution coefficient between the two phases (defined by the quotients between the two phases). Because of differences in pore sizes, there are parts of the gel particle which are available and parts which are not available to the dissolved substance having a molecular weight within the fractionation range limits. In this case, the coefficient indicates the proportion of the stationary phase which is available for the particular substance. The concentration of the substance there is the same as the concentration in the mobile phase.

Two substances with different molecular weights will occupy a different proportion of the stationary phase (the inner volume). If the two substances on a certain gel type have the K_d values K_d' and K_d'', the elution volumes differ by the separation volume V_s according to the equation:

$$V_s = V_e' - V_e'' = (V_o + K_d' \cdot V_i) - (V_o + K_d'' \cdot V_i) \quad (6)$$

$$V_s = (K_d' - K_d'') \cdot V_i \quad (7)$$

Thus, for complete separation of the two materials, the sample volume must not be larger than V_s.

Applications. Among the CHROMATOGRAPHIC methods, gel filtration is the only one which is based primarily on the molecular dimensions of the solutes; it is therefore an important complement to the arsenal of methods available for the purification of PROTEINS and other biological substances. Simplicity of technique, chemically mild and gentle environment, quantitative recoveries, and flexibility are characteristic features of this method. The solvent medium, the degree of cross-linking and the chemical nature of the matrix substance (gel) are factors that affect the distribution of solutes; since these factors may be varied in many ways, gel filtration is an extremely flexible method. Estimation of the molecular weights of the solute substances may be obtained by gel filtration. Gel substances are available which cover molecular weight ranges from a few hundred to several hundred thousand.

References

1. PORATH, J., and FLODIN, P., "Gel Filtration," in "Protides of the Biological Fluids" (PEETERS, H., EDITOR), Vol. 10, Amsterdam, Elsevier, 1963.
2. "Sephadex, Theory and Experimental Technique," published by Pharmacia, Uppsala, Sweden.
3. FLODIN, P., "Dextran Gels and their Applications in Gel Filtration," published by Pharmacia, Uppsala, Sweden.

VELIMIR CUBRILOVIC

GELATIN. See COLLAGENS. The water-insoluble and PROTEOLYTIC ENZYME-resistant collagens are altered or "denatured" by boiling in water, dilute acids, or alkalies, to form soluble, easily digestible gelatins.

GENES

Genes are the physical units of heredity. Precise definitions for them have changed considerably over the years as more has been learned about the chemical nature of the genetic material and its function.

Currently, genes may be defined as segments of genetic material which determine the sequence of amino acids in specific polypeptides, such that there is a one-to-one relation between gene and polypeptide. This definition applies at least to those genes called *structural* genes because they determine the primary structure of proteins. Other kinds of genes may exist as discussed in the following.

Structural Genes. So far as known, structural genes in all organisms are composed of nucleic acids. In the RNA viruses, the genes are RNA (RIBONUCLEIC ACID) only, but in all other organisms, the DNA viruses and the cellular forms which all possess both DNA (DEOXYRIBONUCLEIC ACID) and RNA, the gene material is either known to be DNA, or assumed to be for good reason.

The genes of viruses and bacteria appear to consist of nucleic acid unaccompanied by closely bound protein [see also VIRUSES (COMPOSITION); BACTERIAL GENETICS]. Ordinarily this naked nucleic acid is in the two-stranded condition; exceptions are known among both the RNA and DNA viruses some of which possess single-stranded genetic material. In those organisms with true nuclei, the genetic material is always double-stranded DNA associated with protein ordinarily of the HISTONE type. The function of the protein is not considered to be genetic. It probably controls DNA in its role of determining protein structure. Also it may serve to hold genes together and attached to the CHROMOSOMES of which they are a part.

Structural genes carry out their role of dictating protein structure by producing a messenger RNA (m-RNA) which is a single strand of RNA containing nucleotide bases complementary to one of the strands of the double-stranded DNA of the gene from which it is copied or "transcribed." The evidence is that the same DNA strand of a gene is always transcribed into m-RNA. In this way, only one kind of m-RNA is made for each gene. In the transcription process the C, T, A and G bases of the DNA determine G, A, U and C, respectively, in the m-RNA strand. Transcription effectively constitutes *gene action*. By definition, if a gene is not actively forming m-RNA, it is inactive or "turned off."

Each kind of gene is different from every other gene in its DNA sequence. Hence, as many different kinds of m-RNA are formed as there are different genes in the organism.

After their formation at the gene level, the m-RNA strands attach to RIBOSOMES in the cytoplasm, and the process of PROTEIN BIOSYNTHESIS commences. The significant point to be emphasized here is that the sequence of nucleotide bases, of the "genetic code," in a particular gene is reflected in a specific sequence of amino acids in the polypeptide produced through the protein synthetic mechanism.

The one-to-one relation between gene and polypeptide is a more accurate statement of the situation than the earlier one gene–one enzyme hypothesis. It is now known that a number of proteins are constituted in their functional state of subunits which are polypeptides. When subunits are all identical, the one gene–one protein statement holds with certain exceptions. However, proteins such as vertebrate lactic acid dehydrogenase (LDH) and HEMOGLOBIN are known to be made up of different subunits. For example, the dominant adult hemoglobin in man contains both α and β polypeptides as subunits. These have somewhat different amino acid sequences, and each has been shown to be under the control of a different gene. The genes are not even on the same chromosome. A similar situation has been found for LDH which may be made up of at least two different subunits, each one again under the control of a separate gene.

A term which is currently used by many synonymously with structural gene is *cistron*. Its original definition was based on complementation tests. If two chromosomes bearing the same kinds of genes (homologous chromosomes) are introduced into the same cell, "product interactions" may be observed between the genes of the same type, *i.e.*, genes which control the same kind of polypeptide. If two genes of the same type are mutant, but mutant at different sites, they may *complement* and produce a protein which has an activity comparable to the nonmutant even though each mutation alone or together on the same chromosome, can produce only a mutant, inactive protein. Those mutants which do not complement with the production of an active protein are said to have mutational sites within the same cistron.

Mutation of Genes. A change in the base sequence of the DNA constituting a gene results in an inherited alteration in the code and is called a gene mutation. Changes in base sequence may conceivably result from (1) the deletion or addition of one or more nucleotide pairs in the DNA chain, (2) changes in one or more bases along the chain, or (3) inversion of a segment of the chain.

Good evidence for the occurrence of the first type of mutation exists at least in bacteriophage of the T series which infect *Escherichia coli*. The deletion or addition of a single base pair into a DNA chain of a gene should be expected to cause considerable difficulties in the translation of the code in the derived m-RNA, into an amino acid sequence. For example, if the m-RNA of the nonmutant strain has the sequence:

$$\overline{GCU}\ \overline{AAU}\ \overline{GAA}\ \overline{UUU}\ \overline{AAA}\ \overline{CAU}\ \overline{\cdots}$$

which is read in triplets from left to right to give a particular sequence of amino acids, say ala · asp NH_2 · glu · phe · lys · his, a deletion of a base in the mutant would change the "reading frame" starting at the point of deletion. Thus, the sequence

$$\overline{GCU}\ \overline{AAU}\ \overline{G\underset{A}{\downarrow}AU}\ \overline{UUA}\ \overline{AAC}\ \overline{AU\cdot}\cdots$$

would produce the sequence ala · asp NH_2 · asp · leu · asp NH_2 as one possibility. A similar result would be expected from a duplication, by shifting the reading frame. Such mutations as these should be expected to produce "nonsense" sequences of amino acids after the point of change.

Mutations which are the result of the simple changing of bases, say G $\leftrightarrows$ A, C $\leftrightarrows$ T, or G $\rightleftarrows$ C, or A $\rightleftarrows$ T should obviously cause changes in a single triplet rather than a whole sequence. As a result only a single amino acid change should occur in the polypeptide, if but a single base is changed. Many mutant proteins from a variety of organisms are now known which have but a single amino acid change from the nonmutant, and are therefore presumably the result of a single, or adjacent changes within a single triplet (see TOBACCO MOSAIC VIRUS). The non-occurrence of single mutations causing the substitution of two adjacent amino acids within a chain is evidence that the genetic code is not overlapping. As might be expected, a number of different amino acids may be substituted for the nonmutant acid, but the number of substitutions has been found to be limited for any particular amino acid. This also has connotations for the nature of the genetic code.

Gene mutations of other types such as inversions probably occur in addition to the two discussed above, but techniques have yet to be devised to analyze them.

For the present it is enough to say that a mutation may occur at any point within a gene. Theoretically there should be as many "mutational sites" within a gene as there are nucleotide pairs.

Gene Recombination. When two homologous (*i.e.*, bearing the same kinds of genes) chromosomes are paired in synapsis (as in early meiosis in the nucleated organisms) recombination may occur. Recombination is the exchange, usually equal, of segments of chromosomes. Thus, if a chromosome marked:

A b C D e f g H i J

recombines with one marked:

A b c d E f G h I j

between d and e, the recombinant products will be A b C D E f G h I j and A b c d e f g H i J. This natural process presumably occurs in all organisms both *between* or *within* genes. First, it provides a powerful tool for establishing that the genes are ordered linearly on the chromosome, and in what order, and second, it allows one to establish that there exists a colinearity between the genetic material of a gene and the polypeptide it produces. This has been done by mapping a number of mutational sites for a gene that determines one of the polypeptides (protein A) forming the enzyme tryptophan synthetase in *Escherichia coli.* Each mutant produces a modified protein A which can be shown to differ from the wild type by a single amino acid substitution. When the order of the mutant sites on the *coli* chromosome was compared to the order of amino acids within protein A affected by the mutations, it was found that they were the same. This fundamental finding could only have been possible with the use of a recombination analysis.

Controlling Genes. Genes which do not carry codes for the synthesis of proteins which constitute the enzymes, structural components, etc., of the cell almost certainly exist. These genes may produce proteins, but the proteins presumably act by the regulation of the activity of the structural genes, turning them on and off according to circumstances within the cell (see also METABOLIC CONTROLS).

Examples of such genes are found in *Escherichia coli.* These, termed *regulator genes*, presumably produce substances, possibly proteins, which prevent or repress structural genes from synthesizing m-RNA unless other substances, the inducers, are present to inhibit the repressor substances. Alternatively, repressor substances from other types of regulator genes are active in repression only when certain substances activate the repressor substances. The reason for the existence of these genes would seem to be for the regulation of metabolism by preventing the overproduction of enzymes when their substrates are not present, or of end products such as amino acids. In the latter case, the end product is usually considered to be the substance which activates the repressor produced by the regulator.

References

1. HARTMAN, P. E., AND SUSKIND, S. R., "Gene Action," Englewood Cliffs, N.J., Prentice-Hall, 1965.
2. HAYES, W., "The Genetics of Bacteria and Their Viruses," New York, John Wiley & Sons, 1964.
3. WAGNER, R. P., AND MITCHELL, H. K., "Genetics and Metabolism," New York, John Wiley & Sons, 1964.

R. P. WAGNER

GENETIC CODE

Genetic information stored in the GENES, as a linear sequence of the bases (A, C, G, and T) in DEOXYRIBONUCLEIC ACID molecules, is transcribed into a complementary base sequence (U, G, C, and A, respectively) in the messenger-RNA molecules [see RIBONUCLEIC ACIDS (BIOSYNTHESIS)]; this "coded message" contained in the m-RNA, as a linear sequence or 4-letter "language," is "translated" in the process of PROTEIN BIOSYNTHESIS into a linear sequence of the 20 amino acids within the protein polypeptide chain synthesized. Each nucleotide triplet or "code word" consisting of one of the 64 possible triplet combinations of U, G, C, and A nucleotides) in a messenger-RNA molecule may specify one particular amino acid for incorporation into the polypeptide chain. It appears that certain amino acids may be specified by more than one of the 64 nucleotide triplets; in this respect, the genetic code is said to be "degenerate." A few particular triplet "words" may have special functions, such as to signal polypeptide-chain initiation, or chain termination. The first identification of a particular triplet as the code word for a particular amino acid was the discovery that the sequence UUU (in the form of polyuridylate) appears to be the "code word" specifying incorporation of phenylalanine into a polypeptide, in a cell-free, *in vitro* system containing RIBOSOMES and other required components.[1]

Evidence that a nucleotide *triplet* (and not some smaller or larger run of nucleotides) is the "code word" for incorporation of a specific amino acid has come from studies of the fine structure of genes or DNA of a bacteriophage (virus)[2,3] Many tentative formulations of a "code dictionary" of messenger-RNA triplets, with the corresponding amino acid specified by each triplet, have been proposed, on the basis of both experimental results (primarily those of the Nirenberg group and of the Ochoa group) and theoretical considerations (for example, reference 4). The exact determination of the genetic code, or pattern of correspondence between each possible nucleotide triplet of m-RNA and the amino acid specified by that triplet for incorporation into proteins, has been an active field since about 1961, and current results may be found reviewed in volumes of *Annual Review of Biochemistry*, in the chapters on protein biosynthesis or nucleic acid functions and replication (see also reference 7).

References

1. Nirenberg, M. W., and Matthaei, J. H., "The dependence of cell-free protein synthesis in *E. coli* upon naturally occurring or synthetic polyribonucleotides," *Proc. Natl. Acad. Sci. U.S.*, **47**, 1588–1602 (1961).
2. Crick, F. H. C., "The Genetic Code," *Sci. Am.*, **207**, 66–74 (October 1962).
3. Nirenberg, M. W., "The Genetic Code: II," *Sci. Am.*, **208**, 81–94 (March 1963).
4. Eck, R. V., "Genetic Code: Emergence of a Symmetrical Pattern," *Science*, **140**, 477 (1963).
5. Fraenkel-Conrat, H., "The Genetic Code of a Virus," *Sci. Am.*, **211**, 46–54 (October 1964).
6. Jukes, T. H., "The Genetic Code," *Am. Scientist*, **51**, 227 (1963); "The Genetic Code, II," *ibid.*, **53**, 477 (1965).
7. Crick, F. H. C., "The Genetic Code: III," *Sci. Am.*, **215**, 55 (October 1966).

E. M. Lansford, Jr.

GENETICS, BACTERIAL. See BACTERIAL GENETICS.

GENETICS, BIOCHEMICAL. See BIOCHEMICAL GENETICS.

GENETICS, HUMAN. See BIOCHEMICAL GENETICS; HEMOGLOBINS (IN HUMAN GENETICS).

GENETOTROPHIC PRINCIPLE

This principle can be stated very simply: *The nutritional needs of an organism are determined by its genetic background.*

It is clear that if one gives food that is designed for a green plant to a mammal or fowl, severe malnutrition will result; also, if one gives rat food (which may be quite satisfactory though lacking ascorbic acid) to guinea pigs or monkeys these animals will surely suffer from malnutrition and develop scurvy if the ascorbic acid is absent. By inheritance, guinea pigs and monkeys are not equipped with the metabolic machinery to manufacture ascorbic acid, while rats are so equipped.

The potential importance of the genetotrophic principle lies in the fact that *individuals*, as well as species, differ in their inheritance and, if so, they must *from the quantitative standpoint* have different nutritional needs. This may be the basis of the adage which was stated by Lucretius at least 55 years before Christ: "What is food to one man may be fierce poison to others."

It has been postulated that many diseases such as rheumatoid arthritis, gout, atherosclerosis, dental caries, alcoholism, epilepsy, cataract, acne, mental retardation, multiple sclerosis, muscular dystrophy, schizophrenia, and mental depression may have genetotrophic roots. That is, the individuals who are susceptible to these various diseases (and not all are) may be suffering or may have suffered during early childhood or prenatally, from nutritional deficiencies which arose because of the failure to meet fully their own peculiar individual needs.

Each of the above diseases (and others which may be equally suspect) needs to be investigated specifically to see if the genetotrophic basis can be found. This is, of course, a tremendous task, but since no other hypothesis accounts for many diseases, this one must be followed up.

The possibility that the genetrotophic principle may be operative in the etiology of many diseases of obscure origin is enhanced by the fact that careful determinations of individual needs have been investigated (this is relatively rare), the needs of so-called normal people are found to vary through wide ranges (see, for example, articles on essential amino acids such as LEUCINE, METHIONINE and THREONINE; also BIOCHEMICAL INDIVIDUALITY; ALCOHOLISM; KWASHIORKOR).

The evidence in support of the hypothesis that numerous diseases have a genetotrophic origin is voluminous.

References

1. Williams, R. J., Beerstecher, E. Jr., and Berry, L. J., "The Concept of Genetotrophic Disease," Lancet, 287 (February 18, 1950).
2. Williams, R. J., "Biochemical Individuality," New York, John Wiley & Sons, 1956.

Roger J. Williams

GEOBIOCHEMISTRY

Geobiochemistry deals with the interactions between the biosphere (*i.e.*, the earth's total population of living organisms) and its mineral environment. In practice, geobiochemistry is concerned only with a restricted number of problems, essentially those which do not fall within the scope of any earlier established branch of science, such as coal or petroleum technology, plant nutrition, oceanography, or soil science.

Currently, the principal fields of biogeochemical investigation include the following:

(1) *Contributions of the biosphere*, or the products of its activity, *to the weathering and the building of rocks.* In the "weathering" process, it is now considered possible that lichens and/or the roots of higher plants have a much stronger action than carbon dioxide and water. On the other hand, in the building of limestone formations, the roles played by photosynthesis, and by

changes in the pH of seawater, get greater recognition than formerly.

(2) The mechanisms of *chemical transformations which produced bioliths* such as petroleum and coal, and contemporary phenomena which may fit into such processes.

(3) Studies on the action of the biosphere in modifying the relatively passive geochemical cycles, especially by *concentrating specific elements* at some phase of the cycle, *e.g.*, iodine in sea plants, germanium is some types of vegetation, and uranium and some other metals in peat and other decaying organic material.

(4) The *influence of local variations in a geological substrate on the biosphere* it supports, usually established by observing one or more adverse effects. These adverse effects include: (a) morbidity caused by abnormally high amounts of some less abundant element, such as selenium, or MOLYBDENUM, (b) deficiency diseases caused by a lack of an appropriate amount of some element; *e.g.*, a lack of IODINE contributes to goiter, a lack of MANGANESE causes perosis in fowls, a lack of COBALT or COPPER causes anemia and/or destruction of the nervous system of young cattle. Deficiencies of ZINC, copper, IRON, boron, and molybdenum have been found in plants, mostly in reclaimed land. A balance between several elements may be more important than the absolute amounts of each element present. This aspect of biogeochemistry may usefully be integrated with epidemiological studies.

(5) The geochemical *conditions that have allowed life to appear* at the surface of the earth, including a study (paleobiochemistry) of the organic constituents of fossils, especially amino acids and porphyric pigments.

(6) The action of *heterotrophic bacteria on minerals*, with special emphasis on metallic sulfides. Bacteria are now used for systematically leaching copper from low-grade ore.

(7) The *accumulation of trace elements by specific plants* and their resultant anomalous content of one or more elements that may lead to mutations. This field of study had led to the establishment of new prospecting techniques.

(8) The action of the earth's *radioactive elements* on plants and on people.

Biogeochemistry had its modern beginnings in the USSR in the nineteen twenties, pioneered by Academician V. I. Vernadsky, and in Scandinavia. In these countries, the possibility was conceived of finding buried ore bodies by sampling the nearby vegetal matter. In the decade preceding and during World War II, investigations were carried out for this purpose. By the end of World War II, such biogeochemical investigations by scientific prospectors had extended to great Britain, the United States, and Canada. After a flurry of interest of a decade or so, biogeochemistry became less popular among many geologists than its associated mine-finding science, geophysics. In those countries where biogeochemistry made its headway, for the most part in the USSR, Finland, and Canada, most of the soils were of transported glacial material and were so young that they had poorly developed profiles. In Great Britain, where geologists worked largely on African and Australian problems, and in the United States, most of the areas that were subjected to geochemical studies possessed residual soils with comparatively well-developed soil profiles. Thus, geochemists in the United States and Great Britain tended increasingly to use soil sampling methods rather than geobiochemistry. Biogeochemical techniques are presently accepted in Finland and in the USSR, and to a limited extent in Canada, as having value in prospecting.

Biogeochemistry has much to contribute to other disciplines far removed from geology and mine finding. In the last few decades, investigators in agriculture and forestry, stimulated by the population explosion, have been making desperate efforts to increase both quantitatively and qualitatively the world's supply of food and timber. All have realized the need for better knowledge of many factors, one of the more important of which is the quality and quantity of mineral nutrients in the soil, usually determined by soil sampling. However, in recent years foliar analysis has been found useful in diagnosing the nutrient status of a soil, more useful in some instances than soil analysis. A foliar analysis tells what a tree has been able to absorb; a soil analysis tells only what a tree has available for its use. Two trees growing on identical soil may absorb widely differing amounts of nutrients in that soil; a change, for example, of the pH of a soil may alter the availability of one or more elements in the soil.

The growing importance of biogeochemistry is partly the result of the improved analytical techniques that have largely been developed in the last two decades.

Biogeochemistry, as a branch of geochemistry, offers a great challenge to those interested in wresting new secrets from nature and relating man to the earth on which he lives.

References

1. RANKAMA, K., AND SAHAMA, T. G., "Geochemistry," p. 35, Chicago, University of Chigaco Press, 1949.
2. GINZBURG, I. I., "Principles of Geochemical Prospecting," Intern. Series Monographs on Earth Sciences, Vol. 3, New York, Pergamon Press, 1960 (translated by V. P. Sokoloff).
3. VINOGRADOV, A. P., "The Geochemistry of Rare and Dispersed Chemical Elements in Soils," second edition, New York, Consultants Bureau Inc., 1959.
4. HAWKES, H. E., AND WEBB, J. S., "Geochemistry in Mineral Exploration," New York, Harper and Row.
5. SMALES, A. A., AND WAGER, L. R., "Methods in Geochemistry," New York, Interscience Publishers, 1960.
6. ABELSON, P. H., "Paleobiochemistry," *Sci. Am.*, **195**, 83–92 (July 1956).

H. V. WARREN AND R. E. DELAVAULT

GERMFREE ANIMALS (BIOCHEMICAL ASPECTS)

Pasteur in 1885 stated that life might become impossible in animals deprived of their normal microbial flora. This thinking represented the belief that the bacterium-harboring animal represented the "fittest" form, which consequently evolved from phylogenetic development.

Opposed to the ideas of Pasteur were those biologists who felt that the intestinal flora can be dangerous to the host. Metchnikoff (1901) supposed that toxic products of bacterial metabolism continuously poison the organism. Around the turn of the century, the first attempts were made to test these hypotheses by raising animals which did not harbor *any* bacteria: *i.e.*, germ-free animals. In 1912 Cohendy succeeded in rearing germ-free chickens of remarkable quality, both in terms of growth and general health.

However, a general lack of knowledge in the field of nutrition, and especially of the effect of the necessary sterilization of food hampered early work in this field. Increased insight in nutritional problems enabled Glimstedt to raise germ-free guinea pigs in 1935. Gustafsson reported the raising of germ-free rats in 1946. Starting in the late twenties, Reyniers and associates established a center for germ-free research at the University of Notre Dame, Indiana (the present Lobund Laboratory). They reported the raising of germ-free guinea pigs in 1932, the raising of germ-free rats in 1946, mice in 1956 and rabbits in 1959. Breeding colonies of germ-free rats, mice, guinea pigs and rabbits were subsequently established and are currently in use.

Methods and Apparatus. Germ-free birds are obtained by hatching germicide-treated eggs in presterilized isolators. Smaller germ-free mammals are delivered at term by caesarian operation into a presterilized isolator. Larger germ-free mammals (lambs, goats, swine, mules, cows) are delivered either by direct caesarian operation into the isolator or by hysterectomy and germicidal treatment of the pregnant uterus. The newborn mammals are then handfed compounded milk formulas or purely synthetic diets until they can be weaned to solid food. Guinea pigs can be fed a moist but solid formula directly after birth.

Isolators to house germ-free animals can be

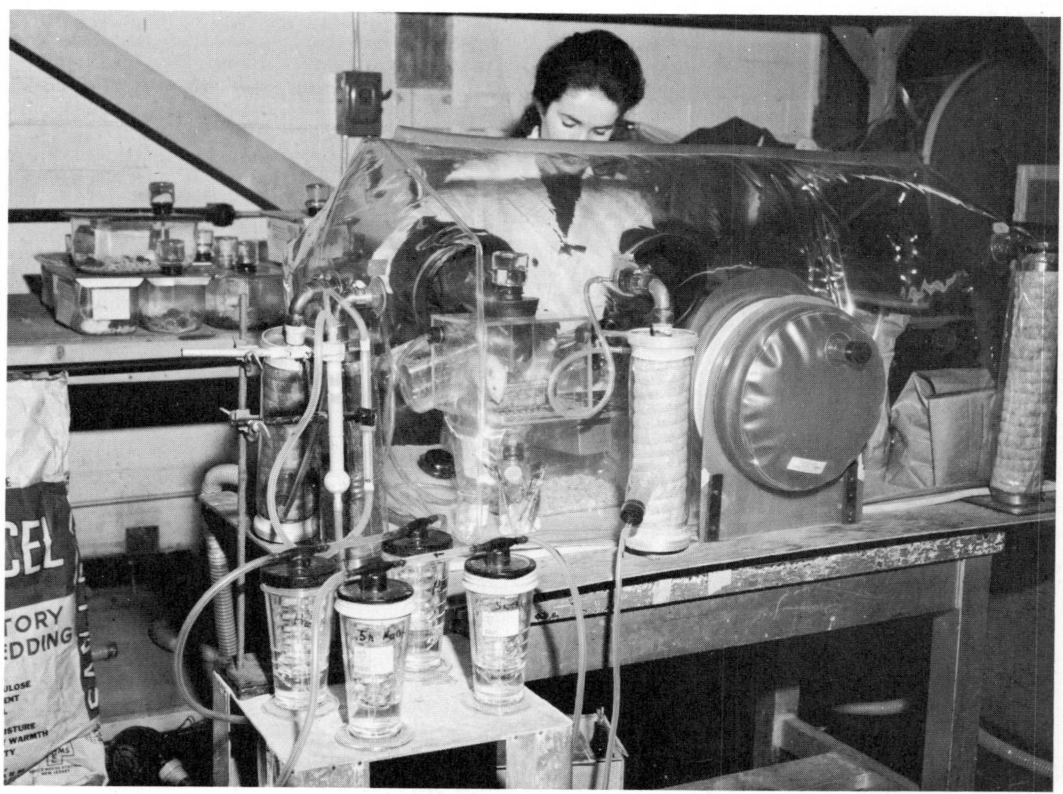

FIG. 1. Oxidative catabolism of cholesterol-26-C^{14} freeing $C^{14}O_2$, measured in germfree rat in metabolism chamber inside plastic isolator. *Left to right:* incoming general air supply via hose and glass-fiber filter. *Foreground:* equipment for washing nitrogen-oxygen mixture flowing through metabolism chamber (filter not visible) and for trapping $C^{14}O_2$ released in metabolism chamber and reaching absorption flasks via filter for outgoing air.

classified as the steel autoclave type (Reyniers), the rigid steel steam-sterilizable type (Gustafsson), and the plastic type (Trexler). The Reyniers-type isolator consists basically of a round autoclave with two or four portholes fitted with gloves for inside manipulation. Glove ports can be closed on the outside with steel covers and the inside of the isolator is then sterilized with pressurized steam. The Gustafsson type requires an autoclave large enough to insert the unit for steam sterilization. The Trexler-type isolator, which consists essentially of a flexible bag made out of pliable vinyl or polyethylene plastic (*e.g.*, 24 × 24 × 48 inches), is sterilized by chemical means (mostly with an atomized 2% solution of peracetic acid). All types are outfitted with glass-fiber filters for incoming air and with similar filters or liquid germicidal traps in the air exhaust. All have sterilizable locks, consisting of a two-door system with a space in between, which permit the introduction of food and other material. After introduction through the outside door, the material in the lock is sterilized with steam or by chemical means, after which the inside door to the main sterile compartment is opened to receive the material.

Characteristics of Germ-free Animals. Germ-free animals are, by definition, free of all detectable microbes. The methods by which germ-free animals are derived do not guarantee, however, that these animals are also free of all viruses. Leukemia and mammary tumor agents have been detected in germ-free mice, and it is possible that other viruses may be found in the future.

Conventional animals develop in an environment where the bacterial microflora is an ever-present challenge. Germ-free animals do not meet this constant challenge, and their adjustment to the less provocative environment is modified accordingly. Externally, germ-free animals resemble their conventional counterparts. On nutritionally adequate sterile diets, their growth is comparable to, or only slightly less than, that of the conventional animal, while the presently available data on rats and mice point to a longer life span in the absence of bacteria. Germ-free rats, mice, rabbits and guinea pigs appear well adjusted, are in good health and reproduce normally.

Internally the germ-free animal shows many signs of the absence of a microflora (see INTESTINAL FLORA). In rodents, the enlargement of the cecum to 5–10 times its normal size is most spectacular. This cecum is filled with contents which are more fluid than usual, indicating changes in the water-household of the animal. It has been demonstrated that in the absence of a "normal" intestinal microflora, certain toxic and musculo-active substances accumulate in the intestinal and cecal contents. These so far unidentified materials may play a role in the abnormal enlargement of the cecum. A further reflection of the absence of bacteria is found in the reduction of the size of spleen and lymph nodes, while in these tissues the concentration of potential antibody-forming cells again is lower than in conventional animals. As a result, the concentration of the gamma (immune) globulins in the serum of germ-free animals is drastically reduced. However, upon contact with bacteria or other antigenic materials, the animal demonstrates that it has not lost its potential of ANTIBODY formation. It responds well to the challenge, albeit somewhat slower and with lower antibody concentrations.

The absence of a microflora is also reflected in a generally thinner small intestine. The intestinal wall proves to be more permeable (passive absorption), but the active transport of digested diet ingredients (simple sugars, amino acid and minerals) is comparable to that of animals harboring a normal population of intestinal bacteria. Oxidation-reduction potentials of the germ-free intestinal contents show values which are on the average 300 mV more positive. Heart and liver have a tendency to be somewhat smaller in the germ-free animal, while studies in rats show that the amount of circulating blood and the cardiac output are reduced. These studies also indicate a reduced work load for the liver, reflected in a much lower blood supply and lower levels of cocarboxylases involved in energy production.

All above-mentioned dissimilarities demonstrate that germ-free animals have come to terms with their environment in ways which are often different from those of their conventional counterparts. As a result, many of the functional aspects of the germ-free animal are different from normal. A knowledge of these deviations from "normal" is of prior and prime importance in the interpretation of results from experiments using germ-free animals to solve specific problems in the life sciences.

Applications in Research. The research potential of the germfree animal is manifold. Basically, not only does this research tool make it feasible to study aspects of systemic metabolism without intestinal microbiota partaking in or at least affecting the metabolic processes, but it opens the possibility of the study of the specific effect of well-defined bacterial species on the metabolic functions of the host.

Germ-free animals generally show a higher concentration of digestive and metabolic enzymes in the mucosal cells of the small intestine (and cecum). This apparently relates to the prolonged life span of these cells in the absence of flora stimulation. The oxidation-reduction potential of the intestinal contents is more positive and in certain cases leads to deficient iron absorption and ANEMIA (rabbits). The germ-free state reveals dietary deficiencies in cases where the animal normally relies on microbial production of certain nutrients, notably vitamins, to cover its requirements (*e.g.*, VITAMIN K and FOLIC ACID). In these cases, where microbial production otherwise covers only part of the animal's requirement, as in the case of thiamine, the germ-free animal reveals dietary deficiency much earlier than its conventional counterpart. Diets marginal in certain amino acids, on the other hand, generally induce better growth in germ-free animals, indicating a certain "consumption" of these amino acids by the microflora of the conventional animals. By the same token, germ-free guinea pigs require less ascorbic acid than conventional guinea pigs. The germ-free animal thus proves ideal to determine systemic nutrient requirements as a base line for practical requirements which are often strongly affected by the presence of an ever-changing microflora. This is especially true since low molecular, water-soluble, chemically defined diets are now available, which will maintain germ-free rats and mice and which will make a precise definition of systemic requirements possible.

In many cases, the intestinal microflora takes an active part in the metabolic cycles of the host. In cholesterol and BILE ACID metabolism, intestinal bacteria increase cholesterol and bile acid turnover by conversion of these compounds to products which are less readily reabsorbed and therefore excreted to a greater extent. Germ-free animals again provide a base-line of systemic turnover, against which the effect of specific bacterial species on cholesterol and bile acid metabolism can be measured by associating the animals with the selected species. It has been shown that the cholesterol-lowering effect of certain lipid fractions was demonstrable in both germfree- and conventional rats, thereby indicating a direct influence on systemic cholesterol anabolism or catabolism. Comparison of results obtained with germ-free and conventional animals has likewise permitted separation of systemic and microbial contributions to the degradation of urea to NH_3 and CO_2. Studies with germ-free rats have established that vitamin K_1 is more efficient than vitamin K_3, on both a prophylactic and a therapeutic basis. This seems to refute the central position of vitamin K_3 as a nucleus from which all forms of vitamin K are produced which function on the cellular level in mammals.

The use of the germ-free animal has revealed syndromes in which bacterial metabolites play an essential role and has made specific studies on biochemical aspects possible. The germ-free animal never shows dental CARIES on diets which strongly affect its conventional counterpart. Association of germ-free rats with strongly acidogenic homofermentative lactic acid bacteria like *Lactobacillus casei* results in only mild cariogenesis in low incidence. *Streptococcus faecalis* as well as certain other streptococci produce rampant caries in these animals though they are slightly less acidogenic than *L. casei*. Their slightly greater heterofermentative character may indicate that acids or products other than lactic acid may be important in the syndrome. On the other hand, heterofermentative *Lactobacillus* strains have not been found to be cariogenic.

Germ-free animals will provide many clues in the study of the complex phenomena of immunity (see IMMUNOCHEMISTRY). Association of these animals with strongly pathogenic forms reveals changes in the concentration of histamine, 5-hydroxytryptamine and possibly of the polypeptides known to play an active role in the inflammatory response. The absence of bacteria, coupled with the feeding of chemically pure diets, makes it possible to exclude bacterial endotoxins and other stimulating macrocomplexes from diet and environment and thus provides an experimental tool for the study of the role of endotoxins in immune response and in the radiation syndrome.

The germ-free animal thus provides an ideal tool to evaluate the many influences of the microbial flora on the biochemical systems which underly the life processes.

References

1. "Germfree vertebrates: Present Status," *Ann. N. Y. Acad. Sci.*, **78**, 1–400 (1959).
2. GORDON, H. A., *J. Exptl. Biol. and Exptl. Medical Res. on Aging*, **3**, 104–114 (1959).
3. "Proceedings III. Symposium on Gnotobiotic Technology," *Lab. Animal Care*, **14**, 569–670 (1963).
4. WOSTMANN, B. S., *et al.*, *Federation Proc.*, **22**, 120–124 (1963); *J. Lipid Res.*, **7**, 77–82 (1966).
5. PLEASANTS, J. R., WOSTMANN, B. S., AND ZIMMERMANN, D. R., *Lab. Animal Care*, **14**, 37–47 (1964).
6. LUCKEY, T. D., "Germfree Life and Gnotobiology," New York, Academic Press, 1963.

BERNARD S. WOSTMANN

GERMINATION (BACTERIAL SPORE).

See SPORES, BACTERIAL.

GERMINATION (PLANT SEED)

The seed plants (*Angiospermae* and *Gymnospermae*) comprise the most familiar part of the earth's vegetation and are important from an economic as well as scientific viewpoint. Most species of seed are dormant when freshly harvested and require varying periods of rest before they become viable, *i.e.*, ready to germinate. Factors which affect this dormancy period and stimulate the rate of seed germination have been widely studied; however, methods of breaking seed dormancy vary in different species. Physical requirements found effective for stimulating germination have included high or low external temperatures, or a daily alternation of temperature; specific light-dark periods; scarification or artificial scratching of the seed coat. In addition, a number of chemical pretreatments of seed have been found to regulate germination; some of these are of natural origin giving rise to symbiotic association of certain plants and microorganisms.

The initial stimulation of germination of a seed is not believed to involve any nutritional factors, since the seed embryo is surrounded by an endosperm (source of stored food) enclosed in the seed coat. However, it is apparent that the imbition of water and oxygen by the seed is required prior to emergence of the hypocotyl and subsequent germination.

For light-sensitive seeds, two photoreactions exist which are reversible, one with a maximum in the "near-red" region (6400–6600 Å) which promotes germination, and the other in the "far red" region (7200–7400 Å) which reverses the light activation; however, some effects may frequently be observed throughout the whole visible range of the spectrum. Temperature-light interactions have also been observed; *e.g.*, lettuce seeds which have been activated by the "near-red" treatment will germinate in the dark at temperatures below about 30°C, but if comparably treated seeds are held at 35°C for short periods they fail to germinate even when placed in an optimum germination temperature (20°C). It is difficult to reduce the multitude of observations on light-temperature relationships in seed germination to a direct biochemical function; however, it does appear that temperature effects are due to biochemical mechanisms other than the light activation.

While auxins (*e.g.*, indoleacetic acid) in general have not been found to induce seed germinations, chemical stimulation of germination has been observed with a number of different compounds including 6-(substituted)-purines (kinetin and its analogues), and several of the gibberellins (see HORMONES, PLANT). In most instances, the effects of these compounds are augmented by the presence of small doses of light, inactive themselves in affecting the rate of germination; thus, the kinins are believed to increase the efficiency with which the photoreaction brings about germination. A synergism has recently been demonstrated between 6-(substituted)-purines and various gibberellins in augmenting the rate of lettuce seed germination.

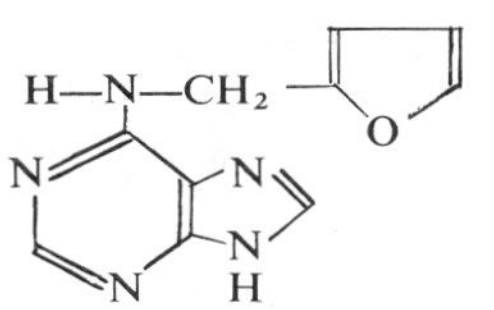

6-Furfurylaminopurine (Kinetin)

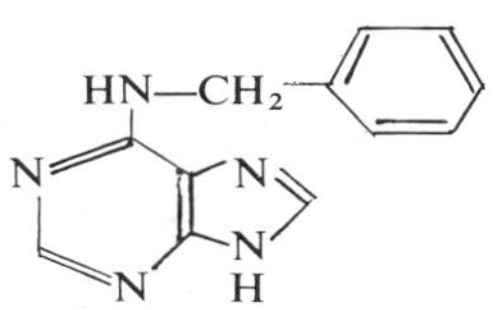

6-Benzylaminopurine

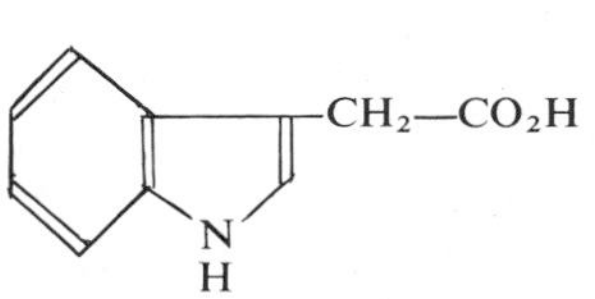

Indoleacetic acid (Auxin)

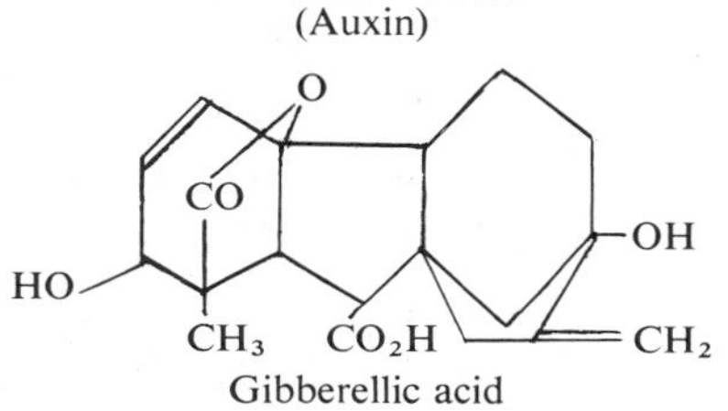

Gibberellic acid

The function of these germinants as well as others such as thiourea, oxygen enrichment, etc., is not clear; however, many investigators consider it reasonable to suppose that these compounds in some manner destroy or bypass naturally occuring inhibitors which normally prevent premature germination. Higher concentrations of these and other comparable reagents actually prevent germination in certain varieties of seed; thus, this phase of plant physiology still remains an important area for future research.

References

1. CROCKER, W., AND BARTON, L. V., "Physiology of Seeds," Waltham, Mass., Chronica Botanica Company, 1957.
2. KOLLER, D., MAYER, A. W., POLJAKOFF-MAYBER, A., AND KLEIN, S., "Seed Germination," *Ann. Rev. Plant Physiol.*, **13**, 437 (1962).
3. AUDUS, L. J., "Plant Growth Substances," London, England, Leonard Hill and Co., 1959.

C. G. SKINNER

GERONTOLOGY (BIOCHEMICAL ASPECTS)

Gerontology is the scientific study of aging. Aging represents the progressive changes which take place in a cell, tissue, organ or organism with the passage of time. The changes which occur after attainment of maturity are of primary interest in gerontology. Geriatrics is the branch of medical science concerned with the prevention and treatment of the diseases of older people and is part of the broader field of gerontology. For the most part, age changes represent a gradual loss in functional capacity which ultimately results in the death of the cell or organism. In fact, for humans and many other animals such as rats, mice, dogs, cats, and even insects, the probability of death increases logarithmically with age.

Although there are many aspects of gerontology, biochemistry must ultimately play a key role in understanding the mechanisms of impairments in cellular function and death which occur even in metazoans such as man. Although biochemists have greatly expanded our knowledge of the mechanisms of energy transformation and protein synthesis, they have given little attention to the effects of age of the cell or tissue on these processes. The determination of such information is of great importance for the science of gerontology and the understanding of aging.

Biochemical theories of aging fall into three major categories, *viz.*, exhaustion of essential materials or accumulation of toxic substances, eversion of complex molecules, and errors in the synthesis of large molecules.

Exhaustion or *accumulation theories* assume that aging results from the exhaustion of some essential material or the accumulation of toxic or deleterious materials in cells. In view of the turnover rates which have been shown for most cellular constituents, it is doubtful whether the exhaustion of any specific material can be the cause of aging. Of course, specific cells in a metazoan may die when they are deprived of their normal source of nutrients by interference with the blood supply. This, however, is usually based on pathological processes, such as arteriosclerosis, and is not a basic mechanism of aging. If there is an impairment in the synthetic mechanisms required for the turnover and replacement of key molecules, exhaustion may occur. However, exhaustion is not the primary factor. The breakdown in the replacement mechanisms is the primary process, and it will be considered later in connection with "error" theories.

The presumption that the accumulation of deleterious substances contributes to aging receives support in the studies of Carrel who found that tissue cultures of chicken heart could not be maintained if serum from old chickens was used in preparing the culture medium. It still remains for biochemists to isolate such a substance from the blood of senescent animals. More recently, it has been found that highly insoluble granules accumulate with advancing age in cells from certain tissues, such as the heart and nervous system. These granules, called "age pigments," are composed of varying proportions of lipid and protein and may occupy a substantial part of the cell at advanced ages. The accumulation of peroxides and free radicals as well as S—S groups in the tissues of old animals has also been proposed as a cause of aging. However, although a slight increase in life span of rats fed various ANTIOXIDANTS (to remove peroxides and free radicals) has been reported, no direct evidence is available on the effect of age on the concentration of peroxides or free radicals in tissues.

The fact that alterations in environmental temperatures can significantly alter longevity in poikilothermic animals has also been interpreted as evidence for the exhaustion or accumulation theory of aging, since it is assumed that changes in temperature will influence the rate of chemical reactions in the animal and hence the rate of utilization of essential materials or the rate of formation and accumulation of deleterious substances. The life spans of *Drosophila* and *Daphnia* are significantly shorter in animals reared at 27°C than in those reared at 15°C. Exposure to high temperature for a short period of time does not influence the mortality curve of the surviving *Drosophila* so that the life shortening effect of the high temperature cannot be attributed to denaturation of essential proteins, but must be related to the rates of chemical reactions in the animal.

In the rat, as well as some other species such as the rotifer, restriction in the amount of food given increases life span. The dietary restriction must begin early in life (at weaning in the rat) and continue throughout life in order to be effective. This treatment prolongs the period of growth, but no firm biochemical explanation for increased longevity can be given although some investigators have assumed that starvation, by prolonging the growth period, delays the accumulation of unknown deleterious products.

Eversion theories of aging are based on the changes in structure and configuration of molecules that take place with the passage of time after they have been formed. Evidence for this theory is based primarily on the changes which take place in connective tissue with advancing age. COLLAGEN from old animals is less readily solubilized than that from young. Its thermal contractility is reduced and in general it attains a more rigid physical and chemical structure. These changes in properties have been attributed to the formation of cross linkages in the collagen molecule which are similar to those induced in the tanning of leather. Some investigators have ascribed these changes to the presence of aldehydes in the body which serve as effective cross-linking agents. Molecular changes also take place in elastin which result in decreased elasticity in many tissues, such as skin, blood vessels, etc., with advancing age. The formation of cross-links in the elastin molecule is regarded as the basic mechanism of these age changes. Thus there is good evidence that molecular changes occur with aging in extra-cellular proteins such as collagen and elastin. Similar changes may also occur in intracellular proteins. Preliminary experiments indicated a significant age difference in the melting temperature of DNA isolated from thymus glands of old and young cattle, which led to the presumption that structural changes in the DNA molecule had taken place with aging. However, subsequent experiments showed that the differences in melting temperatures were due to differences in the histones associated with the DNA rather than to changes in the DNA molecule itself. There is thus some evidence for alterations in an intracellular structural protein or in the DNA-protein complex. Small changes in molecular structure of other intracellular proteins, such as enzymes, might well interfere with their participation in essential biochemical reactions in cells with advancing age.

The *error theory* of aging provides the most attractive formulation to explain most of the facts of aging which are known at the present time, although it is far from proved. The error theory is based on the hypothesis that information with regard to the synthesis of cellular proteins resides in the DNA molecule within the nucleus of the cell. This information is transmitted by messenger RNA from the nucleus to the sites of protein formation in the RIBOSOMES of cells [see also GENES; PROTEINS (BIOSYNTHESIS)]. It is assumed that with increasing age, slightly atypical molecules of messenger RNA are formed so that errors occur in the formation of protein molecules. If the protein molecules which contain errors are enzymes, either they may be completely incapable of participating in the essential chemical reactions in cells, or, they may do so at slower rates. The result of either condition would be an accumulation of the substrates on which the enzymes act. Because of the feedback mechanisms which operate in cells, the accumulation of substrates may stimulate an increased production of messenger RNA and enzymes. When this increase is insufficient to produce adequate amounts of functional enzymes, the cell dies.

Evidence for the error theory stems from a

number of sources. Some studies have indicated the predicted increase in messenger RNA in cells of some tissues as well as decreases in the activities of some enzymes with advancing age in the rat. For example, the succinoxidase activity of cardiac muscle and renal cells of the rat declines by 15–20% between adult (age 10–12 months) and senescent (age 24 months) animals. However, other enzymes such as D-amino acid oxidase and pyrophosphatase remain relatively stable throughout the life span. More detailed studies on isolated mitochondria have failed to show a significant age decrement in specific enzyme activity. It seems probable that any age decrement in mitochondrial enzymatic activity at the cellular level is due primarily to a reduction in the number of mitochondria present in cells.

The error theory also receives support from the observed life-shortening effects of exposure to non-lethal amounts of radiation. Radiation shows its primary biological effects in inducing alterations in DNA which produce mutations in cell lines. These mutations are essentially errors which reduce the viability of the daughter cells. Recently it has been shown that the incidence of atypical chromosomes in liver cells in the rat is increased both by aging and by exposure to radiation.

The important role of DNA in aging is also reflected in genetic differences in life span between different species of animals as well as differences within the same species. The May fly is reported to have a life span of one day in contrast to the potential human life span of 100 years. The range of life spans among different species is much greater than the range of individual life spans within a species and can be regarded as a reflection of differences in DNA. However, selective inbreeding within a species will separate strains with significantly different life spans. In general, inbred strains have shorter life spans than their heterozygous ancestors.

Investigation of the basic mechanisms of senescence is a fertile field for biochemists.

References

1. Barrows, C. H., Jr., Roeder, L. M., and Falzone, J. A., "Effect of Age on the Activities of Enzymes and the Concentrations of Nucleic Acids in the Tissues of Female Wild Rats," *J. Gerontol.*, **17**, 144–147 (1962).
2. Comfort, A., "Ageing: the Biology of Senescence," New York, Holt, Rinehart and Winston, Inc., 1964, xvi, 365pp.
3. Curtis, H. J., "Biological Mechanisms Underlying the Aging Process," *Science*, **141**, 686–694 (1963).
4. de Ropp, R. S., "Man against Aging," New York, St. Martin's Press, 1960, 310pp.
5. Korenchevsky, V., *Physiological and Pathological Ageing*, New York, Hafner Publishing Co., 1961, 514pp.
6. Lansing, A. I. (Editor), "Cowdry's Problems of Ageing: Biological and Medical Aspects," Third Edition, Baltimore, Williams & Wilkins Co., 1952, xxiii, 1061pp.
7. Shock, N. W. (Editor), *Aging: Some Social and Biological Aspects*, American Association for the Advancement of Science, Washington, 1960, viii, 427pp.
8. Strehler, B. L., "Time, Cells, and Aging," New York, Academic Press, 1962, x, 270pp.
9. Verzár, F., "Aging of the Collagen Fiber," in "International Review of Connective Tissue Research," (Hall, D. A., Editor) Vol. 2., pp. 243–300, New York, Academic Press, 1964.
10. von Hahn, H. P., "Age-dependent Thermal Denaturation and Viscosity of Crude and Purified Desoxyribonucleic Acid Prepared from Bovine Thymus," *Gerontologia*, **8**, 123–131, 1963.
11. Wulff, V. J., Quastler, H., Sherman, F. G., and Samis, H. V., "The Effect of Specific Activity of H^3-cytidine on its Incorporation into Tissues of Young and Old Mice," *J. Gerontol.*, **20** (January 1965).

Nathan W. Shock

GIGANTISM

Gigantism, excessive skeletal growth and body weight, is one of the results of an excess of the adenohypophyseal GROWTH HORMONE, also termed *somatotrophin*.

GLOBULINS

Globulins are proteins that are insoluble in water but dissolve readily in aqueous salt solutions (see PROTEINS, section on Classification of Proteins). The term globulins is applied to certain subgroups of the PLASMA PROTEINS.

GLOMERULAR FILTRATION

Glomerular filtration is the process occurring in the kidney of most vertebrates by which the aqueous solution of crystalloidal substances in the blood is separated from the cells and colloidal particles of proteins and lipids. This aqueous solution, the glomerular filtrate, first appears in Bowman's capsular space, or the urinary space. The filter for this separation process, a tuft of capillaries (the glomerulus) with its capsule, is the renal (Malpighian) body or corpuscle. The term glomerulus is sometimes used to include both the tuft of capillaries and the capsule. Recent electron microscopic evidence indicates that the glomerulus and its capsule probably develop as a unit from the blind end of the uriniferous tubule instead of by invagination of the capillary tufts into Bowman's capsule.

Anatomy. Bowman, in 1842, described the gross morphological aspects of this part of the nephron; *i.e.*, an afferent arteriole through which blood enters the tuft of capillaries with the blood leaving through an efferent arteriole. Bowman's original description of the branching of the afferent arteriole to form the glomerular capillary tuft has been confirmed by current studies. As the afferent arteriole penetrates Bowman's capsule, it branches into several primary vessels (Figs. 1

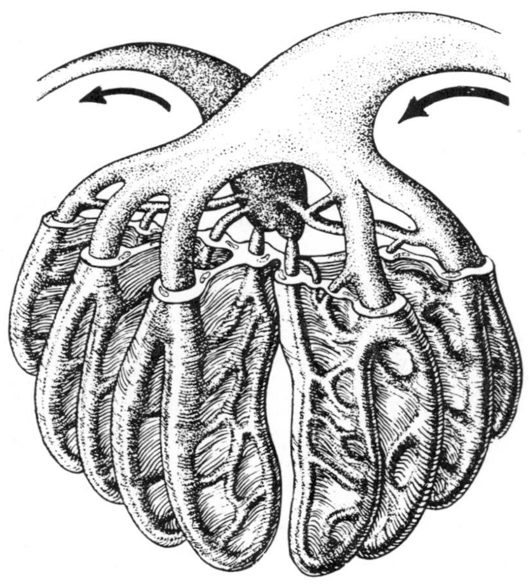

FIG. 1. Schematical stereogram of human renal glomerulus showing capillary lobule around channels arising from afferent arteriole. Arrows indicate blood flow into (afferent arteriole) and out (efferent arteriole) of glomerulus. [From Elias, H., Hassmann, A., Barth, I. B., and Solmor, A., *J. Urol.* **83**, 795 (1960).]

and 5). These primary vessels divide into five to eight secondary through-channels which give rise to capillaries, which in turn anastomose with other capillaries and form a lobule with each secondary channel. The capillaries recombine and join the secondary channel which then becomes the efferent arteriole. The efferent arteriole exits from the capsule adjacent to where the afferent arteriole enters. This arrangement of through-channels with a network of capillaries is structurally suited for the suggested skimming of plasma into the capillaries where filtration occurs. The larger amount of blood containing most of the cells goes directly through the larger through-channels to the efferent arteriole.

The relationship of the glomerulus to the rest of the nephron is depicted in Fig. 2. The two types of nephrons are shown; cortical nephrons with the glomerulus relatively close to the outer cortex and short loops of Henle, and the juxtamedullary nephrons with the glomerulus near the cortico-medullary junction and long loops of Henle dipping down into the inner medulla. In man, the cortical nephrons compose about 80% of the total number and they receive the greater proportion of the blood supply. There are approximately one million nephrons in each human kidney.

Electron microscopic studies have revealed details of the capillary membrane of the glomerulus (Fig. 3). The three main layers are an endothelium, a basement membrane and an epithelium. The schematic representation in Fig. 4 shows the endothelium or the lamina fenestrata

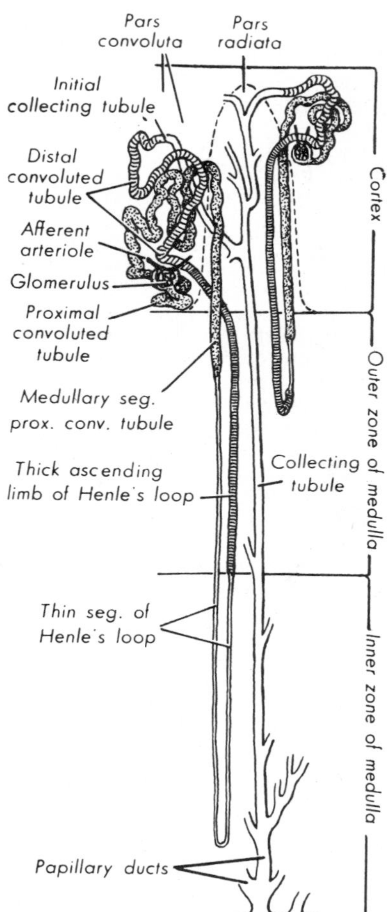

FIG. 2. Diagram of subdivisions of uriniferous tubules to show their relations and locations in a section extending from the outer cortex to the last segments of collecting tubules. (Redrawn and modified from Peter, "Bailey's Textbook of Histology," p. 458, Fifteenth edition, Baltimore, The Williams and Wilkins Co., 1964).

(lamina attenuata) to have pore-like openings, estimated to be from 400–900 Å. This layer probably does not function as a filter for plasma proteins which have effective diameters smaller than these openings. The basement membrane, the lamina densa, situated between an outer and inner cement layer, is believed to be a hydrated gel composed of MUCOPOLYSACCHARIDES. This is considered by many investigators to be the essential filtering membrane. The epithelial layer (visceral layer of Bowman's capsule) is believed by others to be the primary area of separation. The large epithelial cells, podocytes, have extended projections called trabeculae. There are many smaller processes termed pedicels which interdigitate and appear to be anchored in the outer

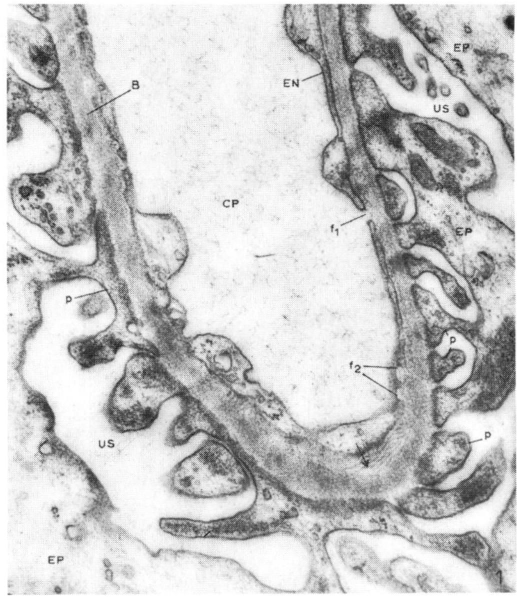

FIG. 3. Electron micrograph of glomerular capillary loop (CP). The three layers of the capillary wall are: the endothelium (EN) with its fenestrae or openings (f_1, f_2), the basement membrane (B), and the epithelium (EP). Bowman's capsular space or urinary space is indicated by US. Several trabeculae of the visceral epithelial cells (podocytes) can be seen at lower right and left, and the interdigitating foot processes (P) are shown. Fine fibrils are indicated (arrow) in the narrow space between the basement membrane and endothelium. [From Farquhar, M. G., Wissig, S. L., and Palade, G. E., "Glomerular Permeability. I. Ferritin Transfer Across the Normal Glomerular Capillary Wall," *J. Exptl. Med.*, **113**, 47–66, Plate 9 (1961).]

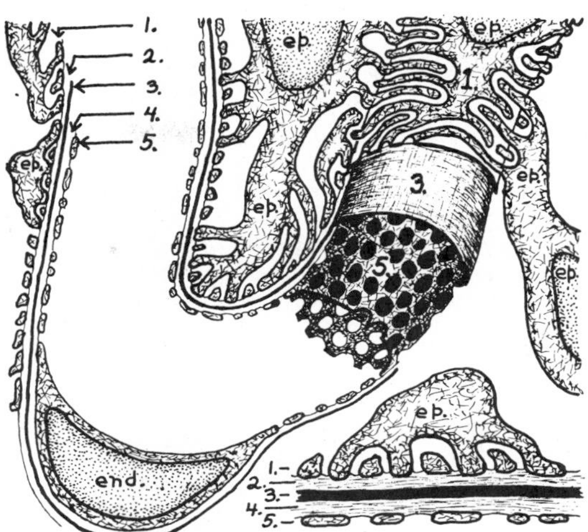

FIG. 4. A schematic illustration of the glomerular capillary. The epithelial cells (ep) or podocytes with processes interdigitating upon capillary surface (layer 1) are shown. A cross section of a podocyte and the other layers are depicted at lower right. The epithelial feet (pedicels) appear to be embedded in the outer cement layer (2). The basement membrane (3) and the inner cement layer (4) and the endothelium (5) are shown. An endothelial cell (end) can be seen at lower left. [From Pease, D. C., "Fine Structures of the Kidney Seen by Electron Microscopy," *J. Histochem.*, **3**, 297, (1955).]

cement layer of the basement membrane. The podocytes with their processes appear to cover the surface of the capillaries. "Pore-slits" between pedicels are found regularly and are estimated to be 100 Å wide.

The space between the visceral and parietal layers of Bowman's capsule (the urinary space) is continuous with the lumen of the proximal tubule, and fluid appearing here drains into the tubule lumen (Fig. 5). The glomerular filtrate successively passing through the rest of the nephron normally is reduced in volume at least 99%. Of the 180 liters of filtrate formed in 24 hours by the two human adult kidneys, only 1–2 liters are excreted as urine.

Mechanism of Filtration. The functional aspects of the glomerulus were described most accurately first by Ludwig in 1844. His idea that the hydrostatic pressure of the blood resulting from the heart beat was the force necessary for glomerular filtration is the generally accepted theory today. The mode of penetration of the filtrate across the capillary wall is still unknown, and a controversy

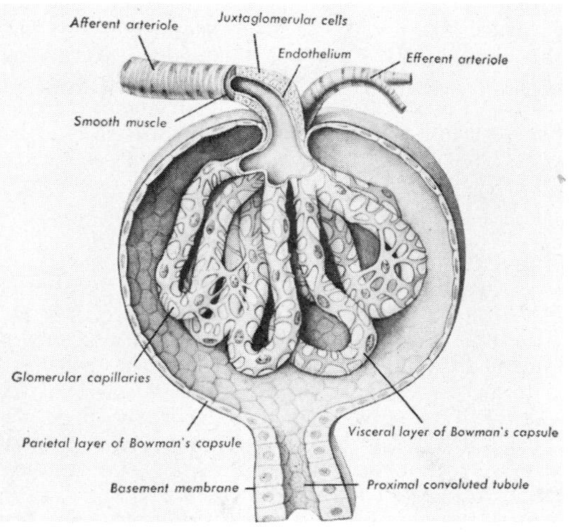

FIG. 5. Diagram of structure of renal corpuscle showing relationship of Bowman's capsule to glomerular capillaries and continuation of "urinary space" with lumen of proximal tubule. (Redrawn and modified from Bargmann, "Bailey's Textbook of Histology," p. 461, Fifteenth edition, Baltimore, the Williams and Wilkins Co., 1964.)

exists as to the best interpretation. The penetration is claimed by some workers to occur by diffusion. Others recently have implicated a pinocytotic process occurring in the podocytes in the formation of the glomerular filtrate. The generally held concept of ultrafiltration based on the idea of pores or channels through which particles of a certain size can pass has not been substantiated by electron microscopic evidence of such pores in the basement membrane. Functionally, however, in dogs the glomerular capillary membrane does behave as though it possesses pores of a diameter from 75–100 Å. Serum ALBUMIN has an effective diameter of about 70 Å. It has been demonstrated in micropuncture experiments that small amounts of serum albumin are filtered which are reabsorbed by the proximal tubules. Particles with larger diameters than albumin are not filtered readily.

The criteria for ultrafiltration in the formation of the glomerular filtrate as given by Pitts are: (1) the ultrafiltrate must be protein free, (2) the crystalloids should be in the same concentration in the filtrate as in the plasma, excepting the slight inequality due to the Gibbs-Donnan effect resulting from the negatively charged proteins in the plasma, and (3) the hydrostatic pressure of the blood should be adequate for the process. Micro puncture studies have confirmed that the filtrate is essentially protein free and that the concentration of crystalloids across the filtering membrane are equal. Pressures of the order of 25–30 mm Hg are adequate for the observed rate of glomerular filtration.

In man, the glomerular capillary hydrostatic pressure (Pg) is about 70% of the mean arterial pressure or equal to 70 mm Hg. Opposing this pressure is the 30 mm Hg colloidal osmotic pressure (Po) due to the proteins in the blood, and the 10 mm Hg tissue or intracapsular pressure (Pi). Therefore, the effective filtering pressure (Pf) is about 30 mm Hg. With mean arterial pressure of 80 mm Hg, filtration decreases; with pressures of 60 mm Hg or below, filtration ceases.

$$Pf = Pg - (Po + Pi)$$

$$30 = 70 - (30 + 10)$$

Clearance Concept. The rate of glomerular filtration can be estimated easily and fairly accurately. The *clearance* of a certain volume of plasma of a substance which meets the following criteria is a measure of glomerular filtration rate (GFR). This substance must be freely filterable, must not be bound to plasma proteins, and must not be reabsorbed or secreted by the tubules, either actively or passively. It must not be metabolized, must have no demonstrable physiological effects, and must be applicable to quantitative analysis in plasma and urine. If the other criteria are met and if the substance is not reabsorbed or secreted by the tubules, that amount that is filtered must be excreted in the urine. Therefore, knowing the concentration (milligrams per milliliter) in plasma (P), the concentration milligrams per milliliter) in urine (U) and the rate of urine excretion or volume (V, milliliters per minute), one can calculate UV, or the amount of the substance excreted per unit time (milligrams per minute). This UV divided by the plasma concentration leads to the volume of plasma which was cleared of the substance per unit time (milliliters per minute). This is not a complete clearing of the volume of plasma determined, but an incomplete clearing of a much larger volume. Since the substance is freely filterable, a volume of fluid equal to the volume of plasma cleared must have appeared as glomerular filtrate in order to contain the amount of substance excreted in the urine.

$$\frac{U(\text{mg/ml})\,V(\text{ml/min})}{P(\text{mg/ml})} = \text{Clearance (ml/min)}$$

Substances which have been used for determination of GFR include: urea, mannitol, creatinine, sodium thiosulfate, and inulin. The clearance of inulin gives the most accurate GFR. Urea and mannitol are reabsorbed, and creatinine is secreted by the tubules of some species. In dogs, *exogenous* creatinine is still used by some workers for measuring glomerular filtration rate as the amount secreted is negligible in relation to the concentrations in urine and plasma. Inulin is a fructose polysaccharide with a molecular weight of 5500 and a diameter of approximately 30 Å. Since it would be acted upon by digestive enzymes if given orally and it is poorly absorbed from subcutaneous injection, it must be given intravenously in clearance studies. In Fig. 6, the direct

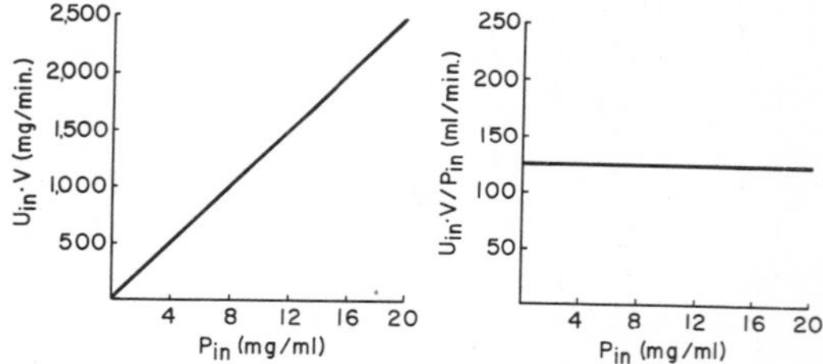

FIG. 6. Relationship of rate of excretion (left) and clearance (right) of inulin to plasma inulin concentration. (From Pitts, R. F., "Physiology of the Kidney and Body Fluids," Chicago, p. 65, Year Book Medical Publishers, 1963.)

linear relationship between plasma concentration and the amount of inulin appearing in the urine is shown. Therefore, the clearance of inulin is independent of plasma concentrations. Inulin clearance and GFR in man is 120 ml/min. By comparing the clearance of other substances to that of inulin, it can be determined if the substance is filtered only like inulin, or if it is filtered and reabsorbed like glucose, or if the substance is filtered and secreted like *p*-aminohippurate.

Substances that are reabsorbed, such as glucose, have clearance rates less than that of inulin, but approach the inulin clearance at high plasma concentrations. Those compounds that are secreted have higher clearances than inulin, but also approach the inulin clearance at high plasma concentrations. This is exemplified in Fig. 7 with clearances of inulin, glucose and *p*-aminohippurate. The latter compound is removed from the plasma by filtration and by such effective secretion by the proximal tubules that its clearance at low plasma levels is a measure of renal plasma flow since essentially all the *p*-aminohippurate is removed in one circuit of the blood (plasma) through the kidney.

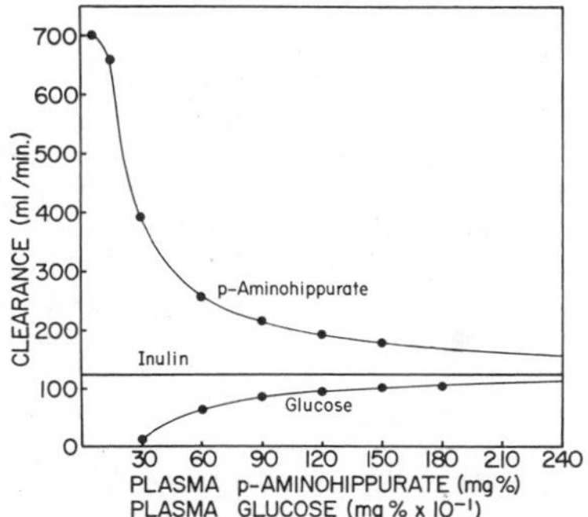

FIG. 7. Relation of clearances to plasma concentrations of *p*-aminohippurate, inulin and glucose in man. (From Pitts, R. F., "Physiology of the Kidney and Body Fluids," Chicago, p. 66, Year Book Medical Publishers, 1963.)

Tests based on clearance concepts are important in an estimation of kidney function in the diagnosis of renal disease and in the testing of drugs for their effects on kidney function.

References

1. METCALFF, J., "Renal Regulation of Body Fluids." Structural and Functional Aspects," *Pediatric Clin. of No. Amer.*, **11**, 833–870 (1964).
2. MUELLER, C. B., "The Structure of the Renal Glomerulus," *Am. Heart J.*, **55**, 304–322 (1958).
3. PITTS, R. F., "Physiology of the Kidney and Body Fluids," Chicago, Year Book Medical Publishers, 1963.
4. SMITH, H. W., "Principles of Renal Physiology," New York, Oxford University Press, 1956.

MARY KATHLEEN CARTER

GLUCAGON

Glucagon is a pancreatic hyperglycemic factor (*i.e.*, a substance that has the effect of raising the blood glucose concentration). Glucagon functions by increasing the rate of glycogenolysis in the liver (see GLYCOGEN AND ITS METABOLISM). Glucagon is a polypeptide reported to have the following structure:

His-Ser-Gln-Gly-Thr-Phe-Thr-Ser-Asp-Tyr-Ser-
Lys-Tyr-Leu-Asp-Ser-Arg-Arg-Ala-Gln-Asp-
Phe-Val-Gln-Try-Leu-Met-Asn-Thr

In raising the blood glucose level, glucagon thus has an overall effect opposite to that of insulin [see INSULIN (FUNCTION)].

Reference

1. WHITE, A., HANDLER, P., AND SMITH, E. L., "The Pancreas," in "Principles of Biochemistry," Third edition, Ch. 50, New York, McGraw-Hill Book Co., 1964.

GLUCONEOGENESIS

(γλυκύς sweet, νέος new, γεννᾶν, to produce.) Gluconeogenesis is the term applied to those processes whereby glucose is biosynthesized from non-glucose precursors. These precursors include the compounds involved in GLYCOLYSIS, the intermediates of the tricarboxylic acid cycle (CITRIC ACID CYCLE), a number of the amino acids (see AMINO ACID METABOLISM), glycerol, propionate, lactate and pyruvate. Gluconeogenesis takes place principally in the liver and kidney and depends importantly upon the balance between several endocrine factors, particularly the adrenal corticoids and insulin. Thus, the description and the evaluation of the mechanisms involved are of considerable significance in the understanding of metabolic diseases.

Early evidence for gluconeogenesis was obtained chiefly from experiments with starved animals and with animals made diabetic by pancreatectomy or by poisoning with phlorhizin. Direct evidence for the conversion of certain amino acids into carbohydrate was obtained by administration to normal, fasted animals and demonstration of an increase, if any, of the GLYCOGEN content of the liver. Since the animals with experimental diabetes excrete glucose in the urine in the absence of dietary carbohydrate, it was concluded that the carbon chains of the glucose were derived from the protein of the diet or, when fasted, from the protein of the tissues. Although the data from these early experiments were not quantitative and permitted only speculation with respect to the mechanisms involved in gluconeogenesis, they did establish certain important biochemical principles. Thus, by the determination of the ratio of the glucose to nitrogen (G : N ratio) in the urine of a phlorhizinized animal, an estimation of the proportion of the protein carbon converted to carbohydrate was made. This ratio was found to be approximately 3.6 from which it was calculated that 55–60% of the protein may be converted to glucose in those conditions where a normal glucose supply is denied the tissues.

The finding that the glucosuria of diabetes could be a result of accelerated gluconeogenesis led to extensive investigations of the effects of other secretions upon the interconversions of the carbon of protein and carbohydrate. Of particular significance was the observation that the removal of the adrenal glands resulted in a diminished capacity for gluconeogenesis whereas the injection of extracts of the adrenal cortex increased the degree of conversion of protein carbon into glucose (see ADRENAL CORTICAL HORMONES). These findings were consistent with clinical observations on patients with adreno-cortical insufficiency or hyperfunction. These facts, together with other related observations, have stimulated extensive research directed toward the definition of the mechanisms of gluconeogenesis and the evaluation of the endocrine regulation of the reactions involved.

Consideration of the now well-known pathways of catabolism of the amino acids and glucose indicates the probable pattern of the reactions involved in gluconeogenesis. These are conveniently discussed under two groups of reactions, those leading from *pyruvate to glucose* and those involved in the conversion of the *carbon of the amino acids into pyruvate*. The reactions whereby pyruvate is formed from glucose involve phosphorylation to glucose-6-phosphate, isomerization to fructose-6-phosphate, and phosphorylation to fructose-1,6-diphosphate, which, through the several reactions of glycolysis, yields pyruvate. The thermodynamic characteristics of the enzymatic reactions involved indicate that there are three reactions in which the degree of reversibility is insufficient to participate in gluconeogenesis. These are (1) the glucokinase reaction whereby glucose reacts with ATP to form the glucose-6-phosphate, (2) the phosphofructokinase reaction whereby fructose-6-phosphate is phosphorylated by ATP to form the fructose-1,6-diphosphate, and (3) the pyruvate kinase reaction whereby phosphoenolpyruvate reacts with ADP to yield ATP and pyruvate. Enzymatic reactions which may circumvent these thermodynamic barriers have been studied extensively in recent years and in each instance new enzymes have been described which appear to function importantly in gluconeogenesis. It has been found that liver and kidney contain a specific glucose-6-phosphatase which promotes the hydrolysis of glucose-6-phosphate to glucose. Quite similarly, these tissues contain a fructose diphosphatase which promotes the hydrolysis of fructose-1,6-diphosphate to fructose-6-phosphate, Thus, these two enzymes bring about a reversal of the reactions (1) and (2) above without involving resynthesis of ATP.

The conversion of pyruvate into phosphoenolpyruvate is a key reaction in gluconeogenesis and has been shown to involve two enzymes not usually mentioned in discussions of pyruvate metabolism. The first of these involves the conversion of pyruvate to oxalacetate through the action of the enzyme, *pyruvate carboxylase*.[1] The reaction may be represented by the following equation:

$$\text{Pyruvate} + CO_2 + \text{ATP} \leftrightarrows \text{Oxalacetate} + \text{ADP} + P_i$$

The reaction as written is an oversimplification since the enzyme contains biotin and displays a small but absolute requirement for acetyl-CoA (see also CARBOXYLATION ENZYMES). The amounts of the enzyme in liver and kidney and the dependence of its concentration upon the physiological state of the tissue, as discussed below, suggest that this enzyme is an important component of the gluconeogenic system.

The conversion of oxalacetate into phosphoenolpyruvate is brought about by the enzyme, *phosphoenolpyruvate carboxykinase*.[2] The reaction is represented by the following equation:

$$\text{Oxalacetate} + \text{GTP (or ITP)} \rightarrow \text{Phosphoenolpyruvate} + CO_2 + \text{GDP (or IDP)}$$

The combined effects of these enzymes constitute a thermodynamically sound pathway from pyruvate to glucose.

With the demonstration of the conversion of oxalacetic acid into phosphoenol pyruvate, the reaction pathway for glucogenesis from the several intermediates of the tricarboxylic acid cycle is clearly defined. It is evident that each of these intermediates can progress through the cycle to oxalacetate and thence to glucose. The reaction pathway from the *glycogenic amino acids* proceeds through these same intermediates. With a few exceptions, the first phase of amino acid catabolism is the formation of the corresponding α-keto acid by transamination or by oxidative deamination. The further metabolism of the keto acid may lead either to pyruvate or to one of the intermediates of the cycle, in which case the pathway to glucose is evident. In those instances where the further metabolism of the keto acid proceeds through acetoacetate, the amino acid is considered to be non-glycogenic.

An important contribution to the mechanisms of the intracellular reactions involved in gluconeogenesis was made by Lardy and his co-workers.[3] These workers were concerned with the fact that the formation of oxalacetate from pyruvate is carried out in the mitochondria and that the rate of release of oxalacetate from these structures into the soluble phase of the cell is insufficient to account for the observed rate of gluconeogenesis. In contrast, aspartate and malate diffuse readily from the mitochondria. Their results indicate that, following the carboxylation of pyruvate to oxalacetate, the latter is transaminated to aspartate or reduced to malate. These compounds diffuse into the extramitochondrial spaces where, through the reoxidation of malate or the transamination of aspartate, oxalacetate is again formed, which may now be converted into phosphoenolpyruvate by the soluble carboxykinase. It should be pointed out that this mechanism not only assigns a role to the transaminases found in both the soluble and the mitochondrial fractions of the cell but also indicates a function for the malate dehydrogenase found in the cytoplasm. Apparently these reactions progress at rates consistent with that demanded for gluconeogenesis.

There is considerable evidence that the tissue concentrations of glucose-1-phosphatase, fructose-1,6-diphosphatase, pyruvate carboxylase and phosphoenolpyruvate carboxykinase depend upon the selective regulatory effects of the adrenal corticoids and INSULIN. Increased levels of these four enzymes were found in rat liver tissues within a few hours after the injection of glucocorticoid hormones.[4-10] These increased quantities of the enzymes are believed to represent *de novo* enzyme syntheses since they may be prevented by typical inhibitors of protein biosynthesis, such as actinomycin and puromycin. The increased tissue levels of the enzymes were not observed when insulin was injected into animals pretreated with the cortical hormones. Conversely, animals made diabetic with alloxan had high levels of these enzymes which could be restored to normal values by control of the diabetes with insulin. These observations are consistent with the theory that the glucocorticoids act as inducers of enzyme biosynthesis and that, in this instance, insulin acts to supress these biosyntheses. This theory is supported by the observation that one of the early effects of administration of adrenal cortical hormone is the stimulation of RNA polymerase activity and messenger RNA synthesis in rat liver nuclei[11,12] [see RIBONUCLEIC ACIDS (BIOSYNTHESIS)]. The rate of turnover of rat liver RNA, as studied by incorporation of radioactive orotic acid, is markedly increased in alloxan diabetic rats and this rate is restored to normal or decreased below normal by administration of insulin. The details of the mechanisms whereby these hormones control the chemical reactions of enzyme biosynthesis must await further research. The topic of control of enzyme reactions has been extensively reviewed in a series of seminars in which a great deal of emphasis is placed upon gluconeogenesis.[13]

References

1. UTTER, M. F., AND KEECH, D. B., *J. Biol. Chem.*, **238**, 2306–2308 (1963).
2. UTTER, M. F., AND KURAHASHI, K., *J. Biol. Chem.*, **207**, 787–802, 821–841 (1954).
3. LARDY, H. A., PAETKAU, V., AND WALTER, P., *Proc. Natl. Acad. Sci.*, **53**, 1410–1415 (1965).
4. WEBER, G., BANERJEE, G., AND BRONSTEIN, S. B., *Am. J. Physiol.*, 202, 137–144 (1962).
5. WEBER, G., SINGHAL, R. L., STAMM, N. B., AND SRIVASTAVA, S. K., *Federation Proc.*, **24**, 745–54 (1965).
6. SHRAGO, E., LARDY, H. A., NORDLIE, R. C., AND FOSTER, D. O., *J. Biol. Chem.*, **238**, 3188–93 (1963).
7. RAY, D. P., FOSTER, D. O., AND LARDY, H. A., *J. Biol. Chem.*, **239**, 3396–3400 (1964).
8. HENNING, H. V., SEIFFERT, I., AND SEUBERT, W., *Biochim. Biophys. Acta*, **77**, 345–348 (1963).
9. ASHMORE, J., WAGLE, S. R., AND UETE, T., *Advan. Enzyme Regulation*, **2**, 101–114 (1964).
10. GREENGARD, O., WEBER, G., AND SINGHAL, R. L., *Science*, **141**, 160–161 (1963).
11. GARREN, L. D., HOWELL, R. R., AND TOMKINS, G. M., *J. Mol. Biol.*, **9**, 100–108 (1964).
12. WEBER, G., SINGHAL, R. L., AND SRIVASTAVA, S. K., *Proc. Natl. Acad. Sci.* **53**, 96–104 (1965).
13. WEBER, G. (EDITOR) *Advan. Enzyme Regulation*, **1**, (1963); **2**, (1964); **3** (1965).

R. L. GARNER

GLUCOSE

D-Glucose (dextrose) has the configuration represented below:

$$\begin{array}{c} \text{CHO} \\ | \\ \text{HCOH} \\ | \\ \text{HOCH} \\ | \\ \text{HCOH} \\ | \\ \text{HCOH} \\ | \\ \text{CH}_2\text{OH} \end{array}$$

Its principal forms are α-D-glucopyranose with a specific rotatory power of about +20°, and β-D-glucopyranose with a rotatory power of about +110° [for structures of these pyranose forms, see CARBOHYDRATES (CLASSIFICATION)]. Exact values are meaningless unless one specifies purity and experimental conditions.

Biochemically, glucose is exceedingly important because it is the predominant fuel used by our bodies. In nature it is outstanding; its general metabolic significance is reflected by the number of carbohydrates, *e.g.*, STARCH, CELLULOSE, GLYCOGEN, maltose, lactose and sucrose, which yield it as a product of hydrolysis. In the case of the first four carbohydrates mentioned, glucose is the sole hydrolytic product.

GLUCURONIDASES

The enzyme β-glucuronidase specifically hydrolyzes aryl, acyl, and alcoholic β-D-glucosiduronic acids. Pyranosiduronic acids are more readily hydrolyzed than the furanosiduronic acids. The tetrasaccharide oligosccharide derived from hyaluronic acid hydrolysate is also a substrate for β-glucuronidase. In addition, the enzyme catalyzes efficiently the transfer of the glucuronyl radical from a donor substrate to an acceptor. The best acceptors are dihydric alcohols. The Fishman Unit is defined as 1 μg of phenolphthalein liberated from 0.001*M* phenolphthalein glucosiduronic acid at pH 4.5 in 1 hour at 38°C. Other substrates which have been employed for assay purposes are the glucosiduronic acids of *p*-nitrophenol, 8-hydroxyquinoline, 6-brom-2-naphthol and umbelliferone.

β-Glucuronidases of mammalian tissue origin usually exhibit pH optima ranging from 3–6 depending on the substrate, in contrast to bacterial β-glucuronidase (*E. coli*) which exhibits a pH optimum at 7. The Michaelis constants range from .01*M* in the case of borneol glucosiduronic acid to $2.2 \times 10^{-5}M$ for naphthol AS–BI β-D-glucosiduronic acid. Saccharo-1,4-lactone is a specific competitive inhibitor of β-glucuronidase. Highly purified mammalian β-glucuronidase dissociates on dilution to inactive products. This dissociation is reversed in the presence of amino compounds such as chitosan, protamine, albumin, DNA and a variety of α,ω-diaminopolymethylenes.

The enzyme is found in all tissues of animal species (mouse, rat, rabbit, dog, man and cat). The pig and guinea pig possess the lowest levels of β-glucuronidase as do several inbred strains of mice (C3H, AKR). The richest tissue in the rat is the preputial gland followed by liver, spleen and tissues of the endocrine system. Body fluids, such as saliva, tears, gastric juice, spinal fluid, urine, amniotic fluid and vaginal fluid all show activity. The enzyme is found in lower forms of life, such as the fish, eel, snails, mollusk, insecta and plants (*Scutellaria baicalensis*). Bacteria also possess this enzyme, particularly certain strains of *E. coli*.

Tissue β-glucuronidase is inherited in inbred strains of mice depending on the complement of the genes, GG, Gg and gg. Mice possessing GG and Gg genes exhibit normal levels of β-glucuronidase, those with gg show marked deficits in enzyme activity.

The purification of the enzyme requires a step to release the enzyme from phospholipid-protein cytoplasmic membranes (flocculation in acetate buffer at 38°C or precipitation by organic solvents). Following the liberation of the enzyme, it can be purified most efficiently by fractionation in ammonium sulfate and in methanol-citrate. The pure enzyme has a specific activity of around 1.5×10^5 Fishman units/mg of protein, and several workers claim to have prepared the enzyme in crystalline form. The enzyme is stable in concentrated solutions.

β-Glucuronidase is widely employed as a tool to hydrolyze glucosiduronic acids of steroids and of drug metabolites which have been isolated from either blood or urine. In particular, the hydrolysis of urinary steroid glucosiduronic acids by β-glucuronidase is preferred to hot mineral acid as the most desirable means of hydrolysing these substances (see BILE AND BILE PIGMENTS; GLYCOSIDES, STEROID).

From cytobiochemical studies, the enzyme appears to be a component of organelles sedimenting in lysosome and MICROSOME fractions. Compared to other acid hydrolases, a relatively higher proportion of the enzyme is found in the microsome fraction.

Modern methods for the enzymorphologic study of β-glucuronidase employ the substrate, naphthol AS-BI β-D-glucosiduronic acid, and a post-incubation coupling technique using either fast Garnet GBC or fast dark Blue R as the diazonium coupling agents. The enzyme is found localized on membranes of the endoplasmic reticulum and of lysosome-like structures.

β-Glucuronidase activity correlates with the action of steroid HORMONES in a number of instances. Thus, castration of female mice leads to a drop in β-glucuronidase activity of the uterus which can be restored to normal by the administration of estrogenic hormone. Other tissues, responding similarly, are adrenal, mammary gland and preputial gland. Recently, the mouse renal β-glucuronidase response to androgens has been the subject of intensive study. Assay conditions have been perfected so that the potency of exogenously administered androgens can be related to testosterone. The response correlates better with the protein anabolic rather than the virilizing property of these steroids. The response can be prevented by prior administration of Actinomycin D which suggests that the synthesis of β-glucuronidase is the consequence of the action of a messenger RNA synthesized through nuclear DNA. The enzyme is excreted in the urine in concentrations which parallel the elevation in the kidney level. *De novo* synthesis of renal β-glucuronidase in androgen-stimulated animals was established in isotope studies. Exogenous gonadotropin produces the response in intact male mice but not in female mice or in castrated male animals.

Serum β-glucuronidase values in normal individuals range from 300–700 Fishman units/100 ml of serum. Higher levels are found in sera of pregnant women and in patients with liver disease. Isoenzymes of β-glucuronidase in human serum have been described. The vaginal fluid β-glucuronidase activity has been found elevated in women with reduced ovarian function or with cancer of the cervix. This level is depressed by the administration of estrogens, and in the menstrual cycle there is a drop in activity at mid-cycle.

Tumors as a rule are rich in β-glucuronidase, and the enzyme has been localized in the cytoplasm of cancer cells as well as in macrophages, histiocytes and leucocytes which are components of many tumor tissues.

References

1. FISHMAN, W. H., *Advan. Enzymol.*, **16**, 361–409 New York (1955).
2. LEVVY, G. A., AND MARSH, C. A., *Advan. Carbohydrate Chem.*, **14**, New York, Academic Press, 1959.
3. FISHMAN, W. H., "Chemistry of Drug Metabolism," Springfield, Illinois, Charles C. Thomas, 1961.

WILLIAM H. FISHMAN

GLUTAMIC ACID

L-Glutamic acid (α-aminoglutaric acid), HOOC—CH_2—CH_2—$CHNH_2$—COOH, is a prominent hydrolysis product of PROTEINS. Many proteins yield 10–20% of their weight of glutamic acid. Some vegetable proteins (*e.g.*, gliadin) may yield as much as 45%.

These high yields are due in part to the presence of GLUTAMINE as a building unit in proteins and the fact that on acid or alkaline hydrolysis glutamine yields glutamic acid.

Glutamic acid enters into metabolism in many ways (see AMINO ACID METABOLISM). It is nonessential nutritionally and thus must be produced endogenously. Its rotatory power in 1.7*N* HCl solution is +31.7°. Monosodium glutamate is extensively used as a condiment to enhance the flavor of foods.

D-Glutamic acid, the optical antipode of the common variety, is yielded as the principal, if not the sole, hydrolysis product of the capsular substance of *Bacillus anthracis*.

See also AMINO ACIDS (OCCURRENCE, STRUCTURE, AND SYNTHESIS).

GLUTAMINE

L-Glutamine, $H_2N—CO—CH_2—CH_2—CHNH_2—COOH$, while not readily produced by protein hydrolysis (since it is hydrolyzed to GLUTAMIC ACID and ammonia by acid or alkali treatment) is one of the more abundant amino acid constituents of PROTEINS.

It is commonly classed as nutritionally nonessential in the diet, yet the fact that it is required by various body cells when grown in TISSUE CULTURE shows that it is important in cellular nutrition. It is unique among the amino acids in that it crosses the BLOOD-BRAIN BARRIER readily and may be metabolized by brain tissue. It, along with glutamic acid (derived from it), is said to represent about 80% of the free amino nitrogen of brain tissue. It is involved in many metabolic processes (see AMINO ACID METABOLISM; PURINE BIOSYNTHESIS; PYRIMIDINE BIOSYNTHESIS).

Glutamine needs to be further explored in the area of nutrition since its use as a nutritional supplement is reportedly of benefit in (1) healing of ulcers, (2) petit mal epilepsy, (3) MENTAL RETARDATION, and (4) ALCOHOLISM.

Glutamine when heated in aqueous solution cyclizes with the formation of ammonia and pyrrolidone carboxylic acid. Its specific rotatory power in water solution is $[\alpha]_D = +6.1°$.

See also AMINO ACIDS (OCCURRENCE, STRUCTURE, AND SYNTHESIS).

GLUTATHIONE

Glutathione (GSH) was isolated, in amorphous state, from yeast, liver, and muscle by F. G. Hopkins in 1921. Its structure was reported to be glutamylcysteine but, after it was obtained in crystalline form in 1929, its structure was established by degradation as γ-glutamylcysteinylglycine. This structure was confirmed by synthesis in 1930:

```
HOOC·CH·CH2·CH2·CO·N·CH·CO·N·CH2·COOH
     |                |  |    |
    NH2               H CH2   H
                         |
                         SH
```

Probably because of the γ-glutamyl peptide link with cysteine, glutathione is resistant to most PROTEOLYTIC ENZYMES. A pancreatic carboxypeptidase does remove glycine but does not attack the γ-glutamyl peptide link. An enzyme found in kidney and in cells of the blood catalyzes the transfer of the γ-glutamyl moiety to water (hydrolysis) or to other amino acids or peptides. The remaining cysteinylglycine is rapidly hydrolyzed by a specific and ubiquitous cysteinylglycinase. A specific enzyme found in pig liver causes cyclization of the glutamyl residue to pyrrolidonecarboxylic acid with simultaneous formation of cysteinylglycine.

As a sulfhydryl compound, glutathione is subject to nonenzymic oxidation in air to the disulfide, oxidized glutathione (GSSG). This reaction is catalyzed by Fe and Cu. Also, it is oxidized by the disulfide (—SS—) groups of other substances such as proteins and cystine and by quinones, hydroperoxides, and iron-porphyrin compounds. The oxidation by —SS— groups occurs in two stages with the formation of mixed disulfides:

$$\text{GSH} + \text{RSSR} \rightleftarrows \text{RSH} + \text{GSSR} \qquad (1)$$

$$\text{GSH} + \text{GSSR} \rightleftarrows \text{GSSG} + \text{RSH} \qquad (2)$$

Metabolic Functions. At the time Hopkins isolated glutathione, sulfhydryl compounds were thought to participate in respiration. Glutathione, because of its presence in all cells, was thought to play an important role in cellular oxidative processes. Now, some 40 years later, it has been relegated to a minor role in the respiration of mammalian tissues. However, it appears to have an important function in respiration in plants. In preparations of germinating pea cotyledons, one-fourth to one-half of the oxygen uptake was found to proceed as follows:

$$\text{Ascorbic Acid} + [\text{O}] \rightarrow \text{Dehydroascorbic Acid (DHA)} \qquad (3)$$

$$\text{DHA} + 2\,\text{GSH} \rightarrow \text{GSSG} + \text{Ascorbic Acid} \qquad (4)$$

$$\text{GSSG} + \text{NADPH} + \text{H}^+ \xrightarrow[\text{reductase, FAD}]{\text{Glutathione}} 2\,\text{GSH} + \text{NADP}^+ \qquad (5)$$

This glutathione-ASCORBIC ACID system may play an important part in the respiration of the nuclei of animal as well as plant cells. The nucleus has substantial concentrations of both glutathione and ascorbic acid but no, or very little, cytochrome pigment. Also, the amount of glutathione reductase in the nucleus is about 20% of the total in the cell. Even though the proportion of the total respiration that proceeds over this route is small, it may be essential for those cellular components with which it is associated. Oxidative PHOSPHORYLATION has not been observed in association with the glutathione-ascorbic acid system.

An important function of glutathione is to protect the erythrocyte against peroxides. Damage by a peroxide sensitizes the erythrocyte to hemolysis in the presence of a wide variety of agents. Glutathione reduces peroxide with the aid of glutathione peroxidase:

$$2\,\text{GSH} + \text{H}_2\text{O}_2 \xrightarrow[\text{peroxidase}]{\text{GSH}} \text{GSSG} + 2\,\text{H}_2\text{O} \qquad (6)$$

GSH is restored by glutathione reductase and the NADPH derived from the hexose monophosphate pathway of glucose metabolism [Eq. (5)]. It has been suggested that the functions of glucose metabolism in the erythrocyte are maintenance of the GSH level as above and production of ATP (glycolytic pathway) for the fresh synthesis of glutathione. Evidence for the first function of glucose metabolism is furnished by the hereditary condition called "primaquine sensitivity" because it was discovered in subjects who were given primaquine for antimalarial treatment. In this condition the erythrocytes have a low activity of glucose-6-phosphate dehydrogenase. Since the action of this enzyme with NAD on glucose-6-phosphate furnishes the NADPH for reduction of GSSG, the deficiency results in a low concentration of GSH and impaired removal of peroxide. Consequently, the erythrocytes are sensitized to hemolysis in the presence of primaquine and other agents.

When it was observed that sulfhydryl compounds, particularly glutathione, activated some enzymes, Waldschmidt-Leitz proposed that the relationship between GSH and GSSG might serve to regulate the activity of such enzymes *in vivo*. In a modified form, this idea is incorporated in some modern concepts. Many hydrolytic, oxidizing-reducing, and transfer enzymes are dependent on one or more intact —SH groups, and an important function of glutathione is considered to be the maintenance of such enzymes in their active conformations.

$$\underset{\text{(active)}}{2\ \text{Enz—SH}} + O_2 \rightarrow \underset{\text{(inactive)}}{\text{Enz—S—S—Enz}} + H_2O_2 \quad (7)$$

$$\text{Enz—S—S—Enz} + \text{GSH} \rightarrow 2\ \text{Enz—SH} + \text{GSSG} \quad (8)$$

It is probable that glutathione in a similar manner maintains cysteine in the sulfhydryl form for incorporation into proteins [Eqs. (*1*) and (*2*), RSSR = cystine, RSH = cysteine].

A reaction similar to Eq. (*5*) serves as the main route for oxidation of the NADPH generated in the brain from carbohydrate metabolism.

Glutathione serves as a cofactor for several enzymes: formaldehyde dehydrogenase, glyoxalase, maleylacetoacetate isomerase, and glyceraldehyde-3-phosphate dehydrogenase. It is a firmly bound component of the last enzyme and does not equilibrate with labeled GSH added to a solution of the enzyme-GSH complex. This is the only instance known of such firm binding of glutathione to a protein. Glutathione is the specific cofactor for glyoxalase. Actually, two enzymes are involved, glyoxalases I and II. The conversion of methylglyoxal to lactic acid proceeds as follows:

$$\begin{array}{c} CH_3 \\ | \\ C{=}O \\ | \\ HC{=}O \end{array} + \text{GSH} \rightarrow \begin{array}{c} CH_3 \\ | \\ C{=}O \\ | \\ HC{-}OH \\ | \\ SG \end{array} \xrightarrow{\text{Glyoxalase I}} \begin{array}{c} CH_3 \\ | \\ HC{-}OH \\ | \\ C{=}O \\ | \\ SG \end{array}$$

$$\xrightarrow{\text{Glyoxalase II}} \begin{array}{c} CH_3 \\ | \\ HC{-}OH \\ | \\ C{=}O \\ | \\ OH \end{array} + \text{GSH} \quad (9)$$

The addition of GSH to the aldehyde group is non-enzymic.

The enzyme, insulinase, which inactivates INSULIN uses GSH as a hydrogen donor for the reaction:

$$\text{Insulin}\langle\begin{array}{c} S \\ | \\ S \end{array} + 2\ \text{GSH} \rightarrow \text{Insulin}\langle\begin{array}{c} SH \\ \\ SH \end{array} + 2\ \text{GSSG} \quad (10)$$

The reduced insulin is inactive.

The *biosynthesis* of glutathione from L-glutamic acid, L-cysteine, and glycine proceeds in two stages, with formation of L-glutamyl-L-cysteine in the first stage. Intact erythrocytes and their stromata are capable of this synthesis. Intact cells can use methionine and serine instead of cysteine, and α-oxoglutaric acid instead of glutamic acid.

Methods for the *determination* of GSH involve titration with a mild oxidant (such as iodine or ferricyanide), amperometric titration, and the glyoxalase reaction. The last is highly specific. Iodimetry tends to give spuriously high values with tissue extracts because of the presence of other reducing agents, but it can be used satisfactorily with erythrocytes. The amperometric method is stated to give accurate results with tissue extracts as well as with erythrocytes.

The first step in the formation of mercapturic acids is conjugation of an aromatic compound with GSH to yield G—S—aryl. After removal of glutamic acid and glycine, in that order, the amino group of the cysteine moiety is acetylated to form the mercapturic acid.

References

1. COLOWICK, S. P., LAZAROW, A., RACKER, E., SCHWARZ, D. R., STADTMAN, D. R., AND WAELSCH, H., "Glutathione," Proceedings of the Symposium held at Ridgefield, Connecticut, November, 1953, New York, Academic Press, 1954, 347pp.
2. CROOK, E. M., "Glutathione," Biochemical Society Symposium Number 17, held at Senate house, University of London, February 15, 1958, Cambridge, England, University Press, 1959, 115pp.
3. KNOX, W. E., "Glutathione," in "The Enzymes," (BOYER, P. D., LARDY, HENRY, AND MYRBÄCK, J., EDITORS), second edition, Vol. 2, pp. 253–294, New York, Academic Press, 1960.
4. PIRIE, N. W., "Glutathione," *Proc. Roy. Soc. London, Ser. B*, **156**, 306–311 (September 18, 1962).

HAROLD L. MASON

GLYCERALDEHYDE

Glyceraldehyde, $HOCH_2$—CHOH—CHO, along with dihydroxyacetone is produced by the mild oxidation of glycerol and may be regarded as a simple sugar even though it does not exist in cyclic forms as do all other sugars. When produced by oxidation of glycerol or by laboratory synthesis, glyceraldehyde is a racemic mixture of the D- and L-forms.

By using the cyanohydrin synthesis (addition of HCN), D-glyceraldehyde can successively be converted to D-tetroses, D-pentoses and D-hexoses including glucose. While all the steps in this complex process have been accomplished, and the configurational relationships among the asymmetric carbons established in this way, the process is not a practical laboratory procedure.

D-Glyceraldehyde has a specific rotatory power of $[\alpha]_D = +21°$. It is an intermediate in CARBOHYDRATE METABOLISM.

Dextrorotatory glyceraldehyde has been arbitrarily chosen as the standard of reference for designation of the optical configuration of other related optically active substances; *i.e.*, it is the prototype of the D-configurational series of carbohydrates [see AMINO ACIDS (OCCURRENCE, STRUCTURE), section on Relative Configurations and Nomenclature; also CARBOHYDRATES (CLASSIFICATION AND STRUCTURAL INTERRELATIONS)].

GLYCEROL

Glycerol (glycerine), $HOCH_2$—CHOH—CH_2OH, is important biochemically in that fats and many other LIPIDS are derivatives (esters) of it, and also because it is related to carbohydrates and can be converted to glucose in the body (see CARBOHYDRATE METABOLISM). It is produced in small amounts by ordinary yeast FERMENTATION. α-Glyceryl phosphate is an intermediate.

Glycerol is a viscous, sweet-tasting liquid which, when pure, boils at 290°C. On heating in the presence of sulfuric acid or $NaHSO_4$, it yields the unsaturated aldehyde acrolein, $CH_2{=}CHCHO$, which has a recognizable irritating odor, and irritates the eyes. This reaction has been used as a test for fats.

GLYCINE

Glycine (aminoacetic acid, glycocoll), CH_2NH_2—COOH, is the simplest protein-derived AMINO ACID. It possesses no asymmetric carbon atom and exists in only one configurational form. It occurs in most PROTEINS. Gelatin is an unusually rich source; it yields about 25% of its weight in glycine on hydrolysis. Silk fibroin yields about 40%. At one time it was supposed that every second amino acid residue in silk fibroin was glycine, but this appears not to be quite true.

This amino acid is classed as nutritionally nonessential for mammals, but for chickens it is nutritionally essential; *i.e.*, it cannot be produced endogenously by the enzymic equipment which chickens possess (see NUTRITIONAL REQUIREMENTS).

Glycine is a sweet-tasting (Greek *glykys*, sweet), neutral substance which like other amino acids exists primarily in the zwitterion form.

GLYCOGEN AND ITS METABOLISM

Structure. Glycogen is a reserve polysaccharide which serves as a storage of carbohydrate in animals just as starch does in plants. It is found in all animal cells (2–10% in the liver, 0.5–1% in the muscle), but also in many other species including bacteria, yeast, fungi and certain higher plants. Being a branched polysaccharide[1] made up of chains of D-glucose units linked by α-1,4-glucosidic bonds with branching points arising from additional α-1,6-linkages (see Fig. 1), it is similar to amylopectin, the branched component of

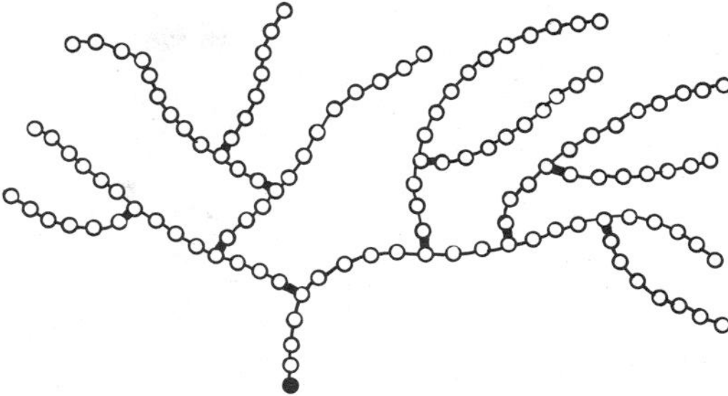

FIG. 1. Schematic representation of glycogen molecule; ●, the reducing group; ○-○, the α-1,4 linkages; ○-○, α-1,6 linkages.

STARCH. Glycogen differs from amylopectin, however, in the following respect: glycogen molecules show a much wider range of molecular weight distribution, characteristic of the physiological state of the animal and, perhaps, of a given species of tissue; it has a much higher content of branching points (8–10% *vs* 4–5%) resulting in a considerably more "bushy" structure; it gives a brown rather than purple-red coloration with iodine (λ max 460 mμ instead of 540 mμ).

Glycogen has been extracted from tissues by the use of hot alkali (KOH), trichloroacetic acid, or water. Usually materials with molecular weights of 4–6 × 10^6 are obtained, although it has been claimed that mild extraction procedures (*e.g.*, cold water, Bueding) yield material with molecular weight up to 500 × 10^6 (sedimentation coefficients upward of 1500S). It is, however, difficult to ascertain whether these giant particles are made up of single molecules or result from non-covalent aggregation.

Degradative studies indicate that inner chains (between two branching points) may be shorter than the average 3–4 glucose units indicated in Fig. 1. There are further indications that the so-called minor components detected in glycogen following enzymatic or chemical hydrolysis (*e.g.*, fructose, maltulose, nigerose, etc.) are artifacts arising from epimerization or transfer reactions.

Biosynthesis. Two enzymes are involved in the biosynthesis of glycogen: *glycogen synthetase*, which synthesizes straight chains of glucose units linked by α-1,4-linkages, and a *branching enzyme* which transfers the terminal portion of this chain onto the chain itself, thus creating a branching point. Glycogen arises from these successive chain elongation and chain transfer reactions.

Glycogen synthetase (UDP glucose: α-1,4-glucan-α-4-glucosyltransferase, EC. 2.4.1.11), discovered by Leloir, catalyzes the reaction:

$$\text{UDPG} + \text{Glycogen}\ (n) \rightarrow \text{UDP} + \text{Glycogen}\ (n+1)$$

where n stands for the number of glucose units in glycogen, and UDPG for uridine diphosphate glucose. This reaction is thermodynamically favored (ΔF°, −7600 calories for the hydrolysis of the UDPG as compared to −4000 calories for the glucosidic bond of glycogen) and, therefore, is essentially irreversible. There is an absolute requirement for a "primer"—an oligo- or polysaccharide acting as a glucosyl acceptor. In this respect, glycogen is far more effective (100 times) than maltotetrose, which is itself far better than maltotriose.

Glycogen synthetase is not entirely specific with respect to the nature of the nucleoside diphosphate glucose donor. ADP-glucose (the best substrate for starch synthesis) is half as effective as UDPG, and pseudo-UDPG and pseudo-TDPG only slightly active. CDPG, ADP-maltose, UDP-galactose and UDP-acetylglucosamine give negative results.

Glycogen synthetase appears to exist in two interconvertible forms, one showing increased activity in the presence of glucose-6-phosphate (dependent, or D-form), the other being independent of glucose-6-phosphate (I-form). Apparently, the I to D interconversion results from phosphorylation and dephosphorylation of the enzyme molecule (Larner) in a manner reminiscent of the interconversion of phosphorylase *b* and *a* (see below). The physiological significance of this interconversion will be discussed later.

Branching enzyme (α-1,4-glucan — α-1,4-glucan-6-glycosyltransferase, EC. 2.4.1.18; also Q enzyme). This enzyme synthesizes branching points by transferring part of the α-1,4-glucan chain from a 4- to a 6-position, therefore converting an "amylose"-type molecule into amylopectin or glycogen. The reaction occurs also with branched polysaccharides when the length of the outer chains exceed 6 glucose residues. Lack of the branching enzyme leads to glycogenosis type IV (Amylopectinosis or Andersen's disease).

Glycogen Degradation. Roughly speaking, glycogen can be degraded either hydrolytically or through a phosphorolytic pathway. However, since phosphorolysis, proceeding from the outer branches, cannot by itself bypass the branching points, an additional transferase and α-1,6-glucosidase would be required for complete degradation.

Hydrolytic Cleavage. Glycogen is hydrolyzed by various types of α-glucosidases, some attacking randomly both inner and outer branches (α-amylase type), others exclusively outer branches. In this case, successive cleavage can involve either every glucosidic bond to produce solely glucose (γ-amylase or glucamylase type) or every other bond to produce maltose (β-amylase type, not found in animals). The physiological significance of the hydrolytic pathway of glycogen catabolism is not really understood, though it is known that lack of one of these α-1,4-glucosidases (acid maltase—Hers) probably results in a severe glycogen storage disease (Type II or Pompe's disease).

The *phosphorolytic cleavage* is undoubtedly the most important metabolic pathway for glycogen utilization. When it was found (Cori) that the phosphorylase reaction was reversible (*i.e.*, formation of polysaccharide from glucose-1-phosphate), it was assumed that this enzyme was responsible also for the biosynthesis of glycogen. This view must now be rejected since it is known that (a) glycogen is synthesized even under physiological conditions favoring glycogen breakdown (high phosphate/glucose-1-phosphate ratio); (b) conditions leading to increased phosphorylase activity (*e.g.*, epinephrine administration) always result in increased glycogen breakdown; and (c) a myopathy in which phosphorylase is missing (McArdle's disease) results in glycogen accumulation rather than depletion.

Phosphorylase (α-1,4-glucan: orthophosphate glucosyl-transferase EC. 2.4.1.1), discovered by Cori and Parnas, catalyzes the reaction:

$$\text{G-1-P} + \text{Glycogen } (n) \rightleftharpoons \text{Glycogen } (n+1) + P_i$$

where G-1-P stands for α-D-glucose-1-phosphate and P_i for inorganic phosphate. The reaction is reversible and equilibrium is achieved when the P_i/G-1-P ratio is approximately 3. As in the case of glycogen synthetase, the enzyme requires an oligo- or polysaccharide primer for activity. Muscle phosphorylase presents many curious structural features that have great bearing on its activity. *First*, it exists in two forms of aggregation: a dimer (phosphorylase *b*, molecular weight 250,000), inactive in the absence of adenosine-5′-phosphate (AMP), and a tetramer (phosphorylase *a*, molecular weight 500,000) active even in the absence of this nucleotide (Cori). There is one AMP binding site per monomer unit. The monomer (molecular weight 125,000) can be produced chemically and is presumably inactive. *Second*, conversion of phosphorylase *b* to *a* results from a phosphorylation of the enzyme by a specific kinase (Krebs and Fischer), and the reverse reaction follows removal of the bound phosphate by a specific phosphatase, according to the following scheme:

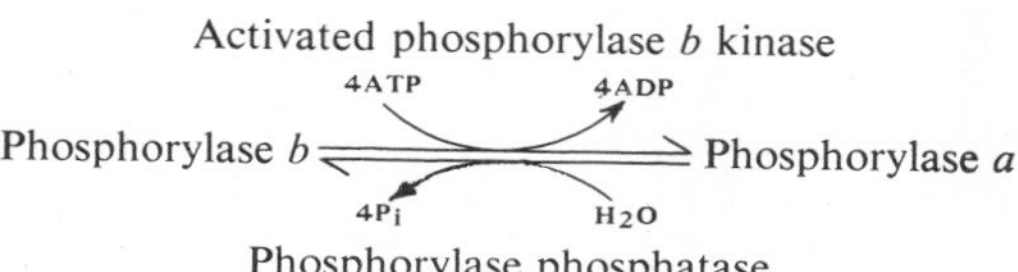

Third, phosphorylase contains pyridoxal-5′-phosphate (PLP) (1 mole/monomer); removal of PLP leads to complete inactivation, but there are several indications that the B_6 derivative acts structurally rather than catalytically (see PYRIDOXINE). Clearly, the activity of the enzyme can be controlled both by the presence of AMP which acts as a structural modifier ("allosteric" effect), or by covalent modification (phosphorylation) of the protein. Phosphorylases from the liver and other tissues have not been as thoroughly investigated; they appear to have slight differences in their structure-to-activity relationship.

Oligo-1,4 → 1,4-*Glucan Transferase and Amylo-1,6-glucosidase*. When a side chain has been partially degraded by phosphorylase, the former enzyme will transfer a fragment generally containing three glucose residues from this side chain to another chain, thus exposing a single glucose residue, the one actually forming the branching point (Brown and Illingworth). This single residue can then be hydrolyzed by amylo-1,6-glucosidase. It is not clear whether or not these two enzymatic activities are associated with a single enzyme. Lack of the 1,6-glucosidase activity leads to Type III glycogenosis (limit dextrinosis or Forbe's disease).

Control of Glycogen Metabolism. Glycogen constitutes an important form of storage of carbohydrate and, therefore, of chemical energy. As a consequence, it exists in a dynamic state of equilibrium determined by its respective rates of synthesis and degradation. These, in turn, are regulated by elaborate control mechanisms in which epinephrine (see ADRENALINE) appears to play a major role. The action of this hormone is mediated by cyclic 3′,5′-adenylic acid (Sutherland) which, itself, affects the activation of some of the regulatory enzymes. A very general and simplified picture of the molecular mechanism of this control is schematically represented in Fig. 2 and Fig. 3.

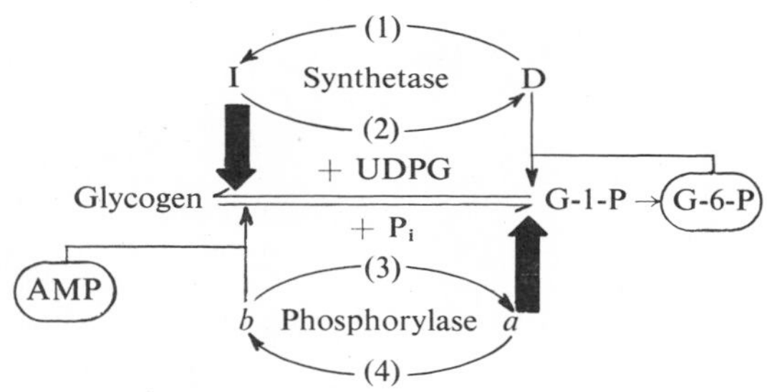

FIG. 2. Glycogen synthesis and breakdown. (1) Activated by insulin (Larner), therefore favoring glycogen synthesis. (2) and (3). Activated by catecholamines (epinephrine), therefore decreasing synthesis and increasing breakdown of glycogen, respectively. (4) Blocked by AMP, significance not as yet understood.

The active forms of both enzymes (I and *a*) appear to result from covalent modification of the structure of the proteins. The inactive or dependent forms are activated by allosteric effectors (G-6-P and AMP, respectively; see also REGULATORY SITES OF ENZYMES). Both these compounds have been shown to affect many other enzymes and, therefore, may have a much more profound role in the regulation of metabolism.

There are good indications that several other key enzymes involved in glycogen metabolism may also be regulated by such a complex cascade of reactions.

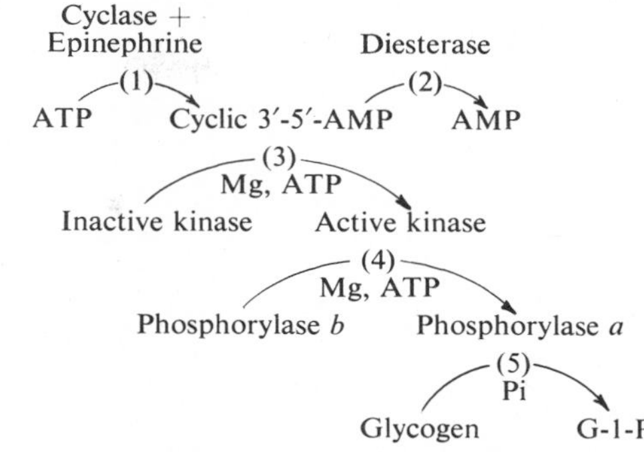

FIG. 3. Control of phosphorylase activity. (1) Activated by catecholamines, inhibited by β-adrenergic blocking agents. (2) Pathway for destruction of cyclic AMP. Blocked by methyl xanthines. (3) Might involve a phosphorylation of kinase, catalyzed by a second enzyme. Active kinase can also be obtained by separate mechanisms, e.g., Ca^{++} and kinase activating factor from muscle, or trypsin. (4) Inhibited by Ca^{++}, reversed by phosphorylase phosphatase. (5) Also catalyzed by phosphorylase *b* in the presence of AMP.

References

1. MEYER, K. H., "The Chemistry of Glycogen," *Advan. Enzymol.*, **3**, 109 (1943).
2. MANNERS, D. J., "Enzymic Synthesis and Degradation of Starch and Glycogen," *Advan. Carbohydrate Chem.*, **17**, 371 (1962).
3. KREBS, E. G., AND FISCHER, E. H., "Molecular Properties and Transformations of Glycogen Phosphorylase in Animal Tissues, *Advan. Enzymol.*, **24**, 263 (1962).
4. "Control of Glycogen Metabolism," Ciba Foundation Symposium, London, Churchill, 1964.

EDMOND H. FISCHER

GLYCOLYSIS

Glycolysis was defined some forty years ago by Otto Warburg[1,p.199] as "the splitting of carbohydrate into lactic acid." This type of lactic acid fermentation was well known to Berzelius, Liebig, Pasteur and Claude Bernard a century ago, as was also alcoholic fermentation. Various kinds of carbohydrates may serve as substrates for glycolysis. Thus, in cancer cells, which can produce lactic acid at a sustained rate probably faster than any other living cell—up to half their own dry weight per hour—glucose is the preferred substrate, although mannose, fructose, and even galactose may be used more slowly. In liver and muscle cells, GLYCOGEN is generally the preferred substrate, partly because such cells have much less glucokinase activity with which to act upon glucose. For the same type of reason, many broken-up cells (homogenates and centrifuged derivatives) use hexosephosphates faster than glucose, if their glucokinase activity is more limiting than the activities of enzymes acting upon hexosephosphates.

In intact living cells, it is probable that most or all of the glycolysis takes place on or in particulate structures (*e.g.*, MITOCHONDRIA, MICROSOMES). However, when such cells are broken up, the glycolytic enzymes rather readily come off the cytoplasmic particulates into the high-speed centrifuge supernatant, especially upon washing the particles even once, thus giving rise to a common fallacy that *cellular* glycolysis is largely non-particulate. Warburg early gave evidence, based on the action of indifferent narcotics, that "glycolysis takes place on the surface of the structural parts of the cell."[1,p.111] This conclusion has been confirmed and extended by other more direct techniques[2,3] with which it was found that glycolysis by cancer cells behaved in all respects tested like glycolysis by mitochondria and microsomes carefully isolated from such cells, whereas glycolysis by the solubilized enzymes in the supernatant fractions did not. Glycolysis by both the cells and their derived particulates was stimulated by exogenous insulin, inhibited by exogenous "anti-insulins" such as sex hormones and corticosteroids and podophyllins, inhibited by prior stress of host animals, inhibited by galactose; also, it used glucose more rapidly than fructose, even more so at low concentrations, and showed very high temperature coefficients, etc. Conceivably, of course, solubilized enzymes from some tissues may yet be found that retain some measure of these types of glucokinase regulation.

It is remarkable that although glycolysis is the sum of a very large number of consecutive intermediate compounds, enzymes, and coenzymes, our present knowledge of all these components, and their sequences, was already acquired nearly a quarter of a century ago. For most animal cells studied, the biochemical sequence from glucose (see CARBOHYDRATE METABOLISM for structures) may be summarized as follows: D-glucose (via glucokinase, ATP, Mg^{++}, insulin: anti-insulin regulators) → D-glucose-6-phosphate (via phosphoglucoisomerase) → D-fructose-6-phosphate (via phosphofructokinase, ATP, Mg^{++}) → D-fructose-1,6-diphosphate (via fructaldolase) → D-glyceraldehyde-3-phosphate (via glyceraldehyde-3-phosphate dehydrogenase, DPN, $HOPO_3^{=}$) → 1,3-diphospho-D-glycerate (via 3-phosphoglycerate kinase, ADP, Mg^{++}) → 3-phospho-D-glycerate (via phosphoglycerate mutase, Mg^{++}) → 2-phospho-D-glycerate (via enolase, Mg^{++}) → phosphoenolpyruvate (via pyruvate kinase, ADP, Mg^{++}) → pyruvate (via pyruvate reductase = lactate dehydrogenase, $DPNH_2$) → L-lactate. The splitting of sugar to lactic acid is thus, to put it in a few words, the shifting of hydrogen by means of the nicotinamide moeity of diphosphopyridine nucleotide (also termed NICOTINAMIDE ADENINE DINUCLEOTIDE). Nicotinamide in DPN takes away 2 atoms of H from phosphorylated carbohydrate, and after dephosphorylation gives back 2H (in $DPNH_2$) to pyruvic acid.

The great biochemical importance of the foregoing sequence of reactions in glycolysis is twofold: (1) Each one of the intermediate compounds formed leads to one or more important possible side reactions also, and these in turn lead to innumerable reactions indispensable to life processes, including respiration. (2) In the entire sequence, and also in some of its parts, comparatively large amounts of free energy are made available, up to a maximum of some 28,000 cal/mole lactate formed under common *in vivo* conditions from one-half mole of glucose.[4–6] This free energy available is considerably larger than the some 9000 calories free energy available from hydrolysis of the so-called high-energy ATP to ADP and inorganic phosphate, although much smaller than the free energy of combustion of a mole of lactate to carbon dioxide and water: some 332,000 calories.[4–6] Whereas the free and heat energies of combustion of lactate are nearly equal, lactic acid fermentation from glucose represents an instance of the relatively rare phenomenon in which the free energy liberated is considerably greater (about 50%) than the heat energy liberated, owing here to the large entropy change involved in the formation of the additional carbonyl (=C=O) bond in two lactates derived from one glucose molecule[4] (see also PHOSPHATE BOND ENERGIES).

The foregoing reaction sequence, commonly called the Embden-Meyerhof pathway after its initial instigators, was in due course worked out in greatest part by Warburg.[7] This pathway is also common to ethyl ALCOHOL FERMENTATION down to the pyruvate stage, which then branches off (via carboxylase) to form acetaldehyde and finally (via alcohol dehydrogenase, $DPNH_2$) ethanol. Alcoholic fermentation is sometimes erroneously referred to as glycolysis, even by those who should know better, but ordinary respiration could also be called glycolysis, since it too shares the common pathway down to pyruvate, and this would indeed lead to confusion! Just as lactate fermentation is the most common fermentation met with in animal cells, so alcoholic fermentation is the most common fermentation met with in plant cells, a distinction most easily observed under anaerobic conditions, and a distinction to which nature has obviously attached considerable evolutionary significance. Even in certain algae where lactate fermention has been observed, D-lactate is formed[8] and not L-lactate as in animal cells.

In view of the fairly exhaustive work extant on the chemistry of glycolysis, the most intriguing and mysterious aspects of glycolysis that remain relatively unsolved are its biological applications, many and varied. Foremost of these is the mechanism of the century-old observation of Pasteur that fermentation is decreased in the presence of air (oxygen), an effect first termed the "Pasteur Effect" by Warburg.[1,p.246] Closely related to this is the "Crabtree Effect," which is the inhibition of respiration (oxygen consumption) caused by addition of glucose. The causes of these effects in various cells under varying particular conditions, as proposed by various investigators, are too numerous to be cited here.[6,8–11] In general, however, a widely prevailing type of cause exists whenever the processes of respiration and fermentation occurring in a given cell (or even subcellular preparation) have to *compete* for some common, limiting intermediate compound, enzyme, or coenzyme: then, obviously, each process will tend to inhibit the other, though not necessarily to the same quantitative or percentage extent. Such competition must take place, however, at parts of the fermentative and respiratory pathways *not* shared in common. As already stated, the pathways of ordinary respiration and fermentation from glucose are common down to pyruvic acid, at which point the pathways separate. In both the oxidation of pyruvate to CO_2 and H_2O, and its reduction to lactate, the nicotinamide moiety is required, and competition for a limiting quantity of DPN + $DPNH_2$ may become involved; this is possibly the most common cause of various observed Pasteur Effects, at least in living or growing cells. In more artificial systems, such as obtain in subcellular preparations, inorganic phosphate and ADP are also experimentally demonstrated instances of competed-for components that can result in Pasteur and/or Crabtree Effects. A large amount of experimental work on glycolysis, the Pasteur Effect, and the Crabtree Effect suffers, however, from lack of due reference back to the living state, or unwarranted application thereto.

Biological applications of glycolysis have concerned, above all else, the metabolism, growth, and chemotherapy of cancer cells. In such cells, Warburg forty years ago first demonstrated the virtually universal occurrence of not only anaerobic but also aerobic glycolysis (*first law of cancer biochemistry*). As recently restated more generally by him, "Shifting of the metabolism to the anaerobic state is the main biochemical difference between the tumor and normal cell." Warburg was also the first to indicate that glycolysis was involved in not only the metabolism but also the *growth* of cancer cells.[1,p.200] All subsequent work, properly interpreted, has confirmed and extended his experimental findings. The most important recent extension (*second law of cancer biochemistry*) has been the finding[3] that the degree of hormonal (insulin: anti-insulin or steroid) regulation of glycolysis at the initial glucokinase stage constitutes a major metabolic- and growth-control mechanism in mammalian cells, be they normal or malignant. In fact, the difference between normal maintenance metabolism or growth (no net increase in cell mass), on the one hand, and increasing, however small, degrees of "uncontrolled" malignant metabolism and growth, on the other hand, appears to depend upon critical increases in the cells' ability to utilize or glycolyze the particular carbohydrate *glucose*.[3]

Recent studies have demonstrated this profound difference between growing cancer cells and normal adult tissues especially clearly and quantitatively in a series of rat hepatomas developed by H. Morris (see also CANCER CELL METABOLISM), in which the growth rates of established lines of hepatomas vary from exceedingly slow to rapid generation or "doubling" times varying from over a year to two weeks. With normal adult liver (no net growth) as a base line, we have found that glucose utilization, and its conversion to lactic acid, increased progressively with growth rate; even more significant, inhibition by "anti-insulins" progressively decreased, as did also the concentration of glucose required for maximum rate of glycolysis (decreased $K_{m \cdot glucose}$). Such phenomena have also been demonstrated for other critical tumor spectra, as in a spectrum of mouse melanomas with different growth rates, and in a spectrum of Sanford-Earle tissue culture sarcoma lines derived originally from the same individual cell, in which the stabilized cell lines grew on host mice at widely varying growth rates.[3]

Finally, one of the most interesting biological aspects of glycolysis is the widespread promoting effect of insulin observable not only with various normal tissues, but especially with malignant tumors taken from host animals or humans. Since a decade ago,[12] many of the factors affecting the quantitative effects of insulin action have been worked out with cancer cells. Prime factors are: the degree of tumor anaplasia; the extent of endogenous anti-insulins resulting from varying degrees of host stress or condition; exogenous anti-insulins added *in vitro*; *in vitro* temperature; concentration and kind of carbohydrate substrate employed. With proper attention to such factors, we have found that the *in vitro* anaerobic glycolysis of virtually every kind of cancer so far tested, taken from a host animal, can be increased by added insulin, under a suitable set of conditions. Thus ascites carcinoma cells taken from the intraperitoneal cavity, often reported in the literature not to have responded to insulin, will indeed readily do so at sufficiently low and maintained glucose concentrations. Other less anaplastic cancers, such as mouse melanomas, which are under far greater endogenous anti-insulin control (at the mitochondrial hexokinase step) show optimal insulin response at glucose concentrations physiological or higher. Brain tissue usually requires still higher glucose concentrations. With certain cancers, larger insulin responses were obtained with fructose than with glucose. Curiously, highly purified endotoxins duplicate many of the insulin responses at the hexokinase step, acting similarly against a variety of anti-insulins that control glucose phosphorylation.[13]

The foregoing account of glycolysis is intended to provide definition and perspective of what the authors regard as major aspects of the three historical phases in the study of glycolysis, pre-Warburg, Warburg, and "post"-Warburg. It is hoped that this account will aid the general reader to wend his way among the mass of innumerable observations and conclusions reported in millions of words on the subject in the literature, commencing this wending, perhaps, with the few references cited below. The book on glycolysis and respiration of tumors by Aisenberg[11] appends a list of 536 references.

References

1. WARBURG, O., "The Metabolism of Tumors," London, Constable, 1930 (translated by F. Dickens).
2. BURK, D., AND WOODS, M., *Radiation Res. Supplement*, **3,** 212–246 (1963).
3. WOODS, M., AND BURK, D., "Hormonal Control of Metabolism in Cancer," in "Control Mechanisms in Respiration and Fermentation," pp. 253–264, Ronald Press, 1963.
4. BURK, D., *Proc. Roy. Soc. London, Ser. B*, **104**, 154–170 (1929).
5. LONG, C. (EDITOR), "Biochemists' Handbook," D. van Nostrand Co., 1961.
6. BURK, D., *Occasional Publ. Amer. Assoc. Adv. Science*, **4,** 121–161 (1937).
7. WARBURG, O., "Wasserstoffübertragende Fermente," Berlin, Verlag Springer, 1948.
8. WARBURG, O., "New Methods of Cell Physiology," New York, Interscience Publishers, 1962.
9. BURK, D., *Cold Spring Harbor Symp. Quant. Biol.*, **7,** 420–459 (1939).
10. DIXON, K. C., Biol. Rev., **12,** 431, 1937.
11. AISENBERG, A. C., "The Glycolysis and Respiration of Tumors," New York, Academic Press, 1961.
12. WOODS, M., HUNTER, J., AND BURK, D., *J. Natl. Cancer Inst.*, **16,** 351–404 (1955).
13. WOODS, M., LANDY, M., BURK, D., AND HOWARD, T., "Effects of Endotoxin on Cellular Metabolism," in "Bacterial Endotoxins," Rutgers, Institute of Microbiology, 1964.

DEAN BURK AND MARK WOODS

GLYCOPROTEINS

Glycoproteins are PROTEINS that contain small amounts of carbohydrate; the term is usually applied to those having less than 4% hexosamine content (see HEXOSAMINES AND HIGHER AMINO SUGARS). Proteins having more than 4% carbohydrate are usually termed mucoproteins. See also PLASMA AND PLASMA PROTEINS.

GLYCOSIDES AND GLYCOSIDASES

Any carbohydrate that forms an internal or cyclic hemiacetal structure can form a linkage, replacing the hydroxyl group of its hemiacetal carbon, to an external alcoholic or hydroxyl group of some other molecule; such a full-acetal linkage is termed a *glycoside* linkage or *glycosidic bond*. A simple example is the linkage from methyl to the glucopyranose ring in methyl glucoside [see CARBOHYDRATES (CLASSIFICATION AND STRUCTURAL INTERRELATIONS)]. The molecule that provides the external hydroxyl group may be another carbohydrate (as well as a non-carbohydrate *aglycone*). Thus the neighboring monosaccharide subunits are linked to each other by glycosidic bonds in homopolysaccharides such as CELLULOSE, in GLYCOGEN, in STARCHES, and in polysaccharides having more than one different repeating subunit, such as MUCOPOLYSACCHARIDES. Enzymes that cleave such glycosidic bonds are glycosidases; examples include lysozyme which attacks BACTERIAL CELL WALLS, various amylases (see DIGESTION), and GLUCURONIDASES.

GLYCOSIDES, STEROID

The pharmacological knowledge of this group considerably antedates the elucidation of their structure. These water-soluble steroid glycosides occur in a great many species of monocotyledons. They are composed of an aglucone or genin and one or more molecules of monosaccharides, such as glucose, galactose, rhamnose, xylose, and such specific 2,6-deoxymonoses as digitoxose and its 3-methyl derivative, cymarose. On acid or enzymatic hydrolysis, they yield water-insoluble aglycones.

The glycosides of *Digitalis lanata*, the digilanides A, B and C (Stoll), contain an acetyl group and yield upon alkaline hydrolysis of this group the *Digitalis purpurea* glycosides A, B and C, which in turn give digitoxin, gitoxin and digoxin, respectively, on hydrolysis by a specific enzyme. Acid hydrolysis liberates 3 moles of digitoxose and the respective genins, all of which are C_{23}-compounds with an unsaturated γ-lactone grouping in the side chain. In scillaridin, the genin of scillaren from squill, we find a δ-lactone and a total of 24 carbon atoms (Stoll).

The members of this group are distinguished by the disposition of various hydroxyl groups in positions, 5, 12, and 14 in addition to the ubiquitous hydroxyl on carbon-3, to which the sugar is attached (see STEROIDS for numbering system). In strophantidin and antiarigenin, we find the angular methyl group at carbon-19 displaced by an aldehyde group. These genins have rings A and B in *cis* position one to another as in the coprostane-cholane series. Only uzarigenin, which is physiologically much less active, displays the *trans* configuration of cholestane and androstane. This great variety of functional groups and isomeric alternatives leads to complicated isomerizations, lactonizations and anhydrizations, which have been studiously explored by Jacobs, Tschesche, Stoll and others.

Many of these substances have been known for ages to the natives of Africa and South America, who used them as arrow poisons. *Digitalis* and *Strophanthus* glycosides have become irreplaceable drugs in the modern therapy of heart disease.

Related cardiac poisons occur in the animal kingdom, as *bufotoxins* in the skin secretion of toads, together with epinephrine and the bicyclic, nitrogenous bufotenin. The genins of the bufotoxins are designated as bufogenins, bufotalin and bufagins, and are conjugated with suberylarginine. Their detailed structure varies from species to species, but they all have a C_{24}-skeleton with the same six-membered δ-lactone ring as scillaridine.

Saponins are water-soluble plant constituents distinguished by their ability to form a soapy foam even in high dilution. One group of them has a triterpenoid structure and need not be discussed here. The other group comprises steroid glycosides which occur in many instances alongside the cardiac glycosides (which are technically saponins too). They have high surface activity and hemolyze red blood corpuscles. They consist of the genins and 1–5 monose molecules, *e.g.*, glucose, galactose, rhamnose, xylose. The corresponding genins have 27 carbon atoms; the carbon skeleton of the side chain is the same as in cholesterol, but carries a virtual keto group on carbon-22 and a hydroxyl group on carbon-26. This keto group exists as a spiroketal, forming one ring with the hydroxyl group on carbon-16 and another one with the hydroxyl group on carbon-26.

Additional secondary hydroxyl groups are found on carbons-2, 6, 12 and 15, besides the OH group on carbon-3 common to all of them. Rings A and B occur in *cis* or *trans* position, and epimerism on the spiro-ring on carbon-22 leads to additional variations. A search for sapogenin-producing plants was stimulated by the need for large quantities of starting material other than cholesterol in the partial synthesis of steroid hormones. Marker, in particular, isolated a great many saponins from tropical plants, established their structure and worked out their degradation to progesterone and related products.

A more potent hemolytic agent than any plant saponins is a recently discovered "animal saponin," a toxic steroid tetrasaccharide, occurring in a tropical sea-cucumber (class *Holothurioideae*, phylum *Echinodermata*). This compound, which we have called Holothurin, differs structurally from the plant saponins and carries also a sulfate ester group, reminiscent of the bile alcohols in the lower vertebrates. Assuming that this compound is paleontologically as old as the animal which produces it, we must marvel at evolution for having first developed a water-soluble surfactant steroid in animals and much later applying the same architectural principle in the elaboration of a plant product.

HARRY SOBOTKA

GLYOXYLATE-MALATE CYCLE. See CITRIC ACID CYCLE, section on the Glyoxylate Shunt.

GOITER

The thyroid gland contains an enzyme system which synthesizes the hormones thyroxine and triiodothyronine from iodine and tyrosine. The chemical pathways leading to the synthesis of thyroid hormones are represented in the following scheme:

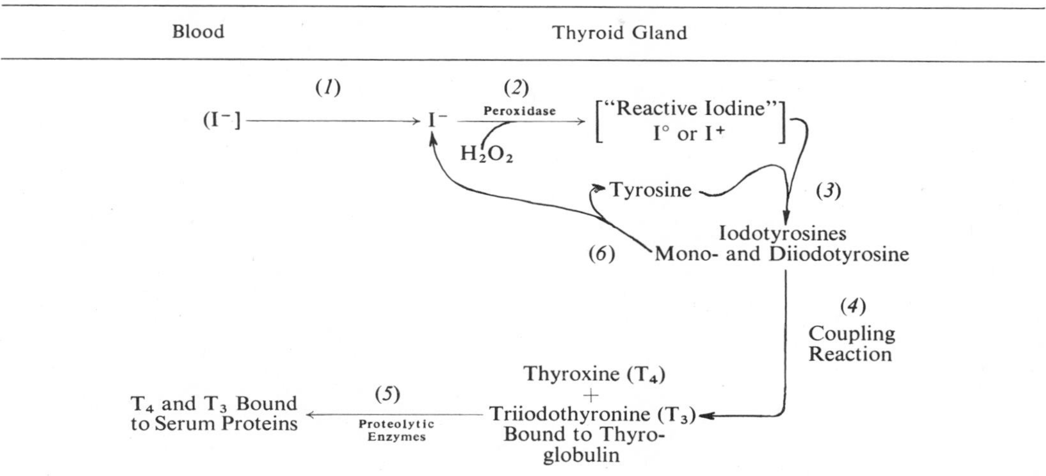

Reaction (*1*) is known as the "iodide trapping mechanism" whereby the thyroid gland removes iodide ion from the bloodstream against a concentration gradient. Reaction (*2*) is catalyzed by a peroxidase that utilizes hydrogen peroxide to convert I^- to "active iodine," with is either I° or the iodinium ion I^+. Reactive iodine then iodinates the amino acid, tyrosine, to form monoiodotyrosine and diiodotyrosine [reaction (*3*)]. This latter reaction may require another enzyme or the peroxidase might possibly serve a dual function by catalyzing it. Thyroxine (T_4) is then formed by the coupling of two diiodotyrosine molecules, as indicated in reaction (*4*). Triiodothyronine (T_3) is formed either by the coupling of monoiodotyrosine and diiodotyrosine or the deiodination of T_4. The coupling reaction apparently occurs while the iodotyrosines are bound to thyroglobulin, a protein that represents 70% of the dry weight of the gland. Free thyroxine is then released into the blood by hydrolysis [reaction (*5*)] of the thyroglobulin and is catalyzed by proteolytic enzymes. Most of the excess free iodotyrosines are deiodinated [reaction (*6*)] by a dehalogenase enzyme to yield iodide ion and free tyrosine, so that the concentration of non-protein bound iodoamino acids remains quite small (see also IODINE; THYROID HORMONES).

Goiter is the abnormal enlargement of the thyroid gland, irrespective of the cause, but many goiters are the result of an inadequate synthesis of the thyroid hormones. A deficiency of hormone synthesis can be a result of genetic enzyme deficiencies in the catalysts that control reactions (*1*), (*2*), (*3*), (*4*) and (*6*). Diets deficient in iodine or the ingestion of thyroid inhibitors can lead to goiters. Goitrogenic materials can interfere with the iodide trapping system (*e.g.*, perchlorate and thiocyanate ions) or inhibit the peroxidase.[6] Peroxidase inhibitors are generally classified as "aromatic" and "thiocarbamide" goitrogens, but another potent inhibitor is aminotriazole.[7] Some goitrogens are found in brassicaceous plants, *e.g.*, turnips, and these foods are potentially goitrogenic. Large amounts of iodide ion can bind the reactive iodine and prevent the iodination of tyrosine.

Other goiters can be caused by infections, neoplasms, and a defective homeostatic mechanism. The latter can lead to an excessive synthesis of thyroid hormones and has been called a toxic goiter.

Thyroid stimulating hormone (TSH) that is produced in the pituitary gland also controls thyroid metabolism. The TSH stimulates growth of the thyroid and reactions (*1*), (*2*), (*3*), (*4*) and (*5*), but the mechanism of this stimulation is unknown (see also THYROTROPIC HORMONE).

References

1. WOLFF, J., AND GOLDBERG, R., in "Biochemical Disorders in Human Disease," (THOMPSON, R. H. S., and KING, E. J., EDITORS), p. 289, New York, Academic Press, 1957.
2. STANBURY, J. B., AND MCGIRR, E. M., *Am. J. Med.*, **22**, 712 (1957).
3. WILLIAMS, R. H., AND BAKKE, J. L., "Textbook of Endocrinology," (WILLIAMS, R. H., EDITOR), Third edition, Chapter 4, p. 96, Philadelphia, Pa., W. B. Saunders Co., 1962.
4. MALOOF, F., AND SOODAK, M., *Pharmacol. Rev.*, **15**, 43 (1963).
5. RAWSON, R. W., SONENBERG, M., AND MONEY, W. L., in "Diseases of Metabolism," (DUNCAN, G. C., EDITOR), Ch. 17, p. 1159, Philadelphia, Pa., W. B. Saunders Co., 1964.
6. ALEXANDER, N. M., *J. Biol. Chem.*, **234**, 1530 (1959).
7. ALEXANDER, N. M., *J. Biol. Chem.*, **234**, 148 (1959).

NICHOLAS M. ALEXANDER

GOLDBERGER, J.

Joseph Goldberger (1874–1929) was an Austrian-born American physician, who distinguished himself by his investigation of pellagra and a convincing demonstration that in humans it is caused by nutritional deficiency. The missing factor was called by him the P-P (pellagra preventing) factor. After his death, it was sometimes called vitamin G in his honor (see NICOTINIC ACID).

GOLGI APPARATUS

The Golgi apparatus is a type of subcellular organelle first reported in 1898 but still of uncertain biochemical significance. The Golgi substance of neurons is discussed in NERVE CELL COMPOSITION AND STRUCTURE.

References

1. DALTON, A. J., "Golgi Apparatus and Secretion Granules," in "The Cell," (BRACHET, J., AND MIRSKY, A. E., EDITORS), Vol. II, New York, Academic Press, 1961.
2. NOVIKOFF, A. B., ESSNER, E., AND QUINTANA, N., "Golgi Apparatus and Lysosomes," *Federation Proc.*, **23**, 1010 (1964).

GOLGI, C.

Camillo Golgi (1844–1926) was an Italian histologist who discovered the use of silver salts to stain cells and tissues. His techniques were particularly useful in studying nervous tisue. The GOLGI APPARATUS still bears his name. He shared the Nobel Prize in medicine and physiology in 1906 with Ramon y Cajal, Spanish histologist, who also used Golgi techniques to study the brain spinal cord and retina, and came up with views opposed to those of Golgi.

GONADOTROPIC HORMONES

The gonadotropins are easily defined as those hormones that stimulate the gonads. In mammals, for which most information is available, the gonadotropins are produced by the *pituitary gland* (adenohypophysis), the *placenta* and the *endometrium* (the latter in the case of the pregnant mare). Recently, some gonadotropic activity has been detected in extracts of hypothalamic tissue, although this material is as yet poorly characterized and its origin is not known. Hormones of the gonadotropin group are *follicle stimulating hormone* (FSH), *luteinizing hormone* (LH), *luteotropic hormone* (LTH), and *chorionic gonadotropin* (HCG—the usual abbreviation, since commonly used preparations are derived from human sources). The biological effects of HCG are similar to those of LH. Preparations from pregnant mare serum (PMS) are also frequently used for an FSH-like gonadotropic activity. *Urinary gonadotropins*, apart from the chorionic gonadotropin in the urine during pregnancy, have proved extremely variable in character both for biological activity and chemical properties. These gonadotropins are in all probability partially degraded pituitary gonadotropins and will not be considered as distinct chemical entities here. This assumption carries the unproved implication that enzymatic degradation products of the gonadotropins can be obtained which still possess biological activity.

Many authors prefer the term *interstitial cell stimulating hormone* (ICH or ICSH) for LH. The different nomenclature arose from differences in bioassay procedures in early studies before it was established that the same hormone was being measured. Luteotropic hormone is also called *prolactin* or *lactogenic hormone*. Although prolactin is the most commonly used name in recent literature, the term luteotropic hormone will be used here since it directs our attention to its function as a gonadotropin. However, it should be noted that a stimulation of corpora lutea secretion by LTH has been demonstrated only in a few rodents.

Early literature on the gonadotropins contained several terms for crude preparations. Preparations containing principally FSH activity have been called prolan A or thyalkentrin; preparations containing principally LH activity have been called prolan B or metakentrin; those containing principally luteotropin have been called mammotropin or galactin. These terms are seldom encountered in current literature.

Although the gonadotropins are readily defined in broad terms, it is very difficult to obtain complete agreement among investigators concerning specific details of the physiological role or mode of action of the gonadotropins. In much of the early literature, points of disagreement were often traced to the degree of purity of the hormone preparations used. The currently available gonadotropin preparations are sufficiently pure that this question is less frequently a factor influencing interpretation of experiments. Investigators are now turning to experiments which include consideration of biological half-life, route or frequency of administration of the gonadotropin, feedback effects in hormonal control, augmentation or inhibitory effects among gonadotropins and other hormones, etc. Although studies delineating the physiological role of gonadotropins are complex, important advances in understanding gonadotropic action are being made.

Chemical and Physical Properties of the Gonadotropins. *Follicle Stimulating Hormone.* Preparations from porcine, ovine, and human pituitary glands have received the most extensive study, although equine FSH has attracted some attention. Bovine pituitaries contain relatively little FSH.

FSH is the most soluble of the pituitary hormones and isolation procedures usually take advantage of its solubility in 50% saturated ammonium sulfate. Final steps in purification utilize a low affinity for carboxyl type ION-EXCHANGE columns at pH near neutrality (conditions under which LH is readily adsorbed) and finally an anion-exchange CHROMATOGRAPHY as with diethylaminoethylcellulose. The best potencies reported for FSH are approximately 15–45 times the potency of the NIH-FSH-S1 standard, the highest reported value being for a sheep FSH preparation. The NIH standard is 2.7 times the Armour standard No. 264–151X, which was used as a reference preparation in FSH studies for several years. Although the reference preparations have proved very stable, as the degree of purification increases, the stability of the FSH decreases considerably. This has proved a major obstacle in the isolation and characterization of FSH. The best preparations have contained mannose, galactose, and fucose. Mannosamine and galactosamine have been reported in some of the preparations. Sialic acid (as much as 5%) has also been reported as a constituent of FSH. Moreover, this seems to be necessary for the biological activity since incubation with neuraminidase rapidly inactivates the hormone.

In view of the purification difficulties, few authors have analyzed the composition of their FSH preparations in detail. Only one report on the amino acid composition of FSH has appeared, this for a pork FSH preparation isolated from a pancreatin-digested pituitary fraction. Most noteworthy were the high content of acidic amino acids (20.8%) and the relatively high cystine content (6.5%). The proline content of this preparation was not remarkably high (4.4%) although probably higher than for most proteins.

The molecular weight of the FSH preparations has not been estimated, but from sedimentation coefficients that have been published the molecular weight is probably in the range 20,000–30,000.

Luteinizing Hormone. Purification methods are available for LH preparations from sheep, human, beef, pork, equine, and even rat pituitary glands. Stability appears much better than reported for FSH which has facilitated LH purification. Most procedures for purification utilize ammonium sulfate fractionation, or extraction in ethanol-acetate buffers in the initial stages, followed by a cation-exchange chromatography, and often a gel filtration.

Preparations from sheep, beef, equine, and rat pituitary glands have achieved potencies of 1–2 times the NIH-LH-S1 standard which is equipotent to the Armour 227–80 LH standard used for several years prior to the availability of the National Institutes of Health (USPHS) reference preparations. Occasional reports of preparations with higher activity have appeared, but these are not readily reproducible. In the case of human and equine LH, biological half-life (in the rat) has been reported as 6–30 times greater than ovine, murine, or porcine LH. Thus, comparison of human or equine LH potencies with a sheep reference standard would not be valid.

The molecular weight of luteinizing hormone from beef, pork, sheep, and human sources is in the range 26,000–30,000. In the case of sheep, pork and beef LH, it has been shown that at pH 1.5 the molecule dissociates into inactive subunits of approximately one-half this molecular weight.

Ovine LH has received the most extensive characterization for amino acid analysis and carbohydrate composition. Noteworthy are a high cystine content (9 residues/mole) a very high proline (25 residues) and an excess of basic amino acids (isoelectric point at pH 7.7). The composition of the beef and pork preparations is very similar although the proline content is lower. All preparations contain mannose, galactose, and fucose, glucosamine and galactosamine, and approximately 2.5% acetyl groups.

Luteotropic Hormone. The preparation from sheep pituitary glands is best characterized. The material has a molecular weight of 23,350. It is a simple protein, has relatively little cystine

(3 residues/mole) but appreciable proline (11 residues). A high content of amino acids with hydrophobic side chains gives LTH a relatively poor water solubility and greater solubility in organic solvents. Most of the isolation procedures take advantage of these properties. Under certain conditions, the monomer readily associates to a rather stable dimeric form. Both forms are biologically active.

Human Chorionic Gonadotropin. This hormone is usually isolated from pregnancy urine by adsorption onto a material such as benzoic acid or kaolin and subsequent fractionation in aqueous ethanol or ethanol acetate solutions. Beyond this point, several techniques have been employed, such as zone ELECTROPHORESIS, COUNTERCURRENT DISTRIBUTION, column chromatography, etc. The best preparations have a potency of 12,000 I.U./mg, a molecular weight of 30,000, and a low isoelectric point (2.95). HCG is glycoprotein (30% carbohydrate) which contains sialic acid (6%) necessary for biological activity (inactivated by neuraminidase). The peptide portion of the molecule is rich in cystine and proline.

Pregnant Mare Serum Gonadotropin. This has been isolated by adsorption or precipitation from pregnant mare serum. Fractional precipitation from aqueous ethanol solution, chromatography, and countercurrent distribution have been used to obtain preparations of 16,000 I.U./mg. Such preparations contain 50% carbohydrate and 40% peptide. The latter was rich in cystine and proline. The material contains 11% sialic acid and is inactivated by neuraminidase.

Physiological Action of the Gonadotropins. Since the hormones controlling gonad function are interrelated in a complex system, elements of which often vary from one species to another, many unresolved points remain. The following is an attempted generalization which of necessity ignores many fine points in our specific knowledge of gonadotropin action. Moreover, the comments apply specifically to mammals although there is a significant literature on gonadotropin function in birds, amphibians, and lower forms.

Growth of the ovarian follicle is stimulated by FSH. As the follicle reaches maturity, LH is apparently required for the final phase of maturation during which time estrogen production increases (see also SEX HORMONES). LH is then required for the rupture of the follicle and release of the ovum. A rise in LH release by the pituitary can be detected prior to ovulation. In some animals, this release of LH is triggered by stimulation of a nerve pathway from the cervix (*e.g.*, induced ovulators such as the rabbit). In spontaneous ovulators (*e.g.*, human, monkey, rat), the triggering mechanism is more subtly controlled but again involves higher nerve centers. Direct control of the pituitary synthesis and/or release of gonadotropins is under hypothalamic control. The hypothalamus has been shown to contain a factor which inhibits the release of luteotropin (PIF or prolactin-inhibiting factor), an FSH-releasing factor (FRF) and an LH-releasing factor (LRF).

The ruptured follicle, under LH stimulation, is converted to a functional corpus luteum. In some species, continued secretion by the corpus luteum requires LTH stimulation. The final role of the gonadotropins is to insure the production of adequate quantities of sex hormones at the proper time. Thus, ultimately, studies of gonadotropin action must deal with steroid synthesis. In this connection, LH has been shown to stimulate conversion of cholesterol to progesterone in slices of beef corpora lutea.

In the male, testicular growth is largely a function of FSH stimulation, while LH stimulates the interstitial cells for the production of testosterone.

Bioassay procedures for gonadotropin have utilized weight increase of the gonads of immature rats or mice or in hypophysectomized animals. Since the weight increase (of either ovary or testes) is largely controlled by FSH-type activity but is somewhat stimulated by an LH-type activity, such an assay is said to measure "total gonadotropin." The most specific bioassays are the Steelman-Pohley method for FSH, which utilizes ovarian weight increase in immature rats which are receiving a high dose of HCG to mask any effects of LH activity in the sample tested. For LH, the Parlow ovarian ascorbic acid depletion test, utilizing ascorbic acid decrease 4 hours after a test dose of LH in pretreated pseudo-pregnant rats, is perhaps the most accurate and specific bioassay. Ventral prostate weight increase in hypophysectomized male rats forms the basis for an LH bioassay that has been used for several years, but is less sensitive. There are a great many other bioassays for gonadotropins, each with their particular advantages and disadvantages.

References

1. COLE, H. H. (EDITOR), "Gonadotropins, Their Chemical and Biological Properties and Secretory Control," San Francisco, W. H. Freeman & Co., 1964.
2. LITVAK, G., AND KRITCHEVSKY, D. (EDITORS), "Actions of Hormones on Molecular Processes," New York, John Wiley & Sons, 1964.

DARRELL N. WARD

GOUT

Gout is a disease involving elevated production and deposition in bodily tissues of a sodium salt of uric acid (see METABOLIC DISEASES; PURINE BIOSYNTHESIS).

GROUND SUBSTANCE

The ground substance is an extracellular matrix in which are imbedded insoluble protein (mainly COLLAGEN) fibers; together these constitute the connective tissue which provides a structural framework for many organs [see BONE FORMATION; PHOSPHATE (IN ANIMAL NUTRITION)].

GROWTH HORMONE, PITUITARY

The early studies of many investigators demonstrated that the growth of an animal is influenced by the anterior lobe of its pituitary gland. The existence of a hormone responsible for this activity was finally proved by the isolation of growth hormone (GH, somatotropin) in highly purified form from bovine pituitaries in 1944. It soon became apparent that although these bovine pituitary growth hormone preparations were potent in promoting growth in the rat, they were without any effect when tested clinically in humans. Many other evidences of species specificity were observed in subsequent studies of the hormone. The rat showed the ability to utilize growth hormone from the widest range of species, but even here, fish pituitary GH, although very active in fish, was inactive in the rat. It has been suggested that the observed species specificity in biological response may be related to molecular variations among the various growth hormone preparations. Recently, with the knowledge that growth hormone of primate origin is active in man, and with the isolation of human growth hormone in 1956, clinical studies on human hypopituitary dwarfs have been initiated. Because of the limited supply of pituitary glands from either monkeys or humans, these studies are of necessity limited, and the real goal for the future is the achievement of laboratory synthesis of a growth hormone with the same structure as that of the primate hormones.

Isolation and Characterization. Growth hormones from the pituitary glands of the following mammalian species have been isolated: ox, sheep, pig, humpback whale, Rhesus monkey, and man. The growth hormones from all the species are purified by a similar fractionation procedure. In general, the pituitaries are ground and extracted with either saline or $Ca(OH)_2$ solution adjusted to pH 10. The growth hormone fraction is then precipitated from the extract by the addition of ammonium sulfate to 0.5 saturation. In the case of the ox, sheep, pig and whale material, the half-saturated ammonium sulfate precipitate is refractionated with ammonium sulfate. For further purification, CHROMATOGRAPHY on "Amberlite" IRC-50 cation-exchange resin is used. All the species of growth hormone tested were found to be adsorbed onto the resin at pH 5.1 in the presence of $0.45M$ $(NH_4)_2SO_4$ and were elutable with a buffer of pH 6. Human or monkey growth hormone, however, could be eluted with water.

For the final purification of the hormone, isoelectric precipitation and alcohol fractionation have been used. The isolation of the porcine hormone required, in addition, a simple ten-transfer COUNTERCURRENT DISTRIBUTION in the solvent system consisting of 2-butanol–0.4% dichloroacetic acid, followed by GEL FILTRATION to remove denatured protein. Gel filtration has also beeen used as the final step in the purification of human growth hormone.

Preparations of growth hormone from all species are homogeneous in free boundary ELECTROPHORESIS. There is a wide variation in isoelectric points. As determined by free boundary electrophoresis in 0.1 ionic strength monovalent buffers, the isoelectric points of the various growth hormone preparations are as follows: ox, 6.85; sheep, 6.8; pig, 6.3; whale, 6.2; monkey, 5.5 and human, 4.9. It is of interest that the bovine and ovine growth hormones have very nearly the same isoelectric points as do the hormones obtained from porcine and whale pituitaries. The two primate hormones, monkey and human, are the most acidic of the hormones studied.

These hormones were also homogeneous when examined by means of zone electrophoresis on starch, and in all cases the biological activity was associated with the protein peak when sections were eluted and assayed.

Molecular weights as obtained from sedimentation measurements in the ultracentrifuge are as follows: ox, 45,000; sheep, 47,800; pig, 41,600; whale, 39,900; monkey, 23,000; human, 21,500. Thus, there is also a considerable range among the species studied. The molecules of ox and sheep growth hormones appear to be the largest at around 45,000, those of pig and whale somewhat smaller at 40,000, and those of the primate hormones the smallest of all, between 21,000 and 25,000.

All the growth hormones isolated have been shown to be simple PROTEINS in that they consist entirely of amino acids and contain no carbohydrate, nucleic acid, lipid, or other prosthetic groups. The ultraviolet absorption spectra are typical of proteins in which the predominant ultraviolet-absorbing moieties are due to tyrosine and tryptophan. As would be expected from the molecular weights, there is a large variation in the number of amino acid residues present in each species of growth hormone. All the common AMINO ACIDS are present. The amino acid content of three species (bovine, monkey and human) of growth hormone (in residues per mole of hormone) is given in Table 1.

In order to determine the amino acid residues at the amino terminus of the various growth hormones, two techniques were utilized; one is the dinitrofluorobenzene method and the other is the phenylisothiocarbamyl method [see PROTEINS (END GROUP ANALYSIS)]. The ox and sheep growth hormones have two NH_2-terminal amino acids, phenylalanine and alanine, indicating that the hormones of these two species either consist of two polypeptide chains or are made up of a branched chain, whereas the other species have but a single amino terminal residue, phenylalanine, and thus are made up of a single polypeptide chain. The amino acids adjacent to the amino or NH_2-terminus were identified as shown in Table 2.

The carboxyl or COOH-terminus of the growth hormone preparations of various species was studied by means of digestion with carboxypeptidase (see PROTEOLYTIC ENZYMES). The results were confirmed by the hydrazinolysis technique. In all cases, phenylalanine was found to be at the COOH-terminus. From kinetic studies with the

TABLE 1. AMINO ACID ANALYSIS OF BOVINE, MONKEY AND HUMAN GROWTH HORMONE

Amino Acids	Number of Residues per Mole of the Hormone		
	Bovine (45,700)	Monkey (23,000)	Human (21,500)
Lysine	22	9	9
Arginine	24	11	10
Histidine	7	3	3
Tyrosine	12	8	8
Tryptophan	3	1	1
Phenylalanine	25	13	13
Aspartic acid	36	21	20
Glutamic acid	49	27	26
Methionine	8	3	3
Half-cystine	8	6	4
Threonine	25	10	10
Serine	27	17	18
Proline	15	10	8
Glycine	23	10	8
Alanine	29	7	7
Valine	14	6	7
Leucine	49	23	25
Isoleucine	13	7	8
Total residues	389	192	188

enzyme, COOH-terminal sequences were proposed as presented in Table 2. It can be noted that although bovine GH has two amino terminal residues, only a single carboxyl terminus is shown for it. This indicates a branched-chain structure.

When the GH preparations of various species were treated with carboxypeptidase so that the entire carboxyl terminal phenylalanine was removed, they were found to be fully active biologically. Thus, the integrity of the carboxyl end of the growth hormone molecule is not essential in order for it to retain its biological potency.

Immunological Studies. When bovine GH was used as an ANTIGEN in the rabbit, antiserum was obtained which contained precipitating ANTIBODIES to the hormone. In the standard test for antigen-antibody reaction, the antiserum was found to react also with ovine growth hormone,

TABLE 2. NH_2- AND COOH-TERMINAL AMINO ACID SEQUENCES OF VARIOUS SPECIES OF GROWTH HORMONE

Species	NH_2-terminal Sequences	COOH-terminal Sequences
Bovine	Phe-Ala-Thr... Ala-Phe-Ala...	...Leu-Ala-Phe-Phe
Sheep	Phe... Ala...	...Ala-Leu-Phe
Pig	Phe-Pro-Ala-Met-Pro...	...Phe-Ala-Phe
Whale	Phe...	...Leu-Ala-Phe
Monkey	Phe...	...Ala-Gly-Phe
Human	Phe-Pro-Thr-Ileu-Pro...	...Phe

but with none of the other purified growth hormone preparations. In a similar study, human and monkey GH both reacted to human GH antiserum, but none of the other species tested showed a reaction. Of special interest is the fact that human GH that has been oxidized with performic acid, thereby converting the cystine residues to cysteic acid, still reacts with antiserum to the native molecule It has been osberved that the oxidized material is devoid of biological activity; thus, the antigenic site of the molecule does not appear to be related to the site responsible for biological activity.

Although all mammalian growth hormones are active in the rat, as noted above, their capacities for eliciting body-weight gain in hypophysectomized rats vary. The non-primate growth hormones (bovine, ovine, porcine, and whale) elicit a continuous increase of body weight in the rat for an apparently unlimited period of time, whereas primate growth hormone exerts this effect for only 10 days. After that time, the animals receiving primate growth hormone become resistant to the hormonal effect, although they still respond to the injection of other non-primate growth hormones.

Biological Properties. From the first demonstration of growth-promoting activity in pituitary extracts, as early as 1921, the chief function of growth hormone was long thought to be the promotion of body growth. This is because such effects are so easily apparent to the clinician. Particularly striking and unmistakable are the widespread changes resulting from the deficiency of this hormone in the body, either from natural disorders of the pituitary or from the experimental removal of this gland from animals. However, we have now come to realize that it is a misconception to think of growth hormone solely as an agent promoting linear and bone growth. We must rather think of it as a major metabolic hormone, whose effects are felt in every process by which an organism functions.

Growth hormone has a marked influence on protein metabolism. For example, a growth-stimulating effect of human growth hormone on human cells *in vitro* has been noted. Human GH has been shown to cause an increase in the nuclear multiplication of human liver cells in tissue culture, and the increase was proportional to the hormonal concentration and length of incubation time. The response of the cells to human GH was species specific, since other proteins, including bovine GH, failed to produce comparable changes. Furthermore, the effect on nuclear multiplication was abolished by the antibody to the hormone. It should also be noted in this connection that bovine growth hormone has been shown to regulate the rate of PROTEIN BIOSYNTHESIS in the rat liver *in vivo*.

In addition to the marked influences of growth hormone on protein metabolism, it is known to play an important role in both fat and carbohydrate metabolism. The percentage of fat in the body weight gained by rats injected with bovine GH is lower than in untreated rats. Recent studies have shown that human GH possesses an intrinsic lipolytic activity as assayed on fat pads from rats, rabbits and guinea pigs. The diabetogenic activity of growth hormone has been demonstrated in cats and dogs with bovine growth hormone and in human subjects with the human hormone.

One of the most characteristic effects of growth hormone is that of biological synergist, enhancing the effects of other hormones when given in conjunction with them. For example, bovine GH enhances the effect of interstitial-cell stimulating hormone (see GONADOTROPIC HORMONES) on the weights of the ventral prostates in hypophysectomized rats. Also, the concurrent administration of growth hormone enhances the effectiveness of ACTH in restoring the adrenal cortex after hypophysectomy.

Growth hormone not only acts in synergism with ACTH but, in its anabolic role in the body, also counteracts some of the effects of ACTH (see ADRENOCORTICOTROPIC HORMONE). This is particularly evident in connection with antibody formation and resistance to infection. ACTH produces a depression of antibody level against a specific antigen when administered during the period of immunization, but growth hormone given simultaneously with the ACTH effectively counteracts this antibody depression. Similarly, ACTH lowers the resistance of animals to bacterial infection, *e.g.*, in studies with *Pasteurella pestis*, and growth hormone counteracts the lowered resistance.

Recent studies with human GH have clearly demonstrated that the hormone possesses intrinsic activities that are similar to those characteristic of prolactin. Human GH promotes pigeon crop-sac growth when administered by either local or systemic procedures, and it induces localized milk secretion when injection with cortisol into rats without pituitaries, ovaries and adrenals. Human growth hormone, of course, does not possess these prolactin-like activities to the same degree as does the ovine lactogenic hormone, but only to an extent of 20% or less of the latter.

The most familiar aspect of growth hormone is, of course, its connection with growth, both normal and abnormal, and the role of the hormone in its latter connection has been under intensive investigation for the last three decades. It was observed early that many tumors developed in the lungs, adrenal medullas, and reproductive organs of normal adult rats that had been treated with bovine growth hormone for many months. The diversity and number of these tumors suggested the possibility of a direct effect of excessive amounts of growth hormone on susceptible tissues. However, similar lesions were found to be absent in growth hormone-treated rats whose pituitaries had been removed. Pituitary removal has been demonstrated to inhibit neoplastic growth, and thus pituitary function appears to be a prime factor in CARCINOGENESIS. On the other hand, in studies of the effects of a carcinogen 9,10-dimethyl-1,2-dibenzanthracene on hypophysectomized rats, animals that were treated with bovine GH in addition to the carcinogen developed more tumors than did those given the carcinogen alone.

References

1. LI, C. H., "Hormones of the Anterior Pituitary Gland," Part I, "Growth and Adrenocorticotropic Hormones," *Advan. Protein Chem.*, **11**, 101 (1956).
2. LI, C. H., "Properties of and Structural Investigations on Growth Hormone Isolated from Bovine, Monkey and Human Pituitary Glands," *Federation Proc.*, **16**, 775 (1957).
3. LI, C. H., "Anterior Pituitary Hormones," *Postgrad. Med.*, **29**, 13 (1961).
4. LI, C. H., AND LIU, W.-K., "Human Pituitary Growth Hormone, VIII," *Experientia*, **20**, 169 (1964).

CHOH HAO LI

GUANINE AND GUANOSINE

Guanine, 2-amino-6-oxypurine, may be represented as follows:

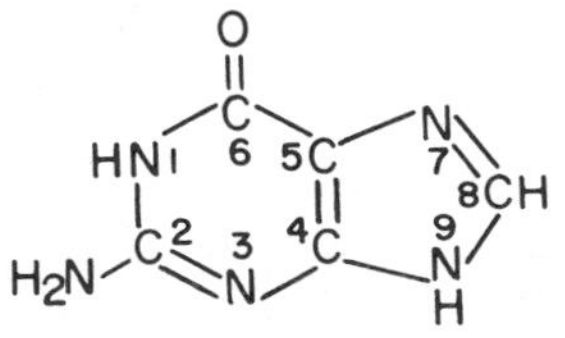

O, C 6, HN 1, 5C, N 7, 8CH, C 2, 3, 4C, N 9, H, H₂N, N

It, like adenine, is a prominent PURINE and occurs in nucleosides and nucleotides, and enters into the make up of nucleic acids (see RIBONUCLEIC ACIDS; DEOXYRIBONUCLEIC ACIDS). Guanine was first discovered as a constituent of guano; hence, its name. It absorbs ultraviolet light readily, as would be anticipated by a glance at its formula.

The structure of the ribonucleoside form, guanosine, is as follows:

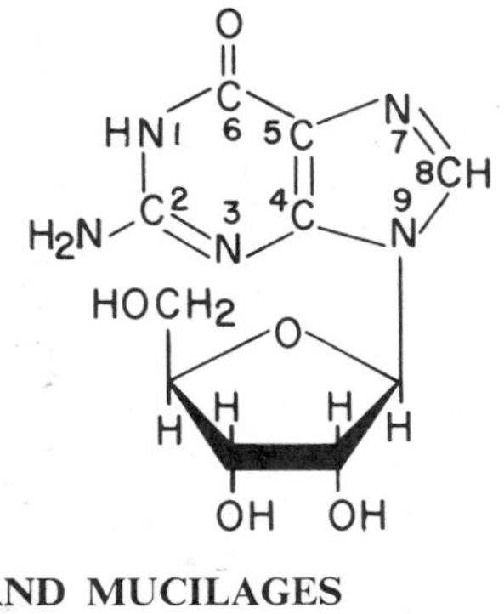

O, C 6, HN 1, 5C, N 7, 8CH, C 2, 3, 4C, N 9, H_2N, N, $HOCH_2$, O, H, H, H, H, OH, OH

GUMS AND MUCILAGES

Gums and mucilages are carbohydrate polymers of high molecular weight obtained from plants.

Reference

1. DUTTON, G. G. S., *et al.*, "Gums and Mucilages," in "The Encyclopedia of Chemistry" (CLARK, G. L., AND HAWLEY, G. G., EDITORS), New York, Reinhold Publishing Corp., 1957.

H

HALLUCINOGENIC DRUGS

There are many substances which will, if taken in the appropriate quantities by normal subjects, produce distortion of perception, vivid images, or hallucinations. Most of these substances will produce powerful peripheral as well as the central effects. Some few agents are characterized by the *predominance* of their actions on mental and psychic functions. This group of drugs has been called hallucinogens, psychotomimetics, psycholytics, and psychedelics, among several ambiguous terms. None of the names which have been suggested to date are adequately descriptive of the compounds. The most commonly used are hallucinogen and psychotomimetic.

The major hallucinogens of current interest[1] may be classed into five groups of chemically distinct compounds: (1) lysergic acid derivatives of which lysergic acid diethylamide (LSD-25) is the prototype; (2) phenylethylamines, the best known of which is mescaline; (3) indolealkylamines, which include psilocybin, psilocin, and bufotenin; (4) piperidyl benzilate esters, typified by Ditran (a 70:30 mixture of N-ethyl-2-pyrrolidymethyl phenylcyclopentylglycolate and N-ethyl-3-piperidyl phenylcyclopentylglycolate), and (5) phenylcyclohexyl piperidines (Sernyl). Figure 1 shows the chemical structures of these compounds.

Drugs representative of the first three groups have been isolated from naturally occurring sources. LSD-25 is a molecular component of ergot, a fungus which infects cereal grains. Mescaline, historically the oldest hallucinogen, was isolated from a Mexican peyote cactus. Psilocybin and psilocin were isolated by A. Hoffman from the Mexican mushroom, *Psilocybe mexicana*. Bufotenin is found in some varieties of toadstools. The indole derivatives are chemically closely related to serotonin (5-hydroxytryptamine), a compound which plays an important, but as yet unknown, role in the central nervous system.

The piperidyl benzilate esters and phenylcyclohexyl piperidines are synthetic compounds, and have not been shown to occur naturally. Some authors[2] do not consider them to be hallucinogens, but active researchers in the field include them among the most active psychotomimetics.[3]

Clinical syndromes from LSD-25, mescaline, and the indoleamines are similar. Somatic symptoms are nausea, dizziness, loss of appetite, blurred vision, paresthesia, weakness, drowsiness, and trembling. These result frequently and are usually associated with sympathomimetic effects, such as increased pulse rate and slight temperature elevation. Perceptual and psychic changes are marked. Visual illusions and vivid hallucinations, decreased concentration, slow thinking, depersonalization, dreamy states, changes in mood, and often, anxiety, are commonly found. Except for severe intoxication, consciousness and judgement remain intact. Experimental subjects are fully aware that their perception is distorted and that the changes are drug induced.

The clinical syndromes from Ditran are different from those produced by the above drugs in some respects. Disorganization of thought, disorientation, confusion, mood changes, and visual and auditory hallucinations are observed. The piperidyl benzilate esters are central anticholinergics, and mental states produced by them are reminiscent of those from other anticholinergics, such as scopolamine.

The effects of phenylcyclohexyl derivatives are also distinctive. Comparatively minor somatic symptoms are evoked. Psychic effects predominate, being typically characterized by feelings of unreality, depression, anxiety, and delusional or illusional experiences. The effects of these drugs are said to be more analogous to natural psychoses than those of the other drugs[1]; however, the same claim has been made for Ditran.[3]

The effects of the hallucinogens on animals vary with the type of drug and the species tested.[2] Generally, the doses necessary to produce behavioral changes in animals are much larger than those which cause symptoms in man. In humans, the effective dose for the various drugs are: (1) mescaline, 5000–10,000 μg/kg of body weight; (2) psilocybin and psilocin, 100–200 μg/kg; (3) Sernyl and Ditran, 150–200 μg/kg; (4) LSD-25, 1–2 μg/kg.

All known hallucinogens are readily absorbed from the intestinal tract, and the effective dose is the same whether administered orally, subcutaneously, or intravenously. Mescaline and LSD-25 are concentrated in the liver, kidney and spleen; neither is found in significant quantities in the brain. The rate of elimination from all organs is rapid. There is no apparent correlation between brain concentration and the observed behavioral changes at different times after drug administration.[2] Psilocybin is metabolized to psilocin, which appears to be the psychoactive agent. After distribution, the highest concentration of this drug is also found in the liver and kidney. None accumulates in the brain. There appears to be no clue to the mechanisms of hallucinogen action in the absorption, distribution, and elimination of the drugs.

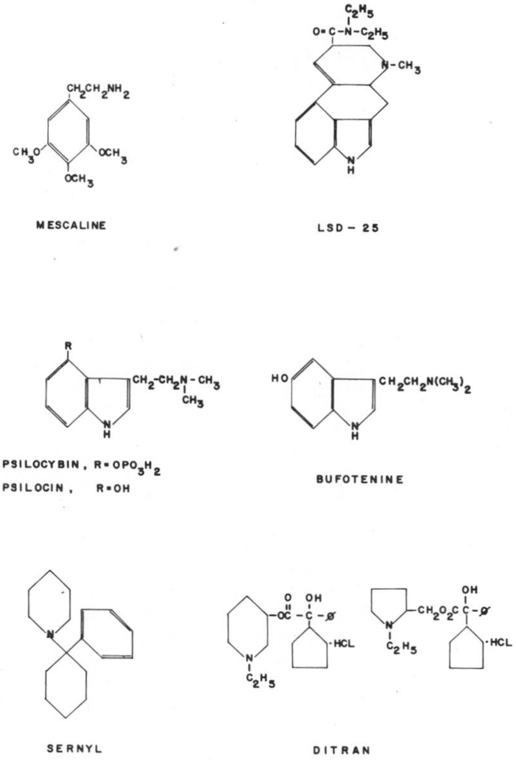

MESCALINE LSD – 25

PSILOCYBIN, R = OPO_3H_2
PSILOCIN, R = OH

BUFOTENINE

SERNYL DITRAN

FIG. 1. Hallucinogenic drugs.

LSD-25 is one of the most powerful drugs known to man. When it was discovered in 1943 by A. Hoffman, its extraordinary potency reawakened interest in the possibility of natural chemical activators in the schizophrenic process. The production of bizarre psychic phenomena, the lack of addictive and toxic properties, and the minimal side effect impressed those investigators who were primarily concerned with mental illness, particularly schizophrenia. From this interest, a large research effort was directed toward known hallucinogens and toward the discovery of new ones. Among the concepts reintroduced was the possibility of using drugs to initiate so-called model psychoses. The use of drugs to produce disturbed mental states has become a popular tool for investigators in many disciplines. The rationale behind these endeavors is as follows: hallucinogenic drugs produce a "schizophrenic-like" state; the physiological and biochemical mechanisms of the drug action may provide clues to mechanisms involved in schizophrenia; drugs counteracting the drug-induced psychoses may provide a therapeutical approach to schizophrenia. A deluge of published work has resulted, much of it lacking scientific discipline and value. Recently, researchers have called for a critical reappraisal[1,4] of the underlying assumption implicit in the use of "model psychoses". These authors decry the lack of careful inquiry which has invested "these agents with an aura of magic, offering creativity to the uninspired, 'kicks' to the jaded, emotional warmth to the cold and inhibited, and total personality reconstruction to the alcoholic or the chronic neurotic."[4] Therapeutically, the hallucinogens have been of little value to psychiatrists; however, recent trends suggest that they may be of clinical value, which properly evaluated under carefully controlled conditions.

Major attention has been focused on the psychotomimetics by some clinicians and lay enthusiasts, who advocate unregulated dispension of these agents, in order that men may "transcend" themselves or "expand their consciousness." Partially as a result of the notoriety, certain elements of the population have been attracted to the use of these agents. Apparently, there are active black markets in major cities. Legally, none of the hallucinating agents can be used, even for investigational use, without prior approval by the Federal Drug Administration.[4,5] They are not available for general prescription purposes, nor are they likely to be in the foreseeable future.

References

1. HOLLISTER, L. E., "Drug Induced Psychoses," *Ann. N.Y. Acad. Sci.*, **96**, 80 (1962).
2. JACOBSEN, E., "The Clinical Pharmacology of the Hallucinogens," *Clin. Pharm. Therap.*, **4**, 4, 480 (1963).
3. GERSHON, S., AND OLARIU, J., "JB-329. A New Psychotomimetic. Its Antagonism by Tetrahydroaminacrin and its Comparison with LSD, Mescaline, and Sernyl," *J. Neuropsychiat.*, **1**, 283 (1960).
4. COLE, J. O., AND KATZ, M. M., "The Psychotomimetic Drugs," *J. Am. Med. Assoc.*, **187**, 759 (1964).
5. BARRON, F., JARVIK, M. E., AND BUNNELL, S., JR., "The Hallucinogenic Drugs," *Sci. Am.*, **211**, #4, 29 (April 1964).

F. CHRISTINE BROWN

HARDEN, A.

Arthur Harden (1865–1940) was an English biochemist who was a pioneer investigator of ALCOHOLIC FERMENTATION. He first found that to carry out fermentation, yeast required a non-dialyzable, heat-labile enzyme(s), *plus* a heat-stable coenzyme of organic nature, *plus* phosphate. In this connection he discovered fructose-1,6-diphosphate. He was awarded the Nobel Prize in chemistry along with HANS VON EULER-CHELPIN in 1929.

HAWORTH, W. N.

Walter Norman Haworth (1883–1950) was a British chemist who distinguished himself in the field of carbohydrate chemistry by establishing the accepted ring formulas for the simple sugars. He also first synthesized ascorbic acid and shared with PAUL KARRER the Nobel Prize in chemistry in 1937.

HEART METABOLISM

The metabolic activities in cardiac muscle reflect the mechanical function of the heart. Because the physiological role of the heart is to perform mechanical work, the most important metabolic events are those concerned with the production of energy in a form that is readily available for use in mechanical activity. Since the heart requires an extraordinary amount of energy, the organization and the function of myocardial metabolism are geared to the production of energy on a large scale. The synthetic processes which occur in heart muscle are largely restricted to those required for sustaining the integrity of the heart itself.

The energy metabolism in heart muscle may be divided into three main phases: (a) energy liberation, (b) energy conservation, and (c) energy utilization. In the phase of energy liberation, the carbon-carbon and carbon-hydrogen bond energy of the substrate is liberated as free energy. Specifically, the processes of GLYCOLYSIS, fatty acid and pyruvic acid oxidation, and the dehydrogenation of the Krebs tricarboxylic acid cycle (see CITRIC ACID CYCLE) occur in this phase. These oxidations result in the conversion of chemical bond energy into the electronic energy of hydrogen, which is transported along the respiratory chain of the MITOCHONDRIA to oxygen. Phase two, that of energy conservation, consists of the process of oxidative PHOSPHORYLATION by which the energy of hydrogen is converted into the terminal bond energy of adenosinetriphosphate (ATP) and, *via* creatine kinase, into CREATINE phosphate. The third phase, that of energy utilization, includes the mechanism by which the terminal high-energy phosphate bond of ATP is channeled into the contractile process which results in mechanical work. In heart muscle, 90% of the energy conserved in ATP is channeled into the contractile process (see CONTRACTILE PROTEINS).

The biochemical activities are localized to a large extent in given compartments of the myocardial cell. The glycolytic reactions which produce pyruvate and lactate from glucose occur in the cytoplasm. The mitochondria have been shown to be the location where a large part of the cell's energy is released by the orderly and rapid oxidation and decarboxylation reactions of the Krebs cycle. The oxidation of fatty acids, pyruvate, acetate, and certain amino acids, such as glutamate, also occurs in the mitochondria. The hydrogen transport enzymes, which transfer hydrogen from DPNH (reduced diphosphopyridine nucleotide) to oxygen, and the associated enzymes which catalyze oxidative phosphorylation are located in the membrane of the mitochondrion. The main reactions of energy production (including energy liberation and energy conservation), therefore, occur in the mitochondria.[1]

The endoplasmic reticulum is closely associated with PROTEIN BIOSYNTHESIS. There is frequently an orderly array of small particles of ribonucleoprotein (RIBOSOMES) over the surface of the endoplasmic reticulum. The ribosomes serve as carriers of messenger RNA or template RNA, which attaches itself to the ribosomes and carries the information of the hereditary pattern for protein synthesis from the deoxyribonucleic acid (DNA) of the chromosomes. The main steps in protein synthesis involve the formation of RNA on a DNA template, and the attachment of mRNA to ribosomes and formation of ribosomal aggregates or polysomes. The amino acid is activated by a specific amino acid activating enzyme and attached to transfer RNA; subsequently, this complex is transported to the mRNA template where the amino acids are polymerized into polypeptide chains.[2]

The contractile events in cardiac muscle occur in the *myofibril*, which may be regarded as another intracellular particle. In cardiac MUSCLE, the myofibrils lie side by side with many mitochondria. The current hypothesis for the shortening of the myofibril provides for a mechanism of ATP-activated sliding of filaments to account for the approximation of the Z-membranes toward one another. This appears to be accomplished by the rapid making and breaking of actomyosin bonds between the thick, myosin-containing and thin, actin-containing filaments.[3,4]

Myocardial Substrate Extraction. The metabolism of the heart, and consequently the extraction of oxygen and substrates, depends on the muscular activity of the heart. When the performance of work increases, the rate of metabolism rises and its energetic efficiency improves to an optimal value. The heart differs, however, from other muscle or other organs with regard to oxygen extraction from the perfusing blood. Peripheral tissues extract about one-third of the oxygen brought by the blood, so that the pooled venous effluent still carries some 13 or 14 volumes per cent of the gas. The myocardial oxygen extraction is, however, more complete, and leaves only 4–7 volumes per cent oxygen in the coronary vein blood. Any further utilization would lead to an undesirable drop in oxygen tension, so that in general, conditions of increased demand are accompanied by an enhanced coronary flow, leaving the oxygen extraction unchanged.

Using the method of coronary sinus catheterization, the extraction and utilization of individual substrates by the heart has been investigated.[5,6] The human heart uses glucose, pyruvate and lactate, and their utilization appears to be a function of their arterial concentration. At normal concentrations, glucose and lactate are used in approximately equal amounts. Pyruvate is utilized by the human heart, but the blood concentration is low and so is its myocardial utilization. Apparently, if complete oxidation of carbohydrates is assumed, the total myocardial carbohydrate metabolism in man and dog accounts for only approximately 35% of the total myocardial oxygen extraction.

The human heart seems to have preference for FATTY ACIDS as a fuel, but the heart is versatile in adapting to the changing nutritional circumstances associated with intermittent feeding. Measurement of the respiratory quotient of the heart and the myocardial oxygen extraction of available substrates has clearly established the importance of lipid as a fuel for respiration, particularly in the fasting state. In the normal heart, the oxygen extraction ratio of fatty acids is, on the average, 67%, which is greater than that of the carbohydrate moiety. Nearly all the fatty acids extracted by the heart are derived from plasma triglyceride or albumin-bound non-esterified fatty acids (NEFA). Myocardial extraction of triglyceride fatty acid appears to be influenced by the activity of a hydrolytic enzyme (lipoprotein lipase) located at or near the cell membrane. The utilization of free fatty acids by the myocardium changes according to quantity and type of fatty acid (carbon chain length and number of double bonds) available in the circulation, the respiratory activity of the heart, and the nutritional state. Fatty acids supplied from the circulation in excess of the needs of the heart for aerobic metabolism are converted to tissue triglyceride, which can subsequently be mobilized as a fuel for respiration. Respiration of fatty acid derived from the circulation and from endogenous myocardial triglyceride exerts a regulatory influence on utilization of carbohydrate substrates (by inhibition of the phosphofructokinase and pyruvate decarboxylation reactions).[7]

It appears to make very little difference what substrate is the source of energy for contraction as long as that substrate is utilized efficiently for the formation of ATP. Marked changes in the pattern of substrate uptake have been noted in man under various conditions without any effect on contractility. For example, in fasting and diabetes, the availability of carbohydrate is decreased and plasma nonesterified fatty acids supply most of the fuel requirements of the heart. There is little evidence that the source of fuel in these conditions is critical, since ATP is the form of energy which ultimately drives the contractile mechanism.

References

1. GREEN, D. E., AND GOLDBERGER, R. F., "Pathways of Metabolism in Heart Muscle," *Am. J. Med.*, **30,** 666 (1961).
2. KONNER, A., "Protein Biosynthesis in Mammalian Tissues. Part I. The Mechanism of Protein Synthesis," in "Mammalian Protein Metabolism," Vol. I (MUNRO, H. N., AND ALLISON, J. B., EDITORS), New York, Academic Press, 1964.
3. HUXLEY, H. E., AND HANSON, J., "Molecular Basis of Contraction in Cross-striated Muscles," in "Structure and Function of Muscle," Vol. I (BOURNE, G., EDITOR), p. 183, New York, Academic Press, 1960.
4. SPIRO, D., AND SONNENBLICK, E. H., "Comparison of the Ultrastructural Basis of the Contractile Process in Heart and Skeletal Muscle," *Circulation Res.*, **15,** II-14 (1964).
5. BING, R. J., "The Metabolism of the Heart," *Harvey Lectures, Series L*, **27** (1954).
6. BING, R. J., "Metabolic Activity of the Intact Heart," *Am. J. Med.*, **30,** 679 (1961).
7. EVANS, J. R., "Importance of Fatty Acid in Myocardial Metabolism," *Circulation Res.*, **15,** II-96 (1964).

S. GUDBJARNASON AND R. J. BING

HEME

Heme is the prosthetic group in the heme proteins: hemoglobin, catalase, PEROXIDASES and cytochrome oxidase (see CYTOCHROMES). Its structure is shown (as structure XIII, Fe(II)-protoporphyrin complex) in the article CHELATION AND METAL CHELATE COMPOUNDS [see also HEMOGLOBINS (COMPARATIVE BIOCHEMISTRY); IRON (IN BIOLOGICAL SYSTEMS); PORPHYRINS].

The chemistry of heme and its functioning is highly complicated and cannot be discussed adequately except in terms of magnetochemistry.[1] The wide distribution of heme in many types of organisms is notable, as is its resemblance to CHLOROPHYLL which contains magnesium. Not all biological pigments which have the heme prefix in their names contain heme. HEMOCYANINS and hemerythrins found in various marine forms contain copper and iron, respectively, but no heme.

Reference

1. SELWOOD, P. W., "Magnetochemistry," Second edition, New York, Interscience Publishers, 1956.

HEMOCYANINS

The hemocyanins represent the only established departure from the use of iron in oxygen-carrying pigments. They are copper PROTEINS, deep blue in color, found dissolved in the serum of a great variety of arthropods and mollusks. The hemocyanins range in molecular weight from about 400,000 (*Palinurus*) to 5,000,000 (*Helix*). In general, the hemocyanins of smallest molecular weight have been found in the crustaceans, those of intermediate size in the cephalopods and in the arachnoids, and the largest in the gastropods. These values are the upper limits of size for the species; the large molecules are capable of reversible dissociation into smaller subunits in a manner which depends upon the pH, ionic strength, protein concentration and nature of the ions in the medium. The subunits are all capable of carrying oxygen. The protein molecular weight per copper is about 36,500 in the arthropods and 25,000 in mollusks. In fact, the smallest subunit found is about twice this size.

Electron microscopy of hemocyanins reveals a variety of regular packing arrangements for the protein subunits in the aggregated states. The subunits of one of the most thoroughly studied, *Helix pomatia*, are arranged in a cylinder possessing tenfold symmetry, about 300 Å in diameter and 335 Å long. This is the largest aggregate observed. Other gastropod hemocyanins appear to be similar, but the arrangement of subunits differs markedly from that found in crustacea such as *Cancer pagurus*, which probably involves eight subunits placed at the corners of a cube. The hemocyanins of a number of species have been crystallized.

There are no heme or other porphyrin groups in hemocyanin. The copper appears to be bound directly to one or more amino acid side chains in the protein, and one oxygen molecule is bound for each two copper ions present. In the deoxygenated state, no cupric ion can be detected either by electron paramagnetic resonance or by specific reagents for cupric copper. Reagents for cuprous copper, on the other hand, react with all the copper ion present. It is generally agreed that the copper ion of deoxygenated hemocyanin is entirely in the cuprous state.

The structure of the active combining site of hemocyanin with oxygen is still not known. On the basis of the stoichiometry of the reaction, it was proposed as early as 1940 by Rawlinson that the oxygen-protein complex involves a pair of copper ions bridged by an oxygen molecule: Protein-Cu-O-O-Cu-Protein. Evidence in favor of this structure, aside from the stoichiometry, is the unusually rapid reaction of deoxyhyemocyanin with small amounts of hydrogen peroxide, which results in destruction of oxygen-carrying capacity and a coordinated conversion of an equivalent of copper to the cupric state for every equivalent of peroxide added. Such behavior is consistent with a simultaneous transfer of two electrons from the peroxide ion to two cuprous ions, a reaction which could take place only if cuprous ions occurred in pairs. There has been much discussion of the valence state of copper in oxyhemocyanin. Chemical methods based upon specific reagents perturb the electron distribution between copper and oxygen in an unpredictable way. Electron paramagnetic resonance studies of oxyhemocyanin reveal no paramagnetism, a result consistent with the oxygen bridge structure, but obviously not conclusive.

Many attempts have been made to produce methemocyanin by oxidation of the protein-bound cuprous ion to the cupric state. Early experiments in which hemocyanin was allowed to react with permanganate or molybdicyanide ions resulted in nonspecific oxidation of protein side chains, often leaving the oxygen-carrying capacity of the protein unchanged. The product of reaction with hydrogen peroxide, however, appears to be a true methemocyanin. Though the methemocyanin of *Limulus polyphemus* cannot be reduced to form active hemocyanin, the oxidation of *Busycon canaliculatum* hemocyanin by hydrogen peroxide is reversed by addition of an excess of hydrogen peroxide, which acts as a reducing agent during the second step.

The copper ions of hemocyanin, though tightly bound, can be dissociated from the protein by dialysis at pH 4, or by addition of low concentrations of sulfide, cysteine, or thiourea. Addition of cyanide ion results in removal of copper from the protein, and quantitative liberation of bound oxygen. Both *Helix* and *Octopus* hemocyanins have been reconstituted from the apoprotein by addition of Cu^{+}, but not Cu^{++}. There is no conclusive evidence concerning the nature of the bonds between copper and protein, though evidence from hydrogen ion titration curves of apohemocyanin suggests that imidazole groups of histidine are involved.

The absorption spectrum of hemocyanin shows bands at 550–560 mμ and 345 mμ which disappear reversibly upon deoxygenation. The protein also possesses optical rotatory properties which are a function of the extent of oxygenation.

The oxygenation curves (per cent oxygenated pigment *vs* partial pressure of oxygen) of the hemocyanins more or less resemble those of the HEMOGLOBINS. Detailed analysis suggests that there is often considerable interaction among oxygen-binding sites, and the oxygenation curves vary greatly with conditions of ionic strength and acidity. It is likely that this variability is related to the change in extent of aggregation. In many cases, increasing acidity tends to decrease the affinity of hemocyanin for oxygen in a manner analogous to, but greater than, the Bohr effect in hemoglobin, but in others the effect is reversed (*Limulus*, *Helix*, *Busycon*) so that increasing acidity increases oxygen affinity. The oxygen affinity of hemocyanins (*Homarus*) under physiological conditions and their total oxygen capacity (*Limulus*, *Helix*) in the blood are so small in some species that it is unlikely they play a major role except in oxygen storage, but in the more active species (*Loligo*), hemocyanin is comparable in efficiency to vertebrate hemoglobin.

Hemocyanin has been reported to have a number of catalytic properties, particularly when denatured; these include activity as a polyphenol oxidase, a catalase, and a lipoxidase. It is not clear that these have any physiological significance. Though the hemocyanins bear superficial resemblance to a number of copper enzymes such as laccase and ascorbic acid oxidase and to proteins such as ceruloplasmin, there is mounting evidence that the active site configurations and copper oxidation states may differ significantly. The hemocyanins remain the only copper proteins known to combine reversibly with oxygen.

References

1. GHIRETTI, F., in "Oxygenases" (HAYAISHI, O., EDITOR), New York, Academic Press, 1962.
2. MANWELL, C., *Ann. Rev. Physiol.*, **22**, 191 (1960).
3. FELSENFELD, G., AND PRINTZ, M. P., *J. Am. Chem. Soc.*, **81**, 6259 (1959).
4. VAN BRUGGEN, E. F. J., SCHUITEN, V., WIEBENGA, E. H., AND GRUBER, M., *J. Mol. Biol.*, **7**, 249 (1963).

GARY FELSENFELD

HEMOGLOBINS (COMPARATIVE BIOCHEMISTRY)

The primary function of the hemoglobin molecule is oxygen transport. The hemoglobin molecules from each species of organism which has been examined differ in the sequence of AMINO ACIDS in their polypeptide chains unless they are very closely related. Chimpanzee and human hemoglobins are apparently identical. Sometimes two or more different kinds of hemoglobin are found simultaneously in the same organism. These structural variations may give rise to differences in the physiological properties which help to determine the efficiency of oxygen transport by the blood from lungs or gills to the tissues. Hemoglobin also plays an important role in CARBON DIOXIDE transport.

Structural Effects of Bound Oxygen. Vertebrate hemoglobins are usually composed of four polypeptide chains of two types, called α and β. The molecules can therefore be described as $\alpha_2\beta_2$. An iron PORPHYRIN moiety, HEME, is associated with each chain. Many different lines of evidence indicate that combination of the heme with oxygen results in structural changes in the protein to which it is bound. Studies of single crystals of horse and human hemoglobins by X-ray diffraction techniques show that the removal of oxygen from the iron atoms of the four hemes results in a separation of the β-chains from one another; the relative positions of the α-chains do not appear to change. Although the molecular basis for this shift is not yet completely understood, the consequences are of considerable importance. It is certain that any change in the mutual relationships of the polypeptide chains will alter the environment of many amino acid residues. These environmental changes are probably responsible for the following properties of hemoglobin: the shape of the curve relating degree of oxygenation to the oxygen pressure and the dependence of the oxygenation reaction upon pH and upon CO_2 concentration.

Typical oxygen equilibrium curves are shown in Fig. 1. The peculiar S-shape may be explained qualitatively on the following basis. Oxygenation of the heme of one chain results in changes in the three-dimensional arrangement or conformation of that chain. This structural change induces a corresponding change in an adjacent chain through certain points of contact in such a way that the oxygen affinity of the heme in the adjacent chain is greatly increased. The physiological importance of this is that hemoglobin must not only be able to combine with oxygen in the lungs or gills but should also be able to release it at high oxygen pressures in the tissues. A hemoglobin with an S-shaped oxygen equilibrium curve will favor delivery at higher oxygen tensions. The "unloading" oxygen pressure of different hemoglobins varies with the organism from which the pigment comes. Among mammals, the smaller the animal, the higher is the metabolic rate (see BASAL METABOLISM) and the higher is the specific rate of oxygen consumption. The hemoglobins appear to be specifically adapted to meet these differing oxygen requirements. In general, the smaller the animal, the higher is the pressure at which its hemoglobin delivers oxygen to the tissues. Thus, referring to Fig. 1, the hemoglobin of a mouse will have an oxygen equilibrium curve to the right of the curve for the hemoglobin from a man or an elephant.

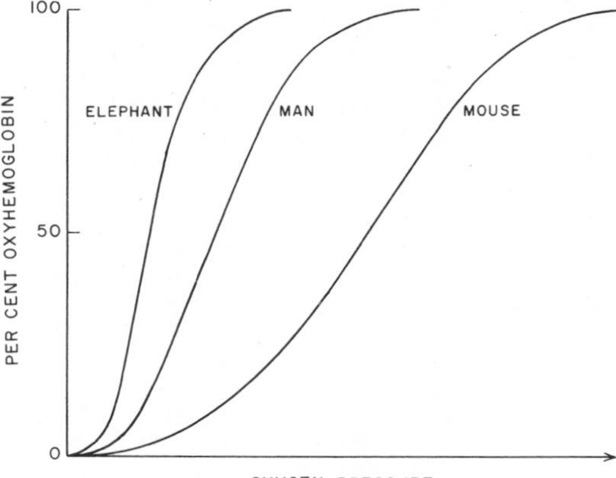

FIG. 1. General appearance of oxygen equilibrium curves of hemoglobins from elephant, man, and mouse.

Effects of Carbon Dioxide and of pH. Hemoglobins are adapted in yet another respect: the dependence of the oxygen equilibrium on pH and CO_2. The lower the pH (within the physiological range), the further to the right is the oxygen equilibrium curve. That is, an increase in the H^+ ion concentration tends to drive oxygen off the hemoglobin. An increase in CO_2 has a similar effect. Thus the CO_2 produced by metabolism in the tissues tends to facilitate the removal of oxygen from the hemoglobin, resulting in a greater availability of oxygen to the tissues. CO_2 does this by two different mechanisms. It lowers the pH of the blood, and it also appears to combine directly with certain amino acid residues in the hemoglobin. This effect of CO_2 on the oxygen affinity is called the "Bohr effect" after its discoverer, the Danish physiologist, Christian Bohr, the father of the atomic physicist, Niels Bohr. The term has gradually been broadened to include not only the CO_2 effect but also the effect of pH.

The Bohr effect not only assists in the release of oxygen in the tissues but also facilitates the removal of CO_2 from the lungs or gills. When oxygen combines with hemoglobin, protons are released. This release of protons facilitates the driving of CO_2 from the blood. Two questions are of interest here to the biochemist. First, what is the molecular mechanism by which oxygenation triggers proton release? Second, how does this process differ in the hemoglobins of various organisms?

The answers to these questions are not yet known with certainty. However, the acid groups which are responsible for the release of protons during oxygenation have been tentatively identified as those of histidyl residues partly because the pH range in which the H^+ ions are released corresponds closely to the acid-base titration curve of the imidazole group of HISTIDINE.

An attempt has been made to determine the location of the acid groups responsible for the Bohr effect. Several possibilities can be considered. One acid group might be associated with each polypeptide chain, and the α- and β-chains might make equal contributions. Several considerations make this hypothesis doubtful. The mammalian fetus possesses a hemoglobin with the same α-chain as in the adult but the other chain is called a γ-chain and differs considerably from the β-chain of the adult. Mutations which alter the amino acid sequence can occur in either the α- or the β-chain of the adult. However, most mutations are deleterious and changes in the α-chain would be more severely selected against in the process of natural selection because any change in the α-chain would affect the sensitive fetus, whereas changes in the β-chain would affect only the adult. This means that evolution tends to favor changes in the β-chain over changes in the α-chain. These considerations indicate that molecular adaptation of hemoglobin—at least in mammals—may involve changes more in the β-chain than in the α-chain. The evolutionary changes associated with the β-chain might then be responsible for adaptations in the Bohr effect, and the acid groups responsible for the acid sensitivity of the oxygen equilibrium might be on the β-chain. One piece of evidence suggests that this is indeed so.

Relative Roles of α- and β-chains. Hemoglobins can be dissociated into their α- and β-subunits. If a mixture of hemoglobins—say from mouse and donkey—is made, the molecules can be made to dissociate, and then to reassociate at random according to the following reaction:

$$\alpha_2{}^D\beta_2{}^D + \alpha_2{}^M\beta_2{}^M \rightleftarrows \alpha_2{}^D\beta_2{}^M + \alpha_2{}^M\beta_2{}^D$$

The superscripts D and M refer to donkey and mouse. Since donkey and mouse hemoglobins possess quite different oxygen equilibria, these "hybrid" molecules offer an opportunity to see if the differences depend on a particular kind of chain. Donkey hemoglobin has a high affinity for oxygen and its oxygen equilibrium has a low sensitivity to changes in pH. In contrast, mouse hemoglobin has a low affinity for oxygen and the equilibrium is much more strongly dependent upon pH. Determination of the properties of the hybrid molecules shows that their properties depend upon the β-chains: $\alpha_2{}^D\beta_2{}^M$ is a hemoglobin very like mouse hemoglobin, whereas $\alpha_2{}^M\beta_2{}^D$ is like donkey hemoglobin.

An adult human hemoglobin called hemoglobin H has been found which lacks α-chains, and consists wholly of β-chains. It might be thought that this hemoglobin would possess an oxygen equilibrium very strongly dependent upon pH, but this is not so. X-ray diffraction studies show that no conformation change occurs during oxygenation. Hemoglobins composed wholly of α-chains have also been studied and lack completely any pH dependence in their oxygen equilibrium. It therefore appears that *linkages between* α- and β-chains are also necessary for a combination with oxygen to result in the release of hydrogen ions from certain acid groups which are at or near the points of contact between the two kinds of chain.

Not only are hemoglobins capable of dissociating into their polypeptide subunits, but certain hemoglobins are also capable of polymerization. Many reptiles and amphibians and certain mice possess hemoglobins which polymerize to form double molecules $(\alpha_2\beta_2)_2$ and sometimes triple or quadruple molecules. In mice, the two molecules are joined together by two disulfide bonds. Many hemoglobins from invertebrate animals have very large molecular weights and are composed of a large number of subunits—as many as 180 in some species. The nature of the forces holding these large aggregates together is not yet known, and neither are the possible homologies of the subunits with vertebrate hemoglobins.

The amino acid sequences of hemoglobins have been extensively altered by mutation during evolution. The idea that knowledge of amino acid sequences from the hemoglobins from various organisms would provide important information about protein evolution and function and animal phylogeny has stimulated many studies. The data now available on the amino acid sequences of the chains from a variety of mammalian and other vertebrate hemoglobins show that the sequence can be varied extensively without drastic change in function. X-ray diffraction studies show that the over-all arrangement ("tertiary structure") of each of the chains of human hemoglobin is remarkably similar to the arrangement of the chain in sperm whale MYOGLOBIN, in spite of substantial differences in amino acid sequence (see Fig. 2). If the α-chains (141 residues) and β-chains (146 residues) from various mammalian hemoglobins are compared with myoglobin (153 residues), only 11 of the corresponding amino acid sites appear invariant. As more studies are completed, this number may shrink further to 2, the number of histidyl residues normally associated with the iron atom of the heme of each chain. This does not mean, however, that there are no restrictions on amino acid substitution. Conservation of the tertiary structure imposes limits on the kinds of substitution. The helical segments (see Fig. 2) must be devoid of proline since the imide linkage in proline destroys helical structure. Any substitutions which would tend to disrupt helical structure probably would be selected against in evolution. Mutations involving a substantial increase in the number of internal bulky residues would tend to alter the tertiary structure. Large changes in the number of external polar residues might greatly change the solubility and interactions with ions. In addition, natural selection would be expected to select against the introduction of those residues into the non-helical regions which would tend to make them helical.

Amino acid residues appear to have four principal structural functions in hemoglobins: (1) "*No function*": Alanine and glycine appear to

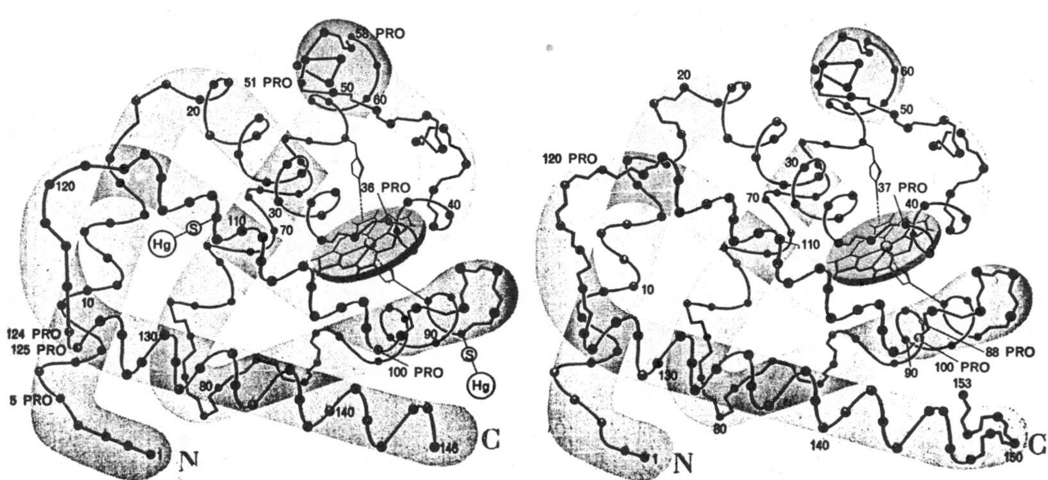

FIG. 2. Comparison of the β-chain of human hemoglobin with myoglobin. Every tenth amino acid is numbered, and the position of prolyl and cysteinyl residues are shown. The heme-linked histidyl residues are also shown. [Reprinted with permission, from M. F. Perutz, "The Hemoglobin Molecule," *Sci. Am.* (November 1964).]

act as fillers and/or spacers and have no interactions to prevent any other interaction by an adjacent residue. (2) *Hydrophobic bonding:* Nonpolar interactions appear to be very important in the interior of the chains. In addition, α-β linkages in the $\alpha_2\beta_2$ unit appear to be largely nonpolar. (3) *Charged group interactions:* Most, if not all, the charged groups are external and are not found in the interior of the polypeptide chain subunits. The β-β and α-α interactions in each $\alpha_2\beta_2$ unit are mostly polar. (4) *Hydrogen bonding:* Threonine and serine may be especially important in forming hydrogen-bonds with their —OH groups [see also PROTEINS (BINDING FORCES)].

There must exist a hierarchy in the functional importance of different parts of a protein. Substitutions in different segments of a polypeptide chain, may, according to the type and position of the substitution, exhibit a spectrum of effects ranging from none detectable to catastrophic. For example, the single substitution of valine for glutamic acid in the sixth-position of the β-chain in human sickle cell hemoglobin results in a large decrease in the solubility of deoxygenated hemoglobin within the red cells. The hemoglobin, by forming a gel, distorts the red cell shape ("sickle") in such a way that flow through the capillaries is retarded. Such drastic consequences do not result if the substitution is lysine rather than glutamic acid (hemoglobin C). Histidine in position 63 of the human β-chain has an essential role stabilizing the ferrous state of the heme iron. Substitution by tyrosine (in hemoglobins "M") results in the loss of this stability because the ferric iron can form a strong linkage with the —OH group of tyrosine. Such a substitution results in a complete loss of the capacity to combine reversibly with oxygen. These are radical substitutions: most effective substitutions appear to be relatively conservative and do not drastically affect the oxygen transport function. Therefore, the number of differences between homologous chains appears to be related not to functional differences, but to the time which has elapsed since the chains diverged from a hypothetical polypeptide ancestor. The mean number of differences between the hemoglobin chains of man, horse, pig, rabbit, and cattle is approximately 11. The common ancestor of these mammals existed some 80 million years ago. Thus approximately 11 effective mutations per chain occurred in 80 million years, or 1 substitution per chain in 7 million years. Zuckerkandl and Pauling, using standard probability theory, have used this figure to estimate the time at which the different human hemoglobin chains (α, β, γ, and δ) are believed to have arisen by gene duplication. These estimates are shown in Table 1.

TABLE 1

Type of Chain Divergence	No. of Differences	Estimated Time since Divergence
β-δ	10	35 million years
β-γ	37	150 million years
β-α	76	380 million years
(α-β)–myoglobin	~135	650 million years

Examination of these estimates indicates that hemoglobins are very old and that it may be possible to find relatives of vertebrate hemoglobins in invertebrate animals. It also suggests that the gene duplication believed to be responsible for the divergence of the α- and β-chains took place in the Devonian period at the time of the appearance of early amphibians and the dominance of fish.

The suggested relationship between numbers of differences and evolutionary time is not wholly secure. It assumes uniformity in the rate of effective amino acid substitution, but this rate may be neither uniform with time nor uniform in different parts of the polypeptide chain. Differences in the rate of effective substitution along the polypeptide chain may be due not only to restrictions imposed by the required tertiary structure but also to differences in the rate at which various parts of the DNA or the GENE mutate. The evolution of hemoglobin may be contrasted with that of cytochrome c (see CYTOCHROMES) in which approximately 50% of the molecule appears to have remained invariant during the time yeast and man have evolved from a common ancestor.

References

1. ANTONINI, E., "Interrelationship between Structure and Function in Hemoglobin and Myoglobin," *Physiol. Rev.*, **45**, 123 (1965).
2. BRAUNITZER, G., HILSE, K., RUDLOFF, V., AND HILSCHMANN, N., "The Hemoglobins," *Advan. Protein Chem.*, **19**, 1 (1964).
3. MANWELL, C., "Comparative Physiology: Blood Pigments," *Ann. Rev. Physiol.*, **22**, 191 (1960).
4. PERUTZ, M. F., "The Hemoglobin Molecule," *Sci. Am.*, 64–76 (November 1964).
5. RIGGS, A., "Functional Properties of Hemoglobins," *Physiol. Rev.*, **45**, 619 (1965).
6. ZUCKERKANDL, E., "The Evolution of Hemoglobin," *Sci. Am.*, **212**, 110–118 (May 1965).
7. ZUCKERKANDL, E., AND PAULING, L., "Evolutionary Divergence and Convergence in Proteins," in "Evolving Genes and Proteins" (VOGEL, HENRY J., EDITOR), New York, Academic Press, 1965.

AUSTEN RIGGS

HEMOGLOBINS (IN HUMAN GENETICS)

Man's knowledge of the relationship between genetics and human hemoglobins began in 1910 when J. B. Herrick reported that the red blood cells of certain persons form bizarre "sickle" shapes when the oxygen pressure is lowered. In subsequent years, it was recognized that persons with these unusual cells almost invariably were of Negro ancestry and that the trait appeared to be familial. Some persons with sickle cells had profound ANEMIA, often leading to death. Others had no anemia and appeared to suffer no ill effects. These two forms became known as sickle cell disease and sickle cell trait, respectively.

The inheritance of sickling was established in 1949 by James V. Neel and E. A. Beet, independently. Persons with the severe disease have two sickle genes; those with the mild trait have one sickle gene and one normal gene. Also in 1949, the first evidence was obtained by L. Pauling, H. A. Itano, S. J. Singer, and I. C. Wells that the defect in sickle cells results from a chemically altered hemoglobin. They showed that persons with sickle disease have a form of hemoglobin which migrates more slowly toward the anode than normal hemoglobin when placed in an electric field under alkaline conditions. Persons with sickle cell trait have both forms of hemoglobin. This was therefore the first established example of a molecular disease.

Three forms of hemoglobin ordinarily occur in persons. The predominant form (>90%) in adults is designated hemoglobin A (Hb A). There is also a small fraction (*ca.* 2.5%) in adults designated Hb A_2. During fetal life, the major portion of hemoglobin is of a different type, fetal hemoglobin (Hb F). The change from Hb F to Hb A occurs about the time of birth. Thereafter, Hb F rapidly diminishes and constitutes less than 1% of the hemoglobin in adults. There is also evidence for an embryonic hemoglobin which precedes Hb F in development.

Hemoglobin has been shown to consist of four polypeptide chains, two of one type and two of another. In the case of Hb A, these were arbitrarily designated α and β. The formula of Hb A can therefore be represented by $\alpha_2\beta_2$. Hemoglobins F and A_2 have the same α-chains combined with different "β-like" chains. The structures are represented by $\alpha_2\gamma_2$ for Hb F and $\alpha_2\delta_2$ for Hb A_2.

The difference in electrophoretic mobility between Hb A and sickle cell hemoglobin hemoglobin (Hb S) could have been due either to differences in composition, Hb S having fewer negatively charged amino acids, or to differences in folding of the polypeptide chains, exposing different amino acids to the surface or changing the shape of the molecule. In 1956, V. M. Ingram resolved the question with an extension of the method successfully used to characterize INSULIN by Sanger. Hemoglobin was digested by trypsin, resulting in small peptides which could be separated from each other by two-dimensional electrophoresis and chromatography. These two-dimensional peptide patterns ("fingerprints") prepared from Hb A and Hb S were identical with one exception. A single peptide of Hb A was replaced by a more electropositive peptide in Hb S. Isolation and analysis of the peptides showed a glutamic acid residue in Hb A to be replaced by a valine residue in Hb S. No other differences were noted.

The demonstration of a single amino acid difference reflecting the action of two different genes permitted the important conclusion that one function of GENES is specifying the primary amino acid sequence of polypeptide chains.

Several dozen variants of human hemoglobin have been reported, some without obvious clinical effect and some causing various degrees of anemia, methemoglobinemia, or other problems. Initially, an attempt was made to name the variants in alphabetical sequence, but this became impractical as the number of variants grew large and when it was demonstrated that variants of similar electrophoretic mobility may have different structures. The geographic place of discovery is now used to distinguish new variants, sometimes in combination with a letter indicating the electrophoretic mobility group (*e.g.*, Hb G Philadelphia) or the formation of methemoglobin (*e.g.*, Hb M Boston).

Peptide analysis has been applied to many of the variants, most of which have been shown to result from single amino acid substitutions as in the case of Hb S. The substitution of valine for glutamic acid in Hb S occurs in the β-chain and therefore affects only adult hemoglobin, which may be written $\alpha_2{}^A\beta_2{}^A$ for normal hemoglobin and $\alpha_2{}^A\beta_2{}^S$ for Hb S. Many other variants also involve substitution in the β-chain, but many are located in the α-chain, thereby affecting Hb F and Hb A_2 as well as adult hemoglobin.

With the complete elucidation of the structure of hemoglobins, it is possible to pinpoint the exact location of each amino acid substitution. Even before this, it had been shown that the Hb S substitution involves the sixth residue from the N-terminus of the β-chain. A more complete symbolism therefore would be $\alpha_2{}^A\beta_2{}^{6\,Val}$. A summary of amino acid substitutions in variant hemoglobins is given in Table 1.

The inheritance of variant hemoglobins in general follows the same rules established for Hb S. Most of the variants are quite rare and are known only in combination with another hemoglobin gene. Occasional families are located in which two variants as well as the usual Hb A occur. This permits analysis of the genetic relationship of the genes to each other. A person has only two homologous genes (alleles) for each unit function. These are located on one of the pairs of chromosomes which comprise the genetic complement. Only one of each chromosome pair is transmitted by a parent to an offspring. If a person has two variant hemoglobins, *e.g.*, Hb S and Hb C, and is married to a person with only Hb A, transmission of both (or neither) Hb S and Hb C to one offspring would be evidence that the genes for these hemoglobin variations are not at homologous positions. They may be on different chromosomes or so distant from each other on the same chromosome that crossing over is likely to occur between them.

In the case of Hb S and Hb C, persons with both always transmit either one or the other to each offspring; the two are never transmitted together. This test of allelism proves that the S and C genes occur at or near homologous positions on the chromosomes. Similar results have been obtained for all combinations of variants involving the β-chain, although the rarity of many has made such tests impossible so far.

Quite different results are obtained when α-chain variants are considered. When they are tested among themselves, the results indicate a single location for the gene. When tested against β-chain variants, there is free assortment, indicating that the gene controlling α-chain structure is separate from the gene controlling β-chain structure and that the two genetic loci are not close together on the same chromosome. They may be on different chromosomes.

These studies yielded another important hypothesis of genetics: Each different polypeptide chain reflects the action of a different gene (locus) in the determination of its structure. Thus there is a Hb_α locus to direct synthesis of the α-chains and a Hb_β locus to direct synthesis of β-chains. The common alleles of these loci are $Hb_\alpha{}^A$ and $Hb_\beta{}^A$, respectively, and a person with only Hb A as the major constituent would have the genetic constitution $Hb_\alpha{}^A Hb_\alpha{}^A / Hb_\beta{}^A Hb_\beta{}^A$. A person with sickle cell trait would be $Hb_\alpha{}^A Hb_\alpha{}^A / Hb_\beta{}^A Hb_\beta{}^S$, forming as products $\alpha_2{}^A\beta_2{}^A$ and $\alpha_2{}^A\beta_2{}^S$. Rare persons have been observed who have two different alleles at each locus, for example, $Hb_\alpha{}^A Hb_\alpha{}^I / Hb_\beta{}^A Hb_\beta{}^S$. Such persons produce four kinds of hemoglobin: $\alpha_2{}^A\beta_2{}^A$, $\alpha_2{}^A\beta_2{}^S$, $\alpha_2{}^I\beta_2{}^A$, and $\alpha_2{}^I\beta_2{}^S$.

Mutations affecting only Hb A_2 are found to

TABLE 1. AMINO ACID SUBSTITUTIONS IN VARIANTS OF HUMAN HEMOGLOBIN

Alpha-chain Variants				Beta-chain Variants			
Position	α^A	Substitution	Name	Position	β^A	Substitution	Name
22	Gly	Asp	J Medellin	6	Glu	Val	S
30	Glu	Gln	G Honolulu	6	Glu	Lys	C
47	Asp	Gly	L Gaslini	7	Glu	Gly	G San Jose
54	Gln	Arg	Shimonoseki	26	Glu	Lys	E
54	Gln	Glu	Mexico	43	Glu	Ala	G Galveston
57	Gly	Asp	Norfolk	63	His	Tyr	M Saskatoon
58	His	Tyr	M Boston	63	His	Arg	Zurich
68	Asn	Lys	G Philadelphia	67	Val	Glu	M Milwaukee-1
87	His	Tyr	M Iwate	121	Glu	Gln	D Punjab
116	Glu	Lys	O Indonesia	121	Glu	Lys	O Arabia

involve alterations in the δ-chain. Although these have not been investigated chemically to the extent of the α- and β-chain variants, it has been possible to show by family studies that the locus directing δ-chain synthesis, Hb_δ, is very close to the Hb_β locus. Variants of the γ-chain in fetal hemoglobin have been observed, but combinations of these variants with α- and β-chain variants which would permit location of the Hb_γ locus with respect to other hemoglobin loci have not been observed.

More precise information on the relationship of the Hb_β and Hb_δ loci comes from the Lepore-type hemoglobin variants. Lepore hemoglobins depart from the normal pattern of single amino acid substitutions. Instead, they have a β-like chain which is structurally similar to a β^A chain in the C-terminal end and to a δ^{A_2}-chain in the N-terminal end. It has been interpreted as a fused δ-β chain resulting from deletion of a chromosome segment containing portions of the Hb_β and Hb_δ genes with fusion of the remaining portions. How this might have happened is illustrated in Fig. 1.

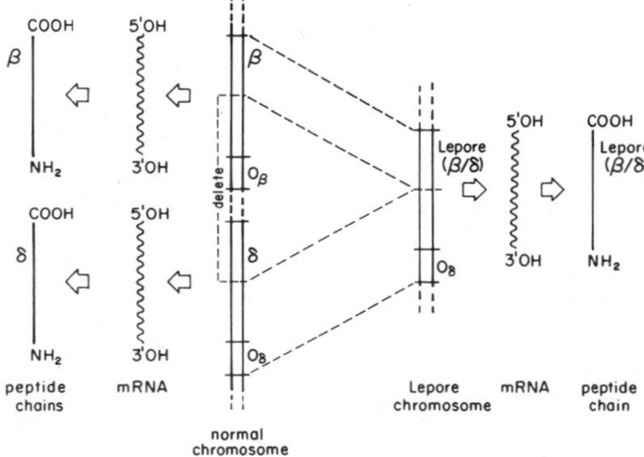

FIG. 1. Diagram showing relationship of Hb_β and Hb_δ loci and the origin of the Lepore chromosome by deletion of complementary segments of these to structural genes. O_β and O_δ are the "operators" from which synthesis of messenger RNA is initiated. From Sutton, H. E., reference 5.

Two additional hemoglobin variants are Hb H and Hb Bart's. These are found in combination with other hemoglobin abnormalities, particularly those involving defects in the α-chain. Structural studies indicate Hb H to consist only of β^A chains, with the formula $\beta_4{}^A$. Hb Bart's is the fetal equivalent, γ_4. The β-like chains have a tendency to form tetramers in the presence of inadequate α-chain synthesis. However, α-chains do not seem to have this property when the supply of β-like chains is limited.

The existence of at least five structurally similar polypeptide chains (α, β, γ, and δ hemoglobin chains and MYOGLOBIN) under the direction of five different structural genes suggests that these genes evolved from a single primitive gene through sequential duplication and divergence. Assuming that greater structural similarity reflects more recent divergence, it has been proposed that the Hb_β and Hb_δ genes are the most recent to separate, these in turn being more closely related to the Hb_γ gene than to Hb_α. All the hemoglobin chains resemble each other more than they do myoglobin.

References

1. BAGLIONI, C., "Correlations between Genetics and Chemistry of Human Hemoglobins," in "Molecular Genetics," Part I (TAYLOR, J. H., EDITOR), pp. 405–475, New York, Academic Press, 1963.
2. INGRAM, V. M., "Hemoglobin and Its Abnormalities," Springfield, Ill., Charles C. Thomas, 1961, 183 pp.
3. LEHMANN, H., HUNTSMAN, R. G., AND AGER, J. A. M., "The Hemoglobinopathies and Thalassemia," in "The Metabolic Basis of Inherited Disease" (STANBURY, J. B., WYNGAARDEN, J. B., AND FREDERICKSON, D. S., EDITORS), pp. 1100–1136, New York, McGraw-Hill Book Co., 1966.
4. RUCKNAGEL, D. L., AND NEEL, J. V., "The Hemoglobinopathies," *Progr. Med. Genet.*, **1**, 158–260 (1961).
5. SUTTON, H. E., "Biochemical Genetics and Man: Accomplishments and Problems," *Science*, **150**, 858–862 (1965).

H. ELDON SUTTON

HEMOPOIESIS

Earliest Stages. In mammals the earliest or prehepatic stage of blood formation takes place in the mesenchyme of the yolk sac, and to some extent in the mesenchyme of the primordia of the umbilical cord and chorion. The yolk sac is evanescent and gives rise to primordia which later take part in the formation of the gut. The mesenchyme forms islands of polymorphic cells, the blood islands, in which the cells are compressed and proliferating rapidly by mitosis between the covering epithelium of squamous cells and the lining columnar epithelium. The columnar cells are of endodermal origin and their structure resembles that of the cells of the acini of the pancreas, with a distal zone of zymogen granules and a basal striated zone of RNA. It is in the mesenchyme, in association with endodermal cells forming the lining cells of the primitive gut, that the primordia of the later hemopoietic centers of liver, thymus and spleen arise. At first, the cells of the blood islands have no specific differences of prognostic value and are totipotent. Soon, however, the peripheral cells of the islands flatten to become endothelial cells, while the cells within round off and assume characteristic nuclear configurations of rapidly changing "blast" cells. The nuclei of these cells have vesicles of denser CHROMATIN, which stain deeply with dyes specific for DNA, and these enclose lighter areas of less dense, less deeply stained chromatin, and tend to have the reaction for RNA. These have been called by some the heterochromatin and the euchromatin. It is thought that this difference in the staining reaction of the chromatin is related to the direction into which these blast cells will differentiate. Also these blast cells have several centers, the nucleoli, which stain specifically for a content of RNA. Staining is abolished in these centers by treatment with RNase. Theoretically they are supposed to be a part of the system for the conversion of inherited elements to their destined fate as expressed by the functional and morphological characteristics of the tissues. Blast cells occur in all tissues and are largest in younger tissues; some are set aside in the remnant of the primitive node, the tail node, to form the vasa and blood cells at the caudal end of the embryo.

The first generation of the blast cells in the blood islands soon develops a deep basophilic cytoplasm, RNA positive; these are called hemocytoblasts (or hemoblasts). The descendants of this type of cell, which may be as large as 15 μ in diameter, are smaller cells with denser nuclei in which the chromatin vesicles become compressed, the nucleolus becomes smaller, and the euchromatin shrinks. It has been suggested that the materials move from the nucleus to the cytoplasm and induce the formation of enzymes necessary for production of the complex substances, such as HEMOGLOBIN, which soon appear within the cytoplasm of some of these cells. Cells such as these are called primitive erythroblasts, and in the electron microscope, they show a Golgi apparatus, a centrosome, a few mitochondria, a little endoplasmic reticulum and ribose granules. These cells are about 10 μ in diameter, and the cytoplasm changes from a polychromatic reaction to an orthochromatic pink reaction as the cells mature; finally they give rise to a generation of primitive erythrocytes, or normoblasts, in which there is a considerable amount of orthochromatic hemoglobin and a small concentrated pycnotic nucleus. It has been suggested that this hemoglobin has a different biochemical structure from the hemoglobin which appears later and that it should be called mesoblast hemoglobin. It must be pointed out that all of these hemoglobiniferous cells can divide by mitosis, and their proliferation accounts for the first population of erythrocytes circulating in the embryo for at least the first month of life.

Some cells in the mesenchyme between the blood islands remain in a primitive condition, but a few of them may differentiate into hemoblasts, which in turn give rise to a few generations of primitive myelocytes, Possibly a few definitive neutrophilic or heterophilic granulocytes may be formed. This relation between the development of the RBCs within the endothelium and the WBCs in the mesenchyme outside of it is a characteristic feature of all of the future hemopoietic centers, including liver, thymus, spleen and bone marrow.

Hepatic Stage. The hepatic stage is so named because of the development of the blood cells in the mesenchyme of the liver. The primitive liver starts at about the 5-mm stage, or beginning of the second month, as a group of liver cords of endodermal origin which grow out into the vitelline vasa, entering the caudal end of the primitive heart and the mesoderm which separates the pericardium from the peritoneum. Some cells between the liver cords and the endothelial sinusoids of the vitelline vasa (future hepatic portal system) differentiate into hemoblasts, the offspring of which may enter the sinusoids and differentiate there into primitive erythroblasts and erythrocytes. This generation gradually fades out, and the erythroblasts give rise to generations of smaller cells or normoblasts which lose the nucleus as the hemoglobin develops and finally become the non-nucleated definitive erythroplastids. It is suggested that the hemoglobin of these erythroplastids is hemoglobin F, and that it should be called hepatic hemoglobin. Very few granulocytes develop in the liver, but as it gets older, they may be seen along with some larger multinucleated cells formed from multiplication of the nuclei without division of cytoplasm in some hemoblasts. These cells are the megakaryocytes, and when they develop in the later organs, by fragmentation, they give rise to the blood platelets. The height of hepatic hemopoiesis is at about the sixth month.

Thymus and Spleen. At about the beginning of the third month, the primordia of the thymus and spleen appear as centers of proliferating mesenchyme around the endoderm of the third branchial pouch and in the dorsal mesentery above the endoderm of the primitive stomach, respectively. Cells from the primitive mesenchyme in the vicinity of the cords of thymic endoderm or stroma round off as small hemoblasts, with the characteristics of the later thymic hemoblasts, and migrate into the stroma where they undergo rapid proliferation, forming generations of larger and smaller hemoblasts, sometimes called thymocytes. Reactive mesenchyme accompanies these cells along the surfaces of the invading vasa. Ordinarily the mesenchyme cannot be distinguished from the endodermal stromal cells, but in the definitive condition when rats are poisoned with nitrogen mustard, the reticulum cell descendants of these mesenchymatous cells become macrophages. The thymus gradually becomes morphologically more compact and vascularized, and cell proliferation becomes limited to larger hemoblasts at the periphery of the organ; because of the functional distribution of the cells, the organ becomes differentiated into lobules each having a dense periphery, the cortex, and a central less dense area, the medulla. In early stages of development of the rat, the proliferation of the cells of the thymus is so great that it produces many more lymphocytes than are used to support its growing size. It is thought by some investigators to be the proliferator of clonal groups of cells which colonize vascular plexuses along the lymphatics at the sites of future lymph nodes. These cells become the parenchymatous primordia of these organs, later differentiating into secondary or germinal centers, and the plasma cells.

There is only a slight proliferation of erythroblasts and primitive erythrocytes in the primitive thymus. These soon leave, and the thymus becomes the site of lymphocyte formation, judging from the tinctorial and histochemical tests, and from supravital observations used for the discrimination of such cells. Many experiments have been carried on with the new techniques of autoradiography to test the life histories and sites of activity of the cells derived from the thymus, which are regarded as important carriers of ANTIBODIES in immunization. These antibodies are absent from lymphocytes or thymocytes of the newly born and do not develop normally until the organism is challenged with foreign proteins. In some animals such as the rat, the thymus continues to grow as long as the rat grows, which is all of its life, although its proliferative activity declines and merely meets the demands of that growth which is correlated with the rest of the body. In other animals where such changes have been investigated, such as the dog, the thymus is active during the youth of the animal, but involutes and becomes a connective tissue remnant with age. Many investigations have been made with foreign materials injected into animals to test the functional activities of the thymus. Also, transplantation of the thymus in relation to definitive hemopoiesis has been tested. Apparently it is an important organ for the production of cells which migrate to various parts of the body to neutralize the effects of adverse metabolic products.

The spleen, after its first appearance as a mesodermal proliferation center in the dorsal mesentery, becomes invaded by a plexus of vasa from the celiac axis which drains into the hepatic portal vein. The mesenchyme around the vasa of this plexus becomes arranged in cords and masses, each of which undergoes varied types of morphological change, the histology of which has not been studied in detail. However, it has been shown that the mesenchyme is capable of differentiating hemoblasts from which arise all of the cellular components found in the circulating blood. Also this organ is supposed to produce an endocrine substance which may stimulate the production of RBCs in the bone marrow. It has been suggested that this may be a site, along the kidney, for the production of the elusive hormone erythropoietin. As the fetus gets older, during the fifth or sixth fetal month, the spleen becomes less of a panhemogenic organ, and its proliferative properties become limited to lymphocytes and monocytes, while the stroma becomes specialized as that part of the reticuloendothelial system in which macrophages develop and digest degenerated WBCs and RBCs. In some animals, the spleen continues to be a panhemogenic organ with complex vascular and histological relations. In some rodents, the spleen is necessary for the protection of the animal against a very widespread rickettsia, *Bartonella*. Blood platelets may develop by fragmentation of megakaryocytes before this occurs in the bone marrow.

Bone Marrow. Finally, starting at about the middle of the third month of fetal life, spaces—primitive medullary cavities—develop within the bones of the axial and appendicular skeleton. These become filled with proliferating mesenchyme and are invaded by vascular plexuses. The mesenchyme differentiates into hemoblasts, reticulum cells, fat cells and osteogenic cells. There is still some controversy as to the fate of the hemoblasts inside and outside of the complicated plexus of wide and narrow endothelial-lined sinuses. According to one view, most of the hemoblasts within the sinusoids which have access to the flowing blood develop into erythroblasts, normoblasts and finally non-nucleated erythroplastids. On the other hand, the surrounding non-endothelial enclosed hemoblasts, without direct access to the flowing blood, differentiate into myeloblasts, myelocytes or the precursors of of the different types of granulocytes and of the megakaryocytes, the precursors of the blood platelets. In the young animals, the so-called red marrow is very extensive, but with age, the central part of the marrow in the diaphysis of the long bones becomes filled with fat. Cells of this tissue have the potency of changing into hemoblasts under certain kinds of metabolic deficiencies. Except for the early stages of development, there is no proliferation of the nucleated blood cells in the circulating blood; although when the blood is cultured for a short time in the new techniques for the study of the chromosomes, these bodies may be observed in cells undergoing mitosis. All normal proliferation of the blood cells occurs in the bone marrow, spleen, thymus and lymphoid tissue. However, since the reticulo-endothelial system extends throughout practically all of the organs of the body, except the nervous system, it has been observed that the primitive totipotency of this tissue can be stimulated to produce extramedullary or hemopoietic proliferation in aberrant sites. Such a condition may be hereditary, and may not appear until relatively late in life. The reader is referred to special books on hematology for the details of development and for the pathological changes in the hemopoietic organs.

References

1. BESSIS, M., "Cytology of the Blood and Blood-forming Organs," New York, Grune and Stratton, 1956 (translated by E. Ponder).
2. BLOOM, W., "Embryogenesis of Mammalian Blood," in "Handbook of Hematology," Vol. 2, p. 865, New York, Hoeber, 1958.
3. PONDER, E., "Hemopoiesis," in "Encyclopedia of Biological Sciences," p. 471, New York, Reinhold Publishing Corp., 1961.
4. MACFARLANE, R. G., AND ROBB-SMITH, A. H. T., "Functions of the Blood," New York, Academic Press, 1961.
5. LOW, F. N., AND FREEMAN, J. A., "Electron Microscopic Atlas of Normal and Leucemic Blood," New York, McGraw-Hill Book Co., 1958.

JAMES E. KINDRED

HEPARIN

Heparin, a carbohydrate, more specifically, an acid MUCOPOLYSACCHARIDE (glycosaminoglycuronoglycan), is a powerful inhibitor of blood coagulation (see BLOOD CLOTTING). It was discovered by McLean. Major contributions to our understanding of its chemical composition, structure and function have been made by Jorpes, Wolfrom, Meyer, Jaques, Winterstein, Jeanloz and their collaborators.

Chemistry. Heparin is obtained from various animal tissues, and procedures have been developed for the preparation of highly active nontoxic fractions for clinical use. This material, however, is not homogeneous according to chemical and physical-chemical criteria, and several of the components exhibit significant anticoagulant activity. This lack of homogeneity causes difficulties in attempts at establishing a definition of the term heparin. A molecular formula of $(C_{12}H_{16}O_{16}NS_2Na_3)_{20}$ and a molecular weight of about 12,000 were recently proposed for sodium heparinate.

The structure of heparin is not entirely established. The molecule consists of a polymer chain of repeating disaccharide units composed of D-glucosamine and a hexuronic acid not yet entirely identified (apparently primarily D-glucuronic acid), with predominantly α-(1 $\rightarrow$ 4)- and perhaps some α-(1 $\rightarrow$ 6)-glycosidic linkages. The amino group and one or more hydroxyl group per disaccharide subunit are substituted with sulfate. A significant portion of native heparin is linked to PROTEIN or peptide. (See also HEXOSAMINES.)

Heparin, a strongly negatively charged polyanion, combines readily with positively charged substances, in particular, with basic proteins such as protamine, and with long-chain aliphatic quaternary ammonium compounds such as cetylpyridinium chloride, yielding water-insoluble complexes. Basic dyes such as Azur A are bound by heparin with a characteristic shift in the dye's absorption spectrum (metachromatic reaction), a reaction utilized for detection of heparin in tissues and on paper chromatograms.

Occurrence. Heparin occurs in many mammalian organs. Thymus, lung, intestine and kidney are relatively rich in heparin, as is dog liver, where heparin was discovered and from which its name was derived. Blood contains only relatively small amounts. Heparin is formed and stored primarily in the mast cells of Ehrlich.

Metabolism. Little is known about the mechanism of heparin biosynthesis. It appears likely that uridine diphosphate derivatives of glucosamine and of a hexuronic acid serve as precursors for the carbohydrate chain of the molecule. The addition of SULFATE is accomplished by means of synthesis of 3′-phosphoadenylyl sulfate from inorganic sulfate and ATP, and transfer of sulfate from this nucleotide to the carbohydrate moiety of heparin.

Heparin appears to be resistant to the action of most known glycosidases. However, adaptive enzymes obtained from a *Flavobacterium* grown in the presence of heparin catalyze its degradation. *In vivo*, exogenous heparin disappears relatively rapidly from the circulation, following intravenous injection. Administration of larger doses results in the excretion of a degradation product in the urine (uroheparin). After injection of heparin labeled with radioactive sulfate, the molecule is desulfated to a certain extent by the animal body, as indicated by the excretion of radioactive inorganic sulfate.

Biological Activities. Heparin is a powerful inhibitor of blood coagulation both *in vivo* and *in vitro*, the most potent known anticoagulant extractable from mammalian tissues. It inhibits thromboplastin generation, it interferes with the conversion of prothrombin to thrombin and with the thrombin-catalyzed conversion of fibrinogen to fibrin, and it decreases platelet agglutination and adhesion.

Furthermore, heparin induces the release of an enzyme (lipoprotein lipase) from tissues into the circulation, which catalyzes the hydrolysis of lipoprotein triglycerides. As a result, chylomicrons (minute fat droplets) and the larger low-density lipoprotein molecules rich in neutral fat disappear from the bloodstream, and lipemia (milky appearance of plasma after fat ingestion) is cleared. The fatty acids and glycerol liberated are then carried to tissues where they are metabolized. This process appears to be a physiological mechanism for the disposal of alimentary blood fat (see also FAT MOBILIZATION; LIPOPROTEINS).

Because of its strong electronegative charge, heparin binds and inactivates many biologically active compounds having basic groups, including enzymes, drugs and various toxic substances.

Clinical Applications. The major clinical application of heparin is based on its anticoagulant action. It is used in venous thrombosis (thromboembolism), in acute myocardial infarction (coronary thrombosis), during vascular and open heart surgery, and in blood transfusion. Another clinical use of heparin in coronary ATHEROSCLEROSIS relates to its ability to stimulate removal of circulating blood lipids.

Most methods for heparin assay are based on its anticoagulant activity. The potency of commercial preparations is usually determined in accord with a procedure in the U.S. Pharmacopeia.

References

1. WOLFROM, M. L., "Heparin and Related Substances," in "Polysaccharides in Biology," *Transactions of Fourth Conference, J. Macy, Jr. Foundation*, 115 (1959).
2. JAQUES, L. B., AND BELL, H. J., "Determination of Heparin," *Methods Biochem. Anal.* (GLICK, D., EDITOR), **7,** 253 (1959).
3. JEANLOZ, R. W., "Heparin," "Comprehensive Biochemistry" (FLORKIN, M., AND STOTZ, E. H., EDITORS), Vol. 5, p. 289, Amsterdam, Elsevier, 1963.
4. ENGELBERG, H., "Heparin, Metabolism, Physiology and Clinical Application," 1963.
5. JORPES, J. E., "Heparin," *Comparative Endocrinology*, (VON EULER, U. S., AND HELLER, H., EDITORS), **2,** 112 (1963).
6. BRIMACOMBE, J. S., AND WEBBER, J. M., "Heparin," in "Mucopolysaccharides," *B.B.A. Library*, **6,** 92 (1964).

7. Muir, H., "Heparin and Heparan Sulfates," in "Chemistry and Metabolism of Connective Tissue Glycosaminoglycans (Mucopolysaccharides)," *International Review of Connective Tissue Research*, **2**, 120 (1964).

Walter Marx

HEXOSAMINES AND HIGHER AMINO SUGARS

The amino sugars are widely distributed in nature and are generally found as components of complex polymers. For example, a polysaccharide called CHITIN is one of the most abundant organic substances in nature, being a major component of the exoskeletal substance in crustacea and insects, and the cell walls of many fungi. When chitin is subjected to vigorous acid hydrolysis, it yields two compounds, a sugar called *glucosamine* and acetic acid. The sugar is a derivative of D-glucose, a six-carbon sugar or hexose, where the alcoholic group on carbon-2 is replaced by an amino group; the derivative is, therefore, 2-amino-glucose, or glucosamine. Glucosamine is a member of the class of sugars called hexosamines (six-carbon amino sugars). Because of the amino group, the hexosamines behave like other primary amines and form salts with acids. In nature, however, particularly in the polymers that contain these sugars, the amino group is coupled to an acyl group, usually acetic acid, and the resulting amide derivative, N-acetylglucosamine, is a neutral substance. One polysaccharide, produced by fungi, is positively charged because some of the amino groups are free.

While chitin illustrates the case of a *homopolysaccharide* where the polymer is formed by condensation of several thousand monomer units of a single type, *i.e.*, N-acetylglucosamine, the amino sugars are also found in much more complex substances. For example, many *heteropolysaccharides* are known that contain one of the N-acetylamino sugars condensed with one or more sugars of another type like the hexoses or uronic acids. Most bacterial and mammalian connective tissue polysaccharides are heteropolysaccharides. *Hyaluronic acid*, a heteropolysaccharide found in skin, umbilical cord, synovial fluid, vitreous humour, streptococci, etc., illustrates this point; the polymer is a long flexible chain composed of alternating units of N-acetylglucosamine and glucuronic acid. Finally, the structures of the sugar-containing polymers become even more complex when glycoproteins and glycolipids are considered. Here, as many as 6 or 7 different sugars may be condensed with each other, and the product, an oligosaccharide or polysaccharide, depending upon the size of the polymer, is covalently linked to protein or to lipid. These substances are widely distributed in mammalian tissues. Most blood proteins are, in fact, glycoproteins, while glycolipids are important constituents of brain, red blood cell stroma, etc.

The replacement of an hydroxyl group in hexose by an amino group to give an amino sugar can theoretically occur at any position in the six-carbon chain, and more than one hydroxyl group could be replaced. In fact, many different amino and diamino sugars are known, and they are not confined to the hexose class of compounds. Most of these sugars are produced by fungi and are frequent components of antibiotics secreted by these molds. If we confine this discussion to the amino sugars found most frequently in the polymers described above, then the 2-amino-hexoses are of prime importance. In addition to glucosamine, the corresponding derivative of another common hexose, D-galactose, also occurs very frequently; the sugar is called *galactosamine* and is usually found in the polymers as its N-acetyl derivative. One other hexosamine is of interest, namely the amino sugar derived from D-mannose, called *mannosamine*. This compound is not a constituent of the polymers, but is an important intermediate in the biosynthesis of a more complex class of amino sugars called the sialic acids.

Sialic Acids. These generally occur in glycoproteins and glycolipids, and only rarely in polysaccharides. They are usually the terminal sugars in the oligo- and polysaccharide chains linked to the protein and lipid units as described above. This class of amino sugars is derived from a parent compound, *neuraminic acid*, that does not occur naturally. Neuraminic acid differs from the hexosamines in several respects. While it is also a linear chain, it is composed of 9 rather than 6 carbon atoms and contains a carboxyl group at carbon-1, a keto group at carbon-2, no hydroxyl group (*i.e.*, two hydrogen atoms) at carbon-3, and an amino group at carbon-5. In fact, it may be considered as an aldol condensation product of pyruvic acid and mannosamine. The sialic acids are derived from neuraminic acid; they all contain an N-acyl group while some also contain O-acyl groups. The N-acyl group is usually acetic acid, but one case is known where it is glycolic acid; the O-acyl group appears to be only acetic acid.

The sialic acids are glycosidically linked to other sugars through the keto group at carbon-2 leaving the carboxyl group free. Each of the sialic acids in a polymer, therefore, contributes a negative charge to the macromolecule. This charge may have important physiological effects; for example, almost all of the negative charge on the surface of erythrocytes results from the sialic acid groups located on this "membrane." When influenza virus attaches to erythrocytes, it is located at or near the sialic acid groups, and a hydrolytic enzyme carried by the virus splits the sialic acid groups from the erythrocytes.

Muramic Acid. One other amino sugar should be mentioned, called muramic acid. This compound is found in a polymer that constitutes the backbone of the rigid structure in BACTERIAL CELL WALLS. It may be regarded as an ether derivative of lactic acid and glucosamine, formed by removing a molecule of water between the hydroxyl group of the lactic acid and that of carbon-3 of the glucosamine. N-Acetylmuramic acid alternates with N-acetylglucosamine units in the cell wall polysaccharide. The carboxyl group of muramic acid is not free in the polymer, but combined with 5 or 6 amino acids to form short polypeptide chains; the amino acids are interesting in that several of them are the "unnatural" amino acids, *i.e.*, the D-configuration.

Biosynthesis and Metabolism. How are the amino sugars formed in nature, and how are they incorporated into the polymers? Much is known concerning the first question, and relatively little concerning the second. The parent compound of galactosamine, mannosamine, muramic acid, and the sialic acids is glucosamine. The steps leading from glucose, the foodstuff of most cells, to glucosamine-6-phosphate have been established. The immediate precursor of glucosamine-6-P is fructose-6-P, the nitrogen atom being derived from the amino acid glutamine. Glucosamine-6-P is then converted to N-acetylglucosamine-6-P; acetyl coenzyme A is the acetyl donor. From this point, different metabolic pathways lead to the other amino sugars described above. An important class of intermediates in these sugar interconversions are *sugar-nucleotides*; these substances are composed of sugars, such as N-acetylglucosamine, glycosidically bound to the phosphate group of a nucleotide such as uridine diphosphate. The sugar-nucleotides not only are important intermediates in sugar interconversions, but serve as a source of sugar units in the building of polysaccharide chains. For example, the transfer of glucose from uridine diphosphate glucose to the end of a growing chain leads to the formation of liver glycogen.

The problem of polymer formation has not been satisfactorily resolved. It appears unlikely that the complex polymers are formed by single-step additions of sugar units from sugar-nucleotides to the end of growing chains, at least in many of the examples cited above. Other intermediates may occur, but have not yet been characterized. In addition, the problem of chain initiation has not yet been solved.

SAUL ROSEMAN

HEXOSE MONOPHOSPHATE SHUNT

This pathway for glucose oxidation, also termed the phosphogluconate oxidative pathway, is discussed in CARBOHYDRATE METABOLISM, section on Other Carbohydrate Interconversions; it is a reaction sequence for glucose oxidation that forms a major alternative to glucose oxidation through the glycolytic sequence (see GLYCOLYSIS) followed by the CITRIC ACID CYCLE. A number of enzymes of the hexose monophosphate shunt are also utilized in the CARBON REDUCTION CYCLE IN PHOTOSYNTHESIS. See also PENTOSES IN METABOLISM.

HEXOSES

Hexoses, or six-carbon sugars, include 16 aldohexoses and 8 ketohexoses, and the various ring forms derived from these fundamental structures. The most common naturally occurring hexoses are D-glucose, D-galactose, D-fructose and D-mannose [see CARBOHYDRATES (CLASSIFICATION AND STRUCTURAL INTERRELATIONS) for the structures of these monosaccharides; CARBOHYDRATE METABOLISM for a discussion of some of the reactions they enter into, in living processes].

HEXURONIC ACIDS

The hexuronic acids are a subclass of the six-carbon sugar acids, in which the terminal carbon, or carbon-6, has been oxidized to a carboxylic acid group. The hexuronic acids thus have the following general formula:

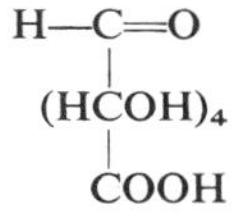

$$\begin{array}{c} H{-}C{=}O \\ | \\ (HCOH)_4 \\ | \\ COOH \end{array}$$

The best-known example is D-glucuronic acid:

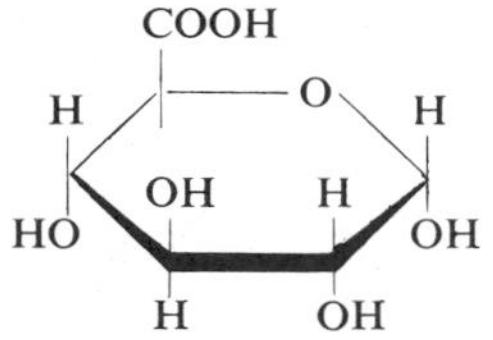

$$\text{[pyranose ring: C-5 bearing COOH; ring substituents H, HO, OH, H, H, OH, H, OH; ring O]}$$

This occurs in the form of glucuronides, bound through a glycosidic linkage to various hydroxylated compounds (aglycones). These glucuronides generally have greater water solubility than the corresponding free aglycones (see, for example, BILE AND BILE PIGMENTS, section on Bile Pigments; GLUCURONIDASES; KETOSTEROIDS).

Uronic acids are formed biosynthetically by oxidation of the corresponding nucleoside diphosphate hexoses. Thus, for example, uridine diphosphate glucose is oxidized (with NICOTINAMIDE ADENINE DINUCLEOTIDE as electron acceptor) to uridine diphosphate glucuronic acid. The latter derivative of glucuronic acid then interacts with an algycone to form the glucosiduronide (glucosiduronic acid). This and other sugar nucleotide reactions are discussed in the article CARBOHYDRATE METABOLISM, section on Other Carbohydrate Interconversions. In many heteropolysaccharides, such as certain MUCOPOLYSACCHARIDES, hexuronic acids occur as repeating structural subunits, usually in alternation with an amino sugar subunit (see HEXOSAMINES AND HIGHER AMINO SUGARS). See also PENTOSES IN METABOLISM, section on Glucuronic Acid Xylylose Cycle.

HIGH-ENERGY PHOSPHATES. See PHOSPHATE BOND ENERGIES; PHOSPHORYLATION, OXIDATIVE.

HISTAMINE

Histamine,

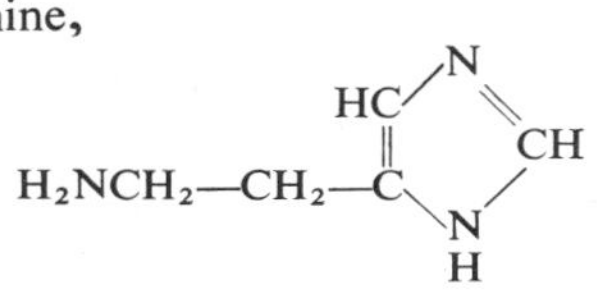

$$H_2NCH_2{-}CH_2{-}\overset{}{C}\begin{array}{l} \!=\!CH{-}N{=}CH \\ \;\;\; {-}NH{-} \end{array}$$

(imidazole ring: HC, N, CH, N–H, C bearing $H_2NCH_2{-}CH_2{-}$)

is a powerful vasodilator which is released in anaphylactic shock (see IMMUNOCHEMISTRY) and occurs in blood and tissues in minute amounts. The injection of 1 μg intravenously in humans, is said to bring about a sharp drop in blood pressure.

Its close relationship to HISTIDINE is emphasized by the fact that the amino acid can be decarboxylated by certain intestinal bacteria, to produce it. Due in part to its powerful action and the minute amounts which occur physiologically, its functions are uncertain.

HISTIDINE

L-Histidine, α-amino-β-imidazolepropionic acid,

$$\begin{array}{c} CH \\ N \quad NH \\ HC{=}C{-}CH_2{-}CHNH_2{-}COOH \end{array}$$

is one of the three more common basic AMINO ACIDS derived from PROTEINS. It occurs widely in proteins. Hemoglobins are a conspicuously rich source and yield about 8.5%. Its specific rotatory power in 6N HCl solution is $[\alpha]_D = +13.3°$

The status of histidine with respect to its essentiality in nutrition is exceptional in that for rats it is indispensable but for humans it appears not to be (see AMINO ACIDS, NUTRITIONALLY ESSENTIAL). The extreme physiological potency of the decarboxylation product of histidine (HISTAMINE) is notable. The amines derived from other amino acids are, as a rule, far less active.

HISTOCHEMICAL METHODS (IN STUDY OF NUCLEIC ACIDS)

Histochemical methods include staining and physical-optical techniques which yield information at the microscope level about the chemical composition of cells. These approaches allow direct observation of the intracellular distribution of nucleic acids, proteins, lipids, and polysaccharides; low molecular weight compounds, like amino acids and sugars, are not ordinarily retained in histological preparations.

The field of histochemistry has grown so rapidly in the post-war era that there are four widely used English-language texts devoted entirely to theory and application. In addition, there are a number of important French and German volumes, a multilingual encyclopedia of histochemistry which is already in its eighth volume, about six international journals devoted exclusively to histochemical techniques and applications, and a proliferating number of reviews and chapters on specialized techniques. It is possible, in this abbreviated article, to describe only the major areas of development in *nucleic acid histochemistry*; references are given at the end of this article to discussions of other constituents.

Interference microscopy may be employed on living cells to yield information about cellular dry mass and protein content. The ultraviolet microscope also has been used with living material to reveal the distribution of nucleic acids by virtue of the natural absorption of constituent nucleic acid bases at 260 mμ. A severe limitation in ultraviolet microscopy of living cells results from radiation damage to the biological system under observation. The need to reduce radiation dosage impinging on cells led to the development of the television and flying spot microscopes. The fact that cells are chemically fixed with acetic acid-ethanol or formalin, prior to histochemical study, is not a severe handicap in the study of nucleic acids since adequate preservation is obtained. In the case of enzyme detection, special methods of preparation are required.

Staining Methods for Nucleic Acids. *The Feulgen Reaction for DNA*. This widely used staining reaction is the most reliable one yet developed for selective detection of DEOXYRIBONUCLEIC ACID or DNA. It is named after Robert Feulgen, a nucleic acid biochemist, who discovered, in 1918, that mild acid treatment of purified DNA released PURINES, leaving a semidegraded product known as apurinic acid (APA). Upon treatment with the aldehyde indicator, Schiff's reagent, APA gave a positive color test. Feulgen soon applied this procedure to tissue sections and announced, in 1923, the successful intracellular localization of DNA in nuclei. By means of this so-called nucleal reaction, he and his collaborators were able to demonstrate, for the first time, that DNA was a nuclear constituent in all plant and animal cells and was not restricted exclusively to animal cells, as has been thought previously. Bacteria were also shown to contain DNA by this procedure. The nucleal reaction immediately became popular and by 1932 there were already several thousand papers published using the so-called Feulgen reaction.

Numerous tests have been carried out with aldehyde blocking reagents and NUCLEASES which confirm the specificity of the reaction for DNA. Indeed, binding is stable enough to allow quantitation of relative DNA content in individual nuclei. A phenomenal number of publications have appeared during the past 15 years based on quantitative applications of the Feulgen reaction to vital questions of DNA metabolism in the cell cycle of normal, malignant, and viral-infected cells. In recent years, other dyes were found to substitute for basic fuchsin in the preparation of Schiff's reagent. These Schiff-type reagents allowed multiple aldehyde staining reactions to be carried out on the same cells using dyes of complementary colors. For example, DNA and polysaccharides could be stained in succession. Some of the new reagents were fluorescent and allowed the Feulgen reaction to be used with increased sensitivity.

Staining by Basic Dyes. During the period of the late nineteenth century, when the chemistry of nucleic acids and associated basic proteins was first enunciated by Friedrich Miescher, A. Kossel and R. Altmann, simultaneous progress was underway in classical histology. Aniline dyes were employed by Paul Ehrlich, A. Pappenheim and others to stain microscopic sections and enhance contrast.

This work started out with a strictly empirical approach by interested anatomists and pathologists. It became apparent that basic dyes, like methylene blue, methyl green, and crystal violet, are bound preferentially to cell nuclei with slight cytoplasmic staining. The nuclear binding property becomes more obvious in the presence of an acidic dye, such as eosin or orange G. With the help of *in vitro* tests, using purified nucleic acids and appropriate dyes, one could demonstrate that positively charged nitrogen groups on dye molecules react primarily in ionic or electrostatic linkage with phosphoric acids of nucleic acids. Other acidic components, like carboxyl groups on PROTEINS and sulfate groups on MUCOPOLYSACCHARIDES and chondroitin sulfate, may compete with nucleic acid phosphate for basic dyes. All of these substances are present in tissues and might present formidable problems were it not for the fact that maximum ionization of these acid groups occurs at different pH values. Optimal staining of sulfuric acid groups occurs at about pH 1.0, while nucleic acids stain in the range of pH 3.5. At higher pH values, protein carboxyl groups are bound. Methods currently employed which utilize basic dyes include methyl green-pyronin, azure B bromide and gallocyanin chromalum.

The methyl green-pyronin dye mixture is a popular one for staining and distinguishing the two nucleic acids. The method was first concocted by Pappenheim near the end of the nineteenth century and uses an unusual combination of two basic dyes to achieve brilliant intracellular staining patterns, namely, red cytoplasm and nucleolus, and green nucleus. The recipe was modified a few years later by Paul Unna and was elevated by Brachet to histochemical status in 1940. The latter found that pretreatment of sections with ribonuclease prevented the red staining of cytoplasm and nucleolus with pyronin. This knowledge of pyronin-RNA staining led immediately to significant findings which implicated RNA as an essential component accompanying protein synthesis during embryonic development. As a result of the studies of Pollister, Leuchtenberger and Kurnick, it is generally accepted that methyl green binds only to highly polymerized DNA in cell nuclei while pyronin stains RNA (and depolymerized DNA if present). When attention is paid to possible protein interference, reliable quantitative measurement from single nuclei may be obtained using methyl green alone. Methyl green-pyronin staining, in combination with RNase controls, continues to serve as an important cytochemical tool for qualitative studies.

Azure B bromide stains DNA and RNA in blue and violet colors respectively at pH 3.8. The cell nucleus appears a pale blue while cytoplasm and nucleolus are stained violet. This metachromatic effect was known for many years, but the property was used by Flax and Himes in 1952 as a basis for carrying out quantitative absorption measurements of nucleic acids at different wavelengths.

Gallocyanin chromalum differs somewhat from other basic dyes mentioned in that a chromalum lake is introduced as a cation to bind nucleic acid phosphate. The dye-lake cation is used at pH 1.6 and binds progressively to cellular elements containing nucleic acids until dye saturation occurs. The method was originally developed by Einarson in 1935. Mainly through the efforts of Sandritter and collaborators, it was demonstrated to be useful for quantitative estimations of DNA following RNase treatment. It is a reliable method for routine staining since the dye complex is not washed out in the dehydrating steps prior to mounting. For quantitative use, the method is receiving increasing attention.

Fluorochroming of Nucleic Acids with Acridine Orange. Recognition of the inherent sensitivity of the fluorescence process over light absorption methods led to a natural interest in fluorescence microscopy as a histo- and cytochemical tool beginning about 1950. There had been extensive studies in pre-war Germany by Strugger which laid the groundwork for future developments. Mainly through the efforts of Schümmelfeder in the post-war period, it was shown that acridine orange (AO) imparted a bicolor fluorescence to cellular regions rich in nucleic acids. RNA appeared in rich shades of orange to red and DNA was bright green. The differential fluorescence was immediately recognized to be a type of metachromasy in which electrostatic binding played a prominent role, as in the case of acid-base staining reactions mentioned previously. The implications of this vastly more sensitive technique were recognized by Armstrong and Niven who employed it to detect nucleic acid alterations in viral-infected cells. Von Bertalanffy applied it clinically to smears of exfoliated cervical cells as a more rapid means of detecting nucleic acid-rich cancer cells. AO fluorochroming has become widely used as a research tool although some workers neglect to use adequate controls, such as nucleases or other stains. This becomes essential in some studies because proteins may induce a green fluorescence similar to that normally produced by DNA. Also, any treatments which affect the physical and chemical state of nucleic acids *in situ* cause an immediate change in color which bears no relation to whether the nucleic acid is DNA or RNA.

Physical-optical Methods for Nucleic Acids at the Cytological Level. *Cytophotometry*. T. Caspersson demonstrated, in 1936, that nucleic acid patterns could be obtained from fixed unstained cells by utilizing the fact that nucleic acids were the prime absorbers of UV radiation at 260 mμ. It was possible to obtain UV absorption curves from cell organelles and to detect marked shifts in nucleic acid in abnormal physiologic states. However, it was almost impossible to obtain valid quantitative data because of protein interference at 280 mμ and nonspecific light losses which were difficult to assess. Fortunately, A. W. Pollister, at Columbia University, with the help of students

(Swift, Moses, Leuchtenberger) circumvented the impasse by demonstrating that the Feulgen reaction for DNA was a useful quantitative tool in the visible light region. Relatively uncomplicated microscopic equipment was employed in the form of a simple microcolorimeter with the stained nucleus taking the place of a cuvette sample. The equipment included a tungsten or mercury light source, narrow-band filters or a monochromator, an iris diaphragm above the microscope to delimit the measured area, and a photomultiplier tube. With appropriate methodology, it was possible to compute the relative DNA content of selected cell nuclei in a heterogeneous cell population, such as liver tissue or testis. The "DNA constancy" hypothesis, which was suggested from the biochemical studies of Boivin and the Vendrelys, was confirmed by the Columbia group. They showed that somatic cell nuclei contained twice as much DNA as mature sperm and that in certain organs, like liver, some cells contained polyploid nuclei with geometric multiples of the basic diploid DNA values, $2N$, $4N$, $8N$. The availability of such a relatively simple quantitative method led to its wide acceptance by many cytologists interested in analyzing the biological role of nucleic acids. Other staining methods became available for quantitation of total proteins, histone proteins, SH-proteins, RNA and DNA. Measuring methods were later devised to allow absorption measurements of irregular-shaped nuclei and cells in mitosis. In recent years, sophisticated equipment has become available through commercial sources to allow rapid measurements to be made of cell populations by mechanical and electronic scanning cytophotometers.

Autoradiography. With the availability of isotopically labeled precursors of nucleic acids, a large number of histo- and cytochemical investigations are making use of autoradiography. The technique involves injecting animals with radioactive compounds, the most popular label being H^3, although C^{14} and P^{32} have also been utilized. In tissues undergoing high levels of RNA synthesis or cell renewal, appropriate precursors, like uridine and cytidine for RNA or thymidine for DNA, are rapidly incorporated. Following sacrifice of the animals, fixed tissue sections are prepared and are coated in the dark with a thin layer of photographic emulsion. After exposure for periods of a few days to several months, the preparation is developed, stained and examined microscopically. Intracellular deposits of silver are observed over cellular areas rich in isotope. A fine degree of localization or resolution is possible when H^3 compounds are used since the average distance traveled by emitted electrons is only 1 μ. This means, for example, that molecules of RNA labeled with H^3-uridine in a small structure like the nucleolus produce silver grains in the emulsion directly above it and not to the sides. The same techniques are applicable to cultured cells with greater experimental sophistication. With the aid of special photographic emulsions, autoradiography has been successfully carried out on ultrathin sections and observations have been made by electron microscopy. This method is being supplemented by other electron histochemical techniques such as indium trichloride and silver solutions for "staining" nucleic acids, and enzyme digestion methods.

The author's work was supported in part by U.S. Public Health Service Research Grants No. CA-07991-02 from the National Cancer Institute and No. 5 RO1-NB-03113-05 from the National Institute of Neurological Diseases and Blindness. Some support was also derived from the U.S. Army Medical Research and Development Command, Department of the Army, under Research Grant No. DA-MD-49-193-65-G138, administered by Dr. Donald E. Rounds.

References

1. Barka, T., and Anderson, P. J., "Histochemistry: Theory, Practice and Bibliography," New York, Paul B. Hoeber, Inc., 1963. (*A recent outline of histochemistry with a comprehensive bibliography of references classified according to chemical components.*)
2. Burstone, M. S., "Enzyme Histochemistry and its Application in the Study of Neoplasms," New York, Academic Press, 1962. (*A comprehensive text which one should consult for information about the growing specialty of enzyme histochemistry.*)
3. Graumann, W., and Neumann, K. (Editors), "Handbuch der Histochemie," Stuttgart, Gustav Fischer Verlag, 1958. (*A modern encyclopedic group of volumes with comprehensive chapters in English, French or German by specialists on histochemical topics. New volumes are added to the series at irregular intervals.*)
4. Lillie, R. D., "Histopathologic Technic and Practical Histochemistry," New York, The Blakiston Co., 1965. (*A reference with concise practical methods of histochemistry. For the novice there are five introductory chapters dealing with the preparation of histological specimens.*)
5. Pearse, A. G. E., "Histochemistry, Theoretical and Applied," Second edition, Boston, Little, Brown and Co., 1960. (*This volume is probably the most thorough text available which describes the theory and practice of histochemistry.*)
6. Sandritter, W., and Kasten, F. H. (Editors), "100 Years of Histochemistry in Germany," Stuttgart, F. K. Schattauer-Verlag, 1964. (*A historical reference with concise summaries of the chemical and physical aspects of histochemistry and biographical sketches of important German histochemists.*)

Frederick H. Kasten

HISTONES

Histones are basic proteins which occur in the nuclei of both plant and animal cells. They are less basic than the protamines, having isoelectric points at about pH 11. Some investigators would restrict the term histone to only those basic proteins anatomically and chemically associated with DNA (DEOXYRIBONUCLEIC ACIDS).

Distribution. Histones appear to be of universal

occurrence in the nuclei of higher plants and animals. However, there have been no systematic studies of histone distribution in invertebrates and microorganisms. While some histone-like proteins have been reported in bacteria and protozoa, these have not been chemically characterized, nor has their close association with DNA been demonstrated. It has been suggested that histones as they exist in mammalian cells, may be limited in distribution to organisms containing polytenic CHROMOSOMES (*i.e.*, chromosomes containing multiple parallel strands of nucleoprotein).

Histones occur in the interphase nucleus in the form of NUCLEOPROTEINS in which histones complex with DNA in such proportions that all or nearly all the negative charges on the phosphate groups of DNA are neutralized by positive charges on histone. Structures in which histone surrounds individual double-stranded DNA helices, as well as those in which histone forms bridges between adjacent coils, have been proposed.

In isolating histones, advantage is taken of the solubility properties of the histone-DNA complex, which is insoluble in dilute salt solutions but soluble at higher ionic strength. Nucleoprotein may also be obtained from some tissues by repeated extractions with distilled water. Calf thymus, which is often used for histone preparation because of its high concentration of nuclei, may be used directly for the preparation of nucleoprotein. With most tissues, however, it is preferable to employ previously isolated cell nuclei as the starting material. The nuclei are first exhaustively washed with dilute salt solutions containing citrate or ethylenediamine tetraacetate (which help disrupt the nuclear membranes) in order to remove soluble proteins. Histones are then extracted from the salt-washed nuclei with dilute acid (0.1–0.3N HCl or H_2SO_4). If the nucleoprotein had been solubilized by high salt concentration, the addition of acid denatures and precipitates DNA leaving histone in solution.

Crude mammalian histone, isolated from a single organ, is a complex mixture of 10–20 individual proteins. They are characterized in their amino acid composition by a high proportion of the basic amino acids LYSINE and ARGININE, by the absence of TRYPTOPHAN, and by the presence of only very small quantities of the sulfur-containing amino acid CYSTEINE. Histone preparations from different species and tissues are very similar in amino acid composition. Small amounts of ε-N-methyllysine, an amino acid previously detected only in the flagellar protein of certain bacteria, are characteristically found in mammalian histone preparations. Acetyl groups occurring principally as the N-terminal group of certain histone fractions are also present.

Purification. Fractionation of histones has been accomplished by a variety of methods; these include selective extraction or precipitation of the crude histone mixture or selective extraction of the salt-washed nuclei and the use of various CHROMATOGRAPHIC techniques. Among the resins employed in the latter procedures are the weak cation exchangers, "Amberlite" IRC-50 and carboxymethylcellulose. The three major fractions obtained by these methods are termed arginine-rich, intermediate or moderately lysine-rich, and lysine-rich histones. These have been defined as having molar lysine-to-arginine ratios of less than one, between one and four, and greater than four, respectively. They differ in amino composition from each other in many respects aside from their lysine to arginine ratio: for example the lysine-rich fraction is much higher in alanine and in proline than are the intermediate and arginine-rich fractions. All of these fractions have been shown to be heterogeneous. Estimates of the molecular weight of the various fractions range from 10,000–20,000. Histones in solution, however, tend to form aggregates of higher molecular weight, a tendency which increases with increasing pH. In calf thymus histone, the N-terminal amino acids are mainly alanine in the arginine-rich fractions, and proline in the intermediate fractions; some of the lysine-rich fractions contain an N-acetyl end group.

Zone ELECTROPHORESIS on starch or polyacrylamide gels which provides the most sensitive analytical method for the identification of histone subfractions has not as yet been successfully employed for preparative purposes.

Biological Functions. The close associations of histones with DNA led some years ago to the hypothesis that histones might play a role in the control of genetic expression at the cellular level. More recently, advances in molecular biology have allowed more detailed mechanisms for such control to be proposed. In particular, it has been suggested that histones, by blocking some areas of the DNA molecule, permit only part of the DNA base sequences to act as templates for the formation of messenger RNA. Thus histones by controlling messenger RNA formation ultimately control PROTEIN BIOSYNTHESIS within the cell. Other investigators believe that the primary role of histone is structural—histone being essential for stabilizing the DNA helix, for the integration of DNA strands into more complex chromosomal structures, and for fixing and maintaining during CELL DIVISION chromosomal changes occurring during differentiation and development. It should be pointed out that these two general ideas about histone function are not mutually exclusive; *i.e.*, it seems possible that histone may fix chromosomal structure in a specific configuration in which the position of the histone molecules could limit RNA formation.

A large number of experiments designed to support the gene regulator hypothesis have shown that in DNA-requiring systems, capable of RNA synthesis in the test tube, histones can act as inhibitors. It is difficult to apply these results directly, however, to the much more complex situation existing within the nucleus of the cell (see METABOLIC CONTROLS).

Among the factors that must be taken into account in considering histone function is the rather small number (10–20) of individual histone species within a given cell. This makes it unlikely for example, that specific gene areas on DNA are controlled by individual histones. In addition there are only slight differences between histones in various organs of the same mammalian species, and tissues of widely differing function such as brain, liver, and kidney, yield histones with indistinguishable electrophoretic patterns. Experiments in which the synthesis and degradation of histones in mammalian organs have been followed employing radioactive amino acids, have shown that histones are replaced at a rate slower than that of any major protein fraction of the cell. It has been suggested that histones may persist for the life of the cell and be replaced only during cell division.

The possibility that histones play a role in genetic mechanisms has led to the suggestion that histone changes may initiate or accompany early cellular changes leading to the formation of tumors. It is not generally agreed at this time whether or not tumor histones differ from those of corresponding normal tissues.

References

1. Phillips, D. M. P., "The Histones," in *Progr. Biophys. Biophys. Chem.*, **12**, 213 (1962).
2. Bonner, J., and Ts'o, P., (Editors), "The Nucleohistones," San Francisco, Holden Day, 1964.
3. Busch, H., "Histones and Other Nuclear Proteins," New York, Academic Press, 1965.
4. Murray, K., "The Basic Proteins of Cell Nuclei," *Ann. Rev. Biochem.*, **34**, 209 (1965).

Amos Neidle and Heinrich Waelsch

HISTORY OF BIOCHEMISTRY

Modern biochemistry, which deals with the chemistry of living things, may be said to have had its beginnings about 1828 when Wohler synthesized urea and it became evident that the same methods that were applied in the area of "mineral" chemistry could be used with profit to investigate the chemistry operating in living organisms. Up to that time "organic chemistry" dealt with the chemistry of organisms but was considered a somewhat hopeless science because of the supposed existence in living things of special vital forces.

No competent biochemist of today would assert that all the mystery has been removed from the process whereby a fertilized hen's egg is transformed into a baby chick, but all biochemists and other scientists agree that scientific exploration, on and on, has yielded extraordinary information about the mechanisms involved, and will continue to do so.

For about seventy-five years after Wohler's discovery, biochemistry, though not traveling under that name, was advancing mainly by building up an increased knowledge of the chemistry of natural substances of living origin: carbohydrates, fats, proteins and other essential constituents of living systems.

About the beginning of the twentieth century, biochemistry began to be called by this name and concurrently to develop along new lines. The organized biochemistry of the later half of the nineteenth century, where it existed, was largely mothered by medical schools; it was called physiological chemistry and dealt predominantly with digestive juices, blood and urine. Advances in the chemistry of CARBOHYDRATES, PROTEINS, etc., were made by organic chemists of whom Emil Herman Fischer was an oustanding leader.

With the turn of the century, there began to emerge the realization that subtle agents, often in minute amounts, were crucially important in biochemistry: enzymes, vitamins, trace elements and hormones. There also came the appreciation that biochemistry has many applications and implications other than with respect to medicine and should be encouraged to stand on its own legs, independent of medical schools.

For about fifty years of the twentieth century, advances in biochemistry centered around the investigation of the nature of many of the agents mentioned above, and something as to how they function. During these years a vast amount of information was accumulated with respect to enzymes—these often involve vitamins and trace elements—and their functioning in metabolism. During these years it came to be appreciated that there is a tremendous biochemical unity in nature inasmuch as the very same amino acids, vitamins, sugars, fat acids, nucleotides and minerals enter into the metabolic functioning of living things of all kinds from the single-celled organisms (or those that appear outwardly to have no cells, like slime molds) up to the most complex multicellular organism—man.

At about the beginning of the last half of the twentieth century, biochemists began to take seriously other subtle entities: VIRUSES, BACTERIOPHAGES and GENES. About the same time, new tools, notably CHROMATOGRAPHY and the ELECTRON MICROSCOPE were developed and were indispensable to the advances which ensued, including the signal advances in our knowledge of intracellular structures, including MULTIENZYME COMPLEXES, and the outstanding advances in BIOCHEMICAL GENETICS.

There are two notable gaps in our biochemical knowledge, namely, (1) how HORMONES function (biochemically) and (2) how (biochemically) the process of differentiation takes place (see DIFFERENTIATION AND MORPHOGENESIS). Definitive information on these matters is almost nonexistent. To cite one instance related to hormones: our knowledge of the chemistry of the SEX HORMONES is relatively advanced; our knowledge as to what they do or how they function biochemically is practically completely lacking!

There is also a developing frontier in biochemistry which has to do with how mental processes, memory, etc., take place. Substantial advances in this area are very much in the future.

There are those in biochemistry who point out, and perhaps decry, our weakness in many areas of *applied* biochemistry. We know about hormones, but we know relatively little about how to use them; we know about the vitamins themselves but we know little about how to help people with them, partly, in both cases, because BIOCHEMICAL INDIVIDUALITY has been explored so little. Further interest in hormones and a resurgent interest in nutrition may be predicted because of the potential value of hormones and nutrients in the promotion of health.

Biochemistry, from the standpoint of its organization and sponsorship, has undergone revolutionary changes in recent years. It is obvious that in many fields which are traditionally *outside* biochemistry—microbiology, botany, entomology, genetics and even psychology and psychiatry—cannot advance without the prominent participation of investigators who are in fact biochemists (versed in the chemistry of living things) whether or not they are so designated. Someone has suggested that all investigators in these areas should in effect have biochemical "hunting licenses." To use a slang expression, biochemistry has, organizationwise, grown "too big for its britches," and something must be (and has been) done about it.

One device is to think up designations which may be more appealing than the term biochemistry, which may, after a time, seem commonplace. Biophysics, molecular biology, biochemical genetics, psychobiochemistry are such terms. It is hoped, however, that the seeming fragmentation of biochemistry will not result in a further fragmentation (departmentalization) in the organization of universities. Something is to be said for departmentalizing teaching operations only and centering every research program around an individual investigator who will investigate as he sees fit.

In any event, in coming decades one can be sure that many developments in the area of biochemistry (the chemistry of living things) will not be *called* biochemistry nor will they be contributed by individuals who call themsleves "biochemists." This is healthy and as it should be.

ROGER J. WILLIAMS

HODGKIN, D. C.

Dorothy Crowfoot Hodgkin (1910–) is a British biochemist who received the Nobel Prize for chemistry in 1964 for her oustanding contributions to the chemistry of VITAMIN B_{12}.

HOMEOSTASIS

Definition. Homeostasis is said to be shown by a physiological system if, given a moderate disturbance that tends to displace the system from its normal values, the parts so react and interact that the harmful effects of the disturbance are much diminished. Here are three examples: (i) When a man is much chilled, the cooling stimulates a mechanism in the base of the brain that sets him shivering. The muscular activity generates heat, which *opposes* the chilling. (ii) Sudden hemorrhage causes a sharp fall in blood pressure. The fall, however, causes the arterioles to constrict, thus lessening the amount of fall. (iii) The tissues need energy, so glucose has to be brought to them continuously by the bloodstream. If the amount of glucose in the blood should fall to a harmfully low level, the fall stimulates the suprarenal glands to secrete epinephrine (ADRENALINE). When epinephrine arrives at the liver, it makes the liver convert some of its store of GLYCOGEN into glucose, which is passed into the blood; so the fall is opposed.

As early as 1878, Claude Bernard[1] had realized that if an organism is to survive against the attacks and disturbances that a harsh environment can inflict on it, the organism *must* react to the disturbance in such a way that its response decreases, rather than augments, the threatening injury. His words have become classic: "Tous les mécanismes vitaux, quelques variés qu'ils soient, n'ont toujours qu'un but, celui de maintenir l'unité des conditions de la vie dans le milieu intérieur." But he spoke essentially as a philosopher, arguing from general principles.

In the years that followed, the experimental physiologists discovered fact after fact, all relating cause to effect (*e.g.*, an injection of epinephrine makes the blood pressure rise; chilling of the skin makes the small blood vessels constrict). The isolated facts accumulated in great numbers; then, in 1932, Walter B. Cannon[2] saw that a unifying theme ran through them. Often these cause-effect actions could be joined into a chain (each effect being itself the cause of the next action), and often these chains were cyclic (having "feedback"[3]), so that if a disturbance were started anywhere in the chain, the sequence of effects would eventually give one that directly affected the initiating disturbance [see METABOLIC CONTROLS (FEEDBACK MECHANISMS)]. He noticed that whenever such chains existed in the living body, the effect evoked by the disturbance always acted in the direction that opposed, or diminished, the disturbance. He called this theme "homeostasis" ("holding at the same, or at a constant, state"). Bernard had predicted that the effects *must* act so; Cannon showed, from the physiologists' own evidence, that the effects did *in fact* do so. Today, the concept is fundamental in providing a unifying principle that runs throughout the organism's vital activities.

Homeostasis Generalized. Cannon's examples were all taken from the vegetative, internal systems of the higher forms, chiefly the mammals, but he saw that the principle must still be applicable to the wide-ranging activities of the free-living animal. It applies, for instance, when the drying animal is forced, by its instinctive reactions, to search for water; it still applies when a man goes through complicated behavior to earn his daily bread. It will also apply at the biochemical level, so far as disturbances at that level threaten survival.

As example, consider the preservation in the tissues of a constant alkalinity (a constant concentration of hydrogen ions)—of the highest importance to many of the tissue's essential biochemical processes. This concentration may be disturbed by many activities and injuries: by carbonic acid produced by muscular activity, by carbonic acid retained by failure of the bloodstream to carry it away, by various other acid-producing or acid-absorbing processes, even by the swallowing of acid poisons. Such disturbances are potential threats to survival, so we are not surprised to discover that the living organism reacts to increasing acidity in ways that ultimately make the acidity decrease. Thus, increasing acidity makes the animal breathe more vigorously, rather than less; then more carbon dioxide is breathed out, and the amount of carbonic acid decreases. Other changes, more complex chemically, join in and also help decrease the acidity. All these processes are homeostatic at the biochemical level.

Positive Feedback. It must not be expected, however, that *every* biochemical event is a direct part of a homeostatic process. Sometimes the organism needs, not constancy but change; muscles, for instance, are useful precisely because they can change suddenly from the slight contraction of resting tonus to the violent activity of contraction at full force. Such a change is helped by *positive* feedback, by the circular chain of causes and effects being such that a little disturbance starts up events that *add* to the disturbance (instead of subtracting from it), making it even larger. An effective explosive must have this property. At the chemical level, the phenomenon of *autocatalysis* is of this type: the formation of a little of the product speeds up the formation of more of it. Such a process is the antithesis of homeostasis.

The two types of response are each seen to be appropriate when one asks: The variable that is affected—does survival demand its constancy or its vigorous change? When its constancy is demanded, then the reaction must be homeostatic. But when its vigorous change is demanded, then the reaction must be of the opposite type. In that case, the reaction will always be found to be acting as a means to an end: the end is constancy of some essential variable (as the animal searching for water is acting to keep its water content constant), but the means may involve inconstancy (as the same animal alternately contracts and relaxes its muscles to move it to the water).

References

1. BERNARD, CLAUDE, "Leçons sur les phénomènes de la vie," Paris, Baillière, 1878.
2. CANNON, WALTER B., "The Wisdom of the Body," New York, Norton, 1939.
3. ASHBY, W. ROSS, "An Introduction to Cybernetics," New York, John Wiley & Sons, 1960.
4. GRODINS, FRED S., "Control Theory and Biological Systems," New York, Columbia University Press, 1963.

W. ROSS ASHBY

HOMOGENTISIC ACID

Homogentisic acid, an intermediate in the metabolism of TYROSINE and PHENYLALANINE, has the following formula:

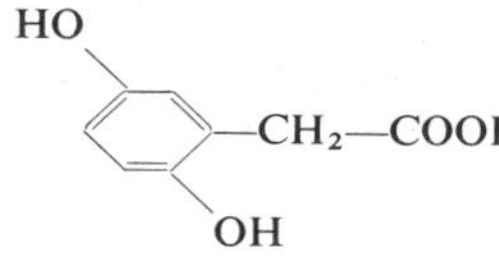

HO

—CH_2—COOH

OH

Like other *p*-diphenols (*e.g.*, hydroquinone) it is extremely easily oxidized in alkaline solution, and when excreted in the urine, it causes the development of a black color by oxidation with oxygen from the air, on standing. The excretion of homogentisic acid takes place in ALKAPTONURIA. The pathway of biosynthesis of homogentisic acid is indicated in the article AROMATIC AMINO ACIDS (BIOSYNTHESIS AND METABOLISM).

HOPKINS, F. G.

Frederick Gowland Hopkins (1861–1947), an English biochemist, was a pioneer in the investigation of nutrition. He first discovered that gelatin was an incomplete protein from the nutritional standpoint because it lacked TRYPTOPHAN. The idea of nutritionally essential AMINO ACIDS stems from his work. He also did pioneer work in the field of vitamins before that term was coined, and postulated in lectures as early as 1906 that rickets (see VITAMIN D) and scurvy (see ASCORBIC ACID) might be deficiency diseases. He shared with CHRISTIAAN EIJKMAN the Nobel Prize in physiology and medicine in 1929.

HOPPE-SEYLER, E. F. I.

Ernst Felix Immanuel Hoppe-Seyler (1825–1895), a German, was one of the earliest biochemists or physiological chemists. His contributions include the crystallization of hemoglobin, the discovery of invertase (α-glucosidase) and lecithin; he did some of the earliest work on what are now known as nucleic acids. He founded the journal which for many years bore his name, Hoppe-Seyler's *Zeitschrift fur Physiologische Chemie.*

HORMONES, ANIMAL

A hormone is a chemical secreted by an endocrine (ductless) gland whose products are released into the circulating fluid. Hormones are regulators of physiological processes by chemical means. Hormones produced by one species usually show similar activity in other species. This is true even though there may be species differences in the chemical composition of the hormones produced. The hormones showing greatest species specificity are PROTEINS or conjugated proteins. However, highly purified preparations of even these hormones are active in a variety of species.

The chemical nature of hormones varies widely. Many are STEROIDS (estrogen, progesterone, aldosterone, cortisone), others are AMINO ACIDS (thyroxine and similar analogues), others are polypeptides (vasopressin), some are proteins of low molecular weight (insulin), and still others are conjugated proteins (follicle stimulating hormone, for example, is a glycoprotein). The polypeptide and protein nature of many hormones has made them difficult to isolate and purify. As mentioned above, these hormones are the ones which show greatest species differences in chemical composition. Amino acid and steroid hormones have been isolated and studied chemically. This has enabled them to be manufactured, and synthetic hormones are generally available for medical purposes.

Hormones are chemicals which help regulate over-all physiological processes such as metabolism, growth, reproduction, metamorphosis, molting, pigmentation, and electrolytic and osmotic balance. The regulation of such gross functions of the endocrine system is accomplished either directly or indirectly. This has been demonstrated in studies of the control of the vertebrate pituitary gland and the control of the molting glands of insects. Immediate effects of the central nervous system on hormonal secretions have been shown in the regulation of pigmentation changes in crustacea and lower vertebrates. Similarly, direct neural control as well as chemical changes in the organs themselves stimulate the production and release of gastric hormones (pancreozymin, gastrin) which, in turn, cause the release of digestive enzymes. Neural control may operate either by direct action of nerve impulses on hormone-producing cells, as in the digestive tract, or more commonly indirectly by formation of substances by cells of the central nervous system. Neurosecretory substances may be transported to a target gland by movement within the axons of the nerve cells. Alternatively, the neurosecretory substances may be transported to the target gland by the circulatory system, and the nervous system itself becomes an endocrine gland. In either case, the adaptive efficiency of both systems is obvious: the nervous system receives the stimulus of the ever-changing environment and coordinates metabolic adjustments within the organism to give adaptive responses to that environment.

Hormones often act on target cells as complex chemical macromolecules formed by combinations with various molecules in plasma or within the target cells themselves. Thus, THYROXINE appears in the plasma bound to PLASMA PROTEINS, and estrogen has been found bound to proteins of liver and uterine cells. The broad metabolic activities of hormones are, therefore, dependent on special chemical combinations for activity at the cellular level. Hormones also may form associations with enzymes and thereby affect the rate of reaction or even the substrate utilized. This appears to be true for the effects of some steroids on some amino acid dehydrogenases.

The wide metabolic effects of hormones are produced by their action on specific biochemical pathways. Not all of the pathways affected are known even for a single hormone, and many hormones have not yet been identified as definitely affecting any pathway. General effects can be observed when a single enzyme in a single pathway is affected by a hormone. However, it is probable that most hormones affect more than a single biochemical reaction and also that they affect more than one biochemical pathway in their target cells. It is also probable that some apparently direct effects on enzyme activity are in reality due to effects on the synthesis by DNA of messenger RNA and, hence, effects on protein synthesis which increase the concentration of a given enzyme. Some of the pathways affected by hormones are summarized below.

Metabolic Areas Influenced by Hormones. Phosphorylation of glucose is influenced by INSULIN and the GROWTH HORMONE. Insulin accelerates phosphorylation while growth hormone depresses it. These hormones also act at the cell membrane on ACTIVE TRANSPORT mechanisms and thereby affect the permeability of the cell to glucose. Insulin increases glucose uptake by cells, growth hormone decreases it. Both hormones also affect other enzymes in carbohydrate pathways, for insulin increases the store of GLYCOGEN in the liver while growth hormone decreases it.

The breakdown and utilization of glycogen are increased in liver and muscle by epinephrine. This hormone increases the concentration of active phosphorylase. Glucagon, a hormone thought to be formed by the alpha cells of the pancreas, has a similar effect but apparently acts on liver phosphorylase but not on muscle phosphorylase. ADRENAL CORTICAL HORMONES increase liver glycogen by increasing the synthesis of glucose from other sources, such as protein. This is correlated with an increased excretion of nitrogen from amino acids.

The activity of succinic dehydrogenase is depressed by cortisone while THYROID HORMONE increases the rate of oxidation of both carbohydrates and fats by increasing the level of cytochrome c. Thyroid hormone also uncouples oxidative PHOSPHORYLATION and electron transfer. This may make substrates of other enzyme systems more readily available for synthetic processes. Thyroid hormone causes swelling of the mitochondria and probably, therefore, can affect all the oxidative enzymes and substrates present in them. Tissues show varying degrees of sensitivity to thyroid substances. Concentrations which stimulate growth in one tissue will depress or even cause regression in another. Thus, concentrations of thyroid hormone which cause METAMORPHOSIS in tadpoles will hasten limb bud formation while at the same time causing tail regression. It is interesting that diethylstibestrol, a synthetic compound with estrogenic activity, also causes mitochondrial swelling while the natural estrogen, estradiol, does not have this effect (see OVARIAN HORMONES).

As indicated above, processes which affect carbohydrate metabolism also affect fat and protein metabolism. Thyroid increases the utilization of fats for energy as do corticosteroids, epinephrine and growth hormone. The synthesis of fats is increased by insulin, while in the absence of this hormone fats are used metabolically for energy and as a source for increased cholesterol formation (see FATTY ACID METABOLISM). These processes are independent of the effects of insulin on glucose utilization.

Adrenal cortical hormones can increase the breakdown of proteins by increasing the amounts of intracellular peptidase formed. Corticosterones probably also inhibit specific pathways in protein synthesis. Thyroid hormone increases incorporation of amino acids and carbohydrates into new proteins presumably by affecting the rate of synthesis of RIBONUCLEIC ACIDS from diphosphate nucleotides. TESTOSTERONE increases protein synthesis in perineal muscle but at the same time it increases protein breakdown in diaphragm muscle. These diverse effects of a single hormone on striated muscle illustrate the insufficiency of terms such as "anabolic" or "catabolic steroid." Growth hormone increases protein synthesis by increasing the rate of amino acid incorporation into RNA-protein. It also increases the activity of L-gulono-γ-lactone hydrolase in liver fractions presumably by its effects on protein synthesis. Insulin affects protein metabolism by its effects on glucose metabolism which provides the necessary energy for synthetic processes. In addition, insulin appears to increase the transport of amino acids into muscle cells.

The above discussion of hormonal reactions on carbohydrate, protein and fat metabolism are not exhaustive but are intended to give a concept of the interrelations between some of the hormones by pointing up their roles in metabolic systems. Hormones can also act to rearrange the metabolism of the cell to form specific substances. These substances may be other hormones or secretions of exocrine (duct) glands. Such substances are not present unless the hormone responsible for their secretion is present. The nervous systems of insects and the pituitary glands of vertebrates both secrete "trophic" hormones. In each case, the target gland produces its special hormone only in the presence of the specific "trophic" hormones. Thus, estrogen is formed in the ovary if follicle stimulating hormone and luteinizing hormone from the anterior pituitary are present (see GONADOTROPIC HORMONES). This seems to indicate that hormones may help initiate synthetic pathways, again, by affecting the formation of messenger RNA from DNA. In a similar fashion, specific secretions may be formed in duct glands (*e.g.*, the seminal vesicle) if the appropriate hormone (*e.g.*, testosterone) is present. Here the hormone either stimulates an existing nonfunctioning metabolic pathway by affecting enzyme activity or helps initiate a new pathway as suggested above. Increases in the intracellular concentration of some enzymes, either by synthesis or by activation of inactive complexes of enzymes, and the changing of metabolic reactions to make special substrates available are possible additional mechanisms of hormone action. Testosterone increases the concentration of succinic dehydrogenase in the seminal vesicle; estrogen activates inactive isocitric dehydrogenase in the placenta; thyroid hormone increases the substrates available for synthesis by uncoupling oxidative phosphorylation and electron transfer. (See also ADRENOCORTICOTROPIC HORMONES; THYROTROPIC HORMONE; SEX HORMONES.)

Electrolyte balance and osmotic equilibrium are related processes in living organisms (see ELECTROLYTE AND WATER REGULATION). In vertebrates, and especially the higher vertebrates, these processes are regulated by hormones. Arthropods also have hormones which affect these two physiological functions particularly at the time of molting when the relatively impermeable exoskeleton is lost. In mammals, sodium reabsorption and POTASSIUM excretion in the tubules of the kidney are increased by aldosterone and deoxycorticosterone. Aldosterone is the most effective in this respect. The increased concentration of sodium thus produced in the tissues causes an increased retention of water there. Thyroid hormone increases phosphorus excretion in the urine. In hyperthyroid organisms, this phosphorus loss is also accompanied by calcium loss. Calcium and phosphorus reabsorption are increased by growth hormone (see CALCIUM; PHOSPHATE IN NUTRITION). PARATHYROID HORMONE increases phosphorus loss in urine while increasing serum calcium by removing calcium phosphate from bone. The loss of glucose in the urine of the diabetic indicates that insulin indirectly affects osmotic balance by affecting glucose utilization. When the serum glucose reaches a certain level, glucose filtered through the glomerulus cannot be reabsorbed by the tubules of the kidney (see RENAL TUBULAR FUNCTIONS). This causes increased osmotic pressure in the tubules which, therefore, decreases the total amount of water that can be reabsorbed.

As can be seen from the foregoing, hormones affect cell permeability. This may occur by facilitating the entrance of certain metabolites, as has been shown in the case of glucose transport. The presence of such metabolites in increased amounts would then increase the metabolism of cell in those pathways in which the concentrations of the metabolites are rate limiting. It is noted that in these cases the hormone is not specifically affecting the metabolic pathway itself. Because hormones affect the resting potential or threshold potential of an irritable tissue, they must therefore affect cell permeability. Such an effect has been demonstrated in uterine smooth muscle where estrogen decreases the threshold and therefore increases the rate of contraction. Oxytocin, a principle isolated from the posterior pituitary, also increases contraction in smooth muscle by decreasing the threshold. Other hormones also affect irritability since irritable tissues are sensitive to ionic concentrations both inside and outside their cells. Thus, muscular weakness is a common symptom of insufficiency of the adrenal cortex. Effects of hormones on cell irritability also may be found in the central nervous system. The display of estrous behavior of females in lower mammalian groups is induced by injections of estrogen in spayed animals. This hormone produces increased motor activity in the brain and spinal cord. Not only do hormones affect the permeability of the cell membrane but they probably change the permeability of intracellular compartments also. The effect of thyroid hormone on mitochondria has been cited.

Hormones may affect capillary permeability as well as cell permeability. Such changes may occur simultaneously with vasomotor changes or they may occur independently. Epinephrine (see ADRENALINE) is a vasoconstrictor in most tissues (not muscle), and estrogen affects both vasomotor control and permeability in connective tissue. It has been found to decrease the spread of local infections.

The growth and metabolism of skeletal tissues are affected by hormones. These effects are obtained by all the mechanisms of activity listed above. In addition, particular systems may be especially sensitive to hormone action. Growth hormone increases bone growth by acting on the epiphyseal plate tissue of developing BONE. Estrogen inhibits growth at the epiphyseal plate and causes fusion of the epiphyses with the diaphyses of the long bones. Thus, after the first estrous cycle there is a gradual decrease in bone growth in female mammalia. There are also strong hormonal influences in skeletal growth and molting in crustacea and insects. Here, the molt of the exoskeleton, the metabolism before, during and after the molt, and the deposition of the new skeleton are controlled by hormones. The changes involved include water uptake, increased oxygen consumption and increased conservation of salts, especially calcium. A summary of hormone names, the organisms that produce them, the glands that produce them, the stimulus for their production and their general effects is presented in Table 1.

TABLE 1. ANIMAL HORMONES

Hormone	Found in	Where Secreted	Stimulus for Production	Gross Effect of Hormone
Growth hormone (STH)	Vertebrates	Adenohypophysis (anterior pituitary)	Hypothalamus secretions carried via pituitary portal blood vessels regulate secretion	Stimulates growth, antagonizes action of insulin
Adrenocorticotrophic hormone (ACTH)	Vertebrates	Adenohypophysis	Same as STH	Stimulates secretion by adrenal cortex
Follicle stimulating hormone (FSH)	Vertebrates	Adenohypophysis	Same as STH	Stimulates growth of ovarian follicles, estrogen secretion
Luteinizing hormone (LH)	Vertebrates	Adenohypophysis	Same as STH	Stimulates ovulation, and corpus luteum growth, estrogen secretion
Luteotrophic hormone (LTH)	Vertebrates	Adenohypophysis	Same as STH	Stimulates secretion by corpus luteum
Thyrotrophic hormone (TSH)	Vertebrates	Adenohypophysis	Same as STH	Stimulates secretion by thyroid
Insulin	Vertebrates	Islets of Langerhans of pancreas, β-cells	Blood sugar changes	Increases glucose metabolism
Glucagon	Vertebrates	Islets of Langerhans, α-cells	Blood sugar changes	Antagonizes action of insulin
Thyroid hormone	Vertebrates	Thyroid gland	Thyrotrophic hormone	Increases metabolic rate
Corticosterones	Vertebrates	Adrenal cortex	ACTH	Antagonizes action of insulin on glucose metabolism
Aldosterone	Vertebrates	Adrenal cortex	ACTH	Electrolyte balance
Epinephrine	Vertebrates	Adrenal medulla	Stress stimuli	Glucose metabolism, vasocon strictor

(TABLE 1—continued)

Hormone	Found in	Where Secreted	Stimulus for Production	Gross Effect of Hormone
Estrogen	Vertebrates	Ovary, follicle	FSH, LH	Growth of uterus
Progesterone	Vertebrates	Ovary, corpus luteum	LH, LTH	Maintenance of pregnancy
Testosterone	Vertebrates	Interstitial cells of testis	LH	Growth and secretion male accessory glands
Parathyroid hormone	Vertebrates	Parathyroid glands	Changes in blood Ca^{++} concentration	Calcium and phosphorus balance
Anti-diuretic hormone (vasopressin)	Vertebrates	Hypothalamus via axons to posterior pituitary (neuro-pophysis)	Osmotic changes cause neuro-secretory cells to form it	Increases water reabsorption in tubules of kidney
Oxytocic hormone	Vertebrates	Hypothalamus via axons to neuropo-physis	Unknown	Causes contraction of smooth muscle, probably helps terminate preg-nancy
Intermedin	Vertebrates	Intermediate lobe of pituitary gland	Neurosecretory cells?	Causes chromato-phore changes, pro-duced lightening and darkening of body color
Gastrin	Vertebrates	Stomach	Neural and chemi-cal changes in stomach	Causes secretion by stomach enzymes
Secretin	Vertebrates	Duodenum	Neural and chemi-cal changes in duodenum	Causes secretion of pancreatic HCO_3^-
Pancreozymin	Vertebrates	Duodenum	Same as secretin	Causes secretion of pancreatic enzymes
Cholecystokinen	Vertebrates	Duodenum	Chemical changes in duodenum (fats)	Causes gall bladder contraction
Corpus cardiacum hormone	Insects	Nervous system via axons to corpus cardiacum	Environmental and internal changes produce neural secretion	Trophic hormone to prothoracic gland, affects molting
Ecdysone	Insects	Prothoracic gland	Trophic hormone or corpus cardiacum	Controls molting, stimulates meta-morphosis
Juvenile hormone	Insects	Corpus allatum gland	Neural stimuli cause secretion	Promotes develop-ment of larval characters; causes ovarian develop-ment in some insects
Sinus gland hormone	Crustacea	Neurosecretory cells via axons to sinus gland	Internal and external environ-mental changes	Inhibit molting, also stimulates ovarian growth
Molting hormone	Crustacea	Y-gland	"Trophic" hormone from neuro-secretory cells	Increases molting rate
Chromatophore lightening and darkening hormones	Crustacea	Eyestalk glands and Postcomissure organs	Neural stimuli cause secretion directly	Cause lightening or darkening of body in response to environment
Androgenic gland	Crustacea	Gland attached to vas deferens	Unknown	Causes develop-ment of testis and male characteristics

References

1. Pincus, G., and Thimann, K. V., "The Hormones," New York, Academic Press, 1955, 5 vols.
2. Gorbman, A., and Bern, H. A., "A Textbook of Comparative Endocrinology," New York, John Wiley & Sons, 1962.
3. Turner, C. D., "General Endocrinology," Philadelphia, Saunders, 1966.
4. Barrington, E. J. W., "An Introduction To General And Comparative Endocrinology," Oxford, Clarendon Press, 1963.
5. Krahl, M. E., "The Action of Insulin on Cells," New York, Academic Press, 1961.
6. Tepperman, J., and Tepperman, H. M., *Pharmacol. Rev.*, **12,** 301–353 (1960).
7. Hoch, F. L., *Physiol. Rev.*, **42,** 605–673 (1962).
8. Randle, P. J., *Ann. Rev. Physiol.*, **25,** 291–324 (1963).
9. Tomkins, G. M., and Maxwell, E. S., *Ann. Rev. Biochem.*, **32,** 677–708 (1963).

Darhl Foreman

HORMONES, PLANT

A plant hormone or "phytohormone" is defined as an organic compound produced naturally in plants, controlling growth or other functions at a site remote from its place of production, and active in minute amounts. Three chemically quite different types of compounds apparently act as plant hormones, the *auxins*, the *gibberellins* and the *kinetins*. In addition, the growth of roots is dependent on VITAMINS of the B group which are synthesized in leaves and transported thence to the roots, thus qualifying as hormones. The status of ethylene as a hormone is discussed below.

Auxins. The best studied hormones are those belonging to the class of *auxins*. These are defined as *organic substances which promote growth along the longitudinal axis, when applied in low concentrations to shoots of plants freed as far as practical from their own inherent growth-promoting substances.* Auxins generally have additional properties but this one is critical. "Low concentrations" is usually taken to mean below $M/1000$. The most widely occurring auxin is indole-3-acetic acid (IAA), which has been isolated in pure form only from fungi and from corn (maize) grains, but its presence has been conclusively demonstrated by biochemical tests or chromatography in a wide variety of flowering plants, including mono- and dicotyledons, as well as in several algae and fungi. The related indole-3-acetaldehyde (IAAld) occurs in a number of etiolated seedlings and in pineapple leaves; indole-3-acetonitrile (IAN) has been isolated pure from cabbage, and its presence indicated in several other plants. These last two act as auxins because they are converted in many plants to indoleacetic acid, the former by an aldehyde dehydrogenase, which occurs widely, the latter by a special enzyme (nitrilase) which has been found only in the leaves of the cereal (grass), cabbage, and banana families. There may be other naturally occurring auxins than indoleacetic acid (IAA), but in spite of many suggestive indications, none has been identified.

Many synthetic auxins have been produced, including 2,4-dichlorophenoxyacetic acid or "2,4-D," naphthalene-1-acetic acid, 2,3,6-trichlorobenzoic acid and others. By definition, these synthetic compounds are not hormones, although they are sometimes loosely referred to as "hormone-type compounds," but they are auxins and their actions resemble those of indoleacetic acid to varying extents. In what follows, the terms "natural auxin" and "IAA" (indole-3-acetic acid) will be used interchangeably, because (a) the identification has been made by chemical or biological tests in a number of instances, (b) the effects of the natural auxin can usually be duplicated by synthetic IAA.

An auxin is formed in *fruits, seeds, pollen, root tips, coleoptile tips*, young *leaves* and especially in *devoloping buds*. It travels away from the site of production in shoots by a special transporting system, dependent on oxygen, which moves it in a predominantly *polar* direction from apex toward base. Movement in the opposite direction, *i.e.*, from base toward apex, takes place to a variable extent depending upon the tissue and the plant; in the *Avena* coleoptile, which is widely used in critical biological work, such anti-polar movement is by diffusion only. In the course of the polar transport, a large part of the IAA becomes bound and is no longer transportable. The transport is rather specifically inhibited by related compounds, especially 2,3,5-triiodobenzoic acid. 2,4-D and other synthetic auxins are transported either more slowly or to a much smaller extent in the polar system. Auxin applied artificially to intact plants, whether IAA or other compounds, can travel rapidly upward by penetrating into the conducting tissues of the wood where it is carried upward in the transpiration stream.

In its normal polar, downward movement the auxin stimulates the cells below the tip to elongate and sometimes to divide also. Specific tissues, notably the cambium, are caused to divide laterally by auxin coming from the developing buds, which accounts for the wave of cell division

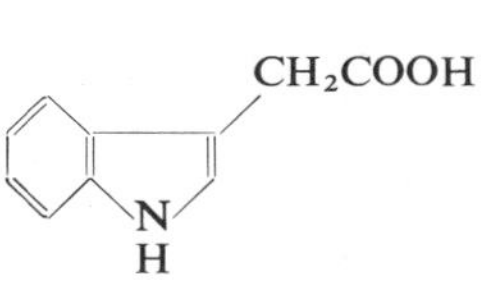

CH_2COOH / N H — IAA

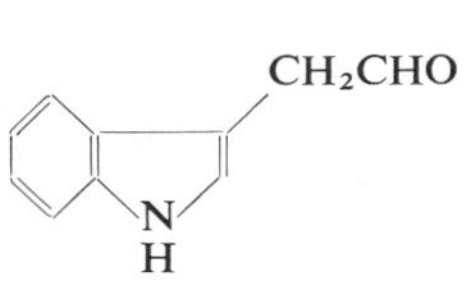

CH_2CHO / N H — IAAld

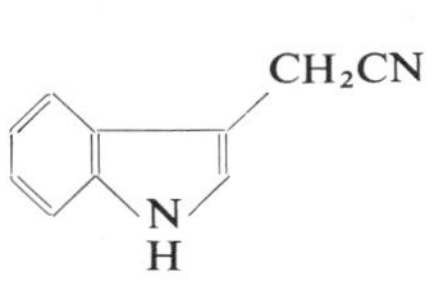

CH_2CN / N H — IAN

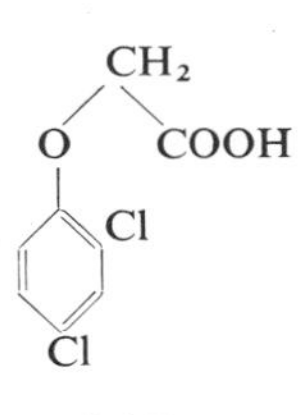

CH_2, O, COOH, Cl, Cl — 2,4-D

occurring in tree trunks in the spring. Stimulation of other stem cells to divide leads to the production of root initials, which grow out as lateral roots. Cells of the young ovary are commonly caused to multiply and enlarge so that an apparently normal fruit is produced without requiring pollination ("parthenocarpic fruit"). This last phenomenon indicates that the growing seeds normally secrete an auxin to which enlargement of the fruit is due, a conclusion which has been directly confirmed by bioassay in several fruit types. Gibberellins can also cause enlargement of fruit (see below). On reaching the lateral buds, however, auxin inhibits their elongation into shoots, and this accounts for "apical dominance," *i.e.*, suppression of the growth of lateral buds by the terminal bud of a shoot. Auxin also inhibits the falling off of leaves or fruits, which normally occurs when they are mature or aged, by the formation of an "abscission layer" of special cells whose walls come apart. That the leaves or fruits do not absciss earlier is due to their steady production of an auxin, which prevents formation of these cells. In the root, auxin inhibits elongation except in very low concentrations, but its level therein is usually low. Auxin can be transported for a short distance from the root apex toward the base, but the transport is not fully polar and in the more basal parts of the root the transport is slight.

When the shoot is placed horizontal, auxin is transported toward the lower side, causing accelerated growth there and hence upward curvature (*geotropism*); in the root, this causes decreased growth on the lower side and hence downward curvature. However, in the downward geotropic curvature of roots other phenomena appear to enter in, and the complexities are not yet resolved. When shoots are illuminated from one side, auxin accumulates on the shaded side and the plant therefore curves toward the light (*phototropism*). Both geotropic and phototropic auxin movements have recently been confirmed with carboxyl labeled C^{14}-IAA. The first observed effect when auxin is applied is the acceleration of the streaming of cytoplasm, but acceleration of growth begins in 7–14 minutes at 23°C. The growth promotion is inhibited by SH reagents like iodoacetate or *p*-chloromercuribenzoate, by cytochrome inhibitors like azide, cyanide or CO in the dark, and by uncouplers like DNP and arsenate. It is, therefore, closely linked to oxidative PHOSPHORYLATION. Recently it has been found also to be inhibited by chloramphenicol, puromycin and actinomycin D, at concentrations identical to those which inhibit the incorporation of labeled amino acids into protein, and this is interpreted to mean that enzymes modifying the cell wall or membrane are synthesized in response to auxin.

In some plants which flower on short days, auxin may inhibit flowering; in plants which flower on long days, however, if close to the transition from the vegetative to the flowering state, auxin may promote flowering. In hemp and some of the squashes, auxin modifies the sexuality of the flowers toward femaleness. In the special case of pineapple, auxin directly causes flowering in an unusually clear-cut and quantitative response.

The principal uses of synthetic auxins are to promote the formation of roots on stem cuttings, to prevent abscission, especially of apples and pears, to induce flowering in pineapples, and occasionally to produce seedless fruits; the largest use, however, is as weed-killers, and depends on the fact that, at concentrations 100–1000 times those occurring naturally, auxins are highly toxic. Monocotyledonous plants, inexplicably, are usually resistant. 2,4-D is the most widely used auxin for this in America, while 2-methyl,4-chlorophenoxyacetic acid ("methoxone") is often used in Europe. Some chemically related compounds antagonize the action of auxins, *e.g.*, relieving the inhibition of root growth caused by 2,4-D. On the other hand, 2,3,5-triiodobenzoic acid synergizes the action, though as noted above it interferes with auxin transport. The auxins are thus a powerful group of tools both for the scientific study of all aspects of plant growth and for the practical production of crops.

Gibberellins. The *gibberellins* were originally isolated from the parasitic fungus *Gibberella fujikuroi*, which causes excessive leaf elongation in rice plants. This and other effects are due to a family of closely related substances which are excreted into the growth medium. Up to now structures have been determined for 11 of them and five have actually been isolated from higher plants. The structure of three fused saturated or nearly saturated rings, with two additional rings perpendicular to them, suggests relationship to the diterpenes, for which there is now strong isotopic evidence; C^{14}-kaurene is readily converted to gibberellic acid (GA_3) by *Gibberella* cultures. The biosynthesis is apparently inhibited by chlorocholine, which is believed to be the basis for the dwarfing action of this compound. GA_7 has the highest activity in most tests. In the following structures, S indicates a saturated ring:

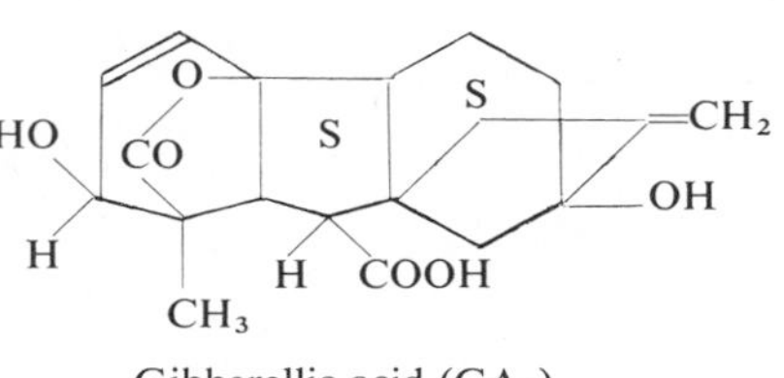

Gibberellic acid (GA_3)

GA$_7$

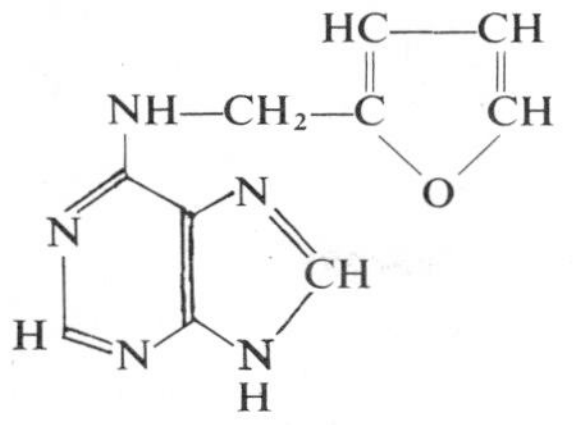

Kinetin, 6-Furfurylaminopurine

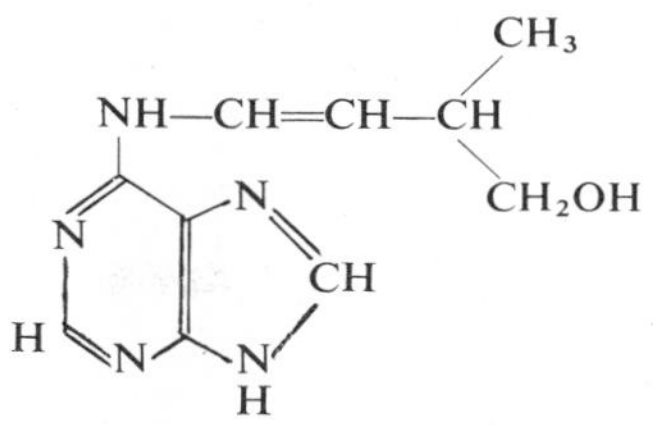

Zeatin, 6-(4′-Hydroxyisopentenyl) aminopurine

Gibberellins cause rapid elongation of shoots; many of the dwarf forms of corn, peas, beans and morning-glory are caused to grow into tall forms indistinguishable from their tall genetic relatives. Many long-day plants are brought into flower in short days by gibberellin, and some biennials, including *Hyoscyamus* (henbane), are made to flower in one year. This process depends on the activation of cell divisions in the shoot apex. Like auxins, gibberellins produce parthenocarpic fruits, especially on tomato; unlike auxins, they do not inhibit lateral bud development, but they inhibit rooting of cuttings and promote the germination of many seeds. Their transport shows no polarity. They are active at concentrations comparable to those of auxins. There is good evidence, indeed, that they act only when auxin is present.

Kinetins. These are less thoroughly known. The first one to be discovered, produced by autoclaving yeast nucleic acid, is 6-furfurylaminopurine, and the other formula shown is that of zeatin, recently isolated from immature corn kernels and believed to occur in other plant materials. The kinetins promote cytokinesis and protein synthesis, thus causing aminoacids to accumulate where kinetins are synthesized (or externally applied) and maintaining the chlorophyll content of yellowing leaves. They antagonize auxin in apical dominance, releasing lateral buds from inhibition by a terminal bud or by applied auxin. Probably through the same mechanism, they promote the development of buds and leaves on tissue cultures. Their action is primarily local, and if there is transport *in vivo* it probably occurs mainly in the transpiration stream (where amino acids are also often found).

Ethylene. The production of ethylene in fruit tissue and in small amounts in leaves may justify its consideration as a new type of hormone, functioning in the gaseous state. *Cherimoyas* and some varieties of pear produce 1000 times the effective physiological concentration. Ethylene formation is closely linked to oxidation and may be centered in the mitochondria. Its effects are to promote cell-wall softening, starch hydrolysis and organic acid disappearance in fruits—the syndrome known as *ripening*. It also decreases the geotropic responses of stems and petioles.

Among the lower plants, hormones controlling the formation of reproductive organs probably occur in the algae *Acetabularia* and *Vaucheria* and perhaps in mushrooms. In the water-mold *Achlya*, the female plants secrete into the medium a substance causing male plants to form antheridia, and these in turn secrete another compound which causes the female to produce oogonia; the oogonia secrete a third compound which attracts the antheridia to them. In *Allomyces* a substance secreted by female plants to attract the male sperms has been isolated and named *sirenin*, $C_{15}H_{24}O_2$. These compounds could be considered "ectohormones."

References

1. Audus, L. J., "Plant Growth Substances," Second edition, London, Leonard Hill, 1959.
2. Crane, J. C., "Growth Substances in Fruit Setting and Development," *Ann. Rev. Plant Physiol.*, **15**, 303–326 (1964).
3. Jones, D. F., MacMillan, J., and Radley, M., *Phytochemistry*, **2**, 307–312 (1963).
4. Leopold, A. C., "Plant Growth and Development," New York, McGraw-Hill Book Co., 1964, 466 pp.
5. Leopold, A. C., "Auxins and Plant Growth," Berkeley, University of California Press, 1955.
6. Pilet, P. E., "Les Phytohormones," Paris, Masson et Cie, 1961.
7. Thimann, K. V., "Plant Growth Substances; Past, Present and Future," in *Ann. Rev. Plant Physiol.*, **14**, 1–18 (1963).
8. Thimann, K. V., in "Proceedings of the Laurentian Hormone Conference of 1964," New York, Academic Press, 1965.
9. Zweig, G., and Rappaport, L., "The Gibberellins: Chemistry and Action," New York, Academic Press, 1960.

Kenneth V. Thimann

HORMONES, STEROID

Steroid hormones are a class of organic compounds containing C, H and O, possessing the cyclopentanoperhydrophenanthrene nucleus. They are produced in endocrine glands such as the adrenal, gonads and placenta, and they exert their action at a distant site. The steroid hormones may be subdivided into androgens, estrogens, progestational substances and corticoids on the basis of their physiological activity.

Androgens are produced in all of the steroid-producing tissues and have the specific function of maintaining the male sex characters of the mammal such as the prostate and seminal vesicles. In man, other secondary sex characters such as facial hair and pitch of voice are controlled by androgens. In the fowl, androgens maintain the comb, wattles and spurs. In addition to sex-specific functions, androgens influence nitrogen metabolism and fat distribution, and to a limited extent, they exert control over electrolyte balance. Testosterone is the most active naturally occurring androgen. Other less active androgens include androst-4-ene-3,17-dione, dehydroepiandrosterone, and 11β-hydroxyandrost-4-ene-3,17-dione. Structurally, androgens possess 19 carbon atoms with oxygen substituents at carbons 3, 11 and 17.

The formation of androgens in the gonads is controlled by the pituitary GONADOTROPIC HORMONES, and the general sequence of biosynthesis in these tissues, the placenta and the adrenal, is (see STEROIDS for numbering system) cholesterol → 20α-hydroxycholesterol (or 22R-hydroxycholesterol) → 20α,22R-dihydroxycholesterol → pregnenolone → progesterone → 17α-hydroxyprogesterone → androst-4-ene-3,17-dione ⇆ testosterone. Only the testis produces significant amounts of testosterone. Adrenal androgens are formed from two additional pathways: pregnenolone → 17α-hydroxypregnenolone → dehydroepiandrosterone → androst-4-ene-3,17-dione and cortisol → 11β-hydroxyandrost-4-ene-3, 17-dione. Androsterone and etiocholanolone are the two principal catabolites of testosterone, androst-4-ene-3, 17-dione and dehydroepiandrosterone. Four 11-oxygenated 17-ketosteroids, 11β-hydroxyandrosterone, 11-ketoandrosterone, 11β-hydroxyetiocholanolone and 11-ketoetiocholanolone are the principal catabolites of 11β-hydroxyandrost-4-ene-3, 17-dione. Androgen biosynthesis may also proceed through pregnenolone-3-sulfate → dehydroepiandrosterone-3-sulfate → dehydroepiandrosterone (see also ANDROGENS; SEX HORMONES; TESTOSTERONE).

Estrogens are a class of C_{18} phenolic steroids where either one or two of the four rings in the structure is aromatic. Estrogens are produced in all steroid-producing tissues. The primary physiological action is maintenance of the female sex characters including growth of the vagina, uterus, mammary glands and Fallopian tubes. Other effects include fat distribution, influence on electrolyte balance (see ELECTROLYTE AND WATER REGULATION), CALCIUM metabolism, and BLOOD CLOTTING.

Estradiol-17β is the principal and most active estrogen and, together with estrone, the corresponding 17-keto derivative, is widely distributed in body fluids. Equilin and equilenin are representative ring B unsaturated estrogens and are associated mainly with the pregnant mare.

Estrogen production in the gonads is controlled by pituitary gonadotropic hormones, while adrenal estrogen formation is regulated by the pituitary ADRENOCORTICOTROPIC HORMONE. The pathway testosterone → 19-hydroxytestosterone → estradiol-17β probably accounts for the major amount of estrogens biosynthesized. Catabolic reactions lead to the formation of a dozen or more products of which estradiol-17β, estrone, and estriol may be considered to be major metabolites. Other catabolites include 2, 6α, 6β, 15α, 15β, 16α, 16β, and 16-keto derivatives of estrone and estradiol-17β.

Progestational hormones include progesterone and its 20-reduced derivatives, 20β-hydroxypregn-4-ene-3, 20-dione and 20α-hydroxypregn-4-ene-3, 20-dione, of which progesterone is the most active. These compounds are produced in all steroid-producing tissues and especially in the corpus luteum and placenta. Progesterone has a specific function on the vaginal and uterine epithelium and mammary glands and, in concert with estrogens, has a role in the maintenance of the female sexual cycle, whether estrus in the rodents or menstrual cycle in the primate. Progestational substances have a unique role in the maintenance of pregnancy. A second primary role of progesterone is to serve as an intermediate in the biosynthesis of all classes of steroid hormones.

The biosynthesis of progesterone in gonadal tissue is regulated by pituitary gonadotropins and by adrenocorticotropin in the adrenal. The pathway of formation is cholesterol → 20α-hydroxycholesterol (or 22R-hydroxycholesterol) → 20α, 22R-dihydroxycholesterol → pregnenolone → progesterone. Progesterone undergoes reductive catabolic reactions of the ketone groups and the nuclear double bond, forming the principal reaction products pregnanediol and pregnanolone.

Corticoids are essential to life and are classified on the basis of their metabolic function. Cortisol and corticosterone protect the organism against stress, form new carbohydrate from protein, influence fat metabolism, cause a dissolution of lymphatic tissue, and have minor influences on electrolyte balance. Aldosterone and deoxycorticosterone are particularly effective in causing sodium retention and POTASSIUM excretion (see also ADRENAL CORTICAL HORMONES).

The biosynthesis of cortisol proceeds by a series of hydroxylations from pregnenolone or progesterone at positions, 11β, 17α, and 21 and this process occurs in the fasciculata and reticularis layers of the adrenal. Corticosterone is produced in all three zones of the adrenal and is formed by 11β- and 21-hydroxylation. Aldosterone is formed exclusively in the glomerulosa and requires the hydroxylation of progesterone at positions 11β, 18, and 21. The 18-hydroxy group is oxidized to an aldehyde function.

The principal but not the only catabolites of cortisol include cortisone, tetrahydrocortisol (3α, 5β), tetrahydrocortisone (3α, 5β), tetrahydroallocortisol (3α, 5α) and the 17-ketosteroids: 11β-hydroxyandrosterone, 11-ketoandrosterone, 11β-hydroxyetiocholanolone, and 11-ketoetiocholanolone. The principal catabolite of aldosterone is the 3α, 5β-tetrahydro derivative.

References

1. DORFMAN, R. I., AND UNGAR, F., "Metabolism of Steroid Hormones," New York, Academic Press, 1965.
2. SOFFER, L. J., DORFMAN, R. I., AND GABRILOVE, J. L., "The Human Adrenal Gland," Philadelphia, Lea and Febiger, 1961.

3. YOUNG, W. C. (EDITOR), "Sex and Internal Secretions," Baltimore, Williams and Wilkins Co., 1961, 2 vols.
4. EICHLER, O., AND FARAH, A., (EDITORS), "Handbuch der Experimentellen Pharmakologie Erganzungswerk," "The Adrenocortical Hormones, Their Origin, Chemistry, Physiology and Pharmacology," Vol. XIV (HELEN WENDLER DEANE, SUBEDITOR), Berlin, Springer Verlag, 1962.

RALPH I. DORFMAN

HYALURONIC ACID AND HYALURONIDASES

Hyaluronic acid is a heteropolysaccharide containing, as repeating subunits in regular linear alternation with each other, D-glucuronic acid (glycosidically linked through its carbon-1 to the carbon-3 of the adjoining glucosamine unit), and N-acetyl-D-glucosamine (glycosidically linked through its carbon-1 to carbon-4 of the adjoining glucuronic acid unit). Enzymes that split the connecting β-glycoside linkages in hyaluronic acid are termed *hyaluronidases* (see HEXOSAMINES; MUCOPOLYSACCHARIDES; SNAKE VENOMS).

HYDRASES

Hydrases are enzymes that catalyze the addition of water to (or its removal from) a substrate. For example, fumarase catalyzes the removal of water from malic acid to form fumaric acid, and the reverse reaction (see CITRIC ACID CYCLE). Other hydrases function in the oxidation of fatty acids (see FATTY ACID METABOLISM).

HYDRATION OF MACROMOLECULES

The first half of the twentieth century may well go down in the history of science as a period in which, among other things, chemists learned how to synthesize a number of useful high polymers. Knowledge acquired by studying these synthetic polymers was quickly applied by investigators to the many problems which nature's own varied macromolecular substances presented to science. The fundamental composition of these macromolecules determined many of the properties and much of the behavior of such materials.

Polymers have been synthesized which are very hydrophobic, *i.e.*, they hate water molecules. As such they find important uses simply because they do not hydrate. On the other hand, synthetic polymers, having water-loving groups of atoms in the longer chains or as part of side chains, are hydrophilic in varying degrees. The long chains of such polymers often hydrogen bond to neighboring chains, giving the polymers unusual but useful properties.

Nature, throughout eons of time, has evolved methods for making the great variety of macromolecules needed by living organisms. This evolution occurred, for the most part, in a water medium or in one having a high water vapor content. So it is no wonder that the majority of the macromolecules involved in the life processes are hydrophilic, in differing degrees. One may find minor exceptions in the cases of certain fats and lipids. It should also be noted that in living organisms, the hydrophilic macromolecules of one kind often join with those of other kinds to produce the useful structures required in the life process, *i.e.*, membranes.

The vast array of PROTEINS all consist of varied polymers of the twenty-one identified AMINO ACIDS that nature has prepared as the building blocks to make these proteins. Nucleic acid polymers of as great a variety exist; on these differing DNA's and RNA's, PROTEIN BIOSYNTHESIS takes place. Since proteins are so hydrophilic, it is not to be wondered that these macromolecules of DNA and RNA are hydrophilic and appear to exist in a number of degrees of hydration. The carbohydrates and polysaccharides (*e.g.*, GLYCOGEN; STARCHES), produced as a product of photosynthesis, all have water-loving groups in their molecular structure. The macromolecular makeup of these particular substances gives materials having varying solubility. Finally the PHOSPHOLIPIDS and the LIPOPROTEINS are hydrophilic to the degree to which they have hydrophilic groups built into their structures. Nature must produce macromolecules that differ in this way in order that living forms may have stability of structure and yet be able to adapt to changing conditions.

Decades ago, biochemists recognized that it was very difficult to remove all of the water from a large number of these macromolecular materials. The term "bound water" was coined to explain the great affinity which many of these materials, especially proteins, exhibited for water. Biochemists were convinced that biological behavior, at least in part, resulted from the amount of "bound water" contained in the macromolecular structure. As an example, water held in plant structures so that it did not freeze in below-freezing temperatures was considered to be bound. At the time these ideas found favor, the method of lyophilization or quick freeze drying had not been perfected. Lyophilization removes the bulk of the water held in biological substances without destroying their structures or their activity.

More recently, lyophilized proteins have been further dried to a constant weight in a high vacuum and then studied as they adsorbed water vapor. The heats of adsorption of the first water vapor molecules were considerably higher than the values obtained as the adsorption approached that of the saturated vapor. These results indicated that the first molecules to be adsorbed were on the most water-loving or active sites. Such adsorbed molecules on the higher-energy sites would be desorbed last in a high vacuum. As part of this same investigation, D_2O (heavy water) vapor was studied for comparison reasons; when the adsorbed D_2O was pumped off in a vacuum, the protein weighed more than it had initially. This could only be due to an exchange reaction between the labile hydrogens of the protein and the adsorbed D_2O. The greatest exchange occurred during the first adsorption-desorption, but several successive adsorption-desorptions were required for a maximum exchange. The whole process could be reversed by adsorbing H_2O on the deuterated protein, but from two to three times as many exchanges were required to remove the deuterium as were needed to deuterate the protein. When the process was complete, the protein again weighed exactly what it did at the beginning of the experiment. The maximum exchanges for all of the proteins studied never exceeded the number of labile hydrogens in a given sample. Only insulin and hemoglobin, of all the substances studied so far, exchanged all of their labile hydrogens. This work seems to be good evidence for the fact that no "bound water" exists in the proteins as dried for this work. These studies were in part confirmed by the findings that the same maxima of exchange resulted when the samples were put into liquid D_2O, each at the pH of its isoelectric point. However, in warm solutions at a higher pH value (more strongly alkaline), all of the labile hydrogens exchanged. The mechanism of the exchange reactions on dry protein surfaces is not fully established, but the evidence strongly suggests that the D_2O molecules are adsorbed in a specific geometric way near a labile hydrogen site before exchange can take place.

Similar studies are under way on DNA, and early results show that it is much more difficult to remove water from DNA than from the proteins. Preliminary exchange values indicate that the maximum exchange can only be accounted for by assuming that a nucleotide pair has three molecules of water in its structure. Thus deuterium exchange studies on dry biological macromolecules are proving to be of great value in the elucidation of the conformational structure and the kind of hydration that may exist within these polymers.

Many biological systems depend in part on their degree of hydration. Most biological membranes are hydrophilic but should the membrane be a multilayered one, the hydration of the different layers may well differ markedly. A great many of the membranes used by living organisms are known to be very selective in what passes through them. They are often called permselective. Membranes are known which pump water into the cells they surround even against high osmotic gradients due to salt concentration. This fascinating field of membrane biophysics is only beginning to receive attention in a quantitative manner. Only as we understand more and more about the hydration of biological macromolecules will we realize the basic importance of water in living systems.

References

1. Eley, D. D., and Leslie, R. B., "The Adsorption of Water on Solid Proteins with Special Reference to Haemoglobin," *Advan. Chem. Phys.*, **7**, "The Structure and Properties of Biomolecules and Biological Systems" (Duchesne, J., Editor) (1964).
2. Reyerson, L. H., and Hnojewyj, W. S., "The Sorption of H_2O and D_2O Vapors by Lyophilized β-Lactoglobulin and the Deuterium-Exchange Effect," *J. Phys. Chem.*, **64**, 811 (1960); *ibid.*, **63**, 1653 (1959); *ibid.*, **65**, 1694 (1961).
3. Thomson, J. F., "The Biological Effects of Deuterium," New York, The Macmillan Company, 1963.
4. Conference on "Forms of Water in Biologic Systems," October 5–8, 1964; *Ann. N.Y. Acad. Sci.*, **125**, Article 2 (August, 1965)

Lloyd H. Reyerson

HYDROCARBON MICROBIOLOGY

Once considered the domain of a few peculiarly gifted microbes, the ability to utilize hydrocarbons as a source of energy and building material has been observed in thousands of examples of yeasts, mold and bacteria. As the known variety of organisms and their hydrocarbon substrates has multiplied, two major areas of interest have developed. (1) *Products*—What do they make from the hydrocarbons? and (2) *Pathways*—How is the oxidation (particularly the first steps) accomplished?

Products. *Catabolic Products.* In many cases, a major portion of the skeletal structure is retained as an acid, alcohol, catechol, etc., which is released into the surrounding medium. The organism must then obtain its energy for the oxidation from the portion removed from the skeleton, from the total oxidation of a percentage of the available hydrocarbon, or from another substrate.

Two classic examples of this type are the production of salicylic acid from naphthalene and fatty acids from normal alkanes.

It is in this category that many observers have come to consider the microbe as a sophisticated catalyst, capable of selective, specific oxidation to produce a desired product from a less desirable substrate. Indeed, the major limitation seems to be not the catabolic ability of the microbe, but rather the ability of the researcher either to select an organism where the catabolic pathway is blocked or limited at the appropriate place, or to manipulate the organism to achieve the same effect.

Cellular Material. To the organism, the biochemistry of converting the hydrocarbon into building blocks for cell constituents is of prime importance. To the hydrocarbon microbiologist, this conversion suggests two interesting possibilities; the removal of certain unwanted hydrocarbons from a crude mixture by substrate-specific organisms, and the production of cell mass suitable for animal feeds or even human consumption.

Both of these possibilities are of interest only if the utilization and growth rates approach or exceed those exhibited by classical carbohydrate fermentors. The fact that hydrocarbon-utilizing microbes are not necessarily more sluggish than their carbohydrate-utilizing cousins is evidenced by the recent flurry of publications (especially patents) in the area of cellular production from hydrocarbons.

There is no evidence to suggest that the biochemistry of these organisms is dramatically different from that observed in carbohydrate utilizers, once the hydrocarbon has been attacked and the catabolic processes are operating.

Biosynthetic Products. Hydrocarbon utilizers have also demonstrated the ability to synthesize materials in excess of their cellular requirements and release them into the media. There seems to be little reason to predict a large number of new products (*i.e.*, not previously demonstrated by carbohydrate utilizers), but the biosynthesis of amino acids, vitamins, antimicrobials, etc., from hydrocarbons as the sole carbon source can be expected.

The most striking example of this is the production of glutamic acid from kerosine in quantities great enough to stimulate interest in the kerosine utilizers as a commercial source of the amino acid.

Pathways. The two categories of oxidation pathways described here are only by way of illustration, since such studies have not included enough genera or species to suggest that these are necessarily the major pathways. In fact, alternate pathways have already been demonstrated in several cases.

Aliphatic Hydrocarbon Oxidation. In general, a pathway closely following chemical oxidation is observed.

$$R—CH_3 \rightarrow [R—CH_2OOH] \rightarrow R—CH_2OH$$
$$\rightarrow RCHO \rightarrow RCOOH$$

The formation of a 1-hydroperoxide early, if not initially, in the oxidation has been strongly implicated by several workers. Cell-free, enzymatic oxidation of a normal alkane to the normal alcohol has been shown to require DPNH and Fe^{++}. After the fatty acid is formed, the oxidation can proceed by the general pathway for fatty acids (see FATTY ACID METABOLISM). Other workers have presented evidence for diterminal oxidation via a similar pathway to yield the corresponding dioic acids, and also for 2-hydroperoxide formation in shorter alkanes to yield methyl ketones.

Aromatic Hydrocarbon Oxidation. The excellent definition of the naphthalene oxidation sequence provides a good example of aromatic oxidations:

H OH
OH
—H
"2–3 Catechol oxidase"
OH
OH
OH COOH
O
O
$CH_3\overset{\|}{C}$ COOH + CHO → COOH
OH OH

Also described are systems which include a "1-2 catechol oxidase." These produce, for example, *cis-cis* muconic acid from benzene by splitting the ring between the adjacent hydroxyl groups. In the case of alkyl aromatics, the attack may proceed either on the ring or on the side chain, depending on the organism's enzymatic ability and the steric relationship of the alkyl substituents.

References

1. DAVIES, J. I., AND EVANS, W. C., *Biochem. J.*, **91**, 251–261 (1964).
2. FOSTER, J. W., in "Oxidases" (HAYAISHI, O., EDITOR), pp. 241–271, New York, Academic Press, 1962.
3. FUHS, G. W., *Arch. Mikrobiol.*, **39**, 374–422 (1961).
4. GHOLSON, R. K., BAPTIST, J. N., AND COON, M. J., *Biochemistry*, **2**, 1155–1159 (1963).

IRA D. HILL

HYDROLYSIS AND HYDROLYZING ENZYMES

A major class of the enzymes (see ENZYME CLASSIFICATION AND NOMENCLATURE) has the ability to catalyze hydrolysis of substrates, *i.e.*, cleavage of the substrate with the introduction of the elements of water. These reactions are in many cases reversible. Hydrolytic enzymes include those which act on ester bonds (esterases, PHOSPHATASES, etc.), on peptide bonds (*e.g.*, PROTEOLYTIC ENZYMES), on thioether bonds, and on amides.

β-HYDROXYBUTYRIC ACID

β-Hydroxybutyric acid,

$$CH_3—CH_2OH—CH_2—COOH,$$

is one of three "acetone substances" (acetoacetic acid and acetone are the others) found in conspicuous amounts in the blood and urine of diabetics, and in starvation. It may be formed (reversibly) by the reduction of acetoacetic acid and, like it, contributes to the acidosis noted in diabetes and starvation. The acetone substances are intermediates formed in FATTY ACID METABOLISM (see also METABOLIC DISEASES). A polymerized form of β-hydroxybutyric acid is produced in certain bacteria.

HYDROXYPROLINE

L-Hydroxyproline, 4-hydroxypyrrolidine-2-carboxylic acid,

```
HO—CH—CH2
  /      \
CH2       CH—COOH
   \     /
     N
     |
     H
```

is one of the commonly accepted AMINO ACIDS derived from PROTEINS, though it, like proline, strictly is an imino acid. It is rarely found in proteins, but gelatin and COLLAGEN, from which gelatin is derived, yield about 14% of hydroxyproline on hydrolysis. Hydroxyproline is nonessential nutritionally. Its specific rotatory power in $1N$ HCl solution is $-47.3°$.

HYDROXYLASES. See OXYGENASES.

HYPERGLYCEMIA

Hyperglycemia is a condition of abnormally elevated glucose concentration in the blood. It may result from overproduction of glucose [*e.g.*, by excessive breakdown of glycogen (see GLYCOGEN AND ITS METABOLISM) or through excessive GLUCONEOGENESIS], or from underutilization of glucose [*e.g.*, in insulin deficiency; see INSULIN (FUNCTION); METABOLIC DISEASES].

Reference

1. WHITE, A., HANDLER, P., AND SMITH, E. L., "Principles of Biochemistry," Third edition, Ch. 21, Section on "Blood Glucose and Its Regulation," pp. 420–426, New York, McGraw-Hill Book Co., 1964.

HYPERTENSION (BIOCHEMICAL ASPECTS)

Hypertension, or high blood pressure, belongs to the group of diseases called "diseases of regulation" because a normal bodily function has become abnormal as a result of disturbance in the equilibria that control arterial pressure. Blood pressure is one link in the mechanisms that control the supply of blood to tissue. Since the total volume of blood is limited, and the demand highly variable, a complicated system of controls is necessary to supply the tissue needs.

There are two broad classes of controls: (1) the nervous, and (2) the chemical. The neural is the more sophisticated class and the chemical more primitive. The body can operate with the chemical controls alone, but the addition of the neural gives much greater flexibility. The two systems are closely interconnected.

To describe the several regulatory systems, the "mosaic theory" of hypertension was proposed by Page. Figure 1 shows an octagon with interconnecting facets indicating equilibria much as in the Phase Rule.

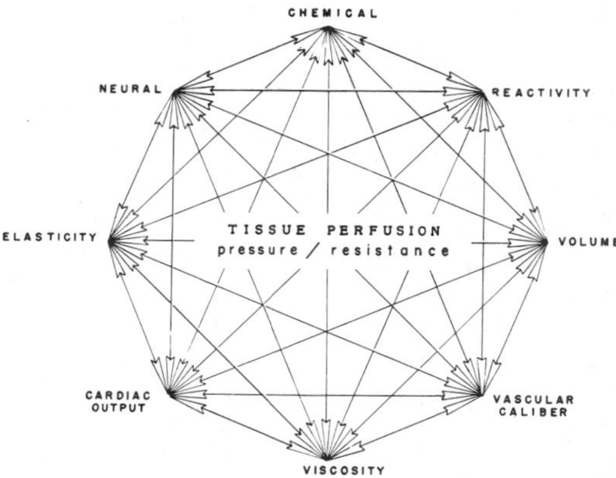

FIG. 1. The mosaic theory which relates the various facets of the regulation of perfusion of tissue with blood. The octagon depicts the equilibria among these facets which maintain blood pressure at relatively fixed levels.

The chemical component currently most widely under study is *angiotensin*, which results from the action of renin, a proteolytic enzyme, on renin substrate to produce angiotensin I. Renin is contained in the juxtaglomerular cells of the kidneys and is released into the blood to act on renin-substrate synthesized by the liver. Angiotensin I is a decapeptide with little physiological activity. Histidyl-leucine is split off it by a "converting enzyme" to form the octapeptide, angiotensin II, which is the most active pressor substance known. Both angiotensin I and angiotensin II have been synthesized. Angiotensin II is further split to inactive peptide fragments by angiotensinase, an aminopeptidase-like enzyme.

Angiotensin II stimulates the adrenal cortex to liberate aldosterone and so provides one link in the regulation of salt metabolism. It also stimulates the smooth muscle of blood vessels to contract and thus raises blood pressure. The efficiency with which the neural transmitter substance, norepinephrine, acts, is increased by angiotensin and increases the response of blood vessels to vasomotor impulses. This effect is believed to be an important reason for the hypertension produced in animals by prolonged infusion of subpressor doses of angiotensin. Both renin and angiotensin can be measured by physical and physiological methods.

Serotonin was first recognized by the appearance of a smooth muscle contracting substance during blood clotting. This contracting activity was found to be due to a substance, liberated from platelets, that proved to be 5-hydroxytryptamine.

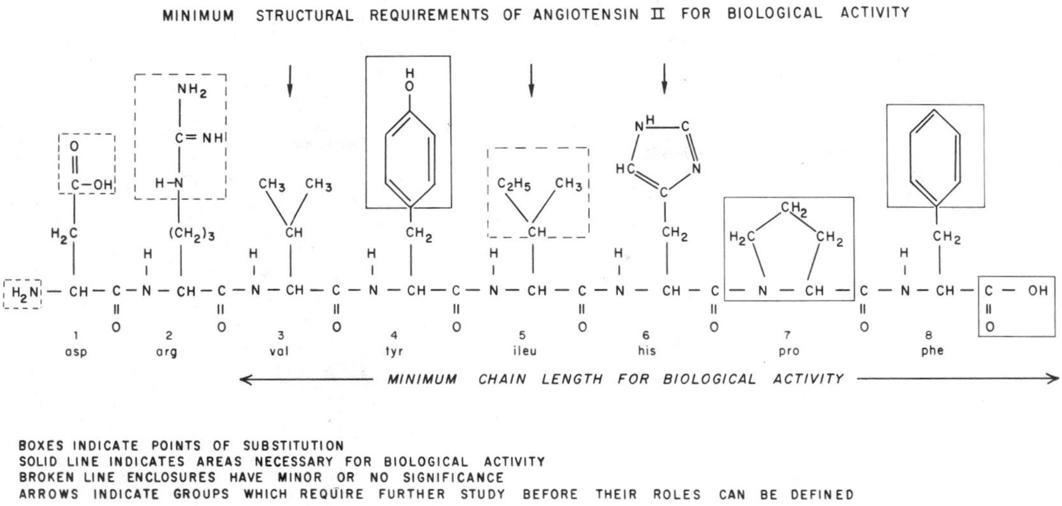

MINIMUM STRUCTURAL REQUIREMENTS OF ANGIOTENSIN II FOR BIOLOGICAL ACTIVITY

1 asp 2 arg 3 val 4 tyr 5 ileu 6 his 7 pro 8 phe

MINIMUM CHAIN LENGTH FOR BIOLOGICAL ACTIVITY

BOXES INDICATE POINTS OF SUBSTITUTION
SOLID LINE INDICATES AREAS NECESSARY FOR BIOLOGICAL ACTIVITY
BROKEN LINE ENCLOSURES HAVE MINOR OR NO SIGNIFICANCE
ARROWS INDICATE GROUPS WHICH REQUIRE FURTHER STUDY BEFORE THEIR ROLES CAN BE DEFINED

FIG. 2. Structure of synthetic angiotension II showing the minimum structural requirements for biological activity.

This compound (also called serotonin) results from decarboxylation of 5-hydroxytryptophan.

The metabolic end product usually measured in the urine is 5-hydroxyindoleacetic acid (see AROMATIC AMINO ACIDS). Important insight into its intermediary metabolism has come from studies on patients suffering carcinoid tumors.

Serotonin has been shown to be widespread in the body. The brain contains much of it and has the enzymatic organization for its synthesis. The intestine also synthesizes large amounts, some of which is taken up and transported by the blood platelets. Serotonin is prevented from being bound, by administration of reserpine. It is believed to have important effects on mentation as well as on gastrointestinal motility. Further, there is some evidence that it participates in the control of the caliber of blood vessels. When blood pressure is abnormally high, serotonin lowers it, and when it is abnormally low, serotonin raises it. For this reason, serotonin has been called an "amphibaric substance." It is believed possible that serotonin can act as a chemical buffer to help regulate blood pressure.

Effective anti-angiotensin, or anti-serotonin compounds, have so far not been found. The remaining substances of recognized importance in the mechanism of blood pressure regulation are norepinephrine and acetylcholine. Norepinephrine is the transmitter substance for sympathetic nerve impulses (see ADRENALINE AND NORADRENALINE), and acetylcholine is the transmitter of both sympathetic and parasympathetic impulses within ganglia (see ACETYCHOLINE AND CHOLINESTERASE). Fortunately, blocking agents for both are known and widely used in the treatment of hypertension.

IRVINE H. PAGE

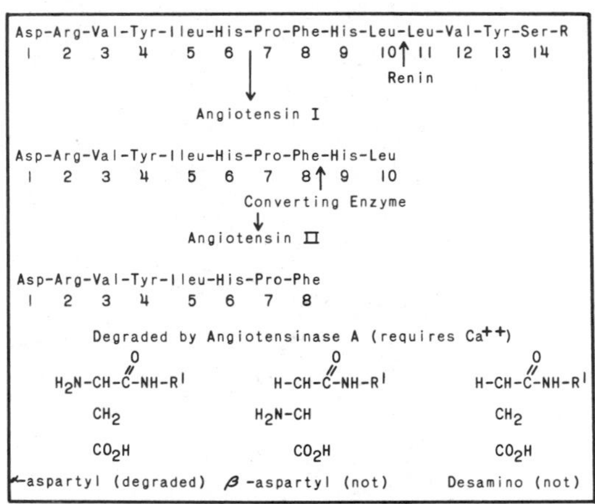

FIG. 3. The metabolism of angiotensin.

HYPOGLYCEMIA

Hypoglycemia, the opposite of HYPERGLYCEMIA, is a condition of lowered, subnormal glucose concentration in the blood. It may be associated with excess insulin activity [see INSULIN (FUNCTION)], with insufficiency of ADRENAL CORTICAL HORMONES, or with starvation (see also GLYCOGEN AND ITS METABOLISM; GLUCONEOGENESIS; METABOLIC DISEASES).

HYPOPHYSIS

The hypophysis, also called the pituitary, is an important ductless or endocrine gland located below the brain. Hormones secreted by the anterior portion, or adenohypophysis, include ADRENOCORTICOTROPIC HORMONE, GONADOTROPIC

HORMONES, GROWTH HORMONE (or somatotropin), and THYROTROPIC HORMONE. The posterior pituitary, or neurohypophysis, secretes the hormones oxytocin and VASOPRESSIN. An intermediate portion of the gland, the pars intermedia, secretes a melanocyte-stimulating hormone (see SKIN PIGMENTATION). The functions of these hormones are summarized and tabulated in the article HORMONES, ANIMAL.

Reference

1. TEPPERMAN, J., "Metabolic and Endocrine Physiology," Chicago, Year Book Medical Publishers, 1962.

HYPOTHALAMUS-HYPOPHYSIS SYSTEM

The hypothalamus is believed to release neurohumoral substances which are transported via hypothalamic-hypophyseal circulation, with the effect of stimulating the adenohypophysis in the secretion of one or more of its hormones (see HORMONES, ANIMAL).

References

1. TEPPERMAN, J., "The Hypothalamo-Hypophysial Relay Systems," in "Metabolic and Endocrine Physiology," Ch. 3, Chicago, Year Book Medical Publishers, 1962.

HYPOXANTHINE

Hypoxanthine, 6-oxypurine, may be formulated:

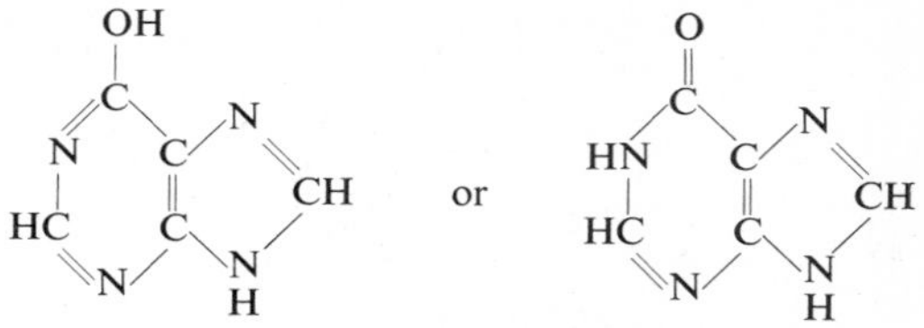

```
       OH                              O
       |                               ‖
      /C\     N                      /C\     N
    N     C/   \\                 HN     C/   \\
    |     ‖      CH      or        |     ‖      CH
   HC     C\   /                  HC     C\   /
     \\N/    N                      \\N/    N
             H                              H
```

Other PURINES, adenine, guanine, xanthine, and uric acid, also are presumed to exist in corresponding tautomeric forms. Hypoxanthine is closely related to adenine and may be produced from it in metabolism. It has been identified as a coenzyme of sulfite oxidase found in dog liver. The ribonucleotide of hypoxanthine (inosinic acid) is the purine initially formed in the metabolic pathway of PURINE BIOSYNTHESIS; this conjugated hypoxanthine moiety is subsequently converted to other purines.

HYPOXIA

Hypoxia is a condition of reduced oxygen (molecular O_2, loosely and reversibly bound to HEMOGLOBIN) concentration in the blood or tissues. It amy result from insufficient oxygen intake to the lungs, or from poisoning by CARBON MONOXIDE (see also CYANIDE POISONING). The brain, which accounts for about one-fourth of the total oxygen consumption of the body at rest, is particularly sensitive to lack of oxygen, and is subject to irreversible damage after brief hypoxia (see BRAIN METABOLISM; GASEOUS EXCHANGE IN BREATHING).

I

IMINO ACIDS

Imino acids are represented primarily by PROLINE and HYDROXYPROLINE, both found in proteins. These do not have in their structure the primary amino (—NH_2) group but rather the secondary amino group, >NH. They react with ninhydrin to yield carbon dioxide and colored condensation products different from those formed from the more typical amino acids.

IMMUNOCHEMISTRY

The Scope of Immunochemistry. Immunochemistry is concerned with the biochemistry and biophysics of the body's defense mechanism. It is particularly concerned with the response of the body to foreign macromolecules (ANTIGENS) and with the chemistry and physics of the interaction between the products of the response (delayed hypersensitivity, ANTIBODY) and the molecules which have elicited them.

The terms "antigen" and "antibody" are defined in a circular manner since an antigen is any agent (foreign macromolecule, virus, cell or tissue) which when introduced into the tissues of an animal, stimulates the formation of antibodies. Antibodies, in turn, are defined as blood PROTEINS whose synthesis is induced by antigens and which have the property of specifically combining with the antigen both *in vitro* and in the living animal.

Contemporary immunochemistry penetrates and enriches many areas of science, basically because of the special techniques it provides for detecting and measuring similarities and differences in the tertiary structure of proteins, nucleic acids and polysaccharides. Thus the microbiologist uses specific antibodies to identify bacteria and viruses by means of their surface antigens, the biochemist is concerned with antibodies as enzyme inhibitors and as models of induced protein synthesis, the biophysicist sees antigen-antibody interactions from the standpoint of specific spatial reactions between macromolecules, the geneticist may use a particular antigen, such as a blood group substance, or allotype, as a marker in studies of inheritance.

The process of blood transfusion depends on proper immunological matching of the blood of donor and recipient. Finally, the success of surgical transplantation of organs such as the kidney from one human to another depends on the extent to which the transplant can be prevented from acting as an antigen in the patient in whom it is implanted.

Defense Mechanisms. Foreign organisms are excluded from the body by skin and mucous membranes. If these barriers are penetrated, the foreign organism may be ingested by phagocytic cells (monocytes, polymorphs macrophages) and subsequently destroyed by cytoplasmic enzymes. In addition to this general constitutive mechanism, special constitutive factors, such as enzymes (lysozyme), destroy certain bacteria. Other processes may limit replication; for instance, fever-inducing substances which are released from polymorphs and, perhaps, lymphocytes increase the body temperature and thus create unfavorable conditions for multiplication of poliomyelitis and possibly some other viruses.

Some time after a foreign macromolecule has entered the body, induced mechanisms come into play, which result in the synthesis of specially adapted molecules (antibodies) able to combine with the foreign substances (antigens) which have elicited them. Most macromolecules (proteins, carbohydrates or nucleic acids) can function as antigens, provided they are different in structure from autologous macromolecules, *i.e.*, from the macromolecules of the responding organism.

Antibodies are proteins with a molecular weight of 150,000–1,000,000 and with the electrophoretic mobility of gamma and beta globulins. The combination between antigen and antibody results in inhibition of the biological activity of the antigen and leads to increased rate of ingestion (opsonization) of the antigen by phagocytic cells. In addition, combination of antigen and antibody results in the activation of a complex chain of interacting constitutive molecules—the complement system—leading to lysis of the cell membranes to which antibody, directed against cellular antigens, is attached.

Specificity. Antibodies can only be elicited by macromolecules (molecular weight greater than 10,000), but if a small molecule such as *p*-diazonium sulfonic acid is covalently linked to a macromolecule, antibody will be formed to it. A molecule which can elicit an antibody only when linked to a macromolecule is called a *hapten*. That part of the structure of an antigen which participates in specific interaction with antibody molecule is referred to as the *determinant* group. Antibodies show a high degree of specificity toward a relatively small portion of the antigen, so that an antibody directed to such a hapten as *p*-diazonium sulfonic acid, reacts more readily with it than with *m*-diazonium sulfonic acid and does not react with *o*-diazonium sulphonic acid. Not only position of substitution in the benzene ring, but also polarity, affects the "fit" between hapten and antibody. An azoprotein, prepared from the methylester of *p*-aminobenzoic acid reacts with antisera to aniline and *p*-toluidine, but only very weakly with antisera to *p*-aminobenzoic acid. Specificity depends on the steric arrangement of the hapten. The three stereoisomeric compounds of tartaric acid (*dextro*, *levo* and *meso* forms) can be distinguished by interaction with antibodies, and so can glucose and galactose which differ only in the spatial arrangement of H and OH on the fourth carbon atom. Each antibody molecule is adapted to a relatively small area, the determinant of the antigen. If any other antigen contains an area of identical or similar conformation, the antibody will react with this other molecule also, and this is called a *cross-reaction*.

Heterogeneity of Antibodies. Antibodies elicited by a macromolecule are heterogeneous with respect to specificity, amino acid sequence, firmness of combination to antigen and size of the area on the macromolecule to which they are adapted.

All antibodies are globulins, but they differ from one another in sedimentation constant (7S, 8S, 11–13S, 19S) and electrophoretic mobility (7S–13S: γG, γD, γA, γE; 19S: γM) (see CENTRIFUGATION, DIFFERENTIAL; ELECTROPHORESIS). In addition to these molecules, other globulins (myeloma proteins, BENCE-JONES PROTEINS) are found, which do not have known antibody activity, but share common antigenic determinants (*i.e.*, common structural features) with the antibodies, and are synthesized by similar cells. The whole group of structurally related proteins is referred to as *immunoglobulins*.

In any one class of antibodies, the structure of a given molecule may not be exactly the same in all individuals of the same species. This can be revealed by immunizing, for example, a rabbit, with the gamma globulin of another rabbit. If the gamma globulin structures of the two rabbits differ, an antibody results which is directed to those parts of the antigen which are distinct from those of the animal synthesizing the antibody. In rabbits, seven differences of this kind, called allotypes, have been recognized which consist of two allelic pairs of three specificities (A1, A2, A3 and A4, A5, A6; see Table 3). An animal, heterozygous with respect to these allotypes, may have four different specificities demonstrable in its gamma globulins. However, not all the specificities are present on all the gamma globulin molecules of a heterozygous individual. It appears that injection with certain antigens may lead to antibody molecules particularly rich in one or the other of the allotypic specificities.

We shall next consider various methods by which antibody can be measured and the properties of the immune systems on which these tests are based.

Assay and Properties of Antibody. All molecular species of antibodies (γG, γA, γD, γE, γM) can neutralize (*i.e.*, inhibit) the biological activity of the antigens which elicited them. Therefore the potency of antibodies can be measured by their ability to neutralize bacterial toxins (antitoxins), enzymes (antienzymes), protein hormones (antihormones) and viruses.

Interactions of this kind can be illustrated by the reaction between enzyme, antibody and substrate. If constant quantities of enzyme are mixed with different quantities of antibody and the mixtures are then added to substrate, the activity of these mixtures decreases with increasing amounts of antibody until a constant residual level of activity is reached which cannot be further decreased by the addition of more antibody (Fig. 1). The residual level of activity depends on the molecular weight of the substrate and is very small for substrates of high molecular weight and relatively large for substrates of low molecular weight. This residual level seems to be due to steric hindrance and more particularly to the fact that some antibodies can inhibit enzyme activity by virtue of the proximity of their combining area to the catalytic area. Other antibody molecules, while not inhibitory (their area of attachment is too far from the catalytic area for steric hindrance, but they may overlap with the area in which inhibiting antibody molecules combine), may compete with inhibitory antibody molecules for the enzyme molecule. If these two types of antibody combine with antigen to give complexes of the same dissociation constant, the competition between inhibiting and non-inhibiting antibodies results in constant residual activity in antibody excess.

The reduction in enzyme activity is much smaller if antigen and antibodies are not preincubated but combine in the presence of substrate. The reason for this "protective" effect of substrate may lie in changes in the shape (conformational changes) of the active site and other sites of the enzyme, as a consequence of combination with substrate. Such conformational changes of enzyme may sometimes also be caused by antibody and may account for the antibodies which augment rather than reduce enzyme activity.

The relative inhibitory capacity of different antibody preparations can only be expressed by a single parameter, if the linear portion of the plot of residual enzyme activity against the logarithm of antibody quantity is the same for all the antibody preparations under examination. If this condition is fulfilled, the relative potency of different antibody preparations can be expressed by finding that quantity of antibody which gives a standard reduction (say 50%) in activity, within the linear range of the relation between residual activity and antibody quantity. It is, however, clear that relative inhibitory capacity must be expressed by at least two parameters if the aforementioned slopes are not parallel. In this case, the relative potency would be different for each level of

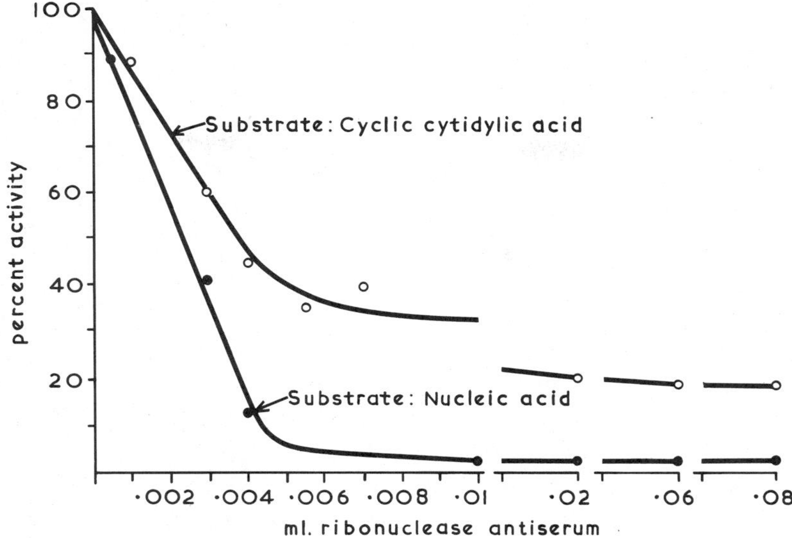

FIG. 1. Neutralization curve. The activity of mixtures of constant quantities of bovine pancreatic ribonuclease and varying quantities of antibody directed against ribonuclease. The activity is determined with a substrate of low molecular weight (cyclic cytidylic acid) and with a substrate of high molecular weight (nucleic acid).[8,24]

activity so that an adequate characterization of relative potency would require at least two parameters.

Antibodies can be assayed by measuring the insoluble product formed when they are mixed with antigen. A large proportion of antibodies of sedimentation constant 7S (*i.e.*, γG) form insoluble precipitates with antigen (*precipitin reaction*), and the nitrogen content of these precipitates can be measured. In the equivalence zone (see Fig. 2), all the antigen and antibody become insoluble, and since the antigen content is known, the antibody content can be determined. The quantitative precipitin reaction will now be discussed in detail.

If relatively small amounts of antigen are added to constant quantities of antibody, all or almost all of the antigen is precipitated in antibody excess (except for horse and certain human antibodies which result in complexes with antigen, that are soluble in antibody excess). In the zone of antibody excess, we can determine the largest number of antibody molecules which can combine with an antigen molecule, *i.e.*, the *valency* of the antigen molecule. This valency is found to be a function of the molecular weight of the antigen, being 3 for ribonuclease (molecular weight 13,400) and about 2000 for tobacco mosaic virus (molecular weight 40,700,000). As the antigen concentration is increased, a relative proportion of antigen and antibody is reached at which all the antibody and all the antigen are precipitated. We call the range of antigen concentration in which this occurs the equivalence zone. Since the quantity of antigen in the precipitate is identical with the total antigen in the system, the amount of antibody in the serum that gives insoluble complexes with antigen can be determined from the difference between nitrogen content of the antigen added and of the precipitate formed at equivalence. If more antigen is added, the zone of antigen excess is reached. Here the complexes between antigen and antibody become soluble in antigen so that only a proportion of the antigen and of the antibody is insoluble.

In the zone of considerable antigen excess, all the antigen-antibody complexes remain soluble (Fig. 2). In this zone we can determine the number of molecules of antigen that can combine with one molecule of antibody and thus determine the valency of the antibody molecule. This can be done by forming antigen-antibody complexes at equivalence, hence of known composition, and dissolving them in antigen excess. This mixture can then be subjected to ELECTROPHORESIS, and the uncombined antigen and antigen-antibody complex can be estimated, since the area under each of the resolved peaks is proportional to the concentration of these three types of compounds. In this way, the molar ratio of antigen to antibody (valency of antibody) in soluble complexes, formed in antigen excess, has been determined to be 2. Another approach to the determination of antibody valency is the technique of equilibrium dialysis. It is based on the use of antibody directed against hapten. In this technique, purified anti-hapten antibody is put into a dialysis bag in the presence on the hapten and is dialyzed against hapten outside the bag. At the end of this dialysis, the concentration of hapten inside and outside the bag would be the same if there were no hapten bound to the antibody molecule. However, antibody does bind hapten, and consequently there is more hapten in the dialysis bag than there is out-

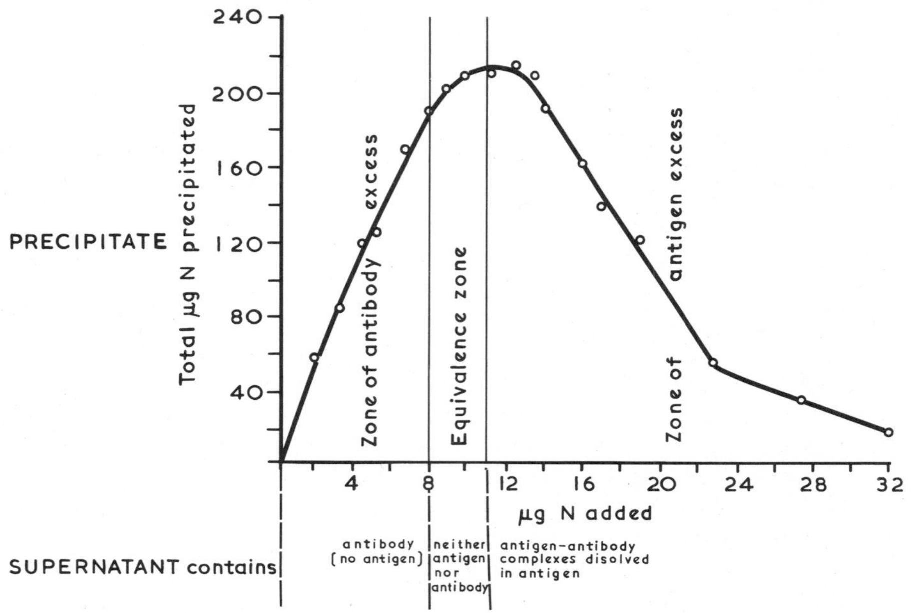

FIG. 2. Quantitative precipitin test. Constant quantities of antibody are incubated with varying quantities of antigen, a precipitate forms, is separated and washed, and its nitrogen content is determined. The total nitrogen of the precipitate is plotted as a function of the quantity of antigen added to the system.[11]

side it. The amount of bound hapten, then, is obtained from the difference between the hapten in the bag (retentate) and the amount of hapten outside the bag (dialysate). Equilibrium dialysis is carried out with varying amounts of hapten, and the largest quantity of hapten that can combine with a given quantity of antibody is so determined. Experiments of this nature have indicated that two molecules of hapten can combine with one molecule of antibody, thus confirming the divalency of antibody molecules.

The precipitin reaction can also be observed by allowing excess antigen to diffuse into agar-gel columns containing antibody (Oudin technique). This results in a moving zone of *antigen-antibody precipitate in the agar* [Fig. 3(a)]. The distance of the zone from the interphase is related to the diffusion constant and concentration of antigen. If several antigens interact with several antibodies, separate zones can be seen for each antigen-antibody interaction. In a modified version of this procedure, cups are cut into agar slabs and are filled with antigen and antibody, respectively, which diffuse and interact in the agar [Ouchterlony technique; Figs. 3(b), 3(c) and 3(d)]. These techniques have the advantage over the quantitative precipitin test in liquid media in that they allow one to discriminate between different antigen-antibody systems because each interacting system of antigen and antibody results in a distinct and separate zone in the agar. Thus, the number of zones, observed in the interaction of antigen and antibody in agar, represents the minimum number of interacting antigens and antibodies in the system. The resolving power of this method can be further increased by first carrying out *electrophoresis of the antigen* in agar, then cutting longitudinal ditches in the agar, parallel to the direction of the current, and introducting the antibody into these ditches. In this way, a series of arcs (zones) can be observed along the direction in which the antigen has been distributed by electrophoresis, each arc corresponding to a separate antigen. By this method, more than thirty distinct serum proteins have been recognized, including some of the immunoglobulins, and the complement components, mentioned in Tables 1 and 2. The method can also be used to detect enzymes since antigen-antibody precipitates in agar retain enzyme activity in the antigen excess portion of the zone. This can be localized by treating the washed agar slide with substrate and converting the product, which reacts with a suitably chosen reagent, to a colored derivative of the product.

All the different molecular species of antibody cause *agglutination*, *i.e.*, clumping of such particulate antigens as red blood cells or bacteria, and the lowest concentration of antibody, resulting in agglutination, is a measure of potency (see BLOOD GROUPS). This assay can be used for more soluble antigens by attaching them to red cells which are then agglutinated by antibody directed against the attached antigens.

Antibodies can also be measured by their interaction with complement, *i.e.*, by the *complement*

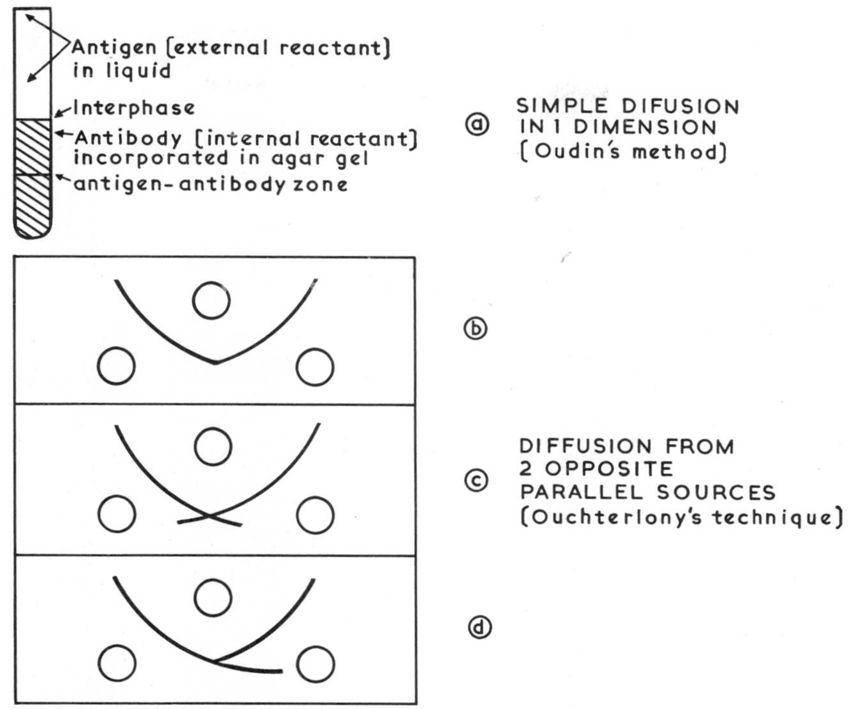

FIG. 3. The reaction between antigen and antibody in agar.
(a) Simple diffusion. A zone of antigen-antibody precipitate has formed and will move toward the bottom of the tube.
(b, c, d) Double diffusion of antibody in the cup, situated at the apex of a triangular arrangement of cups, with two antigens.
(b) The reaction of two identical antigens with antibody.
(c) The reaction with antibody of two structurally completely distinct antigens.
(d) The reaction with antibody of two antigens having some determinants in common (cross-reacting).[15]

fixation test. Red cells, which are sensitized, *i.e.*, combined with antibody, are lysed by complement (usually added as fresh normal guinea pig serum) which consists of a series of substances involved in interlocking reactions. The complex chain of reactions which occurs in this process is shown in Table 1. A given lytic effect of complement is caused by a smaller number of attached macroglobulin antibody molecules (γM, 19S) than of attached 7S antibody. Complement can also interact with complexes between antibody and antigens other than those of the red cells. Since some of the complement factors are removed or inactivated in this process, the hemolytic potency of the added complement is reduced in proportion to the amount of complex added to the complement. This reduction can be measured in terms of the residual hemolytic activity of the complement toward sensitized red cells.

Structure of Antibody. Of all the different molecular forms of antibody, the structure of 7S antibody is most clearly understood. This molecule has an over-all cylindrical shape, with a height of 250 Å and an elliptical base with axes of 41 and 22 Å. The molecule can combine, at its two short ends, with two molecules of antigen, and the specificity of the two combining sites is identical [Figs. 4(a) and 4(b)]. Since the antigen is multivalent and the antibody is divalent, a network, also called a lattice, of cross-linked antigen and antibody can be formed [Fig. 4(c)]. The 7S antibody molecule consists of two equal portions held together by one S—S bond; each of these two portions contains one combining site and consists of one relatively heavy (molecular weight 50,000) and one relatively light (molecular weight 20,000) polypeptide chain. The two polypeptide chains (heavy and light chains) are linked by one S—S bond, and both peptide chains seem to contribute to the combining capacity of the antibody (Fig. 5). The heavy chain of rabbit gamma globulin contains the allotypic specificities A1, A2 and A3, and the light chain contains the allotypic specificities A4, A5 and A6 (cf. Table 3).

All immunoglobulins are built up of heavy and light chains, but different types of immunoglobulins contain different types of chains, and the macroglobulins (γM) contain more than two heavy and two light chains (Table 2; see also ANTIBODIES).

TABLE 1. SEQUENCE AND MECHANISM OF IMMUNE HEMOLYSIS

E = sheep erythrocyte. E* = injured erythrocyte.
A = rabbit antibody directed against sheep erythrocyte.
C′1, C′4, etc. designate serum proteins interacting in a sequence of reactions resulting in immune hemolysis.
C′1a, C′2a, etc. refer to enzymatically active forms of the corresponding component, either known or postulated.
C′1 = macromolecular complex, containing subunits designated C′1q (11S factor), C′1r and C′1s (C′1 proesterase). The intact complex of all these molecules functions as C′1.
Immunoelectrophoretic designation of some components of complement:
C′3 = β_{1C} globulin (converted in free solution to β_{1G} globulin under the influence of the C′4-2a complex).
C′4 = β_{1E} globulin.
C′5 = β_{1F} globulin.

Reaction	Biochemical Event
E + A ⇌ EA	
EA + C′1 $\xrightarrow{Ca^{++}}$ EAC′1a	Attachment of C′1 at a site on C′1q to a receptor on A; activation of a catalytic center at a site on C′1s.
EAC′1a + C′4 → EAC′1a, 4	Enzymatic attack of C′1a on C′4, creating a site on C′4 which attaches to a receptor on E.
EAC′1a, 4 + C′2 $\xrightarrow{Mg^{++}}$ EAC′1a, 4, 2a (↑ Thermal Decay)	Attachment of C′2 to a receptor on C′4; enzymatic attack of C′1a on C′2, creating a catalytic center on C′2; fulfillment of functions of C′1a.
EAC′1a, 4, 2a + C′3 → EAC′1a, 4, 2a, 3a	Activation of C′3 and attachment to a receptor on E.
EAC′1a, 4, 2a, 3a +C′5 + C′6 + C′7 → EAC′1a, 4, 2a, 3a, 5, 6, 7a	Formation of a stable intermediate no longer susceptible to thermal decay: activation of C′5—C′6—C′7.
EAC′1a, 4, 2a, 5, 6, 7a + C′8 + C′9 → E*	Unknown events leading to 80–100 Å holes in the erythrocyte membrane.
E* → Ghost + Hemoglobin	Lysis of erythrocyte and release of hemoglobin.

From: "Antibodies to Biologically Active Molecules," B. CINADER (EDITOR), Oxford, England, Pergamon Press Limited, 1967.

TABLE 2. TYPES OF HUMAN IMMUNOGLOBULIN

Type	Subtype	Heavy Chain	Light Chain	Molecular Formula
γG (or IgG)	γGκ	γ_2	κ_2	$\gamma_2\kappa_2$
	γGL	γ_2	λ_2	$\gamma_2\lambda_2$
γA (or IgA)	γAκ	α_2	κ_2	$\alpha_2\kappa_2$*
	γAL	α_2	λ_2	$\alpha_2\lambda_2$*
γD (or IgD)	γDκ	δ_2	κ_2	$\delta_2\kappa_2$
	γDL	δ_2	λ_2	$\delta_2\lambda_2$
γE (or IgE)	—	—	—	—
	—	—	—	—
γM (or IgM)	γMκ	μ_2	κ_2	$(\mu_2\kappa_2)_n$**
	γML	μ_2	λ_2	$(\mu_2\lambda_2)_n$**

* polymers are also present.
** *n* is 5.
Based on: *Bull. World Health Org.*, **30**, 447 (1964); *J. Exp. Med.*, **121**, 185 (1965); *J. Allerg.*, **37** 185 (1966).

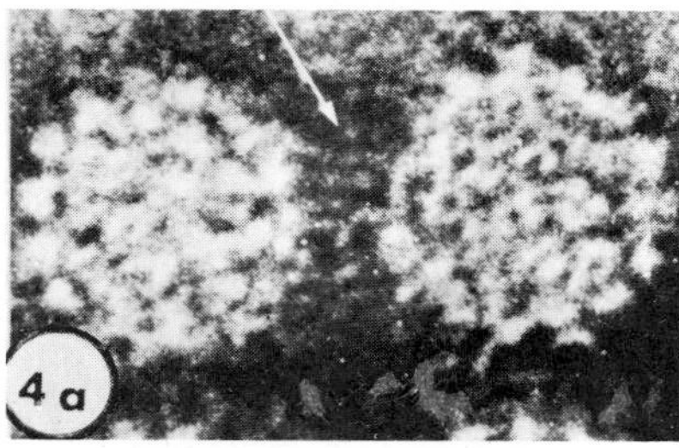

FIG. 4(a) Electron micrograph of antibody from rabbit (see arrows). Two particles of wart virus linked by three distinct antibody molecules, the rod shape of the antibody molecule is clearly visible. Magnification 470,000X.[24]

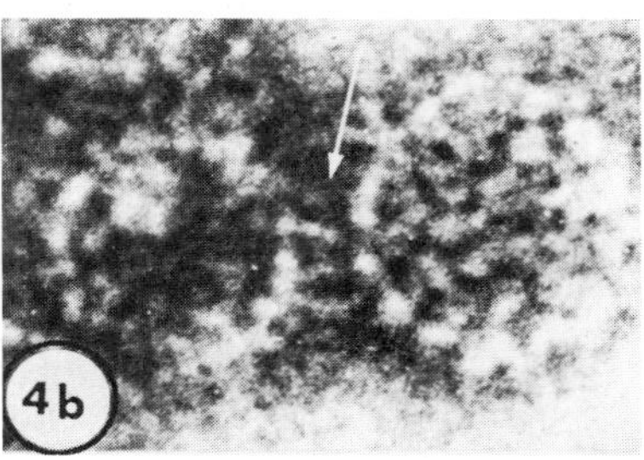

FIG. 4(b) Electron micrograph of antibody from goat (indicated by arrow). Two particles of polyoma virus linked by antibody molecules. Length and width of antibody derived from rabbit and goat showed no appreciable difference. Magnification 479,000X.[24]

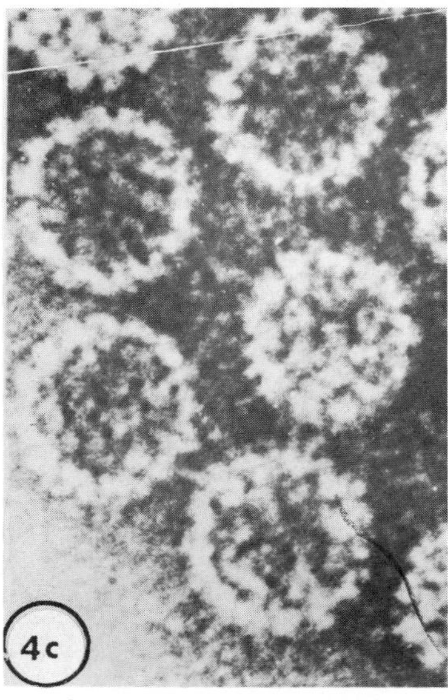

FIG. 4(c) Electron micrograph showing aggregate of polyoma virus particles and antibody. Six particles are linked to one another by two or more antibody molecules. Magnification 459,000X.[24]

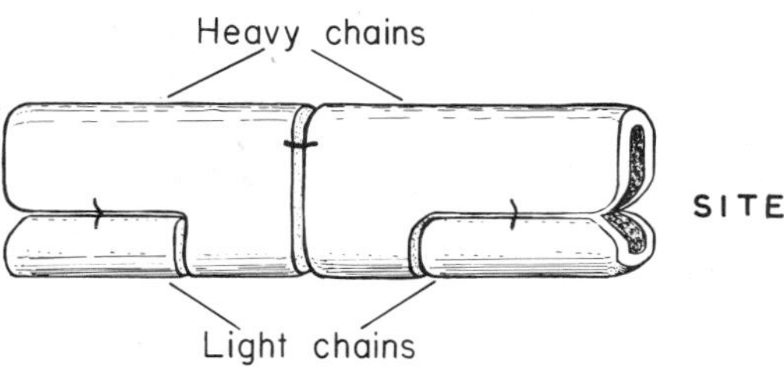

FIG. 5. A model, showing the arrangement of polypeptide chains in a 7S antibody molecule. The combining site can be seen at each end of the divalent molecule, and the S—S bonds, linking the separate parts of the molecule, are indicated by black lines. (Model of Edelman and Gally.)

Having so far dealt with specificity, heterogeneity, assay and structure of antibody, we shall next turn to the changes in quantity and quality of antibody in the course of immunization.

Kinetics of Antibody Formation. After the first intravenous injection of antigen, there is a latent period followed by the appearance of antibody. The antibody reaches a maximum titer two to three days after it has first been detected, and then declines (*primary response*). If antigen is injected once again, months or even years later, antibodies can be detected much earlier than after the first injection. Their quantity increases to a higher level and persists much longer. This *secondary response* has been called the *anamnestic* response because it appears as if the antibody-forming mechanism were remembering previous contact with the antigen. If antigen is injected a third time, the events observed after the second injection are essentially repeated, possibly with a slight increase in the total quantity of antibody formed. Further courses of immunization will soon result in a constant level of antibody production which is not exceeded after repeated immunization and may, ultimately, even decline.

In the primary response, 19S antibody forms first, 7S antibody somewhat later; whereas in the secondary response, both appear simultaneously, but 7S antibody forms the largest part of all the antibody produced. With such antigens as polysaccharides, only antibody of 19S type is formed after a first and even after a second, injection of some species such as horse and man. With virus as antigen, the injection of very small doses of antigen results in a primary response consisting of 19S antibody (γM) only. When large quantities are injected, 19S and 7S antibody are formed. When the primary response results only in the formation of 19S antibody, a second injection of the same antigen does not provoke an anamnestic response. It thus seems as if antigenic stimulation, resulting in the production of 19S antibody only, does not lead to a persisting modification of the antibody-forming system.

TABLE 3. POLYMORPHISM (GENETIC VARIATIONS) OF SERUM PROTEINS

Electrophoretic Designation	Other Designation	Species	Minimum Number of Types of Molecules
Prealbumin	Binds thyroxin	Monkey	3
Albumin		Man	2
		Cattle	5
		Fowl	2
Alpha globulin	Haptoglobin	Man	3*
		Macaca irus	2
		Pig	2
		Seal	2
		Marmot	2
		Ground squirrel	2
	Gcl, Gc2†	Man	2
Beta globulin	Ag(a$^+$), Ag(a$^-$)†† low-density lipoprotein	Man	2
	Transferrin	Man	9
		Mouse	2
		Cattle	3
		Sheep	5
		Goat	2
		Pig	2
		Monkey	7
Gamma globulin	A1, A2, A3 A4, A5, A6, A7	Rabbit	6
	Gml etc.	Man	20
	Inv1, Inv2, Inv3 Inv-		4
	Unsettled	Mouse	9
	GA1, GA2, GA3, GA4	Fowl	4

* Polymerization is a complicating factor in deciding the number of molecules with different determinants.
† Detected by means of hetero-antibodies.
†† Detected by iso-immune serum obtained from a transfused patient. The substance may be a gamma globulin.

During prolonged immunization, various changes in antibody properties are observed. In the course of immunization with bacterial toxins, the antibody obtained after many injections binds the toxin more firmly than does antibody obtained after the first injection. Antibodies of greater inhibitory power, in antibody excess, are formed in the course of prolonged immunization with enzymes. This change may be attributable to the fact that the specificity of the antibodies to a given macromolecule changes in the course of prolonged immunization, and certain determinants (close to the catalytic area), which do not elicit demonstrable antibody in the early phases of an antibody response, elicit antibody molecules during prolonged immunization.

For every antigen and for every individual animal there seems to be a maximum level of antibody production which cannot be exceeded by further immunization. The level of this maximum antibody production is determined by a feedback mechanism which may exist independently for each determinant on a macromolecule. It can, in fact, be shown that the injection of antibody ("passive immunization") into animals already immunized with the antigen to which the injected antibody is directed, can prevent further antibody production to that antigen.

Specific Inhibition of Antibody Formation. Whereas the injection of unpolymerized, soluble antigen into an adult animal usually results in antibody formation, injection of antigen into a newborn animal does not, as a rule, result in antibody formation, but on the contrary, in an unresponsiveness of the animal in later life to the antigen. The more antigen injected at birth, the longer does this unresponsiveness, which is called *acquired immunological tolerance*, persist. The acquired immunological tolerance induced at birth can be prolonged indefinitely by the injection of small doses of antigen, at regular intervals, during adult life.

Acquired immunological tolerance is specific, so that an animal tolerant to one antigen makes a normal antibody response to an antigen structurally unrelated to the macromolecule injected at birth. Tolerance is specific for the determinants of the molecule injected at birth. As a consequence, the injection of a cross-reacting antigen into a tolerant adult animal can result in formation of antibody to the determinants in which the cross-reacting antigen differs from the tolerance-inducing antigen. Prolonged immunization with a cross-reacting antigen results in formation of antibody adapted to a progressively wider area around the determinants, which distinguish the cross-reacting antigen from the tolerance-inducing antigen, so that ultimately an antibody results which can not only react with the cross-reacting antigen but also with the antigen injected at birth.

As a consequence of this change in the antibody, during prolonged immunization, tolerance appears to break down after repeated injection with a cross-reacting antigen. If immunization with the cross-reacting antigen is carried out with antigen incorporated in *Freund's adjuvant* (an emulsion of oil and killed tubercle bacilli, which generally leads to increased antibody production), this breakdown of tolerance can occur in a relatively short time; the greater the difference between the tolerance-inducing and the cross-reacting antigens, the more readily does this breakdown of tolerance occur.

Acquired immunological tolerance to *autologous* proteins, *i.e.*, to proteins of the organism in which antibody formation occurs, is probably responsible for the fact that autoantibodies are, as a rule, not produced. Autologous antigens circulate throughout the body and will consequently induce tolerance during the tolerance-induceable stage in the development of the antibody-forming organs. However, under certain circumstances the body may not be tolerant to some autologous macromolecules. This will occur if the autologous molecules are "secluded" in an organ (for instance, proteins of the lens of the eye and thyroglobulin), will consequently *not* circulate through the body, and will thus never come in contact with the immature antibody-forming organs. It will also occur if synthesis of an autologous macromolecule begins after the tolerance-inducible period is terminated or if an autologous macromolecule is normally produced in such low quantities that it does not interact with a substantial proportion of the antibody-forming apparatus. Furthermore, breakdown of tolerance may occur as the consequence of natural immunization with a macromolecule of bacterial origin, which has structural similarity with a body protein. Finally, it may also occur if autologous antigens undergo some structural alteration as the consequence of a pathological process. In addition to the factors just mentioned, a genetic factor may be involved, which predisposes some individuals toward autoantibody formation.

Acquired immunological tolerance to autologous determinants can help us to understand the great variability in the antibody response to a given antigen in animals of the same species, and the inheritance of the responsiveness to an antigen. It has already been indicated that every animal is tolerant to the determinants on its autologous proteins. It must also be realized that individuals of the same species differ in the determinants of their autologous proteins. This phenomenon is known as *polymorphism*, one instance of it has already been seen in the allotypes of gamma globulin. However, this polymorphism is not confined to this particular class of macromolecules as can be seen from Table 3. As a consequence of this polymorphism, the determinants on a given foreign macromolecule, which can induce antibody formation, will not be the same in different individuals, since they depend on the coincidence between determinants on the foreign protein and on the autologous protein on the injected individual. Since the determinants of autologous proteins are under genetic control, determinants to which antibody cannot be made are also, though indirectly, under genetic control. Thus the responsiveness to the different determinants of an antigen is indirectly controlled by genes, via the direct control of the synthesis of the different polymorphic forms of autologous macromolecules. Since the inheritance of autologous macromolecules is usually dominant or codominant, tolerance-mediated inheritance will be recessive. In experiments with polypeptides, some instances of dominant inheritance of the antibody response have been reported, so that a direct genetic control may also be involved.

Acquired immunological tolerance has been discussed, so far, as a phenomenon encountered in newborn animals; however, it is becoming clear that this process occurs also in adult animals. It appears that tolerance can be induced in adult animals by the injection of some soluble antigens which have been freed of all polymerized material. With large doses, unresponsiveness can even be induced in animals, which have previously responded to the same antigen. Furthermore, the injection with large quantities of polysaccharides, during adult life, results in permanent inhibition of the antibody response to these antigens. This phenomenon is called *immunological paralysis*.

Having now dealt with the primary and secondary antibody response and with the specific inhibition of antibody response, we can turn next to the experimental approaches and results concerning the biosynthesis of antibody.

Biosynthesis of Antibody. Much of the recent information on the mechanisms and cellular origin of antibody formation has been obtained by the application of four techniques:

(1) Fluorescent antibody studies in which a molecule such as gamma globulin, inside a sectioned cell, is allowed to react with an antibody to gamma globulin. The antibody-treated cell is then "flooded" with gamma globulin conjugated with a fluorescent dye. The bright fluorescence of cells containing gamma globulin is studied in a UV microscope.

(2) Studies of the immobilization of flagellated microorganisms in single drop techniques. Single cells from spleen or lymph nodes of an animal, previously immunized with a flagellate organism, are introduced into a microdrop, and the flagellated organism is introduced into the same drop. If the cells make antibody, the microorganism is immobilized by combination of the antibody with its flagellae.

(3) Study of inactivation of phage in single drop technique.

(4) Hemolysis of red cells, in the presence of complement, by antibody secreted from lymphoid cells. Sheep red cells, complement and cells from an animal immunized with sheep red cells are embedded in agar. The lymphnode or spleen cells making antibody will bring about hemolysis of the red cells surrounding them and can thus be identified.

Information obtained from the application of the foregoing techniques may be summarized as follows. The body contains a pool of uncommitted stem cells which differentiate under the influence of antigens. It appears as if antigen were initially being taken up by phagocytic cells from which a molecule not identical with the original antigen passes into a lymphoid type of cell. Whether this is an obligatory step in induction of antibody formation is not clear. At any rate, cells can synthesize antibody to more than one antigen and form 19S antibody as its initial product. After some time the cell undergoes further differentiation and, as a consequence, elaborates both 19S and 7S, and ultimately only 7S, antibody. Antibody of both types seems to exercise an inhibitory effect on the formation of the cells producing 19S and thus on the precursor of the cell producing 7S.

Secondary response seems to involve rapid cell division, and it is quite conceivable that differentiation of a cell forming 19S to a cell forming 7S is preceded by rapid cell division and the formation of clones of antibody (7S) forming cells.

Cellular events in antibody synthesis are primarily, though not necessarily directly, controlled by the antigen and to some extent through antibody in the circulation by a feedback mechanism. However, other biochemical processes must also be implicated. Among these is control of cell differentiation by the thymus and appendix in mammals and by the *Bursa fabricii* and thymus in birds. The effect of these organs is extremely important near the time of birth. If at that time the thymus is extirpated, the cellular apparatus of antibody formation and hypersensitivity is not developed and the animal remains immunologically incompetent. The control by the thymus is, at least in part, mediated through humoral substances.

Hypersensitivity. It has already been mentioned that interaction of antibody with various antigens can trigger reaction chains of constitutive molecules, known as the complement system. However, many other physiological mechanisms can be triggered by the combination of antigen and antibody and the resulting biological phenomena are referred to as specific hypersensitivity. Two types of hypersensitivity or heightened response to an agent can be distinguished: (1) immediate hypersensitivity; (2) delayed hypersensitivity.

Immediate hypersensitivity will be considered first and will be discussed as manifested in anaphylaxis, allergy and in the Arthus phenomenon.

Anaphylaxis, in most animals (but not in the rabbit) depends on the fixation of antibody molecules to tissue cells, the combination of antigen to cell fixed antibody, and the consequent release of pharmacologically active substances. Intravenous injection of an antigen into a guinea pig, pretreated with antigen, usually results in shock and death within minutes. If the animal survives the anaphylactic shock, its reaction to subsequent injection will be progressively less severe; *i.e.*, it is desensitized. An animal can be prepared for anaphylaxis, either actively by immunization, or passively by the injection of antibody. Passive anaphylaxis is useful for the study of the fundamental mechanisms involved in anaphylaxis. If smooth muscle (intestine or uterus) from a sensitized guinea pig is treated *in vitro* with antigen, it will contract (Schultz-Dale reaction). This reaction involves the release of HISTAMINE and can be inhibited by the administration of antihistamines. If antibody is injected into the skin, and antigen, mixed with Evans blue dye, is subsequently injected intravenously, the release of pharmacologically active substances increases the permeability of small blood vessels, resulting in leakage of the blue dye. The severity of the reaction can be measured in terms of the size of the blue area measured on the underside of the skin of the injected animal. The fractionation of antibodies obtained from guinea pigs has shown that only certain types of antibodies cause passive cutaneous anaphylaxis.

Atopy is a term applied to certain types of hypersensitivity states (allergies) of man such as hay fever and some forms of asthma, in which genetic factors play an important role. Atopy is entirely dependent upon non-precipitating antibodies which probably are γE globulins and are called *reagins*. These molecules can attach themselves to cell walls and can then react with antigen (allergen). A genetic factor appears to be responsible for a tendency to synthesize γE globulins in preference to other types of antibody. When serum from an allergic individual is injected intradermally into a normal person and the specific antigen (allergen) is injected into the same site 24 hours later, a wheal and flare reaction of the injected skin develops (Prausnitz-Küstner reaction). If an allergic individual is immunized with the antigen causing the allergy, precipitating antibodies are formed ultimately (γG), which compete for antigen with the non-precipitating antibodies and thus protect the sensitive individual against the consequences of the combination of tissue-fixing (γE) antibody with the antigen.

Other fractions of antibody, though not skin fixing and not inducing anaphylaxis, can nevertheless bring about tissue damage by virtue of combination with antigen. This reaction, called the *Arthus phenomenon*, starts with edema and erythema and ends with hemorrhage and tissue necrosis. The lesion is apparently caused by a chain reaction, triggered by the deposition of antigen-antibody precipitates, causing invasion by neutrophil polymorphs and probably the subsequent release of hydrolytic enzymes (lysosomes) by these cells.

So far, various types of complex pharmacological reactions have been considered, which are mediated by antigen-antibody interaction. There are other types of response which are possibly not mediated by antibody, in the conventional sense, and are known as *delayed-type hypersensitivity*. This reaction develops many hours after the antigen is deposited in or on the tissues. In animals, the reactivity to antigen can be transferred from one individual to another by the transfer of *living* leukocytes. In man, on the other hand, the sensitivity can be transferred even by extracts from killed leukocytes. Furthermore, delayed-type hypersensitivity transferred to a negative recipient can be transferred, in turn, by means of an extract of the now sensitized peripheral blood cells of this first recipient to a second negative recipient.

Delayed hypersensitivity and antibody production may be two related stages of the same process or may be two distinct processes. We cannot distinguish, at present, between the following alternatives:

(1) Delayed-type hypersensitivity is the first response to an antigen, and thus the first manifestation of differentiation of stem-cells to antibody forming cells. The differentiating stem-cell may be able to react with antigen before the cell acquires the capacity to synthesize antibody.

(2) Antibody-producing cells and cells participating in the delayed-type hypersensitivity reactions may be two independent cell lines, which respond in different ways to contact with antigen.

This survey of immunochemistry must now be concluded; it was written in a period during which our understanding has advanced with great speed. Insight into many fascinating problems connected with biosynthesis and structure of antibody is becoming, rapidly, more profound, and the general reader will be well rewarded if he attempts to follow these advances in the occasional articles of such magazines as *Scientific American*.

References

1. Landsteiner, K., "The Specificity of Serological Reactions," Revised edition, New York, Dover, 1962.
2. Boyd, W. C., "Fundamentals of Immunology," Third edition, New York, Interscience Publishers, 1956.
3. Humphrey, J. H., and White, R. G., "Immunology for Students of Medicine," Second edition, Oxford, Blackwell Scientific Publications, 1964.
4. Cohn, M., "Immunochemical Methods for Determining Homogeneity of Proteins and Polysaccharides," *Methods Med. Res.*, **5**, 268, 1952.
5. Williams, R. T. (Editor), "Immunochemistry," *Biochem. Soc. Symp.*, No. 10 (1953).
6. Topley, W. W. C., and Wilson, G. S., "Principles of Bacteriology and Immunity," Fourth edition (Wilson, G. S., and Miles, A. A.), London, E. Arnold, 1955.
7. Burnet, F. M., and Fenner, F., "The Production of Antibodies," Second edition, Melbourne, Macmillan, 1949.
8. Cinader, B. (Editor), "Antibody to Enzymes; A Three-component System," *Ann. N.Y. Acad. Sci.*, **103**, 493 (1963).
9. Race, R. R., and Sanger, R., "Blood Groups in Man," Fourth edition, Oxford, Blackwell Scientific Publications, 1962.
10. Wolstenholme, G. E. W., and Knight, J., (Editors), *Ciba Foundation Symposium on Complement*, 1965.
11. Kabat, E. A., and Mayer, M. M., "Kabat and Mayer's Experimental Immunochemistry," Second edition, Springfield, Ill., Charles C. Thomas, 1961.
12. Campbell, D. H., "Methods in Immunology," New York, W. A. Benjamin, 1963.
13. Ackroyd, J. F., "Immunological Methods," Oxford, Blackwell Scientific Publications, 1964.
14. Holborow, E. J. (Editor), "Antibodies," *Brit. Med. Bull.*, **19**, 169 (1963).
15. Peetoom, F., "Agar-Precipitation Technique," Springfield, Ill., Charles C. Thomas, 1963.
16. Wolstenholme, G. E. W., and Knight, J., (Editors), "The Immunologically Competent Cell," *Ciba Found. Study Group*, No. 16 (1963).
17. Lawrence, H. S. (Editor), "Cellular and Humoral Aspects of the Hypersensitive States," New York, Hoeber-Harper, 1959.
18. Amos, B., and Koprowski, H., (Editors), "Cell-bound Antibodies," Philadelphia, Wister Institute Press, 1963.
19. Bussard, A. (Editor), "Tolérance Acquise et Tolérance Naturelle à l'Égard de Substances Antigéniques Définies" (in English), Paris, Centre National de la Recherche Scientifique, 1963.
20. Wolstenholme, G. E. W., and O'Connor, M., (Editors), "Ciba Foundation Symposium on Cellular Aspects of Immunity," London, J. & A. Churchill, 1960.
21. Good, R. A., and Gabrielsen, A. E., (Editors), "The Thymus in Immunobiology," New York, Hoeber, Medical Division of Harper and Row, 1964.
22. Grabar, P., and Miescher, P., "Mechanism of Cell and Tissue Damage Produced by Immune

Reactions," Basel, Switzerland, Benno Schwabe; New York, Grune and Stratton, 1962.

23. TALIAFERRO, W. H., TALIAFERRO, L. G., AND JAROSLOW, B. N., "Radiation and Immune Mechanisms," New York, Academic Press, 1964.

24. CINADER, B., (EDITOR) "Antibodies to Biologically Active Molecules," Oxford, Pergamon Press Ltd., 1967.

25. CINADER, B., (EDITOR) "Regulation of the Antibody Response," Springfield, Ill., Charles C. Thomas, 1967.

26. DIXON, F. J., JR., AND HUMPHREY, J. H., (EDITORS), *Advan. Immunol.*, **1–4** (1961–1964).

27. Review articles in *Annual Review of Microbiology*, **14–18** (1960–1964).

28. Original articles in *Immunochemistry*, *Immunology*; *Journal of Experimental Medicine*; *Journal of Immunology*.

BERNHARD CINADER

IMMUNODIFFUSION

"Immunodiffusion" is a word which has been used by immunologists during the past few years to describe a form of serologic analysis in which antigen-antibody precipitation, or some related immunologic reaction, occurs within a semisolid anticonvection medium through which ANTIGEN or ANTIBODY, or both, must diffuse to encounter and react with each other In the *single diffusion* variety of immunodiffusion test, first developed in 1946 by J. Oudin, one reactant is caused to diffuse through the supporting medium containing the other. In a second and more versatile version, the *double diffusion* test simultaneously evolved in 1948 by Ö. Ouchterlony and S. D. Elek in different laboratories, the two reactants are separated by anticonvection medium into which they both diffuse toward each other to mix and react. If one of the reactants is electrophoresed before it is permitted to diffuse toward and react with the other, then a third variety of immunodiffusion is being utilized. P. Grabar and C. A. Williams, Jr. generally are given credit for having developed the first practical form of this technique in 1953, which they later named "immunoelectrophoresis."

There are several advantages to performing antigen-antibody precipitin reactions in semisolid, as opposed to fluid, aqueous media. Since antigen and antibody move about each other and intermingle only by diffusion, complex mixtures of antigens and their respective specific antibodies can be analyzed without preliminary purification. Each individual antigen-antibody system precipitates as an opaque band in the semisolid medium usually distinctly separate from bands produced by other systems which, though also present, act independently. This unique characteristic stems from the high immunologic specificity of precipitin reactions coupled with the unlikelihood that any two systems will match each other exactly in concentration, component diffusion rates, and combining characteristics. Hence, each system will occupy a different cross-section of the supporting medium. For example, human serum as a complex mixture of antigens (albumin and the various globulins) will react with its antiserum in liquid medium to form only a single band of precipitate in the classic interfacial precipitin test and later, on standing, only one mass of precipitate. But a double diffusion test readily will resolve human serum into ten or more constituents, and an immunoelectrophoresis test will reveal between twenty and thirty constituents, depending upon the efficiency of the antiserum used, many of which can be identified by appearance, electrophoretic mobility, or auxiliary indicator techniques. In addition to excellent resolution, immunodiffusion techniques offer unique opportunities for the comparison, study, and identification of complex molecules, like proteins and polysaccharides, which purely physical or chemical techniques cannot distinguish (see also ELECTROPHORESIS).

The primary purposes of the semisolid medium used in immunodiffusion are to prevent convection mixing of reactants and to fix antigen-antibody precipitates as and where they form. Therefore, any of several media can be employed which are unreactive with antigens and antibodies, and which will permit diffusion of substances with molecular weights of a few million or less, but will impede diffusion or settling of particles constituted by complexes of antigen and antibody. Those currently favored are, in order of their usage: 1% agar, cellulose acetate paper, 2% gelatin, and 5% polyacrylamide. Of these, only cellulose acetate is not a gelling agent. It is used as a porous, preformed support for liquid, is not transparent like the others, and is utilized by technical adaptation to its own unique characteristics. Thus, antigen-antibody precipitin bands are not seen in this opaque medium until inert ingredients of the original antigen and antibody mixtures have been washed away and the remaining, fixed precipitates have been stained appropriately. This disadvantage severely limits the value of cellulose acetate and similar media for general immunodiffusion usage.

Anticonvection agents usually are employed for immunodiffusion tests in buffered salt solutions of physiologic strength and neutral or slightly alkaline pH. However, for certain purposes they are used with special solutions. For example, immunoelectrophoresis buffers are of only about one-fifth physiologic strength, and they have pH values between 8 and 9. This is to permit a practical speed for the electrophoresis preceding antigen-antibody reaction. As other instances, some varieties of chicken antisera require buffers of five times physiologic strength for adequate antigen-antibody precipitation, and on the other hand, some rabbit antisera against low molecular weight antigens precipitate these antigens better at low than at normal physiologic strength.

Immunodiffusion tests are performed either in tubes or on flat (plate) surfaces such as microscope slides. In either single or double diffusion tube tests, one reactant, usually antibody, is mixed with buffer containing gelling agent and poured into the bottom section of a narrow, glass tube. After this lower layer has gelled, it is topped either with the second reactant (liquid or gel) in the single diffusion test, or for the double diffusion test with buffer-gel mixture followed in turn, after gelling, by the second reactant. In such tube tests, precipitin reactions appear as disks or bands of precipitate which appear to migrate downward in the single diffusion test under immunologic pressure of the diffusing uppermost reactant or, in the double diffusion test, remain stationary as both reactants diffuse into a common area and react with and neutralize each other. Neither single nor double diffusion tube tests can be recommended for general use, because they do not have the resolution or sensitivity of plate tests, they are of little use for comparative qualitative analyses, and artifacts which all immunodiffusion tests sometimes develop cannot readily be identified as such in them.

Plate tests consist of diffusion of reactants from sources cut from, molded in, or resting upon a layer of anticonvection medium so that their areas of diffusion, usually radial, overlap and they form bands or arcs of precipitate in this medium. The shapes of such arcs, their cross-sectional structure, their position between reactant sources, the convergence or divergence of their growing tips, and their interaction when two reactants (*e.g.*, antigens) from adjacent sources are being compared with respect to reaction with a third (*e.g.*, antibody) to provide information on the identity of an antigen, its concentration, its molecular weight, and the nature of its antibody. Primary information of this sort can be supplemented by auxiliary analytic techniques applied to already formed precipitin arcs, such as the use of histochemical indicators to determine whether an antigen is polysaccharide or protein, whether it is an enzyme, or whether it contains a lipid. After immunodiffusion reactions have been completed, immunodiffusion plates can be washed free of inert contaminants, dried, and stained selectively to provide permanent original records of the tests themselves.

Immunoelectrophoresis tests generally are set up on plates. Typically, a complex mixture of antigens, such as serum, is electrophoresed at 1 volt/cm down the center of the long axis of a microscope slide through a 1-mm layer of 1.5% agar in appropriate buffer. Then a trough is cut parallel to and to one side of this axis of migration and charged with antiserum. As antiserum diffuses from this trough across the migration path of the various separated antigens, its antibodies react with these antigens and detect them by forming arcs of precipitate. Since each arc will have a unique position laterally and longitudinally in the gel, exceedingly high resolution of electrophoresed components can be obtained by this immunoelectrophoresis technique.

The chief reagent for immunodiffusion tests is antiserum, and consequently the application of these tests is limited to substances against which antibodies can be induced to form in experimental animals or man. Although precipitating antibodies usually are employed in these tests, there are adaptations for hemagglutinins and complement-fixing reactions.

Immunodiffusion has been used in many fields including, for example, medicine, pathology, microbiology, taxonomy, immunology, genetics, chemistry, agriculture, industry, and criminology.

Detailed information and extensive bibliographies on this subject can be found in the following references.

References

1. Crowle, A. J., "Immunodiffusion," New York, Academic Press, 1961, 333 pp.
2. Ouchterlony, Ö, "Diffusion-in-gel Methods for Immunological Analysis, II," *Progr. Allergy*, **6,** 30–154, 1962.

Alfred J. Crowle

IMMUNOELECTROPHORESIS

See IMMUNODIFFUSION; also, ELECTROPHORESIS; IMMUNOCHEMISTRY.

Reference

1. Williams, C. A., Jr., "Immunoelectrophoresis," *Sci. Am.*, **202,** 130–140 (March 1960).

INBORN ERRORS OF METABOLISM. See BIOCHEMICAL GENETICS. Particular types of inheritable metabolic defect are discussed also in the articles ALKAPTONURIA; GALACTOSE AND GALACTOSEMIA; HEMOGLOBINS (IN HUMAN GENETICS); MENTAL RETARDATION; METABOLIC DISEASES.

INDIVIDUALITY, BIOCHEMICAL. See BIOCHEMICAL INDIVIDUALITY.

INDOLEACETIC ACID

Indole acetic acid

[indole ring]—CH_2—COOH (N–H in ring)

is an intermediate product of tryptophan metabolism and may be formed by intestinal bacteria. It is one of the *auxins* or plant growth hormones which in extremely low concentrations promote cell elongation but not cell multiplication (see HORMONES, PLANT). Through the mediation of auxins, the cells on one side of a plant shoot (or root) may elongate preferentially and cause the shoot to change its direction of growth.

INFORMATION TRANSFER (IN BIOLOGICAL SYSTEMS)

Much biochemical research in recent years has centered around problems concerning the generation, storage, transmission, and "translation" (from one coded molecular structure "language" to another) of biological information, particularly in those areas discussed in this volume in the articles on GENES; PROTEINS (BIOSYNTHESIS); DEOXYRIBONUCLEIC ACIDS (REPLICATION); RIBONUCLEIC ACIDS (BIOSYNTHESIS). General theoretical considerations of coding and information transfer in biological systems are discussed in reference 1. Some early, partly speculative, approaches to the problem of biological coding are discussed in reference 2. See GENETIC CODE for references to experimental approaches to the "coding problem."

References

1. ELIAS, P., "Coding and Information Theory," in "Biophysical Science—A Study Program" (ONCLEY, J. L., EDITOR), Ch. 26, New York, John Wiley & Sons, 1959.
2. LEVINTHAL, C., "Coding Aspects of Protein Biosynthesis," *ibid.*, Ch. 30.
3. QUASTLER, H., "Information Theory in Biology," Urbana, University of Illinois Press, 1963.

INFRARED AND ULTRAVIOLET SPECTROSCOPY

The interaction of electromagnetic energy with matter offers the biochemist a powerful tool for the study of molecular structure and the analysis of complex materials. Instruments for measuring absorption spectra are commercially available for the region from 1600 Å to 300 μ. No one instrument spans this region. Such an instrument would be technically possible but at present would be economically unwise. Too many compromises would be necessary. Figure 1 identifies the regions of the spectrum in two units of wavelength (angstroms and microns) and energy (frequency and electron volts). There is no universal agreement on the bounds of the various spectrosopic regions. The logarithmic scale used here unduly favors the long-wavelength regions. In many respects a linear frequency scale better weights the information content of various spectral regions. Information on sources, windows, regions of atmospheric absorption and absorptivity are included in this Figure.

The absorptivity curve, of course, is only a general representation of the absorption and band width properties of liquids. The unit (cm^{-1}) indicates the inverse of the optical path length of the pure liquid producing an absorbance of 1 on a logarithmic presentation. Thus a log (cm^{-1}) of +6 shows that a sample 10^{-6} cm thick would have an absorbance of 1 or a transmittance of 10%.

The *far-ultraviolet* region between 1600 and 2000 Å has only recently been exploited by analysts. Intense absorption of oxygen and water vapor of the atmosphere necessitate removing (evacuating) or purging (with nitrogen) the optical path of the instrument. The emission from hydrogen (molecular) offers a convenient continuum source for absorption studies. Almost any photoemission detector with a suitably transparent high-purity silica window can serve as a detector. The dispersing element in the monochromator may be a quartz prism or grating. Almost all materials absorb in the far-ultraviolet region. Most compounds exhibit a maximum absorptivity in this region.[8,11,Ch.5] The intense absorption of the carbonyl group is of particular interest to biochemists. Steroids have been studied.[4] Olefins absorb here. Upon conjugation with other unsaturated linkages, the absorption maxima move to longer wavelengths. Absorption bands are made sharper in the vapor phase than

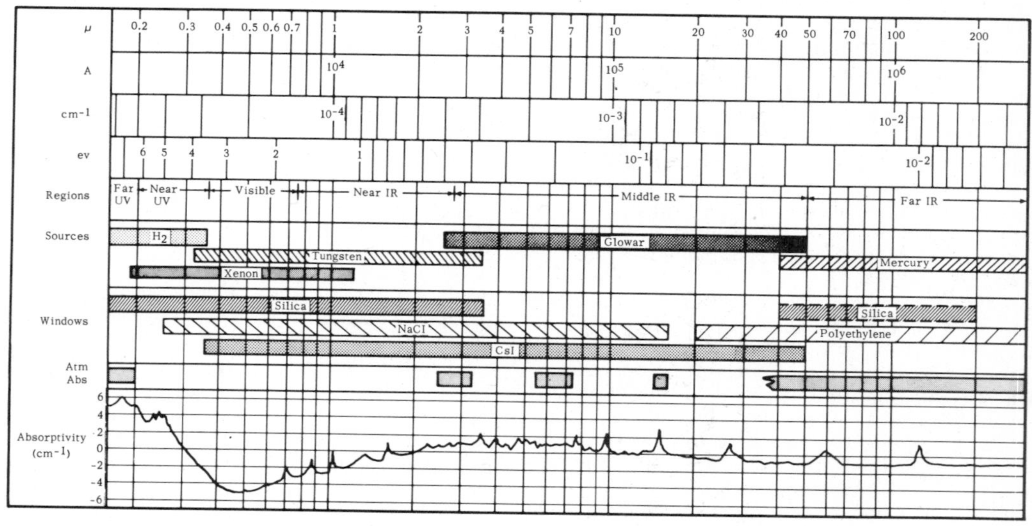

FIG. 1. The electromagnetic spectrum.

in the condensed phases suggesting the utility of combining far-ultraviolet spectroscopy with gas chromatography.[9] To some extent far-ultraviolet spectroscopy of liquids is inhibited by the high absorptivity of most solvents and very thin (0.1 mm) sample cells are usually used. Water, *n*-hexane, trifluoroethanol and acetonitrile are useful solvents.

Far-ultraviolet radiation possesses sufficient energy ($\sim$7 eV) to ionize many compounds; however, the level of intensity in spectrophotometers is not high enough to cause appreciable decomposition in the time required for a spectrum.

The literature of *near-ultraviolet* spectroscopy is very large. The literature is well indexed[22] and collections of spectra have been published.[5] Instrumentation for near-ultraviolet and visible spectroscopy is simple compared to other regions. The hydrogen and tungsten sources are intense emitters. Atmospheric absorption is no problem. Photoemission detectors are very sensitive here. Quartz prisms and gratings are usually used to disperse radiation. Most solvents are transparent except at the short-wavelength end of the near ultraviolet.

A number of spectroscopic methods other than absorption are practiced in the near-ultraviolet region. Excitation spectra are used to obtain the equivalent absorption spectra at concentrations 100 and even 1000 times below that at which conventional absorption spectra can be directly measured.[19] FLUORESCENCE and phosphorescence spectra contribute additional information for the identification of materials.[13,14,21] Proteins exhibit fluorescence because of aromatic constituents.[19] Polarized light is widely used to obtain optical rotatory dispersion spectra. This technique is particularly valuable in identifying the position of substituents on complex molecules such as the steroids.[11,Ch.1] The polarization of exciting and fluorescing light is also valuable for studying the mobility of molecules in solution and for following reactions with "trace" molecules.[23]

The fast response of ultraviolet detectors makes this region ideal for studying the kinetics of reactions. Apparatus for the rapid mixing of chemicals have been developed under the name of "stopped flow" instrumentation.[18] Similar studies are made by rapidly changing the temperature of the solution and are called "temperature jump" studies.[6]

Perhaps the most distinctive property of most organic chemicals in the *visible region* is their high transparency. While our eyes pay particular attention to colored material, the fact is that most compounds are transparent here. Consequently the smaller number of colored compounds is all the more useful for analytical purposes. When a material will complex with a test reagent to produce a stable color, a method of colorimetric analysis is possible.[1] Many clinical procedures are based upon such complexes.

The *near-infrared* spectral region is characterized by overtone and combination absorption bands accompanying the vibrations of hydrogen atoms in organic compounds. Few electronic transitions extend into this region. Almost all organic molecules have low absorptivity at the short-wavelength end of the near infrared and have moderately high absorption at the long-wavelength end.[16] Studies of hydroxyl and amino groups are particularly favored in this region. Studies of hydrogen-bonding phenomenon are made by observing the shift of the OH stretching bands at 2.7 and 1.4 μ.[14]

Instrumentally the near-infrared region is favored because the blackbody emission curves of incandescent sources such as the common tungsten bulb reach a maximum here.[7] The lead sulfide photoconductive detector is usually preferred. Quartz prisms and gratings are usually used to disperse the radiation. Silica cells from 0.01–10 cm in path length are employed. Carbon tetrachloride is the most widely used solvent.

Many molecular groups such as C=O, C—H, C=C, C=N, etc., exhibit characteristic vibrational absorption bands in the *middle-infrared* region.[2] The wavelengths of these absorption bands are often little affected by the remainder of the molecule. As a result, the middle-infrared region is preferred for determination of molecular structure. In addition to these characteristic "group frequency" bands, most molecules exhibit "skeletal" bands whose wavelength and intensity are specific for the total molecule. Consequently, the middle-infrared region is often called the "finger-print" region. Several large collections of spectra are available for comparison with the spectrum of unknowns.[3,17] Microorganisms have been studied.[20]

Polarized infrared radiation can be used to study the orientation of molecules in crystals and films.

Solids are often studied as slurries in mineral oil, as suspensions in solid pressed disks of potassium bromide, or by reflection from an interface of the sample with a transparent material of high refractive index (called attenuated total reflection).[15]

Instrumentation for infrared spectroscopy is more complicated than far-ultraviolet spectroscopy because of the lower emission intensity of sources and the response characteristics of detectors. Windowless incandescent ceramic sources are most widely used. Prisms of NaCl, LiF, CaF_2, KBr, and CsI are used, but gratings are preferred for dispersion of the radiation. Thermocouple detectors are most widely used as detectors, but their low impedance and low output signals require carefully designed amplifiers. The complexity of infrared spectra forces the automatic recording of spectra. Water vapor and carbon dioxide absorb in certain regions of the middle infrared. Purging is desirable but not always necessary. Carbon tetrachloride and carbon disulfide are the most frequently used solvents. Sample cells usually have optical paths from 0.1–1.0 mm.

Instruments for the study of *far-infrared* radiation have only recently become commercially available. Figure 2 is a photograph of the recently introduced Beckman IR-11 spectrophotometer.

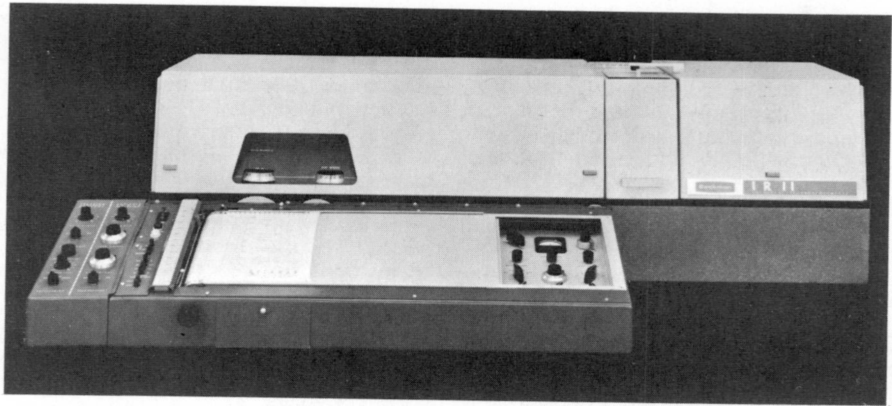

FIG. 2. Infrared spectrophotometer.

Source radiation is very weak in this region, and plasma radiation from a mercury vapor lamp is preferred. All common solid materials absorb in this region. Thin windows of diamond, silica and polyethylene may be used. No prism materials are available, and gratings must be used to disperse the light. Severe filtering is required to remove higher orders of the gratings. Water vapors absorb intensely and must be removed from the optical path. Most samples exhibit lower absorptivity in the far infrared than in the middle infrared. Sample paths from 1–10 mm thickness are usually used.[12]

Certain molecular groups containing bromine, iodine, phosphorus and metalloorganic groups exhibit characteristic absorption bands in the far-infrared region, but most absorption must be classed as skeletal and therefore unique to the total molecule. Among the studies of biochemical interest have been those on amides and esters.[10]

References

1. ALLPORT, N. L., AND KEYSER, J. W., "Colorimetric Analysis," Vol. 1, "Determination of Clinical and Biochemical Significance," London, Chapman & Hall Ltd., 1957.
2. BELLAMY, L. J., "The Infrared Spectra of Complex Molecules," Second edition, London, Methuen & Co. Ltd., 1958.
3. DOBRINER, K., KATZENELLENBOGEN, E. R., AND JONES, R. N., "Infrared Absorption Spectra of Steroids," New York, Interscience Publishers, 1953.
4. FERRARO, J. R., AND ZIOMECH, J. S., (EDITORS), "Developments in Applied Spectroscopy," Vol. 2, New York, Plenum Press, 1963.
5. FRIEDEL, R. A., AND ORCHIN, M., "Ultraviolet Spectra of Aromatic Compounds," New York, John Wiley & Sons, 1951.
6. HAMMES, G. G., AND FOSELLA, P., "A Kinetic Study of Glutamic-Aspartic Transaminase," *J. Am. Chem. Soc.*, **84,** 4644 (1962).
7. KAYE, W., "Near Infrared Spectroscopy. II. Instrumentation and Techniques," *Spectrochim. Acta*, **7,** 181–204 (1955).
8. KAYE, W., "Far Ultraviolet Spectroscopy. II. Analytical Applications," *Appl. Spect.*, **15,** 130 (1961).
9. KAYE, W., "Far Ultraviolet Spectroscopic Detection of Gas Chromatograph Effluent," *Anal. Chem.*, **34,** 287 (1962).
10. MIYAZAWA, T., "Internal Rotational and Low Frequency Spectra of Esters, Monosubstituted Amides and Polyglycine," *Bull. Chem. Soc. Japan*, **34,** 691 (1961).
11. NACHOD, E. C., AND PHILLIPS, W. D., "Determination of Organic Structures by Physical Methods," Vol. 2, New York, Academic Press, 1962.
12. PALIK, E. D., "A Far Infrared Bibliography," *J. Opt. Soc. Am.*, **50,** 1329 (1960).
13. PARKER, C. A., AND HATCHARD, C. G., "The Possibilities of Phosphorescence Measurement in Chemical Analysis; Tests with a New Instrument," *Analyst*, **87,** 664 (1962).
14. PIMENTAL, G. C., AND MCCLELLAND, A. L., "The Hydrogen Bond," San Francisco, W. H. Freeman & Co., 1960.
15. POTTS, W. J., JR., "Chemical Infrared Spectroscopy," Vol. 1, "Techniques," New York, John Wiley & Sons Inc., 1963.
16. REILLEY, C. N., *Advan. Anal. Chem. Instr.*, **1** (1960). New York,
17. "Sadtler Infrared Library," Sadtler Research Laboratory, 1517 Vine St., Philadelphia 2, Pa.
18. SPENCER, T., AND STURTEVANT, J. M., "The Mechanism of Chymotrypsin-Catalyzed Reactions III," *J. Amer. Chem. Soc.*, **81,** 1874 (1959).
19. TEALE, F. W. J., AND WEBER, G., "Ultraviolet Fluorescence of the Aromatic Amino Acids," *Biochem. J.*, **65,** 476 (1957).
20. THOMPSON, H. W. (EDITOR), New York, Interscience Publishers, 1961.
21. UDENFRIEND, S., "Fluorescence Assay in Biology and Medicine," New York, Academic Press, 1962.
22. UNGNADE, H. (EDITOR), "Organic Electronic

Spectral Data," Vols. I and II, Interscience Publishers, 1960.

23. WEBER, G., "Fluorescence-Polarization Spectrum and Electronic-Energy Transfer in Tyrosine, Tryptophan and Related Compounds," *Biochem. J.*, **75**, 335 (1960).

WILBUR KAYE

INHIBITORS, ENZYME. See ENZYMES (ACTIVATION AND INHIBITION).

INOSITOL

Of the various constituents found plentifully in nature, none has remained such an enigma as the inositols. The first of these hexahydroxycyclohexanes or cyclitols was isolated from muscle in 1850 and has subsequently been shown to be the only one of nine isomers to possess full biological activity. *Myo-*, *meso-* or *i*-inositol are synonyms for this cyclic polyalcohol, which is a white, crystalline solid, melting at 225°C, soluble in water, and insoluble in alcohol and ether; its empirical formula is that of glucose, $C_6H_{12}O_6$. Various yeasts and fungi require *i*-inositol for growth, and thus are used for bioassay; bacteria generally need no exogenous source. In the 1940's, several reports showed that certain mammals exhibited improved growth and clearing of alopecia after administration of *i*-inositol. Since it is also present in brewer's yeast, it was included in the vitamin B complex, yet within a decade other investigators, feeding more sophisticated rations, could find no specific symptoms in subjects fed deficient rations or any improved performance after supplementation with inositol. Under defined conditions, it is accepted as a lipotropic factor, mobilizing fat to a lesser extent than choline (see FAT MOBILIZATION).

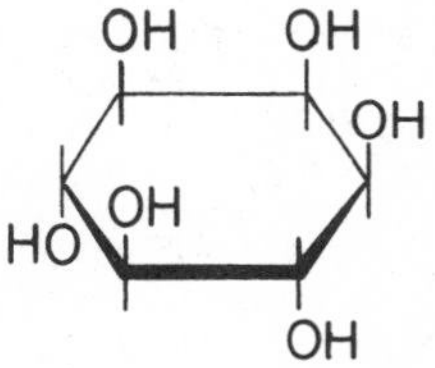

The widespread distribution of inositol in plants and animals has fostered the implication that it must be present functionally. It constitutes up to 2.6% by weight of fresh tissue, secretory glands generally containing highest levels, *e.g.*, thyroid and pituitary. Nerve and cardiac tissue likewise contain significant amounts, while the content of lactating rat mammary glands is elevated 5–6 times over that found during pregnancy. Two sources, dietary and biosynthetic, account for its relatively high levels in tissues, while the intestinal bacteria also produce it. As shown in germ-free rats and mice, among the tissues which most actively synthetize it from labeled glucose is the kidney; this organ also possesses an active enzyme which destroys inositol irreversibly, by converting it to glucuronate. It may also be metabolized to CO_2, yielding energy, like the carbohydrate it resembles. *i*-Inositol has been reported to serve as precursor for other isomeric cyclitols in plants, some being converted to mono- and dimethyl ethers. Cereal grains are the richest plant source, containing inositol as the hexaphosphoric acid ester, phytic acid, which readily forms complexes with polyvalent cations. Within the digestive tract, active phytases are secreted which liberate inositol from these phytates.

In mammalian tissues, biosynthesis of inositol phosphatides (see PHOSPHATIDES AND PHOSPHATIDIC ACIDS) has been demonstrated, in which one, two or three of the hydroxyl moieties of inositol may be esterified with phosphate and linked through one of these to a diglyceride. An important essential for this reaction appears to be cytidine diphosphodiglyceride. Exact function of these inositol phosphatides has not been delineated; with other phospholipids they are present in the membranes of cells, as well as in subcellular organelles such as MICROSOMES. Currently, investigators consider that the inositol phosphatides control permeability of cells or subcellular units to the flux of substances into or out of such structures. That inositol is of prime importance to survival (and reproduction) of the cell was clearly shown by TISSUE CULTURE; of 18 strains of human cells from normal and malignant tissues, each fastidiously required inositol at 10^{-6} *M*.

References

1. WEIDLEIN, E. R., JR., "The Biochemistry of Inositol," Pittsburgh, Pa., Mellon Institute, 1951, 53 pp.
2. Various contributors to "The Vitamins," Vol. II (SEBRELL, W. H., JR., AND HARRIS, R. S., EDITORS), New York, Academic Press, 1954.
3. HAWTHORNE, J. N., "The Inositol Phospholipids," *J. Lipid Res.*, **1**, 255–80 (1960).
4. EAGLE, H., OYAMA, V. I., LEVY, M., AND FREEMAN, A. E., "Myo-inositol as an Essential Growth Factor for Normal and Malignant Human Cells in Tissue Culture," *J. Biol. Chem.*, **226**, 191–206 (1957).
5. FREINKEL, N., AND DAWSON, R. M. C., "The Synthesis of Meso-inositol in Germ-free Rats and Mice," *Biochem. J.*, **81**, 250–4 (1961).
6. HAUSER, G., AND FINELLI, V. N., "The Biosynthesis of Free and Phosphatide Myo-inositol from Glucose by Mammalian Tissue Slices," *J. Biol. Chem.*, **238**, 3224–8 (1963).

E. A. DOISY, JR.

INSECTICIDES

A wide variety of chemical agents are used to minimize the damage of insects to man and animals, and especially in the production of food, feed and fiber. Among the natural insecticidal products of plant origin, some of which have been

TABLE 1. SELECTED EXAMPLES OF INSECTICIDE CHEMICALS

(A) Botanicals

Pyrethrin I

Rotenone

Nicotine

(B) Chlorinated hydrocarbons

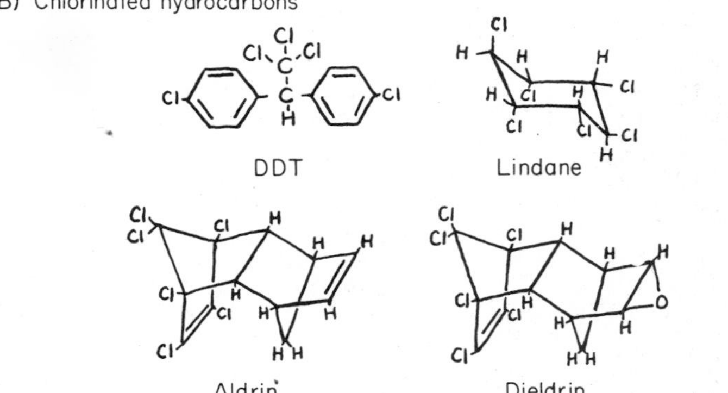

DDT

Lindane

Aldrin

Dieldrin

(C) Organophosphates

Parathion

Ronnel

Malathion

Phorate

(D) Miscellaneous

Carbaryl (carbamate insecticide)

Tepa (chemosterilant)

Piperonyl butoxide (synergist)

used as insecticides for centuries, are the botanicals pyrethrum, rotenone and nicotine (Table 1). The era of synthetic organic insecticides, initiated during World War II, has led to the development and extensive use of chlorinated hydrocarbon (*e.g.*, DDT, lindane, aldrin and dieldrin), organophosphorus (*e.g.*, parathion, ronnel, malathion and phorate) and carbamate (*e.g.*, carbaryl) insecticides (Table 1). Most of the organic insecticides poison insects not only when ingested but also on contact because they readily penetrate the insect cuticle. Some of the organophosphorus insecticides (such as phorate) are systemic, and are absorbed and translocated by plants to kill insects feeding distant from the site of application. When the insecticide is of low toxicity to mammals (such as ronnel), it may be used as a chemotherapeutic agent to control internal insect parasites of livestock. Mites are controlled by many chlorinated hydrocarbons, including analogues of DDT which are rather selective acaricides, and by organophosphates. Many other organic compounds are used as insecticides or acaricides; among them are dinitrophenols and fluoroacetic acid derivatives. Inorganic compounds, containing arsenic and fluorine, are also used as insecticides to a limited, diminishing extent. Insect reproduction can be inhibited by several antimetabolites (antifolics, antipurines, antipyrimidines; see CANCER CHEMOTHERAPY) and by certain alkylating agents (*e.g.*, tepa; see Table 1). Organic compounds are also used to attract and repel insects, some of them being highly potent and selective.

The botanical, chlorinated hydrocarbon, organophosphorus and carbamate insecticides act as nerve poisons. Pyrethrum acts in the central nervous sytem to block NERVE IMPULSE TRANSMISSION by a mechanism which has not been biochemically defined. Rotenone is a selective inhibitor of diphosphopyridine nucleotide-flavin-linked electron transport. Nicotine acts on the cholinergic system in a manner similar to ACETYLCHOLINE, but probably only at the autonomic ganglia. It gives rise to stimulation of voluntary muscles and of the sympathetic innervation of smooth muscle and glands. The chlorinated hydrocarbon insecticides act on the central nervous system resulting in hyperactivity, tremors and convulsions, but the exact mechanism of this action has not been elucidated, although levels of certain nitrogen compounds which may be considered to be of particular importance to the nervous system are altered during poisoning. Organophosphorus and carbamate insecticides act as acetylcholinesterase inhibitors (see ACETYLCHOLINE AND CHOLINESTERASES), an action explaining the biochemical and biophysical lesions implicated in almost all signs of this type of poisoning. Organs innervated by the parasympathetic system are potentiated in response to acetylcholine by these agents, resulting in excessive salivation, lacrimation, myosis, vomiting and other signs. Fibrillations, fasciculations, weakness, ataxia and paralysis appear in the voluntary muscles. Effects on the central nervous system, directly or as reflex responses to peripheral action, lead to asphyxia or, occasionally, to cardiac failure. In the hydrolysis of the normal substrate, acetylcholine, an extremely unstable acetyl derivative of cholinesterase is formed. On reaction with carbamates, a more stable carbamoyl esterase results, and reaction with phosphates yields an even more stable phosphoryl esterase. The enzyme-inhibitor complex formed prior to carbamoylation or phosphorylation may also be important in poisoning, particularly with carbamates. The active site of sensitive esterases, such as acetylcholinesterase, generally consists of the imidazole group of histidine and a serine with the amino-nitrogen linked to the α-carboxyl of a dibasic amino acid residue. It is the serine within this active site that is carbamoylated or phosphorylated. The carbamates may also have a direct effect at the cholinergic receptor sites [see also ENZYMES (ACTIVE CENTERS)].

Selective toxicity is a desirable characteristic of a good insecticide. The biochemistry of the insect nervous system is similar in most respects to that of the mammalian nervous system. However, it appears that insects usually lack cholinergic systems for synaptic transmission at neuromuscular junctions, and the cholinergic SYNAPSES of their central nervous system are more highly protected from penetration of ionized materials than those of mammals. Accordingly, ionized organic nerve poisons are usually more toxic to mammals than insects. Many of the insecticides are much more toxic to insects than to mammals on an equivalent weight basis. Such selectivity may result in part from a slower detoxification of the insecticide in the insect than in the mammal. Malathion is such a compound. In order to act as a cholinesterase inhibitor it must be oxidized *in vivo* from the phosphorodithioate (P=S) form to the active phosphorothiolate (P=O) form, without undergoing other structural modification. The toxicity is completely destroyed by hydrolysis of the carboethoxy groupings, a reaction which occurs very rapidly in mammals. The proportion of the dose converted to the phosphorothiolate toxicant is greater in insects than mammals because of the more rapid hydrolysis of the ester groupings by mammals. In other cases, the susceptible species appear to oxidize phosphorothionates to the active phosphates more rapidly than resistant species. Differences in detoxication rates probably also contribute to the selective toxicity of carbamate insecticides. Insect cholinesterases vary considerably with species in their sensitivity to inhibition by organophosphates and carbamates, an enzyme specificity which may contribute to selective toxicity. Fundamental biochemical differences between insects and mammals might be useful for future development of selective insecticides. Some of these are as follows: different hormones govern the growth, differentiation and reproduction processes; nitrogenous wastes are eliminated as uric acid in insects and as urea in mammals; mammals but not insects synthesize sterols from acetate; α-glycerophosphate dehydrogenase may be more important in insect muscle than in mammalian muscle glycolysis; insects and mammals may differ in the nature of the neuromuscular transmitter substance. The desired degree of selectivity is high, not only affording high toxicity to insects without hazard to man and other higher animals, but also making possible control of the pest without harm to other exposed organisms (including insect predators and parasites). Partial success has been achieved in this respect with plant systemics that are lethal only to the pests feeding on the treated plants.

Resistance Development. The pest population changes on exposure to insecticides for a few to many generations in a manner making control more difficult. Repeated use of insecticides tends to single out as survivors those insects which have the biochemical mechanisms to resist the toxicants and the appropriate gene constitution to transmit this ability. Development of resistance poses a major threat to the continued efficient use of any insecticide, new or old, and has already seriously restricted the areas in which chlorinated hydrocarbons can be used and partially, also, the utility of organophosphates and carbamates. Exposure to one class of insecticides sometimes selects strains resistant to another class of insecticides of entirely different chemical structure and mode of action. Most strains of house flies resistant to DDT have an unusually high level of an enzyme, DDT-dehydrochlorinase, which detoxifies DDT by dehydrochlorination of the trichloroethyl group to yield the dichloroethylene analogue, DDE. Enzymes which hydrolyze organophosphates, or hydroxylate or hydrolyze carbamates, generally are more active in certain strains resistant to these insecticides. An analogue involving only a slight structural modification frequently will still be toxic to the resistant strain. Other compounds may synergize the toxicity of the insecticide by interfering with the detoxification mechanism in the resistant strain. These synergists may be chemicals of the same type as the insecticide, in which case they are usually quite selective in their action, or they may be methylenedioxyphenyl compounds such as piperonyl butoxide (Table 1) and analogues which reduce the detoxification rate of many insecticides and which are commonly used with pyrethrum. Detoxification is not the only resistance mechanism, since penetration rate through the cuticle, direct nerve sensitivity and other factors also vary, in some cases, between resistant and susceptible strains.

Pathways for *biodegradation* of insecticides are similar to those for other compounds containing the same metabolizable groupings. The metabolic pathway for many insecticides is quite well understood because this information is required to evaluate the potential hazards of persisting residues. Toxic metabolites which create special problems in the analysis and interpretation of residue hazards usually arise from oxidative pathways of insecticide metabolism. Oxidative conversions include: phosphorothionates to phosphates; thio ethers to the corresponding sulfoxides and sulfones; N-dealkylation and O-dealkylation; epoxidation, such as in the *in vivo* conversion of aldrin to dieldrin; hydroxylation of N-methyl groups and aromatic rings. Hydrolysis includes attack on all the types of ester groups in organophosphates, and on carbamate and pyrethroid esters. DDT undergoes a variety of metabolic reactions. It is not only dehydrochlorinated to DDE, but sometimes it undergoes reductive dechlorination to the dichloroethane analogue (DDD), and further reactions, including oxidation, to finally yield the acetic acid analogue (DDA).

Use of insecticides on agricultural crops and livestock creates a potential hazard to consumers from residues persisting as food contaminants. Insecticides that are volatile, unstable on exposure to light or weathering, or easily biodegraded pose a lesser problem than the more persistent materials. Generally, the chlorinated hydrocarbons and inorganics are the most persistent insecticides, and organophosphorus and carbamate insecticides are intermediate, and the pyrethroids are the least persistent ones. Of course, the acute and chronic toxicity of the compound, the dosage used, the method of application, and the residue level on the crop are important considerations, too. Final evaluation of the hazards depends on a knowledge of the fate of the compound in the organisms it contacts and, particularly, on the biochemical lesions initiated by the insecticide. (See also TOXINS, PLANT for plant-derived insecticides.)

References

1. HAYES, W. L., JR., "Clinical Handbook on Economic Poisons. Emergency Information for Treating Poisoning," *Public Health Serv. Publ. No.* 476 (1963) 144 pp.
2. HEATH, D. F., "Organophosphorus Poisons. Anticholinesterases and Related Compounds," New York, Pergamon Press, 1961, 403 pp.
3. METCALF, R. L., "Organic Insecticides. Their Chemistry and Mode of Action," New York, Interscience Publishers, 1955, 392 pp.
4. MULLER, P. (EDITOR), "DDT, The Insecticide Dichlorodiphenyltrichloroethane and Its Significance," Vol. I, 1955, 299 pp.; Vol. II, 1959, 570 pp.; Basel, Birkhäuser Verlag.
5. NEGHERBON, W. O., "Handbook of Toxicology," Vol. III, "Insecticides," Philadelphia, W. B. Saunders Co., 1959, 854 pp.
6. O'BRIEN, R. D., "Toxic Phosphorus Esters. Chemistry, Metabolism and Biological Effects," New York, Academic Press, 1960, 434 pp.

JOHN E. CASIDA

INSULIN (FUNCTION IN METABOLISM)

The various carbohydrates which are ingested in foods (STARCH, GLYCOGEN, sugars) are transformed during digestion, absorption and passage through the liver. Ultimately they all form glucose and are metabolized as such. In man and many mammals, the blood serum contains approximately 100 mg of glucose for each 100 ml of fluid. This level is maintained despite the continual utilization by the tissues, because the liver replenishes the blood glucose from its store of glycogen and because the liver can manufacture glucose from non-carbohydrate materials (amino acids, glycerol, lactate, etc.; see GLUCONEOGENESIS). During active absorption of a meal, the glucose level of blood rises, because the influx exceeds the rate at which it can be disposed. This rise in blood glucose seems to serve as a specific stimulus to the Beta Cells of the pancreas, and as a result insulin is secreted. The function of this hormone is to increase the rate at which certain tissues and organs accept the circulating glucose into their cells for storage, oxidation and transformation. As a result of the action of insulin, blood glucose falls and is rapidly deposited as glycogen especially in muscle, transformed to fat primarily in adipose tissue, and oxidized to CO_2 and H_2O in both of these large tissue masses. Insulin produces these effects largely because of its enhancement of the rate by which glucose enters the cell from the surrounding medium. In addition, insulin leads to increased glycogen storage and reduced sugar output by the liver. The hormone also favors the synthesis of cellular proteins from amino acids. Insulin does *not* have any measurable effect on the carbohydrate metabolism of the brain, the gastrointestinal tract or the kidney.

In general, insulin *conserves* foodstuffs, favoring carbohydrate storage, fat deposition and protein synthesis. When the hormone is absent or when its action is impeded, metabolic diabetes results. This disorder is characterized, as expected, by foodstuff wastage. The blood sugar level rises, and much of it is excreted in the urine; fat depots are mobilized, the fatty acids in the blood increase in amount, and the liver produces from these the four-carbon β-keto acids. Protein synthesis is impeded, there is increased protein breakdown, and the resulting amino acids are transformed to sugar. Diabetes is thus characterized by the following abnormalities: hyperglycemia, glucosuria, frequency of urination, thirst, and hunger with loss of weight and strength. When the fatty acid breakdown is severe, the resulting organic acids deplete the alkali reserve of the body producing acidosis and coma. The administration of insulin "pushes" glucose into cells and thus forces the enzymatic machinery into the predominant use of carbohydrate. This in turn inhibits fat and protein breakdown. The blood sugar level falls, the keto acids practically disappear, and glucosuria diminishes sharply. By careful manipulation of dosage, a normal metabolic state can be effectively maintained (see also METABOLIC DISEASES).

If insulin is given in excess so that the blood sugar level is depressed to about 40 mg/100 ml, counterregulatory mechanisms begin to operate, namely secretion of ADRENALINE, glucagon, and pituitary GROWTH HORMONE. Adrenaline and glucagon activate the enzyme systems which facilitate glycogen breakdown, liberating glucose to restore its blood level to normal. The pituitary "growth" hormone, as well as adrenaline, favors the release of fatty acids from stores; the latter in turn lead to inhibition of carbohydrate breakdown. If these countereffects are insufficient to raise the blood sugar level, the central nervous system becomes irritated due to lack of energy fuel. Prolonged hypoglycemia may lead to coma, neurological damage and death. This is not due to an action of insulin on brain tissue; it is a consequence of the low glucose environment created by the action of insulin on muscle and adipose tissue.

Despite the elucidation of the chemical structure of the insulin molecule, its action on a molecular level remains unknown to date. One definite site of action is at the cell membrane where the hormone appears to remove an inhibition to the full, unhampered activity of a sugar transport (carrier?) system. Effects on protein synthesis, on inhibition of sugar output by liver, on fatty acid exit from tissues, etc., are seemingly independent of glucose transport. As yet, the action or actions of insulin cannot be obtained or described except in terms of intact tissues or cells.

The scientific literature concerning insulin is immense. A few references are here listed which would provide more details and further documentation.

References

1. KRAHL, M. E., "The Action of Insulin," New York, Grune & Stratton, 1962.
2. LEVINE, R., "On Some Biochemical Aspects of Diabetes Mellitus," *Am. J. Med.*, **31**, 901 (1961).
3. LEVINE, R., "Concerning the Mechanisms of Insulin Action," *Diabetes*, **10**, 421 (1961).
4. PARK, C. R., *et al.*, "Regulation of Glucose Uptake in Muscle," *Recent Progr. Hormone Res.*, **17**, 493 (1961).

R. LEVINE

INSULIN (STRUCTURE)

Insulin is a polypeptide hormone produced by the β-type islet cells of the pancreas (see HORMONES, ANIMAL; INSULIN (FUNCTION IN METABOLISM)]. The work of Sanger's group on insulin structure was an important biochemical achievement in that it was the first case in which the complete primary structure (amino acid sequence) and disulfide bridge attachment of a protein was established. The methods that were used are discussed in PROTEINS (END GROUP AND SEQUENCE ANALYSIS). The structure of bovine insulin is as follows:

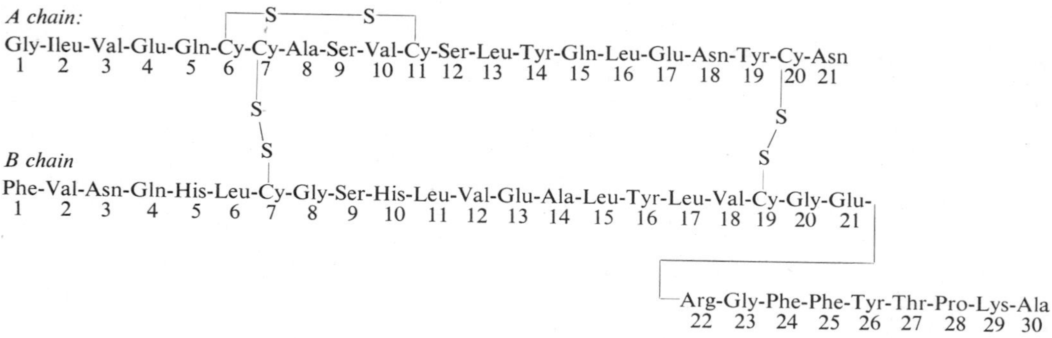

A chain:

Gly-Ileu-Val-Glu-Gln-Cy-Cy-Ala-Ser-Val-Cy-Ser-Leu-Tyr-Gln-Leu-Glu-Asn-Tyr-Cy-Asn
1 2 3 4 5 6 7 8 9 10 11 12 13 14 15 16 17 18 19 20 21

(S—S bridge between A6 and A11; S—S bridges A7–B7 and A20–B19)

B chain

Phe-Val-Asn-Gln-His-Leu-Cy-Gly-Ser-His-Leu-Val-Glu-Ala-Leu-Tyr-Leu-Val-Cy-Gly-Glu-
1 2 3 4 5 6 7 8 9 10 11 12 13 14 15 16 17 18 19 20 21

Arg-Gly-Phe-Phe-Tyr-Thr-Pro-Lys-Ala
22 23 24 25 26 27 28 29 30

Reference

1. Thompson, E. O. P., "The Insulin Molecule," *Sci. Am.*, **193,** 36 (May 1955).

INTERFERONS

Interferons are a class of PROTEINS capable of inhibiting VIRUS REPLICATION in mammalian cells. They are produced by the cells of many kinds of vertebrates. Interferons are of cellular origin and their production may be induced or stimulated by a fairly wide range of materials. They may be produced *in vitro* by cells in tissue culture or *in vivo*. In the latter case, they may be found in the serum and in extracts of tissues from animals after the injection of a proper stimulus for their production.

To be considered an interferon, the viral inhibitor must meet the following criteria. It must inhibit a wide range of viruses, including both ribonucleic acid- and deoxyribonucleic acid-containing viruses. Its biologic activity must be destroyed by the action of PROTEOLYTIC ENZYMES but not by NUCLEASES. It must be small enough so that it is not sedimented by centrifugation at $100{,}000 \times g$ for 2 hours, but it must be large enough to be non-dialyzable. Its anti-viral activity must survive incubation between pH 2 and 10. Its viral inhibitory ability must not reside in any direct action against the viruses inhibited but must result from some intracellular alterations. It must act only on cells from the same species as those from which it was obtained. Some exceptions to this requirement have been reported when cells from closely related species were employed. Interferon activity can be precipitated with ammonium sulfate at a concentration of 75% but not at 50% saturation. It can be quantitatively recovered from the precipitate. Generally, interferons so far tested are stable to precipitation with the heavy metals zinc and cobalt. They are also generally fairly heat stable, resisting incubation at 70°C for 1 hour. Interferons are a heterologous group of proteins as indicated by physicochemical differences between interferons from different species, as well as the species specificity already mentioned. Thus far, it is impossible to say whether heterogeneity also exists among interferons from a single species, although there are indications of this.

Molecular weights have been determined for interferons prepared from chicken cells and mouse cells. The former have molecular weights of 38,000, while the latter are about 26,000. The interferons prepared from chicken cells are more resistant to heat than those from mouse cells. Recently, a material fulfilling the criteria for an interferon has been recovered from the serum of mice stimulated with endotoxin. It was found to be relatively sensitive to heat and to have a molecular weight in excess of 100,000. Too few interferons have been examined to permit a conclusion on the physicochemical characteristics. At least one highly purified interferon from chicken cells was found to be free from nucleic acid and contained only a trace of unidentified carbohydrate. It was slightly basic with an isoelectric point around pH 8.0. The amino acid composition or the sequences of amino acids are not available for any interferons.

Most interferons so far studied were recovered from cells in tissue culture. They have been recovered from cells of the mouse, chicken, human, rabbit, calf and monkey. Various materials were utilized to stimulate their production. Included are: viruses, rickettsia, nucleic acids, nucleotides, and a polyanionic carbohydrate "statolon." Viruses have been most widely used. When a virus stimulus was used, factors such as the type and strain of virus, its virulence for the cell, whether it was active or killed, and the dose added have proved important for each system and quite variable in different laboratories. In addition, environmental factors such as whether the cells were primary cultures or continuous cell lines, the temperature of cultivation, and the pH and composition of the medium have proved important but variable from one cell virus system to another. *In vivo*, the above agents and, in addition, some bacteria and endotoxin have been shown to induce interferon formation. The wide spectrum of stimuli for the production of interferons is of interest, but it must be pointed out that the fundamental mechanism involved is not understood and therefore no generalized statement which might connect these agents is possible.

As mentioned previously, interferons inhibit virus replication by inducing some intracellular changes in sensitive cells. For virus inhibition to become manifest after the addition of interferon, the cells to be affected must continue to synthesize DNA-dependent RNA and protein [see RIBONUCLEIC ACIDS (BIOSYNTHESIS); PROTEINS (BIOSYNTHESIS)]. Thus it appears that interferon acts as an inducer to cause the cells to make new products which are somehow responsible for their resistance to virus infection. The necessity for the synthesis of new cellular products to arise provides a reasonable explanation for the fact that interferon must be present for several hours before its maximum anti-viral effects are realized. If products other than the added interferon are responsible for its activity, this might also explain its great activity. Thus about 0.01 μg of interferon is sufficient to make a million chicken embryo cells completely resistant to any of the effects of equine encephalomyelitis viruses. This *in vitro* activity is greater than that of the antibiotics penicillin and tetracycline assayed with highly sensitive organisms. A considerable amount of work has indicated that viruses susceptible to inhibition by interferon activity are not prevented from adsorbing to the cells or from penetrating to their interior. Reliable evidence indicates that the replication of viral RNA is inhibited. The mechanism by which this replication is prevented remains unknown. The solution of this question will surely impart important knowledge to our understanding of virus replication and its control. It should be pointed out that cells treated with interferon are not necessarily completely incapable of virus synthesis and that cells receiving smaller amounts of interferon may produce smaller amounts of virus and at slower rates. Interferon added after infection of cells has only slight, if any, inhibiting ability. Also, viruses vary considerably in their sensitivity to the inhibitory action of interferon. Thus it is impossible to prevent cell destruction by some viruses despite the absence of new virus synthesis, and still other viruses are practically immune to several hundred times as much interferon as is required to inhibit the most susceptible viruses.

Because interferon inhibited a wide range of viruses, much work has been done to examine its possible use as a prophylactic or therapeutic agent. For such purposes, a substance must be nontoxic and non-immunogenic. Interferon, in the concentrations available, seemed to fulfil both conditions. It is antigenic, however, if given to a proper animal in sufficient doses for long enough periods of time. Evidence from tissue culture work indicates that interferon is a powerful inhibitor when used prophylactically. It, however, has little effect once virus replication has been initiated. The duration of its inhibitory action is also quite transient, gradually diminishing and disappearing over the course of 3–4 days. Likewise, in animals and in man, it has proved valuable when used prophylactically in connection with localized infections. However, disappointing results have been obtained in attempts to spare animals from fatal generalized infections by the parenteral injection of interferon. Perhaps these results are in part due to the unavailability of sufficient amounts of interferon. Sizeable quantities are as yet quite impractical to prepare. However, excretion, destruction, binding at inappropriate tissue sites, etc., also might be responsible, in part, for the less than encouraging results.

More recently, it has been shown that injection of a number of materials (*e.g.*, viruses, endotoxin, statolon, etc.) induced the rapid appearance of what is considered high concentrations of interferon in the serum of animals. Initiation of this interferon response in animals has provided a considerable reduction in the number of deaths resulting from the inoculation of several normally lethal viruses a short time later. Thus it appears that a more successful approach to viral prophylaxis is to induce the animal to make its own interferon. Only a few results of this type are currently available, but on the basis of the successes so far achieved, there is a real need to know the nature of the interferon-inducing mechanism. There is every reason to believe that chemoprophylaxis against a wide range of viruses will be forthcoming.

References

1. Ho, M., "Interferons," *New Engl. J. Med.*, **266,** 1258–1264, 1313–1318, 1367–1371 (1962).
2. Ho, M., "Identification and 'Induction' of Interferon," *Bacterial Rev.*, **28,** 367–379 (1964).
3. WAGNER, R. R., "Cellular Resistance to Viral Infection with Particular Reference to Endogenous Interferon," *Bacterial Rev.*, **27,** 72–86 (1963).
4. HILLEMAN, M. R., "Interferon in Prospect and Perspective," *J. Cellular Comp. Physiol.*, **62,** 337–353.
5. ISAACS, A., "Interferon," *Advan. Virus Res.*, **10,** 1–35 (1963).

ROYCE Z. LOCKART, JR.

INTERSTITIAL CELL STIMULATING HORMONE

This is an alternative term for the luteinizing hormone (LH; see GONADOTROPIC HORMONES; HORMONES, ANIMAL).

INTESTINAL FLORA (IN RELATION TO NUTRITION)

The significance of intestinal microorganisms to the host, particularly to various mammals and birds, has been a subject of continuous interest. One aspect of this relationship is the contribution of these organisms to the nutrition of the host. Certain observations seemingly must be related to the metabolism of the flora in the gastrointestinal tract. There is not, however, adequate information as to the precise contribution of the intestinal flora because of inadequacies in identification and quantification of the complex bacterial populations present in the gastrointestinal tract, in the measurement of the metabolism of a mixed flora in culture, and of appraising the significance of bacterial metabolism in the gastrointestinal tract to the host. This latter difficulty is accentuated by the fact that, except in the ruminant, the principal site of bacterial activity, the cecum and colon, is distal to the site of most active absorption of many materials from the gastrointestinal tract.

In spite of these limitations, advances have been made with several techniques. These approaches include the use of SULFONAMIDES and ANTIBIOTICS to alter the bacterial populations, the study of animals raised GERM FREE or infected with single defined strains of microorganisms, the maintenance of experimental animals under defined "pathogen free" conditions with relatively well-determined intestinal flora, and methods of preventing coprophagy (ingestion of feces) in experimental animals. Of pertinence to this latter point is the realization that in studies of rats, certain dietary requirements have been satisfied in part or totally from ingested feces which contained nutrients of bacterial origin. Unless coprophagy is prevented, it is difficult to determine how much of a nutrient synthesized by intestinal bacteria is absorbed from the lower gastrointestinal tract and what portion has become available only by recycling through the upper gastrointestinal tract where the conditions are favorable for absorption of nutrients and for disruption of bacterial cells (and release of materials to the luminal medium).

Much of experimental nutrition has been concerned with the vitamins. This is also true for research into the nutritional role of intestinal microorganisms. Two types of observations were instrumental initially in focusing the attention of investigators on a possible contribution of the intestinal flora: the difficulty of producing deficiencies of certain vitamins in rats and the existence of markedly negative balances (excretion greater than dietary ingestion) of these vitamins. Rats which were fed diets containing complex carbohydrates, particularly raw starch and dextrin, were often highly resistant to the induction of THIAMINE and other B vitamin deficiencies. This phenomenon was termed refection. The sparing action of feeding raw starch or dextrin (compared to less complex carbohydrate) has often been explained as the result of a greater quantity of undigested carbohydrate reaching the lower gastrointestinal tract to serve as a nutrient source for bacteria. The resulting stimulation of the cecal flora is presumed to result in more active synthesis of the missing nutrients. The types of dietary mono- and disaccharides also influence the makeup of the gastrointestinal flora. A further possible effect of the type of dietary carbohydrate on the flora composition is suggested by the work of Dubos and associates.

The housing of rats on open mesh cage floors through which the feces can drop has been shown to reduce somewhat the practice of coprophagy and to increase the apparent requirement for thiamine and certain other vitamins. Rats maintained under these conditions (which discourage, but do not eliminate, coprophagy) do not generally require an exogenous source of VITAMIN K, FOLIC ACID, or BIOTIN, but they are dependent for optimal growth on a dietary source of the other vitamins. A requirement for vitamin K has been demonstrated in the germ-free rat and in the normal (or infected) rat which is rigorously prevented from ingesting its feces. The prevention of coprophagy in rats fed diets low in folic acid does not induce deficiency signs although such signs appear in the germ-free or sulfonamide treated animal not given folic acid in the diet. This suggests direct absorption from the gastrointestinal tract of the bacterial product. There is, however, no general agreement that rats absorb a substantial amount of any other vitamin synthesized by bacteria unless the feces are recycled. A more detailed comparison of the vitamin requirements of the germ-free rat with those of the animal in which coprophagy is prevented and a reevaluation of the effects on vitamin requirements of feeding diets with different types of carbohydrates to rats in which coprophagy has been prevented may clarify the situation.

Frequently, rats fed diets deficient in a particular vitamin, *e.g.*, thiamine, excrete as much of the vitamin in the feces as rats fed diets containing high levels of the substance in question. Part of the excreted vitamin is inside bacterial cells. It is not surprising that the feeding of diets deficient in a nutrient results in an intestinal bacterial population which does not require that nutrient but can synthesize it.

The destruction or modification of vitamins in the intestinal tract also undoubtedly occurs, although the significance of such activity to the host is not clear. ASCORBIC ACID is often cited as an example of a nutrient readily inactivated by bacteria.

Many other substances have been shown to be altered by bacteria in the gastrointestinal tract. For example, germ-free rats, but not conventional rats (with bacteria in the gastrointestinal tract), excrete active trypsin and invertase in the feces. Apparently bacterial activity, not autodigestion, is the route of degradation of intestinal enzymes. It is quite probable that this is also true of some dietary proteins and peptides which escape digestion in the upper part of the gastrointestinal tract. The further degradation of amino acids to "toxic" products by bacteria has often been considered to have unfavorable effects on the host. The growth-promoting effect of certain antibiotics in some species of experimental and domestic animals has been attributed to a depression of this degradative activity and a consequent reduction in ammonia formed and absorbed. As indicated above in relation to refection, the type of carbohydrate has a profound effect on the composition of the intestinal flora. Bacteria are active in the hydrolysis of polysaccharides as well as in the fermentation of monosaccharides. Little is known of the effect of dietary lipids on bacterial composition or of the effects of bacteria on the dietary lipids that reach the area of the gastrointestinal flora. Bacteria, in general, do not seem to be very active in hydrolyzing ester bonds of complex lipids. Many species do hydrogenate dietary unsaturated fatty acids. ESSENTIAL FATTY ACID deficiency has been shown to be produced more readily in rats that are not allowed to ingest their feces. This may indicate that rats excrete some tissue fatty acids with the ω-6,9-configuration of unsaturation into the feces. Bacterial fatty acids do not appear in significant concentration in the host tissues. Cholesterol has also been shown to undergo conversion to several other compounds but particularly to coprostanol (the saturated, 5β-H compound) in the lower gastrointestinal tract. These conversions almost certainly are mediated by bacteria. Gastrointestinal bacteria seem not to cleave the ring structure of sterols or bile acids.

The effects of bacteria and of diet on BILE ACID metabolism have been rather extensively studied. Hepatic synthesized bile salts are altered in the cecum. The peptide bond between glycine or taurine and the bile acids is hydrolyzed, and the steroid moiety is extensively altered, particularly as regards the constituent hydroxyl groups. One consistent alteration is dehydroxylation of the 7-position of the bile acids. Thus cholic acid is converted to deoxycholic acid, and chenodeoxycholic acid is changed to lithocholic acid. These alterations have been demonstrated in cultures of microorganisms. The feces of germ-free rats contain bile salts which are entirely of the unaltered hepatic type, while the feces of the infected rats contain largely altered bile salts. The turnover times of bile salts are prolonged and the fecal excretion of bile salts is greatly reduced in the germ-free rat. Since the principal site of absorption of bile salts is the terminal ileum, and absorption also occurs from the cecum, it seems possible that the bile salt alterations in some way influence their efficiency of reabsorption. Variations in diet also influence the pattern of intestinal alteration of bile salts and the quantity of bile salts excreted in the feces, perhaps by altering the bacterial flora. Rats and *Cebus* monkeys fed purified diets with simple sugars as the sole carbohydrate source had much lower rates of fecal excretion of bile salts and different patterns of types of bile salts excreted as compared to animals fed "natural" diets based on complex carbohydrates. Interest in the patterns of bile salt excretion has been generated in part by the position of bile salts as products of the hepatic catabolism of cholesterol. Germ-free animals have higher serum cholesterol levels than infected controls, and animals fed synthetic diets including simple carbohydrates have higher serum values than controls fed natural diets based on complex carbohydrates.

The contribution of the intestinal flora to the nutrition of *ruminants* is relatively clear. The rumen and its constituent bacteria are proximal to the site in the gastrointestinal tract of the maximal absorption of nutrients. Thus, the rumen bacteria make a major contribution to the nutrition of the host. The rumen renders it possible for animals to utilize feeds that would be nutritionally inadequate if there were not significant enrichment, particularly of vitamins, by bacteria. Another unusual feature of the rumen bacteria is its action in stabilization of tissue composition even in the face of marked variations in dietary composition. This is particularly well illustrated for the fatty acid composition of the diet and of the tissues. Because of the hydrogenating activity of the rumen bacteria, it is difficult to alter the composition of tissue fatty acids by changing the composition of dietary fatty acids. This is in marked contrast to the non-ruminant (see also RUMINANT NUTRITION).

In summary, gastrointestinal bacteria are responsible for the synthesis and alteration of many compounds of potential importance to the host. It is difficult to determine the significance of this bacterial activity in the non-ruminant because it occurs primarily in a distal portion of the gastrointestinal tract—the cecum. Many observations of the effect of bacteria on nutrition have been made with rats and have failed to evaluate adequately the role of coprophagy. This limitation influences the extent to which the results can be applied to animals (including man) which do not practice coprophagy.

References

1. BARNES, R. H., FIALA, G., AND KWONG, E., "Decreased Growth Rate Resulting from Prevention of Coprophagy," *Federation Proc.*, **22,** 125 (1963).
2. BERGSTRÖM, S., DANIELSSON, H., AND SAMUELSSON, B., "Formation and Metabolism of Bile Acids," in "Lipide Metabolism" (BLOCH, K., EDITOR), New York, Wiley & Sons, 1960.
3. DAFT, F. S., MCDANIEL, E. G., HERMAN, L. G., ROMINE, M. K., AND HEGNER, J. R., "Role of Coprophagy in Utilization of B Vitamins Synthesized by Intestinal Flora," *Federation Proc.*, **22,** 129 (1963).
4. DUBOS, R., SCHAEDLER, R. W., AND COSTELLO, R., "Composition, Alteration, and Effects of the Intestinal Flora," *Federation Proc.*, **22,** 1322 (1963).
5. GUSTAFSSON, B. E., "Vitamin K deficiency in Germ Free Rats," *Ann. N.Y. Acad. Sci.*, **78,** 166 (1959).
6. MICKELSEN, O., "Intestinal Synthesis of Vitamins in the Nonruminant," *Vitamins Hormones*, **12,** 1 (1956).
7. PORTMAN, O. W., "Importance of Diet, Species, and Intestinal Flora in Bile Acid Metabolism," *Federation Proc.*, **21,** 896 (1962).
8. PORTMAN, O. W., "Nutritional Influences on the Metabolism of Bile Acids," *Am. J. Clin. Nutr*, **8,** 462 (1960).

OSCAR W. PORTMAN

INTESTINAL JUICE

Intestinal secretions include many enzymes active in the digestion of various classes of food materials (see DIGESTION AND DIGESTIVE ENZYMES).

INTRINSIC FACTOR

Intrinsic factor is a mucoprotein (or group of mucoproteins) normally produced by the gastric mucosa and essential to absorption of the "extrinsic factor" (vitamin B_{12}) in the diet [see VITAMIN B_{12} (ABSORPTION AND NUTRITIONAL ASPECTS)]. Production of intrinsic factor is deficient in pernicious anemia (see ANEMIAS).

IODINE (IN BIOLOGICAL SYSTEMS)

The most consistent source of iodine in nature at the present time is seawater, in which the element is present as iodide. Because of this, it is not surprising that iodine is found in many plant and animal forms living in or near the ocean, although no specific function can be demonstrated for the iodinated material in many cases. As a historical fact, 3,5-diiodotyrosine (iodogorgoic acid) had been discovered in sponges and corals long before there was any hint concerning thyroid hormone structure even in mammalian forms. Since this time, iodotyrosines have been demonstrated in algae as well as in many animals possessing a horny skeleton. It has been suggested that iodide in the water is activated by peroxidase due to the presence of oxygen in the water and the resulting iodine is accepted by TYROSINE in the protein molecule, similarly to the process in vertebrates. In animals possessing an exoskeleton, the presence of benzoquinones is part of the formation of scleroproteins by a quinone tanning process, and the iodination of tyrosine may be related to this in some way. In protochordates, the endostyle, a structure secreting mucus, has been found capable of iodination of protein present in the mucus which is then secreted into the alimentary canal.

The next phylogenetic development takes place in the vertebrates, in the most primitive members of which there is a structure located in the hypopharynx similar to a thyroid capable of collecting iodine and of forming iodinated protein, which is broken down by a protease, liberating iodinated amino acids. In some of these forms, small amounts of thyroxine are actually found in addition to the iodinated tyrosines. In other vertebrates, culminating in the amphibia and higher vertebrates, a thyroid gland is present in which there is no secretory duct and in which the iodinated protein is held in a storage form known as "colloid." In some, but not all, of these forms, hormonal material liberated by action of proteolytic enzymes is secreted into the bloodstream and plays an essential role in the development of the young animal as well as in the behavior and metabolic activity of the adult. Because of the process known as ontogeny, it is not surprising that the development of the thyroid gland in the human embryo commences with a structure located near the alimentary canal, which then separates and develops its peculiar follicular structure. Not until this type of development has occurred can a genuine function for the iodinated substances be demonstrated.

In these discrete glands, a series of specific chemical reactions can be demonstrated, as shown in the following series:

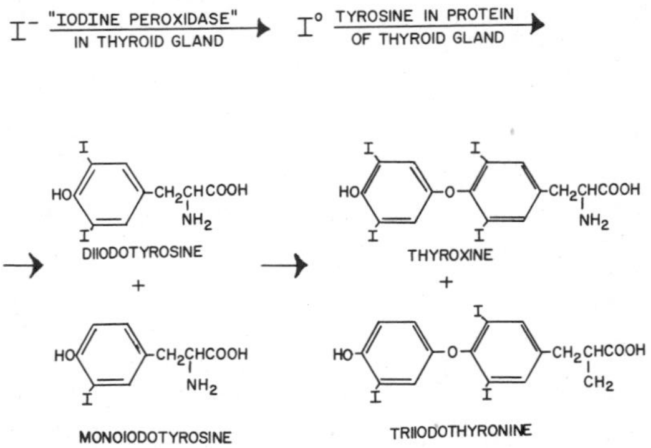

$$I^- \xrightarrow[\text{IN THYROID GLAND}]{\text{"IODINE PEROXIDASE"}} I^0 \xrightarrow[\text{OF THYROID GLAND}]{\text{TYROSINE IN PROTEIN}}$$

→ DIIODOTYROSINE + MONOIODOTYROSINE → THYROXINE + TRIIODOTHYRONINE

As already mentioned, some of these reactions take place in the absence of any apparently specific synthesis of iodotyrosines, and it still is not known how many of these steps require enzymes, even in mammalian forms, although these processes are usually considered as enzymatic. It is possible to iodinate tyrosine in soluble proteins *in vitro* by addition of elemental iodine and under these circumstances thyroxine and triiodothyronine will also be formed, in company with small amounts of iodohistidine.

In amphibian vertebrates undergoing METAMORPHOSIS, thyroxine is known to be essential for this transition from an immature to a mature animal. There is considerable evidence that after metamorphosis has occurred, the thyroid gland is no longer essential, although it may be involved in seasonal changes such as molting.

There is no such clear-cut differentiation as metamorphosis in the mammal, but development is an extremely complex process and has been shown to depend upon the presence of adequate amounts of THYROID HORMONES. Deficient development, especially of the central nervous system, is marked in children suffering from thyroid deficiency early in life, and this inadequacy cannot be overcome completely by medication commenced after the first few weeks. In the adult, thyroxine is important in the maintenance of energy turnover in most of the tissues of the body, such as the heart, skeletal muscle, liver and kidney. Other physiological functions, most notably brain activity and reproduction, are also dependent upon thyroxine, although the metabolic rates of the tissues concerned in these functions do not seem to be altered.

A great deal of work has been done on determining the portions of the thyroxine molecule which are essential to biological activity. As already suggested, the fact that the hormone is an amino acid is almost certainly due to the widespread existence of the excellent iodine acceptor, tyrosine. Thus, the deaminated and decarboxylated metabolic product of thyroxine, tetraiodothyroacetic acid, has been shown to have appreciable biological activity, although quantitatively less than that of thyroxine itself. As for the halogens present on the diphenyl ether portion of the molecule, bromines and chlorines are also active, although diminishing considerably in that order from iodine. With the discovery of 3,5,3′-triiodothyronine as an iodinated material present in the thyroid gland, its greater activity than thyroxine when injected into thyroid-deficient animals has stimulated considerable research. From the illustrations below, it can be concluded that iodine is essential on both the 3- and the 5-position of the benzene ring next to the amino acid side chain, whereas substitution on only the 3′-position of the benzene ring containing the phenol is essential.

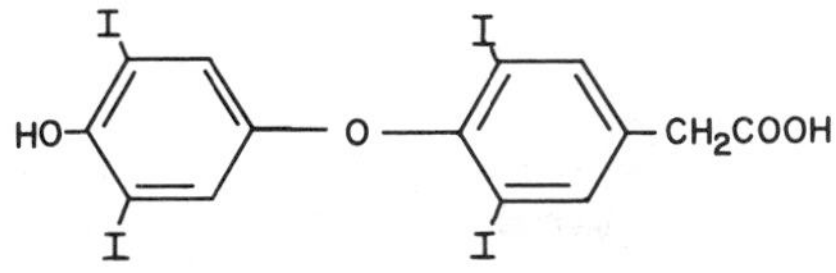

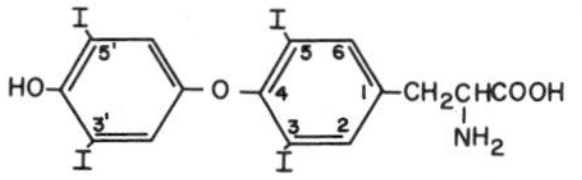

THYROXINE 100%

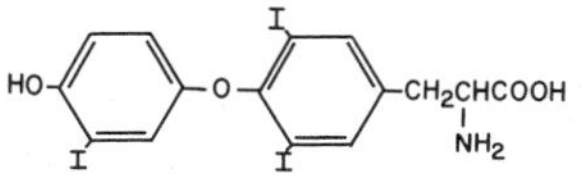

3,5,3′-TRIIODOTHYRONINE 500%

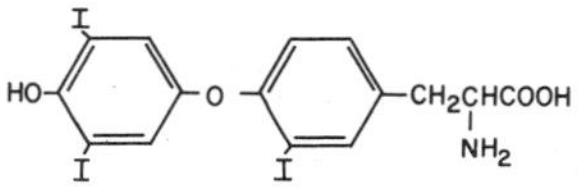

3,3′,5′-TRIIODOTHYRONINE INACTIVE

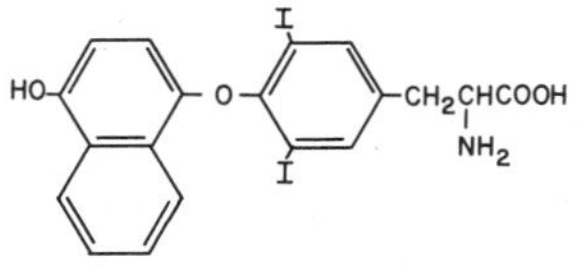

O-(4′-HYDROXY-1′-NAPHTHYL)-3,5-DIIODOTYROSINE 100%

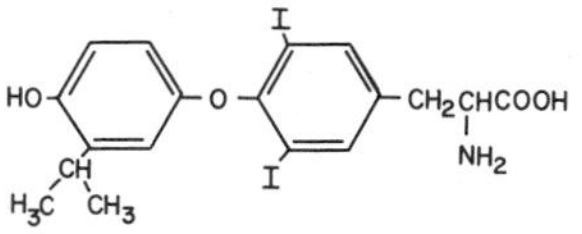

3′-ISOPROPYL-3,5-DIIODOTHYRONINE 750%

Although not a physiological compound, the great activity shown by 3′-isopropyl-3,5-diiodothyronine leads to the conclusion that a bulky substitution is of great value. The high potency of these unsymmetrical compounds has suggested that 5′-deiodination is one step in the process whereby thyroxine influences the functioning of so many different tissues. Confirming evidence is that an inability to deiodinate thyroxine accompanies inactivity of the hormone, although specificity of such deiodination has not yet been proved. Another chemical process which may accompany the onset of activity of thyroxine is the formation of a quinoid type of free radical, requiring both the phenolic hydroxyl and the ether oxygen bridging the two benzene rings:

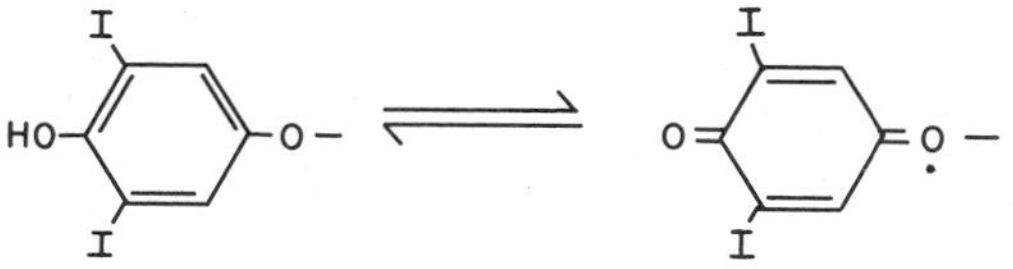

Although evidence for these steps is highly suggestive, it must be considered still quite tentative.

Metabolic Actions of Thyroid Hormones. Assuming that there are some quite definite structural requirements for thyroid hormone activity, it is important to inquire into the specific actions of this materials. It now seems quite clear that thyroxine does not participate directly in any enzyme system, but rather affects the functioning of many systems, presumably by some far more general process. One of the earliest of such demonstrated actions was the uncoupling of oxidation from formation of high-energy phosphate compounds such as adenosine triphosphate. However, such uncoupling is produced by many other substances not showing thyroid hormone-like effects, and more recently it has been shown that thyroxine is actually capable of accelerating coupled reactions under the proper conditions. From this last type of evidence, it has been suggested that the principal role of thyroxine may be the acceleration of enzyme processes ordinarily limiting the level of metabolic turnover. Mitochondria isolated from broken cell preparations by high-speed centrifugation have been shown to swell when placed in contact with thyroxine and similar substances. This has been interpreted as evidence for a membrane function of the hormone, although it is not entirely clear as to how this may alter cellular function in such a specific manner as the hormone does *in vivo*.

More challenging still has been recent work demonstrating an acceleration of protein turnover by thyroxine, implying that the hormone may alter various processes by a specific effect on synthesis of certain key proteins involved in enzymatic reactions. Thus, not only does thyroxine increase the rate of formation of new protein material, but it may also be responsible for the transformation of non-enzymatically active protein into protein with enzymatic activity. The hormone has been shown to be capable of acceleration of the synthesis of urea cycle enzymes (see ARGININE-UREA CYCLE) and probably is essential for the production of a sodium ion transporting mechanism, both of which are essential in the metamorphic transformation of larval forms into mature amphibia. Similar processes may explain many of the actions of thyroxine in higher forms as well. At the present time, it is not possible to say why such a complicated structure should be necessary in order to achieve these results.

References

1. "Symposium on Thyroxine," from eight groups of authors, *Mayo Clinic Proceedings*, **39**, 535–653 (1964).

2. Pitt-Rivers, R., and Trotter, W. R., (Editors), "The Thyroid Gland," London, Butterworths, 1964, 2 vols.
3. "Modern Concepts of Thyroid Physiology," from twenty-two groups of authors, *Ann. N.Y. Acad. Sci.*, **86**, 311–676 (1960).
4. Gorbman, A., "Some Aspects of the Comparative Biochemistry of Iodine Utilization and the Evolution of Thyroidal Function," *Physiol. Rev.*, **35**, 336–346 (1955).

S. B. Barker

IODINE NUMBER

Iodine number is a term applied principally to fats and related compounds. The iodine number of a fat is the per cent of iodine (number of grams of iodine per 100 grams of fat) which the fat will take up. In experimental determinations, iodine monobromide, for example, may be used, but the results are calculated in terms of iodine. Linseed oil, a highly unsaturated fat, may have an iodine number of about 200, while coconut oil may have an iodine number as low as 8 or 9. Completely saturated fats, of course, have an iodine number of 0.

ION EXCHANGE POLYMERS (RESINS)

The biochemical applications of ion exchange resins are varied and numerous. These ionized, insoluble, high molecular weight, branched-chain natural or synthetic polymers possess high resolving power and capacity, a wide range of physical form and chemical specificity, stability, uniformity, and adaptability to a variety of analytical and preparatory techniques, thus providing a unique tool for the chromatographic separation of proteins, peptides and amino acids, polysaccharides and carbohydrates, nucleic acids, nucleotides and their bases, and many other biologically active substances. The principles of ion exchange CHROMATOGRAPHY developed for the separation of inorganic and organic substances apply broadly to biological mixtures. Molecular instability (denaturation) and preservation of biological activity, however, often place restrictions on the freedom of manipulation of the many parameters that control chromatographic separation of conventionally stable compounds. For biological separations, an ion exchanger is selected on the basis of its functional ionic exchange group (strong or weak acid or base), and its physical properties (mesh size, porosity). Equally important are the experimental parameters (buffer, pH, ionic strength, or salt concentration, temperature, flow rate, etc.) and the type of chromatographic operation (batch, stepwise or gradient elution; displacement; frontal development with buffers). Obviously, the exchanger is selected in relation to the nature of the solute, *e.g.*, its pK value, molecular size, etc. Although much practical and theoretical knowledge of adsorption and of ion exchange exists, some measure of trial and error in the design of ion exchange experiments is nevertheless essential.

Natural and Synthetic Polymers. Two types of ion exchange substances are widely used: Monofunctional synthetic polymers with polymethacrylate or polystyrene matrices, and chemically substituted exchangers derived from natural polymers such as cellulose and dextran. In both, the ion exchange properties are determined principally by the kind of functional ionic group introduced into the molecule (Table 1). The matrix of the synthetic polymer is wholly hydrophobic while that of the natural polymer is essentially hydrophilic; this difference also defines to some extent respective spheres of chromatographic performance. The smaller ionic or charged molecular species are readily resolved by the synthetic polymer with its generally higher ionic exchange capacity and wide range of porosity, while macromolecular species, the adsorption of which depends not only on fixed ionic groups but also on adsorptive forces (hydrogen bonding and van der Waals forces), are more successfully resolved by cellulose and dextran derivatives. Resolution is controlled by both ionic and adsorptive forces, often even among smaller molecular weight substances, and is defined in each particular case by the physical and chemical properties of solute, solvent, and the ion exchanger in their equilibrium state. If purified components are available, the physicochemical conditions which control the equilibrium state may be ascertained by calculating the distribution ratio of solute between the mobile and stationary phases according to the plate theory; a selection of conditions which provide preferential elution of one or more components then can be made. Such information is not available ordinarily and solutions of unknown composition usually are examined on a more empirical basis.

The preparation, properties and reactions of synthetic polymer ion exchange resins[1–5] are described in the references; resins often used for biochemical separations are listed in Table 1. Solutes with a molecular weight less than 500 diffuse into ion exchange particles, especially as the porosity is increased (decrease in cross-linkage, decrease in per cent DVB), provided the chemical nature of the solute and resin does not lead to reactions which hinder diffusion. The macromolecular adsorptive capacity of synthetic polymers is relatively low, and such interactions are limited to its surface. In order to increase adsorptive surface, finely powdered resins of mesh size greater than 100 are employed. With increasing mesh size, problems of column operation are created in terms of undesirably low flow rates and column-plugging due to the tight packing of fine particles of resin; conversely, channeling may occur due to shrinkage of the column bed in a changing ionic environment. Upward flow techniques may overcome these difficulties. Uniform sized resins are important for analytical resolution.

The development of cellulose ion exchangers[6,7] introduced an indispensable tool for the fractionation of complex biochemical mixtures. These excellent adsorbents may be prepared in the laboratory or obtained from commercial sources. Such exchangers (Table 1), now derived from both cellulose and dextran, owe their ionic and adsorbent properties to the basic or acidic groups introduced into the biopolymer by synthetic procedures involving the formation of ester or ether linkages with the constituent hydroxyl groups of the molecule, and to the hydrophilic nature of the gel-like matrix. In the polysaccharide molecule the equivalent of low cross-linkage is produced by hydrogen bonding between polyglycose chains. The charged groups are spaced approximately 50 Å apart compared to the closer spacing of 10 Å for synthetic polymers. A limited substitution is essential, in order to preserve an insoluble macromolecular structure. Excessive substitution or degradation imparts undesirable swelling properties which interfere with the hydraulic aspects of column operation, or increase solubility which destroys the solid phase. Stability to strong acids and alkali, relatively little shrinkage or swelling in the pH range of 4–8, and a large surface area represent desirable chromatographic properties. The high degree of resolution achieved with these exchangers is the result of the unique topographical distribution of ionic charges and adsorbent forces on a hydrophilic surface which provides for a multiple point attachment of macromolecules to the adsorbent. The resolution of macromolecular species occurs as a result of rapidly established equilibria on the surface of the adsorbent in contrast to the absorption of small molecules which is characterized by relatively slower diffusion into the interior of the exchanger. While the ionic exchange capacity is much smaller than that of synthetic polymers, the capacity for macromolecules is considerably greater as a result of these properties. A high density of ionic binding sites, however, is not desirable for macromolecules, since very firm binding may alter conformational structure and cause denaturation during the adsorption or elution phases. Dextran exchangers also superimpose a molecular sieving property on that of ion exchange.

Applications. *Amino Acids and Related Compounds.* The synthetic ion exchange resins are admirably suited for the separation of free AMINO ACIDS or amino compounds from urine, serum, and body fluids, and for the separation of the constituent amino acids of hydrolyzed PROTEINS.[1,2] As dipolar ions the four general classes of amino acids (acidic, basic, neutral and aromatic acids) are sufficiently ionized in acid or alkaline solution to undergo ion exchange reactions. Secondary functional groups further differentiate their common ampholytic nature and impart or enhance ionic or nonpolar properties by which certain class separations can be achieved. Under appropriate conditions of pH and ionic strength, the strong cationic ion exchange resins adsorb all amino acids; the weakly ionized cationic resins separate the more basic amino acids (arginine, lysine, histidine) from the less basic amino acids, at the same time allowing neutral and dicarboxylic acids to pass through the column unadsorbed; the anionic strong base resins in the chloride or acetate form retain the dicarboxylic acids (glutamic, aspartic, cysteic acids) while in the hydroxyl or free base form they adsorb neutral and basic amino acids.

The 18 common acids can be resolved on "Dowex" 50-X4* by elution development, with recoveries of 97% or better in 17 of these, in the presence of 50 different solutes, 41 of which are separated in a single column operation. Other elegant fractionation systems separate as many as 148 individual amino acids and related compounds. A complete and quantitative, automated amino acid analysis of protein hydrolysates for the determination of primary protein structure can be performed with a sulfonic acid exchanger ("Amberlite" CG-120, type 111, 8% DVB). The bulk of the amino acids are resolved with 0.2*M* sodium citrate buffer at pH 3.25 and 5.0. A second smaller column with 0.35*M* sodium citrate solution at pH 5.28 separates phenylalanine, tyrosine, lysine, histidine, ammonia and arginine. Other systems have been devised to operate with single or multiple columns, with either "Dowex" 50 × 12, ×8 or ×4 or "Amberlite" IR120 and at more than one temperature. Specific hydraulic fractions of resins having uniform particle size are essential for adequate resolution.

Displacement development techniques[1,2] on cationic or anionic exchangers are used for the preparation of large quantities of amino acids from protein hydrolysates (100–300 grams). The displacement is dependent upon the basicity of the cationic species which are resolved in the order of their pK_1 or pK_2 values; however, recovery is not quantitative. Carrier-displacement development, in which organic acids and bases are employed as intermittent carriers of groups of amino acids also offers a unique and powerful system of fractionation.

Peptides. Ion exchange separation and purification of synthetic or natural peptides, or those obtained from partial acid, alkaline or enzymatic hydrolysis of proteins, are important to studies of protein structure, enzyme action, and the isolation of many biologically or pharmacologically active substances.[1,2,8] Since only the amino and carboxyl end groups are free, peptide separation depends mostly on the interaction of the resin with secondary ionic and nonpolar functional groups. Elution development provides the best resolution and is dependent upon the rapid diffusion of the peptide into the resin particle. Polar attraction predominates; however, nonpolar attraction of aromatic peptides is involved and can be depressed by saturating the resin with aromatic pyridine bases.

Peptides are usually detected by the ninhydrin reaction; obviously ammonium salt buffers must be avoided unless volatile ones are used and removed from the fraction by lyophilization. As the molecular weight of the peptide increases, less ninhydrin color develops and the peptide fraction

* X4 indicates 4% divinylbenzene (DVB).

TABLE 1. ION EXCHANGE POLYMERS

Chemical Designation	Type	Functional Group	Matrix	Manufacturer's Designation*	Ionic Exchange Capacity (meq/g)
		Cationic Synthetic Exchangers			
Sulfonic acid	Strong acid	$R—SO_3^-H^+$	Polystyrene	Amberlite IR112, 120,200 Dowex 50 ZeoKarb 225 Permutit Q	3.6–5.0
Carboxylic acid	Weak acid	$R—COO^-H^+$	Polymethacrylate	Amberlite IRC50 ZeoKarb 226 Duolite CS–101	8.5–9
Iminodiacetic	Weak acid	$R—N—(CH_2—COO^-)_2 2H^+$	Polystyrene	Dowex A 1 Chelex 100	2.9
Phosphonic acid	Intermediate acid	$R—PO_2^-H^+$ $R—PO_3^{=}2H^+$	Polystyrene	Duolite C-62 Biorex C-63	6.6
Methyl sulfonic acid (sulfomethyl)	Strong acid	$R—CH_2—SO_3^-H^+$	Phenolic	Amberlite IR100 ZeoKarb 215 Duolite C-3	2.9
		Anionic Synthetic Exchangers			
Polyamine	Weak base	$R—NH_2$	Polyphenol-formaldehyde	Amberlite IR4 B	5.0
			Polystyrene	Amberlite IR45 Dowex 3	5.0
Trimethylamine	Strong base	$R—N^+{\equiv}(CH_3)_3Cl^-$	Polystyrene	Amberlite IRA400 Dowex 1 DC-Acidite FF	3.4 2–3.6 4.0
Quaternary trimethyl ammonium hydroxide	Strong base	$R—N^+{=}(CH_3)_2Cl^-$ with $CH_2—OH$ on N	Polystyrene	Amberlite IRA410 Dowex 2	5.0 3.7
Pyridinium	Strong base	R—(pyridine ring)N^+Cl^-	Polystyrene	Permutit SK	3.7
		Cationic Natural Polymer Exchangers†			
Carboxymethyl	Weak acid	$R—O—CH_2—COO^-Na^+$	Cellulose	Cellex-CM Selectacel-CM Serva Whatman	0.7 0.7 0.7 0.7
			Dextran	Sephadex-CM	4.5

Table 1 (*continued*)

Chemical Designation	Type	Functional Group	Matrix	Manufacturer's Designation*	Ionic Exchange Capacity (meq/g)
Phosphonic acid Phosphoryl	Intermediate acid	$R—O—PO(O^-Na^+)—O^-Na^+$	Cellulose	Cellex-P	0.85
				Selectacel-P	0.9
				Serva	0.7
				Whatman	0.5–7.4
Sulfoethyl	Strong acid	$R—O—C_2H_4—SO_3^-Na^+$	Cellulose	Cellex-SE	0.2
				Serva	0.2
			Dextran	Sephadex-SE	2.4
Sulfomethyl	Strong acid	$R—O—CH—SO_3^-Na^+$	Cellulose	‡	0.4
Anionic Natural Polymer Exchangers†					
Diethylamino ethyl	Strong base	$R—O—C_2H_4—N^+H{=}(C_2H_5)_2Cl^-$	Cellulose	Cellex-D	0.4–1.34
				Selectacel-DEAE	0.9
				Whatman	0.4–1.0
				Serva	0.5–0.7
			Dextran	Sephadex-DEAE	3.5
Triethylamine	Strong base	$R—O—C_2H_4—N^+{\equiv}(C_2H_5)_3Cl^-$	Cellulose	Cellex-T	0.5
				Selectacel-DEAE	0.3
				Serva	0.5–0.7
Guanidoethyl	Strong base	$R—O—C_2H_4—NH—C(N^+H_2)—N^+H_2$	Cellulose	Cellex GE	0.9
Epichlorhydrin-triethanolamine (Ecteola)	Intermediate base	$R—(NH_2)_x$	Cellulose	Cellex-E	0.3
				Selectacel-Ecteola	0.3
				Whatman	0.3–0.5
				Serva	0.3–0.5
Aminoethyl	Intermediate base	$R—O—C_2H_4—NH_2$	Cellulose	Cellex-AE	0.8
				Whatman	1.0
				Serva	
Polyethyleneimine†	Intermediate base	$R—(CH_2—CH_2—NH)_x$ $R—CH_2—CH_2—NH_2$	Cellulose	Cellex PEI	0.2
Paraaminobenzyl	Weak base	$R—O—CH_2—C_6H_4—NH_2$	Cellulose	Cellex PAB	0.2
				Serva	0.3–0.4

* *Amberlite:* Rohm and Haas Co., Resinous Products Division, Philadelphia, Pa.; *Dowex:* Dow Chemical Co., Midland, Mich.; *Zeo-Karb*, *Deacidite*, *Permutit:* The Permuitt Co., New York, N.Y.; *Duolite:* Chemical Process Co., Redwood City, Calif.; *Chelex*, *Cellex:* Bio-Rad Laboratories, Richmond, Calif.; *Sephadex:* Pharmacia Fine Chemicals, New York, N.Y.; *Serva:* Gallard-Schlesinger Chem. Manufacturing Corp., Carle Place, Long Island, N.Y.; *Selectacel:* Carl Schleicher and Schull, Co., Keane, New Hampshire; Brown and Co., New York, N.Y.; *Whatman:* Reeve Angel, New York, N.Y.[1,2,6,11].

† Mean particle size: 15–20 μ.

‡ Reference 2.

then is detectable only after alkaline hydrolysis.

Best separations have been achieved with cationic sulfonic acid polystyrene resins of sufficiently low cross-linkage to allow rapid equilibrium diffusion. Peptides with as many as 30 amino acid residues have been successfully chromatographed. Discontinuous or convex exponential gradients produced by sodium citrate or sodium acetate solutions have been employed in the range of pH 2.5–8.5 with increasing ionic concentration from 0.2*N*–2.0*N*. As molecular size increases, diffusion becomes a limiting factor and ion exchange interactions occur only on the resin surface. Dipeptides readily diffuse into 8% DVB (divinylbenzene) resins, but larger peptides require resins with 2 or 4% DVB. The isolated peptide fractions must be subjected to rechromatography or studied by other techniques, to demonstrate that a single peptide occupies a single zone.

Anionic exchange resins are particularly effective for some peptide analyses. Peptides with molecular weights in excess of 4000 have been resolved from enzymatic digests of various proteins, (TOBACCO MOSAIC VIRUS, HEMOGLOBIN, COLLAGEN, peptone and corticotropins) on "Dowex" 1 × 2 resins in the acetate cycle with collidine-pyridine-acetic acid buffers at pH 8.2–8.4 as the initial eluent, and with a gradient concentration of acetic acid to 2*N*. The peptides are eluted generally according to their net charge in the following order: basic, neutral, and acidic peptides. The interactions are strictly ionic with little or no side effects due to nonionic adsorption.

Effective and unusual separation of peptides from enzymatic digests and the isolation of a number of biologically active peptides also can be obtained with Ecteola-, CM- or DEAE-cellulose or with DEAE(A50) dextran exchangers.

Proteins. The synthetic ion exchangers have a limited application in the separation of stable, basic proteins of relatively low molecular weight.[1,8] The weak carboxylic acid resin, "Amberlite" IRC-50, was first used successfully to purify CYTOCHROME c, and subsequently, was applied to the purification of other cytochromes, various enzymes, hemoglobins, serum proteins, HISTONES, etc. Quaternary ammonium anionic and sulfonic acid cationic exchangers also have been used for a variety of proteins. Conditions for many separations have been summarized.[1]

Chromatography on "Amberlite" IRC-50 in its cationic form is very dependent on cation concentration and pH. Proteins may be adsorbed and eluted by increasing the cation concentration or even by substituting another eluting cation, *e.g.*, Ca^{++} for Na^{+}. At pH 5–6, neutral proteins, *e.g.*, the hemoglobins, may be chromatographed under rigidly controlled conditions but with a considerable loss of protein by denaturation. Above this pH, proteins bearing positively charged ionic groups react with the exchanger and finite distribution coefficients can be obtained; this is the most useful pH range. Below pH 5–6 excessive and irreversible hydrogen bonding of proteins occurs frequently causing denaturation. The capacity for protein interaction is low at any pH but is increased by employing resins of low cross-linkage and 200–400 mesh ("Amberlite" XE-64). Single buffer systems are generally employed for separation or homogeneity studies; gradient elution with inorganic buffers is difficult due to the large buffering capacity of the resin except in its completely ionized or non-ionized form at high or low pH. Stepwise elution is effective for large-scale preparative protein fractionation. Strong acid cationic sulfonic acid exchangers also are useful; however, about 30% of the protein is irreversibly bound. Proteins also are adsorbed on anionic exchangers, *e.g.*, "Dowex" 2, with tris-buffer at low ionic strength and eluted stepwise with increasing concentration of buffer at a constant pH of 7.2. When protein denaturation does not interfere or dissociation into protein subunits does not occur, IRC-50 may be used in the presence of urea or guanidine buffers which reduce hydrogen bonding and minimize intermolecular complex formation (protein-protein or resin-protein interactions). Low temperatures frequently are of value since they reduce the rate of the denaturation. The alpha, beta, and gamma chains of hemoglobin have been so separated on IRC-50 with a urea gradient; the alpha chains are eluted at 4.9–5.1*M*, the beta chains at 6.2–6.4*M*, and the gamma chains at 5.9–6.1*M* urea.

Cellulose exchangers are used for chromatography of virtually every kind of protein; among these are numerous blood proteins (see PLASMA PROTEINS), ENZYMES, ANTIGENS and ANTIBODIES, HORMONES, VIRUSES and a wide variety of tissue proteins of bacterial, plant and animal origin.[1,2,4–8] The capacity for proteins is large and the resolving power is unique due to the many possibilities for characteristic binding involving only a few of the many available charged sites. Conditions are sought in which primarily ionic bonds bind the proteins and in which hydrophobic and hydrogen bonds become less important. High capacity allows the proteins to be adsorbed in a narrow zone at the top of the column of adsorbent and then to be selectively eluted by altering the electrostatic attraction of the protein for the adsorbent with buffers of increasing or decreasing pH or concentration. This action may be directed against either the adsorbent or the protein component(s) and requires that the adsorbent exist in the ionic state. In non-iozined form (weakly ionized exchangers at high or low pH), hydrogen bonding predominates; this action also is achieved with high concentrations of urea and guanidine.

DEAE-cellulose is the most widely used resin for all proteins because of a basicity that permits unusual resolution. DEAE-dextran has a capacity similar to the cellulose analogue but must be used in the more porous A-50 form, which has combined ionic and molecular sieving action. The less porous A-25 type has low adsorption capacity. Various buffers are employed: most frequently, phosphate over pH range of 5–8, then tris, acetate, bicarbonate, and other buffers. Stepwise and gradient techniques with NaCl or with buffer salts, alone or combined with pH gradient, are particularly effective; buffer mixing devices (Varigrad[7]) are helpful. The more strongly basic TEAE-cellulose is used for acidic proteins, the less basic Ecteola-cellulose for viruses. The cationic CM-cellulose and dextran are widely used for neutral and acidic proteins below pH 7. The remaining derivatives (Table 1) are relatively new but are finding increasing application.

Carbohydrates and Related Compounds. The CARBOHYDRATES are essentially neutral substances, whose hydroxyl groups ionize as extremely weak acids.[1,3,4,10] They react with strong base anionic exchangers in the hydroxyl form only in very alkaline solution in which degradation of the carbohydrate molecule usually occurs; however, stable anionic complexes and derivatives undergo ion exchange reactions. The borate complexes offer the best approach to the resolution of mixtures of various carbohydrates. A completely ionized complex (type 11) is formed in alkaline solution; however, the formation and stability of the complex is dependent upon the orientation of hydroxyl groups and the sterochemical configuration of ring structure. Various mono- and disaccharides with a 1:2 diol grouping in the *cis* position, especially with a planar furanose ring, readily form stable ionic complexes that undergo chromatographic resolution on strong anionic exchange resins. Carbohydrates with hydroxyl groups in the *trans* position or with a pyranose (chair) ring structure are less capable of stable complex formation and do not react as well with anionic exchange resins. However, in some instances, as a result of mutarotation of OH groups or enolization of keto groups, favorable equilibrium conditions can be found for the resolution of such ionic borate complexes. Many hexoses and pentoses and mono-, di-, tri- and tetrasaccharides have been separated in this manner. Strong base exchangers ("Dowex" 1, "Amberlite" IRA-400) are employed on the hydroxy cycle, and solutions of alkali salts of tetraborate are used for equilibration and elution in the pH range of 7–9.

Carbohydrates or their derivatives (carboxyl, sulfate and phosphate esters of simple sugars or complex polysaccharides) can be subjected to chromatography directly on strong or weak base anionic exchangers in the hydroxy, chloride, formate or acetate forms. The dissociation constants of phosphate esters, however, are so similar that these cannot be resolved, except as borate complexes; they obviously can be easily separated from neutral carbohydrates by ion exchange. Borate complexes of sugar alcohols react like those of carbohydrate. The resolution of bisulfite addition compounds of carbohydrates and certain glycosides is dependent more nearly on adsorptive rather than ionic forces. Polysaccharides, such as heparin, chondroitin sulfuric acid, etc., are chromatographed on synthetic as well as cellulose and dextran exchangers. Amino polysaccharides (glucosamine, galactosamine, etc.) are resolved on cationic exchangers. A straightforward but important application of ion exchange resins is the deionization of carbohydrates.

Nucleic Acids and Related Compounds. PURINE and PYRIMIDINE bases, nucleosides and nucleotides are readily separated and resolved on synthetic polymer exchangers; oligonucleotides, polynucleotides and undegraded nucleic acids are preferably separated on cellulose or dextran exchangers.[1,9,10] The ion exchange properties of these compounds ultimately reside in the constituent purine and pyrimidine bases and in the phosphate groups, but chromatographic resolution also is markedly influenced by non-ionic adsorptive forces.

Purine and pyrimidine bases that function as cations in acid solution (cytosine, guanine, adenine) are readily resolved quantitatively on sulfonated polystyrene resins by elution with HCl or NaCl solutions; bases without cationic properties (uracil, thymine) are not adsorbed. The expected order of elution based on dissociation constants of these bases (guanine, adenine, cytosine) is altered by hydrophobic forces. On anionic strong base synthetic polymers ("Dowex" 1, "Amberlite" IRA 400 in chloride or formate systems), the free bases are resolved with ammonium salt buffers at alkaline pH, in an order (cytosine, uracil, thymine, guanine and adenine) that also is not completely predictable from dissociation constant values; thymine with a higher pK is eluted after uracil which has a lower pK value.

Nucleosides are resolved on cationic sulfonated polystyrene resins less satisfactorily on account of the acid lability of the purine ribosides; nevertheless, effective separations are possible by elution with ammonium salt solution from ammonium forms of the exchanger. On anionic synthetic exchangers, however, nucleosides separations similar to those of free bases are obtainable; additionally, cytosine and adenosine must be resolved on a carboxylic acid exchanger. The nucleoside polyphosphates are eluted from DEAE-cellulose or dextran with ammonium bicarbonate buffer at pH 8.6 in the following order: mono-, di-, tri-, tetrapolyphosphates with the purine-containing derivatives being more hydrophobically attracted to the exchangers than the pyrimidine-containing derivatives.

In the nucleotides, the negative phosphate group reduces the basicity of the constituent bases sufficiently to produce a weaker bond with the exchanger. The resultant net charge of these compounds allows effective resolution on cationic sulfonated polystyrene exchangers in a pH range of 2–4 with weak acids or acetate or formate buffers as eluents; the order of elution at a given pH is influenced by both pK values and non-ionic adsorptive forces, the purine compounds being three times as strongly adsorbed as the pyrimidine derivatives. At pH 8, nucleotides are separated from free bases and resolved in the process with a linear concentration gradient of ammonium bicarbonate. On anionic exchangers ("Dowex" 1 and "Amberlite" IRA 400) at pH 2.5–3, cytidylic, adenylic, guanylic acids appear in order of their amino group dissociation constants when eluted with dilute HCl or formic acid–ammonium formate buffer. The free bases and ribosides pass unresolved into the first fraction collected from the column while uridylic acid, which has no NH_2 group, appears between adenylic and guanylic acids. Some isomeric nucleotides are resolved. Ribosyl and deoxyribosyl derivatives are separated as borate complexes.

As with proteins, the chromatography of the macromolecular polynucleotides and oligonucleotides is accomplished best with cellulose and dextran exchangers, although strong base synthetic polymer exchangers (2 or 8% DVB) at acid pH quite satisfactorily resolve mono-, di-, tri- and tetranucleotides by gradient elution with NaCl. The superiority of natural polymers, however, is illustrated by DEAE-cellulose separation of polynucleotides up to dodecanucleotides. Significant shifts in order of elution of polynucleotides occur with slight changes in pH; accordingly, the pH is held constant while salt concentration is changed in stepwise or gradient elution. Volatile buffers at alkaline pH are used for polynucleotides and at neutral or slightly acid pH for oligonucleotides. As the molecular size of these substances increases, multiple site ionic interactions result in irreversible adsorption, and an exchanger of lower milliequivalency becomes necessary. Ecteola-cellulose is useful in this regard for the preparation of RNA and DNA from many normal and malignant organs and tissues.

Miscellaneous Applications. Anionic exchange celluloses (DEAE-, TEAE-, P-, Ecteola-) in nonaqueous solvent systems separate anionic lipids from nonacidic lipids. Special ion exchange resins obtained as acyl chloride derivatives react with antigens to form insoluble exchanger-antigen complexes for the isolation and purification of specific antibodies.[1,7,8,10] PAB-cellulose can be diazotized and coupled with proteins having specific biological activities. Numerous other applications have been made to vitamins, alkaloids, blood and its components, organic acids, therapeutic agents, etc.

Ion exchange resins impregnated in paper function as minuscule columns, and are useful for rapid scanning and selection of resins for column chromatography and for analysis. Fabricated into ion exchange resin membranes which are selectively permeable to inorganic cations or anions and small molecular weight charged organic ions, they are useful for separation of neutral, basic and acidic ions from each other and from macromolecules by electrodialysis.[1,6] Ion exchange resins serve as anti-connectant media for some electrophoretic sparations.[1] Chelating resins are uniquely designed for the analysis of trace metals in biological substances.

References

1. MORRIS, C. J. O. R., AND MORRIS, P., "Separation Methods in Biochemistry," New York, Interscience Publishers (Wiley), 1963.
2. BAILEY, J. L., "Techniques in Protein Chemistry," New York, Elsevier Publishing, 1962.
3. HEFTMAN, E. (EDITOR), "Chromatography," New York, Reinhold Publishing Corp., 1961.
4. CALMAN, C., AND KRESSMAN, T. R. E., "Ion Exchangers in Organic and Biochemistry," New York, Interscience Publishers, 1957.
5. DENKEWALTER, R. G., AND KAZAL, L. A., "Pharmaceutical and Biological Products," in "Ion Exchange Technology" (NACHOD, F. C., AND SCHUBERT, J., EDITORS), New York, Academic Press, 1956.
6. SOBER, H. A., HARTLEY, R. W., JR., CARROLL, W. R., AND PETERSON, E. A., "Fractionation of Proteins," in "The Proteins," Vol. 3, Second edition (NEURATH, H., EDITOR), New York, Academic Press, 1965.
7. PETERSON, E. A., AND SOBER, H. A., in "The Plasma Proteins," Vol. 1 (PUTNAM, F. W., EDITOR), New York, Academic Press, 1960.
8. MOORE, S., AND STEIN, W. H., "Column Chromatography of Peptides and Proteins," *Advan. Protein Chem.*, **11**, 191–239 (1956).
9. COHEN, W. E., "The Separation of Nucleic Acid Derivatives by Chromatography on Ion Exchange Columns," in "The Nucleic Acids," Vol. 1 (CHARGAFF, E., AND DAVIDSON, J. N., EDITORS), New York, Academic Press, 1955.
10. MINOR, R. W., (EDITOR), "Ion Exchange Resins in Medicine and Biological Research," *Ann. N.Y. Acad. Sci.*, **57** (3), 1953.
11. Merck Index, Seventh edition, pp. 1575–85, Rahway, N.J., Merck and Co. Inc., 1960.

LOUIS A. KAZAL

IONIZING RADIATION (BIOCHEMICAL EFFECTS)

During the last decade a tremendous amount of research has been done on the biological effects of ionizing radiation. The basic problem of radiobiology is the site of the primary biochemical lesion and is yet unsolved. At the present time, no one has demonstrated any single biochemical change that is peculiarly and uniquely specific for radiation injury. Ionizing radiations are assumed to affect biological compounds either by direct action or by production of substances in the cell which act indirectly. Some of the biological effects of irradiation are due to ionization, *i.e.*, production of free radicals within the cell, as well as organic peroxides. Organic peroxides are generally very reactive, and in the presence of oxygen, they can elicit chain reactions of peroxidation. Products of organic peroxides within irradiated cells, tissues and organisms have been demonstrated.

Radiation induces breakage of CHROMOSOMES and interferes with CELL DIVISION. The gravity of the genetic effects of radiation is of a different order of magnitude from that of all the other biological effects of this agent in that the genetic effects are essentially irreparable and are therefore, if repeated, cumulative over an unlimited period. Exposure to ionizing radiation reduces life span, regardless of whether the irradiation has been accomplished externally by neutrons, X-rays, or γ-rays, or internally by α-particles, or β-particles.

One of the most noticeable features of radiation experiments is the enormous variation in the doses employed by workers in different fields. Small exposures (50 r) may be adequate to produce chromosome damage or cell death, while on the other hand, very high doses (up to 10^6 r) are required to demonstrate physicochemical changes such as depolymerization or loss of phosphate from nucleic acids. The lethal dose for animals falls into the lower part of the scale. Species differences in susceptibility exist, but the LD_{50} range is not great, extending from 325 r for dogs to 800 r for rabbits.

There have been many investigations of radiation effects on growth, cell division, mitosis and genetic changes. In all probability, some radiation-induced biochemical reaction is involved in these cases. Animals subjected to total-body irradiation developed burns, which are similar to thermal burns. Some of the delayed effects to whole-body irradiation are: alopecia, greying of hair, telangiectasis, skin atrophies, keratosis, ulceration following large doses, cataract formation, and tumor induction; these have been observed in many species.

The site of injury with X-radiation and neutron radiation appears to be directly in the blood-forming organs: the bone marrow, lymph nodes and spleen. The lymphocyte and its precursor are the most sensitive cell in the blood system to irradiation. The ANEMIA which followed radiation from internal and external sources has been found frequently to be a macrocytic type. In mice or rabbits receiving 1000 r, there is a drop in the erythrocyte volume beginning with the third day and reaching a depth at the seventeenth day after irradiation, followed by a slow recovery. There is at first a slight concomitant plasma-volume drop, followed by rise on the tenth day following irradiation.

Whole-body X-irradiation inhibits amino acid transport in the thymus and inhibits amino acid oxidation in many tissues. A decrease in the concentration of serum γ-globulin of fractions II and III has been observed in rats exposed to whole-body radiation. ANTIBODY formation is suppressed by total-body X-irradiation. If the spleen or the appendix of the rabbit is protected by lead shielding during total-body irradiation, the capacity to produce antibodies to an injected antigen is retained to a marked degree even though the lymphatic tissue elsewhere in the body is temporarily destroyed. X-irradiation produces immediate metabolic disturbances in the small intestine of most species studied. Four hours after irradiation, there is an abrupt drop of respiration, diminished absorption of glucose, and a diminished phosphorylation of fructose. There was a linear relation between the inhibition of glucose absorption and the amount of irradiation. The alteration in phosphatase activity of tissues of irradiated animals is confined to radiosensitive tissues. The greatest effect of increased activity is detectable in the spleen and thymus gland. Citrate synthesis and incorporation of P^{32} into ATP are decreased in the spleen. In the thymus, irradiation reduces the viscosity of DNA, and depolymerization results. A loss of potassium occurs from radiosensitive tissues following irradiation. Decreased incorporation of P^{32} and spleen phospholipids has been observed after irradiation. Fatty changes in liver of rats and mice have been observed after high doses of radiation. Fatty acid formation in the bone marrow is inhibited. Irradiation reduces the formation of adrenal steroids. The irradiation of mice with ascites tumor results in a decreased formation of DNA in the tumor; however, at the same time synthesis of several metabolites goes on.

Irradiation inhibits oxidative PHOSPHORYLATION of mitochondria isolated from the liver. Total-body X-irradiation produces in the kidney inhibition of oxidation catalyzed by sulfhydryl enzymes. Potassium excretion is increased in the urine of irradiated animals, and there is a rise in the fecal levels with the onset of diarrhea following large doses of irradiation. There is a decrease in O_2 consumption in the kidney and liver in pyruvate utilization even after 100 r.

Cytochemical methods have shown that X-rays and γ-rays, within the dosage range of 40–4000 r, produce metabolic disturbances in proliferating and undifferentiated cells that are characterized by inhibition of synthesis of deoxyribonucleic acid in the nucleus and accumulation of ribonucleic acid, mainly in the cytoplasm.

References

1. MARGERY, ORD, G., AND STOCKEN, L. A., *Physiol. Rev.*, **33**, 356 (1953).
2. DUBOIS, K. P., AND PETERSEN, D. F., *Ann. Rev. Nucl. Sci.*, **4**, 351 (1954).
3. BARRON, E. S. G., AND ZIRKLE, R. E., (EDITORS), in "Biological Effects of External X and Gamma Radiation," Part I, p. 412, New York, McGraw-Hill Book Co., 1954.
4. WOLSTENHOLME, G. E. W., AND O'CONNOR, G. M., "Ionizing Radiation and Cell Metabolism," *Ciba Found. Symp.*, 1956.
5. LAURENS, H., "Physiological Effects of Radiant Energy," New York, Chemical Catalog Co., 1933.
6. HOLLAENDER, A., "Radiation Biology," Vol. I, New York, McGraw-Hill Book Co., 1954.
7. STANNARD, J. N., AND CASARETT, G. W., *Radiation Res. Supplement* 5, 1964.

W. E. CORNATZER

IRON (IN BIOLOGICAL SYSTEMS)

It has been determined that the iron content of the normal adult is dependent on the size of the individual and the hemoglobin concentration. The distribution of iron in a 70-kg man was estimated by Drabkin[1] and later by Harris[2] as follows:

Compound	Total in Body (g)	Total Iron in Compound (g)	Per Cent of Total Body Iron	Function
Hemoglobin				
Peripheral blood	650.0	2.21	64.0	O_2 transport
Bone marrow	25.0	0.09	2.5	O_2 transport
Myoglobin	40.0	0.14	4.0	O_2 transport and "storage" "storage"
Parenchymal or cellular				
Cytochrome	0.8	0.0034	0.097	O_2 utilization
Catalase	5.0	0.0045	0.13	H_2O_2 destruction
Storage iron				
Ferritin	2.0	0.46	13.0	Fe storage
Hemosiderin	1.5	0.56	16.0	Fe storage
Transport iron	6.5	0.004	0.12	Fe transport
Total iron		3.47		

Iron Absorption. It is generally agreed that iron is absorbed for the most part in its ferrous form directly into the bloodstream. Radioactive iron was shown to be absorbed from any portion of the intestinal tract, but its uptake seemed to be greatest in the duodenum. On the basis of experiments done on the absorption of iron from the intestinal tract of guinea pigs, Granick postulated that iron is taken into the mucosa cells and FERRITIN is formed by a combination of a protein, apoferritin, with iron. After the cell is saturated with ferritin, absorption no longer takes place until the iron of ferritin is transferred to plasma. For a number of years, this concept of a mucosal block was the accepted explanation for iron absorption.

Subsequently, evidence appeared which indicated that there was no absolute block to iron absorption. In man, the ingestion of graded amounts of Fe^{59} as ferrous sulfate was accompanied by a progressive rise in iron uptake. It was found that the absorption of iron in patients and in experimental animals was greater than normal in iron deficiency and in cases where erythropoiesis was accelerated, even when the body iron reserves were elevated. Sufficient evidence is available to conclude that the ferritin concentration in the intestinal mucosa neither controls nor blocks absorption. It has been suggested that an active transport mechanism requiring energy is concerned with iron transfer across the intestinal mucosa.

The factors involved in the absorption of iron in food are more complex than those involving inorganic iron. To obtain Fe^{59}-marked foods, radio iron was injected into hens to obtain labeled eggs and meat; plants were grown in media containing Fe^{59}, and Fe^{59}-enriched bread was prepared. It was shown that iron-deficient subjects absorb more food iron than normal subjects. Absorption from liver, hemoglobin, muscle, and "enriched" bread is greater than from eggs or plants. In all probability, the low absorption from egg yolk is due to the presence of a ferric iron-phosphate complex. Large variations in results were obtained with different subjects.

The type of diet may have a profound effect on absorption. In the presence of a large amount of ASCORBIC ACID, the absorption of iron is appreciably enhanced, which is probably due to the reduction of Fe^{+++} to the Fe^{++} form. In the presence of phosphates, carbonates, and phytates, insoluble iron compounds are formed, thus reducing absorption.

It has been estimated that normal subjects ingesting a mixed diet containing 12–15 mg of iron, retain 5–10% (0.6–1.5 mg), while iron-deficient patients retain 10–20% (1.2–3 mg) iron.

Iron Transportation. After iron enters the bloodstream, it is immediately bound by a specific PLASMA PROTEIN which is a β_1-globulin. This protein, which is appropriately named *transferrin* or siderophilin, has a molecular weight of 90,000 and binds 2 atoms of ferric iron. About 0.25 gram of transferrin in 100 ml of plasma is capable of binding about 300 μg Fe^{+++}, but normally it is only one-third saturated while the remaining two-thirds are unbound reserve. If a small amount of ionized iron is injected intravenously, it is bound by the transferrin which may be completely saturated. If the binding limit is exceeded, ionized iron exhibits toxic effects. The transferrin concentration is increased in iron deficiency and during the latter half of pregnancy; it is decreased during infection and a variety of other disorders.

Electrophoretic studies have shown that there are at least nine genetically controlled variants of human transferrins. They all deliver iron in an equivalent manner for utilization and storage.

Evidence is available to show that iron may be transferred directly to the developing erythroblast. It has been demonstrated that transferrin-bound iron is utilized by reticulocytes for HEMOGLOBIN formation. The transfer of iron is not maximum until 25% of the transferrin is saturated.

Excretion. The total loss of iron from an adult is about 1 mg daily and is distributed in the sweat, feces, hair, and urine. Since approximately 1 mg of iron is normally absorbed daily, the organism is in iron balance. The loss of red cells from the body in normal menstruation would account for 16–32 mg iron, which would amount to an average daily loss of from 0.5–1.0 mg during the 28-day menstrual cycle. Pregnancy would also represent a loss of iron from the body, but this is compensated by the absence of menstruation. During normal hemoglobin catabolism, about 20–25 mg of iron are released per day. The excretion of minute amounts of iron allows the body to conserve and reutilize the iron for the synthesis of hemoglobin. This tenacious conservation has been demonstrated repeatedly by radioactive techniques.

Enzymes. Heme serves as the prosthetic group for catalase, PEROXIDASE, cytochrome oxidase, and the related CYTOCHROMES. Catalase and peroxidase iron are presumably present in the ferric form while the iron of the cytochromes may exist in the reduced or oxidized form. A number of flavoproteins, including succinic dehydrogenase, contain iron in the molecule. Iron appears to act as a coenzyme for aconitase. A number of other enzymes require the presence of iron for their activities (see also FLAVINS AND FLAVOPROTEINS; METALLOPROTEINS).

Storage Iron. Ferritin and hemosiderin represent practically all the iron which is present in the reticulo-endothelial cells of the liver, spleen, and bone marrow and in the parenchymal cells of the liver. Ferritin is an iron protein complex containing up to 23% iron. It is composed of a protein, which has a molecular weight of 450,000, and a colloidal ferric-hydroxide-phosphate complex. Preparations of hemosiderin granules contain up to 40% of iron and are insoluble in water. It appears to be an iron-loaded organelle such as a mitochondrion. The granule contains a small amount of ferritin, but the remaining material is composed of heterogeneous proteins.

References

1. DRABKIN, D. L., "Metabolism of Hemin Chromoproteins," *Physiol. Rev.*, **31**, 345 (1951).
2. HARRIS, J. W., "The Red Cell," Cambridge, University Press, 1963.
3. MOORE, C. V., "Iron Metabolism and Nutrition," *Harvey Lectures*, *Ser.* 55, 67 (1959–1960).
4. BEUTLER, E., FAIRBANKS, V. F., AND FABEY, J. L., "Clinical Disorders of Iron Metabolism," New York, Grune and Stratton, 1963.
5. COONS, C. M., "Iron Metabolism," *Ann. Rev. Biochem.*, **33**, 459 (1964).

ALFRED CHANUTIN

ISOELECTRIC POINT

The isoelectric point is the pH value at which a dipolar ion (such as an amino acid in the zwitterion form, or a protein molecule having many positively and many negatively charged groups) shows no migration in an electric field, *i.e.*, during ELECTROPHORESIS (see PROTEINS, section on Proteins as Acids and Bases; Isoelectric Points).

ISOLEUCINE

L-Isoleucine, α-amino-β-methyl-*n*-valeric acid,

$$CH_3{-}CH_2{-}\underset{\displaystyle H}{\overset{\displaystyle CH_3}{\underset{|}{\overset{|}{C}}}}{-}\underset{\displaystyle H}{\overset{\displaystyle NH_2}{\underset{|}{\overset{|}{C}}}}{-}COOH,$$

is present in most common PROTEINS (except hemoglobins) often in amounts corresponding to about 2–10% of the protein. In its structure may be noted the presence of an isoprene skeleton

$$C{-}C{-}\overset{\displaystyle C}{\overset{|}{C}}{-}C$$

(also present in valine and leucine) and two asymmetric carbon atoms. L-Isoleucine as obtained from proteins is one of four stereoisomers: D- and L-isoleucine and D- and L-alloisoleucine [see AMINO ACIDS (OCCURRENCE STRUCTURE, AND SYNTHESIS)]. Isoleucine is nutritionally essential for rats, chicks, humans, and other higher animal forms. For human beings, the isoleucine requirements based on a small group of men and seven women ranged from 250–700 mg/day. The specific rotatory power in 6*N* HCl is $[\alpha]_D = +40.7°$.

ISOMERASES

Isomerases are enzymes that catalyze an intramolecular rearrangement (see ENZYME CLASSIFICATION AND NOMENCLATURE, group 5, Isomerases). Examples include phosphohexose isomerase, which catalyzes the interconversion of glucose-6-phosphate and fructose-6-phosphate, and phosphoglyceromutase, which catalyzes the interconversion of 3-phosphoglycerate and 2-phosphoglycerate (see CARBOHYDRATE METABOLISM; GLYCOLYSIS).

ISOPRENOID BIOSYNTHESIS

Isoprenoids may be defined as the group of compounds that contain either the repeating unit of

$$C{-}\overset{\displaystyle C}{\overset{|}{C}}{-}C{-}C$$

or are derived by relatively minor structural alterations from compounds that do contain this unit. The isoprenoid compounds then include the terpenes, the STEROIDS, and rubber. In addition, one moiety of a number of complex miscellaneous compounds such as the K vitamins, felinine, and saponins is derived by a biosynthetic pathway common to that of the isoprenoid compounds.

The reaction sequences in isoprenoid biosynthesis are divided into two general groups: reactions common to the biosynthesis of all isoprenoids, and reactions that are required for differentiation and the formation of various specific compounds. The carbon skeleton of the repeating unit is derived from the 2-carbon unit, acetyl coenzyme A (acetyl-CoA), a structure common to all organisms (see COENZYME A). Two molecules of acetyl-CoA condense initially to form an intermediate 4-carbon compound, acetoacetyl-CoA. Acetoacetyl-CoA and a third molecule of acetyl-CoA combine in the presence of condensing enzyme to yield β-hydroxy-β-methylglutaryl-CoA (I) (HMG CoA). HMG CoA is reduced, presumably stepwise, by HMG CoA reductase through the semialdehyde to mevalonic acid (II). Mevalonic acid is phosphorylated in the presence of ATP and mevalonate kinase to form 5-phosphomevalonic acid. Similarly, 5-phosphomevalonic acid is phosphorylated by ATP and 5-phosphomevalonate kinase to 5-pyrophosphomevalonic acid. 5-Pyrophosphomevalonic acid is further phpsphorylated with ATP to yield the hypothetical triphosphate, 3-phospho-5-pyrophosphomevalonic acid. In a spontaneous, concerted process, the phosphate from position-3 is lost as inorganic phosphate and carbon-1 is lost as carbon dioxide yielding isopentenylpyrophosphate (III). Isopentenylpyrophosphate (III) is in equilibrium with its tautomer dimethylallylpyrophosphate (IV). These two compounds, III and IV, may be referred to as the "biosynthetic isoprene units".

After the formation of the biosynthetic isoprene units, many special reactions lead to the formation of the structurally dissimilar isoprenoids. These specialized reactions are treated below (see also CAROTENOIDS; STEROLS).

Formation of New Carbon-Carbon Bonds. The formation of such bonds involves the condensation of the biosynthetic isoprene units. The product of one condensation between isopentenylpyrophosphate and dimethylallylpyrophosphate and subsequent loss of the pyrophosphate group yields a 10-carbon monoterpene, geraniol; similarly sesquiterpenes (C_{15}), diterpenes (C_{20}), triterpenes (C_{30}), etc., are formed from 3,4,6 or more units of III and IV. Condensation may occur either between the carbon atom carrying the pyrophosphoester of IV and the terminal methylene of III (head to tail) or between the two esterified carbon atoms (head to head). The arrangement of the resulting branched compounds may be indicated as follows:

```
    C           C
    |           |
C—C—C—C—C—C—C—C
        ↑   ↑
  Head of IV  Tail of III

    C             C
    |             |
C—C—C—C—C—C—C—C
        ↑   ↑
      Heads of IV
```

In addition to coupling between units, internal coupling and rearrangement may occur as, for example, the following:

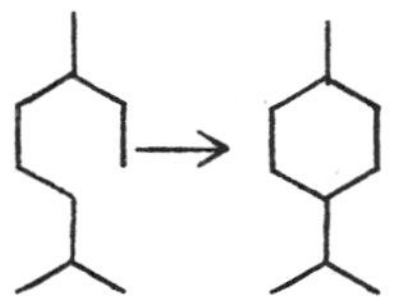

Such coupling and rearrangement reactions may occur to the extent that the products may not appear to have been derived biosynthetically from III and IV. Additional carbon atoms may also become attached to the isoprenoid framework by direct transfer to the molecule from appropriate donors, such as S-adenosylmethionine:

$$R—CH_2—R' \rightarrow R—\underset{}{\overset{CH_3}{\overset{|}{C}}}H—R'$$

Such added carbon atoms are therefore not derived from either III or IV.

Oxidations. Oxidations may be classified either as hydroxylations or dehydrogenations. Hydroxylation of isoprenoid compounds generally involves atmospheric oxygen and reduced pyridine nucleotides:

$$R—CH_2—R' + 1/2O_2 + PNH_2 \rightarrow R—\overset{OH}{\overset{|}{C}}H—R' + PN$$

Dehydrogenation may lead to either the introduction of additional double bonds, the oxidation of alcohols to carbonyls and carboxyls, or the oxidative cleavage of carbon fragments from the parent isoprenoid compound:

$$R—CH_2OH \rightarrow R—CHO \rightarrow R—COOH \rightarrow R—H + CO_2$$

Reductions. Enzymatic reduction or hydrogenation of double bonds, carbonyls, and carbon centers during condensations may occur:

Lanosterol → 24,25-Dihydrolanosterol

Isomerizations. Double bonds may be rearranged enzymatically between carbon centers. These reactions frequently require oxygen, and hydroxylated intermediates may be formed:

Δ^8-Cholestenol → Cholesterol (Δ^5-Cholestenol)

The biosynthesis of *cholesterol* has probably been studied in greater detail than that of any other isoprenoid compound. The biosynthesis of cholesterol illustrates both the reactions common to the biosynthesis of all isoprenoid compounds and a number of the specialized reactions of isoprenoid biosynthesis referred to above. The biosynthetic isoprene units, isopentenylpyrophosphate (III) and dimethylallylpyrophosphate (IV) condense (head to tail) to yield geranylpyrophosphate (C_{10}) which condenses with another isopentenylpyrophosphate to yield the C_{15}-compound, farnesylpyrophosphate (V). Two units of V condense, head to head, to yield squalene (VI). Squalene is attacked by molecular oxygen and a specific enzyme, squalene oxidocyclase, which accomplishes a concerted oxygenation, cyclization, and rearrangement to yield lanosterol (C_{30}) (VII). Four of the six original units of isopentenylpyrophosphate and dimethylallylpyrophosphate are indicated in that part of the lanosterol structure that has not been involved in a rearrangement. Conversion of lanosterol (VII) into cholesterol requires a number of additional

(I) $HOOC{-}CH_2{-}C(H_3C)(OH){-}CH_2{-}C(=O){-}SCoA$

(II) $HOOC{-}CH_2{-}C(H_3C)(OH){-}CH_2{-}CH_2{-}OH$

(III) $CH_2{=}C(H_3C){-}CH_2{-}CH_2{-}O{-}PO_3HPO_3H_2$

(IV) $CH_3{-}C(H_3C){=}CH{-}CH_2{-}O{-}PO_3HPO_3H_2$

(V) $CH_3{-}C(CH_3){=}CH{-}CH_2{-}CH_2{-}C(CH_3){=}CH{-}CH_2{-}CH_2{-}C(CH_3){=}CH{-}CH_2{-}OPO_3HPO_3H_2$

(VI) $H{-}(CH_2{-}C(CH_3){=}CH{-}CH_2)_3{-}(CH_2{-}CH{=}C(CH_3){-}CH_2)_3H$

(VII)

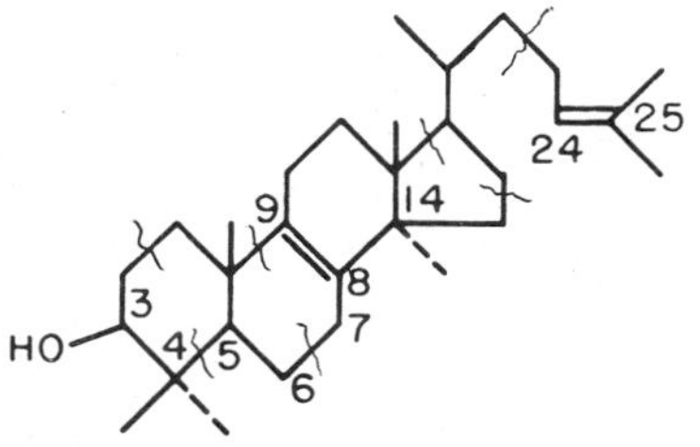

HO 3 4 5 6 7 8 9 14 24 25

enzyme-catalyzed transformations. The methyl groups on positions-4 and 14 are lost oxidatively as carbon dioxide following hydroxylation and dehydrogenation. The double bond in the 8,9-position is isomerized (oxidatively) to yield the 5,6-double bond of cholesterol, and the double bond between carbon atoms 24 and 25 is hydrogenated. Many organisms do not possess all the enzymes essential for sterol biosynthesis and consequently have a nutritive requirement for cholesterol or other sterols. Under these conditions, intermediates in isoprenoid biosynthesis may accumulate.

The control of isoprenoid biosynthesis appears to be accomplished two ways: (1) limitations on the availability of cofactors such as ATP and reduced pyridine nucleotides, and (2) end-product inhibition of mevalonic acid formation particularly at the HMG CoA reductase or mevalonic kinase stages. The first control mechanism illustrates the fact that very large amounts of energy are required for the biosynthesis of isoprenoid compounds. The second process suggests a self-regulatory mechanism in which the accumulation of the products tends to restrict the formation of an essential precursor.

References

1. Wright, L. D., "Biosynthesis of Isoprenoid Compounds," *Ann. Rev. Biochem.*, **30,** 525 (1961).
2. Wolstenholme, G. E. W., and O'Connor, M., (Editors), "Ciba Foundation Symposium on the Biosynthesis of Terpenes and Sterols," Boston, Little, Brown and Co., 1959.
3. Olson, J. A., "The Biosynthesis of Cholesterol," *Ergebnisse der Physiologie*, **56**, 173 (1965).

L. D. Wright and J. L. Gaylor

ISOTOPE DILUTION METHODS

Hevesy and Hofer introduced the concept of isotope dilution analysis in 1934, but it was not until the application of this technique to biochemical research by Rittenburg and Foster in 1940 that the methodology became commonplace. There are basically three ways by which this technique may be used: (1) as an analytical tool; (2) to help identify an intermediary in a metabolic pathway; (3) to prove the incorporation of a precursor in a product.

Analytical Applications. When the isotope dilution method is to be used as an analytical tool, several techniques are possible.

Direct Isotope Dilution. This method permits the quantitative determination of an "inactive" (not radioactive) compound by dilution with an "active" (radioactive) compound. This technique consists of the addition of a *known* amount of a radioactive-labeled compound to a mixture containing an *unknown* amount of the same compound that is not radioactive. The two materials are thoroughly mixed so that there is a uniform distribution. The mixture is then treated in such a way as to isolate the added compound in a pure form. It is essential that the compound isolated be scrupulously pure, but it is not necessary to separate the compound quantitatively from the mixture. Finally, when a representative sample of the pure compound has been isolated and judged to be pure, the radioactivity of the sample is assayed.

EXAMPLE: To a mixture containing an unknown amount of cholesterol is added 1 mg of ^{14}C-cholesterol with 50,000 counts/min. After the materials are mixed well, the mixture is carefully extracted to yield a small amount of pure cholesterol. The cholesterol obtained is then assayed. If the obtained radioactivity is 50 counts/min/mg, it is obvious that the ratio of M $\times$ CPM $= M' \times$ CPM$'$ ($M =$ mass before dilution, $M' =$ total mass after dilution, CPM = counts per minute per milligram of sample before dilution, CPM$'$ = counts per minute per milligram of sample after dilution); therefore, $M = L \times 50000/50 = 1000$, and there are 1000 mg after total dilution and 999 mg of cholesterol present in the original mixture (1000 total -1 added).

Inverse Isotope Dilution. In this method the weight of a known radioactive compound is obtained by dilution of this compound with "inactive" compound of known weight. The materials are mixed, the compound is isolated and purified, and the specific radioactivity is assayed. For instance, an unknown amount of radioactive oleic acid (10000 counts/min/mg) is present in a solution; 250 mg of "cold" oleic acid is added. After the material is well mixed, a small amount of the oleic acid is extracted, purified and assayed, *i.e.*, giving 500 counts/min/mg; therefore, the weight of fatty acid present in the original mixture was:

$$M \times 10000 = (250 + M) \times 500$$

$$M = \frac{(250 + M) \times 500}{10000}$$

$$M = 1.31 \text{ mg}$$

Double Iostope Dilution. In this method of isotope dilution as originally developed by Block and Anker, the knowledge of the specific activity of the unknown substance is not necessary. Two aliquots of the sample to be assayed are removed, and different amounts of carrier $M' + M''$ are added to each. A representative quantity of the substance to be determined is isolated from each sample and carefully purified. The specific activity of each of these is carefully assayed and then by establishing two simultaneous equations and solving for the unknown, the original weight of radioactive substance is obtained.

$$M \times \text{CPM} = (M' - M) \times \text{CPM}'$$

$$M \times \text{CPM} = (M'' - M) \times \text{CPM}''$$

$$(M' - M) \times \text{CPM}' = (M'' - M) \times \text{CPM}''$$

$$M = \frac{\text{CPM}' \cdot M' - \text{CPM}''\, M''}{\text{CPM}'' - \text{CPM}'}$$

EXAMPLE: To an unknown amount of radioactive squalene of unknown radioactivity is added 1000 mg of "cold" squalene. The material is mixed, the squalene is isolated and purified yielding a material of 14.97 counts/min/mg. Then, another identical portion of the first material is mixed with 1500 mg of cold squalene, the material is mixed, and the squalene is isolated and purified yielding a material of 9.99 counts/min/mg.

$$M = \frac{1000 \times 14.97 - 1500 \times 9.99}{14.97 - 9.99} = \frac{15}{4.98}$$

$$M = \text{less than 3 mg}$$

Metabolic Intermediates. Isotope dilution is an aid in the identification of a metabolic intermediary. Into a biological system capable of yielding a radioactive "intermediary" metabolite from a radioactive precursor, a large amount of unlabeled (cold) "intermediary" is added. The materials are then well mixed, and the mixture is subjected to a procedure designed to isolate a representative sample of the "intermediary." After isolation, the intermediary is purified. If a constant specific activity is obtained in the mixed material after repeated purifications and derivatives, the same chemical form as originally present is presumed to have been trapped by the large amount of the known "cold" material added. For instance, in the investigation of cholesterol intermediates, β-hydroxy-β-methylglutaric acid (HMG) was suspected of being an intermediary formed from acetate in the enzymatic system used. To prove this point, to a radioactive incubation of 1 mg acetate in a suitable enzyme system, approximately 100 mg of "cold" HMG were added and the mixture was well mixed. The mixture was then subjected to an extraction procedure to separate HMG, and this compound, when isolated, was carefully purified by repeated crystallization, and several derivatives were made. If all of these compounds gave the same specific radioactivity, these results would establish the radioactive purity of the sample and thus indicate that HMG was an intermediary in the synthesis of cholesterol. The "cold" HMG (in large amounts) was used to trap the "hot" HMG (very slight quantity) formed from the acetate. The trapping is possible because both compounds are identical in all their chemical properties.

Precursor Incorporation. Isotope dilution analysis can be used to prove that a precursor is indeed incorporated in a final product. When a radioactive precursor is suspected of yielding a certain product, the radioactive precursor may be mixed with added amounts of "cold" product. The mixture of these materials containing the original "hot" product and the added "cold" carrier product is then purified by absorption, ion exchange, gas, paper or thin layer chromatography. The "cold" carrier compound, if chemically identical to the final "hot" product, should appear at the same spot and at the same rate in the chromatographic system. Two curves are then plotted: (1) volume of solvent diluted or fraction (tube) number against quantitative chemical composition; (2) volume of solvent diluted or tube number against radioactivity. If coincidence of peaks of the carrier compound and the radioactive compound occurs, this indicates that chemically they are the same. Thus, the product obtained from the precursor is the expected one, and it is identical to the material used as the carrier.

References

1. Hevesey, G., and Hofer, H., *Nature*, **134**, 879 (1939).
2. Rittenberg, D., and Foster, G. L., *J. Biol. Chem.*, **133**, 737 (1940).
3. Bloch, K., and Anker, H. S., *Science*, **107**, 228 (1948).
4. Clingman, W. H., Jr., and Hammen, M. M., *Anal. Chem.*, **32**, 323 (1960).
5. Rabinowitz, J. L., and Gurin, S., *J. Biol. Chem.*, **208**, 307 (1959).
6. Rabinowitz, J. L., *et al.*, *Arch. Biochem. Biophys.*, **113**, 233 (1966).

Joseph L. Rabinowitz

ISOTOPIC TRACERS

In few areas of science have radioactive tracers been employed as widely and as successfully as in biochemistry. In metabolic transformation studies, labeled compounds are injected or fed to the animal or plant. Subsequently, the excretions (solid, liquid or gas) or tissues are analyzed to determine by detection of the radioactive tracer how the original compound has been changed. Labeling of the compound in different positions of the molecule often makes it possible to determine the intermediate steps in the biological processes. In similar fashion, the mechanisms of reaction between enzymes and substrates have been studied in many systems by tracer techniques. Measurements of the rate of metabolic transformations in whole organisms with tracers are more difficult to interpret. The problem of side reactions as well as production of the same compound at various rates by different organs or different parts of the same organ or cell usually allows only comparative rate studies by tracer techniques. An outstanding example of a successful kinetic study was the use of $C^{14}O_2$ in PHOTOSYNTHESIS. Tracer techniques have been used in studying such problems as nucleic acid complementation (C^{14}, P^{32}) and biological macromolecule and protein synthesis (C^{14}, P^{32}). In physiology, by the methods of isotopic dilution analysis, the volume of plasma (I^{131}) and of red cells (Cr^{51}) in the blood, the volume of extracellular fluid (S^{35}), and the water content (H^3) of human or animals are measured. In medicine, radioisotopes are used in the diagnosis of many diseases such as hyperthyroid (I^{131}), anemia and polycythemia (Fe^{59}), pernicious anemia (Co^{60}), brain tumors (P^{32}, I^{131}), gastrointestinal bleeding (Cr^{51}), and kidney and liver malfunctioning (I^{131}).

Many tracers have been used by biochemists, but four are in much more common use: H^3, C^{14}, P^{32} and S^{35}. Unfortunately, the radioactive decay of these four occurs by beta particle emission only, with no accompanying gamma decay. This eliminates the possibility of detection through thick layers of matter, *e.g.*, through test tube walls, which detection of the highly penetrating gamma rays allows. Moreover, the average energy of the emitted electrons is very low for H^3(0.006 meV), C^{14}(0.045 meV) and S^{35}(0.055 meV). Consequently, the biochemist using the most popular tracers, H^3 and C^{14}, must solve some rather difficult counting problems.

Detection Methods. There are two general classes of counters that can be used: one uses a counting gas as the medium sensitive to radiation; the second uses liquids or solids as the sensitive medium. The first class includes *ionization chambers*, *proportional counters* and *Geiger counters*. The biochemical compounds containing C^{14} may be used as obtained or may be converted to solid $BaC^{14}O_3$ samples for counting either by proportional or Geiger counters. It is somewhat more convenient to count such samples with thin window counting tubes rather than by direct insertion in the counting chamber itself. However, the efficiency is less for thin window counting—*e.g.*, transmission through a 1.4 mg/cm² window is 70% for C^{14}. Problems of absorption of the beta particles within the $BaCO_3$ sample itself require careful attention for satisfactory results.

Ionization chambers, Geiger and proportional counting have also been used to count gas samples. The C^{14} compound is oxidized to $C^{14}O_2$ or $C_2^{14}H_2$ which is directly mixed with the counting gas. Similarly, tritium compounds are first converted to HTO, then reduced to HT for gas counting. This technique requires vacuum line equipment for the combustion, purification and transfer of the gas samples into the counting chambers. Disadvantages of gas sample counting include the complexity of the sample handling system, the care necessary to avoid traces of moisture and contaminants which can interfere with the counting efficiency, and the small amount of sample that can be counted at a time.

The second class of counters uses for the sensitive medium a material which emits *scintillations* of light upon passage of radiation. The scintillations are detected and amplified by photomultiplier tubes. A variety of scintillation media are used, but for C^{14} and H^3 the normal system is a solution. The solute such as 2,5-diphenyloxazole (PPO) must be an efficient emitter of light of a wavelength well transmitted through the solution and efficiently detected by the photomultiplier. The solvent such as toluene or xylene must transfer energy well and have a small absorption coefficient for the light emitted by the solute. Commercial liquid scintillation counters are operated at low temperatures to reduce electronic background noise.

The biochemical sample is dissolved, if possible, in the toluene solution. For polar molecules, a dioxane-naphthalene or dioxane-ethanol-toluene solvent may be used. It is necessary that the sample not quench the scintillation. In these cases and in the case of insoluble samples, it may be necessary to use digestion techniques to prepare a soluble, homogeneous system. Some insoluble samples such as peptides, proteins, nucleic acids, CO_2, etc., can be dissolved in toluene by reaction with the hydroxide of hyamine 10-X, a quaternary ammonium base. In some cases of insoluble samples, a finely divided preparation of silicon dioxide (Cab-O-Sil) has been used to form an optically clear thixotropic suspension of the sample in a gel with the scintillating liquid. Scrapings of areas from a thin layer chromatogram have been counted quantitatively for C^{14} and H^3 by this gel technique. Insoluble materials spotted on filter paper as well as sections of paper chromatograms have been counted after immersion in a scintillating solution.

A promising technique is the use of plastic scintillar beads to replace the scintillator solution. Aqueous or alcoholic solutions containing the tracer are allowed to fill the interstices between the beads. The rapidity of sample preparation is a distinct advantage of this technique.

Comparison of the relative sensitivity of liquid scintillation counting and gas sample counting by ionization chamber gives these results: (a) for small, soluble, non-quenching organic-C^{14} samples, the scintillation method is 5 times more sensitive; (b) for larger organic-C^{14} samples, the scintillation method may be as much as 250 times more sensitive; (c) for water-H^3 samples, the two methods have comparable sensitivity; (d) for small, soluble, non-quenching organic-H^3 samples, the scintillation method may be several times more sensitive; (e) for larger organic-H^3 samples the scintillation method may be up to 65 times more sensitive.

Nonradiative Tracers. In principle it is possible to use nonradioactive tracers for the same purposes as radioactive tracers. However, mass spectrometric analysis is required in most cases, and this is usually a more complicated technique than radiation measurement. For oxygen and nitrogen, no suitable radioactive isotopes are known, and tracing must be done with O^{18} and N^{15}. In certain types of experiments, radioactive tracers cannot supply the information that stable tracers can. In radioactive tracing, the gross distribution of activity in several fractions is determined, whereas with mass spectrometric analysis, stable tracing allows observation of specific distributions within a fraction. The study of the cyclization mechanism of squalene can serve as an example. It was uncertain whether a 1,3- or two 1,2-shifts were involved in the methyl migration. It was possible to obtain samples of ethylene from the methyl group and the carbon to which it was attached in the cyclized squalene. Mass spectrometry measured the fraction of ethylene with one C^{13}, with two C^{13} and with no C^{13} atoms. Since only the two 1,2-shift mechanism could yield ethylene with two C^{13} atoms, the presence of this provided the basis of choice of mechanism. With C^{14} tracing, it would be possible only to measure the gross activity of C^{14} in ethylene but not to learn how this activity was distributed on a molecular basis.

References

1. "Packard Technical Bulletins," Packard Instrument Co., Inc., Box 428, LaGrange, Illinois.
2. "Atomlight," New England Nuclear Corp., 575 Albany St., Boston 18, Massachusetts.
3. BELL, C. G., AND HAYES, F. N., (EDITORS), New York, Pergamon Press, 1958.
4. TOLBERT, B. M., AND SIRI, W. E., "Physical Methods of Organic Chemistry (Technique of Organic Chemistry, Part IV, Vol. 1)", (WEISSBERGER, A., EDITOR), New York, Interscience, 1960.
5. COMAR, C. S., "Radioisotopes in Biology and Agriculture," New York, McGraw-Hill Book Co., 1955.
6. ARONOFF, S., "Techniques of Radiobiochemistry," Ames, Iowa, Iowa State College Press, 1956.
7. KAMEN, M. D., "Radioactive Tracers in Biology," Third edition, New York, Academic Press, 1957.

GREGORY R. CHOPPIN

ISOZYMES (ENZYME VARIANTS, ISOENZYMES)

Enzymes are defined in terms of the reactions they catalyze (see ENZYME CLASSIFICATION). Since the same or similar reactions occur in different biological systems, it follows that most enzymes may be isolated from a number of sources. With the development of highly discriminating techniques for the resolution and analysis of PROTEINS, it has become apparent that many different molecular species can exhibit the same biological function. The term "enzyme variants" refers to the family of proteins which catalyze the same or similar reactions. Enzyme variants may be isolated from the same or different species, *i.e.*, there are both intraspecies (ontogenetic) and interspecies (phylogenetic) variants. Studies of both types of enzyme variants are significant from a chemical point of view because they provide an indication of the role of specific functional groups in the determination of the unique chemical and catalytic features of the enzyme molecules. It is possible to define effects of changes in the primary structure (*viz.*, amino acid substitutions, deletions, or additions) on the over-all molecular conformation (tertiary structure) and on the specific catalytic characteristics (affinity for substrates, catalytic efficiency as reflected in reaction velocity, substrate specificity, etc.).

Studies of enzymatic variants are also of interest from a biological point of view. Interspecies variants, in one sense, are a specific (genetic) record of natural history. The amino acid sequence, therefore, can be related to the structure of the nucleic acid of the relevant structural GENE and hence can be used as an evolutionary probe. Intraspecies variants, in some instances, may result simply from metabolic variability, but in other instances they are specifically associated with subtle metabolic adaptations in various tissues. In addition, studies of their distribution and regulation may provide evidence of basic embryological mechanisms.

Homologous and Analogous Variants. Enzyme

variants that exhibit essentially similar molecular structure and catalytic properties may be termed *homologous* variants. It is presumed that such homologous families have arisen by evolutionary modulation (via mutation and natural selection) of the descendants of a common "primitive gene," and hence are "related" in the evolutionary sense. On the other hand, enzyme variants that catalyze the same or similar reactions, but bear little molecular resemblance to each other and appear to have fundamentally different catalytic characteristics, are termed *analogous* variants. These analogous families may have arisen by evolutionarily distinct mechanisms, *i.e.*, from *different* "primitive genes." The presently available evidence allows tentative recognition of a number of instances of analogy and homology.

The most convincingly documented instance of homology is found in the variants of CYTOCHROME C, a component of the electron transport system in BIOLOGICAL OXIDATION. There is striking resemblance in the amino acid sequence of cytochrome c when isolated from phyletically divergent sources. For example, about one-half of the amino acids in yeast cytochrome c are identical with those in similar positions of mammalian cytochrome c! Furthermore, the degree of divergence in primary structure (amino acid sequence) is roughly proportional to the evolutionary divergence. For example, the cytochrome c from insects is structurally intermediate between yeast cytochrome c and mammalian. These points are illustrated in Fig. 1.

Another intriguing example of apparent homology is found in two different pancreatic enzymes isolated from the same species (beef). Trypsin and chymotrypsin are proteins that exhibit different substrate specificity, but have similar molecular and catalytic characteristics. The amino acid sequence of the two molecules is found to be closely related (approximately 40% of the amino acid residues occur in identical positions in the two chains). The two enzymes, therefore, may have resulted from a common primitive gene (perhaps by an evolutionary mechanism involving gene duplication, and subsequent evolutionary modulation of the two genetic copies resulting in different substrate specificity).

It is also possible to perceive analogous and homologous relationships by more gross analyses of structure, coupled with careful studies of catalytic properties. Using this experimental approach, two *analogous* classes of fructose diphosphate aldolase (the enzyme that catalyzes the aldol cleavage of fructose diphosphate in the Embden-Meyerhoff glycolytic pathway; see CARBOHYDRATE METABOLISM) have been detected. Each class has a distinct molecular weight and subunit composition, and specific catalytic characteristics reflecting different structures at the catalytically active site. This point is illustrated in Fig. 2. Members of each class, on the other hand, have very similar properties.

Class I aldolases are found in animals, plants, protozoa, and blue-green algae. Two organisms (*Euglena* and *Chlamydomonas*) contain both Class I and II aldolases. This distribution suggests that the Class I aldolases arose in a progenitor of *Euglena* and *Chlamydomonas* and that subsequently, for a reason not currently appreciated, the Class I enzyme was retained and the Class II enzyme was lost.

Intraorganismic Variants — Isozymes — Isoenzymes. It might be expected from current genetic concepts of mutation (see GENES) that enzymes isolated from divergent species might have different structures. It is, perhaps, less expected that multiple forms of the same enzyme would be discovered within cells of one organism. Such multiplicity could theoretically result by modification of the molecular properties of a single protein molecule, coded by a single gene, or through mechanisms involving synthesis of separate protein species whose structure is defined by more than one gene.

There are a number of possibilities by which different molecular forms may be produced from a single gene, for example:

(1) If the processes involved in protein synthesis are not absolutely specific, then some differences in the amino acid composition of a given protein would result, *i.e.*, a given structural gene might indirectly result in the formation of a heterogeneous population of products. There is at present no evidence for this kind of loose transcriptional and translation specificity within cells. [see PROTEINS (BIOSYNTHESIS)].

(2) There might be multiple stable conformations of the same primary structure, *i.e.*, the same amino acid sequence might assume recognizably different configurations. There are a number of instances where the presence of substrates or other effectors produces demonstrable changes in molecular conformation or state of aggregation. For example, threonine deaminase, the enzyme which catalyzes the conversion of threonine to α-ketobutyrate reversibly dissociates (associates) into a number of molecular species with varying catalytic properties. The degree of dissociation is influenced by a number of factors including the substrate, threonine, as well as ADP, inorganic phosphate, divalent cations, and pH. At low substrate and ADP concentrations, the enzyme is dissociated into subunits having very low catalytic activity. Addition of substrate or ADP causes association with a resultant increase in catalytic activity. The ADP does not react at the catalytic site, but at a different locus, called the *allosteric* site. The multiple forms produced by different degrees of polymerization of a basic subunit or by changes of conformation of a protein whose size remains constant can be termed *allosteric* variants.

Allosteric alterations of protein structure have been implicated as factors in the regulation of metabolic processes (see REGULATORY SITES OF ENZYMES). For example, in the instance mentioned, the levels of threonine and ADP control the rate of degradation of threonine deaminase pathway. At high threonine levels, when the needs for protein synthesis are saturated, threonine deaminase is activated, and activity of the pathway for degradation of this compound is increased.

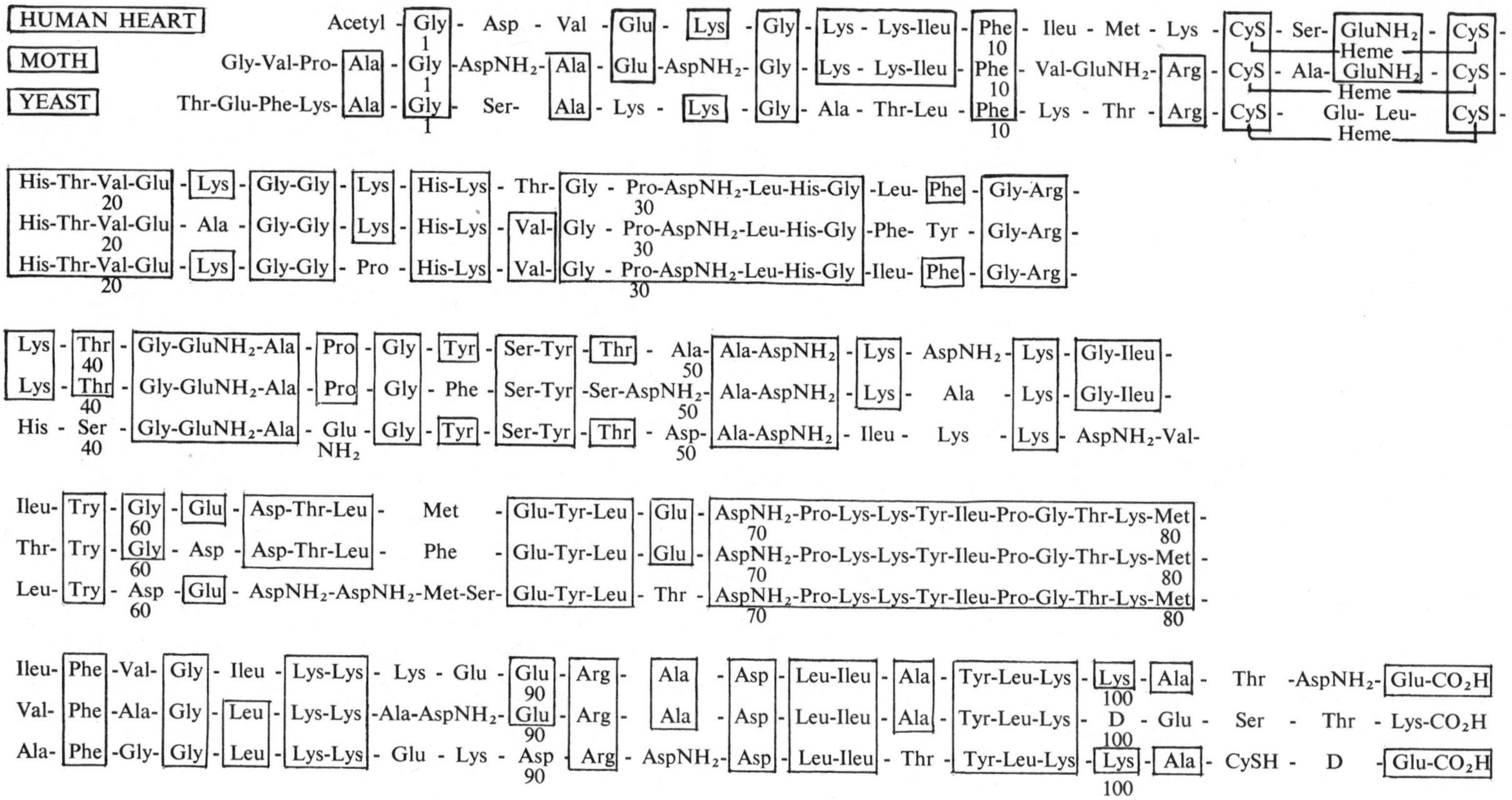

HUMAN HEART Acetyl - Gly - Asp - Val - Glu - Lys - Gly - Lys - Lys-Ileu - Phe - Ileu - Met - Lys - CyS - Ser - GluNH2 - CyS -
(1) (10) (Heme)

MOTH Gly-Val-Pro- Ala - Gly -AspNH2- Ala - Glu -AspNH2- Gly - Lys - Lys-Ileu - Phe - Val-GluNH2- Arg - CyS - Ala- GluNH2 - CyS -
(1) (10) (Heme)

YEAST Thr-Glu-Phe-Lys- Ala - Gly - Ser- Ala - Lys - Lys - Gly - Ala - Thr-Leu - Phe - Lys - Thr - Arg - CyS - Glu- Leu- CyS -
(1) (10) (Heme)

His-Thr-Val-Glu - Lys - Gly-Gly - Lys - His-Lys - Thr- Gly - Pro-AspNH2-Leu-His-Gly -Leu- Phe - Gly-Arg -
(20) (30)

His-Thr-Val-Glu - Ala - Gly-Gly - Lys - His-Lys - Val- Gly - Pro-AspNH2-Leu-His-Gly -Phe- Tyr - Gly-Arg -
(20) (30)

His-Thr-Val-Glu - Lys - Gly-Gly - Pro - His-Lys - Val- Gly - Pro-AspNH2-Leu-His-Gly -Ileu- Phe - Gly-Arg -
(20) (30)

Lys - Thr - Gly-GluNH2-Ala - Pro - Gly - Tyr - Ser-Tyr - Thr - Ala- Ala-AspNH2 - Lys - AspNH2 - Lys - Gly-Ileu -
(40) (50)

Lys - Thr - Gly-GluNH2-Ala - Pro - Gly - Phe - Ser-Tyr -Ser-AspNH2- Ala-AspNH2 - Lys - Ala - Lys - Gly-Ileu -
(40) (50)

His - Ser - Gly-GluNH2-Ala - GluNH2 - Gly - Tyr - Ser-Tyr - Thr - Asp- Ala-AspNH2 - Ileu - Lys - Lys - AspNH2-Val-
(40) (50)

Ileu- Try - Gly - Glu - Asp-Thr-Leu - Met - Glu-Tyr-Leu - Glu - AspNH2-Pro-Lys-Lys-Tyr-Ileu-Pro-Gly-Thr-Lys-Met -
(60) (70) (80)

Thr- Try - Gly - Asp - Asp-Thr-Leu - Phe - Glu-Tyr-Leu - Glu - AspNH2-Pro-Lys-Lys-Tyr-Ileu-Pro-Gly-Thr-Lys-Met -
(60) (70) (80)

Leu- Try - Asp - Glu - AspNH2-AspNH2-Met-Ser- Glu-Tyr-Leu - Thr - AspNH2-Pro-Lys-Lys-Tyr-Ileu-Pro-Gly-Thr-Lys-Met -
(60) (70) (80)

Ileu- Phe -Val- Gly - Ileu - Lys-Lys - Lys - Glu - Glu - Arg - Ala - Asp - Leu-Ileu - Ala - Tyr-Leu-Lys - Lys - Ala - Thr -AspNH2- Glu-CO2H
(90) (100)

Val- Phe -Ala- Gly - Leu - Lys-Lys -Ala-AspNH2- Glu - Arg - Ala - Asp - Leu-Ileu - Ala - Tyr-Leu-Lys - D - Glu - Ser - Thr - Lys-CO2H
(90) (100)

Ala- Phe -Gly- Gly - Leu - Lys-Lys - Glu - Lys - Asp - Arg - AspNH2- Asp - Leu-Ileu - Thr - Tyr-Leu-Lys - Lys - Ala - CySH - D - Glu-CO2H
(90) (100)

FIG. 1. Homologous amino acid sequences in cytochrome c isolated from different species. The following abbreviations for the amino acids are employed: gly = glycine; ala = alanine; val = valine; leu = leucine; ileu = isoleucine; ser = serine; thr = threonine; cys = cysteine; met = methionine; asp = aspartic; aspNH2 = asparagine; glu = glutamic; gluNH2 = glutamine; lys = lysine; arg = arginine; his = histidine; phe = phenylalanine; tyr = tyrosine; try = tryptophan; pro = proline. D implies a presumed deletion of an amino acid has occurred in a particular position (to maximize the homology in the structures). A major difference in the structures occurs at the N-terminus: vertebrates have an acetylated glycine terminus, but "lower species" include instead a number of additional amino acid residues. There are 56 amino acid residues (more than 50%) in common among all three molecules. The cytochrome c's from *human* and the *moth* have 75 residues in common; those from *human* and *yeast* have 65 residues in common.

Similarly, when ATP production is not sufficient to satisfy metabolic demands, the level of ATP falls, and that of ADP rises. High ADP levels result in a relatively greater activity of the enzyme at any level of threonine. Thus, the relative degree of threonine breakdown depends on the ADP available for conversion to ATP, and the substrate available to the system.

(3) An enzyme protein may be modified by other enzyme systems, or by the chemical environment, in such a way as to produce a number of stable products. For example, a variable number of sulfhydryl (—SH) groups in the molecule might be oxidized to form —S—S— linkages with concomitant changes in the orientation of the other groups in the molecule, thus resulting in different

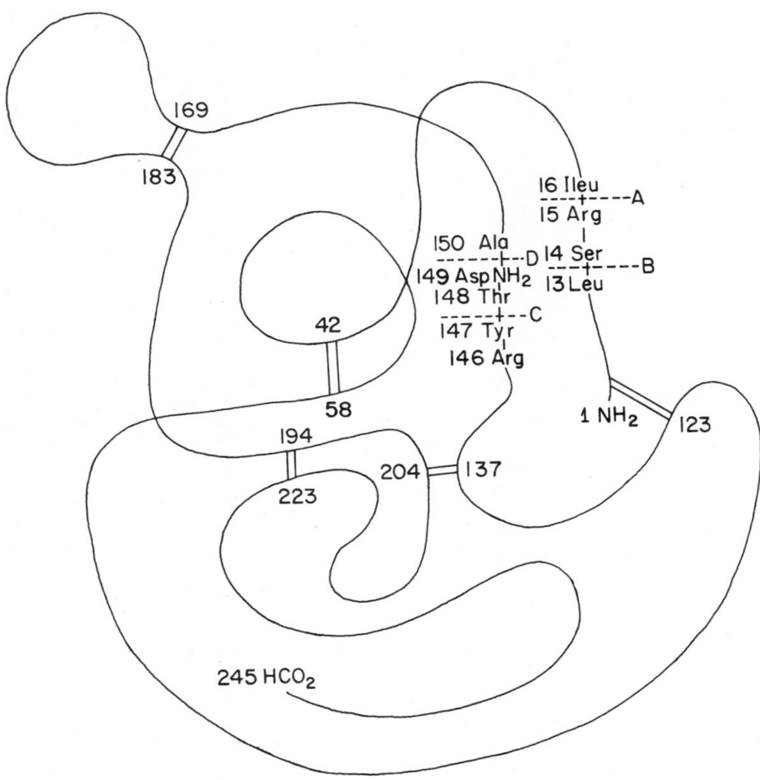

Cleavage at:

A gives rise to Chymotrypsin π

A+B gives rise to Chymotrypsin δ

A,B,C + D gives rise to Chymotrypsin α

FIG. 2. Formation of chymotrypsins α, δ, and π from chymotrypsinogen. Chymotrypsinogen, a single-chain molecule, is synthesized by the pancreas and is enzymatically inert. It is converted to enzymatically active forms by the cleavage of specific bonds by proteinases: cleavage of the arginyl-isoleucine bond at positions 15–16 produces a two-chain enzyme (chymotrypsin π); the additional cleavage of the leucyl–serine bond at positions 13–14 produces a two-chain enzyme (chymotrypsin δ), and cleavage at both of these positions plus the tyrosyl–threonine bond at positions 147–148, and the asparaginyl–alanine bond at 149–150 produces a three-chain enzyme (chymotrypsin α). [Adapted from Hartley *et al.*, *Nature*, **207**, 1157 (1965).]

physical and chemical properties. In addition, enzymes can be partially degraded by proteolytic components of cells to form molecules which should still retain catalytic activity, but which have modified molecular properties. For example, multiple molecular forms of the PROTEOLYTIC ENZYME chymotrypsin are produced by the proteolytic action of trypsin on the inactive protein chymotrypsinogen. Several *different* bonds in the molecule can be cleaved resulting in *different* active chymotrypsins (α-, δ-, π-chymotrypsins). Most of these forms have been isolated and identified. A schematic representation is presented in Fig. 2.

Other intraorganismic variants are the result of formation of multiple polypeptide chains having different amino acid sequences, but having the same function. The chains may be similar in structure and hence homologous, or they may be very different and hence analogous in the evolutionary sense. The general term *isozyme* or *isoenzyme* is usually employed to designate intraorganismic variants *with a distinct genetic origin.* In the simplest case, isozymes may be completely distinct molecules; thus, two different proteins A and B, having different amino acid sequences, may catalyze the same reaction (two separate structural genes are involved). In other cases, the molecular structure may be more complicated. If a dimer is the predominant configuration and A and B are homologous chains existing in the same cell, then three forms (AA, AB, BB) may be expected, purely on the basis of random combination. If trimers were formed, then four multiple forms might be expected: (AAA, AAB, ABB, BBB, etc.). A larger number of non-random arrangements of A and B in heteropolymeric forms might be envisioned. In these instances, two genes cause the formation of more than two different enzymes.

The extensive studies of the structure of variant HEMOGLOBINS have provided a structural and conceptual model for other intraorganismic variants (isozymes). There are three hemoglobin molecules which are known to be present in the human. All three molecules contain four polypeptide chains. Adult hemoglobin A contains two identical α-chains and two identical β-chains, and can be formulated $\alpha_2\beta_2$. Hemoglobin A_2 also contains two chains, but in this molecule, two different polypeptide chains termed δ-chains replace the β-chains, so the molecule can be formulated $\alpha_2\delta_2$. Fetal hemoglobin, like the others, contains the two α-chains but the β-chain is replaced by another peptide chain termed γ, hence, the formulation $\alpha_2\gamma_2$. Thus, there are four different polypeptide chains which unite to form three variant hemoglobins with distinctive molecular characteristics. The amino acid sequence of all four chains (α, β, δ, γ) are similar and, as a matter of fact, resemble MYOGLOBIN, the counterpart of hemoglobin in muscle tissue. Myoglobin exists as a monomer; it does not form a polymer like hemoglobin. The structural similarity has suggested evolutionary homology. It has been proposed that the myoglobin chain, as well as the

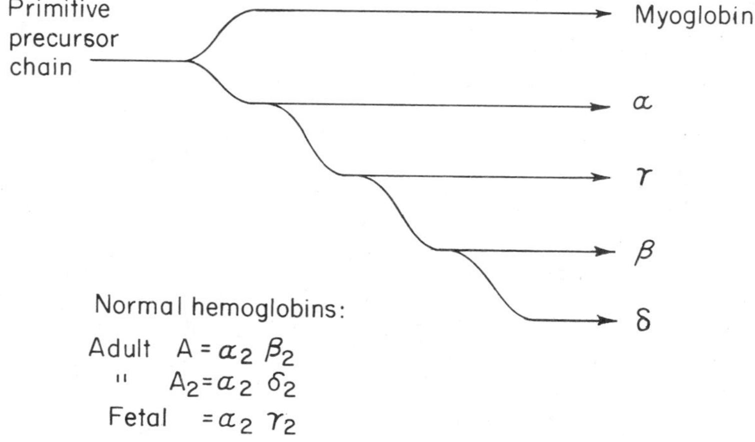

FIG. 3. Possible evolutionary relationships among myoglobin-hemoglobin polypeptide chains. This relationship has been postulated to account for the structural relationships found among the chains; thus, myoglobin $> \alpha > \gamma > \beta > \delta$. This scheme is based on the following postulates: (1) Mutations result in amino acid substitutions, deletions, additions or inversions. (2) At several points in the course of evolution a structural gene for a particular hemoglobin peptide chain undergoes duplication, and the two genes, initially equivalent, evolve independently, each being subject to separate mutational events and the selective pressure of the environment on their protein products. (Adapted from V. Ingram, "The Hemoglobins in Genetics and Evolution," New York, Columbia University Press, 1963.)

hemoglobin chains, were all evolutionarily derived from a common "primitive gene." The degree of structural similarity is employed to construct the pattern of the relationship, as illustrated in Fig. 3.

The specific structure (amino acid sequence) is important in determining the biological activity of the protein. The replacement of a specific glutamyl residue by a valyl residue in a particular position in the β-chain, produces an altered chain which is incorporated into a hemoglobin molecule, termed sickle cell hemoglobin—hemoglobin S. This altered hemoglobin molecule changes the shape of red blood cells at low oxygen tension and indirectly causes the symptoms of sickle cell ANEMIA.

A number of intraorganismic enzyme variants (isozymes) have distinct physiological functions, *i.e.*, the properties of a given isozymic form are modulated to a particular metabolic situation. *Lactic dehydrogenase* and *aldolase* will be discussed as examples. (Studies with the former enzyme led to the term "isozyme.")

Lactic dehydrogenase exists in most animal tissues as five isozymes. Appropriate studies have shown that the molecule is composed of four subunits (polypeptide chains = AAAA). Most animal systems contain two apparently homologous subunits A and B (these have been termed also M and H because vertebrate muscle and heart are rich in M and H subunits, respectively) which interact to form the following hybrid molecules: A_4, A_3B, A_2B_2, AB_3, and B_4. The subunits A and B are immunologically distinct, so that antibodies prepared against A (A_4) react with A_4, A_3B, A_2B_2, AB_3, but not B_4, and antibodies prepared against B (B_4) react against B_4, AB_3, A_2B_2, A_3B, but not A_4. A_4 and B_4, when mixed and treated under appropriate conditions, will dissociate and recombine to form a mixture of A_4, A_3B, A_2B_2, AB_3, and B_4 in the expected proportions for a random process (*i.e.*, 1:4:6:4:1, respectively). In this instance, the two chains A and B are sufficiently different chemically to confer different catalytic characteristics on the isozymes; this is especially noted in the limiting forms A_4 and B_4. The B_4 (H) type is inhibited by its substrate pyruvate at high concentrations; in contrast, the A_4 (M) type is not inhibited by high pyruvate concentrations. Thus, it is possible that the different peptides may be tailored for different physiological roles. The following explanation has been advanced: In the HEART, metabolic energy is derived by complete oxidation of pyruvate and lactate in MITOCHONDRIA. The inhibition by pyruvate of the lactic dehydrogenase reaction favors this pathway. In skeletal muscle, however, energy is supplied by GLYCOLYSIS, and pyruvate is reduced to lactate even in the presence of temporarily high levels of pyruvate.

Another example of functional modulation in isozymes is found with the two homologous (Class I) fructose diphosphate *aldolases* found in vertebrates. Aldolases A and B are each apparently composed of three polypeptide chains. None of the chains in A is identical with those in B. Aldolase A is found in tissues (*e.g.*, muscle, heart and brain) which are predominantly glycolytic (*i.e.*, form glycogen from small molecules) and/or in the metabolism of fructose via fructose-1-phosphate. The specificity of the aldolases is tailored for this function. Aldolase A is very effective in cleaving fructose diphosphate to triose phosphates, the reaction involved in glycolysis, and is a relatively poor catalyst for formation of fructose diphosphate from triose phosphates, or for cleavage of fructose-1-phosphate. Conversely, aldolase B is relatively more effective in the aldol synthesis of fructose diphosphate from tirose phosphates (as occurs in GLUCONEOGENESIS) and in the aldol cleavage of fructose-1-phosphate (as occurs in metabolism of fructose). The distribution of the aldolases in animal tissues suggests that both A and B are not usually synthesized in the same cell. In the early embryonic rudiments of the liver, kidney, and intestine, when the metabolism is primarily glycolytic, the organs contain primarily aldolase A. Later, aldolase A is lost, and the content of aldolase B increases, prior to the development of gluconeogenic activity in the tissue. During this period, there is a shift in the synthesis of A and B in the cells. Other studies suggest that this shift in the synthesis of aldolases A and B may be a fundamental feature of cytodifferentiation in these tissues.

A continued study of enzyme variants is of considerable chemical and biological interest. Further comparative analysis of structure, in relationship to catalytic properties, will yield a better understanding of the chemical basis of the remarkable catalytic properties of proteins. Such comparative studies are also of biological interest. The recognition of homologous and analogous families of enzymes may allow a better understanding of the mechanisms by which functional proteins arise and are perpetuated during evolution.

It is significant that intraorganismic enzyme variants may be associated with specific metabolic functions. Studies of the specific structural and catalytic features of these isozymes and the factors which control their relative concentrations in cells may provide basic information of processes involved in embryological differentiation.

References

1. INGRAM, V., "The Hemoglobins in Genetics and Evolution," New York, Columbia University Press, 1963.
2. SMITH, E. L., AND MARGOLIASH, E., "Evolution of Cytochrome C," *Federation Proc.*, **23,** 1243 (1964).
3. RUTTER, W. J., "Evolution of Aldolase," *Federation Proc.*, **23,** 1248 (1964).
4. MARKERT, C. L., *Science*, **140,** 1329 (1963).
5. KAPLAN, N. O., *Bacteriol. Rev.*, **27,** 155 (1963).
6. WROBLEWSKI, F. (EDITOR), "Multiple Molecular Forms of Enzymes," *Ann. N.Y. Acad. Sci.*, **94,** 655–1030 (1961).

WILLIAM J. RUTTER AND T. V. RAJKUMAR

J

JAUNDICE. See BILE AND BILE PIGMENTS, section on Clinical Significance and Biochemical Disorders.

K

KARRER, P.

Paul Karrer (1889–) is a Swiss chemist who contributed outstandingly to our knowledge of the chemistry of the CAROTENOIDS, VITAMIN A, RIBOFLAVIN and TOCOPHEROLS. Many important compounds in these areas were synthesized by him for the first time. He was awarded, with WALTER NORMAN HAWORTH, the Nobel Prize in chemistry in 1937.

KENDALL, E. C.

Edward Calvin Kendall (1886–), an American biochemist, first distinguished himself by isolating THYROXINE which he named. Later, he further contributed to biochemistry by isolating a large number of steroid HORMONES from the adrenal cortex, among them the hormone eventually called cortisone. Kendall, along with Philip Hench, a physician, and TADEUSZ REICHSTEIN, was awarded the Nobel Prize in physiology and medicine in 1950.

KENDREW, J. C.

John Cowdery Kendrew (1917–), a British biochemist, was awarded a share in the Nobel Prize in chemistry in 1962 for his remarkable work on determining the detailed structure of MYOGLOBIN, a hemoprotein in muscle, using X-ray analysis and electronic computers (see MAX FERDINAND PERUTZ).

KERATIN

Keratin is the structural protein of hair, wool, nails, horn, hoofs, claws, beaks and feathers. Keratin obtained from these sources is sometimes classified as hard keratin, while the structural protein of vertebrate epidermis is designated as soft keratin. The preparation of purified keratin involves drying and pulverizing of these materials, followed by extraction with lipid solvents and digestion of non-keratin proteins by trypsin. Keratin is characterized by a high CYSTINE content providing disulfide cross-links between protein chains. The insolubility, mechanical stability and hardness of keratin is largely due to the cross-linking. If the stabilization by disulfide bridges is made the basis for the definition of keratin, some cross-linked proteins occurring in non-epidermal tissues and in invertebrate organisms must also be designated as keratin.

Chemical Composition. The composition of keratin varies from animal to animal and from site to site in one animal. Even within the keratinized tissue of one site, mixtures of chemical species may be found. Hard keratin contains up to 6% sulfur, while soft keratin contains less than 3% sulfur and correspondingly fewer disulfide cross-links. With the exception of hydroxyproline, all amino acids which occur as constituents of proteins have been found in keratin. It was once thought that hard keratins were characterized by a constant molecular ratio of histidine/lysine/arginine of 1:4:12, while soft keratins were thought to contain these amino acids in the ratio of 1:4:4. Although these ratios are still useful as approximations, more recent analyses have shown that they are not exact integers.

Structure. As in other PROTEINS, the AMINO ACIDS are linked by peptide bonds to form long chains. The chains in turn are linked to each other by disulfide bridges, hydrogen bonds and electrostatic forces. Little is known about the amino acid sequences in the polypeptide chains of keratins. End-group determination [see PROTEINS (END GROUP AND SEQUENCE ANALYSIS)] has shown glycine, threonine, valine, alanine, serine, glutamic acid and aspartic acid to be present in N-terminal positions in wool and human hair keratin. With the exception of valine, the same amino acids were also found as C-terminal groups. Very small amounts of N-terminal group have been found in feather keratin, leading to the suggestion that feather keratin is a cyclic polypeptide. Obviously, keratin is a very complex material and elucidation of its structure or structures by chemical methods is very difficult.

Insolubility and lack of homogeneity are fundamental problems in attempts to determine the chemical structure of keratins. Soluble derivatives, which can be more easily studied, may be obtained by cleavage of disulfide bridges by oxidation (performic acid, peracetic acid) or reduction (thioglycolate), in combination with cleavage of hydrogen bonds by acid, alkali, phenol, urea. Oxidized products are designated as keratoses, reduced products as kerateines. The latter have a strong tendency to reaggregate in the presence of oxygen. This can be prevented by the use of thiol-blocking groups. Fractions with molecular weights ranging from 10,000–80,000 have been obtained from wool by such solubilizing treatments.

Physical methods, particularly X-ray diffraction studies, have produced more detailed information on the structure of keratins. The classical work of Astbury in the 1930's has shown that all mammalian keratins have a typical diffraction pattern. This *α-keratin* pattern is characterized by a sharp meridional reflection corresponding to 5.1 Å (axial repeat spacing) and a strong diffuse reflection around 10 Å (side spacing). In addition, a sharp meridional reflection at 1.5 Å was more recently found. These reflections are not only observed in keratins; myosin and fibrin give typical α-keratin patterns, although they are chemically quite unrelated to the keratins. According to the α-helix model (Pauling and Corey, 1951), the 5.1 Å spacing reflects the pitch of the helix, while the 1.5 Å spacing corresponds to the segment of the helix occupied by one amino acid. The side spacing of 10 Å represents the lateral distance between helices. Lack of exact agreement between observed reflections and spacings expected from a crystal of hexagonally packed parallel α-helices has led to modifications of the original α-helix model. Helices coiled to form a super-helix ("coiled coil"), structures of six helices twisting around a seventh straight helix ("7-strand cable"), of three helices twisted together ("3-strand rope"), and of parallel triplets of helices have been proposed.

When wool or hair is stretched, the X-ray diffraction diagram changes from the α-type to the *β-keratin* pattern which is characterized by equatorial reflections around 4.5 Å (backbone spacing) and 10 Å (side spacing), and meridional reflections about 3.5 Å (axial repeat spacing). Again, this pattern is not limited to keratin materials. Silk fibroin gives a typical β-pattern. Details of the α-to-β transformation are not well understood, but it appears that the 1.5 Å segment occupied by one amino acid in α-keratin is lengthened to 3.5 Å in β-keratin as the structure is changed from the helix conformation to a stretched chain. The equatorial 4.5 Å spacing is attributed to the lateral distance between the chains in the direction in which their approach is not hindered by the amino acid side chains. The equatorial 10 Å spacing reflects the lateral distance between the chains in the approximately perpendicular direction in which their approach is determined by the presence of the amino acid side chains. A "pleated sheet" model with parallel polypeptide chains has been proposed for the β-keratin structure.

The X-ray patterns given by feathers and by reptile scales are much more detailed than those obtained with mammalian keratins. Some of the reflections are typically β, and stretching does not affect the pattern. An axial long spacing of 95 Å has been found, and an arrangement of β-crystallites in a primitive helix with a pitch of 95 Å has been suggested.

Information obtained by X-ray studies at the molecular level of organization has been complemented at the macromolecular and cellular level by electron microscopy and polarized light microscopy. For wool fibers, the most extensively studied keratin fibers, the following (partly hypothetical) picture has emerged: α-helices (diameter 10 Å) are arranged in parallel triplets designated as protofibrils (diameter 20 Å). Protofibrils are grouped in cylinders of about ten to form microfibrils with a diameter of about 70 Å. Numerous microfibrils embedded in a noncrystalline protein matrix make up the macrofibril which has a diameter of one to several microns. The wool fiber (diameter around 20 μ) consists of bundles of macrofibrils. The microfibrils are about 1 μ long, the α-helical units of the protofibril about 200 Å. The keratin of the matrix differs from that of the microfibrils by a higher content of cystine, serine, threonine, and proline. Matrix keratin and microfibrillar keratin are present in about equal proportions.

While in hair and wool the polypeptide chains are oriented in the direction of the fiber axis, a molecular orientation at right angles to the main structural axis, in addition to parallel orientation, has been found in feather, horn, mammalian quills and similar material. This prevents splitting that could more easily occur if the polypeptide chains were laid down exclusively in one direction.

Keratinization. Keratin of vertebrate epidermis is the end product of a process that takes place in a number of histologically distinguishable layers. The innermost or *germinal layer* (adjoining the collagen-containing dermis) consists of mitotically active cells which produce the cells of the *differentiating layer*. Prekeratin, rich in thiol groups, is synthesized in this layer. X-ray analysis of prekeratin shows α-keratin reflections, indicating that the helix conformation is already present. In the *keratinizing layer*, the major portion of the nucleic acids and other non-protein cell constituents is eliminated, dehydration occurs, the SH groups disappear and disulfide cross-links are formed, thus producing the horny cell residues which constitute the *keratinized layer* at the surface. Basically the same sequence of events takes place in the formation of the horny appendages and in the more complex processes of hair or feather growth.

References

1. Alexander, P., and Hudson, R. F., "Wool, Its Chemistry and Physics," New York, Reinhold Publishing Corp., 1964.
2. Fraser, R. D. B., and MacRae, T. P., "An Investigation of the Structure of β-Keratin," *J. Mol. Biol.*, **5**, 457 (1962).
3. Fraser, R. D. B., and MacRae, T. P., "Structural Organization of Feather Keratin," *J. Mol. Biol.*, **7**, 272 (1963).
4. Lundgren, H. P., and Ward, W. H., "Levels of Molecular Organization in α-Keratins," *Arch. Biochem. Biophys.*, *Suppl.* 1, 79–111 (1962).
5. Matoltsy, A. G., "Structural and Chemical Properties of Keratin Forming Tissues," in "Comparative Biochemistry" (Florkin and Mason, Editors), Vol. IV, pp. 343–369, New York, Academic Press, 1962.
6. Mercer, E. H., "Keratin and Keratinization," New York, Pergamon Press, 1961.

J. F. Diehl

KETO ACIDS

Several α-keto acids are important metabolic intermediates; these include pyruvic acid (see CARBOHYDRATE METABOLISM; CITRIC ACID CYCLE), and α-ketoglutaric acid (see CITRIC ACID CYCLE). Keto acids are produced from the corresponding amino acids in transamination reactions (see AMINO ACID METABOLISM; TRANSAMINATION ENZYMES). β-Keto acids in bound form are intermediates in FATTY ACID oxidation [see FATTY ACID METABOLISM (OXIDATION AND BIOSYNTHESIS)].

KETOGENESIS. See AMINO ACID METABOLISM, section on Carbon Chain Conversions.

KETOSTEROIDS

A group of compounds which occur in urine of man and certain higher animals is known as 17-ketosteroids or neutral 17-oxosteroids. These compounds react with *m*-dinitrobenzene to form a chromogen with a characteristic absorption spectrum. In 1931, Butenandt[1] described the isolation of a crystalline steroid from the urine of male individuals which was later identified as *androsterone*. The formulas for androsterone, dehydroisoandrosterone, and etiocholanolone are shown in Fig. 1. These three metabolites comprise approximately 75% of the total urinary ketosteroids in the normal human.[2]

These compounds (Fig. 1) are 17-ketosteroids; they have a keto oxygen in the 17-position of the chemical formula. Essentially none of the other steroid hormones besides ANDROGENS are 17-ketosteroids. The primary androgenic hormones which are precursors of ketosteroids are testosterone, which is secreted by the Leydig cells of the testes, and dehydroisoandrosterone, which is derived from the adrenal cortex.

Because testosterone and dehydroisoandrosterone are metabolized by the tissues and appear in urine as ketosteroids, the rate of secretion of ketosteroids in urine is a very reliable index of the rate of androgen production in the body. Figure 2 shows the ketosteroids which are excreted in the urine following their metabolic conversion from testosterone and dehydroisoandrosterone.[3]

These ketosteroids appear in urine conjugated with either glucuronic or sulfuric acid as the glucuronide or sulfate (see GLUCURONIDASE). A small portion of dehydroisoandrosterone from the adrenal is excreted in urine unchanged, and the remainder is apparently converted to isoandrosterone, androstenedione, etiocholanolone and androsterone. The urinary excretion of dehydroisoandrosterone is increased following the administration of ACTH and decreases after anterior pituitary suppression by hydrocortisone injections. As seen in Fig. 2, some of these ketosteroids are also derived from testosterone and, therefore, urinary ketosteroids are derived from both adrenal and testicular tissues. Metabolic studies following oral administration of hydrocortisone and corticosterone from the adrenal indicate that these substances are also precursors of ketosteroids in urine.[3] However, this is not considered to be major pathway for their metabolism (see also ADRENAL CORTICAL HORMONES).

Between the ages of 4 and 7 years, the average daily excretion of ketosteroids is approximately the same for both male and female. There is a gradual rise in production in both sexes until the age of puberty. Mean values in men between the ages of 20 and 40 are somewhat higher than those in females. A significant decrease in excretion is not normally seen until the age of 50–60 years, after which time there is a marked decrease in excretion of ketosteroids. There is a wide range of the amounts normally excreted in both sexes,

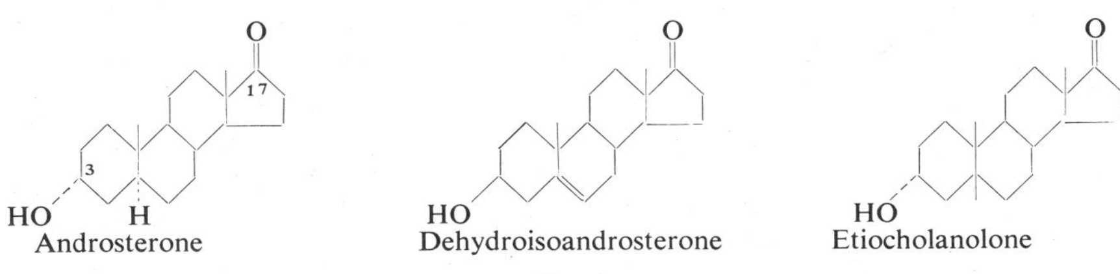

FIG. 1

Tissue — Urine (17-Ketosteroids)

Dehydroisoandrosterone → Dehydroisoandrosterone
Dehydroisoandrosterone → Isoandrosterone
Dehydroisoandrosterone → Androstenedione
Testosterone → Androstenedione
Androstenedione → Etiocholanolone
Androstenedione → Androsterone

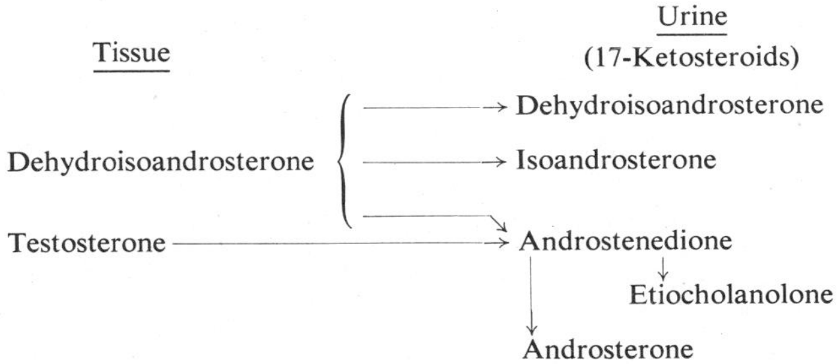

FIG. 2

and it is sometimes difficult to detect a change in patients with an endocrine disturbance. However, a marked increase or decrease in excretion is usually a clue to the nature of the pathological lesion. Males in an age range of 20–40 years excrete approximately 17 mg of ketosteroid per 24 hours, whereas the female in this age range only excretes about 10 mg/day.[3]

An elevation in the level of ketosteroids in urine indicates excessive production of androgens in the testes or adrenals or else an androgen-producing tumor which is sometimes in the ovary. Changes in function of several endocrine organs have been noted to produce variable changes in excretion of ketosteroids in urine.[4] The effect of tumors of the anterior pituitary gland depends on the nature of the lesion and the stage of the disease. In some cases appreciable quantities of these steroids are found in the urine while in others they are virtually absent. An elevation in plasma levels of pituitary ADRENOCORTICOTROPHIC HORMONE as well as the presence of adrenal cortical tumors results in a marked increase in urinary ketosteroids. Interstitial (Leydig) cell tumors of the testes are quite rare, but when such tumors are found, androsterone excretion in urine is found to be markedly increased. Excessive amounts of ketosteroids in urine have been reported in patients with adrenal-like ovarian tumors which have apparently arisen from aberrant adrenocortical cells in the ovary. In cases of "intercurrent amenorrhea," excretion of ketosteroids has been found to increase when the amenorrhea has lasted six months or more.

Following castration in the male, ketosteroid excretion usually falls temporarily, but gradually increases to a level higher than that seen prior to castration. This increase is apparently the result of enhanced production of ketosteroids by the adrenal cortex following castration. Women suffering from Addison's disease consistently show low values for urinary excretion of ketosteroids because the adrenal is the principle source of these steroids in the female. A decreased urinary excretion of ketosteroids is most commonly associated with eunuchoidism and castration in the male and adrenal insufficiency in males and females. However, a decrease in the excretion of these steroids is by no means conclusive, since it also occurs in malnutrition, debilitating diseases, thyrotoxicosis, myxedema and diabetes.

Hirsutism is frequently seen in Cushing's disease, adrenogenital syndrome and an ovarian tumor known as arrhenoblastoma. This abnormal growth of hair is usually closely related to the excretion of ketosteroids. Although total 17-ketosteroids may not always be increased in these conditions, there is usually a definite increase in the urinary excretion of the adrenal steroid dehydroisoandrosterone. Cortisone is usually effective in reducing the excretion of these steroids. The pattern of urinary ketosteroids is extremely important in understanding certain disease states, and the quantity of these substances appearing in urine is a very useful diagnostic aid in the clinical interpretation of changes in testicular and adrenocortical functions. On the other hand, these compounds have little or no therapeutic uses as such. When an androgenic substance is used therapeutically, either testosterone or a synthetic androgenic-anabolic compound is used. Androsterone, the principal ketosteroid normally appearing in urine, has only about one-tenth the biological activity of testosterone, and both etiocholanolone and dehydroisoandrosterone lack androgenic activity.

References

1. BUTENANDT, A., *Z. Angew. Chem.*, **44**, 905 (1931).
2. MOON, H. D. (EDITOR), "The Adrenal Cortex," p. 130, Paul B. Hoeber, Inc., 1961.
3. DRILL, V. A., "Pharmacology in Medicine," Second edition, p. 959, 1958.
4. MASON, L. H., AND ENGSTROM, W. W., *Physiol. Rev.*, **30**, 321 (1950).

JAMES W. FISHER

KIDNEY FUNCTION (BIOCHEMICAL ASPECTS). See GLOMERULAR FILTRATION; RENAL TUBULAR FUNCTIONS.

KOGL, F.

Fritz Kogl (1897–1959) was a Dutch biochemist who is known for his discovery of auxins and of BIOTIN. He became convinced at one time that the "unnatural" D-amino acids played a key role in the cancer problem. This conviction, which has not been corroborated, tended to cast a shadow on his reputation, which on the basis of his isolation of minute amounts of biologically active substances should be substantial.

KORNBERG, A.

Arthur Kornberg (1918–) is an American biochemist. He and SEVERO OCHOA were the joint recipients of the Nobel Prize in physiology and medicine for 1959 for their discoveries of the biological syntheses of DEOXYRIBONUCLEIC ACIDS (DNA) and RIBONUCLEIC ACIDS (RNA), respectively.

KREBS CYCLE. See CITRIC ACID CYCLE.

KREBS, H. A.

Hans Adolf Krebs (1900–) is a German-born British biochemist best known for the discovery of the Krebs Cycle, by which pyruvate is oxidized in animal and plant tissues and in bacteria to yield energy (see CITRIC ACID CYCLE). He has made other important contributions, notably with respect to the mode of formation of urea. He was awarded a Nobel Prize jointly with FRITZ LIPMANN in physiology and medicine in 1953.

KUHN, R.

Richard Kuhn (1900–) is a German chemist of great accomplishment. Much of his investigation has had to do with the chemistry of

CAROTENOIDS, VITAMIN A, RIBOFLAVIN, and PYRIDOXINE. These studies were contemporaneous with those of KARRER, and it is difficult or impossible to assess the proper credit due to each of these outstanding chemists. Kuhn was to be given the Nobel Prize for chemisty in 1938, but he was in a concentration camp and Hitler would not allow him to receive it.

KWASHIORKOR

(Bantu: *Kwashi*—name of a boy born on Sunday; *orkor*—red; *synonyms*: bouffisure d'Annam, cheveux blancs, distrophie des farineux, enfants rouges, Gillan's edema, infantile pellagra, malignant malnutrition, mehlnahrschaden, sindrome pluicarencial infantil.) A nutritional deficiency syndrome, typically of small children, occurring in tropical areas, and clinically characterized by edema, determatoses and growth failure. Gastrointestinal symptoms, behavioral disorders, anemia, and hair changes are frequent concomitants. The most significant biochemical change from a diagnostic standpoint is hypoalbuminemia. It is clinically distinct from marasmus, sprue and pellagra.

Pathogenesis. The etiology is complex and variable, and includes multiple nutritional deficiencies, infection, extreme environmental conditions, and probable systemic disorders. The ultimate limiting factor is undoubtedly a protein deficiency, though the term *protein* is used here in a broad sense. Protein-rich foods such as lean meat, eggs, milk and soya beans always carry, associated with the protein, a substantial assortment of B vitamins. Protein deprivation therefore means also a limitation of the supply of the accompanying vitamins. Along with protein lack is excessive carbohydrate intake. The problem is also complicated by the fact that marked individual metabolic differences clearly operate to cause only *some* of the growth-deficient children in endemic areas to manifest clinical kwashiorkor.

The syndrome is seldom seen outside of the tropical belt, generally occurs between the ages of one and three years, and is identified with cultural groups in which milk and animal products are scarce in contrast to staple grain foods. In a rigid sense, kwashiorkor may be considered to be the result, in a constitutionally susceptible child, of the abrupt change on weaning to a low protein diet, at precisely the time when protein requirements are highest. This concept of the protein-*deprived* child seems more consistent with the etymology of the term than does the reference to skin pigmentation. Riboflavin and niacin deficiency are conspicuous, as they are in the adult population in afflicted areas, but neither vitamin provides effective therapy. Malaria, intestinal parasites and pneumonia are seen in a vast number of the patients.

Pancreatic fibrosis is extremely common, and pancreatic insufficiency as a major conditioning factor in the malnutrition is suggested by the readily observable failure to digest starch and meat, and by the symptoms of steatorrhea, pot-belly and fatty infiltration of the liver. The liver is also fibrotic, a condition that survives throughout the patient's life. The biochemical lesions are thus numerous, variable, and often severe, many patients dying of fluid and electrolyte imbalance, as in the case of other pancreatic disorders.

Symptoms. Edema is invariable but often mild or regional. Its presence, along with frequently excessive subcutaneous fat, may belie the growth failure, which includes decreased height and atrophic musculature. The determatosis is highly variable in appearance and is often described as "crazy pavement" or "flaky paint" dermatosis; it appears on areas of the body not afflicted in pellagra and may resemble any of the recognized vitamin deficiency dermatoses. The hair may become dipigmented or reddish and, on repeated relapses, may show banding. Alopecia is common. Anorexia, vomiting and diarrhea are frequent factors that aggravate the electrolyte imbalance and fluid loss. The ANEMIA is highly variable, but generally unlike that of sprue and nutritional macrocytic anemia. Mental symptoms are predominated by apathy, and hyperirritability is also often seen at the same time. Kwashiorkor is clearly distinguishable from dietary marasmus (severe general malnutrition), in which most of the symptoms listed are absent, but varying intermediate degrees are common. Subclinical forms of kwashiorkor are thought to be widespread.

Therapy. Amino acids, glucose and vitamins are effective therapeutic agents when administered in rational dietary combinations, but a very high protein or amino acid level is most essential. Since most patients receive medical attention only when the disease is well advanced, fluid and electrolyte therapy is an immediate requirement. Attention to concomitant infection is invariably required. Socioeconomic and educational remedies are essential for a permanent cure. The death rate is high, since most cases are not seen until well advanced.

References

1. TROWELL, H. C., DAVIES, J. N. P., AND DEAN, R. F. A., "Kwashiorkor," London, E. Arnold Ltd., 1954.
2. BROCK, J. F., AND HANSEN, J. D. L., in "Clinical Nutrition" (JOLLIFFE, N., EDITOR), pp. 102–107, Second edition, New York, Harper and Bros., 1962.

ERNEST BEERSTECHER, JR.

KYNURENINE

Kynurenine

$$\text{C}_6\text{H}_4(\text{NH}_2)\text{—}\overset{\text{O}}{\overset{\|}{\text{C}}}\text{—CH}_2\text{—CHNH}_2\text{—COOH}$$

is a diamino acid formed as an intermediate in TRYPTOPHAN metabolism both in mammals and in lower forms. In mammalian livers there is an enzyme system which catalyzes the oxidative opening of the pyrrol portion of the indole ring, with the production of kynurenine and formic acid. By complex transformations, kynurenine may be converted into eye pigments in insects; it is also intermediate in the conversion of tryptophan to NICOTINIC ACID in mammals [see AROMATIC AMINO ACIDS (BIOSYNTHESIS AND METABOLISM)].

L

LACTIC ACID

Lactic acid, CH_3—CHOH—COOH, was one of the first biological substances to be investigated from the standpoint of the existence of two optically active forms (see ASYMMETRY). The *dextro* form, which has been called sarcolactic acid, is produced from GLYCOGEN in muscle, contributes the lactic acid found in blood, and can be converted into glycogen in the liver (Cori cycle). It has a configuration corresponding to that of L-alanine (which is also dextrorotatory) and is designated L-lactic acid. The levorotatory form of lactic acid (D-lactic acid) is produced by fermentation by some bacteria. Lactic acids when concentrated tend to form the corresponding lactic anhydrides so that purification as such is generally impractical. Ordinary lactic acid as purchased ("85–90%") is a racemic mixture containing 10–15% of lactic anhydride.

LACTOGENIC HORMONE. See GONADOTROPIC HORMONES; HORMONES, ANIMAL.

LACTOSE

Lactose, a disaccharide, 4-(β-D-galactopyranosyl)-D-glucopyranose, found in the milk of mammals (cows' milk, about 4.5%; human milk about 6.7%) is formulated as follows:

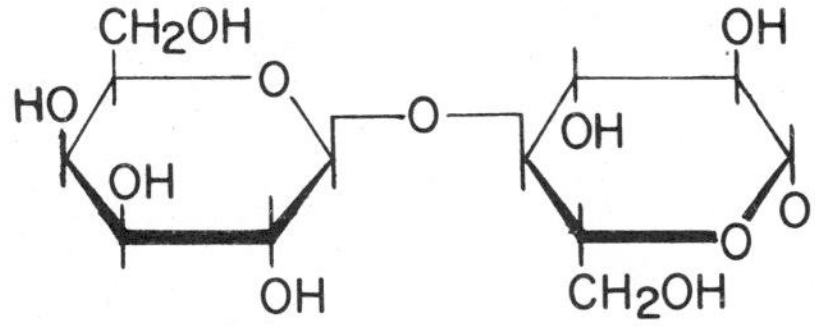

CH2OH, HO, O, OH, OH, —O—, OH, OH, O, OH, CH2OH

It exists in milk at body temperature principally in two tautomeric forms (α and β) in approximately the proportions of 2 to 3. The free aldehyde form (of the glucose portion) is supposed to be in the equilibrium mixture in minute amounts. Lactose is thus a reducing sugar. The β-form is shown in the figure; the α-form (monohydrate) is the commercial purified lactose. Freshly prepared solutions have a rotatory power $[\alpha]_D^{20}$ of $+92.6°$ which changes on standing (mutarotation) to about $+52.3°$. The β-form of lactose pictured above has a rotatory power of about $+34°$ which on standing rises to $+52.3°$. Lactose is not fermented by ordinary yeast. See also CARBOHYDRATES (CLASSIFICATION).

LANDSTEINER, K.

Karl Landsteiner (1868–1943) was an Austrian-born American physician, noted for his discovery of BLOOD GROUPS in 1902. Initially the groups were A, B, AB and O, but later, other factors, M, N and Rh were also discovered by his group. For this pioneering work, he was awarded the Nobel Prize for physiology and medicine in 1930.

LECITHINS. See CEPHALINS AND LECITHINS.

LEDERBERG, J.

Joshua Lederberg (1925–   ) is an American geneticist who is noted for his outstanding contribution to knowledge about bacterial genetics and how bacteriophage virus particles function. He received a share in the Nobel Prize for physiology and medicine in 1958 (see GEORGE WELLS BEADLE; EDWARD LAWRIE TATUM).

LEUCINE

L-Leucine (α-aminoisocaproic acid, 2-amino-4-methylpentanoic acid),

$$(CH_3)_2CH—CH_2—CHNH_2—COOH$$

is present as a structural unit in all the more common PROTEINS. In contrast to isoleucine (which is absent), leucine is present in hemoglobins at a high level (about 15%). It possesses only one ASYMMETRIC carbon atom and thus exists in only two optically active forms. Leucine is nutritionally essential for rats, fowls and human beings. The results obtained in relatively few cases show that the amounts required to maintain human beings in nitrogen equilibrium vary from 170–1100 mg for different individuals (see AMINO ACIDS, NUTRITIONALLY ESSENTIAL). Leucine is ketogenic rather than glucogenic when metabolized in the body. L-Leucine in 6*N* HCl solution has a specific rotatory power $[\alpha]_D = +15.1°$.

LEUKOCYTES AND LEUKEMIA

In spite of many contradictions in the literature on the chemistry of leukocytes, certain areas of agreement have emerged. Before presenting these, however, some of the reasons for contradictions should be considered. (1) Some studies have been performed without taking into account the state of white cell maturation. Thus, comparisons between the polymorphonuclear leukocytes of an exudate and the myeloblasts of acute leukemia must be admitted to be on quite different cell types even though they both belong to the granulocytic series. (2) Methods of preparation of leukocytes for biochemical studies have varied from one laboratory to another. These different techniques impose different degrees of trauma on the cells and thus alter their biochemistry to different degrees. The mitogenic stimulus provided by the bean extracts used for separating peripheral white blood cells (WBC) is an example of a factor obviously influencing lymphocyte chemistry. (3) Normal leukocytes are generally collected from a number of donors and then pooled so that individual variation is averaged out. When comparable studies are performed on leukemia patients, the high leukocyte counts frequently encountered make it possible to use a single donor, and individual variation remains in the sample being compared. (4) Increasingly it is recognized that even cells which are morphologically alike may have quite different biochemistry so that homogeneity of leukocyte populations for comparison of one collection with another may be essentially unattainable.

Chemical studies on leukocytes have been by two principal methods, *cytochemical* and *biochemical*. Cytochemical determinations have the advantage of giving information about the composition and activities of individual cells. However, such results are only qualitative or, in those instances where a scoring procedure is utilized, semiquantitative at best. Biochemical determinations involve isolation of relatively large quantities of leukocytes by various techniques, each of which has certain disadvantges. The results that are obtained, however, are quantitative, although such determinations give average values for the mixture of cell types in a particular cell population.

Carbohydrate Metabolism. Although many studies have been performed on normal leukocyte populations to determine their rates of GLYCOLYSIS and respiration, some question still remains about the role of each. Many of these studies have been directed toward determining whether leukocytes show the kind of metabolism characterized by Warburg as tumor metabolism. This implies that the cell or tissue has a high aerobic glycolysis compared to respiration. Studies on cell homogenates show that normal leukocytes do have a high aerobic glycolysis and glycolysis in such a system is little affected by the presence of respiratory inhibitors. In dealing with a normal cell population, it should be born in mind that the cells are primarily of the granulocytic series. Glycolysis is believed to be the principal energy source for normal leukocytes, and ATP levels drop when glycolysis is inhibited. Homogenates of cells from patients with chronic granulocytic leukemia (CGL) had a lower oxygen consumption, glucose utilization, and lactic acid production than did normal cells. In chronic lymphocytic leukemia (CLL) leukocytes, these values were lower yet. Hexokinase has been implicated as the rate-limiting enzyme of glycolysis in both normal and leukemic leukocytes. Since leukemic cell homogenates are hexokinase deficient, their glycolytic rate is decreased more than that for normal cells. One should be cautious, however, in translating data obtained with homogenates to the situation that may exist in intact cells.

Recent studies with intact cells account for some of the earlier variability that was noted. It now appears that the suspending medium is a factor of considerable importance in determining leukocyte metabolism. It is claimed that aerobic glycolysis of normal white cells is negligible in comparison to oxygen consumption if the cells are suspended in their own citrated plasma. On the other hand, in a balanced salt solution such as Ringers solution, a significant quantity of lactate is produced under aerobic conditions, but without necessarily impairing respiration. The hexose monophosphate shunt pathway of glucose catabolism is also present in leukocytes. In normal cells, less than 10% of the glucose utilized is catabolized by this pathway; for leukemic cell homogenates, the percentage is somewhat higher.

GLYCOGEN is to be found in WBC. Although it is found chiefly in the granulocytes, about 20% of normal lymphocytes give staining reactions (periodic acid Schiff) for glycogen. In lymphocytic leukemia, the number of cells staining positively is considerably increased. While many cells in CGL patients contain glycogen, the level has been estimated as only about half that of normal granulocytes. The blast cells of acute leukemia are devoid of glycogen.

Lipids. The LIPID content of a normal white blood cell population has been estimated to be of the order of 15–20% (dry weight). Most of the common lipids have been observed. These include glycerides, cholesterol, cholesterol esters, lecithin, phosphatidyl ethanolamine, phosphatidyl serine, phosphatidic acid, and sphingomyelin. There is some indication that granulocytes have a higher lipid content than lymphocytes. Thus, in granulocytic leukemia one finds a higher lipid content in the leukocyte population. Leukocytes have an active lipid synthesis as judged by incorporation of labeled acetate into lipid fractions. The observed rate of acetate incorporation is higher when the labeled material is incubated with leukocytes prior to their separation from whole blood. The turnover rate is an inverse function of cell maturity; thus, in acute granulocytic (AGL) or monocytic leukemia (AML) the turnover rate is about four times that of normal cells. It is three times normal levels in CGL, something over twice as high in the cells of ALL, and essentially normal in CLL. In AGL the high turnover rate is particularly evident in the phospholipid fractions. There is some indication that the glyceride fractions of normal leukocytes have the highest turnover rate, but this observation awaits confirmation.

Nucleic Acids. As with most substances to be found in leukocytes, reports concerning nucleic acid content are variable. Many point to some increase in the DNA content of (acute) leukemic cells. There are, however, several studies which indicate that leukocyte DNA is constant irrespective of the cell line and state of maturation. Chromosomal analysis of leukemic leukocytes has shown many aberrations. While there is no characteristic change in acute leukemia, aneuploidy is common, and there is frequently an elevated chromosome number, perhaps accounting for the increased DNA. CGL is exceptional in that a striking and constant chromosomal abnormality manifests itself as a deletion at chromosome 21. RNA is also probably increased in the leukemias, the highest values being found in immature leukocytes. Based on phosphate determinations, the ratio of RNA P to DNA P is about 0.8–1.0 in normal marrow cells. There is not a great deal of information about nucleohistone levels in leukocytes, although there is some evidence of their being increased in patients with lymphatic leukemia. There are also data suggesting that the deoxyribonucleoproteins from normal and leukemic leukocytes exhibit antigenic differences.

A number of studies have been directed toward elucidating physical properties of DNA from normal and leukemic leukocytes. The difficulty of preparing high-quality DNA is a constant problem complicating the interpretation of all such data. It has been reported that the distribution of sedimentation values for DNA prepared by the dodecyl sulfate method varies from one type of leukemia to another. The mean *S* value observed for leukocyte DNA was highest in lymphatic leukemia, intermediate for normal leukocytes, and lowest for granulocytic leukemia. Although differences in chromatographic properties of normal and leukemic leukocyte DNA have been reported, these more likely reflect differences in the method of preparation than they do in the DNA itself. For a number of CGL cases, DNA revealed an absorption peak at 12.9 μ which was not observed in DNA from normal or CLL white cells. Perhaps correlating with this latter finding are potentiometric titration curves of white cell DNA. The denaturation processes responsible for the difference in the back titration curve begin at pH 3 for normal and lymphocytic leukemic DNA. The point of irreversibility for DNA from granulocytic leukemic cells begins at a somewhat higher pH, *i.e.*, 3.8.

Many studies have been concerned with the metabolism of nucleic acids in leukocytes. The labeling of DNA with tritiated thymidine has been a favorite approach to the question of DNA biosynthesis. When peripheral blood is incubated with labeled thymidine, only a few cells of normal blood show thymidine uptake. Considerably higher numbers of cells are generally found in patients with acute leukemia and chronic granulocytic leukemia. The lymphocytes of CLL on the other hand, show a rather poor uptake indicating a relatively long generation time. Estimates of leukocyte life span from such studies are somewhat at variance from different laboratories. Normal granulocytes and lymphocytes have been estimated to have life spans of about 9 and 21 days, respectively. Life spans for their leukemic counterparts are 23 days for granulocytes and 85 days for lymphocytes. While the actual figures may be in for considerable revision, these findings give the surprising picture of the leukemic lymphocyte as a cell that proliferates slowly and has a greatly increased life span. Such has not generally been thought to be the case for malignant cells. In accord with these findings, the turnover rate for RNA and DNA in leukemic granulocutes is about the same, whereas in lymphocytic leukemia the turnover rate of DNA is considerably less than that for RNA, again suggesting that this is a slowly dividing cell. The purine bases may also be directly incorporated into leukocyte DNA. It has been found that as with other tissues the adenine and guanine may be interconverted.

Thymidylate synthetase has been found in the cells of CGL and to a lesser degree in acute leukemia. It has not been observed, however, in either normal leukocytes or in CLL lymphocytes. Thymidylate kinase is present in both normal and chronic leukemic leukocytes. In each instance the enzyme has been shown to be controlled by feedback inhibition by thymidine and deoxycytidine derivatives. The enzymes of PYRIMIDINE BIOSYNTHESIS have been found in leukocytes. These are generally elevated in the granulocytic leukemias, decreasing as a function of cell maturity. Their levels in lymphocytes are apparently more variable. Although the issue is not yet settled, aspartate transcarbamylase, the first enzyme of the biosynthetic sequence, is probably not subject to feedback inhibition as it is in certain bacteria. Most studies of nucleic acid catabolism in leukemia have involved examination of urine specimens. As these shed no light on activities in the leukocytes, they are not included in this discussion. In mouse leukemia, however, there is evidence that the leukemic WBC contain decreased quantities of xanthine oxidase and uricase. This decreased purine metabolism is found in many but not all malignant cells.

Sulfhydryl Compounds. The sulfhydryl compounds of importance in leukocytes are GLUTATHIONE and the cystine-cysteine pair. The importance of these substances in leukocytes is highlighted by the fact that SH compounds have been shown to be essential for cell division and by the fact that within minutes after administration of radioactive L-cystine, it may be found in the circulating leukocytes. Cystine incorporation into white cells is faster in the granulocytic leukemias and slower in CLL than it is in normal individuals. The question of glutathione levels in leukocytes is still being disputed.

Amino Acids. Most of the common AMINO ACIDS have been found free in leukocytes. These include alanine, arginine, glutamic acid, glutamine, histidine, lysine, methionine, ornithine, phenylalanine, proline, threonine, tryptophan, tyrosine, valine, the leucines, serine, glycine, taurine, ethanolamine and β-aminoethylphosphate. The β-aminoethylphosphate is significantly elevated in the WBC of most leukemia patients. The most moderate increases have been observed in CGL. Glutamic acid is also considerably elevated as is proline in leukemic WBC. Ornithine levels have been found to be considerably depressed in the leukocytes of about one-half of leukemia patients.

Phosphatases. The many studies which have been made on leukocyte alkaline PHOSPHATASE point to its being confined essentially to cells of the granulocytic series. Alkaline phosphatase is greatly depressed in CGL granulocytes, and this characteristically low level may be used to distinguish this disorder from leukemoid reactions, which are generally indistinguishable morphologically. In the latter, as well as in cases of acute infection, and in other stress situations, the white cell level of the enzyme rises markedly. In support of the idea that stress might affect leukocyte alkaline phosphatase levels, it was observed that administration of 17-hydroxycorticosteroids results in a significant increase in enzyme activity. It may be that synthesis of alkaline phosphatase is controlled by the Ph^1 chromosome which exhibits a partial deletion in CGL. As these patients achieve remission, alkaline phosphatase levels can return to normal.

Acid phosphatase is also found in the granulocytic series of leukocytes and is present in limited quantities, if at all, in lymphocytes.

Vitamins. Because of the important role played by folic acid antagonists in the therapy (see CANCER CHEMOTHERAPY) of acute leukemia, many studies have been directed toward determining the levels of folic acid and the FOLATE COENZYMES in normal and leukemic WBC. The FOLIC ACID content of leukocytes varies as a function of cell maturity, being highest in the blast cells of acute leukemia, intermediate in the chronic leukemias (somewhat higher in CGL than in the lymphocytes of CLL), and having the lowest values in a normal cell population. As leukemic cells become resistant to the effects of antifolic therapy, the levels of folic acid reductase become considerably increased, presumably permitting the continued conversion of vitamin to active coenzymatic forms. A number of studies have been concerned with vitamin B_{12} which is considerably elevated in the serum of CGL patients. The level of this vitamin in leukocytes of CGL patients, however, appears to be normal. In acute leukemia, the blast cell content is elevated. Studies on the RIBOFLAVIN content of leukocytes indicate that cells in the lymphocytic series contain only 1/3–1/2 as much vitamin as do granulocytes. In contrast to the findings with folic acid, no correlation with cell maturation was observed. Nuclei of lymphocytic cells have a somewhat smaller riboflavin content than granulocytic nuclei, but the major difference lies in the cytoplasmic content of the vitamin. This difference may, however, only reflect the different quantities of cytoplasm in the two cell series. As is characteristic of malignant cells generally, ASCORBIC ACID levels are depressed in leukemic cells. *In vitro* incubation of plasma-suspended WBC with ascorbic acid showed uptake of the vitamin by neutrophils but not by lymphocytes; this same uptake pattern was exhibited by the corresponding leukemic cells. Vitamin E deficiency in monkeys leads to leukocytosis and increased synthesis of bone marrow DNA.

Metal Ions. Studies from a number of laboratories have shown that leukemic leukocytes contain greatly decreased levels of ZINC. In a normal cell population zinc levels of 14–24 $\gamma/10^9$ leukocytes have been reported. In leukemic cells the quantity decreases from about half this value to something approaching only 10%. Data for other metal ions are sparse. Some figures for the leukocytes of CGL and CLL are, respectively, sodium 637 and 409, potassium 1923 and 897; calcium 36 and 12, and magnesium 192 and 84; these are expressed as γ per 10^9 WBC, and the differences may merely reflect differences in cell size. Copper levels for normal WBC may be of the order of 1 $\gamma/10^9$ cells. Leukocytes also contain enzymes for PORPHYRIN synthesis and will incorporate Fe into such a structure. While acute leukemic blast cells incorporate Fe readily to make heme, peripheral WBC from normal individuals or chronic leukemia patients do this very poorly.

Miscellaneous Components. The presence of many enzymes not included in this discussion is implied by what has been said concerning the presence of certain metabolic systems. Others which have been measured are alcohol dehydrogenase, arginase, and phosphorylase. On the basis of limited information, it has been suggested that the phosphorylase level in leukocytes may parallel that in liver, and that this may be useful in diagnosis of GLYCOGEN storage disease characterized by hepatic phosphorylase deficiency. Aryl sulfatase is elevated in the WBC of CGL and depressed in CLL. This enzyme appears to be particularly high in eosinophils. Three proteases with pH optima of 8, 5.5, and 3 have been associated with leukocytes. The pH 8 protease is found in neutrophils and is inhibited by organophosphorus compounds in a manner similar to chymotrypsin inhibition by these substances. The other two proteases are associated primarily with lymphocytes. Leukocytes also carry HISTAMINE. This substance has now been closely associated with basophils and may be completely absent from other types of leukocytes.

Subcellular Organization. A number of studies have been directed toward uncovering the subcellular biochemical organization of leukocytes. While this seems to correspond quite well with that found for many other cells, it is perhaps worth pointing out that the neutrophilic granules resemble liver lysosomes in their enzyme content. In addition to a number of hydrolytic enzymes, cytochrome b has also been associated with the granules. There is also a fair amount of polysaccharide present which is different from glycogen.

Phagocytosis. The vast majority of studies concerning biochemical alterations accompanying phagocytosis has been conducted with polymorphonuclear leukocytes, and most generally with those obtained in exudates. With particle uptake there is increased respiration and glucose oxidation. As this increased activity is not affected by cyanide or dinitrophenol but is inhibited by the glycolytic inhibitor iodoacetate, it has been suggested that the principal energy for phagocytosis comes from glycolysis. When glucose is labeled either at carbon-1 or carbon-6 and the radioactivity of the CO_2 which forms is determined, it is found that phagocytosis leads to much greater stimulation of oxidation of carbon-1 than it does carbon-6. While these findings suggest that there is also stimulation of the hexose monophosphate shunt pathway, there is also evidence that this difference in carbon-1 and carbon-6 oxidation may reflect yet other oxidative mechanisms. The evidence for the latter is as yet sketchy, however. Phagocytosis also results in increased lipid turnover as measured by incorporation of C^{14}-labeled acetate or P^{32}-labeled phosphate into leukocyte lipid. From the P^{32} incorporation it is observed that the phospholipids which have a particular increase in specific activity are phosphatidic acid, inositol phosphatide, and phosphatidyl serine. The more common phospholipids lecithin, phosphatidyl ethanolamine, and sphingomyelin do not show a similar specific activity increase. A number of hydrolytic enzymes associated with the granules of polymorphonuclear leukocytes are released and exhibit increased activity as a result of phagocytosis. Acid phosphatase on the other hand seems to be more resistant to release. Although data for monocytes are limited, it appears that during phagocytosis by these cells there is as much stimulation of glucose carbon-6 oxidation as there is of glucose carbon-1 oxidation.

Abbreviations used: AGL, acute granulocytic leukemia; CGL, chronic granulocytic leukemia; ALL, acute lymphocytic leukemia; CLL, chronic lymphocytic leukemia; WBC, white blood cells.

References

1. Braunsteiner, H., and Zucker-Franklin, D., (Editors), "The Physiology and Pathology of Leukocytes," New York, Grune and Stratton, 1962.
2. Damashek, W., and Gunz, F., "Leukemia," New York, Grune and Stratton, 1964.
3. Hayhoe, F. G. J., "Leukemia, Research and Clinical Practice," Boston, Little, Brown, and Co., 1960.
4. Rebuck, J. W., Bethell, F. H., and Monto, R. W., (Editors), "The Leukemias: Etiology, Pathophysiology, and Treatment," Henry Ford Hospital International Symposium, New York, Academic Press, 1957.
5. Wolstenholme, C. E. W., and O'Connor, M., (Editors), "Biological Activity of the Leukocyte," *Ciba Foundation Study Group*, **10** (1961).

Morton D. Prager

LEVULOSE

Levulose is a name applied in earlier years to D-fructose. This name had the advantage of calling attention to the levorotatory property of one of the two simple sugars derived from sucrose by hydrolysis, in contrast to the dextrorotatory property of the other hydrolysis product which was called "dextrose" (now D-glucose). These names dextrose and levulose are not acceptable chemical names partly because both sugars have optical antipodes for which the names "*levo*-dextrose" or "*dextro*-levulose" would seem unfortunate. See CARBOHYDRATES (CLASSIFICATION).

LIBBY, W. F.

Willard Frank Libby (1908–), American chemist, was awarded the Nobel Prize in chemistry in 1960 for his development of CARBON DATING. This has been particularly invaluable in archeological investigations.

LIGHT ADAPTATION

The VISUAL PIGMENTS of the eye, including the well-known rod pigment employed for dim-light vision, rhodopsin, and three recently discovered cone pigments responsible for color vision in primates, all consist of a chromophore, the 11-*cis* geometrical isomer of VITAMIN A aldehyde (retinene, retinal) bound to an insoluble lipoprotein, opsin. When a dark-adapted animal is placed in the light, the visual pigments, located in projecting outer segments of the retinal rod and cone cells, absorb light, bleach, and then release the all-*trans* isomer of vitamin A aldehyde from the opsin.

Rhodopsin ($\lambda_{max} = 500$ mμ) is the visual pigment which has been studied in greatest detail. In bleaching, it passes through a series of short-lived intermediates before losing its red color. When a molecule of rhodopsin absorbs one quantum of visible light, the geometrical isomerization of the chromophore from the 11-*cis* to the all-*trans* isomer occurs, converting the rhodopsin to pre-luminirhodopsin. This first step in the bleaching process is the only one requiring light. All subsequent reactions can occur in the dark and are thought to involve changes in the conformation of the opsin. Pre-lumirhodopsin ($\lambda_{max} =$ 543 mμ) and the next two intermediates, lumirhodopsin ($\lambda_{max} = 497$ mμ) and metarhodopsin I ($\lambda_{max} = 478$ mμ), are somewhat similar in color to rhodopsin. However, in the following step, the formation of metarhodopsin II ($\lambda_{max} = 380$ mμ), visible bleaching occurs. One of the preceding reactions is in some way geared to trigger the nerve impulses responsible for vision. In the final reaction, metarhodopsin II hydrolyzes, liberating free all-*trans* vitamin A aldehyde.

The all-*trans* vitamin A aldehyde separated from opsin does not accumulate to any great extent in the outer segments of the visual cells as a result of light adaptation. Rather, it is rapidly reduced to vitamin A (retinol) with TPNH by the enzyme ALCOHOL DEHYDROGENASE. In addition to alcohol dehydrogenase, the outer segments contain enzymes capable of metabolizing glucose by way of the glycolytic and hexose monophosphate oxidation pathways (see CARBOHYDRATE

METABOLISM). The latter pathway produces TPNH (also termed NADPH; see NICOTINAMIDE ADENINE DINUCLEOTIDES) and can operate only if it is coupled to reactions which reoxidize TPNH to TPN. Such coupling occurs in the visual cell outer segment when the alcohol dehydrogenase reaction reduces vitamin A aldehyde to vitamin A, employing TPNH supplied by the oxidation of glucose-6-phosphate and 6-phosphogluconate. The hexose monophosphate oxidation pathway is in this manner indirectly activated by light and participates in light adaptation.

Subsequently, the vitamin A diffuses out of the visual cell outer segments, and much of it is converted to vitamin A esters in the retinal MICROSOMES. In the esterification reaction, the fatty acids, largely palmitate, stearate, and oleate, are transferred to vitamin A from an as yet unidentified donor. During prolonged adaptation to bright light, most of the vitamin A esters pass from the retina to the subretinal pigment epithelium where they accumulate.

In *dark adaptation*, the over-all process appears to be reversed, but most of the individual steps have not been described. It is clear that a geometrical isomerization is a key reaction uniquely associated with dark adaptation. In retinal tissue of cattle, an enzyme has been found which catalyzes the interconversion of the all-*trans* and 11-*cis* isomers of vitamin A aldehyde. The final step in dark adaptation appears to be the spontaneous combination of 11-*cis* vitamin A aldehyde with opsin to regenerate the visual pigment (see the list of references following the article VISUAL PIGMENTS).

SIDNEY FUTTERMAN

LIGHT PHASE OF PHOTOSYNTHESIS

Photosynthesis may be defined as the utilization of solar energy by plants and photosynthetic bacteria for the synthesis of organic carbon compounds. The bulk of PHOTOSYNTHESIS on our planet is carried on by green plants in which photosynthesis culminates in the conversion of carbon dioxide into carbohydrates and the evolution of oxygen gas. However, a very small portion of the total planetary photosynthetic product is accounted for by PHOTOSYNTHETIC BACTERIA and a few algal species which carry on a type of photosynthesis in which oxygen is never produced.

The over-all process of photosynthesis can be divided into a light phase and a dark phase. The light phase of photosynthesis, as discussed here, comprises those photochemical reactions which generate the first chemically defined, energy-rich products that are formed *prior* to CO_2 assimilation and are essential for, and utilized in, the conversion of CO_2 into organic compounds. There is general (though not unanimous) agreement among investigators of photosynthesis that in all cells, photosynthetic as well as nonphotosynthetic, the enzymic reactions concerned with conversion of CO_2 into carbohydrates and other organic compounds are dark reactions that are driven by adenosine triphosphate (ATP) and reduced nicotinamide adenine dinucleotide phosphate (NADP) or NICOTINAMIDE ADENINE DINUCLEOTIDE (NAD). In this view, what distinguishes carbon assimilation in photosynthetic cells from carbon assimilation in nonphotosynthetic cells is the manner in which they generate ATP and reduced NAD(P). Nonphotosynthetic cells form them at the expense of chemical energy whereas photosynthetic cells form them at the expense of radiant energy. Thus, our concern here will be with those photochemical reactions in plant photosynthesis that produce, directly or indirectly, ATP and reduced NADP. (NADP rather than NAD is the pyridine nucleotide involved in photosynthetic carbon assimilation by plants.)

The discovery of these early reactions of photosynthesis came not from investigations with intact cells but from studies with subcellular systems, specifically, isolated CHLOROPLASTS. Work with isolated spinach chloroplasts has uncovered two photochemical reactions, independent of CO_2 assimilation, which jointly produce ATP and, indirectly, $NADPH_2$. These photochemical reactions jointly constitute the process of photosynthetic PHOSPHORYLATION which is now subdivided into *cyclic photophosphorylation*, which produces only ATP, and *noncyclic photophosphorylation*, which yields ATP and the reduced form of *ferredoxin*, an iron-containing protein native to chloroplasts, that is neither a heme nor a flavoprotein. With the formation of ATP and reduced ferredoxin, the light phase of photosynthesis comes to an end. All other reactions in photosynthesis (see CARBON REDUCTION CYCLE) are independent of light. The reduction of NADP, formerly considered a photochemical reaction, is now known to be separated from the photoreduction of ferredoxin by two dark, enzymic reactions.

Cyclic Photophosphorylation. Several unique features distinguish cyclic photophosphorylation, discovered in 1954, from the other cellular phosphorylations, *i.e.*, from ATP formation in FERMENTATION and from ATP formation in RESPIRATION (oxidative PHOSPHORYLATION): (1) ATP is formed only in chlorophyll-containing structures and is independent of any other cellular organelles or enzyme systems. (2) No oxygen is consumed or produced. (3) No energy-rich chemical substrate is consumed, the only source of energy being that of the absorbed photons. (4) ATP formation is not accompanied by a measurable electron transport; *i.e.*, there is no net change in any external electron donor or acceptor [Eq. (*1*)].

$$n \cdot \text{ADP} + n \cdot \text{P}_i \xrightarrow{h\nu} n \cdot \text{ATP} \qquad (1)$$

All known cellular phosphorylations occur at the expense of free energy liberated during electron transport from a high-energy electron donor to an electron acceptor. Since there was no direct evidence for this in cyclic photophosphorylation, it was inferred that here ATP formation was coupled to a special type of electron transport hidden in the structure of the chloroplast. The current hypothesis is that a CHLOROPHYLL molecule, on absorbing a quantum of light, becomes excited and promotes an electron to an outer orbital with a higher energy level. This high-energy electron is then transferred to an adjacent electron acceptor molecule with a strongly electronegative oxidation-reduction potential. The transfer of an electron from excited chlorophyll to an adjacent electron acceptor molecule, present in chloroplasts, is considered to be the energy conversion step proper; it constitutes a mechanism for generating a strong reductant at the expense of the excitation energy of chlorophyll. Once the strong reductant is formed, the subsequent electron steps are exergonic. The first electron acceptor molecule in turn transfers an electron to another electron acceptor molecule within the chloroplast with a more electropositive oxidation-reduction potential. The number of electron transfer steps is unknown, but it is envisaged that an electron "bucket brigade" of this kind liberates free energy that is sufficient to form one or more ATP's from ADP and orthophosphate. In the end, the electron originally emitted by an excited chlorophyll molecule returns to the electron-deficient chlorophyll molecule, and the process is repeated. Because of the envisaged cyclic pathway traversed by the emitted electron, the process has been named cyclic photophosphorylation.

The chemical identity of the first electron acceptor in the cyclic electron transport chain has remained in doubt for several years because many different substances of a physiological or a non-physiological character were found to catalyze cyclic photophosphorylation. An example of the former is menadione and of the latter, phenazine methosulfate. Recent findings, however, point to ferredoxin, already mentioned in connection with noncyclic photophosphorylation, as being also the endogenous catalyst of cyclic photophosphorylation in chloroplasts. However, there is no evidence as yet that ferredoxin is also the catalyst of cyclic photophosphorylation in photosynthetic bacteria.

Noncyclic Photophosphorylation. This type of photophosphorylation, discovered in 1957, provided the first, direct experimental evidence for a coupling between light-induced synthesis of ATP and electron transport. Here, in contrast to cyclic photophosphorylation, ATP formation in chloroplasts was stoichiometrically coupled with a light-driven transfer of electrons from water to NADP (or to a nonphysiological electron acceptor such as ferricyanide). More recent evidence has established that illuminated chloroplasts do not react directly with NADP but react with ferredoxin. Photoreduction of ferredoxin is followed by two dark reactions which result in NADP reduction: (a) the reoxidation of reduced ferredoxin by a flavoprotein chloroplast enzyme, ferredoxin-NADP reductase, and (b) the reoxidation of the reduced ferredoxin-NADP reductase by NADP.

The oxidation-reduction of ferredoxin was found to involve the transfer of one electron. This was also evident from the measured stoichiometry of noncyclic photophosphorylation in which the photoreduction of 4 moles of ferredoxin was coupled with the formation of 2 moles of ATP and 1 mole of oxygen [Eq. (*2*)].

$$4Fd_{ox} + 2\,ADP + 2P_i + 2H_2O \xrightarrow{h\nu} 4Fd_{red} + 2ATP + O_2 + 4H^+ \qquad (2)$$

The oxygen produced in noncyclic photophosphorylation by chloroplasts escapes to the atmosphere, whereas ATP and reduced ferredoxin are used for carbon assimilation.

Requirement for Cyclic Photophosphorylation in Photosynthesis. Two moles of reduced ferredoxin are required to produce 1 mole of ATP [Eq. (*2*)]. Since 2 moles of reduced ferredoxin are also required to reduce 1 mole of NADP, noncyclic photophosphorylation gives rise to ATP and $NADPH_2$ in a ratio of 1:1. Were this ratio adequate for carbon assimilation, there would be no need for cyclic photophosphorylation in photosynthesis. However, the carbon reduction cycle appears to have a requirement of 2 $NADPH_2$ and 3 ATP per mole of CO_2, assimilated to the level of glucose. Even if this requirement were reduced to 2 ATP and 2 $NADPH_2$, additional ATP, supplied by cyclic photophosphorylation, would still be necessary to form STARCH, the main product of photosynthesis in leaves. ATP is expended in the formation of ADP-glucose from which the glucosyl moiety is transferred to a starch primer. Moreover, cyclic photophosphorylation may be an important mechanism for providing the large supplies of ATP that are required for protein synthesis and for other endergonic processes in the cell.

Interrelations of Cyclic and Noncyclic Photophosphorylation. Ferredoxin-catalyzed cyclic photophosphorylation and noncyclic photophosphorylation coupled with photoproduction of oxygen by chloroplasts are mutually exclusive. In white light, cyclic photophosphorylation catalyzed by ferredoxin can be unmasked only when photoproduction of oxygen is stopped by means of specific inhibitors. Without the use of inhibitors, this mutually exclusive relation between photoproduction of oxygen and cyclic photophosphorylation can be seen with monochromatic light above 700 mμ. Chloroplasts illuminated in this region of far-red light cannot produce oxygen but can still carry on cyclic photophosphorylation catalyzed by ferredoxin.

The key role of ferredoxin, both in cyclic and in noncyclic photophosphorylation, raises the question about the regulatory mechanism(s) in the cell for switching from noncyclic to cyclic photophosphorylation. The mechanism is still unknown, but one possibility has experimental support: the availability of NADP in the oxidized form. As long as oxidized NADP is available, electrons will flow from water to ferredoxin and thence (via ferredoxin-NADP reductase) to NADP. However, when CO_2 assimilation would temporarily cease for lack of the extra ATP that is contributed by cyclic photophosphorylation, NADP would accumulate in the reduced state and electrons from reduced ferredoxin would begin to "cycle" within the chloroplasts, giving rise to cyclic photophosphorylation. The additional ATP thus generated would reestablish CO_2 assimilation, which, in turn, would oxidize $NADPH_2$ and thereby reestablish noncyclic photophosphorylation. Thus, the operation of the CARBON REDUCTION CYCLE would automatically maintain a balance between cyclic and noncyclic photophosphorylation.

An important distinction between cyclic and noncyclic photophosphorylation is their differential sensitivity to inhibitors. Noncyclic photophosphorylation, but not cyclic, is sensitive to all the inhibitors of oxygen evolution in photosynthesis, the most common of which are *o*-phenanthroline, 3-(4′-chlorophenyl)-1,1-dimethylurea (CMU), and 3-(3′,4′-dichlorophenyl)-1,1-dimethylurea (DCMU). On the other hand, ferredoxin-catalyzed cyclic photophosphorylation is inhibited by low concentrations of antimycin A and 2,4-dinitrophenol, which do not affect noncyclic photophosphorylation. Very recently, another inhibitor, a phlorobutyrophenone derivative, desaspidin, was found which inhibits cyclic photophosphorylation at a very low concentration but does not inhibit noncyclic photophosphorylation.

The view that cyclic and noncyclic photophosphorylation constitute the photochemical energy conversion process in photosynthesis of plants is consistent with recent evidence that fluorescence of spinach chloroplasts is quenched under experimental conditions that give rise to cyclic or noncyclic photophoshporylation.

Cyclic photophosphorylation has been found in every photosynthetic organism tested so far: higher green plants, algae and photosynthetic bacteria. By contrast, noncyclic photophosphorylation, linked with oxygen evolution, is absent from photosynthetic bacteria. Photosynthetic bacteria exhibit a type of photophosphorylation which is noncyclic in the sense that ATP formation is coupled with an open, noncyclic electron flow to NAD from an electron donor other than water. However, in this "bacterial type" of noncyclic photophosphorylation, ATP is formed at the site of cyclic photophosphorylation. Thus, cyclic photophosphorylation emerges as the common denominator of all types of photosyntheses in nature.

Photosynthetic Phosphorylation and the Two Light Reactions of Photosynthesis. Investigations of photosynthesis of intact algal cells in monochromatic light show a drop in quantum efficiency at wavelengths longer than 685 mμ ("red drop"). The addition of shorter wavelengths of light to algal cells exhibiting "red drop" gives a quantum efficiency that is greater than would be expected if the energies of the two beams of light were additive. This phenomenon, known as the "enhancement effect," together with observations of spectral changes in cells illuminated by monochromatic beams of long and short wavelength, gave rise to a hypothesis that photosynthesis in plants requires the interaction of two light reactions: (1) a "short-wavelength" reaction that is associated with photoproduction of oxygen, and (2) a "long-wavelength" reaction that produces reduced NADP and ATP.

A related hypothesis, proposed recently, identifies "noncyclic photophosphorylation" as the "short-wavelength" light reaction and cyclic photophosphorylation as the "long-wavelength" light reaction.

References

1. ARNON, D. I., ALLEN, M. B., AND WHATLEY, F. R., "Photosynthesis by Isolated Chloroplasts," *Nature*, **174**, 394–396 (1954).
2. ARNON, D. I., WHATLEY, F. R., AND ALLEN, M. B., "Photosynthesis by Isolated Chloroplasts. II. Photosynthetic Phosphorylation, the Conversion of Light into Phosphate Bond Energy," *J. Am. Chem. Soc.*, **76**, 6324–6329 (1954).
3. FRENKEL, A. W., "Light Induced Phosphorylation by Cell-free Preparations of Photosynthetic Bacteria," *J. Am. Chem. Soc.*, **76**, 5568–5569 (1954).
4. ARNON, D. I., Cell-free Photosynthesis and the Energy Conversion Process," in *Light and Life*, W. D. MCELROY AND B. GLASS (EDITORS), Baltimore, Johns Hopkins Press, 1961, pp. 489–566.
5. STANIER, R. Y., "Photosynthetic Mechanisms in Bacteria and Plants: Development of a Unitary Concept," *Bacteriological Rev.*, **25**, 1–17 (1961).
6. ARNON, D. I., "The Photochemical Phase of Photosynthesis in Chloroplasts," *Physiol. Rev.*, 1966 (in Press).

DANIEL I. ARNON

LIGHT REACTION

In the over-all process of PHOTOSYNTHESIS, the first stage, or "light reaction," involves the absorption of light by pigments (see CHLOROPHYLLS; CHLOROPLASTS), the energy of the light being utilized for ATP formation (see LIGHT PHASE OF PHOTOSYNTHESIS; PHOSPHORYLATION, PHOTOSYNTHETIC) and for the release (*e.g.*, from water) of hydrogen atoms, which are thereby made available to bring about (in the second stage or "dark reaction") the reduction of carbon dioxide to carbohydrates [see CARBON REDUCTION CYCLE (IN PHOTOSYNTHESIS)].

References

1. ARNON, D. I., "The Role of Light in Photosynthesis," *Sci. Am.*, **203**, 105–118 (Nov. 1960).
2. ARNON, D. I., "Cell-Free Photosynthesis and the Energy Conversion Process," in "Light and Life" (MCELROY, W. D., AND GLASS, B., EDITORS), pp. 489–569, Baltimore, Johns Hopkins Press, 1961.
3. BASSHAM, J. A., "Photosynthesis: Energetics and Related Topics," *Advan. Enzymol.*, **25**, 39–118 (1963).
4. RABINOWITCH, E. I., AND GOVINDJEE, "The Role of Chlorophyll in Photosynthesis," *Sci. Am.*, **213**, 74–83 (July 1965).

LIPIDS (CLASSIFICATION AND DISTRIBUTION)

Lipids are a heterogeneous group of substances which occur ubiquitously in biological materials. They are categorized as a group by their extractability in nonpolar organic solvents such as chloroform, carbon tetrachloride, benzene, ether, carbon disulfide, and petroleum ether. Structural types within the group range from simple straight-chain hydrocarbon molecules to complex ring structures with varying side chains. A useful classification of the lipids is as follows: (1) fatty acids; (2) neutral fats; (3) phosphatides; (4) glycolipids; (5) aliphatic alcohols and waxes; (6) terpenes; (7) steroids.

Fatty Acids. FATTY ACIDS are chains of covalently linked carbon atoms, bearing hydrogen atoms, which terminate in a carboxyl group that is responsible for their properties as acids. The naturally occurring fatty acids are for the most part unbranched molecules but complex structures with branched or cyclic chains do occasionally occur, particularly in lower biological forms. It is noteworthy that most of the natural acids contain an even number of carbon atoms in the chain. Two general types of fatty acids exist, saturated and unsaturated. A saturated fatty acid is one containing no double bonds between carbon atoms and can be represented by the formula for stearic acid:

$$CH_3{-}(CH_2)_{16}{-}COOH$$

Stearic acid

An UNSATURATED FATTY ACID contains one or more double bonds between carbon atoms. A representative member of this group is linoleic acid:

$$CH_3(CH_2)_4CH{=}CHCH_2CH{=}CH(CH_2)_7COOH$$

Linoleic acid

Saturated and unsaturated fatty acids undergo different reactions in biological systems. The unsaturated acids, linoleic, linolenic, and arachidonic, cannot be synthesized by mammalian tissues and are therefore considered essential fatty acids (see FATTY ACIDS, ESSENTIAL). The latter, which are derived primarily from plant sources, have assumed clinical importance in recent years following the observation that a diet rich in unsaturated fatty acids will lower the blood cholesterol while a diet heavy in saturated fatty acids tends to elevate it. Since cholesterol is thought to play a role in the development of coronary ATHEROSCLEROSIS, diets high in unsaturated fatty acids are frequently prescribed in an attempt to minimize the development of atherosclerotic heart disease.

The metabolism of fatty acids in the living organism is a complex process (see FATTY ACID METABOLISM). In general terms, however, two fates can be described. The free fatty acids are either oxidized for energy production or esterified to form more complex lipids which subserve a variety of structural, storage, transport, and hormonal functions.

Neutral Fats. The predominant esterification product of fatty acids from a quantitative standpoint is the neutral fat or *triglyceride* molecule. The latter consists of a molecule of glycerol in ester bond with three molecules of fatty acid. The fatty acids may be the same, but in nature are usually different. A typical triglyceride is depicted below:

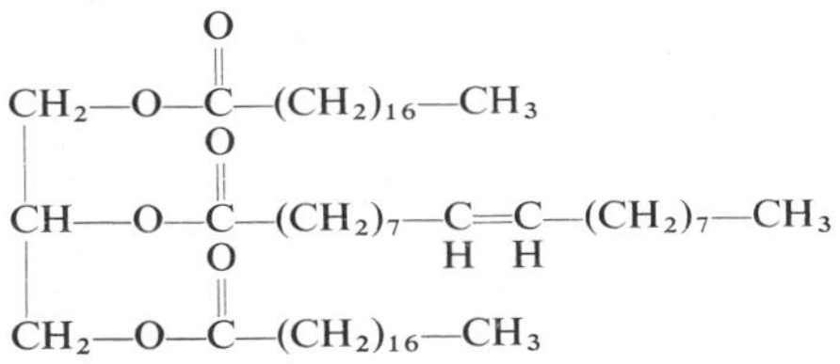

$$\begin{array}{l} CH_2{-}O{-}C({=}O){-}(CH_2)_{16}{-}CH_3 \\ | \\ CH{-}O{-}C({=}O){-}(CH_2)_7{-}CH{=}CH{-}(CH_2)_7{-}CH_3 \\ | \\ CH_2{-}O{-}C({=}O){-}(CH_2)_{16}{-}CH_3 \end{array}$$

Oleodistearin

Triglyceride is the major storage form of fatty acids and comprises almost 100% of adipose tissue. Contrary to popular opinion, it is a potential energy source and can be rapidly mobilized under a variety of circumstances (see FAT MOBILIZATION). Neutral fat is found in all normal diets and can be synthesized by a variety of tissues from both carbohydrate and protein constituents. Indeed, in so far as the caloric intake exceeds the caloric output, net synthesis of triglyceride in adipose tissue occurs.

Phosphatides. Phosphatides, or PHOSPHOLIPIDS, consist of two groups of compounds: PHOSPHATIDIC ACID derivatives and phosphoric acid esters of sphingosine. Phosphatidic acid derivatives are structurally related to neutral fat in that both are esterification products of glycerol. The former differs from the latter in that only two of the hydroxyl groups of glycerol are in ester linkage with fatty acids while the third is esterified to phosphoric acid. Biologically the phosphate ester is synthesized first, the product L-α-glycerophosphate. This compound is an important intermediate of glucose metabolism. The fatty acids are attached subsequently to yield the parent molecule of the group, L-α-phosphatidic acid:

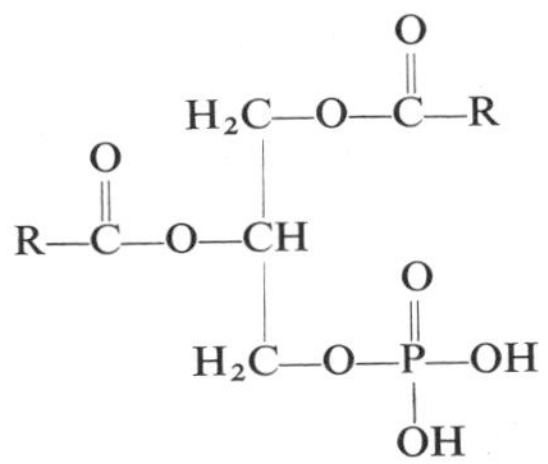

$$\begin{array}{l} \qquad\qquad H_2C{-}O{-}C({=}O){-}R \\ \qquad\qquad\; | \\ R{-}C({=}O){-}O{-}CH \\ \qquad\qquad\; | \\ \qquad\qquad H_2C{-}O{-}P({=}O)(OH){-}OH \end{array}$$

L-α-Phosphatidic acid (R = fatty acid chain)

This compound can be further esterified to yield other members of the phosphatide group, the three most common of which are esters formed between the phosphoric acid portion of the molecule and the nitrogen-containing alcohols, choline, ethanolamine, and L-serine. The formula for phosphatidyl choline, or LECITHIN, is given below:

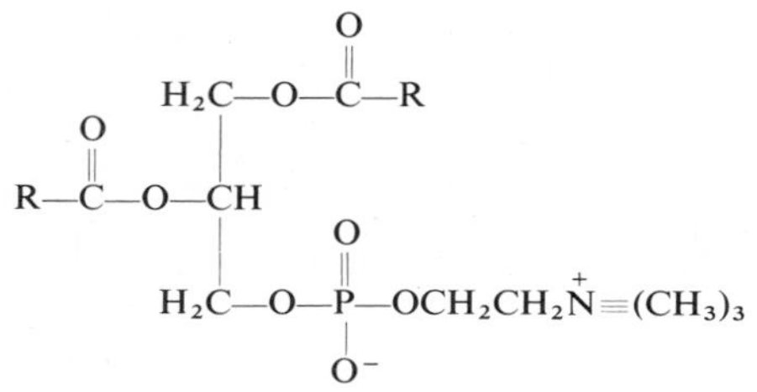

$$\begin{array}{l} \quad\quad\quad\quad\; H_2C-O-\overset{\displaystyle O}{\overset{\|}{C}}-R \\ \quad\; \overset{\displaystyle O}{\overset{\|}{R-C}}-O-\overset{|}{CH} \\ \quad\quad\quad\quad\; \overset{|}{H_2C}-O-\underset{\displaystyle O^-}{\underset{|}{\overset{\displaystyle O}{\overset{\|}{P}}}}-OCH_2CH_2\overset{+}{N}\equiv(CH_3)_3 \end{array}$$

Phosphatidylcholine

Phosphatidylethanolamine and phosphatidylserine are still frequently classified under the generic name of *cephalins* after the original nomenclature of Thudicum. These compounds have the capacity to form mixed micelles and appear to function physiologically as solubilizers of other lipids in body fluids, particularly in combination with protein where they are called LIPOPROTEINS.

Derivatives of sphingosine, the sphingomyelins, are traditionally included as members of the phosphatide group because of their phosphoric acid ester linkage, although some authorities have suggested that all sphingosine-containing compounds should be classified separately as sphingosides. On hydrolysis, sphingomyelins give rise to a fatty acid, choline, phosphoric acid and sphingosine. The following structure is representative:

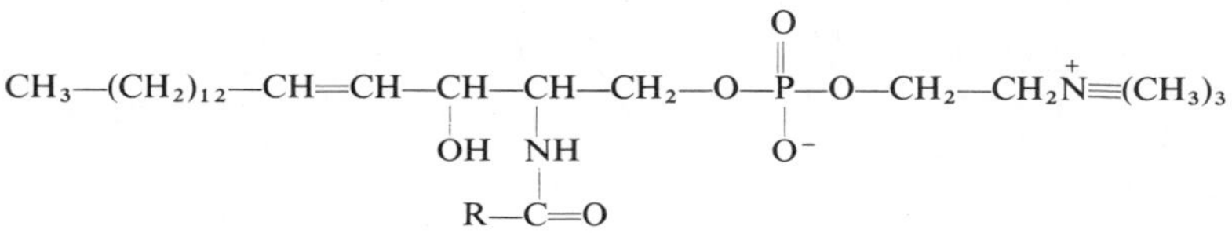

$$CH_3-(CH_2)_{12}-CH=CH-\underset{\displaystyle OH}{\underset{|}{CH}}-\underset{\displaystyle \underset{\displaystyle R-C=O}{\underset{|}{NH}}}{\underset{|}{CH}}-CH_2-O-\underset{\displaystyle O^-}{\underset{|}{\overset{\displaystyle O}{\overset{\|}{P}}}}-O-CH_2-CH_2\overset{+}{N}\equiv(CH_3)_3$$

Sphingomyelin

Glycolipids. Glycolipids are a diverse group of compounds which have the solubility characteristics of lipids but which yield a sugar residue on hydrolysis. The latter is most frequently galactose, but may be glucose, inositol, or other sugars. Phosphoric acid may or may not be present. A large group of galactolipids contain sphingosine and are called CEREBROSIDES because of their abundance in nervous system tissue.

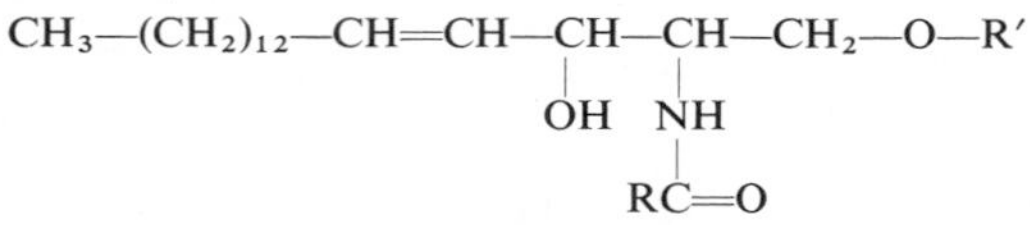

$$CH_3-(CH_2)_{12}-CH=CH-\underset{\displaystyle OH}{\underset{|}{CH}}-\underset{\displaystyle \underset{\displaystyle RC=O}{\underset{|}{NH}}}{\underset{|}{CH}}-CH_2-O-R'$$

Cerebroside (R′ = sugar)

Their precise function is unknown.

Aliphatic Alcohols and Waxes. Long-chain alcohols exist in nature and when esterified with fatty acids are called WAXES. From a quantitative standpoint they represent a minor component of the biologically important lipids. A typical wax, present as a major component of sperm oil, is cetyl laurate:

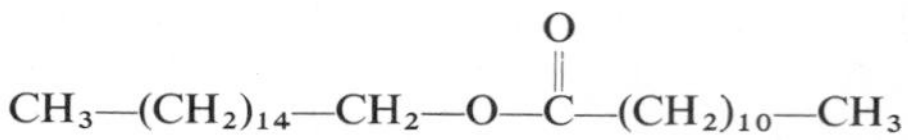

$$CH_3-(CH_2)_{14}-CH_2-O-\overset{\displaystyle O}{\overset{\|}{C}}-(CH_2)_{10}-CH_3$$

Cetyl laurate

Terpenes. Compounds belonging to this class of lipids are characterized by the presence of repeating five-carbon units having a configuration similar to that of isoprene (see ISOPRENOID BIOSYNTHESIS). Members of the group vary from relatively small molecules to complicated polymers and include many substances of biological importance. Vitamin A_1 is a terpene, as is natural rubber (see also CAROTENOIDS). A typical example is the isopentenyl pyrophosphate molecule, an intermediate in cholesterol and steroid hormone synthesis. Its structure is shown below:

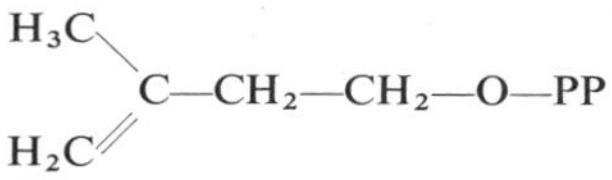

$$\begin{array}{l} H_3C \diagdown \\ \quad\quad\quad C-CH_2-CH_2-O-PP \\ H_2C \diagup\!\!\!\!\diagup \end{array}$$

Isopentenyl pyrophosphate (PP = pyrophosphate)

Steroids. STEROIDS are lipids of complex ring structure which are derived by condensation of intermediates of the terpene class (see STEROL BIOSYNTHESIS). They are of major physiological importance and include the ADRENAL CORTICAL HORMONES and SEX HORMONES. The common characteristic of the group is the possession of the perhydrocyclopentanophenanthrene ring. Traditionally a large subgroup has been separated out under steroids and called sterols. The latter are defined by the presence of an 8–10 carbon chain at position 17 and a hydroxyl group at position 3. At present there seems little advantage in such a subclassification. Cholesterol is a typical member of the group; its structure is shown in the article STEROIDS.

Summary. As is obvious from even this brief and oversimplified classification, the lipids represent a complex group of compounds. They are present in the lowest forms of plant, animal, and microbial life as well as in the most highly developed species, and serve multitudinous functions, some known and many unknown. All are of great interest to the student of biology and biochemistry.

References

1. WHITE, A., HANDLER, P., AND SMITH, E. L., "Principles of Biochemistry," New York, McGraw-Hill Book Co., 1964.

2. Deuel, H. J., Jr., "The Lipids: Their Chemistry and Biochemistry," New York, Interscience Publishers, 1951.
3. Hanahan, Donald, J., "Lipide Chemistry," New York, John Wiley & Sons, 1960.

Ben Bloom and Daniel W. Foster

LIPMANN, F. A.

Fritz Albert Lipmann (1899–) is a German-born American biochemist. He received, with Hans Adolf Krebs, the Nobel Prize in physiology and medicine in 1953, for his discovery of coenzyme a. This coenzyme, which contains pantothenic acid, is of far-reaching significance to all living organisms.

LIPOAMINO ACIDS AND LIPOPEPTIDES

The terms lipoamino acid and lipopeptide describe a variety of complexes and compounds which contain amino acids and lipids in a lipid-soluble, water-insoluble form. This drastic alteration in the solubility characteristics of such highly polar substances as amino acids requires their association with conventional lipoid substances so that the over-all polarity of the combination is drastically reduced. The combination could involve electrostatic attractions between a basic amino acid and a lipid phosphatide. Such an association can be readily formed in a test tube by shaking the two substances together and dissociated by shaking with an aqueous salt solution. General and less well-defined attractive forces of the van der Waals type, as well as dipole interactions, could bond hydrophobic side chains of amino acids and conventional lipids together, resulting in associations which are similar to "molecular compounds." Such complexes are quite stable in nonpolar media and can be fractionated in a manner corresponding to true covalent compounds of aqueous systems. There are also covalently linked amino acids and lipids, such as esters between the carboxyl group of the amino acid and hydroxyl group of glycerol or other alcohol. There are compounds linked by amide bonds between the amino groups of amino acids and the carboxyl groups of fatty acids. Examples of all of the above types of compounds and complexes are known and will be referred to below.

Lipoamino acids and lipopeptides have been found in a wide variety of tissues from microbial, avian, insect, mammalian, and plant sources.[1] The potential significance for lipoamino acids and lipopeptides in cellular physiology is to be found predominantly in the process of amino acid transport (passage from external medium through cell membrane to interior of the cell) and the process of protein biosynthesis. Cells are separated from their environments by membranes which are predominantly lipoidal in nature. For the amino acid to go from its aqueous milieu outside a cell to the water-soluble pool in the cytoplasm, it must negotiate the lipid phase of the membrane. Surprisingly little is known about the manner in which this is accomplished, aside from the fact that the active process is dependent upon energy-yielding metabolism in the cell. Therefore, a lipid-soluble form of the amino acid has potential significance in this process. Studies along these lines have been pursued by several laboratories, as for example, the group at Hannehmann Medical College in Philadelphia. From a variety of tissues, these workers have observed the rapid incorporation of amino acids into a chemically stable, but metabolically dynamic association with phospholipids. The labeling of such substances precedes the buildup of amino acids in the internal soluble pool.[1] (See amino acid pool.)

The principal metabolic fate of amino acids is their incorporation into the various proteins synthesized by living cells. It has been observed in cells which are very active in the synthesis and secretion of relatively large quantities of protein, that the protein-synthesizing unit consists of ribonucleoprotein granules (ribosomes) attached to lipoidal membranes (endoplasmic reticulum; see microsomes). Even in cells where no extensive cytoplasmic network of membranes has been observed and where most of the ribosomes appear to be free either singly (monosomes) or in groups (polysomes) such as in the bacterium *E. coli*, there exists a population of ribosomes attached to the plasma membrane. These attached ribosomes (or polysomes) are the most active in protein synthesis, and furthermore, they are easily torn from the membrane by procedures normally used to split the cell in order to study its components.[2] In gram-positive organisms, internal cytoplasmic membrane systems (mesosomes) are present. Kinetic studies concerned with the metabolism of radioactive amino acids have shown that external amino acids must first pass through the internal cytoplasmic pool of amino acids for oxidation.[3,4] However, for incorporation into protein, in studies with the intact cell, it was found that the amino acid follows a direct route which bypasses the internal pool. Furthermore, newly synthesized protein frequently has been found to be associated with membrane components, so that it has been necessary to use techniques to disrupt membranes in order to release them. These observations would help focus attention on a membrane-bound site for protein synthesis.[3] Kinetic studies with lipoamino acids and lipopeptides further suggest that such materials could be involved in the early stages of protein synthesis.[1] Thus, with intact cells making proteins, it has been observed that amino acids enter more quickly and extensively into association with lipids than with nucleic acids. Similarly, when cells have been pre-labeled with radioactive amino acids and then transferred to fresh unlabeled medium, radioactive amino acids left the lipid phase much more quickly than they left their association with nucleic acids.[3] Agents which attack lipids such as lecithinase A, deoxycholate and lysolecithin are very potent inhibitors of protein synthesis in intact cells.[1] Similarly, agents detrimental to protein synthesis, such as puromycin, chloramphenicol and dinitrophenol, have been found to inhibit the labeling of lipoamino acid and lipopeptide fractions.[1] If future work establishes the role for lipoidal forms of amino acids in PROTEIN BIOSYNTHESIS, it will be necessary to accommodate this information into the scheme of events currently considered to describe the process.

Chemical characterizations for some types of lipopeptide and lipoamino acids have been described in the literature. Thus, Axelrod *et al.* found palmityl amino acids in rat liver.[1] The palmitic acid is linked through an amide bond to the amino group of an amino acid. These investigators isolated and studied an enzyme from liver which was capable of forming the compounds *in vitro*. M. G. Macfarlane has characterized phosphatidyl glycerol amino acid esters from bacterial sources.[5] In this type of compound the amino acid is linked through an ester bond to a hydroxyl group of glycerol. Such compounds, although apparently widespread in bacteria, have not been observed as yet in the tissues of higher organisms. More complex lipopeptides have been thoroughly studied by Lederer and his co-workers on material obtained from microbial sources, principally of the genus mycobacterium.[6] Complex peptide-lipid compounds have also been described by Folch and co-workers as proteolipids[7] and phosphatidopeptides.[8]

In evaluating the metabolic significance of these substances, the question of artifact must be considered. In a worthwhile discussion of this problem, Wren defined those complexes which were formed by merely shaking an amino acid together with lipids in a test tube as artifacts.[1] Obviously, such a combination does not require active metabolism and may have no relation to the metabolic pathways of the cell. Certain of the amino acid-lipid complexes studied will probably fall into this category. However, the fact that some lipids can so readily complex with amino acids and, by so doing, confer completely new solubility characteristics on the amino acid, is not to be totally ignored. Such behavior within a living cell may provide a means for introducing the amino acid into lipid phases for further metabolic utilizations. Therefore, although it is important to distinguish such spontaneously formed complexes from those requiring the active metabolism of the cell, they should not be summarily discarded from any consideration under the term of artifact. Recently, additional evidence for the involvement of membranes and lipid substances in protein biosynthesis has been obtained in other laboratories, and clearly, there is much work that has to be done in this area.

References

1. HENDLER, R. W., "On the Metabolic Importance of Amino Acid-Lipid Complexes," in "Amino Acid Pools," p. 750, Elsevier Publishing Company, 1962.
2. HENDLER, R. W., AND TANI, J., "On the Cytological Unit for In Vivo Protein Synthesis in *E. coli*. Parts I, II and III," *Biochim. Biophys. Acta*, **80**, 279, 294, 307 (1960).
3. HENDLER, R. W., "A Model for Protein Synthesis," *Nature*, **193**, 821 (1962).
4. ROSENBERG, L. E., BERMAN, M., AND SEGAL, S., "The Kinetics of Amino Acid Transport, Incorporation into Protein, and Oxidation in Kidney-Cortex Slices," *Biochim. Biophys. Acta*, **71**, 664 (1963).
5. MACFARLANE, M. G., "Characterization of Lipoamino-Acids as O-Amino-Acid Esters of Phosphatidyl-Glycerol," *Nature*, **196**, 136 (1962).
6. JOLLES, P., SAMOUR, D., AND LEDERER, E., "Analytical Studies on Wax D, A Macromolecular Peptidoglycolipid Fraction from Human Strains of Mycobacterium Tuberculosis," *Arch. Biochem. Biophys. Suppl.*, **1**, 283 (1962).
7. PRITCHARD, E. T., AND FOLCH-PI, J., "Tightly Bound Proteolipid-Phospholipid in Bovine Brain White Matter," *Biochim. Biophys. Acta*, **70**, 481 (1963).
8. FOLCH-PI, J., "The Role of Phosphorus in the Metabolism of Lipids," in "Phosphorus Metabolism" (MCELROY, W. D., AND GLASS, H. B., EDITORS), Vol. II, p. 186, Baltimore, Johns Hopkins Press, 1952.

RICHARD W. HENDLER

LIPOIC ACID

Lipoic acid was discovered independently in several laboratories in the late 1940's as a growth factor and requirement for pyruvate oxidation for certain microorganisms. The trivial names "acetate-replacing factor," "pyruvate oxidation factor," and "protogen" were used to designate the biologically active substance prior to its isolation and identification. The factor was isolated in crystalline form in 1951 from the water-insoluble residue of beef liver by Reed, Gunsalus and a group at the Lilly Research Laboratories. A total of approximately 30 mg of the crystalline substance was obtained from 10 tons of liver residue. These collaborators proposed the trivial name "α-lipoic acid" to designate the isolated compound. Subsequently, a group of investigators at the Lederle Laboratories isolated an oxidation product (sulfoxide) of α-lipoic acid. Chemical analysis by both groups of researchers and synthesis by the latter group established the structure of "α-lipoic acid" as that of 1,2-dithiolane-3-valeric acid. The Lederle investigators proposed the trivial name "6-thioctic acid" to designate this structure. However, the American Society of Biological Chemists has recognized the priority of the name "lipoic acid" and has adopted it as the trivial designation of 1,2-dithiolane-3-valeric acid.

```
        CH2
       /   \
   H2C      CH(CH2)4CO2H
    |        |
    S--------S
```

Lipoic acid (thioctic acid)

Lipoic acid is found in minute amounts, usually in a protein-bound form, in animal and plant tissue and in many microorganisms. It can be released from the bound form by hydrolyzing natural materials with acid or alkali. Most nutritional investigations with higher animals have failed to show a growth response to lipoic acid supplementation. It appears that this biocatalyst can be synthesized by animal tissues. However, there is no doubt that it plays a vital role in animal metabolism, as indicated below. Pharmacological effects of lipoic acid have been noted with a variety of biological systems. Among these extensive studies have been reports of beneficial effects of lipoic acid in hepatic disorders, protection against X-ray damage and against various types of poisoning agents in small vertebrate animals, alleviation of symptoms of ASCORBIC ACID and TOCOPHEROL deficiencies in guinea pigs and rats, respectively, and profound effects on normal processes in several systems undergoing developmental changes, *e.g.*, inhibition of regeneration in hydra and planaria. The concentrations of lipoic acid needed to produce these effects are in most cases several orders of magnitude greater than those usually associated with its vitamin-like activity.

Metabolic Function. At present the only well-defined role of lipoic acid is that of a prosthetic group in MULTIENZYME COMPLEXES which catalyze an oxidative decarboxylation of pyruvic and α-ketoglutaric acids:

$$RCOCO_2H + CoASH + DPN^+ \rightarrow RCOSCoA + CO_2 + DPNH + H^+ \qquad (1)$$

This over-all reaction represents the main pathway of α-keto acid oxidation in animal tissues and is widely distributed among microorganisms. The sequence requires, in addition to lipoic acid ($LipS_2$), the coenzyme forms of four B VITAMINS—thiamine pyrophosphate (TPP), coenzyme A (CoASH) flavin adenine dinucleotide (FAD), and diphosphopyridine nucleotide (DPN; also termed NAD, NICOTINAMIDE ADENINE DINUCLEOTIDE). The available evidence indicates that the over-all reaction [Eq. (*1*)] proceeds via the steps shown in Eqs. (*2*)–(*5*), where the brackets indicate enzyme-bound compounds.

$$RCOCO_2H + [TPP] \rightarrow [RCH(OH)\text{-}TPP] + CO_2 \qquad (2)$$

$$[RCH(OH)\text{-}TPP] + [LipS_2] \rightarrow [RCO\text{-}SLipSH] + [TPP] \qquad (3)$$

$$[RCO\text{-}SLipSH] + CoASH \rightarrow [Lip(SH)_2] + RCOSCoA \qquad (4)$$

$$[Lip(SH)_2] + DPN^+ \xrightarrow{(FAD)} [LipS_2] + DPNH + H^+ \qquad (5)$$

$$R = CH_3,\ HOOC(CH_2)_2\text{—}$$

In its functional form in the *Escherichia coli* pyruvate and α-ketoglutarate dehydrogenase complexes, and presumably in the analogous mammalian complexes as well, the lipoyl moiety is bound in amide linkage to the ε-amino group of a lysine residue. The lipoyl moiety can be released enzymatically with lipoamidase, an enzyme found in extracts of *Streptococcus faecalis*. Reincorporation of the lipoyl moiety is achieved with an enzyme system discovered in extracts of *S. faecalis* and *E. coli*. The latter process requires an energy source, adenosine triphosphate (ATP). The demonstration of inactivation and reactivation of α-keto acid dehydrogenase systems accompanying, respectively, release and reincorporation of the lipoyl moiety, provides direct experimental evidence of lipoic acid involvement in α-keto acid oxidation [Eq. (*1*)].

It should be noted that the lipoyl moiety plays a central role in the sequence represented by Eqs. (*2*)–(*5*), participating in acyl generation, acyl transfer and electron transfer. The decarboxylation reaction [Eq. (*2*)] is visualized as a cleavage of the α-keto acid to form carbon dioxide and an aldehyde moiety linked to thiamine pyrophosphate ("active aldehyde" in biochemical terms). Recent work indicates that the aldehyde moiety is linked to carbon-2 of the thiazolium ring of TPP. The acyl generation reaction [Eq. (*3*)] is visualized as a reductive acylation of the protein-bound lipoyl moiety. Equation (*4*) is visualized as an acyl transfer reaction. The electron transfer reaction [Eq. (*5*)] is catalyzed by a FLAVOPROTEIN.

Some progress has been made in elucidating the biosynthesis of lipoic acid. Tracer experiments with *E. coli* indicate that octanoic acid can function as a precursor of lipoic acid. [1-^{14}C]-Octanoic acid appears to be incorporated into lipoic acid as a unit, carbon-1 of the biosynthesized lipoic acid corresponding to carbon-1 of the octanoic acid.

References

1. REED, L. J., "The Chemistry and Function of Lipoic Acid," *Advan. Enzymol.*, **18**, 319–347 (1957).
2. REED, L. J., "Biochemistry of Lipoic Acid," *Vitamins Hormones*, **20**, 1–38 (1962).
3. REED, L. J., "Chemistry and Function of Lipoic Acid," in "Comprehensive Biochemistry" (FLORKIN, M., AND STOTZ, E. H., EDITORS), Vol. 14, Elsevier Publishing Co., in press.

LESTER J. REED

LIPOPROTEINS

The lipoproteins are a large group of conjugated PROTEINS characterized by their high content of LIPID. They occur in the organism as either soluble complexes, *e.g.*, in serum, or in insoluble complexes such as are present in nervous tissue or in subcellular components of cells. Lipids are ubiquitous in the animal organism and are probably always associated with protein; it is doubtful that any lipid exists in the completely free state.

The soluble lipoproteins in serum constitute the most widely studied group of this class of conjugated proteins. Normal human serum contains 0.5–0.7% lipid. These lipids—cholesterol, cholesterol esters, phosphatides and triglycerides—are either insoluble or only slightly soluble in aqueous solutions, although in serum they are present in solution in relatively high concentration. It is the association of the lipids with protein that enables these water-insoluble materials to exist in a soluble form. The lipoproteins of plasma are soluble in isotonic solution but are insoluble in water, and thus behave like euglobulins (see PLASMA AND PLASMA PROTEINS).

While a relationship between lipids and serum proteins was suggested as early as 1914, it was not until 1929 that the presence of a lipoprotein in serum was first demonstrated. From a horse serum globulin fraction not precipitated by 50% saturated ammonium sulfate, Macheboeuf obtained a material which was later found to be an α_1-globulin and which had a chemical composition similar to that of α_1-lipoproteins isolated from serum of other species.

Purification Methods. Serum lipoproteins have been isolated and fractionated by a number of methods: moving boundary and zone ELECTROPHORESIS, precipitation of low-density lipoprotein by sulfated polysaccharides, ULTRACENTRIFUGATION, and techniques which take advantage of the differential solubility of lipoprotein classes. One example of this latter method is that of Macheboeuf mentioned above. This method of salting out, however, is no longer in general use. Another method of differential solubility is Cohn fractionation of plasma proteins in cold ethanol-water mixtures. This fractionation scheme requires careful regulation of ethanol concentrations, pH, ionic strength, protein concentrations and temperature. The procedure yields two fractions of lipoproteins, one with the electrophoretic mobility of α_1-globulins, and a second with a mobility of β_1-globulins. This technique was the direct result of the large-scale fractionation of plasma proteins carried out at Harvard in Dr. Edwin Cohn's laboratory during World War II. The method is time consuming and exacting, and although it was of great importance in the past, it is not now generally used.

Free electrophoresis was one of the early procedures for lipoprotein fractionation (first used in 1941), but it is not a generally useful method since homogeneous fractions cannot be obtained. On the other hand, zone electrophoresis on paper, starch or silica gel has been widely used for lipoprotein separations and the determination of lipoproteins in serum. By means of these techniques, lipoproteins can be separated into three fractions having electrophoretic mobilities of α_2-, α_1- and β-globulins. However, because of the limited capacity of these techniques, only relatively small amounts of lipoproteins can be separated. Moreover, quantitative elution of lipoproteins from paper and starch is difficult, and on starch the α_1- and β_1-lipoproteins cannot be subfractionated.

The most useful and specific of the methods for lipoprotein separation is ultracentrifugal flotation. Because of their relatively high lipid content, lipoproteins have hydrated densities which are considerably lower than the other serum proteins. If the density of serum is adjusted to 1.21 (accomplished by the addition of appropriate amounts of solid KBr) and the resultant serum is ultracentrifuged at 100,000 $\times$ g for 20–24 hours, all the lipoproteins will float to the top of the tube. By adjusting the original serum to densities intermediate between 1.006 (native density) and 1.21, lipoproteins can be separated from one another. Since the lipoproteins float, the centripetal movement is denoted as a rate of flotation rather than negative sedimentation. The units are expressed as Svedberg units of flotation, S_f, when measured at 26°C in a solution of NaCl whose density is 1.063 g/ml. One S_f unit is a flotation rate of 1×10^{-13} cm/sec/dyne/g at 26°C. The density of 1.063 marks the division point between the dense α_1-lipoproteins ($d > 1.063$) and the low-density β- and α_2-lipoproteins ($d < 1.063$). These two major fractions are the ones isolated by the Cohn fractionation technique mentioned above. A second procedure for separating different classes of lipoproteins is by establishing a density gradient in an ultracentrifuge tube. Upon ultracentrifugation, the various lipoproteins will tend to collect at a level in the tube where the density of the medium corresponds to their own density.

Individual classes of lipoproteins are isolated by ultracentrifugation of serum at different densities. Chylomicrons ($S_f > 400$) are prepared by centrifuging serum at its native density at 26,000 $\times$ g for 30–40 minutes. It has been pointed out that it is the product of g times minutes which is important in the isolation of the chylomicrins, and that ultracentrifugation for 10^6 "g-minutes" will result in separation of the chylomicrons from the other lipoproteins. If serum is centrifuged at its native density at 100,000 $\times$ g for 20–24 hours, lipoproteins from S_f 20–10^5 will float up. Ultracentrifugation of serum at d 1.019 after removal of lipoproteins with S_f to 10^5 will result in the isolation of a fraction with S_f 10–20. If the infranatant is then adjusted to d 1.063 and spun at 100,000 $\times$ g for 20–24 hours, the S_f 0–10 lipoproteins will float up, and the infranatant layer will contain the high-density lipoproteins. These latter lipoproteins may then be ultracentrifuged at d 1.21, or at an intermediate density between 1.063 and 1.21 to obtain two subfractions of high-density lipoproteins.

The designation of lipoproteins is somewhat arbitrary, *e.g.*, some investigators refer to lipoproteins of S_f 10–20 as very low density, while others define very low-density lipoproteins as having S_f values 10–400. Similarly, fraction S_f 0–20 and S_f 0–10 have each been designated as low-density lipoproteins. It is thus important that S_f values, or more commonly, densities or density limits be used to define a given lipoprotein fraction. For large particles, the designation should be "g-minutes" in the ultracentrifuge. Table 1 summarizes some of the physical characteristics of serum lipoproteins including S_f values, diameter, molecular weight and electrophoretic mobility.

In 1955 a number of investigators described the formation of insoluble complexes of sulfated polysaccharides and certain plasma lipoproteins. The polysaccharide-lipoprotein complex is formed at neutral pH and is readily dissociated by

TABLE 1

	S_f	Density Range	Diameter (Å)	Molecular Weight	Electrophoretic Mobility
Chylomicrons	400–100,000	$<$1.006	5×10^3 to 10×10^3		α_2
Very low density	10–400	1.006–1.019	300–700	5.2×10^6	α_2
Low density	0–10[a]	1.019–1.063	200–250	3.2×10^6	β_1
High density	—	1.063–1.21	100–150	435,000	α_1
Very high density	—		$<$100	195,000	α_1

[a] Concentrated in the S_f 3–9 region.

increasing the ionic strength of the medium, or by CHELATION of metal ions which appear to be necessary for the formation of the complex. Several polysaccharides including HEPARIN, dextran sulfate, the sodium salt of sulfated polygalacturonic acid methyl ester methyl glucoside, and corn amylopectin sulfate have been utilized for such isolation of low-density lipoproteins. The interaction of the polysaccharide with the lipoproteins appears to involve the charged amino groups on the phospholipids. The specificity for low-density lipoproteins may be related to the relative amounts of phospholipid and protein in this lipoprotein class. This technique in conjunction with ultracentrifugal separation constitutes an extremely useful method for the isolation of relatively large amounts of pure low-density lipoproteins.

Composition. In humans the β-lipoproteins and α-lipoproteins constitute approximately 5 and 3%, respectively, of the total human plasma proteins. A summary of the composition of several major classes of plasma lipoproteins is presented in Table 2. This Table shows the approximate percentage composition of the plasma lipoproteins. Lipids are removed from the lipoproteins by extraction with solvent systems such as ethanol-ether, 3:1, or chloroform-methanol, 2:1. Ether and other nonpolar solvents will remove primarily neutral lipids from low-density lipoproteins and very little lipid from high-density lipoproteins. The protein moiety of all the low-density lipoproteins in the same, and there appears to be some similarity between other protein of high-density lipoproteins and chylomicrons. There are small amounts of carbohydrate in serum lipoproteins. The proteins of high-density lipoproteins contain about 0.5% hexose and 0.35% HEXOSAMINE. The polypeptide from low-density lipoprotein contains 3.2% hexose (galactose and mannose), 1.2% hexosamine, and 0.35% sialic acid. The major phosphatides of human serum lipoproteins are cephalin (phosphatidyl ethanolamine), 5%; phosphatidyl choline, 68%; lysophosphatidyl choline, 7%; and sphingomyelin, 20%. In addition to the lipids listed in Table 2, free fatty acids and lysolecithin are present, but these lipids are primarily bound to albumin.

Very little is definitely known about the association of lipids and protein. Two types of association may be considered: lipids may be bound to distinct sites on the protein molecule by inclusion between peptide chains of protein, or the protein may combine with the lipid to form a protein-lipid micelle. The type of bonds which link the protein to lipid are not definitely known. However, it is likely that electrostatic forces may be involved between the polar groups of both protein and lipid in addition to bonds between nonpolar groups of lipids and nonpolar amino acid residues. Water molecules are probably essential for the native structure of the lipoprotein since they form hydrogen bonds with various polar groups. If lipoproteins are frozen and thawed, they become insoluble and irreversibly denatured. Free drying also promotes denaturation. In the frozen state the water molecules are

TABLE 2[a]

	S_f	Amount (mg/100 ml plasma)	Approximate Percentage Composition				
			Protein	Phosphatides	Cholesterol Free	Cholesterol Ester	Triglyceride
Chylomicrons	$>$400	100–250	2	7	2	6	83
Very low density	10–400	130–200	9	18	7	15	50
Low density	3–9	210–400	21	22	8	38	10
High density	—	50–130	33	29	7	23	8
Very high density	—	290–400	57	21	3	14	5

[a] Adapted from J. L. Oncley, in "The Lipoproteins: Methods and Clinical Significance" (Homburger, F., and Bernfeld, P., Editors), New York, S. Karger, 1958.

rigidly held in the ice lattice and are not free to form hydrogen bonds with polar groups in the lipoprotein. It is of interest that certain molecules such as glycerol, glucose and sucrose protect lipoproteins against damage. These compounds which are rich in hydroxyl groups may themselves take the place of water molecules which are made unavailable at freezing temperatures.

Low-density β-lipoprotein is thought to have a core which contains nearly all the triglyceride and cholesterol of the lipoprotein. This core is surrounded by a shell of protein, phosphatide, and cholesterol.

The high-density α-lipoprotein contains three peptide chains which may be coated with cholesterol and phosphatide. The cholesterol esters and triglycerides are probably attached to the nonpolar portions of the phosphatides and protein by hydrophobic bonds.

References

1. EDER, H. A., "The Lipoproteins of Human Serum," *Am. J. Med.*, **23**, 269 (1957).
2. GURD, F. R. N., "Association of Lipides with Proteins," in "Lipide Chemistry" (HANAHAN, D., EDITOR), New York, John Wiley & Sons, 1960.
3. GURD, F. R. N., "Some Naturally Occurring Lipoprotein Systems," in "Lipide Chemistry" (HANAHAN, D., EDITOR), New York, John Wiley & Sons, 1960.
4. CORNWELL, D. G., AND HORROCKS, L. A., "Protein-Lipid Complexes," in H. W. SCHULTZ AND A. F. ARYLEMIER, EDITORS, "Proteins and Their Reactions" (SCHULTZ, H. W., AND ARYLEMIER, A. F., EDITORS), Westport, Conn., Avi Publishing Company, 1964.

LEWIS I. GIDEZ

LIVER AND LIVER FUNCTIONS

The blood returning from the intestine to the heart is shunted through a capillary system, the hepatic sinusoids, which are surrounded by epithelial cells arranged in plate forms. These plates cross each other in space at different angles, to permit the greatest possible contact between the blood and these polygonal epithelial cells. The resulting spongelike organ located under the diaphragm and covered by the connective tissue capsule of Glisson is the largest organ of the body. Under normal circumstances, the major part of its blood, between 66 and 75%, comes from the portal vein which drains the splanchnic capillaries, particularly those of intestine, pancreas and spleen. Approximately one quarter to one-third of the hepatic blood comes from the hepatic artery originating from the aorta at the celiac axis. Both hepatic artery and portal vein enter at the hilus of the liver and divide in a dichotomic fashion into parallel running branches. They are surrounded by ramified extensions of Glisson's capsule. The hepatic artery sends branches to the capillary plexus of the portal tracts whereas the bulk of its blood is released into the sinusoids parenchyma as does the portal vein which forms by confluence of superior mesenteric, inferior mesenteric and splenic vein, and receives additional internal radicles from the portal capillary plexus. The sinusoids are blood spaces, normally without the basement membrane otherwise seen in capillaries; they are, therefore, characterized by great permeability for serum protein. Moreover, some of their lining endothelial cells, those which are star-shaped and called Kupffer cells, are part of the reticuloendothelial system. The lining cells form the sinusoidal wall and leave small stomata open through which macromolecular substances pass into a tissue space between liver cell plates and sinusoids and extending between neighboring hepatocytes almost to the bile canaliculus. Tissue fluid is drained toward the lymphatics in either the central canal or, in the human, mainly the portal tract. Arterial and venous blood, mixed to a varying degree, flows toward the tributaries of the hepatic veins which combine to larger veins into which frequently small branches enter at almost right angles. The largest branches enter into the vena cava inferior behind the liver. Vascular sphincter mechanisms in various locations regulate hepatic blood flow and thus function. The portal tracts and the central canals around the tributaries of the hepatic veins cross each other in space and are throughout the liver about 0.3 mm apart. The direction of the blood flow from the portal tracts to the central canals produces the concentric arrangement of the liver cell plates characterizing the liver lobule which conventionally is considered the structural unit of the liver.

The liver forms bile which is released into slits between the liver cells, the bile canaliculi, which are arranged in a chicken-wire-like fashion; the wall of the canaliculi is formed by part of the hepatocellular plasma membrane. The bile canaliculi are drained by small tubes with an independent cuboidal epithelial lining, the ductules or cholangioles. Under normal circumstances hardly any are found within the lobule, the majority being in the periportal zone or in the portal tract. Under abnormal circumstances, they increase in number and are then found deep within the lobule. The ductules continue into the bile ducts located in the portal tracts which unite in dichotomic fashion to finally form the common hepatic duct which leaves the liver where hepatic artery and portal vein enter it. It combines with the cystic duct draining the gallbladder, which concentrates bile by water reabsorption to form the common duct running toward the duodenum. This entrance is controlled by the choledochoduodenal sphincter of Oddi. Bile is produced at an almost constant rate but released from the biliary system in human beings and many animals only if food appears in the duodenum. As a result of this or other mechanisms, the sphincter of Oddi relaxes and the gallbladder contracts. This leads to proper utilization of bile which, while being partly an excretory product, is a secretion essential in intestinal DIGESTION and absorption.

In the liver, several fluid currents exist. Blood and some tissue fluid flow toward the central canal, while bile and most of the tissue fluid (at least in the human) are flowing toward the portal canal. The normal liver consists of approximately 60% hexagonal epithelial cells (hepatocytes), 30% littoral endothelial or Kupffer cells, and about 2% each of bile duct cells, connective tissue and blood vessels. The hepatocytes have three types of borders. Where they are in contact with each other, the border is straight indicating limited, if any, exchange of substance between individual cells. The border toward the tissue space is elongated by narrow extensions of the space between neighboring hepatocytes and particularly by the formation of irregularly shaped finger-like projections in the form of microvilli. This tremendous elongation of the border of the hepatocytes and the preferential location of enzymes in this location reflects structurally the extensive exchange of substances between hepatocytes, tissue space and blood. Much shorter is the border toward the bile canaliculus which is also thrown into microvilli which are far more regular and disappear upon impairment of biliary secretion. Preferential accumulation of ATPase in the villi indicates the intensity of the metabolic processes in bile secretion.

The NUCLEUS is normally vesicular and has conspicuous nucleoli. It varies considerably under normal and pathologic conditions, the majority being tetraploid in adult rodents. Binucleated cells increase in regeneration. The cytoplasm normally contains many and relatively large mitochondria in the matrix, of which the CITRIC ACID CYCLE enzymes and, in the double membrane, the electron transfer enzymes can be demonstrated. RIBOSOMES as ribonuclear protein are arranged around messenger RNA usually in helix form as polysomes. These polysomes as the site of PROTEIN BIOSYNTHESIS may be either free in the cytoplasm or attached to the extensive endoplasmic reticulum (see MICROSOMES) which thus becomes granular and the site of secretion of protein such as serum proteins. The endoplasmic reticulum is also the site of STEROID synthesis, and the smooth endoplasmic reticulum is the site of detoxification and of glucose-6-phosphatase. In addition, one notes the Golgi apparatus responsible for secretion and the perinuclear dark bodies, the lysosomes, which are the site of various hydrolytic enzymes mainly with peak activity in acid medium. They serve to segregate intracellular material after pinocytosis or for storage, secretion and separation of organelles undergoing destruction in the form of autophagic vacuoles. The soluble fraction of the cytoplasm, the hyaloplasm, corresponding to the supernatant fluid in cytochemical analysis, contains proteins and enzymes and cofactors related to CARBOHYDRATE METABOLISM and activation of amino acids and nucleic acids. In addition, in the normal liver, GLYCOGEN and few fat droplets are found as well as some FERRITIN crystals which, under abnormal circumstances, become hemosiderin deposits giving histochemical iron reaction.

The main functions of the liver cells are: (1) secretion of substances into the bloodstream of which the serum proteins particularly ALBUMINS, alpha-globulin, the proteins concerned with BLOOD CLOTTING, haptoglobin and transferrin, as well as some blood enzymes (*e.g.*, esterase), serum cholesterol, and blood glucose are the most important; in contrast to all other tissues which utilize but do not form blood glucose, the liver cells are the main source of the blood glucose (see GLYCOGEN AND ITS METABOLISM) because of a specific phosphatase system; (2) storage of various metabolites particularly glycogen, proteins, fat and vitamins; (3) transformation of various compounds into each other, *e.g.*, fats into carbohydrates and vice versa; (4) detoxification mainly by oxidation or conjugation, the latter mainly for better solubility and urinary excretion; (5) formation of the BILE into which BILE PIGMENT is transmitted by conjugation and BILE ACIDS and cholesterol by transformation.

A variety of sinusoidal cells are seen. Some are flat endothelial cells, others are Kupffer cells with a cytoplasm of varying and irregular outlines and ameboid extensions. They contain few mitochondria but varying inclusions. They are engaged in phagocytosis of circulating exogenous and endogenous macromolecular or corpuscular elements, including bacteria, as well as of hepatocellular breakdown products; they are active in transformation of blood pigment to bile pigment. Other sinusoidal lining cells, rare under normal circumstances, form serum gamma-globulin and correspond to plasma cells. Also fibroblasts can be seen around the sinusoids.

The liver, as a whole, because of its strategic situation near the right heart and because of its sheer bulk, influences circulating blood volume, as well as electrolyte and water metabolism.

The main diffuse reactions of the liver are: (1) regressive changes of the hepatocytes characterized by loss of ribosomes and their disengagement from the endoplasmic reticulum which becomes shattered and dilatated, loss of glycogen, and excess of lysosomes some of which are autophagic vacuoles around focally damaged organelles; if extensive and diffuse enough, the lesion accounts for insufficiency of most of the crucial function of the hepatocytes, particularly those associated with endoplasmic reticulum and mitochondrial function; (2) necrosis of liver cells which also if extensive enough accounts for insufficiency of most of the crucial functions; (3) cholestasis either as a result of extrahepatic biliary obstruction or of intrahepatic processes; this represents an injury of the bile secretory apparatus (biliary microvilli, Golgi apparatus and lysosomes) or the consequence of formation of abnormal bile by the hepatocytes, and is associated with accumulation of biliary substances in the blood such as conjugated bilirubin, bile acids, cholesterol and alkaline phosphatase; (4) permeability change of the hepatocytic plasma membrane directed towards the sinusoidal space; this may be caused by mitochondrial alterations and results in the release of hepatocellular enzymes into the blood, of which the transaminases are the best example; (5) fatty metamorphosis of the hepatocytes which, if not complicated by other injury, interferes with few hepatic functions; (6) inflammation involving intralobular sinusoidal cells and portal tracts; (7) fibrosis which, as such, causes little functional disturbance of the liver but sometimes interferes with portal blood flow to produce portal hypertension; (8) cirrhosis, characterized by regenerative nodules and septa dividing the parenchyma resulting in portal hypertension, diversion of blood from the liver, and damage of hepatic cells with subsequent disturbance of hepatic function; (9) cancer, either primary from the hepatic cells or from bile duct cells, or metastatic from other organs.

References

1. Brauer, R. W. (Editor), "Liver Function. A Symposium on Approaches to the Quantitative Description of Liver Function," Washington, D.C., American Institute of Biological Science, 1958.
2. Child, C. G., III, "The Hepatic Circulation and Portal Hypertension," Philadelphia and London, W. B. Saunders, 1954.
3. Fawcett, D. W., "Observations on the Cytology and Electron Microscopy of Hepatic Cells," *J. Natl. Cancer Inst.*, **15**, 1475 (1955), supplement.
4. Hartman, F. W., LoGrippo, G. A., Mateer, J. G., and Barron, J. (Editors), "Hepatitis Frontiers," Boston and Toronto, Little Brown & Co., 1957.
5. Popper, H., and Schaffner, F., "Liver, Structure and Function," New York, McGraw-Hill Book Co., 1957.
6. Rouiller, C. (Editor), "The Liver, Morphology, Biochemistry, Physiology," Vols. I and II, New York and London, Academic Press, 1963.

Hans Popper

LOEB, J.

Jacques Loeb (1859–1924) was born and educated in Germany but spent the last 33 years of his life in the United States. He was a physiologist, noted for his work on tropisms and their effects on behavior and on artificial parthenogenesis in sea urchins and tadpoles. He was a mechanist and was constantly trying to explain life. He, along with W. J. V. Osterhout, was founder of the *Journal of General Physiology*.

LOEWI, O.

Otto Loewi (1873–1961) was a German-born American physiologist. He is noted for his striking discovery that when the vagus nerve innervating a frog's heart, which is beating in Ringers solution, is stimulated, it not only slows the heartbeat but also releases into the Ringers solution a substance which will slow down a second frog's heart merely by contact with it. This demonstrated the existence of what may be called a neurohormone. In collaboration with Henry Hallet Dale, the active substance was shown to be acetylcholine. Otto Loewi and Henry Hallett Dale were jointly awarded the Nobel Prize in physiology and medicine for this work in 1936.

LUNGS (BIOCHEMICAL FUNCTION). See gaseous exchange in breathing; also the following reference.

Reference

Comroe, J. H., "The Lung," *Sci. Am.*, **214**, 56 (February 1966).

LUTEINIZING HORMONE (LH)

This adenohypophyseal hormone, also termed interstitial cell-stimulating hormone (ICSH), is one of the gonadotropic hormones (see also hormones, animal).

LUTEOTROPIC HORMONE (LTH)

This is another term for prolactin or lactogenic hormone, and is one of the adenohypophyseal gonadotropic hormones (see also hormones, animal).

LYMPH (BIOCHEMICAL ASPECTS)

Lymph may be defined as the fluid in lymph vessels derived from interstitial fluid. Its composition is variable and primarily reflects the activity and permeability of blood capillaries and the consequent pericapillary filtrate of the area drained by a particular lymphatic supply; it secondarily reflects cell metabolism in the area. The exchange between blood and extravascular fluid is by diffusion or by bulk filtration. Water, dissolved electrolytes and small molecules exchange rapidly but at different rates depending upon the size of the ion or molecule and on the relationships between hydrostatic and osmotic pressure in blood and tissue fluids. Macromolecules such as protein and lipids leave the bloodstream more slowly and are found in lymph in lower concentration than in plasma. The lymph protein distribution is similar to that of plasma except in lymph drained from the liver which contains alpha- and beta-globulin fractions not found in lymph from other areas or in plasma. Available evidence from work on the dog suggests that molecules with a weight of approximately 2000 or more are retained almost quantitatively in the lymphatic vessels. Thus, once macromolecules enter the lymphatic system, they recirculate to the blood stream. This recirculation is an important factor in the maintenance of "normal" plasma levels of these constituents. Thus, during the course of a day, 50% or more of total circulating protein escapes from the bloodstream and is returned to it via the lymphatic system.

It has long been known that lymph from all parts of the body clots but, as a rule, less readily than plasma. Recent studies on human thoracic duct lymph confirm and extend studies in animals and show that coagulation substances—prothrombin

V, antihemophylic globulin, hemophilia B factor, fibrinogen and plasminogen—and preactivators are present in thoracic duct lymph in concentrations of about 20–60% of plasma levels (see BLOOD CLOTTING). Trypsin inhibitors have also been found in human lymph. Several of the recently described factors which increase blood capillary permeability have also been described as present in lymph. Again, the recirculation of these substances is an important factor in the maintenance of their plasma levels. Many ANTIBODIES and enzymes are also found in lymph and, in some instances, the lymphatic system appears to be the principal pathway by which these substances travel from their cells of origin to the bloodstream. Among the enzymes found in lymph are alkaline phosphatase, histaminase, transaminase, esterase, amylase, maltase, diastase, catalase, peptidase, saccharase, cholinesterase, and tributyrinase. Recently, angiotensin has been added to the list.

The lymphatic system is particularly concerned with the problem of sterol absorption since the transfer of LIPOPROTEINS to the lacteal lymph vessels as chylomicra is the final step in the absorption of many lipids. Lymph may be the major if not the exclusive agent for the transport of long-chain fatty acids. In contrast, short-chain fatty acids are absorbed directly into the bloodstream. The following tentative mechanism has been suggested to explain *cholesterol* absorption. Free and esterified dietary cholesterol enter the lumen of the small intestine and are mixed with endogenous cholesterol, bile salts, and lipase and cholesterol esterase of the pancreatic juice. Cholesterol esters are hydrolyzed to free cholesterol and fatty acids. The free cholesterol passes into the mucosa along with fatty acids and the products of triglyceride digestion. The free cholesterol becomes mixed with a metabolic pool of free cholesterol in the mucosa. The transfer of cholesterol into the mucosa increases the size of the free cholesterol pool, and there is an increased synthesis of cholesterol esters. These esters and some free cholesterol are then incorporated into chylomicrons and other lipoproteins in the mucosa and finally transferred to the lymph to be carried to the bloodstream for circulation and recirculation.

The lymphatic system has also been shown to be important in the transport of HORMONES from the ovary, testes, adrenal and thyroid and in the transport of lymphocytes. It has been estimated that, in a dog weighing 10 kg, thoracic duct lymph may contain about 10,000 lymphocytes/mm^3. In terms of the rate of lymph flow, this would represent an average output of 200 million/hr or 5 billion/day. Since dogs have about 3000 lymphocytes/mm^3 in circulating blood and, in a 10-kg dog, there is about 1 liter of blood, it can be estimated that the circulating blood of the dog contains not more than 2.5 billion lymphocytes or half the number that enters the blood daily from the thoracic duct. This suggests that blood lymphocytes are replaced at least twice a day in the dog. Evidence is available to suggest that they may be replaced at least 5 times daily in the cat and rabbit. The recirculation of lymphocytes may be significant in the spread of substances, like viruses, which are carried by lymphocytes. For example, although a virus may be too large to gain rapid entrance into the body through the nasal mucosa, it may establish itself and later enter the lymphatics to be widely disseminated to all lymph nodes by the recirculating lymphocytes.

Studies of the composition of lymph under various experimental conditions hold promise of contributing a valuable approach to the study of metabolic patterns. For example, Dietrich and Siegel recently reported an interesting study designed to determine whether nucleotides or nucleotide precursors synthesized in an organ or tissue, *e.g.*, liver, were available to nourish other tissues and organs. The stimulus for their studies arose from observations that certain cell types cannot utilize free bases and must secure the nucleoside containing the base from an external source, apparently other cell types. They argue that if bases and other nucleotide precursors are secreted by a distant organ or cell type, these compounds may be present in both the blood and the lymph which bathes the cell or organ. Blood, however, contains such a mass of living cells that it is difficult to determine whether intermediates found in the plasma are derived from the cells within the blood or from other somatic cells nourished by and yielding their products to the blood. Since the cell population in lymph is insignificant when compared with that of blood, it might be assumed that metabolites found in the lymph would reflect more closely the metabolism of the tissue through which it has passed than that of the lymphocytes. Working on Nembutalized rats, they injected glycine-2-C^{14} and nicotinamide-7-C^{14} and found adenine, guanine (see PURINES), cytosine, uracil (see PYRIMIDINES), and uric acid in measurable amounts in thoracic duct lymph. No detectable quantities of nucleosides were observed. The quantity of acid-soluble nucleotides found was equivalent to that which would be expected from the lymphocytes present in the lymph samples analyzed. Lymph collected for a 45-hour period following the injection of carbon-labeled glycine contained no significant amount of labeled purine derivatives. At the end of this period, liver tissue, however, still contained appreciable quantities of labeled acid-soluble nucleotides. Lymph collected for a similar period of time after the injection of carbon-labeled nicotinamide contained very small amounts of radioactivity. While the results raised many unexplainable questions and suggested the need of further work, they did confirm previous investigations of plasma in indicating that if these compounds are essential for the proper nutrition of certain cell types, these purine derivatives are not transported from sites of synthesis, such as the liver, via the lymphatic ducts.

A second interesting example is in the use of lymph for indicating gas tensions in interstitial fluids and tissues. Lymph from liver, small bowel and muscle yielded CO_2 tensions which were uniformly higher than those of venous blood and O_2 tensions which were considerably lower. The low O_2 lymph tensions, together with high lactate values, suggest that most tissues may operate at O_2 tensions far lower than venous blood and some tissues at tensions close to zero.

References

1. Yoffey, J. M., and Courtice, F. C., "Lymphatics, Lymph and Lymphoid Tissue", Cambridge, Mass., Harvard University Press, 1956.
2. Rusznyàk, I. M., Földi, M., and Szabò, G., "Lymphatics and Lymph Circulation," New York, Pergamon Press, 1960.
3. Mayerson, H. S., "The Physiological Importance of Lymph," in "Handbook of Circulation," Section 2, Vol. II, Washington, American Physiological Society, 1963.
4. Treadwell, C. R., Swell, L., and Vabouny, G. V., "Factors in Sterol Absorption," *Federation Proc.*, **21**, 903–908 (1962).
5. Mayerson, H. S., "The Lymphatic System," *Sci. Am.*, **208**, 80–90 (June 1963).

H. S. Mayerson

LYNEN, F.

Feodor Lynen (1911–) is a German biochemist with a high degree of versatility. He shared with Konrad Bloch the Nobel Prize in physiology and medicine in 1964, for his work related to the biosynthesis of fatty acids.

LYSERGIC ACID DIETHYL AMIDE (LSD). See HALLUCINOGENIC DRUGS.

LYSINE

L-Lysine, α,ε-diaminocaproic acid, $H_2N—CH_2—CH_2—CH_2—CH_2—CHNH_2—COOH$, is an AMINO ACID found in many common PROTEINS, but it is conspicuously low (or lacking) in certain cereal proteins such as gliadin from wheat and zein from corn. Lysine is one of the amino acids which is nutritionally essential for rats, chicks and human beings. Its supplementary use for improving the adequacy of proteins of bread has been advocated and has some justification in the light of the fact that bread is often too prominent a constituent of the diet, especially of those less able to afford more expensive foods. This idea has added merit because, within a relatively small group of adult individuals, the range in needs for maintenance of nitrogen equilibrium has been found to be fourfold, *i.e.*, from 400–1600 mg/day. Some individual children (and adults) with high needs may be severely deficient if their protein source is of poor quality in this respect. (See AMINO ACIDS, NUTRITIONALLY ESSENTIAL.)

L-Lysine like many of the amino acids is dextrorotatory in acid solution. In 6*N* HCl its specific rotatory power $[\alpha]_D$ is $+25.9°$.

LYSOGENY. See BACTERIAL GENETICS, section on Infectious Heredity.

LYSOSOMES

Lysosomes are subcellular organelles which are believed to contain digestive enzymes capable of breaking down many of the cellular constituents. Disruption of the lysosomes and liberation of these enzymes may occur under certain conditions, and can lead to lysis of the cell (see also METAMORPHOSIS, AMPHIBIAN).

References

1. de Duve, C., "The Lysosome," *Sci. Am.*, **208**, 64–72 (May 1963).
2. "Symposium on Lysosomes," *Federation Proc.*, **23**, 1009–1052 (1964).

LYSOSTAPHIN

Lysostaphin is a bacteriolytic enzyme which is specific for members of the genus *Staphylococcus*. The organism which produces this enzyme, discovered by C. A. Schindler and this writer, proved to be a member of the genus *Staphylococcus* and was designated *S. staphylolyticus* n. sp. The staphylolytic enzyme is produced in trypticase-soy broth and similar media. It lyses viable and killed staphylococcal cells and isolated cell walls. The unit of lysostaphin has been designated as the amount required to reduce the optical density of a standard suspension of *S. aureus* FDA 209P 50% in 10 minutes at 37°C.

Lysostaphin has proved to be highly active *in vitro* against all of the more than 500 coagulase-positive isolates of *S. aureus* tested. It has proved inactive against all of 72 viable isolates of 53 species of 21 genera of bacteria other than staphylococci. Coagulase-negative staphylococci are attacked at a more variable rate than coagulase-positive isolates. Efforts are in progress to correlate this variable lysostaphin susceptibility of coagulase-negative staphylococci with their relative virulence.

Lysostaphin is a basic PROTEIN of approximately 30,000 molecular weight, an isoelectric point above 9 and a pH optimum of 7.5. It is relatively labile in solution, but is very stable in the lyophilized form. The rate of staphylococcal cell lysis is dependent upon temperature, pH, lysostaphin concentration and the ionic strength of the reaction medium. The usual reaction medium is 0.05*M* tris-(hydroxymethyl) aminomethane-HCl buffer at pH 7.4 containing 0.145*M* NaCl.

A highly purified (300-fold) preparation of lysostaphin has been reported by the Mead Johnson and Company research biochemists to be a peptidase capable of cleaving glycine and alanine peptide bonds of the glycopeptide components of the staphylococcal cell wall (see BACTERIAL CELL WALLS). They also reported that pentaglycine and other synthetic glycine peptides were hydrolyzed by this purified preparation.

Lysostaphin has been proved effective in therapeutic studies in mice infected with *S. aureus* (Smith). The intraperitoneal therapeutic dose with a partially purified lysostaphin was found to be 0.01 unit (14.55 μg protein/kg) and the therapeutic index was 3000. Additional animal and human therapy and carrier eradication studies are being conducted.

The unique lytic properties of lysostaphin against viable staphylococcal cells provide an antibiotic mechanism which differs strikingly from the mechanism of action of currently used antistaphylococcal ANTIBIOTICS. As such, lysostaphin offers the possibility of a novel approach to the chemotherapy and prophylaxis of staphylococcal disease. Also, because of the unique specificity of the enzyme for members of the genus *Staphylococcus*, lysostaphin should prove to be an additional tool in taxonomic and cell wall composition studies within this genus. Finally, lysostaphin should prove helpful to those who desire to study the intracellular components of the staphylococci without resorting to violent cell disintegration procedures to liberate these components.

V. T. SCHUHARDT

LYSOZYME

Lysozyme is a bacteriolytic enzyme, a glycosidase that hydrolyzes the linkages between acetylmuramic acid and acetylglucosamine (see BACTERIAL CELL WALLS). The binding site of lysozyme has been studied [see ENZYMES (ACTIVE CENTERS)].

References

1. ACKER, R. F., AND HARTSELL, S. E., "Fleming's Lysozyme," *Sci. Am.*, **202**, 132–142 (June 1960).

M

MAGNESIUM (IN BIOLOGICAL SYSTEMS)

The biological functions of magnesium, such as its essential role as a nutrient, its activation of enzyme systems and its pharmacological properties, have been widely investigated. Nevertheless, its critical physiological role remains obscure. Although magnesium plays an important role in plant physiology as an enzyme activator [see ENZYMES (ACTIVATION AND INHIBITION)] and as a constituent of CHLOROPHYLL, this discussion will be restricted to its role in animal systems.

Distribution. Magnesium, primarily an intracellular ion, is distributed among all tissues. It constitutes about 0.05% of the animal body and, of this, 60% occurs in the skeleton and only 1% in the extracellular fluids.

The reported serum magnesium values for most species range from 1.0–3.5 meq/liter with a mean value of about 2. Between 65 and 80% of the plasma magnesium is ultrafiltrable, and most of this exists as the free ion. The nonfilterable portion is reversibly bound to plasma protein. CEREBROSPINAL FLUID contains slightly more than plasma and interstitial fluid is similar to plasma ultrafiltrate.

The magnesium content of soft tissues varies from 0.06–0.13% of dry weight and remains remarkably constant regardless of the magnesium status of the animal. Normally the intracellular concentration is more than 20 times that of the interstitial fluid and the highest concentration occurs in the cell nucleus. Maintenance of such a large concentration gradient across the cell membrane suggests an ACTIVE TRANSPORT mechanism, but unequivocal evidence for such a mechanism is lacking.

The relatively large proportion of magnesium found in the skeleton, which amounts to about 0.6% of dry fat-free bone, serves in part as a body reserve. It occurs largely as Mg^{++} and $MgOH^{+}$ ions held by electrostatic attraction to the apatite crystal surface. During deficiency in young animals, 30% or more of bone magnesium can be mobilized for metabolic functions. Calcium ions appear to replace the magnesium which occupied the original adsorption sites.

Metabolism. The rate of absorption from the intestine exerts an important role in magnesium metabolism, but little is known about the mechanism involved. Whereas *in vitro* studies show that magnesium absorption is positively correlated with the concentration of magnesium, it does not appear to be a purely passive process. Magnesium absorbed in excess of body needs is excreted primarily by way of the kidney. In guinea pigs fed purified diets, approximately 90% of the excreted magnesium is voided in the urine. Urinary excretion is controlled primarily by a filtration-reabsorption mechanism so that magnesium appears in the urine only when GLOMERULAR FILTRATION exceeds tubular reabsorption (see RENAL TUBULAR FUNCTION). However, at least part of the urinary magnesium is secreted directly into the distal tubules. Acute renal failure is accompanied by hypermagnesemia. In some species considerable endogenous magnesium is lost by way of the feces, the amount depending upon the magnesium status of the animal and upon other dietary factors such as the digestibility of the diet. The endogenous fecal magnesium in calves has been estimated at 3.5 mg/kg body weight.

In contrast to the metabolism of calcium, no one endocrine gland exerts a primary regulatory function on magnesium. Thyro-parathyroidectomy in dogs causes only a temporary lowering of plasma magnesium. Adrenalectomy causes a rise, whereas hyper-aldosteronism produces a fall, in the plasma level. Administration of deoxycorticosterone or aldosterone to sheep lowers the magnesium concentration in plasma. Magnesium deficient animals exhibit a higher metabolic rate than normal, and the toxic effect of excess thyroxine is partially overcome by increasing the dietary level of magnesium. The possible relation of these observations to *in vitro* studies of oxidative phosphorylation will be discussed below.

Function. Although magnesium activates isolated enzymes, in most cases an absolute requirement has not been shown because the enzymes are partially active without added magnesium. The stimulating effect is not always specific for magnesium for in some cases MANGANESE or CALCIUM ions also activate the system. There is evidence, although conflicting, that oxidative PHOSPHORYLATION is uncoupled in mitochondria isolated from magnesium-deficient animals as well as in thyrotoxic animals. These effects were overcome by *in vitro* addition of magnesium or by feeding magnesium. If these observations are fully substantiated by future research, it will be one of the rare instances in which an *in vitro* effect of magnesium has an *in vivo* counterpart.

In spite of the inability to correlate physiological function with *in vitro* observations, magnesium must exert a significant role as an intracellular catalyst. It is particularly concerned with enzyme-catalyzed reactions involving the cleavage of phosphate esters and the transfer of phosphate groups. Magnesium ions activate PHOSPHATASES and the phosphorylation reactions involving adenosine triphosphate (ATP). Among the latter group may be mentioned glucokinase, phosphoglucokinase, phosphofructokinase, myokinase, creatine transphosphorylase, arginine transphosphorylase and flavokinase. It has been suggested that an ATP-Mg complex is the active substrate in as much as ATP forms a 1:1 complex with magnesium and maximal activation occurs when the ATP:Mg ratio is 1. Alkaline phosphatases, pyrophosphatases and ATPase are activated by magnesium, as are enolase, certain peptidases and pyruvic oxidase. Since magnesium is tied to ATP utilization, it must be involved in many important metabolic processes such as synthesis of protein, fat and nucleic acids and in the trapping and utilization of energy derived from catabolism of carbohydrate and fat.

Notwithstanding the importance of magnesium in intracellular metabolism, one must look elsewhere for its critical physiological function because there is little change in the magnesium concentration of soft tissues from deficient animals even at the point of death. This fact does not preclude the possibility that a small component of the cell, such as the nucleus or a cell particulate, is deprived of its critical level, but the dramatic drop in extracellular magnesium suggests that a function outside the cell is of greatest significance. It is likely that tetany and convulsions in deficient animals result from a derangement of neuromuscular transmission. Magnesium ion possesses strong pharmacological properties depressing both the central and peripheral nervous systems. These effects are counteracted by CALCIUM. In the presence of normal calcium levels a reduction of extracellular magnesium is believed to increase the release of ACETYLCHOLINE and to decrease the rate of its hydrolysis. Such effects would increase the irritability of the neuromuscular system.

Pathology of Magnesium Deficiency. Although there are numerous clinical symptoms, two cardinal aspects of pathology have been observed in all species of higher animals. These are hyperirritability and soft tissue calcification. While there are species differences as to the dominating syndrome, this determined in part by the severity of the deficiency. Metastatic calcification is more likely to occur in a chronic deficiency in which the animal does not succumb at an early age. Hyperirritability, terminating in convulsions and death, has been observed in the rat, rabbit, pig, calf, chick and duck. Magnesium deficiency in man is characterized by muscle tremors and twitching, often accompanied by delirium and occasionally by convulsions. The guinea pig, calf, dog and cotton rat are prone to metastatic calcification and develop grossly visible deposits in and around joints, along the muscles of the rib cage, and even in the heart, great vessels and other critical organs. Most soft tissues show an elevated ash content and marked histopathology.

The first clinical symptom of magnesium deficiency is a decrease in serum magnesium and in some species this is accompanied by an elevation of inorganic phosphorus. The rat develops a severe hyperemia followed by skin lesions. Muscular dysfunction is a pathological symptom common to many species.

The disease of greatest economic importance is a hypomagnesemia which occurs in cattle and less frequently in sheep and is described by such names as grass tetany, grass staggers, lactation tetany and wheat pasture poisoning. It is observed most frequently when animals are first grazed on lush grass or wheat pastures. The disease is characterized by irritability, tetany and convulsions, and all animals have a subnormal plasma magnesium. Symptoms can be relieved by administration of magnesium salts and can be prevented by providing extra magnesium in the diet. Although the exact cause of the disease is unknown, it seems clear that the magnesium intake is not sufficient to meet physiological needs under the conditions imposed.

Nutritional Requirement. As is true of many mineral nutrients, the requirement for magnesium is affected by other dietary constituents, by the age and species of the animal, and by the criterion of adequacy applied. Tabulated below are reported requirements of various species of young animals fed diets of near normal composition. The

Species	Requirement (mg/100 g diet)
Rat	25
Rabbit	30–40
Guinea pig	80
Pig	40–50
Calf	108
Chick	35
Duck	50

requirement of man has not been determined with precision, but it is believed that the adult requirement for positive balance is about 300 mg/day. The requirement has been reported to be affected by the type of carbohydrate in the diet and the dietary concentration of calcium, phosphorus, potassium, phytate and protein. Of these, calcium and phosphate have by far the most important effect upon magnesium availability (see PHOSPHATE IN METABOLISM). Either of these ions in excess increases the requirement for magnesium, and their effects are additive. Since calcium is known to compete with magnesium pharmacologically, it is reasonable to believe that it also competes with magnesium for absorption sites in the intestine. Probably phosphate decreases magnesium absorption by formation of insoluble magnesium phosphates and excess calcium aggravates the effect by creating a more alkaline intestinal medium. Excess magnesium is toxic, but this effect is largely due to the induction of a calcium deficiency.

References

1. Davis, G. K., in "Nutrition" (Beaton, G. H., and McHenry, E. W., Editors), Vol. I, p. 463, New York, Academic Press, 1964.
2. O'Dell, B. L., *Federation Proc.*, **19**, 648 (1960).
3. Rook, J. A. F., and Storry, J. E., *Nutr. Abstr. Rev.*, **32**, 1055 (1962).
4. Wacker, W. E. C., and Vallee, B. L., *New England J. Med.*, **259**, 431, 475 (1958).
5. Wilson, A. A., *Vet. Rev. Annotations*, **6**, 39 (1960).

B. L. O'Dell

MALIC ACID

Malic acid, HOOC—CH_2—CHOH—COOH (hydroxysuccinic acid), got its name because of its occurrence in unripe apples (Latin *malum*, apple). The L-form is an intermediate in CARBOHYDRATE METABOLISM and, in the CITRIC ACID CYCLE, is convertible into oxaloacetic acid in one direction, and fumaric acid in the other. The rotatory power of malic acid changes from slightly positive to negative when a concentrated solution is diluted.

MALNUTRITION

Malnutrition is a condition characterized by inappropriate quantity, quality, digestion, absorption or utilization of ingested nutrients. Because the specific nutrient requirements of animals are complex, the possibilities for malnutrition are manifold. A nutritious diet for humans must include in adequate amounts:

(1) Water.

(2) A complex mixture of organic compounds. Most of this mixture will consist of "nonessential" substances whose oxidation by the tissues will provide energy as well as starting materials from which the tissues can synthesize other organic components necessary for their structure and function (*i.e.*, fats, carbohydrates, proteins). The organic mixture must also include specific essential compounds which cannot be synthesized by the tissues from other organic compounds (*i.e.*, essential amino acids, essential fatty acids, vitamins).

(3) Inorganic salts needed for osmotic regulation, for catalysis in enzymatic reactions, for incorporation into specific organic compounds, and for the maintenance of a bony skeleton.

If one or more of the approximately 50 essential components of such a diet are not presented in appropriate amounts to the tissues of an individual, that individual is malnourished. Malnourishment at the tissue level is most often attributable to consumption of an inadequate diet. Malnourishment is annually responsible directly or indirectly for thousands of deaths and for incalculable losses in human productivity and well-being. Most vulnerable to the occurrence of malnourishment are groups which must depend on others to supply them with an adequate diet (such as infants and children), those with increased nutrient requirements (growing children, pregnant or lactating women), those unable, because of poverty, to obtain an adequate diet, and those who for one reason or another unnecessarily restrict or distort their dietary intake (alcoholics, food faddists, drug addicts, the aged). Within the latter category, one may include the large numbers of individuals who through ignorance or indifference make no effort to select foods which will assure adequate nutrition. No great effort or detailed knowledge of nutritive values is required. If one consciously selects a *varied* diet (both animal and vegetable foods; both muscle and organ meats; not only leafy vegetables but also roots and seeds) and can allow common sense and the bathroom scales to dictate the amount of food consumed, it is very probable that one will be well nourished.

Types of Malnutrition. *Undernourishment*. Undernourishment results from the consumption of inadequate amounts of food. This type of dietary malnutrition is to be seen in every country of the world today. Its most extreme form—starvation—results from severe and prolonged inadequacy of intake of most or all of the required nutrients. Of three ancient scourges of mankind—war, famine, and pestilence—famine has been the most unrelenting in its recurrence. As recently as 1943, a single famine in Bengal killed 1,500,000 people. However, humans dying from starvation represent only a small fraction of the much vaster reservoir of undernourished individuals whose insufficient diets condemn them to chronic hunger, stunting of growth and development, subnormal mental and physical vigor and endurance, and decreased resistance to disease and other stresses. It is estimated that 10–15% of the people of the world today are undernourished.

Specific Dietary Deficiencies. It is quite possible to suffer from malnourishment while consuming a diet which is quantitatively unrestricted. Deficiency in the diet of one or more specific nutrients results in specific deficiency conditions. For instance a high-energy expenditure relative to CALORIC INTAKE will result in caloric insufficiency. Fats, carbohydrates and proteins can all be oxidized by the tissues to provide usable energy (with yields of approximately 9, 4 and 4 kcal/g, respectively). An adult human expends a minimum of about 1300 kcal/day. Any physical activity increases this amount. Because life can be sustained only by virtue of this continued expenditure of energy, the dietary requirement for calories is in a sense the most urgent of all nutrient requirements. Not only external and internal physical work but also the complex of chemical reactions by which the function and structure of the living organism are maintained exact an energetic toll. The very utilization of the other dietary components is an energy-requiring process. If dietary calories are deficient, the tissues are forced to oxidize for energy other specifically required nutrients and/or internal fat, carbohydrate and protein. Muscle and enzymic protein and other essential tissue constituents will be oxidized, if need be, in order that the daily required caloric expenditure may be maintained. With continued imbalance of caloric intake and expenditure, and with eventual depletion of expendable tissue stores, the level of activity of the organism must decrease until the two are brought into balance or death ensues.

Dietary deficiencies of essential amino acids or of total amino acid nitrogen can occur if an inadequate amount of complete protein or an inadequate mixture of incomplete proteins is eaten [see PROTEINS (NUTRITIONAL QUALITY)]. Of approximately 20 amino acids of which proteins are composed, 8 cannot be synthesized by human tissues and must be contained in the proteins of an adequate diet along with enough "nonessential" amino acids to provide starting material for synthesis and interconversions of the other amino acids by the tissues. Because protein-rich foods are often the most expensive, dietary protein and/or amino acid deficiency is not uncommon, especially among young children in undernourished areas of the world.

Polyunsaturated fatty acids can also be lacking in an otherwise adequate diet resulting, in humans, in a deficiency state which is as yet incompletely defined. Although most tissue lipids can be synthesized from non-lipid dietary components, this is not true of linoleic acid or of the more unsaturated fatty acids for which linoleic acid can serve as a precursor.

Specific dietary vitamin deficiencies cause such diseases as scurvy (vitamin C, see ASCORBIC ACID), pellagra (niacin or NICOTINIC ACID), and rickets (VITAMIN D). Despite the relative abundance of the food supply in the United States, reported deaths due to the above three diseases in 1956 were 7, 70 and 6, respectively. Serious deficiency diseases are also observed in man if the diet is inadequate in other vitamins including VITAMINS A, D, B_6, RIBOFLAVIN and THIAMIN.

Mineral deficiencies can result from inadequate consumption of such inorganic elements as calcium, phosphorus, potassium, sodium, magnesium, iron, zinc, copper, cobalt, chloride, iodine, and probably several others (see MINERAL METABOLISM). Because many of these elements are required in only trace amounts and all are widely distributed in foods, the effects of dietary deficiencies of most of them are observed relatively infrequently in humans and are not clearly defined. CALCIUM, IRON and IODINE represent exceptions to this statement.

Overnutrition. This form of dietary malnutrition is the result of excessive intake of one or more dietary components. For instance, a continued excess of dietary calories results in obesity. The capacity of the body to store excess calories as fat is immense. Normally fat comprises about 12% of the body, but in cases of prolonged caloric excess, the weight of stored fat can far exceed the weight of all of the rest of the body. However, such an overloading of the tissues with fat is accomplished only at the expense of increased demands on the heart and other organs, decreased physical facility, decreased life expectancy and probable increased susceptibility to heart disease.

Dietary excess of some of the vitamins can result in toxic hypervitaminosis. Those vitamins which can be stored in the tissues in large amount—notably vitamins A and D—are potentially dangerous when taken injudiciously in concentrated form as dietary supplements. Excessive intakes of niacin also produce toxic effects. Large dietary amounts of most of the other vitamins are rapidly excreted or metabolized and are not believed to pose toxicity problems.

Secondary Malnutrition. The types of malnutrition discussed above can occur in an otherwise healthy organism. However, the nutritive process is vulnerable at almost any point to disruption by disease. Food allergies, gastric ulcers, anorexia from any of a number of diseases, and other abnormalities can cause inadequate dietary intake. Deficient presentation of required nutrients to the tissues can also be caused by failure of digestive processes. The dietary macromolecules such as STARCHES and PROTEINS must be hydrolyzed to simpler molecules before they can be absorbed from the intestinal tract. The enzymatic reactions by which this DIGESTION is carried out can be depressed by disease-induced decreases in digestive enzyme production or by a too rapid transit of the nutrients through the intestine. Various other abnormalities result in absorptive failures. Specific diseases characterized by failures of intestinal absorption include tropical sprue, pernicious ANEMIA and primary malabsorptive disease. Still other diseases (*i.e.*, diabetes, B_6 dependency) cause impaired utilization or increased requirements of nutrients at the tissue level.

Biochemical Basis for Nutritive Requirements. In each type of malnutrition, the primary result is that the tissues do not have available the chemicals which they require in order to maintain their normal structure and functions. Ultimately, both the necessity for a dietary supply of any given nutrient and the physical consequences of its deficiency will be explainable in terms of its structural function or the individual chemical reactions in which that nutrient participates. To a large extent, these reactions and their obligatory participants are the same or analogous among the many forms of life. In any species, most of the metabolic reactants and products are synthesized by the tissues from other organic compounds. However, the precursors which the organism is able to use for the synthesis of a given substance vary from species to species with the result that different life forms have different nutritive requirements. Thus, because all known forms of life contain protein which in turn contains nitrogen, nitrogen in some form is an essential nutrient for all forms of life. Different species differ however in the complexity of the required source of nitrogen. The NITROGEN-FIXING bacteria can use elemental atmospheric nitrogen. Most plants do not use elemental nitrogen directly for protein synthesis but must instead be supplied with oxidized inorganic nitrogen from the soil or from symbiotic bacteria. Animals in turn generally require a still more complex form of nitrogen: AMINO ACIDS which they must obtain directly or indirectly from plants or bacteria. Even among higher animal species, nutritive amino acid requirements vary: glycine is a specific dietary requirement for the chick but not for man, who can synthesize glycine from other amino acids; adult ruminant animals do not require preformed dietary amino acids if amino acid synthesis by rumen bacteria is adequately supported by other forms of dietary nitrogen.

Most organisms, presumably, require ascorbic acid, but most can synthesize this vitamin. Man is one of the few species which cannot; therefore, for man, ascorbic acid is not only a biochemical necessity but also a dietary one.

Different life forms also exhibit major quantitative differences in nutritive requirements due not merely to differences in size but also to differences in physical characteristics. Thus the required proportion of calcium in the diet will be of a different order of magnitude for those organisms which have a bony skeleton than for those which do not. The relative requirement for cystine is higher in animals with heavy hair cover than for man. In sheep, wool growth accounts for fully one-third of the protein requirement.

For further discussion of nutritional requirements of various organisms, see the separate articles on NUTRITIONAL REQUIREMENTS (AUTOTROPHIC PLANTS); NUTRITIONAL REQUIREMENTS (INVERTEBRATES); NUTRITIONAL REQUIREMENTS (MAMMALS); NUTRITIONAL REQUIREMENTS (VERTEBRATES OTHER THAN MAMMALS); see also GERM-FREE ANIMALS; INTESTINAL FLORA; RUMINANT NUTRITION; TISSUE CULTURE, ANIMAL; TISSUE CULTURE, PLANT.

Conclusion. Only when much more information is available will it be possible to specify in detail and with certainty the exact composition of a diet which will preclude malnutrition in any given individual at any given time. Nevertheless, the application of presently available knowledge would cure the vast majority of cases of malnutrition occurring today. It is clear that progress toward the goal of eradication of human malnutrition must depend heavily on progress in the solution of agricultural, economic and educational problems as well as on the acquisition of new nutritional knowledge.

References

1. DAVIDSON, AND PASSMORE, "Human Nutrition and Dietetics," E. and S. Livingstone Ltd., 1963.
2. JOLLIFFE, N. (EDITOR), "Clinical Nutrition," New York, Harper and Brothers, 1962.
3. MITCHELL, H. H., "Comparative Nutrition of Man and Domestic Animals," New York, Academic Press, 1962.
4. WOHL, M. G., AND GOODHART, R. S., "Modern Nutrition in Health and Disease," Lea and Febiger, 1964.

LUCILE F. ADAMSON

MALTOSE

Maltose, the β-form of which is represented below,

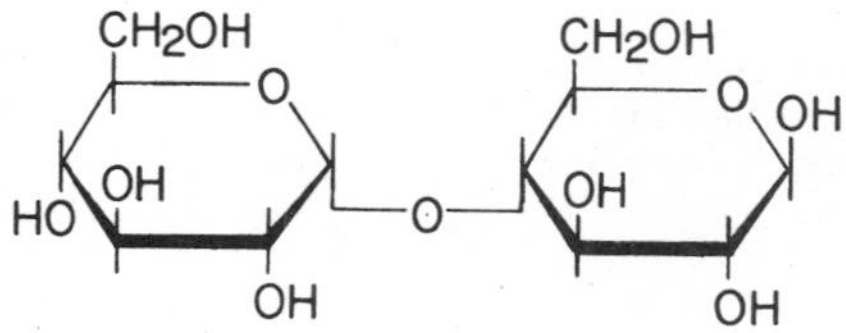

is a glucose glucoside which may be termed 4-(α-D-glucopyranosyl)-D-glucopyranose. It occurs in malted barley (from which it gets its name) and is produced from STARCH by AMYLASES of salivary, pancreatic and plant origins, and also by acid hydrolysis. Its two forms (α and β) are in equilibrium with each other and with a minute amount of the aldehyde form. It possesses all the properties of a reducing sugar and exhibits mutarotation. The linkage between the two glucose residues is of the alpha type so maltose is split under the influence of α-glucosidases and is readily fermented by yeast. It is unaffected by β-glucosidases. Corn syrups which are produced by partial hydrolysis of starch by dilute acids contain appreciable amounts of maltose and other partial hydrolytic products, as well as the principal ingredient glucose. See also CARBOHYDRATES (CLASSIFICATION).

MANGANESE (IN BIOLOGICAL SYSTEMS)

Manganese, the twenty-fifth element of the periodic table, is one of the *essential trace metals*. It occurs in minute concentrations (microgram per gram) within the cells of all living things. Its essentiality has been established for a wide variety of organisms ranging from bacteria through plants to mammals. The evidence for its essentiality rests extensively on the consequences of limiting or curtailing the supply of this metal to various organisms. *Manganese deficiency* has induced in all organisms studied a diminished growth potential and a diminished life expectancy. Additional manifestations depend upon the kind of organism under observation, its age, the degree and duration of manganese deficiency, as well as the coexistence of another deficiency. In *plants*, for example, a striking manifestation is chlorosis in which the leaves become pale or yellow while the veins of the leaves remain green. Manganese plays a significant role in photosynthesis by plants, but it also participates in the regulation of several other enzymic processes. In *chickens*, manganese deficiency causes a different clinical picture when it affects the egg than when it affects the hatched bird; in the first instance, the embrya become swollen and deformed, and their skeletons become defective and fragile ("chondrodystrophy"). Adult birds develop "perosis," a condition in which deformation of bones is so extreme that tendons will slip from their normal position.

The bony deformities seen in the chicken can be induced also in *mammals*. If the deficiency is induced prior to birth, there is a high intra-uterine mortality and whatever young are born alive tend to suffer from an inability to coordinate their muscles (ataxia). These young also have convulsions, delayed growth and defective bone formation. Adequate manganese intake after birth will correct many of these anomalies but not the ataxia. If the deficiency is imposed on adult female mammals, ataxia develops infrequently if at all. Instead there appear anemia, defective bone formation, infertility, a tendency to miscarry, and a tendency to absorb the embrya which die within the uterus. The sickly offspring are jeopardized after birth by a disinterest on the part of the manganese deficient mothers. These avoid nursing their young even when they produce adequate milk. In males, in addition to poor growth, bone deformities and anemia, impotence and infertility develop. Adult animals also develop defects in metabolizing body fat, which are reflected in abnormal amounts and abnormal distribution of body fat. This lipotropic effect of manganese extends also to the metabolism of cholesterol: manganese promotes the synthesis of cholesterol (see STEROL BIOSYNTHESIS) and in this particular role it can be antagonized by VANADIUM. The bone deformities are ascribed primarily to poor synthesis of the MUCOPOLYSACCHARIDES which make up the matrix of the bones. The infertility is a consequence of death of the testicle's germ cells.

All these effects are explained on the basis of the fact that this metal activates various cellular enzymes. A great deal of importance is placed on the particular enzyme systems responsible for oxidative PHOSPHORYLATION. These enzyme systems are indeed important because they determine the generation and utilization of energy from foodstuffs by the cells. Still, manganese activates many other enzymes as well (*i.e.*, arginase, enolase, PEROXIDASES, etc.). Furthermore, it seems to participate in the structure of the nucleic acids which are responsible for the manufacture of enzymes and other proteins. Hence, it is not possible at the present to indicate a single biochemical role for this metal. Whether this metal has one or many biochemical roles, its function cannot be taken over by other metals, and it does not seem possible to replace manganese with other metals in the cells of living animals.

In spite of the small amounts of manganese in mammalian tissues, its concentration seems to be accurately controlled by elaborate mechanisms. As far as has been tested, these mechanisms function primarily by promoting the excretion of excesses of metal from the body rather than by regulating the amounts of metal the body absorbs. These mechanisms are highly developed in *humans*, in whom dietary manganese deficiency has not been described as yet. They are located in the liver and on the mucosa of the gut. Still, in *chronic manganese poisoning* these mechanisms seem to become saturated. This disease afflicts chiefly miners working either in manganese mines or in ore crushing mills. The manganese ore enters the body by inhalation of the dust in the air. Among the many miners exposed throughout the world, some develop brain symptoms. Involvement of the brain manifests itself first in mental aberrations (crying, laughter, sexual anomalies, unprovoked attacks and even homicide). Later, neurological changes occur in the form of trembling, rigidity, salivation, mask-like face, and a general appearance of the patient similar to Parkinson's disease. Chronic manganese poisoning occurs in epidemics and prevails throughout the world, but it has been prevented from occurring in the United States. It is incurable, but not necessarily life-limiting.

From the foregoing it must be apparent that manganese is of interest not only to the biologist, the biochemist and the physiologist, but to the nutritionist, the toxicologist and the industrial physician as well. Furthermore, both its deficiency and its toxicity mimic some spontaneous human diseases whose causes are not known. These considerations have recently provoked the development of new methods applicable to its study. Its small concentrations in tissues necessitate sensitive analytical methods. Therefore, in intact humans and animals, the metabolism of manganese is being studied by means of *artificial radioisotopes*. The quantities of manganese in samples of blood, serum, spinal fluid, etc., are being analyzed by *neutron activation analysis* which utilizes the ability of nuclear reactors to make various elements (including manganese) intensely radioactive. After the natural, nonradioactive metal has become radioactive it can be quantified by measuring the radioactivity it emits. The state of this element in various tissues is being studied with magnetic techniques (nuclear magnetic resonance, electron spin resonance) which reveal the manner in which this metal becomes attached to various tissue receptors.

This work was supported by the U.S. Atomic Energy Commission.

References

1. COTZIAS, G. C., "Manganese" in "Mineral Metabolism: An Advanced Treatise," Vol. 2b (COMAR, C. L., AND BRONNER, F., EDITORS), p. 403, New York, Academic Press, 1962.
2. COTZIAS, G. C., "Trace Metals: Essential or Detrimental to Life," *Brookhaven Lecture Series Number* 26 (April 10, 1963).

GEORGE C. COTZIAS

MANNITOL

D-Mannitol is a hexahydric alcohol which may be derived by reduction of mannose or fructose. It has the following structure:

$$\begin{array}{c} CH_2OH \\ | \\ HO{-}C{-}H \\ | \\ HO{-}C{-}H \\ | \\ H{-}C{-}OH \\ | \\ H{-}C{-}OH \\ | \\ CH_2OH \end{array}$$

[see CARBOHYDRATES (CLASSIFICATION AND STRUCTURAL INTERRELATIONS)]. D-Mannitol occurs naturally in brown algae and in fungi, and is a chief constituent of manna, a dried exudate of *Fraxinus ornus*. D-Sorbitol is the corresponding alcohol derived by reduction of glucose; it occurs in many berries and fruits.

MANOMETRIC METHODS

Many reactions involve a gas or a product or reactant which can be converted into a gas. These reactions can be measured by use of a series of methods called "manometric" since somewhere in the system the pressure is either measured or maintained by way of a manometer. These methods are capable of measuring both the rate and the extent of suitable reactions in a convenient and accurate manner. Their chief disadvantage is that they frequently require larger amounts of expensive substances than do spectrophotometric methods.

These are a group of methods suitable for measuring the rate or the extent of a reaction, providing the reaction involves a gas or involves a reactant or a product convertible into a gas. The methods are based upon the relation, $PV = RT$. For the same amount of gas, $PV/T = P'V'/T'$, but where gas has been used or generated, V' differs from V. If therefore, two of the three variables are held constant, the change in the third is a measure of the change in the gas. In practice, temperature is held constant and either pressure is held constant and change in volume measured, or volume is held constant and change in pressure measured, or both are allowed to vary and both changes are measured. Attention here will be confined to systems in which the volume is held constant and change in pressure is measured and to those in which pressure is held constant and change in volume is measured.

The Constant-volume System; The Warburg Method. In this system the volume is held constant and the change in pressure is measured. The apparatus for this method consists of a flask (detachable), sometimes equipped with one or more sidearms, attached to a manometer containing a liquid of known density. The flask is immersed in a water bath at a constant temperature, and between readings the system is shaken or whirled to promote a rapid gas exchange between the fluid and the gas phase. It is assumed that the temperature of the manometer, which is not immersed, does not differ greatly from that of the flask. The manometer is graduated in millimeters, and has one end open to the atmosphere and the other attached to the flask with the provision of a stopcock so that the flask with manometer may be connected to the external atmosphere or, with the stopcock closed, will constitute a closed system. A given point on the closed (*i.e.*, flask) side of the manometer is chosen, and with the stopcock open to the atmosphere, the fluid in the manometer is adjusted to this fixed point by means of a well, regulated by a screw clamp, at the bottom of the manometer.

When the stopcock is closed and the fluid in the closed arm of the manometer is at a fixed point (*e.g.*, 250 mm), the space within the flask and the manometer to this fixed point is a closed system. If a gas is consumed, by a reaction in the flask, the pressure will be reduced, the fluid will rise in the closed arm and drop in the open arm. (If a gas is generated, the fluid will drop in the closed arm and rise in the open arm.) If one readjusts the fluid in the closed arm to its fixed point (*e.g.*, 250 mm) by means of manipulating the fluid reservoir well, the volume is now the same as it was before (*i.e.*, constant volume), but the pressure has now decreased due to the utilization of gas in the flask. The amount that the pressure has decreased is readily determined by subtracting the reading in the open arm after the gas has been used from the initial reading in the open arm. This decrease in pressure ($-h$; an increase would be $+h$) is related to $-x$, the amount of gas used (or $+x$, produced) by $-x = k(-h)$ where k (a "flask constant") is:

$$k = \frac{V_g \dfrac{273}{T} + V_f \alpha}{P_0}$$

where: V_g = volume of gas in the flask and manometer to the zero point on the manometer,
T = absolute temperature of system,
V_f = volume of fluid,
α = solubility of the gas involved in the fluid used, milliliters of gas dissolved per milliliter of fluid,
P_0 = standard pressure, expressed in terms of the millimeters of manometer fluid corresponding to 760 mm mercury, *i.e.*,

$$P_0 = \frac{760 \times \text{Specific Gravity of Mercury } (=13.60)}{\text{Specific Gravity of the Manometer Fluid}}$$

This is usually chosen so that P_0 comes out to be a whole number. If one uses a salt solution of specific gravity of 1.03, $P_0 = 10{,}000$. Such a solution, containing 50 grams of NaCl and 10 grams of sodium choleate per liter, usually colored with a dye, is called Brodie's solution.

By thus measuring the change in pressure over an interval of time in a system in a known (and constant) volume of both fluid and gas, with a manometer fluid of known density, one may easily calculate the gas used up or generated within the system. Measurement of such a pressure change is very simple, and involves merely recording the reading on the open arm of the manometer at the beginning and at the end of the interval, the level of fluid in the closed arm being adjusted to the same point, and correcting this reading for any change in pressure in the room. For this latter purpose, a flask and manometer system similar to the measurement system (called a thermobarometer), but which does not carry out the reaction, is measured simultaneously. Suppose that the initial reading on an open arm of the manometer was 250 mm in the measurement flask, and after 10 minutes of reaction, it was 227 mm, *i.e.* the pressure has decreased by 23 mm. The simultaneous measurements on the thermobarometer were 250–248 mm, *i.e.*, the pressure in the room had decreased by 2 mm. The decreased pressure due to the reaction in the flask was thus $23 - 2 = 21$ mm, and this times the constant (k) for the system employed will yield the gas uptake in microliters (0.001 ml).

The flask constant contains a factor α, the solubility of the gas in the fluid. Each gas differs in solubility; hence, one must know which gas one is measuring, and under most circumstances, one is able to measure the change in only one gas—not two or three simultaneously.

This type of system, however, is most convenient for measuring oxygen uptake, *i.e.*, respiration, and the process is sometimes called *respirometry*. Of course, in the case of most respiration, CO_2 is liberated simultaneously. If these two gases (CO_2 and O_2) are the only ones involved, one can measure the respiration (O_2 uptake) by absorbing the liberated carbon dioxide in alkali. In the presence of alkali the carbon dioxide pressure in the air is zero within the limits of measurement. The gas exchange caused by the respiration is oxygen absorption plus carbon dioxide liberation. But the alkali keeps the carbon dioxide pressure zero, hence the change noted on the barometer is due solely to the oxygen utilization. The excess of carbon dioxide in solution, of course, continually distills over into the alkali, but it does not affect the observed pressure changes.

The absorption of oxygen by the respiring tissue takes place almost entirely from the oxygen in solution. This is the principal reason for shaking the fluids in the respirometer, *i.e.*, to obtain a fluid phase saturated with the gas phase. But one must, under practical circumstances, take care that the rate of oxygen uptake by the tissue is not greater than can be replaced by the diffusion of oxygen from the atmosphere into the fluid. If the rate of oxygen uptake is so high that the oxygen is used up faster than it can diffuse into fluid, then the rate of respiration observed is dependent upon the rate of which the oxygen diffuses into the fluid and has little to do with the potential rate of the reaction itself.

The rate at which gas diffuses into a liquid is dependent upon the surface layer of the liquid. However, for virtually all respiratory measurements it is sufficient to note that by shaking the flasks a continual new surface is exposed to the gas by virtue of the turbulence of the fluid in the flask. Hence, the greater the rate of shaking, the greater the rate of diffusion of the gas into the liquid, and the greater the rate of respiration one may measure without diffusion errors.

In absorbing CO_2 from the gas phase, the same difficulties are encountered as in the absorption of oxygen. Here, however, because alkali is usually confined to the small center cup, an increased rate of shaking has little effect on increasing the surface. Hence some other method must be employed to increase the surface of the alkali. Usually small rolls or accordion-folded pieces of filter paper are placed in the alkali cup. These should project beyond the side walls of the center cup into the open gas space above. A desirable projection is about 5 mm. Such "KOH papers" are usually prepared in quantity by cutting filter paper into squares with 2-cm sides (the exact dimensions will vary with the depth of the cup employed; this varies from instrument to instrument, but the size need be only approximate). These papers are then folded three or four times, accordion fashion, and when inserted into the center cup, they provide a large surface for the absorption of CO_2.

It is possible, using the constant-volume system (the "Warburg") to make two types of measurements: one in which the *rate* of reaction is determined, the other in which the *amount* of the reaction is determined. For the latter purpose, one usually employs flasks with sidearms. A known quantity of the material to be studied is placed in the sidearm. After equilibration, the rate of oxygen uptake is determined in the absence of substrate (to be certain that it is constant), the material is then tipped in, and the oxygen uptake determined until it again reaches base rate.

It will be noted that from this type of experiment it is possible to obtain both the amount of oxygen used per mole of added substrate and the rate of oxygen uptake. The rate obtained under these conditions may not be, however, the maximum rate possible, since in order to measure the oxygen taken up in a reasonable length of time one may find it necessary to add materials in quantities insufficient to saturate the reaction systems. Normally the reaction systems are considered saturated if one obtains a straight line function with time, but occasionally instances may be found in which the rate of oxygen uptake (or other functions of metabolism) may proceed in a linear manner, yet higher levels of the added materials will increase the rate.

In determining the amount of oxygen taken up per unit of substrate, especially during the metabolism of living tissues, it is frequently a problem to decide whether one should subtract from the oxygen uptake observed in the presence of substrate, the oxygen taken up over the same interval in the absence of substrate. That is, when a substrate is being oxidized at a rapid rate, does the endogenous respiration continue at its constant rate, is it suppressed, or does it increase?

These questions have not yet been answered. Undoubtedly, the response depends upon the tissue involved, and no generalizations can be made. However, it is always good practice to determine the endogenous respiration and to report it, along with the oxidation in the presence of substrate, and to indicate whether or not the endogenous respiration was subtracted from the substrate respiration in calculating the oxygen consumption per mole of substrate.

It is frequently convenient to express the amount of substrate employed in terms of gas produced or absorbed. Since 1 mole of any gas (at standard conditions) occupies 22.4 liters, it is possible to consider any substances in terms of liters with each mole equivalent to 22.4 liters. One may thus speak of adding 11.2 μl of glucose, which means that one has added 0.5 ml of 0.001M glucose solution or 5×10^{-7} mole of glucose.

In this system, in order to calculate the constant, one must know the volume of the system precisely. Since flasks and manometers vary in size, a given combination must be accurately standardized so that the volume of the entire system, including that closed part of the manometer which is filled with gas (to the fixed point), is accurately known. There are several methods for accomplishing this described in the more detailed technical manuals given in the references.

The Constant-pressure System. In essence, this system can use the Warburg apparatus as described earlier, but frequently a more conveniently designed apparatus is used. We shall describe it in terms of the Warburg system ignoring the thermobarometer corrections for the moment. As gas is taken up in the closed system and the fluid in the closed arm rises (because pressure is decreased), a steel or plastic rod can be inserted into the closed capillary of the manometer to just replace the volume of gas that has been used up; thus the pressure will return to its original position. If the inserted rod is of known diameter and attached to a micrometer, the volume of gas taken up is precisely measured. In fact, digital reading micrometers can be used so that they read directly in microliters. Further, small changes in design permit attaching the open arm of the manometer to a large flask (and "open" ends of several manometers may be attached to the same flask) so that changes in room pressure do not affect the system and no thermobarometer is necessary. The only part that needs to be standardized, and this can be done readily during the manufacture, is the micrometer and attached rod; the system then measures microliters without the necessity to standardize flasks, manometers, etc.

Recent Advances. Recently both the constant-volume system and the constant-pressure system have been automated, and instruments are available which will automatically record the microliters absorbed or released. Such instruments promise to have wide use in the future. There are a wide variety of applications of manometric systems to chemical, metabolic, and industrial problems. These are described in more detail in the manuals cited in the references.

References

1. Dixon, M., "Manometric Methods," Third edition, Cambridge, England, University Press, 1951.
2. Perkins, J. J., *Ind. Eng. Chem. Anal. Ed.*, **15**, 61 (1943).
3. Umbreit, W. W., Burris, R. H., and Stauffer, J. F., "Manometric Techniques," Fourth edition, Minneapolis, Minn., Burgess Publishing Company, 1964.
4. Warburg, O., "Uber den Stoffwechsel den Tumoren," Berlin, Springer (English translation by F. Dickens, 1930, Constable, London).

W. W. Umbreit

MARTIN, A. J. P.

Archer John Porter Martin (1910–) is a British biochemist who with Richard Laurence Millington Synge, his collaborator, received the Nobel Prize in chemistry in 1952 for the discovery and development of paper chromatography, a tremendously valuable tool in biochemistry.

MEDAWAR, P. B.

Peter Brian Medawar (1915–) is a British biologist who with Sir Frank Macfarlane Burnet received a share of the Nobel Award in Physiology and Medicine in 1960 for his contributions to the study of antibodies.

MELANINS AND MELANOCYTE STIMULATING HORMONE. See hormones, animal; skin pigmentation.

MEMBRANE EQUILIBRIUM

This term is applied to a special type of osmotic equilibrium, first described in 1911 by Donnan and Harris. They were studying the osmotic pressure of saline solutions of a dye, Congo red, which is the sodium salt of a high molecular weight sulfonic acid. The membranes of their osmometer were permeable to water and to ordinary salts, but impermeable to the dye. After osmotic equilibrium had been attained in their experiments, sodium chloride was present on both sides of the membrane, but its concentration was always higher in the external solution, which contained none of the large ions of the dye. To explain this unequal distribution, Donnan worked out a theory which he expressed in simple equations, based on thermodynamics and the laws of dilute solutions. He showed that diffusible ions tend to be unequally distributed in such a system whenever there is some constraint which prevents at least one kind of ion or charged particle from diffusing freely. Other investigators referred to the unequal distribution of ions as the Donnan effect, and called this type of equilibrium the Donnan equilibrium. After it was pointed out that Donnan might have based his theory on more general equations deduced by Gibbs in

1875–1878, the term Gibbs-Donnan equilibrium came into use.

A simple type of membrane equilibrium may be illustrated by the use of a diagram.

I	II
H_2O	H_2O
$z/n\ R^{n-}$	
$y + z\ Na^+$	$Na^+\ x$
$y\ Cl^-$	$Cl^-\ x$

Here the vertical line represents a membrane impermeable to the ion, R^{n-} but freely permeable to water and to sodium chloride. The molar concentrations of ions, after equilibrium has been reached, are indicated by the small letters. The notation is consistent with the electroneutrality of each solution; z is the equivalent concentration of the anion of valence n and molar concentration z/n. According to Donnan's theory, equilibrium requires an equality of the products of the concentrations of the ions of sodium chloride in the two solutions. This may be expressed by the equation

$$x^2 = y(y + z)$$

which shows at once that x is greater than y, or that the concentration of diffusible salt is greater in the external solution, II. This unequal distribution may be very marked; for example, if z is equal to $100y$, the ratio x/y is 10.05. On the other hand, if y is equal to $100z$, the ratio x/y is only 1.005. It is characteristic of the Donnan equilibrium that a high concentration of any diffusible electrolyte tends to suppress the unequal distribution.

Equilibrium in such a case requires a difference in pressure between the two solutions, and this difference is the difference between their OSMOTIC PRESSURES. The observed difference, although it is often called the colloid osmotic pressure, may be largely due to the unequal distribution of diffusible ions. Donnan pointed out that it would approach that due to the ions of the whole colloidal electrolyte only if x and y were much less than z, while in the opposite extreme, it would approach that due to the colloidal ions alone.

Donnan also deduced the existence of an electric potential difference between the two solutions at equilibrium. Since this is a single potential difference, it cannot be measured directly; the best that can be done is to connect identical electrodes with the solutions on opposite sides of the membrane by way of salt bridges. Many measurements of this sort were made by Loeb in his work on the colloidal behavior of PROTEINS. It was later found that the electromotive force of such cells, of the order of 30 mV, was changed very little by the puncture or removal of the membrane after equilibrium had been reached.

The Donnan equilibrium in non-ideal solutions has been treated mathematically by Overbeek.

The theory of membrane equilibrium has been especially useful in the study of proteins. Biological scientists have found it necessary to consider the Donnan equilibrium in trying to explain differences in ionic concentration, osmotic pressure, and electric potential across cell membranes.

References

1. BOLAM, T. R., "The Donnan Equilibria," London, G. Bell & Sons, 1932.
2. DONNAN, F. G., "The Theory of Membrane Equilibria," *Chem.Rev.*, **1**, 73–90 (1924).
3. HITCHCOCK, D. I., "Proteins and the Donnan Equilibrium," *Physiol. Rev.*, **4**, 505–531 (1924).
4. HITCHCOCK, D. I., "Membrane Potentials in the Donnan Equilibrium. II," *J. Gen. Physiol.*, **37**, 717–727 (1954).
5. LOEB, J., "Proteins and the Theory of Colloidal Behavior," New York, McGraw-Hill Book Co., 1922, 1924.
6. OVERBEEK, J. TH. G., "The Donnan Equilibrium," *Progr. Biophys. Biophys. Chem.*, **6**, 57–84 (1956).

DAVID I. HITCHCOCK

MENDEL, G. J.

Gregor Johann Mendel (1822–1884) was an Austrian monk whose work with peas grown in a monastery garden (1857–) is regarded as the beginning of the science of genetics. Though he was not a brilliant student in the usual sense, having failed examinations for teacher's certificates three times, he applied his genius in his own way and established what are known today as Mendelian laws of inheritance. Sixteen years after Mendel's death his work, which had been carefully written up, was "discovered" by De Vries. Mendel never knew that he would later be regarded as a world-famous scientist.

MENTAL RETARDATION (BIOCHEMICAL ASPECTS)

Only recently have we come to a full realization of the existence of a biochemical component in the etiology of mental retardation. Just as in the psychoses, thinking in this area has been dominated by the Freudian concept of functional disorders. Few would deny that the causes of mental retardation are multifactorial, with both cultural and psychological factors being operative, but recent discoveries have provided compelling arguments for considering biochemical aberrations as basic causes. Especially convincing evidence is at hand from studies of those inborn errors of metabolism which result in intellectual impairment (*e.g.*, phenylketonuria, porphyria).

Many present concepts regarding mental defect are based on knowledge of *phenylketonuria*, an inherited disorder which involves about 1% of the mentally retarded population. Victims of this disease are born without the enzyme, phenylalanine hydroxylase, necessary for the conversion of PHENYLALANINE, an amino acid present in all protein foods, into TYROSINE. This leads to greatly increased concentrations of phenylalanine in the blood, some of which is then converted to phenylpyruvic acid, phenyllactic acid, and *o*-hydroxyphenylacetic acid. All of these metabolites are excreted in the urine of phenylketonurics in higher than normal amounts. Varying concentrations of other unusual metabolites deriving from phenylalanine, tyrosine, and tryptophan are also present. Phenylpyruvic acid gives a characteristic olive-green color in the presence of $FeCl_3$, and its presence in urine is diagnostic of the disorder. Plasma phenylalanine concentration is a confirmatory diagnostic test. (See also AROMATIC AMINO ACIDS; BIOCHEMICAL GENETICS; METABOLIC DISEASES.)

The mental defect in phenylketonuria is usually severe and is apparent by six months of age. Most patients are idiots, a few are imbeciles, and rare individuals have borderline intelligence. Seizures, exema and albinism may be present, and life expectancy is greatly decreased.

That phenylalanine accumulation in the tissues is primarily responsible for the biochemical abnormalities in phenylketonuria is established by demonstrations over the past fifteen years that a low-phenylalanine diet will prevent or reverse these changes. (Some dietary phenylalanine must be provided, since it is an essential amino acid for tissue building and repair.) More important, if such a diet is instituted in the first months of life and continued for three to four years, intellectual impairment can be prevented. Whether the mental defect in older patients can be reversed is less certain, though significant improvements have been reported in a few cases.

The mechanism by which increased levels of phenylalanine affect intellectual development is not definite, but some evidence seems to point to the involvement of serotonin, a hormone known to be important in brain activity. Phenylalanine metabolites, such as phenylpyruvic acid, phenyllactic acid, and phenylacetic acid, inhibit an enzyme which functions in the production of serotonin and another hormone, ADRENALINE, both of which are abnormally low in phenylketonurics. An unresolved problem is the fact that a positive correlation does not always exist between phenylalanine concentration and degree of mental defect.

Parents of phenylketonurics each carry one recessive gene for the condition, but no consistent abnormalities are apparent in such carriers. However, when given a load dose of phenylalanine, carriers show higher plasma phenylalanine concentrations which are sustained over a period of several hours. Also, less tyrosine than normal is formed, indicating a reduction in activity of the phenylalanine hydroxylase enzyme.

The mental retardation which is a clinical feature of the disease *galactosemia* results from an inborn error of carbohydrate metabolism. The congenital absence in the erythrocytes of the enzyme galactose-1-phosphate uridyl transferase prevents the normal conversion of galactose, a sugar found in milk, into glucose. Galactose-1-phosphate accumulates in the red cells. If milk ingestion is continued, the affected individual develops cataracts, enlarged spleen and liver, and mental retardation. Removal of milk, the only food source of galactose, from the diet in infancy prevents the development of these features. The institution of galactose-free diets in older patients usually causes reversal of all symptoms except the mental retardation. Because reversal of intellectual impairment is often not possible, early diagnosis and dietary treatment are most important. Several rapid methods of diagnosis are available. Most patients develop an increased tolerance for galactose as they mature (see also GALACTOSE AND GALACTOSEMIA).

Success in treating phenylketonuria and galactosemia by dietary means has spurred the search for other inborn errors of metabolism among the mentally retarded. More than a dozen such conditions have now been reported, though none has been as thoroughly studied as the two described here. Included are errors in protein, fat and carbohydrate metabolism. Many other inborn errors are known which have no effect on intellectual development.

Mongolism has a clearly defined morphological pattern and accounts for about 20% of the severely retarded. The characteristic physical features—epicanthic fold, flattening at the bridge of the nose, irregularities of palmar creases, and heart malformations—led investigators to believe the condition resulted from some assault to the embryo around the eighth week of development. Biochemical studies disclosed abnormally high excretion levels of β-aminoisobutyric acid, and suggested abnormalities of blood LIPOPROTEIN fractions and absorption of VITAMIN A. Mongoloids are especially susceptible to LEUKEMIA and are often the offspring of older mothers. Genetic analysis has now shown these individuals to possess an extra chromosome at the 21 location (Trisomy 21), and this genetic anomaly is thought to be responsible for the physical, biochemical, and intellectual characteristics of mongolism. Trisomy of other chromosome pairs in severely retarded children has also been reported.

Animal experimentation and epidemiological studies have disclosed a host of *environmental prenatal factors* which produce fetal injury and maldevelopment, often involving the central nervous system. Included are infectious agents, dietary deficiencies, drugs, oxygen deprivation, metallic ions, and ionizing radiations. Virus infections in the young child, especially encephalitis and related diseases, and demyelinating diseases such as MULTIPLE SCLEROSIS often result in severe mental defect. Sparse information is available concerning the specific mechanism of injury and the abnormal metabolites involved in these cases.

Epilepsy is a common occurrence among the mentally retarded and in a portion of the cases appears to be responsible for the retardation. Biochemical findings indicate a decreased ability of epileptogenic brain tissue to bind ACETYLCHOLINE, a chemical transmitter of NERVE IMPULSES, and anticonvulsant drugs have been shown to enhance this capacity. Another suggested mechanism for anticonvulsive agents is that they interfere with the energy supply of neurons by blocking phosphorylation. In any case, the only known means for controlling epileptic seizures is through chemical agents.

Despite the great advances in knowledge of the biochemical and genetic aspects of mental retardation, the causes and means of prevention of most cases of mental defect are still unknown. More must be known about human development, human genetics, human variability, both normal and pathogenic, and the biological basis of intelligence. Recent dramatic advances in these areas enhance the prospect of further progress in the area of mental retardation.

References

1. STANBURY, J. B., WYNGAARDEN, J. B., AND FREDRICKSON, D. S., "The Metabolic Basis of Inherited Disease," New York, McGraw-Hill Book Co., 1960.
2. MASLAND, R. L., SARASON, S. B., AND GLADWIN, T., "Mental Subnormality," New York, Basic Books, 1958.

LORENE L. ROGERS

MESSENGER RIBONUCLEIC ACIDS

Messenger RNA (ribonucleic acid) transfers information for protein structure from a gene to the cytoplasmic site of protein synthesis. Our present concept is that each of the many polypeptides in an organism has its amino acid sequence specified by a different GENE (cistron), and this information, encoded in the nucleotide sequence of the genic DNA (DEOXYRIBONUCLEIC ACIDS), is directly and faithfully transcribed into a messenger RNA according to Watson-Crick base-pairing rules (guanine pairs with cystosine, adenine pairs with thymine of DNA or uracil of RNA). Translation of the transcribed information into a specific amino acid sequence is then accomplished at the RIBOSOMES with the aid of yet another class of RNA's, the transfer RNA's. Every unit of three nucleotides in the messenger RNA selects, again by Watson-Crick rules, a separate transfer RNA which has the appropriate triplet in its structure and also carries a specific amino acid (see SOLUBLE RIBONUCLEIC ACIDS). Thus the sequence of nucleotides in the gene's DNA are translated at a distant site into the remarkably exact sequence of amino acids in PROTEINS.

The conceptual need for a genetic messenger was recognized when it became apparent that the major site of protein synthesis is in the cytoplasm and not at the nuclear chromosomes. In 1961, Jacob and Monod correlated varied genetic and biochemical data to establish that the hypothetical messenger is real and is a particular species of RNA. Bacterial mating experiments, in which the time of appearance of a male gene in the female recipient can be accurately measured, had revealed that the enzyme coded by the gene is synthesized very shortly after the gene enters the female. Furthermore the rate of synthesis is constant and maximal from the start but drops precipitously if the gene is subsequently removed. These experiments suggested that the messenger is a short-lived intermediate that is synthesized and destroyed rapidly. Other experiments with a pyrimidine base analogue, 5-fluorouracil, suggested that the messenger is chemically an RNA. The analogue is rapidly incorporated into RNA and just as rapidly causes alterations in the molecular properties of enzymes synthesized thereafter. A striking feature of the analogue's effect is that the extent of enzyme abnormality is almost immediately maximal and remains constant in the presence of the analogue. If the messenger responsible for information transfer were stable, it would be expected that normal enzyme molecules would continue to be synthesized for some time and the extent of abnormality would increase during exposure to the analogue. The experiments therefore suggested that the messenger is an RNA with a high rate of turnover.

Just such an RNA had been described a few years before in a study of RNA turnover in bacteria infected with bacteriophage. The RNA synthesized after infection is characterized by a high turnover rate in distinction to the major species of RNA which are metabolically quite stable. The unstable RNA's outstanding characteristic was that its base composition reflected the base composition of the phage (not bacterial) DNA. In this particular virus-host system (T2 bacteriophage-*E. coli*), the host DNA is degraded and the cell's synthetic capabilities are subverted to the production of bacteriophage. The genes that are active ("expressed") must then be those of the infecting bacteriophage and the genetic messengers need to be transcribed from the phage DNA. The markedly similar base compositions of unstable RNA and phage DNA indicate that the RNA with high turnover rate was synthesized from the phage DNA template and thereby contained the new genetic information.

In the special virus-host system studied, the only RNA synthesized is messenger RNA and so its properties of high turnover rate and resemblance to DNA's base composition were readily observed. Knowledge of these properties allowed the demonstration of messenger RNA in bacterial and mammalian organisms against the obscuring background of the cell's much larger content of stable RNA. Currently two techniques are widely used for the study of messenger RNA. In one, newly synthesized RNA, labeled by exposure of the cell to a radioactive RNA precursor (such as C^{14} or H^3 uridine or P^{32} inorganic phosphate), can be partially separated from the stable RNA's by centrifuging through a sucrose density gradient. Molecules of different sizes form stable bands that sediment through the gradient at rates proportional to molecular weight. Although messenger RNA represents only 2–5% of the total cell RNA, a radioactive band can be observed and collected. The band is quite diffuse, however, and indicates a wide heterogeneity of molecular weights among messenger RNA's. The diffuseness reflects the variety of polypeptide molecular weights since each amino acid in sequence is coded by a triplet of nucleotides and therefore each messenger RNA must be ten times the molecular weight of the polypeptide to be synthesized.

A second technique useful for detection or separation of messenger RNA depends on the similar base compositions of DNA and messenger RNA. Single-stranded DNA (obtained by heating and quick cooling of native double-stranded DNA) will form a hybrid with its messenger RNA if the two are warmed together. The hybrid is stabilized by hydrogen bonding between the DNA bases and the corresponding (in terms of Watson-Crick rules) RNA bases. If the DNA is fixed in an agar gel or onto nitrocellulose, the technique allows isolation of tangible amounts of messenger RNA because the hybrid can be broken by raising the temperature further. If the DNA is in solution rather than fixed, the hybrid can be separated from the bulk of non-hybrid DNA and RNA by CsCl density gradient centrifugation, a technique which separates molecules according to density (not molecular weight). The hybridization procedure usually leads to the isolation of a mixture of an organism's messenger RNA's, but it can be modified to select a messenger specific to one gene. Basically, the selection requires that hybrid be formed with DNA which differs by one gene (*e.g.*, by genetic deletion) from the DNA that directed the synthesis of the mixed messengers.

The experiments described heretofore have indicated that DNA can direct the synthesis of an RNA so that the DNA's information may be transcribed. To complete the messenger concept there must be proof of participation in protein synthesis. This has come from *in vivo* and *in vitro* experiments. Reinvestigation of the bacteriophage-host experiment described above revealed that the phage-directed messenger RNA becomes associated with the stable ribosomes of the host. Later, with the discovery that polysomes (clusters of ribosomes) are the most active sites of protein synthesis (in microbes to mammals), messenger RNA was found to be concentrated with the polysomes. *In vitro* experiments with purified cell fractions showed that the rate and extent of protein synthesis is dependent on the presence of messenger RNA. When the RNA is added to the *in vitro* system, the ribosomes cluster to form polysomes.

Much of our understanding of messenger RNA has come from a host of *in vitro* experiments. The mechanism of RNA synthesis and information transfer was clarified by work with an extensively purified enzyme that absolutely requires a DNA template in order to synthesize RNA from precursor ribonucleotide triphosphates. The nucleotide sequence in the synthesized RNA was also shown to be directly determined by the sequence in the DNA template. Studies of *in vitro* protein synthesis reveal that specific known proteins are synthesized in systems to which the relevant messenger RNA is added. The striking feature of these experiments is that the nature of the protein synthesized is determined by the messenger and is independent of the biological source of ribosomes and transfer RNA. This supports the concept of a universal code in that a triplet in the messenger from one species can be correctly translated by the transfer RNA of another species. These studies led somewhat unexpectedly to experiments which resulted in a "cracking" of the genetic code. Nirenberg found that synthetic polynucleotides of random sequence could stimulate amino acid incorporation into polypeptide, but the specific amino acid involved depended on the composition of the polynucleotide. Statistical frequencies of different triplets could be calculated and compared to the extent of stimulated incorporation for each amino acid. Comparison of the effects of many different polynucleotides resulted in the elucidation of a consistent genetic code.

Although a great many details of messenger RNA synthesis and action are known, many vexing problems remain. At the same time that the concept of messenger RNA was propounded, Jacob and Monod also suggested a mechanism for the control of messenger synthesis. Their theory implicated a series of specific cytoplasmic repressors which in one state could turn off production of messenger but could also be modified by the cell environment to be ineffective. Thus far the nature of repressor material and the mechanism of its action has not been determined. Another question of interest is whether messenger RNA is directly or indirectly involved in the three-dimensional specific structuring of native proteins. This may be of particular interest in studies of the determinants involved in ANTIBODY synthesis. However, the pace of discovery in this field has been so rapid that we may confidently expect answers to these and other questions in the near future.

References

1. JACOB, F., AND MONOD, J., *J. Mol. Biol.*, **3**, 318–356 (1961).
2. "Synthesis and Structure of Macromolecules," *Cold Spring Harbor Symp. Quant. Biol.*, **28** (1963).

LAZARUS ASTRACHAN

METABOLIC CONTROLS (FEEDBACK MECHANISMS)

The properties of self-replication and growth are inherent in any definition of life. However, without any additional specification of the existence of regulatory mechanisms controlling duplication and growth one might as well be describing the crystalization of lifeless table salt from a brine solution, for each crystal grows in an orderly pattern which it has "inherited" from its "seed" progenitors. The very language itself reflects the similarity to the vital processes, but here the analogy ends. A living organism contains within itself not only the facilities to grow and replicate but also to control these conditions in order to insure a more favorable internal and external environment in which to exist and propagate. The impairment or loss of such control results in a spectrum of effects which may vary from mild and unobtrusive inefficiencies on the one hand, to death on the other.

How is life so organized and regulated that despite constant changes in the environment and in the internal milieu, the cell, the tissue, the organ, the organism itself and even the species all continue to carry out their separate functions efficiently in a manner which, to the untrained observer, seems to be immutable? What are the mechanisms, in biochemical terms, which allow the system to respond to perturbations by making internal corrections to maintain itself on an even keel?

Definition. Feedback control mechanisms are those which take part of the output of a given system and return it to the input of the same system such that the subsequent output is affected. If the effect of the feedback when it is large is to diminish the output, and conversely when it is small to increase the output, one speaks of *negative feedback*, a term which emphasizes the inverse relationship between the magnitudes of the feedback impulse and the subsequent output. Many of the control mechanisms of biologic interest are of the negative feedback type.

Illustration. A mechanical illustration of a negative feedback loop is afforded by a room, a furnace, and a thermostat within the room which can turn the furnace on or off. Fuel consumed in the furnace is converted to heat which elevates the temperature of the room. At some arbitrarily set level, the combination of the output of the system, other sources of heat in the room and all heat losses become such that the thermostat opens and the furnace turns off. When the room loses sufficient heat to lower its temperature, the thermostat closes and the furnace once again supplies heat to the system. The negative feedback here consists of the signal sent from the thermostat to the furnace (in this case an interruption of a current) which modifies the output of the furnace. The term "negative feedback" itself is derived from the lexicon of electronic engineering, a field replete with many examples of feedback loops.

Biological Examples. Within the area of biology, regulatory feedback loops are to be found at all levels of organization. Although, within the past two decades, it has been well established that negative feedback exists at the level of enzyme production and enzyme activity, it would be advisable to first consider some examples of regulation of gross level without necessarily specifying the detailed enzymatic mechanisms.

A culture of microorganisms growing anaerobically on glucose derives its energy, in the form of ATP, from the conversion of glucose to any of a number of two- or three-carbon compounds (lactic acid, acetic acid, alcohol, etc.) depending on the organism. This process yields relatively little energy per glucose molecule utilized. In order to grow and reproduce at a given rate, all other factors being equal, a large amount of glucose is utilized. If, however, oxygen is admitted to the system, glucose can be metabolized aerobically with CO_2 and H_2O being the end products, and much more energy is produced. This is accomplished by means of the enzymes of the CITRIC ACID CYCLE and the associated enzyme systems which are responsible for oxidative PHOSPHORYLATION. If one examines the rate of glucose utilization needed to maintain growth and reproduction at the same rate under aerobic condition as under anaerobic condition, it is found that much less glucose is metabolized. This phenomenon (Pasteur effect) bears a striking resemblance to the furnace-thermostat illustration in that the admission of oxygen having increased energy production satisfies the needs of the organism and allows less fuel to be burned. The mechanism by which the level of ATP regulates the rate of catabolism of glucose is still under investigation.

An example of the general ability of a bacterial system to attempt to maintain a constant environment by a feedback mechanism is provided by the following. If a strain of microorganisms which is capable of growing on protein, hence on amino acids, has the capacity for producing the enzymes L-amino acid decarboxylase and L-amino acid deaminase (oxidase), it can produce more or less of these two enzymes depending on the hydrogen ion concentration (pH) of the medium. Thus, if the pH of the medium drifts to a point which is more acid then the optimum required by the organism, more L-amino acid decarboxylase (and less oxidase) is produced by the cells. As more of the substrate is decarboxylated, CO_2 leaves the medium and more basic amines are produced, the combined effects of which are to increase the pH. Conversely, if the medium tends to become too alkaline for the bacteria, more of the L-amino acid deaminase (and less decarboxylase) is produced by the cells. With increased deamination, ammonia leaves the medium and organic acids are left behind, tending to lower the pH. The cells can therefore finely adjust the amounts of enzyme produced in order to attempt to keep the pH within the narrow region which is optimal for the survival of the organisms. Again, the means by which the pH controls the enzyme-forming mechanism of the cells is not known.

The foregoing example is an illustration of a *negative feedback system* which is *bifunctional* rather than unifunctional. Such systems provide a greater degree of control over the environment than do unifunctional ones. To return to the analogy of the furnace, the thermostat opens at a preset temperature and does not close again till the room cools below this temperature. A lower limit is thus set on the temperature of the room, assuming that the furnace is large enough to overcome any possible heat losses. But, there is no upper limit to temperature and it may increase from external heat sources to uncomfortable levels. If one complements the furnace with an air conditioner and provides the thermostat with an additional switch, thus creating a bifunctional feedback system, it becomes possible to establish an upper limit to the temperature of the room as well.

Feedback mechanisms also control the size of an organism. Unicellular organisms growing under optimal conditions do not continue to increase in size *ad infinitum*. Instead, having reached a given volume, they divide. The same is true of cells in multicellular organisms. Either the cell stops growing (*e.g.*, a muscle cell) or it divides (*e.g.*, a hepatic cell). The same phenomenon is manifested by organs as well as whole organisms in both the animal and the plant kingdoms. The shark or giant sequoia may represent exceptions. Many organs such as the limbs of starfish or liver of almost any animal are capable of regeneration. That is to say, if the organ is reduced in size by an outside influence (accident or surgery), it begins to grow and divide. When the original size is attained, the process stops. The nature of the chemical fact or factors whose concentration determines the rate of growth of the tissue via a feedback loop remains to be discovered. Very similar considerations confound the field of embryology and are discussed elsewhere (see DIFFERENTIATION AND MORPHOGENESIS; METAMORPHOSIS).

Beyond consideration of the needs of the individual organism, complex feedback systems exist which determine the optimal population size and the distribution of individuals within the population based upon the availability of living space and food and on the rate of death due to predation, age and disease. Experimental situations have been devised which demonstrate that the regulatory mechanisms which maintain colony or herd size are complex and involve negative feedback loops.

A tissue in animals which may be thought of as undergoing continual regeneration is bone marrow. At this site hematopoiesis occurs and newly matured erythrocytes are released into the circulation where under normal conditions they survive for a given time and then are destroyed. With unusual demands upon the erythrocyte-forming system, such as may occur after bleeding or in certain anemic states (see ANEMIAS), more red cells are produced. When the optimal level of circulating hemoglobin is attained, the rate of production of red cells by the marrow decreases. It has recently been shown that the control of the rate of erythrocyte production by the marrow is through a negative feedback loop via a substance elaborated by the kidney called *erythropoietin*. Erythropoietin production is elicited under conditions of relative ischemia (low oxygen tension). When oxygenation approaches normal, the amount of erythropoeitin released decreases, allowing the marrow to return to maintenance levels of erythropoiesis.

The process just described in only one of the more recently discovered example of what Claude Bernard in the last century called "the principle of homeostasis." The action of many HORMONES (erythropoietin is a hormone) is to maintain a constant internal milieu via feedback loops. We may cite two examples from the field of endocrinology: (a) The circulating level of glucose in the bloodstream is, among other things, dependent upon the level of INSULIN. Conversely, the amount of insulin secreted by the pancreas, all other factors being constant, is inversely dependent upon the circulating glucose concentration. (b) The hormone corticotropin (ACTH) elaborated by the pituitary gland stimulates the adrenal glands to produce increased levels of circulating corticosteroid hormones. The level of corticosteroids then controls in an inverse manner the amount of ACTH secreted by the pituitary (See HOMEOSTASIS; ADRENAL CORTICAL HORMONES).

Mechanisms. We may now turn our attention directly to the feedback mechanisms which control enzyme activity. These may serve as model systems for more complemented systems. Two distinctly different forms of negative feedback for the control of enzyme systems are distinguishable: (a) feedback inhibition and (b) feedback repression. Only feedback repression involves direct control of enzyme synthesis.

Feedback inhibition of enzyme activity is a mechanism whereby the immediate or distal product of the enzyme activity is capable of inhibiting the action of a constant amount of preformed enzyme. Thus, should the concentration of this product accumulate beyond an optimal amount, it can reduce its own production by inhibiting a key enzyme involved in its own synthesis. When the concentration of the end product is reduced by degradation or utilization, the brake, as it were, is released and the uninhibited enzyme produces more product. As a rule, examples of feedback inhibition involve synthetic metabolic pathways. It is quite often the case that the terminal product of this pathway controls the rate of its own synthesis by being capable of inhibiting a key and often rate-limiting step early in the pathway. As there is usually no chemical similarity between the substrate of the key enzyme and the end product which causes the inhibition, the usual mechanism of competitive inhibition is inapplicable. A new inhibitor site on the enzyme molecule, independent of the substrate site, is required (see REGULATORY SITES OF ENZYMES). Although many examples of feedback inhibition involving *de novo* synthesis of amino acids, purines, and pyrimidines have been delineated in microorganisms, it would be well to give an example from higher animals in order to illustrate the as yet unproved thesis that this mechanism of negative feedback control plays an important role in multicellular organisms as well.

Mammals, including man, are capable of synthesizing cholesterol from acetate by way of a complex series of reactions (see STEROL BIOSYNTHESIS). Cholesterol may also be derived from dietary sources. When the diet is rich in cholesterol, very little is synthesized endogenously; conversely, when the diet is deficient in cholesterol, requirements for this necessary metabolite are met by *de novo* synthesis. It is now known that cholesterol is capable of allosteric inhibition of the enzyme which converts β-methyl-β-hydroxyglutaric acid to mevalonic acid, a key intermediate about one-third of the way on the pathway to cholesterol synthesis. It is of interest that the site of control is at this step, since the pathway between acetate and β-methyl-β-hydroxyglutaric acid is also involved with other metabolic processes. The feedback inhibition occurs at the first step of the pathway which is unique for cholesterol synthesis.

Feedback repression of enzyme synthesis is a mechanism of negative feedback control which allows the level of a given metabolite to regulate the actual amount of enzyme produced without implying any combination of metabolite and enzyme as that postulated for feedback inhibition. Known examples of feedback repression are currently almost exclusively restricted to the field of microbiology (tissue culture of cells from higher organisms included) owing to the difficulties of the experimental techniques. However, as has so often been true in the past, the basic mechanisms which have been delineated from the study of one form of life may generally be applied to other forms of life.

The synthesis of a given protein [see PROTEINS (BIOSYNTHESIS)] is dependent on many factors among which are: (a) the genetic content of the cell (DNA), (b) the formation of messenger ribonucleic acid (mRNA) from DNA, (c) the activation of amino acids by specific enzymes and the transfer of these amino acids to a ribonucleic acid carrier (sRNA), (d) the formation of ribosomes and polysomes and, finally, (e) the transcription of the message on mRNA by the polysomes with concurrent formation of peptide bonds between the activated amino acids to produce a specific polypeptide. The sequence of amino acids in the polypeptide bears a linear relationship to the arrangement of triplets of purine and pyrimidine bases on the structural GENE of the DNA molecule.

The point at which feedback repression occurs is prior to the entire scheme just outlined and has been postulated by Jacob and Monod to occur as follows: Consider one or more structural GENES each specifying the production of an enzyme via the mechanisms described above. If more than one enzyme is involved in the repression-controlled pathway leading to the production of the key end product metabolite (*e.g.*, the amino acid HISTIDINE) then each structural gene for each of these enzymes occupies adjacent spaces on the chromosome. One large mRNA containing the information for the synthesis of all these enzymes is formed from the DNA and proceeds to the cytoplasm where the enzymes are formed. Just prior to the site on the DNA which specifies the first enzyme in the series (*i.e.*, where mRNA synthesis begins) is a region designated as the *operator gene* which does not function to produce mRNA complementary to itself but serves as a switch for the production of mRNA by the structural gene or genes which are adjacent to it on the DNA molecule. The region encompassing the operator gene and its dependent structural genes is termed the operon (see Fig. 1).

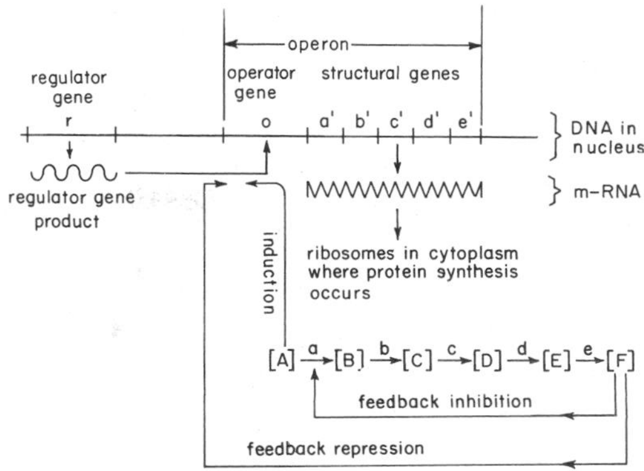

FIG. 1. Mechanisms of action of feedback induction, inhibition and repression. [A], [B], etc., represent the concentration of intermediates in a metabolic pathway; a, b, etc., are the enzymes responsible for the conversion of A to B to C etc.; a′, b′, etc., are the structural genes which specify the formation of enzymes a, b, etc. In this illustration, enzyme "a" may be induced by substrate A and inhibited by end product F, or its synthesis can be repressed by end product F.

At another region on this DNA molecule, or on another DNA molecule, or perhaps even on a DNA molecule in another chromosome, is located a region termed the *regulator gene*. The transcription product of the regulator gene is such that it can bind tightly to the operator gene. Under such circumstances it may be postulated that the operator gene is inactive, and mRNA cannot be formed by the structural gene which comprise the remainder of the operon of which this operator gene is a part. Synthesis of the proteins (enzymes) specified by these affected structural genes would cease in microorganisms since mRNA is known to turn over rapidly in these organisms. In higher animals, the effects of sudden non-production of mRNA would take relatively longer to observe since it is thought that mRNA turns over less rapidly in higher organisms.

It is also conceivable, on the other hand, that the operator gene switch is turned on only when the regulator gene product is firmly bound to the operator gene and that only under these circumstances will enzyme synthesis ensue. Further investigation is needed to establish the nature of the exact mechanism involved, but in either case a system exists which has the requisite properties for metabolic control. Enzyme induction, feedback repression, and the action of embryological ORGANIZER substance are postulated to act by interacting with the regulator gene product in such a manner as to affect its binding to the operator gene. Merely for the sake of illustration let us assume that the operator gene switch is on (*i.e.*, mRNA can be made from the structural gene) only if regulator gene product is attached to the operator gene. Feedback repression would occur when the end product of the synthetic pathway (histidine in our example) would bind the regulator gene product and functionally remove it from the operator gene thus stopping the mRNA synthesis required to form the histidine synthesizing enzymes. Conversely, if we assume that the operator switch is on only when it is free of regulator gene product, then it is conceivable that regulator gene product can only bind to the operator gene when the local concentration of histidine is elevated. If only one structural gene follows the operator gene, only one enzyme is affected by feedback repression. If, however, more than one structural gene for a synthetic pathway follows the operator gene, then feedback repression by the end product will simultaneously cause the lack of synthesis of more than one enzyme; this condition is known as coordinate repression.

A number of compounds are known which will increase the activity of a given amount of enzyme (see ENZYME ACTIVATION) or will increase the amount of enzyme synthesized. Substrates and compounds which closely resemble the substrate may act as *inducers*. One of the most thoroughly studied systems having the property of substrate induction is bacterial β-galactosidase. Organisms grown in the absence of lactose or galactose do not form this enzyme. Within minutes of the addition of either of these compounds to the media, active synthesis of β-galactosidase becomes apparent. In higher animals, one of the actions of certain HORMONES may be to cause specific enzyme induction. By analogy with negative feedback, it is convenient to designate a control mechanism whose primary effect is to increase the rate of synthesis of a given enzyme as an example of *positive feedback*. The term emphasizes the direct relationship between the magnitude of the input and the magnitude of the resultant output. The enzyme tryptophan pyrrolase which is found in certain bacteria and in mammalian liver where it catalyzes the first step in the catabolism of TRYPTOPHAN to NICOTINIC ACID serves as an illustrative example of positive feedback. Increased amounts of the enzyme may be produced by growing the bacteria on tryptophan or treatment of the animal with tryptophan or with hydrocortisone. Both effects are manifest in animals even when they are adrenalectomized. It is thought that the action of hydrocortisone is via interaction at the regulator gene–operator gene loci in a manner opposite to that which occurs in feedback repression. That is to say, hydrocortisone would be capable of turning the operator gene switch on by reacting with regulator gene product. Substrate induction with tryptophan would occur by a similar mechanism. However, although this may be true for bacteria, it is now thought that in mammals tryptophan acts by means of stabilization of the enzyme tryptophan pyrolase, rendering it less subject to degradation. Under conditions of constant synthesis, enzyme accumulates whenever its rate of destruction is decreased.

Aside from the microscopic details of the actual mechanisms involved in feedback regulation (of which we have as yet little knowledge), there is ample evidence that such mechanisms exist and are vitally important in the regulation of the rates of enzyme synthesis. Mutations or deletions of structural genes at times lead to deleterious results for the cell if the reactions catalyzed by the enzyme involved are essential for viability. Mutations of a structural gene may also affect a feedback loop. For example, bacterial mutants which cannot produce the enzyme necessary for the conversion of orotic acid to uracil, a necessary step for RNA synthesis, when grown on limiting amounts of uracil, will accumulate large amounts of orotic acid in the medium (see PYRIMIDINE BIOSYNTHESIS). This situation results from the failure of the concentration of uracil, or other pyrimidines made from it, to attain a high enough level to cause feedback of the enzymes which form orotic acid. The very same situation has been observed in man in the condition known as orotic aciduria. In both the bacterial and the human mutants, the administration of various pyrimidine bases (which the mutant organism could not make) results in the reappearance of negative feedback. Cessation of formation of excessive orotic acid ensues. In the bacterial system, which has been thoroughly investigated, the key end product is the pyrimidine-containing nucleotide cytidine triphosphate (CTP), which reacts with an allosteric site on the enzyme aspartate transcarbamylase, causing feedback inhibition. Aspartate transcarbamylase is the first enzyme on the specific pathway leading to orotic acid and hence to pyrimidine (uracil and CTP) synthesis. Such an example serves as a potent illustration of the basic similarity of a regulation mechanism in species as diverse as *Homo sapiens* and *Escherichia coli*.

Mutations or deletions are known to occur at the level of the operator gene in which case all the enzymes of a pathway controlled by the same operator gene are not synthesized even though no mutation has occurred at the site of the structural genes.

Finally, if mutations occur in the regulator gene, the result may be such that the entire operon is "derepressed" or removed from external control and continues to form mRNA, and hence enzymes, without limit other than the availability of energy and building blocks, *i.e.*, amino acids. It is tempting to speculate that certain tumors may represent mutations of the regulator gene.

References

1. "Cellular Regulatory Mechanism," *Cold Spring Harbor Symp. Quant. Biol.*, **26** (1961).
2. "Synthesis and Structure of Macromolecules," *Cold Spring Harbor Symp.* "Quantitative Biology," **28** (1963).
3. *Journal of Molecular Biology*, 1959 to present.
4. "Advances in Enzyme Regulation," New York, The Macmillan Co., Vol. 1, 1963; Vol. 2, 1964; Vol. 3, 1965.
5. CHANGEUX, J.-P., "The Control of Biochemical Reactions," *Sci. Am.*, **212**, 36–45 (April 1965).

VICTOR H. AUERBACH

METABOLIC DISEASES

Undernutrition. It is self-evident that a lack of any "essential" constituent of food would be considered undernutrition. Therefore, the types of undernutrition are as numerous as the number of essential AMINO ACIDS, essential FATTY ACIDS, essential MINERALS, and VITAMINS. As a rule, each deficiency leads to its own peculiar set of symptoms, which may be ameliorated if the factor which has been lacking in the diet is restored before much damage is done (see also MALNUTRITION; several articles on NUTRITIONAL REQUIREMENTS).

Obesity. Obesity is essentially overweight, which is usually interpreted as 10% or more above the average standard weight for the individual. This may be due simply to overeating or lack of muscular exercise, or both. There are also types of obesity which occur concomitantly with other diseased states but we are not sure whether the obesity is a cause or a result. For example sex organ disturbances are frequently accompanied by obesity and so are diabetes mellitus and hypothyroidism, but this is not invariably the case.

Glycosuria; Diabetes Mellitus. Glycosuria is the term denoting sugar in the urine, but it is usually used for "glucose in the urine," and "melituria" has become the more general term for the appearance of any sugar in the urine.

Glucosuria. Glucose is the physiological sugar and occurs in the blood normally to the extent of about 0.10%. When this figure is exceeded, the excess is eliminated by the kidney. Normally there is practically no glucose in the urine. More than a trace in a 24-hour specimen of urine is abnormal. However, easily detectable amounts may be found in urine voided soon after a high carbohydrate meal because of the temporary rise of the blood sugar; this is "alimentary glycosuria" and may occur in normal individuals. In diabetes mellitus, the blood sugar is usually above 0.10%, the height depending on the severity of the disease and its control. Diabetes mellitus is due to a lack of INSULIN, a hormone secreted by the pancreas. The severity of the disease seems to be related to the amount of insulin produced. Insulin, either the natural form or some modification, is available for clinical use. The severity of the condition cannot be estimated by the blood sugar percentage alone, since this varies also with the fluid intake. It is the actual output of sugar that is important. If the amount excreted is subtracted from that ingested, the remainder will be the amount of sugar utilized. This, together with the blood sugar picture gives the physician an accurate basis for treatment.

Treatment may be (1) a regulation of the diet, (2) the administration of insulin, or (3) the administration of an oral hypoglycemic ("antidiabetic") agent. The two latter methods should also take into account the diet.

Besides diabetes mellitus, there are certain kidney conditions in which glucosuria occurs. These are probably not due to a deranged metabolism but to a fault in the kidney excretory mechanism (see RENAL TUBULAR FUNCTIONS).

Other Meliturias. *Lactosuria* occurs in a considerable proportion of pregnant and lactating women. LACTOSE is a disaccharide and cannot be utilized by the body until digested (hydrolyzed) in the intestinal tract. Its constituent monosaccharides, galactose and glucose, are then absorbed and utilized. Therefore if the mammary gland produces too much milk sugar (lactose) during the process of milk formation, some of it escapes into the bloodstream, and since it cannot be consumed by the woman's own tissues, it is excreted in the urine. It is of no harm, except that its occurrence in urine may be confused with glucose and thus cause difficulty in diagnosis of diabetes mellitus.

Fructosuria is a rather rare condition. Sometimes it occurs in diabetes mellitus along with glucosuria. Alimentary fructosuria sometimes follows the ingestion of large quantities of FRUCTOSE. There is also an "essential fructosuria," which is a specific inability to utilize this sugar, probably as a result of a deficiency of a specific enzyme required for the conversion of fructose to glycogen in the liver. Insulin is of no value in fructosuria, and it is a comparatively harmless phenomenon.

Galactosuria. Galactose is one of the two sugars which are components of lactose. Hence it is understandable that a galactosuria should be tied up with lactose. It occurs in nursing infants suffering from galactosemia (see GALACTOSE AND GALACTOSEMIA). This is a congenital defect in which the infant develops a high blood sugar due chiefly to galactose-1-phosphate, and as a result, galactose appears in the urine. Sometimes it is accompanied by amino aciduria and ketonuria. Enlargement of the liver, cataracts, and MENTAL RETARDATION often accompany this condition. It seems to be due to the congenital absence or deficiency in the enzyme which catalyzes the transformation of galactose-1-phosphate to glucose-1-phosphate. It can be avoided if the infant is maintained on a lactose-free diet.

Pentosuria. Pentoses appear in the urine, as is the case with other sugars, if ingested in very large amounts. This may happen when certain fruits such as prunes, plums, cherries, or grapes are eaten in considerable amounts. This is a temporary condition and is of no moment. A pentose also appears in the urine in cases of morphine addiction (see NARCOTIC DRUGS). There is however a rather rare inherited condition, essential pentosuria, in which the only abnormality is the inability to utilize L-xylulose. It has no other clinical significance, except the possibility that it might be mistaken for glucosuria (see also PENTOSES IN METABOLISM).

Hypoglycemia. The discovery of insulin stimulated the study of blood sugar levels. It was found that insulin lowers these levels, whether they are normal or above or below normal (see INSULIN FUNCTION). Levels above normal are called hyperglycemias and those below normal are hypoglycemias. When they are too low, due to an overdose of insulin, or for some other reason, severe symptoms occur, ranging from weakness and mental confusion to excessive perspiration, cardiac palpitation, loss of memory, diplopia, and even muscular twitching, convulsions, and unconsciousness. Administration of glucose relieves these effects.

An increased activity of the pancreas will, of course, produce more insulin and tend to bring the blood sugar level down. This is seen in adenomas or carcinomas of "islet" tissue of the pancreas. It also occurs in conditions in which there is considerable enlargement of the liver. Such conditions include acute yellow atrophy of the liver, liver necrosis, and carcinoma of the liver. Toxic substances produced by severe burns are known to circulate in the blood and have the same effect. The mechanism is unknown. Newborn infants of diabetic mothers may have hypoglycemia. The reason suggested is that the fetal pancreas, while *in utero*, has been stimulated to provide the mother with insulin in addition to that needed for its own use. Consequently an excess of insulin is present in its blood and has a hypoglycemic effect after the baby is born. There is also a hypoglycemia which appears to be of central nervous origin. This type yields to a high-protein, low-carbohydrate regime.

Glycogen Storage Disease. Several types of derangement of GLYCOGEN METABOLISM in liver and muscle are known to occur. They are relatively rare but the most common, von Gierke's disease, has been known for a number of years. It seems to be due to the absence of an enzyme which converts glucose-6-phosphate to glucose and phosphoric acid. Consequently, glycogen cannot be converted through this intermediate into glucose, and as a result, glycogen piles up in the liver. Hypoglycemia and acidosis are among the consequences. In the cardiac type, glycogen accumulates in the heart and the result is rapidly fatal.

Acidosis and Alkalosis. Carbon dioxide is formed from the products of digestion and metabolism and is chiefly an excretory product, much of it leaving the body via the lungs. When it unites with water it forms carbonic acid. Other acids also arise as a result of the same functions, *viz.*, H_3PO_4, H_2SO_4, uric acid, lactic acid, and others. If fats are incompletely oxidized β-hydroxybutyric acid and acetoacetic acid, among other things, are formed in excessive amounts (see FATTY ACID METABOLISM). All of these acids are neutralized, or buffered, by bases in the blood and tissues. The bases are $NaHCO_3$, $KHCO_3$, proteins, and some organic salts. Ammonia is produced in nitrogen metabolism and is another useful base. The interaction of these various acids and bases makes the blood have a varied pH. However, normally this pH is very constant (7.35–7.45), being kept so largely by the elimination of CO_2, acids, and bases by the lungs, skin and kidneys.

The hydrogen ion concentration of the blood varies with the ratio $B^+HCO_3^-/H_2CO_3$. Now if the relationship between the bicarbonate and the H_2CO_3 of the blood is kept constant, at about 20:1, the pH of the blood will keep within the normal limits of 7.35–7.45, since this "buffer pair" is in equilibrium with all the other blood buffers.

It should be repeated that the acid-base balance depends upon the ratio $B^+HCO_3^-/H_2CO_3$. If more acid is present or less base, the result will be an excess of acid. This is called an acidosis. If more base is present or less acid, there will be an excess of base, or an alkalosis. Either of these is abnormal. However the body has mechanisms ready to "compensate" for either one. It is the uncompensated ones which are most to be feared (see also ACID-BASE REGULATION).

Primary alkali deficit is an acidosis due to overproduction, ingestion or retention of acid. In diabetes mellitus, along with the inability to utilize carbohydrates, there is a disturbance in fat metabolism leading to excessive formation of some organic acids. Excessive intake of HCl to combat gastric disturbances has the same effect. Diseased kidneys may not be able to excrete acids as they normally should. In all of these cases (except the last one), a primarily alkali deficit leads to increased excretion of acid by the kidney as well as an increased formation and excretion of ammonium salts. The body also attempts to get rid of more CO_2 by increasing respiratory movements. If these attempts by the body to get rid of acid are successful, the condition is said to be compensated. If it is uncompensated the patient may go into coma.

Primary alkali excess may be due to overdoses of sodium bicarbonate. More usually, however, loss of acid may have the same effect. This may occur as a result of excessive vomiting, with loss of gastric HCl. The body reacts to this state by increasing the urinary output of base and by lessening the formation and excretion of urinary ammonia. Respiratory loss of CO_2 is depressed and, again, if these mechanisms are successful the alkalosis is compensated. If not, an uncompensated alkalosis with tetany may occur.

Primary CO_2 excess is caused by any depression of respiration with decreased loss of CO_2. It is an acidosis.

Primary CO_2 deficit is an alkalosis caused, usually, by excessive loss of CO_2 by hyperventilation of the lungs. Fever, hot baths and hysteria, are possible causes.

Gout. Gout is related to URIC ACID. This substance is a metabolic product of the PURINES. These occur in all tissues and are characteristic constituents of the NUCLEOPROTEINS. Nucleoproteins are found in the nuclei and cytoplasm of all living tissues, plant and animal. In the breakdown of nucleoproteins, nucleic acids are released, and purines are located in these portions of the nucleoproteins. The purines have as their end product, in man, an oxidized purine, namely uric acid. In addition, there is a pathway for the formation of uric acid which does not involve purines, but which does involve glycine and other simple products.

In gout, there is usually an elevation in the uric acid content of blood serum, but occasionally this does not occur and so it seems possible that some factor other than uric acid sometimes plays a role in the genesis of the disease.

Normally the excretion of uric acid by the kidney keeps pace with its formation from purines of the food, purine metabolism of the tissues and synthesis of uric acid. An elevation of serum uric acid may occur if the kidney cannot eliminate it at a normal rate or if the rate of tissue breakdown is accelerated. This is usually accompanied by rises in other nitrogenous constituents of the blood, *e.g.*, urea, and may not be associated with gout. In gout, the only nitrogenous constituent of serum which rises characteristically is uric acid. Undoubtedly as a result of this increased amount in blood, uric acid precipitates out in various locations in the body. The onset of the first attack is usually a very severe pain in the joint of a finger or toe. The joint becomes red, swollen and extremely tender. Other joints are sometimes affected and frequently more than one finger or toe is involved. It is very variable in its severity; some patients are bedridden for long periods of time, while for others it is relatively mild. Knob-like deformities around the affected joints appear, due to the deposition of uric acid to form "tophi."

Treatment of gout involves the use of drugs, some of which act to relieve pain and others are "uricosuric," *i.e.*, they tend to promote the excretion of uric acid. Dietary procedures seem to be of little avail, probably because uric acid can be synthesized from small-sized molecules.

Porphyrias. Porphyrins are substances which are parts of the HEME molecule; heme, in turn, is an essential constituent of the HEMOGLOBIN of blood, of muscle pigment and of other pigments of the body, including the cytochromes. There are several types and derivatives of the PORPHYRINS and normally the unused portions are metabolized and excreted in small amounts, as coproporphyrin and uroporphyrin, in the feces and urine, respectively. The pigments of the bile are also derived from heme (see BILE AND BILE PIGMENTS).

An abnormality in the conversion of the porphyrins of heme results in "porphyria." A rather rare form is the hereditary one, which is due to the congenital absence of an enzyme necessary for one of the conversions (see also BIOCHEMICAL GENETICS). In this "inborn error of metabolism," there may occur cutaneous photosensitivity, a purplish complexion, gastrointestinal complaints, nervous and mental effects. There may also be a pinkish color of the teeth with fluorescence in ultraviolet light. The urine may also be similarly colored.

Phenylketonuria. Another hereditary anomaly of metabolism is phenylketonuria. Certain feebleminded children were observed to excrete a considerable amount of phenylpyruvic acid in their urine. This was due to the incomplete metabolism of the essential amino acid phenylalanine. A very special diet, containing little phenylalanine, will improve the condition of these patients (see MENTAL RETARDATION).

Alkaptonuria and Tyrosinosis. These two hereditary conditions involve an inability to completely metabolize phenylalanine and tyrosine. The former occurs more frequently in men than in women. A blackish pigment is eliminated in the urine, due to the oxidation of homogentisic acid, one of the normal intermediate products. In tyrosinosis, which is extremely rare, other products are excreted. In both cases, some enzyme probably is lacking. (See ALKAPTONURIA; AROMATIC AMINO ACIDS.)

Cystinuria. Cystine is a "semidispensable" amino acid. That is, it is necessary unless an abundance of another amino acid, methionine, is present. It is found in the urine of certain individuals (see SULFUR METABOLISM). This seems to be a renal condition in which cystine is secreted by the kidney and is not reabsorbed in a normal fashion. The chief danger due to this condition is in the formation of kidney stones composed of cystine.

ISRAEL S. KLEINER

METABOLISM

Metabolism is a very general term, referring to all the chemical changes and energy changes that occur in a living organism (see also ANABOLISM; CATABOLISM). Used in a more special technical sense, BASAL METABOLISM refers to the rate of energy production (due to food combustion) shown by an animal under resting, postabsorptive conditions. The broad term *metabolism* may also be qualified to delimit it to the reactions of substances of a particular structural class, such as the interconversions (as well as the biosynthetic and degradative pathways) of the amino acids, discussed in AMINO ACID METABOLISM; or those of carbohydrates, discussed in CARBOHYDRATE METABOLISM; etc.

METALLOENZYMES. See METALLOPROTEINS.

METALLOPROTEINS

Proteins, especially in solution, readily participate in a greater variety of chemical reactions than any other class of compounds of biological interest. This reactivity is a function primarily of the many polar side chains containing —OH, —COOH, —NH_2, —SH and other groups, all of which can, to varying extents, interact with metal ions. It is not surprising then that in general all PROTEINS can bind metals, some of them very tightly indeed.

For the present purpose, however, it will be important to distinguish between relatively specific and nonspecific binding. A negatively charged protein molecule exerting a nonspecific electrostatic attraction on metal ions would not qualify as metalloprotein. Nor will we consider such insoluble artificial compounds as those obtained by treatment of collagen with trivalent chromium during the tanning process or by precipitation of proteins with $Zn(OH)_2$, even though the action of Cr or Zn may here be fairly specific. Further, ions such as Mg or Mn are often found as important coenzymes, but because the enzyme-metal complexes in these cases are usually very weak they will also be excluded from the present discussion. The term metalloprotein, in agreement with usual practice, will be restricted then to compounds in which under natural conditions a metal ion is relatively specifically and strongly bound directly to a protein molecule in such a way that the compound can be isolated and shown to contain a stoichiometric amount of metal.

It is precisely this stoichiometry which permits the isolation and identification of many metalloproteins. One of the important measures of purity of these compounds is the attainment of a constant metal-to-protein ratio or, in the case of *metalloenzymes*, a constant metal-to-activity ratio. According to this criterion, a considerable number of metalloproteins have been found in nature. We cannot give here an exhaustive list; many of the compounds are well known and are discussed in the present volume under separate headings (*e.g.*, see COPPER; MOLYBDENUM; ZINC; HEMOCYANINS; FERRITIN). A partial list is nevertheless provided here in order to illustrate the following points: (1) a variety of metal ions are found in biologically important metalloproteins; (2) metalloproteins occur in a wide range of biological systems; (3) the function of these metalloproteins, where known, varies widely from one compound to another (see list below).

The chemical properties of the metal in these compounds may be greatly affected by bonding to a protein ligand. The bound metal can play one of many roles. Thus, in an enzyme, the metal ion may permit the formation of a ternary complex between protein, metal and substrate or coenzyme. An instance of this role is provided by the enzyme enolase, which is unable to catalyze the equilibrium between 2-phosphoglycerate and 2-phosphopyruvate in the absence of Mg ions. In other enzymes, the metal may actually participate in electron transport by cyclic oxidation and reduction. Such is probably the case with the Cu in polyphenol oxidase. Finally, the metal may serve primarily for the maintenance of a specific spatial folding of the polypeptide chains in the protein molecule.

A noteworthy omission from the abbreviated list shown here of metalloproteins consists of such compounds as HEMOGLOBIN, the CYTOCHROMES, catalase, PEROXIDASE, CHLOROPHYLL and other metal-PORPHYRIN enzymes. Two arbitrary reasons lead to this omission: (1) Because of the very large amount of information which we possess about these compounds, they are conveniently discussed under separate headings in the present volume. (2) Although the metal does react directly with the apoprotein molecule, as in hemoglobin for instance, it nevertheless primarily forms part of a well-characterized prosthetic group, and a stable metal-porphyrin can in most cases be separated from the protein. It is for these reasons that such compounds are not usually thought of as typical metalloproteins.

The strength of metal-protein bonds in metalloproteins may vary from relatively loose association to very tight binding. When the metal ion is able to dissociate with some ease from the protein, one may look for a single ligand responsible for the metal binding. Such ligand groups are mainly found in the amino acid side chains of the protein molecule (*e.g.*, $—NH_2$ or —OH groups). The interaction of metal and ligand may exhibit strong pH dependence because of competition between metal and hydrogen ions. Of the single ligand groups, by far the strongest is the —SH group in the amino acid cysteine. Even stronger metal bonding to protein may be observed when a divalent or trivalent metal forms chelate complexes with the protein. CHELATION is often indicated not only because of the strength of the bond, but also because of the specificity of the reacting site on the protein molecule for one particular metal. Such a specificity may reflect the coordination requirements of the various metals. The preferred electron donor in the formation of protein-metal coordination compounds is N, such as that of the imidazole nucleus of HISTIDINE, but S and O may also participate in this process. If the protein contains carboxyl or phosphoryl groups, strong

Name	Metal	Source	Function
Hemocuprein	Cu	Erythrocytes	Unknown
Ceruloplasmin	Cu	Serum	? Oxidase
Hepatocuprein	Cu	Liver	Unknown
Polyphenol oxidase	Cu	Mushroom	Enzyme
Hemocyanin	Cu	Mollusks	Respiratory pigment
Tyrosinase	Cu	Mushroom	Enzyme
Metallothionein	Cd + Zn	Kidney	? Na Reabsorption
Xanthine oxidase	Mo	Liver	Enzyme
Carbonic anhydrase	Zn	Erythrocytes	Enzyme
Alcohol dehydrogenase	Zn	Yeast	Enzyme
Ferritin	Fe	Spleen	Fe storage
Transferrin	Fe	Plasma	Fe transport
Conalbumin	Fe	Eggs	Fe storage
Ferredoxin	Fe	Bacteria	Electron transport
DPNH-cytochrome *c* reductase	Fe	Heart muscle	Electron transport
Hemovanadin	V	Tunicates	Respiratory pigment

ionic bonds between metal and protein may be formed. A completely different type of protein-metal interaction is illustrated by the Fe-containing protein FERRITIN. Basically, this compound consists of a coat of protein (apoferritin) surrounding a micelle of hydrated iron hydroxide. The metal can be readily and reversibly removed from the apoprotein.

It is likely that further elucidation of the role of the various trace metals in biological systems will uncover yet other instances of metalloproteins. In many cases the combination of metals with proteins can lead to far-reaching pharmacological effects. The metalloproteins thus make up a group of substances of great academic as well as practical interest.

References

1. VALLEE, B. L., "Zinc and Metalloenzymes," *Advan. Protein Chem.*, **10**, 318–384 (1955).
2. GURD, F. R. N., AND WILCOX, P. E., "Complex Formation between Metallic Cations and Proteins, Peptides and Amino Acids," *Advan. Protein Chem.*, **11**, 311–427 (1956).
3. MALMSTRÖM, B. G., AND ROSENBERG, A., "Mechanism of Metal Ion Activation of Enzymes," *Advan. Enzymol.*, **21**, 131–167 (1959).
4. PASSOW, H., ROTHSTEIN, A., AND CLARKSON, T. W., "The General Pharmacology of the Heavy Metals," *Pharmacol. Rev.*, **13**, 185–224 (1961).

ERNEST C. FOULKES

METAMORPHOSIS, AMPHIBIAN

Amphibian metamorphosis is an important post-embryonic developmental process in which non-reproductive structures of an organism change drastically during a discrete period. It is a classic example of cellular DIFFERENTIATION and offers unique information about COMPARATIVE BIOCHEMISTRY. The spectrum of changes during the transformation of an aquatic larva to a terrestrial frog represents a series of remarkable biochemical and structural adaptations. Amphibian metamorphosis also reflects one of the most dramatic effects of the THYROID HORMONE, or any hormone for that matter. The rapid change in cell type and function are frequently compared to changes occurring in tumor cells. It is now suspected that differentiation and development at any level must be under some sort of genetic control, but there is currently no compelling evidence for a genetic effect during amphibian metamorphosis.

The dramatic nature of morphological transformations during metamorphosis has excited biologists for almost a century. After early descriptive work, the basic hormonal components of this process were established by Gudernatsch and Allen and extended by Etkin and others. Classical anatomical studies culminated in the careful staging procedures of Taylor and Kollros. Cytological studies have been limited but are now in progress in several laboratories. During the past ten years there has been great interest in the biochemical changes which occur during this process and which will be summarized in the subsequent discussion.

Attempts to find a common denominator for the multiplicity of the biochemical effects of the thyroid hormone in initiating amphibian metamorphosis have not been successful to the present time. The impressive contrast in morphological changes in such tissues as the tail, liver, lungs and the developing limbs, must have their counterpart in the contrasting metabolic direction of the biosynthesis of the regulatory macromolecules, particularly nucleic acids and enzymes. It is well known that the protein synthesis occurring in the liver is greatly increased in pace and altered, whereas the reabsorption of the tail is accompanied by an obvious over-all disappearance of tail protein. The conflict in metabolic orientation is most dramatically illustrated in an over-all comparison of RNA and DNA biosynthesis reported by Finamore and Frieden in 1960. A significant increase in radioactive phosphate incorporated into liver RNA and DNA was found one to two days after triiodothyronine (T_3) was injected, in contrast to a decrease in phosphate uptake in tail RNA and DNA. This effect is exactly what would be predicted from the known facts of protein metabolism during metamorphosis.

The genetic control of information transcription and translation is expected to be ultimately reflected in terms of the synthesis of specific protein(s). Experiments on the various aspects of the rate of entry of amino acids into PROTEIN BIOSYNTHESIS have not provided a rationale for the effect of thyroid hormones. Therefore efforts to establish the site of hormone action are now being focused on those steps in the mechanism of protein biosynthesis which involve nucleic acids, particularly the highly metabolically active information-transferring RNA, messenger RNA. (See the work of J. R. Tata and coworkers.) The basic hypothesis is that the cellular differentiation which must proceed in metamorphosis may be the result of a large number of hormonally modified biosynthetic messages. However, at this writing, any of the steps in the temporal sequence of protein biosynthesis outlined below must be regarded as a potential site for thyroid hormone action:

Gene → → DNA → mRNA → → Ribosomes → →

Proteins (Enzymes)→ → →Intracellular Structures

Most of the significant biochemical changes which have been associated with metamorphosis appear to involve modification of proteins or groups of proteins. The metabolic results of these protein changes seem to fulfil the necessary adaptation for the change in environment from water to land. Only several of the more recent developments in this field will be presented in substantial detail.

Nitrogen Metabolism and Excretion: Ammontelism to Ureotelism. In an animal such as the metamorphosing tadpole, in which extensive tissue breakdown and buildup are occurring simultaneously to prepare for a transfer from water to land, it is evident that there will be prodigous changes in the metabolism of nitrogen-containing compounds. Probably the most striking change is the shift from ammontelism (ammonia excretion) to ureotelism (urea excretion). A now well-accepted tenet of comparative biochemistry, developed particularly by Needham and Baldwin, is that the nature of an animal's nitrogenous excretory products is a function of its environment and phylogenetic position. Thus while ammonia is the principal end product of the protein metabolism of aquatic animals, it is supplanted by urea or uric acid in amphibians, birds, reptiles and mammals (see AMMONIA METABOLISM). It is believed that an abundance of water is required for efficient elimination of ammonia because of its greater toxicity. Accordingly, terrestrial animals excrete predominantly urea and, in a few cases, uric acid. Over forty years ago it was demonstrated that anuran tadpoles excrete a large fraction of their nitrogen in the form of ammonia, whereas adult frogs excrete predominantly urea. This has been observed for numerous species of frogs including *R. temporaria*, *R. catesbeiana*, *R. grylio* and several bufo species. This shift in nitrogen excretion to urea has been correlated with the explosive biosynthesis of ARGININE-UREA CYCLE enzymes, now fully documented by Cohen, Brown and associates. These include manifold increases in the activity of enzymes such as arginase, carbamyl phosphate synthetase, ornithine transcarbamylase and arginine synthetase.

Changes in Serum Proteins. The physiological versatility of the serum proteins is well known. Among their more important functions, particularly that of serum ALBUMIN, is the maintenance of blood volume through osmotic regulation. Therefore, it was not unexpected that in the transition from fresh water to land there would be a significant adjustment of the osmotic pressure regulatory mechanisms. Evidence for a remarkable increase in serum albumin concentration during metamorphosis has been found for numerous amphibian species by Herner and Frieden. In fact, it appears that aquatic forms in general have less serum albumin than most land forms. In anuran metamorphosis in such species as *R. grylio* and *R. heckscherii*, there is almost a total absence of serum albumin in the early tadpole. This increases just after metamorphosis to the point where the serum albumin comprises one-half of the total serum proteins. In other species such as *R. catesbeiana*, the typical American bullfrog, and *X. laevis* there is evidence of a small amount of serum albumin in the tadpole and a large increase during metamorphosis. Recently a 10–100 fold increase in ceruloplasmin, the principal copper protein of blood, has been noted during anuran metamorphosis. Significant changes in CARBONIC ANHYDRASE and transferrin have also been noted.

Molecular Changes in Hemoglobin during Metamorphosis. Another most significant adaptive change that occurs during metamorphosis is the alteration in the properties and biosynthesis of HEMOGLOBIN. It was shown earlier by McCutcheon and Riggs that hemoglobin obtained from tadpole red cells has a greater oxygen affinity which is virtually independent of pH, whereas the binding of oxygen by frog hemoglobin is very sensitive to pH (the Bohr effect). This appears to be clearly an adaptive phenomenon since tadpole blood has a large oxygen loading capacity, whereas the blood of the adult frog has a greater unloading capacity, corresponding to the oxygen-poor environment of the tadpole and the oxygen-rich environment of the frog.

These functional differences between tadpole and frog hemoglobins have been studied recently at the molecular level. Both the tadpole and the frog appear to have 3–4 different hemoglobins as shown by different electrophoretic mobility. Tadpole hemoglobins are somewhat unique in having the greatest known electrophoretic mobility of any vertebrate. The widely different amino acid compositions of anuran hemoglobins confirm the fact that there are major changes in the primary structure of these hemoglobins during metamorphosis. The adult hemoglobins are also unique in that they dimerize readily *in vitro* to a molecule weighing 130,000 grams. The ability to form dimers is a function of the stage of metamorphosis and can be correlated with the appearance of highly reactive sulfhydryl groups. In fact, the hemoglobin from young tadpoles has no detectable sulfhydryl groups. During metamorphosis up to eight sulfhydryl groups eventually appear in hemoglobin. The formation of hemoglobin dimer was also correlated with the disappearance of two especially reactive sulfhydryl groups.

Mechanism of Tadpole Tail Resorption. The versatility of the metamorphosing tadpole as a model for the study of differentiation arises not only from the presence of numerous tissues (liver, limbs, lungs) undergoing anabolism, but also from a variety of tissues such as the tail and the gut in which catabolic reactions predominate. It is now accepted that tail resorption is a controlled expression of the activity of intracellular catabolic enzymes. Most tissues seem already to hold the enzymes which are responsible for their own destruction when cells die or assume certain pathological or regulated physiological states. There exist numerous mechanisms, beyond the scope of this discussion, which might account for the control of the catabolic action of these intracellular enzymes.

Virtually every tail enzyme whose over-all impact on the cell is thought to be catabolic undergoes an impressive increase in activity during tail resorption: β-glucuronidase increases 34-fold, cathepsin, 22-fold; collagenase, 10-fold; acid phosphatase, up to 10-fold; several proteinases and peptidases, 6-fold; DNase and phosphatase, several fold. For some of these enzymes it is striking that even as the tissue undergoes degeneration there appears to be an initial synthesis, not just a release, of a particular enzyme which might be involved in tail resorption. Most of these enzymes are thought to be enclosed in cytological particulates known as *lysosomes*. Though no substantial proof is yet available, an obvious convenient hypothesis to account for tail resorption is that the thyroid hormone initiates changes which result in the release of the hydrolytic enzymes of the lysosomes which in turn cause the dissolution of the tail. (See the work of R. Weber.)

Other Biochemical Changes. This brief treatment of the biochemical changes accompanying metamorphosis has not included numerous striking changes in other biochemical systems which may be of less importance in an adaptive sense than those described more fully here. For example, in certain anuran species, metamorphosis is accompanied by a differentiation of VISUAL PIGMENTS from porphyropsin (Vitamin A_2 predominant) to rhodopsin (Vitamin A_1 predominant). Any survival value of a particular visual pigment pattern is not yet appreciated. Quite clearly, there are also significant changes in digestive enzymes as the animal evolves into an exclusive carnivore. Numerous changes in liver enzymes (*e.g.*, glucose-6-phosphatase, ATPase) can also be identified, but the full impact of these changes has not been clearly related to the metamorphic processes. Recently the remarkable temperature sensitivity of this process has been fully documented, and it has been found that the response to thyroid hormones can be arrested at 5°C and resumed 2–3 months later when the animal is returned to 25°C.

Respiration and Metamorphosis. Finally we come to the question of whether metamorphosis is accompanied by an increase in respiration or oxygen uptake. It is paradoxical that we still cannot firmly associate two of the oldest and most important facts regarding the thyroid hormone and anuran metamorphosis. It has been known for over fifty years that the thyroid hormone initiates metamorphosis. It has been realized for an even longer period that the respiration of higher animals is under the profound control of the thyroid hormone. The evidence now at hand suggests that while the injection of thyroid hormones will produce an increase in oxygen uptake in the tadpole, a decrease in respiration probably occurs during spontaneous metamorphosis. These observations are not as contradictory as they might first seem. When triiodothyronine or thyroxine is injected, there is a flood of metabolites which are immediately available for oxidation and/or rapid protein biosynthesis. During normal metamorphosis, metabolites are gradually being used for new protein and other macromolecular reconstruction. While the enzymic machinery involved in oxidation may be poised for greater activity, the absence of metabolites available for oxidative pathways may prevent any increase in oxygen uptake. The elimination of the large surface area of the tail and the extensive modification of skin cells might also contribute to a decrease in oxygen utilization during metamorphosis. Thus, despite a steadily increasing level of thyroid hormone secretion, it is possible to rationalize the increased protein biosynthesis with the lack of increased oxygen uptake in the tadpole during spontaneous metamorphosis.

References

1. FRIEDEN, E., "The Chemistry of Amphibian Metamorphosis," *Sci. Am.*, **209**, 110 (Nov. 1963).
2. BENNETT, T. P., AND FRIEDEN, E., "Metamorphosis and Biochemical Adaptation in Amphibia," in "Treatise on Comparative Biochemistry," Vol. IV B (MASON, H. S., AND FLORKIN, M., EDITORS), pp. 483–556, New York, Academic Press, 1962.
3. BROWN, G. W., JR., "The Metabolism of Amphibia," in "Physiology of the Amphibia" (MOORE, JOHN A., EDITOR), pp. 1–98, New York, Academic Press, 1964.
4. WEBER, R., "The Biochemistry of Amphibian Metamorphosis" in "The Biochemistry of Animal Development," Vol. II, (WEBER, R., EDITOR) New York and London, Academic Press, 1966.
5. FRIEDEN, E., "The Biochemistry of Amphibian Metamorphosis" in "Metamorphosis: A Problem in Developmental Biology," (ETKIN, W. AND GILBERT, L. I., EDITORS) New York, Appleton-Century-Crofts, 1967.
6. ETKIN, W., "How a Tadpole Becomes a Frog," *Sci. Am.*, **214**, 76 (May 1966).

EARL FRIEDEN

METAMORPHOSIS, INSECT

Some aspects of insect metamorphosis that have been correlated with "chromosome puffing" are discussed in the article BALBIANI RINGS.

METHEMOGLOBIN

Methemoglobin (ferrihemoglobin) is chemically the equivalent of hemoglobin except that the iron has been oxidized to the ferric condition; as a result, methemoglobin cannot perform the function of carrying oxygen. It is dark brown in color, with an absorption maximum at 634 mμ. (see HEMOGLOBINS.)

Various oxidizing agents may be used to convert hemoglobin into methemoglobin; methemoglobin is found normally in blood to an extent of 1–2% of the hemoglobin. Its accumulation is prevented by reducing agents in the blood (*e.g.*, GLUTATHIONE). Methemoglobinemia, an excessive amount of methemoglobin in the blood, is usually caused by agents such as nitro compounds, aromatic amines, and drugs such as acetanilide or sulfonamides.

METHIONINE

L-Methionine, α-amino-γ-methylthio-*n*-butyric acid, $CH_3—S—CH_2—CH_2—CHNH_2—COOH$, is one of the protein-derived AMINO ACIDS which is present in most PROTEINS except the simpler protamines and histones. Egg albumin on hydrolysis yields about 5% of its weight of methionine;

most other proteins contain much less. Methionine is nutritionally essential for mammals and fowls. Not only does it furnish a unique skeletal structure but it is also an important methylating agent and facilitates the synthesis of choline in the body. Homocysteine, HS—CH_2—CH_2—$CHNH_2$—COOH (which is not derived from proteins), can serve nutritionally in place of methionine; it is methylated (see METHYLATIONS) in the body to produce methionine. CYSTEINE (or cystine) has a sparing action on methionine; when these other (dispensible) sulfur-containing amino acids are present in the diet, not so much methionine is needed. Methionine serves as a precursor of cysteine and cystine, but the reverse relationship does not exist. The methionine requirement of human beings appears to vary over about a sixfold range from individual to individual. L-Methionine has a specific rotatory power in 3*N* HCl of $[\alpha]_D = +23.4°$.

METHYLATIONS IN METABOLISM

Methyl-containing Compounds and Their Biological Significance. A great variety of methyl-containing compounds occurs in microorganisms as well as in the tissues and body fluids of higher plants and animals. A tentative list of these compounds includes: (a) qualitatively and quantitatively important constituents of cell structures, such as methionine-containing PROTEINS, lecithins and nucleic acids; (b) substances involved in the storage of energy or in certain metabolic reactions, such as CREATINE phosphate and CARNITINE; (c) hormones (*e.g.*, epinephrine) and vitamins (*e.g.*, vitamin B_{12}); (d) compounds which are believed to play special roles (physiological or structural) for certain organisms (such as ALKALOIDS in plants; cyclopropane- or branched fatty acids in bacteria); (e) excretory products, resulting from the catabolism of physiologically active compounds (*e.g.*, metanephrine, methylnicotinamide) or from the detoxication of foreign materials (*e.g.*, methylpyridine).

Some preformed methyl-containing substances are obtained from the diet or from the environment, either directly or after digestion. More generally, methyl groups arise in living bodies by two distinct mechanisms: net synthesis from one-carbon metabolites (methylneogenesis) or transfer from other methylated compounds containing "labile" methyls (transmethylation).

Methylneogenesis. At the present time, the methyl groups of only two compounds, THYMINE and METHIONINE, are known to be formed directly by synthesis. The methyl of thymine (a major component of DNA) originates from one-carbon units, the hydroxymethyl derivative of tetrahydrofolate being the immediate precursor and the uracil moiety of the nucleotide uridine phosphate (deoxy-UMP) being the methyl acceptor. In the reaction

$$\text{5,10-CH}_2\text{OH-folate-H}_4 + \text{Deoxy-UMP} \rightarrow \text{Deoxy-TMP} + \text{Folate-H}_2$$

folate-H_4 functions not only as the carrier, but also as the reducer of the —CH_2OH group (see FOLIC ACID COENZYMES; SINGLE CARBON UNIT METABOLISM).

In the formation of the amino acid methionine (an essential component of dietary and tissue proteins), 5,10-CH_2OH-folate-H_4 is first reduced to methyl folate, $FADH_2$ being the reducing agent. The methyl group is then transferred to homocysteine [$HS\cdot CH_2\cdot CH_2\cdot CH(NH_2)\cdot COO^-$] by pathways which are still not well known and which possibly are different in different species. In *E. coli* and in mammals, the pathway probably involves a vitamin B_{12} enzyme and also catalytic amounts of $FADH_2$ and S-adenosylmethionine (AdMe), and therefore could be represented as follows:

$$\text{5-CH}_3\text{-folate-H}_4 + \text{Homocysteine} \xrightarrow[\text{B}_{12}\text{ enzyme}]{\text{AdMe, FADH}_2} \text{Methionine} + \text{Folate-H}_4$$

Since one-carbon units are formed in the metabolism of many substances, all of these substances (and especially carbon-3 of the amino acid serine) many contribute to the synthesis of methyls (see also COBALAMINS).

Transmethylation. Primarily on the basis of nutritional experiments, it was shown that, unlike most other methyl-containing compounds, methionine and choline contain "labile" methyl groups, *i.e.*, methyls which in the intact animal can be transferred as units to various acceptors and thus give rise to other methylated products. Likewise, the methyl groups of some quantitatively less important natural compounds, such as betaine [$(CH_3)_3N^+\cdot CH_2\cdot COO^-$], dimethylpropiothetin [$(CH_3)_2\cdot S^+\cdot CH_2\cdot CH_2\cdot COO^-$] or S-methylmethionine [$(CH_3)_2\cdot S^+\cdot CH_2\cdot CH_2\cdot CH(NH_2)\cdot COO^-$], and those of certain other substances not yet found in living bodies, such as dimethylacetothetin [$(CH_3)_2\cdot S^+\cdot CH_2\cdot COO^-$] or methyl phosphate ($H_2PO_3\cdot O\cdot CH_3$), seem to be "labile." Actually it appears that one methyl group of choline can be transferred to homocysteine to yield methionine only after oxidation of choline to betaine aldehyde and betaine, as indicated below:

$$\underset{\text{Choline}}{(CH_3)_3\cdot N^+CH_2\cdot CH_2OH} \rightarrow \underset{\text{Betaine aldehyde}}{(CH_3)_3\cdot N^+CH_2\cdot CHO} \rightarrow \underset{\text{Betaine}}{(CH_3)_3N^+CH_2\cdot COO^-} \quad (1)$$

$$\underset{\text{Betaine}}{(CH_3)_3\cdot N^+CH_2\cdot COO^-} + \underset{\text{Homocysteine}}{HS\cdot CH_2\cdot CH_2\cdot CH(NH_2)\cdot COO^-} \rightarrow \underset{\text{Methionine}}{CH_3\cdot S\cdot CH_2CH_2\cdot CH(NH_2)\cdot COO^-} + \underset{\text{Dimethylglycine}}{(CH_3)_2\cdot N\cdot CH_2\cdot COO^-} \quad (2)$$

In the transfer of methyl from betaine (as in the analogous reactions involving thetins as the methyl donors), homocysteine is the obligatory acceptor and therefore methionine is always the methylated product. Instead, in all transmethylations yielding products other than methionine, the immediate donor is AdMe, a compound formed by transferring the adenosyl portion of ATP to the S of methionine:

$$\text{L-Methionine} + \text{ATP} \xrightarrow[\text{Enzyme}]{\text{Mg}^{++},\ \text{Glutathione}} \text{AdMe} + \text{PP}_i + \text{P}_i$$

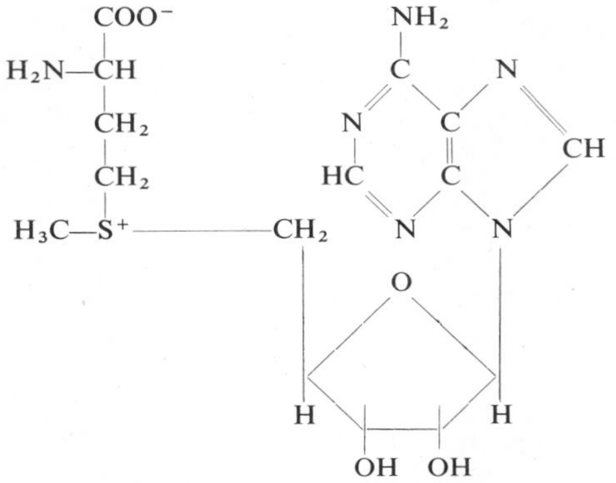

S-Adenosylmethionine (AdMe)

In AdMe the methyl is attached to a sulfonium pole, the energy for the methyl transfer probably being derived from the S atom returning to a divalent state. A similar mechanism has been postulated for the transmethylations from other immediate methyl donors such as the thetins or betaine, in which the methyl is also linked to an onium pole (trivalent S, or tetravalent N). It may be added that the ester bond in methlyphosphate is also probably a high-energy bond.

The enzymes catalyzing the transfer of methyls to various kinds of atoms in the acceptor molecule (N-, O-, S-, or C-methyltransferases) are usually specific for both the methyl donor and the methyl acceptor. On the other hand, the product formed from AdMe in the various transmethylations is adenosylhomocysteine [adenine-ribosyl-$S \cdot CH_2 \cdot CH_2 \cdot CH(NH_2) \cdot COO^-$], from which homocysteine can be easily obtained by hydrolysis. Homocysteine, which is the required acceptor of methyl from 5-methylfolate and betaine, is not a component of dietary or tissue proteins and cannot be synthesized by higher animals. Thus, in these organisms, and with the exception only of the thymine in DNA, methionine is involved in the formation of all methylated compounds; indeed, it is not only the chief dietary source of both homocysteine and preformed methyls, but also the obligatory intermediate in the utilization of the methyls obtained from other sources (synthesis from one-carbon compounds or transmethylation from choline and betaine).

A few examples of the formation of methylated compounds will be discussed here in some more detail.

Methylated Phospholipids. Lecithins are the most abundant PHOSPHOLIPIDS of plant and animal tissues, but are present in only a few bacterial species. Two major pathways for the formation of lecithin are known to occur. One of these pathways utilizes preformed choline (presumably liberated from dietary or tissue phospholipids):

$$\text{Choline} \xrightarrow{\text{ATP}} \text{Phosphorylcholine} \xrightarrow{\text{CTP}} \text{Cytidine Diphosphate Choline} \xrightarrow{\text{1,2-Diglyceride}} \text{Phosphatidyl Choline (Lecithin)}$$

The second pathway involves the stepwise methylation of the ethanolamine moiety of CEPHALINS:

$$\text{Phosphatidyl-EA (Cephalins)} \xrightarrow{\text{AdMe}} \text{Phosphatidyl-MMe} \xrightarrow{\text{AdMe}} \text{Phosphatidyl-DME} \xrightarrow{\text{AdMe}} \text{Phosphatidylcholine (Lecithin)}$$

where EA (ethanolamine, $H_2N \cdot CH_2 \cdot CH_2OH$), MME (monomethylethanolamine, $CH_3 \cdot NH \cdot CH_2 \cdot CH_2OH$), DME (dimethylethanolamine, $(CH_3)_2N \cdot CH_2 \cdot CH_2OH$), and choline [$(CH_3)_3N^+ \cdot CH_2 \cdot CH_2 \cdot OH$] indicate the nitrogenous components of the respective phospholipids. Unequivocal evidence for the occurrence of the second and third methylation steps in particulate liver preparations has been obtained. With these preparations, identification of cephalins as the direct methyl acceptors for the initial methylation is not quite as definite, but recently such a step has been reproduced *in vitro* with a soluble enzyme obtained from *Agrobacterium tume-*

faciens. In animals and higher plants, methylation of cephalins is probably the rate-limiting step, since only minimal amounts of phosphatidyl-MME and phosphatidyl-DME are normally found. It is not yet known whether in these organisms each methylation step is catalyzed by a separate enzyme. This is probably the case in lower forms of life, since mutants of *Neurospora crassa* have been obtained in which phospholipids containing MME or DME accumulate, and *Clostridium butyricum* produces phosphatidyl MME but no phosphatidyl DME or lecithins. At any rate, in microorganisms as well as in animals, AdMe is the donor for all three methylation steps.

The possibility that choline may be synthesized by a route other than the stepwise methylation of the EA moiety of cephalins cannot be excluded, but at the present time no evidence is available for such a possibility.

Little is known about the relative importance of the methylation pathway as compared with the synthesis of lecithin from preformed choline. Preliminary experiments with slices of rat tissues, incubated with relatively large amounts of methionine or choline, indicate that methylation of phospholipids is relevant only in the liver. In all tissues, including the liver, incorporation of choline into the phospholipids is by and large the dominant pathway. It seems possible, however, that lecithin formation by transmethylation may become quite important, if and when the availability of preformed choline is the limiting factor.

Methylated Bases in Nucleic Acids. As mentioned above, the methyl group of thymine, a major constituent of DNA in animals, plants and microorganisms, originates from a one-carbon unit which is then transferred to the free nucleotide deoxy-UMP. In certain types of bacteriophages, hydroxymethylcytosine is a major component of nucleic acids, substituting for cytosine, one of the four main bases in the nucleic acids of living cells. Apparently hydroxymethylcytosine is formed by direct transfer of the —CH_2OH group from 5,10-CH_2OH-folate-H_4. In much smaller proportions, 5-methylcytosine is present in the DNA of animals and plants and 6-methyladenine in the DNA of bacteria. Several methylated bases, both PURINES and PYRIMIDINES, are present also in ribosomal RNA, and more especially in transfer RNA (in which as many as ten different methylated bases have been characterized; see SOLUBLE RIBONUCLEIC ACIDS).

All of the minor methylated components of DNA and RNA are now known to originate by transfer of methyls from AdMe to a carbon or to a nitrogen at the appropriate sites of preformed nucleic acids. Enzymes which catalyze such transmethylations, and which appear to be species specific, have been isolated from a variety of microorganisms and animal tissues. It may be noted that transmethylation from methionine to the uracil moiety of intact RNA yields thymine. Thus the source and the mode of formation of this base in RNA and in DNA are sharply different.

The biological significance of the introduction of methyls in the molecules of DNA and RNA is still obscure, but it seems likely that such processes are quite important in modifying and determining the precise structure and conformation of the various types of nucleic acids [see also RIBONUCLEIC ACIDS (BIOSYNTHESIS)].

Formation and Inactivation of Physiologically or Pharmacologically Active Compounds. Transmethylation from AdMe to the N or the O of certain physiologically active compounds markedly increases or modifies their activity. Thus, phenylethanolamine N-methyltransferase, an enzyme present almost exclusively in the medulla of the adrenals, converts norepinephrine into epinephrine (see ADRENALINE), a hormone with even greater potency and distinctly different physiological effects. The same enzyme further methylates epinephrine to methylepinephrine which is also found normally in the adrenals. N-Methylation of serotonin, a vasoconstrictor substance present in the blood and tissues of most animals, and of tryptamine yields N,N-dimethylserotonin and N,N-dimethyltryptamine, respectively; these two compounds have been shown to cause marked psychotic changes in man. In the pineal gland, O-transmethylation to N-acetylserotonin results in the formation of melatonin (5-methoxy-N-acetyltryptamine), a compound which reverses the darkening effect of the black pigment in the melanophore cells of cold-blooded animals. In plants, a number of pharmacologically active ALKALOIDS, such as morphine, nicotine and ordenine, are formed by N-methylation of the corresponding nor-compounds.

On the other hand, N- or O-methylation can also convert active substances into less active or inactive products. Histamine, a powerful vasodilator substance believed to play a major role in allergic reactions, is converted to the inactive methylhistamine by a specific N-methylferase, present in most mammalian tissues. O-Methylation is the main catabolic pathway for epinephrine, the methylated products [3-methoxyepinephrine (metaepinephrine), 3-methoxy-4-hydroxymandelic acid, and 3-methoxy-4-hydroxyphenylglycol] being excreted in the bile and in the urine, free or conjugated with sulfate or glucuronate. In humans, N-methylnicotinamide is the major catabolic product of nicotinamide, the amide of the vitamin niacin (NICOTINIC ACID) and a characteristic component of the coenzymes DPN and TPN which play fundamental roles in the oxidation and reduction processes of the cell. Methylnicotinamide has no vitamin activity and does not function as a methyl donor.

Methylation is also a mechanism for the detoxication of some foreign substances. N-Methylation of pyridine and quinoline yields products which are less toxic and which are readily excreted. After administration to man of certain phenolic compounds, the corresponding O-methylated products are found in the urine. A system which catalyzes the transfer of methyls from AdMe to non-physiological sulfhydryl compounds, such as BAL or mercaptoethanol, has been demonstrated in the liver of several mammalian species. N- and S-methylations of analogues (thio- or amino-substituted) of natural purines and pyrimidines are catalyzed by enzymes present in many animal tissues and microorganisms.

References

The historical development and the present status of our knowledge are reviewed in references 1 and 2, respectively. The various chapters of reference 3 include presentations and discussions on most types of transmethylation, with the exception only of the methylation of phospholipids. Some research articles on this topic are mentioned in references 4–6.

1. DuVigneaud, V., "A Trail of Research in Sulfur Chemistry and Metabolism and Related Fields," Ithaca, N.Y., Cornell University Press, 1952.
2. Mudd, S. H., and Cantoni, G. L., in "Comprehensive Biochemistry" (Florkin, M., and Stotz, E. H., Editors), Vol. 15, p. 1, Amsterdam, Elsevier, 1964.
3. Shapiro, K. S., and Schlenk, F. (Editors), "Transmethylation and Methionine Biosynthesis," Chicago, University of Chicago Press, 1965.
4. Bremer, J., Figard, P. H., and Greenberg, D. M., *Biochim. Biophys. Acta*, **43**, 477 (1960).
5. Artom, C., and Lofland, H. B., *Biochem. Biophys. Res. Commun.*, **3**, 244 (1960); Artom, C., *ibid.*, **15**, 201 (1964).
6. Artom, C., *Federation Proc.*, **24**, 477 (1965).

Camillo Artom

MEVALONIC ACID

Mevalonic acid, β,δ-dihydroxy-β-methylvaleric acid,

$$HOOC-CH_2-\underset{\displaystyle OH}{\overset{\displaystyle CH_3}{\overset{|}{\underset{|}{C}}}}-CH_2CH_2OH$$

is an intermediate in the biosynthesis of cholesterol and other sterols [see ISOPRENOID BIOSYNTHESIS; STEROLS (BIOGENESIS AND METABOLISM)].

MEYERHOF, O. F.

Otto Fritz Meyerhof (1884–1951) was a German-born biochemist, noted particularly for his discovery of anerobic GLYCOLYSIS in muscle. He was awarded the Nobel Prize in medicine and physiology in 1922, which he shared with A. V. Hill, another outstanding muscle physiologist. His position in Nazi Germany became increasingly uncomfortable so he fled to France and later to the United States where he eventually became a citizen.

MICHAELIS-MENTEN EQUATION. See ENZYMES (KINETICS).

MICROBIOLOGICAL ASSAY METHODS

Microbiological assays have played a significant role in the development of biochemistry, particularly in recent decades.

One of the first examples demonstrating the value of such investigations to biochemistry was that of Mueller, who discovered METHIONINE in 1923, during his attempts to learn the complete nutritional needs of hemolytic streptococci. He investigated a sulfur-containing fraction derived from impure protein which contained an "unknown" required by the organism under investigation. On purification, the sulfur-containing material lost its microbiological activity, but it turned out to be—as a by-product of his research—methionine, which marks a milestone in protein chemistry. In attempts to determine the exact nutritional needs of yeasts, INOSITOL was isolated and found to be a yeast growth substance (Eastcott, 1928). It had been known earlier as a naturally occurring sugar-like compound. In 1936 BIOTIN was first isolated from eggyolk by Kögl, using as a basis its high growth stimulating activity in a microbiological yeast assay. In 1933 the discovery of PANTOTHENIC ACID was announced, and in 1940 its chemical nature was determined (R. J. Williams)—all as a result of application of a microbiological yeast test. Subsequently many discoveries, including those of pyridoxal and pyridoxamine (E. E. Snell), FOLIC ACID, and LIPOIC ACID were made using microbiological assays as a basis. In addition, microbiological assays contributed enormously to the knowledge about numerous other B vitamins (including VITAMIN B_{12}) and related compounds, as well as to protein chemistry.

The principle of microbiological assays is simple: If the complete needs of a microorganism for growth (propagation) consist of a series of recognized nutrients a, b, c, d, ..., the elimination of any single nutrient from the culture medium (when feasible) makes the medium deficient and subject to improvement *only* by adding this missing nutrient. The improvement or lack of improvement of the medium when a mixture of unknown composition is added to it is thus the basis of a *test* or assay for the missing nutrient and can, by the use of suitable standards, be made quantitative. For perfect functioning of a microbiological test, *every* nutritional need of the organism concerned must be known.

In practical use, there are many ramifications and potential difficulties; *e.g.*, toxic, inhibitory and interfering substances must be avoided or compensated for. This is not always easy, though the high dilutions usually involved make it less difficult. Sometimes agents to be assayed stimulate propagation but are not absolutely essential in a long-duration test.

Many microorganisms, including yeasts, lactic acid bacteria and other bacteria have been used for microbiological tests, as have protozoa to a lesser extent. One variant involves the use of *mutant strains* of NEUROSPORA and of bacteria (*e.g.*, *E. coli*) which have, because of artificially induced genetic defects, needs for specific nutrients such as individual amino acids or vitamins.

An outstanding advantage of microbiological assays is that they can readily be used for the determination of minute quantities. By the use of Linderstrom-Lang's microtechniques, it is possible to extend their use (when required) into the ultramicro range. By this means, amino acids can be determined (Agren, 1949) at levels of from 0.066–0.25 μg. Vitamins can be determined at much lower levels. Without resorting to unusual microtechniques, many of the amino acids can be determined in 10–20 mg samples of protein. (Snell, 1948.)

Substances of extremely diverse nature and of fundamental biological importance are capable of being determined microbiologically. These include all the water-soluble vitamins (except possibly vitamin C), all the protein-derived amino acids, many other biologically active substances such as β-alanine, choline, ethanolamine, hematin, individual PURINES, PYRIMIDINES, nucleosides, certain coenzymes and minerals. Not all of the possibilities have been exploited.

Microbiological methods can, of course, be combined with other techniques such as CHROMATOGRAPHY, ELECTROPHORESIS, and ISOTOPIC TRACERS, in a variety of ways. *Bioautographs* are produced, for example, when unknown mixtures are subjected to paper chromatography and the presence of growth-promoting agents is then tested for *in situ* by placing the paper, after the chromatography, in contact with a suitable solidified (*e.g.*, agar-containing) culture medium.

It is largely because of microbiological assays and their applications, that the metabolic unity see (MICROBIOLOGY) in all nature has been elucidated. We now know that all kinds of organisms are composed of the very same amino acids, vitamins, nucleotides and minerals. The pathways in metabolism are complex and intricate, but we know now that there are numerous common pathways which are likely to be used by organisms of the most diverse character. These pathways originated very early, probably even before life began (see ORIGIN OF LIFE).

References

1. *Annual Reviews of Biochemistry*, especially volumes in the decade 1940–1950.
2. JOHNSON, B. CONNOR, "Methods of Vitamin Determination," Minneapolis, Burgess Publishing Co., 1948.
3. HENDLIN, D., "Use of Lactic Acid Bacteria in Microbiological Assays," Part III, in "Symposium on the Lactic Acid Bacteria" (TITTSLER, R. P., *et al.*) *Bacteriol. Rev.*, **16**, 227–260 (1952).

V. CHELDELIN

MICROBIOLOGY (BIOCHEMICAL ASPECTS)

"The most important discoveries of the laws, methods and progress of Nature have nearly always sprung from the examination of the smallest objects which she contains."

J. B. Lamarck
Philosophie Zoologique, 1809

After briefly defining microbiology, it is our intention (1) to discuss how microorganisms have been used as tools for investigating biochemical problems; (2) to point out a few of the unusual substances that appear to be quite unique to either bacteria, molds, yeasts, or viruses; (3) to discuss some of the compounds that are inhibitors of microbial reactions and are being used for chemotherapeutic purposes.

Microbiology is the study of single-celled organisms too small to be observed with the naked eye. Classically, this field has included the study of *algae* and *protozoa*. In this article, however, we have arbitrarily restricted our comments to *bacteria*, *yeasts* and *molds*. Furthermore, we have chosen to include viruses although they are not unicellular. The inclusion of viruses is appropriate because studies of the replication of viruses and their mode of action on the host cell have been among the fastest developing and most exciting areas of biochemical microbiology.

Organisms classified as bacteria form a heterogeneous group which may have widths of 0.2–3 μ and lengths of 0.7–50 μ, depending on the species. Bacteria with spherical shapes have average diameters of 0.8–2.5 μ. Although both yeasts and molds are larger than bacteria, the viruses are at least an order of magnitude smaller, having an average length or diameter of approximately 0.3–0.02 μ [see VIRUSES (COMPOSITION)].

The biochemical study of microorganisms and the use of microorganisms to study biochemical reactions have tended to obliterate the boundary between biochemistry and microbiology. When the cell theory of life and the unitary theory of biochemistry are considered, the fact that such a boundary is no longer clearly defined is not surprising.

Essentially, the cell theory states that all living organisms are composed of units called cells. All cells are basically similar in structure and organization, having similar functional subunits. However, through a process of evolution, some of these subunits have been modified. For example, the protoplasm of all cells is enclosed in a non-rigid membrane. Bacterial cells like plant cells have, in addition, an outermost semirigid cell wall which is lacking in mammalian cells. Another example of a modification is that the genetic material or DEOXYRIBONUCLEIC ACID (DNA) of a *mammalian cell* is present in a clearly defined nucleus surrounded by a nuclear membrane. The bacterial cell also has a nucleus, namely, the region where the DNA is located, but the DNA is not surrounded by a nuclear membrane.

Concept of Comparative Unity of Biochemistry. Just as the cell theory states the over-all structural similarities between various living cells which are recognized by gross or microscopic examination, the unitary theory of biochemistry states that all living processes are due essentially to the same basic chemical reactions common to all cells. The vast variety of different metabolites and structures found in nature only tends to mask these basic chemical reactions common to all cells. All living cells contain nucleic acids, PROTEINS and LIPIDS, and usually contain polysaccharides as well. Certain nucleic acids serve as a means of transferring inherited information from generation to generation and thereby provide each cell with the information needed for the production of proteins which in turn will catalyze the synthesis of macromolecules as well as smaller molecules. Another example of the unitary theory is that the synthesis of DNA requires a DNA template, four deoxynucleoside triphosphates, magnesium and an enzyme, DNA polymerase [see DEOXYRIBONUCLEIC ACIDS (REPLICATION)]. The reaction requires all these components, regardless of whether the DNA is being synthesized in a very small *Escherichia coli* cell, the sea urchin cell or the larger calf thymus cell.

Microorganisms as Tools for Study. Thus, it is not surprising that biochemists studying metabolic pathways quickly turned to bacteria in the 1940's. The techniques for breaking up living cells and their subcellular components in order to study reactions with purified enzymes became well established at that time. These techniques were soon able to clarify the hitherto complex multienzyme systems which make up the metabolism of the cell.

The unsophisticated have often called bacteria "a bag of enzymes." Bacteria are a readily available source of ENZYMES, enzyme substrates and structural components of living systems. An enzymatic reaction found in a bacterial cell usually has a rate many times greater than that of a similar mammalian reaction. Today, microorganisms such as *Escherichia coli* can be purchased in 100-pound quantities. Baker's or brewer's yeast is obtainable in even greater quantitites. Therefore, as a source of living material for biochemical study, it would appear, in many cases, that such microorganisms are easier to obtain than 100 pounds of some rapidly regenerating animal tissues which may consist of more than one type of cell. The rapidity with which large numbers of organisms may be obtained and tested has permitted studies of events which occur at very low frequencies. This is readily seen when one considers that the generation time of *E. coli* is about 20 minutes as compared with 33 years for man.

Microorganisms have been of immeasurable aid in the establishment of the basic concepts of BIOCHEMICAL GENETICS. Foremost among these concepts has been bacterial transformation. The first *in vitro* demonstration of the transfer of genetic information by deoxyribonucleic acid (DNA) was by Avery, MacLeod and McCarty in 1944. They transformed one strain of *Diplococcus pneumoniae* into a second strain of pneumococcus by incubating the former strain with DNA purified from the second strain. This was the first definitive evidence that hereditary information resided in DNA molecules. Work with plant viruses such as TOBACCO MOSAIC VIRUS and, later, with some bacterial and animal viruses showed that in special instances RIBONUCLEIC ACID (RNA) may also carry hereditary information. The classic experiment by Hershey and Chase using bacteriophages whose DNA and proteins were isotopically labeled with P^{32} and S^{35}, respectively, established that intracellular injection of only the DNA of a bacteriophage was necessary to program the production of more bacteriophages. This has been further confirmed by the production of progeny phages in systems where phage DNA, free of its protein coat, is added to bacterial cells capable of taking up DNA molecules [see VIRUSES (REPLICATION)].

The discovery of bacterial recombination by conjugation between two different strains of *E. coli* and bacteriophage-mediated transfer of genetic information from one bacterium to another (transduction) has permitted genetic mapping of the bacterial chromosome (see BACTERIAL GENETICS). This, in turn, has been instrumental in obtaining evidence supporting the concept that a mutation resulting in the change of a single nucleotide along a DNA molecule might cause the substitution of a new amino acid for one normally present in a specific protein molecule. The concept of "messenger RNA" molecules complementary in nucleotide sequence to strands of DNA molecules and their proposed function in directing the synthesis of specific proteins on ribosomes obtained their first support from bacteriophage and bacterial experiments. *In vitro* reactions using purified RNA polymerase enzymes from bacterial sources have lent further support to the idea of an RNA synthesis dependent on a DNA template. These reactions found first in bacterial cells have since been substantiated by studies of mammalian enzymes. RNA synthesis dependent on an RNA template has only been found after infection of a cell with an RNA virus.

The search for the key to a universal GENETIC CODE based on linearly arranged combinations of purines and pyrimidines finally began to look promising with the discovery that enzymes from microbial sources would incorporate particular amino acids into polypeptide chains only in the presence of specific polynucleotides of known base sequence. The genetic code was further deduced to be a sequence of three nucleotides for each amino acid. Universality in transcription of the genetic code was suggested by experiments indicating that viable *animal viruses* may be produced in *bacterial* cells upon introduction of the naked animal virus DNA into the bacteria.

The contribution of microorganisms to the understanding of cellular *regulatory mechanisms* has also been significant. Much of the information on how the synthesis or function of particular enzymes is increased or decreased when the cell is cultured under different conditions has been derived from studies using microbial systems. The rapid growth rate of microorganisms in a wide variety of media, including simple inorganic salts and chemically defined compounds, suggested that the intracellular enzyme composition and concentrations might depend on the substrates available. It appears logical that in a cell of limited size, protein synthesis would be directed toward the making of the enzymes most useful for growth in a particular medium. If this were so, one would expect synthesis of some enzymes to be induced by the presence of the enzyme substrate or related compounds. Such has been the case. Many of the basic concepts of enzyme induction and the genetic control of induction have been developed from studies on the β-galactosidase of *Escherichia coli*. The idea of a genetic regulatory site on the DNA controlling the synthesis of a group of functionally related enzymes has arisen from studies of *E. coli* mutants. Repression, the inhibition of synthesis of a particular enzyme to prevent formation of an unneeded product, has also been extensively studied using microbial systems. Studies on the enzymes necessary for arginine biosynthesis, histidine biosynthesis, and PYRIMIDINE BIOSYNTHESIS in bacteria have suggested that the end product of a sequence of enzyme reactions may inhibit the synthesis of one or more enzymes contributing to its production. Studies on bacterial alkaline phosphatase have shown that the synthesis of this particular enzyme may be repressed by an abundance of inorganic phosphate and may be induced by "starvation" on a low-phosphate medium. (See also METABOLIC CONTROLS.)

Since the excess of a particular product may be deleterious for a cell, another control mechanism is present in bacteria. In this type of control, which is called end product inhibition or feedback inhibition, the end product of the metabolic reaction inhibits the function of the first enzyme of the pathway. The functioning of these regulatory mechanisms either singly or in combination serves to permit cellular growth and tends to maintain a certain constancy of cell composition under a variety of environmental conditions. These initial findings with bacteria pointed out the possibility that identical biochemical regulatory mechanisms exist in all biological cells.

Microorganisms may be used as *analytical* tools. Many microorganisms are unable to synthesize one or more of the compounds they require for growth. Such compounds or nutrients have been called essential growth factors. This requirement of some microorganisms for growth factors has resulted in the use of microorganisms as indicators in MICROBIOLOGICAL ASSAYS for amino acids, vitamins and other compounds. The development of the procedures utilizing microorganisms in such quantitative assays is due largely to the efforts of Snell and Strong.

If a microorganism is unable to synthesize a compound which is required for its growth, and if a basal nutrient medium can be prepared devoid of the growth factor, the organism can often be used as a tool to quantitatively measure the concentration of the compound in a variety of sources. Growth responses of the microorganism to a standard solution containing known amounts of the growth factor are determined and compared with the response to the unknown solution to be analyzed. Inherent in the concept of this type of analysis is that the response must be directly proportional to the amount of the growth factor over a definite concentration range whether it is in the standard or unknown solution. The response should not be influenced by any other factor in the unknown solution. The response measured can be acid production, turbidity, or in the case of fungi, the weight of mycelia.

The microbiological assay not only has been of great aid in the quantitative analysis of nutritive materials of natural origin, but has also resulted in the discovery and elucidation of new biologically active substances of natural as well as synthetic origin. Although the nutritional requirements for THIAMINE, RIBOFLAVIN, PYRIDOXINE and choline (vitamins necessary for the growth and development of organisms) primarily came from animal nutritional studies, the essentiality of pyridoxal, pyridoxamine, *p*-aminobenzoic acid, INOSITOL and PANTOTHENIC ACID as nutritional agents came from studies of the nutrition of bacteria and fungi.

Unique Compounds. Not only have microorganisms been of aid in the study of biochemical problems, but also, biochemistry has helped the understanding of microbiological problems. Some bacterial cells have been modified biochemically to such an extent that they contain compounds unique to their species. Many of these substances are specialized complex pigments. It is not our intention in this brief article to catalogue such substances. Rather, we wish to cite a few examples of compounds unique to microorganisms that are now and will be of significant interest.

The rigid cell wall of bacteria, which has been already mentioned, has, as part of its structure, an unusual amino sugar, *muramic acid* (3-O-carboxyethyl glucosamine). An important part of the cell wall material, especially of Gram-positive organisms, is a mucopeptide which consists of muramic acid, alanine, and glutamic acid, often in the D-configuration, with either L-lysine or diaminopimelic acid also present. Many antibiotics are thought to act by interfering with the synthesis of mucopeptide (see BACTERIAL CELL WALLS).

It is not unusual for the microbial cell wall mucopeptide to contain α,ε-diaminopimelic acid (DAP). This compound has been found only in bacteria and in blue-green algae. Present evidence indicates that either DAP *or* lysine is found in cell wall mucopeptide. DAP also occurs in the walls of SPORES.

Bacterial SPORES also contain another unusual compound, dipicolinic acid (pyridine-2,6-dicarboxylic acid) that appears to be unique to endospores of the genus *Bacillus* and *Clostridium* and of one coccus, *Sporosarcina ureae*. The function of this compound is unknown. The mechanism of "triggering" sporulation, and the explanation for spores being more resistant than vegetative cells to heat and many chemical reagents, are also unknown. It is known, however, that during spore formation there is a spurt of biosynthetic activity. The spore must contain all the genetic information of the bacterium plus new compounds such as dipicolinic acid. The coat proteins of spores are newly synthesized proteins that are not related in composition to the proteins of the vegetative cells.

The modification of biosynthetic apparatus that occurs during spore formation, in many ways, is similar to the changes that occur in the metabolic apparatus of the host after *E. coli* is infected by one of the T-even bacteriophages. The invading phage nucleic acid entering the cell induces the synthesis of new enzymes necessary for making the nucleic acid and proteins of the virus. The four major purine and pyrimidine bases of most DNA's are adenine, thymine, guanine, and cytosine. The DNA of the T-even coliphages do not contain cytosine; the cytosine is replaced by 5-hydroxymethylcytosine (HMC). In nature this unique base, HMC, has only been found to occur in the DNA of T-even coliphages. The T-even phage DNA is also unusual because glucose is attached to a definite percentage of the HMC residues.

The bacteriophages that infect *B. subtilis* may be divided into three groups. The DNA's of the first group contain uracil in place of thymine; the second group has hydroxymethyluracil (HMU) in place of thymine; the third group has thymine. HMU has been found in no other DNA, but it has been found in the growth media of certain fungi, which presumably synthesize the pyrimidine base but do not incorporate it into nucleic acid.

Therapeutic Agents and Other Inhibitors. In 1907 Ehrlich, the father of modern chemotherapy, reported that the dye, trypan red, was both curative and prophylactic for mice infected with *Trypanosoma equinum*, one of the protozoa. This was the first cure of an experimentally produced disease by the administration of a synthetic organic substance of known chemical composition. In 1935 Domagk showed that prontosil, a synthetic dye, cured mice infected with β-hemolytic streptococci, a pathogenic organism. This was the first cure of a *bacterial* disease by an organic compound of known structure. This compound, however, was inactive against the microorganism *in vitro* (in the test tube). Subsequent investigations showed that prontosil was degraded *in vivo* to *p*-aminobenzenesulfonamide and that the latter compound was the active agent. It was soon shown that SULFONAMIDE type drugs block the incorporation of *p*-aminobenzoic acid (PABA) into pteroylglutamic acid (PGA). PGA is required in the metabolism of all living cells (see FOLIC ACID). Most bacteria that are inhibited by the sulfonamide drugs synthesize PGA from PABA. The exposure of such bacteria to sufficient amounts of sulfonamides inhibits the utilization of PABA for PGA synthesis and thus stops bacterial growth. The drug has no effect on the mammalian host since the host does not synthesize PGA from PABA. The host requires preformed PGA in its diet.

The sulfonamide type of drug had been on the organic chemist's shelf for about thirty years before its chemotherapeutic potential was realized. Microorganisms had been manufacturing still another class of extremely potent antibacterial agents for centuries before they were discovered by man. These latter drugs, called ANTIBIOTICS, are substances synthesized and elaborated by one type of microorganism and which inhibit the growth or kill other types of microorganisms. The discovery of the clinical applications of penicillin in the 1940's was a major advance in medical science. Probably no other drug has been of such benefit to mankind in this century.

Most antibiotics have been found to interfere specifically with one of the following essential cell functions: (1) cell wall formation, (2) cell membrane function, (3) protein synthesis or (4) nucleic acid metabolism.

Penicillin, bacitracin, vancomycin, and oxamycin (D-cycloserine) are believed to act by inhibiting the synthesis of cell wall mucopeptide. In particular, penicillin appears to interfere with the transfer of muramyl peptide into polymerized cell wall material. Oxamycin (D-cycloserine) prevents the formation of a precursor of the cell wall by interfering with the formation and utilization of D-alanine.

Some other antibiotics of clinical usefulness are streptomycin, chloramphenicol and the tetracyclines. Chloramphenicol and the tetracyclines are effective agents for blocking the synthesis of proteins. Streptomycin also interferes with protein synthesis but additionally may cause damage to the membrane of cell.

Recently, a group of antibiotics have been found that complex with nucleic acids in the cell. Although these antibiotics are toxic and hence cannot be used clinically they are of value to the laboratory experimenter. Examples of these antibiotics are phleomycin, mitomycin and actinomycin. Phleomycin is found in cultures of *Streptomyces verticillus*. It inhibits DNA biosynthesis but not RNA or protein biosynthesis in *E. coli*. Phleomycin inhibits purified DNA polymerase of *E. coli* by binding of the DNA primer. It has only a slight effect on DNA-dependent RNA polymerase. The higher the adenine plus thymine content of the DNA the greater the inhibition of DNA polymerase by this antibiotic.

Mitomycin interferes with DNA synthesis *in vivo*. This antibiotic *per se* is without effect. In the presence of cell extracts, it is converted to a biologically active agent that causes cross-linking of DNA strands. So although mitomycin itself is inactive in the test tube as an inhibitor of purified DNA polymerase of *E. coli*, it may inhibit DNA polymerase within the cell.

Actinomycin is an inhibitor of DNA-dependent RNA polymerase. It too forms a complex with the DNA template. In contrast to phleomycin, the greater the guanine content of DNA, the greater the inhibitory activity of actinomycin.

We have attempted in this article to present a few of the biochemical aspects of microbiology. It is realized that some of the microbiological aspects of biochemistry have been presented as well. Other articles in this encyclopedia that concern microbial phenomena of scientific or industrial significance include ALCOHOLIC FERMENTATION; ALGAE; FERMENTATIONS; INTESTINAL FLORA (IN RELATION TO NUTRITION); LYSOSTAPHIN; LYSOZYME; NITROGEN FIXATION; OXIDATIVE METABOLISM AND ENERGETICS OF BACTERIA; PHOTOSYNTHETIC BACTERIA; POLYAMINES; RUMINANT NUTRITION AND METABOLISM; TOXINS, ANIMAL AND BACTERIAL; TRANSFORMING FACTORS; VIRUS DISEASES.

References

1. GUNSALUS, I. C., AND STANIER, R. Y. (EDITORS), "The Bacteria," Volume I: Structure; Volume II: Metabolism; Volume III: Biosynthesis; Volume IV: Growth; Volume V: Heredity; New York and London, Academic Press, 1960.
2. HAYES, WILLIAM, "The Genetics of Bacteria and Their Viruses," New York, John Wiley & Sons, 1964, 740 pp.
3. KORNBERG, A., "Enzymatic Synthesis of DNA," *Ciba Lectures Microbial Biochem.*, New York, John Wiley and Sons, 103 pp. (1961).
4. PERKINS, H. R., "Chemical Structure and Biosynthesis of Bacterial Cell Walls," *Bacteriol. Rev.*, **27**, 18–55 (1963).
5. STANIER, R. Y., DOUDOROFF, M., AND ADELBERG, E. A., "The Microbial World," Second edition, Englewood Cliffs, N.J., Prentice-Hall, 1963, 753 pp.

H. VASKEN APOSHIAN AND MUTSUKO NISHIHARA

MICROSOMES AND THE ENDOPLASMIC RETICULUM

The term "microsomes" was originally used by Claude to describe a cellular fraction obtained after high-speed centrifugation of suspensions (in isotonic sucrose) of disrupted cells from which the larger cellular components (largely nuclei, mitochondria, and cellular debris) were removed. Since this high-speed fraction contained cellular components which were not readily visualized with light microscopy they were termed "submicroscopic particles" or microsomes. Present-day usage of the term is restricted to the minute fragments of the *endoplasmic reticulum* (e.r.) with or without the RIBOSOMES attached. Electron microscopy reveals the e.r. as a continuous network of tubules and cisternae within the inner or endoplasmic region of the cytoplasm. Within skeletal MUSCLE the sarcoplasmic reticulum, which is homologous to the e.r., separates the interfibrillar space into two phases. In some cells the e.r. is arranged in a parallel, concentric fashion, while in others the tubules are often tortuous and exhibit branching and anastomoses. Present within the intraluminal space of the e.r. are the electron-dense ribosomes which are small spherical particles 50–200 mμ in diameter. After homogenization in non-electrolyte solutions (*e.g.*, sucrose) the various fragments of the e.r. can be separated into membranous fragments (vesicles) devoid of or containing ribosomes. With the use of surface active agents or by employing the technique of continuous density centrifugation, the microsomal fraction is resolvable into (1) the ribosomes, (2) smooth-surfaced (agranular appearance in the electron microscope) vesicles containing no RNA granules, and (3) rough-surfaced vesicles (granular) with RNA (RIBONUCLEIC ACID) granules. Some investigators have reserved the term "microsome" for the vesicular components (2) and (3), while the term "ribosomes" refers to the RNA granules devoid of membranous components. There are also RNA-containing particles that are smaller in size than ribosomes and are not presumably part of the e.r.

Although from some tissues, particularly liver, it is possible to derive a relatively homogeneous preparation of microsomes, subcellular fractions of other tissues prepared in a similar way may be much more heterogeneous. In a homogenate of brain tissue there are a great many cytoplasmic structures from a variety of cell types with vesicular characteristics and sedimentation properties resembling the e.r. fragments. Included in these are MYELIN, axonal and dendritic fragments, synaptic vesicles, fragmented Golgi reticulum and vesicles, neurofibrils, and lysosomes (see Figs. 1 and 2). Interpretation of biochemical findings on the high-speed "microsomal" fraction derived from such tissues as the brain must be exercised with caution. The granular vesicular fraction is readily identified as microsomal because of the presence of the ribosomes, but the agranular vesicles are considerably more difficult to characterize.

The function of the membranous component of the e.r. is not entirely clear, but it appears to be involved in concentrating and storing the synthesized protein derived from the ribosomes. During embryonic development and rapid cellular proliferation, the membranous components are absent. They appear only after cellular maturation and the elaboration of specific proteins. It has been suggested that in the pancreas, the secretory enzymes, after being synthesized by the ribosomes, are transported through the intraluminar space of the e.r. into the Golgi region of the cell where they aggregate to form the zymogen granules. Other secretory granules may be formed and transported in a similar manner. Electron microscopy also reveals a close functional relationship between the mitochondria and the e.r.

Under various physiological conditions the e.r. can undergo reversible alterations in its arrangements. For example, during fasting the granular e.r. of the chief cells of the stomach or the pancreatic cells, is arranged in the form of closely packed cisternae roughly parallel to the cell border. With feeding, however, the e.r. rearranges into multilayered concentric laminae with a mitochondria, secretory granule or other formed element enclosed within. This structural change may be due to rearrangement in the PHOSPHOLIPID OR LIPOPROTEIN molecules brought about by changes in the degree of hydration.

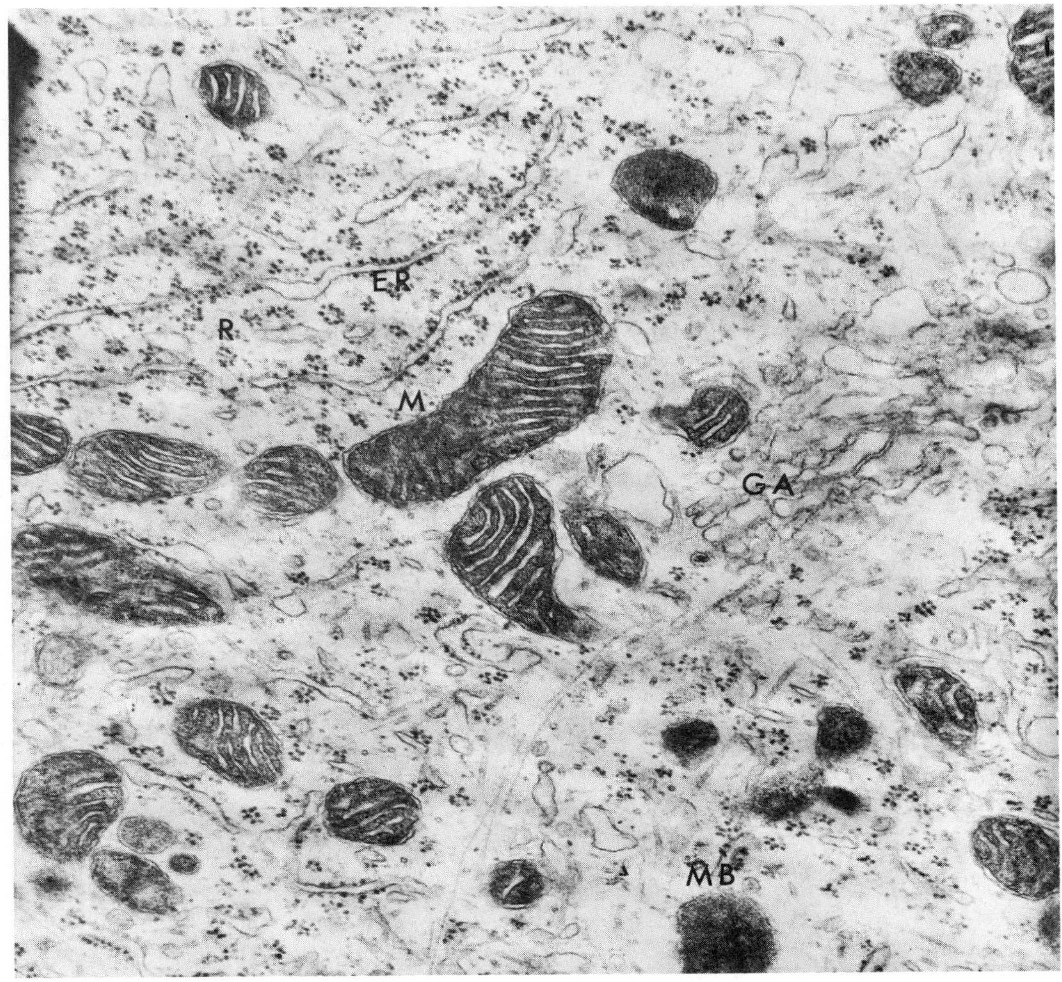

FIG. 1. An electronmicrograph of a motorneutron showing the endoplasmic recticulum (ER), ribosomes (R), Golgi apparatus (GA) with lysosomes, multivesicular bodies (MB), and mitochondria (M), magnification 50,000X. (Courtesy of Drs. K. Barron and Paul Doolin, Neuropathology Research Section, V. A. Hospital, Hines, Illinois.)

The significant biochemical characteristic of the microsomal fraction is its high RNA content, which is largely due to the presence of ribosomes. There remains some question as to whether all the RNA of the e.r. is confined to the ribosomal particles or is partly associated with the membranous components of the e.r. The content of RNA in the microsomal fraction is dependent upon the tissue of origin: it may be as high as 40% of the total cellular RNA in the case of pancreas and 20% in the case of liver. Likewise the content of RNA in the ribosomes is dependent upon the tissue source, but on the average the ribosomes are comprised of equal amounts (by weight) of RNA and protein. Evidently, the protein and RNA are linked together by Mg^{2+}, with the protein functioning in part as a structural matrix for the RNA. The ribosomes are constituted of polynucleotide-protein complexes, with the polynucleotides varying in molecular weight (range of sedimentation units, 20–150 S). About 50% of the total dry weight of the microsomal fraction from such tissues as liver, brain, kidney, and pancreas is comprised of lipids, of which over 80% is phospholipid. Liver microsomes contain almost 60% of the total liver lecithin, 40% of the aminophosphatides, 30% of the phosphoinositides, and 13% of the cardiolipin. Presumably the phospholipids participate in the process of protein synthesis occurring in the microsomes.

Although a number of enzymes are believed to be associated with microsomes, the techniques used for isolation and the inability to precisely

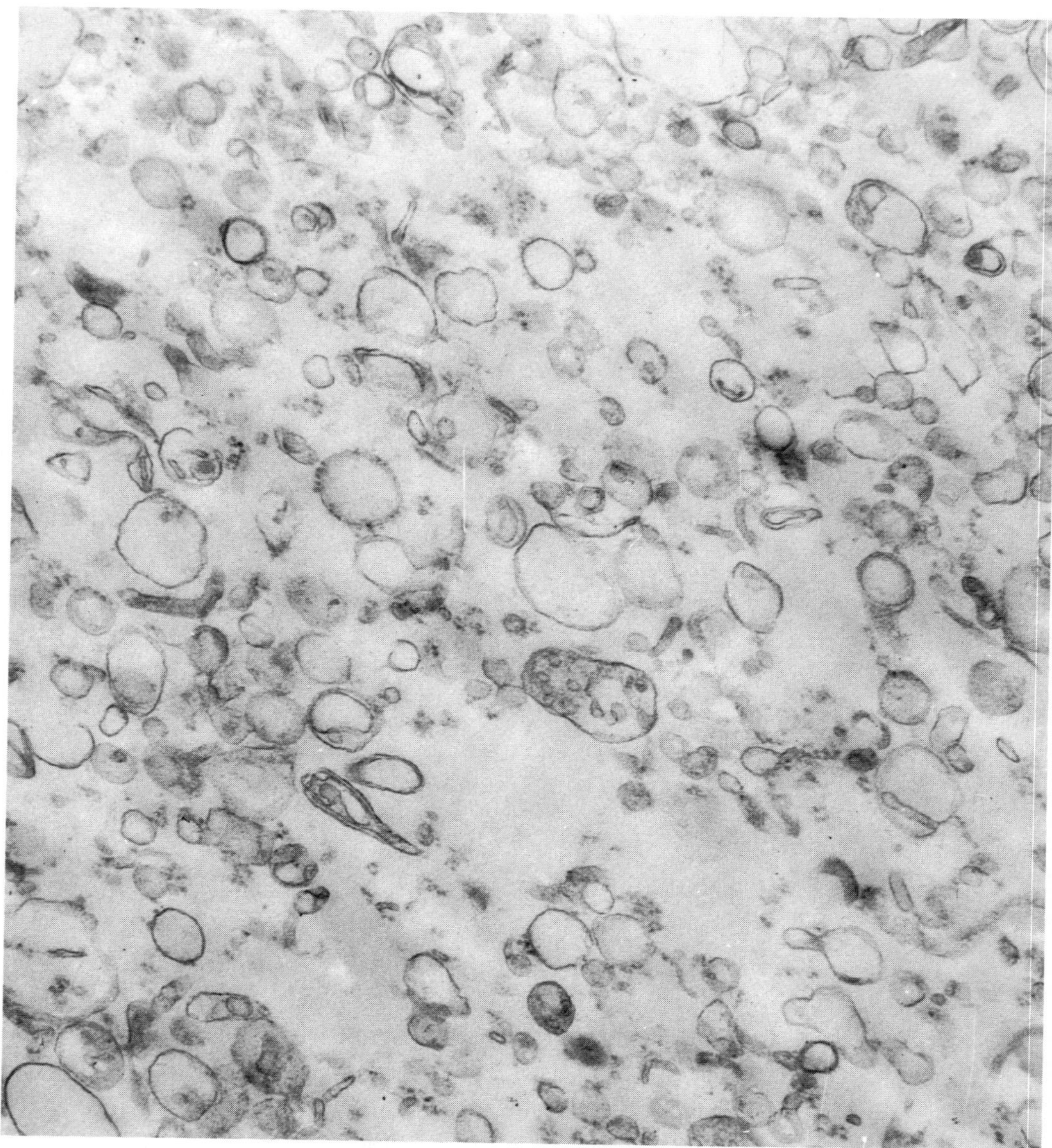

FIG. 2. An electronmicrograph of a crude "microsomal" preparation isolated from rat brain by high-speed centrifugation in 0.25*M* sucrose. The heterogeneity of the preparation is obvious. Besides the ribosomes and e.r, membrane fragments can be seen, fragments of the Golgi apparatus, lysosomes, and multivesicular bodies.

characterize the microsomal fraction often make it difficult to assume that a given enzyme forms an integral part of the e.r. Among the enzymes believed to be associated specifically with the ribosomal particles are PEPTIDASES, ribonuclease, and the amino acid-activating systems (see ACTIVATING ENZYMES) associated with polypeptide synthesis. The amino acid-activating enzymes are not only engaged in peptide synthesis but also catalyze the esterification of the activated amino acid with the specific soluble polynucleotide component. It is believed that the membrane-bound ribosomes are the primary sites of protein synthesis, although the free ribosomes exhibit synthetic activity in the presence of added messenger RNA. Included among the enzymes presumably attached to the membranous components of the e.r. are pyridine nucleotide-cytochrome c reductase, cytochromes m and b_5, glucose-6-phosphatase, adenylate kinase, Mg^{2+}-activated ATPase (also requiring Na^+ and K^+), nucleosidephosphatases (guanosine, inosine, and uridine) and enzymes involved in fatty acid and steroid metabolism. Certain coenzymes such as coenzyme Q_9 and Q_{10} have been reported to be present in liver microsomes. In many of the enzyme localization studies, the homogeneity of the so-called microsomal fraction has not been investigated with the electron microscope; nor is it always possible to establish the endoplasmic origin of the vesicular components in microsomal preparations. HISTOCHEMICAL procedures, employing both light and electron microscopy, have been useful in enzyme localization, and with the refinement of electron microscopy for histochemical localization, more precise biochemical information about the microsomes will be forthcoming.

References

1. ROBERTS, R. B. (EDITOR), "Microsomal Particles and Protein Synthesis," New York, Pergamon Press, 1958.
2. HARRIS, R. J. C. (EDITOR), "The Interpretation of Ultrastructure," New York, Academic Press, 1962.
3. PETERMANN, M. L., "The Physical and Chemical Properties of Ribosomes," New York, American Elsevier Publishing Co., 1964.
4. HAYASHI, T. (EDITOR), "Subcellular Particles," New York, Ronald Press, 1959.

LEO G. ABOOD

MILK (COMPOSITION AND BIOSYNTHESIS)

Milk as the secretion of the mammary gland forms the natural food of the newborn mammal. Although general component characteristics are similar, wide differences exist between species in the gross composition of their milks (Table 1). Fat and protein tend to be higher in the milks of marine and arctic species and those with a rapid growth rate in the young. The emulsified fat and the colloidal calcium phosphoprotein complex present in milk contribute to the white or "milky" appearance. The pH of milk is near neutrality (cow 6.7, human 7.1), and its osmotic pressure is in equilibrium with that of blood.

Most available information on milk concerns that from the dairy cow (*Bos taurus*), and Table 2 and other information will concern that source unless otherwise noted. The triglyceride fatty acid contents shown are typical of the milk of the ruminant species which contains significant quantities of short-chained fatty acids. Nutritionally, milk is an excellent source of most essential nutrients including vitamins, amino acids, and minerals. It is a poor source of iron, copper, and vitamins C and D (D is usually added for human consumption).

Casein (3.2 g/100 ml milk) is the protein complex precipitated from milk by high-speed centrifugation, at pH 4.6, or with the enzyme rennet.[2,3] It is a heterogeneous complex of PROTEINS which separates in an electric field into three major fractions, called the alpha, beta, and gamma components. The largest or alpha component is quite heterogeneous and major fractions isolated from it have included the kappa, $alpha_s$ and other fractions. Upon removal of the casein complex, major proteins remaining in the whey include beta-lactoglobulin (0.30 g/100 ml milk), alpha-lactalbumin (0.15 gram), and several blood proteins including serum albumin (0.03 gram and some "immune" globulins or "lactoglobulins" (0.07 gram). The historic protein "lactalbumin" of milk includes the major entities now recognized as beta-lactoglobulin, alpha-lactalbumin, and serum albumin. The immune lactoglobulin fraction can be as high as 15 g/100 ml, or more, in cow colostrum (first milkings following parturition); this is not true of some other species.[3,4] Casein contains about 0.85% phosphorous esterified to the hydroxy amino acids which is complexed with calcium, magnesium, citrate, more phosphate, etc., to give an intricate equilibrium in the calcium caseinate-phosphate complex present.[2]

Radionuclides naturally present in foods and milk include K^{40}, Ra^{226} in some areas, etc. Others found[5] due to man-made nuclear detonations have included I^{131}, Cs^{137}, Ba^{140}, Sr^{89}, Sr^{90} and C^{14}.

Biosynthesis. The synthesis of milk takes place in specialized secretory cells of the mammary gland under the control of many HORMONES including estrogen and progesterone concerned

TABLE 1. COMPOSITION OF MILKS OF VARIOUS MAMMALS[a]

Species	Total Solids	Fat	Protein	Lactose	Ash
		g/100 g			
Horse	11.0	1.6	2.7	6.1	0.5
Human	12.6	3.8	1.6	7.0	0.2
Goat	12.9	4.1	3.7	4.2	0.8
Cow	13.8	4.4	3.8	4.9	0.7
Dog	21.0	8.5	7.5	3.7	1.2
Rat	30.0	15.0	12.0	3.0	2.0
Reindeer	36.7	22.5	10.3	2.5	1.4
Porpoise	62.0	49.0	11.0	1.3	0.6

[a] From Brody[1] who lists 22 species. Considerable variation within a species exists.

TABLE 2. SOME COMPONENTS PRESENT IN NORMAL COW MILK[2,3]

Amino Acids in Total Proteins (g/100 g protein): Alanine 3.6, arginine 3.5, aspartic acid 7.5, cystine 0.9, glutamic acid 21.7, glycine 2.1, histidine 2.7, isoleucine 6.5, leucine 9.9, lysine 8.0, methionine 2.4, phenylalanine 5.1, proline 9.2, serine 5.2, threonine 4.7, tryptophan 1.3, tyrosine 4.9, valine 6.7.

Vitamins—Water Soluble (μg/100 ml milk): Thiamine 40, riboflavin 150, niacin 70, pyridoxine 70, pantothenic acid 300, biotin 5, folic acid 0.1, choline 15000, B_{12} 0.7, inositol 18000, ascorbic acid 2000.

Minerals (mg/100 ml milk): Ca 125, Mg 10, Na 50, K 150, $PO_4^{\equiv}$ 210, citrate (as acid) 200, Cl 100, HCO_3^- 20, $SO_4^=$ 10, usually traces of Rb, Li, Ba, Sr, Mn, Al, Zn, B, Cu, Fe, Co, I, Mo, and questionable or sometimes traces of Pb, Cs, Cr, Ag, Sn, Ti, V, F, and Si.

Lipids (amt./100 ml milk): Mixed triglycerides 3–5 (g), phospholipides 30 (mg), sterols 10 (mg), traces of waxes, squalene, and free fatty acids, carotenoids 10–60 (μg), vitamin A 10–50 (μg), vitamin D 0.04 (μg), vitamin E 100 (μg), vitamin K (trace).

Triglyceride Fatty Acids (mole % of total fatty acids): Saturated C_4 10, C_6 2, C_8 2, C_{10} 3, C_{12} 4, C_{14} 10, C_{16} 20, C_{18} 10, C_{20} 1; monunsaturated C_{10} 0.2, C_{12} 0.5, C_{14} 1.5, C_{16} 3, C_{18} 30, C_{20} 0.3; polyunsaturated acids about 3.

Enzymes: Catalase, peroxidase, xanthine oxidase, acid and alkaline phosphatases, aldolase, α- and β-amylases, lipases and other esterases, proteases, carbonic anhydrase, etc.

Miscellaneous (mg/100 ml milk): Free glucose 6, free galactose 2; various sugars and metabolic intermediates; N-containing oligosaccharides including acetylglucosamine, neuraminic acid, etc.; ammonia N 0.7; free amino acid (N) 0.35, urea (N) 10, creatine and creatinine (N) 1.5, uric acid 0.7, orotic acid 7 , hippuric acid 5, indican 0.1, dissolved gasses (air exposed) CO_2 10, O_2, 0.8, N_2 1.5.

with mammary development, prolactin with the initiation of lactation, and oxytocin with the release or "let-down" of milk into the collecting passages.[3,4] Milk as normally removed from the mammary gland contains white blood cells (leucocytes), some bacteria, and portions of sluffed-off secretory and other cells lining the intermammary passages.

The mechanisms concerned with the selective assimilation of certain blood components into milk are not known. The secretory cells require all of the usual essential AMINO ACIDS and a proportion of some of the nonessential acids to synthesize the milk proteins.[3] Genetic polymorphs of several of the milk proteins exist in the same species, differing from one another by only a few amino acid residues. The secreting cell is characterized by a high ratio of metabolized glucose proceeding via the pentose phosphate or oxidative "shunt" pathway instead of by glycolysis (see CARBOHYDRATE METABOLISM). One effect is for the $NADPH_2$ generated to enhance reductive synthesis such as takes place in the cells in the formation of the even-chained fatty acids from acetate (See FATTY ACID METABOLISM) by the stepwise condensation of acetyl coenzyme A units, with malonyl coenzyme A also involved.[3] The synthesis of lactose involves the formation of uridine diphosphate glucose from glucose-1-P, its conversion to UDP-galactose and then condensation with glucose to form lactose[3] (see GALACTOSE). In many regards, the constituent ingredients are probably more readily appreciated if milk is considered to have been a part of the actual tissue itself rather than only as a specific product of it.

References

1. BRODY, S., "Bioenergetics and Growth," New York, Reinhold Publishing Corp., 1945.
2. JENNESS, R., AND PATTON, S., "Principles of Dairy Chemistry," New York, John Wiley & Sons, 1959.
3. KON, S. K., AND COWIE, A. T. (EDITORS), "Milk: The Mammary Gland and Its Secretion," New York, Academic Press, 1961, 2 vols.
4. SMITH, V. R., "Physiology of Lactation," Fifth edition, Ames, Iowa State University Press, 1959.
5. "Fallout, Food, and Man" (Symposium), *Federation Proc.*, **22**, 1389 (1963).

B. L. LARSON

MINERAL REQUIREMENTS (VERTEBRATES)

The vertebrate body contains about 4–6% of total mineral matter, the major portion of this consisting of calcium and phosphorus in the skeletal tissues. Interspecies variation in mineral composition of the adult whole body is not great except as a result of differences in relative skeletal size. Thus, the average percentage concentration of minerals in the lean body mass of vertebrates may be listed as follows: Ca 1.1–2.2, P 0.70–1.20, Mg 0.045, K 0.30, Na 0.15, Cl 0.15, Fe 0.008, Zn 0.003, Cu 0.0003. The above minerals are essential to animal life. Other essential minerals which may be found in all animal bodies include S, Mn, Co, I, Mo, Se. A complete analysis of animal tissues will reveal traces of many more mineral elements, some of which may eventually be demonstrated to be essential, some of which are not essential and others which are toxic even in small amounts. Fluorine occupies a peculiar position in being nonessential, toxic, yet distinctly beneficial in preventing tooth decay under conditions that otherwise lead to caries (see FLUORIDE METABOLISM).

Mineral Absorption. Minerals appear in animal tissues mainly as a result of ingestion with food or water. The primary entry into the body is by way of the small intestine from which the minerals are absorbed into the bloodstream and then transported through the body. Some of the minerals are absorbed by simple diffusion processes while others enter into metabolic pathways that enable them to be absorbed against a concentration gradient by energy-requiring processes. Sodium and chlorine are examples of actively absorbed elements. Absorption of calcium is much slower than that of sodium and is at least in part regulated by hormonal and dietary factors. PARATHYROID HORMONE stimulates calcium absorption and thus provides for increased absorption in states of calcium deficiency. VITAMIN D likewise stimulates calcium absorption as does the presence in the diet of relatively slowly absorbed carbohydrate such as lactose. Evidence for active absorption of other divalent cations is scant although there is general agreement that iron absorption often increases in case of increased need by the animal. Mineral absorption is inhibited by dietary factors that tend to reduce solubility of minerals at the pH of the intestinal contents. Excesses of phosphate will inhibit absorption of calcium, iron and magnesium; excesses of calcium will inhibit absorption of phosphate, magnesium and, particularly in presence of phytic acid, zinc.

Mineral Excretion. Excretion of absorbed minerals is by way of the intestinal tract, urinary tract, and in the sweat. Which excretory pathway will be used depends on the mineral in question and upon a variety of dietary, hormonal and environmental factors which act interdependently to exert homeostatic control over the mineral concentrations in the body. Calcium, magnesium, phosphorus and zinc are primarily excreted by way of the intestinal tract. Indeed only small amounts of calcium and zinc are found in urine. Magnesium and phosphorus excretion may be quantitatively greater in the urine, particularly when the cation:anion ratio of the ration is less than 1:1. The monovalent ions, sodium, potassium and chlorine, are primarily excreted in the urine, a process largely regulated by aldosterone, a hormone produced in the adrenal gland. Excretion of absorbed iron by way of urine and feces is very slight. In the human, sweat can be an important pathway of mineral loss, especially for sodium, chlorine and iron.

Metabolic Roles of Minerals. The diversity of roles served by minerals in the body defies simple summarization, and indeed they involve the entire field of nutritional biochemistry. Structural rigidity of the body is largely a function of CALCIUM and phosphorus in bone (see BONE FORMATION; PHOSPHATES). Sulfur is an integral part of a number of organic structures including proteins (the AMINO ACIDS methionine and cysteine), vitamins (BIOTIN, THIAMINE) and a wide variety of sulfated polysaccharides. Further examples of the structural role of minerals include COBALT in vitamin B_{12}; IRON in hemoglobin; IODINE in thyroxine; phosphorus in lecithin, related PHOSPHOLIPIDS and proteins; iron, copper, MOLYBDENUM, ZINC and MAGNESIUM in a diversity of enzymes. In addition to these structural functions, minerals play important roles in a regulatory sense. Sodium, POTASSIUM and chlorine are important in pH regulation and in maintenance of osmotic pressure relationships. These minerals with calcium and magnesium are much concerned with neuromuscular irritability and the transmission of NERVE IMPULSES. Activators of enzymic reactions include calcium for prothrombin synthesis; magnesium for phosphate transfer; magnesium, manganese, copper and zinc for transfer of acyl groups, hydration-dehydration reactions and many others. The exact mechanism whereby the minerals function either as activators of enzyme systems or as constituents of enzymes (forming stoichiometric compounds) is not known, but in the latter class, evidence is accumulating that the metal may contribute stability to the configuration of the protein molecules and also provide sites for ionic attraction and binding of the reactants (see ENZYMES; METALLOPROTEINS). Finally, recent evidence that RIBONUCLEIC ACIDS, which are responsible for the transmission of genetic messages for the synthesis of proteins in tissues, contain significant concentrations of trace metals such as CHROMIUM, nickel and manganese tightly bound in the molecule indicates that these elements may indeed participate in the correct expression of genetic traits.

Quantitative Requirements. Measurements of mineral requirements for the normal growth and maintenance of animals have demonstrated that the quantitative results found are greatly influenced by many factors such as age of the animal, physiological function and dietary constituents; hence it is impossible to speak of "the requirement" for a specific mineral.

In determination of maintenance requirements of adult animals, for example, the conclusion reached will depend on the previous status of the subject. The amount of calcium required to provide for calcium equilibrium may be drastically reduced by using subjects adapted to low-calcium diets. This principle of adaptation of homeostatic mechanisms is not peculiar to mineral nutrients and indeed is a fertile field of investigation in all areas of nutrition and biochemistry.

The mineral requirements for growth are, in general, reflections of the mineral accretions in the total body, modified by the rate of turnover in the tissues and by dietary factors that influence availability to the animal of the mineral in the diet.

In view of the above considerations, it is presumptuous to set forth tables of mineral requirements with the expectation that they will serve as more than guides to practice. Such guides have been published by the NAS-NRC Committees on Animal and on Human Nutrition for example, and they represent our best estimates to date.

The accompanying Tables are included to illustrate the currently available estimates of the mineral requirements of humans (Table 1) and of ruminants, swine, poultry and rats (Table 2).

TABLE 1. ESTIMATED DIETARY MINERAL NEEDS OF ADULT HUMANS, EXPRESSED AS GRAMS PER DAY

Element	Daily Amount	Element	Daily Amount
Ca	0.8–1.4	Mg	0.3
P	0.8–1.4	Fe	0.01–0.015
Na	3	Cu	0.001–0.002
K	0.8	I	0.00025

TABLE 2. ESTIMATED DIETARY MINERAL NEEDS OF RUMINANTS, SWINE, POULTRY AND RATS, EXPRESSED AS PER CENT OF THE RATION

Element	Ruminants	Swine	Poultry	Rats
Ca	0.12–0.5[a]	0.6	1.0–2.8[c]	0.6
P	0.12–0.5[a]	0.4	0.7	0.5
Na	0.2	0.2	0.15	0.05
K		0.3	0.18	0.18
Mg	0.06	0.04	0.05	0.04
Fe		0.008[a]	0.004	0.0025
Cu	0.0006	0.001[a]	0.0004	0.0005
I	0.00008	0.00002	0.000035	0.000015
Zn	0.001	0.001–0.005[b]	0.001–0.005[b]	0.0012
Mn	0.004	0.004	0.005	0.005
Co	0.0001	—	—	—

[a] Higher values are for young animals.
[b] Higher values are for diets containing phytic acid.
[c] Higher values are for laying hens.

On the basis of these guides, recommendations may be made for mineral supplementation of diets or rations consumed by humans and livestock. In general, it is found that otherwise adequate diets for humans, constructed of a variety of animal and vegetable foodstuffs will not need specific mineral supplementation. Farm animals, whose selection of feeds is usually tailored by man to maximize production rate, will ordinarily need to have included in their rations supplements of calcium and phosphorus and sodium. In special cases other supplements will be needed. For example, most poultry rations have manganese added, and in some areas of the world, supplements of cobalt, copper and selenium are beneficial in the rations of grazing animals; iodine supplementation is beneficial to all species of animal life in widespread areas of the world.

All of the minerals may be deleterious to animals if ingested in excessive amounts. The toxic mechanisms are as diverse as the functional roles of the minerals and may range from an effect on absorbability of the element to a displacement of one element by another in an enzyme system or the direct inhibition of a metabolic reaction. Hence the toxicity of mineral elements may be a function not only of the amount present but also of the supply of other minerals in the diet. The toxicity of copper, for example, is enhanced by lowered levels of molybdenum, and molybdenum toxicity is at least in part a reflection of an induced copper deficiency. Both of these interactions are modified by the level of inorganic sulfate in the diet. Zinc tolerance in animals is enhanced by increasing intakes of copper and of iron above normally employed levels. Selenium toxicity may be alleviated by judicious administration of a variety of arsenic compounds. Although effects of this sort may be documented, a mechanistic explanation of them must in most situations await further research.

As the preceding statements indicate, a consideration of the potential hazards of mineral excesses as well as of deficiencies is an integral aspect of proper mineral nutrition.

References

1. UNDERWOOD, E. J., "Trace Elements in Human and Animal Nutrition," Second edition, New York, Academic Press, 1962.
2. MAYNARD, L. A., AND LOOSLI, J. K., "Animal Nutrition," Fifth edition, New York, McGraw-Hill, Book Co., 1962.
3. COMAR, L. C., AND BRONNER, F., "Mineral Metabolism," Vols. 1 and 2, Parts A and B, New York, Academic Press, 1960–1964.
4. MITCHELL, H. H., "Comparative Nutrition of Man and Domestic Animals," Vols. 1 and 2, New York, Academic Press, 1962 and 1964.

5. National Research Council, "Nutrient Requirements of Domestic Animals," Numbers I-X, *Nat. Acad. Sci.—Natl. Res. Council, Publ.*, **296, 331, 464, 504, 579, 827, 989, 912, 990, 1192** (1953-1964).
6. National Research Council, "Recommended Dietary Allowances," *Natl. Acad. Sci.—Natl. Res. Council Publ.*, **1146** (1964).

R. M. FORBES

MITOCHONDRIA

The mitochondrion is a subcellular organelle of variable size and shape bounded by a two-layered membrane. The outer layer of this membrane is considered to be the limiting membrane. The regular invaginations of the internal membrane have been termed *cristae* by Palade [Fig. 1(a)]. The undefined space between cristae has been called the *matrix*. The ELECTRON MICROSCOPIC studies of Palade[1] and Sjöstrand[2], following osmium fixation, established these readily discernible features of mitochondrial ultrastructure. Although these early studies established a base line for correlation of structure with function, the relatively crude preparative techniques and resolution of the electron microscope limited the extent of definitive correlation in the original work. The patterns of sectioned specimens fixed with osmium revealed a structural network, mainly lipoprotein in composition, with a high degree of uniformity, but little indication of enzyme conformation and localization was seen. The lamellar system of the mitochondrion was better characterized functionally than structurally.

The preparative technique involving negative staining, as applied to the electron microscopic study of mitochondria, has made it possible to examine their ultrathin membranes without prior sectioning.[3] In 1961, negative staining, low-temperature methods, and electron microscopy of comparatively high resolution revealed, for the first time, the presence of characteristic, roughly spherical extensions, attached to the continuous basic internal membrane and cristae; these were absent on the external envelope [Fig. 1(b)]. The particles revealed by this method are 80–100 Å in diameter and are attached to the membrane by a visible stalk. The discovery of this repeating particulate unit, designated "elementary particle," as a basic component of mitochondrial membranes established the primary organization of mitochondria into subunits which could be correlated with specific biochemical functions.

Stepwise degradation of the mitochondrion has led to the recognition of three levels of mitochondrial organization: (1) an insoluble network of structural PROTEIN and PHOSPHOLIPID; (2) a particulate respiratory assembly composed of the electron transfer chain and phosphate coupling apparatus; (3) the primary dehydrogenase complexes, *e.g.*, isocitrate dehydrogenase, α-ketoglutarate dehydrogenase, and β-hydroxybutyrate dehydrogenase.

The mitochondrion represents the major site of energy transduction in aerobic animal cells. The mechanism whereby PHOSPHORYLATION is coupled to electron transfer provides energy for metabolic processes and for muscular contraction. The adenosine triphosphate (ATP), generated by the oxidation of reduced diphosphopyridine nucleotide (DPNH) and succinate in the mitochondrion, is used to drive reactions remote from the site of origin. A number of energy-linked functions, in addition to the synthesis of ATP, have been shown to reside in the mitochondrion. These energy-powered reactions are transhydrogenation, reduction of DPN^+ induced by oxidation of succinate or by ATP, swelling and its counterpart contraction which appears to be associated with a contractile protein, and finally ion accumulation.

Energy-driven accumulation of ions, *e.g.*, Ca^{++}, Mg^{++}, Mn^+K^+, $^+$, inorganic phosphate and others, has attracted the interest of many investigators as a potentially important mechanism for the regulation of intracellular ions. This accumulation of ions requires either a substrate capable of promoting oxidative phosphorylation, or ATP itself.

Heart mitochondria contain about 30% of

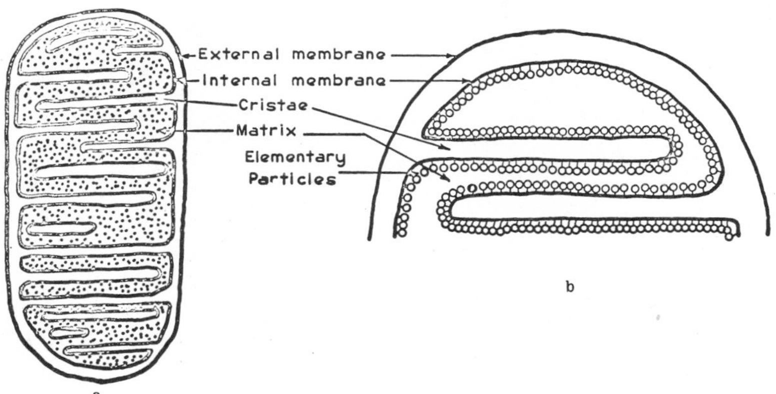

FIG. 1. Diagrammatic sketch of a mitochondrion: (a) fixed in osmium, (b) negatively stained with phosphotungstate.

lipid by weight of which over 90% is phospholipid. Aqueous acetone extraction of mitochondria removes the phospholipid, and electron microscopic observations of such extracted mitochondria, after fixation in osmium tetroxide, present the same appearance as do the non-extracted ones. This evidence strongly suggests that the structure of the mitochondrion is conferred by the protein and not by the phospholipid-protein complex. However, electron transfer activity, as well as the other energy-linked functions, are lost upon the removal of lipid; the electron transfer activity is restored when mitochondrial phospholipid and COENZYME Q are added back.

Criddle *et al.*[4] isolated an apparently inert protein from the membrane fraction of mitochondria which has been designated structural protein. This protein is highly insoluble and binds components of the electron transfer chain, in addition to ATP, and phospholipid. The interactions are thought to be either hydrophobic or electrostatic (or a combination of both) in nature. The monomeric form of structural protein has a molecular weight of approximately 20,000, but it polymerizes readily at neutral pH into aggregates of various sizes. The relationship between structural protein and the contractile protein remains obscure, but it is likely that contractile protein of mitochondria is a contaminant in the structural protein preparations which represent about 50% of the total mitochondrial protein.

The membrane fraction of mitochondria also contains FLAVOPROTEINS, COPPER proteins, nonheme IRON proteins, CYTOCHROMES (heme iron proteins), coupling proteins, and lipids all of which collectively constitute the apparatus of oxidative PHOSPHORYLATION. Some of the primary dehydrogenase enzymes are also bound to the membrane preparations. Fractionation of membrane-bound components invariably requires the use of detergents (bile salts) or alcohols, followed by salt precipitation. Submitochondrial complexes have been obtained, by means of these techniques, which are active and have definite stoichiometry. Four complexes collectively constituting the electron transfer chain have been isolated; they may be represented as shown in Fig. 2.

Slater[5] has defined the criteria which are necessary to designate a new component as an obligatory member of the electron transfer system: (1) It should be present in at least as great concentration as are other members of the electron transfer chain (at least one molecule for each chain). (2) Quantitative extraction of the component should result in loss of enzymic activity (readdition should restore the activity). (3) The component should be oxidizable and reducible at a rate compatible with the over-all electron flux of the chain. In Table 1 are listed the components that may qualify as members of the chain though they have not been shown to meet all of the criteria listed above.

TABLE 1. CONCENTRATIONS OF COMPONENTS OF THE ELECTRON TRANSFER SYSTEM IN BEEF HEART MITOCHONDRIA

Component	Concentration of Component per mg of Protein
Total flavin	0.66 mμmoles
Acid extractable	0.46 mμmoles
Acid-nonextractable	0.20 mμmoles
Total cytochromes	2.65 mμmoles
Cytochrome *a*	1.31 mμmoles
Cytochrome *b*	0.68 mμmoles
Cytochrome c_1	0.21 mμmoles
Cytochrome *c*	0.45 mμmoles
Total iron	9.86 mμmoles
Heme	2.65 mμmoles
Nonheme	7.21 mμmoles
Copper	1.27 mμmoles
Coenzyme Q	4.03 mμmoles
Phospholipid	0.36 mg

Electron Transfer Chain. The chain may be fractionated into four complexes designated I, II, III and IV. Complex I is termed the DPNH-coenzyme Q reductase because it catalyzes the oxidation of DPNH by coenzyme Q. The reaction is inhibited by amytal and rotenone, just as it is in native mitochondria. The complex contains the DPNH flavoprotein, non-heme iron protein, and lipid. The flavin is acid extractable and occurs in the form of flavin mononucleotide; the non-heme iron has been shown to be oxidized and reduced during electron flux through the complex. Complex II, the succinic-coenzyme Q reductase, contains the succinic flavoprotein, non-heme iron protein and cytochrome *b*. This cytochrome *b* is not reduced, as the coenzyme Q is reduced, by this

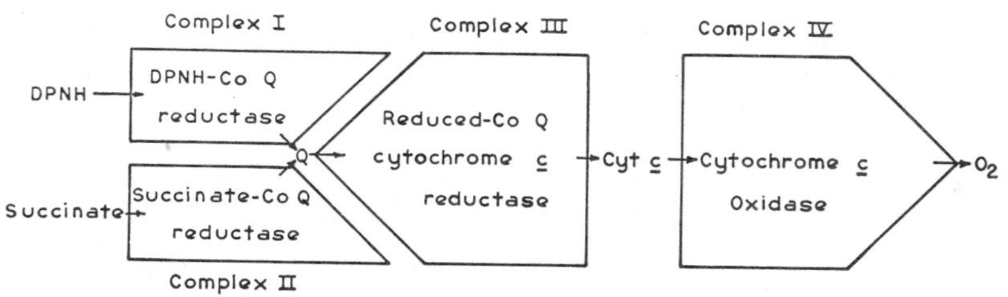

FIG. 2. Schematic representation of the complexes of electron transfer.

complex. The non-heme iron protein appears to differ from that of Complex I. Complex III, the reduced coenzyme Q-cytochrome *c* reductase, catalyzes the reduction of cytochrome *c* by reduced coenzyme Q, and is inhibited by antimycin at concentrations similar to those required for inhibition of the activity in whole mitochondria. It contains about 2 molecules of cytochrome *b* for each molecule of cytochrome c_1 and 2 molecules of non-heme iron. The non-heme iron protein, which again functions as an oxidation-reduction component, has been isolated in relatively pure form. This complex can be separated into a fraction containing cytochromes *b* and c_1. The separation is inhibited by antimycin. Complex IV, cytochrome *c* oxidase, promotes the oxidation of reduced cytochrome *c* by molecular oxygen. The cytochrome *c* oxidase probably consists of 6 molecules of heme *a* protein and 6 molecules of copper protein. A variety of spectral evidence indicates the presence of more than one species of cytochrome in this complex; however, only one heme, heme *a*, has been isolated. The copper is oxidized and reduced at a rate compatible with its participation in the electron flux of the complex. The four isolated complexes can be reconstituted to give a particle which has both DPNH oxidase and succinic oxidase activities.

The active complexes contain lipid in about the same amounts as the mitochondria. Mitochondrial phospholipid exchanges completely with added phospholipid under all conditions of mitochondrial fragmentation which result in the isolation of active complexes. Thus, the phospholipid appears to be involved in the binding of the complexes to each other or to other mitochondrial proteins such as structural protein which in the isolated form contains only about 10% of phospholipid.

Coupling Apparatus. The oxidation of 1 molecule of reduced diphosphopyridine nucleotide leads to the esterification of 3 molecules of inorganic phosphate; *i.e.*, to the formation of three molecules of ATP. Thus, a P/O ratio of 3 is obtained with DPNH as substrate. Succinate yields a P/O of 2; cytochrome *c*, a P/O of 1. This suggests that phosphate esterification can take place at all three sites along the electron transfer chain. Attempts have been made to isolate and to purify the components required in the coupling process. Ideally, removal of a coupling component causes loss of phosphorylative capacity; restoration is observed when the missing component is added to the deficient particle. Protein factors have been described which give partial restoration of phosphate esterification to variously prepared incompetent submitochondrial particles. However, the site specificity of these factors is questionable and in no case has it been possible to show that each coupling component has an additive effect in the presence of the other coupling proteins.

Ancillary Enzymes. The enzymes responsible for feeding substrates (DPNH and succinate) to the electron transfer chain are released from the mitochondrion by several means. Sonication, freezing-thawing, and treatment with phospholipase all lead to the release of some of the enzymes that implement CITRIC ACID CYCLE oxidations, fatty acid oxidation, and β-hydroxybutyrate oxidation, but detectable amounts remain with the particles. It appears, therefore, that the location of these enzyme complexes within the mitochondrion is very much in question. The only certainty is that they are present and it is unlikely that all are to be found in the outer membrane.

Ion Accumulation. The mitochondrion possesses a system capable of moving ions from outside the mitochondrion into the matrix. The accumulated divalent cations are precipitated in the matrix as the metallophosphates. The movement of the ions is an energy-dependent process. Oxidation of substrate is required since the deposition of salts is inhibited by antimycin, cyanide, and anaerobic conditions in the absence of ATP. The ion uptake is essentially stopped by uncouplers of oxidative phosphorylation such as dicumarol and dinitrophenol. Ion accumulation, supported by the oxidation of substrates of electron transfer, is insensitive to oligomycin, an inhibitor of phosphorylation at one of the final steps in the synthesis of ATP. When respiration is inhibited, the accumulation of ions can be supported by ATP, presumably by reversal of the pathway of ATP synthesis, this reaction is inhibited by oligomycin. Thus, a simplified mechanism would show electron transfer generating high-energy intermediates which can (a) produce ATP from ADP and inorganic phosphate, or (b) support another energy-driven reaction or function such as ion accumulation.

Electron micrographs of mitochondria which have accumulated massive amounts of metallophosphates show large dense deposits in the internal space.[6] Some deposits follow the outline of the cristae; therefore, it has been suggested that energy-linked movement of ions occurs from the intra-cristal space into the matrix of the mitochondrion.

Energy-driven Swelling and Shrinking of Mitochondria. The swelling of mitochondria described by Raaflaub[7] is an active change in volume which is dependent upon electron transfer or certain high-energy intermediates produced by the electron flux. It has been postulated that the reversible change in shape is an expression of some form of mechano-enzyme not unlike actomyosin of muscle. This conformational change is an energy-requiring process and may be the primary expression of ion translocation which requires an energy source. Thus, the energized movement of of ions could be a secondary expression of the use of a contractile system present in the mitochondrial membrane or matrix.

Origin of Mitochondria. Three principal theories of mitochondrial origin have been investigated: (1) *de novo* synthesis from precursors present in the cytoplasm; (2) growth and fission of preexisting mitochondria; (3) formation from preexisting nonmitochondrial membranes in the cell by an act of induction. The experiments of Luck have shown quite clearly that growth and division of mitochondria is the most probable mechanism of multiplication of these organelles in *Neurospora* cells.

References

1. Palade, G. E., *Anat. Record*, **114**, 427 (1952).
2. Sjöstrand, F. S., *Nature*, **171**, 303 (1953).
3. Fernandez-Moran, H., *Circulation*, **26**, 1039 (1962).
4. Criddle, R. S., Bock, R. M., Green, D. E., and Tisdale, H., *Biochemistry*, **1**, 829 (1962).
5. Slater, E. C., *Advan. Enzymol.*, **20**, 147 (1958).
6. Brierley, G. P., and Slautterback, D. B., *Biochim. Biophys. Acta*, **82**, 183 (1964).
7. Raaflaub, J., *Helv. Physiol. Pharmacol. Acta*, **11**, 142, 157 (1953).
8. Lehninger, A. L., "The Mitochondrion," New York, W. A. Benjamin Inc., 1964.
9. Chance, B. (Editor), "Energy-Linked Functions of Mitochondria," New York, Academic Press, 1963.
10. Green, D. E., Tzagoloff, A., and Oda, T., in "Intracellular Membranous Structure" (Seno, S., and Cowdry, E. V., Editors), p. 127, 1965.
11. Racker, E., *Advan. Enzymol.*, **23**, 323 (1961).

Paul V. Blair

MOLYBDENUM (IN BIOLOGICAL SYSTEMS)

The importance of molybdenum in animals and plants has been established by studying the effects of a molybdenum deficiency and by studies showing that molybdenum is a constituent of certain metalloflavoprotein enzymes.

Restriction of the molybdenum intake by young rats in a synthetic purified casein diet results in a decreased level of tissue, particularly small intestinal, XANTHINE OXIDASE. The enzyme levels are restored to normal by the inclusion of sodium molybdate and other molybdenum compounds. Sodium tungstate is a competititive inhibitor of molybdate, and dietary intakes of tungstate greatly reduce the molybdenum and xanthine oxidase (by three-fourths) concentrations in rat tissues. Tungstate-administered rats also have a decreased capacity to reduce the nitro group of administered *p*-nitrobenzene sulfonamide. This effect can be prevented by supplying additional dietary molybdate, which suggests some role of molybdenum in the reduction of organic nitro groups.

It has not been possible to completely eliminate molybdenum from the diet, so the absolute requirement for this element in animal nutrition has not been established. Even with tungstate in the diet, rats grow normally, reproduce, and show no changes in the uric acid and allantoin output when the diet contains as little as 0.02 ppm of molybdenum. However, a requirement for molybdenum is indicated by the effects of tungstate in chicks. The uric acid normally excreted is partially replaced by hypoxanthine and xanthine in the excreta when tungstate is added (in a W/Mo ratio of 1000:1 to 2000:1) to low-molybdenum chick diets. Growth and survival are also affected. All of these effects can be prevented by giving the chick additional molybdate in the diet. This species difference may reflect the fact that the end product of protein metabolism in chicks is uric acid.

Xanthine oxidase, obtained from several sources, is a metalloflavoprotein containing molybdenum (in an anion form) and iron in the ratios (flavin:molybdenum:iron) of 2:1:8 from cow's milk, 1:1:8 from chicken liver and 1:1:4 from calf liver. From studies using electron paramagnetic resonance spectroscopy, Bray and co-workers[1] conclude that the sequence of electron transfer in milk xanthine oxidase is from substrate (xanthine → uric acid) to molybdenum (Mo VI → Mo V) to flavin (FAD → FADH) to iron (Fe III → Fe II) to oxygen. Liver aldehyde oxidase, which catalyzes the oxidation of aldehydes to acids, has also been identified as a molybdoprotein.

Molybdenum is of importance in plant and microbial life. Deficiency symptoms in legumes grown in certain soils with low molybdenum availability have established that molybdenum is an essential trace element in the symbiotic NITROGEN FIXATION by leguminous plants. The element is also essential for fungi and higher plants in the process of nitrate assimilation. The role of molybdenum is better defined for nitrate assimilation (the biological conversion of nitrate to the ammonia or amino acid level) than for nitrogen fixation. Nitrate reductase, a molybdoflavoprotein enzyme, catalyzes the first step: TPNH (or DPNH) $+ H^+ + NO_3^+ \rightarrow TPN^+$ (or DPN^+) $+ NO_2^+ + H_2O$. Nitrate reductase functions as an electron carrier as follows: TPNH (or DPNH) → FAD → Mo → NO_3^-. Molybdenum is also essential in plant metabolism for processes other than nitrate reduction and nitrogen fixation.

Legumes, cereal grains and some green leafy vegetables are good sources of molybdenum, whereas fruits, berries and most root or stem vegetables are poor sources. Vertebrate tissues are generally low in molybdenum with concentrations in liver and kidney being higher than in other organs and tissues.

Excess molybdenum intake by cattle causes the disease "teart," characterized by severe diarrhea and "loss of condition." Other animals are also subject to toxicity, but they vary in their degree of tolerance to overdosage. Molybdenum-COPPER interrelationships are of importance since molybdenum is toxic at a lower molybdenum intake when the copper intake is also low; toxic symptoms can be alleviated by copper therapy. The antagonism between molybdenum and copper is affected also by the sulfur content of the diet since sulfate administration results in increased molybdenum excretion in the urine and feces.

References

1. Bray, R. C., Palmer, G., and Beinert, H., *J. Biol. Chem.*, **239**, 2667 (1964).

2. UNDERWOOD, E. J., "Trace Elements in Human and Animal Nutrition," Second edition, p. 100, New York and London, Academic Press, 1962.
3. NASON, A., in "Trace Elements (LAMB, C. A., BENTLEY, O. G., AND BEATIE, J .M. EDITORS), p. 269, New York and London, Academic Press, 1958; KELLER, R. F., AND VARNER, J. E., *ibid.*, p. 297.

DAN A. RICHERT

MONOSACCHARIDES. See CARBOHYDRATES (CLASSIFICATION AND STRUCTURAL INTERRELATIONS).

MORPHOGENESIS (BIOCHEMICAL ASPECTS). See BALBIANI RINGS; CHROMOSOME PUFFS; DIFFERENTIATION AND MORPHOGENESIS; MORPHOGENIC INDUCTION; ORGANIZER (IN CELLULAR DIFFERENTIATION).

MORPHOGENIC INDUCTION

Experimental embryologists have described many instances in which the differentiation of a multicellular region of an embryo is altered following an interaction with another group of cells. This is the basic observation and essentially the definition of embryonic induction.

There are great complexities encountered upon examination of even relatively simple systems. To name one, the capacity of a tissue to induce or to respond to an inductive stimulus is usually limited to specific stages in development. The chemical differences underlying these changes in reactivity are not known.

Two general types of mechanisms are usually considered in attempting to explain how one cell can influence another: (1) by cell-to-cell contact or by the modifying effects of a particular surface on the behavior of a cell, and (2) by the transfer of some substance from one cell to another.

The inhibition of cell movement in TISSUE CULTURE by contact with another cell is a clear example of cellular interaction through mutual contact. The nature of this interaction is not known; it has been postulated that such cell contacts evoke specific alterations of surface patterns.

Considerable evidence has accumulated indicating that many inductions and cellular interactions involve a transfer of material between the reacting systems. That such relationships exist in the adult organism is, of course, amply demonstrated in the activities of the endocrine glands. Effects of HORMONES on differentiative processes have also been clearly demonstrated.

In the years following the classical work of Spemann,[1] much effort was expended in attempting to define the chemical nature of the factor in the specific embryonic tissue responsible for the transformation of the overlying ectoderm into brain, spinal cord, muscle, and associated structures. Unfortunately, it soon became evident that some inducing capacity was present not only in embryonic tissue, but also in extracts of a number of adult organs and, to obscure matters further, in a number of non-biological materials where the common denominator appeared to be a transitory cell injury.

Despite these difficulties, a number of workers have continued the analysis of adult tissue fractions which mimic the normal inductive events, and have reported the purification of proteins which, when allowed to act on ectoderm, will alter the kinds of cells and tissues eventually formed from this ectoderm. Thus, the isolated ectodermal cells may form muscle, kidney, and notochord cells under the influence of one particular protein fraction, and brain structures under the influence of another protein fraction. The exact chemical structure of these factors and the way they affect the target tissues which differentiate are little understood. It is also not known how these factors, isolated from adult tissues, are related to those which presumably exist in the embryo. These aspects of induction have been reviewed by Yamada[2] and by Tiedemann.[3]

The complexities of the classical induction system have led other investigators to simpler systems, in which the inducing tissues are relatively specific and the differentiation of only one type of cell is observed. Many such interacting systems have been described; a good number of them involve the interaction of a layer of epithelial cells with its underlying tissue, resulting in the subsequent differentiation of the epithelium. Examples of such inductions include the formation of differentiated lens, kidney, pancreas and salivary gland cells, and many skin structures such as scales and feathers. Evidence that some of these inductions are the result of the transfer of some chemical substance to the induced tissue has been obtained. One simple, but powerful, method of detecting such a transfer is to separate the two components of an induction system and place them on opposite sides of a filter. This prevents cell contact but permits the diffusion of molecules.

Although inducing agents have been detected in this manner, no one has yet chemically characterized an inducing agent which will duplicate the results observed in the developing animal. Various preparations containing ribonucleic acids, proteins, and subcellular particles have been shown to bring about changes in target tissues; however, their role in development is problematical.

Another embryonic system which has received a great deal of attention is the induction of cartilage cells from somite cells by the action of the embryonic spinal cord and notochord. It has been possible to isolate a cartilage-inducing compound from extracts of the spinal cords and notochords of chick embryos which stimulates the formation of cartilage from somite cells in tissue cultures. The structure of the compound appears to be very complex, since it has been reported to contain a number of amino acids, carbohydrates, and nucleotides.

Complicating aspects are present, however, to cloud the interpretation of these experiments. Namely, it is very difficult in certain experiments to separate nonspecific growth stimulation from specific induction. For example, a cell might be programmed to produce a certain protein or product, but fail to do so, not because it is not "induced," but because it is missing an essential nutrient, hormone or growth factor necessary for cell growth or mitosis. It may be that many of the embryonic inductions involve mechanisms conceptually indistinguishable from a "localized hormone."

Many of the known hormones have morphological as well as biochemical effects in both the embryonic and the adult organism. Hormones are known to trigger the growth and differentiation of a number of organ primordia at the time of pupation during insect development. The well-known effects of thyroxin in precipitating METAMORPHOSIS in amphibia are indicative of the profound effects which hormones may exert during development. Under the influence of thyroxin, certain cells are stimulated to grow while others atrophy.

Still another group of naturally occurring growth-stimulating substances has recently been isolated; these substances exert specific growth-promoting influences on various embryonic target tissues, such as certain nerve cells and epidermal cells. These particular growth factors have been characterized as proteins. When antibodies to the nerve growth factor are injected into newborn animals, the sympathetic nerve cells are almost completely destroyed. It seems probable that many more such hormone-like substances which affect the development of diverse types of cells will be found in the future.

The inducers, hormones and growth factors are not clearly distinct categories, either in terms of chemical composition or biological effects. The response of the target tissues to these substances may be assessed morphologically with the aid of the light microscope and the electron microscope or biochemically in terms of such parameters as chemical composition or enzymatic activities. For example, the induction of cartilage by the chondrogenic factor may be followed by the appearance of chondroitin sulfate and the enzymes necessary for its formation. Similarly, the differentiation of pancreatic epithelium, induced by its underlying mesenchyme, may be traced by examining the fine structure of the cell (*e.g.*, the appearance of ribosomal aggregates, membranes, and zymogen granules) or by the synthesis of a specific enzyme (*e.g.*, amylase).

A major unanswered question concerns the exact mechanisms by which these substances react with their target cells to produce their biochemical and subsequent morphological effects. One possibility, which is under investigation in many laboratories, is that many of these biologically active agents produce their effects by activating, in some as yet unknown manner, the genetic material so that new messenger RNA is synthesized. The protein-synthesizing machinery of the cell (see PROTEIN BIOSYNTHESIS) thus would be directed to make a new protein or enzyme in the differentiating cell. Still further from a solution is the problem of how these biochemical modifications, both within the cell and probably at the cell surface, relate to the morphogenetic movements occurring among groups of cells during development, which play an important role in the shaping of the various tissues and organs of the adult animal.

Original articles reviewing many of the problems discussed here have been collected by Bell.[4]

References

1. SPEMANN, H., "Embryonic Development and Induction," New Haven, Conn., Yale University Press, 1938.
2. YAMADA, T., "A Chemical Approach to the Problem of the Organizer," *Advan. Morphogenesis*, **1**, 1–53 (1961).
3. TIEDEMANN, H., "The Role of Regional Specific Inducers in the Primary Determination and Differentiation of Amphibia," in "Biological Organization at the Cellular and Supercellular Level," (HARRIS, R. J. C., EDITOR), pp. 183–209, New York, Academic Press, 1963.
4. BELL, E., "Molecular and Cellular Aspects of Development," New York, Harper & Row, 1965.

STANLEY COHEN

MUCOPOLYSACCHARIDES

Animal connective tissues contain a group of closely related acidic carbohydrate polymers which are located in the extracellular matrix and are collectively known as mucopolysaccharides.

They are heteropolysaccharides formed by the chain condensation of a pair of monomeric sugar units in an alternating sequence, and as a result, these large polymers are invariably built up from disaccharide repeating units.

One of the monomers is always a HEXOSAMINE (a 2-amino-2-deoxyglycose), being D-glucosamine in the case of *hyaluronic acid* (the viscous polysaccharide component of joint fluids, vitreous humor, and skin) and D-galactosamine in the case of the *chondroitin sulfates* of cartilage. The basic amino group of the hexosamine is always present as the neutral acetamido derivative. The acetylated D-galactosamine of the chondroitin sulfates bears the O-sulfate ester on carbon-4 or 6. The other monomer in the repeating sequence of the polysaccharides is D-glucuronic acid. Structures of the repeating units of hyaluronic acid and chondroitin 4-sulfate are shown in Fig. 1. A stereoisomeric variation of chondroitin 4-sulfate is the polysaccharide dermatan sulfate, originally isolated from skin and containing L-iduronic acid in place of D-glucuronic acid.

The sulfate ester content of the chondroitin sulfates is subject to wide variations, being consistently low in corneal extracts and much higher in most samples obtained from shark connective tissues than from equivalent mammalian sources. A portion of the uronic acid of shark mucopolysaccharides undoubtedly contains sulfate ester groups. It is of interest that phylogenetic studies among various species, while indicating variations

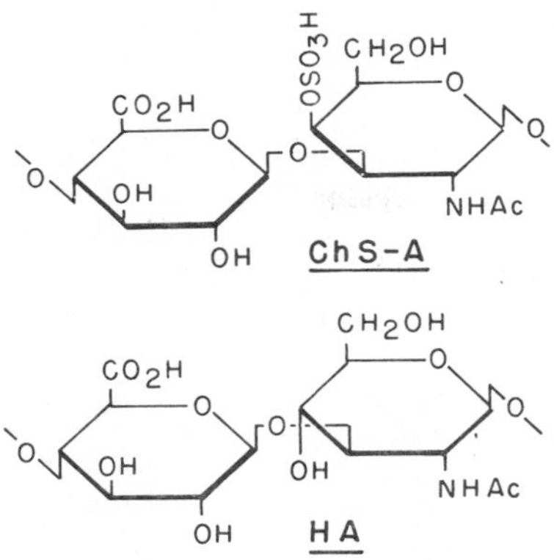

FIG. 1. Repeating units of chondroitin 4-sulfate (ChS-A) and hyaluronic acid (HA). Ac = CH_3CO. A hydrogen substituent on each ring carbon has been omitted for clarity.

in amounts of the polysaccharides, have not disclosed a structurally distinct polysaccharide which could be classified as characteristic for a particular species.

A somewhat different acid mucopolysaccharide is *keratan sulfate*, initially isolated from bovine cornea. The structure of its repeating unit, which contains the monomers D-glucosamine and D-galactose, is shown in Fig. 2. As in the hyaluronic acid molecule, the basic amino group of the glucosamine is present as a neutral acetamido derivative, but in contrast, the glucosamine is present as a 6-O-sulfate ester. The sulfated group converts this otherwise neutral molecule into an anionic polymer; the acetyllactosamine disaccharide repeating unit is found in other carbohydrates, though never as a sulfate ester, notably in blood group substances and milk sugars. An interesting increase of costal cartilage keratan sulfate with aging has been observed.

These polysaccharides are all extracellular components having one or two negatively charged groups per disaccharide repeating unit, the charges being supplied by uronic acid, sulfate monoester, or both. A related intracellular substance found in mast cells is the anticoagulant HEPARIN composed of D-glucosamine, D-glucuronic acid and sulfate. The glucosamine contains a negatively charged sulfamido group rather than the more commonly occurring acetamido derivative. Another remarkable difference is that the glycosidic linkages in heparin are of the α-D-type rather than the β-D- (or α-L-) type found in the extracellular polysaccharides. There appears to be

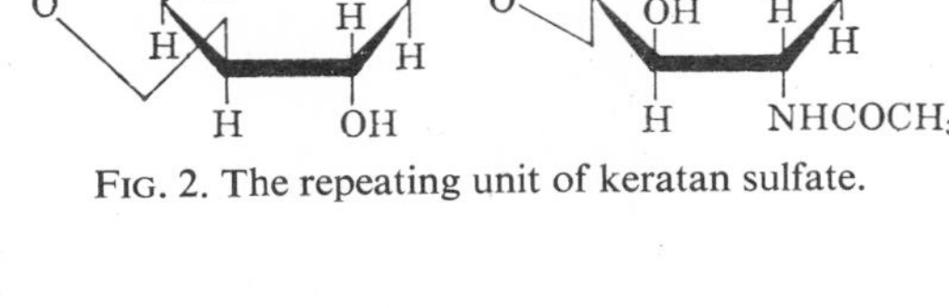

FIG. 2. The repeating unit of keratan sulfate.

an analogy between these and β-D-linked structural polysaccharides such as CELLULOSE and α-D-linked storage polysaccharides such as STARCH or GLYCOGEN.

The mucopolysaccharides are associated with proteins to a greater or lesser degree. The hyaluronic acid which is dissolved in the joint fluids is relatively free of attached proteins and imparts a very high viscosity to the fluid. This is due to the great chain length of the molecule, which has been estimated to contain more than 5000 repeating units within a single chain. The excellent properties of lubrication, resistance to compression, water retention, and prevention of diffusion of the synovial fluids are due to the presence of hyaluronic acid. The decreased efficiency of the fluids in certain pathological joint conditions has been attributed to the presence of degraded hyaluronate molecules of smaller molecular weights.

The sulfated mucopolysaccharides of connective tissues do not appear to attain the great single chain lengths of the hyaluronic acid molecules, having approximately 50 repeating disaccharide units per chain, but the chains are apparently linked covalently to proteins so that giant protein-polysaccharide molecules are present. The polysaccharide chains appear to be linked to the protein by means of a glycosidic attachment to the hydroxyl group of the hydroxyamino acid SERINE. The neutral sugars xylose and galactose appear to be at the end of the chain close to the point of attachment, with xylose probably involved in the linkage. Another type of carbohydrate-to-amino acid linkage found in nature involves the amide group of ASPARAGINE, derived from aspartic acid. This linkage is found in the keratan sulfate isolated from cornea.

The extracellular components, including COLLAGEN, the major fibrous protein of connective tissue, have their origin in specialized cells defined by their site and function and known as fibroblasts, chondroblasts, and osteoblasts. These cells are derived from undifferentiated mesenchymal cells, and the process whereby they become specialized to produce the materials required for the formation and maintenance of a variety of tissues such as skin, blood vessels, tendon, cartilage and bone is little understood.

The biosynthesis of the carbohydrate polymers is mediated by nucleotide (uridine diphospho-) derivates of the monomeric components, but the form in which the intermediates are extruded from the cell, and the influences which determine the disposition of the extracellular substances and the formation, size, and orientation of the fibers in the various tissues have not been defined.

Excess or deficient amounts of hormones or vitamins are known to influence connective tissues in a fundamental manner, not only in the normal metabolism of the tissues, but also in inflammatory responses, wound healing and metamorphoses where enhanced activity is evident. Similarly, the deposition and resorption of bone is under hormonal and vitamin control. In no case, however, has the mechanism of action been satisfactorily outlined. Also the fundamental changes which can be visualized in the connective tissues with aging, the onset of pathological conditions, and deficiency diseases have never been explained on a molecular basis.

References

1. JEANLOZ, R. W., in "Comprehensive Biochemistry" (FLORKIN, M., AND STOTZ, E., EDITORS), Vol. 5, Ch. 7, New York, Elsevier, 1963.
2. JACKSON, S. F., in "The Cell" (BRACHET, J., AND MIRSKY, A. E., EDITORS), Vol. 6, Ch. 6, New York, Academic Press, 1964.
3. MUIR, H., in "International Review of Connective Tissue Research" (HALL, D. A., EDITOR), Vol. 2, pp. 100–145, New York, Academic Press, 1964.
4. ANDERSON, B., HOFFMAN, P., AND MEYER, K., "O-Serine Linkage in Chondroitin 4- or 6-Sulfate Peptides," *J. Biol. Chem.*, **240**, 156 (1965).
5. LINDAHL, U., AND RODÉN, L., "The Linkage of Heparin to Protein," *Biochem. Biophys. Res. Commun.*, **17**, 254 (1964).

PHILIP HOFFMAN

MUCOPROTEINS

Mucoproteins are conjugated PROTEINS in which the protein portion is combined with a relatively large amount of carbohydrate (conventionally more than 4% measured as hexosamine); the carbohydrate portions are complex polysaccharides such as MUCOPOLYSACCHARIDES (see also PLASMA AND PLASMA PROTEINS, section on Mucoproteins).

MULLER, H. J.

Herman Joseph Muller (1890–) is an American geneticist who made many contributions to genetics. He was the first to produce mutations by the use of X rays and for this contribution he was awarded the Nobel Prize in physiology and medicine in 1946.

MULTIENZYME COMPLEXES

A substantial and specific portion of the enzyme complement of most cells is normally found to be more or less firmly bound to subcellular organelles. Until recently it has been tacitly assumed that those enzymes that are not so bound are distributed at random through the liquid phase of the cell. Evidence has begun to accumulate, however, that at least a few of the "soluble" proteins of some cells are organized in specific ways into functionally significant aggregates. It is reasonable to suppose that holding enzymes with related activities close together serves similar ends in relatively simple particles and in the much more elaborate arrays associated with the larger structural elements of the cell, *e.g.*, the respiratory chain of mitochondria.

It should be emphasized at the outset that there are very few multienzyme complexes whose existence has been well-documented. These are the pyruvate and α-ketoglutarate dehydrogenase complexes from several sources, the fatty acid synthetase of baker's yeast, and the tryptophan synthetase of *Escherichia coli*.

Enzyme systems which catalyze the CoA- and DPN-linked oxidative decarboxylation of pyruvic and α-ketoglutaric acids [reaction (*1*)] have been

$$RCOCO_2H + CoA{-}SH + DPN^+ \rightarrow RCO{-}SCoA + CO_2 + DPNH + H^+ \quad (1)$$

isolated as multienzyme complexes with molecular weights of several million, from pigeon breast muscle, pig heart muscle, *E. coli* and beef kidney mitochondria. Two classes of complexes have been obtained, one specific for pyruvate, the other for α-ketoglutarate. The operation of these complexes is represented schematically in Fig. 1, in which the brackets indicate enzyme-bound compounds.

The *E. coli* pyruvate dehydrogenase complex, whose molecular weight is about 4.8 million, has been separated into three enzymes: pyruvate decarboxylase, lipoyl reductase-transacetylase and a flavoprotein (dihydrolipoyl dehydrogenase). The molecular weights of the three enzymes are approximately 183,000, 1.6 million, and 112,000, respectively. The *E. coli* α-ketoglutarate dehydrogenase complex, with molecular weight about 2.4 million, has also been separated into three enzymes, analogous to those obtained from the pyruvate dehydrogenase complex. The three enzymes are α-ketoglutarate decarboxylase, lipoyl reductase-transsuccinylase, and a FLAVOPROTEIN (dihydrolipoyl dehydrogenase). Both complexes have been reconstituted from their respective separate components. The two flavoproteins are functionally interchangeable, but it is not yet certain whether they are identical. The lipoyl moiety, represented as [$LipS_2$] in Fig. 1, is bound by an amide linkage to the ε-amino group of a lysine residue in both transacylases [see LIPOIC ACID for its role in over-all reaction (*1*)].

Lipoyl reductase-transacetylase contains approximately 1 mole of covalently bound lipoic acid per 35,000 grams of protein. The transacetylase is dissociated by treatment with dilute acetic acid into disordered, inactive subunits with a molecular weight about 70,000. There is some evidence that these structures may be dimers of a still smaller subunit. When the acid solution of dissociated subunits is diluted rapidly into phosphate or tris(hydroxymethyl)aminomethane buffers (final pH, 7.0), the subunits reassemble spontaneously to give the enzymatic activity and architecture characteristic of the native aggregate. Both the native and the reconstituted transacetylase combine spontaneously with the decarboxylase and the flavoprotein to produce a large unit which resembles the native pyruvate dehydrogenase complex in composition, enzymatic activities and appearance in the electron microscope. These results support the view that the information required to specify the formation even of very elaborate aggregates resides entirely in the amino acid sequence of the subunits.

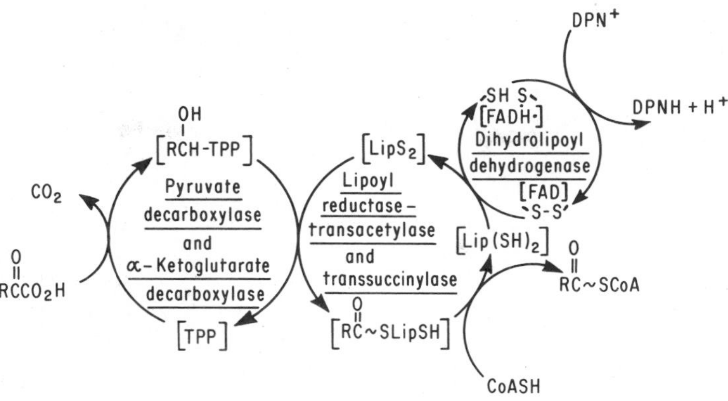

FIG. 1. Schematic representation of the reaction sequence in the CoA- and DPN-linked oxidative decarboxylation of pyruvate and α-ketoglutarate.

On the basis of biochemical evidence, the pyruvate dehydrogenase complex appears to be an organized mosaic of enzymes in which each of the component enzymes is so located as to permit efficient coupling of the individual reactions catalyzed by these enzymes. This concept has been confirmed and extended by electron microscope studies. Electron micrographs of the pyruvate dehydrogenase complex negatively stained with phosphotungstate show a polyhedral structure with a diameter of 300–350 Å [Fig. 2(a)]. The images show four structural elements, arranged as a tetrad, in the center of the polyhedron. Surrounding this tetrad is an orderly array of subunits, 60–90 Å in diameter. Electron micrographs of the isolated lipoyl reductase-transacetylase negatively stained with phosphotungstate show tetrads [Fig. 2(b)] that closely resemble the central tetrad of the native complex [Fig. 2(a)]. Some of the images seen in Fig. 2(b) have the appearance of two parallel strands with a length of 130–160 Å. These results are interpreted as indicating that the subunits of the transacetylase are arranged into four stacks having the appearance of a tetrad (top view) or two parallel strands (side view). The molecules of pyruvate decarboxylase and flavoprotein apparently are disposed in some orderly arrangement around the transacetylase aggregate. On the basis of the biochemical and electron microscope data, tentative models of the pyruvate dehydrogenase complex have been constructed. One such model is shown in Fig. 3.

It has not yet been possible to obtain satisfactory electron micrographs of the intact *E.coli* α-ketoglutarate dehydrogenase complex, since it apparently dissociates during preparation of the specimen for electron microscopy. Electron micrographs of the isolated lipoyl reductase-transsuccinylase have been obtained, however, and these show tetrads similar to the ones found with analogous preparations from the pyruvate dehydrogenase complex [Fig. 2(b)]. The molecules of α-ketoglutarate decarboxylase and of flavoprotein presumably are arranged around this tetrad in the intact α-ketoglutarate dehydrogenase complex.

There are strong indications that both the mammalian pyruvate and α-ketoglutarate dehydrogenase complexes consist of three enzymes analogous to those obtained from the corresponding *E. coli* complexes. This possibility is under active investigation.

The synthesis of long-chain fatty acids is catalyzed by multienzyme systems referred to as "fatty acid synthetases." The over-all reaction may be represented as follows:

$$\text{Acetyl-CoA} + 7\ \text{Malonyl-CoA} + 14\text{TPNH} + 14\text{H}^+ \rightarrow \text{Palmitate} + 14\text{TPN}^+ + 8\text{CoA} + 7\text{CO}_2 + 6\text{H}_2\text{O} \quad (2)$$

This over-all reaction proceeds via a sequence of six or seven consecutive reactions, each of which is probably catalyzed by a different enzyme. Lynen and co-workers have isolated a fatty acid synthetase system from baker's yeast as a structural unit with a molecular weight of approximately 2.3 million. The synthetase has a definite morphology as revealed by electron microscopy. It is visualized by Lynen as a combination of seven different enzymes. There is considerable biochemical evidence to support this conclusion, but attempts to separate the yeast synthetase into enzymatically active components have been unsuccessful The enzymes responsible for fatty acid synthesis in animal tissues may be organized, like the yeast system, into multienzyme complexes. On the other hand, Vagelos and co-workers have made remarkable progress in separating the fatty acid synthesizing system of *E. coli* into individual enzymes. It remains to be seen whether the *E. coli* system occurs in the cell as a multienzyme complex.

The enzyme tryptophan synthetase, as it is isolated from *E. coli* or from *Neurospora crassa*, is

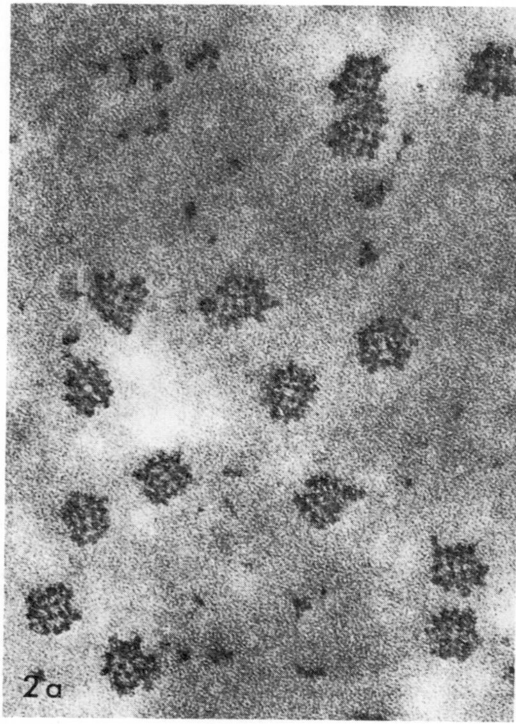

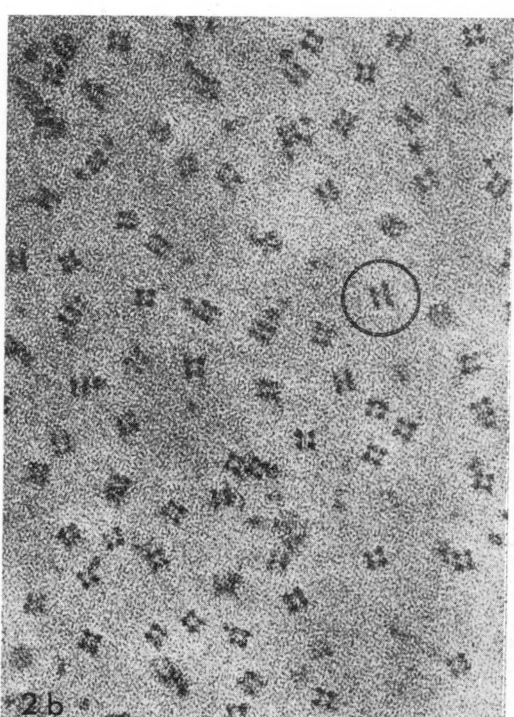

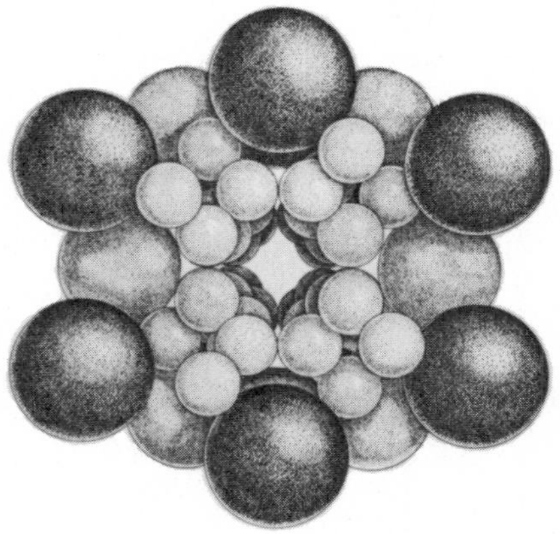

FIG. 2. Electron micrographs of the *E. coli* pyruvate dehydrogenase complex (a) and the lipoyl reductase-transacetylase aggregate (b) negatively stained with 0.25 % potassium phosphotungstate, pH 7.0. Reversed contrast prints 250,000X (from Robert M. Oliver, The University of Texas). Two types of images can be seen in (b): tetrads and parallel strands (circled), presumably corresponding to top and side views, respectively, of the aggregate.

FIG. 3. Tentative model of the *E. coli* pyruvate dehydrogenase complex. The model consists of 12 molecules of pyruvate decarboxylase (large spheres) and 6 molecules of dihydrolipoyl dehydrogenase (medium-size spheres) disposed in an orderly arrangement around the 24 (or 48) subunits comprising the lipoyl reductase-transacetylase aggregate (small spheres).

capable of catalyzing three related chemical reactions:

Indole-glycerol Phosphate + L-Serine →
L-Tryptophan + Glyceraldehyde-3-phosphate (*3*)

Indole-glycerol Phosphate ⇌ Indole +
Glyceraldehyde-3-phosphate (*4*)

Indole + L-Serine → L-Tryptophan (*5*)

Reaction (*3*) probably represents the physiological function of the enzyme. Reactions (*3*) and (*5*) require pyridoxal phosphate as coenzyme, but reaction (*4*) does not. *E. coli* tryptophan synthetase has a molecular weight about 150,000. It can be separated by relatively mild procedures into two components. The A-protein has a molecular weight close to 30,000, that of the B-protein is about 117,000. The A-protein has slight activity in catalyzing reaction (*4*), but no activity in reaction (*3*) or (*5*). The B-protein catalyzes reaction (*5*) but is inactive in reactions (*3*) and (*4*). The intact synthetase apparently consists of one molecule of the A-protein and one of the B-protein. In this sense, it is a multienzyme complex of the simplest type. Tryptophan synthetase from *Neurospora* possesses enzymatic activities very similar to those of the *E. coli* enzyme, and the *Neurospora* enzyme has a molecular weight close to that of the A-B complex from *E. coli*. Although the mold enzyme has been shown to consist of subunits, no active components analogous to the A- and B-proteins have been obtained from it.

Association of the A- and B-proteins profoundly modifies their individual catalytic activities. In this case, at least, the formation of a complex has consequences that go beyond the increased efficiency that might result from simply assembling the enzymes of a metabolic sequence into an organized unit without otherwise altering their structure.

It is appropriate to ask whether there are any obvious advantages a cell might obtain by assembling several of the enzymes concerned with a given metabolic pathway into an aggregate. Advantages of two general types may be suggested: (1) The aggregation of the component polypeptide chains may produce particles which have catalytic properties unlike those of any of the separate chains; *i.e.*, aggregation may itself produce, enhance, or modify the desired activity. (2) If the enzymes which operate in sequence on a given substrate are maintained in proximity to each other, the over-all efficiency of the pathway may be altered, even if the catalytic activities of the individual enzymes are not changed by aggregation.

A survey of the information available concerning multienzyme complexes reveals very few well-documented occurrences of such entities. Any general assessment of the importance of multienzyme complexes in the organization of the cell must inquire whether functional aggregates of enzymes are common or rare. Stated explicitly, the relevant question is: If enzymes are frequently associated with each other, why have so few of the resulting aggregates been found? One possible answer is that multienzyme complexes have been observed rarely because they rarely exist. However, it is also possible to suggest why complexes that do exist may have evaded observation. The chief trend of classical enzymology has been toward breaking down metabolic processes into sequences of reactions, each reaction catalyzed by a single enzyme. A major preoccupation of enzymologists has been the separation and purification of the enzymes involved at each stage of each pathway. To be described as pure, an enzyme must generally be shown to possess the highest possible activity in one reaction and no detectable activity in any other. Judged by this criterion, a multienzyme complex is an impure enzyme. Confronted with such an object in the context of an analytic effort, a worker would proceed, if he could, to remove the "contaminants." Some of the complexes that have been found are known to be unstable under conditions often encountered during the fractionation of proteins. This fact suggests that an attempt to "purify" one component of a complex might very well succeed. The present lack of positive evidence for the general occurrence of multienzyme complexes does not necessarily mean that such entities are rare. With increasing awareness of the fact that many of the classical methods of disrupting cells and extracting and fractionating enzymes may preclude detection and isolation of organized enzyme systems, it is entirely possible that many more such entities will be found.

References

1. Reed, L. J., and Cox, D. J., "Macromolecular Organization of Enzyme Systems," *Ann. Rev. Biochem.*, **35**, 57–84 (1966).
2. Reed, L. J., "Chemistry and Function of Lipoic Acid," in "Comprehensive Biochemistry" (Florkin, M., and Stotz, E. H., Editors), Vol. 14, New York, Elsevier Publishing Co., in press.
3. Sanadi, D. R., "Pyruvate and α-Ketoglutarate Oxidation Enzymes," in "The Enzymes" (Boyer, P. D., Lardy, H., and Myrbäck, K., Editors), Vol. 7, pp. 307–344, New York, Academic Press, 1963.
4. Vagelos, P. R., "Lipid Metabolism," *Ann. Rev. Biochem.*, **33**, 139–172 (1964).

Lester J. Reed and David J. Cox

MULTIPLE SCLEROSIS

In the biochemical makeup of multiple sclerosis (M.S.) two changes are well documented. One is the lipid breakdown of the myelin; the other, increase of gamma globulin in the cerebrospinal fluid (CSF). Brain tissue samples are not taken during life, but after death; therefore, the extent of chemical changes which may occur between the time of death and autopsy remain an undetermined factor. myelin breakdown does not occur in a straight-line fashion, since between the demyelinated areas the tissue may remain intact. Even with these limitations, valid results may be obtained, *i.e.*, when the tissue analyses are compared with that of "normal" patients who died of non-neurological disease.

Several authors have published data on the lipid changes of neural tissue in multiple sclerosis.[1,2] A detailed description of the chemical pathology was given by Lumsden[3]. We wish to restrict our report on the lipid changes mainly to the work of Cumings[4], which is most extensive and representative. He has shown (using chromatography) that the P lipids are reduced greatly with marked losses in phosphatidylcholine, ethanolamine, serine, plasmalogen (phosphatidylethanolamine) and sphingomyelin (see Table 1). In the normal brain, cholesterol is in free

TABLE 1. LIPIDS IN MULTIPLE SCLEROSIS[a]

Substance	Cerebral White Matter[b]	
	Normal	Multiple Sclerosis Plaque
Total lipid phosphorus	890	800
Phosphatidylcholine	195	193
Phosphatidylethanolamine	97	88
Phosphatidylserine	160	153
Phosphoinositide	—	—
Plasmalogens	180	160
Plasmalogens, corrected	361	338
Sphingomyelin	163	161
Unidentified mild alkali and acid-stable phospholipids	100	98
Water, %	73.8	91.0

[a] This Table represents results obtained by Cumings. (From Rose, Augustus S., and Pearson, Carl M., (Editors) "Mechanisms of Demyelination," p. 62, New York, McGraw-Hill Book Co., 1963. Used by permission of McGraw-Hill Book Company.)

[b] Results in mg lipid phosphorus per 100 g dry tissue.

form. Gerstl and co-workers[5] reported reduction of total lipids, plasmalogens and cholesterol. The ester-linked fatty acids were reduced 20–50%. Two findings are significantly outstanding: (1) in other conditions where demyelination occurs as a secondary reaction, the changes differed from those occurring in M.S.; (2) alteration of lipids occurs (or precedes) where no demyelination can be demonstrated by histological methods. The extent of changes is not uniform for all lipids. This might be explained, according to the authors, in that partial resynthesis of the lipids might have taken place. As far as the role of the enzymes, which hydrolyze lipids, is concerned, phospholipase A, ceramidase and phosphoinositidase have been demonstrated in the brain.[5]

Enzymes in M.S. have been studied by Friede.[6] Oxidative (succinic dehydrogenase, DPN diaphorase) and hydrolytic (acid and alkaline phosphatase) enzymes were studied by histochemical procedures. An increase in acid phosphatase and oxidative enzymes were found in the zone of demyelination, indicating a tissue reaction which started in the still myelinated tissue. There was a direct relationship between the process of demyelination and enzyme activity.

As for the action of exogenous enzymes, it was shown that lipase and proteinase, when administered intracerebrally, resulted in demyelination. The myelin lipid complex is actually broken up by the enzyme protease.[7] It may well be that in the long chain of events leading to demyelination, this is the first step.

No detailed studies have, as yet, been carried out on the proteins. The reason probably is that our knowledge of the proteins, even of the normal neural tissue, is rather limited. Gerstl *et al.*[5] determined the protein content of whole white matter and of the myelin prepared by differential centrifugation. In M.S. an increase in the protein content of the myelin fraction was found with concomitant decrease in the supernatant. There was not much difference in the mitochondrial and nuclear fractions.

At present new procedures are being introduced for the fractionation of brain proteins, including column and thin-layer CHROMATOGRAPHY, starch and acrylamide ELECTROPHORESIS. It is hoped that with the use of these techniques, the brain proteins will be separated, purified and ready for physicochemical characterization in the not so distant future. After this, we shall be able to compare well-defined proteins in M.S. and normal conditions.

The second characteristic feature in multiple sclerosis is the significantly increased gamma globulin of the CEREBROSPINAL FLUID. Since the publication of the immunochemical method of Kabat *et al.*,[8] the method was applied to a large number of M.S. cases and an increase in gamma globulin was found in 85% of the cases examined. This was amply confirmed by the use of other techniques, such as electrophoresis on paper, cellulose acetate, agar,[9,10] starch and, more recently, by a quantitative chromatographic technique of Einstein, *et al.*[15] Acrylamide gel electrophoresis, [11,12] which has excellent resolving power for all other proteins, does not separate gamma globulins. The electrophoretic methods have the advantage that other proteins can be measured besides gamma globulin.

Questions connected with the increased gamma globulin in the CSF are: Is the gamma globulin specific, meaning that it has no counterpart in the sera? Is it specific for M.S.? Does it represent antibodies to some of the brain antigens?

In our own studies,[5] rabbits were sensitized with concentrated CSF, then the anti-CSF sera were tested by immunoelectrophoretic (see IMMUNODIFFUSION) technique. The pattern revealed some gamma globulin which remained after absorption with sera of the same patients. However, the major gamma globulin was identical with that of the sera. Hochwald and Thorbecke[13] and MacPherson[14] isolated these specific gamma globulins, named gamma C and gamma trace, respectively. They, however, seem to constitute a minor component when measured quantitatively and, according to MacPherson, were reduced rather than increased in M.S. The major fraction of gamma globulin in M.S. is identical with 7S gamma globulin, present also in normal CSF and sera.

As stated earlier, multiple sclerosis is characterized by breakdown of the myelin with a relative preservation of the axon. Neither its etiology nor its cure is known at present. Several hypotheses have been offered; among them, *autoimmunity* has the strongest support. The argument for autoimmunity is an increased gamma globulin in the CSF, the effect of steroids in lowering the gamma globulin in the CSF, and the work of Bornstein on the demyelinating effect or action of M.S. sera, which indicates the presence of antibodies.[5] Based on this assumption, an experimental model has been developed, called experimental allergic encephalomyelitis (EAE), which is produced by the injection of neural tissue or preparations made from it. The compound responsible for the encephalitogenic activity has been identified as a protein [Kies, Alvord and Roboz (Einstein)]. Several investigators reported their findings on the isolation of the encephalitogenic protein from different species: human, bovine, and guinea pig brain at the 1964 New York Academy of Sciences Symposium.[5] It is a basic protein, and its amino acid composition is similar to that of HISTONE. However, since it is a component of the myelin, which is free of nuclei, it is not part of nucleoprotein. Since the physicochemical characterization of the protein is in progress,[16] it cannot be stated with certainty whether proteins isolated from different species are identical; although, as Helene Rauch has shown with our bovine and human preparations, they cross-react in the passive cutaneous anaphylaxis test. A detailed and scholarly study appeared recently,[17] which describes the experimental allergic encephalomyelitis as a prototype of autoimmune disease. Extensive immunological studies are to be undertaken to test the CSF and blood of M.S. patients, using the encephalitogenic protein(s) to prove that EAE is closely related to M.S. and to give convincing evidence that the underlying mechanism in M.S. is indeed of an autoimmune origin.

References

1. Plum, C. M., and Hansen, S. E., "Studies in Multiple Sclerosis, II," *Acta Psychiat. et Neurol. Scand., Suppl.* 141, **35**, 84 (1960).
2. Majno, G., and Karnovsky, M. L., "A Biochemical and Morphologic Study of Myelination and Demyelination, III," *J. Neurochem.*, **8**, 1 (1961).
3. Lumsden, C. E., "Chemical Pathology of the Lipids in Demyelinating Disease," in "Multiple Sclerosis, A Reappraisal" (McAlpine, Lumsden, and Acheson), Edinburgh, Livingstone, Ltd., 1965.
4. Cumings, J. N., "Some Biochemical Considerations Regarding Different Forms of Demyelination," in "Mechanisms of Demyelination" (Rose and Pearson, Editors), p. 58, New York, McGraw-Hill Book Co., 1963, Cumings, J. N., "Lipid Chemistry of the Brain in Demyelinating Disease," *Brain*, **78**, 554 (1955).
5. N.Y. Academy of Sciences, "Research in Demyelinating Diseases," *Ann. N.Y. Acad. Sci.*, **122**, 1–570 (March 31, 1965).
6. Friede, R. L., "Enzyme Histochemical Studies in Multiple Sclerosis," *Arch. Neurol.*, **5**, 433 (1961).
7. Vogel, F. S., "Demyelinization Induced in Living Rabbits by Means of a Lipolytic Enzyme Preparation," *J. Exptl. Med.*, **93**, 297 (1951).
8. Kabat, E. A., Freedman, D. S., Murray, J. F., and Knaub, V., "A Study of the Crystalline Albumin, Gamma Globulin and Total Protein in the Cerebrospinal Fluid of One Hundred Cases of Multiple Sclerosis and in Other Diseases," *Am. J. Med. Sci.*, **219**, 55 (1950).
9. Lowenthal, A. M., Van Sande, M., and Karcher, D., "The Differential Diagnosis of Neurological Diseases by Fractionating Electrophoretically the CSF Gamma Globulins," *J. Neurochem.*, **6**, 51 (1960).
10. Dencker, S. J., "Studies on Specific Cerebrospinal Gamma Globulin Components," *Acta Neurol. Scand., Suppl.* 29, **4**, 517 (1963).
11. Cunningham, V. R., "Analysis of 'Native' Cerebrospinal Fluid by the Polyacrylamide Disc Electrophoresis Technique," *J. Clin. Pathol.*, **17**, 143–148 (1964).
12. Monseu, G., and Cumings, J. N., "Polyacrylamide Disc Electrophoresis of the Proteins of Cerebrospinal Fluid and Brain," *J. Neurol. Neurosurg. Psychiat.*, **28**, 56 (1965).
13. Hochwald, G. M., and Thorbecke, G. J., "Use of Antiserum against CSF in Demonstrations of Trace Proteins in Biological Fluids," *Proc. Soc. Exptl. Biol. Med.*, **109**, 91 (1962).
14. MacPherson, C. F. C., and Cosgrove, J. B. R., "Immunochemical Evidence for Gamma Globulin Peculiar to CSF," *Canadian J. Biochem. Physiol.*, **39**, 1567 (1961).
15. Einstein, E. R., Richard, K. and Kwa, G. B., "Determination of Gamma Globulin in the Cerebrospinal Fluid by Quantitative Chromatography," *J. Lab. Clin. Med.*, **68**, 120 (1966).
16. Nakao, A., Davis, W. J., and Einstein, E. R., "Basic Proteins from the Acidic Extract of Bovine Spinal Cord, I and II," *Biochim. Biophys. Acta* (in press, 1966).
17. Paterson, P. Y., "Experimental Allergic Encephalomyelitis and Autoimmune Disease," *Advanc. Immunol.* **5**, 131 (1966).

Elizabeth Roboz Einstein

MUSCLE (STRUCTURE AND FUNCTION)

Muscles consist of many fibers held together by connective tissue. Their structure and function vary widely in different organs and animals. On the basis of structure they are divided into smooth and striated muscle. The former are usually found in organs which carry out only sluggish movements. The latter are capable of rapid contractions. Some invertebrates have only smooth muscles, but in arthropods all muscles, including those of the viscera, are striated. Smooth muscles are adapted to many specialized functions. They should not be considered to be primitive; both types of muscles are found side by side even in the lowest forms, such as coelenterates.

Structure. *Smooth Muscle*. One type of smooth muscle has spindle-shaped fibers which are 4–7 μ thick and rarely longer than 0.2 mm. Microscopically they are nearly homogeneous except for a single nucleus, but the electron microscope has shown that they are filled with longitudinal filaments about 80 Å thick [Fig. 1(a)]. This type of muscle is found in the walls of the viscera and blood vessels of vertebrates and in some invertebrates. Another type of smooth muscle, observed only in invertebrates, contains larger fibers with microscopically visible fibrils. Each fibril is made up of many filaments [Fig. 1(b)]. They are often capable of contractions nearly as fast as those of striated muscles. Obliquely striated muscles, which are found in some worms and mollusks, probably are modified striated muscles.

Striated Muscle. Striated muscle fibers contain fibrils which are about 1–2 μ thick. The fibers are made up of alternating bands with different optical properties. The A (anisotropic) bands have a higher refractive index and much greater birefringence than the I (isotropic) bands. The middle of the A bands appears light and is called the H band. In the middle of the I bands is the Z band, a fine network which connects all myofibrils of a fiber transversely and extends to the outer wall of the fibers, the sarcolemma. This structure maintains the A and I segments of all fibrils at nearly the same level, thereby giving the whole fiber its striated appearance [Fig. 1(c)]. It may also help to transmit tension from the fibrils to the outside.

With the ELECTRON MICROSCOPE it has been found that the myofibrils of striated muscle contain a regular array of two types of filaments. Those with a thickness of 130 Å extend through the A band. Thinner filaments pass through the I and most of the A bands, but are interrupted in the middle of the A bands, thereby forming the H band [Fig. 1(d)]. Where the two types of filaments overlap, they are connected by bridges which protrude from the thick filaments.

By selective extraction it has been shown that the thick filaments are mainly composed of the protein MYOSIN, which makes up about 40% of the total muscle protein and has a molecular weight of about 500,000. The thin filaments consist of actin and appear under the electron microscope as two helically wound strands. Each strand is beaded, being made up of molecules of actin G with a molecular weight of about 70,000. Actin G can be obtained in solution and polymerizes easily to form filaments which have the same structure as the natural thin filaments (see also CONTRACTILE PROTEINS).

At rest, most muscles are elastic bodies with rubber-like extensibility. However, the tension of extended muscle is largely, perhaps entirely, due to connective tissue and the supporting structures of the muscle fibers, such as the sarcolemma. Some smooth muscles are plastic within the physiological range of lengths. An external force extends them slowly, as if opposed by a viscous resistance. This resistance probably is caused by the binding between the filaments which slide along each other during extension.

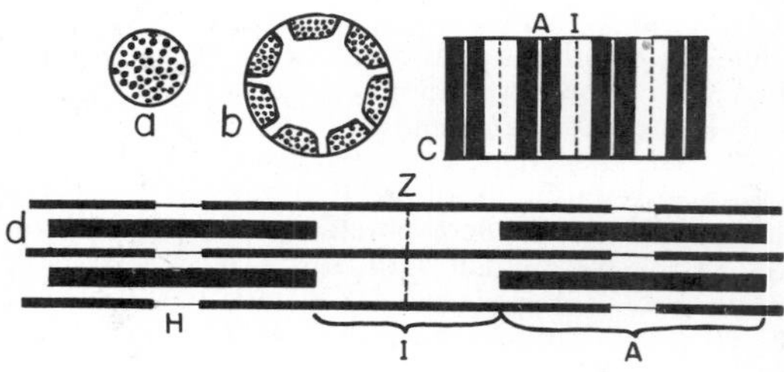

FIG. 1. Diagrammatic representation of structure of smooth and striated muscle. (a) and (b) Cross sections of two common types of smooth muscles. Dots are submicroscopic filaments in cross section. (c) Longitudinal section of striated muscle as seen in a microscope. (d) The same at higher magnification as seen in an electron microscope.

Function. Muscles differ greatly in the speed of their contraction. In vertebrates the response to a single stimulus lasts about 1/40 second in external eye muscles, 0.1–0.3 second in most other skeletal muscles, a few seconds in smooth muscles. There are important differences also between visceral muscles, such as those of the heart and intestine, and skeletal muscle. In the former, contractions are initiated by a spontaneous, slow depolarization of some of the muscle fibers leading up to a conducted response. These muscles are composed of many small fibers, but contractions normally are conducted over the whole muscle, as if it were a single giant fiber, and therefore follow the all or none law.

In skeletal muscles and some smooth muscles, contractions are initiated by motor NERVE IMPULSES. In these muscles, the contraction induced by a nerve impulse generally spreads from the neuromyal junction and thereby activates the whole muscle fiber (see SYNAPSES AND NEUROMUSCULAR JUNCTIONS). Conduction is associated with a brief action potential similar to that of nerve fibers. In some skeletal muscle fibers, a nerve impulse activates only the region near the junction. These fibers have multiple motor innervation. Many visceral muscles of vertebrates and some skeletal muscles of arthropods are also supplied with inhibitory nerve fibers which diminish responsiveness. Some chemical agents induce long continued activity which may not be associated with repetitive action potentials and then are called contractures.

In vertebrate skeletal muscle usually more than

one, sometimes more than 1000 muscle fibers, are innervated by a single motor nerve fiber. Because normally all these fibers respond at the same time, the motor neuron with all the muscle fibers it innervates is called a motor unit. During normal contractions the motor units discharge repetitively at frequencies varying from 5–50 per second. Because the individual responses fuse, a sustained contraction, called tetanus, is produced. Such contractions are stronger than single responses because the mechanical effects of successive responses are added up (summation) if relaxation after each response is incomplete. The strength of contraction increases with the frequency of discharge. Gradation of contraction is also due to the fact that in weak contractions only a small number of motor units participates and the number increases with the height of contraction.

Because a single response is automatically followed by relaxation, a sustained contracted state can be produced only by repeated activation and, therefore, requires continuous energy expenditure. In some muscles, such a state may persist under normal conditions for a long time and is then called tonus. Tonus is generally controlled by the central nervous system through motor nerve fibers, in some smooth muscles also through inhibitory nerve fibers. In slowly contracting muscles, particularly some smooth muscles, tonic contractions can be maintained by a small energy expenditure because it requires infrequent activation of the muscle fibers. Tonus does not represent a distinct physiological mechanism different from that of phasic muscular activity, but is a descriptive term for a sustained, involuntary contraction.

Contraction does not require the presence of oxygen. As in other tissues, energy liberated by the burning of carbohydrate and fat is utilized as PHOSPHATE BOND ENERGY. In skeletal muscles, carbohydrate is the chief source of energy, as shown by the rise in the respiratory quotient during exercise. The active heart, however (see HEART METABOLISM), can take up large amounts of fatty acids and ketonebodies from blood and always has a respiratory quotient below 1. If oxygen supply is insufficient, lactic acid is produced in muscles during activity. In humans this is true only in severe exercise which cannot be sustained for more than an hour. During recovery from exercise, O_2-consumption is increased for a period of time which increases with the severity of the exercise.

The mechanical efficiency of contraction $eff = W/(W + H)$, where W is mechanical work and H heat, can be as high as 40%, but depends greatly on speed of shortening and other factors. The total energy $(W + H)$ liberated during a contraction increases with the extent of shortening (Fenn effect). This has suggested that contraction is a mechanochemical process in which the speed of chemical processes is governed by the tension of the muscle.

Relaxation is essentially a passive process. Heat production during relaxation is accounted for by the conversion of mechanical energy stored in the muscle during contraction into heat. In smooth muscles with slow relaxation the process is governed by the viscous properties of the contractile elements.

Basic Mechanisms. Contractions generally are preceded by depolarization of the cell surface, which occurs normally during the conduction of an impulse; depolarization by KCl solutions or electric current also causes contraction. Activity is conducted rapidly from the surface to the contractile elements in the interior of fibers. This process is much faster than can be accounted for by the diffusion of any material, as shown by the fact that in skeletal muscles heat production reaches a maximum within 2 milliseconds in a single response. Intracellular conduction probably involves one element of the sarcoplastic reticulum, the transverse tubules, which communicate directly with the outside medium and may serve to activate the inside of the fibers. Perhaps electric current flowing in this structure or other disturbances induce contraction by causing the release of Ca from parts of the reticulum near the myofibrils.

The energy for muscular contraction is furnished by the breakdown of adenosine triphosphate (ATP) into adenosine diphosphate (ADP) and inorganic phosphate. This conclusion is supported by the observation that ATP causes contraction of actomyosin gels and of muscle fibers which have been killed and preserved in aqueous solutions of glycerol. However, during a brief contraction the amount of ATP in a muscle does not diminish measurably while inorganic P increases and phosphocreatin disappears, indicating that ATP is turned over many times during a single contraction due to resynthesis by the Lohmann reaction.

ATP has another important function. Disappearance of most of the ATP produces a state of rigor, characterized by stiffness and opacity. Conversely, muscles which have been brought into a state of rigor by extraction in glycerol solution become soft and translucent again by ATP in the presence of Mg. Also inorganic pyrophosphate has this effect. In the extracted fibers a state resembling normal relaxation is produced by the combined action of ATP, Mg^{++} and submicroscopic granules present in muscle extracts. Contraction can be induced in such preparations by the addition of small amounts of Ca ions, subsequent relaxation by removing the Ca ions again by strong chelating agents.

The granular factor producing relaxation consists of small vesicles derived from the sarcoplastic reticulum. The vesicles can accumulate large amounts of Ca in the presence of ATP. The uptake of Ca is associated with a breakdown of ATP and is considered by some to be an active process. The vesicles can be replaced *in vitro* by strong chelating agents such as EDTA and produce their effect by removing Ca from the myofibrils. Thus the contraction-relaxation cycle is controlled by the release and binding of Ca within the muscle fibers.

At the molecular level, contraction of striated

muscle may proceed as follows. With the participation of Ca, a strong bond is formed between the myosin cross-bridges and the actin filaments. By shortening of the bridges or of a larger part of the myosin molecule, the actin filaments are drawn a short distance toward the middle of the myosin filaments causing an over-all shortening of the muscle. The bridges then detach themselves and become longer again. After forming a bond with the actin filaments at a new position, the cycle is repeated, one ATP molecule being split in each cycle. The force responsible for the shortening and lengthening is unknown; it may be due to a change in entropy, the formation of electrostatic, hydrogen or hydrophobic bonds.

References

1. BOZLER, E., "Smooth Muscle" in "Muscle as a Tissue" (RODAHL, K., AND HORVATH, A. M., EDITORS), New York, McGraw-Hill Book Co., 1962.
2. DAVIES, R. E., "A Molecular Theory of Muscular Contraction," *Nature*, **199**, 1068 (1963).
3. EBASHI, S., AND LIPMANN, F., "Adenosine Triphosphate-linked Concentration of Ca Ions in a Particulate Fraction of Rabbit Muscle," *J. Cell. Biology*, **14**, 389 (1962).
4. GERGELY, J. (EDITOR), "The Relaxing Factor of Muscle," *Federation Proc.*, **23**, 885 (1964).
5. GERGELY, J. (EDITOR), "Biochemistry of Muscular Contraction," Boston, Little, Brown Co., 1964.
6. HOYLE, G., "Comparative Physiology of the Nervous Control of Muscular Activity," Cambridge, Cambridge Monographs on Experimental Biology, No. 8, 1957.
7. HUXLEY, H. E., "Electron Microscope Studies on the Structure of Natural and Synthetic Protein Filaments from Striated Muscle," *J. Mol. Biol.*, **7**, 281 (1963).
8. REICHEL, H., "Muskelphysiologie," Berlin, Springer-Verlag, 1960.
9. HUXLEY, H. E., "The Mechanism of Muscular Contraction," *Sci. Am.*, **213**, 18 (December 1965).

EMIL BOZLER

MUSCULAR DYSTROPHY

Muscular dystrophy is characterized by progressive wasting of the voluntary muscles not secondary to any known lesion of the nervous system; in most instances, it is hereditary. In the late stages, most of the muscles of the body may be largely replaced by fat and connective tissue. Syndromes having many features resembling those of muscular dystrophy in man occur in certain animal species as hereditary disease or can be induced experimentally by dietary deficiency, *e.g.*, hereditary muscular dystrophy in mice and chickens, and myopathy of VITAMIN E deficiency in rabbits and other species. These conditions are not identical with the disease in man, but serve as good models for investigations.

As may be expected, a disease in which most of the muscles are lost can be associated with profound biochemical changes. It is important to determine which abnormalities are primary, not merely secondary to other changes, and sufficiently specific to differentiate the disease from other conditions. However, voluntary muscle reacts to noxious circumstances, inborn or exogenous, in a limited number of ways with the result that many of the biochemical alterations in different kinds of muscle wasting are similar.

Urine. The most prominent known urinary alterations are creatinuria, diminution in creatinine and amino aciduria. CREATINE, unlike creatinine, is not a normal constituent of the urine except in infants and children who may excrete appreciable amounts (so-called physiological creatinura) and in women whose small output may be related to the menstrual cycle. In muscular dystrophy of the Duchenne type, with onset in early life, gross creatinuria is common and in the late stages of the disease may greatly exceed the creatinine output. In other forms of muscular dystrophy, creatinuria usually is less pronounced and may be appreciable only after a large number of muscle groups have wasted. The mechanism of creatinuria is not definitely known. A normal man can ingest large amounts of creatine, or its metabolic precursor glycine, without exhibiting creatinuria, but a patient with muscular dystrophy is likely to eliminate greatly increased amounts of creatine. Investigations with N^{15}-labeled glycine indicate that the newly formed creatine is rejected by the muscle or leaks out before it enters the metabolic pool. The defect in retention is unlikely to be related to insufficiency in adenosine triphosphate (ATP) required for phosphorylation because this compound is better maintained in dystrophic muscle than is creatine. Possible causes are increased permeability of the cell membrane or diminished concentrations of creatine kinase. Although creatinuria is a prominent abnormality in muscular dystrophy, it may occur in other diseases that affect voluntary muscle. Creatinine is a normal constituent of urine and is excreted in amounts that are proportional to the total functional mass of muscle. In muscular dystrophy, the reduction in output may serve as an index of the amount of wasting. Amino acids are commonly excreted in the urine, but no characteristic pattern of amino aciduria has been established. In the Duchenne form of the disease, the urinary output of hydroxyproline is reduced.

Muscle Tissue. Interpretations of analyses on wasting muscle are made difficult by lack of a satisfactory basis for comparison with normal muscle. The loss of proteins, changes in fluid and increase in connective tissue and fat make the "non-collagen nitrogen" content of the tissue the usual basis for reference. Intracellular potassium and water are reduced. The rate of glycogenolysis is diminished, and the enzymes aldolase, phosphorylase a and b, and phosphoglucomutase may be reduced to only a fraction of their normal values (see GLYCOGEN AND ITS METABOLISM). Creatine kinase is reduced in proportion to the amount of muscle wasting as are myokinase and pyruvate kinase. There is no agreement whether lactic dehydrogenase is decreased or unchanged. Cytochrome oxidase, hexokinase, succinate dehydrogenase, fumarase, the transaminases and α-glycerophosphate dehydrogenase appear not to be altered.

Of outstanding interest are the gross increases in the cathepsins. The source of these PROTEOLYTIC ENZYMES is not known but is held by many to be the lysosomes. Whether the increased catheptic activity furnishes a mechanism for muscle degeneration or whether it is secondary to tissue destruction is not known. The muscle of the mouse with hereditary muscular dystrophy incorporates labeled amino acids more rapidly than does normal muscle, but the rate of muscle degeneration is even faster, with the result that loss of protein is pronounced. Besides the well-known reduction in the amount of MYOGLOBIN, a qualitative abnormality has been observed. The disproportionately high percentage of the fetal form as compared with the adult form was considered to be evidence that muscular dystrophy is a molecular disease. However, the hypotheses that the adult form of the protein is lost more readily and the fetal form remains in higher relative concentrations, or that the inhibitor normally limiting the RNA production of fetal myoglobin after the age when the adult form is produced is deficient, are equally plausible. Our knowledge of the significance of the different forms of myoglobin in muscular dystrophy is presently of the same order as that concerning the isozymes of lactic acid dehydrogenase and creatine kinase. The activity of 5′-nucleotidase is accelerated and is related to the increased content of connective tissue. Both creatine phosphate and total creatine are reduced, but the ratio of the two remains normal. On the other hand, the energy-rich nucleotides, adenosine diphosphate and adenosine triphosphate, show little variation from normal except in advanced states of muscle atrophy.

Blood. Among the alterations in blood are increased concentrations of calcium, phosphorus, creatine and protein-bound iodine, and a decrease in creatinine. The hyperglycemic response and the decrease in serum potassium and inorganic phosphorus following administration of epinephrine are less than in normal individuals. The only change in serum proteins is an increase of α_2-globulin related to a high level of sialic acid.

Many enzymes may be increased in the blood of patients: phosphohexoisomerase, transaminase, lactic dehydrogenase, aldolase and creatine kinase. Of these, aldolase and creatine kinase (CPK) have been studied most extensively. Inasmuch as CPK normally is confined largely to the muscles, increases in the blood may be assumed to be derived from that organ. In muscular dystrophy, the blood levels may be enormously increased. Characteristically, in the Duchenne type of the disease, the enzyme level in the blood is greatly accelerated even before clinical symptoms are evident, continues at high levels during the early march of the disease, and then falls toward normal when wasting is advanced and the patient is seriously handicapped. The high blood levels are often assumed to be the result of increased permeability of the muscle cell membrane. The membrane of the dystrophic muscle cell has been shown to be abnormally permeable to a number of compounds. Even when the serum levels of CPK are highly elevated, the activity of this enzyme in the muscle may be very little changed.

The Carrier State. When inheritance is autosomal and dominant, the disease in the carrier parent usually is easily recognized, but when transmission is recessive, clinical evidence of disease in the carrier usually is absent or slight. In the Duchenne type, which is inherited by a sex-linked recessive mechanism, boys almost exclusively are affected. Clinical symptoms in girls are comparatively rare, but biochemical alterations may give evidence of the disorder in mild form. A high percentage of carriers of the Duchenne mutant gene have levels of serum CPK that are above normal. More recently, normal relatives of patients were shown to have diminution in total body potassium similar to the patients themselves.

The serum enzyme changes in carriers is explained best by the hypothesis of Lyon that either of the two sex chromosomes in somatic cells may be inactivated, with the result that the muscles of the carriers resemble a mosaic of cells, normal or dystrophic, depending upon the proportion of inactivated X chromosomes containing normal or mutant genes.

References

1. HEYCK, H., LAUDAHN, G., AND LÜDERS, C. J., "Fermentaktivitätsbestimmungen in der gesunden menschlichen Muskulatur und bei Myopathien. II. Mitteilung. Enzymaktivitätsveränderungen in Muskel bei Dystrophia musculorum progressiva," *Klin. Wochschr.*, **41** (10), 500–509 (May 15, 1963).
2. MILHORAT, A. T., "Creatine and Creatinine Metabolism and Diseases of the Neuro-Muscular System," *Proc. Assoc. Research in Nervous and Mental Dis.*, "Metabolic and Toxic Diseases of the Nervous System," **32**, 400–421 (1953).
3. MILHORAT, A. T., AND GOLDSTONE, L., "The Carrier State in Muscular Dystrophy. Identification by Serum Kinase Level," *J. Am. Med. Assn.*, **194**, 130-134 (1965).
4. PEARCE, J. M. S., PENNINGTON, R. J. T., AND WALTON, J. N., "Serum Enzyme Studies in Muscle Disease. Part III. Serum Creatine Kinase Activity in Relatives of Patients with the Duchenne Type of Muscular Dystrophy," *J. Neurol. Neurosurg. Psychiatr.*, **27**, 181–185 (1964).
5. PEARSON, C. M., "Biochemical and Histological Features of Early Muscular Dystrophy," *Rev. Can. Biol.*, **21**, 533–542 (1962).
6. RONZONI, E., BERG, L., AND LANDAU, W., "Enzyme Studies in Progressive Muscular Dystrophy," *Neuromuscular Disorders*, **38**, Res. Publ. Ass. Nerv. Ment. Dis., 721–729 (1961).

A. T. MILHORAT

MUTAGENIC AGENTS

"Spontaneous" mutations are caused by unknown and omnipresent *mutagenic agents* (mutagens). *Induced* mutations are produced at will by subjecting the genetic material to a variety of known mutagens; the comparatively high mutation rates so obtained are superimposed on the spontaneous mutation rates which, as a rule, are rather low. Study of the action of these known mutagens on DNA furnishes some information as to the chemical nature of GENE action in general, and of mutation in particular.

Some of the typical mutagens and their possible effects on DNA are listed in Table 1; a few of these are discussed below as examples. The list is far from complete; in particular, the agents that act on only a few genes or a few species are not listed here. In general, one must realize that for an agent to be effective as a mutagen *in vivo*, its mutagenic action is not the only property required: the agent must also be harmless enough to permit some survival at mutagenic doses. Another requirement is its ease of penetration into the cell. Still another requirement, especially for chemical mutagens, is that they do not react preferentially with the cytoplasm and do not become exhausted on their way to the genes in the nucleus.

Radiation. Although the mutagenic effects of radiation have been recognized for a (comparatively) long time,[1] the mechanisms of their action still eludes us. Various explanations have been offered, mainly on the basis of *in vitro* effects of radiation on DNA. Thus X rays and other *ionizing radiations* are known to produce in aqueous solutions "free radicals" ($HO\cdot$, $HO_2\cdot$), which are short-lived and reactive; these supposedly could react somehow with DNA, causing mutations and, in stronger doses, death of the cell. Since the sequence of bases is believed to carry the genetic information, it was natural to seek a mechanism involving the bases. Thus, in the case of ULTRA-VIOLET RADIATION (*non-ionizing irradiation*), which is not known to produce free radicals, suggestion has been made that dimerization of thymine is responsible for mutagenic and lethal effects;[2] dimerization could cause disturbances when the strands of DNA are ready to replicate.

Heat. While the mechanism of action of the strong mutagens already cited is still far from clear, there is some information as to the manner in which several other mutagens act. One of these is *heat* (thermal oscillations), which is suspected to be a principal mutagen involved in "spontaneous" mutability. Most cells cannot tolerate high temperatures: in the case of dry bacterial cells, spores[3] and dry plant seeds, temperatures up to 130–150°C can be applied for a short time without killing all the cells; the survivors include a large proportion of mutants that are induced (not merely selected out) by heat. If DNA alone is heated under these conditions, an interesting injury can be detected in it: some of the purines become detached (depurination). Among cells surviving such treatment, an injury of this kind

TABLE 1. TYPICAL MUTAGENS AND THEIR POSSIBLE MUTAGENIC EFFECTS ON DNA

Mutagen	Possible Mutagenic Effect on DNA
Radiation:	
X rays	Largely unknown; probably caused by free radicals produced in aqueous solution
Ultraviolet	Largely unknown; dimerization of thymine has been suggested
Auto-oxidizing agents:	
Fe^{++}	Largely unknown; probably caused by free radicals produced in aqueous solution
Heat	Depurination or ionization of guanine at N-1, followed by a change in base pairing
Alkylating and esterifying agents:	Alkylation on N-7 of guanine, possibly followed by depurination or ionization of guanine, and a change in base pairing
Mustards	
Dimethyl- and diethyl sulfate	
β-Propiolactone	
Ethylmethane sulfonate	
Nitrous acid	Deamination of adenine, guanine and cytosine, followed by a change in base pairing
Hydroxylamine	Attachment to the C—C double bond of cytosine, followed by a change in base pairing
Base analogues:	Incorporation into DNA in place of thymine, followed by a change in base pairing
5-Bromouracil	
5-Iodouracil	
5-Chlorouracil	
Deuterium	Replacement of hydrogen, causing an unknown disturbance

would be sufficient to produce loss of a proper base sequence in one or more sites of a DNA molecule, and logically may be offered as an explanation of the observed loss of genetic information (mutation). Thus the first step of the mutational process is an injury which can occur without any replication of DNA (*e.g.*, in dry spores). But this injury cannot be propagated as there is no known biochemical mechanism to utilize a sugar (or a sugar triphosphate) without its base as a building block for DNA synthesis. Thus during the replication of DNA a *second step* must take place: Translation of the primary injury (loss of purines), nonacceptable in the mechanism of DNA replication, into a final acceptable one: replacement of the missing purine by some normal base. Since the replicating strand of DNA has no information as to the missing base, perhaps the incorrect one (the wrong purine or even a pyrimidine) can sometimes be used for the synthesis of the new strand (replication); this represents an error in base sequence that is, by definition, a mutation [see DEOXYRIBONUCLEIC ACIDS (REPLICATION)].

It has recently been suggested that one of the effects of breaking hydrogen bonds by heat or other denaturing agents may be ionization of guanine in position N-1, and that such an ionization may cause guanine to pair with thymine instead of with cytosine, thus contributing to the mutagenic effect of heat.

Alkylating Agents. Passing now to the so-called *chemical mutagens*, one must discuss first the class known as "alkylating agents." To this class belong such substances as mustards, dimethyl sulfate, diethyl sulfate, β-propiolactone, ethylmethane sulfonate, etc. Of these, mustards were the first chemical mutagens discovered and are still used as potent CARCINOGENIC and carcinostatic agents.

The mode of action of all these substances appears to be similar. Under physiologic conditions of temperature and pH they can react with nitrogen atoms of the bases, causing their *alkylation* (methylation, ethylation).

A study[4] has revealed that under physiologic conditions, essentially the only base affected is guanine and the alkyl group becomes attached only to N-7 of this base. Such a reaction produces a positive charge on this nitrogen atom (quaternary ammonium ion) and attaches a bulky group to the base; both changes create serious disturbances and cannot be accepted in the mechanism of replication of DNA. These injuries, then, are the *first step* of the mutational processes (see discussion of the action of heat).

Recently, some evidence has been obtained concerning the second (final) step in which the original injuries are translated into changes acceptable in the mechanism of DNA replication. It appears that guanine with the attached bulky alkyl group simply becomes detached from the sugar, thus causing *depurination*; the subsequent phenomena, leading to the insertion of an incorrect (but natural) base, *i.e.*, mutation, may then be similar to those described for the action of heat.

Another explanation that has been offered recently for the mutagenic effect of alkylating agents is that the positive charge in position N-7 of guanine, produced by alkylation, facilitates the appearance of a negative charge in position N-1 of guanine; the latter ionization would then facilitate pairing of guanine with thymine instead of cytosine, thus leading to an error and a mutational event.

Nitrous Acid. Nitrous acid is known to deaminate adenine, guanine and cytosine slowly. At the low pH used in such experiments, the nitrous acid is so strong that DNA or a living cell, if exposed to it, would be rapidly destroyed. However, conditions have been devised[5] to carry out deamination at a pH not far from neutral, and under these conditions, sufficient survival can be obtained to test the effects of deamination which, although proceeding much more slowly at these pH values, is still easily demonstrable. It has been found that if DNA possessing transforming activity is exposed to such mild deamination conditions, a slow inactivation occurs concomitantly with deamination of only one or a few molecules of adenine and guanine.[5] This observation has been utilized for the demonstration that such sublethal deamination of infective RNA of TOBACCO MOSAIC VIRUS produces viral mutants. Nitrous acid was also found to produce mutants in whole cells, whole bacteriophage, and other viruses, and in DNA having transforming activity.

The production of mutations by treatment of "naked" RNA and DNA *in vitro* with nitrous acid (and, as recently reported, also with other agents) is of particular interest; it clearly indicates that, at least in these cases, no other part of the cell need undergo a primary change in order for the mutational injury to be "registered." The replication of nucleic acids is also not necessary for such "registration" (first step), although it may be necessary for the final step. Moreover, the possibility of subjecting only the nucleic acids and not the whole cell to the mutagenic agents may considerably broaden our experimental possibilities.

According to recent reports, the mutagenic action of nitrous acid on DNA *in vitro* is possible only after the two strands of DNA have been separated (*e.g.*, by heating). One change in nucleic acids produced by nitrous acid and suspected to be mutagenic involves deamination of adenine to hypoxanthine. Ordinarily, on replication of a DNA strand, adenine chooses thymine as a partner in the opposite strand that is being synthesized. However, hypoxanthine may be sufficiently different from adenine to introduce an error in base pairing (cytosine instead of thymine) and such an error may be a mutation. Another possible change is the deamination of guanine to xanthine; however, recent analysis indicates that such a change may be lethal rather than mutagenic. The third possible change is deamination of cytosine to uracil, which can also produce errors in base pairing. Such a change is of particular interest in the case of RNA since the resulting uracil is a normal component of RNA.

References

1. MULLER, H. J., "Artificial Transmutation of the Gene," *Science*, **66**, 84 (1927).
2. BEUKERS, R., AND BERRENDS, W., "The Effect of U.V. Irradiation on Nucleic Acids and Their Components," *Biochim. Biophys. Acta*, **49**, 181 (1961).
3. ZAMENHOF, S., "Effect of Heating Dry Bacteria and Spores on Their Phenotype and Genotype," *Proc. Natl. Acad. Sci. U.S.*, **46**, 101 (1960).
4. REINER, B., AND ZAMENHOF, S., "Studies on the Chemically Reactive Groups of Deoxyribonucleic Acids," *J. Biol. Chem.*, **228**, 475 (1957).
5. ZAMENHOF, S., ALEXANDER, H. E., AND LEIDY, G., "Studies on the Chemistry of the Transforming Activity," *J. Exptl. Med.*, **98**, 373 (1953).
6. ZAMENHOF, S., "Mutations," *Am. J. Med.*, **34**, 609–626 (1963). (A detailed review article with references to original papers.)

STEPHEN ZAMENHOF

MUTATIONS. See MUTAGENIC AGENTS; also BACTERIAL GENETICS; BIOCHEMICAL GENETICS; DEOXYRIBONUCLEIC ACIDS (REPLICATION); GENES; NEUROSPORA (IN BIOCHEMICAL STUDIES); TOBACCO MOSAIC VIRUS.

MYELIN

The Anatomy of Myelin. Myelin is a sheathlike structure which invests the nerve axon circumferentially, somewhat like insulation around a wire. It is formed by a wrapping of the external membrane of the Schwann cell or the oligodendrocyte around the nerve axon. The result is a structure comprised of a series of tightly packed membranes of uniform thickness layered in concentric fashion. A single membrane constituent of this structure, when examined under high-resolution electron microscopy, is seen to be bounded by two dense lines: (1) a thick line, the major dense line, and (2) a thin line, the intraperiod line; there is a clear zone between them (see Fig. 1).

Myelin contains LIPIDS, PROTEINS, polysaccharides, salts and water. The spatial arrangement of these molecules in the myelin membrane has been determined by different procedures. X-ray diffraction analysis of myelin reveals a characteristic pattern for each individual membrane unit, in which two high-density regions (peaks) are separated by a low-density region (trough). The most reasonable interpretation of these curves is that the peripheral regions are occupied by electron-dense polar groups of lipid molecules and adjacent proteins (or polysaccharides) while the central region is occupied by the less dense hydrocarbon tails of the lipid molecules. Hydration studies of myelin in combination with X-ray diffraction analysis and ELECTRON MICROSCOPY substantiate this interpretation by demonstrating that the peripheral regions of the membrane unit are hydrophilic and associated with the presence of polar groups while the central region is hydrophobic and associated with the presence of nonpolar hydrocarbon chains. In addition, the center of the myelin membrane is lightly stained with osmium, indicating that hydrocarbon chains are present here, since lipid hydrocarbon chains do not bind osmium as avidly as their polar groups do. The central region in each membrane unit is approximately 51 Å wide. There is room for two lipid molecules packed tail-to-tail in this region, since the average length of lipid molecules, from phosphate group to hydrocarbon tail, is approximately 26–28 Å. Since this central region fails to expand or shrink when myelin is hydrated or dehydrated, this region must be water free and must be occupied entirely by hydrophobic groups, *i.e.*, the hydrocarbon tails of lipid molecules. These studies indicate that the molecular arrangement in myelin is that of a bimolecular lipid layer bounded by two protein monolayers, with the polar groups of the lipid molecules adjacent to the protein monolayers and the hydrocarbon tails of the lipid molecules extending into the center of the membrane.

The Composition of Myelin. Myelin has a unique chemical composition. It contains more lipid than any other membrane structure analyzed: 78–80% of the dry weight. Myelin also contains high proportions of sphingolipids and PLASMALOGENS which are present in very small amounts elsewhere in the body. When the lipid composition of central myelin was compared to that of gray matter, it was found that myelin contained threefold higher molar proportions of CEREBROSIDE and cerebroside sulfate and lower proportions of ethanolamine glycerophosphatides and choline glycerophosphatides than gray matter (Table 1). A further difference between these topographically related tissues is the fatty acid compositions of the individual lipids. Ethanolamine glycerophosphatides, serine glycerophosphatides, and choline glycerophosphatides in gray matter contain large proportions of polyunsaturated fatty acids (20 and 22 carbon atoms long, containing 4, 5 and 6 double bonds). Myelin glycerophosphatides, however, are much more saturated, they contain only one-sixth as much polyunsaturated fatty acids as the glycerophosphatides in gray matter (Table 2). Another difference is present in the long-chain fatty acids (19–26 carbon atoms) of the sphingolipids from myelin. Myelin sphingolipids, including cerebroside, cerebroside sulfate, sphingomyelin, and ceramide, contain five- to ninefold higher proportions of long-chain fatty acids than the same sphingolipids from gray matter (Table 2). See also PHOSPHOLIPIDS; PHOSPHATIDES; UNSATURATED FATTY ACIDS.

The over-all effect of these differences in lipid composition and structure is as follows. Myelin has one-fifth the molar proportion of lipids containing polyunsaturated fatty acids, compared to gray matter. On the other hand, the molar proportions in myelin of lipids containing very long-chain fatty acids (19–26 carbon atoms) are 10 times

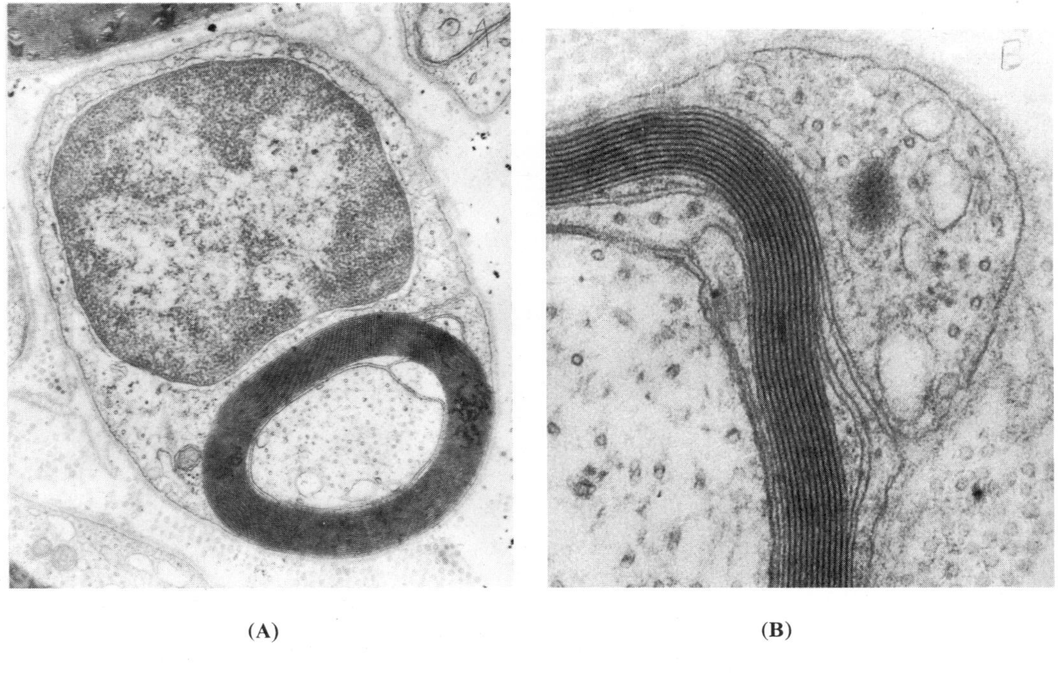

(A) (B)

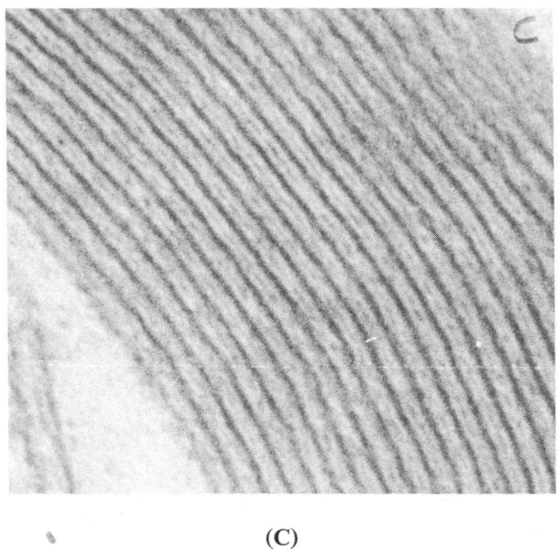

(C)

FIG. 1. Morphology of peripheral nerve myelin. **A.** A single Schwann cell containing a single axon is seen, around which myelin is wrapped. The connection between myelin and the plasma membrane of the Schwann cell (the outer mesaxon) is visible to the right. (22,000X). **B.** Higher magnification (70,000X) of a portion of the myelin sheath demonstrates the layered membranes and the internal and external mesaxons. Axoplasm is shown at the lower left. **C.** Very high (300,000X) magnification of the myelin sheath showing the major dense lines alternating with the intraperiod lines (lighter lines) and the clear zones between them. (*Electron micrographs courtesy of Dr. Humberto Cravioto*).

TABLE 1. LIPID CONTENT OF NORMAL FRONTAL LOBES[a]

Tissue	Total Lipid (% of dry wt)	Individual Lipids (molar percentage of total lipid)								
		Choles-terol	Ethanolamine Glycero-phospha-tides[b]	Serine Glycero-phospha-tides[b]	Choline Glycero-phospha-tides	Sphingo-myelin	Cere-broside	Cerebroside Sulfate	Ceramide	Uncharac-terized[c]
Subject, 10 months										
Gray matter	36.4	35.7	14.9	5.7	24.7	4.3	4.1	1.5	2.5	6.6
White matter	49.0	37.7	16.2	3.6	14.4	3.6	13.6	3.7	2.5	4.7
Myelin	78.0	38.8	15.3	5.1	12.7	4.9	13.8	4.7	1.7	3.0
Subject, 6 years										
Gray matter	35.8	31.1	25.1	7.9	19.6	3.1	2.4	1.4	2.7	6.7
White matter	58.4	37.7	12.3	5.0	11.8	3.8	17.2	3.3	1.6	7.3
Myelin	80.9	42.4	11.4	4.1	9.1	4.3	18.1	3.3	1.3	5.9
Subject, 9 years										
Gray matter	37.6	31.6	22.1	5.8	20.1	6.4	4.2	0.8	1.5	7.5
White matter	66.3	33.5	15.9	6.2	11.2	6.3	12.8	4.3	0.8	9.0
Myelin	78.0	38.7	15.3	5.5	12.8	4.8	13.8	4.6	1.8	2.7
Subject, 55 years										
Gray matter	39.6	31.3	19.6	5.7	19.4	4.2	4.7	1.5	1.6	12.0
White matter	64.6	38.5	11.9	5.2	10.3	6.3	15.3	3.3	1.3	7.9
Myelin	78.0	40.4	11.8	5.3	8.4	4.4	15.7	3.5	1.5	9.0

[a] Reprinted with permission from J. S. O'Brien, *Science*, **147**, 1100 (March 5, 1965).
[b] Includes both diester and plasmalogen forms of these lipids.
[c] Includes inositol glycerophosphatides as principal components, and free fatty acids, gangliosides, phosphatidic acid, and polyglycerophosphatides as minor components. An average molecular weight of 800 was assumed for this fraction.

TABLE 2. NATURE OF LIPID FATTY ACIDS IN BRAIN[a]

Tissue	Percentage of Polyunsaturated Fatty Acids[b] in			Percentage of Fatty Acids with Chains Longer than 18 Carbon Atoms in		
	Ethanolamine Glycerophosphatides	Serine Glycerophosphatides	Choline Glycerophosphatides	Sphingomyelin	Cerebroside	Cerebroside Sulfate
		Subject, 10 months				
Gray matter	46.9	53.0	9.1	2.7	9.4	16.1
White matter	13.9	23.5	3.5	20.3	70.1	71.2
Myelin	12.9	11.5	1.0	23.6	80.1	
		Subject, 6 years				
Gray matter	41.2	36.0	9.5	5.8	15.0	34.0
White matter	24.9	9.0	2.6	50.4	65.7	88.3
Myelin	5.7	2.4	Trace	54.2	86.5	90.0
		Subject, 9 years				
Gray matter	28.4	25.1	7.9	7.9	34.7	32.5
White matter	23.7	12.5	4.3	42.8	85.1	85.6
Myelin	17.4	9.0	3.7	46.0	87.1	89.4
		Subject, 55 years				
Gray matter	41.0	48.5	7.6	25.5	90.2	90.0
White matter	15.5	14.9	2.1	63.2	86.4	82.1
Myelin	4.1	4.4	0.4	59.8	87.5	86.0

[a] Reprinted with permission from J. S. O'Brien, *Science*, **147**, 1102 (March 5, 1965).
[b] Fatty acids containing from 18–22 carbon atoms and from 2–6 double bonds.

the molar proportions in gray matter. In the myelin group of lipids, 1 in 17 fatty acids is polyunsaturated; in the gray matter group, the value is 1 in 5. In the myelin group of lipids, 1 in 5 fatty acids has a chain longer than 18 carbon atoms; for gray matter the value is 1 in 100.

The unique lipid composition and structure of myelin lipids have been proposed as a partial explanation for the unique stability of the myelin membrane.[2,3] Polyunsaturated lipids, which decrease membrane stability, are present in low concentrations in myelin. Long-chain sphingolipids, which enhance membrane stability, are present in high concentrations in myelin. Studies are now in progress on the alterations of myelin lipid composition and structure in diseases in which myelin is unstable to see whether such molecular changes are important in their pathogenesis.

The other quantitatively important components of myelin are proteins. As yet many proteins have been found in central myelin, but their structures, molecular configurations and sizes are not known. An estimate of the average molecular weight of rat central myelin protein has been given as 28,000.[4] The surface charge of a solubilized rat myelin lipoprotein fraction has been found to be electronegative at pH 8.6, with a mobility in an electrical field like that of ALBUMIN.[4] It has also been found that a myelin protein fraction is capable of producing encephalomyelitis when injected into experimental animals.

Two additional characteristics of myelin should be mentioned. One is the remarkable similarity in lipid content and composition of myelin isolated from different species, including rat, ox, guinea pig and man.[5] The second is the similarity in chemical composition between myelin isolated from a baby and that isolated from an adult.[3] It appears that myelin has a rigidly defined composition which varies little between species (at least in the higher mammals studied) and with age. These findings prompt the suggestion that such a composition is essential for myelin formation and that variations in composition outside a limited range may result in faulty myelination.[3]

In spite of the great strides which have been made in understanding the chemistry of myelin, much more information is needed before our knowledge of it is complete. The areas which need exploration include the chemical composition of peripheral nerve myelin, the structure of myelin proteins, the concentration and location of the minor constituents of myelin, including salts, water and polysaccharides, and the nature and significance of the aberrations of myelin composition which occur in human diseases in which myelin is unstable.

References

1. FINEAN, J. B., *Circulation*, **26**, 1151 (1962).
2. VANDENHEUVEL, F. A., *Ann. N.Y. Acad. Sci.*, **122**, 57 (1965).
3. O'BRIEN, J. S., *Science*, **147**, 1099 (1965).
4. GENT, W. L. G., GREGSON, N. A., GAMMACK, D. B., AND RAPER, J. H., *Nature*, **204**, 553 (1964).
5. NORTON, W. T., AND AUTILIO, L. A., *Ann. N.Y. Acad. Sci.*, **122**, 77 (1965).
6. ROBERTSON, J. D., *Progr. Biophys.*, **10**, 356 (1960).
7. O'BRIEN, J. S., AND SAMPSON, E. L., *J. Lipid Res.*, (October 1965).

J. S. O'BRIEN

MYOGLOBINS

Source, Distribution and Isolation. Myoglobins, sometimes called "muscle hemoglobins," are found in muscles of mammals and other orders. They are composed of a PROTEIN moiety, which alone is called apomyglobin, and a prosthetic group, HEME. The heme, iron protoporphyrin IX, is identical to the prosthetic group in HEMOGLOBIN. As in hemoglobin, it is the site of reversible combination with molecular oxygen (see also PORPHYRINS).

Myoglobins are distributed in the cardiac, skeletal and smooth muscles. Much of the red color of fresh muscle is owed to the myoglobin. The protein is particularly abundant in muscles of diving mammals such as whales, dolphins and seals. It is abundant in the flight muscles of birds. In man, it is in highest concentration in cardiac muscle. In the skeletal muscle of the sperm whale, *Physeter catodon*, the myoglobin amounts to about 5% of the wet weight of the tissue.

The myoglobin in a living muscle is retained in the muscle cells, near or combined with the mitochondria. In some rare disease states, some myoglobin may be lost from muscles and appear in the urine. Violent exercise or trauma may have the same effect.

Myoglobin is normally isolated by soaking the minced and ground muscle in distilled water. The myoglobin is isolated from the extract by precipitation with concentrated ammonium sulfate and phosphate buffer or with zinc acetate and alcohol at −8°C. The purity is more than 95–98%. It is readily crystallized from ammonium sulfate or phosphate solutions near pH 6.5–7.0. Last traces of impurities are best removed by ion exchange chromatography.

The isolated myoglobin is very soluble in pure water. In the absence of dissolved salts, it can be dried from the frozen state (lyophilized), stored as a dry powder for long periods in a deep freeze, and reconstituted by adding water or buffer solution.

During isolation the myoglobin is apt to undergo a change of oxidation state of the iron in the heme from +2 (ferro-) to +3 (ferri-). It is the ferromyoglobin that combines reversibly with molecular oxygen. The ferromyoglobin is best isolated by chromatography, a step that may be preceded by the zinc-alcohol purification. The ferrimyoglobin is often called metmyoglobin. Unless oxygen is excluded, the ferromyoglobin will be present in combination with molecular oxygen, a form that is called oxymyoglobin.

The Oxygen-storage Role of Myoglobin. The physiological role of myoglobin is that of the storage of molecular oxygen in muscle. This accords with its high concentration in diving mammals and the report that among men it is elevated in dwellers in the Andes. Oxygen diffusing into the muscle from the blood capillaries combines with deoxygenated ferromyoglobin to form oxymyoglobin. As the consumption of oxygen overtakes the rate of supply from the blood, the oxymyoglobin store is drawn upon. As the oxygen concentration (often expressed as equilibrium pressure) falls, the oxymyoglobin dissociates in a mobile equilibrium. When the supply of oxygen from the blood overtakes the rate of consumption, the reverse process again predominates and the deoxygenated myoglobin is oxygenated once more. The cycle is repeated indefinitely.

The affinity of myoglobin for oxygen is greater than that of hemoglobin. This difference means that the myoglobin reservoir of oxygen can be built up at the expense of the hemoglobin reservoir circulating in the blood. The latter is replenished by exposure to air in the lungs. The oxygen-consuming apparatus (*e.g.*, cytochrome oxidase) in the muscle, on the other hand, is able to reduce the oxygen concentration to such very low levels that the oxymyoglobin storage form can be drawn on efficiently.

Myoglobin contains only one heme group per molecule and thus does not share with hemoglobin the peculiar oxygen dissociation curve that arises from interaction between the four heme groups in the latter. Nor does myoglobin display the Bohr effect [see HEMOGLOBINS (COMPARATIVE BIOCHEMISTRY)].

Methods Used for Structural Studies. Much of the current interest in myoglobins depends on the fact that a myoglobin is the first protein to have its three-dimensional structure nearly completely elucidated. This has been achieved by Kendrew and co-workers by X-ray crystallographic analysis of crystals of sperm whale myoglobin. The detailed interpretation of the results was guided by sequence analysis by Edmundson and Hirs.

The X-ray diffraction analysis was based on the technique of isomorphous replacement. For this technique to be applicable, it is necessary that a number of different substitutions can be made in the structure without disturbing the architecture. Various metal ions and metal complexes were used in the preparation of the substituted molecules in crystalline form. The substituted atoms are of higher atomic number than the other atoms of which the protein is composed, and act somewhat as reference points. It was possible to determine almost the entire structure in this way. An indication of the procedure is given by J. C. Kendrew in the December, 1961 issue of *Scientific American*.

Sperm whale myoglobin contains at least three fractions by ELECTROPHORESIS. The over-all amino acid content is the same in each, the differences are attributed to amide content. In the crystal, the structures of these forms are essentially identical.

Myoglobin from the seal, *Phoca vitulina*, is being studied by the X-ray crystallographic technique. It appears to have a very similar structure to that from the sperm whale.

Structure of Sperm Whale Myoglobin. Figure 1 shows the amino acid sequence of sperm whale myoglobin. The three peptides represent the products of cyanogen bromide cleavage at the two methionine residues. The sequence is shown in sets of three rows. First is the numbering of the residues according to their arrangement in helical

Peptide 3

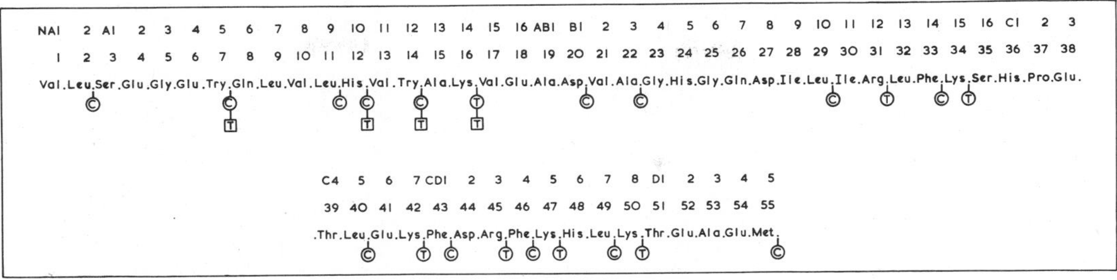

Peptide 2

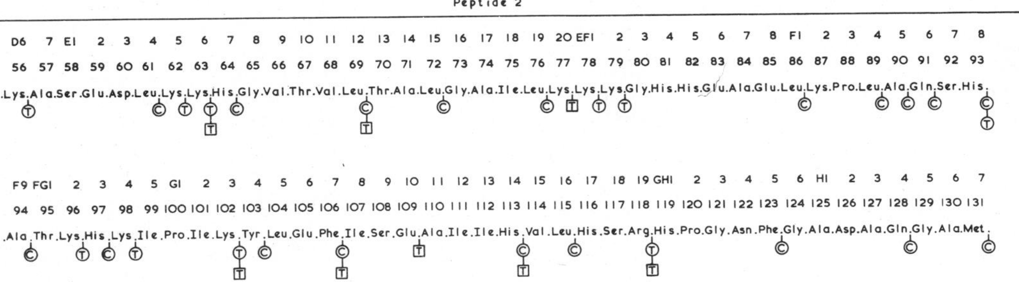

Peptide 1

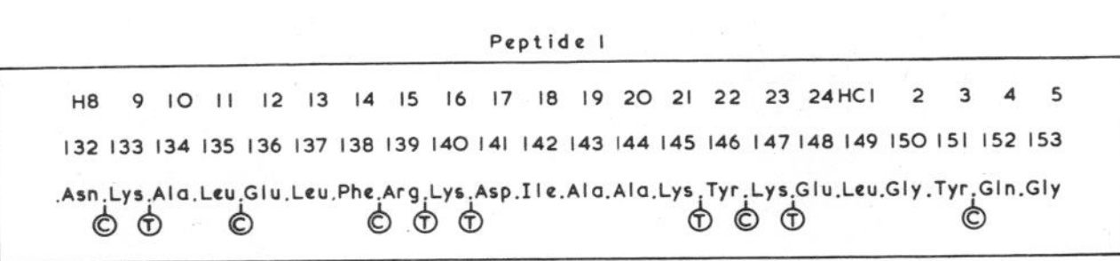

FIG. 1. Amino acid sequence of myoglobin.

and non-helical segments. NA1 and NA2 are the first two residues at the amino terminus and are not in the helical arrangement. The sixteen residues marked A form the A-helix. There are seven other helical sections distinguished by a single letter prefix. Non-helical segments of the chain are named with the letters of both preceding and following helices, *e.g.*, EF. The second row in the set shows the conventional sequential numbering 1 to 153. The third row shows the sequence proper in conventional shorthand. Attention should be called only to the abbreviations Asn and Gln for asparagine and glutamine. The symbols C and T directed between certain residues indicate that the bonds between those residues are sensitive to chymotrypsin and trypsin, respectively. The sequence was constructed by observing overlaps in these cleavages and by stepwise degradation of isolated short peptides by the Edman procedure. The amide distribution corresponds to that of the main component which represents about half of the total myoglobin. [See also PROTEINS (END GROUP AND SEQUENCE ANALYSIS).]

Models of the crystalline sperm whale myoglobin molecule have been constructed, showing that a way is open for the oxygen molecule to reach the iron atom at the center of the nearly planar heme. The protein is folded back on itself to form a compact structure with very few empty spaces inside it. Some side chains protrude noticeably, *e.g.*, the imidazole group of a histidine residue (G17).

A number of observations can be made about the structure of sperm whale myoglobin that may have wide application to other proteins. These are as follows:

(1) As would be expected from electrostatic principles, the charge-bearing side chains protrude into the aqueous solvent and are not buried within the molecule.

(2) The nonpolar side chains, such as those of alanine, valine, leucine, isoleucine, tryptophan, phenylalanine and methionine, are usually buried within the molecule out of touch with the aqueous solvent. Unlike the charged groups, the nonpolar groups assume their preferred arrangement preponderantly but not exclusively.

(3) Polar but uncharged groups such as the side chains of serine, threonine, asparagine and glutamine generally make contact with the solvent.

(4) Summarizing the first three points, one can say that the central region of the molecule is practically exclusively composed of nonpolar side chains in contact with each other. The region on the outside in contact with the solvent is composed of the polypeptide backbone, polar and charged side chains with a few nonpolar side chains as well.

(5) The heme group is arranged so that charged carboxyl groups of the propionic acid side chains are in contact with the solvent and close to charged or polar side chains of the protein moiety. One face of the heme is exposed to solvent at least down to the level of the iron atom. The inner border of the heme, including the methyl and vinyl substituents, is buried in a nonpolar region of the protein.

Questions Opened by the Structure Determination. The following are some examples of questions opened by the structure determination of sperm whale myoglobin:

(1) A major question that hitherto could scarcely be faced is the relation between the crystalline structure of a protein and its structure in solution. This can now be studied. Two promising approaches are low-angle X-ray scattering in solution to measure the radius of gyration, and determination of the relative reactivities of individual side chains to chemical reagents. Preliminary evidence of both kinds points to similarity between crystal and solution structures.

(2) If the preceding question can be settled or the answer closely defined, the protein can be exploited as a model substance in many ways. The reactivity studies could be used to map the effects of steric arrangements and vicinal groups on a scale of subtlety rarely accessible to a chemist. The behavior of individual groups or clusters of groups could be compared with that of smaller model compounds. Optical rotatory dispersion measurements would be interpretable in detail, as would other properties of the solutions. Various tests of stability could be interpreted in terms of binding forces in the secondary and tertiary structure.

(3) The reciprocal role of heme and apoprotein can be studied in detail. The oxygen-binding and other properties of the heme can be studied in terms of the protein environment. Conversely, the influence of the heme on the structure and reactivity of the apoprotein can be studied. The one approach represents an example of a structure-function study of great promise; the other, a step toward understanding allosteric effects.

For references see X-RAY CRYSTALLOGRAPHY.

FRANK R. N. GURD

MYOSINS

General Properties and Isolation. Upon extraction of MUSCLE by means of neutral salt solutions of high ionic strengths at 0°C, a rather viscous solution is obtained which contains the three major proteins of the myofibril—myosin, actin and tropomyosin. By restricting the extraction periods to approximately 10 minutes, a myosin preparation relatively free of the other two proteins can be obtained.

Myosin constitutes a major fraction of the myofibrillar protein material, amounting to approximately 60% in rabbit skeletal muscle, and it has been found in varying amounts in all types of muscle studied thus far. It is readily soluble at ionic strengths above 0.3 and precipitates completely upon dilution at neutral pH to ionic strengths of the order 0.025. This property forms the basis for the preparation of myosin by relatively simple means free of other contaminating proteins. By repeated dissolution (three times) and reprecipitation, preparations which are 95–98% monodisperse may be obtained.

Size and Shape of Molecule. There are a number of difficulties involved in the determination of the size and shape of myosin. It is quite asymmetric as indicated by its intrinsic viscosity of 2.0. In addition, it is heat sensitive and aggregates readily at room temperature. In recent years, however, on the basis of hydrodynamic measurements and from the examination of the molecule in the ELECTRON MICROSCOPE, it has been generally agreed that the myosin molecule is a long rod whose average length is about 1700 Å with a globular portion on one end as seen in Fig. 1. The rod portion is about 20 Å in diameter. The length of the globular region appears to be 100–200 Å in length and about 40 Å in diameter.

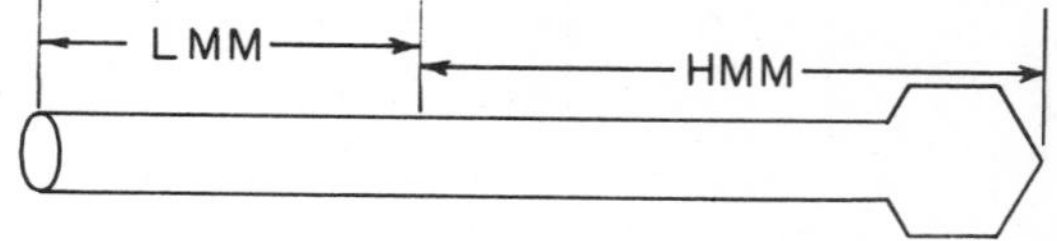

FIG. 1.

The molecular weights reported for myosin have, since 1950, undergone considerable fluctuation. Earlier values were reported as high as 840,000 and have since that time been reported as low as 400,000. In recent years, the values have ranged between 500,000–619,000. On the basis of the latter figure and a weight of 206,000 in 5*M* guanidine, it has been proposed that myosin consists of three polypeptide chains of identical chemical structure wound in the form of a three-stranded rope. However, other experiments have suggested from X-ray data that myosin may have a two-chain structure.

Attempts to delineate the subunit structure of the molecule by the use of enzyme digestion, such as with trypsin, have been fraught with similar difficulties. The finding that short tryptic digestion of myosin produces two large fragments retaining the physical properties of the native molecule was a significant contribution to the understanding of the structure of myosin. These fragments have been called light meromyosin (LMM) and heavy meromyosin (HMM). LMM of molecular weight 140,000 retains the solubility properties of myosin and will depolymerize in urea to smaller molecular weight water-soluble fragments called protomyosins. HMM, of molecular weight 350,000, is soluble in water and retains the ability of myosin to hydrolyze ATP and combine with F-actin. HMM appears to be the globular portion of the molecule shown in Fig. 1.

The two most sensitive biological properties of the myosin molecule are its ability to hydrolyze ATP and to combine with F-actin to form actomyosin. Both of these processes are thought to be of fundamental significance in the mechanism of muscle contraction. The hydrolysis of ATP by myosin according to the following scheme:

$$ATP \rightarrow ADP + P_i$$

is of great interest since myosin is the only known fibrous protein having enzymatic properties. In addition, it appears to provide the chemical energy necessary for the mechanical work of muscle. A voluminous literature exists on the effects of various ions and inhibitors on the above reaction, and these data will be discussed only briefly here.

Studies on the effect of pH on the ATPase activity of myosin have shown that two pH optima exist, a small maximum at pH 6.3 and a much greater activity at pH 9.0. At values above 10, the activity is lost irreversibly so that the activity never reaches a true optimum in the alkaline pH region. These studies have been carried out in the presence of Ca^{++} as activator. An interesting and perhaps significant feature here is that Mg^{++} in low concentrations will inhibit the Ca^{++} activation. The physiologic implication of this effect is not well understood.

Myosin is known to be a sulfhydryl-dependent enzyme, *i.e.*, it appears that 2 sulfhydryl residues/200,000 molecular weight out of 15 must remain intact for ATPase action to occur. By blocking one of these groups with a sulfhydryl inhibitor such as *p*-chloromercuribenzoate or iodoacetamide, a three- to fivefold activation in ATPase takes place. Reaction of the second sulfhydryl residue results in a complete inhibition of enzyme activity.

In addition to its ATPase activity, myosin will interact with F-actin to form a rather viscous complex called actomyosin. This protein is soluble at ionic strengths of around 0.5 and exists in a gel state at ionic strengths of around 0.15. When ATP is added to actomyosin in the sol state there is an immediate fall in the viscosity, and it has been shown by ultracentrifugal studies that a dissociation occurs according to the following scheme:

$$\text{Actomyosin} \underset{}{\overset{\text{ATP}}{\rightleftharpoons}} \text{Myosin} + \text{Actin}$$

When ATP is added to the gel form of actomyosin an immediate flocculation occurs, a process called "superprecipitation" and one believed to mimic an *in vivo* contraction. (See also CONTRACTILE PROTEINS.)

Actomyosin also has the ability to hydrolyze ATP; however, this reaction is activated by Mg^{++} as well as Ca^{++}. Mg^{++}, it will be recalled, inhibited myosin ATPase activity. The mechanism whereby this occurs is not well understood at the present time.

There appear to be two sulfhydryl residues in myosin necessary for the interaction with F-actin. However, these do not seem to be the same groups as those involved in the hydrolysis of ATP.

References

1. Kielley, W. W., "The Biochemistry of Muscle," *Ann. Rev. Biochem.*, **33**, 404 (1964).
2. Gergely, J. (Editor), "Biochemistry of Muscular Contraction," Boston, Little, Brown & Co., 1964.
3. Perry, S. V., "Muscle Proteins in Contraction," Muscle Symposium at the University of Alberta (Paul, W. M., Daniel, E. E., Kay, C. M., and Monckton, G., Editors), New York, Pergamon Press, 1965.
4. Stracher, A., "Characterization of the Sulfhydryl Residues Involved in the Binding of ATP and Actin to Myosin," *ibid.*, 1965.

A. Stracher

MYXOMYCETES. See SLIME MOLDS.

N

NARCOTIC DRUGS

Narcotic drugs are agents which produce sedation or sleep (hypnosis) and relief of pain (analgesia). Although many classes of drugs can cause narcosis, only the addictive analgesics are commonly called narcotics since their use is regulated by the Harrison Narcotic Act of 1914 and its amendments.

Narcotic agents may be classified as natural, semisynthetic and synthetic. Natural narcotics are ingredients of opium, which is the dried exudate of seed capsules of the oriental poppy or *Papaver somniferum*. There are numerous ALKALOIDS in opium, but only two phenanthrene derivatives, *morphine* and *codeine*, are narcotics. Morphine comprises about 10% by weight of opium and was the first of the vegetable alkaloids to be isolated in 1805 by Sertürner. Since the source of the natural alkaloids is opium, all narcotics whose actions resemble those of morphine are sometimes referred to as opiates. Semisynthetic agents are usually made by altering the morphine molecule and include such agents as diacetylmorphine (heroin), ethylmorphine ("Dionin"), dihydromorphinone ("Dilaudid") and methyldihydromorphinone (metopon). Synthetic narcotics include agents with a wide variety of chemical structures. Some of the important synthetic agents are meperidine (piperidine type), levorphanol (morphinan type), methadone (aliphatic type), phenazocine (benzmorphan type) and their derivatives. The structures of the various narcotics are illustrated in Fig. 1.

Since morphine is responsible for the major actions of opium and the actions of all narcotics are qualitatively similar, morphine will be used as a model for discussing narcotic actions. The most prominent effects of morphine in man are on the central nervous system and the gastroenteric tract. The principal central action of morphine is the relief of pain, and this occurs by at least three ways: (1) morphine reduces central perception of pain probably at the thalamic level, (2) it alters the reaction to pain probably at the level of the cerebral cortex, and (3) it elevates the pain threshold by inducing sedation or sleep. In the medulla, morphine depresses the respiratory, cough and vasomotor centers and stimulates indirectly the vomiting center. The nuclei of the occulomotor (III) and vagus (X) nerves are stimulated by sufficient doses of morphine causing miosis (constriction of the pupils), bradycardia (slowing of the heart rate) and increased gastroenteric tone. The over-all effect of morphine on the gastroenteric tract in man is spasmogenic and constipative. Morphine causes the constipative action by several means including increased segmental movement of the large bowels, spastic tonus of the sphincters, decreased defecation reflex and increased reabsorption of water in the large intestines to cause drying of feces.

The metabolic effects of morphine are not marked and are clinically unimportant. The metabolic rate may be decreased slightly due to the lowered activity and tone of the skeletal muscles resulting from the central depression. A rise in blood sugar may be observed after the injection of morphine. The hyperglycemia is due to glycogenolysis in the liver resulting from the release of epinephrine from the adrenal medulla (see GLYCOGEN). The lowering of urine production noted after the administration of the drug is due mainly to the release of antidiuretic hormone from the posterior pituitary gland (see VASOPRESSIN).

Morphine is detoxified or biotransformed mainly in the liver by conjugation with glucuronic acid. Morphine is conjugated by a series of reactions involving the formation of uridine diphosphoglucose (UDP-glucose), the oxidation of carbon-6 of glucose to form uridine diphosphoglucuronic acid (UDP-glucuronic acid) and the transfer of glucuronic acid to morphine to form the morphine glucuronide. The above reactions are diagramed in Fig. 2. The following enzymes catalyze the sequential reactions: reaction (*1*), UDP-glucose pyrophosphorylase; reaction (*2*), UDP-glucose dehydrogenase; reaction (*3*), glucuronyl transferase; reaction (*4*), nucleoside diphosphokinase.

A minor route of metabolism of morphine is N-demethylation which also occurs mainly in the liver. An enzymic system residing in the microsomal fraction of hepatic cells catalyzes the N-dealkylation of morphine, producing normorphine (demethylated morphine) and formaldehyde. O-Dealkylation also occurs in the microsomal fraction, and the formation of morphine (O-demethylated codeine) from codeine has been noted (see OXIDATIVE DEMETHYLATION).

The most serious drawback in the use of narcotic analgesics is their addictive potentiality. The characteristics of drug addiction include

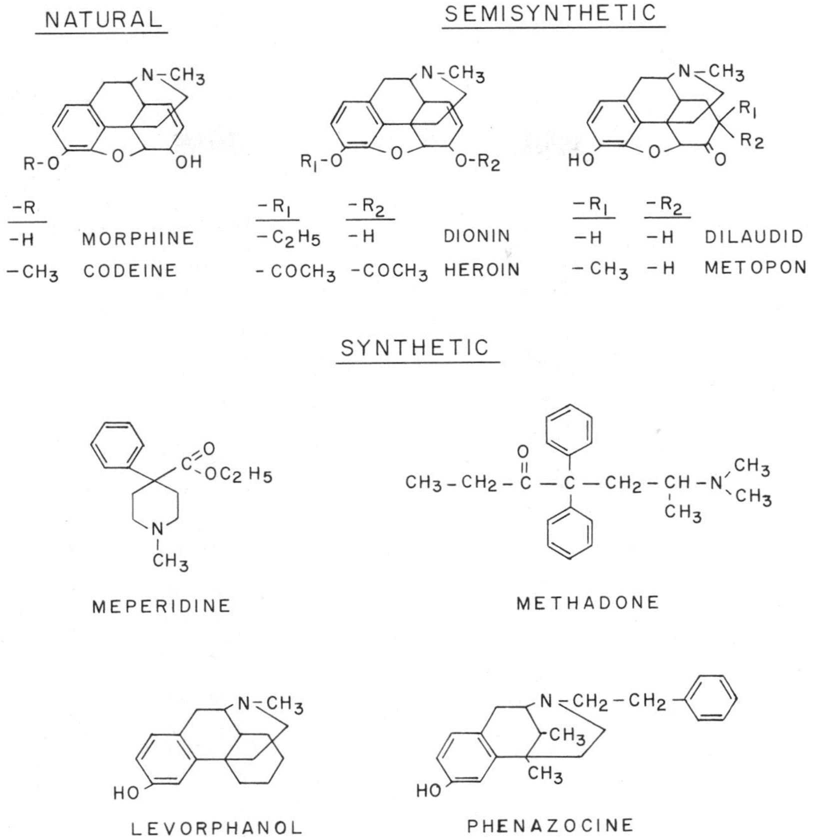

NATURAL

-R	
-H	MORPHINE
$-CH_3$	CODEINE

SEMISYNTHETIC

$-R_1$	$-R_2$	
$-C_2H_5$	-H	DIONIN
$-COCH_3$	$-COCH_3$	HEROIN

$-R_1$	$-R_2$	
-H	-H	DILAUDID
$-CH_3$	-H	METOPON

SYNTHETIC

MEPERIDINE

METHADONE

LEVORPHANOL

PHENAZOCINE

FIG. 1. The chemical structures of various narcotics.

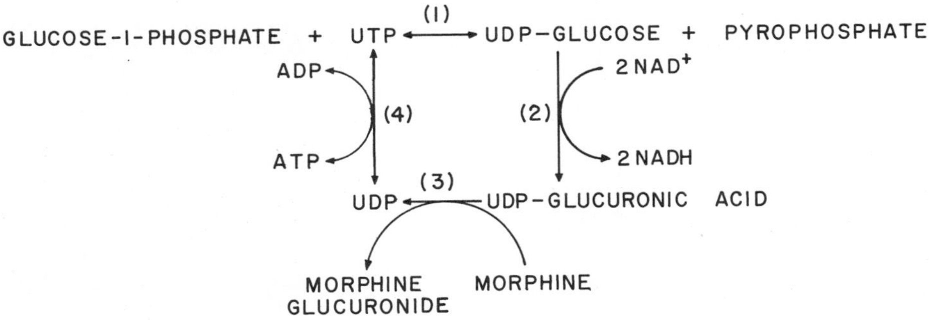

GLUCOSE-1-PHOSPHATE + UTP ⟷(1) UDP-GLUCOSE + PYROPHOSPHATE

(2) $2NAD^+$ → 2NADH

(4) ATP → ADP

UDP ←(3) UDP-GLUCURONIC ACID

MORPHINE → MORPHINE GLUCURONIDE

FIG. 2. The formation of morphine glucuronide. The following abbreviations are used: NAD^+ = nicotinamide adenine dinucleotide, NADH = reduced NAD^+, ATP = adenosine triphosphate, ADP = adenosine diphosphate, UTP = uridine triphosphate, UDP = uridine diphosphate.

psychological need or habituation, tolerance and physical dependence. Habituation consists of an emotional and psychic dependence, and in addiction, the habituation becomes an overpowering desire to take the drug. Tolerance is a phenomenon whereby the dosage of the drug must be continually increased to maintain equivalent pharmacologic effects. Physical dependence develops when the tissues of the body become so adapted to the effects of the drug that the cells of the tissues cannot function normally without the drug in the environment. This is the most vicious characteristic of drug addiction.

The mechanisms underlying the development of tolerance are still unknown. Biochemically, it may be attractive to explain tolerance by decreased absorption, altered distribution, increased biotransformation and/or increased excretion of the drug. However, since these processes have been shown to be unrelated to the development of tolerance, cellular adaptation offers the greatest likelihood for clarifying this phenomenon. Evidence for cellular adaptation is the finding that the respiration of chemically stimulated cortical slices of brain from normal rats is markedly depressed by morphine whereas the respiration of those from rats chronically dosed with morphine is unaffected. The mechanism of this adaptation, however, is still unknown.

Nalorphine, the allyl ($—CH_2—CH=CH_2$) derivative of morphine (N-allylnormorphine), should be mentioned since it has the remarkable property of antagonizing almost all the effects of narcotics. The antagonizing action is specific for the narcotic analgesics. For instance, nalorphine will antagonize the respiratory depression due to morphine or other narcotics but not that caused by other depressants such as hypnotics or anesthetics. This property of nalorphine makes it a particularly useful antidote in cases of acute morphine poisoning. The agent can also precipitate acute withdrawal symptoms if administered to persons addicted to narcotics. The agent has become a useful biochemical tool for studying the mechanism of action of narcotics and tolerance. Since the chemical structures between morphine and nalorphine are so similar, it has been suggested that nalorphine acts by competing with morphine for the receptor site. The antagonistic effect of nalorphine cannot be explained by a simple competitive inhibition if equal affinity for the receptor site with the agonist and antagonist is assumed, because small doses of nalorphine antagonize the effects of much higher doses of the narcotic. Nalorphine also antagonizes the effects of synthetic narcotics of varying chemical structures such as methadone and meperidine. The exact mechanism of action still awaits elucidation.

References

1. Krueger, H., and Sumwalt, M., "The Pharmacology of the Opium Alkaloids," *Public Health Rept., U.S., Suppl.* 165 (1941).
2. Wikler, A., "Sites and Mechanisms of Action of Morphine and Related Drugs in the Central Nervous System," *Pharmacol. Rev.*, **2**, 435 (1950).
3. Woods, L. A., "The Pharmacology of Nalorphine (N-Allylnormorphine)," *Pharmacol. Rev.*, **8**, 175 (1956).
4. Schaumann, O., "Morphin und Morphinähnlich wirkende Verbindugen," in "Handbuch der experimentellen Pharmakologie," Berlin, Springer-Verlag, 1957.
5. Reynolds, A. K., and Randall, L. O., "Morphine and Allied Drugs," Toronto, University of Toronto Press, 1957.
6. Way, E. L., and Adler, T. K., "The Biological Disposition of Morphine and Its Surrogates," *Bull. World Health Organ.*, **25**, 227 (1961); **26**, 51 (1962); **26**, 261 (1962); **27**, 359 (1962).

A. E. Takemori

NATIONAL ACADEMY OF SCIENCES OF THE UNITED STATES

This is a self perpetuating organization of scientists which acts officially as advisor to the Federal government. It was established in 1863 by an Act of Incorporation signed by President Lincoln. Originally it had a membership of 50, but it now numbers about 700 and will doubtlessly be enlarged. Its membership is still highly selective and election carries substantial honor and prestige. A substantial part of the organizational activities entails the selection of new members who measure up to the traditions of the organization.

In 1916 President Wilson called upon the Academy to organize the National Research Council which at the end of the war was, because of its excellent work, perpetuated by executive order. It is made up of about 260 members, many of them representing a large number of scientific organizations. The National Research Council has numerous committees, and many scientific problems of national interest are referred to it.

The *Proceedings of the National Academy of Sciences* is a monthly publication which encompasses many fields. It is recognized at present as one of the leading journals for the publication of biochemical contributions.

NATIONAL INSTITUTES OF HEALTH

The National Institutes of Health, a bureau of the Public Health Service, U.S. Department of Health, Education, and Welfare, is organized to conduct and support medical research. Its laboratories and other facilities occupy thirty brick buildings on a 300-acre tract at Bethesda, Md. Financed by Congressional appropriations, the Institutes expend approximately 15% of their funds in these laboratories and award the remainder for medical research and training in universities, hospitals and other nonfederal institutions.

The Public Health Service has conducted medical research since 1887. Concerned largely with infectious diseases, the Hygienic Laboratory, predecessor of NIH, contributed knowledge and techniques that have helped extend the average life in this country from 47 years in 1900 to over 68 years today. In 1930 the Hygienic Laboratory became the National Institute of Health, and in 1948 the plural form—*Institutes*—was adopted. Early advances against communicable and nutritional diseases include the recognition that pellagra results from a dietary deficiency; the development of a vaccine and serum against Rocky Mountain spotted fever; vaccines for typhus and mumps; the discovery of several diseases, such as ariboflavinosis, Louisiana pneumonitis, and rickettsialpox.

Because more and more people are living longer (to a large extent because of successful assaults against the infectious diseases), the nation's major health problems have shifted from acute infections, particularly those of childhood, to the chronic diseases largely associated with age—heart disease, cancer, arthritis and other metabolic diseases, mental and neurological disorders. In view of this trend, Congress authorized in 1937 the National Cancer Institute, dedicated to research against the disease that had risen from eighth to second place as a cause of death in this country. By 1951 six more National Institutes, some incorporating older laboratories, had been established—Heart, Allergy and Infectious Diseases, Dental Research, Mental Health, Arthritis and Metabolic Diseases, Neurological Diseases and Blindness. In 1962, the National Institute of Child Health and Human Development was established, and the Division of General Medical Sciences was changed to an Institute, making a total of nine Institutes.

In addition to the Institutes, there are now six Divisions. These are concerned with grants, biologics controls, research facilities and research services, regional medical programs, and computer technology. In 1953 the Clinical Center, a 500-bed facility for clinical-laboratory research, was opened, enabling all the Institutes to increase their study of patients.

In fiscal year 1965, NIH employed almost 11,000 scientists and auxiliary workers. The appropriation that year, for both the conduct and support of research, was $1,058,992,000.

The laboratories of NIH apply to the major problems of health and disease a wide range of techniques and scientific specialities. Each of the Institutes approaches the problems in its field through many disciplines, such as chemistry, physics, physiology, pathology and biometrics. As in medical research throughout the world, various chemical approaches—particularly biochemistry, organic chemistry, pharmacology and histochemistry—are paramount. The majority of experiments involve the study of animals or human patients. Results are published by the individual scientists in the journals of their choice.

One type of NIH fellowship, the Visiting Scientist award, has brought many eminent scientists to NIH to conduct research instructive to the permanent staff. Among chemists who have worked at NIH on this basis are Nobel prize winners Warburg and Szent-Györgyi.

Honors to NIH scientists for advances in chemistry have included the Paul-Lewis Laboratories award of the American Chemical Society for four consecutive years, 1951–1954; the Hillebrand prize of the Washington Section of ACS in 1930, 1944, 1949, 1953 and 1954; the award of the Sugar Research Foundation, to C. S. Hudson in 1950; the Joseph Goldberger award for Clinical Nutrition to W. H. Sebrell in 1952 and R. M. Wilder in 1954; the ACS Hillebrand Award to Leon A. Heppel, 1960; ACS Garvan medal to Helen M. Dyer, 1962; ACS Paul-Lewis Award for research in enzyme chemistry to Marshall Nirenberg, 1963; Claude S. Hudson Award for contributions in carbohydrate chemistry to Nelson Richtmyer, 1963; the Eli Lilly Award in Biological Chemistry of ACS to Bruce Ames, 1964.

BRUCE BERMAN

NATIONAL SCIENCE FOUNDATION

The National Science Foundation, an independent agency of the Federal Government, was established by the National Science Foundation Act of 1950 to strengthen basic research and education in the sciences in the United States. The Foundation consists of a National Science Board of 24 members and a Director, each appointed by the President with the advice and consent of the Senate. The Director is the Chief Executive Officer of the Board and serves ex officio as a member of the Board as Chairman of its Executive Committee. The Foundation, located at 1800 G Street, N.W., Washington, D.C., employs a staff of more than 900 scientists and auxiliary personnel. The Foundation's budget for Fiscal Year 1966 is $480,000,000.

Among the activities of the Foundation are:

(1) Award of grants and contracts primarily to universities and other nonprofit institutions in support of basic scientific research. Awards include those made for small and large research projects, for the construction of laboratories or specialized facilities, and for generally strengthening an institution's scientific endeavors. This activity also includes support of concerted research efforts that are planned, coordinated and funded on a national program basis because of the scope of the research being performed and its relationship to national goals. In Fiscal Year 1965 the Foundation made 1859 basic research project grants in the fields of biological and medical sciences, mathematical, physical and engineering sciences, and social sciences.

(2) The support, through contracts, of national centers where large facilities are made available for the use of qualified scientists. At the present time the Foundation is supporting the Kitt Peak National Observatory at Kitt Peak near Tucson, Arizona, the National Radio Astronomy Observatory at Green Bank, West Virginia, and a National Center for Atmospheric Research at Boulder, Colorado.

(3) A program of research and evaluation in the field of weather modification.

(4) The development and dissemination of information relating to scientific resources, including manpower, aimed at facilitating national decisions relating to strengthening the scientific effort of the country.

(5) The award of graduate fellowships in the mathematical, physical, medical, biological, engineering, and social sciences and the provision of support for graduate student traineeship programs at educational institutions. In Fiscal Year 1965, the Foundation supported 4993 fellowships and 2784 traineeships in engineering, mathematical and physical sciences including biochemistry and biophysics.

(6) Programs aimed at improving scientific education in the United States through providing

support for special institutes to improve the competence of teachers of science, mathematics and engineering; projects to modernize materials of instruction and courses of study; projects to afford opportunities for high-ability secondary school and college students to secure added scientific experiences.

(7) A program aimed at improving the coordination of the various scientific information activities within the Federal Government; developing new or improved methods of making scientific information available; fostering the interchange of scientific information among the scientists of the United States and foreign countries; providing support for the translation of foreign scientific information.

HENRY BIRNBAUM

NERVE CELL COMPOSITION AND STRUCTURE

Our concepts about neurons are based upon information drawn from a wide variety of animals and technical procedures, and include morphological, physiological and biochemical data. No attempt can be made here to present either a survey of all the possible neuronal cell types, nor can descriptions of the techniques used to arrive at our present understanding of neurons be included. For extensive bibliographies on neuronal morphology, physiology and biochemistry, reference may be made to Beams *et al.*,[1,2] Eccles,[3] Hodgkin,[4] Nachmansohn,[5] Palay[6] and two recent symposia on nerve cells.[7,8] The following account is based on current knowledge of the fine structure, physiology and biochemistry of neurons from select mammalian, amphibian and invertebrate species which have been the most widely investigated.

The morphological characteristics which distinguish neurons are: long cellular processes called axons, intracellular Nissl substance (granular endoplasmic reticulum), neurofibrillae and cell membrane specializations at synapses and junctional contacts. Other constituents of the cell such as Golgi membranes, cisternae, vesicles, MITOCHONDRIA and NUCLEUS are common to other cell types.

Neurons are relatively large cells due to their elongate, axonal processes which in very large vertebrates may reach several feet. This large size necessitates a relatively high protein synthetic activity, and it has been repeatedly demonstrated that such activity takes place in the perikaryon, the resulting products of which stream into the axonal processes. The morphological substrate for PROTEIN BIOSYNTHESIS activity resides in the ribonucleoprotein(RNP)-containing Nissl substance which consists of scattered masses of granular endoplasmic reticulum (see RIBOSOMES; MICROSOMES). The endoplasmic reticulum of nerve cells is a highly ordered system of membranous cisternae whose membranes are approximately 75 Å thick, on the outer surface of which are situated equally spaced RNP particles which are 150–200 Å in diameter. Proteins manufactured by neurons include a variety of enzymes concerned with cell metabolism.[5] A significant amount of protein found in neurons is in the form of neurofibrillae. Neurofibrillae are long, 50–100 Å thick, protein filaments dispersed in the cytoplasm between Nissl substance, Golgi bodies, mitochondria and vesicles. They appear to traverse the perikaryon in all directions and planes, although loose bundles of filaments are predominantly oriented in one direction in the axon hillock. Neurofibrillae are found as single, 100 Å thick, longitudinally oriented filaments. By polarization optics, it has been shown that at least 10% of the total protein content of axons is in the form of longitudinally oriented filaments. Their function and precise chemical composition are unknown.

Because of glial and Schwann sheaths which surround neurons and the technical obstacles involved in their removal, it has been difficult to isolate single or multiple neurons for biochemical analysis in vertebrates. Therefore, most of our present knowledge about the chemistry of neurons is from invertebrate neurons in which the surrounding sheath material is limited or absent. In freshly prepared neurons of the squid (*Loligo*), the following substances have been found (here listed as millimoles per kilogram wet weight) K 400 mmoles, Na 50 mmoles, Cl 40–150 mmoles, Ca 0.4 mmoles, Mg 10 mmoles, isothionate 250 mmoles, aspartate 75 mmoles, glutamate 12 mmoles, succinate-fumarate 17 mmoles, orthophosphate 2.5–9 mmoles, ATP 0.7–1.7 mmoles, arginine phosphate 1.8–5.7 mmoles, and water 865 g/kg.[4] Neurons also contain relatively large concentrations of choline acetylase, ACETYLCHOLINE, acetylcholinesterase and other nonspecific cholinesterases.

Neuronal cell membranes are composed of an outer, MUCOPOLYSACCHARIDE layer which covers a thin, underlying plasma membrane. It is within this latter membrane that the events of conduction occur, and its structure and biochemistry have been the subject of extensive investigations.[4,5,7,8] The plasma membrane is approximately 75 Å thick as determined by measurements on a variety of neurons. When observed with the ELECTRON MICROSCOPE, the membrane is seen to consist of three alternating electron densities. Plasma membranes from all cell types so far studied are known to consist of a bimolecular layer of LIPOPROTEIN. A variety of direct and indirect evidence shows that the membrane consists of two leaflets of polar, lipid molecules which are oriented with their long axes perpendicular to the surface membrane. The two layers are apposed such that the hydrophobic ends of the molecules face each other, while the hydrophilic ends of the inside layer face inward toward the cell interior and the hydrophilic ends of the outside layer face outward toward the cell exterior. Protein is bound, or adsorbed, to the surface of both hydrophilic layers. Thus a layer of protein is present on the inside and on the outside of the cell. Beyond this, the composition of neuronal plasma membranes includes the usual complement of ATP and enzymes concerned with cell metabolism, particularly those concerned with the transport of ions such as sodium (see ACTIVE TRANSPORT). It has also been demonstrated, by relatively refined homogenization and ultracentrifugation techniques, that neuronal cell membranes contain choline acetylase, acetylcholine and acetylcholinesterase.[5] These substances are believed[5] to be concerned in the conduction process. Nachmansohn[5] has postulated that acetylcholine combines with a receptor protein in the cell membrane which causes a change in the molecular configuration of the receptor protein such as to alter "pore size" and/or "pore charge," thus triggering ionic movement such as the inward flux of sodium during excitation. For convincing arguments against such hypothesis, reference may be made to Eccles[3] and Hodgkin.[4] Sodium, POTASSIUM, CALCIUM and chloride ions are all associated with the neuronal cell membrane. Sodium is most heavily concentrated outside, while potassium and chloride ions are most heavily concentrated inside the cell. Calcium is bound to the protein of the cell membrane. All of these ions have been extensively studied in connection with NERVE IMPULSE CONDUCTION, and much is known about their relative distributions during rest, activity and recovery of neurons. During the nerve cell action potential, there is an inward flux of sodium ions throughout the rising phase of the action potential, which makes the interior relatively less negative than during rest. The movement of sodium and other ions during activity is partly related to their hydrated ion size;[3] thus the concepts of "pore size" and "pore charge" have arisen. For a complete description of ionic movements during excitation, their selective distributions before, during and after the impulse conduction, and membrane properties in many types of neurons, reference may be made to Eccles[3] and Hodgkin.[4]

Structural modifications occur in nerve cell membranes at the level of synaptic and neuroeffector contact areas. These modifications include localized thickenings and increased electron density of presynaptic and postsynaptic membranes. It is here that those molecular events leading to impulse transmission from one cell to another occur. In chemically mediated synapses there is the release of transmitter substance (*e.g.*, acetylcholine) from the presynaptic membrane and its combination with receptor protein in the postsynaptic membrane. These events lead to changes in permeability of the postsynaptic membrane for the particular ion species involved. This results in the so-called synaptic, or junctional, postsynaptic, excitatory potential. In excitatory, acetylcholine-mediated synapses, there is an increase in both sodium and potassium conductances which results in a depolarizing postsynaptic potential. In inhibitory, acetylcholine-mediated synapses, there is an increase in potassium and chloride conductances leading to a hyperpolarizing postsynaptic potential. For a complete description of these events in a variety of SYNAPSES AND NEUROMUSCULAR JUNCTIONS, see reference 3.

On the presynaptic side of synaptic and junctional contacts there is an accumulation of membrane limited vesicles which are 200–300 Å in diameter. Vesicles of this size are found elsewhere in the neuron perikaryon, particularly in the region of the Golgi bodies, and in axonal and dendritic expansions, but they are most heavily concentrated at the synapse. Their presence here may be due to axonal flow of axoplasm distal from the cell body. At least two morphologically distinct types of synaptic vesicles have been described. Those found at acetylcholine-mediated synapses are 300–600 Å in diameter; those found at catecholamine-mediated synapses are 600 1000 Å in diameter and contain an inner electron-dense granule. In some synapses, both types of vesicles have been observed. Both acetylcholine and acetylcholinesterase have been identified in synaptic vesicles. It has been repeatedly speculated both that (1) synaptic vesicles are involved in synaptic and junctional transmission by serving as either precursors or carriers of acetylcholine, and that (2) protein-bound, inactive acetylcholine is transferred from the presynaptic to the intersynaptic space by a process of vesicular and cell membrane fusion. Such a process of cell membrane and vesicular fusion is known to occur in a variety of cells and is the basis for heretofore unrecognized forms of transport. Vesicles at synapses are often seen in contact with synaptic and junctional membranes, but irrevocable proof for their acetylcholine transporting capacity is still wanting.

The Golgi substance (agranular endoplasmic reticulum) of neurons is relatively similar in fine structure to other cell types. It consists of highly oriented, closely packed, flattened cisternal membranes and associated vesicles. Groups of such membranes and vesicles are found circumferentially situated around the nucleus. The cisternal membranes are 100–150 Å thick, and the cisternal lumina have an average width of 200 Å in their undilated portions. Numerous 200–700 Å vesicles are found near the Golgi membranes, particularly at the dilated extremes of the closed cisternal membranes, and some of the membranes of the vesicles are seen to be continuous with the membranes of the cisternae. This leads one to suspect that vesicles are being formed from Golgi membranes. A variety of evidence in many cell types indicates the importance of the Golgi membranes in secretory processes. Such a process in nerve cells is probable. Determinations of the chemical composition of the Golgi substance indicates that it contains phospholipid, polysaccharides, and acid and alkaline phosphatases.

References

1. BEAMS, H. W., *et al.*, "A Correlated Study on Spinal Ganglion Cells and Associated Nerve Fibers with the Light and Electron Microscopes," *J. Comp. Neurol.*, **96**, 249–282 (1952).

2. Beams, H. W., *et al.*, "Studies on the Neurons of the Grasshopper, with Special Reference to the Golgi Bodies, Mitochondria and Neurofibrillae," *La Cellule*, **105**, 293–304 (1953).
3. Eccles, J. C., "The Physiology of Synapses," New York, Academic Press, 1964.
4. Hodgkin, A. L., "The Conduction of the Nervous Impulse," Liverpool, Liverpool University Press, 1964.
5. Nachmansohn, D., "Chemical and Molecular Basis of Nervè Activity," New York, Academic Press, 1959.
6. Palay, S. L., *et al.*, "The Fine Structure of Neurons," *J. Biophys. Biochem. Cytol.*, **1**, 69–88 (1955).
7. Symposium on: "The Submicroscopic Organization and Function of Nerve Cells," *Exptl. Cell Res. Suppl.*, **5**, 1–644 (1958).
8. Symposium on: "Current Problems in Electro-Biology," *Ann. N.Y. Acad. Sci.*, **94**, 339–654 (1961).

James F. Reger

NERVE IMPULSE CONDUCTION AND TRANSMISSION

Communication is a vital and universal need of every organism from the most complicated and multicellular down to the unicellular. The problem is twofold. It involves communication along and within continuous portions of the cell; this is termed *conduction*. It also involves communication across protoplasmic gaps, *i.e.*, from one cell to another; this is called *transmission*. Both processes make use of properties inherent in the protoplasm of all cells.

Impulse Conduction. Among the fundamental properties of protoplasm is conduction as an active process accomplished by the release of stored potential energy, converting it into the kinetic energy of the moving impulse, which is therefore the result of an active process. An equally important passive conduction can be achieved through the process of simple diffusion of chemicals, which then constitute messengers. The specialization of these properties (most developed in nerve cells) has evolved, on the one hand, the self-propagated wave of membrane collapse and restoration which is the conducted nerve, muscle or gland cell impulse and, on the other hand, the diffusion of the chemicals—secreted by cells—across intercellular gaps, constituting chemical or "humoral" transmission.

The possibilities for conduction are facilitated by the creation and storage of potential energy in the form of a concentration battery produced at the cell membranes by the unequal distribution of ions, particularly of the POTASSIUM concentration which is about 27 times higher on the inside than on the outside of the cell, and of the sodium, whose concentration gradient is in the opposite direction because its concentration is 10 times higher outside than inside. The concentration of the anion, Cl^- is about 14 times higher outside than in. These uneven distributions result from the differential diffusion through selectively permeable membranes, which thereby become polarized. The potential from this battery can be recorded between an electrode on the outside and another on the inside of the cell, effecting connections to the two sides of the cell membrane, and reading the voltage which is known as the membrane or resting potential. The measured potential is less than the equilibrium value calculated (by the Nernst equation) for the potential that would be developed by a simple concentration battery formed by the unequal distribution of sodium and potassium ions. The reduced potential appears to be made possible by a special back-pumping, against their concentration and electrochemical gradients, of the sodium (extruded) and potassium (reabsorbed) ions, which goes on simultaneously with, and distorts the diffusion determined by, the membrane semipermeability. The resultant modified potential is the source of the current that eddys into the point of dropped (or even reversed) potential resulting at the site of stimulation, which experiences a transient loss of its restricted permeability permitting temporary free diffusion of ions and a resulting depolarization. The transient free diffusion of ions across the cell membrane carries the current into the cell. The eddy currents in turn act as stimuli to the surrounding membrane, which is then depolarized and draws current from more regions adjacent to the new stimulus sites. These new currents in turn stimulate more distant regions and so on and so on. Meanwhile, the original sites of stimulation and depolarization are repolarizing or recovering. Figure 1 illustrates these events in a diagrammatic and simplified fashion, showing the change of resting, polarized to active, depolarized region and the influence of the subsequent eddy current in creating a progression of the active region from left to right tailed by a region of restored membrane. The latter is temporarily more stable than normal and therefore resistant to restimulation, or refractory for a short period.

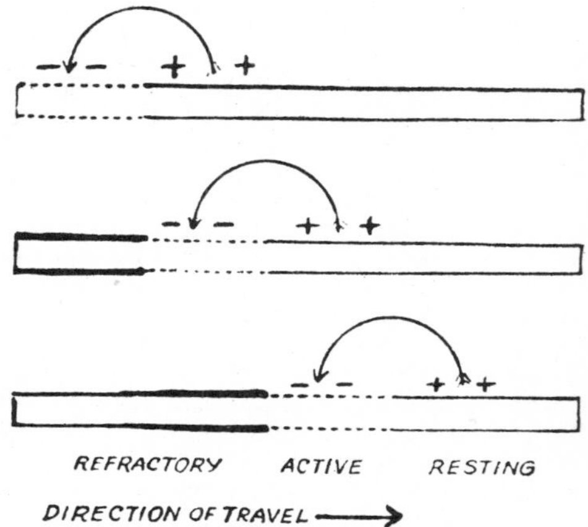

Fig. 1. Spread of active (depolarized) area according to the membrane hypothesis. The local currents cause a repair of the active region and a breakdown of the resting surface beyond.

The eddy of current in this way spreads into a self-propagated wave, which is the conducted impulse whose minute electrical accompaniment can, after suitable amplification, be recorded as the action potential. The growth of electrical charge preceding and culminating in the membrane breakdown, causing the action potential, is evident as the local potential from which the action potential abruptly emerges. The explosive change from a graded disturbance (the accumulation of charge) to the full-fledged self-propagated disturbance, *i.e.*, the impulse, marks the fact that the latter is a yes or no response or an "all or none phenomenon." It is for this reason that computer analogues of nerve networks are built of binary systems.

The mechanism of the transient loss of the selectivity with which ions may permeate the nerve membranes is not known. Neither of the two current ideas is securely established. Because reduction of calcium in a nerve causes it to fire "spontaneously," it has been suggested that the depolarization, which initiates the nerve impulse, removes the calcium ions at the stimulation site or from the ion carrier in the membrane at that site, thus allowing the marked increase in ionic permeability. The presence of ACETYLCHOLINE in many nerves and its importance in transmission from nerve to nerve, as discussed later in this article, have led to acetylcholine being invoked as the agent released by depolarization and acting on a receptor protein in the membrane to cause the increased ionic permeability. However, anticholinesterases, which inhibit the catalysis of the *in situ* destruction of acetylcholine, do not act in a sufficiently characteristic manner on conduction of the nerve impulse to adequately substantiate the claim.

The chemistry underlying the conducted impulse is that of the energy-supplying processes which polarize membranes by segregating ions into unequal concentrations on the two sides of the membrane and of the ionic pumps that supplement and modify the concentration differences and bring about an observed resting potential at variance with the one expected from calculation. The renewal of the potential energy mechanism, as distinguished from the ionic pumps which presumably operate steadily, is accomplished intermittently by oxidative restorative processes, which are characteristically delayed so that a so-called oxygen debt is built up to be paid off during periods of diminished or no call for impulse conduction.

A quite considerable (50% of the activated maximum) steady oxygen consumption that goes on at rest is able to provide the energy that must be expended to counteract the continuous ionic leakage and to operate the postulated pumps. Thus, uncoupling of oxidative PHOSPHORYLATION by dinitrophenol (DNP) and inhibition of oxidation by cyanide and azide or by strophanthin interfere with extrusion of sodium and reabsorption of potassium, so that although the short-circuiting inflow of sodium ions causing the action potential is not immediately affected, the restorative extrusion fails to take place, *i.e.*, the transport energized by adenosine triphosphate (ATP) and phosphorus donors stops (see ACTIVE TRANSPORT).

In keeping with an evident degree of dissociation possible between the diffusion *with* and *against* the concentration gradient, altering the former by changing the external ionic concentration bathing the nerve does not affect the latter process. Likewise, there is a difference in selectivity toward ions as illustrated by the fact the lithium can replace sodium in the "downhill" diffusion with the concentration gradient underlying the action potential mechanism, but is not adequately moved against the concentration gradient by the metabolic sodium pump.

Impulse Transmission. The passage of an impulse across an intercellular gap has been defined as transmission. Examples of this are interneuronal or synaptic and neuroeffector or junctional transmission. The eddy currents and electrical field effects generated here by the arrival of an impulse at the protoplasmic discontinuity are sufficiently attenuated by the passage through the inert conductor offered by the nonpolarized ionic population of the gap (which therefore does not regenerate impulses in the self-propagating manner of a polarized conductor) that they appear generally inadequate *per se* to effect transmission.

A passive conduction mechanism, therefore, becomes an important step in the transmission process at SYNAPSES and neuroeffector junctions. This step starts with the liberation by the nerve impulse of chemical messengers from the ends of the nerves (Fig. 2) where they exist as precursors located in granules and vesicles. The messengers diffuse across the microscopic interspace of about 200 Å and then exercise their specific effects on the cell membrane of the postsynaptic or postjunctional cells. They activate these, generating within them impulses that are conducted throughout the cells in the same manner that conduction took place in the cells originating the messages. Thus the impulse has been relayed from one cell to another by the intervention of a chemical diffusion step which transmits it across the gap. This describes transmission of an *excitatory* impulse. An excitatory messenger chemical that has been identified is acetylcholine; there may be others.

The influence of some cells upon their neighbors is *inhibitory*. In such cases, inhibitory messenger chemicals have been identified, *e.g.*, noradrenaline, ADRENALINE, serotonin and histamine (in order of increasing effectiveness of these biogenic amines); again, there may be others. These messengers, like the excitatory ones, combine with discrete and specific areas or receptors occupying a small fraction of the surface of the responding cells. The receptors are key spots, which initiate, according to their nature, an excitatory or depolarizing process, or an inhibitory or hyperpolarizing process. Some regard the excitatory action of acetylcholine in terms of "punching a hole" in the polarized membrane into which ions can freely diffuse and carry the eddy currents that

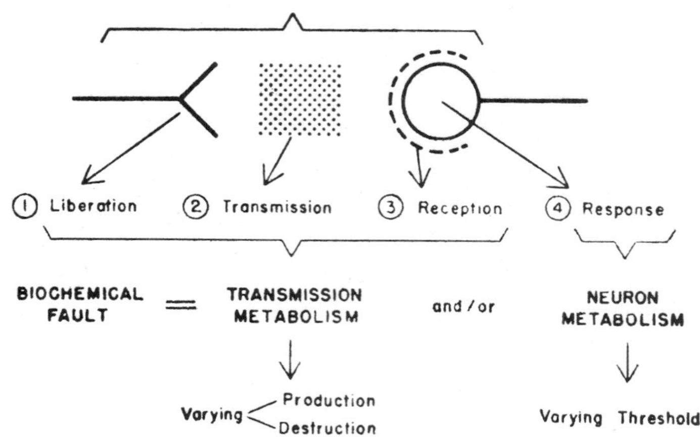

FIG. 2. Potential factors in disturbed synaptic equilibrium.

constitute the first segment of the self-propagating mechanism described above.

The chemical transmitter, excitor or inhibitor, is terminated by the chemical breakdown of the transmitter, accelerated by enzymes strategically located at the gaps where transmission is taking place. This clears the "switchboard" for further messages. [Acetylcholinesterase is the enzyme that has been identified as catalyzing the synthesis and the hydrolysis of acetylcholine, while monoamine oxidase (MAO) and catecholorthomethyl transferase (COMT) are associated with oxidation and degradation metabolism of biogenic amines.]

It seems probable that the simultaneous operation of varying numbers of excitatory and inhibitory fibers, sending messages into the synaptic switchboard, offers the possibility of a finer regulation through a "check and balance" system. An equilibrium between excitatory and inhibitory influences is diagramed in Fig. 2, which summarizes the synaptic mechanisms and suggests the possible derangement or disequilibrium that could occur spontaneously in disease or could be induced experimentally. The points where equilibrium is susceptible to change are both the sites vulnerable to disease and the targets for corrective measures or therapy.

The depolarization process at the receptor is recordable as a local potential (synaptic and motor end plate) from which, again abruptly, rises the much larger action potential, indicating that the self-propagated process, which will invade the whole membrane, has been initiated. A detailed analysis of suitable electrical records can, therefore, identify the complete sequence of events that have been described as conduction, relay or transmission and post-transmission conduction. Individual components can be exhibited with greater clarity or can be isolated by, so to speak, dissecting the various steps through the use of appropriate chemicals, which will enhance or block certain steps, and by the use of polarizing and depolarizing currents to abet or to resist the influences of biological polarization phenomena. Study of the phenomena can likewise be carried on at the level of energy storage and consumption by making use of the usual techniques in the analysis of metabolism.

It takes considerably longer to describe the events in impulse conduction than it does for them to take place. Thus conduction, at its slowest, goes on at the rate of 0.6 m/sec or about 30 yd/min and, at its fastest, at a rate of 120 m/sec or better than 4 miles/min. It is this "flash-like", split-second speed that makes possible the arrival of the impulses from a stubbed toe to the brain and back in time to remove the toe before further injury can take place. This is true despite the relatively longer delays in traversing synaptic switchboards with pauses of half to several microseconds.

The faster impulses are also larger and are conducted in suitably larger and better insulated "cables." These are the large A fibers, which are sheathed with a prominent lipid or MYELIN layer. Small thinly myelinated C fibers conduct the slow, low-amplitude impulses, and the intermediate, moderately myelinated B fibers conduct the intermediate-size impulses that travel at intermediate speeds. A nerve trunk can conduct the several varieties of waves simultaneously, and because of their varying amplitude and speeds, they can be readily, separately identified in the record obtained by "tapping in on the line" with suitable electrodes and amplifying the minute potential changes. The record will display in sequence the faster and larger waves from the A fibers, the waves of intermediate speed and amplitude from B fibers, and lastly the slow, low-amplitude waves from C fibers.

The stream of conducted signals is both amplitude and frequency coded to indicate magnitude of environmental change or stimulus. Thus increased signal strength is accomplished by the increase in the number of fibers participating.

Within the natural frequency range of particular fibers, a mechano-transducer action of sensory receptors also converts increasing stimulus strength to increasing frequency of impulses generated in the individual fibers of the sensory nerve connected to the receptor.

Although a minor degree of interaction can take place between the electrical fields of adjacent fibers, the neural impulse conduction system is essentially a fairly well-insulated system suited to discrete communication. The development of widespread patterns of activity reflects the functioning of interconnections at the synaptic level. It should not be surprising, since transmission is effected by specific chemicals, to find that under special circumstances of great stress, a major depot of such chemicals (adrenaline and noradrenaline) located in the medulla of the adrenal gland can pour these synaptic inhibitory chemicals into the bloodstream. Distributed in this way to all synapses, the adrenal medullary secretion can supply a cutoff influence limiting the massive discharge of impulses that stress is apt to initiate, and thus prevent it from becoming so excessive that it is detrimental. In this way, the neurohumoral transmission mechanism also affords a means of chemical homeostatic regulation.

The vulnerability of impulse transmission to chemical influences may not always serve HOMEOSTASIS; for here also is where the action of poisons like mescaline and lysergic acid diethylamide (see HALLUCINOGENS) takes place. They produce in man a temporary mental derangement or psychosis. Experiments show that these substances inhibit impulse transmission at cerebral synapses and that tranquilizers prevent this effect. Such experiments in disturbed impulse transmission are developing the basis for understanding of chemically induced psychosis and the manner of action of tranquilizers. A sufficient parallelism appears to exist between the experimental laboratory findings and data in clinical psychosis to suggest that the latter may, also, sometimes be a disturbance of impulse transmission in the brain and that tranquilizers tend to restore synaptic equilibrium. In fact, a small polypeptide has been extracted from human blood, and particularly from that of some mentally disturbed patients, which acts on the brain very much in the manner of an inhibitory messenger and like lysergic acid diethylamide.

It seems an inescapable observation that the highly elaborated methods of impulse handling by complex organisms are foreshadowed in the mechanisms found in the unicellular organism or in any single cell.

References

1. MARRAZZI, A. S., "Messengers of the Nervous System," *Sci. Am.*, **196**, 87 (February 1957).
2. ECCLES, J., "The Synapse," *Sci. Am.*, **212**, 56 (January 1965).
3. HODGKIN, A. L., "The Conduction of the Nervous Impulse," Springfield Ill., Charles C. Thomas, 1964.
4. REDICK, T. F., RENFREW, A. G., PIERI, L., AND MARRAZZI, A. S., "A Cerebrally Active Small Moeity from 'Taraxein-Like' Blood Fractions," *Science*, **141**, 646 (1964).
5. BAKER, P. F., "The Nerve Axon," *Sci. Am.*, **214**, 74 (March 1966).

AMEDEO S. MARRAZZI

NERVOUS SYSTEM (BIOCHEMICAL ASPECTS). See BLOOD-BRAIN BARRIER; BRAIN COMPOSITION; BRAIN METABOLISM; CEREBROSPINAL FLUID; MENTAL RETARDATION; MYELIN; NERVE CELL COMPOSITION; NERVE IMPULSE CONDUCTION; SYNAPSES AND NEUROMUSCULAR JUNCTIONS. Classes of substances associated with nervous system tissue include CEPHALINS AND LECITHINS; CEREBROSIDES; GANGLIOSIDES; PLASMALOGENS; SPHINGOLIPIDS.

NEURAMINIC ACID. See HEXOSAMINES AND HIGHER AMINO SUGARS.

NEUROHORMONES

Hormones are "chemical messengers" which are produced by certain specialized cells (nerve cells in the case of neurohormones), and which can then control or regulate the activity of cells elsewhere in the organism. The identities of a number of neurohormones are discussed in NERVE IMPULSE CONDUCTION AND TRANSMISSION, section on Impulse Transmission, and in SYNAPSES AND NEUROMUSCULAR JUNCTIONS (see also NERVE CELL COMPOSITION).

References

1. KATZ, B., "How Cells Communicate," *Sci. Am.*, **205**, 209–220 (September 1961).
2. WURTMAN, R. J., AND AXELROD, J., "The Pineal Gland," *Sci. Am.*, **213**, 50 (July 1965).

NEUROHYPOPHYSIS

The neurohypophysis, or posterior pituitary, produces the hormones VASOPRESSIN and oxytocin (see also HORMONES, ANIMAL; HYPOPHYSIS).

NEUROSPORA (IN BIOCHEMICAL STUDIES)

The main contribution of *Neurospora* to biochemistry was its role in bringing the disciplines of biochemistry and genetics together. Unlike yeast and some other fungi, *Neurospora* was used first, not as a tool of biochemistry, but as a tool for genetic studies. Some of the most exciting discoveries and theories in biology trace their genealogy to the marriage of genetics and biochemistry.

Neorospora was more a witness than a product of this union, called BIOCHEMICAL GENETICS. Professors B. O. Dodge and C. C. Lindegren, at Cal. Tech., ordained *Neurospora* a genetic tool a decade prior to its use in 1940 by the geneticist, G. W. Beadle, and the biochemist, E. L. Tatum.

Genetics made its first contribution to biochemistry as a means for studying intermediary metabolism. One of the basic premises of intermediary metabolism states that low molecular weight metabolites are synthesized in cells in a series of discrete steps each of which is catalyzed by a specific enzyme. Genetics enlarged the premise to state that the specificity of each enzyme is determined by a specific gene.

Mutations of *Neurospora* lacking the capacity to synthesize essential metabolites are usually defective at only one step in the metabolic pathway leading to synthesis of a required metabolite. Correspondingly, *in vitro* activity of the enzyme is absent or altered in extracts of the mutants. This approach to biochemistry made two major contributions: (1) it helped establish the exact route of synthesis of many essential metabolites, and (2) it added impetus to the inherently attractive "one gene: one enzyme" hypothesis.

The one gene: one enzyme hypothesis, after cutting its teeth on *Neurospora*, set off an avalanche of experimentation which spread biochemical genetics in two directions. In one direction, more and more microorganisms were employed, each of which made its own unique contribution, and in the other direction the mechanism of gene-enzyme specificity was attacked. It is with the role of *Neurospora* in attacking this latter problem that this article will deal.

Asexual conidia from two mutant strains, each requiring a different metabolite for growth, when mixed and plated onto unsupplemented medium usually grow at near normal rates. The conidia germinate and their nascent mycelia fuse, permitting entrance of both types of nucleus into a common cytoplasm. The resultant mycelium, with genetically different nuclei, is called a heterocaryon. Growth of inter-allelic heterocaryons is explained as follows: one mutant nucleus is competent, at the GENE level, where the other is incompetent, and *vice versa*. Each nucleus, then, compensates for the deficiency found in the other. Heterocaryosis was used for many years as a test for genetic alleiism on the grounds that nuclei with mutations at the same genetic locus do not complement.

This rationale did not hold for long. When it was discovered that recombination occurs within genes, it became obvious that mutation may alter a gene at many locations (sites) within its borders. *Neurospora* geneticists then found that nuclei mutant in the same gene, but not at the same site within the gene, often complemented when placed in a common cytoplasm. What is involved in this type heterocaryon (called intra-allelic complementation) is as follows: a gene (cistron) specifies the primary structure of a complete polypeptide. If, for example, gene *A* in nucleus 1 is mutant, nucleus 1 will produce no active enzyme α. If gene *A* in nucleus 2 also is mutant, albeit at a different site, nucleus 2 also will fail to yield active enzyme α. If nuclei 1 and 2 are placed into a common cytoplasm, growth of the heterocaryotic mycelium is observed, implying activity of enzyme α. First, it is clear that genetic recombination is not involved since the two nuclei do not fuse or exchange chromosome materials within the cytoplasm. Second, it cannot be a correction in the transcription of messenger RNA since each RNA is synthesized independently while the nuclei remain separate. This leaves essentially two explanations for intra-allelic complementation: either the two mutant messenger RNAs or the two mutant proteins interact in the cytoplasm, after synthesis, to produce active enzyme. One may visualize this as (a) a mutual repair (recombination?) of some sort, or (b) as an aggregation of mutant proteins resulting in some form of hybrid between the two mutant types.

The study of enzyme complementation provided impetus to the study of allosteric proteins in general. The details of this field of study cannot be dealt with here. Suffice it to say that at least three basic structures are now known for enzymes: a single polypeptide chain folded into a tertiary structure, aggregate of two or more identical polypeptide subunits (each coded by the same gene), and aggregates of two or more nonidentical polypeptide subunits (each coded by a different gene). Complementation is explained in either of the latter types by assuming that mutant polypeptides are produced by each of the two nuclei and that when these mutant subunits aggregate in the proper combination in the cytoplasm, restoration of enzyme activity, in some degree, is achieved.

In 1959, D. O. Woodward demonstrated *in vitro* complementation with *Neurospora* extracts. Using mutants void of adenylosuccinase activity, mixtures of extracts of mutant strains capable of *in vivo* complementation showed *in vitro* enzyme activity. This work clearly demonstrates that the phenomenon of complementation is independent of both RNA and protein synthesis and that it must result either from aggregation of mutant polypeptides or from mutual repair of some sort.

In general, heterocaryon enzymes differ from their normal counterparts in temperature sensitivity, pH optima, migration in an electrophoretic field and other physical properties, but they seem always to show the same molecular weight. Examples of these studies are: glutamic dehydrogenase enzyme of *Neurospora*, studied by J. R. S. Fincham; adenylosuccinase enzyme of *Neurospora*, studied by D. O. Woodward; the malic acid dehydrogenase enzyme, studied by K. D. Munkres.

The biochemist, usually with good reason, is skeptical of the biologist who infers chemical reactions or properties of enzymes from biological data. However, the biochemist feels both secure and consistent when extrapolating from the chemical reaction or enzymic phenotype to the biological system. It can be argued that there is a place for deduction and induction; this can be documented with the following experiments conducted with *Neurospora*. These experiments also may serve to describe, in a general way, how

Neurospora has evolved from a genetic to a biochemical tool.

One of the steps in the pathway leading toward synthesis of the amino acid ARGININE involves coupling CARBAMYL PHOSPHATE (CAP) with ornithine to yield citrulline. The reaction is catalyzed by the enzyme ornithine transcarbamylase (OTC). An analogous reaction appears in the pyrimidine pathway where the enzyme aspartate transcarbamylase (ATC) catalyzes the coupling of aspartic acid and CAP. In bacteria CAP is formed by one of several reaction mechanisms, but in all cases studied it forms a common pool for both arginine and PYRIMIDINE BIOSYNTHESIS. In *Neurospora*, however, the evidence points toward a separate synthetic mechanism of CAP for each of the two pathways. This evidence includes the following: mutations at one genetic locus, the *pyr-3* locus, fall into two groups, enzymatically. One group has no *in vitro* ATC activity, and the other group is indistinguishable from wild type with respect to ATC activity. The latter mutants are genetically and enzymatically competent at every step of the pyrimidine pathway subsequent to the ATC step, leaving only the step prior to CAP as a point of genetic block. But since the mutants do not exhibit a dual requirement for both pyrimidine and arginine, as do certain bacterial mutants, the conclusion was questioned. However, the conclusion is reinforced by observing that a double mutant made up of (1) an OTC mutant, incompletely blocked (*i.e.*, the mutant does grow without arginine but only at reduced rates), and (2) an ATC positive, pyrimidine-requiring mutant, is able to grow in the absence of pyrimidine and arginine. (The OTC mutant is called a "suppressor" since its phenotype is the suppression of the *pyr-3* phenotype.) This observation is interpreted to mean that the OTC mutant accumulates CAP synthesized specifically for arginine, and that by metabolic crossfeeding this CAP is made available for pyrimidine synthesis. One of the tentative conclusions derived from this work is that CAP is synthesized in at least two independent ways, one guided by the arginine regulatory mechanism and one by the pyrimidine mechanism.

It is obvious, at the same time, that since a single gene (*pyr-3*) structures the ATC enzyme and since mutants at the *pyr-3* locus require pyrimidine for growth and show ATC activity, the *pyr-3* gene must code a single polypeptide with two functions. One function is postulated to be the synthesis of CAP (for pyrimidines) and the other the coupling of CAP with aspartate to produce carbamyl aspartate (the ATC reaction).

Biological evidence for this conclusion derives from several sources. (1) The *pyr-3* mutants fall into two complementation groups. One group includes all of the ATC positive mutants; these mutants complement with two ATC negative mutants, and both groups fail to complement the remainder of the *pyr-3* mutants, all of which are ATC negative. (2) An arginine-requiring mutant that lacks carbamyl phosphokinase (CPK) activity, the enzyme known to catalyze the formation of CAP (for arginine) in *Neurospora*, does not require pyrimidine for growth. This is a clean-cut demonstration that CAP synthesis for arginine is independent of CAP synthesis for pyrimidine. (3) The two ATC negative mutants which complement the ATC positive mutants are capable of suppressing the arginine requirement of the CPK mutants (suppression by crossfeeding of CAP reciprocal to that described above).

To test the biological interpretations proposed, the following hypothesis is presented: the two functions reside on one polypeptide, or the two functions reside on two separate polypeptides. If the first alternative is correct, one would expect a mutation in the first function (CAP synthesis) to alter the conformational properties of the protein carrying the second function (ATC). This would not be so if the second alternative is correct. Kinetic studies of all mutants suspected of having a mutation at the first function, but not at the second, show that each mutant enzyme exhibits a multiple and varying order of reaction distinct from wild type enzyme, and distinct from other mutant enzymes. This evidence supports the biological interpretation that both functions reside on the same polypeptide. This work was done in our laboratory by Drs. John Hill, K. J. McDougall and the author.

References

1. SHEAR, C. L., AND DODGE, B. O., "Life Histories and Heterothallism of the Red Bread Mold Fungi of the *Monilia sitophila* Group," *J. Agr. Res.*, **34**, 1019 (1927).
2. LINDEGREN, C. C., "The Genetics of *Neurospora*," *Bull. Torrey Botan. Club*, **59**, 85 (1932).
3. BEADLE, G. W., AND TATUM, E. L., "Genetic Control of Biochemical Reactions in *Neurospora*," *Proc. Natl. Acad. Sci., U.S.*, **27**, 499. (1941).
4. BEADLE, G. W., AND COONRADT, V. L., 1944, "Heterocaryosis in *Neurospora crassa*," *Genetics*, **29**, 291.
5. WOODWARD, D. O., "Enzyme Complementation *in vitro* between Adenylosuccinaseless Mutants of *Neurospora crassa*," *Proc. Natl. Acad. Sci. U.S.*, **44**, 1237 (1959).
6. BONNER, D. M., "Gene-enzyme Relationships," Proc. XI Intern. *Congr. Genet., 11th*, 141–149, 1964.
7. SUNDARAM, T. K., AND FINCHAM, J. R. S., "A Mutant Enzyme in *Neurospora crassa* Interconvertible between Electrophoretically Distinct Active and Inactive Forms," *J. Mol. Biol.*, **10**, 423 (1964).

VAL W. WOODWARD

NICOTINAMIDE ADENINE DINUCLEOTIDES

Living organisms carry out a variety of chemical reactions which are catalyzed by ENZYMES, *e.g.*, hydrolytic, isomerization, transfer, synthetic, and oxidation-reduction reactions. In the category of enzymatically catalyzed oxidation-reduction reactions, there are about 200 that are known, and in many cases, electron carriers known as coenzymes are involved. Examples of such coenzymes are nicotinamide adenine dinucleotide (NAD) and a closely related compound nicotinamide adenine dinucleotide phosphate (NADP). Both coenzymes are reduced by the transfer of two hydrogen atoms from substrates to the coenzymes, such transfer being catalyzed by substrate-specific enzymes known as *dehydrogenases*. The reduced products formed are designated NADH and NADPH, respectively.

As the names suggest, the molecules contain a nicotinamide and adenine moiety which are connected together by two molecules of ribose and two phosphate groups as illustrated in Fig. 3 of the article VITAMIN B GROUP.

In the case of NADP, there is an additional phosphate group on the ribose next to the adenine, which is of importance in determining the specificity of combination with different dehydrogenases. However, both of these compounds are functionally similar as far as oxidation and reduction are concerned.

The generalized oxidation-reduction reaction is as follows:

$$\text{Substrate}_1 + \text{NAD}^+ \rightleftarrows \text{Oxidized Substrate}_1 + \text{NADH} + \text{H}^+ \quad (1)$$

This equation represents the transfer of two electrons along with the hydrogen nuclei to form the reduced coenzyme. However, one of the two hydrogens dissociates to release H^+ as shown in Eq. (*1*). The same generalized equation may be applied to reactions involving NADP. Under these conditions, the coenzyme is serving as an electron acceptor. In tissues where other enzymes and electron acceptors are available, the reduced coenzyme may migrate to another enzyme and transfer the electrons to another substrate which becomes reduced as represented in Eq. (*2*).

$$\text{Oxidized Substrate}_2 + \text{NADH} \rightleftarrows \text{Reduced Substrate}_2 + \text{NAD}^+ \quad (2)$$

In this case, the coenzyme is serving as an electron donor with the formation of NAD^+ which may be reduced again. Thus, NAD and NADP are used cyclicly, serving as a carrier of electrons from one substrate to another. In the case of aerobic organisms, the electrons from NADH or NADPH may be transmitted through a series of other electron carriers ultimately ending up on oxygen to form water. In the case of FERMENTATION, the ultimate electron acceptor may be an organic molecule which, after reduction, is not metabolized further and accumulates as a fermentation product in the medium in which the organism is growing.

The substrates for the dehydrogenases utilizing these coenzymes fall into two classes. In the first class are the *primary and secondary alcohols* which are oxidized to aldehydes and ketones, respectively. In the second class of dehydrogenases are those which oxidize *aldehydes* to the corresponding carboxylic acids.

Our knowledge of these coenzymes originated with the work of Warburg, during the 1930's. Since that time, NAD and NADP were found to be ubiquitously distributed in bacteria, fungi, higher plants, invertebrates and vertebrates. In mammals these coenzymes are found in all organs: liver, kidney, heart, brain, lung, spleen, pancreas, testes and others as well. About one hundred enzymes which require these coenzymes have been identified. In Table 1 is a partial list of enzymes which shows the substrate and coenzyme specificity as well as the variety of reactions in which they participate. It will be noted that in the list are enzymes involved in GLYCOLYSIS, FERMENTATION, RESPIRATION, AMINO ACID METABOLISM, and nitrogen metabolism. More reactions are known which utilize NAD, than NADP.

The NAD^+ : NADH system has a redox potential of −0.32 volt. From theoretical considerations, it would be expected that the biochemical compounds which are capable of reducing NAD or NADP would have a more negative redox

TABLE 1. OXIDOREDUCTASES UTILIZING NAD AND NADP

(I) *NAD Enzymes*	*Reaction*
(A) Alcohol dehydrogenase	Alcohol + NAD^+ ⇄ Aldehyde + NADH + H^+
(B) Glycerol phosphate dehydrogenase	Glycerol-3-phosphate + NAD^+ ⇄ Dihydroxyacetone Phosphate + NADH + H^+
(C) Lactic dehydrogenase	Lactate + NAD^+ ⇄ Pyruvate + NADH + H^+
(D) $NADH_2$: oxidized cytochrome *c* reductase	$NADH_2$ + Oxidized Cytochrome *c* → NAD + Reduced Cytochrome *c*
(II) *NADP Enzymes*	*Reaction*
(A) Glucose-6-phosphate dehydrogenase	Glucose-6-phosphate + NADP → Gluconolactone-6-phosphate + NADP + H^+
(B) Isocitric dehydrogenase	Isocitrate + NAD^+ → 2-Oxoglutarate + CO_2 + NADH + H^+
(C) Malate dehydrogenase	Malate + NADP → Pyruvate + CO_2 + NADPH + H^+
(III) *NAD or NADP Enzymes*	*Reaction*
(A) Glutamate dehydrogenase	Glutamate + NAD^+(P) → 2-Oxoglutarate + NH_3 + NADPH + H^+
(B) Quinone reductase	NAD(P)H + Quinone → NAD(P) + Hydroquinone
(C) Nitrate reductase	NAD(P)H + Nitrate → NAD(P) + Nitrite + H_2O

potential. The reduced coenzymes, in turn, would transfer the electrons to other substrates or coenzymes, *e.g.*, FAD (see FLAVINS) having a less negative or more positive potential, according to the conventions used by biochemists (see BIOLOGICAL OXIDATION-REDUCTION).

The equilibrium constant for the general reaction of Eq. (*1*) can be illustrated by a specific example. ALCOHOL DEHYDROGENASE, which has been studied in great detail, catalyzes the following reaction:

$$CH_3CH_2OH + NAD^+ \rightleftharpoons CH_3CHO + NADH + H^+ \quad (3)$$

The equilibrium constant can be obtained by appropriate substitutions as below:

$$K = \frac{[CH_3CHO][NADH][H^+]}{[CH_3CH_2OH][NAD^+]} \quad (4)$$

Since H^+ is a product of the reaction, the equilibrium constant will be affected by pH, and this should be taken into consideration. However, in the cell, the NADH produced would be continuously oxidized and reduced again, therefore equilibrium conditions will not necessarily prevail. Under these conditions, the amount of NAD reduced as compared to the amount oxidized would be related to various factors which would alter the equilibrium conditions. For example, various enzymes would be competing for the coenzymes, and substrate concentrations may vary depending upon the nutritional state of the organism, just to cite two important factors which will lead to what is known as a steady-state condition rather than a condition of equilibrium.

The quantity of these coenzymes present will vary from one organism to another and from one tissue to another. Nevertheless the figures for rat liver are given for purposes of illustration. Rat liver, which has somewhat greater quantities than other tissues, was reported to have 570 μg of NAD and 210 μg of NADP per gram of rat liver. These values represent the sum of both oxidized and reduced forms. Most of these coenzymes are found in the cytoplasm of the cell; however, a significant portion is bound to the mitochondria in which case the electrons removed from the substrates are transferred to the CYTOCHROME system.

In all organs of the rat which have been studied, less than half of the total NAD was found in the reduced form. On the other hand, in most tissues NADP is found in the reduced form to the extent of 90–100%. This difference is considered to be related to the fact that NAD is involved in catabolic reactions while NADP is involved in anabolic reactions. For example, in the synthesis of FATTY ACIDS, the enzymes involved utilize NADPH, while the oxidative enzymes involved in the breakdown of fatty acids utilize NAD (see FATTY ACID METABOLISM).

The reduced form of NAD absorbs ultraviolet light with a peak at 340 mμ which is not associated with the oxidized form of the coenzyme. With the use of an ultraviolet absorption spectrophotometer, this difference in absorption can be used to follow either the reduction of NAD(P) which will result in increased absorbancy, or vice versa for the oxidation of the reduced enzyme. With the use of such spectrophotometric procedures for assay purposes, it has been possible to purify a number of these enzymes to a pure state. Fortunately, some tissues and organisms contain fairly large quantities of these enzymes so that relatively large quantities of pure enzyme can be obtained for detailed studies of the enzyme reaction. For example, various investigators have studied the interaction between pure alcohol dehydrogenase and alcohol, acetaldehyde, NAD and NADH. In such studies, the enzyme is used as a reactant rather than a catalyst. For example, liver alcohol dehydrogenase was found to bind two molecules of NAD. Furthermore, two Zn^{++} were found to be involved in binding NAD to the enzyme. From the point of view of comparative enzymology, it is interesting to note that the enzyme in yeast which catalyzes the same reaction is capable of binding four molecules of NAD by means of four Zn^{++}.

Structural alterations in the coenzymes markedly affect their capability as serving for various dehydrogenases. Although there are different structural requirements for the dehydrogenases, the following points will illustrate the remarkable nature of the specificity. The adenine is required for activity, since experiments have shown that if it is removed, much of the activity is lost. Even in case of a slight change, such as replacing the amino group on adenine with a hydroxyl group, the activity is diminished. The ribose moiety also plays a role, since substitution of deoxyribose for the ribose adjacent to the adenine also reduces activity. Changes in the nicotinamide moiety are also of importance. For example, replacing the $—CONH_2$ with an H eliminates the capacity for this analogue for use as a coenzyme. These as well as many other experiments clearly show that slight structural changes affect the activity of these compounds as coenzymes.

The biosynthesis of NAD is accomplished in about five steps with different enzymes being involved in each step. In brief, the vitamin niacin (or nicotinamide) is required in the diet (see NICOTINIC ACID), and this in turn is conjugated with ribose, then with ATP to form NAD. NADP is formed from NAD by a phosphorylation reaction utilizing ATP. In view of the large number of reactions utilizing these coenzymes, it is readily apparent that a niacin deficiency can be quite serious.

References

1. CONN, E. E., AND STUMPF, P. K., "Outlines of Biochemistry," New York, John Wiley & Sons, 1963.
2. DIXON, M., AND WEBB, E. C., "Enzymes," New York, Academic Press, 1964.
3. For a discussion of the nomenclature and abbreviations of the nicotinamide coenzymes (the NAD-NADP system and the DPN-TPN system, etc.), see DIXON, M., *Science*, **132**, 1548 (1960).

WALTER D. WOSILAIT

NICOTINIC ACID (NIACIN)

Chemical and Physiological Properties. The compound nicotinic acid has been known to the chemist since 1867; however, its role as a vitamin was not recognized until 1937.

The structures of nicotinic acid and the closely related compound nicotinamide are shown in Fig. 1.

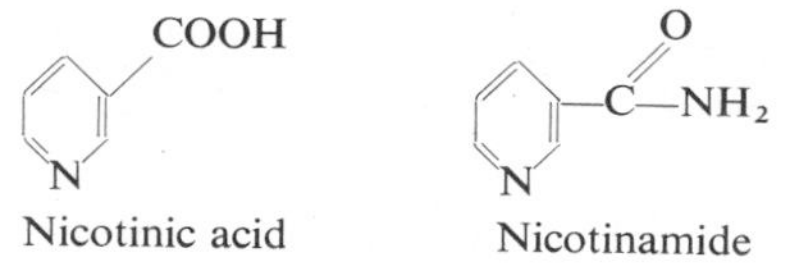

Nicotinic acid (3-COOH pyridine) — Nicotinamide (3-C(=O)—NH_2 pyridine)

Fig. 1.

Nicotinic acid (niacin) is an odorless, white compound which forms needle-like crystals. It is nonhygroscopic and stable in the dry state. It may be autoclaved at 120°C for 20 minutes without destruction and is stable to heating with 1*N* mineral acids and alkali. At 25°C, 1 gram of nicotinic acid will dissolve in 60 ml of water or 80 ml of ethanol; it is more soluble in dilute acid or alkali since it is amphoteric and forms quaternary ammonium or carboxylate salts. It is very soluble in hot water or ethanol, but insoluble in ether.

Nicotinamide (niacinamide) crystallizes from benzene as colorless needles which are soluble in water (1 g/ml) and in ethanol. Unlike nicotinic acid it is also soluble in ether. It is stable in the dry state below 50°C and in aqueous solution up to 120°C for 20 minutes. Heating in the presence of acid or alkali converts it to the acid.

Nicotinic acid and nicotinamide can be estimated colorimetrically through the use of the König reaction in which the pyridine ring is reacted with cyanogen bromide and an aromatic amine to form a yellow color. However, it is somewhat difficult to prepare extracts of biological materials free of interfering substances; hence this chemical method has not been widely used for routine analysis.

Biological assays using either dogs or chicks have been used to measure the niacin activity of foods. However, synthesis of niacin by the test organism from tryptophan in these foods complicates the assay and makes it unreliable.

Microbiological assays of nicotinic acid and nicotinamide have been used successfully. Using *Lactobacillus arabinosus* as the test organism and measuring growth by turbidity or lactic acid production permit the estimation of quantities as small as 0.02 μg/ml medium. This method measures nicotinic acid, its amide, and various combined forms of both, such as nucleosides, nucleotides and the dinucleotides.

Nicotinic acid and related compounds present in urine, blood and other tissues can be measured spectrophotometrically at 260 mμ after removal of protein and separation on "Dowex"-1-formate columns. Paper CHROMATOGRAPHY and chromatography on DEAE-cellulose sheets have been useful techniques for separating small amounts of nicotinic acid and its derivatives.

Nicotinic acid is widely distributed in nature, occurring in all living cells. It is found predominantly in animal tissues in its coenzyme forms, the pyridine nucleotides NAD and NADP (see NICOTINAMIDE ADENINE DINUCLEOTIDES). It can be freed from its bound forms by autoclaving in 1*N* acid. Some plant materials contain biological forms of nicotinic acid which are unavailable to the animal but which may be freed by this treatment. Thus, assay for free nicotinic acid after this treatment may indicate quantities larger than those available to the animal.

The concentration of free nicotinic acid and nicotinamide in fasting human plasma is approximately 0.15 μg/ml; total nicotinic acid and nicotinamide in rat liver is about 175 μg/g fresh tissue. The concentration of pyridine nucleotides in whole blood is about 35 μg/ml, most of this being in the erythrocytes, while the NAD content of human muscle is around 380 μg/g of fresh tissue. Nicotinamide is the predominant free form in animal tissues whereas nicotinic acid is the predominant form in plants. The concentration in leaves ranges from 10–70 μg/g dry weight.

Microorganisms contain relatively large amounts of nicotinic acid and related compounds with concentrations ranging from 100–600 μg/g dry cells. In general, yeasts are higher than most bacteria and molds which have been examined. Some microorganisms, including *Mycobacterium tuberculosis*, seem to produce large quantities of nicotinic acid.

The administration of 50–100 mg of nicotinic acid to a human subject results in a flushing of the face, neck and arms (nitroid reaction). This vasodilation which lasts for an hour or so appears to be a localized phenomenon and may be accompanied by itching, burning, nausea and dizziness. The administration of 30 mg of niacin per day orally (10 mg i.v. or 60 mg i.m.) increases peripheral blood flow and results in an elevated skin temperature. In this case, niacin appears to be acting like histamine and the effect can be counteracted by epinephrine. In contrast, a 500-mg dose of nicotinamide is without visible effect in a human subject.

Dogs, rats and chicks tolerate 2 g/kg of nicotinic acid daily for several months without toxic effects. The LD_{50} of nicotinic acid given subcutaneously to rats and mice is 4–5 g/kg while the LD_{50} for oral administration is 5–7 g/kg. Nicotinamide is approximately twice as toxic as nicotinic acid. Administration is followed by a generalized paralysis, depressed respiration, cyanosis, severe clonic convulsions, and death within 12–36 hours. It may be that the toxicity of nicotinamide reflects the drain of the tissue methyl group supply in attempts to methylate the compound before excretion.

Biosynthesis. Several methods for the chemical synthesis of nicotinic acid have been used; at present, it is synthesized commercially by the hydrolysis of 3-cyanopyridine or by the oxidation of quinoline. The commercial synthesis of nicotinamide is carried out through partial hydrolysis of 3-cyanopyridine in the presence of a quaternary ammonium hydroxide resin.

Niacin is unique among the B vitamins in that it can be synthesized in animal tissues from the amino acid TRYPTOPHAN. Other routes of synthesis exist in plants and certain microorganisms.

The major pathway of tryptophan metabolism in the animal involves its oxidation to kynurenine and subsequent hydroxylation of the benzene ring and cleavage of the 3-carbon side chain to form alanine and 3-hydroxyanthranilic acid. This compound can be further oxidized to carbon dioxide and water or it can be converted to quinolinic acid as shown in Fig. 2. Quinolinic acid

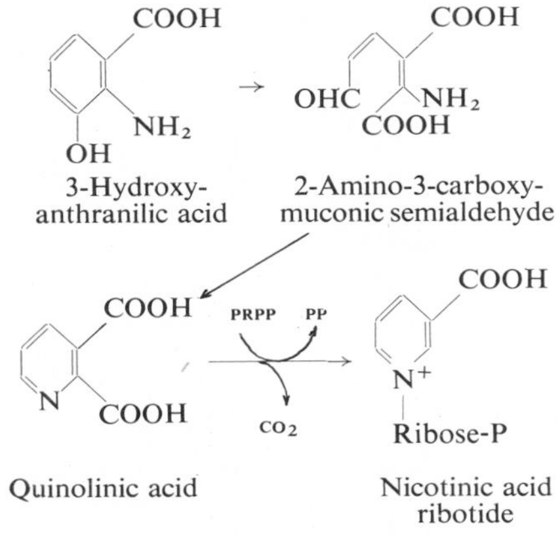

COOH, NH_2, OH — 3-Hydroxyanthranilic acid → COOH, OHC, NH_2, COOH — 2-Amino-3-carboxymuconic semialdehyde

COOH, N, COOH — Quinolinic acid —(PRPP → PP; CO_2)→ COOH, N^+, Ribose-P — Nicotinic acid ribotide

Fig. 2.

condenses with 5-phosphoribosyl-1-pyrophosphate (PRPP) with the simultaneous evolution of carbon dioxide to yield nicotinic acid ribotide. The ribotide can be hydrolyzed to yield free nicotinic acid or it can be converted to NAD or NADP, the coenzyme forms of niacin.

Animals vary considerably in their ability to convert tryptophan to niacin. The intermediate in the conversion of 3-hydroxyanthranilic acid to quinolinic acid (shown in Fig. 2) can be decarboxylated and further oxidized to CO_2 and H_2O. Some animals, such as the cat, decarboxylate all of the 2-amino-3-carboxymuconic semialdehyde formed from tryptophan and do not form quinolinic acid.[1] The dog forms only limited amounts of quinolinic acid.

In animal tissues, nicotinic acid ribotide (shown in Fig. 2) arises from the reaction of either quinolinic acid or nicotinic acid with PRPP. Nicotinic acid ribotide reacts with ATP to yield nicotinic acid-adenine dinucleotide which is aminated to yield NAD. NAD can be phosphorylated to yield NADP.

The tryptophan-to-niacin pathway is operative in *Xanthomonas pruni* and *Neurospora crassa*, but other microorganisms which have been examined appear to synthesize nicotinic acid by another pathway. Quinolinic acid appears to be a precursor of nicotinic acid ribotide in *E. coli*.[2] In *Mycobacterium tuberculosis*, asparagine, aspartic acid, glutamic acid and glycine are reported to serve as sources for the ring nitrogen of nicotinic acid, and synthesis apparently proceeds through quinolinic acid.[3] However, the route of synthesis for nicotinic acid in most microorganisms has not been established.

In most plants which have been examined, tryptophan does not serve as a precursor for nicotinic acid. The best evidence is that plant synthesis of nicotinic acid proceeds through the condensation of 3- and 4-carbon units such as succinate and glycerol to yield quinolinic acid which is converted to nicotinic acid ribotide.[4]

Metabolism and Function. Niacin functions in metabolism as an integral part of the pyridine nucleotides NAD and NADP. These coenzymes act as hydrogen donors and acceptors for BIOLOGICAL OXIDATION-REDUCTION reactions in the metabolism of carbohydrates, amino acids and lipids.

Animal tissues do not retain large amounts of niacin. Excessive amounts are excreted in the urine, in man chiefly as the methylated compounds N^1-methylnicotinamide (N^1-Me) and the 6-pyridone of N^1-Me. Small amounts of free nicotinic acid and nicotinamide are found in human urine, and ingestion of large doses of nicotinic acid results in excretion of nicotinuric acid, the glycine conjugate of nicotinic acid. Dogs, rats, cats and pigs, like man, excrete niacin as N^1-Me; however, the herbivora, rabbits, guinea pigs, sheep and goats excrete other forms. Chickens excrete ornithine derivatives of nicotinic acid. Radioactive carbon dioxide results from the administration of ^{14}C-niacin to mice, rats and dogs. Very little radioactivity is found in the feces of these animals.

Requirements for Niacin and Results of Deficiency. When assessing the dietary requirement for niacin, it is necessary to take into consideration the conversion of tryptophan to niacin. Approximately 60 mg of dietary tryptophan will result in the formation of 1 mg of niacin in human subjects under normal conditions. Therefore, each 60 mg of tryptophan in the diet is recorded as 1 mg of niacin, and the resulting values are called niacin equivalents.

The human requirement for niacin appears to be related to the caloric content of the diet with approximately 4.4 mg of niacin required per 1000 calories. Dextrin in the place of sucrose in the diet of rats seems to decrease the niacin requirement. It may be that dextrin promotes intestinal synthesis, but it is more likely that the fructose portion of sucrose somehow increases the niacin requirement.

The National Research Council's recommended allowance for the adult man is 19 mg and for the adult woman 17 mg of niacin per day. This allowance is increased for pregnancy and lactation. The rat requires approximately 1.5 mg of niacin per 1000 grams of diet, whereas the dog requires 5 mg/100 g of diet or 0.13–0.23 mg/kg body weight per day.

A lack of niacin and tryptophan in the diet results in the disease known as pellagra in man and blacktongue in the dog. *Pellagra* is characterized by dermatitis, glossitis, diarrhea and mental depression. Often the cases of pellagra which are seen in the clinic are complicated by mild deficiencies of other B vitamins. Blacktongue is characterized by a diffuse inflammation of the gums, inner surfaces of the lips, cheeks and areas under the tongue. The tip and margin of the tongue become red, and later dark-bluish patches appear. The mucosa of the entire gastrointestinal tract becomes inflamed.

The incidence of pellagra seems to be much higher in corn-eating populations than in other areas. It was quite prevalent among low-income groups in the south-eastern United States before 1940. Corn contains some niacin, but it is notably low in tryptophan. Pellagra probably results both from the low niacin and tryptophan content of the corn and from other effects of the poor-quality corn protein.

References

1. Ikeda, M., Tsuji, H., Nakamura, S., Ichiyama, A., Nishizuka, Y., and Hayaishi, O., *J. Biol. Chem.*, **240**, 1395 (1965).
2. Andreoli, A. J., Ikeda, M., Nishizuka, Y., and Hayaishi, O., *Biochem. Biophys. Res. Commun.*, **12**, 92 (1963).
3. Konno, K., Oizumi, K., and Oka, S., *Nature*, **205**, 874 (1965).
4. Hadwiger, L. A., Badiei, S. E., Waller, G. R., and Gholson, R. K., *Biochem. Biophys. Res. Commun.*, **13**, 466 (1963).
5. Williams, R. J., Eakin, R. E., Beerstecher, E., Jr. and Shive, W., "The Biochemistry of B Vitamins," New York, Reinhold Publishing Corp., 1950.
6. Hundley, J. M., in "The Vitamins" (Sebrell, W. H., Jr., and Harris, R. S., Editors), Vol. II, New York, Academic Press, 1954.
7. Krehl, W. A., "Vitamins and Hormones, VII," New York, Academic Press, 1949.
8. Goldsmith, G. A., in "Nutrition" (Beaton, G. H., and McHenry, E. W., Editors), New York, Academic Press, 1964.
9. Snell, E. E., in "The Vitamins" (Sebrell, W. H., Jr., and Harris, R. S., Editors), Vol. II, New York, Academic Press, 1954.

Patricia Swan and L. M. Henderson

NINHYDRIN

Ninhydrin (triketohydrindene hydrate),

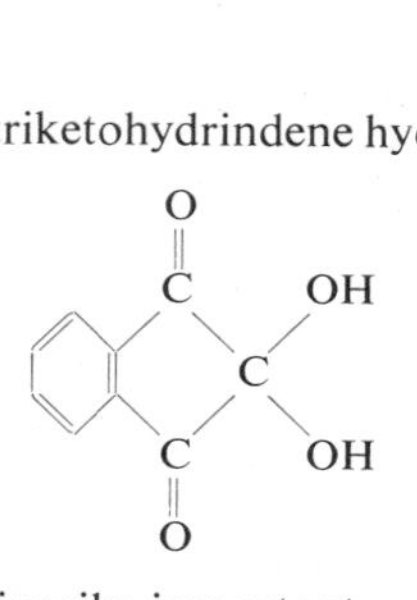

[Structure: benzene ring fused to five-membered ring C(=O)–C(OH)(OH)–C(=O)]

is an extraordinarily important reagent used in the analytical determination of amino acids and related substances.

Typical protein-derived α-AMINO ACIDS are oxidized by ninhydrin quantitatively to produce carbon dioxide, the corresponding aldehyde (*e.g.*, acetaldehyde from alanine), and ammonia. The ammonia formed condenses with unchanged and partially reduced ninhydrin to yield an intensely blue pigment which is formulated as follows:

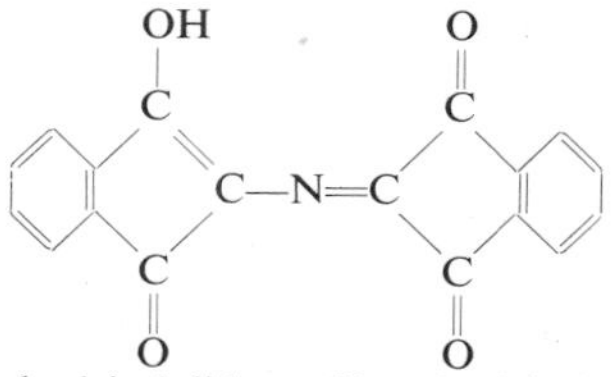

[Structure: two indane-dione units joined by C—N=C; left unit bearing OH and C=O, right unit bearing two C=O]

Diketohydrindylidene-diketohydrindamine

The carbon dioxide, the aldehyde, or the blue pigment produced can be used in the quantitative determination of amino acids or peptides. Of the three, the color reaction is most often used.

NITRATE REDUCTION

Many bacteria and some fungi are able to reduced nitrate to nitrite; some bacteria can reduce nitrate to ammonia or, more commonly, to nitrogen gas (N_2). The element molybdenum is a component of nitrate reductase and is essential for nitrate assimilation [see MOLYBDENUM (IN BIOLOGICAL SYSTEMS)]. Nitrate reductase (at least from some sources) is a molybdoflavoprotein (see FLAVINS AND FLAVOPROTEINS).

Reference

1. Stanier, R. Y., Doudoroff, M., and Adelberg, E. A., "The Principal Types of Microbial Metabolism," in "The Microbial World," Ch. 12, Englewood Cliffs, N.J., Prentice-Hall, 1963.

NITROGEN FIXATION

A positive balance of usable nitrogen on earth depends completely on nitrogen fixation, the process by which atmospheric nitrogen (N_2) is converted by either biological or chemical means to a form of nitrogen (ammonia, NH_3) available for use by plants and other biological agents that are unable to fix their own nitrogen. Of the two processes for converting N_2 to NH_3, the biological one is by far the most significant as far as total amount of N_2 fixed and ease of fixation is concerned. For example, in biological nitrogen fixation, microorganisms either free-living or in symbiosis with plants (mainly in root nodules) reduce N_2 to NH_3 at atmospheric pressure (0.8 atm N_2) and temperatures of 20–37°C compared with the man-conceived chemical process that requires up to 300 atm pressure and temperatures in the range of 200–300°C. The biological process is truly unique.

The importance of biological nitrogen fixation to soil fertility has been known since the late 1800's,[1] and much effort has been expended since that time to obtain chemical knowledge of the process. However, the first real breakthrough toward this goal did not come until 1960 when it was shown that cell-free extracts of the anaerobic bacterium *Clostridium pasteurianum* could be made to fix nitrogen if molecular oxygen (O_2) were rigorously excluded and if pyruvic acid, a source of energy and electrons, was supplied.[2] This finding showed that studies were no longer restricted to whole cells as had been previously indicated and that it should be possible to isolate and chemically identify the components of the nitrogen-fixing system.

The first demonstrable product of cell-free N_2 fixation is ammonia as had been strongly suggested by previous whole-cell studies. Since the reduction of N_2 to $2NH_3$ requires six electrons and since most electron transfer systems known in biochemical pathways involve either a one- or a two-electron transfer, one would expect either six one-electron or three two-electron transfer steps to be involved in nitrogen fixation. This would suggest the existence of nitrogen compounds of valence states (reduction states) intermediate between N_2 and NH_3. No such intermediates have been found even in the systems using cell-free extracts.

Because of the failure to detect intermediates which, if found, would have allowed one to predict the pathway of N_2 fixation, attention was next placed on the mechanism in extracts of *Clostridium pasteurianum* through which electrons were transferred from pyruvic acid to the nitrogen-fixing system. These studies led to the discovery and isolation of the new electron carrier ferredoxin (Fd) which functioned by accepting electrons released during pyruvate oxidation by enzymes present in these clostridial extracts.[3] The electrons from reduced Fd were transferred to a variety of different acceptors as directed by the cell. For example, some of the electrons from reduced Fd were transferred to hydrogenase, an enzyme which combined the electrons with protons (H^+) to produce molecular hydrogen (H_2), a major by-product of this anaerobe. Other electrons from reduced Fd were transferred via a FLAVOPROTEIN carrier to NICOTINAMIDE ADENINE DINUCLEOTIDE phosphate ($NADP^+$) to yield NADPH, a reduced electron carrier shown to be important in the metabolism of all biological agents. It was soon found[4] that electrons from Fd were also required for nitrogen fixation when pyruvate was present as supporting substrate.

A finding of major importance to this discussion was that H_2, through hydrogenase, would act as an electron source for reducing ferredoxin. Thus, in these extracts H_2 could be used to reduce $NADP^+$ to NADPH and NO_2^- to NH_3, and Fd was necessary as an intermediary electron carrier. This is summarized as follows:

Pyruvate —"enzymes"→ Reduced Fd ←Hydrogenase→ H_2

NH_3 $\xleftarrow{NO_2^-}$ Reduced Fd $\xrightleftharpoons{\text{"NADP}^+\text{ reductase"}}$ NADPH

Since Fd is required for pyruvate-supported N_2 fixation, one might expect H_2 to support nitrogen fixation since reduced Fd is readily produced from H_2 in these extracts. Molecular H_2 alone did not support N_2 fixation. This suggested either that a component other than reduced Fd was required, or that H_2, although capable of reducing Fd, was inhibitory to N_2 fixation as previous whole-cell studies had indicated. If an additional component was required, it obviously was produced from pyruvic acid since pyruvic acid supported active N_2 fixation. The products of pyruvate metabolism in *C. pasteurianum* are as follows:

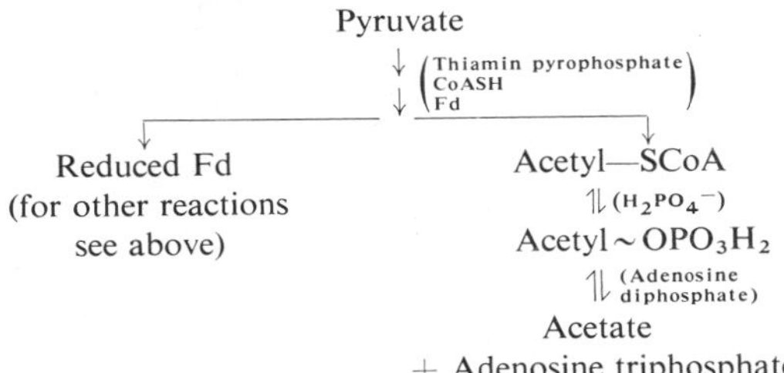

Several attempts were made, without success, to obtain N_2 fixation in extracts to which H_2, N_2 and one of the other products of pyruvate metabolism, ATP, were added. Active N_2 fixation did occur, however, when another product of pyruvate metabolism, acetyl phosphate, was added in addition to H_2 and N_2.[5] When compounds such as ADP were removed from cell extracts by dialysis, no N_2 fixation occurred unless ADP was added together with phosphate, H_2 and N_2. Acetyl phosphate then was acting as a source of ATP. The reason ATP did not work directly was that a continuous supply of ATP was required, and a high concentration of ATP, if added directly to a cell-free extract, was highly inhibitory to N_2 fixation. In whole cells that are fixing N_2, a continuous supply of ATP is made available during sugar metabolism.

Recent research has been directed at determining what function ATP plays in the nitrogen-fixing process. The utilization of ATP has been found to be correlated with the nitrogen-fixing process. For example, both ATP utilization and nitrogen fixation require reduced Fd, and N_2 fixation requires ATP. The utilization of ATP does not require N_2, but it does require the continuous presence of reduced Fd. This suggested that the electrons from reduced Fd were transferred either directly to nitrogenase, the enzyme responsible for activating N_2, or to another protein electron carrier where they were activated by ATP. These highly active electrons would then reduce N_2 if it was available or would combine with protons to form H_2 if N_2 was not available, and probably to some extent even when it was available. This conclusion is supported by the recent discovery that a particle preparation from the aerobic microorganism, *Azotobacter vinelandii*, when supplied with an ATP source and a source of reduced Fd or dithionite ($S_2O_4^{=}$)

will actively fix nitrogen.[6] *Azotobacter*, unlike the clostridia, does not normally evolve H_2 in its metabolism, but it does have an active hydrogenase. When the azotobacter particles were incubated in the presence of ATP and $S_2O_4^=$ in the absence of N_2, H_2 evolution occurred. Thus, in both clostridial and azotobacter extracts, there appears to be an ATP-dependent evolution of hydrogen from a reduced component involved in N_2 fixation. Two possible mechanisms for N_2 fixation are summarized below:

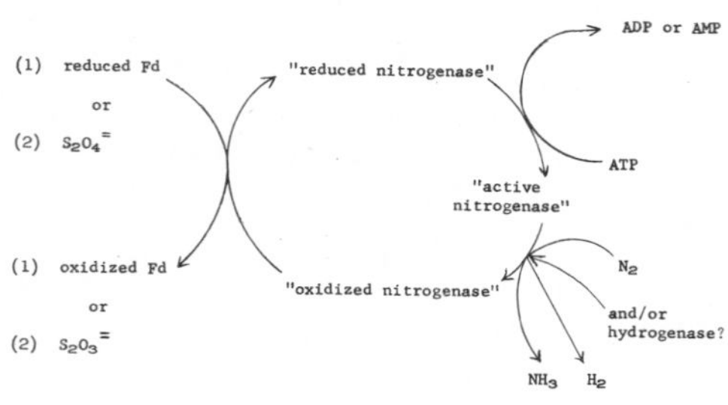

OR

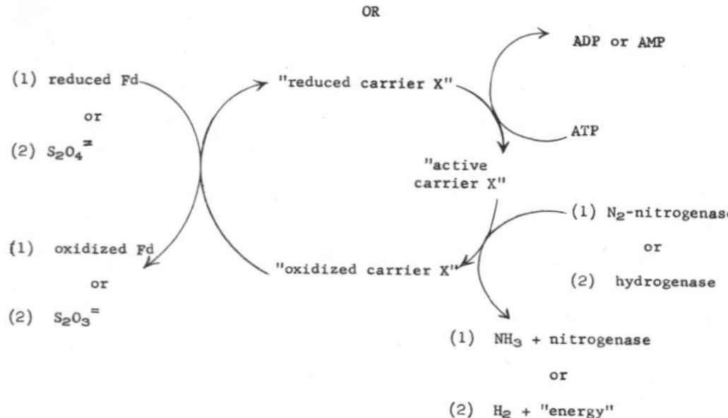

References

1. WILSON, P. W., in "Handbuch der Pflanzenphysiologie" (RUHLAND, W., EDITOR), Berlin, Springer, 1958, vol. 8, p. 9.
2. CARNAHAN, J. E., MORTENSON, L. E., MOWER, H. F. AND CASTLE, J. E., *Biochim. Biophys. Acta*, **44**, 520 (1960).
3. MORTENSON, L. E., *Ann. Rev. Microbiol.*, **17**, 115 (1963).
4. MORTENSON, L. E., *Biochim. Biophys. Acta*, **81**, 473 (1964).
5. MORTENSON, L. E., *Proc. Natl. Acad. Sci. U.S.*, **52**, 272 (1964).
6. BULEN, W. A., BURNS, R. C., AND LECOMTE, J. R., *Biochem. Biophys. Res. Commun.*, **17**, 265 (1964).

L. E. MORTENSON

NITROPRUSSIDE TEST

When CYSTEINE, or peptides or PROTEINS containing free sulfhydryl (SH) groups, are treated with a solution of sodium nitroprusside, $Na_2Fe(CN)_5NO$, a reddish color is produced. Many proteins may react negatively, but after denaturation they react positively, indicating that free SH groups have been formed by the denaturation process. Reducing agents convert cystine, and peptides containing it, into substances that react positively with sodium nitroprusside.

NON-SAPONIFIABLE FRACTION

Non-saponifiable fraction is descriptive of the LIPID material (containing sterols, hydrocarbons, etc.) which remains fat-soluble after a crude fat-containing mixture has been digested with alkali.

When certain vitamins were found to be associated with fats, one of the early questions which had to be answered in each case was whether the biological activity resided in the fat itself. Saponification of the fat and rejection of the water-soluble soaps and glycerol showed that the vitamins in question were to be found in the non-saponifiable residue. Thus VITAMIN A and VITAMIN D, for example, are in the non-saponifiable residue of cod liver oil. VITAMIN E is in the non-saponifiable residue of wheat germ oil. Any tissue will yield at least a small amount of non-saponifiable lipid material.

NORTHRUP, J. H.

John Howard Northrup (1891–) is an American biochemist who first succeeded in obtaining a number of well-known enzymes—pepsin, trypsin and chymotrypsin—in crystalline form. He shared the Nobel Prize in chemistry with JAMES BATCHELLER SUMNER and WENDELL MEREDITH STANLEY in 1946.

NUCLEAR MAGNETIC RESONANCE

The term nuclear magnetic resonance in the chemical and biological sciences usually refers to high-resolution nuclear magnetic resonance to which the following discussion is limited. When a sample of a substance that contains nuclei with a magnetic moment is placed in a very homogeneous magnetic field, the substance can absorb electromagnetic radiation of appropriate frequency. This absorption of electromagnetic radiation is called nuclear magnetic resonance. Typically, a substance containing a "proton" (*i.e.*, a hydrogen atom) in a magnetic field of 10,000 gauss would absorb radiation at about 42.576 megacycles. A larger magnetic field would require a proportionally higher frequency of radiation. A spectrum is produced when the absorption of radiation is plotted as a function of either the magnetic field strength or, less typically, the frequency of irradiation. In the case of protons, the nuclear magnetic resonance (NMR) spectrum for common compounds will extend over a range of no more than 20 ppm of the total applied magnetic field. Typical resolution for protons is a few parts in one hundred million, thus the term high resolution.

There are three basic types of data obtainable from an NMR spectrum that are of use to the chemist in elucidating structure, detecting changes in molecular environment, and so forth. These basic types of data are (a) *the chemical shift*, (b) *the intensity of the absorption bands* and (c) *the coupling constants*.

Chemical shift refers to the observation that similar nuclei in different chemical environments (and thus different magnetic environments) absorb radiation at slightly different magnetic field strengths. Thus, for example, the chemical shift gives us the extremely useful knowledge of how many chemically different types of protons there are in a particular molecule. Consider ethanol in which there are three distinct types of protons, the hydroxyl proton, the methylene protons and the methyl protons. The NMR spectrum of ethanol thus consists of three bands. A trivial example of how this type of information might be used is seen in the isomeric propanols. Isopropyl alcohol has three chemically distinct types of protons and thus exhibits three bands, while *n*-propyl alcohol has four types of protons and hence has four bands.

The *intensity* of a particular band in an NMR spectrum is directly proportional to the number of nuclei represented by that band. In the example of ethanol we find that the areas of the three bands are in the ratio of 1:2:3 corresponding to one hydroxyl proton, two methylene protons, and three methyl protons. The usefulness of relative area data is seen in the case of the isomeric chloronitroethanes. Both 1-chloro-1-nitroethane and 2-chloro-1-nitroethane have a two-band spectra; however, the ratios of band areas are 1:3 and 1:1, respectively.

If we look at the spectrum of ethanol closely, we find that there is fine structure in some of the bands. The band of area three is a triplet, the band of area two is a quartet, while the band of area one is a singlet. The multiplicity of these bands is due to *spin-spin interactions* of the protons. The distance between the individual adjacent components of a multiplet is proportional to the magnitude of the spin-spin interaction and is referred to as the *coupling constant*. A single proton can exist in either of two spin states, parallel or antiparallel. In the case of two identical protons (as the methylene of ethanol), there exist three possible combinations of spin states, parallel-parallel, parallel-antiparallel, and antiparallel-antiparallel. Thus neighboring protons (such as the methyl group of ethanol) will be split into a triplet. The three methyl protons of ethanol can be arranged into four possible combined spin states thus the methylene is split into a quartet. In general, a group of N equivalent protons will split the band due to its neighbor into a $N + 1$ multiplet. Consider patterns due to spin coupling in the isomeric chloronitroethanes. The band of 1-chloro-1-nitroethane with area one will be split into a quartet ($N + 1 = 4$) due to the three β protons, and the band of area three will be split into a doublet ($N + 1 = 2$) due to the single α proton. The spectrum of 2-chloro-1-nitroethane has two bands of equal intensity each of which is a triplet due to the adjacent two equivalent protons. Thus the couplings support the interpretation based on area; and the chemical shifts (by comparison with appropriate correlation charts) also distinguish between these two isomers. This triple check of conclusions about structure gives NMR an inherent reliability that is not generally found in any other analytical method. In molecules with more different types of protons, the splitting may get very complex; however, with the aid of a computer these can sometimes be analyzed. An analysis of spin coupling gives information about the number of nearest neighbors, their equivalence or non-equivalence, and quite often geometrical parameters. For example, the magnitude of coupling between two vicinal protons is a function of the dihedral angle (*i.e.*, the angle between the two carbon-hydrogen vectors).

The vast majority of work to date has been with proton resonance; however, work with other nuclei such as C^{13} or P^{31} shows great promise for the future.

References

Introductory

1. Roberts, J. D., "Nuclear Magnetic Resonance. Applications to Organic Chemistry," New York, McGraw-Hill Book Co., 1959.
2. Dyer, J. R., "Applications of Absorption Spectroscopy of Organic Compounds," Ch. 4, Englewood Cliffs, Prentice-Hall, 1965.

Practical

3. Jackman, L. M., "Applications of Nuclear Magnetic Resonance Spectroscopy in Organic Chemistry," New York, Pergamon Press, 1959.
4. Bhacca, N. S., and Williams, D. H., "Applications of NMR Spectroscopy in Organic Chemistry, Illustrations from the Steroid Field," San Francisco, Holden-Day Inc., 1964.
5. Bhacca, N. S., Johnson, L. F., and Shoolery, J. N., "NMR Spectra Catalog," Vol. I, Palo Alto, Varian Associates, 1962.
6. Bhacca, N. S., Hollis, D. P., Johnson, L. F., and Pier, E. A., "NMR Spectra Catalog," Vol. II, Palo Alto, Varian Associates, 1963.

Theoretical

7. Pople, J. A., Schneider, W. G., and Bernstein, H. J., "High-Resolution Nuclear Magnetic Resonance," New York, McGraw-Hill Book Co., 1959.
8. Roberts, J. D., "An Introduction to Spin-Spin Splitting in High-Resolution Nuclear Magnetic Resonance Spectra," New York, W. A. Benjamin, 1962.

Ben A. Shoulders

NUCLEASES

During the past decade, it has become apparent that a great variety of nucleases exist in nature, many with a high degree of specificity for various features of both the primary and secondary structure of polynucleotides. Perhaps the empirical finding which best emphasizes the importance of these enzymes is that all cells which have been examined carefully contain at least one, and often several, distinct and physically separable nucleases. In reviewing briefly the nucleases which have been described, only those will be considered which have been obtained in reasonably purified form, and even here no attempt will be made to be comprehensive. The emphasis will be on enzymatic specificity.

Exonucleases *versus* Endonucleases. A useful general classification is that introduced by Laskowski which divides nucleases into two categories depending on their mode of attack: (1) endonucleases, which attack polynucleotides at many points within the chain generally producing only a small proportion of mononucleotides; (2) exonucleases, which catalyze a stepwise attack producing exclusively mononucleotides. An exonuclease may initiate its attack at either the 3′- or 5′-hydroxyl end of a polynucleotide (or oligonucleotide). Such enzymes generally require that a free hydroxyl group be available at the site at which attack begins. Substitution of the hydroxyl group with, for example, a phosphoryl or acyl group renders a normally susceptible substrate resistant to enzymatic attack.

Sugar Specificity. Nucleases may show specificity for the pentose moiety of the polynucleotide chain. One category is specific for polynucleotides containing ribose. Among these enzymes (the *ribonucleases*) are pancreatic ribonuclease (and the analogous enzymes from liver and spleen), ribonucleases from plant sources, including pea leaf, tobacco leaf, and rye grass, and from bacterial sources, including *B. subtilis*, *E. coli* and *Aspergillus oryzae* (Takadiastase). The high degree of specificity of these enzymes for polynucleotides containing ribose is readily accounted for by the discovery of Brown, Todd and co-workers that cleavage of internucleotide bonds by pancreatic ribonuclease proceeds first by a very rapid intramolecular transphosphorylation to form the cyclic 2′,3′-phosphate ester followed by a considerably slower hydrolysis of the cyclic ester to form the 3′-phosphomonoester. In one instance (the leaf enzyme), pyrimidine cyclic 2′,3′-phosphates are actually resistant to further hydrolysis. A mechanism involving the obligatory participation of a 2′,3′-cyclic phosphate intermediate ensures the specificity of these nucleases for polyribonucleotides. The ribonucleases cited above are all endonucleases.

A second group of enzymes, the *deoxyribonucleases*, are highly specific for polynucleotides containing 2-deoxyribose. Such nucleases include pancreatic deoxyribonuclease, calf spleen and calf thymus deoxyribonucleases, derived from animal sources, *E. coli* endonuclease I, streptococcal deoxyribonuclease ("streptodornase"), and *E. coli*, exonucleases I, II, and III, from bacteria. The pancreatic, spleen and thymus enzymes as well as *E. coli* endonuclease I are all endonucleases. The other enzymes cited are exonucleases.

A sizable number of enzymes show no specificity with respect to the pentose moiety of the polynucleotide substrate. Enzymes which fall into this category are SNAKE VENOM and spleen phosphodiesterases, both exonucleases, and the nucleases from *Micrococcus pyogenes* (micrococcal nuclease), *Azotobacter agile* and *Neurospora crassa* which are primarily endonucleases.

3′- or 5′-specificity. Nucleases may yield fragments terminated by either 3′- or 5′-phosphomonoester groups. As noted above, ribonucleases invariably yield products with 3′-phosphate groups because of the participation of cyclic 2′,3′-phosphate esters in the process of hydrolysis. In the case of the deoxyribonucleases or enzymes which are nonspecific with respect to the sugar moiety, both 3′- and 5′-phosphate-terminated products may be formed. For example, pancreatic deoxyribonuclease, *E. coli* endonuclease I and "streptodornase" yield oligonucleotides with 5′-phosphoryl groups; the spleen and thymus deoxyribonucleases as well as the micrococcal nuclease form products terminated with 3′-phosphoryl groups. Among the exonucleases, *E. coil* exonucleases I, II, and III and venom phosphodiesterase form nucleoside-5′-phosphates, while spleen phosphodiesterase and *Lactobacillus acidophilus* phosphodiesterase form nucleoside-3′-phosphates. It is of interest that the exonucleases which produce nucleoside-3′-phosphates start their attack at the 5′-hydroxyl end of the polynucleotide; those which yield nucleoside-5′-phosphate start at the 3′-hydroxyl end. As noted above, the exonucleases require a free 3′- or 5′-hydroxyl group depending upon their initial site of attack.

Base Specificity. Of the nucleases which have been described, only two possess an "absolute" base specificity, *i.e.*, they will cleave phosphodiester bonds involving only one nucleic acid base or type of base. The first of these is bovine pancreatic ribonuclease and its counterpart which has been purified from spleen. This enzyme is specific for esters of pyrimidine nucleoside-3′-phosphates; the corresponding purine nucleoside esters are totally resistant. As a result, hydrolysis of ribonucleic acid with pancreatic ribonuclease leads to the formation initially of 2′,3′-cyclic pyrimidine mononucleotides and oligonucleotides composed of runs of purine residues terminated by a pyrimidine 2′,3′-phosphate and, finally, of pyrimidine nucleoside-3′-phosphates and oligonucleotides with terminal pyrimidine nucleoside-3′-phosphates. The capacity of this enzyme to cleave pyrimidine cyclic 2′,3′-phosphates to yield the corresponding nucleoside-3′-phosphate and the spectral shift which accompanies the cleavage form the basis of a precise and relatively specific assay for the enzyme. Pancreatic ribonuclease which is available in crystalline form is the most thoroughly characterized of the nucleases described to date. Its complete amino acid sequence is known and is the subject of much intensive investigation aimed at establishing its active site and mechanism of action. It is a small protein (molecular weight—13,700) with an alkaline isoelectric point; it is stable to boiling at acid pH. Heat stability and relatively low molecular weight are common characteristics of all the ribonucleases which have been characterized physically.

The second base-specific nuclease is the Takadiastase (T_1) ribonuclease, whose specificity is even more highly restricted than pancreatic ribonuclease. The T_1 ribonuclease will attack esters of guanosine-3′-phosphate; all other internucleotide bonds are resistant. Thus, the products of digestion of RNA consist of guanosine-3′-phosphate and oligonucleotides terminated by guanosine-3′-phosphate. As in the case of the pancreatic enzyme, the corresponding cyclic esters appear as intermediates in the course of hydrolysis. A similar enzyme has been found in the smut fungus *Ustilago sphaerogena.*

Both of the base-specific nucleases are ribonucleases. Enzymes of comparable base specificity which attack polydeoxyribonucleotides have not been discovered to date.

Many nucleases display a "partial" base specificity, *i.e.*, certain diester bonds within a polynucleotide are hydrolyzed at a relatively greater rate than others. The cases of "partial" base specificity are too numerous to cite and, in fact, some partial specificity can be attributed to all nucleases thus far examined.

Structural Specificity. By structural specificity is meant a specificity directed toward the secondary structure of a polynucleotide. This type of specificity was first encountered with enzymes attacking DEOXYRIBONUCLEIC ACID, which has a well-defined secondary structure (*i.e.*, the double helix, as described by Watson, Crick and Wilkins).

At this writing, only two enzymes are known which possess an absolute structural specificity. *E. coli* exonuclease I is completely specific for polydeoxyribonucleotides lacking an ordered structure. This enzyme will attack native deoxyribonucleic acid at 1/40,000th the rate of denatured deoxyribonucleic acid. The low level of activity observed in the former case most probably results from the action of the enzymes on denatured, single-stranded ends within the double-helical molecule. *E. coli* exonuclease III, on the other hand, displays a rigid specificity for polydeoxyribonucleotides with an ordered structure. The limited activity observed with denatured deoxyribonucleic acid most probably reflects the activity of the enzyme on small segments of the molecule which have undergone some fortuitous renaturation.

Almost all other nucleases which have been examined show some "partial" structural specificity. Venom and spleen phosphodiesterases, pancreatic ribonuclease, and the micrococcal nuclease attack single-stranded polynucleotides at rates up to 100 times that found with polynucleotides that have assumed an ordered conformation. These enzymes will, however, degrade both types of substrate to completion. On the other hand, pancreatic deoxyribonuclease, *E. coli* endonuclease I, and spleen endonuclease show a distinct preference for polynucleotides with some ordered conformation.

In addition to structural specificity, there are marked effects of polynucleotide size on enzymatic rate. In the case of pancreatic deoxyribonuclease, as well as with other endonucleases, the rate at which diester bonds with a polynucleotide are hydrolyzed diminishes rapidly as the secondary structure is disrupted, and the size of the oligonucleotide fragments produced as a result of nuclease action diminishes.

"Absolute" size specificity has been observed in the case of pancreatic deoxyribonuclease which is unable to attack trinucleotides, and *E. coli* exonuclease I and micrococcal nuclease, both of which are unable to attack dinucleotides.

References

1. KHORANA, H. G., in "The Enzymes" (BOYER, P., LARDY, H., AND MYRBACK, K., EDITORS), Vol. 5, p. 79, New York, Academic Press, 1961.
2. ANFINSEN, C. B., AND WHITE, F. H., JR., *ibid.*, p. 95.
3. LASKOWSKI, M., SR., *ibid.*, p. 123.
4. LEHMAN, I. R., in "Progress in Nucleic Acid Research" (DAVIDSON, J. N., AND COHN, W. E., EDITORS), Vol 2, p. 83, New York, Academic Press, 1963.

I. R. LEHMAN

NUCLEIC ACIDS. See DEOXYRIBONUCLEIC ACIDS (DISTRIBUTION AND STRUCTURE); DEOXYRIBONUCLEIC ACIDS (REPLICATION); NUCLEOPROTEINS; RIBONUCLEIC ACIDS (BIOSYNTHESIS); RIBONUCLEIC ACIDS (STRUCTURE AND FUNCTIONS); SOLUBLE RIBONUCLEIC ACIDS. DNA is a commonly used abbreviation for deoxyribonucleic acids; RNA, for ribonucleic acids. The biological roles of nucleic acids are also discussed in several articles related to BIOCHEMICAL GENETICS, including GENES; GENETIC CODE; MESSENGER RNA; PROTEINS (BIOSYNTHESIS); TRANSFORMING FACTORS.

NUCLEOPROTEINS

Nucleoprotein is a generic term applied to various association products between nucleic acids or polynucleotides and proteins. These compounds were originally called *nucleins* by their discoverer Miescher (1844–1895). In later years, various more specific designations were introduced, such as *ribonucleoprotein* and *deoxyribonucleoprotein*, designating whether the nucleic acid is primarily of the ribose (RNA) or deoxyribose (DNA) type, and *nucleohistone* or *nucleoprotamine*, indicating the nature of the protein in the complex.

An important early criterion used for identifying a nucleoprotein was its phosphorus content, which is very high because of the phosphate in the backbone of the nucleic acid. However, not all phosphorus-containing PROTEINS are nucleoproteins; some, such as the protein vitellin of avian egg yolks, have phosphorus bound directly to the protein, usually to serine residues. Such proteins were occasionally designated as *paranucleins* or *pseudonucleoproteins* but are generally known as *phosphoproteins*. Later tests for nucleoproteins included color reactions for the sugar moieties of the nucleic acids, *i.e.*, ribose and deoxyribose, or analysis for PURINE and PYRIMIDINE bases.

There is a growing tendency not to designate as nucleoproteins those proteins that are covalently linked to small nucleotides or nucleotide-containing structures, such as certain ones found in cell walls or enzymes which require a bound nucleoside phosphate for activity. In general, it is understood that nucleoproteins contain true RNA or DNA. From the conception of a nucleoprotein as a protein with a nucleic acid as a prosthetic group, the emphasis has been shifting toward considering it to be nucleic acid with associated protein. Nucleoproteins are, therefore, not now regarded as molecular substances in the usual sense but as highly specific molecular aggregates. Most early preparations of nucleoproteins were artifacts, in that they resulted from fragmentation, partial dissociation and unnatural reassociation of the components originally present *in vivo*. Nevertheless, most, if not all, of the nucleic acids of cells occur in association with proteins.

Covalent bonds appear to be lacking between the nucleic acid and the protein moieties in a true nucleoprotein complex. The two components are held together mainly by electrostatic bonds between the negative charges of the phosphates in the nucleic acid and the positive charges of basic proteins, as well as by hydrogen bonds between the nucleic acid bases and suitable structures in the protein. Nevertheless, the association between protein and nucleic acid is often very strong indeed, and rather drastic measures were often formerly employed to dissociate the complex. Subsequent procedures are based on the recognized necessity for breaking the electrostatic or the hydrogen bonds, or preferably both. The former are disrupted by strong salts and the latter by detergents, such as sodium dodecyl sulfate (SDS) or strong urea or guanidinium chloride solutions. In addition, it is desirable to separate the protein physically from the nucleic acid following dissociation, which is often accomplished by extraction with liquid phenol in the presence of SDS or certain salts. The protein is partitioned into the phenol while the more polar nucleic acid remains in the aqueous phase. Another method involves precipitation of the nucleic acid with strong salt (for example 2*M* LiCl) under conditions maintaining the protein in solution (4*M* urea). Great care has to be taken in these separations if it is desired to recover both products, since the isolated nucleic acids are exceedingly fragile and the released proteins have a tendency to aggregate irreversibly.

Ribo- and deoxyribonucleoproteins can be separated from each other by differential precipitation with salt (the latter requiring much less salt to be precipitated), as well as by other manipulations, but newer separation techniques depend primarily on differences in size and density. Deoxyribonucleoprotein particles are usually much larger than the corresponding RNA structures and can normally be sedimented out of suspensions in low centrifugal fields, leaving ribonucleoprotein in the supernatant. The latter particles are in turn larger than most free protein and nucleic acid molecules and can be sedimented by high-speed centrifugation. Size separations can also be made effectively by GEL FILTRATION through columns of certain polymers capable of acting as "molecular sieves." A most useful general method for separating deoxyribo- and ribonucleoproteins from each other and from free RNA, DNA and proteins is based on the significant differences in density of these substances. During prolonged centrifugation, starting with a mixture of cell extract and a salt of high density such as cesium chloride, the salt forms a density gradient, and each macromolecular component "bands" at a level where its density is matched. Free RNA has the highest density and free protein the lowest, with intermediate densities for DNA and the various molecular aggregates between the three. (See CENTIFRUGATION.)

DNA is the primary genetic material, existing in the cell as a deoxyribonucleoprotein complex called CHROMATIN. This consists of giant double-stranded DNA molecules [see DEOXYRIBONUCLEIC ACIDS (DISTRIBUTION AND STRUCTURE)], each associated with many different kinds of basic protein (HISTONES), as well as with a number of other types of protein and also some RNA. In higher organisms, chromatin is found in the nucleus of the cell, where it associates to form CHROMOSOMES, the carriers of heredity. The proteins in chromatin have the general function of protecting the DNA from enzymatic attack by nucleolytic enzymes, and the basic histones seem to function in the regulation of the transcription of the information contained in the DNA into "messenger" RNA, which eventually guides PROTEIN BIOSYNTHESIS. Under certain conditions, histones "protect" a portion of the DNA from the enzyme RNA-polymerase and thus prevent the synthesis of RNA. Bacterial chromatin is free in the cytoplasm, since these organisms do not have nuclei.

The principal ribonucleoprotein structures in the cell are the RIBOSOMES. Found mainly in the cytoplasm, they provide the sites of protein synthesis. There are two kinds of ribosomal subunit, a large one ("50S") and a small one ("30S"). Each of these subunits contains a different large molecule of RNA and, again, many different kinds of protein. The large and small subunits are held together by means of Mg^{++} to form potentially active ribosomes capable of binding specifically MESSENGER RNA (see also RIBONUCLEIC ACIDS), aminoacyl-RNA, and the enzymes necessary for assembling the polypeptide chains of proteins.

Most VIRUSES, except for a few large ones, are formed exclusively of nucleic acid and protein, but only a certain class of small plant and bacterial viruses are nucleoproteins in the strict sense, *i.e.*, possess a large nucleic acid molecule tightly and directly associated with proteins. In the larger bacterial and animal viruses, the protein forms an outer protective coat or shell containing the coiled nucleic acid molecule inside. In every case, however, the nucleic acid, which can be double- or single-stranded DNA as well as double- and single-stranded RNA, is the chief functional unit.

Nucleoproteins are thus very important structures in the cell, and information about one or the other kind is still pouring in at a tremendously rapid rate.

References

1. INGRAM, V. M., "The Biosynthesis of Macromolecules," Biology Teaching Monograph Series, New York, W. A. Benjamin, 1965.
2. WATSON, J. D., "Molecular Biology of the Gene," Biology Teaching Monograph Series, New York, W. A. Benjamin, 1965.

I. D. RAACKE AND R. S. BEAR

NUCLEOSIDE DIPHOSPHATE SUGARS

A number of nucleoside diphosphate sugars act as coenzymes, or "carriers" of the particular sugar involved, in metabolic reactions in which the sugar group carried either undergoes some enzymatic conversion (*e.g.*, glucose-galactose epimerization; see GALACTOSE AND GALACTOSEMIA) or serves as the monomeric subunit in some polymerization reaction (*e.g.*, glycogen synthesis, see GLYCOGEN AND ITS METABOLISM; hyaluronic acid synthesis, see MUCOPOLYSACCHARIDES). Either a purine or a pyrimidine base may occur in the nucleoside portion; see, for example, the structure of uridine diphosphate glucose (UDPG) in PYRIMIDINES, section on Pyrimidines in Coenzymes (see also CARBOHYDRATE METABOLISM, section on Other Carbohydrate Conversions).

Reference

1. HASSID, W. Z., "Biosynthesis of Polysaccharides from Nucleoside Diphosphate Sugars," in "The Structure and Biosynthesis of Macromolecules," (BELL, D. J. AND GRANT, J. K., EDITORS), pp. 63-79, Cambridge University Press, 1962.

NUCLEOSIDES AND NUCLEOTIDES

Nucleosides are phosphate-free partial hydrolysis products of nucleotides. Adenosine, guanosine and cytidine are examples of nucleosides. These are constituted of adenine, guanine and cytosine, respectively, linked to ribose. Any purine or pyrimidine base combined with ribose or deoxyribose constitutes a nucleoside. These are sometimes more specifically termed ribonucleosides or deoxyribonucleosides, depending upon the type of sugar contained. Nucleotides are phosphorylated nucleosides; the attached phosphate group may be esterified to the 2′-, 3′-, or 5′-hydroxyl of ribose in the case of ribonucleotides, or to the 3′- or 5′-hydroxyl of deoxyribose in the case of deoxyribonucleotides (for structures of representative compounds see PURINES; PYRIMIDINES). Ribonucleotides are subunits of RIBONUCLEIC ACIDS, and deoxyribonucleotides are subunits of DEOXYRIBONUCLEIC ACIDS.

NUCLEUS AND NUCLEOLUS

Nearly all cells have nuclei which are separated from the cytoplasm by a nuclear envelope. Typical nuclei are spherical bodies of about one-tenth the total volume of the cell, but great variations are found in special types of cells both in size and shape of nuclei. The nucleus typically has a refractive index different from the surrounding cytoplasm and is therefore visible in the living cell. In many cells, the nuclei appear optically empty because of the absence of granules of different refractive index. However, most nuclei contain one or more dense bodies which are easily demonstrated in living cells, especially with the PHASE MICROSCOPE; these are the nucleoli.

When the pH of the medium surrounding the cell is lowered, the nuclei become more conspicuous. When the cell is fixed, especially with acid fixatives, it stands out even more prominently. Basic dyes are readily bound by such nuclei even at the low pH. This is due to the presence of nucleic acids with the phosphoric acid groups available for dye binding. This material has usually been called CHROMATIN because of this property. During the interphase stages between divisions (see CELL DIVISION), the chromatin in fixed cells frequently appears to be a network of threads, sometimes with a few larger clumps. When cells are approaching division, the chromatin becomes condensed into filaments or rods which are distinguishable as individual CHROMOSOMES as prophase advances. The chromosomes are usually visibly double as soon as they can be recognized at prophase. Although chromosomes are visible only during division stages, they are known to have existed in an intact but visually indistinguishable condition all during interphase. They usually are duplicated during middle interphase in rapidly dividing cells. Confidence that they persist intact from one division to the next comes from genetic evidence which shows that genetic loci (GENES) are arranged in a linear order which persists through numerous division cycles. In addition, chromosomes with their DNA labeled by radioactive hydrogen reproduce [see DEOXYRIBONUCLEIC ACIDS (REPLICATION)] and pass through several divisions without losing their label or exchanging it with the newly forming chromosome strands except by an occasional segmental exchange between sister chromatids (daughter chromosomes). This regular sorting of labeled DNA could only occur if the linear subunits of the chromsome containing the DNA remain intact during interphase. Condensation must be accomplished by the folding and coiling of the strand or strands which make up the chromosome. The coiling has been followed by microscopic examination at successive stages of division in cells with large chromosomes. Typically the long prophase chromosomes with many gyres of a coil of a small gyre diameter are transformed into relatively short thick rods with a few large gyres. The detailed changes are difficult to follow, but apparently are accomplished by kinking and formation of a regular coiled coil or by an increase in gyre diameter accompanied by the shortening of the long chromonema, either by coiling at the submicroscopic level or by folding at the molecular level. The coiling cycle is reversed after anaphase, and two interphase nuclei are formed which are usually similar to the parental nucleus in appearance as well as genetic properties. Exceptions occur especially in differentiating tissues in which the appearance of the two daughter nuclei may differ from each other or from the parental nucleus. Whether they differ genetically is much more difficult to answer. In most nuclei the nucleoli disappear during late prophase and reform during late telophase at specific sites on certain chromosomes referred to as nucleolar organizers.

Chemical Components of the Nucleus. The nucleus may contain many of the substances found in the cytoplasm, but relatively little is known of the low molecular weight components, since many of these may leak out or enter the nucleus during its isolation. *In situ* tests can sometimes be utilized, but most of these tests require treatments that would allow movement of small molecules. Nuclei usually do not contain secretory granules and reserves of stored food material, with the exception that some nuclei may contain glycogen.

All of the classes of large molecules, characteristic of cells—protein, RNA (RIBONUCLEIC ACID), DNA (DEOXYRIBONUCLEIC ACIDS), and LIPIDS or LIPOPROTEINS—are present in nuclei. However, the specific proteins and nucleic acids characteristic of the nucleus and cytoplasm may be different. Available evidence indicates that differences do exist. With few exceptions, the DNA appears to be confined to the nucleus and to the chromosomes. Cells with DNA containing viruses or viroids are exceptions, and a DNA-like material is found in the cytoplasm of some eggs where it may be a storage reserve of precursors for the rapid divisions during cleavage.

DNA is present in all types of nuclei so far investigated. The amount of chromosomal DNA present is a characteristic of the species and ordinarily increases only when the cell is about to divide or the cell is becoming polyploid. The increase in DNA is discontinuous even in rapidly dividing cells. Usually the cell spends only about one-third of the cell cycle in the replication of its DNA. Evidence indicates that each molecule replicates only once during each cell division cycle. Extra-chromosomal DNA in viruses or other similar particles in cells are not under such rigid control, but frequently multiply out of sequence with the chromosomal complement.

Although the DNA per nucleus is relatively constant within the species, the variations between species are striking. The nuclei of amphibians and of higher plants of the lily family have the highest amounts of DNA. Man and other mammals have intermediate amounts, and fruit flies and fungi like Neurospora have one-hundredth or less than those with the highest amounts per cell. One of the puzzling features is the significance of the great variations. The genetic information required for growth is probably not vastly different, yet the material for recording it is highly variable. Perhaps the same code is present many times in those with high amounts per cell. However, this would appear to present great problems in evolving changes in such a genetic code.

RNA is found in abundance in both nuclei and cytoplasm of rapidly growing cells, but may be much less abundant in certain non-growing cells. However, in most cells its concentration is low compared to proteins; 5–10% of the dry weight would be considered a very high concentration. Analysis of the purine and pyrimidine bases in RNA from cytoplasm and nuclei indicates differences in some cells. However, there is considerable evidence that a part if not all of the RNA of the cytoplasm is derived from RNA synthesized in the nucleus. Cells with the nucleus removed do not synthesize additional RNA. RNA, labeled with a radioactive isotope in a nucleus transplanted to a homologous, unlabeled cell, moves into the cytoplasm after a few hours. Cells fed labeled intermediates in RNA formation, *e.g.*, cytidine or uridine, incorporate these into the nuclear RNA first. Only after a considerable delay, dependent on the metabolic conditions of the cell, does the labeled RNA appear in the cytoplasm. To account for the differences in cytoplasmic and nuclear RNA, the various RNA's can be assumed to have different rates of destruction in the cytoplasm or to be changed during the transfer, or else the nucleus contains RNA's with different metabolic characteristics and perhaps different functions.

[The metabolic roles of the several types of RNA, particularly in genetic information transfer, are discussed in the articles on RIBONUCLEIC ACIDS (BIOSYNTHESIS); RIBONUCLEIC ACIDS (STRUCTURE AND FUNCTIONS); RIBOSOMES; PROTEINS (BIOSYNTHESIS); SOLUBLE RIBONUCLEIC ACIDS.]

A characteristic class of basic proteins, the HISTONES, are present in most metabolic nuclei; certain bacteria do not appear to have them. They may be present in sperm nuclei, but are sometimes replaced by a more basic class of protein designated protamines, *e.g.*, in fish sperm. Present evidence indicates that the histones are intimately bound to the DNA of the chromosomes, and proper isolation procedures yield nucleohistone, a DNA-histone complex. The histones are present in amounts equivalent to the DNA in some nuclei. They appear to be synthesized simultaneously with DNA replication and to have a low rate of turnover. However, the RIBOSOMES contain basic proteins with amino ratios similar to the histones. Histones have been reported to be present in nucleoli and to be transported to the cytoplasm. However, the cytochemical evidence on which this hypothesis is based has been questioned.

The most abundant proteins in nuclei of metabolically active cells are non-histones. The amount of histones and DNA per nucleus is a species characteristic and does not vary except during synthesis for cell division and during the formation of nuclei that are polyploid (nuclei with multiple sets of chromosomes) or polytene (nuclei with giant multistranded chromosomes). However, the amount of the other proteins may vary greatly with the type of nucleus and its metabolic condition. They appear to be completely absent from sperm nuclei and very abundant in some nuclei of glandular tissues and developing oöcytes. Some of these proteins must be part of the enzymatic equipment of the nucleus, but perhaps a large fraction is synthesized and transferred to the cytoplasm. Proteins have also been shown to enter the nucleus from the cytoplasm.

The presence of high concentrations of proteins and nucleic acids gives nuclei a high viscosity in most cells. The nucleoli are the densest structures in the cell. They have a high concentration of protein and relatively less RNA. Usually no DNA is present. Condensed chromosomes are also rather dense. They contain DNA, histone, some RNA and a residual protein (insoluble in salt solution). Most nuclei have, in addition to the chromosomes and nucleoli, a more fluid phase referred to as karyolymph or nuclear sap. Unlike the cytoplasm, the nucleus is free of membranes except for the nuclear envelope which surrounds it. However, considerable spatial organization is indicated by the fact that chromosomes do not become excessively entangled. The chromosomes have been shown to retain the same position at prophase which they occupied at the preceding telophase.

In summary, then, the nucleus contains most of the cell's genetic information coded in molecules of DNA. This is translated by means of the RNA or perhaps by ribonucleoproteins which are released into the cytoplasm. Much of the protein synthesis and energy-yielding metabolic processes occur in the cytoplasm. However, both protein synthesis and processes that yield ATP (adenosine triphosphate) occur also in the nucleus. Although cells soon die when deprived of a nucleus, nuclei are very dependent on the cytoplasm and cannot survive when freed of their cytoplasm. Contact with fluids other than homologous cytoplasm appears to result in a rapid deterioration of metabolic potential and almost instantaneous loss of ability to function as the genetic apparatus of another cell.

J. HERBERT TAYLOR

NUTRIENT SOLUTIONS AND NUTRITION OF CELLS

Nutrition is the provision, to an organism, of all those elements required for satisfactory maintenance, growth, function and reproduction. In distinguishing the nutrition of *cells* from that of animals or plants, protozoa or bacteria, we are merely recognizing that the cell of an animal or plant is a special sort of organism presenting characteristic limitations. These limitations are less of quality than of condition.

Every cell will require the same basic nutrient constitutents, whether they are brought to it directly, as in a free-living organism, or through other cells within the organized body.

Many free-living single-celled organisms, such as the green algae, are completely *autotrophic*, *i.e.*, they can satisfy all their requirements from simple inorganic sources. The cells of complex plants and animals, on the other hand, are generally *heterotrophic*, being limited in their synthetic capacities. Certain of their constituents must be provided in pre-elaborated forms. This heterotrophy applies especially to three classes of nutrients: (1) energy sources, especially carbohydrates, (2) nitrogenous materials from which proteins are built, and (3) organic catalysts—vitamins, hormones, etc. No animal cells and relatively few cells of higher plants can synthesize carbohydrates from CO_2 and H_2O; these must be provided from external sources. No animal cells and probably no cells of higher plants can fix atmospheric nitrogen. And although many plant cells can utilize inorganic nitrate or ammonia in the synthesis of protein, this is probably true of few if any animal cells; the building blocks for PROTEINS must be supplied in more complex forms, having at least the complexity of simple amino acids. While the requirements of cells for vitamins and hormones may vary widely and some cells may manufacture some of them, little is known as yet of the specific sites of these syntheses. Few cells are known from any organism which do not require an external source of THIAMINE; that all plants get their thiamine presynthesized from the environment is highly improbable—we simply do not as yet know *where*, in the plant, it is generally synthesized.

The nutrients devised for the maintenance of cells of plants and animals are elaborations from those established for complex organisms. Specifically they are, for the most part, elaborations of two such solutions—that of Knop (1864), designed to provide the essential inorganic ions found to be common to all satisfactory soil solutions, which has served ever since as the basis for the formulation of fertilizers and nutrient solutions for plants, and that of Ringer (1886), designed to provide the essential inorganic constituents of blood serum needed to prevent the rapid disintegration of isolated animal organs and tissues. Neither of these solutions was complete, and the history of their elaboration is the history of plant and animal cell nutrition.

Knop's solution consisted of $Ca(NO_3)_2$, KCl, $MgSO_4$, and KH_2PO_4. It contains all of the major inorganic constituents of *plant cells* except carbon. For non-photosynthetic cells, there must first be added a carbohydrate. Dextrose *can* be utilized by practically all cells, plant or animal; it therefore served in the first formula of Gautheret (1934) for plant cells. However, some cells prefer other carbohydrates. Sucrose is a more satisfactory source for many, especially for isolated roots,[8] a preference which Street[6] has explained in terms of the mechanisms of phosphorylation. Other sources are sometimes required—trehalose has proved essential for a small and specialized group of the Myxomycetes. Yet Knop's solution plus dextrose will not support the growth of plant cells. It must be supplemented first with the trace-element catalysts: iron, manganese, zinc, boron, molybdenum and copper. And it must be supplemented with inorganic catalysts as well. Most cells will require the addition of thiamine. Many will not utilize nitrate except in the presence of pyridoxine and must be provided either with pyridoxine plus nitrate or with materials of a still higher level of complexity: amino acid in some form. Many plant cells also require an auxin (see HORMONES, PLANT). For some, indole-acetic acid (IAA) is satisfactory; for others which possess enzyme systems destructive of IAA, another auxin such as 2,4-dichlorophenoxyacetic acid is more effective. Still others may also require a kinin or a gibberellin, or both, although these requirements are less universal than that for auxin.

There have thus resulted certain formulations of which that of White[9] is typical and widely used. This consists of:

	mg/l
$MgSO_4$	360
$Ca(NO_3)_2$	200
Na_2SO_4	200
KNO_3	80
KCl	65
NaH_2PO_4	16.5
$Fe_2(SO_4)_3$	2.5
$MnSO_4$	4.5
$ZnSO_4$	1.5
H_3BO_3	1.5
KI	0.75
Sucrose	20,000
Glycine	3.0
Thiamine	0.1
Niacin	1.0
Pyridoxine	1.0

This nutrient is adequate for most plant cells and organs. For some, one must add 0.1 mg/l of IAA or 2,4-D and/or correspondingly small quantities of biotin, pantothenate, choline, inositol, kinetin, gibberellin, etc. For some, glycine is toxic and must be omitted.

Some plant cells do not grow easily and rapidly in such a nutrient. In the early development of these formulas, a tissue extract such as that of yeast was commonly used as a source of unknown substances. The substitution of thiamine, pyridoxine, glycine, tyrosine, cysteine, inositol, etc., resulted from analyses of these ill-defined complexes. The idea has persisted that when a defined nutrient fails to give satisfactory results, an easy alternative is to resort to these unknowns. Yeast extract, casein digest and coconut milk have been widely used and have, in some cases, given dramatic results. Since, however, all attempts at analysis have brought to light nothing more than complex mixtures of already well-recognized vitamins, amides and other substances, the use of these complexes has contributed little to our knowledge of cell nutrition. Few if any cells used today *require* these complex unknowns for reasonably good maintenance and growth. Their use is therefore to be discouraged though not prohibited. [See also NUTRITIONAL REQUIREMENTS (AUTOTROPHIC PLANTS); TISSUE CULTURE, PLANT].

For *animal cells*, the first solution, *Ringer's solution* (1886), consisted of $CaCl_2$, KCl, NaCl and $NaHCO_3$. This is obviously even less adequate than was that of Knop. It contains, to mention only the inorganic elements, no magnesium, sulfur, phosphorus, nitrogen or iron. It was effective only because, in early work, it served solely as a protective bath, not a nutrient, or was supplemented with massive quantities of serum and/or tissue extract. The inorganic formula has been greatly improved. An excellent replacement widely used today is the formulation of Earle[2] consisting of:

	mg/l
NaCl	6800
KCl	400
$CaCl_2$	200
$MgSO_4$	100
NaH_2PO_4	125
$NaHCO_3$	2200

But this is still no more than a protective bath, ensuring satisfactory conditions of membrane permeability to permit utilization of nutrient materials which must be added. The common energy source is dextrose. Glutamine, although unstable, has proved a very effective source of organic nitrogen for animal as well as for some plant cells. It has usually been supplemented with other amino acids. Waymouth's formula[7] includes 16 additional amino acids, besides hypoxanthine and glutathione. The importance of most of these has not been experimentally established. Even less is known of the role of the vitamins. Eagle includes 8 in his formula,[1] Waymouth 11, but neither essentiality nor optimum concentrations have been established for most of these. It is probable that many may prove to be synthesized by the cells themselves or to be nonessential. On the other hand, it is improbable that the respiratory cofactors: iron, copper, manganese, etc., can be dispensed with. Their omission from nutrient formulas is only made acceptable by the minute quantities required, which are supplied as impurities in other ingredients. One should remember the history of COBALT as a nutrient factor for animals: a "disease" was first traced to the pasturing of animals on certain soils; the defect was attributed to lack of available iron and was shown to be corrected by addition of certain iron salts. However, not all iron salts were effective, and the critical factor was established to be cobalt, present as an impurity in the effective iron salts, and the key element in the molecule of vitamin B_{12}. It is supplied in Waymouth's formula in the form of vitamin B_{12} at a cobalt concentration of one part in 50 million parts of the complete nutrient.

Waymouth's formula is reasonably effective for

many sorts of mammalian cells. It cannot, however, be considered a "general" formula and is still supplemented with 1–5% serum for many purposes.

A similar formula for cells of invertebrates, especially insects, has been devised by Grace[4] which differs from that for vertebrate cells largely in the inclusion of malic, succinic and fumaric acids, and of α-ketoglutarate. How important these are, in the presence of glutamine, seems not to have been established experimentally. (See also NUTRITIONAL REQUIREMENTS, various animal groups; TISSUE CULTURE, ANIMAL).

The above, necessarily brief outline presents the nature of the problem and some of the approaches made to its solution. Nutrients for different purposes will certainly differ widely. Those for maintenance alone will obviously be relatively simple. Wolff[10] has devised special formulas of this sort for the study of differentiation of animal organs without rapid growth. Nutrients planned to support rapid growth, of which Waymouth's formula is an example, are much richer in vitamins, amino acids and energy sources. Nutrients for the support of specific functions must be still different: osteoblasts will multiply with very little calcium or magnesium, but they must have available large quantities of Ca, Mg, PO_4 and CO_3 if they are to build bone; thyroid cells can probably grow without iodine but cannot synthesize thyroxine without this element, etc. Cells must obviously have higher levels of available phosphorus and sulfur for DNA synthesis and cell multiplication than they require for routine maintenance. All of these matters must be taken into consideration in carrying out effective studies of nutritional problems. Ignoring any of them can lead to unsound conclusions.

References

Works Cited

1. EAGLE, H., "Nutrition Needs of Mammalian Cells in Tissue Culture," *Science*, **122**, 501–504 (1955).
2. EARLE, W. R., "Production of Malignancy *in vitro*. IV. The Mouse Fibroblast Cultures and Changes Seen in the Living Cells," *J. Natl. Cancer Inst.*, **4**, 165–212 (1943).
3. GAUTHERET, R. J., "Recherches sur la culture des tissus végétaux: Essais de culture de quelques tissus méristématiques," Paris, François, 1935.
4. GRACE, T. D. C., "The Establishment of Four Strains of Cells from Insect Tissues Grown *in vitro*," *Nature*, **195**, 788–789 (1962).
5. KNOP, W., "Quantitative Untersuchungen über den Ernährungsprozess der Pflanzen," *Landw. Versuchs-Stat.*, **7**, 93–107 (1865).
6. STREET, H. E., "The Role of High-energy Phosphate Bonds in Biosynthesis," *Sci. Prog.*, **38**, 43–66 (1950).
7. WAYMOUTH, C., "Rapid Proliferation of Sublines of NCTC Clone 929 (Strain L) Mouse Cells in a Simple Chemically Defined Medium (MB 752/1)," *J. Natl. Cancer Inst.*, **22**, 1003–1017 (1956).
8. WHITE, P. R., "Potentially Unlimited Growth of Excised Tomato Root Tips in a Liquid Medium," *Plant Physiol.*, **9**, 585–600 (1934).
9. WHITE, P. R., "Nutrient Deficiency Studies and an Improved Inorganic Nutrient for Cultivation of Excised Tomato Roots," *Growth*, **7**, 53–65 (1943).
10. WOLFF, EM., "Analyse des besoins nutritifs d'un organe embryonnaire, le syrinx d'oiseaux, cultivée en milieu synthétique," *Arch. Anat. Microscop. Morphol. Exptl.*, **46**, 407–468 (1957).

General References

11. GAUTHERET, R. J., "La culture des tissus végétaux," Paris, Masson, 1959.
12. PARKER, RAYMOND, R. C., "Methods of Tissue Culture," Third edition, New York, Hoeber, 1961.
13. PAUL, J., "Cell and Tissue Culture," Second edition, Baltimore, Williams & Wilkins, 1960.
14. WHITE, P. R., "A Handbook of Plant Tissue Culture," Lancaster, Pa., Cattell, 1943; "The Cultivation of Animal and Plant Cells," Second edition, New York, Ronald Press, 1963.

PHILIP R. WHITE

NUTRITIONAL REQUIREMENTS (AUTOTROPHIC PLANTS)

Plant nutrients are the inorganic compounds and ions of elements used by autotrophic plants as raw materials in the synthesis of the organic compounds of which cells and tissues of the plant are composed. *Autotrophic plants* differ from heterotrophic ones in being able to survive in a purely inorganic environment and having no requirement for preformed organic compounds. The green plants we see around us, both on land and in aqueous media, are autotrophs. These include all the plants which provide us with food, fiber and shelter as well as those used as ornamentals.

Heterotrophic plants require some preformed organic compounds and often have inorganic nutrient requirements which are qualitatively and quantitatively different from those of the autotrophic plants. Heterotrophic microorganisms, bacteria and fungi are of consequence in plant nutrition because they release mineral nutrients from decaying matter. These mineral nutrients are then available for use by the autotrophs. Some microorganisms convert atmospheric nitrogen to forms useful to autotrophic plants—a process called NITROGEN FIXATION.

Since the human requirement for food, fiber and shelter has been a driving force in experimental work for many years, most of our information on the nutritional requirements of autotrophs has been acquired from experimental work with agronomic plants. Most agronomic plants are annuals or biennials and are small enough to permit experimental activities under controlled conditions such as in a greenhouse or in short-term field experiments. Forested lands support by far the greatest amount of plant growth, but research on the nutritional requirements of these plants has been limited because of many experimental difficulties.

Essential Elements. Although almost every element on the periodic table has been observed, at least in trace amounts, in plant tissues, there is at present little evidence that more than sixteen elements on the periodic table are actually required as nutrients for autotrophic plants. For many years students have used a mnemonic—C. HOPKN'S CaFe, Mighty good (C, H, O, P, K, N, S, Ca, Fe, Mg)—as an aid in remembering ten of the so-called essential nutrients for plants. As a result of research in the last forty years, we must add six more elements; namely boron, chlorine, copper, manganese, molybdenum and zinc.

Three criteria for essentiality for a nutrient for plants are:

(a) the organism cannot complete its life cycle in the absence of the element;

(b) the action of the element must be specific and not replaceable by another element;

(c) the element must have a direct effect on the organism and not merely on some aspect of the environment in which the organism is developing.

While it is convenient to group the essential elements in terms of quantitative requirement, as for example macronutrient elements and micronutrient elements, the distinction is arbitrary and only for the convenience of the physiologist and the chemist, and does not imply relative importance. By definition, the presence in the organism of an essential nutrient in minimal amount is the important feature. An element required by the plant in a few hundredths of a microgram per kilogram amounts of tissue, such as molybdenum, is no less essential for the organism than is one which is required in percentage amounts, *e.g.*, carbon, oxygen, hydrogen, calcium, potassium and others.

Plant Tissue Content of Essential Nutrients. The tissue contents of plants vary rather widely, and the following table is intended to provide only a general idea of the range of concentrations of the elements found in dry plant material. The amounts found depend on the plant part—young or old tissues, fruits, leaves, stems, or roots—selected for analysis. The content of a given plant part, in turn, depends on the supply to the plant, a function of soil fertility and management practices. Often elements not now regarded as essential for the plant may be found in plants and, indeed, may even have an effect on the concentration of essential elements in the tissue.

Nutrient concentration may be expressed either as the weight of nutrient per unit weight of plant tissue (as per cent or as parts per million) or as the relative number of atoms of the nutrient per unit weight of tissue. In the latter expression, it is convenient to use the term milligram-atoms per kilogram of tissue. Because the atomic weights differ widely, concentrations expressed on a weight basis alone give a poor indication of the relative number of atoms present. Thus plant tissues commonly contain about twice as many hydrogen atoms as oxygen atoms, yet the weight of oxygen present is roughly seven times greater than the weight of hydrogen.

In Table 1 the elements are arranged in decreasing order of relative numbers of atoms in the dry plant tissue.

Sources of Nutrients for Plants. Elements required by autotrophic plants are obtained from three sources:

(1) the atmosphere, which supplies carbon from carbon dioxide and, indirectly, nitrogen (through activities of nitrogen-fixing organisms and electrical storms), and hydrogen and oxygen from water and the gases of the atmosphere;

(2) the soil solution, which supplies water and dissolved ions;

(3) the solid phase of the soil—the soil colloids, minerals and organic matter.

Of all the essential elements, carbon is the only one which must be supplied continuously and primarily from the atmosphere. Although carbonates and bicarbonates may play a small role in supplying carbon from the soil, the principal mode of accumulation of carbon is by way of the foliar part of the plant from the minute amounts of carbon dioxide (about 0.03%) in the atmosphere. Plants are capable of obtaining portions of their requirement of other elements through their foliage. Micronutrients are sometimes supplied to agronomic plants by way of foliar sprays. This is often a necessity because some of the micronutrient cations are irreversibly adsorbed (fixed) by soil colloids and organic matter, and rendered unavailable to plants by way of the roots.

The soil solution contains dissolved ions of essential and other elements present as the result of rock and mineral weathering or supplied from commercial fertilizers and irrigation waters. In general, anions are regarded as dissolved in the soil solution. The soil solution also contains cations in solution as hydrated ions and also as organic complexes (chelates).

High proportions of the cations are adsorbed on the clay complex of the soil. These adsorbed cations are in equilibrium with those in soil solution and are the main source of cations to the plant. Weathering of minerals in the soil and decomposition of organic matter in the soil provide an additional source of ions for plants.

The intake of ions by plants is not a simple diffusion process leading to equal concentrations of ions inside and outside the plant cells. The living cells, through the expenditure of metabolic energy, accumulate ions against a concentration gradient, *i.e.*, ions attain a higher concentration within the cell than existed in the external medium. However, under some conditions, plant cells may exclude ions. Accumulation and exclusion processes are not well understood, and intensive research in many laboratories is underway in an attempt to gain insight into this problem (see ABSORPTION; ACTIVE TRANSPORT).

Functions of the Elements. Functions of some elements are still speculative, but for the majority definite functions have been described on a quantitative basis.

Carbon is required in synthesis of all cell and tissue structures, as well as for formation of CARBOHYDRATES (sugars, STARCHES, CELLULOSE,

TABLE 1. THE ESSENTIAL ELEMENTS AND THE APPROXIMATE RANGE OF THEIR AMOUNTS IN PLANTS (ON DRY WEIGHT BASIS)

Element		Per Cent[a]	Milligram-atoms of Element[b] per Kilogram of Plant Tissue
Macronutrients			
Hydrogen	H	5 – 7	50,000 – 70,000
Carbon	C	42 – 44	35,000 – 36,500
Oxygen	O	42 – 44	26,200 – 27,500
Nitrogen	N	1.5 – 3.0	1,070 – 2,140
Potassium	K	0.8 – 2.0	204 – 512
Magnesium	Mg	0.18– 0.25	74.0 – 103
Phosphorus	P	0.20– 0.40	64.5 – 129
Calcium	Ca	0.20– 0.40	50.0 – 100
Sulfur	S	0.10– 0.20	31.2 – 62.4
		Parts per Million[c]	
Micronutrients			
Chlorine	Cl	1000 – 10,000	28.2 – 282
Boron	B	15 – 20	1.38 – 46
Iron	Fe	50 – 250	0.895 – 4.475
Zinc	Zn	10 – 50	0.152 – 0.760
Manganese	Mn	5 – 100	0.091 – 1.820
Copper	Cu	5 – 15	0.079 – 0.236
Molybdenum	Mo	0.05– 5	0.0005 – 0.050

[a] One per cent equals 10,000 ppm.

[b] Milligram-atoms per kilogram equal $\frac{\text{per cent} \times 10{,}000}{\text{atomic weight of element}}$.

[c] One part per million equals 1 mg/kg.

lignin), PROTEIN and LIPIDS. Carbon additionally functions in the energy-collecting mechanism whereby radiant energy from the sun is converted to energy storage compounds, as expressed by the net photosynthetic equation

$$6CO_2 + 6H_2O \leftrightarrows C_6H_{12}O_6 + 6O_2$$

The reverse of this reaction represents the net RESPIRATION reaction whereby energy previously collected is released. It is emphasized that this is a net reaction—the actual mechanisms are very complex (see PHOTOSYNTHESIS).

Hydrogen is also a component of cell and tissue structures. It is a component of all organic compounds in living cells. In addition, hydrogen as part of the water complex plays a very large role in maintaining tissue integrity in the plant and in providing a medium through which ions are absorbed and translocated throughout the tissues of the plant.

Oxygen derived from carbon dioxide, and to a lesser extent from water, is also a constituent of carbohydrates, proteins and many other similar compounds. Oxygen is an essential element of all respiration–energy releasing reactions.

Nitrogen is the characteristic element of proteins along with the previously synthesized carbohydrates. Nitrogen is a component of nucleic acids which play many roles including genetic control in all organisms. The element is a constituent of the PORPHYRIN compounds which are basic units in production of many enzymes, many oxygen transfer systems, and also of the basic unit in CHLOROPHYLL, an important part of the photosynthetic mechanism.

Phosphorus is a constituent of many proteins. It plays an essential role in the energy transfer systems of the plant, both in photosynthesis and respiration. Phosphorus is involved in sugar metabolism. It is an element in many of the enzyme and coenzyme systems.

Sulfur, likewise, is a constituent of many proteins, many enzyme systems and many essential oxidation-reduction systems in the plant by way of sulfhydryl group formation.

Calcium is required in several functions of the plant, as an enzyme activator and in the formation of plant cell walls, probably by way of calcium pectate.

Magnesium is a constituent of chlorophyll and also acts as an activator of some enzyme systems.

Iron, found mainly in the form of porphyrins, is involved in oxidation-reduction systems as, for example, in the CYTOCHROMES, cytochrome oxidases, PEROXIDASES, catalase and probably others.

Manganese functions as a coenzyme in the CITRIC ACID CYCLE system. Manganese also may function as an enzyme activator in other systems.

Copper is a constituent of many enzymes, ascorbic oxidase, polyphenol oxidase, cytochrome oxidase and others.

Zinc, likewise, is a constituent of some enzyme systems, pyridine nucleotide dehydrogenases, carbonic anhydrase and some peptidases. Zinc has a role as a protector of auxins in plant tissues and probably serves a catalytic function in the synthesis of tryptophan which is assumed to be an auxin precursor.

The role of *boron* is poorly understood. It seems to have some function in formation of cell wall structures and may be interrelated with calcium in this function. Boron appears to have a function in carbohydrate translocation.

Molybdenum, another of the micronutrients, is a component of the nitrate reductase system and an inhibitor of acid phosphatase systems.

The role of *chlorine*, the most recently confirmed of the micronutrient elements, is still obscure, but the element appears to have a function in photosynthesis and in water relations in the plant.

Potassium has been regarded as an essential element since the earliest scientific studies in plant nutrition. Still no specific essential function for the element has been described. Virtually every metabolic process in the plant appears to be affected by potassium. Potassium, probably in common with the other macronutrient ions, seems to play a part in mediating ionic and osmotic balance in tissues.

References

1. Bonner, J., and Galston, A. W., "Principles of Plant Physiology," San Francisco, W. H. Freeman & Co., 1955.
2. Chapman, H. D. (Editor), "Diagnostic Criteria for Plants and Soils," University of California Division of Agricultural Sciences, 1966.
3. Russell, E. J., "Soil Conditions and Plant Growth," Ninth edition, New York, John Wiley & Sons, 1961.
4. "Soil," The 1957 Yearbook of Agriculture (Stefferud, A., Editor), Washington, D.C., U.S. Department of Agriculture, 1957.
5. Steward, F. C. (Editor), "Plant Physiology," Vol. III, "Inorganic Nutrition of Plants," New York, Academic Press, 1963.

C. M. Johnson

NUTRITIONAL REQUIREMENTS (INVERTEBRATES)

The nutritional requirements of an animal are not necessarily its natural foodstuffs, but they are the fundamental substances and quantities that the organism must take into itself as prerequisites for performing vital metabolic functions normally. Requirements are inversely proportional to synthesizing abilities. Invertebrates here include animals usually classed inclusively between the Protozoa, which may include photosynthetic numbers, and the Metazoa up to some of the chordates other than vertebrates. However, our knowledge of the nutritional requirements among over a million species distributed into numerous phyla is extremely fragmentary and is not without embarrassing gaps. Moreover, space here permits only a few close views. Consequently, these limitations allow no more than the following broad generalizations on the subject.

Invertebrates have much in common nutritionally with vertebrates, including mammals. The nutrients of both include over 30 elements. Of these, carbon, hydrogen, oxygen, nitrogen and, to a lesser extent, sulfur, phosphorus and others are the constituents of proteins, fats and carbohydrates, and must be obtained for the normal functioning of protoplasm. The complexity of the substances required for this varies with the mode of nutrition.

Several modes are found among invertebrates. With *holozoic* nutrition, solid material is engulfed through temporary or permanent openings, as in most animals, including, for example, insects and many Protozoa. With *saprozoic* nutrition solid food is not ingested, but dissolved substances in food media are absorbed through the body surface, as in cestodes and certain nematodes. *Holophytic* nutrition depends on the presence of chlorophyll, as in typical plants, for photosynthesis of combined substances from carbon dioxide, water and nitrogenous salts, as in certain flagellate protozoans. Some organisms combine modes of nutrition; *e.g.*, some flagellates are saprozoic in darkness and holophytic in sunlight.

Requirements in both holozoic and saprozoic nutrition are essentially for amino acids, sugars, fatty acids and other relatively simple compounds, which in saprozoic nutrition are found in the environment and in holozoic nutrition are obtained by digestive enzymatic dismemberment of large molecules. Holophytic nutrition in invertebrates would seem to require only some inorganic salts, provided one of them is nitrogenous; however, unlike bacteria, holophytic protozoans probably do not rely solely on inorganic material and carbon dioxide. Thus, all invertebrates, except possibly holophytic flagellates, obtain carbon and nitrogen that are already combined into organic compounds. Some species may obtain both only from protein or its degradation products. Moreover, probably all the Metazoa, and possibly most of the Protozoa, require several accessory factors such as vitamins.

Insight into the precise nutritional requirements of invertebrates is largely into a few representatives of Insecta, to a lesser extent some protozoans, and occasionally a few other metazoans. But this understanding of the Insecta and of the Protozoa provides good bases from which we may surmise the nutritional principles of intermediate forms reasonably well on the basis of their mode of nutrition and evolutionary position.

Insects. Despite wide evolutionary separation, insects and mammals generally operate on similar principles of holozoic nutrition, though there are some exceptions. Both have similar enzymes and synthetic abilities (*e.g.*, nonessential amino acids) and often (*e.g.*, Krebs cycle), but not always (*e.g.*, ornithine cycle), similar metabolic mechanisms. Outstanding nutritional differences are that insects, but not mammals, require dietary sterols, and mammals, but not insects, commonly require several fat-soluble vitamins. Nevertheless, insects do not utilize calciferol, or VITAMIN D, but whether certain insects need one or another of the other fat-soluble vitamins currently is being challenged, *e.g.*, VITAMIN A substances. Insect requirements vary with the species and also to some extent with other factors such as variety, sex and age. Generally, however, for growth the insect needs essentially the following: the ten AMINO ACIDS that are needed by the rat—arginine, histidine, isoleucine, leucine, lysine, methionine, phenylalanine, threonine, tryptophan and valine; a number of B vitamins (see VITAMIN B GROUP) usually including biotin, choline, nicotinic acid, pantothenic acid, pteroylglutamic (folic) acid, pyridoxine, riboflavin and thiamine; CHOLESTEROL or structurally similar sterols; a number of inorganic salts, and water.

Lack of some other substances may not necessarily prevent growth, but may decrease its rate or have some other detrimental effect on development. Thus many species need a nonspecific carbohydrate; hexoses, especially glucose, usually suffice. Some species are known that need linoleic acid, oleic acid, ASCORBIC ACID, INOSITOL, CARNITINE or unidentified factors. Diptera need nucleic acids or components such as adenylic acid, cytidylic acid, guanylic acid, certain nucleosides and PURINE bases. The importance of COBALAMIN, or vitamin B_{12}, has not been widely established among insects. Hematin seems necessary only for *Triatoma*, a blood-feeding bug. Sometimes certain substances can be substituted for others, or the presence or absence of one determines the essentiality of another, *e.g.*, the tyrosine-phenylalanine relationship. Apparently the inorganic salts required by insects are similar in variety, but doubtlessly differ in proportions, to those needed by mammals. However, the nutritional value of various proteins, which depends on how well the amino acid composition satisfies the requirements of the organism, is much the same for insects as for mammals [see PROTEINS (NUTRITIONAL QUALITY)]. Proportional relationships are important; *e.g.*, quantitative requirements for several B vitamins in *Drosophila* relate to dietary levels of protein or certain amino acids, and those of riboflavin and thiamine relate to carbohydrate levels.

Metamorphosis, especially where it is complete, with larval, pupal and adult stages is usually attended by changes in food habits and nutrition. In contrast to the somewhat uniform requirements of immature stages, those of adults generally vary widely. Adults of many species may need protein, carbohydrate, and possibly vitamins and inorganic salts; those of some need only sugar and water; those of a few do not feed at all. The differences usually are related to nutrient reserves acquired during the larval feeding stages and to the extent to which growth continues into the adult stage, in particular growth and development of the reproductive organs. As these growth demands may vary with the sex, the nutritional requirements may likewise vary.

Other Metazoa. So far, attempts to cultivate other metazoan invertebrates in a way conducive to determining their actual nutritional requirements have met with indifferent success. The best have not enabled the complete requirements of any of these animals to be elucidated. However, most of these creatures are holozoic, none are holophytic. Evidence is that they commonly need a complete regimen that contains among other things, protein, carbohydrate, fat and other lipid substances, vitamins, and inorganic salts. Moreover, cognizance of the most developed work on these metazoans indicates that the pattern of their nutritional requirements likely conforms closely with the familiar pattern of animal nutrition, as in insects. For instance, it may be provisionally concluded from recent work that all groups of Arthropoda, including Insecta, require a dietary source of sterols such as cholesterol. The clearest general insight, however, into metazoan invertebrates other than insects is limited largely to *nematodes*. The essential amino acids of the free-living *Caenorhabditis* nematode tentatively appear to be arginine, histidine, lysine, methionine, threonine, and valine, probably isoleucine, leucine, and phenylalanine, possibly tryptophan, but perhaps not tyrosine. Moreover, the B vitamins—folic acid, niacinamide, pantothenic acid, pyridoxine, riboflavin and thiamine—are needed, as well as unidentified factors. The parasitic *Neoaplectana* needs a number of amino acids and B vitamins. Purine and pyrimidine derivatives of nucleic acid are beneficial. It seems likely that media of known chemical compositions may be developed before long for some of these nematodes. Then the identity and true significance of various so-called growth factors may be established. On the one hand, the paucity and nebulosity of work on metazoan invertebrates does not warrant further discussion in this brief account.

Protozoa. Nutritional requirements of numerous Protozoa have been investigated, but except in a few species the techniques used usually have not permitted very precise or comprehensive determinations. Nevertheless, common findings that emerged from the work may allow some generalizations, as follows: requirements vary markedly with, and sometimes within, the mode of nutrition; those of holozoic and most saprozoic forms are much the same as for insects, but are simpler for holophytic forms; requirements may vary in some details between closely related species. Moreover, most protozoans require essentially many amino acids, and several forms require fatty acids and sterols, the utilization of which depends on molecular structure and varies with the species. Most protozoans require several B vitamins; flagellates have a rather stereotyped pattern of need for B_{12}, biotin and thiamine, singly or in various combinations. What follows is a brief generalized account of several kinds of protozoans as examples.

The most thoroughly nutritionally understood protozoan is the ciliate *Tetrahymena*, which in nature feeds on particulate material, but is said to be able to absorb some substances through its body surface under certain conditions. It requires essentially the same ten amino acids generally required by insects and the rat—arginine, histidine, isoleucine, leucine, lysine, methionine, phenylalanine, threonine, tryptophan and valine; seven vitamins—nicotinic acid or nicotinamide, pantothenic acid, pteroylglutamic acid, pyridoxal or pyridoxamine, riboflavin, thiamine and thioctic acid; two nucleic acid derivatives—guanine, and uracil or cytidine; inorganic salts—calcium, cobalt, copper, iron, magnesium, manganese, phosphate, potassium and zinc. It utilizes carbohydrates to spare amino acids, but manages well without. Proportional relationships between amino acids are important. Several substances determine the requirement for another; *e.g.*, B_{12} diminished and cholesterol increased the folic acid requirement. Growth is enhanced by acetic and pyruvic acid, and apparently by detergents containing oleic acid. One strain apparently needed serine, and a closely related species needed ethanol or methanol.

Thus the pattern of requirements in *Tetrahymena* is notably like that in insects. Moreover, there is assorted evidence that this pattern for the most part probably persists in other protozoans that are holozoic or saprozoic. Most of these forms require a number of amino acids similar to those required by *Tetrahymena*, though some may or may not need arginine, glycine, proline, serine, tyrosine and others. *Glaucoma*, a holozoic ciliate, needs twelve, but depends on polypeptides to evoke feeding responses, but in *Colpidium*, which responds to large molecules, free amino acids inhibit feeding. Many such species need much the same vitamins as *Tetrahymena*. Several species need B_{12}, *e.g.*, *Paramecium* and *Endamoeba*, but one species of *Paramecium* does not need choline or thioctic acid. Most species require nucleic acid derivatives and utilize one or more nucleotides or purine bases. Many have lipid requirements, *e.g.*, *Paramecium* needs sterols and utilizes oleates and stearates, but not linoleates.

Species that combine holozoic and saprozoic nutrition likewise appear to have extensive requirements. For example, the colorless phytoflagellate *Peranema* requires several amino acids, vitamins, nucleic acid derivatives and cholesterol. *Peranema* and parasitic *Trichomonas* utilize linoleates, and various species of *Trichomonas* require cholesterol.

Among saprozoic forms are many with requirements reminiscent of the typical animal nutrition pattern above. The trypanosome *Herpetomonas* is an example. It is notable that it and some, but not all, trypanosomids require hematin; some others apparently need ascorbic acid. However, free-living saprozoic flagellates seem to have simpler requirements: *Chilomonas* was cultivated in a solution of a few organic and inorganic salts, and *Polytomella* was reared on media supplying ammonium nitrogen, thiamine and acetate. The latter species can use only one of several organic acids and alcohols with no more than five carbons as the sole source of energy.

With holophytic nutrition, *Euglena*, the colorless flagellates which, however, are saprozoic in darkness, need B_{12} and grow best when traces of amino acids are present. One group prefers acetate or related compounds in darkness, but can live photosynthetically without it in sunlight. Most of them can take nitrogen, but no carbon in organic combination, and a few can take both. Most photosynthetic flagellates utilize both nitrates and ammonium ions, but many colorless species cannot use nitrates. Chrysomads, which combine holophytic and saprozoic nutrition, need sugar or glycerol, thiamine and one of several citric acid cycle components, such as citric acid, glutamic acid and histidine; some need biotin.

Thus, we may conclude that the nutritional requirements of metazoan invertebrates and of most protozoans are as complex as those of vertebrates. In fact, there exists a remarkable similarity between the amino acids needed by all of these animals, and apparently the need for riboflavin, thiamine and some other B vitamins is universal. On the other hand, rare and rather specialized requirements are encountered throughout.

References

1. Altman, P. L., and Dittmer, D. S. (Editors), "Biology Data Book," Washington, Federation of American Societies for Experimental Biology, 1964.
2. Dougherty, E. C. (Editor), "Axenic Culture of Invertebrate Metazoa: a Goal," *Ann. N.Y. Acad. Sci.*, **77** (2), 25–406 (1959).
3. Dougherty, E. C., *et al.*, *J. Exptl. Biol.*, **37**, 435–443 (1960).
4. Elliott, A. M., *Ann. Rev. Microbiol.*, **13**, 77–95 (1959).
5. House, H. L., *Exptl. Parasitol.*, **7**, 555–609 (1958).
6. House, H. L., *Ann. Rev. Entomol.*, **6**, 13–26 (1961).
7. House, H. L., *Ann. Rev. Biochem.*, **31**, 653–672 (1962).
8. Hyman, L. H., "The Invertebrates: Protozoa Through Ctenophora," Vol. 1, New York, McGraw-Hill Book Co., 1940.
9. Jackson, G. J., *Exptl. Parasitol.*, **12**, 25–32 (1962).
10. Johnson, W. H., *Ann. Rev. Microbiol.*, **10**, 193–212 (1956).
11. Kidder, G. W., in "Biochemistry and Physiology of Nutrition" (Bourne, G. H., and Kidder, G. W., Editors), Vol. 2, New York, Academic Press, 1953.
12. Provasoli, L., *Ann. Rev. Microbiol.*, **12**, 279–308 (1958).
13. Rogers, W. P., "The Nature of Parasitism," New York, Academic Press, 1962.
14. Rothstein, M., and Tomlinson, G., *Biochim. Biophys. Acta*, **63**, 471–480 (1962).
15. Sang, J. H., *J. Nutr.*, **77**, 355–368 (1962).
16. Trager, Wm., *Physiol. Rev.*, **21**, 1–35 (1941).
17. Zandee, D. I., *Nature*, **202**, 1335–1336 (1964).

H. L. House

NUTRITIONAL REQUIREMENTS (MAMMALS)

Most multicellular animals, through evolutionary processes, have lost the ability to synthesize within their bodies approximately 25–30 relatively simple organic structures such as vitamins, certain amino acids and linoleic acid (or certain related fatty acids). These must be obtained, therefore, from external sources. In contrast, all higher plants, and most lower forms as well, are able to synthesize such compounds by mechanisms not present in animal tissues.

Both plants and animals require water and about 13 mineral elements from external sources, the list of elements required being similar but not identical. If a source of any of these required elements is missing, then growth, life span, productivity and/or reproductive performance are adversely affected. Unlike plants, higher animals require an available source of oxygen for proper energy metabolism and growth, while plants utilize carbon dioxide for this purpose.

Only in recent years, since the discovery of the major vitamins, has it been possible to raise certain mammals on "purified" or "synthetic" diets containing all essential nutrients. Thus, it has now become possible to determine the requirement of an animal species for each organic and inorganic nutrient.

Table 1 summarizes the daily nutrient requirements of growing mammals, based chiefly on data provided by the Committee on Animal Nutrition of the National Academy of Sciences—National Research Council. The dietary requirements of the various species shown in the Table are calculated on a "per kilogram of body weight" basis. The variation seen between species when reported on this basis is primarily a reflection of differences in food consumption, as the smaller species tend to consume more food in relation to their body size.

To convert the values listed in the Table to "per cent of diet" or to "milligrams per kilogram of diet" the following formulas may be used:

When requirement is given in grams:

$$\text{Per Cent of Diet} = \frac{\text{Nutrient (g)} \times 100}{\text{Food Intake (g)}}$$

When requirement is given in milligrams:

$$\text{Milligrams per Kilogram of Diet} = \frac{\text{Nutrient (mg)} \times 1000}{\text{Food Intake (g)}}$$

When requirement is given in micrograms:

$$\text{Milligrams per Kilogram of Diet} = \frac{\text{Nutrient (mcg)}}{\text{Food Intake (g)}}$$

(Nutrient values in these formulas are based on per kilogram of body weight; food intake value in these formulas is given in line 2 of Table 1.)

Many unrelated factors tend to alter the nutrient requirements of mammals. Much of the variation in dietary requirements between and within species can be due to one or more of the following factors:

(*1*) *Metabolic capabilities of the animal:* Different species of animals vary in their ability to synthesize essential nutrients by metabolic means within the body. For instance, ASCORBIC ACID can be synthesized by all mammals except the primates, the guinea pig and the fruit bat; TRYPTOPHAN cannot be converted to niacin (see NICOTINIC ACID) by the cat; again, the cat appears not to be able to convert dietary carotene to VITAMIN A.

(2) *Anatomical differences in the gastrointestinal tract:* Ruminant mammals, although their physiological requirements for nutrients are probably very similar to monogastric mammals, do not show a dietary need for many organic compounds due to microbial synthesis in the anterior portion of their digestive tract (see RUMINANT NUTRITION). The anterior location of this synthetic activity is of importance because of the highly active absorptive areas which follow.

In other mammals, the cecum is the primary site of microbial activity, but due to the less active absorptive state of the lower tract, many of the essential nutrients synthesized are not absorbed in significant amounts in most species.

(3) *Coprophagy:* In some species where coprophagy (consumption of feces) is routinely practiced—as in the rabbit and rat, for instance—dietary requirements for some nutrients are met or substantially reduced (see also INTESTINAL FLORA).

(4) *State of growth, pregnancy, lactation and aging:* The stage and rate of growth, pregnancy and lactation have a profound effect upon total nutrient intake. Also, nutrient requirements may be changed when the product being synthesized has a grossly different chemical composition from the requirement of maintenance or growth, as in the case of high milk production.

The effect of aging upon nutrient requirements is not well understood. However, the adult maintenance requirements for some nutrients—*e.g.*, energy and protein—appear to be substantially lower than the growth requirements.

(5) *Inhibitory and/or nutrient-sparing compounds:* Naturally occurring compounds can alter the nutritive value of a food by such means as inactivating DIGESTIVE ENZYMES (antitrypsin factor of soybeans) or vitamins (AVIDIN of raw egg white inactivates biotin), while others (STARCH) may enhance the nutritive value indirectly by promoting microbial activity in the intestinal tract.

(6) *Disease and parasites:* Disease and parasites may alter the mammal's ability to absorb food, or, in the latter case, may compete for nutrients.

(7) *Environmental temperature:* Environmental temperature may alter energy expenditure and in so doing increase or decrease the requirements for some of the vitamins (*e.g.*, thiamine), the requirements of which are based upon energy intake.

Obviously, all of the above factors do not contribute to the variation seen in Table 1, as the experimental work from which these data were derived was done with young, healthy and actively growing animals (with the exception that human data are for adult maintenance requirements). Nevertheless, these factors do present very real problems in the quantitation of nutritional requirements and, unfortunately, limit their direct application to a practical feeding situation.

TABLE 1. SUMMARY OF DAILY NUTRIENT REQUIREMENTS OF YOUNG, RAPIDLY

	Nutrient	Beef Cattle 454 kg	Cat 0.50 kg	Dairy Heifer 454 kg	Dog 10 kg	Fox 2.3 kg	Guinea Pig 0.10 kg
		1	2	3	4	5	6
1	Metabolizable energy R*, kcal	59	130	40	123	291	196
2	Food consumption R*, g	26	40	22	54	81	80
3	Linoleic acid, g						1.2 R*
4	Protein R*, g	2.6		1.5	9.0	21	16
5	Vitamin A[h], mcg	13 R**	300 R*[i]	16 R**	60 R*	42 R*	418 R*
6	Vitamin D[j], mcg	0.05 R**	1.0 R*	0.03 R*	0.33 R*	NR	3.2
7	Vitamin E, mg	R*	R*	R*	R		5.0 R**
8	Vitamin K, mcg	R*[k]		R*[k]			800 R*
9	Ascorbic acid, mg	NR		NR	NR		16 R**
10	Biotin, mcg		8.0 R		R		R
11	Choline, mg		120 R*		55 R*		120 R**
12	Cobalamin, mcg		R		0.13 R*		R
13	Folic acid, mcg		40 R*		31 R*	16 R*	480 R**
14	Niacin, mg		1.6 R*[n]		0.4 R*	0.8 R*	3.0 R*
15	Pantothenic acid, mcg		200 R*		99 R*	588 R*	1600 R**
16	Pyridoxine, mcg		80 R*		44 R*	91 R	320 R**
17	Riboflavin, mcg		320 R*		88 R*	160 R*	160 R**[l]
18	Thiamine, mcg		160 R*		31 R*	91 R*	160 R*[l]
19	Calcium, mg	51 R**	R	27 R**	528 R*	484 R*	R*
20	Chlorine, mg	77 R	R	28 R**	200 R*	200 R	R*
21	Cobalt, mcg	1.9 R**	R	2.0 R**			R
22	Copper, mcg	114 R**	R	R*	165 R*	R	R
23	Iodine, mcg	10 R*	R	3.0 R**	66 R*	R	R
24	Iron, mg	R	R	R*	1.3 R*	R	R
25	Magnesium, mg	R*	R	20 R**	22 R*	R	R*
26	Manganese, mg	0.2 R*	R	0.44 R**	0.22 R*	R	R
27	Phosphorus, mg	51 R**	R	26 R**	440 R*	484 R*	R*
28	Potassium, mg	R	R	R	440 R*	R	R*
29	Selenium, mcg	R					R
30	Sodium, mg	52 R **	R	18 R	130 R*	160 R	R*
31	Sulfur	R[r]	R[s]	R[r]			
32	Zinc[t], mg	R	R	10 R	0.2 R	R	R*

[a] R = considered to be required, but no substantiating evidence reported; requirement, if given, only an estimate.
[b] R* = experimental evidence indicates a physiological requirement, but quantitative requirement not known, only estimated.
[c] R** = experimental evidence indicates a physiological requirement, and the quantitative requirement is shown.
[d] NR = experimental evidence indicates no dietary requirement.
[e] The growth requirements of mammals represent the minimal value at which no growth impairment occurs.
[f] Blanks indicate no conclusive experimental work done to date.
[g] The values for adult man are for maintenance and include a safety factor of 50% or more above the highest reported requirement.
[h] 0.300 mcg vitamin A alcohol = 1 International Unit of vitamin A.
[i] The cat cannot convert carotene to vitamin A.

GROWING MAMMALS[a–f] (per kg body weight per day unless otherwise indicated)

Hamster 0.076 kg	Horse 454 kg	Lamb 34 kg	Mink 0.72 kg	Monkey 1.0 kg	Mouse 0.016 kg	Pig 17 kg	Rabbit 2.3 kg	Rat 0.075 kg	Adult Man[g] 70 kg
7	8	9	10	11	12	13	14	15	16
284	35	94	330	84	900	182	151	480	31
79	16	44	92	35	250	67	78	134	9.3
				R*	R*	0.53 R**		0.67 R**	0.09 R*
18	1.2	4.7	20	2.5	40	12	13	27	1.0
308 R*	5.5 R*	4.3 R**	47 R	R*	28 R**	35 R**	R*	60 R**	13 R*
NR	0.17	0.14 R*	NR	R*	0.94 R*	0.33 R**	R*	R	R*
2.0 R*	NR	R*	R*	R*	5.0 R*	R	0.3 R	8.0 R**	0.3 R*
R		NR			R*	5.0 R		13 R**	
NR	NR	NR		1.0 R*		NR	NR	NR	1.0 R*
R				10 R	R*		R	NR	2.2 R*
70 R*				R	200 R*	59 R*	R	85 R**	R
R*[l]	NR[m]			1.0 R*	1.25 R**	1.0 R*	R	0.67 R**	0.06 R*
NR			18 R	50 R*	R	R	R	NR	21 R*
R*	R		0.9 R*	1.5 R*	7.5 R	1.2 R*[o]	10 R*	2.0 R**	0.3 R
2910 R*	R		668 R	R*	1875 R**	735 R*[p]	R	990 R**	143 R*
390 R*			101 R	51 R**	250 R**	71 R*[p]	R	132 R**	21 R*
474 R*	44 R*[q]		181 R	30 R**	1000 R**	206 R**[p]	R	530 R*	24 R*
474 R*	55 R*[q]		101 R*	30 R**	625 R**	71 R**[p]	R	167 R**	17 R*
474 R*	31 R*	87 R*	367 R	R	1500 R	435 R**	R*	800 R**	11 R*
R	114 R	152 R*	279 R	R	760 R	59 R*	R	67 R**	53 R*
R	0.8 R**	0.9 R*						NR	R
R	123 R*	200 R*	R	R	R	668 R*	R*	667 R**	29 R*
R	0.2 R*	18 R*	R	R	R	14 R*	R*	20 R**	1.1 R*
R	0.6 R	R	R	R	R*	5.4 R*	R*	3.4 R**	0.17 R*
R	R	26 R*	R	R	R	22 R**	28 R*	53 R**	7.4 R*
R	R	R	R	R	0.5 R	2.6 R*	0.44 R*	6.7 R**	R
276 R*	31 R*	78 R*	367 R	R	1250 R	335 R**	R*	667 R**	11 R*
R	R	R	R	R	500 R**	110 R*	R	240 R**	29 R*
	NR	R				6.7 R		5.3 R**	R
R	75 R	100 R*	180 R	R	500 R	33 R*	R	67 R**	35 R*
	NR	R[r]							
R	R	R*	R	R	1.3 R*	3.3 R*	R	1.6 R**	R

[j] 0.025 mcg vitamin D_3 = 1 International Unit of vitamin D_3.
[k] Required but synthesized by rumen microorganisms.
[l] Briggs *et al.*, unpublished data.
[m] When cobalt is present at 0.05 ppm.
[n] Tryptophan probably not converted to niacin.
[o] In addition to that in a standard corn-soybean meal diet without added tryptophan.
[p] In addition to that in a standard corn-soybean meal diet.
[q] In addition to a poor-quality hay and grain diet.
[r] When low levels of sulfur amino acids are fed.
[s] May serve as source of sulfur in amino acid synthesis.
[t] Requirement may vary with amounts of calcium and phosphorus in the diet.

Despite the difficulties encountered in measuring the dietary requirements of mammals, much progress has been made. As more adequate synthetic, chemically defined diets are developed, more accurate quantitation of the nutrient requirements will be possible.

References

1. ALLISON, J. B., MILLER, S. A., McCOY, J. R., AND BRUSH, M. K., *North Am. Vet.*, **37**, 38 (1956).
2. BECKER, D. E., JENSEN, A. H., AND HARMON, B. G., *Illinois Agr. Expt. Sta. Circ.*, **866**, Urbana, 1963.
3. BRIGGS, G. M., unpublished data, 1965.
4. CAMPBELL, H. A., ROBERTS, W. L., SMITH, W. K., AND LINK, K. P., *J. Biol. Chem.*, **136**, 47 (1940).
5. COOPERMAN, J. M., WAISMAN, H. A., McCALL, K. B., AND ELVEHJEM, C. A., *J. Nutr.*, **30**, 45 (1945).
6. CUNHA, T. J., *Feedstuffs*, **35** (34), 18 (1963).
7. FITCH, C. D., HARVILLE, W. E., DIMMING, J. S., AND PORTER, F. S., *Proc. Soc. Exptl. Biol. Med.*, **116**, 139 (1964).
8. GREENBERG, L. D., AND MOON, H. D., *Arch. Biochem. Biophys.*, **94**, 405 (1961).
9. McCALL, K. B., WAISMAN, H. A., ELVEHJEM, C. A., AND JONES, EDITH S., *J. Nutr.*, **31**, 685 (1946).
10. MILLER, E. R., ULLREY, D. E., ZUTAUT, C. L., BALTZER, B. V., SCHMIDT, D. A., HOEFER, J. A., AND LUECKE, R. W., *J. Nutr.*, **85**, 13 (1965).
11. MORRISON, F. B., "Feeds and Feeding," Twenty-second edition, Ithaca, N.Y., Morrison Publishing Company, 1956.
12. National Research Council, Committee on Animal Nutrition, "Nutrient Requirements of Domestic Animals. Nutrient Requirements of Beef Cattle," *Natl. Acad. Sci.—Natl. Res. Council Publ.*, **1137** (1963).
13. ———, "Nutrient Requirements of Domestic Animals. Nutrient Requirements of Diary Cattle," *ibid.*, **464** (1958).
14. ———, "Nutrient Requirements of Domestic Animals. Nutrient Requirements of Dogs," *ibid.*, **989** (1962).
15. ———, "Nutrient Requirements of Domestic Animals. Nutrient Requirements of Foxes and Minks," *ibid.*, **296** (1953).
16. ———, "Nutrient Requirements of Domestic Animals. Nutrient Requirements of Horses," *ibid.*, **912** (1961).
17. ———, "Nutrient Requirements of Domestic Animals. Nutrient Requirements of Laboratory Animals," *ibid.*, **990** (1962).
18. ———, "Nutrient Requirements of Domestic Animals. Nutrient Requirements of Rabbits," *ibid.*, **331** (1954).
19. ———, "Nutrient Requirements of Domestic Animals. Nutrient Requirements of Sheep," *ibid.*, **1193** (1964).
20. ———, "Nutrient Requirements of Domestic Animals. Nutrient Requirements of Swine," *ibid.*, **1192** (1964).
21. National Research Council, Committee on Food and Nutrition, "Recommended Dietary Allowances," *ibid.*, **1146** (1964).
22. RAMBAUT, P. C., AND MILLER, S. A., *Federation Proc.*, **24**, 373 (1965).
23. REID, MARY E., Pub. 557, Washington, D.C., Human Factors Research Bureau, Inc., 1958.
24. REID, MARY E., AND SALLMANN, L. V., *J. Nutr.*, **70**, 329 (1960).
25. SØRBYE, Ø., KRUSE, INGER, AND DAM, H., *Acta Chem. Scand.*, **4**, 549 (1950).
26. TAPPAN, D. V., AND ELVEHJEM, C. A., *J. Nutr.*, **51**, 469 (1953).
27. WOOD, A. J., AND KENNARD, MARGARET A., *Can. J. Comp. Med.*, **20**, 294 (1956).

JOHN T. TYPPO AND GEORGE M. BRIGGS

NUTRITIONAL REQUIREMENTS (VERTEBRATES OTHER THAN MAMMALS)

The nutritional requirements of birds are quite exacting and include proteins, fats, carbohydrates, minerals and vitamins. The domestication of the wild fowl has increased the size of the egg and egg production and consequently the nutrient requirements. The selection and breeding of chickens and turkeys for specific purposes has led to a further emphasis on nutrient requirements as well as maintenance of the proper balance of nutrients in the diet.

Chickens have been bred and selected specifically for egg production in one direction and for the production of meat in another. The wild turkey has likewise been domesticated and selected for a shorter leg, a wider sternum and a broader back. The intensive selection of turkeys for more efficient meat production has resulted in bird types where natural matings are often difficult. Thus, artificial insemination is practised to a very large degree in commercial operations. In selecting chickens for egg production, the body size of the laying hen has been decreased. This has resulted in a decrease in the body maintenance requirements. The body size of the broiler chick and the commercial turkey has increased by more than 100% over that of the original wild fowl. These birds are grown for meat purposes primarily. Broiler breeder hens which are kept for egg production only and for the purpose of collecting and hatching eggs for incubation purposes have a large body size and a low rate of egg production.

Proteins and Amino Acids. The PROTEIN sources or concentrates which are used in practical chicken feed formulas are soybean oil meal, fish meal, meat and bone scraps, poultry by-product meal, feather meal, cottonseed meal, peanut meal, guar meal and others. All of the protein concentrates listed contain more than 40% protein. The protein content of a vegetable protein concentrate can be increased by sifting out the fibrous material such as the seed hulls. For example, soybean meal as produced in commercial milling operations, contains approximately 44% protein; by removing a portion of the soybean hulls, the protein content is raised to 50%. In commercial practice, the protein content of meat and bone scraps is 50%, poultry by-product meal 55%, and fish meal

60–70%. In addition to the primary protein sources listed above, grains such as corn, milo or grain sorghum, wheat, oats and barley also supply 8–12% protein. The grains are deficient in certain essential amino acids, especially lysine. Fish meal is the only protein concentrate which contains an adequate amount of the ESSENTIAL AMINO ACIDS for the fowl in the proper balance. Soybean meal is deficient in methionine and cystine whereas the grains are fairly rich in methionine; thus, a combination of the grain with protein concentrates tends to make up a feed mixture which contains all of the essential amino acids as well as supplying the nonessential ones.

The growing chicken and turkey require 11 essential amino acids which must be supplied in the diet, either bound in a protein or as the free amino acid, for growth and body maintenance. This list of essential amino acids is: methionine, lysine, glycine, tryptophan, arginine, phenylalanine, histidine, leucine, isoleucine, threonine and valine. The laying hen requires only nine essential amino acids and is able to synthesize glycine and arginine. Methionine can be spared by cystine and phenylalanine by tyrosine. The remaining known amino acids can be synthesized within the body of the bird from the essential ones listed above.

The protein requirement of the chicken is approximately 20% of the total diet. This 20% protein must carry an adequate amount of each of the essential amino acids listed above. The protein requirement of the growing chicken decreases with age, and no more than 15% is required after the birds are 10–12 weeks of age.

Poultry production has become such an intensive business and specialization has progressed to such a degree that a few remarks concerning the feeding of specialized commercial poultry are in order. The broiler chick is fed a high-energy type diet containing approximately 1450 kilocalories of metabolizable energy per pound with a protein level of 24% for the first 5 weeks. The protein level is lowered to 22% for the remaining 3 weeks, and the bird is marketed at an average weight of approximately 3¾ pounds at this time.

The pullet chick which is to be kept for egg production purposes is reared on a diet containing 20% protein for the first 8 weeks after which time the protein level may be lowered to 15% until the bird is 22 weeks of age and egg production is initiated. The laying hen is fed a diet containing 15% protein, balanced in such a manner so as to supply all of the essential amino acids mentioned above except arginine and glycine; the hen can apparently synthesize both arginine and glycine within the body from nonessential amino acids in sufficient amounts to meet the requirements for body maintenance and egg production.

The starting young turkey requires 28–30% protein since this bird grows much faster than the chicken and furthermore attains a heavier weight at maturity. The protein content of the diet may be reduced to 26–28% after the first 4 weeks, then to 24% from the ninth to the twelfth weeks, and to 20–22% from the thirteenth to sixteenth week. In practice, the females are separated from the males at either 12 or 16 weeks of age and fed separately. For example, the females are fed a diet containing not more than 15% protein at the beginning of the seventeenth week and reach maturity at 20–21 weeks of age. The growth of the males is continuing during this period, and the higher level of protein (18–20%) is required from the seventeenth through the twentieth week. After this, the males are fed a lower protein diet (15%) until maturity at 26–28 weeks of age. The turkey female will weigh 12–14 pounds at 20 weeks of age, and the males will weigh 26–28 pounds at 28 weeks of age. The turkey male's weight should equal the age of the bird under optimum conditions up to about 30 weeks of age.

Energy—Carbohydrates. Chickens and turkeys of all ages require a source of energy which may be supplied by adding the principal grains in ground form to the final diet. The grains used in poultry diets are corn, milo, barley, wheat and by-products from the milling of corn and wheat. These include corn gluten meal, wheat middlings and wheat bran. The wheat by-products contain a higher level of CELLULOSE and are not as desirable for use in poultry feeds as are the grains, since the fowl does not digest cellulose. Corn gluten meal contains 41–60% protein and is an excellent source of methionine, in addition to supplying xanthophylls which are necessary for the yellow color in egg yolks and in the skin of the chicken. This yellow color is also obtained from yellow corn and from dehydrated alfalfa meal.

Protein concentrates and grains contain fat which may vary from a level of 4% in corn to as high as 12.5% in poultry by-product meal. In some protein concentrates such as solvent-extracted soybean meal, the fat content has been reduced to less than 1%.

Prior to 1950 it was thought that the fowl did not require fat preformed in the diet. It has now been determined that the fowl has a requirement for fat or, more specifically, for linoleic acid (see FATTY ACIDS, ESSENTIAL). If this fatty acid is completely eliminated from the diet, growth will cease and dermatitis-like lesions will appear around the beak, on the feet, and possibly on other parts of the body. If laying hens are fed a diet completely devoid of linoleic acid, egg production will be decreased and there will be a complete failure of embryonic development if the eggs are incubated.

Vitamins and Cofactors. The chicken and turkey require the following vitamins: A, D, E, K and B_{12}, riboflavin, pantothenic acid, niacin, B_6, choline, folic acid, biotin and thiamine. The first four listed vitamins are fat soluble, and a small amount of fat must be present in the diet for the maximum absorption of these vitamins through the wall of the gastrointestinal tract.

The fowl requires VITAMIN A for growth, maintenance of body tissues, egg production and embryonic development. If a bird is fed on a diet devoid of vitamin A, small, white pustules will appear in the esophagus and on the roof of the mouth. There will be an increased secretion from the eyes which will be followed by the formation of a cheesy exudate in the eye and will finally lead to complete loss of sight or to destruction of the eye itself. Vitamin A may be included in the diet as such, either in the form of the free alcohol or as an ester. It may be also provided by the conversion of the CAROTENES to Vitamin A in the process of absorption and metabolism of the carotenes.

Irradiated cholesterol (D_3) is utilized by the fowl to a much better degree than is irradiated ergosterol (D_2) (see VITAMIN D GROUP). The fowl requires vitamin D_3 for bone formation and the prevention of rickets as well as for growth, egg production and embryonic development.

VITAMIN E is required for the prevention of encephalomalacia, a deterioration of the cerebellum portion of the brain. It also serves as a metabolic ANTIOXIDANT and is necessary for embryonic development.

VITAMIN K is required for the formation of prothrombin in the BLOOD CLOTTING process. In the absence of this vitamin, there is a failure of the blood to clot and a slight injury will result in the bird bleeding to death.

RIBOFLAVIN is necessary for growth and for the prevention of the deterioration of the myelin sheath of the nerves. The outward symptom of riboflavin deficiency in the young chick is "curled toe paralysis." The vitamin is also required for egg production and embryonic development.

PANTOTHENIC ACID is necessary for growth, egg production and embryonic development. Chicks kept on a diet deficient in pantothenic acid develop dermatitis, scaly like lesions, in the area of the beak, on the eyelids and on the feet.

THIAMINE is necessary for the animal to live. This vitamin is stored within the body of the bird to a lesser extent than any other. In its absence, there is a retraction of the head over the back to the rear and the bird assumes a type of paralytic posture.

FOLIC ACID is required for the prevention of cervical paralysis in turkeys and for growth and reproduction in both chickens and turkeys.

BIOTIN is required for growth and for the prevention of dermatitis on the feet and around the beak and eyelids. Hens maintained on a diet deficient in biotin continue to lay eggs, but there is a failure of embryonic development.

VITAMIN B_{12} is required for growth, egg production and embryonic development. Deficiency symptoms other than a failure of growth are not noticeable in the young chick or poult.

A deficiency of Vitamin B_6 causes a hyperirritability, and death results. The vitamin is required for growth, egg production and embryonic development (see PYRIDOXINE).

Niacin, choline, B_{12}, riboflavin, folic acid and biotin are required for the prevention of perosis, characterized by a slipping of the gastrocnemius tendon over the condyles of the tibiotarsal tarsometatarsal hock joint.

It has been possible to rear chickens and turkeys to maturity and have these birds reproduce when all of the above-mentioned vitamins were added to a purified-type diet which did not contain natural food sources. Natural ingredients, such as the oil meals, meat and bone scraps, fish meal, and corn are deficient in certain vitamins mentioned above and thus it is necessary to add the following vitamins to practical formulated poultry feeds: Vitamins A, D, E, K and B_{12}, riboflavin, niacin, choline and pantothenic acid.

The chicken and turkey require the following minerals: calcium, phosphorus, magnesium, manganese, iron, copper, iodine, sodium, potassium, chlorine, zinc, molybdenum, sulfur and selenium. Calcium and phosphorus serve primarily as constituents of the bones and teeth. These are the minerals which give rigidity and strength to the skeleton. These minerals also perform as a part of the organic compounds which make up proteins and lipids, and which function in the muscles, blood and other soft tissues of the body. Magnesium is found in the skeleton and in the soft tissues and is closely associated with calcium and phosphorus in its distribution in metabolism. Manganese is found primarily in the liver, but is also present in the skin, muscle and bones. It is necessary for proper BONE FORMATION and especially for embryonic development. Iron is required as a part of the hemoglobin molecule, the red coloring compound in the blood. Copper is necessary for the formation of the blood. Iodine is required for the formation of thyroxine by the thyroid gland which regulates the metabolism of the bird. Sodium and potassium are closely related. Sodium is found primarily in the extracellular fluids whereas potassium exists primarily within the cell. Chlorine is important for the formation of hydrochloric acid which is required for gastric digestion. Sulfur is required only as a constituent of the sulfur-containing amino acids, methionine and cystine. Zinc is required for bone formation as an enzyme activator and for embryonic development. Selenium is probably necessary primarily as an antioxidant, but it may be also required for the maintenance of the proper equilibrium of the proteins in the blood.

Game Birds. Game birds such as pheasants, quail and others have similar requirements to those listed for the turkey. All research information which has been published for various species of game birds show that the requirements are similar to those of the domestic turkey.

Fish. A limited amount of data is available on the nutritional requirements of fish. Most of the work has been done on rearing channel catfish for commercial purposes. The protein requirement is approximately 40%. Diets are formulated to be relatively low in carbohydrate content or energy and low in fiber. Most of the commercial diets contain approximately the same vitamin and mineral levels as for the broiler chick. In a two-year period, fantastic growth rates have been reported on channel catfish maintained and fed properly in commercial operations. The ability of the catfish to convert feed ingredients to meet his needs is extremely high.

J. R. COUCH

O

OBESITY. See MALNUTRITION; METABOLIC DISEASES.

OCHOA, S.

Severo Ochoa (1905–), a Spanish-born American biochemist, was awarded a share in the Nobel Prize for physiology and medicine in 1959 because of his contributions to the chemistry of RIBONUCLEIC ACID (RNA). ARTHUR KORNBERG received the 1959 award jointly with Severo Ochoa.

OLEIC ACID

Oleic acid is a very widely occurring UNSATURATED FATTY ACID, a component of certain LIPIDS. It is particularly important historically because the determination of its structure was crucial in ascertaining the structure of many other fatty acids. Stearic acid, for example, is produced from it by hydrogenation; if this were not so, the exact chain structure of stearic acid would have been extremely difficult to establish, since tens of thousands of isomers are possible. Oleic acid melts at 13°C; the *trans* stereoisomer of oleic acid is called elaidic acid and melts at 45°C. Oleic acid is converted into elaidic acid by treatment with nitrous acid.

OPSINS. See VISUAL PIGMENTS.

OLIGOSACCHARIDES

Oligosaccharides are composed of two or more (arbitrarily up to about 10) monosaccharides, covalently bonded by glycosidic linkages between the monosaccharides [see CARBOHYDRATES (CLASSIFICATION AND STRUCTURAL INTERRELATIONS)]. Some oligosaccharides occur naturally; these and many others may appear in partial hydrolysates of higher oligosaccharides or polysaccharides.

OPTICAL ACTIVITY AND ISOMERISM

When a light wave passes through matter it becomes modified by the radiation from the electrons and nuclei that have been set in motion. There are two kinds of substances: those whose basic units, whether molecules or unit cells in a crystal, are superimposable on their mirror images and those where this is not the case. Confining our discussion to isotropic media, the effect of the first class of compounds is a change in velocity of the wave with no change in its polarization. On the other hand, the second class of compounds will not only change the velocity but will rotate the plane of polarization as well.

A crude classical picture of the situation may be formulated as follows (see Fig. 1): Consider an electron constrained to move on a helical wire. When a polarized wave traveling toward the reader with its electric vector along the helical axis acts upon the electron, linear motion becomes translated into circular motion, and the electron moves in a circular path as it attempts to follow the direction of $\underline{E}$. This circular motion gives rise to a magnetic field along the direction $\underline{E}$ and perpendicular to $\underline{H}$. The resultant magnetic vector no longer has the same direction as the original, and its direction will be continuously changed as the wave travels through a medium composed of such systems. A more rigorous argument shows that when the effect of the magnetic field is considered, the direction of the electric vector is changed by exactly the same amount, and the two resulting vectors are still mutually perpendicular.

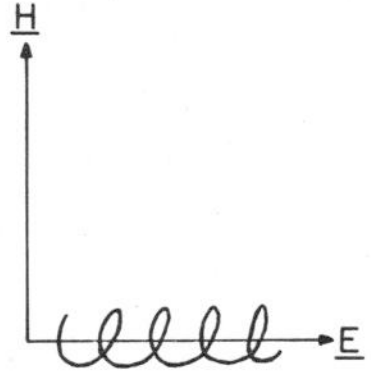

FIG. 1. Light wave with its electric vector polarized along the axis of an helix.

By the convention adopted, a medium is said to be *dextrorotatory* if the plane of polarization is rotated clockwise as the light enters the observer's eye and *levorotatory* if it is rotated counterclockwise. A system of classical electrons moving on right-handed helices will be seen to be dextrorotatory, whereas if left-handed helices are involved, the medium is levorotatory.

One generally measures the rotation, ϕ, in degrees per decimeter. The specific rotation, $[\phi]$, is equal to ϕ divided by the density of the medium in grams per cubic centimeter. Since one is often interested in comparing the rotatory power of single molecules, a quantity, $[M]$, called the molecular rotatory power has been defined by the relation $[M] = [\phi]M/100 = \phi M/100\rho$, where M is the molecular weight of the substance and ρ is its density.

In understanding the two ways in which this phenomenon can provide information on molecular structure, it is necessary to know that only elliptically polarized rays are transmitted in an optically active medium. The left and right elliptically polarized rays are transmitted with different indices of refraction and absorption coefficients. It is the difference in indices of refraction that leads to the rotation of plane polarized light, while the difference in absorption coefficients leads to the phenomenon of circular dichroism. In principle, both rotatory dispersion and circular dichroism yield the same information, for they are mathematically related by a Kronig Kramers transform. In practice, both the dispersion and the dichroism curves are of great help in understanding molecular structure.

The quantum mechanical expression for the rotatory dispersion of a molecule is

$$\phi = \frac{16\pi^2 N_1 \nu^2}{3hc}\left(\frac{n^2+2}{3}\right)\sum_n \frac{I_m(p_{0n}\cdot m_{n0})}{\nu_{0n}^2-\nu^2} \qquad (1)$$

where

N_1 = number of molecules per cubic centimeter,
ν = frequency of incident light,
n = index of refraction of medium,
h = Planck's constant,
c = velocity of light,
p_{0n} = electric dipole moment of the transition from the ground state to state n,
m_{0n} = magnetic dipole moment of the transition from state n to the ground state,
ν_{0n} = frequency of transition between ground state and state n.

When $\nu = \nu_{0n}$ for a particular transition, this expression is seen to become infinite and must therefore be modified in absorption regions. The actual form of a typical dispersion curve, or Cotton effect, is shown in Fig. 2 with the corresponding form of the circular dichroism curve.

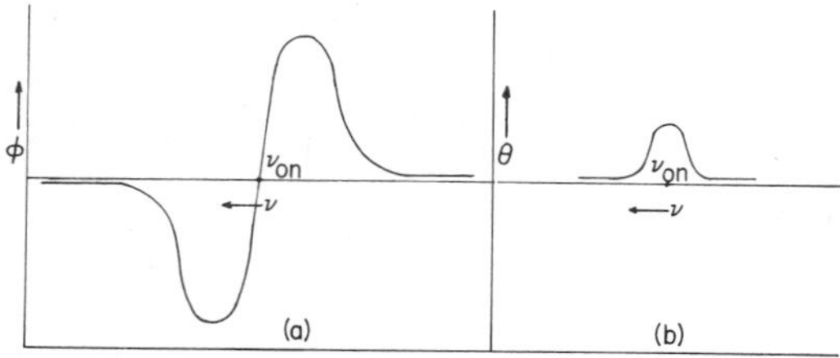

FIG. 2 (a) Rotatory dispersion in absorption region. (b) Circular dichroism in absorption region.

It happens that the dichroism curve tapers off outside the absorption maximum much more rapidly than the dispersion curve, which often extends over several hundred Ångstroms. For this reason, when there are several optically active transitions in a molecule, there will be considerable overlapping in the dispersion curve, and it becomes hard to analyze the data for each individual band. On the other hand, there is rarely such overlapping in the dichroism curve, and quantitative information on individual bands is more readily obtained.

Optical rotation is a measure of the degree of dissymmetry of a molecule [see ASYMMETRY; also AMINO ACIDS (OCCURRENCE), section on Optical Considerations]. In most cases the transitions leading to Cotton effects (cf. Fig. 2) occur in groups which themselves are symmetric, and the other groups in the molecule provide the dissymmetric environment that allows the transition to become active. In optical rotatory dispersion, unlike ordinary absorption, not only a strong electric dipole transition but also a strong magnetic dipole transition can produce a large effect. In the former case, both absorption and rotatory dispersion show strong maxima, while in the second, only the rotatory dispersion curve exhibits a marked effect. The form of the rotatory dispersion curve is particularly sensitive to conformation. Changes in sign are often observed in going from one solvent to another, which may be attributed to a preferred orientation of the groups in one solvent and a different one in the other.

The shape of the rotatory dispersion curve has been found to be particularly sensitive to changes in conformation of individual groups in complex molecules such as steroids; this sensitivity is not exhibited by the dichroism curve, Thus, the dispersion curve of a compound of known conformation can be used as a fingerprint guide in determining the conformations of new compounds (see also POLYPEPTIDES, SYNTHETIC).

DENNIS CALDWELL AND HENRY EYRING

OPTICAL ROTATORY DISPERSION. See OPTICAL ACTIVITY AND ISOMERISM; POLYPEPTIDES; SYNTHETIC.

ORGANIZER (IN CELLULAR DIFFERENTIATION)

The term "organizer" was originally applied to the dorsal lip of the blastopore of the amphibian gastrula. Hans Spemann and his associates, in a masterly analysis of developmental mechanics emanating from the University of Freiburg in the two decades following World War I, identified the germinal region which forms the dorsal lip of the blastopore during gastrulation, destined to become notochord, as an area primarily possessing the ability, when transplanted to the ventral region of an early gastrula, of inducing the formation of a second embryonic axis on the host belly. Such a secondary axis consists of nerve tube with associated sense organs, notochord, somites, pronephros, and even gut. Host and donor cells participate side by side in the structure of any one of these organs. Typically the induced axis parallels that of the host, with brain vesicles, otocysts, etc., occupying identical levels.

The action of the organizer has, in principle, been analyzed into a series or network of simple *inductions*. In embryonic induction, contact or near contact between an active tissue (the inductor) and a receptive (competent) tissue will result in the latter's being *determined* or started in a causal pathway directed toward a specifically differentiated state. In normal development, the primary action of the organizer is exerted on the ectoderm which it comes to underlie during the process of gastrulation; this ectoderm is induced to become neural plate whereas the organizer itself becomes archenteron roof. During neurulation the neural plate is subdivided into the organs of the central nervous system while the adjacent ectoderm gives rise to sensory placodes and other accessory neural structures. The archenteron roof becomes subdivided into notochord, somites, nephros, and lateral plate. These secondary subdivisions depend on adjacent tissues and regions. Thus secondary and tertiary inductions may be thought to follow the original primary induction of the medullary plate in the production of a complete axis. This sequence of events, most clear in amphibian development, appears to hold true, with variations, for all vertebrates. (See also DIFFERENTIATION AND MORPHOGENESIS; MORPHOGENIC INDUCTION.

The dorsal lip or organizer region is by no means homogeneous in itself. Its normal behavior is to roll over the blastopore lip, growing forward beneath the ectoderm to become a lining layer or archenteron roof. The earliest invaginated material thus advances farthest anteriorly and, having passed under the overlying layer progressively from posterior to anterior, comes to underlie the anterior medullary plate (prospective forebrain and midbrain); the last invaginated portion underlies spinal cord. Different organizer levels induce characteristically different axial levels of neural plate when brought into contact with standard receptive ectoderm. Living organizers from different species or even orders can induce effectively.

Experimental analysis of the responding ectoderm has shown that sensitivity to induction is a condition of limited duration and unstable character shared by the whole ectoderm of the gastrula. Once ectoderm has been induced in the neural direction, it itself acquires the capacity to perform neural inductions on sensitive tissues. The response to induction is species specific. Competent ectoderm can form only structures typical of the species, i.e., within its genetic repertory, whatever the source of the inductor to which it is exposed. Furthermore, although early gastrula ectoderm is not capable of acting as a neural inductor when alive, it acquires this capacity when killed by heat, by alcohol fixation, or in various other ways. Indeed, this is true of all early gastrula tissues, ectoderm or not, as well as of tissues from widely different animal and plant groups: killing releases inducing capacity in a great variety of tissues that have no normal relation to organizer activity.

The discovery of inducing ability in killed or extracted tissues gave first impetus to studies of the biochemical machinery involved. The problem is an exceedingly complex one, yielding only reluctantly to analysis at accessible points. The first hypothesis adopted assumed that some substance or substances diffused from the inductor to the reacting cells where a visible result (neural differentiation) was produced in a manner not specified. Chemicals suspected of being active (polysaccharides, carcinogens, nucleoproteins, etc.) have been applied to ventral gastrula ectoderm in carrier substances, such as agar, with some positive results. These results have been equivocal because the only criterion of reaction is an induction, and it is known that the test object itself is capable of self-induction if some of its cells are killed.

However enigmatic the experimental situation may be, there is good evidence that diffusible materials may be active inducing agents. Direct experiments *in vitro*, where amphibian ectoderm cells grown in fluids previously inhabited by organizer tissues have differentiated into neurons or pigment cells, seemingly can be explained only as a response to diffusible material in the culture medium. Various experiments involving the use of tracers have shown that both small and large molecules (*i.e.*, AMINO ACIDS and NUCLEOPROTEINS) can diffuse from organizer to ectoderm. Electron micrographs have shown extensive, apparently temporary cytoplasmic bridges between archenteron roof and overlying ectoderm in some species of amphibian gastrula. Experiments with second- and third-order induction systems have shown that the stimulus can be transmitted across filters, provided the pore size exceeds certain dimensions. These dimensions are of a magnitude to exclude actual cell contact, but not to exclude the intercellular matrix which invariably surrounds living cells, and which has been observed in the pores of such separating filters. The means of passage of materials from one cell layer to another may be through an organized gel rather than by simple diffusion in fluid.

Tests of the inductive action of extracts and fractions from a wide variety of tissue sources in various laboratories agree that a certain specificity of response can be correlated with definite tissues treated in particular ways. In practice, archencephalic (forebrain level), deuterencephalic (hindbrain level), trunk and tail responses can be distinguished. The latter two cases uniformly involve a mesodermal response as well as a neural one. Saxen and Toivonen (1962) hold that the observed antero-posterior spectrum of response can be accounted for by two specific inducing agents, neural and mesodermal, which in pure form would elicit archencephalic and mesodermal responses, respectively, whereas the intermediate inductions would result from varying proportions of the two components. The mechanism by which an active substance elicits neurulation in a competent cell is quite unknown.

Attempts to purify the various active inducing fractions from heterogeneous sources agree in general that neural and mesodermal agents are protein molecules with differing properties, In view of the wide distribution of inductive activity in the animal and plant kingdoms, it has been suggested that a specific inducing activity lies in some regional configuration common to various macromolecules rather than in a single specific protein.

References

1. SAXEN, L., AND TOIVONEN, S., "Primary Embryonic Induction," Englewood Cliffs, N.J., Prentice-Hall, 1962.
2. SPEMANN, H., "Embryonic Development and Induction," New Haven, Yale University Press, 1938.
3. WILLIER, B. H., WEISS, P., AND HAMBURGER, W., (Eds.,) "Analysis of Development," Philadelphia, Saunders, 1955.

DOROTHEA RUDNICK

ORIGIN OF LIFE

Questions relating to the "origin of life" have been of primary concern to mankind from the times when our ancient ancestors first began to think reflectively, but only in the last few decades have observations and experimentation given the information necessary for the construction of general hypotheses which fit into the pattern of modern scientific concepts.

Geologists, studying our own planet, and astronomers, investigating phenomena in the surrounding universe, have established that the formation of the earth as a planetary body took place during a period at least five billion years ago, and since that time, series of occurrences have eventually led to the development on this planet of the millions of species of living organisms existing today. Biologists have been able to reconstruct a rather detailed chronological account of the processes associated with life which have taken place during the latter half of the existence of our planet. This they have done by comparing the morphological relationships existing in modern species and how these structural features are related to those of their earlier ancestors whose fossil remains have been unearthed, and by studying the development of single germ cells into adult organisms which have highly complex functions and structures. More recently, elegant investigations by biochemists and extremely detailed observation by cytologists (made possible by electron microscopy) have established the chemical construction of cells and the dynamic transformations responsible for their living activities, *i.e.*, their metabolism. A large number of different types of cells have been studied, representative of the most diversified species found in modern plants, animals and microorganisms.

Biochemical Unity of Living Organisms. These recent studies show that in spite of the extreme variations in the way in which these different types of life develop, live and contribute to the total economy of nature, all cells living today possess, in common, basic features associated with life. Thus, (a) all cells possess structures and internal compartmentalization that utilize lipid-protein membranes having a common pattern of organization; (b) the activities of all cells are controlled quantitatively and qualitatively by catalysts which are proteins (ENZYMES), each of which is usually associated with a particular inorganic ion, and often with a very specific nonpolymeric organic molecule (a coenzyme); (c) the PROTEINS of all cells are made from the same monomeric units (the twenty or so L-amino acids); (d) the organic coenzymes are identical in all cells; (e) the particular array of enzymes and the intracellular structure is determined in all cells by the same genetic mechanisms which are controlled by nucleic acids; (f) the different nucleic acids which establish the characteristics of each species are formed by all cells from a relatively small number of mononucleotides (see PURINES and PYRIMIDINES) having identical structures; (g) different cells derive the energy needed for enzymatic reactions and other related processes sustaining their living activities from a variety of sources in their environment—sunlight, oxidizable inorganic substances, reduced organic molecules—but in every case the cells convert this energy to the same identical compound, adenosine triphosphate (ATP), in which the energy is conserved in the pyrophosphate bonds; (h) all cells utilize, to a greater or lesser extent, common series of chemically identical reactions for generating ATP, for making the monomeric units needed in forming cell structures, enzymes and genetic material, and for the processes by which the units are condensed into these polymers.

Because of this biochemical unity, most modern scientists interested in the origin of life have accepted the concept that all cells extant today have been derived by evolutionary processes from a dominant colony of primitive type "protocells" having all the structural and functional features enumerated above. Since the oldest fossils found are those of a simple type of algae found in rock formations dated at approximately 1.5–2 billion years ago, the postulated "protocells" must have evolved in geological epochs antedating this time.

The Prebiotic Environment. The question of the origin of life, then, is concerned with events taking place between the time of the formation of the earth as a planetary body and the emergence of these primitive ancestors of modern cells. Most geologists now generally agree as to the nature of the environmental conditions existing in the atmosphere, the hydrosphere, and the lithosphere of this planet when the earth cooled and oceans formed. From several lines of evidence, including the fact that hydrogen is the most common element in the universe, it is believed that the primitive atmosphere of the earth was primarily in a reduced state. Hence, the principal elements forming the organic molecules of living material (C, H, O, N, S) were present in their reduced states (*i.e.*, CH_4, H_2, H_2O, NH_3, H_2S), and these along with molecular nitrogen (N_2) constituted the primitive atmosphere. The oceans contained, in low concentration, the inorganic salts which contribute the ions that are essential in maintaining the external and internal environment of modern cells (Na^+, K^+, Ca^{++}, Mg^{++}, HCO_3^-, Cl^-, $HPO_4^=$, $SO_4^=$) and, in even lower concentrations, compounds containing the transitional elements that are necessary components of modern enzyme systems (*e.g.*, Zn, Fe, Mn, Cu, Co and Mo); there must have been a continual exchange of inorganic substances between the sands, clays, limestones and other minerals of the solid earth and the contacting primordial oceans.

During the 1950's, it was demonstrated experimentally that the simple reduced compounds present in the original atmosphere could be transformed into a mixture of a large number of simple organic compounds, many of which (*e.g.*, succinic acid, glycine, aspartic acid) are known to be essential compounds formed metabolically in all living organisms. In these experimental studies, the energy necessary to disrupt the primitive molecules and allow them to condense into more complex structures was supplied by electrical discharges or ionizing radiations. In the primeval atmosphere, lightning, ultraviolet light, cosmic radiation and decay of radioactive elements in the earth's crust would have similarly affected the formation of organic molecules, and over the eons of time, with no living organisms to consume them, the organic compounds accumulated in the oceans, converting the oceans into what has been termed a "primordial soup."

These oceanic organic compounds would be so dilute and subject to motions disruptive to intermolecular associations, that one cannot visualize their ever having come together to form orderly series of reactions, or to polymerize by dehydrative processes into macromolecules resembling proteins, polysaccharides, nucleic acids and lipids. In the most attractive hypotheses, it is assumed these substances were concentrated in the sands and mud beds bathed by the oceans. The components of the soils (clay, sand, phosphates, carbonates) would act as adsorbents under primordial conditions, just as effectively as they have been used in modern times by biochemists, and would also serve as the surface and matrix upon which reactions could take place. Their surfaces and the adsorbed inorganic ions could function as catalysts, and some of the mineral particles containing anhydride linkages could, by exchange reactions, cause the formation of primitive polymers. Systems developing on such adsorbing surfaces could "assimilate" nutritive material from the surrounding aqueous environment, and at the same time, the adsorptive forces would oppose disintegration of the system due to diffusion.

Some authorities have postulated that in the very early stages in the development of life, the partitioning of the prebiotic systems from their environment was accomplished by the formation of droplets surrounded by membranes formed from oily and proteinaceous substances that arose randomly in the primordial broth—an "exoskeleton" type of structure. Because of the fragility of such organizations, it would seem that the more tenable postulates are those picturing an "endoskeleton"-type framework on the readily available stable adsorbents. One of the most important contributions of ELECTRON MICROSCOPY has been to disclose that living systems depend upon an extensive and highly organized structure of membranes within the cell, and it is possible that this highly reticular ultrastructure found in all modern cells (see MICROSOMES) is a reflection that the original localization of nutrients and the development of the organized systems using the nutrients took place upon the "reticula" of the soil particles. Because of the cohesive pull of adsorption forces, such postulates would not require a "cell membrane" for the original development of orderly series of reactions (*i.e.*, the evolution of a primitive metabolism).

Is it possible by starting with such assumptions concerning the prebiotic environment, to formulate a reasonable pattern of events which led to the emergence of primitive ancestral cells having the properties previously described? Some hypotheses have been advanced which are based upon concepts that at some time there was, through a chance combination, the formation of a specific nucleic acid molecule which could duplicate itself, and that this was the beginning of life. Other hypotheses of a similar nature involve the formation from a random mixture, by chance, of a giant polypeptide having specific enzymatic capacities. In both these types of hypothesis, it is assumed that cellular metabolism—the orderly arrangement of biochemical reactions—was the result rather than the cause of the origin of enzymes and nucleic acids. However, scientists who have given the most critical thought to the problem are in general agreement with the concept that the *development of modern patterns of metabolism antedated* the formation of specific proteins and of the nucleic acids (which became essential at a later stage for the precise genetic mechanisms needed when the organization had reached a degree of complexity such that it could be described as "living").

The Evolution of Metabolism. To what extent can one construct reasonable hypotheses concerning the evolution of metabolism and its relationship to the subsequent development of enzymes and genetic material? Methods used by biologists in reconstructing the probable patterns of the evolution of organisms—investigating fossils, comparing the variations in the anatomy and physiology of modern forms, and observing embryological development—cannot be used for studying the biochemical evolution preceding the appearances of the earliest of "living" organisms. However the concepts established for organic (Darwinian) evolution do provide the basis for forming postulates concerning the evolution of metabolism. In particular, there is the concept that as more and more complex biochemical organization evolved, the development was by small but discrete additions to existing systems—additions which arose by chance but, having increased the efficiency of the system, were maintained by natural selection.

In the prebiotic environment described above, there were undoubtedly countless reactions slowly and randomly taking place in which the low molecular weight organic molecules interacted in reactions catalyzed by the surfaces of the inorganic matrices or by the inorganic ions, or by reactive groups (*e.g.*, sulfhydryl) of the organic molecules themselves. Such reactions in general would lead to no development of organization but would only have created a number of new compounds. However, if a reaction, or a series of reactions, led to the formation of a compound which would serve as a catalyst for its own formation from precursor compounds, then an autocatalytic process would have been set in motion. And as it catalyzed more and more of its own formation, the random transformation of its precursors into other substances would be gradually eliminated and the precursors would be converted exclusively to the autocatalytic compound. In this way, a specific metabolic reaction or series of reactions would become established. If from this initial autocatalytic process there was formed (by chance) another autocatalytic step, a more elaborately directed system would result. If the result were one which led to no further autocatalytis, there would be no further development of an organized pattern of reactions. However, whenever the accretion of a new series of reactions led to the formation of a compound which catalyzed a specific reaction of one of the components in the system thus far organized, then there would be further growth of an integrated system of reactions leading to the formation of more and more specific products from suitable organic substances available to the evolving system. Over the surface of the earth, there must have been hundreds of times when such systems developed to a greater or lesser extent. Different autocatalytic systems, all using the oceanic organic molecules as the source material for the transformations, would develop competition, and those systems which could most efficiently and rapidly transform the organic compounds into substances utilizable in their organized pattern of reaction would become dominant. To maintain their dominance, it was also necessary that they continue to accrete new processes which would yield products that would fit into, and give further direction to, the pattern of metabolism already established and which would give the system new catalytic capabilities (new products either functioning more efficiently as catalysts in reactions previously mediated by more primitive compounds in the system or else catalyzing entirely new reactions), thus expanding the metabolic capabilities and extending the development of the existing systems.

These evolving systems needed not only nutrients for the formation of the chemical compounds but also sources of energy for energy-requiring reactions and for forming polymeric structures from the low molecular weight intermediate products. Advantages accruing to systems forming such *polymers* were that they were less likely to lose to their environment the intermediate monomers, that they created structural material which made them less dependent upon (and finally independent of) the inorganic matrices, and that they were able to form an infinitely greater number of surface configurations than had been furnished by their original inorganic environment. The progressively greater contribution by organic polymers to the catalytic surfaces gave more and more specificity to the surfaces on which particular reactions were catalyzed; thus, primitive "protoenzymes" were evolving. Since all modern organisms universally exploit the energy in pyrophosphate linkages, it is apparent that the system which was eventually to dominate all the rest must have very early acquired the capability of utilizing compounds in the environment containing anhydrides of phosphoric acid, and of developing reactions by which pyrophosphates could be created from inorganic orthophosphates (see PHOSPHATE BOND ENERGIES). A relatively simple biochemical system for this latter reaction, and one found in all living cells, is the anaerobic process of GLYCOLYSIS. It can be visualized that from simple organic molecules (originating abiotically from methane in an inorganic environment) there developed, stepwise, integrated series of reactions forming compounds which, because of their catalytic activity, gave to random reactions direction that resulted in ever-expanding autocatalytic systems, and that the system which became dominant and eventually evolved into the ancestral "protocell" was one which integrated that series of reactions which constitutes the glycolytic process observed in modern cells.

From the compounds formed as intermediate and end products of this pyrophosphate-generating system, many of the protein subunits (AMINO ACIDS) and nucleic acid subunits (nucleotides) can be formed by series of reactions which are considered relatively simple and straightforward by organic chemists, and these processes must have been incorporated stepwise into the evolving glycolysis-based system of metabolism. The mononucleotides are, in modern cells, essential catalysts for the polymerizing mechanisms forming carbohydrates, proteins and lipids. Because certain mononucleotides possess the structural configuration which results in their hydrogen bonding to form complementary pairs, by the process of natural selection they became the components of the ancestral genetic mechanisms which ultimately culminated in the formation of ribonucleic acids and later deoxyribonucleic acids.

As the organic compounds in the primitive environment were gradually used up for structural and energy requirements, it became essential for the surviving system to convert sunlight into utilizable chemical energy for the formation of both (a) ATP and (b) compounds capable of reducing (indirectly) carbon dioxide to organic molecules which were components of the metabolic systems that had previously evolved. Hence it must be assumed that a primitive type of PHOTOSYNTHESIS evolved at an early stage after primordial systems became organized into "cellular units." The compounds (glycine and succinic acid) which are used by modern organisms to form the organic framework for the catalysts necessary for photosynthesis (porphyrins) are two simple molecules which would have been formed (a) abiotically in the primitive atmosphere and (b) later directly by relatively simple reactions from components of the glycolysis system. One of the photosynthetic mechanisms that developed (the one which is, thermodynamically, the most efficient) causes the liberation of oxygen from water, and this photolysis resulted in the accumulation of appreciable concentrations of oxygen in the atmosphere.

The availability of molecular oxygen enabled nonphotosynthetic organisms (which had had to depend upon the highly inefficient anaerobic glycolytic process for generating ATP) to gradually acquire oxidative mechanisms. These mechanisms eventually were organized into the highly efficient phosphorylative-electron transport systems (RESPIRATION) carried out in modern cells by MITOCHRONDRIA (or their bacterial equivalent).

In the epoch during which the biochemical events described above occurred, there evolved a well-organized autocatalytic, self-contained system to which one can, in truth, ascribe the term LIFE.

References

1. OPARIN, A. I., "Life: Its Nature, Origin, and Development," New York, Academic Press, 1961.
2. "The Origin of Life on Earth—Symposium," (Moscow), New York, Pergamon Press, 1959; see BERNAL, J. D., "The Problem of Stages in Biopoesis," p. 38.
3. CALVIN, M., "Chemical Evolution," Eugene, Oregon, University of Oregon Press, 1961.
4. MILLER, S. L., "Production of Some Organic Compounds under Possible Primitive Earth Conditions," *J. Am. Chem. Soc.*, **77**, 2351 (1955).
5. EAKIN, R. E., "An Approach to the Evolution of Metabolism," *Proc. Natl. Acad. Sci. U.S.*, **49**, 360 (1963).

ROBERT E. EAKIN

ORNITHINE

L-Ornithine, α-δ-diamino-*n*-valeric acid, $H_2N—(CH_2)_3—CHNH_2—COOH$, is not one of the amino acids usually obtained by the hydrolysis of PROTEINS, though it enters into the makeup of ARGININE—a urea derivative of ornithine. If arginase (from liver) is present when proteins are being hydrolyzed enzymatically, ornithine (along with urea) is produced from arginine. In the absence of arginase, arginine comes through intact. The exclusion of L-ornithine from the list of protein-derived AMINO ACIDS is thus somewhat arbitrary.

In the "ornithine cycle" (or ARGININE-UREA CYCLE) in which the formation of urea from ammonia and carbon dioxide takes place, ornithine picks up ammonia, carbon dioxide and water to produce citrulline, $H_2N—CO—NH—(CH_2)_3—CHNH_2—COOH$; this picks up ammonia (from aspartic acid) and loses water to form arginine which is then converted into urea and ornithine through the agency of the enzyme arginase. The ornithine formed can then repeat the cycle. Under appropriate conditions, ornithine acts in effect as a catalyst to promote the formation of urea from ammonia and carbon dioxide.

OROTIC ACID

Orotic acid, 6-carboxyuracil,

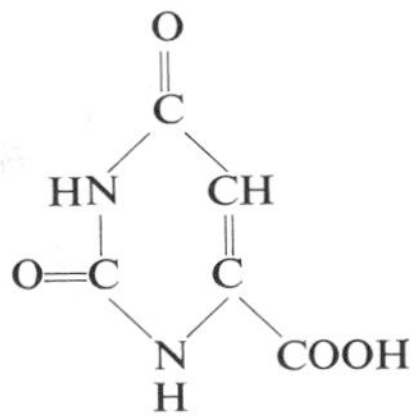

```
        O
        ‖
        C
      /   \
   HN      CH
    |      ‖
  O=C      C
      \   / \
        N     COOH
        H
```

is an intermediate in the synthesis of the PYRIMIDINES which are present both in RNA's and DNA's. Pyrimidines are not included among the nutritional essentials for mammals, but they are essential for cell duplication and, hence, must be produced endogenously (see PYRIMIDINE BIOSYNTHESIS). This is accomplished by a condensation of CARBAMYL PHOSPHATE and aspartic acid to form carbamyl aspartic acid, followed by ring closure and oxidation.

OSMOSIS AND OSMOTIC PRESSURE

If, into the bottom of a jar containing water, a solution of cane sugar is introduced with care so as to avoid mixing, not only will the molecules of cane sugar diffuse into the water but the molecules of water will diffuse into the sugar solution. These processes will go on until the concentration of sugar, and of water, is the same throughout.

If the solution is placed in a container, whose walls are relatively impermeable to the sugar while being permeable to the water, and the container is placed in water, the water will pass from the outside into the container. The term osmosis is usually restricted to the passage of water. If the influx of the water results in an overflow of solution to somewhere other than the surrounding water, this overflow will continue until all the sugar is removed from the container. If the container is closed, water will continue to enter until there is sufficient stress in the stretched walls to cause a pressure on the solution inside; this will eventually stop the influx. Of course, if the walls of the container are not completely impermeable to sugar, then the sugar will be escaping into the water outside the container and this will go on until the concentration of sugar is the same outside and inside. If the walls of the container were impermeable to water but permeable to solute, the latter would escape. The cause of this osmosis, this "pushing," of water into the solution is that the tendency of the water molecules to escape from the pure water is greater than that of the water molecules in the solution. Consider water in contact with a limited volume of air. Of those molecules of water striking the surface some will have sufficient energy to escape into the air and this escape will result in net loss to the air which will continue until the concentration of water vapor molecules there is such that the rate of escape from the air (into the water) equals the rate of escape from the water (into the air). If the volume of the air space is fixed, the pressure will rise. Just as the temperature of all bodies is the same when they are in thermal equilibrium, although their heat content per unit volume varies with their specific heat, so the escaping tendency of the water is the same in all systems when they are in aqueous equilibrium, whether the system is pure water, solution, gas phase, wettable solid, etc. The same concept can be applied to any substance, say mercury in pure mercury, in air containing mercury vapor, and in an amalgam with another metal such as zinc. The term osmosis is usually restricted to the passage of water from a solution where the escaping tendency is higher to a solution where it is lower. Moreover it is usually restricted to the passage through a solid or liquid barrier which prevents the solutions from rapidly mixing. It is not used for the passage of water in the form of vapor through the air from a dilute solution to a stronger solution in the same confined space, although the process is fundamentally the same. It is sometimes restricted to the case where the barrier is semipermeable, *i.e.*, lets through water but not solute.

The escaping tendency of water is lowered by the addition of a solute. If the molecules of the solute have no other effect than to reduce the number of molecules of water in unit volume, then the escaping tendency of the water will be reduced proportionately to the reduction in the mole fraction of water, N_1, the ratio of the moles of water to the sum of the moles of water and solute. Such is a "perfect" solution. If, however, there is some attraction between the solute and water molecules, a smaller fraction of the latter will have energy sufficient to escape—a "nonperfect" solution. The escaping tendency is increased by pressure. Hence a solution in which the water has lower escaping tendency than it has in pure water at the same pressure, P^0, can be brought to water equilibrium by a sufficient increase in the pressure on the solution to a value P. This sufficient increase, $P - P^0$, is the osmotic pressure of the solution. In general we cannot state $P - P^0$, the osmotic pressure, knowing only N_2, the ratio of moles of solute to the sum of the moles of water plus solute, the mole fraction of solute ($N_2 = 1 - N_1$).

What we can say is that if in a solution with a mole fraction N_2 of solute under a pressure P the water has the same escaping tendency as it has in pure water at the same temperature and at a pressure P^0, then $dP/dN_2 = A/B$, where dP/dN_2 is the increase of P relative to increase of N_2 to keep the escaping tendency unchanged, A is the decrease of escaping tendency relative to increase of N_2 when P is unchanged, and B is the increase in escaping tendency relative to increase in P when N_2 is unchanged. For dilute solutions A/B approximates RT/V_1 and so $P - P^0$ approximates N_2RT/V_1, where V_1 is the volume of one mole of water, R is the constant 82.07 cm^3 atm/deg, and T is the absolute temperature. For very dilute solutions, N_2/V_1 approaches n_2/V, the number of moles of solute in a volume V of solution and $P - P^0 = n_2RT/V$ (van't Hoff's equation). This gives an osmotic pressure of 1 atm for one mole of solute in 22.4 liters at 0°C. There is a departure from both these equations for stronger solutions. The fact that one mole of a perfect gas in 22.4 liters at 0°C exerts a pressure of 1 atm, coupled with the above, has led some to say that the osmotic pressure is the bombardment pressure of the solute molecules. It is correct to say that for very dilute solutions, the osmotic pressure of a solution is equal in magnitude to the pressure the solute molecules would exert if they were alone in the same volume and behaved as a perfect gas, but that is another matter.

To measure the osmotic pressure, a semipermeable membrane must be prepared which itself can stand sufficient pressure, or it must be deposited in the walls of a porous pot so that the pressure can be sustained. With the solution inside and water out, pressure is applied to the former until there is no net movement of water.

Observations by Berkeley and Hartley showed that for 3.393 grams of cane sugar per 100 grams of H_2O, the osmotic pressure at 0°C is 2.23 atm while the van't Hoff equation gives 2.17 atm since $n_2/V = 9.72$—10^{-5}. If N_2/V_1 is used instead of n_2/V, the value of 2.22 is obtained. With stronger solutions, the measured osmotic pressure exceeds that calculated: with 33.945 grams of sugar, 24.55 atm is the value measured, while van't Hoff's equation gives 18.41 and the other 21.8 atm. The observed value is given if, in calculating N_2/N_1, it is assumed that each sugar molecule immobilizes five molecules of water.

The solutes in the vacuole of a plant cell are exposed to the inward pressure of the distended cell wall and that of the turgid surrounding cells. Water will pass into the cell vacuole from water outside as long as the total inward pressure on the vacuole falls short of the osmotic pressure of the solution in the vacuole. Passage of water into the vacuole dilutes the contents, lowers the osmotic pressure, increases the inward pressure by distension. The amount by which the inward pressure falls short of the osmotic pressure is called by some the suction pressure.

A substance such as CELLULOSE or gelatin tends to take up water, the tendency decreasing with increase in water content until the stress in the substance causes a sufficient rise in the escaping tendency of the water in the substance. This process, which, like osmosis, is a movement from higher to lower escaping tendency, is called imbibition, and the pressure on the substance sufficient to stop the uptake is the imbibitional pressure. Hence, if a plant cell with a cellulose wall, after coming to equilibrium with a solution, is transferred to water, the wall takes up water by imbibition and the vacuole by osmosis. The latter considers only the over-all movement from outside to vacuole and does not consider the movement from cellulose to vacuole, a process which is the reverse of imbibition. A plant cell in equilibrium with a solution having an osmotic pressure of 25 atm would also be in equilibrium with air about 98% saturated with water vapor. If the cell was transferred to a saturated atmosphere, it would take up water. We lack precise terms for the passage of water from air into the cellulose and into the vacuole. Condensation, which might be used, ranges more widely.

The escaping tendency of water is affected by factors other than concentration of solute and pressure. Increase of temperature increases escaping tendency. This is a complex problem involving not only transfer of water but also of heat. To a minor extent, the passage of water from pure water to a solution involves a heat transfer.

For many naturally occurring membranes, which are not completely semipermeable, i.e., they let solute molecules through slowly, electro-osmosis is important. If the membrane tends to lose negative charges to, or take negative charges from, water or solutions, then the water molecules, in the pores of the membrane, will tend to take on an opposite charge to the membrane. If there is a gradient of electric potential across the membrane, the charged water will move in the appropriate direction. If the potential difference is established by the use of electrodes, this is electro-osmosis.

With some membranes, particularly those containing protein, water may pass from a dilute solution on one side to water on the other—negative osmosis. Under such conditions if the solution was acid the membrane would be positively charged through the uptake of H^+ ions, leaving the water in the pores negative. The greater mobility of the H^+ ions relative to the anions will cause the side of the membrane toward the solution to be negative and so drive the negatively charged water in the pores across the membrane in the opposite direction to normal osmosis.

The rate of osmosis depends not only on the excess of the escaping tendency of the water in the phase from which it moves over that in the phase to which it moves, but also upon the area of surface of interchange and the over-all resistance experienced by the water. The rate of shrinkage of the vacuole of a plant cell when it is placed in a strong solution at first seems surprisingly high. When allowance is made for the fact that the ratio of surface to volume increases as the linear dimension is reduced, then it is realized that when the vacuole of a spherical cell of radius 30 μ shrinks to half its volume in say 5 minutes, the passage of water is only 1 ml per 10,000 cm^2 per minute although the thickness of the layer between vacuole and external solution is of the order of 1 μ in thickness. Under other circumstances, this layer might be said to be relatively impermeable to water. It seems probable that much of the resistance resides not in the cellulose wall or cytoplasm but in the tonoplast which separates the latter from the vacuole.

G. E. Briggs

OVARIAN HORMONES

Physiological Actions. The female sex hormones are steroidal compounds produced chiefly by the ovary in the non-pregnant individual and by the placenta during the later stages of pregnancy. There are two types of female hormones, *estrogens* and *progestins*; these are secreted at different stages of the ovarian cycle.

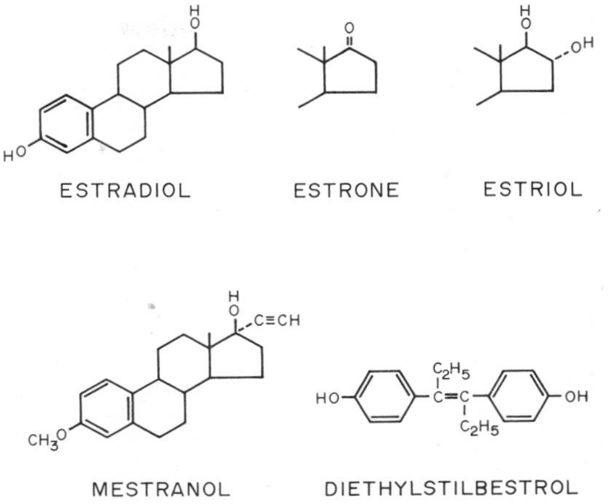

ESTRADIOL ESTRONE ESTRIOL

MESTRANOL DIETHYLSTILBESTROL

FIG. 1. Estrogens.

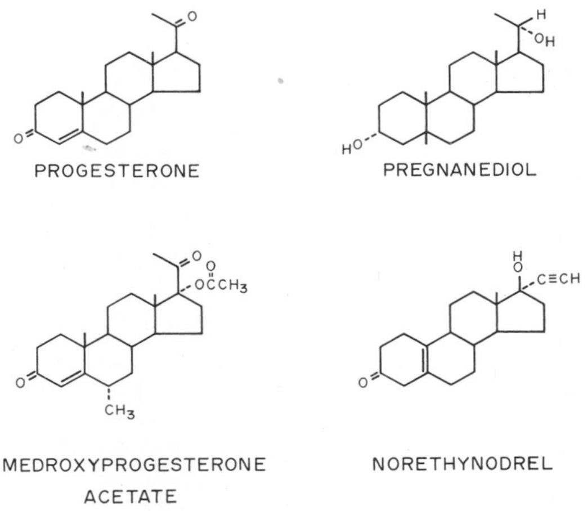

PROGESTERONE PREGNANEDIOL

MEDROXYPROGESTERONE ACETATE NORETHYNODREL

FIG. 2. Progestins.

During the follicular phase, the sac or follicle containing the ripening ovum produces the estrogenic hormone, estradiol (estra-1,3,5(10)-triene-3,17β-diol; see STEROIDS for numbering system), which is primarily responsible for the growth and maintenance of the female reproductive tract and the development in the human of such general feminine characteristics as skin smoothness, distribution of subcutaneous fat and body hair pattern. During the luteal phase, after the mature ovum has been expelled into the Fallopian tube, the wall of the ruptured ovarian follicle forms a yellowish body (corpus luteum) which secretes the progestin, progesterone (pregn-4-ene-3,20-dione). This substance exerts a marked stimulatory effect on the growth and secretion of the uterine lining. In combination with estrogen, it is necessary for the proper implantation of the fertilized ovum, for the maintenance of pregnancy and for the preparation of the mammary gland for lactation.

The activity of an estrogenic hormone usually is evaluated by its ability to promote growth of the uterus or cellular changes in the vaginal epithelium of a castrated or immature rat or mouse. Progestational activity is determined by the extent of proliferation evoked in the uterine endometrium of a rabbit that has been previously stimulated with estrogen.

In addition to their profound effects on the female accessory sex tissues the estrogenic hormones exert an influence on a wide variety of physiological processes. These include the promotion of calcium deposition in BONE, closure of the epiphysis, the retention of sodium and water, the promotion of BLOOD CLOTTING and the retardation in some species of the deposition of cholesterol plaques in the lining of the blood vessels. In some species (ruminants), estrogens enhance general protein anabolism resulting in increased weight of the animal. By their ability to inhibit certain pituitary secretions, estrogens exhibit an indirect inhibitory effect on processes dependent on these pituitary factors.

Biosynthesis and Metabolism. The biochemical precursor of the female sex hormones is cholesterol, which undergoes partial degradation of the side chain to form pregnenolone (3β-hydroxypregn-5-en-20-one). This substance is oxidized to yield progesterone. A combination of 17-hydroxylation and oxidative cleavage of the remaining side chain leads to androstenedione and thence to testosterone, which, by oxidative elimination of the C_{19} methyl group, undergoes A-ring aromatization to form estradiol.

The biosynthesis of steroid hormones in the ovary is stimulated by protein hormones called GONADROPINS which are secreted by the pituitary under regulation by the hypothalamus. Each type of ovarian hormone inhibits the secretion of its gonadotropin by the pituitary, thus effecting a "feedback" control of its own production and a cyclical pattern of ovarian function. During pregnancy, the high levels of sex hormones, produced first by the corpora lutea and later by the placenta, effectively inhibit pituitary gonadotropin secretion, so that follicular development and ovulation do not take place. The use of estrogen-progestin combinations for fertility control is based on this principle.

In the human, as well as many other species studied, estradiol is converted into a large number of metabolites, which are excreted in the urine. In general these transformations involve oxidation of the 17-hydroxyl group, introduction of additional oxygen functions at positions 2 or 6, or 16, or combinations of both. The principal metabolites of estradiol are estrone and estriol; others include 16-ketoestradiol, 16-ketoestrone, 2-methoxyestrone and 16-epiestriol. During pregnancy, when large amounts of estrogen metabolites are excreted in the urine, there is evidence suggesting that some of the estriol may arise from a source other than estradiol. The high urinary level of estriol in the latter stages of pregnancy is an indication of placental function, and a drop in excretion of this steriod is often associated with death of the fetus unless prompt surgical delivery is effected.

The principal urinary metabolite of progesterone is pregnanediol (5β-pregnane-3α,20α-diol), which furnishes a measure of progesterone production in an individual. Since adequate progesterone supply is required for proper gestation, urinary pregnanediol levels usually are checked periodically during pregnancy to determine whether additional progestin need be eliminated to prevent abortion. Other metabolites, which probably represent intermediate stages in the formation of pregnanediol, are 5β-pregnane-3,20-dione, a substance with marked central nervous system depressant activity, and pregnanolone (3α-hydroxy-5β-pregnan-20-one), which elevates the body temperature when administered to humans.

Mechanism of Action. Although little is known about the biochemical mechanism of progesterone action, estrogens have been shown to stimulate a variety of biosynthetic processes in responsive tissues such as uterus within a remarkably short time after administration of the hormone to estrogen-deprived animals. As early as one hour after estrogen treatment, there is an acceleration of the rate of incorporation of labeled precursors into PROTEINS, RIBONUCLEIC ACID, nucleotides and PHOSPHOLIPIDS in rat uterus. Within one to two hours there is an actual accumulation of GLYCOGEN and phospholipid as well as a significant hyperemia and imbibition of fluid by uterine tissue. In the rat and mouse, these early estrogen effects take place without chemical transformation of the estradiol molecule, so that the metabolic conversions of the hormones described above, most of which take place in the liver, do not appear to play any role in the uterotropic process.

The hormone responsive tissues have been shown to possess a component which takes up and retains the steriod; this strong but reversible association with receptor sites appears to be an early if not the initial step of estrogen action, since certain agents which selectively block protein or ribonucleic acid synthesis, and thereby uterine growth response, do not prevent the hormone-receptor interaction. How the formation of the hormone-receptor complex leads to the enhancement of biosynthetic processes remains a mystery.

Clinical and Commercial Uses of Female Sex Hormones. The female sex hormones find a number of important uses both in humans and in certain domestic animals. The most extensive clinical application of ovarian hormones at present is for fertility control in which a combination of estrogen and progestin is employed to prevent follicle maturation and ovulation by inhibiting pituitary gonadotropin secretion (see also REPRODUCTION). Combinations of estrogen and progestin also are used to bring about normal tissue development and endocrine balance in individuals with deficient ovarian function (amenorrhea).

TABLE 1. TYPICAL ORAL PROGESTATIONAL AGENTS AND ANTIFERTILITY PREPARATIONS

Nonproprietary Name ("Trade Name")	Systematic Name	Antiovulatory Product (with Estrogen E or M[a])
17-Hydroxyprogesterone caproate ("Delalutin")	17-Hydroxypregn-4-ene-3,20-dione, caproate	
Medroxyprogesterone acetate ("Provera")	17-Hydroxy-6α-methylpregn-4-ene-3,20-dione, acetate	"Provest" (E)
Megestrol acetate	17-Hydroxy-6-methylpregna-4,6-diene-3,20-dione, acetate	"Volidan" (E)
Chlormadinone acetate	6-Chloro-17-hydroxypregna-4,6-diene-3,20-dione, acetate	"Lutoral," "C-Quens" (M)
Dydrogesterone ("Duphaston")	9β,10α-Pregna-4,6-diene-3,20-dione	
17-Ethynyltestosterone	17-Hydroxy-17α-ethynylandrost-4-en-3-one	
Norethindrone ("Norlutin")	17-Hydroxy-17α-ethynylestr-4-en-3-one	"Norinyl," "Ortho-Novum" (M)
Norethindrone acetate ("Norlutate")	17-Hydroxy-17α-ethynylestr-4-en-3-one, acetate	"Anovlar," "Gestest," "Norlestrin" (E)
Norethynodrel	17-Hydroxy-17α-ethynylestr-5(10)-en-3-one	"Enovid" (M)
Ethynodiol diacetate	17α-Ethynylestr-4-ene-3β,17-diol, diacetate	"Ovulen" (M)
Lynestrenol	17α-Ethynylestr-4-en-17-ol	"Lyndiol" (M)
Dimethisterone ("Secrosterone")	17-Hydroxy-6α-methyl-17α-propynylestr-4-en-3-one	"Ovin," "Oracon" (E)

[a] E = 17-ethynylestradiol; M = mestranol.

Estrogenic hormones are employed for the relief of both physical and psychic symptoms accompanying and following the menopause, for the suppression of lactation after pregnancy and for the treatment of prostatic and some types of breast cancer. There are indications that estrogens may help to prevent or ameliorate ATHEROSCLEROSIS, and their value in this condition is being explored. Estrogens are administered to promote weight gain in sheep and cattle and to impart tenderness to the flesh of poultry.

Since maintenance of pregnancy requires a continued supply of progestin, an important use of these hormones is for the prevention of miscarriage in cases of progesterone insufficiency. Progestins are employed for the treatment of endometriosis and the relief of pain and headache accompanying menstruation (dysmenorrhea). There is some evidence that progestins may alleviate the condition of benign prostatic hypertrophy frequently encountered in older men. Progestins are finding increasing application for the synchronization of estrus in sheep and cattle; by the temporary retardation of ovarian function an entire herd can be brought into estrus at the time when artificial insemination facilities are available.

Since neither estradiol nor progesterone is very effective when given by mouth, certain synthetic derivatives are employed clinically for oral administration. The most widely used oral estrogens are 17α-ethynylestradiol, its 3-methyl ether called mestranol and a simple non-steroidal compound, diethylstilbestrol, which shows all the major physiological actions of estradiol. For animal fattening and tenderizing, either diethylstilbestrol or a closely related analogue, dienestrol diacetate, is commonly employed, either as a feed additive or a subcutaneous implant; estradiol benzoate in combination with either testosterone propionate or progesterone also finds use as a feed additive for cattle.

Because of the recent interest in orally active estrogen-progestin combinations as antifertility agents, a number of oral progestational compounds have been developed. For the most part these are either chemically modified progesterones or 17α-ethynyl derivatives of 19-norandrogens, *i.e.*, substances related to testosterone but lacking the angular methyl group at position 19. Typical oral progestins and estrogen-progestin combinations are listed in Table 1.

ELWOOD V. JENSEN

OXIDASES. See BIOLOGICAL OXIDATION-REDUCTION; OXYGENASES; also ENZYME CLASSIFICATION AND NOMENCLATURE.

OXIDATION-REDUCTION (IN BIOLOGICAL SYSTEMS). See BIOLOGICAL OXIDATION-REDUCTION; OXIDATIVE METABOLISM AND ENERGETICS OF BACTERIA; RESPIRATION. Related topics include ELECTRON TRANSPORT SYSTEMS; OXIDATION-REDUCTION POLYMERS; OXIDATIVE DEMETHYLATION; PHOSPHORYLATION, OXIDATIVE; PHOSPHORYLATION, PHOTOSYNTHETIC; CYTOCHROMES; FLAVINS AND FLAVOPROTEINS; NICOTINAMIDE ADENINE DINUCLEOTIDES.

OXIDATION-REDUCTION POLYMERS

Oxidation-reduction polymers are also termed redox polymers, electron exchange polymers, or electron transfer polymers.

Polymers may be classified, somewhat arbitrarily, as chemically reactive and chemically unreactive types. The latter include most industrial polymers such as polyethylene, polystyrene, nylon, etc., and also CELLULOSE, KERATIN, and a number of inorganic polymers. The former include ion exchangers, redox polymers, enzymes, and other organic and inorganic natural products. Further classification might divide the group of chemically reactive polymers into those that exemplify the acid-base half of chemistry and those that embody the oxidation-reduction half. Ion exexchangers and some enzymes belong in the first subclass, and oxidation-reduction polymers and some other enzymes belong in the second.

Oxidation-reduction polymers are polymeric structures which are capable of undergoing reversible oxidation or reduction (depending upon the initial state) by virtue of the presence of redox functional groups attached to, or as part of the backbone of, the polymeric structure. The redox functional groups that are most commonly found are listed in the following table, where the oxidized and reduced forms are given, and a typical redox polymer is named:

Type of Atom involved	Reduced-oxidized Forms	Polymer
Oxygen	Hydroquinone-quinone	Polyvinylhydroquinone
Sulfur	Thiol-disulfide	Polythiolstyrene mercaptomethylnylon
Nitrogen	Dihydropyridine-pyridinium	Poly(N-vinyl-benzyl nicotinamide) chloride
Iron	Ferrocene-ferricinium	Polyvinylferrocene

Redox polymers may be prepared in linear form as homopolymers and copolymers, and as crosslinked, insoluble materials. They undergo the usual oxidation or reduction reactions, and also show behaviors that depend on the fact that the redox groups are held close to each other, partially fixed in space along the polymer chain. These behaviors model in a simplified way the behaviors of some more complicated biochemically interesting substances. For example, when hydroquinone groups are attached 1,3 to a polymethylene chain, they are in position to interact. Upon half-oxidation, quinone and hydroquinone groups interact to form strongly colored quinhydrone-like charge transfer complexes. The degree of association between quinone and hydroquinone is sensitive to the nature of the solvent, and models the "hydrophobic bond" behavior that is adduced to explain the coiling and denaturation behavior of DNA.[1]

Redox polymers "have been used as simple oxidoreductases, for poising biological systems at a given potential, as virus growth inhibitors, as anti-ulcer agents, in the preparation of hydrogen peroxide, for the removal of oxygen from boiler feedwater, as ANTIOXIDANTS, as semiconductors, as recoverable solid oxidants and reductants, and as electron transfer membranes."[2]

References

1. CASSIDY, H. G., AND KUN, K. A., "Oxidation-Reduction Polymers," New York, Interscience Publishers, 1965.
2. LINDSEY, A. S., *Rev. Pure Appl. Chem.* **14**, 109 (1964).

HAROLD G. CASSIDY

OXIDATIVE DEMETHYLATION

The term "oxidative demethylation" is applied to processes in which a methyl grouping is cleaved from a larger molecule with concomitant oxidation. It appears in general that the methyl moiety is converted to formaldehyde, which distinguishes oxidative demethylation from the methyl transfer reactions (see METHYLATIONS IN METABOLISM).

Examples. At least two types of oxidative demethylation reactions are known. Mackenzie[1] has described preparations derived from MITOCHONDRIA which catalyze the following reactions:

Dimethyglycine → Sarcosine + Formaldehyde

Sarcosine → Glycine + Formaldehyde

In mammalian cells, dimethylglycine can arise from dietrary choline via betaine aldehyde and BETAINE,[2] the last substance being transformed into dimethylglycine in the presence of homocysteine and betaine-homocysteine transmethylase. The glycine derived from complete oxidative demethylation of dimethylygl̃ycine can be transformed into serine or incorporated into PROTEINS and PURINES, while the formaldehyde carbon is presumably conserved by entrance into the folic acid cycle (see SINGLE CARBON UNIT METABOLISM). The enzymes responsible for the oxidative demethylation of dimethylglycine and sarcosine appear to be FLAVOPROTEINS, and purified preparations require the addition of either an artificial electron acceptor such as methylene blue or a protein which couples substrate oxidation to oxygen reduction. No other reactants appear to be necessary for these reactions.

Another type of demethylation reaction is catalyzed by preparations of MICROSOMES from liver cells.[3] In addition to the microsomes, these reactions require the presence of oxygen and reduced NICOTINAMIDE ADENINE DINUCLEOTIDE phosphate; the general over-all reaction appears to be as follows:

$$R{-}CH_3 + O_2 + NADPH + H^+ \rightarrow R{-}H + H_2O + H_2CO + NADP^+$$

Many substances are demethylated in the presence of microsomes, including methylamines, methyl sulfides, methyl ethers and methyl carbamates. Among the compounds so affected are many substances of pharmacological interest such as nicotine, antipyrine and several barbiturates. Also demethylated are a number of INSECTICIDES including "Sevin" and "Isolan." At least for the system observed in rat liver microsomes, this enzymic activity also has two other interesting properties, *viz.*, it is inducible *in vivo* by injection of sublethal amounts of the substrates, and the levels of activity per weight of tissue are consistently higher in males than in females. These observations have led to the proposal that the oxidative demethylation systems present in microsomal preparations are principally detoxification systems used by the organism in handling foreign materials. Arguments for and against this view have recently been summarized in a review by Shuster.[4] (In this connection, it is well to note that microsomal preparations also catalyze steriod ring hydroxylation and fatty acid desaturation, reactions which may turn out to have elements in common with the oxidative demethylation systems.) It is not known at this writing whether or not there is a class of endogenous compounds on which the microsomal oxidative demethylase acts. In the case of the demethylation of methylalkyl anilines, it has been shown that the enzyme complex contains a flavin coenzyme and that the N-oxide of the tertiary amine appears to be an intermediate in the demethylation.[5] *In vivo*, the demethylated substances are presumably conjugated and excreted, and the formaldehyde formed could either undergo the fate described above or be oxidized to carbon dioxide.

Physiological Significance and Distribution. The mitochondrial oxidative demethylation systems appear to be present only in mammalian liver mitochondria and are active in the metabolism of choline. The microsomal oxidative demethylation systems have also been observed mainly in mammalian liver tissue and seem to be largely concerned with the relatively nonspecific detoxification of foreign substances.

References

1. MACKENZIE, C. G., AND HOSKINS, D. D., in "Methods in Enzymology" (COLOWICK, S. P., AND KAPLAN, N. O., Editors), Vol. V, p. 738, New York, Academic Press, 1962.
2. SINGER, T. P., in "The Enzymes" (BOYER, P. D., LARDY, H., AND MYRBÄCK, K., EDITORS), Vol. 7, p. 354, New York, Academic Press 1963.
3. BRODIE, B. B., GILLETTE, J. R., AND LADU, B. N., *Ann. Rev. Biochem.*, **27**, 427 (1958).
4. SHUSTER, L., *Ann. Rev. Biochem.*, **33**, 571 (1964).
5. PETTIT, F. H., ORME-JOHNSON, W., AND ZIEGLER, D. M., *Biochem. Biophys. Res. Commun.*, **16**, 444 (1964).

W. H. ORME-JOHNSON

OXIDATIVE METABOLISM AND ENERGETICS OF BACTERIA

Microoganisms, like other living matter, make use of the ELECTRON TRANSPORT SYSTEMS for the conversion of energy into a chemically utilizable form. The "currency of energy" used for biosynthetic processes is in the form of adenosine triphosphate (ATP). The bacterial enzymes which participate in energy formation are found in highly organized structures which are biochemically analogous to mammalian MITOCHONDRIA. The structure of both the mammalian and the bacterial organelles must be maintained intact in order to generate ATP from oxidative metabolism. In contrast to animal systems which require organic compounds as a source of energy, some microorganisms are capable of utilizing a variety of other chemical substances for this purpose. The heterotrophic (organotrophic) bacterial forms require organic materials for foodstuff, whereas the autotrophic (lithotrophic) organisms can acquire energy from the oxidation of simple inorganic compounds. For example, organisms belonging to the genus *Hydrogenomonas*, an autotrophic microorganism, obtain energy by oxidation of molecular hydrogen. Other autotrophic organisms found in nature are capable of oxidizing hydrogen sulfide, sulfur, ammonia, nitrate, nitrite or ferrous salts as their sole source of energy. Although the energy-forming systems utilize different types of substrates or electron donors (organic *vs* inorganic compounds) and appear to involve different mechanisms, they all have in common a series of oxidative-reductive reactions for the generation of energy-rich phosphate bonds (see PHOSPHATE BOND ENERGIES).

Biological oxidation consists of the stepwise catalytic transfer of hydrogen and electrons from an oxidizable compound to one that can be reduced. Under aerobic conditions, the final transfer of electrons involves the reduction of oxygen. Although the exact sequence of events and carriers is as yet unknown, a diagrammatic scheme for electron flow can be formulated on the basis of the known enzymes and coenzymes found in the isolated intracellular organelles which carry out these reactions, on the basis of the O/R potentials of the isolated coenzymes which act as electron carriers, and by spectrophotometric studies of the interaction of the respiratory chain components (Fig. 1). The sequence of electron flow in some bacterial systems is similar to that described for mammalian systems. It differs, however, from other schemes, in that an additional coenzyme, the naphthoquinone (NQ) is substituted for the benzoquinone usually depicted in schemes of electron transport for mammalian systems (see COENZYME Q). Both benzo- and

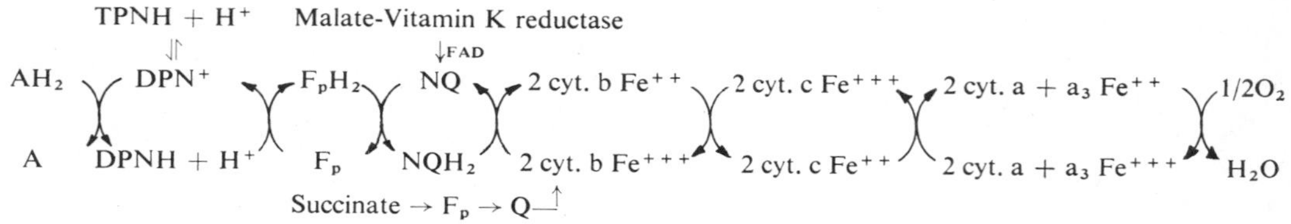

$TPNH + H^+$ Malate-Vitamin K reductase

$\Downarrow\Uparrow$ $\downarrow$FAD

AH_2 — DPN^+ — F_pH_2 — NQ — 2 cyt. b Fe^{++} — 2 cyt. c Fe^{+++} — 2 cyt. a + a_3 Fe^{++} — $1/2O_2$

A — $DPNH + H^+$ — F_p — NQH_2 — 2 cyt. b Fe^{+++} — 2 cyt. c Fe^{++} — 2 cyt. a + a_3 Fe^{+++} — H_2O

Succinate $\rightarrow F_p \rightarrow$ Q$\uparrow$

FIG. 1. Terminal electron transport pathways. The following abbreviations are used: A, AH_2 represent oxidized and reduced substrates (DPN linked); DPN and DPNH, oxidized and reduced diphosphopyridine nucleotide; TPNH, reduced triphosphopyridine nucleotide; F_P and F_PH_2, oxidized and reduced flavoprotein; NQ and NQH_2, oxidized and reduced naphthoquinone; Q, quinone; FAD, flavin adenine dinucleotide.

naphthoquinones have been implicated as coenzymes in electron transport and coupled phosphorylation. The scheme shown in Fig. 1 cannot be applied to all microorganisms since they differ considerably in the nature of their terminal respuratory chains. Microorganisms differ from one another in the nature and type of quinone(s) and in CYTOCHROME composition.

The transfer of two electrons through this series of exergonic reactions usually results in the generation of three molecules of adenosine triphosphate. Although the mechanism of the coupling reactions between electron transport and the generation of phosphate bond energy remains obscure, the loci of the energy "transformers" in the electron transport chain have been described. The P:O ratio (phosphate esterification/2 electrons transferred or μatoms oxygen consumed) has been used as an index of the number of phosphorylative sites associated with the passage of 2 electrons from substrate to oxygen. Three different phosphorylative sites (P:O = 3) have been demonstrated in mammalian mitochondrial systems. Energy-rich phosphate bonds are generated at the DPN level, between cytochrome b and c and between cytochrome a and oxygen. All of the enzymes which carry out the reactions shown in Fig. 1 have been demonstrated in aerobic microorganisms; however, in general the bacterial systems capable of coupling phosphorylation to oxidation differ from intact mitochondria in that they have lower P:O ratios and fail to exhibit respiratory control. The Lower P:O ratios observed with bacterial systems may be a reflection of the lack of one or more energy-generating site(s) in the intact system or to a loss of a phosphorylative site(s) during cell disruption. In addition, bacteria contain nonphosphorylative electron transport "bypass reactions" which lower the P:O ratio. The oxidations carried out by strict aerobic microorganisms (Fig. 1) are referred to as complete oxidation since the end products are H_2O and CO_2 derived from decarboxylation at the substrate level. Some microorganisms, however, lack some of the enzymes necessary for complete oxidation and thus can only carry out a one- or two-step oxidation. This type of oxidation is referred to as incomplete and results in the accumulation of oxidized compounds.

Dehydrogenases. The CITRIC ACID CYCLE appears to be the chief pathway for oxidative metabolism by most microorganisms and serves as the major route for energy formation, synthesis of amino acids and assimilation of carbon compounds. The existence of this pathway in bacteria was difficult to establish because of the lack of permeability to intermediates of the cycle by whole cells. Procedures used to establish the existence of this pathway in animal tissue could therefore not be used with bacteria.

Evidence for this oxidative pathway in bacteria was finally obtained by a number of different methods. The enzymes of the citric acid cycle were demonstrated in cell-free extracts. Analysis of the intracellular pools following the addition of C^{14}-labeled acetate showed that the C^{14} was equally labeled in all the intermediates of the cycle and the distribution of the label in the carbon atoms of the intermediates was that expected from the oxidative reactions of this cycle. Convincing evidence was obtained with mutants of bacteria which were blocked at the condensing enzyme but which retained the ability to form acetyl-CoA. These experiments showed that the condensing enzyme, a vital step in the citric acid cycle, is necessary for acetate oxidation by microorganisms.

The individual enzymes found in this cycle have been isolated from bacteria and studied extensively. They are generally associated with intracellular organelles; however, some of these enzymes can be solubilized from the particles and purified by fractionation procedures. The necessity of intact particles for the coupling of phosphorylation to electron transport can be interpreted as a requirement for a specific spatial orientation of the enzymes and coenzymes involved in these reactions. The enzymes of the citric acid cycle and terminal respiratory chain are associated with the same intracellular structure.

The electrons resulting from the dehydrogenation of isocitrate, α-ketoglutarate and malate are linked to oxygen by the terminal respiratory enzymes shown in Fig. 1. The dehydrogenase involved in the oxidation of malate is DPN^+ linked, whereas isocitric dehydrogenase is TPN^+ linked. Dehydrogenation of α-ketoglutarate differs however, in that it probably involves DPN and LIPOIC ACID. In addition to the DPN^+-linked malate dehydrogenase, some bacteria also oxidize malate by a DPN^+-independent pathway referred to as malate-vitamin K reductase. This FLAVIN enzyme (FAD) links to the DPN^+ chain at the naphthoquinone level of oxidation-reduction.

$$\text{Malate} + \text{DPN}^+ \rightarrow \text{Oxalacetate} + \text{DPNH} + \text{H}^+ \quad \text{(Malic dehydrogenase)} \quad (1)$$

$$\text{Malate} + \text{NQ} \xrightarrow{\text{(FAD)}} \text{Oxalacetate} + \text{NQH}_2 \quad \text{(Malic-vitamin K reductase)} \quad (2)$$

A phosphorylative site at the DPNH level of oxidation in bacterial systems has been demonstrated in *Alcaligenes faecalis* and *Mycobacterium phlei*. Succinate oxidation differs from DPN- and malate-linked pathways in that it is mediated by a different flavoprotein and quinone and linked to the terminal respiratory chain at the cytochrome b level of oxidation-reduction. (DPN is also termed NAD, NICOTINAMIDE ADENINE DINUCLEOTIDE.)

TPNH oxidation is utilized in biosynthetic reactions and for ATP synthesis through the respiratory chain, An enzyme has been found in bacteria which transfers hydrogen from TPNH to DPN.

$$\text{TPNH} + \text{H}^+ + \text{DPN}^+ \rightleftharpoons \text{DPNH} + \text{H}^+ + \text{TPN}^+ \quad (3)$$

This enzyme, pyridine nucleotide transhydrogenase, may permit the oxidation of TPNH to occur via the DPN-linked electron transport pathway. The phosphorylation associated with TPNH oxidation arises from the DPN^+ pathway. The reversal of the transhydrogenase reaction (reduction of TPN from DPNH) was shown to require energy. The energy-dependent reversal of electron transport has been demonstrated in certain microorganisms.

Flavoproteins. There are many flavin enzymes which act as mediators in oxidative reactions. These enzymes differ from one another in the nature of their prosthetic group, in the type of compound oxidized and in their reactions with various electron acceptors. The prosthetic group of most flavoproteins is either riboflavin monophosphate or flavin adenine dinucleotide (for structures, see FLAVINS AND FLAVOPROTEINS). A few electron transport enzymes are known which are activated by riboflavin.

Several flavin enzymes which promote the oxidation of reduced pyridine nucleotide coenzymes have been described. These enzymes differ with respect to the nature of the electron acceptor and the rate at which the acceptor can be reduced. The reaction carried out by each enzyme is generally specific with regard to the electron donor and acceptor. Cytochrome c, oxygen, GLUTATHIONE, nitrate, menadione, VITAMIN K, and various dyes have all been described as electron acceptors for different pyridine nucleotide reductases. The pyridine nucleotide reductases are of particular importance since they form one of the essential links of the terminal respiratory pathway. Most of the reductases, however, utilize as electron acceptors compounds which are not found in biological systems. Some of the flavin enzymes contain metal ions in addition to the flavin prosthetic group. The metal ions (usually non-heme iron) are required for electron transfer to the oxidized acceptor.

Another class of flavoproteins catalyzes the direct transfer of electrons between glucose or amino acids and molecular oxygen. The flavins which react directly with molecular oxygen lead to the formation of hydrogen peroxide. The enzymes which carry out these oxidations are specific with respect to the configuration of the substrate. Thus, the D-glucopyranose configuration is necessary for activity with glucose oxidase and the amino acid oxidases are specific for either the D- or the L-amino acids. Xanthine oxidase differs from glucose or amino acid oxidase in that it is a molybdoflavoprotein like the TPNH-nitrate reductase and hydrogenase.

Quinones. A large variety of benzo- and naphthoquinones (Fig. 2) are found in microorganisms.

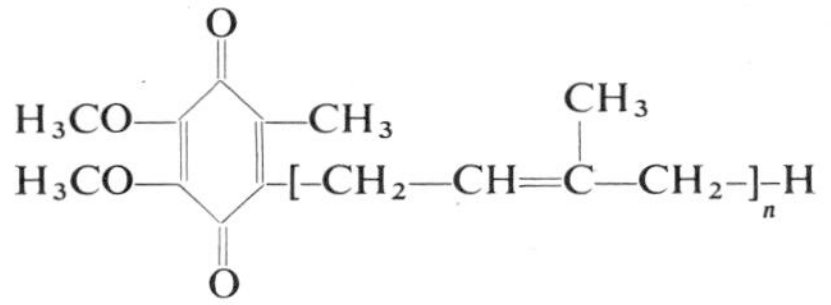

H_3CO, H_3CO, CH_3, $-[-CH_2-CH{=}C(CH_3)-CH_2-]_n-H$

Benzoquinone

CH_3, $-[-CH_2-CH{=}C(CH_3)-CH_2-]_n-H$

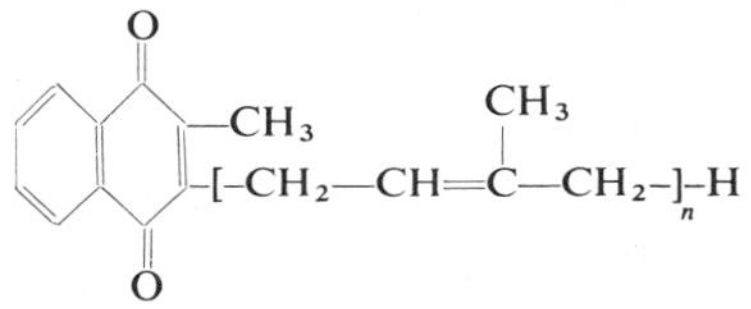

Naphthoquinone

FIG. 2. Naturally occurring quinones.

The naturally occurring quinones differ wirh respect to the type of quinone nucleus and with the length and degree of saturation of the isoprenoid side chain. These lipoidal compounds are localized in the subcellular organelles which carry out electron transport and are required for energy generation. The quinones have been shown to function as coenzymes in electron transport and are required for coupled phosphorylation. Enzymes have been isolated from bacteria which mediate the transfer of electrons from reduced DPN, malate (FAD) or succinate to certain specific quinones. Some microorganisms contain more than one type of quinone and both appear to be functional in electron transport. For example, *E. coli* contains both a benzo- and naphthoquinone; the naphthoquinone serves as a coenzyme on the DPN-linked pathway whereas the benzoquinone mediates electron transport on the succinate-linked pathway. The dependence of electron transport and coupled phosphorylation on this lipoidal coenzyme was demonstrated following removal or destruction of the naturally occurring quinones from the subcellular organelles which carry out coupled phosphorylation. Both oxidation and phosphorylation are lost following such treatment and can only be restored by the addition of quinones with certain specific configurations. It is of interest to note that all of the naturally occurring quinones contain a long side chain adjacent to one of the carbonyl groups with a β-γ unsaturated position (Fig. 2). Quinones which restore both oxidation and phosphorylation contain the β-γ unsaturated side chain. Other quinones are capable of restoring only oxidation. Quinones like menadione, which lack the side chain, restore oxidation by a bypass reaction transferring electrons from DPNH to oxygen or cytochrome b, whereas other quinones like lapachol, which contain substitutions in the 2-position of the naphthoquinone ring, transfer electrons by the same oxidative pathway as the natural quinone. Nevertheless, quinones belonging to the latter group do not restore phosphorylation.

The quinones mediate electron transport between the flavoprotein and cytochrome b on the DPN and succinate-linked pathways. They have been shown to be involved in pyruvate, formate and nitrate oxidation. Quinone-mediated oxidations are inhibited by dicumarol or other naphthoquinone analogues and following destruction of the natural quinone by irradiation with light at 360 mμ.

Terminal Respiratory Pigments. All aerobic forms of life possess iron porphyrin (see structure XIII, in CHELATION AND METAL CHELATE COMPOUNDS) proteins which link the respiratory enzymes to oxygen through a series of one-step electron transfers. Microorganisms, however, differ considerably from one another with regard to the nature of the terminal respiratory pigments. Some aerobic microorganisms contain CYTOCHROME pigments which are similar in absorption characteristics to those found in yeast and mammalian tissues. These microorganisms contain cytochromes whose spectra resemble those described for cytochromes b, c_1, c, a and a_3; however, they usually differ from those found in yeast and mammalian tissues in their sensitivity to Antimycin A, KCN, CO, and light reversibility. Cytochrome a is often missing in some aerobic and facultative microorganisms. This cytochrome, however, is usually replaced by cytochrome a_1 or a_2 or a mixture of the two. In certain facultative microorganisms cytochrome c and b appear to be replaced by one cytochrome pigment, b_1, whereas anaerobic bacteria with few exceptions are completely devoid of cytochromes. Variations in the type and content of the cytochrome pigments can occur by alteration of the degree of aeration, phase of growth, energy source and iron content of the medium employed.

Although the spectral properties of cytochrome pigments among various bacterial species differ, these pigments, nevertheless, must assume the same physiological role in metabolism as that ascribed to the "typical" cytochromes found in yeast and mammalian tissues. Like mammalian tissues, a phosphorylative site has been demonstrated in the span from cytochrome c to oxygen in a few aerobic microorganisms. The differences between the various cytochromes in bacteria merely represents a variation on the same biological theme.

References

1. KAPLAN, N. O., *Bacteriol. Rev.*, **19**, 234 (1952).
2. BRODIE, A. F., *Federation Proc.*, **20**, 995 (1961).
3. DOLIN, M. I., "The Bacteria," Vol. 2, p. 319, New York, Academic Press.
4. BRODIE, A. F., AND RUSSELL, P., JR., *Proc. Intern. Congr. Biochem., 5th, Moscow*, **5**, 89 (1963).
5. SMITH, L., *Bacteriol. Rev.* **18**, 106 (1954).

ARNOLD F. BRODIE

OXIDATIVE PHOSPHORYLATION. See PHOSPHORYLATION, OXIDATIVE.

OXYGEN TRANSPORT. See GASEOUS EXCHANGE IN BREATHING; HEMOCYANINS; HEMOGLOBINS (COMPARATIVE BIOCHEMISTRY); MYOGLOBINS.

OXYGENASES

Definition and Nomenclature. The term "oxygenase" may be defined as a group of enzymes catalyzing the activation of molecular oxygen and the subsequent fixation of oxygen into various substrates. Oxygenases have been found to function in reactions that can be classified into two types. These are schematically shown by Eqs. (*1*) and (*2*), where S stands for a substrate and RH_2 for a reducing agent:

$$S + O_2 \rightarrow SO_2 \qquad (1)$$

$$S + O_2 + RH_2 \rightarrow SO + R + H_2O \qquad (2)$$

When the enzymes catalyze the addition of both atoms of oxygen to a molecule of substrate S as shown in Eq. (*1*), they are designated as "dioxygenases." In Eq. (*2*), one of the oxygen atoms is incorporated into a substrate molecule, and the other atom is reduced to water in the presence of an appropriate electron donor. The enzymes which catalyze the second type of reaction are now called "monooxygenases".[1] They were previously referred to as "mixed function oxidases" by Mason,[2] because these enzymes are bifunctional, carrying out oxidase function on the one hand and oxygenase function on the other. However, it was recently found that anthranilate hydroxylase catalyzed the incorporation of two atoms of oxygen into a molecule of anthranilic acid, followed by the reduction of both atoms of oxygen to form hydroxyl groups of catechol in the presence of DPNH.[1]. This new type of reaction is schematically shown by Eq. (*3*):

$$S + O_2 + RH_2 \rightarrow S\begin{matrix}\diagup OH \\ \diagdown OH\end{matrix} + R \qquad (3)$$

The term hydroxylase has also been used to designate the second type of enzyme. However, these enzymes also catalyze a diverse group of reactions such as dealkylation, decarboxylation, deamination, N- or S-oxide formation and so forth.

Furthermore, not all hydroxylation reactions are catalyzed by the oxygenative hydroxylases. Therefore the term "hydroxylase" is inadequate as a designation for this group of enzymes, although it may still be used for naming individual enzymes.

Methodology. In order to establish whether the suspected enzyme is an oxygenase or not, it is essential to demonstrate the incorporation of molecular oxygen into the reaction product. This is usually carried out by running two parallel experiments differing only in that oxygen 18 is present as a component of water in one case and as atmospheric oxygen in the other. The products are isolated and analyzed for oxygen-18 enrichment by a mass spectrometer. A major technical problem in such experiments is the elimination or minimizing of nonenzymatic exchange of oxygen between the reaction product and water. For details, the readers are referred to an excellent review article by Samuels.[3]

Classification of Oxygenases. Dioxygenases are generally concerned with the ring fission of various aromatic compounds. A large number of enzymes, which contain inorganic iron as the only cofactor, catalyze insertion of two atoms of oxygen into various mono- or di-phenolic compounds. These are, therefore, referred to as phenolytic oxygenases. Pyrocatechase (catechol:oxygen 1,2-oxidoreductase or catechol 1,2-oxygenase) and metapyrocatechase (catechol:oxygen 2,3-oxidoreductase or catechol 2,3-oxygenase) are typical phenolytic oxygenases.[4]

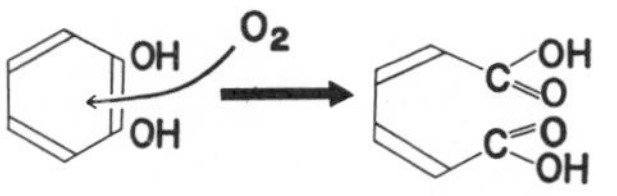

Catechol — O_2 → *cis, cis*-Muconic acid

Pyrocatechase

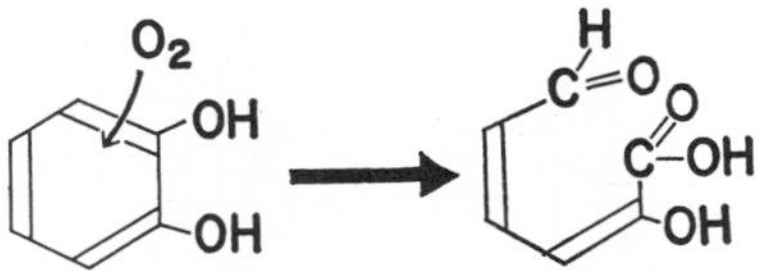

Catechol — O_2 → α-Hydroxymuconic semialdehyde

Metapyrocatechase

The other group of dioxygenases contains HEME instead of inorganic iron. L-Tryptophan oxygenase (pyrrolase) has been isolated from liver and *Pseudomonas*, and D-tryptophan oxygenase was found in intestinal mucosa. L- and D-tryptophan are converted to L- and D-formylkynurenine, respectively.

In most monooxygenase reactions, TPNH appears to be a specific electron donor, but in certain instances DPNH is preferred or serves as the specific electron donor (see NICOTINAMIDE ADENINE DINUCLEOTIDES). ASCORBIC ACID acts as a reductant with phenolase or dopamine β-hydroxylase. Reduced FAD and FMN are reductants for salicylate hydroxylase and diketocamphene monooxygenase, respectively. TetrahydroPTERIDINES are also utilized in the cases of phenylalanine hydroxylase, tyrosine hydroxylase and the conversion of glycerol ether to the corresponding hemiacetal. Although the above-mentioned reactions require the presence of external electron donors, the substrate themselves may serve as internal electron donors for certain monoxygenases. For example, in the case of lactic oxidative decarboxylase, the substrate, L-lactic acid, accepts one atom of oxygen and simultaneously furnishes two atoms of hydrogen to reduce another atom of oxygen by way of enzyme-bound reduced riboflavin-5′-phosphate ($FMNH_2$). Monooxygenases may be classified by the electron donors.[5] Classification of oxygenases and examples are shown in Table 1.

General Properties and Mechanisms of Action. Several dioxygenases have recently been obtained in either crystalline or highly purified forms, which were shown to be homogeneous upon ultracentrifugation and electrophoresis. These include metapyrocatechase, pyrocatechase and 3,4-dihydroxyphenolacetate-4,5-dioxygenases. Experimental evidence obtained with these highly purified enzymes indicates that inorganic iron is the only cofactor involved in such reactions and that iron binds not only with oxygen but also with substrate. It was proposed that in this ternary complex, both oxygen and organic substrate are activated and interact to form an oxygenated product. A similar mechanism has been suggested for tryptophan pyrrolases. For detailed discussions, the readers are referred to a recent review article.[1]

The mechanism of action of monooxygenases has not been fully elucidated. The involvement of heavy metals such as iron, copper, etc., is still a subject for controversy, although there has been an increasing amount of evidence to indicate the participation of heavy metals in monooxygenase-catalyzed reactions. Available evidence indicates that the oxygen is incorporated into a substrate molecule and is simultaneously reduced by the electron donor; the bond between the two oxygen atoms is cleaved, forming a monooxygenated product and water as products.

Physiological Significance and Distribution of Oxygenases in Nature. The oxygenases appear to enjoy a ubiquitous distribution in animals, plants and microorganisms. Their biological role, however, appears to be different from that of the previously known respiratory enzymes. Dehydrogenases and oxidases catalyze transfer of hydrogen atoms and electrons, and are mainly concerned with the generation of ATP. On the other hand, oxygenases catalyze transfer of oxygen and are mainly involved in the transformation of essential metabolites and foreign compounds. They play an important role in the metabolism not only of various aromatic compounds [see AROMATIC

TABLE 1. OXYGENASES

Classification	Cofactors	Examples
Dioxygenase	(1) Inorganic iron	Pyrocatechase, metapyrocatechase, 3-hydroxyanthranilate oxygenase, homogentisate oxygenase
	(2) Heme	L-Tryptophan pyrrolase, D-tryptophan pyrrolase
Monooxygenase	(1) Pyridine nucleotides	Kynurenine hydroxylase, steroid hydroxylases, squalene oxidocyclase
	(2) Flavin nucleotides	Salicylate hydroxylase, diketocamphane monooxygenase
	(3) Cytochromes	Aromatic hydroxylase
	(4) Ascorbic acid	Phenolase, dopamine hydroxylase
	(5) Pteridine derivatives	Phenylalanine hydroxylase, tyrosine hydroxylase, glyceryl-ether hydroxylase
	(6) Substrate itself	Lactic oxidative decarboxylase, inositol oxygenase

AMINO ACIDS (BIOSYNTHESIS AND METABOLISM)] and other cyclic compounds but also of simple aliphatic compounds as well. They also participate in many biosynthetic reactions. The substrates for oxygenases include a wide variety of compounds such as AMINO ACIDS, sugars, fats, hydrocarbons, HEMOGLOBIN, STEROIDS, VITAMINS, HORMONES, toxic and CARCINOGENIC SUBSTANCES and so forth.

References

1. HAYAISHI, O., "Oxygenases," Plenary Volume, VIth International Congress of Biochemistry in New York, I.U.B. Vol. 33, 1964, p. 31.
2. MASON, H. S., "Mechanisms of Oxygen Metabolism," *Science*, **125**, 1185 (1957).
3. SAMUEL, D., "Methodology of Oxygen Isotopes," in "Oxygenases" (HAYAISHI, O., EDITOR), p. 32, New York, Academic Press, 1962.
4. HAYAISHI, O., "Direct Oxygenation by O_2, Oxygenases," in "The Enzymes" (BOYER, P. O., LARDY, H., AND MYRBÄCK, K., EDITORS, Vol. VIII), New York, Academic Press, 1963.
5. HAYAISHI, O., "Biological Oxidation," *Ann. Rev. Biochem.* **31**, 25 (1962).

O. HAYAISHI

OXYTOCIN. See VASOPRESSIN; also HORMONES, ANIMAL.

P

PANTETHEINE

Pantetheine, N-(pantothenyl)-β-amino-ethanethiol, was first discovered as a growth factor for one of the lactic acid bacteria, *Lactobacillus bulgaricus*. All oxygen-using organisms have COENZYME A as part of their metabolic machinery; experimental animals can build it if furnished PANTOTHENIC ACID, or they can utilize various conjugated forms of pantothenic acid, including pantetheine and coenzyme A itself. Some microorganisms are unable to utilize conjugates, probably due to the fact that these do not penetrate their cell walls. *Lactobacillus bulgaricus* is one bacterium that is deficient in the ability to carry out conjugation starting with pantothenic acid itself. Some microorganisms including yeasts are able to build pantothenic acid conjugates when furnished β-alanine as a starting material.

PANTOTHENIC ACID

The first hint as to the existence of what is now called pantothenic acid arose from a publication in 1901 (Wildiers) in which a hypothetical "bios" was found to be a growth-promoting factor essential for yeast. If the authors could have had at the time of their publication a sample of pantothenic acid they would have identified it as "bios" because pantothenic acid alone without other supplements is highly active with respect to the yeast which they were investigating.

The first evidence that a substance answering the general description of pantothenic acid is required by higher animals was gained in 1928 by R. R. Williams and R. E. Waterman, who found a "third factor" in vitamin B in addition to thiamine and a thermostable one. It was destroyed by dry heat, or by long autoclaving, and was essential to weight maintenance and health in pigeons. If pure pantothenic acid had been available to these investigators, they would doubtless have recognized it as their unknown. Pantothenic acid has (infrequently) been designated vitamin B_3.

This vitamin is unique among the vitamin group in that it was one of the first to be isolated using as a basis a microbiological assay method; even more unique is the fact that its structure was largely determined, using a highly quantitative biological yeast test, long before it was isolated or even obtained in concentrated form. Several years before it had been successfully concentrated, the writer and his co-workers described it as an acid with an ionization constant lower than that of an α-hydroxy acid, but about right for a hydroxy acid in which the hydroxyl group was farther removed from the carboxyl group. This was determined by studying the electrical transport of the growth promoting substance, as detected microbiologically. Its molecular weight (*i.e.*, that of the active principle) was found by diffusion techniques to be about 200 (correct, 219); the active principle lacked primary or secondary amino groups (or any group with substantial basic property); it also lacked aldehyde, ketone, olefin or sulfhydryl groups. It did, however, have at least one hydroxyl group in its structure.

Later after the vitamin had been concentrated but not purified, it was found, by the use of microbiological tests, to contain β-alanine in its makeup and to be a condensation product of a lactone and β-alanine. The entire structure is as follows:

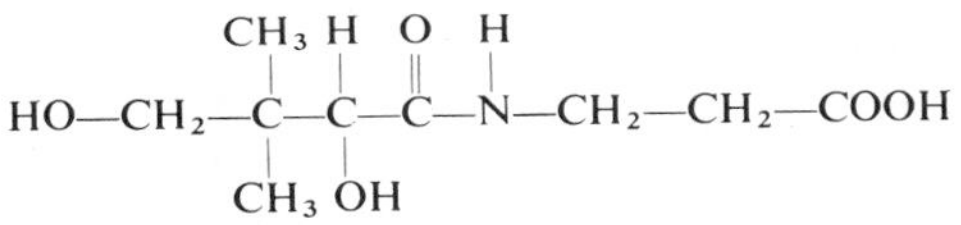

$$HO{-}CH_2{-}C(CH_3)(CH_3){-}C(H)(OH){-}C(=O){-}N(H){-}CH_2{-}CH_2{-}COOH$$

The purification of pantothenic acid from natural sources was accomplished by the writer and his associates only to the extent that about 10 mg of about 90% pure material was obtained from liver in 1938, along with much larger quantities of less potent concentrates. Its highly hydrophilic nature made its isolation difficult. Even its brucine salt is about 1000 times more soluble in water than in chloroform. The acid or its salts have never been obtained in completely pure form from natural sources. The synthetic substance with the D-configuration has in every way exactly the same biological characteristics as the natural vitamin.

Because of the ease of applying the biological yeast test, the whole biological kingdom could be explored for its presence. It was found to be ubiquitous in all kinds of organisms and all types of cells. This led to the choice of the name "pantothenic" which in Greek means *from everywhere*. It was subsequently found to be a dietary essential for chickens, ducks, turkeys, rats, mice, dogs, foxes, pigs, monkeys, cattle and horses, as well as protozoa and insect larvae. While it is synthesized by green plants and many bacteria, it was early shown to be stimulative to alfalfa seedlings, a liverwort, *Ricciocarpus natans*, and pea embryos.

Pantothenol, structurally like pantothenic acid except for the replacement of the carboxyl group by a primary alcohol group, appears to be approximately the nutritional equivalent for animals, indicating that the *in vivo* oxidation of the alcohol is easy.

The manifestations of pantothenic acid deficiency in animals are many, including: dermatitis, ulcers throughout the gastrointestinal tract, intussusceptions, anemia, achromotrichia, depigmentation of tooth enamel, congenital malformations, bowel atony, failure to produce antibodies, hemorrhagic adrenal medulla and cortex, spinal cord lesions, dehydration, fatty liver, thymus involution, kidney damage, heart damage (and sudden death), bone marrow hypoplasia, leucocyte deficiency, spinal curvature, myelin degeneration, uncoordinated gait, etc. The close relationship of pantothenic acid to the reproductive process is suggested by the high content of pantothenic acid in "ROYAL JELLY" and the more recent finding that codfish ovaries are several times richer than "royal jelly," previously the richest known natural source.

The artificial production of any substantial degree of pantothenic acid deficiency in humans may be questioned on humanitarian grounds and from the scientific standpoint; the findings would be scarcely more interesting than those that might be obtained by artificially producing a phosphate deficiency.

The most reasonable interpretation of the extremely diverse effects resulting from pantothenic acid deficiency in various animals (and humans) is based upon the supposition that each tissue in the body is capable of being nourished at various levels of efficiency and that pantothenic acid deficiency, which can potentially cause damage in every tissue, strikes sometimes here and sometimes there, depending upon many factors which reside in the species or in the afflicted individual animal. Despite the absence of any well-defined human deficiency syndrome, there can be no intelligent question regarding the importance of pantothenic acid in human nutrition.

Elsewhere the writer has pointed out that it is unfortunate to assume that since pantothenic acid is ubiquitous, natural deficiencies cannot exist. If ubiquitousness could be interpreted as ruling out the possibility of deficiency, we would be able *a priori* to rule out deficiencies of thiamine, riboflavin, pyridoxine, etc., as well as all individual amino acids and elements such as iodine, calcium or iron.

As with every nutrient, deficiency is determined by *quantitative* considerations: the relative demand and the supply. Because human milk and human muscle are relatively rich in pantothenic acid as compared to other animals, it may be inferred that human needs are relatively high, and therefore deficiencies which manifest themselves in a diffuse way may be common.

Much of the current biochemical interest in pantothenic acid relates to its presence as an integral part of COENZYME A which is discussed in a separate article. PANTETHEINE is a naturally occurring pantothenic acid derivative of interest.

References

1. WILLIAMS, R. J., "Growth-Promoting Nutrilites for Yeasts," *Biol. Rev.*, **16**, 49–80 (1941).
2. WILLIAMS, R. J., "Pantothenic Acid," in "Comprehensive Biochemistry" (FLORKIN, M., AND STOTZ, E. H. EDITORS), Vol. II, Ch. 5, pp. 59–65, Amsterdam, Elsevier, 1963.
3. WILLIAMS, R. J., "Nutrition in a Nutshell," New York, Doubleday and Company, 1962.

ROGER J. WILLIAMS

PAPER CHROMATOGRAPHY. See CHROMATOGRAPHY, PAPER.

PARATHYROID HORMONES

The influence of the parathyroid glands on the regulation of CALCIUM concentrations in the blood of mammals was first recognized by MacCullum and Voegtlin (1909) who reported that removal of these glands produced a marked drop in serum calcium levels causing tetany. That the glands also affected PHOSPHATE was demonstrated by Greenwald (1911) who produced a rapid fall in urinary phosphate by parathyroidectomizing dogs. In 1924 and 1925, Hanson and Collip independently prepared physiologically active extracts of beef parathyroid. Similarly prepared extracts are the only hormone preparations commercially available today.

Recently, several groups of investigators, led by Aurbach, have succeeded in purifying and partially identifying the structure of the hormone, variously called "parathormone," "parathyroid hormone" or "purified parathyroid extract." This is a single-chain peptide hormone with a molecular weight of about 8000. It is expected that very shortly the entire amino acid sequence will be determined. Its partially identified structure is given in Fig. 1.

A second parathyroid hormone, calcitonin, which possesses hypocalcemic properties was postulated by Copp in 1961. The verification of the existence of this hormone has been clouded by the recent discovery by Hirsch[5] of the existence of a similar hypocalcemic principle in thyroid tissue (thyrocalcitonin). Purification of thyrocalcitonin has proceeded rapidly leaving little doubt of its existence. It is generally believed, at the present time, that the hormone postulated by Copp from the parathyroid glands is in actuality the hormone which is now known to be produced by the thyroid gland. However, it is possible that a parathyroid calcitonin may exist in certain species. In both cases however, the only action known is the ability of the extract to lower serum calcium levels, due primarily to inhibition of bone resorption. Little evidence is available at the present time as to the site of action or the part played by the hormone in the physiology of the animal.

$$\textit{NH}_2\textit{-terminus}\quad \text{H}\cdot\text{Ala}\cdot\text{Val}\cdot(\text{Ser,Lys,His,Glu})(\text{Gly,Asp,}\overset{\text{GN}}{\text{AN}}\text{,Ileu,Leu})\text{Met-}$$

$$\left[\text{Phe}\begin{pmatrix}\text{Lys}_6\text{,His}_2\text{,Arg}_5\text{,Asp}_4\text{,Thr}_1\text{,Ser}_3\text{,Glu}_6\text{,}\\ \text{Pro}_3\text{,Gly}_2\text{,Ala}_5\text{,Val}_4\text{,Ileu}_1\text{,Leu}_5\text{,Try}_1\end{pmatrix}\text{Tyr}\right]\text{-}$$

$$(\text{Met,Gly})(\text{His,Phe,Ileu,Leu,Glu,Asp,}\overset{\text{AN}}{\text{GN}}\text{,Lys,Val,Ala})\cdot\text{Ser}\cdot\text{Leu}\cdot\text{OH}\quad \textit{COOH-terminus}$$

FIG. 1. Partial structure of parathyroid hormone. (From POTTS, J. T., JR., and AURBACH, G. D., "The Chemistry of the Parathyroid Hormone," in reference 1.)

The more classical function of parathyroid hormone is concerned with its control of the maintenance of constant circulating calcium levels. For many years this role was attributed to its corollary effect on increased renal phosphate excretion, an effect which is equally demonstrable with both crude and purified hormone preparations and appears to be produced by decreasing renal tubular absorption of phosphate. Also, an opposite effect of the hormone on calcium excretion has been demonstrated in rodents and recently confirmed in man, presumably produced by a hormonally controlled increased tubular reabsorption of calcium. Over the past few years there have also been reports suggesting a positive action of the hormone in producing a stimulation of intestinal absorption of calcium. However, because of conflicting evidence, some with completely contrary results, this function of the hormone cannot be considered proved. As to control, increased secretion of the hormone from the parathyroid glands has been shown to be the result of a direct "feedback" mechanism produced by a lowered ionic calcium level of the blood passing through the gland.

Despite these extraosseous effect of parathyroid hormone, it is obvious that its primary action in controlling calcium concentrations is due to its ability to increase the transfer of calcium from the solid phase of BONE into circulating fluids. The ramifications of this effect on the metabolism of bone are receiving most of the attention of investigators in the field today. A review of a few of the different types of studies in progress is given with the belief that from these, in the near future, the mechanism of action of the hormone will be identified.

(1) Both endogenous and exogenous hormone are capable of increasing the numbers of osteoclasts in bone. Exogenous hormone administration can also produce a decrease in bone COLLAGEN synthesis and a subsequent disappearance of osteoblasts. Additional studies with radiothymidine and radiocytidine add support to the hypothesis that the hormone is capable of influencing the normal modulation of bone cell types and thereby influencing bone remodeling (reshaping).

(2) Parathyroid hormone can be shown to affect citrate metabolism of bone *in vivo*, and to increase both lactate and citrate production of bones incubated in physiological media. These studies have led to the suggestion that solubilization of bone salts is achieved through the acidity and/or the complexing ability of these acids. While it is an acceptable theory, this has not been established.

(3) Numerous studies have demonstrated that crystal breakdown is preceded, accompanied, or immediately followed by breakdown of the components of the matrix of bone; collagenolytic activity in bone has recently been reported. These studies have led to the speculation that the fundamental action of the hormone might be through stimulation of enzymatic breakdown of collagen, MUCOPOLYSACCHARIDES, or both.

(4) Of the many enzymatic processes studied, the most encouraging results pertain to the suppression by the hormone of NADP levels in bone tissue, implicating both the CITRIC ACID CYCLE and areas in the pentose shunt as possible sites of hormonal action.

(5) Several studies have demonstrated hormonal effects on the mitochondrial transport of calcium and/or phosphate. These studies look promising. However, at the present time, they have been limited almost entirely *in vitro* studies with mitochondrial preparation from the liver of rats, using excessive levels of hormonal preparations. This raises some doubts concerning the specificity of the effect.

In summary, progress in this area is rapid. One might hazard the guess that in the near future a single site of hormonal action in bone will be delineated. This must explain both the morphological and the biochemical changes necessary for the proper control of calcium homeostasis, while at the same time allowing for satisfactory growth and remodeling of bone.

References

1. GAILLARD, P. J., TALMAGE, R. V., AND BUDY, A., (EDITORS), "The Parathyroid Glands: Ultrastructure, Secretion, and Function," Chicago, University of Chicago Press, 1965.
2. GREEP, R. O., AND TALMAGE, R. V., "The Parathyroids," Springfield, Ill., Charles C. Thomas, 1961.
3. COPP, D. H., "Parathyroids, Calcitonin, and Control of Plasma Calcium," *Recent Progr. Hormone Res.*, **20**, 59–77 (1964).
4. FIRSCHEIN, H. E., NEUMAN, W. F., MARTIN, G. R., AND MULRYAN, B. J., "Studies on the Mechanism of Action of the Parathyroid Hormone," *Recent Progr. Hormone Res.*, **15**, 427–54 (1959).
5. HIRSCH, P. F., GAUTHIER, G. F., AND MUNSON, P. L., "Thyroid Hypocalcemic Principle and Recurrent Laryngeal Nerve Injury as Factors Affecting the Response to Parathyroidectomy in Rats," *Endocrinology*, **73**, 244–52 (1963).

6. McLean, F. C., and Budy, A. M., "Chemistry and Physiology of the Parathyroid Hormone," *Vitamine Hormone*, **19**, 165–87 (1961).

R. V. Talmage

PASTEUR, L.

Louis Pasteur (1822–1895) was a French scientist of unusual breadth and depth. He received a doctorate in chemistry from Ecole Normale in Paris in 1848. His separation of the *d*- and *l*-forms of tartaric acid, which constituted his thesis, made him famous. One cannot do justice, in so short a space, to his later pioneering contributions. These had to do with fermentation, and later with the diseases of wine, silkworms, sheep and other farm animals, and human beings. He is regarded as the father of immunology if not of microbiology. He did more than any other man to establish (against great resistance) the idea that bacteria cause disease. A biography "The Life of Pasteur," by Rene Vallery-Radet (Doubleday Doran), should be referred to for further information.

PAULING, L. C.

Linus Carl Pauling (1901–) is an American chemist whose fundamental investigations in physical chemistry have been supplemented by highly significant work on protein structure (helical form), antibodies, and the abnormal hemoglobin of sickle cell anemia. For his work on molecular structure, Pauling was awarded the Nobel Prize in chemistry in 1954. In 1963, he received the Nobel Prize in peace.

PAVLOV, I.

Ivan Pavlov (1849–1936) was a Russian physiologist whose work on the physiology of digestion won him the Nobel Prize in medicine and physiology in 1904. He is more widely known, however, for later work in discovering "conditioned reflexes" in dogs. That he found dogs innately to fall into different categories with respect to their diverse responses is rarely mentioned. This clearly indicated individual differences in conditionability, an important observation bearing on fundamental problems in psychology and learning.

PECTINS

Pectins are high molecular weight polysaccharides that occur commonly in plants; they are polygalacturonic acids partially esterified with methanol.

Reference

1. Kertesz, Z. I., "Pectins," in "The Encyclopedia of Chemistry" (Clark, G. L., and Hawley, G. G., Editors), New York, Reinhold Publishing Corp., 1957.

PELLAGRA. See nicotinic acid.

PENTOSES IN METABOLISM

The importance of pentoses in metabolism derives from the fact that (a) D-ribose and D-2-deoxyribose are components of all cells, occurring particularly in nucleic acids and coenzymes, (b) they are intermediates in important metabolic pathways such as the pentose phosphate pathway and the photosynthetic cycle, and (c) the utilization of dietary pentoses and pentitols requires the existence of suitable metabolic reactions for the conversion of the pentoses into useful products.

The Pentose Phosphate Pathway. This pathway,[1,2] also referred to as the *hexose monophosphate shunt*, allows the oxidation of hexose to CO_2 and is important for the production of key metabolites. It is essentially a side pathway which begins and ends with intermediates of the Embden-Meyerhof glycolytic pathway. Figure 1 shows the cyclic nature of the pathway. The most important features to be noted are: (a) carbon-1 of hexose is rapidly lost as CO_2 in an early reaction, but carbon-6 is recycled, converted to trioses of the Embden-Meyerhof pathway, or incorporated in cellular constituents derived from intermediates in the pentose phosphate pathway; (b) the oxidations of glucose-6-phosphate and of 6-phosphogluconate depend upon the availability of NADP (TPN) and readily furnish NADPH (TPNH) for the many biosynthetic reactions that require this reduced coenzyme; (c) the ribose-5-phosphate in the cycle is formed both from direct glucose-6-phosphate oxidation and from non-oxidative transketolase-transaldolase catalyzed reactions beginning with fructose-6-phosphate and triose phosphate. Intermediates used in biosyntheses are: ribose-5-phosphate to supply the pentoses found in nucleic acids, both RNA and DNA; ribulose-5-phosphate for conversion to ribulose-1,5-diphosphate, which reacts with carbon dioxide in photosynthetic organisms to form 3-phosphoglyceric acid and then sugars and other metabolites; erythrose-4-phosphate for the production of aromatic compounds in plants and microorganisms. Biosynthetic contributions of the pentose phosphate pathway are summarized in Fig. 2 [see also carbohydrate metabolism; carbon reduction cycle (in photosynthesis); purine biosynthesis].

The quantitative importance of the pentose phosphate pathway can be evaluated in comparison with other pathways by (a) examining tissues or organisms for the required enzymes, and (b) determining the rate at which carbon-1 of glucose, as compared with carbon-6, is preferentially converted to CO_2. Although quantitative evaluation is sometimes complicated, in general a ratio

$$\frac{^{14}CO_2 \text{ from 1-}^{14}\text{C-glucose}}{^{14}CO_2 \text{ from 6-}^{14}\text{C-glucose}}$$

of unity indicates that glucose is being metabolized exclusively via the Embden-Meyerhof route, a ratio greater than 1 indicates functioning of the pentose phosphate pathway, and a ratio less than 1 indicates that the glucuronate-xylulose

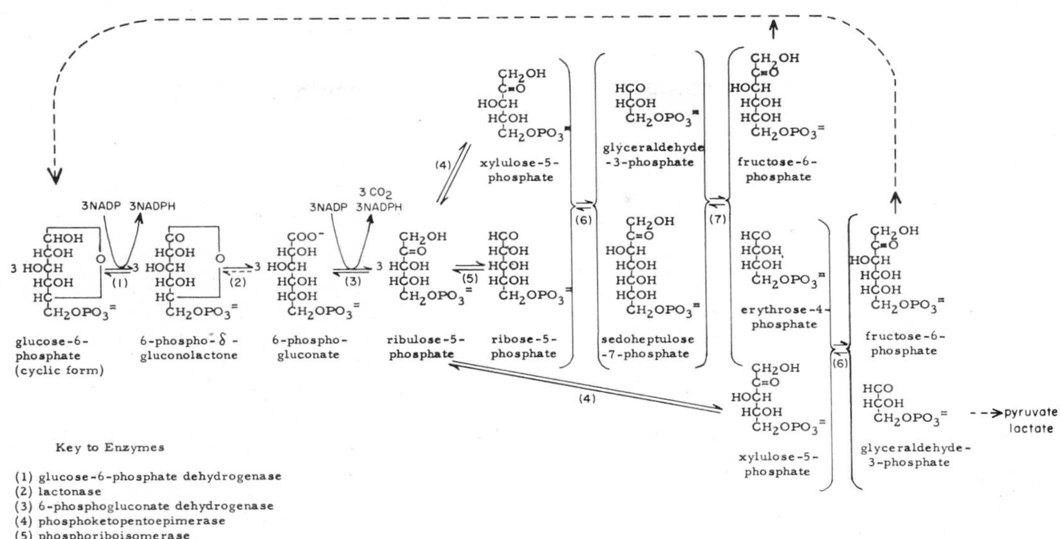

FIG. 1. The pentose phosphate pathway of glucose-6-phosphate oxidation.

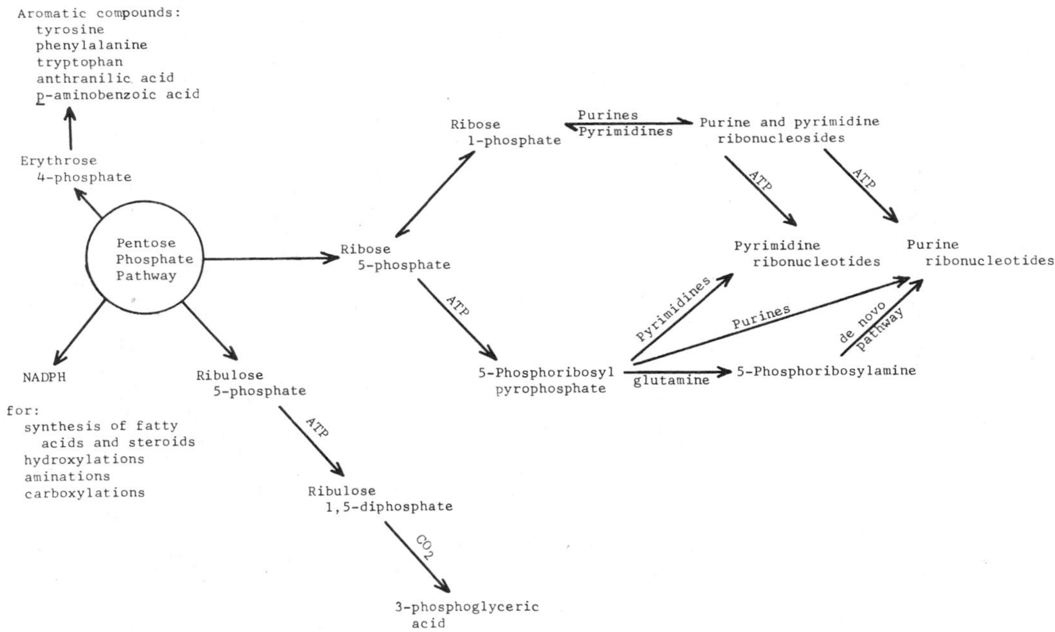

FIG. 2. Biosynthetic reactions involving ribose derivatives.

cycle is operating (see next section). Studies indiacte that the pentose phosphate cycle is widely distributed in nature (mammals, invertebrates, plants, bacteria). In mammals it is a significant pathway for hexose oxidation only in special tissues (cornea, lens, liver, lactating mammary gland). The levels of oxidative enzymes of the pathway may be influenced by nutritional or other factors, and the extent of hexose oxidation through the pathway depends in part on other cellular reactions that regenerate NADP from NADPH. The increased activity of the pathway in lactating mammary glands permits increased synthesis of milk fat by reactions that are dependent upon NADPH (see FATTY ACID METABOLISM). In the human, there are hemolytic diseases that are characterized by deficiency of glucose-6-phosphate dehydrogenase in red blood cells.

The Glucuronic Acid-Xylulose Cycle. This pathway (Fig. 3), which begins with glucose phosphate and terminates with D-xylulose-5-phosphate, a member of the pentose phosphate cycle, leads to the more rapid conversion of carbon-6 of hexose to CO_2 as compared to the conversion of carbon-1 to CO_2. This route[1,3] is operative in mammals, crustaceans and yeast, but it does not appear to be a major pathway for glucose oxidation. Early reactions of the cycle are responsible for the biosynthesis of ASCORBIC ACID (vitamin C) in those animals that can synthesize the vitamin. That some glucose is oxidized through the glucuronic acid–xylulose pathway in the human is suggested by the excretion of gram quantities of L-xylulose by humans with the harmless genetic metabolic defect essential pentosuria, and milligram quantities by normal humans. A member of the cycle, UDP-glucuronic acid, is of great importance as a source of glucuronic acid for the biosynthesis of MUCOPOLYSACCHARIDES and detoxication products. Further investigation is required to elucidate the physiological function of the glucuronate pathway in various species and tissues.

Pentoses in Nucleotides and Nucleic Acids. D-Ribose is found in many coenzymes and is a component of RNA (see RIBONUCLEIC ACIDS). In the synthesis of nucleotides, the ribose is usually attached to a nitrogenous substance via the intermediate 5-phosphoribosyl-1-pyrophosphate (PRPP), although ribose-1-phosphate may be a direct reactant in some transformations (Fig. 2).

2-D-Deoxyribose found in DNA (see DEOXYRIBONUCLEIC ACIDS) is formed directly from ribose after the latter pentose has become a component of nucleotides. For example, an enzymatic system from rat hepatoma and from *Escherichia coli* converts cytidine diphosphate to deoxycytidine diphosphate.[4]

Utilization of Pentoses and Pentitols by Microorganisms.[5,6] In their utilization of pentoses, bacteria generally transform aldopentoses into ketopentoses, which are then phosphorylated. Ribose, however, is usually directly phosphorylated. The utilization of aldopentoses by filamentous fungi and yeasts involves their reduction to pentitols, followed by reoxidation to ketopentoses. Some bacteria (*e.g.*, *Acetobacter*) adapt readily to pentitols as sole carbon sources, forming

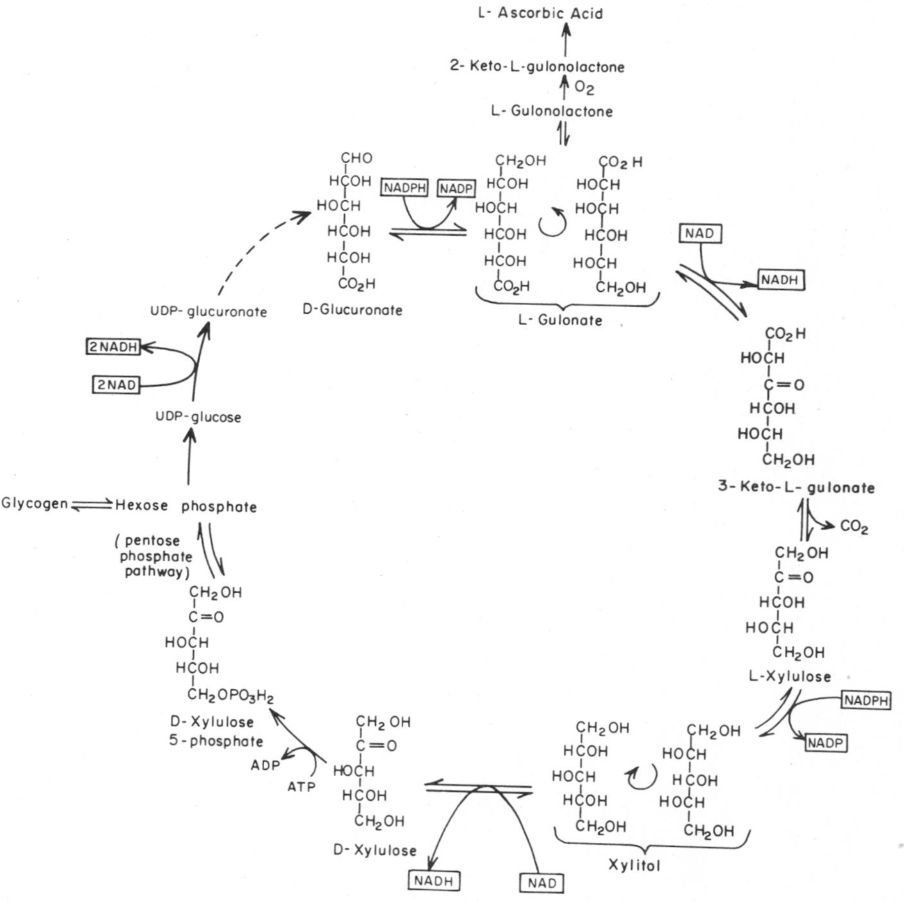

FIG. 3. The glucuronate-xylulose cycle.

pentitol dehydrogenases that yield ketopentoses. Although small amounts of common pentoses can be oxidized by animals, the limited utilization (and consequent urinary excretion) of administered D-xylose by human beings is the basis of a clinical test of intestinal absorption.

Pentoses in Plants. Polysaccharides containing pentoses are very common in plants. The pentoses are usually fed into the polymer synthesizing systems as their nucleotide derivatives. UDP-pentoses can be formed by (a) decarboxylation (*e.g.*, UDP-D-glucuronic acid to UDP-D-xylose) or (b) epimerization (*e.g.*, UDP-D-xylose $\rightleftharpoons$ UDP-L-arabinose).

References

1. HOLLMANN, S., "Non-glycolytic Pathways of Metabolism of Glucose" (TOUSTER, O., translator and reviser), New York, Academic Press, 1964.
2. AXELROD, B., "Other Pathways of Carbohydrate Metabolism," in "Metabolic Pathways" (GREENBERG, D. M., EDITOR), New York, Academic Press, 1960.
3. TOUSTER, O., "Essential Pentosuria and the Glucuronate-Xylulose Pathway," *Federation Proc.*, **19**, 977 (1960).
4. MOORE, E. C., and REICHARD, P., "Enzymatic Synthesis of Deoxyribonucleotides. VI. The Cytidine Diphosphate Reductase System from Novikoff Hepatoma," *J. Biol. Chem.* **239**, 3453 (1964).
5. WOOD, W. A., "Fermentation of Carbohydrates and Related Compounds," in "The Bacteria" GUNSALUS, I. C., AND STANIER, R. Y., EDITORS), New York, Academic Press, 1961.
6. TOUSTER, O., AND SHAW, D. R. D., "The Biochemistry of the Acyclic Polyols," *Physiol. Rev.*, **42**, 18 (1962).

OSCAR TOUSTER

PEPSINS AND PEPSINOGENS

Pepsin is a powerful digestive enzyme in the GASTRIC JUICE of many vertebrates. It catalyzes the hydrolytic breakdown of proteinous foods such as seeds, meats, eggs and milk products. The first recognition of pepsin was made by the German physiologist, Theodor Schwann, in 1836. In 1930 swine pepsin was isolated and identified as a crystalline and solubility pure protein by John H. Northrop of The Rockefeller Institute. Pepsins from cattle, salmon, tuna and whales have also been obtained crystalline pure, whereas shark and chicken pepsins have been only partially purified. Swine pepsin is used commercially in the preparation of hides and for the digestion of proteins used in growth media of microorganisms.

Pepsinogen, the inactive precursor of pepsin, is found in the chief cells of the fundus portion of the stomach mucosa. Nearly a gram of pepsinogen is extractable from the fundus mucosa of one swine stomach. It has also been found in the seminal fluids of humans. This precursor was described in 1892 by Langley in England, but it was not until 1938 that it was isolated, crystallized and studied in some detail by the writer in Dr. Northrop's laboratory. Chicken and human pepsinogen have also been studied.

Properties. *Pepsin.* The catalytic power of pepsin is greatest when the medium is close to pH 2.0, but its action is still observable even at pH 5–6. At 37°C and its pH optimum, pepsin can digest a thousand times its weight of some PROTEINS in an hour. The specificity of its action is broader than that of trypsin or chymotrypsin (see PROTEOLYTIC ENZYMES). Studies of a limited number of simple peptides have indicated that the peptide bond between two aromatic amino acids is particularly susceptible to its action. Despite the easily digestible nature of denatured pepsin, native pepsin shows very little tendency to digest itself at physiological temperatures.

Pepsin is a globular protein with a molecular weight of about 35,000. It is a single peptide chain, the folds of which are held together by two or three cystine disulfide bridges. Pepsin is unusual in its pH stability range, a partial explanation of which may be seen in its amino acid composition and the titration curves. The enzyme is remarkably stable to strong acid, but it is unstable to neutral or weakly alkaline solutions. The low basic amino acid content of pepsin suggests that in strong acid there would be only a small positive charge and no negative charge. The repulsive forces will be small in such a case, and the enzyme should be stable. At pH 8 there is a large negative charge and only a few compensatory positive charges. This produces high electrostatic repulsive forces. It can be surmised, therefore, that above pH 6.5 these repulsive forces exceed the binding forces and this globular protein becomes distorted, *i.e.*, denatured. Van Vunakis *et al.* (1963) have reported that in the presence of anti-pepsinogen serum, swine pepsin is not destroyed at pH 8.

Pepsinogen. Pepsinogen, besides being catalytically inactive, is in most instances distinguishable from the enzyme by being stable to pH 8 and unstable in acid solution. In addition, pepsinogen denaturation at 65°C and pH 7 or room temperature and pH 10.5 is a reversible change. In the latter environment, denaturation follows the ionization of but two groups on the protein. Pepsinogen has a molecular weight of 43,000. It contains considerably more lysine and leucine than does pepsin.

More than one molecular species of pepsinogen was detected in the early studies, and several reports have extended this, especially for human pepsinogen, where three distinguishable species were observed which produced correspondingly different pepsins.

Conversion of Pepsinogen to Pepsin. Pepsinogen is converted to pepsin when it is made more acid than pH 5.5. The reaction is autocatalytic, *i.e.*, pepsin, a product of the reaction, catalyzes the conversion. When the conversion is carried out at pH 4.6, the increase in pepsin follows an S-shaped course with time; the rate increases with time to the midpoint and then decreases as the concentration of precursor becomes limiting. The rate of conversion increases with pepsinogen concentration and with acidity.

Peptides totaling 8000 molecular weight units are liberated from the amino nitrogen end of the precursor peptide chain during its conversion to the enzyme. One of these peptides, rich in lysine and having a molecular weight of 3100, is a dissociable inhibitor of pepsin. It inhibits pepsin only when the pH is 5-6. This inhibition is a simple one-to-one reaction between the inhibitor and the enzyme. Recent studies have reported that human and swine pepsin-inhibitor complexes are stable to pH 7.8, but unstable at pH 9.2. Schlamowitz *et al.* (1963) have partially restored alkali-inactivated pepsin with the inhibitor. In solutions more acid than pH 5, dissociation of the enzyme-inhibitor complex is favored and the inhibitor is then slowly digested by pepsin. The over-all reaction is as follows:

$$\text{Pepsinogen} \xrightarrow{[P]} \text{P-I} + \text{X peptides} \underset{\text{pH} > 5.4}{\overset{\text{pH} < 5.4}{\rightleftharpoons}} \text{P} + \text{I} + \text{X peptides}$$

$$\downarrow$$

$$\text{P} + y + \text{X peptides}$$

P-I = pepsin-inhibitor dissociable complex

Although the conversion of precursor to enzyme appears to produce substantial changes in the protein component, studies by workers in several laboratories with specific antisera suggest that the conformational changes are not particularly striking.

Active Site. Since no structures responsible for the catalytic action of enzymes are well understood, there is some interest in the nature of the protein structures which interact with the substrate and which, for purposes of discussion, may be considered the active site. Acetylation or iodination of the phenolic moieties of the tyrosyl residues and sulfur mustard esterification of carboxyl groups of pepsin resulted in a logarithmic fall in the enzymic activity as the reagent interaction increased. The results suggest that 1-5 tyrosyl and carboxyl groups of pepsin play an essential part in the binding of the substrate.

Pepsin occupies a unique position in some ways among enzymes, by requiring no electrostatic interaction with the substrate. At pH 2.0 virtually all carboxyl groups are un-ionized. The one ionized phosphoester in pepsin does not play an important role, for in 1955 Perlmann showed that its removal produced no change in the enzyme's catalytic action. The few amino groups of the enzyme protein which carry positive charges at pH 2.0 can be discharged by acetylation without loss of catalytic action. A substrate such as acetylphenylalanyl-phenylalanine has no charged groups at pH 2. This leaves only hydrogen or hydrophobic bonds to account for the binding of such a substrate to the enzyme. For the above substrate, carboxyl-carboxyl and phenolic-peptide carboxyl are the obvious possibilities for hydrogen bonding. Hydrophobic bonding might also contribute.

Stepwise reduction of the three disulfide bridges holding the folded peptide chain of pepsinogen together in a relatively stable form produces very different effects. Reduction of the first bridge produces little, if any, effect on the enzyme. The second bridge apparently occupies a critical position for its reduction is followed by a rise in viscosity and a loss of enzymic function. The third bridge is much more resistant to mercaptoethanol reduction. Steiner, Frattati and Edelhoch (1964) have recently found that reduction of pepsinogen until its capacity to form pepsin is nearly lost will recover some of that activity on standing.

Crystalline swine pepsin and pepsinogen are now available commercially, so the next decade or two should see much new light thrown on the structure of this interesting enzyme.

References

1. Northrop, J. H., Kunitz, M., and Herriott, R. M., "Crystalline Enzymes," Second edition, New York, Columbia Press, 1948.

2. Bovey, F. A., and Yanari, S. S., in "The Enzymes" (Boyer, P. D., Lardy, H., and Myrbäck, Editors), Second edition, Vol. 4, p. 63, New York, Academic Press, 1960.

3. Herriott, R. M., *J. Gen. Physiol.*, **45**, Suppl. 2, 57 (1962).

4. Hirschowitz, B., *Physiol. Rev.* **37**, 475 (1957).

Roger M. Herriott

PEPTIDASES

The peptidases are a special group of proteolytic enzymes (proteases) which catalyze the hydrolysis of peptide bonds of PROTEINS or peptides. Historically, the enzymatic breakdown of proteins and peptides has been known for some time, although at first it was not recognized that the hydrolysis observed was the result of the action of several enzymes and not just one. Knowledge of the multiple action soon emerged, and studies of the peptidases have since maintained a prominent place in enzymology. Good reviews of these earlier and, also, more recent studies are available (see references). These studies were extremely valuable in clarifying the problems of protein structure and also in many phases of the chemistry of proteins. There have also been fruitful studies aimed at investigating the problems of specificity and mode of action of enzymes. The peptidases were particularly appropriate for this type of investigation because of the availability of excellent methods for the synthesis of peptides and peptide derivatives which were suitable substrates for the peptidases, but were of much simpler structure than the protein substrates used previously. Thus a large number of synthetic substrates could be made and their structures varied systematically in a known manner. Information obtained in this manner has given real insight into enzymatic specificity and has led to appropriate criteria for classifying the proteases. Customarily a distinction has been made between the PROTEOLYTIC ENZYMES which attack peptide bonds that are not adjacent to a terminal amino or carboxyl group of the peptide chain (endopeptidases or proteinases) and those which only cleave peptide bonds adjacent to such terminal groups (exopeptidases or peptidases). The three principle kinds of peptidases are *carboxypeptidases* which require a free carboxyl group adjacent to the bonds being split, *aminopeptidases* which require a free amino group adjacent to the bond under attack, and lastly the *dipeptidases* which require both free groups and thus act only on dipeptides. The action of each results in the release of an amino acid from the substrate (the molecule being acted on), leaving the remainder of the original peptide chain. The action of either of the first two types can of course be repeated again and again, and if this action occurs, it would leave as final products only free amino acids and one dipeptide. The dipeptide when acted on by a dipeptidase would give two free amino acids. Theoretically, the two enzymes could bring about the complete breakdown of proteins into their composite amino acids. Physiologically, the action of the endopeptidases (spoken of above) supplements the action of the peptidases in bringing about this final result. Thus one of the physiological roles of these enzymes immediately suggests itself, the breaking down of protein or peptide foodstuffs into amino acids which can then be used to synthesize the native proteins of the organism. Also, they can function to mobilize amino acids from the native protein pool when they are needed for the synthesis of some more critically required protein. Peptidases may, however, have another significant physiological function which is less apparent. Theoretically, at least, they can catalyze not only the breakdown of the proteins and peptides to amino acids but also the synthesis of proteins or peptides from amino acids. The importance or extent, *in vivo*, of this latter action is not definitely known. *In vitro*, the equilibrium point of the reaction is such that only the catabolic, or breakdown, function seems important.

Whatever the physiological function(s) of the peptidases, judging by their distribution they seem to be important. Peptidases have been found in almost all tissues examined, in serum and in plants, bacteria and fungi. Just how many individual peptidases there are remains a matter of conjecture. Present evidence indicates a narrowness of specificity, at least for the dipeptidases, which would require a very large number indeed to hydrolyze all the dipeptide possibilities represented by the possible combinations of the 19 amino acids. To date only a few of the peptidases have been obtained in a highly purified form and only one of them, carboxypeptidase, has been studied by kinetic methods.

This class of enzymes has contributed greatly to our knowledge of enzymes in still another area, since it is characteristic of most, if not all, of the peptidases for their catalytic action to be dependent on the presence of a metal ion. These ions are essential for the enzymatic activity and have helped to differentiate various peptidases. Studies of the role of the metal ion in the enzymatic activation of substrates have also been fruitful in examining the mechanism of action of enzymes. An important point that has emerged from such studies is the recognition that in order to explain the optical specificity of the peptidases, *i.e.*, the limiting of their hydrolytic action to peptides composed of amino acids of the L-form, there must be postulated a minimum of three points of attachment of the substrate with the enzyme. Considerable evidence suggests that the metal ion forms a bridge between the substrate and the enzyme involving at least one of these points of attachment and not only is involved with the determination of specificity but also contributes to the "activation" of the susceptible bond of the substrate. Further useful information concerning which types of functional groups of the enzyme and substrate are most probably involved in the active complex has come from studies of various types of inhibitors, particularly the metal-combining agents.

The best example of the types of studies mentioned above have been carried out with some of the more highly purified peptidases. These include leucine aminopeptidases, carboxypeptidases (pancreatic), aminotripeptidases, yeast polypeptidase, glycylglycine dipeptidases, glycyl-L-leucine dipeptidase, carnosinase, prolidase and prolinase. There is also another special class of peptidases, the dehydropeptidases, about which considerable literature is available.

For sources of more specific information or more discussion in detail, the following are suggested as appropriate starting points.

References

1. Bergmann, M., in "Advances in Enzymology," Vol. II, p. 49, New York, Interscience Publishers, 1942.
2. Johnson, M. J., and Berger, J., *ibid.*, Vol. II, p. 69, 1942.
3. Smith, E. L., *ibid.*, Vol. 12, p. 191, 1951.
4. Hoffmann-Ostenhaf, O., *ibid.*, Vol. 14, p. 219, 1953.
5. Malmstrom, B. G., and Rosenberg, A., *ibid.*, Vol. 21, p. 131, 1959.
6. Davis, N. C., and Smith, E. L., in "Methods of Biochemical Analysis," Vol. II, p. 215, New York, Interscience Publishers, 1955.
7. "Biochemists Handbook" (Long, Cyril, Editor), Princeton, N.J., D. Van Nostrand Co., 1961.

Neil C. Davis

PEPTIDE ANALYSIS. See PROTEINS (END GROUP AND SEQUENCE ANALYSIS); also ION EXCHANGE POLYMERS (RESINS), section on Peptides; PROTEINS (AMINO ACID ANALYSIS AND CONTENT).

PERNICIOUS ANEMIA

Pernicious anemia is due to a deficiency of Vitamin B_{12} which, in turn, results from deficient production of INTRINSIC FACTOR in the gastric juice [see VITAMIN B_{12} (ABSORPTION AND NUTRITIONAL ASPECTS); ANEMIAS, section on Defect in Maturation of Red Blood Cells].

PEROXIDASES

Reactions Catalyzed. Peroxidases are enzymes which catalyze the reaction of hydrogen peroxide with a second compound, the latter serving as a hydrogen donor in accordance with the following reaction:

$$AH_2 + H_2O_2 \rightarrow A + 2H_2O$$

AH_2 represents the compound serving as the hydrogen donor for the reaction. A more general classification of this type of reaction under the term "hydroperoxidases" would include the reaction catalyzed by the enzyme, catalase. In this case, the reaction is as follows:

$$H_2O_2 + H_2O_2 \rightarrow O_2 + 2H_2O$$

It is evident that this reaction is simply a special case of the one cited above, with one of the molecules of hydrogen peroxide serving as the hydrogen donor for the reaction.

Peroxidases are widely distributed in plants, and a number of different hydrogen donors have been used for their detection. The most commonly used tests utilize the enzyme-catalyzed reaction of hydrogen peroxide with aromatic polyphenols and polyamines such as *o*-tolidine, guaiacol, pyrogallol, and mesidine. When oxidized, these compounds form colored products, often by reaction of the oxidized compound (*i.e.*, A) with a second molecule of the hydrogen donor (*i.e.*, AH_2). The amount of the colored product which is formed under specified conditions serves as a measure of peroxidase activity. In general, plant peroxidases exhibit a rather broad specificity in regard to the nature of the hydrogen donors which may be used in the enzyme-catalyzed reaction. In addition to the compounds mentioned above, a number of other aromatic phenols and amines, and several enediols, such as reductone and ASCORBIC ACID, will serve as hydrogen donors for the reaction. Peroxidases will also catalyze the breakdown of various organic peroxides such as methyl hydrogen peroxide and ethyl hydrogen peroxide in the presence of suitable hydrogen donors.

In mammalian tissues, peroxidases have been considered to be rare. This may be due, at least partially, to the use of conventional hydrogen donors in testing for peroxidase activity in animal tissues. Some mammalian peroxidases differ markedly from plant peroxidases in regard to their specificity for hydrogen donors. Glutathione peroxidase is active with GLUTATHIONE as hydrogen donor, but is not active with the more conventional hydrogen donors. Other mammalian peroxidases have been noted which utilize the leuko form of the dye sodium 2,6-dichlorobenzenoneindo-3′-chlorophenol as a hydrogen donor, but react much more slowly or not at all when pyrogallol is used. These include lactoperoxidase of milk, myeloperoxidase of leukocytes and a uterine peroxidase which has been studied in rat uterus preparations. It is of interest to note that uric acid may be utilized effectively as a hydrogen donor for the assay of myeloperoxidase activity in leukocytes.

Catalase is found very extensively in mammalian tissues, and has been found in lesser amounts in plants. Catalase from bacterial sources has also been purified and studied. Catalase is an extremely effective enzyme in carrying out its function as a biological catalyst in the presence of adequate amounts of hydrogen peroxide. Even at 0°C, one molecule of enzyme may catalyze the destruction of 40,000 molecules of hydrogen peroxide per second. With very low concentrations of hydrogen peroxide and the presence of a suitable hydrogen donor such as formate or ethanol, catalase may function preferentially as a peroxidase. The low concentration of hydrogen peroxide may be achieved either by a slow enzymatic generation of hydrogen peroxide or by a slow diffusion of hydrogen peroxide into the system. This reaction may be illustrated in the case of ethanol as follows:

$$CH_3CH_2OH + H_2O_2 \rightarrow CH_3CHO + 2H_2O$$

In reactions of this type, the destruction of the hydrogen peroxide is accompanied by the oxidation of the ethanol to acetaldehyde or of the formate to CO_2.

Physiological Significance. Hydrogen peroxide is toxic to animal tissues since it may oxidize a number of essential body components. Furthermore, it is continually being produced in the body as a product of the action of a number of enzymes and may also be produced as a consequence of IONIZING RADIATION. Consequently, there is good reason to believe that the primary function of peroxidases and catalase is to protect the body cells from peroxide damage by preventing an accumulation of peroxide in tissues. Catalase seems to be reasonably effective in this regard, especially in the presence of compounds (*e.g.*, formate or ethanol) which will serve as hydrogen donors for the peroxidatic breakdown of hydrogen peroxide. Studies on intact animals indicate that a considerable amount of formate may be used as a hydrogen donor for the destruction of hydrogen peroxide. A genetically determined deficiency of tissue catalase has been noted in a number of individuals in Japan and also in several Swiss families. Individuals with this deficiency show an increased susceptibility to gangrene as a consequence of infections in the gums adjacent to the teeth, but in other respects they appear to be normal.

Glutathione peroxidase has been shown to have a protective function in liver and erythrocytes. It is the only hydrogen peroxide-destroying enzyme of mammalian tissues which utilizes a hydrogen donor that is normally present in tissues in appreciable amounts. In addition, the oxidized GLUTATHIONE which is formed in the reaction may be rapidly converted back to the reduced form by the reactions illustrated below:

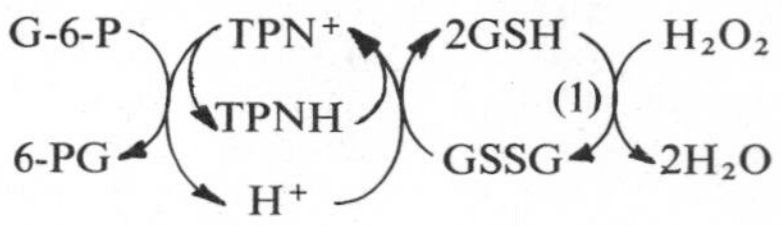

G-6-P → 6-PG; TPN^+ → TPNH + H^+; TPNH + H^+ → TPN^+; GSSG → 2GSH; 2GSH → GSSG (1); H_2O_2 → $2H_2O$

This enzyme-catalyzed sequence of reactions has

been shown to occur readily in liver and erythrocytes. Abbreviations are: G-6-P, glucose 6-phosphate; 6-PG, 6-phosphogluconate; TPN^+ and TPNH, oxidized and reduced triphosphopyridine nucleotide (also termed $NADP^+$ and NADPH); GSSG and GSH, oxidized and reduced glutathione. Reaction (1) is catalyzed by glutathione peroxidase. The over-all physiological significance of the above sequence of reactions, in relation to that of catalase, has not been fully assessed. Iodide peroxidase has a physiological role in the formation of THYROXINE in the thyroid gland. The reaction catalyzed by this enzyme is as follows:

$$2I^- + 2H^+ + H_2O_2 \rightarrow I_2 + 2H_2O$$

The IODINE is subsequently used for the iodination of tyrosine in the initial step of the metabolic pathway leading to the formation of thyroxine.

The specific physiological role of peroxidases in plant tissues is still largely unknown. They probably function in the detoxication of hydrogen peroxide, but the peroxidase-hydrogen peroxide system may also have a physiological function in the oxidation of various aromatic compounds of importance to plant metabolism. Horseradish peroxidase has been proved to be a very useful enzyme in various analytical procedures. It is used extensively in the determination of glucose in blood and urine. In this procedure glucose oxidase specifically oxidizes glucose with hydrogen peroxide as a product. In the presence of horseradish peroxidase and *o*-tolidine (or some other suitable hydrogen donor), the hydrogen peroxide is destroyed and the *o*-tolidine is converted to a colored product. The amount of color serves as a quantitative measure of the glucose in the original sample.

Chemical Nature. Peroxidases and catalases are predominantly PROTEIN in nature with an additional tightly bound compound (prosthetic group), which confers catalytic activity to the molecule. In addition to protein and the prosthetic group, there is an appreciable amount of carbohydrate in some cases that appears to be an integral part of the molecule. Crystallized horseradish peroxidase, for example, contains from 18–25% carbohydrate. Purification of horseradish peroxidase has shown the heterogeneity of this enzyme. Various separative techniques (*e.g.*, paper ELECTROPHORESIS, salt fractionation, or column CHROMATOGRAPHY) have been used to prepare as many as five different active peroxidases from horseradish. The primary difference in these peroxidase isozymes appears to be in the protein with accompanying differences in molecular weights of the different isozymes. In most cases, the prosthetic group of peroxidases or catalases is either HEME or a modified heme. The prosthetic groups of catalase and horseradish peroxidase have been shown to be the same as the prosthetic group of hemoglobin (*i.e.*, protoheme). The different catalytic properties of these protein molecules is due, therefore, to differences in the protein molecule and in the manner of attachment of the protoheme to the protein. In view of the similarities in these different hemoproteins, it is of interest to note that hemoglobin, which is not an enzyme, does exhibit low peroxidase activity.

The specific nature of the prosthetic group in some of the mammalian peroxidases (*e.g.*, lactoperoxidase and myeloperoxidase) is not known. Absorption spectra and susceptibility to inhibition by cyanide and azide provide evidence that the prosthetic groups of these enzymes are iron PORPHYRIN compounds, but they do not appear to be protoheme. The prosthetic group of glutathione peroxidase has not been indentified, but the insensitivity to cyanide and azide indicate that the prosthetic group is not an iron porphyrin. Various properties of glutathione peroxidase support the suggestion that the enzyme may be a FLAVOPROTEIN. The DPNH peroxidase which has been isolated from *Streptococcus faecalis* has been shown to be a flavoprotein. The prosthetic group of this enzyme is flavin adenine dinucleotide.

References

1. MILLS, G. C., "Glutathione Peroxidase and the Destruction of Hydrogen Peroxide in Animal Tissues," *Arch. Biochem. Biophys.*, **86**, 1 (1960).
2. MARTIN, A. P., NEUFELD, H. A., LUCAS, F. V., AND STOTZ, E., "Characterization of Uterine Peroxidase," *J. Biol. Chem.*, **233**, 206 (1958).
3. ALEXANDER, N. M., AND CORCORAN, B. J., "The Reversible Dissociation of Thyroid Iodide Peroxidase into Apoenzyme and Prosthetic Group", *J. Biol. Chem.*, **237**, 243 (1962).
4. NICHOLLS, P., AND SCHONBAUM, G. R., "Catalases," in "The Enzymes" (BOYER, P. D., LARDY, H., AND MYRBÄCK, K., EDITORS), Second edition, Vol. 8, Part B, p. 147, New York, Academic Press, 1963.
5. PAUL, K. G., "Peroxidases", *ibid.*, p. 227.
6. SAUNDERS, B. C., HOLMES-SIEDLE, A. G., and STARK, B. P., "Peroxidase", Washington, D. C., Butterworth, Inc., 1964.

GORDON C. MILLS

PERUTZ, M. F.

Max Ferdinand Perutz (1914–   ), an Austrian-born British biochemist, received with JOHN COWDERY KENDREW a share in the Nobel Prize in chemistry for 1962. Kendrew investigated MYOGLOBIN while Perutz studied HEMOGLOBIN. In both cases, the detailed structures were determined using X-ray diffraction pictures and high-speed computers to analyze the findings.

PESTICIDES. See INSECTICIDES.

pH

The pH value of an aqueous solution is a number describing its acidity or alkalinity. The usual range of pH is from about 1 for 0.1*N* HCl to about 13 for 0.1*N* NaOH. The pH of a neutral solution is 7.2 at 15°C, 7.0 at 25°C, and 6.8 at 35°C.

The pH scale was first used in 1909 by Sørensen. He defined pH as the negative logarithm (base 10) of the concentration of hydrogen ions (equivalents per liter), and he also described an electrometric method for the measurement of pH. Modern chemists have abandoned Sørensen's definition but have retained, in all essentials, his method of measurement.

The approved definition of pH is an operational one.[1] The electromotive force, E_x, of the cell

H_2; solution X: KCl (saturated): reference electrode

and the electromotive force E_s of the cell

H_2; solution S: KCl (saturated): reference electrode

are measured with the same electrodes at the same temperature. The hydrogen electrode is now usually replaced by a glass electrode. The difference in pH between the unknown solution, X, and the standard solution, S, is defined by the equation

$$\text{pH}_x - \text{pH}_s = (E_x - E_s)/k$$

in which k is 2.3026 RT/F or 0.05916 for 25°C if E_x and E_s are expressed in volts.

To complete the definition of pH, values of pH_s have been assigned to standard solutions. The National Bureau of Standards has recommended five solutions as primary standards.[2] One of these, 0.05M potassium hydrogen phthalate, has been adopted as the primary standard of pH in Great Britain and in Japan. For this solution pH_s is 4.00 at 15°C, 4.01 at 25°C, and 4.03 at 40°C.[3]

An interpretation of pH is possible in special cases. The values assigned to the standards have been chosen in such a way as to make pH_S equal to $-\log C_H f_1$, where f_1 is an activity coefficient practically equal to that of sodium chloride. Approximate values of $-\log f_1$ are 0.04, 0.06, 0.09, and 0.11 for ionic strengths of 0.01, 0.02, 0.05, and 0.10, respectively. For solutions of ionic strength not over 0.1 and pH between about 2 and 12, these values may be employed in calculations involving ionic equilibria.

For example, the pH of a "standard acetate" buffer (0.1 M $HC_2H_3O_2$, 0.1 M $NaC_2H_3O_2$) is found to be 4.65. It follows that $-\log C_H$ is 4.65–0.11, or 4.54. For a 1:1 buffer the pH is equal to the pK' of the equation

$$\text{pH} = \text{p}K' + \log C_B/C_A$$

in which C_A is the concentration of the buffer acid and C_B is that of its conjugate base. The apparent constant, K', is not equal to the true constant, K, of the buffer acid because K' lacks a ratio of activity coefficients. In this case it may be assumed that $K = K'f_1$ or $\text{p}K = \text{p}K' - \log f_1$. On this basis p$K$ (or $-\log K$) would be equal to 4.65 + 0.11, or 4.76, which is the accepted value.

A similar calculation may be made for a phosphate buffer in which the acid is $H_2PO_4^-$ and the base is HPO_4^{--}. In this case K_2 is equal to $K_2'f_2/f_1$ where f_2 is the activity coefficient of the bivalent anion. From the Debye-Hückel theory, it follows that $\log f_2 = 4 \log f_1$, and accordingly we have the relation

$$\text{p}K_2 = \text{p}K_2' - 3 \log f_1$$

For a buffer of 0.025 M KH_2PO_4 in 0.025 M Na_2HPO_4, the pH is 6.86 and the ionic strength 0.1. Accordingly, the value of pK_2 should be 6.86 + 0.33, or 7.19. The accepted value is 7.20. Again, the value of $-\log C_H$ is equal to pH + $\log f_1$, being 6.75 for this buffer.

References

1. Bates, R. G., "Electrometric pH Determination, New York, John Wiley & Sons, 1954.
2. Bates, R. G., "Revised Standard Values for pH Measurements from 0 to 95°C," *J. Res. Natl. Bur. Std.*, **66A**, 179–184 (1962).
3. Bates, R. G., and Guggenheim, E. A., "Report on the Standardization of pH and Related Terminology," *Pure Appl. Chem.*, **1**, 163–168 (1960).

David I. Hitchcock

Although the original definition of pH, in terms of the hydrogen ion concentration, has since been modified as Dr. Hitchcock has indicated above, the following table of *approximate* relationship between pH and hydrogen or hydroxyl ion concentration is widely applicable in biological work:

Table 1[a]

	pH Value	Approximate Number of Times H^+ or OH^- Concentration Exceeds That of Pure Water
Acid side (excess of H^+ ions)	1	1,000,000
	2	100,000
	3	10,000
	4	1,000
	5	100
	6	10
Neutrality ⟶	7	
Alkaline side (excess of OH^- ions)	8	10
	9	100
	10	1,000
	11	10,000
	12	100,000
	13	1,000,000

[a] Alexander, J., "Encyclopedia of Chemistry," p. 720, New York, Reinhold Publishing Corp., 1957.

PHASE MICROSCOPY

Definition. The phase contrast microscope is an instrument for observing transparent structures such as living cells; it can also be used for measuring the refractive index of such structures. Since the refractive index of a living cell is related to its solid concentration, the phase contrast microscope is capable of yielding important quantitative cytological information.

Basic Principles. The microscopy of transparent objects has always proved difficult because such objects do not absorb light, and the eye and the photographic plate are only sensitive to variations in light intensity. A light wave may be delayed or speeded-up by passing through a transparent object, but its amplitude is unaffected. In other words, the phase of the light is changed but not its intensity. The problem is to convert these invisible phase changes into visible intensity changes. This was successfully accomplished by the Dutch physicist, F. Zernike, who was awarded the Nobel Prize in 1953. The principle is illustrated in Fig. 1. Wave A represents the light wave incident on a transparent object. If the refractive index of the object is n and that of the surrounding medium is n_0 the light wave undergoes a phase change defined by $\phi = (n - n_0)t$, where t is the thickness of the object. The wave is delayed or advanced depending on whether n is greater or less than n_0. In Fig. 1(a) it is assumed that the wave is delayed, so that the transmitted wave is represented by B. This has the same height as A, so that in the ordinary microscope there is no contrast between the object and the background. It is possible to represent B in a different manner. In Fig. 1(b) the incident wave A is shown together with a dotted wave. If these two waves are added together algebraically, their resultant is wave B. The dotted wave in fact represents the wave scattered or diffracted by the object. The final image is formed by the summation or interference of the direct and diffracted waves. However, since the resultant wave has the same amplitude as the incident wave the image will be invisible. It can be seen from Fig. 1(b) that the incident and diffracted waves are approximately one-quarter of a wavelength out of phase. Zernike pointed out that if the two waves could be separated and their phase relationship altered, the new resultant wave could be made to have a different amplitude. This is shown in Fig. 1(c), in which the phase of the incident wave has been advanced by about one-quarter of a wavelength so that its peak now coincides with the trough of the diffracted wave. The new resultant wave is now lower in amplitude than the incident wave so that the object will appear darker than the background. This system is known as positive phase contrast. We could, however, arrange to make the peak of the two waves coincide so that the resultant wave would have a greater amplitude than the incident wave. The object would now appear brighter than the background, giving negative phase contrast.

Practical Realization. The method of separating the incident and diffracted light is indicated in Fig. 2. In the ordinary microscope, an image of the substage condenser diaphragm is formed at the rear focal plane of the objective lens where it can be seen on removing the eyepiece. In the absence of an object, only the incident light will pass. If an object is introduced, there will be a mixture of incident and diffracted waves, but it will not normally be possible to separate them. If, however, the substage iris diaphragm is replaced by an annular diaphragm D, in the absence of an object an image of this diaphragm will be formed by the incident light at the rear focal plane p. When an object is introduced, the diffracted light fills the aperture of the objective and in addition to the image of the annulus the whole of the rear focal plane will be diffusely illuminated. The incident and diffracted waves will overlap on the image of the annulus itself, but for the most part they are separated. A phase plate is placed at p. This is shown as a transparent disk containing an annular groove which is made to coincide with the image of the substage annulus. Thus, broadly speaking, the direct light passes through the groove whereas the diffracted light passes through the whole of the phase plate. In this way a phase difference, depending on the thickness of the groove, is introduced between the direct and diffracted waves. A layer of absorbing material, such as metal, is usually placed on the groove. The reason for this can be seen in Fig. 1(c). The amplitude of the incident wave is usually much greater than that of the diffracted wave so that even when they interfere they do not produce complete blackness. If the amplitude of the incident wave is reduced by making it pass through an absorbing layer, contrast can generally be improved.

The size of the substage diaphragm D must be matched to the phase plate. The most common arrangement is to use fixed phase plates integral with the objectives and to carry a number of different substage annuli on a rotatable turret in the substage condenser. Centering screws are usually provided in order to make the image of the substage annulus coincide exactly with the phase plate annulus. An auxiliary telescope substituted for the eyepiece and focused on the phase plate makes this adjustment easier.

If it were possible to separate the incident and

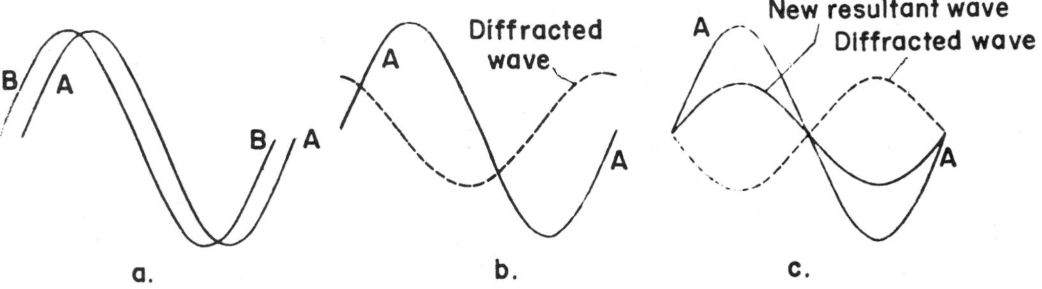

FIG. 1

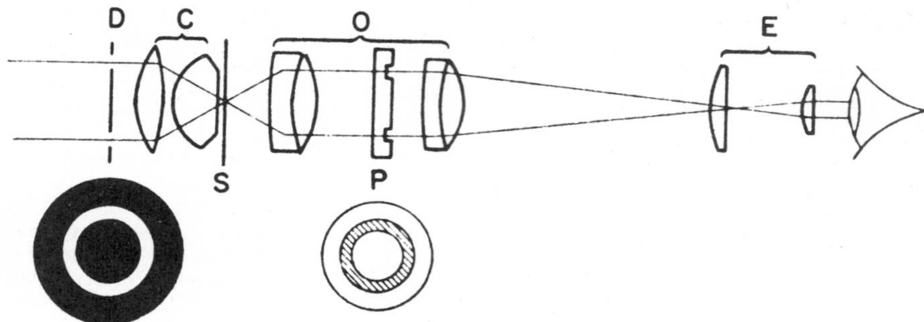

FIG. 2

diffracted waves completely there would be a predictable photometric relationship between the image intensity and the phase change introduced by the object. Since complete separation is not possible in practice, this relationship is disturbed and in general the image is surrounded by a halo of opposite contrast; thus details that appear dark are surrounded by a bright halo. These difficulties are avoided in some forms of interference microscopy. Phase contrast can be regarded as an imperfect form of interference microscopy in which the interfering beams are incompletely separated.

Biochemical Applications. It follows from the definition of ϕ that when the refractive index of the medium is made equal to that of the object, ϕ becomes zero and the object will be invisible. This can be used to determine the refractive index of living cells. Since the refractive index varies linearly with concentration, this provides quantitative cytochemical information. Further details are given in the first two references below.

Phase contrast is frequently used in biochemistry for controlling the isolation of cell particulates[4] and is a valuable aid in correlating biochemical variations with structural changes in cells. Hilz and Eckstein[3], for example, studied changes in yeast cells in relation to metabolic alterations induced by radiation and cytostatic agents.

References

1. BARER, R., "Phase Contrast and Interference Microscopy in Cytology," in "Physical Techniques in Biological Research" (OSTER, G., AND POLLISTER, A. W., EDITORS), Second edition, Vol. 3, New York, Academic Press, 1965.
2. BARER, R., "Phase, Interference and Polarizing Microscopy," in "Analytical Cytology" (MELLORS, R. C., EDITOR), Second edition, New York, McGraw-Hill Book Co., 1959.
3. HILZ, H., AND ECKSTEIN, H., "Teilungssychronisiert Hefezellen," *Biochem. Z.*, **340**, 351, 1964.
4. ROODYN, D. B., "Isolation of Nuclei," *Biochem. Soc. Symp.* **23**, 20 (1963).

R. BARER

PHENYLALANINE

L-Phenylalanine, α-amino-β-phenylpropionic acid,

$$C_6H_5—CH_2—CHNH_2—COOH$$

is present as a structural unit in all the more common PROTEINS (not protamines). Human HEMOGLOBIN is one of the richest proteins; it yields 9.6% phenylalanine. L-Phenylalanine is a nutritionally ESSENTIAL AMINO ACID for mammals and fowls. Adding tyrosine to the diet has a sparing action on the amount of phenylanine required. Part of the phenylalanine, if furnished alone, is used to produce TYROSINE, which is a common constituent of many body proteins. The amount of phenylalanine required daily by adult humans to maintain nitrogen equilibrium in the absence of tyrosine is from 400–1100 mg. In the presence of tyrosine, much less is needed. L-Phenylalanine in *N* HCl solution has a specific rotatory power $[\alpha]_D = -7.7°$. D-Phenylalanine, its antipode, is a structural component of certain ANTIBIOTICS: tyrocidins and gramicidins. The D- form of the amino acid has been administered to human beings whose ability to utilize it and convert it into the L-isomer varies widely from about 3% up to 95% under the conditions used. [See also AMINO ACIDS (OCCURRENCE, STRUCTURE, AND SYNTHESIS); AROMATIC AMINO ACIDS (BIOSYNTHESIS AND METABOLISM).]

PHENYLHYDRAZONES

Phenylhydrazones are derivatives of aldehydes and ketones of principal importance in CARBOHYDRATE chemistry. Aldoses and ketoses condense with phenylhydrazine to form derivatives in which the $C_6H_5NHN=$ group takes the place of the carbonyl oxygen. These derivatives are called phenylhydrazones. Continued treatment with phenylhydrazine causes oxidation of the adjacent alcohol group to a carbonyl group, and a second condensation to yield phenyldihydrazones or osazones. These often are relatively insoluble and crystallize readily. D-Glucose, D-fructose, and

D-mannose all yield a single osazone in common to the three sugars. This fact established their structural relationship as being identical insofar as four carbon atoms are concerned [see CARBOHYDRATES (CLASSIFICATION AND STRUCTURAL INTERRELATIONS)].

PHENYLKETONURIA. See AROMATIC AMINO ACIDS (BIOSYNTHESIS AND METABOLISM); MENTAL RETARDATION; METABOLIC DISEASES.

PHOSPHATASES

Phosphatases are enzymes catalyzing the hydrolysis of compounds of orthophosphoric acid. According to the specificity of individual phosphatases for the different types of linkages of the phosphoryl group(s), they are classified as *phosphomonoesterases*, *phosphodiesterases* (*e.g.*, nucleases, phospholipid phosphodiesterases) and *phosphoric anhydride hydrolases* (*e.g.*, ATPases). While the importance of this classification which was in part proposed in 1932 by S. Uzawa in the laboratory of Akamatsu is illustrated by the usefulness of certain phosphomonoesterases as sensitive and specific analytical tools for phosphomonoester groups, it must be kept in mind that certain phosphatases have overlapping specifities, *e.g.*, acid yeast phosphatase hydrolyzes phosphomonoesters as well as phosphoric anhydrides, but is inactive towards phosphodiester bonds. The DNA-phosphatase-exonuclease of *E. coli* hydrolyzes terminal 3′-phosphomonoester groups as well as internucleotide phosphodiester linkages of DNA (see also NUCLEASES). In analogy to the hydrolysis of carboxylic esters by certain proteases (chymotrypsin), acid prostatic and alkaline intestinal phosphomonoesterases hydrolyze phosphomonoesters as well as aminophosphoric acid.

Phosphomonoesterases. Some phosphomonoesterases (*non-specific phosphomonoesterases*) hydrolyze at similar rates a wide variety of phosphomonoesters, others (*specific phosphomonoesterases*) have highly selective structural requirements for their activity.

A nonspecific phosphomonoesterase (probably a mixture of isozymes) of alkaline pH optimum (pH 9.6) occurs in high concentration in mucosae of the duodenum and jejunum of calves and in dog feces, in considerable but smaller amounts in kidney, mammary gland and bone, and in appreciable quantities in many other tissues of mammals. Most of the mammalian phosphatases of this group are bound to particles and require digestion with trypsin, autolysis or denaturation of associated proteins by butanol (Morton, 1955) for solubilization. They are metalloenzymes and contain a bivalent metal (zinc) as their natural activator. They are inactivated by CN^-, chelating agents and sulfhydryl compounds.

Nonspecific alkaline phosphatase occurs in small concentrations in normal blood plasma. Strongly increased blood levels are observed in certain bone diseases (osteitis fibrosa, osteosarcoma) and during obstruction of the common bile duct. The action of alkaline bone phosphatase is an important mechanism for providing orthophosphate required during BONE FORMATION.[14] Hypophosphatasia[5] is a rare hereditary disease caused by a genetic defect precluding the biosynthesis of alkaline bone phosphomonoesterase. Inhibition of ossification, similar to the symptons observed in rickets, is a characteristic manifestation of this disease.

A nonspecific "alkaline" phosphatase occurs in *E. coli*.[7] It is a soluble enzyme and is predominantly localized in the space between cell membrane and cell wall. Its concentration in the bacteria increases about 40 times when the orthophosphate concentration in the culture medium is kept at the minimal values compatible with growth, owing to the decrease of the repressive effects of orthophosphate on the biosynthesis of this enzyme (derepression). It is the only phosphomonoesterase which has so far been obtained in crystalline form.[12]

Nonspecific phosphomonoesterases of acid pH optima (acid phosphomonoesterases) have been found in many microorganisms and in plant and animal tissues. A soluble phosphomonoesterase of this type occurs in very high concentrations in the mature prostate glands and its secretions of man and anthropoid apes. Acid phosphomonoesterases are strongly and reversibly inhibited by fluoride ions at $10^{-3}M$, but are little affected by chelating agents, cyanide and sulfhydryl compounds.

Acid prostatic phosphomonoesterase is specifically and competitively inhibited by L-tartrate, and may be differentiated from other nonspecific acid phosphomonoesterases on the basis of this property. This is of clinical interest because increased plasma levels of acid prostatic phosphomonoesterase are observed in patients with metastatic carcinoma of the prostate gland. Assays of the blood levels of this enzyme are of some diagnostic and prognostic importance.

The functional role of the high phosphatase activity of the prostate gland and its secretion is as yet unknown.

Nonspecific acid phosphatase of bakers' yeast (pH optimum, 3.6) is a soluble enzyme located near or at the cell surface. Similar to the behavior of the alkaline phosphatase of *E. coli*, its concentration in the cells increases many times when the yeast is incubated in nutrient solutions of very low orthophosphate concentrations.

Specific Phosphomonoesterases. Knowledge of these enzymes began with the discovery in various animal tissues of a 5′-nucleotidase by Reis in 1934.[13] Convenient sources of 5′-nucleotidases which are important analytical tools in studies on nucleic acid structure and metabolism are bull semen and SNAKE VENOMS. The known 5′-nucleotidases act optimally at alkaline pH values and are activated by certain bivalent metal ions (Zn^{++}, Mg^{++}).

More recently, a considerable number of specific phosphomonoesterases were discovered such as glucose-6-phosphomonoesterase of liver whose action is an important intermediary step in the metabolic degradation of GLYCOGEN to glucose, the 3′-nucleotidase of rye grass, phosphoprotein phosphatase, the PR-enzyme which converts phosphorylase a to phosphorylase b by liberation of phosphate, phosphoserine phosphatase, phosphatidic acid phosphomonoesterase, α-glycerophosphate phosphomonoesterase which occurs in yeast and in red blood cells.

Mechanism of Action of Phosphomonoesterases. Cleavage of phosphomonoesters by phosphomonoesterases in O^{18}-enriched water is accompanied by incorporation of O^{18} into the liberated orthophosphate.[17] The data are compatible with the incorporation of one atom of oxygen per molecule of liberated orthophosphate. Thus, the cleavage of phosphomonoesters catalyzed by phosphomonoesterases is a cleavage of the phosphorus-oxygen bond, but not a cleavage of the corresponding carbon-oxygen bond. In analogy, the characteristic, nonenzymatic solvolysis of phosphoric monoesters by water results likewise in rupture of the P—O bond, whereas acid-catalyzed hydrolysis of phosphate esters involves cleavage of the corresponding C—O bond. Nonenzymatic solvolysis of phosphate esters at weakly alkaline reaction is catalyzed by lanthanum salts[3] and proceeds by cleavage of the P—O bond.[4] Possibly, the mechanism of the nonenzymatic lanthanum catalysis and the role of the protein-bound metal in the catalytic action of alkaline phosphomonoesterases are related.

Alkaline phosphatase of *E. coli* forms a phosphate ester of the enzyme when incubated with P^{32}-labeled orthophosphate.[6] Phosphoserine and phosphoserine peptides were isolated from acid hydrolysates of enzyme phosphate ester. Unlike other esterases, however, phosphomonoesterases are not inactivated by diisopropyl fluorophosphate [see also ENZYMES (ACTIVE CENTERS)].

Phosphotransferase Activity of Phosphomonoesterases. Some nonspecific as well as specific phosphomonoesterases were shown to transfer phosphoryl groups to hydroxy compounds other than water, such as glycerol, nucleosides and other aliphatic alcohols and phenols. L-Serine is the preferential receptor for the transfer of phosphoryl groups from L- and D-phosphoryl serine by the action of phosphoserine phosphomonoesterase.

Assay Techniques. In the majority of cases, phosphomonoesterases may be assayed by the colorimetric determination of the liberated orthophosphate in the protein-free filtrate of the assay mixture according to the Fiske-Subbarow technique. Modifications of this technique have been developed for assay mixtures containing acid-labile substrates or compounds forming precipitates with phosphomolybdic acid such as oligonucleotides, protein degradation products, or other organic bases.

Very convenient assay techniques are based on the cleavage of chromogenic aromatic phosphate esters as model substrates. In such cases, the assay may be performed in the spectrophotometer by measurements of the absorption of the liberated phenol. HISTOCHEMICAL visualization of phosphomonoesterase action (introduced by Gomori and refined by Seligman) is based on *in loco* precipitation of the liberated orthophosphate by lead acetate or on *in loco* precipitation of liberated colored, water-insoluble phenol derivatives.

Careful definition of assay conditions is important in the field of phosphomonoesterases because the kinetics of their hydrolytic action are usually complicated by strong inhibitory effects of the liberated orthophosphate as well as by the simultaneous occurrence of phosphoryl transfer reactions.

Phosphodiesterases. Phosphodiester linkages in living organisms occur almost exclusively in nucleic acids, PHOSPHOLIPIDS and in some of their metabolic degradation products. Many phosphodiesterases catalyzing the cleavage of these compounds are specific NUCLEASES (ribonucleases, deoxyribonucleases or lipidphosphohydrolases, respectively. These enzymes do not hydrolyze model substrates such as diphenyl or dinitrophenyl phosphate.

Nonspecific phosphodiesterases hydrolyzing these model substrates as well as phosphodiester linkages of ribo- and deoxyribopolynucleotides have been found in the intestinal mucosa, in spleen, in SNAKE VENOMS and in some microorganisms.

The properties of the "nonspecific" phosphodiesterases resemble those of the nonspecific alkaline phosphomonoesterases with respect to their pH optima, their activation by certain bivalent metals (which might be components of the respective enzyme proteins) and their inhibitors.

The application of these enzymes as analytical tools has been of the greatest importance in the elucidation of the structure of nucleic acids. The discovery of deoxyribose in the laboratory of P. A. Levene,[11] the isolation of the DNA-mononucleotides in the laboratory of S. J. Thannhauser (1931–1934),[10] the complete sequence determination of alanine—s-RNA (see SOLUBLE RIBONUCLEIC ACIDS) in the laboratory of Holley[9] were accomplished by the application of nucleases, phosphodiesterases and phosphomonoesterases to the controlled degradation of nucleic acids.

Assay of Phosphodiesterases. Since the action of phosphodiesterases does not result in the liberation of orthophosphate, the assay of these enzymes is usually more complicated than that of phosphomonoesterases except when aromatic model substrates can be used. In this case, the determination of the liberated phenol can be used as a measure of the enzyme activity. A convenient chromogenic substrate of this type is bis-dinitrophenyl phosphate.

Another useful assay principle is the incubation of substrates of phosphodiesterases with the phosphodiesterase, combined with an excess of a suitable nonspecific phosphomonoesterase, and the subsequent determination of the released orthophosphate.

The phosphoryl groups of phosphomonoesters dissociate in two steps with pK values which are slightly lower than pK_1 and pK_2 of orthophosphoric acid, whereas those of the phosphodiesters have only one dissociable group which is slightly more acidic than pK_1 or orthophosphoric acid. Consequently, the cleavage of phosphodiesters is accompanied by measurable increases of titrable acidity, in contrast to that of phosphomonoesters. While most of the acidimetric techniques are not sufficiently sensitive to permit accurate initial activity measurements the pH-stat has been used successfully to follow the course of phosphodiesterase activity in some instances.

Obviously, these simple routine assay techniques yield only very incomplete (though indispensable) information regarding the activities of the phosphodiesterases. In contrast to the model substrates, the majority of natural substrates are phosphodiesters in which each phosphoryl group is esterified with two different hydroxy compounds. Furthermore, polynucleotides contain in each molecule more than one phosphodiester group (with the exception of dinucleotides or dinucleoside monophosphates).

Phosphoric Anhydride Phosphohydrolases. While some phosphomonoesterases also catalyze the hydrolysis of phosphoanhydride linkages of inorganic pyrophosphate, adenosine di- and triphosphate, and probably of other phosphoric anhydrides, many specific phosphoric anhydride hydrolases have been found.

Inorganic Pyrophosphatase of Bakers' Yeast is the only enzyme of this group which has so far been obtained in crystalline form (Kunitz, 1952).

ATPases (adenosine triphosphate phosphohydrolases) catalyze the hydrolysis of ATP to ADP (adenosine diphosphate) and are in most instances particulate enzymes of rather ubiquitous occurrence in tissues and cells. Of particular interest are the ATPases of MUSCLE tissue in view of the discovery of W. A. Engelhardt (1934) and W. V. Lyubimova (1939) that MYOSIN has ATPase activity. A. Szent-György (1947) found that the contraction induced in actomyosin fibers by ATP is accompanied by liberation of inorganic phosphate, and H. H. Weber (1952) demonstrated a similar phenomenon with glycerin-extracted muscle fibers. The activities of myosin ATPase as well as those of other particulate ATPases are greatly influenced by the ionic environment of the enzyme-substrate system. Most of the ATPases liberate the terminal phosphoryl groups not only from ATP, but also from other nucleoside triphosphates.

Significance of Myosin ATPase. The metabolic utilization of the free energy released during the cleavage of phosphoric anhydride linkages of ATP has been demonstrated in numerous enzyme reactions in cell-free system. Most of these reactions involve the energy-requiring transfer of one of the cleavage products of ATP to a metabolite (*e.g.*, conversion of fatty acids and amino acids to their adenylylcarboxyl anhydrides, conversion of glucose to glucose-6-phosphate). So far, the final manifestation of free energy utilization of the hydrolytic cleavage of ATP by myosin fibers has been the biophysical process of contraction. It seems possible that the enzymatic cleavage of ATP by myosin involves—in analogy to many enzyme reactions—the intermediary formation of an unstable myosin-ATP (enzyme-substrate) compound and that the formation of this compound is responsible for the change of the shape of the myosin molecule which results in contraction.

Apyrases. Enzymes catalyzing the hydrolysis of *both* phosphoric anhydride bonds of ATP with formation of two molecules of orthophosphate and one molecule of adenylate have been purified from potato extracts and from the MITOCHONDRIA of insect muscles.

Nucleotide Pyrophosphatases. An enzyme catalyzing the hydrolysis of deoxycytidine triphosphate to cytidine-5'-monophosphate and inorganic pyrophosphate has been found among the numerous enzymes which are formed in *E. coli* after infection with bacteriophage, but which do not occur in normal *E. coli* cells.

Nucleoside diphosphatases catalyzing the hydrolysis of certain nucleoside diphosphates to inorganic phosphate and mononucleotides were found in muscle and in liver mitochondria.

Acyl Phosphatases. [Enzymes hydrolyzing anhydride bonds between phosphoryl groups and other acyl groups (*e.g.*, carboxyl groups).] Enzymes catalyzing the mixed phosphoric anhydride bonds of acetyl phosphate, acetyl adenylate, 1,3-diphosphoglycerate, and CARBAMYL PHOSPHATE occur in skeletal muscle where they were first detected and in many other mammalian tissues as well as in microorganisms. The muscle enzyme is characterized by its unusually high thermostability which resembles that of myokinase and pancreatic ribonuclease, whereas the liver enzyme is heat labile. The muscle enzyme is inactive toward glycerophosphate and has optimal activity toward acyl phosphates at pH 5.4. Acyl phosphatases catalyze cleavage of the P—O bonds of their substrates, in analogy to the mechanism of action of the phosphomonoesterases.

References

Reviews and Detailed Descriptions of Phosphatases

1. Boyer, P. D., Lardy, H., and Myrbäck, K, (Editors), "The Enzymes," Second edition, Vol. 5. New York and London, Academic Press, 1961.
2. Colowick, S. P., and Kaplan, N. O., (Editors), "Methods in Enzymology," Vols. I–VII, New York and London, Academic Press, 1955–1964.

Articles cited

3. Bamann, E., and Trapmann, H., *Advan. Enzymol.*, **21**, 169 (1959).
4. Butcher, W. W., and Westheimer, F. H., *J. Am. Chem. Soc.*, **77**, 2420 (1955).
5. Cusworth, D. C., *Biochem. J.*, **68**, 262 (1958).
6. Engström, L., *Biochim. Biophys. Acta*, **56**, 606 (1962).
7. Garen, A., and Levinthal, C., *Biochim. Biophys. Acta*, **38**, 470 (1960).
8. Gomori, G., *J. Histochem. Cytochem.*, **4**, 453 (1956).

9. HOLLEY, R. W., APGAR, J., EVERETT, G. A., MADISON, J. T., MARQUISEE, M., MERRILL, S. H., PENSWICK, J. R., AND ZAMIR, A., *Science*, **147**, 1462 (1965).
10. KLEIN, W., AND THANNHAUSER, S. J., *Z. Physiol. Chem.*, **23**, 96 (1935).
11. LEVENE, P. A., AND MORI, T., *J. Biol. Chem.*, **83**, 803 (1929).
12. MALAMY, M. H., AND HORECKER, B. L., *Biochemistry*, **3**, 1893 (1964).
13. REIS, J. L., *Bull. Soc. Chim. Biol.*, **16**, 385 (1934).
14. ROBINSON, R., AND SOAMES, K. M., *Biochem. J.*, **18**, 740 (1924).
15. SELIGMAN, A. M., AND MANNHEIMER, L. H., *J. Natl. Cancer Inst.*, **9**, 427 (1949).
16. SCHWARTZ, J. H., CRESTFIELD, A. M., AND LIPMANN, F., *Proc. Natl. Acad. Sci. U.S.*, **49**, 722 (1963).
17. STEIN, S. S., AND KOSHLAND, D. E., JR., *Arch. Biochem. Biophys.*, **39**, 229 (1952).
18. SZENT-GYÖRGY, A., "Chemical Physiology of Contraction in Body and Heart Muscles," New York, Academic Press, 1953.
19. WEBER, H. H., AND PORTZEHL, H., in "Advances in Protein Chemistry" (ANSON, M. L., BAILEY, K., AND EDSALL, J. T., EDITORS), Vol. VII, p. 161, New York, Academic Press, 1952.

GERHARD SCHMIDT

PHOSPHATE BOND ENERGIES

The Concept of Free Energy Change. To the biochemist the term "phosphate bond energy" refers to the energy released when a particular P—O or P—N bond in a compound is hydrolyzed and the elements of water are used to form new bonds, as for example in the hydrolysis of a phosphate ester:

$$R{-}O{-}PO_3^{=} + HOH \rightarrow ROH + HO{-}PO_3^{=} + \text{Energy} \quad (1)$$

The "phosphate bond energy" is the difference between the "free energy" content of the products (F_P) and the "free energy" content of the reactants (F_R):

$$F_P - F_R = \Delta F \quad (2)$$

The change in "free energy," ΔF, is the *maximum* amount of energy made available to do work as a result of the reaction. If energy is released, then it is obvious that the reactants must have contained more energy than the products and the ΔF of the reaction will be negative. The ΔF of a reaction at any given temperature may be expressed in terms of two other thermodynamic quantities, ΔH and ΔS, where ΔH is the difference in enthalpy (heat content) and ΔS is the difference in entropy:

$$\Delta F = \Delta H - T\Delta S \quad (3)$$

The ΔF of a reaction under a given set of conditions can also be defined in terms of the equilibrium constant (K_{eq}) of the reaction and the displacement of the system from equilibrium, according to Eq. (*4*).

$$\Delta F = -RT \ln K_{eq} + RT \ln \frac{[P_1]^a [P_2]^b \ldots}{[R_1]^c [R_2]^d \ldots} \quad (4)$$

R is the gas constant, T is the temperature in degrees Kelvin, and [P] and [R] are the actual concentrations of products and reactants, respectively, raised to the powers a, b, c, and d which are their respective coefficients in the balanced reaction equation. At 25°C (298°K) Eq. (*4*) can be simplified to Eq. (*5*).

$$\Delta F = -1363 \log K_{eq} + 1363 \log \frac{[P_1]^a [P_2]^b \ldots}{[R_1]^c [R_2]^d \ldots} \quad (5)$$

If a reaction is at equilibrium, then there is no net change of R to P and consequently no ΔF, *i.e.*, $\Delta F = 0$. This can also be seen from Eq. (*5*) where the [P]/[R] term equals K_{eq} at equilibrium. In order to simplify tabulations and comparisons of the ΔF of different reactions, chemists have agreed upon a set of arbitrary "standard-state" conditions in which all reactants and products are at "unit activity." Unit activity is approximated by unit molarity for all solutes in dilute solutions and 1 atm partial pressure for all gases. The solvent, which is always water in biological systems, is arbitrarily taken to be unit activity for dilute solutions even though the concentration of water is actually about 55M. It can be seen from Eq. (*5*) that under standard-state conditions, the [P]/[R] term drops out. The ΔF under standard-state conditions is designated ΔF° and can be directly related to the equilibrium constant of a reaction:

$$\Delta F^\circ = -RT \ln K_{eq} = -1363 \log K_{eq} \text{ at } 25°\text{C} \quad (6)$$

ΔF° then is the change in free energy per mole of reactant that accompanies a reaction under steady-state conditions where the concentration of all reactants and products are maintained at 1M.

Hydrogen ions are frequently products or reactants in biological reactions. Because almost no enzyme-catalyzed reaction can occur at pH 0 (1M or unit activity H^+), biochemists generally define the standard-state free energy change not as ΔF°, but as $\Delta F'$ where all reactants and products *except* H^+ are at unit activity. The actual $[H^+]$ must then be specified and is usually some physiological value (10^{-6}–$10^{-8}M$). The relationship between ΔF° and $\Delta F'$ for a reaction that yields H^+ ions can be calculated from Eq. (*4*):

$$\Delta F = -RT \ln K_{eq} + RT \ln \frac{[P_1]^a [P_2]^b \ldots [H^+]}{[R_1]^c [R_2]^d \ldots}$$

$$\Delta F = \Delta F' = \Delta F^\circ + 1363 \log [H^+], \text{ at } 25°\text{C and unit activity of all P and R except } H^+$$

$$\Delta F' = \Delta F^\circ - 1363 \log \frac{1}{[H^+]}$$

$$\Delta F' = \Delta F^\circ - 1363\text{pH} \quad (7)$$

Even though H^+ ions may not be reactants or products in a given reaction, the free energy changes that are tabulated are still frequently designated as $\Delta F'$ values because they have been calculated from equilibrium or calorimetric data obtained at physiological pH's. It is important to realize that even though tabulated free energy changes have been corrected for physiological hydrogen ion concentrations, the concentrations of all other reactants and products in a living cell are not $1M$ and that the actual ΔF of a reaction *in vivo* may be quite different from the tabulated $\Delta F'$. The true ΔF under *in vivo* conditions can be calculated from Eq. (*8*):

$$\Delta F = \Delta F' + 1363 \log \frac{[P_1]^a [P_2]^b \ldots}{[R_1]^c [R_2]^d \ldots} \qquad (8)$$

As an example, we can estimate the ΔF of hydrolysis of ATP under physiological conditions. In an actively growing cell, the intracellular ATP concentration is maintained at a steady-state level of about $10^{-3}M$. The ATP/ADP ratio is about 10 and the intracellular phosphate level is about $10^{-2}M$. The $\Delta F'$ for the reaction (Table I) is -8 kcal/mole.

$$\Delta F = \Delta F' + 1363 \log \frac{[\text{ADP}]\,[\text{P}_i]}{[\text{ATP}]}$$

$$\Delta F = -8000 + 1363 \log 10^{-3}$$

$$\Delta F = -8000 - 4089$$

$$\Delta F = -12{,}089 \text{ cal/mole}$$

Types of Phosphorus Compounds Found in Cells. The low molecular weight phosphorus compounds found in living cells may be roughly divided into two groups: "energy-rich" (or "high phosphoryl transfer potential") compounds which have a $\Delta F'$ of hydrolysis of -6 to -12 kcal/mole, and "energy-poor" (or "low phosphoryl transfer potential") compounds which have a $\Delta F'$ of hydrolysis of -2 to -4 kcal/mole. Examples of energy-rich compounds include pyrophosphates (*e.g.*, nucleoside tri- and diphosphates, inorganic pyrophosphate and polyphosphates), acyl phosphates (*e.g.*, acetyl phosphate, 1,3-diphosphoglyceric acid), enolic phosphates (*e.g.*, phosphoenolpyruvate), guanidinium phosphates (*e.g.*, creatine phosphate, arginine phosphate), and phosphosulfates (*e.g.*, adenosine phosphosulfate, 3′-phosphoadenosine-5′-phosphosulfate). Thioesters (*e.g.*, acetyl coenzyme A) represent a sixth type of energy-rich compound although the thioester bond contains no phosphorus. Energy-poor compounds include nucleoside monophosphates (*e.g.*, adenosine monophosphate), and most other simple phosphate esters (*e.g.*,glucose-6-phosphate, glycerol phosphate). The dividing line between energy-rich and energy-poor compounds is certainly not a sharp one. The $\Delta F'$ of glucose-1-phosphate hydrolysis is about -5 kcal/mole, and consequently glucose-1-phosphate might be classified as an "intermediate-energy phosphate" compound. Furthermore, it should be reemphasized that the actual ΔF of hydrolysis under *in vivo* conditions may be significantly different from the $\Delta F'$. Unfortunately, the steady-state intracellular levels of many phosphate compounds and their hydrolysis products are unknown and so the actual ΔF cannot be calculated. The $\Delta F'$ values of some important hydrolysis reactions are shown in Table 1.

Factors Affecting the Magnitude of $\Delta F'$. It was originally believed that a large part of the energy in an energy-rich phosphate compound was localized in one or more specific covalent bonds and that it was this energy that was released when the particular P—O or P—N bond was hydrolyzed. For many years the concept of the "high-energy phosphate bond" (abbreviated $\sim$P) was widespread among biochemists. We now know that the magnitude of the $\Delta F'$ of hydrolysis depends on many factors. Some of the more important factors are described below.

Bond Strain Caused by Electrostatic Repulsion. The partial structures of ATP, acetyl phosphate, and phosphoenolpyruvate (PEP) drawn in conventional form are shown below in formulas I, II

TABLE 1. APPROXIMATE $\Delta F'$ VALUES FOR THE HYDROLYSIS OF SOME PHOSPHATE COMPOUNDS

General Type	Compound	Hydrolysis Products	Approximate $\Delta F'$ at pH 6-8 (kcal/mole)
Phosphosulfate anhydride	Adenosine phosphosulfate (APS)	AMP + $SO_4^=$	−18
Pyrophosphate	Inorganic pyrophosphate	2 P_i	− 7
	Adenosine triphosphate (ATP)	ADP + P_i	− 7.7
	Adenosine triphosphate (ATP)	AMP + PP_i	− 8.6
	Adenosine diphosphate (ADP)	AMP + P_i	− 6.4
Acyl phosphate	Acetyl phosphate	Acetic acid + P_i	−10
	1,3-Diphosphoglyceric acid	3-Phosphoglyceric acid + P_i	−12
Thioester[a]	Acetyl-CoA	Acetic acid + CoASH	− 8.2
Enolic phosphate	Phosphoenolpyruvate (PEP)	Keto-pyruvic acid + P_i	−12.8
Guanidinium phosphate	Creatine phosphate	Creatine + P_i	−10.5
Hemiacetal phosphate	α-D-Glucose-l-phosphate	α- and β-D-Glucose + P_i	− 5
Simple phosphate ester	Adenosine monophosphate (AMP)	Adenosine + P_i	− 2
	Glucose-6-phosphate	Glucose + P_i	− 3
	Glycerol-3-phosphate	Glycerol + P_i	− 2.5

[a] Thioesters are energy-rich compounds although the thioester bond contains no phosphorus.

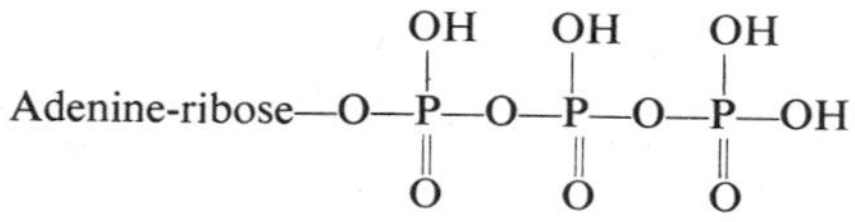

Adenine-ribose—O—P(=O)(OH)—O—P(=O)(OH)—O—P(=O)(OH)—OH

I

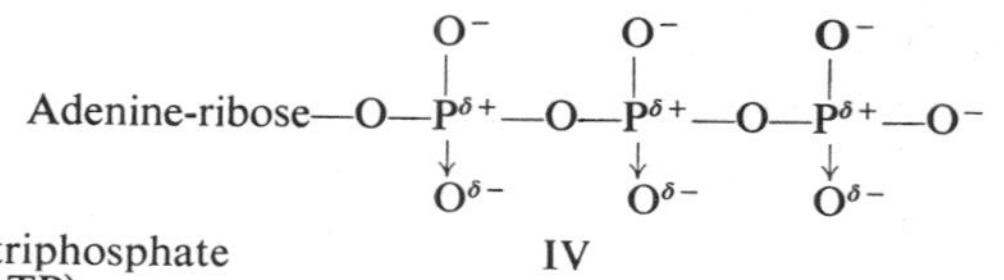

Adenosine triphosphate (ATP)

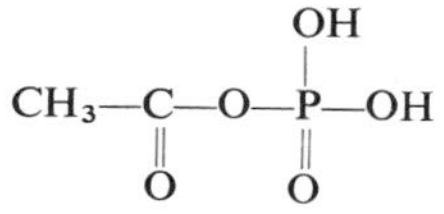

Adenine-ribose—O—$P^{\delta+}$(O^-)($\downarrow O^{\delta-}$)—O—$P^{\delta+}$(O^-)($\downarrow O^{\delta-}$)—O—$P^{\delta+}$(O^-)($\downarrow O^{\delta-}$)—O^-

IV

CH_3—C(=O)—O—P(=O)(OH)—OH

II

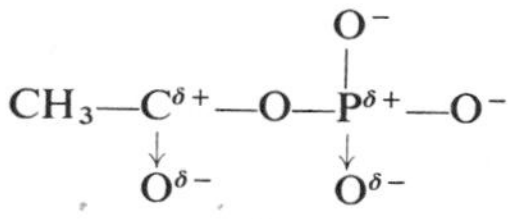

Acetyl phosphate

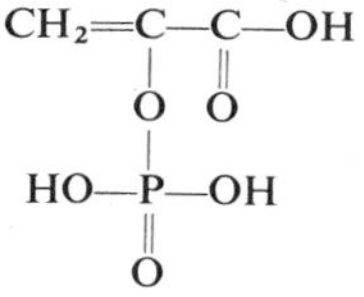

CH_3—$C^{\delta+}$($\downarrow O^{\delta-}$)—O—$P^{\delta+}$(O^-)($\downarrow O^{\delta-}$)—O^-

V

CH_2=C(—O—P(=O)(OH)—OH)—C(=O)—OH

III

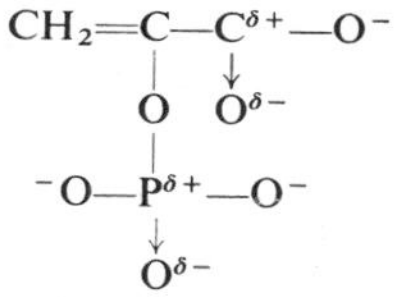

Phosphoenolpyruvate (PEP)

CH_2=C(—O—$P^{\delta+}$($^-$O)($\downarrow O^{\delta-}$)—O^-)—$C^{\delta+}$($\downarrow O^{\delta-}$)—O^-

VI

and III. A more accurate representation of the structures of these energy-rich compounds is shown in formulas IV, V and VI.

Here we recognize that at physiological pH, the acidic hydrogens on the phosphate and carboxyl groups will be ionized, leaving negatively charged oxygens. Furthermore, because oxygen is more electronegative than phosphorus or carbon, the electrons constituting the P=O and C=O double bonds will be drawn much closer to the oxygen, producing an additional partial negative charge on these oxygens and partial positive charges on the P and C atoms. The electrostatic repulsion caused by the close proximity of like charges places severe restrictions on the possible configurations of the ATP, acetyl phosphate and PEP molecules. The most likely configurations are those in which the distances between the repulsive charges are maximized. The fact that the energy-rich molecules remain intact in spite of the repulsive charges means that they must contain sufficient internal energy to overcome their inherent instability. When the anhydride or ester bonds are hydrolyzed, part of the electrostatic energy is released and contributes to the total negative ΔF. The amount of energy contributed by electrostatic repulsion has not actually been measured, but is estimated from theoretical models to be about 5 kcal/mole for ATP and 3–4 kcal/mole for ADP and PEP. Electrostatic repulsion probably contributes significantly to the ΔF of hydrolysis of all pyrophosphate, acylphosphate and enolic phosphate compounds.

Contribution of the ΔF of Isomerization of Hydrolysis Products. Electrostatic repulsion is not the only factor contributing to the high ΔF of PEP hydrolysis. The enol form of pyruvic acid is inherently unstable and spontaneously tautomerizes to the more stable keto form. The $\Delta F'$ of tautomerization has been estimated to be about -6 to -9 kcal/mole. In PEP, the esterified phosphate group prevents the tautomerization and holds the pyruvic acid in the unstable enol form. The actual hydrolysis step to form inorganic phosphate and enol pyruvic acid has a $\Delta F'$ of about -5 kcal/mole (of which 3–4 kcal/mole results from electrostatic repulsion). However, tautomerization increases the over-all $\Delta F'$ to about -13 kcal/mole, making PEP one of the most energy-rich phosphate compounds of biological importance.

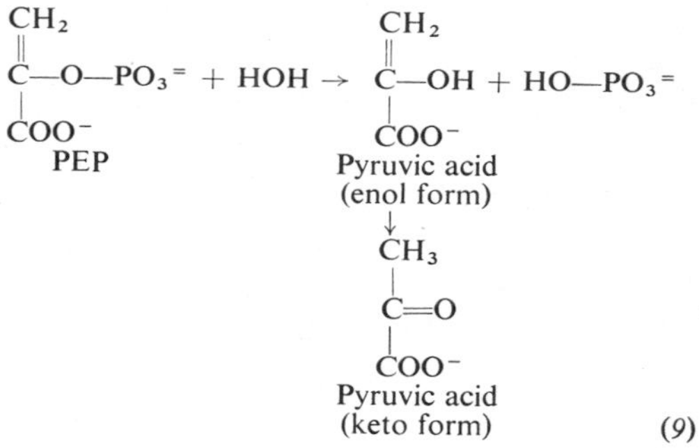

$$CH_2{=}C(COO^-)\text{—O—}PO_3^{=} + HOH \rightarrow CH_2{=}C(COO^-)\text{—OH} + HO\text{—}PO_3^{=} \qquad (9)$$

PEP → Pyruvic acid (enol form) ↓ CH_3—C(=O)—COO^- Pyruvic acid (keto form)

Structural considerations similar to those described above may be used to explain the difference between the $\Delta F'$ of hydrolysis of α-D-glucose-1-phosphate (-5 kcal/mole) and that of glucose-6-phosphate, a typical energy-poor ester (-3 to -4 kcal/mole). In glucose-1-phosphate, the hemiacetal OH group on carbon-1 is esterified and fixed in the α-position. After hydrolysis, the OH group is free to undergo mutarotation to form an equilibrium mixture of α- and β-D-glucose. At equilibrium, the ratio of β/α forms is about 2. Thus the mutarotation contributes about 400 calories of the 1–2 kcal/mole difference. The remainder of the difference is probably the result of the greater interaction of the C-1-phosphate with the OH groups of the ring.

The Effect of pH and Ionization. The relative tendencies of the reactants and products to ionize at the particular reaction pH have a great influence on the $\Delta F'$ of a reaction. The simplest example to consider is the hydrolysis of a compound to produce one or more ionizable groups where none previously existed. Equation (*10*) shows that the hydrolysis of acetyl phosphate produces two new acidic functions—the carboxyl group of acetic acid and a third OH group on inorganic phosphate.

$$CH_3{-}\underset{\underset{O^{\delta-}}{\downarrow}}{C^{\delta+}}{-}O{-}\underset{\underset{O^{\delta-}}{\downarrow}}{\overset{\overset{O^-}{|}}{P^{\delta+}}}{-}O^- + HOH \rightarrow$$

$$CH_3{-}\underset{\underset{O^{\delta-}}{\downarrow}}{C^{\delta+}}{-}OH + HO{-}\underset{\underset{O^{\delta-}}{\downarrow}}{\overset{\overset{O^-}{|}}{P^{\delta+}}}{-}O^- \qquad (10)$$

At physiological pH, the newly created acid group on the phosphate will not ionize because it is an extremely weak acid ($pK_a = 12.7$). Since it did not exist before hydrolysis and is un-ionized after hydrolysis, the formation of this new acidic group has no effect on the $\Delta F'$ of hydrolysis. The newly formed carboxyl group on the other hand, has a pK_a of about 4.8 and so it will spontaneously ionize at physiological pH [Eq. (*11*)]:

$$CH_3{-}\underset{\underset{O^{\delta-}}{\downarrow}}{C^{\delta+}}{-}OH \xrightarrow{pH \gg pK_a} CH_3{-}\underset{\underset{O^{\delta-}}{\downarrow}}{C^{\delta+}}{-}O^- + H^+ \qquad (11)$$

The effect of this ionization is to reduce the concentration of the actual hydrolysis product (un-ionized acetic acid) to a very low level. In fact, for every pH unit above the pK_a, the steady-state concentration of un-ionized acetic acid will decrease by a factor of 10. From Eqs. (*4*) or (*8*) we see that for each decrease in the concentration of any product by a factor of 10, the ΔF will decrease by 1363 cal/mole. The decrease in ΔF caused by the ionization of a newly formed acidic group (ΔF_i) can be calculated directly from Eq. (*12*).

$$\Delta F_i = -1363 \log \left[1 + \frac{K_a}{[H^+]}\right] \qquad (12)$$

It can be seen that the magnitude of the ΔF_i depends on the difference between the pH of the buffered reaction medium and the pK_a of the newly formed ionizable group. The contribution of a new group with a pK_a 1 unit less than the pH is about −1363 cal/mole and the ΔF_i will become more negative by 1363 cal/mole for each pH unit difference.

In addition to the contribution of newly formed ionizable groups, changes in the pK_a of pre-existing ionizable groups must also be considered. For example, the pK_{a1} and pK_{a2} of the terminal phosphate group on ATP are 4 and 6.5. The inorganic phosphate derived from this group has pK_a values of about 2 and 7. The effect of these changes on the $\Delta F'$ of hydrolysis of ATP is discussed in references 3, 5, 6 and 8.

Effect of Metal Ions. Many phosphate compounds form relatively strong metal ion complexes. Just as the pH of the reaction medium influences the ΔF of a reaction when ionizable groups are present, the metal ion concentration will also affect the ΔF if complex-forming groups are present. Consider, for example, the hydrolysis of ATP in the presence and in the absence of Mg^{++}. Both ATP and ADP form magnesium complexes. The effect of the magnesium is to partially neutralize the negatively charged oxygens, thereby diminishing the free energy content resulting from electrostatic repulsion. The binding (or affinity) constant of the $MgATP^{-2}$ complex is about three times stronger than that of the $MgADP^{-1}$ complex, i.e., Mg^{++} is more tightly bound to ATP than to ADP. Thus, ATP loses more free energy content as a result of Mg^{++} binding than does ADP. Consequently, the $\Delta F'$ of ATP hydrolysis in the presence of Mg^{++} is about 1.6 kcal/mole more positive than in its absence.

Resonance Stabilization of Hydrolysis Products. It is apparent from the previous discussion that the ΔF of hydrolysis of a phosphate compound will be highly negative if the products are significantly more stable than the parent compounds. A factor contributing to the greater stability of the energy-rich phosphate hydrolysis products is the delocalization of electrons that constitute the P—O, C—O, and P—N bonds. Before hydrolysis, the electrons have a rather restricted distribution, but after hydrolysis they have, in a sense, a greater "degree of freedom" (*i.e.*, a lower potential or electrostatic energy level). An indication of the degree of freedom of the bond electrons can be obtained by examining the number of permissible resonance forms that can be written for the reactants and products. The fact that we can write several resonance forms for a compound does not mean that the compound is actually present as a mixture of these distinct forms. Rather, the most probable electronic configuration is that which is a hybrid of all the resonance forms. However, the more resonance forms that can be written for a compound, the less localized or restricted the electrons are, and the more stable will the compound be. Formulas VIII, IX and X show three of the resonance forms of ATP. The total number of resonance forms is 12—any combination of 3 for the terminal phosphate and 2 for each of the other phosphates. The products of the reaction, have a total of 18 possible resonance forms—any combination of 3 for the inorganic phosphate (formulas XI, XII, XIII), 3 for the terminal phosphate in ADP, and 2 for the inner phosphate of ADP. The products of ADP hydrolysis (AMP and inorganic phosphate) also possess more resonance forms than the parent compound. Similarly, we can show that acetate and inorganic phosphate possess more resonance forms than their energy-rich precursor, acetyl phosphate.

The energy-rich guanidinium phosphates such as creatine and arginine phosphates possess no obvious electrostatic repulsion forces, nor are there any obvious ionization or tautomerization effects to account for their highly negative ΔF of hydrolysis. Nevertheless, the hydrolysis products are significantly more stable than the guanidinium phosphates. Creatine phosphate possesses 12 possible resonance forms, 3 of which are shown in formulas XIV, XV and XVI. The products of creatine phosphate hydrolysis possess 18 resonance forms—any combination of 3 for the inorganic phosphate and 6 for creatine. Three possible resonance forms of creatine are shown in formulas XVII, XVIII and XIX.

If we examine the hydrolysis of a typical energy-poor phosphate such as glucose-6-phosphate, we find that there is no significant increase in the number of resonance forms as a result of the hydrolysis.

Contribution of Entropy Changes. Not all of the free energy change that occurs upon hydrolysis of an energy rich compound appears as heat. A portion of the energy is retained and utilized in increasing the entropy of the hydrolysis products.

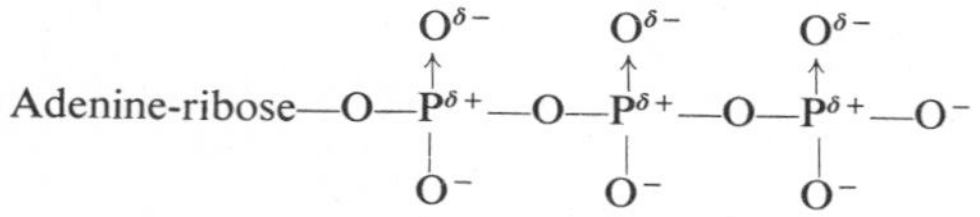

Adenine-ribose—O—$P^{\delta+}$(↑$O^{\delta-}$)(O^-)—O—$P^{\delta+}$(↑$O^{\delta-}$)(O^-)—O—$P^{\delta+}$(↑$O^{\delta-}$)(O^-)—O^-

VIII

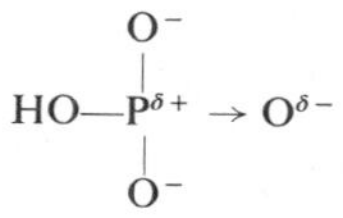

Adenine-ribose—O—$P^{\delta+}$(↑$O^{\delta-}$)(O^-)—O—$P^{\delta+}$(↑$O^{\delta-}$)(O^-)—O—$P^{\delta+}$(O^-)(↓$O^{\delta-}$)—O^-

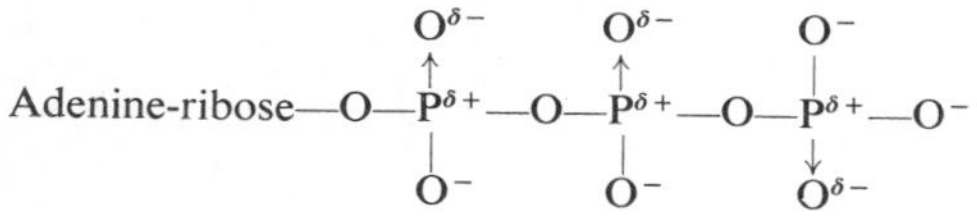

IX

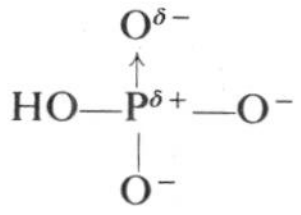

Adenine-ribose—O—$P^{\delta+}$(O^-)(↓$O^{\delta-}$)—O—$P^{\delta+}$(O^-)(↓$O^{\delta-}$)—O—$P^{\delta+}$(O^-)(↓$O^{\delta-}$)—O^-

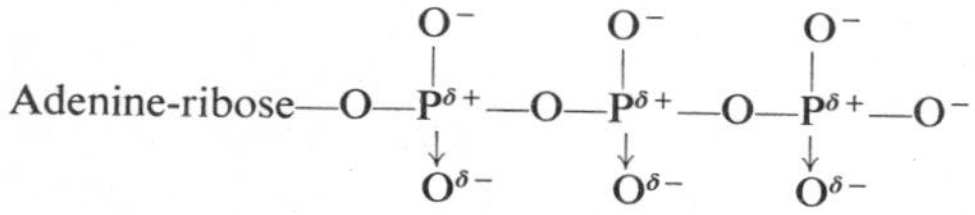

X

Adenosine triphosphate

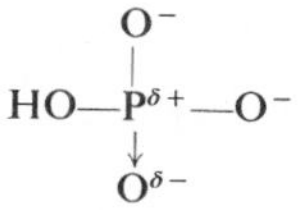

HO—$P^{\delta+}$(O^-)(O^-) → $O^{\delta-}$

XI

HO—$P^{\delta+}$(↑$O^{\delta-}$)(O^-)—O^-

XII

HO—$P^{\delta+}$(O^-)(↓$O^{\delta-}$)—O^-

XIII

Inorganic phosphate

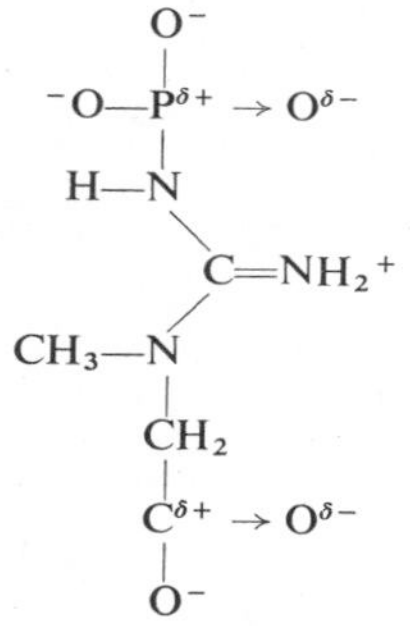

$^-$O—$P^{\delta+}$(O^-) → $O^{\delta-}$; H—N; C=NH_2^+; CH_3—N; CH_2; $C^{\delta+}$ → $O^{\delta-}$; O^-

XIV

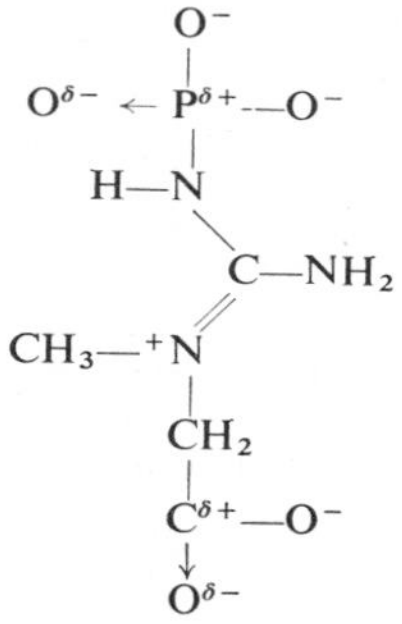

$O^{\delta-}$ ← $P^{\delta+}$(O^-)—O^-; H—N; C—NH_2; CH_3—$^+$N; CH_2; $C^{\delta+}$—O^-; ↓$O^{\delta-}$

XV

$^-$O—$P^{\delta+}$(↑$O^{\delta-}$)—O^-; H—N; C—NH_2; CH_3—$^+$N; CH_2; $C^{\delta+}$—O^-; ↓$O^{\delta-}$

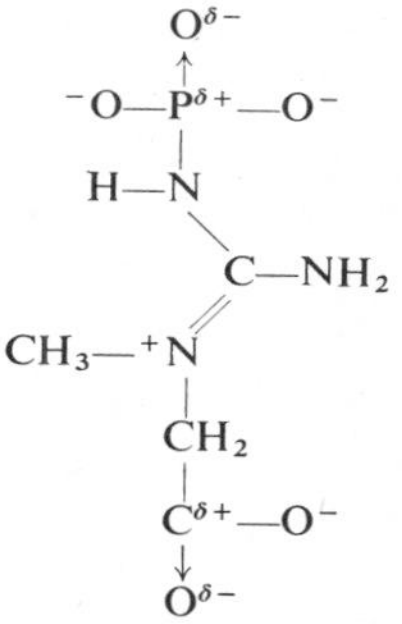

XVI

Creatine phosphate

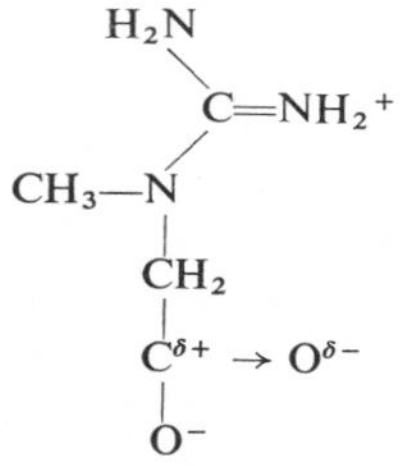

H_2N; C=NH_2^+; CH_3—N; CH_2; $C^{\delta+}$ → $O^{\delta-}$; O^-

XVII

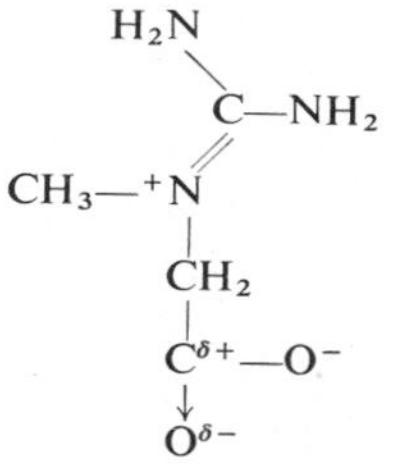

H_2N; C—NH_2; CH_3—$^+$N; CH_2; $C^{\delta+}$—O^-; ↓$O^{\delta-}$

XVIII

H_2N^+; C—NH_2; CH_3—N; CH_2; $C^{\delta+}$ → $O^{\delta-}$; O^-

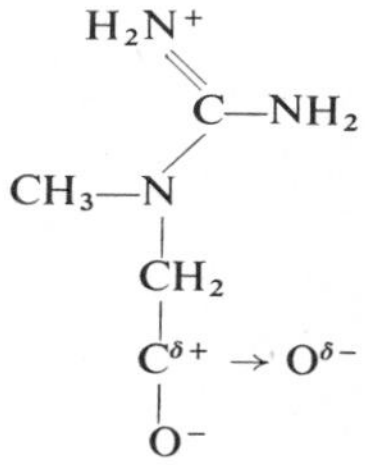

XIX

Creatine

Entropy corresponds to the energy levels of rotation, translation and intramolecular vibration, and may be considered a measure of the degree of randomness or disorder of a system. A quantitative discussion of the entropy change that accompanies a hydrolysis reaction is beyond the scope of this article. However, qualitatively it can be seen [Eq. (*3*)] that as the entropy of the products increases over that of the reactants, the $T\Delta S$ term will become more positive, and the ΔF of hydrolysis will become more negative. Before hydrolysis, energy-rich compounds are maintained in a rather rigid configuration. After hydrolysis, the reaction products are, in a sense, less ordered and thus may have a higher entropy. The contribution of the entropy change depends on the particular energy-rich compound and the conditions of hydrolysis. For the hydrolysis of ATP at physiological pH and temperature, the $\Delta F'$ is about -8 kcal/mole, of which about 5 kcal/mole is released as heat and 3 kcal/mole is retained as the entropy increase.

Biological Functions of Energy-Rich Compounds. While it is convenient to discuss energy-rich compounds in terms of their ΔF of hydrolysis, it should be realized that the energy released upon simple hydrolysis is wasted, and such reactions seldom occur *in vivo*. Rather, the energy in energy-rich compounds is generally used to drive other reactions that are thermodynamically unfavorable. For example, consider the synthesis of glucose-6-phosphate from glucose and inorganic phosphate according to Eq. (*13*):

$$\text{Glucose} + P_i^{-2} \rightarrow \text{Glucose-6-Phosphate}^{-2} + \text{HOH} \quad \Delta F' = +3 \text{ kcal/mole} \quad (13)$$

This reaction is the reversal of the hydrolysis reaction, and in order to proceed, 3 kcal/mole of energy has to be provided. The hydrolysis of ATP [Eq. (*14*)] *yields* 8 kcal/mole, which is more than enough to drive Eq. (*13*).

$$\text{ATP}^{-4} + \text{HOH} \rightarrow \text{ADP}^{-3} + P_i^{-2} + H^+ \quad \Delta F' = -8 \text{ kcal/mole} \quad (14)$$

In the presence of the enzyme hexokinase, Eqs. (*13*) and (*14*) are coupled, and the net reaction of Eq. (*15*) occurs. It can be seen that Eq. (*15*) is the sum of Eqs. (*13*) and (*14*), and because the over-all ΔF is still negative, it will proceed spontaneously.

$$\text{Glucose} + \text{ATP}^{-4} \rightarrow \text{Glucose-6-phosphate}^{-2} + \text{ADP}^{-3} + H^+ \quad \Delta F' = -5 \text{ kcal/mole} \quad (15)$$

Inspection of Eq. (*15*) will show that not only has the ATP provided the energy necessary, it has also provided the phosphate group. There are many other reactions in which the energy rich compound provides the energy needed to drive a reaction as well as a particular transferable group. For example, ATP can act as a phosphorylating agent as in Eq. (*15*), a pyrophosphorylating agent in which the two terminal phosphates are transferred to an acceptor (as in the synthesis of 5′-phosphoribosyl-1′-pyrophosphate) and as an adenylating agent in which the AMP moiety is transferred (as in the formation of an "activated" amino acid-AMP complex). Similarly, acyl phosphates and acyl-CoA compounds can act as phosphorylating and acylating agents, respectively. Phosphosulfates, such as 3′-phosphoadenosine-5′-phosphosulfate (PAPS) act as sulfating agents.

Methods of Measuring the ΔF of Hydrolysis. *Measurements from Equilibrium Data.* The $\Delta F'$ of a reaction may be calculated from Eq. (*8*) if the equilibrium concentrations of reactants or products are not below the limits of detection. Obviously, for the direct hydrolysis of an energy-rich compound such as ATP, the equilibrium level of ATP would be extremely difficult to measure. However, if the hydrolysis can be carried out in several discreet steps, the K_{eq} can be calculated. For example, by employing highly radioactive substrates, the K_{eq} for the hexokinase-catalyzed reaction [Eq. (*15*)] can be determined. Similarly, the K_{eq} for the hydrolysis of glucose-6-phosphate by a phosphatase enzyme [reverse of Eq. (*13*)] can be determined. The sum of these two reactions effectively equals the hydrolysis of ATP and

$$K_{\text{over-all}} = K_{\text{hex}} \times K_{\text{hyd}}$$

and

$$\Delta F'_{\text{over all}} = \Delta F'_{\text{hex}} + \Delta F'_{\text{hyd}}$$

For other methods of calculating or measuring $\Delta F'$, the reader should consult references 2 and 5.

References

1. CONN, E. E., AND STUMPF, P. K., in "Outlines of Biochemistry," Ch. 6, p. 97, New York, John Wiley & Sons, 1963.
2. HUENNEKENS, F. M., AND WHITLEY, F. R., in "Comparative Biochemistry" (FLORKIN, M., AND MASON, H. S., EDITORS), Vol. I, Ch. 4, p. 107, New York, Academic Press, 1960.
3. ATKINSON, M. R., AND MORTON, R. K., in "Comparative Biochemistry" (FLORKIN, M., AND MASON, H. S., EDITORS), Vol. II, Ch. I, New York, Academic Press, 1960.
4. PARDEE, A. B., AND INGRAHAM, L. L., in "Metabolic Pathways" (GREENBERG, D. M., EDITOR), Vol. I, Ch. 1, p. 1, New York, Academic Press, 1960.
5. JOHNSON, M. J., in "The Enzymes" (BOYER, P. D., LARDY, H., AND MYRBÄCK, K., EDITORS), Vol. 3, Ch. 21, p. 407, New York, Academic Press, 1960.
6. GEORGE, P., AND RUTMAN, R. J., in "Progress in Biophysics and Biophysical Chemistry" (BUTLER, J. A. V., AND KATZ, B., EDITORS), Vol. 10, Ch. 1, p. 1, New York, Pergamon Press, 1960.
7. DAVIES, D. D., GIOVANELLI, J., AND REES, T. A., in "Plant Biochemistry" (Vol. 3 of Botanical Monographs, JAMES, W. O., EDITOR), Ch. 2, p. 64, Philadelphia, F. A. Davis Co., 1964.
8. EDSALL, J. T., AND WYMAN, J., in "Biophysical Chemistry," Vol. 1, Ch. 4, pp. 201–217, New York, Academic Press, 1958.

IRWIN H. SEGEL

PHOSPHATE (IN ANIMAL NUTRITION AND METABOLISM)

Phosphorus, as orthophosphate or as the phosphoric acid ester of organic compounds, has many functions in the body, including roles in (a) anabolic and catabolic reactions as exemplified by its essentiality in high-energy bond formation, *e.g.*, ATP, ADP, etc. (see PHOSPHATE BOND ENERGY), and the formation of phosphorylated intermediates in CARBOHYDRATE METABOLISM, (b) the formation of other biologically significant compounds such as the PHOSPHOLIPIDS, important in the synthesis of cell membranes, (c) the synthesis of genetically significant substances, such as DNA (DEOXYRIBONUCLEIC ACID) and RNA (RIBONUCLEIC ACID), (d) contributing to the buffering capacity of body fluids, cells and urine, and (e) the formation of bones and teeth. Like CALCIUM, the majority of the phosphorus in the vertebrate body is contained in the hard tissues; in the adult, approximately 80–86% of the total body phosphorus is contained in the bones and teeth, with the balance found in the soft tissues and body fluids.

Phosphorus is an essential dietary nutrient required for the several biochemical and physiological functions given above. It is an ubiquitous element, found throughout the plant and animal worlds and occurring in all tissues and organs. Because of this, phosphorus deficiencies are rare in humans and most species, but they have been observed in ruminants grazing on forages grown on phosphorus-deficient soils, causing extensive animal losses in various parts of the world. Symptoms of the deficiency are loss of appetite and also a depraved appetite (termed "pica") where the animal chews and consumes extraneous items such as wood, clothing, bones, etc. Vitamin D deficiency may accentuate a marginal lack of phosphorus in the diet.

Experimental phosphorus deficiency can be induced by feeding diets low in this element and by including excesses of Ca^{++}, Sr^{++}, Ba^{++}, Be^{++} and other cations that precipitate phosphates in the intestinal tract. In this situation, BONE FORMATION ceases, and the following histological bone changes have been noted in rats: (a) a thickening of the epiphyseal plate and the formation of a typical rachitic metaphysis, (b) wide osteoid borders of trabecular bone and a considerable rarefaction of the shaft, and (c) irregular or complete cessation of calcification of the zone of provisional calcification of the cartilage matrix. Rickets is often produced in the laboratory by feeding a diet high in Ca and low in P, and containing little or no vitamin D.

The nutrient requirement for phosphorus depends upon the particular species and the physiological status of the animal. During growth, lactation, gestation and egg-laying, a higher P content of the diet is generally required than for the maintained adult. As an example, the National Research Council Report on the Nutrient Requirements of Dairy Cattle (1958) indicates that growing heifers weighing 50, 150, 400 and 800 pounds require dietary phosphorus concentrations of 0.73, 0.44, 0.30 and 0.15%, respectively; the requirement for lactating dairy animals was given as 0.25%. Other reports of the National Research Council are available which deal with other species.

The availability of phosphorus in the diet varies, obviously, with its chemical form and the animal species under question. Diets high in foods of plant origin may contain a considerable portion of phosphorus in the form of phytic acid which is the hexaphosphoric acid ester of INOSITOL. When the acid occurs as salts of Ca, Mg, Na, etc., it is referred to as phytin. Phytate P is usually less available than inorganic phosphate in such species as rats, chickens, dogs, pigs and man. However, a phytase has been shown to be present in the intestine and intestinal secretions of some animals, and the formation of this enzyme is dependent, in part, on the presence of vitamin D. Through the action of phytase, some of the phytate P would be made available for absorption. Experimentation has indicated that under normal dietary conditions and Ca intake, food phytate was of no nutritional concern in humans. The microbial population of the ruminant also elaborates a phytase enzyme which makes phytate P readily available in this class of mammals. Phytates may be of nutritional consequence for another reason: dietary Ca could be bound in an unavailable, insoluble complex, thereby decreasing the absorption of this element.

A number of studies were accomplished in order to determine the availability of phosphorus from other organic and inorganic sources. In chicks, it has been reported that orthophosphates, superphosphates and phosphate rock products were good sources of phosphorus, whereas metaphosphate and pyrophosphate were relatively unavailable. Most organic phosphorus sources such as casein, pork liver, and egg phospholipid, were found to be equally available as inorganic phosphorus. Commonly used phosphorus supplements in human or animal nutrition or both are steamed bone meal, ground limestone, dicalcium phosphate and defluorinated rock phosphates.

Absorption of Phosphate. The phosphate ion readily passes across the gastrointestinal membrane. An ACTIVE TRANSPORT mechanism has recently been proposed from *in vitro* investigations with rat intestine. The intestinal preparation produced a concentration gradient such that serosal $PO_4^{\equiv}$ > mucosal $PO_4^{\equiv}$, and this was inhibited by cyanide or anaerobiosis. Calcium was required and the process was enhanced by the presence of potassium. Flux measurements by others also suggested that a carrier-mediated transport may be involved. This problem is not settled since it was earlier observed that increasing concentrations of phosphate ions in the intestinal lumen linearly enhanced net phosphate absorption with no evidence of saturation, which would be characteristic of a diffusion-like process.

The rate of absorption of phosphate at various intestinal sites in the rat was observed to be most rapid in the duodenum, followed in decreasing order by the jejunum, ileum, colon and stomach.

When transit time was considered, most of the phosphorus that was to be absorbed was done so by the ileum. That is, total absorption by a given segment is a function of both the rate of absorption and the time that the substance spends in a particular segment.

There is no evidence for a direct effect of vitamin D on phosphorus absorption. The absorption of this element will be enhanced by vitamin D, but this effect is secondary to calcium absorption. As noted above, vitamin D may also increase the availability of phosphorus from phytin through the enhanced synthesis of phytase.

Plasma Phosphate. Once absorbed, phosphorus enters the blood, and the majority is present therein as orthophosphate ions. At an ionic strength of 0.165 and at 37°C, calculations showed that the proportional concentrations of the orthophosphate ions in plasma for $H_2PO_4^-$, $HPO_4^=$ and $PO_4^{\equiv}$ were 18.6, 81.4 and 8×10^{-3}%, respectively. The activities in the same order were 0.12×10^{-3}, 0.19×10^{-3} and 5×10^{-6}. Recent estimates show that about 12% of plasma phosphorus is bound to proteins. During egg-laying in birds, the concentration of non-ionized phosphorus compounds in plasma is greatly increased. The administration of diethylstilbestrol to cockerols resulted in the formation of a newly described plasma phosphoprotein which forms relatively firm complexes with Ca. The function of the phosphoprotein appears to be one of phosphorus transport; in laying birds, the phosphoprotein is incorporated into egg yolk.

The approximate average plasma phosphorus levels for several species, in milligrams per 100 ml of plasma, are as follows: pigs, 8.0; sheep, cattle and goats, 6.0; horse, 2.3. Erythrocytes contain considerably more phosphorus than plasma, mostly in the form of organic esters. Some of the latter are acid soluble and hydrolyzable by intracellular enzymes.

Plasma phosphate appears to be homeostatically controlled, but the mechanism is not as well understood as for calcium. The primary organ concerned with phosphate HOMEOSTASIS seems to be the kidney, although the skeleton may also play a part. PARATHYROID HORMONE, by way of its direct action on the kidney and bone, certainly is a significant hormonal factor here.

Phosphate Excretion. The excretion of body phosphorus occurs via the kidney and intestinal tract, the distribution between these pathways varying with species. For example, relatively small amounts of phosphorus are endogenously excreted into the feces of the rat, pig and human, but in the bovine, perhaps 50% or more of the fecal phosphorus may be from endogenous sources. The determination of fecal endogenous phosphorus and net digestibility requires the use of radioisotopes (P^{32}). The former is calculated from the equation:

$$(\text{Average Specific Activity of } P^{32} \text{ in Feces} \div \text{Average Specific Activity of } P^{32} \text{ in Plasma}) \times \text{Daily Fecal P Output, in grams per day}$$

Net digestibility in per cent, is derived from the equation:

$$[(\text{P Intake} - \text{Fecal P} + \text{Endogenous Fecal P}) \times 100] \div \text{P Intake}$$

Typical values for these parameters in a calf weighing 36 kg and receiving phosphorus in the form of casein follow: P intake, 0.81 g/day; total P fecal excretion, 0.20 g/day; fecal endogenous P, 0.15 g/day, yielding an apparent digestibility

$$[(\text{P Intake} - \text{Fecal P}) \times 100] \div \text{P Intake}$$

of 75% and net digestibility of 94%.

The amount of phosphorus excreted in the urine varies with the level of ingested phosphorus and factors influencing phosphorus availability and utilization. It was shown that, in the dog, when plasma phosphate is normal or below, over 99% of the filtered ion is reabsorbed, presumably in the upper part of the proximal tubule (see RENAL TUBULAR FUNCTIONS). As noted by Smith, tubular reabsorption remained essentially complete until the filtered load/T_m ratio reached about 0.75, at which point the urinary excretion of phosphorus began. The tubular reabsorptive process was saturated when the filtered load/T_m ratio was 1.5. Increased plasma concentrations of alanine, glycine and glucose depress phosphate reabsorption, suggesting a common pathway. Phlorizin apparently does not block phosphate reabsorption, but that of glucose is depressed. The mechanism of phosphate reabsorption by the kidney tubule has not been settled. Although some evidence suggests there to be a special mechanism of phosphate reabsorption, recent studies on the Q_{10} and the activation energy indicated that phosphate is reabsorbed by a purely physical process, such as passive diffusion.

Parathyroid hormone and vitamin D influence the urinary excretion of phosphorus. The former depresses tubular reabsorption of phosphorus whereas vitamin D markedly increases the phosphate T_m (in rachitic dogs). Growth hormone also apparently influences phosphate reabsorption directly, resulting in an elevation of serum phosphate levels.

Several studies indicate that phosphate is not secreted across the kidney tubule from plasma to peritubular fluid to lumen although other reports suggest the opposite.

Phosphate of Hard Tissues. As mentioned above, most of the body phosphorus is contained in the intracellular matrix of bone and teeth as some form of hydroxyapatite $[Ca_{10}(PO_4)_6(OH)_2]$, this calcium phosphate salt providing the characteristic hardness of ossified tissue. Phosphate ions are also adsorbed onto the surface of bone crystals and exist in the hydration layer of the crystals. Early theories of calcification placed special emphasis on the role of alkaline PHOSPHATASE and organic esters of phosphoric acid. It was stipulated, as part of the theory, that with the hydrolysis of phosphate restes at the site of calcification, the K_{sp} for bone salt would be exceeded. Although phosphatase may have an important function in bone formation (perhaps in the synthesis of organic matrix), its role as earlier depicted appears now to be untenable. Recent emphasis is given to the specific and characteristic properties of COLLAGEN and other substances such as chondroitin sulfate. This is related to the *local mechanism* of calcification; the other component necessary for calcification is the *humoral mechanism* whereby an adequate supply of calcium, phosphate and other ions is made available to the calcifying site. Collagen supposedly functions by providing suitable chemical groups at required intervals which act as nucleation centers for initiating calcification. It has been proposed either that the functional groups on collagen are anionic, initially binding Ca^{++}, or that the first reaction is with phosphate or phosphorylated intermediates. The first held moiety of bone salt (Ca^{++} or phosphate) subsequently attracts or binds the other component, providing the aggregation or "seed" for subsequent crystal growth. An ATPase-type enzyme has been demonstrated in cartilage, suggesting that ATP may be intimately involved in the calcification mechanism. One proposal is that pyrophosphate is transferred from ATP to free amino groups of collagen, leading to nucleation and followed by combination with calcium and bone salt formation. Another is that ATP provides energy which increases the calcification mechanism. Confusedly, it was also observed that ATP and other pyrophosphates inhibit calcification *in vitro*.

The question of why only part of the body's collagen induces nucleation and calcification is most important. One postulate is that inhibitors of calcification exist in plasma, these being specifically removed or degraded in areas destined to calcify. Such an inhibitor may be inorganic pyrophosphate, recently isolated from human urine, which was shown to inhibit calcium phosphate precipitation *in vitro*. In bone, pyrophosphatase would act locally to degradate the inhibitor, allowing apatite crystals to form. Inhibition of calcification of collagen has also been attributed to other substances occurring in the ground substance, such as the negatively charged MUCOPOLYSACCHARIDES and glycoproteins.

Dietary inorganic phosphates were shown to protect experimental animals, especially rats and hamsters, against dental CARIES. Orthophosphates were effective cariostats but $Na_4P_2O_7$ and $Na_5P_3O_{10}$ were not. Dicalcium phosphate ($CaHPO_4$) was unable to decrease dental caries unless a high level of NaCl was also included in the diet. The mechanism by which this effect is brought about has not been disclosed as yet.

References

1. BARTTER, F. C., "Disturbances of Phosphorus Metabolism," in "Mineral Metabolism" (COMAR, C. L., AND BRONNER, F., EDITORS), Vol. 2A, pp. 315–339, New York, Academic Press, 1964.
2. COMAR, C. L., AND WASSERMAN, R. H., "Macronutrient Metabolism," in "Atomic Energy and Agriculture" (COMAR, C. L., EDITOR), pp. 249–304, Publ. 49 of the American Association for the Advancement of Science, 1957.
3. IRVING, J. T., "Dynamics and Function of Phosphorus," in "Mineral Metabolism" (COMAR, C. L., AND BRONNER, F., EDITORS), Vol. 2A, pp. 249–313, New York, Academic Press, 1964.
4. MAYNARD, L. A., AND LOOSLI, J. K., "Animal Nutrition," pp. 121–143, New York, McGraw-Hill Book Co., 1962.
5. MCLEAN, F. C., "Phosphorus Metabolism," in "Bone as a Tissue" (RODAHL, K., NICHOLSON, J. T., AND BROWN, E. M., EDITORS), pp. 330–336, New York, McGraw-Hill Book Co., 1960.
6. MCLEAN, F. C., AND URIST, M. R., "Bone, an Introduction to the Physiology of Skeletal Tissue," Second edition, Chicago, University of Chicago Press, 1961.
7. NEUMAN, W. F., AND NEUMAN, M. W., "The Chemical Dynamics of Bone Mineral," Chicago, University of Chicago Press, 1958.
8. SMITH, H. W., "The Kidney, Structure and Function in Health and Disease," pp. 113–121, New York, Oxford University Press, 1951.
9. WASSERMAN, R. H., "Calcium and Phosphorus Interactions in Nutrition and Physiology," *Federation Proc.*, **19**, 636–642 (1960).

R. H. WASSERMAN

PHOSPHATIDES AND PHOSPHATIDIC ACIDS

All of the naturally occurring phosphoglycerides appear to de derivatives of L-glycerol-3-phosphoric acid esterified with fatty acids at the 1- and 2-positions of the glycerol. The diacyl compound is called phosphatidic acid. Phosphatidic acid represents a relatively small part of the PHOSPHOLIPIDS of a tissue although many of the more complex phospholipids are phosphatidyl derivatives which contain alcohols esterified to the phosphate. For instance, choline forms phosphatidylcholine which is commonly called lecithin.

```
              O
              ‖
          CH₂OCR₁
   O      |
   ‖      |
R₂COCH
          |    O
          |    ‖
          CH₂OPOH
               |
               OH
```

Phosphatidic acid

```
              O
              ‖
          CH₂OCR₁
   O      |
   ‖      |
R₂COCH
          |    O
          |    ‖
          CH₂OPO(CH₂)₂N(CH₃)₃
               |          +
               O⁻
```

Phosphatidyl choline

The unusually large amounts of phosphatidic acid described in early reports were found to be due to hydrolytic cleavage of the choline from phosphatidylcholine during the isolation process. Nevertheless, most tissues appear to use phosphatidic acid as an intermediate in the biosynthesis of the phosphatides so that it plays a vital role in lipid metabolism without normally accumulating to very high levels.

Other derivatives, such as phosphatidylethanolamine and phosphatidylserine have been isolated from the aminophospholipid (CEPHALIN) fraction of tissue lipids. Phosphatidylinositol occurs in brain and nerve tissue along with its mono- and diphosphate esters and these three lipids have been called mono-, di- and triphosphoinositides, respectively. Glycosides of phosphatidylinositol containing mannose, with properties of both lipids and polysaccharides, have been isolated from mycobacteria. A closely related inositol glycoside has been found attached to ceramide phosphate in plant seeds.[1] Phosphatidylglycerol and diphosphatidylglycerol are additional phosphatides found in many tissues. Appreciable amounts of the choline and ethanolamine phosphoglycerides are found in nearly all animal tissues, and the general distribution of phosphatides in plants and animals was discussed in detail by Dittmer.[4] Orders of Insecta appear to differ among each other in the relative amounts of these two types of phosphatides (*e.g.*, ethanolamine phosphoglycerides predominate in Diptera).[3] In lipids of Gram-negative bacteria, choline is absent and ethanolamine is the principal amine.[2]

The phosphatides, as would be expected of any surface-active agent, are found to a large extent adsorbed in cellular membranes where they may play both structural and functional roles in maintaining the integrity of these membranes. The possibility that phosphatides may modify the surface membrane of a cell has been considered for erythrocytes and microorganisms, and this role may be important in the case of the amino acid esters of phosphatidylglycerol which are found in a variety of microorganisms. The content of the anionic detergent, diphosphatidylglycerol (cardiolipin), in MITOCHONDRIA has been shown to be related to the activity of the cytochrome electron-transport system. The action of cardiolipin as a hapten in the test for the antibodies developed in response to *Treponema pallidum* suggests further roles for this phosphatide. An interesting function of phosphatidic acid as a trans-membrane carrier for cation transport has been extensively investigated and may be significant in specialized salt-excreting glands. Other phosphatides are now recognized as essential cofactors in restoring the activity of various enzymes that have been removed from their normal lipoprotein environment in subcellular membranes. Some enzymes can be treated to show a general requirement for phosphatides (glucose-6-phosphatase and adenosine triphosphatase from the endoplasmic reticulum), while other enzymes require a specific phosphatide for reactivation (β-hydroxybutyrate dehydrogenase from mitochondria).

Other closely related phospholipids that are not phosphatides in a strict sense, are the 1-alkyl and the 1-(1′-alk-1′-enyl) analogues in which the ester at the 1-position is replaced by an ether or alkenyl ether, respectively. The alkyl derivatives are particularly high in erythrocytes of some species, and the alkenyl ether derivatives are major components in the lipids of muscle and nerve tissue (see PLASMALOGENS).

ESSENTIAL FATTY ACIDS are esterified in phosphatides almost entirely at the 2-position and, in general, unsaturated acids (palmitoleate, oleate, linoleate, arachidonate, etc.) are esterified mostly at the 2-position with the saturated acids (palmitate and stearate) predominantly at the 1-position of the phosphatides. Phosphatidylcholine often contains more palmitate than stearate, whereas the reverse is true for phosphatidylethanolamine. Each different phosphatide fraction isolated from a tissue generally contains a wide variety of fatty acids.

Because an individual phosphatide can contain only two of these fatty acids, there are different species for each type of phosphatide, such as 1,2-dioleylglycerol-3-phosphorylcholine, 1-palmityl-2-oleylglycerol-3-phosphorylcholine, or 1-stearyl-2-arachidonylglycerol-3-phosphorylcholine. In mammalian tissues, the palmityl-oleyl species is the predominant one recognized in the choline phosphoglycerides, whereas the stearyl-arachidonyl species is more prevalent among the ethanolamine derivatives.

The interesting problem of separating and identifying the various species of phosphatides is developing in parallel with a similar situation for triglycerides. The existence of these species leads to speculation as to whether or not each species of glycerolipid could have a characteristic function in a certain part of a cell. The manner by which the species of phosphatides originate is discussed in more detail in a recent review.[5]

The rapid development in the last 10 years of such sensitive and facile methods as thin-layer and gas-liquid chromatography for purifying and analyzing lipids is quickly raising the level of sophistication at which phosphatide biochemistry can be discussed, and the reader in this field should expect to check the useful annuals such as *Annual Review of Biochemistry* and *Advances in Lipid Research*.

References

1. CARTER, H. E., JOHNSON, P., AND WEBER, E., "The Glycolipids," *Ann. Rev. Biochem.*, **34**, 109 (1965).
2. KATES, M., "Bacterial Lipids," *Advan. Lipid Res.*, **2**, 17–90 (1964).
3. FAST, P. G., "Insect Lipids: A Review," *Memoirs of the Entomological Society of Canada No.* 37 (1964).
4. DITTMER, J. C., "Distribution of Phospholipids," in "Comparative Biochemistry" (FLORKIN, M., AND MASON, H. S., EDITORS), Vol. 3, pp. 231–264, New York, Academic Press, 1962.

5. LANDS, W. E. M., "Lipid Metabolism," *Ann. Rev. Biochem.*, **34**, 313 (1965).
6. WITTCOFF, H., "The Phosphatides," New York, Reinhold Publishing Corp., 1951.
7. HANAHAN, D. J., "Lipid Chemistry," New York, John Wiley & Sons, 1960.

W. E. M. LANDS

PHOSPHOLIPIDS

The phospholipids belong to the group of fatty acid compounds known as "complex lipids." The simplest are esters of fatty acids with glycerol phosphate and are called PHOSPHATIDIC ACIDS. Others contain the three radicals above plus nitrogen bases such as choline and ethanolamine, the amino acid SERINE or the sugar-like INOSITOL. These are called phosphatidylcholines or lecithins, phosphatidylethanolamines, phosphatidylserines, and phosphatidylinositols, respectively. The latter may have one or more additional phosphate groups attached to the inositol. A similar series also exists containing an aldehyde attached to the 1-position of the glycerol, in the form of an α,β-unsaturated ether. These are commonly referred to as PLASMALOGENS. There are also the polyglycerophosphatides, which contain several molecules of fatty acids, glycerol and phosphoric acid. These have been called cardiolipins. (See also CEPHALINS AND LECITHINS.)

The sphingomyelins contain the nitrogenous alcohol sphingosine instead of glycerol. There are several other complex lipids, such as the CEREBROSIDES and GANGLIOSIDES, which also contain sphingosine. However, since they do not contain phosphoric acid, they will not be discussed here.

Typical formulas of phospholipids are shown below:

With the exception of the more complex sphingosine-containing lipids or the polyglycerol or polyphosphoric acid derivatives, phospholipids such as the phosphatidylcholines, phosphatidylethanolamines, phosphotidylserines and phosphatidylinositols occur in both plant and animal tissues. They are especially abundant in membranes, including the MYELIN sheath of nerve cells, red blood cell membranes and MITOCHONDRIA. Because of technical difficulties in their study, very little is known about either the functions of these compounds or the differences in their metabolism. Unfortunately, for the most part they have been studied and discussed simply as "phospholipids," although it is quite certain that each has its own particular metabolic pattern. However, with present methods of chromatographic separation, progress is being made in differentiating the metabolic function of these important compounds.

The percentage phospholipid content of tissues varies but little under normal physiological conditions. They have, therefore, been referred to as the *élément constant* in contrast to the triglycerides which have been called the *élément variable*. However, under certain conditions, the concentrations of phospholipids have also been known to vary.

In vitro studies with tissue extracts and subcellular particles indicate that phosphatidylcholine, similarly to triglycerides, is derived from acyl coenzyme A and glycerol phosphate, as shown on top of page 653.

The acyl group in the primary position (R) is usually saturated and the one in the secondary position (R′) is usually unsaturated. The current status of the mechanism of biosynthesis of phospholipids has recently been thoroughly reviewed by Ansell and Hawthorne.

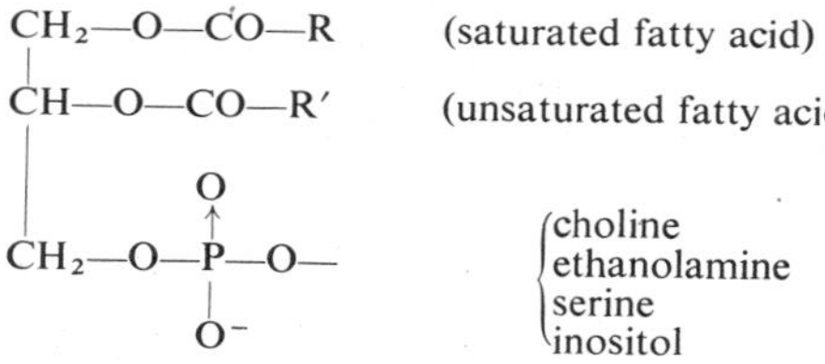

```
CH2—O—CO—R          (saturated fatty acid)
|
CH—O—CO—R'          (unsaturated fatty acid)
|
|        O
|        ↑                {choline
CH2—O—P—O—               {ethanolamine
         |                {serine
         O-               {inositol
```

Phosphatidyl—

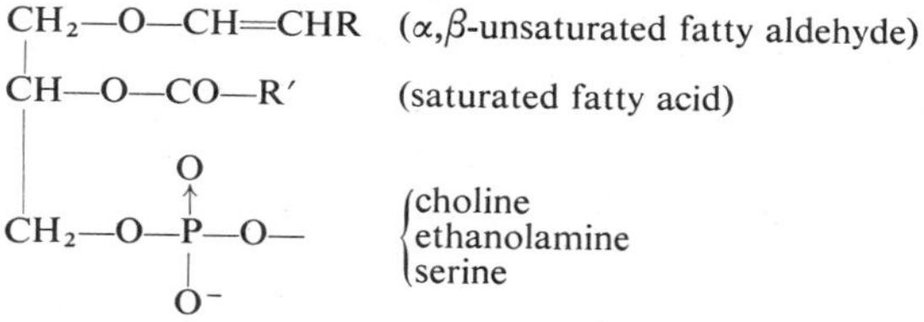

```
CH2—O—CH=CHR   (α,β-unsaturated fatty aldehyde)
|
CH—O—CO—R'     (saturated fatty acid)
|
|       O
|       ↑          {choline
CH2—O—P—O—        {ethanolamine
        |          {serine
        O-
```

Plasmalogens

```
                                              O
                                              ↑
CH3(CH2)12CH=CH—CH—CH—CH2—O—P—O-
                 |    |                       |
                 OH   NH                      O
                      |                       |
                      COR                     CH2—CH2—N+(CH3)3
```

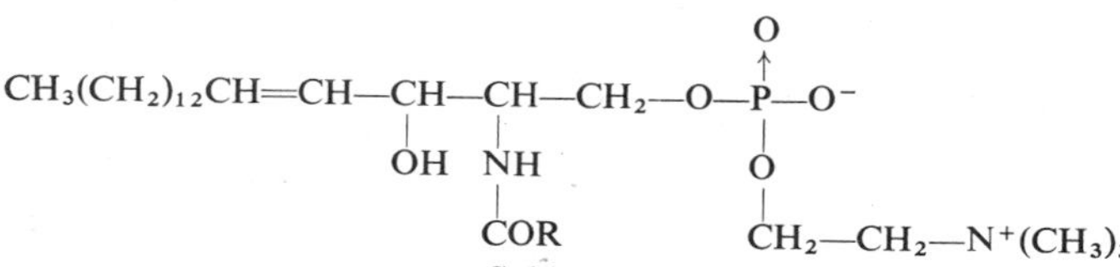

Sphingomyelin

$$
\begin{array}{l}
CH_2\text{—}OH \\
|\\
CH\text{—}OH \\
|\qquad\quad O \\
|\qquad\quad \uparrow \\
CH_2\text{—}O\text{—}P\text{—}O^- \\
\qquad\qquad | \\
\qquad\qquad O^-
\end{array}
\quad + \quad
\begin{array}{l}
R\text{—}CO\text{—}CoA \\
R'\text{—}CO\text{—}CoA
\end{array}
\quad \longrightarrow \quad
\begin{array}{l}
CH_2\text{—}O\text{—}CO\text{—}R \\
|\\
CH\text{—}O\text{—}CO\text{—}R' \\
|\qquad\quad O \\
|\qquad\quad \uparrow \\
CH_2\text{—}O\text{—}P\text{—}O^- \\
\qquad\qquad | \\
\qquad\qquad O^-
\end{array}
$$

1-Glycerol phosphate — Fatty acyl coenzyme A derivatives — Phosphatidic acid

$$
\longrightarrow
\begin{array}{l}
CH_2\text{—}O\text{—}CO\text{—}R \\
|\\
CH\text{—}O\text{—}CO\text{—}R' \\
|\\
CH_2\text{—}OH
\end{array}
\xrightarrow[\text{choline}]{\text{Cytidine diphosphoryl}}
\begin{array}{l}
CH_2\text{—}O\text{—}CO\text{—}R \\
|\\
CH\text{—}O\text{—}CO\text{—}R' \\
|\qquad\quad O \\
|\qquad\quad \uparrow \\
CH_2\text{—}O\text{—}P\text{—}O\text{—}CH_2\text{—}CH_2 \\
\qquad\qquad |\qquad\qquad\qquad | \\
\qquad\qquad O^- \qquad\quad {}^+N(CH_3)_3
\end{array}
$$

1,2-Diglyceride — Phosphatidylcholine

There is evidence that phosphatidylcholine also arises from triglycerides, since the ingestion of glycerol- and fatty acid-labeled triglycerides results in glycerol- and fatty acid-labeled phospholipids in the liver, blood and intestinal mucosa. Since triglycerides are known to be in a dynamic state in which the fatty acids are constantly being removed and replaced from the glycerol moieties, there could well be a diglyceride pool from both endogenous and preformed sources.

Although most blood plasma phospholipids probably originate in the liver, postprandial plasma phospholipid containing dietary fatty acids and glycerol probably arises in the intestinal mucosa.

Phosphatidylethanolamine is less dynamic than phosphatidylcholine which, in turn, is less dynamic than triglycerides. Furthermore, the relative states of incorporation of labeled glycerol as compared to those of fatty acids follow the same order, suggesting that triglycerides may be the precursors of the phosphatidylcholines and these, in turn, the precursors of the phosphatidylethanolamines. That there is at least one other route to the phosphatidylcholines than through 1-glycerol phosphate has been amply demonstrated.

Although the functions of the phospholipids are somewhat obscure, they are probably multiple and different for each phospholipid. Except in pathological states, the sphingosine and carbohydrate containing lipids are found most concentrated in nervous tissues, although not exclusively so. There they appear to serve some insulating function in the MYELIN sheath. A model system describing such function has been formulated.

Since phospholipids have both hydrophilic and hydrophobic parts, they may well function as emulsifying agents to maintain the proper colloidal state of protoplasm. Through their concentration in cell membranes they are involved in the transport of hydrophobic constituents into and out of cells. Since they are also amphoteric, it is likely that this property may be utilized in the maintenance of acid-base balance. The polar groups also give phospholipids the property of association with PROTEINS in the physiologically important LIPOPROTEINS.

In the avian egg, the high phosphatidylcholine level doubtlessly serves to maintain the proper colloidal state in the yolk and also as a source of phosphates for the developing embryo. In this instance, choline, which is a vitamin, is probably used for its vitamin function.

Phospholipids are thought to be involved in the transport of triglycerides through the liver, especially during mobilization from adipose tissue (see FAT MOBILIZATION). Conditions which could be interpreted as interfering with phosphatidylcholine formation, such as a deficiency of choline or its precursors, result in a pronounced increase in liver triglycerides.

Phosphatidylethanolamine has been implicated in the mechanism of BLOOD CLOTTING, probably as a part of the lipoprotein, thromboplastin.

Mitochondrial phospholipids play a role in electron transport and oxidative PHOSPHORYLATION, two mechanisms by which the cell accomplishes the final oxidation of the metabolites to produce energy. Phospholipids have been also implicated in the transport of ions, especially sodium, across membranes. The cardiolipins may serve as complement in immunological reactions.

It has been demonstrated that the fatty acids of the middle carbon atom of phosphatidylcholine are the source of the fatty acids of cholesteryl esters in animal tissues.

References

1. ANSELL, G. B., AND HAWTHORNE, J. N., "Phospholipids," Amsterdam, London, and New York, Elsevier Publishing Co., 1964.
2. HANAHAN, D. J., "Lipid Chemistry," New York, John Wiley & Sons, 1960.
3. WITTCOFF, H., "The Phosphatides," New York, Reinhold Publishing Corp., 1951.

RAYMOND REISER AND NESTOR R. BOTTINO

PHOSPHONIC ACIDS. See AMINO PHOSPHONIC ACIDS.

PHOSPHORYLATION, OXIDATIVE

Oxidative phosphorylation is the enzymic process whereby energy, released from oxidation-reduction reactions during the passage of electrons from substrate to oxygen over the electron transfer chain, is conserved by the synthesis of adenosine triphosphate (ATP) from adenosine diphosphate (ADP) and inorganic orthophosphate (P_i). Since ATP is the major source of energy for biological work, and since most of the net gain of ATP in the animal cell derives from oxidative phosphorylation, the quest for answers to questions regarding the manner in which the cell conserves energy in the form of ATP is one of the most intensive areas of biochemical research today. Although widely differing types of organisms differ as to the components of the electron transfer chain, the mechanism of oxidative phosphorylation is similar in all. Most of our information on this process has come from studies on mammalian systems and thus the description which follows will depend almost entirely on such data.

Oxidative phosphorylation was discovered simultaneously and independently in 1939 by Kalckar in Denmark and by Belitzer in the Soviet Union. It was recognized by these workers that aerobic phosphorylation was different from and independent of phosphorylation supported by GLYCOLYSIS. In addition they found that the stoichiometry of phosphate esterification (ATP synthesis) and oxygen utilized was 2 or more, or that the reduction of 1 atom of oxygen to form water may be accompanied by the "activation" of 2 or more molecules of P_i, thus leading to an expression of the efficiency of the energy conserving system. The efficiency expression is known as the P/O ratio, *i.e.*, the ratio of molecules of P_i esterified per atom of oxygen utilized. The efficiency, or yield, of oxidative phosphorylation is now widely accepted as a maximum of 3 for DPN-linked oxidations, 2 for the oxidation of succinate to fumarate, and 4 for the oxidation of α-ketoglutarate. However, one of the phosphorylations during α-ketoglutarate oxidation is a substrate-level phosphorylation and is not linked, or coupled, directly to the electron transfer process.

The quantitative importance of the ATP synthesized at the expense of energy liberated during electron transfer in the MITOCHONDRION is realized when one follows the conservation of energy during the metabolism of a molecule such as glucose in the cell. The oxidation of 1 mole of glucose to CO_2 and water is accompanied by the release of 673,000 calories. In order to degrade the glucose molecule to a form which can be metabolized further by mitochondrial enzymes, the glycolytic enzymes consume 2 molecules of ATP and also synthesize 2 molecules of ATP in the presence of oxygen, a net energy conservation of zero. The mitochondrion may then degrade the pyruvate supplied by glycolysis to CO_2 and water yielding a net total of 38 molecules of ATP, mostly at the level of the electron transfer process. Thirty eight molecules of ATP per molecule of glucose results in between 260,000 and 380,000 calories conserved, between 39 and 56% of the total energy released in the complete oxidation of glucose, the remainder being released directly as heat. Since the mitochondrion is approximately 50% effective in conserving energy from its major substrate, it is indeed an efficient machine.

Intracellular Localization. The site within the cell where oxidative phosphorylation is catalyzed was not known until 1949 when methods were devised to isolate intact, functional subcellular organelles. It was then found that the carefully isolated mitochondrion could catalyze the complete oxidation of the end product of glycolysis (pyruvate), as well as Krebs cycle (CITRIC ACID CYCLE) intermediates, accompanied by phosphorylation approaching a P/O ratio of 3.0. The P/O ratio of 3.0, long considered the maximum yield of oxidative phosphorylation, has been challenged recently, and the true maximum yield may actually be as high as 6.

Although oxidative phosphorylation is not a stable system and is easily destroyed, methods have been developed to fragment MITOCHONDRIA by both chemical and physical means, yielding small pieces of the inner mitochondrial membrane (cristae) which carry out efficient oxidative phosphorylation. Such preparations, especially those derived from bivine cardiac muscle mitochondria, may be maintained in the frozen state for many days without great loss of phosphorylative capacity and have been the object of probing studies on the mechanism of oxidative phosphorylation. However, it is important to note that to date all preparations capable of catalyzing oxidative phosphorylation have consisted of a membrane structure and appear to contain a rather high degree of structured orientation.

Respiratory Control. One of the basic, and still largely unanswered, questions about oxidative phosphorylation in intact mitochondria involves the linkage, or coupling, between the energy generating (electron transfer) reactions and the energy conserving (phosphorylating) reactions. One of the very significant contributions to this problem was made by Henry Lardy when he and his colleagues demonstrated that the rate at which electron transfer may proceed is severely limited by the availability of ADP to the mitochondrion. This elegant demonstration of the tightness of coupling between oxidation and phosphorylation has important implications to the over-all regulation of cellular energy metabolism since it implies that ATP-utilizing processes in the cell determine the rate of biological oxidation and therefore the rate of energy liberation. Thus, any process requiring the energy contained in ATP, such as muscular contraction, protein synthesis, osmotic work, nerve impulse, etc., would supply ADP and P_i to the mitochondrion and allow electron transfer to proceed at a rate sufficient to supply the demands of the cell for ATP.

Sites of Phosphorylation. Once it was established that three phosphorylations could occur at the level of the electron transfer chain, it became of interest to determine which of the various electron transfer components were specifically involved in the energy conservation process. The known electron transfer components and their relative positions in the electron transfer chain are shown in Fig. 1. Early studies in the laboratories of Lardy and Lehninger demonstrated that 3 sites of energy conservation could be localized. These are called phosphorylation sites I, II, and III preceeding from the level of substrate to that of oxygen, respectively. Site I occurs between DPN and CoQ, site II between CoQ and cytochrome *c*, and site III between cytochrome *c* and oxygen. More detailed localization was achieved by Britton Chance and his colleagues in Philadelphia using an imaginative spectrophotometric technique as follows. In the presence of excess substrate and P_i, mitochondrial electron transfer is severely limited when all of the added ADP has been converted to ATP. During this state of limited electron flow, the electron carriers on the substrate side of the rate limiting or energy conserving step are more reduced than those on the oxygen side of the conservation site. By noting the degree of oxidation or reduction of each of the spectrophotometrically detectable carriers, Chance has determined the sites where the carriers abruptly change from the more reduced to the more oxidized state and has named such sites "crossover points." Crossover points occur between DPN and FP_D, cytochrome *b* and cytochrome *c*, and cytochrome *c* and oxygen, confirming earlier experiments mentioned above. More recently, the possible role of COENZYME Q in phosphorylation has received increased attention. In addition, the recent discovery of non-heme iron in the electron transfer chain may necessitate a reevaluation of the precise position of crossover points. Additional information on phosphorylation sites has been derived from studies of coupling factors and is discussed below.

Inhibitors of Oxidative Phosphorylation. Inhibitors of oxidative phosphorylation fall into three general classifications according to the way in which they accomplish their task. One group inhibits electron transfer directly probably by combining with a specific carrier. Among such compounds are rotenone, cyanide, antimycin A, and 2-heptyl-4-hydroxyquinoline. Since in the presence of such inhibitors electron flow ceases, energy is not released for conservation. Another group of compounds is placed in the catagory of uncouplers and appears to act by catalyzing the immediate breakdown of a very early, non-phosphorylated intermediate in oxidative phosphorylation. In the presence of such uncouplers, electron flow is not reduced, but ATP is not synthesized and all energy from the flow of electrons is released as heat. Some of the uncouplers are substituted phenols (including 2,4-dinitrophenol), dicoumarol, long-chain fatty acids, the antibiotic gramicidin D, and the recently reported substituted carbonyl cyanide phenylhydrazones. The members of a third group are called true inhibitors of oxidative phosphorylation since they appear to combine with one or another of the components of the sequence of enzymes which transfer the energy conserved from electron transfer to the "high-energy" form of ATP, *i.e.*, are involved directly in the energy transfer process. These compounds, including the antibiotics oligomycin, valinomycin and aurovertin, and guanidines, inhibit ATP synthesis and, in well-coupled mitochondria, also result in a limited electron flow.

Related Reactions. In 1951, F. E. Hunter of St. Louis suggested that the mitochondrial ATPase stimulated by 2,4-dinitrophenol (DNP-ATPase) represents a reversal of the reactions leading to the synthesis of ATP during electron transfer phosphorylation. Since then, two additional reactions have been studied in mitochondria which are considered to utilize all or part of the same enzymic sequence. These are the ATP-P_i exchange reaction and the ATP-ADP exchange reaction. That the two exchange reactions, the DNP-ATPase and oxidative phosphorylation are all functionally related is based upon (1) the absolute specificity for the adenine nucleotide in intact mitochondria, (2) the stimulation of the ATPase, the inhibition of the exchanges and the uncoupling of oxidative phosphorylation by DNP, and (3) the fact that all of the reactions are equally sensitive to aging of the mitochondrion.

Reversal of Oxidative Phosphorylation. Mitochondria and submitochondrial particles catalyze an energy-dependent reversal of the reactions involved in oxidative phosphorylation. This phenomenon was first studied in the laboratory of Britton Chance in Philadelphia with intact mitochondria and by Hans Löw of Stockholm in the laboratory of David Green in Wisconsin with submitochondrial particles. Figure 2 shows the pathways of energy transfer and electron transfer for both the forward and reverse directions of oxidative phosphorylation. In the forward, exergonic reaction, ATP is synthesized at the expense of electrons preceeding from reduced DPN, over the various electron carriers shown in Fig. 1, to oxygen. This occurs at energy conservations sites I, II and III by first conserving energy in non-phosphorylated high-energy intermediates and subsequently transferring the energy via the

$$\begin{array}{l} \text{DPNH} \rightarrow \text{FP}_\text{D} \rightarrow \text{FeNH} \\ \qquad\qquad\qquad\qquad \searrow \\ \qquad\qquad\qquad\qquad\quad \text{CoQ} \rightarrow \text{cyt. } b \rightarrow \text{FeNH} \rightarrow \text{cyt. } c_1 \rightarrow \text{cyt. } c \rightarrow \text{cyt. } a \rightarrow \text{cyt. } a_3 \rightarrow \text{O}_2 \\ \qquad\qquad\qquad\qquad \nearrow \\ \text{Succinate} \rightarrow \text{FP}_\text{S} \rightarrow \text{FeNH} \end{array}$$

FIG. 1. The mammalian electron transfer chain. Reduction occurs from left to right. Abbreviations are: FP_D, DPNH dehydrogenase (flavoprotein); FP_S, succinic dehydrogenase (flavoprotein); FeNH, non-heme iron; CoQ, coenzyme Q (ubiquinone); cyt., cytochrome.

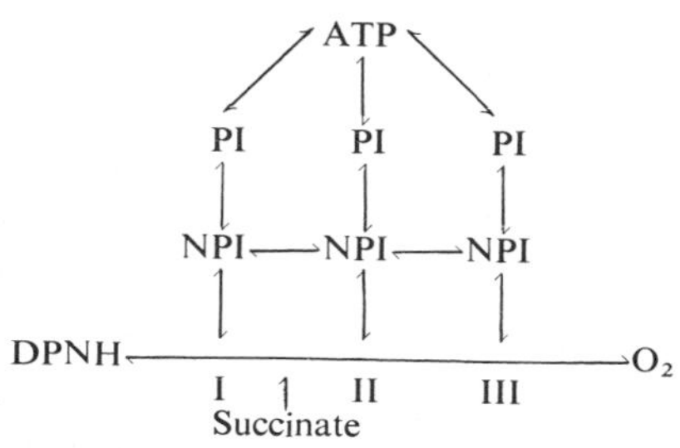

FIG. 2. Pathways of energy and electron transfer. NPI represents non-phosphorylated high-energy intermediate, PI represents phosphorylated high-energy intermediate, and I, II and III identify energy conservation sites I, II and III.

phosphorylated high-energy intermediate to ADP to form ATP. During the reverse, endergonic reaction, energy may be supplied by ATP, electrons or reducing equivalents by succinate, and the production of the reaction is reduced DPN (DPNH). The great advantage of studying the reversal of oxidative phosphorylation in submitochondrial particles is that such particles reduce exogenous DPN while intact mitochondria do not, thus enabling the investigator to easily measure the accumulation of rather large amounts of DPNH. One of the significant findings to be derived from studies of the reversal reaction has been that non-phosphorylated, high-energy intermediates of oxidative phosphorylation may donate energy for the reversal. Lars Ernster in Stockholm has shown that when ATP synthesis is blocked by the presence of the true inhibitor of oxidative phosphorylation, oligomycin, and P_i is omitted from the reaction, non-phosphorylated high-energy intermediates generated at phosphorylation sites II or III may donate energy to reverse electron flow resulting in the reduction of DPN at site I. In addition, Ernster has demonstrated by such experiments (as indicated in Fig. 2) that such intermediates may intercommunicate between phosphorylation sites. The physiological importance of the reversal of oxidative phosphorylation is not clear at present.

"Coupling Factors." In recent years it has been possible to prepare submitochondrial particles from beef heart mitochondria which, although capable of full electron transfer activity, have lost most of the capacity to synthesize ATP. Soluble proteins, isolated from mitochondria, when added to poorly phosphorylating submitochondrial particles restore the ability to converse energy as ATP. It appears that the poorly phosphorylating submitochondrial particles have lost either the enzymes concerned with the NPI or the enzymes concerned with the PI, or both, and that the soluble proteins ("coupling factors") which restore phosphorylative ability to the particles contain the missing enzymes. One "coupling factor," isolated in E. Racker's laboratory in New York, is a large protein, molecular weight of approximately 300,000, which contains an active ATPase and has the unique property of being inactivated by low temperature. Its presence increases the efficiency of oxidative phosphorylation in poorly phosphorylating submitochondrial particles, but it does not appear to participate as an intermediate in phosphorylating reactions. Other "coupling factors" have been isolated and studied in the laboratory of David Green and have been shown to restore phosphorylative capacity to poorly phosphorylating submitochondrial particles at a specific site (*i.e.*, site-specific "coupling factors"). In addition, the site-specific "coupling factors" do appear to participate as high-energy intermediates in oxidative phosphorylation by participating in a reaction which substitutes P_i for the electron transfer component and subsequently transferring the high-energy phosphate to ADP to form ATP. The site-specific "coupling factors" have now been recognized to function as classical synthetases and are termed ATP synthetase I, II or III, according to the phosphorylation site with which they interact. Recent studies, including the isolation of the high-energy phosphorylated form of the ATP synthetases, allow the reactions of oxidative phosphorylation to be written as:

$$\text{ETC} + \text{Synthetase} \rightleftarrows \text{ETC} \sim \text{Synthetase}$$

$$\text{ETC} \sim \text{Synthetase} + P_i \rightleftarrows \text{ETC} + \text{Synthetase} \sim P_i$$

$$\text{Synthetase} \sim P_i + \text{ADP} \rightleftarrows \text{Synthetase} + \text{ATP}$$

in which ETC represents the electron transfer component and $\sim$ indicates a high-energy linkage. Very little information is available on the specific electron transport components which react with the ATP synthetases functioning at each of the energy conservation sites. Cytochrome *c* may serve this purpose at site III in mammalian mitochondria, while DPN may combine with the synthetase at site I in a bacterial system. Kinetic evidence would indicate that cytochrome *b* serves this function at site II, but no direct evidence is available which allows a more specific assignment at this site. Studies on "coupling factors" and ATP synthetases have advanced the understanding of the specific reactions of oxidative phosphorylation and may prove to be important determinants in the final solution of the energy conservation and energy transfer reactions of the cell.

References

1. CHANCE, B. (EDITOR), "Energy-Linked Functions of Mitochondria," New York, Academic Press, 1963.
2. "Symposium on Oxidative Phosphorylation," *Federation Proc.* **22**, No. 4, 1064–1096 (1963).
3. ERNSTER, L., AND LEE, C. P., "Biological Oxidoreductions," *Ann. Rev. Biochem.*, **33**, 729 (1964).
4. LEHNINGER, A. L., "The Mitochondrion," New York, W. A. Benjamin, 1964.
5. LOWEY, A. G., AND SIEKEVITZ, P., "Cell Structure and Function," New York, Holt, Rinehart and Winston, 1963.
6. SANADI, D. R., "Energy-Linked Reactions in Mitochondria," *Ann. Rev. Biochem.*, **34**, 21 (1965).

ROBERT E. BEYER

PHOSPHORYLATION, PHOTOSYNTHETIC

Photosynthetic conversion of light energy into the potential energy of chemical bonds involves an electron transport chain, and the phosphorylation of ADP:

$$\text{ADP} + \text{P} \xrightarrow[\text{Chlorophyll}]{+\text{ Light}} \text{ATP}$$

as intermediate stages. The process of phosphorylation, defined by the above equation, was discovered simultaneously by Arnon and co-workers for green plant chloroplasts and by Frenkel, working with Geller and Lipmann in 1954. For both systems the heart of the mechanism is the creation of a very oxidizing and a very reducing component, utilizing the energy of the photoexcited state of one of the pigment (chlorophyll) molecules. This process will be designated a "photo-act." The redox components are both members of a photosynthetic electron transport chain, bound to the membranes of the CHLOROPLASTS (for green plants) or "chromatophores" (for bacteria). The photo-act can be considered as electron transport *against* the thermochemical gradient—i.e., *away* from the member which is a better electron acceptor (high oxidation-reduction potential), through the excited CHLOROPHYLL, then *to* the member which is a better electron donor (low oxidation-reduction potential). Subsequent steps consist of ordinary, dark electron transport with the thermochemical gradient, and the energy in at least one of these redox reactions in conserved as ATP by a phosphorylation reaction analogous to that found in oxidative phosphorylation by mitochondria.

In *bacteria*, the photo-act proper is accomplished by a special kind of bacteriochlorophyll, amounting to only 3% of the total present. It is unique in having a peak in absorption at 870–890 mμ, or further into the infrared than the remaining 97% of the chlorophyll molecules. It is unique not by virtue of a difference in structure, but because of its "environment"—most probably a close association or complexing with CYTOCHROME molecules. Since its absorption extends to longer wavelengths it is an energy trap, and the function of the bulk of the bacteriochlorophyll is that of capturing light and transmitting it to this active center.

Components of the electron transport chain in bacteria have been shown to include *b*- and *c*-type cytochromes, ubiquinone (fat-soluble substituted quinone, also found in MITROCHONDRIA), ferredoxin (an enzyme containing non-heme iron, bound to sulfide, and having the lowest potential of any known electron carrying enzyme) and one or more FLAVIN enzymes. Of these, a cytochrome (in some bacteria, with absorption maximum at 423.5 mμ, probably c_2) has been shown to be closely associated with the initial photo-act. Chance and Nishimura were able to demonstrate, in *Chromatium*, the oxidation of the cytochrome at liquid nitrogen temperatures, due to illumination of the chlorophyl. At the very least, this implies that the two are bound very closely and no collisions are needed for electron transfer to occur.

Further experiments show that a *b*-type cytochrome is electron donor to the cytochrome c_2, and a flavin enzyme reduces the cytochrome *b*. The over-all pattern of electron flow supporting phosphorylation in bacteria is cyclic, and a generally accepted scheme is shown below:

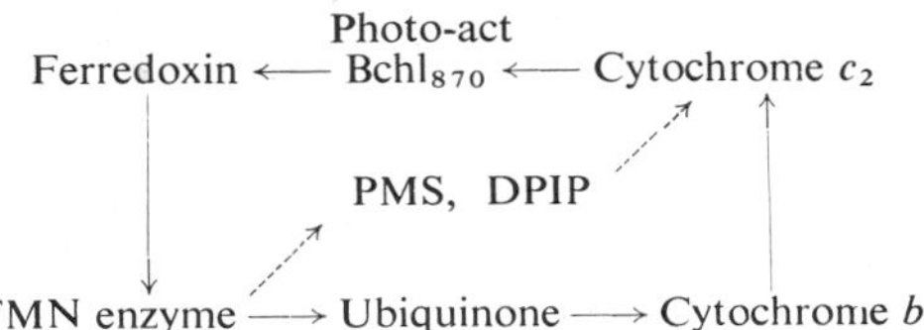

Electron transport can be seen to run from the photoexcited BChl 870 to very negative potential acceptors; then, in thermodynamically downhill reactions, from ferredoxin through flavin enzymes to cytochrome *b* and finally to cytochrome c_2 again.

The site of phosphorylation is partially defined as being at the flavoprotein level, by means of inhibitor and redox-dye bypass experiments. As shown on the diagram, either of the artificial dyes, phenazine methosulfate (PMS) or dichloropheno-lindophenol (DPIP), when added to isolated bacterial chromatophores, permit electron flow to bypass the cytochrome *b*–cytochrome *c* part of the chain. In so doing, the effects of some inhibitors (antimycin A, HOQNO) are circumvented. Nevertheless, phosphorylation continues—suggesting its site must be prior to the point where either of the dyes takes its electrons.

Note that free O_2 is not a requirement for this phosphorylation. Bacterial photophosphorylation was shown to occur under anaerobic conditions from the very start, in striking contrast to oxidative phosphorylation (see PHOSPHORYLATION, OXIDATIVE; also LIGHT PHASE OF PHOTOSYNTHESIS). It is likely to be a more primitive reaction in the evolutionary sense, because photosynthetic systems must have evolved before the atmosphere acquired free oxygen.

The ability of bacterial chromatophores to accomplish photophosphorylation depends strongly on having the correct oxidation-reduction potential in the surrounding milieu. This need for "poising" of the redox potential seems to be explainable by the need to keep electrons flowing within the cycle: an electron sink unrelated to the cycle would eventually deplete the particles of mobile electrons, thereby eliminating (*reduced*) cytochrome c_2, which is the substrate for the light reaction; alternatively, too active an electron donor present might keep all members of the cycle in a completely reduced state, thereby eliminating the other substrate for the light reaction (*oxidized* ferredoxin, according to the above scheme).

An interesting question is the relationship between photosynthetic and oxidative phosphorylation in bacteria (*e.g.*, *Rhodospirillum rubrum*) capable of both. Illumination of the whole cells inhibits respiration, and the isolated subcellular particles appear to be the same for either type of phosphorylation. It may be that a part of the same phosphorylating electron transport chain is used in both the oxidative and the photosynthetic sequence, and the relationship between the two may be governed either by competition of substrates for one or another of the bound catalysts or by other (structural?) considerations.

Non-cyclic electron flow is the rule for photosynthesis by the whole bacteria, which, in the light, effect a net oxidation of a "photosynthetic substrate," and usually carbon dioxide is the ultimate electron acceptor. Isolated chromatophores also accomplish non-cyclic electron flow, against a gradient, as for instance, from $FMNH_2$ to NAD, or from succinate to NAD.

sistent to find that the green plant electron transport requires the cooperation of two distinct photo-acts. One of these functions at the high potential (water) end, the other at the low potential end of the electron transport chain. Components known to participate in electron transport here include Mn^{++}, one or more kinds of plastoquinone (fat-soluble quinones analogous to ubiquinone but unique to photosynthetic systems), plastocyanine (a copper-containing protein), cytochrome *f*, plant ferredoxin (similar to the bacterial enzyme) and one or more flavin enzymes. Participation of cytochrome b_2 has not been proved but may occur. A scheme for the sequence of electron transfer reactions, currently considered to be probable, but not as fully defined as those for oxidative phosphorylation, is shown below:

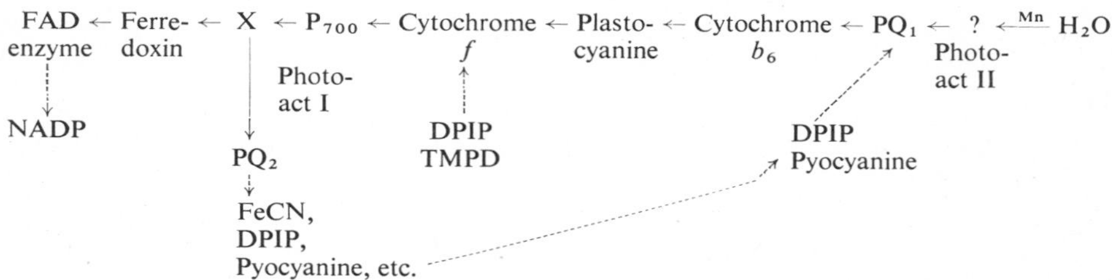

Although phosphorylation may occur simultaneously with non-cyclic electron flow in the light, it is not easy to say that these are coupled because of the difficulty in ruling out the simultaneous occurrence of *cyclic* electron flow, which would be sufficient to account for the measured phosphorylation[1]

An interesting recent suggestion[2] is that the energy-requiring, non-cyclic electron flow (*e.g.*, from succinate to NAD) might be intrinsically a dark process just as in mitochondria, using either ATP or its high-energy precursors, generated during cyclic electron flow, as the driving power. However a dark "reverse electron flow" using the energy of ATP has never been observed in any photosynthetic system, and it seems simpler to depend on the photo-act proper as outlined above, to drive electrons uphill in these preparations (see also PHOTOSYNTHETIC BACTERIA).

Isolated chloroplasts of *green plants* accomplish a non-cyclic electron transport in which water is the electron donor and various added dyes may be the ultimate electron acceptor. Free molecular oxygen is the by-product of this reaction, discovered in 1937 by R. Hill. The most physiological electron acceptor is NADP. The drop in emf going from water to NADP amounts to 1.12 eV and using other electron acceptors a drop of at least 1.25 volts can be shown. This amounts to some 57,000 cal/mole for a 2-electron transfer, or more energy than is found in one quantum of light. It is, of course, a bigger energy requirement than is found with photosynthetic bacteria which use substrates of lower redox potential as electron sources.

In view of this energy requirement, it is con-

In this scheme, dotted arrows indicate components added exogenously.

As in bacteria, the electron transfer specifically associated with the photo-act is a transfer against the thermochemical gradient. Other transfers are exergonic. The component acting at photo-act II is entirely unknown. The active center for photo-act I is known to be a modified chlorophyll (P700), probably complexed with cytochrome *f*, whose absorption at 700 mμ is bleached when oxidation occurs and is regained by subsequent reduction. Its E_0' is $+0.43$ volt when in the ground state; when in the excited state, it is certainly capable of reducing ferredoxin (-0.42 volt) and probably other dyes ranging down to -0.55 volt. The ferredoxin and flavin enzyme components shown are needed for NADP reduction only; oxygen or added heme enzymes as electron acceptors need ferredoxin, but many dyes can be reduced in the absence of these two enzymes which can be removed from the chloroplast membranes. A special platoquinone (PQ_2) functions in the reduction of ferricyanide, but it is not needed for the ferredoxin pathway.

Phosphorylation accompanies this complex scheme of electron transfer. When ferricyanide or NADP are electron acceptors, the ATP formed per two electrons transferred exceeds a ratio of 1.0, under appropriate conditions (high light intensity and the proper pH). This implies that there may be two phosphorylation sites, whose locations are not yet rigorously defined but may be between PQ_1 and cytochrome *f* on the above diagram (*i.e.*, between the two photo-acts).

Phosphorylation can be supported by cyclic electron flow in chloroplasts if a redox dye is added, which can be reduced at the low potential end (after photo-act I), and also re-oxidized at some central portion of the chain (say PQ_1) prior to the phosphorylation step. Effective dyes include pyocyanine, vitamin K, DPIP, and recently an excess of ferredoxin itself.[3] Other conditions must also be adjusted with more or less care depending on the dye; *i.e.*, high light intensities, anaerobiosis, or prior reduction of the dye may be required for or helpful to a cyclic electron flow pattern. This differs from the bacterial system in which the cyclic electron flow pattern is a built-in feature, mediated by firmly bound catalysts. (When cyclic electron flow does occur, the "poising" phenomenon is in evidence just as for bacterial photophosphorylation.)

With carefully prepared chloroplasts, the rate of electron flow is under the control of added ADP as phosphate acceptor. As in the phenomenon designated "respiratory control" in MITOCHONDRIA, this means that the electron transfer reaction coupled to phosphorylation is the rate-limiting step, especially at high light intensities. A large range of chemical compounds uncouple electron flow, so that it may proceed maximally in the absence of simultaneous phosphorylation. While many of these are the same as those effective in mitochondria, a sensitivity to ammonia and organic amines, and a lack of effect of calcium or thyroxine, may distinguish photo- from oxidative phosphorylation.

ATPase is very low in higher plant chloroplasts except under special circumstances. One such is during light-induced electron transport in the presence of Ca ions rather than Mg. Another is an ATPase that survives in the dark (as long as ATP is present), induced by the combination of light, electron flow and high concentrations of —SH containing material (*e.g.* 0.08 *M* cysteine). This ATPase decays rapidly if ATP is removed. These activities probably represent aberrations of the terminal enzyme of photophosphorylation or of its binding so that the hydrolytic function becomes apparent. In *Rhodospirillum rubrum* chromatophores, light induces a dark-persisting P_i-ATP exchange reaction.

By the use of EDTA a protein necessary for photophosphorylation ("coupling factor") can be removed from chloroplast membranes.[4] Incubation of the two fractions together with Mg ions leads to re-binding and partial restoration of activity. The same protein is needed for chloroplast ATP hydrolysis reactions noted above, and antibody to the purified protein inhibits all three reactions.[5] The isolated protein possesses ATPase activity after treatment with trypsin or incubation with dithiothreitol, and is most likely the terminal enzyme of photophosphorylation.

Pre-illumination of chloroplasts gives a transient ability to phosphorylate ADP in the dark immediately afterwards.[6] The amount of ATP that can be formed is 50 times as much as the cytochrome *f* or P700 present, hence the inferred high energy precursor is more likely to be a reservoir than a rapidly turning over chemical compound. Serious attention has been given to the "chemiosmotic" hypothesis of P. Mitchell[7,8] to explain the nature of the intermediate. According to this hypothesis electron transport may be obligately coupled to proton translocation inward, and the high energy state could be a pH differential between the inside and outside of the grana disc membranes. In support of this concept a marked uptake of protons in the light (up to 300 to 400 equivalents per cytochrome *f*) was observed, and related to the phosphorylation mechanism by sensitivity to uncouplers.[6] At the same time cations are excreted.[9] Another relevant observation was the discovery[10] that chloroplasts can make up to 100 moles of ATP per mole of cytochrome *f*, completely in the dark and possibly without electron transport, if they are incubated in acid (pH 4.0) for 15 seconds and then brought rapidly to pH 8.4. The driving force appears to be the pH differential; a minimal pH jump of about 2.8 units is required, to some extent independent of the final pH (*i.e.*, a transition from 4 to 7 is about as effective as one from 5 to 8). In a rather non-specific fashion divalent acids increase the yield of ATP, providing they are at least partly uncharged at pH 4 and fully charged at pH 8—presumably by providing a reservoir of internal acid. Inhibition of ATP formation by uncouplers and a requirement for the same terminal enzyme noted above indicate the usual photophosphorylation mechanism is indeed engaged by the acid-base transition. Some reverse electron transport may also be caused by this treatment because it leads (with very low yield) to fluorescence of system II chlorophyll *a*.[11]

In both bacterial chromatophores and green plant chloroplasts, the existence of photo-induced high energy intermediates or states leads to reversible conformational changes in the structures of the membranes, and to gross swelling and shrinking.[12] These are observed by changes in light scattering, viscosity and sedimentation properties, and by electron microscope observations. The mechanisms may include ion transport followed by water diffusion, internal pH changes leading to conformational changes of proteins, or even something resembling a "contractile protein". The full significance of these changes is not yet established, but it is easy to imagine that the spatial and therefore the functional relationships of some parts of the photosynthetic apparatus may be altered accordingly; indeed the *in vivo* reality of some of these changes has been shown by light scattering studies of whole bacterial cells and by electron microscopy of leaf tissues.

References

Articles Cited Above

1. Bose, S. K., and Gest, H., "Bacterial Photophosphorylation: Regulation by Redox Balance," *Proc. Natl. Acad. Sci. U.S.*, **49**, 337-345 (1963).
2. Gest, H., "Metabolic Aspects of Bacterial Photosynthesis," in "Bacterial Photosynthesis" (Gest, H., San Pietro, A., and Vernon, L. P., Editors), pp. 129–150 Yellow Springs, Ohio, Antioch Press, 1963).

3. ARNON, D. I., TSUJIMOTO, H. Y., AND MCSWAIN, B. D., "Role of Ferrodoxin in Photosynthetic Production of Oxygen and Phosphorylation by Chloroplasts," *Proc. Natl. Acad. Sci. U.S.*, **51**, 174–1282 (1964).

4. AVRON, M., "A Coupling Factor in Photosynthetic Phosphorylation," *Biochim. Biophys. Acta*, **77**. 699–702 (1963).

5. MCCARTY, R., AND RACKER, E., "A Coupling Factor for Photophosphorylation and Hydrogen Ion Transport," in *Brookhaven Symp. Biol.*, **19**, BNL 989 (C-48) (1966).

6. JAGENDORF, A. T., HIND, G., AND NEUMANN, J., "Studies on the Mechanism of Photophosphorylation," in "Photosynthetic Mechanisms of Green Plants," *Natl. Acad. Sci.-Natl. Res. Council, Publ.*, **1145**, 599–610 (1963).

7. MITCHELL, P., "Chemiosmotic Coupling in Oxidative and Photosynthetic Phosphorylation," *Biol. Rev.*, **41**, 445–502 (1966).

8. JAGENDORF, A. T., AND URIBE, E., "Photophosphorylation and the Chemiosmotic Hypothesis," in *Brookhaven Symp. Biol.*, **19**, BNL 989 (C-48) (1966).

9. DILLEY, R., AND VERNON, L. P., "Ion and Water Transport Processes Related to the Light-dependent Shrinkage of Spinach Chloroplasts," *Arch. Biochem. Biophys.*, **111**, 365–371 (1965).

10. JAGENDORF, A. T., AND URIBE, E., "ATP Formation Caused by Acid-base Transition in Spinach Chloroplasts," *Proc. Natl. Acad. Sci. (U.S.)*, **55**, 170–175 (1966).

11. MAYNE, B. C., AND CLAYTON, R. C., "Luminescences of Chlorophyll in Spinach Chloroplasts Induced by Acid-base Transition," *Proc. Natl. Acad. Sci. (U.S.)*, **55**, 494–497 (1966).

12. DILLEY, R. A.; also PACKER, L., DEAMER, D. W., AND CROFTS, A. R.; also IZAWA, S., CONNELLY, T. N., WINGET, G. D., AND GOOD, N. E.—articles in *Brookhaven Symp. Biol.*, **19**, BNL 989 (C-48) (1966).

General Reviews

13. CLAYTON, R. K., "Physical Aspects of the Light Reaction in Photosynthesis," in "Photophysiology" (GIESE, A. C., EDITOR), Vol. I, pp. 155–197, New York, Academic Press, 1964.

14. HOCH, G., AND KOK, B., "Photosynthesis," *Ann. Rev. Plant Physiol.*, **12**, 155–194 (1961).

15. JAGENDORF, A. T., "Biochemistry of Energy Transformation in Photosynthesis," *Surv. Biol. Progr.*, **IV**, 181–344 (1962).

16. VERNON, L. P. "Bacterial Photosynthesis," *Ann. Rev. Plant Physiol.*, **15**, 73–100 (1964).

17. WHATLEY, F. R., AND LOSADA, M., "The Photochemical Reactions of Photosynthesis," in "Photophysiology" (GEISE, A. C., EDITOR), Vol. I, pp. 111–154, New York, Academic Press, 1964.

ANDRE T. JAGENDORF

PHOTOREACTIVATION. See BACTERIAL GENETICS, section on DNA Synthesis.

PHOTOSYNTHESIS*

Photosynthesis is the process by which green plants harness the energy of sunlight absorbed by chlorophyll to build organic compounds from carbon dioxide and water. This reaction is often referred to as assimilation or fixation of carbon.

Fundamental Role of Photosynthesis. For their subsistence, growth and multiplication, all living beings, plants and animals alike, need organic food. The organic foodstuffs are utilized by living cells as building stones and as sources of energy. The energy stored in the foodstuffs is released mainly by RESPIRATION, the reaction in which organic matter combines with the oxygen of the air yielding carbon dioxide and water as final products. The rate at which living beings die and consume each other is so high that they would all disappear from the earth within the lifetime of a human generation if there were not a process providing for the re-formation of organic matter.

While there are many ways by which organic substances are decomposed in respiration and similar reactions, there is only one reaction, photosynthesis, which for millions of years has counterbalanced death and decomposition. In the course of photosynthesis, the hydrogen of water is used to transform carbon dioxide into carbohydrate; simultaneously the oxygen of the water is liberated as free oxygen gas. Among the many pigments appearing in the plant kingdom, only CHLOROPHYLL and, to a lesser extent, CAROTENOIDS and phycobilins are known to convert sunlight into chemical energy. Chlorophyll is the green dye whose color is so characteristic of meadows and forests. The green plants may transform the carbohydrates produced by photosynthesis into fats, proteins and many other substances. Thus, directly or indirectly, all the organic food of plants or animals depends upon the photosynthetic process. In the cycle of synthesis and death, the basic materials, carbon dioxide and water, can be used over and over again, while the energy released upon the destruction of organic matter is lost forever; it is dissipated as heat in space and must constantly be replaced by the sun's radiation.

Origin and Evolution of Life. The chlorophyll system is a highly complex structure that ceases to function whenever its protein components are damaged. Obviously, therefore, life cannot have started with the particular synthetic reaction that is its sole support at present. Life probably began with organic substances produced by ultraviolet radiation or by electrical discharges when the earth's atmosphere consisted mainly of hydrogen, methane, ammonia and water vapor. Experiments have shown that such a gas mixture, if subjected to electrical discharges, yields many aliphatic and amino acids. Contemporary living cells still use these simple compounds as building stones, not only for proteins and fats but for the synthesis of even more intricate organic molecules. The most famous example is the formation of the PORPHYRIN ring, the parent structure of chlorophyll and of hemin, the red pigment in blood. It is formed by a succession of a few simple condensation steps. This synthesis starts with glycine and acetic acid, compounds now assumed to have been present in large quantities at the beginning of organic evolution on earth. It is likely, therefore, that porphyrins and hence their iron and magnesium derivatives, *i.e.*, hemin and chlorophyll, appeared early and took part in the emergence of the first living things (see also ORIGIN OF LIFE).

Thus the complex mechanism of photosynthesis may have evolved as a part of the first organisms. Even at the initial stages of life the simplest photochemical reactions (reactions initiated by light) known to occur with porphyrins must have been instrumental in speeding up and selecting special types of synthetic reactions. Because the process of photosynthesis as found in present-day plants can be broken up into separate sets of partial reactions, it is assumed that, during evolution, these were added one by one, with the light-absorbing chlorophyll complex as the original center. The last of these evolutionary steps was the capacity to release free oxygen. In this way, photosynthesis first furnished and then maintained the supply of oxygen in the atmosphere, which made the Darwinian evolution of respiring organisms possible. The preceding developmental step, the photochemical reduction of carbon dioxide at the expense of compounds other than water, seems never to have risen above the level of unicellular microorganisms.

Rate of Carbon Dioxide Turnover. In bright sunlight the rate of the photosynthetic reaction is about 15–30 times the respiratory rate. A sunflower leaf may gain 9% of its dry weight per hour. Considering that respiration proceeds continuously while photosynthesis occurs only during the hours of light, the net rate of the latter is about five times greater. This means an excess of carbohydrate, which is partly stored and partly used for the growth of the plant. Non-green plants and animals, which constitute the heterotrophic organisms, depend upon this excess for their organic food.

Land plants, which utilize the carbon dioxide present in the air, grow faster if supplied with more carbon dioxide than they can now find in nature. This averages only 0.03% of the gas volume at the earth's surface. Aquatic plants thrive on the carbon dioxide dissolved as gas or as carbonates in the water. The best estimate for the rate of the total carbon dioxide reduction on earth is a turnover of 10^{10} tons of carbon per year, more than two-thirds of which is contributed by the flora of the oceans. An amount of carbon dioxide equal to the total readily available reservoir passes through the life cycle in about 350 years. The cycle of carbon requires also one of oxygen and of hydrogen. Since both these elements are available in larger quantities, it takes about 2000 years to renew all the oxygen in the air and 2,000,000 years to decompose all water on earth. Probably there was more carbon dioxide in the atmosphere of the earth at earlier times. The enormous deposits of coal and oil, the so-called fossil fuels, are the remnants of once living plants and microorganisms. These products of ancient photosynthesis (which, since the beginning of the Industrial Revolution around 1750, man is returning to the air in ever increasing amounts as carbon dioxide) have been out of circulation for many millions of years.

Efficiency of Photosynthesis. In land plants, between 1 and 3% of the light falling upon a green plant is transformed into chemical energy. Important as this answer may be for problems of agriculture, it is of no value in respect to the theoretical understanding of the process of photosynthesis. Most of the incident radiation is lost by reflection, transmission and ineffective absorption by pigments other than chlorophyll. Only a fraction is absorbed by the active pigments. The efficiency with which this latter fraction is utilized in the course of the photochemical process is of great theoretical interest. It is measured by the minimum number of light quanta necessary to bring one molecule of carbon dioxide to the energetic level of a carbohydrate. From the mid-1920s to the early 1960's, a prodigious amount of work was devoted to the determination of this number. It is certain that eight quanta of red light totaling 320 kilogram-calories (kgcal) suffice to store the equivalent of 112 kgcal in terms of organic matter. The net efficiency of photosynthesis, therefore, is about 35%. Much higher efficiencies are alleged to have been found in certain experiments where plants were kept under unusual conditions. In the unicellular green alga *Chlorella*, the quantum yield remains essentially constant over the entire range of the chlorophyll absorption spectrum. This means that, expressed in terms of calories, the efficiency increases in the direction from blue to red in the spectrum, because the energy content of the light quanta decreases according to the laws of physics. Chlorophyll in the living cell obeys these laws exactly as do the chlorophyll-sensitizing photochemical oxidations in vitro.

Artificial Food Production and Space Travel. Because the earth's human population is increasing so rapidly, with no corresponding increase in food production, scientists have wondered how much more food would be available if full use could be made of the great efficiency of photosynthesis, which in laboratory experiments has been found to approach 35% of the intensity of the radiation absorbed by the active pigments. The average acre of land under cultivation yields hardly more than 1 ton of total dry organic matter per year because most of the light falling on the surface is not trapped by chlorophyll and thus is lost for photosynthesis. If agriculture were superseded by the continuous cultivation of tiny aquatic algae in thin layers of fertilized water, the same area could produce between 30 and 50 times more edible material. Before such a project can become practical, two other problems must be solved: how to maintain, easily and cheaply over large areas, conditions for optimal growth of algae and how to induce people to eat them. A technological photochemistry for the synthesis of selected chemicals is certainly practicable. By contrast, a completely artificial manufacture of food, imitating photosynthesis, does not appear to be a sensible aim to strive for. One cannot conceive of an artificial device as small and as efficient as a green plant cell for the purpose of producing all necessary nutrients.

The problem of how to nourish human beings

and supply them with oxygen during long-lasting expeditions into outer space can be solved on the basis of natural photosynthesis. Oxygen for respiration would be provided by illuminated plants, which in turn would use the exhaled carbon dioxide and other waste products of man's metabolism as the raw material (carbon source, vitamins and mineral nutrients) to grow and produce the necessary food. In short, the carbon cycle of the organic world on earth would have to be duplicated in miniature. The light energy to keep it running could be obtained either directly from the sun or indirectly from the power that promotes the space rocket.

A third problem, the conversion of light energy into electricity for industrial purposes, is solved in principle on the basis of purely physical reactions found in "solar batteries," and a solution may be found for photochemical reactions also, yet neither method of utilizing light resembles the photochemistry of chlorophyll in plants.

Energy Storage (Decrease in Entropy). Light energy may be used in several ways: (1) to speed up reactions that would proceed in the dark in the same way but at much slower rates (autoxidation, the bleaching of colors, for example); (2) for reactions that will not occur unless light energy is available (for instance, the release of oxygen from water at normal temperatures); (3) for a reaction in which the new products still hold as potential chemical energy a part of the light energy originally absorbed. In the course of organic evolution, photosynthesis reached the third stage and achieved an over-all molecular efficiency of 35%. This efficiency is due to the incorporation of water as a reactant into the photochemical system and to subsequent release of free oxygen arising from water.

There are many other mechanisms in nature in which light plays a role (vision, for instance) and some in which carbon dioxide is transformed into organic matter without an over-all gain in energy. The latter fall in two classes, one where light is not needed at all and one where it is needed but does not produce a net gain in free energy. A few species of autotrophic bacteria are known to grow in the absence of light with carbon dioxide as the sole source of carbon. They utilize the energy derived from the oxidation of inorganic compounds such as ferrous iron salts, sulfur, thiosulfate, hydrogen sulfide and molecular hydrogen. This fact was generally ignored in earlier attempts to understand the mechanism of photosynthesis. The amount of organic matter formed by this means was considered insignificantly small as compared with photosynthesis, and the reduction of carbon dioxide in the dark appeared to be fundamentally different from photochemical reduction. In the decade between 1930 and 1940, however, it became evident that many organisms and tissues, from propionic acid bacteria to pigeon liver, include carbon dioxide among their metabolic substrates (see CARBOXYLATION ENZYMES). In these cases a specific enzymatic system provides for the proper coupling between an energy-yielding reaction, respiration or fermentation, and a reaction in which hydrogen is transferred to carbon dioxide or its equivalent derivative. The result is called a carboxylation. Since the very life of autotrophic organisms depends on catching what little there is of carbon dioxide in the air, their carboxylation enzymes are usually more efficient than the same enzymes found in heterotrophs. The decisive difference between both classes of organisms lies, however, not in the way in which the carbon dioxide molecule is attacked but in the energy balance. No ordinary metabolic synthesis can end with an over-all gain and storage of chemical energy, because an excess of combustible material, previously synthesized, has to be sacrificed to make the reaction go.

In photosynthesis it is the energy of light that in the course of the very reaction furnishes the necessary combustible material or hydrogen donor by the "splitting" of water. The utilization of light energy, however, is not sufficient to guarantee a successful accumulation of chemical energy. This was shown by studies on the metabolism of purple bacteria, strongly colored yellowish, red or purple unicellular microorganisms (see PHOTOSYNTHETIC BACTERIA). Because their growth depends on light, they were first believed to photosynthesize as do the green plants. Investigations proved that purple bacteria are unable to produce molecular oxygen and are unable to reduce carbon dioxide simply with water. In addition to radiant energy, they need energetically valuable hydrogen donors. According to the specificity of each species, the purple bacteria require for the assimilation of a certain amount of carbon dioxide an equivalent amount of either sulfur, hydrogen sulfide, thiosulfate, molecular hydrogen, aliphatic acids or alcohols. In purple bacteria, therefore, the energy of light is not as successfully stored as in the green plants but is partially wasted. The light serves here only as a promoter, as it were, of reactions that might as well be carried on thermally in the dark. The overall gain consists quite visibly in the appearance of complex living forms in place of the simple molecules of "food". This conversion means a decrease in entropy within the system under observation.

In 1940 it was found that the photochemical reaction proper is the same in purple bacteria and in green plants. Certain species of unicellular green ALGAE change to a type of metabolism similar to that of some purple bacteria if they are incubated during a few hours in pure hydrogen gas. Upon subsequent moderate illumination, they reduce carbon dioxide as usual, but they do not evolve oxygen. Instead they absorb the equivalent amount (twice the volume) of hydrogen. Once adapted to the use of hydrogen the same algae can reduce carbon dioxide, as do the "Knallgas" bacteria, in the dark, if some oxygen is supplied instead of light. Under the influence of intense irradiation the algae revert to normal photosynthesis with the production of oxygen. This reversible separation of carbon dioxide reduction from oxygen evolution or from the absorption of light demonstrated what had long been postulated—that photosynthesis consists of a series of complicated reactions that must be studied separately before the process as a whole is clearly understood.

After 1940 all students of the problem agreed that at least three sets of partial reactions may be distinguished: (1) decomposition of water with the aid of light absorbed by chlorophyll, leading, on the one hand, to compounds having a surplus of hydrogen and serving as intermediary hydrogen donors, and, on the other hand, to compounds having a surplus of oxygen; (2) transfer of hydrogen from the intermediary donors to carbon dioxide; (3) disposal of the compounds having a surplus of oxygen. The permanent conversion of radiant energy into chemical energy is possible only because green plants accomplish the task of removing the surplus oxygen by liberating molecular oxygen and not by consuming hydrogen donors, as do, for instance, the purple bacteria. This picture of photosynthesis requires, among other things, that all the oxygen evolved be derived from water and not from the reduced carbon dioxide molecule.

The separate phases of photosynthesis represented by the three sets of partial reactions just mentioned are discussed in more detail in the articles LIGHT PHASE OF PHOTOSYNTHESIS; PHOSPHORYLATION, PHOTOSYNTHETIC; CARBON REDUCTION CYCLE (IN PHOTOSYNTHESIS).

References

1. *Handbuch der Pflanzenphysiologie*, Vol. V, "CO_2 Assimilation," (PIRSON, A., EDITOR), Berlin, Springer-Verlag, 1960.
2. RABINOWITCH, E., "Photosynthesis and Related Processes," 3 vols., New York, Interscience Publishers, 1945–56.
3. *Annual Review of Plant Physiology*, 1950 and later issues.
4. GAFFRON, H., "Energy Storage: Photosynthesis," in "Plant Physiology" (STEWARD, F. C., EDITOR), Vol. IB, pp. 1–277, 1960.

HANS GAFFRON

PHOTOSYNTHETIC BACTERIA

The photosynthetic bacteria comprise a morphologically diverse group of pigmented bacteria which grow anaerobically in the light. They are unable to grow anaerobically in the dark, *i.e.*, cannot grow by FERMENTATION, but some species can grow aerobically in the dark, *i.e.*, by aerobic RESPIRATION. In their natural habitats, it is probable that they all grow photosynthetically. Their pigment systems containing a number of CAROTENOIDS (which are, in most cases, responsible for their color) and a single CHLOROPHYLL, are particularly adapted to the absorption of light in the red region of the spectrum where green plant photosynthesis is inefficient. Consequently, photosynthetic bacteria may frequently be found on the under surface of leaves in stagnant waters. In certain situations, photosynthetic bacteria may occur in nature in high concentrations, giving rise to purple or red "blooms." Most photosynthetic bacteria are motile and are phototactic, showing a characteristic response ("fright-movement"), moving away by a reversal of swimming direction when passing from an illuminated to a dark region. As a result, photosynthetic bacteria may be concentrated in an illuminated area.

Unlike green plants and algae, the photosynthetic bacteria do not evolve oxygen. Photosynthetic bacteria contain a single species of chlorophyll, although several distinct *bacteriochlorophylls* have been identified in different photosynthetic bacteria. The bacteriochlorophylls are similar in structure to the plant chlorophylls; they are magnesium chlorin complexes which are more reduced than the plant chlorophylls and have different substituents. All photosynthetic bacteria require an accessory reductant (hydrogen donor) for carbon dioxide fixation. Different photosynthetic bacteria may use sulfide, thiosulfate, sulfur, hydrogen gas or a number of simple organic compounds as reductants.

The major groups of photosynthetic bacteria are as follows:

(1) The green sulfur bacteria (*Chlorobacteriaceae*) includes the genus *Chlorobium* which is strictly anaerobic, strictly autotrophic, and strictly photosynthetic, using inorganic sulfur compounds as reductants for carbon dioxide fixation. *Chlorobium* species are non-motile, short bacilli. More recently, a motile green photosynthetic bacterium *Chloropseudomonas ethylicum* has been isolated which is also able to grow anaerobically in the presence of a limited number of simple organic compounds including ethanol and acetate.

(2) The red sulfur bacteria (*Thiorhodaceae*) are a morphologically heterogeneous group which are strictly anaerobic and strictly photosynthetic, which can grow autotrophically or heterotrophically. Reduced sulfur compounds are commonly used as reductants for autotrophic growth, but hydrogen gas may be used. *Chromatium* species may be large motile bacilli, but motile spiral organisms (*Thiospirillum*) are known. These organisms accumulate elementary sulfur when grown with reduced sulfur compounds. Some species require VITAMIN B_{12}.

(3) The non-sulfur purple and brown bacteria (*Athiorhodaceae*) include motile rods (*Rhodopseudomonas*) and spirilla (*Rhodospirillum*) most of which grow photosynthetically under anaerobic conditions in the presence of organic compounds, and also aerobically in the dark on the same organic compounds. Some species have been reported to grow autotrophically in the light. One or more vitamins of the VITAMIN B GROUP are required as growth factors.

Some of the properties of the major genera of photosynthetic bacteria are summarized in Table 1.

Rhodomicrobium vannielii is an exceptional photosynthetic bacterium both physiologically and morphologically. It is a strict anaerobe which grows on organic compounds, but does not require growth factors. The cells are oval and

TABLE 1. PROPERTIES OF THE PHOTOSYNTHETIC BACTERIA

Group	Main Genera	Vitamin Requirements	Aerobic/Dark Growth	Hydrogen Donors used for Growth
Green sulfur bacteria (Chlorobacteriaceae)	*Chlorobium*	None	No	Inorganic
	Chloropseudomonas	None	No	Inorganic or organic
Purple sulfur bacteria (Thiorhodaceae)	*Chromatium*	Some require B_{12}	No	Inorganic or organic
	Thiospirillum	B_{12}	No	
Purple non-sulfur bacteria (Athiorhodaceae)	*Rhodopseudomonas*	Some B vitamins	Yes	Organic, but some species can use inorganic H-donors
	Rhodospirillim		Yes (some)	
(Hyphomicrobiaceae)	*Rhodomicrobium*	None	No	

commonly reproduce by extruding a filament, which may branch. New cells are formed by "budding" from the ends of the filaments. Characteristic non-motile clusters of cells joined by filaments are thus formed. However, under certain growth conditions single motile cells with peritrichous flagella are formed.

Unlike plants and eucaryotic algae, photosynthetic bacteria do not contain CHLOROPLASTS. However, the photosynthetic pigments are localized within the bacterial cell, and pigment-bearing particles call "chromatophores" can be isolated from disrupted cell preparations. Under suitable conditions, these "chromatophores" show limited photosynthetic activity, namely the ability to catalyze photosynthetic PHOSPHORYLATION and certain photooxidations and photoreductions. "Chromatophores" unlike chloroplasts are incomplete photosynthetic units and are unable to catalyze carbon dioxide fixation. It appears that the "chromotophores" of most photosynthetic bacteria are artifacts derived by the disruption of the bacterial cytoplasmic membrane system, which in most cases is seen in electron micrographs to contain vesicles. Like cytoplasmic membrane preparations of non-photosynthetic bacteria, "chromatophores" contain electron transport components including CYTOCHROMES and quinones, non-heme iron, and appreciable amounts of PHOSPHOLIPIDS. The structure of the cytoplasmic membrane is not identical in all the photosynthetic bacteria, and even within a single species *Rhodospirillum rubrum* the structure of the membrane systems is influenced by cultural conditions, especially the light intensity during growth. *Rhodomicrobium vannielii* has a system of peripheral lamellae which contains the photosynthetic pigments and in this respect resembles the procaryotic blue-green algae. The green sulfur bacteria appear to contain distinct "chromatophores" which are oblong membrane-bounded vesicles underlying the cytoplasmic membrane and arranged peripherally.

All photosynthetic bacteria fix carbon dioxide, and the mechanism is believed to be the same as in green plants and non-photosynthetic chemoautotrophic bacteria. The primary pathway of carbon dioxide fixation proceeds by the carboxylation of ribulose-1,5-diphosphate to yield 3-phosphoglyceric acid and subsequently other phosphorylated carbohydrates (see CARBON REDUCTION CYCLE). Other CARBOXYLATION reactions do occur and may be at least of equal importance with photosynthetic bacteria growing on organic compounds.

The thorough investigations of Van Niel led to the important unifying concept of photosynthesis represented by the reaction:

$$CO_2 + 2H_2X \xrightarrow{\text{Light}} (CH_2O) + H_2O + 2X$$

in which H_2X represents the reductant for the fixation of carbon dioxide to "cell material" at the oxidation level of carbohydrate (CH_2O). In plant photosynthesis, water is the reductant and so the oxidized product is oxygen which is a characteristic product of plant photosynthesis. Photosynthetic bacteria cannot use water as a reductant for carbon dioxide fixation, as a result of which oxygen is not produced. Accessory reductants need to be provided, as with the green sulfur bacteria which can use hydrogen sulfide and produce elementary sulfur. The close analogy between photosynthesis in green plants and green sulfur bacteria is clear from the following reactions:

Green plants: $CO_2 + 2H_2O \rightarrow (CH_2O) + H_2O + O_2$

Green sulfur bacteria:

$$CO_2 + 2H_2S \rightarrow (CH_2O) + H_2O + 2S$$

When photosynthetic bacteria grow with organic compounds, these serve not only as reductants for carbon dioxide fixation but also as important carbon sources for growth. Such bacteria assimilate organic compounds in the light and use them as a major source of cell carbon. They are thus able to use organic compounds extremely efficiently, and high cell yields may be obtained under optimal growth conditions. Even the strict autotroph *Chlorobium* assimilates low concentrations of some simple organic compounds such as acetate although this is not a major source of cell carbon for the green sulfur bacteria of this genus. It is clear that additional carboxylation reactions are involved in the photo-assimilation of organic compounds. Washed suspensions of *Rhodospirillum rubrum* can assimilate acetate in the light in the absence of carbon dioxide to form a storage product: a polyester of β-hydroxybutyric acid which has been found in many bacteria. Carbon dioxide is required for the conversion of poly-β-hydroxybutyrate to other cell constituents. Under certain conditions, photosynthetic bacteria fix nitrogen gas. Greatest biochemical versatility is displayed by the Athiorhodaceae which can grow either aerobically in the dark or anaerobically in the light. They thus share the advantages of two ways of life—photosynthesis and respiration—and can adapt from one condition to the other with remarkable facility. Aerobic dark grown cells have a very low concentration of chlorophyll and carotenoid pigments, but on transfer to anaerobic conditions in the light, pigments are formed and are localized within the membrane system after a "lag period" during which there is a rapid synthesis of the enzymes required to form the photosynthetic pigments. Light itself is not essential for pigment formation since pigment formation will take place in the dark at low oxygen concentrations.

Photosynthetic bacteria have contributed greatly to our understanding of photosynthesis in its broadest sense, and their diverse metabolic activities still pose many challenging problems to biochemists.

References

1. Gest, H., and Kamen, M. D., "The Photosynthetic Bacteria" in "Encyclopedia of Plant Physiology" (Ruhland, W., Editor), Vol. 5, Pt. 2., Berlin, Springer-Verlag, 1960.
2. Gest, H., San Pietro, A., and Vernon, L. P., (Editors), "Bacterial Photosynthesis," Yellow Springs, Ohio, Antioch Press, 1963.
3. Kondrateva, E. N., "Photosynthetic Bacteria," translated from Russian and published for National Science Foundation, Washington, D.C., Israel Program for Scientific Translations, Ltd., Cat. No. 1214, 1965.
4. Larsen, H., "On the Microbiology and Biochemistry of the Photosynthetic Green Sulfur Bacteria," *Kgl. Norske Videnskab. Selskabs Skrifter*, **1953**, No. 1.
5. Schlegel, H. G., and Pfenning, N., "Die Anreichungskultur einiger Schwefelpurpurbakterien," *Arch. Mikrobiol.*, **38**, 1 (1961).
6. Stanier, R. Y., "The Organization of the Photosynthetic Apparatus in Purple Bacteria," in "The General Physiology of Cell Specialisation," New York, McGraw-Hill Book Co., 1963.
7. Niel, C. B. van, "The Culture, General Physiology, Morphology and Classification of the Non-Sulfur Purple Bacteria," *Bacteriol. Rev.*, **8**, 1 (1944).
8. Niel, C. B. van, "The Comparative Biochemistry of Photosynthesis," in "Photosynthesis in Plants" (Frank, J., and Loomis, W. E., Editors), Ames, Ia., Iowa State College Press, 1949.

D. S. Hoare

PHOTOSYNTHETIC PHOSPHORYLATION. See PHOSPHORYLATION, PHOTOSYNTHETIC.

PHOTOTROPISM

When cylindrical plant organs are exposed to unilateral light, or to an asymmetric light field, they usually respond by curving toward or away from the direction of higher light intensity. This curvature, due to unequal growth on the two sides of the organ, is called *phototropism*. Stems generally curve toward light and roots away from the light. In the cylindrical coleoptiles (leaf sheaths) of grass seedlings, the unequal growth which produces curvature is related to an asymmetry in the distribution of the plant growth HORMONE (also called *auxin* and chemically indole-3-acetic acid). Light seems to cause the migration of auxin from the light to the dark side of the coleoptile, and the final result is that there is roughly twice as much auxin on the dark side as on the light side. It is this asymmetry in auxin which causes the asymmetry in growth, and in turn leads to the curvature toward the light. In roots, all these phenomena are the same, but roots curve away from the light because the higher auxin content inhibits, rather than promotes the growth of the cells on the dark side.

The amount of curvature depends on the amount of light. In general, it does not matter whether a bright flash is given for a short period, or whether a dim source is applied for a much longer period: if the total light energy administered is the same, then the curvature will be the same.

The effect of light varies with its wavelength, or color. Blue light (ca. 450 mμ) is most effective in producing curvature, and near ultraviolet (ca. 370 mμ) is also effective. Wavelengths longer than 500 mμ (green) are essentially ineffective, although certain specialized plant organs do respond to red light. If a careful study is made of the relation between wavelength and curvature induced, the results indicate that a yellow pigment, probably either a CAROTENOID or a RIBOFLAVIN derivative, is the light receptor. The carotenoid absorption spectrum seems to fit the action spectrum for phototropism a little better in the visible part of the spectrum, but the riboflavin-type spectrum seems to fit better in the ultraviolet. Certain mutant plants have been obtained which are low in carotenoids, normal in riboflavin and close to normal in phototropism. However, since they still do have some carotenoid present, and since only a little may be needed, this experiment does not establish with certainty the nature of the receptor pigment.

It can be shown that the lighted side of a unilaterally exposed coleoptile becomes electronegative with respect to the darkened side. This induced transverse electrical polarity of about 100 mV may help explain the lateral movement of auxin in the coleoptile. At the pH of the cell, the auxin molecule would tend to ionize, yielding a positive hydrogen ion and a negative indoleacetate ion. The latter might then be caused to migrate toward the positive (dark) side of the coleoptile. In support of this theory, it has been found possible to induce curvature by properly applied electrical currents. Depending on the sign of the applied electrodes, electricity can be made either to reinforce or to annul the curvature induced by light. How the absorption of light energy is converted into an electrical potential is unknown, but it is surmised that the normal electron-transport systems of respiration are involved.

Certain unicellular fungal organs, such as the erect sporangiophores of *Phycomyces* and *Pilobolus*, are highly sensitive to phototropic and light stimuli. The action spectrum for fungal phototropism almost exactly matches that for higher plant phototropism, indicating that both contain the same photoreceptor. Here, however, the similarity ends, for since fungus hyphae do not respond to auxin, it cannot be the intermediary in their response to light.

Symmetrical light causes an acceleration of growth of these fungus hyphae, instead of a depression of growth, as in grass colepotiles. Immersion of the fungus hyphae in paraffin oil leads to a reversal of the direction of their curvature. This is believed to be due to the cylindrical lens action of the sporangiophore. When light enters the sporangiophore from an air medium, the cylindrical surface of the hyphae focuses it on the "shaded" side, thus inducing the abovementioned growth stimulation, which in turn leads to the curvature toward the light. But when light passes from the oil, with its higher index of refraction, into the hypha, it is diverged, rather than converged. Thus, the full light stimulus is given to the lighted side, which, stimulated in its growth, leads to curvature away from the light. This interpretation may need revision in view of the recent finding that grass coleoptiles (which probably do *not* act as cylindrical lenses) still reverse their direction of curvature under paraffin oil.

The positive phototropic behavior of stems and the negative phototropic behavior of roots have obvious survival value for the plant. Petioles of leaves are also phototropically sensitive and serve to produce a mosaic of leaves which effectively absorbs all useful light impinging on the plant.

ARTHUR W. GALSTON

PIGMENTS, ANIMAL. See SKIN PIGMENTATION.

PIGMENTS, PLANT

Nature affords no wider variety of colors than those found in the world of plants. The prevailing color of plants is the green of CHLOROPHYLL, often followed, in some parts of the world, by the yellow, orange, and red of autumn foliage. The yellow colors of leaves are due principally to the group of fat-soluble plastid pigments known as the CAROTENOIDS.

The much more various colors of flowers and fruits range from dead white, through ivory, yellow, orange, red and finally violet and blue. Combinations of these can give rise to nearly the whole of the visible spectrum of colour, and, in certain combinations and concentrations, the pigments responsible for the primary colors can produce brown and even what may appear to the eye as black colorations. Carotenoids, which will not be considered in detail in this section, are often involved in the pigmentation of yellow flowers and fruits; however, certain non-carotenoid yellow and orange pigments are often found to be the principal cause of yellow petal colors.

White flower petals are more often pale ivory than true white. The latter color is not common and is the result of a total lack of pigmentation. Most white flowers contain one or more representatives of a class of pigments known as flavones (I, see also FLAVONOIDS AND RELATED COMPOUNDS). These compounds, usually bearing hydroxyl groups at three or more of the positions 3′, 4′, 3, 5, 6, 7, 8 (and most commonly at 3′, 4′, 5 and 7), are themselves pale yellow in color and usually impart little yellow color to the plant parts in which they occur. It is believed that in some cases their combination with metal ions in the cell sap may lead to flavone-metal complexes sufficiently deeper in color than the free flavones to be capable of imparting yellow colors to what would otherwise be white petals. The presence of flavone derivatives in white petals can usually be demonstrated by treating the petal with an alkaline solution: the deeply yellow salt of the flavone can be recognised by the change in color of the petal.

Yellow flowers and fruits are nearly always carotenoid-pigmented. However, certain flowers (*Coreopsis*, *Antirrhinum*, *Oxalis* and some others) contain pigments belonging to the chalcone (II) and benzalcoumaranone (or aurone) (III) classes. The tri- and tetrahydroxy-II and -III are deep-yellow to orange compounds and are capable of imparting lively yellow colors to flower petals. Exposure of a yellow *Coreopsis* petal to alkali causes the color to change to the orange-red color of the salts of chacone and aurone pigments present in the flower.

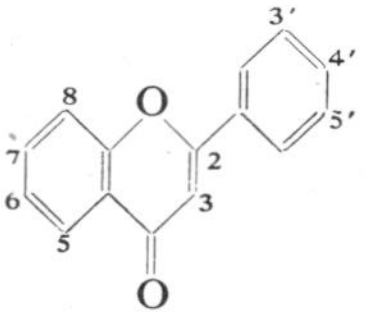

I

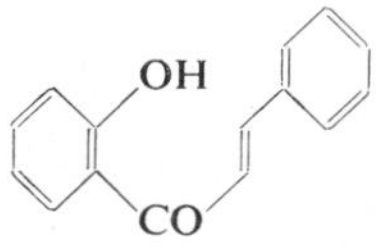

II

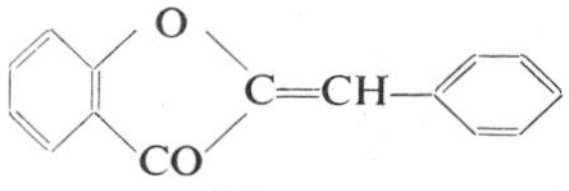

III

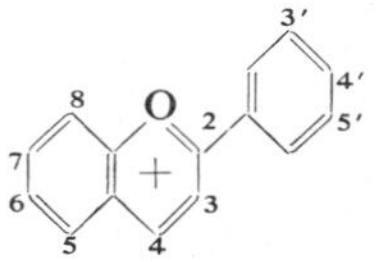

IV

Orange flowers may be pigmented with chalcones or aurones, but in most cases they owe their colors to mixtures of yellow carotenoid and red anthocyanin pigments. The orange of snapdragons is the result of the presence of aurone and anthocyanin pigments, but this is an unusual example. Orange autumn leaves are pigmented with carotenoid and anthocyanin mixtures.

Red colors in flowers, fruits and leaves are nearly always due to a group of pyrilium salts called anthocyanins (IV), and pigments of this class are also responsible for the *blue* shades of color as well. By a surprisingly limited number of structural alterations, coupled with alteration of the cell-sap environment brought about by pH changes, the presence of metallic ions and other constituents, the extraordinary gamut of colors ranging from pale salmon, through pink, scarlet, lavender, crimson, violet, purple and finally to blue and blue-black are produced.

The fundamental structure IV appears in most of the natural anthocyanins in three modifications; all of them bear hydroxyl groups at 3, 5 and 7, and differ only in the presence of additional hydroxyl groups at 4′; 3′, 4′; or 3′, 4′, 5′. These three anthocyanidins are called respectively, pelargonidin, cyanidin and delphinidin. In nature, they bear sugars, linked to form β-glycosides, at 3 or 5 or 3, 5. The glycosylated anthocyanidins are the anthocyanins. Pelargonidin glycosides are characteristic of scarlet, those of cyanidin of crimson to magenta, those of delphinidin of violet to blue flowers and fruits. In some cases one or more (usually 3′ or 3′, 5′) of the hydroxyl groups may be methylated. Thus, a rather limited number of rather superficial structural changes can give rise to an almost endless number of subtle color changes in the mature plant tissues in which these pigments are found.

Most of the pigments mentioned in the above discussion occur, as do the anthocyanins, not as the free phenolic substances, but combined with sugars in the form of glycosides. The water solubility thus imparted to what would otherwise be rather insoluble compounds permits their occurrence in solution in the aqueous contents of the cell.

Recent studies have disclosed at least a part of the pathway by which compounds of the above classes (known collectively as flavonoid compounds) are formed in the plant. The basic carbon skeleton is derived by combination of the C_6—C_3 unit shown on the right in the formulas with three two-carbon ("acetate") fragments to give a precursor which, by eventual cyclization and alteration in the oxidation level of the C_3 portion of the molecule, leads to the final compounds. Little is known about the exact sequence of events by which two such compounds as quercetin (I, with OH at 3, 3′, 4′, 5, 7) and cyanidin (IV, with OH at 3, 3′, 4′, 5, 7) are linked. The relationship between quercetin and cyanidin is formally that of a single two-electron reduction of the flavone to the anthocyanidin, but existing evidence does not support the view that this process actually takes place in the cell.

T. A. Geissman

PITUITARY HORMONES. See HORMONES, ANIMAL; HYPOPHYSIS.

PLASMA AND PLASMA PROTEINS

Introduction. *Definition.* If one were to obtain blood from the circulation of any of the higher orders of animals and, by centrifuging, remove from it the formed elements (erythrocytes, leukocytes and platelets), there would remain a fluid portion called plasma. This clear fluid can be visualized as the extracellular matrix of whole blood which may be regarded as a somewhat atypical tissue.

Plasma is an aqueous solution of low and high molecular weight components of diverse composition and function. If whole blood is allowed to coagulate, the formed elements aggregate in a mesh of fibrin derived from one of these soluble components, fibrinogen (see BLOOD CLOTTING). Upon standing, this clot retracts, leaving a clear, straw-colored fluid, serum, which may be obtained by centrifugation. The operational difference between *plasma* and *serum* is the absence from the latter of fibrinogen which remains in the clot in the form of fibrin.

General Physiological Functions. With the exception of the respiratory, gas-exchange function of hemoglobin in the erythrocytes, the formed elements are not directly concerned in the over-all metabolism of the animal. In general, metabolic interest has been predominantly focused on the plasma. This fluid is the medium of transport for all nutrients from absorbing to utilizing tissues, of metabolic intermediates from one tissue to another, of metabolic end products from tissues of origin to those of excretion, and of hormones from endocrine to target tissues. The interstitial fluid of the tissues is in equilibrium with the plasma across the capillary walls, permitting the assumption that the composition of plasma is an indication of the composition of extracellular fluids in general. Easily obtained, plasma is employed in studies of normal and abnormal metabolism.

Many of the low molecular weight components of plasma, particularly inorganic anions and cations, are important pH and osmotic buffers in addition to their specific functions as nutritional and cofactor materials.

The plasma proteins also figure significantly in the regulation of pH and colloid osmotic pressure. Quantitatively, they represent a large and metabolically labile segment of the organism's reserve of nitrogen compounds.

Composition. *General.* Dissolved and suspended in mammalian plasma are solutes and insoluble materials equivalent to about 10% of the plasma volume. Inorganic salts represent about 0.9% of the plasma. A variety of low molecular weight organic materials represent about 2%, while the plasma proteins total some 7% of the plasma volume.

The Plasma Proteins. The plasma proteins comprise the majority of the high molecular weight, nitrogenous materials which are suspended or dissolved in plasma. The study of their characteristics has been dependent upon the development in the last 25 years of methods of separation and isolation that take advantage of their individual physical and chemical characteristics. Among these characteristics are: their polyelectrolyte nature, their high molecular weight with their corresponding large particle size, and their density. Probably the most widely applied method of separation is ELECTROPHORESIS, which depends on the ionic nature of proteins. Conventionally, mammalian serum proteins are brought to a pH of about 8.6 and are caused to migrate at characteristic rates by the application of an electrical field across the solution, in which case they separate into five or six major groups. The protein which migrates furthest toward the anode is albumin, followed by one or two peaks representing the α-globulins. There is then a very sharp boundary which represents fibrinogen, and one or more β-globulin peaks followed closely by a broad γ-globulin region. The more recently developed technique of gel electrophoresis permits the separation of several of these groups into sub-components, differentiated on the basis of particle size by the molecular sieve characteristics of the separating medium. The relative quantities of these constitutents are shown in Table 1.

TABLE 1. CONCENTRATION RANGES OF MAJOR PROTEIN COMPONENTS OF HUMAN PLASMA

Component	Range (g/100 ml)
Albumin	2.8–4.5
Globulins (total)	3.0–3.5
α_1-Globulins	0.3–0.6
α_1-Lipoproteins	0.3–0.5
α_2-Globulins	0.4–0.9
β-Globulins	0.6–1.1
β_1-Lipoproteins	0.3–0.5
Transferrin	0.4
Fibrinogen	0.3
γ-Globulins	0.7–1.5

The individual serum proteins were first made available for study following the development of the precise salting-out methods of Cohn and his co-workers. This method also takes advantage of the electrolyte nature of the proteins. Raising the salt concentration or the ionic strength of protein solutions in a stepwise fashion permits the fractionation of the total protein components of plasma into five or six fractions, one of which represents albumin, and others various overlapping or separate α-, β-, and γ-globulin constituents.

Upon subjecting whole plasma to a method of separation that is based on a different physical characteristic, *i. e.*, molecular weight or sedimentation rate, using the ultracentrifuge, two major broad peaks are usually seen, one representing a group of proteins ranging in molecular weight from 150,000–200,000 and one representing proteins ranging from 55,000–70,000, along with several smaller components.

Quite recently, methods for separating plasma proteins by chromatographing on ion-exchange materials have been developed, and these have permitted extensive purification of the individual proteins.

Individual Plasma Proteins. *γ-Globulins.* The γ-globulin fraction has been defined as the plasma protein component with the lowest electrophoretic mobility at pH 8.6. Plasma proteins of the γ-globulin group have some common physical and immunochemical characteristics but are heterogeneous by other criteria, including a wide range of electrophoretic mobility and their diffuse spread on ion-exchange chromatography.

Upon ultracentrifugation the γ-globulins contain two classes of proteins. Members of the first class, representing 85–90% of the γ-globulins have a sedimentation constant approximating 7 S, corresponding to a molecular weight of about 150,000. The second group, representing from 10–15% of the total γ-globulin, has a sedimentation constant of 19S, or a molecular weight of about 900,000. Chemically, the 7 S γ-globulins are thought to consist of four peptide chains linked by disulfide bonds. The smaller two of the chains are called B chains and have an average molecular weight approximating 20,000 and the A chains have an average molecular weight of approximately 55,000. The 19 S γ-globulins have not been as extensively studied and would appear not to be simple aggregates of 7 S proteins although there is evidence that some of their peptide chains are similar.

The main specific function of the γ-globulin appears to be ANTIBODY activity. When foreign macromolecules, ANTIGENS, are injected into an animal, these specific combining proteins are formed. The antibodies found in human blood are mainly associated with the γ-globulins, though some are found in the β-globulin fraction.

Plasma is the medium of transport for antibodies functioning as a defence against foreign macromolecules. The γ-globulins, and hence the antibodies, are synthesized in several tissues, most probably by the widely spread cells of the reticuloendothelial system (see also IMMUNOCHEMISTRY).

β- and α-Globulins. The β- and α-globulins comprise a whole spectrum of proteins of varied physical, chemical and physiological characteristics. As groups they seem to have nothing in common but their electrophoretic mobility. Many of the plasma proteins concerned with lipid, vitamin or metal transport fall into this electrophoretic class.

(1) *Lipoproteins.* Lipids do not occur as free entities in plasma but are firmly though reversibly associated with proteins. A small quantity of fatty acid is present as fatty acid-albumin complexes. The remainder of the total plasma lipid, mostly in the form of esterified fatty acids (triglycerides, cholesterol esters, phospholipids, etc.), is present in complex form as LIPOPROTEINS. The lipids are

bound to the protein chains by non-covalent linkages. Lipoproteins are usually separated from serum by taking advantage of the fact that they are less dense than other plasma constituents and float to the top of solutions of carefully adjusted densities following high-speed centrifugation.

The lipoproteins present in serum range in molecular weight from 200,000 to approximately 5,000,000 and contain from 5–97% lipid. The lipoproteins are usually divided into two classes, depending upon their electrophoretic mobility. The β-lipoproteins represent those which migrate in the β-globulin range. They contain relatively more lipid, and these lipid moieties would appear to be metabolically more labile. The major portion of the β-lipoproteins is apparently synthesized in the liver and represents the main vehicle for carrying alimentary lipid, which has been accumulated in the liver, out toward the tissues for further metabolism. The α-lipoproteins have the electrophoretic mobility of α-globulins, appear to be mainly synthsized in the intestine, have in general a lower lipid content, and would appear to be more stable with regard to their metabolic function.

(2) *Mucoproteins.* Although many of the γ- and β-globulins, especially the immune globulins, contain varying but small amounts of carbohydrate, there is also a group of proteins migrating largely in the α- and β-range that have carbohydrate contents ranging from 40% on up. Carbohydrates present include various monosaccharides and their amino and acidic derivatives linked in oligosaccharides of various sizes and sequence, which are in turn covalently linked to the protein moieties. Among the mucoproteins are orosomucoid, which is an acidic mucoprotein, and haptoglobin, an α_2-globulin which can combine with hemoglobin to form a weak peroxidase.

(3) *The Metal-binding Proteins.* β- and α-globulins capable of combining with various metals have been isolated from plasma. Albumin also combines with many metal ions. There is a crystalline β_1-globulin specifically involved in complexes with iron, copper and zinc. This protein represents approximately 3% of the total plasma protein, has a molecular weight of approximately 90,000, and is capable of binding two atoms of metal per molecule of protein. Apparently, the main physiological function of this metal-binding protein, which has been called transferrin, is to transport iron from its intestinal absorption site to the hematopoietic tissues.

A blue, copper-containing protein, ceruloplasmin, has also been isolated from serum. At least 90% of the total serum copper is transported complexed to this protein. Ceruloplasmin is an α_2-globulin, has a molecular weight of about 160,000, and binds eight atoms of copper per protein molecule.

Fibrinogen. Fibrinogen is a protein with a molecular weight of about 330,000, important because of its role in BLOOD CLOTTING. It is synthesized in the liver. The action of thrombin on fibrinogen results in the liberation of two peptides which together have a molecular weight of approximately 9000. The protein remaining is called fibrin monomer, and in the presence of calcium and perhaps other plasma factors, it polymerizes to form a network which is then the matrix of the blood clot described above. Electrophoretically, fibrinogen is seen in plasma as a sharp peak somewhere in the β-region.

Plasma Albumin. Plasma ALBUMIN is the largest single plasma protein species. It has two major functions. Because it represents approximately 40–50% of the total protein in most mammalian plasma, it is important as both an osmotic and a pH buffer and, indeed, is largely responsible for maintaining these two physiological parameters. A second major physiological function is the transport of numerous kinds of small molecules. A large proportion of the free metal ions, such as calcium, magnesium and manganese, which circulate in the plasma, are bound reversibly to the albumin molecules. In a like fashion, a small proportion of the total free fatty acid and several other large organic molecules are also carried or bound in a more or less specific fashion to this protein.

Chemically, mammalian plasma albumins tend to have molecular weights of the order of 70,000. All of them, as far as is known, consist of a single polypeptide chain composed solely of amino acids with no covalently linked carbohydrate lipid or other organic group. Plasma albumin appears to be synthesized solely in the liver.

References

General Reference

1. PUTNAM, F. W., in "The Proteins" (NEURATH. H., EDITOR), Second edition, Vol. III, Ch. 14, New York, Academic Press, 1965.

Analytical Techniques

2. PENNELL, R. B., "The Plasma Proteins" (PUTNAM, F. W., EDITOR), Vol. I, Ch. 2, New York, Academic Press, 1960.
3. SMITHIES, O., *Advan. Protein Chem.*, **14**, 65 (1959).
4. SOBER, H. A., HARTLEY, R. W., JR., CARROLL, W. R., AND PETERSON, E. A., in "The Proteins" (NEURATH, H., EDITOR), Second edition, Vol. III, Ch. 12, New York, Academic Press, 1965.

MELVIN FRIED

PLASMALOGENS

Plasmalogens are a class of LIPIDS that generate higher fatty aldehydes on hydrolysis with mild acid or acidic reagents, particularly mercuric chloride. They are present in most animal tissues but are not found in significant concentrations in plants or bacteria, although within the last several years two strains of obligate anaerobes, *Clostridium butyricum* and *Bacteroides succinogenes*, were found to contain appreciable amounts. Plasmalogens are usually PHOSPHATIDES, and only very small amounts have been detected in the neutral lipid fraction, in particular in neutral lipids of butter and of the digestive gland of the sea star. Although the aldehydogenic property is found in phosphatides containing ethanolamine, serine and choline, it is distributed in tissues most widely with phosphatides containing ethanolamine. The specific molecule has been called phosphatid*al* ethanolamine to indicate the aldehydogenic property as well as a chemical structure very similar to that of phosphatid*yl* ethanolamine. Whereas phosphatidyl ethanolamine has its two hydrocarbon chains linked to the glycerol residue of glycerylphosphorylethanolamine through acyl ester linkages, phosphatidal ethanolamine has only the β-chain so linked. The α-chain is attached to the glycerol through an α', β'-unsaturated ether linkage (enol ether or 1′-alkenyl ether), and it is the hydrolysis of this linkage that gives rise to higher fatty aldehyde.

$$R—\underset{\delta^-}{CH}=CH—\ddot{O}—R' \xrightarrow{H^+}$$

$$[R—CH_2—\overset{+}{C}H—O—R'] \xrightarrow{H_2O}$$

$$R—CH_2—CHO + R'—OH + H^+$$

Among mammalian organs, brain contains the highest concentration of plasmalogen and liver the lowest (see BRAIN COMPOSITION). Other organs are intermediate. The concentration in human plasma is low. Among invertebrates, mollusk (bivalve) tissues have rather high concentrations, and these are highest in the gill. Concentrations are appreciably different in the same organ of different mammalian species. Some representative values in rat and rabbit tissues are as follows. In the rat, the concentrations in brain, heart, lung, skeletal muscle, and liver (in milligrams per gram of fresh tissue) are 11.2, 2.1, 3.1, 1.3 and 0.9, respectively, corresponding to 24, 13, 15, 13 and 4% of the total phosphatide in the organ. In the same tissues of the rabbit, the values are 12.4, 5.7, 2.7, 1.4 and 0.6 mg/g fresh tissue, corresponding to 27, 32, 14, 22 and 2%, respectively, of the total phosphatide content. In larger animals such as man, the proportion of the total phosphatide found as plasmalogen tends to increase. This is particularly true of the heart and other muscular tissues, in which substantial quantities of phosphatidal choline are found in addition to phosphatidal ethanolamine. Appreciable quantities of phosphatidal choline are found in mammalian sperm and in certain specialized organs of invertebrates. For example, in the posterior gills of *Eriocheir sinensis* (Chinese crab), its presence has been considered to have functional significance in relation to the capacity of this tissue to serve as a salt-concentrating organ.

Plasmalogens were discovered in 1924 by Feulgen through an accident of histologic technique. He observed that the cytoplasm of cells would stain purple with fuchsin-sulfurous acid, but that this occurred only after pretreatment with acidic fixatives such as mercuric chloride. It was quickly perceived that the precursor was a lipid, but it was not until 1939 that Feulgen established that these compounds were phosphatides containing glycerylphosphorylethanolamine. To account for the generation of fatty aldehyde and the stability to alkali, he formulated their structure as an acetal of fatty aldehyde linked to the primary and secondary hydroxyl groups of glycerol (I).

This structure did not account either for the chemical reactivity of the molecule in acid medium or the retention of "lipid phosphorus" after mild acidic hydrolysis. In 1954, independent observations were reported by Klenk and by Rapport in which catalytic hydrogenation resulted in the complete loss of the aldehydogenic property. Analysis of the character of this hydrogenation by Rapport showed that it was simply hydrogen addition and involved no hydrogenolysis. These observations were correctly interpreted to indicate the presence of an α,β-unsaturated ether structure (II). Confirming evidence was quickly obtained by means of an independent, stoichiometric reaction involving addition of iodine. Since the unsaturated ether linkage is an activated double bond, it will add iodine under conditions where no such addition is observed with ordinary olefinic unsaturation. The enol ether structure, which is not found in other natural products, appears to account for all of the special properties of plasmalogens including the generation of fatty aldehyde, the loss of this property on hydrogenation with the formation of a stable ether linkage, the catalytic effect of mercuric salts on hydrolysis, the specific addition of iodine, the stability to alkali, and the infrared absorption band at 6.0–6.05 mμ (1670–1650 cm^{-1}). The double bond has been shown to have the *cis* configuration.

Analytical methods for the specific structural feature of plasmalogens (as contrasted with general properties of phosphatides) are based on three properties: the generation of aldehydes, the addition of iodine, and the measurement of "lipid phosphorus" that is stable under mild alkaline conditions but unstable to mild acid treatment. Of the methods for determining aldehydes, formation of the *p*-nitrophenylhydrazone is superior to that based on fuchsin-sulfurous acid. Measurement of aldehydes as their dimethyl

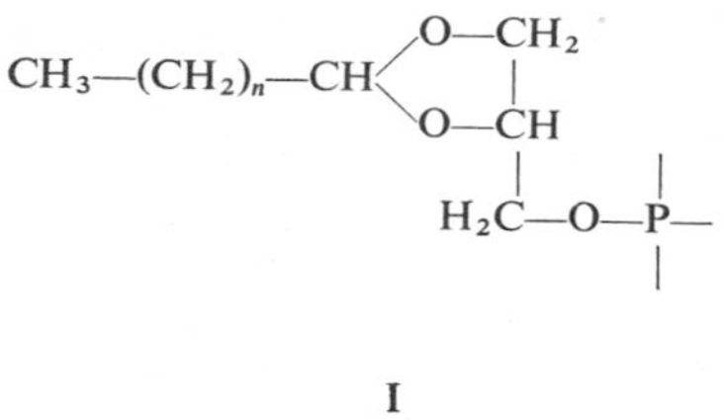

$$CH_3—(CH_2)_n—CH\begin{matrix}\diagup O—CH_2 \\ \quad | \\ \diagdown O—CH \\ \quad | \\ H_2C—O—\underset{|}{\overset{|}{P}}—\end{matrix}$$

I

$$\begin{matrix} CH_3—(CH_2)_n—CH=CH—O—CH_2 \\ | \\ CH_3—(CH_2)_m—\underset{\underset{O}{\|}}{C}—O—CH \\ | \\ H_2C—O—\underset{|}{\overset{|}{P}}— \end{matrix}$$

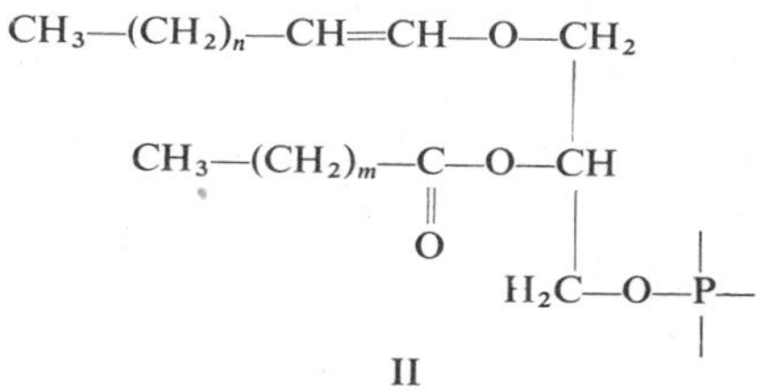

II

acetals by gas-liquid CHROMATOGRAPHY is also feasible. This latter method shows that in phosphatidal ethanolamine and phosphatidal choline from various sources, C_{16} and C_{18} saturated chains predominate, followed by a branched C_{15} saturated chain in phosphatidal choline of ox spleen and in both phosphatidal ethanolamine and phosphatidal choline of ox liver.

Despite their unique chemical reactivity which suggests some parallel with high-energy intermediates of metabolism, no specific function has yet been demonstrated for these compounds. Two observations suggest that the unsaturated ether group, as part of the polar portion of this amphipathic molecule, may be important in relation to interaction with protein, and thus be relevant for the structural stability (and function) of cell membranes. These two observations are: (1) that phosphatidal choline is attacked very much more slowly by rattlesnake venom than phosphatidyl choline, a difference which is relatively insignificant with cobra venom; (2) that lysophosphatidal choline, in contrast to lysophosphatidyl choline, cannot be reacylated enzymatically by liver microsomal enzyme.

Reference

1. RAPPORT, M. M., AND NORTON, W. T., "Chemistry of the Lipids," *Ann. Rev. Biochem.*, **31**, 104–122 (1962).

MAURICE M. RAPPORT

POLYAMINES

Spermidine, spermine and related amines including putrescine, cadaverine and propanediamine are the bases in biological systems which are classified as the polyamines. The chemical structures reveal the multiple amino (—NH_2) and imino (—NH—) groups characterizing these molecules:

$NH_2(CH_2)_3NH_2$	1,3-Propanediamine
$NH_2(CH_2)_4NH_2$	Putrescine
$NH_2(CH_2)_5NH_2$	Cadaverine
$NH_2(CH_2)_3NH(CH_2)_4NH_2$	Spermidine
$NH_2(CH_2)_3NH(CH_2)_4NH(CH_2)_3NH_2$	Spermine

These basic nitrogen groups are responsible for the tendency of the polyamines to react with acidic molecules in biological systems. For example, putrescine and spermidine neutralize about 40% of the acidic phosphate groups of the nucleic acid of certain bacterial viruses—the T-even *Escherichia coli* bacteriophages. Many microorganisms contain putrescine and spermidine, and evidence has been obtained for bacterial nucleic acid–polyamine interactions which appear to be especially significant in Gram-negative bacteria. These organisms contain substantial quantities of the bases as contrasted with the Gram-positive bacteria which contain little or no detectable polyamine.

Spermine forms an insoluble salt with many phosphates, and spermine phosphate crystals, which precipitated from semen, were described by Leeuwenhoek in 1677 in a famous study with microscopic lenses. The name "spermine" was applied in 1888 because of the unusually high concentration of the base in semen. The Rosenheims and Dudley in England, and Wrede and co-workers in Germany established the presence of spermine in various mammalian tissues, determined the correct chemical structure, and confirmed the structure by synthesis during the period 1923–1926. Spermidine was isolated and named by the English workers shortly thereafter. Putrescine and cadaverine were described in numerous investigations on decomposing animal material, and the names of these compounds were derived from their characteristic association with materials undergoing bacterial decomposition.

The most effective methods of separating and identifying the polyamines in tissues involve (a) extraction into trichloroacetic acid, (b) partial purification by extraction into *n*-butyl alcohol after the addition of a Na_2SO_4-Na_3PO_4 salt mixture, (c) removal of the butanol by evaporation and separation of the polyamines in the concentrated extract by ion-exchange chromatography, paper chromatography or paper electrophoresis. Quantitative analysis is accomplished by spectrophotometric analysis of yellow dinitrophenyl-polyamine derivatives in column eluates or purple ninhydrin-polyamine reaction products in eluates from filter paper.

Spermine and spermidine are present in many mammalian tissues, but there is considerable variability in the concentration in different organs and also a great species variability. For example, human semen contains approximately 15 μmoles of spermine per gram and only traces of spermidine; bull semen, on the other hand, contains only a trace of spermine and no detectable spermidine. Data on the distribution of putrescine, propanediamine and cadaverine in animal tissues are very incomplete since the concentrations are low and analytical results are inaccurate. The presence of polyamines in microorganisms and bacterial viruses has already been cited. Polyamines also occur in plant tissue including seeds and seedlings, orange juice, Chinese cabbage, and mushrooms. The nitrogenous component of the ALKALOID pithecolobine is identical with spermine, and the alkaloids palustrin and lunarine have recently been characterized as spermidine derivatives.

Research on the polyamines was stimulated by the discovery in 1949 that putrescine is an essential growth factor for *Hemophilus parainfluenzae* 7901. Spermidine, spermine and 1,3-propanediamine also show comparable growth factor activity, and several other microorganisms (*Aspergillus nidulans* pu_1, *Neisseria perflava*, *Pasteurella tularensis*, *Lactobacillus casei*) require a polyamine for maximum growth. Very recently (1964) the polyamines were identified as growth factors for a Chinese hamster cell line in culture. Spermine is also an antimicrobial factor and is toxic when added to a variety of mammalian and chick cell lines in culture. These toxic effects are due to aldehyde products of the enzymatic oxidation of the polyamine.

Biochemical Functions. Polyamines affect a great variety of biological systems. In many instances, these effects can be attributed to the reaction of polyamine cations with negatively-charged particulates or molecules. Reactions or systems involving nucleic acids have been most actively investigated ,and polyamines are reported to participate in the following: (1) formation and stabilization of large ribosomal particles which are essential in PROTEIN BIOSYNTHESIS; (2) release of pancreatic ribosomal amylase, ribonuclease and chymotrypsinogen; (3) protection of DNA and RNA against thermal denaturation; (4) inhibition of the degradation of RNA and DNA in whole cells or tissue preparations; (5) stimulation of RNA synthesis by RNA polymerase. Spermidine and spermine are most active in these systems.

Spermine also affects the activity of several enzyme systems which do not involve nucleic acids as reactants. For example, dihydrofolate reductase, tetrahydrofolic acid formylase, uridine-diphosphogalactose-4-epimerase, muscle phosphorylase and bovine testicular hyaluronidase are activated by spermine. On the other hand, lysozyme, hexokinase and pyruvic carboxylase are inhibited by spermine, and the polyamine is effective in preventing the dilution inactivation of β-glucuronidase and prostatic acid phosphatase.

Since the reactions are diverse—they involve different reactants and exhibit no common reaction mechanism—it seems unlikely that the polyamine effects in these systems are specific. In several instances, evidence has been obtained for a polyamine enhancement of enzyme-substrate interaction. If a common mechanism of action of these basic molecules is to be derived from the experimental evidence, a role in the modifications of charges and charge effects, either on or between particulates or large molecules and small molecules, seems plausible.

Biosynthesis of Polyamines. The biosynthesis of spermidine in *E. coli* occurs by the following reaction:

$$\text{Putrescine} + \text{S-Adenosylmethionine} \xrightarrow{-CO_2} \text{Spermidine} + \text{Methylthioadenosine}$$

Two enzymes involved in this reaction, which occurs in two stages, have been isolated from *E. coli* extracts and partially purified. The formation of spermidine and spermine from C^{14}-putrescine and C^{14}-methionine has been demonstrated in developing chick embryos, and tritiated putrescine, administered intraveneously to rats, is incorporated into liver spermidine. The enzymes and the reaction mechanisms have not been clarified in these animal systems. The diamines, putrescine and cadaverine, are formed by the action of ornithine decarboxylase and lysine decarboxylase on their respective substrates. These reactions have been studied in bacteria and plants, but animal systems have not been investigated. 1,3-Propanediamine is formed by the oxidation of spermidine by a polyamine oxidase present in several bacteria. The enzyme, purified from *Serratia marcescens*, catalyzes the following reaction:

$$\underset{\text{Spermidine}}{NH_2(CH_2)_3NH(CH_2)_4NH_2} + O_2 + H_2O \rightarrow$$

$$[NH_2CH_2CH_2CH_2CHO] + \underset{\text{1,3-Propanediamine}}{NH_2(CH_2)_3NH_2} + H_2O_2$$

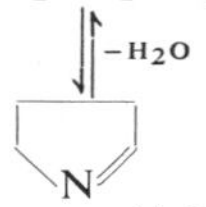

$$\downarrow -H_2O$$

Δ^1-Pyrroline

A polyamine oxidase from beef plasma has also been purified (see also AMINE OXIDASES). The products of polyamine degradation by the enzyme are aldehydes which are very toxic in various bacterial and tissue culture systems. Animal and plant diamine oxidases which oxidize putrescine to Δ^1-pyrroline, ammonia and hydrogen peroxide have been isolated and purified.

References

1. GUGGENHEIM, M., "Die biogenen Amine," Fourth edition, Basel, S. Karger, 1951.
2. HERBST, E. J., AND SNELL, E. E., "Putrescine and Related Compounds as Growth Factors for *Hemophilus parainfluenzae* 7901," *J. Biol. Chem.*, **181**, 47 (1949).
3. TABOR, H., *et al.*, "The Biochemistry of Polyamines: Spermidine and Spermine," *Ann. Rev. Biochem.*, **30**, 579 (1961).
4. TABOR, H., AND TABOR, C. W., "Spermidine, Spermine, and Related Amines," *Pharmacol. Rev.*, **16**, 245 (1964).

EDWARD J. HERBST

POLYNUCLEOTIDES, SYNTHETIC

The DNA and RNA synthesizing enzymes can, in the absence of a nucleic acid template, catalyze the formation of polynucleotides of very simple base sequence. The resulting polymers are of high molecular weight and resemble the natural nucleic acids in many ways (see articles on DEOXYRIBONUCLEIC ACIDS and RIBONUCLEIC ACIDS).

Polydeoxyribonucleotides.[1] The polydeoxyribonucleotides that have so far been synthesized by DNA polymerase have an ordered structure resembling natural DNA and exhibit, at high temperature or extremes of pH, a helix–random coil transition. The polymers formed in unprimed reactions have one of two types of base sequence. In one type of sequence, two bases alternate in each strand (alternating copolymers), and the following polymers have so far been synthesized with this sequence, $\overline{dAT}$:$\overline{dAT}$, $\overline{dAU}$:$\overline{dAU}$, $\overline{dABU}$:$\overline{dABU}$ and $\overline{dAIU}$:$\overline{dAIU}$. The helical structure of these polymers appears to be stabilized, at least in part, by hydrogen bonding between complementary bases in each strand in the same way as with natural DNA. The base sequence in the other class of polydeoxyribonucleotide is such that each strand is homopolymeric, the bases in one strand being complementary (in terms of their hydrogen bonding ability) to the bases in the other strand. The homopolymer pairs that have been synthesized by DNA polymerase are: dG:dC, dG:d$\overline{BC}$, dI:dC, dI:d$\overline{BC}$.

Preparation of Synthetic polydeoxyribonucleotides.[1] The synthetic DNA polymers are prepared by incubating DNA polymerase with the two required deoxyribonucleoside triphosphates in the presence of Mg^{++}. The synthesis may be followed conveniently by the decrease in optical density (at 260 mμ) that accompanies the conversion of the deoxyribonucleoside triphosphates into helical polymer. The increase in viscosity or the production of acid-insoluble products are alternative methods of determining the extent and the rate of synthesis. The reaction may be terminated by heating (to inactivate the enzyme) or by increasing the ionic strength. Unlike a primed synthesis, the unprimed reaction exhibits a lag period (which can be many hours) before DNA synthesis can be detected by the methods outlined above. The polymers resulting from an unprimed synthesis may be used as templates, in a primed synthesis, to produce polymers with a similar base sequence.

The mechanism whereby the alternating copolymer and homopolymer base pair sequences are first generated in an unprimed synthesis is not understood at this time.

The homopolymer pair, dA:dT, can be synthesized by DNA polymerase if the reaction is primed by an rA:rU template.[2] Ribonucleoside triphosphates can be incorporated during enzymatic synthesis of DNA if Mn^{++} is present.[3] A DNA very similar to the dAT:dAT copolymer has been isolated from *Cancer corealis* (crab) testis.

Properties of Polydeoxyribonucleotides. (a) Alternating Copolymers.[1] dAT:dAT and dA$\overline{BU}$:dA$\overline{BU}$ have been studied extensively, and both polymers exhibit a helix–random coil transition when solutions are heated or adjusted to extremes of pH. The X-ray pattern obtained from the lithium salt of dAT:dAT indicates that the polymer has the same helical dimensions as natural DNA. The copolymers have high viscosities and high sedimentation coefficients. Both polymers, by virtue of their alternating base sequence, can form intra-stranded helical structures, but an inter-stranded hybrid, dAT:dA$\overline{BU}$, has also been formed both enzymatically and physically.

Electron micrographs of dA$\overline{IU}$:dA$\overline{IU}$ indicate that although short lengths of this material are indistinguishable from natural DNA, the synthetic polymer, unlike natural DNA, has a branched appearance.

(b) Homopolymer Pairs.[1,4] Studies on the homopolymer pairs indicate that these polymers also exist as ordered or helical structures. The dG:dC polymer is more stable toward heat than is dAT:dAT, and this is similar to natural DNA's which exhibit a helical stability which is dependent on the base composition. When solutions of the homopolymer pairs are adjusted to high pH, their helical structure is destroyed, the two homopolymeric strands dissociate and can be physically separated by sedimentation in a density gradient. In this way, the homopolymers dI, dC, d$\overline{BC}$ and dG have been prepared. Studies have been made on dI, dC and d$\overline{BC}$, and in each case conditions can be found where ordered structures exist. The individual homopolymers dI and dC or dI and d$\overline{BC}$ can be mixed together to reform the homopolymer pairs dI:dC or dI:d$\overline{BC}$, and further, under certain conditions, more complex structures are favored which are presumed to be three-stranded complexes of the type dI:dC:dI and dI:d$\overline{BC}$:dI.

Polyribonucleotides.[5] A number of synthetic polyribonucleotides have been synthesized as a result of the discovery and purification of the enzymes polynucleotide phosphorylase and *E. coli* RNA polymerase.

Polynucleotide phosphorylase has been used to synthesize the homopolymers rA, rU, rI, rC, rT and rG and a variety of random copolymers (*e.g.*, rA,U and rA,G,U,C). The synthesis requires, in addition to the enzyme, the nucleoside diphosphate and Mg^{++}.

Recently the DNA dependent *E. coli* RNA polymerase has been used to catalyze the synthesis of polyribonucleotides. If dAT:dAT is used as primer, then the alternating copolymer rAU:rAU can be formed.[6] Similarly a polymer with a repeating triplet sequence has been synthesized (rAAG) from a nonanucleotide template $d(TTC)_3$.[7] The requirements for the RNA polymerase reaction are ribonucleoside triphosphates, Mg^{++} and Mn^{++}.

Properties of Polyribonucleotides.[7] The polyribonucleotides have been studied by a variety methods; it is known that the homopolymers rA, rI, rC and rG can form ordered structures with themselves, which are stable above room temperature. X-ray diffraction patterns indicate that ordered rA and rI are helical. The ordered structures arise from base pairing via hydrogen bonds, and the structures can be converted to the random coil form by heat in the same way as with the DNA-like polymers.

Interaction of one homopolymer with another has been investigated, and structures such as the homopolymer pairs, rA:rU, rG:rC, rI:rA and rI:rC, and the more complex interactions rI:rA:rI and rA:rU:rU and rG:rC:rG, have been demonstrated.

The synthetic polyribonucleotides have proved to be very useful in elucidating how the four bases in RNA can specify a particular sequence of 20 different amino acids in PROTEIN BIOSYNTHESIS. For example, it has been found that rU can promote the synthesis of polyphenylalanine in a system containing *E. coli* supernatant and ribosomes.[8]

Abbreviations. In this article the following abbreviations have been used:

(1) The prefix r or d indicates that the polymer contains ribo- or deoxyribonucleotides.

(2) The contents of the polymers are given by the following abbreviations: A for adenine, T for thymine, G for guanine, C for cytosine, U for uracil, I for hypoxanthine or inosine, BU for 5-bromouracil and BC for 5-bromocytosine.

When the abbreviation for a base contains more than one letter, a bar above the abbreviation shows which letters belong together.

(3) The order of bases in the chain is given by the sequence of the abbreviations for the bases (thus dAT stands for an alternating sequence of A and T). If the sequence in a chain is not specified, the symbols have commas between them (thus rA,G,U,C).

(4) A colon shows that two chains are hydrogen bonded together (thus dG : dC and d$\overline{\text{ABU}}$: d$\overline{\text{ABU}}$).

References

1. An extensive list of references to the work of Kornberg and co-workers relating to the preparation and properties of the synthetic polydeoxyribonucleotides can be found in Kornberg, A., "Enzymatic Synthesis of DNA," C.I.B.A. Lectures in Microbiology and Biochemistry, New York, John Wiley & Sons, 1961.
2. Lee-Huang, S., and Cavalieri, L. F., *Proc. Natl. Acad. Sci. U.S.*, **50**, 1116 (1963).
3. Berg, P., Fancher, H., and Chamberlain, M., in "Informational Macromolecules" (Vogel, H. J., Bryson, V., and Lampen, J. O., Editors), p. 467, New York, Academic Press, 1963.
4. Inman, R. B., and Baldwin, R. L., *J. Mol. Biol.*, **8**, 452 (1964).
5. An extensive list of references to the research on polyribonucleotides can be found in Steiner, R. F., and Beers, R. F., "Polynucleotides," New York, Elsevier Publishing Co., 1961.
6. Chamberlain, M., Baldwin, R. L., and Berg, P., *J. Mol. Biol.*, **7**, 334 (1963).
7. Nishimura, S., Jacob, T. M., and Khorana, H.G., *Proc. Natl. Acad. Sci. U.S.*, **52**, 1494 (1964).
8. Nirenberg, M. W., and Matthaei, J. H., Fifth International Congress of Biochemistry, Moscow (August 1961).

Ross B. Inman

POLYPEPTIDES, SYNTHETIC

The term, polypeptide, is usually given to any compound having the repeating unit,

$$[-\underset{\text{H}}{\underset{|}{\text{N}}}-\underset{\text{R}}{\underset{|}{\text{CH}}}-\overset{\text{O}}{\overset{\|}{\text{C}}}-]_n,$$

formally obtained by the dehydration of a series of α-amino acids.

$$\text{H}-\underset{\text{H}}{\underset{|}{\text{N}}}-\text{CHR}-\overset{\text{O}}{\overset{\|}{\text{C}}}-\text{OH}+\text{H}-\underset{\text{H}}{\underset{|}{\text{N}}}-\text{CHR}'-\overset{\text{O}}{\overset{\|}{\text{C}}}-\text{OH}\rightarrow$$

$$\underset{\text{H}}{\underset{|}{\text{HN}}}-\text{CHR}-\overset{\text{O}}{\overset{\|}{\text{C}}}-\underset{\text{H}}{\underset{|}{\text{N}}}-\text{CHR}'-\overset{\text{O}}{\overset{\|}{\text{C}}}\text{OH}+\text{H}_2\text{O} \qquad (1)$$

Within this broad classification are the naturally occurring biologically important PROTEINS and ENZYMES. Here we shall consider synthetic polypeptides, prepared by the chemist, which may or may not be identical to natural polypeptides. The synthesis, structure and applications of synthetic polypeptides in biochemistry will be outlined briefly within this entry.

The stepwise synthesis of a polypeptide may be illustrated by the synthesis of a specific dipeptide having a single peptide bond. L-Alanyl-alanine has been chosen for simplicity. The use of blocking groups, fundamental to successful stepwise peptide synthesis, will be illustrated by the carbobenzyloxy group introduced by M. Bergmann:

$$\text{C}_6\text{H}_5\text{CH}_2\text{O}\overset{\text{O}}{\overset{\|}{\text{C}}}\text{Cl}+\text{NH}_2-\overset{\text{CH}_3}{\overset{|}{\text{C}}}\text{HCOOH}\rightarrow\text{C}_6\text{H}_5\text{CH}_2\text{O}\overset{\text{O}}{\overset{\|}{\text{C}}}\text{NH}\overset{\text{CH}_3}{\overset{|}{\text{C}}}\text{HCOOH}+\text{HCl} \qquad (2)$$

$$\text{NH}_2\overset{\text{CH}_3}{\overset{|}{\text{C}}}\text{HCOOH}+\text{CH}_3\text{OH}\rightarrow\text{NH}_2\overset{\text{CH}_3}{\overset{|}{\text{C}}}\text{HCOOCH}_3+\text{H}_2\text{O} \qquad (3)$$

$$\text{C}_6\text{H}_5\text{CH}_2\text{O}\overset{\text{O}}{\overset{\|}{\text{C}}}\text{NH}-\overset{\text{CH}_3}{\overset{|}{\text{C}}}\text{HCOOH}+\text{NH}_2\overset{\text{CH}_3}{\overset{|}{\text{C}}}\text{HCOOCH}_3\rightarrow\text{C}_6\text{H}_5\text{CH}_2\text{O}\overset{\text{O}}{\overset{\|}{\text{C}}}\text{NH}\overset{\text{CH}_3}{\overset{|}{\text{C}}}\text{H}\overset{\text{O}}{\overset{\|}{\text{C}}}-\text{NH}\overset{\text{CH}_3}{\overset{|}{\text{C}}}\text{HCOOCH}_3+\text{H}_2\text{O} \qquad (4)$$

$$\text{C}_6\text{H}_5\text{CH}_2\text{O}\overset{\text{O}}{\overset{\|}{\text{C}}}\text{NH}\overset{\text{CH}_3}{\overset{|}{\text{C}}}\text{H}\overset{\text{O}}{\overset{\|}{\text{C}}}\text{NH}\overset{\text{CH}_3}{\overset{|}{\text{C}}}\text{HCOOCH}_3\xrightarrow[\text{steps}]{\text{two}}\text{NH}_2\overset{\text{CH}_3}{\overset{|}{\text{C}}}\text{H}-\overset{\text{O}}{\overset{\|}{\text{C}}}\text{NH}-\overset{\text{CH}_3}{\overset{|}{\text{C}}}\text{H}-\text{COOH} \qquad (5)$$

Alanylalanine—a dipeptide

In Eq. (*5*), the ester group is removed by mild hydrolysis and the carbobenzyloxy group,

$$C_6H_5CH_2O\text{—}\overset{\overset{\displaystyle O}{\|}}{C}\text{—}$$

by hydrogenolysis. While Eq. (*4*) depicts the dehydration necessary for the formation of the peptide link, this reaction does not proceed spontaneously and has been the subject of continuing research. Great effort has been expended to find methods of forming the peptide bond without causing racemization of the optically active center in the amino-blocked alanine derivative in Eq. (*4*). The problem has been reviewed[1a–d]. The conversion of the acid component to an acid azide, $\overset{\overset{\displaystyle O}{\|}}{C}\text{—}N_3$, the *p*-nitrophenyl ester,

$$\text{—}\overset{\overset{\displaystyle O}{\|}}{C}\text{—O—}C_6H_4\text{—}NO_2$$

and the use of a diimide are among the more favored methods of forming the peptide bond. A further advance in the stepwise synthesis of polypeptides was described recently by R. B. Merrifield who employed a modified resin as the blocking group and was able to synthesize bradykinin, a nonapeptide, without the isolation of the various peptide intermediates, in 32% yield. Commercial synthesis of natural polypeptides of importance may soon become a reality.

Polypeptides can also be obtained by less laborious techniques, *i.e.*, the polymerization of an activated unit to polymer with a molecular weight distribution. The most succesful technique utilizes the N-carboxyanhydride (NCA) of an α-amino acid, a method first reported in 1906 by Leuchs. The anhydrides are best made by the phosgenation procedure:

$$NH_2\text{—CHR—COOH} \xrightarrow[\text{Inert solvent}]{COCl_2} \begin{array}{c}\text{NH——CHR}\\ | \qquad\quad |\\ \text{O=C} \qquad \text{C=O}\\ \diagdown\ \ \diagup\\ \text{O}\end{array} + 2HCl \qquad (6)$$

$$n\ \begin{array}{c}\text{NH——CHR}\\ | \qquad\quad |\\ \text{O=C} \qquad \text{C=O}\\ \diagdown\ \ \diagup\\ \text{O}\end{array} + NH_2R' \rightarrow NH_2\text{—CHR—}\overset{\overset{\displaystyle O}{\|}}{C}\text{—[NH—CHR—}\overset{\overset{\displaystyle O}{\|}}{C}\text{—]}_n\text{—NHR}' + nCO_2 \qquad (7)$$

Although Leuchs reported the synthesis of a polymeric material from glycine-NCA, the general applicability of the N-carboxyanhydrides to the formation of polypeptides is more recent. The careful purification and subsequent initiation of the polymerization of the NCA of γ-benzyl-L-glutamate, for example, has led to a polymer with a molecular weight of over one million. The polypeptides of most amino acids have been obtained by this technique. Reviews have appeared in the literature[2a,b].

The synthesis of a polypeptide with a repeating sequence of amino acids has been less succesful. Materials with average molecular weights of about 5000–15,000 have been obtained by the polymerization in solution of a free tripeptide *p*-nitrophenyl ester. The bulk polymerization of the tripeptide alkyl esters at elevated temperatures has been far less effective. Much work is needed to devise syntheses of polypeptides with ordered sequences without entailing the work demanded by the step-by-step syntheses.

Over a period of years, investigations of the shape of proteins have revealed a profound relationship between structure and biological application. Synthetic polypeptides have been used as models for obtaining information about the shape of the natural polypeptides. While several physical measurements may be made, X-ray crystallography, INFRARED SPECTRA and optical rotary dispersion have been most widely used.

X-ray crystallography studies on natural and synthetic polypeptides provide a method of "looking" at the coiling of the polymer. Several different three-dimensional shapes have been uncovered. In the α-helix, described by Pauling, the hydrogen bonding of amide carbonyl to amide hydrogen occurs internally, making a thirteen-membered ring. There are about 3.6 amino acid residues per turn of helix. Extended β-structures have also been described. These sheet-like arrays of polypeptide chains are formed by hydrogen bonding the amide carbonyl to the amide hydrogen in another chain, residue by residue. While the muscle proteins and poly-L-alanine assume the α-helical conformation, certain of the KERATINS exist in the β-form[2b]. Studies on poly-L-proline have revealed its similarity in three-dimensional structure to COLLAGEN[2c]. Other synthetic polypeptides have been examined[3] [see PROTEINS (BINDING FORCES IN SECONDARY AND TERTIARY STRUCTURE)].

Infrared spectra are caused by changes in the vibrational energy of the molecules after absorbing radiation. While different frequencies can be assigned to various groupings, little can be determined about the shape of the polypeptide without resorting to infrared dichroism. In this technique, the polymer is oriented and the direction of orientation is determined. If the polypeptide on orientation assumes an α-structure, *e. g.*, an α-helix, the carbonyl absorption maximum at about 1650 cm^{-1} will exhibit parallel dichroism.

Thus the 1650 cm^{-1} peak will be of greater intensity when light polarized parallel to the axis of orientation is used for scanning the spectrum than when light perpendicularly polarized is used. The synthetic polypeptides, poly-γ-benzyl-L-glutamate and poly-L-methionine show parallel dichroism and exist in the α-form.

Other natural polypeptides show perpendicular dichroism, the carbonyl groupings in the peptide links being perpendicular to the orientation axis. Several synthetic polypeptides exhibit an α-structure, but certain exceptions have been noted, for example, poly-L-cysteine exhibits perpendicular dichroism. While a detailed discussion of the field is beyond this entry, reference 2b contains an extensive account.

Optical rotary dispersion (ORD) data have been used with great success to provide a means for estimating the helical content of optically active polypeptides. W. Moffitt provided a theoretical basis for the technique. Optical activity was assigned to the helix beyond that attributable to the optically active residues. This contribution to the total optical activity of the polymer is due to the dissymmetry of the helix. By analyzing the changes in optical rotation through much of the ultraviolet and visible spectrum, the "per cent helix" of synthetic polypeptides and proteins may be derived. In a study of E. Blout, the screw sense of the helix of poly-γ-benzyl-L-glutamate was shown to be opposite in sense to that of poly-β-benzyl-L-aspartate. The percentage coiling of the synthetic polypeptides is often directly related to the amino acid composition. The examination of the globular proteins by ORD reveals a smaller percentage helix than the muscle proteins. Other phenomena have been uncovered by studying changes in optical rotation. The mutarotation of poly-L-proline has been compared to that of the protein, collagen, and similarities in structure have been noted. The use of ORD in the study of the structure of polypeptides has been adequately reviewed[4a,b].

In conclusion, the biochemical applications of synthetic polypeptides shall be considered. Employing the techniques of peptide synthesis, du Vigneaud reported the synthesis of the natural peptide hormone, oxytocin, in 1953 and shortly thereafter announced the total synthesis of the VASOPRESSINS. Many other naturally occurring polypeptides have since been synthesized. The complete biological activity of the ADRENOCORTICOTROPIC HORMONE, ACTH, for example, has been obtained by K. Hoffman from a synthetic polypeptide having the first 23 amino acids of the natural material.

The potential biological properties of the poly-α-amino acids obtained from N-carboxy anhydrides have also been studied. Many of these synthetic polypeptides are substrates for the PROTEOLYTIC ENZYMES and have been used to investigate the mode of action of the enzymes. Immunological data have been obtained by studies on the antigenicity of water-soluble polypeptides. A review of the studies on the biological properties of poly-α-amino acids has appeared[5].

References

1a. GOODMAN, M., AND KENNER, G. W., *Advan. Protein Chem.*, **12**, 465–638 (1957).
1b. ALBERTSON, N. F., *Org. Reactions*, **12**, 157–355 (1962).
1c. BOISSONNAS, R. A., *Advan. Org. Chem.*,**3**, 159–190 (1963).
1d. SCHRÖDER, E., AND LÜBKE, K., "The Peptides, Vol. I, Methods of Peptide Synthesis," New York, Academic Press, 1965.
2a. KATCHALSKI, E., AND SELA, M., *Advan. Protein Chem.*, **13**, 243–492 (1958).
2b. BAMFORD, C. H., ELIOTT, A., AND HANBY, W. E., "Synthetic Polypeptides," New York, Academic Press, 1956.
2c. RAMACHANDRAN, G. N., "Treatise on Collagen, Vol. I, Chemistry of Collagen," London, Academic Press (in press).
3. RAMACHANDRAN, G. N., "Aspects of Protein Structure," pp. 13–137, New York, Academic Press, 1963.
4a. BLOUT, E. R., in "Optical Rotatory Dispersion—Applications to Organic Chemistry" (DJERASSI, C., EDITOR), pp. 238–272, New York, McGraw-Hill Book Co., 1960.
4b. URNESS, P., AND DOTY, P., *Advan. Protein Chem.*, **16**, 402–544 (1961).
5. SELA, M., AND KATCHALSKI, E., *Advan. Protein Chem.*, 392–477 (1959).

STANLEY M. BLOOM

POLYRIBOSOMES

Polyribosomes are aggregates of RIBOSOMES. These aggregates have been demonstrated by sedimentation studies and by electron microscopy, and are believed by many investigators to be held together by a thread of RIBONUCLEIC ACID, of the particular type termed "messenger-RNA." During the period of attachment of each ribosome to an mRNA strand, a growing polypeptide chain or protein molecule (also attached to the ribosome) lengthens progressively as the ribosome advances along the length of the mRNA strand, each amino acid entering the polypeptide chain in sequence being determined by the specific nucleotide triplet ("codon") adjacent momentarily to the ribosome [see GENES; GENETIC CODE; PROTEINS (BIOSYNTHESIS)].

Reference

1. RICH, A., "Polyribosomes," *Sci. Am.*, **209**, 44–53 (December 1963).

POLYSACCHARIDES. See CARBOHYDRATES (CLASSIFICATION AND STRUCTURAL INTERRELATIONS); also separate articles on certain types of polysaccharides, including CELLULOSE; CHITIN; GLYCOGEN; MUCOPOLYSACCHARIDES; STARCHES.

PORPHYRIAS. See ANEMIAS; METABOLIC DISEASES.

PORPHYRINS

The porphyrins are intensely colored macrocylic tetrapyrrole compounds widely distributed in animals, plants, and microorganisms. Iron porphyrins known as hemes (Fe^{2+}) and hemins (Fe^{3+}) occur in heme proteins. Magnesium porphyrins are present as intermediates in the pathways of the biosynthesis of CHLOROPHYLLS. The chlorophylls are magnesium complexes of chlorins (dihydroporphyrins). The occurrence of small quantities of copper, zinc and manganese porphyrins in nature has been reported, but as biologically significant functions have not been found for these compounds, it has been suggested that such compounds may simply arise spontaneously (non-enzymatically) from the combination of the porphyrins with available metal ions. Metal-free porphyrins have no known biological function other than as biosynthetic intermediates and, under normal conditions, only occur in small amounts in nature. Somewhat larger amounts are found in animals with porphyria diseases, heavy metal poisoning or certain ANEMIAS. Porphyrin accumulation has also been noted in certain microorganisms.

Protoporphyrin IX, the porphyrin found in several heme proteins, including HEMOGLOBINS, MYOGLOBINS, b-type CYTOCHROMES, catalases and PEROXIDASES, was the first natural porphyrin to have its structure fully elucidated. Crystalline protoporphyrin IX iron (III) chloride ($C_{34}H_{32}N_4FeCl$) or hemin, as it is usually called, was found many years ago to be readily obtained when blood was treated with hot acetic acid and sodium chloride. After many years of research chiefly by Nencki, Kuster, Willstätter, and finally by Hans Fischer, the structure of hemin was elucidated and confirmed by synthesis in Fischer's laboratory in 1930 (see Fig. 1). Recent studies by INFRARED and NUCLEAR MAGNETIC RESONANCE spectroscopy and by X-ray crystallography fully support Fischer's structure. Porphyrins were thus found to be distinguished by a macrocylic structure where four pyrrole-like rings are linked at alpha positions by carbon atoms. The four carbon bridge positions are known as *meso* positions and are designated as alpha, beta, gamma and delta positions. The eight other peripheral positions available for substitution are called *beta* positions and are designated (Fig. 1) by the numbers 1,2,...,8.

Other heme proteins have porphyrins structually related to protoporphyrin IX. Thus in *c*-type CYTOCHROMES the porphyrin has been found to be identical with protoporphyrin IX except for ethyl groups at the 2- and 4-positions with thioether linkages to cysteine residues of the peptide chain. Porphyrin *a* of cytochrome *c* oxidase (cytochromes *a* and a_3) differs from protoporphyrin IX in having formyl and long-chain alkyl (about 17 carbons) groups, presumably at the 8- and 2-positions, respectively. Chlorocruoro or spirographis porphyrin has a 2-formyl group in place of the 2-vinyl group.

The structural similarities found among the naturally occurring porphyrins are expected in view of the fact that, as far as is now known, the same biosynthetic pathways are used by all forms of life. The pathways used for iron porphyrins and for chlorophylls are the same up to the stage of protoporphyrin IX. The ability to synthesize porphyrins from simple precursors is nearly universal among living forms; only a very few organisms require preformed tetrapyrrole compounds as growth factors.

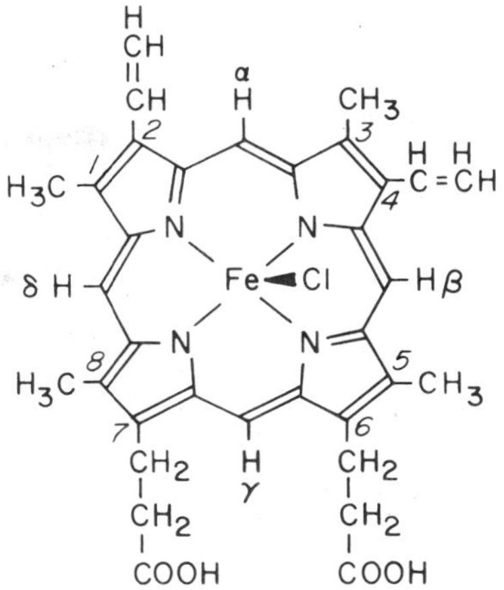

Early studies in the use of isotopically labeled compounds in biological systems, particularly the studies of Shemin and his colleagues at Columbia University, showed acetate and glycine were incorporated into hemin in humans and rats. Glycine was able to provide all four nitrogen atoms, the *meso* carbons, and one carbon of each pyrrole and of each alpha position (next to the ring) for substituents in positions 2,4,6 and 7. These and further studies demonstrated that δ-aminolevulinic acid (ALA) and a monopyrrole, porphobilinogen (PBG), were early heme precursors. The pathway to form protoporphyrin IX appears to be as follows: ALA to PBG to uroporphyrinogen III to coproporphyrinogen III to protoporphyrinogen IX to protoporphyrin IX. Iron is inserted, presumably enzymatically, into protoporphyrin IX to form heme. *Porphyrinogens* are colorless reduced porphyrins where the four pyrrole rings are separated by saturated —CH_2— (methylene) bridges. Thus six reducing equivalents are required to convert porphyrins to porphyrinogens which are readily autoxidized back to porphyrins. Evidently, uroporphyrin III and coproporphyrin III, though observed in small concentrations, are not directly on the biosynthetic pathway. The uro III compounds have acetic acid and propionic acid groups at positions 1,3,5 and 8 and positions 2,4,6 and 7, respectively; the copro III compounds are of similar structure except for methyl groups at positions 1,3,5 and 8. The III designation for the uro and copro compounds and IX designation for the proto compounds serve to indicate the positions of substitutents on the porphyrin ring. Uro I

FIG. 1. Protoporphyrin IX iron (III) chloride; hemin.

and copro I isomers are symmetrically substituted with propionic acid groups at positions 1,3,5 and 7, and although these isomers are apparently not useful as biosynthetic intermediates, they are, nevertheless, occasionally found in presumably abnormal situations.

Synthesis and Reactivity. A large number of differently substituted porphyrins have been synthesized, in large measure as a result of the brilliant investigations carried out during the 1920's and 1930's in Hans Fischer's laboratory in Munich In these syntheses, dipyrryl intermediates (dipyrrylmethenes and dipyrrylmethanes) have been used extensively whereas monopyrroles and tetrapyrroles (bilanes) have been used less frequently. The great majority of condensation reactions which produce the porphyrin ring system proceed in very low yield. Yields of 1–3% are very common, and the desired product is frequently obtained in a yield of a few tenths of a per cent Moreover, such condensations often produce several different porphyrins from which the desired product must be separated. When it is considered that the pyrrole intermediates required for porphyrin synthesis must themselves be prepared by several step syntheses, the difficulty frequently encountered in preparing even small amounts of synthetic porphyrins becomes readily apparent. By far the most widely used intermediates for porphyrin ring synthesis, the alpha-bromo-alpha-methyldipyrrylmethenes and alpha-bromo-alpha(bromomethyl)dipyrrylmethenes, have proved of great importance in establishing through synthesis the structure of naturally occurring porphyrins and their degradation products. Other important intermediates are alpha-unsubstituted dipyrryllmethanes which often give porphyrins when condensed with formic acid. Reported syntheses where porphyrins have been obtained directly from monopyrrole intermediates include the reaction of pyrrole itself with certain aldehydes to give alpha, beta, gamma, delta-tetrasubstituted porphyrins, frequently in excellent yield. The alpha-(hydroxymethyl) pyrroles and alpha- (dialkylaminomethyl) pyrroles have also been convenient intermediates. Syntheses of porphyrins from monopyrryl intermediates which have two different beta-substituent groups suffer from the disadvantage that a mixture of isomeric porphyrins must be anticipated. In certain porphyrin syntheses, both from monopyrryl and dipyrryl intermediates, the presence of metallic ions, such as copper, zinc or magnesium, has been observed to increase the yield of porphyrin, and the metal may be subsequently removed by treatment with acid.

Porphyrins with free ring positions undergo electrophilic substitution reactions typical of an aromatic system. Acylation, halogenation, sulfonation and chloromethylation reactions can be carried out at *beta* positions smoothly and in good yield. Nitration has been shown to take place at *meso* positions. In acid solution, deuterium exchange for hydrogen has been shown to take place both at *meso* and at *beta* positions. Porphyrins undergo stepwise chemical or photochemical reduction to dihydroporphyrins (chlorins result when reductions take place at *beta* positions; phlorins are obtained when reduction takes place at *meso* positions), tetrahydroporphyrins and hexahydroporphyrins. Chlorins are green and phlorins are blue in organic solvents. Tetrahydroporphyrins are purple. Drastic reduction with hydriodic acid cleaves a porphyrin into monopyrroles. Vigorous oxidation of a porphyrin yields substituted maleimides. Both the pyrroles from reduction and the maleimides from oxidation have proved valuable in elucidating the structure of naturally occurring porphyrins.

Coordination Compounds. Coordination compounds of porphyrins with a large number of metal ions have been prepared. With alkali metal ions (Li^+, Na^+, K^+) and with Cu^+ and Ag^+, two metal ions can be bound to one porphyrin molecule to give complexes which exhibit low stability under weakly acidic conditions. Mg^{2+}, Zn^{2+}, Cd^{2+} and transition metal ions form one-to-one complexes. With certain complexes (Pt^{2+}, Pd^{2+}, Ni^{2+} and Cu^{2+}), highly acidic conditions (*e.g.*, concentrated sulfuric acid) may be required to displace the metal ion. With others, such as those with Mg^{2+}, Zn^{2+}, and Cd^{2+}, the metal ion can be readily displaced under mildly acidic conditions. Additional (non-porphyrin) ligands may also bind with the metal ion. The ability of iron porphyrins to bind additional ligands such as oxygen, hydrogen peroxide, carbon monoxide, histidine, etc., and to assume different oxidation and spin states is of particular biochemical importance. Oxygen and carbon monoxide bind to low-spin (diamagnetic) Fe^{2+} complexes; here pi-bonding, with iron acting as the pi-donor, contributes importantly to the bonding. Fe^{2+} complexes are frequently very readily autoxidized. The explanation of those structural features of the hemoglobin and myoglobin molecules which permit these proteins to bind oxygen reversibly without undergoing oxidation remains an intriguing biochemical question.

Crystallographic studies with Fe^{3+} chloride and methoxide porphyrins (and also metmyoglobin) have revealed the iron atom to be out of the plane of the porphyrin ring by nearly 0.5Å in a square pyramidal arrangement where the porphyrin nitrogens constitute the base of the pyramid with the anion (chloride or methoxide) at the apex. With Ni^{2+} complexes, X-ray data have shown an essentially square-planar arrangement. In the low-spin Fe^{2+} complexes, such as the dipyridine complexes (hemochromogens) or with monopyridine-monocarbon monoxide complexes, the iron atom is also expected to be inplane; however in high-spin Fe^{2+} complexes, *e.g.*, deoxyhemoglobin, the iron atom is most likely appreciably out-of-plane.

The aromatic character of the porphyrin ring system as demonstrated by substitution reactions is also supported by the high ring current field effects observed in both nuclear magnetic resonance spectra and diamagnetic susceptibility values. Also consistent with a high degree of electron delocalization are the quite extensive theoretical treatments of the highly characteristic (and most useful) electronic spectra of porphyrin derivatives. The porphyrin system conforms to the Huckel rule requirements for aromaticity with an "inner ring" of 18 pi-electrons and an "outer ring" of 18 or 22 pi-electrons for the diprotonated or dianion species, respectively. That despite the aromatic character of the porphyrin system, appreciable deviations from planarity can readily occur was first shown by the preparation of N-alkylated porphyrins: it is sterically impossible to accommodate an alkyl group at a central nitrogen atom without the over-all out-of-plane distortion clearly evident in NMR spectra of N-ethyl and N-methyl etioporphyrins. X-ray data for *meso* substituted porphyrins (alpha, beta, gamma, delta-tetraphenylporphyrins) have also revealed appreciable deviations from planarity. However, with naturally occurring porphyrins where the substituents are only at the *beta* positions, only minor deviations from planarity by carbon and nitrogen atoms of the porphyrin ring have been observed.

Differences in peripheral substitution can markedly affect the properties of porphyrins. In general, it has been found that the more electron-withdrawing the peripheral substituents, the lower is the basicity of the central nitrogen atoms, the less strongly will pi-acceptor ligands such as oxygen and carbon monoxide be bound to low-spin Fe^{2+} complexes, the more strongly are (sigma) donor ligands such as amines bound to central metal ions, the less extensive is intermolecular association (presumably through pi-pi interactions between porphyrin molecules) in a solvent such as chloroform, the lower is the rate of autoxidation of Fe^{2+} porphyrins, and the more stable is the Fe^{2+} compared with Fe^{3+} oxidation state. Electronic spectra, characterized by very strong absorption near 400 mμ (the Soret band) and by weaker bands at both longer and shorter wavelengths, reflect differences in both the type and the relative positions of substituent groups. Frequently, but by no means always, the more electron-withdrawing the substituent groups, the longer are the wavelengths of the absorption maxima. Different metal ions may result in characteristically different electronic spectra.

The increased insight into the effects of structure and medium on the properties and reactions of porphyrin compounds has stimulated, and promises to continue to stimulate, attempts to interpret structure-function relationships among heme proteins.

References

1. Fischer, H., and Orth, H., "Die Chemie Des Pyrrols," Vol. II (two parts), Leipzig, Akademische Verlagsgesellschafts M.B.H., 1937 and 1940.
2. Lemberg, R., and Legge, J. W., "Haematin Compounds and Bile Pigments," New York, Interscience Publishers, 1949.
3. Falk, J. E., "Porphyrins and Metalloporphyrins," New York, Elsevier Publishing Co., 1964.
4. Lascelles, J., "Tetrapyrrole Biosynthesis and its Regulation," New York," New York, W. A. Benjamin Co., 1964.
5. Caughey, W. S., Alben, J. O., and Beaudreau, C. A., Structure and Medium Effects on the Reaction of Iron (II) Porphyrins with Oxygen and Carbon Monoxide," in "Oxidases and Related Redox Systems" (King, T. E., Mason, H. S., and Morrison, M., Editors), New York, John Wiley & Sons, 1965.
6. Caughey, W. S., Deal, R. M., Weiss, C , and Gouterman, M., "Electronic Spectra of Substituted Metal Deuteroporphyrins," *J. Mol. Spectry.*, **16**, 427 (1965).

Winslow S. Caughey and J. Lyndal York

POTASSIUM AND SODIUM (IN BIOLOGICAL SYSTEMS)

Sodium (Na) is essential to higher animals which regulate the composition of their body fluids and to some marine organisms, but it is dispensible for many bacteria and most plants, except for the blue-green algae. Potassium (K), on the other hand, is essential for all, or nearly all, forms of life. The importance of these cations for all forms of life has been related to the predominance of Na and K in the ocean where primitive forms of life are thought to have originated and developed. During most of the period of evolution of living organisms, there has been little change in the Na and K content of seawater, either as to proportion or total amount. The body fluids of sea animals are, in most instances, similar to seawater in Na and K level and ratio. In freshwater and terrestrial animals, the Na and K level of body fluids is usually somewhat lower, and the ratio is likely to vary from the 40:1 ratio of seawater. Most "fresh" waters contain small and variable amounts of Na and K, usually in a ratio of from about 1:1 to 4:1.

In spite of the higher level of Na in natural water, K is universally the characteristic cation found within both plant and animal cells. While Na is not an absolute requirement for most plants and bacteria, it is found in these organisms and is essential to higher animals where it is the principal cation of the extracellular fluids (see MINERAL REQUIREMENTS). Na and K are important constituents of both intra- and extracellular fluids. It may be said that, generally, the best external and internal medium for function of cells not adjusted to low salt levels is a medium involving a balance of Na and K.

Beyond the osmotic effects depending on the sum of the concentration of the ions in the solution, Ringer found in 1882 that to maintain the contractillity of isolated frog heart, it was necessary to perfuse it with a medium containing Na, K and Ca ions in the proportion of seawater (see NUTRIENT SOLUTIONS). It has since been recognized that the normal life activities of tissues and cells may depend on a proper balance among the inorganic cations to which they are exposed. Sodium is required for the sustained contractility of mammalian muscle while K has a paralyzing effect; a balance is necessary for normal function. Other investigators have found that the antagonism among univalent and divalent cations observed by Ringer is demonstrable with various simpler or more complicated organisms or biological systems.

Excessive salt in soil, such as soils recently soaked with seawater, is toxic to most plants, though there are many plants, *e.g.*, those of the salt marshes and the sea, which are adapted to a high salt concentration.

Ingestion of seawater by man as the only source of water is eventually fatal because of the inability of his body to eliminate salt at a concentration comparable to that of seawater. This results in accumulation of salt, with severe toxic effects and eventually fatal results.

Potassium is a usual but variable constituent of most soils, but available K may be depleted through loss due to leaching by rain water, or through loss due to accumulation in plants grown on the soil. Potassium is thus an important major constituent of the fertilizers usually added to soils for improvement of plant growth.

It is probable that K is absorbed by the plant roots from the soil by an active transport mechanism which carries it through the cell wall structure. Similarly, K and Na if required, are accumulated by animals also by ACTIVE TRANSPORT. The actual cellular content of K and Na is likewise controlled by transport mechanisms which specifically move K in and Na out of the cell against the concentration gradient. The energy for this is derived from the metabolic processes of the cell. The nature of these transport mechanisms has not been fully determined.

Potassium differs from most other essential constituents of plant and animal cells in that it is not built into the cell as a part of an organic compound, but is rather an ion from a soluble inorganic or organic salt. Potassium ions may chelate with cellular constituents such as polyphosphates. The ion is of the correct size to fit into the water lattice adsorbed to the proteins in the cell. In general, the K and Na ions are attracted to protein or other colloidal or structural units having a negative charge. MUCOPOLYSACCHARIDES within the cell, on the cell surfaces and of the intercellular structures, are of particular importance in holding cations such as K and Na. Active centers or other configurational features of the PROTEINS in the cell may be affected or altered by the K held by electrostatic or covalent binding. There are several enzyme systems which are activated by K.

An appreciable proportion of the Na in animals is not readily exchangeable with the Na in body fluids, as it is held in body tissues, particularly in bone and connective tissues. In general, most of Na and K in the animal is in a dynamic state, exchanging between different parts of the cell, between the cell and the extracellular fluid, and intermixing with ingested Na and K in body fluids.

Most cellular constituents do not selectively bind K in preference to Na; MYOSIN of muscle fibers, for example, will bind either. On the other hand, the MITOCHONDRIA and RIBOSOMES are organized cellular organelles able to selectively take up or extrude K; this accounts for only a part of the K held in the cell.

In blue-green algae and some yeasts, Na may in part replace cellular K. While K is usually the principal cation concerned with maintenance of the OSMOTIC PRESSURE within the cell, Na contributes appreciably to the total, and amino acids and other organic compounds may help make up any deficit, particularly in marine invertebrates.

The Na content of the body extracellular fluids of marine invertebrates from the coelenterate through the arthropod phyla is approximately that of seawater (89–113% of the 440 meq/liter of seawater). In freshwater and terrestrial invertebrates, the Na of body fluids varies over a wide range and there is considerable variation among vertebrates. The Na content of the body fluid of Mammals is closely regulated at about 145 meq/liter (see ELECTROLYTE AND WATER REGULATION). There are both fish and crustaceans which are so highly adaptable that they are able to live in either fresh or salt water.

The regulation of the osmotic pressure within the cell and the control of passage of water into or out of the cell is dependent to a considerable extent on the control of the K and Na in the cell by the transport systems of the cell wall. The cell wall itself is of protein-lipid composition and is in general relatively impermeable to passage of water and inorganic salts. Recent studies of the cell walls with electron microscopes and with the use of other investigative techniques indicate that the cell wall contains pores connecting the cell contents with the extracellular fluid, or in some plants, with other cells. In cells having an endoplasmic reticulum, the intracellular vacuolar system may have openings through the cell wall communicating with the extracellular fluid. The ease with which water passes in or out of the cell in response to changes in external or internal osmotic pressure varies over an extreme range, from easy passage to rigid control, depending on the cell and its functions.

Phagocytosis and pinocytosis may bring salts and water, as well other substances, into the cell.

In some unicellular organisms, osmotic equilibrium may be maintained by a contractile vacuole which collects water; in other organisms, water may be excreted through the cell wall. The kidney and sweat glands of higher animals, gills of fish and salt glands of birds serve to excrete salt. Most animals, through control of Na and K excretion and loss, are able to adapt to a wide range of intake.

The importance of NaCl in nutrition has been recognized from the beginning of history. Agricultural populations who lived on cereal grains, nuts, berries and other vegetable foods poor in Na, experienced a hunger for salt which led them to go to great lengths to obtain this mineral. This was particularly true if they lived in a hot climate with the attendant increased loss of salt in perspiration. Similarly, herbivorous animals will travel long distances to supply their need for additional salt. On the other hand, peoples or animals subsisting on meat, milk and other foods receive quite appreciable amounts of Na salts in the diet, and experience no special desire or hunger for salt.

In plants the meristematic tissues in general are particularly rich in K, as are other metabolically active regions such as buds, young leaves and root tips. Potassium deficiency may produce both gross and microscopic changes in the structure of plants. Effects of deficiency reported include leaf damage, high or low water content of leaves, decreased photosynthesis, disturbed carbohydrate metabolism, low protein content and other abnormalities. The necessity for supplying K in fertilizers to compensate for soil deficiency to insure adequate plant growth and function was previously mentioned.

Since K is found abundantly present in most natural foods consumed by animals, deficiency is ordinarily no problem. With prolonged maintenance through parenteral (intravenous) feeding when normal oral feeding is not possible, K must be supplied: 2–3 meq of K per 100 calories metabolized or 40–60 meq of K (3.0–4.5 grams KCl) for an individual using 2000 calories is considered adequate.

Experimental K deficiency in rats results in stunted growth, loss of chloride with hypochloremic acidosis, loss of K and increase of Na in muscle. In man, disease of the gastrointestinal tract, involving loss of secretions through vomiting or diarrhea, may result in serious loss of both Na and K. Trauma, surgery, anoxia, ischemia, shock and any damage to or wasting away of tissues may result in loss of cellular K to the extracellular fluid and plasma, and the loss from the body through kidney excretion. Recovery with rapid uptake of K by the tissues may result in low plasma levels. Low extracellular K concentration may cause muscular weakness, changes in cardiac and kidney function, lethargy and even coma in severe cases. In man, the cells normally contain 3500 meq, or more, of K while less than 100 meq can be found in the extracellular fluids.

Growth of higher animals on low Na diets is limited although no special and specific functions can be attributed to Na. It is obvious of primary importance to animals, since depletion results in serious symptoms and even death. As NaCl is sought by, or supplied to animals, and used as a condiment by man, deficiency is seldom encountered except where loss is excessive. Serious deficiency of Na may result when losses exceed dietary intake because of sweating or loss of body fluid, excessive excretion due to kidney abnormalities, or lack of adrenal cortical hormones. The resulting Na deficiency produces symptoms of thirst, anorexia and nausea. There is a lassitude and loss of energy, muscle cramps, and there are mental disturbances. Replacement of lost salt and water corrects the abnormalities. (See articles on NUTRITIONAL REQUIREMENTS).

A dietary intake of 2.5–3.5 grams of salt per day for an adult is generally considered adequate unless loss or excretion is high. There are no reserve stores of either Na or K in the animal body, so any loss beyond the amount of intake comes from the functional supply of cells and tissues.

The kidney is the key regulator of the Na and K content of higher animals and makes possible adaptation to wide variations of intake (see RENAL TUBULAR FUNCTIONS.)

In the glomerulus of the kidney nephron (or individual unit), an ultrafiltrate containing the smaller molecules of plasma is normally produced. As this ultrafiltrate passes down the kidney tubule, 97.5% or more of the Na is actively resorbed along with nearly all of the K. The remaining 2.5% of the Na is sufficient to account for even the maximum sodium excretion. Potassium is added to the filtrate in the distal tubule through exchange for Na. Control of this exchange appears to be the principal mode of action of aldosterone, which thus exerts a final control over Na excretion. Aldosterone is a steroid hormone from the adrenal cortex (see ADRENAL CORTICAL HORMONES), secretion of which seems to result from lowering of the Na/K ratio in the blood. Water is passively resorbed with the electrolytes along the length of the tubule.

Water excretion is further controlled by the antidiuretic hormone from the posterior pituitary gland which acts to increase water resorption in the kidney through making the collecting tubule permeable to water for additional resorption beyond what took place in the tubule. The posterior pituitary gland secretes the hormone as a rapid and sensitive response to a rise in the osmotic pressure of the extracellular fluid. The osmotic pressure of the extracellular fluid is, of course, principally due to its NaCl content.

With low intake of Na, excretion is reduced to a very low level to conserve the supply in the body. Potassium is not so efficiently conserved.

The kidney regulates the acid-base balance of the body by control over resorption of Na ions which may exchange for hydrogen ions in the kidney tubule. Since most dietaries are of acid-ash, the urine is usually more acid than the original plasma filtrate and much of the phosphate excreted is thus changed to the acid monosodium salt, Within the range of normal variability, with an alkaline ash diet, the urine may become alkaline, and in extreme instances, some sodium bicarbonate may be excreted.

The salts of the buffer pairs responsible for control of the pH of plasma and extracellular fluid involve Na as the principal cation, while the cellular buffers involve K salts (see ACID-BASE REGULATION).

References

1. Comar, C. L., and Bronner, F., "Mineral Metabolism," Vols. IA, IB and IIB, New York, Academic Press, 1960–1962.
2. Florkin, M., and Mason, H. S., "Comparative Biochemistry," Vol. II, pp. 403–518; Vol. IV, pp. 677–720, New York, Academic Press, 1960.

3. JOLIFFE, N. "Clinical Nutrition," pp. 398–474, New York, Harper and Brothers, 1962.
4. PITTS, R. F., "Physiology of the Kidney and Body Fluids," Chicago, Year Book Medical Publishers, 1963.
5. STEWARD, F. C., "Plant Physiology," Vol. III, "Inorganic Nutrition of Plants," New York, Academic Press, 1963.

W. KNOWLTON HALL

PRECIPITIN REACTION. See IMMUNOCHEMISTRY, section on Assay and Properties of Antibody.

PROGESTERONE. See HORMONES, ANIMAL; HORMONES, STEROID; OVARIAN HORMONES; SEX HORMONES.

PROLACTIN. See GONADOTROPIC HORMONES; HORMONES, ANIMAL. Prolactin is also termed lactogenic hormone, or luteotropic hormone (LTH).

PROLINE

L-Proline, α-pyrrolidinecarboxylic acid,

```
  H2C—CH2
   /     \
H2C       HC—COOH
   \     /
     N
     H
```

is very widespread in PROTEINS including protamines. COLLAGEN, and gelatin, its degradation product, yield about 15%. These proteins also have a high content of hydroxyproline. L-Proline is nonessential nutritionally for mammals and fowls. In $0.5N$ HCl solution, its specific rotatory power $[\alpha]_D$ is —52.6° [see also AMINO ACIDS (OCCURRENCE)].

PROTAMINES

Protamines are strongly basic PROTEINS (more strongly basic than HISTONES). Like histones, protamines occur in association with nucleic acids, as for example salmine, a protamine that occurs in sperm cells of the salmon. The basic properties of protamines depend upon their high content of basic AMINO ACIDS such as arginine. (See also NUCLEOPROTEINS).

PROTEIN BIOSYNTHESIS. See PROTEINS (BIOSYNTHESIS).

PROTEINS

Proteins are large molecules found universally in the cells of living organisms, or in such biological fluids as blood plasma. They invariably contain carbon, hydrogen, oxygen and nitrogen, almost invariably contain sulfur, and sometimes contain phosphorus. They are specifically characterized by yielding a mixture of alpha-AMINO ACIDS when hydrolyzed by means of acids, alkalies or certain enzymes (see PROTEOLYTIC ENZYMES).

Proteins are exceedingly diverse in properties and functions. Some are relatively inert fibres, such as the KERATINS of wool, hair or horn, or the COLLAGENS of tendon and connective tissue, which play an important structural role in animal organisms. Others are readily soluble in water or in dilute salt solutions, such as ovalbumin or egg white, serum albumin of blood plasma, or HEMOGLOBIN of red blood cells, and their molecules are not very far from spherical in shape. These are often called globular or corpuscular proteins, in contradistinction to the fibrous proteins. Many of them can be obtained from water in crystalline form, and X-ray studies have shown that protein crystals are highly ordered systems—true crystals in every respect. All known ENZYMES, the essential catalysts of biological systems, are proteins—many being present in solution in cytoplasm or cellular secretions, others more or less firmly anchored to larger cellular structures. A number of HORMONES such as insulin and several of the hormones of the pituitary gland, are also proteins, as are all the ANTIBODIES which are called forth in immunological reactions. Protein foods are essential to the nutrition of all animals. Proteins are thus of prime importance in the functioning of all living organisms.

Amino Acids Derived from Proteins. The AMINO ACIDS which proteins yield on hydrolysis are all α-amino acids. The simplest is glycine, with the formula $^+H_3N \cdot CH_2 \cdot COO^-$ or $H_2N \cdot CH_2 \cdot COOH$ (Either of the two formulas may be used to represent glycine; the latter is the more usual, but the former represents the structure more accurately.)

All but two of the amino acids derived from proteins may be represented by a generalization of this formula, as $^+H_3N \cdot CHR \cdot COO^-$. There are more than twenty possible groups which can occupy the position denoted by R. All the amino acids found in proteins, except glycine, are optically active, because of the ASYMMETRY of the α-carbon atom which is surrounded by H, R, $—COO^-$ and $—NH_3^+$ groups. The relative configuration of all these groups is the same for all amino acids derived from proteins. These amino acids are all known as L-amino acids, their enantiomorphs as D-amino acids. A number of D-amino acids have been found in nature; they are present in certain ANTIBIOTICS and in various other compounds. So far as is known, however, they are not found in proteins (see also AMINO ACIDS, NON-PROTEIN).

Many methods have been used in the past for determining the content of amino acid residues in proteins. The most widely used techniques today involve acid hydrolysis, followed by separation of the amino acids on an ion-exchange resin, or by paper chromatography. In the procedure of W. H. Stein and S. Moore, the protein is generally first hydrolyzed by heating in a strongly acid solution—such as $6N$ HCl at 100°C for 24 hours—and the amino acids in the hydrolysate are then separated by passage of the solution over a suitable ion-exchange resin, and collected in a series of fractions. The separated amino acids are then treated with ninhydrin, which produces colored derivatives which may then be determined colorimetrically. The process has been developed so that the data are registered automatically on a recorder. By this method, a virtually complete amino acid analysis may be carried out on a few milligrams of protein. Tryptophan is largely destroyed on acid hydrolysis, and must be determined separately on an alkaline or enzymatic hydrolysate. Serine and threonine are also partially destroyed, and this destruction must be corrected for. [See PROTEINS (AMINO ACID ANALYSIS AND CONTENT).]

Simple and Conjugated Proteins. Many proteins have been obtained which yield only α-amino acids, and no other substances, on hydrolysis—*e.g.*, the hormone insulin and the enzyme pepsin. These are known as *simple* proteins. Others yield additional compounds beside the amino acids; these are known as *conjugated* proteins. There are numerous classes of conjugated proteins: they include the glycoproteins and mucoproteins, which contain carbohydrate groups; LIPOPROTEINS, which contain fatty acids, cholesterol and phospholipids; heme proteins, such as hemoglobin and several oxidative enzymes, which contain iron-porphyrin (heme) groups; NUCLEOPROTEINS, in which proteins are associated with nucleic acids, and many others. Many of the simpler viruses, such as tobacco mosaic or tomato bushy stunt virus, are nucleoproteins; all viruses are constituted largely of proteins and nucleic acids.

Classification of Proteins. In the past, proteins have commonly been classified according to their solubility in various solvents; for instance, *albumins*, which are readily soluble in pure water; *globulins*, which are insoluble in water but dissolve readily in aqueous salt solutions; *prolamines*, which are soluble in alcohol-water mixtures but insoluble in either pure alcohol or pure water, and various other classes. This terminology is largely arbitrary, although still used in the naming of many proteins (see also PLASMA AND PLASMA PROTEINS).

Proteins which are known to exert a specific activity are often given names which distinguish this activity. This is particularly true of the enzyme proteins, which catalyze specific chemical reactions, and the protein hormones, which arouse specific physiological responses (see HORMONES, ANIMAL).

Proteins as Acids and Bases; Isoelectric Points. Even the simplest of the amino acids, glycine, contains one free amino and one carboxyl group. Thus on addition of acid it acquires a positive charge:

$$^{+}H_3N \cdot CH_2 \cdot COO^- + H^+Cl^- \rightleftharpoons \; ^{+}H_3N \cdot CH_2 \cdot COOH + Cl^- + H_2O$$

On addition of alkali it acquires a negative charge:

$$^{+}H_3N \cdot CH_2 \cdot COO^- + Na^+OH^- \rightleftharpoons H_2N \cdot CH_2 \cdot COO^- + Na^+ + H_2O$$

Therefore, when placed in an electric field, glycine in acid solution migrates as a positively charged ion to the cathode; in alkaline solution, it migrates as a negatively charged ion to the anode. At an intermediate pH value, near neutrality, the positive and negative charges on the molecule just balance one another, and the molecule does not move in either direction when placed in an electric field. The pH at which this occurs is known as the *isoelectric point* of the molecule. For glycine the isoelectric point is near pH 6, not far from neutrality; for glutamic acid, with two carboxyl groups and one amino group, it is more acid, near pH 3; for lysine, with two amino groups and one carboxyl, it lies at an alkaline pH value, near 10.

Proteins behave similarly to amino acids, except that one protein molecule may contain scores or hundreds of free carboxyl groups (from aspartic and glutamic acid residues), amino groups (from lysine residues), imidazole groups (from histidine residues) and guanidine groups (from arginine residues). According to the amino acid composition of the protein, and the pH of the medium, the state of charge on the protein molecule can be varied from a maximum positive value in strongly acid solution to a maximum negative value in strongly alkaline solution. When the solution is placed in an electric field (see ELECTROPHORESIS), the protein migrates as an anion or a cation, depending on the pH of the solution, the speed of migration increasing with the net charge of the protein. Each protein has a characteristic isoelectric point, depending on the relative number of acid and basic side chains in the molecule. Some proteins bind other ions than hydrogen ions—cations like calcium, for instance, or anions such as chloride—so that the net charge on the protein, and hence the isoelectric point, is somewhat dependent on the amount of these other ions present in the medium. Approximate isoelectric points of a few proteins are listed below.

Protein	Source	Isoelectric Point (pH)
Bushy stunt virus	Tomato plant infection	4.1
β-Lactoglobulin	Cow's milk	5.2
Carboxypeptidase	Pancreas	6.0
Hemoglobin	Human red cells	6.7
Growth hormone	Anterior pituitary	6.8
Ribonuclease	Pancreas	9.5
Cytochrome *c*	Horse heart	10.7
Lysozyme	Egg white	10.7

Peptide Chains in Proteins. Decisive evidence indicates that the amino acids in proteins are joined by peptide (—CO·NH—) linkages. These are formed, with elimination of water, by the combination of two amino acid residues. The equilibrium in such reactions generally lies well to the left; free energy must be provided from some coupled reaction in order to synthesize a peptide bond. The study of the biosynthesis of peptide bonds, and of protein molecules, is now being vigorously pursued [see PROTEINS (BIOSYNTHESIS); RIBOSOMES].

Peptide chains may be very short, or they may consist of several hundred amino acid residues linked together. Their existence in proteins is shown by the action of proteolytic enzymes, which readily break proteins down into smaller molecules, and which have been shown by their action on simpler substrates to attack no linkages other than peptide bonds.

The amino acid cystine contains two amino and two carboxyl groups, all of which can enter into peptide linkage. Thus cystine can act as a bridge between two peptide chains:

$$\begin{array}{c}\cdots R_1CH{-}CO{-}HN\cdot CH\cdot CO{-}NH{-}CHR_2\cdots \\ | \\ CH_2 \\ | \\ S \\ | \\ S \\ | \\ CH_2 \\ | \\ \cdots R_3CH\cdot CO{-}HN\cdot CH\cdot CO{-}NH{-}CHR_4\cdots\end{array}$$

Such cross-links appear to be frequent in fibrous proteins of the keratin class, which contain large amounts of cystine residues. They have been definitely shown to exist, and their positions exactly located, in the hormone INSULIN from beef pancreas. In other cases, a single peptide chain may be folded back on itself, and fastened by a disulfide bridge so as to form a loop in the chain. One such loop has been shown to occur in the insulin molecule, and some proteins, such as serum albumin or ribonuclease, contain a considerable number of such folds and loops. The exact position of these cross-links has now been determined in several enzymes, including ribonuclease and lysozyme.

Some proteins, such as α- and β-casein of milk, contain phosphorus in the form of phosphate esters of hydroxyamino acids. These may in some cases be diesters, or pyrophosphate esters, which act as bridges between two peptide chains. Such esters may also serve to form a loop in a single peptide chain as with a disulfide bridge.

Determination of the order in which the amino acid residues are arranged in the peptide chains of a protein is discussed in the article on PROTEINS (END GROUP AND SEQUENCE ANALYSIS). The spatial configuration of peptide chains is discussed in the article PROTEINS (BINDING FORCES IN SECONDARY AND TERTIARY STRUCTURE).

Molecular Weights of Proteins. Proteins are large molecules; even the smaller proteins generally have molecular weights over 10,000, while molecular weights of many millions are not uncommon. Molecular weights may be determined by such methods as osmotic pressure, light scattering, sedimentation equilibrium in the ultracentrifuge, sedimentation velocity and diffusion, and other procedures. Information regarding molecular size and shape is also obtained from study of viscosity, double refraction of flow, dispersion of the dielectric constant, and depolarization of fluorescent light from protein solutions. Some typical values for the molecular weights of a few proteins are given below:

Protein	Molecular Weight
Insulin	5,733
Lysozyme (egg white)	14,307
Ribonuclease A (beef pancreas)	13,383
Chymotrypsinogen	21,000
Ovalbumin	43,000
Hemoglobin	64,450
Serum albumin	66,000
Gamma globulin	150,000
Edestin	310,000
Fibrinogen	340,000
Thyroglobulin	640,000
Myosin	500,000
Hemocyanin (octopus)	2,800,000
Hemocyanin (snail)	8,900,000
Tomato bushy stunt virus	7,600,000
Tobacco mosaic virus	40,000,000

Denaturation. Protein molecules are complex and delicate structures. Most globular proteins readily undergo a change known as denaturation. Denatured proteins are usually far less soluble, near their isoelectric points, than the native proteins from which they were derived. Proteins which crystalline readily in the native state become incapable of crystallization after denaturation, and other changes are observed. Denaturation can be produced by heating the protein solution briefly to temperatures near the boiling point or by exposure to acid and alkaline solutions, to concentrated solutions of urea or guanidine hydrochloride, or to various detergents. Precipitation by reagents such as alcohol or acetone, at room temperature, may produce denaturation; although the same reagents at lower temperatures may be employed to precipitate the protein without denaturation. Denaturation often involves no change in the molecular weight of the protein, although sometimes the native protein may be split into two or more subunits on denaturation. True denaturation does not involve the splitting of peptide bonds; apparently it involves a breaking of many weak internal bonds within the native protein molecule, such as hydrogen bonds; the result is that the highly ordered structure of the native protein molecule is replaced by a much looser and more random structure. Denaturation is probably not a single unique process; different denaturing agents, applied in different ways and for different times, may give a whole series of different denatured products. For some proteins, such as egg albumin, denaturation appears to be an essentially irreversible process; for others, such as serum albumin or hemoglobin, it is possible to restore the protein to a state at least approaching its native configuration by readjusting pH, temperature, and other conditions to those favoring the native state of the protein.

Recent work indicates indeed that many, perhaps most proteins will return completely to the native state, even after exposure to such solutions as 8*M* urea. For instance, C. B. Anfinsen and his associates have broken the disulfide bonds of ribonuclease by reduction in 8*M* urea; this drastically disarranges the original pattern of the peptide chain and all enzymatic activity vanishes. On removing the urea, and reforming the disulfide bonds by oxidation in air in neutral aqueous solution, the enzyme activity returns, and the resulting protein is indistinguishable from native ribonuclease by numerous searching tests. Similar experiments have now been carried out successfully on a number of other proteins, including lysozyme, trypsin, myoglobin, taka-amylase A, and alkaline phosphatase.

References

1. Haurowitz, F., "The Chemistry and Function of Proteins," Second edition, New York, Academic Press, 1963.
2. Neurath, H. (Editor), "The Proteins," Second edition, Vols. I, II, and III, New York, Academic Press, 1963, 1964 and 1965.
3. Perutz, M. F., "Proteins and Nucleic Acids," New York, Elsevier Publishing Co., 1962.
4. *Advances in Protein Chemistry*, New York, Academic Press, published annually.

John T. Edsall

PROTEINS (AMINO ACID ANALYSIS AND CONTENT)

The 24 amino acids recognized as occuring naturally in proteins may be classified according to their structure [see AMINO ACIDS (OCCURRENCE, STRUCTURE, AND SYNTHESIS)]. The usual classification is made according to the number of acidic and basic groups that are present. Thus, the neutral amino acids contain one acidic and one basic group. These include the aliphatic series—glycine, alanine, valine, leucine, isoleucine, serine and threonine; the aromatic series—phenylalanine, tyrosine and tryptophan; the sulfur-containing series, cystine, cysteine and methionine; and the only three containing a secondary amino group—proline, 4-hydroxyproline and 3-hydroxyproline. Another group, the acidic amino acids, have one extra acidic group and include aspartic acid, asparagine, glutamic acid and glutamine. Asparagine and glutamine are the β- and γ-amide derivatives, respectively, of aspartic and glutamic acids and, while actually being neutral, they are usually listed in this group to show their relationship to the corresponding acidic amino acids. The remaining basic amino acids each contain an extra basic group and include histidine, arginine, lysine and hydroxylysine. All of the above amino acids have been shown to occur in proteins although three of these, 4-hydroxyproline, 3-hydroxyproline and hydroxylysine, have been found thus far only in the structural proteins COLLAGEN and gelatin. The last of these amino acids to be recognized as existing in proteins, 3-hydroxyproline, has only recently been added to this listing and is the only new one to be thus identified in the past 35 years. New amino acids of very limited occurrence may yet be discovered, but because of numerous past and present investigations, the likelihood of this happening becomes increasingly smaller.

The rapid and accurate analysis of amino acids has become very important in many areas of biochemical research and has become of prime importance in determining the composition and structure of proteins. Indeed, the quantitative determination of amino acids is related to our understanding of proteins in nearly the same way that our knowledge of carbon, hydrogen and nitrogen is related to our understanding of simpler organic compounds.

To accomplish the task of analyzing a purified protein for its constituent amino acids, one must first liberate these amino acids by breaking the peptide bonds which bind them together. Acid hydrolysis is the method almost universally used. One must then carry out an analysis which will determine what amino acids are present and the concentration of each.

Over the years several methods have been used for the analysis of amino acids. The older colorimetric and gravimetric methods, and even the more recent microbiological methods, with isolated exceptions for a few individual amino acids, have all given way to the newer and much more powerful chromatographic and electrophoretic techniques. These latter are partition CHROMATOGRAPHY on paper, ELECTROPHORESIS on paper, gas chromatography, and ion-exchange chromatography with columns of synthetic resin (see ION EXCHANGE POLYMERS).

Paper chromatography is probably the most widely used method for the analysis of amino acids and owes its popularity in part to the relative simplicity of paper chromatographic procedures and to the relatively inexpensive materials and equipment needed. One-dimensional separations with a variety of solvent systems provide rapid, qualitative estimations of relatively simple amino acid mixtures. Where all of the amino acids in a protein hydrolysate are desired, either several different one-dimensional systems must all be used or a two-dimensional technique must be employed. The high resolution provided by paper chromatography makes it an attractive method for quantitative determination. However, when quantitative recoveries are desired, the method immediately becomes more complex, and more instrumentation and equipment are needed. Attention must be paid to the proper conditioning and storage of the paper that is used, and somewhat more elaborate containers are needed to carry out the chromatographic separations. The amount of total effort involved also increases because the separated amino acids, after visualization on the paper sheet, must be quantified. This is done either by measuring the spots with a densitometer or by cutting out the spots, eluting the amino acids and estimating their concentration in a spectrophotometer following color formation. The sensitivity of the method depends on the individual amino acid and on experimental conditions, particularly the freedom from ammonia contamination, but in general, the lower limit of detection is about 0.01 μmole; for most identifications, 0.1 μmole is usually a convenient amount. Paper chromatographic procedures have been satisfactory for certain applications, but for most other they can hardly compete with ion-exchange chromatography, which now has about the same rapidity and even greater sensitivity and which greatly exceeds the precision of paper chromatographic techniques. It is interesting to note that the first complete amino acid sequence of a protein, INSULIN, was determined almost entirely using paper chromatographic techniques.

In *paper electrophoresis*, separations are carried out with the paper strip held vertically in a large volume of organic solvent or placed horizontally between two cooled metal or glass plates. This technique has been used mainly in the field of peptide separations, but the application of high voltages has recently been shown to be useful for the separation and analysis of amino acids. To provide rapid separations, high voltages and current densities must be employed, and rather elaborate cooling equipment is needed to dissipate the resulting heat efficiently and evenly. In general, this method suffers from the same limitation as all methods wherein a paper support is used, *i.e.*, it is more difficult to obtain reproducible quantitative results than is possible with a column effluent.

Gas chromatography has become an extremely valuable analytical tool in recent years, particularly in the petroleum industry, and although the technique is better suited to more volatile compounds, it has also been applied to amino acids. The amino acids, as a group, have such low vapor pressures that it has been necessary to convert them to chemical derivatives by acetylating the free amino groups and esterifying the free carboxyl groups. No single system so far studied has yielded completely satisfactory results on all the common amino acids, and the quantitative formation of the amino acid derivative has not been demonstrated on a scale comparable in amount to that which is needed for an analysis. The potential advantages lie in its high sensitivity and short analysis time. Present detectors and capillary columns make it possible to detect amino acid derivatives in the 10^{-10} to 10^{-9} mole range. The sensitivity of these detectors varies significantly toward equimolar quantities of the various derivatives, but they probably provide the method as a whole with a five- to ten-fold increase in sensitivity compared to current ion-exchange methods.

The most accurate and precise method of amino acid analysis, in use today is that of chromatography on columns of ion-exchange resin, followed by reaction with ninhydrin as used in the *automatic amino acid analyzer*. This analyzer first described and introduced in 1958, is now produced commercially by two companies in the United States and by others abroad; it has rapidly become an indispensable item in any laboratory studying protein chemistry. The analyzer and associated accessories are not simple, and their high cost is justified only by the performance which they provide. The resin employed for the exchange process is a copolymer of polystyrene and divinylbenzene which has been sulfonated to attach strong acidic groups to the polymerized material. On application of a sample of hydrolyzed protein, the constituent amino acids are attracted to the resin through an ionic interaction between the basic amino group of the amino acid and the acidic group on the resin. Through proper control of the available parameters, the amino acids are bound to the resin with varying degrees of attraction such that when the eluting buffer is pumped down through the column, each amino acid moves at an individual and independent rate, emerging from the bottom of the column in succession and without overlap. With the acidic eluents employed, the basic amino acids are retarded the most strongly and the acidic amino acids are retarded the least. Thus the acidic amino acids are first to come off the columns, followed by the neutrals and finally the basics. Mechanical, positive displacement, piston-type pumps are employed to pump sodium citrate buffers through the columns at relatively high, constant rates and against quite high back pressures. Most analyzers make use of two separate columns for each complete analysis (although a one-column system is also used.) The basic amino acids are analyzed with a small column, and the acidics and neutrals from a longer intermediate column. The use of two columns, while requiring two aliquots of sample, lends considerabe versatility to the method and allows the analysis to be completed in a shorter time than is possible with a single column. The stream emerging from the column is combined with an additional stream of ninhydrin reagent provided by a separate pump, and the mixture then flows through a long coil of tubing immersed in a boiling water bath where color formation takes place. The colored stream then flows through the colorimeter. Here, the extent of color formation is measured and printed on a strip chart recorder as a series of separate peaks, each corresponding to an individual amino acid in the sample mixture. The use of three photometer units in the colorimeter allows three separate channels to be plotted by the recorder. The first gives the absorption at 570 mμ wavelength of the blue color resulting from nearly all amino acids. The yellow color given by proline and the hydroxyprolines is measured by the second photometer at 440 mμ. The final channel shows the absorption at 570 mμ from a cuvette which has one-third the normal depth. The operation of the system as a continuous flow technique, the protection of the effluent stream from atmospheric contaminants, and the constancy of the pumps all combine to result in a constant and highly stable base line, permitting an increase in precision, accuracy and sensitivity over earlier manual methods.

Since the automatic amino acid analyzer was described in 1958, improvements in equipment and technique of usage have drastically shortened the length of time required for a complete amino acid analysis of a protein hydrolyzate and have increased the sensitivity almost tenfold. Procedures currently being used in the author's laboratory and elsewhere allow a complete analysis to be obtained in 4 hours, allowing three complete analyses to be obtained in a normal 9-hour work day. Using newly developed long flow path cuvettes and smaller columns of improved resin preparations, an optimal analysis can be carried out on 0.05–0.1 μmole of each amino acid. The level of detectibility is about 5×10^{-10} mole. A precision of $100 \pm 3\%$ can be obtained for concentration levels ranging from 0.01–2 μmole. For optimal loads and under ideal operating conditions, the precision can be increased to $100 \pm 1\frac{1}{2}\%$. And finally, with one of the new, very recently developed electronic integrators attached to the automatic analyzer, the effort and time required for calculating the results of an analysis are significantly curtailed. This production item, in effect, is a small computer and automatically measures the area under a peak as the peak material passes through the colorimeter. At the conclusion of the peak, the area is printed, presenting a digital figure corresponding to the concentration of the corresponding amino acid. If desired, a dual channel integrator is available which continuously integrates any two of the analyzer channels.

References

1. Spackman, D. H., Stein, W. H., and Moore, S., "Automatic Recording Apparatus for Use in the Chromatography of Amino Acids," *Anal. Chem.*, **30**, 1190 (1958).
2. Light, A., and Smith, E. L., "Amino Acid Analysis of Peptides and Proteins," in "The Proteins" (Neurath, H., Editor), Second edition, Vol. 1, Ch. 1, New York, Academic Press, 1963.
3. Piez, K. A., and Saroff, H. A., "Chromatography of Amino Acids and Peptides," in Chromatography" (Heftmann, E., Editor), Ch. 14, New York, Reinhold Publishing Corp., 1961.
4. Spackman, D. H., "Ion Exchange Chromatography of Protein Hydrolyzates. A. The Automatic Amino Acid Analyzer and Accelerated Procedures," in "Serum Proteins and the Dysproteinemias" (Sundermann, F. W., and Sunderman, F. W., Jr., Editors), Ch. 19, Philadelphia, Lippincott, 1964.

Darrel H. Spackman

PROTEINS (AMINO ACID SEQUENCES). See PROTEINS (END GROUP AND SEQUENCE ANALYSIS).

PROTEINS (BINDING FORCES IN SECONDARY AND TERTIARY STRUCTURES)

Theoretically, a polypeptide chain can assume a large number of conformations, *i.e.*, spatial arrangements of its constituent atoms, due to the possibility of rotation around the single bonds of the amino acid residues (bonds b and c in Fig. 1A). The spatial arrangements of the atoms in a protein are described in terms of structures of various levels. The sequence of constituent amino acids is termed the *primary* structure (Fig. 1A). A regular or semiregular folding of the polypeptide chain which may occur, and which is usually constrained by hydrogen bonds between the N—H and C═O groups of the peptide backbone, is called the *secondary* structure (Fig. 1B). The arrangement of the polypeptide chain in space, as determined by various types of non-covalent interactions between the amino acid side chains, is termed the *tertiary* structure (Fig. 1C). This nomenclature has been proposed by K. U. Linderstrøm-Lang in 1952. The aggregation of protein molecules (as "subunits") into oligomers is called *quaternary* structure (Fig. 1D), following a suggestion of J. D. Bernal.

The primary structure is determined by the chemical bonding. The other three levels of structure are largely a resultant of the non-covalent interactions of the constituent groups of a protein with each other and with the solvent. Recent evidence suggests that under a given set of physical conditions (solvent composition, pH, ionic strength, temperature, presence of other solutes), a given polypeptide chain possesses a unique, most stable conformation (or a small number of them). If this rule holds, then secondary, tertiary and quaternary structures are determined uniquely by the primary structure and by the physical environment of the protein molecule.

Since the terms secondary structure and tertiary structure both refer to the spatial folding of the protein, and since both structures are determined by forces of a similar nature, it is not always possible to consider the two levels of structure separately. A distinction may be artificial at times, and the more general term *protein conformation*, referring to both levels of structure, is preferable.

Hydrogen Bonds. These are interactions between a hydrogen atom attached to an electronegative atom (the *donor*, usually O or N) and a second electronegative atom (the *acceptor*). Their strength arises mainly from the electrostatic interaction between the electron clouds of the acceptor and the relatively exposed hydrogen nucleus, together with some covalent-type interaction. Because of the latter, the formation of stable hydrogen bonds usually requires that the three atoms involved be collinear or nearly so. Hydrogen bonding groups found in proteins are listed in Table 1. Usually, only one hydrogen bond is formed between groups, for example, in a tyrosyl-carboxyl bond,

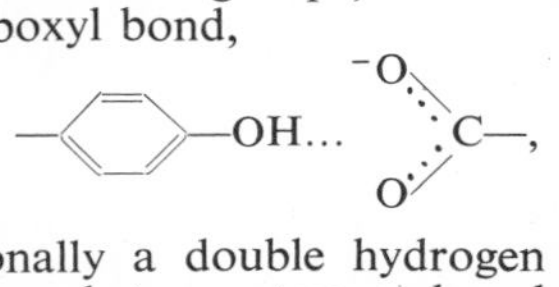

—C₆H₄—OH... $^{-}$O/O═C—,

but occasionally a double hydrogen bond can form, such as between two carboxyl groups:

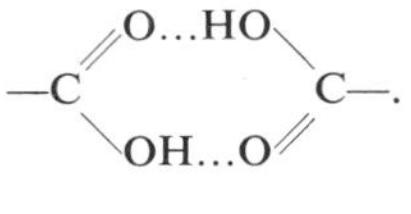

—C(═O...HO)(—OH...O═)C—.

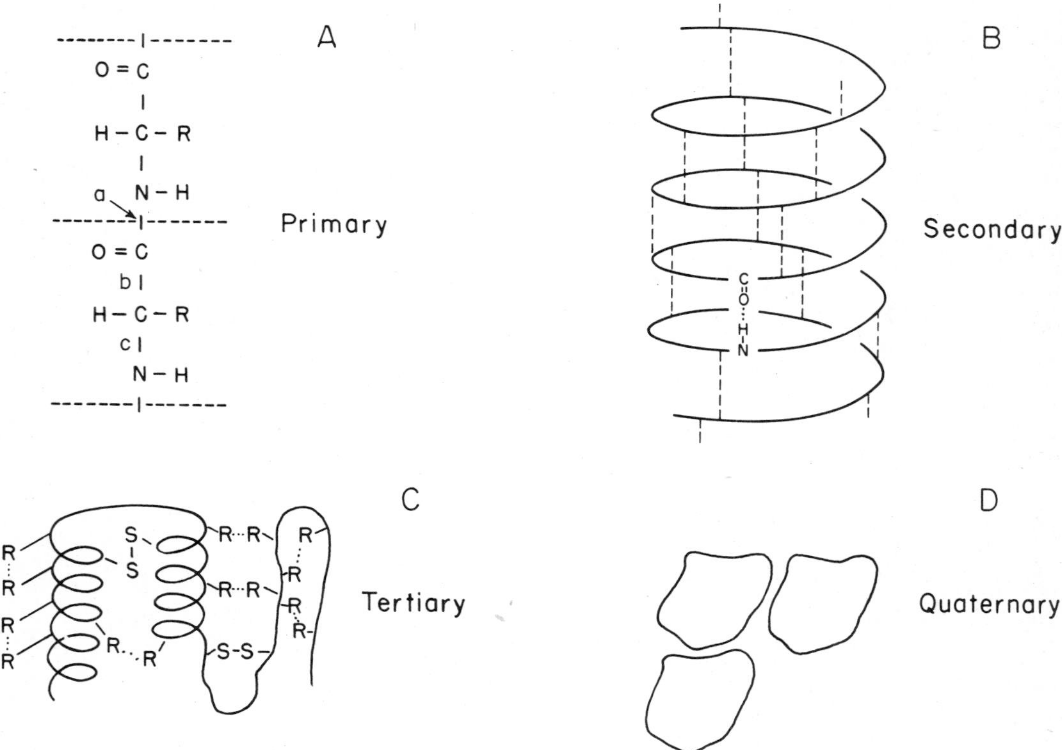

FIG. 1. Schematic representation of the structural levels in proteins. (A) Primary structure. Rotation around bond *a* (the peptide bond) is usually restricted to *cis* and *trans* forms. Less restricted rotation is possible around bonds *b* and *c*. (B) Secondary structure. (C) Tertiary structure. Amino acid side chains are denoted by R, non-covalent interactions by ... . (D) Quaternary structure.

TABLE 1. AMINO ACID RESIDUES PARTICIPATING IN VARIOUS TYPES OF NON-COVALENT BONDS

Hydrogen Bond		Hydrophobic Bond	
Donor Groups	Acceptor Groups	Nonpolar Side Chains	Side Chains Having Nonpolar Parts
Lysyl	Tyrosyl	Alanyl	Arginyl
Seryl	Aspartyl	Valyl	Lysyl
Threonyl	Glutamyl	Leucyl	Tyrosyl
Tyrosyl	Asparagyl	Isoleucyl	Tryptophyl
Aspartyl	Glutaminyl	Phenylalanyl	Glutamyl
Glutamyl		Methionyl	Glutaminyl
Asparagyl		Cysteyl	
Glutaminyl		Prolyl	
Peptide N—H	Peptide C=O		Peptide $C^{\alpha}H$, $C^{\alpha}H_2$ (glycyl)

The secondary structure of a protein is usually determined by the peptide hydrogen bonds, N—H...O═C, between backbone groups. Regular repeating arrangements of peptide hydrogen bonds result in the formation of helical structures if donor and acceptor groups belong to nearby residues of the peptide chain, or the formation of pleated sheets if the hydrogen bonds join adjacent chains. An example of the former is the α-helix (Fig. 2), first postulated by Pauling and Corey in 1951, and observed later in the crystal structures of hemoglobin and MYOGLOBIN. Side chain groups can form hydrogen bonds with each other and with backbone peptide groups. The contribution of hydrogen bonds to the stabilization of structures in water is very small because water molecules can also form hydrogen bonds with both donor and acceptor groups of proteins. By being present in large excess in dilute solutions, water is a very effective competitor to internal hydrogen bonding. Hydrogen bonds play an important stabilizing role only when the groups involved are inaccessible to water, such as by being "buried" in nonpolar regions inside a protein molecule, or in the case of fibrous proteins, or in nonaqueous solutions of proteins. The formation of a hydrogen bond is an exothermic process. Therefore, the stability of any protein structure maintained primarily by hydrogen bonds decreases at elevated temperatures.

Hydrophobic Interactions. Approximately 35–50% of the amino acid residues in many proteins have hydrocarbon-like or similarly nonpolar side chains. Some other side chains are partly nonpolar (Table 1). In aqueous solution, these side chains will tend to come into contact with each other, diminishing the extent of contact with water (Fig. 3), thus forming hydrophobic bonds (between two side chains) or hydrophobic regions (involving several side chains).

The source of the stability of hydrophobic bonds is the same as the reason for the low solubility of hydrocarbons in water or for the formation of micelles, namely the occurrence of structural changes in water. In liquid water, a majority of the possible hydrogen bonds between water molecules is maintained. The solubility behavior of nonpolar substances indicates that water molecules next to these solutes take up a more ordered structure, resulting in the formation of more hydrogen bonds. This ordering, described thermodynamically as a decrease of entropy, is the main contribution to the unfavorable free energy of solution of these solutes. When nonpolar groups in proteins form hydrophobic bonds, the extent of contact with water is diminished and thereby the ordering of water structure is decreased, resulting in a favorable free energy change and a stabilization of the bond. Attractive van der Waals forces between the nonpolar groups make a relatively small contribution. The over-all process is endothermic (below about 55°C) because energy is required to break the hydrogen bonds. As a result, hydrophobic bonds become stronger with increasing temperature. The addition of organic solvents, such as alcohols, to aqueous protein solutions weakens hydrophobic bonds, because nonpolar group-solvent hydrophobic interactions act as a competition.

Lyophobic Bonding. In nonaqueous but strongly polar solvents (such as N, N-dimethylformamide) without a hydrogen-bonded structure like water, the solvent undergoes no structural changes.

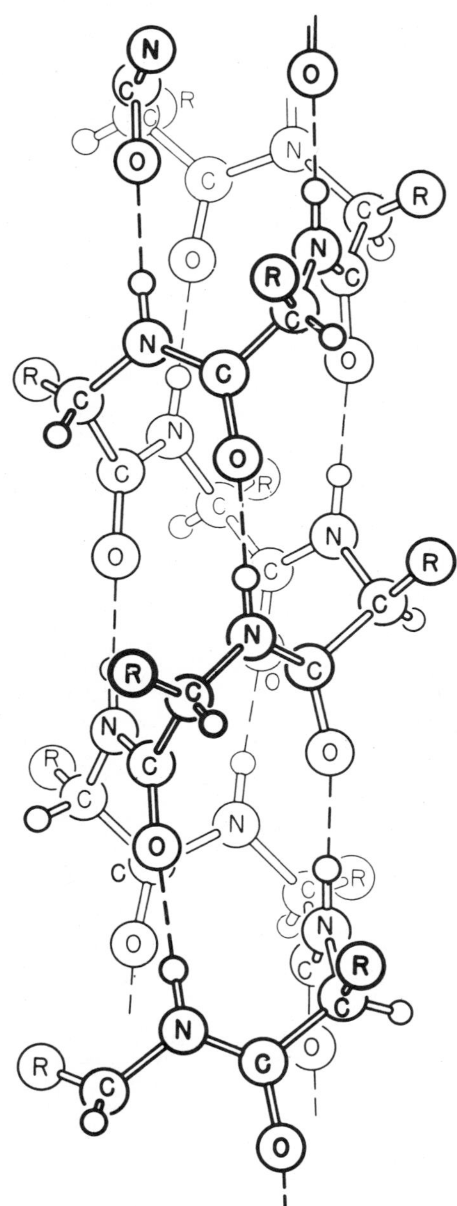

FIG. 2. Secondary structure: right-handed α-helix. The side chains (denoted by R) correspond to L-amino acids. (Reprinted with permission, from L. PAULING, R. B. COREY, and H. R. BRANSON, *Proc. Natl. Acad. Sci. U.S.*, **37**, 205 (1951).

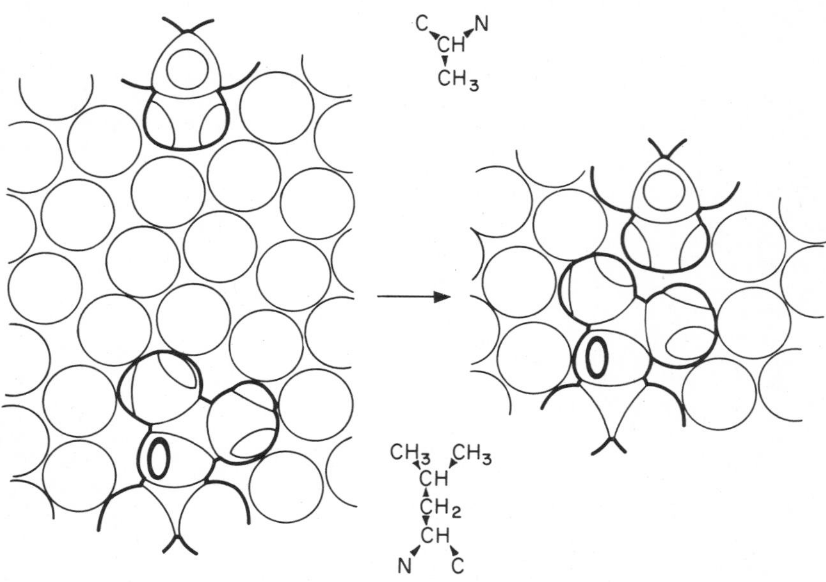

FIG. 3. Schematic representation of the formation of a hydrophobic bond between an alanyl and a leucyl side chain. The two side chains are drawn corresponding to the structural formulae shown. Water molecules are indicated by circles, with no attempt to show the structure of the liquid. (After Némethy and Scheraga. Reprinted with permission from the *Journal of Physical Chemistry*, **66**, No. 10, p. 1774, October 1962; copyright 1962 by the American Chemical Society.)

Nonpolar groups will still tend to cluster in preference to being exposed to solvent, but this is a result of unfavorable van der Waals interaction energies between solvent and solute molecules. Lyophobic bonds are weaker than hydrophobic bonds and their strength decreases at high temperatures.

Ionic Interactions. Since proteins contain charged groups at most pH values of the solution, opposite charges can attract each other and may contribute to the stabilization of certain conformations. Such interactions, sometimes also called *salt linkages*, can arise, for example, between NH_3^+ and COO^- or —⟨benzene ring⟩—O^- groups. When these groups are close together, the distinction between such an ion pair and a hydrogen bond becomes arbitrary. Ion pairing does not occur in water between exposed charged groups on the surface of a protein, but it may occur in nonaqueous solvents or when the ion pair is buried inside the protein. Divalent metal ions (such as Mg^{++} or Zn^{++}) can form ionic bonds by interacting simultaneously with two anionic groups. The presence of many nearby charges of like sign on a protein (at low or high pH) destabilizes secondary and tertiary structures by electrostatic repulsion.

Disulfide Bridges. Cystine residues form part of two peptide chains and thus restrict the conformational freedom of adjacent portions of the protein. Although their presence is properly described in the primary structure of the protein, they exert a strong restriction on secondary and tertiary structure.

Steric Interactions. (a) Rotation around single bonds is not completely free. In the absence of other, stronger interactions, substituents attached to two adjacent atoms tend to take up staggered positions in preference to eclipsing each other. This results in the preferment of certain conformations. (b) Due to the finite size of atoms, not all possible conformations of a polypeptide chain are allowed. All conformations are excluded in which atoms would approach each other more closely than certain distances (usually of the order of 3–4Å). These two restrictions can diminish the number of allowed conformations by a large factor.

Relative Importance. The strengths of various types of interactions are compared in Table 2. In aqueous solution, hydrophobic bonds presumably make the numerically largest contribution to the stabilization of protein conformations because of their large number and because water diminishes the strength of hydrogen bonds. On the other hand, hydrophobic bonds have low specificity, because any number and kind of nonpolar groups can form them. Hydrogen bonds and ionic interactions, where they can form at all, pose much more specific requirements for secondary and tertiary structure.

TABLE 2. STRENGTH OF VARIOUS NON-COVALENT INTERACTIONS IN PROTEINS AT 25°C
(Approximate values of magnitudes of effects)

Type of Interaction	Free Energy of Formation,[a] $\Delta F°$ (kcal/mole)	Heat of Formation, $\Delta H°$ (kcal/mole)	Entropy, $\Delta S°$ (cal/deg mole)
Hydrogen bond			
in aqueous solution	−0.5 to +1	−1.5 to 0	−4 (backbone) −3 to −5 (side chain)
in nonaqueous environment	−5 to −4	−6 to −5	Same
Hydrophobic bond[b]	−1.5 to −0.2	+0.3 to +1.8	+1.7 to +11
Lyophobic interactions[b]	−1 to 0	−1 to 0	0
Ionic interactions			
in aqueous solution	>0		
in nonaqueous environment	−2 to −1	+0.5 to +1.0	+5 to +10

[a] The equilibrium constant for the formation of the bond is given by $K = \exp(-\Delta F°/RT)$.
[b] The numbers refer to the interaction between two side chains only. If several side chains interact, the contribution becomes correspondingly larger.

References

1. KAUZMANN, W., *Advan. Protein Chem.*, **14**, 1 (1959).
2. MARTIN, R. B., "Introduction to Biophysical Chemistry," Ch. 16, New York, McGraw-Hill Book Co., 1964.
3. SCHELLMAN, J. A., AND SCHELLMAN, C., in "The Proteins" (NEURATH, H., EDITOR), Second edition, Vol. II, Ch. 7, New York, Academic Press, 1964.
4. SCHERAGA, H. A., in "The Proteins" (NEURATH, H., EDITOR), Second edition, Vol. I, Ch. 6, New York, Academic Press, 1963.
5. PAULING, L., AND COREY, R. B., *Proc. Natl. Acad. Sci. U.S.*, **37**, 235–282 (1951).

GEORGE NÉMETHY

PROTEINS (BIOSYNTHESIS)

The synthesis of protein has long been a question that has stimulated biochemical research, and it is only in recent years that a clear insight into the problem has been obtained. As a scientific study, protein synthesis presents a complex problem because this area of research encompasses a number of pertinent biological phenomena, ranging from the *biochemical* nature of the process, *e.g.*, number and nature of enzymes involved, energy requirements, site of synthesis, to the matter of the *specificity* of protein structure, *e.g.*, why it is that a cow produces its own particular molecular species of protein and not that of a mouse. This latter question imposes the genetic considerations that were emphasized by the one gene–one enzyme hypothesis (see BIOCHEMICAL GENETICS) proposed by Beadle and Tatum in the 1940's. The hypothesis, in essence, states that the function of a gene is to determine the specificity of a particular enzyme produced by the cell. Since enzymes are proteins, the question of GENE action is, therefore, intimately associated with that of protein synthesis. Recent studies have succeeded in furnishing a lucid and coherent picture of the mechanisms involved in these complex metabolic processes, and this area of research is an outstanding example of work being done in modern biology.

In a simple chemical description, the synthesis of protein is accomplished by the linking of a series of amino acids, one to another, by means of peptide bonds (see PROTEINS). The polymer formed is called a protein or polypeptide. Although approximately 100 naturally occurring AMINO ACIDS have been identified, only 20 are involved in protein synthesis and have been referred to as the "magic 20." The term "magic" stems from the evidence that the utilization of these following amino acids is universal: glycine, alanine, valine, isoleucine, leucine, serine, threonine, aspartic acid, glutamic acid, asparagine, glutamine, lysine, arginine, cysteine, methionine, tryptophan, phenylalanine, tyrosine, histidine and proline.

The incorporation of amino acids into a biologically active protein is now known to proceed by a series of biochemical events that involve both the cytoplasm and nucleus. In the cytoplasm, events are initiated by the activation of amino acids. This activation can be viewed as a two-step reaction and is catalyzed by a group of ACTIVATING ENZYMES called amino acyl RNA synthetases. Each of the 20 amino acids is activated by a specific amino acyl RNA synthetase, *e.g.*, glycine is acted upon only by the glycine-activating enzyme. The first step involves the reaction of an amino acid with energy-rich adenosine triphosphate (ATP) to produce an amino acid adenylate (AMP-amino acid) and inorganic pyrophosphate (PP). The carboxyl group of the amino acid is linked to the phosphate of adenosine-5′-monophosphate by means of an anhydride bond, a high-energy bond that preserves the energy contained in the original ATP molecule. The AMP-amino acid is unstable and exists as an enzyme-bound intermediate, *i.e.*, it is not found free in the cell, and is utilized (with its energy) in the second step which is the transfer of the amino acid from the adenylate to a molecule of SOLUBLE RIBONUCLEIC ACID (sRNA), a low molecular weight nucleic acid. Each amino acid is transferred to a specific type of sRNA; *e.g.*, glycine is transferred to a sRNA that accepts only glycine, and the resultant compound is called an amino acyl RNA. The amino acid is linked to the sRNA through a stable ester linkage between the carboxyl group of the amino acid and a *cis*-hydroxyl group (whether 2′ or 3′ is still unknown) of the terminal nucleotide (an adenylate) of the sRNA. The reaction for the activation process can be written:

(1) ATP + Amino Acid →
AMP-Amino Acid + PP

(2) AMP-Amino Acid + sRNA →
sRNA-Amino Acid + AMP

Overall: ATP + Amino Acid + sRNA →
sRNA-Amino Acid + AMP + PP

The high degree of specificity that is maintained is what would be required under physiological conditions to insure that an amino acid is linked to the proper sRNA.

A second type of RIBONUCLEIC ACID associated with protein synthesis is produced in the nucleus and is called informational or messenger RNA (mRNA). This species of RNA is responsible for carrying the genetic information from the GENES to the cytoplasm and serves as the template for protein synthesis. The interrelationship between gene action and protein synthesis is expressed in the "Central Dogma": DNA → RNA → Protein, which states that genetic information is *transcribed* to mRNA, from which it is then *translated* into protein structure. Genetic material is composed of double-helical DEOXYRIBONUCLEIC ACID, and the formation of mRNA, catalyzed by the enzyme RNA polymerase, involves the production of a strand of RNA that is complementary in base sequence to one of the DNA strands of the double helix, as shown:

DNA: {TTTTTGCCCAAAGTGAAA—etc.
AAAAACGGGTTTCACTTT—etc.

mRNA: UUUUUGCCCAAAGUGAAA—etc.

In this example, the base composition in the mRNA complements that of the strand of DNA that commences with the sequence AAA, with uracil (U) of RNA replacing the T of DNA.

The linear arrangement of the four different types of nucleotides in the DNA, *i.e.*, base sequence, is the key to gene action. The transcribed genetic information in the mRNA molecule that dictates the specificity of protein structure (linear arrangement of the amino acids in the polypeptide chain) is obtained from the sequential order of the four bases within the DNA that composes a gene. A particular sequence of three bases (triplet codon) serves to code for a given amino acid. From a fixed starting point within the gene, this code is "read off" in triplets, each triplet serving as a code for one amino acid. This reading refers to the production of mRNA, and the "message," therefore, contains instructions for the protein that is to be made.

Evidence concerning the identity of the triplets in mRNA that code for the 20 amino acids was obtained by using enzymatically prepared RNAs that serve as mRNA in the *in vitro* system for protein synthesis. The base ratios of each synthetic mRNA and the amino acid composition of the polypeptide synthesized under its direction provided data that allowed for an assignment of codons. Of the possible 64 triplet combinations that can be obtained from the four bases, three-fourths of them have been shown to serve as codons; this is accepted as evidence that the code is "degenerate," *i.e.*, more than one triplet can code for a given amino acid. Below is a list of some of the amino acids and their assigned codons. The base sequence of each codon was determined by the pairing on ribosomes of specific sRNAs with triplets of known sequence and by the use of mRNAs of known composition.

arginine (arg): CGU, CGC, CGA, CGG
glycine (gly): GGU, GGC, GGA, GGG
lysine (lys): AAA, AAG
phenylalanine (phe): UUU, UUC
valine (val): GUU, GUC, GUA, GUG
proline (pro): CCC, CCU, CCG, CCA

In the cytoplasm, the following three components, excluding enzymes, are necessary for the final step of synthesis: ribosomes, mRNA and sRNA(s)-amino acids(s). RIBOSOMES (70 S) are considered "factories" or "workbenches" for polypeptide formation, and each is composed of a 30 S and a 50 S component. The mRNA becomes associated with the 30 S component of the ribosome and the sRNA-amino acid molecules with the 50 S component. Translation occurs on the ribosomes and relies on recognition that occurs between codons of the mRNA and specific sRNA-amino acid molecules. In addition to this type of recognition, which is thought to occur by base pairing between the two types of RNA, there is ionic bonding (between the sRNA molecules and the ribosome) which insures the spatial relationships that allow the two amino acids attached to adjacent sRNA molecules to undergo covalent bonding. The following diagrams illustrate how the mRNA "tape" is "read" by the ribosome (the latter represented schematically as a rectangle):

mRNA: UUUGUUCCCAAAGGUAAA—etc.
Ribosome
sRNA   sRNA
| |
H_2N—phe—CO—NH—val—COOH

In this example, recognition takes place on the ribosome between the triplet UUU of mRNA and the sRNA carrying phenylalanine (phe) and between GUU and sRNA-valine (val). Synthesis commences at the N-terminal end of the polypeptide and proceeds by peptide bond formation between the carboxyl group of the first amino acid (phe) and the amino group of the amino acid (val) on the adjacent sRNA. The carboxyl group of this second amino acid will undergo bonding with the amino group of the amino acid (pro) attached to the next sRNA recognized in the sequence. Following is a second example:

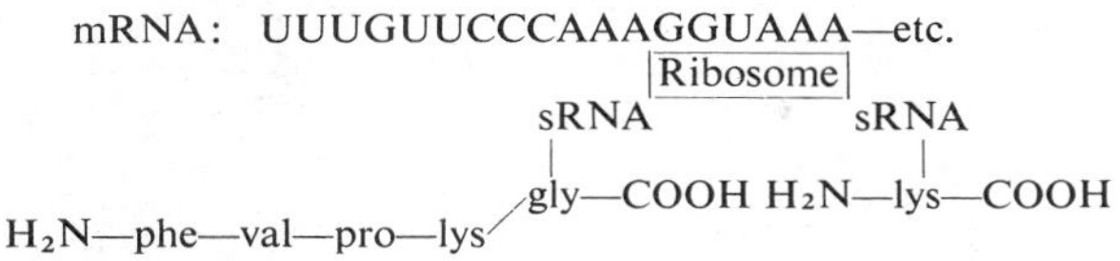

```
mRNA:  UUUGUUCCCAAAGGUAAA—etc.
                   |Ribosome|
                 sRNA         sRNA
                   |            |
                  /gly—COOH H₂N—lys—COOH
H₂N—phe—val—pro—lys/
```

The second illustration depicts the growth of the polypeptide chain as the ribosome "rolls along the tape" and synthesis continues. The chain is attached to the sRNA-glycine that will undergo peptide bond formation with lysine of the next sRNA. Once a sRNA has its amino acid incorporated into the polypeptide structure, it is released from the ribosome and becomes available again for the amino acid activation process. Synthesis proceeds along the mRNA "tape" until the message has been read, *i.e.*, to C-terminal ending; the nascent polypeptide is then released from the ribosome. Because of the specificity involved in each phase of synthesis, it is evident that a mRNA molecule, as the cytoplasmic agent of a gene, always directs the synthesis of a specific protein.

A mRNA molecule can accommodate a number of ribosomes simultaneously, and the length of an individual mRNA determines the number of ribosomes that can be associated with it. These complexes, often observed as aggregates, are called polyribosomes or polysomes. The polyribosome responsible for the synthesis of hemoglobin contains five ribosomes, and polyribosomes containing as many as 115 ribosomes have been reported.

Many features of protein synthesis are now elucidated, but various important aspects remain unclear, *e.g.*, details on factors controlling the physical relationships between mRNA, sRNA and ribosomes, enzymes and the exact nature of the reactions occurring on the polyribosome [see also TRANSFER FACTORS (IN PROTEIN SYNTHESIS)]. The answers to these and other uncertainties rely on future research.

The above description has emphasized synthesis as it occurs on ribosomes in the cytoplasm, and although this pathway is considered to be the major route of protein synthesis in the cell, it is likely that other modes of synthesis will be uncovered for certain proteins and peptides, *e.g.*, proteins of BACTERIAL CELL WALLS.

References

1. SCHWEET, R., AND BISHOP, J., "Protein Synthesis in Relation to Gene Action," in "Molecular Genetics," Part I (TAYLOR, J. Herbert, EDITOR), Ch. 8, New York, Academic Press, 1963.
2. WAGNER, R. P., AND MITCHELL, H. K., "Genetics and Metabolism," Second edition, New York, John Wiley & Sons, 1964.
3. MCELROY, W. D., "Cell Physiology and Biochemistry," Foundations of Modern Biology Series, Second edition, Englewood Cliffs, N.J., Prentice-Hall, 1964.
4. HARTMAN, P. E., AND SUSKIND, S. R., "Gene Action," Foundations of Modern Genetics Series, Englewood Cliffs, N.J., Prentice-Hall, 1965.

FRANK B. ARMSTRONG

PROTEINS (CONTRACTILE). See CONTRACTILE PROTEINS; MYOSINS.

PROTEINS (DETERMINATION OF MOLECULAR WEIGHT). See PROTEINS, section on Molecular Weights of Proteins.

PROTEINS (END GROUP AND SEQUENCE ANALYSIS)

The suggestion that proteins consist of the general structure shown following:

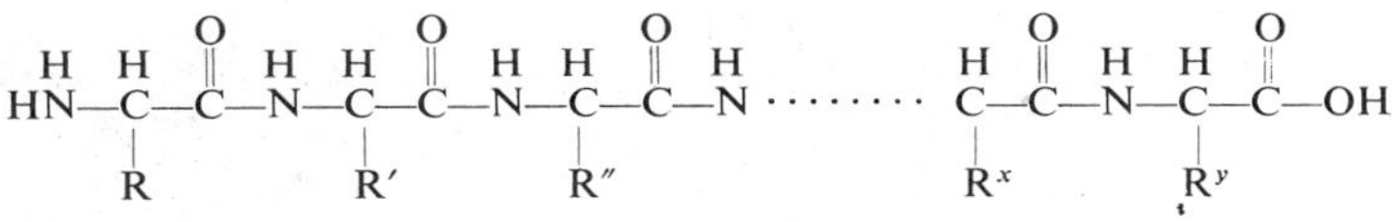

```
      H  H  O   H  H  O   H  H  O   H                H  O   H  H  O
           ||        ||        ||                        ||        ||
     HN—C—C—N—C—C—N—C—C—N ········ C—C—N—C—C—OH
         |          |          |                  |          |
         R          R′         R″                 Rˣ         Rʸ
```

was made independently about the turn of the century by Fischer and Hofmeister. Only within the last decade, however, has it been possible to determine, with increasing facility, in what sequence the AMINO ACID residues (an amino acid residue is —N(H)—C(H)(R)—C(=O)—) are arranged in a protein. The relatively small protein INSULIN was the first to yield its sequence to the efforts of Frederic Sanger who was awarded the Nobel Prize for his investigations.

If we consider the above structure and ignore any reactive groups that may be on R, R′, etc., it will be seen that one end has a free α-amino group and the other a free α-carboxyl group. These are defined as the *N-terminal* and *C-terminal* ends, respectively, and are attached to the N-terminal and C-terminal residues. These termini are unique in any polypeptide chain (*i.e.*, any sequence of amino acids joined as above); if they can be qualitatively identified and quantitatively determined, their kind will tell something about sequence and their number will tell how many chains are in the protein (if the molecular weight is known).

Two methods are commonly used for the *determination of N-terminal residues.* One, due to Sanger, is based upon these chemical reactions:

$$\underset{\text{(I)}}{F\text{-}C_6H_3(NO_2)_2} + \underset{\text{(II)}}{HN\text{—}CH(R)\text{—}CO\text{—}NH\text{—}CH(R')\text{—}CO\text{—}NH\text{—}CH(R'')\text{—}CO\cdots} \rightarrow \underset{\text{(III)}}{(NO_2)_2C_6H_3\text{—}NH\text{—}CH(R)\text{—}CO\text{—}NH\text{—}CH(R')\text{—}CO\text{—}NH\text{—}CH(R'')\text{—}CO\cdots}$$

$$+ HF \rightarrow \underset{\text{(IV)}}{(NO_2)_2C_6H_3\text{—}NH\text{—}CH(R)\text{—}C(=O)\text{—}OH} + \underbrace{HN\text{—}CH(R')\text{—}C(=O)\text{—}OH + HN\text{—}CH(R'')\text{—}C(=O)\text{—}OH + \cdots}_{\text{(V)}}$$

When 2,4-dinitro-1-fluorobenzene (I) reacts with a protein (II), the dinitrophenyl (DNP) protein (III) is formed. After isolation of the DNP-protein, the N-terminal DNP-amino acid (IV) and free amino acids (V) are released by refluxing in hydrochloric acid which hydrolyzes the peptide bonds ($\text{—}C(=O)\text{—}NH\text{—}$) but not the DNP-to-nitrogen bond. The DNP-amino acid may then be qualitatively and quantitatively determined.

The second method is the Edman degradation:

$$\underset{\text{(VI)}}{C_6H_5\text{—}N{=}C{=}S} + \underset{\text{(II)}}{NH\text{—}CH(R)\text{—}CO\text{—}NH\text{—}CH(R')\text{—}CO\text{—}NH\text{—}CH(R'')\text{—}CO\text{—}NH\cdots} \rightarrow$$

$$\underset{\text{(VII)}}{C_6H_5\text{—}NH\text{—}C(=S)\text{—}NH\text{—}CH(R)\text{—}CO\text{—}NH\text{—}CH(R')\text{—}CO\text{—}NH\text{—}CH(R'')\text{—}CO\text{—}NH\cdots} \xrightarrow{H^+}$$

$$\underset{\text{(VIII)}}{\text{PTH-amino acid: } C_6H_5\text{—}N\text{—}C(=S)\text{—}NH\text{—}CHR\text{—}C(=O)\text{ (ring)}} + \underset{\text{(IX)}}{HN\text{—}CH(R')\text{—}CO\text{—}NH\text{—}CH(R'')\text{—}CO\text{—}NH\cdots}$$

When phenyl isothiocyanate (VI) reacts with a protein (II), the phenylthiocarbamyl (PTC) protein (VII) is formed. In the presence of acid, the N-terminal amino acid is removed as the phenylthiohydantoin (VIII) (the PTH-amino acid as it is commonly called); the remaining protein (IX) has one less residue. The Edman method is particularly advantageous because the N-terminal residue alone is removed, a new N-terminus is formed, and repeated cycles of degradation can be carried out to determine sequences.

Thus, it is usually possible by these or another method to determine the kind and number of N-termini in a protein. In some instances, no N-terminal residue can be detected; when this is so, the N-terminal amino group may, for example, have an acetyl group attached to it and no longer be reactive toward the various reagents.

Methods for the *determination of C-terminal residues* are far less satisfactory than those for N-termini. Thus, if a protein is treated with anhydrous hydrazine (N_2H_4), all residues except the C-terminal will form the corresponding hydrazide

(HN—CHR—CO—NH—NH); the C-terminal amino acid will be free and therefore unique. Another method requires the use of the enzymes carboxypeptidases A and B. These enzymes (with certain specificity) split only C-terminal residues from a polypeptide chain. The rate at which free amino acids appear must be determined because, as soon as one residue is removed, another is opened to the enzyme. Potentially, one can determine the sequence of several residues near the C-terminus. Unfortunately, neither of these methods has been applied successfully in very many instances.

If the N-terminal determination has shown that the protein contains more than one type of chain, then it is advantageous to separate the types of chains before proceeding to a determination of total sequence. Suppose, then, that we have a protein with a single type of chain or have isolated a single kind from a protein with multiple chains. How does one approach the determination of sequence? Let us represent the polypeptide chain by a single line with its N- and C-termini:

N________________________________C

In many instances, the number of residues thus depicted will be of the order of 150 more or less. The use of N-terminal and C-terminal methods may identify a few residues from each end, but most of the determination will require other methods. The most generally used approach is to hydrolyze the protein first with the enzyme trypsin. Trypsin is very specific in its cleavage of proteins in that it breaks only those peptide bonds that are associated with the carboxyl groups of the basic amino acids lysine and arginine. Trypsin, then, might break up our protein as follows:

N 1 | 2 | 3 | 4 | 5 | 6 | 7 | 8 C

How many peptides may we expect from a tryptic hydrolysis? Because of the specificity of trypsin, each peptide (*i.e.*, combination of two or more residues) will contain C-terminal lysine or arginine except perhaps the one that derived from the C-terminus of the protein itself. Therefore, the number of peptides should not exceed one more than the sum of lysyl and arginyl residues. If the complete amino acid analysis of our example had revealed 6 lysyl and one arginyl residues, then the 8 peptides in our example are the maximum that we can expect. The number of tryptic peptides can be indicative of identical polypeptide chains in a protein. Suppose the protein is found to have 4 N-terminal residues and a total of 50 lysyl and arginyl residues but only 25 tryptic peptides can be detected; one is justified in concluding that the protein is made of identical halves and that each half has two different chains.

The peptides in a tryptic hydrolysate can help us little unless the mixture can be separated. This task usually falls to CHROMATOGRAPHY. If relatively large amounts (of the order of 10–20 μmoles) are to be separated, chromatography on columns of ION-EXCHANGE RESINS is an effective means of separation. Smaller amounts may be handled by paper chromatography or ELECTROPHORESIS. After each peptide has been isolated, the first necessity is a determination of the amino acid composition. This is a rapid and accurate procedure with the automatic amino acid analyzers that are now available [see PROTEINS (AMINO ACID ANALYSIS)]. When the amino acid analyses have been completed, it will generally be evident, as our example also shows, that the peptides vary greatly in the number of residues—from free lysine or arginine which may arise in certain instances to peptides with 30 or 40 residues. If the (internal) sequence of the individual peptides can be determined, much will have been accomplished. Frequently, the entire sequence of peptides as large as decapeptides (10 residues) can be determined by repeated application of the Edman procedure that has been described above. If the Edman procedure cannot completely determine the sequence, it is then necessary to split the peptide further by other enzymes or by chemical means, to separate these fragments, to determine the amino acid composition, and to apply the Edman degradation again.

Although the entire sequence of each tryptic peptide may have been determined, we still have no basis for concluding that peptide No. 4 of our example should be attached to No. 5 rather than to No. 8. However, if a few residues have been placed in sequence at the N-terminus of the intact protein, say, by use of the Edman degradation, then the tryptic peptide with this sequence can be placed at the N-terminus; likewise, if one peptide contains no lysine or arginine, it may be positioned at the C-terminus. To arrange the tryptic peptides in their correct order requires a breaking of the polypeptide chain in another way. This may be done effectively by another enzyme such as chymotrypsin which has a different action from trypsin; it splits most efficiently those bonds that are associated with the carboxyl groups of tyrosine, phenylalanine, tryptophan, and leucine. Our polypeptide chain might then be broken by chymotrypsin in this way

N I | II | III | IV | V | VI | VII | VIII | IX C

instead of in this way

N 1 | 2 | 3 | 4 | 5 | 6 | 7 | 8 C

as it was with trypsin. These chymotryptic peptides must then be separated, analyzed, and have their sequences determined. If a chymotryptic peptide contains arginine or lysine, it is a useful "overlap" to determine the order of two tryptic peptides because part comes from one tryptic peptide and part from another. Suppose that the sequence of chymotryptic peptide No. VIII was found to be A-C-F-B-L-G-L-D-M-O where "L" represents a lysyl residue and the other letters some other residues. (By convention, the N-terminus of any sequence is written at the left.) From our presumed knowledge of the tryptic peptides, we recognize that G-L is tryptic peptide No. 7, that A-C-F-B-L is a unique sequence at the C-terminal end of tryptic peptide No. 6, and that D-M-O is the N-terminal portion of tryptic peptide No. 8 which contained no arginine or lysine and was at the C-terminus. Peptide No. VIII, therefore, links together Nos. 6, 7, and 8 in that order. Likewise, II would link 1 and 2, III would overlap 2, 3 and 4, etc. Such chymotryptic peptides as I, V, VI, and IX, would be of no help in bridging from one to another tryptic peptide, but a redetermination of their sequence would give added assurance to the previous determination of sequence by means of the tryptic peptides.

The ideal results that we have obtained with our model protein are hardly to be expected in actual practice. Each protein presents new problems that test the ingenuity of the investigator. Despite this fact, the determination of sequence in many proteins continues apace. This knowledge of sequence in combination with other data is telling more and more how these vital substances carry out their functions.

References

1. "The Proteins" (Neurath, H., Editor), Second edition, New York, Academic Press.
2. *Advances in Protein Chemistry*—recent volumes.
3. *Annual Reviews of Biochemistry*—recent volumes.

W. A. Schroeder

PROTEINS (NUTRITIONAL QUALITY)

It has been known for many years that PROTEINS vary markedly in their ability to support growth of animals. The fundamental studies of Osborne and Mendel (1914) led to a realization that the nutritive value of various proteins is dependent upon their content of the ESSENTIAL AMINO ACIDS required for growth and maintenance. The proteins of cereals are deficient primarily in lysine, but some are almost equally deficient in tryptophan (corn) or in threonine (rice, wheat). Animal proteins such as those of egg, meat and milk, contain more ideal amounts of each of the essential amino acids. Tissue proteins cannot be synthesized if any one of the essential amino acids is absent; the extent of protein synthesis appears to be governed by the concentration of the limiting amino acids (*e.g.*, lysine in cereals). A useful method for estimating nutritional quality of proteins involves determining the "chemical score," by measuring the amount of each amino acid in the test protein and expressing each value as a per cent of the corresponding amount of each of these amino acids in egg. Egg is chosen for reference purposes because of its nearly ideal balance of essential amino acids. The lowest value or "score" is given by the limiting amino acid for growth and is a quantative expression of the nutritive value of the protein. Numerous studies with young growing animals have shown excellent correlations between chemical scores and nutritive value of food proteins as judged by animal growth. Nevertheless some highly processed foods yield lower values by animal growth than predicted by chemical score; this is apparently due to poor digestibility or to inadequate availability of one or more of the essential amino acids therein. It follows, therefore, that procedures for evaluating nutritional quality of proteins must take into account the content of "available" amino acids.

Despite the undoubted value of chemical score, animal growth tests remain the most satisfactory yardstick for assessing the quantitative pattern of available amino acids in a protein.

Nutritionists generally agree that the growing rat and man utilize proteins similarly and that protein quality of human diets may be evaluated by rat growth tests. The most widely used rat growth method is the protein efficiency ratio (P.E.R.) method. In this procedure, the grams of weight gain per gram of protein consumed are determined in young growing rats eating a 10% protein diet. P.E.R. values range from zero for certain processed cereal proteins to about 4.0 for the protein mixture of whole egg. The P.E.R. method has been criticized because it does not give any credit to a protein for its amino acid contribution to maintenance. A young growing animal requires amino acids for growth and for maintenance, whereas a mature animal needs amino acids only for maintenance. The quantitative pattern of amino acids for maintenance and for growth are different. For example, much less lysine is required for maintenance than for rapid growth. For this reason, cereal proteins which are deficient in lysine are of greater value for maintenance than for growth.

In the body, amino acids may be considered as entering a dynamic "pool" contributed to by both dietary and tissue protein. Since nitrogen balance is the sum of the gains and losses of all tissue proteins of the body, nitrogen balance may be used as a measure of the value of dietary protein. The fraction of absorbed nitrogen retained in the body has been defined as the "biological value" of the protein. Although methods of protein evaluation based upon nitrogen retention measure the value of a protein for both maintenance and growth, such methods are time consuming and are not generally applicable to the routine evaluation of foods. Nevertheless, the term "biological value" has become synonymous with protein quality. Egg has a biological value of close to 100, whereas white flour has a value of approximately 50. Apparently, protein requirements may be satisfied by x grams of a protein with a biological value of 100 or $2x$ grams of a protein with a biological value of 50. Thus in order to express the true protein value of a food, both quality and quantity must be considered; the product of the two factors is a convenient way of doing this. Most peoples of the world obtain their protein needs by eating adequate amounts of relatively poor-quality proteins; the amino acids which are not utilized for protein synthesis are metabolized for energy. Human diets contain mixtures of proteins, and a deficiency of an amino acid in one food may be corrected by an excess of that amino acid in another protein in the diet. Staple combinations of foods such as breakfast cereal and milk, macaroni and cheese, corn and beans, rice and legumes, etc., are good examples of supplementary mixtures of proteins; the protein quality of the mixture may be better than either protein alone. In the final analysis, therefore, it is quantity and quality of the mixture of proteins eaten that are important.

References

1. Block, R. J., and Mitchell, H. H., *Nutr. Abstr. & Rev.*, **16**, 249 (1946).
2. Morrison, A. B., in "Symposium on Foods—Proteins and Their Relations" (Schultz, H. W., and Anglemier, A. F., Editors), Westport, Conn., Avi Publishing Co., 1964.
3. Allison, J. B., *Physiol. Rev.*, **35**, 664 (1955).
4. "Protein and Amino Acid Nutrition" (Albanese, A. A., Editor), New York and London, Academic Press, 1959.

J. M. McLaughlan

PROTEOLYTIC ENZYMES (AND THEIR ZYMOGENS)

The proteolytic enzymes are a large group of PROTEINS which share a common ability to catalyze the hydrolysis of the amide, or "peptide," bond between two AMINO ACIDS in polypeptides, proteins and closely related compounds.

$$\mathrm{R{-}C({=}O){-}N(H){-}C(H)(R'){-}C({=}O){-}N(H){-}C(H)(R''){-}C({=}O){-}N(H){-}R'''} + \mathrm{H_2O}$$

$$\xrightarrow{\text{Proteolytic enzyme}} \mathrm{R{-}C({=}O){-}N(H){-}C(H)(R'){-}C({=}O){-}OH}$$

$$+$$

$$\mathrm{H_2N{-}C(H)(R''){-}C({=}O){-}N(H){-}R'''}$$

These enzymes have been obtained from a great diversity of tissues, and they are differentiated on the basis of detailed aspects of their chemical structure and of the specificity that each shows in catalyzing hydrolysis of peptide bonds in which certain specific amino acids are involved. Thus, for example, *trypsin* obtained from the mammalian pancreas has a molecular weight of 24,000, a defined amino acid sequence in its single peptide chain, and catalyzes the hydrolysis of peptide bonds in peptides or proteins only where arginine or lysine contributes the acyl function $(\mathrm{{-}N(H){-}C(H)(R'){-}C({=}O){-}})$. *Leucine amino peptidase*, on the other hand, is obtained from swine kidney, has a molecular weight of 300,000, an as yet undetermined amino acid sequence, an absolute requirement for Mg^{++} or Mn^{++} for activity, and catalyzes the hydrolysis of the peptide bond in peptides and proteins between the amino acid, preferably leucine, at the end of the peptide chain which has its amino group free (the N-terminal amino acid) and the adjacent amino acid. Although it has been useful in the past to classify proteolytic enzymes as to source and substrate specificity, in recent years our knowledge of the *structure* and at least some aspects of the *chemical mechanisms* through which these enzymes catalyze peptide hydrolysis has developed to a point where classification on the basis of these more fundamental characteristics has been attempted by Hartley. Thus, we have (a) the serine proteases, in which a single specific serine residue in the protein has been shown to be directly involved in catalysis, (b) the thiol proteases, in which enzyme activity is dependent upon the presence in the proteins of the reduced form of specific cysteine residues, (c) the metallo proteinases, which require the presence of specific metal ions for catalytic activity and (d) the acid proteases which function catalytically near pH 1-3. Representative examples of each class are given in Table 1. A brief description of the most thoroughly investigated member of each class will serve to indicate the current status of our knowledge of the structure and function of the proteolytic enzymes.

Serine Proteases. The most intensively studied proteolytic enzyme has been α-chymotrypsin, which is commonly obtained in crystalline form from beef or pork pancreas extracts. This enzyme and trypsin, a similar enzyme from the same sources, occur chiefly in their inactive or *zymogen* forms, chymotrypsinogen A and trypsinogen. Before the characteristic catalytic properties of the enzymes can be observed, the zymogens must be converted to the active form. This process involves a limited hydrolysis of the zymogen and will be described somewhat later, after the structure and mechanism of the enzymes have been described.

Mechanism of Action. In 1949 it was discovered by Jansen and Balls that diisopropyl fluorophosphate (DFP), a potent inhibitor of the enzyme acetylcholinesterase, would also inhibit α-chymotrypsin and trypsin. In this reaction, only one mole of DFP reacts with one mole of the enzyme to form an inactive mono-(di-isopropyl phosphoryl)chymotrypsin (DIP-cht) and one mole of HF. The physical chemical characteristics of DIP-cht, including the molecular weight of 25,100, are identical with those of the fully active chymotrypsin, which suggests that no change in the

TABLE 1

Class	Representative Enzymes or Groups of Enzymes
Serine proteases	Trypsin—pancreas; chymotrypsin—pancreas; thrombin—serum; elastase—pancreas; subtilisin—*B. subtilis*
Thiol proteases	Papain—papaya latex; ficin—fig latex; bromelain—pineapple stems; some mammalian intracellular proteases (cathepsins)
Metallo proteases	Carboxypeptidase A—pancreas; leucine amino peptidase—kidney; prolinase—kidney; prolidase—kidney; dipeptidase—various mammalian tissues
Acid proteases	Pepsin—stomach; rennin—stomach

protein has occurred except for the phosphorylation of some specific site which is important in the catalytic mechanism. Subsequently, Oosterbaan and his associates were able to show that phosphorylation occurred on the hydroxyl group of a single specific serine in the peptide chain of chymotrypsin,

gly.asp.ser-DIP.gly.gly.pro.leu

It was later shown in Neurath's laboratory that the site of reaction of DFP with trypsin was serine in the sequence:

...cySO_3.glu.gly.asp.ser-DIP.
gly.gly.pro.val.val.cySO_3...

Other studies have shown that a ...gly.asp.ser.gly... or ...gly.glu.ser.ala... sequence is characteristic of irreversibility of the DFP inhibition, however, the acetyl group is lost by spontaneous hydrolysis if the pH of the solution is raised toward pH 7-8. Again Oosterbaan, Cohen and their associates showed that the site of acetylation was the hydroxyl group of the same serine residue which had been implicated in the DFP reaction.

Since chymotrypsin had been shown by Neurath and Schwert to catalyze the hydrolysis of amino acid esters as well as the more typical peptide bond, the suggestion was made by a number of investigators, notably Wilson, that normal catalysis proceeded through an acyl enzyme intermediate, and that the DFP and NPA reactions were closely related processes in which the intermediate phosphoryl or acyl enzyme could be distinguished:

$$\mathrm{E} + \underset{\substack{\text{(a) Amino acid ester}\\ \text{(b) } p\text{-Nitrophenyl}\\ \text{acetate (pH 8)}}}{\mathrm{R{-}\overset{\overset{\displaystyle O}{\|}}{C}{-}O{-}R'}} \rightleftharpoons \underset{\substack{\text{Michaelis-}\\ \text{Menten}\\ \text{complex}}}{\mathrm{E\cdot R{-}\overset{\overset{\displaystyle O}{\|}}{C}{-}OR'}} \rightarrow \underset{\substack{\text{Transient}\\ \text{acyl}\\ \text{enzyme}}}{\mathrm{[E\cdot \overset{\overset{\displaystyle O}{\|}}{C}{-}R]}} + \mathrm{R'OH} \xrightarrow{+H_2O} \mathrm{E} + \mathrm{R{-}\overset{\overset{\displaystyle O}{\|}}{C}{-}OH} \quad (1)$$

$$\mathrm{E} + \underset{\substack{\text{Diisopropyl}\\ \text{fluorophosphate}}}{\mathrm{(RO)_2{-}\overset{\overset{\displaystyle O}{\|}}{P}{-}F}} \rightarrow \underset{\text{DIP-enzyme}}{\mathrm{E{-}\overset{\overset{\displaystyle O}{\|}}{P}{-}(OR)_2}} + \mathrm{HF} \quad (2)$$

$$\mathrm{E} + \underset{\substack{p\text{-Nitrophenyl}\\ \text{acetate (pH 5)}}}{\mathrm{CH_3{-}\overset{\overset{\displaystyle O}{\|}}{C}{-}O{-}R}} \rightarrow \underset{\substack{\text{Stable mono-}\\ \text{acetyl enzyme}}}{\mathrm{E{-}\overset{\overset{\displaystyle O}{\|}}{C}{-}CH_3}} + \mathrm{ROH} \quad (3)$$

most, though not all, of the serine proteases, and suggest that this serine residue plays a key role in a catalytic mechanism which may be common to all of the serine proteases. Between 1954 and 1956 Hartley and Kilby, and Balls and his co-workers found that one mole of *p*-nitrophenyl acetate (NPA) would also react with chymotrypsin at pH values near 5 to give an inactive monoacetyl chymotrypsin. In marked contrast with the

The direct involvement of histidine in the catalytic mechanism was recognized early as a result of studies of pH dependence which showed that catalytic activity varied in a manner which implied the participation of an uncharged imidazole side chain. Other studies have shown that photooxidation, or the reaction of a specific alkylation reagent [L-1-tosylamido-2-phenylethyl chloromethyl ketone (TPCK)] with one of the two histidine residues of chymotrypsin yields an inactive enzyme [see ENZYMES (ACTIVE CENTERS)]. This type of evidence led Cunningham to suggest that the catalytically active center of chymotrypsin and other serine proteases contained hydrogen bond-linked serine and histidine residues in which the increased nucleophilic property of the serine oxygen and electrophilic character of the imidazole nitrogen produced the catalytic hydrolysis of sensitive substrates through a transient intermediate in which an amino acid or peptide acylates the serine hydroxyl. Other mechanisms have been suggested, but it would appear at present that the weight of evidence strongly favours the mechanism derived from the acyl enzyme postulate. Important changes and improvements in this mechanism have been suggested by Bender, Bruice, Sturtevant, Jencks, and others as a result of detailed kinetic analyses of chymotrypsin-catalyzed reactions.

Substrate Specificity. It has already been noted that trypsin preferentially catalyzes the hydrolysis of peptide bonds in which lysine and arginine contribute the acyl portion. Another serine protease, thrombin, exhibits similar specificity. Chymotrypsin has been, however, the most widely studied protease from the point of view of specificity as well as mechanism. This enzyme preferentially hydrolyzes peptide bonds in which tryptophan, tyrosine and phenylalanine contribute the acyl portion, and in addition, though at a slower rate, similar bonds involving methionine, leucine, valine and other amino acids. In addition to the peptide bond, chymotrypsin and other serine proteases will catalyze hydrolysis of a variety of other derivatives at an appreciable rate if a "specific" amino acid furnishes the acyl portion. Thus chymotrypsin will hydrolyze N-acetyl-L-tyrosine ethyl ester to N-acetyl-L-tyrosine and ethanol even more rapidly than it will hydrolyze the N-acetyl-L-tyrosinamide or N-acetyl-L-tyrosyl glycinamide. Other derivatives of this amino acid hydrolyzed by this enzyme include N-acetyl-L-tyrosine hydroxamide, N-acetyl-L-tyrosine hydrazide and N-acetyl-L-tyrosine methyl amide. All of these, as well as other substances relatively unrelated to specific amino acid substrates, have been widely studied as tools for the exploration of the structural foundation of enzyme specificity. Among the latter are *p*-nitrophenyl acetate, cinnamic acid methyl ester, cinnamoyl imidazole, diethyl-α-acetamidoglutarate, and methyl-β-phenyllactate. The recent studies of Niemann and his associates and of S. G. Cohen and his associates have again focused attention on the importance of the net interaction of all four constituents about the asymmetric α-carbon of the amino acid derivatives, with the corresponding four aspects of the stereochemical environment on the enzyme surface near the catalytic center, as the controlling factors in influencing the strength and specificity of enzyme-substrate interaction, including sterospecificity.

Amino Acid Sequence. The determination of the amino acid sequence of chymotrypsin has been the goal of a number of investigations carried out over a ten-year period since 1954. Figure 1 includes the most recent and complete report, that of Hartley, as well as a recent progress report by Walsh and Neurath on the sequence of trypsin. In both cases, the sequence given is that of the inactive precursor or zymogen. Gaps in the sequence of trypsinogen have been left, following Walsh and Neurath, so as to emphasize numerous similarities over many sections of the peptide sequence of these two proteases from bovine pancreas. Of special interest are the very similar regions from residue 190–198 in chymotrypsinogen and residue 178–186 in trypsin which include the DFP-reactive serines, 195 and 183 respectively, and the regions from 37–58 in chymotrypsin and 26–47 in trypsin which include both histidines, including the one, 57, which reacts with TPCK. Other similarities include the facts that six of the seven proline residues of chymotrypsin, occur in the same position in trypsin, and that all four tryptophan residues of trypsinogen are in identical positions in chymotrypsinogen. Although these similarities are in accord with the deduction from studies of the mechanisms of action of these enzymes that an identical catalytic process is involved, the differences of amino acid composition and sequence must also be of considerable significance in view of the markedly different substrate specificities of these two enzymes.

Activation of Zymogens. The inactive precursor of trypsin, trypsinogen, may be activated by an enzyme, enterokinase, or autocatalytically by trypsin itself. Activation involves hydrolysis between residues 6 and 7 of trypsinogen (Fig. 1). The liberation of the acid peptide 1–6 into the solution then permits internal structural rearrangements which in turn generate the specific three-dimensional array of amino acids of the catalytic center. The structural features which control specificity however, are already largely present in the zymogens, since several investigators including Doherty and Vaslow have shown that the zymogen forms do bind rather strongly the various compounds known to be substrates for the active enzyme. Activation of chymotrypsinogen A occurs as a result of sequential tryptic and chymotryptic hydrolysis of bonds 15–16, 13–14, 146–147 and 148–149, the liberation of the dipeptides ser.arg and thr.asn, and internal structural rearrangements. Various intermediates in this process possess catalytic activity and may be isolated, but the over-all process described yields α-chymotrypsin, the most widely studied form.

Metallo Proteases. Although a large number of proteolytic enzymes require specific metal ions for activity, they do not necessarily share a common catalytic mechanism. Our knowledge of this class of enzymes is very incomplete, but two representatives have been examined in detail and a summary of their properties will illustrate some of the similarities and differences.

Carboxypeptidase A. This enzyme was originally prepared by Anson from pancreas extracts. It is an exopeptidase in the classification of Bergman; *i.e.*, it catalyzes the hydrolysis only of the carboxylterminal amino acid of a peptide chain, in

1 2 3 4 5 6 7 8 9 10 11 12 13 14 15 16 17 18 19
Chymotrypsinogen cys- gly- val- pro- ala- ile- gln- pro- val- leu- ser- gly- leu- ser- arg- ILE-VAL-GLY- asp-
Trypsinogen val- asp- asp- asp- asp- lys- ILE-VAL-GLY- gly-
1 2 3 4 5 6 7 8 9 10

20 21 22 23 24 25 26 27 28 29 30 31 32 33 34 35 36 37 38 39 40 41
glu-glu- ala- val- pro- gly- ser- trp- PRO-trp- GLN- VAL-SER- LEU-gln- asp- lys-thr- GLY- phe- HIS- PHE-
tyr-thr- cys- gly- ala- asn- thr- val- PRO-tyr- GLN- VAL-SER- LEU-asn- -ser- GLY- tyr- HIS- PHE-
11 12 13 14 15 16 17 18 19 20 21 22 23 24 25 26 27 28 30 30

42 43 44 45 46 47 48 49 50 51 52 53 54 55 56 57 58 59 60 61
CYS- GLY- GLY- SER- LEU-ILE- ASN-glu- asn- TRP-VAL-VAL-thr- ALA-ALA-HIS- CYS-gly- val- thr-
CYS- GLY- GLY- SER- LEU-ILE- ASN-ser- gln-TRP-VAL-VAL-ser- ALA-ALA-HIS- CYS-tyr- lys- ser-
31 32 33 34 35 36 37 38 39 40 41 42 43 44 45 46 47 48 49 50

62 63 64 65 66 67 68 69 70 71 72 73 74 75 76 77 78 79 80 81 82 83 84 85
thr-ser- asp- VAL-val- val- ala- gly- glu- phe-asp- gln- gly- ser- ser- ser- glu- lys- ile- gln- lys-leu- lys- ile-
gly-ile- gln- VAL-arg- leu- gly- glu- asp- asn- ile- asn- val- val- glu- gly- asp- glu- gln- phe- ile- ser- ala- ser-
51 52 53 54 55 56 57 58 59 60 61 62 63 64 65 66 67 68 69 70 71 72 73 74

86 87 88 89 90 91 92 93 94 95 96 97 98 99 100 101 102 103 104 105 106 107
ala-lys- val- phe-lys- asn- SER- lys-TYR-ASN-ser- leu- thr- ile- ASN-ASN-asn- ILE- thr- LEU-leu- LYS-
lys-ser- ile- val- his- pro- SER- -TYR-ASN(pro, leu, thr, asn) ASN-ASN-asp- ILE- met-LEU-ile- LYS-
75 76 77 78 79 80 81 82 83 84 85 86 87 88 89 90 91 92 93 94 95

108 109 110 111 112 113 114 115 116 117 118 119 120 121 122 123 124 125 126 127 128 129
LEU-ser- thr- ALA-ALA-SER-phe-ser- gln- thr- VAL-ser- ala- val- cys- LEU-PRO-ser- ala- ser- asp-asp-
LEU-lys- ser- ALA-ALA-SER-leu-asn-ser- arg- VAL-ala- ser- ile- ser- LEU-PRO-thr- ser- cys-
96 97 98 99 100 101 104 103 104 105 106 107 108 109 110 111 112 113 114 115

130 131 132 133 134 135 136 137 138 139 140 141 142 143 144 145 146 147 148 149
phe- ala- ALA-GLY- THR- thr- CYS-val- thr- thr- GLY- TRP- GLY- leu- THR- arg- tyr- thr- asn- ala-
ala- ser- ALA-GLY- THR- gln- CYS-leu- ile- ser- GLY- TRP- GLY- asn- THR- lys- ser- ser- gly- thr-
116 117 118 119 120 121 122 123 124 125 126 127 128 129 130 131 132 133 134 135

150 151 152 153 154 155 156 157 158 159 160 161 162 163 164 165 166 167 168 169 170 171
asn- thr- PRO- ASP- arg- LEU-gln- gln- ala- ser- leu- PRO-leu- LEU-SER-asn- thr- asn- CYS-LYS-lys- tyr-
ser- tyr- PRO- ASP- val- LEU-lys- cys- leu- lys- ala- PRO-ile- LEU-SER-asp-ser- ser- CYS-LYS-ser- ala-
136 137 138 139 140 141 142 143 144 145 146 147 148 149 150 151 152 153 154 155 156 167

172 173 174 175 176 177 178 179 180 181 182 183 184 185 186 187 188 189 190 191
trp- gly- thr- lys- ILE- lys- asp- ala- MET- ile- CYS-ALA-GLY- ala- ser- gly- val- ser- -SER-CYS-
tyr- pro- gly- gln- ILE- thr- ser- asn- MET- phe-CYS-ALA-GLY- tyr- leu- glu- gly- gly- lys-asn- SER-CYS-
158 159 160 161 162 163 164 165 166 167 168 169 170 171 172 173 174 175 176 177 178 179

192 193 194 195 196 197 198 199 200 201 202 203 204 205 206 207 208 209 210 211 212
met-GLY-ASP-SER-GLY- GLY- PRO-leu- VAL-CYS-lys- lys- asn- gly- ala- trp- thr- leu- val- GLY-ILE-
gln- GLY-ASP-SER-GLY- GLY- PRO-val- VAL-CYS-ser- gly- lys- leu- gln- - GLY-ILE-
180 181 182 183 184 185 186 187 188 189 190 191 192 193 194 195 196

213 214 215 216 217 218 219 220 221 222 223 224 225 226 227 228 229 230 231 232 233 234
VAL-ser-SER-TRP-GLY-SER-ser- thr- cys- ser- thr- ser- thr- PRO-GLY-VAL-TYR- ala- arg- VAL-thr-ala-
VAL -SER-TRP-GLY-SER-gly- cys- ala- gln- lys- asn- lys- PRO-GLY-VAL-TYR- thr- lys- VAL-cys-asn-
197 198 199 200 201 202 203 204 205 206 207 208 209 210 211 212 213 214 215 216 217

235 236 237 238 239 240 241 242 243 244 245 246
leu-VAL-asn- TRP-val- gln- GLN-THR-leu- ALA- ala- ASN
tyr-VAL-ser- TRP-ile- lys- GLN-THR-ile- ALA- ser- ASN
218 219 220 221 222 223 224 225 226 227 228 229

FIG. 1. Amino Acid Sequence of Trypsinogen and Chymotrypsinogen.

contrast with the endopeptidase, chymotrypsin, which catalyzes hydrolysis at all susceptible bonds regardless of their position in a peptide chain. The crystalline enzyme has a molecular weight of about 33,000 based on various physical studies. The amino acid composition is known, but very little of the sequence has been determined. Much of our recent knowledge of the structure and mechanism of action of this enzyme is due to the work of Vallee and of Neurath and their associates. The enzyme appears to be a single polypeptide chain with single N-terminal and C-terminal asparagines. It contains one ZINC atom per mole which can be removed only by dialysis below pH 5.5 or at neutral pH by powerful chelating agents. The resulting zinc-free protein is stable but completely inactive. It has been shown that the zinc is bound to the protein by CHELATION with a specific sulfhydryl group and an amino group.

The substrate specificity of carboxypeptidase has been defined on the basis of comparative studies of a large number of compounds. They fall into two groups, illustrated by the following general formulas:

(a) Peptides or amides

$$R-\overset{O}{\overset{\|}{C}}-\overset{H}{\overset{|}{N}}-\underset{R'}{\underset{|}{\overset{H}{\overset{|}{C}}}}-\overset{O}{\overset{\|}{C}}-\ \vdots\ -\overset{H}{\overset{|}{N}}-\underset{R''}{\underset{|}{\overset{H}{\overset{|}{C}}}}-\overset{O}{\overset{\|}{C}}-OH$$

$$R-\overset{O}{\overset{\|}{C}}-\ \vdots\ -\overset{H}{\overset{|}{N}}-\underset{R''}{\underset{|}{\overset{H}{\overset{|}{C}}}}-\overset{O}{\overset{\|}{C}}-OH$$

(b) Ester analogues

$$R-\overset{O}{\overset{\|}{C}}-\overset{H}{\overset{|}{N}}-\underset{R'}{\underset{|}{\overset{H}{\overset{|}{C}}}}-\overset{O}{\overset{\|}{C}}-\ \vdots\ -O-\underset{R''}{\underset{|}{\overset{H}{\overset{|}{C}}}}-\overset{O}{\overset{\|}{C}}-OH$$

$$R-\overset{O}{\overset{\|}{C}}-\ \vdots\ -O-\underset{R''}{\underset{|}{\overset{H}{\overset{|}{C}}}}-\overset{O}{\overset{\|}{C}}-OH$$

Details of the specificity requirements may be found in the accompanying references, but it should be emphasized that a free terminal carboxyl group is required, that the free α-amino group of a dipeptide is inhibitory, that phenylalanine, tyrosine, tryptophan, leucine, methionine and isoleucine side chains in the R'' position make the best substrates, and that arginine, lysine and proline are the poorest substrates for this enzyme. It is special interest to note that ester analogues of these compounds, such as benzoylglycyl-β-phenyl-L-lactic acid are hydrolyzed, although carboxypeptidase A, in contrast with chymotrypsin, is not inhibited by the esterase inhibitor DFP.

Vallee and his associates have shown that metal ions other than zinc may be bound to carboxypeptidase and that these in turn lead to changes in the catalytic properties of the protein. Thus Co^{++} and Ni^{++} form a more active peptidase than Zn^{++} itself, while Hg^{++} and Cd^{++} cause activation of the esterase activity and complete inhibition of the peptidase activity. Other chemical treatments of the Zn^{++}-enzyme such as limited acetylation and iodination also lead to a decline in peptidase activity coincident with increased esterase activity.

Carboxypeptidase A, as well as the closely related enzyme, carboxypeptidase B, have been employed widely as aids to the determination of the chemical nature of the C-terminal amino acids of various peptides and proteins. Carboxypeptidase B utilizes basic amino acids as substrates much more effectively than does carboxypeptidase A.

Like chymotrypsin and trypsin, carboxypeptidase A is also derived from an active zymogen precursor, procarboxypeptidase A. The activation process is exceedingly complex, however, as procarboxypeptidase A has been shown to have a molecular weight of 87,000 and to consist of three discrete subunits. In the course of the activation process, trypsin first activates an endopeptidase activity within the zymogen, and only somewhat later does the typical peptidase activity appear. Each activity is associated with only one of three subunits.

Leucine Aminopeptidase. This metallo peptidase is perhaps more representative of the large group of tissue peptidases. It has been prepared from intestinal mucosa and kidney, and has been detected in several other tissues. So far as is known, it is not derived from a zymogen precursor. Full activity of the enzyme requires that it be incubated for several minutes in the presence of an excess of Mg^{++} or Mn^{++}. The metal binding is not so strong as in carboxypeptidase A, but apparently one mole of bound metal ion per mole of enzyme (molecular weight 300,000) is sufficient for activation. The enzyme specifically hydrolyzes the N-terminal amino acid of peptide chains and closely related compounds. It acts preferentially on derivatives containing leucine, but it will hydrolytically release a very broad spectrum of N-terminal amino acids. A useful synthetic substrate has been L-leucinamide. Slow hydrolysis of L-leucine *n*-butyl ester has been observed, although the enzyme is not inhibited by DFP. Much of our knowledge of this and other peptidases has been due to E. L. Smith and his associates, who were the first to postulate a role for ternary metallo-protein-substrate complexes in proteolysis. In leucine aminopeptidase, metal ion chelation by the protein is considerably more reversible than in the case of carboxypeptidase A. Subsequent interaction of the leucyl (or other) side chain, followed by chelation of the protein-bound metal ion with the nitrogen of the free N-terminal amino group and the nitrogen of the peptide bond, allows electron displacement into the metal ion and thereby labilizes the susceptible bond to nucleophilic attack by water.

Thiol Proteases. It has been widely observed that, despite the resistance of serine proteases and many metallo proteases to inhibition by sulfhydryl reagents such as *p*-chloromercuribenzoate and idoacetate, the proteolytic activity of extracts of many animal and plant tissues is markedly decreased by these reagents. A typical sulfhydryl-dependent protease which has been isolated in crystalline form and characterized by many chemical and physical techniques is papain. Although no single enzyme can be considered to be completely characteristic of any of these classifications, a consideration of papain will serve as an introduction to the properties of the thiol proteases.

Crystalline papain, obtained from papaya latex, has a molecular weight of about 21,000. The amino acid composition is known, and E. L. Smith and his associates have reported a large portion of the amino acid sequence. Full proteolytic activity can be obtained only if the enzyme is incubated with a combination of a mercaptan, such as cysteine, and the metal binding agent, ethylenediaminetetraacetic acid. Papain is inhibited by iodine, hydrogen peroxide, prolonged standing in air, Hg^{++}, or *p*-chloromercuribenzoate. It is also inhibited by reaction with idioacetic acid, and subsequent hydrolysis leads to the recovery of one mole of S-carboxymethylcysteine. Thus a single cysteine residue in papain appears to be an obligatory participant in catalysis.

Papain is an endopeptidase of rather broad specificity though it does show a preference for peptide bonds in which arginine and lysine contribute the acyl portion. Like the other proteases it will catalyze the hydrolysis of low molecular weight synthetic substrates which conform to its minimal specificity requirements. Examples include α-benzoyl-L-argininamide, carbobenzoxy-L-glutamic acid diamide, benzoylglycine amide, α-benzoyl-L-arginine ethyl ester, and benzoylglycine methyl ester. No evidence for a zymogen precursor has been reported, and the enzyme is not susceptible to DFP.

Detailed studies of the kinetics of the papain-catalyzed hydrolysis of several synthetic substrates as a function of pH have led to the suggestion that the carboxyl group of an aspartic or glutamic acid residue also plays a key role in the catalytic mechanism of papain. Some evidence for the participation of an amino group has also been reported. Most suggested mechanisms include the transient formation of an acyl-thiol enzyme intermediate. Other plant proteases, such as ficin, appear to be very similar to papain, but the relationship to a large group of tissue thiol-cathepsins is less certain.

Acid Proteases. This final classification of proteases is best exemplified by pepsin, a protease of the gastric secretion of many species. Like the pancreatic proteases, chymotrypsin, trypsin and carboxypeptidase A, pepsin is derived from a zymogen. Pepsinogen is autocatalytically converted to pepsin at low pH as a result of a controlled proteolysis which liberates several peptides including one which can be shown to act as a reversible competitive inhibitor of pepsin. Crystalline pepsin itself is essentially homogenous and has a molecular weight of about 36,000. The amino acid composition is known but relatively few sequence studies have been reported. Unusual properties of this protein are its very low isoelectric point and its instability toward denaturation above pH 6.

The substrate specificity of pepsin is rather broad. It has been established by examining both the rates of hydrolysis of synthetic peptides and the sites of peptic hydrolysis in various proteins. Among the most susceptible peptide linkages are those in which two hydrophobic amino acids participate, *e.g.*, phe.leu, phy.phe, and phe.tyr. No amino acid esters have been found susceptible to pepsin-catalyzed hydrolysis. (Further details are presented in the separate article PEPSINS AND PEPSINOGENS.)

References

1. Cunningham, L. W., in "Comprehensive Biochemistry" (Florkin and Stotz, Editors), Amsterdam, Elsevier Publishing Co., in press.
2. Desnuelle, P., in "The Enzymes" (Boyer, Lardy and Myrbäck, Editors), Vol. 4, pp. 93, 119, New York, Academic Press, 1960.
3. Green, N. M., and Neurath, H., in "The Proteins" (Neurath, H., and Bailey, K., Editors), Vol. 2, Pt. B, p. 1057, New York, Academic Press, 1954.
4. Walsh, K. A., and Neurath, H., *Proc. Natl. Acad. Sci. U.S.*, **52**, 884 (1964).
5. Hartley, B. S., *Nature*, **201**, 1284 (1964).
6. Bovey, F. A., and Yanari, S. S., in "The Enzymes" (Boyer, Lardy and Myrbäck, Editors), Vol. 4, p. 63, New York, Academic Press, 1960.
7. Neurath, H., "Protein-Digesting Enzymes," *Sci. Am.*, **211**, 68 (December 1964).

Leon Cunningham

PROTOPLASTS. See BACTERIAL CELL WALLS.

PTERIDINES

The pteridine ring system is formally related to the hydrocarbon naphthalene (it is a 1,3,5,8-tetraazanaphthalene). More profitably, it can be related to the monocyclic nitrogen heterocycles, pyrimidine and pyrazine (it is a fused pyrimido-4,5-b-pyrazine), or to the purine ring system (from which it can formally be derived by addition of one carbon to the five-membered ring). The ring is now generally numbered as shown in I, although it was originally numbered (II) to conform with the purine system and this numbering is still encountered.

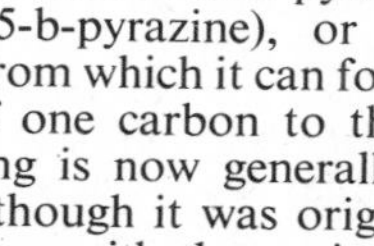

I

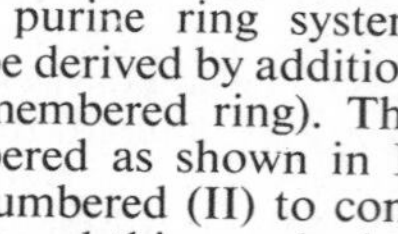

II

The parent compound, pteridine, has never been isolated from natural sources. Indeed, with only two exceptions (as yet), all natural compounds are 2-amino-4-hydroxy-pteridines (for which residue the name "pterin" has been resurrected, but is not widely accepted). The large number of such compounds from natural sources differ only in the substituents in, and the state of oxidation of, the pyrazine ring. The two known 2,4-dihydroxy-pteridines are intimately concerned with RIBOFLAVIN biosynthesis (riboflavin itself being a fused benzpteridine).

Structure and Occurrence. Pteridines were first isolated from natural sources over 75 years ago by Hopkins. The material was butterfly wings (hence, the derivation of the name from the Greek, *pteron*, a wing), but the correct structures of these wing pigments were not established until 1940. The reason for this long delay between discovery and recognition is to be found in the properties of the naturally occurring compounds. They are difficult to combust for elementary analysis, they have no melting point, they retain water tenaciously but non-stoichiometrically, and until the advent of paper chromatography, reliable criteria of purity were lacking.

The three best-known compounds in this group of wing pigments are xanthopterin (2-amino-4,6-dihydroxy-pteridine), isoxanthopterin (2-amino-4,7-dihydroxy-pteridine and leucopterin (2-amino-4,6,7-trihydroxy-pteridine).

It should be pointed out that these chemical names, are not strictly correct, since recent work has shown that as a general rule (and certainly with regard to the compounds described in this article) hydroxy groups—which are always on a carbon adjacent to a ring nitrogen—do not exist as such, but rather as oxo groups, the structures then being true cyclic amides. Thus 2-amino-4-hydroxy-pteridine is strictly, 2-amino-3,4-dihydro-4-oxo-pteridine. However, with this reservation, the "hydroxy" nomenclature is retained here for convenience and simplicity.

None of the above compounds has been shown, with certainty, to have biological activity (with the possible exception of xanthopterin). They appear to be metabolic end products. Other compounds which have been isolated and identified at about the same time (some later) and can be included in this group are the yellow chrysopterin (2-amino-4,6-dihydroxy-7-methylpteridine), the orange-red erythropterin [3′-(2-amino-4,6-dihydroxy-7-pteridinyl)pyruvic acid (III)], and the red bicyclic pterorhodin (IV).

Two other colorless compounds, found in the meal moth *Ephestia kuhniella*, are closely related to erythropterin. They are ekapterin [(III) with the pyruvic acid residue replaced by lactic acid] and lepidopterin [(III) with the —C=O group of the pyruvic acid residue replaced by —$CHNH_2$].

In contrast to the above biologically inert compounds, a new group of pteridines was discovered in the 1940's *because of* their biological activities as growth factors for a variety of organisms, particularly microorganisms. Many such compounds have been isolated, but generically they are all similar, being based on pteroic acid [2-amino-4-hydroxy-6-(*p*-carboxyanilinomethyl)pteridine]. The first of these was FOLIC ACID (isolated from spinach leaves, hence the derivation of the name from the Latin *folia*, a leaf), and this has now become a general designation for compounds containing the pteroyl radical showing biological activity.

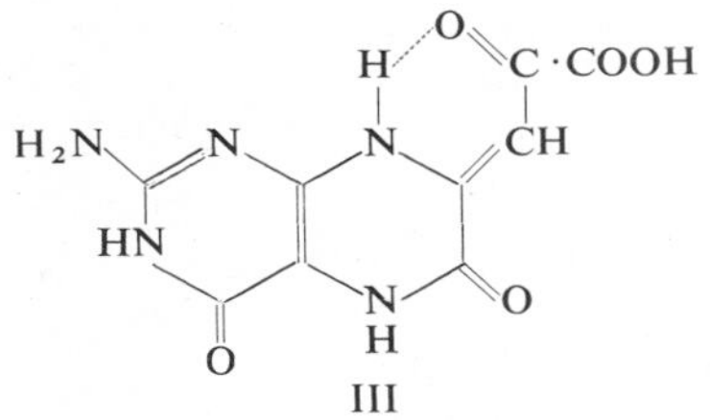

III

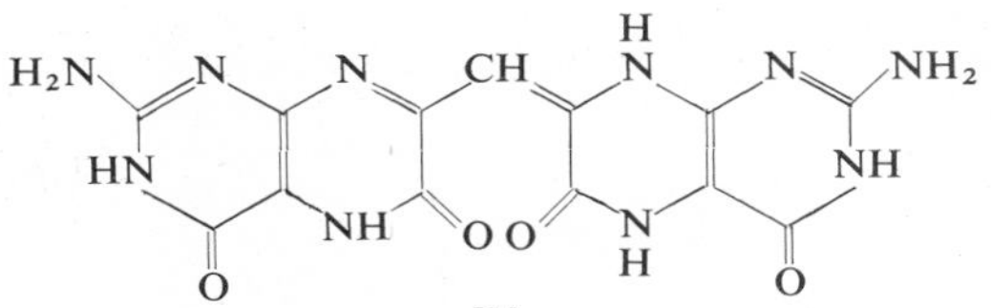

IV

A third group of closely related compounds contains those associated with biopterin [2-amino-4-hydroxy-6-(L-*erythro*-1′,2′-dihydroxypropyl)pteridine] which was first isolated from human urine and, simultaneously, from the fruit fly *Drosophila melanogaster* in 1955, and which has growth factor activity for the flagellate *Crithidia fasciculata*. It is also said to be required for the transformation of a bee larva into a queen. It has since been isolated from *Ephestia*, amphibia, fish and red ants. Naturally occurring compounds related to this material are: ichthyopterin (7-hydroxybiopterin) found in the skin of carp and other fish; neopterin [2-amino-4-hydroxy-6-(D-*erythro*-trihydroxypropyl)-pteridine], isolated from bee larvae, and its corresponding 3′-phosphate isolated from *Escherichia coli*; biopterin glycosides (in which a variety of sugar residues are attached to the 1′- or the 2′-position of the side chain), isolated from blue-green algae; the dihydro-pteridines sepiapterin (2-amino-7,8-dihydro-4-hydroxy-6-lactylpteridine) and isosepiapterin (2-amino-4-hydroxy-6-propionylpteridine) both isolated from *Drosophila melanogaster*, and the latter in considerable quantity from blue-green algae; lastly, the drosopterins (drosopterin, isodrosopterin and neodrosopterin) the red pigments in *Drosophila* eyes. The drosopterins also occur in the skin of some fish and in the skin folds of various reptiles. The structures of these latter compounds are not known with certainty; they may be dipteridyl derivatives. One tetrahydro compound related to biopterin has been partially characterized, although its structure is not well established. This "tetrahydrobiopterin" has been isolated from the skin of amphibians and fish, and a closely related compound alleged to have a ribosyl and a phosphate attached to it has been obtained from the blue-green alga (*Anacystis nidulans*) and partially characterized.

Various other pteridines have been isolated from natural sources. 2-Amino-4-hydroxypteridine occurs in *Drosophila melanogaster* and can be obtained from a variety of sources. It is known to be the immediate precursor of isoxanthopterin. 2-Amino-4-hydroxypteridine-6-carboxylic acid is probably a breakdown product, an artifact of isolation of a number of the compounds described above, as is 2,6-diamino-4-hydroxypteridine, first isolated from blue-green algae and *Drosophila*. 2-Amino-4-hydroxy-6-hydroxymethylpteridine has been found in one species of blue-green algae, and more recently in a *Pseudomonas* species; 2-amino-7,8-dihydro-4-hydroxy-6-lactyl-8-methylpteridine has been isolated from the silk worm, and finally 2-amino-4-hydroxypteridine-6-sulfate occurs in considerable quantity in the bacterium, *Azotomonas insolita*.

It should be mentioned—as can be deduced from the above presentation—that the distinction between a naturally occurring compound and an artifact of isolation tends to be rather fine in this group of compounds. The corollary of this, of course, is that the naturally occurring forms of some of the compounds described above remain to be isolated. This becomes particularly obvious when one looks for biological activities of these compounds. Most of them are biologically inert, probably because they are not in the correct state of oxidation for activity, and there is no direct biological route to that state; alternatively, they may have been changed chemically during isolation.

Biosynthesis of Pteridines. This subject is dealt with in some detail under FOLIC ACID. In brief, there seems to be little doubt that pteridines are derived from PURINES which are probably in the form of ribonucleotides (see PURINE BIOSYNTHESIS). Thus the purine ribonucleotide (*e.g.*, guanylic acid) is thought to cleave at carbon-8, the resulting pyrimidine ribosyl phosphate undergoing an Amadori rearrangement and cyclizing with incorporation of two of the carbon atoms of the ribotyl residue into the newly formed pyrazine ring and the remaining three forming a side chain at the 6-position of the ring. From the resulting compound, which can be oxidized to the (2-amino-4-hydroxy-6-pteridinyl)glycerol phosphate described above, folic acid is derived by an as yet unknown mechanism. Indeed many of the fine details in the observed conversion of guanylic acid into a structure containing the pteridine ring are matters for speculation. Unambiguous evidence of the nature of the intermediates is lacking.

A unifying scheme for the biosynthesis of members of the folic acid group and of the biopterin group has been proposed. In this, the three-carbon phosphate side chain is cleaved and a one-carbon unit (for folic acid) or a three-carbon unit (for biopterin) is added to the resulting 2-amino-4-hydroxypteridine derivative. Many of the compounds in the biopterin group can be related to one another by known chemical reactions, although their biological interconversions again remain uncertain.

Biological Functions. As already mentioned, the "wing pigment" group of pteridines appear to be metabolic end products. Xanthopterin has some surprising effects on specific cells, causing them to divide and proliferate to a remarkable extent. The biochemical basis for this effect is unknown. A good deal is known about the metabolic activities of the folic acid group of pteridines (see FOLIC ACID COENZYMES).

The biological functions of the biopterin group are slowly becoming clear. Biopterin is a growth factor for the flagellate *Crithida fasciculata*, but the biochemical basis for this is not known. The conversion of phenylalanine into tyrosine requires a pteridine cofactor whose structure is believed to be that of a dihydrobiopterin. A few other oxidation reactions, *e.g.*, the oxidation of long-chain alkyl ethers of glycerol to the corresponding fatty acid and glycerol, and the conversion of L-tyrosine to DOPA, require pteridine cofactors which are probably of the same type. It seems likely that more reactions of this general type will also be shown to have pteridine cofactor requirements. The mechanism by which the pteridine functions is not known, but it appears to involve a redox couple between a tetrahydro compound (presumably the 5,6,7,8-tetrahydro) and a dihydro compound [possibly the 5,6-dihydro or a "quinonoid" dihydro derivative (V)].

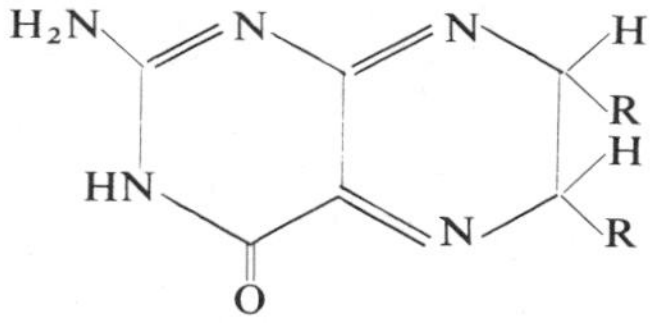

H_2N N N H R H HN N R O

V

References

1. GATES, M., *Chem. Rev.*, **41**, 63 (1947).
2. ALBERT, A., *Quant. Rev.*, **6**, 197 (1952).
3. ALBERT, A., *Adv. Chem. Org. Natural Products*, **11**, 350 (1954).
4. PFLEIDERER, W., *Angew. Chem., Intern. Ed.*, **3**, 114 (1964).
5. WOLSTENHOLME, G. E. W., AND CAMERON, M. P. (EDITORS), "Chemistry and Biology of Pteridines," *Ciba Found. Symp.* (1954).
6. FORREST, H. S., in "Comparative Biochemistry," (FLORKIN, M., AND MASON, H. S., EDITORS), **IVB**, New York, Academic Press, 1964, p. 615.
7. PFLEIDERER, W., AND TAYLOR, E. C. (EDITORS), "Pteridine Chemistry," Proceedings of the Third International Symposium, Oxford, Pergamon Press, 1964.

HUGH S. FORREST

PTEROYLGLUTAMIC ACID

This is an alternative name for FOLIC ACID, one member of the VITAMIN B GROUP. Pteroylglutamic acid is one of a variety of naturally occurring

PTERIDINES. Reduced forms of pteroylglutamic acid, namely dihydropteroylglutamic acid (or dihydrofolic acid) and tetrahydropteroylglutamic acid (or tetrahydrofolic acid), and certain of their derivatives, are members of the group of FOLIC ACID COENZYMES (FOLATE COENZYMES), and play catalytic roles in SINGLE CARBON UNIT METABOLISM.

PURINE BIOSYNTHESIS AND METABOLISM

Derivatives of the PURINES have essential functions in all living things in the forms of (a) nucleotides and COENZYMES, which participate in nearly all of the biosynthetic processes of cells as carriers of amino acids, carbohydrates, phosphate, sulfate, and acyl groups, as well as what might be regarded as "chemical energy," and (b) nucleic acids, which serve both as the repository for genetic information and as the intermediates necessary for transcribing this information into protein structure. In addition, purines are excreted by certain animals as the major product for the elimination of superfluous dietary nitrogen.

Structure and Nomenclature. We will consider here five purine bases which have general biological importance (Fig. 1). The purines are usually metabolically active as nucleosides, compounds containing a five-carbon sugar (D-ribose or D-2-deoxyribose), or as nucleotides, the phosphate esters of nucleosides [see PURINES (STRUCTURE AND OCCURRENCE) for structures, *e.g.*, of adenosine-5′-phosphate; also for conventional numbering of the ring systems]. The nucleotides of the purines in Fig. 1 are called inosinic acid (inosine monophosphate), adenylic acid (adenosine monophosphate), guanylic acid (guanosine monophosphate), and xanthylic acid (xanthosine monophosphate), and sometimes abbreviated IMP, AMP, GMP and XMP. The nucleotide of uric acid does not occur naturally. The prefix *deoxy-* is used for the compounds containing deoxyribose. Nucleotides also function metabolically as diphosphates (abbreviated, for example, as ADP, adenosine diphosphate) and triphosphates [as ATP, adenosine triphosphate; see structure in PURINES (STRUCTURE AND OCCURRENCE)].

Biosynthesis. Almost all species of living things have the ability to synthesize purine compounds from small-molecule precursors. In higher animals, it appears that only the liver and certain rapidly proliferating tissues such as the spleen, intestinal epithelium and bone marrow can carry out this process to a significant extent. Purine synthesis occurs particularly rapidly in avian liver, since birds excrete a purine (uric acid) as the major end product of nitrogen metabolism. Experiments with isotopic tracers, first in intact pigeons, and later with enzyme systems isolated from pigeon liver, established the origin of each of the atoms of the purine ring system. These precursors are glycine, carbon dioxide, a fragment related to formic acid, and nitrogen atoms derived from aspartic acid and glutamine. The initial studies with the enzymes systems from pigeon liver showed that the *first purine compound to be formed* is the ribonucleotide of hypoxanthine, *inosinic acid*, and demonstrated, furthermore, that all of the intermediate compounds in the biosynthetic sequence are 5-phosphoribosyl derivatives. In Fig. 2 is shown the sequence of reactions by which inosinic acid is constructed from small-molecule precursors. Each reaction is catalyzed by a specific enzyme. It may be noted that many of the reactions involve the synthesis of carbon-nitrogen bonds by the abstraction of the elements of water from the reactants by ATP. Adenylic acid and guanylic acid are derived from inosinic acid by the reactions of amination and oxidation shown in Fig. 3. It has been found subsequently that all organisms so far studied synthesize purine nucleotides essentially by the pathways shown in Figs. 2 and 3.

Dietary Sources of Purine Derivatives. In addition to their synthesis from small molecules, purine compounds may also be derived from the nucleic acids and nucleotides in the diet. In the

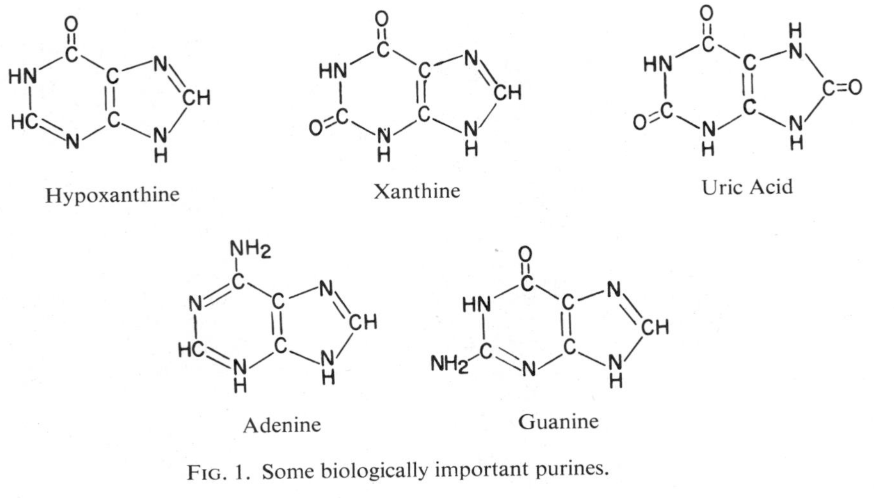

Hypoxanthine

Xanthine

Uric Acid

Adenine

Guanine

FIG. 1. Some biologically important purines.

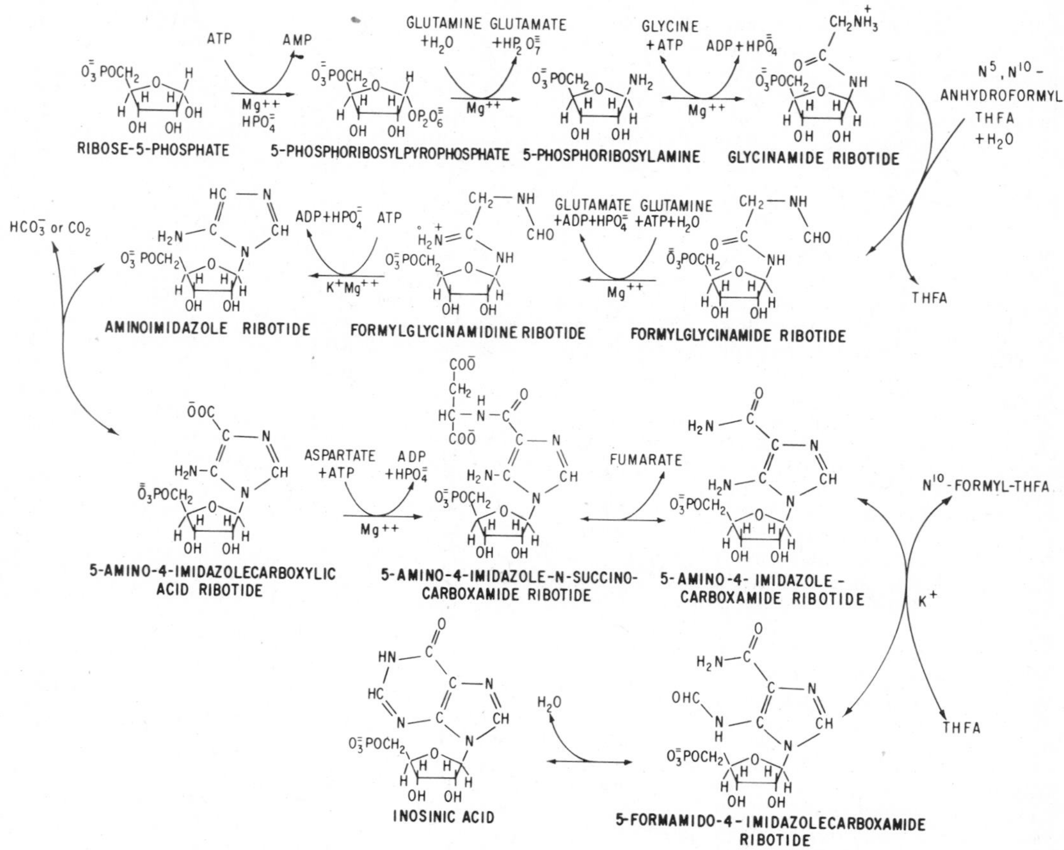

FIG. 2. Biosynthesis of inosinic acid. (THFA denotes tetrahydrofolic acid.)

digestive tract of higher animals, enzymes called NUCLEASES and PHOSPHATASES hydrolyze nucleic acids to nucleotides and then to nucleosides. The nucleosides may be further degraded by enzymes found in certain tissues by phosphorolytic cleavage to the free purines and ribose-1-phosphate. Resynthesis of the purine nucleotides in cells may occur by two pathways. One involves the reversal of the last reaction, the combination of a purine base with ribose-1-phosphate to form the nucleoside, followed by phosphorylation of the nucleoside by ATP to yield the nucleoside-5′-phosphate and ADP. Many organisms depend largely upon a second, one-step, process for the conversion of purines to the metabolically active nucleotides. In this reaction, the purine base is condensed with 5-phosphoribosyl-1-pyrophosphate, to yield directly the nucleotide and pyrophosphate.

Control of Purine Synthesis. In most organisms the biosynthesis of purine nucleotides is reduced when an adequate supply of purine derivatives is available. Control of purine synthesis may be exerted by the intracellular levels of the nucleotide derivatives by the mechanisms of FEEDBACK inhibition and enzyme repression. The first enzyme of the pathway, which effects the reaction between glutamine and phosphoribosyl pyrophosphate, is inhibited by purine nucleotides, so that the flow of compounds through the pathway is reduced when substantial levels of the end products are present. Also, the synthesis of the early enzymes, at least through the first four steps, is repressed by sufficient levels of the purine nucleotides. In addition, both of these mechanisms of control have been observed in the steps shown in Fig. 3 by the respective end products of each sequence of reactions.

Degradation and Excretion of Purines. In higher animals, purine compounds derived from the diet and by synthesis in amounts in excess of those required by the various tissues are degraded and excreted. The nucleotides and nucleosides are converted, chiefly in the liver and kidneys, to the free bases by the reactions mentioned previously. The amino groups of adenine and guanine are removed by hydrolysis to yield hypoxanthine and xanthine, respectively. Enzymatic oxidation of the latter two compounds produces uric acid. The subsequent metabolism of uric acid varies from species to species. In many animals, uric acid is

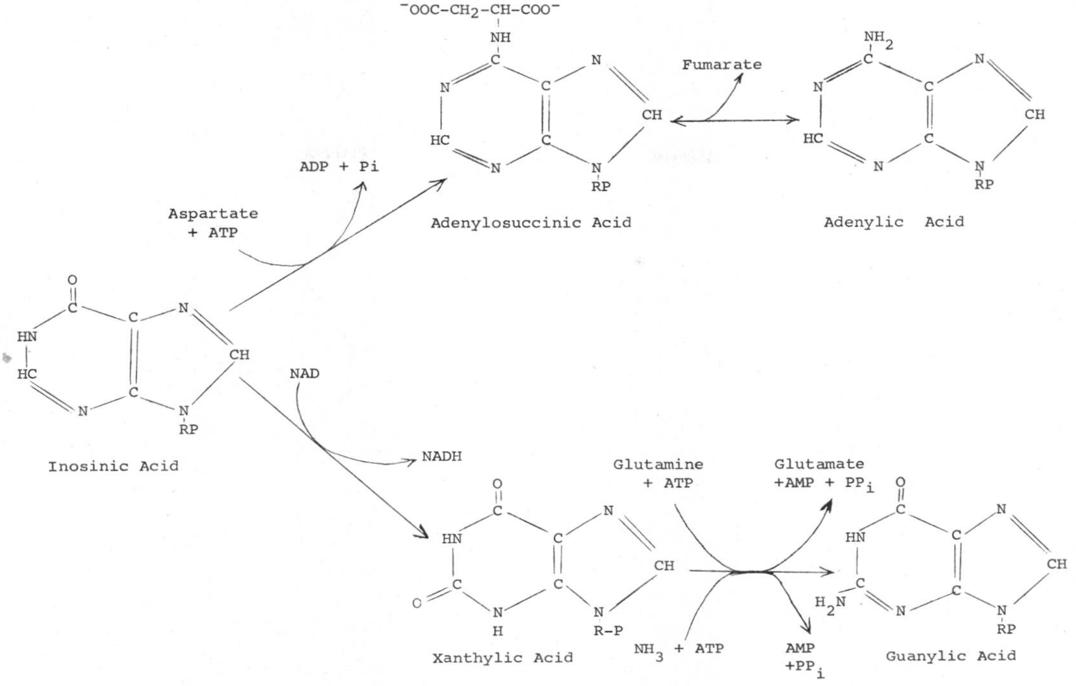

FIG. 3. Biosynthesis of adenylic and guanylic acids. (RP denotes 5-phosphoribosyl, NAD is nicotinamide adenine dinucleotide, and PP_i is pyrophosphate.)

converted to allantoin by the enzyme uricase and is excreted as such. Most marine animals further degrade allantoin to allantoic acid, urea or ammonia. Some species, including man, higher primates and birds, lack uricase, so that the uric acid is excreted intact, either in solid form (birds) or in soluble form in the urine. In birds and reptiles, the excretion of uric acid as the end product of nitrogen metabolism is an important mechanism for the conservation of water. The low solubility of uric acid allows it to be eliminated in solid form with very little loss of water. On the other hand, the low solubility of uric acid and its salts may present problems in other species, such as humans, which must excrete their uric acid in solution form. Some individuals produce more purines than can normally be excreted as urinary uric acid. The uric acid content of the blood and tissues increases to the point where salts of this purine may crystallize out, particularly in the joints. These solid deposits produce an arthritic condition known as tophaceous gout.

Metabolic Functions of Purine Derivatives. The purine (and pyrimidine) nucleoside phosphates are widely involved in intermediary metabolism in the form of anhydride derivatives with carboxylic, phosphoric, and sulfuric acids. Such compounds are both chemically and metabolically reactive and are often termed "energy-rich" compounds. Probably the most important anhydride compound in metabolism is ATP, which contains two "pyrophosphate" bonds. ATP is formed from ADP and orthophosphate during the fermentative and oxidative degradation of many foodstuffs. The chemical energy of the foodstuffs is thereby stored in the ATP, from which the energy may be recovered to drive many biosynthetic reactions of the sort shown in Figs. 2 and 3. One of the pyrophosphate bonds of ATP is cleaved in such biosynthetic reactions, to form either ADP and orthophosphate, or AMP and pyrophosphate. ATP may also serve as a phosphorylating agent: typical examples are the conversion of ribose-5-phosphate to phosphoribosyl pyrophosphate, of nucleosides to nucleoside-5′-phosphates, and of the nucleoside monophosphate sequentially to the corresponding diphosphates and triphosphates.

Biological macromolecules such as proteins, nucleic acids, polysaccharides, and complex lipids are built up by the condensation of activated subunits. The activation is achieved by formation of an anhydride derivative between the subunit and a nucleoside triphosphate, usually with the splitting out of pyrophosphate. Thus, amino acyl adenylates are the activated intermediates in protein synthesis, and carbohydrates are transferred in polysaccharide formation from nucleoside diphosphate glycosyl derivatives. The nucleic acids are formed by direct condensation of nucleoside (or deoxynucleoside) triphosphates. [See CARBOHYDRATE METABOLISM; DEOXYRIBONUCLEIC ACIDS (REPLICATION); PROTEINS (BIOSYNTHESIS); RIBONUCLEIC ACIDS (BIOSYNTHESIS)].

SULFATE is introduced into organic compounds by transfer from 3′-phosphoadenosine-5′-phosphosulfate, a mixed anhydride biosynthesized from ATP and sulfate. The metabolically active compound for the transfer of methyl groups is S-adenosyl methionine. This sulfonium derivative is formed from ATP and methionine.

Several vitamins of the B group, including nicotinamide (niacin), riboflavin, pantothenic acid, and B_{12} may occur as nucleotide derivatives, in which form they are active in enzyme systems. These COENZYME forms of the vitamins participate in biological reactions of hydride, electron, acyl and alkyl group transfer. In addition to these biosynthetic roles of the nucleoside triphosphates, the chemical energy stored in such compounds may also be utilized for mechanical and electrical work, such as in muscular contraction and nerve action.

References

1. DAVIDSON, J. N., "The Biochemistry of Nucleic Acids," Fourth edition, New York, John Wiley & Sons, 1959.
2. BUCHANAN, J. M., AND HARTMAN, S. C., "Enzymatic Reactions in the Synthesis of the Purines," *Advan. Enzymol.*, **21**, 199 (1959).
3. KORNBERG, A., "Pyrophosphorylases and Phosphorylases in Biosynthetic Reactions," *Advan. Enzymol.*, **17**, 191 (1957).
4. WHITE, A., HANDLER, P., AND SMITH, E. L., "Principles of Biochemistry," Third edition, Chs. 10, 28, New York, McGraw-Hill Book Co., 1964.

STANDISH C. HARTMAN

PURINES (STRUCTURE AND OCCURRENCE)

Purine is a planar heterocyclic compound composed of a pyrimidine and an imidazole ring with two carbon atoms in common. The parent compound, as well as a relatively large number of oxo and amino derivatives, have been isolated from every type of organism: animal, plant and microbial cells, and from viral nucleoproteins.

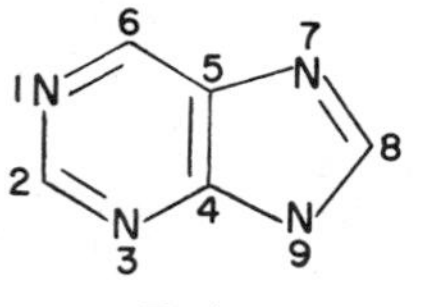

Purine

Purines (and pyrimidines) are building units of nucleic acids, and they play the important role in the storage of genetic information in the processes of cell division, reproduction and the transmission of hereditary factors.

Purines, when substituted on position 9 with a sugar, form what are known as purine *nucleosides.* Further esterification of such nucleosides with phosphoric acid, results in *nucleotides*, *e.g.*, adenine (6-aminopurine) is a purine base, adenosine (9-β-D-ribofuranosyladenine) is a purine nucleoside and adenylic acid b (9-β-D-ribofuranosyl-adenine-3′-phosphate or adenosine-3′-phosphate) is a purine nucleotide. Another nucleotide, adenosine-5′-phosphate, has the structure shown.

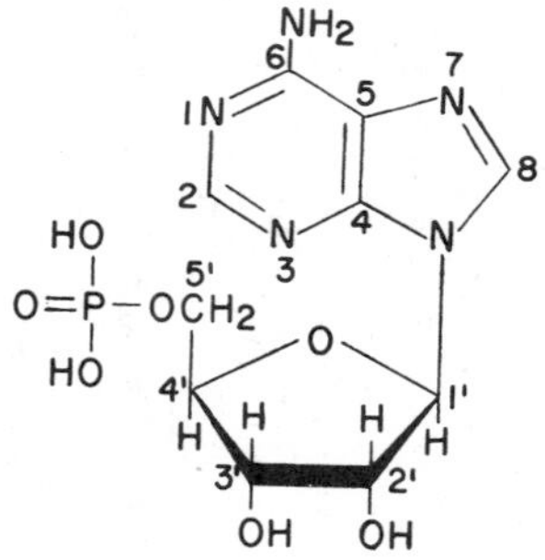

Adenosine-5′-phosphate

In nucleic acids, purine (and pyrimidine) nucleotides are joined together by phosphodiester bonds through the 3′- and 5′-positions. Nucleic acids containing ribose as the sugar component are designated RIBONUCLEIC ACIDS or RNA, whereas those containing 2-deoxyribose are designated DEOXYRIBONUCLEIC ACIDS or DNA. The most common of the purine bases, *i.e.*, *adenine* and *guanine* (2-amino-6-purine-one) (see PURINE BIOSYNTHESIS, Fig. 1, for structures), are always present in both RNA and DNA. Adenine also occurs in the free form in human urine and feces, and guanine is found in the excreta of birds. The iridescence of fish scales and the white, shiny appearance of the skin of many amphibia and reptiles are also due to the presence of crystals of guanine in these tissues.

Methylated and Oxo Derivatives. Besides adenine and guanine several other purines have been identified as minor components of RNA; *e.g.*, N^1-methyl-adenine, 2-methyladenine and N^6, N^6-dimethyladenine have been found in soluble RNA from animal, plant and bacterial cells. Also N^1-methylguanine, N^2-methylguanine, N^2,N^2-dimethylguanine and 7-methylguanine have been isolated from similar sources. Several oxo derivatives of purine, *i.e.*, *hypoxanthine* (6-purine-one), *xanthine* (2-6-purinedione) and *uric acid* (2,6,8-purinetrione) (see PURINE BIOSYNTHESIS, Fig. 1, for structures), are of great metabolic importance. Uric acid is the chief end product of purine (and protein) metabolism in many animal species.

Xanthines methylated at one or more of the nitrogen atoms are important natural products. 1,3-Dimethylxanthine (*theophylline*) occurs in tea leaves; 3,7-dimethylxanthine (*theobromine*) is the most important alkaloid of the cocoa bean, and 1,3,7-trimethylxanthine (*caffeine*) occurs in different plants all of which are sources of stimulant beverages (*e.g.*, coffee beans).

Purines in Coenzymes. Several important COENZYMES are derivatives of purine nucleotides.

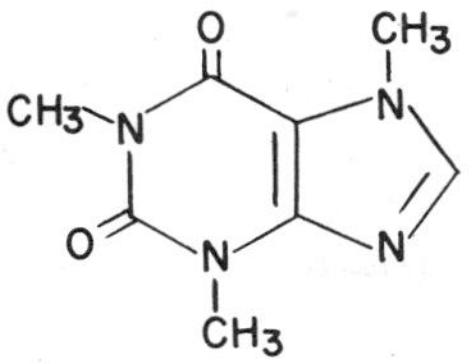

1,3,7-Trimethylxanthine (caffeine)

Coenzymes containing adenine, guanine and hypoxanthine bring about a great many of the metabolic tranformations which take place in the cell. All cells contain a number of these "free" nucleotides. The mono-, di- and tri-5′-phosphates

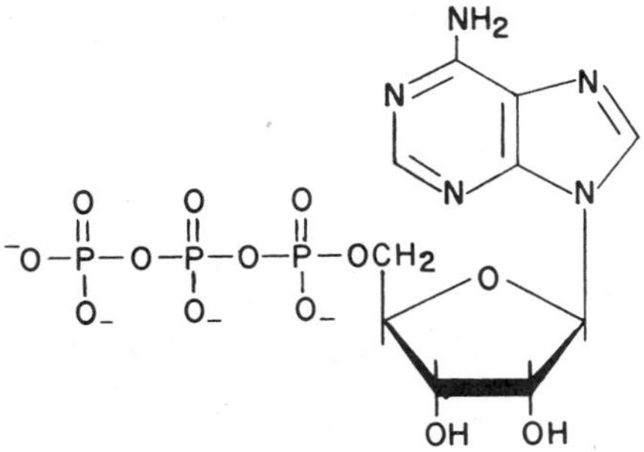

Adenosine-5′-triphosphate (ATP)

of adenosine (AMP, ADP, ATP) are involved in transphosphorylation and energy transfer reactions. In addition, anhydrides of fatty acids or amino acids and adenosine polyphosphates represent the "biochemically activated" forms of those acids (see PHOSPHATE BOND ENERGIES). Also S-adenosylmethionine is involved in transmethylation reactions (see METHYLATIONS IN METABOLISM). Other adenine coenzymes widely distributed in nature are: NICOTINAMIDE ADENINE DINUCLEOTIDE (NAD), nicotinamide adenine dinucleotide phosphate (NADP), flavin adenine dinucleotide (FAD) (see FLAVINS), and COENZYME A (CoA). NAD and NADP occur in plant and animal tissues, where they form the prosthetic group of a large number of enzymes, primarily dehydrogenases. The presence of FAD with its reduced form ($FAD \cdot H_2$) has been demonstrated in a variety of material including microorganisms and several different animal tissues, where it is involved in oxidation-reduction reactions. Coenzyme A is readily acetylated on the terminal mercapto group and thus effects the transfer of acetyl groups in the biological systems. The word formulas for the above coenzymes are the following:

Guanosine polyphosphates (GTP and GDP) have been found in many sources including rat brain, muscle and liver, rabbit muscle, yeast, etc. Large amounts of GTP are found in embryonic tissue. Also, a guanosine polyphosphate derivative (GDP-mannose) has been isolated from yeast. It seems likely that GTP plays an important role in PROTEIN BIOSYNTHESIS.

The natural occurrence of inosine polyphosphates has been demonstrated in a few cases. ITP has been found in frog muscle and mitochondrial nucleotides. The biological importance of inosine polyphosphates as coenzymes is not clear.

Purines in Antibiotics. A number of ANTIBIOTICS containing purine or related groups have been isolated. These are the following: *Pathocidin* produced by *Streptomyces albus* var. *pathocidicus*. It was identified as 8-azaguanine. *Nebularine* (9-β-D-ribofuranosyl purine) was isolated from the mushroom *Agaricus* (*Clitocybe*) *nebularis* and from the streptomycete *Streptomyces* (strain B 34) *yokosukaensis*. It is active against *Mycobacterium tuberculosis*, *Mycobacterium phlei* and *Candida albicans*. *Nucleocidin* is a derivative of adenine (adenine-9-$C_6H_{10}O_5$-OSO_2NH_2) found in cultures of *Streptomyces calvus* which possesses antitrypanosomal and broad spectrum antibacterial properties. *Tubercidin* is produced by

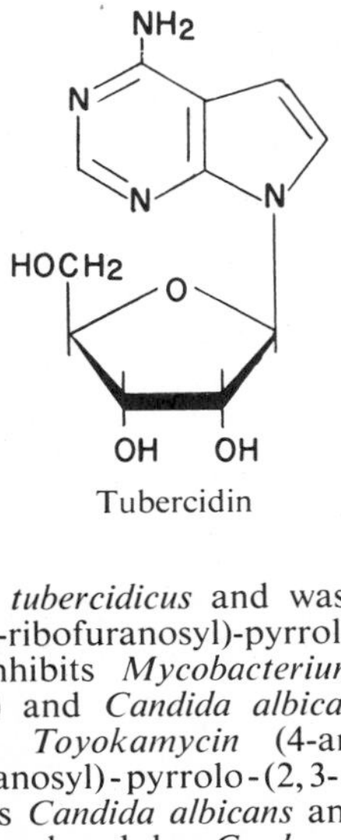

Tubercidin

Streptomyces tubercidicus and was identified as 4-amino-7-(D-ribofuranosyl)-pyrrolo-(2,3-d)-pyrimidine. It inhibits *Mycobacterium tuberculosis* (strain BCG) and *Candida albicans*. A related antibiotic is *Toyokamycin* (4-amino-3-cyano-7-(D-ribofuranosyl)-pyrrolo-(2,3-d)-pyrimidine which inhibits *Candida albicans* and other fungi. *Cordycepin* produced by *Cordyceps militaris* is 9-(D-3-deoxyribofuranosyl)adenine. *Angustmycin A* (decoyinine) and *Angustmycin C* (psicofuranine) are two related adenine antibiotics isolated from

Adenine-1′-ribose-5′-P-P-5′-ribose-1′-nicotinamide (NAD)

Adenine-1′-ribose-5′-P-P-5′-ribose-1′-nicotinamide (NADP)
|
2′P

Adenine-1′-ribose-5′-P-P-5′-ribitol-1′-flavin (FAD)

Adenine-1′-ribose-5′-P-P-pantothenate-aminoethylthiol (CoA)
|
3′-P

Streptomyces hygroscopicus. Other adenine derivatives are *Septacidin* obtained from *Fusarium bulbigenum*, *3′-amino-3′-deoxyadenosine* isolated from *Helminthosporium* and *Homocitrullylaminoadenosine* produced by *Cordyceps militaris*. *Puromycin* isolated from the actinomycete *Streptomyces alboniger* consists of an aminonucleoside linked to the amino acid *p*-methoxyphenylalanine.

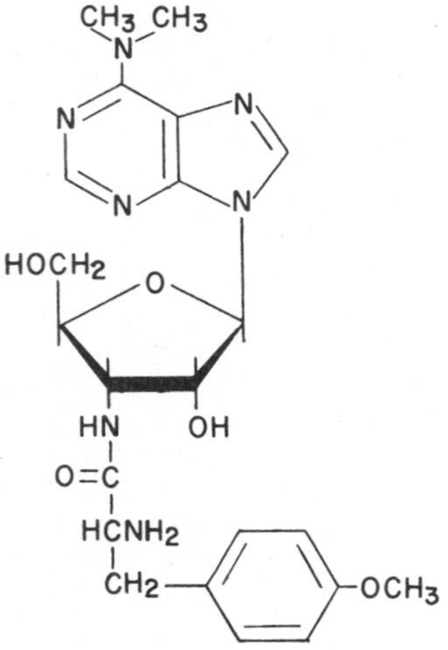

Puromycin

It inhibits protein synthesis by interfering with the transfer of amino acids by soluble RNA to the ribosomes of bacteria and animal cells.

References

1. Bendich, A., in "The Nucleic Acids" (Chargaff, E., and Davidson, J. N., Editors), pp. 81–136, New York, Academic Press, 1955.
2. Michelson, A. M., "The Chemistry of Nucleosides and Nucleotides," New York, Academic Press, 1963.
3. Bock, R. M., in "The Enzymes" (Boyer, P. D., Lardy, H., and Myrbäck, K., Editors), Vol. 2, p. 3, New York, Academic Press, 1960; Utter, M. F., *ibid.*, Vol. 2, p. 75.
4. Umezawa, H., in "Recent Advances in Chemistry and Biochemistry of Antibiotics," p. 167, Microbial Chemistry Research Foundation, Tokyo, 1964.

John D. Fissekis

PYRIDINE NUCLEOTIDES. See nicotinamide adenine dinucleotides. Diphosphopyridine nucleotide (DPN) is an older term for nicotinamide adenine dinucleotide (NAD); triphosphopyridine nucleotide (TPN) in an older term for nicotinamide adenine dinucleotide phosphate (NADP).

PYRIDOXINE FAMILY OF VITAMINS

By the late 1920's it had become clear that there was more than one water-soluble of "B" vitamin, but it was not until 1934 that the component of the "B-complex" known as vitamin B_6 was clearly defined. The absence of this new factor from the diet of the rat led to a characteristic dermatitis sometimes referred to as "rat pellagra." This condition developed even when supplements of both vitamin B_1 and riboflavin were added to the diet. In subsequent years, rapid progress was made in the study of the new vitamin, and it was isolated in crystalline form in 1938 from such natural materials as rice bran and yeast independently by five different research groups. The structure was quickly established and confirmed through synthesis, and the name "pyridoxine" was proposed for this new alcohol. More recently the name "pyridoxol" has been adopted (Fig. 1).

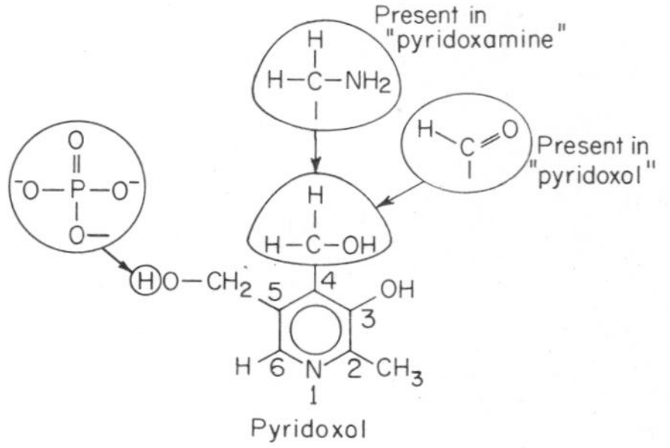

Present in "pyridoxamine"

Present in "pyridoxol"

Pyridoxol

Fig. 1. Structure of the pyridoxine family of vitamins.

On the basis of nutritional studies with animals, a single form of vitamin B_6 was recognized, but studies of the growth requirements of the lactic acid bacteria showed that there must be additional forms. Not only do these bacteria require the vitamin, but the growth of certain strains is promoted much more by the new forms than by pyridoxol. In 1944 the aldehyde, "pyridoxal," and the amine, "pyridoxamine," (Fig. 1), were established as forms of vitamin B_6. These, together with the alcohol, pyridoxol, make up the "pyridoxine family" of vitamins.

Pyridoxine Vitamins in Nutrition. The biological significance of the pyridoxine vitamins far transcends that suggested by its role in preventing "rat pellagra." Although 1 or 2 mg is an adequate daily intake for man, the need for the vitamin is absolute. A most striking demonstration lies in the development of convulsive seizures among infants receiving inadequate dietary pyridoxine. Other deficiency symptoms in man include dermatitis, anemia and a derangement of tryptophan metabolism. Studies of the distribution of the vitamin in nature, together with its known function in enzymes, show conclusively that the requirement for the vitamin is not limited to a few specialized tissues but is universal. As far as we know, all living cells contain the vitamin and make use of it in their metabolism. Accurate estimates of the content of the vitamin in tissues is difficult. The total content of pyridoxine in rat liver, a rich source, has been estimated as 10 mg/kg, or about 6×10^{-5} moles/liter as the average tissue concentration. The predominant form in liver and other animal tissues is pyridoxal. Pyridoxamine is also present as a substantial fraction of the total, but pyridoxol is almost absent. Plant tissues usually contain a higher proportion of pyridoxol. In all cases, the majority of the vitamin exists in the form of esters with phosphoric acid (Fig. 1). The two forms, pyridoxal phosphate and pyridoxamine phosphate, are of particular significance as they are the functioning *coenzyme forms* of the vitamin. All three free forms and the coenzyme forms of the vitamin are interconvertible in living cells. The vitamin is slowly lost from the animal body, in part through oxidation at the 4-position to a carboxylic acid, 4-pyridoxic acid, which is excreted in the urine—hence the need for a regular dietary intake.

Biochemical Functions. A major function of pyridoxal and pyridoxamine phosphates is found in their participation in the catalysis of biological "transamination" between α-amino acids and keto acids, as was first proposed by Esmond E. Snell in 1944. In 1945 it was shown that pyridoxal phosphate also serves as the coenzyme for bacterial decarboxylases acting upon amino acids. Within ten years about two dozen other enzymes, all catalyzing reactions of AMINO ACIDS, had been shown to depend upon the same coenzyme, and it became clear that the action of pyridoxal phosphate lies at the very center of nitrogen metabolism (see AMINO ACID METABOLISM). Surprisingly, the same coenzyme is found in glycogen phosphorylase, an enzyme that does not act on amino acids, but its mode of action in this case is not known. Other evidence suggests still another function for vitamins B_6 in fat metabolism.

The principal chemical reactions of the pyridoxine family are those of the 3-hydroxypyridine nucleus and of the additional functional groups in the 4- and 5-positions. A property of great practical value for biochemical investigations is the strong ultraviolet light absorption whose characteristics are sensitively dependent upon alterations in the 3-position and upon the presence or absence of a double bond in the 4-substituent group. Pyridoxol itself absorbs light maximally at 291 mμ at low pH where the nitrogen atom of the pyriding ring is protonated. In neutral solutions, it absorbs at 324 mμ, and the shift can be shown to result from dissociation of the 3-hydroxyl group ($pK_a = 5.0$). Thus, in neutral solution a dipolar ionic structure predominates (*e.g.*, see Fig. 2). In alkaline solution, the proton dissociates from the ring nitrogen ($pK_a = 9.0$) and the absorption maximum shifts again to 310 mμ. Introduction of an aldehyde group in the 4-position, as in pyridoxal phosphate, shifts the principal absorption band to 390 mμ in neutral solution. Conversion to a "Schiff base" by reaction with an amino acid (Fig. 2) leads to an additional shift, typically to about 415 mμ in neutral solution.[5] The spectrum of the coenzyme attached to various enzymes can also be observed, and the study of the changes in the spectrum during the action of an enzyme is an important method of investigation of the biological role of the vitamin B_6 coenzymes.

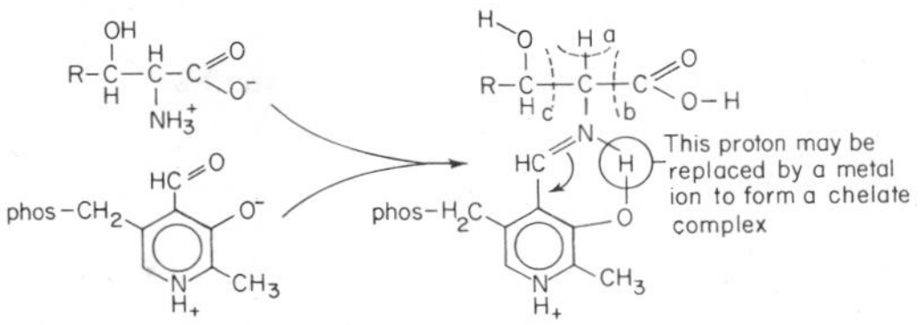

This proton may be replaced by a metal ion to form a chelate complex

FIG. 2. Schiff base formation between pyridoxal phosphate and an amino acid. The Schiff base (right) may be cleaved at any one of positions a, b or c in subsequent reactions. The arrows indicate the direction of electron movement into the pyridine ring.

The 3-hydroxypyridine nucleus of pyridoxol is highly substituted, but the 6-position is free and reactive in electrophilic substitution reactions. The colored products formed by reaction with diazonium ions or with 2,6-dichloroquinone chlorimide form the basis for qualitative and quantitative tests for the pyridoxine vitamins. Other tests of value for specific members of the group utilize the characteristic bright orange color of pyridoxamine with ninhydrin and the formation of hydrazones of pyridoxal or pyridoxal phosphate.

Pyridoxol is resistant to oxidation by air, but it is readily converted to pyridoxal by the action of manganese dioxide. Pyridoxal may in turn be converted to pyridoxamine by a transamination reaction with an amino acid. Pyridoxol, by reaction with acetone, is converted to an "isopropylidene" derivative in which the 3- and 4-positions are "blocked." This leaves the 5-position free for chemical alterations and provides the basis for the synthesis of a large number of analogues of the pyridoxine vitamins.[6]

Pyridoxal and pyridoxal phosphate react reversibly with amino acids in aqueous solutions to yield Schiff bases (imines) as shown in Fig. 2. These in turn may react with metal ions to form stable chelates (see CHELATION). The Schiff bases, and especially their metal chelates, are chemically reactive and break down to a variety of products depending upon the conditions and upon the nature of the amino acid involved. One of these reactions is "transamination." Others include the breakdown of serine to pyruvic acid and of threonine to acetaldehyde and glycine, and the racemization of optically active amino acids. These reactions parallel remarkably the reactions catalyzed by pyridoxal phosphate-dependent enzymes as described in the following section.

Both the enzymic transformations and the nonenzymic reactions of amino acids which are mediated by pyridoxal phosphate may all be vizualized as proceeding by the common initial step of Schiff base formation (Fig. 2), as was first suggested in 1952 by Braunstein and Shemyakin. The reaction occurring after Schiff base formation depends upon the structure of the amino acid and the specificity of the enzyme. The various reactions stem from the possibility of cleaving the bound amino acid at each of three different bonds surrounding the α-carbon atom, as indicated in Fig. 2. The electron-attracting power of the N-protonated pyridine ring is vizualized as a primary factor in facilitating the cleavage. Cleavage at position a (Fig. 2) leads to removal of the α-hydrogen atom as a proton. In the case of a racemase, a proton is added to the resulting intermediate (Fig. 3) at position d to give a Schiff base which is identical to the original, but which consists of a mixture of both possible steric configurations at the α-carbon atom. Following the hydrolytic cleavage of the Schiff base, the pyridoxal phosphate and the free, racemic amino acid are formed. In the case of a transaminase, a proton is added at position e to yield a new Schiff base (Fig. 3), which upon hydrolysis produces pyridoxamine phosphate and an α-keto acid derived from the original amino acid. In a second transamination step (just the inverse of that described above), the pyridoxamine phosphate, still attached to the enzyme, reacts with a molecule of a different α-keto acid to yield a new amino acid and to regenerate the pyridoxal phosphate. The net reaction catalyzed by a typical transaminase, glutamic-oxaloacetic transaminase, is as follows:

$$^{-}OOC{-}CH_2{-}CH_2{-}\underset{\underset{NH_3^+}{|}}{CH}{-}COO^-$$
Glutamate

$$+\ ^{-}OOC{-}CH_2{-}\underset{\underset{O}{\|}}{C}{-}COO^- \rightleftarrows$$
Oxaloacetate

$$HOOC{-}CH_2{-}CH_2{-}\underset{\underset{O}{\|}}{C}{-}COO^-$$
α-Ketoglutarate

$$+\ ^{-}OOC{-}CH_2{-}\underset{\underset{NH_3^+}{|}}{CH}{-}COO^-$$
Aspartate

This reaction plays a central role in the nitrogen metabolism of cells.

The intermediate pictured in Fig. 3 can react in still another way if a suitable group, Y, exists in the β-position of the amino acid. In this case the group may be eliminated as an anion Y^-. The group, Y, may be —OH,—SH, an indole ring or various others. The unsaturated intermediate which remains is hydrolyzed to pyridoxal phosphate, and in several steps to an α-keto acid and ammonia. Thus the amino acid serine yields pyruvic acid and ammonia while cysteine yields hydrogen sulfide as well.

Returning to Fig. 2 we shall consider briefly some other types of reactions. Cleavage at position b yields carbon dioxide and, after addition of a proton at the α-carbon atom and hydrolysis of the Schiff base, a primary amine. The bacterial amino acid decarboxylases have been mentioned previously. In the animal body, histamine arises from histidine in the same manner.

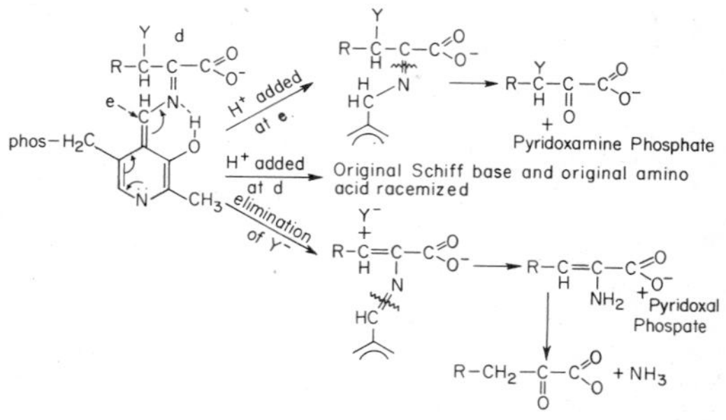

FIG. 3. Three types of reaction of intermediate formed by loss of the α-hydrogen atom of an amino acid-pyridoxal Schiff base.

Several enzymes are known which cause the cleavage of the carbon-carbon bond to the side chain of an amino acid (position c, Fig. 2). Thus threonine and serine (Y = OH) are cleaved to glycine and acetaldehyde or formaldehyde. Lack of space prevents a more detailed consideration of these and several other types of reactions. The reader is referred to the review by Braunstein for a detailed treatment.

In recent years, several of the pyridoxal phosphate-dependent enzymes have been crystallized or highly purified and they are now being studied intensively in order to learn exactly how nature has combined the remarkably versatile molecule of pyridoxal phosphate with specific enzyme proteins to accomplish the many different metabolic reactions which depend upon vitaim B_6.

References

1. BRAUNSTEIN, A. E., in "The Enzymes" (BOYER, P. D., LARDY, H., AND MYRBÄCK, K., EDITORS), Vol. 2, Ch. 6, pp. 113–184, Second edition, New York, Academic Press, 1960.
2. WAGNER, A. F., AND FOLKERS, K., "Vitamins and Coenzymes," pp. 160–193, New York, John Wiley & Sons, 1964.
3. SNELL, E. E., FASELLA, P. M., BRAUNSTEIN, A., AND ROSSI-FANNELLI, A., (EDITORS), "Chemical and Biological Aspects of Pyridoxal Catalysis," New York, The Macmillan Co., 1963.
4. "International Symposium on Vitamin B_6," in "Vitamins and Hormones" (HARRIS, R. S., WOOL, I. G., AND LORAINE, J. A., EDITORS), Vol. 22, New York, Academic Press, 1964.

5. METZLER, D. E., AND SNELL, E. E., *J. Am. Chem. Soc.*, **77**, 2431 (1955); METZLER, D. E., *J. Am. Chem. Soc.*, **79**, 485 (1957).
6. KORYTNYK, W., AND PAUL, B., *J. Heterocyclic Chem.*, **2**, 144 (1965).

DAVID E. METZLER

PYRIMIDINE BIOSYNTHESIS AND METABOLISM

The central role of the polynucleotides RNA and DNA in the maintenance and continuity of life in all organisms lends importance to consideration of pathways of biosynthesis and degradation of the polynucleotides. The key compounds in the biosynthesis of the polynucleotides are the nucleotides; this article will be restricted to biosynthesis of the pyrimidine nucleotides [see PURINE BIOSYNTHESIS AND METABOLISM; DEOXYRIBONUCLEIC ACIDS (REPLICATION); RIBONUCLEIC ACIDS (BIOSYNTHESIS)]. The pyrimidine nucleotides also function as coenzymes in the metabolism of carbohydrates and phospholipids [see PYRIMIDINES (STRUCTURE AND OCCURRENCE)].

Formation of Ribonucleotides *de novo*. The biosynthesis of pyrimidine ribonucleotides is shown in Fig. 1 (see reference 1, pp. 137–140, 158–163). The first major feature to be noted is that the initial precursors are intermediates in CARBOHYDRATE METABOLISM and AMINO ACID METABOLISM: aspartic acid, ammonia, carbon dioxide, ribose-5-phosphate and glutamine. Thus most organisms are capable of the complete synthesis of nucleic acids without requirement for nucleic acid derivatives in diet or environment. The second feature of note is that the biosynthesis is a sequence of enzymatic steps entirely unrelated to the steps in the degradation of nucleic acids; in other words, the major pyrimidines (uracil, cytosine and thymine), which were first discovered as products of chemical and enzymatic degradation of nucleic acids, are not intermediates in the biosynthesis. Here biosynthesis or formation *de novo* is distinguished from utilization of preformed pyrimidines, which is discussed in a later paragraph. The first pyrimidine intermediate in the *de novo* sequence is the unique compound, orotic acid, which is not a component of the nucleic acids or (as far as is known) of any other biochemical system. Orotic acid is condensed with 5-phosphoribosyl-1-pyrophosphate (PRPP) to form a nucleotide-like compound, orotidine-5′-phosphate (OMP). PRPP, synthesized from ribose-5-phosphate by transfer of pyrophosphate from ATP, is also the donor of ribose-5-phosphate in the biosynthesis of purines. OMP is decarboxylated in an irreversible step to uridine-5′-phosphate (UMP) which is readily phosphorylated to uridine diphosphate (UDP) and uridine triphosphate (UTP) by the action of ATP-phosphotransferases. UTP is converted to cytidine triphosphate (CTP) by an amination step; the amino donor is glutamine with ATP as a reactant and guanosine triphosphate (GTP) as an activator of the enzyme.[2] Apparently all of these nucleotide reactions require the presence of magnesium ion. UTP also serves as the donor of the uridine phosphate moiety in the synthesis of RNA and in the synthesis of the uridine nucleotide-carbohydrate coenzymes (see reference 3, pp. 36–82). CTP serves as donor of the cytidine phosphate moiety

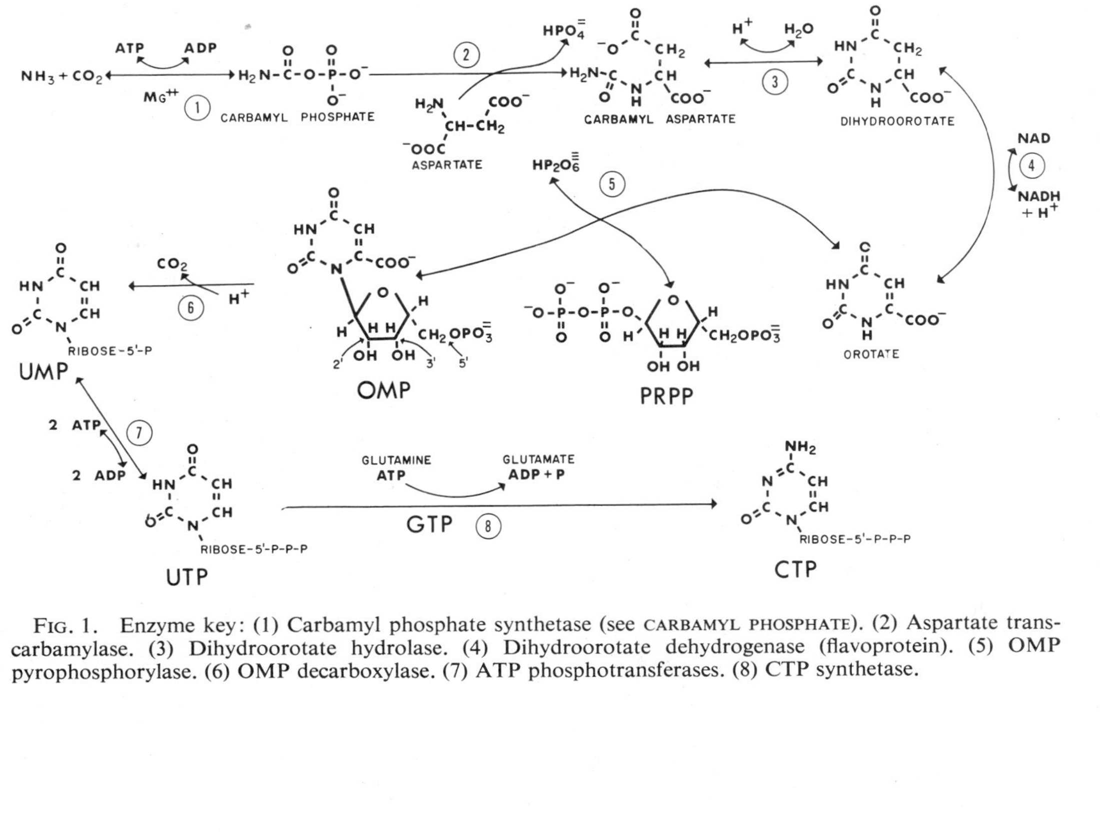

FIG. 1. Enzyme key: (1) Carbamyl phosphate synthetase (see CARBAMYL PHOSPHATE). (2) Aspartate transcarbamylase. (3) Dihydroorotate hydrolase. (4) Dihydroorotate dehydrogenase (flavoprotein). (5) OMP pyrophosphorylase. (6) OMP decarboxylase. (7) ATP phosphotransferases. (8) CTP synthetase.

in the synthesis of RNA and in the synthesis of CDP-choline and related compounds (see reference 3, pp. 36–82), and is interconverted with CDP and CMP by the ATP phosphotransferases.

The formation of 5-ribosyluracil-5′-phosphate ("pseudo UMP") from uracil and ribose-5-phosphate by enzymes from *Tetrahymena* (see reference 3, pp. 19–20), and the incorporation of this nucleotide from "pseudo-UTP" into RNA by RNA polymerase have been demonstrated;[4] whether this is the primary pathway of biosynthesis of the trace component of RNA, 5-ribosyluracil, is not yet established. The other minor components of RNA, 5-methyluracil, 5-methylcytosine, N^3-methyluracil and N^3-methylcytosine, are known to be formed by enzymatic methylations of uracil and cytosine groups at specific locations in the polynucleotide structure.[5]

Formation of Deoxyribonucleotides *de novo*. The biosynthesis of the pyrimidine deoxyribonucleotides is shown in Fig. 2; they appear to be derived from the ribonucleotides rather than synthesized by pathways parallel to ribonucleotide synthesis.

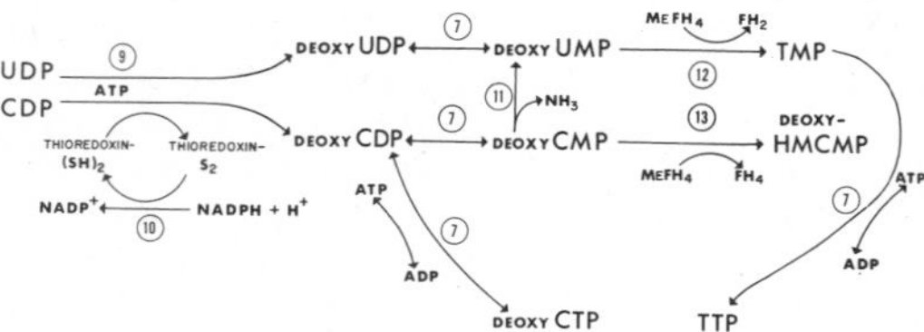

FIG. 2. Enzyme key: (9) Ribonucleoside diphosphate reductases. (10) Thioredoxin reductase. (11) Deoxycytidylate deaminase. (12) Thymidylate synthetase. (13) Deoxycytidylate hydroxymethylase.

The known condensation of acetaldehyde and glyceraldehyde phosphate by an aldolase-like reaction to form deoxyribose-5-phosphate has an equilibrium unfavorable to synthesis under physiological conditions, and no deoxyribose donor corresponding to PRPP has yet been found.

The ribonucleotides of uracil and cytosine (as well as of adenine and guanine) are converted at the diphosphate level to the deoxyribose derivatives by removal of an oxygen atom from the 2′-position of the ribose. The reducing agent is a disulfhydryl polypeptide, thioredoxin-$(SH)_2$, which is regenerated by reaction with the reduced pyridine nucleotide NADPH. This mechanism has been established for *E. coli* and rat tissues;[6] in extracts of *Lactobacillus*, a reduction of CTP involving vitamin B_{12} as a cofactor has been demonstrated.[7] The exact mechanism of the reduction step and the fate of the oxygen atom are not yet known. DeoxyUMP is methylated to thymidylic acid (5-methyldeoxyuridylic acid, TMP); the methyl group is derived from the "one-carbon pool" via methylenetetrahydrofolic acid ($MeFH_4$) (see reference 8, pp. 136–137; see also SINGLE CARBON UNIT METABOLISM). TTP and deoxyCTP are the direct donors of the thymine and cytosine nucleotides, respectively, in the replication of DNA via DNA polymerase. 5-MethyldeoxyCTP, the precursor of the trace component methylcytosine in DNA, is possibly formed by analogous reactions starting with deoxyCMP. The 5-hydroxymethylcytosine (HMC) of *E. coli* T-even phages is known to be formed by a non-reductive reaction of deoxyCMP with methylenetetrahydrofolic acid.[9]

Degradation of Nucleotides and Polynucleotides. Enzymes present in cells and digestive secretions are capable of complete degradation of the nucleic acids (see reference 1, pp. 219–234). The ribonucleases and deoxyribonucleases depolymerize RNA and DNA, respectively, by rupture of the pentose-phosphate bonds to yield short-chain polynucleotides. Less specific phosphodiesterases further attack these products, yielding nucleotides of either the 2′(3′)-phosphate or the 5′-phosphate class, depending on the enzyme (see NUCLEASES). A group of specific and nonspecific PHOSPHATASES hydrolyze these mononucleotides to nucleosides and phosphate. Specific nucleoside hydrolases and nucleoside phosphorylases split the nucleosides at the glycoside bond; the latter enzyme liberates the free pyrimidine and the pentose-1-phosphate. Deaminase enzymes attack deoxyCMP, deoxycytidine and cytidine to liberate ammonia and produce the corresponding uracil derivatives. The free pyrimidines, uracil and thymine, are reduced to the 4,5-dihydropyrimidines by pyridine nucleotide-dependent reactions and cleaved to CO_2, NH_3 and β-alanine or β-aminoisobutyric acid, respectively.

Utilization of Nucleic Acid Degradation Product. Most organisms are capable of synthesizing the 5′-nucleotides from free bases and nucleosides if these are available from nutritional sources or from intracellular degradations. The reversibility of the nucleoside phosphorylase reaction permits uracil and thymine to be converted to uridine and thymidine by condensation with the appropriate pentose-1-phosphate; cytosine does not appear to be readily converted. Enzymes from some microorganisms are also capable of conversion of uracil directly to UMP by reaction with PRPP, in analogy with the utilization of several free purines. The nucleosides uridine, cytidine, thymidine and deoxycytidine are readily taken up by cells and phosphorylated to the nucleoside-5′-phosphates by ATP-phosphotransferases. These reactions are frequently used in research applications for the controlled introduction of isotope-labeled bases and nucleosides into the nucleic acids of living organisms. If suitable free pyrimidines or nucleosides are available as nutrients, the rates of the *de novo* biosynthesis are suppressed by regulatory mechanisms, both at the gene expression level (synthesis of enzymes) and the enzymatic level (activity of enzymes) [see METABOLIC CONTROLS (FEEDBACK MECHANISMS)].

Control of Nucleotide Biosynthesis. The biosynthesis of the nucleotides is regulated at the enzymatic level by the concentrations of certain nucleotides which are terminal products and which are also activators or inhibitors of enzymes acting early in the biosynthetic pathway[10] [see

ENZYMES (ACTIVATION AND INHIBITION); REGULATORY SITES OF ENZYMES]. For example, CTP is an inhibitor of the enzyme aspartate transcarbamylase; thus elevation of the concentration of CTP, whether derived from exogenous pyrimidines or by *de novo* synthesis, will diminish the rate of *de novo* pyrimidine synthesis, thereby conserving energy and intermediary metabolites. In addition, deoxynucleoside triphosphates inhibit the ribonucleotide reductase system strongly (in a complex pattern of mutual activations and inhibitions) with the result that the intracellular concentrations of the deoxynucleotides are normally maintained at low levels but are rapidly restored if drawn upon for the synthesis of DNA.

Nucleic acid function can be deranged by inhibition of the biosynthetic steps or by inclusion of analogues which are incorporated into the nucleic acid (see reference 8, pp. 167–189). Examples are: azauridine, which as azaUMP inhibits orotidylic acid decarboxylase, hence pyrimidine biosynthesis; the glutamine analogue, diazo-oxo-norleucine, which inhibits the conversion of UTP to CTP; fluorouracil, which on conversion to fluorouridylic acid inhibits thymidylate synthetase, hence DNA synthesis; bromouracil, which is incorporated into DNA as an analogue of thymine but does not permit normal replication of the DNA (see POLYNUCLEOTIDES, SYNTHETIC).

References

1. POTTER, V. R., "Nucleic Acid Outlines. Vol. I, Structure and Metabolism," Minneapolis, Burgess Publishing Co., 1960.
2. CHAKRABORTY, K. P., AND HURLBERT, R. B., "Role of Glutamine in The Biosynthesis of Cytidine Nucleotides in *E. coli*," *Biochim. Biophys. Acta*, **47**, 607–609 (1961).
3. HUTCHINSON, D. W., "Nucleotides and Coenzymes," London, Methuen and Co., 1964.
4. GOLDBERG, I. H., AND RABINOWITZ, M., "Comparative Utilization of PseudoUTP and UTP by RNA Polymerase," *J. Biol. Chem.*, **238**, 1793–1800 (1963).
5. BOREK, E., "The Methylation of Transfer RNA: Mechanism and Function," *Cold Spring Harbor Symp. Quant. Biol.*, **28**, 139–148 (1963).
6. LAURENT, T. V., MOORE, E. C., AND REICHARD P., "Enzymatic Synthesis of Deoxyribonucleotides IV," *J. Biol. Chem.*, **239**, 3436–3444 (1964).
7. ABRAMS, R., "CTP as the Precursor of Deoxycytidylate in *Lactobacillus leichmannii*," *J. Biol. Chem.*, **240**, PC 3697 (1965).
8. BROCKMAN, R. W., "Mechanisms of Resistance to Anticancer Agents," *Advan. Cancer Res.*, **7**, 129 (1963).
9. COHEN, S. S., "Virus-Induced Acquisition of Metabolic Function," *Federation Proc.*, **20**, 641-649 (1961).
10. MONOD, J., CHANGEUX, J.-P., AND JACOB, F., "Allosteric Proteins and Cellular Control Systems," *J. Mol. Biol.*, **6**, 306–329 (1963).

ROBERT B. HURLBERT

PYRIMIDINES (STRUCTURE AND OCCURRENCE)

The parent compound, *pyrimidine* (also known as 1,3-diazine or meta-diazine) is a planar heterocyclic ring containing four carbon and two nitrogen atoms. The pyrimidine ring is also found

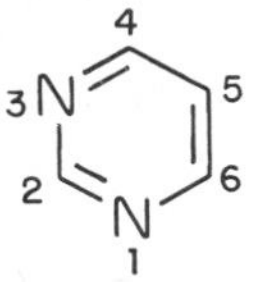

Pyrimidine

structurally in other fused ring heterocyclic systems such as PURINES, PTERIDINES and quinazolines. Though pyrimidine itself has not been found to occur naturally, some of its oxo and amino derivatives are widely distributed in nature—especially in cell and viral nucleic acids, vitamins, and coenzymes. Accordingly, pyrimidines play a vital role in many biological processes. Many synthetic pyrimidines are important as pharmacological and chemotherapeutic agents.

Nucleic Acid Pyrimidines. Pyrimidines, as well as purines, are found in all nucleic acids as component subunits (*nucleosides*, *e.g.*, uridine) in which a sugar (ribose or 2-deoxyribose) is substituted on position 1 of the pyrimidine ring.

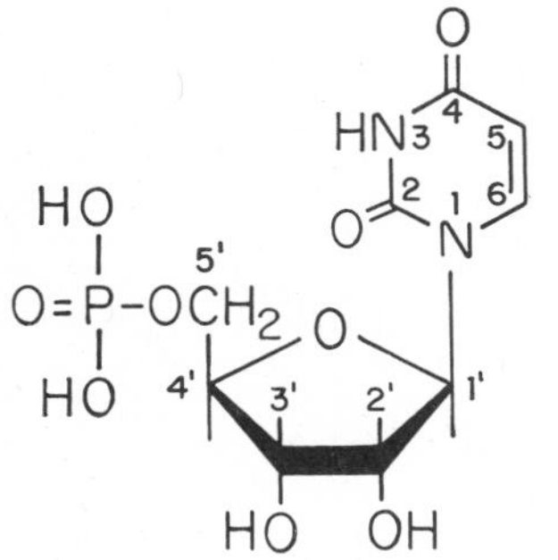

Uridine 5'-phosphate
(a nucleotide)

The phosphoric esters of the nucleosides are called *nucleotides* (*e.g.*, uridine-5′-phosphate) which by 3′→ 5′ phosphodiester bonds form the long-chain polymers (*polynucleotides*) comprising the nucleic acids. Chemical and/or enzymatic hydrolyses of nucleic acids can yield the free pyrimidines, their nucleosides or nucleotides.

RIBONUCLEIC ACIDS (RNA) contain *uracil* (I, R = H) (2,4-pyrimidinedione) and *cytosine* (II, R = H) (4-amino-2-pyrimidinone) as the main pyrimidine components of their polynucleotide sequences. *Thymine* (5-methyluracil) and 5-*methylcytosine* as well as N^3-methylated uracil and cytosine have also been discovered as minor components in the hydrolysates of transfer (soluble) ribonucleic acids. A 5-ribosylated uracil (III, *pseudouridine*) is also present in small amounts in the polynucleotide sequence of transfer RNA. More recently, the ribonucleotide of 5,6-*dihydrouracil* (IV) has been found in crude yeast transfer RNA.

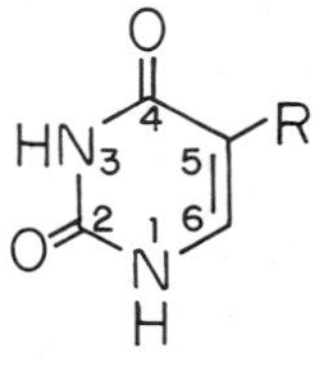

Uracil (R=H)
Thymine ($R=CH_3$)

I

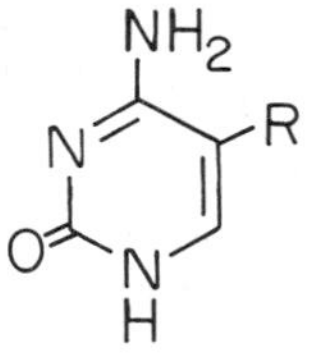

Cytosine (R=H)

II

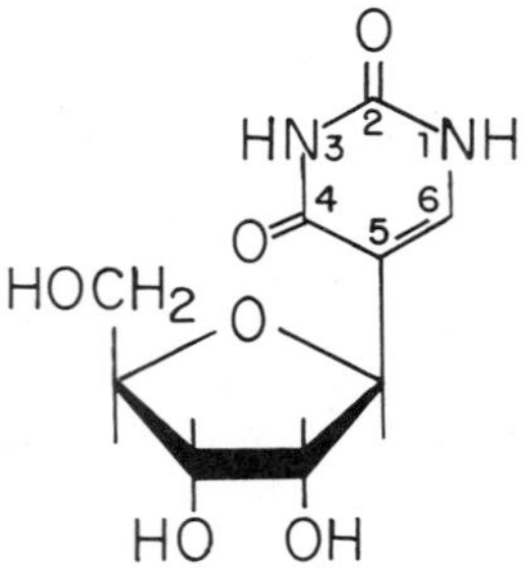

Pseudouridine

III

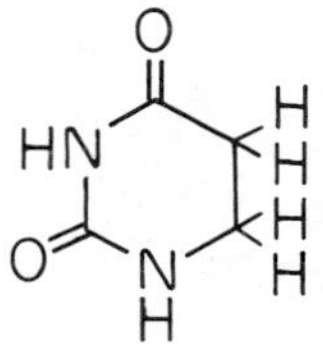

Dihydrouracil

IV

Thymine and *cytosine* are the main pyrimidine components in the DEOXYRIBONUCLEIC ACIDS (DNA) of cells. 5-*Methylcytosine* also occurs as a minor component in DNA from several sources (*e.g.*, wheat germ DNA). The T-even coliphages contain 5-*hydroxymethylcytosine* (II, R = CH_2OH) instead of cytosine in their DNA. 5-*Hydroxymethyluracil* (I, R = CH_2OH) has been found in the DNA of a *Bacillus subtilis* bacteriophage, and uracil has replaced thymine in another.

Pyrimidines in Coenzymes. *Uracil*, *cytosine* and *thymine* occur as components of several "nucleotide coenzymes" or "nucleotide anhydrides" which are of considerable biochemical significance. Uridine diphosphate glucose (UDPG), a coenzyme of the system which catalyzes the conversion of galactose-1-phosphate to glucose-1-phosphate, is one example of this class (see GALACTOSE AND GALACTOSEMIA).

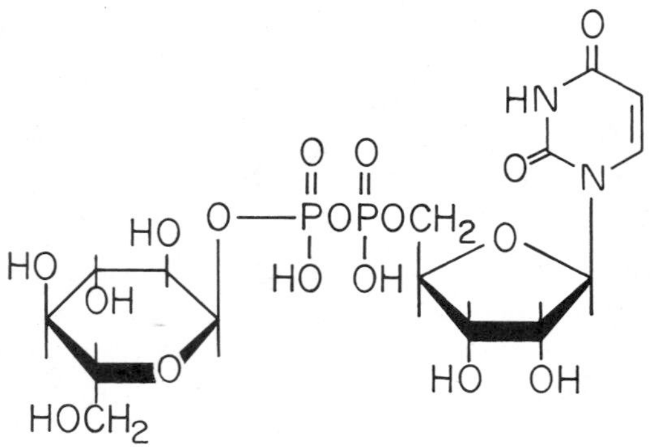

UDPG

Many other naturally occurring UDP derivatives have been isolated in which the uridine diphosphate is bound to galactose, xylose, arabinose, glucuronic acid, galacturonic acid, N-acetylglucosamine, N-acetylgalactosamine, N-acetylmuramic acid, sugar-peptides, etc. These have been shown to be of considerable significance in sugar metabolism, the biosynthesis of polysaccharides, and cell wall synthesis in microorganisms (see BACTERIAL CELL WALLS; CARBOHYDRATE METABOLISM).

Several derivatives of cytidine diphosphate (CDP) have also been discovered in nature. CDP derivatives containing ribitol, glycerol and ethanolamine have been isolated from a number of bacteria. These cytidine coenzymes are important in the biosynthesis of PHOSPHOLIPIDS. Derivatives of 2′-deoxycytidine diphosphate (*d*CDP) and thymidine diphosphate (TDP) are also present in certain bacteria, of which TDP-glucose, TDP-rhamnose, TDP-mannose and TDP-ribose are some examples. In addition, 2′-deoxycytidine diphosphate (*d*CDP) has recently been found in nature linked to choline and ethanolamine.

A number of uracil nucleotide derivatives containing sugar-peptides, which are precursors of peptides in cell walls, have been isolated recently from microorganisms. It is expected that many new nucleotide anhydrides will be discovered in the future.

Pyrimidine Nucleoside Antibiotics. Five pyrimidine nucleoside antibiotics have been isolated thus far from cultures of *Streptomyces*. These are: Amicetin (from *Streptomyces vinaceus-drappus* and *Streptomyces plicatus*), Bamicetin and Plicacetin (*Streptomyces plicatus*), Blasticidin S (from *Streptomyces griseochromogenes*) (a very effective agent against rice blast disease), and Gougerotin (from *Streptomyces gougerotii*). All of these antibiotics contain the pyrimidine, *cytosine*, in their complex structures along with a 4-aminohexose and amino acids. It is quite likely that additional pyrimidine nucleoside antibiotics will be discovered in the next decade.

Other Pyrimidines in Nature. *Orotic acid* (uracil-6-carboxylic acid), occurs in the whey of cow's milk, and it also accumulates during the growth of *Neurospora crassa* mutants. Arabinofuranosyl-uracil and -thymine nucleosides (*spongouridine* and *spongothymidine*) occur in *Cryptotethia crypta*, a sponge found in the Florida-Bimini Island region. *Vicine* (4-amino-2,5,6-trihydroxypyrimidine) and the 5-substituted glucoside have been isolated from vetch seeds. The anti-beriberi vitamin (B_1, aneurin, THIAMINE) found in rice bran, yeast and other sources, contains a 4-amino-2,5-dimethylpyrimidine derivative in its structure.

References

1. BROWN, D. J., "The Pyrimidines," New York, Interscience Publishers, 1962.
2. BENDICH, A., in "The Nucleic Acids" (CHARGAFF, E., AND DAVIDSON, J. N., EDITORS), pp. 81–136, New York, Academic Press, 1955.
3. FOX, J. J. and WEMPEN, I., *Advan. Carbohydrate Chem.*, **14**, 283–360 (1959)
4. KENNER, G. W., AND TODD, A. R., in "Heterocyclic Compounds" (ELDERFIELD, R. C., EDITOR), Vol. 6, pp. 234–323, New York, John Wiley & Sons, 1957.
5. FOX, J. J., WATANABE, K. A., AND BLOCH, A., in "Progress in Nucleic Acid Research and Molecular Biology," DAVIDSON, J. N., AND COHN, W. E. (EDITORS), pp. 251–313, New York, Academic Press, 1966.
6. MICHELSON, A. M., "The Chemistry of Nucleosides and Nucleotides," New York, Academic Press, 1963.

JACK J. FOX

PYROPHOSPHATES

Pyrophosphates are derivatives of pyrophosphoric acid, which is sometimes formulated

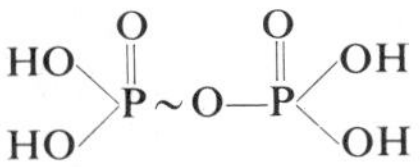

$$(HO)_2P(=O){\sim}O{-}P(=O)(OH)_2$$

using the symbol ~ to indicate the presence of an anhydride ("high-energy") phosphate bond (see PHOSPHATE BOND ENERGIES). Adenosine diphosphate (ADP), adenosine triphosphate (ATP; for structures see PURINES), NICOTINAMIDE ADENINE DINUCLEOTIDES, and thiamine pyrophosphate (for structure see VITAMIN B GROUP) are among the better known pyrophosphate derivatives. 5-Phosphoribosyl-1-pyrophosphate (PRPP) is a key phosphoribosylating agent in PURINE BIOSYNTHESIS and in PYRIMIDINE BIOSYNTHESIS.

PYRUVIC ACID

Pyruvic acid, $CH_3{-}C(=O){-}COOH$, is an extremely important and versatile intermediate in CARBOHYDRATE METABOLISM. GLYCOLYSIS in animal tissues leads to the formation of lactic and pyruvic acids (which are interconvertible). In the oxidation of pyruvic acid, it is converted into acetyl coenzyme A, which in turn is oxidized in the CITRIC ACID CYCLE. Pyruvic acid can also be decarboxylated and reduced to alcohol as in ALCOHOLIC FERMENTATION. Its amination leads to the formation of alanine. In muscle, the synthesis of glycogen takes place by way of the phosphorylation of pyruvate to produce phosphoenolpyruvate,

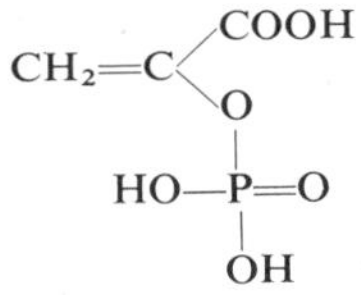

$$CH_2{=}C(COOH){-}O{-}P(=O)(OH){-}OH$$

and thence the reversal of the glycolysis process.

Q

QUANTASOMES

This term has been applied to particles observed in electron micrographs, arrayed in a regular two-dimensional arrangement within each of the lamellae, stacks of which occur within CHLOROPLASTS, the subcellular organelles or plastids that carry out PHOTOSYNTHESIS (see also LIGHT PHASE OF PHOTOSYNTHESIS; PHOTOSYNTHETIC PHOSPHORYLATION).

Reference

1. RABINOWITCH, E. I., AND GOVINDJEE, "The Role of Chlorophyll in Photosynthesis," *Sci. Am.*, **213**, 74–83 (July 1965).

R

RADIATION EFFECTS. See BACTERIAL GENETICS, section on DNA Synthesis; IONIZATION RADIATION (BIOCHEMICAL EFFECTS); MUTAGENIC AGENTS; ULTRAVIOLET RADIATION (BIOCHEMICAL EFFECTS).

RADIOACTIVE TRACERS. See ISOTOPE DILUTION METHODS; ISOTOPIC TRACERS. Particular applications of radioactive tracers are discussed also in CARBON (RADIOCARBON) DATING; HISTOCHEMICAL METHODS, section on Autoradiography.

RADIOAUTOGRAPHY

See HISTOCHEMICAL METHODS, section on Autoradiography. Examples of the use of radioautographs in cytological studies appear in the following references.

References

1. BASERGA, R., AND KISIELESKI, W. E., "Autobiographies of Cells," *Sci. Am.*, **209**, 103–110 (August 1963).
2. DAVIDSON, E. H., "Hormones and Genes," *Sci. Am.*, **212**, 36–45 (June 1965).
3. CAIRNS, J., "The Bacterial Chromosome," *Sci. Am.*, **214**, 36–44 (January 1966).

REDOX POTENTIALS. This is a condensed term for oxidation-reduction potentials (see BIOLOGICAL OXIDATION-REDUCTION).

REGULATORY SITES OF ENZYMES

The enzymes contained in any living cell possess a wide range of catalytic activities. These activities must be regulated in a complex and interdependent way; otherwise the cell would rapidly degenerate into a non-living aggregation of organic compounds participating in a variety of uncoordinated and mainly degradative reactions. Metabolic regulation is at present very poorly understood, but control mechanisms of two major types are known to occur. These mechanisms control (a) the rate of synthesis of specific enzymes and (b) the catalytic behavior of preformed enzymes. This article deals with the processes through which regulation of the second type is effected.

The pronounced effect of adenosine-5′-monophosphate (AMP) on glycogen phosphorylase was one of the earliest examples of this type of regulation to be discovered (see GLYCOGEN AND ITS METABOLISM). This same compound (AMP) has recently been observed to affect the kinetics of several other enzymes, including a threonine dehydrase, phosphofructokinase, and DPN-specific isocitrate dehydrogenase.

A physiological role for such effects was first suggested in 1956 independently by Umbarger and by Yates and Pardee. Umbarger's finding that one of the threonine dehydrases of *Escherichia coli* is inhibited by isoleucine and Pardee's observation that aspartate transcarbamylase is inhibited by cytosine triphosphate led both of these workers to the very important concept of feedback inhibition of biosynthetic sequences. Threonine dehydrase catalyzes an early step in isoleucine biosynthesis, and aspartate transcarbamylase is similarly involved in the biosynthesis of cytosine triphosphate and other pyrimidines. Control of the level of a metabolite by feedback control of an early step in its synthesis is now recognized as a general phenomenon (see METABOLIC CONTROLS). Similar effects have been reported for a number of enzymes that participate in degradative metabolism, and it appears that the kinetic characteristics of many key enzymes are strongly affected by the concentrations of specific metabolites. Such effects seem to play important roles in keeping the multitude of reactions in a living cell in balance.

Several sets of terms are used in the description of such regulatory proteins. The small molecule that modulates behavior of the enzyme is usually termed an *effector*, *modifier* or *modulator*. Presumably its effects are exerted as a consequence of its binding at a specific site on the enzyme, the *regulatory*, *modifier* or *control* site (as distinguished from the *catalytic* site at which the reaction occurs). The term *allosteric inhibitor* was suggested by Monod and Jacob to emphasize that there need be no steric resemblance between the substrate and the effector. However, the adjective *allosteric* has lost this original meaning and is now applied rather loosely to proteins, enzymes, sites, effectors, conformational transitions, and reactions with the general implication that a protein and a small molecule interact in some other relationship than that of enzyme and substrate.

Although the mechanism by which an effector modulates the behavior of an enzyme is not understood in any case, the kinetics of most regulatory enzymes have a number of features in common. The curve of velocity as a function of substrate concentration is usually sigmoid (the slope first increases, reaches a maximum, and then decreases as substrate concentration is increased—see Fig. 1). *Positive effectors* move this curve to the left (a smaller concentration of substrate is required for half-maximal reaction in the presence of the effector than in its absence) and *negative effectors* have the opposite effect. The effector causes changes primarily in the apparent affinity of the enzyme for substrate; in many cases, the maximal velocity attainable is changed very little by the effector.

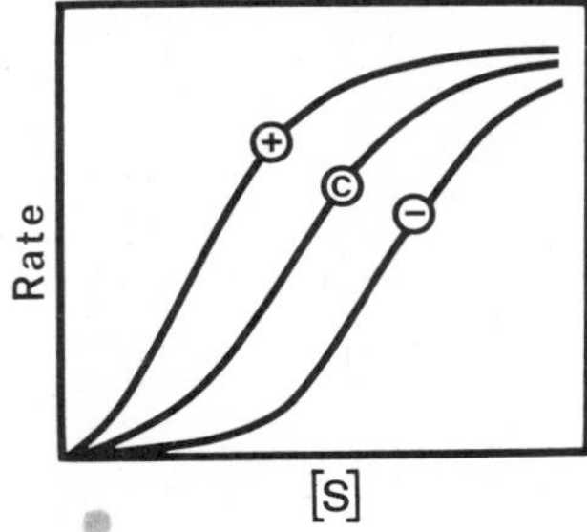

Fig. 1. Generalized plot of rate as a function of substrate concentration for a reaction catalyzed by a regulatory enzyme. Curve identifications: *c*, control; −, negative effector added; +, positive effector added. In addition to the characteristic changes in apparent affinity for substrate that are indicated in the Figure, effectors may also alter the maximal velocity obtainable at saturating levels of substrate.

The sigmoid rate curve that seems characteristic of regulatory enzymes differs markedly from the more common hyperbolic (Michaelis) enzyme response curve [see ENZYMES (KINETICS)]. A sigmoid response can be obtained only if more than one molecule of substrate participates in the rate-determining step or steps; *i.e.*, the order of the reaction with regard to substrate must be greater than one. The simplest assumption is that each regulatory enzyme molecule contains several sites capable of binding substrate. These sites need not all be catalytic, but they must bind cooperatively—*i.e.*, the presence of substrate at one site must enhance the affinity of other sites for substrate molecules. The order of the reaction could not exceed one, regardless of the number of catalytic or other sites on each enzyme molecule, if each site functioned independently. The mechanism by which the effector increases or decreases the apparent affinity of the enzyme for substrate is unknown. In some cases the apparent kinetic order of the reaction with regard to substrate is changed by the effector, which suggests that interactions between substrate-binding sites are enhanced or weakened by the effector, but in other cases no such effect is seen. The belief that the effector binds at a specific regulatory site distinct from the catalytic site was based originally on the lack of steric resemblance between substrate and effector. This belief has been strengthened by subsequent study of effector action, and especially by the recognition of positive effectors, which enhance substrate binding and thus cannot occupy a substrate-binding site. Direct evidence for distinct sites has as yet been obtained only in the case of aspartate transcarbamylase. This enzyme has been dissociated (Gerhart and Schachman) into two types of subunits, one of which bears the catalytic site (and is catalytically active), whereas the other binds the negative effector, cytosine triphosphate. The catalytic subunit is unaffected by cytosine triphosphate, but regulatory behavior is regained when the subunits are allowed to recombine.

Most regulatory enzymes, like many others, probably contain several subunits. There is little evidence at present in support of an early suggestion that effectors exert their action by changing the state of aggregation of their target enzymes. It seems necessary to assume, however, that the primary action of an effector on its enzyme is to cause some conformational change (probably in most cases much less drastic than dissociation). The observed kinetic effects are then considered to result from the conformational alteration. There is little direct physical evidence as yet for such changes under conditions approaching those used for *in vitro* assays, and evidence regarding conformational changes *in vivo* seems at present unattainable.

The life of a cell must depend as strictly on the regulatory properties of key enzymes as on their catalytic activity as such. Both the high steric specificity of regulatory sites for their effectors and the specificity of the kinetic consequences of effector binding are evidence for stringent evolutionary selection.

References

1. Umbarger, H. E., *Science*, **145**, 674 (1964).
2. Gerhart, J. C., and Pardee, A. B., *J. Biol. Chem.*, **237**, 891 (1962).
3. Monod, J., Changeux, J. P., and Jacob, F., *J. Mol. Biol.*, **6**, 306 (1963).
4. Monod, J., Wyman, J., and Changeux, J. P., *J. Mol. Biol.*, **12**, 88 (1965).
5. Atkinson, D. E., Hathaway, J. A., and Smith, E. C., *J. Biol. Chem.*, **240**, 2682 (1965).
6. Gerhart, J. C., and Schachman, H. K., *Biochemistry*, **4**, 1054 (1965).
7. Atkinson, D. E., *Ann. Rev. Biochem.*, **35**, 85 (1966).
8. Stadtman, E. R., *Adv. Enzymol.*, **28**, 41 (1966).

D. E. Atkinson

REICHSTEIN, T.

Tadeusz Reichstein (1897–) is a Polish-Swiss chemist. He was one of the first to synthesize ASCORBIC ACID. His later work involved isolating and studying the chemistry of the steroid HORMONES of the adrenal cortex. With Edward Calvin Kendall and Philip Hench, a physician, Reichstein received the Nobel Prize in physiology and medicine in 1950.

RELAXIN

The separation of the pubic bones of the females of several species of animals during pregnancy and parturition is under hormonal control. One of the substances which is involved is a protein to which the name *relaxin* has been given. Relaxin was first discovered in the serum of pregnant rodents (guinea pigs and rabbits); subsequently it was found to be present in high concentration in the ovaries of pregnant females of several non-rodent species, including pigs, cattle, and sheep. In addition, its isolation from rooster testis has been reported.

Assay. Relaxin assays may be performed in guinea pigs or mice. The subcutaneous or intramuscular injection of the hormone into estrogen-treated females of either species is followed by relaxation of the symphyseal ligaments; the separation can be detected by manual palpation in guinea pigs, or visually, by transillumination of the dissected pelves in mice. About 6–10 hours are required for maximal response of guinea pigs following administration of relaxin in aqueous solution. The response in mice is slower, requiring 18–24 hours; for maximum effect, relaxin must be given in combination with a vehicle which delays absorption.

Isolation and Purification. For the isolation of relaxin in large quantity, the most convenient starting material is the ovary of the pregnant sow. Methods for extraction include the use of butanol saturated with dilute aqueous trichloroacetic acid,[2] hot glacial acetic acid,[3] and acid acetone.[4] These methods afford preparations which contain from 100–300 GPU (guinea pig units, see reference 1) per milligram. Further purification may be achieved by ION-EXCHANGE chromatography on DEAE-cellulose or IRC-50 and by GEL FILTRATION on sephadex G-50.[5] By the application of such procedures, preparations which appear to be nearly homogeneous when subjected to ELECTROPHORESIS on starch gel at pH 8.0 have been obtained. As little as 0.3–0.6 μgm of the purified material will cause a detectable response in guinea pigs or mice, corresponding to an activity of 1500–3000 GPU/mg.

Chemical Properties. As judged by its diffusion and sedimentation properties, as well as its behavior on analytical sephadex columns, relaxin has a molecular weight of 10,000. At pH 8 on starch gel, the principal active component of purified preparations moves toward the cathode about one-half as rapidly as does synthetic α-MSH. Another, slower moving, active component can also be observed in some preparations. Analytical data indicate the probable absence of non-amino acid constituents in the relaxin molecule. Activity is rapidly lost upon exposure to mild reducing agents, to acid methanol and to proteolytic enzymes. Relaxin is stable at elevated temperature (100°C) for short periods in neutral or slightly acid solution. Relaxin has been found to be antigenic in guinea pigs and rabbits; anti-porcine relaxin will inhibit the effects of relaxin from other species of animals.

Physiological Properties. In addition to their effects upon the symphyseal ligaments of guinea pigs and mice, relaxin preparations will inhibit spontaneous contractions of excised rat or mouse uterine segments, will inhibit decidual reactions in rats and mice, and have been reported to initiate cervical softening in several species of animals (including humans). Whether these phenomena are intrinsic properties of relaxin, or are ascribable to contaminants, has not yet been ascertained.

References

1. For a review of earlier literature, see FRIEDEN, E. H., AND HISAW, F. L., *Recent Progr. Hormone Res.*, **8**, 333 (1953).
2. FRIEDEN, E. H., AND LAYMAN, N. W., *J. Biol. Chem.*, **229**, 569 (1957).
3. COHEN, H., AND KLEINBERG, W., *Arch. Biochem. Biophys.*, **91**, 17 (1960).
4. GRISS, G., KECK, J., ENGELHORN, R., AND TUPPY, H., *Experentia*, **18**, 545 (1962).
5. FRIEDEN, E. H., *Trans. N.Y. Acad. Sci.*, **25**, 331 (1963).

EDWARD H. FRIEDEN

RENAL TUBULAR FUNCTIONS (AND URINE COMPOSITION)

Details of kidney structure should be reviewed before functional aspects are considered (see GLOMERULAR FILTRATION and its illustrations).

Nature and Composition of the Urine. About 1250 ml of urine are excreted in 24 hours by normal man, with specific gravities usually between 1.018 and 1.024 (extremes, 1.003–1.040). Flow ranges from 0.5–20 ml/min with extremes of dehydration and hydration. Maximum osmolar concentration is 1400, compared to plasma osmolarity of 300 mOsm/liter. In diabetes insipidus, characterized by inadequate *anti-diuretic hormone* (ADH) production (see VASOPRESSIN), volumes of 15–25 liters/day of dilute urine are formed.

In addition to substances listed in Table 1, there are trace amounts of purine bases and methylated purines, glucuronates, the pigments urochrome and urobilin, hippuric acid and amino acids. In pathological states, other substances may appear: proteins (nephrosis); bile pigments and salts (biliary obstruction); glucose, acetone, aceto-acetic acid, and β-hydroxybutyric acid (diabetes mellitus).

The U/P ratios of the substances in Table 1 vary widely because of differential handling by the kidney. Quantitative knowledge of glomerular filtration, tubular reabsorption, and secretion of these requires an understanding of the concept of *renal plasma clearance*.

Excretion Rates and Plasma Clearance. The rate at which a substance (X) is excreted in the urine is the product of its urinary concentration, U_X (mg/ml), and the volume or urine per minute, V. The rate of excretion ($U_X V$) depends, among other factors, on the concentration of X in the plasma,

TABLE 1. COMPOSITION OF 24-HOUR URINE IN THE NORMAL ADULT[a]

	Amount		U/P[b]
Urea	6.0 –18.0 g N		60.0
Creatinine	0.3 – 0.8 g N		70.0
Ammonia	0.4 – 1.0 g N		—
Uric acid	0.08– 0.2 g N		20.0
Sodium	2.0 – 4.0 g	(100.0–200.0 meq)	0.8– 1.5
Potassium	1.5 – 2.0 g	(35.0– 50.0 meq)	10.0–15.0
Calcium	0.1 – 0.3 g	(2.5– 7.5 meq)	—
Magnesium	0.1 – 0.2 g	(8.0– 16.0 meq)	—
Chloride	4.0 – 8.0 g	(100.0–250.0 meq)	0.8–2.0
Bicarbonate	—	(0.0– 50.0 meq)	0.0–2.0
Phosphate	0.7 – 1.6 g P	(2.0– 50.0 mmoles)	25.0
Inorganic sulfate	0.6 – 1.8 g S	(40.0–120.0 meq)	50.0
Organic sulfate	0.06– 0.2 g S		—

[a] From White, Handler, Smith and Stetten, "Principles of Biochemistry," New York, McGraw-Hill Book Co., 1959.
[b] U/P = ratio of urinary to plasma concentration.

P_X (mg/ml). It is therefore reasonable to relate $U_X V$ to P_X, and this is called the clearance ratio: $(U_X \cdot V)/P_X$, or more generally, UV/P. This has the dimensions of volume and is in reality the smallest volume from which the kidneys can obtain the amount of X excreted per minute. It must be understood that the kidneys do not usually clear the plasma completely of X, but clear a larger volume incompletely. The clearance is therefore not a real but a *virtual* volume. When substances are being cleared simultaneously, each has its own clearance rate, depending on the amount absorbed from the glomerular filtrate or added by tubular secretion. The former will have the lower clearance, the latter the higher. Those cleared only by glomerular filtration will be intermediate, and their clearance will in effect measure the rate of *glomerular filtration* in milliliters per minute.

The best-known substance which can be infused into the blood to provide a clearance equal to glomerular filtration rate is *inulin*, a polymer of fructose containing 32 hexose molecules (molecular weight, 5200).

Strong evidence exists that it is neither reabsorbed nor secreted, is freely filterable, is not metabolized, and has no physiological influences. Its clearance in man is 120–130 ml/min. This is taken to be the glomerular filtration rate (GFR) or C_F (amount of plasma water filtered through glomeruli per minute). Besides inulin in the dog and other vertebrates, creatinine, thiosulphate, ferrocyanide and mannitol also fulfill these requirements.

Knowing the glomerular filtration rate permits quantitation of the amount of any substance freely filtered [C_F (ml/min) $\times$ P_X (mg/ml)]. Subtracting from this one minute's excretion, $U_X V$, would give the amount *reabsorbed* in milligrams per minute. A classical example is the glucose mechanism. At normal plasma concentrations, none or a trace appears in the urine. When plasma glucose is elevated to about 180–200 mg% (the "threshold"), the amount appearing in the urine begins to increase. As concentration is raised more, the nephrons become progressively saturated until the rate of reabsorption becomes constant and maximal. This indication of saturation of the transport system is referred to as the T_m, "tubular maximum" (here, T_{mG}). In humans T_{mG} has the value of 340 mg/min. Absorption occurs in the proximal convoluted tubules.

Tubular Reabsorption of Organic Substances. *Sugars.* Xylose, fructose and galactose when introduced into the blood are reabsorbed by the same transfer system as glucose, but much less completely. When glucose load is increased, their reabsorption is blocked (*competitive inhibition*). The glycoside, *phlorizin* (contained in the bark of apple, pear and other fruit trees), blocks reabsorption of all sugars with resulting increase in urinary excretion.

Amino Acids. The total clearance is low in man (1–8 ml/min). Much of the characterization of mechanisms has been done in the dog. Several amino acids (glycine, arginine and lysine) demonstrate relatively poor reabsorption, with small T_m's. Other amino acids are so efficiently reabsorbed that saturation is not achieved by concentrations which do not cause physiological disturbances.

Ascorbic Acid. This has a T_m of 2.0 ± 0.19 mg/1.73 SQ.M.S.A. in man (or *ca.* 1.5 mg/100 ml of filtrate). In the dog, net reabsorption is 0.5 mg/100 ml of filtrate.

Urea. This major product of protein metabolism is filtered and reabsorbed to varying degrees (40–70%), inversely related to the rate of urine production. Reabsorption is largely a process of back-diffusion in man, dog, rabbit and chicken, although active mechanisms operate in the kidneys of Elasmobranchii. In amphibian (Anuran) kidneys, the tubules secrete urea.

Uric Acid. In mammals, this appears as a consequence of metabolism of purine bases. In most mammals, it is oxidized to allantoic acid, but not in primates or the Dalmatian coach dog. It has been generally assumed that a T_m characterizes its reabsorption, average 15 mg/min in man, a value so high compared to the amounts existing in the plasma that saturation should not occur normally. Under conditions of injections of an uricosuric drug [G-28315 (Sulfinpyrazone)] and vigorous mannitol diuresis, excreted urate/filtered urate ratios up to 1.23 have been observed in man, demonstrating tubular secretion. A three-component system of filtration, reabsorption and secretion is suggested, so that the amount excreted is the *net effect* of these operations. Definite evidence for tubular secretion has been found in birds and reptiles.

Creatine. This is a product of muscle metabolism which disappears from the urine of humans after adolescence. It is reabsorbed in concentrations below 0.5 mg%. At higher concentrations, reabsorption is incomplete and excretion is enhanced. No T_m has been demonstrated. At higher plasma levels, the creatine/C_F ratio becomes constant at 0.8.

Tubular Secretion. Evidence has accumulated that the proximal tubule is the site of active secretion of some physiologically occurring substances as well as certain foreign substances when injected into the circulation.

Tubular Secretion of Foreign Substances. *p-Aminohippuric Acid* (*PAH*). At low plasma concentrations, this substance is almost completely cleared from the plasma by a combination of glomerular filtration and efficient tubular secretion. Hence, its clearance measures 90% of total plasma flow through the kidneys. In man, the clearance is 600–700 ml/min and, corrected for hematocrit, gives the *effective renal blood flow*.

When plasma levels are elevated in the range of 30–50 mg%, the excretory mechanism becomes saturated and a T_m can be discerned. The quantity of the substance filtered is $P_{PAH} \times C_F$; the total excreted in the urine is $U_{PAH} \times V$, with T_{PAH} designated as the tubular contribution. Then T, $T_{mPAH} = U_{PAH} \cdot V - U_{PAH} \cdot G_F$. PAH in plasma is largely freely filterable. Some substances, *e.g.*, diodrast and phenol red, are bound to protein, and a correction factor needs to be introduced. T_m for PAH in man is *ca.* 77 mg/min/1.73 M.S.A. In the dog, it is 19.1 mg/min/SQ.M.S.A.

Others. Additional foreign substances, secreted by the tubules are *diodrast*, *phenol red* and *penicillin*. *Carinamide* and *benemid* block the secretion of the above. The blocking is competitive, for these substances are also secreted by the tubules.

Tubular Secretion of Physiological Substances. *Creatinine*. This is derived from creatine in muscles. Exogenous creatinine is cleared by glomerular filtration plus tubular secretion in man; the T_m is small. It is also secreted by tubules of certain teleosts, alligator, chicken, goat and rat. Glomerular filtration alone is the mechanism in the dog, rabbit, sheep, seal, cat, frog and turtle.

N-Methylnicotinamide. This metabolic derivative of nicotinic acid has a clearance up to three times the C_F in dogs.

Ammonium. This is synthesized in the tubular epithelium from glutamine and other amino acids and secreted into the distal tubular urine.

Potassium. Its clearance is usually well below that of inulin in man and dog, suggesting fairly complete reabsorption. Under certain circumstances, *e.g.*, giving large amounts of K salts of foreign anions, the clearance of K arises above C_F. It appears that a three-component system operates, with proximal reabsorption and distal secretion.

Hydrogen Ions. The distal tubule is able to generate and secrete H^+ as a means of acidifying the urine by the aid of the enzyme, CARBONIC ANHYDRASE:

$$CO_2 + H_2O \overset{C.A.}{\rightleftharpoons} H_2CO_3 \begin{matrix} \nearrow HCO_3^- \\ \searrow H^+ \end{matrix}$$

Acid-base Regulation. All three (NH_4^+, K^+, and H^+) are secreted into the distal tubular urine to be exchanged for Na (of Na_2HPO_2, $NaHCO_3$ and NaCl), conserving valuable base and ridding the body of metabolic acids.

Excretion of Electrolytes. *Cations*. Sodium: About 99% of the filtered sodium is reabsorbed by active mechanisms, 80–85% in the proximal tubule and 14–19% in the distal tubule and collecting duct. In man, *ca.* 200 μeq/min are excreted. *Aldosterone*, secreted by the adrenal cortex, is necessary for efficient reabsorption (see ADRENAL CORTICAL HORMONES).

The distal absorption is concerned with the acid-base regulatory mechanism. No T_m for Na can be discerned. Increased loading is followed by increase in total reabsorption, but with decreased efficiency, so that urinary excretion is increased.

Calcium: Its excretion is complicated by the fact that a significant part of its plasma content is combined with plasma proteins. That which is filtered is in complexes poorly ionized; hence it is not handled as Ca^{++}. Urinary excretion is low (*ca.* 8.5 μeq/min) suggesting efficient tubular reabsorption.

Magnesium: That not bound by plasma proteins is filtered and reabsorbed. About 5–6 μeq/min are excreted in man.

Anions. Chloride: This renal mechanism is similar to sodium, being the chief "indifferent" anion that accompanies Na^+ through the kidney.

Phosphate: In the plasma about 80% occurs as HPO_4^- and 20% as $H_2PO_4^-$, both usually combined with Na. The ultimate ratio of $H_2PO_4^-/HPO_4^-$ in urine is determined by the pH. PO_4 T_m in the dog is 0.10–0.15 mmoles/min, and in man, 0.13 mmoles/min. Excretion in man is 7–20 μmole/min.

Sulfate: This is actively reabsorbed in dog and man, and shows a well-defined, although small T_m. In the dog, it is 0.05 mmole/min. With slight increases in plasma concentration, excretion is rapidly increased.

Excretion of Water. Mechanisms which concern tubular reabsorption of water are intimately tied up with the handling of osmotic constituents, primarily sodium salts. Another factor is the action of antidiuretic hormone (ADH), elaborated by the supra-optic and paraventricular nuclei of the hypothalamus (see VASOPRESSIN). Finally, the composite mechanisms are currently integrated in the light of a *countercurrent diffusion multiplier system*, operating particularly in nephrons which project long loops of Henle and vasa recta into the papillary zones of the renal medulla, as exemplified by the golden hamster and kangaroo rat.

Since the fluid in the proximal convoluted tubule has been found to be isosmotic by direct puncture studies, it is assumed that water follows Na in on a passive osmotic obligatory basis. Evidence from micro-cryoscopic methods shows the osmotic pressure in the cortex equal to that of the plasma, but that it is stratified in increasing concentrations proceeding from the cortex to the tip of the papillae where it is 3–4 times that of plasma.

Recently, this has been verified by micropuncture of the loops of Henle and the accompanying vasa recta, both arranged in the principle of a "hairpin" countercurrent system. Finally, the osmotic concentration in the collecting tubules has been found to parallel the concentration in the loops as the tubules pass from the cortex to their point of exit at the tip of the papillae.

The principle of the countercurrent system as it applies to the medullary loop of Henle system is as follows: Sodium, by an active mechanism, and chloride, as the result of an electrochemical gradient thus established, are believed to be transported out of the relatively water-impermeable ascending limb of the loop of Henle into the interstitium of the medulla until a gradient of *ca.* 200 mOsm kg H_2O has been established. This single effect is multiplied as the fluid of the thin descending limb comes into osmotic equilibrium with the interstitial fluid by the diffusion of water out (and probably by the diffusion of some NaCl into the descending limb), thus raising the osmolarity of the fluid rounding the hairpin loop into the ascending limb. The increased concentration here, also raising that in the interstitium, now favors further movement of fluid out of the descending limb, further increasing concentration and so on.

In this fashion, an increasing osmotic gradient is established in the direction of the tip of the papillae, and yet at no level is there a large osmotic difference between the luminal and interstitial fluid. The collecting ducts in the presence of ADH are believed to be water permeable and somewhat Na impermeable (net transport small, although there may be diffusion in and active transport out). This results in diffusion of water out of the collecting ducts into the hyperosmotic medullary institium, and ultimately into the vasa recta to be carried away until the fluid in the ducts becomes corresponding concentrated.

The view is favored that ADH acts in a permissive fashion to let water diffuse out, perhaps altering the size of "pores" in the base of the cells of the tubular epithelium. In addition, the membrane of the opposite end of the cell presents an obstacle more readily passed by water than by hydrophilic solutes.

The active hormone is bound to a carrier neurosecretory material (NSM) as it is formed in the supra-optic and paraventricular nuclei of the hypothalamus. This substance is supposed to flow via the supra-optic–hypophyseal tract to the posterior pituitary lobe, which functions as a storage and release center.

The change in effective osmotic pressure in the blood to the centers appears to be the effective stimulus, hypertonicity causing increased release, and urinary concentration. The possibility exists that certain *volume receptors* respond to isotonic fluid expansion to inhibit ADH action and produce diuresis.

References

1. SELKURT, E. E., "Kidney, Water and Electrolyte Metabolism," *Ann. Rev. Physiol.*, **21**, 117–150 (1959).
2. SMITH, H. W., "Principles of Renal Physiology," Oxford, The University Press, 1956.
3. SPERBER, I., "Secretion of Organic Anions in the Formation of Urine and Bile," *Pharmacol. Rev.*, **11**, 109–134 (1959).

EWALD E. SELKURT

REPRODUCTION, FERTILITY, AND STERILITY

Much has been learned concerning the biochemistry of mammalian reproduction, fertility and sterility in recent years. However, many aspects of the complicated processes involved in reproduction in man and higher animals still need clarification and explanation. Knowledge of the chemicals, enzymes and hormones involved in the intricate biological reactions of reproduction has accumulated largely since the turn of the century. In many instances, discovery and understanding has paralleled or followed the identification of substances involved in controlling, or essential to, the reproductive processes. Greater emphasis and interest in controlling reproduction has been stimulated periodically. The economic pressures brought on by the need to raise greater numbers of high-producing farm livestock and to control certain diseases, especially resulting during and after recent wars, have been an important factor. The urging and desires of childless couples have repeatedly spurred interest and efforts to understand reproductive phenomena. Controlling reproduction for family planning has for many years brought attention to the reproductive processes of man. More recently and probably for some years to come, interest and efforts to understand and control reproduction will continue because of the threat that population growth may exceed the food supply and living space available to man.

The Male Role. The male's part in the reproductive process in man and higher animals is primarily that of producing spermatozoa. Spermatozoa carry the sample half of the CHROMOSOMES needed to restore a full complement of chromosomes and to initiate the development of an ovum at the time of fertilization. The male role usually includes also the deposition of spermatozoa into the female tract at the time of mating except in instances where artificial insemination is used.

The paired testes of the male produce millions of spermatozoa that pass into the epididymis where they mature. During this passage, the relationship of the minerals, particularly sodium and potassium change and a reduction in the level of potassium and an increase in sodium, is thought to be conducive to the initiation of motility upon ejaculation. The testes also play a hormonal role by producing the ANDROGENIC hormones (testosterone and androstenedione) which promote maleness and sex drive.

Many factors are known to affect the fertility of the male. In most mammals where the testes are carried in a scrotum outside and away from the body, any condition which brings the testes to body temperature or higher usually impairs spermatogenesis, reduces fertility, and may cause temporary sterility.

Spermatozoa are cells that display metabolic activity similar to many body cells. This metabolism, which usually involves the breakdown of glucose or fructose either by anaerobic or aerobic pathways, yields energy in the form of adenosine triphosphate which supports spermatozoan motility. Spermatozoa in the testes and in storage in the epididymides do not have an abundant supply of metabolizable substrates and thus tend to remain immobilized. At the time of ejaculation, they are mixed with secretions from the accessory glands which supply liberal quantities of fructose, citric acid and many other chemical substances which permit the generation of energy needed for motility.

The head of the sperm cell consists largely of nuclear material including the DEOXYRIBONUCLEIC ACID (DNA) which forms the basis for starting the development of a new individual when a spermatozoon fertilizes an ovum.

The Female Role. The paired ovaries of the female produce ova. In mammals, ova are usually from 125–175 μ in diameter. The cytoplasm, in addition to carrying the CHROMATIN material, usually has many small fat globules present. MUCOPOLYSACCHARIDES are also present in small quantities. The clear band of material surrounding the cytoplasm is called the zona pellucida and consists of neutral or weakly acidic mucoprotein. When first shed from the ovary, numerous cumulus cells cling to the ovum. These cells are held together by an intracellular substance reported to be mainly hyaluronic acid, which is soon dissolved as the freed ovum passes down the oviduct.

Two predominant female hormones are produced in the ovary. *Estrogens* (mainly estradiol) are produced as follicles are developing. Estrogens promote the female characteristics and induce estrus. After an ovarian follicle has ruptured and the ovum has been shed, the cavity becomes filled by a rapid growing yellowish tissue, called luteal tissue, forming a corpus luteum. This tissue produces another sex steroid hormone called *progesterone*. This hormone is essential in the preparation of the uterine tissue for implantation of a fertilized ovum and, in many species, is required for the maintenance of pregnancy (see also OVARIAN HORMONES; SEX HORMONES).

Spermatozoa are usually deposited in the vagina of the female at the time of mating or into the cervix or uterus by artificial insemination. Oxytocin is released due to the stimulus of mating and produces contractions in the female reproductive tract. These contractions aid in transporting spermatozoa to the oviducts where fertilization usually occurs.

If spermatozoa are present in the female reproductive tract when an ovum is released, fertilization is brought about by penetration of one sperm cell into the cytoplasm of the ovum. Within the cytoplasm of the ovum, the male and female pronuclei fuse and CELL DIVISION soon begins. The biochemical phenomena involved in this process are not yet well understood. Metabolic activity of the ovum is greatly increased soon after ferilization. Within 3 or 4 days, the fertilized ovum migrates from the oviduct into the uterus. The developing embryo is nourished at first by the materials present in the ovum and by substances absorbed from the uterine environment in which it is located. Soon extra-embryonic membranes are developed, circulation is established to and from the embryo, and the fetal blood exchanges waste products and nutrients across the membranes of the placenta. No direct exchange of blood occurs between the fetus and the mother.

As the fetus nears the age of normal gestation for the species, the endocrine balance begins to change, usually with the estrogen levels rising and the progesterone levels falling, and this together with other factors such as the size and pressures of the fetus initiate parturition. Oxytocin is released at this time and stimulates uterine muscle contractions essential to the expulsion of the fetus.

Estrous Cycle Control. In recent years, great strides have been made toward artificially controlling the estrous cycle, particularly in the farm animals. By synchronizing the estrous cycles of groups of female farm animals, a number of females can be bred so that the calf, lamb or pig crop arrives within a few days, thus facilitating farm operations and producing a uniform age group of young animals. Estrous cycle control is brought about by administering hormones that influence the pituitary release of follicle stimulating hormone (FSH) and luteinizing hormone (LH) which cause follicle growth and induce ovulation (see GONADOTROPIC HORMONES). Progestrone or substances with progestational activity, which inhibit estrus by suppressing pituitary release of gonadotropic hormones, given by injection or orally have been quite effective in bringing about estrous cycle control. Estrus is suppressed over a period of two weeks or more by administering the hormone with progestational activity, then upon withdrawal a high percentage of the animals come into estrus within 3 to 6 days. The fertility level of animals with synchronized estrous cycles has not been quite as high as with control animals, although much progress has been made in the last two years by the combined use of progestational agents and follicle stimulating substances.

Control of Fertility by Drugs. Progestationally active substances that are orally effective have become the basis of the "pill" method of preventing conception in the human female. A number of compounds have been developed which inhibit ovulation if the "pill" is taken regularly. In some cases, estrogenic substances have been added that also help to inhibit ovulation (see Table in article on OVARIAN HORMONES). Recent findings indicate that some of the compounds not only inhibit ovulation but may also cause the disappearance and degeneration of ova if they are shed. Most reports indicate a highly effective preventative level when the drugs are taken according to prescription. Side effects which were noted with some of the compounds used in early tests now seem to be minimal. The return to fertility when the "pill" is discontinued seems to be normal. In some instances, proper administration of some of these same hormones can help to restore fertility in individuals with temporary sterility.

References

1. BISHOP, D. W. (EDITOR), "Spermatozoan Motility," Washington, D.C., American Association for the Advancement of Science, Publ. No. 72, 1962.
2. COLE, H. H., AND CUPPS, P. T. (EDITORS), "Reproduction in Domestic Animals," New York, Academic Press, 1959.
3. HARTMEN, C. G. (EDITOR), "Mechanisms Concerned with Conception," New York, The Macmillan Co., 1963.
4. MANN, T., "The Biochemistry of Semen," New York, John Wiley & Sons, 1964.
5. NALBANDOV, A. V., "Reproductive Physiology," San Francisco, W. H. Freeman and Co., 1964.
6. SALISBURY, G. W., AND VANDEMARK, N. L., "Physiology of Reproduction and Artificial Insemination of Cattle," San Francisco, W. H. Freeman and Co., 1961.
7. WILLIAMS, W. W., "Sterility: The Diagnostic Survey of the Infertile Couple," Springfield, Mass., W. W. Williams, 1964.

N. L. VANDEMARK

RESPIRATION

Respiration is primarily an energy-yielding dissimilation process and a phenomenon that is exhibited by all living cells. In this process, high-energy containing substances generally carbohydrates (*e.g.*, starch, glycogen, sucrose, glucose) or lipids are broken down in a stepwise manner, under enzymatic control, to simpler substances of lower energy content. Chemical or free energy is liberated at certain specific stages in the form of high-energy phosphate bonds (~P) which are trapped by the adenylic acceptor system (adenosine diphosphate, ADP) and stored in the pyrophate bonds of adenosine triphosphate (ATP; see PHOSPHATE BOND ENERGIES). The over-all thermodynamic efficiency is estimated at 60–70% which is high in comparison to man-made machines (steam or combustion engines, etc.).

Respiration may be either aerobic or anaerobic (FERMENTATIVE) depending on the organism or tissue concerned, and also on the oxygen tension in the atmosphere or at the sites of respiration.

Aerobic respiration is essential for the survival of the vast majority of terrestrial and aquatic organisms, although some fungi (yeasts) and, for example, the enteric bacteria are facultative anaerobes and other bacteria (Clostridia) are obligate anaerobes. In animals, fish and insects, aerobic respiration includes breathing (pumping air in or out of air sacs and lungs or water over gills). This is referred to as external respiration (*vide infra*) in contrast to internal or cellular (tissue) respiration.

Respiration is a catabolic process and as well as being the least complex it is also the best understood of all physiological processes. However, all other synthetic (anabolic) processes which are essential for growth and reproduction of organisms are in one way or another completely dependent on its normal operation.

Thus, in obligately aerobic organisms, if respiration is inhibited by heavy metal poisons (*e.g.*, cyanide and carbon monoxide) that interfere with the iron porphyrin oxygen acceptor systems, death ensues rapidly. Similarly in organisms that "breathe" in their over-all aerobic respiration, considerable discomfort and death may be experienced if the oxygen tension of the atmosphere (21%) falls below a critical point (14%).

Aerobic Cellular Respiration. The aerobic respiration of glucose involves the consumption of oxygen, the release of an equivalent volume of carbon dioxide, the liberation of energy and the formation of water according to the following over-all equation:

$$C_6H_{12}O_6 + 6O_2 \rightarrow 6CO_2 + 6H_2O$$

$$\Delta F = -688 \text{ kg cal/mole}$$

The volume ratio of CO_2/O_2 is known as the respiratory quotient (RQ) and is indicative of the substrate being respired. Thus the RQ is unity for hexoses and varies from <1 for fats and to >1 when organic acids are being preferentially respired.

The equation for aerobic respiration does little to indicate either its mechanism or its special relationship with fermentation or anaerobic respiration. It is now well established that the aerobic respiration of carbohydrates takes place in three main phases, of which the first two are shared with fermentation.

Phase I—Mobilization of Reserve Carbohydrates. In Phase I, the reserve carbohydrates are mobilized by phosphorylations involving inorganic phosphate and ATP to glucose-6-phosphate and fructose-1,6-diphosphate in which the esterified phosphate linkages are "energy poor" (—P *ca.* 3 kg cal/mole) in comparison with the "energy-rich" pyrophosphate bonds of ATP ($\sim$P *ca.* 12 kg cal/mole). No oxidation occurs in this phase and energy is lost from the ATP reservoir.

Phase II—The Embden-Meyerhof-Parnas Sequence of Glycolysis. In Phase II the process of GLYCOLYSIS occurs in which this hexose diphosphate is cleaved into two triose phosphates. One of these, 3-phosphoglyceraldehyde, undergoes a coupled oxidation and substrate phosphorylation involving inorganic phosphate and diphosphoridine nucleotide (DPN, or NAD, NICOTINAMIDE ADENINE DINUCLEOTIDE). Sufficient energy is liberated by the oxidation of the carbonyl to the carboxyl group to reduce DPN and to form an unstable $\sim$P in 1,3-diphosphoglyceric acid which is trapped by the adenylic acceptor system and stored as a pyrophosphate bond in ATP. A second $\sim$P is formed in the intramolecular transfer of the low-energy ester phosphate of 3-phosphoglyceric acid in the 2-position accompanied by a dehydration with the formation of phosphoenolpyruvic acid. This phosphate is transferred to ATP with the formation of pyruvic acid.

The second triose phosphate arising from hexose diphosphate is converted to 3-phosphoglyceraldehyde and is carried through the same series of reactions. Thus, per mole of hexose diphosphate, two moles of pyruvic acid are formed, two moles of DPN are reduced, and four moles of ATP are synthesized from ADP according to the following equation:

$$C_6H_{12}O_6\text{—(2P)} + 2DPN + 4ADP \rightarrow$$

$$2CH_3COCOOH + 2DPNH_2 + 4ATP$$

and this ends the phase of glycolysis.

Phase III—The Krebs Tricarboxylic Acid Cycle, or CITRIC ACID CYCLE. By far the major part of the energy of the glucose molecule is liberated in the steps that lead to the ultimate aerobic fate of pyruvic acid into carbon dioxide and water. The enzymes of the Krebs cycle, in contrast to glycolytic enzymes, are bound up in discrete cellular organelles called MITOCHONDRIA. Entrance to the cycle is gained by acetyl coenzyme A which arises from the oxidative decarboxylation of pyruvic acid involving COENZYME A and the reduction of DPN. Acetyl-CoA condenses with one molecule of oxaloacetic acid to form citric acid which undergoes successive decarboxylations and oxidations involving primarily DPN, but also TPN (triphosphopyridine nucleotide, coenzyme II, NADP), and also coenzyme A to form another molecule of oxaloacetic acid which condenses with acetyl-CoA arising from the second molecule of pyruvic acid. Thus, for two complete revolutions of the cycle, the carbon skeleton of glucose is entirely released as carbon dioxide in these decarboxylations.

The remaining available energy of the glucose molecule withheld in two molecules of pyruvic acid is liberated by processes of oxidative PHOSPHORYLATION which accompany the passage of H^+ and electrons along hydrogen and electron carrier systems involving the pyridine nucleotides, flavoproteins, cytochromes and cytochrome oxidase to molecular oxygen.

Inorganic phosphate is trapped and stored in the pyrophosphate bonds of ATP by the energy liberated at the successive steps in the transfer of H^+ and electrons to oxygen where they react to form water. Similarly, reduced DPN arising from triose phosphate dehydrogenation in Phase II (glycolysis) would also release its energy by oxidative phosphorylations of this type. It is estimated that per mole of hexose diphosphate between 30 and 40 pyrophosphate bonds are synthesized by the steps of oxidative phosphorylation in which oxygen is an essential factor, as compared to the four that are produced by substrate phosphorylations in anaerobic glycolysis.

The primary function of respiration is to release chemical energy, but this would be to no avail if there were no energy-conserving systems such as the adenylic acceptor system. Likewise, if ATP were not utilized for synthetic reactions in growth the adenylic acceptor system would become saturated and would operate as a FEEDBACK control that would regulate the rate of respiration. Thus the highest respiration rates are found at the sites of most active synthesis.

The Pentose Phosphate Pathway. Many, if not all organisms have an alternative aerobic pathway for the oxidation of glucose, and one of widespread occurrence has been variously referred to as the "hexose monophosphate shunt," the "direct oxidation pathway" or the "pentose phosphate cycle." The starting point is glucose-6-phosphate which undergoes two successive dehydrogenations involving TPN, followed by a decarboxylation of carbon-1 to pentose phosphate. For futher details, see CARBOHYDRATE METABOLISM; CARBON REDUCTION CYCLE (IN PHOTOSYNTHESIS); PENTOSES IN METABOLISM.

Anaerobic Cellular Respiration. Anaerobic respiration is synonymous with FERMENTATION. The latter is often the preferred term and connotes in Pasteur's original definition "life without oxygen," or more specifically, the degradation of carbohydrate into two or more simpler molecules by processes not requiring molecular oxygen. In fermentations, the carbon skeleton of glucose is never completely released as carbon dioxide and in some it may not appear at all. The term aerobic fermentation may be used to designate fermentations occurring in aerobic atmospheres along with a component of aerobic respiration.

The classical fermentation is the ALCOHOLIC FERMENTATION of glucose by yeast, according to the equation:

$$C_6H_{12}O_6 \rightarrow 2CO_2 + 2C_2H_5OH$$

$$\Delta F = -56.1 \text{ kg cal/mole}$$

Since one of the products of this fermentation

cannot escape from the cells, it eventually becomes toxic and yeasts vary widely in their capacity to withstand the accumulation of alcohol. Thus Baker's yeast which has been selected for the rapid production of carbon dioxide can only withstand a concentration of 3–4% alcohol. Brewer's yeasts ferment glucose more slowly and can produce as much as 10% alcohol, and wine yeasts can withstand concentrations of 14–16% alcohol. When yeasts are transferred to an aerobic environment, they continue to ferment (aerobic fermentation), but they also carry out a small aerobic respiration.

Certain body tissues, *e.g.*, muscle and liver, when respiring may support components of both aerobic respiration and a lactic acid fermentation according to the equation:

$$C_6H_{12}O_6 \rightarrow 2C_3H_6O_3 \quad \Delta F = -47.4 \text{ kg cal/mole}$$

The relative contribution of each process to the total repiration is determined by the supply of oxygen to the muscle tissue. Thus, lactic acid fermentation is augmented by anoxia brought about by heavy exercise, and the accumulation of lactic acid can cause stiffening of the muscles.

In the tissues and organs of green plants, aerobic respiration is the normal process, although some organs, *e.g.*, apple fruits, etc., can endure long exposure (100 hours or more) of anaerobic conditions without harmful effects. In such cases, fermentations take place exclusively. In between a completely anaerobic environment and one that contains a critical volume of oxygen (*ca.* 5%), components of both aerobic respiration and aerobic fermentation contribute to the over-all process. This critical oxygen tension is known as the "extinction point" of aerobic fermentation, above which aerobic respiration is exclusively dominant. When fermentations occur in plants, the products are generally alcohol and carbon dioxide as in yeast fermentation, *e.g.*, apples, carrot roots, grapes, pea seedlings, but in potato tubers and other tissues, lactic acid and carbon dioxide are the main products.

Many microorganisms ferment carbohydrates to a variety of products including formic, acetic, propionic, butyric and succinic acids, etc.

Both alcoholic and lactic acid fermentations share a common pathway with aerobic respiration leading to the production of pyruvate in glycolysis (Phases I and II). The primary function of these fermentative processes is to release energy in the form of $\sim$P which is trapped and stored in the pyrophosphate bonds of ATP for subsequent use in synthetic growth processes. Pasteur recognized that they could only be substitute energy-yielding processes in normally aerobic organisms, but a comparison of the free energy changes (ΔF) indicates that they would have to utilize about 12 times as much glucose per unit time in order to yield the same amount of energy as aerobic respiration and this is never achieved in nature.

In many normally aerobic animal and plant tissues, the actual rate of carbohydrate destruction increases on exposure to anaerobic conditions, but this rarely exceeds a two- to fourfold increase. Conversely, on transfer from anaerobic to aerobic conditions the rate of carbohydrate utilization decreases. It is quite obvious why aerobic respiratory processes do not occur in the absence of oxygen, but it is not so clear why fermentations are extinguished in an oxygen atmosphere above a critical oxygen tension (extinction point). Oxygen, therefore, plays a positive role in suppressing or decreasing fermentation and the products of anaerobic metabolism. Much more subtle to the organism is the action of oxygen in diminishing carbohydrate destruction; in other words, oxygen conserves carbohydrates and prevents the inefficient drain on foodstuffs that would occur if fermentations operated independently and unlimited. This is known as the "Pasteur effect" which is considered to be operating when it can be shown that the rate of carbohydrate utilization under anaerobic conditions exceeds that under aerobic conditions at oxygen tensions above the extinction point of anaerobic respiration. The mechanism of the "Pasteur effect" and the role of oxygen have been the subject of considerable investigation and much controversy.

The lower efficiency of alcoholic and lactic acid fermentations as compared to aerobic respiration can be readily explained by the fact that in the absence of oxygen, the processes of oxidative phosphorylation which release the major part of the energy of the glucose molecule and which accompany the transfer of H^+ and electrons to molecular oxygen are inoperative. Since these processes are irrevocably linked with the Krebs tricarboxylic acid cycle which is responsible for the aerobic degradation of pyruvate, this substance is forced into anaerobic channels of metabolism where it is reduced directly or indirectly by reduced DPN produced in the oxidation of 3-phosphoglyceraldehyde in glycolysis (Phase II).

External Respiration. The maintenance of respiration within cells demands an external system to facilitate gaseous exchange. In microorganisms, aqueous diffusion is sufficient, but in multicellular organisms, more elaborate mechanisms are required to maintain a suitable gaseous environment at the cellular level. In land plants this takes the form of gasous diffusion paths in the intercellular spaces, controlled at the epidermis, of leaves, for example, by stomata. Insects also rely upon diffusion and air sacs, but in higher animals where much of the actively respiring tissue is far from the surface, a more complex two-stage system is found which involves the mechanical pumping (breathing) of air or water (in fish) into organs that effect the attainment of gaseous equilibria (lungs or gills) and the transport of respiratory gases in a circulatory system.

External respiration involves expiration (breathing out) of carbon dioxide and inspiration (breathing in) of oxygen of the air or water transported to and from the lungs or gills by HEMOGLOBIN of the venous and arterial blood. Breathing is involuntary, but it may be controlled voluntarily. The rate of involuntary breathing is adjusted by nerve centers of the brain so that 20 liters of fresh air are inspired for every liter of carbon dioxide expired. Normally a man at rest breathes 7 liters of air per minute. When oxygen is in short supply, the need for oxygen takes precedence over the elimination of carbon dioxide. The most important organ of respiration in higher vertebrates is the lung where the exchange of gases occurs. The lung has no muscle tissue, but during inspiration, the diaphragm contracts and flattens, extending the chest cavity downward, and the outer muscles of the ribs, etc., raise the ribs and increase the diameter of the chest. This enlargement decreases the lung air pressure resulting in a flow of air through the mouth, trachea, bronchus and bronchioles. Expiration is a passive movement in which the reverse process takes place. The bronchioles end in thin walled sacs (alveoli) which allow oxygen and carbon dioxide to diffuse through. It is here that the exchange of gases occurs (see GASEOUS EXCHANGE IN BREATHING). The transport of oxygen and carbon dioxide by blood affords the body tissue ample oxygen to carry out the processes of aerobic cellular respiration with the removal of carbon dioxide. However, under conditions of anoxia, the fermentative mechanisms, of muscle tissue for example, provide a substitute source of energy.

References

1. BEEVERS, H., "Respiratory Metabolism in Plants," Evanston, Ill., Row-Peterson and Co., 1961.
2. KREBS, H. A., AND KORNBERG, H. L., "Energy Transformations in Living Matter," Berlin, Springer-Verlag, 1957.
3. MCELROY, W. D., "Cell Physiology and Biochemistry," Second edition, Englewood Cliffs, N.J., Prentice-Hall, 1964.
4. SCHMIDT-NIELSEN, K., "Animal Physiology," Second edition, Englewood Cliffs, N.J., Prentice-Hall, 1964.
5. SPECTOR, W. S. (EDITOR), "Handbook of Biological Data," Philadelphia, Saunders, 1956.

E. R. WAYGOOD

RESPIRATORY CHAIN. See BIOLOGICAL OXIDATION-REDUCTION; ELECTRON TRANSPORT SYSTEMS; MITOCHONDRIA, section on Electron Transfer Chain.

RESPIRATORY QUOTIENT. See RESPIRATION, section on Aerobic Cellular Respiration; GASEOUS EXCHANGE IN BREATHING.

RETINAL (RETINENE). See LIGHT ADAPTATION; VISUAL PIGMENTS; VITAMIN A.

Rh FACTOR. See BLOOD GROUPS.

RIBOFLAVIN

Riboflavin is synthesized by green plants and by most bacteria, molds, fungi and yeasts. Higher animals require the substances as a growth factor (for review, see reference 1). The biogenesis of riboflavin has been examined in a large number of microorganisms;[2] some produce it in large amounts, *e.g.*, *Clostridium acetobutylicum*,[3] *Mycobacterium smegmatis*,[4] *Eremothecium ashbyii*,[5,6] *Candida sp.*,[7,8] *Mycocandida riboflavina*,[9] and *Ashbya gossypii*.[10] Conditions leading to maximal production of riboflavin have been studied in particular detail with the ascomycetes *Ashbya gossypii* and *Eremothecium ashbyii* (for review, see references 11–14).

Investigations with *A. gossypii* have shown that rings B and C of riboflavin can be derived from the same simple precursors which are utilized in the biosynthesis of purines, while the *o*-xylene portion (ring A) is derived from two-carbon precursors.[15] The precursor role of purine derivatives was indicated by the finding of Maclaren[16] that adenine, guanine and xanthine increased the production of riboflavin by *E. ashbyii*. More definitive evidence for a precursor-product relationship between purines and riboflavin was provided by the demonstration that uniformly ^{14}C-labeled adenine is converted to riboflavin by growing cultures of *E. ashbyii* while 8-^{14}C-adenine does not contribute significant radioactivity to riboflavin.[17] The nitrogen atoms of the pyrimidine and the imidazole rings of the purine are transferred intact to riboflavin.[18] The obligatory participation of a purine derivative in riboflavin synthesis has been shown in a purineless strain of *E. coli*.[19]

The origin of the ribityl group of riboflavin is uncertain. It has been found that label from the ribosyl moiety of uniformly labeled guanosine is not incorporated into the side chain of riboflavin. This suggested that the ribityl group does not arise by reduction of the ribose moiety of guanosine.[20] Also, purine ribonucleosides are not superior to the free bases in the stimulation of riboflavin production.[17,21] Studies on the incorporation of various labeled substances into riboflavin by intact cells of *A. gossypii* indicate that the side chain of riboflavin in this organism is formed from glucose by pathways involving the transketolase and aldolase reactions.[22]

A number of fluorescent compounds other than flavin derivatives have been isolated from the mycelium of *E. ashbyii* and *A. gossypii*.[23–27] A green fluorescent material has been identified by chemical synthesis as 6,7-dimethyl-8-D-ribityllumazine,[27] and a blue fluorescent substance as 6-methyl-7-hydroxy-8-D-ribityllumazine.[28,29] Extracts capable of converting 6,7-dimethyl-8-ribityllumazine to riboflavin have been obtained from a number of microorganisms known to grow in media without added riboflavin.[26,30–33] *Lactobacillus casei* which requires riboflavin for growth[34] does not appear to possess conversion activity.[14]

Riboflavin synthetase which catalyzes the reaction:

6,7-DIMETHYL-8-RIBITYLLUMAZINE → RIBOFLAVIN + 4-RIBITYLAMINO-5-AMINO-2,6-DIHYDROXYPYRIMIDINE

has been purified from extracts of *E. coli*, *A. ashbyii* and baker's yeast. The stoichiometry of the reaction has been established with such enzyme preparations by measurement of substrate disappearance and product formation, and the demonstration that labels from 6,7-dimethyl-^{14}C-8-ribityllumazine can be recovered in the methyl groups and carbon atoms 6 and 9 of riboflavin.[35] Similar results were obtained with extracts from *Candida flareri* using 6,7-dimethyl-8-ribityllumazine labeled in carbon atoms 6 and 7 and in the methyl substituents.[36] The reaction also occurs in green plants.[37] 4-Ribitylamino-5-amino-2,6-dihydroxypyrimidine has been identified as the second product of the riboflavin synthetase reaction.[38]

Riboflavin synthetase from *A. gossypii* possesses a marked degree of substrate specificity. A number of analogues of 6,7-dimethyl-8-ribityllumazine modified in positions 2,6,7 or 8 have been synthesized[39,40] and tested.[41] Of these, only 6,7-dimethyl-8-[1′-5′-deoxy-D-ribityl)]lumazine is converted to the corresponding isoalloxazine, 5′-deoxyriboflavin. However, a number of analogues, *e.g.*, 6,7-dimethyl-8-D-xylityllumazine, 6-methyl-7-hydroxy-8-D-ribityllumazine, and 6,7-dihydroxy-8-D-ribityllumazine, are inhibitors of the conversion of 6,7-dimethyl-8-ribityllumazine to riboflavin.

Riboflavin synthetase has been purified about 2000-fold from extracts of baker's yeast. The enzyme forms stoichiometric complexes with 6-7-dimethyl-8-D-ribityllumazine, riboflavin or certain lumazine derivatives. Only those substances are bound which have a kinetic effect on the enzyme. The substrate is probably bound in the complex at the site which leads to the donation of a four-carbon moiety from one molecule of 6,7-dimethyl-8-ribityllumazine to another forming riboflavin.[42,43]

The coenzyme forms of riboflavin and their metabolic functions are discussed in the article FLAVINS AND FLAVOPROTEINS.

References

1. WILLIAMS, R. J., EAKIN, R. E., BEERSTECHER, E., AND SHIVE, W., "Biochemistry of the B Vitamins," p. 669, Reinhold Publishing Corp., 1950.
2. PETERSON AND PETERSON, *Bacteriol. Rev.*, **9**, 49 (1945).
3. YAMASAKI, I., *Biochem.*, 307, **431** (1940).
4. MAYER, R. L., AND RODBART, M., *Arch Biochem.* 11, **49** (1946).
5. GUILLIERMOND, A., *Rev. Mycol.*, **1**, 115 (1936).
6. RAFFY, A., *Comptes. Rendus.*, **209**, 900 (1939).
7. BURKHOLDER, R. R., *Proc. Natl. Acad. Sci., U.S.*, **29**, 166 (1943).
8. TANNER, F. W., JR., VOJNOVICH, C., AND VANLANEN, J. M., *Science*, **101**, 180 (1945).
9. MCCLARY, J. E., U. S. Patent No. 2537148 (1951).
10. WICKERHAM, L. J., FLICKINGER, M. H., AND JOHNSTON, R. M., *Arch. Biochem.*, **9** 95 (1946).
11. PRIDHAM, T. G., *Econ. Botany*, **6**, 185 (1952).
12. VAN LANEN, J. M., AND TANNER, F. W. JR., *Vitamins Hormones*, **6**, 163 (1948).
13. HICKEY, R. J., in "Industrial Fermentations" (UNDERKOFLER, L. A., AND HICKEY, R. J., EDITORS), Vol. 2, p. 157, New York, Chemical Publishing Company, 1954.
14. PLAUT, G. W. E., in "Metabolic Pathways" (GREENBERG, D. M., EDITOR), Vol. II, p. 673, New York, Academic Press, 1961.
15. PLAUT, G. W. E., *J. Biol. Chem.*, **208**, 513 (1954); *ibid.*, **211**, 111 (1954).
16. MCCLAREN, J., *J. Bacteriol.*, **63**, 233 (1952).
17. MCNUTT, W. S., *J. Biol. Chem.*, **210**, 511 (1954); *ibid.*, **219**, 365 (1956).
18. MCNUTT, W. S., *J. Am. Chem. Soc.*, **83**, 2305 (1961).
19. HOWELLS, D. J., AND PLAUT, G. W. E., *Biochem. J.*, **94**, 755 (1965).
20. MCNUTT, W. S., AND FORREST, H. S. *J. Am. Chem. Soc.*, **80**, 951 (1958).
21. GOODWIN, T. W., AND PENDLINGTON, S., *Biochem. J.*, **57**, 631 (1954).
22. PLAUT, G. W. E. AND BROBERG, P. L., *J. Biol. Chem.*, **219**, 131 (1956).
23. MASUDA, T., *Pharm. Bull.* (*Tokyo*) 4, **375** (1956).
24. MASUDA, T., KISHI, T., AND ASAI, M., *Pharm. Bull.* (*Tokyo*) 5, **598** (1957).
25. FORREST, H. S., AND MCNUTT, W. S., *J. Am. Chem. Soc.*, **80**, 739 (1958).
26. MALEY, G. F., AND PLAUT, G. W. E., *Federation Proc.*, **17**, 268 (1958).
27. MALEY, G. F., AND PLAUT, G. W. E., *J. Biol. Chem.* **234**, 641 (1959).
28. PLAUT, G. W. E., AND MALEY, G. F., *Arch. Biochem. Biophys.*, **80**, 219 (1959).
29. PLAUT, G. W. E., AND MALEY, G. F., *J. Biol. Chem.* **234**, 3010 (1959).

30. Maley, G. F., and Plaut, G. W. E., *J. Am. Chem. Soc.*, **181**, 2025 (1959).
31. Kuwada, S., Masuda, T., Kishi, T., and Asai, M. *Chem. and Pharm. Bull.* (*Tokyo*), **6**, 618 (1958).
32. Katagiri, H., Takeda, I., and Imai, K., *J. Vitamin* (*Osaka*), **4**, 211 (1958); *ibid*, **4**, 278 (1958).
33. Korte, F., and Aldag, H. V., *Anal. Chem. Liebigs*, **628**, 144 (1959).
34. Snell, E. E., and Strong, F. M., *Ind. Eng. Chem. and Ed.*, **11**, 346 (1939).
35. Plaut, G. W. E., *J. Biol. Chem.*, **238**, 2225 (1963).
36. Goodwin, T. W., and Horton, A. A., *Nature*, **191**, 772 (1961).
37. Mitsuda, H., *Vitamins*, **28**, 466 (1963).
38. Wacker, H., Harvey, R. A., Winestock, C. H., and Plaut, G. W. E., *J. Biol. Chem.*, **239**, 3493 (1964).
39. Winestock, C. H., and Plaut, G. W. E., *J. Org. Chem.*, **26**, 4456 (1961).
40. Davoll, J., and Evans, D. D., *J. Chem. Soc.*, 5041 (1960).
41. Winestock, C. H., Aogaichi, T., and Plaut, G. W. E., *J. Biol. Chem.*, **238**, 2866 (1963).
42. Plaut, G. W. E., and Harvey R. A., *Federation-Proc.*, **24**, 481 (1965).
43. Plaut, G. W. E., *Abstracts of the Sixth International Congress Biochemistry V*, 382 (1964).

G. W. E. Plaut

RIBONUCLEASE

See nucleases, section on Sugar Specificity; proteins, section on Denaturation; see also antigens, section on Proteins. Ribonuclease (from bovine pancreas) was the first protein possessing enzymatic activity for which the complete amino acid sequence (primary structure) was determined, as well as the locations of intramolecular disulfide cross-linkages. The complete primary structure of ribonuclease has been widely reported, for example, in the following reference.

Reference

1. Stein, W. H., and Moore, S., "The Chemical Structure of Proteins," *Sci. Am.*, **204**, 81–92 (February 1961).

RIBONUCLEIC ACIDS (BIOSYNTHESIS)

The genetic apparatus of the cell consists of a mechanism whereby the information contained in the cellular DNA is expressed in the form of proteins which ultimately direct all biological processes. The pathway from DNA to the synthesis of proteins proceeds by two steps (Fig. 1). DNA is a polymer

$$\text{DNA} \xrightarrow[\text{(1)}]{\text{transcription}} \text{RNA} \xrightarrow[\text{(2)}]{\text{translation}} \text{Protein}$$

Fig. 1

consisting of four distinct monomeric units. The linear sequence of these monomers in DNA makes up the genetic code (see also genes). In the first step, this code is faithfully copied (*transcribed*) enzymically by RNA polymerase into the linear sequence of the four monomeric units of RNA. In the second step, the sequence of monomers in RNA is *translated* into the sequence of amino acids in proteins. The species of RNA which carries the genetic code from DNA to the sites of protein synthesis has been termed messenger RNA. This article on RNA biosynthesis is primarily directed to an explanation of the transcription process. For a description of the events taking place in translation, see proteins (biosynthesis).

Polynucleotide Phosphorylase. The first enzyme which was known to catalyze the synthesis of polyribonucleotides is polynucleotide phosphorylase [Eq. (*1*)]:

$$n\ \text{N-R-P-P} \underset{\text{Polynucleotide phosphorylase}}{\overset{\text{Mg}^{++}}{\rightleftharpoons}} (\text{N-R-P})_n + n\ \text{Pi} \qquad (1)$$

Here N is a purine (adenine or guanine) or pyrimidine (cytosine or uracil) base, N-R a nucleoside consisting of a base covalently linked with R the five-carbon sugar ribose, N-R-P-P the 5′-nucleoside diphosphate, and $(\text{N-R-P})_n$ a RNA polymer containing n monomeric units of the nucleoside monophosphate N-R-P, joined by 3′,5′-phosphodiester linkages. The reaction proceeds with a release of inorganic phosphate equivalent to the incorporation of nucleoside-phosphate units into RNA and is readily reversible. When two or more nucleoside diphosphates are included in the system, their nucleoside phosphates are incorporated into RNA in a random sequence to an extent proportional to the relative concentrations of the diphosphates. Although polynucleotide phosphorylase synthesizes RNA *in vitro*, it is currently believed to have a degradative function *in vivo*. The mechanism of this reaction has been recently reviewed (see reference 2, pp. 93–133).

RNA Polymerase. There is general agreement that the synthesis of cellular RNA is catalyzed by RNA polymerase:

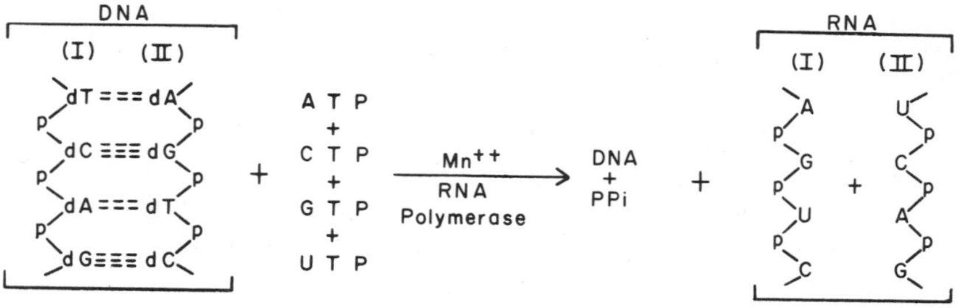

DNA (I) (II): dT===dA, p, dC≡≡≡dG, p, dA===dT, p, dG≡≡≡dC + ATP + CTP + GTP + UTP $\xrightarrow[\text{RNA Polymerase}]{\text{Mn}^{++}}$ DNA + PPi + RNA (I): A, p, G, p, U, p, C + (II): U, p, C, p, A, p, G (*2*)

The RNA polymerase reaction [Eq. (*2*)] has an absolute requirement for a DNA "template" which directs the assembly of the polyribonucleotide from the nucleoside triphosphate precursors (ATP, CTP, GTP, UTP). One mole of pyrophosphate is liberated from the nucleoside triphosphates for each mole of nucleoside phosphate incorporated into RNA. Analysis of the RNA formed in the reaction shows that its average base composition is *complementary* to the base composition of the DNA template. This is the result of a base-pairing mechanism where the bases of the nucleosides dA, dC, dG and dT in DNA direct the incorporation of U, G, C and A, respectively, into RNA. The pairs formed by the purine and pyrimidine bases of the nucleosides dA-U, dC-G, dG-C and dT-A, are termed complementary base pairs. Thus, the transcription of DNA strand (I) by complementary base pairing would give rise to RNA strand (I), and that of DNA strand (II) to RNA strand (II). This process can further be illustrated by substituting in Eq. (*2*) an unnatural DNA containing dT as the only nucleoside component. The utilization of this DNA as a template would result in the synthesis of a RNA containing only A. More extensive treatments of this aspect of the transcription process have appeared recently (see reference 1, pp. 61–66; reference 2, pp. 59–92; reference 3, pp. 91–100).

Similarities between Naturally Occurring RNA and That Synthesized *in vitro* by RNA Polymerase. DNA normally exists as a helical, double-stranded structure (*i.e.*, DNA:DNA complex). When DNA is heated in solution, the helical structure collapses, and the two strands, which are complementary in base composition [as shown in Eq. (*2*)] become separated from one another. When the heated DNA preparations are "annealed" by incubation at a temperature slightly below that where strand separation occurs, a great deal of the DNA duplex structure is restored. This process is illustrated in Fig. 2. The reformation of DNA:DNA complexes is quite specific, and occurs only when heated DNA from the same organism or closely related organisms is annealed. For example, when heated DNA from *E. coli* and the bacterial virus T2 are annealed, T2 DNA:*E. coli* DNA hybrid complexes are not formed. This high degree of specificity required for the formation of a double-strand structure is an indication of the existence of complementary base sequences in the two strands.

It is also possible to form DNA:RNA hybrids using the annealing procedure. When *E. coli* is infected with T2 bacterial virus, the synthesis of *E. coli* RNA ceases, and an unstable species of RNA (messenger RNA) is formed which has a base composition different from that of *E. coli* DNA, but similar to that of T2 DNA. When this "T2 RNA" is subjected to annealing, DNA:RNA hybrids are formed with T2 DNA, but not with other DNA. The T2 RNA synthesized *in vitro* in a RNA polymerase system using a T2 DNA template exhibits the same DNA:RNA hybridization specificity as the T2 RNA formed *in vivo*. The results of these experiments not only show that naturally occurring RNA and the RNA polymerase product have similar properties when annealed with DNA, but are also an indication that they contain base *sequences* complementary to those of DNA (see reference 2, pp. 231–300).

Stable species of RNA, *e.g.*, ribosomal RNA and sRNA, are found in all organisms and account for most of the cellular RNA. By refining the DNA:RNA hybridization technique, it has been shown that *E. coli* ribosomal RNA and sRNA form hybrids with *E. coli* DNA, and that the hybridized regions of the DNA account for approximately 0.2 and 0.02%, respectively, of the total DNA genome. It has thus been concluded that the base sequences of these polyribonucleotides arise from complementary sequences of DNA (see reference 4, pp. 162–164). A similar approach using the technique of RNA:RNA hybridization allows the conclusion that RNA containing the base sequences of ribosomal RNA and sRNA is formed *in vitro* in the RNA polymerase reaction. It can be seen in Eq. (*2*) that both strands of DNA are transcribed by RNA polymerase *in vitro*. When the RNA product of the polymerase reaction (cRNA) is separated from the template DNA and annealed by itself, RNA:RNA hybrids are formed. RNA:RNA hybrids are also produced when *E. coli* cRNA (the RNA polymerase product obtained by using an *E. coli* DNA template) is annealed with an excess of either *E. coli* ribosomal or sRNA (Fig.

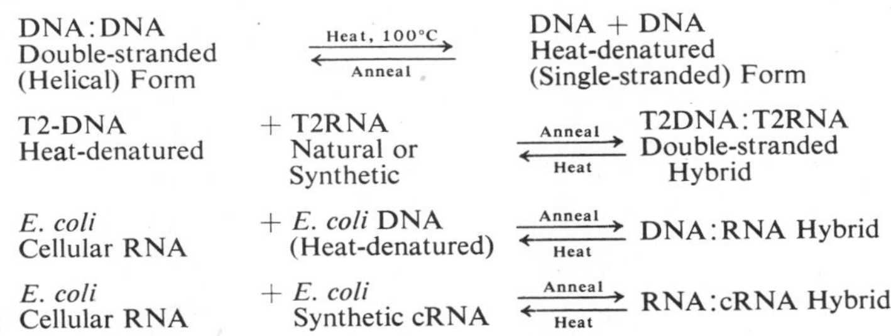

FIG. 2. Summary of polynucleotide hybridizations. *Note:* Natural T2 RNA refers to RNA produced and isolated after T2 phage infection of bacteria. T2 and *E. coli* synthetic RNA (cRNA) refer to RNA made *in vitro* with RNA polymerase using either T2 or *E. coli* DNA as the template.

2), and the extent of RNA:RNA hybrid formation is similar to that which was given previously for the DNA:RNA hybrids. Since *E. coli* ribosomal RNA, for example, can hybridize with its complement present in the RNA polymerase product (Fig. 3), and since both DNA strands are copied in the polymerase reaction, we can conclude that a RNA containing the base sequences of ribosomal RNA is present in the RNA polymerase product. The same argument holds for sRNA.

Another line of evidence suggesting that RNA polymerase is responsible for the synthesis of RNA *in vivo* comes from experiments with Actinomycin D. This ANTIBIOTIC, which inhibits the *in vivo* synthesis of RNA almost completely, also inhibits the *in vitro* RNA synthesis by RNA polymerase at Actinomycin concentrations consistent with those effective *in vivo* (see reference 2, pp. 82–84).

Asymmetric Transcription of DNA. The two complementary strands of a viral DNA have been isolated from one another. When these two DNA preparations are annealed separately with the RNA produced after virus infection, DNA:RNA hybrid formation is observed with only *one* of the two DNA strands (see reference 4, pp. 174–178, pp. 191–204). This suggests that only one of the two DNA strands is transcribed. Although both strands are normally transcribed by RNA polymerase *in vitro*, recent investigations show that single-strand transcription can occur *in vitro* under carefully controlled conditions. Although the mechanism by which single-strand transcription occurs is not yet known, it appears that the transcription of both strands which has been observed *in vitro* may well be an artifact.

A current hypothesis asserts that specific regions in DNA may serve as initiation points for the transcription of cistronic regions by RNA polymerase, and may provide a mechanism for selection of the proper DNA template strand as well.

Messenger RNA. The average base composition of the total cellular RNA of bacteria differs considerably from that of the bacterial DNA. This is not surprising since about 90% of cellular RNA is composed of the metabolically stable sRNA (10–15%) and ribosomal RNA (75–80%) fractions. This stable RNA arises from only about 0.2% of the total DNA genome (*i.e.*, the total chromosomal DNA) and has an average base composition quite different from that of DNA. The other 10% of the RNA, however, is quite unique. It has a rapid turnover rate and resembles the cellular DNA in base composition (see reference 3, pp. 75–90, pp. 101–132). The existence of this RNA fraction which resembles the previously mentioned T2 virus messenger RNA in its instability, was actually predicted on the basis of experiments on the induction and repression of enzyme synthesis (see reference 3, pp. 193–211). It was known that the synthesis of certain bacterial enzymes is rapidly induced when their substrates

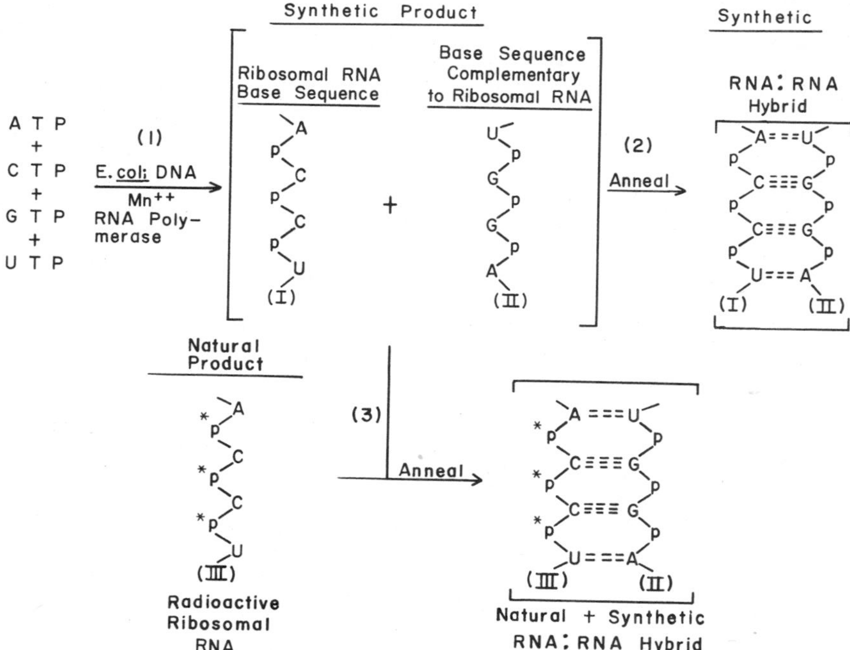

FIG. 3. Hybridization of ribosomal RNA with the RNA polymerase product. *Note*: The formation of natural + synthetic RNA:RNA hybrids is detected by utilizing naturally occurring RNA labeled *in vivo* with ^{32}P. Unhybridized RNA is selectively degraded by ribonucleases, and the extent of hybrid formation is detected by recovering the undegraded, hybridized RNA and assaying it for radioactivity.

are added to the growth medium. The withdrawal of the inducers results in a cessation of the production of new enzymes. It was concluded that the presence of the inducer would result in the synthesis of a RNA which would carry the information for the synthesis of the induced enzymes from DNA to the sites of protein synthesis, thus the term *messenger* RNA. Since withdrawal of the inducer causes a cessation of new enzyme synthesis, it was further hypothesized that this RNA fraction would turn over rapidly.

The DNA: RNA hybridization technique provided an adequate tool for showing the existence of the total cellular messenger RNA produced by bacteria. The demonstration of a RNA carrying the information for the expression of a specific genetic region proved more difficult since a single messenger RNA is greatly diluted by the total messenger fraction. This difficulty was overcome by incorporating a region of a bacterial genome carrying the information for the synthesis of an inducible enzyme system into the genome of a bacterial virus (see reference 4, pp. 171–174, pp. 363–372). When such viral DNA is used for annealing with messenger RNA samples isolated from induced and noninduced bacteria, less of the interfering total messenger RNA fraction is annealed, and the new RNA species can readily be observed in the induced cells. Other experiments have suggested that at least some messenger RNA molecules are *polycistronic*, *i.e.*, they contain information for the synthesis of more than one protein.

Replication of Viral RNA. Many viruses which infect bacterial, plant and animal cells contain a single-stranded RNA, rather than DNA, as the primary genetic material. After infection of the host cell, the viral RNA first serves as a messenger for the production of proteins including the enzyme(s) which serve as the RNA-directed RNA polymerase system (RNA synthetase) for the synthesis of new viral RNA (see reference 4, pp. 95–104). This process is called replication [see VIRUSES (REPLICATION)]. Replication first proceeds with the synthesis of a double-stranded *replicative form*. This consists of the original viral RNA strand and a second strand, complementary to the original in base sequence. Synthesis of the replicative form presumably proceeds by complementary base pairing (U substituting for T) similar to that shown in Eq. (*2*). Once the replicative form has been assembled, it serves as a template for the synthesis of new viral RNA, *i.e.*, that strand of the replicative form *identical* in base composition and base sequence to the original viral RNA strand.

This model for RNA virus replication is based on experiments which show that a portion of the $^{32}PO_4$-viral RNA is incorporated into a ribonuclease resistant form following infection of the host cells. It is supported by results obtained with a RNA virus temperature-sensitive mutant which does not yield nuclease-resistant viral RNA after infection at high temperature. If cells growing at low temperature are infected with this viral mutant, labelled with $^{32}PO_4$ in the RNA, and shifted to a higher temperature, the nuclease resistant RNA formed initially is rapidly lost. On the other hand the RNA of the parental virus is not found in the progeny. Nor has evidence for the existence of a double-stranded RNA replicative intermediate been obtained in experiments where the synthesis of infectious viral RNA has been demonstrated in a cell free system. Thus where certain lines of evidence strongly suggest the involvement of a double-stranded RNA replicative form in the cycle of infection, others indicate that the replication of at least some single-stranded RNA viruses may proceed by a different mechanism. Additional evidence is required to resolve these discrepancies.

Methylation of Preformed Polyribonucleotides. A stable fraction of the total cellular RNA, *i.e.*, sRNA, contains a number of methylated purine and pyrimidine bases. The mechanism by which these bases are methylated was demonstrated after the discovery that certain mutant bacterial strains, which require methionine for growth, produce "methyl-poor" sRNA when deprived of the essential nutrient. When methyl-poor sRNA is incubated with S-adenosyl methionine and the RNA methylating enzymes, the methyl groups are transferred to the bases of sRNA [Eq. (3); see reference 4, pp. 139–159]:

$$\text{S-Adenosyl Methionine} + \text{Methyl-poor sRNA} \xrightarrow{\text{RNA methylase}} \text{Methyl-sRNA} + \text{S-Adenosyl Homocysteine} \quad (3)$$

A number of different RNA methylases have been isolated, and each is specific for the production of a distinctly different methylated base. Although the mechanism of synthesis of the methylated bases in sRNA has been well established, the physiological importance of this process is not known (see also SOLUBLE RIBONUCLEIC ACIDS).

The abbreviations used are: DNA, deoxyribonucleic acid; RNA, ribonucleic acid, a polyribonucleotide; Pi, inorganic orthophosphate, PPi, inorganic pyrophosphate; A, C, G, U, the ribonucleosides containing adenine, cytosine, guanine and uracil, respectively; dA, dC, dG, dT, the deoxyribonucleosides containing adenine, cytosine, guanine and thymine, respectively; ATP, CTP, GTP, UTP, the 5′(pyro)-triphosphates of the nucleosides adenosine, cytidine, guanosine and uridine; sRNA, soluble RNA; cRNA, complementary RNA.

References

1. VOGEL, H. J., BRYSON, V., AND LAMPEN, J. O. (EDITORS), "Information Macromolecules," New York, Academic Press, 1963.
2. DAVIDSON, J. N., AND COHN, W. E. (EDITORS) "Progress in Nucleic Acid Research," Vol. I, New York, Academic Press, 1963.
3. "Cellular Regulatory Mechanisms," *Cold Spring Harbor Symp. Quant. Biol.* (1961).
4. "Synthesis and Structure of Macromolecule," *Cold Spring Harbor Symp. Quant. Biol.* (1963).

C. FRED FOX

RIBONUCLEIC ACIDS (STRUCTURE AND FUNCTIONS)

Ribonucleic acid (RNA) is a generic term for a group of polymers containing ribonucleotides as the fundamental repeating unit. The ribonucleic acids contain ribonucleotides of the heterocyclic bases adenine, guanine, uracil and cytosine, as well as a number of minor ribonucleotide constituents with rare bases such as thymine, methylated purines and dihydrouracil, and unusual ribonucleotides such as pseudouridine (see PURINES; PYRIMIDINES). Minor ribonucleotide constituents may also have methylated sugars. The ribonucleotide units are linked together through 3′,5′-phosphodiester bonds on the ribose moieties, and naturally occurring RNA's may contain from 70–5000 ribonucleotide units. Some RNA's possess a secondary structure resulting from hydrogen bonding between complementary base pairs.

RNA occurs naturally in several subclasses, the better known being described below. All known classes are involved in the function of the genetic apparatus, either as secondary "messages" transcribed from a primary DEOXYRIBONUCLEIC ACID structure in complex living systems, or as primary genetic material in simpler replicating structures, as in certain VIRUSES. Thus, the ribonucleic acids are substances that have a functional genetic specificity, and this specificity resides in the sequence of bases along the polyribonucleotide chain. Any *class* of RNA may therefore be heterogeneous with respect to the sequences existing in the different molecules within the class. Only certain classes have been fractionated into homogeneous subclasses, for example, certain transfer RNA's. Viral RNA's are probably homogeneous in sequence within the class.

The relation between the various classes of RNA and their functions may be described as follows:

Messenger RNA. Messenger RNA is synthesized in the living cell by the action of an enzyme, RNA polymerase, that carries out the polymerization of ribonucleotides on a DNA template region which carries the information for the primary sequence of amino acids in a structural protein. The messenger RNA is a ribonucleotide copy of the deoxynucleotide sequences in the primary genetic material [see RIBONUCLEIC ACIDS (BIOSYNTHESIS)]. If three nucleotides, a triplet, are required for the coding of one amino acid, then messenger RNA's contain 180–900 nucleotides for the synthesis of primary peptide chains containing 60–300 amino acids (molecular weight 6000–30,000 g/mole). If messenger RNA's are polycistronic they will be even larger—up to 2 million—and will contain sequence information for more than one polypeptide chain; in addition they must contain termination information in the form of triplets that code for no amino acid, permitting gaps in the synthesis to produce several separate and different peptide chains. Natural messenger RNA's have been isolated and partially characterized, but most information about the messenger RNA function has been deduced from model experiments with synthetic "messengers" such as polyuridylic acid and polyadenylic acid which lead to the formation of polyphenylalanine and polylysine, respectively [see also PROTEINS (BIOSYNTHESIS); MESSENGER RIBONUCLEIC ACIDS].

Ribosomal RNA. This exists as a part of a functional unit within living cells called the RIBOSOME, a particle containing protein and ribosomal RNA in roughly 1:2 parts by weight, having a particle weight of about 3 million. The ribosome plays an essential role in protein synthesis, but the function of the ribosomal RNA is poorly understood. The 70 S ribosome has two parts, a 30 S and 50 S subunit, and the molecular weights of their respective RNA's are about 0.6 and 1.2 million. Messenger RNA combines with ribosomes to form polysomes containing several ribosome units, usually 5 (*e.g.*, during hemoglobin synthesis), complexed to the messenger RNA molecule. This aggregate structure is the active template for protein biosynthesis.

Transfer RNA. This is the smallest and best charactertized RNA class. Transfer RNA molecules contain only about 80 nucleotides per chain. Within the class of transfer RNA molecules there are probably at least 20 separate kinds, correspondingly related to each of the 20 amino acids naturally occurring in proteins. Some of the transfer RNA's have been separated into homogeneous subclasses related only to one amino acid. The primary structure of one such transfer RNA, alanine transfer RNA, has been determined and is shown in Fig. 1. Transfer RNA is distinguished as a class by having a relatively high content of unusual bases. Transfer RNA must have at least two kinds of specificity:

(1) It must recognize (or be recognized by) the proper amino acid ACTIVATING ENZYME in order that the proper amino acid is transferred to its free 2′- or 3′-OH group.

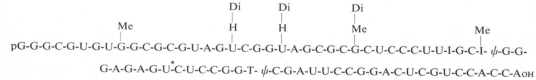

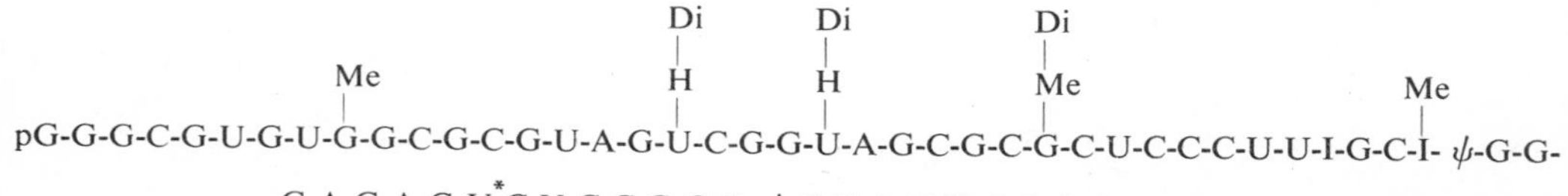

FIG. 1. Structure of an alanine transfer RNA from yeast. G = guanine, C = cytosine, U = uracil, A = adenine, ψ = pseudouridine, DiHU = dihydrouracil, U* = mixture of DiHU and U, I = hypoxanthine, DiMeG = N^2-dimethyl guanine, MeI = 1-methylhypoxanthine, MeG = 1-methyl guanine, and T = thymine. [Reprinted with permission, from R. W. Holley, *et al.*, *Science*, **147**, 1462 (1965).]

(2) It must recognize the proper triplet on the messenger RNA-ribosome aggregate.

Having these properties, the transfer RNA accepts or forms an intermediate transfer RNA-amino acid (aminoacyl RNA) that finds its way to the polysome, complexes at a triplet coding for the activated amino acid, and allows transfer of the amino acid into peptide linkage [see also SOLUBLE RIBONUCLEIC ACIDS].

Viral RNA. Viral RNA is isolated from a number of plant, animal and bacterial viruses and may be thought of as a polycistronic messenger RNA. It has been shown to have molecular weights of 1 or 2 million. Generally speaking, there is one molecule of RNA per infective virus particle. The RNA of RNA virus can be separated from its protein component and is also infective, bringing about the formation of complete virus [see TOBACCO MOSAIC VIRUS; VIRUSES (REPLICATION)].

References

1. PETERMAN, M. L., "The Physical and Chemical Properties of Ribosomes," Amsterdam, Elsevier Publishing Co., 1964.
2. SPIRIN, A. S., "Macromolecular Structure of Ribonucleic Acids," New York, Reinhold Publishing Corp., 1964.
3. SPIRIN, A. S., in "Progress in Nucleic Acid Research (DAVIDSON, J. N., AND COHN, W. E. (EDITORS), Vol. I, pp. 301–345, New York, Academic Press, 1963.

F. J. BOLLUM

RIBOSE

D-Ribose when it occurs as a free sugar is largely in the pyranose form (six-membered ring), but as a constituent of nucleosides, nucleotides and RIBONUCLEIC ACIDS it is present in the furanose form:

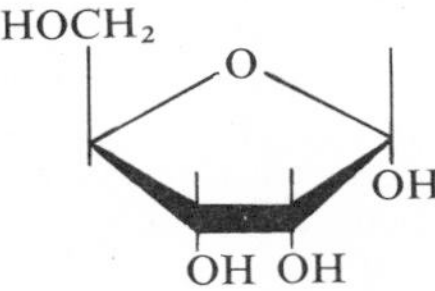

Ribose is not nutritionally essential and can presumably be biosynthesized, if necessary (see PENTOSES IN METABOLISM).

RIBOSOMES

Ribosomes, or ribonucleoprotein particles, the smallest organized structures in the cell, have been found in all organisms as yet examined, including bacteria, fungi, algae, protozoa, and all the cells of the multicellular organism, with the possible exception of the mature erythrocyte. However, their disposition and abundance within different cell types do vary. In most unicellular organisms, they fill the cytoplasm lying freely or in clusters. In muscle cells, for example, they are extremely scarce, lying freely in the cytoplasm here and there. In some cells, as in the reticulocyte, they exist freely, unattached to membranes (Fig. 1), while in others, as in glandular cells, and in those cells which manufacture protein for export,

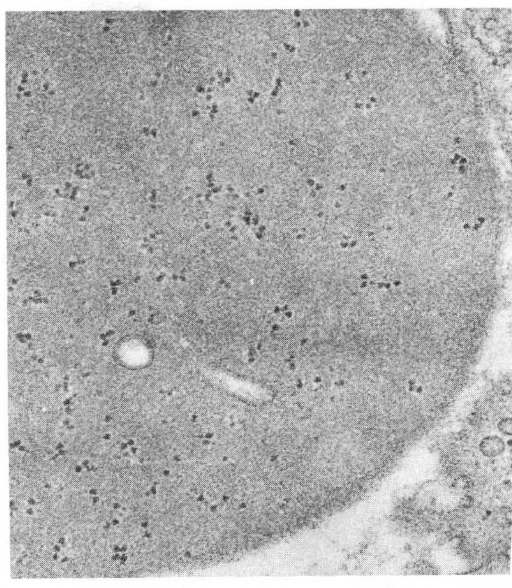

FIG. 1. Electron micrograph showing reticulocyte ribosomes.

as pancreas acinar and hepatic parenchyma cells, most of the ribosomes are found attached to the membranes of the endoplasmic reticulum (E.R.) (cf. Fig. 2—fibroplast). In fact, the division of

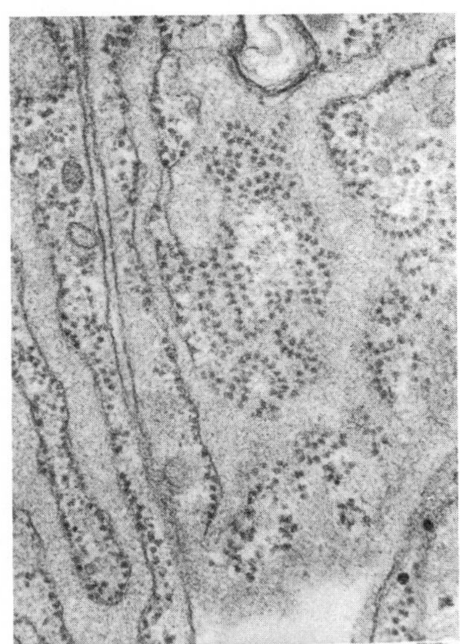

FIG. 2. Electron micrograph showing fibroblast ribosomes.

cell types is so good that we can say that in cells engaged in manufacturing proteins only for their own use, as for protein replacement and for protein manufacture preparatory to cell division, the ribosomes are unattached to membranes, while in those cells which are engaged in synthesizing protein for export, the ribosomes lie on the membranes of the E.R. It is thought that protein synthesized by the ribosomes are disengaged from them by passage across or through the membranes of the E.R., on their passageway outside the cell. Within the cell itself, ribosomes can be seen in, and have been isolated from, nuclei and chloroplasts; the evidence for their existence in MITOCHONDRIA is as yet meager (see also MICROSOMES AND THE ENDOPLASMIC RETICULUM).

The ubiquitous nature of these particles reflects their utmost importance for the economy of the cell; they are the centers of protein synthesis by the cell [see PROTEINS (BIOSYNTHESIS)]. As far as we know at present, all protein synthesis of the cell occurs on the surfaces of ribosomal particles, whether these are in the nucleus, in the cytoplasm, or in various organelles within the cytoplasm, as chloroplasts and possibly mitochondria. This involvement has been shown by *in vivo* experiments: when radioactive amino acid is injected into an animal, or if whole organisms are incubated in the presence of radioactive amino acid, the first appearance or radioactive protein is on the surface of the ribosomal particles. In addition, these particles can be isolated (cf. below) and incubated *in vitro*: it has been found, in the presence of suitable additions, that the ribosomes can incorporate radioactive amino acids into proteins bound to them, and in fact, isolated ribosomes, as reticulocyte ribosomes, can synthesize a discrete protein, in this case hemoglobin.

These particles have been isolated in various ways, dependent on the nature of the cells. In the case of unicellular organisms, the cells are ground with sand or alumina, or broken in a French pressure cell. The suspension of cellular material is centrifuged at low speed; the supernatant from this centrifugation is then spun at high speed, and the ribosomal particles come down in the form of a pellet. Some cells (*e.g.*, the reticulocytes) can be broken by osmotic shock, while tissue culture cells are broken by mild sonication or by hand grinding. In the case of the soft tissues, in most cases the cells are homogenized in an isoosmotic medium (sucrose being the solution of choice). In all cases, the pellet which has resulted from the high-speed centrifugation ($\sim 100{,}000 \times g$ for 60–120 minutes) contains the ribosomes. However, as mentioned above, in some cell types, as liver, pancreas and glandular tissue, the ribosomes are attached to membranes; thus in these cases, the high-speed pellet will contain the particles still attached to membranes (the so-called MICROSOME fraction see Fig. 3). To disengage the particles from the membranes, the use of detergents, either ionic (*e.g.*, deoxycholate) or non-ionic (*e.g.*, "Tween," "Lubrol," or "Triton") have been used; all these "solubilize" the membranes, allowing the ribosomes to be re-sedimented as a high-speed pellet freed of membranous material (Fig. 4). However, another ionic detergent, dodecyl sulfate, has been used to free the RNA, in good form, from the protein of the ribosome. A word about the term "microsome": it is used nowadays interchangeably for ribosomes and microsomes. It really only describes the pellet resulting when the supernatant from a low-speed mitochondria or chloroplast spin is then spun at high speed. Thus, in the case of cells where ribosomes are attached to membranes, the resulting pellet will consist of these attached structures; the ribosomes must then be obtained from this "microsome" fraction. In the case of those cells where no membranes are visible, as in practically all bacteria, the high-speed pellet will only contain ribosomes; thus in the latter case, "microsomes" and "ribosomes" are synonymous terms, describing the same structure.

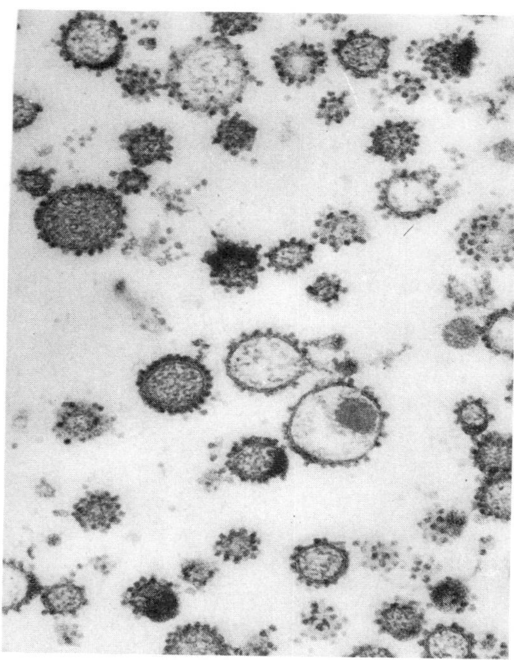

FIG. 3. Electron micrograph of microsomes (pancreas), showing attached ribosomes.

As visualized with the ELECTRON MICROSCOPE, the ribosomes are oblate spheroids, measuring 150–200 Å in diameter if viewed in positive contrast, and 200–300 Å if viewed in negative contrast. Almost the only internal structure which can be seen clearly, even in negative stained preparations, is the groove between the large and small subunits (cf. below). It can be calculated that in reticulocytes there are about 10^5 free ribosomal particles per cell, in *E. coli* about 6×10^3 free particles per cell, and about 5×10^5 free ribosomes in the yeast cell; in all cases, this value comes out to a concentration of $10^{-5}M$ to $2 \times 10^{-6}M$, a concentrated 1–4% solution.

Ribosomes contain only protein and RNA (see

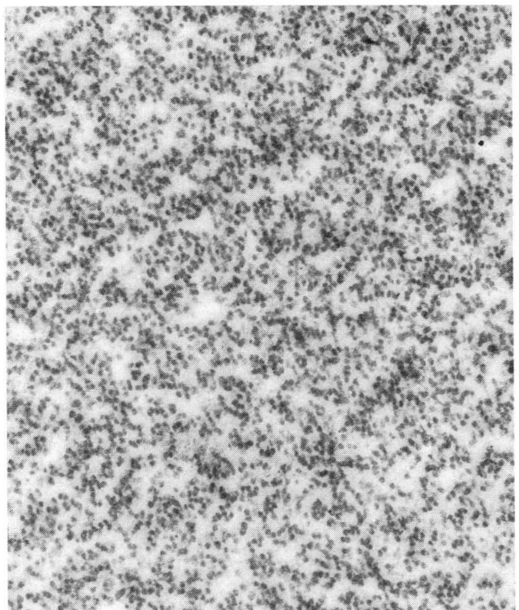

FIG. 4. Electron micrograph of isolated (pancreas) ribosomes.

RIBONUCLEIC ACIDS) macromolecules, although some (~1%) PHOSPHOLIPID is found in mammalian ribosomes. However, as isolated, the ribosomes are highly porous, containing water to about 2–4 times their mass. In bacterial ribosomes, up to 60% of the dry mass is RNA; in mammalian ribosomes, this goes down to 45–55%. Ribosomes, because of this high RNA value, contain 80–90% of the RNA of the whole cell. The base composition of this RNA varies somewhat from cell type to cell type, and generally does not follow the Watson-Crick DNA formulation of G = C and A = U; also the proportion of guanine in ribosomal RNA is always high, varying from 30–40% of the total bases. There are also small amounts of methylated bases and 5-ribosyluracil in ribosomal RNA, though not as much as in soluble (transfer) RNA. This RNA is high in molecular weight and, as far as is known, is composed of two strands—a 23 S (*E. coli.*) or 28 S (mammalian cells) RNA found in the large subunit and a 16 S (*E. coli.*) or 18 S (mammalian cells) RNA found in the small subunit. Both of these RNA species are double-stranded through most of the molecule but not in the form of a perfect helix. However, it has not been ruled out that within the ribosomes *in situ* the large strands of RNA are really composed of tightly intercomplexed smaller RNA species.

The protein composition of the ribosomes is not very clearly known; there are contradictory statements as to the number of discrete proteins present in the particles, varying from 2–30. A possible reason for this reported range is the "stickiness" of ribosomes; many proteins, mostly cationic ones, adsorb strongly to the ribosomes and are not easily washed off. However, the amino acid composition of the ribosome proteins, all of which are highly insoluble, suggests a basic character, being preponderant in the basic and dicarboxylic amino acids, with the former being predominant, very similar to the situation in HISTONES. Molecular weight determinations of the proteins again vary widely, ranging from 12,000–25,000.

Other important components of the particles are Mg and Ca ions, and the POLYAMINES such as spermine and putrescine. All ribosomes so far examined have been found to contain Mg^{++}, ranging from 1–4 moles Mg^{++} per 10 moles RNA-phosphate. Ca^{++} has been found in some cases at about one-tenth the amount of Mg^{++}. The polyamines have been found in bacterial and mammalian ribosomes at about one-tenth to one-twentieth the amount of Mg^{++}. All of these components are thought to aid in binding the ribosomal subunits (cf. below) to each other via the free phosphate groups of the RNA.

The physical properties of ribosomes has been greatly studied. The corrected sedimentation coefficients of ribosomes from all sources range from 70 S (*E. coli.*) to 83 S (clover), but it is not certain whether ribosomes from different classes of organisms differ from each other in this respect. The particle weight determinations again vary widely, from $\sim 3 \times 10^6$ for *E. coli.* to $\sim 5 \times 10^6$ for liver ribosomes; the reason for this large difference is not known at present. The charge on the outside of the particles is strongly negative, reflecting a preponderance of the secondary phosphate groups of the RNA over basic groups of the protein. Just where and how the RNA and protein are intertwined within the particles is not known; the surmise is that the RNA is covered by the protein, except for a few spots on the outside. Since the sedimentation properties of the ribosomes can be changed by changes in ribosomal concentration and in the ionic strength of the medium, the ribosomes behave like polyelectrolyte systems. A singular property of all ribosomes so far examined is the ease with which the 70 S or 83 S particle (monomer) can be dissociated into subunits. In the case of *E. coli* particles, the 70 S monomer can be broken up by simple resuspension in Mg^{++}-free medium into a large (50 S) and a small (30 S) subunit. In mammalian ribosomes, the Mg^{++} seems to be more firmly held, for it takes treatment with EDTA to dissociate the monomer; the 78 S liver particle breaks up into a 47 S and a 32 S particle upon treatment with EDTA. These subunits can be almost completely (*E. coli.*) or partially (liver) reconverted back to the monomer upon addition of Mg^{++}. Thus it seems clear that Mg^{++} is very instrumental in the reversal of dissociation of the ribosomes. The RNA/protein ratios of the subunits are very similar to each other, and the base composition of the RNA in each of the particles (cf. above) seems to be the same. A small variety of subparticles of other sizes (as determined by sedimentation coefficients) have been described, ranging from 30–60 S; it is thus thought by some that the dissociation is more complex than a simple one of a monomer into large and small subunits; for example, the monomer might be composed of six subunits.

The meaning of the division of ribosomes into subunits is made clear by experiments involving protein synthesis. First, new data make it appear that in those cells where the ribosomes are attached to the membranes of the E.R., it is the large subunit which is attached to the membrane; the small subunit is then bound to the large. We now know of species of RNA molecules other than ribosomal RNA which are involved in protein synthesis; namely, "messenger" RNA and soluble (transfer) RNA. It now seems that "messenger" RNA is bound to the small subunit while soluble RNA is bound to the large subunit. Even from the fragmentary data we have at present, we can assume that the positioning of the soluble RNA on the ribosomes with regard to the messenger RNA must be exact. The distance between the anti-codon site (messenger RNA recognition site) on the soluble RNA and the —C—C—A—amino acid end of the molecule must be rigidly fixed so that subunit structure of the ribosomes may provide the exact framework upon which this can occur. However, the exact functioning, or even meaning, of ribosomal RNA within the ribosomes is not known; it could act as a specific structural framework of the particles or as a modifier of the coding function of messenger RNA. Little is known of the localization of the other components of the protein-synthesizing system. The activating enzymes certainly seem to be soluble, not bound to ribosomes. However, there are at least two other enzymatic factors involved in protein synthesis—one needs GTP as cofactor, and the other is probably the peptide-bond forming enzyme; we have only little knowledge whether these enzymes are ribosomal bound in the cell, but one or both of these factors can be removed from ribosomes *in vitro*.

Ribosomes as isolated have been found to contain nascent protein on them, and it is believed that these ribosomal-bound proteins are the precursors to the soluble proteins. It is not known whether these nascent proteins on the ribosomes are in their tertiary configuration (fully active if they are enzymes) or whether they are released from the ribosomes as a primary chain, and secondarily folded thereafter. Finally, all ribosomes so far examined have been found to contain small amounts of ribonuclease tightly bound to them, but the enzyme activity is latent, having to be activated by EDTA; the meaning of these findings for the function of ribosomes is uncertain.

Recent observations have indicated that ribosomes active in protein synthesis are held together in chains or clusters in so-called polyribosomes. Electron microscopic observations of many cell types have indicated that many ribosomes exist in the cell in the form of clusters (cf. Fig. 1) or as chains, like beads on a string, the chains showing up very clearly in the case of those ribosomes on membranes (cf. Fig. 2). These polysomal structures can be isolated as heavy aggregates containing from 4–20 ribosomes, separate from the monomeric ribosomes, by means of sucrose-density gradient centrifugation. When amino acid incorporation experiments are performed, it is seen, with preparations from various cell types, that most of the newly synthesized protein is in the region of the polyribosomes rather than in the region of the monomers. Cells active in protein synthesis yield more polyribosomes than those less active. Furthermore, *in vitro* amino-acid incorporation is more potent with polyribosomes than with ribosomes. Because in many cases the polysome structure can be disengaged by addition of RNase, the conjecture at present is that these structures consist of ribosomes held together by a chain of messenger RNA. Thus, the thought is that either the ribosomes move along the messenger, or the messenger moves along the ribosomes, so that messenger RNA, as it moves along, codes for amino acid incorporation by a whole series of ribosomes engaged simultaneously in synthesizing the same specific protein. However, other interpretations of the experiments are possible, and there are some data which do not fit the theory, so that at present we can only say that ribosomes having newly synthesized proteins bound to them are held together in chains or clusters, and that messenger RNA may be one of the agents holding them together.

The question of the biosynthesis of the structural protein and of the RNA of the ribosomes remains largely unanswered. There is some thought that the monomer is synthesized via stages from smaller particles having higher proportions of RNA than the mature particles, protein later being accrued to the particles. The turnover of ribosomal RNA is much less than that of the other RNA species of the cell; recent evidence suggests that the half-life of liver ribosomal RNA is about 5 days. However, this RNA, like the soluble and "messenger" RNA's does seem to be formed via certain segments of the DNA genome in the nucleus, or nuclear region, of the cell.

In conclusion, I must emphasize the tentative nature of a good deal of what has been written, for it is only within a span of a few years that we have been dealing with this complex macromolecular functioning structure called a ribosome.

References

1. McQuillen, K., *Progr. Biophys. Biophys. Chem.*, **12**, 69 (1961).
2. T'so, P. O. P., *Ann. Rev. Plant Physiol.* **13**, 45 (1962)
3. Petermann, M., "Physical and Chemical Properties of Ribosomes," New York, Elsevier Press, 1965.

Philip Siekevitz

ROBINSON, R.

Robert Robinson (1886–   ), a British chemist, is a prominent investigator in the area of alkaloids, and also did chemical work concerned with steroids and flavones. He received the Nobel Prize in chemistry in 1947.

ROYAL JELLY

Royal jelly is a mixture secreted by bees, for feeding those young larvae which are to become queens. All the female larvae are the same, and a special interest in royal jelly derives from the fact that receiving it causes larvae to develop into fertile queens which live for several years. Larvae that do not get this special food develop into infertile workers which have a relatively short life span.

Royal jelly contains an assortment of amino acids and the various B vitamins, including relatively large amounts of pantothenic acid and biotin. Biopterin (see PTERIDINES) has been reported to be one of its active constituents. It also contains a relatively high amount of zinc. A somewhat unique constituent is 10-hydroxy-Δ^2-decenoic acid. The physiological significance of these various constituents has not been elucidated. Some compounds in royal jelly are labile, and during storage royal jelly loses its ability to produce queens. There is no reason to suppose that royal jelly as such, especially in small amounts, has any special nutritional value for mammals.

RUBBER, NATURAL

Natural rubber is a high molecular weight polymer of isoprene units; in rubber all the double bonds have the *cis* arrangement. It is produced by coagulating the milk juice (latex) of certain tropical trees, chiefly *Hevea brasiliensis*. Its biosynthesis presumably results from the coupling of "biosynthetic isoprene units" (see ISOPRENOID BIOSYNTHESIS, section on Formation of New Carbon-Carbon Bonds).

References

1. FISHER, H. L., "Rubber," *Sci. Am.*, **195**, 75–88 (November 1956).
2. NATTA, G., "Precisely Constructed Polymers," *Sci. Am.*, **205**, 33–41 (August 1961).

RUMINANT NUTRITION AND METABOLISM

Ruminants are a particular suborder (ruminantia) of the class of mammals; fall within the categories of undulates (hoofed) and actiodactyles (even-toed) and are characterized by a four-chambered stomach. The most well-known and intensively studied genera are: *Bos* (cattle), *Ovis* (sheep) and *Capra* (goats) probably because of the commercial importance of their products the world over.

Ruminants have assumed their economic importance because they are well adapted to the consumption and utilization of large quantities of fibrous plant material (roughages) which are not well utilized by simple-stomach animals. Ruminant tissues *per se* are not capable of digesting the celluloses and hemicelluloses present in these roughages. It is rather the masses of microorganisms that inhabit the anterior portion of the ruminant gut which convert the complex nutrients of the roughages into simple compounds that are readily metabolized by the ruminant tissues.

In addition to releasing the energy of the complex carbohydrates, these microorganisms convert both protein and non-protein nitrogen into microbial protein and synthesize vitamins for the benefit of the host. The ruminant, in turn, provides for the microorganisms by: (1) eating frequently to provide nutrients and substrates, (2) absorbing the end products of digestion to prevent their accumulation from inhibiting the fermentive and synthetic activities of the organisms, and (3) maintaining a relative constancy of temperature, pH and moisture.

The nutritionist is as concerned with providing the proper nutrients and conditions for activity of the microorganisms as he is with providing the requirements of the ruminant *per se*. The first two stomach portions are the reticulum and the rumen which together form a functional unit, the "fermentation vat." Although the correct anatomical term for them is the rumeno-reticulum, the processes occurring there are commonly referred to as rumen function, and the study of them as rumenology.

The rumeno-reticulum largely accounts for the differences in digestive physiology between ruminants and simple-stomach animals. These differences are attributable to the end products of digestion that are formed in, and absorbed from, the rumeno-reticulum. Actually, absorption of some materials occurs from the third portion, the omasum, in addition to water absorption which appears to be its primary function. The way in which the material entering the abomasum (true stomach) and traversing the remainder of the digestive tract is digested does not differ markedly from other animals. True, much of this material is in the form of micro-organisms *per se* and their products. Further, the physiological requirements of the ruminant tissues for nutrients are not believed to differ substantially from other animals. It is the lot of the nutritionist then to provide for the physiological requirements of the animal by supplementing natural feedstuffs in such a way that the combined end products of rumen and of lower-tract digestion furnish the necessary nutrients. Extensive biochemical research in recent decades has demonstrated that the final metabolic pathways of nutrient utilization (intermediary metabolism) are virtually invariant among animals.

The fact that ruminants derive a larger share of their energy (49–75%) from short-chain volatile fatty acids (VFA) rather than from glucose largely distinguishes them from simple-stomach animals. Neonatal ruminants approach non-ruminants in digestive physiology, and gradually levels of blood glucose decrease and those for VFA increase. This changeover takes place within a few weeks after birth, and its rapidity is related to how soon liquid feeding (milk) is replaced by solid feed. The reticulum, rumen and omasum are considerably smaller than the abomasum in the newborn and increase rapidly within gestion of solid food to a point where the rumen alone accounts for 80% of the total stomach volume. Also, prior to this changeover, there is little vitamin synthesis taking place and the young ruminant is as dependent on a dietary supply as the dog, rat, etc.

Rumen Digestion. Ruminants consume their food rapidly with only enough mastication to moisten the dry material for swallowing, after which it is stored in the rumen to await rumination. This process includes the mechanical factors of digestion in which the feed is regurgitated, remasticated, reinsalivated and reswallowed. This series of reflex acts is not essential for life. It is largely dependent on the physical nature of the diet. Ruminants on high roughage diets often spend one-third of their time ruminating. A single cycle of rumination may consume about 1 minute, and an animal may indulge in this practice during 10 or 15 separate periods in 24 hours. These movements are under afferent and efferent vagal control (parasympathetic) and are mediated by the hypothalamus.

While in the rumen, feed undergoes mixing, maceration and fermentation. A particular portion of roughage may be subjected to several rumination cycles. Passage of feed residues into the omasum and abomasum depends mainly on particle size. Forage consumed in the long form may undergo fermentation for several days before the residue is passed along to the lower tract. The time required to eliminate 80% of the undigested residue from long forage is approximately 48 hours in the goat and 72 hours in the cow.

In addition to the physical action that feed undergoes, other factors contribute to its degradation. Although saliva contains no digestive enzymes in adult ruminants, it has several important functions in the digestive process. Copious quantities are produced (approximately 8 liters/day for sheep and 100 liter/day for cattle). The sublingual and superior molar glands secrete continuously as do the parotids which are stimulated to much greater output by feeding and somewhat greater output by rumination. Submaxillary glands, in the cow at least, produce mucus-rich saliva during feeding but not in response to rumination. The mechanical importance of saliva for swallowing is apparent when it is noted that much of the feed consumed by ruminants may contain less than 15% moisture. Chemically its most important function is that of providing buffers (primarily bicarbonate and phosphate) for the large amounts of acid produced in the fermentation process to control pH and permit the microorganisms to continue to operate effectively. In addition, saliva provides the organisms with a small but continuous supply of phosphate, nitrogen and sulfur, and its ionic composition appears to be controlled to help regulate the electrolyte balance of the entire animal.

Having discussed the mechanical aspects of digestion and the importance of saliva in maintaining the constancy of rumen environment, a discussion of the microorganisms is pertinent. Since no digestive enzymes are secreted by tissues of the rumen, reticulum or omasum, the removal of 70–85% of the digestible dry matter of the ingesta prior to the abomasum depends entirely on the microflora.

Prior to 1900, it was known that organisms inhabited the ruminant digestive tract and "assisted" the animal in the digestion of fiber. Only since the birth of rumenology, about 1940, have extensive studies been made of the kinds of organisms present and the specific reactions carried out by each. Total viable counts of rumen bacteria published over a 10-year period rose from 10^6 to 10^{10} or 10^{11}/g due, of course, to progress in counting technique. Total bacterial numbers vary with the substrate, especially available carbohydrate and nitrogen, as do the types of organisms present. Numerous morphological types, Gram positive and negative, capsulated and non-capsulated are present. More than a hundred species of protozoa have been found in the rumen in total numbers up to 10^6/g. Chemical defaunation of the rumen failed to show any marked depression of feed digestibility, but this does not mean that the protozoa do not normally contribute appreciably to efficient nutrietn utilization.

The symbiotic relationship between the rumen microorganisms and the host ruminant has been pointed out. Also, innumerable examples of symbiosis among the organisms inhabiting the rumen have been demonstrated or suggested.

Metabolism in the Rumen. The fate of various dietary constituents and the resulting end products provided for absorption and utilization by the ruminant tissues will be discussed in various categories.

Carbohydrates. Investigators in the late 1800's suggested that end products of cellulose digestion were carbon dioxide, methane, hydrogen, and acetic, butyric, propionic and other volatile fatty acids. It remained for the extensive work since 1940 to elaborate the proportions of these end products under various feeding situations, their absorption and their value to the animal. Breakdown of sugars and STARCHES takes place rapidly and that of CELLULOSES and hemicelluloses less rapidly in the rumen. Although glucose can be absorbed slowly through the rumen wall, it normally gets metabolized so rapidly by rumen organisms that no appreciable amount ever enters the blood from the rumen. Some readily available carbohydrates may escape conversion to VFA in the rumen by being incorporated into capsular material of some bacteria or into starch-digesting protozoa only to be absorbed as glucose from the lower tract following the digestion of the organisms. The primary products of carbohydrate breakdown are the VFA, and the following are their approximate proportions: acetic, 60%; propionic, 22%; butyric, 10%; longer chain, 8%. Diets rich in sugar and starch favor more propionic acid production possibly with lactic acid as an intermediate, whereas roughage diets generally favor acetic and butyric acids with glucose and cellobiose as intermediates. Several species of bacteria may contribute to the several-stage process of cellulose digestion, but no protozoa are known with certainty to contribute.

The VFA are almost completely absorbed from the first three stomach compartments. Acetic acid and butyric acid, which is largely converted to β-hydroxybutyric during absorption, are primarily used for lipogenesis. The propionic acid is glucogenic.

Nitrogen Compounds. The bulk of nitrogen in normal feeds is in the protein form, but appreciable amounts may occur as non-protein nitrogen (NPN). The possibility of conversion of NPN to protein by rumen organisms suggested prior to 1900 has been conclusively demonstrated in recent years. Feeding up to one-third of the nitrogen required by ruminants in an NPN form is a common and economical practice. This means of meeting the protein requirement is actually an extension of a natural process since enough urea enters the rumen via the saliva to provide for about 15% of the protein requirement of a ruminant on a low dietary intake. The only known requirement of the rumen organisms for nitrogen is as ammonia. This does not mean that amino acids and peptides are not used efficiently by rumen organisms to eventually meet the animal's requirements. The total microbial protein formed daily (2/3 bacterial and 1/3 protozoa) equals two to three times that required to maintain the animal. Proteins from the two classes of organisms appear to be of equal biological value, but the protozoal fraction is more digestible.

The main rumen end product of most feed protein and NPN compounds is ammonia. Proteolytic activity is always strong, and the peptide and amino acids liberated are rapidly deaminated. The conversion of feed protein to microbial protein occurring in the rumen is estimated to range from 40% for zein to 90% for casein. Too rapid liberation of feed nitrogen as ammonia contributes to poor utilization because much of the ammonia is absorbed, converted to urea and eliminated in the urine. Therefore, rate of liberation of its nitrogen is probably a better criterion of protein quality for ruminants than is amino acid composition—the one used for simple-stomach animals [see PROTEINS (NUTRITIONAL QUALITY)]. Although amino acid deficiency is unlikely in ruminants, individual amino acids may be limiting on certain rations for particular production functions. For example, high-producing dairy cows on high-corn diets may require additionally sine, and fast-growing lambs on high-urea diets may require methionine supplementation or additional inorganic sulfur.

Lipids. Although lipid consumption by ruminants is usually low, a cow could consume enough lipid (500 g/day) to provide about one-fourth of her energy requirement for maintenance. Forage LIPIDS contain a large percentage of unsaturated C_{18} acids (especially linoleic), as do many of the common concentrates. In spite of this, ruminant depot fats are characteristically rather hard and high in stearic acid.

It is now well established that rumen organisms rapidly hydrolyze glycerides and PHOSPHOLIPIDS so that over half of the lipids present in the rumen often consist of long-chain fatty acids in the free form. The glycerol portion is also rapidly fermented to propionic acid. Further, rumen bacteria and probably ciliate protozoa rapidly hydrogenate the unsaturated C_{18} acids to stearic acid. Microbial action also results in the formation of appreciable quantities of *trans* and small quantities of *iso* forms of the long-chain unsaturated acids which are absent in forages. This rumen activity causes the depot and milk fats of ruminants to be more constant in composition than those of the non-ruminant which reflect the dietary fatty acid composition. However, when excessive quantities of oils are fed (exceeding the capacity of the microorganisms) or are injected intravenously (to bypass the rumen), fatty acid composition of both depot and milk fats is altered markedly.

The microorganisms *per se* contain lipids that become available in the lower tract. However, if the adult ruminant requires "essential" FATTY ACIDS, it is probable that they must originate in the diet and escape rumen hydrogenation. Details of lipid absorption by ruminants are not available, but it is likely that long-chain acids from feed and microbial sources are absorbed via blood and lymph and that synthesis into glycerides and cholesterol esters occurs in the intestinal mucosa.

Because the ether extract from roughages has more waxes, resins and essential oils, it is less well utilized than that from concentrates which contain a greater percentage of true fats. Low levels (5% or less) of supplemental animal or vegetable fat may be well utilized by ruminants. High levels (10% or more) lower feed digestibility, especially the fiber, probably due to a depressing effect on cellulose digesting organisms. In the case of a lactating animal, the total feed utilization may be severely reduced by excess fat because of the marked depression in milk fat production.

Minerals. The macroelements Na, K, Ca, Mg, Cl, P and S are essential in the ruminant, contribute to osmotic pressure in body fluids, and are varied in concentration to maintain acid-base equilibrium. Na, K, Cl and S are excreted mainly in the urine, whereas Ca, Mg and P leave by way of the feces. An intestinal "pump" makes absorption of Na very efficient even against a considerable gradient (see ACTIVE TRANSPORT). Normal forages have an excess of cations (Ca in legumes and K in grasses), whereas concentrates may have an acid excess (especially P). Urine pH values are normally alkaline, but may range from 8.3 to 5. Mineral excesses in urine are usually balanced by bicarbonate or ammonium ions.

In addition to the minerals named above, I, Mn, Co, Cu, Fe and Zn complete the list of minerals believed to be required by all animals for digestion, respiration, muscular activity and specific metabolic roles, as well as skeletal and soft tissue composition. Fluorine, selenium and molybdenum excesses are all associated with specific maladies of ruminants and yet each may be necessary in small quantities.

Sodium and Cl both appear to be absorbed from the rumen against concentration gradients; K appears to diffuse freely. Phosphate can pass rumen epithelium in either direction, whereas it is relatively impermeable to Ca and Mg. Activity of rumen organisms (cellulose digestion, vitamin synthesis, etc.) can be affected by specific mineral concentrations. It has been previously mentioned that phosphate and sulfur reenter the rumen via saliva, thereby minimizing fluctuations in their concentrations. The most dramatic effect of a mineral lack on ruminant performance results from vitamin B_{12} deficiency which occurs when Co is not provided in adequate amounts for synthesis of the vitamin by rumen microorganisms (see COBALT).

Large quantities of Ca and P are required for high levels of milk production, and Co, Cu and S levels are important for wool growth. Cobalt and IODINE deficiencies are likely to be problems in certain geographical areas. Calcium, P, Na and Cl are needed in large enough amounts so that sources of each are normally provided *ad libitum* to ruminants [see also MINERAL REQUIREMENTS (VERTEBRATES)].

Vitamins. Ruminants are generally conceded to be independent of an exogenous source of VITAMIN K and those of the VITAMIN B GROUP (B complex). Thiamine, riboflavin, B_6, nicotinic acid, pantothenic acid, biotin, folic acid and B_{12} have been shown to be synthesized in the rumen. Vitamins entering the rumen in combined forms in feed are liberated in the digestive processes. Some of those liberated from feed sources and those synthesized and secreted into rumen fluid can probably be absorbed directly from the rumen. Those escaping to the lower tract are absorbed along with those available from the digested microorganisms. Exact requirements for these vitamins by either the organisms or the ruminants are not known. Individual bacterial species have known requirements for certain of these vitamins which probably are provided by other bacteria.

Fat-soluble vitamins A, D and E or their precursors are required in ruminant diets. The feed β-carotene, part of which is converted to the vitamin in the intestine, provides the required VITAMIN A. A number of sterols of natural feedstuffs have VITAMIN D activity, and numerous tocopherols are also present to provide ruminants with VITAMIN E activity. In general, exposure of forages to the sun increases vitamin D and reduces vitamin A and E activities. These three vitamins have the same metabolic roles for ruminants as for other animals, except that there is no evidence of E deficiency causing reproductive failure in cattle, sheep or goats.

Special Nutritional Considerations. The single most important factor limiting animal production is energy. The desired level of energy intake varies with the species and productive purpose intended, to wit: Maintenance or sub-maintenance level for a beef cow on range, 1.2–2 times maintenance for a growing animal, 1.5–3 times for fattening, and 2–4 times for lactation. The improved genetic potential of animals together with economic conditions has resulted in increased amounts o. concentrates being fed to sustain high production Common rations for ruminants range from 100% roughage to almost 100% concentrates.

Actually, ruminants have no physiological requirement for roughage except to maintain normal milk fat production. Low milk fat can result from low roughage intake, feeding ground or pelleted roughage or grazing lush pasture; it appears that high concentrate feeding exaggerates the effect. Also, milk fat composition may be altered by a reduction in short-chain fatty acids and an increase in unsaturated ones. These anomalies can be traced to rumen function in that the relative or absolute production of acetic acid (an important milk fat precursor) is reduced and propionic acid production is increased. Feeding acetate, or sodium or potassium bicarbonate, at least partially corrects for the low milk fat. More subtle changes also occur in other milk constituents which appear to be associated with specific VFA changes in the rumen and with energy intake.

Feeding acetate or bicarbonate salts of Na and K permits consumption and utilization of purified high-concentrate diets that are otherwise unsatisfactory. In general, high levels of acetic acid production are associated with a high wastage of energy as heat in ruminants. Based on some of these observations, diets with ground pelleted hay and steamed flaked corn were developed which increased propionic acid production as predicted and resulted in improved weight gains and feed efficiency for growing ruminants.

In the previous century, many of the observations of ruminant performance on various rations were not understood or appreciated until the intensive work in recent decades unfolded some of the mysteries of rumen metabolism. The examples given above serve to point out the necessity for understanding rumen digestion and metabolism in order to control or regulate the end products available to the animal. The nutritionist draws on this knowledge to plan, improve, provide or supplement rations for ruminants which will provide optimally for the desired production function. The application of knowledge to date suggests that beginnings have been small compared with the progress possible with additional understanding. It remains for the combined efforts of physicist, chemist, microbiologist, biochemist, physiologist, biomathematician and nutritionist to provide additional information basic to such progress.

References

1. ANNISON, E. E., AND LEWIS, D., "Metabolism in the Rumen," New York, John Wiley & Sons, 1959.
2. DOUGHERTY, R. W., ALLEN, R. D., BURROUGHS, WISE, JACOBSON, N. L., AND MCGILLIARD, A. D., "Physiology of Digestion in the Ruminant," London, Butterworths, 1965.
3. DUKES, H. H., "The Physiology of the Domestic Animals," Ithaca, N.Y., Comstock Publ. Assoc., 1955.
4. LEWIS, D., "Digestive Physiology and Nutrition of the Ruminant," London, Butterworths, 1961.
5. MAYNARD, L. A., AND LOOSLI, J. K., "Animal Nutrition," Fifth edition, New York, McGraw-Hill Book Co., 1962

RICHARD D. MOCHRIE

S

SALIVA. See DIGESTION AND DIGESTIVE ENZYMES, section on Carbohydrate Digestion; also RUMINANT NUTRITION AND METABOLISM, section on Rumen Digestion.

SANGER, F.

Frederick Sanger (1918–) is a British biochemist whose outstanding contribution was the determination of the complete amino acid sequence in the protein hormone INSULIN. Many years before he accomplished this he developed the use of 2,4-dinitrofluorobenzene (Sanger's reagent) which was an invaluable tool [see PROTEINS (END GROUP AND SEQUENCE ANALYSIS)]. In 1948 he was awarded the Nobel Prize for chemistry.

SAPONINS. See GLYCOSIDES, STEROID, section on Saponins.

SARCOPLASMIC RETICULUM AND T SYSTEM

The sarcoplasmic reticulum and the T (for transverse) system are complex networks of tubules and enclosed spaces within fibers of MUSCLE. Tubules of the sarcoplasmic reticulum appear to run parallel to the myofibrils in the spaces between, as well as transversely around the myofibrils; the transverse portions occur at the levels of the Z line and of the H zone. The transverse channels of the T system appear to form an independent network having no open continuity with the sarcoplasmic reticulum channels. It has been suggested that the T system channels may conduct electrical stimuli to the interior regions of the myofibril and that the contents of the sarcoplasmic reticulum may have a metabolic function in producing adenosine triphosphate (ATP) additional to that produced in the muscle mitochondria.

Reference

1. PORTER, K. R., AND FRANZINI-ARMSTRONG, C., "The Sarcoplasmic Reticulum," *Sci. Am.*, **212**, 72–80 (March 1965).

SCURVY. See ASCORBIC ACID (VITAMIN C) AND SCURVY.

SECRETIN. See HORMONES, ANIMAL.

SEDIMENTATION. See CENTRIFUGATION, DIFFERENTIAL; ULTRACENTRIFUGATION.

SELENIUM (IN BIOLOGICAL SYSTEMS). See ANTIOXIDANTS; MINERAL REQUIREMENTS (VERTEBRATES); NUTRITIONAL REQUIREMENTS (MAMMALS), Selenium is a constituent of the so-called Factor 3, which has biological activity similar to that of tocopherols (see VITAMIN E). Excess selenium is poisonous to animals (see TOXINS, PLANT, section on Minerals).

References

1. "Symposium on Nutritional Significance of Selenium (Factor 3)," *Federation Proc.*, **20**, 665–702 (1961).
2. ROSENFELD, I., AND BEATH, O. A., "Selenium: Geobotany, Biochemistry, Toxicity, and Nutrition," New York, Academic Press, 1964.

SERINE

L-Serine, α-amino-β-hydroxypropionic acid, $HOCH_2—CHNH_2—COOH$, is found widely as a constituent of common PROTEINS; it often constitutes 5–10% of the protein. It is nutritionally nonessential for mammals since it can be synthesized from glycine, but has unique functions in that it enters into the composition of phosphatidyl serines (see PHOSPHOLIPIDS) and is esterified (phosphoserine) in phosphoproteins. Serine is decarboxylated to form ethanolamine which enters into the makeup of phosphatidylethanolamines, another type of phospholipid. Choline, in lecithins, may be formed from ethanolamine by methylation. The specific rotation of L-serine in 1*N* HCl solution is $[\alpha]_D = +15°$. See also AMINO ACIDS (OCCURRENCE).

SEROTONIN

Serotonin (5-hydroxytryptamine) arises biosynthetically by decarboxylation of 5-hydroxytryptophan (see AROMATIC AMINO ACIDS, section on Tryptophan). It is a vasoconstrictor occurring, among other locations, in brain and in blood platelets. It appears to play a role in blood pressure regulation (see HYPERTENSION).

SERUM AND SERUM PROTEINS

When whole blood is allowed to coagulate (see BLOOD CLOTTING), the formed or cellular elements (see BLOOD, CELLULAR COMPONENTS) aggregate in a network of fibrin derived from one of the soluble components, fibrinogen. The clot may be separated by centrifugation, leaving a clear, straw-colored fluid which is termed serum. Thus, blood *serum* differs from blood *plasma* (the fluid, non-cellular portion of unclotted whole blood) essentially in the absence from serum of fibrinogen, which is present in plasma but remains in the clot in the form of fibrin when clotting has occurred (see PLASMA AND PLASMA PROTEINS).

SEX HORMONES

The hormones capable of stimulating the development of accessory reproductive organs and secondary sex characteristics are termed "sex hormones." The sex hormones may be divided into three groups in accordance with their major biological activities: androgens (Greek *andros* = male and *gennao* = produce; *i.e.*, masculinizing hormones), estrogens (Greek *oistros* = gadfly; hence: sting, frenzy; *i.e.*, feminizing hormones) and gestagens (Latin *gestatio* = carrying; *i.e.*, hormones required for maintenance of pregnancy).

Sources. Both testes and ovaries are secretory organs for the three groups of sex hormones, the quantitative relationship differing with species, sex, age and season. In addition, extra-gonadal tissue may contribute significantly to the total body concentration of sex hormones in mammals. The adrenal cortices (see ADRENAL CORTICAL HORMONES) of certain mammalian species (*e.g.*, man, rat, guinea pig, cattle, mole, dog) produce androgens, and estrogens have been isolated from human and bovine adrenals. Sex hormone secretion by the adrenal cortex may be common to all mammals. The placenta is a source of gestagens and estrogens (see Hormones of Reproductive Cycles, below). Non-endocrine tissues convert corticosteroids to androgen in small amounts, and possibly androgens to estrogens. The role of the adrenal and of other tissues in sex hormone elaboration in non-mammalian species is obscure (see also HORMONES, STEROID).

Structure. The sex hormones of vertebrates are STEROIDS and are included among the "steroid hormones." Most androgens, and their metabolites, are C_{19} steroids; similarly, most estrogens are aromatic C_{18} steroids. The adrenal androgens, unlike those of the gonads, also include C_{19} steroids with an oxygen function at carbon-11, *e.g.*, 11β-hydroxy-Δ^4-androstene-3,17-dione. The configurations of some of the principal sex hormones are shown in Fig. 1. (See also ANDROGENS; OVARIAN HORMONES.)

Biosynthesis. In every species investigated, gonads and adrenals synthesize sex steroids from cholesterol, a product of lipid synthesis and a constituent of diet. As discussed elsewhere in detail (see STEROLS, BIOGENESIS), the enzymatic reactions leading to the formation of steroidal sex hormones are (a) cholesterol to 3β-hydroxy-Δ^5-pregnene-20-one, (b) 3β-hydroxy-Δ^5-pregnene-20-one via at least two possible pathways to C_{19} steroid androgens, and (c) androgen transformation to C_{18} steroid estrogens. One sequence of reactions under (b) includes the formation of the gestagen progesterone: 3β-hydroxy-Δ^5-pregnene-20-one $\rightarrow$ progesterone $\rightarrow$ C_{19} steroids. Evidence exists for other pathways of biosynthesis. Also, one gland may form steroid hormones from steroidal precursors elaborated by another gland. This is exemplified in the pregnant woman by the conversion in the placenta to estrogens of circulating C_{19} steroids, arising from fetal and maternal adrenals.

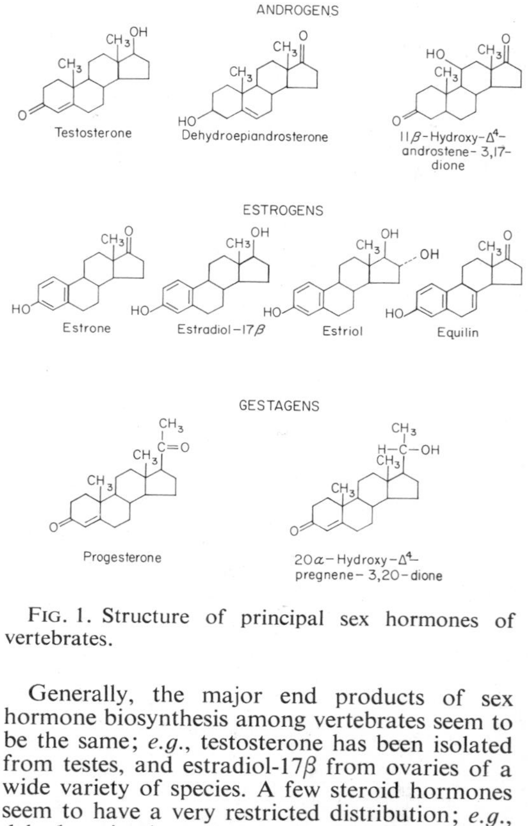

FIG. 1. Structure of principal sex hormones of vertebrates.

Generally, the major end products of sex hormone biosynthesis among vertebrates seem to be the same; *e.g.*, testosterone has been isolated from testes, and estradiol-17β from ovaries of a wide variety of species. A few steroid hormones seem to have a very restricted distribution; *e.g.*, dehydroepiandrosterone and estriol in primates, and equilin in the equine species.

It should be noted that intermediates in the sequence of reactions leading to steroid hormone formation may have hormonal activities of their own (*e.g.*, progesterone formed as an intermediate during testosterone biosynthesis). Intermediates are usually secreted by the organ in small amounts but may, through excessive synthesis or reduced metabolism, become major secretory products, with ensuing profound biological effects. These facts have been important in explaining the results found in experimental studies and in human aberrant steroid-endocrine function.

Metabolism. The sex hormones are secreted into the circulation and are present in blood, associated with the serum proteins. After being metabolized to varying degrees by different tissues, including the "target organs," the metabolites appear in the circulation as steroid conjugates (see below). Many, but not all, metabolites are inactive as sex hormones. A principal site of metabolism is the liver. The major routes of excretion are via the urine and the feces.

In general, but with notable species differences, androgens and gestagens are metabolized by reduction of ring A of the steroid nucleus and of one or more carbonyl groups attached to the steroid nucleus. Estrogen metabolism proceeds mainly by hydroxylation of the nucleus, which also occurs to some extent with other steroid hormones. Most of the metabolites are excreted coupled with glucuronic or sulfuric acid; such compounds are collectively designated as "steroid conjugates." (An interesting steroid conjugate is dehydroepiandrosterone sulfate, which is the chief form in which dehydroepiandrosterone is secreted by the human adrenal gland.) Examples of principal metabolites are shown in Fig. 2.

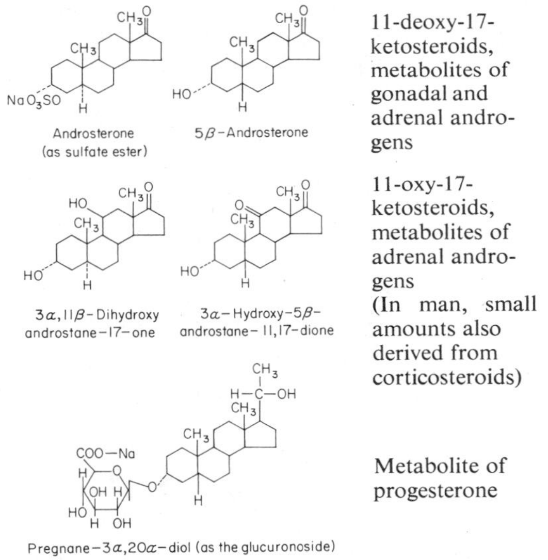

Androsterone (as sulfate ester) — 5β-Androsterone — 11-deoxy-17-ketosteroids, metabolites of gonadal and adrenal androgens

3α,11β-Dihydroxy androstane-17-one — 3α-Hydroxy-5β-androstane-11,17-dione — 11-oxy-17-ketosteroids, metabolites of adrenal androgens (In man, small amounts also derived from corticosteroids)

Pregnane-3α,20α-diol (as the glucuronoside) — Metabolite of progesterone

FIG. 2. Structure of principal metabolites of sex hormones in man. (Principal metabolites produced in most vertebrates are not known.) *Note:* Major metabolites of estrogens are conjugated estrone and estriol (see Fig. 1). Ovarian estrone and estradiol-17β are partially converted to estriol by the human liver.

The metabolites of steroid hormones present in human urine and blood are used as indices of gonad or adrenal gland (and indirectly of the anterior pituitary) functioning. For example, urinary pregnanediol levels reflect ovarian progesterone secretion and are a measure of corpus luteum function, while urinary 17-KETOSTEROIDS —a group of metabolites derived largely from C_{19} steroids—reflect adrenal cortical and gonadal secretory activity. Blood testosterone determinations have been used to assess ovarian and adrenal production of androgens.

Physiological Activity. Androgens and estrogens stimulate and partially regulate the development and growth of the reproductive organs and of secondary sex characteristics, including mating behavior. These sex hormones also exert generalized effects on the body and must be considered as general, as well as specific, growth factors. In many of their biological activities, the sex hormones interact with each other, either antagonistically or synergistically. The degree of their influence and the nature of the response vary with different species, with age, with sex, and with the general physiological milieu. The rate and amount of hormone secreted or administered, the route of administration, the nature of the circulating hormones, all play important roles in biological effectiveness. Some of these considerations are illustrated in Table 1.

The mechanisms of action of androgenic and estrogenic hormones at the subcellular and molecular level appear to involve a stimulation of those processes regulating enzyme activity and metabolism (see METABOLIC CONTROLS; also OVARIAN HORMONES, section on Mechanism of Action). The alterations in metabolism ultimately result in both the gross and the subtle changes implied by the terms "masculinization" and "feminization."

Sexual maturation and reproductive activity occur following the onset of gonadal secretion of sex hormones, under the stimulus of pituitary gonadotrophin (see GONADOTROPIC HORMONES). Removal of the gonads (gonadectomy) of the immature animal generally prevents sexual maturation, *i.e.*, the development of accessory sex organs and of secondary sex characteristics, and changes in those metabolic activities influenced or regulated by sex hormones. Gonadectomy performed after sexual maturation generally causes the same changes as those in the immature animal, but with much less intensity. Conversely, sex hormone administration prevents or reverses many of the effects of gonadectomy and, in the immature intact animal, causes precocious sexual maturation.

Androgens. In the development of the embryo and fetus, androgenic hormones, arising from the embryonic testes, control or influence male differentiation. In lower vertebrates, *e.g.*, amphibians, androgens may cause sex reversal; in mammals, their effect is to partially masculinize female fetuses and, in some species, to form sterile intersexes possessing both testicular and ovarian tissue.

Androgens stimulate the growth of the male reproductive tract (sperm duct, seminal vesicles and associated structures). Spermatogenesis is enhanced by androgens. Secondary sex characteristics develop under the influence of androgens,

TABLE 1. RELATIVE ACTIVITY OF SOME ANDROGENS ON SEXUAL AND NON-SEXUAL END POINTS

End Point	Testo-sterone	Methyl-testo-sterone	Δ^4-Andro-stene-3,17-dione	Dehydro-epiandro-sterone	Andro-sterone
(1) Seminal vesicle and prostate gland growth in rats and mice	1[a]	1	2	3–4	2–3
(2) Comb growth in capon	1[b]	1	2–3[c]	3	2–3[c]
(3) Comb growth in chicken	2	—	—	3	1
(4) Pituitary gonadotrophin inhibition in rats	1	1	—	2	2
(5) N_2 Retention in humans	1	1	3	4	4
(6) Creatinuria in humans	neg.[d]	1	neg.	neg.	—
(7) Restoration of spermatogenesis in hypophysectomized rats	neg.	neg.	neg.	2	—

Activities ranked as 1 to 4, with 1 showing greatest activity
[a] Maximum effect 3 days after injection. Maximum effect occurs at 11 days after injection if testosterone is injected as propionate ester.
[b] Inhibited by estradiol-17β or progesterone.
[c] More active by local application than by systemic administration.
[d] neg. = no activity.
Adapted from reference 2.

e.g., pigmentation of fish and of the plumage of birds, size of the thumb pad of frogs, size and color of the comb and wattles in chickens, antler growth of deer, and voice change in man. The male sex hormones hasten ossification of the epiphyses of long bones and are responsible for a "male distribution" of body fat and hair. Estrogens antagonize many of these androgenic activities.

Metabolic effects of androgens on the sex organs include increased fructose production by seminal vesicles, with concomitant enhancement of the activities of several enzymes. In the rat prostate, amino acid incorporation into protein and the synthesis of citric acid and fatty acids are stimulated by androgens. The activity of the enzyme alkaline PHOSPHATASE in prostatic and renal tissue responds to androgens.

Androgens exert a striking stimulatory or anabolic influence on nitrogen metabolism, causing nitrogen retention and a limited increase in muscle strength and development. This anabolic effect has been termed "myotropic" activity, which is not necessarily correlated with androgenic activity. Indeed, attempts have been made to dissociate these two biological properties of androgens for therapeutic purposes (see Sex Hormone Analogues, below).

In the female, androgens can cause the development of male secondary sex characteristics and of the remnants of male sex structures. Normal female structures show diverse responses to androgens; in mammals, androgens exert a gestagenic effect on the uterus, while in birds and lizards the effect on the oviducts is estrogenic. Ovarian function eventually ceases in response to prolonged androgen administration.

Estrogens. The development of the female reproductive tract (oviducts, mammalian uterus and associated structures) is stimulated by estrogens. Estrogens exert a stimulatory effect on uterine nucleotide and protein synthesis, on enzymatic activity (*e.g.*, alkaline phosphatase, lactic acid dehydrogenase) and on fluid accumulation. The sum of these effects leads to increased uterine weight. Uterine contractility and responsivity to oxytocin (see VASOPRESSIN) and the uterine blood supply are increased. The estrogens increase mitotic activity and induce characteristic changes in the cell lining of the oviducts, uterus, cervix and vagina. The oviducts of some species respond to estrogen administration with secretory activity.

Estrogens are responsible for the development of the secondary sex characteristics of the female, including distribution of body hair, alterations in body contour and closure of the epiphyses of the long bones. They affect the plumage of certain species of birds, and in mammals, they exert a proliferative action on the mammary gland. As with androgens, estrogens exert biological effects which are not confined to the reproductive organs. They have a profound action on CALCIUM and phosphorous metabolism, and probably play a role in normal BONE metabolism. Lipid metabolism is affected by estrogens: in humans, estrogens affect fat deposition and decrease blood lipid concentrations. In birds, blood lipids, calcium and phosphate are increased by these hormones. The estrogens influence the concentration and mode of transport of ADRENAL CORTICAL HORMONES in blood, and may exert an action on the pituitary production of ADRENOCORTICOTROPHIC and GROWTH hormones. Androgens counteract many of the effects of estrogens.

In the male, chronic estrogen intake suppresses testicular function.

Gestagens. In mammals, gestagenic action is intimately connected with uterine development in pregnancy and progesterone has often been called "the hormone of pregnancy." Gestagens act on the uterine endometrium, previously prepared by estrogens, to cause glycogen and mucus secretions. These secretions are indispensable for the implantation of the fertilized ova. The depressing effects of progesterone on uterine contractility and response to oxytocin are thought to maintain the uterus quiescent during pregnancy. Other effects of gestagens are (a) a stimulation of mammary gland development in conjunction with estrogens and other hormones, (b) a rise in body temperatures in women, (c) an inhibition of ovulation (in women and rabbits), and (d) in large doses, the promotion of salt and water retention. In mammals, gestagens inhibit pituitary luteinizing hormone production; in chickens and ducks, the effect is one of stimulation.

Hormones of Reproductive Cycles. Periodicity is a prominent characteristic of reproductive processes. However, the periods, rhythms and cycles constituting this periodicity are rarely the same in any two species of vertebrates. In the majority of vertebrates, an environmental stimulus is required to initiate the reproductive process.This stimulus is frequently provided by climatic, seasonal or population changes. Among some vertebrates, particularly in mammals, "inherent cycles" of physiological changes exist which determine sexual activity and which are more or less independent of environmental stimuli. In general, internal and environmental stimuli activate the hypothalamic-pituitary-gonadal mechanisms of both sexes; this activation results in physiological, anatomical and behavioral changes in the animal. The net effect of these changes is to prepare the sexually mature male and female of the species for successful breeding.

The sections to follow will be confined to the endocrine modifications and interrelationships occurring among mammalian females possessing inherent sexual cycles. Other aspects of the endocrinology of vertebrate reproduction (*e.g.*, spontaneous and induced ovulation, delayed implantation, the effects of seasonal changes, and of age) are beyond the scope of this brief presentation.

Two GONADOTROPHIC HORMONES of the pituitary, follicle-stimulating hormone (FSH) and luteinizing hormone (LH), act on the ovary to cause follicular maturation and to increase estrogen secretion. Estrogen secretion results in uterine proliferation (see Estrogens, above). The rising titers of estrogens have a depressing action on the pituitary ("feedback" mechanism), probably via the hypothalamus, to inhibit FSH secretion and, presumably, to stimulate LH secretion. Ovulation ensues as a result of the discharge of LH. (A discharge of LH may also be induced by mating and other stimuli which act via neurogenic pathways on the hypothalamic-pituitary mechanism). Concomitant with or before ovulation, estrogen secretion diminishes. The postovulatory follicle, now luteinized (corpus luteum), persists for variable periods of time depending upon the species. In mammals with short sexual cycles (*e.g.*, rat, hamster), corpora lutea degenerate rapidly. In mammals with longer sexual cycles (*e.g.*, guinea pig, dog, man), and in short-cycle mammals after mating, the corpora lutea persist and secrete progesterone under the stimulus of a luteotrophic hormone released by the pituitary. Progesterone acts on the estrogen-prepared uterus to cause physiological changes necessary for successful implantation (see Gestagens, above) and exerts a negative "feedback" action to inhibit pituitary LH release. In man (other species ?), the corpus luteum also secretes estrogens. The degeneration of corpora lutea and diminishing hormone secretion signal the end of one sexual cycle.

In the event of pregnancy, the life of the corpus luteum is prolonged. Luteotrophic hormones continue to be secreted, by both the pituitary and the placenta. There is continued and increasing progesterone secretion; in many species, the placenta replaces the ovary as the major source of progesterone during pregnancy. Progesterone is essential for the maintenance of pregnancy and for mammary development. Estrogen production from the ovary or placenta increases during pregnancy in a few species. In man, the placenta secretes relatively enormous amounts of estriol before birth. With the termination of pregnancy, the circulating titer of sex hormones decreases. The diminution of sex hormone concentrations removes their inhibitory influence exerted on the pituitary and permits initiation of lactation.

An exquisite balance exists among the gonadal and extra-gonadal secretions of the several sex hormones. Within the context of other physiological activities, this balance regulates reproductive processes. An acute and massive, or a chronic, alteration in this equilibrium will be reflected in the capacity for, and performance in, reproduction.

Other Substances with Sex Hormone Activity. *Sex Hormone Analogues.* A number of synthetic steroidal and non-steroidal compounds have been prepared which have marked sex hormone-like activities. Such compounds include the androgen methyltestosterone, the gestagens 17α-ethynyl-19-nortestosterone and 6α-methyl-17α-acetoxyprogesterone, and the estrogens 17α-ethynylestradiol and stilbestrol (see Fig. 3). Stilbestrol is among the oldest of the synthetic estrogens and is a non-steroidal estrogen. The 19-nortestosterone compounds are derivatives of testosterone which lack the carbon-19 methyl group, and they possess gestagenic activity. Some of the sex hormone analogues exhibit higher potency when given orally than do steroidal sex hormones of natural origin. Others exhibit a high potency in a particular biological activity; for instance, 19-nortestosterone is 20 times as active as testosterone as an anabolic agent though much less active as an androgen. The sex hormone analogues have found wide clinical application in problems of fertility and sterility, adolescence, the post-menopausal period, ATHEROSCLEROSIS, cancer of the breast and prostate, and muscle wasting conditions.

Substances with Sex Hormone Activity from Invertebrates and Plants. Substances with the biological activities of some of the steroidal sex hormones of vertebrates are found in invertebrates and plants.

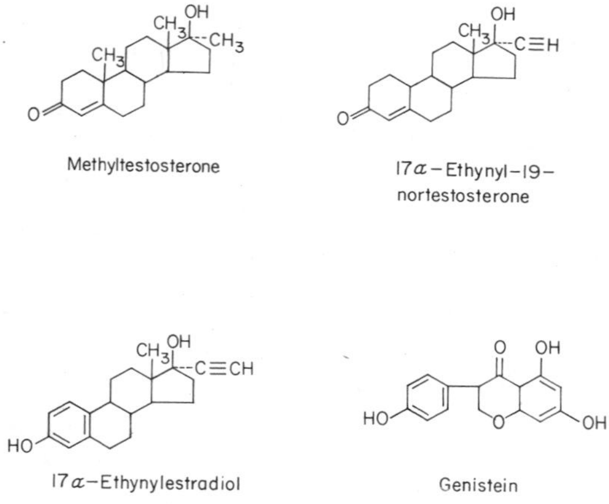

Methyltestosterone

17α-Ethynyl-19-nortestosterone

17α-Ethynylestradiol

Genistein

FIG. 3. Structure of some sex hormone analogues and of genistein. The structures of some others, including 6α-methyl-17α-acetoxyprogesterone (medroxyprogesterone acetate), and diethylstilbestrol, are illustrated in the article OVARIAN HORMONES.

In studies with invertebrates, steroidal estrogens have been isolated from mollusks and starfish. The hormones regulating growth, moulting and metamorphosis in insects appear to be lipid or steroid in nature. The juvenile hormone and the prothoracic gland hormone (ecdysone) of insects and the sex attractant of the American cockroach are such substances. It is possible that the steroid sex hormones of vertebrates evolved from sterol and structually related sex hormones of invertebrates. Several groups of plants contain compounds with sex hormone simulating activities. Most of the active substances have not been identified; one that has is genistein (see Fig. 3). Genistein, contained in the subterranean clover of Australia, has led to permanent estrus and sterility in grazing cattle.

References

1. HEFTMANN, E., AND MOSETTIG, E., "Biochemistry of Steroids," New York, Reinhold Publishing Corp., 1960.
2. DORFMAN, R. I., AND SHIPLEY, R. A., "Androgens," New York, John Wiley & Sons, 1956.
3. YOUNG, W. C., "Sex and Internal Secretions," Third edition, Baltimore, Williams and Wilkins Co., 1961, 2 vols.
4. VON EULER, U. S., AND HELLER, H., "Comparative Endocrinology," New York, Academic Press, 1963, 2 vols.
5. ZUCKERMAN, S., "The Ovary," New York, Academic Press, 1962, 2 vols.
6. CAREY, H. M., "Human Reproductive Physiology," Washington, D.C., Butterworth, 1963.
7. RYAN, K. J., "Hormones of the Placenta," *Am. J. Obstet. Gynecol.*, **84**, 1695 (1962).

ERIC BLOCH

SHERMAN, H. C.

Henry Clapp Sherman (1875–1955) was an American biochemist, noted for his contributions to nutrition. His animal experiments involved assaying for specific nutrients and devising diets on which animals (rats) could propagate for generations. His work emphasized the importance of calcium as a nutrient. He demonstrated that rickets could be induced by an improper balance between calcium and phosphorus, as well as by a lack of vitamin D.

SICKLE CELL ANEMIA. See ANEMIAS; HEMOGLOBINS (IN HUMAN GENETICS).

SINGLE CARBON UNIT METABOLISM

The single carbon units at various oxidation states are familiar as the chemical compounds carbon dioxide, formate, formaldehyde, methanol and methane. The radicals derived from formate, formaldehyde and methanol are interrelated in metabolic systems. These are the formyl (—CHO) or methenyl (—CH═) radicals derived from formate, the hydroxymethyl (—CH_2OH) or methylene (—CH_2—) radicals derived from formaldehyde, and the methyl (CH_3—) radical derived from methanol. They occur as derivatives of the vitamin FOLIC ACID, N-[4-{[(2-amino-4-hydroxy-6-pteridyl)methyl]amino}-benzoyl] glutamic acid or more briefly pteroylglutamic acid [Fig. 1(A)]. The functional forms of the one carbon units are derivatives of 5,6,7,8-tetrahydrofolate or the polyglutamate derivative of tetrahydrofolate [Fig. 1(B)], in which the one carbon radicals occur as substitutions in the 5- and/or 10-positions of the pteridine ring. Other PTERIDINE derivatives have been isolated that do not contain

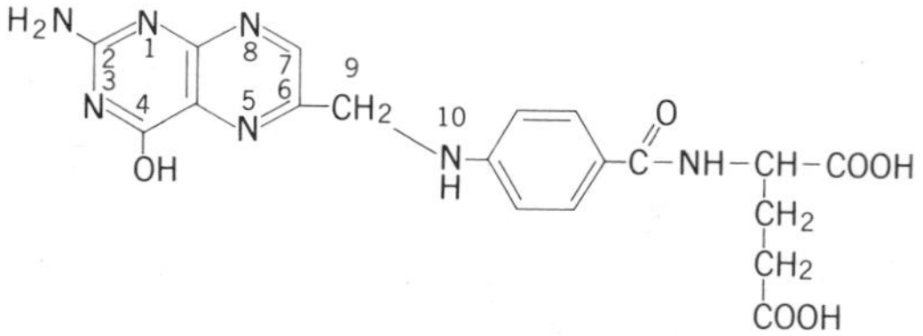

FIG. 1a

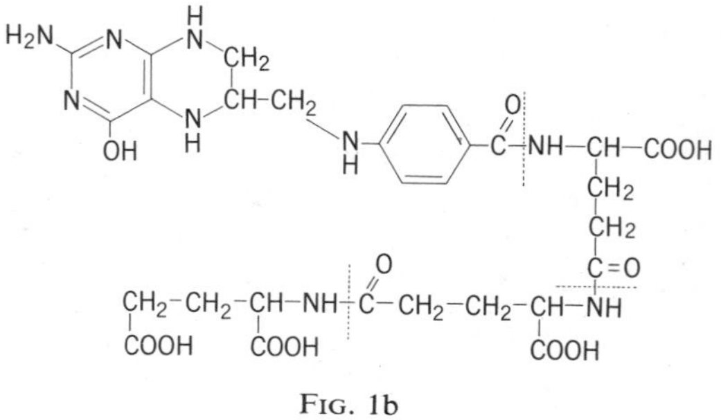

FIG. 1b

the benzoylglutamyl moiety, such as 2-amino-4,6-dihydroxypteridine (xanthopterin) and 2-amino-4-hydroxy-6[1,2-dihydroxypropyl-(L-*erythro*)]-pteridine (biopterin). Although these compounds do not function in one carbon unit metabolism, they may function in other metabolic reactions. Dihydrobiopterin, for example, is the coenzyme in the hydroxylation of phenylalanine to form tyrosine.

Formyltetrahydrofolate in Purine Biosynthesis. The activation of formate to a derivative that serves as a source of one carbon units in biosynthetic reactions is catalyzed by the enzyme formyltetrahydrofolate synthetase:

Formate + *l*(L)-Tetrahydrofolate + ATP

$\rightleftharpoons$ 10-Formyltetrahydrofolate + ADP + P_i

The reaction is reversible and may also serve to generate ATP under certain conditions. The enzyme has wide distribution. It has been obtained in crystalline form from a bacterial source. The product, 10-formyltetrahydrofolate, is in chemical equilibrium with 5,10-methenyltetrahydrofolate and with the 5-formyl isomer:

5,10-Methenyltetrahydrofolate

$OH^{\ominus}$ $H^{\oplus}$ H_2O $H^{\oplus}$

$OH^{\ominus}$

5-Formyltetrahydrofolate $\longleftarrow$ 10-Formyltetrahydrofolate

Both the 10-formyl and the 5,10-methenyl derivatives function as donors of the single carbon unit in specific enzymic reactions involved in the biosynthesis of the purine derivative, inosinic acid (see PURINE BIOSYNTHESIS). 10-Formyltetrahydrofolate serves as the source of the C-2 atom of the purine ring in a reaction with 5′-phosphoribosyl-5-amino-4-imidazolecarboximade to form the 5-formamido derivative. 5,10-Methenyltetrahydrofolate serves as the source of the C-8 atom of the purine ring in the formylation of 5′-phosphoribosylglycineamide to form the N-formylglycineamide derivative. Inosinic acid serves as a precursor of the purine derivatives found in nucleic acids such as adenine and guanine.

Formiminotetrahydrofolate and Histidine and Purine Degradation. The N-formimino derivatives of glutamate and glycine have been encountered as intermediates in the degradation metabolism of histidine and purines, respectively. Animals in a state of folic acid deficiency excrete large amounts of formiminoglutamate in their urine. Tetrahydrofolate is required for the further metabolism of these formimino derivatives. In the presence of the specific enzyme, the formimino group (—CH=NH) is transferred from the amino acid to tetrahydrofolate to yield 5-formiminotetrahydrofolate and the free amino acid, which is then further metabolized.

l(L)-Tetrahydrofolate + Formiminoglycine $\rightleftharpoons$

5-Formiminotetrahydrofolate + Glycine

The formiminotetrahydrofolate is converted to 5,10-methenyltetrahydrofolate and ammonia by the enzyme formiminotetrahydrofolate cyclodeaminase. There is no evidence that this is a reversible reaction.

5-Formiminotetrahydrofolate + H^+ $\rightarrow$

5,10-Methenyltetrahydrofolate$^+$ + NH_3

The cyclic product is hydrolyzed enzymically or non-enzymically to yield 10-formyltetrahydrofolate.

Methylenetetrahydrofolate in Serine Metabolism. 5,10-Methenyltetrahydrofolate can be reduced enzymically by a dehydrogenase in the presence of TPNH (NADPH) to yield 5,10-methylenetetrahydrofolate.

5,10-Methenyltetrahydrofolate$^+$ + TPNH $\rightleftharpoons$

5,10-Methylenetetrahydrofolate + TPN^+

The same product may be found non-enzymically by the condensation of formaldehyde with tetrahydrofolate. This compound functions as the donor of the one carbon unit to glycine in the presence of the enzyme serine hydroxymethyltransferase to form L-serine. This is a pyridoxal protein (see PYRIDOXINE FAMILY). The reaction is reversible, and the enzyme can therefore also function to generate, from serine, metabolically active one carbon units at the level of oxidation of formaldehyde.

5,10-Methylenetetrahydrofolate + Glycine $\rightleftharpoons$

Tetrahydrofolate + L-Serine

5-Hydroxymethylcytosine occurs in the DNA of the T-even phages. Infection of *E. coli* by these phages is accompanied by the appearance of an enzyme that catalyzes the transfer of the hydroxymethyl group of 5,10-methylenetetrahydrofolate to deoxycytidylate, resulting in the formation of 5-hydroxymethyldeoxycytidylate.

5,10-Methylenetetrahydrofolate + dCMP $\rightleftharpoons$

Tetrahydrofolate + 5-Hydroxymethyl-dCMP

A pyridoxal enzyme that catalyzes the cleavage of glycine to methylenetetrahydrofolate and carbon dioxide has been detected in several microorganisms.

Tetrahydrofolate + Glycine $\rightarrow$

5,10-Methylenetetrahydrofolate + CO_2 + NH_3

Methylenetetrahydrofolate in Thymine Biosynthesis. Thymidylate is formed from deoxyuridylate by the transfer of the one carbon unit from 5,10-methylenetetrahydrofolate in the presence of the enzyme thymidylate synthetase (see also PYRIMIDINE BIOSYNTHESIS). The reduction of the one carbon unit from the oxidation level of formaldehyde to that of methanol that occurs in this reaction is accompanied by the oxidation of the tetrahydrofolate nucleus to 7,8-dihydrofolate. This reaction involves the transfer of the hydrogen atoms from the reduced pyrazine ring of the pteridine ring to the methyl group of thymine.

5,10-Methylenetetrahydrofolate + dUMP →

Dihydrofolate + dTMP

Dihydrofolate formed as the product in this reaction can be reduced to tetrahydrofolate by the enzyme dihydrofolate reductase with TPNH or DPNH, depending on the pH of the reaction mixture.

7,8-Dihydrofolate + TPNH + H^+ ⇄

l(L)-Tetrahydrofolate + TPN^+

The enzyme may also reduce folate to dihydrofolate. This enzyme is strongly inhibited by 4-amino-4-deoxyfolate (aminopterin) or other 4-amino-4-deoxyfolate derivatives.

5-methyltetrahydropteroylmono- or tri-glutamate and a vitamin B_{12}-containing enzyme (see also COBALAMINS; METHYLATIONS IN METABOLISM).

5-Methyltetrahydropteroyltriglutamate +

L-Homocysteine →

Tetrahydropteroyltriglutamate + L-Methionine

S-adenosylmethionine, the activated form of methionine, rather than the methyl derivative of tetrahydrofolate, functions as the substrate in the transmethylation reactions involved in the formation of methylated purine and pyrimidine bases, and in the formation of choline and its derivatives.

The relationships among the single carbon units involved in the reactions discussed are summarized in Fig. 2.

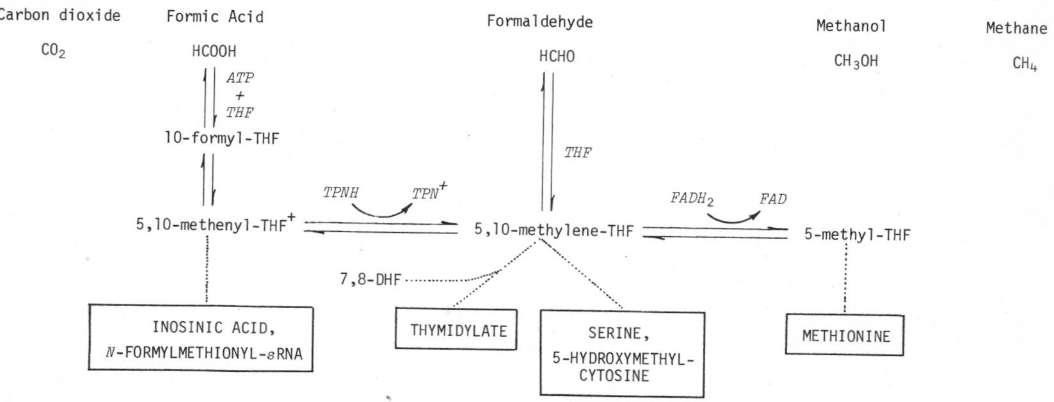

FIG. 2

Methyltetrahydrofolate in Methionine Biosynthesis. 5,10-Methylenetetrahydrofolate may be further reduced to yield 5-methyltetrahydrofolate by the FLAVOPROTEIN, 5,10-methylenetetrahydrofolate oxidoreductase.

5,10-Methylenetetrahydrofolate + $FADH_2$ ⇄

5-Methyltetrahydrofolate + FAD

A flavin reductase is required to catalyze the formation of the $FADH_2$ from FAD and DPNH that is required in the reaction. 5-Methyltetrahydrofolate may also be formed by the chemical reduction of 5,10-methylenetetrahydrofolate with sodium borohydride.

The methyltetrahydrofolate derivative serves as the methyl donor in the biosynthesis of METHIONINE from homocysteine. Two systems have been described that carry out this synthesis. One system, found in bacterial strains normally able to synthesize methionine, requires 5-methyltetrahydropteroyltriglutamate as the methyl donor and shows no activity with the monoglutamyl derivative. The second system found in methionineless, vitamin B_{12}-requiring bacterial mutants and in mammalian systems, can utilize either

References

1. STOKSTAD, E. L. R., in "The Vitamins, Chemistry, Physiology, Pathology" (SEBRELL, W. H., JR., AND HARRIS, R. S., EDITORS), Vol. III, p. 89, New York, Academic Press, 1954.
2. HUENNEKENS, F. M., AND OSBORN, M. J., *Advan. Enzymol.* **21**, 369 (1959).
3. RABINOWITZ, J. C., in "The Enzymes" (BOYER, P. D., LARDY, H., AND MYRBÄCK, K., EDITORS), Second edition, Vol. II, p. 185, New York, Academic Press, 1960.
4. FRIEDKIN, M., *Ann. Rev. Biochem.*, **32**, 185 (1963).

JESSE C. RABINOWITZ

SKIN PIGMENTATION

Melanin pigments are the substances usually considered when one comments on the color of skin, hair and eyes. However, many other pigments are present in these tissues, and in normal and abnormal states they can modify the color produced by melanin. For example, if the hemoglobin content of the blood is reduced, the skin becomes light in color. On the other hand if the hemoglobin content remains constant or even decreases when peripheral vasodilatation occurs, the skin becomes red in color. BILE pigments give a yellow color to skin. Chemicals such as atabrine also produce yellowing whereas silver salts leave a deep blue color. The KERATIN of the epidermis may also impart a yellowish cast to the skin. In this brief discussion only variations in skin color resulting from changes in melanin pigmentation will be described.

Melanin is produced from the enzymic oxidation of the amino acid TYROSINE through several steps to 5,6-dihydroxyindole. This indole is in turn oxidized and polymerized to form a brown pigment. *In vivo* the quinones arising from the oxidation process, *e.g.*, indole-5,6-quinone are readily bound to peptides and proteins. The final pigment is a melano-protein, and these pigments appear to be different in the skin, hair and eyes of a given individual. Furthermore, the melanin polymer may exist in different states of oxidation. In a mildly oxidized form melanin is darker than when it is in a reduced state. Vigorous oxidation with hydrogen peroxide will result in the destruction of melanin. Thus the nature of the melanin-protein complex and the state of oxidation of the melanin itself are two of the factors that are important in determining the color of skin, hair and eyes.

The enzymic formation of melanin occurs in the cytoplasm of cells called *melanocytes*. These cells are derived embryologically from the neural crest. In many ways they behave as do nerve cells. Eventually melanocytes come to rest in the skin at the epidermal-dermal junction and about hair bulbs, in the eyes along the uveal tract and in the central nervous system along the leptomeninges. It is the copper-containing enzyme tyrosinase, present in the cytoplasm of these cells, that catalyzes the oxidation of tyrosine to dihydroxyphenylalanine (dopa) and then the further oxidation of dopa. Subsequent rearrangements result in the formation of indoles, and these in turn form melanin. Ordinarily, the copper-containing tyrosinase exists in an inhibited state. Activation steps are required to initiate pigment formation. For example, the tanning that occurs after exposure to sunlight results from at least two actions. (1) Sunlight produces a direct oxidation of melanin already present in the skin to make it darker. (2) The light acts to increase markedly the activity of tyrosinase so that more pigment can form. If, for any reason, tyrosinase is not present or is inhibited, the skin, eyes and hair will be light in color. Thus in phenylketonuria, inhibition of tyrosinase by phenylalanine appears to be responsible for the decreased coloring of those subjects. In albinism, no functioning tyrosinase is present. It is not known at what state the defect in tyrosinase occurs. On the other hand, releasing normal inhibitors of tyrosinase results in a darkening of skin. This occurs not only with sunlight but also with other forms of radiant energy. In addition darkening occurs after the ingestion of arsenic, an agent that combines with sulfhydryl groups which may be normal inhibitors of tyrosinase.

Of the numerous factors that control the melanin color of skin, hair and eyes, we have thus far mentioned only three, *viz.*, the amount of melanin, the state of oxidation of melanin and the nature of the melano-protein. Another important factor relates to the state of dispersion of pigment granules and perhaps of tyrosinase as well, within the cytoplasm of the melanocytes. The *melanocyte-stimulating hormones*, α- and β-MSH, produced in the intermediate lobe of the pituitary gland, are peptides that can bring about a rapid darkening of skin of human subjects, guinea pigs, frogs, tadpoles and other animals. In marine animals, the darkening results from a dispersion of pigment granules within the cytoplasm of the pigment cells. The mechanism of darkening for mammalian melanocytes is less clearly defined. However, we do know that MSH brings about an increase in tyrosinase activity within the melanocytes. The melanocyte-stimulating hormones are responsible for the hyperpigmentation seen in patients with adrenocortical insufficiency. They probably are contributing factors to the darkening of skin that occurs during pregnancy. Some patients with tumors of the pituitary gland secrete both MSH and ACTH so that there is not only darkening of skin but also the production of Cushing's syndrome. ACTH is related chemically to MSH, and it has about 1% of the darkening activity of MSH on frog skin. A few patients have been observed to have metastatic carcinoma originating from tumors of the pancreas or lung with the production of intense hyperpigmentation. These individuals produce sufficient MSH-like peptides from their tumors to bring about darkening of the skin. ACTH-like peptides may also be produced, and the patients may have adrenocortical hypertrophy.

Our knowledge of changes in skin color is based on variations in activity of tyrosinase and the output of MSH from the pituitary gland. However, we know relatively little about the lightening processes involving the graying of hair or the production of vitiligo (the development of white spots in the skin). There is good evidence to suggest that a neurochemical agent, either a combination of ACETYLCHOLINE and noradrenaline, or an unidentified substance, is responsible for the lightening of pigment cells in these abnormal states. For frogs, there are three potent lightening agents. All inhibit the action of MSH but work differently. Melatonin is 5-methoxy-N-acetyltryptamine. It is produced by the pineal gland and is a potent lightening agent. Melatonin will always lighten MSH-darkened skin and the action is reversible. Noradrenaline behaves in the same way as does melatonin, although it is considerably less potent. It is made at the peripheral nerve endings. However, noradrenaline differs from melatonin in that sympathetic blocking agents such as ergotamine can prevent the action of noradrenaline, but not that of melatonin. Acetylcholine is less active than noradrenaline and only works on about one-third of the frogs tested; two-thirds are non-reactors. Also the lightening of skin by acetylcholine is not reversible. None of these agents has been shown to bring about lightening of human skin. This phase of the melanin problem is under active investigation at the present time.

References

1. Lerner, A. B., "Hormones and Skin Color," *Sci. Am.*, **205**, 99–108 (July 1961).
2. Wurtman, R. J., and Axelrod, J., "The Pineal Gland," *Sci. Am.* **213**, 50–60 (July 1965).

Aaron B. Lerner

SLEEP

The biochemical aspects of sleep have received comparatively little detailed attention. There are many ancillary problems which have been investigated but the central one—what is accomplished biochemically each day by long hours of sleep—has been attacked only a few times and with no marked success. The symptomatic treatment of insomnia by the use of sleep-inducing drugs is largely on an empirical basis. Knowledge of the fundamental reasons behind our need to sleep is the sort that can be obtained only by biochemical investigation.

The physiology of sleep is discussed in the following reference.

Reference

1. Kleitman, N., "Sleep and Wakefulness as Alternating Phases in the Cycle of Existence," Chicago, University of Chicago Press, 1939.

SLIME MOLDS

The Myxomycetes (Mycetozoa, Plasmodial slime molds) are organisms characterized by a naked, multinucleate, free-living, amoeboid assimilative stage—the *plasmodium*—which, under favorable environmental conditions, usually becomes converted into a large number of fruiting bodies which bear spores.

Biochemical studies in the Myxomycetes are almost entirely confined to *Physarum polycephalum* which has been used as a tool by biochemists and biophysicists to investigate protoplasmic streaming, the mitotic cycle, the nutrition of the plasmodium, the conditions which induce sporulation, and the biochemical changes which accompany sporulation.

The most significant discovery relative to the problem of protoplasmic streaming in the plasmodium centers around at least two contractile proteins (myxomyosin and myosin B) which have been found in the plasmodium of *P. polycephalum*. Both show a reversible change in viscosity when ATP is added. The presence of KCl appears to be necessary in the response to ATP of at least one of the two proteins (myxomyosin). It is interesting in this connection to point out that K^+ seems to play an important role in plasmodial migration, an activity which is inextricably linked with protoplasmic streaming in the plasmodium.

The plasmodium of *P. polycephalum* is the only one which has been grown in axenic culture on a chemically defined liquid medium. Such a medium contains inorganic salts, glucose, citric acid, various amino acids, vitamins and hematin. Nutritional requirements appear to be specific for different slime molds inasmuch as this medium will not support the growth of other species which have been studied.

During the growth phase, plasmodial nuclei undergo synchronous division, mitotic synchrony in the young plasmodia being nearly absolute under controlled conditions. When several plasmodia of *P. polycephalum*, with different mitotic rhythms, are permitted to coalesce, a single rhythm is established throughout the whole mass within 6–7 hours, about half the usual interphase time. In connection with the above studies, it has been demonstrated that the DNA in the plasmodial nuclei is synthesized only within a short period (1–2 hours) following mitosis, but that RNA synthesis continues at a more or less even rate throughout the mitotic cycle. Spectrophotometric studies indicate the DNA is synthesized in the same manner, in various stages of the life cycle, in four other species. Of interest in this connection is the discovery that in two of the three heterothallic species studied, the gamete nuclei of the two compatible mating types (fusion of which is prerequisite to plasmodial formation) have different amounts of DNA.

The conditions which bring about sporulation have been critically studied since Gray showed in 1938 that pigmented plasmodia require light in order to fruit. The conditions which bring about sporulation in *P. polycephalum*, which has a yellow plasmodium, relate to: (1) an optimal age; (2) availability of niacin, niacinamide or tryptophan in the post-growth sporulation medium; (3) a period of exposure to light of 310–500 mμ wavelength following a dark incubation period of 4 days. The biochemical changes which take place during sporulation are not adequately known. It has been found, however, that there is about 3 times as much cytochrome oxidase activity in the spores as in the plasmodia and about 6 times as much ascorbic acid oxidase activity in the plasmodia as in the spores. Nothing is known about the chemical composition of the plasmodial pigments which act as photoreceptors. Published results on this topic are completely contradictory. There is strong evidence, on the other hand, that these pigments are indeed closely linked with the photoreceptor mechanism which induces sporulation.

References

1. Alexopoulos, C. J., "Myxomycetes II," *Botan. Rev.*, **29**, 1–78 (1963).
2. Allen, R. D., and Kamiya, N., (Editors), "Primitive Motile Systems in Cell Biology," pp. 69–136, New York, Academic Press, 1964.
3. Prescott, D. M. (Editor), "Methods in Cell Physiology," pp. 3–54, New York, Academic Press, 1964.

4. THERRIEN, C. D., "Microspectrophotometric Analysis of Nuclear Deoxyribonucleic Acid in Some Myxomycetes," University of Texas, Ph.D. Dissertation, 1965.
5. WARD, J. M., "Shift of Oxidases with Morphogenesis in the Slime Mold *Physarum polycephalum*," *Science*, **127**, 596 (1958).

C. J. ALEXOPOULOS

SNAKE VENOMS

Snake venoms present the biochemist with an interesting array of research problems in protein chemistry. Unlike the toxic materials of many plants and some animal (*e.g.*, toad poisons), snake venoms are largely mixtures of individual proteins; in fact, proteins have been reported to comprise 90% or more of the dry weight of some snake venoms.[1]

Poisonous snakes are found over an exceedingly large geographical area of the world, although certain areas are free of them.[1] Traditionally, the venoms of the Elapidae (cobras, coral snakes and their allies) have been classified as neurotoxic, while venoms of the vipers and crotalid snakes have been considered to be hemolytic. This generalization does not imply that the elapid venoms contain no hemolytic principles or that the viperid and crotalid venoms contain no neurotoxic components, however.

Snake Venoms as Sources of Enzymes. Only in comparatively recent times have the components of snake venoms been investigated systematically, and it is now known that venoms contain large numbers of catalytic proteins, *i.e.*, enzymes. Much of the biochemical research on venoms has been concentrated on the study of these enzymes, some of which have served as useful tools in the investigation of other biochemical problems.

Inasmuch as venoms are considered to have evolved from the digestive tract, it is not surprising that many of the enzymes found in them are hydrolytic in their action. It has long been recognized that animals killed by snake bite decompose more rapidly than those that die from other causes. Moreover, it appears that the digestive and toxic functions of snake venoms may be closely interrelated. Some of the enzymes that have been found in snake venoms are briefly discussed below. In addition to the trivial names of these enzymes, the systematic names and code numbers recommended by the Commission on Enzymes of the International Union of Biochemistry are given for most of them.

Phospholipase A, Formerly Called Lecithinase A (phosphatide acyl-hydrolase, 3.1.1.4). These enzymes are found in venoms of all three groups of poisonous snakes. They act by removing one of the fatty acids of phosphatidycholine (lecithin), phosphatidylethanolamine, phosphatidylserine or phosphatidylinositol (see CEPHALINS). In the past, considerable confusion has risen concerning the point of cleavage by the phospolipases A, and this is the subject of current research interest. Recent work indicates that hydrolysis normally occurs at the β-position, which is usually occupied by an unsaturated fatty acid residue. The mode of action of phospholipase A may be represented thus:

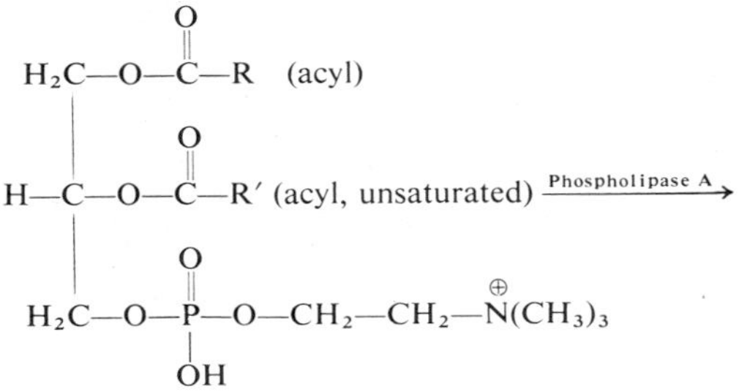

$$\begin{array}{l} H_2C-O-\overset{\overset{\displaystyle O}{\|}}{C}-R \quad \text{(acyl)} \\ \;\;| \\ H-C-O-\overset{\overset{\displaystyle O}{\|}}{C}-R' \quad \text{(acyl, unsaturated)} \xrightarrow{\text{Phospholipase A}} \\ \;\;| \\ H_2C-O-\overset{\overset{\displaystyle O}{\|}}{\underset{\underset{\displaystyle OH}{|}}{P}}-O-CH_2-CH_2-\overset{\oplus}{N}(CH_3)_3 \end{array}$$

A lecithin

$$\begin{array}{l} H_2C-O-\overset{\overset{\displaystyle O}{\|}}{C}-R \\ \;\;| \\ H-COH \\ \;\;| \\ H_2C-O-\overset{\overset{\displaystyle O}{\|}}{\underset{\underset{\displaystyle OH}{|}}{P}}-O-CH_2-CH_2-N^{\oplus}(CH_3)_3 \end{array} \quad + \; R'-COOH$$

Unsaturated fatty acid

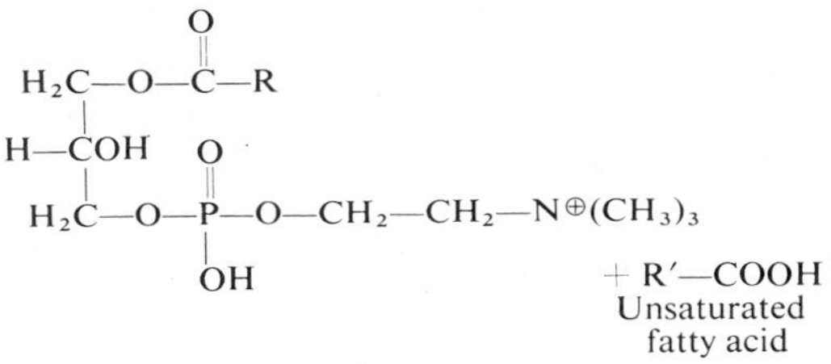

A lysolecithin

The compounds formed by removal of one fatty acid residue from a phospholipid are called lysolecithins (or lysocephalins), and are potent agents for hemolyzing red blood cells.

Phospholipases A have been isolated in highly purified form from a number of snake venoms including a species of South American rattlesnake. In this venom, the phospholipase forms a conjugated protein with croactin, a protein of pronounced toxic effects.[2]

Amino Acid Oxidase (L-amino acid: O_2 oxidoreductase, deaminating, 1.4.3.2). The venoms of more than fifty species of snakes have been found to contain amino acid oxidases specific for the L-enantiomorphs of AMINO ACIDS.[3]

Some of these enzymes have been highly purified and are available from commercial sources. These enzymes, which contain flavin adenine dinucleotide (FAD) as a prosthetic group, catalyze the following type of reactions:

$$\text{Enzyme—FAD} + R-\underset{\underset{\displaystyle NH_2}{|}}{\overset{\overset{\displaystyle H}{|}}{C}}-COOH \rightarrow \text{Enzyme—FADH}_2 + \left[\begin{array}{c} R-\underset{\underset{\displaystyle NH}{\|}}{C}-COOH \end{array}\right] \qquad (1)$$

$$\left[\begin{array}{c} R-\underset{\underset{\displaystyle NH}{\|}}{C}-COOH \end{array}\right] + H_2O \rightarrow R-\underset{\underset{\displaystyle O}{\|}}{C}-COOH + NH_3 \qquad (2)$$

The reduced flavoprotein is then reoxidized by reaction with oxygen:

$$\text{Enzyme—FADH}_2 + O_2 \rightarrow \text{Enzyme—FAD} + H_2O_2$$

In the absence of catalase (which is not present in venoms), the hydrogen peroxide decarboxylates the α-keto acid oxidatively:

$$R\text{—}\underset{\underset{O}{\|}}{C}\text{—COOH} + H_2O_2 \rightarrow R\text{—COOH} + CO_2 + H_2O$$

If catalase is added to the reaction mixture *in vitro*, the hydrogen peroxide is decomposed, and the α-keto acid may be recovered. This procedure has been utilized for the preparation of α-keto acids from their analogous amino acids.

The L-amino acid oxidase of *Crotalus adamanteus* has been isolated in crystalline form.[4] It has a molecular weight of 130,000 and contains two moles of FAD per mole of enzyme. It has been proposed that each FAD is only one-half reduced by the substrate:

$$\text{Enzyme}\left\langle\begin{matrix}\text{FAD}\\ \text{FAD}\end{matrix}\right. + R\text{—}\underset{NH_2}{\overset{H}{C}}\text{—COOH} \rightarrow \text{Enzyme}\left\langle\begin{matrix}\text{FADH}\\ \text{FADH}\end{matrix}\right. + \left[R\text{—}\underset{\underset{NH}{\|}}{C}\text{—COOH}\right]$$

$$\left[R\text{—}\underset{\underset{NH}{\|}}{C}\text{—COOH}\right] + H_2O \rightarrow R\text{—}\underset{\underset{O}{\|}}{C}\text{—COOH} + NH_3$$

The L-amino acid oxidases of snake venoms are more active than the D-amino acid oxidases and L-amino acid oxidases found in other animal tissues. The enzymes in different snake venoms evidently differ somewhat from one another and appear to be a group of homologous enzymes.[3]

Snake venom L-amino acid oxidases have been used not only for the preparation of α-keto acids, but to detect the presence of L-amino acids in an excess of the D-isomer, to determine quantitatively L-amino acids, and to prepare D-amino acids from a racemic mixture. All of these applications are based on what appears to be an absolute antipodal specificity of these L-amino acid oxidases. (See also AMINE OXIDASES.)

Hyaluronidase (hyaluronate glycanohydrolase 3.2.1.d; nomenclature and code number cited by Dixon and Webb[11]). Hyaluronidase, formerly referred to as "spreading factor" hydrolyzes the hyaluronic acid gel in the spaces between cells and fibers of connective tissues. Hyaluronidases are widely distributed in snake venoms, where their function apparently is to reduce the viscosity of the ground substance and thus to accelerate the spread of the venom. The venom of the common krait has been reported to possess an unusually high hyaluronidase activity.[1] Snake venom hyaluronidases attack hyaluronic acid with the liberation of tetrasaccharides (80–85%) and disaccharides (10–15%). Their action is that of an endohexosaminidase.[5] (See also MUCOPOLYSACCHARIDES.)

Proteinases (peptide hydrolases, 3.4). Enzymes capable of hydrolyzing proteins have long been known to exist in the venoms of many snakes. The viper venoms are generally more proteolytic than those of the Elapidae, and the venoms of pit vipers of the western hemisphere (Crotalidae) are particularly rich in proteolytic activity.

Endopeptidases (PROTEOLYTIC ENZYMES capable of attacking protein substrates at internal peptide bonds) have been found in a wide variety of venoms. Relatively few studies have been performed with synthetic substrates, however, and only recently have purified preparations been made. It is evident that different proteinases exist not only among the various species, but also within the same venom, as research workers have recently separated three proteinases from the venoms of an Asiatic moccasin[6] and the Western Diamondback Rattlesnake.[7] The venoms of several vipers have been reported to hydrolyze ester and amide substrates for the pancreatic proteinases, trypsin and chymotrypsin. Recent work shows that the hydrolysis of these substrates is carried out by enzymes other than the proteinases, however, and the substrate specificities of snake venom proteinases remain to be elucidated.

Exopeptidases (proteolytic enzymes which liberate the N-terminal or C-terminal amino acids from peptide chains) appear to be rare in snake venoms; only in the venom of *Echis carinatus* has appreciable exopeptidase activity been observed.[3]

Acetylcholinesterase (acetylcholine acetyl-hydrolase, 3.1.1.7). The venoms of many, if not all, elapid snakes contain cholinesterases; these enzymes appear to be absent from viperid venoms. The cholinesterases found in snake venoms are of the acetylcholinesterase type, in that they hydrolyze β-methylacetylcholine, but not benzoylcholine, and are inhibited by high substrate concentration (see ACETYLCHOLINE).

Enzymes that Hydrolyze Phosphate Bonds. The presence in snake venoms of several enzymes which attack phosphate bonds has been known for more than a quarter of a century. Certain of these enzymes are extremely useful in determinations of the structures of nucleic acids and of certain other types of molecules. Earlier nomenclature has undergone various changes, but the presence in snake venoms of at least four types of phosphate bond-hydrolyzing enzymes has been established. Other enzymes of a similar nature may occur in venoms, but the four discussed below have been at least partially purified (see also PHOSPHATASES; NUCLEASES).

Phosphodiesterase (orthophosphoric diester phosphohydrolase, 3.1.4.1). Phosphodiesterases are enzymes that hydrolyze bonds in diesters of phosphoric acid. Venom phosphodiesterase has proved useful in the study of nucleic acid structures, as it degrades both DEOXYRIBONUCLEIC ACID (DNA) and RIBONUCLEIC ACID (RNA). The attack on nucleic acids and oligonucleotides occurs from the end possessing a free hydroxyl on carbon-3′ of the sugar, with the liberation of 5′-nucleotides. This enzyme is therefore classified as an exonucleotidase. Venom phosphodiesterase rapidly attacks polynucleotides which possess a monoesterified phosphate on carbon-5′, and substrates without the 5′-phosphate are attacked more slowly. Nucleotides with a monoesterified 3′-phosphate are much more resistant to hydrolysis than those containing a free hydroxyl on carbon-3′.

Snake venom phosphodiesterases also hydrolyze a number of synthetic substrates such as di-*p*-nitrophenyl phosphate, *p*-nitrophenyl-uridine-5′-phosphate, thymidyl-(5′-3′)-thymidine, and *p*-dinitrophenyl-thymidine-5′-phosphate.

The purification of a phosphodiesterase from *Bothrops atrox* venom has recently been described.[8]

Endonuclease. The presence in snake venom of an enzyme which splits phosphate bonds in the interior of nucleic acids has recently been established.[9] This enzyme has been purified from *Bothrops atrox* venom, and was found to degrade both DNA and RNA. The enzyme forms 3′-monophosphate esters of relatively uniform length. It thus acts in a completely different manner from that of the phosphodiesterase described above, which attacks the end of nucleotide chains.

5′-Nucleotidase (5′-ribonucleotide phosphohydrolase, 3.1.3.5). A number of snake venoms contain an enzyme capable of hydrolyzing nucleotides with the phosphate esterified in the 5′-position to yield a nucleoside and inorganic phosphate. A 5′-nucleotidase, recently purified from *Bothrops atrox* venom,[10] showed activity toward a number of nucleotides, including the 5′-phosphates of adenosine, inosine, uridine, guanosine and cytidine; the deoxy analogues were also attacked. In addition to the 5′-phosphate, this enzyme requires a nitrogenous base on carbon-1′ of the pentose and a free hydroxyl in the 3′-position.

Alkaline Nonspecific Phosphatase (orthophosphoric monoester phosphohydrolase, 3.1.3.1). An enzyme capable of hydrolyzing a wide variety of phosphates, including adenosine triphosphate, flavin mononucleotide, ribose-5-phosphate, and 3′-adenosine monophosphate, has also been found in certain snake venoms, and has been partially purified.[10]

The Toxic Nature of Snake Venoms. The role of enzymes in the toxicity of snake venoms is a debatable topic, but there is no doubt that the enzymes in venoms exert an adverse effect on snake bite victims. For example, the production of lysolecithins by phospholipase A results in hemolysis. Proteases are evidently responsible for considerable necrosis of tissue; moreover, some of them retard blood coagulation, while others accelerate it. Hyaluronidase probably promotes the spread of the venom through tissue, and it appears likely that venom acetylcholinesterase can affect nerve action.

Nevertheless, not all enzymes found in snake venoms have obvious function in poisoning, and not all of the toxic agents have been shown to possess enzymatic activity. Several protein components with pronounced toxic action have been identified, and some have been purified. Among these are neurotoxin, cardiotoxin and respiratory toxin from cobra venom, and crotamine and croactin from a species of South American rattlesnake. Croactin constitutes one of the two peptide chains of crotoxin, a protein isolated by Slotta and Fraenkel-Conrat, and shown by exacting criteria to be homogeneous. The other peptide chain of crotoxin is a phospholipase A; thus, there occurs a complex protein composed of an enzyme and a toxin.[2]

It appears reasonable to expect that the newer procedures for protein fractionation now available will yield increased information concerning the enzymatic and toxic components of snake venoms.

References

1. Buckley, E. E., and Porges, N. (Editors), "Venoms," Publication No. 44, American Association for the Advancement of Science, 1956.
2. Slotta, K. H., in "The Enzymes" (Boyer, P. D., Lardy, H., and Myrbäck, K., Editors), Second edition, Vol. 4, Part A, p. 551, New York, Academic Press, 1960.
3. Zeller, E. A., in "The Enzymes" (Sumner, J. B., and Myrbäck, K., Editors), Vol. 1, p. 986, New York, Academic Press, 1950.
4. Wellner, D., and Meister, A., *J. Biol. Chem.*, **235**, 2013 (1960).
5. Meyer, K., Hoffman, P., and Linker, A., in "The Enzymes" (Boyer, P. D., Lardy, H., and Myrbäck, K., Editors), Second edition, Vol. 4, Part A, p. 447, New York, Academic Press, 1960.
6. Murata, Y., Satake, M., and Suzuki, T., *J. Biochem.*, **53**, 431 (1963).
7. Pfleiderer, G., and Sumyk, G., *Biochim. Biophys. Acta*, **51**, 482 (1961).
8. Bjork, W., *J. Biol. Chem.*, **238**, 2487 (1963).
9. Georgatsos, J. G., and Laskowski, Sr., M., *Biochemistry* **1**, 288 (1962).
10. Sulkowski, E., Bjork, W., and Laskowski, Sr., M., *J. Biol. Chem.*, **238**, 2477 (1963).
11. Dixon, M., and Webb, E. C., "Enzymes," Second edition, p. 744, New York, Academic Press, 1964.

J. M. Prescott

SODIUM (IN BIOLOGICAL SYSTEMS). See POTASSIUM AND SODIUM (IN BIOLOGICAL SYSTEMS); also ACTIVE TRANSPORT; ELECTROLYTE AND WATER REGULATION; MINERAL REQUIREMENTS (VERTEBRATES).

SOLUBLE RIBONUCLEIC ACIDS

The soluble ribonucleic acids (sRNA) are the smallest of the nucleic acids for which a biological function is known. sRNA's combine with activated

Di Di Di
Me H H Me Me
pG-G-G-C-G-U-G-U-G-G-C-G-C-G-U-A-G-U-C-G-G-U-A-G-C-G-C-G-C-U-C-C-C-U-U-I-G-C-I-Ψ-G-G-G-A-G-A-G-U-*C-U-C-C-G-G-T-Ψ-C-G-A-U-U-C-C-G-G-A-C-U-C-G-U-C-C-A-C-C-AOH

FIG. 1. Nucleotide sequence of an alanine sRNA from yeast. A-, G-, C-, and U- represent the major nucleotides adenosine-, guanosine-, cytidine- and uridine-3′-phosphates. Minor nucleotides are DiHU-, 5,6-dihydrouridine-3′-phosphate; Di MeG-, N^2-dimethylguanosine-3′-phosphate; MeG-, 1-methylguanosine-3′-phosphate; I-, inosine-3′-phosphate; MeI-, 1-methylinosine-3′-phosphate; T-, ribothymidine (5-methyluridine) 3′-phosphate; ψ-, 5-ribosyluracil (pseudouridine) -3′-phosphate. The "p" at the left represents a phosphate group attached to the terminal G-. The "OH" at the right shows that the adenosine is not phosphorylated in the 2 or 3 carbon of ribose.

amino acids and act as adaptors for carrying the amino acids to RIBOSOMES, the site of protein synthesis [see PROTEINS (BIOSYNTHESIS)]. The sRNA's interact there with another RNA—"messenger RNA" (mRNA)—through complementary base triplets, and position themselves in a specific sequence on the surface of the ribosome. The amino acids are then polymerized into peptides in the sequence specified by the mRNA, and the sRNA's return to the cytoplasm to be recharged with amino acids. sRNA as isolated may contain inactive material. The fraction of sRNA that is involved in amino acid transport is often referred to as "transfer RNA" (tRNA).

Distribution. sRNA's are distributed mainly in the cytoplasm of the cell, although small amounts may be found associated with cellular organelles. The name "soluble ribonucleic acid" originally arose from the observation that a ribonucleic acid fraction remained soluble in the supernatant after high-speed centrifugation of broken cell suspensions had precipitated organelles and the bulk of the cellular RNA. In the analytical ULTRACENTRIFUGE, sRNA exhibits a sedimentation constant of about 4 Svedberg units. It constitutes, depending upon the source, 5–20% of the total cellular RNA.

Structure. Preparations of sRNA are obtained by differential centrifugation, phenol extraction or by various chromatographic procedures. Mixtures of the various amino acid transfer RNA's are usually isolated. Preparations of sRNA, whatever the source, appear to be of approximately the same size, with a molecular weight of about 25,000, and 70–80 nucleotides in length per single-stranded chain.

sRNA is characterized by a relatively high content of "minor" nucleotides. The most common of these is 5-ribosyluracil phosphate (pseudouridine phosphate). The structure of pseudouridine is shown in the article PYRIMIDINES. In addition, numerous base-methylated derivatives of the major nucleotides have been identified in sRNA. Methylation of the bases occurs after synthesis of the sRNA, and is carried out by RNA methylases, a group of enzymes which are specific both for nucleic acid, and purine or pyrimidine base [see RIBONUCLEIC ACIDS (BIOSYNTHESIS), section on Methylation]. Other minor nucleotides are present in lesser amounts. The functional significance of the minor constituents is not presently known.

A recent noteworthy achievement is the purification of alanine sRNA from yeast and the determination of its primary structure. This is shown in Fig. 1. On the right is the —C—C—A amino acid acceptor end. Other sites which are recognized as being present in this and other sRNA's are (a) the aminoacyl synthetase site, which recognizes a specific activated amino acid–enzyme complex (see ACTIVATING ENZYMES), and (b) the anticodon, or codon-recognition site, which is specific for a particular charged sRNA. This latter site recognizes coding triplets of mRNA, probably by means of complementary bases. A ribosome binding site, by which the sRNA interacts with the ribosome, may also be present. The locations of these sites in sRNA (other than the amino acid acceptor site) are uncertain.

Conformation. sRNA possesses considerable secondary structure, and the various site activities are probably dependent upon appropriate conformation of the chain. There is good evidence that a strand of sRNA doubles back upon itself and forms helical double-stranded regions in its structure by complementary base pairing of the type seen in DNA. Possible conformations of the sRNA of Fig. 1 are shown in Fig. 2; parallel double strands represent helical regions of base pairing. It should be emphasized that these schematic representations are speculative. The structures of additional sRNA's, when known, may provide evidence for the location of various active sites and may show common features of structure which will indicate their three-dimensional shapes.

Mechanism of Action. With the structure of a sRNA before us, let us now consider the means by which an amino acid is activated and eventually incorporated into the peptide chain. The initial step is a reaction involving the amino acid, ATP and an enzyme known as an aminoacyl-RNA synthetase, which is specific for the amino acid:

$$\mathrm{R{-}\underset{\underset{NH_2}{|}}{\overset{\overset{H}{|}}{C}}{-}COOH} + \mathrm{ATP} \xrightarrow{\text{Enzyme}} \left[\mathrm{R{-}\underset{\underset{NH_2}{|}}{\overset{\overset{H}{|}}{C}}{-}\overset{\overset{O}{\|}}{C}{-}O{-}\underset{\underset{OH}{|}}{\overset{\overset{O}{\|}}{P}}{-}O{-}adenosine}\right]\text{—enzyme} + \text{Inorganic Pyrophosphate} \qquad (1)$$

A high-energy, mixed anhydride is produced, and forms a strong complex with the enzyme. An aminoacyl-RNA synthetase (ACTIVATING ENZYME) exists for each of the twenty or so amino acids. The complex then interacts through the recognition site on sRNA with a sRNA specific for this complex. At least one sRNA exists for each of the twenty amino acids. The reaction may be formulated as follows:

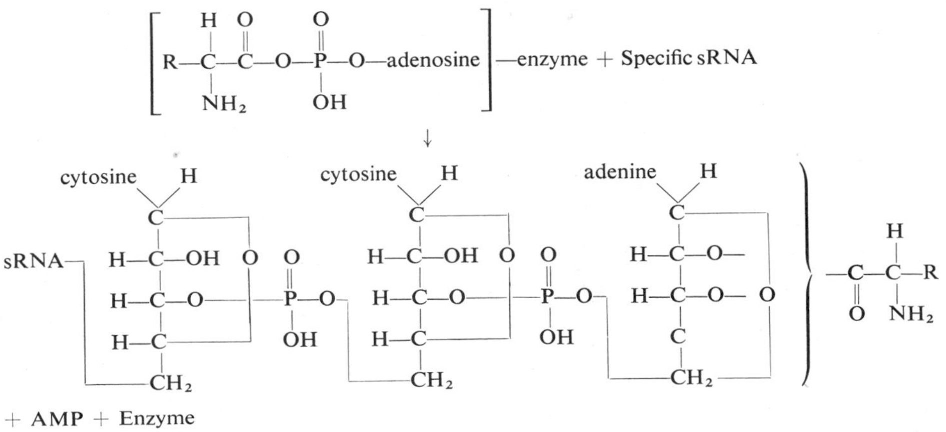

$$\left[R{-}\underset{NH_2}{\overset{H}{C}}{-}\overset{O}{\overset{\|}{C}}{-}O{-}\underset{OH}{\overset{O}{\overset{\|}{P}}}{-}O{-}\text{adenosine} \right]{-}\text{enzyme} + \text{Specific sRNA}$$

$$\downarrow \qquad (2)$$

$$\text{sRNA}{-}[\text{cytosine ribose}]{-}O{-}\underset{OH}{\overset{O}{\overset{\|}{P}}}{-}O{-}[\text{cytosine ribose}]{-}O{-}\underset{OH}{\overset{O}{\overset{\|}{P}}}{-}O{-}[\text{adenine ribose}]{-}O{-}\overset{O}{\overset{\|}{C}}{-}\underset{NH_2}{\overset{H}{C}}{-}R$$

$$+ \text{AMP} + \text{Enzyme}$$

The terminal trinucleotide sequence

(—C—C—A)

of the amino acid acceptor end, shown in detail above, appears to be common to all transfer RNA's. The amino acid is attached to the terminal adenosine (amino acid acceptor site) via an ester bond involving the 2′- or 3′-hydroxyl group of the terminal adenosine. This linkage is a high-energy bond, and the amino acid remains in an activated form. Hydrolysis of this ester bond can provide sufficient energy for the formation of the peptide bond.

Peptide Bond Formation. The placement of amino acids in proper sequence in the peptide chain is the result of interaction between the sRNA adaptors and the mRNA with its coded message. The amino acid itself plays no part in its specific, accurate placement in the peptide. After attachment of an amino acid to sRNA, the amino acid side chain can, in fact, be chemically modified, and the modified amino acid will then be incorporated into the peptide in the same position as the unmodified amino acid.

mRNA is held on the surface of the ribosome and may bind several ribosomes in a polyribosome ("polysome") structure. The sRNA's move to the ribosome and align themselves via their coding sites to matching triplets on the mRNA. Each ribosome binds two sRNA molecules and is involved in forming a single peptide chain. Enzymes known as transferases (see TRANSFER FACTORS IN PROTEIN SYNTHESIS) effect the cleavage of the ester bond between the sRNA and the amino acid, and the simultaneous formation of a peptide bond between the amino acid and the growing peptide chain. This is shown in Fig. 3.

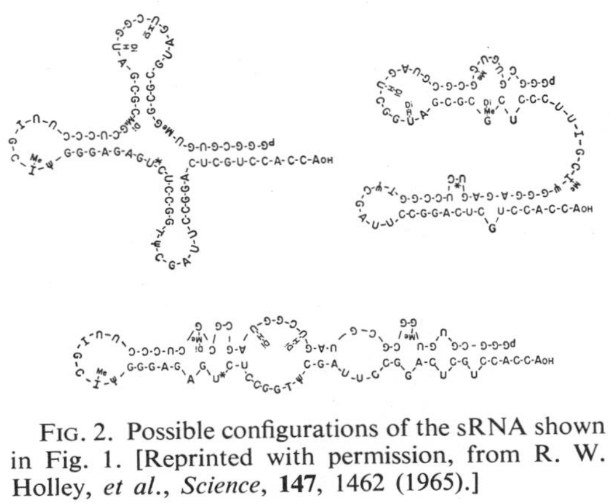

$$H_3N^+{-}[AA_1]{-}[AA_2]{-}[AA_3]{-}\overset{O}{\overset{\|}{C}}{-}O{-}\text{sRNA}_a \;+\; H_3N^+{-}[AA_4]{-}\overset{O}{\overset{\|}{C}}{-}O{-}\text{sRNA}_b$$

$$H_3N^+{-}[AA_1]{-}[AA_2]{-}[AA_3]{-}[AA_4]{-}\overset{O}{\overset{\|}{C}}{-}O{-}\text{sRNA}_b + \text{sRNA}_a$$

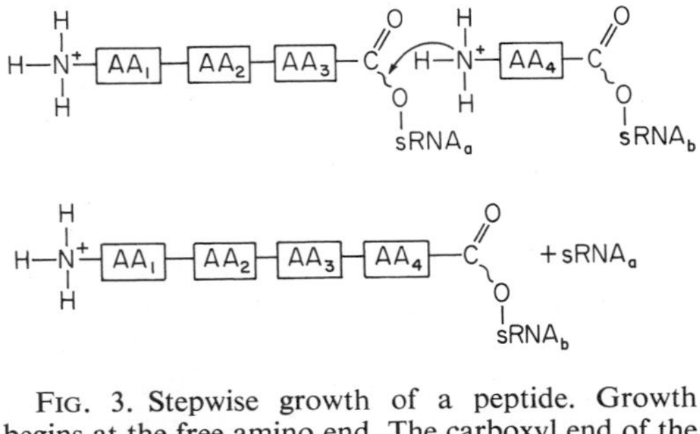

FIG. 3. Stepwise growth of a peptide. Growth begins at the free amino end. The carboxyl end of the incomplete chain terminates in an sRNA. [Reprinted with permission, from J. D. Watson, "Molecular Biology of the Gene," New York, W. A. Benjamin, 1965.]

FIG. 2. Possible configurations of the sRNA shown in Fig. 1. [Reprinted with permission, from R. W. Holley, *et al.*, *Science*, **147**, 1462 (1965).]

The sRNA then returns to the cytoplasm to accept another amino acid, and the mRNA moves over the ribosome surface so as to bring the next coding triplet in the sequence into position. Guanosine triphosphate is required for peptide synthesis, though its mode of action is not clearly established.

Species Specificity of sRNA. The GENETIC CODE appears to be largely universal. This suggests that sRNA from one source might be employed with mRNA, ribosomes and transferases from another source to carry out *in vitro* protein synthesis. This has been done. A cell-free system containing bacterial sRNA charged with amino acids, GTP, mRNA, ribosomes and enzymes from reticulocytes has been shown to synthesize hemoglobin peptides. Hemoglobin is the protein commonly synthesized by these reticulocytes and is not produced in bacteria. The ability to substitute sRNA's in this manner indicates that major structural similarities exist among the sRNA's of various species.

Code Degeneracy and Multiple Forms of sRNA. Many of the amino acids are coded for by more than one base triplet. This is termed "degeneracy" of the code. Leucine, for example, is coded for by four different triplets. Multiple forms of leucine sRNA have been isolated, and specificity for specific codons has been shown. Degeneracy of the code and multiple forms of a given sRNA may be related to regulatory mechanisms of the cell.

References

1. WATSON, J. D., "The Molecular Biology of the Gene," New York, W. A. Benjamin, 1965.
2. INGRAM, V. M., "The Biosynthesis of Macromolecules," New York, W. A. Benjamin, 1965.
3. STEINER, R. F., "The Chemical Foundations of Molecular Biology, Princeton, N.J., D. Van Nostrand Co. 1965.
4. HOAGLAND, M. B., STEPHENSON, M. L. SCOTT, J. F., HECHT, L. I., AND ZAMECNIK, P. C., "A Soluble Ribonucleic Acid Intermediate in Protein Synthesis," *J. Biol. Chem.*, **231**, 241 (1958).
5. HOLLEY, R. W., APGAR, J., EVERETT, G. A., MADISON, J. T., MARQUISEE, M., MERRILL, S. H., PENSWICK, J. R., AND ZAMIR, A., "Structure of a Ribonucleic Acid," *Science*, **147**, 1462 (1965).
6. HOLLEY, R. W., "The Nucleotide Sequence of a Nucleic Acid," *Sci. Am.*, **214**, 30 (February 1966).

FRANK F. DAVIS

SPECTROPHOTOMETRY AND SPECTROSCOPY.

See FLAME PHOTOMETRY AND ATOMIC ABSORPTION SPECTROSCOPY; INFRARED AND ULTRAVIOLET SPECTROSCOPY.

SPHINGOLIPIDS AND SPHINGOSINE

Sphingolipids are LIPIDS that contain sphingosine, $CH_3(CH_2)_{12}CH{=}CH{-}CHOH{-}CHNH_2{-}CH_2OH$, or a closely related base such as dihydrosphingosine. A major class of sphingolipids are the *sphingomyelins*, in which the nitrogen of sphingosine is joined through an amide linkage to a fatty acid residue, and the terminal hydroxyl group of sphingosine is esterified to the phosphoryl portion of phosphorylcholine (for structure of a sphingomyelin, see LIPIDS, section on Phosphatides.) The sphingomyelins are thus a subclass of the PHOSPHOLIPIDS. Another class of sphingolipids, containing no phosphoric acid, includes the CEREBROSIDES, in which the terminal hydroxyl group of the fatty acyl sphingosine is glycosidically linked to a sugar, and the glycosphingolipids, e.g., GANGLIOSIDES, in which a polysaccharide is bound to the fatty acyl sphingosine. The N-fatty acyl sphingosine portion of the molecule is sometimes termed the ceramide portion; thus, the gangliosides and certain other glycosphingolipids may be termed ceramide polysaccharides, or ceramide oligosaccharides.

Sphingolipids are found primarily in brain and central nervous system tissues (see BRAIN COMPOSITION), but they occur also in other animal tissues and in plants (see also MYELIN; NERVE CELL COMPOSITION AND STRUCTURE).

References

1. CELMER, W. D., AND CARTER, H. E., "Chemistry of Phosphatides and Cerebrosides," *Physiol. Rev.* **32**, 167 (1952).
2. KLENK, E., AND DEBUCH, H., "The Lipides," *Ann. Rev. Biochem.*, **28**, 39 (1959).
3. CARTER, H. E., JOHNSON, P., AND WEBER, E. J., "Glycolipids," *Ann. Rev. Biochem.*, **34**, 109 (1965).

SPIES, T. D.

Tom Douglas Spies (1902–1960) was an American physician who was a pioneer in the clinical use of vitamins in the treatment of deficiency diseases.

SPORES, BACTERIAL

John Tyndall was the first to detect the existence of unique microscopic forms of life resistant to boiling. In 1871 Ferdinand Cohn, a German botanist, correctly described these entities as bacterial forms which are morphologically distinguishable and possess remarkable powers of resistance. This resistance character of bacterial spores provides the basis for most present-day STERILIZATION and food preservation processes.

Spores are formed by all members of the *Bacillus* (aerobic) and *Clostridium* (anaerobic) and by certain *Sporosarcina* and *Spirillum*. Medically important species include *Bacillus anthracis*, the causative agent of anthrax; *Clostridium tetani*, the causative agent of tetanus; *Clostridium botulinum*, which produces the most potent toxin known. Some assume great importance as food spoilage organisms, and others such as *Bacillus thuringiensis* are pathogenic for insects and are employed as bacterial insecticides. Most, however, are of little known practical importance.

The Dormant Spore—A Resting Cell. Spores are formed within vegetative cells, hence their designation as endospores. Spore-forming vegetative cells are called sporangia. Though spores may be formed at various sites within the cell, the intracellular location is constant for a given species.

ELECTRON MICROSCOPE studies of thin sections reveal that the bacterial spore has a complex anatomy which differs markedly from that of the progenitor vegetative cell. Spores are characterized by one or more outer envelopes or "coats." Directly underlying these is the "cortex," a complex laminated enveloping structure of prominent size as viewed in thin sections but probably highly compacted in the native state. The central body, often oval and commonly referred to as the "core," is bounded by a limiting membrane which appears to be in intimate association with the innermost surface of the enveloping cortex.

In a number of respects, spores bear little resemblance chemically to vegetative cells. The outermost coats are predominantly proteinaceous, and the high cystine content of spores appears to be localized in these structures.[1] Recent evidence[2] suggests that the cortex is the site of the spore mucopeptide which, though its composition may vary somewhat between species, usually contains glucosamine, muramic acid, alanine, D-glutamic acid, and α,ε-diaminopimelic acid as primary constituents. The metal ion content of spores is higher than that of vegetative cells. This is particularly true of calcium[3] which may exceed 2% of the spore dry weight. The most unusual spore constituent is dipicolinic acid.[4] This compound (pyridine-2,6-dicarboxylic acid) is present apparently in all bacterial spores and may account for as much as 5–15% of the spore solids. Whether it exists in spores as the free acid or as the calcium chelate is not known; neither has its cellular localization or its functional role been clarified.

The resting spore, viewed under the light microscope, is a refractile cell which does not stain with bacteriological dyes at ordinary temperatures. Available evidence indicates that it is a highly condensed body with a low water content. This relative anhydrous state, in combination with the high coat cystine content, may account in large measure for its extraordinary resistance to heat, desiccation, ionizing radiation and bactericidal agents. Although dormant spores have been shown, upon rupture, to contain a normal complement of enzymes, the intact spore is metabolically quite inert. Recent studies indicate a Q_{O_2} as low as 0.001, a value approximately 1/10,000 that of corresponding vegetative cells. This is compatible with the remarkable longevity of dormant spores which are known to remain viable 70 years or more and probably survive several hundred years in appropriate environments.

Germination—The Breaking of Dormancy. Marked cellular changes occur when the dormant spore is induced to germinate. The cell loses refractility, the optical density of suspensions is reduced greatly, the cell becomes readily stainable, it swells somewhat due probably to imbibition of water, its resistance to heat and other deleterious environments vanishes almost instantly, and it resumes normal metabolic activity.

Chemical alterations and structural changes accompany these gross manifestations of germination. The spore loses 30% or more of its solids in the form of a soluble germination exudate which contains, as major constituents, approximately 90% of the spore calcium, all of the dipicolinic acid, and spore mucopeptide in amounts governed by the germination conditions. In addition, smaller quantities of various inorganic materials, small peptides and some free amino acids may be found in the exudate. The calcium of the exudate is almost invariably detected in association with the dipicolinic acid as a chelate complex. Associated with this discharge of spore materials, the coats become flaccid or may in other cases rupture. In either event, the cortical mucopeptide is solubilized, and a complete discharge or emergence of the central body from its enveloping coat structures ensues.

For convenience, the germination process is often divided into two sequential steps: (1) the period of primary events or activation, and (2) the secondary outgrowth phase of synthetic activity which ultimately leads to a vegetative cell and division.

In the course of the primary events of germination, the chemical and structural changes described earlier, which collectively characterize the event, transpire. This is a period of breakdown of existing spore structures. It may be induced in various ways and by a variety of chemicals. Prior exposure of spores to elevated nonlethal temperatures[5] predisposes many types for rapid germination in appropriate chemical environments. Complex biological materials which contain appropriate ions, amino acids and ribosides are often effective. Germinative compounds include L-alanine,[6] inosine,[7] glucose, potassium nitrate[8] and other inorganic and organic salts, and surfactants.[9] Since the primary germinative events may occur (a) in the presence of various enzyme inhibitors, (b) as a consequence of exposure of spores to surfactants at markedly elevated temperatures, (c) in the absence of an external energy source, and (d) as a consequence of mechanical abrasion under conditions designed to preclude enzymatic activity,[10] there is good evidence that these initial germinative changes are not a consequence of the synthetic capacities of the cell.

In contrast, the secondary period of outgrowth is characterized by active metabolism and a marked synthetic capacity which leads to the formation of new proteins, nucleic acids, cell wall structures (see BACTERIAL CELL WALLS) and, ultimately, the new vegetative cell. The nutritional requirements for outgrowth do not appear to differ markedly from the requirements for ordinary vegetative growth.

Sporogenesis—Formation of the Spore. Bacterial spores are formed within genetically competent mother cells, but vegetative growth of such cells does not lead invariably to spore formation; sporulation, like germination, is affected greatly by environmental factors, and so long as the proper nutritional conditions exist vegetative growth will continue. Induction of sporulation is not well understood, but the process may be initiated by competent cells in environments depleted of the nutritional requirements for continued growth; indeed it appears that sporangial cells, once irreversibly committed to sporulation, may complete the process in the absence of external nutrients.[11]

Cytological evidence[12] indicates that sporogenesis commences with the enclosure of a portion of the nuclear material of the mother cell and a small portion of the cytoplasm by a thin septum growing from a site near the inside of the cell wall. The spore body so formed undergoes further morphogenesis and maturation within the mother cell; the cortical structure is formed peripheral to the septum of the spore body and the outermost coat layers then appear. Coincident with these cytological events, a marked uptake of calcium occurs, dipicolinic acid and spore mucopeptide are synthesized, the cystine-rich proteins are formed, and the features which collectively characterize the dormant spore rapidly emerge. Finally, the mature spore is liberated from its enclosure within the mother cell to its own independent existence as a free spore.

Whereas formation of a spore does not provide reproductive advantage since each vegetative cell produces only one spore, survival advantage in certain environments is acknowledged, but other organisms survive in nature without spores.

References

1. VINTER, V., *Nature (London)*, **183**, 998 (1959).
2. WARTH, A. D., OHYE, D. F., AND MURRELL, W. G., *J. Cell Biol.* **16**, 593 (1963).
3. CURRAN, H. R., BRUNSTETTER, B. C., AND MYERS, A. T., *J. Bacteriol.*, **45**, 484 (1943).
4. POWELL, J. F., *Biochem. J.*, **54**, 210 (1953).
5. EVANS, F. R., AND CURRAN, H. R., *J. Bacteriol.*, **46**, 513 (1943).
6. HILLS, G. M., *Biochem. J.*, **45**, 353 (1949).
7. POWELL, J. F., *J. Appl. Bacteriol.*, **20**, 349 (1957).
8. HYATT, M. T., AND LEVINSON, H. S., *J. Bacteriol.*, **81**, 204 (1961).
9. RODE, L. J., AND FOSTER, J. W., *J. Bacteriol.*, **81**, 768 (1961).
10. RODE, L. J., AND FOSTER, J. W., *Proc. Natl. Acad. Sci. (U.S.)*, **46**, 118 (1960).
11. HARDWICK, W. A., AND FOSTER, J. W., *J. Gen. Physiol.*, **35**, 907 (1952).
12. YOUNG, I. E., AND FITZ-JAMES, P. C., *J. Biophys. Biochem. Cytol.*, **6**, 467 (1959).

L. J. RODE

STANLEY, W. M.

Wendell Meredith Stanley (1904–), an American biochemist, made the remarkable discovery that a virus (TOBACCO MOSAIC VIRUS) could be crystallized in fine needles and was thus a chemical substance rather than a living entity. For this accomplishment, which has been followed by many more, he was awarded the Nobel Prize in chemistry in 1946, along with JAMES BATCHELLER SUMNER and JOHN HOWARD NORTHRUP.

STARCHES

Starches are widely distributed throughout the plant world as reserve polysaccharides, and are stored principally in seeds, fruits, tubers, roots and stem pith [see CABOHYDRATES (ACCUMULATION)]. They usually occur as discrete particles or granules of 2–150 μ in diameter. The physical appearance and properties of granules vary widely from one plant to another, and may be used to classify starches as to origin. Some are round, some elliptical and some polygonal. Many have a spot, termed the hilum, which is the intersection of two or more lines or creases. The granules are anisotropic and show strong birefringence with two dark extinction lines extending from edge to edge of the granule with their intersection at the hilum. Some granules show a series of "striations" arranged concentrically around the hilum.

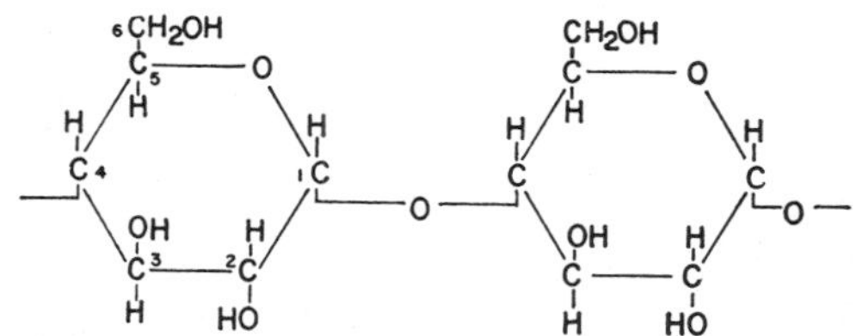

FIG. 1. Basic repeating unit of amylose and amylopectin.

Although starches are hydrolized only to D-glucose, they are not single substances but, except in very rare instances, are mixtures of two structurally different glucans. One, termed amylose, is a linear chain (Fig. 2, structure A) of D-glucose units joined by $\alpha(1 \rightarrow 4)$ links, and the other, known as amylopectin, is a bush-shaped structure (Fig. 2, structure B) of $(1 \rightarrow 4)$ linked α-D-glucose units with $\alpha(1 \rightarrow 6)$ links at the branch points which occur about every 25–27 sugar units. Most starches contain 22–26% amylose and 74–78% amylopectin. Amylose may be selectively precipitated from a hot starch dispersion by the addition of substances such as butanol, fatty acids, various phenols and nitrocompounds such as nitropropane and nitrobenzene. On cooling slowly, the amylose combines with the fractionating agent to form a complex which separates as microscopic crystals. Disruption of the complexes occurs when they are dissolved in hot water or extracted with ethanol.

Structure A Structure B

FIG. 2.

Corn starch is the most important of the starches manufactured in the United States. Approximately 120 million bushels of corn are processed annually. Waxy corn starch is made from waxy corn. Most waxy starches are entirely free of amylose, and therefore consist solely of branched, amylopectin molecules.

Starch may be prepared from white potatoes by either batch or continuous processes. Commercial wheat starch is usually prepared by the Martin process. Starches from other sources (cassava, tapioca, etc.) are prepared in a manner similar to that for corn or potato starch.

Several different crystalline forms of starches may be obtained. Native cereal starches occur as the "A" form and tuber starches as the "B" form. Intermediates are designated as "C" types. In general, the B type is obtained by evaporation of pastes at room temperature or by precipitation by freezing or retrogradation, the C type at higher temperatures, and the A type at 80–90°C. At higher temperatures, amorphous starch is usually obtained. Precipitation of starch pastes with alcohols or some other precipitating agents gives a "V" type of starch with a more symmetrical arrangement of molecules.

One of the most important properties of starch granules is their behavior on heating with water. On heating, water is at first slowly and reversibly taken up and limited swelling occurs, although there are no perceptible changes in viscosity and birefringence. At a temperature characteristic for the type of starch, the granules undergo an irreversible rapid swelling and lose their birefringence, and the viscosity of the suspension increases rapidly. Finally, at higher temperatures starch diffuses from some granules and others are ruptured leaving formless sacs. Swelling may be induced at room temperature by numerous chemicals such as formamide, formic acid, chloral, strong bases and metallic salts. Magnesium sulfate impedes gelatinization.

When aqueous starch solutions are allowed to stand under aseptic conditions, they become opalescent and finally undergo precipitation, known as retrogradation, to give a starch with a "B" X-ray pattern. Amylose molecules retrograde more readily than amylopectin molecules. In certain cases, such as white potato starch, there is an initial preferential precipitation of amylose. Generally, retrogradation is increased as the starch concentration is increased, and as the temperature is decreased toward 0°C.

Starch solutions have a high positive optical rotation because of the presence of α-D-glucosidic linkages. The values range from +180° to +220° for aqueous alkaline solutions. Osmotic pressure measurements on acetylated amyloses and amylopectins from several starches give values for which the molecular weights of the unacetylated components have been calculated to be 100,000–210,000 for amyloses and 1,000,000–6,000,000 for amylopectins. At present there is little information on the shape of the molecules although amylose molecules have been shown to be very flexible.

Starches behave as polyhydroxy alcohols and, in the presence of an impelling agent, are capable of ether formation with alkyl and acyl halides and alkyl sulfates, and of ester formation with both inorganic and organic acids. For the preparation of acyl derivatives, the impelling agent such as pyridine or an acid anhydride is necessary to absorb the water formed during acylation. Alkalies, especially sodium hydroxide, are normally used as impelling agents for etherification. The fully methylated ethers of amylose and amylopectin have been utilized extensively in structural determinations.

The enzymes which hydrolyze starch may be classified as: (1) liquefying (dextrogenic) or α-amylases, (2) saccharogenic or β-amylases, and (3) phosphorylases. α-Amylases rapidly decrease the solution vicosity and cause rapid and extensive degradation. The main products of degradation are malto-dextrins of about six D-glucose units. Malt α-amylase hydrolyzes starches to fermentable sugars in about 90% of the theoretical yield. The stage at which the solution is no longer colored by iodine is known as the *achroic point* and the reducing value calculated as maltose is termed the *achroic R-value*. In contrast to α-amylases, β-amylases cause only a slow change in the viscosity of starch solutions. With whole starches, the hydrolysis proceeds rapidly until about 50–53% of the theoretical amount of maltose is produced, and then very slowly until a limit of about 61–68% is reached. The unhydrolyzed residue is called a β-amylase limit dextrin and may be likened to a pruned amylopectin molecule. Phosphorylases such as P-enzyme bring about the reversible hydrolysis of starch to α-D-glucose-1-(dihydrogen phosphate) (see GLYCOGEN).

Starches are hydrolyzed for the manufacture of syrups and D-glucose. In the manufacture of these products, a starch-water slurry is treated at 30–45 pounds of steam pressure with hydrochloric acid at pH 1.5–2.0. For the preparation of syrups, the hydrolysis is stopped at 42% conversion of starch to D-glucose. About one-third of the total corn starch and 95% of the total corn syrup is sold for food purposes.

Roy L. Whistler and William M. Corbett

STAUDINGER, H.

Hermann Staudinger (1881–) is a German chemist whose work with high polymers is largely extraneous to biochemistry yet it is rooted in, and impinges on, the chemistry of polysaccharides and proteins. He received the Nobel Prize in chemistry in 1953.

STERILIZATION

This term has two related common usages.

In its broadest sense it is applied to higher forms of life and denotes treatments which deprive individual living organisms of the ability to produce offspring. It can be achieved by a wide variety of methods such as surgery or treatment with various radiations or chemicals.

A more restricted sense of the term is in wider use however, and this will be discussed exclusively in the following text. It denotes the process of treating a defined system, such as a solid article or a container filled with liquid or gas, in a manner which effectively destroys or removes all living organisms; the organisms concerned are usually the microorganisms, bacteria, yeasts and molds. It must be emphasized that this is an exact meaning, namely that the system which has been sterilized contains *no* living microorganisms. A degree of treatment which can be shown to sterilize one sample, however, will not necessarily sterilize all samples of a large batch, since there is a statistical probability of survival of the last organism in any one sample. Increasing the severity of treatment will increase the certainty that all samples of the batch are sterile, and in selecting a process for use in practice it is necessary to define the degree of certainty required for the particular end use concerned. The canning industry, for example, requires an extremely high degree of certainty that every processed container will be free from viable pathogenic organisms capable of growth in the product.

The term sterilization is also sometimes used loosely in circumstances where although the treatment is insufficient to destroy all microorganisms present, those which survive are of types which can be tolerated for particular applications. For example, "sterilization" of equipment in boiling water for short periods will not inactivate heat-resistant organisms; this process is, however, adequate for many purposes since the numbers of organisms remaining will often be so small as not to constitute a significant hazard. In the canning industry, a term "commercial sterilization" is applied to processes which inactivate all organisms which are able to grow in the treated product. Use of these loose terms will not be discussed further.

Physical Methods of Sterilization. *Heat treatment* is the most common form of sterilization. Yeasts, molds and vegetative bacteria are inactivated fairly readily; exposure to a temperature of 80°C for 10 minutes will kill most of them. Bacterial spores, however, can be exceptionally resistant and may require temperatures well above the boiling point of water for inactivation in practical, convenient times. The heat is usually applied by a hot gas or liquid, but direct flaming can be used, or indirect methods such as radiofrequency heating.

Moist heat is much more lethal than dry heat, and processing in steam under pressure in an autoclave is used extensively. The precise treatment may vary according to the chemical nature of the sample to be sterilized and the types of microorganisms present, but the aim is usually to heat the organisms for 20–30 minutes at 121°C (15 pounds steam pressure), after allowing for delay in reaching the sterilizing temperature at all points in the sample. Penetration is much slower in solid products than in liquids where convection is possible, and the necessary correction factor is usually obtained by direct determination of the rate of heat penetration. Longer exposures to temperatures lower than 120°C may be used with thermally unstable products, or alternatively temperatures up to 150°C have been used with products which can be heated and cooled rapidly. Treatment times of the order of seconds are adequate at the highest temperatures and can produce a given lethal action with relatively little chemical damage to the product. This process is used commercially for a number of fluid foods such as milk; the treated product which shows minimal heat damage is packed aseptically into a previously sterilized can, bottle or plastic container.

Dry heat is also used for sterilization, but higher temperatures are required. A typical process is 1–2 hours at 160°C, but again the precise figures depend on the product concerned. Inactivation is believed to be due to oxidative processes, in contrast with the denaturing effect of moist heat.

Much interest has been aroused in recent years in a completely different physical method of sterilization, namely, by using the lethal action of *ionizing radiations* such as gamma rays, cathode rays, X rays and beta rays. Their main practical advantages are that they cause little rise in temperature and freely penetrate most materials (within limits which can be closely calculated). Gamma rays for example can penetrate many inches of products. Most microorganisms, including bacterial spores, can be inactivated by doses of the order of 3 Mrads, but in food products where there is a danger of the growth of *Clostridium botulinum*, a dose of approximately 5 Mrads is usually regarded as necessary to obtain an adequate degree of certainty that no spore will survive. (The *rad* is the unit of dose and corresponds to absorbing radiation energy equivalent to 100 ergs/g). Large electrical generators or radioactive sources (usually cobalt 60) are needed for treating appreciable quantities of product and must be used with proper safety precautions, but the method is economically acceptable for many products and is gaining increasing acceptance particularly in the medical field; heat-sensitive products, such as plastic equipment, sterilized in sealed containers are especially suitable. Application is limited in some cases by destructive action of the radiation on chemical constituents of the product itself, and the treatment of each product must be studied directly.

Other physical procedures include the use of *ultraviolet* and *ultrasonic radiations*. Although these are highly lethal to microorganisms under suitable conditions, their effectiveness in practice is often limited. The penetration of ULTRAVIOLET RADIATION is much reduced by the presence of organic matter, and this radiation is best used for the treatment of relatively clean gases or liquids, or smooth surfaces. The effectiveness of ULTRASONIC VIBRATIONS depends on factors such as the viscosity of the medium, and the method has been little used in practice for full sterilization.

Liquid and gaseous products can also be sterilized by *filtration*, and this method is much used in the manufacture of heat-sensitive biochemicals. Porcelain, sintered glass, diatomaceous earth and asbestos pads are in common use, and pore size can be selected for removing microorganisms of specific sizes. The only real problem is to avoid clogging the fine pores of the filter with large inert particles; these are best removed by a prior stage of centrifugation or filtering through a coarse filter.

Chemical Methods. The most versatile chemical procedures utilize highly reactive gases. *Formaldehyde* has been employed for this purpose for many years at concentrations of 1–2 mg/liter of air and with exposure time of 1–2 hours. Penetration tends to be limited and the gas can polymerize in the treated article with subsequent difficulty of removal. *Ethylene oxide* is superior in both respects and its use is increasing. Owing to inflammability, it is often used in admixture with carbon dioxide (10% ethylene oxide) or fluorinated hydrocarbons, and in such form there is no explosion hazard. Penetration can be assisted by prior evacuation of the system, the gas then being left in contact with the product for several hours. Plastic packages constitute little barrier to the gas, and it is commercially practicable to sterilize the contents of sealed packages. *Propylene oxide* has similar application and is preferred where a less volatile agent is required. Other bactericidal gases, including *ozone*, find limited use for sterilization.

Many chemicals are available for inactivating microorganisms in the liquid state, and these are highly effective under favorable circumstances. In most practical conditions, however, their efficiency seldom approaches that of true sterilization since the presence of organic material tends to protect the organisms. Chemical agents which can be used include: strong acids and alkalies; phenols of various types; halogen-containing compounds (particularly chlorine in the form of hypochlorite); peroxides and other oxidizing agents (*e.g.*, permanganates); alcohols and aldehydes; surface active agents including quaternary ammonium compounds; heavy metal salts; various dyes. The activity and suitability of these compounds vary widely, but in general their action is more one of sanitization than full sterilization.

A recent interesting development has been to use active agents which subsequently decompose into harmless residues. Hydrogen peroxide is a well-established agent of this type, and a more recent example is diethyl pyrocarbonate which decomposes slowly in the presence of water into ethyl alcohol and carbon dioxide.

Combination Procedures. There is a considerable field for further exploration into the merits of combining two or more of the foregoing procedures. For example, heat treatment may be combined with any of the others. It is also possible to increase the lethal effect by simple manipulations or time sequences. For example, in the process known as Tyndallization, a heat treatment may be broken up into a succession of short treatments; during the intervening periods of time, bacterial SPORES will tend to germinate and become susceptible to subsequent heating. In this way, a given total heat treatment produces a greater total lethal effect than a single treatment. There is also particular interest in finding procedures for inducing the germination of bacterial spores since a successful procedure would permit sterilization at much lower levels of treatment than at present. No fully reliable method is available however.

Viruses. Increasing attention is being paid to the ability of sterilization procedures to inactivate viruses. These are sensitive to heat and to the other treatments discussed, but the necessary processing procedures have only been worked out in selected cases. Experimental techniques in this field tend to be complex, and in most cases, acceptable levels of residual infection must still be established.

References

1. REDDISH, G. F. (EDITOR), "Antiseptics, Disinfectants, Fungicides and Sterilization," London, Kimpton, 1957.
2. SYKES, G., "Disinfection and Sterilization," London, Spon, 1965.
3. RUBBO, S. D., AND GARDNER, J. F., "A Review of Sterilization and Disinfection," London, Lloyd-Luke, 1965.
4. WHITTET, T. D., HUGO, W. B., AND WILKINSON, G. R., "Sterilization and Disinfection," London, Heinemann, 1965.
5. GOLDBLITH, S. A., JOSLYN, M. A., AND NICKERSON, J. T. R., "Introduction to Thermal Processing of Foods," Westport, AVI, 1961.

R. S. HANNAN

STEROID GLYCOSIDES. See GLYCOSIDES, STEROID.

STEROIDS

Steroids are defined as isocyclic products of plant or animal origin with a tetracyclic hydroaromatic ring system (perhydrocyclopentanophenanthrene). It consists of one five-membered and three six-membered rings in angular annelation; it usually carries two angular methyl groups, C_{18} on C_{13}, and C_{19} on C_{10}, and a side chain of 2 to 11 carbon atoms on C_{17}. Hydroxyl and carbonyl groups are the most important functions occurring in the cyclic and acyclic portions of the molecule, while a carboxyl group in the side chain characterizes an important subdivision, the BILE ACIDS. Additional variety is caused by the presence in the cyclic and acyclic portions of double bonds which, in the estrogenic hormones, are accumulated in one or two rings, giving rise to benzene and naphthalene systems. An immense number of further isomeric possibilities are offered by the steric configurations on 4 to 10 asymmetric carbon atoms, but remarkably enough, only a small number of them are realized in natural products. This selectivity of nature may be responsible for the great stability of the steroid skeleton, which survives postmortally all other organic constituents of the animal body, *i.e.*, carbohydrates, proteins and fats; it is found as "adipocere" under certain conditions in skeletonized cadavers and has also been linked to the origin of mineral oil.

TABLE 1. STEROIDS

No. of C Atoms	Animal Steroids	Plant Steroids
29	—	Sterols of higher plants
28	—	Sterols of fungi
27	Cholesterol, other zoosterols; scymnol	Sapogenins; steroid alkaloids
24	Bile acids; toad poisons	Squill aglycones
23	—	Cardiac aglycones
21	Progestogens; adrenocortical hormones	—
19	Androgens	—
18	Estrogens	—

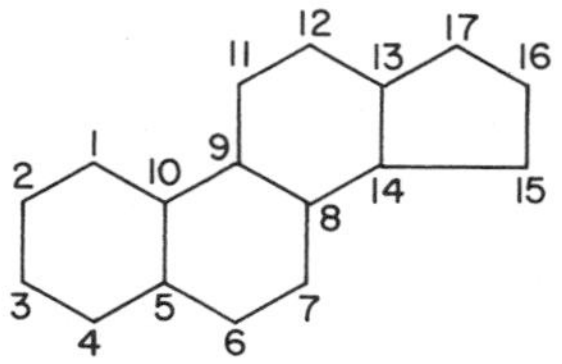

Table 1 shows animal and plant representatives of steroids arranged according to number of carbon atoms. One principal difference between these two series is the preponderance of tertiary hydroxyl groups in plant steroids, but of secondary hydroxyl groups in animal steroids. As steroids are mostly of hydrophobic character, nature endeavors to render them water soluble for various physiological functions by conjugation with hydrophilic groups. This is accomplished in the plant (see GLYCOSIDES, STEROID) by glycosidic linkages to hexoses and certain deoxyhexoses in the neutral saponins, the cardiac glycosides and certain steroid alkaloids, and in the animal world by conjugation with the amino acids glycine and taurine in the case of the BILE ACIDS, with sulfuric acid in scymnol in shark's bile, and with suberylarginine in toad poison.

The early history of steroid chemistry forms a fascinating chapter of organic chemistry. Mauthner, Windaus, Wieland, Jacobs, Diels and others accumulated observations on the products of oxidation, aromatization, and other reactions of cholesterol, bile acids and plant glycosides. The interrelationship between sterols and bile acids was soon recognized. Shortly after the steroid character of the female and male sex hormones had been established, X-ray crystallography demonstrated the impossibility of certain structural proposals by Windaus and Wieland. Rosenheim and King, on the basis of work with monomolecular layers and of the X-ray evidence of Bernal, formulated the cyclopentanophenanthrene system (first as an alternative to chrysene), which was quickly and generally accepted (see Fig. 1).

Once the basic skeleton of the steroids was established, one of the next tasks was to elucidate the steric relationships. Leaving the stereochemistry of the functional groups aside for the moment, one finds up to 9 asymmetric carbon atoms in the steroid skeleton: C_5, C_{10}, C_9, C_8, C_{14}, C_{13} in the ring system; in addition C_{17}, where the side chain is attached, C_{20} in the side chain, and C_{24} likewise in the side chain of the phytosterols. The relative configuration of C_5 and C_{10},

FIG. 1. Cyclopentanophenanthrene skeleton.

of C_9 and C_8, and of C_{14} and C_{13}, determines whether the junctions between rings A/B, B/C, and C/D, respectively, are *trans* or *cis*. Knowledge of the relationship of C_{10} to C_9, and of C_8 to C_{14}, often described by the alternative of "syn" or "anti," is essential for the description of the steric position of ring A in relation to ring C and of ring B in relation to ring D. X-ray crystallographic observations supported the intuitive attribution of the "*trans*-anti-*trans*-anti-*trans*" configuration to cholesterol, uzarigenin, allocholane, allopregnane and androstane, and of the "*cis*-anti-*trans*-anti-*trans*" configuration to coprostane, digitoxigenin, cholane, pregnane and etiocholane derivatives. According to an arbitrary convention, one designates the substituent groups as "α" and "β" depending on whether they are situated below or above the plane of the molecule, when depicted in a certain way. Figure 2 illustrates one of the two most important configurations of steroid skeletons which nature has selected among 64 possibilities. The side chain is usually attached in β-position to C_{17}. The configurations on C_{20} and C_{24} have likewise been determined and are known to produce steric isomerisms.

The sterols, from which the name of the entire group is derived, are monovalent alcohols with a secondary hydroxyl group on C_3 usually in β-position. The best known representative is cholesterol. It forms esters with a great variety of acids. Both the free and the esterified sterols accompany the neutral fat and the phosphatides in most animal and plant fat. Upon alkaline hydrolysis, the other LIPID constituents form fatty acid soaps; the fraction which remains insoluble in aqueous alkaline solution is called the "unsaponifiable" and consists primarily of sterols. The variations in the cholesterol content of blood in animals, particularly in man, are of great significance for the diagnosis of various diseases, but nothing is known about their purpose in the economy of animals or plants, except for the function of 7-dehydrosterols as provitamins D. Ergosterol (ergostatriene-5,7,22-ol-3), a sterol abundant in yeast, undergoes a series of reactions upon ultraviolet irradiation, leading to lumisterol, tachysterol (scission of link C_9—C_{10}) and culminating in the formation of vitamin D_2. The natural vitamin D, known as vitamin D_3, found for example in codliver oil, is a homologue of vitamin D_2, derived from 7-dehydrocholesterol. The vitamins D are not steroids in the proper sense of the word, as their skeleton consists of a six-membered ring, connected by 2 carbon atoms C_6 and C_7 with a

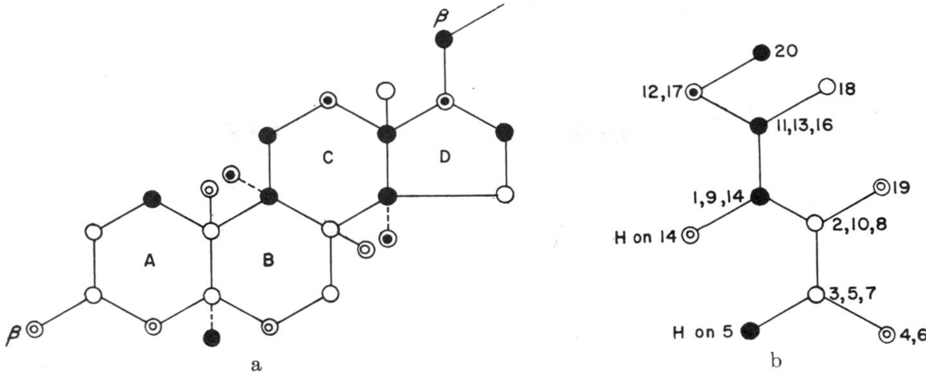

FIG. 2. Constellation of a saturated β-sterol; ◉ atoms in bottom plane, ● in second, ○ in third, ◎ in top plane. b represents a lateral view of the molecule.

hydrindene system. Recently a saponin-like protosteroid was isolated from a sea cucumber (*Echinodermata*), and its structure was established. This toxic surfactant is a tetrasaccharide and at the same time a sulfate ester of a steroid, which still carries the 3 methyl groups characteristic for the lanosterol series.

The sterols occurring in yeasts and fungi mostly contain 28 carbon atoms, for example cerevisterol and fungisterol, but zymosterol from yeast is a cholestadiene-8-ol-3 with 27 carbon atoms. The phytosterols of algae and phanerogamia usually contain 29 carbon atoms and 2 double bonds, like fucosterol from brown algae, spinasterol from spinach and stigmasterol from *Physostigma venenosum*. The various sitosterols from wheat (Greek *sitos*) and other plants are monounsaturated; brassicasterol from *Brassica rapa* is an ergostadiene-5,22-ol with but 28 carbon atoms.

Cholesterol is the principal sterol of all vertebrate animals; it is also found in some mollusks and in crustaceans, where it may be of alimentary origin. The so-called sterols from wool fat with 30 carbon atoms have been recognized as pentacyclic triterpenes by Ruzicka. Sponges, echinoderms (starfish) and marine mollusks, such as clams and oysters, contain a variety of mono- and disaturated C_{28} and C_{29} sterols, which are related to various phytosterols. Their isolation and the elucidation of their steric configuration are owed to W. Bergmann and others.

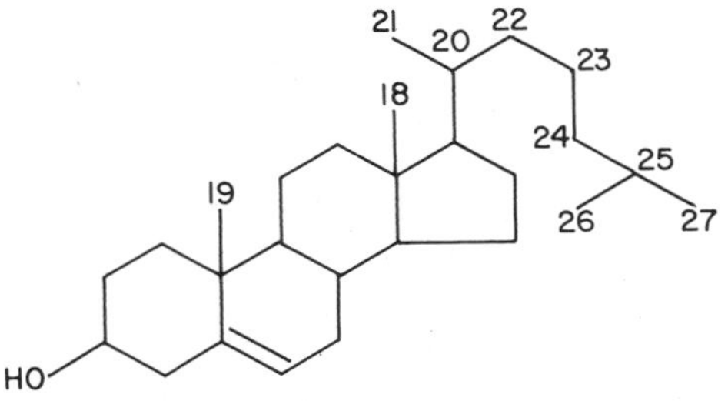

FIG. 3. Cholesterol.

Biogenesis. The finding that ingestion of radioactively labeled acetic acid leads to the synthesis of radioactive cholesterol was the first step in the elucidation of sterol biosynthesis. Actually a growth factor for *Lactobacilli*, replaceable by acetic acid, was found to have the structure $HOCH_2 \cdot CH_2 \cdot C(CH_3)(OH) \cdot CH_2 \cdot CO_2H$; it was called mevalonic acid (see ISOPRENOID BIOSYNTHESIS) and its close relationship to a trimer of acetic acid is evident. Six molecules of this C_6-acid polymerize, losing their carboxyl groups, to the linear isoprenolog squalene, a hydrocarbon occurring in nature. Twelve of the carbon atoms (marked by circles in Fig. 4) originate from the

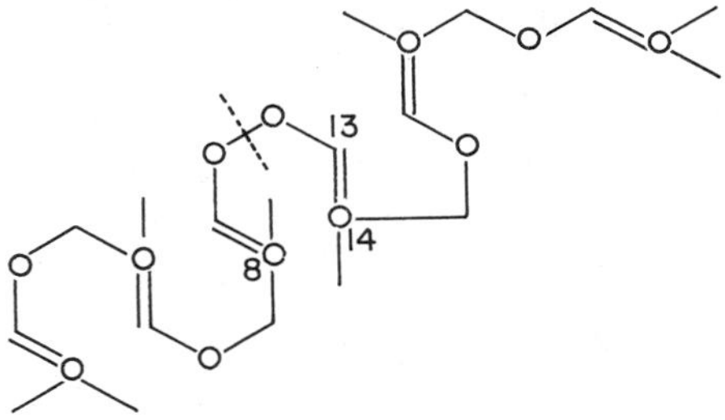

FIG. 4. Squalene.

carboxyl groups of the original acetic acid, the remaining 18 from the methyl groups. Squalene folds in the manner indicated in the formula and yields (with a two-step rearrangement of the methyl group from C_8 to C_{13}) lanosterol (for structure see STEROLS, BIOGENESIS), a "protosterol." This "protosterol," found in wool, fat, loses 3 methyl groups in positions 4, 4 and 14 in the course of biosynthesis, yielding zymosterol, found in yeast, which is convertible into cholesterol. The gradual oxidative degradation of the side chain in cholesterol to bile acids and subsequently to the various steroid hormones in animals is well established and has also been confirmed by C^{14} tracer studies. Many of the enzymes operative during these hormone syntheses in the insertion of hydroxyl groups on individual carbon atoms have been separated and even localized in various cell constituents. (See HORMONES, STEROID.)

Analytical Chemistry of the Steroids. The general analytical methods of organic chemistry have been employed in the steroid field such as colorimetry, ultraviolet and INFRARED SPECTROSCOPY, NUCLEAR MAGNETIC RESONANCE, mass spectroscopy, and anomalous rotation dispersion. A variety of color reactions have been described for cholesterol. The most important is the Liebermann-Burchard reaction with acetic anhydride and a few drops of concentrated sulfuric acid which are added to a chloroform solution of cholesterol and produce a series of colors ending with a reasonably stable green. Replacement of acetic anhydride by acetyl chloride or benzoyl chloride gives an amber color (Bernouilli). Rosenheim describes a red color with trichloroacetic acid for $\Delta_{4,5}$-sterols.

Sterols as well as some less complicated hydroaromatic secondary alcohols form ethanol-insoluble complexes with digitonin (Windaus). These have been used for their gravimetric determination and especially for the separation of free sterols from their esters, which do not react with digitonin. The reaction is in general specific for sterols carrying the hydroxyl group on C_3 in β-position.

HARRY SOBOTKA

STEROLS (BIOGENESIS AND METABOLISM)

Sterols are found in all organisms except the bacteria and the blue-green algae, which are in many respects more closely related to bacteria than to plants. Although the most common or best known sterol, cholesterol, bears little apparent structural similarity to the non-cyclic terpenes, it has been clearly established that cholesterol is derived from a non-cyclic triterpene squalene. Thus, one may view the metabolism of sterols as consisting of two phases: the first phase being the synthesis of the triterpene squalene, and the second phase being the formation of the sterol nucleus and subsequent transformations. These two phases differ from each other in many fundamental respects. (1) Although the second phase does not take place ordinarily in bacteria, most of the first phase is present also in bacteria and serves the important function of providing the terpenoid side chain of the ubiquinones or napthoquinones. (2) The first phase leads to an increase in the size of the molecules, whereas the "synthetic" activities of the second phase are accompanied by a decrease in the size of the sterol molecules. (3) The first phase is reductive and can take place under anaerobic conditions whereas the second phase is oxidative and is characterized by the direct participation of molecule oxygen in many, if not most, reactions.

Each of these two phases can be subdivided into various stages or reaction sequences. These are briefly summarized below and titled stages 1, 2, etc., with the structures of key compounds illustrated in Fig. 1.

Stage 1. Synthesis of the Biological Isoprene Units. Three molecules of acetyl-CoA condense to form β-hydroxy-β-methylglutaryl-CoA which is reduced by TPNH to mevalonic acid. Mevalonic acid is sequentially phosphorylated by ATP to

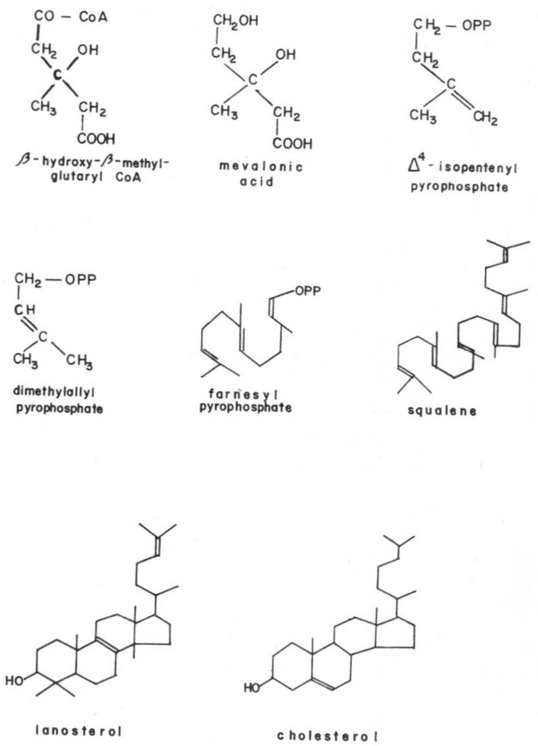

β-hydroxy-β-methylglutaryl CoA; mevalonic acid; Δ^4-isopentenyl pyrophosphate; dimethylallyl pyrophosphate; farnesyl pyrophosphate; squalene; lanosterol; cholesterol

FIG. 1.

phosphomevalonic acid, pyrophosphomevalonic acid and an as yet un-isolated triphosphate which undergoes simultaneous decarboxylation and elimination of phosphate to yield Δ^4-isopentenyl pyrophosphate. This is one of the two biological isoprene units and can be isomerized to the Δ^3-isomer, dimethylallyl pyrophosphate. This sequence of reactions converts a member of the common metabolite pool, acetate, into special compounds which feed into a relatively limited number of specialized synthetic pathways (see also ISOPRENOID BIOSYNTHESIS).

Stage 2. Head-to-tail Condensation of the Biological Isoprene Units. This condensation is based on the reactivity of the allylic pyrophosphate structure in dimethylallyl pyrophosphate and of the double bond in the Δ^4-isopentenyl pyrophosphate. The condensation yields geranyl pyrophosphate which is another allyl pyrophosphate and can in turn condense with another molecule of Δ^4-isopentenyl pyrophosphate to form the C_{15} allylic pyrophosphate, farnesyl pyrophosphate. The important feature of this type of condensation is that a reactive allylic pyrophosphate is always regenerated, thus permitting continuous condensation and lengthening of the chain, such as in the formation of the C_{20} halves of the carotenes (see CAROTENOIDS) and of the polyisoprenoid side chains of the ubiquinones, plastoquinones (see COENZYME Q), VITAMIN E and VITAMIN K. These reactions are thus general in nature and applicable to the synthesis of practically all the biological polyisoprenoid derivatives.

Stage 3. Symmetrical Reductive Condensation of Farnesyl Pyrophosphate to Squalene. This reaction again utilizes TPNH as the reductant and, by a mechanism not yet fully understood, forms the symmetrical triterpene squalene which is the direct precursor of a large number of polycyclic triterpenes including lanosterol, the precursor of all sterols.

Stage 4. Oxidative Cyclization of Squalene to Lanosterol. This enzyme is an interesting OXYGENASE which utilizes molecular oxygen and TPNH to achieve the complex oxidative cyclization of squalene to lanosterol. The reaction marks the beginning of the "aerobic phase" of sterol metabolism. The direct insertion of one atom of molecular oxygen into substrate with the simultaneous reduction of the other oxygen atom at the expense of TPNH represents a type of reaction (oxygenase, or mixed function oxygen transferase) which is particularly prevalent in the metabolism of the sterols. Indeed, nearly all the oxygen atoms that one finds in the many and varied sterols are derived from such oxygenase-type reactions.

Stage 5. Conversion of Lanosterol to Cholesterol. This transformation involves the reduction of the side chain double bond by TPNH, the oxidative removal of the three methyl groups on carbons 4 and 14 by reactions involving oxygenases and the "migration" of the double bond in the nucleus. There is evidence that these reactions may occur in different orders in various organisms and/or tissues, thus leading to the formation of a large number of intermediates which can only be accounted for by the presence of alternative pathways between lanosterol and cholesterol. In the plant kingdom, C_{24} methylation or ethylation takes place with some of these side chain unsaturated sterols.

Stage 6A. Esterification of Cholesterol. Esterification of cholesterol takes place in many tissues, the acid being derived from either acyl-CoA or phospholipids. Although the enzyme cholesterol esterase can catalyze both the hydrolysis and the esterification, it probably acts *in vivo* only as a hydrolytic enzyme.

Stage 6B. Formation of Bile Alcohols and Acids. By a large number of oxygenase reactions, together with some other reactions, various animals convert cholesterol to polyalcohols, polyhydroxy acids or conjugates of polyhydroxy acids with amino acids which share the same property of being surface active agents and thus aid the absorption of lipids. Comparative studies show that the evolution of animals is accompanied by a shift from 27-carbon bile alcohols to 27-carbon BILE ACIDS to 24-carbon bile acids and their conjugates to glycine or taurine.

Stage 6C. Conversion of Cholesterol to the Steroid Hormones. The biochemistry of the steroid HORMONES is reviewed elsewhere. It will, therefore, only be stated here that the sequence cholesterol → C_{21} corticosteroids → C_{19} androgens → C_{20} estrogens is again characterized by the wide occurrence of oxygenase-type reactions.

Stage 6D. Plant Alkaloids. Little is known about the biogenesis of those ALKALOIDS which are undoubtedly derived from cholesterol.

Stage 6E. Various microorganisms are known to be able to carry out limited structural modifications or total degradation of sterols. Some of these are of practical pharmaceutical importance.

The above brief summary emphasizes the dynamics of cholesterol metabolism. It should be pointed out that in certain stable membrane structures, such as adult nervous systems, the cholesterol is metabolically inert and serves a structural function. There is also evidence that in animals at least, cholesterol also serves as structural unit for less stable membrane structures. The rate of cholesterol synthesis, at least in mammals, is also under homeostatic control mechanism (see METABOLIC CONTROLS) although this feedback mechanism may be somewhat sluggish or not too finely tuned such that the serum cholesterol level can be significantly altered by dietary means. Little information is available concerning the role of sterols in plants although nutritional studies have shown that sterols constitute essential nutrients for yeast grown under anaerobic conditions. The universal presence of sterols throughout the animal and plant kingdoms (with the exception of the blue-green algae if they are considered as plants) and their almost complete absence in bacteria suggest the evolutionary closeness between plants and animals as compared with these two kingdoms and the bacteria. In contrast, the universal presence of isoprenoids suggests that the synthesis of mevalonic acid and the biological isoprene units must be more primitive or basic processes from the evolutionary viewpoint. It should also be noted that bacteria and blue-green algae do not have endoplasmic reticulum (see MICROSOMES) and that it is possible that this may be related to the absence of sterols.

References

1. RICHARDS, J. H., AND HENDRICKSON, J. B., "The Biosynthesis of Sterols, Terpenes and Acetogenins," New York, W. A. Benjamin, 1964.
2. "Biosynthesis of Terpenes and Sterols," Boston, Little, Brown, 1959.
3. WRIGHT, L. D., "Biosynthesis of Isoprenoid Compounds," *Ann. Rev. Biochem.*, **30**, 525 (1961).

T. T. TCHEN

SUCCINIC ACID

Succinic acid, butanedioic acid, HOOC—CH_2—CH_2—COOH, derives its name from the Latin name for amber (*succinum*) from which fossil resin it was first obtained by dry distillation. It occurs widely as an intermediate in metabolism since it functions, for example, in the CITRIC ACID CYCLE.

SULFATE AND ORGANIC SULFATES

Inorganic sulfate ion ($SO_4^=$) occurs widely in nature. It is not surprising, therefore, that this ion can be put to a number of uses by biological systems. These uses can be divided primarily into two categories—the formation of sulfate esters and the reduction of sulfate to a form that will serve as a precursor of the amino acids CYSTEINE and METHIONINE. These uses are discussed separately below. Certain specialized bacteria use sulfate to oxidize carbon compounds and thus reduce sulfate to sulfide, while other specialized bacterial species derive energy from the oxidation of inorganic sulfur compounds to sulfate. While the intermediary metabolism of these two bacterial groups has been clarified by the outstanding work of Peck,[1] these specialized systems will not be considered in detail here.

Sulfate Esters. Among the variety of sulfate esters formed by living cells are the sulfate esters of phenolic and steroid compounds excreted by animals, sulfated polysaccharides, and simple esters such as choline sulfate. The key intermediate in the formation of all of these compounds has been shown to be 3′-phosphoadenosine-5′-phosphosulfate (PAPS). As will be seen below, this nucleotide also serves as an intermediate in sulfate reduction. Essentially, PAPS can be considered as an "activated" form of sulfate since the phosphate-sulfate anhydride reacts much more readily with alcohols or reducing agents than inorganic sulfate.

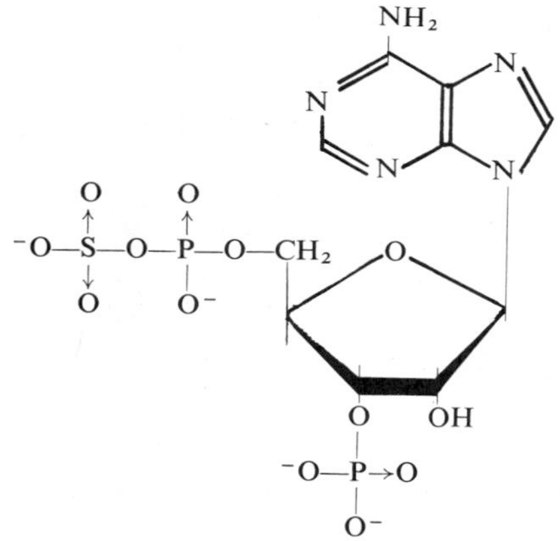

PAPS

In all systems that have been investigated so far, PAPS is formed from sulfate and ATP (cf. adenosine triphosphate) in two sequential enzymatic steps.[2] The reactions are:

$$\text{ATP} + \text{SO}_4^= \rightleftarrows \text{APS} + \text{PP}_i$$
$$\text{APS} + \text{ATP} \rightarrow \text{PAPS} + \text{ADP}$$
$$\text{Sum: } 2\text{ATP} + \text{SO}_4^= \rightarrow \text{PAPS} + \text{ADP} + \text{PP}_i$$

The first step involves the displacement of the terminal pyrophosphate group of ATP by sulfate, and the second step is the phosphorylation of the 3′-ribose hydroxyl group of APS by ATP. Since the first step is thermodynamically unfavorable in the direction of APS formation, the second phosphorylation step as well as the hydrolysis of inorganic pyrophosphate (cf. pyrophosphatase) aids in the activation process. The system for PAPS formation has been shown to occur in a great variety of tissues and organisms including liver, chick embryo, snails, fungi, algae and bacteria. Thus PAPS is justifiably referred to as "active sulfate."

"Sulfokinase" is a general term used to refer to enzymes that catalyze the transfer of sulfuryl groups from PAPS to alcoholic hydroxyl groups, phenolic hydroxyl groups or organic amines. In addition to the sulfated product, PAP (3′-phosphoadenosine-5′-phosphate) is formed by the reaction. A great many sulfokinases have been described in the biochemical literature, and the reader should consult reviews and individual papers for details. A few leading references follow:

	Reference No.
Sulfokinases—general review	3
Phenol sulfokinases	4
Steroid sulfokinases	5
Other "detoxification" sulfokinases	6
Choline sulfokinase	7
Polysaccharide sulfokinases	8

Sulfate Reduction. In organisms that utilize sulfate as a source of sulfur for the synthesis of cysteine and methionine, the first step in the reduction process is the formation of PAPS. This is not surprising since the direct reduction of sulfate ion itself is an extremely difficult chemical process. It is known that the reduction of esters and anhydrides occurs much more readily than the reduction of the corresponding anions. Following activation, the sulfuryl group of PAPS is reduced to sulfite ion ($SO_3^=$) by reduced triphosphopyridine nucleotide (TPNH) and a complex enzyme system. This system seems to include a low molecular weight polypeptide that has been highly purified.[9] Following the reduction of PAPS to sulfite, additional reduction steps readily produce hydrogen sulfide, which appears to be a direct precursor of the amino acid cysteine.

Sulfatases. In addition to the biosynthetic reactions discussed above, mention should also be made of the sulfatases, a widely distributed group of enzymes that hydrolyze simple sulfate esters to inorganic sulfate. A number of these enzymes from plant, animal and bacterial sources have been studied in detail. Table 1, taken from the review by Robbins,[10] provides a summary of the characteristics of some of these enzymes.

TABLE 1. REPRESENTATIVE SULFATASES

Enzyme	Source	Characteristics
Arylsulfatase A	Most animal tissues	Inhibited by SO_4^{2-}, HPO_4^{2-}, F^-. Nitrocatechol sulfate is the usual substrate. Anomalous kinetics.[11-13]
Arysulfatase B	Most animal tissues	Similar to A, but normal kinetics.[12] Lower affinity toward nitrocatechol sulfate.
Arysulfatase C	Most animal tissues	Microsomal enzyme, inhibited by CN^-. *p*-Nitrophenyl sulfate or *p*-acetylphenylsulfate is the usual substrate.[15]
Arysulfatase	Microbial (*Alcaligenes metalcaligenes*)	Shows anticompetitive inhibition by CN^- and hydrazine.[16]
Arysulfatase	Taka-Diastase (*Aspergillus oryzae*)	Similar to mammalian arylsulfatase C.[17,18]
Arylsulfatase	Marine mollusks	Inhibited by SO_4^{2-} and HPO_4^{2-}.[19-21]
Steroid sulfatase	Marine mollusks	Specific for 3β-OH sulfate esters of 5α or $\triangle^5$ steroids.[22]
Steroid sulfatase	Mammalian liver	Microsomal enzyme.[23,24] Same substrate specificity as mollusk enzyme.
Chondrosulfatase	*Proteus vulgaris*	Hydrolyzes sulfate from oligosaccharides obtained by degradation of chondroitin sulfate. Does not act on native chondroitin sulfate.[25]
Glycosulfatase	*Littorina littorea*[26] and other sources.	Hydrolyzes glucose-6-sulfate and adenosine-5′-sulfate.[27]
Myrosulfatase	Mustard seed	Specific enzyme for the hydrolysis of sinigrin.[28,29]

References

1. PECK, H. J., *Proc. Natl. Acad. Sci. U.S.*, **45**, 701 (1959); *ibid.*, **46**, 1053 (1960).
2. ROBBINS, P. W., in "The Enzymes" (BOYER, P. D., LARDY, H., AND MYRBÄCK, K., EDITORS), Second edition, Vol. 6, p. 469, New York, Academic Press, 1962.
3. GREGORY, J. D., AND ROBBINS, P. W., *Ann. Rev. Biochem.*, **29**, 347 (1960).
4. GREGORY, J. D., AND LIPMANN, F., *J. Biol. Chem.*, **229**, 1081 (1957).
5. NOSE, Y., AND LIPMANN, F., *J. Biol. Chem.*, **233**, 1348 (1958).
6. WILLIAMS, R. T., in "Biogenesis of Natural Compounds" (BERNFELD, P., EDITOR), p. 427, Oxford, Pergamon Press, 1963.
7. KAJI, A., AND GREGORY, J. D., *J. Biol. Chem.*, **234**, 3007 (1959).
8. ADAMS, J. B., *Biochim. Biophys. Acta*, **83**, 127 (1964).
9. WILSON, L. G., ASAHI, T., AND BANDURSKI, R. S., *J. Biol. Chem.*, **236**, 1822 (1961); ASAHI, T., BANDURSKI, R. S., AND WILSON, L. G., *J. Biol. Chem.*, **236**, 1830 (1961).
10. ROBBINS, P. W., in "The Enzymes" (BOYER, P. D., LARDY, H., AND MYRBÄCK, K., EDITORS), Second edition, Vol. 6, p. 363, New York, Academic Press, 1962.
11. ROY, A. B., *Experientia*, **13**, 32 (1957).
12. BAUM, H., AND DODGSON, K. S., *Biochem. J.*, **69**, 573 (1958).
13. ANDERSON, S. O., *Acta Chem. Scand.*, **13**, 120, 884, 1671 (1959).
14. DODGSON, K. S., AND WYNN, C. H., *Biochem. J.*, **68**, 387 (1958).
15. DODGSON, K. S., SPENCER, B., AND WYNN, C. H., *Biochem. J.*, **62**, 500 (1956).
16. DODGSON, K. S., SPENCER, B., AND WILLIAMS, K., *Nature*, **177**, 432 (1956).
17. ROBINSON, D., SMITH, J. N., SPENCER, B., AND WILLIAMS, R. T., *Biochem. J.*, **51**, 202 (1952).
18. BOYLAND, E., MANSON, D., SIMS, P., AND WILLIAMS, D. C., *Biochem. J.*, **62**, 68 (1956).
19. SODA, T., in "Die Methoden der Fermentforschung" (BAMANN, E., AND MYRBÄCK, K., EDITORS), Vol. II, p. 1695, Leipzig, Georg Thieme, 1941.
20. DODGSON, K. S., LEWIS, J. I. M., AND SPENCER, B., *Biochem. J.*, **55**, 253 (1953).
21. DODGSON, K. S., AND POWELL, G. M., *Biochem. J.*, **73**, 666, 672 (1959).
22. ROY, A. B., *Biochem. J.*, **62**, 41 (1956).
23. GIBIAN, H., AND BRATFISCH, G., *Z. Physiol. Chem.*, **305**, 265 (1956).
24. ROY, A. B., *Biochem. J.*, **66**, 700 (1957).
25. DODGSON, K. S., AND LLOYD, A. G., *Biochem. J.*, **66**, 532 (1957).
26. DODGSON, K. S., AND SPENCER, B., *Biochem. J.*, **57**, 310 (1954).
27. EGAMI, F., AND TAKAHASHI, N., *Bull. Chem. Soc. Japan*, **28**, 666 (1955).
28. ETTLINGER, M. G., AND LUNDEEN, A. J., *J. Am. Chem. Soc.*, **78**, 4172 (1956).
29. BAUM, H., AND DODGSON, K. S., *Nature*, **179**, 312 (1957).

PHILLIPS W. ROBBINS

SULFHYDRYL COMPOUNDS

Sulfhydryl compounds, *i.e.*, those containing the SH group (thiol group), are widespread and exceedingly important in biochemistry. Among these are CYSTEINE and all PROTEINS and peptides that have this unit in their structures, including many enzymes, MYOSIN, HEMOGLOBIN and GLUTATHIONE. LIPOIC ACID in reduced form possesses two sulfhydryl groups. Sulfhydryl groups are notably reactive in that they undergo oxidation easily to form disulfides, and combine with heavy metals. COENZYME A contains a reactive thiol group, essential to its function as a carrier of acyl groups bound to the coenzyme A through a "high-energy" thioester linkage. Certain of the PROTEOLYTIC ENZYMES (the thiol proteases) require the presence of at least one thiol group as an obligatory participant in catalysis. (See also SULFUR METABOLISM.)

SULFONAMIDES

The discovery that the azo dye "Prontosil" had chemotherapeutic activity against streptococcal infections[5] can be regarded as the most important breakthrough in the history of medicinal chemistry. It destroyed a stalemate in chemotherapy of nearly 30 years duration, led to conceptual advances of broad application and, on the practical side, gave rise to a variety of new pharmacodynamic drugs as well as the antimicrobial agents that were evolved directly from sulfanilamide (Fig. 1).

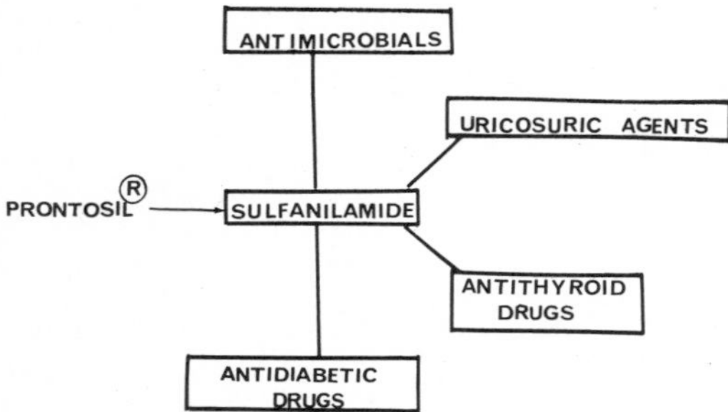

FIG. 1. Pharmacodynamic agents derived from sulfanilamide.

The recognition of sulfanilamide as the active moiety of "Prontosil",[6] and the discovery that *p*-aminobenzoic acid (*p*AB) would nullify its effects,[7] provided the basis for the metabolite: ANTIMETABOLITE concept and the prediction that this would result in a generalized approach to chemotherapy.[8] Within a few years, FOLIC ACID had been isolated and *p*AB was found to be an intrinsic component of this vitamin (Fig. 2).

It is thus with the *synthesis* of this vitamin that sulfanilamide and its congeners interfere, and the effects of sulfonamides on the metabolism of susceptible microorganisms are explainable in terms of deficiencies in folate-containing coenzymes (see FOLIC ACID COENZYMES).

The nature of the mutual antagonism between *p*AB and sulfonamides was investigated exhaustively before folic acid had been characterized. These studies, carried out in the main with growing cultures, showed this antagonism to be competitive, *i.e.*, growth in mixtures of metabolite and antimetabolite depended only on the ratio of the two, and not on the absolute concentration. Similarly, the biosynthesis of folates in resting

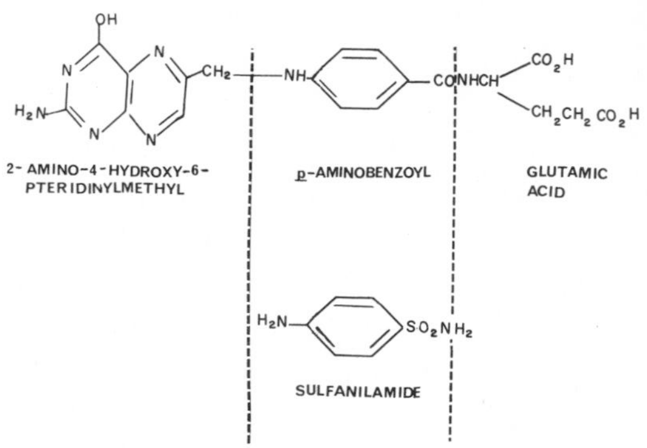

2-AMINO-4-HYDROXY-6-PTERIDINYLMETHYL | p-AMINOBENZOYL | GLUTAMIC ACID

SULFANILAMIDE

FIG. 2. Folic acid showing structural relationship of sulfanilamide with *p*-aminobenzoic acid.

cell suspensions of suitable bacteria exhibited the same competitive relationship. Tracer studies with radioactive *p*AB and sulfanilamide showed that sulfonamides excluded *p*AB from the cells, but were not themselves incorporated. Later, in studies of (dihydro) folate biosynthesis at the enzymatic level, it was found that sulfonamides not only block the condensation of 2-amino-4-hydroxy-7,8-dihydropteridine-6-methanol with *p*AB, but appear to condense with the pteridine themselves to form spurious folates.[9] Some of the more active participants in this lethal synthesis are relatively inactive with whole cell preparations. This suggests that the transport across cell membranes may be more important to the activity of a sulfonamide than its relative affinity for the dihydrofolate synthetase within the cell.

The therapeutic effectiveness of the sulfonamides depends on two differences between parasitic microorganisms and host cells in the biogenesis of tetrahydrofolate-containing coenzymes. The latter are formed in microorganisms by the condensation of *p*AB with a dihydropteridine, then with glutamate, reduction to tetrahydrofolate and its union with appropriate "one-carbon" metabolites. This process is short-circuited in the mammal, which lacks the enzymatic equipment to produce folate by biosynthesis and is dependent on exogenous sources of this vitamin. The reaction inhibited by sulfonamides is thus present in the parasite and absent from the host, a metabolic difference ideal for the purposes of chemotherapy. But this, in itself, is insufficient to account for the difference in response. Since the host possesses the coenzymatic folates in quantities ample for cellular growth, there is no thoeretical reason why parasitic cells should not avail themselves of the opportunity to take these, preformed, from the host's cellular and tissue fluids, and thus in effect eliminate the necessity for their biosynthesis. That microorganisms are unable to do this reflects the lack of appropriate transport mechanisms. In cells that use preformed folate, its assimilation appears to be an active, energy-requiring process. Many microorganisms do possess mechanisms for the utilization of the ultimate end products of the folate system, purines, thymine, methionine and serine, and for this reason, sulfonamides lose effectiveness in the presence of extensive tissue damage where such end products are readily available.

Since the sulfonamides block a reaction that is present in the parasite but absent from the host, there is no theoretical reason why an essentially nontoxic sulfonamide should not be attainable. In practice, the drugs become involved in many pharmacodynamic activities which are quite irrelevant to their mode of action as antimicrobials. Indeed it was careful observation of these "side effects" which led to the creation of new antithyroid, diuretic, uricosuric and antidiabetic drugs. On the other hand, the objective of the medicinal chemist, whose purpose is optimal antimicrobial effects, has been to eliminate, or minimize, these pharmacodynamic activities, and to provide drugs with suitable pharmacokinetic properties.

The useful sulfonamides carry heterocyclic substituents on the amide nitrogen and have pK_a values close to the normal serum pH. Earlier derivatives, sulfanilamide, sulfapyridine and sulfathiazole, had deleterious side effects such as methemoglobin formation and an extensive conversion to acetyl derivatives which not only inactivated the antibacterial but created problems of renal excretion. These problems largely have been overcome, and newer derivatives have been selected, in the main, on the basis of pharmacokinetic properties. This is not to say that sulfonamides do not differ among themselves in their spectra of antibacterial activities. However, the ultimate clinical usefulness of a sulfonamide is determined by the integral of all its antibacterial, pharmacodynamic and pharmacokinetic properties, and the best sulfonamide for a given purpose represents a favorable balance with respect to all of these. Speed of absorption, lipid: water distribution, acetylation, protein binding, tissue: plasma distribution, "detoxicating" mechanisms and renal clearance are factors, quite apart from its primary antimicrobial activity, which determine its ultimate utility.

Until 1957, the more useful sulfonamides (sulfadiazine, sulfisoxazole) were those possessing relatively short clearance times. In that year, sulfamethoxypyridazine (half-life, 40 hours) was introduced and it was soon followed by sulfadimethoxine (half-life, 36 hours); soon the whole gamut from 4–120 hours had been provided. The longer-acting derivatives depend on low acetylation rates, high protein binding and perhaps other factors, for their long survival. At first glance, the chemotherapeutic effectiveness of a sulfonamide might be expected to depend primarily on the concentration of free, unbound drug, but in fact, many of the derivatives which are extensively bound to serum proteins have high therapeutic effectiveness. In fact, it is the relative binding to the microbial acceptor sites, and freedom of transport across cellular membranes, which are important.

With the advent of antibiotics, the sulfonamides are no longer unique, or predominant, among antimicrobial chemotherapeutic agents. Nevertheless, they maintain a sphere of usefulness, especially in urinary tract infections, that is reflected in the continued writing of 15–20 million new prescriptions per year in the United States alone. The development of selective inhibitors of dihydrofolate reductase, used as potentiators of sulfonamides, may further extend their utility.[4]

Utility aside, the impact of the sulfonamide story on biochemistry and chemotherapy has had repercussions of immeasurable magnitude and duration.

References

1. Northey, E. H., "Sulfonamides and Allied Compounds," New York, Reinhold Publishing Corp., 1948.
2. Jukes, T. H., and Broquist, H. P., "Sulfonamides and Folic Acid Antagonists" in "Metabolic Inhibitors" (Hochster, R. M., and Quastel, J. H., Editors), Vol. I, p. 481, New York, Academic Press, 1963.
3. Zbinden, G., "Molecular Modification in the Development of Newer Antiinfective Agents," in "Molecular Modification in Drug Design" (Gould, R. F., Editor), p. 25, Washington, D.C., American Chemical Society, 1964.
4. Hitchings, G. H., and Burchall, J., "Inhibition of Folate Biosynthesis and Function as a Basis for Chemotherapy," in "Advances in Enzymology" (Nord, F. F., Editor), Vol. 27, New York, Interscience Publishers, 1965, p. 417.
5. Domagk, G., *Deut. Med. Wochschr.*, **61**, 250 (1935).
6. Fourneau, E., Trefouel, J., Trefouel, J., Nitti, F., and Bovet, D., *Compt. Rend. Soc. Biol.*, **122**, 652 (1936).
7. Woods, D. D., *Brit. J. Exptl. Pathol.*, **21**, 74 (1940).
8. Fildes, P., *Lancet*, **1**, 955 (1940).
9. Brown, G. M., *J. Biol. Chem.*, **237**, 536 (1962).

George H. Hitchings

SULFUR METABOLISM

Sulfur in some form is required by all living organisms. It is utilized in various oxidation states including sulfide, elemental sulfur, sulfite, sulfate and thiosulfate by lower forms and in organic combination by all. The more important sulfur-containing organic compounds include: the amino acids, CYSTEINE, cystine and METHIONINE which are components of PROTEIN; the vitamins, THIAMINE and BIOTIN; the cofactors, LIPOIC ACID and COENZYME A; certain complex lipids of nerve tissue, the sulfatides; components of MUCOPOLYSACCHARIDES, the sulfated polysaccharides; various low-molecular weight compounds such as GLUTATHIONE and the hormones, VASOPRESSIN and oxytocin; many therapeutic agents such as the SULFONAMIDES and penicillins (see ANTIBIOTICS) as well as most of the oral hypoglycemic agents used in the treatment of diabetes mellitus. Sulfhydryl groups of the cysteine residues in enzyme proteins and related compounds such as HEMOGLOBIN, play a key role in many biocatalytic processes; sulfhydryl-disulfide interchange reactions involving the cysteine residues of proteins are critical events in the immune processes, in transport across cell membranes and in BLOOD CLOTTING. The so-called S—S bridges between these residues play a key role in the maintenance of the tertiary structure of most proteins [see PROTEINS (BINDING FORCES)].

The electronic structure of sulfur is such that a variety of oxidation states are readily obtainable. Indeed, a sulfur cycle exists in nature:

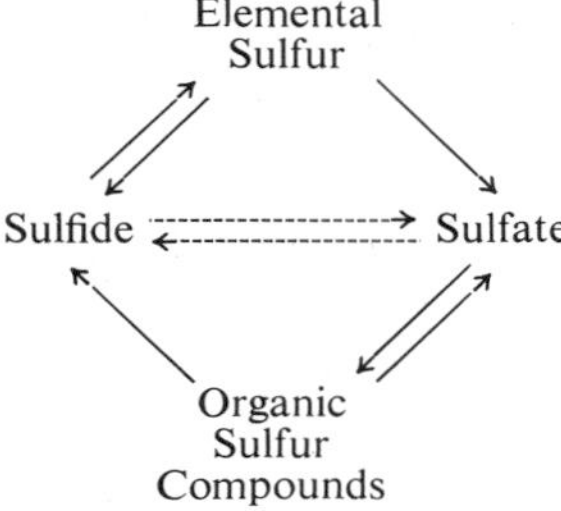

The oxidation and reduction of elemental sulfur and sulfide occur in different species of bacteria, *e.g.*, the oxidation of sulfides via elemental sulfur to sulfate takes place in *Chromatia*, the alternative oxidation to sulfate in *Thiobacilli*. The reduction of sulfate to sulfide occurs in *Desulfovibrio*. The biosynthesis of organic sulfur compounds from SULFATE takes place mainly in plants and bacteria, and the oxidation of these compounds to sulfate is characteristic of animal species and of heterotrophic bacteria.

The amino acids *cysteine* and *cystine* are interconverted by oxidation-reduction reactions as follows:

$$\begin{array}{ccccc} S & \text{—} & S & & SH \\ | & & | & & | \\ CH_2 & & CH_2 & & CH_2 \\ | & & | & & | \\ CHNH_2 & & CHNH_2 & + 2H \rightleftharpoons 2 & CHNH_2 \\ | & & | & & | \\ COOH & & COOH & & COOH \\ & \text{Cystine} & & & \text{Cysteine} \end{array}$$

Cystine was first isolated from a urinary calculus by Wollaston in 1805. It was shown to be a component of protein by Morner in 1899 and independently by Embden in 1900; proof of its structure was provided by Friedman in 1902.

Methionine ($CH_3SCH_2CH_2CH(NH_2)COOH$) is the methyl thioether of the next higher homologue of cysteine. It was discovered in casein by Mueller in 1921, and its structure was elucidated by Barger and Coyne in 1928. All higher species studied, from protozoa to man have a nutritional requirement for methionine. It has been shown that the sulfur of methionine and the carbon and nitrogen of the amino acid serine are utilized for the biosynthesis of cysteine. Du Vigneau and his colleagues demonstrated that the intermediate cystathionine is formed in the process, and Binkley and others found that a pyridoxal phosphate enzyme, cystathioninase, is involved in the cleavage to form cysteine and homoserine. It is of interest that methionine will not replace the requirement for cystine for the growth of various mammalian cells in TISSUE CULTURE.

The oxidation of the sulfur-containing amino acids was studied early by H. B. Lewis and his students and by Medes and Toennies and their colleagues. Cyst(e)ine undergoes a variety of oxidative reactions in mammalian liver (in some cases transamination of intermediates may be involved) to yield ultimately pyruvate, NH_3 and sulfate. Sulfite is an intermediate in the formation of sulfate; its conversion to sulfate is mediated by a flavin enzyme containing a heme-like group. In other reactions, sulfide and thiosulfate are produced. Cysteine also undergoes oxidation and decarboxylation to form taurine ($SO_3HCH_2CH_2NH_2$) which is utilized for the formation of the BILE ACID, taurocholic acid. Cysteine is also the precursor of a component of coenzyme A, β-mercaptoethylamine.

Humans with the hereditary disorder, cystinuria, have a defect in membrane transport of a group of amino acids which includes (in addition to cystine) lysine, arginine and ornithine. There is a failure of reabsorption of these amino acids in the proximal tubules of the kidney (see RENAL TUBULAR FUNCTIONS) and an accompanying enhanced excretion in the urine. The presence of excessive amounts of cystine in the urine of these individuals is readily detected by a simple test for cystine, which involves the addition of a solution of sodium cyanide (to reduce the cystine to cysteine) and of sodium nitroprusside to yield a magenta color. Due to the limited solubility of cystine at the pH of the urine, beautiful hexagonal plates of cystine are often seen upon microscopic examination of the urine of such patients. Cystine calculi make up about 1% of all urinary calculi which are encountered in the human. Cystinuria is distinctly different from another complex hereditary disorder sometimes referred to as "cystinosis." As part of the complex pathologic findings in the latter disorder, there is a deposition of needle-like cystine crystals in the spleen and characteristically in the cornea of the eye.

The outline of the many reactions involved in the interconversion of the methyl groups of methionine and of choline was elucidated in large part by du Vigneaud and his colleagues. The active form of methionine, S-adenosylmethionine, was demonstrated by Cantoni to be a key compound in a host of biological METHYLATION reactions. Included in the latter, for example, are those leading to the methylation of norepinephrine and its metabolites and the methylation of transfer RNA. Although the animal organism has the potential to synthesize methyl groups, dietary methionine serves as the principal source thereof.

The major end product of sulfur metabolism in the animal organism is sulfate. Some three-fourths of the urinary sulfur is inorganic sulfate. A small fraction of sulfate is esterified with a variety of phenolic compounds and represents an important function of sulfate in detoxication of many substances including various metabolites as well as foreign organic compounds ingested by the host. The active form of sulfate was shown by the researches of De Meio and Lipmann and their collaborators to be 3′-phosphoadenosine-5′-phosphosulfate (see SULFATE AND ORGANIC SULFATES). In addition to an inactivation function (*e.g.*, of adrenal steroids) and a detoxication function, biological sulfatation plays a role in the biosynthesis of certain of the polysaccharides such as the chondroitin sulfate moiety of mucopolysaccharides and in the formation of the tyrosine-O-sulfate residues in fibrinogen. Considerable strides by workers such as Dziewiatkowski have been made in our understanding of formation of cartilage and connective tissue by the utilization of sulfur-35 labeled sulfate to tag the sulfated polysaccharides (S^{35} is a weak beta emitter—0.18 MeV—and is well suited for radioautographic work in the hands of careful investigators).

An interesting compound, ergothioneine, the betaine of 2-thiolhistidine, was isolated from ergot by Tanret in 1909. It was later (1925–1926) isolated by G. Hunter and Eagles and by Benedict from porcine erythrocytes. From the more recent studies of Melville and Heath and their co-workers, ergothioneine appears to be synthesized only by fungi such as *Claviceps purpurea* and *Neurospora crassa*. Ergothioneine has been isolated from oats; however, it is assumed that the compound is produced by the fungi which are present on most cereal grains. The relatively high concentrations found in mammalian erythrocytes can be accounted for by the intake of foodstuffs containing ergothioneine; it is incorporated into the red cell during erythropoiesis. Its function, if any, in the animal organism is unknown.

The tripeptide GLUTATHIONE (γ-glutamylcysteinylglycine) was first isolated by Hopkins from red cells, but it is present in highest concentratino in the lens of the eye—some 600 mg/100 g in the cortex. In the lens are found small amounts of analogues of glutathione in which cysteine is replaced by α-aminobutyric acid (ophthalmic acid) or by S-sulfocysteine or by S-(α,β,dicarboxyethyl)-cysteine. The enzyme glutathione reductase, a FLAVOPROTEIN which catalyzes the interconversion of the oxidized and reduced forms of glutathione by interaction with the coenzyme NADP (2GSH $+ NADP^+ \rightleftharpoons GSSG + NADPH + H^+$), has been found in bacteria, plant and animal tissues. Decreased levels of glutathione are found in the erythrocytes of humans who have the sex-linked genetic disorder referred to as "primaquine sensitivity." These persons commonly develop a hemolytic crisis and anemia following ingestion of certain drugs such as the antimalarial primaquine, acetanilide, etc. It has been found that they lack the enzyme glucose-6-phosphate dehydrogenase which catalyzes a reaction leading to formation of the reduced cofactor, NADPH, which in turn is required to maintain glutathione in the reduced state as shown above. The reduced glutathione has been proposed as a cofactor for a PEROXIDASE which protects hemoglobin from oxidation by peroxides in the red blood cell. Glutathione also serves as a cofactor in the action of certain enzymes such as glyceraldehyde phosphate dehydrogenase according to Racker. Glutathione apparently plays a role in a variety of transhydrogenase reactions involved in the reduction of S—S bonds to the sulfhydryl form and the conversion of the latter to the S—S form. Important roles have been ascribed by Mazia to the formation and reduction of S—S bonds in the molecular processes involved in CELL DIVISION.

Perhaps the most dramatic example of detoxication in the animal organism is the conversion of CYANIDE to thiocyanate in the liver as catalyzed by the enzyme rhodanese (3 mg HCN/kg of body weight will kill a rabbit, but some 500 mg/kg of NaCNS is required). The sulfur may be furnished by elemental sulfur or thiosulfate; thiosulfate is the more likely precursor of the sulfur used as it is normally excreted in the urine. Mercaptopyruvate ($SHCH_2\overset{\overset{O}{\|}}{C}COOH$), formed by transamination of cysteine, and cysteine sufinate ($SO_2CH_2CHNH_2$ COOH), and intermediate in the oxidation of cysteine, react to form pyruvate and alanine thiosulfate ($^-SSO_2CH_2CHNH_2COOH$) which in turn yields thiosulfate. There is some evidence however that *in vivo*, mercaptopyruvate may also transfer its thiol group to a sulfhydryl group on the enzyme to form a rhodanese-SSH complex which furnishes the sulfur for the formation of ^-SCN from ^-CN.

There is increasing evidence that the disulfide peptide hormones such as nonapeptide of the neurohypophysis VASOPRESSIN (pitressin) exert their action by exchange reaction between sulfhydryl groups on membranes and the S—S groups of the hormones. In the case of vasopressin, it has been shown that the capacity of the urinary bladder of the toad to form S—S bonds between the hormone and SH groups of the membrane parallels the capacity to allow a passive, non-energy dependent transfer of water between regions of different osmotic pressure. INSULIN, which is composed of two peptide chains held together by two S—S bridges, may act similarly in facilitating the transport of glucose across the cell membrane. It is of interest that vasopressin, oxytocin and insulin are some 100 times as effective as are simple disulfides like oxidized glutathione in inducing swelling of mitochondria *in vitro*, which involves the transport of water and other substances into the mitochondrion.

The compound β-mercaptoethanol, which is formed enzymatically from β-mercaptopyruvate in yeast, has proved to be a most important tool for the reduction or protection of certain enzymes during purification. This compound, introduced as a protective agent by Meister and his colleagues, made it possible for Nirenberg to prepare stable enzyme preparations for use in his studies of PROTEIN BIOSYNTHESIS in the elucidation of the genetic nucleotide code.

References

1. BLACK, S., "The Biochemistry of Sulfur-containing Compounds," *Ann. Rev. Biochem.*, **32**, 399 (1963).

2. du Vigneaud, V., "A Trail of Research in Sulfur Chemistry and Metabolism," Ithaca, N.Y., Cornell University Press, 1952.
3. Dziewiatkowski, D., "Sulfur," in "Mineral Metabolism" (Comar, C. L., and Bronner, F., Editors), Vol. 2B, New York, Academic Press, 1962.
4. Greenstein, J. P., and Winitz, M., "Chemistry of the Amino Acids," New York, Academic Press, 1961, 3 vols.
5. White, A., Handler, P., and Smith, E. L., "Principles of Biochemistry," Third edition, New York, McGraw-Hill Book Co., 1964.
6. Young, L., and Maw, G. A., "The Metabolism of Sulfur Compounds," London, Metheun & Co., New York, John Wiley & Sons, 1958.

Robert C. Baldridge

SUMNER, J. B.

James Batcheller Sumner (1887–1955), an American biochemist, has the distinction of being the first to obtain an enzyme in crystalline form. The enzyme was urease and was obtained from jack bean meal. He shared with John Howard Northrup and Wendell Meredith Stanley the Nobel Prize in chemistry in 1946.

SUPERSONIC VIBRATIONS. See ultrasonic vibrations.

SVEDBERG, T.

Theodor Svedberg (1884–), a Swedish chemist, is noted for his development of the ultracentrifuge and its use in the study of proteins and other macromolecules. For this work, he received the Nobel Prize for chemistry in 1926.

SWEAT

Sweat Formation. The skin as an organ of excretion is concerned with the loss of water, electrolytes, and organic substances, predominantly waste products. Water is lost in part as *insensible perspiration* at the rate of 30 ml/hr in man, two-thirds of which is lost through the skin and one-third through the respiratory tract. The origin of this water is the interstitial fluid of the dermis, which diffuses through the skin without wetting it. *Sensible perspiration* is supplied by some 2½ million sweat glands in man, innervated by cholineric fibers of the sympathetic nervous system. There are two kinds of sweat glands; *apocrine*, in which the secretion end of the gland when filled is pinched off (these occur largely in axillae and nipples); and *eccrine* (exocrine) glands, by far the most numerous covering the other areas of the skin. The surface film of skin is an emulsion of fatty and aqueous materials; the lipid component is derived mainly from *sebum*, the excretory produce of sebaceous glands. Formation of sweat is not a simple filtration process, for the sweat glands may develop a pressure of 250 mm Hg, considerably higher than arterial blood pressure. The stimulus to secretion is a rise in blood temperature, activating centers in the hypothalamus, or reflex afferents, *e.g.*, gustatory nerve stimulation.

Volume of Sweat. The volume of production varies widely depending largely on environmental temperature and amount of physical activity. This is illustrated in Fig. 1. The volume lost may easily reach 1–2 liters/hr and with strenuous exercise in the sun, 4 liters/hr.

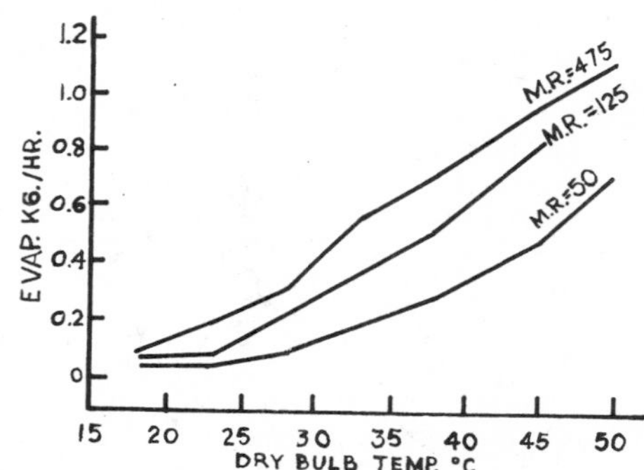

Fig. 1. Effects of increasing room temperature on rates of sweating at three different metabolic rates: at rest (M.R. 50 Cal/m²/hr); walking on treadmill at 4.5 km/hr (M.R. 125 Cal/m³/hr); walking at 5.6 km/hr up a 2.5% grade (M.R. 175 Cal/m²/hr). (*Courtesy of S. Robinson*[2].)

Composition of Sweat. Sweat is about 99% water, with a specific gravity of about 1.003. The pH is usually acid, from 5–7.5. Sweat is hypotonic, the solid content averaging about 680 mg%, of which about two-thirds is ash, and one-third organic substance.

Inorganic Content. NaCl: Its concentration is very variable, but is usually in the range of 18–97 meq/liter. The Na/Cl ratio averages 1:11. Men working in a hot environment, losing 12 liters of sweat per day with a concentration of NaCl of 100 meq/liter, would lose 70 grams of salt, seven times the daily intake.

K: Its concentration averages 4.5 meq/liter, and varies inversely with Na. The Na/K ratio is elevated to 15 in a hot environment without acclimatization; in about 5 days of adaptation it drops to 5, due to an increase in K concentration, and a 50% fall in Na. This is taken as an indication of increased adrenal cortical activity.

Other Minerals: Ca ranges from 1–8 mg%, with a decrease in concentration as sweating proceeds. *Mg:* 0.04–04 mg%; *P:* 0.003–0.04 mg%; SO_4: 4–17 mg%; traces of *Cu*, *Fe* and *Mn*.

Organic Content. Urea Nitrogen: 12–39 mg%; the sweat urea/plasma urea equals 1.92 ± 0.48, regardless of the concentration in the plasma or the rate of sweating.

NH_3 *Nitrogen:* Its concentration in eccrine sweat ranges from 5–9 mg% (50–200 times the blood content).

Creatinine: Averages 0.4 mg% (0.1–1.3).

Uric Acid: is less concentrated than in the blood (0–1.5 mg%).

Lactic Acid: 4–40 meq of acid per liter of sweat (4–40 times that in the blood).

Pyruvic Acid: 0.1–0.8 meq/liter (1.3–10 times that of blood).

Miscellaneous Organic: Glucose in insignificant amounts; amino acids, vitamins, phenol, and histamine. *Drugs* such as nicotine, atabrine, morphine, sulfanilamide, and alcohol appear in sweat after administration.

References

1. Lorincz, A. L., and Stoughton, R. B., "Specific Metabolic Processes of Skin," *Physiol. Rev.*, **38**, 481–502 (1958).
2. Robinson, S., and Robinson, A. H., "Chemical Composition of the Sweat," *Physiol. Rev.*, **34**, 202–220 (1954).

Ewald E. Selkurt

SYMBIOSIS

Symbiosis is a close association between two or more different species, for all or part of their lives, and as such is identifiable only on the basis of recognition of the constancy of the species relationship. Parasitoids are excluded from this discussion since, by common usage, symbiosis implies mutual benefit. This understanding by itself reveals nothing of the biochemical aspects, if indeed there are any. Numerous marine animals have symbiotic relationships founded in stereotyped behavior. Several functions are served to the mutual advantage of both partners, and yet individuals of either species can, and frequently do, live alone at irregular intervals.

Plant Symbioses. There are more biochemical processes operative in plant symbioses, where behavior is not an element. In lichens, consisting of a saprophytic fungus and a photosynthetic alga, the hyphae are closely appressed to, or penetrate into, the algal cells, forming a functional unit, the thallus. The entire thallus has a respiration rate, under optimum conditions, which is comparable to that of mature angiosperm leaves, per unit of surface area. However, the rate of PHOTOSYNTHESIS is much lower because there is less photosynthetic pigment than in leaves, so that the net rate of carbon dioxide assimilation is lower, maximum values so far recorded ranging between 0.2–3.2 mg carbon dioxide per 50 cm^2 per hour, in the light. Rates of both respiration and photosynthesis are dependent upon the water content of the thallus, which varies with environmental conditions.

It is usually postulated that carbon compounds produced in photosynthesis by the alga move to the fungus, but this hypothesis has been difficult to demonstrate. In intact thalli, when $H_2C^{14}O_3$ was supplied in the light, labeled carbon first appeared in organic compounds in the algal zone. With either of the symbiotes isolated in pure culture, there is no clear-cut evidence of requirements for growth factors, such as B vitamins. In one species, it has been demonstrated that the lichen alga excretes a variety of substances into the medium, including vitamins, polysaccharides and nitrogenous compounds, on which the lichen fungus can exist. However, since both the morphology and the reproductive behavior of symbiotes in pure culture are different from that in the thallus, the physiology in isolated culture may not be typical. In most healthy lichen thalli, dead algal cells are rare, indicating that the symbiosis does not involve digestion of algal cells by the fungus.

All saprophytic higher plants, some pteridophytes and bryophytes, and a number of monocotyledons, dicotyledons, and conifers form associations of their roots with fungi, the complex being a mycorrhiza. Basidiomycetes are especially common in mycorrhizas with forest trees, over 40 species being known for *Pinus silvestris* alone. Prior to initial infection, the plant rootlets secrete materials which are essential to germination and growth of the fungus. The exudates contain amino acids, nucleic acid constituents, B vitamins, enzymes, and simple sugars. (The fungi cannot compete with saprophytic fungi in utilizing cellulose.) The most active growth-promoting constituent of the root exudates is a complex metabolite known as "M-factor," which contains at least two substances, one of which is diffusible through plasma membranes and can be replaced in its effect on hyphal growth by DPN (diphosphopyridine nucleotide). The second substance is bound within the root cells but is taken up by the fungi enzymically.

Once the fungi have been stimulated, they appear to produce auxin, inducing the plant rootlets to increase in width and branch profusely. Thus the rootlets are more efficient soil-nutrient absorbing organs than are uninfected roots.

Legume root-nodule symbiosis is obviously mutualistic, the host benefiting from nitrogen fixed in the nodule while the microorganisms in the nodule acquire nutrition and a favorable environment (see also NITROGEN FIXATION). About 10^8 tons of nitrogen are biologically fixed, mainly by symbiotic sources, annually in the world. (This discussion omits any consideration of the numerous asymbiotic systems which fix nitrogen.) Although nine genera other than legumes possess nitrogen-fixing nodules in soils deficient in fixed nitrogen, only in the legumes is much known of the biochemistry of the reactions. Prior to infection, the plant root exudes trytophan, which is converted by the *Rhizobium* bacteria to indoleacetic acid. Indoleacetic acid causes a characteristic deformation or curling of the root hairs. The bacteria then produce an extracellular polysaccharide slime, which in turn induces the plant to secrete the enzyme, polygalacturonase. The actual function of this enzyme is unknown, but it seems to act with indoleacetic acid to affect the plasticity of the primary wall of the young root hair and thus assist penetration of the infection thread. The latter is a hypha-like structure consisting largely of cellulose, laid down by the host cell, and containing the bacteria. The thread grows only at its tip, and when it reaches a preformed tetraploid cell in the cortex, this cell and its diploid neighbors are stimulated to repeated divisions, which form the nodule. Vesicles in the infection thread rupture, releasing bacteria into the plant cell, where they multiply rapidly. Folds of protoplasmic membrane of the host cell then arise and form a complex, producing numerous membrane envelopes, each envelope surrounding a small group of bacteria. The bacteria are then transformed into swollen and sometimes branched forms known as *bacteroids*; included in the cytological transformations is a prominent development of respiratory MITOCHONDRIA. The bacteroids do not divide within the nodule and have not been cultivated on artificial media. Concomitantly with bacteroid transformation, hemoglobin appears in the nodule and nitrogen is fixed. Eventually the hemoglobin is replaced by bile pigments, the bacteroids lyse, and the nodule either degenerates or resumes new growth.

Beyond that, largely because crushed nodules readily lose their capacity to fix nitrogen and bacteroids taken from nodules are inactive, little is known of the biochemistry. Fractions of crushed nodules have been washed and the plant tissues separated from the bacteroids by differential centrifugation. With N^{15} analysis, it was found that N^{15} enrichment occurs first in the plant membranes, not in the bacteroids, so that these membranes are the likely site of nitrogen fixation.

The current hypothesis of symbiotic nitrogen fixation in legume nodules is as follows: Nitrogen (in the absence of oxygen) oxidizes hemoglobin to hemiglobin; nitrogen is thus activated and reduced to ammonia in the plant membranes. Bacteroid respiration reduces some of the hemiglobin to hemoglobin, and the hemoglobin-hemiglobin system acts as a carrier for electron transport. At the same time, the host is supplying carbon compounds which are partially oxidized by the bacteroids and thus contribute electrons for the reduction of the activated nitrogen on the membranes. These incompletely oxidized carbon substrates then accept ammonia through the activity of the bacteroids. Thus amino acids are made available to the host plant. The first compound containing fixed nitrogen still remains to be discovered.

The nature of the links in the electron transport chain between bacteroid respiration and hemiglobin is unknown. Effective bacteroids have only cytochromes *b* and *c*. Although enzymes for PORPHYRIN synthesis are active in both crushed nodules and isolated bacteroids, the site of synthesis of hemoglobin is unknown. MOLYBDENUM and COBALT are both involved, but their specific functions have not been elucidated.

Plant—Animal Symbioses. As to plant-animal relationships, it is well known that algae live symbiotically, intracellularly, in numerous aquatic invertebrates, mostly marine. Either the hosts are small and light-transparent, or else the algae live in parts of the body near enough to the surface to permit light to penetrate. As a result of the photosynthetic activities of the algae, oxygen is excreted, but this is not necessary for the animals, since there is always sufficient light for a net excess of photosynthesis over respiration by the plants in the community. Similarly, there is an abundance of carbon dioxide in both seawater and fresh water, so that the algae are amply provided. The one aspect of animal respiration which seems to be involved is calcification, especially in reef corals. The coral skeleton is secreted by ectodermal cells, controlled in part by the enzyme carbonic anhydrase in the tissues, catalyzing the reaction $H_2CO_3 \rightleftharpoons H_2O + CO_2$. The deposition of calcium carbonate (as aragonite) from calcium and bicarbonate ions results in the production of hydrogen ions, as in the reaction

$$Ca^{2+} + HCO_3^- \rightleftharpoons CaCO_3 + H^+.$$

In large accumulations of skeleton-building animals, the hydrogen ions produced could effectively lower the pH of the water to such an extent that the precipitation of calcium carbonate would be inhibited. However, if carbonic anhydrase is uninhibited, H^+ should be removed by the formation of $CO_2 + H_2O$. This reaction is favored by removal of the CO_2 by the photosynthetic algae, which helps to maintain the animals' tissue pH. Corals containing symbiotic algae have a 10:1 ratio between calcium deposition in the light and that in the dark.

Nitrogen and phosphorus are excreted by the animals in forms such as orthophosphate, nucelotide phosphate, uric acid and urea. The algae can directly utilize these compounds, although apparently they are not adequate to supply the whole demand of the algae. A great deal of organic matter, vitamins, and other micronutrients may pass between algae and animal tissues, but little is known of the significance of these exchanges because of lack of precise knowledge of the nutritional requirements, particularly of the animal.

Microorganism—Invertebrate Symbioses. Insects, ticks and mites harbor symbiotic microorganisms covering the range of protozoa, bacteria, yeast, rickettsiae and other bizarre forms not even identified. In general, a symbiotic insect deprived of its microorganisms suffers from impaired growth and fecundity, but the loss of symbiotes can be at least partially compensated for by feeding the insect large supplements of yeast. Many of the microorganisms are intracytoplasmic, included either in the gut epithelium or in more interior organs. None of these has been cultivated axenically *in vitro* and practically nothing is known of their metabolism. Mycetocytes, which contain symbiotic bacteroids, have been isolated from American cockroach tissues, and several enzymic reactions have been identified in them, such as glutamate-aspartate transamination, alanine-pyruvate transamination, succinic dehydrogenase and cytochrome oxidase. These reactions also proceed, to a lesser extent, in non-symbiotic tissues, so that their presence in the symbiotic tissue does not explain the dependence of the insect on the microorganisms. Mycetocytes are highly modified, with numerous membranous structures and minute rod-shaped organelles, and the extent to which these may be active, as in legume nodules, is a matter of conjecture.

In cockroaches, the symbiotes are inherited as extrachromosomal factors via the egg, but numerous factors are able to interrupt the transmission. Several antibiotics, mineral imbalances, and lipid and sterol deficiencies in the female insect's nutrition cause a loss of symbiotes in the offspring.

Intestinal microorganisms of insects are somewhat more amenable to analysis, and a number of them have been cultured *in vitro*. Even though these cultured organisms may synthesize and excrete certain B vitamins into the culture substrate, it has not been a simple matter to demonstrate a need of the insect host for the excreted vitamins, probably because the optimum concentrations have not been established. Yeasts in the intestines of Anobiid beetles produce five B vitamins and a sterol, which are deficient in the natural food of the insects.

Flagellated protozoa in the intestines of termites and wood-eating roaches have not been cultured axenically for any length of time, but surviving, washed suspensions of them have been maintained long enough to establish the biochemical step essential to their hosts. The flagellates decompose CELLULOSE to glucose, but the glucose as such is not recovered in the surrounding medium. Instead glucose is anaerobically broken down to acetic acid, hydrogen, carbon dioxide and other unidentified volatile acids. The intestinal epithelium is permeable to acetic acid, which is probably taken up by the hemolymph and thus is the carbon source of energy utilized by the insects.

The growth and differentiation hormone (ecdysone) of the insects is responsible for initiating fundamental internal reorganizations leading to sexual reprodution in the flagellates, which otherwise reproduce asexually.

Microorganism—Vertebrate Symbioses. The rumen, or first chamber of the complex stomach of even-toed, hoofed grazing mammals, the Ruminantia, provides a highly specialized environment, in which the physical conditions are held relatively constant; in it, numerous species of bacteria and protozoa have become adapted. By simulating the rumen environment, cultivation of the rumen bacteria has been achieved and their reactions have been studied. Rumen bacteria consist of a number of species which attack the components of forage, each digesting sugars, cellulose, starch, hemicellulose, xylan or pectin, and producing a number of intermediate organic acids, alcohols or gases. These compounds, succinate, lactate, formate, ethanol, carbon dioxide, hydrogen or methane, are in turn attacked by other species which produce the final products: acetic, propionic and butyric acids, carbon dioxide, and methane. The salts of the volatile fatty acids (acetate, propionate and butyrate) are converted to carbohydrate and utilized by the ruminant. Methane, however, is excreted. Thus the usefulness of rumen fermentation is a function of the degree to which the fermentation products of value are not all converted to methane. (See also RUMINANT NUTRITION.)

In contrast to the substrate specificity of the carbohydrate digesters, nitrogen conversion is widely distributed among the rumen bacteria. Most of the protein is rapidly digested and broken down to ammonia. Much of the ammonia is synthesized into microbial nitrogenous components, and only a low concentration of amino acids is present in the rumen fluid. Excess ammonia is absorbed from the rumen and converted into urea, which in turn can be hydrolyzed to a small degree by many species of microorganisms. In all, about 46% of the total rumen nitrogen is in the bacterial cells, while another 21% is in the protozoa, 26% is in the plant tissues, and 7% is soluble. Of the energy substrates, between 8 and 12% is converted into microbial cells. Ultimately, not only the products of microbial fermentation, but the cells themselves, are absorbed and utilized by the animal.

The flagellates and ciliates in the rumen have not been cultured azenically *in vitro*. Much of the difficulty of clarifying protozoan metabolism is due to the fact that there are bacteria inside of the gastric sacs, and their contributions cannot be estimated.

The nitrogen requirements of the ciliates may be satisfied by ammonia, which is obtained from the rumen fluid, and by a few amino acids, which may be obtained by hydrolysis of the engulfed bacteria. The protozoal bodies contribute up to 20% of the host's nutritional requirements under favorable conditions [see also INTESTINAL FLORA (IN RELATION TO NUTRITION)].

In summary, it can be said that the currently known biochemical aspects of symbiosis are mainly nutritional.

Reference

1. "Symbiotic Associations," Thirteenth Symposium of the Society for General Microbiology, London, Cambridge University Press, 1963, 365pp.

MARION A. BROOKS

SYNAPSES AND NEUROMUSCULAR JUNCTIONS

The area of functional contact between two nerve cells (synapse) and between an axon terminal and a muscle fiber (neuromuscular junction) are regions of unique anatomical, physiological and biochemical specializations which are concerned with information transfer. Since there are a wide variety of types of synapses and neuromuscular junctions, the following discussion will be limited to a description of those from select mammalian, amphibian and invertebrate species which have been the most extensively investigated. For a detailed description of the physiology of synapses and neuromuscular junctions from a large number of animals, reference may be made to Eccles.[2]

Synapses and neuromuscular junctions have been classified on a variety of morphological, functional and biochemical bases. Anatomically, synapses have been classified as being either central (brain and spinal cord) or peripheral (autonomic ganglia, peripheral sensory ganglia and effector junctions) in position and/or axo-dentritic, axo-somatic or axo-axonic in contact relationships. Physiologically, synapses have been classified as either excitatory or inhibitory in action and/or as chemically or electrically mediated. Biochemically, synapses have been classified as either releasing acetylcholine, epinephrine, norepineprine or substances for which the precise chemical composition is not yet fully known. Neuromuscular junctions are classified in the following manner: anatomically as within the peripheral nervous system; functionally as excitatory, inhibitory or sensory; biochemically as releasing acetylcholine.

When synapses and neuromuscular junctions are observed with the electron microscope, their fine structures are seen to be basically similar. The similarities include increased electron density of the presynaptic and postsynaptic membranes, the presence presynaptically of large numbers of MITOCHONDRIA and synaptic vesicles, and an intersynaptic cleft space of 200–300 Å. Synaptic vesicles are 300–600 Å in diameter in synapses where acetylcholine is the mediator, and 600–1000 Å in diameter in synapses where epinephrine or norepinephrine are the mediators. The accumulation of mitochondria and synaptic vesicles on the presynaptic side occurs as a result of the distal migration of materials from the cell body into the axon. Differences in the morphology of central nervous system synapses have been reported,[6] and various functional attributes have been alluded to.[2,6] These include differences in presynaptic and postsynaptic interspace distances, postsynaptic membrane electron densities, and amounts and electron density of interspace substance. A Type I synapse[6] has a synaptic cleft width of approximately 300 Å and a greater electron density in the postsynaptic membrane and interspace substance adjacent to it relative to other synapses. A Type II synapse has a synaptic cleft width of approximately 200 Å and relatively equal electron densities in presynaptic and postsynaptic membranes. On the basis of limited evidence,[2,6] it has been suggested that the morphology of the Type I synapse is typical of all excitatory synapses and the morphology of the Type II synapse is typical of all inhibitory synapses. For other types of morphological variations for which various functional attributes have been assigned, reference may be made to Whittaker *et al.*[6] Typical mammalian and amphibian neuromuscular junctions consist of the apposition of presynaptic axolemma to a highly infolded, postsynatic sarcolemma in which the synaptic cleft width is approximately 200–300 Å. Neuromuscular junctions of teleosts and certain invertebrates do not have an infolded, posysynaptic sarcolemma. Other slight differences in the fine structure of neuromuscular junctions exist which are probably of little functional significance.

Synapses transmit either electrically or chemically, and they may be either excitatory or inhibitory in action. Most synapses and neuromuscular junctions are chemically mediated and excitatory in action. Types of chemically mediated synapses include those from the mammalisn and amphibian central nervous system, sympathetic and parasympathetic ganglia, and sympathetic innervation of the adrenal chromaffin cells. The three most widely known transmitter substances are acetylcholine, epinephrine and norepinephrine. Their chemical formulas (see also ACETYLCHOLINE; ADRENALINE) are:

$$CH_3{-}\overset{O}{\overset{\|}{C}}{-}O{-}CH_2{\cdot}CH_2{\cdot}\overset{+}{N}(CH_3)_3$$

Acetylcholine

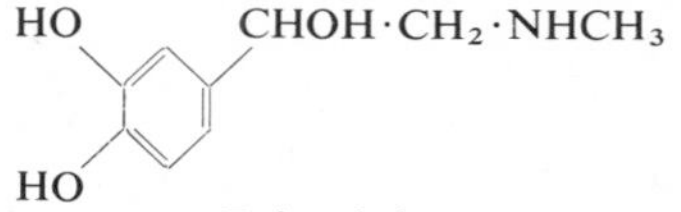

HO CHOH·CH$_2$·NHCH$_3$

HO

Epinephrine

HO CHOH·CH$_2$NH$_2$

HO

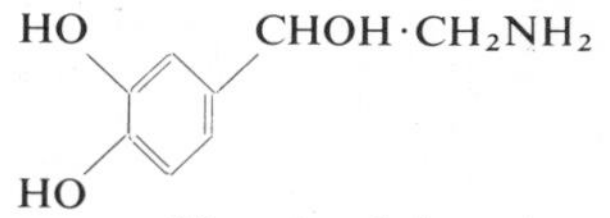

Norepinephrine

There are a variety of other substances which serve as chemical mediators. For a list and description of these substances, reference may be made to McLennan.[4] It has been known for many years that acetylcholine is released from the axon terminals of secondary neurons in the parasympathetic nervous system and that epinephrine is released from the axon terminals of secondary neurons in the sympathetic system during synaptic activity. Since this early work, a variety of other types of synapses and all neuromuscular junctions examined from both vertebrates and invertebrates have been found to utilize acetylcholine as the transmitter substance.

The first evidence for the role of acetylcholine in synaptic activity was the finding that acetylcholine was present in large concentrations from perfusates of heart muscle following synaptic activity.[4] Since the early work, more sophisticated procedures have been utilized to study the mechanisms of synaptic activity more precisely.[3] As conduction reaches the nerve terminals, acetylcholine is released into the narrow, synaptic interspace. It has been postulated by several investigators[2,3] that the presynaptic vesicles are units of, or contain, acetylcholine which is released from the presynaptic membrane into the synaptic cleft. The acetylcholine then combines with receptor protein of the postsynaptic membrane resulting in changes in postsynaptic membrane permeability, thus triggering the excitatory, postsynaptic potential. The precise mechanism of acetylcholine binding to receptor protein is unknown and as Eccles[2] has stated: "The concept of receptor sites is conveniently used in all discussions relating to chemical transmission, yet practically nothing is known about their structure or distribution on the postsynaptic side of the synaptic cleft." Following acetylcholine-receptor combination there is an increase in sodium and potassium conductances giving rise to a depolarizing, excitatory, postsynaptic potential. This is due to an inward flux of sodium ions during the rising phase of the excitatory, postsynaptic potential and an outward flux of potassium ions during the descending phase of the excitatory, postsynaptic potential. The excess acetylcholine present in the synaptic cleft either diffuses from the synaptic cleft or is immediately hydrolyzed to acetate and choline by the acetylcholinesterase present in the postsynaptic membrane. Explanations offered to explain the time course of synaptic delay include the hydrolysis by acetylcholinesterase and/or the rate of diffusion of acetylcholine from the synaptic cleft.

Because of technical limitations, it has been difficult to isolate single, or multiple, synapses and/or neuromuscular junctions for precise biochemical analyses, but beginnings have been made following relatively refined homogenization and ultracentrifugation procedures.[6] It is now known that both acetylcholine and acetylcholinesterases are present in high concentrations in those synapses and neuromuscular junctions where chemical mediation is by acetylcholine. This information has been derived from both biochemical and histochemical techniques.[1,6] It is interesting to note here that the data available show that acetylcholine and acetylcholinesterase are present on both the presynaptic and postsynaptic side of synapses and along the entire length of neurons. These findings lend support to the hypothesis of Nachmansohn[5] which suggests that conduction along neuronal cell membranes involves processes in which choline acetylase, acetylcholine and acetylcholinesterase are intimately linked along the whole neuron and not just at the synapse. Nachmansohn[5] postulates that as a result of acetylcholine's combination with receptor protein in the cell membrane, there is a change in the molecular configuration of the receptor protein such as to change "pore size" or "pore charge" and thus trigger ionic fluxes appropriate to the occasion. For convincing arguments against such an hypothesis, reference may be made to Katz[3] and Eccles.[2]

Chemically mediated, inhibitory synapses are also found in the central and peripheral nervous systems. Inhibitory transmitter substances include: acetylcholine in the case of parasympathetic innervation of the vertebrate heart, and epinephrine or norepinephrine in certain vertebrate sympathetic ganglion cells. Many other synaptic inhibitors are known. For a list of these and their actions, reference may be made to McLennan.[4] It is interesting to note here that inhibition is accomplished by exactly the same substances which cause excitation in other types of synapses. There have been several explanations offered as to why an identical substance, chemically, can act as an inhibitor in one type of synapse and as an exciter in another type of synapse. For a description of all of the explanations, reference may be made to Eccles.[2] Two possible explanations underlie synaptic inhibition; either the transmitter released by inhibitory impulses competes with excitatory transmitter for receptor sites and changes the conductance characteristics of the membrane, or inhibitory impulses depress the release of some excitatory transmitter substance in addition to the specific action of the transmitter substance on the postsynaptic membrane. In most inhibitory synapses, hyperpolarization rather than depolarization occurs in the postsynaptic membrane. This results from the fact that there is an increase in chloride and potassium conductances such as to make the inside of the postsynaptic cytoplasm relatively more negative.

While most synapses and all neuromuscular junctions studied transmit chemically, by acetylcholine or other substances, a few synapses have been found to transmit electrically.[2] Some electrotonic synapses are excitatory and others are inhibitory in action. Of particular interest is whether there is unidirectional or bidirectional transmission. Most electrotonic synapses studied thus far transmit in both directions, whereas chemically mediated synapses transmit in only one direction. Only two cases of unidirectional, electrotonic transmission are known. These are the synapse between the giant axon and giant motor fibers of the crayfish and the Mauthner cell of the goldfish.[2] In those electrotonic, transmitting synapses which have so far been studied with the electron microscope, it was seen that there is a fusion between the outer, lipoprotein membrane leaflets of the presynaptic and postsynaptic membranes.

References

1. Couteaux, R.,"Principaux critères morphologiques et cytochimiques utilisábles aujourd hui pour dé finir les divers types de synapses," *Actualités Neurophysiologie*, **3**, 145–173 (1961).
2. Eccles, J. C., "The Physiology of Synapses," New York, Academic Press, 1964; "The Synapse," *Sci. Am.*, **212**, 56 (January 1965).
3. Katz, B., "The transmission of Impulses from Nerve to Muscle, and the Subcellular Unit of Synaptic Action," *Proc. Roy. Soc. London, Ser. B*, **155**, 455–479 (1962).
4. McLennan, H., "Synaptic Transmission," Philadelphia, W. B. Saunders Co., 1963.
5. Nachmansohn, D., "Chemical and Molecular Basis of Nerve Activity," New York, Academic Press, 1959.
6. Whittaker, V. P., *et al.*, "The Synapse: Biology and Morphology," *Brit. Med. Bull.*, **18**, 223–228 (1962).

James F. Reger

SYNGE, R. L. M.

Richard Laurence Millington Synge (1914–) is a British biochemist who with his collaborator, Archer John Porter Martin, discovered and developed paper chromatography. This is such an important tool in biochemistry that these two investigators were awarded the Nobel Prize in chemistry in 1952.

SZENT-GYORGYI, A.

Albert Szent-Gyorgyi (1893–) is a Hungarian-born chemist who has been in the United States since 1947. He has worked broadly, dealing with the functions of the adrenal glands, the thymus gland and the chemical mechanisms of muscular activity. He is noted for his work with what came to be known as ASCORBIC ACID. He isolated "hexuronic acid" from adrenal glands and suspected, upon hearing of the work of Charles Glen King, that it might be vitamin C. He was awarded the Nobel Prize in physiology and medicine in 1937.

T

TASTE RECEPTORS

Receptor cells especially sensitive to chemicals are found in virtually all animals. By convention, those receptors normally excited by contact with chemicals in liquid phase at relatively high concentrations are termed *taste* or *gustatory* receptors, although the distinctions between taste and smell are not critical at cellular or molecular levels. Since the taste receptors play a central role in food intake and regulation of nutrition, they are the focus of much interest among biologists. Biochemists find that the mechanisms of taste receptor stimulation are ideally suited for analyzing the basic problem of relationships between the structures of molecules and their physiological actions.

Chemical aspects of taste receptor functions are most directly studied by recording the patterns of electrical potentials in receptor cells while the cells are being stimulated with pure chemicals of known structures and properties. Since 1955, when this method was first successfully applied to single taste receptor cells, using receptors on the mouthparts of a fly, many earlier theories of taste stimulation have been revised. This is true of the popular concept of four different types of taste receptor cells, each sensitive to a single type of chemical stimulus (and giving rise to sweet, salt, acid or bitter sensations). Intracellular recordings from taste cells of rat and hamster show that even primary receptor cells are sensitive to 3 or 4 of the so-called basic taste modalities. Consequently, it is generally held that a variety of different receptor *sites* commonly exist on the receptor membrane of any one receptor *cell*. Biochemical characterization of events at receptor sites has progressed furthest in analyzing electrolyte and carbohydrate stimulation.

Electrolyte stimulation is chiefly a function of monovalent cations in all animals which have been adequately studied. Consequently, the receptor sites are thought to be anionic. The pH relationships of stimulation also indicate that strongly acidic (*e.g.*, $PO_4^{\equiv}$ or $SO_4^{=}$) receptor groups are involved. Calculations of free energy changes of the reaction between salt and receptor site give values between 0 and -1 kcal/mole; and low ΔF values suggest that the reaction involves only weak physical forces. The reaction occurs extremely rapidly, since typical nerve impulses can be recorded within 1 msec after stimulating electrolytes are applied. In blow-flies, 0.004*M* NaCl, which produces 1 impulse per second, represents the threshold for behavioral responses; thresholds are slightly higher in man.

No one type of receptor site or reaction can account for the extreme structural specificities, yet curious assortments of molecules, which elicit "sweet" sensations. Early studies on a variety of organisms demonstrated that ring structures and D-isomers were more stimulating in polyol compounds than straight-chains and L-isomers. Thus, inositol (Fig. 1) with its ring structure, stimulates; the straight-chain polyhydric alcohols, sorbitol, dulcitol (Fig. 2), and mannitol, do not. Possession of an α-D-glucopyranoside linkage generally increases stimulating capacity of sugars. Maltose, with a 1,4-linkage, turanose with a 1,3-linkage, and the non-reducing sugars, all stimulate. Lactose, with its 1,4-linkage, and melibiose, with a 1,6-linkage, both lack the α-link and are relatively non-stimulating.

Conformation, as well as configuration, is important in determining the stimulating power of sugar molecules. Glucose, which exists in solution almost entirely in an aldopyranose "chair" conformation, has derivatives of both 1C and C1 conformations (Figs. 3 and 4). Those of the C1 type are much more stimulating. The hydroxyl groups attached to C3 and C4, and inclined 19° above and 19° below the adjacent plane of the molecule, appear to be necessary for the critical linkage at the receptor site. Lack of effects by metabolic inhibitors (azide, fluoride, iodoacetate, etc.), or of temperature effects upon the initial excitatory process, suggests that this step depends upon specific physical, rather than chemical, reactions.

The nature of other polyol receptor sites and the molecular basis for genetic and species differences in taste capabilities remain largely unknown. Saccharin (*o*-sulfobenzimide; Fig. 5) exemplifies both puzzles, since its molecule does not fit any known sugar receptor site, yet it is confused with sugar stimuli by man and other primates, but not by non-primate animals. The substitution of other groups for one hydrogen (dotted lines in Fig. 5) renders saccharin tasteless. The genetic basis of taste has been most thoroughly studied with phenylthiocarbamide (PTC). A strong bitter taste of PTC depends upon the chemical components indicated by dotted lines in Fig. 6, and upon possession of a dominant "taster" gene in man. Curiously, a small change

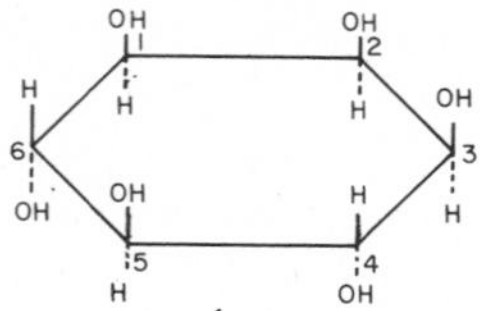

FIGS. 1–7. Molecules illustrating relationships between structure and effects on taste receptors (for explanations, see text).

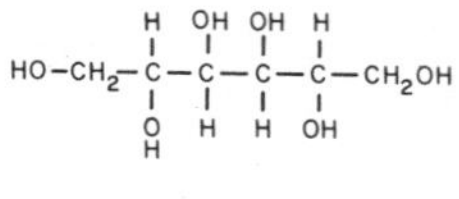

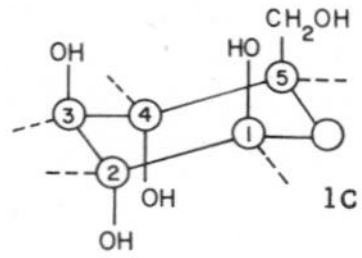

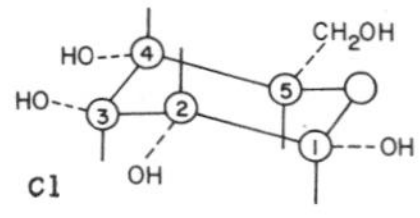

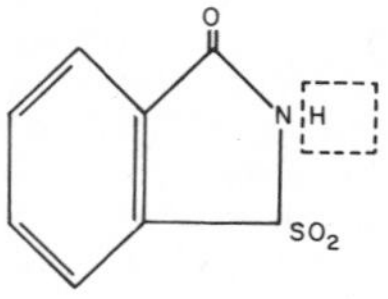

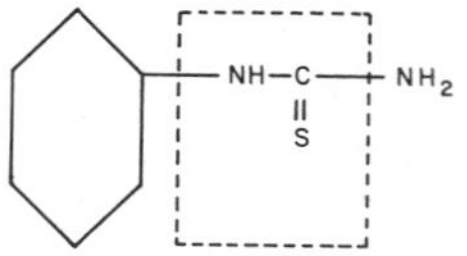

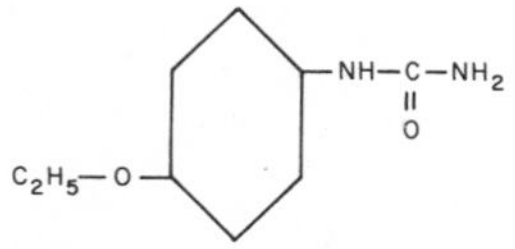

in this molecule (Fig. 7) yields a product 250–300 times sweeter than sugar, and used as a commercial sweetening agent (dulcin).

Taste receptors for water have been reported to occur on mouth parts of mammals and invertebrates. Specialized amino acid and amine receptors are found on the legs of various arthropods. The mechanisms by which adequate stimuli initiate nerve impulses in these cells offer a rich field for investigation. Some stimuli, especially long-chain hydrocarbons, are known to act in the opposite manner—*i.e.*, by decreasing, rather than increasing, the output of receptor impulses. Their effects resemble the actions of narcotics. Taste sensations, as ultimately perceived, therefore, probably result from a complex coded pattern of augmented or depressed frequencies of nerve impulses, originating in the different cells of a heterogeneous population of taste receptors.

References

1. BEIDLER, L. M., "Physiology of Olfaction and Gustation," *Ann. Otol., Rhinol. & Laryngol.*, **69**, 398 (1960).
2. HODGSON, E. S., "Chemoreception," in "Physiology of Insecta" (ROCKSTEIN, M., EDITOR), Vol. I, Ch. 9, New York, Academic Press, 1964, 640pp.
3. MONCRIEFF, R. W., "The Chemical Senses," New York, John Wiley & Sons, 1946, 424pp.
4. PFAFFMANN, C., "The Sense of Taste," in "Handbook of Physiology" (FIELD, J., EDITOR), Washington, D.C., American Physiological Society, 1959.
5. ZOTTERMAN, Y. (EDITOR), "Olfaction and Taste," New York, The Macmillan Co., 1963, 418pp.

EDWARD S. HODGSON

TATUM, E. L.

Edward Lawrie Tatum (1909–) is an American biochemist who, in collaboration with GEORGE WELLS BEADLE, did pioneer work in BIOCHEMICAL GENETICS. Using *Neurospora*, these investigators demonstrated for the first time the close connection between genes and enzymes. They jointly received a share in the Nobel Prize in physiology and medicine in 1958. JOSHUA LEDEBERG was the other recipient that year.

TAURINE

Taurine, aminoethylsulfonic acid,

$$H_2N—CH_2CH_2—SO_3H,$$

is one of the biologically important AMINO ACIDS which does not have a protein origin. It may be produced by the oxidation of CYSTEINE to cysteic acid, $H_2N—CH(COOH)—CH_2SO_3H$, the decarboxylation of which yields taurine. Taurine is present in small amounts in urine, but was early found to be constituent of certain BILE ACIDS.

TERPENES

Terpenes are naturally occurring compounds (hydrocarbons, in the strict use of the term terpene) that are built up from isoprenoid units (see ISOPRENOID BIOSYNTHESIS). A monoterpene (or monoterpenoid such as geraniol, derived by hydrolytic loss of pyrophosphate from geranyl pyrophosphate) contains 10 carbons (2 isoprene units); a sesquiterpene, 15 carbons (3 isoprene units); a triterpene, 30 carbons (6 isoprene units), etc. The CAROTENOIDS are tetraterpenoids, consisting of 8 isoprenoid residues. Many plant essential oils, and many pigments of plants and of microorganisms, are terpenoids. Natural rubber and guttapercha are polyterpenes.

Reference

1. SCHUERCH, C., "Terpenes," in "The Encyclopedia of Chemistry" (CLARK, G. L., AND HAWLEY, G. G., EDITORS), New York, Reinhold Publishing Corp., 1957.

TESTOSTERONE

This steroid (Δ^4-androstene-17β-ol,3-one; $C_{19}H_{28}O_2$; melting point 155°C)

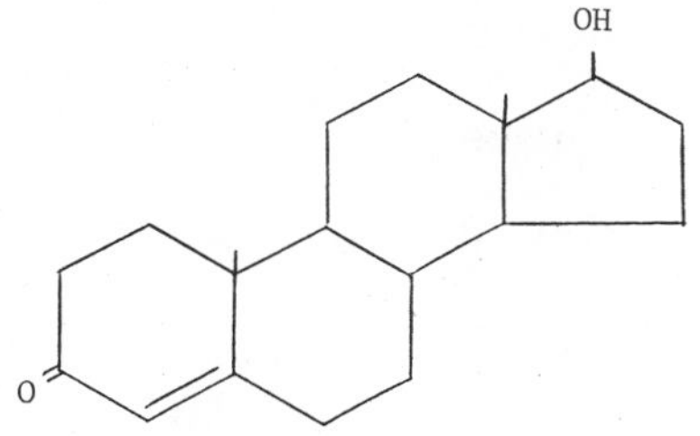

OH

O

Testosterone

is considered to be the primary androgenic hormone elaborated by the mammalian testis and was first isolated in 1935 by Laqueur from steer testis.

The route of its biosynthesis is by way of acetate units to cholesterol then through pregnenolone to testosterone. Acetate is a proven precursor *in vivo*, *in vitro* and in perfusion experiments involving testis, adrenal tumors and ovary. In this regard, progesterone and pregnenolone have been considered by many to be more immediate substrates for androgen biosynthesis since they yield in testis and ovary homogenate systems the products androst-4-ene-3,17-dione and testosterone. The dione is metabolically readily convertible to testosterone and vice versa. The mechanism of cleavage of the side chain of cholesterol and progesterone and the introduction of oxygen to yield the 17β-hydroxy group of testosterone is unsettled. Corticosteroid C_{21} compounds are precursors *in vivo* of urinary 17-ketosteroids.

A related steroid, 5α-androst-16-ene-3α-ol, originally isolated from testis has now been detected in human urine as the glucosiduronic acid. Men excrete approximately 1.3 mg/day in contrast to 0.45 mg/day for women. The significance of this substance as a precursor or a metabolite of testosterone remains to be ascertained.

The mean free testosterone level in normal men is given as 0.56 μg/100 ml of serum, and in normal women, 0.12 μg/100 ml of serum.

The catabolism of testosterone proceeds by oxidation to androstenedione, saturation of the 4,5-double bond gives the two androstanediones, reduction at carbon-3 yields the four corresponding hydroxyketones, and reduction of the 17-keto group leads to androstane-3,17-diol. The very first metabolic reaction of testosterone is its conversion to the corresponding glucosiduronic acid in which form the steroid moiety undergoes metabolic change. The unconjugated testosterone may follow a different sequence of metabolic reactions. Hydroxylation occurs at positions 2,6,11 and 16. The 17α-hydroxy epimer of testosterone, epitestosterone, is also a catabolite of testosterone. The substantial portion of the urinary 17-ketosteroids determined by the Zimmermann reaction, however, is attributed to steroid catabolites of the adrenal—a major component being dehydroisoandrosterone.

A most significant route of metabolism is the conversion *in vivo* of testosterone to estradiol, estrone and estriol via the oxidative removal of the angular methyl group at carbon-19.

The biological effects of testosterone are its stimulation of growth of sexual and nonsexual tissues. In the male, testosterone stimulates the growth of seminal vesicles, prostate, vas deferentia, epididymis, penis, coagulating glands and preputial glands, and in birds it stimulates the growth of comb and wattles. It stimulates the process of spermatogenesis. In the female, growth effects are noted in mammary gland and clitoris, and the hormone induces ovulation in birds and rodents. Moreover, testosterone inhibits the pituitary gland through a feedback control mechanism effective on the secretion of GONADOTROPINS (interstitial cell stimulating hormone). Also, testosterone is responsible for the mating behavior of birds and mammals.

Effects on non-sex tissues include protein anabolism (nitrogen retention, increase in mass of levator ani muscle), increase in vascularity and flushing of the skin, increased production of red blood cells and hemoglobin, sebaceous secretion and acne, pigmentation, bone growth and maturation of epiphyseal centers and the renotrophic effect (see also ANDROGENS; HORMONES, STEROID; SEX HORMONES).

A number of anzymes increase their activity in response to testosterone *in vivo*. In the mouse kidney, arginase, D-AMINO ACID OXIDASE and β-GLUCURONIDASE respond; in humans, androgens will elevate the serum acid phosphatase in metastatic prostatic cancer.

The bioassay of androgens is performed by several procedures. In the capon comb test, the order of biological activity is testosterone > androsterone > dehydroisoandrosterone > epiandrosterone. A somewhat similar order of activity is found in the rodent assays in which the weight of seminal vesicles and prostate gland is measured in castrated animals. The rat levator ani muscle is considered to be a measure of protein anabolic activity, and the results of this assay correlate well with the mouse renal β-glucoronidase response. Analogues of testosterone have been prepared, and these show enhanced biological activity, *e.g.*, the propionate and cyclopentyl propionate esters of testosterone, various enol ethers and also 17-alkyl substituted testosterone. The absence of the 19-methyl group, as in 19-nortestosterone reduces classical androgenic activity but not the extent of protein anabolism. Other steroids possessing a preponderance of protein anabolic activity are methylandrostenediol, halotestin and androstanazole.

In therapy, testosterone and its derivatives are used in correcting the deficit in hypogonadal men and in the palliation of metastatic breast carcinoma.

The mechanism of action of testosterone cannot be explained by its serving as a substrate for reversible transhydrogenases or by its participation as a coenzyme in a key enzyme of metabolism. It appears from recent work more likely that the site of action of the androgen in the responsive cell is nuclear DNA and that this DNA directs the production of specific messenger RNA, leading to the synthesis of specific enzymic and non-enzymic proteins. The evidence for this hypothesis rests on the ability of prior treatment by Antinomycin D to prevent a number of responses to testosterone administration.

References

1. Dorfman, R. I., Forchielli, E., and Gut, M., "Androgen Biosynthesis and Related Studies," *Recent Progr. Hormone Res.*, **19**, 251–274 (1963).
2. Fieser, L. F., and Fieser, M., "Steroids," New York, Reinhold Publishing Corp., 1959.
3. Baulieu, E. E., Corpechot, C., Dray, F., Embiozzi, E., Lebeau, M. C., Mauvais-Jarvis, P., and Robel, P., "An Adrenal-secreted 'Androgen'; Dehydroandrosterone Sulfate, its Metabolism and a Tentative Generalization on the Metabolism of other Steroid Conjugates in Man," *Recent Progr. Hormone Res.*, **21**, 411 (1965).

William H. Fishman

THEORELL, A. H. T.

Axel Hugo Teodor Theorell (1903–) is a Swedish biochemist whose outstanding contribution, for which he received the Nobel Prize in physiology and medicine in 1955, was the elucidation of the chemistry and functioning of flavoprotein enzymes.

THIAMINE AND BERIBERI

Beriberi is an ancient disease, being mentioned in the Chinese Neiching believed to date from 2697 B.C. Its primary focus has always been in Asia, but it has been repeatedly reported on every continent and in scores of countries, as well as on hundreds of ships propelled by sail, and later by steam. Presumably it became more prevalent after the advent of mechanical power for the milling of rice. It has been caused primarily by the consumption of excessive proportions of white rice, but several extensive epidemics, *e.g.*, in Africa and in Labrador, were associated with diets including little or no rice.

The early history of beriberi is covered best by two books; one by W. L. Braddon (1907), the other and better one by E. B. Vedder (1913). The latter book was the first to set forth the modern and now universally accepted theory of the cause of beriberi, namely a deficiency of a necessary component of the diet now generally called vitamin B_1 or thiamine. The first symptom of beriberi to appear is usually numbness of the extremities especially the lower part of the legs. It is often possible to diagnose beriberi by exploring the calves and shins with the point of a pin. One can often draw blood without the patient being aware of the pin pricks. Such anaesthesia is frequently accompanied by quite marked sensitiveness to pressure on the calves which may induce quite acute pain. The common practice now is to use loss of the patellar reflexes or ankle jerks as a leading symptom of beriberi. The latest book on the subject which reflects our present full understanding of the disease is cited below as reference 1.

During the first decade of this century, before its fundamental cause was understood, there were thousands of deaths from beriberi, and its incidence and the mortality from it were conspicuously greatest among men doing hard labor. Later when the cause, first suggested by Christian Eijkman in Java, had been clearly demonstrated by Fraser and Stanton in Malaya to be a deficiency in white rice, the disease became less deadly and is now largely limited to mothers and babies in whom the symptoms of the disease are less conspicuous and less well defined.

Edema may frequently appear among the early symptoms. Sometimes edema appears only later and sometimes not at all. It can not be too strongly emphasized that there is great variability in the appearance of beriberi. In infacts the disease usually begins with incessant crying which in a few hours, or at most a few days, turns into a characteristic whine. The baby is restless, cyanotic around the lips and there is a generalized pallor of the body. There is a very low output of urine. The baby is not comatose but is very indifferent to its surroundings. In some cases, there is a more or less complete loss of voice and the baby may cry for hours almost noiselessly.

At autopsy, the adult and infant forms exhibit much the same abnormalities. In both, autopsy reveals very widespread and extensive changes which show that most tissues of the body are affected. The heart, especially the right ventricle, is dilated and engorged with blood. There is some clear fluid in the pericardium also often in the lungs. Hyperaemic spots are found widely in the intestines and other internal organs. There is often considerable edema of the genital organs and especially the lower part of the legs, but this edema may extend throughout the body. All the principal nerves have degenerated with loss of the myelin sheath. There may be some degeneration of the spinal cord. Beriberi is clearly primarily a nerve disorder, but many other functions are seriously deranged. Beriberi has been widely referred to as wet or dry depending on whether or not edema is a conspicuous symptom. Death from heart failure occurs in both cases very frequently.

Beriberi was very prevalent in Japan from 1870 to 1910 or 1915, and the Japanese have made numerous contributions to our knowledge of the disease. Takaki, a surgeon general of the Japanese Navy, first called emphatic attention to the dietary cause of the disease. He liberalized and diversified the Navy ration and thus brought about a dramatic decrease in the incidence of beriberi among the sailors. Later it was Hirota who first recognized and described the infantile form of the disease. Many of the early Japanese publications were in Japanese and never have been made available in a western language. This writer, therefore, asked the Japanese Committee on Vitamins to summarize all Japanese contributions to the subject in the English language, and this summary has been scheduled for publication.

When I first began to study beriberi in 1910, the greatest single problem which it presented was a chemical one, namely the identity of the essential substance lacking in white rice and present in the bran coats. I began to try to isolate the substance at that time but for many years had little success. The first success in isolating it was reported by Jansen and Donath in Java in 1926. They obtained the substance in pure form but in very low yields so they made little progress in studying its chemistry. Even its empirical formula remained unknown for they overlooked the presence of sulfur in the molecule, and their microanalyses, therefore, made little sense.

I was unable to secure any crystals by the Jansen and Donath procedure and continued to try to develop procedures which would give better yields. I was at last successful in 1933, just 23 years after my first efforts. We soon established the composition as $C_{12}H_{18}N_4OSCl_2$. In our earliest experiments we quantitatively split this substance into two fragments in accordance with the following equation:

$$\underset{}{C_{12}H_{18}N_4OS} + H_2SO_3 \rightarrow \underset{\text{I}}{C_6H_9N_3SO_3} + \underset{\text{II}}{C_6H_9NOS}$$

Product I was judged from its composition and its ultraviolet absorption to be a pyrimidine sulfonic acid. There were, however, theoretically possible, 15 substances of this description, and none of them were described in the literature. After many other endeavors which proved to be in vain, we reduced product I with sodium in liquid ammonia and thus obtained 2,5-dimethyl-6-aminopyrimidine which proved identical with that substance which we had produced synthetically by methods which predetermined its structure. We could therefore, conclude that product I had the structure:

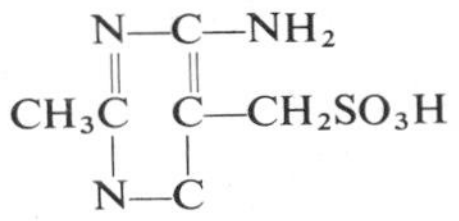

```
      N—C—NH2
      ‖   ‖
  CH3C   C—CH2SO3H
      |   |
      N—C
```

Product II was more easily identified. On oxidation with nitric acid, product II gave a substance $C_5H_5NSO_2$, which clearly was 4-methylthiazole-carboxylic acid as recorded by Wohrman in 1890. Product II was also shown (by heating it in a sealed tube with hydrochloric acid) to contain a hydroxyl in the side chain, for the product of this reaction had the composition $C_6H_8NSCl \cdot HCl$. Product II was, therefore, believed to be

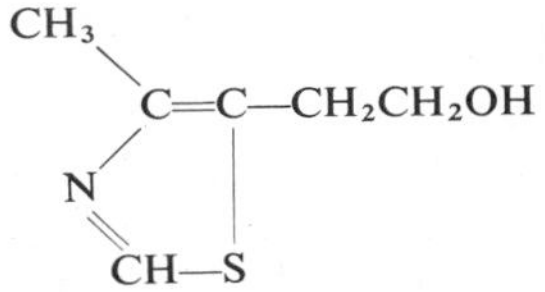

```
  CH3
     \
      C=C—CH2CH2OH
     /   |
    N    |
     \\  |
      CH—S
```

and this was proved by synthesis from brominated acetopropyl alcohol plus thioformamide.

When we gently heated 2-methyl-6-amino-5-bromomethylpyrimidine hydrobromide with product II, we obtained fairly abundant quantities of the vitamin itself, as was proved by its ability to cure polyneuritis in rats and pigeons when given is doses of 5 μg or thereabouts. Hundreds of tons of the substance have since been manufactured and sold in the United States as the chloride hydrochloride or the nitrate for enrichment of bread, flour and corn meal. It is only necessary to treat the bromide hydrobromide described above with silver chloride to obtain the chloride hydrochloride which has the following structure:

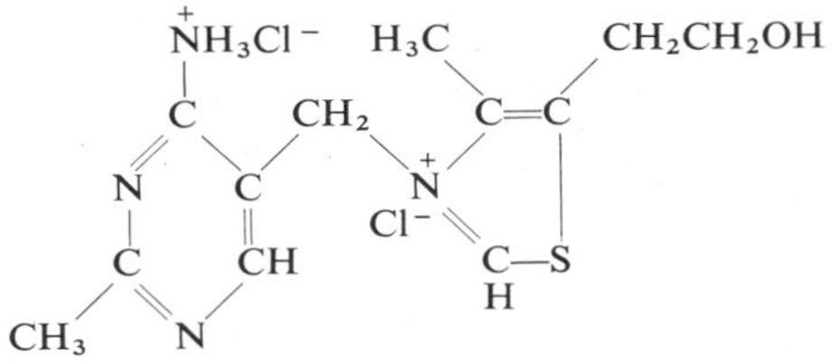

```
        +
        NH3Cl⁻   H3C          CH2CH2OH
        |           \         /
        C      CH2    C=C
      //  \   /   \  /    |
     N     C       N+     |
     |     ‖  Cl⁻ \\      |
     C     CH       C—S
    / \\  /         H
 CH3    N
```

Thiamine chloride hydrochloride

The Philippines is the only country in the world which has systematically kept track of the deaths from beriberi ever since 1904 (except the years of the Japanese occupation). There are about 25,000 deaths a year from the disease in the Philippines. Salcedo and his associates clearly showed in Bataan that beriberi in human beings can be prevented by including 2 mg of thiamine in each pound of rice consumed. So far the Philippine government has not taken steps to enforce the rice enrichment law enacted in 1952 prohibiting the sale of white rice without vitaminization. We still hope that the Philippines will adopt such measures, for we believe their example will be very effective throughout Asia and thus will help stamp out a great scourge of that continent.

Reference

1. Williams, R. R., "Toward the Conquest of Beriberi," Cambridge, Harvard University Press, 1961.

Robert R. Williams

THREONINE

L-Threonine (α-amino-β-hydroxy-n-butyric acid),

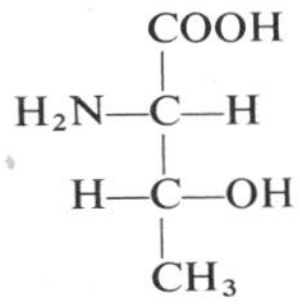

```
        COOH
        |
   H2N—C—H
        |
     H—C—OH
        |
        CH3
```

is one of the AMINO ACIDS that is common to many PROTEINS. Since it has two ASYMMETRIC carbon atoms, the name given above in parentheses is not definitive and applies equally well to the three other optical isomers of L-threonine: D-threonine and D- and L-allothreonine (see DIASTEREOISOMERS).

L-Threonine is a nutritionally ESSENTIAL AMINO ACID and was discovered by Rose because of this fact. Rats fed the amino acids known before the discovery of threonine lacked something essential, which was finally identified as L-threonine. The individual human needs, as determined by the amounts required to maintain nitrogen equilibrium, vary from about 100–500 mg/day. L-Threonine in water solution has a specific rotatory power $[\alpha]_D = -28.3°$.

THYMINE AND THYMIDINE

Thymine (5-methyluracil; 2,4-hydroxy-5-methylpyrimidine), is one of the PYRIMIDINE bases which is commonly found in DEOXYRIBONUCLEIC ACIDS (DNA). In RIBONUCLEIC ACIDS (RNA), uracil is present rather than thymine. Thymine gets its name from "thymus nucleic acid" where it was first found. The structure of thymidine (1-β-D-deoxyribofuranosylthymine) is also shown.

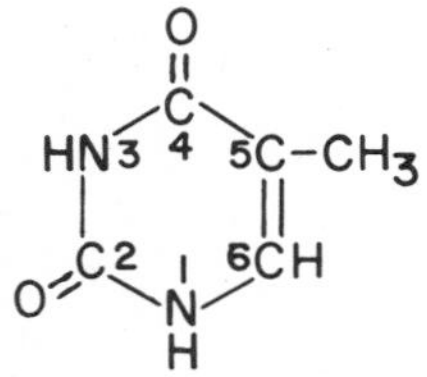

Thymine

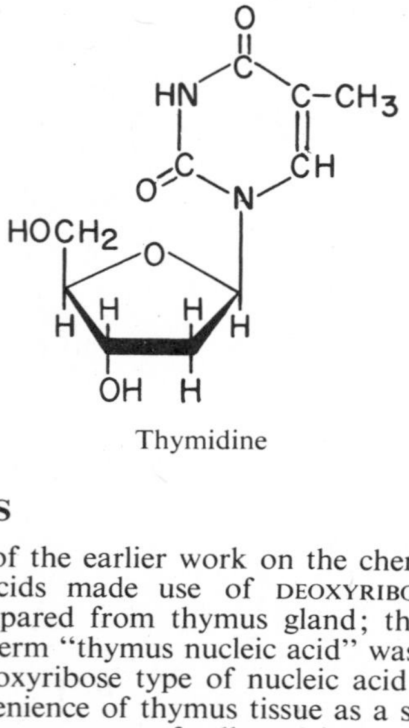

Thymidine

THYMUS

Much of the earlier work on the chemistry of nucleic acids made use of DEOXYRIBONUCLEIC ACIDS prepared from thymus gland; thus for a time the term "thymus nucleic acid" was applied to the deoxyribose type of nucleic acid (DNA). The convenience of thymus tissue as a source of DNA (a component of cell nuclei) depended on the fact that the cells of the thymus are predominantly lymphocytes, which have an unusually high volume ratio of nucleus to cytoplasm.

The biochemical role of the thymus is closely involved with the production of ANTIBODIES, the protective immunological agents of the body (see IMMUNOCHEMISTRY). In higher animals, the thymus appears to function by producing, and stocking the rest of the body with, lymphocytes; relatively early in life this process is finished, and the thymus decreases in size, leaving only a remnant in the adult animal.

References

1. BURNET, M., "The Thymus Gland," *Sci. Am.*, **207**, 50–57 (November 1962).
2. LEVEY, R. H., "The Thymus Hormone," *Sci. Am.*, **211**, 66–77 (July 1964).

THYROID HORMONES

The thyroid hormones are unique biological compounds is that they contain IODINE. The principal component found in the thyroid gland and circulated in the plasma of most animals is L-thyroxine, shown in Fig. 1. Other iodinated compounds isolated from the gland include L-3-iodotyrosine, L-3,5-diiodotyrosine (Fig. 2), L-3,3′-diiodothyronine, L-3,5,3′-triiodothyronine (Fig 3), and L-3,3′,5′-triiodothyronine. Of these other five, only L-3,5,3′-triiodothyronine exhibits appreciable thyroxine-like activity.

Activity requires the L-configuration and diphenyl ether grouping, and is reduced by changing the length of the three-carbon side chain or substitution of the iodine by bromine or chlorine.

***In Vivo* Effects of Thyroid Hormone.** Two types of effects produced by thyroxine must be distinguished—physiological and pharmacological. The latter is produced by tissue levels greatly in excess of those normally found. The peripheral tissue content of L-thyroxine has been estimated to be of the order of $5 \times 10^{-8}M$. Many of the current concepts regarding the physiological role of thyroxine stem from work done with animals given relatively high doses of thyroid hormone, and in only a few cases has there been an attempt to do a broad time study of various parameters with minimal doses. It is therefore difficult to determine whether many of the effects reported are of a primary or secondary nature.

The most clearly established effect of thyroid hormones is on cellular oxidative processes. A hyperthyroid animal at rest shows an increased rate of oxygen consumption (basal metabolic rate, BMR), and tissue slices prepared from its liver, kidney and muscle also show increased oxygen consumption. The increase in oxygen consumption does not appear to occur in all tissues because brain, spleen and gonads are unaffected. The reverse situation holds with hypothyroid animals. Thyroxine can only produce an increase in oxygen consumption when administered *in vivo*, *i.e.*, the addition of thyroxine to tissue slices prepared from normal animals does not increase oxygen consumption.

The increase in cellular oxidation is a reflection of increased production of heat and of energy for cellular functions. Thyroid hormones are thought to play some role in heat production, but the mechanism remains undefined.

Thyroxine also has an effect on growth that may be unrelated to its action on oxidative processes. Thyroidectomized animals are subject to a type of dwarfism characterized by an inhibition of bone

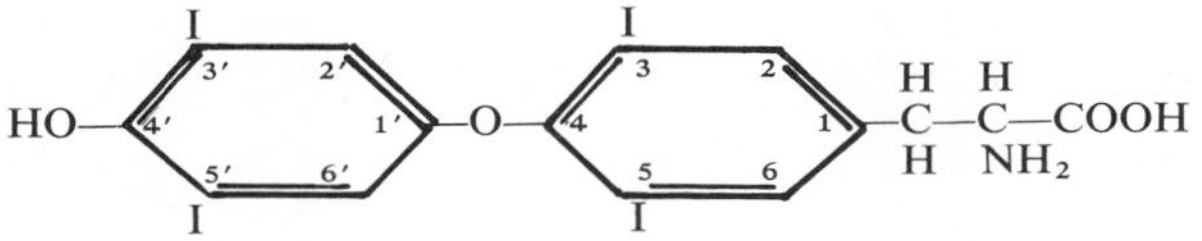

FIG. 1.

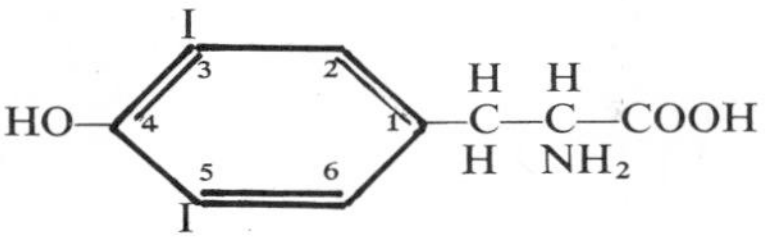

FIG. 2.

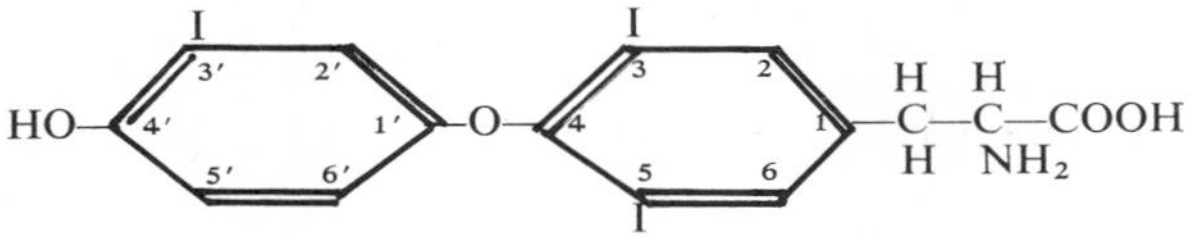

FIG. 3.

ossification and the maintenance of an infantile type bone structure. It is not possible at present to say whether this is caused by a direct role of thyroxine on growth processes or whether it is a manifestation of an indirect effect on GROWTH HORMONE.

Thyroxine produces a marked acceleration of METAMORPHOSIS in amphibia. Thyroidectomized tadpoles continue to grow to a relatively enormous size but do not undergo metamorphosis unless they receive thyroid hormone. The mechanism of this effect is not understood.

In vivo effects on water and electrolyle metabolism, on nitrogen and lipid metabolism, on central nervous system activity, on lactation, and on pulse rate, cardiac output and respiration also have been well documented, but the basis of these effects is poorly understood.

***In Vitro* Effects of Thyroxine.** Since thyroxine has more than one physiological effect, it may have more than one target site. Three are currently being considered, a direct effect on one or more key enzymes, an effect on membrane permeability, and an interaction with metal ions.

It is unlikely that the mechanism(s) of action of thyroxine involves a direct effect on one or more rate-limiting enzymes. A wide variety of different enzymes have been reported to be affected, and in addition there appears to be a poor correlation between the chemical structure required for *in vivo* action and that producing an *in vitro* effect. The lack of correlation with *in vivo* structural requirements may however be a reflection of differences in distribution and metabolism of thyroxine analogues *in vivo*. The amount of hormone required to affect enzyme reactions *in vitro* is high, thereby raising the question of whether this represents a physiological action. Finally, there is no obvious correlation between the myriad of *in vitro* effects observed and the established physiological action of thyroxine.

The second proposed target site, *i.e.*, membrane permeability can more readily lead to an explanation of differing physiological effects. A structural change in the membrane could affect the availability of reactants and products, of ions and also the activity of membrane-bound enzymes. The most useful study system has been the MITOCHONDRIA. Concentrations as low as $1 \times 10^{-8}M$ produce an increase in the volume of normal mitochondria *in vitro*. Mitochondria from hyperthyroid rats show increased membrane fragility whereas those from hypothyroid animals are more stable than normal. There is also an interesting parallelism between thyroxine-induced swelling of mitochondria from different tissues of normal rat and the increased oxygen consumption of hyperthyroid animals. Mitochondria from liver and kidney swell rapidly on exposure to thyroxine, whereas those from brain, spleen and testis do not. Mitochondrial enzyme reactions that are affected *in vitro* by high concentrations of thyroxine, such as oxidative PHOSPHORYLATION and ATPase, are unaffected when submitochondrial fragments are tested, suggesting that the primary effect in intact mitochondria is on the mitochondrial membrane. However, thyroxine analogues such as D-thyroxine having little or no *in vivo* effect can also produce mitochondrial swelling *in vitro*.

Thyroxine can form insoluble complexes with a variety of divalent cations. Many enzymes require divalent cations for their activity, and it has been suggested that some of the *in vitro* effects of thyroxine can be explained on this basis. However, there is no evidence at present that a direct reaction can occur between thyroxine and a metalloenzyme either *in vivo* or *in vitro*.

References

1. LITWACK, G., AND KRITCHEVSKY, D. (EDITORS), "Actions of Hormones on Molecular Processes," New York, John Wiley & Sons, 1964.

2. Pitt-Rivers, R., and Tata, J. R., "The Thyroid Hormones," New York, Pergamon Press, 1959.
3. Pitt-Rivers, R., and Trotter, W. R., "The Thyroid Gland," Washington, D.C., Butterworths, 1964.
4. Tapley, D. F., and Hatfield, W. B., *Vitamins Hormones*, **20**, 251 (1962).
5. Tata, J. R., Ernster, L., Lindberg, O., Arrhenius, E., Pedersen, S., and Hedman, R., *Biochem. J.*, **86**, 408 (1963).

Cecil Cooper

THYROTROPIC HORMONE

Thyrotropic hormone (TSH) is a protein hormone, secreted by certain basophilic cells in the anterior pituitary gland, which stimulates the release of THYROID HORMONE (thyroxine) from the thyroid gland as well as the synthesis of new hormone from iodide and tyrosine.

The rate of TSH secretion is controlled both by the circulating level of thyroxine not bound to specific carrier proteins ("free thyroxine") and by a peptide thyrotropin releasing factor (TRF) from the anterior hypothalamus. Thyroxine shuts off TSH production largely by a direct effect on the thyrotrophic cells of the pituitary, although an additional action on hypothalamic centers has not been ruled out. TRF may not be essential for a basal level of TSH secretion, but it is needed to permit the maximal TSH secretion required for thyroid gland hypertrophy.

Chemical purification of TSH has yielded a protein of molecular weight 28,000. Highly purified TSH is unstable and loses two-thirds of its activity rapidly even at very low temperatures. This instability may be due to contamination by proteolytic enzymes in the very early stages of extraction. Even the best TSH preparations are inhomogeneous in starch gel ELECTROPHORESIS and resolve into 6 bands all of which have biological activity and very similar amino acid and carbohydrate content. Amino acid sequences have not yet been determined. Since all TSH preparations have luteinizing hormone activity, there may be some structural overlap between the two hormones (see GONADOTROPIC HORMONES).

Eel TSH differs markedly from beef or other mammalian TSH in its biological activity. Mammalian thyrotropins show moderate immunologic cross-reaction. TSH is not responsible for human thyrotoxicosis (Graves' disease). However, the "long-acting thyroid stimulator" which appears in the circulation of many hyperthyroid patients, and seems responsible for both thyroid overactivity and ophthalmoplegia, may be either an immune γ-globulin duplicating a significant portion of the TSH structure or else a TSH fragment bound to a γ-globulin carrier.

The mechanism by which TSH affects the functioning of the thyroid gland is still in doubt. Enhanced permeability of thyroid cells or accelerated movement of colloid droplets to the sites of proteolytic breakdown of thyroglobulin to thyroxine seem most likely. Exogenous TSH has not been shown to concentrate in the thyroid and is eliminated largely by the kidneys. The activity of TSH disappears from human serum with a half-life of about 35 minutes. Human serum TSH levels are very low (equivalent to 1 μg of the best pituitary TSH preparations in 10 liters of plasma), and even in primary thyroidal failure, the grossly elevated TSH levels are still only of the order of a few micrograms per liter. Chemical purification of plasma TSH has, therefore, not been attempted.

References

1. Purves, H. D., "Control of Thyroid Function" in "The Thyroid Gland" (Pitt-Rivers, R., and Trotter, W. R., Editors), Vol. 2, pp. 1–38, Washington, Butterworths, 1964.
2. Carsten, M. E., and Pierce, J. G., "Starch Gel Electrophoresis and Chromatography in the Purification of Beef Thyrotropic Hormone," *J. Biol. Chem.*, **235**, 78–84 (1960).
3. Morris, C. J. O. R., "Chemistry of Thyrotropin" in "Thyrotropin" (Werner, S. C., Editor), pp. 209–215, Springfield, Ill., C. C. Thomas, 1963.
4. Greer, M. A., "The Participation of the Nervous System in the Control of Thyroid Function," *Ann. N.Y. Acad. Sci.*, **86**, 667–675 (1960).

Warner H. Florsheim

THYROXINE. See THYROID HORMONES; also AMINO ACIDS (OCCURRENCE, STRUCTURE, AND SYNTHESIS); GOITER; IODINE (IN BIOLOGICAL SYSTEMS).

TISELIUS, A. W. K.

Arne Wilhelm Kaurin Tiselius (1902–), a Swedish biochemist, is noted for his painstaking and ingenious development of ELECTROPHORESIS as a practical method for separating and studying PROTEINS. He received the Nobel Prize in chemistry in 1948.

TISSUE CULTURE, ANIMAL

The first phase of the classical science of biochemistry was concerned with study of the chemistry of products of living organisms; with the effects of diet, drugs and other environmental factors upon chemical processes in whole organisms; and with analyses of the composition of organisms and their parts (organs, whole blood, serum, etc.). When the groundwork of basic chemistry of the proteins, carbohydrates, and fats had been laid, and some details of the importance of vitamins, hormones, and enzymes became clear, the second phase began. In this, the emphasis shifted to the study of biochemical pathways, their mediation by enzymes, and their interrelationships in metabolism. In this second phase, the elucidation of specific pathways depended heavily upon the study of enzymatically controlled systems in tissue minces and breis, upon the isolation of parts of cells for the identification of specific enzymes carried upon them, and finally upon the isolation and purification of the enzymes themselves. We are today in the phase where many pathways of metabolism and biosynthesis, and ways in which they can be influenced, are recognized. It is now also well understood that the coordination of the multitude of chemical reactions which proceed simultaneously in the cell is to a great extent controlled by molecular arrangements in space, particularly upon inter- and intracellular surfaces. This phase of biochemical investigation therefore demands the use of whole, living cells, in conditions as close as possible to those obtaining *in situ* in the natural state.

The methods of tissue culture are designed for just this kind of situation. The aim is to isolate and maintain living, metabolizing cells, tissues or organs in artificial media, so that each population of cells, organ, or fragment of tissue can be studied in a controlled environment, independent of the fluctuating and incompletely known influences (hormones and all the products of metabolism of diverse other cells) which play upon them in the organism as a whole. Cells in tissue culture, in contrast to homogenates, retain the structural integrity which is important for coordination of function; in addition, they are in experimentally controllable environments, in contrast to similar cells *in vivo*.

It is thus possible, using cells in artificial culture, to investigate the nutrition and metabolism of muscle, liver, bone, mammary tissue, etc., independently of the effects of other tissues. Certain limitations are inherent in the method, one of which is the size of the organ or fragment which can be successfully cultivated. Central necrosis occurs when the thickness of tissue is too great to permit adequate penetration of oxygen and other essential nutrients. However, this difficulty can to some extent be overcome by the use of thin, strip-shaped explants. Cell cultures are generally grown as monolayers on glass or plastic surfaces or, with proper attention to aeration, in suspension. It is now possible to grow massive cultures, from which grams of cells can be regularly harvested, in very large monolayers, *e.g.*, in 5 liter bottles, or in aerated suspensions. Such cultures are generally made from cells of a single type, or even from clones of cells derived from single cells. On the other hand, interesting observations can be made on the interactions of one tissue upon another, by deliberate juxtaposition of unlike tissues, throwing light on the biochemical mechanisms underlying organogenesis.

The development of chemically defined nutrient media for cells has markedly increased the degree of definition of the environment and has made possible the identification of nutrients which are necessary for cell survival and proliferation. Most cells can grow satisfactorily in media containing only salts, carbohydrate, amino acids and a few water-soluble vitamins, in the right proportions. The presence of protein may be beneficial, but is usually not essential, particularly for cells which themselves make, and discharge into the medium, protein from simpler components. Fats and fat-soluble vitamins are not necessary, though interesting modulations of certain cells are produced by the addition or exclusion of vitamin A (mucous metaplasia or keratinization of skin). Hormones have quite specific pharmacological effects upon certain cell types, but are only needed for the survival and growth of certain specialized tissues, *e.g.*, mammary gland.

Some differences are to be noted between short-term (primary) and long-term (established) cultures. It is usual in biochemical and physiological practice to use tissue slices suspended in balanced salt solutions (see NUTRIENT SOLUTIONS). Such surviving tissues are essentially in a state of delayed death. The short-term culture, providing more adequate nutrition to the tissue, bridges the gap between such surviving tissues and long-term cultures, in which cells may with proper care be carried along indefinitely in potentially unlimited proliferation. The gap is not a purely temporal one. Organ cultures fall into the first classification, of limited systems. Maintenance of the inter-tissue relationships proper to an organ is, *in vitro* as in life, generally incompatible with progressive and unrestricted increase in the cell aggregate. By no means is every primary culture capable of progressing to grow as an established cell culture. The reasons for this are poorly understood. They may be largely lack of knowledge of correct conditions permitting "establishment," *i.e.*, conditions under which cells of a primary culture may be induced to remain in, or (if they have not this capacity) enter into, a state in which cell proliferation is possible. It is known that many enzyme activities commonly diminish early in culture life, and the failure of many cells to pass the threshold from primary to established cultures may be associated with these losses. It has also been demonstrated that certain synthetic functions of cells may proceed at too slow a rate to supply enough of the end product to maintain cell growth. Serine, cystine and inositol which many cells can synthesize from precursors, but at too slow a rate to meet the needs of the cells, are examples of nutrients whose addition, preformed, may permit survival and growth. Such supplements are particularly important for cultures of small numbers of cells in a large volume of medium, where slowly formed metabolites diffuse away and become unavailable at the cell sites where they are needed. Cells which do survive and establish long-term cultures are those which, by either selection or mutation, are capable of growing under the restrictions imposed upon them by particular culture conditions. This fact places some limitation upon their use, because the behavior of cells in established culture is to a significant extent a function of the culture conditions and may differ in many respects from the behavior of freshly explanted cells. Long-term cultures are nevertheless of great value for studying the general features of growing cells.

Long-term cultures under optimum conditions grow in a manner similar to cultures of micro-organisms. Cells placed in fresh medium (provided the inoculum is large enough) enter first a *lag phase*, during which they adjust to the new environment and carry on the synthetic processes leading to cell division. Growth takes place by logarithmic increase in cell number during the so-called *log phase*. Due to exhaustion of limiting nutrients, the culture then enters a *stationary phase*, until supplied with new medium (Fig. 1). During the *log*

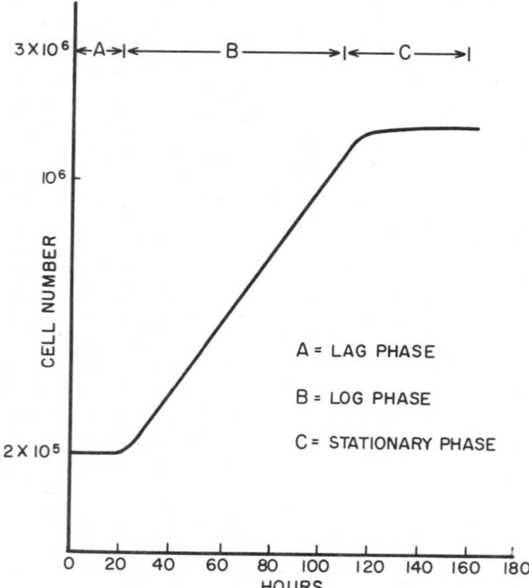

FIG. 1. Growth curve for cells in tissue culture.

phase, the increase in cell number can be approximately represented by the growth equation:

$$\log_{10} N = \log_{10} N_0 + \log_{10} 2 \cdot Kt$$

where

N = cell number at time t,

N_0 = cell number at time 0,

K = regression coefficient,

t = time.

Populations of like cells (in cell culture) or organized groups of cells (in organ culture) may be treated as systems for biochemical study, if one always bears in mind that the environment, which includes the medium, the substrate (usually glass or plastic, or a protein-covered surface), and the gas phase, as well as other cells, are integral parts of the system. Analyses of changes in the medium in which cells have grown, especially when a chemically defined medium has been employed, may be at least as informative as analyses of the cells themselves and have the additional advantage that they can be carried out serially without destruction of the cells.

From the viewpoint of cell nutrition, cultures are used for examination of the utilization of particular nutrients, *e.g.*, AMINO ACIDS, VITAMINS (an example is shown in Fig. 2), PURINES, PYRIMIDINES and CARBOHYDRATES, and of the interrelationships between them. Nutritional variants (*e.g.*, cells requiring asparagine or INOSITOL) can be induced

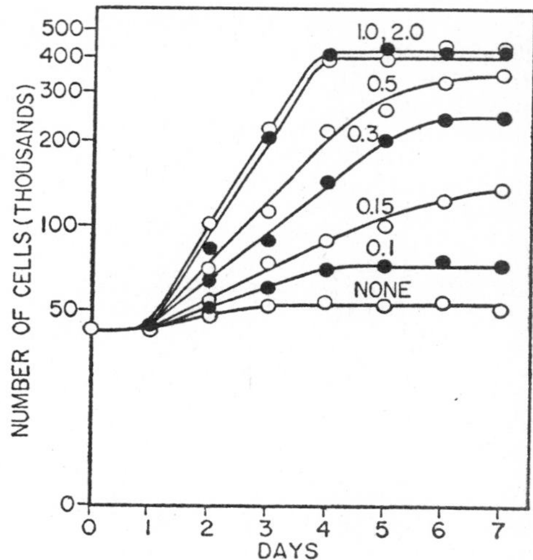

FIG. 2. The effect of different concentrations of folic acid on the growth of liver cells. The values represent millimicromoles of folic acid per milliliter. [Reprinted with permission from I. LIEBERMAN AND P. OVE, *J. Biol. Chem.*, **235**, 1120 (1960).]

by radiation, hormones or drugs. The effects of specific nutritional deficiencies or of specific antagonists to known nutrients can be explored. Biosynthetic pathways can be followed, *e.g.*, those leading to protein synthesis (including the synthesis of specific enzymes and immunologically specific proteins), those leading to nucleic acid synthesis, and those involved in the energy-producing systems of the cell. Substrate-induced or adaptive enzyme formation can be followed in cultures (Fig. 3 illustrates an example of this) as well as loss of enzyme activities under particular conditions of culture. The use of radioactive isotopes facilitates studies of metabolic pathways and

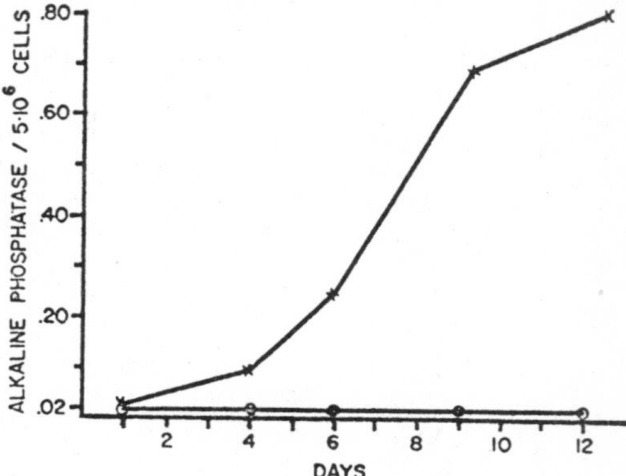

FIG. 3. Alkaline phosphates activity of replicate cultures grown with (×) and without (○) 5 m*M* phenylphosphate (human skin fibroblast strain Mas). [Reprinted with permission from R. P. COX and G. PONTECORVO, *Proc. Natl. Acad. Sci. U.S.*, **47**, No.6 841 (1961).]

of the turnover and incorporation of labeled components into cellular proteins and nucleic acids. This technique can conveniently be combined with autoradiography (see HISTOCHEMICAL METHODS) for the identification of the sites of turnover and incorporation in the cell, *e.g.*, in following the time course of protein and nucleic acid synthesis in chromosomes and other cell organelles during the mitotic cycle.[5] For such experiments, cultures of cells dividing synchronously are convenient. The high degree of synchrony possible in cultures of microorganisms has not been achieved with somatic cells, but approximations to it have been made, *e.g.*, by alternate cycles of cooling and warming, or by blocking a synthetic pathway and later releasing it again, as by addition and withdrawal of an excess of thymidine.

Comparisons of biochemical pathways in normal and tumor cells appear to offer hope for insight into differences in biochemical behavior of cells which have acquired malignant properties (see also CANCER CELL METABOLISM). This approach, however, has revealed more similarities between normal and cancer cells, and more differences within these classes, then differences clearly attributable to the malignant transformation alone. A time disconnection of nuclear protein synthesis and ribonucleic acid synthesis from deoxyribonucleic acid synthesis may be a true difference in biosynthetic behavior associated with tumor cells.[5]

From the use of cells growing in culture has come much information about the cell as a living unit, its capacities for independence, its degrees of dependence upon external anabolites, and its interdependence with other cells. The effects of HORMONES, growth stimulants and inhibitors, radiation protective agents, radiomimetic MUTAGENS, and CARCINOGENS, at the cellular level, throw light not only upon the inherent properties of the cell *per se* but also upon the effects of these agents in the physiological and pathological processes of the organism as a whole.

References

1. COX, R. P., AND PONTECORVO, G., "Induction of Alkaline Phosphatase by Substrates in Established Cultures of Cells from Individual Human Donors," *Proc. Nat. Acad. Sci., U.S.*, **47**, 839–845 (1961).
2. LASNITZKI, I., "Carcinogens, Hormones, Vitamins and Organ Cultures," *Int. Rev. Cytol.*, **7**, 79–121 (1958).
3. LEVINTOW, L., AND EAGLE, H., "Biochemistry of Cultured Mammalian Cells," *Ann. Rev. Biochem.*, **30**, 605–640 (1961).
4. LIEBERMAN, L., AND OVE, P., "Control of Growth of Mammalian Cells in Culture with Folic Acid, Thymidine, and Purines," *J. Biol. Chem.*, **235**, 1119–1123 (1960).
5. SEED, J., "Studies of Biochemistry and Physiology of Normal and Tumour Strain Cells. Synthesis of Ribonucleic Acid, Deoxyribonucleic Acid and Nuclear Protein in Normal and Tumour Strain Cells," *Nature*, **198**, 147–153 (1963).
6. WILLMER, E. N. (EDITOR), "Cells and Tissues in Culture: Methods, Biology and Physiology," 3 vols., New York and London, Academic Press, 1965.

CHARITY WAYMOUTH

TISSUE CULTURE, PLANT

Plant tissues, cells, and organs can be grown in test tubes under sterile conditions and used for studies on the biochemistry of plant growth and function. Many different plants and a wide variety of techniques have been used. The basic techniques are relatively simple, if one chooses to work with a tissue for which the physical and nutritional requirements are known. Good examples are the tissues derived from tobacco stems, carrot roots, corn endosperm and tomato fruit, all of which can be grown continuously on defined media for many years. It is only necessary to transplant the tissues to fresh nutrient and maintain them in the same manner as microbial cultures. Some tissues may be grown in liquid media as suspended cells and groups of cells, and these can be plated out on agar or dispensed by quantitative methods. Through the use of the appropriate material and techniques, plant tissue culture offers several advantages for experimental work in plant biochemistry: (1) the material is relatively uniform; (2) the conditions are controlled; (3) the nutrient medium can be defined; (4) sterility is part of the system.

Plant Products. Plants are known to produce many types of compounds, and it is only natural that a search should be made for the occurrence of these compounds in tissue cultures. The identification of natural products in tissue cultures could lead to investigations of their biosynthesis and their role in plant metabolism. To cite some examples, excised roots and tissues of tobacco are known to produce nicotine; up to 29 μg/mg dry weight of the ALKALOID was found in isolated tobacco roots grown in culture and 7μg/mg in tobacco tissue cultures. Atropine from belladonna, hyosine from *Datura*, lycorine and haemanthamine from floral primordia of *Narcissus*, squalene from tobacco, FLAVONOIDS from *Citrus*, and terpenoids from *Rosa* are all reported as being synthesized by plant tissues grown on nutrient media. This is an impressive list, but it is only the beginning of work on plant products. Commercial production of such chemicals may be possible in the future.

Growth. The rate of growth of some tissue cultures is rapid enough to facilitate biochemical studies while the rate of growth of others is very slow. An example of a fast-growing tissue is tobacco which will double in cell population in 2–4 days when grown in a liquid medium. Cultures which yield 2×10^5 cells/ml (60 mg/ml fresh weight and 3 mg/ml dry weight) have been reported. In such studies, the amount of inoculum is important in decreasing the lag phase of growth; apparently, the dividing cells are nourished partly by the medium and partly by substances provided by the nondividing cells of the culture.

Growth Substances. Growth-promoting substances such as the auxins, kinins and gibberellins play such an important role in plant growth and development that much work has been done with these compounds as they affect tissue cultures. Kinetin, for example, has been shown to decrease glucose-6-phosphate dehydrogenase and transketolase activity, inhibit oxygen uptake, increase shikimic acid levels, and induce more lignin to form in tobacco tissue cultures. It is postulated that the Embden-Meyerhof pathway and RNA metabolism are both affected by kinetin. What the primary action is remains to be determined (see HORMONES, PLANT).

Plant Cell Walls. The wall material of plant cells is one of their distinguishing characteristics. As a result, lignin, CELLULOSE and other wall constituents have been studied in many plant tissue cultures. Phenylpropanoids, for example, have been shown to be precursors of lignin formation in white pine, *Sequoia*, lilac, rose, carrot, and geranium tissue cultures. Moreover, the biosynthesis of lignin has been shown to be affected by kinetin, boron and major elements such as calcium. Also, it has been established that the cellulosic walls of plants contain a protein which is rich in the amino acid hydroxyproline. This has been shown for carrot, tobacco and sycamore tissues. Hydroxyproline in the wall accounts for as much as 90% of the total of this amino acid in the cell; hence it can be used as an index of wall growth. This hydroxyproline is formed by the hydroxylation of proline *in situ*; the oxygen of the hydroxyl group comes from atmospheric oxygen. It is postulated that the special wall protein (containing hydroxyproline) participates as an enzyme, "extensin," involved in wall growth.

Polysaccharides are deposited in cell walls and also outside of the cell. This has been shown for galacturonates of cells of sycamore maple. The methyl esterification of the galacturonic acids is thought to reduce the possibility of cross-linkages and thus to account for the great friability of cells of this tissue. It is suggested that the methyl donor may be methyltetrahydrofolic acid and that methylation may occur at the sugar-nucleotide in the cell.

Enzymes. Plant tissue cultures are being investigated for their enzyme content and activity. For example, isolated MITOCHONDRIA from normal and crown gall tumor tissues of tomato have been used to study respiratory and phosphorylative enzymes; the changes in oxidative enzymes have been studied during tumor induction in carrot tissue; the secretion of enzymes by plant cells (such as α-amylase by *Rumex* virus tumor tissue or peroxidase, indoleacetic acid oxidase, or phosphatase by other tissues) has been shown; changes in nitrate reductase levels in orchid embryos have been correlated with changes in development.

Biochemical Genetics. The relationship between biochemistry and genetics is strong, as is obvious from work on microbial systems. However, at the present time, very little is being done with the genetic control of the metabolism of plant tissue cultures. A step in this direction is evident from some work on tobacco. Scopoletin, a glycoside, is found in both the tissue and the medium of tobacco tissue cultures, and the level of the glycoside can be regulated by plant growth substances (auxin and kinetin). Recent work suggests that certain interspecific hybrids of tobacco form tumors and produce more scopoletin than nontumor hybrids. This begins to relate the genetics of tumor formation to the biochemistry of the plant, and further work in this direction can be expected.

Techniques for the selection of clones, strains and mutants of plant tissue cultures are available. Their use in studies on somatic variants will undoubtedly become a part of the biochemical genetics of higher plant cells.

Metabolism. Investigations on nitrogen metabolism illustrate the type of work being done on plant tissue cultures. The rate of protein turnover during active growth of primary explants of carrot phloem, proline and arginine metabolism in Jerusalem artichoke, protein synthesis by nuclei isolated from tobacco tissue, and experiments on the ornithine cycle in tissues of *Pinus* are examples of the type of work that is being done.

A few intensely green, autotrophic higher plant tissue cultures are available for biochemical studies, but more are needed. These will be useful for work on PHOTOSYNTHESIS, photoperiodism, light effects on cell growth and development, as well as many of the areas mentioned above. In the same way, one can predict that cell organelles (nuclei, mitochondria, plastids, etc.) will be studied to find out the mechanisms for their biochemical function. Ultimately, the goal of these studies on plant tissue cultures is to explain, control and direct plant functions *in vivo*. The future should find these techniques contributing to this goal.

References

1. RIKER, A. J., AND HILDEBRANDT, A. C., "Plant Tissue Cultures Open a Botanical Frontier," *Ann. Rev. Microbiol.*, **12**, 469–490 (1958).
2. STABA, E. J., "The Biosynthetic Potential of Plant Tissue Cultures," *Develop. Ind. Microbiol.*, **4**, 193–198 (1963).
3. STEWARD, F. C., WITH MAPES, M. O., KENT, A. E., AND HOLSTEN, R. D., "Growth and Development of Cultured Plant Cells," *Science*, **143**, 20–27 (1964).
4. TULECKE, W. AND NICKELL, L. G., "Methods, Problems, and Results of Growing Plant Cells under Submerged Conditions," *Trans. N.Y. Acad. Sci.*, **22**, 196–206 (1960).
5. WHITE, P. R., "A Handbook of Plant Tissue Culture," Second edition, New York, Ronald Press, 1963.

WALTER TULECKE

TOBACCO MOSAIC VIRUS

Tobacco mosaic virus (TMV) is a rod-shaped plant virus which causes the mosaic disease of tobacco. It was the first virus discovered and the first to be purified, and it has been investigated more intensively than has any other virus. TMV is particularly suitable for biochemical and physicochemical studies on viruses, because it is quite small and simple in structure, is available in large quantities, is relatively easy to isolate, and is extremely infectious and stable.

Isolation, Structure and Properties of TMV. Viruses, the simplest structures that exhibit some of the properties of life, cannot reproduce themselves outside living cells. However, upon invading a living host cell they become engaged in converting the cell's normal metabolic processes to the function of synthesizing virus. One of the hosts for the wild strain (vulgare) of TMV is the tobacco variety Samsun. Symptoms produced in the plant are usually streaking or mottling of the leaves, some blistering and distortion of the upper leaves, and a general stunting of growth. Symptoms of the disease appear about seven days after inoculation of young plants.

TMV is isolated from infected plant leaves by standard methods of virus purification. Juice is extracted from the leaves and the virus is purified by alternate cycles of low-speed and high-speed centrifugation. From 1 kg of diseased leaf tissue, as much as 5 grams of TMV (about 10^{17} virus particles) can be isolated.

Although too small to be observed in an ordinary microscope, the virus particles may be seen in an ELECTRON MICROSCOPE, which can magnify them up to several hundred thousand times. Figure 1 shows a typical electron micrograph of a preparation of TMV. The isolated virus consists of uniform rods of length 3000 Å (an angstrom is one ten-thousandth of a micron) and diameter of 180 Å. Each rod is hollow with a central hole of diameter 40 Å extending the full length of the particle. The virus has a molecular weight of 40 million atomic mass units (amu).

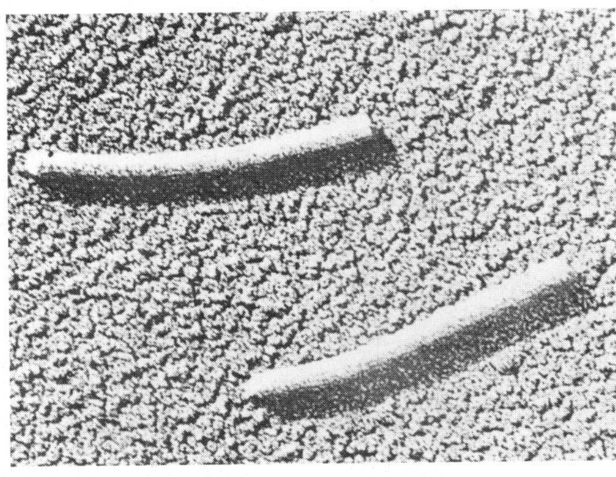

FIG. 1. Electron micrograph of tobacco mosiac virus particles. (Magnification *ca.* 130,000 ×.)

Chemical analysis showed that the virus is a nucleoprotein containing 95% protein and 5% ribonucleic acid (RNA). PROTEINS are large chain-like molecules made out of some 20 different kinds of chemical units called amino acids. The amino acids are held together by chemical bonds called peptide bonds; thus proteins are often called polypeptides. The protein of TMV consists of identical subunits of molecular weight 17,500 amu, and there are 2130 subunits in each virus particle. The RNA (see RIBONUCLEIC ACIDS) consists of a long single strand of chemical links called nucleotides, of which there are four kinds, containing the PURINE and PYRIMIDINE bases adenine, guanine, cytosine and uracil; each nucleotide also includes a ribose group and a phosphate group. The RNA of TMV contains 6400 nucleotides. It was found that the viral RNA itself is the essential infectious agent.

The structure of TMV has been determined by combining the results from X-ray diffraction, electron microscopy, ultracentrifuge sedimentation, viscosity, diffusion, light scattering and other physical and chemical studies. X-ray diffraction studies showed that the virus has a helical structure (see Fig. 2). The protein subunits, which are roughly ellipsoidal in shape, are closely packed in a helical pattern about the long axis of the virus rod. About 16 subunits form a single turn of the helix, which has a pitch of 23 Å. The total number of turns in the rod is 130. The RNA chain is deeply embedded in the protein subunits. It is held in a helical configuration in the intact virus and follows the pitch of the protein subunits. The RNA does not extend into the central hole of the virus rod but is completely surrounded by its outer coat of protein.

FIG. 2. Model of a portion of tobacco mosaic virus. The protein subunits are schematically illustrated in a helical pattern about the long axis of the particle. Part of the RNA chain is shown without its supporting framework of protein.

TMV is exceedingly stable as compared with its component parts. It remains native and infectious over a period of years at room temperature and for very short periods at temperatures up to 80°C. However, the isolated protein subunits may be easily denatured and the isolated RNA is quite thermolabile and susceptible to attack by ribonucleases (enzymes which degrade RNA).

In the intact virus, the RNA is resistant to these enzymes due to the protection offered by the protein coat. The thermal stability of the intact virus is the result of interactions between the RNA and protein parts. Hydrophobic bonds (bonds in which chemical groups, such as nonpolar side chains of amino acids, attract each other more than they do water molecules) play a significant role in holding the protein subunits together. Salt linkages occur between the phosphate groups of the RNA and the basic groups of the protein [see PROTEINS (BINDING FORCES)].

Isolation and Properties of RNA. TMV was the first virus from which infectious RNA was isolated. The RNA can be isolated by mixing the virus with phenol, buffer and bentonite, an acidic silicate clay which binds and inactivates nucleases. The protein becomes denatured and dissolves in the phenol layer while the RNA remains in the aqueous buffer layer. Bentonite is removed by centrifugation. The isolated RNA alone can initiate the infectious process and the production of complete progeny virus. It is, however, less efficient than the complete nucleoprotein particle. It was found that continuity of the single RNA chain is essential for its infectivity. The split of one of its 6400 chemical bonds results in loss of infectivity.

The molecular weight of TMV RNA, 2 million amu, was determined by physicochemical methods such as light scattering and analytical ULTRACENTRIFUGATION coupled with measurements of viscosity or diffusion. The isolated RNA does not maintain the regular helical form which it assumes in the virus particle. Its configuration in solution depends markedly upon the conditions of solution such as ionic strength, temperature and metal ion content. In dilute neutral salt at low temperature, portions of the chain are bound by hydrogen bonds to other portions. By studying the effect of temperature on the optical rotation of the RNA, it was shown that the hydrogen bonding occurs in definite areas and results in helical segments. About 50–60% of the nucleotides are involved in helical regions.

The biological specificity of the RNA depends on the sequence in which the nucleotides occur along the chain. This sequence contains the genetic information which enables the host cell to synthesize more virus coat protein, RNA and enzymes required for the synthesis [see PROTEINS (BIOSYNTHESIS)]. Methods have not been worked out for determining the complete sequence of the nucleotides. The end groups of TMV RNA have been determined by carefully controlled alkaline hydrolysis and by cleavage with specific enzymes. Both ends of the chain contain adenine. Neither end carries a phosphate group. Additional information about nucleotide sequences between the two ends has been obtained by splitting the chain with specific NUCLEASES followed by fractionation of the products and quantitative spectrophotometry. The sequences of many pieces (oligonucleotides) of the RNA have been determined.

Free single-stranded TMV RNA not yet enclosed in viral protein shells has been isolated from infected plant cells. In addition, a virus-specific double-stranded RNA occurring in very small amounts has been isolated. It consists of the parental-type TMV RNA strand and a second RNA strand whose base sequence is complementary [see DEOXYRIBONUCLEIC ACIDS (DISTRIBUTION AND STRUCTURE)] to the parental-type strand. This double-stranded RNA is believed to be the replicative form or template involved in production of progeny RNA strands in the replicative process.

Isolation, Structure and Properties of Protein. TMV is the first virus whose protein was shown to consist of subunits and the first whose complete amino acid sequence of protein has been determined. Each of the identical protein subunits has 158 amino acids. The amino acid composition of the protein subunit was determined for several TMV strains and is shown in Table 1.

TABLE 1. AMINO ACID COMPOSITION OF FOUR STRAINS OF TOBACCO MOSAIC VIRUS

Amino Acid	Strain			
	Vulgare	Dahlemense	G-TAMV	Holme's Ribgrass
Alanine	14	11	18	18
Arginine	11	9	8	11
Aspartic acid	18	17	22	17
Cysteine	1	1	1	1
Glutamic acid	16	19	16	22
Glycine	6	6	4	4
Histidine	0	0	0	1
Isoleucine	9	7	8	8
Leucine	12	13	11	11
Lysine	2	2	1	2
Methionine	0	1	2	3
Phenylalanine	8	8	8	6
Proline	8	8	10	8
Serine	16	16	10	13
Threonine	16	17	19	14
Tryptophan	3	3	2	2
Tyrosine	4	5	6	7
Valine	14	15	12	10
Total	158	158	158	158

It is possible to obtain the protein subunits by a variety of methods, namely by treatment of the virus with 67% acetic acid, 30% pyridine, or dilute alkali, or by the following method: the virus is dialyzed at 4°C against a buffer of pH 10.4, whereby the virus particles are split into so-called A-protein consisting of aggregates of several subunits. The RNA is removed. By high dilution (protein concentration $< 0.01\%$) at 4°C, the

A-protein is disaggregated to subunits. The molecular weight of the subunit, 17,500 amu, was calculated from a combination of results of analytical sedimentation and diffusion studies. This value is in very good agreement with the value determined from amino acid analysis.

The subunits may be reaggregated under certain conditions of ionic strength, pH and temperature to virus-like rods of various lengths with the same helical structure as that of the virus. The rods, which contain no RNA, are non-infectious, and are considerably unstable to heat as compared with the whole virus. The serological and ELECTROPHORETIC properties of these reaggregated particles are identical to those of the virus particles, and differ from those of the A-protein and protein representing intermediate aggregation stages between A-protein and rods. Evaluation of electrophoretic measurements leads to the conclusion that those positions of the subunits which lie on the surface of the rods have less positive and more negative charges than what corresponds to a homogeneous distribution of charge.

Native virus-like protein consisting of polymers of various sizes of the protein subunit can be isolated along with mature particles from infected plants. This so-called X-protein represents TMV protein made in excess in the infected cell.

By recombining A-protein and RNA, complete infectious virus particles can be reconstituted. If the RNA from one TMV strain and protein from a different strain are combined, the progeny to which they give rise has the genetic characteristics contributed by the RNA.

For determination of the amino acid sequence, the virus is split into protein and RNA by treatment with phenol, acetic acid or pyridine. After removal of the RNA the protein is cleaved with the enzyme trypsin. The resulting 12 tryptic peptides are isolated by ion-exchange chromatography or isoelectric precipitation and purified by paper chromatography and electrophoresis. The tryptic peptides are cleaved into smaller peptides by treatment with other protein-splitting enzymes (chymotrypsin, pepsin, papain, subtilisin). These peptides are then separated from each other, and their structures are analyzed by chemical and enzymatic stepwise degradation procedures (Edman degradation, DNP method, hydrazinolysis, carboxypeptidase A and B). After establishing the sequences of amino acids in the 12 tryptic peptides, their proper order within the peptide chain is determined by isolation and analysis of so-called bridge peptides, which possess the partial sequence of two neighboring tryptic peptides. The complete amino acid sequence determined in this manner for the TMV strain vulgare is shown in Fig. 3 [see also PROTEINS (END GROUP AND SEQUENCE ANALYSIS)].

Mutants. Mutants of TMV are produced by treatment of TMV RNA with a mutagenic agent such as nitrous acid, hydroxylamine, or alkylating substances, or by incorporation of 5-fluorouracil in place of uracil. The mutagenic effect of these substances is based on alteration of one or more of the bases in the RNA leading to a base exchange or exchanges in RNA during replication. The following mutagenic changes take place by treatment with nitrous acid: cytosine is converted to uracil and adenine is replaced by guanine. The change of one of the 6400 nucleotides in TMV RNA is sufficient to produce a mutation. The mutants can be distinguished from the parent strain by altered symptoms (color of leaf mottling, degree of leaf deformation, systemic virus propagation or local lesion formation, etc.), by amino acid exchanges in the virus coat protein, or by altered symptoms *and* amino acid exchanges.

The exact positions of the amino acid exchanges for about 50 chemically induced and spontaneously occurring mutants have been localized. Many mutants have only one exchange, several have two exchanges and only very few have three exchanges. Quantitative evaluation of data concerning the relationship between the number of nucleotide conversions and the number of amino acid exchanges for these mutants produced by nitrous acid treatment leads to the conclusion: 55% of

1 2 3 4 5 6 7 8 9 10 11 12 13 14 15 16 17 18 19 20 21
Acetyl Ser-Tyr-Ser-Ileu-Thr-Thr-Pro-Ser-GluN-Phe-Val-Phe-Leu-Ser-Ser-Ala-Try-Ala-Asp-Pro-Ileu-

22 23 24 25 26 27 28 29 30 31 32 33 34 35 36 37 38 39 40 41 42
Glu-Leu-Ileu-AspN-Leu-Cys-Thr-AspN-Ala-Leu-Gly-AspN-GluN-Phe-GluN-Thr-GluN-GluN-Ala-Arg-Thr-

43 44 45 46 47 48 49 50 51 52 53 54 55 56 57 58 59 60 61 62 63 64 65
Val-Val-GluN-Arg-GluN-Phe-Ser-GluN-Val-Try-Lys-Pro-Ser-Pro-GluN-Val-Thr-Val-Arg-Phe-Pro-Asp-Ser-

66 67 68 69 70 71 72 73 74 75 76 77 78 79 80 81 82 83 84 85 86 87 88
Asp-Phe-Lys-Val-Tyr-Arg-Tyr-AspN-Ala-Val-Leu-Asp-Pro-Leu-Val-Thr-Ala-Leu-Leu-Gly-Ala-Phe-Asp-

89 90 91 92 93 94 95 96 97 98 99 100 101 102 103 104 105 106 107 108 109 110
Thr-Arg-AspN-Arg-Ileu-Ileu-Glu-Val-Glu-AspN-GluN-Ala-AspN-Pro-Thr-Thr-Ala-Glu-Thr-Leu-Asp-Ala-

111 112 113 114 115 116 117 118 119 120 121 122 123 124 125 126 127 128 129 130 131
Thr-Arg-Arg-Val-Asp-Asp-Ala-Thr-Val-Ala-Ileu-Arg-Ser-Ala-Ileu-AspN-AspN-Leu-Ileu-Val-Glu-

132 133 134 135 136 137 138 139 140 141 142 143 144 145 146 147 148 149 150 151 152 153 154
Leu-Ileu-Arg-Gly-Thr-Gly-Ser-Tyr-AspN-Arg-Ser-Ser-Phe-Glu-Ser-Ser-Ser-Gly-Leu-Val-Try-Thr-Ser-

155 156 157 158
Gly-Pro-Ala-Thr

FIG. 3. Sequence of the 158 amino acids in the protein subunit of the common strain (vulgare) of tobacco mosaic virus.

the cytosine or adenine conversions in the region (cistron) of the RNA containing information for production of the virus coat protein lead to amino acid exchanges in the protein of the surviving virus particles, 25% lead to lethal mutations and 20% to the production of virus with no amino acid exchanges in the coat protein, in which case no change in coat protein information had resulted from the nucleotide conversions on account of degeneration of the GENETIC CODE (see also GENES).

The finding that there are many mutants which show no exchanges in the coat protein in spite of marked changes in symptoms demonstrates the polycistronic nature of TMV RNA, *i.e.*, only a part of the RNA is necessary for the determination of the virus coat protein. How many and what additional proteins (probably enzymes which are required for virus synthesis) are determined by the TMV RNA is not known. Attempts to synthesize TMV protein in cell-free systems have so far been negative.

Another finding which shows that in addition to the cistron for the coat protein there exist still other cistrons in TMV RNA is the following: there are heat-sensitive mutants which, in contrast to the wild strain, cannot reproduce at high temperatures (>30°C); in this case, the defect is not caused by amino acid exchanges in coat proteins, as no changes were observed. The fact that for other heat-sensitive mutants the heat sensitivity is caused by altered coat proteins can be shown by the following experiment: these mutants grown at low temperatures (20–25°C) are split into RNA and A-protein. If reaggregation of the A-protein is attempted at low and at high temperatures, the aggregation to rods is successful only at low temperatures. The A-protein of the wild strain aggregates to rods at high temperatures as well as low. Analyses of the coat proteins of the mutants show one (or at most two) amino acid exchanges as compared with the wild strain. This illustrates that the type and position of amino acid exchange in the protein chain and not the number of exchanges is decisive for hindrance of the aggregation process. A further illustration is the occurrence of a mutant whose protein can aggregate neither *in vivo* nor *in vitro* under all conditions investigated, although it differs from the wild strain vulgare by only two amino acids.

There are TMV strains known which were isolated in nature and whose coat proteins show normal aggregation behavior in spite of large differences in amino acid exchange. The complete amino acid sequence of two such strains has been determined. A comparison with vulgare shows that only 70% of all amino acid positions among the three TMV strains is identical. The number of amino acids is the same in all cases, namely 158.

References

1. CASPAR, D. L. D., "Assembly and Stability of the Tobacco Mosaic Virus Particle," *Advan. Protein Chem.*, **18**, 37 (1963).
2. KNIGHT, C. A., "The Chemistry of Viruses," "Protoplasmatologia, Handbuch der Protoplasmaforschung," Vol. IV, No. 2, Vienna, Springer-Verlag, 1963.
3. WITTMAN, H. G., AND WITTMANN-LIEBOLD, B., "Tobacco Mosaic Virus Mutants and the Genetic Coding Problem," *Cold Spring Harbor Symp. Quant. Biol.*, **28**, 1963.
4. ANDERER, F. A., "Recent Studies on the Structure of Tobacco Mosaic Virus," *Advan. Protein Chem.*, **18**, 1 (1963).
5. MARKHAM, R., "The Biochemistry of Plant Viruses," in "The Viruses" (BURNET AND STANLEY, EDITORS) Vol. 2, p. 33, New York, Academic Press, 1959.
6. FRAENKEL-CONRAT, H., "The Genetic Code of a Virus," *Sci. Am.*, **211**, 46 (October 1964).

H. G. WITTMANN AND L. E. BOCKSTAHLER

TOCOPHEROLS. See ANTIOXIDANTS; VITAMIN E.

TODD, A. R.

Alexander Robertus Todd (1907–), is a British biochemist whose contributions have been diverse and distinguished. His outstanding work, for which he received the Nobel Prize in chemistry in 1957, had to do with a long-range and thorough study of the nucleotides. He was the first to synthesize adenosine diphosphate (ADP) and adenosine triphosphate (ATP) as well as many other nucleotides. Lord Todd is very active and influential in governmental affairs related to science.

TOXICOLOGY

The benefits arising from modern chemical technology have brought in their train some undesirable by-products, of which an important one is the harmful effect of chemicals. While toxic effects of certain naturally occurring substances have been known for a very long time, such effects of man-made chemicals or of other substances which man has learned to use beneficially now appear with greater frequency and intensity. Primitive societies, while without the benefits of modern western technology, were free of the threat that their food, atmosphere, and water would be polluted by chemicals which might be harmful. One may say that when the savage dies at an early age of exposure, starvation or disease, his tissues unlike those of his longer-lived civilized fellow-man, are free of potentially undesirable chemicals. To assist in the retention and further expansion of the benefits of modern chemical technology by recognition and control of its harmful concomitants is the special function of the science of toxicology.

Toxicology. Toxicology is the science which strives to understand the toxic, including fatal, action on living systems of *substances acting chemically*. The definition, therefore, excludes action by physical forces such as electrocution, concussion (explosives), missiles, and heat, and broadly speaking includes all living systems to which the substances are applied in small amounts. Such substances are commonly called poisons. More narrowly considered, toxicology concerns itself with the harmful effects to man and to other living forms, exerted by chemical substances which are useful to man in some way. Toxic chemicals may gain access to the living system by ingestion, inhalation, or contact with skin and other body surfaces. Deliberate usé of toxic action against living forms which are inimical to man in some way is an important product of the science of toxicology (*e.g.*, pesticides, antiseptics; see INSECTICIDES). While the body of toxicological knowledge may be divided in many ways, it has been very useful to do so on a combined physiological and biochemical basis, *i.e.*, substances exerting harmful action on the nervous system, the respiratory system, blood-forming and circulatory system, anti-metabolite action, etc., of higher living forms. Such a division or classification will not be made here; rather, three major topics will be discussed which are fundamental to a proper perspective for the non-toxicologist. These are concepts of *toxicity* and of *hazard*, and the subject of mechanisms of toxic action.

Toxicity. The term toxicity refers to the relative capacity of a substance for harmful chemical action to a living system. Arbitrarily, substances which are harmless when ingested by a human adult in average health in quantities of 50 grams or more are considered to have no toxicity; substances which are harmful in much smaller amounts only to rare individual systems somehow specifically sensitized to them (ALLERGIC reactions) may fall into this category also. Realization that all substances, including water, may be toxic under certain conditions makes precise definition of toxicity or poison difficult. Consequently, practical guidelines for safe handling and use do not readily follow from the definition. This has led to the ancillary concept of hazard, which, while it does not solve the problem of definition, permits the adoption of guidelines.

Hazard. The term hazard as applied to a potentially harmful chemical substance refers to the likelihood of toxic injury to a living system under circumstances of intended use of the substance. It can be readily seen that if, as an example, the amount of a substance needed to keep a slice of bread soft approaches closely the amount which is toxic, the substance is highly (and perhaps prohibitively) hazardous. In common parlance, the substance is "too toxic" for the purpose. It is to be noted that hazard is not dependent solely on the inherent toxicity or the absolute amount needed to cause injury, but is rather dependent jointly on the amount technologically needed *and* the toxicity.

Mechanisms of Toxicity. In most instances, little or nothing is known concerning the details of mechanism of toxic action. Often it is even difficult to give a reasonably specific description of the signs of toxicity by which to characterize the action and thereby identify the poison. Many poisons appear to act on a wide variety of tissues, cells and processes, so that in such cases an action on some fundamental property common to all cells is to be reasonably suspected (see also CYTOTOXIC CHEMICALS). However, there are poisons whose actions have been sufficiently characterized biochemically and localized physiologically to serve as illustrations of a broad biochemical classification. Such basis for understanding is limited by methodology and by biochemical processes yet unknown.

(1) *Corrosion.* Harmful action by this means is readily understood—the biological system or important parts thereof coming into contact with the poison are literally destroyed. Strong acids, alkalies, and oxidizing and reducing agents come readily to mind as examples. Such action is often termed massive or severe necrosis and is an obvious morphological lesion.

(2) *Suppression of Biochemical Processes.* This type of harmful action is likewise readily understood. The common characteristic is the chemical combination of the poison with some member comprising a chain or cycle of chemical entities whose interaction executes a vital biochemical fuction (*e.g.*, enzymatic), the resulting combination constituting a block to the function. In most of the cases which are reasonably well understood, catalytic systems, *i.e.*, enzymes, are the targets of the poisons. Numerous examples of this basis for toxicity exist; a few are given below. Such poisons may be said to produce biochemical lesions. (See also BERYLLIUM AND BERYLLIOSIS.)

(a) The cyanide ion combines with a tissue enzyme, cytochrome oxidase, so firmly that vital tissue oxidation is blocked, with obvious consequences (see CYANIDE POISONING).

(b) CARBON MONOXIDE combines with hemoglobin, the oxygen-carrying blood pigment, thus reducing its oxygen-carrying capacity with resultant injury.

(c) Fluoroacetate enters the Krebs' CITRIC ACID CYCLE in place of acetate, producing a fluorinated member of the cycle which interferes with the operation of the cycle. Function of tissues dependent upon operation of the cycle is seriously impaired or ceases.

(d) Arsenic-containing compounds combine with sulfhydryl groups borne and required by some enzymes, thus interfering with their normal enzymatic action and causing reduction or cessation of the function which they carry out.

(e) Mustard gases combine with cell nucleic acids in such a way as to interfere with the formation of new nucleic acids which are needed for replacement of cells in repair and for normal cellular turnover, or for new cells in growth and differentiation. The tissue or animal is soon deprived of the services of cells which cannot be replaced without a source of appropriate nucleic acids (see MUTAGENIC AGENTS).

(f) Arrow poison (curare) and similarly acting poisons block the effects of NERVE IMPULSES to various muscles and other structures, thus interfering with normal action. For example, paralysis of muscles which ventilate the lungs results in death of the organism by suffocation.

(3) *Exaggeration of a Vital Process.* Indiscriminate or uncontrolled action (overactivity) in a vital process can produce serious disruption or death of an organism. One of the best examples is that of strychnine (see ALKALOIDS), the most obvious effects of which are convulsions which may be fatal. Events leading to convulsions arise from heightened sensitivity of nerve cells causing uncoordinated, indiscriminate, massive discharge of nervous impulses to skeletal muscles, thus abolishing necessary reciprocal muscle actions. The resulting interference with normal rhythmic respiratory muscle activity leads to suffocation of the organism. Another example is that of poisons which interfere with the enzymatic removal of the chemical concomitant or transmitter (an ester) of the nerve impulse, the physiological action of which is a discontinuous event; before another impulse is possible, the preceding charge of transmitter must be removed by an esterase (see ACETYLCHOLINE AND CHOLINESTERASE). The transmitter being a choline ester, the enzyme is cholinesterase, and the poisons are called anticholinesterases, many of which are esters of phosphoric acids (pesticides and nerve gases). While they exaggerate the action of choline esters, the end result is interference with the over-all function ending in paralysis and death of the organism. It is apparent that this example may also be regarded as one producing a biochemical lesion.

(4) *Cancerogenesis.* That certain chemicals produce cancers or act jointly with others to produce cancers has been a well-recognized fact for some time (see CARCINOGENIC SUBSTANCES). While by some definition this type of action is not called toxicity, it is nevertheless an obvious harmful effect induced chemically. Cancer being essentially an uncontrolled proliferation of cells characterizing the cancer, often with derangements in differentiation, cancerogenic action must, at least in some cases be centered on the normal controls and restraints to proliferation and differentiation (see CANCER CELL METABOLISM). Unfortunately we do not understand the mechanism of chemical cancerogenesis at the biochemical level in the sense of that in (2) and (3) *supra*, almost entirely because we lack knowledge of the normal controls themselves. Examples of known carcinogens are benzpyrene, β-naphthylamine and tobacco tars.

(5) *Radioactivity.* Radioactive substances gaining access to the body may be said to be toxic because injury results if the radiation lasts long enough and if ionization of tissue components occurs (see IONIZING RADIATION). Some similarities with the effects of mustard gases are known, such as interfering with cell proliferation and proper tissue repair. There is evidence that abnormal production of peroxides in exposed tissues may be the fundamental biochemical lesion involved. Radiation applied externally, such as X rays used by radiologists in medical diagnostic work, may be cancer producing; here the biochemical lesion is unknown but may well comprise injury to the tissue mechanisms for control of normal cell proliferation and differentiation akin to that discussed in (4).

References

1. CHENOWETH, M. B., "Monofluoracetic Acid and Related Compounds," *Pharmacol. Rev.*, **1**, 383 (1949).
2. HOMBURGER, F., AND FISHMAN, W. C., "Physiopathology of Cancer," New York, P. B. Hoeber, 1959.
3. JOHNSON, J. M., AND BERGEL, F., "Biological Alkylating Agents," in "Metabolic Inhibition" (HOCHSTER AND QUASTEL, EDITORS), Vol. II, New York, Academic Press, 1963.
4. NACHMANSOHN, D., "Chemical and Molecular Basis of Nerve Activity," New York, Academic Press, 1959.
5. SOLLMANN, T., "A Manual of Pharmacology," Philadelphia, W. B. Saunders Co., 1957.
6. STOTZ, E., ALTSCHUL, A. M., AND HOGNESS, T. R., "The Cytochrome c-Cytochrome Oxidase Complex," *J. Biol. Chem.*, **124**, 745 (1938).

STEPHEN KROP AND ARNOLD J. LEHMAN

TOXINS, ANIMAL AND BACTERIAL

Toxins may be defined as high molecular weight poisons of plant, animal or bacterial origin that are highly toxic to particular species of animals including man. Most toxins are antigenic (see ANTIGENS), and when injected into animals in sublethal doses or in a detoxified form (toxoid), they elicit the formation of ANTITOXINS which in many cases, but not all, confer specific protection against the particular toxin in question.

Among toxins of animal origin, the most studied have been those secreted by the venom glands of certain snakes and other poisonous reptiles. However, antigenic proteins of high toxicity are also produced by certain species of fish, such as the stingrays, and by many predatory arthropods including scorpions, spiders and wasps. The lethal principle in SNAKE VENOMS is probably a neurotoxin in most cases. However, reptilian venoms, irrespective of species, generally contain a variety of enzymes that may contribute indirectly to the toxicity by breaking down permeability barriers and facilitating the spread of the true toxin. Such enzymes include phospholipases that produce hemolysis, numerous types of phosphoesterases, nucleotidases and hyaluronidases. Venoms may contain proteolytic enzymes that either enhance or inhibit blood coagulation. *Crotoxin*, the neurotoxin from the rattlesnake *Crotalus terrificus* venom, was isolated as a crystalline protein some years ago and that from the cobra (*Naja*) has also been obtained in a purified form.

Bacterial Toxins. Because of their role in the pathogenesis of disease, it is the toxins produced by certain pathogenic bacteria that have been most extensively investigated. Toxins formed by *Clostridium botulinum*, Type A, *Clostridium tetani* and *Corynebacterium diphtheriae* have each been isolated as heat-labile, crystalline proteins. The injection of minute doses of these toxic proteins into susceptible animals faithfully reproduces most of the signs and symptoms of the diseases botulism, tetanus and diphtheria, respectively. Botulinus and tetanus toxins are so toxic that it can be estimated that 1 mg of either protein would be sufficient to kill more than 1000 tons of a susceptible animal (such as guinea pig or man). Some properties of purified toxins are summarized in Table 1.

specific O-antigens as part of their structure. These O-antigens are highly toxic lipopolysaccharides and have been termed *endotoxins* (see also BACTERIAL CELL WALLS). They differ markedly in their properties from the soluble "exotoxin" proteins discussed above.

TABLE 1. PROPERTIES OF CRYSTALLINE TOXINS

	Crotoxin	Diphtheria	Tetanus	Botulinus Type A
Toxicity[a] for				
guinea pig	—	8×10^{11}	8×10^{13}	1.1×10^{15}
mouse	3×10^{7}	8×10^{8}	1.3×10^{13}	0.5×10^{15}
Molecular weight	30,000	72,000	70,000	900,000
Sedimentation constant	—	4.6 S	4.5 S	17.3 S

[a] Toxicity expressed as LD50 per kilo animal per mole toxin.

When diphtheria, botulinus or tetanus toxins are treated with dilute formalin at slightly alkaline pH and at ordinary temperatures, they become completely detoxified and converted to toxoid without losing either their antigenicity or their ability to react with their homologous antitoxin. The diseases diphtheria and tetanus have been virtually eradicated from those countries that have introduced universal immunization with toxoids (see also ANTITOXINS).

Relatively little is known about the nature and mode of action of toxins. The composition of the three bacterial toxins cited in Table 1 may be completely accounted for in terms of the usual amino acids. Many pathogenic bacteria produce a number of extracellular enzymes, which resemble those found in venoms. In some cases, these enzymes may be toxic, as for example, the hemolytic and cytotoxic alpha-toxin of *Clostridium welchii* which cleaves lecithin to yield phosphoryl choline and a diglyceride. Numerous strains of hemolytic streptococci of Group A and of *Bacillus pestis* produce toxic enzymes that hydrolyze pyridine nucleotides. Other exo-enzymes break down permeability barriers and permit the rapid spread of bacteria and the lethal toxins released by them into the tissues. Botulinus, Shiga dysentery and tetanus toxins all act selectively on nerve tissue. The fixation of tetanus toxin by the central nervous system is due to a specific interaction of this toxin with the GANGLIOSIDE moiety of a cerebroside-ganglioside complex. The role of ganglioside in the action of the toxin is not known. Diphtheria toxin, on the other hand, is not restricted to the nervous system in its action, and its mode of action is entirely different. Purified diptheria toxin is a powerful inhibitor of protein synthesis. Low concentrations of this toxin strongly inhibit amino acid incorporation into protein both *in vivo* by mammalian cell lines growing in tissue culture and in cell-free extracts from sensitive animals.

Cell walls of *Escherchia coli*, various *Salmonella* species and other Gram-negative bacteria contain

References

1. WELSH, J. H., "Composition and Mode of Action of Some Invertebrate Venoms," *Ann. Rev. Pharmacol.*, **4**, 293 (1964)
2. BUCKLEY, E. E., AND PORGES, N., (EDITORS), "Venoms," Publication No. 44, Washington, D.C., American Association for the Advancement of Science.
3. HOWIE, J. W., AND O'HEA, A. J., (EDITORS), "Mechanisms of Microbial Pathogenicity," Cambridge, The University Press, 1955.
4. VAN HEYNINGEN, W. E., AND ARSECULERATNE, S. N., "Exotoxins," *Ann. Rev. Microbiol.*, **18**, 195 (1964).

A. M. PAPPENHEIMER, JR.

TOXINS, PLANT

Poisonous plants may be classified biochemically according to their toxic principles, whether natural constituents or substances absorbed from the soil. Some plants may have more than one toxic principle. The commonest in the higher plants are alkaloids, amines, essential oils, glycosides, minerals, oxalates, photodynamic substances, phytotoxins or toxalbumins, polypeptides, resins and resinoids, and tannins. This classification, naturally, excludes species that are mechanically injurious to animals.

Alkaloids are basic, nitrogenous organic compounds almost invariably physiologically active; they poison through the nervous system, although some have a direct effect on the liver and other organs. Although no alkaloids have been found in algae or bryophytes and few in ferns and gymnosperms, they are widespread in the plant kingdom. In the Angiosperms, they are commoner in certain families (*Apocynaceae*; *Amaryllidaceae*; *Leguminosae*; *Liliaceae*; *Menispermaceae*; *Rubiaceae*; *Solanaceae*) than others. Appreciable amounts have been isolated from nearly 4000 species, but it is suspected that alkaloids occur in 10% of all higher plants. Alkaloids in the major toxic plants belong primarily to the following configurations: *guinolizidine* (*e.g.*, sparteine and isosparteine in Scotch broom; more than a dozen, mainly lupinine, in lupine; cytisine in mescal bean); *indole* (harmine in *Peganum Harmala* and *Banisteriopsis*; physostigmine in calabar bean; rauwolfine and related compounds in snake root; yohimbine in yohimbe bark; amides of lysergic acid, chanoclavine and clymoclavine in certain morning glories; several alkaloids related to strychnine in *Gelsemium*; strychnine in many species of *Strychnos*; a strangely hallucinogenic indole structure with a phosphylated chain—psilocybine—in *Psilocybe*, *Panaeolus* and other mushrooms); *isoquinoline* (morphine, codeine, papaverine and 20 others in opium poppy; mescaline and seven others in peyote cactus; sanguinarine and several others in bloodroot); *diterpinoid alkaloids* (delphinine in larkspur; aconitine in monkshood); *pyridine* (nicotine in tobacco; coniine in poison hemlock); *pyrrolizidine* (heliotrine and lassiocarpine in heliotrope; senecionine in *Senecio* spp.; monocrotaline in rattle-box); *steroidal alkaloids* of two kinds—*Solanum* type (solanidine in tomato, potato) and *Veratrum* type (veratramine in false hellebore); *tropane* (atropine in belladonna; atropine, hyoscyamine, scopolamine in *Datura* spp.). *Purines*, sometimes considered as alkaloids, are methylxanthines, weakly basic nitrogenous compounds, often toxic but (as caffeine in coffee, tea, guaraná, maté, yoco, kola; theobromine in cacao) occasionally with strong stimulating effects (see also ALKALOIDS; PURINES).

Amines, simpler natural bases derived from amino acids, cause toxicity in some plants (phenylethylamine in mistletoe; N-methyl-beta-phenylethylamine in an *Acacia*). Certain species of *Lathyrus*, including the garden sweet pea, have toxic substances related to beta-cyano-L-alanine that cause stunting of growth, skeletal deformity and paralysis in livestock. Amines often give plants a fetic odor. Toxic amines are not abundant in higher plants. The alkaloids in *Claviceps* are present with amines, and part of ergot-poisoning may be due to amines. The toxicity of some mushrooms is due, in part at least, to amines (choline and muscarine in *Amanita*).

Essential oils, either free or in combination with glycosides in the plant, are the toxic agent in certain species (oil of parsley, juniper, rue, wormwood, nutmeg, tansy, etc., are a few examples). Essential oils are widely distributed in higher plants, but those with poisonous action are common in certain families (*Labiatae*, *Lauraceae*, *Myristicaceae*, *Piperaceae*, *Umbelliferae*).

Glycosides, although more widespread than alkaloids, are second to them in importance as toxic principles. When hydrolyzed with acids or enzymes they yield a sugar (glucose, pentoses, hexoses or rarer sugars) or a closely allied carbohydrate and an aglycone component (acids, alcohols, aldehydes, phenols) that is normally responsible for the toxicity. Glycosides are of several types. *Cyanogenic glycosides*, themselves usually innocuous, may have toxic effects due to their hydrocyanic acid component; common in the *Rosaceae* (amygdaline from almond; others in cherries, apples, etc.), cyanogenic glycosides occur throughout the higher plants, but their concentration may vary in a given plant with climate, rainfall, season, maturity, etc. One of the most interesting is that occurring in tapioca (*Manihot esculenta*) that South American natives learned to remove by leaching in preparing an edible starchy meal from the root; seeds of rubber (*Hevea*) are likewise employed as food in the Amazon after removal of the glycoside. Other well known examples are phaseolunatine (flax, lima beans) and gynocardine (seeds of *Gynocardia odorata* or false chaulmoogra). A number of grasses cultivated for grain and fodder are poisonous because of cyanogenic glycosides: *Sorghum* (dhurrine), *Glyceria*, *Holcus*, *Triglochin* and *Zea Mays*. Some legumes (white clover, vetch) and other well known plants (hydrangea, *Stillingia*) fall into this toxic category. Ruminants are more susceptible to cyanic poisoning than man and most other mammals. *Goitrogenic glycosides* of two kinds—thiocyanates and thiooxazolidone—have been found in soybean, flax, sundry species of *Brassica* (turnip, kohlrabi, mustard seeds, kale, broccoli, cabbage, Brussels sprouts). This toxin causes hyperthyroidism resulting in sickness and death in stock animals. *Mustard oil glycosides*, peculiar to cruciferous seeds (horseradish, mustards, charlock, fanweed, wild radish), yield upon hydrolysis a vesicant sulfurous isothiocyanate or mustard oil. Most of the toxic species contain the mustard oil allyl isothiocyanate, extremely poisonous to cattle. Hydrolysis of ranunculine, the glycoside in certain *Ranunculaceae* (anemone, buttercup, marsh marigold), yields an irritant oil, protoanemoanine. *Coumarin glycosides*, with the aglycone component related to coumarin, are present in horse chestnut, *Daphne*, etc.). *Steroidal* and *triterpenoidal glycosides*, with aglycones of complex carbon chains, are numerous and widespread, acting frequently like saponines. The aglycone component may itself be physiologically active. Of special interest are those steroidal glycosides stimulating to heart muscle—the *cardiac glycosides*, of which 400 are known. *Digitalis*, containing more than 10 cardiac glycosides the aglycones of which are cyclopentanophenanthrene derivatives, is the source of a widely prescribed medicine, digitoxin. Cardiac glycosides are characteristic of the *Scrophulariaceae*, *Apocynaceae* and *Liliaceae*; especially noteworthy are *Apocynum*, *Nerium*, *Thevetia* and *Urginea*; lily-of-the-valley, with convallarine and related glycosides, may often poison children who eat the conspicuous fruits from garden plants. *Saponines*, known in 400 species from 50 families, are amorphous steroidal or triterpenoidal glycosides that form a non-alkaline colloidal solution or foam in water; their aglycones are sapogenines. Physiologically, saponines vary greatly but are toxic usually only when associated substances cause some injury permitting the saponine to enter the blood. Noteworthy are pokeweed, alfalfa, English ivy, beech, soap berry, soapwort and the tropical *Entada scandens* and *Barringtonia*. Many are piscicidal, some saponines killing fish in concentrations of 1/200,000 (see also GLYCOSIDES, STEROID).

Certain **minerals,** accumulated or selectively concentrated by plants, may lead to poisoning, especially in livestock. Many species of crop plants (oats, rye, corn, sorghum, celery, rape, squash, carrots, alfalfa) and weeds (especially *Amaranthaceae*, *Chenopodiaceae*, *Compositae*, *Cruciferae*, *Solanaceae*) may concentrate nitrates to levels causing abortion, faulty lactation, intestinal upsets, and hypothyroidism in cattle and sheep. Other types of nitrogen poisoning may result from nitrites or gaseous oxides of nitrogen in silo fermentation. Sundry species, especially *Leguminosae*, take up *selenium* from the soil, frequently causing severe loss where cattle raising is a major industry. Plants may be obligate selenium accumulators (*Astragalus* spp.) requiring this element and not growing where it is lacking, or they may facultatively accumulate selenium when it occurs, yet be capable of thriving without it (*Aster*, *Atriplex*). Symptoms vary with the animal, but selenium poisoning often includes depression, dyspnea, respiratory and myocardial abnormalities, hemorrhage, enteritis, and liver and kidney damage. "Blind staggers" and "alkali disease" are syndromes attributable to selenium toxicity. Certain species of *Astragalus* and *Oxytropis*, responsible for selenium-induced sickness in cattle, are called "loco weeds." In poultry, selenium poisoning affects hatchability and produces malformed chicks. Numerous forage plants, especially legumes, sometimes accumulate *molybdenum* and poison livestock, especially ruminants, several months after ingestion, producing anemia, reproductive difficulty, alteration of coat color and other symptoms, including even death. Fluorosis, locally common (as in the Punjab of India), occurs in man and livestock when excessive amounts of *fluorine* in water or concentrated by plants from the soil are ingested.

Oxalic acid or oxalates are accumulated sufficiently by some plants to cause poisoning; their concentration may vary with season and environment, some species being most dangerous in autumn. Precipitation in the blood of ionic calcium may be correlated with absorption of oxalates, but the toxicity of oxalates is due also to precipitation of crystals in renal tubules and resulting epithelial necrosis. Oxalates derive the tissue of calcium through precipitation. Purslane, rhubarb, dock, sorrel, beet are temperate species of high oxalate content. Calcium oxalate, causing mechanical irritation of mucous membranes, is common in aroids, such as skunk cabbage, jack-in-the-pulpit; this family is much more oxalate-toxic in its tropical species, such as *Xanthosoma*, *Colocasia* and *Monstera*.

Photodynamic substances are usually pigments that poison through photosensitization, in which an animal experiences erythema and pruritis, edema and dermal necrosis. Proteins make up probably the basic element that suffers oxidation on exposure to light: photosensitization is not known in animals with pigmented skin. The biochemistry of photosensitization, however, is not yet clearly understood. Several kinds of photosensitization are recognized: they are classified as primary when the photosensitizing substances are contained in the plant and hepatogenic when digestion frees a breakdown product to be eliminated through the liver. Photosensitizing plants can be classed as icterogenous (*Tribulus*) and non-icterogenous (buckwheat, St. Johnswort), according to their ability to damage the liver or not. In buckwheat and St. Johnswort, the photosensitizing agents are fagopyrine and hypericine, respectively, napthrodiathrone derivatives. Buckwheat poisonings in human beings are suspected to be allergic reactions, not photosensitizations as in many animals. Hepatogenic photosensitization is due to liver dysfunction as the result of a pigment such as phylloerythrine, the toxic agent in devil's thorn (*Tribulus*); in such plants as *Lippia* and *Lantana*, the toxin is a polycyclic triterpene, but pyrrolizidine alkaloids in *Senecio* as well as other vegetal substances may cause hepatic lesions. *Tribulus terrestris* causes the "big head" disease of cattle in South Africa; the Mexican *Agave Lecheguilla* and *Nolina texana* of southwestern United States, as well as the cultivated rape, horsebush and some species of *Panicum* cause hepatogenic photosensitivity in livestock. The trefoil (*Medicago denticulata*), through sensitization, causes the "aphis" disease of cattle in Australia, but the toxic principle is not known. Other kinds of photosensitizers are found in blue-green algae, some molds, *Sorghum*, vetches, clover, alfalfa, oats and *Polygonum*.

Phytotoxins or **toxalbumins** are proteinaceous compounds; they sometimes resemble bacterial toxins. Since they are antigenic, they cause the production of ANTIBODIES and often themselves offer an immunization mechanism. Susceptibility varies with the animal and the individual. Some, like ricine (from castor oil bean), which has hemagglutinative properties, may be toxic in extremely minute amounts. Readily absorbed through the intestinal wall, a few are gastrointestinal irritants and find therapeutic employment as laxatives. Not many species produce phytotoxins, but some that do are unusually dangerous to livestock. Jequirity, black locust and tung are some of the most poisonous. The famous sand-box tree and the physic nut are euphorbiaceous species that contain, among other toxic substances, dangerous phytotoxins. An African member of the passion fruit family (*Adenia digitata*) owes its toxicity to a toxalbumin (modeccine) as well as a cyanogenic glycoside.

A few **polypeptides** are highly toxic. The akee, certain fungi and algae (especially blue-green algae, such as *Microcystis aeruginosa*) act through poisonous peptides. In addition to several alkaloids and amines, ergot sclerotia contain polypeptides.

Resins and **resinoids**, chemically very heterogeneous, are soluble in organic solvents, insoluble in water; they irritate physiologically by direct action on muscle or nerve tissue. Many are fatally toxic. Common resin- or resinoid-toxic plants are marijuana, water hemlock, mountain laurel and milkweeds. The genus *Rhus* is perhaps the best known contact poison, acting through catechal compounds with unsaturated side chains. Poison ivy (*Rhus Toxicodendron*) causes an exceptionally severe dermatitis primarily because of 3-*n*-pentadecylcatechol in the phloem; a similar resin-like compound causes the occupational dermatitis characteristic of workers and artisans in the Chinese lacquer (*Rhus verniciflua*) industry; certain tropical species of the same family (*Anacardiaceae*) —mango, pistachio, cashew nut, *Metopium* (poison wood) and guao (*Comocladia*)—are toxic, due presumably to similar substances that have blistering effects upon the skin. The South African tree known as Cape mahogany (*Trichilia emetica*) owes its toxicity to a resin and a tannin. Many of the resin-containing species have properties of purgatives: examples are sundry members of the morning glory family. Resins with insecticidal and other poisonous properties are present in some legumes (*Derris*, *Tephrosia*) and in the pantropical strand-weed *Calotropis*.

Tannins are phenolic compounds, non-nitrogenous and often glycosidal in nature; they are generally astringent. Some (as in certain acorns and leaves of *Quercus*) have been known to poison livestock severely.

A few plants owe their toxicity to principles (aldehydes, esters, ketones, etc.) that cannot be accommodated in the foregoing categories of chemical compounds; for many others (such as andromedotoxine from *Rhododendron* and *Kalmia*), the exact nature of the principles is still not known. Pyrethrines (insecticidal principles in *Chrysanthemum*) are esters. Rotenone, from *Derris* and *Lonchocarpus*, one of the most important insecticides, is a ketone. A miscellany of other compounds would include such principles as plumbaginine, a naphthaquinone derivative from several species of *Plumbago*.

Poisonous plants are better known in our temperate zones, because the floras are limited and better studied. The tropics of both hemispheres are exceedingly rich in toxic plants, too many of which are still biochemically uninvestigated. Even such well known and virulently toxic species as manchineel (*Hippomane Mancinella*) lack definitive studies, and the complete understanding of their poisoning is still not available. Furthermore, many species with poisonous properties still remain to be discovered in the tropics.

Poisonous plants have had a far-reaching and neglected role in cultural and political history. They have plagued man from earliest time, but they have often been bent to his use as medicines, arrow poisons, fish poisons, in witchcraft and magic, to administer justice, as abortifacients, as narcotics and in political intrigue; modern man counts some of them as important allies in his fight against insects, rodents and other pests and as medicines. Primitive man early acquired discerning knowledge of poisonous plants: his earliest attempts to classify plants must have separated the useful from the harmful ones, and his knowledge of the toxic flora must have been necessary for his very health and survival.

An ancient use of poisoning, surviving in Africa and Madagascar, is determination of guilt or innocence through ordeal: poisons being administered in the belief that their good spirits will punish guilt by death, sparing the innocent. The important ordeal poisons are *Leguminosae*; bark of *Erythrophleum*, *Parkia*, *Detarium*; seeds of *Physostigma venenosum*. *Erythrophleum* contains a cardiac glycoside, erythrophleine; *Physostigma venenosum*, an alkaloid, physostigmine, sedative of the spinal cord causing death by asphyxiation and an important medicine in modern ophthalmology. Minor ordeal poisons, almost all alkaloidal, belong to the *Apocynaceae*, *Asclepiadaceae*, *Euphorbiaceae*, *Loganiaceae* and *Sapotaceae*.

Fish poisoning is worldwide and ancient. At least 154 species in 68 families are used, most (109) in South America. Piscicides owe their toxicity to a variety of principles, especially to alkaloids and to saponine and cyanogenic glycosides. Usually crushed leaves, stems or roots are thrown into still or sluggish water. Few actually kill fish, merely stupefying and generally interfering with respiration. Many belong to the *Acanthaceae*, *Amaryllidaceae*, *Araceae*, *Ebenaceae*, *Flacourtiaceae*, *Lecythidaceae*, *Solanaceae and Taxaceae*; the most important families are the *Compositae*, *Euphorbiaceae*, *Leguminosae* and *Sapindaceae*. *Derris* (southeastern Asia) and *Lonchocarpus* (South America) are outstandingly important; they contain rotenone, a ketone, the most effective of modern contact insecticides. Sundry species of *Tephrosia*, native to both hemispheres and owing their effects mainly to the alkaloid tephrosine, are employed as piscicides in the Americas, Africa, Asia and Australia. Many species of *Serjania* and *Paullinia*, acting probably through saponines, are used locally in the tropics. *Clibadium* and *Phyllanthus* are important South American fish poisons; the saponine-rich *Yucca* was used in North America. In Europe, *Verbascum*, *Cyclamen* and *Taxus* were employed.

Other than *Derris* and *Lonchocarpus*, few plants have provided insecticides. The most notable are tobacco, hellebore and pyrethrum daisies (*Chrysanthemum*). Chrysanthemum flower heads were first used insecticidally in ancient Persia. Pyrethrum is exceptionally valuable against household insects, since it is not toxic to man. Used in oil sprays, the two active principles (pyrethrine I and pyrethrine II), both esters, are now extremely important commercially. Most pyrethrum daisies are cultivated in Japan and Brazil.

The lily, *Urginea maritima*, employed as a rodenticide in the Mediterranean from classical times, has a glycoside that poisons almost no other animals than rodents. It became important commercially in the 1930's, only to be displaced by more effective synthetic poisons.

The use of arrow poisons is likewise worldwide and ancient, and many botanical families enter their preparation: *Araceae*, *Dioscoriaceae* (East Indies); *Amaryllidaceae*, *Leguminosae*, *Rubiaceae* (Africa); *Ranunculaceae* (Europe, India); *Annonaceae* (South America); *Celastraceae*, *Rutaceae* (Philippines); *Sapindaceae* (West Indies); *Ascle-*

piadaceae (North America). The notable families are *Apocynaceae*, *Loganiaceae*, *Menispermaceae* and *Moraceae*. The active principles are primarily alkaloids and glycosides. Historically interesting is *Antiaris toxicaria* (upas tree), the basis of arrow poisons of southeastern Asia; the latex contains two glycosides, for which effective antidotes are still not known. First reported in Europe in 1300, this poison wrought great slaughter among the Portuguese in capturing Malacca in 1511. The botanist Rumpf identified the source plant in 1750. *Strophanthus* is a major African arrow poison plant. South American arrow poisons (*curares*) employ the greatest number and diversity of species. Although many different plants, varying from tribe to tribe, enter into their preparation, most curares owe their activity either to loganiaceous (*Strychnos* spp.) or menispermaceous (*Abuta*, *Chondrodendron*, *Sciatodenia*) species. Bark, leaves and roots are boiled to a syrup and sun-dried to a paste that is applied to darts for blow guns, less frequently to arrows. Forty-two alkaloids are known from these curares: curarine and related compounds are valuable therapeutically as muscle relaxants.

One danger from poisoning lies in unfamiliar, introduced ornamentals. This is true especially of tropical ornamentals (*Datura*, *Nerium*, *Plumeria*, etc.), but certain temperate and subtropical species, when ingested, may be toxic: *Aconitum*, *Colchium*, *Convallaria*, *Delphinium*, *Digitalis*, *Lupinus*, *Hedera*, *Narcissus*, *Taxus* and many more.

Of the hundreds of poisonous species of native floras, only a few have been historically important. Ergot (*Claviceps purpurea*), a fungal parasite on rye, occasionally causes fatal mass poisoning in Europe when infected grains, accidentally passing through the mill, contaminate bread flour. Superstitions have often been associated with toxic plants: the frenzy of witches in the Middle Ages was due often to unguents of mandrake and other solanaceous poisons. Poison hemlock (*Conium maculatum*) of Europe, now a widely naturalized weed, was the source of Socrates' death potion; its alkaloids, chiefly coniine, act on the sensory and phrenic nerves, killing by asphyxiation from paralysis of the diaphragm. Water hemlock (*Cicuta*) has similar effects due to the resinoid, cicutine. When mixed with hay, the horse-tail (*Equisetum*) may slowly weaken cattle through a nerve poison. The wood, bark and leaves of yew (*Taxus*) contain taxine, poisonous to man and browsing animals. Some of our commonest temperate plants may have toxic effects with susceptible individuals at certain seasons: oaks (*Quercus*), lady slipper (*Cypripedium*), some hollies (*Ilex*), mountain laurel (*Kalmia*) and other ericaceous species, milkweeds (*Asclepias*), etc. Toxic plants abound in the *Compositae*, especially in warmer, drier areas, and many legumes of both temperate and tropical climates are suspect.

The use of toxic plants has influenced affairs in many cultures and lands far beyond the extent that is usually suspected. Undoubtedly the most intensively diabolic use of poisonous plants grew up during the Middle Ages in Europe, when the Italians developed the study of criminal and political poisoning into a cult; the name of the Borgia family is inextricably linked with this phase of European history. The *hashishin* of ancient Asia Minor were political murderers, excited to their nefarious work by ingesting hasheesh (*Cannabis sativa*): from this Arabic term comes our word *assassin*. It has been suggested that the Viking *berserker*, who went on periodic frenzies of lust and murder, were maddened for their frightful purpose by eating *Amanita muscaria*.

Most of our plant medicines and narcotics are poisons. The difference between a poison on the one hand and a medicine on the other is often only one of dosage. Thus, digitoxin (*Digitalis*), one of our most valuable cardiac medicines, and reserpine (*Rauwolfia*), the valuable tranquilizer, can be fatally toxic substances. All of our narcotics, many employed in primitive culture in magico-religious ceremonies, may similarly be classified as poisons: peyote (*Lophophora*); ololiuqui (*Ipomoea*, *Rivea*); toloache (*Datura*); red bean (*Sophora*); teonanacatl (*Psilocybe* and other mushrooms); yopo (*Anadenanthera*); caapi or ayahuasca (*Banisteriopsis*, *Tetrapterys*); yakee (*Virola*), coca (*Erythroxylon*), and many others. Opium and tobacco may likewise be counted among our plants that are employed for their toxic principles but for reasons other than outright poisoning.

References

1. Blohm, H., "Poisonous Plants of Venezuela," 1962.
2. Chopra, R. N., Badhwar, R. L., and Ghosh, S., "Poisonous Plants of India," 1949.
3. Gardner, C. A., and Bennetts, H. W., "The Toxic Plants of Western Australia," 1956.
4. Kingsbury, J. M., "Poisonous Plants of the United States and Canada," 1964.
5. Muenscher, W. C., "Poisonous Plants of the United States," revised edition, 1960.
6. Steyn, D. G., "The Toxicology of Plants in South Africa," 1934.
7. Watt, J. M., and Breyer-Brandwijk, M. S., "Medicinal and Poisonous Plants of Southern and Eastern Africa," Second edition, 1962.

Richard Evans Schultes

TRACE ELEMENTS. See mineral requirements (vertebrates); the several articles on nutritional requirements; articles on a number of individual elements that are nutritionally required, or occur in tissues, at low concentration levels, including chromium, cobalt, copper, fluoride, iodine, manganese, molybdenum, selenium, vanadium, and zinc.

TRACER METHODS. See isotopic tracers; isotope dilution methods.

TRANSALDOLASES. See carbohydrate metabolism; carbon reduction cycle (in photosynthesis); pentoses in metabolism.

TRANSAMINATION ENZYMES

The principal biological mechanism for both the biosynthesis and the degradation of amino acids is a transfer of the amino group involving the corresponding keto acid (see AMINO ACID METABOLISM). The enzymes are called transaminases or aminotransferases. The individual transaminases are mostly called by trivial names since no satisfactory nomenclature has yet been developed. This may be misleading, for even a well-characterized purified enzyme is referred to by a variety of names. The variability in substrate specificity requirements, between otherwise comparable enzymes from different sources, precludes the interchangeable use of a restricted number of names.

Guirard and Snell[2] have compiled a table of many partially characterized enzymes, listing their properties, approved names and some appropriate references. Assay procedures may be derived from these references or from the review by Aspen and Meister.[1] The optimum pH is commonly around pH 8. Since there has been continuing improvement in assay procedures, current papers should be consulted where possible.

The Nature of the Reaction. The physiological amino donor is most commonly another α-amino acid, but enzymes are known which transfer the β-amino group of β-alanine, the γ-amino group of γ-aminobutyric, the δ-amino group of ornithine, and even the amino groups of putrescine. It was originally thought that the reaction involved a direct transfer from such donors to the keto acid, for such a reaction was already known to occur non-enzymatically. Subsequently it was discovered that the key step in the reaction, the removal of the α-hydrogen atom or conversion of the amino group to an imino group, involved protein-bound 5′-phosphorylated derivatives of the aldehyde and amino forms of vitamin B_6 (see PYRIDOXINE FAMILY OF VITAMINES). Three likely enzyme-bound tautomeric intermediates are shown in Fig. 1.

Although the transamination occurs in effect with protein-bound vitamin B_6, the details of the reaction mechanism are still largely unknown due to the fact that purification of the enzymes concerned has only recently been undertaken. Table 1 shows the spectroscopic properties of the pyridoxal forms of those transaminases which have been purified. A general feature of all the enzymes, and for that matter of all vitamin B_6 enzymes so far investigated, is that the aldehyde is not free but rather bound to the protein either as an imine or an imine derivative. It is interesting that marked differences occur between enzymes catalyzing the same type of reaction, but possessing different substrate specificities.

The binding of the aldehyde group of the pyridoxal phosphate does not prevent the formation of imine derivatives with the substrates but rather accounts for the rapidity of their formation. Thus the bound pyridoxal phosphate reacts with "carbonyl" reagents much more rapidly than free pyridoxal phosphate.

Model transamination reactions between amino acids and pyridoxal phosphate have been extensively studied. Metal ions were found to stimulate the reaction in aqueous solutions but not in non-aqueous media. The enzymes do not contain metal ions.

Activation by Cofactors. Although it appears that they do not function as freely dissociable coenzymes, with tissue extracts and purified enzyme preparations it is generally found that pyridoxal phosphate and pyridoxamine phosphate are necessary for activity or cause a marked stimulation. Preincubation of the enzyme with the cofactor is often necessary for the maximum effect; the time of preincubation is very dependent upon the conditions. Occasionally pyridoxal phosphate has been found to inhibit when a large excess is used presumably because, being a very reactive compound, it reacts with other groups in the protein. The pyridoxal form appears to be most stable, and this stability may be conferred by the addition of keto acid substrates. Additional stability may be obtained by the further addition of carboxylic acid substrate analogues and pyridoxamine phosphate. The pyridoxal phosphate is usually bound too tightly to be removed by dialysis and some enzymes are in fact very resistant to resolution.

Two types of enzymatic reaction have been investigated:

(1) *Exchange transamination*: In the presence of an amino acid and its corresponding keto acid, transaminases will catalyze the formation of radioactive keto acid from labeled amino acid and,

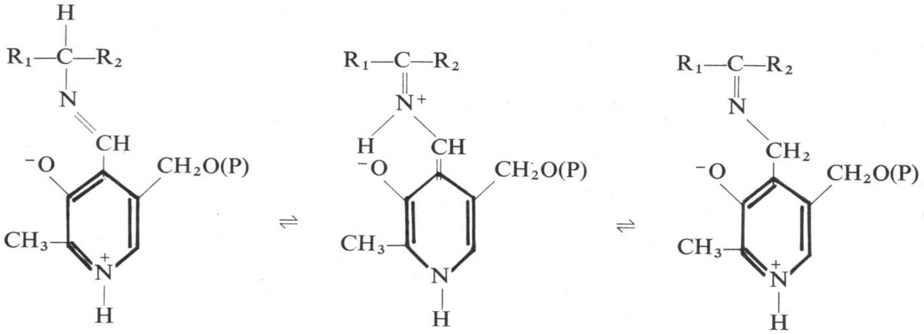

FIG. 1.

TABLE 1. SPECTRAL PROPERTIES OF TRANSAMINASES AND PYRIDOXAL DERIVATIVES

Transaminase	Source	pH	λ_{max} (mμ) Pyridoxal Enzyme	*p*K
Glutamic-aspartic	Pig heart supernatant	4.6 8.4	430 362	6.3
Glutamic-aspartic	Pig heart mitochondria	5.0 8.0	435 355	
Glutamic-aspartic	Beef liver supernatant	5.0 8.0	430 362	6.2
Glutamic-aspartic	Beef liver mitochondria	5.0 8.0	435 355	6.2
Glutamic-alanine	Pig heart supernatant	4.5 8.5	426 Multiple 330–450	7.3
Leucine-isoleucine	Pig heart supernatant	4.8–10.5	414,326	None observed
D-Amino acid	*Bacillus subtilis*	5–9	330,415	None observed
Pyridoxamine-pyruvate	*Pseudomonas* sp.		415	(Bound pyridoxal)
Aspartic-β-decarboxylase[b]	*Achromobacter* sp; *Clostridia* sp.	3–8	360	None observed
Serine[b] Transhydroxymethylase	Rabbit liver	5–9.7	430	None observed
Pyridoxal phosphate		7	388,330	
Pyridoxal		7	318,390	
Pyridoxal-valine imine		7 12	324,425 361	10.9

[a] The spectra of enzyme-substrate complexes depend on both the enzyme and the particular substrate. The maxima are again characteristic of imines but for a characteristic maxima at about 495 mμ which was assigned to the quinoid intermediary binary complex.

[b] Aspartic-β-decarboxylase and serine transhydroxymethylase will both react slowly with certain amino acids to yield the corresponding keto acid.

at an even greater rate, the labilization of the α-hydrogen of the amino acid. Because the concentrations of the substrates do not change, this system is in a steady state and is by far the easiest for enzymological analysis. It has been found in such systems that not only do the two distinct forms of the enzyme (E_1) and (E_2) exist, but there are three types of binary complexes (Fig. 2).

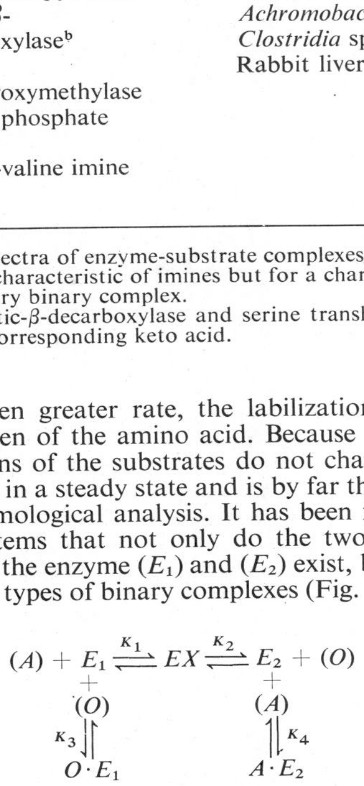

$$\begin{array}{ccccc} (A) + E_1 & \overset{K_1}{\rightleftharpoons} & EX & \overset{K_2}{\rightleftharpoons} & E_2 + (O) \\ + & & & & + \\ (O) & & & & (A) \\ K_3 \updownarrow & & & & \updownarrow K_4 \\ O \cdot E_1 & & & & A \cdot E_2 \end{array}$$

FIG. 2.

The formation of the "abortive" binary complexes of keto acid and pyridoxal enzyme ($O \cdot E_1$) and between amino acid and pyridoxamine enzyme ($A \cdot E_2$) complicates the analysis unless the ratio of amino acid to keto acid is maintained constant $(A)/(O) = R$. In such a case, one can readily obtain an apparent dissociation constant (K') for (A) for the formation of a mixture of binary complexes where K' is related to the individual dissociation constants by the expression:

$$K' = [K_1 + R \cdot K_2]/[1 + K_1/(R \cdot K_3) + (R \cdot K_2)/K_4]$$

(a) Spectroscopic analysis with high enzyme concentrations: The spectral changes (D_1 to D_2) at any wavelength caused by increasing the concentrations of both substrates the same relative amount (A_1 to A_2) are simply related to the apparent dissociation constant K' by the practical equation:

$$\frac{1 - (A_1)/(A_2)}{D_1 - D_2} = \frac{[K' + (A_1)][1 + (A_2)/K']}{[e_1' - e_2'][(E_0)]}$$

where $e_1'E_0$ would be the absorbance in the theoretical limit as the concentration approaches zero and $e_2'E_0$ would be the other limit as the substrate concentrations approached infinity. The apparent dissociation constant K' is determined by making all measurements relative to a particular density (D_2) obtained with a known substrate concentration A_2. When the left-hand side of the equation is plotted against the known values of A_1, a straight line results, cutting the abcissa at $-K'$. From the intercept on the ordinate may be derived e_1' and hence the important ratio K_1/K_2 for

$$e_1'[1 + R \cdot K_2/K_1] = e_1 + e_2 \cdot R \cdot K_2/K_1$$

Values for K_1/K_2 have commonly been found to be greater than 10 which shows that the transamination reaction with the bound vitamin B_6 lies far in favor of the amino acid and pyridoxal form of the enzyme.

(b) Kinetic analysis with low enzyme concentrations: The rate of radioactive exchange obeys the theoretical equation:

$$v \cdot t \cdot [1/(A) + 1/(O)] = \ln [(C_0 - C_e)/(C_t - C_e)]$$

where v is the velocity of exchange, t the time, and C_0, C_t, C_e are the observed counts in any reactant initially, at time t, and at the final isotopic equilibrium. When the ratio of amino acid to keto acid is maintained constant, this equation may readily be combined with that for (EX) to yield the practical equation below, from which the apparent dissociation constant K' again may be derived:

$$\frac{(A)}{v} = \frac{(1 + R)t}{\ln [(C_0 - C_t)/(C_t - C_e)]} = \frac{K' + (A)}{Ve}$$

Ve is the maximum rate of exchange for particular ratio of amino acid to keto acid.

Although the spectroscopic and kinetic methods employ much different enzyme concentrations, they have been found to yield comparable results with the glutamic aspartic transaminase.

(2) *Mixed transaminations of the type*

$$(A_1) + (O_2) \rightleftharpoons (A_2) + (O_1):$$

Such reactions are perhaps of more physiological interest but their analysis is much more complicated. It has been customary to avoid substrate concentrations which form appreciable amounts of the abortive complexes. The usual approach is to measure the variation in the initial velocity as a function of the concentration of one of the substrates holding the concentration of the cosubstrate constant. The analysis is based upon the general equation for two-substrate systems which has the form:

$$Vf/v = 1 + K_a/(A) + K_o/(O) + K_{ao}/(A)(O)$$

If the reaction is merely the summation of two exchange transaminations K_{ao} is zero. It is however very difficult to prove that K_{ao} is zero, particularly because it would be expected to have a minimum at the optimum pH where it is most convenient to carry out the analysis. It appears likely that the binary complexes which play such a key role in the reaction might be formed from existing binary complexes rather than necessarily being formed from the free enzyme. In such a case it is possible that measurable values of K_{ao} will be found.

Reversibility. If equal concentrations of an amino acid and keto acid are mixed and transamination is allowed to proceed, the reaction will reach the thermodynamic equilibrium in the majority of cases at around 50% completion. The rates of the initial enzymatic reactions in the forward and reverse directions are not, however, necessarily comparable. The rat liver glutamine transaminase, for example, will use many keto acids as acceptors but will not use the corresponding amino acids as donors unless very high concentrations are employed. Certain reactions, especially in crude systems, may appear to be irreversible because one product is removed by some subsequent reaction. Thus glutamine yields ketoglutaramate which is rapidly hydrolyzed, ornithine yields glutamic semialdehyde which is both oxidized and cyclizes, kynurenine yields a keto acid which cyclizes to kynurenic acid, cysteine sulfinate yields a very unstable keto acid which decomposes to pyruvate, etc.

It should be noted that removal of a single product does not yield a zero-order reaction because of competitive inhibition between the remaining product and its corresponding reactant.

Inhibitors. Transaminase inhibitors are of a variety of different types. Carbonyl reagents, especially hydroxylamine, for example are found to react only with the pyridoxal form of the enzyme, and hence compete solely with the amino acid. Carboxylic acid substrate analogues on the other hand compete with both amino acid and keto acid substrates although not necessarily to the same extent. There has been much effort devoted to devising specific inhibitors for certain transaminases with limited success. Cycloserine and aminoxyacetic acid are potent specific inhibitors of certain transaminases utilizing alanine. A not uncommon finding is that the substrates themselves inhibit; presumably this is due to the formation of abortive binary complexes. Most of the enzymes have been shown to be very sensitive to thiol reagents with the notable exception of the D transaminase of *Bacillus subtilis*.

Influence of Hormones. Transaminases play such a key role in protein nutrition that there is much interest in possible mechanisms of their hormonal control. The studies have been of two kinds: (1) The effects of prior hormone administration on the different enzyme levels in different tissues and relative to one another in a single tissue: Two enzymes, the glutamic-alanine and tyrosine transaminases of rat liver, which increase after administration of steroid hormones, have been investigated in some detail. An interesting finding in this work has been that the effect is localized to a particular isozyme.[3] This finding emphasizes that it may be misleading to compare over-all percentage increases in activities determined with crude extracts. The net rise for the glutamic-alanine transaminase has been found to be due to a stimulation of both *de novo* synthesis and degradation of the enzyme protein.[5] (2) The effects of hormones on the *in vitro* enzymatic assay activity: It has been found that certain steroid conjugates in extremely low concentration will prevent the formation of holoenzyme from pyridoxal phosphate and apoenzyme. Of particular interest is the finding that estrogen and synthetic estrogen derivatives possess comparable activities (see OVARIAN HORMONES).

Isozymes. Two or more enzymes which apparently possess the same catalytic activity are termed ISOZYMES. The occurrence of isozymes appears to be a characteristic feature of transaminases. Thus few transaminase negative mutants are known for they cannot be isolated by the standard procedures because the loss of a particular transaminase does not necessarily lead to a phenotypic block at that point in amino acid synthesis. Many transaminases, for example, are capable of synthesizing glutamate with ketoglutarate. There are also genetically distinct transaminases with virtually identical substrate specificity requirements present in a single cell, albeit in different intracellular locations. Such enzymes have been compared and found to have similar but yet distinct spectral and catalytic properties.[6] It has been suggested that the MITOCHONDRIAL enzymes might have some more specialized function than amino acid breakdown/synthesis associated with control of respiration through the Krebs CITRIC ACID CYCLE. There is some indication for minor variants also in ternary and quaternary structure.

Clinical Diagnosis. Plasma transaminase levels have been widely used as indicators of tissue necrosis particularly after myocardial infarction and in hepatic diseases.

References

1. ASPEN, A. J., AND MEISTER, A., in "Methods of Biochemical Analysis," Vol. 6, New York, Interscience Publishers, 1958.
2. GUIRARD, B. M., AND SNELL, E. E., "Vitamin B_6 Function in Transamination and Decarboxylation Reactions," in "Comprehensive Biochemistry" (FLORKIN, I. M., AND STOTZ, E. H., EDITORS), Vol. 15, p. 138, Amsterdam, Elsevier, 1964.
3. JAKOBY, G. A., LADU, B. N., "Studies on the Specificity of Tyrosine–α-Ketoglutarate Transaminase," *J. Biol. Chem.*, **239**, 419, 1964.
4. MEISTER, A., in "The Enzymes" (BOYER, P. D., LARDY, H., AND MYRBÄCK, K., EDITORS), Second edition, Vol. 6, p. 193, New York, Academic Press, 1962.
5. SEGAL, H. L., AND KIM, Y. S., "Glucocorticoid Stimulation of the Biosynthesis of Glutamic Alanine Transaminase," *Proc. Natl. Acad. Sci. U.S.*, **50**, 912 (1963).
6. SNELL, E. E., FASELLA, P. M., BRAUNSTEIN, A. E., AND ROSSI-FANELLI, A., (EDITORS), "Symposium on Chemical and Biological Aspects of Pyridoxal Catalysis," Oxford, Pergamon Press, 1963.

W. T. JENKINS

TRANSDUCTION. See BACTERIAL GENETICS, section on Infectious Heredity.

TRANSFER FACTORS (IN PROTEIN SYNTHESIS)

In protein synthesis, the initial amino acid-activation reaction is followed by the aminoacylation of SOLUBLE RNA (sRNA), and subsequently, in the presence of ribonucleoprotein particles (RIBOSOMES) and the factors discussed below, by the transfer of the aminoacyl moiety to protein [see PROTEINS (BIOSYNTHESIS)]. Hoagland and co-workers reported in 1957[1] that C^{14}-amino acid-labeled "activating enzymes" preparations, isolated from incubation of C^{14}-amino acid, ATP and rat liver "pH 5-insoluble activating enzymes" fraction, transferred significant amounts of C^{14}-amino acid to protein. The incorporation into protein required microsomes, ATP, GTP and a nucleoside triphosphate-generating system. Incubation of microsomes with sRNA-amino acid-C^{14}, isolated from the labeled "pH 5-insoluble activating enzymes" fraction by extraction with phenol, also led to the transfer of amino acid to protein; GTP, ATP, a nucleoside triphosphate-generating system and small amounts of "pH 5-insoluble" fraction were required. Subsequently, the transfer of sRNA-bound amino acids to microsomal protein was found to be catalyzed by a soluble enzyme fraction present in that portion of the $105{,}000 \times g$ supernatant of rat liver that does not precipitate at pH 5, although it is also present to a small extent in the "pH 5-insoluble" fraction. Studies with partially purified aminoacyl-transferring preparations indicate that GTP is the only nucleotide required in this reaction and that ATP serves to regenerate GTP; in addition, a sulfhydryl compound such as glutathione is also required with purified preparations. Aminoacyl transfer to polypeptide is also observed with purified ribonucleoprotein particles (ribosomes), and recent data indicate that the active component in protein synthesis consists of ribosomal clusters (polyribosomes) held together by messenger RNA [see RIBONUCLEIC ACIDS (STRUCTURE), and MESSENGER RIBONUCLEIC ACIDS].

Purification of the soluble enzyme system that catalyzes this reaction, from several sources, has revealed the presence of at least two protein factors. Resolution of two complementary fractions has been obtained from rat liver by salt fractionation,[2] from rabbit reticulocytes by chromatography on calcium phosphate gel,[3] and from *E. coli* by chromatography on DEAE-Sephadex.[4]

Aminoacyl-transferring enzyme preparations exhibit considerable species specificity with respect to ribosomes. However, the soluble factors involved in aminoacyl transfer do not appear to be specific with respect to the amino acid incorporated; indeed, the simultaneous presence and transfer of all sRNA-bound amino acids with native polyribosomes appears to be essential and is consistent with the synthesis of protein. With artificial polyribosomes, formed from ribosomes and synthetic polynucleotides, only the sRNA-bound amino acids "coded" by the artificial messenger are transferred to a peptide-bound form. Thus, the sequence of amino acids in a polypeptide is determined by the nucleotide sequences in messenger RNA and sRNA but apparently not by the other soluble factors involved in the transfer reaction.

A mechanism involving the movement of ribosomes along the messenger RNA strand, coordinated with peptide chain synthesis, has been postulated. Interaction of sRNA with polyribosomes leads to the binding of aminoacyl sRNA at specific positions adjacent to the carboxyl-terminal end of a nascent peptide chain whose carboxyl group is esterified to sRNA. A new peptide bond is formed between the amino group of the incoming aminoacyl sRNA and the ester carboxyl of the terminal amino acid in peptidyl sRNA, releasing the sRNA that was previously at the end of the growing peptide chain. Thus, as this process is repeated, the polypeptide chain is extended from its amino-terminal to its carboxyl-terminal end by the sequential addition of amino-acyl sRNA's; the new carboxyl-terminal residue of the peptide is bound to its corresponding sRNA on polyribosomes.

The roles of the aminoacyl-transferring enzymes and of GTP in the reactions between aminoacyl sRNA and polypeptide have not been established. Experiments with *E. coli* preparations[5,6] indicate that the binding of aminoacyl sRNA to polyribosomes, prior to peptide bond synthesis, is a non-enzymatic reaction; it occurs at relatively low temperatures and only requires magnesium and potassium or ammonium ions. Subsequent incubation of the polyribosomal-aminoacyl sRNA complex with both transferring enzymes and GTP results in peptide bond formation. However, experiments with reticulocyte preparations[3] indicate that the binding reaction between aminoacyl sRNA and polyribosomes requires one of the two transferring enzymes and GTP, while the peptide bond-synthesizing reaction requires the second enzyme. Whether these data actually reflect different mechanisms in these two different systems remains to be determined. The exact mechanism of the aminoacyl transfer reaction involving the participation of possible intermediates and of various transfer factors, in a series of events whereby amino acids are polymerized into a protein, remains one of the more complicated and interesting problems in biochemistry presently under investigation.

References

1. Hoagland, M. B., Stephenson, M. L., Scott, J. F., Hecht, L. I., and Zamecnik, P. C., *J. Biol. Chem.*, **231**, 241 (1958).
2. Fessenden, J. M., and Moldave, K., *J. Biol. Chem.*, **238**, 1479 (1963).
3. Arlinghaus, R., Shaeffer, J., and Schweet, R., *Proc. Natl. Acad. Sci. U.S.*, **51**, 1291 (1964).
4. Allende, J. E., Munro, R., and Lipmann, F., *Proc. Natl. Acad. Sci. U.S.*, **51**, 1211 (1964).
5. Cannon, M., Krug, R., and Gilbert, W., *J. Mol. Biol.*, **7**, 360 (1963).
6. Spyrides, G. J., *Proc. Natl. Acad. Sci. U.S.*, **51**, 1220 (1964).

Kivie Moldave

TRANSFER RIBONUCLEIC ACIDS (TRANSFER RNA). See SOLUBLE RIBONUCLEIC ACIDS; also RIBONUCLEIC ACIDS (BIOSYNTHESIS); RIBONUCLEIC ACIDS (STRUCTURE AND FUNCTIONS).

TRANSFORMING FACTORS

Bacterial transformation is defined as the transfer of a genetic trait from one organism to another one via free DEOXYRIBONUCLEIC ACID (DNA) molecules, *i.e.*, not requiring cell-to-cell contact (conjugation) or a virus carrier (transduction). The active materials have been variously called transforming "principles," "factors" or "agents," as well as DNA. "Transformation" of mammalian cells in culture due to viral infection is a different phenomenon, not to be confused with this one.

A typical example is the transfer of drug resistance. Given a pure culture of streptomycin-resistant *Pneumococcus*, the DNA may be isolated in pure form and a small amount of it added to a streptomycin-sensitive culture of the same species. Under the right conditions, a fraction of the sensitive cells, up to a few per cent, acquire permanently the trait of streptomycin resistance, and the DNA of their progeny can now be used to repeat the transformation of other sensitive cells.

To meet all the criteria for transformation, the DNA must come from bacteria of different genetic constitution than the recipient cells, with regard to the genes transferred; the recipient cells must be altered *hereditarily—i.e.*, their progeny carry the new property in their genes and can donate it; the phenomenon must be dependent on free macromolecular DNA in solution, subject to degradation by pancreatic deoxyribonuclease. These requirements rule out mutations induced by nonspecific nucleic acids or their degradation products, temporary physiological effects on the recipient cells, and viral transduction.

This phenomenon provided the most direct evidence that genes are made of DNA. Transformation was discovered in 1928 by Griffith, who found that live, non-virulent, pneumococci were converted to virulence by dead cells of a different genetic type, namely one governing the kind of polysaccharide capsule surrounding the cell. It was established in the 1930's that extracts of the donor cells were effective, and in their classic 1944 paper, Avery, MacLeod and McCarty identified the active principle as high molecular weight DNA. Their evidence was that the activity fractionated with the DNA at all steps, was not affected by enzymes attacking protein or RNA, sedimented with DNA in a centrifuge, and was rapidly inactivated by agents known to attack DNA, especially pancreatic deoxyribonuclease as purified by Kunitz. There was some controversy as to whether a resistant protein or perhaps some polysaccharide might be the true agent, particularly since the only genetic trait known then to be transferred involved polysaccharide properties. Over the next few years, however, these investigators and others, particularly Rollin Hotchkiss, succeeded in eliminating most rational doubts. Hotchkiss found he could transfer other traits, such as those for drug resistance, and lowered the protein level in his preparations to a maximum of 1 part protein in 5000 parts of DNA. This DNA was fully active.

Zamenhof then showed that the viscosity of DNA declined sharply at the same temperature where transforming activity was lost, and in further work the correlation of DNA properties and biological activities has been extensively confirmed and used as a tool for study of related problems. Many kinds of genetic properties have now been transformed, the limitation being only whether one has an assay available.

The conditions under which bacteria are susceptible to transformation are relatively stringent, even in the best cases, and relatively few species have been definitely proved to transform at all. Three systems are most widely used at present, *Pneumococcus*, *Hemophilus influenza* (1951) and *Bacillus subtilis* (1958). In addition, work is being done with *Neisseria*, *Streptococcus* and other species of *Hemophilus*. Because of the mass of other genetic information on *Escherichia coli*, there have been many attempts to transform it. While there have been reports of limited success, it appears that no one has a working *E. coli* system now, with the exception of a specialized case in which lambda phage DNA can be made to carry in a restricted set of genes associated with it.

Effort has also gone into transformation of mammalian cells in TISSUE CULTURE via free DNA, with one report of limited success by Szybalka and Szybalski in 1962. The most dramatic reports are perhaps those of Benoit in France, who attempted to transform ducks and reported an apparent positive result in 1957. Work continues on this line, although no further successes have yet been published.

Following the initial proof of DNA's role in bacterial transformation, many questions arose. Does the DNA actually penetrate the cell wall, or does it act from without? Is a GENE one molecule, and what is a molecule of DNA anyhow? Is the new information added *to* or substituted *for* that already in the cell? Is there a regular ordering of DNA in the bacterial chromosome? What properties of DNA are truly essential to its genetic roles? The answers to these queries have opened further ones.

In order to get at these problems, it was necessary to establish quantitative features of the system. First, the recipient cells are "competent" to take up the DNA only at certain stages late in the growth cycle of the culture, and then only under certain growth conditions not well understood. At present, the routine production of competent cells in the laboratory is still a cookbook art, with small differences in details of handling the cultures of major practical importance. Fortunately, cells in this state may be frozen and stored for weeks to months, to be thawed and used as needed.

Given competent cells, their reaction with DNA is describable in terms at least formally similar to enzyme kinetics. The reaction may be stopped by adding excess deoxyribonuclease (DNase). A curve of transformed cells *vs* time of exposure to DNA rises linearly, after a short lag of 1–3 minutes, for upwards of 20 minutes, depending to some extent on whether the cells are exhausting the available DNA, as they can in the most competent systems. At low DNA levels (below 0.01 μg/ml), the velocity of reaction, in terms of transformed cells per milliliter per second, is proportional to DNA concentration, or its activity, in experiments on damage to the DNA. At levels above 1 μg/ml, the velocity saturates as though some feature of the cells is limiting. Foreign or inactive DNA shows competitive inhibition at this level, implying a binding site. P^{32}-labeled DNA binds transiently and then permanently to the cells, and this binding is inhibited proportionately to transformation by foreign DNA. Not all studies show complete saturation, however, and the nature of the binding site and the competitive step is subject to continuing study. Clearly, it is here that one looks for the detailed molecular mechanism.

At the lower end of the titration curve, where DNA is limiting, a highly competent culture will exhibit 10^{13}–10^{14} transformed cells per gram DNA added (*e.g.*, 3×10^3 transformants per 10^{-10} gram DNA). In the most favorable situations, one transformed cell is obtained per cell equivalent of donor DNA taken up by the culture. This means that the process is a highly efficient one, not a rare event. Since the average donor cell probably has two or more chromosomal copies, the exact efficiency is somewhat uncertain, but high.

Once the DNA has been taken up by the cell, some period of time must elapse for phenotypic expression of the new gene, *i.e.*, transcription of its message and synthesis of whatever new molecules are controlled by it. In some cases, the new property begins to be expressed within minutes after the DNA has been added, and in others it is delayed for longer times. These phenomena depend on the physiological state of the cells, the molecular properties being altered, and on the molecular mechanism of incorporation of the incoming DNA and its transcription.

None of the phenomena above definitely proved that the DNA molecule *per se* penetrated the cell, although the radioactive label became permanently bound. The first really strong evidence on this point arose from studies of linked genes, which also established other facts as to the relation of the gene to the "molecule" and the chromosome. "Linkage" in the transformation sense refers to the tendency of occasional pairs or groups of genes to be co-transferred to one recipient cell, independently of the DNA concentration. At saturation DNA levels, two or more particles of DNA may well penetrate one cell and transform it for two genes. If the genes are always on separate DNA particles, however, the frequency of double transformants declines as the square of the DNA dilution factor, whereas if they are on one molecule they dilute linearly, as do the single transformants. Such linked markers were found by Marmur and Hotchkiss in 1954. Using them, they established (a) that there is a definite position on the chromosome for each gene, and (b) that the incoming gene replaces the one previously there, rather than just being added to it. In 1961, two groups demonstrated that physical linkage of incoming and recipient genes occurred rapidly, within minutes after addition of the DNA, and in the absence of measurable DNA synthesis. This implied very strongly that the DNA did physically penetrate, and was strong evidence against the "copy-choice" hypothesis for genetic recombination.

More recent studies using density and radioactive labels, as well as genetic linkage, plus CsCl density gradient fractionation, have confirmed that the donor DNA does react directly with the recipient cell's DNA. Details of the molecular recombination process are under study at present.

The existence of linked transforming activities has also helped clarify the nature of the DNA "molecule." The typical DNA preparation from bacteria is heterogeneous in "molecular weight," and this is now thought to be the result of random shearing of the chromosome during extraction. Linked genetic markers may be separated by mild shear forces, such as flow through a needle, without substantial damage to the single markers themselves. The degree of linkage is strongly subject to the shear attendant in preparation of the DNA, and there is no evidence for discrete molecular units corresponding to single genes. The physical size of the minimum piece of DNA capable of transforming a cell has a molecular weight of less than 300,000, compared to mean particle weights of many millions in the typical preparation. The present view is that a functional gene is a region of a DNA molecule, *i.e.*, the *cistron*, and it is presumably set off from other genes only by the base sequence of the DNA at its ends. The usual preparation of bacterial DNA is a collection of fragments of the chromosome, which behaves as though in the cell it is one continuous structure of particle weight $1\text{–}3 \times 10^9$.

The recombination process in transformation may take place *within* a gene, just as in other genetic systems. Lacks and Hotchkiss first clearly demonstrated this point by transforming cells unable to make functional amylomaltase with DNA from another mutant strain also unable to make the enzyme.

Many further studies have been done both as to the phenomenon of transformation itself and using it as an assay for properties of the DNA. One of the most significant of these was the finding of *renaturation* of denatured DNA by Marmur and Lane in 1960. In this phenomenon, the separated strands of heat-denatured DNA were shown to be capable of pairing spontaneously and reforming a helical structure, by standing for a while at 20°C below the melting temperature of the helix. Transforming activity returned from a few per cent up to 50% or more of that of the original native DNA. This result and later experiments showing that there is intrinsic activity in the denatured DNA itself, even in fractionated single strands, have confirmed completely the concept that the *information* content of the DNA resides in the base sequence alone, not in the secondary structure of the helix, which, however, is of great importance for the *mechanisms* of action of the nucleic acids.

References

1. Ravin, A. W., in *Advan. Genet.*, **10**, 61 (1961).
2. Hayes, W., "The Genetics of Bacteria and Their Viruses," Ch. 20, New York, John Wiley & Sons, 1964.

WALTER R. GUILD

TRANSGLYCOSYLATION REACTIONS. See CARBOHYDRATE METABOLISM.

TRANSKETOLASES. See CARBOHYDRATE METABOLISM; CARBON REDUCTION CYCLE (IN PHOTOSYNTHESIS); PENTOSES IN METABOLISM.

TRANSMETHYLATION REACTIONS. See METHYLATIONS IN METABOLISM.

TRANSPEPTIDATION REACTIONS. See PEPTIDASES.

TRANSPHOSPHORYLATION REACTIONS

Transphosphorylation is the transfer of phosphoryl groups from one molecule to another. Enzymically catalyzed transphosphorylations are of central importance in biochemical energy storage, transfer and utilization. Chemical bond energy (released from oxidation reactions) is stored in living systems in the form of "high-energy" compounds, principally adenosine triphosphate [ATP; see PHOSPHATE BOND ENERGIES; also PURINES (STRUCTURE AND OCCURRENCE), section on Coenzymes, for structure of ATP]. Once ATP has been synthesized, the bond energy stored in it can be transferred to other structures by transphosphorylation reactions, in which a phosphoryl group (in some reactions a pyrophosphoryl group or an adenylyl group) is transferred to a substrate molecule. Such phosphorylations, at the expense of ATP, occur widely in many metabolic areas in the course of biosynthetic reactions [see, for example, ACTIVATING ENZYMES; CARBOHYDRATE METABOLISM; DEOXYRIBONUCLEIC ACIDS (REPLICATION); PURINE BIOSYNTHESIS; PYRIMIDINE BIOSYNTHESIS; RIBONUCLEIC ACIDS (BIOSYNTHESIS)].

The formation of ATP by phosphorylation of adenosine diphosphate (ADP) is an energy-consuming reaction, coupled to the energy-releasing processes of GLYCOLYSIS and RESPIRATION, and to the light-absorption processes of PHOTOSYNTHESIS. (See also BACTERIAL METABOLISM; LIGHT PHASE OF PHOTOSYNTHESIS; PHOSPHORYLATION, OXIDATIVE; PHOSPHORYLATION, PHOTOSYNTHETIC.)

TRICARBOXYLIC ACID CYCLE. See CITRIC ACID CYCLE.

TRIGLYCERIDES

Triglycerides are the principal constituents of natural fats (see LIPIDS). The resolution of natural fats into individual pure triglycerides is not usually feasible, but evidence indicates that the natural fats are complex mixtures of triglycerides in which mixed glycerides (*i.e.*, those containing more than one kind of fatty acid radical per molecule) predominate over the simple glycerides such as tripalmitin (glyceryl tripalmitate) or triolein (glyceryl trioleate).

TRYPSINS AND TRYPSINOGENS. See PROTEOLYTIC ENZYMES.

TRYPTOPHAN (TRYPTOPHANE)

L-Tryptophan, α-amino-β-indolepropionic acid,

$$\text{(indole ring)}\text{—C(=CH—NH)—}CH_2\text{—}CHNH_2\text{—}COOH$$

was early recognized as one of the AMINO ACIDS derived from PROTEINS, because by condensation reactions with various aldehydes it yields highly colored products which can serve as means of identification and estimation. It is unevenly distributed in common proteins and often at a low level (less than 2%). Tryptophan was one of the first amino acids for which essentiality in the diet was demonstrated. Nutritionally, it is essential as an amino acid; it is also important because it can serve as a raw material for the production of niacinamide (see NICOTINIC ACID). Diets poor in tryptophan are thus conducive to the production of pellagra. Individual human needs for tryptophan (nitrogen equilibrium studies) vary from 82–250 mg/day. L-Tryptophan in 0.5*N* hydrochloric acid solution has a specific rotatory power $[\alpha]_D = +2.4°$.

TURNOVER NUMBER. See ENZYMES (KINETICS), section on Application of Reaction Rates.

TYROSINE

L-Tyrosine (*p*-hydroxyphenylalanine),

$$HO\text{—}C_6H_4\text{—}CH_2\text{—}CHNH_2\text{—}COOH$$

has long been recognized as one of the AMINO ACIDS yielded by the hydrolysis of PROTEINS. Millon's reaction (the development of red color by heating in the presence of a mixture of mercuric nitrate and mercuric nitrite) is used to detect the presence of tyrosine or tyrosine-containing proteins. Tyrosine is present in most proteins, often at levels between 1 and 6%. INSULIN and papain are among the richest proteins in this regard; they yield about 13%.

Tyrosine is nutritionally nonessential, although it has a sparing action on phenylalanine needs. If phenylalanine alone is in the diet, it must supply the phenylalanine needs and also act as a precursor for whatever tyrosine is needed for protein building. If tyrosine is supplied along with phenylalanine, the need for the latter is decreased. The specific rotatory power of L-tyrosine in acid solution is $[\alpha]_D = -8.6°$.

U

ULTRACENTRIFUGATION

Preparative Ultracentrifugation. *Differential centrifugation* is probably the most used preparative technique in which a mixture of particles is separated by differences in sedimentation rate of the components present. Both high and low centrifugal forces are commonly used, and in many cases, repeated cycling of high and low forces is used advantageously. (The principles involved will be found under CENTRIFUGATION, DIFFERENTIAL.) The equipment necessary to carry out this technique effectively must be capable of high-speed rotation in order to produce sufficiently great centrifugal forces. For a great many years, the electrically driven ultracentrifuge in use as a standard throughout the world has been the Beckman Model L Ultracentrifuge. This instrument is capable of spinning a 12-tube rotor at 50,000 rpm, producing a relative centrifugal force of 226,400 × gravity. Two types of rotors are available, fixed-angle rotors and swinging-bucket rotors. A variety of sizes of each type of rotor can be selected to meet capacity requirements. The larger and heavier the rotor, of course, the lower is the maximum speed. The rotor chamber is refrigerated to keep the rotor and its contents cold, and this chamber is also evacuated by the instrument's vacuum pump in order to avoid heating due to friction between air molecules and the rapidly spinning rotor. A higher-capacity instrument was introduced several years ago having a larger rotor chamber, a rotor stabilizer and several new controls for which higher-capacity rotors are available. This instrument is the Beckman Model L-2. Most recently available is the Beckman Model L2-65, which is capable of 65,000 rpm, for which high-speed rotors are available. The vacuum in this instrument is enhanced by having a diffusion pump backed up by a mechanical pump. The types of tubes used in these instruments vary according to the needs of the work to be accomplished. Transparent cellulose nitrate tubes with tube caps are most commonly employed, but a variety of requirements are met by polyallomer, stainless steel, quartz and polycarbonate.

Density gradient centrifugation is a technique used to measure the density heterogeneity of a macromolecular system. A gradient can be formed from sucrose, cesium chloride, or other materials, through which macromolecules are sedimented. When a mixture of molecular species is layered on top of the gradient column, each component will form its own band or zone, separated from each other by distances related to their specific sedimentation rates. These bands can be drawn off for further analysis, permitted to drip out of a punctured tube and collected in fractions, or made to flow through a continuous-flow monitoring device. The gradient for this method, known as zone centrifugation, must be quite steep in order to minimize convective stirring. The isopycnic gradient centrifuge technique separates mixtures of particles which differ in density. The gradient, in this case, must encompass the densities of every particle in the mixture. A centrifugal force is then applied continuously until each species of molecule reaches the position in the gradient corresponding to its own density. A great amount of consideration has been given to this latter method as a means for determining molecular weights.

The equipment used for density gradient centrifugation is essentially the same as described for differential centrifugation.

Important applications include the isolation of a variety of VIRUSES from both plant and animal material, the ability to distinguish specific DNA molecules, determination of sedimentation rates of many viruses, separation and isolation of cellular components such as MITOCHONDRIA and RIBOSOMES, purification of LIPOPROTEINS, and sedimentation characteristics of enzymes, ANTIGENS and ANTIBODIES.

Zonal ultracentrifugation is now distinguished from density gradient centrifugation mainly in terms of instrumentation which avoids the reorienting movements of swinging buckets during acceleration and deceleration of the rotor. The centrifuge itself need not be different from the Model L except that the vacuum chamber is extended in height to accommodate the longer zonal rotors. These rotors are long, narrow, essentially hollow cylinders divided internally into sector-shaped compartments by septa. Rotor stability is maintained by attaching an upper shaft to a bearing assembly, so that, in this case, the rotor is coupled at both the top and the bottom. In addition, a high-speed rotating seal permits the introduction and expulsion of fluids from the rotor while it is spinning. A density gradient solution is introduced into such a rotating rotor at the periphery of the rotor, light end first. When the rotor is completely full, it is displaced from the rotor through the center core line. The flow is then halted

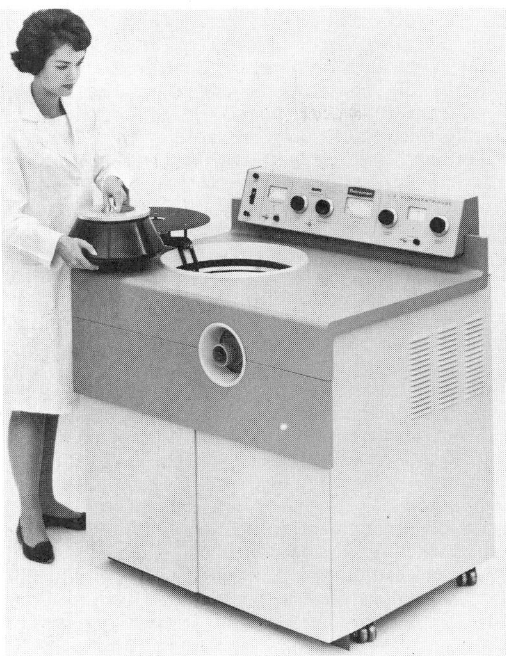

FIG. 1. The Beckman model L-2 preparative ultracentrifuge.

and sample is introduced through the center so that it rests on the lightest edge of the gradient. The rotor is then accelerated to high speed (up to 40,000 rpm). After zones of particles are formed, the entire contents of the rotor are displaced through the center line by pumping in very dense solution through the edge line. This displaced solution is collected in fractions for monitoring by various means such as spectrophotometry.

While zonal ultracentrifugation can be considered as still in its infancy, a number of very significant applications have already been explored. Extensive studies have been made on separation and isolation of cellular components such as ribosomes, polysomes, mitochondria, nuclei and membranes. It has been possible to isolate T2 and T3 bacteriophages, polio virus,

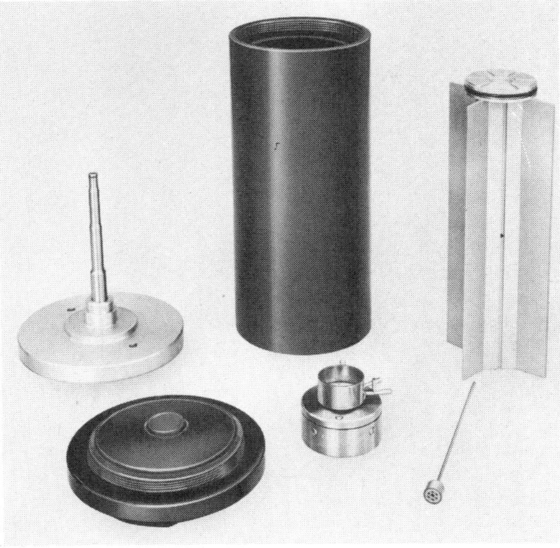

FIG. 2. The Beckman B-IV zonal ultracentrifuge rotor disassembled.

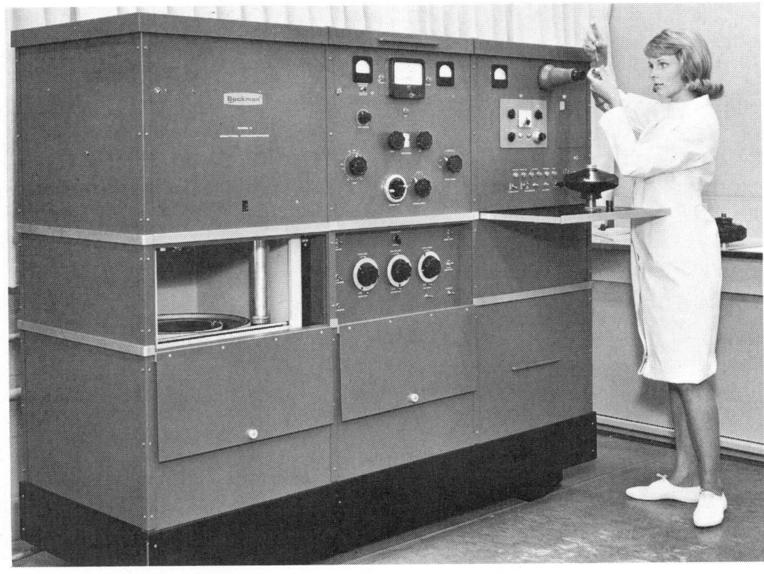

FIG. 3. The Beckman model E analytical ultracentrifuge.

bromegrass mosaic virus, adenovirus 2, echo 28, and the Rauscher mouse leukemia virus. About twenty fractions of different size and sedimentation rate GLYCOGEN particles have been separated from liver tissue. It has been possible to show that a particle responsible for the relaxing factor in muscle can be separated from soluble tissue proteins as well as from mitochondria. Studies are in progress to detect and isolate trace amounts of virus which may be present in tumors.

FIG. 4. An analytical rotor.

Special rotors for the Beckman Model L4-CF Zonal Ultracentrifuge enable one to use the principles of zonal centrifugation in continuous flow work. For example, large liquid volumes may be run through the continuous-flow rotor in order to remove virus particles with 95–99% efficiency, at a rate of 10 liters/hr.

Analytical Ultracentrifugation. The electrically driven analytical ultracentrifuge (Beckman Model E) is based in large part on the early centrifuges developed by Svedberg which were either oil or air driven. The Model E which is in use today is capable of speeds over 65,000 rpm spinning a 13-pound eliptical rotor coupled by a flexible piano-wire shaft, permitting the rotor to seek its own axis of rotation resulting in a self-balancing system. The cells which fit the rotor are sector-shaped and come with a variety of centerpieces, allowing numerous techniques to be applied. The rotor operates in a temperature-controlled vacuum chamber and becomes in effect a part of the optical system employed. The instrument's optical systems enable the visualization of sedimenting materials by either schlieren patterns, Rayleigh interferometric fringes or ultraviolet absorption. A monochromator accessory can be used for absorbtion studies at a variety of wavelengths, and an ultraviolet scanner scans the rotating cell and plots on a recorder the optical density as it varies with distance in the cell. Photographs can be taken automatically through two optical systems at preselected time intervals and exposure times.

There are three main applications of analytical ultracentrifugation: sedimentation velocity, sedimentation equilibrium and the approach to sedimentation equilibrium. In the sedimentation velocity method, the ultracentrifuge is operated at high speeds which cause the initially randomly distributed particles to migrate in the centrifugal direction through the solvent. A boundary is produced when a portion of the solvent is cleared of solute molecules and a layer of uniform concentration results from sedimentation. This boundary is then the region between the cleared solvent and the layer of uniform concentration and consists of a concentration gradient in which the concentration varies with distance in the cell. The movement of this boundary with time is recorded photographically through the optical system as changes in refractive index. This method provides the data necessary for calculation of sedimentation coefficients as well as diffusion coefficients, which can then be used for molecular weight determinations.

For the method known as sedimentation equilibrium, the ultracentrifuge is operated at lower speeds. At these lower centrifugal forces there is a relatively greater effect of the forces of diffusion which tend to redistribute the particles at random. The diffusion forces which are of concern are those which are in the centripetal direction. At some point in time the concentration variations with distance in the cell are stabilized and undergo no further change, *i.e.*, a state of equilibrium is

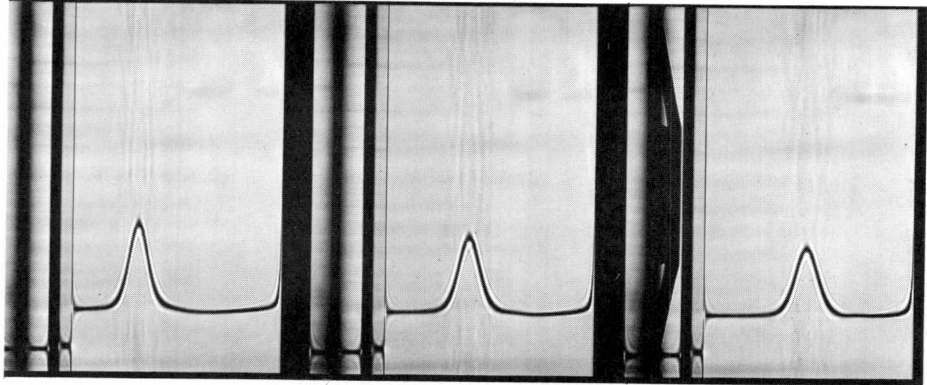

FIG. 5. Schlieren patterns obtained in the analytical ultracentrifuge.

FIG. 6. Ultraviolet absorption photographs showing the sedimentation of UV absorbing material with time.

achieved. Molecular weights can be calculated directly from the measurement of the concentration distribution at equilibrium.

The approach to sedimentation equilibrium requires only the measurement of the net transport of solute molecules across a given plane in the cell. In contrast to waiting for an equilibrium to be established, this method takes considerably less time for an experiment to be completed.

In addition to the previously mentioned sedimentation and diffusion coefficients and molecular weight determinations, analytical ultracentrifugation is used for shape analysis, concentration determinations, homogeneity studies, and association-dissociation investigations in an attempt to characterize protein-protein and protein-small molecule complexes.

Buoyant density gradient techniques have been adapted to the analytical ultracentrifuge in a highly successful manner, and the continued development of specialized cell centerpieces and other accessories lends significantly to its already very versatile capabilities.

References

1. ANDERSON, N. G., *J. Phys. Chem.*, **66**, 1984 (1962).
2. ANDERSON, N. G., *Fractions*, No. **1**, 2 (1965).
3. BRAKKE, M. K., *J. Am. Chem. Soc.*, **73**, 1847 (1951).
4. MESELSON, M., STAHL, F. W., AND VINOGRAD, J., *Proc. Natl. Acad. Sci. U.S.*, **43**, 581 (1959).
5. SCHACHMAN, H. K., "Ultracentrifugation in Biochemistry," New York and London, Academic Press, 1959.
6. SVEDBERG, T., AND PEDERSEN, K. O., "The Ultracentrifuge," London and New York, Oxford University Press; New York, Johnson Reprint Corporation, 1940.
7. TRAUTMAN, R., "Instrumental Methods of Experimental Biology" (NEWMAN, D. W., EDITOR), New York, The Macmillan Co., 1964.
8. WILLIAMS, J. W., "Ultracentrifugal Analysis in Theory and Experiment," New York and London, Academic Press, 1963.

MANUEL J. GORDON

ULTRASONIC VIBRATIONS

Sound consists of the mechanical vibrations of molecules or atoms of a gas, liquid or solid about the equilibrium positions of the particles. The sound waves with which man is most familiar, and the only kind that can exist in air, consist of alternate high- and low-pressure volumes along the lines of propagation. The human ear can detect those vibrating in a range between 10,000 and 20,000 cycles of vibration per second. Thus, those vibrations above 20,000 cycles/second (20 kc/sec) are considered ultrasonic (or supersonic). However, the term ultrasonics is usually reserved for frequencies of 500,000 cycles/sec and higher. Most of the research involving plants and animals has been conducted at those levels.

The first ultrasonic vibrations were produced over forty years ago by means of the electric spark and the singing arc. A dependable source of controllable frequency and intensity arose from the underwater signaling efforts of Langevin in 1912. While Langevin hinted at the possibilities of the piezoelectric effects of quartz on biological material, nothing was published until 1926. The range of ultrasonic wavelengths in liquids is from 6 cm to approximately 2×10^{-6} cm. The extremely short wavelengths have made possible the discovery of many interesting phenomena that are not ordinarily associated with acoustics.

In order to produce ultrasonic vibrations it is necessary to have a power source (a generator for a source of energy), a converter (transforms potential energy into a form suitable for activation), and finally a transducer (actually generates the ultrasonic vibrations). A number of different kinds of transducers are available, *e.g.*, quartz crystals or those made from ceramics, such as barium titanate.

During the past fifty years, ultrasound as a physical force has been investigated as it applies to every branch of science. In 1927, several investigators introduced ultrasound to the biological sciences primarily to investigate its destructive force upon biological organisms and living tissues. Today, it has progressed from its use as a destructive agent to a diagnostic tool in visualizing tumours within an intact skull in the form of a sonogram.

Ultrasound exerts its effect by pressure alternations and cavitations. In addition to these mechanical effects, various thermal effects are observed due to the absorption of acoustical energy. Cavitation occurs with explosive violence, thereby resulting in the dissociation of molecules. Pecht and Anbar[6] investigated the formation of hydrogen peroxide induced by ultrasonic waves and suggested it was a gas phase reaction occurring in the cavitation bubbles. If ultrasound is applied in large dosages, it is not unlike X rays and atomic radiations. In smaller dosages, it has been found useful as a therapeutic agent (as in arthritis).

Rammler *et al.*[7] have described the release of sulfatase by sonic disruption while studying the sulfate metabolism of aerogenes. Ultrasound is frequently employed to disrupt cells that have resisted other methods in an effort to release their components for biochemical analysis. Ultrasound accelerates the decalcification of bone in the presence of acid.[8] Biochemical differences between the ATPase activity and the actin-binding ability of both L-myosin and meromyosin are indicated by treatment with ultrasound.[3] While studying the effects of sonic vibrations of the ATPase activities of MYOSIN and actomysin, it was learned by several Japanese investigators that greater activation resulted for F-actin ATPase than for myosin.[1,2]

The denaturating effect of ultrasound on PROTEINS has been demonstrated (Grabar, 1951; Griffin, 1952). Dietrich[4] inactivated egg-white lysozyme by ultrasound and indicated that the reaction follows not first-order kinetics but a more complicated pattern. Fenn and Belcastro[5] found that treating several surface active agents with ultrasound did not affect their stability and that the amount of hydrogen peroxide formed during sonication was not sufficient to disturb stability. Therefore, they suggested that thermal, mechanical or other factors must be involved.

Ultrasonic irradiation has been used with some success in increasing the permeability of cells to a variety of chemical compounds and is currently being investigated as a tool in combination with them in chemotherapy. Its effects upon nerves, blood and chromosome behavior are also being studied.

While some controversial results are obtained in the use of ultrasound, it appears that they may be due to the difficulty in reproducing experiments in which exact parameters of intensity are being compared.

References

1. Asakura, S., Taniguchi, M., and Oosawa, F., "The Effect of Sonic Vibration on Exchangeability and Reactivity of the Bound Adenosine Diphosphate of F-actin," *Biochim. Biophys. Acta*, **74**, No. SC2256, 140–142 (1963).
2. Asakura. S., Taniguchi, M., and Oosawa, F., "The Effect of Sonic Vibration on Myosin and Actomyosin Adenosine Triphosphate Activities," *Biochim. Biophys. Acta*, **74**, No. SC2255, 142–144 (1963).
3. Barany, M., Barany, K., and Oppenheimer, H., "Effect of Ultrasonics on the Adenosine Triphosphatase Activity and Actin-binding Ability of L-Myosin and Heavy Meromyosin," *Nature*, **199**, No. 4894, 694–695 (August 17, 1963).
4. Dietrich, F. M., "Inactivation of Egg-White Lysozyme by Ultrasonic Waves and Protective Effect of Amino-Acids," *Nature*, **195**, No. 4837, 146–148 (July 14, 1962).
5. Fenn, G. D., and Belcastro, P. F., "Effect of Ultrasonic Waves on the Stability of Selected Surface-Active Agents, Sulfonamides, and *p*-Aminobenzoic Acid," *J. Pharm. Sciences*, **52**, No. 4, 340–342 (April 1963).
6. Pecht, I., and Anbar, M., "On the Mechanism of Hydrogen Peroxide Formation in Water Induced by Ultrasonic Radiation," *Israel Journal of Chemistry*, **1**, No. 3a, 283 (1963).
7. Rammler, D. H., Grado, C., and Fowler, L. R., "Sulfur Metabolism of *Aerobacter aerogenes*," *Biochemistry* (February 1964).
8. Thorpe, E. J., Bellomy, B. B., and Sellers, R. F., "Ultrasonic Decalcification of Bone," *J. Bone Joint Surg.*, 45-A, No. 6, 1257–1259 (September 1963).

Dale C. Braungart

ULTRAVIOLET RADIATION (BIOCHEMICAL EFFECTS)

We can think of biochemical effects as the action of ultraviolet light (UV) on substances of biological interest or/and as explanation of the action of UV on biological systems in biochemical terms. Among the most obvious or well known are sunburn, production of VITAMIN D in milk, sterilization of foodstuffs and utensils, and production of mutations in microorganisms, pollen, and other lower organisms (see MUTAGENIC AGENTS).

A conceptually clear approach to such phenomena is to measure an *action spectrum* for the effect of ultraviolet light. The system is exposed to the same dose of UV at each of several selected wavelengths and the amount of effect is measured. These amounts are plotted (logarithmically) *vs* the wavelengths, and hopefully the plot will bear a resemblance to the absorption spectrum of some known pigment or submolecular chromophoric group. If it does, one can say that the substance absorbing and initiating the biological action has been identified. This approach was employed by Gates who suggested that nucleic acid was the vital target in the killing of bacteria; this interpretation was later given for the production of mutations in corn pollen. In some examples, the action spectrum implicates protein as well.[2]

It is now known that one of the more important photochemical steps which can take place during irradiation of nucleic acid is the formation of dimers between two THYMINE bases. This reaction constitutes a lesion; in some instances, this lesion can be repaired by one of several "tricks" available to the cell or to the photobiologists. The cell may substitute thymine for the dimer in the nucleic acid (dark recovery); with the aid of visible light, a photosensitive enzyme may cleave the dimer (photorecovery). The dimer may also be split by UV of a shorter wavelength. Such repair mechanisms may restore a cell to a normal reproduction state.

Photoreversal of short wavelength UV damage by visible light was first observed by Hausser and von Oehmcke in 1933,[1] with banana skin; how general this protective phenomenon is in the plant kingdom remains to be determined.

By suitable choice of wavelength it is possible to inactivate nucleic acids in some viruses, *e.g.*, poliomyelitis, without appreciable effect on the protein moiety; with such inactive viruses, ANTIBODIES may be prepared which have medicinal value.

The question of the mechanism of UV-induced

mutations is still largely a mystery, but one possibility stems from the observation by Grossman that irradiation at 2537 Å of polyuridylic acid, —UUU—, which normally codes for phenylalanine [see PROTEINS (BIOSYNTHESIS)] leads to a modified polymer coding for serine. The mechanism may involve the formation of a uridine hydrate (H) in the polymer which may code for serine as UUH. Such a mechanism could be involved in the synthesis of modified enzymes, which in turn could alter the dynamics of cellular growth and differentiation.

Steenboch found that rickets could be prevented by irradiation of some foods or of the patient with UV of less than 3100 Å. During irradiation, ergosterol is converted to vitamin D_2.

Ultraviolet erythema is generally thought to be caused by the action of UV on the epidermis, the vasodilation in the dermis being effected by one or more mediator substances diffusing from the epidermis. These substances have yet to be identified with certainty, and the same may be said for the casual factor in UV-induced skin cancer. In both phenomena, NUCLEOPROTEINS could be the absorber.

References

1. SELIGER, H. H., AND MCELROY, W. D., "Light: Physical and Biological Action," New York, Academic Press, 1965.
2. GIESE, A. C., "Photophysiology," Vols. I and II, New York, Academic Press, 1964.
3. HOLLAENDER, A., "Radiation Biology," Vol. II, New York, McGraw-Hill Book Co., 1955.
4. MCLAREN, A. D., AND SHUGAR, D., "Photochemistry of Proteins and Nucleic Acids," Oxford, Pergamon Press, 1964.
5. ELLIS, C., WELLS, A. A., AND HEYROTH, F. F., "The Chemical Action of Ultraviolet Rays," New York, Reinhold Publishing Corp., 1941.
6. *Photochemistry and Photobiology*, An international journal, Oxford, Pergamon Press (Volume I, 1962, etc.).

A. D. MCLAREN

ULTRAVIOLET SPECTROPHOTOMETRY. See INFRARED AND ULTRAVIOLET SPECTROSCOPY.

UNSATURATED FATTY ACIDS

Unsaturated fatty acids are aliphatic carboxylic acids which contain one or more carbon-carbon double (olefinic) bonds or triple (acetylenic) bonds. Although fatty acids containing triple bonds occur with relative rarity in nature, fatty acids of the olefinic type (particularly of the *cis* configuration) appear to be universal constituents of living cells.

Biosynthesis and Comparative Biochemistry. The often bewildering differences in unsaturated fatty acid composition, both quantitatively and qualitatively, to be found among various organisms is due in part to the assimilation of unsaturated fatty acids from diets of different composition, but they may also reflect fundamental differences both in the mode of *in toto* biosynthesis of olefinic acids and in the type of transformations (chain elongation, further desaturation) which the organism may carry out on unsaturated fatty acids assimilated from the diet. An understanding of the biosynthetic pathways involved in the formation and transformation of unsaturated fatty acids, as they operate in various life forms, will often allow us to distinguish between assimilation products and biosynthetic products and thus permit us to make meaningful comparisons based on fatty acid composition.

Monounsaturated Fatty Acids. Monounsaturated fatty acids (fatty acids containing one carbon-carbon double bond in the chain) can apparently be synthesized by all cellular organisms with the possible exception of certain obligate parasites. In some bacteria and, with rare exceptions, in all higher organisms, the predominant monounsaturated acids synthesized are palmitoleic (*cis*-9-hexadecenoic) and oleic (*cis*-9-octadecenoic) acids, *i.e.*, 16- and 18-carbon acids with the double bond in the 9-position counting from the carboxyl carbon. Among the bacteria, the anaerobes, facultative anaerobes and some aerobes synthesize *cis*-vaccenic (*cis*-11-octadecenoic) acid rather than oleic acid although palmitoleic remains the major C_{16} acid. At least two distinctly different pathways for the biosynthesis of monounsaturated fatty acids have been demonstrated. One pathway, found only in bacteria, produces C_{16} and C_{18} monounsaturated fatty acids by chain elongation of *cis*-3-unsaturated intermediates. The over-all process can proceed under anaerobic conditions. The second pathway, which is found in many aerobic bacteria and probably in all higher organisms, involves the direct, oxygen-dependent desaturation of long-chain saturated fatty acids to the *cis*-9-unsaturated derivatives. The two mechanisms seem to be mutually exclusive; no organism has yet been found which can utilize both pathways.

The *anaerobic pathway* is presumably the more primitive and, except for one step (the formation of the *cis*-3-double bond), is identical with the pathway for saturated fatty acid synthesis. The details of this pathway are illustrated in Fig. 1.

It should be noted that the representation of intermediates as free acids is a simplification since they are actually bound to enzymes by thioester linkages. Thus, although a number of discrete steps are shown, they are probably all carried out on the same enzyme complex without the release of intermediates. The key step in the control of unsaturated fatty acid synthesis by the anaerobic mechanism is the dehydration of β-hydroxydecanoate, a normal intermediate in saturated fatty acid synthesis (see FATTY ACID METABOLISM). If the dehydration is α,β (as it is almost exclusively for longer and shorter chain-length β-hydroxy acids), the resulting *trans*-2-double bond is reduced to the saturated analogue which is further chain-elongated to palmitic and stearic acids. If the dehydration is β,γ, the resulting *cis*-3-double bond is not reduced, and elongation then results

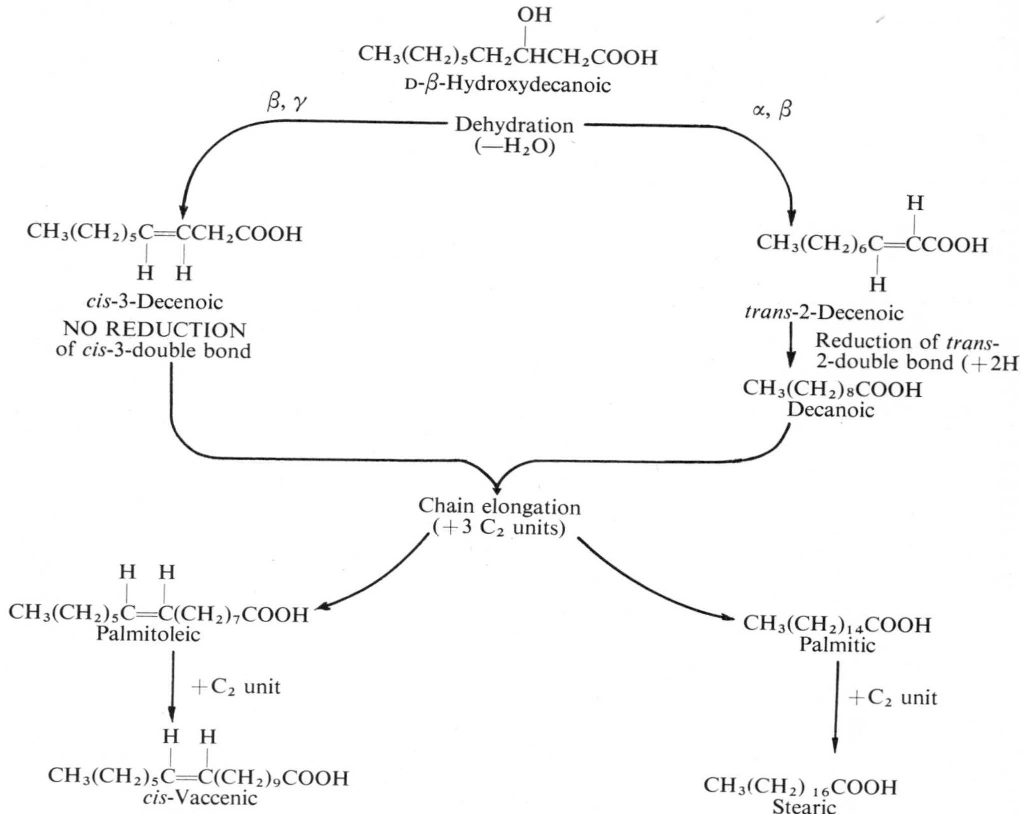

FIG. 1. The anaerobic pathway for monounsaturated fatty acid biosynthesis.

in palmitoleic and vaccenic acids. Thus, the percentage of unsaturated acids formed would ultimately depend upon the activity of the β,γ-dehydrase.

The *oxygen-dependent* mechanism was first demonstrated in yeast in 1958 and since then has been found to operate in many other organisms from bacteria to mammals. Unlike the anaerobic pathway, the O_2-dependent pathway is completely separate from the fatty acid synthesizing system. In this reaction, two hydrogens are removed from a saturated fatty acid (usually the coenzyme A ester of stearic or palmitic acid) to form a *cis*-double bond in the 9-position. The reaction requires molecular oxygen and a reduced pyridine nucleotide (usually TPNH). In one enzyme system (from the bacterium *Mycobacterium phlei*), the additional requirements for ferrous ion and a flavin (either FAD or FMN) have been demonstrated. There is some evidence which suggests that ferrous ion and flavin may be involved in the O_2-dependent desaturating systems of other organisms as well. Experiments utilizing the bacterium *Corynebacterium diphtheriae* have demonstrated that the 9D and 10D hydrogens are stereospecifically removed from stearic acid to form the *cis*-double bond of oleic and that the abstraction of the 9D hydrogen is probably the rate-limiting step. The exact function of the various cofactors involved in the desaturation process is not known.

Other Pathways. The formation of *cis*-9-monounsaturated acids in plants is O_2-dependent, and under certain circumstances, a direct desaturation of saturated acids can be demonstrated. However, there is evidence that another pathway is generally utilized by higher plants and green algae which requires molecular oxygen, yields palmitoleic and oleic acids but does not seem to involve direct desaturation of the saturated analogues.

There are also several instances known where the direct, O_2-dependent desaturation of palmitic or stearic acid does not produce a *cis*-9-double bond. In the bacterium *Bacillus megaterium*, *cis*-double bonds in the 5-positions result, while in *Mycobacterium phlei*, the major product of the desaturation of palmitic acid is *cis*-10-hexadecenoic acid.

Polyunsaturated Fatty Acids. Although polyunsaturated fatty acids (fatty acids containing two or more double bonds) cannot be synthesized by bacteria, almost all higher organisms can produce them. The only pathway known for their formation is further oxygen-dependent desaturation of preexisting unsaturated acids.

The new double bonds are introduced to yield (with rare exceptions) all *cis*-1,4-polyenes, *i.e.*, fatty acids in which successive double bonds are isolated by single methylene groups. Oleic acid, for example, is readily converted to linoleic (*cis,cis*-9,12-octadecadienoic) acid in many fungi, algae and higher plants by an O_2-dependent desaturation at the 12-13 positions. Linoleic, in turn, can be converted to α-linolenic (*cis,cis,cis*-9,12,15-octadecatrienoic) acid by removal of the 15-16 hydrogens. In this series, the new double bonds are introduced between the terminal methyl groups and the preexisting double bond(s). Most fungi, algae and modern higher plants do not form significant amounts of polyunsaturated fatty acids by desaturation in the opposite direction (*i.e.*, introduction of double bonds between the carboxyl group and preexisting double bonds). Certain primitive plants (ferns, mosses) and a number of protozoa, however, can desaturate in either direction. Thus, many amoebas and ciliates can convert oleic acid to γ-linolenic (*cis,cis,cis*,-6,9,12-octadecatrienoic) acid by introducing methylene-interrupted double bonds on either side of the 9 double bond. Higher animals are incapable of desaturating toward the terminal methyl group. Thus, in the rat or man, oleic could give rise to 6,9-octadecadienoic but not to linoleic acid. Dietary linoleic acid can be transformed to give γ-linolenic acid but not α-linolenic. Chain elongation of γ-linolenic followed by further desaturation toward the carboxyl group gives rise to the 20-carbon acid arachidonic (*cis,cis,cis, cis*-5,8,11,14-eicosatetraenoic) which is considered a typical "animal" fatty acid, even though its precursor (linoleic) cannot be synthesized by the animal. Certain protozoa and primitive plants also form arachidonic acid, but it is virtually absent in more highly evolved plants, algae and fungi.

Polyunsaturated Acids Containing Isolated Double Bonds. Although polyunsaturated fatty acids containing isolated double bonds (double bonds separated from other double bonds by two or more methylene groups) are rare in nature, they have been found to occur in a number of apparently unrelated organisms, but especially in certain primitive plants (*Ginkgo biloba, Equisetum*) and in the slime mold *Dictyostelium discoideum*. These unique acids contain an isolated *cis*-double bond in the 5-position with other double bonds occurring in the more typical 1,4-sequence elsewhere in the chain.

Taxonomy. Enough is now known about the comparative biosynthetic pathways for the formation of unsaturated fatty acids so that we can begin to logically group organisms according to the way in which they form or transform unsaturated fatty acids. Although these groupings do not necessarily coincide with taxonomic groupings based on morphological, physiological and other biochemical characteristics, they do not conflict with these groupings and someday may be valuable in establishing or clarifying certain evolutionary relationships (see also CHEMOTAXONOMY; COMPARATIVE BIOCHEMISTRY).

Metabolism and Function. Fatty acids generally do not occur naturally in the free state but rather appear as integral parts of more complex lipids such was WAX esters, triglycerides, PHOSPHOLIPIDS, glycolipids, sphingolipids and other large molecules. Wax esters and triglycerides may serve as energy reserves or as protective or insulating materials, and in these capacities the presence or absence of double bonds in the fatty acid components would seem to be of secondary importance. It is known, for example, that unsaturated fatty acids can be readily degraded to acetate by the same β-oxidation systems which catabolize saturated fatty acids (see FATTY ACID METABOLISM). We are more concerned here, however, with the unique functions of unsaturated fatty acids in which the number and positions of double bonds are of primary importance.

For many higher animals, linoleic acid and its transformation products, γ-linolenic and arachidonic acids, are considered "ESSENTIAL FATTY ACIDS" since if at least one of these acids is not included in the diet, a disease state results, especially in young, rapidly growing animals. It is probable that the essential fatty acids and particularly arachidonic acid, are specifically required as structural components of animal cell membranes. In addition, they serve as precursors to the mammalian hormones, the prostaglandins and are utilized in the transport of cholesterol (as cholesterol esters) in the bloodstream. It is probable that polyunsaturated fatty acids are also important to the normal function of the mitochondrial electron transport system, but their exact role is unknown.

In green plants and algae, α-linolenic acid is usually a major constituent, and recent experiments from several laboratories have demonstrated a definite relationship between the photosynthetic efficiency of a plant tissue and the concentration of linolenic acid in the CHLOROPLASTS. From other evidence, it seems that α-linolenic acid, in the form of galactoglycerides, is utilized as a "structural cement" in the assembly of chlorophyll molecules and other components into photosynthetically active units.

It has been observed repeatedly in organisms ranging from anaerobic bacteria to mammals that the proportion of unsaturated to saturated fatty acids of a tissue increases, within limits, with decreasing environmental temperature. Since this effect appears to be independent of the mechanism involved for the biosynthesis of unsaturated fatty acids and seems to occur among all phyla, there must be survival advantages associated with increased unsaturation at lower temperatures.

Again, the reasons for this phenomenon are unknown, but it is noteworthy that the melting points of unsaturated fatty acids and their derivatives are invariably lower than the saturated analogues. It is probable that many more examples will be discovered in the near future where the configurations and positions of double bonds within a fatty acid molecule confer upon it unique properties which enable it to perform vital and highly specific roles within the cell.

References

1. Erwin, J., and Bloch, K., "Biosynthesis of Unsaturated Fatty Acids in Microorganisms," *Science*, **143**, 1006–1012 (1964).
2. Bloch, K., "The Biological Synthesis of Unsaturated Fatty Acids," in "The Control of Lipid Metabolism," *Biochem. Soc. Symp.*, No. 24 (1965).
3. Mead, J. F., "Synthesis and Metabolism of Polyunsaturated Acids," *Federation Proc.*, **20**, 952–955 (1961).
4. Stoffel, W., "Metabolism of the Unsaturated Fatty Acids. I. Biosynthesis of Highly Unsaturated Fatty Acids," *Z. Physiol. Chem.*, **333**, 71–88 (1963).
5. Fulco, A. J., and Bloch, K., "Cofactor Requirements for the formation of Δ^9-Unsaturated Fatty Acids in Mycobacterium Phlei," *J. Biol. Chem.*, **239**, 993–997 (1964).
6. Schlenk, H., and Gellerman, J. L., "Arachidonic, 5,11,14,17-Eicosatetraeonic and Related Acids in Plants—Identification of Unsaturated Acids," *J. Am. Oil Chemists' Soc.*, **42**, 504-511 (1965).

Armand J. Fulco

URACIL AND URIDINE

Uracil is one of the naturally occurring PYRIMIDINES; its structure is shown at left. It is one of the bases attached to the ribose phosphate chain in RIBONUCLEIC ACIDS. Its nucleoside form (uridine, uracil ribonucleoside, or 1-β-D-ribosyluracil) has the structure shown at right (see also PYRIMIDINE BIOSYNTHESIS).

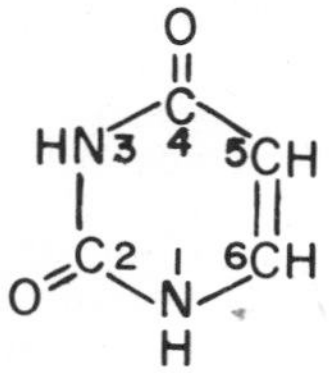

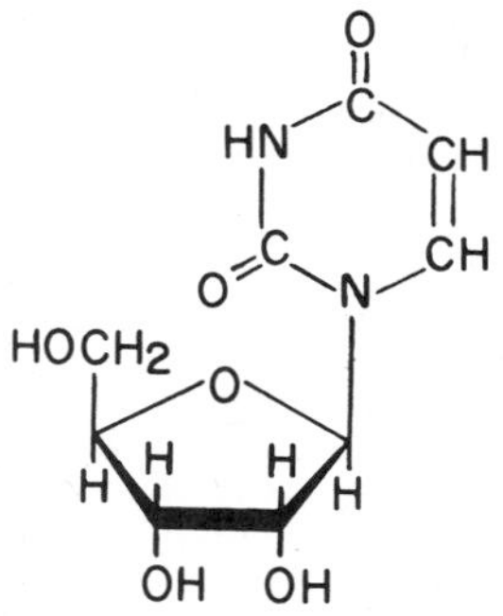

O
HN3 4 5CH
O C2 1 6CH
N
H

HOCH2 O H H H H OH OH
HN C CH CH O C N

UREA AND UREA-ARGININE CYCLE. See AMMONIA METABOLISM; ARGININE-UREA CYCLE; RENAL TUBULAR FUNCTIONS, section on Tubular Reabsorption.

UREASE

In 1926, J. B. Sumner announced the crystallization of the enzyme urease. It was the first enzyme to be isolated as a crystalline protein; this accomplishment confirmed the then growing belief that ENZYMES, the biological catalysts, were indeed from the chemical standpoint PROTEIN molecules. Urease catalyzes the cleavage of urea to ammonia and carbon dioxide.

URIC ACID

Uric acid, 2,6,8-trihydroxypurine (for structure, see PURINE BIOSYNTHESIS, Fig. 1) is the principal form in which nitrogen (from ingested protein) is excreted in reptiles, birds and Dalmatian dogs. In human beings, uric acid is excreted in small amounts (of the order of 0.7 g/day) and is the principal end product of purine metabolism (see PURINE BIOSYNTHESIS AND METABOLISM). On a low purine diet the uric acid excretion is about the same as in starvation (endogenous purines), but on a high purine diet it rises. Uric acid in the blood of healthy individuals is present at the level of 2–6 mg/100 ml. In gout the levels are high, but factors other than the uric acid level of the blood determine whether or not insoluble salts of uric acid are deposited in the tissues and joints (see also METABOLIC DISEASES). High rates of synthesis or uric acid or low rates of urinary excretion are, however, conducive to gout.

URINE COMPOSITION. See RENAL TUBULAR FUNCTIONS (AND URINE COMPOSITION).

URONIC ACIDS. See HEXURONIC ACIDS.

U.S. FOOD AND DRUG ADMINISTRATION

The Food and Drug Administration, a constituent unit of the U.S. Department of Health, Education, and Welfare, was created to enforce the Federal Food and Drugs Act of 1906 and certain related statutes. The Act of 1906 was the first national law regulating the interstate distribution of foods and drugs. Its passage was in no small measure due to the personal campaigning of Dr. Harvey W. Wiley, then chief of the Bureau of Chemistry of the U.S. Department of Agriculture. He gave widespread publicity to the findings of the Bureau on then current practices in the adulteration and debasement of foods and the worthless and frequently dangerous character of many of the nostrums and patent medicines of that day. Such disclosures and the public reaction to them led Congress to pass the original "Pure Food and Drug Law" of 1906. Enforcement was first vested in the Bureau of Chemistry, headed by Dr. Wiley, and in 1927 the Food and Drug Administration was established as a separate unit to handle the regulatory functions of the Bureau. A governmental reorganization in June 1940 effected the transfer of the Food and Drug Administration from the Department of Agriculture to the Federal Security Agency, and the latter agency became the Department of Health, Education, and Welfare in 1953.

Administrative experience and judicial interpretations demonstrated certain deficiencies in the original Act, and a stronger and more inclusive law was passed in 1938, the present Federal Food, Drug, and Cosmetic Act. It too has been amended as weaknesses have been revealed by court decisions or changing conditions.

The broad purpose of this Act is to insure that foods, drugs, cosmetics and therapeutic devices shipped in interstate commerce are safe and suitable for their intended use and that they are truthfully and informatively labeled.

The food provisions ban products containing poisonous or deleterious substances, except that provision is made for the establishment of the amount and kind of pesticides that may remain on raw agricultural products, and of additives to processed food, when such articles are shipped in interstate commerce. Preparation under insanitary conditions or the presence of filth or decomposition is likewise prohibited. The omission or removal of a valuable constituent of a food, the substitution of a cheaper substance for one more valuable, or the concealment of inferiority or damage all constitute prohibited acts. The law also authorized promulgation of regulations fixing reasonable standards of identity, quality and fill of container for foods. Except in the case of foods so standardized, the label must list the ingredients.

Among the most important drug provisions of the Act are those which require informative labeling as to composition, adequate directions for use, and warnings against misuse where necessary for the protection of consumers. The Act prohibits any false or misleading statements in the labeling of a drug. Drugs listed in the United States Pharmacopeia and the National Formulary are required to meet the standards set forth in these compendia. A new drug may not be marketed until proof of its safety and effectiveness acceptable to the Food and Drug Administration has been submitted in the form of a new drug application. Predistribution certification is required for insulin, antibiotics, and certain colors used in foods, drugs, and cosmetics.

The Act of 1938 for the first time brought therapeutic devices and cosmetics under Federal regulation. The most important provisions relating to devices are the requirements that they be safe to use in the manner recommended and that they have the properties claimed for them in their labeling. In practical effect, the most important provision relating to cosmetics is the prohibition against the use of dangerous ingredients.

It should be understood that with a few exceptions, such as those noted, effective enforcement of the statute depends upon factory inspection, sampling and analysis, rather than upon predistribution controls. Substantial penalties of fine and/or imprisonment are provided for refusal to permit inspections or for the shipment of violative products. Seizure of the offending merchandise and injunction to halt a continuing violation are also authorized under the statute.

To carry out its enforcement operations, the Administration has eighteen field districts in strategically located cities throughout the country, equipped with laboratories where most of the examinations of foods, drugs, and cosmetics are made. Establishments engaged in the production of articles subject to the law are inspected for sanitary conditions, raw materials used, and controls exercised in compounding, processing, packaging, and labeling products destined for interstate shipment. Surveillance is maintained over such commodities in interstate commerce through warehouse inspections, sampling, and analysis. When domestic violations are found, the facts are reported to the Department of Justice with a recommendation for seizure, criminal prosecution, or injunction actions in the Federal courts. Imports are refused admission if they do not comply with the requirements of applicable laws.

The central headquarters of the Food and Drug Administration in Washington, D.C., exercises general supervision over the field work, controls broadly the allocation of time to the various projects, and determines the type of legal proceedings to be brought for violations. In addition, the staff laboratories in Washington undertake fundamental investigations to form the basis for enforcement policy and to develop new or improved methods of analysis. These laboratories also make certain specialized types of analyses for which the field laboratories are not equipped. Other functions of the Washington staff and laboratories are the evaluation of new-drug applications; the establishment of safe limits for pesticide residues, food additives and colors; the formulation of food standards; and the certification of insulin, antibiotics, and colors.

It is apparent that effective enforcement of the Federal Food, Drug, and Cosmetic Act is dependent upon the analyst, who as a result of his examination testifies as to the composition, presence or absence of constituents or ingredients, or other factors significant in the application of the provisions of the law. Even though the composition of the product may not be directly at issue, the findings and testimony of the chemist may form the basis for the action. For example, the physician who is testifying in a case alleging false or misleading therapeutic claims may base his conclusions as to the effect of the drug on prior testimony of the chemist about its composition.

The Food and Drug Administration is thus vitally concerned with the development of adequate methods of analysis and has always taken an active part in the work of the Association of Agricultural Chemists. This Association has as its primary purpose the development of methods of analysis through collaborative studies engaged in by laboratories of industrial concerns as well as those in state and federal regulatory agencies. Analytical methods so developed are published in the "Official Methods of Analysis of the Association of Official Agricultural Chemists." While these methods have no specific recognition in the Federal Food, Drug, and Cosmetic Act, they have quite generally found acceptance in the courts. The A.O.A.C. methods are incorporated in many of the food standards promulgated under authority of the Act.

N. E. Cook

V

VALINE

L-Valine, α-aminoisovaleric acid, $(CH_3)_2CH$—$CHNH_2$—COOH, is one of the AMINO ACIDS that is found widely distributed in all common PROTEINS; it is usually found at 3–10% levels. It is related structurally to isoprene. Valine is nutritionally essential for mammals and fowls. The human requirement is from about 400 to 800 mg/day. The specific rotatory power of L-valine in 6*N* HCl solution is $[\alpha]_D = +28.8°$.

VAN SLYKE METHOD

This is historically an important method for quantitative determination of free amino nitrogen. In a special apparatus, the sample is treated with nitrous acid and the evolved nitrogen is measured. Since 1 mmole of an AMINO ACID yields 22.4 ml of nitrogen measured under standard conditions, the sample size need not be large. The Van Slyke Method can be used, for example, to demonstrate readily that as a PROTEIN is hydrolyzed the free amino groups increase to levels many times that in the original protein.

VANADIUM (IN BIOLOGICAL SYSTEMS)

The effect of vanadium on a number of biological activities has been demonstrated.

Vanadium has been shown to be necessary for the optimal growth of certain green algae, being concerned with photosynthesis under strong light. Otherwise, notwithstanding its almost universal presence in plant and animal tissues, it has not been proved to serve an essential role.

The vanadium content of plant and animal tissues varies from less than 1 ppm in the tissues of some plants and animals to up to 4% in the blood of a tunicate, the pleobranch ascidians. Adult man contains about 25 mg of vanadium in his body. Most of the vanadium in man is present in the fatty tissues, the serum, and the bones and teeth.

Many of the described actions of vanadium in mammalian tissues have been related to lipid metabolism. Mammalian liver suspensions show increased oxidation of PHOSPHOLIPIDS from brain, liver and soybeans in the presence of small amounts of vanadium. The hepatic synthesis of cholesterol and fatty acids in rats and rabbits decreases in the presence of trace amounts of vanadium. The site of vanadium inhibition of cholesterol biosynthesis has been localized at the enzyme, squalene synthetase [see STEROLS (BIOGENESIS)]. The hepatic content of COENZYME A in the intact rat is decreased by vanadium injection. Functional levels of ATP in crude liver homogenates are reduced by the presence of vanadium salts. In experimental ATHEROSCLEROSIS, addition of small amounts of vanadium to an atherogenic diet produces a decrease in the aortic content of cholesterol in rabbits. Dietary vanadium has also been shown to lower hepatic, plasma and tissue phospholipids and cholesterol in rabbits. The hepatic synthesis of phospholipids has been demonstrated to be decreased, and their oxidation to be increased, by vanadium. Vanadium has been shown to decrease the total body content of cholesterol in young men under controlled conditions, but no effect has been demonstrated on the serum cholesterol levels of older subjects ingesting a regular diet.

In a number of *in vitro* systems, the effects of vanadium on lipid synthesis and oxidation are counteracted by other transition elements. This antithetic action has been especially demonstrated with MANGANESE, CHROMIUM and COBALT.

In a similar fashion to MOLYBDENUM, vanadium increases the fixation of atmospheric nitrogen by Azobacter (see NITROGEN FIXATION). Vanadium at 5 ppm inhibits the growth of tubercle bacilli, and dietary vanadium decreases the mortality of tuberculous mice and the size of pulmonary lesions of tuberculous rabbits.

Vanadium-deficient rats and guinea pigs have a higher incidence of dental CARIES. Hamsters given a cariogenic diet develop less dental caries if they receive oral or subcutaneous vanadium.

The activity of guinea pig liver monoaminoxidase is increased by the presence of a vanadium salt in the incubating buffer.

Although vanadium has been shown to have a number of biological effects, final proof of its essentiality as a trace metal for mammals is still lacking.

References

1. CURRAN, G. L., in "Metal Binding in Medicine" (SEVEN, M. J., EDITOR), p. 216, Philadelphia, Lippincott, 1960.
2. SCHROEDER, H. A., BALASSA, J. J., AND TIPTON, I. H., *J. Chronic Diseases*, **16**, 1047, 1963.

G. L. CURRAN

VASOPRESSIN

Vasopressin is a neurohypophyseal polypeptide hormone of relatively simple structure containing eight amino acids. The brilliant work of du Vigneaud and his collaborators made possible the knowledge concerning the chemical structure of this physiologically active material whose structure may be represented as follows:

$$\overline{\text{CyS—Tyr—Phe—Glu(NH}_2\text{)—Asp(NH}_2\text{)—CyS}}\text{—Pro—Arg—Gly(NH}_2\text{)}$$
(1) (2) (3) (4) (5) (6) (7) (8) (9)

Another neurohypophyseal polypeptide hormone with different physiological properties, named *oxytocin*, has also been isolated, characterized and synthesized. Its chemical structure is the same as that of vasopressin with the exception that isoleucine and leucine are substituted in positions (3) and (8), respectively. With respect to species variation, it is of interest that vasopressin isolated from horse, beef or human pituitaries contains arginine at position (8), whereas pig vasopressin contains lysine at this position. In the lower forms of animals such as teleost fish, amphibia, reptiles and birds, the ring structure of oxytocin is found in combination with the side chain of vasopressin resulting in the formation of a physiologically active material known as arginine-vasotocin. The disulfide bridge as well as the amide groups are necessary for full activity of vasopressin. Replacement of the aromatic side chain in (3) by an aliphatic one decreases the antidiuretic and vasopressor activity and enhances the oxytocic activity. The presence of a basic group in position (8) is essential for antidiuretic and vasopressor activity, the degree being dependent on the basicity of the substituent. The name antidiuretic hormone (ADH) for vasopressin indicates that this certainly represents the main physiological action of vasopressin, the effect being brought about by an increase in the amount of water reabsorbed in the renal tubules (see RENAL TUBULAR FUNCTIONS). Contraction of muscle in the walls of blood vessels causes a rise in blood pressure, the first effect to be demonstrated for this active material and thus the origin of the name. Vasopressin also acts on the smooth muscle of the gut and increases peristalsis, but only in large and unphysiological doses. Current data support the neurosecretory view of the hypothalamico-hypophyseal system for vasopressin as well as for oxytocin. Vasopressin is made in the cells of the supraoptic and paraventricular nuclei and passes along the axonic process to the neurohypophysis where it is stored in a form bound to particulate components. The secretion of vasopressin by the neurohypophysis is believed to be under reflex nervous control as well as osmotic regulation, the former being demonstrable under conditions such as coitus or suckling, and the latter being related to the osmolar concentration of the blood perfusing the hypothalamico-neurohypophyseal areas. *Diabetes insipidus* is the disease state associated with a lack of vasopressin in which a hypotonic urine is produced with a volume of 10–20 times that of a normal individual. Under normal conditions about 7.5–50 milli units of vasopressin are secreted per hour.

A variety of body fluids and tissue extracts have been found to enzymatically inactivate oxytocin; however, there is some controversy regarding the identity or lack of it with regard to vasopressinase and oxytocinase. A highly purified preparation of vasopressinase from human placenta has been extensively studied which demonstrated its difference from oxytocinase. The most sensitive quantitative method for the assay of vasopressin is the antidiuretic effect on unanesthetized hydrated rats or dogs. Another method utilizes the pressor activity in the anesthetized rat. In the course of quantitative assays, natural or synthetic preparations of vasopressin must be compared with either the Third International Standard for Oxytocic, Vasopressor and Antidiuretic Substances or the U.S.P. Posterior Pituitary Reference Standard in which one unit is equal to 0.5 mg of the dry standard powder.

References

1. DU VIGNEAUD, V., RESSLER, C., AND TRIPETT, S., *J. Biol. Chem.*, **205**, 959 (1953).
2. DU VIGNEAUD, V., RESSLER, C., SWAN, J. M., ROBERTS, P. O., KATSOYANNIS, P. G., AND GORDON, S., *J. Am. Chem. Soc.*, **75**, 4879 (1953).
3. DU VIGNEAUD, V., LAWLER, H. C., AND POPENSE, E. A., *J. Am. Chem. Soc.*, **75**, 4880 (1953).
4. SACHTER, M., "Polypeptides which Affect Smooth Muscles and Blood Vessels," New York, Pergamon Press, 1960.
5. RUDINGER, J., "Symposium on Oxytocin, Vasopressin, and their Analogues," New York, Pergamon Press, 1964.

CLYDE G. HUGGINS

VIRTANEN, A. I.

Artturi Ilmari Virtanen (1895–), a Finnish biochemist, received the Nobel Prize in chemistry in 1945 for his investigations in agricultural and nutritional chemistry, particularly relating to fodder preservation.

VIRUS DISEASES

Until little more than a decade ago, most of the known viruses were named for the diseases which they caused, and they were identified and quantitated by their pathological effects in susceptible animals or plants. Thus we have viruses called tobacco mosaic, yellow fever, smallpox, etc. The tools and methods for working with viruses have been improved and many new viruses have been isolated. Large quantities of many viruses can now be obtained *in vitro* in cell cultures, and better methods of virus purification have enabled investigators to carry out chemical tests which were hitherto impossible.

Many of the biological tests for viruses, such as complement fixation, hemagglutination, or cytopathic changes in cell cultures and serum-virus neutralization tests in cell cultures or animals, are extremely sensitive, and these methods are used routinely for viral identification. There are a few viruses, however, whose identification depends on biochemical tests and still others for which biochemical tests in conjunction with biological tests are useful aids for identification.

All the viruses which have been studied in detail have a core of nucleic acid surrounded by a protein coat or coats. The viral nucleic acids are of two types, ribose (RNA) or deoxyribose (DNA). They represent the genetic material necessary for the replication of intact virus progeny [see VIRUSES (REPLICATION)]. Nucleic acid extracted from purified virus using phenol or dodecyl sulfate is easily destroyed by the homologous NUCLEASES which are present in normal sera or tissues. DNA is destroyed by the enzyme deoxyribonuclease, and RNA by ribonuclease, and this is one means of identifying the type of nucleic acid. The intact virus is not affected by these enzymes.

Without a protein coat, the nucleic acid is ordinarily incapable of entering a cell. Entry can be accomplished by washing the cells to remove any serum and by pretreating the warmed cells with hypertonic salt solution. On entry into the cell, the viral nucleic acid induces the synthesis of a new protein coat from the host material. Some viral nucleic acids are capable of replicating at least one virus cycle in hosts which do not permit replication by the whole virus. Poliovirus nucleic acid has been known to give rise to whole virus in embryonic chick cells, a host which is not susceptible to infection with poliovirus; polyoma virus nucleic acid has produced whole virus in *B. subtilis*, a bacterium. In general, the DNA viruses multiply in the nucleus of the host cell and RNA viruses in the cytoplasm. Some of the RNA viruses appear to emerge as buds from the cell membrane as seen in the electron microscope.

The nucleic acids and the proteins which they control exist in an orderly fashion. In the RNA viruses, the nucleic acid may be in either a cubical or a helical form; the DNA viral nucleic acids occur only in a cubical form. The cubical nucleic acids are enclosed in a container or capsid made up of similar protein subunits or capsomers in crystalline array, usually as an icosahedron. Certain groups of viruses are consistent in that they are the same size, and the number and arrangement of the capsomers are the same. In some of the RNA viruses the nucleic acid is arranged as a double- or single-helical structure surrounded by protein subunits. The combination of nucleic acid and protein may constitute the whole virus, or there may be additional outer membranes which contain lipids. Viruses which have lipid membranes can usually be inactivated by ether, choloroform or bile salts.

The two groups of viruses RNA and DNA are further divided according to size, morphology, and biological and chemical properties. Thus the cubical RNA viruses which are ether stable are divided into the picornaviruses and the reoviruses. The name picornavirus comes from "pico" meaning "very small" and "rna" indicating the type of nucleic acid. Included in the group are the enteroviruses: polio, Coxsackie, and foot-and-mouth, etc. The reoviruses cause inapparent infection in man and other animals, and their relationship to spontaneous disease is uncertain. They are morphologically similar to the wound-tumour virus of clover, and a small cross activity with this virus by means of complement fixation has been reported.

The arboviruses are those which undergo a biological cycle in both arthropods and vertebrate hosts. Not all of these viruses have been studied for fundamental chemical characteristics. Those which have been studied were found to be cubical RNA viruses, sensitive to the action of ether and relatively unstable.

The myxoviruses and rabies are examples of RNA helical viruses, which contain an outer, ether-labile, lipid-containing membrane. The myxoviruses include the influenza, parainfluenza, mumps, newcastle, simian SV5 and SV11, and possibly also measles, rinderpest and dog distemper viruses. Most of the myxoviruses have an affinity for mucins. They adhere to the surface of erthyrocytes causing them to agglutinate, and the substance responsible for the adhesion is neuraminic acid (see HEXOSAMINES AND HIGHER AMINO SUGARS). Adsorption to the cell surface is inhibited by mucoproteins which are present in sera and other biological fluids. *Vibrio cholera* produces an enzyme known as receptor destroying enzyme (RDE), which when added to erythrocytes prior to adding virus, inhibits cell attachment and prevents hemagglutination. Less is known about the biochemical aspects of measles, dog distemper, rinderpest or rabies viruses.

Information is lacking concerning the symmetry of a large number of RNA ether-sensitive oncogenic viruses. The fowl leucosis, murine leukemia and mouse mammary carcinoma viruses make up this group.

All of the known DNA viruses have cubical symmetry. Two groups which are ether stable are the papovaviruses and the adenoviruses. There are also two groups of DNA viruses which are generally ether sensitive: the herpes viruses and the pox viruses. All of the herpes viruses are ether labile, but ether stability varies within the pox virus group.

The chemical nature of many viruses, which either do not grow well or do not lend themselves to purification, is unknown. The Riley lactic dehydrogenase virus is a non-pathogenic virus which is recognized only by an increase in lactic dehydrogenase in the blood of infected mice. A lipovirus described by Chang causes marked degradation of infected cells and releases a lypogenic toxin dissociable from infectivity, which is capable of inducing fatty degeneration in other uninfected cells. A marked increase in the gamma-globulin fraction of blood serum of mink infected with Aleutian mink disease is an indication of infection with a virus which causes a color change in the fur and often sickness and death.

The chemical characteristics of viruses are not only an aid in their classification and identification but are also useful in understanding the nature of known viral or viral-induced host ANTIGENS and possibly in the control of disease. The adenoviruses, for example, contain at least three protein moieties, and certain types are capable of inducing one or more new host antigens—the chemistry of which is unknown at present. The viral proteins can be separated by gel diffusion and correlated with complement fixation. One moiety is the toxic protein which causes the host cell to degenerate; another corresponds to the group antigen common to all the 31 types of adenoviruses; the third is the type specific protein.

A number of viruses induce the host cell to elaborate a substance known as INTERFERON which will inhibit the growth of many viruses in cells of the same species but not in cells from other animal species. Interferons produced by an influenza virus and also a herpes virus were both active against the vesicular stomatitis virus. They both proved to be proteins of low molecular weight, trypsin sensitive, and stable at a temperature of 70°C and over a wide range of pH (1–11).

There are a number of chemical compounds which in some way interfere with the formation or function of viral RNA or DNA or the proteins they sensitize, or which prevent viral attachment to the cell. Some of these have been used to prevent or treat disease in animals and man. Certain of the thiosemicarbazones prevent formation of certain soluble pox virus antigens and of complete virus. They can be given by mouth and are reported to be effective in the prophylaxis of smallpox.

The halogenated thymidine analogues, bromo-, chloro- and iododeoxyuridine, interfere with the incorporation of thymidine in DNA and prevent the formation of complete virus. They cause serious side effects on injection, but they are used topically in treatment of herpes eye infections.

Cytosine arabinoside inhibits cell division and reproduction of some of the DNA viruses but allows some of the viral protein to be synthesized.

Puromycin and parafluorophenylalanine arrest the production of viral capsid protein and RNA.

The synthesis of viral RNA and antigenic protein of polioviruses is inhibited by guanidine, but the drug does not inactivate poliovirus or prevent its entry into the cell.

Examples of drugs which have no effect on the virus particle but prevent attachment of the virus to the cell are the synthetic sulfated polysaccharides which inhibit the arbo-, picorna- and herpes viruses, and adamantamine hydrochloride which inhibits influenza viruses. The latter is active when given by mouth.

The correlation of the chemical properties of viruses with their biological activity and morphology is a difficult and complex problem. The metabolic functions of viruses and host cells are interrelated so that it is difficult to study single reactions or pathways. Many biochemical properties of viruses are known, and the armamentarium of drugs now at hand should enable investigators to gain further insight into the fundamental chemistry of viral infections, and eventually their control.

References

1. ANDREWES, C. H., "Viruses of Vertebrates," pp.1–392, Baltimore, Williams and Wilkins Company, 1964.
2. LAMPSON, G. P., TYTELL, A. A., NEMES, M. M., AND HILLEMAN, M. R., "Characterization of Chick Embryo Interferon Induced by a DNA Virus," *Proc. Soc. Exptl. Biol. Med.*, **118**, 441–448 (1965).
3. "Basic Mechanisms in Animal Virus Biology," *Cold Spring Harbor Symp. Quant. Biol.*, **27**, 1–525 (1962).
4. CHANG, R. S., "An Immunologic Study of the 'Lipovirus'," *J. Immunol.*, **92**, 305–312 (1964).
5. POLLARD, M. (EDITOR), "Perspectives in Virology," Vol. 3, pp. 1–281, New York, Hoeber Medical Division, Harper and Row, Publishers Inc., 1962.
6. "Viruses, Nucleic Acid and Cancer," *Symp. Fundamental Cancer Res. 17th*, Baltimore, Williams and Wilkins Company, 1963, pp. 1–639.

BERNICE E. EDDY

VIRUS, TOBACCO MOSAIC. See TOBACCO MOSAIC VIRUS.

VIRUSES (COMPOSITION AND STRUCTURE)

Viruses are considered to be the smallest infectious agents capable of replicating themselves in living cells. The majority of these extremely small infectious particles fall within a size range of about 0.02–0.25 μ (1 μ = 0.001 mm), and can only be visualized directly in the ELECTRON MICROSCOPE. The precise mechanism of how a virus transfers its nucleic acid to the host cell, and how the normal function of the cell is directed toward the production of progeny virus, remains obscure. It is, in many respects, one of the most fundamental problems in molecular biology [see VIRUSES (REPLICATION)].

Most of the physical and chemical techniques employed in virus research have been aimed at elucidating the characteristics of the isolated infectious virus particle and its components. These studies have clearly demonstrated that the genetic information is carried in a limited amount of nucleic acid located in the core of the virus particle. An outer shell or coat of protein surrounds and protects the nucleic acid.

Investigations carried out by the application of X-ray diffraction, electron microscopy and high-speed ULTRACENTRIFUGATION techniques, have revealed the protein shells to be constructed according to a precise, geometrical and aesthetically pleasing pattern. It is not the shape but the symmetrical packing of the protein macromolecules forming the shell which is considered to be of structural importance. Some viruses have been described as having "spherical" shapes, but recent research has shown that the protein coats of roughly "spherical" viruses possess icosahedral symmetry. This symmetrical arrangement is well illustrated in the electron micrograph of human adenovirus shown in Fig. 1A. The macromolecules form a shell with 20 equilateral triangular faces and 12 vertexes. At each of the vertexes, a central macromolecule can be seen surrounded by five neighbors. On the faces and edges, each central macromolecule is surrounded by six neighbors. These features are consistent with a body possessing icosahedral symmetry. The model of an icosahedron constructed from 252 spheres is identical to the particle shown in the electron micrograph, Fig. 1B.

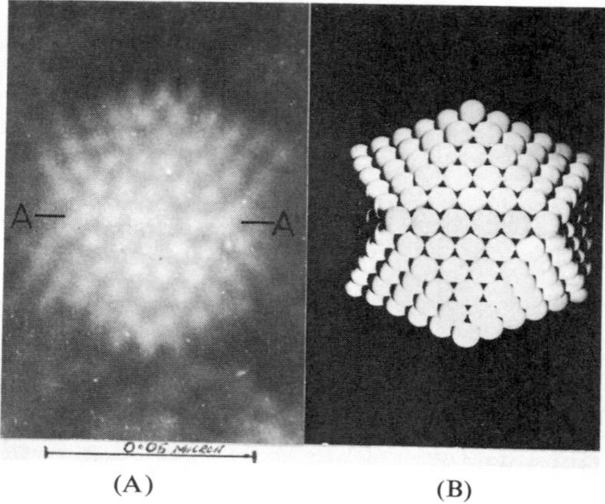

(A) (B)

FIG. 1A. The electron micrograph shows a single adenovirus particle with a protein shell composed of macromolecules arranged in accordance with icosahedral symmetry. The two of the 12 vertexes indicated at A are surrounded by 5 units.

FIG. 1B. A model of an icosahedron constructed from 252 spheres photographed in the same orientation as the particle in Fig. 1A. From R. W. Horne, *et al.*, *J. Molec. Biol.*, **1**, 84 (1959). (Reprinted with permission from Academic Press.)

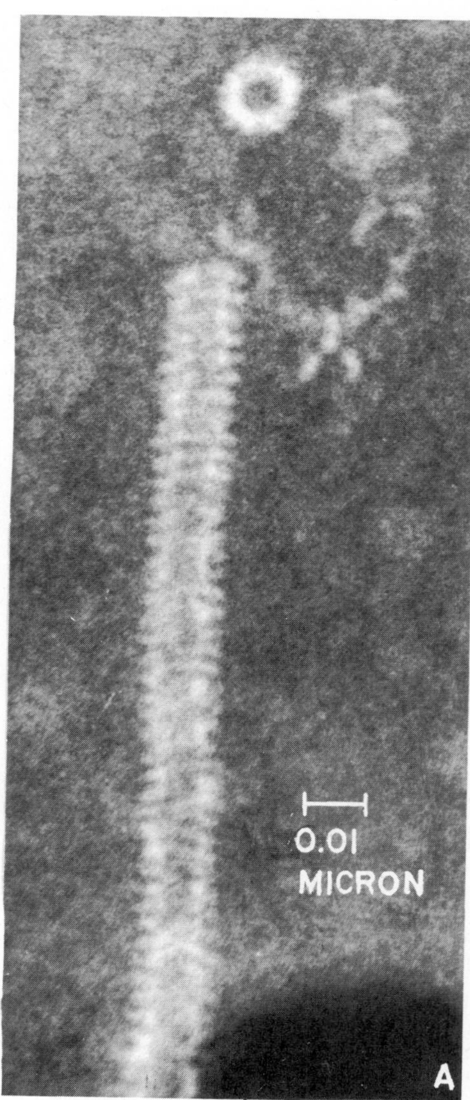

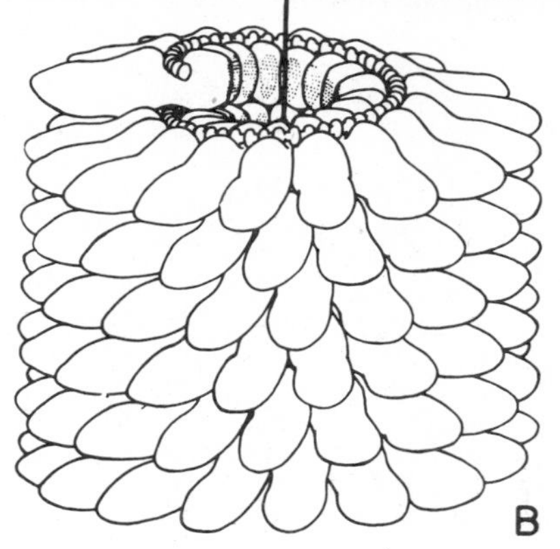

FIG. 2A. Electron micrograph of part of a tobacco mosaic virus rod. The electron dense material surrounding the rod reveals the hollow central region of the virus as well as the grooves arranged along the axis of the virus. The small doughnut-shaped particle is a fragment of the virus seen end on.

FIG. 2B. The diagram derived from X-ray diffraction data illustrates a small portion of a rod of tobacco mosaic virus. The helical pattern, grooves and hollow central region resulting from the packing of the elongated protein molecules are shown. Part of the coiled nucleic acid is shown exposed for diagramatic purposes. (Reprinted by permission of D. L. D. Casper and A. Klug.)

A relatively large number of the so-called spherical viruses have been studied in recent years including, poliovirus, herpes virus, varicella virus, polyoma virus, several plant viruses and certain small bacterial viruses. These and other "spherical" viruses have been shown to be constructed according to the same basic icosahedral plan. The size of the viruses and the number of macromolecules forming their shells have been observed to vary considerably, but their icosahedral symmetry is basically the same.

Other viruses which cause diseases such as influenza and mumps in man have their protein coats arranged in a helical pattern. One of the most extensively studied viruses is TOBACCO MOSAIC VIRUS (TMV), which is responsible for a serious disease in tobacco plants. This virus appears in the form of long slender rigid rods and, when examined in the electron microscope, shows the structural features illustrated in Fig. 2A. The electron-dense material surrounding the particle also reveals an axial hollow central region as well as a regular periodicity along the rod axis. The periodicity results from a series of grooves produced by the helical packing of the protein molecules as illustrated in the diagram, Fig. 2B, which was based on X-ray diffraction analysis of TMV virus crystals. The nucleic acid appears to follow the same helical pattern and is deeply embedded in the protein structure.

The inner component located in influenza virus, mumps and related viruses (myxoviruses) has the same basic helical arrangement observed in TMV, but is more flexible as shown in the electron micrograph, Fig. 3A. Myxoviruses, when fully assembled, are more complicated due to additional components associated with their surfaces. The myxoviruses seem to acquire an outer LIPOPROTEIN envelope when they are being released at the host cell surface. They also possess a series of spike-like projections located at the outer virus surface, which has been identified with the haemagglutinating properties of the virus (Fig. 3B). The possible arrangement of the various components in myxoviruses is illustrated in the model shown in Fig. 4.

There remains a third, and more sophisticated group of viruses whose protein components are assembled in a complex symmetrical pattern. A typical example of such a complex structure and relationship to function is well illustrated in the coli T_2 bacterial virus shown in Fig. 5. The head of the tadpole-like virus has the form of a bipyramidal hexagonal prism which contains the

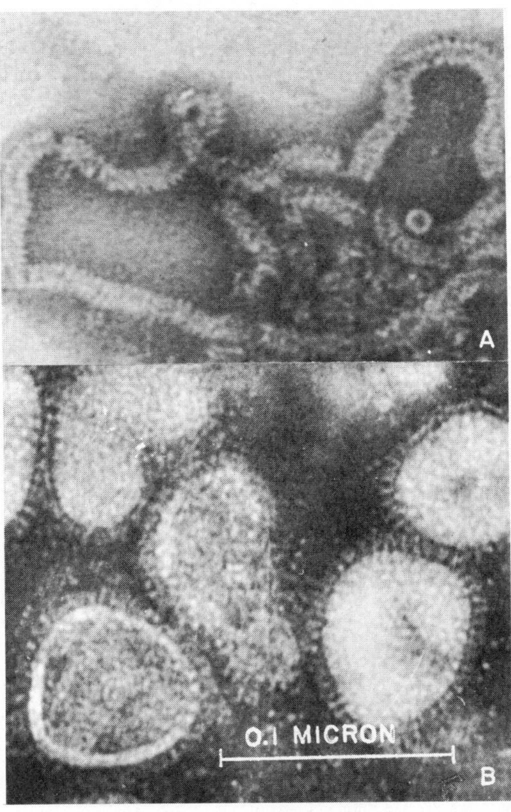

FIG. 3A. The inner nucleoprotein component present in the myxoviruses has the same basic helical pattern seen in TMV rods, but the structure appears to be more loosely assembled and is flexible.

FIG. 3B. An electron micrograph of influenza virus showing the irregular shape of the particle. The nucleoprotein helical component is enclosed in the outer spiked envelope, and is difficult to resolve unless the virus is disrupted.

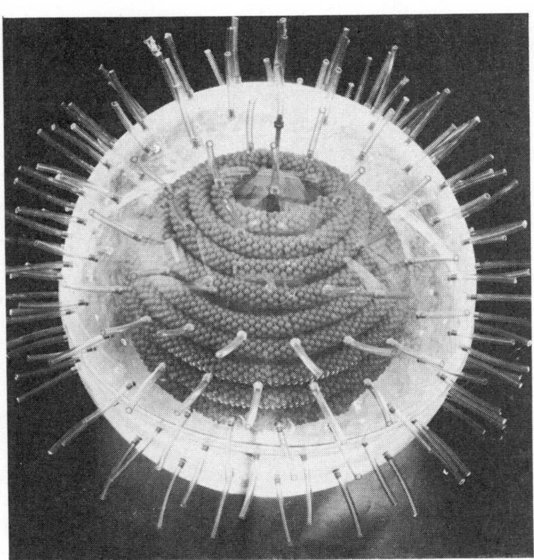

FIG. 4. The model of myxovirus has been constructed to show a possible arrangement of the various components observed in the particles. (From R. W. Horne *et al.* Reprinted by permission of *Virology*, Academic Press.)

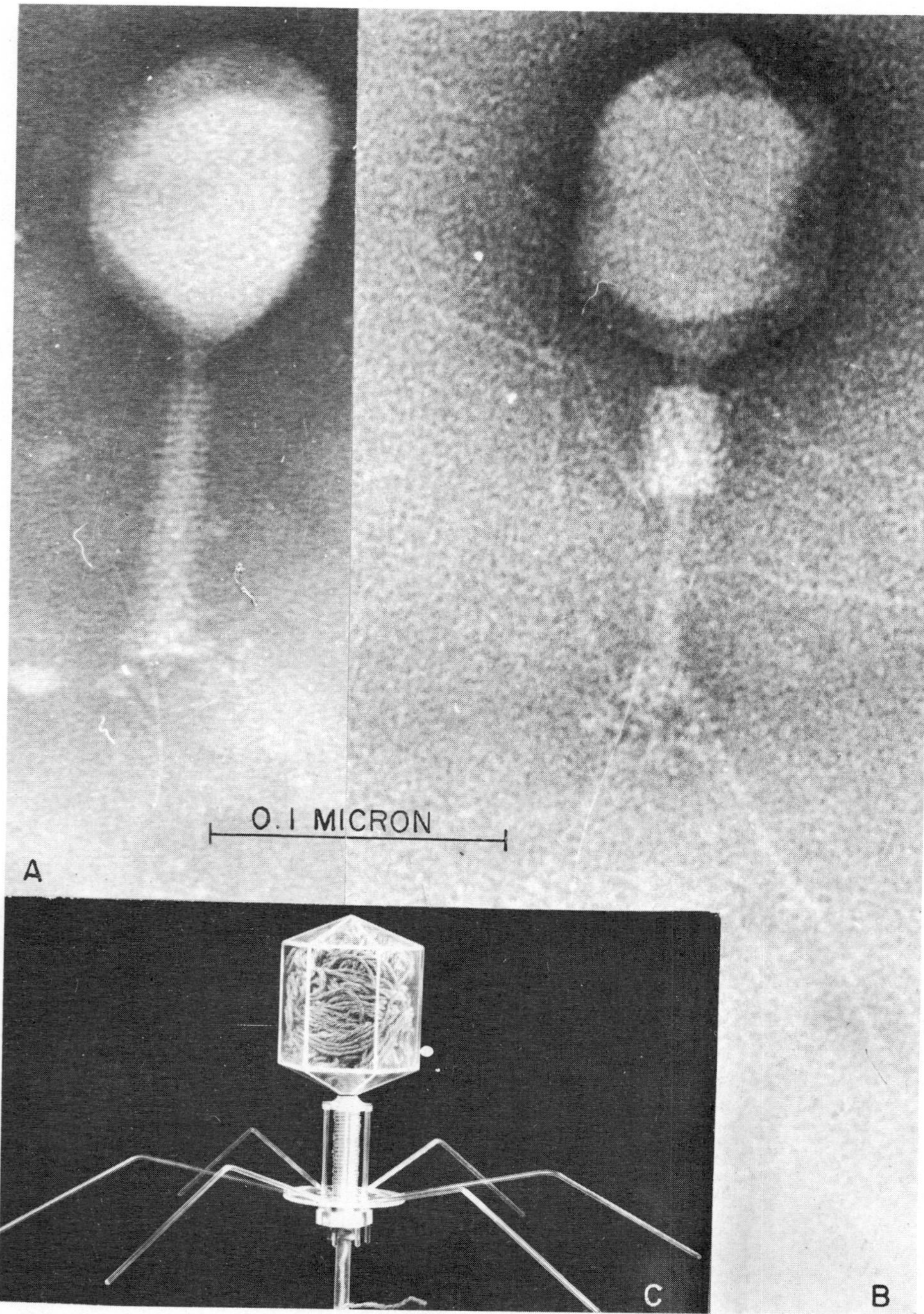

FIG. 5A. The electron micrographs reveal the sophisticated structure of the *E. coli* T_2 bacterial virus. The six tail fibers attached to the base plate are normally placed very close to the tail structure.

FIG. 5B. Contraction of the tail sheath, shown in a chemically "triggered" phage, exposes the core and tail fibers.

FIG. 5C. The model illustrates the relationship and function of the tail components in the T_2 bacteriophage. (From S. Brenner, R. W. Horne *et al.* Reprinted by permission of *Journal of Molecular Biology*, Academic Press.)

viral nucleic acid. Attached to one end of the head at a point of sixfold symmetry, is a tail structure composed of several protein components. These consist of an outer contractile sheath, hollow central core, hexagonal plate and 6 tail fibers.

The problem is how the virus transfers the nucleic acid from the bacteriophage head through the relatively thick bacterial cell wall. It appears that the virus is initially attached to the host by the hexagonal plate located at the extreme end of the tail. Electron micrographs have indicated that the 6 tail fibers splay out and also make contact with the cell wall. The helical tail sheath contracts with the hexagonal plate, which results in the penetration of the hollow core into the cell wall. This absorption mechanism acts not unlike a microsyringe system, but the mechanism for transferring the nucleic acid from the virus head through the hollow core still remains obscure.

Once absorption and release of the nucleic acid has occurred, the normal function and synthesis within the host cell ceases and is switched to the assembly of new virus particles. It is a remarkable fact that within about 20 minutes following a single infection, a large number of complete T_2 progeny virus particles are assembled and released, finally resulting in the death of the cell.

Other structurally complex viruses have been studied with the aid of the electron microscope. The large particles forming the pox group of

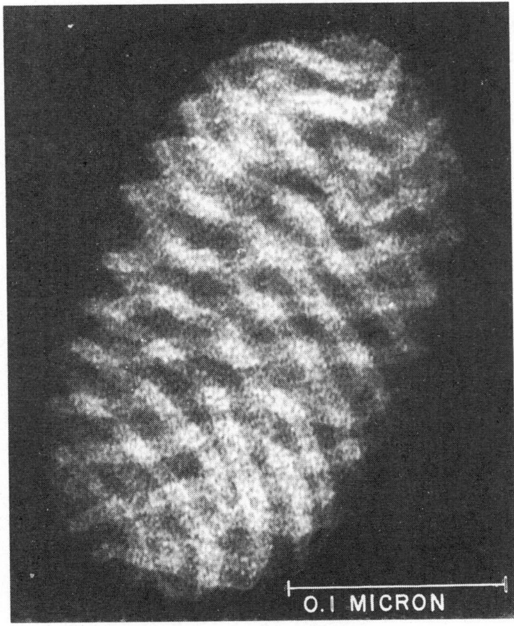

FIG. 6. Contagious dermatitis virus belongs to the pox group of viruses, and the complex structural features are shown in the electron micrograph. (From J. Nagington and R. W. Horne. Reprinted by permission of *Virology*, Academic Press.)

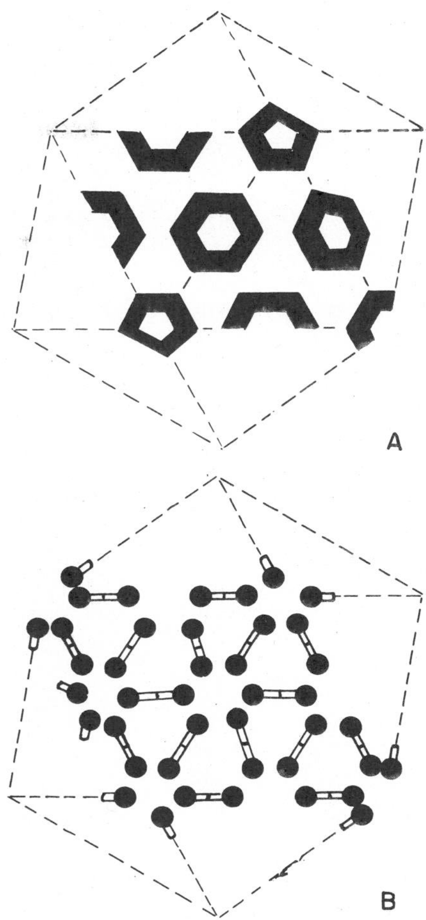

FIG. 7. The diagram illustrates how the morphological units seen in the electron microscope (A) can be constructed from a number of identical structure units arranged on the surface of an icosahedron (B).

viruses have not been found easy to analyze by morphology alone. Contagious pustular dermatitis, for instance, is a member of the pox virus group and presents an intriguing structural pattern as shown in the electron micrograph in Fig. 6.

The criss-cross pattern results from the superimposition of the top and bottom of a helical pattern. It is known from histological studies that the coiled or helical component is only one of several layers forming the infective virus, and the function of these multilayers still requires further study.

From the few examples discussed above, it has been shown that viruses are constructed according to three basic symmetrical groups: those with protein shells possessing icosahedral symmetry, helical symmetry and complex symmetry. The reasons for the symmetrical packing of the protein forming the virus shells are twofold. (1) The amount of nucleic acid contained in a virus particle is limited, and it is only capable of "coding" for a small number of proteins. (2) The most economical way to pack identical units in the form of macromolecules is to arrange them in some form of symmetrical pattern. The most likely patterns originally suggested on theoretical grounds and from X-ray studies have now been demonstrated in the electron microscope.

It should be emphasized that the electron microscope shows only part of the macromolecular organization of the virus shell. It is likely that the morphological units resolved on the surfaces of a number of virus particles are composed of smaller components or structure units. The diagram shown in Fig. 7 illustrates how identical structure units could be arranged on the surface of an icosahedral virus to form the macromolecules visualized in the electron microscope.

Although a great deal of research has been devoted to viruses, we need to know more about how viruses enter their host cells and deliver up their nucleic acid, and the sequence of events following infection. There are other important problems concerned with how viruses are assembled within the infected cells, and how under some circumstances "infected" cells are capable of producing virus without undergoing serious pathological changes.

References

1. Crick, F. H. C., and Watson, J. D., *Ciba Found. Symp. Nature Viruses*, 5 (1957).
2. Caspar, D. L. D., and Klug, A., *Cold Spring Harbor Symp. Quant. Biol.*, **27**, 1 (1962).
3. Horne, R. W., and Wildy, P., *Advan. Virus Res.*, **10**, 101 (1963).
4. Waterson, A. P., "Introduction to Animal Virology," Cambridge, The University Press, 1962.
5. "Viruses, Nucleic Acids and Cancer," *Symp. Fundamental Cancer Res., 17th* (1963).

R. W. Horne

VIRUSES (REPLICATION)

The viruses of plants, animals, bacteria or insects are biochemically and morphologically diverse groups. The following concerns some generalizations about replication which justify the classification virus.

Transmissible forms of virus (virion) consist of one type of NUCLEIC ACID, either DNA or RNA and one or more antigenically specific PROTEINS. Some viruses also contain LIPIDS and others may have incorporated host cell components. With few exceptions, no enzyme activity is detected. In the absence of a host cell, the virion does not replicate, mutate or function metabolically. These manifestations of independence or life only follow upon integration of the virus with a cell.

The initial stages of this integration entail an interaction of viral and cellular surfaces, resulting in exposure of the viral nucleic acid to an intracellular environment wherein it is reactive. The initiation of infection has been detailed most finely for the T-phages of *E. coli*[1] and the myxoviruses,[2] and has been subdivided into two stages. The first, a second-order reaction of virus and cell which obeys pseudo-first-order kinetics, under experimental conditions, is ionic, reversible, nonenzymatic and involves a cell-virus complex in which the virus remains sensitive to the neutralizing action of viral immune serum. The following stage is sensitive to temperature, appears enzymatic, is irreversible and involves a cell-virus complex resistant to immune serum.[2]

Despite differences in viral structures and cell surfaces, this two-stage cell-virus interaction may be a broadly encountered feature. However, the mechanisms and chemical components involved to produce these kinetics differ markedly with the virus system observed, as illustrated by the following brief outline of the initiation of infection by T_2 phages and myxoviruses.

Strains of myxoviruses differ one from another in the pattern and density of electric charges on the cell surface with which they react. The charge pattern results from a distribution of neuraminic acid residues (see HEXOSAMINES) bound terminally by α-glucoside linkages to the polysaccharide of surface glycoproteins. The reactivity of the cell surface to individual strains of influenza virus can be selectively modified by progressive alteration of the charge by the action of soluble neuraminidase or virus containing the enzyme which splits off negatively charged residues. Interaction of all myxoviruses with the cell surface may be excluded by sufficiently intense enzyme action. In the initial stages of binding, the virus is held to the cell by two types of linkages: one sensitive to this action of neuraminidase and a second that can be blocked by α-amino-*p*-methoxyphenyl methane sulfonic acid.[3] The latter appears to act at the same site as the viral inhibitor adamantanamine.

The T_2 phage of *E. coli* is asymmetrical, and its cell receptors are localized in special fibers at the end of the tail. Ionic attachment is followed by a series of reactions which involve alteration of the tip of the phage tail, contraction of the tail sheath (a reaction likened to muscle contraction), enzymatic weakening of the cell wall, penetration of the cell membrane by the tail core protein and injection of the DNA into the cell.[1] Most of the phage protein remains exterior to the cell wall, but may not be without effect. Phage ghosts, protein coats without DNA, may react with susceptible cells to cause cellular death or lysis from without. In contrast, the infecting vaccinia virus has been visualized intact early in the intracellular state.

The nucleic acid of several plant, animal and bacterial virions has been isolated in the laboratory and introduced with low efficiency into insusceptible, as well as normally susceptible cells, whereupon the normal sequence of infection ensues and a yield of complete virus with all usual components is produced. For example, the RNA of poliovirus (normally replicating in only human or simian tissue) has been used to infect the chick, mouse, rabbit, guinea pig and hamster.[4] From such experiments it is clear that the protein of the virion has little function in the phase of viral replication and that the characteristic host-virus specificity is determined by the compatibility of the interacting surfaces rather than the suitability of the interacting biochemical milieu.

Once the early stages are completed and the viral structure disarranged, the infecting principle, viable in that cell, is no longer extractable and transmissible to other cells with any meaningful efficiency. This is called the eclipse phase and persists until the replicating form reaches a mature state.

The intracellular replicating form of the virus, in contrast to the transmissible form, is more complex and may consist of additional functional proteins, *e.g.*, enzymes, and viral specific RNA may be found in cells infected with DNA viruses. Under control of the viral genome, the viral proteins and nucleic acids are synthesized *de novo* from amino acids and nucleotides. These are derived from the intracellular pool or from extracellular sources. With T phages, where dissolution of the cellular DNA occurs, nucleotides may be salvaged, whereas cellular proteins are poor sources of amino acids.[5] While the nucleic acid of the virion is internally located in a protein coat, the components of the replicating form are not confined by a viral specific membrane and replication is not a consequence of growth and binary fission. No generalizations can be made concerning the intracellular localization or distribution of viral components.

Replication proceeds from a sequence of biosynthetic reactions which produce the viral components in a fixed order.[6] This sequential development finds expression in a time-dependent sensitivity of viral replication to inhibition by metabolic antagonists, in a time-ordered appearance of enzyme activities, in a sequential appearance of viral specific antigens, and in morphologic components identifiable with the electron microscope. The viral nucleic acid is probably a single molecule, and at present it is unknown how the genetic information is sequentially utilized in this time-ordered series of biosynthetic reactions.

With a DNA virus, the following biochemical sequence has been elucidated. Firstly, there is an expression of the viral nucleic acid in synthesis of a complementary or messenger RNA, mediated presumably through the action of a DNA-primed RNA polymerase of cellular origin. This RNA has been particularly identified in the T_2 phage-*E. coli* system. Next appears a number of proteins which either act upon the host cell (*e.g.*, DNAase, deoxycytidine triphosphatase, basic histones) or function in the synthesis of viral nucleic acid (nucleotide kinases, deoxyhydroxymethylcytidine monophosphate kinase, glucosyl transferases, DNA polymerase, etc.). The DNA of the T even phages contains hydroxymethyl cytosine rather than cytosine. This base does not occur in the normal cell, and it is presumed that DNA replication must wait upon the function of DNA to produce the enzymes necessary for synthesis of the unique base. Subsequent to polymerization of the nucleotide triphosphate, the hydroxymethylcytosine group sof the DNA polymer are partially glucosylated.[7] Proceeding along with the replication of DNA is the synthesis of the structural proteins of the virion. A pool of structural elements accumulates prior to formulation of the infectious virus. The mechanism of the latter process is not well understood, but since TOBACCO MOSAIC VIRUS, for example, can be reconstituted *in vitro* from isolated RNA and protein subunits, the ultimate structure is believed to be a consequence of the intrinsic properties of the subunits from which it is constructed and organizes spontaneously. There may follow a distinct intracellular phase of accumulated mature virus prior to an abrupt lysis of the cell, or the intact virus may have a transient intracellular existence being organized at the cell surface just prior to release, as with influenza virus where maturation and release occur over many hours.

The molecular weight of most RNA of viruses is approximately 2×10^6. This relatively small molecule can, according to current theory, code information for only 5 to 7 proteins. One of these is reported (poliovirus) to be an RNA primed RNA polymerase which is responsible for the independent replication of the RNA virus [see RIBONUCLEIC ACIDS (BIOSYNTHESIS)].

Even with the DNA viruses, which contain far more nucleic acid, the induced complement of viral specific enzymes (which may not be com-completely defined as yet for any virus) is inadequate for viral replication, and many essential reactions are catalyzed by host cell enzymes. In general, there are no viral specific glycolytic enzymes, or enzymes of the citric acid cycle. High-energy phosphate needed for synthetic reactions is produced by the persisting cellular activity. In extreme cases, termed satellite viruses, the deficient requirements of the replicating virus cannot be supplied by the host cell, and its replication occurs only in cells already infected with a different virus.

Generalizations concerning the cytopathic effect of viruses cannot be made. However, the morphologic and biochemical lesions at the cellular level have been carefully detailed for numerous individual examples. The range of effects includes dissolution of the cell nucleus and complete lysis of the cell to a selective alteration of host metabolism which allows survival of the cell with viral replication. Of particular interest are the induced biochemical alterations of tumorogenic viruses leading to malignant transformation of the host cell.

References

1. GAREN, A., AND KOZLOFF, L. M., "The Initiation of Bacteriophage Infection," *The Viruses* (BURNET, F. M., AND STANLEY, W. M., EDITORS), Vol. 2, p. 203, New York, Academic Press, 1959.

2. Ishida, N., and Ackermann, W. W., "Growth Characteristics of Influenza Virus. Properties of the Initial Cell-virus Complex," *J. Exptl. Med.*, **104**, 501 (1956).
3. Ackermann, W. W., "Cell Surface Phenomena of Newcastle Disease Virus," in "Newcastle Disease Virus: An Evolving Pathogen," p. 153, Madison, Wisc., The University of Wisconsin Press, 1964.
4. Holland, J. J., McLaren, L. C., and Syverton, J. T., "Mammalian Cell-virus relationship. III. Poliovirus Production by Non-primate Cells Exposed to Poliovirus Ribonucleic Acid," *Proc. Soc. Exptl. Biol. Med.*, **100**, 843 (1959).
5. Evans, E. A., "Biochemical Studies of Bacterial Viruses," Chicago, The University of Chicago Press, 1952.
6. Cohen, S. S., "Biochemistry of the Virus Infected Cell," in "Viral and Rickettsial Infections of Man" (Rivers, T. M., and Horsfall, F. L., Editors), Third edition, p. 49, I. B. Lippincott Co., 1959.
7. Kornberg, A., "Enzymatic Synthesis of DNA," New York, John Wiley & Sons, 1961.

W. Wilbur Ackermann

VISUAL PIGMENTS

The eyes of most vertebrates contain two kinds of light receptor: rods, for vision in dim light; cones, for vision in bright light and color vision. Each of these organs contains a photosensitive pigment that bleaches on exposure to light. Some aspect of this process triggers a nervous excitation which, transmitted from neuron to neuron along the optic pathways to the brain, ends in exciting visual sensations.

Constituents of the Pigments. All known visual pigments are conjugated proteins, in which VITAMIN A aldehyde is bound as chromophore to a characteristic type of protein found in the outer segments of the rods and cones, called opsin.

Vitamin A aldehyde was first discovered in the retina and was named *retinene* before its structure was appreciated. It is now proposed to call vitamin A *retinol*, and its aldehyde *retinal* or *retinaldehyde*. The corresponding 3-dehydro molecules, containing one more conjugated double bond in the ionone ring, and called heretofore vitamin A_2 and retinene$_2$, will be referred to hereafter as *retinol*$_2$ and *retinal*$_2$:

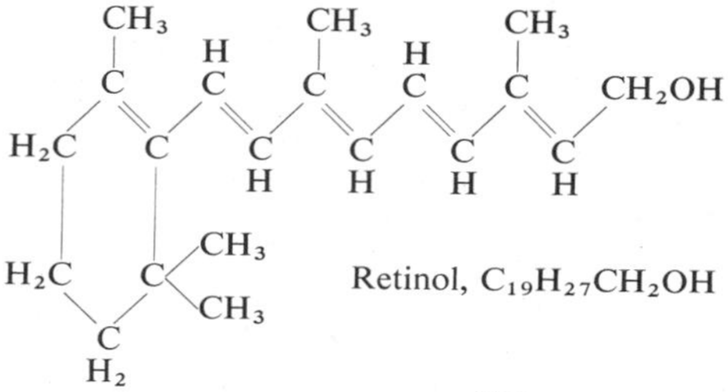

Retinol, $C_{19}H_{27}CH_2OH$

Retinal, $C_{19}H_{27}CHO$:

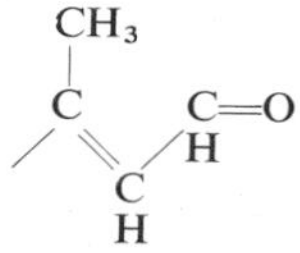

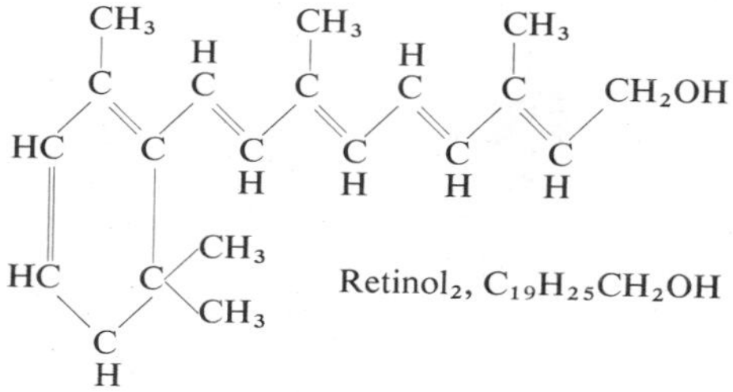

Retinol$_2$, $C_{19}H_{25}CH_2OH$

Retinal$_2$, $C_{19}H_{25}CHO$:

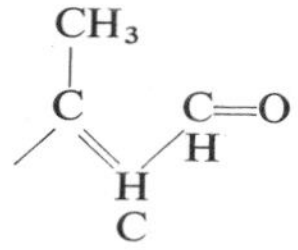

Two great families of opsins are found in vertebrate eyes, those of the rods and those of the cones. These combine with retinal$_1$ and retinal$_2$ to yield the four major pigments of vertebrate vision:

				Usual λ_{max} ($m\mu$)
Retinol$_1$ $\underset{\text{DPN–H}}{\overset{\text{DPN+}}{\rightleftharpoons}}$ Retinal$_1$	+ Rod Opsin	$\overset{\text{light}}{\rightleftharpoons}$	Rhodopsin	500
(Alcohol dehydrogenase)	+ Cone Opsin	$\overset{\text{light}}{\rightleftharpoons}$	Iodopsin	562
Retinol$_2$ $\underset{\text{DPN–H}}{\overset{\text{DPN+}}{\rightleftharpoons}}$ Retinal$_2$	+ Rod Opsin	$\overset{\text{light}}{\rightleftharpoons}$	Porphyropsin	522
	+ Cone Opsin	$\overset{\text{light}}{\rightleftharpoons}$	Cyanopsin	620

In all the visual pigments so far analyzed, the chromophores are found in a particular, sterically hindered *cis* configuration of retinal 1 or 2, the 11-*cis* isomer. This isomer is twisted as well as bent at the *cis* linkage, through conflict between the H on C_{10} and the methyl group on C_{13}. As a result, it is thermodynamically unstable, and probably for this reason also it is photoisomerized to all-*trans* with the highest quantum efficiency of all the *cis* isomers of retinal.

Effect of Light. The only action of light in vision is to isomerize the chromophore of a visual pigment from the 11-*cis* configuration to all-*trans*. All other effects are "dark" consequences of this initial photoreaction. This isomerization is shown by the following structures:

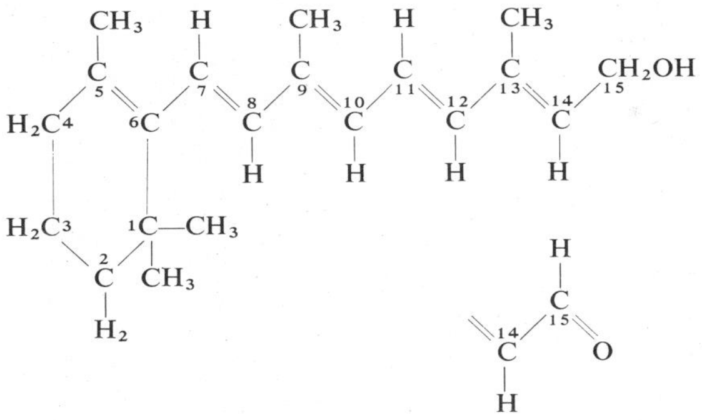

All-*trans*-Retinol (Vitamin A)

Retinaldehyde

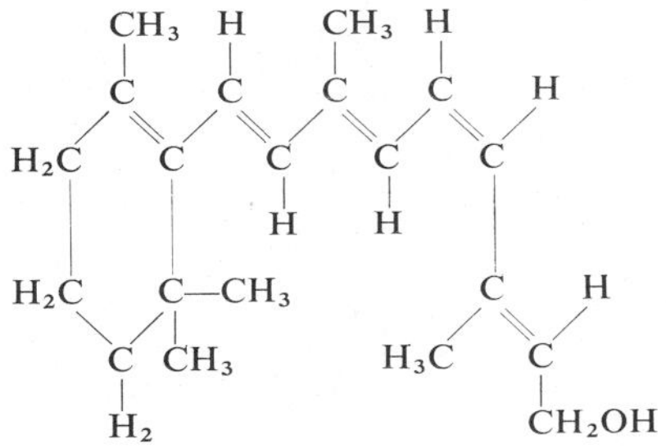

11-*cis* Retinol

The photoisomerization of the chromophore is followed by a spontaneous, stepwise rearrangement of the configuration of opsin (Fig. 1). The first product of irradiation (pre-lumirhodopsin, pre-lumiiodopsin) consists of all-*trans* retinal bound at the chromophoric site to as yet unchanged opsin. This is stable at liquid nitrogen temperature, but on progressive warming in the dark changes over a series of intermediates (*e.g.*, lumirhodopsin, metarhodopsins I and II), hydrolyzing finally to a mixture of opsin and all-*trans* retinal. During this stepwise rearrangement of opsin, new groups are exposed, including two —SH groups per molecule, and one H^+-binding group with pK about 6.6 (imidazole?).

From the point of view of visual excitation, the critical event may be this exposure of new groups on opsin. It is known that a dark-adapted rod can be excited by absorbing one photon, presumably in one molecule of rhodopsin. Some very large amplification must intervene between this unimolecular event and neural excitation. It has been suggested that the visual pigments may be proenzymes or zymogens, in which the catalytic

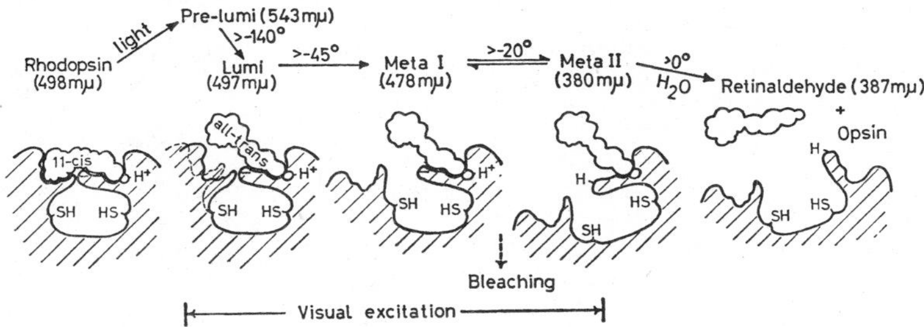

FIG. 1. Stages in the bleaching of rhodopsin. Rhodopsin has as chromophore 11-*cis* retinaldehyde, which fits closely a section of the opsin structure. The only action of light is to isomerize retinaldehyde from the 11-*cis* to the all-*trans* configuration (pre-lumirhodopsin). Then the structure of opsin opens progressively (lumi- and the metrahodopsins), ending in the hydrolysis of retinaldehyde from opsin. Bleaching occurs in going from metarhodopsin I to II; and visual excitation must have occurred by this stage. The opening of opsin exposes new chemical groups, including two —SH groups and one H^+-binding group. The absorption maxima shown are for prelumirhodopsin at $-190°C$, lumirhod opsin at $-65°C$, and the other pigments at room temperature.

center is covered by the chromophore. Light, by isomerizing the chromophore, or through the subsequent opening up of opsin structure, might uncover the catalytic center, activating the enzyme. Such an enzyme, capable of turning over many molecules of substrate, would constitute one stage of amplification. If its substrate were a second proenzyme, activated by opsin, that would provide a second stage of amplification. A chain of such enzyme activations arranged in series would be in effect a biochemical photomultiplier, capable of deriving a large and rapid effect from the action of light on a single molecule of visual pigment.

In the visual pigments, retinaldehyde is anchored to an amino group of opsin (apparently an ϵ-amino group of lysine) through a Schiff base linkage:

$$\underset{\text{Retinal}}{C_{19}H_{27}HC{=}O} + \underset{\text{Opsin}}{H_2N{-}\text{opsin}} \rightarrow$$

$$\underset{\text{Visual pigment}}{C_{19}H_{27}HC{=}\overset{H}{N^+}{-}\text{opsin}} + H_2O$$

An ordinary Schiff base of retinal would have λ_{max} about 365 mμ. To account for the large shift of spectrum that accompanies the formation of a visual pigment (to 500 mμ in rhodopsin or 562 mμ in iodopsin), it is assumed that the Schiff base N binds a proton, so producing the conjugate acid, which ordinarily would have λ_{max} about 440 mμ, and that in addition, side-chain interactions between retinal and a negatively charged group on opsin tend to draw the positive charge on retinal up into the conjugated system, thus greatly stimulating resonance.

All these features of structure and behavior run parallel in the A_2 series of visual pigments. These are found characteristically in freshwater vertebrates—freshwater fish—and occur also among vertebrates that spawn in fresh water—such anadromous fish as the salmonids, larval lampreys and some amphibian tadpoles.

Besides vertebrates, two other animal phyla have developed complex, image-resolving eyes: the mollusks (particularly in such cephalopods as squid and octopus) and the arthropods (insects, crustacea). (A few annelid worms have well-formed eyes, but nothing is yet known of their chemistry.) In all invertebrates yet examined, the visual pigments are "rhodopsins" based upon retinal$_1$. The ALCOHOL DEHYDROGENASE system is not an intrinsic component of these visual systems. Hence, the action of light on them ends with the production of metarhodopsins or, at most, the liberation of retinal.

A recent exploration of the visual systems of primates has shown that in the rhesus monkey and man, typical rhodopsins occur in the rods, but the cones contain three different visual pigments, which serve as the basis of color vision. Direct microspectrophotometry of human foveas and single human cones has revealed the difference spectra of three color vision pigments with λ_{max} about 440, 540 and 570 mμ. All three pigments appear to have the same chromophore, 11-*cis* retinal, attached to three different cone opsins.

References

1. BRINDLEY, G. S., "Physiology of the Retina and the Visual Pathways," London, Arnold, 1960.
2. DARTNALL, H. J. A., "The Visual Pigments," London, Methuen, 1957.
3. MORTON, R. A., AND PITT, G. A. J., *Progr. Chem. Org. Nat. Prod.*, **14**, 244 (1957).
4. WALD, G., in "Comparative Biochemistry" (FLORKIN, M., AND MASON, H. S., EDITORS), Vol. 1, p. 311, New York, Academic Press, 1960.
5. WALD, G., *Vitamins Hormones*, **18**, 417 (1961).
6. WALD, G., in "Life and Light" (MCELROY, W. D., AND GLASS, B., EDITORS), p. 724, Baltimore, Johns Hopkins Press, 1961.
7. WALD, G., BROWN, P. K., AND GIBBONS, I. R., *J. Opt. Soc. Am.*, **53**, 20 (1963).
8. WALD, G., AND BROWN, P. K., *Cold Spring Harbor Symp.*, **30**, 345 (1965).
9. MACNICHOL, E. F., JR., "Three-Pigment Color Vision," *Sci. Am.*, **211**, 48–56 (1964).

GEORGE WALD

VITAMIN A

Vitamin A is found in all vertebrates, but only crustacea of the invertebrates contain the vitamin. Animals either ingest the vitamin preformed from other animal sources or synthesize it from various provitamin CAROTENOIDS of the plant kingdom. Most animal species store appreciable amounts of the vitamin in their livers, have low concentrations in the blood and undetectable quantities in most other tissues. A deficiency of the vitamin produces a variety of symptoms, the most uniform being eye lesions, nerve degeneration, bone abnormalities, membrane keratinization, reproductive failure and congenital abnormalities. Toxic symptoms from large doses of vitamin A are readily produced in animals and in man.

According to the International Union of Pure and Applied Chemistry in coordination with the International Biochemistry Union, the form of vitamin A which predominates in higher animals and marine fish shall be called *retinol* (vitamin A_1 alcohol, also axerophthol, $C_{20}H_{30}O$). This molecule has the all-*trans* configuration. The corresponding vitamin A aldehyde shall be called *retinal* (formerly retinene$_1$). Oxidation of retinal leads to vitamin A_1 acid, to be called retionic acid. In freshwater fish, vitamin A_2 ($C_{20}H_{28}O$, dehydroretinol) predominates. Considerable amounts of mono- and di-*cis* isomers of retinol are found in fish liver, whereas in higher animals, except for a specific stereoisomerization in the retina, *cis* isomers are absent. Rat liver has been shown to convert various *cis*-retinols to all-*trans*-retinol *in vitro*.

Retinol is most readily determined in unsaponifiable lipid extracts by its reaction with antimony trichloride or trifluoracetic acid (λ_{max} = 620 mμ,

$E^{1\%}_{1cm} = ca.\ 5000$). Separation of retinol, retinol esters and retinal can be accomplished by column, paper or thin-layer CHROMATOGRAPHIC techniques.

Vitamin A appears to have at least two different biochemical functions, one clearly elucidated in the retina and one unknown but thought to involve many tissues, particularly those with a mucous epithelium.

In the rods of the retina, retinal is found combined with the protein opsin, the complex being called rhodopsin (visual purple). Although the entire series of reactions involved in dark vision has not entirely worked out, the major steps in the cycle are quite clear.

All-*trans*-retinol from the blood is oxidized by ALCOHOL DEHYDROGENASE (with NADP) to retinol which in turn is isomerized in the retina to 11-*cis*-retinal. This combines with opsin to form rhodopsin (the all-*trans* isomer will not combine with opsin). On exposure to light, rhodopsin undergoes a sequence of changes with the eventual splitting off of retinal, which now has the all-*trans* configuration. This presumably can be reutilized in the retina by isomerization, or it can be reduced to retinol by alcohol dehydrogenase and returned to the circulation either as the free alcohol or as an ester. (See also LIGHT ADAPTATION; VISUAL PIGMENTS.)

Retinal has been shown to be present in trace amounts ($<1\ \mu g/g$) in porcine and bovine liver, while in hen's blood 10–15% of the total vitamin A can be in this form. Eggs and ovaries of a variety of species, particularly birds, fish and reptiles, contain retinal which can be equal to or greater than the amount of retinol. Retinal, but not retinol, has been found in the heads of honeybees and house flies. The equilibrium for liver alcohol dehydrogenase lies far to the side of reduction (retinol), but when an acceptor molecule (protein) is present to trap the aldehyde then detectable quantities of retinal accumulate.

The relatively recent observation that retinoic acid can replace retinol or retinal for normal growth of animals has given rise to new concepts in the biochemistry of vitamin A. Although retinoic acid cannot be demonstrated to be present normally in animal tissues, its formation by liver aldehyde dehydrogenase (NAD) and aldehyde oxidase has been accomplished so that the molecule must be considered in the general scheme of vitamin A metabolism. When retinoic acid is given to animals as the only form of vitamin A, growth is normal, but they eventually become sterile and also blind. This has led to the consideration that vitamin A may have at least three independent functions: one for growth, one for vision and one for reproduction. The reversal of the oxidative pathway of vitamin A (retinol → retinal → retinoic acid) does not occur in the body. When retinoic acid is fed to animals, even in relatively large doses, there is no storage and, in fact, the molecule is rapidly metabolized and cannot be found several hours after administration. The metabolic products have not been identified; several fractions from liver or intestine, isolated after administering retinoic acid marked with carbon 14, have been shown to have biological activity. Such observations have given rise to the concept that there is an "active metabolite" which is the biochemically effective form of vitamin A in the generalized functions related to growth.

Attempts to implicate vitamin A in specific biochemical reactions (other than in vision) have not been definitive, although a variety of effects on tissues or cells have been reported. Among these may be mentioned (1) the release of proteolytic enzymes from lysosomes treated with vitamin A; (2) an effect on SULFATE incorporation into MUCOPOLYSACCHARIDES in colon homogenates; (3) an effect on sulfation reactions generally; (4) formation of corticosterone from cholesterol in adrenal homogenates. In addition, implications have been made for a direct function of vitamin A in protein utilization and in spermatogenesis.

References

1. MOORE, T., "Vitamin A," Amsterdam, Elsevier Publishing Co., 1957.
2. "Symposium on Vitamin A and Metabolism," in "Vitamins and Hormones" (HARRIS, R. S., AND INGLE, D. J., EDITORS), Vol. 18, pp. 291–571, New York, Academic Press, 1960.
3. OLSON, J. A., "The Biosynthesis and Metabolism of Carotenoids and Retinol (Vitamin A)," *J. Lipid Res.*, **5**, 281–299 (1964).

JOHN G. BIERI

VITAMIN B GROUP (B COMPLEX)

A number of vitamins are commonly grouped together as "the vitamin B complex." They are so designated for historical reasons, and also because of similarity in solubility, distribution in natural sources, and biocatalytic function.

In 1912, Casimir Funk, a Polish biochemist, proposed the existence of "vitamines," substances required in trace amounts in the diet to prevent nutritional deficiency diseases such as scurvy, beriberi, and rickets. Later, in 1915, McCollum and Davis separated two vitamin fractions from egg yolk and called them "fat-soluble A" and "water-soluble B."

Further studies revealed that water-soluble vitamin B could be separated into a heat-labile vitamin B_1, which prevented polyneuritis in rats and birds, and heat-stable vitamin B_2, which promoted growth. After vitamin B_1 and vitamin B_2 were successfully purified, isolated and chemically synthesized, it was found that these pure vitamins did not completely substitute for natural vitamin sources such as liver or yeast in the diets of experimental animals. Intensive study of various fractions of natural foods led to the current knowledge that there are many components to "water-soluble B." These components, which are organic compounds required in small amounts in the diet, constitute the vitamins of the B complex.

It is now possible to prepare diets for experimental animals from protein, cabohydrate, fat and small amounts of chemically synthesized vitamins. Deletion of one of the vitamins from the diet often results in characteristic deficiency symptoms. Study of biochemical lesions of the vitamin deficiencies can yield valuable clues to the metabolic function of the vitamin. The vitamins of the B complex are characterized by their functioning as *coenzymes* in numerous metabolic reactions.

The following discussion will briefly outline the structure and function of eight B vitamins: PANTOTHENIC ACID, NICOTINIC ACID, RIBOFLAVIN, THIAMINE, BIOTIN, PYRIDOXINE, FOLIC ACID and COBALAMIN. Choline is also discussed in this section, since it is required in the diet under certain conditions and is present in many natural sources of the vitamin B complex. However, choline is not strictly regarded as a vitamin because it does not function as a coenzyme in any enzyme reaction. Other compounds often discussed in relation to B vitamins are LIPOIC ACID (thioctic acid), INOSITOL, and *para*-aminobenzoic acid. All are growth factors for certain microorganisms. However, since these compounds are not essential animal dietary requirements, they will not be considered in this discussion of the B complex. The information in this article is condensed and admittedly incomplete. Interested readers are directed to the more detailed sections on individual vitamins in this Encyclopedia and to the references listed at the end.

B Vitamins and Coenzymes. At this point it may be helpful to define a coenzyme. A coenzyme is a non-protein substance that is closely associated with or bound to the protein component (apoenzyme) of an enzyme. Together the coenzyme and apoenzyme form the complete enzyme known as the holoenzyme. The presence of a coenzyme is necessary for enzyme activity.

An "Activating" Coenzyme. Pantothenic acid is a constituent of COENZYME A (CoA), which participates in numerous enzyme reactions. CoA (Fig. 1) was discovered as an essential cofactor for the acetylation of sulfanilamide in the liver and of choline in the brain. Now CoA is known to be involved in many biochemical reactions in the body as an "activator" of normally less reactive carbon fragments and a "transferer" of these fragments to different molecules.

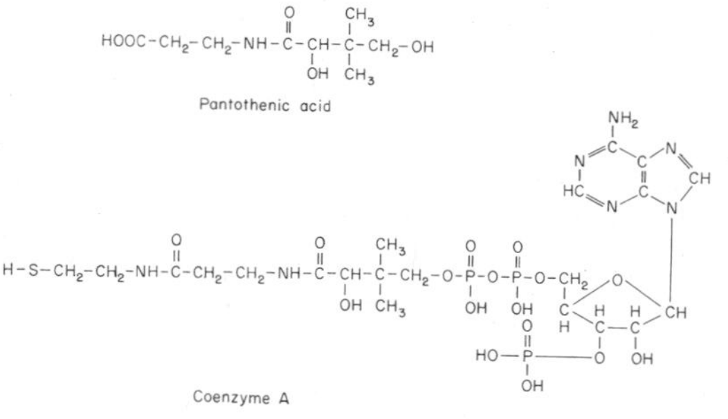

Pantothenic acid

Coenzyme A

FIG. 1. Structures of pantothenic acid and coenzyme A.

CoA is particularly important in the initial reaction of the CITRIC ACID CYCLE of carbohydrate metabolism and energy production. After oxidative decarboxylation of pyruvic acid (see MULTIENZYME COMPLEXES), CoA combines with the two-carbon acetate fragment to form acetyl-CoA or "active" acetate.

$$\underset{\text{Pyruvic acid}}{CH_3—\overset{O}{\overset{\|}{C}}—COOH} + CoA \rightarrow \underset{\substack{\text{Acetyl-CoA}\\ \text{"active acetate"}}}{CH_3—\overset{O}{\overset{\|}{C}}—CoA} + CO_2$$

Oxidation (actually dehydrogenation) and decarboxylation are performed by other B vitamin coenzymes that will be discussed subsequently. "Active acetate" is very reactive and is essential for the proper functioning of the citric acid cycle.

Coenzyme A is necessary for the activation, synthesis and degradation of fatty acids. Synthesis of cholesterol and ultimately the production of steroid hormones are also coenzyme A dependent (see FATTY ACID METABOLISM).

Oxidation-reduction Coenzymes. Many metabolic reactions involve either oxidation or reduction of the substrate. This is usually achieved by the enzymatic removal of hydrogen (oxidation) or the reverse, donation of hydrogen (reduction). A highly simplified diagram of this oxidation-reduction system is seen in Fig. 2. The first two components of this system, NAD and FAD, are B vitamin coenzymes.

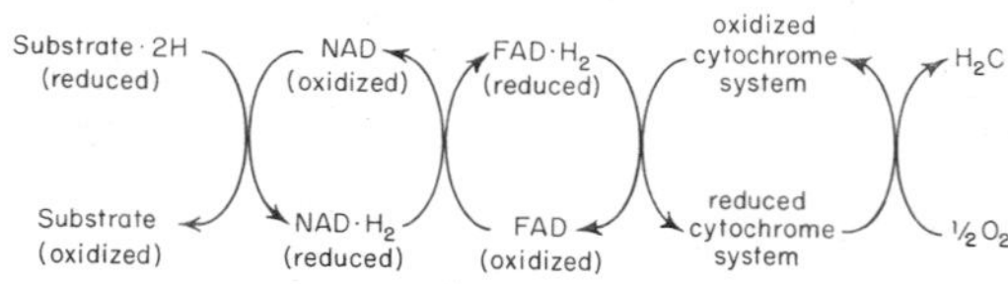

FIG. 2. Simplified diagram of the biological oxidation-reduction system.

(A) *Nicotinic Acid Coenzymes.* Nicotinic acid can be converted to nicotinamide in the body and, in this form, is found as a component of two oxidation-reduction coenzymes (Fig. 3): NICOTINAMIDE ADENINE DINUCLEOTIDE (NAD) and nicotinamide adenine dinucleotide phosphate (NADP). The nicotinamide portion of the coenzyme transfers hydrogens by alternating between an oxidized quaternary nitrogen and a reduced tertiary nitrogen (Fig. 4).

Enzymes that contain NAD or NADP are usually called dehydrogenases. They participate in many biochemical reactions of lipid, carbohydrate and protein metabolism. An example of an NAD-requiring enzyme is lactic dehydrogenase, which catalyzes the conversion of lactic acid to pyruvic acid. Over fifty NAD-dependent enzyme systems are known to exist.

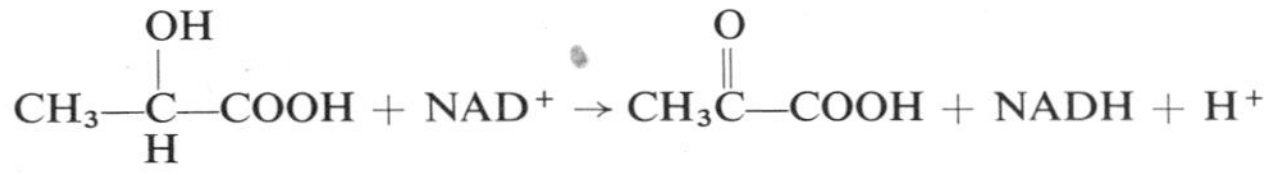

$$CH_3\text{—}\underset{H}{\overset{OH}{C}}\text{—}COOH + NAD^+ \rightarrow CH_3\overset{O}{\overset{\|}{C}}\text{—}COOH + NADH + H^+$$

Lactic acid Pyruvic acid

NADP is an essential coenzyme for glucose-6-phosphate dehydrogenase which catalyzes the oxidation of glucose-6-phosphate to 6-phosphogluconic acid. This reaction initiates metabolism

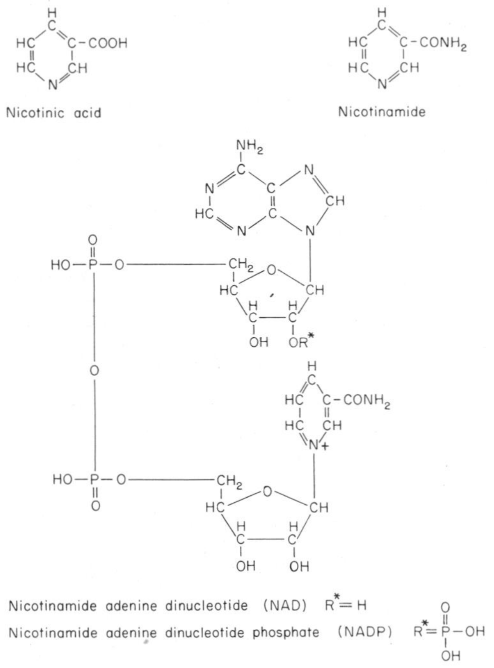

Nicotinic acid

Nicotinamide

Nicotinamide adenine dinucleotide (NAD) R* = H

Nicotinamide adenine dinucleotide phosphate (NADP) R* = P(=O)(OH)–OH

FIG. 3. Structures of nicotinic acid, nicotinamide and nicotinamide coenzymes.

of glucose by a pathway other than the citric acid cycle. The alternate route is known as the phosphogluconate oxidative pathway or the hexose monophosphate shunt; its first step follows:

In the biological oxidation-reduction system, reduced NAD (*i.e.*, NADH) is reoxidized to NAD by the riboflavin-containing coenzyme FAD (see Fig. 2).

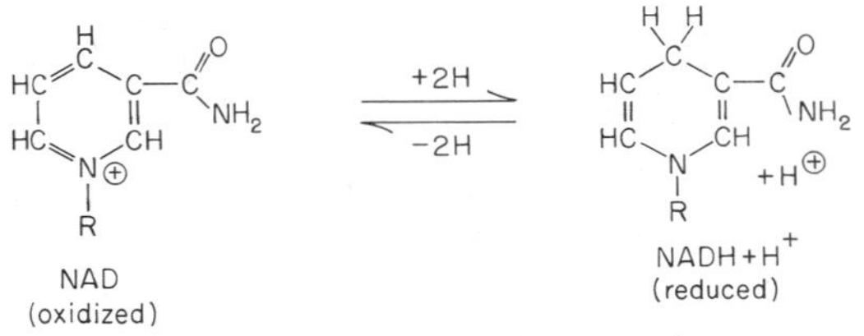

NAD (oxidized) ⇌ (+2H / −2H) NADH + H^+ (reduced)

FIG. 4. Oxidized and reduced states of nicotinamide coenzymes. R represents the rest of the coenzyme as seen in Fig. 3.

(B) *Riboflavin Coenzymes.* Riboflavin has been shown to be a constituent of two coenzymes (Fig. 5): flavin mononucleotide (FMN) and flavin adenine dinucleotide (FAD).

FMN was originally discovered as the coenzyme of an enzyme system that catalyzes the oxidation of the reduced nicotinamide coenzyme, NADPH, to NADP. Most of the many other riboflavin-containing enzymes contain FAD. FAD is an integral part of the biological oxidation-reduction system (Fig. 2) where it mediates the transfer of hydrogen ions from NADH to the oxidized CYTOCHROME system. FAD can also accept hydrogen ions directly from a metabolite and transfer them to either NAD, a metal ion, a heme derivative or molecular oxygen. The various mechanisms of action of FAD are probably due to differences in the protein apoenzymes to which it is bound (see FLAVINS AND FLAVOPROTEINS). The oxidized and reduced states of the flavin portion of FAD are seen in Fig. 6.

Decarboxylation Coenzymes. Thiamine, biotin and pyridoxine (vitamin B_6) coenzymes are grouped together because they catalyze similar phenomena, *i.e.*, the removal of a carboxyl group (—COOH) from a metabolite. However, each

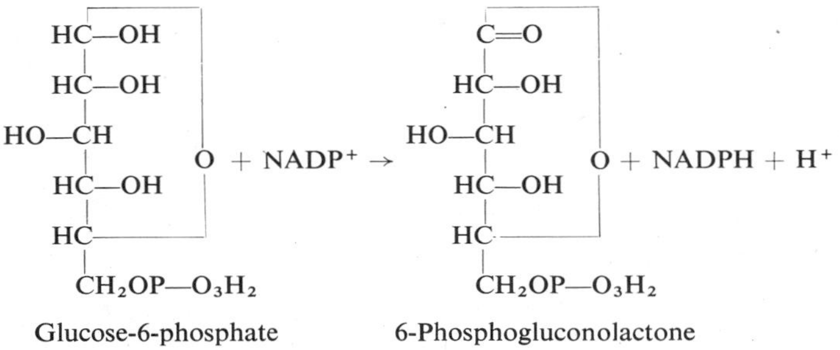

Glucose-6-phosphate + $NADP^+$ → 6-Phosphogluconolactone + NADPH + H^+

(HC—OH, HC—OH, HO—CH, HC—OH, HC—O (ring), $CH_2OP\text{—}O_3H_2$ → C=O, HC—OH, HO—CH, HC—OH, HC—O (ring), $CH_2OP\text{—}O_3H_2$)

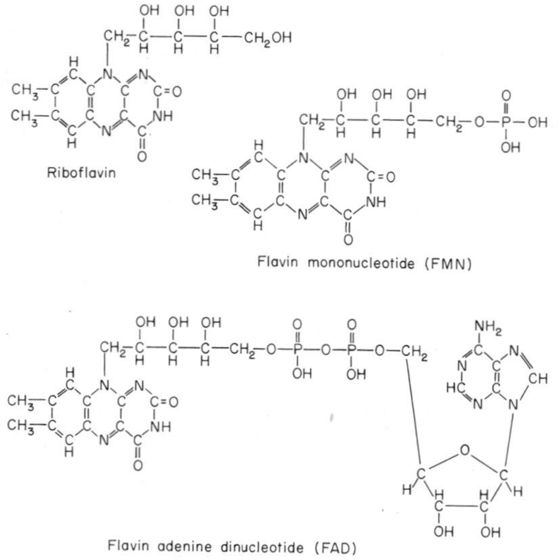

Riboflavin

Flavin mononucleotide (FMN)

Flavin adenine dinucleotide (FAD)

FIG. 5. Structures of riboflavin and flavin coenzymes.

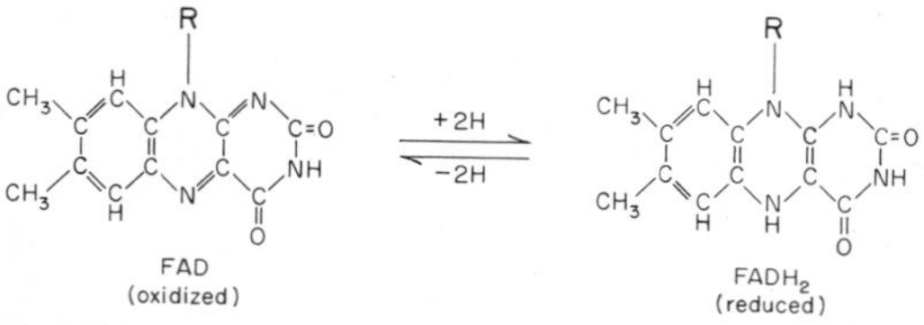

FAD (oxidized) $\underset{-2H}{\overset{+2H}{\rightleftharpoons}}$ $FADH_2$ (reduced)

FIG. 6. Oxidized and reduced states of flavin coenzymes. R represents the rest of the coenzyme as seen in Fig. 5.

require different specific circumstances. Thiamine coenzyme decarboxylates only α-keto acids ($R{-}\overset{\overset{\large O}{\|}}{C}{-}COOH$), is frequently accompanied by dehydrogenation, and is mainly associated with CARBOHYDRATE METABOLISM. Biotin enzymes do not require the α-keto configuration, are readily reversible, and are concerned primarily with lipid metabolism. Pyridoxine coenzymes perform non-oxidative decarboxylation and are closely allied with AMINO ACID METABOLISM.

(A) *Thiamine Coenzyme*. Thiamine, also known as vitamin B_1 and the anti-beriberi factor, is metabolically active as thiamine pyrophosphate (TPP) (Fig. 7). (See THIAMINE AND BERIBERI).

TPP functions as a coenzyme which participates in decarboxylation of α-keto acids. As previously mentioned in the discussion of coenzyme A, dehydrogenation and decarboxylation must precede the formation of "active acetate" in the initial reaction of the citric acid cycle:

$$\underset{\text{Pyruvic acid}}{CH_3{-}\overset{\overset{O}{\|}}{C}{-}COOH} + NAD^+ + CoA \xrightarrow{(FAD,\ TPP)} \underset{\substack{\text{Acetyl-CoA} \\ \text{"active acetate"}}}{CH_3{-}\overset{\overset{O}{\|}}{C}{-}CoA} + CO_2 + NADH + H^+$$

This reaction is a good example of the interrelationship of vitamin B coenzymes. Four vitamin coenzymes are necessary for this one reaction: thiamine (in TPP) for decarboxylation, nicotinic acid (in NAD) and riboflavin (in FAD) for dehydrogenation, and pantothenic acid (in CoA) for activation of the acetate fragment.

TPP also mediates the oxidative decarboxylation of α-ketoglutaric acid, another intermediate of carboxydrate metabolism in the citric acid cycle. The nutritional requirement for thiamine increases as dietary carbohydrate increases because of a greater demand for TPP.

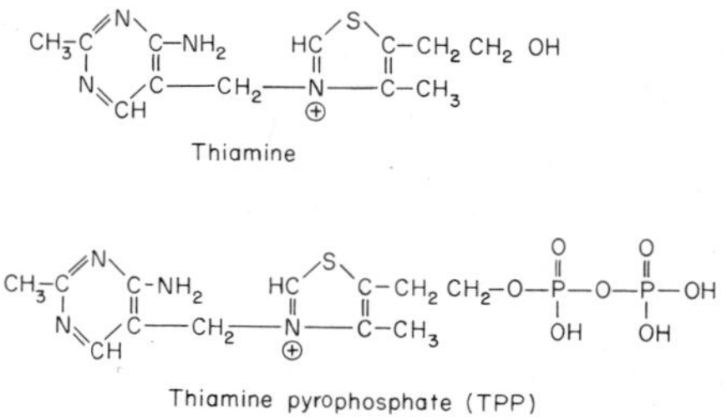

Thiamine

Thiamine pyrophosphate (TPP)

FIG. 7. Structures of thiamine and thiamine pyrophosphate.

(B) *Biotin Enzymes*. Biotin reacts with an oxidized carbon fragment (denoted as CO_2) and an energy-rich compound, adenosine triphosphate (ATP) to form carboxy biotin (Fig. 8), which is "activated carbon dioxide." Biotin is firmly bound to its enzyme protein by a peptide linkage.

Biotin enzymes are believed to function primarily in reversible CARBOXYLATION-decarboxylation REACTIONS. For example, a biotin enzyme mediates the carboxylation of propionic acid to methylmalonic acid which is subsequently con-

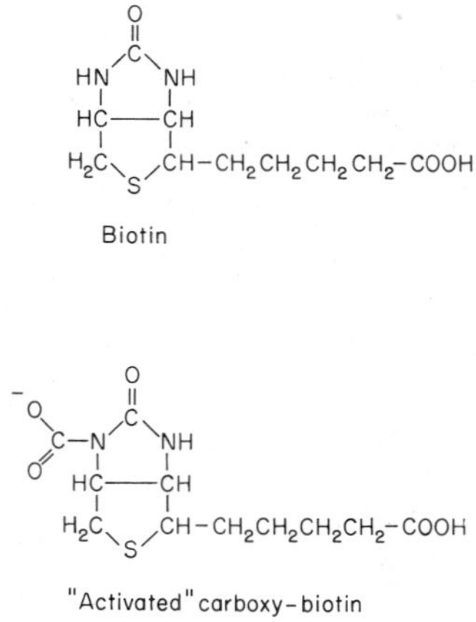

Biotin

"Activated" carboxy-biotin

FIG. 8. Structures of biotin and carboxy-biotin.

verted to succinic acid, a citric acid cycle intermediate. A vitamin B_{12} coenzyme and coenzyme A are also essential to this over-all reaction, again pointing out the interdependence of the B vitamin coenzymes. The reaction is as follows:

$$\text{“}CO_2\text{”—biotin} + CH_3CH_2\overset{\overset{\displaystyle O}{\|}}{C}\text{—CoA} \rightarrow CH_3CH\text{—}\overset{\overset{\displaystyle O}{\|}}{\underset{\underset{\displaystyle COOH}{|}}{C}}\text{—CoA} + \text{Biotin}$$

Propionyl-CoA Methylmalonyl-CoA

Another biotin enzyme-mediated reaction is the formation of malonyl-CoA by carboxylation of acetyl-CoA ("active acetate"). Malonyl-CoA is now believed to be a key intermediate in fatty acid synthesis. The reaction follows:

$$\text{“}CO_2\text{”—biotin} + CH_3\text{—}\overset{\overset{\displaystyle O}{\|}}{C}\text{—CoA} \rightarrow HOOC\text{—}CH_2\text{—}\overset{\overset{\displaystyle O}{\|}}{C}\text{—CoA} + \text{Biotin}$$

Acetyl-CoA Malonyl-CoA

(C) *Vitamin B_6 (Pyridoxine) Coenzyme.* Various forms of vitamin B_6 are pyridoxine, pyridoxal and pyridoxamine. All can be converted in the body to the metabolically active form, pyridoxal phosphate (Fig. 9).

Pyridoxal phosphate enzymes mediate the non-oxidative decarboxylation of amino acids. This mechanism is of primary importance in bacteria, but it may also be essential to proper functioning of the nervous system in man by providing a pathway for the synthesis of a nerve impulse inhibitor, γ-amino-butyric acid from glutamic acid:

$$HOOC\text{—}CH_2CH_2\underset{\underset{\displaystyle NH_2}{|}}{CH}\text{—}COOH \xrightarrow{\text{Pyridoxal phosphate}} HOOC\text{—}CH_2CH_2CH_2NH_2 + CO_2$$

Glutamic acid γ-Aminobutyric acid

Pyridoxal phosphate is also a cofactor for TRANSAMINATION reactions. In these reactions, an amino group is transferred from an amino acid to an α-keto acid, thus forming a new amino acid and a new α-keto acid:

$$\left\{\begin{matrix} HOOC\text{—}CH_2CH_2\overset{\overset{\displaystyle NH_2}{|}}{CH}\text{—}COOH \\ \text{Glutamic acid} \\ + \\ CH_3\overset{\overset{\displaystyle O}{\|}}{C}\text{—}COOH \\ \text{Pyruvic acid} \end{matrix}\right\} \xrightarrow{\text{Pyridoxal phosphate}} \left\{\begin{matrix} HOOC\text{—}CH_2CH_2\overset{\overset{\displaystyle O}{\|}}{C}\text{—}COOH \\ \alpha\text{-Ketoglutaric acid} \\ + \\ CH_3\overset{\overset{\displaystyle NH_2}{|}}{CH}\text{—}COOH \\ \text{Alanine} \end{matrix}\right\}$$

Transamination reactions are important for the synthesis of amino acids from non-protein metabolites and for the degradation of amino acids for energy production. Since pyridoxal phosphate is intimately involved in amino acid metabolism, the dietary requirement for vitamin B_6 increases as the protein content of the diet increases.

"One-carbon Fragment" Transferring Coenzymes. Folic acid functions as a coenzyme in enzyme reactions which involve the transfer of one-carbon fragments at various levels of oxidation. Vitamin B_2 (cobalamin) may be interrelated with folic acid in these reactions. Folic acid and vitamin B_{12} are also considered together since certain clinical ANEMIAS can be correctd by administration of either of the two vitamins (see also SINGLE CARBON UNIT METABOLISM).

(A) *Folic Acid Coenzymes.* The coenzyme forms of folic acid are derivatives of tetrahydrofolic acid (FH_4, Fig. 10; see also FOLIC ACID COENZYMES).

One-carbon fragments in various oxidation states are: (1) formyl (—CHO), (2) hydroxymethyl (—CH_2OH), and (3) methyl (—CH_3). The coenzyme forms of folic acid have one of these groups attached to either the N-5 or N-10 of tetrahydrofolic acid. One folic acid coenzyme, methyltetrahydrofolate (CH_3—FH_4) transfers its methyl group to homocysteine to yield methionine, in a reaction which also requires a vitamin B_{12} coenzyme:

$$\underset{\text{Homocysteine}}{HS\text{—}CH_2CH_2\overset{\overset{NH_2}{|}}{C}HCOOH} + CH_3\text{—}FH_4 \xrightarrow{B_{12}\text{ coenzyme}} \underset{\text{Methionine}}{CH_3\text{—}S\text{—}CH_2CH_2\overset{\overset{NH_2}{|}}{C}HCOOH} + FH_4$$

Other folic acid enzymes are involved in the transfer of one-carbon fragments in the synthesis of serine and degradation of histidine. They also participate in the biosynthesis of PURINES and PYRIMIDINES—important constituents of nucleic acids, the genetic material in the chromosomes.

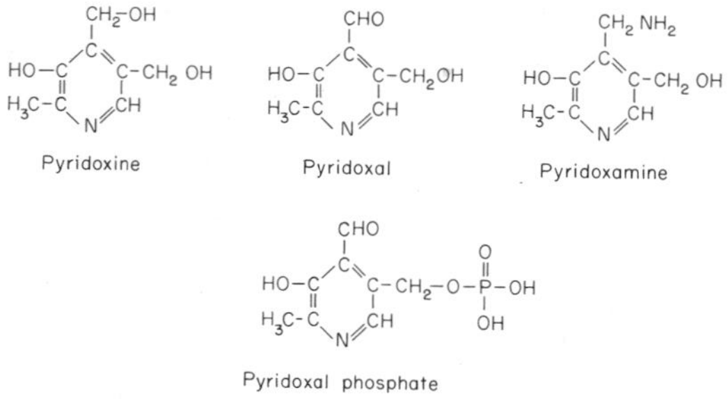

Pyridoxine

Pyridoxal

Pyridoxamine

Pyridoxal phosphate

FIG. 9. Structures of some forms of vitamin B_6 and pyridoxal phosphate.

(B) *Cobamide Coenzymes.* Vitamin B_{12} or cyanocobalamin is a highly substituted corrin structure attached to dimethylbenzimidazole ribotide (Fig. 11). COBALT is covalently and coordinately bound in the center of the corrin configuration.

Cobalamin is also biologically active if the cyanide (—CN) moiety is replaced by chloride, hydroxyl, nitrate or sulfate. In the body, vitamin B_{12} is converted to its coenzyme form. The coenzymes (cobamide coenzymes) so far established have a deoxyadenosyl group in place of cyanide. The B_{12} coenzyme most prevalent in animals and man is dimethylbenzimidazole

Folic acid (PGA)

Tetrahydrofolic acid (H_4-PGA)

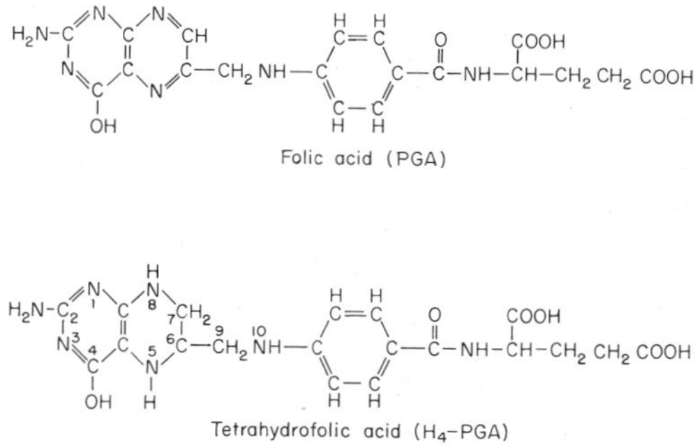

FIG. 10. Structures of folic acid and tetrahydrofolic acid.

cobamide (Fig. 11). Other cobamide coenzymes contain a different heterocyclic base in place of dimethylbenzimidazole.

The mechanism of action of cobamide coenzymes has not yet been established. However, B_{12} coenzymes are essential cofactors for isomerization of glutamic acid to β-methylaspartic acid and of succinyl-CoA to methylmalonyl-CoA in certain microorganisms:

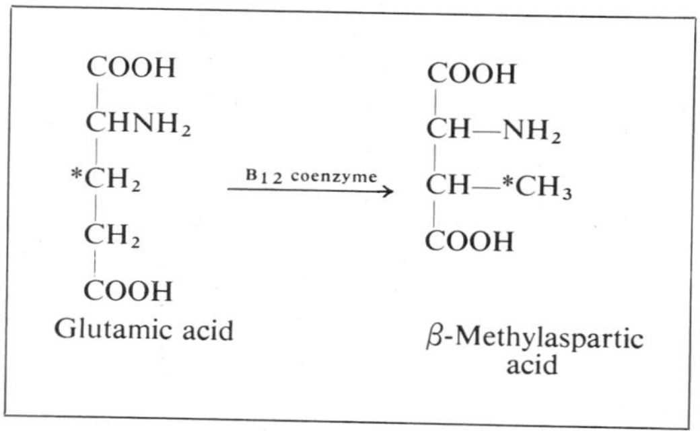

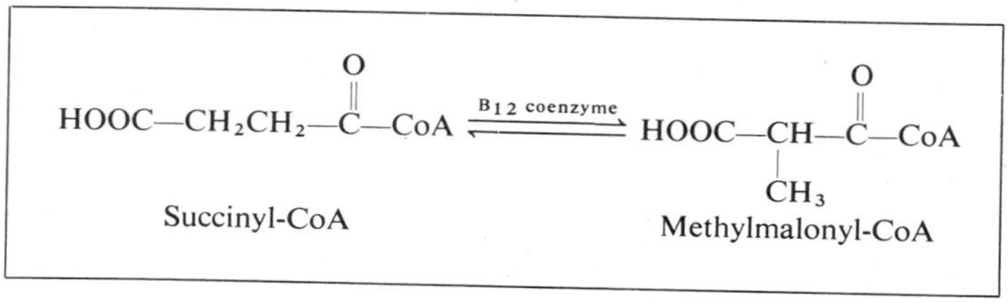

$$\underset{\text{Glutamic acid}}{COOH\text{—}CHNH_2\text{—}{}^*CH_2\text{—}CH_2\text{—}COOH} \xrightarrow{B_{12}\text{ coenzyme}} \underset{\beta\text{-Methylaspartic acid}}{COOH\text{—}CH(NH_2)\text{—}CH({}^*CH_3)\text{—}COOH}$$

$$\underset{\text{Succinyl-CoA}}{HOOC\text{—}CH_2CH_2\text{—}\overset{\overset{O}{\|}}{C}\text{—}CoA} \underset{}{\overset{B_{12}\text{ coenzyme}}{\rightleftharpoons}} \underset{\text{Methylmalonyl-CoA}}{HOOC\text{—}\underset{CH_3}{\underset{|}{C}H}\text{—}\overset{\overset{O}{\|}}{C}\text{—}CoA}$$

As previously mentioned, a B_{12} coenzyme is required by some animals and microorganisms for the synthesis of methionine from CH_3—FH_4 and homocysteine.

In man, vitamin B_{12} is required for growth and normal blood formation. Pernicious anemia is a disease which results from failure to absorb cyanocobalamin from the intestine.

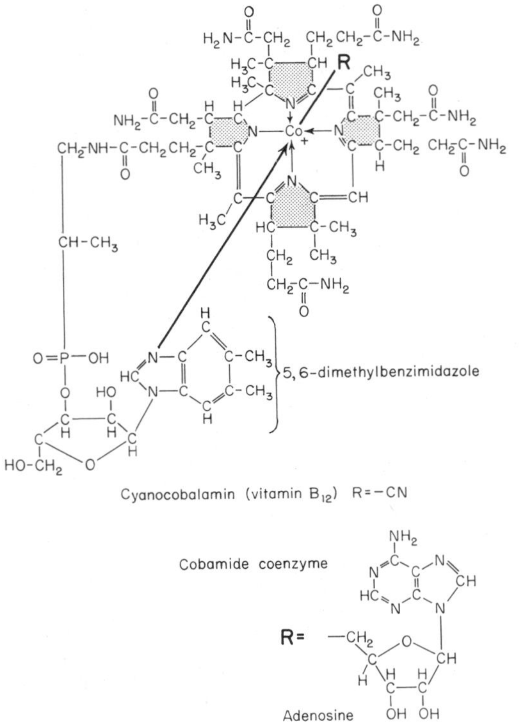

Cyanocobalamin (vitamin B_{12}) R=−CN

Cobamide coenzyme

R= Adenosine

FIG. 11. Structures of vitamin B_{12} and a cobamide coenzyme.

A Lipotropic Agent. Choline does not function as a coenzyme, but it is usually included with the B vitamins since it is water soluble, is required in the diet under certain conditions, and is found in many of the same sources as other B vitamins.

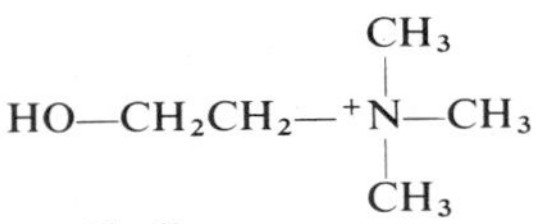

$$HO{-}CH_2CH_2{-}{}^{+}N(CH_3)_3$$

Choline

Choline functions as a lipotropic agent, *i.e.*, it aids in the transport of fat and thus prevents fatty infiltration of the liver (see also FAT MOBILIZATION). Choline is a component of many PHOSPHOLIPIDS, such as lecithin. ACETYLCHOLINE, a nerve transmitter substance, is synthesized by acetylation of choline by acetyl coenzyme A.

References

Books

1. HARRIS, L. J., "Vitamins in Theory and Practice," Cambridge, The University Press, 1955.
2. ROBINSON, F. A., "The Vitamin B Complex," New York, John Wiley & Sons, 1951.
2. SEBRELL, W. H., JR., AND HARRIS, R. S., (EDITORS), "The Vitamins," 3 vols., New York, Academic Press, 1954.
4. WAGNER, A. F., AND FOLKERS, K., "Vitamins and Coenzymes," New York, John Wiley & Sons, 1964.
5. WILLIAMS, R. J., EAKIN, R. E., BEERSTECHER, E., JR., AND SHIVE, W., "The Biochemistry of B Vitamins," New York, Reinhold Publishing Corp., 1950.

Review Articles

6. BROWN, G. M., "Biosynthesis of Water Soluble Vitamins and Derived Coenzymes," *Phys. Rev.*, **40**, 331 (1960).
7. BROZEK, J., AND VAES, G., "Experimental Investigations on the Effect of Dietary Deficiencies on Animal and Human Behavior," *Vitamins Hormones*, **19**, 43 (1961).
8. BURNS, J. J., AND CONNEY, A. H., "Water Soluble Vitamins (Ascorbic Acid, Nicotine Acid, Vitamin B_6, Biotin and Inositol)," *Ann. Rev. Biochem.*, **29**, 413 (1960).
9. COATES, M. E., AND PORTER, J. W. G., "Water Soluble Vitamins (Vitamin B_{12}, Folic Acid, Ascorbic Acid, Biotin and Vitamin B_6)," *Ann. Rev. Biochem.*, **28**, 439 (1959).
10. DINNING, J. S., "Water Soluble Vitamins (Vitamin B_{12}, Folic Acid, Thiamine, Riboflavin, and Pantothenic Acid)," *Ann. Rev. Biochem*, **29**, 437 (1960).
11. FRIEDKIN, M., "Enzymatic Aspects of Folic Acid," *Ann. Rev. Biochem.*, **32**, 185 (1963).
12. HORWITT, M. K., "Water Soluble Vitamins (Thiamine, Riboflavin, Pantothenic Acid, Nicotinamide, Lipoic Acid)," *Ann. Rev. Biochem.*, **28**, 411 (1959).
13. "International Symposium on Vitamin B_6 in Honor of Professor Paul Gyorgy," *Vitamins Hormones*, **22**, 361 (1964).
14. JAENICKE, L., "Vitamin and Coenzyme Function —Vitamin B_{12} and Folic Acid," *Ann. Rev. Biochem.*, **33**, 287 (1964).
15. MISTRY, S. P., AND DAKSHINAMURTI, K., "Biochemistry of Biotin," *Vitamins Hormones*, **22**, 1 (1964).
16. STOKSTAD, E. L. R., "The Biochemistry of the Water Soluble Vitamins," *Ann. Rev. Biochem.*, **31**, 451 (1962).
17. STOKSTAD, E. L. R., AND OACE, S. M., "Folic acid, Biotin and Pantothenic Acid," in "Newer Methods of Nutritional Biochemistry" (ALBANESE, A., EDITOR), Vol. 2, New York, Academic Press, 1965.

SUSAN M. OACE AND E. L. R. STOKSTAD

VITAMIN B_{12} (ABSORPTION, AND NUTRITIONAL ASPECTS)

Absorption.[1–3] Vitamin B_{12} exists in food in various peptide-bound coenzymatically active forms (see COBALAMINS AND COBAMIDE COENZYMES). A variable percentage (perhaps 5–10%, on the average) of the vitamin B_{12} content of food is freed from its peptide bonds by acid and enzymes in the stomach and also by enzymes in the small intestine, and thus becomes available for absorption.

There are two separate and distinct mechanisms for absorption of vitamin B_{12}. One mechanism is active, the other passive; both operate simultaneously. The active process is physiologically more important, since it is operative primarily in the presence of the small (1–2 μg) quantities of vitamin B_{12} made available for absorption from the average meal. This special mechanism, perhaps uniquely necessary for vitamin B_{12} because of its large size and polar properties, operates as follows: the normal gastric mucosa secretes a substance called the *intrinsic factor of Castle*, which combines with free vitamin B_{12}; the complex travels down the intestine to the ileum, where, in the presence of calcium and pH above 6, it attaches to "receptors" lining the wall of the ileal mucosa. Vitamin B_{12} is then freed from intrinsic factor via a "releasing factor" mechanism of unknown nature, operating either at the surface of or within the ileal mucosal cell, and passes into the bloodstream. Thus, important requirements for normal absorption of vitamin B_{12} from food are: (1) the vitamin must be freed from its peptide bonds in food; (2) the gastric mucosa must secrete an adequate quantity of intrinsic factor; (3) the ileal mucosa must be sufficiently normal both structurally and functionally so that vitamin B_{12} may be absorbed across it.

Intrinsic factor has not yet been isolated in pure form. It is probably a glycoprotein or MUCOPOLYSACCHARIDE with a molecular weight in the range of 50,000 and an end group conformation akin to that of partly degraded blood group substance. The sole known role of intrinsic factor is to facilitate the transport of the large (molecular weight = 1355) vitamin B_{12} molecule across the wall of the ileal mucosa and into the bloodstream. How it performs that function, and its fate afterward, are almost totally unknown. It is an interesting fact that antibody to intrinsic factor exists in the serum of approximately half of all patients with pernicious anemia.

The second mechanism for vitamin B_{12} absorption is operative primarily in the presence of quantities of vitamin B_{12} greater than those made available for absorption from the average diet (*i.e.*, quantities greater than about 30 μg). This mechanism is a passive one, probably diffusion, and most likely occurs along the entire length of the small intestine. It operates when patients with pernicious anemia (vitamin B_{12} deficiency due to inadequate or absent intrinsic factor secretion of unknown cause) are treated with large quantities (500 μg or more daily) of oral vitamin B_{12}. Such treatment is probably better than treatment with oral hog intrinsic factor, to which refractoriness often develops, but it is not as certain as treatment with monthy injections of vitamin B_{12}.

Nutritional Roles.[2,4] Vitamin B_{12} is required for DNA synthesis and, therefore, is necessary in every reproducing cell in man for maintenance of the ability to divide. It functions coenzymatically in the methylation of homocysteine to methionine (see METHYLATIONS IN METABOLISM). It is important in several isomerization reactions (see COBALAMINS AND COBAMIDE COENZYMES), and as a reducing agent, and is probably of special importance in enzymatic reduction of ribosides to deoxyribosides. It is involved in protein synthesis, partly via its role in the conversion of homocysteine to methionine; in fat and carbohydrate metabolism, partly via its role in the isomerization of succinate to methylmalonate (which then may be decarboxylated to propionate), and in folate metabolism (see SINGLE CARBON UNIT METABOLISM). Where these two vitamins interrelate, vitamin B_{12} appears to serve as a coenzyme and folate as a substrate; such is true in the vitamin B_{12}-mediated transfer of a methyl group from N^5-methyltetrahydrofolic acid to homocysteine, which is thereby converted to methionine.

Vitamin B_{12} is one of the most potent nutrients known; the minimal daily requirement for absorption by the normal adult is probably in the range of 0.1 μg. This equals, for example, one five-hundredth of the minimal daily adult folate requirement, which is in the range of 50 μg.

As with all nutritional deficiencies, deficiency of vitamin B_{12} may arise from inadequate ingestion, absorption or utilization, and from increased requirement or increased excretion. Deficiency of vitamin B_{12} produces megaloblastic (large germ cell) ANEMIA, damage to the alimentary tract (glossitis being the most striking feature), and neurologic damage. It is not yet known whether the anemia and alimentary tract damage are directly due to the vitamin B_{12} deficiency or arise partly from the secondary folate deficiency, due to inability to adequately utilize folate without the coenzymatic action of vitamin B_{12}. However, nerve damage does appear to be directly due to vitamin B_{12} deficiency, since it is observable under the microscope only with vitamin B_{12} deficiency, and not with folate deficiency. The most classic neurologic sign of vitamin B_{12} deficiency is decreased ability to perceive the vibration of a tuning fork pressed against the ankles. This finding is associated with damage to the posterior and lateral columns of the spinal cord, and also with damage to the peripheral nerves. This damage occurs because vitamin B_{12} deficiency results in gradual deterioration of the MYELIN sheath, which is followed by deterioration of the axon. These processes occur slowly over a period of months to years, and during this stage are reversible by treatment with vitamin B_{12}. However, when the nerve nucleus finally deteriorates, the neurologic damage becomes irreversible.

Because of its vital role as a nutrient of very high potency, vitamin B_{12} has been endowed with magical qualities by hucksters peddling vitamins for a multiplicity of problems from poor growth in children, through reduced sexual potency in adults, to poor appetite in the elderly. The double irony of this advertising is not only that there is not a sliver of evidence that vitamin B_{12} treats anything but deficiency of vitamin B_{12}, but also that the vitamin in the quantities usually supplied by hucksters is not absorbed when eaten by the very subjects who really need it—elderly people who have vitamin B_{12} deficiency due to inability to absorb the vitamin due to inadequate or absent secretion of intrinsic factor. In these people, the treatment of choice is the vitamin given by injection.

References

1. HERBERT, V., STREIFF, R. R., AND SULLIVAN, L. W., "Notes on Vitamin B_{12} Absorption; Autoimmunity and Childhood Pernicious Anemia; Relation of Intrinsic Factor and Blood Group Substance," *Medicine* (1964).
2. HERBERT, V., "Drugs Effective in Megaloblastic Anemias," in "The Pharmacological Basis of Therapeutics" (GOODMAN, L. S., AND GILMAN, A., EDITORS), Third edition, New York, The Macmillan Co., 1965.
3. HERBERT, V., AND CASTLE, W. B., "Intrinsic Factor," *New Engl. J. Med.*, **270**, 1181–1185 (1964).
4. SMITH, E. L., "Vitamin B_{12}," Second edition, New York, John Wiley & Sons, 1965.

VICTOR HERBERT

VITAMIN D GROUP

While the term "vitamin D" (singular) is often used in nutritional discussions, it is objectionable when used in a strict biochemical context, on the grounds that there are probably at least 10 different substances each of which is capable of performing the nutritional function of "vitamin D," namely that of promoting growth (including BONE growth) and preventing rickets in young animals.

The most important or at least the best known members of the family of D vitamins are vitamin D_2 (calciferol), which has the structure indicated in abbreviated form below and can be produced by ultraviolet irradiation of ergosterol, and vitamin D_3, which may be produced by the irradiation of 7-dehydrocholesterol.

H_2C

HO

Vitamin D_2

H_2C

HO

Vitamin D_3

It should be noted that subscript numerals have a different connotation here than they have in connection with the B vitamins. Vitamins B_1, B_2, B_6, B_{12}, etc., represent individual substances which have little or no chemical resemblance to each other and perform different metabolic functions. The various vitamin D's, however, have very similar structures, differing only in the side chains, and perform the same functions.

There are a number of unique features exhibited in connection with the D vitamins. First, they are not required nutritionally at all if the organism has access to ultraviolet light (which is present in sunlight). Some animals, kept away from ultraviolet light, require so little vitamin D that the need cannot be demonstrated using ordinary diets. Rats, for example, exhibit a need for vitamin D when the calcium/phosphorus ratio in the diet is about 5:1 but not when it is the more usual 1:1. Chickens, on the other hand, exhibit a need even when the calcium/phosphorous ratio is "normal" (1.5:1).

Different species of animals respond distinctively to the different members of the vitamin D family. The most striking example of this is the fact that vitamin D_2 (calciferol) has practically no vitamin D activity for chickens. Rats respond about equally to D_2 and D_3. Human beings respond both to D_2 and D_3. Information as to how various animals react to the other less known forms of vitamin D is largely lacking and for practical reasons is not sought after.

Members of the vitamin D family are extremely difficult to isolate and identify in pure form from any source. Fish liver oils are rich sources, and vitamins D_2 and D_3 have been isolated from them. Most ordinary foods are such poor sources in terms of amounts present, that the presence of D vitamins in them has not been demonstrated. Sterols which can be converted into some form of vitamin D by ultraviolet light are, however, widespread, and it may be inferred that D vitamins are often present even when their presence has never been demonstrated.

The requirements of animals for vitamin D in terms of actual weight are extremely small. It is estimated that human beings need about 400 international units of vitamin D per day. Since an international unit of vitamin D corresponds to 0.025 μg of crystalline vitamin D, this means that the daily human requirement is about 0.01 mg. Foods can contain as little as 0.02 parts per million of vitamin D and yet furnish an ample supply on the basis of the above estimate.

Especially when vitamin D first became available at low cost without prescription, it was sometimes used indiscriminately on the supposition that "if a little is good, more is better." This resulted in cases of excessive calcification and severe damage (hypervitaminosis). The complete story of vitamin D dosage is an interesting one in that it has been observed that some "susceptible" children do not respond to the usual doses but require 5000–10,000 units per day to keep them free from rickets. There are other children that are afflicted with "vitamin D-resistant rickets" who do not respond even to these high doses but may do so when doses of the order of 500,000–1,000,000 units are administered. Though the picture is not clear, it would seem that in some individuals the vitamin D has difficulty getting through to where it is needed. Whether and to what extent this same difficulty exists with other nutrients is largely unknown.

Reference

1. Rosenberg, H. R., "Chemistry and Physiology of the Vitamins," New York, Interscience Publishers, 1945.

Roger J. Williams

VITAMIN E (THE TOCOPHEROLS)

In 1920 Matil and Conklin found evidence of a factor necessary for rats. Later Evans demonstrated that it was not vitamins A, B, C or D. Eventually, the tocopherols were identified as naturally occurring oily substances and were characterized as alpha, beta and gamma forms. They had biological activity in descending order. Vitamin E is necessary for the normal growth of animals. Without it, they develop infertility, abnormalities of the central nervous system, and myopathies involving both skeletal and cardiac muscle. The tocopherols exert an antioxidant effect chemically, the magnitude of which is in reverse order to that of its vitamin activity. Muscular tissue taken from a deficient animal has an increased rate of oxygen utilization. The tocopherols are so widely distributed in natural foods that a spontaneous deficiency does not occur unless diseases of the gastrointestinal or biliary systems hinder absorption.

The evidences for such deficiency have rested largely upon the findings of a low level of tocopherol in the serum and upon an abnormal hemolysis of erthrocytes by hydrogen peroxide.

Although nearly every vitamin has been used unwisely in the treatment of human diseases, perhaps no other substance has aroused a greater degree of controversy among clinicians than vitamin E. Because deficient animals develop a form of myopathy, it was natural to test the therapeutic efficacy of vitamin E in various forms of progressive muscular dystrophy and in diseases of the reproductive system. Many enthusiastic claims have been refuted by investigators whose methods were meticulous and objective. At present there is no recognized indication for the administration of vitamin E except in patients with a malabsorption syndrone. Recent evidence indicates that diets which contain stripped corn oil without other fats my result in a deficiency of vitamin E.

Despite the lack of therapeutic indication, many pharmaceutical preparations contain the tocopherols. As yet there is no evidence that an excess of vitamin E produces toxic reactions. On the basis of animal studies, it has been estimated that normal persons require approximately 30 mg of naturally occurring tocopherols daily, but more than this amount is supplied by the average diet. Vitamin E is certainly an essential for man yet much remains to be learned about it (see also ANTIOXIDANTS, and references in that article).

Robert E. Hodges

VITAMIN K GROUP

Occurrence. The quinones that make up the group of K vitamins are quite ubiquitous. They are present in animal tissues, microorganisms and plants. Although these naphthoquinone derivatives constitute a very small portion of the naphthoquinones found throughout nature, they have received a great deal of attention from biochemists because they are required for normal metabolism and they are chemically related to VITAMIN E and ubiquinone (see COENZYME Q). Isolation of vitamin K from animal tissues has not been achieved *per se*, but its presence in them has been reported by the use of C^{14}-labeled compounds or by feeding the dried tissue to vitamin K-deficient chicks. The hemorrhagic disease developed by chicks on a diet essentially free of vitamin K is characterized by slow clotting of blood. Until recently, the bioassay for vitamin K involved supplementing the vitamin-deficient chick or rat and measuring clotting time or prothrombin time. A recent report describes a naphthoquinone-dependent strain of *Bacteroides melaninogenicus* which may replace or supplement the former assay. Extraction and separation of the several K vitamins from microorganisms or plants can be achieved by using several solvent systems and subjecting the extracts to standard column, paper or thin layer chromatographic techniques.

Structure. The vitamin K (Koagulations-vitamin) that was isolated from alfalfa was designated vitamin K_1, since it was the first of the group recognized. The one isolated a few years later (which was actually of microbial origin) from putrified fish meal was called vitamin K_2. Vitamin K_1 was found to be (−)2-methyl-3-phytyl-1,4-naphthoquinone, and K_2, 2-methyl-3-farnesyl-digeranyl-1,4-naphthoquinone. With the increasing number of vitamins isolated and characterized, several unofficial conventions to name them have been introduced: (1) by the number of carbon atoms in the 3-substituent of the quinone ring, and (2) by the number of isoprenoid units in the side chain, and in both instances also according to the degree of unsaturation of this substituent. Thus, vitamin K_1 from alfalfa has been referred to as $K_{1(20)}$ and $K_{1(4)}$; K_2 from fish meal, $K_{2(35)}$ and $K_{2(7)}$, and so on. Since these are unofficial nomenclatures, other notations will be found. Other derivatives of vitamin K which show anti-hemorrhagic activity, although not homologues, have retained their trivial names and notations: vitamin K_3 (menadione), K_4 (menadiol), K_5 (4-amino-2-methyl-1-naphthol hydrochloride), K_6 (1,4-diamino-2-methylnaphthalene dihydrochloride), and K_7 (4-amino-3-methyl-1-naphthol hydrochloride). Similar chemical properties and structures of vitamin K, vitamin E and ubiquinone have resulted in speculation proposing similar roles of the quinones in biological systems. All are *para*-quinones, completely substituted, and possess a side chain in the 3-position with basically the same carbon skeleton. Although vitamin K and ubiquinone possess unsaturated open side chains and vitamin E is a ring structure, it is well-known that the two side chain structures, open and ring, are readily interconvertible chemically.

Function. In general, when vitamin K is absent or lacking in the diet of animals, including man, a hemorrhagic disease appears. Young fowls that are allowed to continue on a deficient diet for extended periods will die of internal hemorrhage or from extensive bleeding from small external wounds. The hemorrhagic condition is a direct result of a decrease in concentration of one or more proteins necessary for normal BLOOD CLOTTING. Fowls experience difficulty in absorbing vitamin K from the intestine, whereas humans, rats and dogs absorb it readily and normally obtain their requirement of vitamin K from intestinal bacteria (see INTESTINAL FLORA) without need of dietary supplementation. If, however, bacterial synthesis is inhibited by the use of sulfa drugs or certain antibiotics, the disease will develop unless the diet is supplemented with some form of vitamin K. When there is a decrease in the amount of BILE salts in the intestine, as in obstructive jaundice, vitamin K is absorbed in such small amounts that the disease will also ensue. It has been recommended by the American Academy of Pediatrics that the prophylactic use of vitamin K (0.5–1.0 mg I. M., or 1–2 mg *per os*) to control and prevent the disease in premature babies be adopted universally. Vitamin K_1 is also able to reverse the hemorrhagic condition resulting from administration of dicumarol to animals.

It has been reported that vitamin $K_{1(4)}$ and several of the vitamin K_2 homologues are capable of restoring electron transport in solvent-extracted or irradiated bacterial and MITOCHONDRIAL preparations. Other reports suggests that vitamin K is concerned with the phosphorylation reactions accompanying oxidative PHOSPHORYLATION. The capacity of these compounds to exist in several forms, *i.e.*, quinone, quinol, chromanol, etc., seems to strengthen the proposal linking them with oxidative phosphorylation. Recently, data have been presented suggesting that vitamin K acts to induce prothrombin synthesis. Since prothrombin has been shown to be synthesized only by liver parenchymal cells, in the dog, it would appear that the proposed role for vitamin K is not specific for only prothrombin synthesis but applicable to other proteins. Although our knowledge of the mechanism of action of "fat-soluble" vitamins is much less than that of "water-soluble" vitamins, it seems reasonable to assume that they also may be concerned with enzymes and function as coenzymes.

Biosynthesis. Our knowledge of the biosynthesis of vitamin K is very limited. In recent years, the biosynthesis of fatty acids and STEROLS from acetate units has been established and has led to speculation that vitamin K and other quinones could be synthesized by plants and microorganisms by cyclization of polyacetic acid precursors. The synthesis of compounds containing an aromatic ring by various microorganisms, on the other hand, has been shown to require a cyclohexane precursor, dehydroquinic acid (which perhaps was formed from smaller aliphatic compounds).

The isoprenoid side chain of vitamin K has been shown to be formed according to the principle of farnesol synthesis, extended by several C_5 units (see ISOPRENOID BIOSYNTHESIS). By the use of isotope labeling, it has been shown that menadione is converted to vitamin $K_{2(4)}$ in chicks and rats. It has also been shown that dietary vitamin $K_{1(4)}$ or $K_{2(6)}$ is efficiently converted to vitamin $K_{2(4)}$ by either the chick, pigeon or rat. *In vitro* experiments strongly suggest that vitamin K of dietary origin is converted to vitamin $K_{2(4)}$ in liver, where the majority of the vitamin appears to be stored.

See reference 14 for tentative rules of nomenclature for compounds in the vitamin K, vitamin E, and coenzyme Q groups, recently adopted by the IUPAC-IUB Commission on Biochemical Nomenclature.

References

1. American Acad. Pediatrics, Inc., "Standards and Recommendations for Hospital Care of New Born Infants," 1960.
2. ANDERSON, G. F., AND BARNHART, M. I., *Am. J. Physiol.*, **206**, 929 (1964).
3. ROSENBERG, H. R., "Chemistry and Physiology of the Vitamins," New York, Interscience Publishers, 1945.
4. CHEN, L. H., AND DALLAM, R. D., *Arch. Biochem. Biophys.* (1965).
5. DALLAM, R. D., *Am. J. Clin. Nutr.*, **9**, 104 (1961).
6. DALLAM, R. D., AND HAMILTON, J. H., *Arch. Biochem. Biophys.*, **105**, 630 (1964).
7. GIBBONS, R. J., AND ENGLE, L. P., *Science*, **146**, 1307 (1964).
8. GUSTAFSSON, B. E., *Ann. N.Y. Acad. Sci.*, **78**, 166 (1959).
9. MURTHY, P. S., AND BRODIE, A. F., *J. Biol. Chem.* **239**, 4292 (1964).
10. THOMSON, R. H., "Naturally Occurring Quinones," New York, Academic Press, 1957.
11. OLSON, R. E., *Science*, **145**, 926 (1964).
12. EDDY, W. H., "Vitaminology," Baltimore, The Williams and Wilkins Co., 1949.
13. WAGNER, A. F., AND FOLKERS, K., "Vitamins and Coenzymes," New York, Interscience Publishers, 1964.
14. FOLKERS, K., *et al.*, "Nomenclature of Quinones with Isoprenoid Side-chains," *Biochem. Biophys. Acta*, **107**, 5–10 (1965).

R. DUNCAN DALLAM

VITAMINS (GENERAL)

The historical background of the discoveries of the various vitamins, and their general biochemical functions, are discussed in the article "Vitamins" by R. J. Williams in "The Encyclopedia of Chemistry" (G. L. Clark and G. G. Hawley, Editors), New York, Reinhold Publishing Corporation, 1957; Second Edition, 1966.

In this volume, the vitamins are discussed in more detail in separate articles. The water-soluble or B vitamins are surveyed, and their structures shown, in the article VITAMIN B GROUP. The B vitamins are each discussed also in separate articles emphasizing the simplest nutritionally active form of the vitamin, as well as in articles emphasizing the coenzyme form (or functional aspects and enzymology) of the same vitamin. These are as follows (with the article on the coenzyme or bound form named in parentheses): BIOTIN; FOLIC ACID (also FOLIC ACID COENZYMES); LIPOLIC ACID (also MULTIENZYME COMPLEXES); NICOTINIC ACID (also NICOTINAMIDE ADENINE DINUCLEOTIDES); PANTOTHENIC ACID (also COENZYME A); PYRIDOXINE FAMILY OF VITAMINS (also TRANSAMINATION ENZYMES); RIBOFLAVIN (also FLAVINS AND FLAVOPROTEINS); THIAMINE AND BERIBERI; VITAMIN B_{12} (also COBALAMINS AND COBAMIDE COENZYMES). For Vitamin C, see ASCORBIC ACID AND SCURVY.

The fat-soluble vitamins are discussed in the articles: COENZYME Q; VITAMIN A (see also VISUAL PIGMENTS); VITAMIN D GROUP; VITAMIN E; VITAMIN K GROUP. See also the various articles on NUTRITIONAL REQUIREMENTS.

WAKSMAN, S. A.

Selman Abraham Waksman (1888–), a Russian-born American microbiologist, is noted for his discovery and isolation of streptomycin. For his work in the area of ANTIBIOTICS (a term he coined), he received the Nobel Prize in physiology and medicine in 1952.

WARBURG, O.

Otto Warburg (1883–), a German biochemist, is noted for his studies involving gaseous exchange in tissue slices and for the widely used Warburg apparatus for carrying out this technique (see MANOMETRIC METHODS). His contributions to enzyme chemistry and the study of respiration are many and cannot be described in a short space (see GLYCOLYSIS). In 1931 he received the Nobel Prize for physiology and medicine.

WATER METABOLISM. See ELECTROLYTE AND WATER REGULATION in this Enyclopedia, and the article "Water Metabolism of Vertebrates" by R. M. Chew in "The Encyclopedia of the Biological Sciences" (Gray, P., Editor), New York, Reinhold Publishing Corp., 1961.

References

1. BARTHOLOMEW, G. A., AND CADE, T. J., "The Water Economy of Land Birds," *Auk*, **80**(4), 504–539 (1963).
2. BLACK, V. S., "Excretion and Osmoregulation," in "The Physiology of Fishes" (BROWN, M. E., EDITOR), Vol. 1, pp. 163–205, New York, Academic Press, 1957.
3. CHEW, R. M., "Water Metabolism of Mammals," in "Physiological Mammalogy" (MAYER, W. V., AND VAN GELDER, R. G., EDITORS), Vol. 2, pp. 44–178, New York, Academic Press, 1965.
4. DEYRUP, I. J., "Water Balance and Kidney," in "Physiology of the Amphibia" (MOORE, J. A., EDITOR), pp. 251–328, New York, Academic Press, 1964.
5. SCHMIDT-NIELSEN, K., "Desert Animals, Physiological Problems of Heat and Water," London, Oxford University Press, 1964.

WATSON, J. D.

James Dewey Watson (1928–) is an American biochemist who with FRANCIS HARRY COMPTON CRICK received a share in the Nobel Prize in physiology and medicine in 1962, for contributing a double-helix model for the structure of DNA molecules. Maurice Wilkins, a British physicist, received a share of the same award.

WAXES

The English term *wax* is derived from the Anglo-Saxon *weax*, which was the term applied to the material gleaned from the honeycomb of the bee. Beeswax is a somewhat unctuous, though sticky, yellow substance, which melts readily when heated, and is not subject to ordinary oxidation or change. Its unique physical and stable characteristics led the Egyptians in 4200 B.C. to find numerous and varied uses for the wax.

The ancient Egyptians made use of the wax to preserve mummies. The word mummy is derived from the Persian *mumiai*, wax. The fibrous wrapping material which encased the corpse was first dipped in molten beeswax, and wax was also used in sealing the coffin. A resin was used in conjunction with the wax. The Egyptians are also known to have made square wax (or wax and resin) tablets that could be rubbed down and reused. Several tablets were often fastened together with fiber; these so-called wax tablets were the forerunner of books.

The custom of first making a model in wax of what was later to be cut from stone or cast in bronze, dates back to the thirteenth century B.C. Some of the earliest Egyptian and Greek bronzes were so produced. In the Renaissance (1475–1610) in Spain, beautiful wax figures of saints, distinguished in form and coloring, were achieved in the realm of religious art. *Cire perdue*, or the "lost wax art," an ancient craftsmanship by goldsmiths, was revived by Benvenuto Cellini in the Italian Renaissance, in the design and sculpture of his many marvellous works of art such as the gold medallion of "Leda and the Swan." In modern times, "cire perdue" has been adopted in the jewelry trade for duplicating rings, bracelets and brooches of intricate design. The models of Cellini were most probably sculptured in beeswax.

Quantity production of many precision parts for electrical equipment, ordnance turbine blades, and aircraft instruments is now dependent upon the art of precision casting which is a slight modification of the lost wax art process. Beeswax, however, has given way to the use of other waxes for the purpose.

Lipid Waxes. At one time, the term *wax* was synonomous with the present term "lipid wax." It excluded petroleum hydrocarbons such as paraffin. In physical characteristics and chemical structure, a natural wax differs from that of a neutral fat. A neutral fat is comprised of glyceryl esters of FATTY ACIDS, chiefly palmitic, stearic, oleic and linoleic. When a neutral fat is saponified, the glycerol of these triglycerides splits off and dissolves, leaving the insoluble fatty acids, which can be easily separated. Of the derived lipids, both palmitic and stearic acids are wax-like. When a lipid wax is saponified, both alcohols and fatty acids remain as residues. The alcohols, monatomic, have a high carbon range (C_{12}–C_{32}) contrasted with glycerol, $C_3H_5(OH)_3$. Included in the classification of lipid waxes are those of the esters of cyclic alcohols (*e.g.*, cholesteryl palmitate) and the free sterols. Cholesterol and lanosterol, as well as their fatty acid esters, are the principal constituents of *woolwax* derived from the hair of sheep. Sterols found in plants, known as phytosterols, differ little from the animal sterols, zoosterols. Woolwax is much better known as "anhydrous lanolin." The latter is comprised of fatty acid esters of both aliphatic alcohols and sterols, with small amounts of free sterols and fatty acids. Peculiarly, the acid components of the esters, which early investigators termed "lano acids" and "agno acids," have now proved to be iso-acids and ante-iso acids.

Waxes which are animal in origin may be grouped as (a) insect waxes, (b) land animal waxes and (c) marine animal waxes. The two principal subgroups of the wax-producing insects are (1) the *Apidae*, of which honey bees are the outstanding members, and (2) the *Coccidae* to which the *Coccus ceriferus* (source of Chinese insect wax) belong, and the *Carteria lacca*, source of stick lac from which shellac and shellac wax is derived.

Insect Waxes. The most important insect wax from an economic standpoint is *beeswax*, secreted by the hive-bee, in many parts of the world. Wax scales are secreted by eight wax glands on the underside of the abdomen of the worker bee. These wax wafers are used by the bee in building its honeycomb. Comb foundations are provided for hive-bees so as not to waste honey; 1½–3 pounds of wax can be obtained from ten combs when they are scraped. The crude wax must be rendered and refined before it can be sold as "yellow beeswax." Yellow beeswax, when further refined either by sun bleaching or by means of peroxide chemicals, is known as "white beeswax."

The chemical components of beeswax are alkyl esters of monocarboxylic acids (71–72%), cholesteryl esters (0.6–0.8%), lactone (0.6%), free alcohols (1–1½%), free wax acids (13.5–14.5%), hydrocarbons (10.5–11.5%), coloring matter (0.3%), moisture and mineral impurities (0.9–2%). Myricyl palmitate ($C_{46}H_{92}O_2$) is the principal constituent of the alkyl esters (49–53%, wax basis); other esters include myricyl palmitoleate and ceryl hydroxypalmitate. The free wax acid component is "cerotic acid," and the principal hydrocarbon is hentriacontane ($C_{31}H_{64}$). Earlier investigators reported the free wax acids of beeswax as all having an odd number of carbons and labeled them as neocerotic ($C_{25}H_{50}O_2$), carbocerotic ($C_{27}H_{54}O_2$), montanic ($C_{29}H_{58}O_2$), and melissic ($C_{31}H_{62}O_2$). From our advanced knowledge of the melting points of synthetic carboxylic acids, those given for beeswax correspond very well with those of *iso-acids*.

Of the insect waxes, other than beeswax, two are of economic importance, namely, shellac wax and Chinese insect wax (China wax). Shellac wax is obtained as a by-product recovered from the dewaxing of shellac spar varnishes. Shellac is derived from the lac insect, a parasite that feeds on the sap on the lac tree indigenous to India. Native lac wax melts at 72–80°C, whereas commercial shellac wax melts at 80–84.5°C. The high melting point and good dielectric properties favor its use in the electrical industry for insulation.

China wax resembles spermaceti in whiteness and crystalline appearance, but it is of greater hardness and friability. It is of considerable value in China and Japan as a coating material for tallow candles, in treating silk and cotton fabrics to give them a sheen, in the sizing and glazing of papers, and in medicaments.

Land Animal Waxes. The land animal waxes are either solid or liquid. Anhydrous lanolin, derived from the wool of the sheep, is of great economic value. It is of a stiff, unctuous consistency. The unsaponifiables of anhydrous lanolin are known as "wool wax alcohols" and belong in the class of derived lipids. They are in considerable demand by cosmetic and pharmaceutical industries. Woolwax has a great affinity for water, of which it will absorb 25–30%. Refined woolwax is kneaded with water to produce a water-white, colorless ointment, known as hydrous lanolin or "lanolin USP." Anhydrous lanolin is widely used in cosmetic creams, since it is readily absorbed by the skin. It is also used in leather dressings and shoe pastes, as a superfatting agent for toilet soap, as a protective coating for metals, etc. The only representative of liquid land animal wax is "mutton bird oil" obtainable from the stomach of the mutton bird.

Marine Animal Waxes. The marine animal waxes are either solid or liquid. *Spermaceti* is derived from a concrete obtained from the head of the sperm whale. The sperm whale is 60–80 feet in length, with an enormous head, 30 feet in circumference, in which there is a large hollow on the upper surface of the skull, filled with a peculiar fatty tissue. When the spongy mass is removed from the head, the oil is allowed to separate by draining. Spermaceti is in the oil in a dissolved state while the animal is living, but tends to concrete after its death. The resultant mass, after draining the oil, is boiled in a 2–3% lye solution to clean it; it is then washed free from alkali, and the wax is melted and molded into cakes. Sperm oil itself is in demand as a lubricant for the spindles of cotton and wool mills and for delicate mechanisms of all kinds. Its limpidity, freedom from tendency to "gum" and become rancid, and ability to retain its viscosity at relatively high temperatures make it preferable to other fixed oil lubricants.

Spermaceti may be described as a white, somewhat translucent, slightly unctuous mass of a scaly crystalline fracture, pearly in luster, with a faint odor and a bland mild taste. The refractive index ($n_D{}^{70}$) of spermaceti is 1.4397, almost identical with that of cetyl palmitate, its principal constituent. A spermaceti wax is also obtained by hydrogenating the blubber oil of the sperm whale; the resulting wax is a trifle harder and higher in melting point (46–50°C) than the natural spermaceti (43–47°C). Since spermaceti has in the past been adulterated with paraffin, considerable attention has been given by cosmetic and pharmaceutical authorities to the promulgation of suitable standards to define its purity. As an example, for a cosmetic grade the saponification value limits are set at 108–134, and iodine value at 3–4. There is a fairly large and increasing demand for the derived lipid wax *cetyl alcohol*, yielded from spermaceti wax by saponification; it is used in the manufacture of lipstick, shampoo and other cosmetics.

Plant Waxes. Waxes are formed in the "protective" cellulose walls of plants, as distinguished from the several other cellulose walls. The wax components consist chiefly of alkyl esters produced by the esterification of ethanoid alcohols with acids, both of high molecular weight. The esters are usually accompanied by some free alcohol and/or free acid, more frequently the latter, and by end residues of hydrocarbons high in molecular weight, *e.g.*, hentriacontane, $C_{31}H_{64}$.

The biosynthesis of the long hydrocarbon chains (containing the methylene group, $—CH_2—$) of the fatty acids is discussed in the article FATTY ACID METABOLISM (OXIDATION AND BIOSYNTHESIS). Alcohols with an even number of carbons become esterified by the acids. Any surplus of alcohols of even carbons remain free. In esterification, the free fatty acid combines with another mole of alcohol to form an ester by the elimination of one mole of water, namely

$$HR_m \cdot OH + HR'_nCO \cdot OH \rightarrow HR'_nCO \cdot OR_mH + H_2O$$

In the metamorphosis of plant material, glycerides in the form of fat or oil, built up by a reductive process and carboxylation, apparently are transformed into new chemical entities, such as high molecular weight alcohols, acids and hydrocarbons, noticeably so in the arid plants which produce wax in abundance. Hydrocarbons are formed by the decarboxylation of esters and always have an *odd* number of carbons as might be expected; *e.g.*, cetyl palmitate, $C_{15}H_{31}CO \cdot CO_{16}H_{32}$ yields hentriacontane, $C_{31}H_{64}$, on decarboxylation. Cetyl palmitate itself is never a constituent in arid plant waxes because of its decarboxylation, whereas hentriacontane is abundant.

The leaves of several species of palm trees furnish wax of great economic importance. This is particularly true of the product furnished by harvesting the leaves of the carnauba palm, *Copernicis cerifera* Mart.; the industry centers at Parahyba, Brazil. The wax is removed from the leaves by sun drying, trenching, threshing and beating; the powdered wax is melted in a clay or iron pot over a fire, strained, cast into blocks, and broken into chunks for export shipment from Bahia on the Atlantic Seaboard, most of it going to the United States. The various grades of *carnauba wax* carry provincial names as well as commercial designations; *e.g.*, *Flor fina* is Yellow No. 1, *Primeira* is Yellow No. 2, *Gordurosa* is North Country No. 3, etc. Carnauba and candelilla (a stem wax) are the hardest of all natural waxes. It is the high content of esters of hydroxylated saturated carboxylic acids that gives carnauba its extreme hardness, whereas the esters of the hydroxylated unsaturated fatty acids produce its outstanding luster. These properties can be imparted to waxes employed in polishes, carbon paper coatings, etc., even though the carnauba content is less than 10%.

Ouricury, carandá and raffia are commercial palm leaf waxes of lesser importance. Ouricury wax has a very high content of esters of hydroxylated carboxylic acids and is used as a substitute for carnauba in carbon paper coatings. Carandá and raffia waxes have very low contents of these acids and make unsatisfactory substitutes for carnauba.

The most important waxes obtainable from the stems of plants are *candelilla wax* and sugarcane wax. They are entirely different in their composition; candelilla contains far more hydrocarbons (50–51%) than any other known natural wax, whereas sugarcane wax contains 78–82% wax esters but only 3–5% hydrocarbons. The candelilla is an arid plant that grows in Mexico and the southwestern United States. To recover the wax, the plant stalks are pulled up by the roots and boiled in acidulated water. On cooling, the congealed wax is removed from the surface of the water in the tank. The crude wax is given an additional refinement before it is placed on the market. Candelilla wax is brownish in color and melts at 66–78°C. It is often used in conjunction with carnauba in leather dressings, floor waxes, etc. It is also used in sound records, electrical insulators, candle compositions, etc.

Because of the enormous tonnage of sugarcane processed in many parts of the world, it is possible to recover an appreciable tonnage of *sugarcane wax* as a by-product. The crude wax contains about one-third each of wax, resin and oil, and hence needs considerable refinement by selective solvents before it can become of value for industrial use. The refined wax is dull yellow in color, melts at 79–81°C, and is hard and brittle. Some of the esters in its composition are sterols—sitosterol and stigmasterol—combined with palmitic acid, and are responsible for the good emulsification properties of the wax itself in the preparation of polishes and the like.

Of waxes obtained from fruits, *japanwax* is the only one of great economic importance, particularly in the Asiatic countries. The wax occurs as a greenish coating on the kernels of the fruit of a small sumac-like tree. Japanwax is actually a vegetable tallow, since it is comprised of 90–91% glycerides, including those of dicarboxylic acids, *e.g.*, japanic acid, $(CH_2)_{19}\cdot(COOH)_2$, with an odd number of carbons. The textile industries in the past have been large users of japanwax since it is a source of emulsifying softening agents.

Waxes from grasses include bamboo leaf wax, esparto wax, and hemp fiber wax. Esparto wax is a hard, tough wax, with a melting point of 73–78°C, and is the most important grass wax. Esparto, or Alfa grass, is shipped from Mediterranean countries to Scotland, where it is dewaxed so that it can be made into paper. The accumulated dustings, flailed from the grass, are extracted with a light petroleum solvent by percolation and the wax is recovered by boiling off the solvent.

Mineral Waxes. The mineral waxes comprise fossil (lipobiolithic) waxes—peat waxes and montan waxes, lignite paraffins, and earth waxes. The peat deposits in Eire and in Chatham Islands, New Zealand, cover vast areas. Peat will yield 9.5–10.5% wax when extracted by azeotrope-like mixtures of solvents. Peat waxes are still in the development stage. Montan waxes, on the other hand, have been extensively developed, particularly the montan wax, "bergwachs," refined at Halle-Riebeck, Germany. The recovery of wax from the bituminous lignite is 10–18%. The selective solvents used are benzene and unrefined wood alcohol in the proportion of 85:15, and at a temperature of 100°C. The alcohol acting on the cell walls of the crushed lignite frees the wax which is readily dissolved by the benzene. The wax solution is separated, and the solvent distilled off, leaving the crude wax as a residue. The crude wax must be further refined to obtain the "deresinified," "yellow refined" and "white bleached" grades.

Crude montan wax has an approximate composition of 50–55% esters of wax acids, 16–18% of free wax acids with a melting point of 81.5–83°C, 1–2% free alcohols, 3–6% ketones, 20–23% resin, and 3% asphaltic material. The free wax acids are isoacids, ranging from C_{21} to C_{31}, with the C_{29} acid, isononacosanoic, prevailing. There are a number of industrial uses for montan wax: electrical insulation, leather finishes, polishes, carbon papers, etc. Much of the "deresinified" grade was shipped to Gersthofen to produce modified ester waxes, known as "I.G. waxes" but now as "Gersthofen waxes." These modified ester waxes, which include long chain esters produced from montanic acid, are of great value as adjuncts in wax compounding, as are also the glycol esters of montanic acid, when used in preparing wax emulsions.

Lignite paraffins are obtained from the destructive distillation of brown coal or lignite, and are like the lower-melting petroleum waxes in composition. In the dry-distillation process, the resultant lignite tar is worked up into oils and paraffin wax, particularly in the Saxon-Thüringian district. Paraffin waxes have also been obtained from the crude shale oil produced by a batch process at the Bureau of Mines Oil-Shale Demonstration Plant, Rifle, Colorado.

The earth waxes are naturally occurring mineral waxes consisting of hydrocarbons of isoparaffinic structure and very high molecular weight. These native mineral waxes or wax bitumens are associated with earth deposits of a sedimentary nature belonging to the Tertiary period. The earth wax of great economic importance is *ozocerite*, which was originally called "ceresin." Ozocerite is mined at Boryslaw, Poland, but shipped elsewhere to be refined, to rid it of the earth impurities. The pure ozocerite is microcrystalline, has a melting point of 74.4–75°C, and is entirely free from brittleness. When fused with high-melting paraffins or other waxes, it tends to break down their macrocystalline structure and make them much less liable to fracture in various wax coatings, etc. The electrical industry has always been a large consumer of ozocerite, crude or refined, as it is an excellent insulator. Pure ozocerite is used in the manufacture of pharmaceutical and cosmetic preparations. However, its importance in the arts has now ceded to the petroleum microcrystalline waxes.

Petroleum is the largest single source of hydrocarbon waxes. The production in the United States is estimated at well over $1\frac{1}{2}$ billion pounds per annum, of which about one-fourth is exported. The petroleum waxes may be roughly divided between *paraffin waxes*, which are macrocrystalline in structure (needles and plates), and *microcrystalline waxes*. Petroleum waxes, as well as a number of types of synthetic waxes, are discussed in reference 2.

References

1. WARTH, A. H., "Waxes: Physical and Chemical Characteristics," in "Biology Data Book" (ALTMAN, P. L., AND DITTMER, D. S., EDITORS), p. 382, Washington, D.C., Federation of American Societies for Experimental Biology, 1964.
2. WARTH, A. H., "Waxes," in "The Encyclopedia of Chemistry" (CLARK, G. L., AND HAWLEY, G. G., EDITORS), New York, Reinhold Publishing Corp., 1957.

ALBIN H. WARTH

WIELAND, H. O.

Heinrich Otto Wieland (1877–1957) was a German organic chemist whose work was strongly oriented toward biochemistry inasmuch as it dealt with BILE ACIDS, STEROIDS and BIOLOGICAL OXIDATIONS. In opposition to OTTO WARBURG, Wieland emphasized the importance of dehydrogenation in biological oxidations. He received the Nobel Prize in chemistry in 1927 primarily on the basis of his contributions to steroid structure.

WILLSTÄTTER, R.

Richard Willstätter (1872–1942) was a German chemist whose leanings were strongly toward biochemistry. He received the Nobel Prize in chemistry in 1915 for his work on plant pigments, notably CHLOROPHYLL. One of his main interests was enzymes which he thought were elusive substances *associated with* colloidal proteins. He was reluctant to abandon this erroneous view. He had to flee Germany to Switzerland at the beginning of World War II.

WINDAUS, A.

Adolf Windaus (1876–1959) was a German organic chemist who first synthesized HISTAMINE, but is chiefly noted for his researches on the structure of cholesterol which made possible the elucidation of the structure of VITAMIN D (also Windaus' work). His investigations are basic to all steroid studies. In 1928 he received the Nobel Prize in chemistry.

X

XANTHINE

Xanthine, 2,6-dihydroxypurine (for structure, see PURINE BIOSYNTHESIS, Fig. 1), is a purine base, which may be formed by the oxidation of hypoxanthine or by deamination of guanine. It, in turn, may be oxidized to uric acid. The enzyme required for this oxidation, XANTHINE OXIDASE, contains 0.03% molybdenum, which element is therefore nutritionally essential in minute amounts.

XANTHINE OXIDASE

Uric acid, which is the end product of purine catabolism in primates, the major nitrogenous excretory product of birds, and an intermediate in the catabolism of purines by other life forms (see PURINE BIOSYNTHESIS AND METABOLISM), is produced in all cases by the action of xanthine oxidase. This enzyme, which has been the subject of many hundreds of research reports has attracted a great deal of interest, not primarily because of its metabolic importance but rather because its complexity and catalytic versatility have challenged and intrigued enzymologists for over half a century. Thus, with a molecular weight of close to 300,000, a xanthine oxidase molecule is composed of approximately 2,500 amino acid residues. In addition, it is known to contain two molecules of flavin adenine dinucleotide, two atoms of molybdenum and eight atoms of iron per molecule of enzyme. Xanthine oxidase is capable of catalyzing the oxidation of a large number of different PURINES, aldehydes, PTERIDINES and other heterocyclic compounds by a variety of electron acceptors, including molecular oxygen, nitrate and nitro compounds, quinones, dyes, ferricyanide and cytochrome C. Indeed, its ability to catalyze such diverse oxidation-reduction reactions misled early investigators into attributing the aldehyde and purine oxidizing activities of whole milk to separate enzymes. A great deal of excellent experimentation was then directed at proving that a single enzyme was responsible for catalyzing the oxidation of both classes of compounds. The wide substrate specificity of xanthine oxidase poses the possibility that in addition to oxidizing purines *in vivo*, it may also play a role in various aldehyde oxidations, such as that of VITAMIN A aldehyde to vitamin A acid. Xanthine oxidase has definitely been implicated in catalyzing the pteridine oxidations which produce the pigments of *Drosphila*. In a similar vein, the wide electron acceptor specificity of xanthine oxidase leaves its *in vivo* acceptor a matter for conjecture. When acting on its substrates in the presence of oxygen, xanthine oxidase has also been shown to generate reactive radicals, presumably oxygen radicals, whose presence can be detected by their ability to initiate free radical chain reactions, such as the aerobic oxidation of sulfite, and by their chemiluminescent interaction with luminol.

Xanthine oxidase is found primarily in the livers of animals although it has also been reported to be present in the kidney, blood and intestinal mucosa of certain species. The level of xanthine oxidase activity in mammalian liver has been found to be remarkably responsive to dietary influences. Thus, nutritional deficiencies of MOLYBDENUM, IRON, RIBOFLAVIN or ESSENTIAL AMINO ACIDS have all been seen to cause sharp decreases in the measurable xanthine oxidase of liver. Indeed, it was such studies of the dietary compounds required for the maintenance of high levels of liver xanthine oxidase which first led to an appreciation of the essentiality of molybdenum in the diet and its role in the action of xanthine oxidase.

For reasons which have not yet been elucidated, xanthine oxidase is also abundantly present in milk where it occurs as part of the protein membrane surrounding and thus stabilizing the glovules of fat. In fact, most of the iron in cows' milk can be accounted for as the iron of the xanthine oxidase in that milk. The level of this enzyme in milk is apparently species dependent. Thus, human milk contains less than 10% the xanthine oxidase activity of cow's milk. It may be proposed that newborn mammals cannot synthesize adequate xanthine oxidase to meet their metabolic needs and that the xanthine oxidase in milk is absorbed by the suckling to meet this need. The observation that newborn rats contain no xanthine oxidase in their livers is in accord with this speculation. Because it is so readily available, the xanthine oxidase of whole, unpasteurized, cows' milk, has been the subject of most of the studies of the physical and catalytic properties of this enzyme. The few studies which have been performed on the enzyme from mammalian liver indicate that it is similar to the milk enzyme, whereas the enzyme purified from avian liver appears to contain more iron and to be unable to transfer electrons directly to molecular oxygen.

Milk xanthine oxidase has been purified and crystallized by a procedure that involves the precipitation of casein by treatment with the mixed PROTEOLYTIC ENZYMES of pancreas (pancreatin) followed by column chromatography on calcium phosphate. Final crystallization was from an ethanolic phosphate buffer. The crystalline material appeared to be largely homogeneous in the ultracentrifuge. However, a more sensitive criterion, the constant solvent solubility test, clearly indicated heterogeneity. Furthermore, it has been demonstrated that treatment of milk xanthine oxidase with pancreatin generates several components separable by column chromatography, all of which are enzymically active. This was undoubtedly due to limited proteolysis of the primary form of the enzyme. The enzyme as isolated from milk by the usual procedure has also been found to contain several small peptides which are not dialyzable from the native enzyme but which are readily liberated after heat denaturation. Crystallization of the enzyme from milk by a procedure which avoids treatment with proteolytic agents is a highly desirable but as yet unrealized goal. The purified enzyme is quite stable when stored at deep freeze temperatures as a precipitate under saturated ammonium sulfate solution and may be kept for many months in this condition with only minor losses of activity. The enzyme is usually assayed spectrophotometrically by exploiting the differences in the absorption spectra of its substrates and their products. However, FLUOROMETRIC, MANOMETRIC, photometric and tracer methods have also been employed to advantage.

The oxidation of hypoxanthine to xanthine to uric acid and the oxidation of aldehydes to carboxylic acids are formally hydroxylation reactions, in that a hydrogen atom attached to carbon is replaced by a hydroxyl group. Studies with H_2O^{18} have demonstrated that the oxygen thus introduced into the subtrate has its origin in the oxygen of water. Oxidation, as catalyzed by xanthine oxidase, is in reality the removal of an electron pair and a hydrogen nucleus, perhaps in combination as a hydride ion, followed by their replacement by a hydroxyl ion from the solvent. The electron pair removed from the substrate passes onto the enzyme and is transferred by way of its components to the final electron acceptor in the reaction mixture. If this is oxygen, then it is reduced to hydrogen peroxide. Recent studies by the technique of electron spin resonance spectrometry have indicated that electrons from the substrate make their way through the components of the enzyme in the sequence molybdenum → flavin → iron and then to oxygen. Xanthine oxidase thus contains an electron transport chain analogous to that operating in MITOCHONDRIA, to which it has, in fact, been likened.

The concentration of xanthine required to achieve half the maximal rate of uric acid production (referred to as the K_m for xanthine) is very much a function of pH, changing from $1 \times 10^{-6}M$ at pH 7.0 to $3 \times 10^{-4}M$ at pH 11.0. At any given pH, the K_m for xanthine is also a function of the concentration of the electron acceptor. Thus, at pH 10.0 as the oxygen concentration is varied from zero to infinity, the apparent K_m for xanthine rises from zero to a limiting value of $1.5 \times 10^{-4}M$. Similarly, under these same conditions, as the concentration of xanthine is varied from zero to infinity, the apparent K_m for molecular oxygen rises from zero to a limiting value of $8 \times 10^{-5}M$. Such behavior indicates that the product of xanthine oxidation, *i.e.*, urate, dissociates from the enzyme before the electron acceptor, which is oxygen in this case, reacts with the reduced enzyme. Failure to appreciate the profound effects of pH on K_m for xanthine, and of xanthine concentration on K_m for oxygen undoubtedly accounts for many of the seeming disagreements in the literature concerning the magnitudes of these constants.

Concentrations of xanthine well above the K_m level are inhibitory to the enzyme. This phenomenon of excess substrate inhibition has been extensively studied. It has been suggested that the enzyme contains two binding sites for xanthine—at one site xanthine acts as a substrate, whereas at the other it competes with the electron acceptor and thus inhibits. This suggestion appears to be reasonable. Thus, it has been reported that in the absence of electron acceptors, xanthine oxidase is capable of catalyzing the dismutation of a mixture of hypoxanthine and uric acid to xanthine. Inhibition by excess xanthine was also observed to be more pronounced at low levels of electron acceptor (methylene blue) than at high concentrations thereof. Like the K_m for xanthine, the sensitivity to inhibition by excess xanthine is strikingly affected by pH. For this reason, investigations of the effect of pH on the activity of this enzyme which have often been conducted at a single arbitrary substrate concentration have yielded results which are misleading and difficult to interpret, since the observed pH optima are then due to fortuitous balances between the effects of Ph on K_m, excess substrate inhibition, and V_m.

Xanthine oxidase can transfer electrons to CYTOCHROME C. Oxygen which might, *a priori*, have been expected to act as a competing electron acceptor is, in fact, absolutely required for cytochrome C reduction and probably functions as an electron conducting bridge from the enzyme to the cytochrome C. The observation that MYOGLOBIN or the native globin thereof is a potent competitive inhibitor of cytochrome C reduction but does not inhibit the reduction of oxygen, quinones or dyes, undoubtedly accounts for much of the confusion in the literature concerning the relative ability of cytochrome C to act as an electron acceptor from xanthine oxidase, since cytochrome C preparations are frequently contaminated with myoglobin. The ability of myoglobin to inhibit the reduction of cytochrome C by xanthine oxidase is very much a function of the species from which the myoglobin was isolated. This suggests that the interaction of xanthine oxidase with myoglobin is a highly specific protein-protein interaction which could be used in studies of the conformation of these proteins.

Xanthine oxidase exhibits an activity-dependent sensitivity to heavy metals. Thus, in the absence of substrates, the enzyme is unaffected by the traces of metals ordinarily present in distilled water and in buffer salts, whereas in the presence of substrate, these same metal contaminants are capable of inactivating the enzyme within a few minutes. This phenomenon which may be due to the exposure, by reduction, of a sulfhydryl group on the catalytic site, necessitates the use of ethylene diamine tetraacetate or other chelating agents to protect the enzyme during routine assays. Bovine serum albumin has occasionally been used to serve this same purpose.

Xanthine oxidase is inhibited by cyanide in a manner which is quite unlike the rapid and reversible combination of cyanide with metal complexes. Thus, cyanide inhibition of xanthine oxidase is a moderately slow and irreversible reaction. Substrates, various reducing agents, and competitive inhibitors act to protect the enzyme against cyanide inhibition. Xanthine oxidase is inhibited reversibly and competitively by purine analogues such as pyrazolo-pyrimidines, 6-chloropurine and 6-mercaptopurine. Borate, urea, s-triazines, guanidinium, nitroguanidine and nitrosoguanidine are also competitive with respect to xanthine. The most effective competitive inhibitor described to date is 2-amino-4-hydroxy-6-formylpteridine for which a K_I of less than $10^{-9}M$ has been reported. Salicylate which is also an inhibitor has been found to stabilize the enzyme and has found use in prolonging the shelf life of purified xanthine oxidase. Competitive inhibitors of xanthine oxidase have been used with some success in attempts to control gout.

The reddish color of concentrated solutions of purified xanthine oxidase was noted by early investigators. Since simple flavoenzymes are yellow, a non-flavin chromogen was proposed. It has been shown that the iron in xanthine oxidase accounts for its red color. Precipitation of the enzyme with methanol liberated the FLAVIN while leaving the iron on the protein. The resultant iron-protein exhibited a reddish color and had an absorption spectrum similar to those reported for the non-heme iron-containing proteins isolated from mitochondria and from plant chloroplasts. Aldehyde oxidase from rabbit liver and dihydroorotic dehydrogenase from *Zymobacterium oroticum* which also contain non-heme iron and flavin exhibited absorption spectra similar to that of xanthine oxidase. In all cases, incubation of the enzymes with their substrates under anaerobic conditions results in rapid bleaching of that part of the spectrum attributable to the flavin and also of that absorbancy associated with the iron. These changes are rapidly reversed by admission of oxygen. These observations accord with those made by electron spin resonance spectrometry in assigning an electron transfer role to the iron of these enzymes.

In summary, much has been learned about the microcosm of xanthine oxidase, but much more which remains yet hidden from view will, when exposed, certainly enhance our understanding of the chemical architecture of enzymes and of life.

References

1. DIXON, M., *Enzymologia*, **5**, 198 (1938).
2. DERENZO, E. C., *Advan. Enzymol.*, **17**, 293 (1956).
3. BRAY, R. C., in "The Enzymes" (BOYER, P. D., LARDY, H., AND MYRBÄCK, K., EDITORS), Vol. 7, p. 533, New York, Academic Press, 1963.
4. RAJAGOPALAN, K. V., AND HANDLER, P., *J. Biol. Chem.*, **239**, 1509 (1964).
5. PALMER, G., BRAY, R. C., AND BEINERT, H., *J. Biol. Chem.*, **239**, 2657 (1964).
6. BRAY, R. C., PALMER, G., AND BEINERT, H., *J. Biol. Chem.*, **239**, 2667 (1964).

IRWIN FRIDOVICH

X-RAY CRYSTALLOGRAPHY

Structural studies with X rays have made important contributions to present knowledge of the molecular architecture of many complex substances produced by living organisms. Examples include the recognition of the α-helix structure in polypeptide chains of proteins [see PROTEINS (BINDING FORCES IN SECONDARY AND TERTIARY STRUCTURE)], the double-helical structure of DNA [see DEOXYRIBONUCLEIC ACIDS (DISTRIBUTION AND STRUCTURE)], the three-dimensional structure of MYOGLOBIN (reference 3) and of lysozyme (reference 5), and the porphyrin-like structure of Vitamin B_{12} (see COBALAMINS).

References

1. RICH, A., AND GREEN, D. W., "X-Ray Studies of Compounds of Biological Interest," *Ann. Rev. Biochem.*, **30**, 93–132 (1961).
2. KRAUT, J., "Structural Studies with X-Rays," *Ann. Rev. Biochem.*, **34**, 247–268 (1965).
3. KENDREW, J. C., "The Three-Dimensional Structure of a Protein Molecule," *Sci. Am.*, **205**, 96–110 (December 1961).
4. PERUTZ, M. F., "The Hemoglobin Molecule," *Sci. Am.*, **211**, 64 (November 1964).
5. PHILLIPS, D. C., "The Three-dimensional Structure of an Enzyme Molecule," *Sci. Am.*, **215**, 78-90 (November, 1966).

X-RAYS (BIOCHEMICAL EFFECTS). See IONIZING RADIATION (BIOCHEMICAL EFFECTS); MUTAGENIC AGENTS.

Y

YEASTS

The scientific study of yeasts, which have been used since antiquity in baking, brewing and wine making, belongs in the domain of plant biology or in microbiology. The influence of these organisms on biochemistry, however, has been profound and far reaching. The fact that the word *enzyme* comes from the Greek meaning *in yeast* is some indication of the importance of these organisms.

ALCOHOLIC FERMENTATION was of prime interest to biochemists for decades, and investigations in this area paved the way for the study of all enzymes, for the elucidation of the functioning of the B vitamins as coenzymes, and for an appreciation of the metabolic unity in nature which is now often taken for granted.

Early nutritional studies on animals made extensive use of yeast as a convenient source of "vitamin B" before the existence of several vitamins belonging in this category was known. The nutrition of yeast itself was used as a basis for discovering two B vitamins—PANTOTHENIC ACID and BIOTIN—as well as something about the biological importance of INOSITOL.

The yeasts which have received the principal attention of biochemists are the conveniently available "brewers" and "bakers" yeasts which are known by the name *Saccharomyces cerevisiae.* The use of this term has been misleading at times because there are many varieties of yeast falling in this classification, which may be morphologically indistinguishable, but nevertheless are distinct in their nutritional needs and in their responses to vitamins and other nutrilites.

Reference

1. ROSE, A. H., "Yeasts," *Sci. Am.*, **202**, 136–146 (February 1960).

Z

ZINC (IN BIOLOGICAL SYSTEMS)

Zinc has been recognized as a "trace element" ever since it was identified first as a growth factor for *Aspergillus niger* (Raulin, 1869). Since then, zinc has also been shown to be essential for normal growth and development in higher organisms. Evidence for a specific biochemical role of zinc was first obtained by Keilin and Mann in 1940, when the metal was shown to be a stoichiometric component of bovine CARBONIC ANHYDRASE. The discovery of many other zinc-containing enzymes has since revealed a wide diversity in its biochemical function.

Zinc (molecular weight 65.38) occurs as element 30 just after the first transition metal series. With the *d*-orbitals fully occupied, zinc has only one valency state available, and its salts are colorless. Chemically, zinc resembles both the preceding transition elements and the other group IIB elements Cd and Hg. Zinc is measured analytically by a variety of sensitive techniques among which colorimetry with dithizone, emission spectrography and atomic absorption spectroscopy are most prominent. The radioactive isotopes ^{65}Zn (256 days) and ^{63}Zn (38 minutes) have been applied to biological studies.

The high affinity of zinc for nitrogenous and sulfur-containing ligands seems chiefly responsible for the occurrence of zinc in a wide variety of biological compounds such as PROTEINS, AMINO ACIDS, NUCLEIC ACIDS and PORPHYRINS. The association of zinc with enzymes has been studied most thoroughly. Operationally the enzymes which are affected by zinc can be considered in two groups: Zinc metalloenzymes and zinc metal-enzyme complexes. Zinc metalloenzymes incorporate zinc so firmly in the protein matrix that they can be thought of as an "entity" in nature. Under reasonably mild conditions, the metal and protein moiety are isolated together and exhibit an integral stoichiometric relationship. On the same basis, a strict correlation is preserved between metal content and enzymatic activity, allowing the inferential identification of a specific biological function of zinc *in vivo*. Zinc metal-enzyme complexes on the other hand comprise enzymes which are activated *in vitro* by the addition of zinc ions. The loose association and the relative lack of metal ion specificity render it difficult in such cases to assign specific biological significance to zinc *in vivo*.

Zinc Metalloenzymes. Thus far some 20 zinc metalloenzymes belonging to widely different classes of enzymes have been characterized (Table 1). In all of them zinc is a selective stoichiometric constituent and essential for catalytic activity. It is often present in numerical correspondence to the number of active enzymatic sites, coenzyme binding sites, or enzyme subunits. Removal of zinc results in complete loss of activity. Inhibition by metal complexing agents is a characteristic feature of all zinc metalloenzymes. However, no direct relationship holds between the inhibitory effectiveness of these agents and their affinity for ionic zinc. Inhibition may occur either through formation of mixed enzyme-Zn-inhibitor complexes (*i.e.*, in the alcohol dehydrogenases) or through removal of zinc from the enzyme (*i.e.*, carboxypeptidase A). The zinc-protein bonds are very firm, implying chemically specific and structurally distinct binding sites. Apparent association constants of 3×10^{10} and 10^{12} have been measured in carboxypeptidase A and carbonic anhydrase, respectively. In carboxypeptidase A, zinc is bound to a sulfur and a nitrogen ligand. Evidence for binding to cysteine residues has also been obtained for horse liver alcohol dehydrogenase. Sulfur is not involved in binding zinc to bovine carbonic anhydrase, however. The metal seems located at the active enzymatic center. In bovine carboxypeptidase A, zinc is bound near the substrate binding site. In the alcohol dehydrogenases of yeast and of horse and human liver, one atom of zinc is located at or near each coenzyme binding site. In these enzymes, zinc is also essential for the maintenance of the quaternary structure. Its removal results in dissociation into four subunits. Zinc may thus participate in substrate and coenzyme binding, in the stabilization of a critical subunit configuration, as well as in the catalytic event. Though zinc is the only constituent of zinc metalloenzymes *in vivo*, it can be replaced by other metals *in vitro*. Substitution by metals such as Co, Ni, Fe, Mn, Cd, Hg, Pb has been achieved in the carboxypeptidases, the carbonic anhydrases, alkaline phosphatase of *E. coli* and others. The resulting new metalloenzymes exhibit specific differences in catalytic competence.

Zinc Metalloproteins. A number of zinc metalloproteins have been isolated for which no enzymatic function has been identified as yet. They include zinc metalloproteins from human serum, a protein from human leukocytes which

TABLE 1. ZINC METALLOENZYMES[a]

Enzyme	Molecular Weight	Zn (g at./mole)	Coenzyme (moles/mole)	Source
Oxidoreductases				
Alcohol dehydrogenase	150,000	4	4 NAD	Yeast
Alcohol dehydrogenase	84,000	3–4[b]	2 NAD	Equine liver
Alcohol dehydrogenase	87,000	4[b]	2 NAD	Human liver
Lactate dehydrogenase			NAD	Rabbit muscle
Malate dehydrogenase	70,000	2	2 NAD	Porcine heart
D-Glyceraldehyde-3-phosphate dehydrogenase	137,000	2	3 NAD	Rabbit muscle, bovine muscle, yeast
Glutamate dehydrogenase	1,000,000	2–4	*n* NAD	Bovine liver
D(–)Lactate dehydrogenase	50,000	2	1 FAD	Yeast
Spermine oxidase				Bovine plasma
Hydrolases				
Alkaline phosphatase	80,000	2	—	*E. coli*
Alkaline phosphatase			—	Mammalian kidney
Alkaline phosphatase			—	Human leukocytes
Carboxypeptidase A	34,300	1	—	Bovine pancreas
Carboxypeptidase A	34,800	1	—	Porcine pancreas
Carboxypeptidase B	34,000	1	—	Bovine pancreas
Carboxypeptidase B	34,300	1	—	Procine pancreas
Lyases				
Aldolase	65–75,000	1	—	Yeast
Aldolase	50,000	1	—	*Asp. niger*
Carbonic anhydrase	30,000	1	—	Bovine erythrocytes
Carbonic anhydrase	34,000	1	—	Human erythrocytes

[a] For bibliography, see reference 3. [b] Unpublished observations.

contains 0.3% zinc, and metallothionein from human and equine kidney cortex containing 2.6% and 2.2% zinc, respectively. The latter proteins also contain 4.2% and 5.8% cadmium, and as much as 8.5% sulfur. Both zinc and cadmium atoms are bound to three cysteine residues each.

Zinc Metal-enzyme Complexes. Zinc-activated enzymes are numerous and occur among various classes of enzymes. Generally zinc is but one of several activating metal ions. Because of the weak association, little is known about the stoichiometry and the mode of interaction of the metal with these enzymes.

Zinc in Biological Material. Zinc is a ubiquitous component of animal and plant tissue. In vertebrates, most organs including pancreas contain 20–30 μg of zinc per gram of wet tissue. Liver, voluntary muscle and bone hold about double this amount. Zinc contents ranging from 100–1000 μg/g wet weight were measured in islet tissue of certain teleost fish. Correlation between zinc content and INSULIN storage suggested a parallelism, but direct evidence for the existence of zinc-insulin complexes *in vivo* is lacking. Insulin is known to interact with zinc and other bivalent metals *in vitro*. High zinc contents have been noted in seminal vesicles, epididymis, the prostate and semen. The dorsalateral prostate of the adult rat holds about 180 μg of zinc per gram of wet tissue. Dried human semen contains 1–2 μg/g. Outstanding in zinc content are the tissues of the eye. The retina, iris and choroid of fish, frogs and carnivores contain up to 1000, 6000 and 30,000 μg/g dry weight, respectively. The highest zinc content—up to 150,000 μg/g—is found in the *tapetum lucidum cellulosum* of adult fox and seals. Zinc is bound as a zinc-cysteine-monohydrate complex. Its functional significance is unknown.

Human blood contains 7–8 μg zinc/ml. About 12% is present in serum, 3% in leukocytes and 85% in erythrocytes. In all three compartments, zinc occurs as part of zinc proteins and zinc metalloenzymes. In erythrocytes, zinc is correlated to carbonic anhydrase activity.

Zinc Deficiency. The enzymatic consequences of zinc deprivation have been studied in microorganisms, plants and animals. Experimental zinc deficiency has been generated in rat, mouse, turkey, chick and hog. Diets containing 22 μg zinc reduce the growth rate of rats and induce changes in carbohydrate, protein and nucleic acid metabolism. A number of pathological states in higher plants have been attributed specifically to lack of zinc. Parakeratosis in hogs has recently been recognized as a calcium-conditioned deficiency of zinc. Supplements of zinc induce recovery of these conditions. In humans, primary zinc deficiency was noted in chronic beriberi. It has also been suggested to be a factor in post-alcoholic cirrhosis.

References

1. VALLEE, B. L., "Biochemistry, Physiology and Pathology of Zinc," *Physiol. Rev.*, **39**, 443 (1959).
2. VALLEE, B. L., "Metal and Enzyme Interactions: Correlation of Composition, Function, and Structure," in "The Enzymes" (BOYER, P. D., LARDY, H., AND MYRBÄCK, K., EDITORS), Vol. 3, New York, Academic Press, 1960.
3. VALLEE, B. L., AND COLEMAN, J. E., "Metal Coordination and Enzyme Action," in "Comprehensive Biochemistry" (FLORKIN, M., AND STOTZ, E. H., EDITORS), Vol. 12, Amsterdam, Elsevier Publishing Co., 1964.

JEREMIAS H. R. KÄGI

ZYMOGENS. See PROTEOLYTIC ENZYMES (AND THEIR ZYMOGENS); PEPSINS AND PEPSINOGENS.

INDEX